Horst Ibelgaufts

Dictionary of Cytokines

Distribution:
VCH, P.O. Box 10 11 61, D-69451 Weinheim (Federal Republic of Germany)
Switzerland: VCH, P.O. Box, CH-4020 Basel (Switzerland)
United Kingdom and Ireland: VCH (UK) Ltd., Wellington Court, Cambridge CB1 1HZ (England)
USA and Canada: VCH, 220 East 23rd Street, New York, NY 10010-4606 (USA)
Japan: VCH, Eikow Building, 10-9 Hongo 1-chome, Bunkyo-ku, Tokyo 113 (Japan)

ISBN 3-527-30042-2

Horst Ibelgaufts

Dictionary of Cytokines

Weinheim · New York
Basel · Cambridge · Tokyo

Horst Ibelgaufts
Institute of Biochemistry
University of Munich (LMU)
at the Institute of Molecular Biology
Würmtalstraße 221
D-81375 München
Federal Republic of Germany
email (Internet): Ibelgaufts@vms.biochem.mpg.de

Published jointly by
VCH Verlagsgesellschaft mbH, Weinheim (Federal Republic of Germany)
VCH Publishers, Inc., New York, NY (USA)

Editorial Director: Dr. Hans-Joachim Kraus
email (Internet): Kraus@vchgroup.de
Production Manager: Dipl.-Ing. (FH) Hans Jörg Maier

Library of Congress Card No. Applied for.

A catalogue record for this book is available from the British Library.

Deutsche Bibliothek Cataloguing-in-Publication Data:

Ibelgaufts, Horst:
Dictionary of cytokines / Horst Ibelgaufts. – Weinheim ; New York ; Basel ; Cambridge ; Tokyo: VCH, 1994
ISBN 3-527-30042-2
NE: HST

Printed on acid-free and chlorine-free paper.

Converted by: Filmsatz Unger & Sommer GmbH, D-69469 Weinheim. Printing: Zechnersche Buchdruckerei GmbH & Co. KG, D-67330 Speyer. Bookbinding: IVB Industrie- und Verlagsbuchbinderei Heppenheim GmbH, D-64646 Heppenheim.

Printed in the Federal Republic of Germany.

This book is dedicated to PAULA *who always reminds me that the real world begins beyond the screen of my word processor.*

"Life: the ability to communicate"
(Arnold de Loof, Int. J. Biochem. 25: 1715-21 (1993))

"If it's secreted, we like it"
(read in a message I received through the INTERNET)

"You have but to know an object by its proper name for it to lose its dangerous magic."
(Elias Canetti, Auto da Fè)

"As they had the same name, when you saw them together, you could never tell one from the other."
(Eugene Ionesco, The Bald Prima Donna)

„Eliminate all the other factors, and the one which remains must be the truth."
(Sherlock Holmes, The Sign of the Four)

„...Would you think me impertinent if I were to put your theories to a more severe test?"
(Dr. Watson, The Sign of the Four)

Some Personal Remarks

"*It might be very interesting if one could demonstrate that under certain circumstances certain differences were observable between different reactions taking place in various systems.*" This sentence once found on a laboratory blackboard door most fittingly describes my feelings during the compilation of this dictionary.

Indeed, cytokine research is haunted by an abundance of phenomena and conditions, suitable or unsuitable systems, differences, subtleties, and redundancies: all sorts of cells appear to produce and secrete, upon all sorts of stimuli, all sorts of molecules into the growth medium, which then induce or suppress all sorts of reactions in all sorts of other cells, either by themselves or in combination with each other. The most frequently cited metaphor in cytokine research is that of a finely meshed cytokine network. This is, of course, true to the extreme: connect everything with everything in a diagram of 100 cell types and 500 factors – and add some extra lines. When compiling this dictionary my private metaphor was that of a zoo of factors in a jungle of interactions surrounded by deep morasses of acronyms and bleak deserts of synonyms.

This dictionary is my attempt to map my ways through the jungles, morasses, and deserts of cytokine-land. I started with this subject as a complete and innocent newcomer and I have, therefore, tried to include everything that I felt newcomers to the field would want to find in a work of reference. Many entries originate from my searches for new material for a lecture series on genetic engineering held each term at Munich university but the book later developed a life of its own. Cytokines have provided just the right mixture of phenomena and factors which are not only exciting (in several ways) and interesting to deal with but which also beg and offer themselves to be analyzed in terms of gene structure, gene function and regulatory mechanisms by employing modern techniques of genetic engineering. The increasing number of cytokine genes that have been identified, characterized, and cloned in this way are magnificent examples of the power of this technology. Recent advances in creating transgenic animals overexpressing certain cytokines or carrying null mutations of cytokine genes as well as first attempts of gene therapy based on the transfer of cytokine genes further underscores its importance. Incidentally, even a cursory examination of many learned medical journals demonstrates that they abound with articles heavily relying on the applications of concepts and techniques of molecular biology and genetic engineering; now, if that is not a pleasing prospect for hard-boiled molecular biologists.

Three years ago I was naive enough not to waste a second thought before venturing to put my own ignorance into alphabetical order. After having struggled for some time with the sequelae of informational overload I must admit that I have begun to collect cytokines and information about related matters like other people collect stamps. Compiling such a dictionary and stamp collecting truly have much in common, including, for example, the joy of the hunt, the desire to be complete, and the satisfaction of being in possession of a quite rare specimen. Yet, while even uninitiated stamp collectors can easily spot their items of interest in one of the many stamp catalogues the situation is not as easy as that for newcomers in cytokine research.

Severe acronymania seems to be some sort of occupational disease if you are dealing with people who deal with cytokines. The panoply of names alone is truly forbidding: there are factors that go by two dozens of different names. And then, of course, most of the original names are misnomers. NGF, the "nerve growth factor" affects all sorts of tissues; "fibroblast" growth factors have an exceptional range of target cells, and "platelet-derived" growth factor is produced by many cells other than platelets. There are growth factors that do anything but promoting growth, interferons that mainly function as immunomodulators, interleukins that have profound effects on many non-immune cells, inhibitors that promote growth or induce differentiation, and chemotactic proteins with lots of unrelated functions, to name just a few inconsistencies. The state of the nomenclature in this field is summed up brilliantly by a statement once overheard in a laboratory: "*The light at the end of the tunnel is the light of the on-coming train.*"

There positively *is* some charm – not to mention the necessities of mnemonics – in giving factors names related to their functions but, alas, there are usually too many functions associated with one and the same cytokine. There is also an understandable reluctance to rename a pet factor that may have survived several waves of doctoral and postdoctoral fellows under a specific name. Of course, finding out that several factors with different names referring to their various activities are actually identical begs for a consolidating nomenclature and there have been many attempts to rectify this situation. Now we have things neatly numbered, with heparin-binding growth factors (HBGF), for example, up to 8, fibroblast growth factors (FGF) up to 9 (and overlapping with HBGFs), and interleukins even up to 15. Thank Goodness, nobody has so far come up with the idea of numbering consecutively the 30 or so chemokines that a while ago still were intercrines and a little longer ago featured as small inducible genes (*sig*) and small inducible cytokines (*scy*).

While numbering may introduce an element of precision I wonder who will remember easily what interleukin 23 will be good for, once we will have come that far. Sometimes I can't even remember the differences between *two* numbers such as type I and type II membrane proteins. What about entirely new names for them to please the scientific community, to further widen the gap between the ordinary lay person and the insiders, and to provide other dictionary-makers with some more entries? I herewith proudly propose that they be called Entons/Cetons (or maybe Enteins/Ceteins?) or even better, Ntons/Ctons (Nteins/Cteins). (Note added in proof: When I wrote this paragraph some months ago I was quite sure that Entons should be those membrane proteins having their N-terminal end on the outside of the cell. Now, while rereading this, I wonder whether it was not meant the other way round).

My feeling is that there will be no easy remedy for inadequacies of, and overkill by, nomenclature. We will never be able to drain completely the morasses of names in spite of all sorts of nomenclature meetings, reports, and recommendations. As a dictionary maker I will not be out of work for a long time. There will always be these familiar weary feelings when, in the title of a recently published article, an acronym, *A*, crops up that is long considered to be obsolete for factor *B*: are these people really talking about *B* – sometimes they are – or is that just again one of these cases of identical acronyms for two different factors, one old, one new? With modern cloning strategies yielding clones and sequence information galore we will always be confronted with the situation that the newly isolated factor *B* is found out to be identical with factor *A* of long standing and reputation. It would be silly in this case to dream up a new name, *C*, but I have encountered some instances of something previously called *A* having been renamed *C*, even by the very same people who initially described *A*, without *B* coming into question at all. This *is* downright silly and just adds a new twist to what unnerved editors all over the world know as EUAs (extra unnecessary acronyms).

Maybe a word of caution is justified with respect to the very many "synonyms" for individual factors found in this dictionary. Very often they are not synonyms in the grammatical sense. Upon close examination of protein or nucleic acid sequences many of the synonymous factors turn out to be slightly altered variants of one parent factor. They may arise by differential use of alternatively spliced transcripts or by post-translational processing yielding, for example, N or C terminal truncations or extensions, or microheterogeneous products with slight sequence alterations. In some instances it has been shown that these altered factors also possess some altered biological activities, but in most instances this has not been investigated and is not known. It appears most likely, though, that factors and their variants will have different biological functions because cells in general do not waste things. Some of the observed redundancies between cytokine activities are probably not redundancies at all but just reflections of our incomplete knowledge of what is really going on. Of course there are true redundancies as cellular safeguards, so to speak. Recent experiments involving the directed inactivation of cytokine functions by employing strategies of antisense oligonucleotides, cell ablation, or targeted gene deletion attest to it. These experiments also show more clearly than ever that observations made *in vitro* must be treated with caution in that they do not necessarily allow conclusions to be drawn about *in vivo* activities.

One point needs to be stressed also: cytokine mediators, although of considerable importance, represent one side of the coin only. The ultimate fate and the functional activities of cells are influenced in a variety of ways by an intricate intracellular signaling machinery and by other mediators, including, for example, cellular interaction molecules, cell surface proteins, components of the extracellular matrix and the basement membrane, and a plethora of other low-molecular weight compounds, many of which are not even proteins or peptides. They all influence the regulated expression of soluble cytokines and, in turn, are themselves controlled by them. In addition, many cytokines are now known to exist also in membrane bound forms and many of their cellular receptors have been described to yield bioactive soluble variants. Likewise, "classical" membrane bound molecules, some CD antigens, for example, have also been shown to exist as soluble forms and to behave like cytokines. This effectively blurs all distinctions sometimes imposed by nomenclature between classical diffusible messengers and resident molecules.

I should add that, being a humble molecular biologist who has been brought up with genes and their products, I have left out entries specifically dealing with mediators that are not proteins or peptides derived thereof. We all have our little idiosyncrasies but after all most researchers in this field will agree that cytokines *are* regulatory protein/peptide factors. I knowingly present only this one side of the coin although I refer to these other non-protein mediators often enough in the text entries.

Time must tell whether some of the included factors are just filing cards out of date – the German expression is much more colorful and literally translates as "filing card *corpses*". There are many activities once reported that then sometimes live on as autocitations never to be cited by anyone else outside the reporting lab or, worse, disappear entirely from the realms of published material for ever after – published *and* perished. I always wonder whether these modern *fata morganas* are an indication of dwindling or dried-up funds, students having changed the subject, postdocs having left the lab for good to turn to other things, or heads of departments having lost their interest or having acquired that blissful status of *emeritus*? Constant confrontation with these *fata morganas* probably is the price we have to pay for having access to omniscient electronic databases with roughly 100 million articles that they never forget!

I could have added many more factors had it not been for the fact that many researchers obviously feel inclined to report activities *without* giving them a proper name – a veritable nightmare for a dictionary-maker who thrives on lexemes. Hence Canetti's citation above and hence my plea: if you isolate anything consult data banks to avoid hackneyed abbreviations or acronyms, *give it a proper name,* mention this name in titles or abstracts – and drop me a line!

Before coming to a close I feel compelled to add at least one observation made in the course of compiling material for this dictionary. Once having decided to provide full references I have come to thoroughly *hate* journals that do allow citations without titles and last page numbers. Is that not most annoying to all of us? I once saw a cartoon of a poor *schlemiel* standing in front of his poster presentation sporting a large placard "*Get a free drink if you read my poster.*" The artist certainly knew something about the tricks of the trade. Are incomplete references, I ask you, a scheme to induce you to look up cited articles which, perhaps, you would not have done after having read the title? There is so much reading to do and titles can often be an enormous help in creating one's own personal selection of what is considered essential, noteworthy, or superfluous in a given context. Missing full references are time sinks and are definitely not due to editorial space restrictions (one of the things that publishers would like you to believe that they exist): the majority of journals that leave out full bibliographical information prefer to grace you with half-empty columns or pages after each article.

Having got this off my chest I hope that this dictionary will be a little bit more than a mere *hortus siccus* representing an assiduously established collection of exotic or outré mediators that reflects my and cytokine researcher's craze for factors. This is not meant to be deprecatory, by the way: whoever is engaged in analyzing growth, differentiation, and functional capacities of cells at the molecular level will be confronted with a factor sooner or later. Incidentally, if the factor isolated by putting a lot of money, a lot of time, and a lot of work into the project, is really brand-new the joy is great; the enthusiasm is subdued if it turns out that the newly isolated factor A with its action B on the cell type C is identical with, or even only related to, another published factor D that shows an action E on the cell type F. But such is life.

In spite of more than 14,000 cited references it is needless to say that comprehensive coverage of the literature has been impossible. Books the size of this dictionary have been written already on some of the factors included herein. Although I have made every attempt to trace the origins of names and to provide the proper citations in many instances I have cited review articles or recently published material rather than first publications. The references found in this dictionary should be considered as "first-aid sets" providing, if necessary, easy access to further articles.

I also hope that there are a minimum of omissions and inaccuracies and that these can be rectified in a later edition. Maybe this work will be of some use in backing up readers for the dog-fight with genuine publications found in learned journals (we all have our little dreams, why not admit it). I would be pleased if this book were of some assistance in knocking some holes into existing language barriers between and among different scientific disciplines, getting things straightened out, for example, if one encounters a semantic masterpiece like the following, which completely boggled me when I started my work and tends to leave a feeling of stoic equanimity when I read it now:

"In contrast to BCDFγ and BCDFε activities of BSF-1, BCGFII could function as BCDFα ..."

Acknowledgments

I am truly grateful to *Dr. Karin Thalmeier*, GSF-Hämatologikum München, for critically reading some passages of the manuscript and for making many valuable useful suggestions and encouraging comments. She also kindly performed some PCR-based cytokine analyses and provided some of the photographs on page 160.

Dr. *Anna Mayer*, Dept. of Virology, Johannes Gutenberg University, Mainz, kindly provided the microphotographs of colonies of hematopoietic cells on page 159.

I would also like to accentuate the benevolent support received by many of the members of the Institute of Biochemistry who patiently sat through a term of cytokine lectures that I talked them into. They gave me the opportunity to piece together the information now scattered over many entries (see: KWN). In sometimes desperately trying to make a coherent story of what I have learned in compiling this dictionary they induced me to read even more material, in particular stuff that could not be included here due to space restrictions (*sic*!). These initial trial lectures and the relaxed atmosphere provided by my colleagues were most helpful in allowing me to organize and reorganize my thoughts and to propagate the idea that cytokines, hematology, immunology, and endocrinology can thrill you to bits.

Thanks are also due to *Dr. Walter Behrens* of Hoffmann-La Roche AG, Grenzach-Wyhlen who was most helpful in providing me with literature in the initial stages of my work.

Thanks and acknowledgments would be incomplete if I did not mention the very valuable services provided by the various BIOSCIENCE interest groups that I connected to many times during the preparation of this book. The INTERNET network is really a marvelous tool and I would wish more funds and more publicity to those who provide world-wide communication within the scientific community by organizing the network. For further information write to biosci @ net.bio.net.

Thanks are also due to Dr. Hans-Joachim Kraus of VCH Publishers who competently got things organized at the publisher's side. I apologize for the slight (?) shock, cytokine-mediated of course, he must have suffered in view of a manuscript that met the deadline but was much too long and could not be shortened. You can't have the cake and eat it. I also wish to acknowledge the work of Hans Jörg Maier of VCH Publishers who was responsible for the technical side of the production of the book.

Last but not least I would like to pay credit to the Scottish bag pipers whose music, unfortunately from records only, helped to soothe me during and after the many long hours spent in front of my word processor, ear-deep in printed material, or lost in data bank space.

München, Spring 1994

Contents

How to Use this Dictionary

Main entries and corresponding references: Abbreviations/acronyms usually constitute the main entries, with long forms cross-referenced to point to them. Many cytokines have been described under different names and/or have been renamed several times. Relevant information concerning the activities that gave rise to these alternative names, including references allowing the origins of these names to be traced, are listed under the appropriate acronyms/abbreviations. These references are not included in references of main entries.

Cross-references: "see:" or "see also:" are used to point to other entries that may be of interest in the current context. "»" within the text is used solely to indicate that the following term has its own entry, and I hope the "»" is unobtrusive enough.

Although I have tried to write entries that can be read without being caught in a maze of cross-references (see: KWN) I am afraid that this is a dictionary; so I felt that I could not present data for a factor in a single entry, leaving, in some cases, as mere cross-references 20 alternative names/acronyms representing at least 20 different activities. It may therefore be necessary after all to enter the maze. I have also used **non-lexemes** to point to relevant entries and I hope that this will make it easier to find things.

Is anyone out there who uses a dictionary for **browsing**? If someone feels inclined to do so, this may lead him/her on a journey truly reflecting all the intricacies of the cytokine network referred to in the preface. In this case I suggest that the journey be started by reading the entries "Cytokines", "Hematopoiesis", and "Neuroimmune network" first and see how far you get.

The alphabet is structured as follows: Brackets, hyphens, and chemical position indicators (ortho, o; meta, m; para, p; numbers) are ignored. **Names remain together and are not permuted:** Colony stimulating factors, Human interleukin for DA cells, T cell replacing factors etc.

The **order of entries with prefixed/suffixed numbers or Greek letters** is exemplified in the following list:

......

.3T3

5q$^-$

6-26

26 kDa Protein

......

Alpha-Interferon

......

Beta-CSF

......

CSF

CSF-1

CSF-1R

CSF-2

CSF-2α

CSF-2β

CSF-3

CSF-309

CSF-α

CSF-β

CSF-Box

......

IFN-β

IFN-β2

......

Abbreviations

To save space the following abbreviations/acronyms are used throughout the text without explanation. They all have entries of their own.

aFGF	acidic fibroblast growth factor
bFGF	basic fibroblast growth factor
CSF	colony-stimulating factor(s)
EGF	epidermal growth factor
Epo	erythropoietin
FGF	fibroblast growth factor
G-CSF	granulocyte colony-stimulating factor
GM-CSF	granulocyte/macrophage colony-stimulating factor
IFN	Interferon
IFN-α	Alpha-Interferon
IFN-β	Beta-Interferon
IFN-γ	Gamma-Interferon
IGF	insulin-like growth factor (IGF-I and IGF-II)
IL	Interleukin (IL1 to IL14)
LIF	leukemia inhibitory factor
M-CSF	macrophage colony-stimulating factor
NGF	nerve growth factor
PDGF	platelet-derived growth factor
TGF	transforming growth factor
TGF-α	Alpha-TGF
TGF-β	Beta-TGF
TNF	tumor necrosis factor
TNF-α	Alpha-TNF
TNF-β	Beta-TNF

Amino acids one and three letter codes

A	Ala	Alanine
C	Cys	Cysteine
D	Asp	Aspartic acid
E	Glu	Glutamic acid
F	Phe	Phenylalanine
G	Gly	Glycine
H	His	Histidine
I	Ile	Isoleucine
K	Lys	Lysine
L	Leu	Leucine
M	Met	Methionine
N	Asn	Asparagine
P	Pro	Proline
Q	Gln	Glutamine
R	Arg	Arginine
S	Ser	Serine
T	Thr	Threonine
V	Val	Valine
W	Trp	Tryptophan
Y	Tyr	Tyrosine

Journal citations

To save space the following abbreviations are used for the most frequently cited journals:

Advances in Experimental Medicine and Biology	AEMB
Advances in Cancer Research	ACR
Advances in Immunology	AI
Analytical Biochemistry	AB
Annals of the New York Academy of Science	ANY
Annual Review of Biochemistry	ARB
Annual Review of Cell Biology	ARC
Annual Review of Genetics	ARG
Annual Review of Immunology	ARI
Annual Review of Medicine	ARM
Annual Review of Neuroscience	ARN
Annual Review of Pharmacology and Toxicology	ARPT
Annual Review of Physiology	ARP
Annual Review of Respiratory Disease	ARRD
Anticancer Research	ACR
Archives of Biochemistry & Biophysics	ABB
Biochemical Biophysical Research Communication	BBRC
Biochemical Journal	BJ
Biochemistry	B
Biochemistry International	BI
Biochimica Biophysica Acta	BBA
BioTechnology	BT
British Medical Bulletin	BMB
Cancer Cells	CC
Cancer Genetics Cytogenetics	CGC
Cancer Research	CR
Cellular Immunology	CI
Cold Spring Harbor Symposium on Quantitative Biology	CSHSQB
Current Opinion in Cell Biology	COCB
Current Opinion in Immunology	COI
Current Topics in Developmental Biology	CTDB
Current Topics in Membranes and Transport	CTMT
Current Topics in Microbiology and Immunology	CTMI
EMBO Journal	EJ
Endocrinology	E
European Cytokine Network	ECN
European Journal of Biochemistry	EJB
European Journal of Clinical Investigation	EJCI
European Journal of Immunology	EJI
Experimental Cell Research	ECR
Experimental Hematology	EH
FASEB Journal	FJ
FEBS Letters	FL
Genes & Development	GD
Growth Factors	GF
Immunological Reviews	IR
Immunology Today	IT
International Journal of Cancer	IJC
International Journal of Cell Cloning	IJCC
International Journal of Immuno-pharmacology	IJI
Japanese Journal of Cancer Research	JJCR
Journal of Biochemical & Biophysical Methods	JBBM
Journal of Biochemistry	JB
Journal of Biological Chemistry	JBC
Journal of Cancer Research and Clinical Oncology	JCRCO
Journal of Cellular Biochemistry	JCBc
Journal of Cell Biology	JCB
Journal of Cell Science	JCS
Journal of Cellular Physiology	JCP
Journal of Clinical Investigation	JCI
Journal of Experimental Medicine	JEM
Journal of Immunological Methods	JIM
Journal of Immunology	JI
Journal of Steroid Biochemistry and Molecular Biology	JSBMB
Journal of the National Cancer Institute	JNCI
Lymphokine Research	LR
Methods in Enzymology	MiE
Molecular Cellular Biology	MCB
Molecular Endocrinology	ME
Nature (London)	N
New England Journal of Medicine	NEJM
Nucleic Acids ResearchNAR	
Oncogene	O
Oncogene Research	OR
Proceedings of the National Academy of Science (USA)	PNAS
Proceedings of the Society for Experimental Biology and Medicine	PSEBM
Progress in Clinical and Biological Research	PCBR
Progress in Growth Factor Research	PGFR
Science (Wash.)	S

Trends in Biochemical Sciences (Elsevier)	TIBS	Trends in Neurological Sciences (Elsevier)	TINS
Trends in Biotechnology (Elsevier)	TibTech	Trends in Pharmacological Sciences (Elsevier)	TIPS
Trends in Genetics (Elsevier)	TIG		

Introduction

In the 60ies analysis of the specific functions of B and T cells led to the conclusion that the maturation of B cells into immune-competent antibody-secreting cells could not be explained solely on the basis of antigens interacting with B cell receptors. Moreover, it was felt that the very complex reactions of immune responses and their regulation could not be explained sufficiently either by assuming the existence of specific and genetically determined recognition proteins mediating specific interactions between these and other cells.

It has been known for some time now that immune cells secrete a plethora of soluble factors which act in a very complex and highly specific manner and are responsible for the maintenance of an intricate communication network between these cells. These factors, designated lymphokines and interleukins, are responsible for their survival, proliferation, development, and differentiation and thus influence and control their individual biological activities. They can easily form concentration gradients, reach distant parts of an organism, address multiple target cells, and are easily subject to degradation or neutralization. The unique possibilities offered by these mechanisms of temporal and spatial regulation of factor activities by far exceed those provided by simple direct specific or unspecific cell-to-cell contacts

The real break-through that rang in the age of factors, so to speak, came with the development of techniques allowing T lymphocytes to be kept indefinitely in vitro provided their growth media was supplemented with sterile conditioned media in which other stimulated T cells had been grown previously. The growth-promoting activity of T cells was soon attributed to a specific T cell growth factor (TCGF) that was present in the conditioned medium. This factor which was soon purified and characterized biochemically was renamed in 1979 and is now known as Interleukin 2 (IL2).

Initially immunologists were surprised and even frustrated by the observation that similar, if not identical, supernatants of cells were capable of eliciting a number of different biological responses, depending entirely on the test system employed. In a relatively short time the biochemical analysis of these phenomena resulted in a rather voluminous catalogue of stimulatory and inhibitory factors, many of which acted either synergistically with, or antagonized the activities of, other factors. The sheer size of this catalogue made it difficult even for insiders not to lose track of things and the problem was certainly not lessened by the resulting (and still rising) flood of exotic acronyms.

The many lymphokines, monokines, interleukins, and activating or inhibitory factors, growth factors, and hormones are now generally and collectively known as cytokines. The original nomenclature at most reflects the context in which these factors had been identified initially. In many instances the old names refer to bioactivities rather than biochemically well-defined distinct protein factors. In some instances these names suggest relationships or functions that do not exist or do not reflect main activities. Latest additions to the growing number of terms, such as the recently discovered intercrines (and even more recently renamed chemokines) exemplify the teething troubles and pains of nomenclature.

Soon after the identification of the first soluble cellular mediator substances it was found that their action was not restricted to cells of the immune system; cytokines affect many different types of cells and may, in fact, act upon all known cell types. Many cytokines initially identified in the immune system are also produced by a variety of non-immune cells.

As a rule the synthesis of cytokines is inducible although some factors are known to be produced constitutively. Endogenous rhythms in the expression of cytokines and age-related alterations have also been observed. Appropriately activated cell usually synthesize many different factors at the same time. Most of these cells also express specific receptors either inducibly or constitutively so that they can react to a wide spectrum of factors. The initial concept of "one producer, one cytokine, one target cell" has certainly turned out to be wrong. Even those few factors that now appear to act specifically with regard to function and cell type will most certainly be found to have other activities as well.

Our knowledge of the physiological, cell biological, biochemical, and molecular biological properties and actions of cytokines and their receptors has grown tremendously over the last decade. What has begun in the 60ies as a break-through in deciphering the language of immunology, leading within a decade to a virtual "immunological tower of Babel", has developed to such an extent that individual words of the biological regulatory language and also some structures of its grammar are now before our eyes.

Many factors the existence of which was only conjecture in the early days of cytokine research, have now been identified as proteins or peptides. The use of molecular biological methods and the application of powerful biochemical separation techniques have made possible the cloning of genes encoding many of these factors so that they are now available in large quantities as highly purified recombinant proteins. The use of these modern techniques has also enlarged considerably our knowledge of the corresponding cytokine receptors. The value of these techniques becomes apparent if one considers that most of these factors are active at nano- or pico- to attomolar concentrations. Even today their isolation from blood, tissues, or conditioned culture media is not a mean task and represents a problem that stretches a biochemist's knowledge to the limits.

In many instances the discovery, isolation, cloning, and analysis of cytokines and other factors in non-human organisms has been the incentive to search for, and to find, similar or identical factors in humans. This is not surprising if one considers the overwhelming evidence of functional and structural relationships between cytokines from different species that has emerged during the past several years. It is therefore no longer surprising that even research on yeasts can provide important clues about cytokines in higher organisms that have direct implications for clinical and therapeutical matters. It is pleasing to record that the field is converging.

I believe that, in view of the increasing potentials of cytokines for medical research and application, cytokine research offers an exceptional example of the very important role of basic research: the long way from the vaults and basements of basic research into the first floor of therapeutic application has become markedly shorter.

Cytokine research has benefited tremendously from many fields of modern biological sciences. The use of molecular biological techniques now allows the analysis of these factors at a level of sophistication seemingly unattainable a decade ago. Gene cloning, protein engineering, and directed mutagenesis not only allow cytokines to be produced in large quantities but also to construct "streamlined" cytokines and receptors and also hybrid cytokines with novel and sometimes improved biological, pharmacological or clinically useful activities. The use of transgenic animals, gene targeting techniques, cell ablation procedures, and antisense techniques is Watson's severe test cited at the beginning of the preface: not only do these techniques allow individual factors and their activities to be identified with precision but also their biological activities to be modulated at will. Future developments may include the very interesting possibilities of synthesizing specific antagonists for individual cytokines.

Further developments which have influenced cytokine research to a large extent and will continue to do so include improved techniques of tissue culture such as the development of chemically defined media, the possibilities offered by cell-type specific monoclonal antibodies, which allow individual populations of cells to be typed specifically, and also the development of cell lines whose growth depends entirely on the exogenous supply of specific growth factors.

The introduction of the techniques and concepts mentioned previously has already had a tremendous impact on our understanding of mechanisms and specific cause-and-effect relationships of disease. After all, our first attempts of using cytokines in a clinical setting have already yielded some successes although much remains to be learned. It should be kept in mind that the entire spectrum of indications for cytokine therapies has not yet been recognized or realized. The use of cytokines and specific cytokine inhibitors will permit the specific manipulation of almost all functions of the immune system and non-immune cells. Since most immune and non-immune defects involve the action of one or more cytokines in one way or another clinicians of many different medical disciplines will be offered a plethora of entirely new ways of diagnosis and treatment of illnesses.

The importance of cytokines in medicine is perhaps best illustrated by two points. One is the observation that many types of tumor cells also produce one or the other cytokine and use them to enhance their growth and to influence metastatic processes, including tumor angiogenesis. The other point refers to recent findings of how viruses subvert the host immune system by mimicking or blocking cytokine activities.

Some people may think that, in spite of all the progress that has been made, the current situation is still characterized best by describing it as a mixture of superficial knowledge and terribly precise incompetence. Although we are only just beginning to comprehend the full implication of cytokines in both normal and pathological processes one of the triumphs of cytokine research is certainly the provision of a rationale for clinical trials now under way to treat some human malignancies by cytokine gene transfer.

Clarification of how distinct cytokines contribute to the problem and/or solution of pathological situations are expected to guide efforts to diagnose and treat pathological conditions in ways oriented to the molecular and cellular pathophysiology of individual diseases to an extent previously thought to be impossible. With several cytokines in clinical use and others being in various stages of clinical trials the prospects look good and the future will be exciting.

Entries with Number Prefixes

IaIF: *MHC class II (Ia) inducing factor* This poorly characterized factor of ≈ 50 kDa is produced by activated lymphoma cells (A20.1 B cell lymphoma) following cross-linking of membrane IgG2a or IgD. It induces the expression of major histocompatibility class II (Ia) cell surface antigens on B cells. At least one factor with IaIF activity is identical with » IL4.

Justement LB et al Production of multiple lymphokines by the A20.1 B cell lymphoma after cross-linking of membrane Ig by immobilized anti-Ig. JI 143: 881-89 (1989); **Kim KJ & Finnegan A** Induction of IgG2a secretion from mIgG2a$^+$ B lymphoid tumor cells by BCDF present in several antigen-specific T helper clones. CI 110: 149-62 (1987); **Lee F et al** Isolation and characterization of a mouse interleukin cDNA clone that expresses B cell stimulatory factor 1 activities and T cell- and mast cell-stimulating activities. PNAS 83: 2061-5 (1986); **Stuart PM et al** Induction of class I and class II MHC antigen expression on murine bone marrow-derived macrophages by IL4 (B cell stimulatory factor 1). JI 140: 1542-7 (1988)

Ia-inducing factor: see: IaIF.

1xN/2b: A murine stromal cell line obtained from long-term Whitlock-Witte-type bone marrow cultures (see: LTBMC) that is absolutely dependent on » IL7 for growth and viability. The cells are used to assay this factor (see also: Bioassays). IL7 activity is determined by measuring the incorporation of ^{3}H thymidine into the newly synthesized DNA of proliferating cells. Cell proliferation can be determined also by employing the » MTT assay. An alternative and entirely different method of detecting IL7 is » Message amplification phenotyping.

Namen AE et al Stimulation of B cell progenitors by cloned murine interleukin-7. N 333: 571-73 (1988); **Namen AE et al** B cell precursor growth-promoting activity: purification and characterization of a growth factor active on lymphocyte precursors. JEM 167: 988-1002 (1988); **Park LS et al** Murine interleukin 7 (IL7) receptor. Characterization of an IL7-dependent cell line. JEM 171: 1073-89 (1990); **Whitlock CA et al** Bone marrow stromal cell lines with lymphopoietic activity express high levels of a pre B neoplasia-associated molecule. Cell 48: 1009-14 (1987)

2B11-BCGF: *2B11 B cell growth factor* This poorly characterized B cell growth factor (see: BCGF) of ≈ 20 kDa stimulates the proliferation of rat astrocytes.

Benveniste EN et al Human B cell growth factor enhances proliferation and glial fibrillary acidic protein gene expression in rat astrocytes. Int. Immunol. 1: 219-28 (1989) [erratum in Int. Immunol. 1: 555]

2D9: A subclone of the human glioblastoma cell line 86HG39. Expression of indoleamine 2,3-dioxygenase, which leads to the production of N-formyl kynurenine from L-tryptophan, has been used as a sensitive bioassay for » IFN-γ.

Däubener W et al A new, simple bioassay for human IFN-γ. JIM 168: 39-47 (1994)

2E8: (ATCC TIB-239 2E8) 2E8 is a B lymphocyte cell line derived from the bone marrow of BALB/c-xid strain mice (see also: Immunodeficient mice). The cells express complete surface IgM molecules and class I MHC antigens. A proportion of cells also express class II MHC (59.0%) and BP-1 (72.5%) antigens. 2E8 cells depend on » IL7 for growth and survival and can be used as an indicator cell line for detecting and quantitating this factor. 2E8 cells respond to human or murine IL7, but the latter may be preferable for routine passage and maintenance. The growth of 2E8 cells in methylcellulose assays has been shown to be unaffected by » IFN-γ, which partially or completely inhibits the replication of other lymphoid lines.

Pietrangeli CE et al Stromal cell lines which support lymphocyte growth: characterization, sensitivity to radiation and responsiveness to growth factors. EJI 18: 863-72 (1988); **Gimble JM et al** Modulation of lymphohematopoiesis in long-term cultures by γ interferon: direct and indirect action on lymphoid and stromal cells. EH 21: 224-30 (1993); **Ishihara K et al** Stromal-cell and cytokine-dependent lymphocyte clones which span the pre-B to B cell transition. Dev. Immunol. 1: 149-61 (1991)

3B6-IL1: This factor is produced by an Epstein-Barr virus-positive B lymphoblastoid cell line (721 LCL; subclone 3B6). It has a biological activity similar to that of » IL1 and was originally believed to be identical with » IL1. It has been shown now that it is identical with » ADF (adult T cell leukemia-derived factor) and hence with » Thioredoxin.

Rimsky L et al Purification to homogeneity and NH2-terminal amino acid sequence of a novel interleukin 1 species derived from a human B cell line. JI 136: 3304-10 (1986); **Wakasugi H et al** Epstein-Barr virus-containing B cell line produces an interleukin 1 that it uses as a growth factor. PNAS 84: 804-8 (1987); **Wollman EE et al** Cloning and expression of a cDNA for human thioredoxin. JBC 263: 15506-12 (1988)

3CH134: An » ERG (early response gene) whose transcription is rapidly and transiently stimulated by serum growth factors. It is expressed predominantly in murine lung and encodes a protein of 367 amino acids with no significant homology to other known proteins. The protein is also found in serum-stimulated Balb/c » 3T3 cells. The gene is also expressed following cross-linking of membrane immunoglobulin on B lymphocytes in several B lymphoma cell lines.

3CH134 encodes a protein tyrosine phosphatase. 3CH134 is the murine homologue of the human » CL100 gene.

Charles CH et al cDNA sequence of a growth factor-inducible immediate early gene and characterization of its encoded protein. O 7: 187-90 (1992); **Charles CH et al** The growth factor-inducible immediate-early gene 3CH134 encodes a protein-tyrosine-phosphatase. PNAS 90: 5292-6 (1993); **Mittelstadt PR & De Franco AL** Induction of early response genes by cross-linking membrane Ig on B lymphocytes. JI 150: 4822-32 (1993)

3T3: An aneuploid murine cell line with fibroblast-like morphology. The name 3T3 derives from very early studies of the behavior of primary murine cell cultures which were found to grow much better and longer if subcultivated at low cell density (3×10^5 cells/dish) every three days (3T3 protocol). Accordingly, „3T3“ cell lines have been established from primary cultures of murine embryos of different strains, e. g. BALB/c and Swiss mice

3T3 cells have been used as a standard cell line for many biochemical and cell biological studies, including studies of growth factor, because they have a wide range of receptors and second messenger systems. Many variant sublines of these cells are now available. 3T3 cells show a strong contact inhibition *in vitro*, i. e. under standard conditions they cease growing as soon as cell-to-cell contacts are established (see also: Cell culture). 3T3 cells are used, for example, because they are very sensitive towards transforming growth factors (see: TGF-α and TGF-β). In contrast to the parent cells which require a solid matrix for growth transformed 3T3 cells can be grown as colonies in soft agar or other semisolid media, e. g. those containing methyl cellulose.

3T3 fibroblasts are also used to assay » PDGF; the cells respond equally well to all three molecular forms of PDGF. Heparin-binding growth factors (see: HBGF) induce proliferation of 3T3 fibroblasts almost as good as PDGF. The cells also respond to other growth factors, but to a much lesser extent than to PDGF. Stimulation of proliferation of 3T3 cells by » EGF or TGF-β is ≈ 20 % that of serum. Synergistic effects between EGF and TGF-β have been observed.

A very sensitive line of 3T3 cells, known as ***NIH3T3*** and established at the National Institutes of Health from Swiss mice. This cell line is easily transformed by the introduction of functional » oncogenes and is used frequently for studies of the transforming activities of oncogenes and their relationship to growth factor/growth factor receptor expression and signal transduction pathways.

The induction of cell proliferation following treatment with » PDGF is the basis of a bioassay for this factor.

Aaronson SA & Todaro GJ Development of 3T3-like lines from Balb/c mouse embryo cultures: transformation susceptibility to SV40. JCP 72: 141-48 (1968); **Heidaran MA et al** Transformation of NIH3T3 fibroblasts by an expression vector for the human epidermal growth factor precursor. O 5: 1265-70 (1990); **Pruss RM & Herschman HR** Variants of 3T3 cells lacking mitogenic response to epidermal growth factor. PNAS 74: 3918-21 (1977); **Terwilliger E & Herschman HR** 3T3 variants unable to bind epidermal growth factor cannot complement in coculture. BBRC 118: 60-4 (1984);**Todaro GJ & Green H** Quantitative studies of the growth of mouse embryo cells in culture and their development into established lines. JCB 17: 299-313 (1963); **Wigler M et al** Biochemical transfer of single-copy eukaryotic genes using total cellular DNA as donor. Cell 14: 725-31 (1978); **Van Zoelen EJJ et al** Transforming growth factor-β and retinoic acid modulate phenotypic transformation of normal rat kidney cells induced by epidermal growth factor and platelet-derived growth factor. JBC 261: 5003-9 (1986); **Van Veggel JH** et al PC13 embryonal carcinoma cells produce a heparin-binding growth factor. ECR 169: 280-6 (1987); **Van Zoelen EJJ et al** PDGF-like growth factor induces EGF-potentiated phenotypic transformation of normal rat kidney cells in the absence of TGFβ. BBRC 141: 1229-35 (1986)

3-10C: Name of a cDNA the expression of which is induced at least 10-fold in peripheral human blood leukocytes by staphylococcal enterotoxin A. It encodes a protein of 99 amino acids that resembles human β-thromboglobulin (42 % homology) and shows extensive homology to members of the » chemokine family of proteins. 3-10C is strongly related to » LUCT (lung carcinoma-derived chemotaxin). The protein encoded by 3-10C is identical with » MDNCF (monocyte-derived neutrophil chemotactic factor) and » NAF (neutrophil-activating factor) and hence identical with » IL8.

Schmid J & Weismann C Induction of mRNA for a serine protease and a β-thromboglobulin-like protein in mitogen-stimulated human leukocytes. JI 139: 250-6 (1987); **van Damme J et al** A novel NH_2-terminal sequence-characterized human monokine possessing neutrophil chemotactic, skin-reactive, and granulocytosis-promoting activity. JEM 167: 1364-76 (1988); **Zipfel PF et al** Mitogenic activation of human T cells induces two closely related genes which share structural similarities with a new family of secreted factors. JI 142: 1582-90 (1989)

4-HC: *4-hydroperoxycyclophosphamide* The mutagenic oxazaphosphorine compound ***Cyclophosphamide*** is a bifunctional alkylating agent and a potent animal teratogen inducing a variety of malformations. Cyclophosphamide-induced DNA damage has been implicated as a primary mechanism underlying teratogenesis . Cyclophosphamide must be metabolically converted into the active 4-hydroxylated intermediate 4-hydroxycyclophosphamide by liver microsomal monooxygenases to exert its effects. 4-HC is a preactivated derivative of cyclophosphamide that hydrolyzes spontaneously in aqueous solution to form 4-hydroxycyclophosphamide. ***Mafosfamide*** (ASTA Z 7557) is a cyclophosphamide derivative that rapidly generates 4-HC after aqueous dissolution.

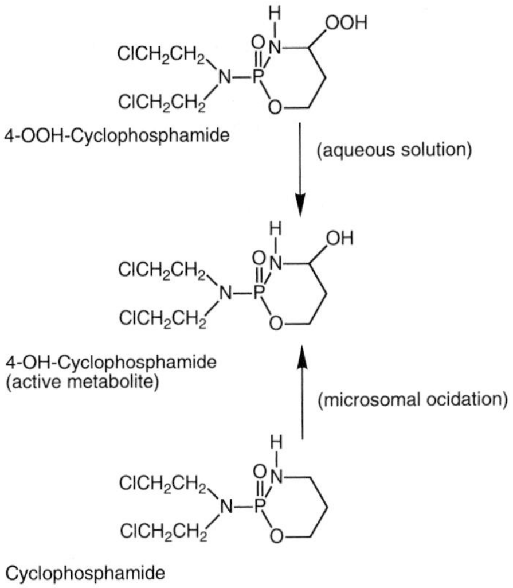

4-OOH-Cyclophosphamide

(aqueous solution)

4-OH-Cyclophosphamide (active metabolite)

(microsomal ocidation)

Cyclophosphamide

Pretreatment of bone marrow cells with 4-HC or other compounds such as » fluorouracil diminishes or totally inhibits a variety of hematopoietic progenitor cells (see: Hematopoiesis) and appears to spare cells which are in a slowly- or non-cycling state (see also: Cell cycle). Cellular aldehyde dehydrogenases, known to catalyze the oxidation of mafosfamide metabolites to relatively nontoxic compounds, have been implicated as one determinant of differential sensitivity of cells towards 4-HC. Inhibitors of these enzymes such as diethyldithiocarbamate, diethylaminobenzaldehyde, or cyanamide potentiate the cytotoxic action of mafosfamide towards hematopoietic progenitors, i. e. they prevent protection. Preincubation with » IL1 and TNF-α, known to stimulate expression of aldehyde dehydrogenases, can protect early progenitor cells from phenylketophosphamide, an analog of 4-HC which is resistant to inactivation by aldehyde dehydrogenase. This protective effect must be due, therefore, to other, as yet unidentified, mechanisms. A reduction of glutathione levels has also been suggested to play a role in the differential sensitivity of cells towards exposure to 4-HC.

Cells surviving 4-HC treatment can be assayed in long-term bone marrow cultures (see: LTBMC), by » limiting dilution analysis, and/or » colony formation assays and these studies show that sensitivity to the drug can be used to distinguish primitive progenitor cells (measured, for example as » CFU-S or as » CD34-positive cells; see also: hematopoietic stem cells) from those cells committed to any of the hematopoietic lineages. In the mouse treatment with 4-HC leads to a dose-dependent killing and differential sensitivity , with » BFU-E being most sensitive and CFU-GEMM being most resistant cell types. It has been shown *in vitro* that treatment of human marrow cells with 4-HC also spares, to some extent, early hematopoietic progenitors such as » CFU-GEMM and » BFU-E . Although treated marrow cells show marked reduction in committed progenitor cell frequency, and often total absence of detectable progenitors, there is no significant loss of marrow reconstituting ability (see: MRA). The examination of hematopoietic progenitors such as » BFU-E, » CFU-GEMM, and also » CFU-GM has shown that these assayable cell types do not predict pluripotent stem cell survival *in vitro* and that detection of their presence does not reflect marrow reconstituting ability. Transplantable stem cells from human cord blood expressing the early hematopoietic marker antigen » CD34 have been shown to be resistant to treatment with 4-HC; they also do not give rise to colonies when plated in clonogenic assays. Nevertheless, the availability of enriched population of primitive hemopoietic progenitors provides an opportunity to study the interactions between these cells, cytotoxic drugs, and purified hematopoietic growth factors (see also: Hematopoietins). Several » cytokines are known to protect early hematopoietic cells from both irradiation and chemotherapy damage. The mechanism of action of these cytokines and which cells are protected is not known. Growth of 4HC-resistant (unresponsive) highly immature cells in LTBMC certainly involves as yet undefined interactions with marrow stroma in addition to known hematopoietic growth factors. The cytokines » IL1, » TNF-α, » TGF-β3, » IL3, and » IL6 can protect hematopoietic cells capable of repopulating irradiated long-term bone marrow stromal cultures from 4-HC. These » LTRCs (long-term culture-initiating cells) are closely related to marrow reconstituting stem cells (see also: MRA, marrow repopulating ability for an assay system). Since dose intensity is emerging as a crucial determinant of success in cytotoxic cancer therapy and myelosuppression presents as one of its major complications the use of cytokine administration in combination with cytotoxic drug treatment may help to overcome this problem.

Cyclophosphamide derivatives are widely used at high doses for *ex vivo* chemoseparation or ***bone marrow* purging**. Suitable bone marrow is taken from the patient at a propitious time in the history of the disease and usually cryopreserved. This purging strategy is aimed to eliminate residual occult tumor cells from marrows used for autologous bone marrow transplantation (ABMT) after ablative chemotherapy and/or total-body irradiation (TBI), while sparing normal hematopoietic stem cells. Autologous bone marrow transplants with preserved hemopoietic support capacity can repopulate the hematological system of patients but treatment of the bone marrow graft to eliminate residual tumor cells prior to reinfusion is not always effective and can be associated also with delayed engraftment and prolonged cytopenias. Although cytotoxic drug exposure usually leads to a reversible reduction of the size of the stem cell compartments with recovery of compartment size resulting from a transiently increased proliferative activity of surviving stem cells an irreversibly decreased proliferative potential of pluripotent stem cells has been observed after some cytotoxic agents, in particular after repeated exposure. Delays in the return of peripheral blood elements are assumed to result from damage to, or loss of, » hematopoietic stem cells responsible for hematological recovery.

Abboud M et al Study of early hematopoietic precursors in human cord blood. EH 20: 1043-7 (1992); **Beran M et al** Regrowth of granulocyte-macrophage progenitor cells (GM-CFC) in suspension cultures of bone marrow depleted of GM-CFC with 4-hydroperoxycyclophosphamide (4-HC). Eur. J. Haematol. 39: 118-24 (1987); **Bhalla K et al** Effect of combined treatment with interleukin-3 and interleukin-6 on 4-hydroperoxycyclophosphamide-mediated reduction of glutathione levels and cytotoxicity in normal and leukemic bone marrow progenitor cells. Leukemia 6: 814-9 (1992); **Gale RP** Antineoplastic chemotherapy myelosuppression: mechanisms and new approaches. EH 16: s3-7 (1985); **Gordon MY et al** 4-Hydroperoxycyclophosphamide inhibits proliferation by human granulocyte-macrophage colony-forming cells (GM-CFC) but spares more primitive progenitor cells. Leuk. Res. 9: 1017-21 (1985); **Jones RJ et al** Purging with 4-hydroperoxycyclophosphamide combinations. PCBR 377: 1-9 (1992); **Kohn FR & Sladek NE** Effects of aldehyde dehydrogenase inhibitors on the *ex vivo* sensitivity of murine late spleen colony-forming cells (day-12 CFU-S) and hematopoietic repopulating cells to mafosfamide (ASTA Z 7557). Biochem. Pharmacol. 36: 2805-11 (1987); **Komatsu N et al** Survival of highly proliferative colony forming cells after treatment of bone marrow cells with 4-hydroperoxycyclophosphamide. CR 47: 6371-6 (1987); **Lemoli RM et al** TGF-β 3 protects normal human hematopoietic progenitor cells treated with 4-hydroperoxycyclophosphamide *in vitro*. EH 20: 1252-6 (1992); **Mirkes PE et al** Identification of cyclophosphamide-DNA adducts in rat embryos exposed *in vitro* to 4-hydroperoxycyclophosphamide. Chem. Res. Toxicol. 5: 382-5 (1992); **Moreb J et al** Role of aldehyde dehydrogenase in the protection of hematopoietic progenitor cells from 4-hydroperoxycyclophosphamide by interleukin 1 β and tumor necrosis factor. CR 52: 1770-4 (1992); **Moreb J & Zucali JR** The therapeutic potential of interleukin-1 and tumor necrosis factor on hematopoietic stem cells. Leuk. Lymphoma 8: 267-75 (1992); **Mortensen BT et al** Mafosfamide (ASTA-Z-7654) *in vitro* treatment of bone marrow does not eradicate Philadelphia-positive cells in chronic myeloid leukemia. Eur. J. Haematol. 41: 218-22 (1988); **Porcellini A et al** Effect of two cyclophosphamide derivatives on hemopoietic progenitor cells and pluripotential stem cells. EH 12: 863-6 (1984); **Rowley SD et al** Human multilineage progenitor cell sensitivity to 4-hydroperoxycyclophosphamide. EH 13: 295-8 (1985); **Rowley SD et al** CFU-GM content of bone marrow graft correlates with time to hematologic reconstitution following autologous bone marrow transplantation with 4-hydroperoxycyclophosphamide-purged bone marrow. Blood 70: 271-5 (1987); **Rowley SD et al** Hematopoietic precursors resistant to treatment with 4-hydroperoxycyclophosphamide: requirement for an interaction with marrow stroma in addition to hematopoietic growth factors for maximal generation of colony-forming activity. Blood 82: 60-5 (1993); **Sahovic EA et al** Role for aldehyde dehydrogenase in survival of progenitors for murine blast cell colonies after treatment with 4-hydroperoxycyclophosphamide *in vitro*. CR 48: 1223-6 (1988); **Siena S et al** Effects of *in vitro* purging with 4-hydroperoxycyclophosphamide on the hematopoietic and microenvironmental elements of human bone marrow. Blood 65: 655-62 (1985); **Slott VL & Hales BF** Role of the 4-hydroxy intermediate in the *in vitro* embryotoxicity of cyclophosphamide and dechlorocyclophosphamide. Toxicol. Appl. Pharmacol. 92: 170-8 (1988); **Winton EF & Colenda KW** Use of long-term human marrow cultures to demonstrate progenitor cell precursors in marrow treated with 4-hydroperoxycyclophosphamide. EH 15: 710-4 (1987); **Zucali JR et al** Protection of cells capable of reconstituting long-term bone marrow stromal cultures from 4-hydroperoxycyclophosphamide by interleukin 1 and tumor necrosis factor. EH 20: 969-73 (1992)

4q cytokine family: A name sometimes given to the » chemokine family of cytokines which, among other factors, comprises a variety of factors with chemotactic (see: Chemotaxis) and pro-inflammatory activities (see: Inflammation) encoded by genes located on human chromosome 4q.
Rollins BJ et al Assignment of the human small inducible cytokine A2 gene, SCYA2 (encoding JE or MCP-1) to 17q11.2-12 – evolutionary relatedness of cytokines clustered at the same locus. Genomics 10: 489-92 (1991)

5c8: This $CD4^+$ T cell-restricted surface activation protein, also called ***T-Bam***, has been shown to be a component of the contact-dependent helper signal to B cells. 5c8 is the ligand for » CD40.
Lederman S et al Molecular interactions mediating T-B lymphocyte collaboration in human lymphoid follicles. Roles of T cell-B cell-activating molecule (5c8 antigen) and CD40 in contact-dependent help. JI 149: 3817-26 (1992); **Lederman S et al** Identification of a novel surface protein on activated $CD4^+$ T cells that induces contact-dependent B cell differentiation (help). JEM 175: 1091-101 (1992)

5-Fluorouracil: see: Fluorouracil.

5-FU: see: Fluorouracil.

5q⁻ syndrome: The 5q⁻ syndrome is a chromosomal aberration characterized by deletions of parts of the long arm of human chromosome 5. Affected patients develop myelodysplastic syndromes (MDS; refractory anemia), a heterogeneous group of therapeutically refractory pre-leukemic disorders resulting from a clonal hemopoietic stem cell disorder often associated with cytogenetic abnormalities and leading to varying degrees of blast cell accumulation (monocytosis in the blood and myelomonocytic dysplasia in the bone marrow) and even transformation to leukemia. Deletions of genes in the 5q region, initially described as the hallmark of a unique type of MDS with refractory anemia subsequently was demonstrated to occur in 30% of patients with MDS, in 50% of patients with acute myelogenous leukemia arising secondary to MDS or to prior chemotherapy, in 15% of *de novo* acute myelogenous leukemias, and in 2% of *de novo* acute lymphocytic leukemias.
Bone marrow cultures (see also: LTBMC, long-term bone marrow culture) from patients with 5q⁻ syndrome display an abnormal pattern of colony formation *in vitro* with reduced production or even a complete lack of various early hematopoietic progenitor cell types such as » BFU-E, » CFU-GM and » CFU-Meg (see also: colony formation assay).
Chromosomal aberrations of the long arm of chromosome 5 are of particular interest in cytokine research because several genes encoding cytokines or their receptors have been mapped to this region. Loci include genes for » aFGF, » GM-CSF, » IL3, » IL4, » IL5, IL9, IL12 (both subunits), the gene encoding IRF1 (interferon regulatory factor 1; see: IRS interferon response sequence) at 5q31.1, and the genes encoding the receptor for » M-CSF (see: *fms* oncogene) and » PDGF, and the FGF receptor, FGFR-4. Some of these genes lie in close vicinity to each other. The genes encoding IL4 and IL5 are located on a DNA fragment of ≈ 100-170 kb. The genes encoding GM-CSF and IL3 are only 9 kb apart and have the same orientation. The receptor gene for M-CSF is separated from the receptor gene for PDGF by only 800 bp. Both genes map in the vicinity of another receptor gene with intrinsic tyrosine-specific protein kinase *flt*4 (see: *flt*). A similar close vicinity of the receptor genes for M-CSF and PDGF is also seen in mouse chromosome 18 which is syntenic with human chromosome 5.
It has been postulated that some of the genes deleted in 5q⁻ deletions, involving, for example, the genes coding for GM-CSF, IL3, aFGF, and the receptors for M-CSF and PDGF, may somehow be connected with the various disorders of the hematopoietic system in the affected patients. However, the importance of these deletions is still controversial because many of these patients still retain a second unaffected chromosome.
Bartram CR Molecular genetic aspects of myelodysplastic syndromes. Hematol. Oncol. Clin. North. Am. 6: 557-70 (1992); **Boultwood J et al** Loss of both CSF1R (*fms*) alleles in patients with myelodysplasia and a chromosome 5 deletion. PNAS 88: 6176-80 (1991); Growth and characteristics of bone marrow colonies from patients with 5q⁻ syndrome. Blood 66: 463-65 (1985); **Bunn HF** 5q- and disordered hematopoiesis. Clin. Haemat. 15: 1023-35 (1986); **Eccles MR** Genes encoding the platelet-derived growth factor (PDGF) receptor and colony-stimulating factor 1 (CSF-1) receptor are physically associated in mice as in humans. Gene 108: 285-88 (1991); **Frolova EI et al** Linkage mapping of the human CSF2 and IL3 genes. PNAS 88: 4821-24 (1991); **Huebner K et al** The human gene encoding GM-CSF is at 5q21-32, the chromosome region deleted in the 5q⁻-anomaly. S 230: 1282-5 (1985); **Le Beau MM et al** The interleukin 3 gene is located on human chromosome 5 and is deleted in myeloid leukaemias with a deletion of 5q. PNAS 84: 5913-17 (1987); **Le Beau MM et al** Interleukin 4 and interleukin 5 map to human chromosome 5 in a region encoding growth factors and receptors and are deleted in myeloid leukaemias with a del(5q). Blood 73: 647-50 (1989); **Le Beau MM et al** Evidence for the involvement of GM-CSF and *fms* in the deletion (5q) in myeloid disorders. Science 231: 984-7 (1986); **Le Beau MM et al** Cytogenetic and molecular delineation of the smallest commonly deleted region of chromosome 5 in malignant myeloid diseases. PNAS 90: 5484-8, (1993); **Mathew P et al** The 5q-syndrome: a single-institution study of 43 consecutive patients.

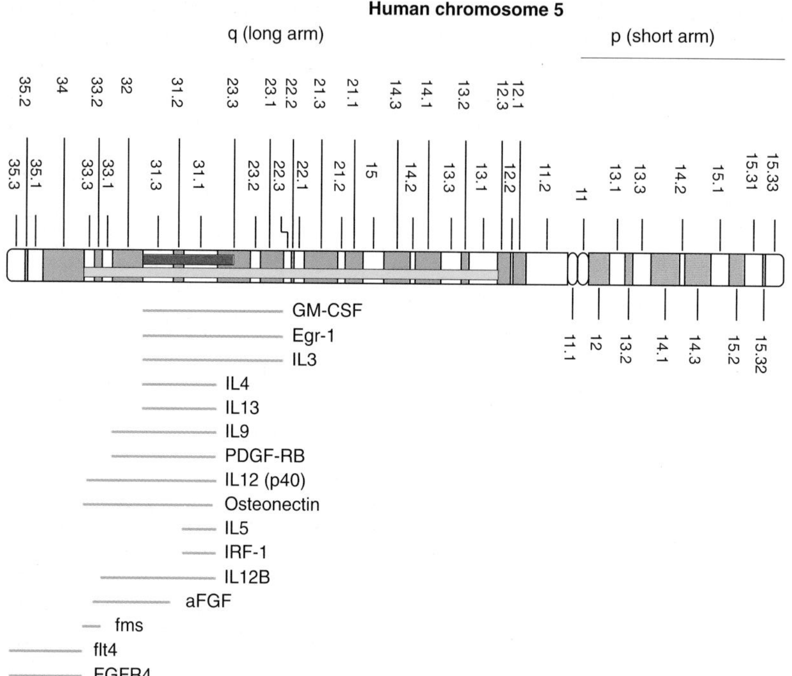

Genetic map of human chromosome 5 with special reference to cytokine genes.
At least 15 genes encoding cytokines or cytokine receptors map to the long arm of human chromosome 5. Interstitial deletions within the long arm of chromosome 5 removing some cytokine genes are a frequent cytogenetic abnormality in patients with a unique type of myelodysplastic syndrome with refractory anemia. These deletions have been shown affect one or both alleles. The proximal and distal breakpoints vary from patient to patient. The smallest commonly deleted segment is band 5q31 (indicated by a red bar). However, deletions can cover the region from 5q13 to 5q33 (indicated by a green bar).
The order of some genes on chromosome 5q has been determined in greater detail.
The known order of genes is: Tel IL9 - - - GM-CSF - IL3 - - - - IRF-1 - IL5 - IL4 - IL13 - Cen. The distance between IRF-1 and IL5 is ≈ 200 kb. The genes encoding IL4 and IL5 are located on a DNA fragment of ≈ 110-180 kb. The distance between IL4 and IL13 is ≈ 50 kb. The genes encoding GM-CSF and IL3 are separated by 9 kb apart and have the same orientation. The GM-CSF/IL3 loci are separated from IL4/IL5 by a minimum of 600 kb. The receptor genes for M-CSF and PDGF are separated by 800 bp.

Blood 81: 1040-5 (1993); **Nagarajan L et al** Molecular anatomy of a 5q interstitial deletion. Blood 75: 82-7 (1990); **Nienhuis AW et al** Expression of the human c-*fms* proto-oncogene in hematopoietic cells and its deletion in the 5q syndrome. Cell 42: 421-8 (1985); **Nimer SD & Golde DW** The 5q⁻ abnormality. Blood 70: 1705-12 (1987); **Pettenati MJ et al** Assignment of CSF-1 to 5q32. 1: evidence for clustering of genes regulating hematopoiesis and for their involvement in the deletion of the long arm of chromosome 5 in myeloid disorders. PNAS 83: 2970-74 (1986); **Sokal G et al** A new hematologic syndrome with a distinct phenotype: the 5q⁻ chromosome. Blood 46: 519-33 (1975); **Stock W et al** Characterization of yeast artificial chromosomes containing interleukin genes on human chromosome 5. Cytogenet. Cell Genet. 61: 263-5 (1992); **Sutherland GR et al** Interleukin 5 is at 5q31 and is deleted in the 5q⁻ syndrome. Blood 71: 1150-52 (1988); **Tanaka K et al** Parallel loss of c-*fms* and GM-CSF genes in myeloid leukemias with 5q-chromosome. Rinsho Ketsueki 32: 931-37 (1991); **Thangavelu M et al** A physical and genetic linkage map of the distal long arm of human chromosome 5. Cytogenet. Cell Genet. 59: 27-30 (1992); **Thornton DE et al** Characterization of the 5q⁻ breakpoint in an acute non-lymphocytic leukemia patient using pulsed-field gel electrophoresis. Am. J. Med. Genet. 41: 557-69 (1991); **Tinegate H et al** The 5q-syndrome: an underdiagnosed form of macrocytic anemia. Brit. J. Haemat. 53: 103-10 (1983); **Van der Berghe H et al** The 5q⁻-anomaly. CGC 17: 189-255 (1985); **van Leeuwen BH et al** Molecular organization of the cytokine gene cluster involving the human IL3, IL4, IL5, and GM-CSF genes on human chromosome 5. Blood 73: 1142-48 (1989); **Warrington JA et al** A radiation hybrid map of 13 loci on the long arm of chromosome 5. Genomics 11: 701-8 (1991); **Westbrook CA et al** Report of the chromosome 5 workshop. Genomics 10: 1105-09 (1991); **Willman CL et al** Deletion of IRF-1, mapping to chromosome 5q31.1, in human leukemia and preleukemic myelodysplasia. S 259: 968-711(1993); **Wong KF et al** Clonal evolution in primary 5q-syndrome. Cancer 70: 100-3 (1992); **Yates P & Potter MN** Eosinophilic leukemia with an abnormality of 5q31, the site of the IL5 gene. Clin. Lab. Haematol. 13: 211-15 (1991)

6-26: A gene specifically expressed in cells in response to human interferon. The protein encoded by this gene is identical with human thymosin β 4. See: Thymic hormones.

Clauss IM et al Human thymosin-β 4/6-26 gene is part of a multigene family composed of seven members located on seven different chromosomes. Genomics 9: 174-80 (1991)

7S NGF: *7S nerve growth factor* see: NGF.

7TD1: A murine hybridoma cell line formed by fusion of the mouse myeloma cell line Sp2/0-Ag14 with spleen cells from a C57BL/6 mouse immunized with *Escherichia coli* lipopolysaccharide three days before fusion. The cell line is dependent on » IL6 for its growth and is used to assay this factor (see also: Bioassays and HGF, hybridoma growth factor). IL6 activity is determined by measuring the incorporation of ^{3}H thymidine into the newly synthesized DNA of proliferating cells and blocking of incorporation following treatment with specific anti-IL6 antibodies. Cell proliferation can be determined also by employing the » MTT assay. An alternative and entirely different method of detecting IL6 is » Message amplification phenotyping.

Balkwill FR & Burke F The cytokine network. IT 10: 299-07 (1989); **Everson MP et al** Interleukin-6 and granulocyte-macrophage colony-stimulating factor are candidate growth factors for chronic myelomonocytic leukemia cells. Blood 74: 1472-76 (1989); **Van Damme J et al** Identification of the human 26-kD protein, interferon β2 (IFN-β2), as a B cell hybridoma/plasmacytoma growth factor induced by interleukin 1 and tumor necrosis factor. JEM 165: 914-19 (1987); **Van Snick J et al** Purification and NH_2-terminal amino acid sequence of a T cell-derived lymphokine with growth factor activity for B cell hybridomas. PNAS 83: 9679-83 (1986); **Van Snick J** From hybridoma and plasmacytoma growth factors to interleukin 6. in: Del Guerico P & Cruse JM (eds): B lymphocytes: Function and Regulation. Contrib. Microbiol. Immunol. Karger, Basel, Vol. 11, pp. 73-95 (1989)

9E3: 9E3 is identical with » CEF-4 and is the avian homologue of the human GRO genes (see: MGSA, melanoma growth stimulatory activity). Sequence homology studies have shown that 9E3 is a member of the » chemokine family of cytokines which comprises many factors with chemotactic (see: Chemotaxis) and pro-inflammatory activities (see: Inflammation).

The biological functions of 9E3 are unknown. *In vivo* 9E3 is expressed at high levels in connective tissue cells and is also transiently overproduced after injuries. Expression is highest in those regions in which new blood vessels are formed (see also: Angiogenesis factors). The synthesis of 9E3 is induced strongly by » aFGF, » bFGF, » TGF-α, » TGF-β, less well by » PDGF and not at all by » EGF. *In vitro* 9E3 is expressed at high levels by proliferating or serum-stimulated cells. It is not expressed by confluent cells. 9E3 is synthesized constitutively by Rous sarcoma virus-transformed fibroblasts derived from normal embryonic chicken tissues.

Transcriptional activation of 9E3 expression in Rous sarcoma virus-transformed fibroblasts is mediated by constitutively expressed transcription factors (see also: Gene expression) which bind to a *src*-responsive sequence (SRU) in the promoter region of the 9E3 gene that is composed of three elements (AP-1, see: *jun*; PRD II/kappa B; TAACGCAATT). Signaling involves the action of the » *fes* » oncogene.

Alexandropoulos K et al Evidence that a G-protein transduces signals initiated by the protein-tyrosine kinase v-Fps. JBC 266(24): 15583-6 (1991); **Dehbi M et al** Transcriptional activation of the CEF-4/9E3 cytokine gene by pp60v-*src*. MCB 12: 1490-99 (1992); **Martins-Green M et al** Wound-factor-induced and cell cycle phase-dependent expression of 9E3/CEF-4, the avian *gro* gene. Cell Regul. 2: 739-52 (1991); **Spangler R et al** Evidence that v-src and v-fps gene products use a protein kinase C-mediated pathway to induce expression of a transformation-related gene. PNAS 86: 7017-21 (1989); **Sugano S et al** Transformation by Rous sarcoma virus induces a novel gene with homology to a mitogenic platelet protein. Cell 49: 321-28 (1987)

10K protein: This protein of 10 kDa is expressed from early stages in the embryonic chicken lens. Chicken 10K and its homologues from mouse and human lenses have been identified by protein sequencing and cloning to be identical with » MIF (macrophage migration inhibitory factor), originally identified in activated human T cells. The expression of 10K/MIF strongly correlates with cell differentiation in the developing chicken lens. Northern blot analysis shows that 10K/MIF is widely expressed in mouse tissues.

Wistow GJ et al A macrophage migration inhibitory factor is expressed in the differentiating cells of the eye lens. PNAS 90: 1272-5 (1993)

12 kDa BCGF: *B cell growth factor 12 kDa* see: LMW-BCGF (low molecular weight B cell growth factor). See also: BCGF (B cell growth factors).

16C8: see: EPA (erythroid promoting activity).

16K I-factor: see I factor.

17q cytokine family: A name sometimes given to the » chemokine family of cytokines which, among other factors, comprises a variety of factors

with chemotactic (see: Chemotaxis) and pro-inflammatory activities (see: Inflammation) encoded by genes located on human chromosome 17q.
Rollins BJ et al Assignment of the human small inducible cytokine A2 gene, SCYA2 (encoding JE or MCP-1) to 17q11.2-12 – evolutionary relatedness of cytokines clustered at the same locus. Genomics 10: 489-92 (1991)

19 kDa vaccinia protein: This factor is identical with » VVGF (vaccinia virus growth factor).

20α-hydroxysteroid dehydrogenase-inducing factor: This factor induces the synthesis of 20α-hydroxysteroid dehydrogenase, a specific enzyme marker of mature T cells, in murine spleen cells. The term » IL3 has been proposed for this factor.
Ihle JN et al Regulation of T cell differentiation: *in vitro* induction of 20 α-hydroxysteroid dehydrogenase in splenic lymphocytes from athymic mice by a unique lymphokine. JI 126: 2184-89 (1981)

20K-BCGF: *20 kDa B cell growth factor* This factor is also known as » BSF-1, » BCGF 1, » EL4-BCGF, and » BRF. It is identical with » IL4. See also: BCGF (B cell growth factors).
Content J et al Induction of a 26 kDa protein mRNA in human cells treated with an interleukin 1-related leukocyte-derived factor. EJB 152: 253-57 (1985); **Okada M et al** B cell growth factor (BCGF) and B cell differentiation factor from human T hybridomas. Two distinct kinds of BCGFs and their synergism in B cell proliferation. JEM 157: 583-90 (1983); **Yoshizaki K et al** Characterization of human B cell growth factor (BCGF) from a cloned T cell or mitogen-stimulated T cells. JI 130: 1241-46 (1983)

22 kDa factor: This factor is produced by mitogen-stimulated mononuclear cells from human peripheral blood. It induces the production of human » IFN-β in cultured human fibroblasts, thereby rendering these cells resistant to virus infection. The factor is identical with one of the biologically active forms of » IL1 (IL1β).
Bunning RA et al Homogeneous interferon-β-inducing 22 k factor (IL1β) has connective tissue cell stimulating activities. BBRC 139: 1150-57 (1986) **Content J et al** Induction of a 26 kDa protein mRNA in human cells treated with an interleukin 1-related leukocyte-derived factor. EJB 152: 253-57 (1985); **Van Damme J et al** Homogeneous interferon-inducing 22K factor is related to endogenous pyrogen and interleukin-1. N 314: 266-68 (1985)

26 kDa protein: This protein, also called IL1-inducible 26k factor, was initially detected as a factor induced in human fibroblasts following their stimulation with » IL1, » TNF, » PDGF or poly(I)-poly(C). The protein was also immunoprecipitated by antiserum raised against partially purified human » IFN-β. The factor was found to be identical with » HGF (hybridoma growth factor), » PCT-GF (plasmacytoma growth factor), » IFN-β2, or » HPGF (hybridoma/plasmacytoma growth factor). The new name suggested for the 26 kDa protein is » IL6.
Haegeman G et al Structural analysis of the sequence coding for an inducible 26 kDa protein in human fibroblasts. EJB 159: 625-32 (1986); **Poupart P et al** B cell growth modulating and differentiating activity of recombinant human 26-kd protein (BSF-2, HuIFN-β 2, HPGF). EJ 6: 1219-24 (1987); **Van Damme J et al** Identification of the human 26-kD protein, interferon β2 (IFN-β2), as a B cell hybridoma/plasmacytoma growth factor induced by interleukin 1 and tumor necrosis factor. JEM 165: 914-19 (1987)

32D: An immortalized cell line originally derived from long-term cultures of murine bone marrow supplemented with » WEHI-3-conditioned media. It is used to assay » IL3 and » G-CSF (see also: Bioassays). The activities of the two factors are determined by measuring the incorporation of ^{3}H thymidine into the newly synthesized DNA of the proliferating factor-dependent T cells and blocking of incorporation following treatment with specific anti-IL3 and anti-G-CSF antibodies, respectively. Cell proliferation can be determined also by employing the » MTT assay. An alternative and entirely different method of detecting these factors is » Message amplification phenotyping. 32D cells require ≈ 10-30 times more G-CSF to reach the same level of proliferation as the » NFS-60 cell line. The cells secrete » IL6. Dibutyryl cyclic adenosine monophosphate has been shown to retard the rapid loss of viable cells seen in the absence of IL3.
32D cells are characterized by a lack of extracellular matrix components and they can be propagated in suspension cultures (see also: Cell culture). Cells can be kept indefinitely as undifferentiated cells as long as they are stimulated by IL3. Removal of IL3 arrests them in the G_0/G_1 phase of the » cell cycle and cell growth is synchronized upon renewed addition of IL3. Prolonged removal of IL3 leads to cell death. 32D cells harboring a constitutively expressed » *myc* oncogene immediately die from » apoptosis if IL3 is removed, suggesting that apoptosis may be an important mechanism in the elimination of hematopoietic cells harboring mutations, such as constitutive c-*myc* expression, which leads to an imbalance of normal cell cycle regulatory controls.
In the presence of » G-CSF and suboptimal concentrations of IL3 32D cells undergo terminal dif-

ferentiation into granulocytes. The expression of an » oncogene, designated *Evi*-1, a nuclear DNA-binding protein of 145 kDa, blocks granulocytic differentiation in response to G-CSF; expression of » *myb* has the same effect and creates cells that proliferate indefinitely in response to G-CSF. Treatment of the cells with » GM-CSF inhibits cell proliferation and the cells die. The incorporation of a functional c-*fms* gene, which encodes the receptor for G-CSF, yields cell variants that still respond to IL3 and » M-CSF but that cannot be differentiated into granulocytes by treatment with G-CSF. The expression of a functional *src* oncogene in 32 D cells creates cell variants that no longer differentiate into mature granulocytes following treatment with G-CSF. These variants are tumorigenic in nude mice (see also: Immunodeficient mice) and growth factor-independent cells are recoverable from the tumors.

The growth of 32D cells and IL3-independent *abl*- and *src*-transformed sublines is inhibited by TGF-β. The oncogene-transfected cells express a greater number of TGF-β1 receptors than the parental cell line and responds to TGF-β1 with increased sensitivity.

The cell line has been used also to produce a number of sublines differing in their factor dependency, sublines with growth requirements for IL3 (32D clone 3), G-CSF (32D G), and also » Epo (32D Epo), and » GM-CSF (32D GM) only and that express the corresponding receptors on their cell surface have been described. The introduction of » oncogenes such as » *src* or » *abl* by means of retroviral vectors yields cell variants, 32D-*src*, and 32D-*abl*, respectively, that have lost their requirements for IL3. At least in some cases the functional substitution of IL3 by an expressed oncogene is due to constitutive activation of proteins involved in IL3 signal transduction.

Anderson SM et al Abrogation of IL3 dependent growth requires a functional v-*src* gene product: evidence for an autocrine growth cycle. O 5: 317-25 (1990); **Askew DS et al** Constitutive c-*myc* expression in an IL3-dependent myeloid cell line suppresses cell cycle arrest and accelerates apoptosis. O 6: 1915-22 (1991); **Berridge MV et al** Cyclic adenosine monophosphate promotes cell survival and retards apoptosis in a factor-dependent bone marrow-derived cell line. EH 21: 269-76 (1993); **Birchenall-Roberts MC et al** Transcriptional regulation of the transforming growth factor β 1 promoter by v-*src* gene products is mediated through the AP-1 complex. MCB 10: 4978-83 (1990); **Birchenall-Roberts MC et al** Differential expression of transforming growth factor-β 1 (TGF-β 1) receptors in murine myeloid cell lines transformed with oncogenes. Correlation with differential growth inhibition by TGF-β 1. JBC 266: 9617-21 (1991); **Gascan H et al** Response of murine IL3-sensitive cell lines to cytokines of human and murine origin. LR 8: 79-84 (1989); **Greenberger JS et al** Demonstration of permanent factor-dependent multipotential (erythroid/neutrophil/basophil) hematopoietic progenitor cell lines. PNAS 80: 2931-5 (1983); **Hagel AJ et al** Different colony-stimulating factors are detected by the interleukin-3-dependent cell lines FDC-P1 and 32D cl-23. Blood 64: 786-90 (1984); **Hültner L et al** Interleukin 3 mediates interleukin 6 production in murine interleukin 3-dependent hemopoietic cells. GF 2: 43-51 (1989); **Jirik FR et al** Transfection of a factor-dependent cell line with the murine interleukin-3 (IL3) cDNA results in autonomous growth and tumorigenesis. Leuk. Res. 11: 1127-34 (1987); **Kato J & Sherr CJ** Human colony-stimulating factor 1 (CSF-1) receptor confers CSF-1 responsiveness to interleukin-3-dependent 32DC13 mouse myeloid cells and abrogates differentiation in response to granulocyte CSF. Blood 75: 1780-87 (1990); **Kreider BL et al** Induction of the granulocyte-macrophage colony-stimulating factor (CSF) receptor by granulocyte CSF increases the differentiative options of a murine hematopoietic progenitor cell. MCB 10: 4846-53 (1990); **Kreider BL et al** The immediate early gene response to a differentiative stimulus is disrupted by the v-*abl* and v-*ras* oncogenes. O 7: 135-40 (1992); **Krüger A & Anderson SM** The v-*src* oncogene blocks the differentiation of a murine myeloid progenitor cell line and induces a tumorigenic phenotype. O 6: 245-56 (1991); **Laneuville P et al** Expression of the chronic myelogenous leukemia-associated p210bcr/*abl* oncoprotein in a murine IL3 dependent myeloid cell line. O 6: 275-82 (1991); **Mavilio F et al** Alterations of growth and differentiation factors response by Kirsten and Harvey sarcoma viruses in the IL3-dependent murine hematopoietic cell line 32D cl 3(G). O 4: 301-8 (1989); **Metcalf D** Multi-CSF-dependent colony formation by cells of a murine hemopoietic cell line: specificity and action of multi-CSF. Blood 65: 357-62 (1985); **Metcalf D** Mechanisms responsible for size differences between hemopoietic colonies: an analysis using a CSF-dependent hemopoietic cell line. IJCC 10: 116-25 (1992); **Migliaccio G et al** Selection of lineage-restricted cell lines immortalized at different stages of hematopoietic differentiation from the murine cell line 32D. JCB 109: 833-41 (1989); **Migliaccio G et al** Response to erythropoietin in erythroid subclones of the factor-dependent cell line 32D is determined by translocation of the erythropoietin receptor to the cell surface. PNAS 88: 11086-90 (1991); **Morishita K et al** Expression of the Evi-1 zinc finger gene in 32Dcl3 myeloid cells blocks granulocytic differentiation in response to granulocyte colony-stimulating factor. MCB 12: 183-9 (1992); **Morrissey PJ et al** The response of the IL6-dependent cell line, B9, to IL6 is not augmented by recombinant derived purified IL2 JIM 147: 141-3 (1992); **Otani H et al** Interleukin (IL)-2 and IL3 induce distinct but overlapping responses in murine IL3-dependent 32D cells transduced with human IL2 receptor β chain: involvement of tyrosine kinase(s) other than p56lck. PNAS 89: 2789-93 (1992); **Patel G et al** v-*myb* blocks granulocyte colony-stimulating factor-induced myeloid cell differentiation but not proliferation. MCB 13: 2269-76 (1993); **Redner RL et al** Variable pattern of *jun* and *fos* gene expression in different hematopoietic cell lines during interleukin 3-induced entry into the cell cycle. O 7: 43-50 (1992); **Rovera G et al** Alteration of the program of terminal differentiation caused by oncogenes in the hemopoietic progenitor cell line 32D C13 (G). ANY 567: 154-64 (1989); **Ruggiero M et al** Mitogenic signal transduction in normal and transformed 32D hematopoietic cells. FL 291: 203-7 (1991); **Shimada Y et al** Erythropoietin-specific cell cycle progression in erythroid subclones of the interleukin-3-dependent cell line 32D. Blood 81: 935-41 (1993); **Valtieri M et al** Cytokine-dependent granulocyte differentiation. Regulation of proliferative and differentiative responses in a murine progenitor cell line. JI 138: 3829-35 (1987); **Watson JD et al** Purification to homogeneity of a human hematopoietic growth factor that stimulates the growth of a murine interleukin-3-dependent cell line. JI 137: 854-57 (1986); **Whetton AD & Dexter TM** Myeloid hematopoietic growth factors. BBA 989: 111-32 (1989)

38K gene of Cowpox virus: see: IL1β-Convertase.

50K-BCGF: *50 kDa B cell growth factor* This factor is also known as » BCGF 2 (see also: BCGF (B cell growth factors). It is identical with » EDF (eosinophil differentiation factor) which, in turn, is identical with » IL5.

Swain SL & Dutton RW Production of a B cell growth-promoting activity (DL)BCGF, from a cloned T cell line and its assay on the BCL1 B cell tumor. JEM 156: 1821-34 (1982); **Swain SL et al** Evidence for two distinct classes of murine B cell growth factors with activities in different functional assays. JEM 158: 822-35 (1983); **Vasquez A et al** Different human B cell subsets respond to interleukin 2 and to a high molecular weight B cell growth factor (BCGF). EJI 16: 1503-7 (1986); **Yoshizaki K et al** Characterization of human B cell growth factor (BCGF) from cloned T cells or mitogen-stimulated T cells. JI 130: 1241-46 (1983)

60A: see: *Drosophila* 60A.

60K-BCGF: *60 kDa B cell growth factor* see: HMW-BCGF (high molecular weight BCGF). See also: BCGF (B cell growth factors).

86HG39: see: 2D9.

160 SF: *160 hybridoma-derived suppressor factor* This poorly characterized factor (≈ 50-70 kDa) is constitutively produced by a human T-T cell hybridoma generated by fusion of Con A-activated human peripheral blood lymphocytes from a normal donor with cells of the Jurkat tumor T cell line. A similar but distinct factor is also produced by the Jurkat parent line.

160 SF inhibits antibody production by human and mouse lymphocytes. 160 SF arrests phytohemagglutinin-induced proliferation of peripheral blood mononuclear cells in the G0/G1 phase of the » cell cycle. The factor also inhibits the growth, but not the viability, of cells from the following human tumor cell lines: A673 sarcoma cell line, SK-LC-6 and SK-LC-14 lung cell lines, SB, Raji, and Daudi lymphoblastoid cell lines, and FARR malignant melanoma cell line. It does not affect the growth of murine L1210 cells and FS-4 normal human diploid fibroblasts. 160 SF is not identical with » IFN-γ, » IFN-α, » TNF, or » TGF-β.

Fox FE et al Human hybridoma-derived suppressor factor 160 and transforming growth factor β are different molecules. Lymphokine Cytokine Res.11: 307-15 (1992); **Kunicka JE et al** Hybridoma-derived human suppressor factors: inhibition of growth of tumor cell lines and effect on cytotoxic cells. Hum. Antibodies Hybridomas. 2: 160-9 (1991); **Kunicka JE et al** Human suppressor factors constitutively produced by T-T cell hybridomas: functional and biochemical characterization. Hybridoma 8: 127-51 (1989)

446-BCDF: *446 B cell differentiation factor* This poorly characterized factor of ≈ 32 kDa and a pI of 6.0 is produced and secreted by anti-CD3-stimulated T cells. It regulates terminal B cell differentiation, enhancing the synthesis of immunoglobulins by SAC-activated B cells by a factor of 10-100. This activity is not inhibited by antibodies directed against » IL6 and is enhanced after passage over an anti-IL6 affinity column.

Mayer L et al Cytokines regulating human B cell growth and differentiation. Clin. Immunol. Immunopathol. 6c1: S28-S36 (1991)

464. 1: Name of a DNA clone encoding MIP-α. See: MIP (macrophage inflammatory protein).

Irving SG et al Two inflammatory mediator cytokine genes are closely linked and variably amplified on chromosome 17q. NAR 18: 3261-70 (1990); **Obaru KM et al** A cDNA clone used to study mRNA inducible in human tonsillar lymphocytes by a tumor promoter. JB 99: 885-94 (1986); **Zipfel PF et al** Mitogenic activation of human T cells induces two closely related genes which share structural similarities with a new family of secreted factors. JI 142: 1582-90 (1989)

744. 1: Name of a DNA clone encoding MIP-β. See: MIP (macrophage inflammatory protein).

see: 464. 1.

1788: see: RPMI 1788.

5637: (ATCC: HTB 9 = subline of 5637) Human cell line with epithelial-like characteristics established from a bladder carcinoma of a 68 year old male patient. 5637 cells are used in cytokine research because they are constitutive producers of a number of growth factors, including » GM-CSF, » M-CSF, » G-CSF (See also: CSF for general information about colony stimulating factors), » IL1, » IL 6, » IL7, » IL8, » LIF, » TNF-α, and » TNF-β. The expression of » IL3, » IL4, » IL5 and » IL9 is not observed. The culture medium (see also: CM, conditioned medium) is used, among other things, to isolate factors and to provide an (undefined) growth factor cocktail for other cell studies.

5637 cells also secrete a poorly characterized factor, termed ***SF-1*** (synergistic factor 1), distinct from » G-CSF, » GM-CSF, and » IL6, which synergizes with » IL1 in stimulating primitive bone marrow progenitor cells, termed high proliferative-potential colony-forming cells (HPP-CFC), in the presence of an optimal dose of » M-CSF.

Gordon MY et al An *in vitro* model for the production of committed hemopoietic progenitor cells stimulated by exposure to single and combined recombinant growth factors. Bone Marrow Transplant. 4: 353-58 (1989); **Kaashoek JGJ et al** Cytokine production by the bladder carcinoma cell line 5637: Rapid analysis of mRNA expression levels using a cDNA-PCR procedure. Lymphokine and Cytokine Research 10: 231-35 (1991); **McNiece IK et al** Studies on the myeloid synergistic factor from 5637: comparison with interleukin-1 α. Blood 73: 919-23 (1989); **Morioka E et al** Purification of a granulocyte colony-stimulating factor from the conditioned medium of a subclone of human bladder carcinoma cell line 5637, HTB9. Res. Exp. Med. Berl. 190: 229-38 (1990); **Pfluger KH et al** Karyotype and ultrastructure of a colony stimulating factor (CSF) producing cell line (5637) originated from a carcinoma of the human urinary bladder. Blut 53: 89-100 (1986); **Sorg R et al** Rapid and sensitive mRNA phenotyping for interleukins (IL1 to IL6) and colony-stimulating factors (G-CSF, M-CSF, and GM-CSF) by reverse transcription and subsequent polymerase chain reaction. EH 19: 882-7 (1991)

5637-derived factor: This factor, secreted, among others, by the » 5637 cell line, was originally detected as a factor promoting the proliferation of the murine 32D cell line. Comparison of its amino acid sequence shows that this factor is identical with » G-CSF.

Watson JD et al Purification to homogeneity of a human hematopoietic growth factor that stimulates the growth of a murine interleukin-3-dependent cell line. JI 137: 854-57 (1986)

a: (a) as a prefix "acidic" such as in aFGF (acidic FGF.

A20: An » ERG (early response gene) encoding a » zinc finger protein whose transcription is rapidly and transiently stimulated by » TNF-α in a variety of cells. Increased expression of A20, which encodes a » zinc finger protein of 790 amino acids, has been observed in TNF-resistant human breast carcinoma cells. A20 negatively regulates its own expression. Constitutive expression of A20 after stable transfection of » 3T3 fibroblasts and WEHI 164 cells results in significant, but partial, resistance to TNF cytotoxicity. A20 protein expression in lymphocytic and monocytic cells lines correlates with lymphocyte activation and monocyte differentiation. A20 expression is also induced in Jurkat T cells expressing the human T cell leukemia virus type I Tax protein. The gene is also induced by » Phorbol esters.
A20 Has been found to be constitutively expressed in Epstein-Barr virus (EBV)-immortalized B cells. Transfection experiments demonstrate that the EBV latent membrane protein LMP1, a fundamentally important viral transforming protein, induces A20 expression.

Krikos A et al Transcriptional activation of the tumor necrosis factor α-inducible zinc finger protein, A20, is mediated by kappa B elements. JBC 267: 17971-6 (1992); **Laherty CD et al** Human T cell leukemia virus type I Tax and phorbol 12-myristate 13-acetate induce expression of the A20 zinc finger protein by distinct mechanisms involving nuclear factor kappa B. JBC 268: 5032-9 (1993); **Laherty CD et al** The Epstein-Barr virus LMP1 gene product induces A20 zinc finger protein expression by activating nuclear factor kappa B. JBC 267: 24157-60 (1992); **Opipari AW Jr et al** The A20 zinc finger protein protects cells from tumor necrosis factor cytotoxicity. JBC 267: 12424-7 (1992)

A375: A human melanoma cell line that is used in cytokine research to assay IL1α or IL1β (see: IL1). The growth of these cells is inhibited by both factors at concentrations of 10-30 pg/mL. The assay is not influenced by prostaglandin E2, plant lectins, bacterial » endotoxins and cytokines such as » IL2, » TNF, interferons (see: IFN) or colony-stimulating factors (see: CSF). The growth of A375 cells is inhibited by » Oncostatin M and IL6 but not » LIF or » G-CSF.

Bruce AG et al Oncostatin M is a differentiation factor for myeloid leukemia cells. JI 149: 1271-5 (1992); **Nakai S et al** A simple, sensitive bioassay for the detection of interleukin 1 using human melanoma A375 cell line. BBRC 154: 1189-96 (1988)

A431: (ATCC CRL 1555) Human cell line established from an epidermoid carcinoma of a 85 year old female patient. The cells grow into colonies when cultivated in soft agar or on lawns of human fibroblasts and develop into subcutaneous tumors following subcutaneous injection into mice treated with anti-thymocyte antiserum.
The cell line is frequently used for the biochemical analysis of » EGF physiology. It expresses an extremely high number of EGF receptors on its cell surface (3×10^6/cell; corresponding to ≈ 0.2 % of total cell protein), due, at least in part, to the amplification of EGF receptor DNA sequences (30-fold). This is in marked contrast to normal human fibroblasts which have a much more reduced EGF receptor density (1×10^5/cell).
In spite of the very high EGF receptor density the growth of A431 cells is inhibited by EGF even in the presence of serum. Growth inhibition by EGF therefore serves as a » bioassay for this factor, with factor activity being determined by measuring the reduction of incorporation of ^{3}H thymidine into DNA. Factors such as » IGF or HBGF (heparin-binding growth factors) do not interfere with this assay. The detection limit of this assay is on the order of 0.1-0.2 ng of EGF. A431 cell variants have been described that respond positively to EGF.
EGF has been shown to have a biphasic effect on the growth of A431 cells, being a stimulator of growth at picomolar and inhibited at nanomolar concentrations of EGF. It has been shown that the growth inhibitory effect is due to the induction of terminal differentiation in these cells.
Under low serum conditions » TGF-β can enhance the proliferation of A431 cells. The factor known as » SCSGF (Sertoli cell secreted growth factor) is a very strong mitogen for A431 cells even in the

presence of serum. Growth and DNA synthesis of A431 cells are also stimulated by » HGF (hepatocyte growth factor). HGF also enhances the motility of these cells. A431 cells also respond to » MIS (Müllerian inhibiting substance).

Barnes DW Epidermal growth factor inhibits the growth of A431 human epidermoid carcinoma in serum-free culture. JCB 93: 1-4 (1982); **Kawamoto T et al** Growth stimulation of A431 cells by epidermal growth factor: identification of high-affinity receptors for epidermal growth factor by an anti-receptor monoclonal antibody. PNAS 80: 1337-41 (1993); **Konger RL & Chan TCK** Epidermal growth factor induces terminal differentiation in human epidermoid carcinoma cells. JCP 156: 515-21 (1993); **Lee L et al** Reciprocal effects of epidermal growth factor and transforming growth factor β on the anchorage-dependent and -independent growth of A431 epidermoid carcinoma cells. ECR 173: 156-62 (1987); **Merlino GT et al** Amplification and enhanced expression of the epidermal growth factor receptor gene in A431 human carcinoma cells. S 224: 417-19 (1984); **Rizzino A et al** Isolation and characterization of A431 cells that retain high epidermal growth factor binding capacity and respond to epidermal growth factor by growth stimulation. CR 48: 2377-81 (1988); **Shimizu N et al** Genetic analysis of hyperproduction of epidermal growth factor receptors in human epidermoid carcinoma A431 cells. Somatic Cell. Mol. Genet. 10: 45-53 (1984); **Stoscheck C et al** Biology of the A431 cell: a useful organism for hormone research. JCBc 23: 191-202 (1983); **Sunada H et al** Monoclonal antibody against epidermal growth factor receptor is internalized without stimulating receptor phosphorylation. PNAS 83: 3825-29 (1986); **Tajima H et al** Regulation of cell growth and motility by hepatocyte growth factor and receptor expression in various cell species. ECR 202: 423-31 (1992); **Weber W et al** Production of an epidermal growth factor receptor-related protein. S 224: 294-97 (1984)

A23187: see: Calcium ionophore.

ABAE cell growth-inhibitory activity: *adult bovine aortic endothelial cell growth-inhibitory activity* This factor was originally isolated from bovine pituitaries and found to inhibit the growth of adult aortic endothelial cells. It is identical with » LIF.

Ferrara N et al Pituitary follicular cells secrete an inhibitor of aortic endothelial cell growth: identification as leukemia inhibitory factor. PNAS 89: 698-702 (1992)

aBCGF: *autostimulatory B cell growth factor* abbrev. also: aBGF. This poorly characterized factor of 16 kDa is secreted constitutively by Epstein Barr virus-transformed human B cell lines. It exists as monomeric dimeric, and tetrameric forms. aBCGF acts as an » autocrine growth factor for Epstein Barr virus-transformed B cells. aBCGF is not identical with » IL1 or » IL6 that have been described also to possess growth-promoting activity for Epstein Barr virus-transformed B cells. See also: BCGF (B cell growth factors).

Buck J et al Purification and biochemical characterization of a human autocrine growth factor produced by Epstein-Barr virus-transformed cells. JI 138: 2923-28 (1987)

aBGF: *autostimulatory B cell growth factor* see: aBCGF.

abl: This » oncogene is found in the genome of the acutely transforming replication-defective retrovirus, Abelson murine leukemia virus (Ab-MuLV). The cellular homologue, c-*abl*, located on human chromosome 9q34.1, is a member of the tyrosine kinase family of proteins. Its biological functions are still unknown but amino acid sequences of proteins from various species are highly conserved. the protein encoded by cellular *abl* has been shown to bind to specific DNA sequences in the nucleus and to possess the unique function of binding to actin filaments in the cytoplasm. c-*abl* also becomes hyperphosphorylated during the metaphase of the » cell cycle, which results in an inhibition of its DNA binding activity.

Translocation of the c-*abl* proto-oncogene to the BCR (breakpoint cluster region) gene located on chromosome 22 produces a hybrid protein of 210 kDa (p210BCR/ABL) with constitutively overexpress an abnormal tyrosine kinase activity. This elevated kinase activity is essential for transformation. The c-*abl* translocation leading to the expression of the aberrant fusion protein has been observed in 90% of CML (chronic myelogenous leukemia) cases examined, leading to the hallmark of chronic myelogenous leukemia (CML), the Philadelphia chromosome. The detection of the bcr/*abl* fusion is also used to monitor residual Philadelphia chromosome-positive cells in patients with CML. Using » antisense oligonucleotides it has been possible to inhibit the proliferation of bcr/*abl*-containing cell lines.

v-*abl* and the cellular » *myc* oncogene act synergistically to transform mature B cells with high efficiency as shown by the rapid induction of plasmacytomas in mice infected with a retrovirus expressing both oncogenes.

c-*abl* contains a so-called » *src* homology region, a sequence of ≈ 50 amino acids found in many non-receptor tyrosine kinases and other proteins. Deletion of the SH3 region activates the *abl* proto-oncogene, thereby suggesting the participation of the SH3 region in the negative regulation of transformation.

Ab-MuLV infection can render myeloid and lymphoid cells independent of the growth factors »

IL3 and » GM-CSF. v-*abl* can also relieve » IL2 dependence in T cells. Temperature-sensitive mutations of the viral v-*abl* kinase result in a temperature-sensitive phenotype of growth factor dependence. Transformation of cells induced by expression of *abl* can be reverted by treatment of cells with » Herbimycin A.

■ **Transgenic/Knock-out/Antisense Studies:** Introduction of a mutated c-*abl* gene encoding a shortened protein that does no longer bind to DNA but retains its tyrosine kinase activity into the mouse germline via targeted gene disruption of embryonic stem cells (see: ES cells) shows that mice homozygous for this mutation display increased perinatal mortality, runtedness, and abnormal spleen, head, and eye development. An examination of the immune system reveals major reductions in B cell progenitors in the adult bone marrow, with less dramatic reductions in developing T cell compartments. The same observations have been made with a mutant mouse in which the *abl* gene function was completely disrupted. These results suggest that, among other things, expression of c-*abl* may be critical for the proliferation of lymphocytes.

● **Reviews: Wang JY et al** *abl* tyrosine kinase in signal transduction and cell-cycle regulation. Curr. Opin. Genet. Dev. 3: 35-43 (1993); **Witte ON** Role of the BCR-ABL oncogene in human leukemia: fifteenth Richard and Hinda Rosenthal Foundation Award Lecture. CR 53: 485-9 (1993)

Chen W et al T cell immunity to the joining region of p210BCR-ABL protein. PNAS 89: 1468-72 (1992); **Cleveland JL et al** Tyrosine kinase oncogenes abrogate interleukin-3 dependence of murine myeloid cells through signaling pathways involving c-*myc*: conditional regulation of c-*myc* transcription by temperature-sensitive v-*abl*. MCB 9: 5685-95 (1989); **Cook WD et al** Abelson virus transformation of an interleukin 2-dependent antigen-specific T cell line. MCB 7: 2631-5 (1987); **Fabegra S et al** Polymerase chain reaction: a method for monitoring tumor cell purge by long-term culture in BCR/ABL positive acute lymphoblastic leukemia. Bone Marrow Transplant. 11: 169-73 (1993); **Han XD et al** Chronic myeloproliferative disease induced by site-specific integration of Abelson murine leukemia virus-infected hemopoietic stem cells. PNAS 88: 10129-33 (1991); **Holland GD et al** Conservation of function of *Drosophila melanogaster abl* and murine v-*abl* proteins in transformation of mammalian cells. J. Virol. 64: 2226-35 (1990); **Jhanwar SC et al** Localization of the cellular oncogenes ABL, SIS, and FES on human germ-line chromosomes. Cytogenet. Cell. Genet. 38: 73-5 (1984); **Kipreos ET & Wang JYJ** Reversible dependence on growth factor interleukin-3 in myeloid cells expressing temperature sensitive v-*abl* oncogene. OR 2: 277-84 (1988); **Kipreos ET & Wang JYJ** Cell cycle-regulated binding of c-abl tyrosine kinase to DNA. S 256: 382-5 (1992); **Klinken SP** Erythroproliferation *in vitro* can be induced by *abl*, *fes*, *src*, *ras*, *bas*, *raf*, *raf/myc*, *erb* B and *cbl* oncogenes but not by *myc*, *myb* and *fos*. OR 3: 187-92 (1988); **Largaespada DA et al** A retrovirus that expresses v-*abl* and c-*myc* oncogenes rapidly induces plasmacytomas. O 7: 811-9 (1992); **Martiat P et al** Retrovirally transduced antisense sequences stably suppress P210BCR-ABL expression and inhibit the proliferation of BCR/ABL-containing cell lines. Blood 81: 502-9 (1993); **Miyamura K et al** Long persistent bcr-*abl* positive transcript detected by polymerase chain reaction after marrow transplant for chronic myelogenous leukemia without clinical relapse: a study of 64 patients. Blood 81: 1089-93 (1993); **Owen PJ et al** Cellular signaling events elicited by v-*abl* associated with growth factor independence in an interleukin-3-dependent cell line. JBC 268: 15696-703 (1993); **Rayter SI** Expression and purification of active *abl* protein-tyrosine kinase in *Escherichia coli*. MiE 200: 596-604 (1991); **Rovera G et al** Effect of Abelson murine leukemia virus on granulocytic differentiation and interleukin-3 dependence of a murine progenitor cell line. O 1: 29-35 (1987); **Van Etten RA** Malignant transformation by *abl* and BCR/ABL. Cancer Treat. Res. 63: 167-92 (1992); **Westbrook, CA et al** Clinical significance of the BCR-ABL fusion gene in adult acute lymphoblastic leukemia: a Cancer and Leukemia Group B Study (8762) Blood 80: 2983-90 (1992)

● **Transgenic/Knock-out/Antisense Studies: Harris AW et al** Lymphoid tumorigenesis by v-*abl* and BCR-v-*abl* in transgenic mice. CTMI 166: 165-73 (1990); **Rosenbaum H et al** An E μ-v-*abl* transgene elicits plasmacytomas in concert with an activated *myc* gene. EJ 9: 897-905 (1990); **Rosti V et al** c-*abl* function in normal and chronic myelogenous leukemia hematopoiesis: *in vitro* studies with antisense oligomers. Leukemia 6: 1-7 (1992); **Schwartzberg PL et al** Mice homozygous for the abl^{mt} mutation show poor viability and depletion of selected B and T cell populations. Cell 65: 1165-75 (1991); **Tybulewicz VLJ et al** Neonatal lethality and lymphopenia in mice with homozygous disruption of the c-*abl* proto-oncogene. Cell 65: 1153-63 (1991)

ABMT: abbrev. for autologous bone marrow transplantation.

abnormal wing discs: abbrev. awd. See: nm23.

Acetylcholine receptor inducing activity: see: ARIA.

Acetyl-N-Ser-Asp-Lys-Pro: see: AcSDKP.

Acidic fibroblast growth factor: see: aFGF.

Acidic astroglial growth factor: see: AGF-1.

AcNPV: abbrev. for *Autographa californica* nuclear polyhedrosis virus. See: Baculovirus expression system.

AcSDKP: AcSDKP is a tetrapeptide (acetyl-N-Ser-Asp-Lys-Pro) originally isolated from fetal calf bone marrow. It is produced constitutively by bone marrow cells in long-term culture (see also: LTBMC). AcSDKP can be generated from thymosin β 4, one of the » thymic hormones, by a one-step enzymatic cleavage *in vitro* and *in vivo*. Cleavage is probably mediated by mammalian endoproteinase Asp-N. Thymosin β 4 has been shown to be induced during » GM-CSF-induced differentiation of bone marrow cells. Degradation of AcSDKP in serum has been shown to be the result of some plasma metalloproteases which can be inhibited by protease inhibitors.

AcSDKP substantially inhibits entry into the » cell cycle of murine pluripotent » hematopoietic stem cells (see: CFU-S) *in vitro* and *in vivo*. It plays an essential role in maintaining CFU-S in a physiological quiescent state in normal mice since treatment of mice with polyclonal antisera directed against the tetrapeptide leads to a dramatic increase in the percentage of CFU-S in DNA synthesis. AcSDKP has a marked protective effect in mice during anticancer chemotherapy with phase-specific drugs. AcSDKP appears to be inactive on actively cycling cells and is only capable of maintaining quiescent cells in the G0 phase of the cell cycle.

AcSDKP enhances binding of hemopoietic cell to stromal cell layers *in vitro*, probably through the activation of stromal cells. AcSDKP is able to inhibit the *in vitro* growth of human bone marrow progenitor cells (see also: Hematopoiesis), in particular » CFU-GM, CFU-E, and » BFU-E, and to decrease their proportion in cell cycle at nanomolar concentration. The peptide appears to be inactive on the » FDCP-2 hematopoietic cell line and the promyelocytic leukemia cell line K562.

AcSDKP also decreases mixed lymphocyte reaction intensities when H-2 incompatible allogeneic spleen cells are used as stimulators and has been suggested to be of potential therapeutic value for clinical bone marrow transplantation. AcSDKP appears to be ineffective on human leukemic cells and therefore, by acting selectively on normal progenitors, may represent a potent therapeutical agent for the protection of normal bone marrow progenitors during chemotherapy. For other hematopoietic inhibitory peptides see also: pEEDCK, EIP (epidermal inhibitory pentapeptide).

Aizawa S et al Biological activities of tetrapeptide AcSDKP on hemopoietic cell binding to the stromal cell *in vitro*. EH 20: 896-9 (1992); **Bogden AE et al** Amelioration of chemotherapy-induced toxicity by cotreatment with AcSDKP, a tetrapeptide inhibitor of hematopoietic stem cell proliferation. ANY 628: 126-39 (1991); **Bonnet D et al** The tetrapeptide AcSDKP, an inhibitor of the cell-cycle status for normal human hematopoietic progenitors, has no effect on leukemic cells. EH 20: 251-5 (1992); **Frindel E & Monpezat JP** The physiological role of the endogenous colony forming units-spleen (CFU-S) inhibitor acetyl-N-Ser-Asp-Lys-Pro (AcSDKP). Leukemia 3: 753-4 (1989); **Frindel E et al** Inhibitory effects of AcSDKP on the mixed lymphocyte reaction (MLR). Part I. MLR with mouse spleen cells. Leukemia 6: 1043-4 (1992); **Grillon C et al** Involvement of thymosin β 4 and endoproteinase Asp-N in the biosynthesis of the tetrapeptide AcSerAspLysPro a regulator of the hematopoietic system. FL 274: 30-4 (1990); **Grillon C et al** Optimization of cell culture conditions for the evaluation of the biological activities of the tetrapeptide N-Acetyl-Ser-Asp-Lys-Pro, a natural hemoregulatory factor. GF 9: 133-8 (1993); **Guigon M et al** Inhibition of human bone marrow progenitors by the synthetic tetrapeptide AcSDKP. EH 18: 1112-5 (1990); **Lauret E et al** Further studies on the biological activities of the CFU-S inhibitory tetrapeptide AcSDKP. II. Unresponsiveness of isolated adult rat hepatocytes, 3T3, FDCP-2, and K562 cell lines to AcSDKP. Possible involvement of intermediary cell(s) in the mechanism of AcSDKP action. EH 17: 1081-5 (1989); **Lavignac C et al** Inhibitory effects of AcSDKP on the mixed lymphocyte reaction (MLR). Part II. Human whole blood cells. Leukemia 6: 1045-7 (1992); **Lenfant M et al** Inhibitor of hematopoietic pluripotent stem cell proliferation: purification and determination of its structure. PNAS 86: 779-82 (1989); **Lenfant M et al** Formation of Acetyl-Ser-Asp-Lys-Pro, a new regulator of the hematopoietic system, through enzymatic processing of thymosin β4. ANY 628: 115-25 (1991); **Monpezat JP & Frindel E** Further studies on the biological activities of the CFU-S inhibitory tetrapeptide AcSDKP. I. The precise point of the cell cycle sensitive to AcSDKP. Studies on the effect of AcSDKP on GM-CFC and on the possible involvement of T lymphocytes in AcSDKP response. EH 17: 1077-80 (1989); **Moscinski LC et al** Identification of a series of differentiation-associated gene sequences from GM-CSF-stimulated bone marrow. O 5: 31-7 (1990); **Pradelles P et al** Negative regulator of pluripotent hematopoietic stem cell proliferation in human white blood cells and plasma as analyzed by enzyme immunoassay. BBRC 170: 986-93 (1990); **Wdzieczak-Bakala J et al** AcSDKP, an inhibitor of CFU-S proliferation, is synthesized in mice under steady-state conditions and secreted by bone marrow in long-term culture. Leukemia 4: 235-7 (1990)

ACT-2: Act-2 (abbrev. for immune activation gene), also known as » G26, » pAT 744, » SCY A4, and MIP-1β (macrophage inflammatory protein see: MIP), is a factor secreted by T cells, B cells, monocytes and various tumor cell lines, following their stimulation by mitogens such as phytohemagglutinin, bacterial » endotoxins or *Staphylococcus aureus*. The synthesis of ACT-2 is stimulated also by » IL2, while glucocorticoids, » TGF-β and » CsA (cyclosporin A) inhibit the synthesis. ACT2 is encoded by a gene located on human chromosome 17q21-q23. Studies of sequence homology and genome organization have shown that Act-2 belongs to the » chemokine family of cytokines. In rats the subcutaneous injection of the factor precipitates local inflammatory reactions characterized by the infiltration of neutrophils.

Leonard WJ et al Structure, function, and regulation of the interleukin-2 receptor and identification of a novel immune activation gene. Phil. Trans. R. Soc. Lond. (Biol.) 327: 187-92 (1990); **Lipes MA et al** Identification, cloning, and characterization of an immune activation gene. PNAS 85: 9704-08 (1988); **Miller MD et al** A novel polypeptide secreted by activated human T lymphocytes. JI 143: 2907-16 (1989); **Modi WS et al** Chromosomal localization of the ACT-2 cytokine. Cytogenet. Cell Genet. 58: 2008 (1991); **Napolitano M et al** The gene encoding the Act-2 cytokine. Genomic structure, HTLV-I/Tax responsiveness of 5´ upstream sequences, and chromosomal localization. JBC 266: 17531-36 (1991); **Napolitano M et al** The gene encoding the Act-2 cytokine – chromosomal localization and identification of 5´ regulatory sequences. Clin. Res. 38: A302 (1990); **Napolitano M et al** Identification of cell surface receptors for Act-2. JEM 172: 285-89 (1990); **Zipfel PF et al** Mitogenic activation of human T cells induces two closely related genes which share structural similarity with a new family of secreted factors. JI 142: 1582-90 (1989)

ACTH: *adrenocorticotropic hormone* see: POMC (proopiomelanocortin).

Activated cells: see: Cell activation.

Activating factor: This factor is identical with » ILI.

Activating transcription factor: abbrev. ATF. See: CRE (cyclic AMP-responsive element).

Activation-induced cell death: abbrev. AICD. See: Apoptosis.

Activation-inducing molecule: see: AIM.

Activator protein 1: abbrev. AP-1. AP-1 is the gene product of the » *jun* oncogene and functions as a transcription factor (see: Gene expression).

Activin A:

■ **ALTERNATIVE NAMES:** EDF (erythroid differentiation factor); FRP (FSH-(follicle stimulating hormone) releasing protein; WEHI-MIF (WEHI mesoderm-inducing factor). See also: individual entries for further information.

■ **SOURCES:** Activins have been found in the testis and in limited areas of the adult brain. Activin A is synthesized by normal bone marrow cells and also by a number or myelomonocytic cell lines. Its synthesis in monocytes is enhanced by » GM-CSF and to a lesser degree by » IFN-γ. Its synthesis in bone marrow stromal cells is induced by » TNF-α, IL1α (see: IL1), and bacterial » endotoxins.

■ **PROTEIN CHARACTERISTICS:** There are three different activins, designated A, B, and A-B. Activin A is identical with the homodimer of the βA subunit of » Inhibin. It has been suggested that the 2 forms of the inhibin β subunit be referred to as β-A (see also: EDF, erythroid differentiation factor) and β-B. The βB dimer of inhibin, which stimulates FSH secretion, should be called ***activin B*** and the heterodimer consisting of 1 β-A and 1 β-B subunit should be termed ***activin AB***. Sequence analysis reveals that activin A belongs to a large family of proteins related to » TGF-β, showing ≈ 40 % of sequence homology in the carboxyterminal region.

■ **GENE STRUCTURE:** For mapping data see: Inhibins.

■ **RELATED FACTORS:** A homologue of activin A is » XTC-MIF isolated from » *Xenopus laevis.*

■ **RECEPTOR STRUCTURE, GENE(S), EXPRESSION:** Activin A receptors are expressed in murine erythroleukemic cells with a density of 3200-3500 receptors/cell. Binding studies have revealed the existence of three receptor types with molecular masses of 140 kDa, 76 kDa, and 67 kDa, respectively. There also seems to exist a class of TGF-β receptors that binds TGF-β, activin A, and » inhibins. Two human receptors, designated TSR-I and AcrR-I, have been cloned and found to encode transmembrane serine/threonine kinases distantly related to known mammalian » TGF-β and activin receptors and forming heterodimeric complexes with them.

One of the receptors cloned from a murine corticotrophic cells encodes a protein of 494 amino acids with an intracellular domain functioning as a serine/threonine-specific protein kinase. Activin A receptor-related receptors have been isolated (see: ALK).

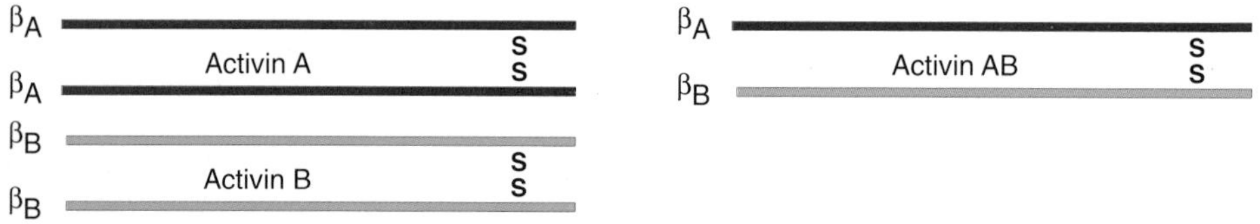

Molecular structure of activins.
There are three different activins, designated A, B, and A-B. Activin A is a homodimer of the βA subunit of Inhibin A. Activin B is a dimer of the βB subunit of inhibin B. Activin A-B is the heterodimer consisting of β-A and β-B. Sequence analysis reveals that activin A belongs to a large family of proteins related to TGF-β, showing ≈ 40 % of sequence homology in the carboxyterminal region.

■ **Biological Activities:** No functional differences between activin A and activin AB have yet been identified. Originally activin A was described as a protein stimulating *in vitro* the release of FSH (follicle stimulating hormone). Purified recombinant activin B has been shown to possess similar potency as activin A in its ability to stimulate FSH production in an *in vitro* pituitary assay. However, Activin A and also the related » inhibins show a variety of other biological activities. An involvement of these in the induction of the mesoderm during embryonic development has been suggested. It has been found, however, that the corresponding genes are expressed during embryogenesis at a time that precludes their activity as physiological inducers of mesodermal development.

Activin A controls the expression and secretion of a number of hormones such as FSH (follicle stimulating hormone), » prolactin, ACTH (corticotropin), and also the secretion of oxytocin. Activin A suppresses basal ACTH secretion and » POMC (proopiomelanocortin) mRNA accumulation in pituitary cell lines *in vitro*. Activin A also stimulates the synthesis of gonadotropin-releasing hormone receptors and inhibits the release of » Growth hormone.

Activin A acts as a neuronal survival factor (see also: neurotrophins) and inhibits neuronal differentiation. It also influences the development of granulosa cells. Activin A also supports the proliferation of a number of cells lines.

In rodents Activin A enhances the production of normal erythroid precursor cells *in vitro* and *in vivo*. The activity as an erythroid differentiation factor, which can be observed also in human bone marrow stromal cells, suggests that activin A is a regulatory factor of » hematopoiesis. It has been observed that activin A stimulates the production of colonies of » BFU-E and » CFU-E which is probably due to the increased secretion by bone marrow stromal cells of other enhancing factors (see also: BPA, burst promoting activities). It has been suggested that activin A may also be involved in controlling megakaryocytopoiesis. The promyelocytic cell line » HL-60 differentiates into monocytes and macrophages following exposure to activin A. EDF can induce *in vitro* the differentiation of human and murine erythroleukemic cell lines, inducing the development of hemoglobin-positive cells.

Activin A has been shown to modify the growth of vascular smooth muscle cells by complex mechanisms involving » autocrine production of » IGF-1 and modification of the action of IGF-1.

The biological activities of activin A can be inhibited specifically by » follistatin.

■ **Interactions:** Activin A stimulates the synthesis of » ET (endothelin). In murine erythroleukemic cells activin A induces the synthesis of hemoglobin and thromboxane A_2 synthase (TX synthase).

■ **Assays:** Activin A can be detected in a protein

Activin and Inhibin: Functional diversity by differential subunit association.

	Activin (ββ)	Inhibin (αβ)
Pituitary		
Basal follicle stimulating hormone	↑	↓
Basal ACTH	↓	
Gonadotropin releasing hormone	↑	↓
Growth hormone releasing hormone	↓	↑
Gonads		
Luteninizing hormone	↓	↑
Follicle stimulating hormone	↑	↓
Placenta		
Chorionic gonadotropin	↑	↓
Bone marrow		
Erythropoiesis	↑	↓
Brain		
Oxitocin	↑	

binding assay involving the use of a specific binding protein, » Follistatin. It is also detected by its activity as » EDF (erythroid differentiation factor). Immunoassays are also available. For further information see also subentry "Assays" in the reference section.

■ **Clinical Use & Significance:** Activin A may have clinical significance as a growth inhibitory and an erythroid differentiation factor in the treatment of erythroleukemia (see also: Inhibins).

● **Reviews: Burger HG et al** Inhibin: definition and nomenclature, including related substances. J. Clin. Endocr. Metab. 66: 885-6 (1988); **DePaolo LV et al** Follistatin and activin: A potential intrinsic regulatory system within diverse tissues. PSEBM198: 500-12 (1991); **Hillier SG & Miro F** Inhibin, activin, and follistatin. Potential roles in ovarian physiology. ANY 687: 29-38 (1993); **Ling N et al** Inhibins and activins. Vitamins & Hormones – Advances in Research & Applications, pp. 1-46, Academic Press, San Diego 1988; **Vale W et al** The inhibin/activin family of hormones and growth factors. In: Sporn MA & Roberts AB (eds) Peptide growth factors and their receptors. Handbook of Experimental Pharmacology, Vol. 95: 211-48, Springer, Berlin 1990

● **Biochemistry & Molecular Biology: Chertov OY et al** Mesoderm-inducing factor from bovine amniotic fluid: purification and N-terminal amino acid sequence determination. Biomed. Sci. 1: 499-506 (1990); **Mason A et al** Complementary DNA sequences of ovarian follicular fluid inhibin show precursor structure and homology with transforming growth factor β. N 318: 659-63 (1985); **Smith JC et al** Identification of a potent *Xenopus* mesoderm-inducing factor as a homologue of activin A. N 345: 729-31 (1990); **Vale W et al** Purification and characterization of an FSH releasing protein from porcine ovarian follicular fluid. N 321: 776-77 (1986)

● **Receptors: Attisano L et al** Identification of human activin and TGFβ type I receptors that form heterodimeric kinase complexes with type II receptors. Cell 75: 671-80 (1993); **Donaldson CJ et al** Molecular cloning and binding properties of the human type II activin receptor. BBRC 184: 310-6 (1992); **Hardie DG** Cell signaling: reaction with activin. The receptor for activin, a peptide messenger possibly involved in early differentiation steps in the embryo, may be ligand-activated serine kinase. Curr. Biol. 1: 321-2 (1991); **Hino M et al** Characterization of cellular receptors for erythroid differentiation factor on murine erythroleukaemia cells. JBC 264: 10309-14 (1989); **Kondo S et al** Identification of the two types of specific receptors for activin/EDF expressed on Friend leukemia and embryonal carcinoma cells. BBRC 161: 1267-72 (1989); **Mathews LS & Vale WW** Expression cloning of an activin receptor, a predicted transmembrane serine kinase. Cell 65: 973-82 (1991); **Mathews LS et al** Molecular cloning and binding properties of the human type II activin receptor. S 255: 1702-5 (1992); **Matzuk MM & Bradley A** Structure of the mouse activin receptor type II gene. BBRC 185: 404-13 (1992); **Matzuk MM & Bradley A** Cloning of the human activin receptor cDNA reveals high evolutionary conservation. BBA 1130: 105-8 (1992); **Nishimatsu S et al** Multiple genes for *Xenopus* activin receptor expressed during early embryogenesis. FL 303: 81-4 (1992)

● **Biological Activities: Frigon NL jr et al** Regulation of globin gene expression in human K562 cells by recombinant activin A. Blood 79: 765-72 (1992); **González-Manchón C & Vale W** Activin A, inhibin, and TGF-β modulate growth of two gonadal cell lines. E 125: 1666-72 (1989); **González-Manchón C et al** Activin-A modulates gonadotropin-releasing hormone secretion from a gonadotropin-releasing hormone-secreting neuronal cell line. Neuroendocrinol. 54: 373-77 (1991); **Hashimoto M et al** Activin/EDF as an inhibitor of neural differentiation. BBRC 173: 193-200 (1990); **Kitaoka M et al** Inhibition of growth hormone secretion by activin A in human growth hormone-secreting tumor cells. Acta Endocrinol. (Copenh.) 124: 666-71 (1991); **Kojima I et al** Modulation of growth of vascular smooth muscle cells by activin A. ECR 206: 152-6 (1993); **LaPolt PS et al** Activin stimulation of inhibin secretion and messenger RNA levels in cultured granulosa cells. ME 3: 1666-73 (1989); **Ling N et al** Novel ovarian regulatory peptides: inhibin, activin, and follistatin. Clin. Obstet. Gynecol. 33: 690-702 (1990); **Ling N et al** Pituitary FSH is released by a heterodimer of the β-subunits from the two forms of inhibin. N 321: 779-82 (1986); **Mitrani E et al** Activin can induce the formation of axial structures and is expressed in the hypoblast of the chick. Cell 63: 495-501 (1990); **Murata T & Ying SY** Transforming growth factor-β and activin inhibit basal secretion of prolactin in a pituitary monolayer culture system PSEBM198: 599-605 (1991); **Schubert D & Kimura H** Substratum-growth factor collaborations are required for the mitogenic activities of activin and FGF on embryonal carcinoma cells. JCB 114: 841-46 (1991); **Schubert D et al** Activin is a nerve cell survival molecule. N 344: 868-70 (1990); **Shao LE et al** Regulation of production of activin A in human marrow stromal cells and monocytes. EH 20: 1235-42 (1992); **Shao LE et al** Effect of activin A on globin expression in purified human erythroid progenitors. Blood 79: 773-81 (1992); **Thomsen G et al** Activins are expressed early in *Xenopus* embryogenesis and can induce axial mesoderm and anterior structures. Cell 485-93 (1990); **Ueno N et al** Activin is a cell differentiation factor. PGFR 2: 113-24 (1990); **Xiao S & Findlay JK** Interactions between activin and follicle-stimulating hormone-suppressing protein and their mechanisms of action on cultured rat granulosa cells. Mol. Cell. Endocrinol. 79: 99-107 (1991); **Yamashita T et al** Activin A/erythroid differentiation factor induces thromboxane A2 synthetic activity in murine erythroleukaemia cells. JBC 266: 3888-92 (1991); **Yamashita T et al** Expression of activin A/erythroid differentiation factor in murine bone marrow stromal cells. Blood 79: 304-07 (1992)

● **Assays: Demura R et al** Competitive protein binding assay for activin A/EDF using follistatin determination of activin levels in human plasma. BBRC 185: 1148-54 (1992); **Shintani Y et al** Radioimmunoassay for activin A/EDF. Method and measurement of immunoreactive activin A/EDF levels in various biological materials. JIM 137: 267-74 (1991); **Wong WLT et al** Monoclonal antibody based ELISAs for measurement of activins in biological fluids. JIM 165: 1-10 (1993); for additional information see also: references in entries referring to the alternative names of activin A.

● **Clinical Use & Significance: Broxmeyer HE et al** Selective and indirect modulation of human multipotential and erythroid hematopoietic progenitor cell proliferation by recombinant activin and inhibin. PNAS 85: 9052-6 (1988); **Nakao K et al** Effects of erythroid differentiation factor (EDF) on proliferation and differentiation of human hematopoietic progenitors. EH 19: 1090-95 (1991); **Shiozaki M et al** Evidence for the participation of endogenous activin A/erythroid differentiation factor in the regulation of erythropoiesis. PNAS 89: 1553-56 (1992); **Shiozaki M et al** *In vivo* treatment with erythroid differentiation factor (EDF/activin A) increases erythroid precursors (CFU-E and BFU-E) in mice. BBRC 165: 1155-61 (1989); **Shiozaki M et al** Differentiation-inducing and growth-inhibitory activities of erythroid differentiation factor (EDF/activin A) toward mouse erythroleukemic cells *in vivo*.

Activin A knock-outs: see addendum: Activin A.

IJC 45: 719-23 (1990); **Yu J et al** Importance of FSH-releasing protein and inhibin in erythrodifferentiation. N 330: 765-67 (1987); **Yu J et al** Specific roles of activin/inhibin in human erythropoiesis *in vitro* ANY 628: 199-211 (1991)

Activin AB: see: Activin A.

Activin B: see: Activin A.

Activin-binding protein: The protein that binds » Activin A is identical with » Follistatin.

Activin receptor-like kinases: see: ALK.

Acute phase reaction: The term acute phase response summarizes a number of very complex neurological, endocrine, and metabolic changes observed in an organism, either locally or systemically, a short time after injuries or the onset of infections, immunological reactions, and inflammatory processes (see also: Neuroimmune network). Each form of injury or tissue disorder that precipitates an inflammatory response (see also: Inflammation) inevitably also causes an acute phase reaction.

An acute phase reaction is characterized, among other things, by fever, and an increase in the num-

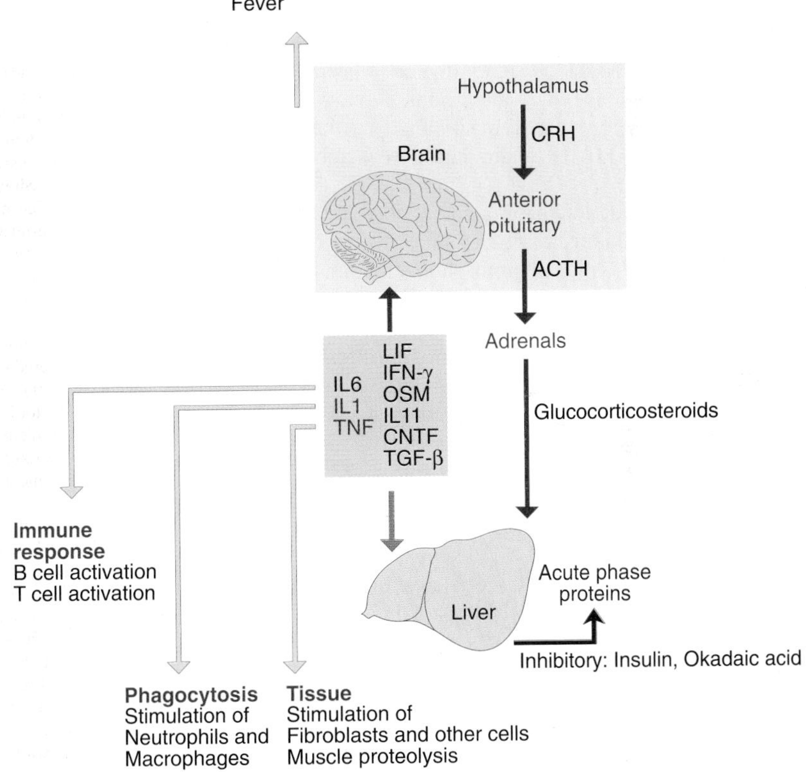

Regulation of Acute phase reactions and synthesis of Acute phase proteins.
Inflammatory cytokines such as IL6, IL1, TNF, and others such as TGF, IFN, and LIF are produced by inflammatory cells. They induce local and systemic reactions. Among other things these mediators are involved in activation of leukocytes, fibroblasts, endothelial cells, and smooth muscle cells, inducing the synthesis of further cytokines. These mediators also have direct actions in hepatocytes of the liver.
Activities are enhanced indirectly by activation of the pituitary/adrenal gland axis which involves synthesis of ACTH and subsequent production of cortisol. Cortisol can enhance expression of IL6 receptors in liver cells and thus promotes IL6-mediated synthesis of acute phase proteins.
Negative regulatory loops can involve inhibition of synthesis of IL6, IL1, and TNF by cortisol and inhibition of the synthesis of IL1 and TNF in monocytes by IL6.
Of all mediators participating in the induction and regulation of acute phase protein synthesis IL6 appears to induce the broadest spectrum of acute phase proteins whereas IL1 and TNF only induce the synthesis of subsets of these proteins.

Examples of acute phase protein expression levels in various disease states.

Strongly elevated expression	Negligible expression	Unaltered expression
Bacterial infections	Arthralgia	Colitis ulcerosa
Fractures	Myalgia	Leukemia
Juvenile chronic arthritis	Irritable colon	Mixed connective tissue disease
Tumors	Back pain	Osteoarthritis
Crohn's disease		Sclerodermia
Surgery		Systemic lupus erythematodes
Polymyalgia rheumatica		
Rheumatoid arthritis		
Systemic vasculitis		
Burns		

bers of peripheral leukocytes, in particular an increase in the numbers of circulating neutrophils and their precursors. At the same time one observes cellular and biochemical alterations, in particular the coordinated synthesis of so-called » acute phase proteins (APP) or acute phase reactants (APR) in the liver.

The acute phase reaction is initiated and mediated by a number of inflammatory » cytokines secreted by a variety of cell types (polymorphonuclear leukocytes, fibroblasts, endothelial cells, monocytes, lymphocytes etc.). The cascades of inflammatory cytokines in different tissues represent amplification and regulatory pathways controlling the development of acute phase responses *in vivo*. Therefore, this reaction is a direct consequence of the biological activities of an organism's own mediator substances and not the result of intrinsic properties of the infectious and/or inflammatory agents *per se*.

The main mediator of the acute phase reaction is » IL6, which, in turn, is regulated by » IL1. In cultures of hepatocytes IL6 induces a spectrum of biochemical alterations which more or less resemble the patterns also observed in the serum during an acute phase reaction *in vivo*. Like IL6 » LIF and » IL11 also induce an almost complete spectrum of physiological changes. A much more restricted pattern of alterations is found with » IL1 and » TNF with each of them showing almost identical activities. Some acute phase proteins are also induced by » CNTF (ciliary neurotrophic factor).

Type, duration and degree of cellular activities and changes in these activities are influenced by the doses and the combinations of the different cytokines and also by the order in which individual factors act upon a particular target cell.

A variety of low molecular weight substances normally acting on the cells also play a major role in the sequence of events. Glucocorticoids, for example, are of particular importance for the progression of the acute phase reaction because their synthesis can be influenced by a number of cytokines, and they also, in turn, influence the synthesis of cytokines. In hypophysectomized individuals one observes a pronounced reduction of the synthesis of » acute phase proteins.

Akira S & Kishimoto T IL6 and NF-IL6 in acute-phase response and viral infection. IR 127: 25-50 (1992); **Andus T et al** Effects of cytokines on the liver. Hepatology 13: 364-75 (1991); **Baumann H & Gauldie J** The acute phase response. IT 15: 74-80 (1994); **Heinrich PC et al** Interleukin 6 and the acute phase response. BJ 269, Suppl. 51-66 (1990); **Kushner I et al** The acute phase response is mediated by heterogeneous mechanisms. ANY 557: 19-30 (1988); **Kushner I et al** The acute phase response: (an overview). MiE 163: 373-83 (1988); **Pepys MB** (edt) Acute phase proteins and the acute phase response. Springer, Heidelberg, 1989); **Sehgal PB et al** Regulation of the acute phase and immune responses: interleukin-6. ANY Vol. 557 (1989); **Thompson D et al** Insulin modulation of acute phase protein production in a human hepatoma cell line. Cytokine 3: 619-26 (1991); for further information see also: Acute phase proteins, » Inflammation, and » Septic shock.

Acute phase protein inducing factor: see: APPIF.

Acute phase proteins: (abbrev. APP) or acute phase reactants (APR) is the generic name given to a group of ≈ 30 different biochemically and functionally unrelated proteins. The levels of acute phase proteins in the serum are either increased (positive APPs) or reduced (negative APPs) approximately 90 minutes after the onset of an inflammatory reaction (see: Inflammation). The more important acute phase proteins are usually glycoproteins. Exceptions are C-reactive protein (CRP) and serum amyloid A protein (SAA).

Some acute phase proteins of inflammation and their function.

Protein	Function
α1 acid glycoprotein	interaction with collagen promotion of fibroblast growth binding of certain steroids
α1-antichymotrypsinogen	protease inhibitor
α1-antitrypsin	protease inhibitor resolution of emphysema
α2 antiplasmin	modulation of coagulation cascade
Antithrombin III	modulation of coagulation cascade
C1 inhibitor	negative control of complement cascade
C2	complement component
C3	complement component
C4	complement component
C4 binding protein	complement component
C5	complement component
C9	complement component
C-reactive protein	binding to membrane phosphorylcholine with complement activation and opsonization interaction with T and B cells
Ceruloplasmin	copper transport protein reactive oxygen scavenger ?
Factor VIII	clotting formation of fibrin matrix for repair
Factor B	complement component
Ferritin	iron transport protein
Fibrinogen	clotting formation of fibrin matrix for repair
Fibronectin	fibrin clot formation
Haptoglobin	hemoglobin scavenger
Heme oxygenase	heme degradation
Hemopexin	heme binding and transport protein
Heparin cofactor II	proteinase inhibitor
Kallikrein	vascular permeability and dilatation
LPS binding protein	macrophage activation
Manganese superoxide dismutase	copper zinc binding protein formation of reactive oxygen species
Mannose-binding protein	serum lectin
Plasminogen	proteolytic activation of complement, clotting, fibrinolysis
Plasminogen activator inhibitor I	protease inhibitor
Prothrombin	clotting formation of fibrin matrix for repair
Serum amyloid A	cholesterol and HDL scavenger
Serum amyloid P component	formation of IgG immune complexes
von Willebrand factor	coagulation protein

Acute phase proteins are synthesized predominantly in the liver with each hepatocyte possessing the capacity to produce the entire spectrum of these proteins. Following stimulation of single hepatocytes within individual lobules one observes a stimulation of further hepatocytes and this process continues until almost all hepatocytes produce these proteins and release them into the circulation. The various acute phase proteins differ markedly in the rise or decline of their plasma levels and also in their final concentrations. Nevertheless, acute phase responses generate a characteristic serum protein profile which may, however, differ widely from species to species.

Acute phase proteins regulate immune responses, function as mediators and inhibitors of inflammatory processes, act as transport proteins for products generated during the inflammatory process (the heme-binding protein hemopexin, and » haptoglobin), and/or play an active role in tissue

Acute phase protein expression: some interspecies differences.

Protein	Human	Rabbit	Mouse	Rat
α1 acidic glycoprotein	++	?	+++	+++
α1 proteinase inhibitor	++	?	?	++
α1 antichymotrypsin	++	?	+	–
α2 macroglobulin	0	++	?	+++
C-reactive protein	+++	+++	+	+
Ceruloplasmin	+	?	?	+
Fibrinogen	++	++	++	++
Haptoglobin	++	++	++	++
Hemopexin	+	?	+	++
Serum amyloid A protein	+++	+++	A	+++
Serum amyloid P protein	0	?	++	0
Transferrin	–	+	–	?

A: absent in plasma; 0: no significant change; ?: unknown; +++: elevation 100-fold or more;
++: elevation 2-5-fold; +: elevation less than 2-fold.

repair and remodeling. Some of the acute phase proteins function as cytokines. C-reactive protein, for example, activates macrophages (see also: MAF, macrophage activating factor) while some other acute phase proteins influence the chemotactic behavior of cells (see: MIF, migration inhibition factor). Some acute phase proteins possess antiproteolytic activity and presumably block the migration of cells into the lumen of blood vessels thus helping to prevent the establishment of a generalized systemic inflammation. A failure to control these processes, i. e., an uncontrolled acute phase reaction, eventually has severe pathological consequences (see also: septic shock).

The elevated serum concentrations of certain acute phase proteins are of diagnostic relevance and also of prognostic value. Their measurement, for example, allows inflammatory processes to be distinguished from functional disturbances with similar or identical clinical pictures. Under normal circumstances an acute phase response is not observed with functional disturbances that are not the result of an inflammatory process, thereby allowing the differentiation between failure of funcOtion and organic disease.

Some acute phase reactions are also observed in chronic disorders such as rheumatoid arthritis and chronic infections while malignant diseases are almost invariably associated with acute phase responses and therefore the determination of acute phase protein levels cannot be used for differential diagnoses in these instances. There are many diseases in which the rise in the synthesis of acute phase proteins parallels the degree and progression of the inflammatory processes.

The co-ordinated expression of many acute phase proteins as a direct consequence of the activities of several cytokine stimuli can be explained, at least in part, by the fact that the regulatory sequences of the genes encoding these acute phase proteins contain so-called cytokine response elements (see also: IL6-RE as an IL6-specific element). These elements are recognized specifically by transcription factors that mediate the activity of these genes in a cell- and/or tissue-specific manner (see also: Gene expression).

Additive, synergistic, co-operative, and antagonistic effects between cytokines and other mediator substances do occur and have been observed in almost all combinations. IL1 and also » IFN-γ reduce some of the effects of IL6. Some of the effects of » IL2 and IL6 are antagonized by » TGF-β. The combined action of two or even more cytokines may produce effects that no factor on its own would be able to achieve. In cultured HepG2 hepatoma cells » IL1, » IL6, » TNF-α and » TGF-β induce the synthesis of antichymotrypsin and at the same time repress the synthesis of albumin and AFP (α-Fetoprotein). The synthesis of fibrinogen is induced by » IL6 and this effect is, in turn, suppressed by IL1α (see: IL1), » TNF-α or » TGF-β1. The increased synthesis of » haptoglobin mediated by IL6 is suppressed by TNF-α. Insulin inhibits the synthesis of some negative acute phase proteins (prealbumin, » transferrin, and fibrinogen, in HepG2 hepatoma cells.

Baumann H The electrophoretic analysis of acute phase plasma proteins. MiE 163: 566-94 (1988); **Campos SP & Baumann H** Insulin is a prominent modulator of the cytokine-stimulated expression of acute-phase plasma protein genes. MCB 12: 1789-97 (1992); **De Simone V & Cortese R** The transcriptional regulation of liver-specific gene expression. In: McLean N (edt) Oxford Surveys on Eukaryotic Genes, Vol. 5, pp. 51-90, (1988); **Dofferhoff ASM et al** Patterns of cytokines, plasma endotoxin, plasminogen activator inhibitor, and acute phase proteins during the treatment of severe sepsis in humans. Crit. Care Med. 20: 185-92 (1992); **Ikawa M & Shozen Y** Quantification of acute phase proteins in rat serum and in the supernatants of a cultured rat hepatoma cell line and cultured primary hepatocytes by an enzyme-linked immunosorbent assay. JIM 134: 101-6 (1990); **Isshiki H et al** Reciprocal expression of NF-IL6 and C/EBP in hepatocytes: possible involvement of NF-IL6 in acute phase protein gene expression. The New Biologist 3: 63-70 (1991); **Koj A** Definition and classification of acute phase proteins. In: Gordon AH & Koj A (eds) The acute phase response to injury and infection, V. 10. Elsevier, Amsterdam, pp. 139-232 (1985); **Koj A et al** Role of cytokines and growth factors in the induced synthesis of proteinase inhibitors belonging to acute phase proteins. Biomed. Biochem. Acta 50: 421-5 (1991); **Liao WSL & Stark GR** Cloning of rat cDNAs for eight acute phase reactants and kinetics of induction of mRNAs following acute inflammation. Adv. Inflamm. Res. 10: 220-22 (1986); **Mackiewicz A et al** Transforming growth factor β1 regulates production of acute phase proteins. PNAS 87: 1491-5 (1990); **Mackiewicz A et al** Effects of cytokine combinations on acute phage protein production in two human hepatoma cell lines JI 146: 3032-37 (1991); **Mackiewicz A et al** Mechanisms regulating glycosylation of human acute phase proteins. Arch. Immunol. Ther. Exp. Warsz. 39 365-73 (1991); **Oldenburg HSA et al** Cachexia and the acute phase protein response in inflammation are regulated by interleukin 6. EJI 23: 1889-94 (1993); **Perlmutter DH et al** Cachectin/tumor necrosis factor regulates hepatic acute phase gene expression. JCI 78: 1349-54 (1986); **Richards C et al** Cytokine control of acute phase protein expression. ECN 2: 89-98 (1991); **Rienhoff HY Jr** Molecular and cellular biology of serum amyloid A. Mol. Biol. Med. 7: 287-98 (1990); **Steel DM & Whitehead AS** The major acute phase reactants: C-reactive protein, serum amyloid P component and serum amyloid A protein. IT 15: 81-8 (1994); **Thompson D et al** Insulin modulation of acute phase protein production in a human hepatoma cell line. Cytokine 3: 619-26 (1991); **Thompson D et al** The value of acute phase protein measurements in clinical practice. Ann. Clin. Biochem. 29: 123-31 (1992); **Whicher JT et al** Acute phase proteins. Mod. Meth. Pharmacol. 5: 101-28 (1989); for further information see also: Acute phase reaction, » Inflammation, and » Septic shock.

Acute phase response: see: Acute phase reaction.

ADCC: *antibody-dependent cellular cytotoxicity*
ADCC is an antibody-dependent immune reaction mediated by LGL cells (large granular lymphocytes), i. e. killer cells and NK lymphocytes. The ADCC reaction is very sensitive. It plays an important role in hypersensitivity and occurs at antibody concentrations which are well below those leading to complement lysis.
The importance of ADCC *in vivo* is difficult to evaluate because the activities of NK cells and also the direct effects of antibodies cannot be separated from the ADCC reaction.
NK cells play a major role as effector cells in the spontaneous destruction of tumor cells and virus-infected cells without having been previously sensitized antigenically. The activity of the NK cells is not coupled to functionally to the expression of major histocompatibility complex. This means that tumor cells and virus-infected cells not expressing MHC can be attacked also. The reaction is characterized by humoral antibodies being used as target cell antigens.
The lysis of the cells occurs after a specific contact between the Fc receptors of the K cells and IgG antibodies located on the target cells (see also: CD16, CD32).
The K cell receptor has a low affinity for free IgG and the cells therefore can also be active *in vivo* in the presence of a large excess of non-target cell-specific serum antibodies.

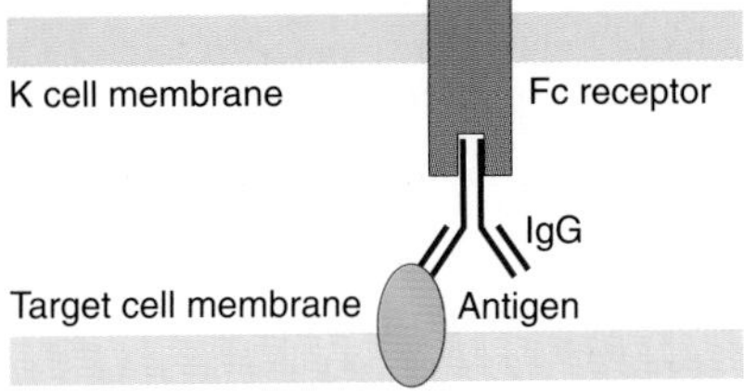

Antibody-dependent cellular cytotoxicity (ADCC). The specific contact between the killer cell and a target cell presenting an antigen is mediated by antibodies anchored via their Fc domain to Fc receptors on the K cell. Cell-to-cell contacts most likely involve interactions with other adhesion molecules and their counterreceptors.

The ADCC reaction can be induced in macrophages and granulocytes by several » hematopoietins (hematopoietic growth factors) including » M-CSF. It is also influenced by interferons (see: IFN for general information about interferons, and IFN-α, IFN-β, IFN-γ), » TNF-α, » TNF-β and » IL4.
ADCC defects in combination with a reduced activity of NK cells and anomalous granulocyte granules are observed in patients with autosomal recessive Chediak-Steinbrinck-Higashi syndrome. Afflicted patients show disturbances of the granulopoiesis and the neutrophil leukocytes contain abnormally large lysosomal granules. These patients die early in childhood mainly due to severe bacterial infections and lymphoproliferative diseases.

Eisenthal A & Rosenberg SA The effect of various cytokines in the *in vitro* induction of antibody-dependent cellular cytotoxicity in murine cells. JI 142: 2307-13 (1989); **Fan S et al** Activation of macrophages for ADCC *in vitro* Effects of IL4, TNF, interferons-α/β, interferon-γ, and GM-CSF. CI 135: 78-87 (1991); **Ravetch JV & Kinet JP** Fc receptors. ARI 9: 457-92 (1991)

Adenosine-uridine binding factor: abbrev. AUBF. See: ARE (AU-rich element).

ADF: *ATL (adult T cell leukemia)-derived factor*
■ **ALTERNATIVE NAMES:** 3B6-IL1; 3B6-IL1-like factor; SASP (surface-associated sulfhydryl protein); TCGF(IL2)-receptor inducing factor; TIA (IL2R/p55 (Tac) inducing factor). Two activities described as » BSF MP6 (B cell stimulatory factor) and » ECEF (eosinophil cytotoxicity-enhancing factor) are very closely related to, or identical with, ADF. See also: individual entries for further information.
■ **SOURCES:** ADF is constitutively produced and secreted by T cells. The factor is also produced by HTLV-I-transformed T cells. Cell lines producing ADF also appear to produce another factor, » HGI (hepatocyte growth inhibitory factor), although the biological significance of this finding is unclear. ADF/thioredoxin also appear to exist in a membrane-associated form.
■ **PROTEIN CHARACTERISTICS:** ADF is an acidic protein of 11.6 kDa not related to any other known lymphokine. ADF contains a redox-active disulfide (-Cys-Gly-Pro-Cys-) and is one of the homologues of the oxidoreductase » thioredoxin (104 amino acids).
■ **GENE STRUCTURE:** The human thioredoxin (Trx) gene extends over 13 kb and consists of five exons. The promoter region is very G + C rich and does not contain a classical TATA or CCAAT box, but has three consensus sequences for high-affinity Sp1 binding (see also: Gene expression). Southern analysis demonstrates the presence of several Trx genes in the human genome.
■ **BIOLOGICAL ACTIVITIES:** ADF induces the synthesis and expression of high affinity receptors for » IL2 in various lymphoid cell types. It thus indirectly promotes the » autocrine growth of IL2-dependent T cells. ADF regulates the binding of transcription factor NFκB (see also: Gene expression) to the enhancer region of the IL2R/p55 gene (IL2 receptor).
The growth-promoting effect of ADF is correlated with the thiol group-dependent proton transfer activity of thioredoxin, a protein with dithiol-disulfide oxidoreductase activity. ADF synergizes with » IL1 and IL2 and allows virus-infected cells to respond to suboptimal doses of growth factors. ADF can protect cells against the cytotoxic actions of » TNF. ADF also plays a role in the immortalization of lymphocytes by viruses such as HTLV-I and EBV. ADF acts as a growth factor for hepatoma cell lines. ADF has been shown to regulate eosinophil migration.
■ **CLINICAL USE & SIGNIFICANCE:** 13-cis-retinoic acid (RA), which is a competitive inhibitor of thioredoxin reductase, suppresses the expression of IL2 receptors on HTLV-I (+) cells and reduces the cell number, viability, and proliferation of ADF-producing lymphoid cells. The inhibition of the ADF/thioredoxin system may therefore be a new therapeutic approach for retrovirus-related disorders. A loss of ADF-producing cells has been observed to occur in patients undergoing HIV infections. Down-regulation of ADF production by HIV-1 may play a role in the pathophysiology of HIV-infected individuals.
● **BIOCHEMISTRY & MOLECULAR BIOLOGY: Fujii S et al** Immunohistochemical localization of adult T cell leukemia-derived factor, a human thioredoxin homologue, in human fetal tissues. Virchows Arch. [A], 419: 317-26 (1991); **Forman-Kay JD et al** Determination of the positions of bound water molecules in the solution structure of reduced human thioredoxin by heteronuclear three-dimensional nuclear magnetic resonance spectroscopy. J. Mol. Biol. 220: 209-16 (1991); **Jacquot JP et al** Human thioredoxin reactivity-structure/function relationship. BBRC 173: 1375-81 (1990); **Martin H & Dean M** Identification of a thioredoxin-related protein associated with plasma membranes. BBRC 175: 123-8 (1991); **Rosen A et al** A T helper cell x Molt 4 human hybridoma constitutively producing B cell stimulatory and inhibitory factors. LR 5: 185-204 (1986); **Teshigawara K et al** Adult T leukemia cells produce a lymphokine that augments interleukin 2 receptor expression. J. Mol. Cell. Immunol. 2: 17-26 (1985); **Tonissen KF & Wells JR** Isolation and characterization of human thioredoxin-encoding genes. Gene 102: 221-8 (1991); **Wollman EE et al** Cloning and expression of a cDNA for human thioredoxin. JBC 263: 15506-12 (1988)
● **BIOLOGICAL ACTIVITIES: Hori K et al** Regulation of eosinophil migration by adult T cell leukemia-derived factor. JI 151: 5624-30 (1993); **Matsuda M et al** Protective activity of adult T cell leukemia-derived factor (ADF) against tumor necrosis factor-dependent cytotoxicity on U937 cells. JI 147: 3837-41 (1991); **Mitsui A et al** Reactive oxygen-reducing and protein-refolding activities of adult T cell leukemia-derived factor/human thioredoxin. BBRC 186: 1220-6 (1992); **Nakamura H et al** Expression and growth-promoting effect of adult T cell leukemia-derived factor. A human thioredoxin homologue in hepatocellular carcinoma. Cancer 69: 2091-7 (1992); **Okada M et al** TCGF(IL 2)-receptor inducing factor(s). II. Possible role of ATL-derived factor (ADF) on constitutive IL 2 receptor expression of HTLV-I(+) T cell lines. JI 135: 3995-4003 (1985); **Tagaya Y et al** ATL-derived factor (ADF), an IL2 receptor/Tac inducer homologous to thioredoxin; possible involvement of dithiol-reduction in the IL2 receptor induction. EJ 8: 757-64 (1988); **Tagaya Y et al** Role of ATL-derived factor (ADF) in the normal and abnormal cellular activation: involvement of dithiol related

reduction. Mol. Immunol. 27: 1279-89 (1990); **Wakasugi N et al** Adult T cell leukemia-derived factor/Thioredoxin produced by both HTLV-1 and EBV-transformed lymphocytes acts as an autocrine growth factor and synergises with IL1 and IL2. PNAS 87: 8282-86 (1990); **Wollman EE et al** Cloning and expression of a cDNA for human thioredoxin. JBC 263: 15506-12 (1988); **Yamamoto S et al** Induction of Tac antigen and proliferation of myeloid leukemic cells by ATL-derived factor: comparison with other agents that promote differentiation of human myeloid or monocytic leukemic cells. Blood 67: 1714-20 (1986); **Yamauchi A et al** Lymphocyte transformation and thiol compounds: the role of ADF/thioredoxin as an endogenous reducing agent. Mol. Immunol. 29: 263-70 (1992); **Yodoi J et al** IL2 receptor and Fc ε R2 gene activation in lymphocyte transformation: possible roles of ATL-derived factor. Int. Symp. Princess Takamatsu Cancer Res. Fund 19: 73-86 (1988); **Yodoi J & Tursz T** ADF, a growth-promoting factor derived from adult T cell leukemia and homologous to thioredoxin: involvement in lymphocyte immortalization by HTLV-I and EBV. ACR 57: 381-411 (1991)

● **CLINICAL USE & SIGNIFICANCE: Masutani H et al** ADF (adult T cell leukemia-derived factor)/human thioredoxin and viral infection: possible new therapeutic approach. AEMB 319: 265-74 (1992); **Masutani H et al** Dysregulation of adult T-cell leukemia-derived factor (ADF)/thioredoxin in HIV infection: loss of ADF high-producer cells in lymphoid tissues of AIDS patients. AIDS Res. Hum. Retroviruses 8: 1707-15 (1992)

ADH: *antidiuretic hormone* The hormone is identical with » Vasopressin.

Adherence-promoting factor: This factor is released by cultured human lung fragments and promotes the adhesion of polymorphonuclear leukocytes to cultured human vascular endothelial cells. The activity is neutralized by anti-human IL1 antisera and appears to be a mixture of the two molecular forms of » IL1.

Bochner BS et al Interleukin 1 production by human lung tissue. I. Identification and characterization. JI 139: 2297-302 (1987)

Adherent cell growth: see: colony formation assay. See also: Cell culture.

Adherent lymphokine-activated killer cells: abbrev. ALAK. See: LAK cells (lymphokine-activated killer cells).

ADIF: *autocrine differentiation-inhibiting factor* A poorly characterized factor secreted by tsAEV-LSCC HD3 chicken erythroid cells transformed by the avian erythroblastosis virus. ADIF blocks the differentiation of the chicken erythroid cells that synthesize and secrete it into the culture medium without affecting proliferation (see also: Autocrine stimulation). ADIF also prevents dimethylsulfoxide-induced synthesis of hemoglobin by murine Friend erythroleukemia cells and also blocks erythroid differentiation in normal human and murine bone marrow, selectively affecting early » BFU-E (burst-forming) erythroid precursor cells, but not the more advanced » CFU-E erythroid precursors. The biological activities of AGIF resemble those of » DIP (differentiation-inhibiting protein).

Krsmanovic V et al Autocrine differentiation-inhibiting factor (ADIF) from chicken erythroleukemia cells acts on human and mouse early BFU-E erythroid progenitors. BBRC 157: 762-9 (1988)

Adipocyte differentiation associated protein P422: see: MDGI (mammary-derived growth inhibitor).

Adipocyte growth factor: abbrev. AGF. This factor is identical with » bFGF.

Adipogenesis inhibitory factor: see: AGIF.

Adiuretin: This factor is identical with » Vasopressin.

Adoptive immunotherapy: abbrev. AIT. Name given to a form of therapy in which immune cells are transferred to tumor-bearing hosts. The immune cells have antitumor reactivity and can mediate direct or indirect antitumor effects. In human patients the treatment protocol involves the intravenous injection of » IL2 for several days, followed by leukopheresis (collection of leukocytes), cultivation of isolated leukocytes *in vitro* in the presence of high concentrations of IL2, and subsequent re-infusion of the leukocytes. This *in vitro* treatment leads to the proliferation and expansion of a subpopulation of peripheral mononuclear blood cells known as » LAK cells (lymphokine-activated killer cells; see also: Cell activation).

Adoptive immunotherapy is used in the treatment of various tumors including melanoma, renal carcinoma, and other tumor targets. It can also be employed for maintenance therapy when the tumor burden is low, or after bone marrow transplantation to eliminate minimal residual disease, or in early relapse. Adoptive immunotherapy with LAK cells may also be useful in preventing GVHD in human bone marrow transplant recipients due to the ability of LAK cells to inhibit generation of allospecific cytotoxic T cells.

The IL2/LAK therapy is very demanding, both for the patient and for the clinician, and due to its toxicity usually has to be carried out in an intensive care unit. Adverse side effects of the IL2 treatment include fever, diarrhea, hypotony, myocardial infarction, confusion, insomnia, coma, thrombocytopenia, and pulmonary edemas due to the interstitial incorporation of water caused by a capillary leakage syndrome. IL2 interferes with blood-brain-barrier functions and the integrity of the endothelium of brain vessels, and this may, at least in part, explain the neuropsychiatric side effects (sleepiness, disorientation, depression) associated with IL2 therapy.

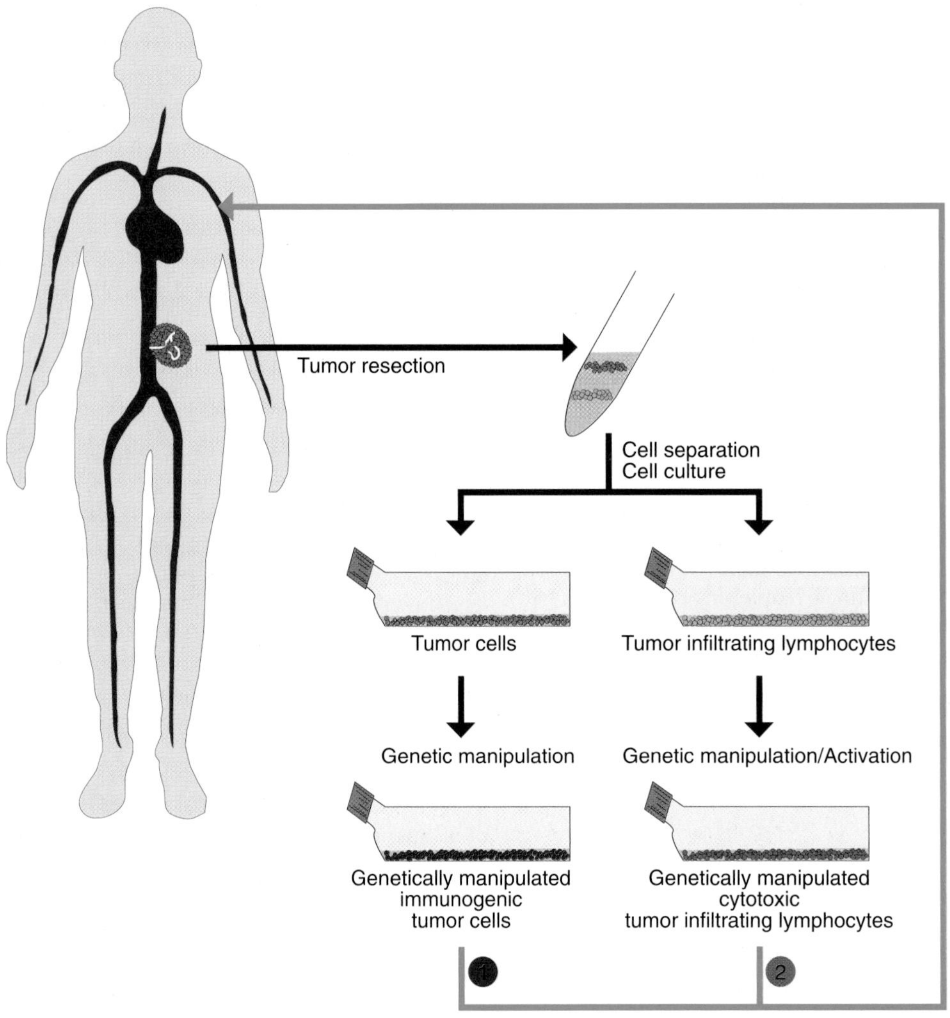

Two strategies of adoptive immunotherapy.
Following resection of the tumor the cells are separated to obtain pure cultures of tumor cells. Alternatively tumor tissues are used to collect tumnor infiltrating lymphocytes.
Tumor cells can be propagated and manipulated in vitro. Genetic manipulation is used to introduce foreign genes into the tumor cell genome. One example is the introduction of a gene encoding a particular cell surface marker that is expressed at very low levels by the unmanipulated tumor cells. Overexpression of this cell surface marker by manipulated tumor cells renders them immunogenic. The rationale behind reinfusion of the manipulated tumor cells into the patient is to provide a trigger stimulus to the patients's immune system which will then also recognize the same marker on unmanipulated residual tumor cells (strategy 1).
Another strategy (strategy (2) employs isolation of tumor infiltrating lymphocytes. These cells can be grown and activated/manipulated in vitro to render them more cytotoxic to tumor cells. Following reinfusion into the patients the manipulated/activated cells home in on tumor tissues and can be used to destroy selectively any residual tumor masses.

Many of the side effects resemble those also observed as a reaction to » TNF, and it has been assumed that they may be caused by IL2 causing the secretion of TNF and other cytokines, and also low molecular weight substances such as prostaglandins. Drastic overstimulation of LAK cells by IL2 may cause them to attack non-malignant cells with the toxic effects of an LAK/IL2 treatment then resembling those also observed during graft-versus-host reactions.

For a number of tumors refractory to conventional chemo- or radiotherapy, such as progredient hypernephromas, malignant melanomas, and colorectal carcinomas, IL2/LAK treatment may result in a complete or partial remission if more than 10^{10}-10^{11} LAK cells are reinfused into the patient. Approximately one third of the remissions in melanoma and hypernephroma patients are complete and of long duration. Partial remissions, defined as a reduction of the total tumor mass by more than 50 % of the initial values, have been observed to be short-lived in melanoma and hypernephroma patients.

It has been observed that tumors responding to IL2/LAK treatment appear to express certain MHC class II antigens (HLA-DR antigens) while non-responders do not express them. Another reason for unsatisfactory results obtained *in vivo* may be due to the preferential migration of transfused LAK cells to the lung, liver, and spleen, thus effectively bypassing the tumor. In addition, several inhibitors of LAK cell activity may also play an important role.

One way to enhance the efficacy of cytokine therapy and at the same time to reduce toxicity may be the simultaneous administrations of several synergising cytokines. Synergistic effects have been observed, for example, with combinations of IL1 and » IFN-α or » TNF-α in the treatment of tumor-bearing mice. The mechanism of action of these combinations is still a matter of controversy and does not allow the definition of optimal regimens (see also: LAK cells).

Improvements of adoptive immunotherapy have been reported with the use of adherent » LAK cells (so-called ALAK cells) or » TILs (tumor-infiltrating lymphocytes) isolated from tumor tissue and activated *in vitro*. Regional administration of lymphokine-activated killer cells, if possible, can be superior to intravenous application in some instances. It has been observed that » CIK (cytokine-induced killer cells) can be generated which are even more cytotoxic than LAK cells.

Lately IL2 was shown to be capable of inducing LAK cells *in vivo*, thus obviating the need of a demanding, time-consuming, and expensive cell transfer process (see: LAK cells) while at the same time reducing toxicity without affecting efficacy. For a novel approach in cancer therapy involving the use of genetically engineered cells see: Cytokine gene transfer.

● **REVIEWS: Andreesen R & Hennemann B** Adoptive immunotherapy with autologous macrophages: Current status and future perspectives. Pathobiology 59: 259-63 (1991); **Atzpodien J & Kirchner H** Cancer, cytokines, and cytotoxic cells: interleukin-2 in the immunotherapy of human neoplasms. Klin. Wochenschr. 68: 1-11 (1990); **Bertagnolli MM et al** Approaches to immunotherapy of cancer: Characterization of lymphokines as second signals for cytotoxic T cell generation. Surgery 110: 459-68 (1991); **Dullens HFJ et al** Cancer treatment with interleukins 1, 4, and 6 and combinations of cytokines: a review. *In vivo* 5: 567-70 (1991); **Hillman GG et al** Adoptive immunotherapy of cancer: biological response modifiers and cytotoxic cell therapy. Biotherapy 5: 119-29 (1992); **Kaplan G et al** Rational immunotherapy with interleukin 2. BT 10: 157-62 (1992); **Kolitz JE & Mertelsmann R** The immunotherapy of human cancer with interleukin 2: Present status and future directions. Cancer Invest. 9: 529-42 (1991); **Melief CJ** Tumor eradication by adoptive transfer of cytotoxic T lymphocytes. ACR 58: 143-75 (1992); **Parkinson DR** Interleukin 2 in cancer therapy. Semin. Oncol. 15; 10-26 (1988); **Rosenberg SA** The development of new immunotherapies for the treatment of cancer using interleukin-2. Ann. Surg. 208: 121-35 (1989); **Rosenberg SA** Immunotherapy and gene therapy of cancer. CR 51: s5074-s9 (1991); **West WH** IL2 and adoptive cellular therapy of renal carcinoma and other malignancies. Cancer Treatm. Rev. Suppl. June, 83-9 (1989)

● **CLINICAL USE & SIGNIFICANCE: Den Otter W et al** Effective immunotherapy with local low doses of interleukin-2. *In vivo* 5: 561-6 (1991); **Dillman RO et al** Continuous interleukin-2 and lymphokine-activated killer cells for advanced cancer: A National Biotherapy Study Group trial. J. Clin. Oncol. 9: 1233-40 (1991); **Fortis C et al** Recombinant interleukin-2 and lymphokine-activated killer cells in renal cancer patients: II. Characterization of cells cultured *ex vivo* and their contribution to the *in vivo* immunomodulation. Cancer Immunol. Immunother. 33: 128-32 (1991); **Gautier Hermann G et al** Recombinant interleukin-2 and lymphokine-activated killer cell treatment of advanced bladder cancer: Clinical results and immunological effects. CR 52: 726-33 (1992); **George RE et al** *In vitro* cytolysis of primitive neuroectodermal tumors of the posterior fossa (medulloblastoma) by lymphokine-activated killer cells. J. Neurosurg. 69: 403-9 (1988); **Koretz MJ et al** Randomized study of interleukin 2 (IL2) alone vs IL2 plus lymphokine-activated killer cells for treatment of melanoma and renal cell cancer. Arch. Surg. 126: 898-903 (1991); **Lauria F et al** Continuous infusion of interleukin-2 in two relapsed high grade non-Hodgkin lymphoma patients Effectiveness and tolerability. Eur. J. Cancer 27: 521-2 (1991); **Riddell SR et al** Phase I study of cellular adoptive immunotherapy using genetically modified $CD8^+$ HIV-specific T cells for HIV seropositive patients undergoing allogeneic bone marrow transplant. The Fred Hutchinson Cancer Research Center and the University of Washington School of Medicine, Department of Medicine, Division of Oncology. Hum. Gene Ther. 3: 319-38 (1992); **Rosenberg SA et al** A progress report on the treatment of 157 patients with advanced cancer using lymphokine-activated killer cells and interleukin-2 or high-dose interleukin-2 alone. NEJM 316: 889-97 (1987); **Rubin JT et al** Immunohistochemical correlates of response to recombinant interleukin-2-based immunotherapy in humans. CR 49: 7086-92 (1989); **Walewski J et al** Evaluation of natural killer and lymphokine-activated killer (LAK) cell activity *in vivo* in patients treated with high dose interleukin-2 and adoptive transfer of autologous LAK cells. JCRCO 115: 170-4 (1989); **Weiss GR et al** A randomized phase II trial of continuous infusion interleukin-2 or bolus injection interleukin-2 plus lymphokine-activated killer cells for advanced renal cell carcinoma. J. Clin. Oncol. 10: 275-81 (1992); **West WH et al** Constant-infusion recombinant interleukin-2 in adoptive immunotherapy of advanced cancer. NEJM 316: 898-905 (1987); **Whitehead RP et al** Phase II study of intravenous bolus recombinant interleukin-2 in advanced malignant melanoma: Southwest Oncology Group Study. JNCI 83: 1250-2 (1991); for further information see also: CIK (cytokine-induced killer cells), LAK cells (lymphokine-activated killer cells), TILs (tumor-infiltrating lymphocytes).

Adrenal growth factor: see: AGF.

Adrenocorticotropic hormone: abbrev. ACTH. See: POMC (proopiomelanocortin).

Adult bovine aortic endothelial cell growth-inhibitory activity: See: ABAE cell growth-inhibitory activity.

Adult T cell leukemia-derived factor: abbrev. ATL-derived factor. See: ADF.

AEF: *allogeneic effect factor* This poorly characterized factor consists of two subunits (40 kDa, 10-12 kDa) and is secreted by T cells activated *in vivo* (see also: Cell activation) following their stimulation by a second specific antigen *in vitro*. AEF functions as a T cell growth factor and enhances the generation of cytotoxic lymphocytes.
Katz DH & Zuberi RI Allogeneic effect factor. MiE 116: 428-40 (1985)

AFBP: *amniotic fluid binding protein* This protein is identical with one of the insulin-like growth factor binding proteins (see: IGF-BP).

AFC: *antibody-forming cell.*

aFGF: *acidic fibroblast growth factor*
■ **ALTERNATIVE NAMES:** AGF-1 (astroglial growth factor 1); aHDGF (anionic hypothalamus-derived growth factor); BDGF-A (brain-derived growth factor A); Brain FGF; CLAF (corpus luteum angiogenic factor); ECGF ((hypothalamus-derived) endothelial cell growth factor); ECGF-α (endothelial cell growth factor α); EDGF 2 (eye-derived growth factor 2); EK-AF-1 (embryonic kidney-derived angiogenesis factor 1); ESC-CEB derived endothelial cell growth factor; FGF-1 (fibroblast growth factor 1); FGF-α (fibroblast growth factor α); GMF (glial maturation factor); HBGF-1 (heparin-binding growth factor 1); HDGF (heart-derived growth factor); HDGF-A (hypothalamus-derived growth factor A); HGF-α (heparin-binding growth factor α); HGF-β (heparin-binding growth factor β); OGF (ovarian growth factor); Pituitary FGF; PrDGF (prostate-derived growth factor); prostate epithelial cell growth factor; prostatropin; RDGF-α (retina-derived growth factor α). See also: individual entries for further information.
The recommended name for aFGF is ***FGF-1*** (see: Nomenclature Meeting Report and Recommendation, January 18, 1991, reprinted in: ANY Vol. 638, pp. xiii-xvi).
■ **SOURCES:** One of the best sources of aFGF is brain tissue. aFGF is produced by a similar spectrum of cells of mesodermal and neuroectodermal origin that also produces » bFGF.
■ **PROTEIN CHARACTERISTICS:** aFGF is a 16 kDa protein of 140 amino acids (pI = 5,5-6). Variants with a length of 134 and 155 amino acids have also been described. aFGF does not contain disulfide bonds and is not glycosylated. aFGF is derived by posttranslational processing of ECGF-β (endothelial cell growth factor).
At the protein level aFGF and » bFGF display 55 % homology which is limited, however, to a few functional domains. The structure of aFGF, revealed by X-ray crystallography, shows a similar folding structure as » IL1. Bovine and human aFGF only differ from each other by 11 amino acid exchanges. aFGF appears to be less conserved than » bFGF.
aFGF, like » bFGF, » CNTF (ciliary neuronotrophic factor), and » PD-ECGF (platelet-derived endothelial cell growth factor), does not possess a » signal sequence that would allow secretion of the factor by classical secretion pathways (endoplasmatic reticulum/Golgi system). The mechanism underlying the release of aFGF is unknown.
A truncated variant of aFGF generated by alternative splicing and removal of the entire second coding exon is only 60 amino acids long (6.7 kDa) elicits only minimal fibroblast proliferation and antagonizes the effects of full-length aFGF when added exogenously to or when coexpressed with aFGF in » 3T3 fibroblasts. It has been suggested that the truncated variant may provide a unique mechanism for endogenous regulation of responses to aFGF.

■ **Gene Structure:** The aFGF gene extends over more than 19 kb and is located on human chromosome 5q31. 3-33. 2 (see also: $5q^-$ syndrome). The gene contains three exons separated by two very large introns and produces a single transcript of 4.1 kb. Different cDNA species differentially expressed in human kidney and brain tissues have also been isolated.

■ **Related Factors:** aFGF is a member of an evolutionary closely conserved family of proteins the prototype of which is » bFGF. Members of this so-called FGF family display a remarkable affinity to heparin and are therefore also called » HBGF (heparin-binding growth factors). See also: FGF-1 to FGF-7 and HBGF-1 to HBGF-8.

■ **Receptor Structure, Gene(s), Expression:** FGF receptors are encoded by a gene family consisting of at least four receptor tyrosine kinases that transduce signals important in a variety of developmental and physiological processes related to cell growth and differentiation. A unique feature of FGF receptors is the multitude of structural variants and a high degree of cross-reactivity between these receptors and their various ligands. Some receptor subtypes have been described that preferentially bind aFGF, » bFGF, and/or FGF-related proteins. The receptor known as FGFR-2 binds aFGF with ≈ 1000-fold higher affinity than » bFGF. The receptor known as FGFR-3 binds aFGF and bFGF while the FGFR-4 receptor binds aFGF with highest affinity, followed by » K-FGF/*hst*-1 and bFGF. There are several indications that receptor variants that differentially bind FGFs with different affinities may arise by differential splicing of messages encoding the extracellular receptor domains.

A binding protein for aFGF, designated » HBp17, has been described to modulate aFGF activities.

■ **Biological Activities:** The wide spectrum of biological activities is revealed by the plethora of different names. Because aFGF binds to the same receptor as » bFGF aFGF displays more or less the same spectrum of activities. It is, however, generally ≈ 50-100-fold less active than bFGF in similar assays.

The glycosaminoglycan heparin found in the extracellular matrix (see: ECM) of cells has been shown to enhance the biological activities to levels reaching those shown by » bFGF. In addition, heparin appears to protect aFGF from inactivation by proteases, acids, and heat, thus increasing its biological half life (see also: HBGF, heparin-binding growth factors).

A multifunctional role of aFGF is suggested by the many different receptor phenotypes expressed in various cell types. *In vitro*, aFGF participates in various cellular functions including proliferation, differentiation, angiogenesis, and cell migration, but *in vivo*, the physiologic role of this growth factor is still not clearly defined. aFGF is a potent mitogen *in vitro* for many cells of ectodermal and mesodermal embryonic origin including skin-derived epidermal keratinocytes, dermal fibroblasts and vascular endothelial cells. aFGF is an » angiogenesis factor and promotes » wound healing. Myocyte-derived aFGF has been implicated in cardiomyocyte maturation and ventricular remodeling. aFGF has been implicated in an » autocrine system by which calcium regulates parathyroid cell growth. It has been demonstrated that the expression of aFGF is highest during the late stages of hepatic morphogenesis in newborn rats as well as during hepatic differentiation in adult liver. Several findings are consistent with various neurotrophic, mitogenic, and neuromodulatory functions associated with aFGF in the mammalian central nervous system. aFGF, like » bFGF, is a potent mitogen for glial precursor cells *in vitro*. Both growth factors inhibit the differentiation of astrocyte or oligodendrocyte precursors. FGFs thus may play a critical role in gliogenesis and the timing of glial differentiation in the brain.

aFGF has been shown to provide positive growth signals for megakaryocyte progenitor cells (see also: Hematopoiesis). aFGF significantly stimulates the formation of megakaryocyte colonies *in vitro* in mice, synergising with » IL3. The activity of aFGF is indirect and is completely abrogated by a monoclonal antimouse IL6 antibody.

■ **Transgenic/Knock-out/Antisense Studies:** Direct gene transfer into porcine arteries of an aFGF gene genetically engineered to allow secretion of aFGF has demonstrated that aFGF stimulates intimal hyperplasia *in vivo* and also induces formation of new blood vessels in the expanded intima.

■ **Assays:** bFGF activities can be assayed in a proliferation assay using CCL39 hamster fibroblast cells, human » SW-13 cells or bovine » FBHE cells. aFGF can be detected by means of a sensitive enzyme immunoassay. An alternative and entirely different detection method is » Message amplification phenotyping. For further information see also subentry "Assays" in the reference section.

■ **Clinical Use & Significance:** The intravenous application of aFGF has shown that the factor promotes the regeneration of the endothelium following arterial intravascular injuries. aFGF appears to block the thickening of the intima normally observed after these processes (see also: bFGF). Overexpression of aFGF and bFGF in pancreatic cancers has been found to be associated with a more advanced tumor stage. aFGF has been found to be expressed in the epithelial cell compartment of bladder transitional cell carcinoma and appears likely to be involved in the biology of these cancers.

A possible physiological function for aFGF in the normal support of hippocampal CA1 pyramidal cells and neurons in some other brain regions is suggested by the observation that aFGF prevents neuronal death following experimental ischemia. Recombinant aFGF has been shown to induce both rapid angiogenesis and neurogenesis through a synthetic conduit across a 15 mm surgical gap in the peripheral nerve of the rat. aFGF may thus serve as an important mediator of controlled growth during peripheral nerve regeneration, inducing a return of motor function in 57 % of animals 24 weeks after surgery.

It has been shown recently that chimeric toxins composed of aFGF fused to mutant forms of *Pseudomonas* exotoxin are cytotoxic to a variety of tumor cell lines with FGF receptors. aFGF-toxin fusion proteins have specific *in vitro* cytotoxic, antiangiogenic, and antimigratory effects on microvascular endothelia (see also: Cytokine fusion toxins).

● **Reviews: Baird A & Klagsbrun M** The fibroblast growth factor family. CC 3: 239-43 (1991); **Burgess WH & Maciag T** The heparin-binding (fibroblast) growth factor family of proteins. ARB 58: 575-606 (1989); **Goldfarb M** The fibroblast growth factor family. Cell Growth Differ. 1: 439-45 (1990); **Gospodarowicz D et al** Structural characterization and biological functions of fibroblast growth factor. Endocrinol. Rev. 8: 95-113 (1987); **Gospodarowicz D et al** Fibroblast growth factor: structural and biological properties. JCP (Suppl.) 5: 15-26 (1987); **Klagsbrun M** The affinity of fibroblast growth factors (FGFs) for heparin; FGF-heparan sulfate interactions in cells and extracellular matrix. COCB 2: 857-63 (1990); **Sensenbrenner M** The neurotrophic activity of fibroblast growth factors. Prog. Neurobiol. 41: 683-704 (1993); **Sullivan DE & Storch TG** Tissue- and development-specific expression of HBGF-1 mRNA. BBA Gene Struct. Expression 1090: 17-21 (1991); **Wagner JA** The fibroblast growth factors: an emerging family of neural growth factors. CTMI 165: 95-118 (1991)

● **Biochemistry & Molecular Biology: Alterio J et al** Characterization of a bovine acidic FGF cDNA clone and its expression in brain and retina. FL 242: 41-6 (1988); **Arakawa T et al** Production and characterization of an analog of acidic fibroblast growth factor with enhanced stability and biological activity. Protein Eng. 6: 541-6 (1993); **Burgess WH et al** Structure-function studies of heparin-binding (acidic fibroblast) growth factor-1 using site-directed mutagenesis. JCBc 45: 131-8 (1991); **Chiu IM et al** Alternative splicing generates two forms of mRNA coding for human heparin-binding growth factor 1. O 5: 755-62 (1990); **Esch F et al** Primary structure of bovine brain acidic fibroblast growth factor (FGF). BBRC 133: 554-62 (1985); **Esch F et al** Primary structure of bovine pituitary basic fibroblast growth factor (FGF) and comparison with the amino-terminal sequence of bovine brain acidic FGF. PNAS 82: 6507-11 (1985); **Gautschi-Sova P et al** Amino acid sequence of human acidic fibroblast growth factor. BBRC140: 874-80 (1986); **Gimenez-Gallego G et al** Brain-derived acidic fibroblast growth factor: complete amino acid sequence and homologies. S 230: 1385-8 (1985); **Gospodarowicz D et al** Heparin protects acidic and basic FGF from inactivation. JCP 128: 475-84 (1986); **Halley C et al** Nucleotide sequence of a bovine acidic FGF cDNA. NAR 16: 10913 (1988); **Jaye M et al** Human endothelial cell growth factor: cloning, nucleotide sequence, and chromosomal localization. S 233: 541-5 (1986); **Linemeyer DL et al** Expression in *Escherichia coli* of a chemically synthesized gene for biologically active bovine acidic fibroblast growth factor BT 5: 960-5 (1987); **Mergia A et al** The genes for basic and acidic fibroblast growth factors are on different human chromosomes. BBRC 138: 644-51 (1986); **Myers RL et al** Gene structure and differential expression of acidic fibroblast growth factor mRNA: identification and distribution of four different transcripts. O 8: 341-9 (1993); **Payson RA et al** Cloning of two novel forms of human acidic fibroblast growth factor (aFGF) mRNA. NAR 21: 489-95 (1993); **Thomas KA et al** Structural modifications of acidic fibroblast growth factor alter activity, stability, and heparin dependence. ANY 638: 9-17 (1991); **Wang WP et al** Cloning and sequence analysis of the human acidic fibroblast growth factor gene and its preservation in leukemia patients. O 6: 1521-9 (1991); **Yu YL et al** An acidic fibroblast growth factor protein generated by alternate splicing acts like an antagonist. JEM 175: 1073-80 (1992); **Zhu X et al** Three-dimensional structures of acidic and basic fibroblast growth factors. S 251: 90-3 (1991)

● **Receptors: Dell KR et al** A novel form of fibroblast growth factor receptor 2. Alternative splicing of the third immunoglobulin-like domain confers ligand binding specificity. JBC 267: 21225-9 (1992); **Johnson DE et al** Diverse forms of a receptor for acidic and basic fibroblast growth factors. MCB 10: 4728-36 (1990); **Kan M et al** Identification and assay of fibroblast growth factor receptors. MiE 198: 158-74 (1991); **Partanen J et al** FGFR-4, a novel acidic fibroblast growth factor receptor with a distinct expression pattern. EJ 10: 1347-54 (1991); **Payson RA** Cloning of two novel forms of human acidic fibroblast growth factor (aFGF) mRNA. NAR 21: 489-95 (1993); **Sakaguchi K et al** Identification of heparan sulfate proteoglycan as a high affinity receptor for acidic fibroblast growth factor (aFGF) in a parathyroid cell line. JBC 266: 7270-8 (1991); **Vainikka S et al** Fibroblast growth factor receptor-4 shows novel features in genomic structure, ligand binding and signal transduction. EJ 11: 4273-80 (1992); **Wang WP et al** Cloning and sequence analysis of the human acidic fibroblast growth factor gene and its preservation in leukemia patients. O 6: 1521-9 (1991); **Werner S et al** Differential splicing in the extracellular region of fibroblast growth factor receptor 1 generates receptor variants with different ligand-binding specificities. MCB 12: 82-8 (1992)

● **Biological Activities: Bjornsson TD et al** Acidic fibroblast growth factor promotes vascular repair. PNAS 88: 8651-5 (1991); **De Vito WJ et al** Effect of *in vivo* administration of recombinant acidic fibroblast growth factor on thyroid function in the rat: induction of colloid goiter. Endocrinology 131: 729-

35 (1992); **Engele J & Bohn MC** Effects of acidic and basic fibroblast growth factors (aFGF, bFGF) on glial precursor cell proliferation: age dependency and brain region specificity. Dev. Biol. 152: 363-72 (1992); **Engelmann GL et al** Acidic fibroblast growth factor and heart development. Role in myocyte proliferation and capillary angiogenesis. Circ. Res. 72: 7-19 (1993); **Han ZC et al** Recombinant acidic human fibroblast growth factor (aFGF) stimulates murine megakaryocyte colony formation *in vitro*. Int. J. Hematol. 55: 281-6 (1992); **Jouanneau J et al** Secreted and non-secreted forms of acidic fibroblast growth factor produced by transfected epithelial cells influence cell morphology, motility, and invasive potential PNAS 88: 2893-7 (1991); **Marsden ER et al** Expression of acidic fibroblast growth factor in regenerating liver and during hepatic differentiation. Lab. Invest. 67: 427-33 (1992); **Sakaguchi K** Acidic fibroblast growth factor autocrine system as a mediator of calcium-regulated parathyroid cell growth. JBC 267: 24554-62 (1992)

● **TRANSGENIC/KNOCK-OUT/ANTISENSE STUDIES: Nabel EG et al** Recombinant fibroblast growth factor-1 promotes intimal hyperplasia and angiogenesis in arteries *in vivo*. N 362: 844-6 (1993)

● **ASSAYS: Ishikawa R et al** Developmental changes in distribution of acidic fibroblast growth factor in rat brain evaluated by a sensitive two-site enzyme immunoassay. J. Neurochem. 56: 836-41 (1991); **Kardami E et al** Biochemical and ultrastructural evidence for the association of basic fibroblast growth factor with cardiac gap junctions. JBC 266: 19551-7 (1991); **van Zoelen EJ** The use of biological assays for detection of polypeptide growth factors. Prog. Growth Factor Res. 2: 131-52 (1990)

● **CLINICAL USE & SIGNIFICANCE: Klein-Soyer C et al** Opposing effects of heparin with TGF-β or aFGF during repair of a mechanical wound of human endothelium. Influence of cAMP on cell migration. Biol. Cell. 75: 155-62 (1992); **Mellin TN et al** Acidic fibroblast growth factor accelerates dermal wound healing. GF 7: 1-14 (1992); **Merwin JR et al** Acidic fibroblast growth factor-Pseudomonas exotoxin chimeric protein elicits antiangiogenic effects on endothelial cells. CR 52: 4995-5001 (1992); **Ravery V et al** Immunohistochemical detection of acidic fibroblast growth factor in bladder transitional cell carcinoma. Urol. Res. 1992; 20: 211-4 (1992); **Sasaki K et al** Acidic fibroblast growth factor prevents death of hippocampal CA1 pyramidal cells following ischemia. Neurochem. Int. 21: 397-402 (1992); **Walter MA et al** Enhanced peripheral nerve regeneration by acidic fibroblast growth factor. Lymphokine Cytokine Res. 12: 135-41 (1993); see also: bFGF as a prototype of fibroblast growth factors.

AG126: see: Tyrphostins.

AG213: see: Tyrphostins.

AG370: see: Tyrphostins.

AG490: see: Tyrphostins.

Agammaglobulinemia tyrosine kinase: abbrev. atk. See: Immunodeficient mice, subentry xid.

AGF: *adipocyte growth factor* This factor was originally detected as an activity stimulating the proliferation of adipocyte precursors. It is identical with » bFGF.

Lau DC et al Purification of a pituitary polypeptide that stimulates the replication of adipocyte precursors in culture. FL 153: 395-8 (1983); **Rowe JM et al** Purification and characterization of a human pituitary growth factor. B 25: 6421-5 (1986)

AGF: *adrenal growth factor* This factor is isolated from bovine adrenal glands. It is identical with aFGF or bFGF.

Gospodarowicz D et al Isolation of fibroblast growth factor from bovine adrenal gland: physicochemical and biological characterization. Endocrinol. 118: 82-90 (1986)

AGF: *autocrine growth factor* This poorly characterized factor is produced by a subclone of the EBV-negative Burkitt-like lymphoma cell line BJAB. It supports the long-term culture of this subclone without added cytokines. The factor is not identical with » IL2, » IL3, » IL4, » IL5, » IL6, » IL7, » GM-CSF, » TNF-α, » TNF-β, » IFN-γ, and » TGF-β. AGF shows some IL1-like activities; these are, however, neutralized by anti-IL1-α antibodies without affecting the growth-promoting activities of BJAB cells.

The acronym AGF is sometimes also used as a general term for an autocrine growth factor activity (see also: Autocrine stimulation).

Lahm H et al Autocrine growth factors secreted by the malignant human B cell-line BJAB are distinct from other known cytokines. ECN 1: 41-6 (1990)

AGF-1: *astroglial growth factor 1* AGF-1 induces a characteristic morphological change in cultured rat astroglial cells and stimulates their proliferation. This factor is identical with » aFGF. See also: AGF-2.

Courty J et al [*In vitro* properties of various growth factors and *in vivo* effects] Biochimie 66: 419-28 (1984); **Lagente O et al** Isolation of heparin-binding growth factors from dogfish (Mustela canis) brain and retina. FL 202: 207-10 (1986); **Pettmann B et al** Purification of two astroglial growth factors from bovine brain. FL 189: 102-8 (1985); **Sensenbrenner M et al** Neuronal-derived factors regulating glial cell proliferation and maturation. J. Physiol. Paris 82: 288-90 (1987); **Unsicker K et al** Astroglial and fibroblast growth factors have neurotrophic functions for cultured peripheral and central nervous system neurons. PNAS 84: 5459-63 (1987); see also: references for » AGF-2.

AGF-2: *astroglial growth factor 2* AGF-2 induces a characteristic morphological change in cultured rat astroglial cells and neuroblasts and stimulates their proliferation. This factor is identical with » bFGF. See also: AGF-1.

Gensburger C et al Effect of basic FGF on the proliferation of rat neuroblasts in culture CR Acad. Sci. III 303: 465-468 (1986); **Lagente O et al** Isolation of heparin-binding growth factors from dogfish (Mustela canis) brain and retina. FL 202: 207-10 (1986); **Latzkovits L** Sodium and potassium uptake in primary cultures of rat astroglial cells induced by long-term exposure to the basic astroglial growth factor (AGF-2). Neurochem. Res. 14: 1025-1030 (1989); see also: references for » AGF-1.

AGIF: *adipogenesis inhibitory factor* This factor of 178 amino acids is produced as a precursor of 199 amino acids. Purified AGIF inhibits the process of adipogenesis in mouse 3T3-L1 preadipocytes. AGIF inhibits the differentiation of bone marrow pre-adipocytes into adipocytes. The cDNA has been isolated from a library established from a human bone marrow stromal cell line (KM-102) and is identical to » IL11.

Kawashima I et al Molecular cloning of cDNA encoding adipogenesis inhibitory factor and identity with interleukin 11. FL 283: 199-202 (1991); **Ohsumi J et al** Adipogenesis inhibitory factor. A novel inhibitory regulator of adipose conversion in bone marrow. FL 288: 13-16 (1991)

AGM-1470: O-(Chloroacetylcarbamoyl)-Fumagillol. See: Fumagillin.

AGP7: see: myeloblastin.

aHDGF: *anionic hypothalamus-derived growth factor* This factor is closely related to other growth factors isolated from retina (see: RDGF-α, RDGF-β). AHDGF is identical with » aFGF.

D'Amore PA & Klagsbrun M Endothelial cell mitogens derived from retina and hypothalamus: biochemical and biological similarities. JCP 99: 1545-9 (1984); **Klagsbrun M & Shing Y** Heparin affinity of anionic and cationic capillary endothelial cell growth factors: analysis of hypothalamus-derived growth factors and fibroblast growth factors. PNAS 82: 805-9 (1985)

AIC2A: *activation-induced* This murine protein is identical with the IL3-binding β subunit of the murine receptor for » IL3. The murine protein designated AIC2B is a homologous but separate gene, displaying 91 % sequence identity at the amino acid level with AIC2A, and does not bind IL3.

AIC2B is a subunit required for the generation of high affinity receptors for » IL5 and » GM-CSF. AIC2B and AIC2A are homologues of the β c (β common) subunit of the human cytokine receptors for IL3, GM-CSF and IL5.

Both murine genes contain 14 exons. They have a length of 28 kb and have been located to murine chromosome 15 in the vicinity of the » *sis* oncogene locus. Several transcripts of different lengths have been reported to originate by differential splicing.

Devos R et al Molecular basis of a high affinity murine interleukin-5 receptor. EJ 10: 2133-7 (1991); **Duronio V et al** Tyrosine phosphorylation of receptor β subunits and common substrates in response to interleukin-3 and granulocyte-macrophage colony-stimulating factor. JBC 267: 21856-63 (1992); **Fung MC et al** Distinguishing between mouse IL3 and IL3 receptor like (IL5/GM-CSF receptor converter) mRNAs using the polymerase chain reaction method. JIM 149: 97-103 (1992); **Gorman DM et al** Cloning and expression of a gene encoding an interleukin 3 receptor-like protein: identification of another member of the cytokine receptor gene family. PNAS 87: 5459-63 (1990); **Gorman DM et al** Chromosomal localization and organization of the murine genes encoding the β subunits (AIC2A and AIC2B) of the interleukin 3, granulocyte/macrophage colony-stimulating factor, and interleukin 5 receptors. JBC 267: 15842-8 (1992); **Hara T & Miyajima A** Two distinct functional high affinity receptors for mouse interleukin-3 (IL3). EJ 11: 1875-84 (1992); **Itoh N et al** Cloning of an interleukin-3 receptor gene: a member of a distinct receptor gene family. S 247: 324-7 (1990); **Kitamura T et al** Reconstitution of functional receptors for human granulocyte/macrophage colony-stimulating factor (GM-CSF): evidence that the protein encoded by the AIC2B cDNA is a subunit of the murine GM-CSF receptor. PNAS 88: 5082-6 (1991); **Mui AL et al** Purification of the murine interleukin 3 receptor. JBC 267: 16523-30 (1992); **Ogorochi T et al** Monoclonal antibodies specific for low-affinity interleukin-3 (IL3) binding protein AIC2A: evidence that AIC2A is a component of a high-affinity IL3 receptor. Blood 79: 895-903 (1992); **Park LS et al** Cloning of the low-affinity murine granulocyte-macrophage colony-stimulating factor receptor and reconstitution of a high-affinity receptor complex. PNAS 89: 4295-9 (1992); **Takaki S et al** Identification of the second subunit of the murine interleukin-5 receptor: Interleukin-3 receptor-like protein, AIC2B is a component of the high affinity interleukin-5 receptor. EJ 10: 2833-8 (1991); **Wang HM et al** Structure of mouse interleukin 3 (IL3) binding protein (AIC2A). Amino acid residues critical for IL3 binding. JBC 267: 979-83 (1992)

AIC2B: see: AIC2A.

AICD: *activation-induced cell death* Cell death occurring after, and induced by, » cell activation. See: Apoptosis.

aIFN: abbrev. for alpha-interferon. See: IFN-α. See also: IFN for general information about interferons.

AIGF: *androgen-induced growth factor* This factor is secreted by a murine androgen-dependent mammary carcinoma cell line (SC-3). It has a length of 215 amino acids and displays a 30-40% homology to other factors belonging the fibroblast growth factor family (see: FGF). With the cloning of the ninth member of the FGF family (see: FGF-9) AIGF has been referred to as FGF-8.

The synthesis of AIGF is induced by testosterone. It functions as an » autocrine growth factor for SC-3 cells in the absence of the hormone.

Tanaka A et al Cloning and characterization of an androgen-induced growth factor essential for the androgen-dependent growth of mouse mammary carcinoma cells. PNAS 89: 8928-32 (1992)

AIM: *activation-inducing molecule* This poorly characterized protein is produced by activated T lymphocytes (see also: Cell activation). It serves as a secondary signal mediating the maximal expression of the T cell growth factor » IL2.
Cebrián M et al Triggering T cell proliferation through AIM, an activation inducer molecule expressed on activated human lymphocytes. JEM 168: 1621-37 (1988)

AIM: *astroglia-inducing molecule* This glycoprotein factor of ≈ 50 kDa (and multimers thereof) is a constituent of fetal bovine serum and is also found in human serum. Treatment with 6 M guanidine hydrochloride generates an AIM with a molecular mass 12-18 kDa. AIM induces bipotential glial precursors known as oligodendrocyte-type 2 astrocyte (O-2A) progenitors to become type 2 astroglia rather than oligodendroglia. AIM may be a novel differentiation factor as it has little effect on type 1 astroglia and none of the known growth factors that have been tested to date mimics its effects.
Levison SW & McCarthy KD Characterization and partial purification of AIM: a plasma protein that induces rat cerebral type 2 astroglia from bipotential glial progenitors. J. Neurochem. 57: 782-94 (1991)

AIT: abbrev. for » Adoptive Immunotherapy.

AKR-2B: An established murine cell line which is used in cytokine research to quantitate » TGF-α and » TGF-β (see also: Bioassays). Both factors induce colony formation in soft agar or other semi-solid media such as methylcellulose.
Moses HL et al Transforming growth factor production by chemically transformed cells. CR 41: 2842-8 (1981); **Rizzino A** Soft agar growth assays for transforming growth factors and mitogenic peptides. MiE 146: 341-53 (1987).

ALAK: *adherent lymphokine-activated killer cells* see: LAK cells (lymphokine-activated killer cells).

Albumin: see: LGF (liver growth factor).

ALK: *activin receptor-like kinase* Human cDNA clones encoding four novel putative transmembrane protein serine/threonine kinases, denoted ALK-1, -2, -3 and -4 have been isolated. The cDNA clones for ALK-1, -2 and -3 encode complete proteins of 503, 509 and 532 amino acids respectively. The ALK-4 cDNA is incomplete and the predicted protein of 383 amino acids has a truncated extracellular domain. The ALKs share similar domain structures and sequence identities (60-79%) among each other. They also have approximately 40% sequence identity to » activin receptors, » TGF-β type II receptor and *Daf*-1 in the kinase domains. ALK-1 to -4 probably are receptors that bind ligands that are members of the TGF-β superfamily. ALK-2 and ALK-4 show ubiquitous tissue expression patterns, whereas the distribution of ALK-1 and ALK-3 varies strongly between different tissues with more restricted expression patterns.
ten Dijke P et al Activin receptor-like kinases: a novel subclass of cell-surface receptors with predicted serine/threonine kinase activity. O 8: 2879-87 (1993)

ALLFIT: see: Bioassays.

Allogeneic effect factor: see: AEF.

Alpha-BCDF: *B cell differentiation factor α* see: BCDF-α.

Alpha-Chemokines: see: Chemokines.

Alpha-CSF: *colony stimulating factor α* See: CSF-α.

Alpha-ECGF: *endothelial cell growth factor* See: ECGF-α.

Alpha-ENAP: *endothelial cell neutrophil activating peptide* see: ENAP.

Alpha-Endorphin: see: POMC (proopiomelanocortin).

Alpha-FGF: *fibroblast growth factor α* see: FGF-α.

Alpha-Intercrines: see: Chemokines.

Alpha-Interferon: see: IFN-α. See also: IFN for general information about interferons.

Alpha 1-Microglobulin: see: ITI (Inter-alpha-trypsin inhibitor).

Alpha 2-Macroglobulin: abbrev. α_2M. α_2M is a glycoprotein of 725 kDa encoded by a gene that spans ≈ 48 kb and consists of 36 exons. The protein is found in serum at concentrations of 2-4 mg/ml. It consists of two identical subunits which themselves are made up of two peptide chains covalently linked by disulfide bonds. α_2M is an » acute phase protein (see also: Acute phase reaction). Its gene promoter contains an IL6 responsive element (see also: Gene expression) and its expression is enhanced by » IL6 by a factor of up to several 100-fold through the action of the human transcription factor » NF-IL6.

Native α_2M does not bind to a receptor. However, it undergoes a conformational change upon reaction with either small primary amines such as methylamine or with protein ligands to form activated α_2M. As a result, a receptor recognition site is exposed on each subunit of the molecule enabling it to bind to its receptors on macrophages. The receptor for α_2M, first described as the low-density lipoprotein receptor-related protein (LRP), is encoded by a gene on human chromosome 12q13-q14. α_2MR/LRP consists of two polypeptides, 515 and 85 kDa, that are noncovalently associated. A 39 kDa polypeptide, termed the receptor-associated protein (RAP), interacts with the 515-kDa subunit after biosynthesis of these molecules and remains associated on the cell surface. This molecule regulates ligand binding of α_2MR/LRP. α_2MR/LRP expression in macrophages can be markedly decreased by bacterial lipopolysaccharides and by » IFN-γ.

α_2M is a natural inhibitor of several serum proteases such as cathepsin, chymotrypsin, collagenase, elastase, kallikrein, plasmin, factor Xa, thrombin, and trypsin, which it inhibits by irreversible complex formation. These complexes include unusual forms in which more than one of the four identical subunits of α_2M are cross-linked by amide bonds to more than one lysyl amino group of the bound protease. Bound proteases retain residual biological activity and preferentially react with low molecular weight substrates. Plasma levels of α_2M are reduced in patients with sepsis (see also: Septic shock) and have been described to be associated with fatal outcome in some studies.

The interactions between α_2M and cytokines are thought to modulate the expression of acute phase proteins in liver hepatocytes. α_2M inhibits the growth of several tumors *in vitro* and *in vivo* (rat). The conditioned medium (see also: CM) of astroglial cells contains α_2M, and this protein has been shown to promote the outgrowth of neurites. The main fraction of » TGF-β in the serum is complexed with α_2M. α_2M inhibits the binding of isoforms of » TGF-β (TGF-β1) to cell surface receptors. It does not block TGF-β1 inhibition of » CCL-64 cells but reduces this activity of TGF-β2 $\approx$ 10-fold. α_2M neutralizes the inhibition by TGF-β of the IL6-induced release of C-reaction protein by human hepatoma cells. In spite of the fact that IL6 is also complexed with α_2M induction of the synthesis of C-reactive protein by IL6 is not influenced. α_2M has been found to be the major protein secreted by bovine adrenocortical cells in primary culture and its synthesis is stimulated by TGF-β. α_2M/TGF-β appears to be biologically active on adrenocortical cells.

Many other cytokines, including » Activin A, » Inhibins, the γ subunit of » NGF, » IL1, » IL2, » IL6, » PDGF, » aFGF, » bFGF, » VEGF (vascular endothelial growth factor), » transferrin, and » defensins have also been reported to bind to α_2M. In some instances the bound cytokine may be a transport form and complexation with α_2M may be a way of increasing the biological half life. In other instances binding of a ligand to α_2M has been shown to inhibit some of the biological activities of the bound ligand. Although it has been suggested that cytokines bound to α_2M may be destined for degradation and do not represent potentially bioactive molecules the peptide binding function of α_2M probably is a mechanism that allows targeting of biologically active peptides to different cell types expressing the α_2MR/LRP receptor.

Interactions between α_2M and various cytokines are not only important parameters to consider in view of » Bioassays and other » Cytokine assays as they may influence generation, release interaction and stability or mask immunogenic epitopes; they are also important parameters to consider in view of the therapeutical uses of cytokines as they may affect biodistribution, bioavailability, stability, clearance or cytokine induction.

● **Reviews: Borth W** α 2-macroglobulin, a multifunctional binding protein with targeting characteristics. FJ 6: 3345-53 (1992); **Gonias SL** α 2-macroglobulin: a protein at the interface of fibrinolysis and cellular growth regulation. EH 20: 301-11 (1992); **James K** Interactions between cytokines and α2-macroglobulin. IT 11: 163-6 (1990)

Bonner JC et al Reversible binding of platelet-derived growth factor-AA, -AB, and -BB isoforms to a similar site on the "slow" and "fast" conformations of α 2-macroglobulin. JBC 267: 12837-44 (1992); **Borth W & Teodorescu M** Inactivation of human interleukin-2 (IL2) by α 2-macroglobulin-trypsin complexes. Immunology 57: 367-71 (1986); **Chen BJ et al** Structure of α 2-macroglobulin-protease complexes. Methylamine competition shows that proteases bridge two disulfide-bonded half-molecules. B 31: 8960-6 (1992); **de Boer JP et al** Alpha-2-Macroglobulin functions as an inhibitor of fibrinolytic, clotting, and neutrophilic proteinases in sepsis: studies using a baboon model. Infect. Immun. 61: 5035-43 (1993); **Graziadei I et al** The hepatic acute-phase proteins α 1-antitrypsin and α 2-macroglobulin inhibit binding of transferrin to its receptor. BJ 290:

109-13 (1993); **Hall SW et al** Binding of transforming growth factor-β 1 to methylamine-modified α 2-macroglobulin and to binary and ternary α 2-macroglobulin-proteinase complexes. BJ 281: 569-75 (1992); **Hilliker C et al** Assignment of the gene coding for the α 2-macroglobulin receptor to mouse chromosome 15 and to human chromosome 12q13-q14 by isotopic and nonisotopic *in situ* hybridization. Genomics 13: 472-4 (1992); **James K et al** The effect of α2 macroglobulin in commercial cytokine assays. JIM 168: 33-7 (1994); **Keramidas M et al** Inhibition of adrenocortical steroidogenesis by α 2-macroglobulin is caused by associated transforming growth factor β. Mol. Cell. Endocrinol. 84: 243-51 (1992); **Koo PH & Liebl DJ** Inhibition of nerve growth factor-stimulated neurite outgrowth by methylamine-modified α 2-macroglobulin. J. Neurosci. Res. 31: 678-92 (1992); **LaMarre J et al** Regulation of macrophage α 2-macroglobulin receptor/low density lipoprotein receptor-related protein by lipopolysaccharide and interferon-γ. JCI 91: 1219-24 (1993); **Matthijs G et al** Structure of the human α-2 macroglobulin gene and its promoter. BBRC 184: 596-603 (1992); **Saitoh S et al** *De novo* production of α-2-macroglobulin in cultured astroglia from rat brain. Brain Res. Mol. Brain Res. 12: 155-61 (1992); **Poller W et al** Cloning of the human α 2-macroglobulin gene and detection of mutations in two functional domains: the bait region and the thiolester site. Hum. Genet. 88: 313-9 (1992); **Saitoh S et al** *De novo* production of α 2-macroglobulin in cultured astroglia from rat brain. Brain Res. Mol. Brain Res. 12: 155-61 (1992); **Schroeter JP et al** Three-dimensional structures of the human α 2-macroglobulin-methylamine and chymotrypsin complexes. J. Struct. Biol 109: 235-47 (1993); **Soker S et al** Vascular endothelial growth factor is inactivated by binding to α 2-macroglobulin and the binding is inhibited by heparin. JBC 268: 7685-91 (1993); **Stouffer GA et al** Activated α2 macroglobulin and transforming growth factor β1 induce a synergistic smooth muscle cell proliferative response. JBC 268: 18340-44 (1993); **Taylor AW & Mortensen RF** Effect of α-2-macroglobulin on cytokine-mediated human C-reactive protein production. Inflammation 15: 61-70 (1991); **Vaughan JM & Vale WW** α 2-macroglobulin is a binding protein of inhibin and activin. Endocrinology 132: 2038-50 (1993); **Wolf BB** Reaction of nerve growth factor γ and 7S nerve growth factor complex with human and murine α 2-macroglobulin. B 32: 1875-82 (1993)

Alpha-MSH: *melanocyte stimulating hormone* see: POMC (proopiomelanocortin).

Alpha-N-peptide: see: Inhibins.

Alpha-Pluripoietin: See Pluripoietin-α.

Alpha-pregnancy-associated endometrial globulin: This protein is identical with one of the insulin-like growth factor binding proteins (see: IGF-BP).

Alpha-RDGF: *retina-derived growth factor α* See: RDGF-α.

Alpha-TGF: see: TGF-α.

Alpha-TNF: see: TNF-α.

Alpha-trypsin-inhibitor: see: ITI (Inter-α-trypsin inhibitor).

Alveolar macrophage chemotactic factor: see: AMCF.

Alveolar macrophage-derived growth factor: abbrev. AMDGF. See: MDGF (macrophage-derived growth factor).

AMCC: *adherent marrow cell culture* see: LTBMC (long-term bone marrow culture).

AMCF: *alveolar macrophage chemotactic factor* A factor of 10 kDa isolated from porcine alveolar macrophages. AMCF is a very strong chemoattractant for neutrophils activated by bacterial » endotoxins (see also: Cell activation). Two different and apparently unrelated factors have been described. The factor designated ***AMCF-I*** (cDNA of 1466 nucleotides) is also a chemoattractant for human neutrophils. AMCF-I shares 74% identity with human » IL8 and 84% identity with rabbit IL8, and probably is the porcine homologue of IL8. The second factor, ***AMCF-II***. appears to be species-specific. It shares 53% identity with human » NAP-2 (neutrophil activating peptide 2), 61 % identity with rat » CINC, and 67 % identity with » ENA-78. Both porcine factors belong to the » chemokine family of cytokines.

Goodman RB et al Identification of two neutrophil chemotactic peptides produced by porcine alveolar macrophages. JBC 266: 8455-63 (1991)

AMDGF: *alveolar macrophage-derived growth factor* see: MDGF (macrophage-derived growth factor).

Ames dwarf mouse: see: Dwarf mice.

AMF: *autocrine motility factors* s A general term for » motogenic cytokines that influence the motility of those cells that secrete them. See also: Autocrine stimulation.

AMF: *autocrine motility factor*

■ **ALTERNATIVE NAMES:** Autotaxin; B16-F1 melanoma autocrine motility factor; TAMF (tumor autocrine motility factor). See also: individual entries for further information. The abbreviation AMF has also been used as a generic name for factors influencing the motility of cells that produce it (see also: Autocrine).

■ **SOURCES:** AMF is produced constitutively by various melanomas, hepatomas, leukemic cell lines, and some metastasizing tumors.
■ **PROTEIN CHARACTERISTICS:** AMF is a 55 kDa unglycosylated protein. It contains internal disulfide bonds essential for biological activity.
■ **RECEPTOR STRUCTURE, GENE(S), EXPRESSION:** The receptor is a 78 kDa glycoprotein (gp78) with a cDNA-derived length of 323 amino acids. The cDNA shows strong homology (50 %) to a tumor suppressor gene designated p53 (see: oncogenes). The biological relevance of this observations is unclear.
The receptor is expressed by various cell types. A hydrophobic region has been detected between positions 111 and 137, indicating that the receptor may contain a transmembrane segment. The receptor is phosphorylated after binding of its ligand. The activity of AMF can be blocked by pertussis toxin, indicating the involvement of a pertussis toxin-sensitive G protein in signal transduction.
Treatment of B16-F1 melanoma cells with an antibody directed against gp78 increases the capacity of these cells to colonize lung tissues following intravenous injection of these cells into mice. The cellular changes observed in the melanoma cells correlate with their altered metastasizing potential and alterations in the extent of O-glycosylation of gp78.
■ **BIOLOGICAL ACTIVITIES:** AMF enhances the directed chemotactic and random (chemokinetic) motility (see also: Chemotaxis) of those cells secreting it (see also: Autocrine stimulation). This process is accompanied by an increased generation of pseudopodia and an increased methylation of phospholipids. The factor also stimulates the lung-colonizing ability of tumor cells *in vivo*. AMF acts synergistically with IGF-1 (see: IGF) or » GM-CSF.
AMF purified from HT-1080 human fibrosarcoma cells has been described to stimulate the growth and motility of 3T3-A31 fibroblasts at a concentration of 0.1 ng/ml or less. AMF also stimulates the healing of experimentally wounded, density-arrested A31 monolayer cultures. The growth factor activities of AMF suggest a possible role in normal tissue regeneration and tumor cell dissemination.
■ **ASSAYS:** Activities of this factor are measured by motility tests employing a Boyden chamber (see: Chemotaxis).

■ **CLINICAL USE & SIGNIFICANCE:** A study of bladder carcinoma patients has shown that the concentration of urinary AMF correlates with the degree of invasive growth of the tumors.

● **REVIEWS: Nabi IR et al** Autocrine motility factor and its receptor: role in cell locomotion and metastasis. Cancer Metastasis Rev. 11: 5-20 (1992); **Stoker M & Gherardi E** Regulation of cell movement: the motogenic cytokines. BBA 1072: 81-102 (1991)
● **BIOCHEMISTRY & MOLECULAR BIOLOGY: Evans CP et al** An autocrine motility factor secreted by the Dunning R-3327 rat prostatic adenocarcinoma cell subtype AT2.1. IJC 49: 109-13 (1991); **Liotta LA et al** Tumor cell autocrine motility factor. PNAS 83: 3302-6 (1986); **Silletti S et al** Purification of B16-F1 melanoma autocrine motility factor and its receptor. CR 51: 3507-11 (1991); **Watanabe H et al** Purification of human tumor cell autocrine motility factor and molecular cloning of its receptor. JBC 266: 13442-48 (1991)
● **RECEPTORS: Lotan R et al** Suppression of melanoma cell motility factor receptor expression by retinoic acid. CR 52: 4878-84 (1992); **Nabi IR et al** B16-F1 melanoma autocrine motility-like factor receptor. CR 50: 409-14 (1990); **Nabi IR et al** Tumor cell autocrine motility factor receptor. Experientia Suppl. 59: 163-77 (1991)
● **BIOLOGICAL ACTIVITIES: Atnip KD et al** Chemotactic response of rat mammary adenocarcinoma cell clones to tumor-derived cytokines. BBRC 146: 996-1002 (1987); **Kohn EC et al** Granulocyte-macrophage colony-stimulating factor induces human melanoma-cell migration. IJC 53: 968-72 (1993); **Lotan R et al** Suppression of melanoma cell motility factor receptor expression by retinoic acid. CR 52: 4878-84 (1992); **Nabi IR & Raz A** Cell shape modulation alters glycosylation of a metastatic melanoma cell-surface antigen. IJC 40: 396-402 (1987); **Silletti S & Raz** Autocrine motility factor is a growth factor. BBRC 194: 446-57 (1993); **Stracke ML et al** Insulin-like growth factors stimulate chemotaxis in human melanoma cells. BBRC 153: 1076-83 (1988); **Stracke ML et al** Cell motility, a principal requirement for metastasis. Experientia Suppl. 59: 147-62 (1991); **Watanabe H et al** Expression of autocrine motility factor receptor in serum- and protein-independent fibrosarcoma cells: implications for autonomy in tumor-cell motility and metastasis. IJC 53: 689-95 (1993)
● **CLINICAL USE & SIGNIFICANCE: Guirguis R et al** Detection of autocrine motility factor in urine as a marker of bladder cancer. JNCI 80: 1203-11 (1988); **Javadpour N & Guirguis R** Tumor collagenase-stimulating factor and tumor autocrine motility factor as tumor markers in bladder cancer – an update. Eur. Urol. 21: s1-4 (1992)

AMH: *anti-Müllerian hormone* This hormone is identical with » MIS (Müllerian inhibiting substance).

AML 193: A human monocytic leukemia cell line that is used for the detection of » IL3 (see also: Bioassays). IL3 activity is determined by measuring the incorporation of ^{3}H thymidine into the newly synthesized DNA of proliferating cells. Cell proliferation can be determined also by employing the » MTT assay. The assay is not specific because these cells also respond to » GM-CSF and » G-

CSF. The specificity must be checked therefore by using appropriate antibodies neutralizing the unwanted activities. An alternative and entirely different detection method is » Message amplification phenotyping.

Lange B et al Growth factor requirements of childhood acute leukemia: establishment of GM-CSF-dependent cell lines. Blood 70: 192-9 (1987); **Saito K et al** Bioassays of recombinant human granulocyte colony stimulating factor – evaluations of the measurement of biological activity and correlations among *in vitro* colony-forming, NFS-60, AML-193, and *in vivo* CPA-mouse methods. Jap. Pharmacol. Therapeutics. 18 (Suppl.) 9: 31-41 (1990)

Amniotic fluid binding protein: abbrev. AFBP. This protein is identical with one of the insulin-like growth factor binding proteins (see: IGF-BP).

Amphiregulin: see: AR.

AMR: abbrev. for amphiregulin. See: AR.

ANAP: *anionic neutrophil-activating peptide* ANAP was originally isolated from psoriatic scales as a protein that activates neutrophils (see also: Cell activation) and that was not identical with » IL1, initially believed to be responsible for this activity. ANAP is identical with » MDNCF (monocyte-derived neutrophil chemotactic factor). The factor is identical with » IL8 and therefore belongs to the » chemokine family of cytokines.

Schröder JM et al Identification of C5ades arg and an anionic neutrophil-activating peptide (ANAP) in psoriatic scales. J. Invest. Dermatol. 87: 53-58 (1986)

Anchorage-dependent cells: A term used to characterize cells requiring a solid substrate for growth, i. e. the solid surface of a culture dish, and which, therefore, do not grow in suspension cultures or semisolid soft agar (see also: colony formation assay). See also: Cell culture.

Androgen-induced growth factor: see: AIGF.

ANF: *atrial natriuretic factor* This factor is a member of the natriuretic peptide family of proteins. Other members are ***BNP*** (brain natriuretic peptide) and ***CNP*** (C-type natriuretic peptide). ANP and BNP are produced mainly by atrial heart muscle cells (ANF) and ventricle cells (BNP). They are stored within granules of atrial cardiocytes. ANF is also found in the central nervous system and the kidney but at concentrations which are orders of magnitude lower than in cardiac tissues. CNP is found mainly in the central nervous system and can be produced by vascular endothelial cells.

ANF is produced as a preproprotein of 151 amino acids. In humans and rats the main form of ANF is a processed form of the prepro-protein, called ***cardionatrin IV*** (126 amino acids), also called ***γ-ANP*** (atrial natriuretic peptide). The circulating form of ANF is a peptide of 28 amino acids (***cardionatrin I*** or ***α-ANP***). A variant of ANF is » Urodilatin.

The release of ANF is stimulated by adrenaline, arginine » vasopressin, acetylcholine. Specific receptors for ANF have been found in blood vessels, kidney, in the thymic cortex, medulla, and splenic white pulp, and on astrocytes.

ANF is a potent natriuretic and hypotensive protein, and inhibits the secretion of renin and aldosterone. ANF also inhibits the production and secretion of » ET (endothelins) by cultured endothelial cells.

Apart from its pharmacological activities ANF also functions as a cytokine. It functions as an anti-growth factor for vascular smooth muscle cells and endothelial cells and also has significant anti-proliferative effects on astrocytes. ANP is also a suppressor of chondrocyte proliferation. It also appears to enhance natural killer cell activity and has neuromodulatory effects. ANF also inhibits the production and secretion of » ET (endothelins) by cultured endothelial cells. Synthesis of CNP by vascular endothelial cells has been shown to be inducible by » IL1 and » TNF-α. CNP inhibits the growth of vascular endothelial cells and counteracts the growth-promoting activities of IL1 and TNF-α.

■ **TRANSGENIC/KNOCK-OUT/ANTISENSE STUDIES:** The role of ANF in long-term cardiovascular regulation has been studied in a » transgenic mouse model involving lifelong elevated plasma ANF levels. These mice are chronically hypotensive. The mice adequately compensate for the renal effects but not the hemodynamic effects of the hormone as no obvious natriuretic or diuretic phenotype was observable.

Ballermann BJ A highly sensitive radioreceptor assay for atrial natriuretic peptide in rat plasma. Am J. Physiol. 254: F159-63 (1988); **Burgisser E et al** Human cardiac plasma concentrations of atrial natriuretic peptide quantified by radioreceptor assay. BBRC 133: 1201-9 (1985); **Cahill PA & Hassid A** Clearance receptor-binding atrial natriuretic peptides inhibit mitogenesis and proliferation of rat aortic smooth muscle cells. BBRC 179: 1606-13 (1991); **Capper SJ et al** Specificities compared for a radioreceptor assay and a radioimmunoassay of atrial natriuretic peptide. Clin. Chem. 36: 656-8 (1990); **Drewett JG et al** Neuromodulatory effects of atrial natriuretic peptides correlate with

ANF transgenics: see addendum: ANF.

an inhibition of adenylate cyclase but not an activation of guanylate cyclase. J. Pharmacol. Exp. Ther. 260: 689-96 (1992); **Furuya M et al** C-type natriuretic peptide is a growth inhibitor of rat vascular smooth muscle cells. BBRC 177: 927-31 (1991); **Gutkowska J et al** Radioreceptor assay for atrial natriuretic factor. AB 168: 100-6 (1988); **Halevy O et al** Epidermal growth factor receptor gene expression in avian epiphyseal growth-plate cartilage cells: effect of serum, parathyroid hormone and atrial natriuretic peptide. Mol. Cell. Endocrinol. 75: 229-35 (1991); **Hu RM et al** Atrial natriuretic peptide inhibits the production and secretion of endothelin from cultured endothelial cells. JBC 267: 17384-9 (1992); **Itoh H et al** Atrial natriuretic polypeptide as a novel antigrowth factor of endothelial cells. Hypertension 19: 758-61 (1992); **Kanagawa K & Matsuo H** Purification and complete amino acid sequence of a-human atrial natriuretic polypeptide. BBRC 118: 131-9 (1984); **Levin ER & Frank HJ** Natriuretic peptides inhibit rat astroglial proliferation: mediation by C receptor. Am. J. Physiol. 261: R453-7 (1991); **Moss RB & Golightly MG** *In vitro* enhancement of natural cytotoxicity by atrial natriuretic peptide fragment 4-28. Peptides 12: 851-4 (1991); **Mukoyama M et al** Brain natriuretic peptide as a novel cardiac hormone in humans – evidence for an exquisite dual natriuretic peptide system, atrial natriuretic peptide, and brain natriuretic peptide. JCI 87: 1402-12 (1991); **Nakao K et al** Molecular biology and biochemistry of the natriuretic peptide system I. Natriuretic peptides. H. Hypertens. 10: 907-12 (1992); **Ogawa Y et al** Human C-type natriuretic peptide – characterization of the gene and peptide. Hypertension 19: 809-13 (1992); **Ong H et al** A highly specific radioreceptor assay for the active circulating form of atrial natriuretic factor in human plasma. Clin. Chem. 34: 2275-9 (1988); **Porter JG et al** C-type natriuretic peptide inhibits growth factor-dependent DNA synthesis in smooth muscle cells. Am. J. Physiol. 263: C1001-6 (1992); **Sudo T et al** C-type natriuretic peptide (CNP): a new member of natriuretic peptide family identified in porcine brain. BBRC 168: 863-70 (1990); **Suga S et al** Endothelial production of C-type natriuretic peptide and its marked augmentation by transforming growth factor β – possible existence of vascular natriuretic peptide system. JCI 90: 1145-9 (1992); **Suga S et al** Cytokine-induced C-type natriuretic peptide (CNP) secretion from vascular endothelial cells – evidence for CNP as a novel autocrine/paracrine regulator from endothelial cells. Endocrinology 133: 3038-41 (1993)

● **TRANSGENIC/KNOCK-OUT/ANTISENSE STUDIES: Koh GY et al** Atrial natriuretic factor and transgenic mice. Hypertension 22: 634-9 (1993)

Angiogenesis factors: ***Angiogenesis*** is the preferred term for processes leading to the generation of new blood vessels through sprouting from already existing blood vessels. Blood vessel growth occurs in the embryo and rarely in the adult with exceptions such as the female reproductive system, » wound healing, and pathological processes such as cancer.

The term ***vasculogenesis*** is mainly used for the *de novo* generation of blood vessels occurring during embryogenesis. ***Vascular expansion*** is the enlargement of small or occluded vessels which is frequently observed during the generation of collateral blood vessels.

Under normal conditions all processes involving the new formation or the remodeling of existing of

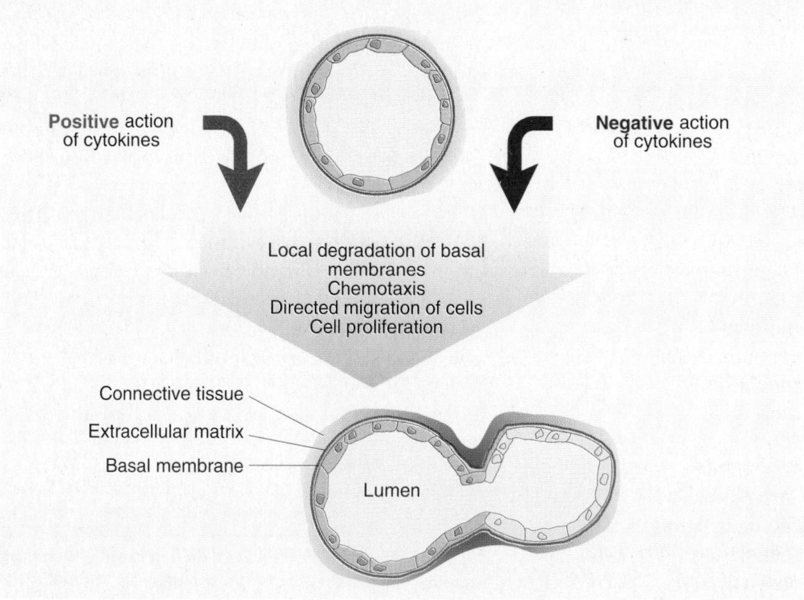

Cytokines in Angiogenesis.
Many different cytokines regulate processes involved in the generation of new blood vessels, including local degradation of basal membrane, reorganization of the extracellular matrix, directed influx of new cells. See text and table for details.

new blood vessels, which normally is a self-limiting process, require the controlled and concerted growth of various specific cell types in order to avoid the unwanted overgrowth. During tumor angiogenesis one observes the directed sprouting of new blood vessels into the direction of the solid tumor mass. This process guarantees the blood supply of the tumor and in its absence tumor cells would die by necrosis.

A plethora of experiments have suggested that tissues secrete angiogenic factors which promote angiogenesis under conditions of poor blood supply during normal and pathological angiogenesis processes. The first indication of the existence of such diffusible substances was gleaned from filtration experiments demonstrating that tumor cells separated from underlying tissues by filters that do not allow passage of cells are nevertheless capable of supporting vessel growth in these tissues. The formation of blood vessels is initiated and maintained by a variety of factors secreted either by the tumor cells themselves or by accessory cells.

Some of these factors are protein factors initially detected due to some other biological activities and later shown to promote angiogenesis. The list of angiogenically active protein factors includes fibroblast growth factors (see: FGF), » angiogenin, » EGF, » VEGF, » TNF-α, TGF-β, » PD-ECGF, » PDGF, » IGF, and » IL8. Fibrin fragment E has also been shown to have angiogenic activity. Angiogenic activities also include a number of other compounds such as prostaglandins E_1 and E_2, steroids, heparin, 1-butyryl glycerol (monobutyrin) secreted by adipocytes, and many undefined derivatives of the arachidonic acid metabolism. The biologically active principle extracted from some carcinoma cells and identified as nicotinic amide is also a potent angiogenic compound in several bioassays although its mechanism of action remains to be elucidated.

These factors and compounds differ in cell specificity and also in the mechanisms by which they induce the growth of new blood vessels. Not all compounds that are active *in vivo* show the same spectrum of activities for endothelial cells *in vitro*. Many of these factors are pleiotropic and, among other things, may induce the migration and proliferation of endothelial cells, the production of collagenase and plasminogen activator. Some of these factors, however, are neither mitogenic nor chemotactic for endothelial cells. They elicit their effects probably by attracting other cells to the site of growth, by activating these cells (see also: Cell activation), and by inducing them to secrete angiogenic factors.

As many angiogenic factors are mitogenic and chemotactic for endothelial cells their biological activities can be determined *in vitro* by measuring the induced migration of endothelial cells or the effect of these factor on endothelial cell proliferation (for further information see also subentry "Assays" in the reference section). However, there are also a number of bioassays which allow direct determination of angiogenic activities. One assay employs the use of the non-vascularized rabbit eye (***cornea pocket assay***) which has the advantage that new blood vessels are easily detected and essentially must be newly formed blood vessels. The use of the chicken chorioallantoic membrane (***CAM assay***) is inexpensive but more problematic because of the expertise required to differentiate pre-existing from newly formed blood vessels.

The immediate reaction following exposure of tissue to angiogenic factors is the dissolution of the basal membrane of pre-existing blood vessels. The next step is the migration and proliferation of endothelial cells. The immature new vessels is elongated, develops a lumen, and branches. The final stage of capillary development is the formation of a new basal membrane (see also: ECM, extracellular matrix). In some instances one observes the formation of a layer of perivascular cells (pericytes).

Since the development of new blood vessels is to be avoided under normal conditions it can be assumed that the synthesis of angiogenically active factors is subject to constitutive repression in normal tissues. Many control mechanisms have been described to be active. Some angiogenic factors are produced only if the cells have become activated during a process requiring the new formation of blood vessels. Other factors such as » TGF-β are produced by normal cells as inactive proforms that can be activated in a controlled manner. Factors such as » aFGF and » bFGF are not secreted but stored in depots of the basal membrane while » TNF-α is capable of promoting angiogenesis only from without a blood vessels but not if it is released in the lumen.

In tumor cells these tight regulatory mechanisms have been corrupted in most cases by the ability of tumor cells to synthesize angiogenic growth factors constitutively. Some tumor cells also produce slightly altered factors that are directly mitogenic for endothelial cells or they elaborate hydrolytic enzymes capable of releasing angiogenesis factors from their membrane depots or of activating inactive angiogenic factors.

Angiogenic factors and also their inhibitors are of clinical significance because they could be used to directly interfere with angiogenic processes involved, for example, in » wound repair. Angiogenesis inhibitors should be of value in treatment of diseases of pathogenic neovascularization such as Kaposi's sarcoma, diabetic retinopathy, and malignant tumor growth. One example of compounds, also known as ***angioinhibins***, which inhibit angiogenesis, is » Fumagillin. Other factors negatively influencing angiogenesis by inhibiting the proliferation of endothelial cells include » PF4 (platelet factor 4), » IFN-α and » IFN-γ, » thrombospondin isolated from platelets, a proteinase inhibitor of cartilage known as » CDI (cartilage-derived inhibitor), and » ChDI (chondrocyte-derived inhibitor). A nonanticoagulating derivative of heparin, heparin adipic hydrazide (HAH), covalently linked by an acid-labile bond to the anti-angiogenic steroid, cortisol has been shown to have potent anti-angiogenesis and antitumour activity in a mouse model.

● **REVIEWS: Billington DC** Angiogenesis and its inhibition: potential new therapies in oncology and non-neoplastic diseases. Drug Des. Discov. 8: 3-35 (1991); **Blood CH & Zetter BR** Tumor interactions with the vasculature: angiogenesis and tumor metastasis. BBA 1032: 89-118 (1990); **Bouck N** Tumor angiogenesis: the role of oncogenes and tumor suppressor genes. CC 2: 179-85 (1990); **Denekamp J** Review article: angiogenesis, neovascular proliferation and vascular pathophysiology as targets for cancer therapy. Br. J. Radiol. 66: 181-96 (1993); **Eisenstein R** Angiogenesis in arteries: Review. Pharmacol Ther. 49: 1-19 (1991); **Folkman J & Klagsbrun M** Angiogenic factors. S 235: 442-6 (1987); **Folkman J** What is the evidence that tumors are angiogenesis dependent? JNCI 82: 4-6 (1990); **Folkman J & Ingber D** Inhibition of angiogenesis. Semin. Cancer Biol. 3: 89-96 (1992); **Folkman J & Shing V.** Control of angiogenesis by heparin and other sulfated polysaccharides. AEMB 313: 355-64 (1992); **Hom DB & Maisel RH** Angiogenic growth factors: their effects and potential in soft tissue wound healing. Ann. Otol. Rhinol. Laryngol. 101: 349-54 (1992); **Ingber D** Extracellular matrix and cell shape: Potential control points for inhibition of angiogenesis. JCBc 47: 236-41 (1991); **Klagsbrun M & D'Amore PA** Regulators of angiogenesis. ARP 53: 217-39 (1991); **Leibovitch SJ et al** Macrophages, wound repair and angiogenesis. PCBR 266: 131-45 (1988); **Liotta LA et al** Cancer metastasis and angiogenesis: An imbalance of positive and negative regulation. Cell 64: 327-36 (1991); **Maciag T** Molecular and cellular mechanisms of angiogenesis. In: DeVita VT et al (eds) Important advances in oncology, pp. 85-100, JB Lippincott, Philadelphia, 1990; **Mitchell MA & Wilks JW** Inhibitors of angiogenesis. Ann. Rep. Medicinal Chem. 27: 139-48 (1992); **Ribatti D et al** Angiogenesis under normal and pathological conditions. Haematologica 76: 311-20 (1991); **Risau W** Angiogenic growth factors. PGFR 2: 71-9 (1990); **Schweigerer L & Fotsis T** Angiogenesis and angiogenesis inhibitors in pediatric diseases. Eur. J. Pediatr. 151: 472-6 (1992); **Schultz GS & Grant MB** Neovascular growth factors. Eye 5: 170-80 (1991); **Vlodavsky I et al** Extracellular matrix-resident growth factors and enzymes: possible involvement in tumor metastasis and angiogenesis. Cancer Metastasis Rev. 9: 203-26 (1990); **Weidner N et al** Tumor angiogenesis and metastasis – correlation in invasive breast carcinoma. NEJM 324: 1-8 (1991)

● **BIOCHEMISTRY & MOLECULAR BIOLOGY: Crum R et al** A new class of steroids inhibit angiogenesis in the presence of heparin or a heparin fragment. S 230: 1375-8 (1985); **Dobson DE et al** 1-Butyryl-glycerol: a novel angiogenesis factor secreted by differentiating adipocytes. Cell 61: 223-30 (1990); **Maione TE & Sharpe RJ** Development of angiogenesis inhibitors for clinical applications. TIPS 11: 457-61 (1990); **Masferrer JL et al** 12(R)-hydroxyeicosatrienoic acid, a potent chemotactic and angiogenic factor produced by the cornea. Exp. Eye. Res. 52: 417-24 (1991); **Maxwell M et al** Expression of angiogenic growth factor genes in primary human astrocytomas may contribute to their growth and progression. CR 51: 1345-51 (1991); **Odedra R & Weiss JB** Low molecular weight angiogenesis factors. Pharmacol. Ther. 49: 111-24 (1991); **Polverini PJ et al** Assay and purification of naturally occurring inhibitor of angiogenesis. MiE 198: 440-50 (1991); **Sato E et al** Ovarian glycosaminoglycans potentiate angiogenic activity of epidermal growth factor in mice. Endocrinology 128: 2402-6 (1991); **Sunderkotter C et al** Macrophage-derived angiogenesis factors. Pharmacol. Therapeutics 51: 195-216 (1991); **Taylor CM et al** Matrix integrity and the control of angiogenesis. Int. J. Radiat. Biol. 60: 61-4 (1991); **Thompson WD et al** Angiogenic activity of fibrin degradation products is located in fibrin fragment E. J. Pathol. 168: 47-53 (1992); **Thorpe PE et al** Heparin-steroid conjugates: new angiogenesis inhibitors with antitumor activity in mice. CR 53: 3000-7 (1993); **Wilkison WO et al** Biosynthetic regulation of monobutyrin, an adipocyte-secreted lipid with angiogenic activity. JBC 266: 16886-91 (1991)

● **ASSAYS: Auerbach R et al** A simple procedure for the long-term cultivation of chicken embryos. Dev. Biol. 41: 391-94 (1974); **Castellot JJ Jr et al** Heparin potentiation of 3T3-adipocyte stimulated angiogenesis: mechanisms of action on endothelial cells. JCP 127: 323-9 (1986); **Gaudric A et al** Quantification of angiogenesis due to basic fibroblast growth factor in a modified rabbit corneal model. Ophthalmic. Res. 24: 181-8 (1992); **Greenblatt M & Shubik P** Tumor angiogenesis: transfilter diffusion studies in the hamster by the transparent chamber technique. JNCI 41: 111-24 (1968); **Hayek A et al** The use of digital image processing to quantitate angiogenesis induced by basic fibroblast growth factor and transplanted pancreatic islets. Microvasc. Res. 41: 203-9 (1991); **Mourad MM et al** An immunoradiometric assay for von Willebrand factor antigen to assess angiogenesis. Br. J. Dermatol. 123: 21-8 (1990); **Okamura K et al** A model system for tumor angiogenesis: involvement of transforming growth factor-α in tube formation of human microvascular endothelial cells induced by esophageal cancer cells. BBRC 186: 1471-9 (1992); **Olivo M et al** A comparative study on the effects of tumor necrosis factor-α (TNF-α), human angiogenic factor (h-AF) and basic fibroblast growth factor (bFGF) on the chorioallantoic membrane of the chick embryo. Anat. Rec. 234: 105-15 (1992); **Passaniti A et al** A simple, quantitative method for assessing angiogenesis and antiangiogenic agents using reconstituted basement membrane, heparin, and fibroblast growth factor. Lab. Invest. 67: 519-28 (1992); **Peek MJ et al** The chick chorioallantoic membrane assay: an improved technique for the study of angiogenic activity. Exp. Pathol. 34: 35-40 (1988); **Polverini PJ et al** Assay and purification of naturally occurring inhibitor of angiogenesis. MiE 198; 440-50 (1991); **Sato N et al** Actions of TNF and IFN-γ on angiogenesis *in vitro*. J. Invest. Dermatol. 95: 85S-89S (1990); **Shahabuddin S et al** A study of angiogenesis factors from five different sources using

Some cytokines with angiogenic or anti-angiogenic activity and their actions on several cell types.

Factor	endothelial cells	smooth muscle cells	fibroblasts
9E3	chemotactic		
Activin A		modulatory †	
aFGF	mitogenic/chemotactic		mitogenic
Amphiregulin			mitogenic
ANF	inhibitory	inhibitory	
Angiogenin	not mitogenic, not chemotactic		
Angiotensin II		modulatory/mitogenic ¶	
Angiotropin	chemotactic, inhibits proliferation	mitogenic	
Betacellulin		mitogenic	
Beta2 microglobulin			mitogenic
Beta-Thromboglobulin			chemotactic
bFGF	mitogenic/chemotactic		mitogenic/chemotactic
Bombesin			mitogenic
CLAF **	mitogenic/chemotactic		
CDI	inhibits proliferation		
ChDI	inhibits proliferation		
CTAP III			mitogenic/chemotactic
ECDGF ‡	?	mitogenic	mitogenic
EDMF	chemotactic		
EGF #	mitogenic, not chemotactic		
ELMER			mitogenic
Endothelins		mitogenic	mitogenic
Epithelins			mitogenic/inhibitory ¬
f-ECGF	mitogenic		
FGF-6 ‡	?		
Fibrin	not mitogenic, chemotactic		
GSM ‡	?		
HAF ‡	not mitogenic		
Haptoglobin ‡	?		
HB-EGF	Not mitogenic	mitogenic	mitogenic
HBNF ‡	mitogenic		mitogenic
HGF ‡	mitogenic		
IFN-α	Inhibitory, not chemotactic		Inhibitory
IFN-γ	Inhibitory, not chemotactic	Inhibitory	
IGF	mitogenic, not chemotactic		
IL1			stimulatory
IL8	mitogenic, chemotactic		
KAF ‡	?		
K-FGF	mitogenic		mitogenic
MDGF		stimulatory	stimulatory
MGSA			mitogenic
Neuromedin B		mitogenic	mitogenic
Neuromedin C		mitogenic	mitogenic
Oncostatin M	mitogenic	mitogenic	mitogenic
PAF ***	mitogenic/chemotactic		
PD-ECGF	mitogenic/chemotactic	not chemotactic/mitogenic	not chemotactic/mitogenic
PDGF	Not mitogenic	mitogenic ∞	
PF4 ‡‡	Inhibitory		chemotactic/mitogenic
PIGF	mitogenic		
Prolactin 16 kDa ‡‡	Not mitogenic		
SDGF #1	mitogenic	mitogenic	
SDGF #2		mitogenic	
SDGF #3			mitogenic
SDMF		chemotactic/not mitogenic	
TGF-α	mitogenic, chemotactic		
TGF-β ‡	inhibitory	mitogenic, inhibitory ††	mitogenic, inhibitory
Thrombospondin	Inhibitory, not chemotactic	mitogenic	mitogenic
TNF-α ‡	Inhibitory, chemotactic		stimulatory
TNF-β	Inhibitory, not chemotactic		stimulatory
VEGF/VPF *	mitogenic, chemotactic		

‡ angiogenic *in vivo*
‡‡ inhibitory *in vivo*
# potentiated by thrombin
* synergizes with bFGF
** aminoterminally truncated form of bFGF
*** aminoterminally extended bFGF
† involving » autocrine production of IGF-1 and modulation of its activity
†† TGF-β1 in synergy with α2 macroglobulin; inhibitory for smooth muscle cells at low concentrations and mitogenic at higher concentrations; bimodal activity mediated by PDGF
¶ stimulates hypertrophied growth response in vascular smooth muscle cells (increases in cell size and protein synthesis, but not cell number); promotes growth of vascular smooth muscle cells in vitro via the autocrine production of growth factors such as PDGF, bFGF, HB-EGF, and » TGF-β.

¬ Epithelin 1 stimulates the proliferation of fibroblasts and Epithelin 2 inhibits Epithelin 1-induced cell proliferation.
∞ PDGF-AA, in contrast to PDGF-AB and PDGF-BB, is a poor mitogen for vascular smooth muscle cells. Together with bFGF, which upregulates PDGF-A receptors, it acts synergistically on DNA synthesis of these cells.
#1 schwannoma-derived growth factor; enhances proliferation of these cells induced by PDGF.
#2 smooth muscle cell derived growth factor
#3 spleen-derived growth factor

a radioimmunoassay. IJC 35: 87-91 (1985); **Splawinski J et al** Angiogenesis: quantitative assessment by the chick chorioallantoic membrane assay. Methods Find. Exp. Clin. Pharmacol. 10: 221-6 (1988); **Tanaka NG et al** Inhibitory effects of anti-angiogenic agents on neovascularisation and growth of the chorioallantoic membrane (CAM). The possibility of a new CAM assay for angiogenesis inhibition. Exp. Pathol. 30: 143-50 (1986); **Vlodavsky I et al** Extracellular matrix-resident basic fibroblast growth factor: implication for the control of angiogenesis. JCBc 45: 167-76 (1991); **Vu MT et al** An evaluation of methods to quantitate the chick chorioallantoic membrane assay in angiogenesis. Lab. Invest. 53: 499-508 (1985); **Wilting J et al** A modified chorioallantoic membrane (CAM) assay for qualitative and quantitative study of growth factors. Studies on the effects of carriers, PBS, angiogenin, and bFGF. Anat. Embryol. 183: 259-71 (1991)

Angiogenic factor: This glioma-derived endothelial cell growth factor (see also: ECGF) is identical with » aFGF.

Libermann TA et al An angiogenic growth factor is expressed in human glioma cells. EJ 6: 1627-32 (1987)

Angiogenin:

■ **SOURCES:** Angiogenin is synthesized and secreted by human and murine carcinoma cells and fibroblasts and is found circulating in the serum. Angiogenin mRNA is expressed in a wide spectrum of cells and is not correlated to a particular cell phenotype.

■ **PROTEIN CHARACTERISTICS:** Angiogenin is a basic protein of 14.4 kDa. Angiogenin shows a 33 % homology to ribonuclease A. The homology of human and bovine angiogenin is 65 % at the protein level.

■ **GENE STRUCTURE:** The human angiogenin has been located to chromosome 14q11 proximal to the α/δ T cell receptor. The murine angiogenin gene maps to chromosome 14.

■ **RELATED FACTORS:** Angiogenin belongs to a family of proteins, designated RNAse superfamily, which also include eosinophil cationic protein, eosinophil-derived neurotoxin, sialic-acid binding lectin and anti-tumor protein p30.

■ **RECEPTOR STRUCTURE, GENE(S), EXPRESSION:** A 49 kDa receptor protein has been described. A 42 kDa angiogenin binding protein, found as a dissociable cell-surface component of calf pulmonary artery endothelial cells and a transformed bovine endothelial cell line, GM7373, has been identified as a cell surface actin. The interaction between this actin and angiogenin has been suggested to constitute an essential step in the angiogenesis process induced by angiogenin.

■ **BIOLOGICAL ACTIVITIES:** Angiogenin displays limited ribonuclease activity towards wheat-germ RNA, yeast RNA, poly(C) and poly(U). This activity can be blocked by ribonuclease inhibitors, including a specific inhibitor of mammalians, designated ***RAI*** (ribonuclease/angiogenin inhibitor), which is encoded by a gene on human chromosome 11p15.5, approximately 90 kb away from the Harvey *ras* proto-oncogene. Angiogenin specifically cleaves 18S and 28S RNA in rabbit reticulocyte lysates.

Several bioassays shown that angiogenin is a potent » angiogenesis factor. Blocking of the ribonuclease activity concomitantly also blocks the angiogenic activity of angiogenin. Mutations that enhance ribonuclease activity also enhance the angiogenic activity. Very likely the biological activity of angiogenin is due to the activation of accessory cells induced to secrete a cytokine that activates endothelial cells. Angiogenin by itself is neither a mitogen nor a chemoattractant for endothelial cells. Angiogenin also promotes the adhesion of endothelial cells and fibroblasts.

In endothelial cells from capillaries and the umbilical chord angiogenin induces the secretion of prostacyclin without having any effect on the prostaglandins E. Angiogenin activates phospholipase C and induces the rapid incorporation of fatty acid into cholesterol esters in vascular smooth muscle cells.

■ **CLINICAL USE & SIGNIFICANCE:** Angiogenin has been used to induce neovascularisation within the poorly vascularized meniscal fibrocartilage, and to improve the results of meniscal repair in New Zealand white rabbits.

● **REVIEWS: Fox EA & Riordan JF** The molecular biology of angiogenin. In: Chien S (edt) Molecular biology of the cardiovascular system, pp. 139-54, Lea & Fabiger, Philadelphia 1990; **Riordan JF & Vallee BL** Human angiogenin, an organogenic protein. Brit. J. Cancer 57: 587-90 (1988)

● **BIOCHEMISTRY & MOLECULAR BIOLOGY: Acharya KR et al** Crystallization and preliminary X-ray analysis of human angiogenin. J. Mol. Biol. 228: 1269-70 (1992); **Allemann RK et al** A hybrid of bovine pancreatic ribonuclease and human angiogenin: an external loop as a module controlling substrate specificity? Prot. Engin. 4: 831-5 (1991); **Blanco FJ et al** The homologous angiogenin and ribonuclease N-terminal fragments fold into very similar helices when isolated. BBRC 182: 1491-8 (1992); **Bond MD et al** Replacement of residues 8-22 of angiogenin with 7-21 of RNAse A selectively affects protein synthesis inhibition and angiogenesis. B 29: 3341-9 (1990); **Bond MD & Vallee BL** Isolation and sequencing of mouse angiogenin DNA. BBRC 171: 988-95 (1990); **Bond MD et al** Characterization and sequencing of rabbit, pig and mouse angiogenins: discernment of functionally important residues and regions. BBA 1162: 177-86 (1993); **Curran TP et al** Alteration of the enzymatic specificity of human angiogenin by site-directed mutagenesis. B 32: 2307-13 (1993); **Fett JW et al** Isolation and characterization of angiogenin, an angiogenic protein from human carcinoma cells. B 24: 5480-6 (1985); **Hallahan TW et al** Importance of asparagine-61 and asparagine-109 to the angiogenic activity of human angiogenin. B 31: 8022-9 (1992); **Harper JW & Vallee BL** Human angiogenin is a blood vessel inducing protein whose primary structure displays 33 % identity to that of bovine pancreatic ribonuclease A (RNase A). B 28: 1875-84 (1989); **Harper JW & Vallee BL** Mutagenesis of aspartic acid-116 enhances the ribonucleolytic activity and angiogenic potency of angiogenin. PNAS 85: 7139-43 (1988); **Harper JW & Vallee BL** A covalent angiogenin/ribonuclease hybrid with a fourth disulfide bond generated by regional mutagenesis. B 28: 1875-84 (1989); **Kamiya Y et al** Amino acid sequence of a lectin from Japanese frog (Rana japonica) eggs. J. Biochem. Tokyo 108: 139-43 (1990); **Kurachi K et al** Sequence of the cDNA and gene for angiogenin, a human angiogenesis factor. B 24: 5494-9 (1985); **Lapidus AL et al** Expression of the hAng gene in *Escherichia coli*; isolation and characterization of human recombinant Ser-(-1) angiogenin. Biomed. Sci. 1: 597-604 (1990); **Lee FS & Vallee BL** Structure and action of mammalian ribonuclease (angiogenin) inhibitor. Prog. Nucleic Acid Res. Mol. Biol. 44: 1-30 (1993); **Maes P et al** The complete amino acid sequence of bovine milk angiogenin. FL 241: 41-5 (1988); **Moore F & Riordan JF** Angiogenin activates phospholipase C and elicits a rapid incorporation of fatty acid into cholesterol esters in vascular smooth muscle cells. B 29: 228-33 (1990); **Rybak SM et al** Angiogenin mRNA in human tumor and normal cells. BBRC 146: 1240-8 (1987); **Rybak SM et al** C-terminal angiogenin peptides inhibit the biological and enzymatic activities of angiogenin. BBRC 162: 535-43 (1989); **Schneider R & Schweiger M** The yeast RNA1 protein, necessary for RNA processing, is homologous to the human ribonuclease/angiogenin inhibitor (RAI). Mol. Gen. Genet. 233: 315-8 (1992); **Shapiro R et al** Role of lysines in human angiogenin: chemical modification and site-directed mutagenesis. B 28: 1726-32 (1989); **Shapiro R et al** Site-directed mutagenesis of histidine-13 and histidine-114 of human angiogenin. Alanine derivatives inhibit angiogenin-induced angiogenesis. B 28: 7401-8 (1989); **Steinhelper ME & Field LJ** Assignment of the angiogenin gene to mouse chromosome 14 using a rapid PCR-RFLP mapping technique. Genomics. 12: 177-9 (1992); **Strydom DJ et al** Amino acid sequence of human tumor-derived angiogenin. B 24: 5486-94 (1985); **Weremowicz S et al** Localization of the human angiogenin gene to chromosome band 14q11, proximal to the T cell receptor α/δ locus. Am. J. Hum. Gene. 47: 973-81 (1990)

● **Receptors: Chamoux M et al** Characterization of angiogenin receptors on bovine brain capillary endothelial cells. BBRC 176: 833-9 (1991); **Hu GF et al** An angiogenin-binding protein from endothelial cells. PNAS 88: 2227-31 (1991); **Hu GF et al** Actin is a binding protein for angiogenin. PNAS 90: 1217-21 (1993)

● **Biological Activities: Badet J et al** *In vivo* and *in vitro* studies of angiogenin – a potent angiogenic factor. Blood Coagul. Fibrinolysis. 1: 721-4 (1990); **Bicknell R & Vallee BL** Angiogenin stimulates endothelial cell prostacyclin secretion by activation of phospholipase A2. PNAS 86: 1573-7 (1989); **Fett JW et al** Induction of angiogenesis by mixtures of two angiogenic proteins, angiogenin and acidic fibroblast growth factor, in the chick chorioallantoic membrane. BBRC 146: 1122-31 (1987); **Saxena SK et al** Angiogenin is a cytotoxic, tRNA-specific ribonuclease in the RNase A superfamily. JBC 267: 21982-6 (1992); **Soncin F** Angiogenin supports endothelial and fibroblast cell adhesion. PNAS 89: 2232-6 (1992); **St Clair DK et al** Angiogenin abolishes cell-free protein synthesis by specific ribonucleolytic inactivation of 40S ribosomes. B 27: 7263-8 (1988)

● **Clinical Use & Significance: King TV & Vallee BL** Neovascularisation of the meniscus with angiogenin. An experimental study in rabbits. J. Bone Joint Surg. Br. 73: 587-90 (1991)

Angioinhibins: Generic name given to substances that negatively influence the generation of new blood vessels (see also: Angiogenesis factors). One of the most active angioinhibins known is » Fumagillin.

Angiopeptin: see: Somatostatin.

Angiotensin: Angiotensin is a decapeptide (DRVYIHPFHL) originally found to be produced by kidney-derived renin from an α-2 hepatic globulin. It is mainly known for its potent pharmacological activities. Angiotensin elevates blood pressure through its direct vasoconstrictor, sympathomimetic, and (through release of aldosterone) sodium-retaining activities.

The original angiotensin (***Angiotensin I***) gives rise to ***Angiotensin II*** by removal of a C-terminal dipeptide. This reaction is mediated by angiotensin converting enzyme in plasma, liver and nerve tissues. Further removal of the aminoterminal amino acid from angiotensin II by an aspartate amino peptidase yields ***Angiotensin III***. Angiotensin II and Angiotensin III have been shown to be even more potent with respect to modulation of peripheral blood pressure and may more readily traverse biological membranes. All of the substrates and enzymes needed for angiotensin conversion have been found to be present in the brain. Several studies also indicate that various cells of the immune and the nervous system containing and/or responding to one or another of these peptides, suggesting a role in modulating functions within the » Neuroimmune network.

Apart from its pharmacological activities angiotensin II also has cytokine-like activities. It has been described as a growth-promoting factor in several cellular systems and shares many signal transduction mechanisms with growth factors. The introduction of a functional angiotensin II receptor into Chinese hamster ovary cells has been shown to confer growth properties to angiotensin II for these cells.

Angiotensin II acts as a growth factor in the cardiovascular system and has been implicated in angiogenesis. It is a mitogen for neonatal rat cardiac fibroblasts and has also been shown to be a hypertrophogenic cytokine for proximal tubular cells of the kidney.

Angiotensin II appears to be a bifunctional vascular smooth muscle cell (VSMC) growth modulator. It stimulates a hypertrophied growth response in vascular smooth muscle cells characterized by

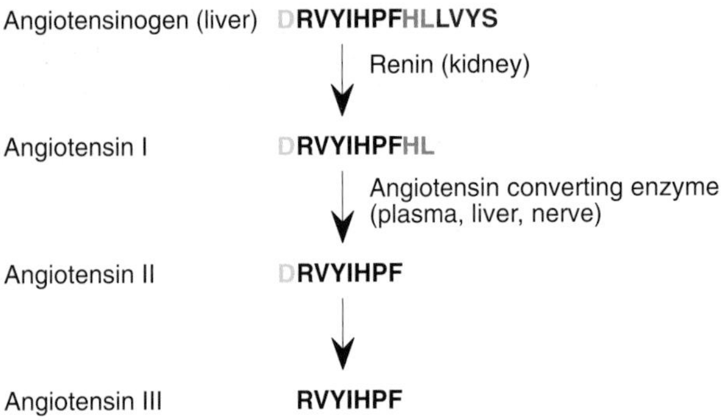

The angiotensin family.
Angiotensins are derived from angiotensinogen by removal of amino acids from the C-terminal end. Angiotensin III is a naturally occurring N-des-Asp Angiotensin II.

increases in cell size and protein synthesis, but not cell number. Angiotensin II also promotes growth of vascular smooth muscle cells *in vitro* via the » autocrine production of growth factors such as » PDGF, » bFGF, » HB-EGF, and » TGF-β. Experimental studies have demonstrated that angiotensin infusion can enhance smooth muscle proliferation after balloon injury *in vivo*. Angiotensin II thus plays a key role as a mediator of vascular neointima formation. It is probably one of the factors responsible for an increase in the thickness of the arterial medial smooth muscle cell layer as one of the major consequences of hypertension.

Angiotensin II has been shown to be mitogenic for bovine adrenocortical zona fasciculata/reticularis cells. Angiotensin II is a growth factor for human SH-SY5Y neuroblastoma cells. Insulin at high concentrations and » IGF at low concentrations enhance the proliferative response of these cells to angiotensin II. Locally produced Angiotensin II can act in an » autocrine or » paracrine fashion to alter the growth of human mesangial and neuroblastoma cells.

Angiotensin III has been found to be a chemotactic factor for human polymorphonuclear neutrophils.

Chen LI et al The interaction of insulin and angiotensin II on the regulation of human neuroblastoma cell growth. Mol. Chem. Neuropathol. 18: 189-96 (1993); **Clyne CD et al** Angiotensin II stimulates growth and steroidogenesis in zona fasciculata/reticularis cells from bovine adrenal cortex via the AT1 receptor subtype. Endocrinology 132: 2206-12 (1993); **Cook JL et al** The use of antisense oligonucleotides to establish autocrine angiotensin growth effects in human neuroblastoma and mesangial cells. Antisense Res. Dev. 2: 199-210 (1992); **Costerousse O et al** Angiotensin I-converting enzyme in human circulating mononuclear cells: genetic polymorphism of expression in T-lymphocytes. BJ 290: 33-40 (1993); **Dubey RK et al** Culture of renal arteriolar smooth muscle cells. Mitogenic responses to angiotensin II. Circ. Res. 71: 1143-52 (1992); **Ferrario CM et al** Pathways of angiotensin formation and function in the brain. Hypertension 9: 1-11 (1990); **Ganong WF** Angiotensin II in the brain and pituitary: contrasting roles in the regulation of adenohypophyseal secretion. Horm. Res. 31: 24-31 (1989); **Gibbons GH et al** Vascular smooth muscle cell hypertrophy vs. hyperplasia. Autocrine transforming growth factor-β 1 expression determines growth response to angiotensin II. JCI 90: 456-61 (1992); **Kunert-Radek J & Pawlikowski M** Angiotensin II stimulation of the rat pituitary tumoral cell proliferation *in vitro*. BBRC 183: 27-30 (1992); **Laporte S & Escher E** Neointima formation after vascular injury is angiotensin II mediated. BBRC 187: 1510-6 (1992); **Nakahara K et al** Identification of three types of PDGF-A chain gene transcripts in rabbit vascular smooth muscle and their regulated expression during development and by angiotensin II. BBRC 184: 811-8 (1992); **Naftilan AJ** The role of angiotensin II in vascular smooth muscle cell growth. J. Cardiovasc. Pharmacol. 20: S37-40 (1992); **Schorb W et al** Angiotensin II is mitogenic in neonatal rat cardiac fibroblasts. Circ. Res. 72: 1245-54 (1993); **Stouffer GA & Owens GK** Angiotensin II-induced mitogenesis of spontaneously hypertensive rat-derived cultured smooth muscle cells is dependent on autocrine production of transforming growth factor-β. Circ. Res. 70: 820-8 (1992); **Teutsch B et al** A recombinant rat vascular AT1 receptor confers growth properties to angiotensin II in Chinese hamster ovary cells. BBRC 187: 1381-8 (1992); **Veale D et al** Production of angiotensin converting enzyme by rheumatoid synovial membrane. Ann. Rheum. Dis. 51: 476-80 (1992); **Wolf G & Neilson EG** Angiotensin II as a hypertrophogenic cytokine for proximal tubular cells. Kidney Int. Suppl. 39: S100-7 (1993); **Wolf G et al** Angiotensin II-induced proliferation of cultured murine mesangial cells: inhibitory role of atrial natriuretic peptide. J. Am. Soc. Nephrol. 3: 1270-8 (1992); **Yamamoto Y et al** Angiotensin III is a new chemoattractant for polymorphonuclear leukocytes. BBRC 193: 1038-43 (1993)

Angiotropin: This factor, also known as ***MAT*** (Monocyto-Angiotropin), can be isolated from activated peripheral blood monocytes and macrophages. Angiotropin is a 4.5 kDa polyribonucleoprotein consisting of a peptide chain of 38 amino acids an a ribonucleic acid of 43 bases. In several bioassays angiotropin is found to be a very powerful » angiogenesis factor and a strong chemoattractant (see also: Chemotaxis) for endothelial cells the proliferation of which it does not stimulate. The activity as an angiogenesis factor appears to be linked to the presence of the RNA moiety (see also: Angiogenin). *In vivo* angiotropin has vasodilatory activity. In cultures of confluent endothelial cells angiotropin induces the generation of three-dimensional networks of cells which morphologically resemble capillaries.

Hockel M et al Purified monocyte-derived angiogenic substance (angiotropin) induces controlled angiogenesis associated with regulated tissue proliferation in rabbit skin. JCI 82: 1075-90 (1988); **Hockel M et al** Purified monocyte-derived angiogenic substance (angiotropin) stimulates migration, phenotypic changes, and "tube formation" but not proliferation of capillary endothelial cells *in vitro*. JCP 133: 1-13 (1987); **Hockel M et al** Angiotropin treatment prevents flap necrosis and enhances dermal regeneration in rabbits. Arch. Surg. 124: 693-8 (1989)

Anionic neutrophil-activating peptide: See: ANAP.

ANP: *atrial natriuretic polypeptide* see: ANF (atrial natriuretic factor).

antennapaedia: see: Homeotic genes.

anti-ACTH peptides: see: Corticostatins.

Antibody-dependent cellular cytotoxicity: see: ADCC.

AntiCD3-induced cytokine-related syndrome: see: OKT3-induced cytokine-related syndrome.

Antigen-induced interferon: This interferon is identical with » IFN-γ. See also: IFN for general information about interferons.

Antigen non-specific inhibitor: see: nsINH.

Anti-Müllerian hormone: abbrev. AMH. This hormone is identical with » MIS (Müllerian inhibiting substance).

Anti-oncogenes: see: Oncogene.

Antiproliferative protein: see: Neural antiproliferative protein.

Antisense oligonucleotides: Antisense RNA is complementary to the mRNA transcribed from a gene. These anti-mRNAs, or shorter oligonucleotides derived from the sequence of the mRNA under study, can form complexes with the primary RNA transcript of a gene. The interaction of the antisense gene with its target is only seen when the target gene is expressed. The efficiency with which antisense RNA or oligonucleotides can be used to inhibit endogenous gene expression primarily not only depends on the synchronous or overlapping expression of the antisense RNA and the endogenous gene but usually also on a relative excess of the antisense transcripts.

Sense RNA/antisense RNA complexes effectively and specifically block the normal process of » gene expression without affecting the expression of other genes because correct processing of the target gene transcript, its transport, or its translation by ribosomes either become impossible or are severely impaired. In some biological systems antisense RNAs have been found to be natural methods of gene control.

The introduction of antisense transcripts into living cells can be used to effectively block the activity of a gene, thus making it possible to analyze its function and the physiological consequences associated with the selective elimination of a particular protein. The use of antisense oligonucleotides is often used as an alternative to » Knock-out mice that carry null mutations of the gene under study and are more difficult to produce (for a related approach see also: Genetic ablation).

Antisense oligonucleotides can also be used to interfere with the expression of growth factors, growth factor receptor genes and/or » oncogenes (and of course, any desirable gene) at the level of transcription. Antisense-induced reduction in the expression of several oncogenes, including » *myb*, *myc*, *ras* and *fos*, has been shown to inhibit growth and transformed phenotypes in various cell lines. Antisense oligonucleotides have been used also to block the proliferation of tumor cells by functionally inactivating a cytokine that serves as an » autocrine growth factor for these cells. The directed introduction of antisense oligonucleotides into tumor cells may therefore also be of potential clinical interest if the growth requirements of a tumor are known.

Peptides encoded in the antisense strand of DNA (***Antisense peptides***) have been found to bind to sense peptides and proteins with significant selectivity and affinity. Understanding the mechanism of sense-antisense peptide recognition may also be helpful to learn how to design agents that can be

sense transcription ⟶ 5′CUACGAUCCCACGGAU 3′

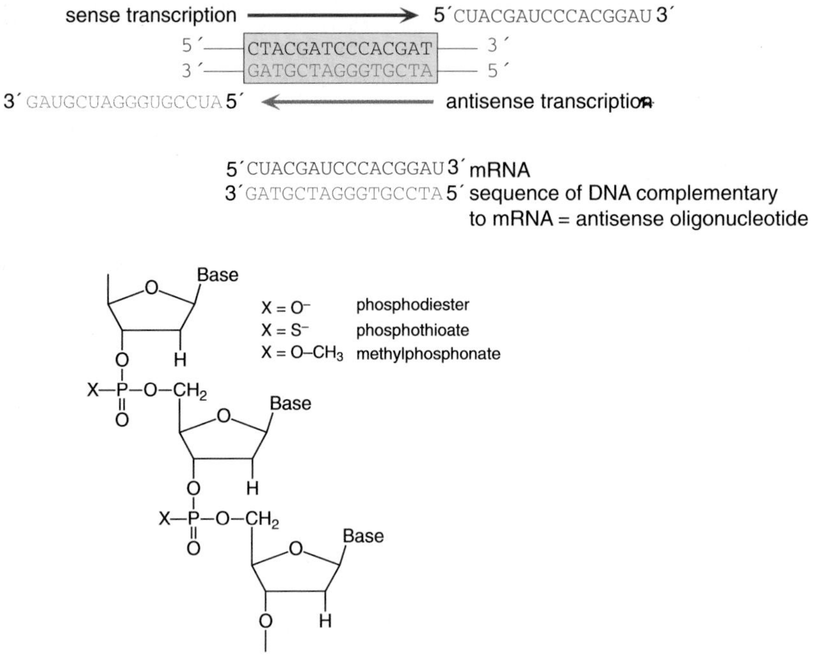

5′— CTACGATCCCACGAT — 3′
3′— GATGCTAGGGTGCTA — 5′

3′ GAUGCUAGGGUGCCUA 5′ ⟵ antisense transcription

5′CUACGAUCCCACGGAU 3′ mRNA
3′GATGCTAGGGTGCCTA 5′ sequence of DNA complementary to mRNA = antisense oligonucleotide

X = O^- phosphodiester
X = S^- phosphothioate
X = $O–CH_3$ methylphosphonate

Base; O; H; X–P–O–CH_2; O

Principle of antisense oligonucleotides.
Sense transcription of a given DNA sequence yields a mRNA with the sequence as indicated. Antisense transcription would yield an antisense RNA that could form complexes with this mRNA thus preventing its translation. It has been observed that this type of interaction is used in some systems to control gene expression.
Antisense oligonucleotides used for therapeutic purposes essentially consist of chemically synthesized DNA complementary to mRNA. Chemical modifications such as phosphothioates or methylphosphonates serve to produce molecules that are more resistant to cleavage by endogenous enzymes than normal DNAwithin an organism.

useful for separation, diagnostic and therapeutical purposes.

A listing of antisense patents, 1971-1991. Antisense Res. Dev. 1: 219-26 (1991); An indexed bibliography of antisense literature, 1991. Antisense Res. Dev. 1992 Spring; 2: 63-107 (1992) **Calabretta B et al** Antisense oligonucleotides. Semin. Cancer Biol. 3: 391-8 (1992); **Calabretta B et al** Prospects for gene-directed therapy with antisense oligodeoxynucleotides. Cancer Treat. Rev. 19: 169-79 (1993); **Carter G & Lemoine NR** Antisense technology for cancer therapy: does it make sense? Br. J. Cancer 67: 869-76 (1993); **Chaiken I** Interactions and uses of antisense peptides in affinity technology. J. Chromatogr. 597: 29-36 (1992); **Dzau YJ & Pratt RE** Antisense technology to block autocrine growth factors. J. Vasc. Surg. 15: 934-5 (1992); **Erickson RP et al** The use of antisense approaches to study development. Dev. Genet. 14: 251-7 (1993); **Ghosh MK & Cohen JS** Oligodeoxynucleotides as antisense inhibitors of gene expression. Prog. Nucl. Acid Res. Mol. Biol. 42: 79-126 (1992); **Leonetti JP et al** Cell delivery and mechanisms of action of antisense oligonucleotides. Prog. Nucleic Acid Res. Mol. Biol. 44: 143-66 (1993); **Malcolm AD** Uses of antisense nucleic acids – an introduction. Biochem. Soc. Trans. 20: 745-6 (1992); **Munir MI et al** Antisense RNA production in transgenic mice. Somat. Cell. Mol. Genet. 16: 383-94 (1990); **Murray JAH** (edt) Common antisense. Wiley-Liss, New York, 1992; **Neckers L et al** Antisense inhibition of oncogene expression. Crit. Rev. Oncog. 3: 175-231 (1992); **Stein CA & Cheng YC** Antisense oligonucleotides as therapeutic agents – is the bullet really magical? S 261: 1004-12 (1993); **Tropsha A et al** Making sense from antisense: a review of experimental data and developing ideas on sense-antisense peptide recognition. J. Mol. Recognit. 5: 43-54 (1992)

Antisense peptides: see: Antisense oligonucleotides.

Anti-TAC: Name given to an antibody directed against the cell surface antigen » CD25 (see also: CD antigens), the low affinity receptor for » IL2.

Antril: Tradename of the recombinant L-1 receptor antagonist » IL1ra (Synergen Inc. Boulder, Colorado).

Aortic endothelial cell growth-inhibitory activity: This factor is isolated from bovine pituitaries and is identical with » LIF.

AP-1: *activator protein 1* AP-1 is the gene product of the » *jun* oncogene that acts as a transcription factor. See also: Gene expression.

APC: *antigen-presenting cell*

Apo-1: This antigen was originally found to be expressed on the cell surface of activated human T and B lymphocytes and a variety of malignant human lymphoid cell lines. Cross-linking of the antigen by specific antibodies induces » apoptosis. Signals ultimately leading to apoptosis can be blocked by enforced expression of the » *bcl*-2 » oncogene). The induction of anti-Apo-1-mediated apoptosis appears to be correlated with *bcl*-2 mRNA down-regulation.
The Apo-1 antigen is a transmembrane glycoprotein of 48 kDa encoded by a gene on human chromosome 10q24.1. Apo-1 is identical with ***Fas antigen***, a cysteine-rich transmembrane protein of 335 amino acids with significant similarity to the members of the TNF/NGF receptor superfamily, originally defined by a mouse monoclonal antibody. The Apo/Fas antigen has been renamed as ***CD95*** (see also: CD Antigens).
A ligand for Apo-1/Fas has been cloned recently. Fas ligand is a membrane-bound protein of 31 kDa (278 amino acids) that belongs to the family of TNF-related proteins, which also includes ligands for » CD27, CD30, and » CD40. Overexpression of the Fas ligand on the cell surface has been shown to cause shedding of the protein, and the soluble form of the Fas ligand can actively trigger » Apoptosis by binding to Fas.
Mice carrying the » *lpr* (lymphoproliferation) mutation are prone to systemic lupus erythematosus. The underlying cause of this defect is a mutation in the Fas antigen gene, which is caused by the insertion of an early transposable element in intron 2 of the Fas antigen gene.
Chronically activated lymphocytes die from apoptosis after anti-Fas treatment while non-activated lymphocytes and those activated for short periods (less than 4 days) are refractory to antibody-mediated cell killing although they express the Fas antigen. Fas may therefore prove to be a target for immunosuppressive intervention. It has been shown that a variety of EBV-positive lymphoblastoid cell lines express Apo-1. The sensitivity of these cells to anti-Apo-1-mediated apoptosis may therefore open a new therapeutic approach for the treatment of EBV-induced lymphoproliferative lesions in immunocompromised individuals. Selective targeting of anti-Apo-1 antibodies to tumors cells may also be an efficient antitumor mechanism.

Adachi M et al Aberrant transcription caused by the insertion of an early transposable element in an intron of the Fas antigen gene of lpr mice. PNAS 90: 1756-60 (1993); **Dhein J et al** Induction of apoptosis by monoclonal antibody anti-APO-1 class switch variants is dependent on cross-linking of APO-1 cell surface antigens. JI 149: 3166-73 (1992); **Falk MH et al** Expression of the APO-1 antigen in Burkitt lymphoma cell lines correlates with a shift towards a lymphoblastoid phenotype. Blood 79: 3300-6 (1992); **Inazawa J et al** Assignment of the human Fas antigen gene (Fas) to 10q24.1. Genomics 14: 821-2 (1992); **Itoh N et al** The polypeptide encoded by the cDNA for human cell surface antigen Fas can mediate apoptosis. Cell 66: 233-43 (1991); **Mapara MV et al** APO-1 mediated apoptosis or proliferation in human chronic B lymphocytic leukemia: correlation with *bcl*-2 oncogene expression. EJI 23: 702-8 (1993); **Oehm A et al** Purification and molecular cloning of the APO-1 cell surface antigen, a member of the tumor necrosis factor/nerve growth factor receptor superfamily. Sequence identity with the Fas antigen. JBC 267: 10709-15 (1992); **Owen-Schaub LB et al** DNA fragmentation and cell death is selectively triggered in activated human lymphocytes by Fas antigen engagement. CI 140: 197-205 (1992); **Suda T et al** Molecular cloning and expression of the Fas ligand, a novel member of the tumor necrosis factor family. Cell 75: 1169-78 (1993); **Watanabe-Fukunaga R et al** Lymphoproliferation disorder in mice explained by defects in Fas antigen that mediates apoptosis. N 356: 314-7 (1992); **Watanabe-Fukunaga R et al** The cDNA structure, expression, and chromosomal assignment of the mouse Fas antigen. JI 148: 1274-9 (1992)

Apoptosis: *apoptosis* Apoptosis is a term referring to the cytologically observable changes associated with ***programmed cell death*** (abbrev. ***PCD***), a process of active cellular self-destruction observed in all eukaryotes. Unlike necrosis, which is a passive process, the induction of apoptosis is an active genetically regulated process and, like other gene-directed processes such as differentiation, requires the co-ordinated expression of many genes. This process, once set in motion, is essentially irreversible. Programmed cell death can be triggered by cytotoxic lymphocytes, glucocorticoids, and various cytolytic cytokines, e. g. TNF-α. Removal of growth factors from factor-dependent cells also induces programmed cell death. Accordingly, the process is also called ***activation-induced cell death*** (***AICD***). In hematopoietic cells (see also: Hematopoiesis), suppression of apoptosis is primarily due to the continuous presence of » CSFs (colony-stimulating factors). Apoptosis can also be initiated by cross-linking of the » Apo-1 cell surface antigen. Signals ultimately leading to apoptosis can be blocked by enforced expression of the » *bcl*-2 » oncogene).
Apoptosis allows selective elimination of cells

Apo-1: see addendum: Fas ligand.

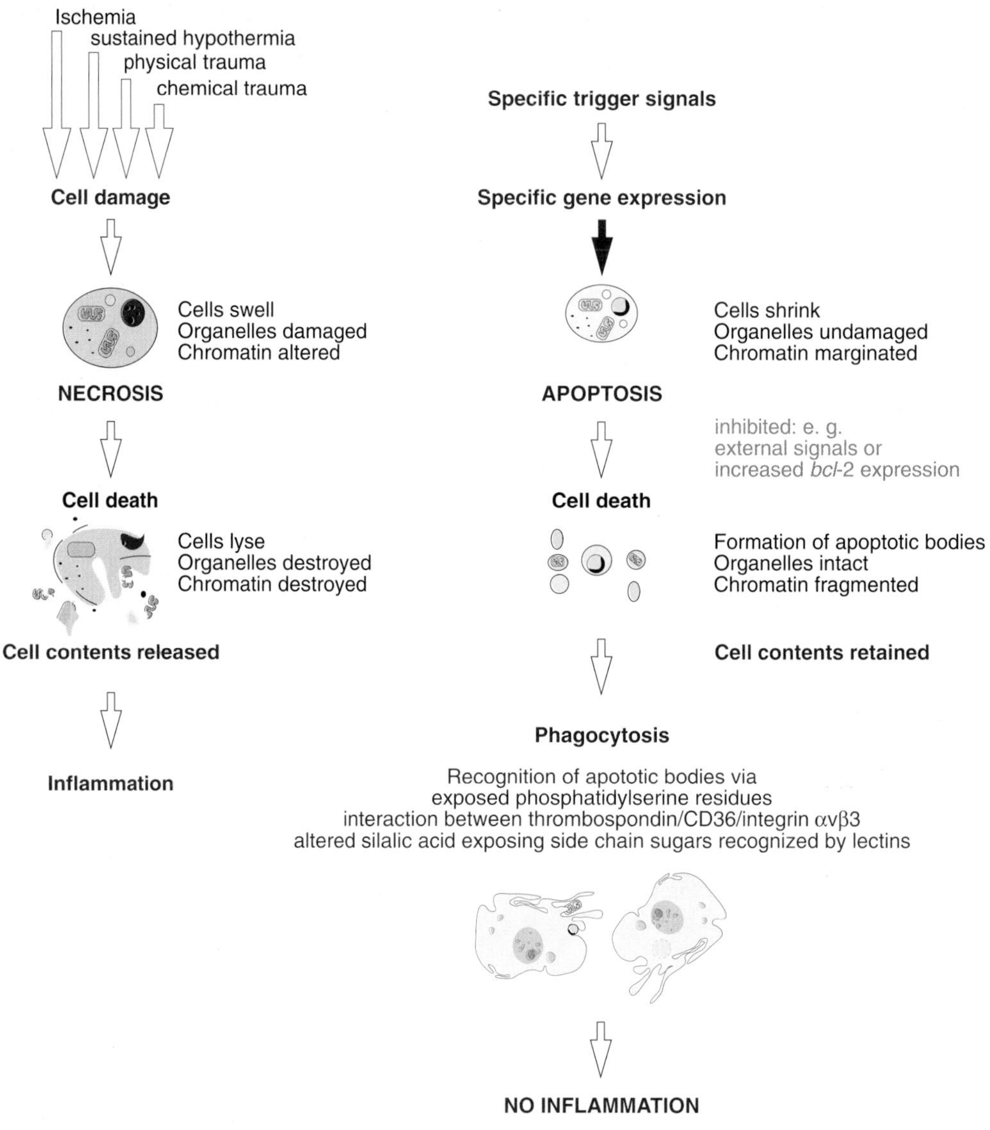

Ischemia
sustained hypothermia
physical trauma
chemical trauma

Cell damage

Cells swell
Organelles damaged
Chromatin altered

NECROSIS

Cell death

Cells lyse
Organelles destroyed
Chromatin destroyed

Cell contents released

Inflammation

Specific trigger signals

Specific gene expression

Cells shrink
Organelles undamaged
Chromatin marginated

APOPTOSIS

inhibited: e. g.
external signals or
increased *bcl*-2 expression

Cell death

Formation of apoptotic bodies
Organelles intact
Chromatin fragmented

Cell contents retained

Phagocytosis

Recognition of apotopic bodies via
exposed phosphatidylserine residues
interaction between thrombospondin/CD36/integrin $\alpha v\beta 3$
altered silalic acid exposing side chain sugars recognized by lectins

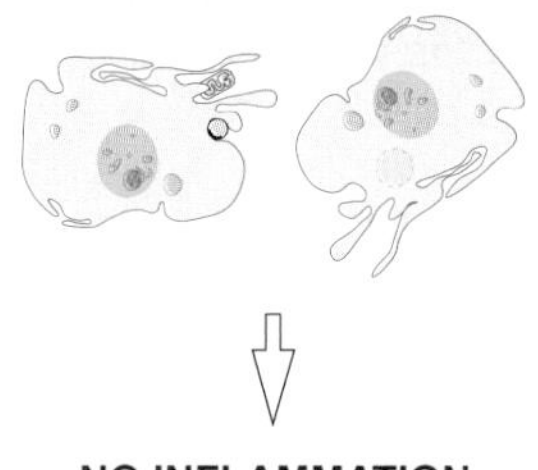

NO INFLAMMATION

Comparison of cell death by necrosis and apoptosis.
The two processes differ in various aspects. Cellular necrosis is usually caused by cell damage and does not require further gene activity. The membrane is the major site of damage and, among other things, loses its ability to regulate osmotic pressure. Eventually cell contents are released and elicit inflammatory reactions.
Apoptotic cell death requires gene activity. It can be prevented by increased expression of some genes or by external signals. Apoptotic cells eventually break up into apoptotic cell bodies. Cell contents are not released and there is no inflammation.

from a proliferating cell population. Apoptotic processes are observed, for example, during embryonal development, during metamorphosis, in endocrine tissue atrophy, during the normal turnover of tissues, and during tumor regression. They play a decisive role in the elimination of self-recognizing T lymphocytes in the thymus. In thymocytes apoptosis can be induced by treatment with glucocorticoids or irradiation. In certain hematopoietic cell types the growth of which depends on the continuous presence of growth factors (see also: Hematopoiesis), growth factor withdrawal also leads to apoptotic cell death rather than a cessation of cell growth (see also: Factor-dependent cell lines). It has been assumed that apoptotic processes are the controlling factors for the limitation of the population of immature hematopoietic progenitor cells in the absence of growth factors. The cell number in higher organisms therefore can be regulated in different organ systems by the availability of limited amounts of survival factors and the competition of cells for these factors (see also: neurotrophins). The disruption of normal processes leading to apoptosis results in illegitimate cell survival and can cause developmental abnormalities and facilitate cancer development.

Cell death by apoptosis differs from necrosis which may be a consequence, for example, of injuries, complement attacks, severe hypoxia, hypothermia, lytic virus infections or exposure to a number of toxins and which eventually leads to cell lysis.

Unlike necrosis, apoptosis requires a functional energy-producing system. The earliest indications of apoptotic cell death are morphological alterations of the cells such as chromatin condensation, disappearance of the nucleolus, and alterations of the cell surface characterized by the occurrence of blebs. These signs are followed by a margination of the chromatin at the inner surface of the nuclear membrane. Eventually the activation of specific intracellular calcium-dependent endonucleases (see also: Calcium ionophore) leads to the fragmentation of DNA, generating fragments of the size of nucleosomes. In contrast to cell death by necro-

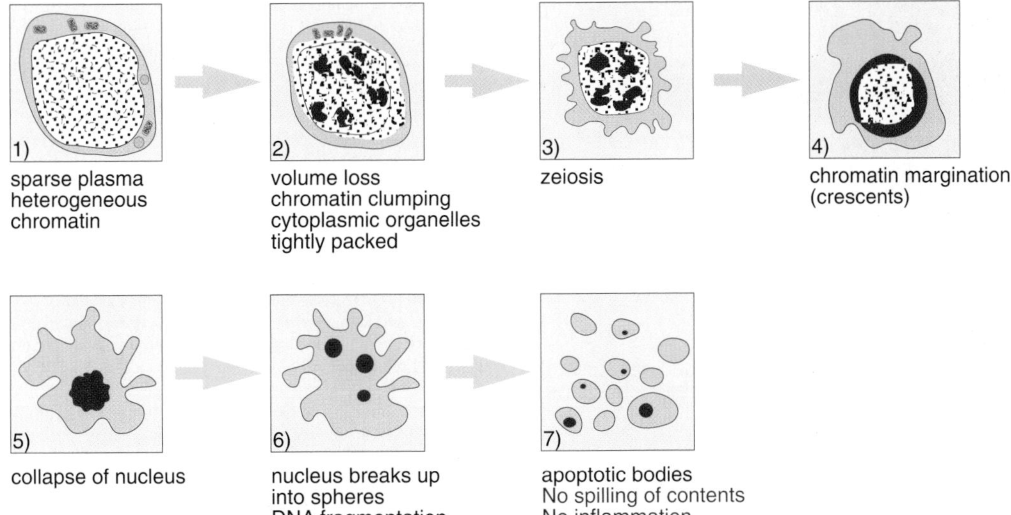

Schematic representation of different stages of apoptosis in lymphocytes.
(1) Normal cell have a volume of ≈ 90 pL. They show a sparse plasma and heterogeneous chromatin. (2) Chromatin begins to clump. Cytoplasmic organelles are tightly packed and the cell has lost some volume. These cells already show membrane changes and can be recognized by phagocytes as apoptotic cells. (3) The plasma membrane becomes ruffled and blebbed (zeiosis). (4) Chromatin marginates along the nuclear envelope, showing a typical crescent-like shape. The volume is now ≈ 70 pL. (5) The nucleus collapses. (6) The nucleus breaks up into spheres. Often the DNA becomes fragmented into pieces of the size of nucleosomes. (7) The cell is fragmented into many small apoptotic bodies that can be shown to exclude vital dyes for some time. Contents are retained and thus their is no inflammatory reaction.

sis, cells dying by apoptosis shrink and eventually break up into apoptotic bodies. Since intracellular contents are not released from apoptotic cells and their fragments this process is not accompanied by inflammatory processes (see also: Inflammation). Being an active process apoptotic cell death can be regarded as a differentiation event. The process can be inhibited by inhibitors of RNA and protein synthesis, suggesting that a number of specific proteins are required for its initiation and progression. Some of these proteins, appropriately called ***death proteins***, have already been identified. The sequences of some of them suggest that they are DNA binding proteins which, in turn, regulate the expression of other proteins (see also: Gene expression). Other death proteins appear to encode membrane proteins and a class of genes known as tumor suppressor genes (see: oncogenes). One particular tumor suppressor gene, designated p53 (see: oncogenes), has been found not to be expressed in certain myeloleukemic cell lines (see: M1 as an example). The introduction of a functional p43 gene into these cells causes apoptotic cell death which, in turn, can be prevented by the presence of » IL6 which induces the differentiation of these cells into monocytes.

The genetic analysis of some genes of the nematode *Caenorhabditis elegans* has revealed the existence of some genes involved in the control of apoptosis. The expression of these genes inducing apoptotic cell death is subject to very complex regulatory circuits involving the products of other genes inhibiting the expression of the apoptosis genes. Similar genes have also been detected in humans (see: *bcl*-2, IL1β convertase).

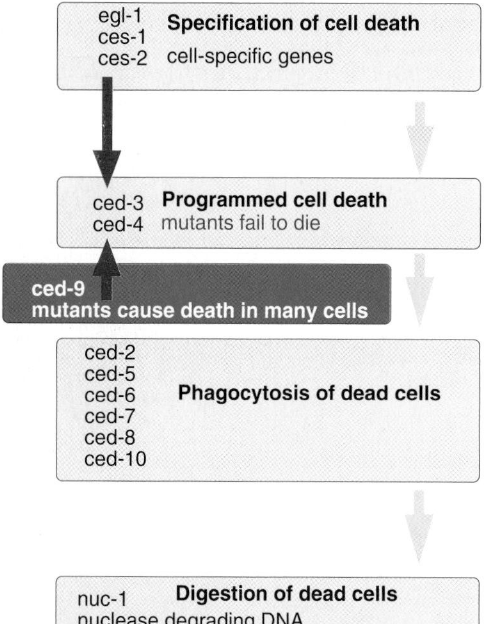

Cell death in *Caenorhabditis elegans.*
Cells in Caenorhabditis elegans undergo "true" (non-stochastic) programmed cell death, i. e. the fate of each individual cell can be predicted.
131 known somatic cells out of 1090 generated in the normal lineage die.
These include 105 cells from neuronal lineage and 26 non-neuronal cells (epithelial and muscle cells). As shown by mutational analysis various genes specify which cells are destined to die. Others regulate phagocytosis of dead cells and digestion of phagocytized cells. Mutations in some genes have been shown to prevent cell death while others cause death in many cells.

● **Reviews: Clarke P** Developmental cell death: morphological diversity and multiple mechanisms. Anat. Embryol. 181: 195-213 (1990); **Cohen JJ** Programmed cell death in the immune system. AI 50: 55-85 (1991); **Collins MKL & Rivas AL** The control of apoptosis in mammalian cells. TIBS 18: 307-9 (1993); **Ellis RE et al** Mechanisms and functions of cell death. ARC 7: 663-698 (1991); **Cotter TG et al** Cell death via apoptosis and its relationship to growth, development, and differentiation of both tumor and normal cells. ACR 10: 1153-9 (1990); **Fawthrop DJ et al** Mechanisms of cell death. Arch. Toxicol. 65: 437-44 (1991); **Johnson EMR jr. & Deckwerth TL** Molecular mechanism of developmental neuronal death. ARN 16: 31-46 (1993); **Kerr JFR et al** Apoptosis: a basic biological phenomenon with wide-ranging implications in tissue kinetics. Br. J. Cancer 26: 239-57 (1972); **Lee S et al** Apoptosis and signal transduction: clues to a molecular mechanism. Curr. Opin. Cell Biol. 5: 286-91 (1993); **Oppenheim RW** Naturally occurring cell death during neural development. TINS 8: 487-93 (1985); **Oppenheim RW** Cell death during development of the nervous system. ARN 14: 453-501 (1991); **Peitsch MC et al** The apoptosis endonucleases: cleaning up after cell death? Trends Cell Biol. 4: 37-41 (1994); **Raff MC** Social controls on cell survival and cell death. N 356: 397-400 (1992); **Schwartz LM et al** The role of cell death genes during development. Bioessays 13: 389-95 (1991); **Schwartz LM & Osborne BA** Programmed cell death, apoptosis and killer genes. IT 14: 582-90 (1993); **Tomei LD & Cope FO** (eds) The molecular basis of cell death. Cold Spring Harbor Laboratory, Cold Spring Harbor, New York 1991); **Ucker DS** Death by suicide: one way to go in mammalian cellular development? The New Biologist 3: 103-9 (1991); **Walker NI et al** Patterns of cell death. Methods Achiev. Exp. Pathol. 13: 18-54 (1988); **Waring P et al** Apoptosis or programmed cell death. Med. Res. Rev. 11: 219-36 (1991); **Williams GT** Programmed cell death: Apoptosis and oncogenesis. Cell 65: 1097-8 (1991); **Williams, GT & Smith CA** Molecular regulation of apoptosis: genetic controls on cell death. Cell 74: 777-9 (1993); **Wyllie AH et al** Cell death: the significance of apoptosis. Int. Rev. Cytol. 68: 251-305 (1980)

● **Biochemistry & Molecular Biology: Compton MM** Development of an apoptosis endonuclease assay. DNA Cell Biol. 10: 133-41 (1991); **Facchinetti A et al** An improved method for the detection of DNA fragmentation. JIM 136: 125-31 (1991); **Fesus L** Apoptosis fashions T and B cell repertoire. Immunol Lett. 30: 277-82 (1991); **Gaido ML & Cidlowski JA** Identification, purification, and characterization of a calcium-dependent endonuclease (NUC18) from apoptotic rat thymocytes. JBC 266: 18580-5 (1991); **Hengartner MO et al** *Caenorhabditis elegans* gene *ced*-9 protects cells from programmed cell death. N 356: 494-9 (1992); **Hughes FM Jr & Gorospe WC** Biochemical identification of apoptosis (programmed cell death) in granulosa cells: Evidence for a potential mechanism underlying follicular atresia. E 129: 2415-22 (1991); **Martin DP et al** Inhibitors of protein synthesis and RNA synthesis prevent neuronal death caused by nerve growth factor deprivation. JCB 106: 829-44 (1988); **Oppenheim RW** Naturally occurring and induced neuronal death in the chick embryo *in vivo* requires protein and RNA synthesis: evidence for the role of cell death genes. Dev. Biol. 138: 104-13 (1990); **Owens GP et al** Identification of mRNAs associated with programmed cell death in immature thymocytes. MCB 11: 4177-88 (1991); **Sachs L & Lotem J** Control of programmed cell death in normal and leukemic cells: new implications for therapy. Blood 82: 15-21 (1993); **Schwartz LM et al** Gene activation is required for developmentally programmed cell death. PNAS 87: 6594-8 (1990); **Schwartz LM et al** Do all programmed cell deaths occur via apoptosis? PNAS 90: 980-4 (1993); **Shi Y et al** Cyclosporin A inhibits activation-induced cell death in T cell hybridomas and thymocytes. N 339: 625-6 (1991); **Smith CA et al** Antibodies to CD3/T cell receptor complex induce death by apoptosis in immature T cells in thymic cultures. N 337: 181-4 (1989); **Wadewitz AG & Lockshin RA** Programmed cell death: dying cells synthesize a co-ordinated unique set of proteins in two different episodes of cell death. FL 241: 19-23 (1988); **Williams GT et al** Haemopoietic colony stimulating factors promote cell survival by suppressing apoptosis. N 343: 76-9 (1990); **Yonish-Rouack E et al** Wild-type p53 induces apoptosis of myeloid leukaemic cells that is inhibited by interleukin-6. N 352: 345-7 (1991)

APP: abbrev. for » acute phase protein(s). See also: Acute phase reaction.

APPIF: *acute phase protein inducing factor* This factor induces the synthesis of various » acute phase proteins (see also: Acute phase reaction) such as » haptoglobin, the heme-binding protein hemopexin, and ceruloplasmin. It is identical with one of the molecular forms of » IL1.

Matsushima K et al Purification of human interleukin 1 from human monocyte culture supernatants and identity of thymocyte comitogenic factor, fibroblast proliferation factor, acute phase protein-inducing factor, and endogenous pyrogen. CI 92: 290-301 (1985); **Suga T et al** Macrophage-mediated acute phase transport protein production induced by lentinan. IJI 87: 691-9 (1986)

APR: *acute phase reactant* see: Acute phase proteins. See also: Acute phase reaction.

Aqueous humor lymphocyte inhibitory activity: This factor with immunosuppressive activity, isolated from aqueous humor, is identical with » TGF-β.

Cousins SW et al Identification of transforming growth factor-β as an immunosuppressive factor in aqueous humor. Invest. Ophthalmol. Vis. Sci. 32: 2201-11 (1991)

AR: *amphiregulin* also abbrev. AMR or AREG.

■ **Alternative Names:** CRDGF (colorectum cell-derived growth factor), KAF (keratinocyte-derived autocrine factor). See also: individual entries for further information.

■ **Sources:** Amphiregulin can be isolated from placenta, testes, ovaries, and a number of tumor cell lines.

■ **Protein Characteristics:** Amphiregulin is an extremely hydrophilic glycoprotein with a length of 84 amino acids. A truncated form of 78 amino acids has also been described. Amphiregulin contains 6 cysteine residues engaged in the formation of disulfide bonds (positions 46/59; 54/70; 72/81) which are essential for its biological activity. The carboxyl-terminal from residues 46 to 84 exhibit striking homology to the » EGF family of proteins.

The factor is synthesized as a membrane-bound precursor of 252 amino acids. It contains putative nuclear targeting sequences (NTS; see: signal sequences) and it has been shown recently that nuclear targeting is involved in AR-mediated growth responses.

■ **Gene Structure:** The human amphiregulin gene extends over ≈ 10.2 kb, comprises six exons, and is located on human chromosome 4q13-21.

■ **Related Factors:** The carboxyterminal amino acids of amphiregulin between positions 46-84 show a strong homology to » EGF and contain EGF-like sequence domains. Amphiregulin is a member of the EGF family of proteins. A 76 % homology is observed with » SDGF (schwannoma-derived growth factor).

■ **Receptor Structure, Gene(s), Expression:** Amphiregulin binds to the same receptors as » EGF and » TGF-α, albeit with lower affinity.

■ **Biological Activities:** Amphiregulin inhibits the growth of various human carcinoma cell lines *in vitro*. The proliferation of some tumor cell lines and also of fibroblasts and other normal cells is enhanced by amphiregulin. Amphiregulin can functionally replace » EGF and » TGF-α in murine keratinocytes.

■ **Clinical Use & Significance:** Amphiregulin acts as an » autocrine growth factors for some human, colon carcinomas. There are some indications that amphiregulin mRNA expression is markedly elevated in epidermal biopsies derived from human psoriatic lesions.

Cook PW et al Amphiregulin messenger RNA is elevated in psoriatic epidermis and gastrointestinal carcinomas. CR 52: 3324-7 (1992); **Disteche CM et al** Mapping of the amphiregulin and the platelet-growth factor receptor α genes to the proximal long arm of chromosome 4. Cytogenet. Cell Genet. 51: 990 (1989); **Johnson GR et al** Response to and expression of amphiregulin by ovarian carcinoma and normal ovarian surface epithelial cells: nuclear localization of endogenous amphiregulin. BBRC 180: 481-8 (1991); **Johnson GR et al** Autocrine action of amphiregulin in a colon carcinoma cell line and immunocytochemical localization of amphiregulin in human colon. JCB 118: 741-51 (1992); **Johnson GR et al** Amphiregulin induces tyrosine phosphorylation of the epidermal growth factor receptor and p185*erb*B2. Evidence that amphiregulin acts exclusively through the epidermal growth factor receptor at the surface of human epithelial cells. JBC 268: 2924-31 (1993); **Johnson GR et al** Characterization of high and low molecular weight forms of amphiregulin that differ in glycosylation and peptide core length. Evidence that the NH2-terminal region is not critical for bioactivity. JBC 268: 18835-43 (1993); **Modrell B et al** The interaction of amphiregulin with nuclei and putative nuclear localization sequence binding proteins. GF 7: 305-14 (1992); **Plowman GD et al** The amphiregulin gene encodes a novel epidermal growth factor-related protein with tumor-inhibitory activity. MCB 10: 1969-81 (1990); **Saeki T et al** Differential immunohistochemical detection of amphiregulin and cripto in human normal colon and colorectal tumors. CR 52: 3467-73 (1992); **Shoyab M et al** Amphiregulin: a bifunctional growth-modulating glycoprotein produced by the phorbol 12-myristate 13-acetate-treated human breast adenocarcinoma cell line MCF-7. PNAS 85: 6528-32 (1988); **Shoyab M et al** Structure and function of human amphiregulin: a member of the epidermal growth factor family. S 243: 1074-6 (1989)

ARE: *AU-rich element* A pentanucleotide sequence element consisting of one or more AUUUA tandem repetitions found in the 3´untranslated regions of certain mRNAs. The presence of these elements shortens mRNA half-lifes and the introduction of such elements into mRNAs that do not contain them leads to a rapid decay of these engineered mRNAs *in vivo*.

The modulation of mRNA half-life by selective degradation of mRNA appears to be one of the mechanisms allowing transient expression of genes (see also: Gene expression; see also: ERG, early response genes). On the other hand, selective transcript stabilization, probably mediated by cytoplasmic inhibitors of selective ribonucleases, also occurs and may be a common ingredient of most inductive stimuli.

AU-rich sequence motifs have been found in many transcripts encoding growth factors, growth factor-like substances, » oncogenes, and other proteins with growth-regulating activity. These motifs are observed, for example, in the genes encoding » IL1, » IL3, » IL8, » GM-CSF, » IFN-γ, » PF4 (platelet factor 4), » TNF-α, and » *sis*, » *jun*, *ets*, *raf*, » *myc*, » *fos*.

A number of AU-binding factors have recently been discovered, suggesting that specific regulation may occur through specific protein-mRNA interaction(s). The mechanisms by which these proteins influence ARE-directed mRNA turnover are presently unclear but it is thought that may participate in the assembly of larger protein/mRNA complexes that then mediate mRNA degradation. ***AUBF*** (adenosine-uridine binding factor) is a cytosolic phosphoprotein that binds to, and stabilizes, four AUUUA tandem repetitions found in the 3´ untranslated regions of certain mRNAs. ***AUF1*** is a recently cloned protein of 37 kDa that binds specifically to the 3´untranslated AU-rich regions of mRNAs encoding » *myc* and » GM-CSF.

Akashi M et al Role of AUUU sequences in stabilization of granulocyte-macrophage colony-stimulating factor RNA in stimulated cells. Blood 78: 2005-12 (1991); **Bagby GC et al** Interleukin 1 stimulation stabilizes GM-CSF mRNA in human vascular endothelial cells: preliminary studies on the role of the 3´ AU-rich motif. PCBR 352: 233-9 (1990); **Bagby GC et al** Vascular endothelial cells and hematopoiesis: regulation of gene expression in human vascular endothelial cells. Hematol. Pathol. 5: 93-9 (1991); **Bickel M et al** Binding of sequence-specific proteins to the adenosine- plus uridine-rich sequences of the murine granulocyte/macrophage colony-stimulating factor mRNA. PNAS 89: 10001-5 (1992); **Bohjanen PR et al** An inducible cytoplasmic factor (AU-B) binds selectively to AUUUA multimers in the 3´ untranslated region of lymphokine mRNA. MCB 11: 3288-95 (1991); **Gillis P & Malter JS** The adenosine-uridine binding factor recognizes the AU-rich elements of cytokine, lymphokine, and oncogene mRNAs. JBC 266: 3172-7 (1991); **Iwai Y et al** Identification of sequences within the murine granulocyte-macrophage colony-stimulating factor mRNA 3'-untranslated region that mediate mRNA stabilization induced by mitogen treatment of EL-4 thymoma cells. JBC 266: 17959-65 (1991); **Kruys V et al** Translational control mediated by UA-rich sequences. Enzyme 44: 193-202 (1990); **Malter JS et al** Adenosine-uridine binding factor requires metals for binding to granulocyte-macrophage colony-stimulating factor mRNA. Enzyme 44: 203-13 (1990); **Müller WE et al** Association of AUUUA-binding protein with A+U-rich mRNA during nucleocytoplasmic transport. JMB 226: 721-33 (1992); **Ross HJ et al** Cytokine messenger RNA stability is enhanced in tumor cells. Blood 77: 1787-95 (1991); **Savant-Bhonsdale S & Cleveland DW** Evidence for instability of mRNAs containing AUUUA motifs mediated through translation-dependent assembly of a >20S degradation complex. Genes Dev. 6: 1927-39 (1992); **Stephens JM et al** Tumor necrosis factor α-induced glucose transporter (GLUT-1) mRNA stabilization in 3T3-L1 preadipocytes. Regulation by the adenosine-uridine binding factor. JBC 267: 8336-41 (1992); **Tonouchi N et al** Deletion of 3' untranslated region of human BSF-2 mRNA causes stabilization of the mRNA and high-level expression in mouse NIH3T3 cells. BBRC 163: 1056-62 (1989); **Zhang W et al** Purification, characterization, and cDNA cloning of an AU-rich element RNA-binding protein, AUF1. MCB 13: 7652-65 (1993)

Arginine vasopressin: abbrev. AVP. See: Vasopressin.

ARIA: *acetylcholine receptor inducing activity* ARIA is a 42 kDa glycoprotein that binds moder-

ately to heparin (see also: HBGF, heparin-binding growth factors). It was purified from chicken brain on the basis of its ability to increase the synthesis of acetylcholine receptors in chicken myotubes. ARIA also regulates different receptor subunits that appear early and late during the course of nerve-muscle synapse formation. It also increases the number of muscle voltage-gated sodium channels.
Chicken ARIA is homologous to rat » NDF (*neu* differentiation factor) and human » heregulin, ligands for the receptor tyrosine kinase encoded by the » *neu* proto-oncogene. ARIA also appears to be identical with » GGF (glial growth factor). These *neu* ligands are collectively known as » neuregulins.
The biological activities of ARIA are mediated by a glycoprotein receptor of 185 kDa that is phosphorylated on tyrosine residues after addition of ARIA to intact cells. It has been suggested that members of the ARIA protein family promote the formation and maintenance of chemical synapses.
Corfas G et al ARIA, a protein that stimulates acetylcholine receptor synthesis, also induces tyrosine phosphorylation of a 185 kDa muscle transmembrane protein. PNAS 90: 1624-8 (1993); **Corfas G & Fischbach GD** The number of Na^+ channels in cultured chick muscle is increased by ARIA, an acetylcholine receptor-inducing activity. J. Neurosci. 13: 2118-25 (1993); **Falls DL et al** ARIA, a protein that stimulates acetylcholine receptor synthesis, is a member of the *neu* ligand family. Cell 72: 801-15 (1993); **Harris DA et al** Acetylcholine receptor-inducing factor from chicken brain increases the level of mRNA encoding the receptor α subunit. PNAS 85: 1983-7 (1988); **Martinou JC et al** Acetylcholine receptor-inducing activity stimulates expression of the ε-subunit gene of the muscle acetylcholine receptor. PNAS 88: 7669-73 (1991); **Usdin TB & Fischbach GD** Purification and characterization of a polypeptide from chick brain that promotes the accumulation of acetylcholine receptors in chick myotubes. JCB 103: 493-507 (1986)

ARRE: *antigen receptor response element* A short nucleotide sequence, designated ARRE-2, is found in the promoter region of the » IL2 gene. It contains the binding site for a transcription factor (see also: Gene expression), known as NFAT-1, which is expressed exclusively in activated T cells (see also: Cell activation) and appears to modulate the expression of the IL2 gene.
Durand DB et al Characterization of antigen receptor response elements within the interleukin-2 enhancer. MCB 8: 1715-1724 (1988)

ASC: abbrev. for antibody secreting cells.

ASRA: *autocrine serum-replacing activity* This poorly characterized factor of ≈ 60 kDa is found in the culture medium of confluent murine hematopoietic » FDC-P1 cells. It is also produced by human skin fibroblasts. It permits the serum-free survival of FDC-P1 cells even at low density and allows fibroblasts to grow in serum-free medium (see also: SFM) in the absence of » EGF. The factor is not identical with » EGF.
Bohmer RM & Burgess AW An autocrine activity capable of substituting for serum in cell cultures. GF 1: 347-56 (1989)

ASH: see addendum.

Astacin: see: BMP (bone morphogenetic protein).

ASTA Z 7557: (mafosfamide) see: 4-HC (Hydroperoxycyclophosphamide).

AST-CF: *astrocyte cytotoxic factor* This poorly characterized factor is secreted by astrocytes and is cytotoxic for oligodendrocytes. It is probably identical or related with » TNF.
Robbins DS et al Production of cytotoxic factor for oligodendrocytes by stimulated astrocytes. JI 139: 2593-8 (1987)

Astrocyte cytotoxic factor: see: AST-CF.

Astroglia-inducing molecule: see: AIM.

Astroglial growth factor 1: see: AGF-1.

Astroglial growth factor 2: see: AGF-2.

Astrostatine: This poorly characterized 17 kDa protein is secreted by normal neurons. Two neuronal cell lines, the » PC12 rat pheocromocytoma and the neuro 2A (N2A) murine neuroblastoma also release such an activity when the cells are at high density and show a low proliferation rate. The protein does not react with antibodies directed against » TGF-β or » EGF receptors. It inhibits the proliferation of type I astrocytes. The factor may play a role in the development of astrogliosis.
Moonen G Physiopathology of neural death and of gliosis. Bull. Mem. Acad. R. Med. Belg. 145: 110-8 (1990); **Rogister B et al** Cultured neurons release an inhibitor of astroglia proliferation (astrostatine). J. Neurosci. Res. 25: 58-70 (1990)

AT464: see: pAT464.

AT744: see: pAT744.

ATF: *activating transcription factor* One of the transcription factors (see also: Gene expression) that bind to the » CRE (cyclic AMP-responsive) element found in the enhancer region of various cytokine and cytokine-regulated genes.

ATH8: This human HTLV-1-infected T helper cell line is used instead of murine » CTLL-2 cells in » bioassays detecting » IL2. The cells express the IL2 receptor and depend entirely on IL2 for their growth.

Antonen J et al The use of an HTLV-I-infected human T cell line (ATH8) in an interleukin 2 bioassay. JIM 99: 271-5 ((1987)

atk: *agammaglobulinemia tyrosine kinase* see: Immunodeficient mice, subentry xid.

ATL-derived factor: see: ADF.

Atrial natriuretic polypeptide: see: ANF (atrial natriuretic factor).

AtT20-ECGF: *AtT20 endothelial cell growth factor* This poorly characterized factor is a homodimeric glycoprotein (2 × 23 kDa) secreted by the pituitary cell line AtT20. The factor is a strong mitogen for vascular endothelial cells. It is not related to other endothelial cell growth factors (see also: ECGF).

Plouet J et al Isolation and characterization of a newly identified endothelial cell mitogen produced by AtT20 cells. EJ 8: 3801-6 (1989)

ATX: see: Autotaxin.

AUBF: *adenosine-uridine binding factor* see: ARE (AU-rich element).

AU-rich element: see: ARE.

Autacoids: Generic name given to a group of chemically diverse substances produced by various tissues in the body with » autocrine or » paracrine activity. These substances, sometimes also called ***local hormones*** or ***autopharmacologic agents***, have varied pharmacological activities. Notable other autacoids include, among others, prostaglandins, leukotriens, kinins, and a variety of low molecular weight compounds (lipids, nitrogen bases, reactive oxygen species, nitric oxide, platelet-activating factor). In many tissues the actions of » cytokines are indirectly mediated through the production of autacoids or other cytokines.

Autocrine differentiation-inhibiting factor: see: ADIF.

Autocrine: A term referring to a mechanism of growth control, also called autostimulatory growth control, in which a cell secretes a soluble factor (***AGF***; autocrine growth factor) that binds to its receptor, which is also expressed by the factor-producing cell. This effectively creates an autogenous loop in which a product acts back on the cells that produce it.

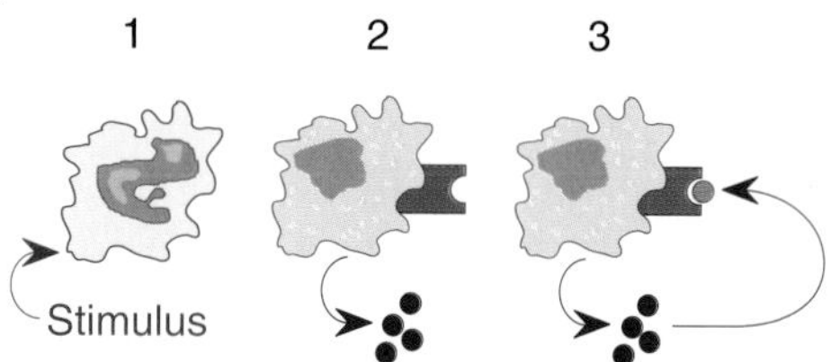

Simple autocrine growth control loop.
The stimulation of cells (1) induces the expression and secretion of a mediator (2) and its membrane-bound receptor. By interacting with its receptor (3) the mediator can elicit further responses in the cell by which it is produced.

The uncontrolled synthesis of such autocrine growth factors by cells that also express the corresponding receptors is a mechanisms allowing, for example, tumor cells to regulate their own growth and to become independent of exogenous growth control mechanisms. In some cases activated » oncogenes are responsible for the constitutive synthesis of aberrant proteins that take over the role of authentic growth factors. The frequently observed regulated expression of growth factor receptors appears to be a safeguard against a catastrophic spread of cell proliferation by a growth factor stimulus. Again, the uncontrolled expression of receptor genes subverts this mechanism of orderly growth.
The existence of such autocrine loop in tumor cells has been demonstrated by *in vitro* experiments showing that factor-specific neutralizing antibodies are capable of inhibiting further proliferation of the factor-producing cells. The immunization of mice with an autocrine growth factor effectively blocks the *in vivo* growth of the tumor cells expressing and secreting this growth factor (see: *hst*).
Examples for the endogenous production of tumor-associated growth factors are, among many others, » IL3, » EGF, » bFGF, » IGF and » PDGF. In some instances autocrine loops have been shown to be very complicated mechanisms involving more than a single factor and its receptor gene (see figure).
The establishment of autocrine growth control loops is not restricted to regulatory peptide factors. It has been observed, for example, that Epstein-

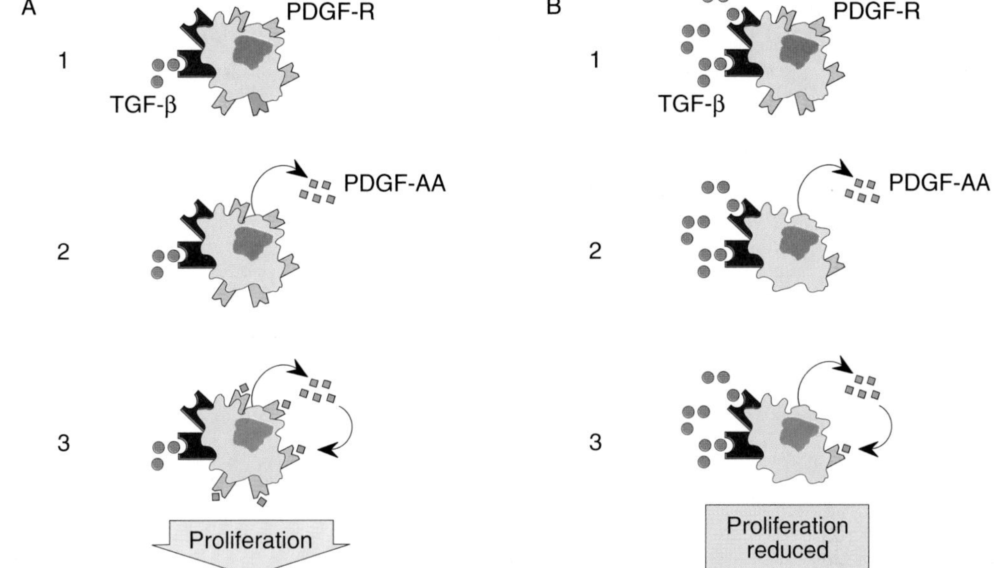

Concentration-dependent complex control of an autocrine loop.
A) At low concentrations binding of TGF to its membrane receptor (1) induces the production of PDGF-AA (2). PDGF-AA then binds to its own receptor which is also expressed by the cell and thus can stimulate cell proliferation. B) High concentrations of TGF (1) down-regulate expression of receptors for PDGF-AA (2). Biological responses elicited by PDGF-AA are thus diminished due to a paucity of PDGF-AA receptors. The effect is reduced cell proliferation.
Concentration-dependent differential actions involving an interplay of several factors and down-regulation of receptors have been observed also for a number of other cytokines.

Barr virus-transformed B lymphocytes secrete protein factors but that the main mediator allowing autocrine growth is lactate. The neurotransmitter serotonin appears to be an autocrine growth factor for small cell lung carcinomas (see also: Bombesin).

The autocrine growth control model is the basis for a number of therapeutical concepts that, at least in principle, should allow direct interference with the growth control mechanisms of tumor cells.

A reduction in the tissue and plasma concentrations of growth factors can be achieved by the administration of antibodies directed against the growth factor. At least experimentally this has been shown to inhibit the growth of factor-dependent tumor cell lines and also of human tumor cells transplanted into nude mice (see also: Immunodeficient mice).

The reverse, i. e. the provision of additional growth factors for the purpose of inhibiting tumor cell growth, is based on the observation that many growth factors induce the down-modulation of receptor expression (see also: receptor transmodulation). It may be envisaged that increased doses of growth factors may induce resting tumor cells to enter the » cell cycle and hence to become vulnerable to chemotherapy. A similar concept is used, for example, in the therapy of mammary carcinomas which may involve low-dose treatment with estrogens before chemotherapy.

Another way of blocking tumor cell growth might be a blockade of the growth factor receptors by specific monoclonal antibodies. Anti-EGF receptor antibodies have been shown to lead to tumor regression of transplanted epithelial carcinomas in the nude mouse model (see also: Immunodeficient mice).

Yet another possibility of exploiting the autocrine growth control concept might be the use of » antisense oligonucleotides that can be used to interfere with the expression of growth factors and growth factor receptor genes at the level of transcription (see also: Gene expression).

For other mechanisms of growth control see also: Intracrine, Juxtacrine, Paracrine, Retrocrine.

Avis IL et al Preclinical evaluation of an anti-autocrine growth factor monoclonal antibody for treatment of patients with small-cell lung cancer. JNCI 83: 1470-6 (1991); **Bajzer Z & Yuk-Pavlovic S** Quantitative aspects of autocrine regulation in

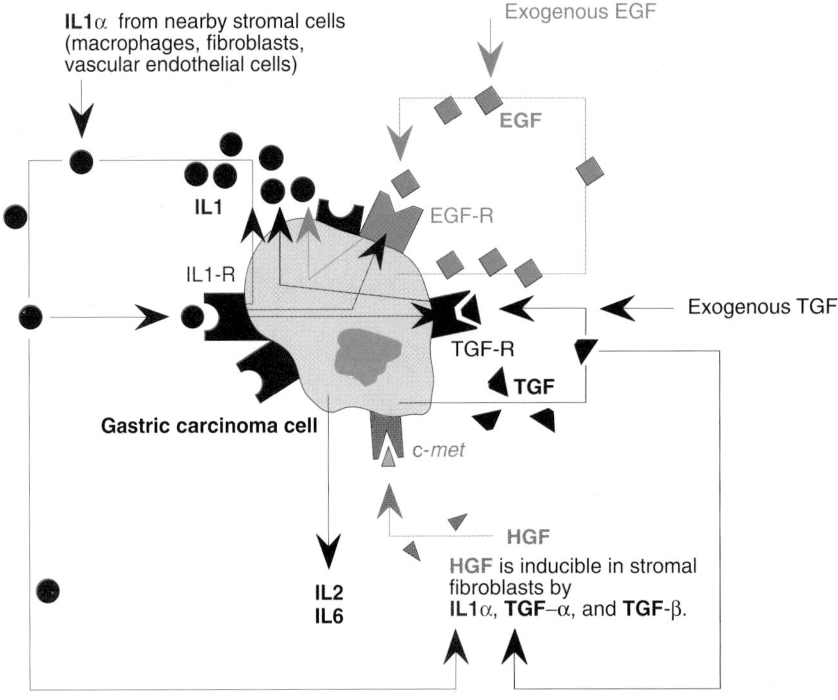

Example of complex multiple autocrine growth control loops in gastric carcinoma cells.
Gastric carcinoma cells secrete large amounts of IL1α and express significant amounts of the IL1 receptor. IL1α can enhance expression of itself and is provided also by nearby stromal cells such as macrophages, fibroblasts, and vascular endothelial cells. In addition, production of IL1α by gastric carcinoma cells is induced also by EGF and TGF-α. Autocrine growth of gastric carcinoma cells mediated by an interplay between IL1 and its receptor is demonstrated by the fact that antibodies against IL1α and also IL1 receptor antagonist (IL1ra) significantly suppress cell growth.
Gastric carcinoma cell lines also produce EGF and TGF-α and their receptors. IL1α can induce expression of EGF and TGF-α receptors. EGF and TGF-α can induce expression of IL1α.
EGF and TGF-α can induce the synthesis of their own receptors and also of stromelysin and collagenase. This sets in motion a cascade of events leading to extracellular matrix degradation, which is convenient for tumor progression and invasion.
HGF (hepatocyte growth factor) can be produced by stromal fibroblasts and its synthesis by these cells can be enhanced by TGF-α, TGF-β, and IL1α. HGF can also stimulate growth of gastric carcinoma cells which express the HGF receptor c-*met*. In addition, gastric carcinoma cells can also secrete IL2 and IL6.

tumors. Crit. Rev. Oncogen. 2: 53-73 (1990); **Battegay EJ et al** TGF-β induces bimodal proliferation of connective tissue cells via a complex control of an autocrine PDGF loop. Cell 63: 515-24 (1990); **Becker D et al** Proliferation of human malignant melanomas is inhibited by antisense oligodeoxynucleotides targeted against basic fibroblast growth factor. EJ 8: 3685-91 (1989); **Dunn AR & Wilks AF** Contributions of autocrine and non-autocrine mechanisms to tumorigenicity in a murine model for leukemia. Ciba Found. Symp. 148: 145-55 (1990); **Forsten KE & Lauffenburger DA** Autocrine ligand binding to cell receptors. Mathematical analysis of competition by solution "decoys". Biophys. J. 61: 518-29 (1992); **Forsten KE et al** Interrupting autocrine ligand-receptor binding: comparison between receptor blockers and ligand decoys. Biophys. J. 63: 857-61 (1992); **Garbers DL** Guanylyl cyclase receptors and their endocrine, paracrine, and autocrine ligands. Cell 71: 1-4 (1992); **Gordon J & Cairns JA** Autocrine regulation of normal and malignant B lymphocytes. ACR 56: 313-34 (1991); **Heldin CH & Westermark B** Autocrine stimulation of growth of normal and transformed cells. In: Habenicht A (edt) Growth factors, differentiation factors, and cytokines, pp. 267-278, Springer, Berlin 1990; **Jasmin C et al** Autocrine growth of leukemic cells. Leuk. Res. 14: 689-93 (1990); **Lang RA & Burgess AW** Autocrine growth factors and tumorigenic transformation. IT 11: 244-9 (1990); **Leof EB et al** Induction of c-*sis* mRNA and activity similar to platelet-derived growth factor by transforming growth factor-β: a proposed model for indirect mitogenesis involving autocrine activity. PNAS 83: 2453-7 (1986); **Mulshine JL et al** Autocrine growth factors as therapeutic targets in lung cancer. Chest (Suppl.) 96: 31s-4s (1989); **Nicolson GL** Paracrine/autocrine growth mechanisms in tumor metastasis. Oncol. Res. 4: 389-99 (1992); **Nicolson GL** Cancer progression and growth: relationship of paracrine and autocrine growth mechanisms to organ preference of metastasis. ECR 204: 171-80 (1993); **Pike SE et al** The role of lactic acid in autocrine B cell growth stimulation. PNAS 88: 11081-5 (1991); **Rodan GA et al**

Autocrine/paracrine regulation of osteoblast growth and differentiation Lab. Invest. 64: 593-5 (1991); **Schüller HM** Receptor-mediated mitogenic signals and lung cancer. Cancer Cells 3: 496-503 (1991); **Schwab G et al** Characterization of an interleukin-6-mediated autocrine growth loop in the human multiple myeloma cell line, U266. Blood 77: 587-93 (1991); **Sporn MB & Roberts AB** Autocrine growth factors and cancer. N 313: 745-7 (1985); **Sporn MB & Roberts AB** Autocrine secretion: 10 years later. Ann. Intern. Med. 117: 408-14 (1992); **Sporn MB & Todaro GJ** Autocrine secretion and malignant transformation of cells. NEJM 3093: 878-80 (1980); **Trepel JB et al** A novel bombesin receptor antagonist inhibits autocrine signals in a small cell lung carcinoma cell line. BBRC 156: 1383-9 (1988); **Yayon A & Klagsbrun M** Autocrine regulation of cell growth and transformation by basic fibroblast growth factor. Cancer Metastasis Rev. 9: 191-202 (1990); **Walsh JH et al** Autocrine growth factors and solid tumor malignancy. West. J. Med. 155: 152-63 (1991)

Examples of factors modulating growth, differentiation or functional activities of cells in an autocrine fashion.

Factor	Cell type	Reference
5-hydroxy tryptamine	human pancreatic carcinoid cells	JCP 150: 1 (1992)
Activin A	growth inhibitor in rat hepatocytes	JCI 92: 1491 (1993)
	modulator of Sertoli cell function	E 132: 975 (1993)
ADF	EBV- or HTLV-1-transformed human lymphoid cells	AEMB 319: 265 (1992)
aFGF	human nasopharyngeal carcinoma cells	IJC 54: 807 (1993)
α-fetoprotein	human colon carcinoma cells	IJC 49: 425 (1991)
	normal/malignant peripheral blood mononuclear cells	Leukemia 7: 1807 (1993)
AGF	malignant human B cell lines	ECN 1: 41 (1990)
AMF	murine fibroblasts	BBRC 194: 446 (1993)
Angiotensin II	vascular smooth muscle cells	JCI 90: 456(1992)
		JCI 91: 2268 (1993)
		Am. J. Physiol. 264: C390 (1993)
	cardiomyocytes	JCI 92: 398 (1993)
AR	normal human mammary epithelial cells	JCP 153: 103 (1992)
	human colon carcinoma cells	JCB 118: 741 (1992)
BCGF-12KDB	cell chronic lymphocytic leukemia cells	EJI 22: 1927 (1992)
β galactoside binding protein	autocrine cytostatic factor and cell growth regulator	Cell 64: 91 (1991)
bFGF	autocrine modulator of thyroid follicular cell growth	E 130: 2363 (1992)
	aortic endothelial cells	JCP 157: 279 (1993)
	human glioma cell line	J. Neurosurg. Sci. 34: 189 (1991)
	human neuroectodermal tumors	Neurosurg. 33: 679 (1993)
	human nasopharyngeal carcinoma cells	IJC 54: 807 (1993)
Bombesin	small cell lung cancer	PCBR 354A: 193 (1990)
CD23	EBV-transformed B cells from Wiskott-Aldrich syndrome	Clin. Exp. Immunol. 91: 43 (1993)
CGRP	F9 teratocarcinoma cells	BBRC 187: 381 (1992)
Cystatin C	glomerular mesangial cells	BBRC 182: 1082 (1992)
EGF	trophoblast function	J. Clin. Endocrinol. Metab. 74: 981 (1992)
	human prostate cancer cells	Prostate 20: 151 (1992)
	human prostatic tissue	J. Steroid Biochem. Mol. Biol. 46: 463 (1993)
	some human skin squamous cell carcinomas	J. Dermatol. Sci. 2: 161 (1991)
	human transitional cell carcinoma	CR 53: 5300 (1993)
	human esophageal carcinoma cells	CR 51: 1898 (1991)
	kidney epithelial cells in polycystic kidney disease	Eur. J. Cell Biol. 61: 131 (1993)
Epo	some erythroleukemias	JCI 88: 789 (1991)
ET-1	some human cancer cell lines	JCI 87: 1867 (1991)
	renal tubular cells	Am. J. Physiol. 265: F542 (1993)
	ovarian granulosa cells	J. Endocrinol. Invest. 16: 425 (1993)
	rat testis	Am. J. Physiol. 265: E267 (1993)
	rat mesangial cells	Am. J. Physiol. 265: C188 (1993)
	cardiomyocytes	JCI 92: 398 (1993)
FGF-3	mammary epithelial cells	JNCI 84: 887 (1992)
Gastrin	pancreatic tumor cells	Br. J. Cancer 66: 32 (1992)
	cultured colon tumor cells	ECR 186: 15 (1990)
	gastric cancer cells	Mol. Cell. Endocrinol. 93: 23 (1993)
G-CSF	some acute myeloblastic leukemia (AML)	Blood Rev. 6: 149 (1992)
GM-CSF	some acute myeloblastic leukemia (AML)	Leukemia 6: 562 (1992)
		Blood 81: 3068 (1993)
	autocrine differentiating factor in EoL-1 cells	Int. Arch. Allergy Immunol. 100: 240 (1993)
GRO/CINC	normal rat kidney cells	JCP 156: 412 (1993)
GRP	human small cell lung cancer	JNCI 83: 1470 (1991)
HGF/SF	human normal bronchial epithelial cells	Cell Growth Differ. 4: 571 (1993)
	some lung carcinoma cells	
IFN-β	human vascular smooth muscle cells	Lab. Invest. 66: 715 (1992)
IGF-1	human thyroid papillary cancer cell lines	Eur. J. Cancer 28A: 1904 (1992)
	human midgut carcinoid tumors	IJC 51: 195 (1992)
	human colon cancer cell differentiation and growth	Cancer Lett. 62: 23 (1992)
	glioma cells	E 130: 2683 (1992)

Factor	Cell type	Reference
	human esophageal carcinoma cells	CR 51: 1898 (1991)
	small cell lung cancer	CR 50: 2511 (1990)
	endothelial cells	Am. J. Physiol. 265: C801 (1993)
	human cardiac myocytes	ECR 207: 348 (1993)
	P19 embryonic carcinoma cells	Dev. Genet. 14: 194 (1993)
	human embryonic lung fibroblast WI38	Mol. Endocrinol. 7: 171 (1993)
IGF-2	human rhabdomyosarcomas	CR 52: 1830 (1992)
	human intestinal epithelium	E 131: 1359 (1992)
	skeletal myoblasts	JBC 266: 15917 (1991)
	human breast cancer	Mol. Endocrinol. 5: 709 (1991)
		J. Clin. Endocrinol. Metab. 77: 963 (1993)
	human neuroblastomas	PCBR 366: 257 (1991)
		JCP 155: 290 (1993)
	P19 embryonic carcinoma cells	Dev. Genet. 14: 194 (1993)
IL1	human multiple myelomas	Jpn. J. Clin. Oncol. 21: 22 (1991)
	some T helper type 2 cells	JI 146: 3849 (1991)
	negative growth regulator for some renal cancer cells	Nippon Hinyokika Gakkai Zasshi 84: 1470 (1993)
	some human gastric carcinoma cells	CR 53: 4102 (1993)
	human thyroid carcinoma cell line NIM1	J. Clin. Endocrinol. Metab. 76: 127 (1993)
IL2	normal T lymphocytes	
	B cells	Cytokine 2: 272 (1990)
IL4	rat pre-leukemic cells	IJC 50: 481 (1992)
	Reed-Sternberg cells of Hodgkin's lymphoma	Blood 79: 191 (1992)
IL5	EBV-transformed lymphocytes	Blood 79: 1763 (1992)
IL6	some acute myeloblastic leukemia (AML)	Blood Rev. 6: 149 (1992)
	myelomas	Semin. Cancer Biol. 3: 17 (1992)
	human pleomorphic adenoma of parotid gland	Cancer 70: 559 (1992)
	mesangial cell growth	Kidney Int. 41: 604 (1992)
	human megakaryocytic leukemia cells	Br. J. Haematol. 77: 32 (1991)
	negative growth regulator in some human lung cancers	CR 53: 4175 (1993)
	some human cervical carcinoma cells	Gynecol. Oncol. 50: 15 (1993)
	some malignant human melanoma cells	JCB 120: 1281 (1993)
IL8	some human malignant melanoma cells	JI 151: 2667 (1993)
IL10/BCRF1	some AIDS lymphomas	EJI 22: 2937 (1992)
IL11	human megakaryoblastic cells	Blood 81: 889 (1993)
KAF	human keratinocytes	MCB 11: 2547 (1991)
K-fgf	human Kaposi sarcomas	Cell Growth Differ. 1: 63 (1990)
Luteinizing hormone (LH)	autocrine co-mitogen for human thymocytes	BBRC 187: 1187 (1992)
NGF	human keratinocytes	JBC 268: 22838 (1993)
Parathyroid hormone	some tumor cells	BBRC 167: 1134 (1990)
	autocrine growth inhibitor for adult T cell leukemia and lymphocytes	BBRC 166: 1088 (1990)
PDGF	human cytotrophoblasts of placenta	GF 6: 219 (1992)
	vascular smooth muscle cells	JCB 123: 741 (1993)
	some subtypes of human Kaposi sarcomas	PNAS 89: 7046 (1992)
	various human tumors	O 7: 1355 (1992)
	human gliomas	CR 52: 3213 (1992)
	human meningiomas	IJC 49: 398 (1991)
	human neuroectodermal tumors	Neurosurg. 33: 679 (1993)
Prolactin	lymphoproliferation	PNAS 89: 7713 (1992)
	pituitary cells	E 131: 595 (1992)
SCF	subgroups of AML	Nouv. Rev. Fr. Hematol. 35: 285 (1993)
	megakaryoblastic leukemia cell line	Leukemia7: 235 (1993)
Substance P	autocrine modulator of macrophage functions	Immunology 71: 52 (1990)
TAMF	autocrine motility of tumor cells, metastasis formation	Cancer Metastasis Rev. 11: 5 (1992)
TCSF	autocrine motility of bladder cancer cells	PCBR 370: 393 (1991)
TGF-α	ovarian carcinoma cell lines	CR 52: 341 (1992)
	some thyroid papillary carcinoma cells	ECR 207: 348 (1993)
	human esophageal carcinoma cells	Cancer 71: 2902 (1993)
	squamous epithelium of cholesteatoma	Am. J. Otolaryngol. 14: 82 (1993)
	benign placental cytotrophoblasts	Cell Growth Differ. 4: 387 (1993)
TGF-β	autocrine stabilizer for rhythmic beating of cardiac myocytes	JCI 90: 2056 (1992)
	growth modulator for vascular smooth muscle cells	JCI 90: 456 (1992)
	negative growth factor for astrocytes	J. Neuroimmunol. 39: 163 (1992)
	negative growth regulator for human colon carcinoma	Cell Growth Differ. 4: 115 (1993)
		Cancer Lett. 68: 33 (1993)
	autocrine regulator of adrenal cortex functions	Endocr. Res. 17: 267 (1991)
	low-grade and malignant human gliomas	IJC 49: 129 (1991)
	negative growth regulator of HL-60 cells	Blood 77: 1248 (1991)
	negative growth regulator of human breast cancer cells	Cell Growth. Differ. 1: 367 (1990)

Factor	Cell type	Reference
Thrombin	vascular smooth muscle cells	Am. J. Physiol. 265: C806 (1993)
TNF-α	autocrine growth inhibitor for human papillomavirus type 16-harboring human keratinocytes	JI 149: 2702 (1992)
	human chronic lymphocytic leukemia cells	Blood 80: 1299 (1992)
	human neuroblastomas	CR 52: 3194 (1992)
	autocrine growth regulator during macrophage differentiation	PNAS 89: 4754 (1992) PNAS 90: 4763 (1993)
	negative growth regulator in low grade Ki-1-positive malignant lymphoma cells	Am. J. Pathol. 140: 709 (1992)
	human ovarian cancer cells	CR 53: 1939 (1993)
TNF-β	some EBV-infected B cell lines	JEM 177: 763 (1993)
Transferrin	promyelocytic leukemia HL-60 cells	BBRC 176: 473 (1991)
TSTGF	growth promoter of T cell proliferation	J. Leukoc. Biol. 47: 121 (1990)
Vasopressin	aortic smooth muscle cells	Am. J. Physiol. 264: C390 (1993)
VIP	human neuroblastoma cells	Regul. Pept. 37: 213 (1992) Arch. Surg. 128: 591 (1993)
Wnt-1	mammary hyperplasia	CR 52: 4413 (1992)

Autocrine motility factor: see: AMF.

Autocrine serum-replacing activity: see: ASRA.

***Autographa californica* nuclear polyhedrosis virus:** abbrev. AcNPV. See: Baculovirus expression system.

Autopharmacological agents: see: autacoids.

Autophosphorylation: This term refers to the phosphorylation, i. e. the transfer of phosphate groups, to amino acids within a protein possessing the activity of a protein kinase.
Autophosphorylation reactions are frequently observed following the binding of a ligand to a receptor with intrinsic protein kinase activity. The binding of a ligand first induces the formation of receptor dimers. An autophosphorylation therefore is really a transphosphorylation reaction in which one receptor subunit of the dimer phosphorylates the other subunit. The induction of tyrosine phosphorylation is usually transient and returns to basal level within 20 to 30 minutes. Dephosphorylation is mediated by (specific) protein tyrosine phosphatases (see, for example: HCP). It has been observed that inhibitors of such phosphatases can transiently substitute for growth factors and induce a mitogenic response (see: Okadaic acid).
The majority of the » hematopoietin (hematopoietic growth factor) receptors do not possess intrinsic tyrosine kinase activity in their intracellular domain. These receptors have been shown to become phosphorylated by other proteins (see, for example: *lck*, Janus kinases) that interact with the intracellular domains of such receptors via » *src* homology domains. In addition, ligand binding to such receptors also leads to the phosphorylation of other cellular substrate molecules.
Autophosphorylation reactions appear to play an important role in the regulation of the biological activities of many cytokine receptors. It has been observed that the intrinsic protein kinase activity of these receptors is significantly enhanced after having become phosphorylated. If the process of receptor phosphorylation is blocked by suitable inhibitors, receptor-mediated biological responses are usually also blocked. A critical role for tyrosine phosphorylation in growth regulation is suggested by the observation that inhibitors of tyrosine kinases inhibit cell growth, that inhibitors of tyrosine phosphatases can stimulate cell proliferation in the absence of a suitable receptor ligand and that mutant receptors show a direct correlation between the ability to induce tyrosine phosphorylation and the ability to support ligand-induced cell growth. Compounds that have been described to block receptor phosphorylation and other phosphorylation reactions include » Genistein, » H8, » Herbimycin, » Lavendustin A, » Staurosporine, » Suramin and » tyrphostins.

Autostimulatory B cell growth factor: see: aBCGF.

Autotaxin: abbrev. ATX. A 125 kDa factor secreted by human A2058 melanoma cells. Partial sequence analysis reveals that the factor does not have any significant homology to other known growth factors or motility factors (see also: Motogenic cytokines). The factor stimulates both random and directed migration of the cells that produce it. Pretreatment of the cells with pertussis toxin abolishes the response, indicating that a pertussis toxin-sensitive G protein is involved in signal transduction.

Another factor that has also been called Autotaxin is » AMF (autocrine motility factor) but these two factors do not seem to be identical.

Stracke ML et al Identification, purification, and partial sequence analysis of autotaxin, a novel motility-stimulating protein. JBC 167: 2524-9 (1992)

AVLPR-IL8: see: FDNCF (fibroblast-derived neutrophil chemotactic factor).

AVP: *arginine vasopressin* see: Vasopressin.

awd: *abnormal wing discs* see: nm23.

Azurocidin: see: CAP 37.

B

b: (a) as a prefix "basic" such as in bFGF (basic FGF; (b) depending upon context a prefixed b may also denote "of bovine origin".

B2: see: CD21.

B6SUt-A: A multipotential murine hematopoietic progenitor cell line cloned from nonadherent cells of virus-infected long-term marrow cultures (see also: LTBMC). The cells depend on murine » IL3 for their growth. Murine » GM-CSF is as potent as murine IL3 in stimulating these cells. B6SUt-A is considered to be a » hematopoietic stem cell line and has been used also to assay » murine » IL3 and » GM-CSF. Sodium orthovanadate, an inhibitor of phosphotyrosine phosphatase, and 12-O-tetradecanoyl-phorbol-13-acetate (TPA), a known activator of protein kinase C, also stimulates DNA synthesis in these cells.
B6SUtA cells undergo erythroid differentiation with limited proliferation after addition of » Epo. Clonal sublines of B6SUtA form mixed colonies containing erythroid cells, neutrophil-granulocytes, and basophil/mast cells in semisolid medium containing » Epo and conditioned medium from pokeweed mitogen-stimulated spleen cells.
Broudy VC et al Dynamics of erythropoietin receptor expression on erythropoietin-responsive murine cell lines. Blood 75: 1622-6 (1990); **Enver T et al** Erythropoietin changes the globin program of an interleukin 3-dependent multipotential cell line. PNAS 85: 9091-5 (1988); **Greenberger JS et al** Demonstration of permanent factor-dependent multipotential (erythroid/neutrophil/basophil) hematopoietic progenitor cell lines. PNAS 80: 2931-5 (1983); **Sorensen PH et al** Interleukin-3, GM-CSF, and TPA induce distinct phosphorylation events in an interleukin 3-dependent multipotential cell line. Blood 73: 406-18 (1989)

B7: Name of a B cell activation antigen (see also: CD antigens) that is the natural ligand of the cell surface protein » CD28 (see also: Cell activation).

B9: A subclone of the murine hybridoma cell line B13.29. B9 cells require » IL6 for its survival and proliferation *in vitro*. The cells are used to assay IL6 (see also: Bioassays) which at very low concentration stimulates the proliferation of B9 cells. The cells respond to murine, rat and human IL6. The cells also respond to » IL11. Since many plasma or serum samples kill B9 cells by a complement-mediated mechanisms such samples must be heated to 56 °C for 30 minutes to prevent this. IL6 activity is determined by measuring the incorporation of ^{3}H thymidine into the newly synthesized DNA of proliferating cells. 1 unit is approximately equal to 1 pg/ml of a recombinant (r)IL6 standard. Cell proliferation can be determined also by employing the » MTT assay. An alternative and entirely different method of detecting IL6 is » Message amplification phenotyping.
Aarden LA et al Production of hybridoma growth factor by monocytes EJI 17: 1411-6 (1987); **Borinaga AM et al** Interleukin-6 is a cofactor for the growth of myeloid cells from human bone marrow aspirates but does not affect the clonogenicity of myeloma cells *in vitro*. Br. J. Haematol. 76: 476-83 (1990); **Burger R & Gramatzki M** Responsiveness of the interleukin (IL)-6-dependent cell line B9 to IL11. JIM 158: 147-8 (1993); **Helle M et al** Functional discrimination between interleukin 6 and interleukin 1. EJI 18: 1535-8 (1988); **Hermann E et al** Correlation of synovial fluid interleukin 6 (IL6) activities with IgG concentrations in patients with inflammatory joint disease and osteoarthritis. Clin. Exp. Rheumatol. 7: 411-4 (1989); **May LT et al** High levels of "complexed" interleukin-6 in human blood. JBC 267: 19696-704 (1992); **Morris DD et al** Endotoxin-induced production of interleukin 6 by equine peritoneal macrophages *in vitro*. Am. J. Vet. Res. 53: 1298-301 (1992); **Morrissey PJ et al** The response of the IL6-dependent cell line, B9, to IL6 is not augmented by recombinant derived purified IL2 JIM 147: 141-3 (1992); **Ruef C et al** Interleukin 6 is an autocrine growth factor for mesangial cells. Kidney Int. 38: 249-57 (1990); **Straneva JE et al** Is interleukin 6 the physiological regulator of thrombopoiesis? EH 20: 47-50 (1990)

B12: An » ERG (early response gene) encoding a protein of 316 amino acids whose transcription is rapidly and transiently stimulated by » TNF-α in human umbilical vein endothelial cells. The human gene, located on chromosome 17q22-q23 appears to be highly conserved. Expression appears to be developmentally regulated in a tissue-specific manner, with B12 being differentially expressed in the heart and liver during mouse embryogenesis.
Wolf FW et al Characterization of a novel tumor necrosis factor-α-induced endothelial primary response gene. JBC 267: 1317-26 (1992)

B13: A factor-dependent murine Ly-1-positive, CD5-positive pre-B cell line which is used for the detection of murine and human » IL5 and murine » IL3 (see also: Bioassays). The cell line does not respond to human IL3. Factor activity is determined by measuring the incorporation of ^{3}H thymidine into the newly synthesized DNA of proliferating cells. Cell proliferation can be determined also by employing the » MTT assay. An alternative and entirely different method of detecting these factors is » Message amplification phenotyping.
The proliferation of B13 cells in response to IL5 is inhibited by addition of sphingosine, staurosporine, or H-7.
Introduction of a functional » IL5 gene by means of a retroviral expression vector into B13 cells creates an » autocrine loop, generating IL5-independent cells that are tumorigenic *in vivo*.
Barry SC et al Analysis of interleukin 5 receptors on murine eosinophils: a comparison with receptors on B13 cells. Cytokine 3: 339-44 (1991); **Blankenstein T et al** Retroviral interleukin 5 gene transfer into interleukin 5-dependent growing cell lines results in autocrine growth and tumorigenicity. EJI 20: 2699-705 (1990); **Li VS et al** IL5-induced IgA synthesis by LPS-stimulated mouse B cells is prevented by protein kinase C inhibitors. ECN 3: 103-7 (1992); **Rolink AG et al** Monoclonal antibodies reactive with the mouse interleukin 5 receptor. JEM 169: 1693-1701 (1989)

B13.29: see: B9.

B15-TRF: *B15 T cell replacing factor* see: B151-TRF. For some aspects of nomenclature see also: TRF (T cell replacing factors).

B15R: B15R is a vaccinia virus gene. It encodes a secreted glycoprotein of 33 kDa functioning as a soluble receptor for one of the molecular forms of murine » IL1. B15R protein specifically binds to IL1β, but not to IL1α or to the IL1 receptor antagonist known as » IL1ra.
Viruses with a deleted B15R gene have been found to be much more virulent than wild type viruses. Following intranasal injection of mice they cause symptoms much faster and the mortality rate of these much is also much enhanced.
Blockade of the synthesis of IL1β by the viral protein is regarded as a viral strategy allowing systemic antiviral reactions elicited by IL1 to be suppressed or diminished. Binding proteins effectively blocking the functions of IL1 with similar activity as B15R have also been found to be encoded by genes of the cowpox virus. For other viral strategies allowing subversion of host defense mechanism see also: viroceptor.
Alcami A & Smith GL A soluble receptor for interleukin-1 β encoded by vaccinia virus: a novel mechanism of virus modulation of the host response to infection. Cell 71: 153-67 (1992); **Spriggs MK et al** Vaccinia and cowpox viruses encode a novel secreted interleukin-1-binding protein. Cell 71: 145-52 (1992)

B16-F1 melanoma autocrine motility factor: This factor was isolated from the B16-F1 melanoma cell line which has a low metastasizing potential. The factor which enhances the motility of these cell is identical with » AMF (autocrine motility factor). See also: Motogenic cytokines.
Silletti S et al Purification of B16-F1 melanoma autocrine motility factor and its receptor. CR 51: 3507-11 (1991)

B61: An » ERG (early response gene) encoding a secreted protein of 25 kDa (187 amino acids) whose transcription is rapidly and transiently stimulated by » TNF-α in human umbilical vein endothelial cells. The protein has no significant homology to other proteins and appears to be evolutionarily conserved.
Holzman LB et al A novel immediate-early response gene of endothelium is induced by cytokines and encodes a secreted protein. MCB 10: 5830-8 (1990)

B151K12 TRF: see: B151-TRF.

B151-TRF: *B151 T cell replacing factor* This factor is produced by the T cell hybridoma B151K12. It also known as ***B15-TRF***, and ***B151K12 TRF***. It is secreted constitutively by the murine T cell hybridoma B151K12. It induces the synthesis of IgM in the leukemic B cell line » BCL_1. The factor functions as a B cell growth and differentiation factor; it has » BCGF 2 (B cell growth factor 2) activity and is identical with » IL5. For some aspects of nomenclature see also: TRF (T cell replacing factors).
B151K12 cells have been found to secrete two different T cell replacing factors, designated ***B151-TRF1*** and ***B151-TRF2***. B151-TRF1 induces growth and differentiation of antigen-activated B cells into antigen-specific plaque-forming cells (PFC) and is identical with » IL5.
B151-TRF2 is a poorly characterized factor that induces polyclonal differentiation of unstimulated B cells into IgM-secreting cells without concomitant stimulation of antigen, mitogen, or anti-Ig antibody. In a murine model of systemic lupus erythematosus with polyclonal B cell activation associated with the T cell hyperfunction B151-TRF2 appears to be one of the B cell differentiation factors responsible for polyclonal B cell activation leading to autoantibody production.

Dobashi K et al Polyclonal B cell activation by a B cell differentiation factor, B151-TRF2. III. B151-TRF2 as a B cell differentiation factor closely associated with autoimmune disease. JI 138: 780-7 (1987); **Harada N et al** BCGF II activity on activated B cells of a purified murine T cell-replacing factor (TRF) from a T cell hybridoma (B151K12). JI 134: 3944-51 (1985); **Harada N et al** Production of a monoclonal antibody to and its use in the molecular characterization of murine T cell-replacing factor (TRF) and B cell growth factor II (BCGF II). PNAS 84: 4581-85 (1987); **Howard M et al** B cell growth and differentiation factors. IR 78: 185-210 (1984); **Kinashi T et al** Cloning of complementary DNA encoding T cell-replacing factor and identity with B cell growth factor 2. N 324: 70-3 (1986); **Matsumoto M et al** Role of T cell-replacing factor (TRF) in the murine B cell differentiation: induction of increased levels of expression of secreted type IgM mRNA. JI 138: 1826-33 (1987); **Nakanishi K et al** Both interleukin 2 and a second T cell derived factor in EL4-supernatant have activity as differentiation factors in IgM synthesis. JEM 160: 1605-21 (1984); **Nakanishi K et al** Soluble factors involved in B cell differentiation: identification of two distinct T cell-replacing factors (TRF). JI 130: 2219-24 (1983); **Nakanishi K et al** Ig RNA expression in normal B cells stimulated with anti-IgM antibody and T cell-derived growth and differentiation factors. JEM 160: 1736-51 (1984); **Nakanishi K et al** Both B151-T cell replacing factor 1 and IL5 regulate Ig secretion and IL2 receptor expression on a cloned B lymphoma line. JI 140: 1168-74 (1988); **Ono S et al** Identification of two distinct factors B151-TRF1 and B151-TRF2 inducing differentiation of activated B cells and small resting B cells into antibody-producing cells. JI 137: 187-96 (1986); **Ono S et al** Ia-restricted B-B cell interaction. I. The MHC haplotype of bone marrow cells present during B cell ontogeny dictates the self-recognition specificity of B cells in the polyclonal B cell activation by a B cell differentiation factor, B151-TRF2. JI 139: 3213-23 (1987); **Takahashi T et al** Structural comparison of murine T cell (B151K12)-derived T cell-replacing factor (IL5) with rIL5: dimer formation is essential for the expression of biological activity. Mol. Immunol. 27: 911-20 (1991); **Takatsu K et al** Antigen-induced T cell replacing factor (TRF). III. Establishment of T cell hybridoma continuously producing TRF and functional analysis of released TRF. JI 125: 2646-53 (1980); **Takatsu K et al** Purification and physicochemical characterization of murine T cell replacing factor (TRF). JI 134: 382-9 (1985); **Tominaga A et al** Establishment of an assay system for the detection of translated materials of T cell-replacing factor mRNA in *Xenopus* oocytes. Microbiol. Immunol. 30: 789-98 (1986)

B151-TRF1: *B151 T cell replacing factor* see: B151-TRF.

B151-TRF2: *B151 T cell replacing factor* see: B151-TRF.

BAC1. 2F5: A mouse macrophage cell line derived from splenic adherent cells from a (BALB/c X A.CA) F1 mouse and generated by transfection with SV40 DNA mutated in the origin of replication. BAC1 cells produce lysozyme, collagenase, and esterase, bear Fc receptors, express Ia antigens, and engage in Fc-mediated phagocytosis. The cells can be induced to secrete » IL1. The growth of these cells depends upon » M-CSF and » GM-CSF and it can, therefore, be used to detect these factors (see also: Bioassays). Alternative methods of detecting these factors are » colony formation assays and » Message amplification phenotyping. See also: CSF for general information about colony stimulating factors.

Morgan C et al Isolation and characterization of a cloned growth factor dependent macrophage cell line, BAC1.2F5. JCP 130: 429-7 (1987); **Schwarzbaum S et al** The generation of macrophage-like cell lines by transfection with SV40 origin defective DNA. JI 132: 1158-62 (1984); **Whetton AD & Dexter TM** Myeloid hematopoietic growth factors. BBA 989: 111-132 (1989)

Bacterial endotoxins: see: Endotoxin.

Bactericidal/permeability-increasing protein: abbrev. BPI. See: Endotoxin.

Baculovirus expression system: Baculoviruses are insect-pathogenic and -specific viruses belonging to the group of ***nuclear polyhedrosis viruses***. The most prominent member of this group is ***Autographa californica* nuclear polyhedrosis virus (*AcNPV*)**. AcNPV infects more than 30 different species of insects of the lepidoptera family.
The genome of AcNPV is a circular closed double-stranded DNA with a length of ≈ 129 kb. Free viruses consist of a viral capsid that is surrounded by a lipid-containing membrane. Several nucleocapsids can aggregate to form superstructures. Following productive infection of susceptible insect cells (e.g. ***Spodoptera frugiperda*** or ***Trichoplusia ni*** cells) one observes free extracellular virus particles and intracellular forms of the virus that are easily distinguished microscopically as dense viral inclusion bodies surrounded by proteins.
The infection cycle of AcNPV is biphasic. Initially infectious virus particles infect gut cells of an insect. It migrates to the cell nucleus and sheds its protein coat. After ≈ 6 hrs one observes first signs of viral DNA replication. After ≈ 12 hrs new infectious viruses are released by budding. Viruses are taken up by other cells either by endocytosis or by fusion. The production of extracellular virus particles reaches its maximum ≈ 36-48 hrs post infection.
≈ 18-24 post infection inclusion bodies with up to 100 embedded virus particles are formed in the nuclei of infected cells. These bodies consist of polyhedrin, a virus-encoded protein of 29 kDa. Inclusion bodies accumulate until the cell lyses 4-5 days post infection. Viral inclusion bodies are an

important component of the infection cycle. Polyhedrin crystals are relatively inert and resistant against environmental influences. They are released into the environment after the death of an infected insect and are then taken up by other insects with their food. Inclusion bodies lyse in the alkaline milieu of the insect gut and then release their infectious virus particles, thus closing the infection cycle.

As the polyhedrin gene is neither required for virus infectivity nor intracellular multiplication it has been used to design vectors allowing the expression of heterologous proteins in infected insect cells. Polyhedrin has been shown to accumulate in virus-infected cells. As it can make up 20 % of the total protein content of a cell yields are on the order of 1 gram of foreign protein per liter of infected cell cultures. (1 mg of protein per 1-2 $\times 10^6$ of infected cells). Cloning vectors exploit the extraordinary strength of the polyhedrin gene promoter and are constructed in a way allowing the foreign gene to be expressed at high levels under the control of the polyhedrin gene promoter.

Due to its size the virus genome itself cannot be used as a cloning vector. Instead, a so-called intermediate or transfer vector is used which allows al cloning operations to be carried out in bacteria. The foreign gene to be expressed in insect cells is inserted between 5´ and 3´ sequences of the polyhedrin gene which is located on the transfer vector. Infectious recombinant baculoviruses are usually obtained by recombination *in vivo*, following simultaneous infection of insect cells with wild-type viruses and the engineered cloning vector carrying the foreign gene.

Recombinant baculoviruses form plaques and are easily distinguished from wild-type viruses still carrying an intact polyhedrin gene. Recombinant baculoviruses can be used also to infect insect larvae to increase the yields of recombinant viruses/proteins.

Polyhedrin has been shown to accumulate in virus-infected cells. As it can make up 20 % of the total protein content of a cell yields are on the order of 1 gram of foreign protein per liter of infected cell cultures. (1 mg of protein per 1-2 $\times 10^6$ of infected cells). Cloning vectors exploit the extraordinary strength of the polyhedrin gene promoter and are constructed in a way allowing the foreign gene to be expressed under the control of the polyhedrin gene promoter.

One particular advantage of the use of baculovirus vectors in insect cell systems is the fact that insect cells process and modify heterologous proteins correctly. The foreign proteins have antigenic and other immunological properties resembling those of genuine proteins produced in their native cell systems. This is a distinct advantage over heterologous gene expression in bacteria which cannot process eukaryotic proteins correctly (see also: *Escherichia coli*). The baculovirus/insect system has therefore been used to produce recombinant » cytokines and other proteins of interest since it is known that correct post-translational processing of heterologously expressed proteins is often required for full bioactivity as a pharmacological compound. Insect cells, however, have been shown to lack the capacity to convert a high mannose core to complex N-linked oligosaccharides during glycosylation. If this causes problems, mammalian cell systems must be used to express the recombinant protein (see: BHK cells, CHO cells, COS cells, Namalwa cells).

● BASICS: **Cameron IR et al.** Insect cell culture technology in baculovirus expression systems. Trends Biotechnol. 7: 66-70 (1989); **Davis TR et al** Comparative recombinant protein production of eight insect cell lines. In Vitro Cell. Dev. Biol. 29A: 388-90 (1993); **Doerfler W & Böhm P** (edt) The molecular biology of baculoviruses. Curr. Top. Microbiol. Immunol. 131: 51-68 (1986); **Fraser MJ** The baculovirus-infected insect cell as a eukaryotic gene expression system. CTMI 158: 131-72 (1992); **Friesen PD & Miller LK** The regulation of baculovirus gene expression. Curr. Top. Microbiol. Immunol. 131: 31-49 (1986); **Kang CY** Baculovirus vectors for expression of foreign genes. Adv. Virus Res. 35: 177-192 (1988); **Keddie BA et al** The pathway of infection of *Autographa californica* nuclear polyhedrosis virus in an insect host. Science 243: 1728-1730 (1989); **Luckow VA & Summers MD** Trends in the development of baculovirus expression vectors, Bio/Technology 6: 47-55 (1988); **Luckow VA** Baculovirus systems for the expression of human gene products. Curr. Opin. Biotechnol. 4: 564-72 (1993); **Maiorella B et al** Large-scale insect cell culture for recombinant protein production. Bio/Technology 6: 1406-1410 (1988); **Matsuura Y et al** Baculoviruses expression vectors: the requirements for high level expression of proteins, including glycoproteins. J. Gen. Virol. 68: 1233-1250 (1987); **Miller LK** Baculoviruses as gene expression vectors. Ann. Rev. Microbiol. 42: 177-179 (1988); **Miller LK** Baculoviruses: high-level expression in insect cells. Curr. Opin. Genet. Dev. 3: 97-101 (1993); **Summers MD & Smith G** A manual of methods for baculovirus vectors and insect cell culture procedures. Texas Agricultural Experiment Station Bulletin No. 1555. Texas Agricultural Experiment station, College Station, Texas, 1987; **Summers MD & Smith GE** Genetic engineering of the genome of the *Autographa californica* nuclear polyhedrosis virus. In: Fields, B. et al. (edt) Banbury Report 22: Genetically altered viruses and the environment, pp. 319-339, Cold Spring Harbor Laboratory, Cold Spring Harbor, New York, 1985); **Vaughn JL et al** The establishment of two cell lines from the insect *Spodoptera frugiperda (lepidoptera, noctuidae)*. In Vitro 13: 213-217 (1977)

● EXPRESSION OF CYTOKINES AND OTHER PROTEINS: **Alcover A et al** A soluble form of the human CD8 α chain expressed in the baculovirus system: biochemical characterization and binding to MHC class I. Mol. Immunol. 30: 55-67 (1993); **Barnett J et al** Physicochemical characterization of recombinant human nerve growth factor produced in insect cells with a baculovirus vector. J. Neurochem. 57: 1052-61 (1991); **Brown SA & Friesel R** Production of recombinant *Xenopus* fibroblast growth factor receptor-1 using a baculovirus expression system. BBRC 193: 1116-22 (1993); **Cohen T et al** High levels of biologically active vascular endothelial growth factor (VEGF) are produced by the baculovirus expression system. GF 7: 131-8 (1992); **Fiebich BL et al** Synthesis and assembly of functionally active human vascular endothelial growth factor homodimers in insect cells. EJB 211: 19-26 (1993); **Grabenhorst E et al** Biosynthesis and secretion of human interleukin 2 glycoprotein variants from baculovirus-infected Sf21 cells. Characterization of polypeptides and posttranslational modifications. EJB 215: 189-97 (1993); **Graber P et al** Purification and characterization of biologically active human recombinant 37 kDa soluble CD23 (sFc ε RII) expressed in insect cells. JIM 149: 215-26 (1992); **Guisez Y et al** Expression, purification and crystallization of fully active, glycosylated human interleukin-5. FL 331: 49-52 (1993); **Ingley E et al** Production and purification of recombinant human interleukin-5 from yeast and baculovirus expression systems. EJB 196: 623-9 (1991); **Kang XQ et al** Production, purification, and characterization of human recombinant IL8 from the eukaryotic vector expression system baculovirus. Protein Expr. Purif. 3: 313-21 (1992); **Kunimoto DY et al** High-level production of murine interleukin-5 (IL5) utilizing recombinant baculovirus expression. Purification of the rIL5 and its use in assessing the biologic role of IL5 glycosylation. Cytokine 3: 224-30 (1991); **Lindqvist C et al** Expression of human IL2 receptor α- and β-chains using the baculovirus expression system. Scand. J. Immunol. 38: 267-72 (1993); **Lowe PN et al** Expression of polyisoprenylated *Ras* proteins in the insect/baculovirus system. Biochem. Soc. Trans. 20: 484-7 (1992); **Matsuura Y et al** Biological function of recombinant IL6 expressed in a baculovirus system. Lymphokine Cytokine Res. 10: 201-5 (1991); **Morishita K et al** Expression and characterization of kinase-active v-*erb*B protein using a baculovirus vector system. Jpn. J. Cancer Res. 83: 52-60 (1992); **Peakman TC et al** Expression of the mouse c-*abl* type IV proto-oncogene product in the insect cell baculovirus system. BBA 1138: 68-74 (1992); **Quelle DE et al** Phosphorylatable and epitope-tagged human erythropoietins: utility and purification of native baculovirus-derived forms. Protein Expr. Purif. 3: 461-9 (1992); **Ragona G et al** The transcriptional factor Egr-1 is synthesized by baculovirus-infected insect cells in an active, DNA-binding form. DNA Cell Biol. 10: 61-6 (1991); **Rahn HP et al** Expression of human salivary-gland kallikrein in insect cells by a baculovirus vector. Agents Actions Suppl. 38: 66-73 (1992); **Ramer SE et al** Purification and initial characterization of the lymphoid-cell protein-tyrosine kinase p56*lck* from a baculovirus expression system. PNAS 88: 6254-8 (1991); **Smith, G. E. et al.** Production of human β interferon in insect cells infected with a baculovirus expression vector. MCB3: 2156-65 (1983); **Smithgall TE et al** Construction of a cDNA for the human c-fes proto-oncogene protein-tyrosine kinase and its expression in a baculovirus system. B 31: 4828-33 (1992); **Vranovsky K et al** Production of v-*Myb* and c-*Myb* in insect cells infected with recombinant baculoviruses. Gene 117: 233-5 (1992); **Watts JD et al** Purification and initial characterization of the lymphocyte-specific protein-tyrosyl kinase p56*lck* from a baculovirus expression system. JBC 267: 901-7 (1992); **Yee CJ et al** Expression and characterization of biologically active human hepatocyte growth factor (HGF) by insect cells infected with HGF-recombinant baculovirus. B 32: 7922-31 (1993)

BAF: *B cell activating factors* Generic name given to factors that activate B cells (see also: Cell activation). Some of these factors have been identified (BAF = IL1α) while other remain uncharacterized (see: BCAF). For problems of nomenclature see also: BSF (B cell stimulating factors).

BAF: *B cell activating factor* This factor is identical with one of the molecular forms of » IL1 (IL1α). For some aspects of nomenclature see also: BCDF (B cell differentiation factors).

Aarden LA et al Revised nomenclature for antigen-nonspecific T cell proliferation and helper factors. CI 48: 433-36 (1979); **Koopman WJ et al** Evidence for the identification of lymphocyte-activating factor as the adherent T cell-derived mediator responsible for enhanced antibody synthesis by nude mouse spleen cells. CI 35: 92-8 (1978); **Wood DD & Cameron PM** Stimulation of the release of a C cell-activating factor from human monocytes. CI 21: 133-45 (1976); **Wood DD et al** Resolution of a factor that enhances the antibody response of T cell-depleted murine splenocytes from several other monocyte products. CI 21: 88-96 (1976); **Wood DD** Purification and properties of human B cell-activating factor. I. Comparison of the plaque-stimulating activity with the thymocyte-stimulating activity. JI 123: 2395-99 (1979)

BaF3: An immortalized murine bone marrow-derived » IL3-dependent pro-B cell line. When cultured in the presence of IL3, or other growth promoting factors, BaF3 cells are highly resistant to X-irradiation and the cytotoxic drugs etoposide and cisplatin. BaF-3 cells undergo » apoptosis upon cytokine withdrawal. In the absence of IL3, cell death can be prevented by the » calcium ionophores A23187 (1 microM) and ionomycin (0.5 microM) which maintain cell viability while reversibly arresting the cell cycle.

The cells have been used as a model to study the functional reconstitution of growth factor receptors for » IL3, IL5, and » GM-CSF by introduction of cloned and expressed receptor components common to all three cytokine receptors.

The cell line has been used to create cells that grow in response to » Epo by introduction of an expression vector encoding functional Epo receptors. These altered cells have been used to study the pathogenicity and mitogenicity of » SFFV (spleen focus-forming virus).

Collins MK et al Interleukin 3 protects murine bone marrow cells from apoptosis induced by DNA damaging agents. JEM 176: 1043-51 (1992); **Daley GQ & Baltimore D** Transformation of an interleukin 3-dependent hematopoietic cell line by the chronic myelogenous leukemia-specific P210bcr/abl protein. PNAS 85: 9321-6 (1988); **Kitamura T et al** Reconstitution of functional receptors for human granulocyte/macrophage colony-stimulating factor (GM-CSF): evidence that the protein encoded by the AIC2B cDNA is a subunit of the murine GM-CSF receptor. PNAS 88: 5082-6 (1991); **Rodriguez-Tarduchy G et al** Regulation of apoptosis in interleukin-3-dependent hemopoietic cells by interleukin-3 and calcium ionophores. EJ 9: 2997-3002 (1990); **Sakamaki K et al** Critical cytoplasmic domains of the common β subunit of the human GM-CSF, IL3 and IL5 receptors for growth signal transduction and tyrosine phosphorylation. EJ 11: 3541-9 (1992); **Wang Y et al** Erythropoietin receptor (EpoR)-dependent mitogenicity of spleen focus-forming virus correlates with viral pathogenicity and processing of env protein but not with formation of gp52-EpoR complexes in the endoplasmic reticulum. J. Virol. 67: 1322-7 (1993); **Yoshimura A et al** Mutations in the Trp-Ser-X-Trp-Ser motif of the erythropoietin receptor abolish processing, ligand binding, and activation of the receptor. JBC 267: 11619-25 (1992)

Baker's yeast: See: *Saccharomyces cerevisiae.*

BALM-4: This human B cell line is used to assay » IL4 (see also: Bioassays). The activity of the factor is determined by measuring the incorporation of ^{3}H thymidine into the newly synthesized DNA of proliferating cells. An alternative and entirely different detection method is » Message amplification phenotyping.
Callard RE The marmoset B lymphoblastoid cell line (B95-8) produces and responds to B cell growth and differentiation factors: role of shed CD23 (sCD23). Immunology 65: 379-84 (1988); **Shields JG et al** The response of selected human B cell lines to B cell growth and differentiation factors. EJI 17: 535-40 (1987)

Basic astroglial growth factor: see: AGF-2.

Basic fibroblast growth factor: see: bFGF.

Bax: see: *bcl*-2.

Bb: see: HMW-BCGF (high molecular weight B cell growth factor).

B-BCDF: *B cell differentiation factor.* This B cell differentiation factor (see also: BCDF for some aspects of nomenclature) is secreted by the B lymphoblastoid cell line » CESS-2. It induces the synthesis of IgG in activated B cells. The biological activity makes it very likely that it is identical with another B cell differentiation factor, designated » BCDF-γ. See also: BCDF 1. This factor is probably not identical with another factor also abbrev. B-BCDF.
Yoshizaki K et al Isolation and characterization of B cell differentiation factor (BCDF) secreted from a human B lymphoblastoid cell line. JI 132: 2948-54 (1984)

B-BCDF: *B cell-derived B cell growth factor* This poorly characterized factor is produced by some B cell lines and normal activated B cells. B cell lines, immortalized by Epstein-Barr virus, constitutively secrete 10-fold higher level of B-BCGF compared with normal activated B cells. B-BCGF appears to be distinct from » IL1, » IL 2, » IL4 (see: BSF 1; B cell stimulatory factor); » IL5 (see: BCGF-II), » IFN-γ, or » TGF. B-BCGF acts on activated B cells, but not on resting G_0 phase B cells (see also: Cell cycle) to induce proliferation. This factor is probably not identical with another factor also abbrev. B-BCDF.
Muraguchi A et al B cell-derived BCGF functions as autocrine growth factor(s) in normal and transformed B lymphocytes. JI 137: 179-86 (1986)

B-BPA: *B lymphocyte-derived burst promoting activity* See: E-CSF (erythroid colony-stimulating factor).

BCAF: *B cell activating factor* see: BAF.

BCAF: *B cell activating factor* This poorly characterized factor of 30 kDa found in supernatants of anti-CD2 monoclonal antibody-stimulated human T cell clones induces resting B cells to enter the S phase of the » cell cycle. Its biological activity makes it unlike that it is identical with » IL1, » IL2, » IFN or other known B cell stimulating factors (see also: BSF, B cell stimulating factor), including » IL4 which also acts on resting B cells.
Bowen DL et al Identification and characterization of a B cell activation factor (BCAF) produced by a human T cell line. JI 136: 2158-63 (1986); **Diu A et al** Activation of resting human B cells by helper T cell clone supernatant: characterization of a human B cell activating factor. PNAS 84: 9140-4 (1987); **Diu A et al** Further evidence for a human B cell activating factor distinct from IL4. CI 125: 14-28 (1990); **Fevrier M et al** Human B cell activating factor (BCAF): production by a human T cell tumor line. LR 8: 59-67 (1989); **Leclercq L et al** Supernatant from a cloned helper T cell line stimulates most small resting B cells to undergo increased I-A expression, blastogenesis, and progression through the cell cycle. JI 136: 539-45 (1986)

BCDC: *bladder carcinoma-derived cytokine* This poorly characterized factor of 22 kDa is secreted by a human bladder carcinoma cell line, J82. It induces the expression of tissue thromboplastin in human umbilical chord endothelial cells.
Noguchi M et al Identification and partial purification of a novel tumor-derived protein that induces tissue factor on cultured human endothelial cells. BBRC 160: 222-7 (1989)

BCDF: *B cell differentiation factors*
A normal immune response requires the co-ordinated interaction and also the specific contact between T and B cells. Some of the helper functions

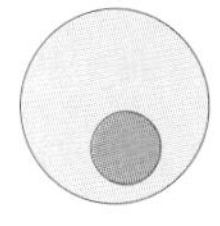

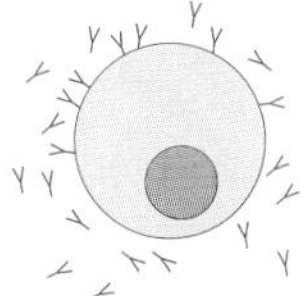

B cell differentiation factors.
B cell differentiation factors promote development of activated B cells into antibody-producing cells. They do not support clonal expansion. Factors with BCDF activity have also been described as BMF (B cell maturation factors) or BSF (B cell stimulation factors). Some TRFs (T cell replacing factors) also possess activities as BCDF/BMF/BSF. Some factors, designated BRMF (B cell replication/maturation factors) or BGDF (B cell growth/differentiation factors) promote development of activated B cells into antibody-producing cells and clonal expansion.

of T cells are due to the continuous production and secretion of specific growth factors. These T cell-derived B cell differentiation factors play a critical role in the regulation of B cell differentiation.
The term BCDF was coined at the beginning of cytokine research to distinguish T cell-derived factors (see also: TRF, T cell replacing factor) mediating the (terminal) differentiation of B cells into immunoglobulin-secreting B cells from other factors that induce B cell proliferation (see also: BCGF, B cell growth factors). A generic name for factors, possessing both types of activities is » BSF (B cell stimulating factor) or » BRMF (B cell replication and maturation factors). For an assay system see also: JR-2(82) cells.
In many instances BCDF is an operational definition of biological activities and does not signify biochemically identified distinct factors. These activities have occasionally been described also as » ***BMF*** (B cell maturation factors). BCDF activity is detected by employing various cell lines (see, for example CESS; HFB-1; SKW6-C14).
The pleiotropic actions of many growth factors (see also: Cytokines) demonstrates that the strict distinction of differentiation-inducing and proliferation-inducing factors is not very useful. Among the factors that influence the growth and differentiation of B cells are » IL1, » IL2, » IL4, » IL5, » IL6, » IFN-γ, » TNF-α and » TNF-β. The term BCDF is still used, however, to signify that a newly isolated factor acts on B cells and promotes their maturation. The term is used in particular to describe less well defined activities obtained, for example, from culture supernatants or from serum. Occasionally a suffix serves to indicate a particular activity. BCDF-IgG, for example, signifies that the factor promotes the differentiation of B cells into cells secreting IgG.

Brooks K et al B cell growth and differentiation factors. MiE 116: 372-9 (1985); **Callard RE** Human B cell responses to cytokines. In: Balkwill FR (edt) Cytokines – A practical approach, pp. 121-50, IRL Press, Oxford 1991; **Callard RE et al** Assay for human B cell growth and differentiation factors. in: Clemens MJ et al (eds) Lymphokines and Interferons. A practical Approach, pp. 345-64, IRL Press, Oxford 1987; **Callard RE** The marmoset B lymphoblastoid cell line (B95-8) produces and responds to B cell growth and differentiation factors: role of shed CD23 (sCD23). Immunology 65: 379-84 (1988); **Hamaoka T & Ono S** Regulation of B cell differentiation: interactions of factors and corresponding receptors. ARI 4: 167-204 (1986); **Howard M et al** B cell growth and differentiation factors. IR 78: 185-10 (1984); **Kishimoto T** Factors affecting B cell growth and differentiation. ARI 3: 133-57 (1985); **Leanderson T et al** B cell growth factors: distinction from T cell growth factor and B cell maturation factor. PNAS 79: 7455-9 (1982); **Melchers F et al** H-2-unrestricted polyclonal maturation without replication of small B cell induced by antigen-activated T cell help factors. EJI 10: 679-85 (1980); **O'Garra A et al** B cell factors are pleiotropic. IT 9: 45-54 (1988); **Okada M et al** B cell growth factors and B cell differentiation factor from human T hybridomas. Two distinct kinds of B cell growth factor and their synergism in B cell proliferation. JEM 157: 583-90 (1983); **Romagnani S** T cell-derived B cell growth and differentiation factors. Ric. Clin. Lab. 17: 181-97 (1987); **Shields JG et al** The response of selected human B cell lines to B cell growth and differentiation factors. EJI 17: 535-40 (1987)

BCDF: *B cell differentiation factor* This specific B cell differentiation factor is identical with » IL1. For some aspects of nomenclature see also: BCDF (B cell differentiation factors).

Jandl RC et al Interleukin-1 stimulation of human B lymphoblast differentiation. Clin. Immunol. Immunopharmacol. 46: 115-21 (1988)

BCDF: *B cell differentiation factor* This specific B cell differentiation factor is identical with » BSF p2 (B cell stimulating factor p2) which, in turn, is identical with » IL6. For some aspects of nomenclature see also: BCDF (B cell differentiation factors).

Harigai M et al Stimulation of interleukin 6-like B cell differentiation factor production in human adherent synovial cells by recombinant interleukin 1. Scand. J. Immunol. 29: 289-97 (1989); **Hirano T et al** Purification to homogeneity and characterization of human B cell differentiation factor (BCDF or BSFp-2). PNAS 82: 5490-4 (1985); **Sehgal PB et al** Human β-2-interferon and B cell differentiation factor are identical. S 235: 731-2 (1987)

BCDF-1: *B cell differentiation factor* This B cell differentiation factor (see also: BCDF for some aspects of nomenclature) induces the synthesis of IgG in activated B cells. The biological activity makes it very likely that it is identical with another B cell differentiation factor, designated » BCDF-γ. See also: B-BCDF.
Hirano T et al Human helper T cell factor(s). III. Characterization of B cell differentiation factor I (BCDF 1). JI 132: 229-34 (1984); **Hirano T et al** Human helper T cell factor(s). IV. Demonstration of a human late-acting B cell differentiation factor acting on *Staphylococcus aureus* Cowan I-stimulated B cells. JI 133: 798-802 (1984)

BCDF-A20: *A20 B cell differentiation factor* This poorly characterized factor is produced by several murine T cell clones and induces secretion of IgG2a in membrane IgG2a (mIgG2a)-bearing BALB/c B lymphoid tumor cells, A20. BCDF-A20 activity is not due to » IL1, » IL2, various » TRFs (T cell replacing factors), » BMFs (B cell maturation factors) or » IL4 (see: BSF 1).
Kim KJ & Finnegan A Induction of IgG2a secretion from mIgG2a$^+$ B lymphoid tumor cells by BCDF present in several antigen-specific T helper clones. CI 110: 149-62 (1987)

BCDF α: *B cell differentiation factor α* This B cell differentiation factor induces the expression of IgA receptors on B cells which is signified by the suffix α. One specific factor with BCDF α activity is » BCGF 2 (B cell growth factor). In addition, there are other (unidentified factors) detected in culture supernatants of various cell lines possessing the same activity. See also: BCDF, BCGF (B cell growth factors), and » BSF (B cell stimulating factors) for some aspects of nomenclature.
Kanowith-Klein S et al Constitutive production of B cell differentiation factor-like activity by human T and B cell lines. EJI 17: 593-8 (1987); **Kawanishi H et al** Autoreactive T cell hybridoma-derived B cell stimulatory factor(s) governing IgA isotype immunoglobulin production by murine Peyer's patch B cells. Immunol. Invest. 17: 587-613 (1988)

BCDF ε: *B cell differentiation factor ε* This B cell differentiation factor induces the expression of FceR II (IgE receptor; see also: CD23) and enhances IgE production on B cells. One specific factor with BCDF ε activity is » BSF 1 (B cell stimulating factor) which is identical with » IL4. In addition, there are other (unidentified factors) detected in culture supernatants of various cell lines possessing the same activity. See also: BCDF, BCGF (B cell growth factors), and » BSF (B cell stimulating factors) for some aspects of nomenclature.
Coffman RL et al B cell stimulatory factor-1 enhances the IgE response of lipopolysaccharide-activated B cells. JI 136: 4538-41 (1986); **Hudak SA et al** Murine B cell stimulatory factor 1 (interleukin 4) increases expression of the Fc receptor for IgE on mouse B cells. PNAS 84: 4606-10 (1987); **Kanowith-Klein S et al** Constitutive production of B cell differentiation factor-like activity by human T and B cell lines. EJI 17: 593-8 (1987); **Lebrun P & Spiegelberg HL** Concomitant immunoglobulin E and immunoglobulin G1 formation in *Nippostrongylus brasiliensis*-infected mice. JI 139: 1459-65 (1987); **Moller E & Strom H** Biological characterization of T cell-replacing factor in the synovial fluid of rheumatoid arthritis patients. Scand. J. Immunol. 27: 717-24 (1988); **Snapper CM et al** Differential regulation of IgG1 and IgE synthesis by interleukin 4. JEM 167: 183-96 (1988); **Vercelli D et al** Human recombinant interleukin 4 induces Fc ε R2/CD23 on normal human monocytes. JEM 167: 1406-16 (1988)

BCDF γ: *B cell differentiation factor γ, B cell differentiation factor for IgG1* This B cell differentiation factor induces the expression of IgG by activated B cells which is signified by the suffix γ. The factor is identical with » BSF-1 (B cell stimulating factor 1) and hence identical with » IL4. See also: BCDF, BCGF (B cell growth factors), and » BSF (B cell stimulating factors) for some aspects of nomenclature.
Layton JE et al Clonal analysis of B cells induced to secrete IgG by T cell-derived lymphokine(s). JEM 160: 1850-63 (1984); **Lebrun P & Spiegelberg HL** Concomitant immunoglobulin E and immunoglobulin G1 formation in *Nippostrongylus brasiliensis*-infected mice. JI 139: 1459-65 (1987); **Pure E et al** T cell-derived B cell growth and differentiation factors. Dichotomy between the responsiveness of B cells from adult and neonatal mice. JEM 157: 600-12 (1983); **Tesch H et al** Lymphokines regulate immunoglobulin isotype expression in an antigen-specific immune response. JI 136: 2892-5 (1986); **Vitetta E et al** Serological, biochemical, and functional identity of B cell-stimulatory factor 1 and B cell differentiation factor for IgG1. JEM 162: 1726-31 (1985); **Yuan D et al** Activation of the γ 1 gene by lipopolysaccharide and T cell-derived lymphokines containing a B cell differentiation factor for IgG1 (BCDF γ). JI 135: 1465-9 (1985)

BCDF-IgG: Generic name given to B cell differentiation factors (see also: BCDF) inducing the synthesis of IgG in activated B cells. One factor with this activity is » BCDF-γ. Another factor with this activity is » IL2. See also: BCDF, BCGF (B cell growth factors), and » BSF (B cell stimulating factors) for some aspects of nomenclature.
Kanowith-Klein S et al Constitutive production of B cell differentiation factor-like activity by human T and B cell lines. EJI 17: 593-8 (1987); **Kikutani H et al** Effect of B cell differentia-

tion factor (BCDF) on biosynthesis and secretion of immunoglobulin molecules in human B cell lines. JI 134: 990-5 (1985); **Kim KJ & Finnegan A** Induction of IgG2a secretion from mIgG2a$^+$ B lymphoid tumor cells by BCDF present in several antigen-specific T helper clones. CI 110: 149-62 (1987); **Kinashi T et al** Cloning of complementary DNA encoding T cell replacing factor and identity with B cell growth factor II. N 324: 70-3 (1986); **Kishi H et al** Induction of IgG secretion in a human B cell clone with recombinant IL 2. JI 134: 3104-7 (1985); **Lau VL et al** Epstein-Barr-virus-transformed lymphoblastoid cell lines derived from patients with X-linked agammaglobulinaemia and Wiskott-Aldrich syndrome: responses to B cell growth and differentiation factors. Clin. Exp. Immunol. 75: 190-5 (1989); **Mulero J et al** Spontaneously increased B cell growth factor and B cell differentiation factor activities in the synovial fluid of patients with rheumatoid arthritis. Ann. Rheum. Dis. 48: 400-5 (1989); **Warrington RJ** The characterization of a human B cell line utilizable for the assay of B cell growth factors. JIM 100: 117-22 (1987); **Yabuhara A et al** Giant lymph node hyperplasia (Castleman's disease) with spontaneous production of high levels of B cell differentiation factor activity. Cancer 63: 260-5 (1989)

BCDF μ: *B cell differentiation factor μ* This B cell differentiation factor is found in T cell supernatants of phorbol-12-myristate 13-acetate (PMA)-induced » EL-4 cells or concanavalin A (Con A)-induced 7.1.1a cells, It induces the expression of IgM by activated B cells which is signified by the suffix μ. The factor is identical with » IL5. See also: BCDF, BCGF (B cell growth factors), and » BSF (B cell stimulating factors) for some aspects of nomenclature.

Brooks K et al Lymphokine-induced IgM secretion by clones of neoplastic B cells. N 302: 825-6 (1983); **Kikutani H et al** Effect of B cell differentiation factor (BCDF) on biosynthesis and secretion of immunoglobulin molecules in human B cell lines. JI 134: 990-5 (1985); **Nakanishi K et al** Both interleukin 2 and a second T cell derived factor in EL4-supernatant have activity as differentiation factors in IgM synthesis. JEM 160: 1605-21 (1984); **Pure E et al** Induction of B cell differentiation by T cell factors. Stimulation of IgM secretion by products of a T cell hybridoma and a T cell line. JI 127: 1953-8 (1981); **Pure E et al** T cell-derived B cell growth and differentiation factors. Dichotomy between the responsiveness of B cells from adult and neonatal mice. JEM 157: 600-12 (1983); **Vitetta ES et al** T cell-derived lymphokines that induce IgM and IgG secretion in activated murine B cells. IR 78: 137-57 (1984)

BCDF-Na1: This B cell differentiation factor secreted in relatively high amounts by the human T cell line » Na1 (TCL Na1). It is identical with » BSF 2. See also: BCDF, BCGF (B cell growth factors), and » BSF (B cell stimulating factors) for some aspects of nomenclature.

Hirano T et al Purification to homogeneity and characterization of human B cell differentiation factor (BCDF or BSFp-2). PNAS 82: 5490-4 (1985); **Shimizu K et al** Immortalization of BCDF (BCGFII)- and BCDF-producing T cells by human T cell leukemia virus (HTLV) and characterization of human BDGF (BCGFII). JI 134: 1728-33 (1985)

B cell activating factor: see: BAF, BCAF.

B cell-derived B cell growth factor: see: B-BCDF.

B cell-derived enhancing factor: abbrev. BEF. See: Interleukin B.

B cell-derived growth enhancing factor: see: BGEF.

B cell-derived growth enhancing factor 2: see: BGEF-2.

B cell-derived leukocyte (migration) inhibitory factor: abbrev. B-LIF. See: LIF (leukocyte (migration) inhibitory factor).

B cell-derived T cell growth factor: see: B-TCGF.

B cell differentiation factor α: see: BCDF α.

B cell differentiation factor ε: see: BCDF ε.

B cell differentiation factor for IgG1: see: BCDF γ:

B cell differentiation factor γ: see: BCDF γ.

B cell differentiation factor μ: see: BCDF μ.

B cell differentiation factors: see: BCDF for some aspects of nomenclature.

B cell growth factor 1: see: BCGF 1.

B cell growth factor 2: see: BCGF 2.

B cell growth factor γ: see: BCGF γ.

B cell growth factors: For some aspects of nomenclature see: BCGF (B cell growth factors) and BCDF (B cell differentiation factors).

B cell growth inhibitory factor: see: BIF.

B cell inducing factor: see: BIF.

B cell interferon: This factor is identical with » IFN-α. See also: IFN for general information about interferons.

B cell maturation factors: see: BMF.

B cell precursor growth-promoting activity: This factor is identical with » IL7.

Namen AE et al B cell precursor growth-promoting activity: purification and characterization of a growth factor active on lymphocyte precursors. JEM 167: 988-1002 (1988)

B cell progenitor kinase: abbrev. bpk. See: Immunodeficient mice, subentry xid.

B cell proliferation factor: see: BCPF.

B cell replication and maturation factor: see: BRMF.

B cell replication factors: see: BRF.

B cell stimulating factor 1: see: BSF 1.

B cell stimulating factor 2: see: BSF 2.

B cell stimulating factor p1: see: BSFp1.

B cell stimulating factor p2: see: BSFp2.

B cell stimulatory factor 2: see: BCSF 2.

BCGF: *B cell growth factor* Human » IL2 is released by T lymphocytes on stimulation with antigen or mitogen and functions as a T cell growth factor (TCGF) by inducing proliferation of activated T cells. It is generally accepted that resting or activated B cells do not respond directly to IL2 but require for their proliferation other T cell-derived lymphokines usually referred to as B cell growth factors (BCGFs). For an assay system see also: JR-2(82). In fact, clonal expansion of B lymphocytes and their differentiation into antibody-producing cells are regulated by a complex series of soluble products. Some of them are not specific for B cells, but act as competence factors or competence cofactors for different hemopoietic cell lines (see also: hematopoiesis).

The term BCGF was coined at the beginning of cytokine research to distinguish T cell-derived factors mediating the proliferation of B cells from other factors that induce B cell develop into immunoglobulin-secreting B cells (see also: BCDF, B cell differentiation factors). In many instances BCGF is an operational definition of biological activities and does not signify biochemically identified distinct factors.

Some factors that influence the growth and differentiation of B cells are » » CD23, IL1, » IL2, » IL4, » IL5, » IL6, » IFN-γ, » TNF-α and » TNF-β. In combination with » IL5 and other interleukins adrenocorticotropic hormone, ACTH (see: POMC), behaves as a late-acting B cell growth factor, thus establishing a link between stress axis and immune system.

It has been observed that differences exist for one and the same factor in different species with regard to its activity as a B cell growth factor. In analogous » bioassays murine » IL5, for example, but not human IL5, acts as a B cell growth factor.

The pleiotropic actions of many growth factors (see also: Cytokines) demonstrates that the strict distinction of differentiation-inducing and proliferation-inducing factors is not very useful. Some factors, designated » BGDF (B cell growth and differentiation factors), induce proliferation and differentiation of B cells. The term BCGF is still used, however, to signify that a newly isolated factor acts on B cells and promotes their maturation. The term is used in particular to describe less well defined activities obtained, for example, from culture supernatants or from serum. For activities that induce proliferation without differentiation see: BCPF (B cell proliferation factor).

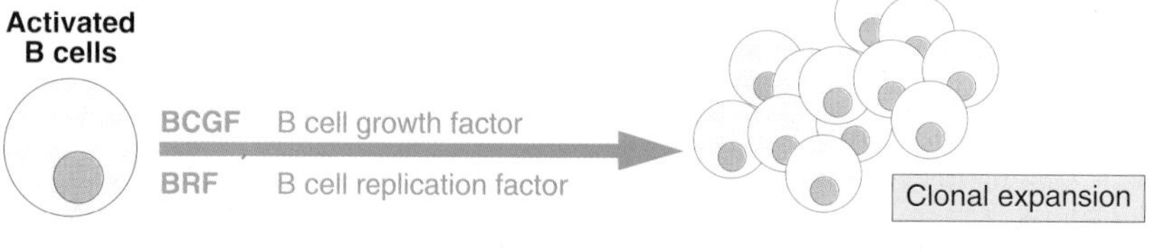

Activated
B cells

BCGF B cell growth factor

BRF B cell replication factor

Clonal expansion

B cell growth factors.
B cell growth factors promote clonal expansion of activated B cells. They do not support development into antibody-producing cells. Factors with BCGF activity have also been described as BRF (B cell replication factors). Some factors, designated BRMF (B cell replication/maturation factors) or BGDF (B cell growth/differentiation factors) promote development of activated B cells into antibody-producing cells and clonal expansion.

A frequently used assay for factors with BCGF activity is to assay the proliferative responses of T cell-depleted normal B cells activated with either anti-IgM, » phorbol esters, or *S. aureus* Cowan I (SAC) preparations. Since these cells may still contain subpopulations of B cells with different activation requirements, selected indicator cell lines are employed.

Benjamin D et al Heterogeneity of B cell growth factor receptor reactivity in healthy donors and in patients with chronic lymphatic leukemia: relationship to B cell-derived lymphokines. CI 103: 394-408 (1986); **Brooks KH** Adrenocorticotropin (ACTH) functions as a late-acting B cell growth factor and synergizes with interleukin 5. J. Mol. Cell. Immunol. 4: 327-35 (1990); **Butler JL et al** Development of a human T-T cell hybridoma secreting B cell growth factor. JEM 157: 60-8 (1983); **Callard RE** Human B cell responses to cytokines. In: Balkwill FR (edt) Cytokines – A practical approach, pp. 121-50, IRL Press, Oxford 1991; **Callard RE et al** Assay for human B cell growth and differentiation factors. in: Clemens MJ et al (eds) Lymphokines and Interferons. A practical Approach, pp. 345-64, IRL Press, Oxford 1987; **Callard RE** The marmoset B lymphoblastoid cell line (B95-8) produces and responds to B cell growth and differentiation factors: role of shed CD23 (sCD23). Immunology 65: 379-84 (1988); **Clutterbuck E et al** Recombinant human interleukin 5 is an eosinophil differentiation factor but has no activity in standard human B cell growth factor assays. EJI 17: 1743-50 (1987); **Defrance T et al** Human interferon-γ acts as a B cell growth factor in the anti-IgM antibody co-stimulatory assay but has no direct B cell differentiation activity. JI 137: 3861-7 (1986); **Howard M et al** B cell growth and differentiation factors. IR 78: 185-210 (1984); **Kishimoto T** Factors affecting B cell growth and differentiation. ARI 3: 133-57 (1985); **Kumar A et al** Human BCGF-12kD functions as an autocrine growth factor in transformed B cells. ECN 1: 109-13 (1990); **Leanderson T et al** B cell growth factors: distinction from T cell growth factor and B cell maturation factor. PNAS 79: 7455-9 (1982); **Mehta SR et al** Purification of human B cell growth factor. JI 135: 3298-302 (1985); **Mingari MC et al** B cell growth factor activity of immunoaffinity-purified and recombinant human interleukin 2. EJI 15: 193-6 (1985); **Mingari MC et al** Heterogeneity of B cell growth factor (BCGF)-producing T cells in humans. Clonal analysis of BCGF-producing cells within T4$^+$ and T8$^+$ subsets and evidence for the involvement of different growth factors in different BCGF assays. Ric. Clin. Lab. 16: 23-8 (1986); **O'Garra A et al** B cell factors are pleiotropic. IT 9: 45-54 (1988); **Okada M et al** B cell growth factors and B cell differentiation factor from human T hybridomas. Two distinct kinds of B cell growth factor and their synergism in B cell proliferation. JEM 157: 583-90 (1983); **Romagnani S** T cell-derived B cell growth and differentiation factors. Ric. Clin. Lab. 17: 181-97 (1987); **Sekhsaria S & Sahasrabuddhe CG** Structural homology of interferon γ with B cell growth factor and its proliferative effect on long term B cell lines. LR 8: 47-57 (1989); **Sharma S et al** Molecular cloning and expression of a human B cell growth factor gene in *Escherichia coli*. S 235: 1489-92 (1987); **Shields JG et al** The response of selected human B cell lines to B cell growth and differentiation factors. EJI 17: 535-40 (1987); **Storek J et al** A novel B cell stimulation/proliferation assay using simultaneous flow cytometric detection of cell surface markers and DNA content. JIM 151: 261-7 (1992); **Swain SL et al** Evidence for two distinct classes of murine B cell growth factors with activities in different functional assays. JEM 158: 822-35 (1983); **Tosato G et al** Monocyte-derived human B cell growth factor identified as interferon-β 2 (BSF-2, IL6). S 239: 502-4 (1988); **Tosato G & Pike SE** A monocyte-derived B cell growth factor is IFN-β/BSF-2/IL6. ANY 557: 181-90 (1989); **Tucci A et al** Effects of eleven cytokines and of IL1 and tumor necrosis factor inhibitors in a human B cell assay. JI 148: 2778-84 (1992); **Yoshizaki K et al** Characterization of human B cell growth factor (BCGF) from cloned T cells or mitogen-stimulated T cells. JI 130: 1241-6 (1983)

BCGF: *B cell growth factor* This particular B cell growth factor, subsequently renamed » BCGF-1, is identical with » IL4. For some aspects of nomenclature see also: BCGF (B cell growth factors) and BCDF (B cell differentiation factors).

Swain SL et al Evidence for two distinct classes of murine B cell growth factors with activities in different functional assays. JEM 158: 822-35 (1983)

BCGF 1: *B cell growth factor 1* This B cell growth factor, also known as 20 K-BCGF, » EL4-BCGF, » BSF-1 (B cell stimulating factor 1), (B cell stimulating factor 1), or BRF (B cell replication factor) was originally identified in supernatants of the » EL4 thymoma cell line. The factor was then renamed as » BSF-p1. It is identical with » IL4. For some aspects of nomenclature see also: BCGF (B cell growth factors) and BCDF (B cell differentiation factors).

Farrar J et al Biochemical and biophysical characterization of mouse B cell growth factor: a lymphokine distinct from interleukin 2. JI 131: 1838-42 (1983); **Howard M et al** Interleukin 2 induces antigen-reactive T cell lines to secrete BCGF-1. JEM 158: 2024-39 (1983)

BCGF 2: *B cell growth factor 2* This B cell growth factor causes proliferation of » BCL_1 cells. It is identical with » EDF (eosinophil differentiation factor) and one of the factors described as » TRF (T cell replacing factor). It has also been described as » BGDF (B cell growth and differentiation factor). All factors are now known as » IL5. Another factor described in the older literature, also designated BCGF 2, is not identical with IL5 but identical with » IL4 (see also: IgG1 induction factor, EL 4-BCGF, and B151-TRF). The factor secreted by WEHI cells (WEHI-231 BCGF 2) is probably also identical with IL4.

The BCGF 2 that serves as an » autocrine T cell growth factor for murine D10 cells is probably identical with the activity found in supernatants of D 10 cells (BCGF 2B) and is probably identical with » IL4. The BCGF 2 activity isolated from stimulated human peripheral blood T lymphocytes that has been named BCGF 2A (human BCGF) is probably » IL5.

For some aspects of nomenclature see also: BCGF

(B cell growth factors) and BCDF (B cell differentiation factors).

Campbell HD et al Isolation, structure, and expression of cDNA and genomic clones for murine eosinophil differentiation factor. EJB 174: 345-52 (1988); **Ennist DL et al** Activity of a partially purified human BCGF on murine assays for B cell stimulatory factors. I. BCGF II-like activity of human BCGF. CI 110: 77-94 (1987); **Ennist DL et al** Activity of a partially purified human BCGF on murine assays for B cell stimulatory factors. II. Synergy between two distinct BCGF II-like factors in promoting B cell differentiation. JI 139: 1525-31 (1987); **Hara Y et al** LR 4: 239-43 (1985); **Harada N et al** BCGFII activity on activated B cells of a purified murine T cell replacing factor (TRF) from a T cell hybridoma (B151k12). JI 134: 3944-51 (1985); **Harada N et al** Production of a monoclonal antibody useful in the molecular characterization of murine T cell-replacing factor/B cell growth factor II. PNAS 84: 4581-5 (1987); **Kinashi T et al** Cloning of a complementary DNA encoding T cell replacing factor and identity with B cell growth factor II. N 324: 70-3 (1986); **Kupper T et al** Autocrine growth of T cells independent of interleukin 2: identification of interleukin 4 (IL 4, BSF-1) as an autocrine growth factor for a cloned antigen-specific helper T cell. JI 138: 4280-7 (1987); **McKenzie D et al** Purification and partial amino acid sequence of murine B cell growth factor II (BCGFII). JI 139: 2661-8 (1987); **Nakajima K et al** Physicochemical and functional properties of murine B cell-derived B cell growth factor II (WEHI-231-BCGF-II). JI 135: 1207-12 (1985); **Sanderson CJ et al** The production of lymphokines by primary alloreactive T cell clones: a co-ordinate analysis of 233 clones in seven lymphokine assays. Immunology 56: 575-84 (1985); **Sanderson CJ et al** Eosinophil differentiation factor also has B cell growth factor activity: proposed name interleukin 4. PNAS 83: 437-40 (1986); **Swain SL & Dutton RW** Production of a B cell growth promoting activity (DL) BCGF, from a cloned T cell line and its assay on the BCL 1 B cell tumor. JEM 156: 1821-34 (1982); **Swain SL et al** Evidence for two distinct classes of murine B cell growth factors with activities in different functional assays. JEM 158: 822-35 (1983); **Yokota T et al** Isolation and characterization of lymphokine cDNA clones encoding mouse and human IgA-enhancing factor and eosinophil colony-stimulating factor activities: relationship to interleukin 5. PNAS 84: 7388-92 (1987)

BCGF-12kD: *B cell growth factor 12 kDa* see: LMW-BCGF (low molecular weight B cell growth factor).

BCGF costimulation assay: see: BCGF (B cell growth factors).

BCGFγ: *B cell growth factor γ* This B cell differentiation factor induces the expression of IgG by activated B cells which is signified by the suffix γ. The factor is identical with » IL4. See also: BCDF γ (B cell differentiation factor γ, B cell differentiation factor for IgG1). See also: BCGF and BCDF (B cell differentiation factors) for some aspects of nomenclature.

Sideras P et al Partial biochemical characterization of IgG1-inducing factor. EJI 15: 593-8 (1985); **Vitetta E et al** Serological, biochemical, and functional identity of B cell-stimulatory factor 1 and B cell differentiation factor for IgG1. JEM 162: 1726-31 (1985); **Vitetta ES et al** T cell-derived lymphokines that induce IgM and IgG secretion in activated murine B cells. IR 78: 137-57 (1984)

BCGF-H: see: HMW-BCGF (high molecular weight B cell growth factor).

$BCGF_{high}$: see: HMW-BCGF (high molecular weight B cell growth factor).

BCGF-L: see: LMW-BCGF (low molecular weight B cell growth factor).

$BCGF_{low}$: see: LMW-BCGF (low molecular weight B cell growth factor).

BCL_1: A cell line established from a spontaneous lymphoma obtained from a female BALB/c mouse. This murine B lineage leukemia has many features of human CLL (chronic lymphocytic leukemia). Cells are passaged as spleenic tumors *in vivo* and isolated from the tumor by centrifugation in Ficoll gradients and depletion of T cells with monoclonal antibodies.

In vitro the cell line can be induced by bacterial » endotoxins to produce IgM. It expressed IgM and also IgD on its cell surface. The growth of this cell line depends upon » IL5 and the cells are therefore used to assay this factor (see also: Bioassays). IL5 activity is determined by measuring the incorporation of ^{3}H thymidine into the newly synthesized DNA of the proliferating factor-dependent cells. Cell proliferation can be determined also by employing the » MTT assay. An alternative and entirely different method of detecting IL5 is » Message amplification phenotyping.

The lymphokines » IL1, » IL2, » IL3, and » IL6 have no effect on the growth of BCL_1. BCL_1 cells are ≈ 1000-fold less sensitive for human IL5. The human factor is therefore detected by employing the » TF-1 cell line. BCL_1 cells also respond to murine » IL4 and » GM-CSF. Anti-GM-CSF antibodies inhibit the proliferation of BCL1 cells induced by IL4 and IL5, and GM-CSF. Anti-IL6 antibodies is also a very effective inhibitor of BCL1 proliferation induced by either IL4 or IL5 but not by GM-CSF. The IL5-induced proliferation of the cells is abrogated by » TGF-β and to a lesser extent by » IFN-γ. BCL_1 cells also constitutively secrete » IL10.

Brooks KH et al Cell cycle-related expression of the receptor for a B cell differentiation factor. JI 134: 742-7 (1985); **Chelstrom LM et al** Treatment of BCL-1 murine B cell leukemia with recombinant cytokines. Comparative analysis of the antileukemic potential of interleukin 1 β (IL1 β), interleukin 2 (IL2), interleukin-6 (IL6), tumor necrosis factor α(TNF α), granulocyte colony stimulating factor (G-CSF), granulocyte-macrophage colony stimulating factor (GM-CSF), and their combination. Leuk. Lymphoma. 7: 79-86 (1992); **Clutterbuck EJ et al** Recombinant human interleukin 5 is an eosinophil differentiation factor but has no activity in standard human B cell growth factor assays. EJI 17: 1743-50 (1987); **Mosman TR & Fong TAT** Specific assays for cytokine production by T cells JIM 116: 151-8 (1989); **Ni K & O'Neill HC** Proliferation of the BCL1 B cell lymphoma induced by IL4 and IL5 is dependent on IL6 and GM-CSF. Immunol. Cell. Biol. 70: 315-22 (1992); **O'Garra A et al** The BCL1 B lymphoma responds to IL4, IL5, and GM-CSF. CI 123: 189-200 (1989); **O'Garra A et al** Ly-1 B (B-1) cells are the main source of B cell-derived interleukin 10. EJI 22: 711-7 (1992); **Slavin S & Strober S** Spontaneous murine B cell leukemia. N 272: 624-6 (1977); **Swain SL & Dutton RW** Production of a B cell growth promoting activity (DL) BCGF, from a cloned T cell line and its assay on the BCL 1 B cell tumor. JEM 156: 1821-34 (1982)

***bcl-2*:** (B cell lymphoma/leukemia 2) An » oncogene located on human chromosome 18q21. The gene contains three exons and extends over at least 370 kb. It codes for a protein of 239 amino acids that is located in the inner mitochondrial membrane. A shorter form of *bcl*-2 lacking the hydrophobic transmembrane domain within its carboxyl terminus responsible for anchoring the protein in the mitochondrial membrane can be obtained by alternative splicing of the *bcl*-2 gene. This protein does not block apoptosis. *bcl*-2 has also been localized to other membranes, including the endoplasmic reticulum and nuclear membrane.

The overexpression of *bcl*-2 prevents B and T cell death by » apoptosis and counteracts apoptotic mechanisms elicited by cross-linking of the » Apo-1 cell surface protein. Human *bcl*-2 has been shown to reduce the number of programmed cell deaths in the nematode *Caenorhabditis elegans*, suggesting that the mechanisms of apoptosis are the same in humans and in the nematode.

The induction of anti-Apo-1-mediated apoptosis appears to be correlated with *bcl*-2 mRNA down-regulation. The *bcl*-2 proto-oncogene is activated by translocation in a variety of B lymphoid tumors and synergizes with the » *myc* oncogene in tumor progression. The majority of follicular lymphoma cells carry a typical chromosome translocation 14;18, which juxtaposes the *bcl*-2 gene to the immunoglobulin heavy-chain (IgH) gene and leads to the deregulated expression of an unaltered *bcl*-2 gene product. Variant translocations of the *bcl*-2 gene to the Ig λ or Ig κ gene have been found in ≈ 10% of chronic lymphocytic leukemia.

bcl-2-mediated inducible cell survival may be a regulatory process in normal homeostasis and morphogenesis in many fetal tissues and structures. Enforced expression of *bcl*-2, for example, can prevent apoptotic cell death induced by serum or growth factor withdrawal but it does not do so in targets of cell-mediated killing. *bcl*-2 prevents apoptosis of hematopoietic cell lines dependent on » IL3, » IL4, or » GM-CSF but fails to prevent apoptosis in other cell lines following » IL2 or » IL6 deprivation. *bcl*-2 also fails to prevent antigen receptor-induced apoptosis in some B cell lines. *bcl*-2 confers growth and survival advantage to» IL7-dependent early pre-B cells which become factor independent by a multistep process in culture. Sensory neurons that depend for survival on one or more neurotrophic factors (» NGF, » BDNF, and » NT-3) are rescued by *bcl*-2, whereas » CNTF-dependent ciliary neurons are not. The failure of *bcl*-2 to prevent all types of apoptotic deaths suggests the existence of multiple independent intracellular mechanisms of apoptosis, some of which are *bcl*-2-independent.

The existence of other genes regulating *bcl*-2 functions has recently been described. ***Bax***, a protein of 21 kDa which has extensive amino acid homology with *bcl*-2 in some protein domains homodimerizes and forms heterodimers with *bcl*-2. Overexpression of Bax accelerates apoptotic death induced by cytokine deprivation in an IL3-dependent cell line and counters the repressor activities of *bcl*-2. It appears that the ratio of *bcl*-2 to Bax determines survival or death following an apoptotic stimulus. ***bcl-x*** is a *bcl*-2-related gene that functions as a dominant regulator of apoptotic cell death. *bcl*-x inhibits cell death almost as well as *bcl*-2 in some instances. A shorter splice variant of *bcl*-x appears to encode a protein that inhibits the ability of *bcl*-2 to enhance the survival of growth factor-deprived cells. While the longer form of *bcl*-x is expressed in tissues containing long-lived postmitotic cells the shorter form appears to predominate in tissues that undergo a high rate of turnover. Epstein-Barr virus has recently been shown to contain an open reading frame that appears to be the viral functional homologue of *bcl*-2 (see: BHRF1).

■ **TRANSGENIC/KNOCK-OUT/ANTISENSE STUDIES:** Transgenic mice overexpressing the *bcl*-2 protein are characterized by a long-lasting persistence of immunoglobulin-secreting cells and an increased life span of memory B cells, a finding which implicates the *bcl*-2 protein in regulating the life span of maturing thymocytes and in the antigenic selection process. An overexpression of this oncogene is seen in the majority of human follicular B cell lymphomas with a t(14q32;18q21) translocation. These tumors are also observed in » transgenic animals overexpressing the oncogene in their lymphocytes.

Some human disorders associated with extensive cell death such as neurodegenerative diseases, may be associated with functional disturbances in genes such as *bcl*-2. The overexpression of *bcl*-2 may be one factor leading to prolonged survival of acute lymphoblastic leukemia cells. *bcl*-2 overexpression has also been reported to block chemotherapy-induced apoptosis in a human leukemia cell line, thus creating an opportunity for re-initiation of cell growth upon drug withdrawal.

IL4 may play an essential role in the pathogenesis of chronic lymphocytic leukemia disease, by preventing both the death and the proliferation of the malignant B cells. It protects chronic lymphocytic leukemic B cells from death by apoptosis and upregulates *bcl*-2 expression.

Transgenic mice in which expression of the *bcl*-2 gene has been disrupted by targeted inactivation of the *bcl*-2 gene in » ES cells have been generated to study the function of *bcl*-2. *bcl*-2-deficient mice complete their embryonic development but are growth-retarded and die early. Thymus and spleen undergo massive apoptotic involution. *bcl*-2-negative mice initially show a normal differentiation of lymphocytes into phenotypically mature cells. T and B cells that do not express *bcl*-2 disappear from the bone marrow, thymus, and periphery by 4 weeks of age, indicating that the gene product is dispensible for lymphocyte maturation but is required for a stable immune system after birth. Mature T cells lacking *bcl*-2 have also shorter life spans *in vitro* and show increased sensitivity to glucocorticoids and γ-irradiation.

Allsopp TE et al The proto-oncogene *bcl*-2 can selectively rescue neurotrophic factor-dependent neurons from apoptosis. Cell 73: 295-307 (1993); **Baffy G et al** Apoptosis induced by withdrawal of interleukin-3 (IL3) from an IL3-dependent hematopoietic cell line is associated with repartitioning of intracellular calcium and is blocked by enforced *bcl*-2 oncoprotein production. JBC 268: 6511-9 (1993); **Bagg A & Cossman J** *bcl*-2: physiology and role in neoplasia. Cancer Treat. Rep. 63: 141-66 (1992); **Boise LH et al** *bcl*-x, a *bcl*-2-related gene that functions as a dominant regulator of apoptotic cell death. Cell 74: 597-608 (1993); **Borzillo GV et al** *bcl*-2 confers growth and survival advantage to interleukin 7-dependent early pre-B cells which become factor independent by a multistep process in culture. O 7: 869-76 (1992); **Campana D et al** Prolonged survival of B lineage acute lymphoblastic leukemia cells is accompanied by overexpression of *bcl*-2 protein. Blood 81: 1025-31 (1993); **Cleary ML et al** Cloning and structural analysis of cDNAs for *bcl*-2 and a hybrid *bcl*-2/immunoglobulin transcript resulting from the 5(14;18) translocation. Cell 47: 19-28 (1986); **Elefanty AG et al** *bcr-abl*, the hallmark of chronic myeloid leukemia in man, induces multiple hematopoietic neoplasms in mice. EJ 9: 1069-78 (1990); **Garcia I et al** Prevention of programmed cell death of sympathetic neurons by the *bcl*-2 proto-oncogene. S 258: 302-4 (1992); **Gribben JG et al** Bone marrows of non-Hodgkin's lymphoma patients with a *bcl*-2 translocation can be purged of polymerase chain reaction-detectable lymphoma cells using monoclonal antibodies and immunomagnetic bead depletion. Blood 80: 1083-9 (1992); **Hengartner MO et al** *Caenorhabditis elegans* gene ced-9 protects cells from programmed cell death. N 356: 494-9 (1992); **Hockenberry DM** The *bcl*-2 oncogene and apoptosis. Semin. Immunol. 4: 413-20 (1992); **Jacobson MD et al** *bcl*-2 blocks apoptosis in cells lacking mitochondrial DNA. N 361: 365-9 (1993); **Korsmeyer SJ** *bcl*-2: an antidote to programmed cell death. Cancer Surv. 15: 105-18 (1992); **LeBrun DP et al** Expression of *bcl*-2 in fetal tissues suggests a role in morphogenesis. AM. J. Pathol. 142: 743-53 (1993); **Mah SP et al** The proto-oncogene *bcl*-2 inhibits apoptosis in PC12 cells. J. Neurochem. 60: 1183-6 (1993); **Mapara MV et al** APO-1 mediated apoptosis or proliferation in human chronic B lymphocytic leukemia: correlation with *bcl*-2 oncogene expression. EJI 23: 702-8 (1993); **Monaghan P et al** Ultrastructural localization of the *bcl*-2 protein. J. Histochem. Cytochem. 40: 1819-25 (1992); **Nunez G et al** *bcl*-2 maintains B cell memory. N 353: 71-4 (1991); **Miyashita T & Reed JC** *bcl*-2 oncoprotein blocks chemotherapy-induced apoptosis in a human leukemia cell line. Blood 81: 151-7 (1993); **Oltvai ZN et al** *bcl*-2 heterodimerizes *in vivo* with a conserved homolog, Bax, that accelerates programmed cell death. Cell 74: 609-19 (1993); **Perlmutter RM & Ziegler S** Proto-oncogene expression following lymphocyte activation. Curr. Top. Membr. Transp. 35: 571-86 (1990); **Reed JC** *Bcl*-2 and the regulation of programmed cell death. JCB 124: 1-6 (1994); **Sentman CL et al** *bcl*-2 inhibits multiple forms of apoptosis but not negative selection in thymocytes. Cell 67: 879-88 (1991); **Tanaka S et al** Structure-function analysis of the *bcl*-2 oncoprotein. Addition of a heterologous transmembrane domain to portions of the *bcl*-2 β protein restores function as a regulator of cell survival. JBC 268: 10920-6 (1993); **Tsujimoto Y et al** Characterization of the protein product of *bcl*-2, the gene involved in human follicular lymphoma. O 2: 3-7 (1987); **Tsujimoto Y et al** Cloning of the chromosome breakpoint of neoplastic B cells with the t(14;18) chromosome translocation. S 226: 1097-99 (1984); **Vaux DL et al** *bcl*-2 gene promotes hemopoietic cell survival and co-operates with c-*myc* to immortalize pre B cells. N 335: 440-2 (1988); **Vaux DL et al** *bcl*-2 prevents death of factor-deprived cells but fails to prevent apoptosis in targets of cell mediated killing. Int. Immunol. 4: 821-4 (1992); **Vaux DL et al** Prevention of programmed cell death in Caenorhabditis elegans by human *bcl*-2. S 258: 1955-7 (1992)

● **Transgenic/Knock-out/Antisense Studies: Katsumata M et al** Differential effects of *bcl*-2 on T and B cells in transgenic mice. PNAS 89: 11376-80 (1992); **McDonnell TJ et al** *bcl*-2-immunoglobulin transgenic mice demonstrate extended B cell survival and follicular lymphoproliferation. Cell 57: 79-88 (1989); **McDonnell TJ et al** Deregulated *bcl*-2-immunoglobulin transgene expands a resting but responsive immunoglobulin M and D-expressing B cell population. MCB 10: 1901-7 (1990); **Nakayama KI et al** Disappearance of the lymphoid system in *bcl*-2 homozygous mutant chimeric mice. S 261: 1584-8 (1993); **Siegel RM et al** Inhibition of thymocyte apoptosis and negative antigenic selection in *bcl*-2 transgenic mice. PNAS 89: 7003-7 (1992); **Strasser A et al** Abnormalities of the immune system induced by dysregulated *bcl*-2 expression in transgenic mice. CTMI 166: 175-81 (1990); **Strasser A et al** Novel primitive lymphoid tumors induced in transgenic mice by cooperation between *myc* and *bcl*-2. N 348: 331-3 (1990); **Strasser A et al** *bcl*-2 transgene inhibits T cell death and perturbs thymic self-censorship. Cell 67: 889-99 (1991); **Veis DJ et al** *bcl*-2-deficient mice demonstrate fulminant lymphoid apoptosis, polycystic kidneys, and hypopigmented hair. Cell 75: 229-40 (1993)

bcl-x: see: *bcl*-2.

BCPF: *B cell proliferation factor* This term refers to poorly characterized factors active on B cells initially found to be secreted by human T cell hybridomas. BCPF acts on B cells to induce proliferation without differentiation and is distinct from conventional B cell growth factors (see also: BCGF). BCPF acts like an antigen *in vivo* or like anti-μ *in vitro*, pre-activating B cells and rendering them responsive to B cell growth factors. BCPF induces cells to proliferate without pre-activation (see also: Cell activation).

Mayer L et al Regulation of B cell activation and differentiation with factors generated by human T cell hybridomas. IR 78: 119-35 (1984); **Mayer L** Terminal maturation of resting B cells by proliferation-independent B cells differentiation factors. JEM 164: 383-92 (1986); **Stohl W & Mayer L** Inhibition of T cell-dependent human B cell proliferation and B cell differentiation by polyspecific monomeric IgG. Clin. Exp. Immunol. 70: 649-57 (1987)

BCRF-1: abbrev. for *Bam* HI C fragment rightward reading frame. This is a gene found in the genome of Epstein-Barr virus. It encodes a protein of 145 amino acids that specifically inhibits the synthesis of » IFN-γ. BCRF-1 closely resembles » IL10 and is therefore also called ***viral IL10*** (***vIL10***).

Baer R et al DNA sequence and expression of the B95-8 Epstein-Barr virus genome. N 310: 207-11 (1984); **Hudson GS et al** The short unique region of the B95-8 Epstein-Barr virus genome. Virology 147: 81-98 (1985); **Moore KW et al** Homology of cytokine synthesis inhibitory factor (IL10) with the Epstein-Barr virus gene BCRF1. S 248: 1230-4 (1990), [erratum in S 250: 494]

BCSF: *B cell stimulating factor* see: BSF (B cell stimulating factor).

BDF: *B cell differentiation factor* This factor is identical with » BAF (B cell activating factor). It is therefore identical with » IL1. For some aspects of nomenclature see also: BCDF (B cell differentiation factors).

Aarden LA et al Revised nomenclature for antigen-nonspecific T cell proliferation and helper factors. CI 48: 433-36 (1979); **Hoffmann MK et al** Macrophage factor controlling differentiation of B cells. JI 122: 497-501 (1979)

BDGF: *brain-derived growth factor* The biological activities of these factors resemble that of » aFGF isolated from brain, » ECGF (endothelial cell growth factor), and some growth factors isolated from retina (see: RGDF, retina-derived growth factor; EDGF, eye-derived growth factor) hypothalamus (see: HDGF, hypothalamus-derived growth factor), and pituitary (see: Pituitary FGF).

Purification of BDGF by heparin-sepharose affinity chromatography (see also: HBGF, heparin binding growth factor) yields two forms, designated ***BDGF-A*** and ***BDGF-B***, which are mitogenic for many different cell types, including endothelial cells, chondrocytes, osteoblasts, epithelial cells, glial cells, fibroblasts, smooth muscle cell. The two factors are also chemotactic for fibroblasts and astroglial cells.

In general, many of the factors described in the older literature, either called BDGF according to the source, and also BDGF-A or BDGF-B, are most likely identical with » aFGF or » bFGF.

Courty J et al [*In vitro* properties of various growth factors and *in vivo* effects] Biochimie 66: 419-28 (1984); **Huang JS et al** Bovine brain-derived growth factor. Purification and characterization of its interaction with responsive cells. JBC 261: 11600-7 (1986); **Huang SS et al** Transforming growth factor activity of bovine brain-derived growth factor. BBRC 139: 619-25 (1986); **Jaye M et al** Human endothelial cell growth factor: cloning, nucleotide sequence, and chromosomal localization. S 233: 541-5 (1986); **Senior RM et al** Brain-derived growth factor is a chemoattractant for fibroblasts and astroglial cells. BBRC 141: 67-72 (1986)

BDGF: *bone derived growth factor(s)* Generic name given to growth factors isolated from bone and cartilage which stimulate the growth of bone and cartilage cells. See also individual entries; see also: BMP (bone morphogenetic protein).

Hauschka PV et al Growth factors in bone matrix. Isolation of multiple types by affinity chromatography on heparin-sepharose. JBC 261: 12665-74 (1986); **Hauschka PV et al** Polypeptide growth factors in bone matrix. Ciba Found. Symp. 136: 207-25 (1988); **Watrous DA & Andrews BS** The metabolism and immunology of bone. Semin. Arthritis Rheum. 19: 45-65 (1989)

BDGF 0.45: *bone-derived growth factor 0.45* This growth factor, isolated from demineralized bone material, is probably identical with » PDGF. The number refers to the molarity of the NaCl solution used to elute the factor from heparin-sepharose affinity columns (see also: HBGF, heparin-binding growth factor).

Hauschka PV et al Growth factors in bone matrix. Isolation of multiple types by affinity chromatography on heparin-sepharose. JBC 261: 12665-74 (1986)

BDGF-1: *bone-derived growth factor 1* This poorly characterized growth factor of ≈ 20-30 kDa

is isolated from chondrocyte cultures. It stimulates the proliferation of bone and cartilage cells.
Kato Y et al Effect of bone-derived growth factor on DNA, RNA, and proteoglycan synthesis in cultures of rabbit costal chondrocytes. Metabolism 31: 812-5 (1982)

BDGF 1.1: *bone-derived growth factor 1.1* This growth factor, isolated from demineralized bone material, is identical with » bFGF. The number refers to the molarity of the NaCl solution used to elute the factor from heparin-sepharose affinity columns (see also: HBGF, heparin-binding growth factor).
Hauschka PV et al Growth factors in bone matrix. Isolation of multiple types by affinity chromatography on heparin-sepharose. JBC 261: 12665-74 (1986)

BDGF 1.7: *bone-derived growth factor 1.7* his growth factor, isolated from demineralized bone material, is identical with » bFGF. The number refers to the molarity of the NaCl solution used to elute the factor from heparin-sepharose affinity columns (see also: HBGF, heparin-binding growth factor).
Hauschka PV et al Growth factors in bone matrix. Isolation of multiple types by affinity chromatography on heparin-sepharose. JBC 261: 12665-74 (1986)

BDGF 2: *bone-derived growth factor 2* This factor of 11.6 kDa was initially isolated from bone material and found to induce collagen synthesis and DNA synthesis in bone cells. The factor is synthesized by osteoblasts and its production is enhanced by » IFN-γ. The factor is identical with the plasma protein » $Beta_2$ microglobulin.
Canalis E et al Stimulation of DNA and collagen synthesis by an autologous growth factor in cultured fetal rat calvaria. S 210: 1021-3 (1980); **Canalis E et al** A bone-derived growth factor isolated from rat calvariae is β_2-microglobulin. E 121: 1198-1200 (1987); **Canalis E & Centrella M** Isolation of a non-transforming bone-derived growth factor from medium conditioned by fetal rat calvariae. E 118: 2002-8 (1986); **Evans DB et al** Immunoreactivity and proliferative actions of β 2 microglobulin on human bone-derived cells *in vitro*. BBRC 175: 795-803 (1991); **Kato Y et al** Effect of bone-derived growth factor on DNA, RNA, and proteoglycan synthesis in cultures of rabbit costal chondrocytes. Metabolism 31: 812-5 (1982)

BDGF-A: *brain-derived growth factor A* see: BDGF.

BDGF-B: *brain-derived growth factor B* see: BDGF.

BDNF: *brain-derived neurotrophic factor*
■ **Sources:** BDNF is found in neurons of the central nervous system. It is expressed predominantly in hippocampus, cortex, and synapses of the basal forebrain. The synthesis of BDNF is subject to regulation by neuronal activity and specific transmitter systems. BDNF expression is also switched on in Schwann cells following peripheral nerve lesion. BDNF is also expressed in muscles and its expression is upregulated in denervated muscles.
■ **Protein Characteristics:** BDNF is a basic protein (pI = 9,99) of 252 amino acids that is synthesized as a precursor with a 18 amino acid hydrophobic » signal sequence and a prosequence of 112 amino acids. The proteins isolated from various mammals are almost identical in their sequences and also display a conserved tissue distribution. Some protein domains of BDNF are identical with those of » NGF and another neurotrophic factor, designated NT-3 (neurotrophin 3); polyclonal antibodies raised against murine NGF cross-react with both other proteins and totally or partially block the bioactivities of the other proteins. At the protein level these three proteins show 50 % total homology. The variable domains are believed to control the specificity of expression in certain types of neurons.
■ **Gene Structure:** The human BDNF gene is located on chromosome 11p15.5-p11.2 between the loci FSHB and HVBS1 (which covers a region of ≈ 4 Mb). The murine gene is located on chromosome 2.
■ **Receptor Structure, Gene(s), Expression:** The biological activity of BDNF is mediated by a receptor that belongs to the » *trk* family of receptors encoding a tyrosine-specific protein kinase. BDNF only binds weakly to the $gp140^{trk}$ receptor (to which NGF binds with high affinity), and it binds to the NGF receptor known as LNGFR (see: NGF). It has been possible, by combination of structural elements from » NGF, BDNF and » NT-3, to engineer a multifunctional pan-neurotrophin that efficiently activates all » *trk* receptors and displays multiple neurotrophic specificities.
■ **Biological Activities:** BDNF selectively supports the survival of primary sensory neurons and retinal ganglia. The factor supports survival and differentiation of certain cholinergic neurons and also some dopaminergic neurons *in vitro*. BDNF has been shown recently to prevent death of cultured embryonic rat spinal motor neurons at picomolar concentrations and to rescue substantial numbers of motor neurons that would otherwise die after lesioning of the neonatal sciatic or facial nerve. BDNF inhibits the normal cell death of embryonic chick motor neurons.

BDNF does not appear to act on sympathetic ganglia. In specific neurons of the central nervous system located in the hippocampus and the cortex the synthesis of BDNF is influenced by neuronal activity either positively (glutamate transmitter system) or negatively (GABA transmitter system). BDNF has been shown to rapidly potentiate the spontaneous and impulse-evoked synaptic activity of developing neuromuscular synapses in culture and thus appears to be involved in the regulation of functions of developing synapses.

The biological activities of BDNF and NT-3 (neurotrophin-3) are additive, and BDNF also interacts with » LIF.

The tissue distributions and also the neuronal specificities of BDNF, NGF and NT-3 differ markedly from each other. BDNF, and also NT-3 act on some cells that do not respond to NGF. It has been possible to create chimaeric proteins consisting of various BDNF and NGF domains and these hybrid proteins have been found to show simultaneous some of the BDNF/NGF activities.

Glucocorticoid hormones, which are important regulators of brain development and aging and which can impair the capacity of hippocampal neurons to survive various neurological insults have been shown to prevent activity-dependent increases of BDNF mRNA in cultures of rat hippocampal neurons. It has been suggested, therefore, that the known ability of glucocorticoids to exacerbate neuronal injury following ischemia and other metabolic insults may be due to antagonism of regulatory mechanisms governing levels of BDNF and other » neurotrophins in the brain.

■ **TRANSGENIC/KNOCK-OUT/ANTISENSE STUDIES:** A targeted deletion of the BDNF gene has been generated by homologous recombination in » ES cells. Mice lacking BDNF expression have severe deficiencies in several sensory ganglia including the vestibular ganglion. The few remaining vestibular axons fail to contact the vestibular sensory epithelia. They terminate in the adjacent connective tissue. Survival of sympathetic, midbrain dopaminergic and motor neurons is not affected.

■ **ASSAYS:** Specific ELISA assays are available. The bioactivity of BDNF can be monitored by a survival assay of embryonic chicken sensory and sympathetic neurons.

■ **CLINICAL USE & SIGNIFICANCE:** Since BDNF supports the survival of sensory neurons, retinal ganglion cells, basal forebrain cholinergic neurons, and mesencephalic dopaminergic neurons *in vitro* it may be of use in local treatment of nerve degeneration disorders such as Parkinson's disease. The effects of BDNF on motor neurons raise the possibility that it may be useful in treating patients with motor neuropathies and amyotrophic lateral sclerosis.

● **REVIEWS: Barde YA** The nerve growth factor family. PGFR 2: 237-48 (1990); **Thoenen H** The changing scene of neurotrophic factors. TINS 14: 165-70 (1991); see also literature cited for » neurotrophins.

● **BIOCHEMISTRY & MOLECULAR BIOLOGY: Acklin C et al** Recombinant human brain-derived neurotrophic factor (rHuBDNF). Disulfide structure and characterization of BDNF expressed in CHO cells. Int. J. Pept. Protein Res. 41: 548-52 (1993); **Barde YA et al** Purification of a new neurotrophic factor from mammalian brain. EJ 1: 549-53 (1982); **Götz R et al** Brain-derived neurotrophic factor is more highly conserved in structure and function than nerve growth factor during vertebrate evolution. J. Neurochem. 59: 432-42 (1992); **Hamel W et al** Neurotrophin gene expression by cell lines derived from human gliomas. J. Neurosci. Res. 34: 147-57 (1993); **Hanson IM et al** The human BDNF gene maps between FSHB and HVBS1 at the boundary of 11p13-p14. Genomics 13: 1331-3 (1992); **Hofer M et al** Regional distribution of brain-derived neurotrophic factor mRNA in the adult mouse brain. EJ 9: 2459-64 (1990); **Ibáñez CF et al** Chimaeric molecules with multiple neurotrophic activities reveal structural elements determining the specificities of NGF and BDNF. EJ 10: 2105-10 (1991); **Isackson PJ et al** Comparison of mammalian, chicken and *Xenopus* brain-derived neurotrophic factor coding sequences. FL 285: 260-4 (1991); **Jones KR & Reichardt LF** Molecular cloning of a human gene that is a member of the nerve growth factor family. PNAS 87: 8060-4 (1990); **Leibrock J et al** Molecular cloning and expression of brain-derived neurotrophic factor. N 341: 149-52 (1989); **Maisonpierre PC et al** Gene sequences of chicken BDNF and NT-3. DNA Seq. 3: 49-54 (1992); **Matsuoka I et al** Differential regulation of nerve growth factor and brain-derived neurotrophic factor expression in the peripheral nervous system. ANY 633: 550-2 (1991); **Özcelik T et al** Chromosomal mapping of brain-derived neurotrophic factor and neurotrophin 3. Genomics 10: 569-75 (1991); **Narhi LO et al** Comparison of the biophysical characteristics of human brain-derived neurotrophic factor, neurotrophin-3, and nerve growth factor. JBC 268: 13309-17 (1993); **Phillips HS et al** Widespread expression of BDNF but not NT-3 by target areas of basal forebrain cholinergic neurons. S 250: 290-4 (1990); **Radziejewski C et al** Dimeric structure and conformational stability of brain-derived neurotrophic factor and neurotrophin-3. B 31: 4431-6 (1992); **Rosenthal A et al** Primary structure and biological activity of human brain-derived neurotrophic factor. E 129: 1289-94 (1991); **Shintani A et al** Characterization of the 5'-flanking region of the human brain-derived neurotrophic factor gene. BBRC 182: 325-32 (1992); **Thoenen H et al** The synthesis of nerve growth factor and brain-derived neurotrophic factor in hippocampal and cortical neurons is regulated by specific transmitter systems. ANY 640: 86-90 (1991); **Timmusk T et al** Multiple promoters direct tissue-specific expression of the rat BDNF gene. Neuron 10: 475-89 (1993); **Wetmore C et al** Brain-derived neurotrophic factor: subcellular compartmentalization and interneuronal transfer as visualized with anti-peptide antibodies. PNAS 88: 9843-7 (1991); **Wetmore C et al** Localization of brain-derived neurotrophic factor mRNA to neurons in the brain by *in situ* hybridisation. Exp. Neurol. 109: 141-52 (1990); **Zafra F et al** Interplay between glutamate and γ-aminobutyric acid transmitter systems

BDNF knockouts: see addendum BDNF.

in the physiological regulation of brain-derived neurotrophic factor and nerve growth factor synthesis in hippocampal neurons. PNAS 88: 10037-41 (1991)
● RECEPTORS: **Escandon E et al** Characterization of neurotrophin receptors by affinity crosslinking. J. Neurosci. Res. 34: 601-13 (1993); **Glass DJ et al** *trk*B mediates BDNF/NT-3-dependent survival and proliferation in fibroblasts lacking the low affinity NGF receptor. Cell 66: 405-13 (1991); **Klein R et al** The *trk*B tyrosine protein kinase is a receptor for brain-derived neurotrophic factor and Neurotrophin-3. Cell 66: 395-403 (1991); **Rodriguez-Tebar A et al** Binding of neurotrophin-3 to its neuronal receptors and interactions with nerve growth factor and brain-derived neurotrophic factor. EJ 11: 917-22 (1992); **Soppet D et al** The neurotrophic factors brain-derived neurotrophic factor and neurotrophin-3 are ligands for the *trk*B tyrosine kinase receptor. Cell 65: 895-903 (1991)
● TRANSGENIC/KNOCK-OUT/ANTISENSE STUDIES: **Ernfors P et al** Mice lacking brain-derived neurotrophic factor develop with sensory deficits. Nature 368: 147-50 (1994)
● ASSAYS: **Murphy RA et al** Immunological relationships of NGF, BDNF, and NT-3: recognition and functional inhibition by antibodies to NGF. J. Neurosci. 13: 2853-62 (1993)
● BIOLOGICAL ACTIVITIES: **Alderson RF et al** Brain-derived neurotrophic factor increases survival and differentiated functions of rat septal cholinergic neurons in culture. 5: 297-306 (1990); **Ceccatelli S et al** Expanded distribution of mRNA for nerve growth factor, brain-derived neurotrophic factor, and neurotrophin 3 in the rat brain after colchicine treatment. PNAS 88: 10352-6 (1991); **Cosi C et al** Glucocorticoids depress activity-dependent expression of BDNF mRNA in hippocampal neurons. Neuroreport 4: 527-30 (1993); **Hamel W et al** Neurotrophin gene expression by cell lines derived from human gliomas. J. Neurosci. Res. 34: 147-57 (1993); **Henderson CE et al** Neurotrophins promote motor neuron survival and are present in embryonic limb bud. N 363: 266-270 (1993); **Hofer M et al** Brain-derived neurotrophic factor prevents neuronal death *in vivo*. N 331: 261-2 (1988); **Hyman C et al** BDNF is a neurotrophic factor for dopaminergic neurons of the substantia nigra. N 350: 230-2 (1991); **Knüsel B et al** Promotion of central cholinergic and dopaminergic neuron differentiation by brain-derived neurotrophic factor but not neurotrophin-3. PNAS 88: 961-5 (1991); **Koliatsos VE et al** Evidence that brain-derived neurotrophic factor is a trophic factor for motor neurons *in vivo*. Neuron 10: 359-67 (1993); **Lohof AM et al** Potentiation of developing neuromuscular synapses by the neurotrophins NT-3 and BDNF. N 363: 350-3 (1993); **Maisonpierre PC et al** NT-3, BDNF, and NGF in the developing rat nervous system: parallel as well as reciprocal patterns of expression. 5: 501-9 (1990); **Meyer M et al** Enhanced synthesis of brain-derived neurotrophic factor in the lesioned peripheral nerve: different mechanisms are responsible for the regulation of BDNF and NGF mRNA. JCB 119: 45-54 (1992); **Oppenheim RW et al** Brain-derived neurotrophic factor rescues developing avian motoneurons from cell death. N 360: 755-7 (1992); **Phillips HS et al** BDNF mRNA is decreased in the hippocampus of individuals with Alzheimer's disease. 7: 695-702 (1991); **Schecterson LC & Bothwell M** Novel roles for neurotrophins are suggested by BDNF and NT-3 mRNA expression in developing neurons. 9: 449-63 (1992); **Sendtner M et al** Brain-derived neurotrophic factor prevents the death of motoneurons in newborn rats after nerve section. N 360: 757-9 (1992); **Sieber-Blum M** Role of the neurotrophic factors BDNF and NGF in the commitment of pluripotent neural crest cells. 6: 949-55 (1991); **Suter U** NGF/BDNF chimaeric proteins: analysis of neurotrophin specificity by homologue-scanning mutagenesis. J. Neurosci. 12: 306-18 (1992); **Widmer HR et al** BDNF protection of basal forebrain cholinergic neurons after axotomy: complete protection of p75NGFR-positive cells. Neuroreport 4: 363-6 (1993)

BEF: *B cell-derived enhancing factor* see: Interleukin B.

Beige mouse: see: Immunodeficient mice.

Beige-Nude-Xid mouse: abbrev. BNX. See: Immunodeficient mice.

***bek*:** A gene isolated from the genomes of chicken, mice and humans that encodes a high-affinity transmembrane receptor for » aFGF, » bFGF, and » K-FGF. The *bek* receptor is also called FGFR-2 and arises by differential splicing of the FGFR-2 receptor gene (see: bFGF).
Binding of FGFs to the *bek* receptor is significantly enhanced by heparin (see also: HBGF, heparin-binding growth factors) and requires the interaction with heparan sulfate proteoglycans of the extracellular matrix (see also: ECM).
bek is a homologue of the human » K-*sam* gene and is related to a gene designated » TK-14. *bek* is also related to the » KGF (keratinocyte growth factor) receptor gene and arises as a differentially spliced gene product from a gene that also encodes the KGF receptor. The human *bek* gene is located on chromosome 10q25.3-q26.
Champion-Arnaud P et al Multiple mRNAs code for proteins related to the *bek* fibroblast growth factor receptor. O 6: 979-87 (1991); **Dionne CA et al** Cloning and expression of two distinct high-affinity receptors cross-reacting with acidic and basic fibroblast growth factors. EJ 9: 2685-92 (1990); **Dionne CA et al** *bek* , a receptor for multiple members of the fibroblast growth factor (FGF) family, maps to human chromosome 10q25.3-q26. Cytogenet. Cell. Genet. 60: 34-6 (1992); **Gilbert E et al** Control of *bek* and K-*sam* sites in alternative splicing of the fibroblast growth factor receptor 2 pre-mRNA. MCB 13: 5461-8 (1993); **Mansukhani A et al** Characterization of the murine *bek* fibroblast growth factor (FGF) receptor: activation by three members of the FGF family and requirement for heparin. PNAS 89: 3305-9 (1992); **Mattei MG et al** Assignment by *in situ* hybridisation of a fibroblast growth factor receptor gene to human chromosome band 10q26. Hum. Genet. 87: 84-6 (1991); **Mansukhani A et al** Characterization of the murine *bek* fibroblast growth factor (FGF) receptor: Activation by three members of the FGF family and requirement for heparin. PNAS 89: 3305-9 (1992); **Miki T et al** Determination of ligand-binding specificity by alternative splicing: two distinct growth factor receptors encoded by a single gene. PNAS 89: 246-50 (1992); **Orr-Urtreger A et al** Developmental expression of two murine fibroblast growth factor receptors, *flg* and *bek* . Development 113: 1419-34 (1991); **Sato M et al** Tissue-specific expression of two isoforms of chicken fibroblast growth factor receptor, *bek* and Cek3. Cell Growth Differ. 3: 355-61 (1992); **Yayon A et al** A confined variable region confers ligand specificity on fibroblast growth factor receptors: implications for the origin of the immunoglobulin fold. EJ 11: 1885-90 (1992)

Ber H2 antigen: see: CD30.

Beta$_2$-ANAP: anionic neutrophil-activating peptide β_2 see: ANAP.

Beta$_2$ microglobulin: abbrev. β_2M. β_2M is a polypeptide with a length of 99 amino acids which is found in the serum of normal individuals and in the urine in elevated amounts in patients with Wilson disease, cadmium poisoning, and other conditions leading to renal tubular dysfunction. It is produced by many cell types β_2M has been shown to be identical with » BDGF 2 (bone-derived growth factor 2), CRG-8 (see: CRG, cytokine-responsive genes), and » Thymotaxin. The expression of β_2M is inducible by interferons and » TNF. In human hepatoblastoma and hepatoma cells the synthesis of β_2M is stimulated by » IL6. β_2M has been shown to enhance the proliferation of fibroblasts. β_2M is identical with the non-MHC-encoded β chain of MHC class I molecules. β_2M associates with and directs the intracellular transport of major histocompatibility complex class I molecules. The lack of expression of HLA class I antigens in the human melanoma cell line FO-1 has been shown to be the result of a defect in the β_2Mgene. β_2M and components of the major histocompatibility complex interact with hormone and other growth factor receptors (possibly » IGF receptors). β_2M therefore acts as a growth factor and also modulates the binding of other growth factors to their receptors. β_2M also binds to the Fc receptor in neonatal intestine and controls migration of hematopoietic cells from bone marrow.
Until recently, β_2M had not been implicated as an etiologic factor in any disease. A protein that accumulates in amyloid-laden tissue obtained from chronic hemodialysis patients with carpal tunnel syndrome has been identified as β_2M. Depositions of β_2Mhave also been observed in hemodialysis-related amyloidosis, a form of systemic amyloidosis with a predilection for the synovium and bone that occurs with a high frequency among patients on long-term hemodialysis. The clinical features include carpal tunnel syndrome, erosive arthropathy, spondyloarthropathy, lytic bone lesions, and pathologic fractures.

■ **Transgenic/Knock-out/Antisense Studies:** Mice homozygous for a β2-microglobulin gene disruption (created by gene targeting in » ES cells) express little if any functional MHC class I antigen on the cell surface yet are fertile and apparently healthy. They show no mature CD4-negative CD8-positive cells and are defective in CD4-negative/CD8-positive T cell-mediated cytotoxicity. These deficient mice have been shown to suffer high parasitemias and early death when infected with the obligate cytoplasmic protozoan parasite *Trypanosoma cruzi*. Despite this increased susceptibility, the β2-microglobulin-deficient mice are more responsive than their normal litter mates in terms of lymphokine production, making higher levels of both IL2 and » IFN-γ in response to mitogen stimulation. The deficient mice show essentially no inflammatory response in parasite-infected tissues. These results demonstrate the importance of CD8-positive T cells in immune protection in *Trypanosoma cruzi* infection and also implicate CD8-positive T cells and/or class I MHC molecules in regulation of lymphokine production and recruitment of inflammatory cells.

Arce-Gomez B et al The genetic control of HLA-A and B antigens in somatic cell hybrids: requirements for β-2-microglobulin. Tissue Antigens 11: 96-112 (1978); **Casey TT et al** Tumoral amyloidosis of bone of β-2-microglobulin origin in association with long-term hemodialysis: a new type of amyloid disease. Hum. Path. 17: 731-8 (1986); **Cunningham BA et al** The complete amino acid sequence of β-2-microglobulin. B 12: 4811-21 (1973); **Drew PD et al** Regulation of MHC class I and β2-microglobulin gene expression in human neuronal cells: factor binding to conserved cis-acting regulatory sequences correlates with expression of the genes. JI 150: 3300-10 (1993); **D'Urso CM et al** Lack of HLA class I antigen expression by cultured melanoma cells FO-1 due to a defect in B(2)m gene expression. JCI 87: 284-92 (1991); **Gejyo F et al** A new form of amyloid protein associated with chronic hemodialysis was identified as β-2- microglobulin. BBRC 129: 701-6 (1985); **Gejyo F & Arakawa M** β 2-microglobulin-associated amyloidoses. J. Intern. Med. 232: 531-2 (1992); **Gorevic PD et al** Polymerization of intact β-2-microglobulin in tissue causes amyloidosis in patients on chronic hemodialysis. PNAS 83: 7908-12 (1986); **Homma T et al** Secretion of β-2-microglobulin from human hepatoblastoma and hepatoma cells on stimulation with interleukin-6. Eur. Surg. Res. 24: 204-10 (1992); **Jennings JC et al** β2-Microglobulin is not a bone cell mitogen. E 125: 404-9 (1989); **Lonergan M et al** A regulatory element in the β2-microglobulin promoter identified by *in vivo* footprinting. MCB 13: 6629-39 (1993); **McClure J et al** Carpal tunnel syndrome caused by amyloid containing β-2-microglobulin: a new amyloid and a complication of long term hemodialysis. Ann. Rheum. Dis. 45: 1007-11 (1986); **Parnes JR & Seidman JG** Structure of wild type and mutant mouse β2-microglobulin genes. Cell 29: 661-9 (1982); **Severinsson L & Peterson PA** β2-Microglobulin induces intracellular transcription of human class I transplantation antigen heavy chains in *Xenopus laevis* oocytes. JCB 99: 226-32 (1984); **Simister NE & Mostov KE** An Fc receptor structurally related to MHC class I antigens. N 337: 184-7 (1989)

● **Transgenic/Knock-out/Antisense Studies: Chamberlain JW et al** Tissue-specific and cell surface expression of human major histocompatibility complex class I heavy (HLA-B7) and light (β2-microglobulin) chain genes in transgenic mice. PNAS 85: 7690-4 (1988); **Flynn JL et al** Major histocompatibility complex class I-restricted T cells are required for resistance to *Mycobacterium tuberculosis* infection. PNAS 89:

12013-7 (1992); **Koller BH & Smithies O** Inactivating the B2 microglobulin locus in mouse embryonic stem cells by homologous recombination. PNAS 86: 8932-5 (1989); **Koller BH et al** Normal development of mice deficient in B2-M, MHC class I proteins, and $CD8^+T$ cells. S 248:1227-30 (1990); **Pereira P et al** Blockade of transgenic γ δ T cell development in β 2-microglobulin deficient mice. EJ 11: 25-31 (1992); **Tarleton RL et al** Susceptibility of β 2-microglobulin-deficient mice to Trypanosoma *cruzi* infection. N 356: 338-40 (1992); **Zijlstra M et al** β 2-microglobulin deficient mice lack $CD4^-8^+$ cytolytic T cells. N 344: 742-6 (1990), Comment in: N 344: 709-11 (1990)

Betacellulin: abbrev. BTC. A growth factor purified from the conditioned media of mouse pancreatic β cell tumor cells. The BTC gene is expressed in several mouse tissues, including kidney, liver, and a mouse β tumor cell line.
The human BTC cDNA, cloned from the human breast adenocarcinoma cell line MCF-7, encodes a protein of 32 kDa (178 amino acids), including a putative signal sequence. Betacellulin appears to be processed from a larger transmembrane precursor by proteolytic cleavage.
The amino acid sequences of human and murine BTC precursor proteins exhibit 79% similarity. The carboxyl-terminal domain of betacellulin has 50 % sequence similarity with that of rat » TGF-α. BTC is a member of the » EGF family of proteins. Betacellulin is a potent mitogen for retinal pigment epithelial cells and vascular smooth muscle cells.
Sasada R et al Cloning and expression of cDNA encoding human betacellulin, a new member of the EGF family. BBRC 190: 1173-9 (1993); **Shing Y et al** Betacellulin: a mitogen from pancreatic β cell tumors. S 259: 1604-7 (1993)

Beta-Chemokines: see: Chemokines.

Beta-CSF: *colony-stimulating factor* see: CSF-β. See also: CSF for general information about colony stimulating factors.

Beta-ENAP: *endothelial cell neutrophil activating peptide β* see: ENAP.

Beta-Endorphin: see: POMC (proopiomelanocortin).

Beta-FGF: *fibroblast growth factor β* see: FGF-β.

Betaglycan: A proteoglycan protein component of the extracellular matrix (see: ECM), which functions as a receptor for » TGF-β.
Andres JL et al Membrane-anchored and soluble forms of betaglycan, a polymorphic proteoglycan that binds transforming growth factor-β. JCB 109: 3137-45 (1989)

Beta-GMF: see: GMF-β (glial maturation factor).

beta-IG-M1: TGF-β-inducible gene M1 see: CTGF (connective tissue growth factor).

Beta-Intercrine: see: Chemokines.

Beta-Interferon: see: IFN-β. See also: IFN for general information about interferons.

Beta-Lipotropin: see: POMC (proopiomelanocortin).

Beta-Microseminoprotein: see: IBF (immunoglobulin binding factors).

Beta-MSH: *melanocyte stimulating hormone* see: POMC (proopiomelanocortin).

Beta-NGF: see: NGF.

Beta-Pluripoietin: see: pluripoietin-β.

Beta-RDGF: *retina-derived growth factor* see: RDGF-β.

Beta-TG: see: Beta-Thromboglobulin.

Beta-TGF: see: » TGF-β.

Beta-Thromboglobulin: abbrev. β-TG.
■ **Alternative Names:** Pro-Platelet basic protein (abbrev. PPBP).
■ **Sources:** β-TG is stored in the α-granules of platelets and released in large amounts after platelet activation.
■ **Protein Characteristics:** β-TG is a protein of 8.85 kDa. It is synthesized in the cells as a biologically inactive 15 kDa precursor called ***PBP*** (platelet basic protein). Removal of the nine aminoterminal amino acids from PBP by proteolytic cleavage yields a protein known as » CTAP-3 (connective tissue-activating protein) or LA-PF4 (low-affinity platelet factor 4). Removal of three amino acids from the aminoterminus of CTAP-3 yields β-TG. CTAP-3 is the predominant form of the protein in platelets. The removal of 11 aminoterminal amino acids from β-TG by Cathepsin G yields another protein with neutrophil-activating activities, known as » NAP-2 (neutrophil activating peptide 2).

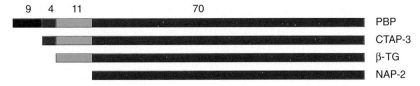

Platelet basic protein yields β-thromboglobulin and other active cytokines.
Successive removal of aminoterminal amino acids from platelet basic protein (PBP) yields connective tissue activating protein 3 (CTAP-3), Beta-thromboglobulin, and neutrophil activating protein 2 (NAP-2).

■ **GENE STRUCTURE:** The human gene encoding PBP, CTAP-3, and β-TG maps to chromosome 4q12-q13. It comprises three introns and has a length of 1.139 kb. A gene duplication of the β-TG gene, designated β-TG2, has been described. The β-TG gene is in close proximity (≈ 7 kb) of the gene encoding » PF4 (platelet factor 4). The β-TG is also located in a region in which several genes encoding proteins belonging to the » chemokine family of growth factors have been found. The genes encoding » IL8, » MGSA (melanoma growth stimulating activity) and β-TG are located within a stretch of 700 kb.
Southern blot analysis of genomic DNA suggests that there are at least two copies of the β-TG gene, designated TGB1 and TGB2, in the human genome closely situated to each other. Only one gene /TGB1) has been shown to be expressed in a megakaryocyte-specific fashion.

■ **RELATED FACTORS:** Similarities of the protein sequences of β-TG and its precursor demonstrate that β-TG belongs to the » chemokine family of cytokines.

■ **BIOLOGICAL ACTIVITIES:** β-TG is a strong chemoattractant for fibroblasts and is weakly chemotactic for neutrophils (see also: Chemotaxis). It stimulates mitogenesis, extracellular matrix synthesis, glucose metabolism, and plasminogen activator synthesis in human fibroblast cultures. β-TG, its precursor, and its cleavage products influence the functional activities of neutrophil granulocytes. β-TG affects the maturation of human megakaryocytes and thus could play a role in the physiological regulation of platelet production by megakaryocytes.

■ **ASSAYS:** β-TG is assayed by ELISA and other immunological tests.

■ **CLINICAL USE & SIGNIFICANCE:** Serum plasma levels of β-TG (often in conjunction with » PF4) and also concentrations of β-TG in urine are used as an index of platelet activation in a variety of clinical settings.

● **REVIEWS: Kaplan KL & Niewiarowski S** Nomenclature of secreted platelet proteins – report of the working party on secreted platelet proteins of the subcommittee on platelets. Thromb. Haemost. 53: 282-4 (1985)

● **BIOCHEMISTRY & MOLECULAR BIOLOGY: Begg GS et al** Complete covalent structure of human β-Thromboglobulin. B 17: 1739-44 (1978); **Castor CW et al** Structural and biological characteristics of connective tissue activating peptide (CTAP-III), a major human platelet derived growth factor. PNAS 80: 765-9 (1983); **Cohen AB et al** Generation of the neutrophil-activating peptide-2 by cathepsin G and cathepsin G-treated human platelets. Am. J. Physiol. 263: L249-56 (1992); **Doi T et al** Structure of the rat platelet factor 4 gene: a marker for megakaryocyte differentiation. MCB 7: 898-904 (1987); **Green CJ et al** Identification and characterization of PF4*var*1, a human gene variant of platelet factor 4. MCB 9: 1445-51 (1989); **Holt JC et al** Characterization of human platelet basic protein, a precursor form of low-affinity platelet factor 4 and β-thromboglobulin. B 25: 1988-96 (1986); **Holt JC & Niewiarowski S** Platelet basic protein, low-affinity platelet factor 4, and β-thromboglobulin: purification and identification. MiE 169: 224-33 (1989); **Majumdar S et al** Characterization of the human β-thromboglobulin gene. Comparison with the gene for platelet factor 4. JBC 266: 5785-9 (1991); **Niewiarowski S et al** Identification and separation of secreted platelet proteins by isoelectric focusing. Evidence that low-affinity platelet factor 4 is converted to β-thromboglobulin by limited proteolysis. Blood 55: 453-6 (1980); **Tunnacliffe A et al** Genes for β-thromboglobulin and platelet factor 4 are closely linked and form part of a cluster of related genes on chromosome 4. Blood 79: 2896-2900 (1992); **Van Damme J et al** Purification of granulocyte chemotactic peptide/interleukin-8 reveals N-terminal sequence heterogeneity similar to that of β-thromboglobulin. EJB 181: 337-44 (1989); **Villanueva GB et al** Circular dichroism of platelet factor 4. ABB 170: 1745-50 (1989); **Wenger RH et al** Human platelet basic protein/connective tissue activating peptide- III maps in a gene cluster on chromosome 4q12-q13 along with other genes of the β-thromboglobulin superfamily. Hum. Genet. 87: 367-8 (1991)

● **BIOLOGICAL ACTIVITIES: Abgrall JF et al** Inhibitory effect of highly purified human platelet β-thromboglobulin on *in vitro* human megakaryocyte colony formation. EH 19: 202-5 (1991); **Allegrezza-Giulietti A et al** β-Thromboglobulin and polymorphonuclear leukocytes activation. (Effects on chemiluminescence, release of membrane bound calcium, NADPH-oxidase activity and membrane fluidity). BI 24: 273-9 (1991); **Allegrezza-Giulietti A et al** Platelet release products modulate some aspects of polymorphonuclear leukocyte activation. JCBc 47: 242-50 (1991); **Castor CW et al** Connective tissue activation. XXIX. Stimulation of glucose transport by connective tissue activating peptide-III. B 24: 1762-1767 (1985); **Han ZC et al** Negative regulation of human megakaryocytopoiesis by human platelet factor 4 and β thromboglobulin: comparative analysis in bone marrow cultures from normal individuals and patients with essential thrombocythemia and immune thrombocytopenic purpura. Br. J. Haematol. 74: 395-401 (1990); **Senior RM et al** Chemotactic activity of a platelet α granule protein for fibroblasts. JCB 96: 382-5 (1983); **Van Damme J et al** The neutrophil-activating proteins interleukin 8 and β-thromboglobulin: *in vitro* and *in vivo* comparison of NH2-terminally processed forms. EJI 20: 2113-8 (1990)

● **ASSAYS: Donlon JA et al** Effect of proteases on the β-thromboglobulin radioimmunoassay. Life Sci. 36: 525-31 (1985); **Fiskerstrand CE et al** Radioimmunoassay of β-thromboglobulin. False high values from frozen plasma samples. Am. J. Clin. Pathol. 90: 610-2 (1988); **Kerry PJ & Curtis AD** Standardization of β-thromboglobulin (β-TG) and platelet factor 4 (PF4): a collaborative study to establish international standards for β-TG and PF4. Thromb. Haemost. 53: 51-5 (1985); **Lane DA et al** Detection of enhanced *in vivo* platelet α-granule release in different patient groups—comparison of β-thromboglobulin, platelet factor 4 and thrombospondin assays. Thromb. Haemost. 52: 183-7 (1984); **Sander HJ et al** Immunocytochemical localization of fibrinogen, platelet factor 4, and β thromboglobulin in thin frozen sections of human blood platelets. JCI 72: 1277-87 (1983); **Scarabin PV et al** Reliability of a single β-thromboglobulin measurement in a diabetic population: importance of PGE1 in anticoagulant mixture. Damad Study Group. Thromb. Haemost. 57: 201-4 (1987); **Stenberg PE et al** Optimal techniques for the immunocytochemical demonstration of β-thromboglobulin, platelet factor 4, and fibrinogen in the α granules of unstimulated platelets. Histochem. J. 16: 983-1001 (1984); **Strauss W et al** Serial determinations of PF4 and β TG: comparisons between multiple venipunctures vs a catheter infusion system. Thromb. Haemost. 59: 491-4 (1988); **Takahashi H et al** Measurement of platelet factor 4 and β-thromboglobulin by an enzyme-linked immunosorbent assay Clin. Chim. Acta 175: 113-4 (1988); **Van Oost BA et al** Increased urinary β-thromboglobulin excretion in diabetes assayed with a modified RIA kit-technique. Thromb. Haemost. 49: 18-20 (1983); **Van Wyk V et al** A formula for correcting for the *in vitro* release of platelet β-thromboglobulin. Thromb. Res. 46: 659-68 (1987) **Walkowiak B & Cierniewski CS** Kinetics of the β-thromboglobulin release from α-granules of blood platelets activated by ADP. Thromb. Res. 46: 727-36 (1987)

● **CLINICAL USE & SIGNIFICANCE: Adamides S et al** A study of β-thromboglobulin and platelet factor-4 plasma levels in steady state sickle cell patients. Blut 61: 245-7 (1990); **Benedetti-Panici P et al** Elevated plasma levels of β-thromboglobulin in breast cancer. Oncology 43: 208-11 (1986); **De Caterina R et al** Platelet activation in angina at rest. Evidence by paired measurement of plasma β-thromboglobulin and platelet factor 4. Eur. Heart J. 9: 913-22 (1988); **Fabris F et al** Clinical significance of β-thromboglobulin in patients with high platelet count. Acta Haematol. 71: 32-8 (1984); **Fenandez-Vigo J et al** Platelet function in diabetic retinopathy: levels of β-thromboglobulin and platelet factor 4. Metab. Pediatr. Syst. Ophthalmol. 15: 5-8 (1992); **Ffrench P et al** Comparative evaluation of plasma thrombospondin β-thromboglobulin and platelet factor 4 in acute myocardial infarction. Thromb. Res. 39: 619-24 (1985); **Gavaghan TP et al** Increased plasma β-thromboglobulin in patients with coronary artery vein graft occlusion: response to low dose aspirin. J. Am. Coll. Cardiol. 15: 1250-8 (1990); **Gomi T et al** Plasma β-thromboglobulin to platelet factor 4 ratios as indices of vascular complications in essential hypertension. J. Hypertens. 6: 389-92 (1988); **Hopper AH et al** Urinary β-thromboglobulin correlates with impairment of renal function in patients with diabetic nephropathy. Thromb. Haemost. 56: 229-31 (1986); **Kaplan KL & Owen J** Plasma levels of β-thromboglobulin and platelet factor 4 as indices of platelet activation *in vivo*. Blood 57: 199-202 (1981); **Kubisz P et al** Relationship between platelet aggregation and plasma β-thromboglobulin levels in arterio-vascular and renal diseases. Atherosclerosis 55: 363-8 (1985); **McRoyan DK et al** β-Thromboglobulin and platelet factor 4 levels in a case of acute thrombotic thrombocytopenic purpura. South. Med. J. 78: 745-8 (1985); **Musumeci V et al** Urine β-thromboglobulin concentration or β-thromboglobulin/creatinine ratio in single voided urine samples cannot be reliably used to estimate quantitative β-thromboglobulin excretion. Thromb. Haemost. 55: 2-5 (1986); **Nieszpaur M et al** Plasma β-thromboglobulin in acute myocardial infarction with ventricular arrhythmias. Pol. J. Pharmacol. Pharm. 42: 89-92 (1990); **Pechan J et al** Circadian rhythm of plasma β-thromboglobulin in healthy human subjects. Blood Coagul. Fibrinolysis 3: 105-7 (1992); **Rapold HJ et al** Plasma levels of plasminogen activator inhibitor type 1, β-thromboglobulin, and fibrinopeptide A before, during, and after treatment of acute myocardial infarction with alteplase. Blood 78: 1490-5 (1991); **Rinder HM & Snyder EL** Activation of platelet concentrate during preparation and storage. Blood Cells 18: 445-56 (1992); **Seibold JR & Harris JN** Plasma β-thromboglobulin in the differential diagnosis of Raynaud's phenomenon. J. Rheumatol. 12: 99-103 (1985); **Taomoto K et al** Usefulness of the measurement of plasma β-thromboglobulin (β-TG) in cerebrovascular disease. Stroke 14: 518-24 (1983); **Toth L et al** Elevated levels of plasma and urine β-thromboglobulin or thromboxane-B2 as markers of real platelet hyperactivation in diabetic nephropathy. Hemostasis 22: 334-9 (1992); **Woo E et al** β-thromboglobulin in cerebral infarction. J. Neurol. Neurosurg. Psychiatry 51: 557-62 (1988); **Yamauchi K et al** Plasma β-thromboglobulin and platelet factor 4 concentrations in patients with atrial fibrillation. Jpn. Heart J. 27: 481-7 (1986)

Beta-TNF: see: TNF-β.

Beta-Urogastrone: This factor is identical with » EGF.

Gregory H Isolation and structure of urogastrone and its relationship to epidermal growth factor. N 257: 325-7 (1975)

BF: *blastogenic factor* This factor is found in the supernatant of a mixed leukocyte culture. It restimulates cells primed in allogeneic mixed cultures independently of the original stimulator cells. In addition to stimulating proliferation of human cytotoxic lymphocytes BF can also cooperate in a mouse thymocyte costimulator assay and sustain mouse and human IL2-dependent cell lines. The factor is identical with » IL2.

Kasakura S & Lowenstein L A factor stimulating DNA synthesis derived from the medium of leukocyte cultures. N 208: 794-8 (1965); **Rode H et al** Characterization of a monoclonal antibody raised to the human lymphokine blastogenic factor. In: Interleukins, Lymphokines, and Cytokines. Academic Press, pp. 11-6, New York 1983

BF: *blocking factor* For a specific blocking factor see: TNF-BF (tumor necrosis factor blocking factor).

BFA: *burst feeder activity* see: BFU (burst promoting activities).

bFGF: *basic fibroblast growth factor*
■ **ALTERNATIVE NAMES:** AGF (adipocyte growth factor); AGF-2 (astroglial growth factor 2); BDGF 1. 7 (bone-derived growth factor 1. 7); BDNF (brain-derived neurotrophic factor); CDGF 1. 7 (cartilage-derived growth factor 1. 7); CDGF-1 (cartilage-derived growth factor 1); CDGF, CGF (chondrosarcoma growth factor); CGF (colonic growth factor); cementum mitogenic factor; ChDGF, CHSA-GF (chondrosarcoma-derived growth factor); CLAF (corpus luteum angiogenic factor); CMGF (chicken muscle growth factor); ECGF (embryonic carcinoma-derived growth factor); EDGF 1 (eye-derived growth factor 1); EK-AF-2 (embryonic kidney-derived angiogenesis factor 2); FGF-2 (fibroblast growth factor 2); FGF-β (fibroblast growth factor β); GGF (glial growth factor); GPA (growth-promoting activity); HBGF (heparin-binding growth factor); HBGF-2 (heparin-binding growth factor 2); HDGF-C (hypothalamus-derived growth factor); HDGF (hepatoma-derived growth factor); HDGF-2 (hepatoma-derived growth factor 2); HGF (hepatocyte growth factor); HGF (hepatoma growth factor); HGF (hypothalamic growth factor); HPGF (human pituitary growth factor); human placenta purified factor; KAF (kidney angiogenic factor); MDGF (macrophage-derived growth factor; bFGF is at least one component of MDGF, the major component being PDGF); MDGF (melanoma-derived growth factor; MeGF (melanocyte growth factor); MGF (macrophage growth factor); MGF (myogenic growth factor); pituitary-derived chondrocyte growth factor; MTGF (mammary tumor-derived growth factor); OGF (ovarian growth factor); PAF (placental angiogenic factor); PGF (pituitary growth factor); POF (prostatic osteoblastic factor); PrGF (prostatic growth factor); RDGF-β (retina-derived growth factor); TAF (tumor angiogenesis factor); UDGF (uterine-derived growth factor). See also: individual entries for further information.

The recommended name for bFGF is ***FGF-2*** (see: Nomenclature Meeting Report and Recommendation, January 18, 1991, reprinted in: ANY Vol. 638, pp. xiii-xvi).

■ **SOURCES:** bFGF is found in almost all tissues of mesodermal and neuroectodermal origin and also in tumors derived from these tissues. Endothelial cells produce large amounts of this factor. Some bFGF is associated with the extracellular matrix (see: ECM) of the subendothelial cells. Many cells express bFGF only transiently and store it in a biologically inactive form. The mechanism by which the factor is released by the cells is not known. It is released after tissue injuries and during inflammatory processes (see: Inflammation) and also during the proliferation of tumor cells. The expression of bFGF and related factors seems to be regulated differentially, depending on cell type and developmental age.

■ **PROTEIN CHARACTERISTICS:** bFGF is an 18 kDa protein with a length of 155 amino acids and an isoelectric point of 9.6. The factor does not contain disulfide bonds and is not glycosylated. Shorter variants with a length of 131 and 146 amino acids, respectively, have been described. The 18 kDa protein is obtained by using the classical AUG initiation codon. Some higher molecular weight forms of bFGF (22, 23, 24, 25 kDa) have also been described. They arise by translation beginning at a CUG codon. The smallest form (18 kDa) occurs predominantly in the cytosol, while the higher molecular weight forms (22, 22.5, 24 kDa) are associated with the nucleus and ribosomes.

bFGF is related to another member of the FGF family, » aFGF and shows a homology of 55 % which is limited, however, to a few functional domains. The structure of bFGF, revealed by X-ray crystallography, shows a similar folding structure as » IL1.

The sequences of bovine and human bFGF differ in only 2 amino acids. Human and ovine bFGF are identical. A homologue of human bFGF isolated from » *Xenopus* shows an overall sequence homology of 84 %.

bFGF, like » aFGF, » CNTF (ciliary neuronotrophic factor), and » PD-ECGF (platelet-derived endothelial cell growth factor), does not possess a » signal sequence that would allow secretion of the factor by classical secretion pathways (endoplasmatic reticulum/Golgi system). The mechanism underlying the release of aFGF is unknown. The higher molecular weight forms differ from the 18 kDa form of bFGF in their N-terminal ends. These ends contain a » signal sequence known as nuclear targeting sequence which is responsible for the localization of the factor in the cell nucleus (see also: Signal sequences). A mechanism of action, called » intracrine growth control, has been proposed to account for some of the actions of bFGF.

■ **GENE STRUCTURE:** The bFGF gene has a length of ≈ 38 kb and contains three exons separated by three large introns. The first intron separated codons 60 and 61, the second intron codons 94 and 95. The bFGF gene codes for a total of four polypeptides, three of which using initiation from non-AUG codons. The human bFGF gene is located on chromosome 4q25. Some cells appear to express two polyadenylated different transcripts of 3.7 and 7 kb.

■ **RELATED FACTORS:** bFGF is the prototype of a large family of proteins, designated FGF family. Its members display a remarkable affinity to heparin and are therefore also called » HBGF (heparin-binding growth factors). See also: FGF-1 to FGF-7 and HBGF-1 to HBGF-8.

FGF proteins are evolutionary closely conserved. Some » oncogenes isolated from gastric, mammary, and bladder carcinomas and Kaposi sarcomas (see: *int*-2, *hst*, K-FGF) encode biologically active bFGF-related growth factors which show an overall sequence homology of 30-45 %. A factor designated keratinocyte growth factor (see: KGF = FGF-7) and two additional factors encoded by oncogenes, designated » FGF-5 and » FGF-6, are also members of the FGF family.

■ **RECEPTOR STRUCTURE, GENE(S), EXPRESSION:** FGF receptors are encoded by a gene family consisting of at least four receptor tyrosine kinases that transduce signals important in a variety of developmental and physiological processes related to cell growth and differentiation. bFGF and also » aFGF glycoprotein membrane receptors of 125-130 kDa and 145-165 kDa are expressed on the cell surfaces of various FGF-sensitive cells at densities of $2 \times 10^3 - 10^5$/cell. The intracellular domains of these receptors encode a tyrosine-specific protein kinase. Binding of the ligand to the receptor leads to » autophosphorylation of the receptor protein.

The genes encoding bFGF receptors are members of a gene family, and homologues binding either bFGF, aFGF, or related factors have also been isolated from other species. The gene known as » *flg* (*fms*-like gene) encodes a 145 kDa receptor protein which is the human homologue of the corresponding chicken gene (*Cek*1). The official new nomenclature for *flg* is ***FGFR-1***. This gene is also related to another gene isolated from humans, mice, and chicken (***Cek*3**), designated » ***bek*** (125 kDa), a human gene called **TK-14**, and a gene encoding the receptor for » KGF (keratinocyte growth factor). The official nomenclature for this type of bFGF receptor which binds aFGF equally well is ***FGFR-2***, and alternative splicing of the FGFR-2 gene has been demonstrated to yield the splice variants » *bek* and » K-*sam*. Yet another type of bFGF receptor, designated ***FGFR-3***, also binds aFGF and bFGF equally well. This receptor exists as a membrane-bound form of 125 kDa. A fourth receptor, mapping at 5q33-qter (see also: $5q^-$-syndrome) and designated ***FGFR-4***, binds » aFGF with highest affinity, followed by » K-FGF/*hst*-1 and bFGF.

Differential splicing in the extracellular region of FGFR-1 has been shown to generate receptor variants with different ligand-binding specificities. One variant binds bFGF with a 50-fold lower affinity.

Binding of bFGF to one of its receptors requires the interaction with heparan sulfate and heparan sulfate proteoglycans (Syndecan) of the extracellular matrix (see also: ECM) before full functional activity is obtained. This is also demonstrated by the ability of heparitinase to inhibit receptor binding and biological activity of bFGF. It has been observed that the membrane-associated factor is phosphorylated by a protein kinase also located on the cell surface, which may additionally alter its activity and bioavailability. Heparin has been shown to protect bFGF from inactivation by proteases, acids, and heat. It also improves its capacity to bind to the receptors and hence potentiates the biological activities of bFGF. This feature may be of physiological importance because mast cells, for example, contain a lot of heparin which could be released during degranulation. In addition, heparin also increases the biological half life of bFGF.

Binding of bFGF (and also aFGF) to receptor proteins can be inhibited by » Suramin, and also by protamine. At least one receptor type has been described to be utilized as a receptor for Herpes simplex virus (HSV-1).

■ **BIOLOGICAL ACTIVITIES:** The wide spectrum of biological activities is revealed by the plethora of different names. A multifunctional role of bFGF is suggested by the many different receptor phenotypes expressed in various cell types. bFGF stimulates the growth of fibroblasts, myoblasts, osteoblasts, neuronal cells, endothelial cells, keratinocytes, chondrocytes, and many other cell types. In capillary endothelial cells bFGF acts as an » autocrine growth factor. The mitogenic action of bFGF for endothelial cells can be potentiated by thrombin. Transferrin and HDL (high density lipoprotein) also support the activity of bFGF on endothelial cells.

bFGF promotes the maturation and maintenance of cholinergic neurons and acts as a mitogen for chomaffin cells. It also influences the proliferation, differentiation, and function of astrocytes and oligodendrocytes. For some cell lines, for example » PC12, bFGF is a neurite outgrowth-promoting factor with an activity that is on the same order of magnitude as » NGF. For some cholinergic, dopaminergic, and GABAergic neuronal cells bFGF acts as a differentiation factor promoting outgrowth of neurites and promoting survival. In some cell types this activity is as pronounced as that of » CNTF (ciliary neuronotrophic factor). In some cells bFGF appears to induce the expression of certain neuronal-specific genes (e. g. SCG-10) the synthesis of which is also induced by » NGF.
Rat hippocampal and human cortical neurons have been shown to be protected by bFGF against induced damage induced by iron, which is believed to contribute to the process of cell damage and death resulting from ischemic and traumatic insults by catalyzing the oxidation of protein and lipids. bFGF has been shown to allow long-term culture of rat primary hippocampal neurons in serum-free culture media, yielding continuous untransformed cell lines that can be passaged.
Immortalized rat fibroblasts, genetically altered to secrete bFGF, have been shown to decrease excitotoxic lesion size by 30 % after implantation in rat brain near the striatum 7 days before striatal infusion of excitotoxic quantities of an NMDA-receptor agonist.
bFGF has been shown to be a promoting or inhibitory modulator of cellular differentiation also for other cell types. bFGF is not only a mitogen for chondrocytes but also inhibits their terminal differentiation. In early embryos bFGF functions as a differentiation factor that induces tissues destined to produce ectodermal structures to differentiate into mesodermal tissues. In cultures of rat retina cells bFGF induces photoreceptor differentiation. The influence of bFGF on differentiation processes are probably the result of a complex interaction with other factors. Elevated levels of bFGF are observed during early embryonic development and it has been suggested that bFGF, in combination with other factors showing similar activities, e. g. TGF-β, may be involved in the development of the mesoderm.
There are many instances in which bFGF retards senescence of cells, thus allowing maintenance of cells *in vitro* which would normally lose their differentiated phenotype in long-term cultures.
In the pituitary bFGF regulates the secretion of » thyrotropin and » prolactin. bFGF also acts as an ovarian hormone and differentially regulates the expression of steroids. bFGF modulates the proliferation and differentiation of granulosa cells and inhibits induction of receptors for » luteinizing hormone mediated by FSH (follicle stimulating hormone). bFGF also blocks the synthesis of enzymes the expression of which is induced by FSH. bFGF stimulates the synthesis of progesterone. Some of these effects of bFGF are, in turn, blocked by » TGF-β, suggesting that the ratio of bFGF/TGF-β is used for fine control of these processes.
bFGF also has direct central actions and inhibits food intake in rats after intracerebroventricular injection for at least 12 hours. bFGF has been implicated as an embryonic mesoderm-inducing factor and there is also some evidence for an organ- and sex-specific role of FGF in the development of the fetal mammalian reproductive tract.
bFGF and also aFGF also acts as an » angiogenesis factor by controlling the proliferation and migration of vascular endothelial cells. The expression of plasminogen activator and collagenase activity by these cells is enhanced by bFGF. bFGF is probably one of the factors responsible early in development for the growth of new capillary blood vessels colonizing the mesencephalon and telencephalon which at this time are essentially free of blood vessels. bFGF and/or closely related factors may also be engaged in tumor angiogenesis, facilitating the invasive growth and metastasis of tumors by inducing the synthesis of proteases. Tumors expressing certain » oncogenes encoding bFGF-like factors constitutively have been shown to support unlimited angiogenesis.
bFGF also plays an important physiological role in tissue regeneration and » wound healing.
If bFGF that lacks a normal secretory signal is furnished with a hydrophobic signal sequence the introduction of this recombinant genes into cells leads to malignant transformation and yields cells that are tumorigenic *in vivo*. Monoclonal antibodies directed against bFGF inhibit the growth of tumor cells. Treatment of these cells with low and nontoxic concentrations (0.5-2.5 μg/ml) of negatively charged, nonsulfated aromatic compounds (e.g., aurin tricarboxylic acid, 4-hydroxyphenoxyacetic acid) result in restoration of their normal proliferative rate, morphological appearance, and adhesion properties.

■ **Interactions:** The synthesis of bFGF by vascular smooth muscle cells and skin fibroblasts is induced by » IL1. In human endothelial cells bFGF reduces the expression of cyclo-oxygenase. The same inhibitory action on bFGF activities is also shown by » IFN-γ which is itself induced by FGFs. In endothelial cells bFGF reduces the expression of the receptor for TGF-β (see also: Receptor transmodulation), thus effectively modulating the inhibitory action of TGF-β on endothelial cells. The activity of bFGF (and also of aFGF) on the vascular epithelium is inhibited strongly by » TGF-β. TGF-β may, however, also have positive actions on bFGF-mediated processes in other cell types.

The release of bFGF by astrocytes is significantly enhanced by » IL1, » IL6 or » EGF. Other lymphokines, and also » NGF, have no effect. In microglia the release of bFGF is considerably reduced by » IL3, » EGF and » NGF.

■ **Transgenic/Knock-out/Antisense Studies:** A dominant-negative FGF receptor mutant has been used to block FGF function in suprabasal keratinocytes of » transgenic mice. Expression of the mutant receptor, which blocks signal transduction in cells when co-expressed with wild-type receptors, disrupts the organization of epidermal keratinocytes and induces epidermal hyperthickening, showing that FGF is essential for the morphogenesis of suprabasal keratinocytes and for the establishment of the normal program of keratinocyte differentiation.

Treatment of various human tumor cell lines (melanomas, glioblastomas) with » Antisense oligonucleotides directed against bFGF has been shown to inhibit the proliferation of these cells and their ability to form colonies in soft agar (see also: Colony formation assay).

■ **Assays:** bFGF activities can be assayed in a proliferation assay using CCL39 hamster fibroblast cells, human » SW-13 cells or bovine » FBHE cells. bFGF can be assayed by a sensitive enzyme immunoassay. BFGF can also be detected by a modification of the » cell blot assay. An alternative and entirely different method of detecting IL7 is » Message amplification phenotyping. For further information see also subentry "Assays" in the reference section.

■ **Clinical Use & Significance:** Some preclinical studies have been performed with recombinant bFGF. It is assumed that bFGF will be of potential interest as a factor promoting » wound healing and revascularization of tissues. Animals experiments with bFGF have shown that it promotes endosteal, but not periosteal, bone formation and bFGF may thus be a potential agent for treatment osteoporosis which may increase bone mass without causing outward deformation of the skeletal bones.

Another possible use of bFGF is suggested by its neurotrophic activities (see also: neurotrophins). bFGF and also related factors may be useful to support regeneration of tissues after brain, spinal chord, and peripheral nerve injuries. Animal experiments have shown that the infusion of bFGF positively influences the growth of retinal ganglion cells and also nerve regeneration. Injections of bFGF maintain photoreceptors which would normally degenerate, for example, in cases of inherited retinal dystrophy. bFGF and related factors may also play a role in neurodegenerative disorders such as Alzheimer's disease, Parkinson's disease, and Huntington's disease.

bFGF is produced in most gliomas and is involved in tumorigenesis and malignant progression as an » autocrine growth factor that may play an important role also in tumor neovascularization as a » paracrine angiogenic factor (see also: Angiogenesis factors). Overexpression of bFGF and aFGF in pancreatic cancers has been found to be associated with a more advanced tumor stage. bFGF has also been found to be overexpressed in gastric and renal cell carcinomas. Very high levels of bFGF have been found in the serum and/or urine of patients with many types of cancer. bFGF has also been implicated in the pathology of juvenile nasopharyngeal angiofibromas.

Since the migration and proliferation of endothelial cells play a pivotal role in various vascular diseases and bFGF plays a significant role as an » autocrine endothelial growth factor specific ways to block its expression may have considerable therapeutic implications. By employing » antisense oligonucleotides specifically blocking the expression of bFGF it has been possible to suppress endothelial cell proliferation.

bFGF has been shown to improve cardiac systolic function, to reduce infarct size, and to increase the number of arterioles and capillaries in the infarct. Thus, the angiogenic action of bFGF might lead to a reduction in infarct size in a canine experimental myocardial infarct model.

The observation that HSV-1 uses one of the bFGF receptors as an entry port suggests that it may be possible to use specific receptor antagonists to block virus entry, thus preventing later complications of the infection. For further uses of bFGF exploiting the specific interaction of bFGF with its receptor see also: Mitotoxins, Saporin.

● Reviews: **Baird A & Klagsbrun M** The fibroblast growth factor family. CC 3: 239-43 (1991); **Burgess WH & Maciag T** The heparin-binding (fibroblast) growth factor family of proteins. ARB 58: 575-606 (1989); **Goldfarb M** The fibroblast growth factor family. Cell Growth Differ. 1: 439-45 (1990); **Gospodarowicz D et al** Structural characterization and biological functions of fibroblast growth factor. Endocrinol. Rev. 8: 95-113 (1987); **Klagsbrun M** Mediators of angiogenesis: the biological significance of basic fibroblast growth factor (bFGF)-heparin and heparan sulfate interactions. Semin. Cancer. Biol. 3: 81-7 (1992); **Logan A & Berry M** Transforming growth factor β1 and basic fibroblast growth factor in the injured CNS. TIPS 14: 337-43 (1993); **Schweigerer L** Basic fibroblast growth factor: properties and clinical implications. In: Habenicht A (edt) Growth factors, differentiation factors, and cytokines, pp. 42-66, Springer, Berlin 1990; **Sensenbrenner M** The neurotrophic activity of fibroblast growth factors. Prog. Neurobiol. 41: 683-704 (1993); **Wagner JA** The fibroblast growth factors: An emerging family of neural growth factors. CTMI 165: 95-118 (1991); **Westerman R et al** Basic fibroblast growth factor (bFGF), a multifunctional growth factor for neuroectodermal cells. JCS 13: s97-s117 (1990)

● Biochemistry & Molecular Biology: **Abraham JA et al** Human basic fibroblast growth factor: nucleotide sequence, genome organization, and expression in mammalian cells. CSH-SQB 51: 657-68 (1986); **Baird A et al** Receptor- and heparin-binding domains of basic fibroblast growth factor. PNAS 85: 2324-8 (1988); **Baldin V et al** Translocation of bFGF to the nucleus is G_1 phase cell cycle specific in bovine aortic endothelial cells. EJ 9: 1511-7 (1990); **Bashkin P et al** Basic fibroblast growth factor binds to subendothelial extracellular matrix and is released by heparitinase and heparin-like molecules. B 28: 1737-43 (1989); **Bugler B et al** Alternative initiation of translation determines cytoplasmic and nuclear localization of basic fibroblast growth factor. MCB 11: 573-7 (1991); **Burgess WH** Structure-function studies of acidic fibroblast growth factor. ANY 638: 89-97 (1991); **Eriksson AE et al** Three-dimensional structure of human basic fibroblast growth factor. PNAS 88: 3441-5 (1991); **Esch F et al** Primary structure of bovine pituitary basic fibroblast growth factor (FGF) and comparison with the amino-terminal sequence of bovine brain acidic FGF. PNAS 82: 6507-11 (1985); **Florkiewicz RZ et al** Basic fibroblast growth factor gene expression. ANY 638: 109-26 (1991); **Fukushima Y et al** The human basic fibroblast growth factor gene (FGFB) is assigned to chromosome 4q25. Cytogenet. Cell Genet. 54: 159-60 (1990); **Gospodarowicz D et al** Heparin protects acidic and basic FGF from inactivation. JCP 128: 475-84 (1986); **Heath WF et al** Mutations in the heparin-binding domains of human basic fibroblast growth factor alter its biological activity. B 30: 5608-15 (1991); **Hebert JM et al** Isolation of cDNAs encoding four mouse FGF family members and characterization of their expression patterns during embryogenesis. Dev. Biol. 138: 454-63 (1990); **Imamura T et al** Recovery of mitogenic activity of a growth factor mutant with a nuclear translocation sequence. S 249: 1567-70 (1990); **Imamura T et al** Identification of a heparin-binding growth factor-1 nuclear translocation sequence by deletion mutation analysis. JBC 267: 5676-9 (1992); **Klagsbrun M** The affinity of fibroblast growth factors (FGFs) for heparin; FGF-heparan sulfate interactions in cells and extracellular matrix. COCB 2: 857-63 (1990); **Kurokawa T et al** Cloning and expression of cDNA encoding human basic fibroblast growth factor. FL 213: 189-94 (1987); genes. Mammalian Genome 2: 135-7 (1992); **Mergia A et al** The genes for basic and acidic fibroblast growth factors are on different human chromosomes. BBRC 138: 644-51 (1986); **Mignatti P et al** Basic fibroblast growth factor, a protein devoid of secretory signal sequence, is released by cells via a pathway independent of the endoplasmic reticulum-Golgi complex. JCP 151: 81-93 (1992); **Mignatti P & Rifkin DB** Release of basic fibroblast growth factor, an angiogenic factor devoid of secretory signal sequence: A trivial phenomenon or a novel secretion mechanism? JCB 47: 201-7 (1991); **Prats H et al** High molecular mass forms of basic fibroblast growth factor are initiated by alternative CUG codons. PNAS 86: 1836-40 (1989); **Presta M et al** Biologically active synthetic fragments of human basic fibroblast growth factor (bFGF): Identification of two Asp-Gly-Arg-containing domains involved in the mitogenic activity of bFGF in endothelial cells. JCP 149: 512-24 (1991); **Prestrelski SJ et al** Binding of heparin to basic fibroblast growth factor induces a conformational change. ABB 293: 314-9 (1992); **Quarto N et al** The NH2-terminal extension of high molecular weight bFGF is a nuclear targeting signal. JCP 147: 311-8 (1991); **Saksela O & Rifkind DB** Release of basic fibroblast growth factor-heparan sulfate complexes from endothelial cells by plasminogen activator-mediated proteolytic activity. JCB 110: 767-75 (1990); **Saksela O et al** Endothelial cell-derived heparan sulfate binds fibroblast growth factors and protects it from proteolytic degradation. JCB 107: 743-51 (1988); **Shimasaki J et al** Complementary DNA cloning and sequencing of rat ovarian basic fibroblast growth factor and tissue distribution of its mRNA. BBRC 157: 256-63 (1988); **Thompson SA et al** Cloning, recombinant expression, and characterization of basic fibroblast growth factor. MiE 198: 96-116 (1991); **Vilgrain I & Baird A** Phosphorylation of basic fibroblast growth factor by a protein kinase associated with the outer surface of a target cell. Mol. Endocrinol. 5: 1003-12 (1991); **Vlodavsky I et al** Extracellular sequestration and release of fibroblast growth factor: a regulatory mechanism? TIBS 16: 261-8 (1991); **Vlodavsky I et al** Sequestration and release of basic fibroblast growth factor. ANY 638: 207-20 (1991); **Yoshitake Y et al** Derivation of monoclonal antibody to basic fibroblast growth factor and its application. MiE 198: 148-57 (1991); **Zhang J et al** Three-dimensional structure of human basic fibroblast growth factor, a structural homologue of interleukin 1β. PNAS 88: 3446-50 (1991); **Zhu X et al** Three-dimensional structures of acidic and basic fibroblast growth factors. S 251: 90-3 (1991)

● Receptors: **Armstrong E et al** Localization of the fibroblast growth factor receptor-4 gene to chromosome region 5q33-qter. Genes Chromosom. Cancer 4: 94-8 (1992); **Avivi A et al** A novel form of FGF receptor-3 using an alternative exon in the immunoglobulin domain III. FL 330: 249-52 (1993); **Coltrini D et al** Biochemical bases of the interaction of human basic fibroblast growth factor with glycosaminoglycans. New insights from trypsin digestion studies. EJI 214: 51-8 (1993); **Dionne CA et al** Cloning and expression of two distinct high-affinity receptors cross-acting with acidic and basic fibroblast growth factors. EJ 9: 2685-92 (1990); **Eisemann A et al** Alternative splicing generates at least five different isoforms of the human basic FGF receptor. O 6: 1195-1202 (1991); **Johnson DE et al** Diverse forms of a receptor for acidic and basic fibroblast growth factors. MCB 10: 4728-36 (1990); **Johnston DL et al** The human fibroblast growth factor receptor genes: a common structural arrangement underlies the mechanisms for generating receptor forms that differ in their third immunoglobulin domain. MCB 11: 4627-34 (1991); **Kaner RJ et al** Fibroblast growth factor receptor is a portal of entry for herpes simplex virus type 1. S 248: 1410-3 (1990); **Keegan K et al** Isolation of an additional member of the fibroblast growth factor receptor family, FGFR-3. PNAS 88: 1095-9 (1991); **Keegan K et al** Structural and biosynthetic characterization of the fibroblast growth factor receptor 3 (FGFR-3) protein. O 6: 2229-36 (1991); **Kiefer MC et al** The molecular biology of heparan sulfate fibroblast growth factor receptors. ANY 638: 167-76 (1991); **Kiefer MC et al** Molecular cloning of a human basic fibroblast growth factor receptor cDNA and expression of a biologically active extracellular domain in a baculovirus system. GF 5: 115-27 (1991); **Lee PL et al** Purification and complementary DNA cloning of a receptor for basic fibroblast growth factor. S 245: 57-60 (1989); **Moscatelli D et al** Interaction of basic fibroblast growth factor with extracellular matrix and receptors. ANY 638: 177-81 (1991); **Olwin BB & Rapraeger A** Repression of myogenic differentiation by aFGF, bFGF, and K-FGF is dependent on cellular heparan sulfate. JCB 118: 631-9 (1992); **Ornitz DM et al** Heparin is required for cell-free binding of basic fibroblast growth factor to a soluble receptor and for mitogenesis in whole cells. MCB 12: 240-7 (1992); **Partanen J et al** Diverse receptors for fibroblast growth factors. PGFR 4: 69-83 (1992); **Rapraeger AC et al** Requirement of heparan sulfate for bFGF-mediated fibroblast growth and myoblast differentiation. S 252: 1705-8 (1991); **Stark KL et al** FGFR-4, a new member of the fibroblast growth factor receptor family, expressed in the definitive endoderm and skeletal muscle lineages of the mouse. Development 113: 641-51 (1991); **Vainikka S et al** Fibroblast growth factor receptor-4 shows novel features in genomic structure, ligand binding and signal transduction. EJ 11: 4273-80 (1992); **Wennstrom S et al** cDNA cloning and expression of a human FGF receptor which binds acidic and basic FGF. GF 4: 197-208 (1991); **Werner S et al** Differential splicing in the extracellular region of fibroblast growth factor receptor 1 generates receptor variants with different ligand-binding specificities. MCB 12: 82-8 (1992); **Yayon A et al** Cell surface, heparin-like molecules are required for binding of basic fibroblast growth factor to its high affinity receptor. Cell 64: 841-8 (1991)

● **Biological Activities: Alarid ET et al** Evidence for an organ- and sex-specific role of basic fibroblast growth factor in the development of the fetal mammalian reproductive tract. E 129: 2148-54 (1991); **Araujo DM & Cotman CW** Basic FGF in astroglial, microglial, and neuronal cultures: characterization of binding sites and modulation of release by lymphokines and trophic factors. J. Neurosci. 12: 1668-78 (1992); **Benezra M et al** Reversal of basic fibroblast growth factor-mediated autocrine cell transformation by aromatic anionic compounds. CR 52: 5656-62 (1992); **Davidson JM & Broadley KN** Manipulation of the wound-healing process with basic fibroblast growth factor. ANY 638: 306-15 (1991); **Faktorovich EG et al** Photoreceptor degeneration in inherited retinal dystrophy delayed by basic fibroblast growth factor. N 347: 83-6 (1990); **Frim DM et al** Effects of biologically delivered NGF, BDNF and bFGF on striatal excitotoxic lesions. Neuroreport 4: 367-70 (1993); **Gallicchio VS et al** Basic fibroblast growth factor (B-FGF) induces early- (CFU-s) and late-stage hematopoietic progenitor cell colony formation (CFU-gm, CFU-meg, and BFU-e) by synergizing with GM-CSF, Meg-CSF, and erythropoietin, and is a radioprotective agent *in vitro*. Int. J. Cell. Cloning 9: 220-32 (1991); **Gospodarowicz D** Biological activities of fibroblast growth factors. ANY 638: 1-8 (1991); **Gay CG & Winkles JA** Interleukin 1 regulates heparin-binding growth factor 2 gene expression in vascular smooth muscle cells. PNAS 88: 296-300 (1991); **Gray CW & Patel AJ** Characterization of a neurotrophic factor produced by cultured astrocytes involved in the regulation of subcortical cholinergic neurons. Brain Res. 574: 257-65 (1992); **Hla T & Maciag T** Cyclo-oxygenase gene expression is down-regulated by heparin-binding (acidic fibroblast) growth factor-1 in human endothelial cells. JBC 266: 24059-63 (1991); **Hicks D & Courtois Y** Fibroblast growth factor stimulates photoreceptor differentiation *in vitro*. J. Neurosci. 12: 2022-33 (1992); **Ishikawa K et al** Neurotrophic effects of fibroblast growth factors on peptide-containing neurons in culture from postnatal rat hypothalamus. Neuroendocrinology 55: 193-8 (1992); **Kan M et al** High and low affinity binding of heparin-binding growth factor to a 130 kDa receptor correlates with stimulation and inhibition of growth of a differentiated human hepatoma cell. JBC 11306-13 (1988); **Logan A et al** Basic fibroblast growth factor and central nervous system injury. ANY 638: 474-6 (1991); **Plata-Salaman CR** Food intake suppression by growth factors and platelet peptides by direct action in the central nervous system. Neurosci. Lett. 94: 161-6 (1988); **Ray J et al** Proliferation, differentiation, and long-term culture of primary hippocampal neurons. PNAS 90: 3602-6 (1993); **Sato Y et al** Autocrinological role of basic fibroblast growth factor on tube formation of fascular endothelial cells *in vitro*. BBRC 180: 1098-1102 (1991); **Thomas KA** Transforming potential of fibroblast growth factor genes. TIBS 13: 327-8 (1988); **Unsicker K et al** Basic fibroblast growth factor in neurons and its putative functions. ANY 638: 300-5 (1991); **Vlodavsky I et al** Extracellular matrix-resident basic fibroblast growth factor: implication for the control of angiogenesis. JCBc 45: 167-76 (1991); **Zhang Y et al** Basic FGF, NGF, and IGFs protect hippocampal and cortical neurons against iron-induced degeneration. J. Cereb. Blood Flow Metab. 13: 378-88 (1993)

● **Transgenic/Knock-out/Antisense Studies: Becker D et al** Proliferation of human malignant melanomas is inhibited by antisense oligodeoxynucleotides targeted against basic fibroblast growth factor. EJ 8: 3685-91 (1989); **Morrison RS** Suppression of basic fibroblast growth factor expression by antisense oligodeoxynucleotides inhibits the growth of transformed human astrocytes. JBC 266: 728-34 (1991); **Murphy PR et al** Phosphorothioate antisense oligonucleotides against basic fibroblast growth factor inhibit anchorage-dependent and anchorage-independent growth of a malignant glioblastoma cell line. Mol. Endocrinol. 6: 877-84 (1992); **Werner S et al** Targeted expression of a dominant-negative FGF receptor mutant in the epidermis of transgenic mice reveals a role of FGF in keratinocyte organization and differentiation. EJ 12: 2635-43 (1993)

● **Assays: Baird A et al** Radioimmunoassay for fibroblast growth factor (FGF): release by the bovine anterior pituitary *in vitro*. Regul. Pept. 10: 309-17 (1985); **el Husseini AE et al** PCR detection of the rat brain basic fibroblast growth factor (bFGF) mRNA containing a unique 3' untranslated region. BBA 1131: 314-6 (1992); **Feige JJ & Baird A** Phosphorylation and identification of phosphorylated forms of basic fibroblast growth factor. MiE 198: 138-47 (1991); **Guthridge M et al** Detection of FGF-β mRNA in chondrosarcoma cells by a new *in situ* hybridization technique with synthetic oligonucleotide probes. JBBM 22: 279-88 (1991); **Ii M et al** Improved enzyme immunoassay for human basic fibroblast growth factor using a new enhanced chemiluminescence system. BBRC 193: 540-5 (1993); **Ishihara M et al** A cell-based assay for evaluating the interaction of heparin-like molecules and basic fibroblast growth factor. AB 202: 310-5 (1992); **Joseph-Silverstein J et al** The development of a quantitative RIA for basic fibroblast growth factor using polyclonal antibodies against the 157 amino acid form of human bFGF. The identification of bFGF in adherent elicited murine peritoneal macrophages. JIM 110: 183-92 (1988); **Kajio T et al** Quantitative colorimetric assay for basic fibroblast growth factor using bovine endothelial cells and heparin. J. Pharmacol. Toxicol. Methods 28: 9-14 (1992); **Kan M et al** Identification and assay of fibroblast growth factor receptors. MiE 198: 158-71 (1991); **Kardami E et al** Biochemical and ultrastructural evidence for the association of basic fibroblast growth factor with cardiac gap junctions. JBC 266: 19551-7 (1991); **Kurobe M et al** Fluorometric enzyme immunoassay of basic fibroblast growth factor with monoclonal antibodies. Clin. Chem. 38: 2121-3 (1992);

Nagasaka M et al [The determination of basic fibroblast growth factor (bFGF) in human placental tissue and the changes in bFGF during pregnancy] Nippon-Sanka Fujinka Gakkai Zasshi. 44: 329-35 (1992); **Schmitt JF et al** A new quantitative polymerase chain reaction-high performance ion exchange liquid chromatographic method for the detection of fibroblast growth factor-β (FGF-β) gene amplification. JBBM 24: 119-33 (1992); **Schulze-Osthoff K et al** *In situ* detection of basic fibroblast growth factor by highly specific antibodies. Am. J. Pathol. 137: 85-92 (1990); **Songsakphisarn R & Goldstein N** The use of a novel immune complex to isolate neutralizing antibodies to basic fibroblast growth factor. Hybridoma 12: 343-8 (1993); **van Zoelen EJ** The use of biological assays for detection of polypeptide growth factors. PGFR 2: 131-52 (1990); **Watanabe H et al** A sensitive enzyme immunoassay for human basic fibroblast growth factor. BBRC 175: 229-35 (1991)

● CLINICAL USE & SIGNIFICANCE: **Aspenberg P & Lohmander LS** Fibroblast growth factor stimulates bone formation. Acta Orthop. Scand. 60: 473-6 (1989); **Chodak GW et al** Increased levels of fibroblast growth factor-like activity in urine from patients with bladder or kidney cancer. CR 48: 2083-8 (1988); **Cuevas P et al** Basic fibroblast growth factor (FGF) promotes cartilage repair *in vivo*. BBRC 156: 611-8 (1988); **Eguchi J et al** Gene expression and immunohistochemical localization of basic fibroblast growth facto in renal cell carcinoma. BBRC 183: 937-44 (1992); **Fiddes JC et al** Preclinical wound-healing studies with recombinant human basic fibroblast growth factor. ANY 638: 316-28 (1991); **Fujimoto K et al** Increased serum levels of basic fibroblast growth factor in patients with renal cell carcinoma. BBRC 180: 386-92 (1991); **Hebda PA et al** Basic fibroblast growth factor stimulation of epidermal wound healing in pigs. J. Invest. Dermatol. 95: 626-31 (1990); **Itoh H et al** Specific blockade of basic fibroblast growth factor gene expression in endothelial cells by antisense oligonucleotide. BBRC 188: 1205-13 (1992); **Khoury RK et al** The effect of basic fibroblast growth factor on the neovascularisation process: skin flap survival and staged flap transfers. Br. J. Plast. Surg. 44: 585-8 (1991); **Lindner V & Reidy MA** Proliferation of smooth muscle cells after vascular injury is inhibited by an antibody against basic fibroblast growth factor. PNAS 88: 3739-43 (1991); **Lindner V et al** Role of basic fibroblast growth factor in vascular lesion formation. Circ. Res. 68: 106-13 (1991); **Mayahara H et al** *In vivo* stimulation of endosteal bone formation by basic fibroblast growth factor in rats. GF 9: 73-80 (1993); **Mazué G et al** Preclinical and clinical studies with recombinant human basic fibroblast growth factor. ANY 638: 329-40 (1991); **Mignatti P et al** Basic fibroblast growth factor released by single, isolated cells stimulates their migration in an autocrine manner. PNAS 88: 11007-11 (1991); **Nguyen M et al** Elevated levels of an angiogenic peptide, basic fibroblast growth factor, in the urine of bladder cancer patients. JNCI 85: 241-2 (1993); **Robson MC et al** The safety and effect of topically applied recombinant basic fibroblast growth factor on the healing of chronic pressure sores. Ann. Surg. 216: 401-6 (1992); **Schiff M et al** Juvenile nasopharyngeal angiofibroma contain an angiogenic growth factor: basic FGF. Laryngoscope 102: 940-5 (1992); **Takahashi JA et al** Correlation of basic fibroblast growth factor expression levels with the degree of malignancy and vascularity in human gliomas. J. Neurosurg. 76: 792-8 (1992); **Tanimoto H et al** Expression of basic fibroblast growth factor in human gastric carcinomas. Virchows Arch. B Cell Pathol. 61: 263-7 (1991); **Tsuboi R & Rifkin DB** Recombinant basic fibroblast growth factor stimulates wound healing in healing-impaired db/db mice. JEM 172: 245-51 (1990); **Yanagisawa-Miwa A et al** Salvage of infarcted myocardium by angiogenic action of basic fibroblast growth factor. S 257: 1401-3 (1992); **Yamanaka Y et al** Overexpression of acidic and basic fibroblast growth factors in human pancreatic cancer correlates with advanced tumor stage. CR 53: 5289-96 (1993)

BFU-E: *burst forming unit, erythroid* Term used to describe the earliest known erythroid precursor cells that eventually differentiate into erythrocytes (see also: Hematopoiesis). BFU-e produce large colonies of erythroid cells that consist of "bursts" of smaller colonies. These colonies appear after 10 to 15 days in culture.

The factor now known as » IL3 was initially isolated as a burst promoting activity (see: BPA) stimulating the growth of BFU-E. BFU-E entirely depends on the continuous presence of IL3. IL3 appears to be a priming factor *in vitro* and *in vivo* that renders progenitor cells sensitive for the action of other cytokines.

The growth of BFU-E is also promoted by » EDF (eosinophil differentiation factor), a factor now known to be identical with » IL5. This activity probably depends also upon other factors secreted by adherent stromal cells stimulated by EDF. Other factors that stimulate growth of BFU-E are » IL4, » IL6, and » IL11. Survival of BFU-E cells is also promoted by » IL1. This factor has also been shown to inhibit the growth of BFU-E. This activity is mediated by IFN-γ and the effect is, in turn, reversible by » Epo. In the presence of optimal concentrations of Epo and IL3 their effects on murine and human BFU-E are synergised by » bFGF. The growth of BFU-E is inhibited by » TGF-β1. It has been shown that » IGF (I and II) also influence the growth of BFU-E via a direct mechanism involving the type I IGF receptor. It has been shown that » Activin A acts as a commitment factor and/or a promoter of erythroid progenitors that influences further development of BFU-E. BFU-E colony formation is enhanced by activin A only when the cells are exposed to IL3 plus Epo, but not when exposed to Epo or Epo plus SCF. The growth of BFU E is inhibited by » Inhibin. Colony formation of BFU-E stimulated by GM-CSF plus SCF is suppressed by several » chemokines, including » MIP-1α, MIP-2α, » PF4, » IL8, and » MCAF. MIP-1β blocks the suppressive effects of MIP-1α. MIP-2β or GRO-α block the suppressive effects of IL8 and PF4. BFU-E is inhibited by » TNF-α and » TNF-β.

BFU-E is characterized by the expression of several cell surface markers such as » CD33, » CD34 (see also: CD antigens) and HLA-DR. The cells do not yet express glycophorin A.

BFU-E can be assayed and identified in a » colony formation assay by the specific morphology of the cells. The growth of BFU-E is independent of » Epo. The cells, however, respond to Epo by developing hemoglobinized cell clusters during *in vitro* culture. This effect is synergised by » IL9 which also supports survival and the early stage of proliferation of BFU-E. Under certain culture conditions » IGF-I can entirely replace Epo although at least a 100-fold higher molar concentration than that of Epo is required to reach maximal stimulation. BFU-E also responds to » uteroferrin and » E-CSF (erythroid colony-stimulating factor). BFU-E is also inhibited by the tetrapeptide » AcSDKP but not by the pentapeptide » pEEDCK. Retinoic acid also suppresses BFU-E formation.

The cell surface antigen CD45, which encodes a protein tyrosine phosphatase, has been implicated in the stimulation of early human myeloid progenitor cells by various growth factors. Treatment of cells with various » antisense oligodeoxynucleotides the CD45 gene, or with monoclonal anti-CD45, significantly decreases GM-CSF-, IL3, and SCF-enhanced Epo-dependent BFU-E colony formation, but had little or no effect on BFU-E colony formation stimulated by Epo alone.

Further development of BFU-E requires stem cell factor (see: SCF). In suspension cultures of BFU-e containing no » Epo a population of cells that contain globin within 1 week can be obtained by treatment with » SCF (stem cell factor) in synergy with » IL3. The effects of SCF on BFU-E, but not those of Epo, can be mimicked by » vanadate. DNA synthesis in BFU-E is also inhibited by » NRP (negative regulatory protein).

A reduced production of BFU-E colonies *in vitro* or a complete absence is observed in patients with myelodysplastic syndromes (MDS). In mice mutations in the receptor for stem cell factor » SCF (*W*-Locus) also lead to a reduction in BFU-E cell counts. Within the erythropoietic stem cell lineage BFU-E precedes the more mature » CFU-E (colony-forming units erythroid). Maturation of the cells is accompanied by a progressive increase of Epo receptors on BFU-E and the cells become progressively more dependent upon Epo. It has been shown that among various hematopoietic progenitors BFU-E is the most sensitive to treatment with » 4-HC (4-hydroperoxycyclophosphamide).

Aglietta M et al Interleukin-3 *in vivo*: kinetic of response of target cells. Blood 82: 2054-61 (1993); **Aglietta M et al** Granulocyte-macrophage colony stimulating factor and interleukin 3: target cells and kinetics of response *in vivo*. Stem Cells Dayt. 11: s83-7 (1993); **Akahane K et al** Pure erythropoietic colony and burst formations in serum-free culture and their enhancement by insulin-like growth factor I. EH 15: 797-802 (1987); **Bot FJ et al** Stimulating spectrum of human recombinant multi-CSF (IL3) on human marrow precursors: importance of accessory cells. Blood 71: 1609-14 (1988); **Broxmeyer HE et al** CD45 cell surface antigens are linked to stimulation of early human myeloid progenitor cells by interleukin 3 (IL3), granulocyte/macrophage colony-stimulating factor (GM-CSF), a GM-CSF/IL3 fusion protein, and mast cell growth factor (a c-*kit* ligand). JEM 174: 447-58 (1991); **Broxmeyer HE et al** Growth characteristics and expansion of human umbilical cord blood and estimation of its potential for transplantation in adults. PNAS 89: 4109-13 (1992); **Broxmeyer HE et al** Comparative analysis of the human macrophage inflammatory protein family of cytokines (chemokines) on proliferation of human myeloid progenitor cells. Interacting effects involving suppression, synergistic suppression, and blocking of suppression. JI 150: 3448-58 (1993); **Correa PN et al** Production of erythropoietic bursts by progenitor cells from adult human peripheral blood in an improved serum-free medium: role of insulin-like growth factor 12. Blood 78: 2823-33 (1991); **Dai CH et al** Human burst-forming units-erythroid need direct interaction with stem cell factor for further development. Blood 78: 2493-7 (1991); **Dai CH & Krantz S** Vanadate mimics the effect of stem cell factor on highly purified human erythroid burst-forming units *in vitro*, but not the effect of erythropoietin. EH 20: 1055-60 (1992); **Dainiak N et al** Primary human marrow cultures for erythroid bursts in a serum-substituted system. EH 18: 1073-9 (1985); **Deldar A et al** Canine BFU-e progenitors: adaptation of a reproducible assay and anatomical distribution. IJCC 9: 579-93 (1991); **Del-Rizzo DF et al** Interleukin 3 opposes the action of negative regulatory protein (NRP) and of transforming growth factor β (TGF-β) in their inhibition of DNA synthesis of the erythroid stem cell BFU-E. EH 18: 138-42 (1990); **Dowton LA & Ma DD** A method for enriching myeloid (CFU-GM) and erythroid (BFU-E) progenitor cells from human cord blood by accessory cell depletion. Pathology 24: 291-5 (1992); **Ferrajoli A et al** Analysis of the effects of tumor necrosis factor inhibitors on human hematopoiesis. Stem Cells Dayt. 11: 112-9 (1993); **Gallicchio YS et al** Basic fibroblast growth factor (B-FGF) induces early- (CFU-s) and late-stage hematopoietic progenitor cell colony formation (CFU-gm, CFU-meg, and BFU-e) by synergizing with GM-CSF, Meg-CSF, and erythropoietin, and is a radioprotective agent *in vitro*. IJCC 9: 220-32 (1991); **Gebbia V et al** Analysis of human dysplastic hematopoiesis in long-term bone marrow culture. Blood 59: 442-8 (1989); **Gratas C et al** Retinoid acid supports granulocytic but not erythroid differentiation of myeloid progenitors in normal bone marrow cells. Leukemia 7: 1156-62 (1993); **Gregory CJ et al** Three stages of erythropoietic progenitor cell differentiation distinguished by a number of physical and biologic properties. Blood 51: 527 (1978); **Hangoc G et al** Influence of IL1 α and -1 β on the survival of human bone marrow cells responding to hematopoietic colony-stimulating factors. JI 142: 4329-34 (1989); **Hangoc G et al** Effects *in vivo* of recombinant human inhibin on myelopoiesis in mice. EH 20: 1243-6 (1992); **Hangoc G et al** *In vivo* effects of recombinant interleukin-11 on myelopoiesis in mice. Blood 81: 965-72 (1993); **Heath DS et al** Separation of the erythropoietin-responsive progenitors BFU-E and CFU-E in mouse bone marrow by unit gravity sedimentation. Blood 47: 777-92 (1976); **Iscove VW & Sieber F** Erythroid progenitors in mouse bone marrow detected by macroscopic colony formation in culture. EH 3: 32-43 (1975); **Kanfer EJ et al** The *in vitro* effects of interferon-γ, interferon-α, and tumor-necrosis factor-α on erythroid burst-forming unit growth in patients with non-leukaemic myeloproliferative disorders. Eur. J. Haematol. 50: 250-4 (1993); **Keller JR et al** Transforming growth factor β: possible roles in the regulation of normal and leukemic hematopoietic cell growth. JCBc 39: 175-84 (1989);

Means RT Jr & Krantz SB Inhibition of human erythroid colony-forming units by γ interferon can be corrected by recombinant human erythropoietin. Blood 78: 2564-7 (1991); **Means RT Jr et al** Inhibition of human erythroid colony-forming units by interleukin-1 is mediated by γ interferon. JCP 150: 59-64 (1992); **Merchav S et al** Comparative studies of the erythroid-potentiating effects of biosynthetic human insulin-like growth factors-I and -II. J. Clin. Endocrinol. Metab. 74: 447-52 (1992); **Migliaccio AR et al** Direct effects of IL4 on the *in vitro* differentiation and proliferation of hematopoietic progenitor cells. Biotechnol. Ther. 1: 347-60 (1989-90) **Mitjavila MT et al** Effects of five recombinant hematopoietic growth factors on enriched human erythroid progenitors in serum-replaced cultures. JCP 138: 617-23 (1989); **Mizuguchi T et al** Activin A suppresses proliferation of interleukin-3-responsive granulocyte-macrophage colony-forming progenitors and stimulates proliferation and differentiation of interleukin-3-responsive erythroid burst-forming progenitors in the peripheral blood. Blood 81: 2891-7 (1993) **Muta K & Krantz SB** Apoptosis of human erythroid colony-forming cells is decreased by stem cell factor and insulin-like growth factor I as well as erythropoietin. JCP 156: 264-71 (1993); **Nakamura K et al** Effect of erythroid differentiation factor on maintenance of human hematopoietic cells in co-cultures with allogenic stromal cells. BBRC 194: 1103-10 (1993); **Nakao K et al** Effects of erythroid differentiation factor (EDF) on proliferation and differentiation of human hematopoietic progenitors. EH 19: 1090-5 (1991); **Ottmann OG & Pelus LM** Differential proliferative effects of transforming growth factor β on human hematopoietic progenitor cells. JI 140: 26561-5 (1988); **Pantel K & Nakeff A** The role of lymphoid cells in hematopoietic regulation. EH 21: 738-42 (1993); **Papayannopoulou T et al** *kit* ligand in synergy with interleukin-3 amplifies the erythropoietin-independent, globin-synthesizing progeny of normal human burst-forming units-erythroid in suspension cultures: physiologic implications. Blood 81: 299-310 (1993); **Peschle C et al** c-*kit* ligand reactivates fetal hemoglobin synthesis in serum-free culture of stringently purified normal adult burst-forming unit-erythroid. Blood 81: 328-36 (1993); **Schaafsma MR et al** Interleukin-9 stimulates the proliferation of enriched human erythroid progenitor cells: additive effect with GM-CSF. Ann. Hematol. 66: 45-9 (1993); **Serke S et al** Analysis of CD34-positive hemopoietic progenitor cells from normal human adult peripheral blood: flow-cytometrical studies and in-vitro colony (CFU-GM, BFU-E) assays. Ann. Hematol. 62: 45-53 (1991); **Shibuya T & Mak TW** Isolation and induction of erythroleukemic cell lines with properties of erythroid progenitor burst-forming cell (BFU-E) and erythroid precursor cell (CFU-E). PNAS 80: 3721-5 (1983); **Snoeck HW et al** Differential regulation of the expression of CD38 and human leukocyte antigen-DR on $CD34^+$ hematopoietic progenitor cells by interleukin-4 and interferon-γ. EH 21: 1480-6 (1993); **Sonoda Y et al** Erythroid burst-promoting activity of purified recombinant human GM-CSF and interleukin 3: studies with anti-GM-CSF and anti-IL3 sera and studies in serum-free cultures. Blood 72: 1381-6 (1988); **Testa U et al** Cascade transactivation of growth factor receptors in early human hematopoiesis. Blood 81: 1442-56 (1993); **Tong J et al** *In vivo* administration of recombinant methionyl human stem cell factor expands the number of human marrow hematopoietic stem cells. Blood 82: 784-91 (1993); **Vainchenker W et al** Fetal hemoglobin synthesis in culture of early erythroid precursors (BFU-E) from the blood of normal adults. JCP 102: 297-303 (1980); **van Bockstaele DR** Direct effects of 13-cis and all-trans retinoic acid on normal bone marrow (BM) progenitors: comparative study on BM mononuclear cells and on isolated $CD34^+$ BM cells. Ann. Hematol. 66: 61-6 (1993); **Wisniewski D et al** Enrichment of hematopoietic progenitor cells (CFUC and BFUE) from human peripheral blood. EH 10: 817-29 (1982)

BFU-Mk: *burst forming unit, megakaryocyte* See: CFU-MEG.

BGDF: *B cell growth and differentiation factor* Collective term used to describe factors that pro-

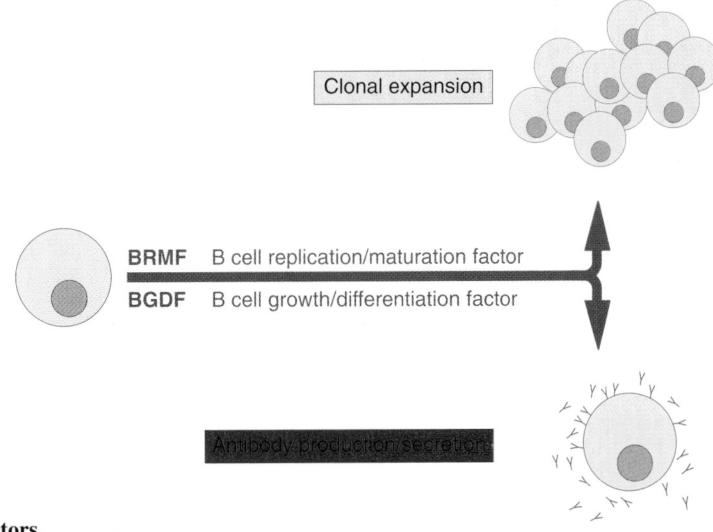

B cell growth/differentiation factors.
B cell growth/differentiation factors support development of activated B cells into antibody-producing cells and promote clonal expansion. Factors with BGDF activity have also been described as BRMF (B cell replication/maturation factors). The activities os some factors are more restricted. They either support only clonal expansion or only differentiation into antibody-expressing cells but not both.

mote the proliferation and the differentiation of B cells. For problems of nomenclature see: BCGF (B cell growth factors), BCDF (B cell differentiation factors), and » BSF (B cell stimulating factor). One particular factor with BGDF activity that induces both proliferation and IgM secretion in the mouse leukemic B cell line, BCL_1, is » BCGF 2 and is therefore identical with » IL5. Some BGDF activity may be due to » IL4 (see also BCDF). Another cytokine with BGDF activity is » IL13.

Callard RE et al Assay for human B cell growth and differentiation factors. in: Clemens MJ et al (eds) Lymphokines and Interferons. A practical Approach, pp. 345-64, IRL Press, Oxford 1987; **Nakajima K et al** Physicochemical and functional properties of murine B cell-derived B cell growth factor II (WEHI-231-BCGF-II). JI 135: 1207-12 (1985); **Pike BL et al** A high-efficiency cloning system for single hapten-specific B lymphocytes that is suitable for assay of putative growth and differentiation factors. PNAS 82: 3395-9 (1985); **Shimizu K et al** Immortalization of BGDF (BCGF II)- and BCDF-producing T cells by human T cell leukemia virus (HTLV) and characterization of human BGDF (BCGF II). JI 134: 1728-33 (1985)

BGEF: *B cell derived growth enhancing factor 2* This poorly characterized factor of 60-65 kDa (gel filtration) is secreted by a human B cell line established from the peripheral blood of a patient with rheumatoid arthritis. BGEF enhances IL1-induced proliferation of peanut agglutinin nonagglutinated thymocytes and IL1-induced production of (IL2 by thymocytes and a human T cell clone, HSB.2 C5B2. BGEF alone does not induce the production of IL2. BGEF differs from » BGEF-2.

Kang H et al Characterization of a B cell-derived growth-enhancing factor produced by a human B cell line established from a patient with rheumatoid arthritis. JI 139: 1154-60 (1987)

BGEF-2: *B cell derived growth enhancing factor 2* This poorly characterized factor(s) of 15-20 kDa (gel filtration) was initially identified and purified as an IL2 enhancing factor enhancing the IL2-dependent proliferation of CTLL A/J cells. The activity is produced by B cells of patients with rheumatoid arthritis and systemic lupus erythematosus only when they are in the active stage of the disease. BGEF-2 enhances IL2-dependent growth of peripheral blood T cells from patients with active rheumatoid arthritis but does not enhance the growth of T cells from healthy volunteers.

BGEF-2 itself does not display » IL2 or » IL1 activity and » IFN-α, » IFN-γ, » TNF, » IL4, » IL5, » IL6 do not display BGEF-2 activity. BGEF-2 also appears to be different from another B cell-derived growth enhancing factor, » BGEF.

Tomura K et al IL2 enhancing factor(s) in B cell supernatants from patients with rheumatoid arthritis or systemic lupus erythematosus. Tohoku J. Exp. Med. 159: 171-83 (1989); **Tomura K et al** Correlations between IL2 enhancing activity and clinical parameters in patients with rheumatoid arthritis and systemic lupus erythematosus. Tohoku J. Exp. Med. 163: 269-77 (1991)

BGF: *bone growth factor* This growth factor is isolated from normal and cancerous rat prostates. It is identical with » bFGF. See also: PGF (prostatic growth factor)

Matsuo Y et al Heparin binding affinity of rat prostatic growth factor in normal and cancerous prostates: partial purification and characterization of rat prostatic growth factor in Dunning tumor. CR 47: 188-92 (1987)

BHK: *Baby hamster kidney* This cell line, established from pooled kidneys of Syrian or Golden hamster (*Mesocricetus auratus*) in the early 60s has been used as one of the standard cell lines for many biochemical investigations including studies of vaccine production, cellular metabolism, » gene expression, » cell cycle, mutagenesis, and use of » antisense oligonucleotides. Many different sublines of the original cell line are available. BHK cells are used, among other things, for producing recombinant » cytokines and other proteins. Recombinant proteins produced in BHK cells usually undergo correct post-translational processing which may be required for full bioactivity of the factor when used as a pharmacological compound. For other expression systems used in cytokine research see also: CHO cells, COS cells, Namalwa cells, Baculovirus expression system, *Escherichia coli*).

Brill G et al BHK-21-derived cell lines that produce basic fibroblast growth factor, but not parental BHK-21 cells, initiate neuronal differentiation of neural crest progenitors. Development 115: 1059-69 (1992); **Hayakawa T et al** *In vivo* biological activities of recombinant human erythropoietin analogs produced by CHO cells, BHK cells and C127 cells. Biologicals 20: 253-7 (1992); **Nimtz M et al** Structures of sialylated oligosaccharides of human erythropoietin expressed in recombinant BHK-21 cells. EJB 213: 39-56 (1993)

BHRF1: This is an open reading frame encoded by the human herpesvirus Epstein-Barr virus. The protein is expressed in the early lytic cycle but is not essential *in vitro* either for virus-induced B cell transformation of lymphoblastoid cell lines or full virus replication. BHRF1 appears to be the viral homologue of » *bcl*-2 and resembles the *bcl*-2 gene both in its subcellular localization and in its capacity to enhance survival of B cells. BHRF1 may serve to delay cell death by » apoptosis and thereby maximize virus production. For another virus-encoded protein increasing survival by suppressing antiviral immune responses see: BCRF1 (viral IL10; see: IL10).

Henderson S et al Epstein-Barr virus-coded BHRF1 protein, a viral homologue of *bcl*-2, protects human B cells from programmed cell death. PNAS 90: 8479-83 (1993)

BIF: *B cell growth inhibitory factor* This poorly characterized factor of ≈ 80 kDa was found in culture supernatants of Con A-activated human peripheral blood mononuclear cells. It was found to be an inhibitor of antigen-stimulated B cells proliferation and clonal expansion induced by a B cell growth factor (see: BCGF). BIF does not inhibit the proliferation of murine » 3T3 fibroblasts, murine myeloma cell line NS-1, human lymphoid cell lines (MOLT-4, HSB-2, and Daudi), or a human myeloid cell line (K-562).
Higher levels of BIF activity appear to found in patients with multiple myeloma but not in patients with benign monoclonal gammopathy.
Kawano M et al Identification and characterization of a B cell growth inhibitory factor (BIF) on BCGF-dependent B cell proliferation. JI 134: 375-81 (1985); **Kawano M et al** Altered cytokine activities are related to the suppression of synthesis of normal immunoglobulin in multiple myeloma. Am. J. Hematol. 30: 91-6 (1989)

BIF: *B cell inducing factor* This factor of ca 20 kDa is produced by human peripheral blood T cells. BIF is assayed by induction of IgM-, IgG-, and IgA-secreting cells in peripheral blood B (non-T) cells stimulated into proliferation with Staphylococcus aureus bacteria strain Cowan I (Sac), and in the IgM cell line » SKW6.4. BIF production by T cells is induced by dexamethasone. BIF is separable from » IL2 and is probably identical with one of the factors described as » BSF (B cell stimulating factors).
Jeong G et al Independent regulation of B cell inducing factor and IL2 production by T lymphocytes, and direct and indirect promotion of immunoglobulin secretion by glucocorticosteroid. CI 103: 199-206 (1986); **Ralph P et al** IgM and IgG secretion in human B cell lines regulated by B cell-inducing factors (BIF) and phorbol ester. Immunol. Lett. 7: 17-23 (1983); **Ralph P et al** Stimulation of immunoglobulin secretion in human B lymphocytes as a direct effect of high concentrations of IL 2. JI 133: 2442-5 (1984); **Ralph P et al** Human B cell-inducing factor(s) for production of IgM, IgG and IgA: independence from IL 2. JI 132: 1858-62 (1984)

bIFN: abbrev. for Beta-interferon. See: IFN-β. See also: IFN for general information about interferons.

Bikunin: Bikunin is the smallest of three subunits of » ITI (Inter-α-trypsin inhibitor). Bikunin is identical with » ECGF-2b (endothelial cell growth factor 2b).

Bilirubin: see: LGF (liver growth factor).

Binding proteins 25, 26, 28, or **29:** These proteins are identical with some of the insulin-like growth factor binding proteins (see: IGF-BP).

Bioassays: Assays in which concentrations of biologically active cytokines are quantitated either in absolute terms or in relation to a standard by the magnitude of a biological effect that these cytokines have in sensitive isolated cells and sometimes also in tissues, or organisms.
Bioassays exploit the many different biological activities of cytokines and measure, for example, cytokine-induced cell proliferation, cytotoxicity, capacity to induce colony formation (see also: colony formation assay; see also: Limiting dilution analysis), cellular degranulation, or the induction of secretion of further cytokines or other compounds. Generally cytokine-induced effects show saturation kinetics which can be used to quantitate their amounts from dose-response curves. It should be noted that cytokine assays frequently measure only one single aspect of the many biological activities of a given cytokine and that these *in vitro* assay also involve the use of unphysiological stimuli such as lectins or bacterial lipopolysaccharides. These assays frequently involve the use of primary cell cultures and established cell lines that depend upon the presence of (a) particular cytokine(s) for their growth or survival (see also: » factor-dependent cells) or that respond to a cytokine in a particular way. One should be aware of the fact that at this moment in time the relevance of many *in vitro* activities of cytokines to their endogenous functions within an intact organism is not clearly defined (see also: Transgenic animals; see also: subentry "Transgenic/Knock-out/Antisense studies" for individual cytokines) and that these assays based on function really only serve one purpose, namely to measure factor concentrations.
In many instances bioassays measure the effects of a given factor directly. However, there are also assays, known as ***conversion assays***, which involve the use of two different cell lines. In these assays the activity of a particular factor is determined indirectly by measuring, in another cell line, the activity of a second factor induced by the first factor.
A variety of immunoassays (see: Cytokine assays) allow determination and quantification of many cytokines by their immunoreactivities (for refe-

rences see individual factors). Bioassays specifically detect bioactivities, i. e. they specifically measure concentrations of biologically active cytokines. This may, of course, complicate the interpretation of quantitative measurements considerably if one considers, for example, to correlate global serum levels of different cytokines with their possible involvement in a particular disease situation.

In general it is advisable to use several complementary assay systems to avoid drawbacks and pitfalls of individual bioassays. In many instances the cell lines employed in these assays respond to more than one particular cytokine. In some cases specificity for a particular factor can be obtained by using antibodies specifically neutralizing known interfering factors and even receptor-blocking antibodies; however, this may not always be possible or even complicate analysis (see also, for example: Autocrine growth control). Potential complex interactions among cytokines must also be considered when interpreting data obtained from cytokine bioassays. Different cytokines may act through the same receptor and therefore induce similar biological responses that may only differ in the doses required for half-maximum effects. If available, monoclonal cytokine-specific antibodies can be used to confirm that the signal in the assay is due to the correct cytokine. Bacterial endotoxins are known to be good inducers of cytokines and hence their presence in biological fluids to be assayed may lead to false results. Adhesion of cells has also been shown to stimulate cytokine production.

Inhibitory factors or otherwise blocking activities (see: BP, binding proteins) in the biological samples assayed may also interfere with bioassays and may yield values that are much too low. A classical example is the downregulation by prostaglandin E2 of thymocyte proliferation induced by IL1. Some of the problems can be overcome by the use of serial dilutions of the unknown and standard preparations, a careful examination of dose-response curves, and the use of neutralizing antibodies. (Useful computer programs for non-linear fitting of raw response data are ***ALLFIT*** and ***FLEXFIT*** (obtainable in DOS and Macintosh formats from Dr. PJ Munson, Laboratory of Theoretical and Physical Biology, NICHHD, NIH, Bethesda, MD 20892, USA)).

Another drawback of bioassays may be the use of supernatants or sera because biologically active cytokines bound to the cell surface (see, for example: Paracrine) cannot be detected in this way.

Some bioassays do not differentiate between different molecular species of a particular cytokine or unrelated factors binding to the same cell surface receptor. IL1α and IL1β (see: IL1), for example, are structurally unrelated factors that show almost identical biological activities because they bind to the same receptor. An alternative and entirely different technique that may be used to detect levels of cytokines even in single cells by measuring the corresponding mRNAs is » Message amplification phenotyping.

Callard RE et al Assay for human B cell growth and differentiation factors. in: Clemens MJ et al (eds) Lymphokines and Interferons. A practical Approach, pp. 345-64, IRL Press, Oxford 1987; **Coligan JE et al** Current protocols in immunology. Grene and Wiley-Interscience, New York 1991); **Dotsika EN** Assays for mediators affecting cellular immune functions. Curr. Opin. Immunol. 2: 932-5 (1989); **Feldmann M et al** Cytokine assays: role in evaluation of the pathogenesis of autoimmunity. IR 119: 105-123 (1991); **Hamblin AS & O'Garra A** Assays for interleukins and other related factors. In: Lymphocytes, a practical approach, Klaus GGB (edt), pp. 209-28, IRL Press, Oxford, (1987); **Laska EM & Meisner MJ** Statistical methods and applications of bioassay. Annu. Rev. Pharmacol. Toxicol. 27: 385-97 (1987); **Mosman TR & Fong TAT** Specific assays for cytokine production by T cells JIM 116: 151-8 (1989); **Newton RC & Uhl J** Assays relevant to the detection and quantitation of cytokines and their inhibitors. Modern Methods in Pharmacol. 5: 83-99 (1989); **Thorpe R et al** Detection and measurement of cytokines. Blood Rev. 6: 133-48 (1992); **van Zoelen EJ** The use of biological assays for detection of polypeptide growth factors. Prog. Growth Factor Res. 2: 131-52 (1990); **Winstanley FP** Cytokine bioassay. In: Gallagher G et al (eds) Tumor Immunobiology, A practical Approach. Oxford University Press, pp. 179-303 (1993); **Wadha M et al** Quantitative biological assays for individual cytokines. In: Balkwill FR (edt) Cytokines, A practical approach. Oxford University press, pp. 309-330 (1991); for further information see also subentry "Assay" for many cytokines listed in this dictionary, which contains further cross-references to factor-dependent cell lines.

Bioleukin™: Trademark for recombinant IL2 (Biogen, Geneva).

Biological response modifiers: abbrev. BRM. See: Immunomodulation.

Bipotent stem cells: abbrev. BPSC. See: Hematopoiesis.

bithorax: see: Homeotic genes.

BL2M3-GPF: *BL2M3 cell growth promoting factor* see: LMW-BCGF (low molecular weight B cell growth factor).

Bladder carcinoma-derived cytokine: see: BCDC.

Blast2: Alternative name used previously for the cell surface antigen » CD23 (see also: CD antigens for general information).

Blast colony-forming cells: A type of early hematopoietic progenitors (see: Hematopoietic stem cells) in normal bone marrow that yields colonies of blast cells after extended incubation. The colonies also contain progenitor cells and produce secondary colonies on replating that contain multilineage hematopoietic cells (see also: Hematopoiesis). Most blast colonies contain some early hematopoietic stem cells that are defined as » CFU-S in a spleen colony formation assay. Blast colony-forming cells may even be pre-CFU-S cells. It has been suggested that blast CFCs resemble the most primitive high-proliferative potential colony-forming cells (see: HPP-CFC).

Leary AM & Ogawa M Blast colony assay for umbilical blood and adult bone marrow progenitors. Blood 69: 953-6 (1987); **Nakahata T & Ogawa M** Clonal origin of murine hematopoietic colonies with apparent restriction to granulocyte-macrophage-megakaryocyte (GMM) differentiation. JCP 111: 239-46 (1982); **Wright EG & Pragnell IB** The stem cell compartment: assays and negative regulators. CTMI 177: 136-49 (1992)

Blastogenic factor: see: BF.

B-LIF: *B cell-derived leukocyte (migration) inhibitory factor* see: LIF (leukocyte (migration) inhibitory factor).

***blk*:** *B lymphoid kinase* A complementary DNA specifying a polypeptide of 55 kDa has been isolated from murine lymphocytes. *blk* (on murine chromosome 14; > 30 kb, 13 exons) encodes a protein with tyrosine kinase activity that is related to other *src*-related protein kinases. *blk* is specifically expressed in the B cell lineage. Expression of *blk* is regulated during B-cell development: *blk* RNA is expressed in all pro-B-, pre-B-, and mature B-cell lines examined, but is absent from plasma cell lines.

The tyrosine kinase encoded by *blk* may function in a signal transduction pathway that is restricted to B lymphoid cells. *blk* is among a number of *src*-related protein tyrosine kinases expressed after stimulation of resting B lymphocytes with antibodies to surface immunoglobulins (sIgD, sIgM). *blk* and other *src* family tyrosine kinases including» *fyn*, » *lyn*, and perhaps » *lck*, are activated upon engagement of the B cell antigen receptor complex. These kinases then act directly or indirectly to phosphorylate and/or activate effector proteins including p42 (microtubule-associated protein kinase, MAPK), phospholipases C-γ 1 and C-γ 2, phosphatidylinositol 3-kinase (PI 3-K), and p21*ras*-GTPase-activating protein (GAP).

Burkhardt AL et al Anti-immunoglobulin stimulation of B lymphocytes activates *src*-related protein-tyrosine kinases. PNAS 88: 7410-4 (1991); **Cambier JC & Campbell KS** Membrane immunoglobulin and its accomplices: new lessons from an old receptor. FJ 6: 3207-17 (1992); **Dymecki SM et al** Specific expression of a tyrosine kinase gene, *blk*, in B lymphoid cells. S 247: 332-6 (1990); **Dymecki SM et al** Structure and developmental regulation of the B-lymphoid tyrosine kinase gene *blk*. JBC 267: 4815-23 (1992); **Law DA et al** Examination of B lymphoid cell lines for membrane immunoglobulin-stimulated tyrosine phosphorylation and *src*-family tyrosine kinase mRNA expression. Mol. Immunol. 29: 917-26 (1992); **Lin J & Justement LB** The MB-1/B29 heterodimer couples the B cell antigen receptor to multiple *src* family protein tyrosine kinases. JI 149: 1548-55 (1992); **Pleiman CM et al** Mapping of sites on the *src* family protein tyrosine kinases p55*blk*, p59*fyn*, and p56*lyn* which interact with the effector molecules phospholipase C-γ 2, microtubule-associated protein kinase, GTPase-activating protein, and phosphatidylinositol 3-kinase. MCB 13: 5877-87 (1993); **Yao XR & Scott DW** Expression of protein tyrosine kinases in the Ig complex of anti-μ-sensitive and anti-μ-resistant B cell lymphomas: role of the p55*blk* kinase in signaling growth arrest and apoptosis. IR 132: 163-86 (1993)

Blood progenitor activator: abbrev. BPA. This factor is identical with » IL3.

B lymphocyte-derived burst promoting activity: abbrev. B-BPA. See: E-CSF (erythroid colony-stimulating factor).

BM: *bone marrow* see: Hematopoiesis. See also: LTBMC (long-term bone marrow culture).

BMC: *bone marrow culture; bone marrow cells* see also: LTBMC (long-term bone marrow culture). See also: Hematopoiesis.

BMCM: *bone marrow conditioned medium* Conditioned medium (see: CM) obtained after cultivation of bone marrow cells.

BMF: *B cell maturation factor* Generic name given to factors that promote the maturation of resting B cells into Ig-secreting cells without having an effect on cell proliferation. Differentiation (or terminal differentiation) of B cells is the process leading to antibody secretion. The differentiation process is often divided into an *induction phase*, when class switching, i. e. expression of a different class of membrane-bound immunoglobulin occurs, and *maturation* in which the soluble form of this immunoglobulin is secreted by the cell.

BMFs induce the pre-B-like 70Z/3 tumor cell line to express complete Ig molecules on its cell surface, and cause the mature B cell-like WEHI-279 tumor cell line to increase its ratio of secretory to membrane μ production, begin high rate Ig secretion, and then die.
Some known factors with BMF activity are » IL2, » IL5 (see: BCGF 2, B cell growth factor 2), and » IFN-γ. For some aspects of nomenclature see also: BCDF (B cell differentiation factors) and BCGF (B cell growth factors). The aberrant expression of a B cell maturation factor is responsible for the severe immunodeficiency in » motheaten mice.
Karasuyama H et al Recombinant interleukin 2 or 5 but not 3 or 4 induces maturation of resting mouse B lymphocytes and propagates proliferation of activated B cell blasts. JEM 167: 1377-90 (1988); **Leanderson T et al** B cell growth factors: distinction from T cell growth factor and B cell maturation factor. PNAS 79: 7455-9 (1982); **Sherris DI & Sidman CL** Distinction of B cell maturation factors from lymphokines affecting B cell growth and viability. JI 136: 994-8 (1986); **Sidman CL et al** γ interferon is one of several direct B cell maturing lymphokines. N 309: 801-4 (1984); **Sidman CL et al** B cell maturation factor (BMF): a lymphokine or family of lymphokines promoting the maturation of B lymphocytes. JI 132: 209-22 (1984); **Sidman CL et al** B cell maturation factor: effects on various cell populations. JI 132: 845-50 (1984)

BMP: *bone morphogenetic proteins* BMP is the generic name of a family of proteins, originally identified in extracts of demineralized bone that were capable of inducing bone formation at ectopic sites (for further information see also subentry "Assays" in the reference section). BMPs are found in minute amounts in bone material (≈ 1 μg/kg dry weight of bone).
This family of proteins comprises ***BMP-1*** (730 amino acids, incl. 22 amino acids encoding a » signal sequence; related to » BP10), ***BMP-2A*** (renamed BMP-2; 68 % homology with *Drosophila » dpp*), ***BMP-2B*** (renamed BMP-4), ***BMP-3*** (identical with » osteogenin), ***BMP-4*** (identical with BMP-2B = DVR-4, see: DVR gene family; 72 % homology with *Drosophila » dpp*), ***BMP-5***, ***BMP-6*** (= DVR-6; vegetal-specific-related-1; see also: Vg-1), ***BMP-7*** (identical with » OP-1, osteogenic protein 1; see also: Vg-1), and ***BMP-8*** (identical with » OP-2). Some of these proteins exist as heterodimers, for example OP1/BMP2A).
With the exception of BMP-1, which is probably a Zinc-containing metalloproteinase and a homologue of ***astacin***, a protease isolated from crayfish (*Astacidae*), the proteins of the BMP family are members of the transforming growth factor family of proteins (see: TGF-β), based on primary amino acid sequence homology, including the absolute conservation of seven cysteine residues between TGF-β and the BMPs. From a high degree of amino acid sequence homology (≈ 90 %), BMP-5, BMP-6, and BMP-7 are recognized as a subfamily of the BMPs. Sequence analysis suggests that the » *Drosophila* 60A gene is the dipteran homologue of the BMP subfamily
The human genes encoding BMP-1, BMP-2A and BMP-3 map to chromosomes 8, 20p12, and 4p13-q21, respectively. They lie in the vicinity of the genes involved in the formation of cartilage and bone tissues. The genes encoding BMP-5 and BMP-6 map to human chromosome 6. The gene encoding BMP-7 maps to human chromosome 20. In the mouse the BMP-2A gene is located close to the Tsk (tight skin) locus, and it has been suggested that this gene may be the site for this mutation.
BMPs can be isolated from demineralized bones and osteosarcoma cells. They have also been shown to be expressed in a variety of epithelial and mesenchymal tissues in the embryo. Some BMPs (e. g. BMP-2 and BMP-4) have been shown to elicit qualitatively identical effects (cartilage and bone formation) and to have the ability to substitute for one another.
BMPs are proteins which act to induce the differentiation of mesenchymal-type cells into chondrocytes and osteoblasts before initiating bone formation. They promote the differentiation of cartilage- and bone-forming cells near sites of fractures but also at ectopic locations. Some of the proteins induce the synthesis of alkaline phosphatase and collagen in osteoblasts. Some BMPs act directly on osteoblasts and promote their maturation while at the same time suppressing myogenous differentiation. Other BMPs promote the conversion of typical fibroblasts into chondrocytes and are also capable of inducing the expression of an osteoblast phenotype in non-osteogenic cell types. In addition, some BMPs and also some related factors may be involved in embryonic development. Osteogenin and related BMPs also promote additional successive steps in the endochondral bone formation cascade by functioning as potent chemoattractants for circulating monocytes and by inducing, among other things, the synthesis and secretion of » TGF-β1 by monocytes. Monocytes stimulated by TGF-β secrete a number of chemotactic and mitogenic cytokines that recruit endothelial and mesenchymal cells and promote their synthesis of collagen and associated matrix constituents.

Limb development and outgrowth depends on epithelial-mesenchymal interactions. It is regulated by a combination of stimulatory and inhibitory signals secreted by the apical ectodermal ridge, a specialized epithelium at the limb tip, responsible for the stimulation of proliferation of the underlying mesenchyme. Proliferation of mesenchyme in the early mouse limb-bud is stimulated by » FGF-4 produced by the apical ectodermal ridge, and BMP-2 has been shown to inhibit limb growth.

It has been shown that mutations at the classic mouse locus ***short ear*** (se) on chromosome 9, associated with a specific spectrum of morphologic alterations in the ear and many internal skeletal structures, disrupt the mouse homologue of the BMP-5 gene. The mutant animals also have a defect in bone fracture repair and show soft tissue abnormalities including lung cysts, liver granulomas, and hydrotic kidneys.

The activity of BMPs is antagonized *in vivo* and *in vitro* by a factor called » OIP (osteogenesis inhibitory protein) while some activities of BMPs may be enhanced by » Activin A or » TGF-β.

The clinical use of these factors is still in its infancy. Relatively impure preparations have been used for the treatment of bone fractures. Some studies suggest that factors, such as osteogenin, must be combined with a bone-derived matrix in order to initiate bone differentiation. Osteogenin, in combination with insoluble collagenous bone matrix, has been used to induce local endochondral bone differentiation in calvarial defects of adult primates. It also appears that the osteoinductive potential of BMP preparations bound to porous β-tricalcium phosphate or the use of BMP combined with true bone ceramic as a bone grafting material are superior to treatment with BMPs alone for the treatment of bone tissue defects and the promotion of new bone formation.

BMP-2A has been suggested as a reasonable candidate for the human condition fibrodysplasia (myositis) ossificans progressiva (FOP).

■ **TRANSGENIC/KNOCK-OUT/ANTISENSE STUDIES:** The biological activities of BMP-4 have been studied in » transgenic animals (mice) expressing BMP-4 under the control of a bovine cytokeratin IV promoter. These animals show a defect in their fur and whiskers which is associated with progressive balding. It has been observed that the expression pattern of cytokeratin markers is disturbed in some transgenic hair follicles. In response to transgene expression outer root sheath cells below the stem cell compartment and hair matrix cells around the dermal papilla cease proliferation. Hyperproliferative responses in dermal fibroblasts have not been observed in these animals.

● **REVIEWS: Aldinger G et al** Bone morphogenetic protein: a review. Int. Orthop. 15: 169-77 (1991); **Canalis E et al** Growth factors and cytokines in bone cell metabolism. ARM 42: 17-24 (1991); **Hauschka PV et al** Polypeptide growth factors in bone matrix. Ciba Found. Symp. 136: 207-25 (1988); **Joyce ME et al** Role of growth factors in fracture healing. PCBR 365: 391-416 (1991); **Luyten FP et al** Advances in osteogenin and related bone morphogenetic proteins in bone induction and repair. Acta Orthop. Belg. 58: 263-7 (1992); **Ripamonti U & Reddi AH** Growth and morphogenetic factors in bone induction: role of osteogenin and related bone morphogenetic proteins in craniofacial and periodontal bone repair. Crit. Rev. Oral Biol. Med. 1992; 3: 1-14 (1992); **Rosen V & Thies RS** The BMP proteins in bone formation and repair. TIG 8: 97-102 (1992); **Wang EA** Bone morphogenetic proteins (BMPs): therapeutic potential in healing bony defects. TibTech 11: 379-83 (1993); **Watrous DA & Andrews BS** The metabolism and immunology of bone. Semin. Arthritis Rheum. 19: 45-65 (1989); **Wozney JM et al** Growth factors influencing bone development. JCS 13(suppl): 149-56 (1990)

● **BIOCHEMISTRY & MOLECULAR BIOLOGY: Aldinger G et al** Bone morphogenetic protein: a review. Int. Orthop. 15: 169-77 (1991); **Celeste AJ et al** Identification of transforming growth factor β family members present in bone-inductive proteins purified from bovine bone. PNAS 87: 9843-7 (1990); **Dickinson ME et al** Chromosomal localization of seven members of the murine TGF-β superfamily suggests close linkage to several morphogenetic mutant loci. Genomics 6: 505-20 (1990); **Dumermuth E et al** The astacin family of metalloendopeptidases. JBC 266: 21381-5 (1992); **Gopal Rao VVN et al** The gene for bone morphogenetic protein 2A (BMP2A) is localized to human chromosome 20p12 by radioactive and nonradioactive *in situ* hybridization. Hum. Genet. 90: 299-302 (1992); **Hahn GV et al** A bone morphogenetic protein subfamily: chromosomal localization of human genes for BMP5, BMP6, and BMP7. Genomics 14: 759-62 (1992); **Israel DI et al** Expression and characterization of bone morphogenetic protein-2 in Chinese hamster ovary cells. GF 7: 139-50 (1992); **Nishimatsu S et al** Genes for bone morphogenetic proteins are differentially transcribed in early amphibian embryos. BBRC 186: 1487-95 (1992); **Plessow S et al** cDNA sequence of *Xenopus laevis* bone morphogenetic protein 2 (BMP-2). BBA 1089: 280-2 (1991); **Shimell MJ et al** The *Drosophila* dorsal-ventral patterning gene *tolloid* is related to human bone morphogenetic protein 1. Cell 57: 469-81 (1991); **Stocker W et al** Implications of the three-dimensional structure of astacin for the structure and function of the astacin family of zinc-endopeptidases. EJB 214: 215-31 (1993); **Tabas JA et al** Bone morphogenetic protein: chromosomal localization of human genes for BMP1, BMP2A, and BMP3. Genomics 9: 283-9 (1991); **Takaoka K et al** Gene cloning and expression of a bone morphogenetic protein derived from a murine osteosarcoma. Clin. Orthop. 294: 344-52 (1993); **Wang EA et al** Purification of other distinct bone-inducing factors. PNAS 85: 9484-8 (1988); **Wozney JM et al** Novel regulators of bone formation: molecular clones and activities. S 242: 1528-34 (1988); **Wozney JM** The bone morphogenetic protein family and osteogenesis. Mol. Reprod. Dev. 32: 160-7 (1992)

● **RECEPTORS: Paralkar VM et al** Identification and characterization of cellular binding proteins (receptors) for recombinant human bone morphogenetic protein 2B, an initiator of bone differentiation cascade. PNAS 88: 3397-401 (1991)

● **BIOLOGICAL ACTIVITIES: Bentz H et al** Transforming growth factor-β2 enhances the osteoinductive activity of a bovine bone-derived fraction containing bone morphogenetic protein-2 and 3. Matrix 11: 269-75 (1991); **Chen TL et al** Bone morphogenetic protein 2b stimulation of growth and osteogenic phenotypes in rat osteoblast-like cells: comparison with TGF-β1. J. Bone Miner. Res. 6: 1387-93 (1991); **Chen P et al** Stimulation of chondrogenesis in limb bud mesoderm cells by recombinant human bone morphogenetic protein 2B (BMP-2B) and modulation by transforming growth factor β 1 and β 2. ECR 195: 509-15 (1991); **Cunningham NS et al** Osteogenin and recombinant bone morphogenetic protein 2B are chemotactic for human monocytes and stimulate transforming growth factor β 1 mRNA expression. PNAS 89: 11740-4 (1992); **Fujimori Y et al** Heterotopic bone formation induced by bone morphogenetic protein in mice with collagen-induced arthritis. BBRC 186: 1362-7 (1992); **Hammonds RG et al** Bone-inducing activity of mature BMP-2b produced from a hybrid BMP-2a/2b precursor. ME 5: 149-55 (1991); **Harrison ET Jr** Osteogenin promotes reexpression of cartilage phenotype by dedifferentiated articular chondrocytes in serum-free medium. ECR 192: 340-5 (1991); **Harrison ET Jr et al** Transforming growth factor-β: its effect on phenotype reexpression by dedifferentiated chondrocytes in the presence and absence of osteogenin. In Vitro Cell Dev. Biol. 28A: 445-8 (1992); **Hiraki Y et al** Bone morphogenetic proteins (BMP-2 and BMP-3) promote growth and expression of the differentiated phenotype of rabbit chondrocytes and osteoblastic MC3T3-E1 cells. J. Bone Miner. Res. 6: 1373-85 (1991); **Jones CM et al** DVR-4 (bone morphogenetic protein-4) as a posterior-ventralizing factor in *Xenopus* mesoderm induction. Development. 115: 639-47 (1992); **Kaplan FS et al** Fibrodysplasia ossificans progressiva: a clue from the fly? Calcif. Tissue Int. 47: 117-25 (1990); **Katagiri T et al** The non-osteogenic mouse pluripotent cell line C3H10T1/2 is induced to differentiate into osteoblastic cells by recombinant human bone morphogenetic protein 2. BBRC 172: 295-9 (1990); **Kingsley DM et al** The mouse short ear skeletal morphogenesis locus is associated with defects in a bone morphogenetic member of the TGF β superfamily. Cell 71: 399-410 (1992); **Koster M et al** Bone morphogenetic protein 4 (BMP-4), a member of the TGF-β family, in early embryos of *Xenopus laevis*: analysis of mesoderm inducing activity. Mech. Dev. 33: 191-9 (1991); **Luyten FP et al** Natural bovine osteogenin and recombinant human bone morphogenetic protein 2B are equipotent in the maintenance of proteoglycans in bovine articular cartilage explant cultures. JBC 267: 3691-5 (1992); **Lyons KM et al** Patterns of expression of murine *Vgr*-1 and BMP-2a RNA suggest that transforming growth factor β-like genes co-ordinately regulate aspects of embryonic development. GD 3: 1657-68 (1989); **Lyons KM et al** Organogenesis and pattern formation in the mouse: RNA distribution patterns suggest a role for bone morphogenetic protein-2A (BMP-2A). Development 109: 833-44 (1990); **Niswander L & Martin GR** FGF-4 and BMP-2 have opposite effects on limb growth. N 361: 68-71 (1993); **Ogawa Y et al** Bovine bone activin enhances bone morphogenetic protein-induced ectopic bone formation. JBC 267: 14233-7 (1992); **Paralkar VM et al** Interaction of osteogenin, a heparin-binding bone morphogenetic protein, with type IV collagen. JBC 265: 17281-4 (1990); **Takuwa Y et al** Bone morphogenetic protein-2 stimulates alkaline phosphatase activity and collagen synthesis in cultured osteoblastic cells, MC3T3-E1. BBRC 174: 96-101 (1991); **Thies RS et al** Recombinant human bone morphogenetic protein 2 induces osteoblastic differentiation in W-20-17 stromal cells. Endocrinology 130: 1318-24 (1992); **Vainio S et al** Identification of BMP-4 as a signal mediating secondary induction between epithelial and mesenchymal tissues during early tooth development. Cell 75: 45-8 (1993); **Vukicevic S et al** Stimulation of the expression of osteogenic and chondrogenic phenotypes *in vitro* by osteogenin. PNAS 86: 8793-7 (1989); **Yabu M et al** Ultramicroscopic aspects of the conversion of fibroblasts to chondrocytes in the mouse dorsal subfascia by bone morphogenetic protein (BMP). Arch. Histol. Cytol. 54: 95-102 (1991); **Yamaguchi A et al** Recombinant human bone morphogenetic protein-2 stimulates osteoblastic maturation and inhibits myogenic differentiation *in vitro*. JCB 113: 681-7 (1991); **Yu YM et al** Changes in the gene expression of collagens, fibronectin, integrin, and proteoglycans during matrix-induced bone morphogenesis. BBRC 177: 427-32 (1991)

● **TRANSGENIC/KNOCK-OUT/ANTISENSE STUDIES: Blessing M et al** Transgenic mice as a model to study the role of TGF-β related molecules in hair follicles. Genes Dev. 7: 204-15 (1993)

● **ASSAYS: Kataoka H & Urist MR** Transplant of bone marrow and muscle-derived connective tissue cultures in diffusion chambers for bioassay of bone morphogenetic protein. Clin. Orthop. 286: 262-70 (1993); **Nogami H et al** Bioassay of chondrocyte differentiation by bone morphogenetic protein. Clin. Orthop. 258: 295-9 (1990); **Urist MR & Hudak RT** Radioimmunoassay of bone morphogenetic protein in serum: a tissue-specific parameter of bone metabolism. PSEBM 176: 472-5 (1984); **Urist MR et al** Preparation and bioassay of bone morphogenetic protein and polypeptide fragments. MiE 146: 294-312 (1987); **Wozney JM et al** Novel regulators of bone formation: molecular clones and activities. S 242: 1528-34 (1988)

● **CLINICAL USE & SIGNIFICANCE: Gerhart TN et al** Healing segmental femoral defects in sheep using recombinant human bone morphogenetic protein. Clin. Orthop. 293: 317-26 (1993); **Heckman JD et al** The use of bone morphogenetic protein in the treatment of non-union in a canine model. J. Bone Joint Surg. 73: 750-64 (1991); **Horisaka Y et al** subperiosteal implantation of bone morphogenetic protein adsorbed to hydroxyapatite. Clin. Orthop. 268: 303-12 (1991); **Johnson EE et al** Repair of segmental defects of the tibia with cancellous bone grafts augmented with human bone morphogenetic protein. A preliminary report. Clin. Orthop. 236: 249-57 (1988); **Johnson EE et al** Distal metaphyseal tibial non-union. Deformity and bone loss treated by open reduction, internal fixation, and human bone morphogenetic protein (hBMP). Clin. Orthop. 250: 234-40 (1990); **Johnson EE et al** Resistant nonunions and partial or complete segmental defects of long bones. Treatment with implants of a composite of human bone morphogenetic protein (BMP) and autolyzed, antigen-extracted, allogeneic (AAA) bone. Clin. Orthop. 277: 229-37 (1992); **Kaplan FS et al** Fibrodysplasia ossificans progressiva: a clue from the fly? Calcif. Tissue Int. 47: 117-25 (1990); **Katoh T et al** Osteogenesis in sintered bone combined with bovine bone morphogenetic protein. Clin. Orthop. 287: 266-75 (1993); **Reddi AH & Cunningham NS** Recent progress in bone induction by osteogenin and bone morphogenetic proteins: challenges for biomechanical and tissue engineering. J. Biomech. Eng. 113: 189-90 (1991); **Ripamonti U** Bone induction in non-human primates. An experimental study on the baboon. Clin. Orthop. 269: 284-94 (1991); **Ripamonti U** The induction of bone in osteogenic composites of bone matrix and porous hydroxyapatite replicas: an experimental study on the baboon (*Papio ursinus*). J. Oral. Maxillofac. Surg. 49: 817-30 (1991); **Ripamonti U & Reddi AH** Growth and morphogenetic factors in bone induction: role of osteogenin and related bone morphogenetic proteins in craniofacial and periodontal bone repair. Crit. Rev. Oral Biol. Med. 3: 1-14 (1992); **Ripamonti U et al** Osteogenin, a bone morphogenetic protein, adsorbed on porous hydroxyapatite substrata, induces rapid bone differentiation in calvarial defects of adult primates. Plast. Reconstr. Surg. 90: 382-93 (1992); **Ripamonti U et al** Induction of bone in composites of osteogenin and porous hydroxyapatite in baboons. Plast. Reconstr. Surg. 89: 731-9 (1992); **Ripamonti U** Calvarial regeneration in primates with autolyzed antigen-extracted allogeneic bone. Clin. Orthop. Sept. 282: 293-303 (1992); **Ripamonti U et al** Initiation of bone regeneration in adult baboons by osteogenin, a bone morphogenetic protein. Matrix 12: 369-80 (1993); **Ripamonti U et al** Reconstruction of the bone-bone marrow organ by osteogenin, a bone morphogenetic protein, and demineralized bone matrix in calvarial defects of adult primates. Plast. Reconstr. Surg. 91: 27-36 (1993); **Takaoka K et al** Telopeptide-depleted bovine skin collagen as a carrier for bone morphogenetic protein. J. Orthop. Res. 9: 902-7 (1991); **Toriumi DM et al** Mandibular reconstruction with a recombinant bone-inducing factor. Functional, histologic, and biomechanical evaluation. Arch. Otolaryngol. Head Neck Surg. 117: 1101-12 (1991); **Wang EA et al** Recombinant human bone morphogenetic protein induces bone formation. PNAS 87: 2220-4 (1990); **Wu CH et al** Enhanced osteoinduction by intramuscular grafting of BMP-β-TCP compound pellets into murine models. Arch. Histol. Cytol. 55: 97-112 (1992); **Xiang W et al** The effect of bone morphogenetic protein on osseointegration of titanium implants. J. Oral. Maxillofac. Surg. 51: 647-51 (1993); **Yasko AW et al** The healing of segmental bone defects, induced by recombinant human bone morphogenetic protein (rhBMP-2). A radiographic, histological, and biomechanical study in rats. J. Bone Joint Surg. Am. 74: 659-70 (1992)

BMP-1 to **BMP-7:** see: BMP (bone morphogenetic proteins).

BMPG: *bone marrow proteoglycan* This immunoregulatory factor with a length of 206 amino acids is produced by a T cell hybridoma. The factor has three types of sugar chains and shows a marked homology with animal lectins, including the human asialoglycoprotein receptor, chicken hepatic lectin and the homing receptor of lymphocytes.

Yoshimatsu K et al Purification and cDNA cloning of a novel factor produced by a human T cell hybridoma: sequence homology with animal lectins. Mol. Immunol. 29: 537-46 (1992)

BMT: abbrev. for bone marrow transplantation.

BN: abbrev. for » Bombesin.

BNP: *brain natriuretic peptide* see: ANF (atrial natriuretic factor).

Bnx mouse: abbrev. for beige/nude/xid. See: Immunodeficient mice.

Bo: *bovine* E. g. as a prefix such as BoEGF.

Bombesin: abbrev. BN.

■ **Sources:** Bombesin is a neuropeptide found in the central and peripheral nervous system of amphibians (*Bombina bombina*). Bombesin-like peptides are produced by neurons of the central and peripheral nervous system, and many other neuroendocrine cell types.

■ **Protein Characteristics:** Bombesin is a small peptide of 14 amino acids with the sequence pEQRLGNQWAVGHLM-NH_2.

■ **Related Factors:** Bombesin-like peptides are grouped into three families, based upon sequence identities in the eight carboxyterminal amino acids which are also responsible for the biological activities and receptor binding. These families are the Bombesin family, Ranatensin family, and the Phyllolitorin family. The factors known as neuromedins (see also: Tachykinins) are members of the Ranatensin subfamily. Human » GRP (gastrin-releasing peptide) is a member of the Bombesin subfamily and is considered to be the human homologue of the amphibian bombesin.

■ **Receptor Structure, Gene(s), Expression:** The biological activity of bombesin is mediated by specific receptors that also bind » GRP (gastrin-releasing peptide) and Bombesin-like factors. These receptors are expressed in many different cell types and tissues, including » 3T3 cells, rat pituitary adenomas, pancreatic cells, gastrin-producing cells, brain.

■ **Biological Activities:** Bombesin and Bombesin-like factors show a wide spectrum of biological activities. These include regulation of the contraction of smooth muscle cells, induction of the secretion of neuropeptides and hormones. In human mammary carcinomas »GRP (gastrin releasing peptide, mammalian bombesin) induces the synthesis of » ET (Endothelin) which acts as a » paracrine growth factor for stromal cells of the mammary gland. Bombesin is synthesized after electric stimulation of nerve cells and induces the release of gastrin and cholecystokinin in the intestines and the pancreas.

Bombesin	MLHGVAWQNGLRQE
GRP (Rat)	MLHGVAWHSGRPYMKALVTGGGAGTSVPA
GRP (Human)	MLHGVAWHNGRPYMKTLVTGGGAPLPV
NMB (Rat)	MFHGTAWLNGRPHVRIKSARSRPEPLDWSFPT
Ranatensin	MFHGVAWQPVE

Sequence comparison of Bombesin and Bombesin-like peptides.
Amino-terminal amino acid positions required for biological activity and receptor binding are conserved (blue shading). NMB = neuromedine B; GRP = Gastrin releasing peptide.

BNP transgenics: see addendum: BNP.

Apart from the classical role of neurohumoral hormone bombesin also acts as a growth factor and therefore shows cytokine-like activities although it is not classified as a cytokine due to its small size. In serum-free medium (see also: SFM) bombesin stimulates DNA synthesis and proliferation of murine fibroblasts (3T3 cells) in the absence of other growth factors. Several analogs of » SP (substance P) competitively block the binding of bombesins to their receptor and all the events leading to mitogenesis. The growth of human mammary carcinoma cells is also enhanced by bombesin. The growth-promoting activities of bombesin are potentiated by insulin. The biological activity of bombesin as a mitogen is coupled to the activation of $Ca2^+$-mobilizing G proteins. In human alveolar macrophages bombesin-like factors modulate the synthesis of » IL1.

Small cell lung cancers (SCLC) have been shown to secrete many bombesin-like peptides. It has been suggested that these peptides act as » autocrine growth factors and this view is supported by the inhibition of tumor cell growth following transplantation into the nude mouse (see also: Immunodeficient mice) and treatment with monoclonal antibodies directed against bombesins. Some transplanted tumors escaping this treatment and developing into progressively growing tumors may be the result of mutations making these cells independent of the growth factor.

■ **CLINICAL USE & SIGNIFICANCE:** The inhibition of the interaction of these factors with their receptors may be of importance for the treatment of SCLC (small cell lung cancers). It has been observed that bombesin can be used to increase the sensitivity of cisplatin-sensitive and resistant variants of human ovarian carvinoma cell lines to cisplatin.

● REVIEWS: **Bevis CL & Zasloff M** Peptides from frog skin. ARB 59: 395-414 (1990); **Castiglione R de & Gozzini L** Nonmammalian peptides: structure determination synthesis, and biological activity. Chimica*oggi* April 1991, pp. 9-15; **Rozengurt E** Neuropeptides as cellular growth factors: Role of multiple signaling pathways. EJCI 21: 123-34 (1991); **Schüller HM** Receptor-mediated mitogenic signals and lung cancer. Cancer Cells 3: 496-503 (1991); **Woll P & Rozengurt E** Neuropeptides as growth regulators. BMB 45: 492-505 (1989)

● BIOCHEMISTRY & MOLECULAR BIOLOGY: **Lévesque A et al** Synthesis of bombesin analogs by the Fmoc method. ACR 11: 2215-22 (1991); **Shipp MA et al** CD10/neutral endopeptidase 24. 11 hydrolyses bombesin-like peptides and regulates the growth of small cell carcinomas of the lung. PNAS 88: 10662-6 (1991)

● RECEPTORS: **Battey J & Wada E** Two distinct receptor subtypes for mammalian bombesin-like peptides. TINS 14: 524-8 (1991); **Battey J et al** Molecular cloning of the bombesin/gastrin-releasing peptide precursor from Swiss 3T3 cells. PNAS 88: 395-9 (1991); **Corjay MH et al** Two distinct bombesin receptor subtypes are expressed and functional in human lung carcinoma cells. JBC 266: 18771-9 (1991); **Houben H & Denef C** Bombesin receptor antagonists and their use in the evaluation of paracrine and autocrine intercellular communication. Front. Horm. Res. 19: 176-95 (1991); **Kane MA et al** Isolation of the bombesin/gastrin-releasing peptide receptor from human small cell lung carcinoma NCI-H345 cells. JBC 266: 9486-93 (1991); **Rozengurt E & Sinnett-Smith J** Bombesin stimulation of fibroblast mitogenesis: specific receptors, signal transduction, and early events. Phil. Trans. R. Soc. Lond. Biol. 327: 209-21 (1990); **Rozengurt E et al** Mitogenic signaling through the bombesin receptor: Role of a guanine nucleotide regulatory protein. JCS13: s43-s56 (1990); **Shapira H et al** Distinguishing bombesin receptor subtypes using the oocyte assay. BBRC 176: 79-86 (1991); **Zachary I & Rozengurt E** High affinity receptors for peptides of the bombesin family in Swiss 3T3 cells. PNAS 82: 7616-20 (1985); **Zachary I et al** Bombesin, vasopressin, and endothelin rapidly stimulate tyrosine phosphorylation in intact Swiss 3T3 cells. PNAS 88: 4577-81 (1991)

● BIOLOGICAL ACTIVITIES: **Carney DN et al** Bombesin is an autocrine growth factor for human small cell lung cancer cell lines. Proc. Am. Fed. Clin. Res. 31: 404 (1983); **Cuttitta F et al** Bombesin-like peptides can function as autocrine growth factors in human small-cell lung cancer. N 316: 823-6 (1985); **Endo T et al** Bombesin and bradykinin increase inositol phosphates and cytosolic free Ca^{2+}, and stimulate DNA synthesis in human endometrial stromal cells. J. Endocrinol. 131: 313-8 (1991); **Isonishi S et al** Modulation of cisplatin sensitivity and growth rate of an ovarian carcinoma cell line by bombesin and tumor necrosis factor-α. JCI 90: 1436-42 (1992); **Johnson TC & Sharifi BG** Abrogation of the mitogenic activity of bombesin by a cell surface sialoglycopeptide growth inhibitor. BBRC 161: 468-74 (1989); **Lemaire I** Bombesin-related peptides modulate interleukin-1 production by alveolar macrophages. Neuropeptides 20: 217-23 (1991); **Narayan S et al** A potent bombesin receptor antagonist inhibits bombesin-stimulated growth of mouse colon cancer cells *in vitro*: absence of autocrine effects. Cell Growth Diff. 3: 111-8 (1992); **Nelson J et al** Bombesin stimulates proliferation of human breast cancer cells in culture. Br. J. Cancer 63: 933-6 (1991); **Radulovic S et al** Inhibition of growth of HT-29 human colon cancer xenografts in nude mice by treatment with bombesin/gastrin releasing peptide antagonist (RC-3095). CR 51: 6006-9 (1991); **Rozengurt E & Sinnett-Smith J** Bombesin stimulation of DNA synthesis and cell division of Swiss 3T3 cells. PNAS 80: 2936-40 (1983); **Schrey MP et al** Bombesin and glucocorticoids stimulate human breast cancer cells to produce endothelin, a paracrine mitogen for breast stromal cells. CR 52: 1786-90 (1992); **Sehti T et al** Growth of small cell lung cancer cells: stimulation by multiple neuropeptides and inhibition by broad spectrum antagonists *in vitro* and *in vivo*. CR 52: 2737s-42s (1992); **Willey JC et al** Bombesin and the C-terminal tetradecapeptide of gastrin-releasing peptide are growth factors for human bronchial epithelial cells. ECR 153: 245-8 (1984); **Woll PJ & Rozengurt E** Bombesin and bombesin antagonists: studies in Swiss 3T3 cells and human small cell lung cancer. Br. J. Cancer 57: 579-86 (1988); **Woll PJ & Rozengurt E** Two classes of antagonist interact with receptors for the mitogenic neuropeptides bombesin, bradykinin, and vasopressin. GF 1: 75-83 (1988); **Yano T et al** Stimulation by bombesin and inhibition by bombesin/gastrin-releasing peptide antagonist RC-3095 of growth of human breast cancer cell lines. CR 52: 4545-7 (1992)

Bone-derived growth factor 1.7: See: BDGF 1.7.

Bone-derived growth factor 2: see: BDGF 2.

Bone growth factor: see: BMP (bone morphogenetic protein).

Bone marrow natural suppressor cell-derived suppressor factor: see: NS suppressor factor.

Bone marrow proteoglycan: see: BMPG

Bone marrow purging: see: 4-HC (Hydroperoxycyclophosphamide).

Bone morphogenetic proteins: see: BMP.

Bone resorption-stimulating factor: see: BRSF.

Bov: *bovine* E. g. as a prefix such as bov-GM-CSF.

Boyden chamber: see: Chemotaxis.

BP: *binding protein* General name for proteins which bind ligands through non-covalent interactions. Many binding proteins bind other proteins. Some binding proteins which are important for cytokine research circulate in the serum and bind » cytokines. In many instances cytokines that are bound to such carrier proteins are biologically inactive. Apart from their presumed role as mere transport proteins these carriers may act also as physiological antagonists of cytokines. Carrier proteins have been described, for example, for » EGF, » IGF and » TGF (see also: TGF-BF, TNF blocking factor). It has been described that a complexed cytokine often has greater plasma half life than an uncomplexed factor (see also: Cytokine inhibitors).
Special binding proteins are those that either specifically or unspecifically bind to DNA. Many of these DNA binding proteins are transcription factors that positively or negatively affect » gene expression.

BP: *burst promoting (activity)* This factor is identical with » IL3.

BP10: This factor has been inferred from a cDNA clone isolated from a sea urchin library. The gene encodes a secreted 64 kDa protein with protease activity that is related with BMP-1 (see: BMP, bone morphogenetic protein). BP10 appears to be involved in the development of the ectoderm. Structurally and functionally BP10 is also related with the *tolloid* gene (see: *dpp*; *decapentaplegic*) identified in *Drosophila*.
Lepage T et al Spatial and temporal expression pattern during sea urchin embryogenesis of a gene coding for a protease homologous to the human protein BMP-1 and to the product of the *Drosophila* dorsal-ventral patterning gene *tolloid*. Development 114: 147-63 (1992)

Bp50: see: CD40.

BPA: *blood progenitor activator* This factor is identical with » IL3.

BPA: *burst promoting activities* Also abbrev. BP. Collective name given to a group of hematopoietic growth factors (see: Hematopoietins) that stimulate the growth of early immature erythroid progenitor cells (see: BFU-E) from bone marrow (see also: Hematopoiesis). These cells give rise to large erythroid multicentric colonies, called ***bursts***, in » Colony formation assays.
The term is frequently used to describe more or less undefined activities in growth media, sera and cell culture supernatants. An equivalent term is ***BFA*** (*burst feeder activity*). In some older references BPA appears to be used as the murine equivalent of activities otherwise known as » EPA (erythroid potentiating activity).
Some known factors with BPA activity are » E-CSF (erythroid colony-stimulating factor), » GM-CSF, » G-CSF, » IL3, » IL4, » IL9, Epo, » HILDA, and » SCGF (stem cell growth factor. Other factors with BPA activity have only been poorly characterized. For a membrane-bound form of a factor with BPA activity see: E-CSF (erythroid colony-stimulating factor).
Correa PN et al Production of erythropoietic bursts by progenitor cells from adult human peripheral blood in an improved serum-free medium: role of insulin-like growth factor 12. Blood 78: 2823-33 (1991); **Dukes PP et al** Measurement of human erythroid burst-promoting activity by a specific cell culture system. EH 13: 59-66 (1985); **Feldman L et al** Purification of a membrane-derived human erythroid growth factor. PNAS 84: 6775-9 (1987); **Feldman L & Dainiak N** B lymphocyte-derived erythroid burst-promoting activity is distinct from other known lymphokines. Blood 73: 1814-20 (1989); **Gauwerky CE et al** Human leukemia cell line K562 responds to erythroid-potentiating activity. Blood 59: 300-5 (1982); **Porter PN & Ogawa M** Characterization of human erythroid burst-promoting activity derived from bone marrow conditioned medium. Blood 59: 1207-12 (1982); **Sasaki H et al** Three quantitative assays for human erythroid burst-promoting activity of recombinant growth factors and of omentum-conditioned medium. EH 18: 84-8 (1990); **Skettino S et al** Selective generation of erythroid burst-promoting activity by recombinant interleukin 2-stimulated human T lymphocytes and natural killer cells. Blood 71: 907-14 (1988); **Sonoda Y et al** Erythroid burst-promoting activity of purified recombinant human GM-CSF and interleukin-3: studies

with anti-GM-CSF and anti-IL3 sera and studies in serum-free cultures. Blood 72: 1381-6 (1988);**Tsuda E et al** Factor with erythroid burst-promoting activity in human urine unlike other hematopoietic growth factors. Int. J. Hematol. 54: 363-9 (1991)

BPA: *burst promoting activity* One factor with this activity was originally isolated from human stromal bone marrow cells. It is identical with » GM-CSF. Another factor with the same name and the same activity produced by endothelial cells stimulated with » IL1 is also identical with GM-CSF.

Kohama T et al A burst-promoting activity derived from the human bone marrow stromal cell line KM-102 is identical to the granulocyte-macrophage colony-stimulating factor. EH 16: 603-8 (1988); **Segal GM et al** Erythroid burst-promoting activity produced by interleukin-1-stimulated endothelial cells is granulocyte-macrophage colony-stimulating factor. Blood 72: 1364-7 (1988); **Sieff CA et al** Human recombinant granulocyte-macrophage colony-stimulating factor: a multilineage hematopoietin. S 230: 1171-3 (1985); **Sonoda Y et al** Erythroid burst-promoting activity of purified recombinant human GM-CSF and interleukin-3: studies with anti-GM-CSF and anti-IL3 sera and studies in serum-free cultures. Blood 72: 1381-6 (1988)

BPA: *burst promoting activity* This factor was originally described as an activity promoting the growth of early hematopoietic precursors. It is identical with » IL3.

Dexter TM et al The role of hemopoietic cell growth factor (interleukin 3) in the development of hemopoietic cells. Ciba Found. Symp. 116: 129-47 (1985); **Lee JC et al** Constitutive production of a unique lymphokine (IL3) by the WEHI-3 cell line. JI 128: 2393-8 (1982); **Sonoda Y et al** Erythroid burst-promoting activity of purified recombinant human GM-CSF and interleukin-3: studies with anti-GM-CSF and anti-IL3 sera and studies in serum-free cultures. Blood 72: 1381-6 (1988)

BPI: *bactericidal/permeability-increasing protein* see: Endotoxin.

bpk: *B cell progenitor kinase* see: Immunodeficient mice, subentry xid.

BPSC: *bipotent stem cells* see: Hematopoiesis.

Bradykinin:

■ **Alternative Names:** Kinin 9, Kallidin

■ **Sources:** Bradykinin is the final product of the kinin system and is split from a serum α2-globulin precursor by the enzyme kallikrein and also by trypsin or plasmin.

■ **Protein Characteristics:** Bradykinin is a short nonapeptide with the sequence RPPGFSPFR.

■ **Biological Activities:** Bradykinin reduces blood pressure by dilating blood vessels. In bronchial smooth muscles and also in the intestines and the uterus bradykinin leads to muscle contraction. Bradykinin is also one the most potent known substances inducing pain. Bradykinin stimulates the synthesis of prostaglandins and this activity is potentiated by » Haptoglobin.

Apart from the classical role of a hormone bradykinin also acts as a growth factor and therefore shows cytokine-like activities although it is not classified as a cytokine due to its small size. In serum-free medium (see also: SFM) bradykinin in nanomolar concentrations stimulates DNA synthesis and proliferation of murine fibroblasts in the absence of other growth factors. It is also a weak mitogen for human fibroblasts. The growth-promoting activities of bradykinin are potentiated by insulin. The bradykinin-induced influx of calcium can be potentiated by » NGF (see also: Calcium ionophore).

Bradykinin is a » paracrine growth factor for small cell lung cancers. It inhibits the growth of mammary stromal cells. Bradykinin receptor antagonists potentiate the » IL1-induced synthesis of prostaglandins in fibroblasts. Bradykinin induces the synthesis of » IL6 and synergizes with IL1 in the IL1-induced resorption of bones.

● **Reviews: Bevis CL & Zasloff M** Peptides from frog skin. ARB 59: 395-414 (1990); **Castiglione R de & Gozzini L** Non-mammalian peptides: structure determination synthesis, and biological activity. Chimica*oggi* April 1991, pp. 9-15; **Rozengurt E** Neuropeptides as cellular growth factors: Role of multiple signaling pathways. EJCI 21: 123-34 (1991); **Schüller HM** Receptor-mediated mitogenic signals and lung cancer. Cancer Cells 3: 496-503 (1991); **Woll P & Rozengurt E** Neuropeptides as growth regulators. BMB 45: 492-505 (1989)

● **Receptors: Burch RM & Kyle DJ** Recent developments in the understanding of bradykinin receptors. Life Sci. 50: 829-38 (1992); **Hess JF et al** Cloning and pharmacological characterization of a human bradykinin (BK-2) receptor. BBRC184: 260-8 (1992); **Phillips E et al** Expression of functional bradykinin receptors in *Xenopus* oocytes. J. Neurochem. 58: 243-9 (1992)

● **Biological Activities: Bunn PA Jr et al** Neuropeptide signal transduction in lung cancer: Clinical implications of bradykinin sensitivity and overall heterogeneity. CR 52: 24-31 (1992); **Bush AB et al** Nerve growth factor potentiates bradykinin-induced calcium influx and release in PC12 cells. J. Neurochem. 57: 562-74 (1991); **Endo T et al** Bombesin and bradykinin increase inositol phosphates and cytosolic free Ca^{2+}, and stimulate DNA synthesis in human endometrial stromal cells. J. Endocrinol. 131: 313-8 (1991); **Francel PC** Bradykinin and neuronal injury. J. Neurotrauma 9: S27-S45 (1992); **Godin C et al** Bradykinin stimulates DNA synthesis in competent Balb/c 3T3 cells and enhances inositol phosphate formation induced by platelet-derived growth factor. Biochem. Pharmacol. 42: 117-22 (1991); **Lerner UH** Bradykinin synergistically potentiates interleukin-1 induced bone resorption and prostanoid biosynthesis in neonatal mouse calvarial bones. BBRC 175: 775-83 (1991); **Lerner UH & Modéer T** Bradykinin B1 and B2 receptor agonists synergistically potentiate interleukin-1-induced prostaglandin biosynthesis in human gingival fibroblasts. Inflammation 15: 427-36 (1991); **Patel KV & Schrey MP** Inhibition of DNA synthesis and growth in human breast stromal cells by Evidence for independent roles of B1 and B2 receptors in the respective control of cell growth and phospholipid hydrolysis. CR 52: 334-40 (1992); **Sethi T & Rozengurt E** Multiple neuropeptides stimulate clonal growth of small cell lung cancer: effects of bradykinin, vasopressin, cholecystokinin, galanin, and neurotensin. CR 51: 3621-3 (1991); **Skidgel RA et al** Metabolism of substance P and bradykinin by human neutrophils. Biochem. Pharmacol. 41: 1335-44 (1991); **Woll PJ & Rozengurt E** Two classes of antagonist interact with receptors for the mitogenic neuropeptides bombesin, bradykinin, and vasopressin. GF 1: 75-83 (1988); **Vandekerckhove F et al** Bradykinin induces interleukin-6 and synergizes with interleukin-1. Lymphokine Cytokine Res. 10: 285-9 (1991)

Brain-derived growth factor: See: BDGF.

Brain-derived growth factor A: abbrev. BDGF-A. See: BDGF.

Brain-derived growth factor B: abbrev. BDGF-B. See: BDGF.

Brain-derived neurotrophic factor: see: BDNF. See also: Neurotrophins for general information.

Brain FGF: This factor is identical with » aFGF.
Thomas KA et al Purification and characterization of acidic fibroblast growth factor from bovine brain. PNAS 81: 357-61 (1984)

Brain natriuretic peptide: abbrev. BNP See: ANF (atrial natriuretic factor).

BRF: *B cell replication factors* This is a generic name for factors that promote proliferation of B cells, which leads to clonal expansion. See also: BCGF (B cell growth factors).

BRF: *B cell replication factor* This factor is identical with those later named » BSF 1 (B cell stimulating factor) and » BCGF 1 (B cell growth factor). These factors are identical with » IL4. See also: BCGF (B cell growth factors).
Leonhard W et al T cell hybridomas which produce B lymphocyte replication factors only. N 300: 355-7 (1982)

BRM: *biological response modifiers* see: Immunomodulation.

BRMF: *B cell replication and maturation factor* Generic name for factors that induce B cell replication (proliferation) and maturation (i. e. terminal differentiation; see: BMF, B cell maturation factor). See also: BCDF (B cell differentiation factors).
Melchers F & Corbel C Studies on B cell activation *in vitro*. Ann. Immunol. Paris 134D: 63-73 (1983); **Schimpl A** Lymphokines active in B cell proliferation and differentiation. Springer Semin. Immunopathol. 7: 299-310 (1984)

BRMP Unit: *Biological Response Modifier Program unit* An activity standard of » interleukins and other growth factors with immunomodulatory activities created by the working party on immunomodulators within the biological response modifier program initiated by the National Institutes of Health (NIH).
It is frequently very difficult to compare the biological activities of immunomodulatory factors because producers and also the National Cancer Institute (NCI) and WHO (World Health Organization) use their own reference standards (see also: Cytokine assays, Bioassays). Some of the confusion has been resolved by introducing international standards (IS). The correlations for IL2, for example, are 2.3 Cetus units = 1 BRMP unit = 6 international units.
One Cetus unit is defined as the amount of IL2/mL that induces half-maximal incorporation of tritium-labeled thymidine after a 24 Hr incubation by IL2-dependent murine T cell lines (i. e. an EC_{50} unit).
Oldham RK Biological response modifiers program and cancer chemotherapy. Int. J. Tissue React. 4: 173-88 (1982); **Rossio JL et al** The BRMP IL2 reference reagent. LR 5 (Suppl. 1) S13-18 (1986)

BRSF: *bone resorption-stimulating factor* This poorly characterized factor (> 3.5 kDa) has been found to be produced by peripheral blood leukemic lymphocytes from patients with adult T cell leukemia (ATL). The factor has bone resorption activity in a bioassay employing fetal mouse forearm bones in organ culture. It may be involved in hypercalcemia observed in patients with ATL.
Fujihira T et al Evidence of bone resorption-stimulating factor in adult T cell leukemia. Jpn. J. Clin. Oncol. 15: 385-91 (1985)

Bruton's tyrosine kinase: abbrev. btk. See: Immunodeficient mice, subentry xid.

Bryostatins: Bryostatins are macrocyclic lactones isolated from the marine invertebrates *Bugula neritina*. More than 13 different but structurally closely related compounds have been isolated. Bryostatins are powerful activators of protein kinase C but may also in some cases inhibit members of the PKC family. Bryostatins only elicit some of those effects normally observed with » phorbol esters. TPA, for example, is a potent growth inhibitor for human A549 cells for which bryostatin 1 has no effect at all.

Bryostatins are used, among other things, to test whether binding of a cytokine to its receptor involves protein kinase C in signal processing (for other agents used to dissect cytokine-mediated signal transduction pathways see: Calcium ionophore, Calphostin C, Genistein, H8, Herbimycin A, K-252a, Lavendustin A, Phorbol esters, Okadaic acid, Staurosporine, Suramin, Tyrphostins, Vanadate).

Bryostatins have significant antitumor activity against murine lymphomas, leukemia cells and melanoma cells. Their cytostatic activities generally markedly increase with decreasing concentrations. *In vitro* they induce the release of » IFN-γ, interleukins, and » hematopoietins (hematopoietic growth factors) from accessory cells and also activate T and B cells. In neutrophils bryostatins induce degranulation. Bryostatins potentiate the radioprotective effect of » GM-CSF.

Bryostatin1 has been shown to support the *in vitro* growth of multipotent human hematopoietic progenitors, probably through an indirect mechanism involving accessory cells. Bryostatin 1 potentiates the ability of recombinant » GM-CSF or » IL3 to support the formation of » CFU-GM by » CD34-positive human bone marrow mononuclear cells. The compound selectively increases the formation of pure and mixed neutrophil and macrophage colonies in response to IL3 or GM-CSF, while inhibiting the growth of eosinophilic cells. It has also been shown to stimulate normal erythropoiesis in human bone marrow progenitor assays.

Bryostatin 1 lacks tumor promoting activity and is able to induce differentiation in maturation-arrested leukemia cells. It has a potential therapeutic role as a B cell differentiating agent and has been shown to modulate differentiation in a number of human non-Hodgkin's lymphoma cell lines.

● Reviews: **Pettit GR** The bryostatins. Fortschr. Chem. Org. Naturst. 57: 153-95 (1991)

al-Katib A et al Bryostatin 1-induced hairy cell features on chronic lymphocytic leukemia cells *in vitro*. EH 21: 61-5 (1993); **Drexler HG et al** Bryostatin 1 induces differentiation of B chronic lymphocytic leukemia cells. Blood 74: 1747-57 (1989); **Gebbia V et al** The effects of the macrocyclic lactone bryostatin-1 on leukemic cells *in vitro*. Tumori 78: 167-71 (1992); **Gescher A** Towards selective pharmacological modulation of protein kinase C: opportunities for the development of novel antineoplastic agents Br. J. Cancer 66: 10-9 (1992); **Grant S et al** Effect of bryostatin 1 on the *in vitro* radioprotective capacity of recombinant granulocyte-macrophage colony-stimulating factor (rGM-CSF) toward committed human myeloid progenitor cells (CFU-GM). EH 20: 34-42 (1992); **Hornung RL et al** Preclinical evaluation of bryostatin as an anticancer agent against several murine tumor cell lines: *in vitro* versus *in vivo* activity. CR 52: 101-7 (1992); **Jones RJ et al** Bryostatin 1, a unique biologic response modifier: anti-leukemic activity *in vitro*. Blood 75: 1319-23 (1990); **Kennedy MJ et al** Differential effects of bryostatin 1 and phorbol ester on human breast cancer cell lines. CR 52: 1278-83 (1992); **Klein SB et al** Regulation of TGF-α expression in human keratinocytes: PKC-dependent and -independent pathways. JCP 151: 326-36 (1992); **Leonard JP et al** Regulation of hematopoiesis IV. The role of interleukin 3 and bryostatin 1 in the growth of erythropoietic progenitors from normal and anemic W/Wv mice. Blood 72: 1492-6 (1988); **Levine BL et al** Response of Jurkat T cells to phorbol ester and bryostatin. Development of sublines with distinct functional responses and changes in protein kinase C activity. JI 147: 3474-81 (1991); **Li F et al** Bryostatin 1 modulates the proliferation and lineage commitment of human myeloid progenitor cells exposed to recombinant interleukin-3 and recombinant granulocyte-macrophage colony-stimulating factor. Blood 80: 2495-502 (1992); **McCrady CW et al** Effect of pharmacologic manipulation of protein kinase C by phorbol dibutyrate and bryostatin 1 on the clonogenic response of human granulocyte-macrophage progenitors to recombinant GM-CSF. Br. J. Haematol. 77: 5-15 (1991); **McCrady CW et al** Modulation of the activity of a human granulocyte-macrophage colony-stimulating factor/interleukin-3 fusion protein (pIXY321) by the macrocyclic lactone protein kinase C activator bryostatin 1. EH 21: 893-900 (1993); **Mohammad RM et al** Differential effects of bryostatin 1 on human non-Hodgkin's B lymphoma cell lines. Leuk. Res. 17: 1-8 (1993); **Pettit GR et al** Structural modifications of bryostatin 2. Anticancer Drug Res. 7: 101-14 (1992); **Schaufelberger DE et al** The large-scale isolation of bryostatin 1 from *Bugula neritina* following current good manufacturing practices. J. Nat. Prod. 54: 1265-70 (1991); **Sharkis SJ et al** The action of bryostatin on normal human hematopoietic progenitors is mediated by accessory cell release of growth factors. Blood 76: 716-20 (1990); **Steube KG & Drexler HG** Differentiation and growth modulation of myeloid leukemia cells by the protein kinase C activating agent bryostatin-1. Leuk. Lymphoma 9: 141-8 (1993); **Trenn G et al** Immunomodulating properties of a novel series of protein kinase C activators. The bryostatins. JI 140: 433-9 (1988); **Tuttle TM et al** Bryostatin 1 activates T cells that have antitumor activity. J. Immunother. 12: 75-81 (1992); **Workman P** Signal transduction inhibitors as novel anticancer drugs: where are we? Ann. Oncol. 3: 527-31 (1992)

BSC-1 cell growth inhibitor: The inhibitor, occasionally also referred to as » polyergin, was initially isolated from the growth medium (see also: CM, conditioned medium) of the established African Green Monkey kidney cell line, BSC-1.

The inhibitor, TGF-β2 (see: TGF-β) from human platelets, and CIF-B (cartilage inducing factor; see: CIF-A) are related, with only 32 amino acid exchanges out of 112 amino acids observed for the inhibitor and TGF-β2.
The inhibitor and TGF-β2 have nearly identical biological activity and compete for binding to the same cell membrane receptor. The inhibitor is extremely active as a growth inhibitor with some cells, but not with others. Approximately 50% inhibition of thymidine incorporation is observed with CCL64 cells at 0.05 ng/mL and with BSC-1 cells at 1 ng/ml. The growth inhibitor stimulates colony formation in soft agar by » AKR-2B cells.
Hanks SK et al Amino acid sequence of the BCS-1 cell growth inhibitor (polyergin) deduced from the nucleotide sequence of the cDNA. PNAS 85: 79-82 (1988); **Holley RW et al** A growth regulatory factor that can both inhibit and stimulate growth. Ciba Found. Symp. 116: 241-52 (1985); **Holley RW et al** Isolation of the BSC-1 monkey kidney cell growth inhibitor. MiE 146: 163-73 (1987); **McPherson JM et al** The growth inhibitor of African green monkey (BSC-1) cells is transforming growth factors β 1 and β 2. B 28: 3442-7 (1989); **Tucker RF et al** Growth inhibitor from BSC-1 cells closely related to platelet type β transforming growth factor. S 226: 705-7 (1984)

BSF: *B cell stimulating factors* BSF, and also BCSF (B cell stimulating factor), in particular in older references, is the generic name given to B cell-specific growth and differentiation factors involved in the T cell-dependent activation of B cells into antibody-secreting cells. stimulating factors. These factors are also referred to as either » BCDF (B cell differentiation factors) or » BCGF (B cell growth factors).
Factors with BSF activity that have been identified specifically are » IL1, » BSF 1 (now known as » IL4), » BCGF 2 (B cell growth factor, now known as » IL5), » BSF 2 (now known as » IL6), and » IL10, and » VIP. An unidentified factor with BSF activity found in the synovial fluid of patients with rheumatoid arthritis (RA-SF; see: TRF (T cell replacing factors)).
Go NF et al Interleukin 10, a novel B cell stimulatory factor: unresponsiveness of X chromosome-linked immunodeficiency B cells. JEM 172: 1625-31 (1990); **Howard M et al** B cell growth and differentiation factors. IR 78: 185-210 (1984); **Kishimoto T** B cell stimulatory factors (BSFs): molecular structure, biological function, and regulation of expression. J. Clin. Immunol. 7: 343-55 (1987); **Kishimoto T & Hirano T** Molecular regulation of B lymphocyte response. ARI 6: 485-512 (1988); **Paul WE** Proposed nomenclature for B cell stimulating factors. IT 4: 332 (1983); **Rosen A et al** A T helper cell x Molt4 human hybridoma constitutively producing B cell stimulatory and inhibitory factors. LR 5: 185-204 (1986); **Schwarting R et al** Biochemical characterization and purification of human B cell stimulatory factor (BSF). EJI 15: 632-7 (1985); **Smeland E et al** Characterization of two murine monoclonal antibodies reactive with human B cells. Their use in a high-yield, high-purity method for isolation of B cells and utilization of such cells in an assay for B cell stimulating factor. Scand. J. Immunol. 21: 205-14 (1985)

BSF 1: *B cell stimulating factor 1* This factor is a T cell-derived factor purified originally from » EL-4 murine thymoma cells. It is required for entry into the S phase of the » cell cycle by B cells stimulated with low concentrations of anti-IgM antibodies. It acts directly on resting B cells to prepare them to synthesize DNA more promptly on subsequent exposure to competent stimuli and to strikingly enhance their expression of class II molecules of the major histocompatibility complex.
The factor, initially named BCGF-1 (B cell growth factor 1), was found to be physically and functionally different from » IL1, » IL2 and » IL3 and identical with » BCDF γ (B cell differentiation factor γ). It is identical with » IL4. See also: BCGF (B cell growth factors).
Grabstein K et al Purification to homogeneity of B cell stimulating factor. A molecule that stimulates proliferation of multiple lymphokine-dependent cell lines. JEM 163: 1405-14 (1986); **Howard M et al** Identification of a T cell derived B cell growth factor distinct from interleukin 2. JEM 155: 914-23 (1982); **Lee F et al** Isolation and characterization of a mouse interleukin cDNA clone that expresses B cell stimulatory factor 1 activities and T cell- and mast-cell-stimulating activities. PNAS 83: 2061-5 (1986); **Mosmann TR et al** T cell and mast cell lines respond to B cell stimulatory factor 1. PNAS 83: 5654-8 (1986); **Noelle R et al** Increased expression of Ia antigen on resting B cells: an additional role for B cell growth factor. PNAS 81: 6149-53 (1984); **Ohara J et al** Partial purification of murine B cell stimulatory factor (BSF)-1. JI 135: 2518-23 (1985); **Ohara J & Paul WE** Production of a monoclonal antibody to and molecular characterization of B cell stimulatory factor-1. N 315: 333-6 (1985); **Paul WE** Living with lymphocytes. Int. Arch. Allergy Appl. Immunol. 77: 7-12 (1985); **Rabin EM et al** B cell stimulatory factor 1 activates resting B cells. PNAS 82: 2935-9 (1985); **Rabin EM et al** B cell stimulatory factor 1 (BSF-1) prepares resting B cells to enter S phase in response to anti-IgM and lipopolysaccharide. JEM 164: 517-31 (1986); **Vitetta E et al** Serological, biochemical, and functional identity of B cell-stimulatory factor 1 and B cell differentiation factor for IgG1. JEM 162: 1726-31 (1985); **Yokota T et al** Isolation and characterization of a human interleukin cDNA clone, homologous to mouse B cell stimulatory factor 1, that expresses B cell and T cell-stimulating activities. PNAS 83: 5894-8 (1986)

BSF 2: *B cell stimulating factor 2* This factor induces the final maturation of B cells into immunoglobulin-secreting cells and also mediates the generation of cytotoxic T lymphocytes (see: CDF, cytotoxic differentiation factor; BCDF, B cell differentiation factor). The factor, which is also secreted by the human bladder carcinoma cell line T24 (see: T24 BCDF), is identical with » IL6.

Billiau A BSF-2 is not just a differentiation factor N 324: 415 (1986); **Gauldie J et al** Interferon β2/B cell stimulatory factor 2 shares identity with monocyte derived hepatocyte stimulatory factor and regulates the major acute phase protein response in liver cells. PNAS 84: 7251-5 (1987); **Hirano T et al** Complementary DNA for novel human interleukin (BSF-2) that induces B lymphocytes to produce immunoglobulin. N 324: 73-6 (1986); **Hirano T et al** Human B cell differentiation factor defined by an anti-peptide antibody and its possible role in autoantibody production. PNAS 84: 228-31 (1987); **Hirano T et al** Absence of antiviral activity in recombinant B cell stimulatory factor 2 (BSF-2). Immunol. Lett. 17: 41-5 (1988)

BSF MP6: *B cell stimulating factor MP-6* This factor, produced by a T cell hybridoma of the same name is identical with the redox-active enzyme thioredoxin and hence identical with » ADF (adult T cell leukemia-derived factor).

Carlsson M et al Interleukin-2 and a cell hybridoma (MP6) derived factor act synergistically to induce proliferation and differentiation of human B chronic lymphocytic leukemia cells. Leukemia 3: 595-601 (1989); **Carlsson M et al** Interleukin-4 (IL4) enhances homotypic adhesion of activated B chronic lymphocytic leukemia (B-CLL) cells via a selective up-regulation of CD54. Scand. J. Immunol. 37: 515-22 (1993); **Ericson ML et al** Secretion of thioredoxin after *in vitro* activation of human B cells. Lymphokine Cytokine Res. 11: 201-7 (1992)

BSF p1: *B cell stimulating factor p1* This factor is identical with » IgG1 induction factor and hence identical with » IL4.

Howard M et al Identification of a T cell derived B cell growth factor distinct from interleukin 2. JEM 155: 914-23 (1982); **Oliver K et al** B cell growth factor (B cell growth factor I or B cell-stimulating factor, provisional 1) is a differentiation factor for resting B cells and may not induce cell growth. PNAS 82: 2465-7 (1985)

BSF p2: *B cell stimulating factor p2* BSF p2 induces the final differentiation of B cells into high-rate Ig-secreting cells. The factor is also known as » BCDF (B cell differentiation factor). It is identical with » IL6.

Hirano T et al Purification to homogeneity and characterization of human B cell differentiation factor (BCDF or BSFp-2). PNAS 82: 5490-94 (1985)

BSF-TC: *TC-1-derived B cell stimulating factor* This poorly characterized factor is produced by the murine stromal cell line TC-1. The factor stimulates proliferation of B cells previously activated by anti-Ig (anti-Ig blasts). Proliferation of anti-Ig blasts is not induced by purified cytokines known to be produced by TC-1 (» M-CSF, » GM-CSF, or G-CSF) or by » IL1, » IL2, » IL3, » IL4, » IL5, or » IL6. The supernatant of TC-1 cultures also enhances proliferation of B cells that were co-cultured with bacterial lipopolysaccharides, anti-Ig, or dextran sulfate. Proliferation of low, but not high, density B cells isolated from spleen is also directly stimulated by TC-1 culture supernatants.

Simpson L et al Detection and characterization of a B cell stimulatory factor (BSF-TC) derived from a bone marrow stromal cell line. JI 142: 3894-900 (1989)

BT-20: A human mammary epithelial cell line established from an estrogen-independent estrogen receptor-negative adenocarcinoma of the breast. The cells respond to » bFGF and express high-affinity bFGF receptors. The cells also express receptors for » M-CSF, and this factor increases the invasive growth of the cell in an *in vitro* invasion model. BT-20 cells express a high number of » EGF receptors. BT-20 cells secrete an » IGF binding protein (IGFBP2; see: IGFBP).
Recombinant human » GM-CSF and » IL3 enhance colony formation by BT-20 cells in a soft agar clonogenic assay (see also: Colony formation assay). HLA class I and class II antigens are induced in BT-20 cells by » IFN-α, IFN-γ, and » IL1 (IL1α).
BT-20 cells are used to detect natural and recombinant human and murine » TNF-α and » TNF-β activities which induce » apoptosis. Human IFN-γ appears to be the only other cytokine showing greater than additive activity in combination with recombinant TNF-α.

Bellomo G et al Tumor necrosis factor α induces apoptosis in mammary adenocarcinoma cells by an increase in intranuclear free Ca^{2+} concentration and DNA fragmentation. CR 52: 1342-6 (1992); **Clemmons DR et al** Insulin-like growth factor binding protein secretion by breast carcinoma cell lines: correlation with estrogen receptor status. Endocrinology 127: 2679-86 (1990); **Filderman AE et al** Macrophage colony-stimulating factor (CSF-1) enhances invasiveness in CSF-1 receptor-positive carcinoma cell lines. CR 52: 3661-6 (1992); **Kim I et al** Identification and regulation of insulin-like growth factor binding proteins produced by hormone-dependent and -independent human breast cancer cell lines. Mol. Cell. Endocrinol. 78: 71-8 (1991); **Nachbaur D et al** Stimulation of colony formation of various human carcinoma cell lines by rhGM-CSF and rhIL3. Cancer Lett. 50: 197-201 (1990); **Peyrat JP et al** Basic fibroblast growth factor (bFGF): mitogenic activity and binding sites in human breast cancer. J. Steroid Biochem. Mol. Biol. 43: 87-94 (1992); **Sedlak J et al** Cytokine (IFN-α, IFN-γ, IL1-α, TNF-α)-induced modulation of HLA cell surface expression in human breast cancer cell lines. Neoplasma 39: 269-72 (1992); **Yen J & Kramer SM** A rapid *in vitro* cytotoxicity assay for the detection of tumor necrosis factor on human BT-20 cells. J. Immunother. 20: 174-81 (1991)

B-TCGF: *B cell-derived T cell growth factor* This factor is secreted by a B cell lymphoma line and has a molecular mass of 27-50 kDa. It does not stimulate the proliferation of thymocytes on its own but stimulates the growth of fetal and adult thy-

mocytes in the presence of » IL2, » IL4 and » IL7. Antibodies directed against » TNF-α, » GM-CSF, or » IL6 do not neutralize the biological activity of this factor. The factor is identical with » IL10.
MacNeil IA et al IL10, a novel growth cofactor for mature and immature T cells. JI 145: 4167-73 1990); **Suda T et al** Identification of a novel thymocyte growth-promoting factor derived from B cell lymphomas. CI 129: 228-40 (1990); **Willoughby PB et al** Analysis of a murine B cell lymphoma, CH44, with an associated non-neoplastic T cell population. I. Proliferation of normal T lymphocytes is induced by a secreted product of the malignant B cells. Am. J. Pathol. 133: 507-15 (1988)

BTG1: This human gene encodes a protein with antiproliferative functions. It is identical with TIS-21 (see: TIS genes).

B-TGF: Alternative name of » TGF-β, omitting Greek lettering.

Btk: *Bruton's tyrosine kinase* See: Immunodeficient mice, subentry xid.

Buffy coat interferon: This factor is identical with » IFN-α isolated from the leukocyte-containing fractions (buffy coat) of processed stored blood. See also: IFN for general information about interferons.

Burst feeder activity: abbrev. BFA. See: BFU (burst promoting activities).

Burst forming unit, erythroid: see: BFU-E.

Burst promoting activities: see: BPA.

C

c: In combinations with gene symbols a prefixed c signifies "cellular", for example c-*sis* (the cellular homologue of the » *sis* » oncogene).

C3b: see: HSE-MSF (hepatic sinusoidal endothelial cell-derived migration stimulating factor).

C7: This murine protein is expressed in murine peritoneal macrophages after induction with bacterial lipopolysaccharides or interferons. It is also called CRG-2 (see: CRG, cytokine responsive gene). It is closely related to the interferon-induced protein » IP-10, showing ≈ 77 % homology if conservative amino acid changes are taken into account, and 67 % identity.
Narumi S & Hamilton TA Dexamethasone selectively regulates LPS-inducible gene expression in murine peritoneal macrophages. Immunopharmacology 19: 93-101 (1990); **Ohmori V & Hamilton TA** A macrophage LPS-inducible early gene encodes the murine homologue of IP-10. BBRC 168: 1261-7 (1990)

C9-related protein: This protein, which is the ninth component of complement, is related to perforin 1 (see: Perforins).
Shinkai Y et al Homology of perforin to the ninth component of complement (C9). N 334: 525-7 (1988); **Trapani JA et al** Genomic organization of the mouse pore-forming protein (perforin) gene and localization to chromosome 10: similarities to and differences from C9. JEM 171: 545-57 (1989)

C10: C10 is one of several mRNAs that are acutely expressed in myelopoietic mouse bone marrow cultures stimulated by » GM-CSF. It is also strongly elevated during the induction of neutrophilic differentiation of IL3-dependent » 32D (clone 3) cells by granulocyte colony-stimulating factor.
C10 is related to MIP-1 (see: MIP, macrophage inflammatory protein). Its protein sequence demonstrates that C10 belongs to the » chemokine family of cytokines (β-Chemokines). The C10 gene contains a second exon not found in other chemokines.
Berger MS et al The gene for C10, a member of the β chemokine family, is located on mouse chromosome 11 and contains a novel second exon not found in other chemokines. DNA Cell Biol. 12: 839-47 (1993); **Orlofsky A et al** Novel expression pattern of a new member of the MIP-1 family of cytokine-like genes. Cell. Regul. 2: 403-12 (1991)

Cachectin: A protein, initially found to be secreted by macrophages, that *in vitro* suppressed the expression of enzymes of lipid metabolism in adipocytes and induced a cachectic state *in vivo* caused by systemic suppression of the enzyme lipoprotein lipase (LPL). Cachectin is identical with » TNF-α.
A recently described cachexia-inducing factor of 25 kDa which is produced by the human melanoma cell line SEKI and also inhibits lipoprotein lipase expression in murine fibroblasts may be a distinct factor not related to TNF-α.
Beutler B et al Identity of tumor necrosis factor and the macrophage-secreted factor cachectin. N 316: 552-4 (1985); **Beutler B et al** Purification of cachectin, a lipoprotein lipase-suppressing hormone secreted by endotoxin-induced RAW 264.7 cells. JEM 161: 984-95 (1985); **Beutler & Cerami** The history, properties, and biological effects of cachectin. B 27: 7575-82 (1988); **Kawakami M et al** Suppression of lipoprotein lipase in 3T3-L1 cells by a mediator produced by SEKI melanoma, a cachexia-inducing human melanoma cell line. J. Biochem. Tokyo 109: 2481-8 (1991)

Cachexia-inducing factor: see: Cachectin.

CAF: *chondrocyte activating factor* see: CGF (chondrocyte growth factor).

CAFC: *cobblestone area-forming cells* This primitive cell type has been observed in Dexter-type long-term bone marrow cultures of murine cells (see: LTBMC) utilizing primary stromal cell layers reseeded with hematopoietic cells. These cells are detected in » limiting dilution assays and resemble cells from the so-called cobblestone area morphologically defined as active zones of » hematopoiesis *in vivo*. Such cells and/or close relatives can be enriched and studied further by treatment with such agents as » fluorouracil or » 4-HC (4-hydroperoxycyclophosphamide).
The frequency of precursors forming cobblestone areas on day 28 after reseeding has been proposed to be a measure of » hematopoietic stem cells capable of repopulating irradiated bone marrow *in*

vivo (see: MRA, marrow-repopulating ability for an assay system) and therefore constitute one primitive type of hematopoietic stem cells or very early progenitors of myeloid hematopoietic stem cells. The frequency of day 28 CAFCs appears to closely correlate with that of » CFU-S cells obtained at day 12.

Bartelmez SH et al Uncovering the heterogeneity of hematopoietic repopulating cells EH 19: 861-2 (1991); **Deryugina EI & Müller-Sieburg CE** Stromal cells in long-term cultures: keys to the elucidation of hematopoietic development? Crit. Rev. Immunol. 13: 115-50 (1993); **Neben S et al** Quantitation of murine hematopoietic stem cells *in vitro* by limiting dilution analysis of cobblestone area formation on a clonal stromal cell line. EH 21: 438-43 (1993); **Ploemacher RE et al** An *in vitro* limiting dilution assay of long-term repopulating hematopoietic stem cells in the mouse. Blood 74: 2755-63 (1989); **Ploemacher RE & van der Sluijs JP** *In vitro* frequency analysis of spleen colony forming and marrow-repopulating hemopoietic stem cells in the mouse. J. Tiss. Cult. Meth. 13: 63-8 (1991); **Ploemacher RE et al** Murine hemopoietic stem cells with long-term engraftment and marrow repopulating ability are more resistant to γ-radiation than are spleen colony forming cells. Int. J. Radiat. Biol. 61: 489-99 (1992); **Ploemacher RE et al** Wheat germ agglutinin affinity of murine hemopoietic stem cell subpopulations is an inverse function of their long-term repopulating ability *in vitro* and *in vivo*. Leukemia 7: 120-30 (1993)

Calcitonin gene-related peptide: see: CGRP.

Calcium ionophore: Ionophores are compounds that increase the permeability of cellular membrane barriers to ions by functioning as mobile ion carriers or channel formers. They contain hydrophobic regions conferring lipid solubility and hydrophilic ion-binding regions which delocalise the charge of the ion to shield it from the hydrophobic regions of the membrane lipid bilayer.

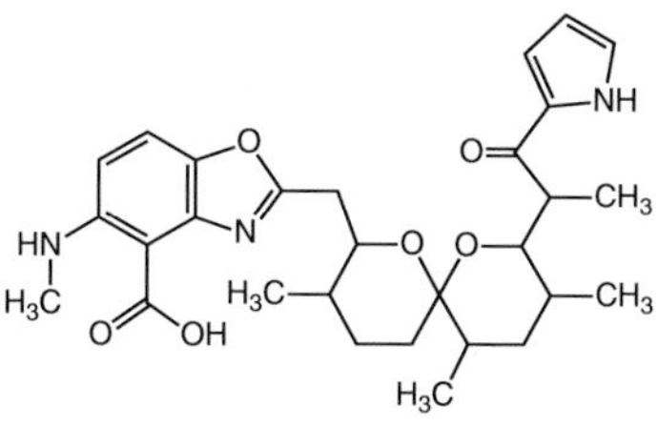

Calcium ionophore A23187

Calcium ionophore ***A23187*** is an artificial mobile iron carrier that normally acts as an ion-exchange shuttle molecule transporting one divalent calcium ion into the cell in exchange of two H^+. As intracellular calcium levels can be monitored by a variety of fluorescent probes use of A23187 provides information about the involvement of elevated levels of cytosolic free calcium (and indirect information about associated secondary messenger systems) in » cytokine responses following cytokine receptor-mediated signal transduction processes. A postulated participation of calcium in the process under study can be confirmed by employing calcium-specific chelators such as EGTA to produce very low intracellular calcium levels which should then block the response.

A23187 can cause » Cell activation, differentiation, or proliferation and thus mimics cellular processes normally observed in response to » cytokines. It can therefore be used to probe functional capacities of cells and to dissect complex processes into a series of discrete stages at the molecular level, for example the ability of cells to produce and release » cytokines, to express » ERGs (early response genes), » oncogenes, differentiation antigens (see also: CD antigens), intracellular adhesion molecules, to undergo » apoptosis, to go through the » cell cycle, or to inhibit these processes. Simultaneous treatment of cells with calcium ionophores and other agents (cytokines, drugs, hormones) can be used to investigate whether any of these agents affect (enhance or reverse) any of the elicited responses. For other agents used to dissect cytokine-mediated signal transduction pathways see: Bryostatins, Calphostin C, Genistein, H8, Herbimycin A, K-252a, Lavendustin A, Phorbol esters, Okadaic acid, Staurosporine, Suramin, Tyrphostins, Vanadate.

Balasubramanian SV et al Bilayers containing calcium ionophore A23187 form channels. BBRC 189: 1038-42 (1992); **Berridge MJ** Inositol trisphosphate and calcium signaling. N 361: 315-25 (1993); **Hepler PK** Calcium and mitosis. Int. Rev. Cytol. 138: 239-68 (1992); **Scharf O & Foder B** Regulation of cytosolic calcium in blood cells. Physiol. Rev. 73: 547-82 (1993)

Calgranulin A: see: Calgranulins.

Calgranulin B: see: Calgranulins

Calgranulins: Calmodulins are intracellular calcium-binding proteins consisting of at least two different peptide chains designated Calgranulin A (abbrev. CAGA) and Calgranulin B (abbrev. CAGB). These proteins are biochemically related to » S100. CAGA and CAGB map to human chromosome 1q12-q21.

The two calgranulins are also known as ***myeloid-associated proteins***. Calgranulin A (98 amino acids, 11 kDa) is identical with ***Cystatin A***, ***Stefin A*** or ***MRP 8*** (MIF-related protein; see also: MIF, migration inhibition factor). Calgranulin A forms tight complexes with papain and the cathepsins B,

H, and L. Calgranulin is identical with *L1*, *Calprotectin*, *MRP 14* (MIF-related protein; see: MIF, migration inhibition factor). The genes encoding the two calgranulins have been located to human chromosome 3q21.

The two calgranulins are synthesized by peripheral blood neutrophil granulocytes and monocytes and also by squamous cell epithelia and keratinocytes of patients with inflammatory dermatoses. They are not produced by tissue macrophages. They function as inflammatory cytokines (see also: Inflammation). Production of Calgranulins ceases with progressing differentiation of the monocytes. Elevated amounts of both calgranulins are found in the serum of patients with cystic fibrosis and clinically normal heterozygous carriers.

MRP 8 and MRP 14 occur as non-covalently linked complexes (heterodimers, trimers, and tetramers). They belong to the » S100 family of proteins. Mouse MRP8 and MRP14 proteins share 59% identity with their human counterparts, but they are more divergent than the other members of the S100 protein family.

MRP8 and MRP14 are recognized by some monoclonal antibodies directed against MIF activities. The expression of these proteins is specific for cells of myeloid origin, namely granulocytes, monocytes and macrophages, and is observed in blood granulocytes and monocytes but not in normal tissue macrophages. MRP8/MRP14 expression and complexation are characteristic for granulocytes and distinct stages of macrophage differentiation in mice. In all inflammation models tested so far the cells arriving first at the lesion are MRP8- and MRP14-positive. MRP8/14 is also found in body fluids in inflammatory conditions and thus may be considered as a very sensitive inflammation marker.

MRP8/14 complexes have also been described as ***L1 complex***. They are identical with the ***cystic fibrosis antigen*** (CFAG) found to be elevated in the serum of cystic fibrosis homozygotes and heterozygous carriers and to be inducible in the promyelocytic cell line » HL-60. CFAG consists of CAGA and CAGB. MRP8/14 complexes have been shown to be a specific inhibitor of casein kinase I and II. This complex is expressed during the late stages of terminal differentiation induced in human promyelocytic HL-60 leukemia cells by 1 α,25-dihydroxyvitamin D3 (but not by » TNF-α or » IFN-γ), and in human monocytic THP-1 leukemia cells by phorbol 12-myristate 13-acetate (see also: Phorbol esters). At concentrations of 5-15 nM, the purified complex inhibits the growth of HL-60 and THP-1 cells, as well as other cell types belonging to different cell lineages.

MRP8 has been shown to be expressed also outside the cells of the immune system and to be identical with a protein from chicken, human, and mouse lens (see: 10K protein).

Chilosi M et al Multimarker immunohistochemical staining of calgranulins, chloroacetate esterase, and S100 for simultaneous demonstration of inflammatory cells on paraffin sections. J. Histochem. Cytochem. 38: 1669-75 (1990); **Clark BR et al** Calgranulin expression and association with the keratinocyte cytoskeleton. J. Pathol. 160: 25-30 (1990); **Dorin JR et al** Related calcium-binding proteins map to the same subregion of chromosome 1q and to an extended region of synteny on mouse chromosome 3. Genomics 8: 420-6 (1990); **Dorin JR et al** A clue to the basic defect in cystic fibrosis from cloning the CF antigen gene. N 326: 614-7 (1987); **Goebeler M et al** Expression and complex assembly of calcium-binding proteins MRP8 and MRP14 during differentiation of murine myelomonocytic cells. J. Leukoc. Biol. 53: 11-8 (1993); **Hsieh WT et al** Mapping of the gene for human cysteine proteinase inhibitor stefin A, STF1, to chromosome 3cen-q21. Genomics 9: 207-9 (1991); **Kelly SE et al** Calgranulin expression in inflammatory dermatoses. J. Pathol. 159: 17-21 (1989); **Lagasse E & Weissman IL** Mouse MRP8 and MRP14, two intracellular calcium-binding proteins associated with the development of the myeloid lineage. Blood 79: 1907-15 (1992); **van Kelly SE et al** Morphological evidence for calcium-dependent association of calgranulin with the epidermal cytoskeleton in inflammatory dermatoses. Br. J. Dermatol. 124: 403-9 (1991); **Heyningen V & Dorin J** Possible role for two calcium-binding proteins of the S-100 family, co-expressed in granulocytes and certain epithelia. AEMB 269: 139-43 (1990); **Odink K et al** Two calcium-binding proteins in infiltrate macrophages of rheumatoid arthritis. N 330: 80-2 (1987); **Roth J et al** Expression of calcium-binding proteins MRP8 and MRP14 is associated with distinct monocytic differentiation pathways in HL-60 cells. BBRC 191: 565-70 (1993); **Sorg C et al** The calcium binding proteins MRP8 and MRP14 in acute and chronic inflammation. Behring Inst. Mitt. 91: 126-37 (1992); **Strauss M et al** Chemical synthesis of a gene for human stefin A and its expression in *E. coli*. Biol. Chem. Hoppe-Seyler 369: 1019-30 (1988); **Tsui FWL et al** Molecular characterization and mapping of murine genes encoding three members of the stefin family of cysteine proteinase inhibitors. Genomics 15: 507-14 (1993); **Warner-Bartnicki AL et al** Regulated expression of the MRP8 and MRP14 genes in human promoyelocytic leukaemic HL60 cells treated with the differentiation-inducing agents mycophenolic acid and 1a,25-dihydroxyvitamin D3. ECR 204: 241-6 (1993); **Wilkinson MM et al** Expression pattern of two related cystic fibrosis-associated calcium-binding proteins in normal and abnormal tissues. JCS 91: 221-30 (1988); **Zwadlo G et al** Two calcium-binding proteins associated with specific stages of myeloid cell differentiation are expressed by subsets of macrophages in inflammatory tissues. Clin. Exp. Immunol. 72: 510-5 (1988)

Calphostin C: A secondary metabolite of the fungus *Cladosporium cladosporioides*. It is an inhibitor of protein kinase C that binds to the diacylglycerol regulatory binding site of the enzyme. It can be used to study whether this enzyme is involved in cytokine-mediated signal transduction processes. Calphostin C is highly toxic to tumor cell lines but it is not known whether this is related to its ability to inhibit protein kinase C. Calphostin C has been shown to inhibit succinate-dependent mitochondrial electron transport by a mechanism that does not involve protein kinase C.

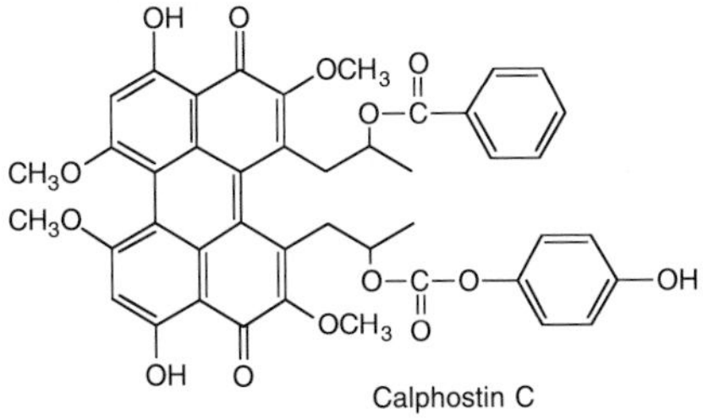

Calphostin C

For other agents used to dissect cytokine-mediated signal transduction pathways see: Bryostatins, Calcium ionophore, Genistein, H8, Herbimycin A, K-252a, Lavendustin A, Phorbol esters, Okadaic acid, Staurosporine, Suramin, Tyrphostins, Vanadate.

● REVIEWS: **Casnellie JE** Protein kinase inhibitors: probes for the functions of protein phosphorylation. Adv. Pharmacol. 22: 167-205 (1991); **Tamaoki T** Use and specificity of staurosporine, UCN-01, and calphostin C as protein kinase inhibitors. MiE 201: 340-47 (1991)

Berridge MY & Tan AS calphostin C, inhibits succinate-dependent mitochondrial reduction of MTT by a mechanism that does not involve protein kinase C. BBRC 185: 806-11 (1992); **Minana MD et al** Differential effects of the protein kinase C inhibitors H7 and calphostin C on the cell cycle of neuroblastoma cells. Brain Res. 596: 157-62 (1992)

Calprotectin: see: Calgranulins.

Calyculin A: see: Okadaic acid.

CAM assay: *chorioallantoic membrane assay* A bioassay to determine the biological activity of » angiogenesis factors.

cAMP responsive element: see: CRE

cAMP responsive element binding protein: abbrev. CREB. See: CRE (cyclic AMP-responsive element).

CAM-RF: cellular adhesion molecule regulatory factor This factor of 50 kDa is secreted by the human CCRF-CEM T lymphoma cell line initially isolated from a child lymphoma patient with central nervous system involvement. The cells grow invasively in a monolayer invasion assay system *in vitro*. CAM-RF strongly upregulates the expression of ICAM-1 (intercellular adhesion molecule 1, CD54), ELAM-1 (endothelial leukocyte adhesion molecule 1, E-selectin), adhesion molecules on vascular endothelial cells and also the expression of VCAM-1 (vascular cell adhesion molecule, INCAM-110) to a lesser extent. CAM-RF also promotes the cellular binding between CEM cells and vascular endothelial cells *in vitro* and it is thought that it may promote lymphoma invasion through upregulation of adhesion molecules.

Totsuka T et al Invasive human T lymphoma cells produce a novel factor that upregulates expression of adhesion molecules on endothelial cells. EH 21: 1544-9 (1993)

Cancer vaccination: see: Cytokine gene transfer.

CAP 37: *cationic antimicrobial protein* CAP37 is found in azurophil granules, specialized lysosomes of the neutrophil. It is a glycoprotein of 222 amino acids. It is identical with ***Azurocidin***. The gene maps to the short arm of human chromosome 19. The cDNA-derived amino acid sequence of CAP37 shows 44, 42, and 32% homology at the amino acid level to neutrophil elastase, myeloblastin, and cathepsin G, respectively, suggesting that CAP37 is a member of the serine protease gene family that mediate various functions in inflammation. CAP37 does not possess serine protease activity due to mutations in the active site residues serine and histidine.

CAP37 is expressed in undifferentiated » HL-60 cells but not in mature neutrophils (see also: myeloblastin). It is produced and packaged into azurophil granules in large quantities during neutrophil differentiation together with other serine proteases. It is a specific chemoattractant for monocytes and possesses antibacterial activity. CAP37 is not chemotactic for neutrophils and lymphocytes. CAP37 released from neutrophils during phagocytosis and degranulation may mediate recruitment of monocytes in the second wave of inflammation.

Almeida RP et al Complementary DNA sequence of human neutrophil azurocidin, an antibiotic with extensive homology to serine proteases. BBRC 177: 688-95 (1991); **Campanelli D et al** Azurocidin and a homologous serine protease from neutrophils: differential antimicrobial and proteolytic properties. JCI 85: 904-15 (1990); **Morgan JG et al** Cloning of the cDNA for the serine protease homologue CAP37/azurocidin, a microbicidal and chemotactic protein from human granulocytes. JI 147: 3210-4 (1991); **Pereira HA et al** CAP37, a human neutrophil-derived chemotactic factor with monocyte specific activity. JCI 85: 1468-76 (1990); **Pohl J et al** Amino acid sequence of CAP37, a human neutrophil granule-derived antibacterial and monocyte-specific chemotactic glycoprotein structurally similar to neutrophil elastase. FL 272: 200-4 (1990); **Wilde CG et al** Characterization of two azurophil granule proteases with active-site homology to neutrophil elastase. JBC 265: 2038-41 (1990); **Zimmer M et al** Three human elastase-like genes coordinately expressed in the myelomonocyte lineage are organized as a single genetic locus on 19pter. PNAS 89: 8215-9 (1992)

CArG box: see: SRE (serum response element).

Cartilage-derived factor: see: CDF.

Cartilage-derived growth factor 1: see: CDGF-1.

Cartilage-derived growth factor 1. 7: abbrev. CDGF. This factor is identical with » bFGF.

Cartilage-derived inhibitor: see: CDI.

Cartilage-inducing factor A: see: CIF-A.

Cartilage-inducing factor B: see: CIF-A.

Catabolin: This factor was initially isolated as a protein causing the resorption of cartilage tissue and the resorption of proteoglycans in the extracellular matrix (see: ECM). Catabolin is identical with one of the molecular forms of » IL1 (IL1β).
Pilsworth LM & Saklatvala J The cartilage-resorbing protein catabolin is made by synovial fibroblasts and its production is increased by phorbol myristate acetate. BJ 216: 481-9 (1983); **Saklatvala J et al** Pig interleukin-1. Purification of two immunologically different leukocyte proteins that cause cartilage resorption, lymphocyte activation, and fever. JEM 162: 1208-22 (1985); **Saklatvala J et al** Characterization of catabolin, the major product of pig synovial tissue that induces resorption of cartilage proteoglycan *in vitro*. EJB 199: 705-14 (1981); **Saklatvala J et al** Identification of catabolin, a protein from synovium which induces degradation of cartilage in organ culture. BBRC 96: 1225-31 (1980); **Saklatvala J et al** Purification to homogeneity of pig leukocyte catabolin, a protein that causes cartilage resorption *in vitro*. BJ 215: 385-92 (1983)

CAT development factor: *choline acetyltransferase development factor* see: CDF.

Cationic antimicrobial protein: see: CAP 37.

CBGF: *chick brain-derived growth factor* CBGF is a poorly characterized protein of ≈ 1.5 kDa that was isolated from chicken brain and acts as a strong mitogen for astrocytes.
Carlone R et al Purification of a chick brain-derived growth factor by reversed-phase high-performance liquid chromatography. J. Neurosci. 7: 2163-7 (1987)

***cbl*:** An » oncogene detected initially as the transforming gene in the genome of the murine Cas NS-1 retrovirus. This virus induces pro-B, pre-B and myeloid tumors in mice inoculated at birth. Sequencing of c-*cbl* has revealed that the viral oncogene has been generated by a large truncation that removes 60% of the C-terminus of the corresponding protein. The truncated protein is a DNA-binding transcription factor.
The human cellular homologue of v-*cbl* is located on chromosome 11q23.3, which is in the region of translocation breakpoints in a range of acute leukemias. The human gene, which is expressed in a range of hemopoietic lineages, encodes a cytoplasmic protein of 120 kDa (906 amino acids).
Blake TJ et al The sequences of the human and mouse c-*cbl* proto-oncogenes show v-*cbl* was generated by a large truncation encompassing a proline-rich domain and a leucine zipper-like motif. O 6: 653-7 (1991); **Blake TJ & Langdon WY** A rearrangement of the c-*cbl* proto-oncogene in HUT78 T-lymphoma cells results in a truncated protein. O 7: 757-62 (1992); **Blake TJ et al** The truncation that generated the v-*cbl* oncogene reveals an ability for nuclear transport, DNA binding and acute transformation. EJ 12: 2017-26 (1993); **Langdon WY et al** The localization of the products of the c-*cbl* and v-*cbl* oncogenes during mitosis and transformation. CTMI 182: 467-74 (1992)

CCF: *crystal-induced chemotactic factor* This poorly characterized 8.4 kDa factor is synthesized by polymorphonuclear leukocytes (PMN-CCF). It was initially detected in the culture media of neutrophils phagocytizing urate and calcium pyrophosphate crystals. The factor is a chemoattractant (see also: Chemotaxis) for segmented neutrophil granulocytes and monocytes. The factor is possibly involved in the pathogenesis of arthritic processes.
Bhatt AK & Spilberg I Monoclonal antibody to the crystal-induced chemotactic factor from human neutrophils detects similar immunoreactive and biologically active protein in promyelocytic HL-60 cells. Hybridoma 8: 323-30 (1989); **Spilberg I & Mehta J** Demonstration of a specific neutrophil receptor for a cell-derived chemotactic factor. JCI 63: 85-8 (1979); **Spilberg I & Bhatt AK** Crystal-induced chemotactic factor. MiE 162: 193-7 (1988)

C-C-Intercrines: A name sometimes given to some members of the » chemokine family of cytokines which, among other factors, comprises a variety of factors with chemotactic (see: Chemotaxis) and pro-inflammatory activities (see: Inflammation). The C-C refers to the spacing of two neighboring cytosine residues which is conserved in members of this group.

CCSF: *chondrocyte colony-stimulating factor* see: CDF (cartilage-derived factors).

CCL39: A Chinese hamster lung fibroblast cell line. It displays the properties characteristic of normal secondary cultures of Chinese hamster fibroblasts including reversible G0 growth arrest (see also: cell cycle), anchorage dependence (see also: cell culture, and high serum-growth factor dependence. Injection of CCl39 cells, or anchorage-independent variants, into nude mice (see also: Immunodeficient mice) leads to tumor formation. Proliferation of these cells is stimulated by α-thrombin and also by » bFGF or » aFGF which are as potent as α-thrombin to reinitiate DNA synthesis in G0-arrested cells. Protamine and polylysine (at 5 microM) inhibit the mitogenic effect induced by bFGF.

CCL39 cells require several growth factors (insulin, α-thrombin, EGF) for their continuous growth.

Chadwick DW & Lagarde AE Coincidental acquisition of growth autonomy and metastatic potential during the malignant transformation of factor-dependent CCL39 lung fibroblasts. JNCI 80: 318-25 (1988); **Dauchel MC et al** Modulation of mitogenic activity and cellular binding of basic fibroblast growth factor by basic proteins. JCBc 39: 411-20 (1989); **Magnaldo I et al** The mitogenic signaling pathway of fibroblast growth factor is not mediated through polyphosphoinositide hydrolysis and protein kinase C activation in hamster fibroblasts. JBC 261: 16916-22 (1986)

CCL-64: (ML-CCL64; ATCC CCL-64 = Mv1Lu) An epithelial cell line established from mink lung. The cell line is used to assay » TGF-β which inhibits the proliferation of these cells even under serum-containing conditions (see also: Bioassays). This assay is an order of magnitude more sensitive than a similar assay employing » NRK (normal rat kidney) cells. Of the three isoforms, TGF-β1, -β2, and -β3, TGF-β3 is somewhat more potent (ED_{50} = 0.5 pM versus 2 pM) than TGF-β1 and TGF-β2 as a growth inhibitor. TGF-β1 inhibits the proliferative response of CCL64 to serum mitogens, to insulin, and to » EGF and this response can be inhibited by prior exposure of the cells to nanogram concentrations of pertussis toxin. Inhibition of CCL-64 cell proliferation by TGF-β2, but not by TGF-β1, is reduced ≈ 10-fold by » Alpha-2-Macroglobulin. Levels of TGF are underestimated if TGF-β production is studied in mononuclear cells, due to the presence of platelets. Retinoic acid and the tumor promoter tetradecanoyl phorbol acetate (see also: Phorbol esters) have been shown to interfere with the TGF-β growth inhibition assay, causing themselves a high degree of cell proliferation inhibition.

The TGF-β assay with CCL-64 is not affected by » PDGF, » IGF, or heparin-binding growth factors (see: HBGF). Although » EGF has a small positive effect of cell proliferation, it does not antagonize the inhibitory effect of TGF-β.

TGF-resistant cell lines developed from mutagenized Mv1Lu cells do not express TGF-β receptors. Growth and DNA synthesis of CCL-64 cells are also stimulated by » HGF (hepatocyte growth factor).

An alternative and entirely different detection method is » Message amplification phenotyping.

Cheifetz S et al Distinct transforming growth factor-β (TGF-β) receptor subsets as determinants of cellular responsiveness to three TGF-β isoforms. JBC 265: 20533-8 (1990); **Danielpour D et al** Immunodetection and quantitation of the two forms of transforming growth factor-β (TGF-β and TGF-β 2) secreted by cells in culture. JCP 138: 79-86 (1989); **Danielpour D et al** Differential inhibition of transforming growth factor β 1 and β 2 activity by α 2-macroglobulin. JBC 265: 6973-7 (1990); **Laiho M et al** Concomitant loss of transforming growth factor (TGF)-β receptor types I and II in TGF-β-resistant cell mutants implicates both receptor types in signal transduction. JBC 265: 18518-24 (1990); **Merino J et al** The measurement of transforming growth factor type β (TGF β) levels produced by peripheral blood mononuclear cells requires the efficient elimination of contaminating platelets. JIM 153: 151-9 (1992); **Slootweg MC et al** Characterization of growth factor activity produced by fetal mouse osteoblasts. J. Endocrinol. 124: 301-9 (1990); **Su HC et al** A role for transforming growth factor-β 1 in regulating natural killer cell and T lymphocyte proliferative responses during acute infection with lymphocytic choriomeningitis virus. JI 147: 2717-27 (1991); **Tajima H et al** Regulation of cell growth and motility by hepatocyte growth factor and receptor expression in various cell species. ECR 202: 423-31 (1992); **Tucker RF et al** Growth inhibitor from BSC-1 cells closely related to platelet type β transforming growth factor. S 226: 705-7 (1984); **Van Zoelen EJJ et al** PDGF-like growth factor induces EGF-potentiated phenotypic transformation of normal rat kidney cells in the absence of TGFβ. BBRC 141: 1229-35 (1986)

CD2: This cell surface antigen (see also: CD antigens) is also known as ***T11***, ***LFA-2*** (lymphocyte function-associated), ***Leu-5***, ***OKT11*** and ***SRBC receptor*** (sheep red blood cell receptor). CD2 is found on ≈ 90 % of mature peripheral T cells and 70 % of thymocytes with the exception of very immature forms. It is also expressed on NK cells. This marker is not expressed on B lymphocytes.

CD2 is a 50 kDa transmembrane protein belonging to the immunoglobulin superfamily. CD2 consists of 327 amino acids with 185 amino acids representing the extracellular domain, and 117 amino acids for the proline-rich, basic intracellular domain. The human CD2 gene maps to chromosome 1p13 near the gene encoding CD58. It is ≈ 12 kb in length and comprises five exons.

CD2 functions in normal human immune responses include recognition involving T helper cells and antigen-presenting cells, cytolytic effector

functions of natural killer cells and cytotoxic T lymphocyte. CD2 is also involved in the interaction of thymic epithelial cells and thymocytes.
The extracellular domain of CD2 is a receptor for a 55-70 kDa protein called ***CD58*** or ***LFA-3*** which is expressed on many different cell types. The LFA-3 gene lies in close vicinity of the CD2 gene. Binding of LFA-3 to CD2 allows the establishment of very co-ordinated cell-to-cell-contacts as are observed, for example, between antibody-presenting cells or cytotoxic lymphocytes with their target cells or the interactions of maturing thymocytes with epithelial cells of the thymus. In addition, CD59 (= protectin, = MIRL, = HRF20, = MACIF, = P18, = H19) appears to be a ligand for CD2, synergising with CD58 in mediating T cell adhesive interactions and activation. There is another, as yet unidentified, CD2 ligand identified in the K562 erythroleukemic cell line, which binds CD2 with a 10-fold lower affinity than CD58. CD48 (= murine sgp-60 = human Blast-1, rat OX-45 = murine BCM1) has also been shown to function as a low affinity ligand for CD2. Monoclonal antibodies directed against CD48 have been shown to inhibit IL2 production, IL2 receptor expression, and T cell proliferation in murine cells.
The cytoplasmic domain of CD2 is considerably conserved in humans, rats and mice. It functions as a signaling domain. Binding of the ligand to CD2 activates the metabolism of inositol phospholipids, increases intracellular calcium levels (see also: Calcium ionophore), and activates protein kinase C, eventually leading to the expression of the T cell growth factor » IL2 and its receptors.
The synthesis and secretion of several cytokines and also cell proliferation can also be induced by anti-CD2 antibodies binding to CD2. Some of these antibodies are capable of blocking the interaction of the T cell receptor (TCR) with its ligand CD3, thus effectively blocking T cell activation (see also: Cell activation).

Abraham DJ et al Function and regulation of the murine lymphocyte CD2 receptor. Leukocyte Biol. 49: 329-41 (1991); **Antonio RN et al** A soluble multimeric recombinant CD2 protein identifies CD48 as a low affinity ligand for human CD2: divergence of CD2 ligands during the evolution of humans and mice. JEM 177: 1439-50 (1993); **Beyers AD et al** Autonomous roles for the cytoplasmic domains of the CD2 and CD4 T cell surface antigens. EJ 10: 377-85 (1991); **Bierer BE et al** Expression of the T cell surface molecule CD2 and an epitope-loss CD2 mutant to define the role of lymphocyte function-associated antigen 3 (LFA-3) in T cell activation. PNAS 85: 1194-8 (1988); **Bierer BE et al** Synergistic T cell activation via the physiological ligands for CD2 and the T cell receptor. JEM 168: 1145-56 (1988); **Cabrero JG et al** Identification, by protein sequencing and gene transfection, of sgp-60 as the murine homologue of CD48. PNAS 90: 3418-22 (1993); **Cerdan C et al** IL1α is produced by T lymphocytes activated via the CD2 plus CD28 pathways. JI 146: 560-4 (1991); **Chang HC et al** Dissection of the human CD2 intracellular domain. Identification of a segment required for signal transduction and interleukin 2 production. JEM 169: 2073-83 (1989); **Clayton LK et al** Murine and human T11 (CD2) cDNA sequences suggest a common signal transduction mechanism. EJI 17: 1367-70 (1987); **Clayton LK et al** The gene for T11 (CD2) maps to chromosome 1 in humans and to chromosome 3 in mice. JI140: 3617-21 (1988); **Clipstone NA & Crumpton MJ** Stable expression of the cDNA encoding the human T lymphocyte-specific CD2 antigen in murine L cells. EJI 18: 1541-5 (1988); **Danielian S et al** The tyrosine kinase activity of p56*lck* is increased in human T cells activated via CD2. EJI 21: 1967-70 (1991); **Deckert M et al** CD58 and CD59 molecules exhibit potentializing effects in T cell adhesion and activation. JI 148: 672-7 (1992); **Diamond DJ et al** Exon-intron organization and sequence comparison of human and murine T11 (CD2) genes. PNAS 1615-9 (1988); **He Q et al** A role in transmembrane signaling for the cytoplasmic domain of the CD2 T lymphocyte surface antigen. Cell 54: 979-84 (1988); **Kingsmore SF et al** Physical linkage of genes encoding the lymphocyte adhesion molecules CD2 and its ligand LFA-3. Immunogenetics 30: 123-125 (1989); **Makgoba MW et al** The CD2-LFA3 and LFA1-ICAM pathways: relevance to T cell recognition. IT 10: 417-21 (1989); **Moingeon P et al** The structural biology of CD2. IR 111: 111-44 (1989); **Parish CR et al** Detection of a glycosylation-dependent ligand for the T lymphocyte cell adhesion molecule CD2 using a novel multimeric recombinant CD2-binding assay. JI 150: 4833-43 (1993); **Richardson NE et al** Adhesion domain of human T11 (CD2) is encoded by a single exon. PNAS 85: 5176-80 (1988); **Sayre PH et al** Molecular cloning and expression of T11 cDNAs reveal a receptor-like structure on human T lymphocytes. PNAS 84: 2941-5 (1987); **Seed B & Aruffo A** Molecular cloning of the CD2 antigen, the T cell erythrocyte receptor by a rapid immunoselection procedure. PNAS 84: 3365-9 (1987); **Sewell WA et al** Molecular cloning of the human T lymphocyte surface CD2 (T11) antigen. PNAS 83: 8718-22 (1986) (Erratum in PNAS 84: 7256 (1987); **Sewell W A et al** The murine homologue of the T lymphocyte CD2 antigen: molecular cloning, chromosome assignment and cell surface expression. EJI 17: 1015-20 (1987); **Yagita H et al** Molecular cloning of the murine homologue of CD2. Homology of the molecule to its human counterpart T11. JI 140: 1321-6 (1988)
● **Transgenic/Knock-out/Antisense Studies: Lang G et al** The structure of the human CD2 gene and its expression in transgenic mice. EJ 7: 1675-82 (1988); **Monostori E et al** Human CD2 is functional in CD2 transgenic mice. Immunology 74: 369-72 (1991)

CD4: This cell surface antigen (see also: CD antigens) is also known as ***T4***, ***Leu-3***, ***OKT4*** or ***L3T4*** (murine). CD4 is a 55 kDa transmembrane glycoprotein belonging to the immunoglobulin superfamily. In humans CD4 is expressed of peripheral T cells, thymocytes, on macrophages and granulocytes. It is also expressed in a developmentally regulated manner in specific regions of the brain. The human CD4 gene is on chromosome 12pter-p12.
T cell progenitor cells in the thymus initially do not express CD4 and another antigen called » CD8

and develop into mature T cells in several steps. CD4-positive cells are T helper cells which, upon activation (see also: Cell activation) produce a number of different cytokines and play a role in the activation and/or proliferation of B cells, cytotoxic T lymphocytes, and macrophages.

CD4 acts as a cellular adhesion molecule binding class II MHC molecules. It stabilizes the interaction of class II MHC-restricted T cells and antigen-presenting cells expressing an antigen in combination with a class II MHC molecule. Interactions of CD4 with MHC class II molecules are crucial during thymic development and subsequently for the function of single-positive $CD4^+CD8^-$ T lymphocytes. CD4 also acts as a receptor for HIV. Another ligand of CD4 is » LCF (lymphocyte chemoattractant factor). The expression of CD4 in brain suggest that it may play a more general role in mediating cell recognition events than merely those of cellular immune responses.

The expression of » IL4 in » transgenic animals blocks the maturation of T cells and considerably reduces the population of immature $CD4^+8^+$ thymocytes and peripheral T cells while at the same time greatly enhancing the numbers of CD8-positive mature thymocytes. In CD4-negative mouse strains the development of CD8-positive cells proceeds normally although the activity of helper cells is greatly reduced. The proliferation of thymocytes with the marker spectrum $CD4^-8^-$, $CD4^+8^-$, $CD4^-8^+$ is stimulated by » IL7 which is therefore an important developmental factor for functionally different subpopulations of T lymphocytes.

$CD4^+8^-$ thymocytes produce factors such as » IL4, » IL5, » IL10 and » IFN-γ (see also: TH1/TH2 cytokines).

CD4 functions as an effector molecule for intracellular signal transduction. In T lymphocytes CD4 has been found to be associated with tyrosine kinase $p56^{lck}$ (see: *lck*) which phosphorylates a number of other cell surface proteins (including CD3) and is probably involved in transducing the intracellular signal when these proteins are bound to their receptors (see also: HT-2 cell line).

Monoclonal antibodies directed against CD4 inhibit several cellular functions *in vitro* and have been used, among other things, to treat autoimmune diseases in experimental animals. At present, these antibodies, and also antibody-toxin chimeras are tested as drugs against infections with HIV. One particular class of genetically engineered proteins, known as immunoadhesins, are fusion proteins combining the constant Fc region of IgG with the CD4 receptor domain acting as a binding site for pg120 protein of HIV. Such fusion proteins therefore contain an additional functional constant region of the IgG antibody and can be used to eliminate cells and viruses expressing the corresponding antigens.

Pluripotent murine » hematopoietic stem cells have recently been shown also to express the CD4 antigen. The CD4-positive cells isolated from mouse marrow repopulate all hematopoietic lineages in several assays (see: LTBMC, long-term bone marrow culture; MRA, marrow repopulating ability), indicating that this population contains primitive stem cells with extensive repopulation capacity. This also suggests that CD4 may also play an important role in lineage definition in early hematopoietic differentiation (see also: Hematopoiesis).

● REVIEWS: **Barcley AN et al** CD4 and the immunoglobulin superfamily. Phil. Trans. R. Soc. Lond. B 342: 7-12 (1993); **Janeway CA et al** $CD4^+$ T cells: specificity and function. IR 101: 39-80 (1988); **Littman DR** The structure of the CD4 and CD8 genes. ARI 5: 561-84 (1987); **Olive D & Mawas C** Therapeutic applications of anti-CD4 antibodies. Crit. Rev. Ther. Drug Carrier Syst. 10: 29-63 (1993); **Parnes JR** Molecular biology and function of CD4 and CD8. AI 44: 265-311 (1989)

● BIOCHEMISTRY & MOLECULAR BIOLOGY: **Carr SA et al** Protein and carbohydrate structural analysis of a recombinant soluble CD4 receptor by mass spectrometry JBC 264: 21286-95 (1989), Erratum in JBC 265: 3585 (1990); **Davis SJ et al** High level expression in Chinese hamster ovary cells of soluble forms of CD4 T lymphocyte glycoprotein including glycosylation variants. JBC 265: 10410-8 (1990); **Dialynas DP et al** Characterization of the murine antigenic determinant, designated L3T4a, recognized by monoclonal antibody GK1.5: expression of L3T4a by functional T cell clones appears to correlate primarily with class II MHC antigen-reactivity. IR 74: 29-56 (1983); **Isobe M et al** The gene encoding the T cell surface protein T4 is located on human chromosome 12. PNAS 83: 4399-4402 (1986); **König R et al** Overexpression and biosynthesis of CD4 in Chinese hamster ovary cells: co-amplification using the multiple drug resistance gene. PNAS 86: 9188-92 (1989); **Kwong PD et al** Molecular characteristics of recombinant human CD4 as deduced from polymorphic crystals. PNAS 87: 6423-7 (1990); **Maddon PJ et al** The isolation and nucleotide sequence of a cDNA encoding the T cell surface protein T4: a new member of the immunoglobulin gene family. Cell 42: 93-104 (1985); **Maddon PJ et al** Structure and expression of the human and mouse T4 genes. PNAS 84: 9155-9 (1987); **Sleckman BP et al** Expression and function of CD4 in a murine T cell hybridoma. N 328: 351-3 (1987); **Webb NR et al** Cell-surface expression and purification of human CD4 produced in baculovirus-infected insect cells. PNAS 86: 7731-5 (1989)

● TRANSGENIC/KNOCK-OUT/ANTISENSE STUDIES: **Barzaga-Gilbert E et al** Species specificity and augmentation of responses to class II major histocompatibility complex molecules in human CD4 transgenic mice. JEM 175: 1707-15 (1992); **Gillespie FP et al** Tissue-specific expression of human CD4 in transgenic mice. MCB 13: 2952-8 (1993); **Lores P et al** Expression of human CD4 in transgenic mice does not confer sensitivity to human immunodeficiency virus infection. AIDS Res. Hum.

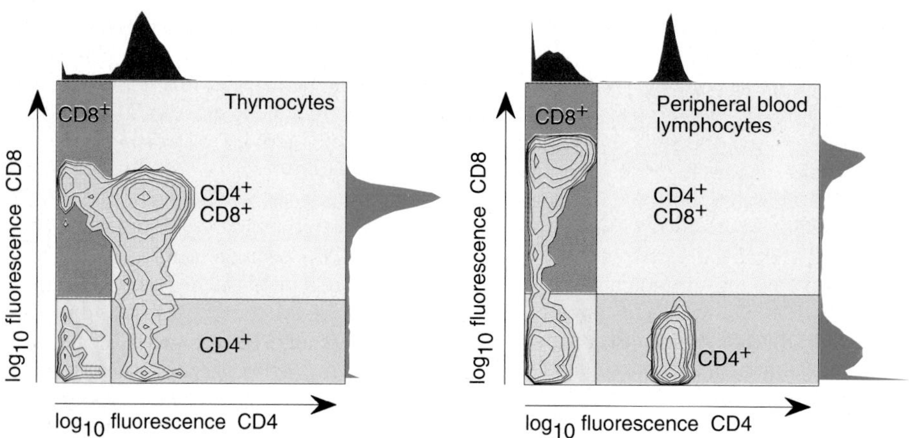

FACS analysis of thymocytes and peripheral blood lymphocytes.
This graphic shows the results obtained by analyzing fluorescence intensity of thymocytes or peripheral blood lymphocytes labelled with fluorescent antibodies directed against CD4 or CD8. Thymocyte populations contain double-negative (yellow) and double-positive (light blue) cell types while peripheral blood lymphocytes contain no double-positive cells but only single positive or double-negative cells. Arrows indicate increasing fluorescence intensity (log10 fluorescence).

Retroviruses 8: 2063-71 (1992); **Rahemtulla A et al** Normal development and function of CD8$^+$ cells but markedly decreased helper cell activity in mice lacking CD4. N 353: 180-4 (1991); **Robey EA et al** Expression of CD4 in transgenic mice alters the specificity of CD8 cells for allogeneic major histocompatibility complex. PNAS 88: 608-12 (1991); **Seong RH & Parnes JR** Alteration of T cell lineage commitment by expression of a hybrid CD8/CD4 transgene. AEMB 323: 79-87 (1992); **Weber S et al** Constitutive expression of high levels of soluble mouse CD4 in transgenic mice does not interfere with their immune function. EJI 23: 511-6 (1993); **Yu SH et al** Functional human CD4 protein produced in milk of transgenic mice. Mol. Biol. Med. 6: 255-61 (1989)

● Ligands: **Barber EK et al** The CD4 and CD8 antigens are coupled to a protein-tyrosine kinase (p56*lck*) that phosphorylates the CD3 complex. PNAS 86: 3277-81 (1989); **Burgess KE et al** Biochemical identification of a direct physical interaction between the CD4: p56*lck* and Ti(TcR)/CD3 complexes. EJI 21: 1663-8 (1991); **Rudd CE et al** Molecular analysis of the interaction of p56*lck* with the CD4 and CD8 antigens. AEMB 292: 85-96 (1991); **Vega MA et al** Structural nature of the interaction between T lymphocyte surface molecule CD4 and the intracellular protein tyrosine kinase *lck*. EJI 20: 453-56 (1990); **Veillette A et al** Signal transduction through the CD4 receptor involves the activation of the internal membrane tyrosine-protein kinase p56*lck*. N 338: 257-9 (1989)

● Clinical Use & Significance: **Auffray C et al** CD4-targeted immune intervention: A strategy for the therapy of AIDS and autoimmune disease. TibTech 9: 124-31 (1991); **Capon DJ et al** Designing CD4 immunoadhesins for AIDS therapy. N 337: 525-31 (1989); **Clampham P et al** Soluble CD4 blocks the infection of diverse strains of HIV and SIV for T cells and monocytes but not for brain and muscle cells. N 337: 368-70 (1989); **Emmrich J et al** Treatment of inflammatory bowel disease with anti-CD4 monoclonal antibody. Lancet 338: 570-71 (1991); **Gorman SD et al** Reshaping a therapeutic CD4 antibody. PNAS 88: 4181-5 (1991); **Reiter C et al** Treatment of rheumatoid arthritis with monoclonal CD4 antibody M-T151: Clinical results and immunopharmacologic effects in an open study, including repeated administration. Arthritis Rheum. 34: 525-6 (1991); **Suzuki H et al** CD4 and CD7 molecules as targets for drug delivery from antibody bearing liposomes. ECR 193: 112-19 (1991); **Symons JA et al** Soluble CD4 in patients with rheumatoid arthritis and osteoarthritis. Clin. Immunol. Immunopathol. 60: 72-82 (1991); **Ward RHR et al** Prevention of HIV-1 IIIB infection in chimpanzees by CD4 immunoadhesin. N 352: 434-6 (1991); **Wineman JP et al** CD4 is expressed on murine pluripotent hematopoietic stem cells. Blood 80: 1717-24 (1992)

CD8: This cell surface antigen (see also: CD antigens) is also known as ***T8***, ***Leu-2***, ***Lyt-2*** or ***OKT8***. CD8 belongs to the immunoglobulin superfamily. CD8 occurs either as a disulfide-linked homodimer or homomultimer of two 34 kDa subunits (***CD8α*** = Lyt-2) or as a heterodimer complexed with another protein named ***CD8β*** (Lyt-3) of which there are multiple forms arising by alternative splicing of its mRNA. Alternative splicing may also lead to the synthesis of a secreted form of CD8α which may also form homodimers. The human genes encoding CD8α (≈ 8 kb, 6 exons) and CD8β map to chromosome 2p12, lying in close proximity to each other. These genes are also closely linked to the immunoglobulin kappa light chain cluster.

T cell progenitor cells in the thymus initially do not express CD8 and another antigen called » CD4 and develop into mature T cells in several steps. Immature thymocytes later coexpress CD8 and » CD4, and these cells give rise to mature CD4$^+$CD8$^-$ and CD4$^-$CD8$^+$ T cells. Those T cells that recognize self-MHC are selected to mature by a process known as positive selection in which class I MHC generates an instructive signal that directs differentiation to a CD8 lineage. The functional consequences of CD4 expression in a population of class I-selected CD8⁺ lymphocytes have been investigated in a transgenic mouse line in which a » CD4 transgene is expressed on a significant fraction of the mature CD8⁺ lymphocytes but not in the thymus. CD8⁺ lymphocytes expressing the CD4 transgene proliferate in response to allogeneic class I and class II major histocompatibility complex, whereas CD8⁺ cells from control animals proliferate only to allogeneic class I gene products. The ability of a T cell population to react with class II allogeneic major histocompatibility complex is therefore determined by the presence of » CD4.

CD8-positive cells are cytotoxic T lymphocytes which are capable of lysing target cells by direct cell contact. These cells play a role in the elimination of virus-infected cells and tumor cells and are also involved in transplant rejection processes.

CD8 is required for the development of cytotoxic T cells but not for the development of T helper cells. CD8 functions as an adhesion molecule that binds class I MHC molecules. Antibodies against CD8 can block the killing activity of cytotoxic cells by inhibiting target cell adhesion.

In CD4-positive cells the expression of CD8 can be induced by » IL4. The expression of » IL4 in » transgenic animals blocks the maturation of T cells and considerably reduces the population of immature $CD4^{+}8^{+}$ thymocytes and peripheral T cells while at the same time greatly enhancing the numbers of CD8-positive mature thymocytes. The proliferation of thymocytes with the marker spectrum $CD4^{-}8^{-}$, $CD4^{+}8^{-}$, $CD4^{-}8^{+}$ is stimulated by » IL7 which is therefore an important developmental factor for functionally different subpopulations of T lymphocytes.

CD8 functions as an effector molecule for intracellular signal transduction. In T lymphocytes CD8 has been found to be associated with tyrosine kinase $p56^{lck}$ (see: *lck*) which phosphorylates a number of other cell surface proteins (including CD3) and is probably involved in transducing the intracellular signal when these proteins are bound to their receptors (see also: HT-2 cell line).

Mice lacking CD8 have been generated by homologous recombination in embryonal stem cells (see: ES cells) and bred with the experimental allergic encephalomyelitis (EAE)-susceptible PL/JH-2u to investigate the role of CD8⁺ T cells in this model of multiple sclerosis. The disease onset and susceptibility were similar to those of wild-type mice. However, the mutant mice had a milder acute EAE, reflected by fewer deaths, but more chronic EAE, reflected by a higher frequency of relapse. This suggests that CD8⁺ T lymphocytes may participate as both effectors and regulators in this animal model (for a role of CD8-positive cells in parasitemia see also: BDGF 2 (bone-derived growth factor 2).

● **Reviews: Littman DR** The structure of the CD4 and CD8 genes. ARI 5: 561-84 (1987); **Parnes JR** Molecular biology and function of CD4 and CD8. AI 44: 265-311 (1989)

● **Biochemistry & Molecular Biology: de Totero D et al** Heterogeneous immunophenotype of granular lymphocyte expansions: differential expression of the CD8 α and CD8 β chains. Blood 80: 1765-1773 (1992); **Giblin P et al** A secreted form of the human lymphocyte cell surface molecule CD8 arises from alternative splicing. PNAS 86: 998-1002 (1989); **Johnson PA** human homologue of the mouse CD8 molecule, Lyt-3: genomic sequence and expression. Immunogenetics 26: 174-7 (1987); **Kavathas P et al** Isolation of the gene encoding the human T lymphocyte differentiation antigen Leu-2 (T8) by gene transfer and cDNA subtraction. PNAS 81: 7688-92 (1984); **Littman DR et al** The isolation and sequence of the gene encoding T8: a molecule defining functional classes of T lymphocytes. Cell 40: 237-46 (1985); **Moebius U et al** Expression of different CD8 isoforms on distinct human lymphocyte subpopulations. EJI 21: 1793-1800 (1991); **Nakauchi H et al** Molecular cloning of Lyt-2, a membrane glycoprotein marking a subset of mouse T lymphocytes: molecular homology to its human counterpart, Leu-2/T8, and to immunoglobulin variable regions. PNAS 82: 5126-30 (1985); **Nakauchi H et al** Molecular cloning of Lyt-3, a membrane glycoprotein marking a subset of mouse T lymphocytes: molecular homology to immunoglobulin and T cell receptor variable and joining regions. PNAS 84: 4210-4 (1987); **Norment AM & Littman DR** A second subunit of CD8 is expressed in human T cells. EJ 7: 3433-9 (1988); **Norment AM** Alternatively spliced mRNA encodes a secreted form of human CD8 α. Characterization of the human CD8 α gene. JI 142: 3312-9 (1989); **Panaccio M et al** Molecular characterization of the murine cytotoxic T cell membrane glycoprotein Ly-3 (CD8). PNAS 84: 6874-8 (1987); **Shiue L et al** A second chain of human CD8 is expressed on peripheral blood lymphocytes. JEM 168: 1993-2005 (1988); **Spurr NK et al** CD8B, the human equivalent of the mouse Ly-3 gene is localized on chromosome 2. Immunogenetics 27: 70-2 (1988); **Sukhatme VP et al** Gene for the human T cell differentiation antigen Leu-2/T8 is closely linked to the kappa light chain locus on chromosome 2. JEM 161: 429-34 (1985); **Sukhatme VP et al** The T cell differentiation antigen Leu-2/T8 is homologous to immunoglobulin and T cell receptor variable regions. Cell 40: 591-7 (1985); **Weichhold GM et al** The CD8-α locus is located on the telomere side of the immunoglobulin-kappa locus at a distance of 2 Mb. Genomics 16: 512-4 (1993)

● **Transgenic/Knock-out/Antisense Studies: Koh DR et al** Less mortality but more relapses in experimental allergic encephalomyelitis in CD8-/- mice. S 256: 1219-3 (1992); **Lee et al** CD8 surface levels alter the fate of α/β T cell receptor-expressing thymocytes in transgenic mice. JEM 175: 1013-25 (1992); **Robey EA et al** Expression of CD4 in transgenic mice alters the specificity of CD8 cells for allogeneic major histocompatibility complex. PNAS 88: 608-12 (1991); **Robey EA et al** Thymic selection in CD8 transgenic mice supports an instructive model for commitment to a CD4 or CD8 lineage. Cell 64: 99-107 (1991); **Russell JH et al** Evidence for CD8-independent T cell maturation in transgenic mice. JI 144: 3318-25 (1990); **Seong RH & Parnes JR** Alteration of T cell lineage commitment by expression of a hybrid CD8/CD4 transgene. AEMB 323: 79-87 (1992) ● **LIGANDS: Barber EK et al** The CD4 and CD8 antigens are coupled to a protein-tyrosine kinase (p56*lck*) that phosphorylates the CD3 complex. PNAS 86: 3277-81 (1989); **Chalupny NJ et al** Association of CD8 with p56*lck* is required for early T cell signaling events. EJ 10: 1201-1207 (1991); **Norment AM et al** Cell-cell adhesion mediated by CD8 and MHC class I molecules. N 336: 79-81 (1988); **Rudd CE et al** Molecular analysis of the interaction of p56*lck* with the CD4 and CD8 antigens. AEMB 292: 85-96 (1991)

CD16: This cell surface antigen (see also: CD antigens) is also known as ***neutrophil antigen NA*** and ***Leu-11***. It is identical with hFcRIII = $F_{c\gamma}$RIII, an Fc-IgG receptor also expressed on NK cells and neutrophils. The human CD16 gene maps to chromosome 1q23. CD16 is part of a multimeric complex consisting of three functionally and biochemically distinct proteins that are membrane-spanning subunits mediating assembly and signal transduction. The complex is associated with a protein tyrosine kinase, » *lck*, which is also involved in other signal transduction pathways.

CD16 is one of the three subtypes of Fc-IgG receptors. These receptors are widely distributed on many cells of the immune system and contribute to the pathogenesis of immune complex- and auto-antibody-mediated disorders such as vasculitis, rheumatoid arthritis, idiopathic thrombocytopenic purpura or autoimmune neutropenia.

Among other things this receptor is involved in establishing cell-to-cell interactions mediating antibody-dependent cellular cytotoxicity (see: ADCC). CD16 is also secreted as a soluble protein (see also: IBF, immunoglobulin binding factor). The expression of CD16 can be induced by TGF-β. This effect is antagonized by » IL4.

A case of neonatal isoimmune neutropenia in which the mother was completely lacking CD16 but had normal » CD32 on natural killer lymphocytes has been described. The mother had isoantibodies CD16 antigen, apparently produced during pregnancy and responsible for the neutropenia in her child. Her parents were heterozygous for the neutrophil-FcRIII deficiency caused by a deletion of the FcRIII-1 (CD16) gene.

Anderson P et al Fc-γ receptor type III (CD16) is included in the zeta NK receptor complex expressed by human natural killer cells. PNAS 87: 2274-8 (1990); **Gergely J & Sarmay G** The two binding-site models of human IgG binding Fcγ receptors. FJ 4: 3275-83 (1990); **Grundy HO et al** The polymorphic Fc-γ receptor II gene maps to human chromosome 1q. Immunogenetics 29: 331-9 (1989); **Huizinga TW et al** The PI-linked receptor FcRIII is released on stimulation of neutrophils. N 333: 667-9 (1988); **Huizinga TWJ et al** Maternal genomic neutrophil FcRIII deficiency leading to neonatal isoimmune neutropenia. Blood 76: 1927-32 (1990); **Lanier LL et al** Functional and biochemical analysis of CD16 antigen on natural killer cells and granulocytes. JI 141: 3478-85 (1988); **Le Coniat M et al** The human genes for the α and γ subunits of the mast cell receptor for immunoglobulin E are located on human chromosome band 1q23. Immunogenetics 32: 183-6 (1990); **Mouawad R et al** Antibodies raised against peptide fragments of CD16 reacting with membrane and soluble form of CD16. I: Production and characterization. Hybridoma 11: 447-59 (1992); **Ory PA et al** Characterization of polymorphic forms of Fc receptor III on human neutrophils. JCI 83: 1676-81 (1989); **Ory P et al** Sequences of complementary DNAs that encode the NA1 and NA2 forms of Fc-γ receptor III on neutrophils. JCI 84: 1688-91 (1989); **Peltz GA et al** Human Fc γ RIII: cloning, expression, and identification of the chromosomal locus of two Fc receptors for IgG. PNAS 86: 1013-7 (1989); **Perussia B & Ravetch JV** FcγRIII (CD16) on human macrophages is a functional product of the FcγRIII-2 gene. EJI 21: 425-9 (1991); **Procopio ADG et al** GTP-binding proteins transduce signals generated via human FCγ receptor IIIA (CD16). JI 146: 3550-6 (1991); **Ravetch JV & Kinet JP** Fc receptors. ARI 9: 457-92 (1991); **Salcedo TW et al** Physical and functional association of p56lck with FcγRIIIA (CD16) in natural killer cells. JEM 177: 1475-80 (1993); **Simmons D & Seed B** The Fc γ receptor of natural killer cells is a phospholipid-linked membrane protein N 333: 568-70 (1988) [erratum in N 340: 662]; **Unkeless JC et al** Structure and function of human and murine receptors for IgG. ARI 6: 251-81 (1988); **Unkeless JC** Function and heterogeneity of human Fc receptors for immunoglobulin G. JCI 83: 355-61 (1989); **Wong HL et al** IL4 antagonizes induction of FcγRIII (CD16) expression by transforming growth factor-β on human monocytes. JI 147: 1843-8 (1991)

CD21: This cell surface antigen (see also: CD antigens) is also called ***B2***. It identical with the complement fragment C3d receptor known as ***CR2***. The human CR2 gene maps to chromosome 1q32. CD21 is expressed on mature B cells. It is a 145 kDa receptor protein that is involved in the activation of B cells (see also: Cell activation). On B lymphocytes CD21 also binds » IFN-α.

CD21 is also a ligand for another cell surface antigen, » CD23, and regulates the synthesis of IgE. CD21 also functions as a receptor for Epstein-Barr virus. The virus receptor found on human epithelial cells is identical with CD21 found on B lymphocytes. CD21/CD23 together with LFA-1/ICAM-1 appear to be the major molecules involved in homotypic aggregation of human B cells.

Ahearn JM & Fearon DT Structure and function of the complement receptors, CR1 (CD35) and CR2 (CD21). AI 46: 183-219 (1989); **Aubry JP et al** CD21 is a ligand for CD23 and regulates IgE production. N 358: 505-7 (1992); **Björck P et al** CD23 and CD21 function as adhesion molecules in homotypic aggregation of human B lymphocytes. EJI 23: 1771-5 (1993); **Delcayre AX et al** Epstein Barr virus/complement C3d receptor is an interferon α receptor. EJ 10: 919-26 (1991); **Kieff E** Characterization of an Epstein-Barr virus receptor on human epithelial cells. JEM 176: 1405-14 (1992); **Moore MD et al** Molecular cloning of the cDNA encoding the Epstein-Barr virus/C3d receptor (complement receptor type 2) of human B lymphocytes.

PNAS 84: 9194-8 (1987); **Rayhel EJ et al** Characterization of the human complement receptor 2 (CR2, CD21) promoter reveals sequences shared with regulatory regions of other developmentally restricted B cell proteins. JI 146: 2021-6 (1991); **Weis JJ et al** Identification of a 145,000 M(r) membrane protein as the C3d receptor (CR2) of human B lymphocytes. PNAS 81: 881-5 (1984); **Weis JH et al** A complement receptor locus: genes encoding C3b/C4b receptor and C3d/Epstein-Barr virus receptor. map to 1q32. JI138: 312-5 (1987); **Yefenof E et al** Surface markers on human B and T lymphocytes. IX. Two color immunofluorescence studies on the association between EBV receptors and complement receptors on the surface of lymphoid cell lines. IJC 17: 693-700 (1976)

CD23: This cell surface antigen (see also: CD antigens) is also known as ***OKT10*** or ***Blast2***. The CD23 gene maps to human chromosome 19p13.3. CD23 is expressed on antigen-activated T and B cells (see also: Cell activation). Constitutive expression of CD23 is a characteristic feature of many EBV-transformed B lymphblasts and the factor is therefore also called ***EBVCS*** (EBV cell surface antigen).

CD23 is a 45 kDa protein. The nucleotide sequence of the cDNA predicts a polypeptide with 321 amino acids and a molecular mass of 36.3 kDa. Proteolytic cleavage which may occur autocatalytically yields instable 33 kDa and 37 kDa variants of CD23 which are further processed into end products of 25 kDa. These soluble forms of CD23 are called ***sCD23*** (soluble CD23. They are identical with » ***IgE-BF*** (IgE binding factors) found in serum and may also account for some activities have been described as » EK-derived basophil promoting activity.

In B cells and monocytes the expression of CD23 and sCD23 is induced by » IL4 (see also: Ramos) and » IL13. In phytohemagglutinin-activated T cells these factors are induced by » IL2 and » IL4. In monocytes they are induced by » IL3, » IL4, and » GM-CSF, and IFN-γ. In IL4-stimulated B cells the synthesis of sCD23 is inhibited by» IFN-γ. IFN-γ also promotes the release of sCD23 in epidermal Langerhans cells. The synthesis of sCD23 mediated by » IFN-γ or » IL4can be inhibited by » IFN-α.

The membrane form of CD23 functions as a low-affinity receptor for IgE (FcεR II) and also binds » CD21.

A variety of observations suggest that CD23 secreted by B cells is a multifunctional cytokine that modulates immune responses. CD23 and also sCD23 function as B cell growth factors (see also: BCGF) and are capable of inhibiting the growth of normal and transformed human B cells. The EBV-producing marmoset B cell line B95-8 proliferates in response to purified human sCD23. It is at least 1000 times more sensitive in this assay than human EBV-transformed lymphoblastoid cell lines.

A stable 14 kDa fragment of CD23 also functions as a B cell differentiation factor (see also: BCDF) A 16 kDa sCD23 variant obtained by cleavage of 160 N-terminal amino acids and processing of the c-terminal end binds IgE and blocks ongoing synthesis of IgE and also synthesis of IgE induced by » IL4. sCD23 also inhibits the migration of monocytic cell lines such as » U937 and therefore functions as a migration inhibition factor (see: MIF).

In bone marrow progenitors sCD23 potentiates the release of histamines induced by » IL3 although it is not a histamine releasing factor on its own. sCD23 synergizes with » IL1 in the maturation of normal and leukemic hematopoietic progenitor cells and the maturation of immature thymocytes. The IL3-dependent growth of CD34-positive AML cells (acute myeloid leukemia) is inhibited by sCD23. Antibodies against soluble CD23 have been shown to decrease the total numbers of hematopoietic cells and early progenitors » CFU-GM in long-term bone marrow cultures.

The intracellular domain of CD23 is associated with a tyrosine-specific protein kinase called » *fyn* (p59 *fyn*) that mediates post-receptor signaling processes.

Alderson MR et al Regulation of human monocyte cell-surface and soluble CD23 (Fc ε RII) by granulocyte-macrophage colony-stimulating factor and IL3. JI 149: 1252-7 (1992); **Arock M et al** Soluble CD23 increases IL3 induction of histamine synthesis by human bone marrow cells. Int. Arch. Allergy Appl. Immunol. 96: 190-2 (1991); **Bertho JM et al** Synergistic effect of interleukin 1 and soluble CD23 on the growth of human $CD4^+$ bone marrow-derived T cells. EJI 21: 1073-6 (1991); **Bieber T & Delespesse G** γ-interferon promotes the release of IgE-binding factors (soluble CD23) by human epidermal Langerhans cells. J. Invest. Dermatol. 97: 600-3 (1991); **Bonnefoy JY et al** Activation of normal human B cells through their antigen receptor induces membrane expression of IL1α and secretion of IL1β. JI 143: 864-9 (1989); **Cairns JA & Gordon J** Intact, 45 kDa (membrane) form of CD23 is consistently mitogenic for normal and transformed B lymphoblasts. EJI 20: 539-43 (1990); **Callard RE et al** The marmoset B lymphoblastoid cell line (B95-8) produces and responds to B cell growth and differentiation factors: role of shed CD23 (sCD23). Immunology 65: 379-84 (1988); **Chretien I et al** Regulation of human IgE synthesis. I. Human IgE synthesis *in vitro* is determined by the reciprocal antagonistic effects of interleukin 4 and interferon-γ. EJI 20: 243-51 (1990); **Delespesse G et al** Influence of recombinant IL4, IFN-α, and IFN-γ on the production of human IgE-binding factor (soluble CD23). JI 142: 134-8 (1989); **Delespesse G et al** Human IgE-binding factors. IT 10: 159-164 (1989); **Fischer A & König W** Regulation of CD23 expression, soluble CD23 release and immunoglobulin synthesis of peripheral blood lym-

CD23 knockouts: see addendum: CD23.

phocytes by glucocorticoids. Immunology 71: 473-9 (1990); **Flores-Romo L et al** Soluble fragments of the low-affinity IgE receptor (CD23) inhibit the spontaneous migration of U937 monocytic cells: neutralization of MIF activity by a CD23 antibody. Immunology 67: 547-9 (1989); **Fourcade C et al** Expression of CD23 by human bone marrow stromal cells. Eur-Cytokine-Netw. 3: 539-43 (1992); **Gordon J et al** CD23: a multifunctional receptor/lymphokine? IT 10: 153-7 (1989); **Gordon J et al** Role of membrane and soluble CD23 in lymphocyte physiology. Monogr. Allergy 29: 156-68 (1991); **Graber P et al** Purification and characterization of biologically active human recombinant 37 kDa soluble CD23 (sFc ε RII) expressed in insect cells. JIM 149; 215-26 (1992); **Kawabe T et al** Induction of Fc ε RII/CD23 on phytohemagglutinin-activated human peripheral blood T lymphocytes. I. Enhancement by IL2 and IL4. JI 147: 548-53 (1991); **Kikutani H et al** Molecular structure of human lymphocyte receptor for immunoglobulin E. Cell 47: 657-66 (1986); **Lecron JC et al** Soluble CD23 displays T cell growth enhancing activity. Immunology 74: 561-3 (1991); **Letellier M et al** Mechanism of formation of human IgE-binding factors (soluble CD23): III. Evidence for a receptor (Fc ε RII)-associated proteolytic activity. JEM 172: 693-700 (1990); **Ludin C et al** Cloning and expression of the cDNA coding for a human lymphocyte IgE receptor. EMBO J. 6: 109-14 (1987); **Mossalayi MD et al** Soluble CD23 (FcεR II) and interleukin 1 synergistically induce early thymocyte maturation. JEM 171: 959-64 (1990); **Mossalayi MD et al** Effect of CD23 on purified human hematopoietic cells. Bone. Marrow. Transplant. 9 Suppl 1: 50-53 (1992); **Mossalayi MD et al** Inhibition of interleukin-3-dependent growth of CD34$^+$ acute myelogenous leukemia cells by recombinant soluble CD23. ANY 628: 362-7 (1991); **Pene J et al** Modulation of IL4-induced human IgE production *in vitro* by IFN-γ and IL5: the role of soluble CD23 (s-CD23). JCBc 39: 253-64 (1989); **Ravetch JV & Kinet JP** Fc receptors. ARI 9: 457-492 (1991); **Rose K et al** Partial characterization of natural and recombinant human soluble CD23. BJ 286: 819-24 (1992); **Sarfati M et al** Native and recombinant soluble CD23 fragments with IgE suppressive activity. Immunology 76: 662-7 (1992); **Sugie K et al** *Fyn* tyrosine kinase associated with Fc ε RII/CD23: possible multiple roles in lymphocyte activation. PNAS 88: 9132-5 (1991); **Swendeman S & Thorley-Lawson DA** The activation antigen BLAST-2, when shed, is an autocrine BCGF for normal and transformed B cells. EJ 6: 1637-42 (1987); **Trask B et al** Fluorescence *in situ* hybridization mapping of human chromosome 19: cytogenetic band location of 540 cosmids and 70 genes or DNA markers. Genomics 15: 133-45 (1993); **Vercelli D et al** Human recombinant interleukin 4 induces Fc ε R2/CD23 on normal human monocytes. JEM 167: 1406-16 (1988); **Wendel-Hansen V et al** The gene encoding CD23 leukocyte antigen (FCE2) is located on human chromosome 19. Somat. Cell Molec. Genet. 16: 283-6 (1990); **Yanagihara Y et al** Establishment of a sensitive radioimmunoassay for the detection of human IgE-binding factor (soluble CD23). Int. Arch. Allergy Immunol. 98: 189-99 (1992); **Yokota A et al** Two species of human Fc ε receptor II (Fc ε RII/CD23): tissue-specific and IL4-specific regulation of gene expression. Cell 55: 611-8 (1988)

CD25: This cell surface antigen (see also: CD antigens) is also known as » TAC antigen. It is expressed on antigen-activated T and B cells (see also: Cell activation). CD25 is identical with the low affinity receptor for » IL2 (IL2 receptor α chain).

Leonard WJ et al Structure of the human interleukin-2 receptor gene. S 230: 633-9 (1985); **Uchiyama T et al** A monoclonal antibody (anti-TAC) reactive with activated and functionally mature human T cells. I. Production of anti-TAC monoclonal antibody and distribution of TAC8$^+$) cells. JI 126: 1393-7 (1981)

CD25 expression inhibitory protein: This factor of 29 kDa is secreted by promonocytic cells chronically infected with HIV. It inhibits the expression of » CD25 in phytohemagglutinin-activated human monocytes. A factor with a similar biological activity and a similar molecular mass is » IL2R α chain inhibitory activity. At present it is unclear whether these two factors are identical.

Celerier J et al Purification and identification of a CD25 expression inhibitory protein from cell-free supernatants of chronically HIV-infected promonocytic cells. BBA 1139: 300-6 (1992)

CD27: This cell surface antigen is also known as 1A4, Tp55 (human), or S152 (murine). The human CD27 gene maps to chromosome 12p13. It encodes a transmembrane glycoprotein of 55 kDa (Tp55) which forms a disulfide-linked homodimer on resting T cells. Upon activation cells may also express a 55 kDa monomer and a 32 kDa variant. A soluble form of CD27 (sCD27; 32 kDa) derived from the transmembrane form is detected in the supernatant of T cells activated with anti-CD3 or combinations of anti-CD2 antibodies and found in both serum and urine from healthy donors. sCD27 levels are significantly elevated in cerebrospinal fluid (CSF) of multiple sclerosis (MS) patients and of patients suffering from other inflammatory neurological diseases.

CD27 belongs to a recently characterized family of cysteine-rich receptors with known ligands include » NGF, » TNF-α and » TNF-β. Structural similarities suggest that CD27 belongs to a lymphocyte-specific subgroup of the family, comprised of the B cell antigen » CD40, the lymphoid activation antigen CD30, the rat T cell subset antigen OX40, the mouse T cell activation antigen 4-1BB, the surface antigen Fas, and the pox virus gene product T2. Several members of this family play a role in cell differentiation, proliferation, and survival.

CD27 is present on a large subset (75 %) of peripheral T lymphocytes and most medullary thymocytes. Membrane expression of CD27 strongly increases after T cell activation via the TCR/CD3 complex or the CD2 molecule. The expression of CD27 is decreased by » phorbol esters. CD27-negative cells can be induced to express the CD27 antigen through stimulation with immobilized anti-CD3 antibodies. Some anti-CD27 antibodies appear to increase T cell proliferation while others inhibit T cell activation. The expression of CD27 on activated T cells is rapidly down-regulated by treatment with some anti-CD27 antibodies.

CD27-positive T cells are present in both the $CD4^+$ and $CD8^+$ subpopulations. CD27 is preferentially expressed on the $CD45RA^+$CD45RO-CD29low subset of CD4 cells. CD27 is also expressed on a subpopulation of the normal human B cell lineage which is absent from cord blood but present in tonsils and in the peripheral blood of adult individuals. It is also expressed by some B lymphoblastoid cell lines. CD27 on B cells can be induced selectively by the combination of Staphylococcus aureus plus » IL2, but not by either treatment alone. $CD27^+$ B lymphocytes express high levels of the adhesion structures CD11a (LFA-1), CD54 (ICAM-1), CD58 (LFA-3) and of the lymphocyte homing receptor CD44.

After *in vitro* stimulation, $CD27^+$ but not CD27- B cells secrete large amounts of both IgM and IgG. Three functionally different B cell subsets representing distinct stages of B cell differentiation have been observed: CD27- IgD^+ B cells do not secrete appreciable Ig; $CD27^+IgD^+$ B cells exclusively secrete IgM; $CD27^+$IgD- B cells produce IgG.

Some resting peripheral blood natural killer cells also express CD27 and this expression is up-regulated by » IL2. The cytolytic activity of IL2-activated, but not resting, NK cells is inhibited by an anti-CD27 monoclonal antibody (anti-1A4) so that it can be assumed that CD27 also plays an important role in the regulation of activated NK cells.

CD27 has been shown recently to be the receptor for a ligand (***CD27L***; new designation: ***CD70***, see also: CD Antigens) which is a transmembrane glycoprotein of 194 amino acids with an extracellular carboxyterminal domain (type II transmembrane protein). It displays homology with CD40 ligand (see: TRAP, TNF-related activation protein) and the ligand for » CD30. The gene encoding CD27L maps to human chromosome 19p13. CD27L enhances the generation of cytolytic cells in the presence of suboptimal concentrations of a mitogen and may therefore be involved in T cell maturation. CD40L also induces proliferation of costimulated T cells in the presence of antisera directed against » IL2. Since CD27L is a ligand expressed on the cell surface it may be involved in direct cell-to-cell interactions between T cells.

Bigler RD et al S152 (CD27). A modulating disulfide-linked T cell activation antigen. JI 141: 21-8 (1988); **Borst J et al** Alternative molecular form of human T cell-specific antigen CD27 expressed upon T cell activation. EJI 19: 357-64 (1989); **Camerini D et al** The T cell activation antigen CD27 is a member of the nerve growth factor/tumor necrosis factor receptor gene family. JI 147: 3165-9 (1991); **De Jong R et al** Regulation of expression of CD27, a T cell-specific member of a novel family of membrane receptors. JI 146: 2488-2494 (1991); **De Jong R et al** The CD27- subset of peripheral blood memory $CD4^+$ lymphocytes contains functionally differentiated T lymphocytes that develop by persistent antigenic stimulation *in vivo*. EJI 22: 993-9 (1992); **de Rie MA et al** Quantitation of soluble CD27, a T cell activation antigen, and soluble interleukin-2 receptor in serum from patients with psoriasis. Arch. Dermatol. Res. 283: 533-4 (1991); **Goodwin RG et al** Molecular and biological characterization of a ligand for CD27 defines a new family of cytokines with homology to tumor necrosis factor. Cell 73: 447-56 (1993); **Hintzen RQ et al** A soluble form of the human T cell differentiation antigen CD27 is released after triggering of the TCR/CD3 complex. JI 147: 29-35 (1991); **Hintzen RQ et al** Elevated levels of a soluble form of the T cell activation antigen CD27 in cerebrospinal fluid of multiple sclerosis patients. J. Neuroimmunol. 35: 211-7 (1991); **Loenen WA et al** The CD27 membrane receptor, a lymphocyte-specific member of the nerve growth factor receptor family, gives rise to a soluble form by protein processing that does not involve receptor endocytosis. EJI 22: 447-55 (1992); **Loenen WA et al** Genomic organization and chromosomal localization of the human CD27 gene. JI 149: 3937-43 (1992); **Martorell J et al** CD27 induction on thymocytes. JI 145: 1356-63 (1990); **Maurer D et al** CD27 expression by a distinct subpopulation of human B lymphocytes. EJI 20: 2679-84 (1990); **Maurer D et al** IgM and IgG but not cytokine secretion is restricted to the $CD27^+$ B lymphocyte subset. JI 148: 3700-5 (1992); **Sugita K et al** The 1A4 molecule (CD27) is involved in T cell activation. JI 147: 1477-1483 (1991); **Sugita K et al** Participation of the CD27 antigen in the regulation of IL2-activated human natural killer cells. JI 149: 1199-203 (1992); **Sugita K et al** CD27, a member of the nerve growth factor receptor family, is preferentially expressed on $CD45RA^+$ CD4 T cell clones and involved in distinct immunoregulatory functions. JI 149: 3208-16 (1992); **Van Lier RA et al** Tissue distribution and biochemical and functional properties of Tp55 (CD27), a novel T cell differentiation antigen. JI 139: 1589-96 (1987); **Van Lier RA et al** Anti-CD27 monoclonal antibodies identify two functionally distinct subpopulations within the $CD4^+$ T cell subset. EJI 18: 811-6 (1988)

CD28: This cell surface antigen (see also: CD antigens) is also known as ***T90/44*** antigen or ***Tp44***. The CD28 gene contains 4 exons and maps to human 2q33-q34. CD28 is a 44 kDa homodimeric highly glycosylated protein which is expressed on most human T cells. Almost all CD4-positive cells, ≈ 50 % of the CD8-positive cells, and all CD4-CD8-positive cells express this marker.

The biological functions of this protein are unknown. It has been shown, however, that CD28 is a receptor for costimulatory proteins acting on T cells. Posttranslational processing of CD28 appears to result in variant isotypes with potentially different physiologic roles on the cell surface. The natural ligand of CD28 is a 44-54 kDa glycoprotein, called ***B7 or BB1*** (new designation: ***CD80***, see also: CD Antigens) which is primarily expressed on activated B cells and other antigen-presenting cells. Binding of B7 to CD28 on T cells delivers a costimulatory signal that triggers T cell proliferation by stimulating a transcription factor (see also: Gene expression) that, in turn, induces the synthesis and secretion of IL2 and other cytokines. It has been shown that CD28 engagement by B7 induces transient down-regulation of CD28 synthesis and prolonged unresponsiveness to CD28 signaling.

B7 binds to another protein structurally related to CD28, called ***CTLA-4***. CTLA-4 is expressed in low copy number by T cells only after activation, but it binds B7 with approximately 20-fold higher affinity than CD28. A soluble form of the extracellular domain of CTLA-4 has been shown to bind B7 with high avidity and to suppress T cell-dependent antibody responses *in vivo*. Large doses of this soluble protein also suppress responses to a second immunization.

The human gene encoding B 7 has six exons, spans at least 32 kb and maps to chromosome 3q13.3-q21. Trisomy of chromosome 3 is a recurrent chromosome change seen in various lymphomas and lymphoproliferative disorders and chromosomal defects involving 3q21 have been described in leukemia and myelodysplastic states.

In T cells activated via the T cell receptor antibodies directed against CD28 leads to an increased synthesis and secretion of » IL2, » IL13, » TNF-α, » GM-CSF, » IFN-γ and » TNF-β.

In vitro experiments have demonstrated that signals transduced by the CD28 receptor can determine whether the occupancy of the T cell receptor results in a productive immune response or clonal anergy. If the costimulation by B7 is blocked, humoral responses are suppressed effectively. It has been shown that this may lead to long-term acceptance of tissue xenografts.

Aruffo A & Seed B Molecular cloning of a CD28 cDNA by a high-efficiency COS cell expression system. PNAS 84: 8573-7 (1987); **Buonavista N et al** Molecular linkage of the human CTLA4 and CD28 Ig- superfamily genes in yeast artificial chromosomes. Genomics 13: 856-61 (1992); **Damle NK et al** Differential regulatory effects of intercellular adhesion molecule-1 on costimulation by the CD28 counter-receptor B7. JI 149: 2541-8 (1992); **Damle NK et al** Differential costimulatory effects of adhesion molecules B7, ICAM-1, LFA-3, and VCAM-1 on resting and antigen-primed $CD4^+$ T lymphocytes. JI 148: 1985-92 (1992); **deBoer M et al** Generation of monoclonal antibodies to human lymphocyte cell surface antigens using insect cells expressing recombinant proteins. JIM 152: 15-23 (1992); **Fraser JD et al** Regulation of Interleukin 2 gene enhancer activity by the T cell accessory molecule CD28. S 251: 313-6 (1991); **Freeman GJ et al** The gene for B7, a costimulatory signal for T cell activation, maps to chromosomal region 3q13.3-3q21. Blood 79: 489-94 (1992); **Gimmi CD et al B cell** surface antigen B7 provides a costimulatory signal that induces T cells to proliferate and secrete interleukin 2. PNAS 88: 6575-9 (1991); **Gross JA et al** The murine homologue of the T lymphocyte antigen CD28. Molecular cloning and cell surface expression. 144: 3201-10 (1990); **Harding FA et al** CD28-mediated signaling co-stimulates murine T cells and prevents induction of anergy in T cell clones. N 356: 607-9 (1992); **Lafage-Pochitaloff M et al** Human CD28 and CTLA-4 Ig superfamily genes are located on chromosome 2 at bands q33-q34. Immunogenetics 31: 198-201 (1990); **Lee KP et al** The genomic organization of the CD28 gene: implications for the regulation of CD28 mRNA expression and heterogeneity. JI 145: 344-52 (1990); **Lenschow DJ et al** Long-term survival of xenogeneic pancreatic islet grafts induced by CTLA4lg. S 257: 789-92 (1992); **Lesslauer W et al** Purification and N- terminal amino acid sequence of the human T90/44 (CD28) antigen. Immunogenetics 27: 388-91 (1988); **Linsley PS et al** T cell antigen CD28 mediates adhesion with B cells interacting with activation antigen B7/BB-1. PNAS 87: 5031-5 (1990); **Linsley PS et al** Immunosuppression *in vivo* by a soluble form of the CTLA-4 T cell activation molecule. S 257: 792-5 (1992); **Linsley PS et al** CD28 engagement by B7/BB-1 induces transient down-regulation of CD28 synthesis and prolonged unresponsiveness to CD28 signaling. JI 150: 3161-9 (1993); **Linsley PS & Ledbetter JA** The role of the CD28 receptor during T cell responses to antigen. ARI 11: 191-212 (1993); **Linsley PS et al** Coexpression and functional cooperation of CTLA-4 and CD28 on activated T lymphocytes. JEM 176: 1595-604 (1992); **Norton SD et al** The CD28 ligand, B7, enhances IL2 production by providing a costimulatory signal to T cells. JI 149: 1556-61 (1992); **Schwartz RH** Costimulation of T lymphocytes: the role of CD28, CTLA-4, and B7/BB1 in interleukin-2 production and immunotherapy. Cell 71: 1065-8 (1992); **Selvakumar A et al** Genomic organization and chromosomal location of the human gene encoding the B lymphocyte activation antigen B7. Immunogenetics 36: 175-81 (1992); **Turka LA et al** T cell activation by the CD28 ligand B7 is required for cardiac allograft rejection *in vivo*. PNAS 89: 11102-5 (1992); **Verweij CL et al** Activation of interleukin-2 gene transcription via the T cell surface molecule CD28 is mediated through an NF-kB-like response element. JBC 266: 14179-82 (1991)

CD30: This cell surface antigen (see also: CD antigens) is also known as ***Ki-1*** or ***Ber-H2 antigen***. It is expressed on mitogen-activated B and T cells but not on resting lymphocytes or monocytes. CD30 is also expressed on Hodgkin and Sternberg-Reed cells of Hodgkin's lymphomas and on most Hodgkin-derived cell lines. CD30 is also expressed on a subset of non-Hodgkin's lymphomas, including Burkitt's lymphomas, and on several virus-transformed cell lines

CD30 is a glycoprotein of 120 kDa that shows sequence homology to members of the » TNF receptor superfamily. The gene encoding human CD30 maps to chromosome 1p36. Soluble forms of CD30 have been found in the serum of some patients with adult T cell leukemia or other CD30-

positive lymphoma patients. Highest levels of soluble CD30 are found in the serum of acute patients.

A ligand for murine CD30, designated CD30L, has recently be described to be a type II membrane protein of 239 amino acids. Its carboxyterminal domain shows significant homology to » TNF-α, TNF-β, and the ligands for CD40 (see: TRAP, TNF-related activation protein) and » CD27. Human CD30L is 72 % identical with murine CD30L at the amino acid level. The human CD30L gene maps to chromosome 9q33.

CD30L enhances proliferation of CD3-activated T cells. It induces cell death in several CD30-positive lymphoma-derived cell lines.

Cotelingam JD Phenotype of atypical CD30-positive cells in infectious mononucleosis. Reply. Am. J. Clin. Pathol. 96: 146-146 (1991); **Delsol G et al** Antibody BNH9 detects red blood cell-related antigens on anaplastic large cell ($CD30^+$) lymphomas. Br. J. Cancer 64: 321-326 (1991); **Dürkop H et al** Molecular cloning and expression of a new member of the nerve growth factor receptor family that is characteristic for Hodgkin's disease. Cell 68: 421-7 (1992); **Gause A et al** Clinical significance of soluble CD30 antigen in the sera of patients with untreated Hodgkin's disease. Blood 77: 1983-1988 (1991); **Pohl C et al** Idiotype vaccine against Hodgkin's lymphoma: Generation and characterization of an anti-idiotypic monoclonal antibody against the Hodgkin-associated (anti-CD 30) monoclonal antibody HRS-3. Anticancer Res. 11: 1115-1124 (1991); **Schwab U et al** Production of a monoclonal antibody specific for Hodgkin and Sternberg-Reed cells of Hodgkin's disease and a subset of normal lymphoid cells. N 299: 65-7 (1982); **Smith CA et al** CD30 antigen, a marker for Hodgkin's lymphoma, is a receptor whose ligand defines an emerging family of cytokines with homology to TNF. Cell 73: 1349-60 (1993)

CD32: This cell surface antigen (see also: CD antigens) is identical with hFcRII = $F_{c\gamma}RII$, the most frequently utilized receptor for the Fc portion of immunoglobulin IgG. It plays an essential role in the removal of antigen-antibody complexes from the circulation (see also: IBF, immunoglobulin binding factor).

The human gene encoding CD32 maps to chromosome 1q23-q24. CD32 is a transmembrane glycoprotein of 40 kDa that is expressed on almost all cells of the immune system and also on some epithelial cells. CD32 is one of the three subtypes of Fc-IgG receptors. These receptors are widely distributed on many cells of the immune system and contribute to the pathogenesis of immune complex- and auto-antibody-mediated diseases such as vasculitis, rheumatoid arthritis, idiopathic thrombocytopenic purpura or autoimmune neutropenia. CD32 is involved in cell-to-cell interactions mediating antibody-dependent cellular cytotoxicity (see: ADCC). Binding of antibody-antigen complexes to CD32 activates the non-receptor protein kinase » *fgr*,

Brooks DG et al Structure and expression of human IgG FcRII(CD32). Functional heterogeneity is encoded by the alternatively spliced products of multiple genes. JEM 170: 1369-85 (1989); **Gergely J & Sarmay G** The two binding-site models of human IgG binding Fcγ receptors. FJ 4: 3275-83 (1990); **Ravetch JV & Kinet JP** Fc receptors. ARI 9: 457-92 (1991); **Sammartino L et al** Assignment of the gene coding for human FcRII (CD32) to bands q23q24 on chromosome 1. Immunogenetics 28: 380-1 (1988); **Stuart SG et al** Isolation and expression of cDNA clones encoding a human receptor for IgG (Fc γ RII). JEM 166: 1668-84 (1987); **Stuart SG et al** Human IgG Fc receptor (hFcII; CD32) exists as multiple isoforms in macrophages, lymphocytes, and IgG-transporting placental epithelium. EJ 8: 3657-66 (1989); **Unkeless JC et al** Structure and function of human and murine receptors for IgG. ARI 6: 251-81 (1988); **Unkeless JC** Function and heterogeneity of human Fc receptors for immunoglobulin G. JCI 83: 355-61 (1989)

CD33: This cell surface antigen (see also: CD antigens) is an integral 67 kDa membrane glycoprotein that is used a surface marker for very early bone marrow stem cells. It is expressed on » BFU-E, » CFU-Eo, » CFU-G, » CFU-GEMM, » CFU-GM, » CFU-M, and » CFU-MEG (see also: Hematopoiesis). CD33 is expressed by myeloid precursors after » CD34. CD33 is also expressed on peripheral monocytes. CD33 is also expressed on the leukemic cells of most patients with acute myeloid leukemia. It is a marker used to distinguish myelogenous leukemia cells from lymphoid or erythroid leukemias. The human gene encoding CD33 maps to chromosome 19q13.3.

Bernstein ID et al Blast colony-forming cells and precursors of colony-forming cells detectable in long-term marrow culture express the same phenotype ($CD33^-$ $CD34^+$). EH 19: 680-2 (1991); **Bühring HJ et al** Sequential expression of CD34 and CD33 antigens on myeloid colony-forming cells. Eur. J. Haematol. 42: 143-9 (1989); **Peiper SC et al** Molecular cloning, expression, and chromosomal localization of a human gene encoding the CD33 myeloid differentiation antigen. Blood 72: 314-21 (1988); **Simmons D & Seed B** Isolation of a cDNA encoding CD33, a differentiation antigen of myeloid progenitor cells. JI 141: 2797-2800 (1988)

CD34: This cell surface antigen (see also: CD antigens), formerly known as hemopoietic progenitor cell antigen 1 (***HPCA1***), is also called ***MY10***. The human CD34 gene, which maps to chromosome 1q32, spans 26 kb and has 8 exons. CD34 is a 67 kDa transmembrane glycoprotein that is selectively expressed on human hematopoietic progenitor cells. The biological function of CD34 is still unknown. It is used a surface marker for very early » hematopoietic stem cells. It is expressed on » BFU-E, » CFU-Eo, » CFU-GEMM, » CFU-GM, » CFU-MEG (see also:

Hematopoiesis). CD34 is found on ≈ 30 % of the blast cells of patients with acute myeloid and lymphocytic leukemia. Recent studies have shown that the CD34 gene is also expressed on endothelial cells of small blood vessels.

Cells expressing CD34 respond to » SCF (stem cell factor) and » LIF in the presence of colony-stimulating factors (see: CSF) and these factors can induce myeloid differentiation in these cells. The effects of some growth factors on CD34-positive cells are counteracted by the IFN-γ-inducible protein » IP10. Interleukin 11 stimulates The proliferation of human CD34-positive cells is also stimulated by » IL11, and IL11 synergises with SCF, » IL3, and » GM-CSF.

Allogeneic CD34-positive marrow cells devoid of detectable mature and immature T and B lymphocytes can engraft and reconstitute stable long-term myelopoiesis and lymphopoiesis in lethally irradiated baboons, suggesting that these cells comprise a population of pluripotent » hematopoietic stem cells capable of fully reconstituting lympho-hematopoiesis in the transplanted host.

Andrews RG et al $CD34^+$ marrow cells, devoid of T and B lymphocytes, reconstitute stable lymphopoiesis and myelopoiesis in lethally irradiated allogeneic baboons. Blood 80: 1693-701 (1992); **Bartelmez SH et al** Uncovering the heterogeneity of hematopoietic repopulating cells EH 19: 861-2 (1991); **Bender JG et al** Identification and comparison of CD34-positive cells and their subpopulations from normal peripheral blood and bone marrow using multicolor flow cytometry. Blood 77: 2591-6 (1991); **Bernstein ID et al** Recombinant human stem cell factor enhances the formation of colonies by $CD34^+$ and $CD34^+lin^-$cells, and the generation of colony-forming cell progeny from $CD34^+lin^-$ cells cultured with interleukin-3, granulocyte colony-stimulating factor, or granulocyte-macrophage colony-stimulating factor. Blood 77: 2316-21 (1991); **Broxmeyer HE et al** Comparative analysis of the human macrophage inflammatory protein family of cytokines (chemokines) on proliferation of human myeloid progenitor cells. Interacting effects involving suppression, synergistic suppression, and blocking of suppression. JI 150: 3448-58 (1993); **Bühring HJ et al** Sequential expression of CD34 and CD33 antigens on myeloid colony-forming cells. Eur. J. Haematol. 42: 143-9 (1989); **Civin CI et al** Positive stem cell selection – basic science. PCBR 333: 387-401 (1990); **Egeland T et al** Myeloid differentiation of purified $CD34^+$ cells after stimulation with recombinant human granulocyte-monocyte colony-stimulating factor (CSF), granulocyte-CSF, monocyte-CSF, and interleukin-3. Blood 78: 3192-9 (1991); **Ema H et al** Multipotent and committed $CD34^+$ cells in bone marrow transplantation. JJCR 82: 547-52 (1991); **Greaves MF et al** Molecular features of CD34: a hemopoietic progenitor cell-associated molecule. Leukemia 6: 31-6 (1992); **He XY et al** Isolation and molecular characterization of the human CD34 gene. Blood 79: 2296-2302 (1992); **Howell SM et al** Localization of the gene coding for the hemopoietic stem cell antigen CD34 to chromosome 1q32. Hum. Genet. 87: 625-7 (1991); **Issaad C et al** A murine stromal cell line allows the proliferation of very primitive human $CD34^+/CD38^-$ progenitor cells in long-term cultures and semisolid assays. Blood 81: 2916-24 (1993); **Janssen WE et al** The $CD34^+$ cell fraction in bone marrow and blood is not universally predictive of CFU-GM. EH 20: 528-30 (1992); **Janssen WE et al** Use of $CD34^+$ cell fraction as a measure of hematopoietic stem cells in bone marrow and peripheral blood: comparison with the CFU-GM assay. PCBR 377: 513-21 (1992); **Kerst JM et al** Combined measurement of growth and differentiation in suspension cultures of purified human CD34-positive cells enables a detailed analysis of myelopoiesis. EH 20: 1188-93 (1992); **Kirshenbaum AS et al** Effect of IL3 and stem cell factor on the appearance of human basophils and mast cells from $CD34^+$ pluripotent progenitor cells. JI 148: 772-7 (1991); **Lansdorp PM & Dragowska W** Long-term erythropoiesis from constant numbers of $CD34^+$ cells in serum-free cultures initiated with highly purified progenitor cells from human bone marrow. JEM 175: 1501-9 (1992); **Leary AG et al** Growth factor requirements for survival in G0 and entry into the cell cycle of primitive human hemopoietic progenitors. PNAS 89: 4013-7 (1992); **Lemoli RM et al** Interleukin 11 stimulates the proliferation of human hematopoietic $CD34^+$ and $CD34^+CD33^-DR^-$ cells and synergises with stem cell factor, interleukin-3, and granulocyte-macrophage colony-stimulating factor. EH 21: 1668-72 (1993); **McNiece I et al** Action of Interleukin-3, G-CSF, and GM-CSF on highly enriched human hematopoietic progenitor cells: synergistic interaction of GM-CSF plus G-CSF. Blood 74: 110-4 (1989); **Ottmann OG et al** Regulation of early hematopoiesis in serum-deprived cultures of mafosfamide-treated and untreated CD34-enriched bone marrow cells. EH 19: 773-8 (1991); **Sato N et al** Purification of human marrow progenitor cells and demonstration of the direct action of macrophage colony-stimulating factor on colony-forming unit-macrophage. Blood 78: 967-74 (1991); **Satterthwaite AB et al** Structure of the gene encoding CD34, a human hematopoietic stem cell antigen. Genomics 12: 788-94 (1992); **Silvestri F et al** The CD34 hemopoietic progenitor cell associated antigen: biology and clinical applications. Haematologica 77: 265-73 (1992); **Simmons DL et al** Molecular cloning of a cDNA encoding CD34, a sialomucin of human hematopoietic stem cells. JI 148: 267-71 (1992); **Suda J et al** Two types of murine CD34 mRNA generated by alternative splicing. Blood 79: 2288-95 (1992); **Sutherland DR et al** Structural and partial amino acid sequence analysis of the human hemopoietic progenitor cell antigen CD34. Leukemia 2: 793-803 (1988); **Tenen DG et al** Chromosome 1 localization of the gene for CD34, a surface antigen of human stem cells. Cytogenet. Cell Genet. 53: 55-7 (1990); **Terstappen LWMM et al** Sequential generations of hematopoietic colonies derived from single nonlineage-committed $CD34^+CD38^-$ progenitor cells. Blood, 77: 1218-27 (1991); **Traweek ST et al** The human hematopoietic progenitor cell antigen (CD34) in vascular neoplasia. Am. J. Clin. Pathol. 96: 25-31 (1991); **Udomsakdi C et al** Separation of functionally distinct subpopulations of primitive human hematopoietic cells using rhodamine-123. EH 19: 338-42 (1991); **Williams DA** Human $CD34^+$ $HLA\text{-}DR^-$ bone marrow cells contain progenitor cells capable of self-renewal, multilineage differentiation, and long-term *in vitro* hematopoiesis. Commentary: In search of the self-renewing hematopoietic stem cell. Blood Cells, 17: 296-300 (1991)

CD40: This cell surface antigen (see also: CD antigens) is also called ***Bp50***. It is a transmembrane glycoprotein with a length of 277 amino acids (48 kDa). CD40 is a phosphoprotein and can be expressed as a homodimer. A soluble form of CD40 (28 kDa) has also been described.

The protein is expressed on all B lymphocytes during various stages of development, activated T cells and monocytes, follicular dendritic cells, thymic epithelial cells, and various carcinoma cell lines. It is expressed on most mature B cell malignancies and on some early B cell acute lymphocytic leukemias. Induction of CD40 mRNA and enhancement of cell surface protein expression in primary human monocytes is observed after treatment with » GM-CSF, » IL3, or » IFN-γ. The human CD40 gene maps to chromosome 20.
CD40 has been proposed to play a role in the development of memory cells. It also plays a role in » cell activation, functioning as a competence factor and progression factor. Crosslinking of the CD 40 antigen (in combination with cytokines such as » IL4 and » IL5) leads to B cell proliferation and induces immunoglobulin class switching from IgM to the synthesis of IgG, IgA, and IgE in the absence of activated T cells. CD40 is one of the obligatory signals required for commitment of naive B cells to IgA secretion; the mechanism of IgA induction requires the cooperation of » IL10 and » TGF-β. Soluble CD40 inhibits T-dependent B cell proliferation.
Monoclonal antibodies against CD40 mediate a variety of effects on B lymphocytes, including induction of intercellular adhesion (via CD11a/CD18 (LFA-1)), short- and long-term proliferation, differentiation and enhanced tyrosine phosphorylation of proteins. Germinal center centrocytes are prevented from undergoing » apoptosis by activation through CD40 and receptor for antigen.
In human resting B cells expression of CD40 is induced by » IL4. Treatment of human B cells with » IL6 leads to the phosphorylation of the intracellular CD40 domain. CD40 does not, however, function as a receptor for IL6. In activated human B cells the synthesis of IL6 is induced by treatment of the cells with monoclonal antibodies directed against CD40, suggesting that CD40 participates in IL6-dependent signal transduction mechanisms. Some limited sequence homologies have been found with receptors for » NGF and »TNF-α and » CD27 and it has been assumed that CD40 may also be involved in modulating the biological activity of these and other cytokines. Human CD 40 has been found to be the receptor for a protein (CD40L; see: TRAP, TNF-related activation protein) involved in an X-linked immunodeficiency syndrome. The CD40 ligand is identical with a T cell-restricted cell surface activation antigen designated 5c8. The murine ligand of CD40 is » gp39.
The induction of a state of sustained proliferation in human resting B lymphocytes by CD40-mediated activation *in vitro* in the presence of » IL4 (a potent activator of B cell proliferation *in vitro*) has been demonstrated to be a significant technological advance in the preparation of human monoclonal antibodies. This CD40 system, which involves cross-linking of the CD40 receptor with an anti-CD40 antibody, allows the expansion of human B lymphocyte populations and the isolation and differentiation of individual B lymphocyte clones, thus facilitating the formation of stable human antibody-secreting heterohybridomas after fusion with suitable mouse myeloma cell lines, or the establishment of antibody-secreting human cell lines following their immortalization with EBV.

Alderson MR et al CD40 expression by human monocytes: regulation by cytokines and activation of monocytes by the ligand for CD40. JEM 178: 669-74 (1993); **Banchereau J et al** Long-term human B cell lines dependent on Interleukin 4 and antibody to CD40. S 251: 70-72 (1991); **Banchereau J & Rousset F** Growing human B lymphocytes in the CD40 system. N 353: 678-9 (1991); **Banchereau J et al** Long-term human B cell lines dependent on interleukin 4 and antibody to CD40. S 251: 70-72 (1991); **Banchereau J et al** Role of cytokines in human B lymphocyte growth and differentiation. Nouv. Rev. Fr. Hematol. 35: 61-6 (1993); **Barrett TB et al** CD40 signaling activates CD11a/CD18 (LFA-1)-mediated adhesion in B cells. JI 146: 1722-9 (1991); **Björck P et al** Expression of CD40 and CD43 during activation of human B lymphocytes. Scand. J. Immunol. 33: 211-8 (1991); **Clark EA & Shu G** Association between IL6 and CD40 signaling. IL6 induces phosphorylation of CD40 receptors. JI 145: 1400-6 (1990); **Clark EA** CD40: a cytokine receptor in search of a ligand. Tissue Antigens 35: 33-6 (1990); **Clark EA et al** CDw40 and BLCa-specific monoclonal antibodies detect two distinct molecules which transmit progression signals to human B cells. EJI 18: 451-7 (1988); **Darveau A et al** Efficient preparation of human monoclonal antibody-secreting heterohybridomas using peripheral B lymphocytes cultured in the CD40 system. JIM 159: 139-43 (1993); **deBoer M et al** Generation of monoclonal antibodies to human lymphocyte cell surface antigens using insect cells expressing recombinant proteins. JIM 152: 15-23 (1992); **Defrance T et al** Interleukin 10 and transforming growth factor β cooperate to induce anti-CD40-activated naive human B cells to secrete immunoglobulin A. JEM 175: 671-82 (1992); **Defrance T et al** Proliferation and differentiation of human $CD5^+$ and CD5- B cell subsets activated through their antigen receptors or CD40 antigens. EJI 22: 2831-9 (1992); **Fanslow WC et al** Soluble forms of CD40 inhibit biologic responses of human B cells. JI 149: 655-60 (1992); **Galy AH & Spits H** CD40 is functionally expressed on human thymic epithelial cells. JI 149: 775-82 (1992); **Gascan H et al** Anti-CD40 monoclonal antibodies or $CD4^+$ T cell clones and IL4 induce IgG4 and IgE switching in purified human B cells via different signaling pathways. JI 147: 8-13 (1991); **Grimaldi JC et al** Genomic structure and chromosomal mapping of the murine CD40 gene. JI 149: 3921-6 (1992); **Inui S et al** Identification of the intracytoplasmic region essential for signal transduction through a B cell activation molecule, CD40. EJI 20:

CD40: see addendum.

1747-53 (1990); **Jabara HH et al** CD40 and IgE: synergism between anti-CD40 monoclonal antibody and interleukin 4 in the induction of IgE synthesis by highly purified human B cells. JEM 172: 1861-4 (1990); **Loetscher H et al** Molecular cloning and expression of the human 55 kd tumor necrosis factor receptor Cell 61: 351-9 (1990); **Maliszewski CR et al** Recombinant CD40 ligand stimulation of murine B cell growth and differentiation: cooperative effects of cytokines. EJI 23: 1044-9 (1993); **Marshall LS et al** The molecular basis for T cell help in humoral immunity: CD40 and its ligand, gp39. JCI 13: 165-74 (1993); **Ramesh N et al** Chromosomal localization of the gene for human B-cell antigen CD40. Somat. Cell. Mol. Genet. 19: 295-8 (1993); **Rousset F et al** Cytokine-induced proliferation and immunoglobulin production of human B lymphocytes triggered through their CD40 antigen. JEM 173: 705-10 (1991); **Schall TJ et al** Molecular cloning and expression of a receptor for human tumor necrosis factor Cell 61: 361-70 (1990); **Smith CA et al** A receptor for tumor necrosis factor defines an unusual family of cellular and viral proteins. S 248: 1019-23 (1990); **Splawski JB et al** Immunoregulatory role of CD40 in human B cell differentiation. JI 150: 1276-85 (1993); **Stamenkovic I et al** A B lymphocyte activation molecule related to the nerve growth factor receptor and induced by cytokines in carcinomas. EJ 8: 1403-10 (1989); **Torres RM et al** Differential increase of an alternatively polyadenylated mRNA species of murine CD40 upon B lymphocyte activation. JI 148: 620-6 (1992); **Uckun FM et al** Stimulation of protein tyrosine phosphorylation, phosphoinositide turnover, and multiple previously unidentified serine/threonine-specific protein kinases by the pan-B cell receptor CD40/Bp50 at discrete developmental stages of human B cell ontogeny. JBC 266: 17478-85 (1991); **Zhang K et al** CD40 stimulation provides an IFN-γ-independent and IL4-dependent differentiation signal directly to human B cells for IgE production. JI 146: 1836-42 (1991)

CD40L: *CD40 ligand* see: CD40. For the ligand see: TRAP (TNF-related activation protein).

CD58: This cell surface antigen (see also: CD antigens) of 55 kDa is also known as LFA-3. CD58 is expressed on hematopoietic and various non-hematopoietic cells, including endothelium. It is also expressed on macrophages, germinal center B cells, medullary thymocytes, and memory T cells. The CD58 gene maps to human chromosome 1p13 in the same region as the gene for its receptor. It is a ligand of another cell surface marker called » CD2 and thus mediates cell adhesion. Binding of CD58 to CD2 has been shown to enhance antigen-specific T cell activation. CD58 also stimulates the release of » IL1 from thymic epithelial cells and monocytes.

Barbosa JA et al Gene mapping and somatic cell hybrid analysis of the role of human lymphocyte function- associated antigen-3 (LFA-3) in CTL-target cell interactions. JI136: 3085-91 (1986); **Quillet-Mary A et al** Target lysis by human LAK cells is critically dependent upon target binding properties, but LFA-1, LFA-3 and ICAM-1 are not the major adhesion ligands on targets. IJC 47: 473-9 (1991); **Rivoltini L et al** The high lysability by LAK cells of colon-carcinoma cells resistant to doxorubicin is associated with a high expression of ICAM-1, LFA-3, NCA and a less-differentiated phenotype. IJC 47: 746-54 (1991); **Wallner BP et al** Primary structure of lymphocyte function-associated antigen 3 (LFA-3): the ligand of the T lymphocyte CD2 glycoprotein. JEM 166: 923-32 (1987)

CD64: This cell surface antigen (see also: CD antigens) is a 72 kDa glycoprotein. It is identical with hFcRI = $F_{c\gamma}$RI, the high affinity receptor for the Fc portion of immunoglobulin IgG. CD64 is mainly expressed on mononuclear phagocytes. Its expression is induced by » IFN-γ.

CD64 is one of the three subtypes of Fc-IgG receptors. It is the only IgG receptor which binds monomeric ligands. These receptors are widely distributed on many cells of the immune system. CD64 plays a central role in antibody-dependent cytotoxicity and clearance of immune complexes; it contribute to the pathogenesis of immune complex- and auto-antibody-mediated diseases such as vasculitis, rheumatoid arthritis, idiopathic thrombocytopenic purpura or autoimmune neutropenia. The receptor can be released by cells and circulates in biological fluids as » IBF (immunoglobulin binding factors).

Gergely J & Sarmay G The two binding-site models of human IgG binding Fcγ receptors. FJ 4: 3275-83 (1990); **Pearse RN et al** Characterization of the promoter of the human gene encoding the high-affinity IgG receptor: transcriptional induction by γ-interferon is mediated through common DNA response elements. PNAS 88: 11305-9 (1991); **Ravetch JV & Kinet JP** Fc receptors. ARI 9: 457-92 (1991); **Repp R et al** Neutrophils express the high affinity receptor for IgG (FcγRI, CD64) after *in vivo* application of recombinant human granulocyte colony-stimulating factor. Blood 78: 885-9 (1991); **Van de Winkel JGJ et al** Gene organization of the human high affinity receptor for IgG, FcγRI (CD64). Characterization and evidence for a second gene. JBC 266: 13449-55 (1991); **Unkeless JC** Function and heterogeneity of human Fc receptors for immunoglobulin G. JCI 83: 355-61 (1989)

CD70: New designation for the ligand of » CD27.
see: references for » CD Antigens (Schlossman et al).

CD80: New designation for the B7 antigen, a ligand of » CD28.
see: references for » CD Antigens (Schlossman et al).

CD95: New designation for the Fas antigen. See: Apo-1.
see: references for » CD Antigens (Schlossman et al).

CD115: New designation for the receptor of » M-CSF.
see: references for » CD Antigens (Schlossman et al).

CD117: New designation for the receptor of » SCF (see: *kit*).
see: references for » CD Antigens (Schlossman et al).

CD40 ligand: see addendum: CD40. **CD70:** see addendum.

CD120a: New designation for the 55 kDa receptor of » TN-α.
see: references for » CD Antigens (Schlossman et al).

CD120b: New designation for the 55 kDa receptor of » TN-α.
see: references for » CD Antigens (Schlossman et al).

CD122: New designation for the 75 kDa receptor of » IL2.
see: references for » CD Antigens (Schlossman et al).

CD126: New designation for the receptor for » IL6.
see: references for » CD Antigens (Schlossman et al).

CD antigens: *cluster of differentiation, cluster of determinants*
Generic name given to a large collection of cell surface proteins selectively expressed as differentiation antigens on leukocytes. The number that follows is arbitrarily assigned.
Many monoclonal antibodies directed against these differentiation antigens have been produced to study the development and functional activities of human lymphocytes. The old names of these cell surface proteins, frequently derived from the names of the antibodies used to identify and detect them, are often used as synonyms for these CD antigens although many have been assigned a CD number (see also: individual entries). A small *w* before the number designation stands for "workshop". It indicates that the CD assignment is tentative.
In many instances these CD antigens are expressed only at certain stages of development and the biological functions of many of the identified proteins are still unknown. Unlike the morphological criteria used in classical hematology for the description of specific leukocyte developmental stages, the use of monoclonal antibodies allows the objective and precise analysis and standardized typing of mature and immature normal and malignant cells of all lineages. They also help to delineate the biologic traits that distinguish normal immune and hematopoietic cells from their malignant counterparts which is of fundamental importance in understanding hematological malignancies.
Some of the known CD antigens are identical with cytokine receptors (see: CD25), carry out cytokine receptor-like functions (see: CD27, CD30, CD40), or are involved in modulating the biological activities of cytokines (see: CD4, CD28, CD40). The expression of many CD antigens is also influenced by cytokines. In some cases binding of ligands to CD antigens has been shown to modulate cytokine expression.
In some cases the cell specificity of the monoclonal antibodies directed against specific CD antigens has been exploited in the immunotherapy of several cancers.
Barcley AN et al (eds) The Leukocyte Antigen Facts Book. Academic Press 1993; **Bernhard A et al** (eds) IVth Internatl. Workshop on Human Leukocyte Differentiation Antigens. Oxford Univ. Press, Oxford 1990; **Horejsi W & Bazil V** Surface proteins and glycoproteins of human leukocytes. BJ 253: 1-26 (1989); **Knapp W et al** (eds) Leukocyte Typing IV. White cell differentiation antigens. Oxford University Press, Oxford 1989; **Schlossman SF et al** (eds) Leucocyte typing V: White cell differentiation antigens. Oxford University Press, Oxford 1994); **Zaleski MB** Cell-surface molecules in the regulation of immune responses. Immunol. Invest. 20: 103-31 (1991)

CDF: *cartilage-derived factor* A growth factor of 11 kDa isolated from fetal cartilage that stimulates the sulfation of glycosaminoglycans and proteoglycans in chondrocyte cultures (see also: ECM). The factor is closely related to, or identical with » IGF.
Two other proteins of 16 kDa and 22-26 kDa have also been described in the literature as CDF and these are probably distinct factors unrelated to IGF. The 26 kDa protein has also been called ***CCSF*** (chondrocyte colony-stimulating factor). All three CDFs stimulate the growth of chondrocytes in soft agar cultures.
Canalis E et al Effect of cartilage-derived factor on DNA and protein synthesis in cultured rat calvariae. Calcif. Tissue Int. 36: 102-7 (1984); **Hamerman D et al** A cartilage-derived growth factor enhances hyaluronate synthesis and diminishes sulfated glycosaminoglycan synthesis in chondrocytes. JCP 127: 317-22 (1986); **Hiraki Y et al** Combined effects of somatomedin-like growth factors with fibroblast growth factor or epidermal growth factor in DNA synthesis in rabbit chondrocytes. Mol. Cell. Biochem. 76: 185-93 (1987); **Kato Y et al** Purification of growth factors from cartilage. MiE 198: 416-24 (1991); **Shimomura Y & Suzuki F** Cultured growth of cartilage cells. Clin. Orthop. 184: 93-105 (1984); **Suzuki F et al** Preparation of cartilage-derived factor. MiE 146: 313-20 (1987)

CDF: *CAT (or ChAT) development factor, choline acetyltransferase development factor* This 22 kDa anionic factor is produced by muscle cells and induces the synthesis of choline acetyl transferase in several human neuroblastoma cells lines which show cholinergic and adrenergic characteristics. CDF synergizes with » bFGF which has also been shown to possess CDF activity. The factor does not influence the expression of tyrosine hydroxylase and is not a mitogen for these cells. CDF also promotes the survival of motor neurons *in vitro*. Some of the activity of CDF may be due to » IL6.

Hama T et al Interleukin-6 as a neurotrophic factor for promoting the survival of cultured basal forebrain cholinergic neurons from postnatal rats. Neurosci. Lett. 104: 340-54 (1989); **McManaman JL & Crawford FG** Skeletal muscle proteins stimulate cholinergic differentiation of human neuroblastoma cells. J. Neurochem. 57: 258-66 (1991); **McManaman JL et al** Rescue of motor neurons from cell death by a purified skeletal muscle polypeptide: effects of the ChAT development factor, CDF. Neuron 4: 891-8 (1990); **Rabinovsky ED et al** Differential effects of neurotrophic factors on neurotransmitter development in the IMR-32 human neuroblastoma cell line. J. Neurosci. 12: 171-9 (1992)

CDF: *cholinergic differentiation factor* This factor, originally purified from cardiac and skeletal muscle cell-conditioned medium, is also called » CNDF (cholinergic neuronal differentiation factor). The enhanced survival of the motoneurons, which would have died between days 1 and 4 of culture, is accompanied by a 4-fold increase in choline acetyltransferase (ChAT) activity. CDF also increases ChAT activity in dorsal spinal cord cultures, but has no detectable effect on ChAT levels in septal or striatal neuronal cultures. CDF has been shown to alter neurotransmitter production as well as the levels of several neuropeptides in cultured rat sympathetic neurons. It induces mRNA for choline acetyltransferase (EC 2.3.1.6), » Somatostatin, » SP (substance P), and » VIP (vasoactive intestinal polypeptide) while lowering mRNA levels of tyrosine hydroxylase (EC 1.14.16.2) and neuropeptide Y (NPY).

CDF is identical with » LIF and the factor is therefore also named CDF/LIF.

Martinou JC et al Cholinergic differentiation factor (CDF/LIF) promotes survival of isolated rat embryonic motor neurons *in vitro*. Neuron 8: 737-44 (1992); **Nawa H et al** Recombinant cholinergic differentiation factor (leukemia inhibitory factor) regulates sympathetic neuron phenotype by alterations in the size and amounts of neuropeptide mRNAs. J. Neurochem. 56: 2147-50 (1991); **Rao MS & Landis CC** Cell interactions that determine sympathetic neuron transmitter phenotype and the neurokines that mediate them. J. Neurobiol. 24: 215-32 (1993)

CDF: *cytolytic differentiation factor for T lymphocytes* also abbrev. CTL differentiation factor (cytolytic T lymphocyte differentiation factor). This is an operational definition of biological activities secreted by stimulated mononuclear cells which promote the development of thymocyte precursor cells into CD8-positive killer cells. One of the known factors with CDF activity is identical with IL6. » IL6 and there may be others showing the same activity.

Cernetti C et al Identification of a 24 kDa cytokine that is required for development of cytolytic T lymphocytes. PNAS 85: 1605-9 (1988); **Cernetti C et al** Bioassays for the detection of a cytokine that enhances the development of cytolytic T lymphocytes. CI 109: 148-58 (1987); **Ming JE et al** Interleukin 6 is the principal cytolytic T lymphocyte differentiation factor for thymocytes in human leukocyte conditioned medium. J. Mol. Cell. Immunol. 4: 203-211 and 211-2 (1989)

CDF: *cytotoxic differentiation factor* This factor promotes the development of cytotoxic T lymphocytes. It is identical with » IL6 and probably the same factor as » CDF (cytolytic differentiation factor for T lymphocytes, CTL differentiation factor).

Takai Y et al B cell stimulatory factor 2 is involved in the differentiation of cytotoxic T lymphocytes. JI 140: 508-12 (1988); **Takai Y et al** Requirement for three distinct lymphokines for the induction of cytotoxic T lymphocytes from thymocytes. JI 137: 3494-3500 (1986)

CDGF: *cartilage-derived growth factor* These growth factors isolated from cartilage are identical either with » aFGF or » bFGF.

Courty J et al [*In vitro* properties of various growth factors and *in vivo* effects] Biochimie 66: 419-28 (1984); **Davidson JM et al** Accelerated wound repair, cell proliferation, and collagen accumulation are produced by a cartilage-derived growth factor. JCB 100: 1219-27 (1985); **Klagsbrun M & Beckoff MC** Purification of cartilage-derived growth factor. JBC 255: 10859-66 (1980); **Nichols WK et al** Differential effects of cartilage-derived growth factor stimulation of collagen secretion by bovine aortic and microvascular endothelial cells. Coll. Relat. Res. 6: 505-14 (1987)

CDGF: *chondrosarcoma-derived growth factor* see: CGF.

CDGF 1. 7: *cartilage-derived growth factor* This factor is identical with » bFGF.

CDGF-1: *cartilage-derived growth factor* This factor is identical with » bFGF.

Courty J et al [*In vitro* properties of various growth factors and *in vivo* effects] Biochimie 66: 419-28 (1984); **Sullivan R & Klagsbrun M** Purification of cartilage-derived growth factor by heparin affinity chromatography. JBC 260: 2399-2403 (1985)

CDGF: *CEF-derived growth factor* (CEF = chicken embryo fibroblast) CDGF is a 32 kDa heterodimeric factor consisting of a 15 kDa and a 17 kDa subunit linked by disulfide bonds. The factor is isolated from the serum-free culture supernatants of primary cultures of confluent chicken embryo fibroblasts. It acts as a strong mitogen for fibroblasts. The physicochemical characteristics of this factor make it very likely that it is a new growth factor.

Geistlich A & Gehring H Isolation and characterization of a novel type of growth factor derived from serum-free conditioned medium of chicken embryo fibroblasts. EJB 207: 147-53 (1992); **Geistlich A & Gehring H** CDGF (chicken embryo fibroblast-derived growth factor) is mitogenically related to TGF-β and modulates PDGF, bFGF, and IGF-I action on sparse NIH/3T3 cells. ECR 204: 329-35 (1993)

CDI: *cartilage-derived inhibitor* This factor is isolated from cartilage. It is an endogenous inhibitor of metalloproteinases (collagenases) which inhibits the proliferation of endothelial cells and also inhibits angiogenesis *in vivo* (see also: Angiogenesis factors).
Moses MA & Langer R A metalloproteinase inhibitor as an inhibitor of neovascularisation. JCBc 47: 230-5 (1991)

cDNA: see: Gene libraries.

cDNA libraries: see: Gene libraries.

cDNA-PCR: reverse transcriptase polymerase chain reaction see: Message amplification phenotyping.

cDNA subtraction: see: Gene libraries.

CDw121a: Tentative new designation for the type I receptor of » IL1.
see: references for » CD Antigens (Schlossman et al).

CDw121b: Tentative new designation for the type II receptor of » IL1.
see: references for » CD Antigens (Schlossman et al).

CDw124: Tentative new designation for the receptor for » IL4.
see: references for » CD Antigens (Schlossman et al).

CDw127: Tentative new designation for the receptor for » IL7.
see: references for » CD Antigens (Schlossman et al).

CDw128: Tentative new designation for the receptor for » IL8.
see: references for » CD Antigens (Schlossman et al).

CEF-4: This protein is identical with another cytokine called » 9E3.
Gonneville L et al Complex expression pattern of the CEF-4 cytokine in transformed and mitogenically stimulated cells. O 6: 1825-33 (1991)

CEF-10: see: CTGF (connective tissue growth factor).

CEF-derived growth factor: (CEF = chicken embryo fibroblast). See: CDGF.

Cell ablation: see: Genetic ablation.

Cell activation: This term describes the extent of stimulation of cellular processes initiated as a response to external stimuli reaching the cell. More strictly this term refers to the transition of cells from the G_0 phase to the G_1 phase of the eukaryotic » cell cycle.
The concept of activated cells was developed from observations of cells cultured *in vitro* which reacted to a variety of substances by altering their biochemical and functional activities and hence the term activation refers to new specific activities observed in these cells. As a rule the activation process is never a all-or-nothing reaction but rather a continuous spectrum of alterations in comparison to an unactivated cell.
Electron microscopic studies reveal an increase in the numbers of intracellular vesicles as the first indication of the activation of some cellular processes. Biochemically different activation states may be identified and classified by the phase-specific expression of novel antigens on the surfaces of activated cells. In addition, cell activation is normally associated with the rapid induction of the expression of a number of new genes, including those encoding transcription factors (see also: Gene expression), » oncogenes, cytokine genes, cell surface molecules (see, for example: CD anti-

Properties of activated neutrophils

Enhanced phagocytotic activity
Increased cell migration
Increased chemotaxis
Increased myeloperoxidase activity
Elevated oxidative metabolism
Enhanced microbicidal activity
Enhanced cytotoxicity
Release of lysosomal enzymes

Properties of activated macrophages/monocytes

Reduced motility
Expression of new surface antigens
Increased Ia expression
Synthesis of plasminogen activator
Enhanced cytotoxicity against tumor cells
Increased ability to kill intracellular parasites
Increased production and release of cytokines
Increased synthesis of prostaglandins/leukotrienes
Increased production of reactive oxygen intermediates
Increased production of reactive nitrogen intermediates
Increased procoagulatory activity

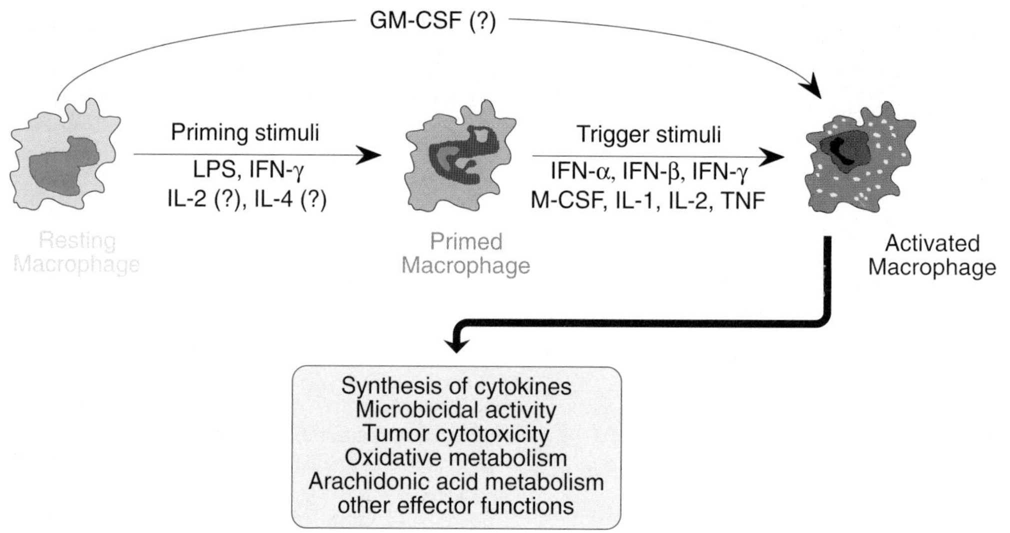

Cytokines involved in macrophage activation.
Several cytokines are known to function as pre-activating factors (priming factors). The priming process is associated with the expression of various cytokine receptors and hence priming factors render macrophages sensitive to other cytokines (trigger factors). The pre-activated "primed" macrophage can react with a cascade of metabolic reactions including the synthesis and release of other cytokines.

gens), adhesion molecules, and many other genes the biological functions of which are as yet unknown.

Stimuli that activate cells and control functional specialization are usually binding of a ligand, for example a cytokine, to its receptor and also the binding of an antigen to its membrane receptor. Such stimuli are also of physiological importance while many other biochemical stimuli known to activate cells and to induce functional alterations *in vitro* may not be relevant *in vivo*.

Binding of a ligand to its receptor induces conformational changes which serve to transmit a signal into the interior of the cell. This process involves a number of secondary messengers and a complicated cascade of reactions eventually leading to a modulation of gene activity in the nucleus. Intracellular protein phosphorylation mediated by intrinsic protein kinase activities of receptors or by non-receptor protein kinases is thought to be the initial step in cell activation. The rapid and transient redistribution of Ca^{2+} from intracellular membrane-bound compartments (stores) is also a key event of cell activation.

Factors that activate cells can be functionally distinguished into so-called ***competence factors*** and ***progression factors***. Competence factors are ***priming factors*** that pre-activate cells. Many of these factors are encoded by so-called early or immediate early response genes (see also: Gene expression). They play an important role for the transition from G_0 to G_1 but these factors are not capable of promoting entry into the S phase of the cell cycle. The simultaneous or successive activity of progression factors provides a trigger signal required for entry into the S phase. The expression or repression of genes may later initiate cell proliferation and/or differentiation (for a technique allowing characterization of such genes by genomic difference cloning see: Gene libraries).

An indication of how subtle the processes of cell activation may be is the observation that mere adhesion of circulating cells to the surfaces of blood vessels can activate them. This process is mediated, at least in part, by numerous cell adhesion molecules expressed on the surfaces of the interacting cells.

In a certain way the available repertoire of external stimuli leading to cell activation is redundant because identical functional activities can be induced by completely different stimuli (see also: Cytokines). On the other hand, activation processes can also be reversed in that some factors are capable of deactivating pre-activated cells (see for example: MDF, macrophage deactivation factor). The directed inhibition of cell activation by cytokine antagonists, antibodies or drugs is used as a therapeutic approach to immunointervention in a variety of clinical and experimental settings (see: Cytokine inhibitors). There is also a growing list of genetic defects associated with impaired activation of cells and these contribute to the understanding of cellular functions *in vivo*.

Abu-Ghazaleh RI et al Eosinophil activation and function in health and disease. Immunol. Ser. 57: 137-67 (1992); **Adams DO & Hamilton TA** The cell biology of macrophage activation. ARI 2: 283-318 (1984); **Amaiz-Villena A et al** Human T-cell activation deficiencies. IT 13: 259-65 (1992); **Cantrell DA et al** T lymphocyte activation signals. Ciba Found. Symp. 164: 208-18 (1992); **Crabtree GR** Contingent genetic regulatory events in T lymphocyte activation. S 243: 355-61 (1989); **Damjanovich S et al** Dynamic physical interactions of plasma membrane molecules generate cell surface patterns and regulate cell activation processes. Immunobiology 185: 337-49 (1992); **Desiderio SV** B cell activation. Curr. Opin. Immunol. 4: 252-6 (1992); **Di Persio JF & Abboud CN** Activation of neutrophils by granulocyte-macrophage colony-stimulating factor. Immunol. Ser. 57: 457-84 (1992); **Gardner P** Calcium and T lymphocyte activation. Cell 59: 15-20 (1989); **Geppert TD et al** Accessory cell signals involved in T cell activation. IR 117: 5-66 (1990); **June CH** Analysis of lymphocyte activation and metabolism by flow cytometry. Curr. Opin. Immunol. 4: 200-4 (1992); **Mills GB** Activation of lymphocytes by lymphokines. CTMT 35: 495-535 (1990); **Nathan C** Mechanism and modulation of macrophage activation. Behring Inst. Mitt. Feb 88: 200-7 (1991); **Parker DC** T cell-dependent B cell activation. ARI 11: 331-60 (1993); **Perlmutter RM** *in vivo* dissection of lymphocyte signaling pathways. Clin. Immunol. Immunopathol. 67: S44-9 (1993); **Revillard JP & Berard-Bonnefoy N** CD4, CD8, and other surface molecules regulating T cell activation. Immunol. Ser. 59: 41-61 (1993); **Tobias PS et al** Participation of lipopolysaccharide-binding protein in lipopolysaccharide-dependent macrophage activation. Am. J. Respir. Cell Mol. Biol. 7: 239-45 (1992); **West MA** Role of Cytokines in leukocyte activation: phagocytic cells. CTMT 35: 537-70 (1990)

Cell blot assay: This assay utilizes biologically active growth factors retained on nitrocellulose membranes after transfer from polyacrylamide gels. Cells cultured on blotted lanes can respond to growth factors and this assay is therefore particularly useful for a precise determination of apparent molecular masses of active factors in crude tissue extracts. The cell blot assay has been adapted to detect » CNTF (ciliary neurotrophic factor), » FGF, and » NGF. For other immunoassays see also: Cytokine assays.

Carnow TB et al Localized survival of ciliary ganglionic neurons identifies neuronotrophic factor bands on nitrocellulose blots. J. Neurosci. 5: 1965-71 (1985); **Hayman EG et al** Cell attachment on replicas of SDS polyacrylamide gels reveals two adhesive plasma proteins. JCB 95: 20-3 (1982); **Kendall ME & Hymer WC** Measurement of hormone secretion from individual cells by cell blot assay. MiE 168: 327-38 (1989); **Manthorpe M et al** Detection and analysis of growth factors affecting neural cells. Neuromethods 23: 87-137, Boulton A et al (edt.) The Humana Press Inc. (1992); **Pettmann B et al** Biological activities of nerve growth factor bound to nitrocellulose paper by Western blotting. J. Neurosci. 8: 3524-42 (1989); **Rudge J et al** An examination of ciliary neuronotrophic factors from avian and rodent tissue extracts using a blot and culture technique. Dev. Brain Res. 32: 103-110 (1987)

Cell culture: Cell culture is a technique allowing growth and propagation of cells of higher organisms under controlled conditions.

Culture dishes may be of glass or plastic and may also have pre-treated surfaces, coated, for example with collagen, polylysine, or components of the extracellular matrix (see: ECM) to facilitate the attachment of the cells. Some sophisticated techniques also employ entire layers of adherent cells, known as » feeder cells, which are used to support the growth of cells with more fastidious growth requirements.

In order to mimic the normal physiological environment for the cells they are kept in incubators with a CO_2 atmosphere using semi-synthetic growth media. These media are buffered and contain, among other things, amino acids, nucleotides, salts, vitamins, and also a supplement of serum such as fetal calf serum (FCS) or newborn calf serum (NCS). Some fastidious cells are also grown in media supplemented with horse serum or even human serum. Although the addition of serum has been instrumental in developing modern cell culture techniques, its use also has disadvantages. Commercially available serum is obtained from pooled sources and, despite rigorous quality controls, individual lots may vary considerably in their biological composition and may contain varying amounts of proteases, selective inhibitors, bacterial toxins, lipids, and other factors interfering with the growth of cells and their functional activities.

Currently available serum replacements contain growth factors and inhibitors such as hormones, transferrin, insulin, selenium, and attachment factors. Their use may reduce the necessary amounts of serum normally required for optimal cell growth by 50-80 %. Some chemically defined media, so-called ***serum-free media*** (see: SFM) and even absolutely protein-free media have been formulated for a number of cell lines. These media are particularly well suited to investigate the growth requirements of cells and the effects of » cytokines and growth factors. They facilitate the detection of these factors and also the enrichment and the analysis of factors secreted by the cultured cells. It is known, however, that cell variants obtained, for example, from one parent line, may vary considerably in their growth requirements when grown in defined serum-free media. In these cases it may be necessary either to formulate new medium compositions or to resort to serum-reduced media. Maintenance of new cell variants may even then be difficult to achieve. In some instances the growth of some fastidious cells can be supported by using conditioned medium (see: CM) which essentially is the sterile filtered used growth medium which normally is discarded.

Cells grown *in vitro* do not organize themselves into tissues. Instead, they grow as monolayers (or in some instances as multilayers) on the surface of tissue culture dishes. The cells usually multiply until they come into contact with each other, form a so-called ***confluent monolayer***. Normal cells stop growing when they come into contact with each other and this phenomenon is known as ***contact inhibition***.

Anchorage-dependent cells show the phenomenon of adherence, i. e. they grow and multiply only if attached to the inert surface of a culture dish or another suitable support. Such cells cannot normally be grown without a solid support. Many cells do not require this solid surface and show a phenomenon known as ***anchorage-independent growth*** (abbrev. AIG). Accordingly, one variant of growing these cells in culture is the use of ***spinner cultures*** (or ***suspension cultures***) in which single cells float freely in the medium and are kept like this by constant stirring. This technique is particularly useful for growing large amount of cells. Anchorage-independent cells are usually also capable of forming colonies in semisolid media (see also colony formation assay). Some techniques have been developed that can be used also to grow anchorage-dependent cells in spinner cultures. They make use of microscopically small positively-charged dextran beads to which these cells can attach (***micro-carrier cultures***).

The starting material for the establishment of a cell culture are organs or tissues of a suitable donor obtained under sterile conditions. These tissues are minced and frequently treated with proteolytic enzymes such as trypsin, collagenase of dispase to obtain a single cell suspension that can be used to inoculate a culture dish. In some cases dispersion of tissue is also effectively achieved by treatment with buffers containing EDTA. A particular form of initiating a cell culture is the use of tiny pieces of tissues (***explant cultures***) from which cells may grow out *in vitro*.

Cell cultures freshly initiated from tissues or organ pieces are called ***primary cell cultures***. These are characterized by a normally limited life span of the cells which may die after a number of cell divisions. This process is called senescence. A ***cell line*** may be obtained from the primary culture by renewed disruption of the cell layers with trypsin or any other suitable agent and fresh inoculation of a limited number of cells into new dishes containing fresh medium (see also: 3T3 cells). This process, which is repeated normally every three to four days, is called ***cell passage*** or ***subcultivation***. After approximately 70 subcultivation steps the cell line is called an **established cell line**.

Primary cell cultures maintained for some passages may undergo a process referred to as ***crisis***. This is usually associated with alterations of the properties of the cells and may proceed quickly or extend over many passages. Loss of contact inhibition is frequently an indication of cells having lost their normal characteristics. These cells then growth as multilayers in tissue culture dishes. The most pronounced feature of abnormal cells is the alteration in chromosome numbers, with many cells surviving this process being aneuploid. The switch to abnormal chromosome numbers is usually referred to as ***cell transformation*** and this process may give rise to cells that can then be cultivated for indefinite periods of time by serial passaging. Such cells give rise to so-called ***continuous cell line***.

Continuous cell lines and various culture techniques play an important role in cytokine research. They are employed to obtain » recombinant cytokines and other proteins of interest. So-called » factor-dependent cell lines are used to study the bioactivities of cytokines and other growth factors.

Barnes, D. & Sato, G. Methods for growth of cultured cells in serum-free medium. Anal. Biochem. 102: 255-270 (1980); **Continuous Cell Lines** – An International Workshop on Current Issues. Proceedings. Bethesda, Maryland, March 20-22, 1991. Dev. Biol. Stand. 76: 1-343 (1992); **Barnes, D. & Sato, G.** Serum-free cell culture: a unifying approach. Cell 22: 649-656 (1980); **Butler M & Dawson M** (eds) Cell culture Lab Fax. Bios Scientific Publ. Oxford 1992; **Crowe, R. et al.** Experiments with normal and transformed cells. A laboratory manual for working with cells in culture. Cold Spring Harbor Laboratory, Cold Spring Harbor, New York, 1983; **Fischer, G. et al.** (eds) Hormonally defined media. Springer Verlag, Berlin 1983; **Freshney RI** (edt.) Animal cell culture, a practical approach. IRL Press Oxford 1986; **Freshney, R. I.** Culture of animal cells. A manual of basic technique. A. R. Liss, New York, 1983; **Jacoby, W. B. & Pastan, I. H.** (edt.) Cell culture. Methods Enzymol. 58, Academic Press, Inc. 1979; **Kruse, P. F. et al.** Tissue culture, methods and applications. Academic Press, New York, 1973; **Potten CS & Hendry** (eds) Cell clones: Manual of Mammalian Cell Techniques. Churchill Livingstone, Edinburgh 1985

Cell cycle: This term collectively refers to the different discrete successive stages that can be discerned in mitotically active eukaryotic cells. The well-organized sequence of events yields two identical daughter cells. Four phases are usually distinguished. In the presynthetic G_1 phase (Gap 1) the cell prepares itself for subsequent DNA synthesis. Enzymes and proteins required for initiating and carrying out DNA synthesis are synthesized late in G_1 and early in S. The G_1 phase is followed by the S phase (synthetic phase) in which the cell replicates its DNA. The postsynthetic G_2 phase (Gap 2) is the phase after completion of DNA synthesis in which the cells controls whether DNA replication has been complete and prepares for cell division by synthesizing molecules required in mitotic operation. Mammalian cells do not initiate mitosis in the presence of either unreplicated or damaged DNA, but some drugs are known, e. g. » staurosporine, that uncouple mitosis from the completion of DNA replication. G_1, S, and G_2 are often collectively referred to as the interphase, i. e. the period between successive cell divisions. The M phase (mitotic phase) following G_2 is characterized by the disappearance of nuclear membranes and nucleoli, appearance of the spindle apparatus (prophase), condensation of chromatin into chromosomes, parallel alignment of chromosomes in the equatorial plane of the spindle (metaphase), separation of chromosomes into pairs, and simultaneous movement of chromosomes to opposite poles (anaphase), the reassembly of two nuclei (telophase), and cytoplasmic division (cytokinesis), i. e. the segmentation and separation of the cytoplasm, resulting in the formation of two separate cells.

Depending upon the cell type and also on growth conditions, the various phases may be of different lengths. S-, G_2 and M phase usually last ≈ 10 hours while the lengths of the G_1 phases may vary considerably.

Recent studies on the cooperation between different growth factors for mitogenesis have shown that multiple requirements exist for a cell to proceed through the entire cycle and that the cycle proceeds through multiple check-points at which the cell's progress is monitored to assure that previous events are completed before the next step is initiated. The individual stages of the cycle are under strict control and the exact and unperturbed temporal order of associated gene activities is an absolute prerequisite for successful completion of the cycle. In addition, in multicellular organisms cell division must be co-ordinated during development for organogenesis.

A plethora of so-called CDC (cell cycle control) genes have been identified. Mutations in any of these genes lead to the arrest of the cell at a stage requiring the presence of the corresponding gene product, and in many instances genes from higher eukaryotes, including humans, have been shown to be capable of complementing the defects in cell cycle mutants of yeast. Cell cycle regulatory proteins also include a special family of proteins, designated cyclins, and associated proteins such as protein kinases (see: ERK, extracellular signal-regulated protein kinases) and protein phosphatases. The name cyclins for these internal regulators of the cell cycle derives from the observation that in all cells these proteins accumulate periodically in a cell cycle phase-specific manner. The concentrations of cyclins A and B, for example, are highest at the transition point between G_2 and M phase and these proteins are relatively rapidly degraded with entry of cells into the M phase. Some

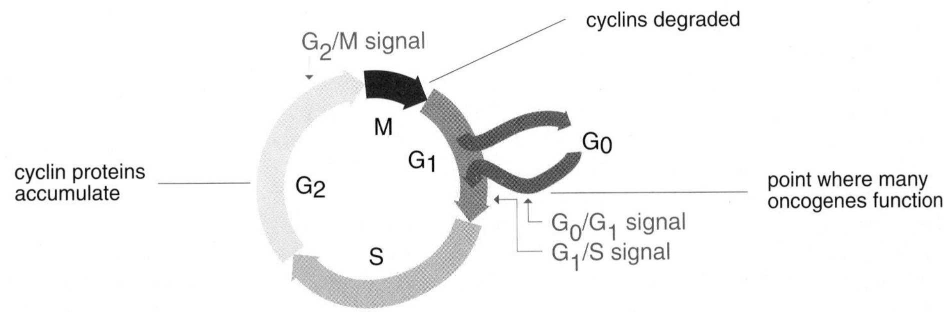

The cell cycle phases and some control points.
See text for details.

of these proteins also show alterations in the degree of phosphorylation in different cell cycle phases. To date, human cyclins have been grouped into five types, A through E, which share ≈ 16-32 % amino acid identity and ≈ 42-55 % similarity. In addition, several proteins, known as ***cyclin-dependent kinases*** (CDKs) bind to a cyclin to become active protein kinases that phosphorylate a plethora of intracellular target proteins. cdc2 has been shown to be the major regulator of the cell cycle. In all eukaryotes entry into M phase critically depends on the activation of a cdc2/cyclin B complex. The rapid destruction of the cyclins (by the ubiquitin pathway) also leads to rapid loss of activity of their kinase companions. A number of mitogen-regulated proteins (MRP) of largely unknown function are synthesized after stimulation of cells by mitogenic growth factors. Transcriptional changes and transient increases in mRNAs corresponding to various » oncogenes (e. g. retinoblastoma gene product and p53), transcription factors (see also: Gene expression), and » cytokines are observed during all phases of the cell cycle. Resting (quiescent) cells, i. e. non-dividing cells, are usually called G_0 cells (they are also said to be in the plateau phase). Normally these cells are arrested somewhere before the G_1 phase before the initiation of DNA synthesis. Some epithelial cells may also be arrested in the G_2 phase Important examples of resting cells are » Hematopoietic stem cells, which are spared from the cytotoxic activities of such agents as » 4-HC or » fluorouracil, and stromal cells of the hematopoietic microenvironment (see: Hematopoiesis). The transition of cells from G_0 to G_1 following a mitogenic stimulus provided by cytokines and other factors is a very complex process involving the selective and co-ordinated induction and expression of several hundred different genes and probably also the simultaneous repression of other genes (see also: ERG, early response genes). Resting cells in G_0 express only a limited number of receptors for growth factors and other ligands. After stimulation of these cells by ligands binding to their respective receptors (initiation phase) they become "competent" to respond to further factors following the expression of suitable receptors on the cell surface. Binding of these new ligands to their corresponding receptors mediates a signal for the further transit through the cell cycle (progression phase). Such factors are generally also known as ***competence factors*** and ***progression factors***, respectively. Some cytokines, such as » IL2, are known to act only if their corresponding receptors are occupied for a longer period of time in order to provide a continuous signal for progression.

Most of the known growth factors act on G_0 cells and stimulate them to re-enter the cell cycle (see also: Cell activation). Resting 3T3 fibroblast cells, for example, become competent for entering the G_1 phase following their stimulation by » PDGF. Withdrawal of » IL3 from the factor-dependent cell lines such as » 32D or » FDC-P1 arrests the cells in the G_0/G_1 phase of the cell cycle and cell growth is synchronized upon renewed addition of IL3. Some other progression factors such as IGF-1 (see: IGF) and also » EGF are required as progression factors to reach the S phase. The proliferation of macrophages requires the presence of » M-CSF during the entire length of the G_1 phase to guarantee cell survival and entry into the S phase. Many cells, including a variety of tumor cell types, are known to produce » autocrine growth factors that keep them in a cycling mode. A variety of negative stimuli are known to keep cells such as » hematopoietic stem cells in G_0 (see, for example: AcSDKP, SCI, stem cell inhibitor) while others, for example » IL11, specifically shorten the G_0 period of the cell cycle of stem cells.

Information about cell cycle stages, cycle times, and their regulation has been acquired by employing a variety of methods, including treatment with diverse cell cycle inhibitors (e. g. 5-azacytidine, cycloheximide, puromycin, doxorubicin, 5-fluorouracil, hydroxyurea), use of (temperature-sensitive) cell mutants that are blocked at specific points in the cell cycle. Special cloning strategies such as subtractive hybridisation allow isolation of genes specifically expressed or induced at different stages of the cell cycle (see, for example: GOS, G_o/G_1 switch genes). If cells are labeled with radioactive DNA precursors (tritiated thymidine) for a short period of time immediate autoradiographic analysis of fixed cells can be used to measure the fraction of cells in S phase. Phase cycle times and phase durations can be determined by labeling cells for a short period and by measuring the number of labeled mitotic cells at different intervals and comparing these with the total number of mitotic cells observed in a sample (percent labeled mitoses). By comparing the number of mitotic cells with the total number of cells in a sample it is possible to obtain the mitotic index. High mitotic indices indicate short cell cycle times (rapid cell proliferation) and the presence of a large fraction of cycling cells in a population. Flow cytometry employing fluorescent DNA-specific stains can be used to measure DNA contents. The resultant histograms show G_1 and G_2 + M cells as discrete populations with G_2 + M cells having twice the DNA content of the G_1 population. Cells in S phase form a broad distribution and the technique can also be used to discriminate G_0 cells from cells in G_1. This type of analysis is particularly useful to gain immediate information about the effects of cytotoxic drug or cytokine treatment on tumor cell populations.

The more or less exact knowledge of processes regulating the cell cycle has provided the prerequisites for a type of tumor therapy known as cell cycle therapy. It is essentially based on the treatment of resting cells with cytokines to induce their entry into the S phase thus effectively rendering them vulnerable to the action of S phase-specific drugs. In addition, this knowledge can also be used to design treatment regimens that specifically protect non-cycling stem cells.

Bravo R Genes induced during the G0/G1 transition in mouse fibroblasts. Semin. Cancer Biol. 1: 37-46 (1990); **Cross F et al** Simple and complex cell cycles. ARC 5: 341-395 (1989); **Dasso M et al** DNA replication and progression through the cell cycle. Cibas Found. Symp. 170: 180-6 (1992); **Denhardt DT et al** Gene expression during the mammalian cell cycle. BBA 865: 83-125 (1986); **Draetta G** Cell cycle control in eukaryotes: molecular mechanisms of cdc2 activation. TIBS 15: 378-83 (1990); **Ducommun B** From growth to cell cycle control. Semin. Cell. Biol. 2: 233-41 (1991); **Freeman RS & Donoghue DJ** Protein kinases and proto-oncogenes: biochemical regulators of the eukaryotic cell cycle. B 30: 2293-2302 (1991); **Hanley-Hide J** Cyclins in the cell cycle: an overview. CTMI 182: 461-6 (1992); **Hunt T** Cyclins and their partners: from a simple idea to complicated reality. Semin. Cell. Biol. 2: 213-22 (1991); **Kimchi A** Cytokine triggered molecular pathways that control cell cycle arrest. JCBc 50: 1-9 (1992); **Kirschner M** The cell cycle then and now. TIBS 17: 281-5 (1992); **Lewin B** Driving the cell cycle: M phase kinase, its partners, and substrates. Cell 61: 743-52 (1990); **Maller JL** *Xenopus* oocytes and the biochemistry of cell division. B 29: 3157-66 (1990); **Masui Y** Towards understanding the control of the division cycle in animal cells. Biochem. Cell. Biol. 70: 920-45 (1992); **Motokura T & Arnold A** Cyclins and oncogenesis. BBA 1155: 63-78 (1993); **Müller R et al** Signals and genes in the control of cell cycle progression. BBA 1155: 151-79 (1993); **Nasmyth KA** FAR-reaching discoveries about the regulation of START. Cell 63: 1117-20 (1990); **Nigg EA** Targets of cyclin-dependent protein kinases. Curr. Opin. Cell Biol. 5: 187-93 (1993); **Nigg EA** Cellular substrates of p34cdc2 and its companion cyclin-dependent kinases. Trends Cell Biol. 3: 296: 301 (1993); **Pagano M & Draetta G** Cyclin A, cell cycle control and oncogenesis. PGFR 3: 267-77 (1991); **Pardee AB** G1 events and regulation of cell proliferation. S 246: 603-8 (1989); **Pines J** Cyclins and cyclin-dependent kinases: take your partners. TIBS 18: 195-7 (1993); **Rittling SR & Baserga R** Regulatory mechanisms in the expression of cell cycle dependent genes. ACR 7: 541-52 (1987); **Schneider C et al** Genes specifically expressed at growth arrest of mammalian cells. Cell 54: 787-93 (1988); **Seuwen K & Pouyssegur J** G protein-controlled signal transduction pathways and the regulation of cell proliferation. ACR 58: 75-94 (1992); **Sherr CJ** Mammalian G1 cyclins. Cell 73: 1059-65 (1993); **Solomon MJ** Activation of the various cyclin/cdc2 protein kinases. Curr. Opin. Cell Biol. 5: 180-6 (1993); **Tam SW & Schlegel R** Staurosporine overrides checkpoints for mitotic onset in BHK cells. Cell Growth Differ. 3: 811-7 (1992); **Weinberg RA** The retinoblastoma gene and cell growth control. TIBS 15: 199-202 (1990)

Cell growth-inhibitory factor: see addendum: CGIF.

Cell cycle therapy: see: Cell cycle.

Cell lines: see: individual entries. Cell lines (see also: Cell culture) have been instrumental in elucidating the bioactivities of » cytokines (see: Factor-dependent cell lines). For selected cell lines used in cytokine research see: subentry "Assays" for each cytokine.

Cell scattering factors: see: Motogenic cytokines.

Cell passage: see: Cell culture.

Cellular adhesion molecule regulatory factor: see: CAM-RF.

Cellular oncogenes: see: Oncogenes.

Cementum-derived growth factor: see: CGF.

Cementum mitogenic factor: This factor is identical with » bFGF. See also: CGF (cementum-derived growth factor)

Nakae H et al Isolation and partial characterization of mitogenic factors from cementum. B 30: 7047-52 (1991)

CEMF: *corneal endothelium modulation factor* This poorly characterized factor is secreted by polymorphonuclear leukocytes. It affects a population of rabbit corneal endothelial cells with polygonal cell shape by modulating them to fibroblast-like cells and by further stimulating their growth, leading to the formation of colonies of multilayered modulated cells. It also induces the synthesis of type I collagen (fibrillar collagen) and leads to a dramatically reduced amount of type IV collagen (basement membrane collagen) in the endothelial cells.

CEMF significantly increases the production of bFGF in corneal endothelial cells. The same effects are observed after treatment of corneal endothelial cells by »bFGF, but the modulating effects of bFGF are further augmented by CEMF.

Kay EP et al Corneal endothelium modulation factor released by polymorphonuclear leukocytes. Partial purification and initial characterization. Invest-Ophthalmol. Vis. Sci. 31: 313-22

(1990); **Kay EP et al** Corneal endothelial modulation: a factor released by leukocytes induces basic fibroblast growth factor that modulates cell shape and collagen. Invest-Ophthalmol. Vis. Sci. 34: 663-72 (1993)

CESS: A human lymphoblastoid CD43-positive (leukocyte sialoglycoprotein sialophorin) B cell line established from peripheral blood cells of a patient with myelomonocytic leukemia and immortalized by transformation with Epstein Barr virus. It is used to assay (see also: Bioassays) B cell differentiation factors (see: BCDF) and T cell replacing factors (see: TRF). The activities of these factors are determined by measuring the induction of the synthesis of immunoglobulins which is only observed after the stimulation of the cells with one of these factors.

The cells constitutively secrete » IL6 and secrete IgG in response to exogenous IL6 without any requirement for proliferation. This response is also used as a bioassay of IL6. Histamine increases the IL6 binding by CESS cells. IL6 also increases the phosphorylation of CD40 in CESS cells. The cells do not respond to B cell growth factors (see: BCGF). CESS cells also produce » IL12. Proliferation of these cells is inhibited by treatment with antibodies directed against insulin or insulin receptor.

Baier TG et al Influence of antibodies against IGF-I, insulin or their receptors on proliferation of human acute lymphoblastic leukemia cell lines. Leuk. Res. 16: 807-14 (1992); **Bradley TR et al** Cell lines derived from a human myelomonocytic leukemia. Br. J. Haematol. 5: 595-604 (1982); **Callard RE** The marmoset B lymphoblastoid cell line (B95-8) produces and responds to B cell growth and differentiation factors: role of shed CD23 (sCD23). Immunology 65: 379-84 (1988); **Clark EA & Shu G** Association between IL6 and CD40 signaling. IL6 induces phosphorylation of CD40 receptors. JI 145: 1400-6 (1990); **Hutchins D et al** Production and regulation of interleukin 6 in human B lymphoid cells. EJI 20: 961-8 (1990); **Kikutani H et al** Effect of B cell differentiation factor (BCDF) on biosynthesis and secretion of immunoglobulin molecules in human B cell lines. JI 134: 990-5 (1985); **Meretey K et al** Histamine influences the expression of the interleukin-6 receptor on human lymphoid, monocytoid and hepatoma cell lines. Agents Actions 33: 189-91 (1991); **Muraguchi A et al** T cell replacing factor-induced IgG secretion in a human B blastoid cell line and demonstration of acceptors for TRF. JI 127: 412-6 (1981); **Peest D et al** Suppression of polyclonal B cell proliferation mediated by supernatants from human myeloma bone marrow cell cultures. Clin. Exp. Immunol. 75: 252-7 (1989); **Poupart P et al** B cell growth modulating and differentiating activity of recombinant human 26-kd protein (BSF-2, HuIFN-β 2, HPGF). EJ 6: 1219-24 (1987); **Raynal MC et al** Interleukin 6 induces secretion of IgG1 by coordinated transcriptional activation and differential mRNA accumulation. PNAS 86: 8024-8 (1989); **Shields JG et al** The response of selected human B cell lines to B cell growth and differentiation factors. EJI 17: 535-40 (1987); **Stevens C et al** The effects of immunosuppressive agents on *in vitro* production of human immunoglobulins. Transplantation 51: 1240-4 (1992); **Valiante NM et al** Role of the production of natural killer cell stimulatory factor (NKSF/IL12) in the ability of B cell lines to stimulate T and NK cell proliferation. CI 145: 187-98 (1992)

CETAF: *corneal epithelial cell-derived thymocyte activating factor* The term CETAF is more or less an operational definition of biological activities identified in thymocyte co-stimulation assays, which greatly stimulates the growth of murine thymocytes. The main CETAF activity appears to be due to » IL1.

Grabner G et al Corneal epithelial cell-derived thymocyte-activating factor (CETAF). Invest. Ophthalmol. Vis. Sci. 23: 757-63 (1982); **Grabner G et al** Biological properties of the thymocyte-activating factor (CETAF) produced by a rabbit cell line (SIRC). Invest. Ophthalmol. Vis. Sci. 24: 589-95 (1983); **Luger TA et al** A corneal cytokine triggers the *in vivo* synthesis of serum amyloid A by hepatocytes. Ophthalmic Res. 15: 121-5 (1983)

CF: *chemotactic factor* see: Chemotaxis. CF also appears in abbreviations of other factors, e. g. LDCF (leukocyte-derived chemotactic factor). (b) cytostatic factor.

CF: *cytostatic factor* This poorly characterized factor of 50 kDa is purified from the conditioned medium of the LPS-treated murine myelomonocytic leukemia cell line » WEHI-3. CF is a potent reversible growth inhibitor for » CHO cells. It does not inhibit murine » L929 cells. CF also inhibits Con A-induced mitogenesis in mouse and rat spleen cells.

Du XT et al A cytostatic protein isolated from the conditioned medium of mouse monocytic leukemia WEHI-3 cell cultures. Immunol. Lett. 23: 119-24 (1989)

CF: *cytotoxic factor* This poorly characterized factor is released from rat basophilic leukemia cells sensitized with anti-ovalbumin mouse serum after incubation with anti-ovalbumin. It is identical with » TNF-α.

Ohno et al The production of tumor necrosis factor-α by rat basophilic leukemia cells with triggering IgE receptor. Tohoku J. Exp. Med. 156: 209-10 (1988)

CFC: *colony forming cells* A general term referring to bone marrow stem cells which, unlike pluripotent stem cells of the marrow, can be shown to be committed in their development to a particular lineage. These cells can give rise to individual colonies when grown in soft agar (see also: colony formation assay; see also: Limiting dilution analysis) following treatment with colony-stimulating factors (see: CSF). Growth and final differentiation of these progenitor cells depend on the continuous presence of these CSFs and is not seen in their absence. The particular type of differentiated cells to which the CFCs are committed is usually indicated by a suffix. ***CFC-Eo*** is committed to the development into erythrocytes. ***CFC-G*** is committed to the development into granulocytes. ***CFC-GM*** is committed to the development into granulocytes and macrophages. ***CFC-M*** is committed to the development into macrophages. The term » ***CFC-Mix*** refers to a human multipotential hematopoietic progenitor cell type which gives rise to colonies containing all types of mature myeloid cell types. ***HPP-CFC*** (high proliferative potential colony-forming cells) give rise to large colonies in soft agar cultures; they are not considered to be a measure of true pluripotential stem cells but, rather, of their slightly more differentiated progeny (see LTBMC (long-term bone marrow culture).

see: references for » Colony formation assay.

CFC-Eo: *eosinophil colony forming cells* see: CFC.

CFC-G: *granulocyte colony forming cells* see: CFC.

CFC-GM: *granulocyte-macrophage colony forming cells* see: CFC.

CFC-M: *macrophage colony forming cells* see: CFC.

CFC-Mix: (colony-forming cells, mixed) The term refers to a human multipotential hematopoietic progenitor cell type which gives rise to colonies containing all types of mature myeloid cell types, including erythrocytes, neutrophils, macrophages, eosinophils, mast cells, megakaryocytes, when grown in soft agar (see also: colony formation assay). This cell type is more or less ill defined because some of the more mature forms may be missing in individual colonies. However, comparison with similar cell types in mice demonstrate that there is an overlap between CFC-Mix and » CFU-S in mice and that measuring CFC-Mix may be a suitable system for the analysis of human multipotential stem cells (see LTBMC (long-term bone marrow culture).

Johnson GR & Metcalf D Pure and mixed erythroid colony formation *in vitro* stimulated by spleen conditioned medium with no detectable erythropoietin. PNAS 74: 3879-82 (1977); **Johnson GR** Colony formation in agar by adult bone marrow multipotential hemopoietic cells. JCP 103: 371-83 (1980); **Hara H & Ogawa M** Murine hemopoietic colonies in culture containing normoblasts, macrophages, and megakaryocytes. Am. J. Hematol. 4: 23-4 (1978); **Humphries, RK et al** Characterization of a primitive erythropoietic progenitor found in mouse marrow before and after several weeks in culture. Blood 53: 746-63 (1979); **Humphries RK et al** CFU-S in individual erythroid colonies *in vitro* derived from adult mouse marrow. N 279: 718-20 (1979); **Metcalf D** Hemopoietic colonies: *in vitro* cloning of normal and leukaemic cells. Recent Res. Cancer Res. 61: 1-227 (1977); **Metcalf D et al** Colony formation in agar by multipotential hemopoietic cells JCP 98: 401-12 (1979); **Humphries RK et al** Self-renewal of hemopoietic stem cells during mixed colony formation *in vitro*. PNAS 78: 3629-33 (1981)

CFCS: *cell-free culture supernatants* see: CM (conditioned medium).

CFS: used sometimes as a suffix to denote cell-free supernatant, as in U937-CFS. See also: CM (conditioned medium).

CFU: *colony forming unit in vitro* clonable colony-forming cells. This term was coined at a time when a granulocytic cell type was the only one that could be grown in semisolid media (see: Colony formation assay). They are the same cell types described by other authors as » CFU-GM. It was found subsequently that other cells types can also be grown in semisolid culture media. CFU is now used as a general term to describe distinct precursor cell population types in the bone marrow. The major cell type observed in the colonies obtained is added as a suffix. For a description of individual types see: CFU-Bas, CFU-E, CFU-Eo, CFU-G, CFU-GEMM, CFU-GM, CFU-M, CFU-MEG. See also: Hematopoiesis.

Metcalf D Hemopoietic colony stimulating factors. In: Handbook of Experimental Pharmacology Vol. 57: 343-84 (1981)

CFU-A: (colony forming units) These cells are detected in an assay system using conditioned medium from a rat fibroblast line (AD-1-19T, which produces a cocktail of cytokines, mainly » GM-CSF, M-CSF, » SCF, small amounts of » IL1, » IL6, but no » IL3. Normal bone marrow cells give rise to large colonies in the presence of the cytokine cocktail with high cloning efficiency. The CFU-A assay detects multipotent » hematopoietic stem cells that are also detected by the *in vivo* spleen colony formation assay (see: CFU-S). The clonogenic CFU-A cells also closely resemble the most primitive type of » HPP-CFC. CFU-A cell growth is inhibited by » TGF-β and also by » SCI (stem cell inhibitor) and this activity has been used to monitor purification of the latter. CFU-A survive treatment with » Fluorouracil, which eliminates later precursors.

Maltman J et al Transforming growth factor β: is it a downregulator of stem cell inhibition by macrophage inflammatory protein 1 α? JEM 178: 925-32 (1993); **Quesniaux YF et al** Use of 5-fluorouracil to analyze the effect of macrophage inflammatory protein-1 α on long-term reconstituting stem cells *in vivo*. Blood 81: 1497-504 (1993); **Wright EG & Pragnell IB** The stem cell compartment: assays and negative regulators. CTMI 177: 136-49 (1992)

CFU-C: *colony forming unit (in culture)* A term that refers to cells obtained in an *in vitro* culture system (see: Colony formation assays) allowing clonal growth of early » hematopoietic stem cells. CFU-C are granulocytic precursors and do not represent the earliest hematopoietic stem cells. This term was coined at a time when this cell type was the only one that could be grown in culture. They are the same cell types described by other authors as » CFU-GM.

Bradley TR & Metcalf D The growth of mouse bone marrow cells *in vitro*. Aust. Exp. Biol. Med. Sci. 44: 287 (1966); **Kirshner JJ & Goldberg J** Types of colonies formed by normal human bone marrow, peripheral blood and umbilical cord blood CFUc. EH 8: 1202-7 (1980); **Mertelsmann R et al** Diagnostic and prognostic significance of the CFU-c assay in acute nonlymphoblastic leukemia. CR 41: 4844-8 (1981); **Pike BL & Robinson BA** Human bone marrow colony growth in agar gel. JCP 76: 77 (1970); **Pluznik DH & Sachs L** The induction of clones of normal mast cells by a substance from conditioned medium. ECR 43: 553 (1966); **Yourmo J et al** Rapid *in situ* cytochemical classification of CFU-C colonies in soft agar culture: permanent whole culture preparations. J. Histochem. Cytochem. 28: 991-6 (1980)

CFU-E: *colony forming unit, erythroid* Term used to describe very early erythroid precursor cells that eventually differentiate into erythrocytes (see also: Hematopoiesis). CFU-E which arise from » BFU-E can be assayed and identified in a » colony formation assay by the specific morphology of the cells. The cells are characterized by expressing the cell surface marker CD36.

Of all erythroid progenitor cells CFU-E are the ones expressing the highest amounts of receptors for » Epo and their growth is absolutely dependent on the presence of Epo.

Treatment of the cells with Epo leads to several rounds of cell division eventually culminating in terminal differentiation and the loss of the capacity to undergo cell division. The growth of CFU-E is enhanced by IGF-1 (see: IGF) and inhibited by EDF (eosinophil differentiation factor = » IL5) which induces differentiation. Another inhibitor which inhibits the capacity of CFU-E to form colonies in soft agar is » TNF-α. CFU-E also responds to » E-CSF (erythroid colony-stimulating factor). Human CFU-E is also inhibited by the tetrapeptide » AcSDKP but not by the pentapeptide » pEE-DCK. Retinoic acid suppresses CFU-E colony formation. It has been shown that » Activin A acts as a commitment factor and/or a promoter of CFU-E.

Boyer SH et al Roles of erythropoietin, insulin-like growth factor 1, and unidentified serum factors in promoting maturation of purified murine erythroid colony-forming units. Blood. 1992 Nov 15; 80: 2503-12 (1992); **Cotton EW et al** Quantitation of insulin-like growth factor-1 binding to highly purified human erythroid colony-forming units. EH 19: 278-81 (1991); **Eaves CJ et al** *In vitro* characterization of erythroid precursor cells and the erythropoietic differentiation process. In: Stamatoyanopoulos G & Nienhuis AW (eds) Cellular and molecular regulation of hemoglobin switching. Grune and Stratton, New York, pp 251-73 (1979); **Gratas C et al** Retinoid acid supports granulocytic but not erythroid differentiation of myeloid progenitors in normal bone marrow cells. Leukemia 7: 1156-62 (1993); **Heath DS et al** Separation of the erythropoietin-responsive progenitors BFU-E and CFU-E in mouse bone marrow by unit gravity sedimentation. Blood 47: 777-92 (1976); **Iscove VW & Sieber F** Erythroid progenitors in mouse bone marrow detected by macroscopic colony formation in culture. EH 3: 32-43 (1975); **Iscove NN et al** Complete replacement of serum in primary cultures of erythropoietin-dependent red cell precursors (CFU-E) by albumin, transferrin, iron, unsaturated fatty acid, lecithin and cholesterol. ECR 126: 121-6 (1980); **Means RT et al** Inhibition of human colony-forming-unit erythroid by tumor necrosis factor requires accessory cells. JCI 86: 538-41 (1990); **Merchav S et al** Comparative studies of the erythroid-potentiating effects of biosynthetic human insulin-like growth factors-I and -II. J. Clin. Endocrinol. Metab. 74: 447-52 (1992); **Nakamura K et al** Effect of erythroid differentiation factor on maintenance of human hematopoietic cells in co-cultures with allogenic stromal cells. BBRC 194: 1103-10 (1993); **Nakao K et al** Effects of erythroid differentiation factor (EDF) on proliferation and differentiation of human hematopoietic progenitors. EH 19: 1090-5 (1991); **Ottmann OG & Pelus LM** Differential proliferative effects of transforming growth factor β on human hematopoietic progenitor cells. JI 140: 26561-65 (1988); **Sakata S & Enoki Y** Improved microbioassay for plasma erythropoietin based on CFU-E colony formation. Ann. Hematol. 64: 224-30 (1992); **Sawada K et al** Human colony-forming units-erythroid do not require accessory cells, but do require direct interaction with insulin-like growth factor I and/or insulin for erythroid development. JCI 83: 1701-9 (1989); **Shao L et al** Regulation of production of activin A in human marrow stromal cells and monocytes. EH 20: 1235-42 (1992); **Shibuya T & Mak TW** Isolation and induction of erythroleukemic cell lines with properties of erythroid progenitor burst-forming cell (BFU-E) and erythroid precursor cell (CFU-E). PNAS 80: 3721-5 (1983); **Spivak JL** The mechanism of action of erythropoietin. IJCC 4: 139-66 (1986)

CFU-Eo: *colony forming unit eosinophil* Term used to describe a particular developmental type of blood-forming cell. CFU-Eo can be identified in a » colony formation assay by the specific morphology of the cells. The cells are characterized by expressing the cell surface markers CD13, » CD33, » CD34 (see also: CD antigens) and HLA-

DR. See also: Hematopoiesis, LTBMC (long-term bone marrow culture).

Eosinophil colony stimulation is stimulated by » IL5 (see also: EDF, eosinophil differentiation factor). Stimulation of eosinophil colony formation is also stimulated by » IL7 (synergising with » IL3 and » GM-CSF); this supportive effect on eosinophil precursors is mediated by the endogenous release of IL5. Eosinophilia observed in patients undergoing therapy with » IL2 also appears to be due to the induction of IL5.

Bjornson BH et al *In vitro* culture of circulating CFUEOS from normal donors. EH 10: 271-6 (1982); **Bot FJ et al** Stimulating spectrum of human recombinant multi-CSF (IL3) on human marrow precursors: importance of accessory cells. Blood 71: 1609-14 (1988); **Clutterbuck EJ & Sanderson CJ** Regulation of human eosinophil precursor production by cytokines: a comparison of recombinant human interleukin-1 (rhIL1), rhIL3, rhIL5, rhIL6, and rh granulocyte-macrophage colony-stimulating factor. Blood 75: 1774-9 (1990); **Enokihara H et al** Effect of human recombinant interleukin 5 and G-CSF on eosinophil colony formation. Immunol. Lett. 18: 73-6 (1988); **Macdonald D et al** Interleukin-2 treatment-associated eosinophilia is mediated by interleukin-5 production. Br. J. Haematol. 76: 168-73 (1990); **Schrezenmeier H et al** Interleukin-5 is the predominant eosinophilopoietin produced by cloned T lymphocytes in hypereosinophilic syndrome. EH 21: 358-65 (1993); **Vellenga E et al** The supportive effects of IL7 on eosinophil progenitors from human bone marrow cells can be blocked by anti-IL5. JI 149: 2992-5 (1992); **Warren DJ & Moore MA** Synergism among interleukin 1, interleukin 3, and interleukin 5 in the production of eosinophils from primitive hemopoietic stem cells. JI 140: 94-9 (1988)

CFU-Eo GSF: *colony forming unit eosinophil growth stimulating factor* This T cell-derived factor was found in the conditioned media prepared from E rosette forming cells, treated with » IL2, of patients with reactive eosinophilia. The factor is identical with » IL5.

Enokihara H et al Interleukin 2 stimulates the T cells from patients with eosinophilia to produce CFU-Eo growth stimulating factor. Br. J. Haematol. 69: 431-6 (1988); **Enokihara H et al** Effect of human recombinant interleukin 5 and G-CSF on eosinophil colony formation. Immunol. Lett. 18: 73-6 (1988); **Enokihara H et al** T cells from eosinophilic patients produce interleukin-5 with interleukin-2 stimulation. Blood 73: 1809-13 (1989)

CFU-G: *colony forming unit granulocyte* Term used to describe a particular developmental type of blood-forming cell. CFU-G can be identified in a » colony formation assay by the specific morphology of the cells. The cells are characterized by expressing the cell surface markers CD13, CD15, » CD33 (see also: CD antigens) and HLA-DR. See also: Hematopoiesis, LTBMC (long-term bone marrow culture). » GM-CSF strongly synergizes with » G-CSF in the formation of granulocytic colonies with respect to number and size and enhances the *in vitro* survival of CFU-G. » IL3 does not stimulate granulocytic colony formation by itself but appears to regulate the survival and proliferative rate of granulocytic progenitors. G-CSF-stimulated growth of CFU-G is further stimulated by » IL4. G-CSF-stimulated growth of CFU-G is inhibited by » IFN-γ, and this effect is counteracted by IL4. IL4 inhibits colony formation induced by other colony-stimulating factors. Retinoic acid has been shown to stimulate CFU-G. Unstimulated lymphocytes co-cultured with normal bone marrow cells release a high molecular weight glycoprotein that inhibits day 7 CFU-G in culture. Lymphocytes treated with » IL2 produce and release » TNF-α and » IFN-γ, which suppress CFU-G.

Sera obtained from patients with acute leukemia in remission undergoing intensive chemotherapy or sera of patients with malignant lymphomas in the recovery phase contain a factor, designated ***CFU-G inhibitory factor***, which completely inhibits the growth of allogenous and autologous bone marrow cells *in vitro*. This poorly characterized factor inhibits the growth of cells responding to » G-CSF. The growth of other hematopoietic cell types such as » CFU-E, » BFU-E, or cells responding to » GM-CSF is not influenced by this activity.

Bot FJ et al Human granulocyte-macrophage colony-stimulating factor (GM-CSF) stimulates immature marrow precursors but no CFU-GM, CFU-G, or CFU-M. EH 17: 292-5 (1989); **Bot FJ et al** Synergistic effects between GM-CSF and G-CSF or M-CSF on highly enriched human marrow progenitor cells. Leukemia 4: 325-8 (1990); **Bot FJ et al** Effects of human interleukin-3 on granulocytic colony-forming cells in human bone marrow. Blood 73: 1157-60 (1989); **Iizuka Y et al** The influence of the supernatant obtained from a culture of IL2-activated lymphocytes on human hematopoietic progenitors. EH 18: 119-23 (1990); **Pantel K & Nakeff A** The role of lymphoid cells in hematopoietic regulation. EH 21: 738-42 (1993); **Snoeck HW et al** Differential regulation of the expression of CD38 and human leukocyte antigen-DR on $CD34^+$ hematopoietic progenitor cells by interleukin-4 and interferon-γ. EH 21: 1480-6 (1993); **Snoeck HW et al** Effects of interleukin-4 on myelopoiesis: localization of the action of IL4 in the $CD34^+HLA\text{-}DR^{++}$ subset and distinction between direct and indirect effects of IL4. EH 21: 635-9 (1993); **Snoeck HW et al** Interferon-γ and interleukin-4 reciprocally regulate the production of monocytes/macrophages and neutrophils through a direct effect on committed monopotential bone marrow progenitor cells. EJI 23: 1072-7 (1993); **Snoeck HW et al** Effects of interleukin-4 (IL4) on myelopoiesis: studies on highly purified $CD34^+$ hematopoietic progenitor cells. Leukemia 7: 625-9 (1993); **Urase F et al** Inhibition of CFU-G formation by human serum during granulopoiesis after chemotherapy: purification of CFU-G inhibitory factor and its activities. Nat. Immun. Cell Growth Regul. 10: 94-104 (1991); **van Bockstaele DR** Direct effects of 13-cis and all-trans retinoic acid on normal bone marrow (BM) progenitors: comparative study on BM mononuclear cells and on isolated $CD34^+$ BM cells. Ann. Hematol. 66: 61-6 (1993)

CFU-G inhibitory factor: see: CFU-G (colony forming unit granulocyte).

CFU-GEMM: *colony forming unit granulocyte, erythrocyte, monocyte, macrophage* Term used to describe a pluripotent precursor cell type in the lineage of blood-forming cells (see also: Hematopoiesis) which is capable to differentiate into these particular fully differentiated cell types. CFU-GEMM can be identified in a » colony formation assay (see also: LTBMC, long-term bone marrow culture) by the specific morphology of the cells. The cells are characterized by expressing the cell surface markers » CD33, » CD34 (see also: CD antigens) and HLA-DR. IL3 and GM-CSF as single factors are equally active in stimulating CFU-GEMM, but the combination of both factors produces additive stimulatory effects upon CFU-GEMM. CFU-GEMM also responds to » uteroferrin. The growth of CFU-GEMM is stimulated by » SCF. SCF has been found also to synergize with » GM-CSF, » IL6, » IL3, » IL11 or » Epo to increase the numbers of CFU-GEMM. The growth of CFU-GEMM is inhibited by » TGF-β and » Inhibin. It has been shown that CFU-GEMM is relatively resistant to treatment with » 4-HC (4-hydroperoxycyclophosphamide). Colony formation of CFU-GEMM stimulated by GM-CSF plus SCF is suppressed by several » chemokines, including » MIP-1α, MIP-2α, » PF4, » IL8, and » MCAF. MIP-1β blocks the suppressive effects of MIP-1α. MIP-2β or GRO-α block the suppressive effects of IL8 and PF4.

The cell surface antigen CD45, which encodes a protein tyrosine phosphatase, has been implicated in the stimulation of early human myeloid progenitor cells by various growth factors. Treatment of cells with various » antisense oligodeoxynucleotides the CD45 gene, or with monoclonal anti-CD45, significantly decreases GM-CSF-, IL3-, and SCF-enhanced Epo-dependent CFU-GEMM colony formation, but had little or no effect on CFU-GEMM colony formation stimulated by Epo alone.

Ash RC et al Studies of human pluripotential hemopoietic stem cells (CFU-GEMM) *in vitro*. Blood 58: 309-16 (1981); **Bot FJ et al** Stimulating spectrum of human recombinant multi-CSF (IL3) on human marrow precursors: importance of accessory cells. Blood 71: 1609-14 (1988); **Bot FJ et al** Human granulocyte-macrophage colony-stimulating factor (GM-CSF) stimulates immature marrow precursors but no CFU-GM, CFU-G, or CFU-M. EH 17: 292-5 (1989); **Broxmeyer HE et al** CD45 cell surface antigens are linked to stimulation of early human myeloid progenitor cells by interleukin 3 (IL3), granulocyte/macrophage colony-stimulating factor (GM-CSF), a GM-CSF/IL3 fusion protein, and mast cell growth factor (a c-*kit* ligand). JEM 174: 447-58 (1991); **Broxmeyer HE et al** Growth characteristics and expansion of human umbilical cord blood and estimation of its potential for transplantation in adults. PNAS 89: 4109-13 (1992); **Broxmeyer HE et al** Comparative analysis of the human macrophage inflammatory protein family of cytokines (chemokines) on proliferation of human myeloid progenitor cells. Interacting effects involving suppression, synergistic suppression, and blocking of suppression. JI 150: 3448-58 (1993); **Erickson-Miller CL et al** Megakaryocytopoiesis and platelet production: does stem cell factor play a role? Stem Cells Dayt. 11: s163-9 (1993); **Fauser AA & Messner HA** Proliferative state of human pluripotent hemopoietic progenitors (CFU-GEMM) in normal individuals and under regenerative conditions after bone marrow transplantation. Blood 54: 1197-1200 (1979); **Hangoc G et al** Effects *in vivo* of recombinant human inhibin on myelopoiesis in mice. EH 20: 1243-6 (1992); **Hangoc G et al** *In vivo* effects of recombinant interleukin-11 on myelopoiesis in mice. Blood 81: 965-72 (1993); **Kanz L et al** Identification of human megakaryocytes derived from pure megakaryocytic colonies (CFU-M), megakaryocytic-erythroid colonies (CFU-M/E), and mixed hemopoietic colonies (CFU-GEMM) by antibodies against platelet associated antigens. Blut 45: 267-74 (1982); **Keller JR et al** Transforming growth factor β: possible roles in the regulation of normal and leukemic hematopoietic cell growth. JCBc 39: 175-84 (1989); **Keller JR et al** Transforming growth factor β directly regulates primitive murine hematopoietic cell proliferation. Blood 75: 596-602 (1990); **Li ML et al** Co-stimulatory effects of steel factor, the c-kit ligand, on purified human hematopoietic progenitors in low cell density culture. Nouv. Rev. Fr. Hematol. 35: 81-6 (1993); **Messner HA** The role of CFU-GEMM in human hemopoiesis. Blut 53: 259-77 (1986); **Ottmann OG & Pelus LM** Differential proliferative effects of transforming growth factor β on human hematopoietic progenitor cells. JI 140: 26561-5 (1988); **Pantel K & Nakeff A** The role of lymphoid cells in hematopoietic regulation. EH 21: 738-42 (1993); **Testa U et al** Cascade transactivation of growth factor receptors in early human hematopoiesis. Blood 81: 1442-56 (1993); **Tong J et al** *In vivo* administration of recombinant methionyl human stem cell factor expands the number of human marrow hematopoietic stem cells. Blood 82: 784-91 (1993)

CFU-GM: *colony forming unit granulocyte, macrophage* Term used to describe a pluripotent precursor cell type in the lineage of blood-forming cells. CFU-GM can be identified in a » colony formation assay by the specific morphology of the cells. The cells are characterized by expressing the cell surface markers CD13, » CD33, » CD34 (see also: CD antigens) and HLA-DR. See also: Hematopoiesis, LTBMC (long-term bone marrow culture). Assays to detect CFU-GM are employed to provide information about the presence of » hematopoietic stem cells in bone marrow used for transplantation.

A reduced production of CFU-GM colonies *in vitro* or even a complete lack is observed in patients with myelodysplastic syndromes (MDS). IL3 appears to be a priming factor *in vitro* and *in vivo* that renders progenitor cells sensitive for the action of other cytokines. The growth of CFU-GM is inhibited by » TGF-β, » Activin A, and » Inhibin. In the presence of maximal concentrations of colony-stimulating factors (see also: CSF) TGF-β enhances proliferation of CFU-GM while the growth of other progenitor cell types (» CFU-GEMM, » CFU-E, » BFU-E) remains inhibited The survival of CFU-GM is also promoted by » IL1, » IL4, and » IL6. CFU-GM also responds to » IL9, » IL11, » IL12, » uteroferrin and » bFGF. Antibodies against soluble » CD23 have been shown to decrease the total numbers of CFU-GM. CFU-GM differentiation is promoted by retinoic acid. CFU-GM is inhibited by the tetrapeptide » AcSDKP and the pentapeptide » pEEDCK. 5-lipoxygenase metabolites have been shown to be physiologically important in regulating proliferation of CFU-GM, with inhibitors of lipoxygenase activity leading to an increased recovery of CFU-GM in liquid cultures. A subpopulation of CFU-GM has been shown to be stimulated by choline. Colony formation of CFU-GM stimulated by GM-CSF plus SCF is suppressed by several » chemokines, including » MIP-1α, MIP-2α, » PF4, » IL8, and » MCAF. MIP-1β blocks the suppressive effects of MIP-1α. MIP-2β or GRO-α block the suppressive effects of IL8 and PF4. CFU-GM is inhibited by » TNF-α and » TNF-β.

The cell surface antigen CD45, which encodes a protein tyrosine phosphatase, has been implicated in the stimulation of early human myeloid progenitor cells by various growth factors. Treatment of cells with various » antisense oligodeoxynucleotides the CD45 gene, or with monoclonal anti-CD45, significantly decreases CFU-GM colony formation stimulated with » GM-CSF, » IL3, and GM-CSF + » SCF, but not with » G-CSF or » M-CSF. Human » IL3 and » GM-CSF have been reported to induce the expression of » IL2 receptors on a small population of normal human bone marrow cells immunophenotypically related to CFU-GM.

Aglietta M et al Interleukin-3 *in vivo*: kinetic of response of target cells. Blood 82: 2054-61 (1993); **Bot FJ et al** Stimulating spectrum of human recombinant multi-CSF (IL3) on human marrow precursors: importance of accessory cells. Blood 71: 1609-14 (1988); **Bot FJ et al** Human granulocyte-macrophage colony-stimulating factor (GM-CSF) stimulates immature marrow precursors but no CFU-GM, CFU-G, or CFU-M. EH 17: 292-5 (1989); **Broxmeyer HE et al** Human umbilical cord blood as a potential source of transplantable hematopoietic stem/progenitor cells. PNAS 86: 3828-32 (1989); **Broxmeyer HE et al** CD45 cell surface antigens are linked to stimulation of early human myeloid progenitor cells by interleukin 3 (IL3), granulocyte/macrophage colony-stimulating factor (GM-CSF), a GM-CSF/IL3 fusion protein, and mast cell growth factor (a c-*kit* ligand). JEM 174: 447-58 (1991); **Broxmeyer HE et al** Comparative analysis of the human macrophage inflammatory protein family of cytokines (chemokines) on proliferation of human myeloid progenitor cells. Interacting effects involving suppression, synergistic suppression, and blocking of suppression. JI 150: 3448-58 (1993); **Broxmeyer HE et al** Growth characteristics and expansion of human umbilical cord blood and estimation of its potential for transplantation in adults. PNAS 89: 4109-13 (1992); **Cooper S & Broxmeyer HE** Clonogenic methods *in vitro* for the enumeration of granulocyte-macrophage progenitor cells (CFU-GM) in human bone marrow and mouse bone marrow and spleen. J. Tiss. Cult. Meth. 13: 77-82 (1991); **Dowton LA & Ma DD** A method for enriching myeloid (CFU-GM) and erythroid (BFU-E) progenitor cells from human cord blood by accessory cell depletion. Pathology 24: 291-5 (1992); **Du XX et al** Effects of recombinant human interleukin-11 on hematopoietic reconstitution in transplant mice: acceleration of recovery of peripheral blood neutrophils and platelets. Blood 81: 27-34 (1993); **Estrov HM et al** Suppression of chronic myelogenous leukemia colony growth by interleukin-4. Leukemia 7: 214-20 (1993); **Ferrajoli A et al** Analysis of the effects of tumor necrosis factor inhibitors on human hematopoiesis. Stem Cells Dayt. 11: 112-9 (1993); **Fourcade C et al** Expression of CD23 by human bone marrow stromal cells. Eur-Cytokine-Netw. 3: 539-43 (1992); **Gallicchio YS et al** The influence of choline chloride on murine hematopoiesis *in vivo* and *in vitro*. Int. J. Cell Cloning 1: 451-63 (1983); **Gallicchio YS et al** Basic fibroblast growth factor (B-FGF) induces early- (CFU-s) and late-stage hematopoietic progenitor cell colony formation (CFU-gm, CFU-meg, and BFU-e) by synergizing with GM-CSF, Meg-CSF, and erythropoietin, and is a radioprotective agent *in vitro*. IJCC 9: 220-32 (1991); **Gazzola MV et al** Recombinant interleukin 3 induces interleukin 2 receptor expression on early myeloid cells in normal human bone marrow. EH 20: 201-8 (1992); **Gebbia V et al** Analysis of human dysplastic hematopoiesis in long-term bone marrow culture. Blood 59: 442-8 (1989); **Gerhartz HH & Schmetzer H** Detection of minimal residual disease in acute myeloid leukemia. Leukemia 4: 508-16 (1990); **Gooding RP et al** An assessment of myeloid colony-forming-cell generation in liquid human bone marrow cultures: influence of accessory cells and cytokines IL1 α, β and IL6. Int. J. Exp. Pathol. 74: 173-80 (1993); **Gratas C et al** Retinoid acid supports granulocytic but not erythroid differentiation of myeloid progenitors in normal bone marrow cells. Leukemia 7: 1156-62 (1993); **Haldar C et al** Response of CFU-GM (colony forming units for granulocytes and macrophages) from intact and pinealectomized rat bone marrow to murine recombinant interleukin-3 (rIl-3), recombinant granulocyte-macrophage colony stimulating factor (rGM-CSF) and human recombinant erythropoietin (rEPO). Prog. Brain Res. 91: 323-5 (1992); **Hangoc G et al** Influence of IL1 α and -1 β on the survival of human bone marrow cells responding to hematopoietic colony-stimulating factors. JI 142: 4329-34 (1989); **Hangoc G et al** Effects *in vivo* of recombinant human inhibin on myelopoiesis in mice. EH 20: 1243-6 (1992); **Hangoc G et al** *In vivo* effects of recombinant interleukin-11 on myelopoiesis in mice. Blood 81: 965-72 (1993); **Hestdal K et al** Increased granulopoiesis after sequential administration of transforming growth factor-β1 and granulocyte-macrophage colony-stimulating factor. EH 21: 799-805 (1993); **Janssen WE et al** The CD34[+] cell fraction in bone marrow and blood is not universally predictive of CFU-GM. EH 20: 528-30 (1992); **Janssen WE et al** Use of CD34[+] cell fraction as a measure of hematopoietic stem cells in bone marrow and peripheral blood: comparison with the CFU-GM assay. PCBR 377: 513-21 (1992); **Keller JR et al** Transforming growth factor β: possible roles in the regulation of normal and leukemic hematopoietic cell growth. JCBc 39: 175-84 (1989); **Kerst JM et al** Combined measurement of growth and differentiation in suspension cultures of purified human CD34-positive cells enables a detailed analysis of myelopoiesis. EH 20: 1188-93 (1992); **Kerst JM et al** Interleukin-6 is a survival factor for committed myeloid progenitor cells. EH 21: 1550-7 (1993); **Li ML et al** Co-stimulatory effects of steel factor, the c-kit ligand, on purified human hematopoietic progenitors in low cell density culture. Nouv. Rev. Fr. Hematol. 35: 81-6 (1993); **Lu L et al** Characterization of adult human marrow hematopoietic progenitors highly enriched by two-color sorting with My10 and major histocompatibility (MHC) class II monoclonal antibodies. JI 139: 1823-9 (1987); **Migliaccio AR et al** Direct effects of IL4 on the *in vitro* differentiation and proliferation of hematopoietic progenitor cells. Biotechnol. Ther. 1: 347-60 (1989-90) **Mizuguchi T et al** Activin A suppresses proliferation of interleukin-3-responsive granulocyte-macrophage colony-forming progenitors and stimulates proliferation and differentiation of interleukin-3-responsive erythroid burst-forming progenitors in the peripheral blood. Blood 81: 2891-7 (1993) **Ottmann OG & Pelus LM** Differential proliferative effects of transforming growth factor β on human hematopoietic progenitor cells. JI 140: 26561-5 (1988); **Pantel K & Nakeff A** The role of lymphoid cells in hematopoietic regulation. EH 21: 738-42 (1993); **Pasquale D & Chikkappa G** Lipoxygenase products regulate proliferation of granulocyte-macrophage progenitors. EH 21: 1361-5 (1993); **Piacibello W et al** Responsiveness of highly enriched CFU-GM subpopulations from bone marrow, peripheral Blood and cord blood to hemopoietic growth inhibitors. EH 19: 1084-9 (1991); **Ploemacher RE et al** Autocrine transforming growth factor β 1 blocks colony formation and progenitor cell generation by hemopoietic stem cells stimulated with steel factor. Stem Cells Dayt. 11: 336-47 (1993); **Schaafsma MR et al** Interleukin-9 stimulates the proliferation of enriched human erythroid progenitor cells: additive effect with GM-CSF. Ann. Hematol. 66: 45-9 (1993); **Serke S et al** Analysis of CD34-positive hemopoietic progenitor cells from normal human adult peripheral blood: flow-cytometrical studies and in-vitro colony (CFU-GM, BFU-E) assays. Ann. Hematol. 62: 45-53 (1991); **Smith C et al** Purification and partial characterization of a human hematopoietic precursor population. Blood 77: 2122-8 (1991); **Snoeck HW et al** Effects of interleukin-4 (IL4) on myelopoiesis: studies on highly purified $CD34^+$ hematopoietic progenitor cells. Leukemia 7: 625-9 (1993); **Testa U et al** Cascade transactivation of growth factor receptors in early human hematopoiesis. Blood 81: 1442-56 (1993); **Tong J et al** *In vivo* administration of recombinant methionyl human stem cell factor expands the number of human marrow hematopoietic stem cells. Blood 82: 784-91 (1993); **Williams DE et al** Characterization of hematopoietic stem and progenitor cells. Immunol. Res. 6: 294-304 (1987); **Williams DE et al** Purification of murine bone marrow-derived granulocyte-macrophage colony-forming cells and partial characterization of their growth and regulation *in vitro*. EH 15: 243-50 (1987)

CFU-M: *colony forming unit macrophage* Term used to describe a particular developmental type of blood-forming cells. CFU-M can be identified in a » colony formation assay by the specific morphology of the cells. These cells express the cell surface markers CD13, CD15, » CD33, CD38 (see also: CD antigens) and HLA-DR. The cells respond to » IL3, » IL4, » IFN-γ, » SCF, and colony stimulating factors (see: CSF). CFU-M is inhibited by retinoic acid and under certain conditions also by » IL4. Growth of CFU-M in response to colony-stimulating factors is further stimulated by » IFN-γ and this effect is reversed by IL4. See also: Hematopoiesis, LTBMC (long-term bone marrow culture).

Bot FJ et al Stimulating spectrum of human recombinant multi-CSF (IL3) on human marrow precursors: importance of accessory cells. Blood 71: 1609-14 (1988); **Bot FJ et al** Human granulocyte-macrophage colony-stimulating factor (GM-CSF) stimulates immature marrow precursors but no CFU-GM, CFU-G, or CFU-M. EH 17: 292-5 (1989); **Geissler D et al** Clonal growth of human megakaryocytic progenitor cells in a micro-agar culture system: simultaneous proliferation of megakaryocytic, granulocytic, and erythroid progenitor cells (CFU-M, CFU-C, BFU-E) and T lymphocytic colonies (CFU-TL). IJCC1: 377-88 (1983); **Kanz L et al** Identification of human megakaryocytes derived from pure megakaryocytic colonies (CFU-M), megakaryocytic-erythroid colonies (CFU-M/E), and mixed hemopoietic colonies (CFU-GEMM) by antibodies against platelet associated antigens. Blut 45: 267-74 (1982); **Nakeff A et al** *In vitro* colony assay for a new class of megakaryocyte precursor: colony forming unit megakaryocyte (CFU-M). PSEBM 151: 587-90 (1976); **Sato N et al** Purification of human marrow progenitor cells and demonstration of the direct action of macrophage colony-stimulating factor on colony-forming unit-macrophage. Blood 78: 967-74 (1991); **Snoeck HW et al** Differential regulation of the expression of CD38 and human leukocyte antigen-DR on $CD34^+$ hematopoietic progenitor cells by interleukin-4 and interferon-γ. EH 21: 1480-6 (1993); **Snoeck HW et al** Effects of interleukin-4 (IL4) on myelopoiesis: studies on highly purified $CD34^+$ hematopoietic progenitor cells. Leukemia 7: 625-9 (1993); **Snoeck HW et al** Effects of interleukin-4 on myelopoiesis: localization of the action of IL4 in the $CD34^+HLA\text{-}DR^{++}$ subset and distinction between direct and indirect effects of IL4. EH 21: 635-9 (1993); **Snoeck HW et al** Interferon-γ and interleukin-4 reciprocally regulate the production of monocytes/macrophages and neutrophils through a direct effect on committed monopotential bone marrow progenitor cells. EJI 23: 1072-7 (1993); **van Bockstaele DR** Direct effects of 13-cis and all-trans retinoic acid on normal bone marrow (BM) progenitors: comparative study on BM mononuclear cells and on isolated $CD34^+$ BM cells. Ann. Hematol. 66: 61-6 (1993)

CFU-Mast: A circulating mast cell progenitor, previously believed to be a monocyte or basophil. The cells are $CD34^+$, kit^+, Ly^-, $CD14^-$, $CD17^-$. Mast cells are produced in response to » SCF.

Agis H et al Monocytes do not make mast cells when cultured in the presence of SCF. Characterization of the circulating mast cell progenitor as a c-kit^+, $CD34^+$, Ly-, CD14-, CD17-, colony-forming cell. JI 151: 4221-7 (1993)

CFU-MEG: *colony forming unit megakaryocyte* Also abbrev. CFU-Mk. Term used to describe an early precursor cells in the developmental lineage giving rise to megakaryocytes, which is more developed than the precursor type known as BFU-Mk (burst forming unit, megakaryocyte). CFU-MEG can be identified in a » colony formation assay by the specific morphology of the cells. These cells express the cell surface markers » CD34 (see also: CD antigens) and HLA-DR. See also: Hematopoiesis, LTBMC (long-term bone marrow culture).

A reduced production of CFU-MEG colonies *in vitro* or even a complete lack is observed in many patients with myelodysplastic syndromes (MDS). Some studies indicate that megakaryocyte progenitors in myeloproliferative disorders form spontaneous megakaryocyte colonies without the addition of megakaryocyte colony-stimulating factor (Meg-CSF). This spontaneous colony formation appears to be due to a hypersensitivity to » IL3.

IL3 appears to be a priming factor *in vitro* and *in vivo* that renders progenitor cells sensitive for the action of other cytokines. For factors influencing the growth and development of CFU-MEG and BFU-Mk see also: MEG-CSF (megakaryocyte colony-stimulating factor), Mk-potentiator, and TPO (thrombopoietin). The interleukin » IL6 has been reported to act as a maturation and differentiation factor for CFU-MEG. SCF has been found to synergize with » GM-CSF, » IL6, » IL3, » IL11 or » Epo to increase the numbers of CFU-Meg. Choline also stimulates megakaryocyte colony formation. Proliferation of CFU-MEG is also stimulated by » bFGF. Growth of CFU-MEG is inhibited by » TGF-β.

Aglietta M et al Interleukin-3 *in vivo*: kinetic of response of target cells. Blood 82: 2054-61 (1993); **Aglietta M et al** Granulocyte-macrophage colony stimulating factor and interleukin 3: target cells and kinetics of response *in vivo*. Stem Cells Dayt. 11: s83-7 (1993); **Briddell RA et al** Characterization of the human burst-forming unit-megakaryocyte. Blood 74: 145-51 (1989); **Bruno E et al** Recombinant GM-CSF/IL3 fusion protein: its effect on *in vitro* human megakaryocytopoiesis. EH 20: 494-99 (1992); **Bruno E et al** Basic fibroblast growth factor promotes the proliferation of human megakaryocyte progenitor cells. Blood 82: 430-5 (1993); **Cairo MS et al** Effect of interleukin-11 with and without granulocyte colony-stimulating factor on *in vivo* neonatal rat hematopoiesis: induction of neonatal thrombocytosis by interleukin-11 and synergistic enhancement of neutrophilia by interleukin-11 + granulocyte colony-stimulating factor. Pediatr. Res. 34: 56-61 (1993); **Diochot S et al** Effects of endothelins on the human megakaryoblastic cell line MEG-01. Eur. J. Pharmacol. 227: 427-31 (1992); **Erickson-Miller CL et al** Megakaryocytopoiesis and platelet production: does stem cell factor play a role? Stem Cells Dayt. 11: s163-9 (1993); **Gallicchio YS et al** The influence of choline chloride on murine hematopoiesis *in vivo* and *in vitro*. Int. J. Cell Cloning 1: 451-63 (1983); **Gebbia V et al** Analysis of human dysplastic hematopoiesis in long-term bone marrow culture. Blood 59: 442-8 (1989); **Gewirtz AM & Hoffman R** Human megakaryocytes production: cell biology and clinical considerations. Hematol. Oncol. Clin. North Am. 4: 43-63 (1990); **Ishibashi T et al** Thrombopoietic effects of interleukin-6 in long-term administration in mice. EH 21: 640-6 (1993); **Kanz I et al** Human megakaryocytic progenitor cells. Klin. Wochenschr. 65: 297-307 (1987); **Kimura H et al** Interleukin 6 is a differentiation factor for human megakaryocytes *in vitro*. EJI 20: 1927-31 (1990); **Kobayashi S et al** Circulating megakaryocyte progenitors in myeloproliferative disorders are hypersensitive to interleukin-3. Br. J. Haematol. 83: 539-44 (1993); **Long MW et al** *In vitro* differences in responsiveness of early (BFU-Mk) and late (CFU-Mk) murine megakaryocyte progenitor cells. In: Levine RI (edt) Megakaryocyte development and function. Alan R. Liss, New York, pp. 179-86 (1986); **Messner HA et al** The growth of large megakaryocyte colonies from human bone marrow. JCP (Suppl.) 1: 45-51 (1982); **Murphy MJ jr & Ogata K** *In vitro* cloning of murine megakaryocyte progenitors (CFU-Meg). J. Tiss. Cult. Meth. 13: 83-8 (1991); **Nakeff A & Dicke KA** Stem cell differentiation into megakaryocytes from mouse bone marrow cultured with the thin layer agar technique. EH 22: 58 (1972); **Pantel K & Nakeff A** The role of lymphoid cells in hematopoietic regulation. EH 21: 738-42 (1993); **Tong J et al** *In vivo* administration of recombinant methionyl human stem cell factor expands the number of human marrow hematopoietic stem cells. Blood 82: 784-91 (1993); **Zauli G et al** Presence and characteristics of circulating megakaryocyte progenitor cells in human fetal blood. Blood 81: 385-90 (1993)

CFU-Mk: see: CFU-MEG.

CFU-S: *colony forming unit spleen* Term used to describe very early murine pluripotent » hematopoietic stem cells derived from the bone marrow (see also: Hematopoiesis) that give rise to discrete nodules in the spleens of lethally irradiated animals within eight days. On histologic section, some of these nodules contain erythroid, granulocytic, and megakaryocytic elements. The progeny of a single spleen colony has been found to be able to repopulate the entire hematopoietic organ in lethally irradiated recipients and forms new colonies upon retransplantation.

Spleen colony formation by CFU-S in irradiated mice is used as a short-term quantitative *in vivo* assay for pluripotent hematopoietic stem cells. This ***Spleen colony assay*** is used as an alternative to the time-consuming assay of hematopoietic stem cells to reconstitute lympho-and myelopoiesis when transferred into irradiated hosts, which may take several months before full and lasting reconstitution with donor cells can be unequivocally demonstrated. It has been shown that day 8 CFU-S have a poor repopulating capacity and the spleen cell assay is now considered to measure slightly more differentiated progeny of true stem cells (see also: LTBMC, long term bone marrow culture) and may be equivalent to human » CFU-Mix cell assays. A cell type which is more primitive than » CFU-S (***pre-CFU-S***) is considered to possess long-term » MRA (marrow repopulating ability).

Day 12 CFU-S are more potent in terms of repopulating ability than day 8 CFU-S. Day 12 CFU-S have been subdivided further by staining with ***rhodamine-123***, with one actively proliferating subpopulation staining positively with the dye and another fraction showing dull staining. Cell resistant to the cell cycle drug » fluorouracil share properties with rhodamine-123 dull cells in being relatively unresponsive to » IL3 and » M-CSF. Rhdodamine-123 dull cells form long-lasting cobble-stone areas (see: CAFC, cobblestone area-forming cells). C57BL/6 mice have only 2.6 % stem cells that are in active » cell cycle whereas DBA/2 mice have 24 % CFU-S in S phase.

The most primitive spleen colony-forming cells are multipotent and can undergo extensive self-renewal. They produce mature cells representative of all the myeloid cell lineages. Normally, only a minority of CFU-S undergo DNA synthesis, but the proportion of replicating cells is increased dramatically during recovery from chemotherapy or following bone marrow transplantation.

These cells do not respond very well to a factor known as » SCF (stem cell factor). However, the effect of SCF is greatly enhanced by » IL1 and » IL3. IL1 in combination with M-CSF has been shown to stimulate the formation of colonies from a cell population highly enriched for CFU-S. CFU-S have also been shown to respond to » bFGF *in vitro*. They are expanded *in vivo* by treatment with » IL11.

One factor, designated » MIP (macrophage inflammatory protein), prevents colony formation by CFU-S. Some low molecular weight compounds also appear to be inhibitory regulators of stem cells (see: AcSDKP). Generation of these cells in long-term bone marrow cultures is also suppressed by antibodies directed against » IL6. Colony formation of CFU-S induced by IL3 can be inhibited by » TGF-β.

Baines P et al Physical and kinetic properties of hemopoietic progenitor cell populations from mouse marrow detected in five different assay systems. Leuk. Res. 6: 81-8 (1982); **Barker JE et al** Advantages of gradient vs. 5-fluorouracil enrichment of stem cells for retroviral-mediated gene transfer. EH 21: 47-54 (1993); **Bartelmez SH et al** Uncovering the heterogeneity of hematopoietic repopulating cells EH 19: 861-2 (1991); **Bearpark AD & Gordon MY** Adhesive properties distinguish subpopulations of hemopoietic stem cells with different spleen colony-forming and marrow repopulating capacities. Bone Marrow Transplant. 4: 625-8 (1989); **Bertoncello I et al** Multiparameter analysis of transplantable hemopoietic stem cells. I. Separation and enrichment of stem cells homing to marrow and spleen on the basis of rhodamine-123 fluorescence. EH 13: 999-1006 (1985); **Bodine DM et al** Effects of hematopoietic growth factors on the survival of primitive stem cells in liquid suspension culture. Blood 78: 914-20 (1991); **Chen MG & Schooley JC** A study of the clonal nature of spleen colonies using chromosome markers. Transplantation 6: 121-6 (1968); **Corso A et al** A protocol for the enrichment of different types of CFU-S from fetal mouse liver. Haematologica 78: 5-11 (1993); **Curry JL & Trentin JJ** Hemopoietic spleen colony studies. I. Growth and differentiation. Dev. Biol. 15: 395-413 (1967); **Deryugina EI & Müller-Sieburg CE** Stromal cells in long-term cultures: keys to the elucidation of hematopoietic development? Crit. Rev. Immunol. 13: 115-50 (1993); **de Vries P et al** The effect of recombinant mast cell growth factor on purified murine hemopoietic stem cells. JEM 173: 1205-11 (1991); **Du XX et al** Effects of recombinant human interleukin-11 on hematopoietic reconstitution in transplant mice: acceleration of recovery of peripheral blood neutrophils and platelets. Blood 81: 27-34 (1993); **Gallicchio VS et al** Basic fibroblast growth factor (B-FGF) induces early- (CFU-s) and late-stage hematopoietic progenitor cell colony formation (CFU-gm, CFU-meg, and BFU-e) by synergizing with GM-CSF, Meg-CSF, and erythropoietin, and is a radioprotective agent *in vitro*. Int. J. Cell. Cloning 9: 220-32 (1991); **Hampson J et al** The effects of TGFβ on hemopoietic cells. GF 1: 193-202 (1989); **Harris RA et al** An antigenic difference between cells forming early and late hemopoietic spleen colonies. N 307: 638-641 (1984); **Harrison DE** Evaluating functional abilities of primitive hematopoietic stem cell populations. CTMI 177: 13-30 (1992); **Harrison DE et al** Primitive hemopoietic stem cells: direct assay of most productive populations by competitive repopulation with simple binomial, correlation and covariance calculations. EH 21: 206-19 (1993); **Heyworth CM et al** The response of hemopoietic cells to growth factors: developmental implications of synergistic interactions. JCS 91: 239-47 (1988); **Hodgson GS et al** *In vitro* production of CFU-S and cells with erythropoiesis repopulating ability by 5-fluorouracil treated mouse bone marrow. Int. J. Cell Coning 1: 49-56 (1983); **Hudak S et al** Anti-IL6 antibodies suppress myeloid cell production and the generation of CFU-c in long-term bone marrow cultures. EH 20: 412-7 (1992); **Humphries RK et al** CFU-S in individual erythroid colonies *in vitro* derived from adult mouse marrow. N 279: 718-20 (1979); **Inoue T et al** Separation and concentration of murine hematopoietic stem cells (CFUS) using a combination of density gradient sedimentation and counterflow centrifugal elutriation. EH 9: 563-72 (1981); **Jones RJ et al** Separation of pluripotent haematopoietic stem cells from spleen colony-forming cells. N 347: 188-9 (1990); **Jones RJ** Spleen colony-forming unit: a myeloid stem cell. CTMI 177: 75-82 (1992); **McCulloch EA & Till JE** Proliferation of hematopoietic colony-forming cells transplanted into irradiated mice. Radiation Res. 22: 383-9 (1964); **Molineux G et al** Development of spleen CFU-S colonies from day 8 to day 11: relationship to self-renewal capacity. EH 14: 710-3 (1986); **Nijof W & Wierenga PK** Thiamphenicol as an inhibitor of early red cell differentiation. Hoppe Seylers Z. Physiol. Chemie 361: 1371-9 (1980); **Orlic D & Bodine DM** Pluripotent hematopoietic stem cells of low and high density can repopulate W/Wv mice. EH 20: 1291-5 (1992); **Pantel K & Nakeff A** The role of lymphoid cells in hematopoietic regulation. EH 21: 738-42 (1993); **Ploemacher RE et al** Separation of CFU-S from primitive cells responsible for reconstitution of the bone marrow hemopoietic stem cell compartment following irradiation: evidence for a pre-CFU-S cell. EH 17: 263-6 (1989); **Ploemacher RE & Brons RHC** Separation of CFU-S from primitive cells responsible for reconstitution of the bone marrow hemopoietic stem cell compartment following irradiation: evidence for a pre-CFU-S cell. EH 17: 263-6 (1989); **Ploemacher RE & van der Sluijs JP** *In vitro* frequency analysis os spleen colony forming and marrow-repopulating hemopoietic stem cells in the mouse. J. Tiss. Cult. Meth. 13: 63-8 (1991); **Robinson S & Riches A** Haematopoietic stem cell proliferation regulators investigated using an *in vitro* assay. J. Anat. 174: 153-62 (1991); **Rosendaal M et al** Haemopoietic stem cells: possibility of toxic effects of 5-fluorouracil on spleen colony formation. Blood Cells 7: 561-74 (1981); **Rosendaal M et al** Hematopoietic stem cells are organized for use on the basis of their generation age. N 264: 68-9 (1976); **Schofield R & Lajtha LG** Effect of isopropylmethane sulfonate (IMS) on hemopoietic colony-forming cells. Br. J. Haematol. 25: 195-202 (1973); **Siminovitch L et al** The distribution of colony-forming cells among spleen colonies. J. Cell. Comp. Physiol. 62: 327-36 (1963); **Spangrude GJ et al** Purification and characterization of mouse hematopoietic stem cells. S 241: 58-62 (1988); **Till JE & McCulloch EA** Hemopoietic stem cell differentiation. BBA 605: 431-59 (1980); **van der Sluijs JP et al** Marrow repopulating cells, but not CFU-S, establish long-term *in vitro* hemopoiesis on a marrow-derived stromal layer. EH 18: 893-6 (1990); **Visser JWM et al** Isolation of murine pluripotent hemopoietic stem cells. JEM 50: 1576-90 (1984); **Wolf NS et al** *In vivo* and *in vitro* characterization of long-term repopulating primitive hematopoietic cells isolated by sequential Hoechst 33342-rhodamine 123 FACS selection. EH 21: 614-22 (1993); **Worton RG et al** Physical separation of hemopoietic stem cells differing in their capacity for self renewal. JEM 130: 91-103 (1969); **Zipori D** The renewal and differentiation of hematopoietic stem cells. FJ 6: 2691-7 (1992)

CFU-s: *colony forming unit spleen* This factor which supports the formation of colonies by hematopoietic precursors isolated from the spleen is identical with » IL3.

Schrader JW & Clark-Lewis I A T cell-derived factor stimulating multipotential hemopoietic stem cells: molecular weight and distinction from T cell growth factor and T cell-derived granulocyte-macrophage colony-stimulating factor. JI 129: 30-35 (1982)

CFU-SA: *colony forming unit stimulating activity* This factor was initially found to be produced by spleen cells and also by certain ConA-stimulated T cell hybridomas. It induces the generation of multipotent » hematopoietic stem cells in cultures of murine bone marrow cells. The factor appears to have the same activity as » PCSF (P cell stimulating activity) and is identical with » IL3.

Cerny J et al Stimulation of bone marrow hemopoietic stem cells by a factor from activated T cells. N 249: 63-66 (1974); **Schrader JW & Clark-Lewis I** A T cell-derived factor stimulating multipotential hemopoietic stem cells: molecular weight and distinction from T cell growth factor and T cell-derived granulocyte-macrophage colony-stimulating factor. JI 129: 30-35 (1982)

CFU-s inhibitory activity: This factor is identical with » SCI (stem cell inhibitor).

CFU-s SF: *colony forming unit spleen stimulating factor* This factor is identical with » IL3.

CGF: *cementum-derived growth factor* This factor of 23 kDa is isolated from human cementum. It is mitogenic for a number of cell types and this activity is potentiated by » EGF. The factor is not identical with » EGF or » PDGF. Another factor, also designated CGF, has a molecular mass of 22 kDa. It is not clear whether these two factors are identical. Another factor, also isolated from human cementum (see: cementum mitogenic factor) is identical with » bFGF.

Nakae H et al Isolation and partial characterization of mitogenic factors from cementum. B 30: 7047-52 (1991); **Yonemura K et al** Isolation and partial characterization of a growth factor from human cementum. Bone Miner. 18: 187-98 (1992)

CGF: *chondrocyte growth factor* This factor (or factors) is isolated from human pituitaries by virtue of the ability to bind to heparin. CGF is identical with » bFGF. TGF-β has also been shown to possess CGF activity and the possibility of other factors having the same activity cannot be excluded. Some of these factors have been described as CAF (chondrocyte activating factor).

Bandara G et al The synovial activation of chondrocytes: evidence for complex cytokine interactions involving a possible novel factor. BBA 1134: 309-18 (1992); **Baratz ME et al** Studies on the autocrine activation of a synovial cell line. J. Orthop. Res. 9: 651-7 (1991); **Kasper S & Friesen HG** Human pituitary tissue secretes a potent growth factor for chondrocyte proliferation. J. Clin. Endocrinol. Metab. 62: 70-6 (1986); **Kato Y et al** Purification of growth factors from cartilage. MiE 198: 416-424 (1991); **Webber RJ & Sokoloff L** *In vitro* culture of rabbit growth plate chondrocytes. 1. Age-dependence of response to fibroblast growth factor and "chondrocyte growth factor". Growth 45: 252-68 (1981); **Pepper MS et al** Plasminogen activator inhibitor 1 is induced in microvascular endothelial cells by a chondrocyte-derived transforming growth factor-β. BBRC 176: 633-8 (1991); **Too CK et al** Further purification of human pituitary-derived chondrocyte growth factor: heparin binding and cross-reactivity with antiserum to basic FGF. BBRC 144: 1128-34 (1987)

CGF: *chondrosarcoma growth factor* abbrev. also: CDGF. This factor is isolated from cultured chondrosarcoma cells. It is mitogenic for capillary endothelial cells. The factor is identical with » bFGF.

Shing Y et al Heparin affinity: purification of a tumor-derived capillary endothelial cell growth factor. S 223: 1296-9 (1984)

CGF: *colonic growth factor* This factor, isolated from epithelial tissues of the colonic mucosa, is a mitogen for epithelial cells. It is an N-terminal extended form of » bFGF. Other factors also possessing CGF activity include » EGF, IGF-1 (see: IGF) and also » IL1.

Nice EC The major cell mitogen extractable from colonic mucosa is an N terminally extended form of basic fibroblast growth factor. JBC 266: 14425-30 (1991); **Whithead RH et al** Detection of colonic growth factors using a human colonic carcinoma cell line (LIM1215). IJC 46: 858-63 (1990)

CGF: *cytokine growth factor* A general term used occasionally as a synonym for » cytokines.

Ford RJ et al Human B cell lymphomas: *in vitro* and *in vivo* studies on growth factors and cell growth. Leuk. Lymphoma 10: s51-6 (1993)

CGF: *cytotoxic cell-generating factor* This poorly characterized factor of 28 kDa is produced constitutively by certain rat T cell hybridomas and promotes the generation of cytotoxic T cells. The factor may be identical with » CDF (cytolytic differentiation factor for T lymphocytes, CTL differentiation factor). In certain T cell hybridomas CGF also induces the expression of » IL2 receptors. The factor is not identical with » IL1, » IL2, » IL3, » IL4, » IL5, » IL6 or » IFN-γ.

Murakami T et al Functional analysis of mononuclear cells infiltrating into tumors. IV. Purification and functional characterization of cytotoxic cell-generating factor. JI 141: 4235-42 (1988); **Uede T et al** A soluble factor distinct from interleukin 2 is involved in the generation and differentiation of cytotoxic T cells. JJCR 76: 871-9 (1985); **Uede T et al** Establishment of rat-mouse T cell hybridomas that constitutively produce a soluble factor that is needed for the generation of cytotoxic cells: biochemical and functional characterization. JI 135: 3252-7 (1985); **Yamaki T et al** Cellular mechanisms of tumor rejection in rats. Nat. Immun. Cell. Growth Regul. 9: 1-25 (1990)

CGRP: *calcitonin gene-related peptide* CGRP is a hypothalamic neuropeptide of 37 amino acids that has been identified in numerous tissues including immune organs. CGRP is a potent vasodilator with a long duration of action. Calcitonin (32 amino acids) and CGRP are encoded by a single gene, the CALC-I gene. They are expressed by alternative splicing of the pre-messenger RNA derived from the CALC-I gene. Calcitonin mRNA is the predominant product in parafollicular thyroid C cells, whereas CGRP mRNA is the main product in neurons of the central and peripheral nervous systems. A variety of N-truncated CGRP fragments have been described that have different pharmacological activities and may also behave as competitive antagonists of CGRP in certain bioassays. Cleavage of CGRP with endopeptidase 24.11 yields a tetrapeptide, Val32-Gly-Ser-Glu35, known to function as an eosinophil granulocyte chemotactic factor.

Apart from the classical role of neurohumoral hormone CGRP also shows cytokine-like activities although it is not classified as a cytokine due to its small size. CGRP inhibits the proliferation of spleen cells. It induces a dose-dependent cAMP accumulation in IL2-producing TH1 cells and inhibits their production of IL2 by decreasing IL2 mRNA accumulation. These effects are prevented by CGRP8-37, a CGRP antagonist that is missing the first 7 amino acids. CGRP also inhibits the accumulation of mRNA coding for » TNF-α, » TNF-β, and » IFN-γ. CGRP also appears to be an » autocrine growth factor for murine F9 teratocarcinoma cells and may be a growth factor during early embryogenesis. Calcitonin has been found to inhibit the growth of the human gastric carcinoma cell line KATO III.

In folliculo-stellate cells, one of the nonhormone secreting pituitary cells, CGRP induces the synthesis of » IL6. CGRP may also be involved in macrophage immune responses since specific receptors for calcitonin gene-related peptide (CGRP) have been identified on plasma membranes from the IL1 secreting murine macrophage-like cell line P388 D1. CGRP has also been shown to be a potent activator of neutrophils and also promotes neutrophil adherence to venular endothelium. CGRP is chemotactic for human T cells. CGRP, as a consequence of its sustained vasodilator activity can have prolonged potentiating effects on neutrophil accumulation and edema formation in inflammatory conditions, mediated, for example, by » IL1.

Abello J et al Characterization of calcitonin gene-related peptide receptors and adenylate cyclase response in the murine macrophage cell line P388 D1. Neuropeptides 19: 43-9 (1991); **Adema GJ & Baas PD** Deregulation of alternative processing of Calcitonin/CGRP-I pre-mRNA by a single point mutation. BBRC 178: 985-92 (1991); **Boudard F & Bastide M** Inhibition of mouse T cell proliferation by CGRP and VIP: effects of these neuropeptides on IL2 production and cAMP synthesis. J. Neurosci. Res. 29: 29-41 (1991); **Buckley TL et al** Inflammatory edema induced by interactions between IL1 and the neuropeptide calcitonin gene-related peptide. JI 146: 3424-30 (1991); **Davies D et al** Endopeptidase-24.11 cleaves a chemotactic factor from α-calcitonin gene-related peptide. Biochem. Pharmacol. 43: 1753-6 (1992); **Emeson RB et al** Posttranscriptional regulation of calcitonin/CGRP gene expression. ANY 657: 18-35 (1992); **Foster CA et al** Calcitonin gene-related peptide is chemotactic for human T lymphocytes. ANY 657: 397-404 (1992); **Maggi CA et al** Human α-calcitonin gene-related peptide-(8-37) as an antagonist of exogenous and endogenous calcitonin gene-related peptide. Eur. J. Pharmacol. 192: 85-8 (1991); **Mimeault M et al** Comparative affinities and antagonistic potencies of various human calcitonin gene-related peptide fragments on calcitonin gene-related peptide receptors in brain and periphery. J. Pharmacol. Exp. Ther. 258: 1084-90 (1991); **Nakamura A et al** Calcitonin inhibits the growth of human gastric carcinoma cell line KATO III. Regul. Pept. 37: 183-94 (1992); **Richter J et al** Calcitonin gene-related peptide (CGRP) activates human neutrophils - inhibition by chemotactic peptide antagonist BOC-MLP. Immunology 77: 416-21 (1992); **Segond N et al** Calcitonin gene-related peptide: an autocrine growth factor with regulatory activity *in vitro*. BBRC 187: 381-8 (1992); **Sung CP et al** CGRP stimulates the adhesion of leukocytes to vascular endothelial cells. Peptides 13: 429-34

CGIF: see addendum.

(1992); **Tatsuno I et al** Neuropeptide regulation of interleukin-6 production from the pituitary: stimulation by pituitary adenylate cyclase activating polypeptide and calcitonin gene-related peptide. Endocrinology 129: 1797-804 (1991); **Wang F et al** Calcitonin gene-related peptide inhibits interleukin 2 production by murine T lymphocytes. JBC 267: 21052-7 (1992); **Wimalawansa SJ** CGRP radioreceptor assay: a new diagnostic tool for medullary thyroid carcinoma. J. Bone. Miner. Res. 8: 467-73 (1993); **Zimmerman BJ et al** Neuropeptides promote neutrophil adherence to endothelial cell monolayers. AM. J. Physiol. 263: 678-82 (1992)

ChAT development factor: *choline acetyltransferase development factor* see: CDF.

cHDGF: *cationic hypothalamus-derived growth factor* This factor is identical with » bFGF.
Klagsbrun M & Shing Y Heparin affinity of anionic and cationic capillary endothelial cell growth factors: analysis of hypothalamus-derived growth factors and fibroblast growth factors. PNAS 82: 805-9 (1985)

ChDGF: *chondrosarcoma-derived growth factor* abbrev. also CHSA-GF. This factor of ≈ 18 kDa was originally isolated from a rat chondrosarcoma. It is mitogenic for murine » 3T3 fibroblasts and capillary endothelial cells and is also an » Angiogenesis factor. It is identical with » bFGF. See also: CGF which is most likely identical.
Shing Y et al Angiogenesis is stimulated by tumor-derived endothelial cell growth factor. JCBc 29: 275-87 (1985); **Sullivan RC et al** Use of size-exclusion and ion-exchange high-performance liquid chromatography for the isolation of biologically active growth factors. J. Chromatogr. 266: 301-11 (1983)

ChDI: *chondrocyte-derived inhibitor* This factor of 35 kDa is obtained from the conditioned medium (see: CM) of chondrocytes. The factor strongly inhibits neovascularization of tissues *in vivo* (see also: Angiogenesis factors) and inhibits the proliferation and migration of FGF-stimulated capillary endothelial cells *in vitro*. ChDI also acts as an inhibitor of collagenases.
Moses MA et al Isolation and characterization of an inhibitor of neovascularization from scapular chondrocytes. JCB 119: 475-82 (1992)

Chemokine receptor R1: see: CKR-1.

Chemokines: New generic name given to a family of activation-inducible cytokines previously referred to as members of » ***SIS family***, » ***sig family***, » ***scy family*** (scy = small cytokine), ***platelet factor 4 superfamily*** or » ***intercrines***. These proteins are mainly chemotactic (hence the name which is derived from *chemo*tactic cyto*kines*; see also: Chemotaxis).

Chemokines have molecular masses of 8-10 kDa and show ≈ 20-50 % sequence homology among each other at the protein level. The proteins also share common gene structures and tertiary structures. All chemokines possess four conserved cysteine residues involved in intramolecular disulfide bond formation.

According to the chromosomal locations of individual genes two different subfamilies are distinguished. Members of the ***α-chemokine*** or ***4q family*** map to human chromosome 4q12-21. The first two cysteine residues are separated by a single amino acids and these proteins are therefore also called ***C-X-C-chemokine***. This subfamily includes » IL8, » PF4 (platelet factor 4), » Beta-Thromboglobulin, » IP-10, » KC, » MIP-2 (see: MIP, macrophage inflammatory protein), » MGSA (melanoma growth stimulatory activity), » 9E3, and » ENA-78.

Members of the ***β-chemokine*** or ***17q family*** map to human chromosome 17q11-32 (murine chromosome11). The first two cysteine residues are adjacent and these proteins are therefore also called ***C-C-chemokine***. This subfamily includes » LD78, » ACT-2, » FIC (fibroblast-inducible cytokine), » I-309, » LAG-1 (lymphocyte activation gene 1), » RANTES, » JE, » HC-21, » MIP (macrophage inflammatory protein), FIC, » MCAF, and » MARC protein.

The biological activities of intercrines are mediated by specific receptors and also by receptors that bind several of these proteins (see: CKR-1). MGSA and MIP-2 (see: MIP), for example, compete with » IL8 for binding to the same receptor in » U937 cells. This receptor does not bind MIP-1 (see: MIP), CTAP-III (see: Beta-Thromboglobulin), LD78, ACT-2 or PF-4. The » CKR-1 receptor has been shown to be identical with the Duffy blood group antigen that mediates binding of *Plasmodium vivax* to erythrocytes. For viral chemokine receptor homologues see also: ECRF3, EBI 1 (EBV-induced gene 1).

Chemokines are multipotent cytokines that localize and enhance » inflammation by inducing » Chemotaxis and activation of different types of inflammatory cells typically present at inflammatory sites. Chemokines and other mediators are also secreted by these cells. Chemokine have been shown to exert their effects on distinct subsets of cells. C-X-C chemokines, for example appear to attract neutrophils but not macrophages, while C-C- chemokines preferentially induce migration of macrophages. Some chemokines have been shown

```
ENA78     AGPAAAVLRELRCVCLQT-TQGVHPKMISNLQVFAIGPQCSKVEVVASLKN-GKEICLDPEAPFLKKVIQKILDGGNKEN
GCP-2      GPVSAVLTELRCTCLRVTLR...
GRO α         ASVATELRCQCLQT-LQGIHPKNIQSVNVKSPGPHCAQTEVIATLKN-GRKACLNPASPIVKKIIEKMLNSDKSN
GRO β         APLATELRCQCLQT-LQGIHLKNIQSVKVKSPGPHCAQTEVIATLKN-GQKACLNPASPMVKKIIEKMLKNGKSN
GRO γ         ASVVTELRCQCLQT-LQGIHLKNIQSVNVRSPGPHCAQTEVIATLKN-GKKACLNPASPMVQKIIEKILNKGSTN
IL8        AVLPRSAKELRCQCIKTYSKPFHPKFIKELRVIESGPHCANTEIIVKLSD-GRELCLDPKENWVQRVVEKFLKRAENS
IP10          VPLSRTVRCTCISISNQPVNPRSLEKLEIIPASQFCPRVEIIATMKKKGEKRCLNPESKAIKNLLKAVSKEMSKRSP
NAP-2             AELRCMCIKT-TSGIHPKNIQSLEVIGKGTHCNQVEVIATLKD-GRKICLDPDAPRIKKIVQKKLAGDESAD
PF4          EAEEDGDLQCLCVKT-TSQVRPRHITSLEVIKAGPHCPTAQLIATLKN-GRKICLDLQAPLYKKIIKKLLES

HC14         QPDSVSIPITCCFNVINRKIPIQRLESY-TRITNIQCPKEAVIFK-TKRGKEVCADPKERWVRDSMKHLDQIFQNLKP
I309        SKSMQVPFS-RCCFSFAEQEIPLRAILCY--RNTSSICSNEGLIFK-LKRGKEACALDTVGWVQRHRKMLRHCPSKRK
MCP-1        QPDAINAPVTCCYNFTNRKISVQRLASY-RRITSSKCPKEAVIFK-TIVAKEICADPKQKWVQDSMDHLDKQTQTPKT
MCP-3        QPVGINTSTTCCYRFINKKIPKQRLESY-RRTTSSHCPREAVIFK-TKLDKEICADPTQKWVQDFMKHLDKKTQTPKL
MIP-1α       ASLAADTPTACCFSYTSRQIPQNFIADY--FETSSQCSKPGVIFL-TKRSRQVCADPSEEWVQKYVSDLELSA
MIP-1β       APMGSDPPTACCFSYTARKLPRNFVVDY--YETSSLCSQPAVVFQ-TKRSKQVCADPSESWVQEYVYDLELN
RANTES       SPYSSDT-TPCCFAYIARPLPRAHIKEY--FYTSGKCSNPAVVFV-TRKNRQVCANPEKKWVREYINSLEMS
```

Structure of chemokines.
Alignment of some chemokine sequences showing the typical arrangement of cysteine residues in C-X-C (top) and CC (bottom) chemokines. Cysteine residues are shown in yellow. The ELR (Glu-Leu-Arg) sequence (red) is common to all C-X-C chemokines that activate neutrophil leukocytes.

to induce selective migration of leukocyte subsets. It is now assumed that the combinatorial effects of multiple chemokines and other mediators are responsible for the cellular composition at inflammatory sites. In addition, many chemokines also directly activate cells. Some of them activate granulocytes and/or monocytes and cause respiratory bursts, degranulation, and the release of lysosomal enzymes. Others prime immune cells to respond to sub-optimal amounts of other inflammatory mediators. Yet others have been shown to be potent histamine releasing factors for basophils. It has been proposed that erythrocytes through their promiscuous chemokine receptor play an important role in regulating the chemokine network. Chemokines bound to the erythrocyte receptor are known to be inaccessible to their normal target cells. This appears to provide a sink for superfluous chemokines and limits the systemic effects of these mediators without disrupting localized processes taking place at the site of inflammation. Many of these proteins are strongly expressed during the course of a number of pathophysiological processes including autoimmune diseases, cancer, atherosclerosis, and chronic inflammatory diseases.

Baggiolini M et al Interleukin 8 and related chemotactic cytokines – CXC and CC chemokines. Adv. Immunol. 55: 97-179 (1994); **Broxmeyer HE et al** Comparative analysis of the human macrophage inflammatory protein family of cytokines (chemokines) on proliferation of human myeloid progenitor cells. Interacting effects involving suppression, synergistic suppression, and blocking of suppression. JI 150: 3448-58 (1993); **Irving SG et al** 2 inflammatory mediator cytokine genes are closely linked and variably amplified on chromosome 17q. NAR 18: 3261-70 (1990); **Miller MD & Krangel MS** Biology and biochemistry of the chemokines: a family of chemotactic and inflammatory cytokines. Crit. Rev. Immunol. 12: 17-46 (1992); **Mukaida N et al** Regulation of human interleukin 8 gene expression and binding of several other members of the intercrine family to receptors for interleukin-8. AEMB 305: 31-8 (1991); **Sager R** GRO as a cytokine. In: Oppenheim JJ et al (eds) Molecular and cellular biology of cytokines. pp. 327-32, Wiley-Liss, New York, 1990; **Oppenheim JJ et al** Properties of the novel proinflammatory supergene "intercrine" cytokine family. ARI 9: 617-48 (1991); **Rollins BJ et al** Assignment of the human small inducible cytokine A2 gene, SCYA2 (encoding JE or MCP-1) to 17q11.2-12 – evolutionary relatedness of cytokines clustered at the same locus. Genomics 10: 489-92 (1991); **Stoeckle MY & Barker KA** Two burgeoning families of platelet factor 4-related proteins: mediators of the inflammatory response. New Biol. 2: 313-23 (1990); **Wenger RH et al** Human platelet basic protein/connective tissue activating peptide- III maps in a gene cluster on chromosome 4q12-q13 along with other genes of the β-thromboglobulin superfamily. Hum. Genet. 87: 367-8 (1991); **Widmer U et al** Genomic structure of murine macrophage inflammatory protein-1 α and conservation of potential regulatory sequences with a human homologue, LD78. JI 146: 4031-40 (1991); **Wilson SD et al** Clustering of cytokine genes on chromosome 11. JEM 171: 1301-14 (1990); **Wolpe SD et al** Macrophage inflammatory protein 1 and 2: members of a novel superfamily of cytokines. FJ 3: 2565-73 (1989)

Chemokinesis: see: Chemotaxis.

Chemokinetic inhibitory factor: see: CIF.

Chemotactic cytokines: see: chemokines.

Chemotactic index: abbrev. CI. See: Chemotaxis.

Chemotactic monokine: This factor is identical with » IL8.

Chemotaxin: This protein is secreted by the myeloid leukemia cell line ML-1 when stimulated with phorbol-12-myristate 13-acetate (see also: Phorbol esters). N-terminal sequence analysis reveals the identity with » IL8.
The term chemotaxin (or chemoattractant) is also used as a generic name for substances promoting » chemotaxis of cells and thus applies to a variety of cytokines/proteins or non-protein molecules, including proteins of the complement cascade, fibronectin, » fMLP, leukotrienes, prostaglandins, platelet activating factor, etc.
Suzuki K et al Isolation and amino acid sequence of a chemotactic protein, LECT/interleukin 8, from a human myeloid leukemia cell line, ML-1. Immunol. Lett. 36: 71-81 (1993)

Chemotaxis: General term used to describe the directed cell locomotion in concentration gradients of soluble extracellular agents. Substances that induce a chemotactic response (CTF = chemotactic factors) are also known under the general name of ***cytotaxins***, ***chemotaxins***, or ***chemoattractants*** Cells showing ***positive chemotaxis*** move towards areas with higher concentrations of these agents, those showing ***negative chemotaxis*** move away from these areas. This behavior is in marked contrast to the reactions elicited by some other compounds, known as ***chemokinesis***, essentially characterized by an undirected movement of cells (see also: motogenic cytokines). The dose-dependent migration of cells induced by gradients of substratum-bound substances has been called ***haptotaxis***.
A variety of methods have been used to measure chemotaxis, including migration of cells under layers of agarose, phagokinetic tract analysis, cell orientation assays, time-lapse cinematography. The chemotactic behavior of cells can be assayed in a so-called ***Boyden chamber*** in which the cells are separated from a test substance by a membrane. Chemotactically active compounds induce the migration of cells through the membrane into the compartment containing the chemotactic agent. Agarose gel clots have been used in similar assays to measure the activity of cells in the presence of chemotactic compounds or the inhibition of migration by inhibitors.
A quantitation of the chemotactic response of cells has been made possible by calculating the ***chemotactic index*** (abbrev. CI). It is calculated by measuring the lengths that cells migrate in the presence and the absence of a chemotactic factor as follows: CI = (length migrated in the presence of the factor – length migrated in the absence of the factor)/(length migrated in the absence of the factor).
Chemotactically active factors do not constitute a homogeneous class of compounds and include such diverse substances as component C5a of the complement system, N-formylated oligopeptides of bacterial origin, intermediates of lipid metabolism such as arachidonic acid and leukotriene B4 (LTB4), and many » cytokines. Chemotactically active factors are active in the nano- to picomole region. They are important mediators of a number of processes including, for example, immune responses, » inflammation, » wound healing, and general systemic reactions following tissue or organ injuries.
The biological activity of chemotactically active factors is mediated by specific cell surface receptors the expression of which can be modulated positively or negatively by almost all cytokines. Some components of the extracellular matrix (see: ECM), cellular adhesion molecules, and cytoskeletal elements also play an important role in the chemotactic responses of cells.
Most of the chemotactically active cytokines that lead to a local accumulation of cells at the site of secretion merely influence the migration of cells at low concentrations. At higher concentrations (10-100-fold) these cytokines also induce a series of co-ordinated biochemical processes and alter ion fluxes through the cellular membrane, remodeling of the cytoskeleton, alterations in lipid metabolism, activation of protein kinases, production of superoxide ions, release of lysosomal enzymes etc. (see also: Cell activation).
Agrawal PK & Reynolds DL Evaluation of the cell-mediated immune response of chickens vaccinated with Newcastle disease virus as determined by the under-agarose leukocyte-migration-inhibition technique. Avian Dis. 35: 360-4 (1991); **Anonymous** Chemotactic cytokines. Biology of the inflammatory peptide supergene family. Proceedings of the Second International Symposium on Chemotactic Cytokines. AEMB 305: 1-186 (1991); **Bignold LP et al** Studies of chemotactic, chemotactic movement-inhibiting, and random movement-inhibiting effects of interleukin 1 α and β, tumor necrosis factors α and β, and interferon γ on human neutrophils in assays using "sparse-pore" polycarbonate (Nucleopore) membranes in the Boyden chamber. Int. Arch. Allergy Appl. Immunol. 91: 1-7 (1990); **Bignold LP** Kinetics of chemo-attraction of polymorphonuclear leukocytes towards N-formyl peptide studied with a novel polycarbonate (Nucleopore) membrane in the Boyden chamber. Experientia 44: 518-21 (1988); **Camp R et al** Chemotactic cytokines in inflammatory skin disease. AEMB 305: 109-18 (1991); **Haddox JL et**

al A visual assay for quantitating neutrophil chemotaxis in a collagen gel matrix. A novel chemotactic chamber. JIM 141: 41-52 (1991); **Hughes FJ & McCulloch CA** Quantification of chemotactic response of quiescent and proliferating fibroblasts in Boyden chambers by computer-assisted image analysis. J. Histochem. Cytochem. 39: 243-6 (1991); **Jensen P & Kharazmi A** Computer-assisted image analysis assay of human neutrophil chemotaxis *in vitro*. JIM 144: 43-8 (1991); **Lauffenburger D et al** Measurement of leukocyte motility and chemotaxis parameters with a linear under-agarose migration assay. JI 131: 940-7 (1983); **Maderazo EG et al** Inhibitors of chemotaxis. MiE 162: 223-35 (1988); **Michna H** Induced locomotion of human and murine macrophages: a comparative analysis by means of the modified Boyden chamber system and the agarose migration assay. Cell Tissue Res. 255: 423-9 (1989); **Pike MC** Chemoattractant receptors as regulators of phagocytic cell function. CTMT 35: 19-43 (1990); **Pilaro AM et al** An improved *in vitro* assay to quantitate chemotaxis of rat peripheral blood large granular lymphocytes (LGL). JIM 135: 213-23 (1990); **Rice JE & Bignold LP** Chemotaxis of polymorphonuclear leukocytes in whole blood in the 'sparse-pore' polycarbonate (Nuclepore) membrane/Boyden chamber assay. JIM 149: 121-5 (1992); **Schell-Frederick E et al** Identification and quantification of human peripheral blood lymphocytes and monocytes following migration into nitrocellulose filters. JIM 139: 25-30 (1991); **Shi Y et al** A rapid, multiwell colorimetric assay for chemotaxis. JIM 164: 149-54 (1993); **Somersalo K et al** A simplified Boyden chamber assay for neutrophil chemotaxis based on quantitation of myeloperoxidase. AB 185: 238-42 (1990); **Terpstra GK & Houben LA** Directional cellular movement of cell populations: its description by chemotactic assays. Agents Actions. 26: 224-6 (1989); **Thomson MK & Jensen AL** Reassessment of two Boyden chamber methods for measuring canine neutrophil migration: the leading front and the lower surface count assays. Vet. Immunol. Immunopathol. 29: 197-211 (1991); **Thoren-Tolling K** Chemiluminescence and chemotaxis assay of porcine polymorphonuclear cells. A methodological study. Zentralbl. Veterinärmed. A. 37: 174-85 (1990); **Transquillo RT et al** Measurement of the chemotaxis coefficient for human neutrophils in the under-agarose migration assay. Cell Motil. Cytoskeleton. 11: 1-15 (1988); **Watanabe K et al** Very rapid assay of polymorphonuclear leukocyte chemotaxis *in vitro*. J. Pharmacol. Methods 22: 13-8 (1989); **Wilkinson EC** Micropore filter methods for leukocyte chemotaxis. MiE 162: 38-50 (1988)

Chick brain-derived growth factor: see: CBGF.

Chicken embryo fibroblast-derived growth factor: (CEF-derived growth factor). See: CDGF.

Chicken muscle growth factor: see: CMGF.

Chicken myelomonocytic growth factor: see: cMGF.

O-(Chloroacetylcarbamoyl)-fumagillol: AGM-1470. See: Fumagillin.

CHO: *Chinese hamster ovary* This cell line, established from Chinese hamster (*Cricetulus griseus*) ovary cells in the late 50s has been used as one of the standard cell lines for many biochemical investigations. It has been used as one of the standard cell lines for many biochemical investigations including studies of vaccine production, cellular metabolism, » gene expression, » cell cycle, mutagenesis, and use of » antisense oligonucleotides. Many different sublines are available. CHO cells are used, among other things, for producing recombinant » cytokines and other proteins. Unlike bacterially derived recombinant cytokines (see also: *Escherichia coli*) factors produced in CHO cells usually undergo correct post-translational processing which may be required for full bioactivity of the factor when used as a pharmacological compound. For alternatives see also: BHK cells, COS cells, Namalwa cells, Baculovirus expression system.

Acklin C et al Recombinant human brain-derived neurotrophic factor (rHuBDNF). Disulfide structure and characterization of BDNF expressed in CHO cells. Int. J. Pept. Protein Res. 41: 548-52 (1993); **Fukushima K et al** N-linked sugar chain structure of recombinant human lymphotoxin produced by CHO cells: the functional role of carbohydrate as to its lectin-like character and clearance velocity. ABB 304: 144-53 (1993); **Hayakawa T et al** *In vivo* biological activities of recombinant human erythropoietin analogs produced by CHO cells, BHK cells and C127 cells. Biologicals 20: 253-7 (1992); **Israel DI et al** Expression and characterization of bone morphogenetic protein-2 in Chinese hamster ovary cells. GF 7: 139-50 (1992); **Langley KE et al** Purification and characterization of soluble forms of human and rat stem cell factor recombinantly expressed by *Escherichia coli* and by Chinese hamster ovary cells. ABB 295: 21-8 (1992); **Lu HS et al** Post-translational processing of membrane-associated recombinant human stem cell factor expressed in Chinese hamster ovary cells. ABB 298: 150-8 (1992); **Malik N et al** Amplification and expression of heterologous oncostatin M in Chinese hamster ovary cells. DNA Cell Biol. 11: 453-9 (1992); **Nagao M et al** Production and ligand-binding characteristics of the soluble form of murine erythropoietin receptor. BBRC 188: 888-97 (1992); **Rice KG et al** Quantitative mapping of the N-linked sialyloligosaccharides of recombinant erythropoietin: combination of direct high-performance anion-exchange chromatography and 2-aminopyridine derivatization. Anal. Biochem. 206: 278-87 (1992); **Schmelzer CH et al** Purification and partial characterization of recombinant human differentiation-stimulating factor. Protein Expr. Purif. 1: 54-62 (1990); **Schmelzer CH et al** Biochemical characterization of recombinant human nerve growth factor. J. Neurochem. 59: 1675-83 (1992); **Shima N et al** Tumor cytotoxic factor/hepatocyte growth factor from human fibroblasts: cloning of its cDNA, purification and characterization of recombinant protein. BBRC 180: 1151-8 (1992); **Sun XJ et al** Expression and function of IRS-1 in insulin signal transmission. JBC 267: 22662-72 (1992); **Suzuki A et al** Biochemical properties of amphibian bone morphogenetic protein-4 expressed in CHO cells. BJ 291: 413-7 (1993); **Tressel TJ et al** Purification and characterization of human recombinant insulin-like growth factor binding protein 3 expressed in Chinese hamster ovary cells. BBRC 178: 625-33 (1991)

Cholecystokinin B: see: Gastrin.

Cholinergic differentiation factor: abbrev. CDF or » CNDF (cholinergic neuronal differentiation factor). This factor is identical with » LIF.

Cholinergic neuronal differentiation factor: see: CNDF.

Choline acetyltransferase development factor: abbrev. CAT development factor. See: CDF.

Chondrocyte activating factor: abbrev. CAF. See: CGF (chondrocyte growth factor).

Chondrocyte colony-stimulating factor: abbrev. CCSF. See: CDF (cartilage-derived factors).

Chondrocyte growth factor: see: CGF.

Chondrosarcoma-derived growth factor: also abbrev. CHSA-GF. See: ChDGF.

Chondrosarcoma growth factor: see CGF.

Chorioallantoic membrane assay: abbrev. CAM assay. See: Angiogenesis factors

CHSA-GF: *chondrosarcoma-derived growth factor* see: » ChDGF.

CI: *chemotactic index* see: Chemotaxis.

CIF: *chemokinetic inhibitory factor* A poorly characterized glycoprotein factor of ≈ 30 kDa secreted by human B chronic lymphocytic leukemia (CLL) cells *in vitro* and found also in the serum of CLL patients. Resting normal B lymphocytes from blood and spleen have the capacity to produce CIF spontaneously. CIF receptors have been detected on neutrophils and CIF inhibits human neutrophil migration.

Siegbahn A et al The chemokinetic inhibitory factor (CIF) in serum of CLL patients: correlation with infection propensity and disease activity. Scand. J. Haematol. 35: 80-7 (1985); **Siegbahn A et al** Production of chemokinetic inhibitory factor (CIF) by normal blood and spleen B lymphocytes. Leuk. Res. 10: 179-86 (1986); **Siegbahn A et al** Specific binding of B-CLL cell-derived chemokinetic inhibitory factor (CIF) to human polymorphonuclear leukocytes. Eur. J. Haematol. 39: 172-9 (1987)

CIF: *colony inhibitory factor* This factor has been found in crude extracts of normal human neutrophils. It inhibits the production of » CSA (colony stimulating activity) by monocytes and macrophages *in vitro*. CIF is identical with » Lactoferrin.

Broxmeyer HE et al Identification of lactoferrin as the granulocyte-derived inhibitor of colony-stimulating activity production. JEM 148: 1052-67 (1978)

CIF: *contact inhibitory factor* A poorly characterized factor derived from confluent cultures of a contact-inhibited hamster melanocytic line that reversibly restores density-, anchorage-, and serum-dependent growth of malignant hamster, human, and murine melanoma cells and some other malignant cell lines. CIF also restores several other normal phenotypic characteristics of these cells, including pigment cell differentiation in amelanotic cells. It also influences the distribution and content of specific cytoskeletal elements in some cell lines.

Treatment of the cells with this factor also induces resistance to lysis by natural killer cells without affecting cell viability or cytotoxicity of the NK cells. CIF did not induce the production of interferon in cytotoxicity assays.

Fass E et al Liposome-entrapped contact inhibitory factor: transfer of capacity for density-dependent growth to melanoma cells. J. Invest. Dermatol. 87: 309-12 (1986); **Lipkin G et al** Vitiligo-related pigment cell differentiation antigens are expressed on malignant melanoma cells following phenotypic reversion induced by contact inhibitory factor. Differentiation 30: 35-9 (1985); **Lipkin G et al** Contact inhibitory factor also restores anchorage and serum dependence to hamster melanoma cells. J. Invest. Dermatol. 87: 305-8 (1986); **Nabi ZF et al** Susceptibility to NK cell lysis is abolished in tumor cells by a factor which restores their contact inhibited growth. Cancer 58: 1461-5 (1986)

CIF-A: *cartilage-inducing factor A* This factor is isolated from demineralised bovine bones. It is a dimer consisting of two 13 kDa subunits.

Another form of this factor found during purification and designated ***CIF-B*** is related to CIF-A but differs from it in its aminoterminal sequence. The known sequence of the first 30 aminoterminal amino acids and also some of the biological and biochemical characteristics of both factors suggest that CIF-A is identical with » BSC-1 cell growth inhibitor which, in turn, is identical with TGF-β1 (see: TGF-β) isolated from platelets. CIF-B is identical with TGF-β2 (see: TGF-β).

CIF-A induces the generation of chondrocytes in cultures of fetal mesenchymal rat cells. In soft agar cultures of rat kidney cells CIF-A, in combination with » EGF or » TGF-α, leads to a loss of contact inhibition.

Bonewald LF et al Role of transforming growth factor-β in bone remodeling. Clin. Orthop. 250: 261-76 (1990); **Ellingworth LR et al** Antibodies to the N-terminal portion of cartilage-inducing factor A and transforming growth factor β. JBC 261: 12362-7 (1986); **Hauschka PV et al** Growth factors in bone matrix. JBC

Chromogranin A–C: see addendum: Granins. **Chromostatin:** see addendum: Granins.

261: 12665-74 (1986); **Higgins PJ et al** Contact-inhibitory factor induces alterations in the distribution and content of specific cytoskeletal elements in an established line of rat hepatic tumor cells. IJC 40: 792-801 (1987); **Seyedin SM et al** Cartilage inducing factor B is a unique protein structurally and functionally related to transforming growth factor-β. JBC 262: 1946-9 (1987); **Seyedin SM et al** Cartilage inducing factor A: Apparent identity to transforming growth factor-β. JBC 261: 5693-5 (1986)

CIF-B: *cartilage-inducing factor* B see: CIF-A.

CIK: *cytokine-induced killer cells* These cells are generated *in vitro* by culturing human peripheral blood mononuclear cells in the presence of » IFN-γ, anti-CD3 monoclonal antibody, » IL1, and » IL2. The timing of IFN-γ treatment is critical and optimal results are obtained when IFN-γ is added before IL2 treatment. These cells are highly efficient cytotoxic effector cells capable of lysing tumor cell targets by a mechanism that is non-major histocompatibility antigen (MHC)-restricted. The cells have been used to study the growth behavior of human B cell lymphoma cells in severe combined immune deficient (SCID) mice (see also: Immunodeficient mice). The animals injected with bone marrow contaminated with human lymphoma cells had enhanced survival if the bone marrow was treated with CIK cells before infusion. It has been observed that CIK cells proliferate more rapidly than » LAK cells (lymphokine-activated killer cells) and that they also show a greatly enhanced cytotoxicity.
The majority of CIK effector cells have a T cell phenotype and resemble tumor-infiltrating lymphocytes (see: TIL). Unlike TILs CIK cells do not require contact with tumor cells for stimulation and proliferation. CIK cells express CD3, CD56 and TCRa/b. The cytotoxic effect of these cells against tumor targets is blocked by antibodies directed against lymphocyte function-associated antigen (LFA-1) and its counterreceptor, ICAM-1 (intercellular adhesion molecule 1).
Schmidt-Wolf IGH et al Use of a SCID mouse/human lymphoma model to evaluate cytokine induced killer cells with potent anti-tumor activity. JEM 174: 139-49 (1991); **Schmidt-Wolf IGH et al** Phenotypic characterization and identification of effector cells involved in tumor cell recognition of cytokine-induced killer cells. EH 21: 1673-9 (1993)

Ciliary neuronotrophic factor: see: CNTF.

Ciliary neurotrophic factor: see: CNTF.

CIM6PR: *cation-independent mannose 6-phosphate receptor* see: MPR.

CINC: *cytokine-induced neutrophil chemoattractant* CINC is a chemotactic rat protein that specifically induces the infiltration of neutrophils. It is the rat homologue of » MGSA (melanoma growth stimulatory activity). Unlike MGSA, for which there are three genes, rat CINC is encoded by a single gene. The protein belongs to the » chemokine family of cytokines. Rat CINC shows 61 % sequence identity with » AMCF II (alveolar macrophage-derived neutrophil chemotactic factor).
Hirasawa N et al Induction of neutrophil infiltration by rat chemotactic cytokine (CINC) and its inhibition by dexamethasone in rats. Inflammation 16: 187-96 (1992); **Iida M et al** Level of neutrophil chemotactic factor/gro, a member of the interleukin 8 family, associated with lipopolysaccharide induced inflammation in rats. Infect. Immun. 60: 1268-72 (1992); **Nakagawa H et al** Production of an interleukin-8-like chemokine by cytokine-stimulated rat NRK-49F fibroblasts and its suppression by anti-inflammatory steroids. Biochem. Pharmacol. 45: 1425-30 (1993); **Watanabe K et al** Effect of rat CINC/*gro*, a member of the interleukin-8 family, on leukocytes in microcirculation of the rat mesentery. Exp. Mol. Pathol. 56: 60-9 (1992); **Watanabe K et al** Level of neutrophil chemotactic factor CINC/*gro*, a member of the interleukin-8 family, associated with lipopolysaccharide-induced inflammation in rats. Infect. Immun. 60: 1268-72 (1992); **Watanabe K et al** Rat CINC, a member of the interleukin-8 family, is a neutrophil-specific chemoattractant *in vivo*. Exp. Mol. Pathol. 55: 30-37 (1991); **Wittwer AJ et al** High-level expression of cytokine-induced neutrophil chemoattractant (CINC) by a metastatic rat cell line: purification and production of blocking antibodies. JCP 156: 421-7 (1993); **Zagorski J & DeLarco JE** Rat CINC (cytokine-induced neutrophil chemoattractant) is the homologue of the human GRO proteins but is encoded by a single gene. BBRC 190: 104-10 (1993)

CK: *cytokine* abbrev. for » Cytokines. This abbreviation should not be confused with the abbrev. used in clinical chemistry where CK stands for creatinin kinase.

CK-1: see: Cytokine 1.

c-*kit* ligand: An alternative name for » SCF (stem cell factor) the receptor of which is encoded by » *kit*.

CKR-1: *chemokine receptor R1* This protein has been isolated as a cDNA that encodes a seven trans-membrane-spanning G protein-coupled receptor of 39 kDa with homology to other chemoattractant receptors. CKR-1 acts as an erythrocyte receptor for the C-C family of » chemokines. It binds human and murine macrophage inflammatory protein 1 α (see: MIP), human monocyte chemotactic protein 1 (see: MCP-1), and » RANTES with varying affinities. Another RANTES/MIP-1α receptor has been cloned independently. Both receptors appear

to be related or homologues of » US28, a gene product encoded by cytomegalovirus.
This particular receptor has been shown recently to act also as a receptor for the malarial parasites *Plasmodium vivax* and *Plasmodium falciparum*. It appears to be identical with a previously described ***Duffy blood group antigen***. The lack of expression of this antigen in a high percentage of African Americans is responsible for their resistance to invasion by *Plasmodium*. It has been shown that » IL8 binds minimally to Duffy-negative erythrocytes and that monoclonal antibodies directed against the Duffy blood group antigen block binding of IL8 and other » chemokines to Duffy positive erythrocytes. Two chemokines, » MGSA and IL8 also block the binding of the parasite ligand and the invasion of human erythrocytes by *P. knowlesi*.

Gao JL et al Structure and functional expression of the human macrophage inflammatory protein 1 α/RANTES receptor. JEM 177: 1421-7 (1993); **Horuk R et al** Purification, receptor binding analysis, and biological characterization of human melanoma growth stimulating activity (MGSA). Evidence for a novel MGSA receptor. JBC 268: 541-6 (1993); **Horuk R et al** A receptor for the malarial parasite *Plasmodium vivax*: the erythrocyte chemokine receptor. S 261: 1182-4 (1993); **Horuk R et al** The human erythrocyte inflammatory peptide (chemokine) receptor. Biochemical characterization, solubilization, and development of a binding assay for the soluble receptor. B 32: 5733-8 (1993); **Neote K et al** Molecular cloning, functional expression, and signaling characteristics of a C-C chemokine receptor. Cell 72: 415-25 (1993)

CL100: An » ERG (early response gene) whose transcription is rapidly and transiently stimulated in fibroblasts exposed to oxidative/heat stress and growth factors. The gene is the human homologue of the murine » 3CH134 gene. CL100 encodes a protein tyrosine/threonine phosphatase or 39.3 kDa with significant amino acid sequence similarity to a Tyr/Ser-protein phosphatase encoded by the late gene H1 of vaccinia virus. CL100 protein rapidly and potently inactivates MAP kinase (see: ERK, extracellular signal-regulated kinases) by dephosphorylation and suppresses its activation by the activated *ras* oncogene.

Alessi DR et al The human CL100 gene encodes a Tyr/Thr-protein phosphatase which potently and specifically inactivates MAP kinase and suppresses its activation by oncogenic *ras* in *Xenopus* oocyte extracts. O 8: 2015-20 (1993); **Keyse SM & Emslie EA** Oxidative stress and heat shock induce a human gene encoding a protein-tyrosine phosphatase. N 359: 644-7 (1992)

CLAF: *corpus luteum angiogenic factor* This factor with the activity of an » angiogenesis factor is an aminoterminally truncated form of brain fibroblast growth factor » bFGF. The biological activity of CLAF is indistinguishable from that of pituitary or brain FGF. It is highly mitogenic for cultured bovine vascular endothelial cells. CLAF also stimulates the proliferation of a wide variety of mesoderm- and neuroectoderm-derived cells, including vascular smooth muscle cells, granulosa and adrenal cortex cells, rabbit costal chondrocytes, and corneal endothelial cells.

Baird A & Hsueh AJ Fibroblast growth factor as an intraovarian hormone: differential regulation of steroidogenesis by an angiogenic factor. Regul. Pept. 16: 243-50 (1986); **Gospodarowicz D et al** Corpus luteum angiogenic factor is related to fibroblast growth factor. Endocrinol. 117: 2383-91 (1985)

Class switching of immunoglobulin heavy chains: see: Isotype switching.

CLE: *conserved lymphokine element* This sequence element, also known as CSF Box due to its presence in the promoters of colony stimulating factors, consists of a highly conserved decanucleotide (5′-GGAGATTCCA-3′) found at positions -99 to -108 in the promoter regions of many lymphokine genes, including, for example, the genes encoding » IL2, » IL3, » IL4 » IL5, » IFN-γ, » G-CSF and » GM-CSF. CLE serves as a binding site for specific DNA binding proteins that function as transcription factors and influence the expression of those genes containing a CLE (see also: Gene expression). Optimum binding is stimulated by phorbol 12-myristate 13 acetate (see also: Phorbol esters) and » calcium ionophores (A23187). For a related conserved element see also: Cytokine 1.

Arai KI et al Cytokines: co-ordinators of immune and inflammatory responses. ARB 59: 783-6 (1990); **Masuda ES et al** The granulocyte-macrophage colony-stimulating factor promoter cis-acting element CFL0 mediates induction signals in T cells and is recognized by factors related to AP1 and NFAT. MCB 13: 7399-7407 (1993); **Miyatake S et al** Characterization of the mouse granulocyte-macrophage colony-stimulating factor (GM-CSF) gene promoter: nuclear factors that interact with an element shared by three lymphokine genes – those for GM-CSF, Interleukin-4 (IL4), and IL5. MCB 11: 5894-901 (1991); **Shannon MF et al** Nuclear protein interacting with the promoter region of the human granulocyte/macrophage colony-stimulating factor gene. PNAS 85: 674-8 (1988)

CLMF: *cytotoxic lymphocyte maturation factor* This factor is isolated from a B lymphoblastoid cell line following their stimulation with » phorbol ester and » calcium ionophore. The factor stimulates the proliferation of human phytohemagglutinin-activated lymphoblasts and acts in synergy with suboptimal doses of » IL2. The factor has been renamed » IL12.
Another activity, described as ***CTL maturation factor*** (abbrev. TcMF) which is not identical with

» IL1, » IL2, » IL4, » IFN-α or » IFN-γ also appears to be identical with » IL12.

Gubler U et al Coexpression of two distinct genes is required to generate secreted bioactive cytotoxic lymphocyte maturation factor. PNAS 88: 4143-7 (1991); **Stern AS et al** Purification to homogeneity and partial characterization of cytotoxic lymphocyte maturation factor from human B lymphoblastoid cells. PNAS 87: 6808-12 (1990); **Wong HL et al** Characterization of a factor(s) which synergises with recombinant interleukin 2 in promoting allogeneic human cytolytic T lymphocyte response *in vitro*. CI 111: 39-54 (1988)

Cloning of cytokine genes: see: Gene libraries.

Clonogenic assays: see: Colony formation assay.

CM: *conditioned medium* Occasionally also referred to as CFCS (cell-free culture supernatants). Cm is the general term to describe media in which cells have already been grown. CM is obtained by sterile filtration of these used media and is added to fresh culture media for up to one third of the final volume.

These media are used for the cultivation of particularly fastidious cells and cell lines because they contain many mediator substances secreted by the other cells, that may promote the growth of new cells and may help them to "take".

The biological activity of conditioned media obtained from cultures of different cell lines has been the first indication of the existence of soluble mediators now known under the collective term of » cytokines. The analysis of the effects of conditioned media on different cell types and the biochemical analysis of the constituents responsible for a particular biological activity is still one of a number of ways pursued to detect novel secreted growth factors. The intricacies of this approach are illustrated best by the observation that it may make a great difference if conditioned media are obtained from short-term or long-term cultures of cells, both of which may contain entirely different secreted factors. The effects of conditioned media may be particularly noticeable if chemically defined culture media devoid of serum additions are employed (see: SFM, serum-free medium).

cMG1: An » ERG (early response gene) whose transcription is rapidly and transiently stimulated epithelial cells following mitogenic stimulation. The gene encodes a protein of 338 amino acids and functions as a transcription factor. cMG1 is rapidly induced by » EGF, » Angiotensin, serum, and insulin in intestinal epithelial cells. The human homologue of cMK1 is » Erf-1 or » CL100, both of which are identical with the gene encoding transcription factor » GATA-1.

Gomperts M et al The nucleotide sequence of a cDNA encoding an EGF-inducible gene indicates the existence of a new family of mitogen-induced genes. O 5: 1081-3 (1990); **Gomperts M et al** Mitogen-induced expression of the primary response gene cMG1 in a rat intestinal epithelial cell-line (RIE-1). FL 306: 1-4 (1992)

CMGF: *chicken muscle growth factor* This factor is identical with » bFGF.

Kardami E et al Myogenic growth factor present in skeletal muscle is purified by heparin affinity chromatography. PNAS 82: 8044-7 (1985); **Kardami E et al** Heparin inhibits skeletal muscle growth *in vitro*. Dev. Biol. 126: 19-28 (1988)

cMGF: *chicken myelomonocytic growth factor* abbrev. also MGF (myelomonocytic growth factor).

■ **Sources:** cMGF is isolated from retrovirus-transformed chicken myeloblasts.

■ **Protein Characteristics:** cMGF is a 27 kDa factor containing intramolecular disulfide bonds. The cDNA-derived protein sequence has a length of 189 amino acids. The protein is synthesized with a hydrophobic secretory » signal sequence of 23 amino acids. cMGF is extensively glycosylated and this accounts for the more than 20 different forms separable by two-dimensional gel electrophoresis. These different forms of the protein may also show some varying biological activities.

■ **Gene Structure:** The expression of the cMGF gene involves the interaction of transcription factors AP-1 (see: *jun*) and NF-M, a transcription factor specifically expressed in myeloid cells (see also: Gene expression).

■ **Related factors:** cMGF shows Limited protein homologies to murine and human » G-CSF (35-38 %) and to human » IL6 (18-23 %).

■ **Receptor Structure, Gene(s), Expression:** The biological activity of cMGF is mediated by a specific receptor. This protein of 120 kDa is expressed with a density of ≈ 60-100/ in factor-dependent myeloid cell lines. It is not expressed on other hematopoietic cell lineages.

■ **Biological Activities:** The biological activities of cMGF resemble those of colony-stimulating factors » GM-CSF and » M-CSF (see also: CSF for general information about colony stimulating factors). The transformation of myeloid cells by » oncogenes » *myc* or » *myb* gives rise to cells that do not longer require cMGF for their growth. Dependency of cells on cMGF can be abolished by introducing a functional oncogene into these cells expressing an intracellular domain encoding a tyrosine-specific protein kinase (e. g. *erb*B; see: *neu*). cMGF then acts as an » autocrine growth factor for myeloid cells of different developmental stages.

Adkins B et al Autocrine growth induces by *src*-related oncogenes in transformed chicken myeloid cells. Cell 39: 439-45 (1984); **Gausepohl H et al** Molecular cloning of the chicken myelomonocytic growth factor (cMGF) reveals relationship with interleukin 6 and granulocyte colony stimulating factor EJ 8: 175-81 (1989); **Leutz A & Graf T** Chicken myelomonocytic growth factor. In: Habenicht A (edt) Growth factors, differentiation factors, and cytokines, pp. 215-31, Springer, Berlin 1990); **Leutz A et al** Molecular cloning of the chicken myelomonocytic growth factor (cMGF) reveals relationship to interleukin 6 and granulocyte colony stimulating factor. EJ 8: 175-81 (1989); **Leutz A et al** Purification and characterization of cMGF, a novel chicken myelomonocytic growth factor. EJ 3: 3191-7 (1984); **Sterneck E et al** Autocrine growth induced by kinase type oncogenes in myeloid cells requires AP-1 and NF-M, a myeloid-specific, C/EBP-like factor. EJ 11: 115-26 (1992); **Sterneck E et al** Structure of the chicken myelomonocytic growth factor gene and specific activation of its promoter in avian myelomonocytic cells by protein kinases. MCB 12: 1728-35 (1992)

CNDF: *cholinergic neuronal differentiation factor* abbrev. also CDF (cholinergic differentiation factor). This factor is isolated from rat heart cell cultures. CNDF can induce previously noradrenergic neurons to synthesize acetylcholine and form cholinergic synapses. This change in phenotype occurs without alteration in neuronal survival or growth.

CNDF induces the synthesis of choline acetyltransferase, » Somatostatin, » SP (substance P) and » VIP (vasoactive intestinal peptide) and inhibits the synthesis of tyrosine hydroxylase and neuropeptide Y. The factor is identical with » LIF.

Fukada K Purification and partial characterization of a cholinergic neuronal differentiation factor. PNAS 82: 8795-9 (1985); **Fukada K et al** Immunoaffinity purification and dose-response of cholinergic neuronal differentiation factor. Dev. Brain Res. 62: 203-14 (1991); **Nawa H et al** Recombinant cholinergic differentiation factor (leukemia inhibitory factor) regulates sympathetic neuron phenotype by alterations in the size and amounts of neuropeptide mRNAs. J. Neurochem. 56: 2147-50 (1991); **Rao MS et al** The cholinergic neuronal differentiation factor from heart cell conditioned medium is different from the cholinergic factors in sciatic nerve and spinal cord. Dev. Biol. 139: 65-74 (1990); **Yamamori T et al** The cholinergic neuronal differentiation factor from heart cells is identical to leukemia inhibitory factor. S 246: 1412-6 (1989); Erratum in S 247: 271 (1990); **Yamamori T et al** CDF/LIF selectively increases c-*fos* and *jun*-B transcripts in sympathetic neurons. Neuroreport 2: 173-6 (1991)

CNP: *C-type natriuretic peptide* see: ANF (atrial natriuretic factor).

CNTF: *ciliary neuronotrophic factor*

■ **Alternative Names:** ciliary neurotrophic factor (CNTF); MANS (membrane-associated neurotransmitter-stimulating factor).

■ **Sources:** CNTF is found predominantly in peripheral nerve tissues. The main source appears to be myelin-associated Schwann cells in peripheral nerves and astrocytes in the central nervous system. CNTF is present at very high concentrations within intraocular tissues that contain the same muscle cells innervated by ciliary ganglionic neurons *in vivo*. A substantial portion of the cholinergic differentiation and ciliary neurotrophic activities present in » MANS (Membrane-associated neurotransmitter stimulating factor) preparations can be attributed to CNTF or a CNTF-like molecule.

■ **Protein Characteristics:** CNTF is an acidic cytosolic protein of ≈ 24 kDa. CNTF does not display any homology to other neurotrophic factors such as » NGF and » neurotrophins. At the protein level CNTF from rabbits and humans show ≈ 76 % sequence identity. Rat CNTF and human CNTF show 84 % homology.

CNTF, like » aFGF, » bFGF, and » PD-ECGF (platelet-derived endothelial cell growth factor), does not possess a » signal sequence that would allow secretion of the factor by classical secretion pathways (endoplasmatic reticulum/Golgi system). The mechanism underlying the release of aFGF is unknown.

■ **Gene Structure:** The human gene encoding CNTF maps to chromosome 11q12.2.

■ **Receptor Structure, Gene(s), Expression:** CNTF receptors are expressed exclusively in the nervous system and skeletal muscle. The α subunit of the receptor is anchored to the cell membrane by a glycosyl-phosphatidylinositol linkage. One possible function of this type of linkage is to allow for the regulated release of this receptor component. Potential physiological roles for the soluble CNTF receptor α subunit are suggested by its presence in cerebrospinal fluid and by its release from skeletal muscle in response to peripheral nerve injury.

The other receptor subunit, known as » gp130, is also a component of receptors for » LIF, » Oncostatin M, » IL6 and » IL11. This explains, at least in part, some of the very similar biological activities of these factors which are otherwise unrelated. The dimeric receptor can be converted into a trimeric receptor form containing another protein (LIF binding protein; see: LIF)

CNTF-induced stimulation of fibrinogen gene expression in hepatocytes occurs, at least in part, by CNTF interacting with the IL6 receptor.

Components of the signaling pathway that couples the CNTF receptor to » gene expression have been identified as p91, a protein involved in signaling through the » IFN receptor (see: IRS: interferon response sequence), and two p91-related proteins.

■ **BIOLOGICAL ACTIVITIES:** In contrast to other » neurotrophins, CNTF appears to be expressed relatively late during ontogenesis. CNTF has been proposed to be a lesion factor that is released after nerve injuries and that, in combination with other factors, promotes the survival and the regeneration of neurons. There are some indications, however, that the synthesis of CNTF may also decrease following nerve injuries.

In chicken embryos CNTF prevents the cell death by » apoptosis of motor neurons during the development of the nervous system. It also protects oligodendrocytes against natural and » TNF-induced death (» Apoptosis) but not against complement-induced death (necrosis). In newborn rats local administration of CNTF prevents the degeneration of motor neurons following axotomy. Progressive degeneration of motor neurons in pmn (progressive motor neuronopathy) mice, which carry an autosomal recessive mutation associated with progressive degeneration of the caudo-cranial motor neurons, can be prevented partially by treatment with CNTF. This treatment also prolongs considerably the survival of the animals.

In vitro CNTF promotes the growth of parasympathetic neurons and sympathetic, sensory, and spinal motor neurons. CNTF also plays an important role in the differentiation of O-2A progenitor cells into oligodendrocytes and type 2 astrocytes. In ganglion cells CNTF induces the synthesis of » SP (substance P; see also: tachykinins), » VIP (vasoactive intestinal peptide), » Somatostatin. *In vivo* CNTF stimulates the synthesis of choline acetyltransferase and of a low-affinity receptor for NGF in sympathetic neurons. CNTF has been shown to act synergistically with a Schwann cell-derived factor to provide trophic support to neonatal sympathetic neurons in rats and to down-regulate the responsiveness of those neurons to » NGF.

CNTF induces the hepatic » acute phase protein genes » haptoglobin, α1-antichymotrypsin, α2-macroglobulin and β-fibrinogen in human hepatoma cells (HepG2) and in primary rat hepatocytes with a time course and dose-response comparable with that of » IL6. CNTF has also been shown to function as an endogenous pyrogen (see also: EP) and also induces additional endogenous pyrogenic activities.

■ **ASSAYS:** A human neuroblastoma cell line, » NBFL, responds to CNTF by elevated levels of » VIP (vasoactive intestinal protein) and this has been used to assay this factor. CNTF can also be detected by a modification of the » cell blot assay. A monoclonal antibody directed against human CNTF has been produced recently.

■ **TRANSGENIC/KNOCK-OUT/ANTISENSE STUDIES:** A targeted deletion of the CNTF gene has been created in » ES cells (embryonic stem cells) by homologous recombination. These cells were used subsequently to generate » transgenic mice lacking CNTF. Embryos and animals during the first postnatal weeks do not show any functional or morphological changes. With increasing age the animals display a progressive atrophy and loss of motor neurons showing that CNTF plays an essential role in the maintenance of motor neuron functions in the postnatal period.

■ **CLINICAL USE & SIGNIFICANCE:** One possible use of CNTF is suggested by its neurotrophic activities (see also: neurotrophins). Since CNTF promotes the survival of motor neurons it may be possible to use the factor in the treatment of amyotrophic lateral sclerosis (ALS), a neurodegenerative disorder characterized by a progredient degeneration of upper and lower motor neurons in the spinal chord.

● **REVIEWS: Davis S & Yancopoulos GD** The molecular biology of the CNTF receptor. Curr. Opin. Cell Biol. 5: 281-5 (1993); **Ip NY & Yancopoulos GD** Ciliary neurotrophic factor and its receptor complex. PGFR 4: 139-55 (1992); **Patterson PH** The emerging neuropoietic cytokine family: first CDF/LIF, CNTF and IL6; next ONC, MGF, GCSF? Curr. Opin. Neurobiol. 2: 94-7 (1992); **Rao MS & Landis CC** Cell interactions that determine sympathetic neuron transmitter phenotype and the neurokines that mediate them. J. Neurobiol. 24: 215-32 (1993); **Thoenen H** The changing scene of neurotrophic factors. TINS 14: 165-70 (1991); see also literature cited for » neurotrophins.

● **BIOCHEMISTRY & MOLECULAR BIOLOGY: Barbin G et al** Purification of the chick eye ciliary neuronotrophic factor. J. Neurochem. 43: 1468-78 (1984); **Giovannini M et al** Chromosomal localization of the human ciliary neurotrophic factor gene (CNTF) to 11q12 by fluorescence *in situ* hybridization. Cytogenet. Cell Genet. 63: 62-3 (1993); **Gupta SK et al** Preparation and biological properties of native and recombinant ciliary neurotrophic factor. J. Neurobiol. 23: 481-90 (1992); **Kaupmann K et al** The gene for ciliary neurotrophic factor (CNTF) maps to murine chromosome 19 and its expression is not affected in the hereditary motor neuron disease 'wobbler' of the mouse. Eur. J. Neurosci. 3: 1182-6 (1991); **Lam A et al** Sequence and structural organization of the human gene encoding ciliary neurotrophic factor. Gene 102: 271-6 (1991); **Lam A et al** Expression cloning of neurotrophic factors using *Xenopus* oocytes. J. Neurosci. Res. 32: 43-50 (1992); **Lin LFH et al**

CNTF knockouts: see addendum: CNTF.

Purification, cloning, and expression of ciliary neurotrophic factor. S 246: 1023-5 (1989); **Manthorpe M et al** Purification of adult rat sciatic nerve ciliary neuronotrophic factor. Brain Res. 367: 282-6 (1986); **Masiakowski P et al** Recombinant human and rat ciliary neurotrophic factors. J. Neurochem. 57: 1003-12 (1991); **McDonald JR et al** Expression and characterization of recombinant human ciliary neurotrophic factor from *Escherichia coli*. BBA Gene Struct. Expression 1090: 70-80 (1991); **Negro A et al** Cloning and expression of human ciliary neurotrophic factor. EJB 201: 289-94 (1991); **Negro A et al** Synthesis, purification, and characterization of human ciliary neuronotrophic factor from *E. coli*. J. Neurosci. Res. 29: 251-60 (1991); **Stockli KA et al** Molecular cloning, expression, and regional distribution of rat ciliary neurotrophic factor. N 342: 920-3 (1989); **Watters D et al** Monoclonal antibody that inhibits biological activity of a mammalian ciliary neurotrophic factor. J. Neurosci. Res. 22: 60-4 (1989)

● **RECEPTORS: Davis S et al** The receptor for ciliary neurotrophic factor. S 253: 59-63 (1991); **Davis S et al** Released form of CNTF receptor α component as a soluble mediator of CNTF responses. S 259: 1736-9 (1993); **Davis S et al** LIFRβ and gp130 as heterodimerizing signal transducers of the tripartite CNTF receptor. S 260: 1805-8 (1993); **Ip NY et al** CNTF and LIF act on neuronal cells via shared signaling pathways that involve the IL6 signal transducing receptor component gp130. Cell 69: 1121-32 (1992); **Nesbitt JE et al** Ciliary neurotrophic factor regulates fibrinogen gene expression in hepatocytes by binding to the interleukin-6 receptor. BBRC 190: 544-50 (1993)

● **ASSAYS: Cazzola F et al** Production and characterization of monoclonal antibodies to human ciliary neurotrophic factor with defined epitope recognition. Hybridoma 12: 259-70 (1993)

● **BIOLOGICAL ACTIVITIES: Arakawa Y et al** Survival effect of ciliary neurotrophic factor (CNTF) on chick embryonic motor neurons in culture: comparison with other neurotrophic factors and cytokines. J. Neurosci. 10: 3507-15 (1990); **Burnham P et al** Effects of ciliary neurotrophic factor on the survival and response to nerve growth factor of cultured rat sympathetic neurons. Dev. Biol. 161: 96-106 (1994); **Clutterbuck RE et al** Ciliary neurotrophic factor prevents retrograde neuronal death in the adult central nervous system. PNAS 90: 2222-6 (1993); **Fann MJ Patterson PH** A novel approach to screen for cytokine effects on neuronal gene expression. J. Neurochem. 61: 1349-55 (1993); **Friedman B et al** Regulation of ciliary neurotrophic factor expression in myelin-related Schwann cells *in vivo*. Neuron 9: 295-305 (1992); **Gurney ME et al** Induction of motor neuron sprouting *in vivo* by ciliary neurotrophic factor and basic fibroblast growth factor. J. Neurosci. 12: 3241-7 (1992); **Hagg T et al** Ciliary neurotrophic factor prevents neuronal degeneration and promotes low affinity NGF receptor expression in the adult rat CNS. Neuron 8: 145-58 (1992); **Hagg T & Varon S** Ciliary neurotrophic factor prevents degeneration of adult rat substantia nigra dopaminergic neurons *in vivo*. PNAS 90: 6315-9 (1993); **Hughes SM et al** Ciliary neurotrophic factor induces type-2 astrocyte differentiation in culture. N 335: 70-3 (1988); **Ip NY et al** Ciliary neurotrophic factor enhances neuronal survival in embryonic rat hippocampal cultures. J. Neurosci. 11: 3124-34 (1991); **Lillien LE et al** Extracellular matrix-associated molecules collaborate with ciliary neurotrophic factor to induce type-2 astrocyte development. JCB 111: 635-44 (1990); **Lillien LE et al** Type 2 astrocyte development in rat brain cultures is initiated by a CNTF-like protein produced by type 1 astrocytes. Neuron 1: 485-94 (1988); **Louis JC et al** CNTF protection of oligodendrocytes against natural and tumor necrosis factor-induced death S 259: 689-92 (1993); **Magal E et al** Effect of CNTF on low-affinity NGF receptor expression by cultured neurons from different rat brain regions. J. Neurosci. Res. 30: 560-6 (1991); **Oppenheim RW et al** Control of embryonic motor neuron survival *in vivo* by ciliary neurotrophic factor. S 251: 1616-8 (1991); **Rabinovsky ED** Peripheral nerve injury down-regulates CNTF expression in adult rat sciatic nerves. J. Neurosci. Res. 31: 188-92 (1992); **Rao MS et al** The cholinergic neuronal differentiation factor from heart cell conditioned medium is different from the cholinergic factors in sciatic nerve and spinal cord. Dev. Biol. 139: 65-74 (1990); **Rao MS et al** Effects of ciliary neurotrophic factor (CNTF) and depolarization on neuropeptide expression in cultured sympathetic neurons. Dev. Biol. 150: 281-93 (1992); **Saadat S et al** Ciliary neurotrophic factor induces cholinergic differentiation of rat sympathetic neurons in culture. JCB 108: 1807-16 (1989); **Sendtner M et al** Ciliary neurotrophic factor prevents the degeneration of motor neurons after axotomy. N 345: 440-1 (1990); **Sendtner M et al** Effect of ciliary neurotrophic factor (CNTF) on motoneuron survival. JCS 1991; 15: s103-s9 (1991); **Sendtner M et al** Ciliary neurotrophic factor prevents degeneration of motor neurons in mouse mutant progressive motor neuronopathy N 358: 502-4 (1992); **Seniuk N et al** Decreased synthesis of ciliary neurotrophic factor in degenerating peripheral nerves. Brain Res. 572: 300-2 (1992); **Shapiro L et al** Ciliary neurotrophic factor is an endogenous pyrogen. PNAS 90: 8614-8 (1993); **Symes AJ et al** Ciliary neurotrophic factor coordinately activates transcription of neuropeptide genes in a neuroblastoma cell line. PNAS 90: 572-576 (1993); **Unsicker K et al** Stimulation of neuron survival by basic FGF and CNTF is a direct effect and not mediated by non-neuronal cells: evidence from single cell cultures. Brain Res. Dev. Brain Res. 65: 285-8 (1992); **Wewetzer K et al** CNTF rescues motor neurons from ontogenetic cell death *in vivo*, but not *in vitro*. Neuroreport 1: 203-6 (1990)

● **TRANSGENIC/KNOCK-OUT/ANTISENSE STUDIES: Masu Y et al** Disruption of the CNTF gene results in motor neuron degeneration. N 365: 27-32 (1993)

Cobblestone area: Term used to describe morphologically distinct regions of bone marrow which are zones of active » hematopoiesis. See also: LTBMC (long-term bone marrow culture).

Cobblestone area-forming cells: see: CAFC. See also: LTBMC (long-term bone marrow culture).

Coley's toxin: Name given to the active principle responsible for the antitumor activity observed in terminally ill cancer patients treated with a preparation of heat-killed Gram-positive or Gram-negative bacteria. The study of this phenomenon led to the identification of bacterial lipopolysaccharides (see: LPS) as an active component of the toxins. The active principle ultimately responsible for the antitumor responses is now known as » TNF-β, the synthesis of which is induced by bacterial endotoxins.

Coley WB The treatment of malignant tumors by repeated inoculations of erysipelas: with a report of ten original cases. Am. J. Med. Sci. 105: 487-511 (1893); **Starnes CO** Coley's toxins in perspective. N 357: 11-2 (1992)

Colon cancer cell growth inhibitor: This inhibitor is identical with a binding protein for » IGF known as HT29-IGFBP (HT29 insulin-like growth factor binding protein), isolated from the human colon carcinoma cell line HT29. See: IGF-BP (insulin-like growth factor binding protein).

Culouscou JM & Shoyab M Purification of a colon cancer cell growth inhibitor and its identification as an insulin-like growth factor binding protein. CR 51: 2813-9 (1991)

Colonic growth factor: see: CGF.

Colony forming cells: see: CFC.

Colony formation assay: In this test cells are grown *in vitro* in soft agar (tissue culture medium containing agar as a gelling agent; see also: Cell culture) or other highly viscous media, containing, for example, methylcellulose, plasma gel or fibrin clots. These semisolid media reduce cell movement and allow individual cells to develop into cell clones that are identified as single colonies. These assays are also generally referred to as ***clonogenic assays***. Semisolid media are used to grow transformed cells and hematopoietic cells.

Many normal cells show the phenomenon of adherence, i. e. they grow and divide only if attached to a solid inert support, as is provided, for example, by the glass or plastic surfaces of tissue culture dishes. This requirement is frequently lost during prolonged cell culture and the development of transformed cells. It is indeed one of the phenotypically recognizable characteristics of cell transformation. Transformed cells acquire the ability to grow as colonies in semisolid media (anchorage-independent growth; abbrev. AIG). Several cytokines (see, for example: TGF) also induce the substrate-independent growth of cells. Soft agar colony formation assays are also used to measure the sensitivity of human tumors to anticancer drugs (see: HTCA, human tumor clonogenic assay).

The generation of hematopoietic colonies in a colony formation assay is absolutely dependent on the continuous presence of so-called colony-stimulating factors (see also: CSF for a general discussion). The activity of CSFs is assayed by their capacity to induce hematopoietic progenitor cells (see also: Hematopoiesis) of the bone marrow to develop colonies of differentiated cells after a culture period of 7 – 14 days, consisting of more than 40 cells. These colonies are the progeny of single cells called ***colony-forming cells*** (abbrev. CFC) or ***colony forming units*** (abbrev. CFU). The nature of a particular colony-stimulating factor can sometimes be inferred after staining of these colonies and close examination of the distinct morphological cell types that have developed. Colony formation assays therefore allow the study of the influences of a given factor on the determination of the lineage along which colony forming cells differentiate (see also: Factor-dependent cell lines).

The specific activity of a particular colony-stimulating factor can be derived from dose-response curves obtained by plotting the number of colonies against the concentrations of the factor under examination. These activity curves demonstrate that many of the colony-stimulating factors are active at concentrations in the picomolar range. For a liquid culture assay alternative to classical colony formation assays see: Limiting dilution analysis.

Barr RD et al Growth of erythroid colonies in agar cultures of normal human bone marrow. Blut 50: 179-83 (1985); **Bradley ER & Metcalf D** The growth of mouse bone marrow cells *in vitro*. Aust. J. Exp. Biol. Med. 44: 287-300 (1966); **Broxmeyer HE** Colony assays of hematopoietic progenitor cells and correlations with clinical situations. CRC Crit. Rev. Oncol./Hematol. 1: 227-57 (1983); **Dainiak N et al** Primary human marrow cultures for erythroid bursts in a serum-substituted system. EH 18: 1073-9 (1985); **Dresch C et al** Eosinophil colony formation in semisolid cultures of human bone marrow cells. Blood 49: 835-44 (1977); **Eaves AC & Eaves CJ** Erythropoiesis in culture. In: McCullock EA (edt) Cell culture techniques – clinics in hematology. WB Saunders, Eastbourne, pp 371-91 (1984); **Fauser AA & Messner HA** Identification of megakaryocytes, macrophages, and eosinophils in colonies of human bone marrow containing neutrophilic granulocytes and erythrocytes. Blood 53: 1023-7 (1979); **Iscove VW et al** Erythroid colony formation in cultures of mouse and human bone marrow: analysis of the requirements for erythropoietin by gel formation and affinity chromatography on agarose-concanavalin A. JCP 83: 309-20 (1974); **Iscove VW & Sieber F** Erythroid progenitors in mouse bone marrow detected by macroscopic colony formation in culture. EH 3: 32-43 (1975); **Metcalf D** Hemopoietic colonies. *In vitro* cloning of normal and leukemic cells. Springer, Verlin 1977; **Metcalf D** Mechanisms responsible for size differences between hemopoietic colonies: an analysis using a CSF-dependent hemopoietic cell line. IJCC 10: 116-25 (1992); **Nakeff A et al** *In vitro* colony assay for a new class of megakaryocyte precursor: colony forming unit megakaryocyte (CFU-M). PSEBM 151: 587-90 (1976); **Ogawa M et al** Human marrow erythropoiesis in culture. I. Characterization of methyl cellulose assay. Blood 48: 407-17 (1976); **Pike BL & Robinson WA** Human bone marrow colony growth in agar gel. JCP 76: 77-84 (1970); **Pluznik DH & Sachs L** The cloning of normal mast cells in tissue culture. JCP 66: 319-24 (1965); **Rizzino A** Soft agar growth assays for transforming growth factors and mitogenic peptides. MiE 146: 341-53 (1987); **Rosendaal M et al** Assessing cultured colonies automatically. Leuk. Res. 10: 539-47 (1986); **Stephenson JR et al** Induction of colonies of hemoglobin synthesizing cells by erythropoietin *in vitro*. PNAS 68: 1542-6 (1971); **Testa NG et al** Assays for hematopoietic growth factors. In: Balkwill FR (edt) Cytokines, a practical approach, pp 229-44; IRL Press Oxford 1991

Colony forming unit eosinophil growth stimulating factor: see: CFU-Eo GSF.

Colony forming units: see: CFU.

Colony forming unit spleen: see: CFU-s.

Colony forming unit stimulating activity: see: CFU-SA.

Colony forming unit stimulating factor: abbrev. CFU-SF. This factor is identical with » IL3.

Colony inhibitory factor: see: CIF.

Colony promoting activity: see: CPA.

Colony promoting factor: abbrev. CPF. See: CPA (colony promoting activity).

Colony stimulating activity: abbrev. CSA. See: CSF (colony stimulating factor).

Colony stimulating factors: see: CSF for general information about colony stimulating factors.

Colony stimulating factor 1: see: CSF-1.

Colony stimulating factor 2: see: CSF-2.

Colony stimulating factor 2α: see: CSF-2α.

Colony stimulating factor 2γ: see: CSF-2γ.

Colony stimulating factor 3: see: CSF-3.

Colony stimulating factor 4: see: CSF-4.

Colony stimulating factor 309: see: CSF-309.

Colony stimulating factor α: see: CSF-α.

Colony stimulating factor β: see: CSF-β.

Colorectum cell-derived growth factor: see: CRDGF.

Committed stem cells: see: Hematopoiesis.

Competence factors: see: Cell activation.

Competence genes: see: Gene expression.

Competitive repopulating unit: abbrev. CRU. See: MRA (marrow repopulating ability).

Competitive repopulation assay: see: MRA (marrow repopulating ability).

Complementary DNA: abbrev. cDNA. See: Gene libraries.

Examples of hematopoietic colonies formed by human hematopoietic progenitor cells. ➧
(a) granulocyte-macrophage colony derived from CFU-GM following stimulation with 10 % GCT-CM (giant cell tumour-conditioned medium) containing G-CSF and GM-CSF. Colonies were screened after growth in soft agar for 14 days.
(b) erythroid colony obtained from BFU-E following stimulation with Epo and PHA-LCM (phytohaemagglutinin-stimulated leukocytes-conditioned medium) which contains G-CSF, GM-CSF and IL3. Cells were grown for 14 days in methyl cellulose.
(c) mixed cell coloy obtained from pluripotent CFU-GEMM following stimulation with Epo and PHA-LCM (phytohaemagglutinin-stimulated leukocytes-conditioned medium). Cells were grown in methyl cellulse. Epo was added at day 4 of incubation.
(d) like (a) at lower magnification.
(e) erythroid colony obtained from CFU-E following stimulation with Epo. Cells were grown in methyl cellulose for 7 days.
(f) same as (c) at lower magnification.
(g) multicentric BFU-E colony obtained after stimulation with SCF and 5637-conditioned medium. Colonies were screened after growth in methyl cellulose for 14 days.
(h) same as g, showing unicentric compact cell cluster.
(i) same as g and h showing multicentric and unicentric cell clusters.
(j) same as k.
(k) colonies of CFU-GM obtained after stimulation with SCF and 5637-conditioned medium. Colonies were screened after growth in methyl cellulose for 14 days.
(l) CFC-blast cells showing stromal cell-adherent progenitor cells on primary stromal cell layer.

Photographs a-f courtesy of Dr. A. Mayer, Dept. of Virology, Johannes Gutenberg University, Mainz. Photographs g-l courtesy of Dr. Karin Thalmeier, GSF Hämatologikum, München.

Colony formation assay

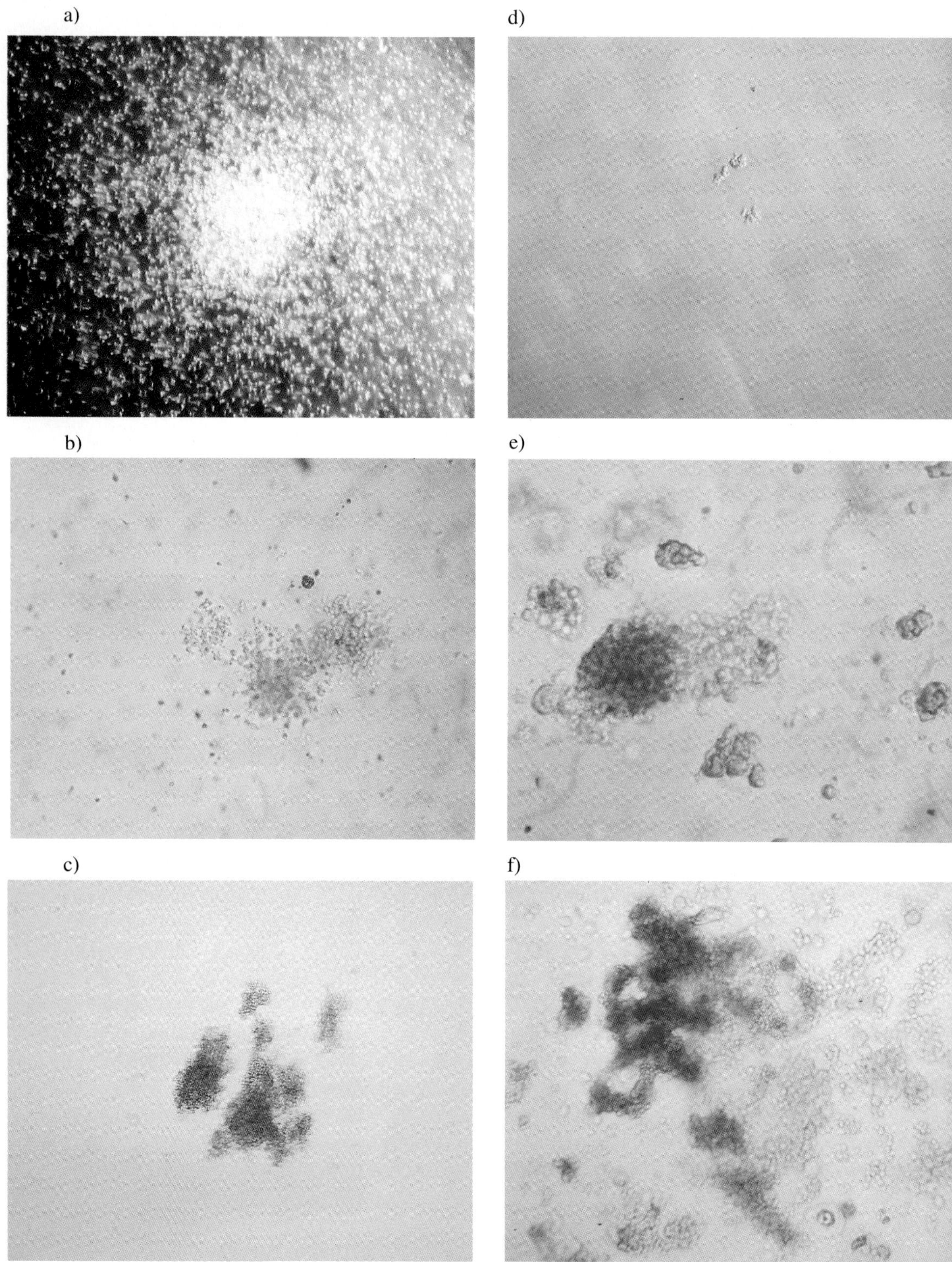

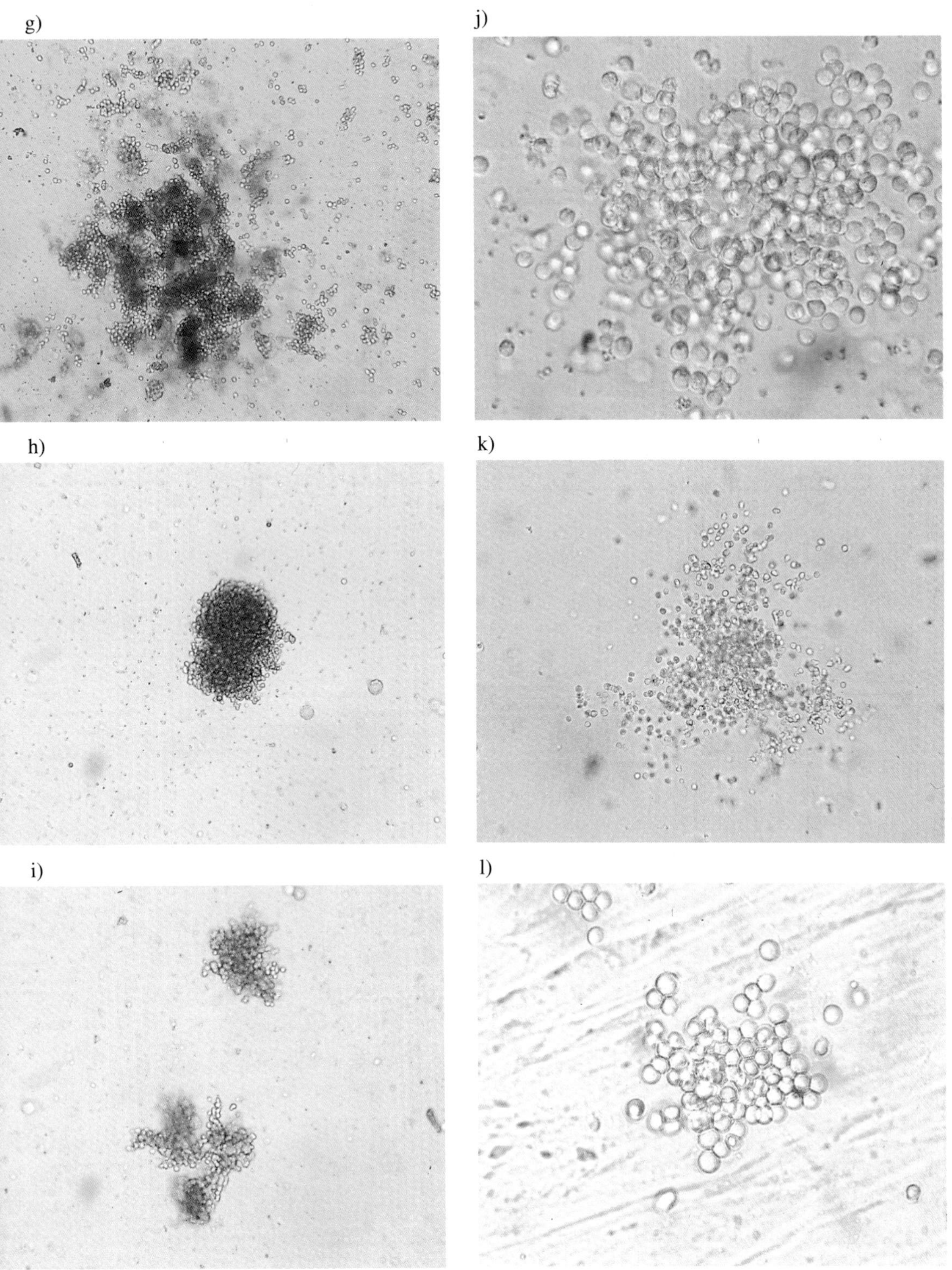
g)
j)
h)
k)
i)
l)

Complement factor Bb: see: HMW-BCGF (high molecular weight B cell growth factor).

Complement S protein: This factor is identical with » vitronectin.

Conditioned medium: see: CM.

Connective tissue activating peptide-3: see: CTAP-3. See also: Beta-Thromboglobulin.

Connective tissue growth factor: see: CTGF.

Consensus interferon: Recombinant consensus » IFN-α (IFN-α Con1) is derived from a synthetic gene comprising the most frequently occurring nucleotide positions at sites differing between individual subtypes of IFN-α. Earlier studies have shown that minor differences in primary structure among the IFN-α protein family are reflected in their potency in selected biological assays and the hope was to obtain an interferon with unique properties. Some studies have shown that the consensus interferon is biologically active and possesses antitumor activity. Some studies suggest that IFN-Con1 may be more effective at lower protein concentrations in clinical applications than other available IFNs.

Klein ML et al Structural characterization of recombinant consensus interferon-α. J. Chromatogr. 454: 205-15 (1988); **Neidhart JA et al** Phase I study of recombinant methionyl human consensus interferon (r-metHuIFN-Con1). J. Biol. Response Mod. 7: 240-8 (1988); **Ozes ON et al** A comparison of interferon-Con1 with natural recombinant interferons-α: antiviral, antiproliferative, and natural killer-inducing activities. J. Interferon. Res. 12: 55-9 (1992)

Conserved lymphokine element: see: CLE.

Constitutive hematopoiesis: see: Hematopoiesis.

Contact inhibitory factor: see: CIF.

Continuous cell lines: see: Cell culture.

contra-IL1: An inhibitor of » IL1. See: IL1ra (Interleukin 1 receptor antagonist).

Conversion assays: see: Bioassays.

Corneal endothelium modulation factor: see: CEMF.

Corneal epithelial cell-derived thymocyte activating factor: see: CETAF.

Corpus luteum angiogenic factor: see: CLAF.

Corticostatins: abbrev. CS; also called: anti-ACTH peptides. A family of arginine-rich cysteine-rich proteins structurally and functionally more or less identical with leukocyte-derived » defensins. Four corticostatins have been isolated from rabbit lung extracts and peritoneal neutrophil extracts and one from human neutrophils. These proteins are expressed also in the pituitary, adrenal medulla, and small intestine.

Some, but not all, corticostatins compete with corticotropin (see: POMC, proopiomelanocortin) for occupancy of the corticotropin receptor, inhibiting adrenocortical ACTH-induced corticosterone production. They show a high degree of specificity in that they do not inhibit the action of » angiotensin II in the adrenal cortex. Of the human defensins only HNP-4 shows significant corticostatic activity, and it is less potent than some rabbit defensins.

Corticostatins may modify the hypothalamic-pituitary-adrenal axis (see: Neuroimmune network) in an endocrine or » paracrine manner in response to infection.

Belcourt D et al Purification of cationic cystine-rich peptides from rat bone marrow. Primary structures and biological activity of the rat corticostatin family of peptides. Regul. Pept. 40: 87-100 (1992); **Singh A et al** Structure of a novel human granulocyte peptide with anti-ACTH activity. BBRC 155: 524-9 (1988); **Tominaga T et al** Effects of corticostatin -I on rat adrenal cells *in vitro*. J. Endocrinol. 125: 287-92 (1990); **Tominaga T et al** Distribution and characterization of immunoreactive corticostatin in the hypothalamic-pituitary-adrenal axis. Endocrinology 130: 1593-8 (1992); **Zhu QZ et al** The corticostatic (anti-ACTH) and cytotoxic activity of peptides isolated from fetal, adult and tumor-bearing lung. J. Steroid. Biochem. 27: 1017-22 (1987); **Zhu QZ et al** Isolation and structure of corticostatin peptides from rabbit fetal and adult lung. PNAS 85: 592-6 (1988); **Zhu QZ et al** Isolation and biological activity of corticostatic peptides (anti-ACTH). Endocrin. Res. 15: 129-49 (1989); **Zhu QZ & Solomon S** Isolation and mode of action of rabbit corticostatic (antiadrenocorticotropin) peptides. Endocrinology 130: 1413-23 (1992)

Corticotropin: see: POMC (proopiomelanocortin).

Corticotropin-releasing factor: abbrev. CRF. See: POMC (proopiomelanocortin).

COS cells: Established cell lines obtained by transformation of African green monkey kidney cells (CV-1 cell line) with Simian virus SV40 that had been mutated so that it could no longer replicate its DNA (SV40*ori*-defective). COS cells constitutively express SV40 large T antigen , an early viral gene product that binds to the SV40 origin of DNA replication and allows host cell DNA polymerases to carry out multiple cycles of viral DNA replication. When the viral SV40 ori sequence is inserted into cloning vectors (for example plasmids of bacterial origin) these altered plasmids can replicate in high copy numbers in monkey cells infected with SV40 helper viruses providing SV40 large T antigen. COS cells are permissive for the lytic growth of wild-type SV40 and support the growth of SV40 mutants encoding non-functional large T antigens; therefore, their use eliminates the need for helper viruses to achieve replication of plasmids carrying the SV40 ori.

COS cells are easy to maintain in culture and a large number of SV40-based cloning vectors are available. These cells are used widely to study » Gene expression, to clone genes by expression, and to test the efficacy of mammalian expression constructs. COS cells are often used as a general purpose mammalian expression system to produce small quantities of recombinant proteins for structural and functional studies. For other expression systems see: Baculovirus expression system, BHK cells, CHO cells, Namalwa cells, *Escherichia coli*.

Edwards CP & Aruffo A Current applications of COS cell based transient expression systems. Curr. Opin. Biotechnol. 4: 558-63 (1993)

Costimulator: This factor was initially described as an activity which was essential for concanavalin A-induced mitogenesis of murine thymocytes. The factor by itself was not mitogenic. Costimulator is identical with » IL2.

Aarden LA et al Revised nomenclature for antigen-nonspecific T cell proliferation and helper factors. CI 48: 433-36 (1979); **Paetkau V et al** Proliferation of murine thymic lymphocytes *in vitro* is mediated by the concanavalin A-induced release of a lymphokine (costimulator). JI 117: 1320-24 (1976); **Paetkau V et al** Cellular origins and targets of Costimulator (IL2). IR 51: 157-75 (1980); **Shaw J et al** Partial purification and molecular characterization of a lymphokine (costimulator) required for the mitogenic response of mouse thymocytes *in vitro*. JI 120: 1967-73 (1978)

Cowpox Virus 38K gene: see: IL1β-Convertase.

CP-10: This factor of 10.3 kDa (76 amino acids) is isolated from supernatants of activated murine spleen cells. CP-10 shows 55% sequence homology with » S100 with pronounced amino acid sequence similarities within the putative N- and C-terminal Ca^{2+}-binding sites.

CP-10 has chemotactic activity for murine polymorphonuclear cells and macrophages. CP-10 has maximal chemotactic activity for neutrophils at 10^{-13} M. The protein mediates localized accumulation of neutrophils and mononuclear cells over 24 hours in delayed-type hypersensitivity reactions.

Lackmann M et al Purification and structural analysis of a murine chemotactic cytokine (CP-10) with sequence homology to S100 proteins. JBC 267: 7499-504 (1992); **Lackmann M et al** Identification of a chemotactic domain of the pro-inflammatory S100 protein CP-10. JI 150: 2981-91 (1993)

CPA: *colony promoting activity* The terms CPA or ***CPF*** (colony promoting factor) have been used as a generic term for (unidentified) factors with colony-stimulating activities (see: CSF).

One distinct CPA activity is isolated from the supernatants of long-term bone marrow cultures. It enhances granulocyte-macrophage colony formation in the presence of colony-stimulating factors. CPA stimulates proliferation and differentiation of more immature cells to CSF-responding granulocytes-macrophage progenitors (GM-CFC). The addition of » IL3 to a GM colony assay system suppresses GM colony formation. CPA does not stimulate proliferation of IL3 dependent » DA cells but facilitates the proliferation of IL6-dependent » MH60.BSF2 cells. CPA is different from » IL1, IL3 or » GM-CSF but similar to or the same as » IL6.

Another poorly characterized CPA activity (> 14 kDa) has been found to be produced by adherent marrow macrophages in long-term murine bone marrow cultures (see also: LTBMC). It does not stimulate granulocyte-macrophage colony formation (see also: CFU-GM) by itself but increases the number of colonies induced by other colony-stimulating factors. Colony formation by marrow cells treated with » fluorouracil is induced by CPA plus » GM-CSF but not by GM-CSF alone. This CPA differs from IL1, IL3, » IL4, IL6, TNF-α, TNF-β, and » SCF and resembles » GM-EF, a human granulomonopoietic enhancing factor.

Chen YR et al Cell source and biological characteristics of murine bone marrow-derived colony promoting activity EH 21: 1219-26 (1993); **Izumi H et al** Role of humoral factors in granulopoiesis: colony promoting factor (CPF) and its target 'pre-CFU-c' in long-term bone marrow culture. Leuk. Res. 7: 155-65 (1983); **Izumi H et al** Characterization of possible receptor of colony-promoting activity (CPA) by comparison with that of colony-stimulating factor (CSF). EH 12: 231-6 (1984); **Kashiwakura I et al** Comparative studies of the colony-promoting activity of porcine kidney extract with several interleukins and colony-stimulating factors. Chem. Pharm. Bull. Tokyo 39: 1495-8 (1991); **Kashiwakura I et al** Properties of colony promoting activity in porcine kidney extract. Chem. Pharm. Bull. Tokyo 39: 425-7 (1991); **Sugimoto K et al** Colony promoting activity: differences from interleukin-3 and similarities to interleukin-6. Biomed. Pharmacother. 44: 135-40 (1990); **Tsurusawa M et al** Colony promoting activity (CPA) produced in long-term culture of murine bone marrow cells. Leuk. Res. 8: 377-86 (1984); **Wang SY et al** Effect of lipopolysaccharide on the production of colony-stimulating factors by the stromal cells in long-term bone marrow culture. EH19: 122-7 (1991)

CPF: *colony promoting factor* see: CPA (colony promoting activity).

CPH: abbrev. for cyclophilin. See: CsA (cyclosporin A).

CR-1: see: CRIPTO.

CR2: Alternative name of the cell surface protein » CD21.

CRBP: *cellular retinoid-binding protein* see: MDGI (mammary-derived growth inhibitor).

CRDGF: *colorectum cell-derived growth factor* This 25 kDa factor is isolated from the human colon carcinoma cell line HT29. It is a homologue of » AR (amphiregulin) and is therefore a member of the EGF-like growth factors (see also: EGF). CRDGF is detected in a Western blot by AR-specific antibodies. The factor, like AR, stimulates the phosphorylation of the EGF receptor.

Culouscou JM et al Colorectum cell-derived growth factor (CRDGF) is homologous to amphiregulin, a member of the epidermal growth factor family. GF 7: 195-205 (1992)

CRE: *cyclic AMP-responsive element* cAMP second messenger pathways provide a chief means by which cellular growth, differentiation, and function can be influenced by extracellular signals. CREs are a common feature of all cAMP-responsive gene promoters and were first identified as an inducible enhancer of genes that can be transcribed in response to increased cAMP levels. This sequence element is an octanucleotide motif (TGACGTCA) which is characterized by a conserved core sequence, 5´-TGACG-3´. CRE serves as a binding site for specific DNA binding proteins that function as transcription factors and influence the expression of those genes containing a CRE. These regulatory factors are generally named ***CREB*** (cAMP responsive element binding proteins). They form a protein family of related factors which bind with different affinities to the DNA either as homodimers or as heterodimers.

At least ten different CREB protein genes have been cloned, including CREB = CREB1, CREB2 = CRE-BP1 = ATF2, ATF (activating transcription factor), ATF-43 = ATF-1 = TREB 36. The CREB1 gene maps to human chromosome 2q32.3-q34. The human CREB2 gene maps to 2q24.1-q32, a site very close to that of the CREB1 gene. CREB proteins are activated by phosphorylation by cAMP-dependent protein kinase A. The physiological consequences of CREB protein phosphorylation have been studied in transgenic animals (mice) expressing a transcriptionally inactive mutant of CREB, which cannot be phosphorylated, in cells of the anterior pituitary. These mice exhibit a dwarf phenotype with atrophied pituitary glands markedly deficient in somatotroph but not other cell types. The normal development of a highly restricted cell type therefore critically depends on the transcriptional activation of CREB.

Cole TJ et al The mouse CREB (cAMP responsive element binding protein) gene: structure, promoter analysis, and chromosomal localization. Genomics 13: 974-82 (1992); **Diep A et al** Assignment of the gene for cyclic AMP-response element binding protein 2 (CREB2) to human chromosome 2q24.1-q32. Genomics 11: 1161-3 (1991); **Flint KJ & Jones NC** Differential regulation of three members of the ATF/CREB family of DNA-binding proteins. O 6: 2019-26 (1991); **Hoeffler JP et al** Cyclic AMP-responsive DNA-binding protein: structure based on a cloned placental cDNA. S 242: 1430-3 (1988); **Hurst HC et al** Identification and functional characterization of the cellular activating transcription factor 43 (ATF-43) protein. NAR 19: 4601-19 (1991); **Karpinski BA et al** Molecular cloning of human CREB-2: an ATF/CREB transcription factor that can negatively regulate transcription from the cAMP response element. PNAS 89: 4820-4 (1992); **Maekawa T et al** Leucine zipper structure of the protein CRE-BP1 binding to the cyclic AMP response element in brain. EJ 8: 2023-8 (1989); **Montminy MR & Bilezikjian LM** Binding of a nuclear protein to the cyclic AMP response element of the somatostatin gene. N 328: 175-8 (1975); **Montminy MR et al** Identification of a cyclic-AMP-responsive element within the rat somatostatin gene. PNAS 83: 6682- 6 (1986); **Taylor AK et al** Assignment of the human gene for CREB1 to chromosome 2q32.3-q34. Genomics 7: 416-21 (1990); **Zhu Z et al** Purification and characterization of a 43-kDa transcription factor required for rat somatostatin gene expression. JBC 264: 6550-6 (1989)

● **TRANSGENIC/KNOCK-OUT/ANTISENSE STUDIES: Struthers RS et al** Somatotroph hypoplasia and dwarfism in transgenic mice expressing a non-phosphorylatable CREB mutant. N 350: 622-4 (1991)

CREB: *cAMP responsive element binding protein* see: CRE (cyclic AMP-responsive element).

CRG: *cytokine responsive genes* General term for a number of genes the expression of which is greatly enhanced after treatment of a murine macrophage cell line (RAW 264.7) with » IFN-γ or the conditioned medium of a mitogen-stimulated spleen cell culture.

Some of these genes encode transcription factors (see also: Gene expression). Other genes are identical with known cytokine genes or genes known

CREB knockouts: see addendum. **CRF knockouts:** see addendum: CRF.

to be expressed after cytokine-treatment of cells (see also: ERG, early response genes). ***CRG-1*** is identical with transcription factor IRF-1 (see: IRS, interferon response sequence) which modulates the activity of interferon genes. ***CRG-2***, which is also called » C7, encodes a protein of 98 amino acids. Its synthesis is induced by » IFN and bacterial lipopolysaccharides. It is identical with, or related to, » IP-10, a protein that belongs to the » chemokine family of cytokines. ***CRG-3*** is identical with a recently identified transcription factor, zif/268. ***CRG-4*** is identical with metallothionein II, the expression of which is known to be induced by interferons. ***CRG-5*** is identical with a transcription factor also known as LRF-1. ***CRG-6*** encodes an unknown interferon-inducible protein. ***CRG-8*** is identical with » Beta-2-Microglobulin the expression of which is also induced by interferon treatment. ***CRG-9*** is a homologue of prostaglandin synthase. ***CRG-10*** is identical with » *mig* (monokine induced by γ interferon).

Farber JM A collection of mRNA species that are inducible in the RAW 264.7 mouse macrophage cell line by γ interferon and other agents. MCB 12: 1535-45 (1992); **Vanguri P & Farber JM** Identification of CRG-2: an interferon-inducible mRNA predicted to encode a murine monokine. JBC 265: 15049-57 (1990)

CRGF: *Cripto growth factor* see: Cripto.

CRIPTO: This protein is produced in large amounts by human undifferentiated teratocarcinoma cells (NTERA-2 clone D1) and the murine F9 teratocarcinoma cell line. CRIPTO is also known as ***CRGF*** (Cripto growth factor) or ***TDGF-1*** (teratocacinoma-derived growth factor). CRIPTO transcripts are detected only in undifferentiated cells and disappear after cell differentiation induced by retinoic acid treatment.

CRIPTO is a member of the family of EGF-related growth factors (see: EGF) and has a length of 188 amino acids. The human gene, designated CR-1, contains six exons and maps to chromosome 3. A related gene, CR-3, with characteristics of a functional retroposon, is on human chromosome Xq21-q22. Southern blot analysis of murine and human DNA reveal the existence of several other CRIPTO-like sequences in the genome.

CRIPTO can function as a dominantly acting oncogene since its overexpression in murine NIH3T3 fibroblasts and mammary epithelial cells can lead to transformation *in vitro*.

CRIPTO is preferentially expressed in colorectal tumors, functioning as an » autocrine or » paracrine growth factor for these cells. It does not appear to be expressed in normal colonic tissues and may therefore be useful as a tumor marker.

■ **TRANSGENIC/KNOCK-OUT/ANTISENSE STUDIES:** Inhibition of CRIPTO expression by the use of » Antisense oligonucleotides has recently been shown to inhibit the growth of human colon carcinoma cell lines expressing this factor. Cells expressing the antisense message have also been found to be less tumorigenic following transplantation into nude mice (see also: Immunodeficient mice) as demonstrated by the smaller size and the longer latency periods of the transplanted tumor cells.

Ciardiello F et al Expression of cripto, a novel gene of the epidermal growth factor gene family, leads to *in vitro* transformation of a normal mouse mammary epithelial cell line. CR 51: 1051-4 (1991); **Ciccodicola A et al** Molecular characterization of a gene of the 'EGF family' expressed in undifferentiated human NTERA2 teratocarcinoma cells. EMBO J. 8: 1987-91 (1989); **Dono R et al** Isolation and characterization of the CRIPTO autosomal gene and its X-linked related sequence. Am. J. Hum. Genet. 49: 555-65 (1991); **Kuniyasu H et al** Expression of cripto, a novel gene of the epidermal growth factor family, in human gastrointestinal carcinomas. Jpn. J. Cancer Res. 82: 9069-73 (1991); **Saeki T et al** Differential immunohistochemical detection of amphiregulin and cripto in human normal colon and colorectal tumors. CR 52: 3467-73 (1992)

● **TRANSGENIC/KNOCK-OUT/ANTISENSE STUDIES: Ciardiello F et al** Inhibition of CRIPTO expression and tumorigenicity in human colon cancer cells by antisense RNA and oligodeoxynucleotides. O 9: 291-98 (1994)

crmA: *cytokine response modifier A* see: IL1β-Convertase.

CRU: *competitive repopulating unit* see: MRA (marrow repopulating ability).

Cryptdins: see: Defensins.

Crystal-induced chemotactic factor: see: CCF.

CSA: *colony stimulating activity* A general term used in the older literature (around 1981 and before) to describe colony-stimulating factor activities (see: CSF), i. e. factors supporting the clonal growth of hematopoietic cells in a » Colony formation assays. This term was used as a generic term in particular to refer to factors or activities that were believed to be different from other colony stimulating factors known at this time. Specific activities are indicated by a prefix indicating the target cell type responding (see: G-CSF, GM-CSF, M-CSF, Multi-CSF.

The acronym CSA is also in current use for unidentified or poorly characterized activities behaving like colony-stimulating factors.

CSA: *colony stimulating activity* This factor is identical with » M-CSF. See also: CSF for general information about colony stimulating factors.

Dexter TM et al The role of hemopoietic cell growth factor (interleukin 3) in the development of hemopoietic cells. Ciba Found. Symp. 116: 129-47 (1985)

CsA: also abbrev. CyA. Abbrev. for cyclosporin. A water-insoluble cyclic oligopeptide with a molecular mass of 1202.63. isolated from the fungus imperfectus *Tolypochladium inflatum* Gams. The compound contains some unusual components such as L-α-Aminobutyric acid, D-alanine, and the C9-amino acid 3-Hydroxyl-4-Methyl-2-methylamino-6-octenic acid. Several amino acids are N-methylated. Many structurally related metabolites of cyclosporin (cyclosporins B-Z) showing identical or similar pharmacological activities have been described.
Cyclosporins have antifungal, antiparasitic and anti-inflammatory activities. CsA inhibits the growth of smooth muscle cells of the arterial intima and may therefore be an inhibitor for vascular overreactions, as seen, for example, in the development of stenoses after arterial injuries. The most important activity of CsA from a clinical viewpoint is probably the suppression of humoral and cell-mediated immune responses observed in all species tested so far.
The immune-suppressive activity of cyclosporin is the result of the inhibition of T helper cells and cytotoxic T cells. T suppressor cells are less affected. CsA selectively blocks many immunoregulatory functions of activated T cells (see also: Cell activation) and, among other things, inhibits the release of » IL2, » IL3, » IL4, » IL5, » IFN-γ, » GM-CSF and » TNFα. CsA does not act directly but exerts its effects by inhibiting certain lymphocyte-specific DNA-binding transcription factors which, in turn, control the expression of many cytokine genes by binding to regulatory sequences contained in their promoter regions (see also: Gene expression).
The exact mechanism of action of cyclosporin is still unknown. CsA completely inhibits the activation of transcription of the IL2 gene which is modulated by transcription factors NF-AT, NFIL2A and NF-IL2B (see also: Figure showing the promoter structure of » IL2). Cyclosporin also partially inhibits transcription factor NFκB-mediated transcriptional activation. Destabilization of cytokine mRNAs may also be involved in CsA activities.
Cyclosporin binds with high affinity to a protein found in large amounts in the cytosol but also observed in a membrane-bound form. This 17 kDa protein, known as ***Cyclophilin*** (CYPH), constitutes ≈ 0.1-0.4 % of all cytosolic proteins. The analysis of human and rat genomic DNA has shown that cyclophilin is a member of a multigene family. Cyclophilin is identical with ***Peptidyl-prolyl cis-trans-Isomerase*** (PPIase), an enzyme normally catalyzing the isomerisation of proline-containing peptide bonds. This enzyme is also thought to be essential for the correct sterical folding of a number of enzymes. It has been proposed that the isomerisation of some of the substrates of cyclophilin may expose DNA binding sites required for the activation of some genes encoding cytokines. Two related human cyclophilin genes have been cloned. One, encoding cyclophilin B, maps to human chromosome 15.
A homologue of cyclophilin, designated ***sp18***, has been observed to be released by murine LPS-activated macrophages. This protein also acts as a

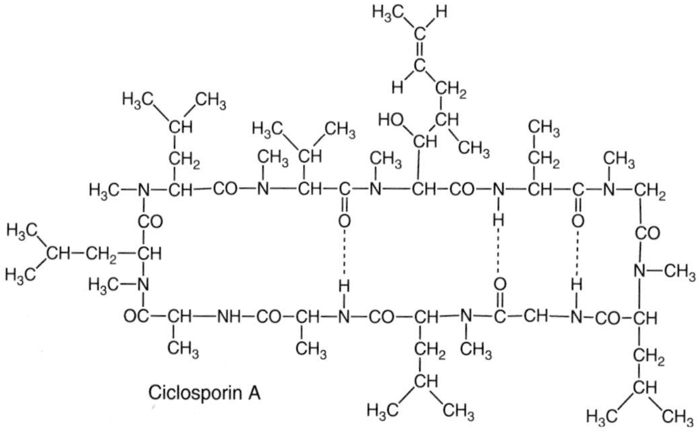

Ciclosporin A

PPIase, is a potent chemoattractant (see also: Chemotaxis) for human leukocytes and monocytes and has pro-inflammatory activity (see also: Inflammation). Like cyclophilin sp18 binds CsA which also inhibits its chemotactic activity. The pro-inflammatory activity of secreted sp18 suggests that it is a cytokine. In addition, these observations also suggest that some of the immunosuppressive actions of CsA may be due to the interactions with extracellular forms of cyclophilin and that these proteins act on different *in vivo* targets. A cyclophilin-like secreted protein of 21 kDa with similar activities, designated SCYLP (cyclophilin-like protein), has been found in human milk.

The capacity of cyclosporin to prevent T cell activation (see also: Cell activation) can be abolished by exogenous » IL2. This demonstrates that the inhibition of IL2 synthesis plays a pivotal role for the biological activities of CsA.

It has been shown that the measurement of IL2 receptor levels may allow discrimination of CsA-mediated hepatotoxicity and a beginning graft-versus-host reaction.

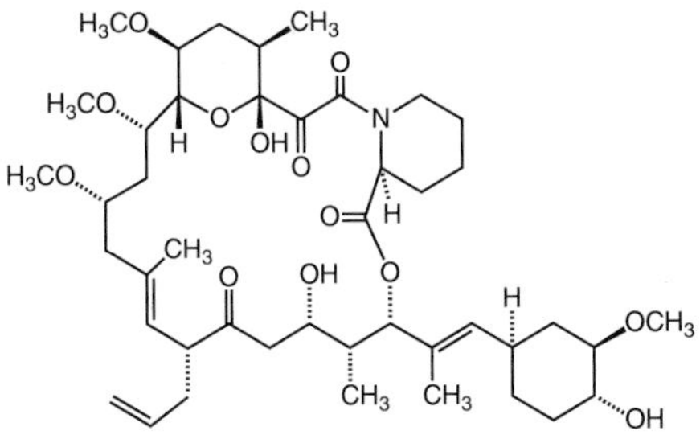

FK506

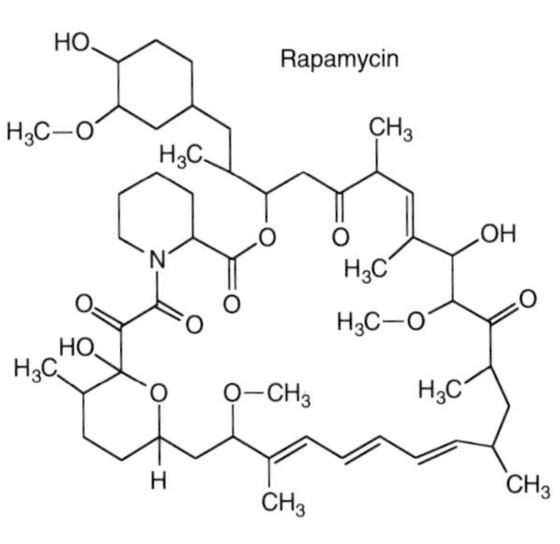

Rapamycin

Another compound showing biological activities similar to those of CsA is the structurally unrelated macrolide compound ***FK506***. FK506 and its structural analog, ***Rapamycin***, also acts by binding to intracellular proteins and suppresses T cell activation processes and the expression of lymphokines at concentrations ≈ 10-100-fold lower than CsA. The intracellular FK506 binding protein, FKBP, which is encoded by a gene located on human chromosome 20, is also a PPIase which is, however, not identical with cyclophilin. It is most likely identical with an endogenous inhibitor of protein kinase C.

● **REVIEWS: Foxwell BM & Ruffel B** The mechanisms of action of cyclosporin. Cardiol. Clin. 8: 107-17 (1990); **Hohman RJ & Hultsch T** Cyclosporin A: new insights for cell biologists and biochemists. New Biol. 2: 663-73 (1990); **McKeon F** When worlds collide: immunosuppressants meet protein phosphatases. Cell 66: 823-6 (1991); **Schumacher A & Nordheim A** Progress towards a molecular understanding of cyclosporin A-mediated immunosuppression. Clin. Investig. 70: 773-9 (1992)

● **BINDING PROTEINS: Danielson PE et al** p1B15A cDNA clone of the rat mRNA encoding cyclophilin. DNA 7: 261-7 (1988); **Dilella AG** Chromosomal assignment of the human immunophilin FKBG-12 gene. BBRC 179: 1427-33 (1991); **Flanagan WM et al** Nuclear association of a T cell transcription factor blocked by FK-506 and cyclosporin A. N 352: 803-7 (1991); **Friedman J & Weissman I** Two cytoplasmic candidates for immunophilin action are revealed by affinity for a new cyclophilin: one in the presence and one in the absence of CsA. Cell 66: 799-806 (1991); **Goebl MG** The peptidyl-prolyl isomerase, FK506-binding protein, is most likely the 12 kd endogenous inhibitor 2 of protein kinase C. Cell 64: 1051-2 (1991); **Haendler B & Hofer B** Characterization of the human cyclophilin gene and of related processed pseudogenes. EJB 190: 477-82 (1990); **Harding MW et al** A receptor for the immunosuppressant FK506 is a *cis-trans* peptidyl-prolyl isomerase. N 341: 758-60 (1989); **Hasel KW et al** An endoplasmatic reticulum-specific cyclophilin. MCB 11: 3484-91 (1991); **Jin YJ et al** Molecular cloning of a membrane-associated human FK506- and rapamycin-binding protein, FKBP-13. PNAS 88: 6677-81 (1991); **Liu J et al** Cloning, expression, and purification of human cyclophilin in *Escherichia coli* and assessment of the catalytic role of cysteines by site-directed mutagenesis. PNAS 87: 2304-8 (1990); **Maki N et al** Complementary DNA encoding the human T cell FK506-binding protein, a peptidylprolyl cis-trans isomerase distinct from cyclophilin. PNAS 87: 5440-3 (1990); **Michnick SW et al** Solution structure of FKBP, a rotamase enzyme and receptor for FK506 and rapamycin. S 252: 836-8 (1991); **Price ER et al** Human cyclophilin B: a second cyclophilin gene encodes a peptidyl-prolyl isomerase with a signal sequence. PNAS 88: 1903-7 (1991); **Sherry B et al** Identification of cyclophilin as a proinflammatory secretory product of lipopolysaccharide-activated macrophages. PNAS 89: 3511-5 (1992); **Spik G et al** A novel secreted cyclophilin-like protein (SCYLP). JBC 266: 10735-8 (1991; **Standaert RF et al** Molecular cloning and overexpression of the human FK506-binding protein FKBP. N 346: 671-4 (1990); **Takahashi N et al** Peptidyl-prolyl *cis-trans* isomerase is the cyclosporin A-binding cyclophilin. N 337: 473-5 (1989); **Van Duyne GD et al** Atomic structure of FKBG-FK506, an immunophilin-immunosuppressant complex. S 252: 839-42 (1991)

● **BIOLOGICAL ACTIVITIES: Andersson J et al** The effects of FK 506 on cytokine production are dependent on the mode of cell activation. Transplant. Proc. 23: 2916-9 (1991); **Andersson J et al** Effects of FK506 and cyclosporin A on cytokine production studied *in vitro* at a single-cell level. Immunology 75: 136-42 (1992); **Banerji SS et al** The immunosuppressant FK-506 specifically inhibits mitogen-induced activation of the interleukin-2 promoter and the isolated enhancer elements NFIL2A and NF-

AT1. MCB 11: 4074-7 (1991); **Baumann G et al** Cyclosporin A and FK-506 both affect DNA binding of regulatory nuclear proteins in the human interleukin-2 promoter. The New Biologist 3: 270-8 (1991); **Brabletz T et al** The immunosuppressive FK506 and cyclosporin A inhibit the generation of protein factors binding to the two purine boxes of the interleukin 2 enhancer. NAR 19: 61-7 (1991); **Emmel EA et al** Cyclosporin A specifically inhibits function of nuclear proteins involved in T cell activation. S 246: 1617-20 (1989); **Ferns GA et al** Vascular effects of cyclosporin A *in vivo* and *in vitro*. Am. J. Pathol. 137: 403-13 (1990); **Hanke JH et al** FK506 and rapamycin selectively enhance degradation of IL2 and GM-CSF mRNA. Lymphokine Cytokine Res. 11: 221-31 (1992); **Hatfield SM & Roehm NW** Cyclosporine and FK506 inhibition of murine mast cell cytokine production. J. Pharmacol. Exp. Ther. 260: 680-8 (1992); **Johansson A & Moller E** Evidence that the immunosuppressive effects of FK506 and cyclosporin are identical. Transplantation 50: 1001-7 (1990); **Jonasson L et al** Cyclosporin A inhibits smooth muscle proliferation in the vascular response to injury. PNAS 85: 2303-6 (1988); **Kallen J et al** Structure of human cyclophilin and its binding site for cyclosporin A determined by X-ray crystallography and NMR spectroscopy. N 353: 276-9 (1991); **Krönke M et al** Cyclosporin A inhibits T cell growth factor gene expression at the level of mRNA transcription. PNAS 81: 5214-8 (1984); Li, B et al Inhibition of interleukin 2 receptor expression in normal human T cells by cyclosporine. Demonstration at the mRNA, protein, and functional levels. Transplantation 53: 146-51 (1992); **Mattila PS et al** The actions of cyclosporin A and FK506 suggest a novel step in the activation of T lymphocytes. EJ 9: 4425-33 (1990); **Randak C et al** Cyclosporin A suppresses the expression of the interleukin 2 gene by inhibiting the binding of lymphocyte-specific factors to the IL2 enhancer. EJ 9: 2529-36 (1990); **Shi Y et al** Cyclosporin A inhibits activation-induced cell death in T cell hybridomas and thymocytes. N 339: 625-6 (1991); **Simpson MA et al** Sequential interleukin 2 and interleukin 2 receptor levels distinguish rejection from cyclosporin toxicity in liver allograft recipients. Arch. Surg. 126: 717-20 (1991); **Thomson AW** The immunosuppressive macrolides FK-506 and rapamycin. Immunol. Lett. 29: 105-11 (1991); **Tocci MJ et al** The immunosuppressant FK506 selectively inhibits expression of early T cell activation genes. JI 143: 718-26 (1989); **Yoshimura N et al** Effect of a new immunosuppressive agent, FK-506, on human lymphocyte responses *in vitro*. II. Inhibition of the production of IL2 and γ-IFN, but not B cell stimulating factor α. Transplantation 47: 356-9 (1989)

CSBP-1: This transcription factor is identical with » GATA-1.

Gumucio DL et al Phylogenetic footprinting reveals a nuclear protein which binds to silencer sequences in the human γ and ε globin genes. MCB 12: 4919-29 (1992)

CSF: *cell scattering factors* see: Motogenic cytokines.

CSF: *colony stimulating factor; colony stimulating activities* also abbrev. CSA.

This term goes back to experiments carried out in the 60ies which attempted to grow colonies of bone marrow cells immobilized in soft agar or methyl cellulose. It was found that » feeder cell layers and also the supplementation of the growth media with conditioned medium (see: CM) considerably promoted colony growth, suggesting the existence of colony-stimulating factors secreted by the cells. While hematopoietic progenitor cells could be maintained only for short periods of time in the absence of such factors, their presence allowed the development of colonies containing erythroid cells, neutrophils, eosinophils, macrophages, and megakaryocytes (see also: Hematopoiesis). The biochemical analysis of various colony-stimulating activities supporting the growth and development of these cell types revealed that there existed many different and distinct factors of this sort.

Colony-stimulating factors belong to the group of regulatory proteins and peptides known as » cytokines. Many of these factors are either N- or O-glycosylated. Glycosylation has been shown to enhance the solubility, stability and resistance to proteolytic enzymes. It does not appear to be required for the full spectrum of biological activities of these factors. The genes encoding many of the human colony-stimulating factors have been cloned and mapped. Some of the genes are in close vicinity but they do not show great homology among each other with the exception of some conserved regions.

Colony-stimulating factors are produced by many different cell types, including, for example, B lymphocytes, epithelial cells, fibroblasts, endothelial cells, macrophages, stromal cells, T lymphocytes. They are synthesized as precursor molecules containing a classical hydrophobic secretory » signal sequence of ≈ 25-32 amino acids (see also: Gene expression). The secreted factors have an extremely high specific biological activity ($10^8 - 10^9$ units/mg protein) and are, therefore, active at very low concentrations (1-100 pM). These factors are absolutely required for the proliferation of hematopoietic progenitor cells. The concentrations required for the mere maintenance of viability are usually orders of magnitude lower than those required to induce cell proliferation or to elicit specific functional activities of the cells.

The names of the individual factors usually indicate the cell types that respond to these factors. The classical colony-stimulating factors include » M-CSF (macrophage-specific), » G-CSF (granulocyte-specific), » GM-CSF (macrophage/granulocyte-specific), » IL3 (multifunctional) and » MEG-CSA (megakaryocyte-specific). G-CSF and M-CSF are lineage-specific while GM-CSF and IL3 are multifunctional hematopoietic factors acting on earlier stages of differentiation of hema-

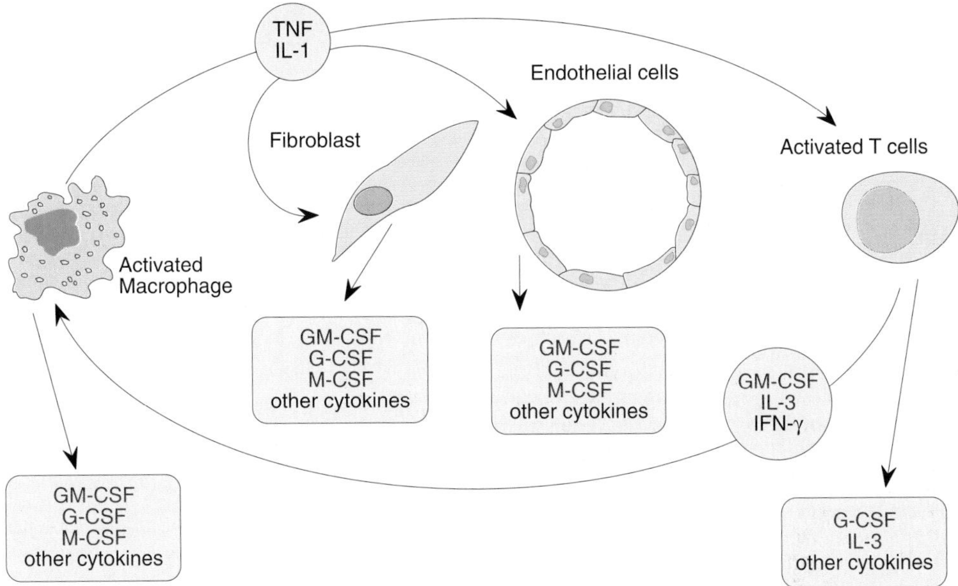

Sources of colony-stimulating factors.
Colony-stimulating factors are constituents of a complicated cytokine network.
They are produced together with other cytokines by a variety of cell types and their synthesis is influenced by many different factors.

topoietic progenitor cells (see also: Hematopoiesis).

Some of the biological activities of colony-stimulating factors are concentration-dependent, i. e. different biological activities may be observed at low and high concentrations. In addition, the activities of many of these factors can be synergised by other factors.

Factors such as » Epo which is involved in erythropoiesis, but also most of the factors known as » interleukins, are not classified as colony-stimulating factors. In combination with colony-forming factors they constitute the group of so-called » hematopoietins (hematopoietic growth factors) (see also: Hematopoiesis).

The synthesis of colony-stimulating factors can be stimulated by immunological reactions and products of infectious agents. In some cell types their synthesis can be stimulated by a factor of up to 1000. The induction of synthesis may be direct or a consequence of the activity of other hematopoietic factors. The synthesis of » IL1, for example, is induced by bacterial » endotoxins or foreign antigens, and IL1, in turn, is capable of stimulating the synthesis of various colony-stimulating factors.

Colony-stimulating factors may act as differentiation factors initiating the irreversible terminal differentiation of progenitor cells. Colony-stimulating factors also induce the synthesis of many other growth factors and play an important role in the maintenance of membrane integrity and cell viability. The proliferation of mature cells is usually not influenced by these factors, they may, however, prolong their life span. Apart from their actions on a number of hematopoietic cell types colony-stimulating factors also influence other cell types.

The biological activities of colony-stimulating factors are mediated by highly specific membrane receptors. As a rule receptor-ligand complexes are internalized by the cells. This process is followed by a complex cascade of more or less biochemically poorly characterized reactions which eventually culminate in cell proliferation or cell differentiation. Some colony-stimulating factor receptors are bifunctional molecules, i. e., they contain a growth factor binding site in the extracellular domain and possess the activity of protein kinases in their intracellular domain. These protein kinases are responsible for the » autophosphorylation of receptor proteins which is an important event signaling that a ligand has bound to the receptor. Post-receptor events include a rapid and transient transition of cytoplasmic protein kinase C into the membranes and the activation of the Na^+/H^+ antiport system which leads to an intracellular alkalization. Withdrawal of the factors induces, among other things, a reduction of glucose transport and a rapid decline in intracellular ATP concentration. It can therefore be assumed that these factors also play an important role in processes governing the primary metabolism of their target cells.

The stimulation of hematopoietic cells in particular by colony-stimulating factors is the basis of their clinical use in patients suffering from diverse forms of hematopoietic disorders. The use of molecular biological techniques, culminating in the cloning and expression of the genes encoding the colony-stimulating factors has been instrumental in these development. Most of these factors are now available in highly purified form and in sufficient amounts allowing an assessment of their therapeutical efficacy and their use in the clinic.

The use of these mediators opens ways of reconstituting aberrant hematopoietic processes in a physiological manner. This includes more or less all varieties of disorders which are characterized either by an insufficient maturation of blood cells or by a reduced production of leukocytes. The fact that different cell types vary in their responses to individual colony-stimulating factors can be used to expand differentially defined populations of progenitor cells. On the one hand, these factors can be used to correct chemotherapy-induced cytopenias. On the other hand colony-stimulating factors can be used to counteract cytopenia-related predisposition to infections and hemorrhages. Since these factors are also capable to influence the growth cycle of many malignant cell types, many of them showing at least tumoricidal activities, colony-stimulating factors may also be used in the directed treatment of various cancers either alone or in combination with cytotoxic drugs.

The major advantage of treatments involving the use of colony-stimulating factors is certainly the more or less physiological reconstitution of hematopoietic processes. So far there are no indications that treatments with these factors lead to an exhaustion of the normal stem cell population in the bone marrow or to the proliferation on non-myeloid tumors.

A novel concept is the so-called peripheral stem cell separation. This technique involves the expansion of peripheral stem cells by *in vivo* pretreatment with suitable colony-stimulating factors and their subsequent collection. These cells can later be re-infused to normalize blood parameters, following, for example, extensive courses of drug treatment.

In summary the available colony-stimulating factors can be used in many instances and the entire spectrum of clinical indications is only just now emerging.

Behring Institute Mitteilungen Vol. 90: Colony stimulating factors; **Brugger W et al** Clinical role of colony stimulating factors. Acta Haematol. 86: 138-47 (1991); **Clark SC & Kamen R** The human hematopoietic colony-stimulating factors. S 236: 1229-37 (1987); **Dexter TM et al** (eds) Colony stimulating factors: Molecular and cellular biology. Marcel Dekker, New York 1990; **Habenicht A** (edt) Growth factors, differentiation factors, and cytokines. Springer, Berlin, 1990; **Kelso A & Metcalf D** T lymphocyte-derived colony-stimulating factors. AI 48: 69-105 (1990); **Metcalf D** The colony stimulating factors. Discovery, development, and clinical applications. Cancer 10: 2185-95 (1990); **Metcalf D et al** Synergistic suppression: anomalous inhibition of the proliferation of factor-dependent hemopoietic cells by combination of two colony-stimulating factors. PNAS 89: 2819-23 (1992); **Moore MAS** The clinical use of colony stimulating factors. ARI 9: 159-91 (1991); **Nakayama N et al** Colony-stimulating factors, cytokines, and hematopoiesis. COI 2: 68-77 (1989); **Negrin RS & Greenberg PL** Therapy of hematopoietic disorders with recombinant colony-stimulating factors. Adv. Pharmacol. 23: 263-296 (1992); **Neidhart JA** Hematopoietic colony-stimulating factors. Uses in combination with standard chemotherapeutic regimens and in support of dose intensification. Cancer 70: s913-20 (1992); **Nicola NA** Receptors for colony-stimulating factors. Brit. J. Haematol. 77: 133-8 (1991); **Pimentel E** Colony-stimulating factors. Ann. Clin. Lab. Sci. 20: 36-55 (1990); **Talmadge JF et al** Protective, restorative, and therapeutic properties of recombinant colony stimulating factors. Blood 79: 2093-2103 (1989); **Parry DAD et al** Conformation homologies among the cytokines: interleukins and colony-stimulating factors. J. Mol. Recog. 1: 107-11 (1988); **Polli EE** The role of hematopoietic growth factors in bone marrow transplantation. Acta Haematol. 86: 155-61 (1991); **Williams GT et al** Haemopoietic colony stimulating factors promote cell survival by suppressing apoptosis. N 343: 76-9 (1990)

CSF-1: *colony stimulating factor 1* This factor was initially described as a macrophage growth factor promoting the development of macrophage-containing colonies in a » colony formation assay. It is identical with » M-CSF. A splice variant of CSF-1 has been named ***CSF-4***. See also: CSF for general information about colony stimulating factors.

Ladner MB et al Human CSF-1: gene structure and alternative splicing of mRNA precursors. EJ 6: 2693-8 (1987); **Pimentel E** Colony-stimulating factors. Ann. Clin. Lab. Sci. 20: 36-55 (1990); **Ralph P et al** Molecular and biological properties of human macrophage growth factor, CSF-1. CSHSQB 51: 679-83 (1986); **Stanley ER et al** CSF-1 – a mononuclear phagocyte lineage-specific hemopoietic growth factor. JCBc 21: 151-9 (1983)

CSF-2: *colony stimulating factor 2* This factor was initially described as an activity promoting the development of macrophage- and granulocyte-containing colonies in a » colony formation assay.

It is isolated from cultured murine lung cells and human squamous cells line CHU-2. The factor is identical with » GM-CSF. See also: CSF for general information about colony stimulating factors.
Bowlin TL et al Biologic properties of interleukin 3. II. Serologic comparison of 20-α -SDH-inducing activity, colony-stimulating activity, and WEHI-3 growth factor activity by using an antiserum against IL 3. JI 133: 2001-6 (1984); **Hapel AJ et al** Different colony-stimulating factors are detected by the "interleukin-3"-dependent cell lines FDC-Pl and 32D cl-23. Blood 64: 786-90 (1984); **Nicola NA et al** Purification of a factor inducing differentiation in murine myelomonocytic leukemia cells. JBC 258: 9017-23 (1983); **Nomura H et al** Purification and characterization of human granulocyte colony-stimulating factor (G-CSF). EJ 5: 871-6 (1986)

CSF-2α: *colony stimulating factor 2α* This factor was initially described as an activity secreted by » LBRM-33 and » WEHI-3 cells, stimulating the growth of bone marrow cells in the granulocyte, macrophage, megakaryocyte, mast cell, and erythrocyte lineages, as well as the growth of the CSF-dependent cell line » FDC-P2 in a » colony formation assay. It is identical with » IL3. See also: CSF for general information about colony stimulating factors.
Conlon PJ et al Generation of anti-peptide monoclonal antibodies which recognize mature CSF-2 α (IL 3) protein. JI 135: 328-32 (1985); **Park LS** Characterization of the cell surface receptor for a multi-lineage colony-stimulating factor (CSF-2α). JBC 261: 205-10 (1986); **Prestidge RL et al** Biochemical comparison of murine colony-stimulating factors secreted by a T cell lymphoma and a myelomonocytic leukemia. JI 133: 293-8 (1984)

CSF-2β: *colony stimulating factor 2γ* This factor was initially described as an activity secreted by » LBRM-33 and » WEHI-3 cells, stimulating the growth of bone marrow cells in the granulocyte, macrophage, megakaryocyte, mast cell, and erythrocyte lineages, as well as the growth of the CSF-dependent cell line » FDC-P2 in a » colony formation assay. It is identical with » IL3. See also: CSF for general information about colony stimulating factors.
Prestidge RL et al Biochemical comparison of murine colony-stimulating factors secreted by a T cell lymphoma and a myelomonocytic leukemia. JI 133: 293-8 (1984)

CSF-2γ: *colony stimulating factor 2β* This factor was initially described as an activity secreted by » LBRM-33 cells, stimulating the growth of bone marrow cells in the granulocyte and macrophage lineages. It does not support the growth of the CSF-dependent cell line » FDC-P2 in a » colony formation assay. It is identical with » GM-CSF. See also: CSF for general information about colony stimulating factors.
Prestidge RL et al Biochemical comparison of murine colony-stimulating factors secreted by a T cell lymphoma and a myelomonocytic leukemia. JI 133: 293-8 (1984)

CSF-3: *colony stimulating factor 3* This factor was initially described as an activity promoting the development of granulocyte-containing colonies in a » colony formation assay. It is isolated from cultured murine lung, human T cell lines, human Hodgkin tumor cell lines, and human fibrous histiocytoma cell lines. The factor is identical with » GM-CSF. See also: CSF for general information about colony stimulating factors.
The name CSF-3 has also been proposed for » G-CSF.
Burgess AW et al Purification and properties of colony-stimulating factors from mouse lung-conditioned medium. JBC 252: 1998-2003 (1977); **Burgess AW & Nice EC** Murine granulocyte-macrophage colony-stimulating factor. MiE 116: 588-600 (1985); **Burgess AW et al** Granulocyte-macrophage colony-stimulating factor from mouse lung-conditioned medium. BJ 235: 805-14 (1986); **Byrne PV** Human granulocyte-macrophage colony-stimulating factor purified from a Hodgkin´s tumor cell line. BBA 874: 266-73 (1986); **Erickson-Miller EL et al** Purification of GCT cell-derived human colony-stimulating factors. EH 16: 184-89 (1988); **Pimentel E** Colony-stimulating factors. Ann. Clin. Lab. Sci. 20: 36-55 (1990)

CSF-4: *colony stimulating factor 4* see: CSF-1.

CSF 309: This factor is identical with » IL6. See also: CSF for general information about colony stimulating factors.
Garman RD et al B cell-stimulatory factor 2 (β 2 interferon) functions as a second signal for interleukin 2 production by mature murine T cells. PNAS 84: 7629-33 (1987)

CSF-α: *colony stimulating factor α* This factor was initially described as an activity promoting the development of macrophage- and granulocyte-containing colonies in a » colony formation assay. It is identical with » GM-CSF. See also: CSF for general information about colony stimulating factors.
Dessein AJ et al Enhancement of human blood eosinophil cytotoxicity by semi-purified eosinophil colony-stimulating factor(s). JEM 156: 90-103 (1982); **Lopez AF & Vadas MA** Stimulation of human granulocyte function by monoclonal antibody WEM-G1. PNAS 81: 1818-21 (1984); **Vadas MA et al** Activation of antibody-dependent cell-mediated cytotoxicity of human neutrophils and eosinophils by separate colony-stimulating factors. JI 130: 795-9 (1983); **Vadas MA et al** Eosinophil activation by colony-stimulating factor in man: metabolic effects and analysis by flow cytometry. Blood 61: 1232-41 (1983)

CSF-β: *colony stimulating factor β* This factor was initially described as an activity promoting the development of macrophage- and granulocyte-containing colonies in a » colony formation assay. It is identical with » GM-CSF. See also: CSF for general information about colony stimulating factors.

Nicola NA et al Identification of the human analog of a regulator that induces differentiation in murine leukaemic cells. N 314: 625-8 (1985); **Vadas MA et al** Activation of antibody-dependent cell-mediated cytotoxicity of human neutrophils and eosinophils by separate colony-stimulating factors. JI 130: 795-9 (1983)

CSF box: colony stimulating factor box see: CLE (conserved lymphokine element).

CSF-Eo: *colony-stimulating factor specific for eosinophils* see: Eo-CSF. See also: CSF for general information about colony stimulating factors.

CSF-G: *colony-stimulating factor specific for granulocytes* see: G-CSF. See also: CSF for general information about colony stimulating factors.

CSF-HU: *human urinary colony stimulating factor* see: Urinary colony-stimulating factor.

CSF-GM: *colony-stimulating factor specific for granulocytes-macrophages* see GM-CSF. See also: CSF for general information about colony stimulating factors.

CSF-M: *colony-stimulating factor specific for macrophages* see: M-CSF. See also: CSF for general information about colony stimulating factors.

CSIF: *cytokine synthesis inhibitory factor* CSIF was initially identified as a factor produced by murine TH2 helper cells (see: TH1/TH2 cytokines). CSIF can inhibit the production of » IL2, » IL3, » TNF-β, » IFN-γ, and » GM-CSF by TH1 helper cells responding to antigen and antigen-presenting cells, but TH2 cytokine synthesis is not significantly affected. CSIF is identical with » IL10.

Fiorentino DF et al Two types of mouse T helper cell. IV. Th2 clones secrete a factor that inhibits cytokine production by Th1 clones. JEM 170: 2081-95 (1989)

CT.4S: a cell line obtained after extended culture of CT.EV cells (a subline of » CTLL cells) in » IL4 and subsequent mutagenesis of these cells with ethylmethane sulfonate and selection for lack of expression of the p55 chain of the » IL2 receptor. The cells are hyporesponsive to IL2 but retain their full sensitivity to IL4. CT.4S cells respond poorly to IL2 at concentrations of 100 U/ml or less although they respond vigorously to higher IL2 concentrations. CT.4S cells give measurable responses to 3-10 U/ml (approximately 15-50 pg/ml) of IL4. CT.4S cells do not respond to » IL1, » IL3, » IL6, » IL7, » IL10, » IL12, » GM-CSF, » G-CSF, » MCSF or IFN-γ. and are used as a sensitive and specific bioassay for IL4. IL4-induced proliferation of CT.4S cells is highly sensitive to inhibition by » TNF-α. Less than 500 pg/mL is capable of blocking greater than 95 % of the response to 1 ng/mL of IL4.

Variants of CT.4S harboring a retroviral vector allowing the constitutive synthesis of IL4 have been constructed. These cells grow continuously in the absence of exogenous IL4 but do not form tumors if transplanted into mice. This is in contrast to many other factor-dependent cell lines that become growth factor-independent and tumorigenic after transfer of an expression vector carrying the gene for the cytokine that these cells require for their growth.

Blankenstein T et al Retroviral interleukin 4 gene transfer into an interleukin 4-dependent cell line results in autocrine growth but not in tumorigenicity. EJI 20: 935-8 (1990); **Hu-Li J et al** Derivation of a T cell line that is highly responsive to IL4 and IL2 (CT.4R) and of an IL2 hyporesponsive mutant of that line (CT.4S). JI 142: 800-7 (1989); **Li WQ et al** Lack of tumorigenicity of interleukin 4 autocrine growing cells seems related to the anti-tumor function of interleukin 4. Mol. Immunol. 27: 1331-7 (1990); **Ramsdell F et al** TNF abrogates the response of the CT.4S cell line to IL4. JIM 167: 299-301 (1994); **Seder RA et al** Increased frequency of interleukin 4-producing T cells as a result of polyclonal priming. Use of a single-cell assay to detect interleukin 4-producing cells. EJI 21: 1241-7 (1991); **Wang H et al** Evidence for the existence of IL4 and IFN γ secreting cells in the T cell repertoire of naive mice. Immunol. Lett. 31: 169-75 (1992)

CT6: A murine cytotoxic/suppressor T cell line the viability and proliferation of which depends on » IL2 and which is used to assay this factor. The proliferative effect of IL 2 is attenuated by simultaneous exposure to prostaglandin E2. The cells also proliferate in response to porcine IL2, murine » IL4 and » TNF-α but not in response to human TNF-α. The cells have also been shown to proliferate in response to » IL7 to the same extent as to » IL2.

CT6 cells are highly responsive to phorbol myristate acetate (see also: Phorbol esters) and to a lesser extent but significantly responsive to lipopolysaccharide (LPS) in short-term proliferation assays. PMA does not support a long-term culture of CT6 cells. Incubation of CT-6 cells with mixed bovine brain gangliosides results in a dose-dependent inhibition of cell proliferation without influencing cell viability. The inhibition is totally reversed when the gangliosides are removed.

CSK: see addendum.

Treatment of G-arrested cultures of CT6 cells with IL 2 induces the rapid sequential expression of the cellular » oncogenes » *fos*, » *myc*, and » *myb*. CT6-derived factor-independent lines show no mitogenic response to IL 2, yet binding of IL 2 with its receptor in the cells is capable of inducing the expression of *fos* and *myc*. In factor-independent cell lines, *myc* is expressed constitutively at high levels, suggesting that it abrogates growth factor requirements of these cells.

Picomolar concentrations of TGF-β inhibit S-phase progression stimulated by IL2 or IL4. TGF-β pretreatment does not decrease the expression of high affinity IL2R on CT6 cells but markedly reduces the IL2-stimulated transferrin receptor expression. TGF-β also significantly reduces early (1 to 2 h) increases in cellular » *myc* mRNA levels stimulated by IL2 or IL4.

Elevation of intracellular cyclic adenosine 3': 5' monophosphate (cAMP) and also exposure to stable cAMP-derivatives (8-bromoadenosine 3': 5'-monophosphate) inhibit IL2-stimulated proliferation of these cells.

Beckner SK & Farrar WL Interleukin 2 modulation of adenylate cyclase. Potential role of protein kinase C. JBC 261: 3043-7 (1986); **Cleveland JL et al** Role of c-*myc* and other genes in interleukin 2 regulated CT6 T lymphocytes and their malignant variants. JI 138: 3495-504 (1987); **Farrar WL et al** Effects of anti-proliferative cyclic AMP on interleukin 2-stimulated gene expression. JI 139: 2075-80 (1987); **Ho SN et al** Differential bioassay of interleukin 2 and interleukin 4. JIM 98: 99-104 (1987); **Merritt WD et al** Inhibition of interleukin-2-dependent cytotoxic T lymphocyte growth by gangliosides. CI 89: 1-10 (1984); **Mukaida N et al** Interleukin 2-dependent T cell line acquires responsiveness to phorbol myristate acetate and lipopolysaccharide in the course of long-term culture. Immunol. Commun. 13: 475-86 (1984); **Ranges GE et al** Tumor necrosis factor-α as a proliferative signal for an IL2-dependent T cell line: strict species specificity of action. JI 142: 1203-8 (1989); **Ruegemer JJ et al** Regulatory effects of transforming growth factor-β on IL2- and IL4-dependent T cell-cycle progression. JI 144: 1767-76 (1990); **Willcocks JL et al** The murine T cell line CT6 provides a novel bioassay for interleukin-7. EJI 23: 716-20 (1993)

CTAP-3: *connective tissue activating peptide-3* CTAP-3 is also known as » LA-PF4 (low-affinity platelet factor 4). The factor is identical with » PF4 (platelet factor 4). It is generated by proteolytic removal of the first nine aminoterminal amino acids of a biologically inactive 15 kDa precursor protein called ***PBP*** (platelet basic protein) and is therefore also called pro-platelet basic protein (abbrev. PPBP or Pro-PBP). Removal of further four aminoterminal amino acids from CTAP-3 yields » Beta-Thromboglobulin. CTAP-3 is the predominant form of Beta-Thromboglobulin-like proteins in platelets. Cleavage of CTAP-3 with cathepsin G yields the neutrophil-activating peptides » NAP-2. CTAP-3 and its cleavage products belong to the » chemokine family of cytokines. The human gene encoding PBP and CTAP-3 maps to chromosome 4q12-q13.

CTAP-3 is a multifunctional protein. It is a chemoattractant for fibroblasts, neutrophils, and monocytes (see also: Chemotaxis). It is mitogenic for fibroblasts and also induces the synthesis and secretion of hyaluronic acid, glycosaminoglycans, proteoglycan core protein, plasminogen activator and prostaglandin E_2 by these cells. In basophils CTAP-3 induces the release of histamines. It also induces marked stimulation of 2-deoxy-glucose uptake in cultures of human synovial cells, chondrocytes, and dermal fibroblasts.

Car BD et al Formation of neutrophil-activating peptide 2 from platelet-derived connective-tissue-activating peptide III by different tissue proteinases. BJ 275: 581-4 (1991); **Castor CW et al** Structural and biological characteristics of connective tissue activating peptide (CTAP-3), a major human platelet-derived growth factor. PNAS 80: 765-9 (1983); **Castor CW et al** Connective tissue activation. XXII. A platelet growth factor (connective tissue activating peptide III) in human growth-hormone-deficient patients. J. Clin. Endocrinol. Metab. 52: 128-32 (1981); **Castor CW et al** Preparation and bioassay of connective tissue activating peptide III and its isoforms. MiE 198: 405-16 (1991); **Castor CW et al** Connective tissue activation. XXXV. Detection of connective tissue activating peptide-III isoforms in synovium from osteoarthritis and rheumatoid arthritis patients: patterns of interaction with other synovial cytokines in cell culture. Arthritis Rheum. 35: 783-93 (1992); **Cohen AB et al** Generation of the neutrophil-activating peptide-2 by cathepsin G and cathepsin G-treated human platelets. Am. J. Physiol. 263: L249-56 (1992); **Han ZC et al** Negative regulation of human megakaryocytopoiesis by human platelet factor 4 (PF4) and connective tissue-activating peptide (CTAP-III). IJCC 8: 253-9 (1990); **Holt JC & Niewiarowski S** Platelet basic protein, low-affinity platelet factor 4, and β-thromboglobulin: purification and identification. MiE 169: 224-33 (1989); **MacCarter DK et al** Connective tissue activation. XXIII. Increased plasma levels of a platelet growth factor (CTAP III) in patients with rheumatic diseases. Clin. Chim. Acta 115: 125-34 (1981); **Reddigari SR et al** Connective tissue-activating peptide-III and its derivative, neutrophil-activating peptide-2, release histamine from human basophils. J. Allergy Clin. Immunol. 89: 666-72 (1992); **Tanabe N et al** The effect of low affinity platelet factor 4 (LAPF4) secreted by human megakaryoblastic cell line (MEG-01) upon human bone marrow fibroblasts. Haematologica 76: 193-9 (1991); **Wenger RH et al** Cloning of cDNA for connective tissue activating peptide 3 from a human platelet-derived λgt11 expression library. Blood 73: 1498-1503 (1989); **Wenger RH et al** Human platelet basic protein/connective tissue activating peptide-III maps in a gene cluster on chromosome 4q12-q13 along with other genes of the β-thromboglobulin superfamily. Hum. Genet. 87: 367-8 (1991)

CTF: *chemotactic factor* see: Chemotaxis.

CTGF: *connective tissue growth factor* This factor is produced by human umbilical vein endothelial cells. The protein is predicted from its cDNA to be a 38 kDa cysteine-rich secreted protein related to » PDGF. It can be isolated after chromatography using antisera directed against » PDGF. This factor is the major activity of PDGF-like factors secreted by human vascular endothelial cells.

A locus on human chromosome 6q23.1 proximal to the cellular » *myb* » oncogene sharing homology with the nov proto-oncogene overexpressed in avian nephroblastoma has been described to correspond to the CTGF gene. This location is of interest since chromosomal abnormalities involving this region have been associated with different human tumors including Wilms' tumors.

At the protein level CTGF shows a homology of 45 % to the translation product of a gene, designated ***CEF-10***, which is induced by the activity of the viral *src* » oncogene in chicken embryo fibroblasts.

cyr61 is a growth factor-inducible immediate early gene initially identified in serum-stimulated mouse fibroblasts encoding a member of an emerging family of cysteine-rich secreted proteins that includes CTGF. Cyr61 is identical with ***βIG-M1*** (TGF-—inducible gene M1), a gene transcription of which is found to be increased after treatment of murine cells with » TGF-β. Cyr61 is expressed in the developing mouse embryo and extraembryonic tissues. ***βIG-M1*** encodes a 379 amino-acid protein which is 81% homologous to CEF-10. ***βIG-M1*** shows 50 % homology with another protein of 348 amino acids, designated β-IG-M2, also induced by TGF-β. Cyr61 transcripts are found in mesenchymal cells of both mesodermal and ectodermal origin during their differentiation into chondrocytes.

Bork P The modular architecture of a new family of growth regulators related to connective tissue growth factor. FL 327: 125-30 (1993); **Bradham DM et al** Connective tissue growth factor: a cysteine-rich mitogen secreted by human vascular endothelial cells is related to the *src*-induced immediate early gene product CEF-10. JCB 114: 1285-94 (1991); **Brunner A et al** Identification of a gene family regulated by transforming growth factor-β. DNA Cell Biol. 10: 293-300 (1991); **Igarashi A et al** Connective tissue growth factor. J. Dermatol. 19: 642-3 (1992); **Joliot V et al** Proviral rearrangements and overexpression of a new cellular gene (nov) in myeloblastosis-associated virus type 1-induced nephroblastomas. MCB 12: 1021 (1992); **Martinerie C et al** Physical mapping of human loci homologous to the chicken nov proto-oncogene. O 7: 2529-34 (1992); **O'Brien TP & Lau LF** Expression of the growth factor-inducible immediate early gene cyr61 correlates with chondrogenesis during mouse embryonic development. Cell Growth Differ. 3: 645-54 (1992)

CTIL: *cytotoxic tumor infiltrating lymphocytes* see: TILs (tumor-infiltrating lymphocytes).

CTL44: A subline derived from » CTLL cells showing rapid and non-adherent growth. It is used in » bioassays of murine » IL4 and has been described to be hyoresponsive to » IL2, in contrast to its parent cell line.

Favre N & Erb P Use of the CTL44 cell line, a derivative of CTL/L cells, to identify and quantify mouse interleukin 4 by bioassay. JIM 164: 213-20 (1993)

CTL differentiation factor: see: CDF (cytolytic differentiation factor for T lymphocytes).

CTLL: A murine cytotoxic T cell line derived from C57/Bl/6 inbred mice. This cell line constitutively expressed IL2 receptors and depends entirely on the presence of » IL2 for its growth. It is therefore used to assay this factor (see also: Bioassays). The cells appear to be capable of growing on a matrix containing immobilized IL2.

The activity of IL2 is determined by measuring the incorporation of ^{3}H thymidine into the newly synthesized DNA of the proliferating cells. Cell proliferation can be determined also by employing the » MTT assay. The CTLL assay detects IL2 in the concentration range of ≈ 50 pg/mL to 50 ng/mL. Treatment of CTLL cells with» TGF-β inhibits IL2-dependent growth, induces morphological alterations, causes an increased adherence, and induces the *de novo* expression of CD8α (see also: CD8). Vitamin E has been found to stimulate IL2 dependent cellular growth, glycosylation, and expression of certain glycoproteins in CTLL cells. Micellar gangliosides, in particular highly sialylated gangliosides, are potent inhibitors of the IL2-induced proliferation of CTLL cells. They abolish both DNA and protein synthesis, and depressed cellular expansion, without affecting viability. These effects are reversible for at least 12 hr following ganglioside treatment. Ganglioside micelles and lipid vesicles containing gangliosides are able to bind IL2 and prevent binding of IL2 to high-affinity receptors on the lymphocyte surface. Inhibition of DNA synthesis by gangliosides can be reversed partially by high concentrations of exogenous IL2.

Glycophorin A, the major sialoglycoprotein of the human erythrocyte membrane, inhibits IL2-stimulated proliferation of CTLL cells and the IL2-dependent cell line » HT-2 in a dose-dependent manner.

CTLL-2 cells also respond to murine » IL4. Neutralizing anti-IL4 antibodies can be employed to differentiate between IL2 and IL4 activities.

In many instances CTLL cells are also used for the detection of other factors in an indirect assay. This approach is called conversion assay (see also: Bioassays). The amount of » IL1, for example, is inferred by measuring in the CTLL assay the amounts of IL2 that are induced by IL1 in another cell line. In some cases this assay can be performed by cocultivating CTLL cells with another cell line (see: EL4 for the detection of IL1).
An alternative and entirely different method of detecting these factors is » Message amplification phenotyping.

Chu JW & Sharom FJ Effect of micellar and bilayer gangliosides on proliferation of interleukin-2-dependent lymphocytes. CI 132: 319-38 (1991); **Chu JW & Sharom FJ** Glycophorin A interacts with interleukin-2 and inhibits interleukin-2-dependent T lymphocyte proliferation. CI 145: 223-39 (1992); **Conlon PJ** A rapid biologic assay for the detection of interleukin-1. JI 131: 1280-2 (1983); **Eskandari MK et al** Effects of arachidonic acid metabolites and other compounds on the CTLL assay for interleukin 2. JIM 10: 85-9 (1989); **Gearing AJH & Thorpe R** The international standard for human interleukin-2. Calibration by international collaborative study. JIM 114: 3-9 (1988); **Gillis S et al T cell** growth factor: parameters of production and a quantitative microassay for activity. JI 120: 2027-31 (1978); **Gogu SR & Blumberg JB** Vitamin E increases interleukin-2 dependent cellular growth and glycoprotein glycosylation in murine cytotoxic T-cell line. BBRC 193: 872-7 (1993); **Hämmerling U et al** Development and validation of a bioassay for interleukin-2. J. Pharm. Biomed. Anal. 10: 547-53 (1992); **Horwitz JI et al** Immobilized IL2 preserves the viability of an IL2 dependent cell line. Mol. Immunol. 30: 1041-8 (1993); **Igietseme JU & Herscowitz HB** A modified in situ enzyme-linked immunosorbent assay for quantitating interleukin-2 activity employing monoclonal anti-IL2 receptor antibody. JIM 108: 145-52 (1988); **Inge TH et al** Immunomodulatory effects of transforming growth factor-β on T lymphocytes. Induction of CD8 expression in the CTLL-2 cell line and in normal thymocytes. JI 148: 3847-56 (1992); **Sladowski D et al** An improved MTT assay. JIM 157: 203-7 (1993); **Watson J** Continuous proliferation of murine antigen-specific helper T lymphocytes in culture. JEM 150: 1510 (1979)

CTL maturation factor: abbrev. TcMF. See: CLMF (cytotoxic lymphocyte maturation factor).

CTX: *cytotoxin* This factor is identical with » TNF-α or » TNF-β. This term is also used in a more general sense, referring to both species of » TNF.

Kuhl FC & Cuatrecasas Macrophage cytotoxin. In: Interleukins, Lymphokines, and Cytokines. Academic Press, pp. 511-9, New York 1983

C-type natriuretic peptide: abbrev. CNP See: ANF (atrial natriuretic factor).

C-X-C-Intercrine: A name sometimes given to some members of the » chemokine family of cytokines which, among other factors, comprises a variety of factors with chemotactic (see: Chemotaxis) and pro-inflammatory activities (see: Inflammation). The C-X-C refers to the spacing of two cytosine residues separated by another amino acid, which is conserved in members of this group.

CyA: abbrev. for cyclosporin. See: CsA.

Cyclic AMP-responsive element: see: CRE.

Cyclic AMP responsive element binding protein: abbrev. CREB. See: CRE (cyclic AMP-responsive element).

Cyclophilin: see: CsA (cyclosporin).

Cyclophosphamide: see: 4-HC (Hydroperoxycyclophosphamide).

Cyclosporin: see: CsA.

Cyr61: see: CTGF (connective tissue growth factor).

CySF: This poorly characterized factor (≈ 10 kDa or less) is isolated from dialysable human extract as three different species, designated CySF-L1, CySF-L2, and CySF-L3.
CySFs function as a cytotoxicity-stimulating factor and activate natural killer cytotoxicity against NK-sensitive and insensitive tumor cells (K562; Daudi; Raji; MOLT4) when preincubated with effector cells for 72 h. CySF-L2 is also capable of activating NK cytotoxicity of highly purified CD56-positive CD3-negative NK cells and monocytes. CySF-L2 induces the release of » IFN-γ from CD3-positive T cells and CD56-positive CD3-negative NK cells and of » TNF-α and prostaglandin E2 from monocytes.
Induction of IFN-γ release appears to be the crucial step during CySF-L2-mediated NK cytotoxicity activation since enhancement of NK activity is completely blocked when anti-IFN -γ antibodies are present during treatment of human peripheral blood mononuclear cells. Anti-IFN-α, anti-TNF-α, anti-IL1, and anti-IL2 antibodies show no blocking effect.

Doelker I & Anderer FA The CySF-L2 factor from dialysable human extract activates natural killer cytotoxicity by induction of interferon γ. Cancer Immunol. Immunother. 34: 299-305 (1992); **Neidlinger AC & Anderer FA** Dialysable factors from human leukocyte extracts activating human NK cytotoxicity against tumor cells: identification of factor-producing leukocyte subpopulations. Anticancer Res. 11: 2023-8 (1991)

Cystatin A: see: Calgranulins.

Cystic fibrosis antigen: see: MIF: (macrophage migration inhibitory factor).

Cytokine 1: abbrev. CK-1. CK-1 is the binding site for a transcription factor called NF-GMa. which is found in the promoter region of the gene encoding » G-CSF. This transcription factor is responsible for the increased expression of G-CSF observed in fibroblasts after treatment with » TNF-α or IL1β (see: IL1). An identical NF-GMa-binding sequence has also been found in the promoter region of the gene encoding » GM-CSF. However, in this case » TNF-α and IL1β do not induce transcription of the GM-CSF gene. For a related conserved element see also: CLE (conserved lymphokine element).

Kuczek ES et al A granulocyte-colony-stimulating factor gene promoter element responsive to inflammatory mediators is functionally distinct from an identical sequence in the granulocyte-macrophage colony-stimulating factor gene. JI 146: 2426-33 (1991); **McCaffrey PG et al** A T cell nuclear factor resembling NF-AT binds to an NF-kappaB site and to the conserved lymphokine promoter sequence "cytokine-1". JBC 267: 1864-71 (1992); **Shannon MF et al** Three essential promoter elements mediate tumor necrosis factor and interleukin-1 activation of the granulocyte-colony stimulating factor gene. GF 7: 181-93 (1992)

Cytokine antagonists: see: Cytokine inhibitors.

Cytokine assays: Many different assay systems are available for quantifying cytokine concentrations or for detecting cells that express them (see also: BRMP unit). Immunoassays generally measure immunoreactivities, i. e. they also detect biologically inactive cytokine molecules or fragments of these factors. They are therefore useful indicators of the presence of a cytokine but they reveal nothing about the biological activity of the detected molecules. This is particularly important for cytokines that are produced in latent forms, for example » TGF-β or » HGF (hepatocyte growth factor) which require extracellular activation by proteolytic processing.
Immunoassays such as radioimmunoassays (***RIA***), immunoradiometric assays (***IRMA***), and enzyme-linked immunosorbent assays (***ELISA***) require cytokine-specific antibodies and/or labeled cytokines or their receptors or labeled antibodies. It has been shown for some cytokines that the same ELISA used in different laboratories yields comparable results, whereas different ELISAs usually detect the highest levels in the same samples, but yield different absolute values. Such observations highlight the necessity of establishing international standards for all immunoassays.
Immunological assays are (commercially) available for many cytokines (for references see the subentry "Assays" for individual cytokines; for the determination of cytokine receptors see also: Scatchard plots). Some of these assays may be influenced by the presence of cytokine binding proteins such as » Alpha 2 macroglobulin. Radioreceptor assays (***RRA***) measure cytokine concentrations by displacing ligands from cell-bound receptors.
The ***RHPA*** (reverse hemolytic plaque assay) is an adaptation of a plaque assay initially established to detect immunoglobulin-secreting cells and can be used to detect individual cytokine-secreting cells and to determine the amounts of a particular cytokine secreted by this cell. The » ***cell blot assay*** also allows visualization of cytokine release by producer cells.
Alternative cytokine assays frequently employ » factor-dependent cell lines or cell lines responding in a particular way to individual cytokines (see: Bioassays for a general discussion; references for individual cell lines used in such assays are found under the appropriate entry for these cell lines). A technique allowing the determination of cytokine-specific mRNA is » Message amplification phenotyping.

Capper SJ Immunoassays for growth factors. In: McKay I & Leigh I (eds). Growth factors, a practical approach, IRL Press, pp. 181-199, Oxford (1992); **Hamblin AS & O'Garra A** Assays for interleukins and other related factors. In: Klaus GGB (edt) Lymphocytes, a practical approach, pp 209-28. IRL Press, Oxford 1987; **James K et al** The effect of α2 macroglobulin in commercial cytokine assays. JIM 168: 33-7 (1994); **Lewis CE et al** Cytokine release by single immunophenotyped human cells: use of the reverse hemolytic plaque assay. IR 119: 23-39 (1991); **Meager A** RIA, IRMA, and ELISA assays for cytokines. In: Balkwill FR (edt) Cytokines, a practical approach, pp. 299-307, IRL Press, Oxford 1991; **Parker CW** Immunoassays MiE 182: 700-18 (1990); **Perret G & Simon P** [Radioreceptor assay: principles and applications to pharmacology] J. Pharmacol. 15: 265-86 (1984); determination in human synovial fluids: a consensus study of the European Workshop for Rheumatology Research. The Cytokine Consensus Study Group of the European Workshop for Rheumatology Research. Clin. Exp. Rheumatol. 10: 515-20 (1992); **Walker MR et al** Enzyme-labeled antibodies in bioassays. Methods Biochem. Anal. 36: 179-208 (1992)

Cytokine cascades: see: Cytokine network.

Cytokine-dependent cell lines: see: Factor-dependent cell lines.

Cytokine expression profiles: see: Message amplification phenotyping.

Cytokine families: see: Gene families.

Cytokine fusion toxins: Cytotoxic chimaeric proteins derived from a number of biologically active cytokines. These toxins are obtained by fusion of interleukin cDNAs with cDNAs containing the coding regions of toxic protein molecules.

The cytokine portion of the fusion proteins can interact with the high affinity binding receptors. These receptors are rapidly internalized after ligand binding and the toxic protein is activated inside the cell.

The toxic moieties of most chimaeric proteins consist of diphtheria toxin (DT) or Pseudomonas endotoxin (PE), both of which irreversibly inhibit protein biosynthesis in eukaryotic cells by inhibiting elongation factor 2.

Constructs containing » IL2, » TGF-α, » IL4 and » IL6 coupled to either DT or PE have been described. Single chain antigen binding proteins, such as the variable heavy and light regions of the monoclonal antibody anti-Tac have also been used in constructing these chimeric toxins. These toxic chimaeric proteins are cytotoxic *in vitro* and also after transplantation into experimental animals for a number of human tumor cell lines expressing the corresponding growth factor receptors. These chimeric toxins also kill fresh tumor cells from patients and display anti-tumor activity toward human malignant tumors in nude mice (see also: Immunodeficient mice). In some instances it has been possible to increase the cytotoxic potency and binding affinity to receptor bearing cells of the fusion toxins by employing genetic engineering techniques such as site-directed mutagenesis.

Such toxic fusion proteins may be useful for the targeted destruction of receptor-positive lymphoma cells. They may also be suitable to prevent graft-versus-host reaction in transplant recipients due to their cell-specific toxicity. These constructs have also been used to selectively remove HIV-1-infected cells from heterogeneous cell populations containing infected and uninfected cells.

Toxic constructs containing factor moieties such as » FGF could be employed in certain situations for preventing angiogenesis under certain conditions (see also: Angiogenesis factors).

Another strategy for the generation of toxic protein chimeras may be the combination of receptor antibodies with a toxic compound. A monoclonal antibody directed against the IL2 receptor has been coupled with ricin A, for example. This antibody chimera is capable of selectively destroying cells expressing the IL2 receptor. This concept has been exploited at least experimentally to selectively remove leukemic cells. For other cytotoxic constructs see: immunotoxins, saporin, mitotoxins.

The general applicability of the toxic chimera approach may be hampered by the fact that many normal cells also express receptors for cytokines used in the construction of these fusion proteins. Clinical trials are now beginning with some of these agents.

Banker DE et al An epidermal growth factor-ricin A chain (EGF-RTA)-resistant mutant and an epidermal growth factor-*Pseudomonas* endotoxin (EGF-PE-)-resistant mutant have distinct phenotypes. JCP 139: 51-7 (1989); **Case JP et al** Chimaeric cytotoxin IL2-PE40 delays and mitigates adjuvant-induced arthritis in rats. PNAS 86: 287-91 (1989); **Fitzgerald D et al** Generation of chimeric toxins. Targeted. Diagn. Ther. 7: 447-62 (1992); **Debinski W et al** Substitution of foreign protein sequences into a chimaeric toxin composed of transforming growth factor α and *Pseudomonas* exotoxin. MCB 11: 1751-3 (1991); **Finberg RW et al** Selective elimination of HIV-1-infected cells with an interleukin-2 receptor-specific cytotoxin. S 252: 1703-5 (1991); **Heimbrook DC et al** Biological activity of a transforming growth factor α-*Pseudomonas* exotoxin fusion protein *in vivo* and *in vitro*. J. Indust. Microbiol. 7: 203-7 (1991); **Heimbrook DC et al** Transforming growth factor α-*Pseudomonas* exotoxin fusion protein prolongs survival of nude mice bearing tumor xenografts. PNAS 87: 4697-701 (1990); **Kelley VE et al** Interleukin 2-diphtheria toxin fusion protein can abolish cell-mediated immunity *in vivo*. PNAS 85: 3980-4 (1988); **Kiyokawa T et al** Protein engineering of diphtheria-toxin-related interleukin-2 fusion toxins to increase cytotoxic potency for high-affinity IL2-receptor-bearing target cells. Prot. Engin. 4: 463-8 (1991); **Kreitman RJ et al** Targeting growth factor receptors with fusion toxins. IJI 14: 465-72 (1992); **Kronke M et al** Adult T cell leukemia: a potential target for ricin A chain immunotoxins. Blood 65: 1416-21 (1985); **Kronke M et al** Selective killing of human T lymphotropic virus-1 infected leukaemic T cells by monoclonal anti-interleukin 2 receptor antibody-ricin A chain conjugates. CR 46: 3295-8 (1986); **Lakkis F et al** Phe496 and Leu497 are essential for receptor binding and cytotoxic action of the murine interleukin-4 receptor targeted fusion toxin DAB389-mIL4. Protein Eng. 5: 241-8 (1992); **Lappi DA & Baird A** Mitotoxins: growth factor-targeted cytotoxic molecules. PGFR 2: 223-36 (1990); **Lorberboum-Galski H et al** Cytotoxic activity of an interleukin 2-*Pseudomonas* exotoxin chimaeric protein produced in *Escherichia coli*. PNAS 85: 1922-6 (1988); **Meneghetti CM & LeMaistre CF** Initial clinical experiences with an interleukin-2 fusion toxin (DAB486-IL2). Targeted Diagn. Ther. 7: 395-401 (1992); **Merwin JR et al** Acidic fibroblast growth factor-Pseudomonas exotoxin chimeric protein elicits antiangiogenic effects on endothelial cells. CR 52: 4995-5001 (1992); **Murphy JR et al** Genetic construct, expression, and melanoma-selective cytotoxicity of a diphtheria toxin-related a-melanocyte-stimulated hormone fusion protein. PNAS 83: 8258-62 (1986); **Pai LH et al** Antitumor activity of a transforming growth factor α-*Pseudomonas* exotoxin fusion protein (TGF-α-PE4G). CR 51: 2808-12 (1991); **Pastan I & Fitzgerald D** Recombinant toxins for cancer treatment. S 254: 1173-7 (1991); **Perentesis JP et al** Protein toxin inhibitors of protein synthesis. Biofactors 3: 173-84 (1992); **Rose JW et al** Chimeric cytotoxin IL2-PE40 inhibits relapsing experimental allergic encephalomyelitis. J. Neuroimmunol. 32: 209-17 (1991); **Shapiro ME et al** *In vivo* studies with chimeric toxins. Interleukin-2 fusion toxins as immunosuppressive agents. Targeted Diagn. Ther. 7: 383-93 (1992); **Siegall CB et al** Cytotoxic activity of chimeric proteins composed of acidic fibroblast growth factor and *Pseudomonas* exotoxin on a variety of cell types. FJ 5: 2843-9 (1991); **Strom TB & Kelley VE** Toward more selective therapies to block undesired immune responses. Kidney Int. 35: 1026-33 (1989)

Cytokine gene transfer: Cytokines are important modulators of host antitumor responses. Antitumor cytokine therapies are currently limited by important systemic and often life-threatening toxicities. With the availability of cloned cytokine genes and recombinant cytokines and a better, albeit still limited, understanding of tumor/host interactions the tools and concepts of gene therapy are now being applied to the development of effective new cancer treatments. Cytokine gene therapy will probably be a useful complement to conventional treatments (surgery, chemotherapy, radiotherapy) in the near future and may allow the prophylactic or therapeutical effective vaccination against cancer under safe and clinically feasible conditions, which has been an elusive goal for years. Cytokine gene transfer techniques are based on the use of highly efficient and targeted gene delivery vectors (e. g. replication-deficient retroviruses, HSV vectors, or adenovirus vectors). These vectors are used to introduce and stably express cloned cytokine genes (or the corresponding receptor genes) into the genomes of tumor cell lines, immune cells, or hematopoietic cells, yielding cells that overexpress the transferred cytokine gene in a

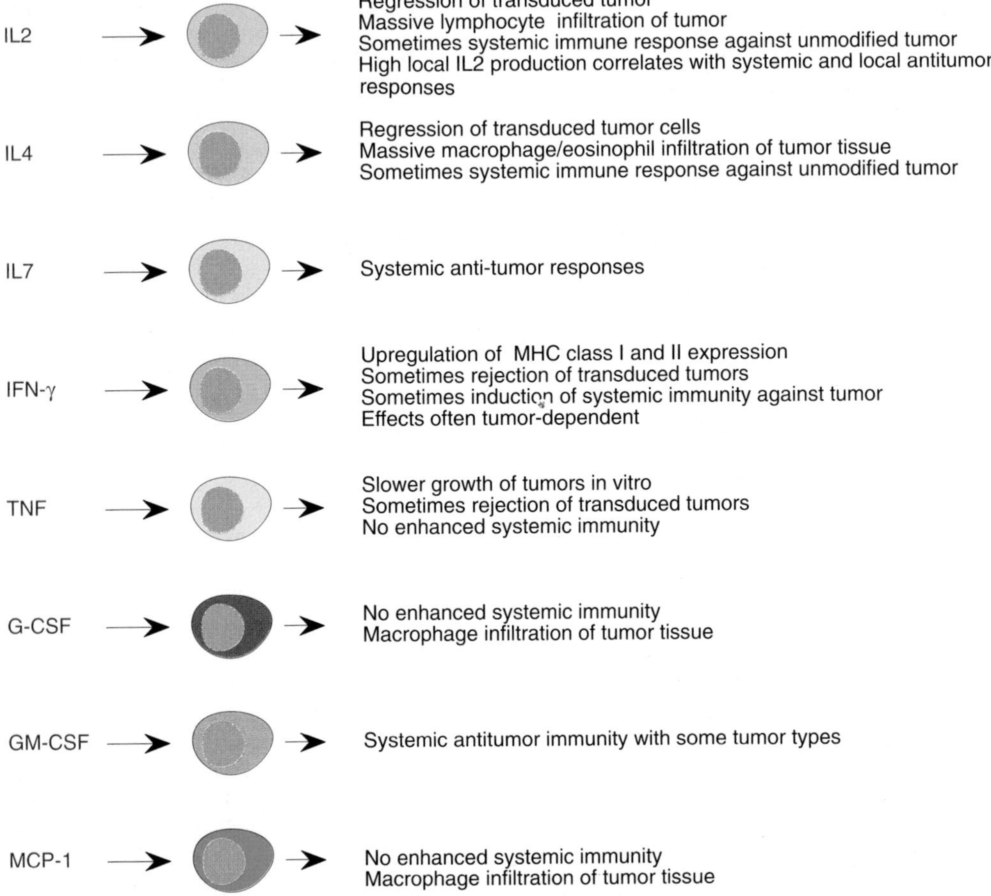

Some examples of cytokine gene transfer in the development of cancer vaccines.
A variety of cytokine genes have been transferred into murine tumors. Some of the manipulated tumor cells elicit systemic immune responses also against unmanipulated tumor cells. The efficacy of some of these strategies is now being investigated in human clinical trials.

controlled manner (for a reverse process, i. e. cytokine gene inactivation, see: Knock-out mice).
Cytokine gene transfer into tumor cells or infiltrating peri-tumoral cells is aimed at guaranteeing the prolonged and elevated release of cytokines in the tumor microenvironment, in order to achieve local efficacy and to circumvent systemic toxicities. This approach mimics the physiologic release of cytokines at the effector-target sites in the immediate vicinity of a tumor.
Recent investigations have demonstrated that the *in vivo* growth of weakly immunogenic tumors in syngeneic hosts can be inhibited by genetic manipulations that enable these tumor cells to secrete a variety of cytokines, including » IL1, » IL2, » IL4, » IL6, » IL7, » JE, » GM-CSF, » G-CSF, » TNF-α, » IFN-α, » IFN-γ. It has also been shown that vaccination with cells expressing IFN-γ, IL2, IL4, IL6, IL7 or TNF-α increases systemic immunity since mice vaccinated with transduced cells also reject a subsequent challenge of non-transduced cells, and, in some cases, of preexisting tumors. Since most human tumors fail to elicit a detectable host immune response the introduction of a func-

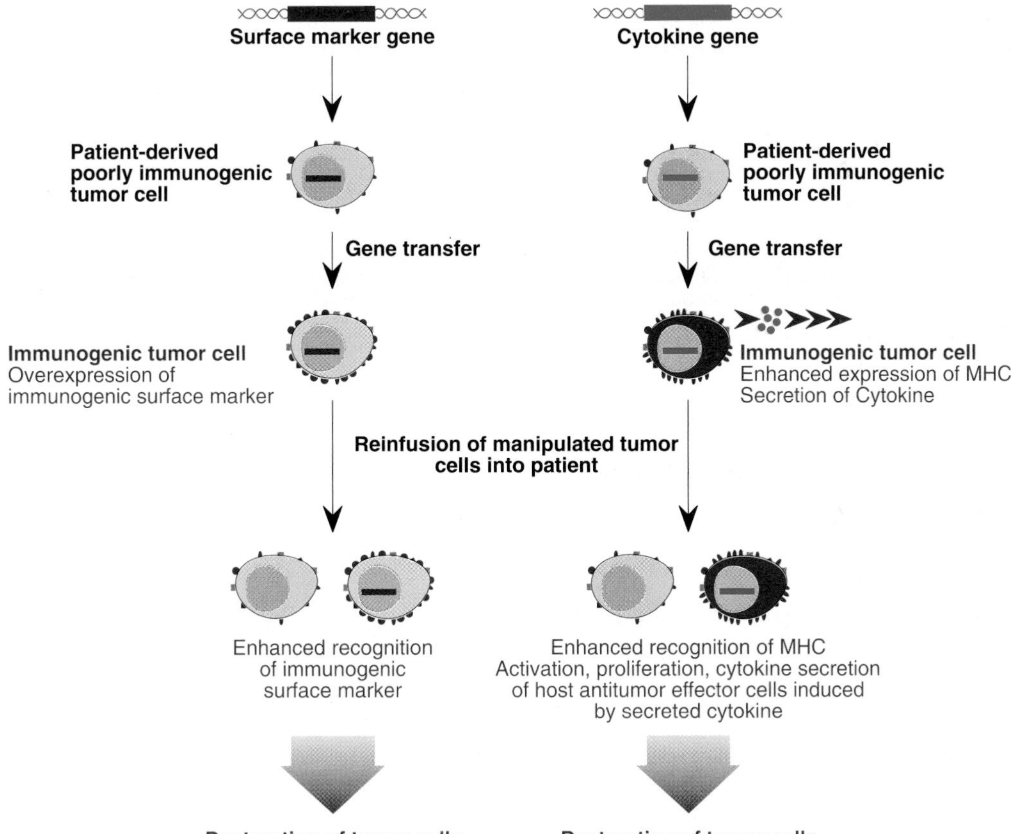

Anticancer mechanisms following cytokine gene transfer into tumor cells.
Tumor cells manipulated by cytokine gene transfer have been shown to elicit systemic immune responses also against unmanipulated tumor cells following their reinfusion into the cancer bearing organism. Gene transfer into poorly immunogenic tumor cells can be used to induce overexpression of cell surface markers expressed only at low levels on the surface of the poorly immunogenic tumor cells, including expression of MHC antigens. Overexpression of the surface marker triggers immune functions and the immune system then also attacks tumor cells with low level expression of this surface marker.
In principle this approach to cancer treatment involves direct transfer of a gene encoding an immunogenic cell surface marker (left, red circles) or of a gene encoding a cytokine indirectly mediating overexpression of such markers or of MHC molecules (right, purple ovals). Tumor cell-targeted cytokine gene transfer also promotes activation, proliferation, and cytokine secretion of host effector cells infiltrating the tumor and actively participating in tumor cell killing (cytotoxic T lymphocytes, LAK cells, neutrophils, eosinophils, basophils, B cells).

tional cytokine gene provides a strategy with potential application for the development of immunotherapies for nonimmunogenic tumors. The tumor targeting nature of » TILs (tumor-infiltrating lymphocytes) in particular creates the possibility of using these cells as a vehicle to deliver gene product specifically to tumor tissue.

Experimental studies have demonstrated the feasibility of cytokine gene transfer into explanted tumor cells and suggest a means for generating autologous or HLA-matched allogeneic tumor cell vaccines for the treatment of patients with various tumors. It has been demonstrated that the enhanced expression and secretion of cytokines by engineered tumor cells enhances specific immune responses, for example by inducing the activation of T cells, and thus provide a modality for the treatment of these tumors and their metastases (see also: Genetic ablation). Many problems remain to be solved for a better understanding of the functioning of transfected genes; nevertheless, the first human clinical trials in adenosine deaminase deficiency, which causes a form of severe combined immunodeficiency. renal cell carcinomas, and metastatic melanoma have recently demonstrated the feasibility and safety of using gene transduction for human gene and cancer therapy. At present patients eligible for cytokine gene transfer tumor therapy are those with cancer that has failed all standard effective treatment and for which no other effective treatment options are available.

It has also been observed that some features of the tumor phenotype can be suppressed *in vitro* through the restoration of expression of tumor suppressor genes (see: oncogenes) or the transfer of other genes, such as cell adhesion molecules that affect the differentiation or the metastatic potential of malignant cells in several *in vitro* model systems. There are also several examples of suppression of the malignant phenotype in certain cases *in vivo* in nude mice (see also: Immunodeficient mice). The experimental evidence for tumor suppression by restored gene expression and the pivotal role played by these genes in the regulation of cell replication suggests that the restored expression of some tumor suppressor genes in some tumor cells may also eventually play a role in cancer gene therapy. The use of » antisense oligonucleotides targeted to oncogenes has also shown that it is possible to alter the proliferative potential of malignant cells *in vitro*, and their tumorigenicity or metastatic potential *in vivo*.

The anti-tumor activity of murine » TILs against modified and unmodified tumor cells has been shown to be increased after introduction of MHC class I antigens into tumor cells, and this approach may also be of potential significance as an immunomodulating strategy.

In addition, gene transfer of genes that alter neuronal physiology into neural cells in the adult mammalian brain also has great promise both for elucidating neuronal physiology and brain mechanisms, and for gene therapy of neurological diseases.

Andersen JK et al Gene transfer into mammalian central nervous system using herpes virus vectors: extended expression of bacterial lacZ in neurons using the neuron-specific enolase promoter. Hum. Gene Ther. 3: 487-99 (1992); **Bandara G et al** Gene transfer to synoviocytes: prospects for gene treatment of arthritis. DNA Cell Biol. 11: 227-31 (1992); **Belldegrun A et al** Human renal carcinoma line transfected with interleukin-2 and/or interferon α gene(s): implications for live cancer vaccines. JNCI 85: 207-16 (1993); **Bottazzi B et al** Monocyte chemotactic cytokine gene transfer modulates macrophage infiltration, growth, and susceptibility to IL2 therapy of a murine melanoma. JI 148: 1280-5 (1992); **Bradley A** Embryonic stem cells: proliferation and differentiation. Curr. Opin. Cell Biol. 2: 1013-7 (1992); **Bregni M et al** Human peripheral blood hematopoietic progenitors are optimal targets of retroviral-mediated gene transfer. Blood 80: 1418-22 (1992); **Bubenik J et al** Use of IL2 gene transfer in local immunotherapy of cancer. Cancer Lett. 62: 257-62 (1992); **Bubenik J et al** Utilization of interleukin-2 gene transfer in local immunotherapy of cancer. J. Cancer Res. Clin. Oncol. 119: 253-6 (1993); **Chambers CA et al** Long term expression of IL4 *in vivo* using retroviral-mediated gene transfer. JI 149: 2899-905 (1992); **Colombo MP et al** Cytokine gene transfer in tumor cells as an approach to antitumor therapy. Int. J. Clin. Lab. Res. 21: 278-82 (1992); **Cournoyer D et al** Gene therapy of the immune system. ARI 11: 297-329 (1993); **Douvdevani A et al** Reduced tumorigenicity of fibrosarcomas which constitutively generate IL1 α either spontaneously or following IL1 α gene transfer. IJC 51: 822-30 (1992); **Federoff HJ et al** Expression of nerve growth factor *in vivo* from a defective herpes simplex virus 1 vector prevents effects of axotomy on sympathetic ganglia. PNAS 89: 1636-40 (1992); **Ferrantini M et al** α 1-interferon gene transfer into metastatic Friend leukemia cells abrogated tumorigenicity in immunocompetent mice: antitumor therapy by means of interferon-producing cells. CR 53: 1107-12 (1993); **Foa R et al** IL2 treatment for cancer: from biology to gene therapy. Br. J. Cancer 66: 992-8 (1992); **Friedmann T et al** Gene therapy of cancer through restoration of tumor-suppressor functions? Cancer 70: s1810-7 (1992); **Gastl G et al** Retroviral vector-mediated lymphokine gene transfer into human renal cancer cells. CR 52: 6229-36 (1992); **Gansbacher B et al** Retroviral gene transfer induced constitutive expression of interleukin-2 or interferon-γ in irradiated human melanoma cells. Blood 80: 2817-25 (1992); **Geller AI** Herpesviruses: expression of genes in postmitotic brain cells. Curr. Opin. Genet. Dev. 3: 81-5 (1993); **Hillman GG et al** Adoptive immunotherapy of cancer: biological response modifiers and cytotoxic cell therapy. Biotherapy 5: 119-29 (1992); **Hoogerbrugge PM et al** Treatment of patients with severe combined immunodeficiency due to adenosine deaminase (ADA) deficiency by autologous transplantation of genetically modified bone marrow cells. Hum. Gene Ther. 3: 553-8 (1992); **Hughes PF et al** Retroviral gene transfer to primitive normal and leukemic hematopoietic cells using clinically applicable procedures. JCI 89: 1817-24 (1992); **Karp SE et al** Cytokine secretion by genetically modified nonimmunogenic murine fibrosarcoma. Tumor inhibition by IL2 but not tumor necrosis factor. JI 150: 896-908 (1993); **Kawaja MD et al** Somatic gene transfer of nerve growth factor promotes the survival of axotomized septal neurons and the regeneration of their axons in adult rats. J. Neurosci. 12: 2849-64 (1992); **Lotze MT et al** New biologic agents come to bat for cancer therapy. Curr. Opin. Oncol. 4: 1116-23 (1992); **McBride WH et al** Genetic modification of a murine fibrosarcoma to produce interleukin 7 stimulates host cell infiltration and tumor immunity. CR 52: 3931-7 (1992); **Mizuno M et al** Growth inhibition of glioma cells by liposome-mediated cell transfection with tumor necrosis factor-α gene – its enhancement by prior γ-interferon treatment. Neurol. Med. Chir. Tokyo 32: 873-6 (1992); **Nolta JA et al** Retroviral vector-mediated gene transfer into primitive human hematopoietic progenitor cells: effects of mast cell growth factor (MGF) combined with other cytokines. EH 20: 1065-71 (1992); **Obiri NI et al** Expression of high affinity interleukin-4 receptors on human renal cell carcinoma cells and inhibition of tumor cell growth *in vitro* by interleukin-4. JCI 91: 88-93 (1993); **Oldham R** Gene therapy and cancer: is it for everybody? Cancer Invest. 10: 607-9 (1992); **Pardoll DM et al** Molecular engineering of the antitumor immune response. Bone Marrow Transplant. 9: s182-s6 (1992); **Pardoll D** New strategies for active immunotherapy with genetically engineered tumor cells. Curr. Opin. Immunol. 4: 619-23 (1992); **Porgador A et al** Antimetastatic vaccination of tumor-bearing mice with two types of IFN-γ gene-inserted tumor cells. JI 150: 1458-70 (1993); **Restifo NP et al** A nonimmunogenic sarcoma transduced with the cDNA for interferon γ elicits $CD8^+$ T cells against the wild-type tumor: correlation with antigen presentation capability. JEM 175: 1423-31 (1992); **Rosenberg SA** Karnofsky Memorial Lecture. The immunotherapy and gene therapy of cancer. J. Clin. Oncol. 10: 180-99 (1992); **Rosenberg SA** Gene therapy of cancer. Important Adv. Oncol. 1992: 17-38; **Roth C et al** IL2 gene transduction in malignant cells: applications in cancer containment. Bone Marrow Transplant. 9: s174-5 (1992); **Schendel DJ & Gansbacher B** Tumor-specific lysis of human renal cell carcinomas by tumor-infiltrating lymphocytes: modulation of recognition through retroviral transduction of tumor cells with interleukin 2 complementary DNA and exogenous α interferon treatment. CR 53: 4020-5 (1993); **Sivanandham M et al** Prospects for gene therapy and lymphokine therapy for metastatic melanoma. Ann. Plast. Surg. 28: 114-8 (1992); **Tsai SC et al** Induction of antitumor immunity by interleukin-2 gene-transduced mouse mammary tumor cells versus transduced mammary stromal fibroblasts. JNCI 85: 546-53 (1993); **Uchiyama A et al** Transfection of interleukin 2 gene into human melanoma cells augments cellular immune response. CR 53: 949-52 (1993); **van Beusechem VW et al** Long-term expression of human adenosine deaminase in rhesus monkeys transplanted with retrovirus-infected bone-marrow cells. PNAS 89: 7640-4 (1992); **Watanabe Y** Transfection of interferon-γ gene in animal tumors – a model for local cytokine production and tumor immunity. Semin. Cancer Biol. 3: 43-6 (1992); **Zimmer A** Manipulating the genome by homologous recombination in embryonic stem cells. ARN 15: 115-37 (1992)

Cytokine-induced killer cells: see: CIK.

Cytokine-induced neutrophil chemoattractant: see: CINC.

Cytokine inhibitors: Cytokine inhibitors are cytokine-specific substances that inhibit the biological activities of specific cytokines in a number of different ways. These proteins probably function to buffer or limit the effects of cytokines as part of a regulatory network.

In principle such a specific inhibitor could be an antagonist binding to the same receptor and competing for receptor binding with the genuine cytokine. This type of cytokine inhibitor is exemplified by the IL1-specific receptor antagonists (see: IL1ra, IL1 receptor antagonist) which bind to IL1 receptors without having agonist effects themselves.

Another class of more or less specific cytokine inhibitors consists of specific soluble binding proteins (see: IGF-BP, TNF-BP, uromodulin as examples). These binding proteins do not themselves bind to cytokine receptors. However, they prevent the genuine cytokine to interact with its receptor by complexing the cytokine.

In many instances cytokine-binding inhibitors are soluble receptor variants which are generated either by proteolytic cleavage of membrane-bound receptors or by translation of alternatively spliced receptor RNAs (see also: Receptor shedding). Many of these soluble cytokine receptors lack a transmembrane and a cytoplasmic domain normally found in membrane-bound forms of the receptor.

As a rule cytokine binding proteins show a very high affinity towards the respective cytokines, which may exceed the affinity of cytokine-specific antibodies by a factor of 100-1000. These proteins may therefore be valuable drugs that might be used to block, for example, the » autocrine activities of some of the cytokines. In addition these physiological inhibitors do not contain structures (such as the Fc portion of antibodies) that would allow interaction with other cells. They should therefore not be recognized by the immune system.

Another class of cytokine inhibitors of cytokine action are non-specific binding proteins that do not interfere with binding of the cytokines to their respective receptors. One protein capable of binding unspecifically to several cytokines is » Alpha-2-Macroglobulin (α_2M). Some drugs also interfere with cytokine actions (see, for example: CsA, Cyclosporin A; Pentamidine, Pentoxifylline).

Cytokine knock-out studies: see: Knock-out mice.

Cytokine MAPPing: see: Message amplification phenotyping.

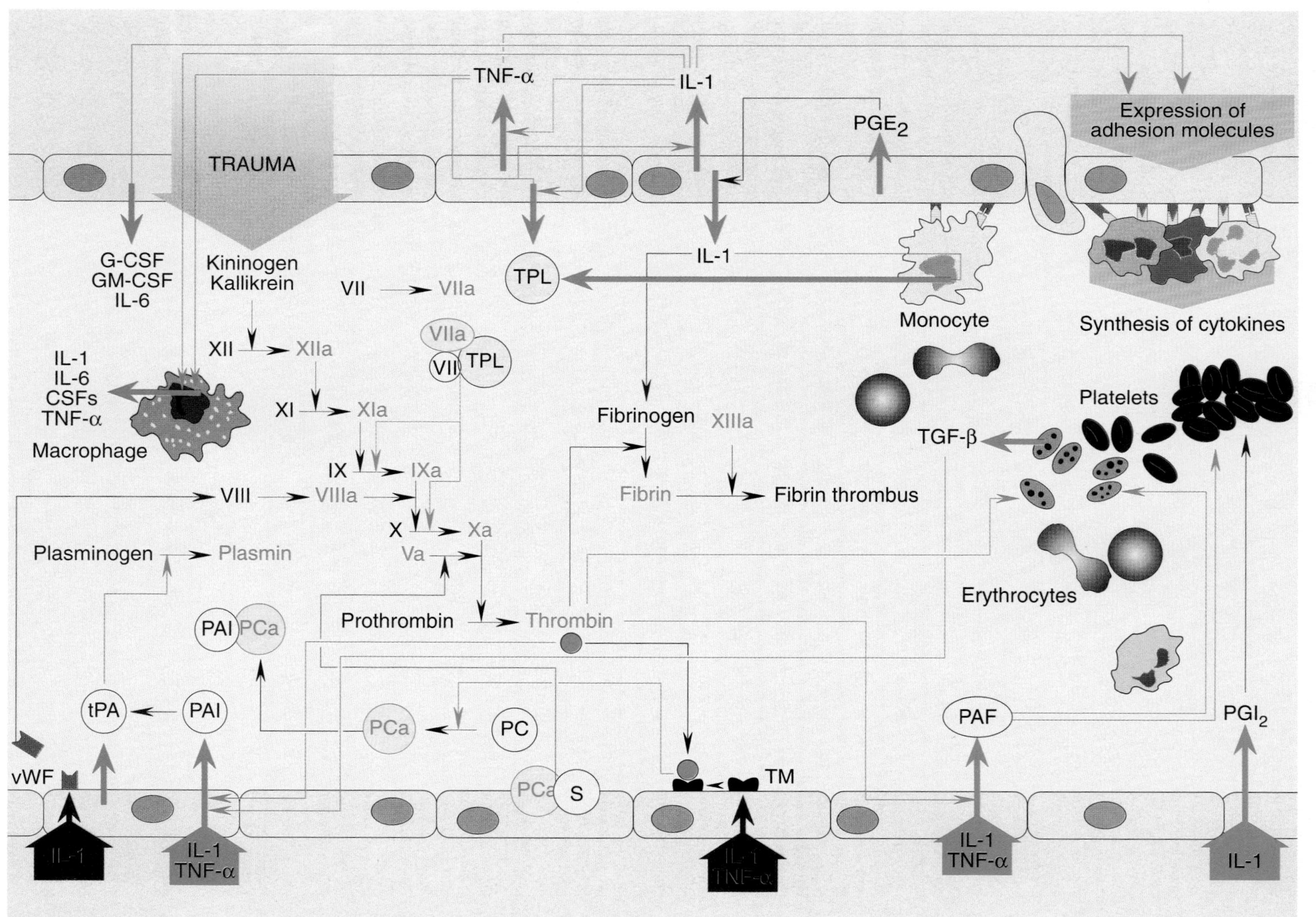

TNF-α
IL-1
PGE$_2$
Expression of adhesion molecules
TRAUMA
IL-1
G-CSF
GM-CSF
IL-6
Kininogen
Kallikrein
VII
VIIa
TPL
Monocyte
Synthesis of cytokines
VIIa
VII
TPL
XII
XIIa
IL-1
IL-6
CSFs
TNF-α
Macrophage
XI
XIa
Fibrinogen
XIIIa
Platelets
TGF-β
IX
IXa
VIII
VIIIa
Fibrin
Fibrin thrombus
X
Xa
Plasminogen
Plasmin
Va
Erythrocytes
PAI
PCa
Prothrombin
Thrombin
tPA
PAI
PCa
PC
PAF
PGI$_2$
vWF
PCa
S
TM
IL-1
IL-1
TNF-α
IL-1
TNF-α
IL-1
TNF-α
IL-1

PAF platelet activating factor
PAI plasminogen activator inhibitor
PC protein C
PGE_2 prostaglandin E_2
PGI_2 prostaglandin
S protein S
TM thrombomodulin
tPA tissue plasminogen activator
TPL tissue thromboplastin
vWF von Willebrandt factor
a small a at the end indicates the activated state

→ synthesis of
→ synthesis is inhibited
→ induced or activated
→ inhibits or reduced

Simplified view of the many interactions of cytokines and other mediators as exemplified by the bioactivities of IL1 and TNF-α on endothelial cells and components of the coagulation cascades.

Apart from being involved in the regulation of capillary permeability and vessel tonus the endothelium of blood vessels plays a decisive role in maintaining a non-thrombogenic surface by providing activators and inhibitors of coagulation and fibrinolysis. The endothelium is also involved in modulation of various immunological processes.

In endothelial cells IL1 induces, among other things, the synthesis of various colony-stimulating factors, IL6, TNF-α, prostaglandins, platelet activating factor (PAF), plasminogen activator inhibitor (PAI). An important endogenous regulator of IL1 is PGE2 which inhibits secretion of IL1 and TNF-α.

TNF-α , among other things, induces the synthesis of tissue thromboplastin (TPL) which participates in the formation of the factor X activator complex that catalyses generation of thrombin. The complex also activates factor IX. TNF-α also induces the synthesis of IL1 by endothelial cells and this IL1 can also promote production of TPL by endothelial cells. The synthesis of TNF-α can also be induced by IL1.

TNF-α and IL1 also induce the synthesis of cell surface antigens. The expression of adhesion molecules increases adhesion of lymphocytes and leukocytes at the vessel wall and thus facilitates transendothelial migration of these cells. TNF-α and IL-1 down-regulate the Protein C inactivation system by inhibiting the synthesis of thrombomodulin. Thrombomodulin complexed with thrombin activates protein C which can then form complexes with membrane-bound protein S. These complexes inhibit factor Va.

Activated protein C also neutralizes plasminogen activator inhibitor (PAI by complexation. TNF-α and also IL1 thus reduce inactivation of factor Va. TNF-α , IL1 and thrombin inhibit the formation of PAI.

Cytokine network: This term essentially refers to the complex interactions of » cytokines by which they induce or suppress their own synthesis or that of other cytokines or their receptors, and antagonize or synergise with each other in many different ways. These interactions are often ***cytokine cascades*** with one cytokine initially triggering the expression of one or more other cytokines that, in turn, trigger the expression of further factors and create complicated feedback regulatory circuits.

The cytokine network which has arisen during evolution allows considerable flexibility, depending on which part of the network is activated and the ready amplification of response to a particular stimulus. A network may also be necessary to stabilize the whole system and compensate for the lack of one component (see, for example: Knock-out mice); for some quantitative studies of cytokine gene expression *in vivo* allowing proper description of the cytokine network see: Message amplification phenotyping.

The cytokine network can be either protective or damaging by counteracting or supporting pathophysiologically important functions (see, for example, Angiogenesis factors; Autocrine; Cell activation; Hematopoiesis; Neuroimmune network, Septic shock; inflammation; wound healing).

The intricacies of the cytokine network are particularly relevant to the therapeutical uses of cytokines. Cytokine treatment may be associated, for example, with cell type- or cytokine-specific hematological alterations that profoundly influence treatment modalities and the interpretation of the effects. Examples are the transient lymphopenias induced by » IFN-α, » IFN-γ, » IL2 and » TNF-α, transient monocytopenias induced by » IFN-γ, and » TNF-α, and neutrophilias induced by » IL2, » IFN-α and » TNF-α). Treatment regimes have demonstrated that maximally tolerated doses of a biologically active cytokine may by no means be identical with therapeutically optimal doses. A study of the activation of the cytokine network by individual cytokines or other agents provides a rationale for devising cytokine combination and cytokine antagonist treatments in a variety of clinical settings.

It is almost mandatory that the important role of cytokines be seen holistically, i. e. within the entire framework of a living organism, not only taking into account specific actions of specific cytokines on specific target cells but also other features such as the actions of cell adhesion molecules, cell surface antigens (see: CD antigens), components of the extracellular matrix (see: ECM), and the plethora of low-molecular mass compounds that have long been a domain of endocrinologists and immunologists. All these factors interact with cytokines and determine the functional activities of individual cells, tissues and organs. They also establish bidirectional communication between the immune and the nervous system (see: Neuroimmune network).

Balkwill FR & Burke F The cytokine network IT 10: 299-304 (1989); **Olsson I** The cytokine network J. Intern. Med. 233: 103-5 (1993); see also references under "Cytokines".

Cytokine receptor superfamily: The notable amino acid homology among some mammalian cytokine receptors has led to speculation that these receptors are derived from a common evolutionary precursor and that these receptors can be grouped into a cytokine receptor superfamily. This superfamily of cytokine receptors comprises receptors for » Epo, » IL2 (β-subunit), » IL3, » IL4, » IL5, » IL6, » IL7, » GM-CSF, » G-CSF, » LIF, » CNTF, and also the receptors for » Growth hormone and » Prolactin. Common structural features of members of this superfamily are the absence of a tyrosine-specific kinase domain in the intracellular region, the presence of a cysteine motif (four aligned cysteine residues, generally found in the N-terminal region of the extracellular domain), and a Trp-Ser-X-Trp-Ser motif (***WSXWS***) located proximal to the transmembrane domain. The four cysteines appear to be critical to the maintenance of the structural and functional integrity of the receptors. The WSXWS consensus sequence is thought to serve as a recognition site for functional protein-protein interaction of cytokine receptors. Peptides encompassed within the WSXWS region have been shown to inhibit the priming effect of » IL3, » IL5, and » GM-CSF on the synthesis of leukotriene C4 induced by chemotactic peptides in a basophil mediator release assay. Mutational analysis of the » Epo receptor within the WSXWS motif indicates that it is critical for protein folding, ligand-binding, and signal transduction. The cytoplasmic domains of the family members are less well conserved, but some sequence similarities have been reported.

The gene ***CRFB4*** (cytokine receptor family II, member 4), a receptor belonging to the same family with an as yet unidentified ligand, encodes protein ***CRF2-4***. The gene spans more than 30 kb and maps to human chromosome 21 less than 35 kb away from the » IFN-α receptor gene.

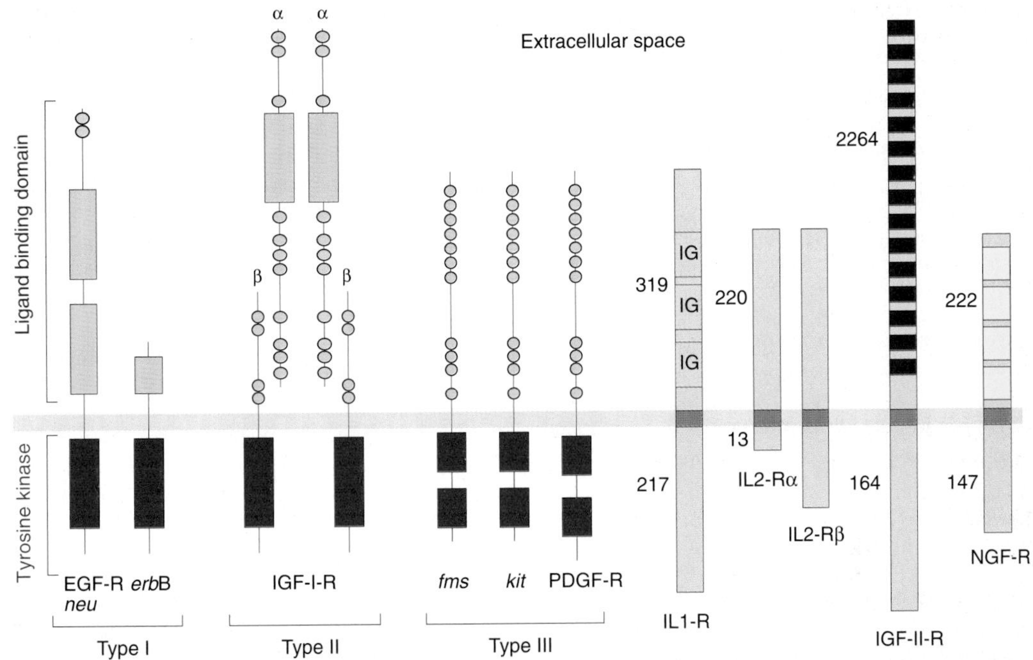

Principle structures of cytokine receptors.
Two groups of cytokine receptors can be distinguished. One group is characterized by large intracellular domains possessing intrinsic protein tyrosine kinase activity (red rectangles). The other group comprises cytokine receptors with intracellular domains that do not encode a tyrosine kinase.
Apart from differences in their extracellular domains the three types of group I cytokine receptors, designated type I, type II, and type III, differ from each other in the structures of their tyrosine kinase domains. Green rectangles or circles indicate various short homology domains. Type I receptors have large tyrosine kinase domains. Type two receptors consist of subunits. Type III receptors contain tyrosine kinase domains which are split into two domains by a short sequence region generally referred to as the kinase insert.
The extracellular portion of the IL1 receptor contains three IgG-like sequence domains. The IGF-II receptor contains 15 repetitive sequence domains of 150 amino acids each. The NGF receptor contains four cysteine-rich sequence domains. Transmembrane regions are indicated by small blue rectangles. Numbers to the left of the receptors indicate lengths in amino acids.

Most of the receptors belonging to the cytokine receptor superfamily consist of subunits which are responsible for the high-affinity binding of the respective ligands and also for post-receptor signaling processes. Some receptor subunits have been found to be part of several receptor complexes. The possibility to differentially regulate the synthesis of individual receptor subunits may be one way to increase the spectrum of cellular responses elicited after binding of a cytokine to the receptor. Many of the cytokine receptors exist in membrane-bound and soluble forms. The latter may arise by proteolytic cleavage of transmembrane receptors or by utilizing alternatively spliced receptor mRNAs. They may act as » Cytokine inhibitors or in a » Retrocrine fashion.

A subgroup of the cytokine receptor superfamily, called ***hematopoietin receptor superfamily***, comprises receptors for » IL4, » IL6, » IL7, » Epo, » GM-CSF, » prolactin, » growth hormone, and the β subunit of the » IL2 receptor.

Bazan JF Structural design and molecular evolution of a cytokine receptor superfamily. PNAS 87: 6934-8 (1990); **Bischoff SC et al** Peptide analogs of consensus receptor sequence inhibit the action of cytokines on human basophils. Lymphokine Cytokine Res. 11: 33-7 (1992); **Cosman D et al** A new cytokine receptor superfamily. TIBS 15: 265-70 (1990); **Fernandez-Botran R** Soluble cytokine receptors: Their role in immunoregulation. FJ 5: 2567-74 (1991); **Kishimoto T et al** Cytokine signal transduction. Cell 76: 253-62 (1994); **Lutfalla G et al** A new member of the cytokine receptor gene family maps on chromosome 21 at less than 35 kb from IFNαR. Genomics 16: 366-73 (1993); **Miyajima A et al** Cytokine receptors and signal transduction. ARI 10: 295-331 (1992); **Miyajima A et al** Common subunits of cytokine receptors and the functional redundancy of cytokines. TIBS 17: 378-82 (1992); **Miyazaki T et al** The integrity of the conserved 'WS motif' common to IL2 and other cytokine receptors is essential for ligand binding and signal transduction. EJ 10: 3191-7 (1991); **Murakami M et al** Critical cytoplasmic region of the interleukin 6 signal transducer gp130 is conserved in the cytokine receptor family. PNAS 88: 11349-53 (1991); **Olsson I et al** The receptors for regulatory molecules in hematopoiesis. Eur. J. Haematol. 48: 1-9 (1992); **Schreurs J et al** Cytokine receptors: a new superfamily of receptors. Int. Rev. Cytol. 137B: 121-55 (1992); **Sprang SR & Bazan JF** Cytokine structural taxonomy and mechanisms of receptor engagement. Curr. Biol. 3: 815-27 (1993); **Yoshimura A et al** Mutations in the Trp-Ser-X-Trp-Ser motif of the erythropoietin receptor abolish processing, ligand binding, and activation of the receptor. JBC 267: 11619-25 (1992)

Cytokine-related syndrome: see: OKT3-induced cytokine-related syndrome.

Cytokine release syndrome: see: OKT3-induced cytokine-related syndrome.

Cytokine response modifier A: abbrev. *crmA*. See: IL1β-Convertase.

Cytokine responsive genes: see: CRG.

Cytokines: The term cytokine, or immunocytokines, was used initially to separate a group of immunomodulatory proteins, also called immunotransmitters, from other growth factors that modulate the proliferation and bioactivities of non-immune cells. This clear-cut distinction, however, cannot be maintained and may not be meaningful altogether. Some cytokines are produced by a rather limited number of different cell types while others are produced by almost the entire spectrum of known cell types. The initial concept "one producer cell – one cytokine – one target cell" has been falsified for almost every cytokine investigated more closely. A definition of these factors on the basis of their producer or target cells is therefore also problematic. The same applies to classifications based upon the biological activities of cytokines (see, for example: BCDF, B cell differentiation factors; BCGF, B cell growth factors, and TRF, T cell replacing factors).
Today the term cytokine is used as a generic name for a diverse group of soluble proteins and peptides which act as humoral regulators at nano- to picomolar concentrations and which, either under normal or pathological conditions, modulate the functional activities of individual cells and tissues. These proteins also mediate interactions between cells directly and regulate processes taking place in the extracellular environment (see also: Autocrine, Paracrine, Juxtacrine).

In many respects the biological activities of cytokines resemble those of classical hormones produced in specialized glandular tissues. Some cytokines also behave like classical hormones in that they act at a systemic level (see, for example: Septic shock, Acute Phase Reaction). In general cytokines act on a wider spectrum of target cells than hormones. The perhaps major feature distinguishing cytokines from true hormones is the fact that, unlike hormones, cytokines are not produced by cells that are organized in specialized glands, i. e. there is not a single organ source for these mediators.
Cytokines normally do not possess enzymatic activities although there is a growing list of exceptions. Some cytokines have been found, upon determination of their primary structures, to be identical with classical enzymes (see, for example: ADF (adult T cell leukemia-derived factor), nm23, PD-ECGF (platelet-derived endothelial cell growth factor), neuroleukin).
In the more restricted sense cytokines comprise » interleukins, initially thought to be produced exclusively by leukocytes, » lymphokines, initially thought to be produced exclusively by lymphocytes, » monokines, initially thought to be produced exclusively by monocytes, interferons (see: IFN), colony-stimulating factors (see: CSF). and a variety of other proteins. It has been suggested that the generic term ***regulatory peptide factors*** (abbrev. PRF) be used for all these factors to avoid the general difficulties with the nomenclature. This term has the advantage that it includes also a number of low molecular mass peptides which are generally not regarded as cytokines although they have many cytokine-like activities. Some of these low molecular weight proteins and peptides have indeed been referred to as » ***mini-cytokines***. Most cytokines are structurally unrelated although some can be grouped into families (see: Gene families, Cytokine receptor families).
Most cytokines are glycoproteins which are secreted by cells using classical secretory pathways (see also: Signal sequences). Membrane-bound forms have been described for many cytokines, and some may also be associated with the extracellular matrix (see: ECM). It is likely that the switching between soluble and membrane forms of cytokines is an important regulatory event (see also: Autocrine, Paracrine, Juxtacrine, Retrocrine). In some cases membrane forms of a cytokine have been found to be indispensable for normal development, with soluble forms being unable to entirely substitute for them.

Most cytokines are generally not stored inside cells (exceptions are, for example » TGF-β and » PDGF which are stored in platelets). The expression of most cytokines is strictly regulated, i. e. these factors are usually produced only by activated cells (see also: Cell activation) in response to an induction signal. Expression is normally transient and can be regulated at all levels of » Gene expression (see also: ARE (AU-rich element)). However, constitutive expression has also been observed. The expression of many cytokines also seems to be regulated differentially, depending on cell type and developmental age.
Most cytokines were initially detected in functional tests *in vitro* as biochemically undefined activities or as distinct factors with distinct biological activities. This also explains, at least in part, the plethora of different names for some of the cytokines. In many instances these activities were named after a particular biological activity observed in an *in vitro* assay (see also: Bioassays and Cytokine assays for alternatives) or after cells that were found to elaborate these factors (for techniques allowing identification of cytokine genes, cytokine receptor genes, and other relevant genes without prior knowledge of their activities see: Gene libraries). One should be aware of the fact that at this moment in time the relevance of many *in vitro* activities of cytokines to their endogenous functions within an intact organism is not clearly defined.
Almost all cytokines are pleiotropic effectors and their biological activities can overlap considerably. One of the consequences of this functional overlap is the observation that one factor may frequently functionally replace another factor altogether or at least partially compensate for the lack of another factor. Since most cytokines have ubiquitous biological activities, their physiologic significance as normal regulators of physiology is often difficult to assess. Studies of gene functions in experimental animals in which a cytokine gene has been functionally inactivated by gene targeting

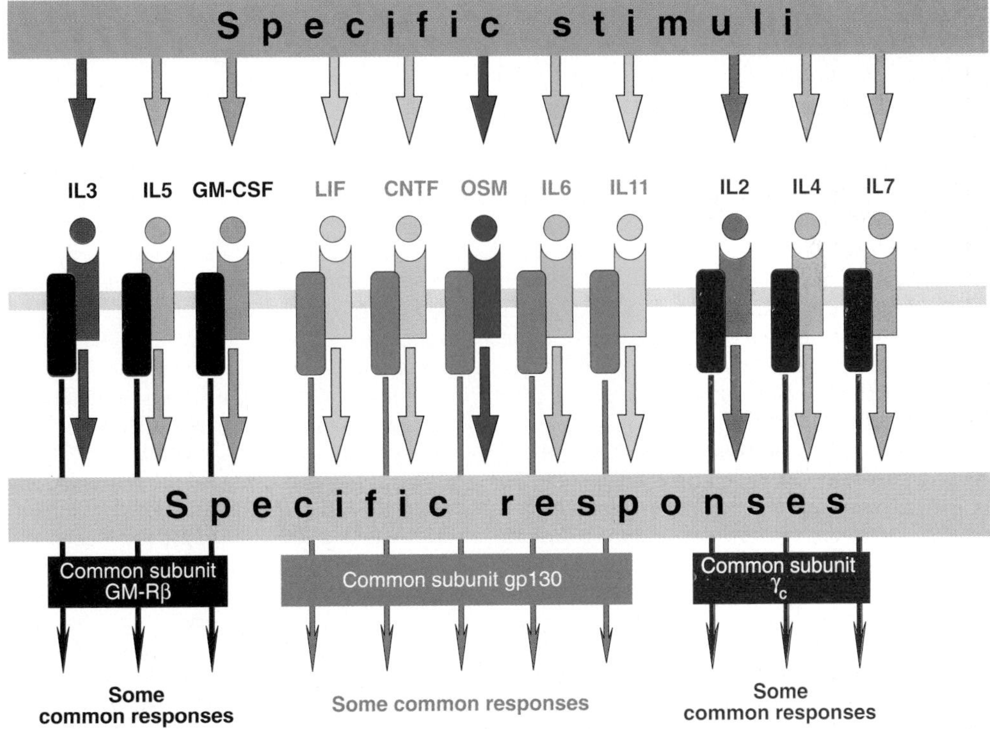

Comparative structures of some cytokine receptor systems.
A variety of cytokine receptors have been shown to consist of a cytokine-specific domain which binds the ligand and a signalling subunit that is shared between cytokines. This set-up allows for responses common to all cytokines utilizing the common signalling molecules and the corresponding cytokines. Precisely how utilization of common signalling subunits eventually also leads to cytokine-specific responses is unknown.

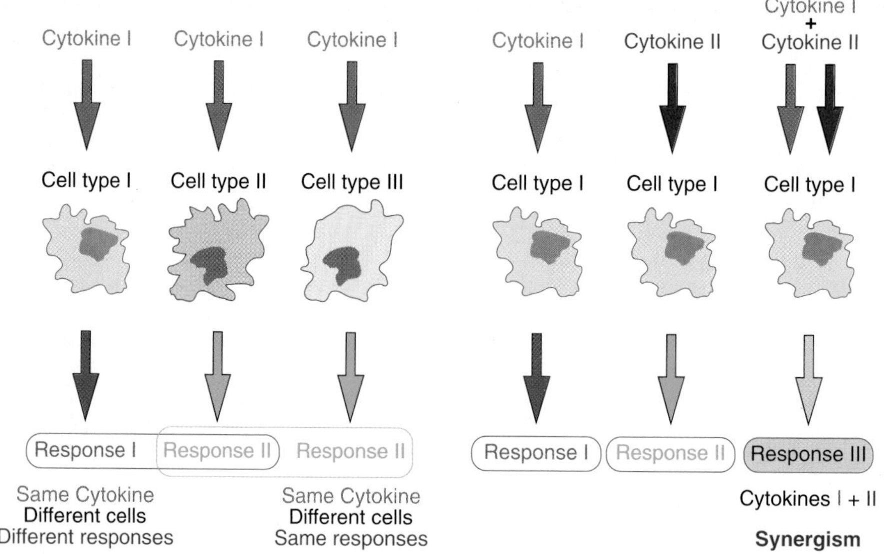

Cytokine synergism.
Cytokines can elicit quite different responses, depending upon the cell types and the presence of other cytokines. Combinations of cytokines can elicit responses that are different from those invoked by either cytokine acting on its own.

(see also: Transgenic animals) are of particular importance in cytokine research because, unlike *in vitro* studies, they provide information about the true *in vivo* functions of a given cytokine by highlighting the effects of their absence. In many instances these studies have shown that null mutations of particular cytokine genes do not have the effects *in vivo* expected from their activities *in vitro*. If information about knock-out studies is available for a given cytokine or its receptor it can be found in this dictionary as a special subentry (Transgenic/Knock-out/Antisense studies) for each particular cytokine.

Many cytokines show stimulating or inhibitory activities and may also synergise or antagonize the actions of other factors. A single cytokine may also under certain circumstances elicit reactions which are the reverse of those shown under other circumstances. The type, the duration, and also the extent of cellular activities induced by a particular cytokine can be influenced considerably by the micro-environment of a cell, depending, for example, on the growth state of the cells (sparse or confluent), the type of neighboring cells, cytokine concentrations, the combination of other cytokines present at the same time, and even on the temporal sequence of several cytokines acting on the same cell. The fact that every cell type may have different responses to the same growth factor can be explained, at least in part, by different spectrums of genes expressed in these cells and the availability and levels of various transcription factors that drive » Gene expression. The responses elicited by cytokines are therefore contextual and the "informational content", i. e. the intrinsic activities of a given cytokine may vary with conditions. It has been observed, for example, that » bFGF is a strong mitogen for fibroblasts at low concentrations and a chemoattractant at high concentrations. bFGF has also been shown to be a biphasic regulator of human hepatoblastoma-derived HepG2 cells, depending upon concentration. The interferon » IFN-γ can stimulate the proliferation of B cells prestimulated with Anti-IgM, and inhibits the » IL4-induced activities of the same cells. On the other hand, IL4 activates B cells and promotes their proliferation while inhibiting the effects induced by » IL2 in the same cells. The activity of at least two cytokines (IL1α and IL1β) is regulated by an endogenous receptor antagonist, the IL1 receptor antagonist (see: IL1ra). Several cytokines, including » TNF, » IFN-γ, » IL2 and » IL4, are inhibited by soluble receptors (see also: Receptor shedding, Cytokine inhibitors, Retrocrine). Several cytokines, including » IL10 and » TGF-β, act to inhibit other cytokines.

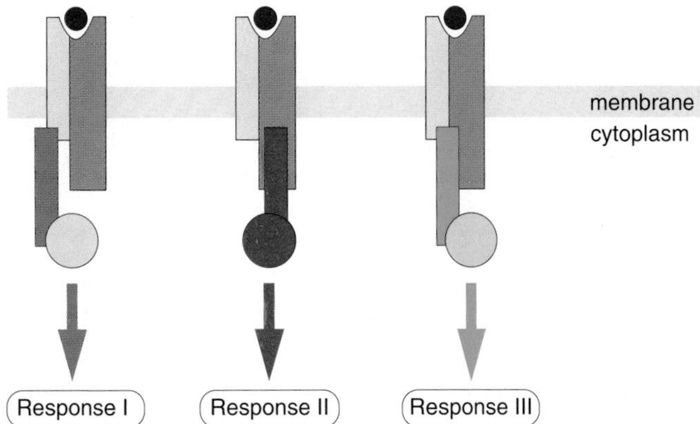

A possible structural basis for the pleiotropic activities of some cytokines.
Many cytokines have been shown to exert a broad range of biological effects, some of which are also elicited by other cytokines. Some of the similar activities of different cytokines can be explained by common receptor-associated signaling subunits. Some cytokine receptors can be associated also with different signaling moieties thus allowing a single cytokine to elicit many different effects.

The processes responsible for the regulation of cytokines are not well understood. Cells utilize distinct biochemical pathways converging on mediator release and these can be probed, among other things, by employing a variety of substances mimicking or inhibiting the actions of cytokines (see, for example: Bryostatins, Calcium ionophore, Genistein, H8, Herbimycin A, K-252a, Lavendustin A, Phorbol esters, Okadaic acid, Staurosporine, Suramin, Tyrphostins, Vanadate). Frequently one observes a hierarchical order of cytokine actions with some early cytokines preactivating cells so that they then can respond to late-acting cytokines (see also: Cell activation). Many cytokines induce the synthesis of novel gene products once they have bound to their respective receptors (see: ERGs, early response genes). Some of the novel products are themselves cytokines (see: Chemokines, for example).

Cytokine mediators can be transported quickly to remote areas of a multicellular organism. They can address multiple target cells and can be degraded quickly. Concentration gradients can be used to elicit specific responses. These possibilities by far exceed the possibilities provided by mere cell-to-cell contacts within a multicellular organism. It can be assumed that cytokines play a pivotal role in all sorts of cell-to-cell communication processes although many of the mechanisms of their actions have not yet been elucidated in full detail. A close examination of the physiological and pathological effects of the regulated or deregulated (see: Transgenic animals) expression of cytokines in complex organisms has shown that these mediators are involved in virtually all general systemic reactions of an organism, including such important processes as the regulation of immune responses (see, for example: BCDF, B cell growth and differentiation factors; BCGF, B cell growth factors; TRF, T cell replacing factors; Isotype switching), inflammatory processes (see: Inflammation), » hematopoiesis (see also: Hematopoietins), and » wound healing. Cytokines are important mediators involved in embryogenesis and organ development and their activities in these processes may differ from those observed postnatally. In addition they play a key role in neuroimmunological, neuroendocrinological, and neuroregulatory processes (see: Neuroimmune network). It has been shown recently that a number of infectious agents exploit the cytokine repertoire of organisms to evade immune responses of the host (see: Viroceptors).

The biological activities of cytokines are mediated by specific membrane receptors which can be expressed on virtually all cell types known. Their expression is also subject to several regulatory mechanisms (see: Receptor transmodulation) although some receptors are also expressed constitutively.

Unlike the cytokines themselves, which are rarely related among each other in terms of primary sequences (although they appear to have some common three-dimensional features), cytokine receptor proteins have been shown to share a number of characteristics. Many receptors are members of a » cytokine receptor superfamily. Many receptors are multi-subunit structures that bind ligands and at the same time possess functions as signal transducers due to their intrinsic tyrosine kinase activity (see also: Autophosphorylation). Some of these subunits are shared by several receptors. Many receptors are associated with special signal transducing proteins in the interior of the cell (see, for example Janus kinases). Some receptors may bind more than one cytokine. Several cytokine receptors have been shown to be converted into soluble binding proteins that regulate ligand access to the cell by specific proteolytic cleavage of receptor ectodomains.

The many specific activities of individual cytokines have been the basis for current concepts of therapeutical intervention, in particular of the treatment of hematopoietic malfunctions and tumor therapy. Applications involve the support of chemo- and radiotherapy, bone marrow transplantation, and general immunostimulation (see also: Adoptive immunotherapy, LAK cells, TIL, Cytokine gene transfer). Although some cytokines are now in clinical use (see also: Recombinant cytokines) one must be aware of the fact that current knowledge is still limited and that the *in vivo* modulation of the activity of one factor may not have the desired effect (see: Cytokine network).

Nevertheless it can be stated that our new (and growing) understanding of the biological mechanisms governing cytokine actions are an important contribution to medical knowledge. The biochemistry and molecular biology of cytokine actions explain some well-known and sometimes also some of the more obscure clinical aspects of diseases. Knowledge that cytokines create regulatory hierarchies and provide independent and/or interrelated regulatory mechanisms that can confer distinct and interactive developmental functions lays a solid, albeit rather complicated foundation, for current and future clinical experiences.

● **Books: Balkwill FR** (edt) Cytokines – A practical approach, IRL Press, Oxford 1991; **Barnes D et al** (eds) Peptide growth factors, MiE Vol 198, Academic Press 1991; **Callard RE** (edt) Cytokines and B lymphocytes. Academic Press, London, 1991; **Clemens MJ** Cytokines βIOS Scientific Publishers Oxford 1991; **Dawson MM** Lymphokines and interleukins. Open University Press, Milton Keynes, 1991; **Dinarello CA et al** (eds) The physiological and pathological effects of cytokines. Liss, New York, 1990; **Habenicht A** (edt) Growth factors, differentiation factors, and cytokines, Springer, Berlin 1990; **Nilsen-Hamilton M** (edt) Growth factors and development. (Current Topics in Developmental Biology, Vol. 24, Academic Press 1990; **O'Dorisio MS & Panerai A** (eds) Neuropeptides and immunopeptides: messengers in a neuroimmune axis. ANY 594: 1-503 (1990); **Oppenheim JJ et al** (eds) The molecular and cellular aspects of cytokines and pathophysiological and therapeutic roles of cytokines. A. R. Liss, New York 1990; **Westwick SL et al** (eds) Chemotactic cytokines. Plenum Press, New York 1991); **Waterfield MD** (edt) Growth factors. BMB Vol. 45, 1989

● **Reviews: Adamson ED** Growth factors and their receptors in development. Dev. Genet. 14: 159-64 (1993); **Aggarwal BB & Pocsik E** Cytokines: From clone to clinic. ABB 292: 335-59 (1992); **Anderson TD** Cytokine-induced changes in the leukon. Toxicol. Pathol. 21: 147-57 (1993); **Arai KI et al** Cytokines: coordinators of immune and inflammatory responses. ARB 59: 783-836 (1990); **Balkwill FR & Burke F** The cytokine network. IT 10: 299-307 (1989); **Bateman A et al** The immune-hypothalamic-pituitary-adrenal axis. Endocrine Rev. 10: 92-112 (1989); **Beck G & Habicht GS** Primitive cytokines: Harbingers of vertebrate defense. IT 12: 180-3 (1991); **Bonneau RH** Stress-induced modulation of the immune response. ANY 594: 253-69 (1990); **Borden EC & Sondel PM** Lymphokines and cytokines as cancer treatment: immunotherapy realised. Cancer 65 (Suppl. 3) 800-14 (1990); **Cairo MS** Cytokines: A new immunotherapy. Clinics in Perinatology 18: 343-59 (1991); **Carding SR et al** Cytokines in T cell development. IT 12: 239-45 (1991); **Cournoyer D & Caskey CT** Gene therapy of the immune system. ARI 11: 297-329 (1993); **Cross M & Dexter TM** Growth factors in development, transformation, and tumorigenesis. Cell 64: 271-80 (1991); **Dainiak N** Cell membrane family of growth regulatory factors. PCBR 352: 49-61 (1990); **Deuel TF** Polypeptide growth factors: roles in normal and abnormal cell growth. ARC 3: 443-92 (1987); **Farrar WL et al** The molecular basis of immune cytokine action. CRC Crit. Rev. Ther. Drug. Carr. Syst. 5: 229-61 (1989); **Fernandez-Botran R** Soluble cytokine receptors: Their role in immunoregulation. FJ 5: 2567-74 (1991); **Green JB & Smith JC** Growth factors as morphogens: do gradients and thresholds establish body plan? TIG 7: 245-50 (1991); **Hall AK & Rao MS** Cytokines and neurokines: related ligands and related receptors. TINS 15: 35-7 (1992); **Hermus ARMM & Sweep CGJ** Cytokines and the hypothalamic-pituitary-adrenal axis. JSBMB 37: 867-71 (1990); **Holán V et al** Immunological tolerance and lymphokines. CRC Crit. Rev. Immunol. 10: 481-93 (1991); **Imura H et al** Cytokines and endocrine function: An interaction between the immune and neuroendocrine system. Clin. Endocrinol. 35: 107-15 (1991); **Iversen OH** The hunt for endogenous growth-inhibitory and/or tumor suppression factors: Their role in physiological and pathological growth regulation. ACR 57: 413-53 (1991); **Jessell TM & Melton DA** Diffusible factors in vertebrate embryonic induction. Cell 68: 257-70 (1992); **Kennedy RL & Jones THJ** Cytokines in endocrinology: Their roles in health and in disease. E 129: 167-78 (1991); **Kroemer G & Martínez AC** Cytokines and autoimmune disease. Clin. Immunol. Immunopathol. 61: 275-95 (1991); **Kroemer G et al** Immunoregulation by cytokines. Crit. Rev. Immunol. 13: 163-91 (1993); **Lee B & Ciardelli TL** Clinical applications of cytokines for immunostimulation and immunosuppression. Prog. Drug. Res. 39: 167-96 (1992); **Lotze MT et al** New biologic agents come to bat for cancer therapy. Curr. Opin. Oncol. 4: 1116-23 (1992); **Lübbert M et al** Animal models for the biological effects of continuous high cytokine levels. Blut 61: 253-57 (1990); **Massagué J & Pandiella A** Membrane-anchored growth factors ARB 62: 515-41 (1993); **McDonald NG & Hendrickson WA** A structural superfamily of growth factors containing a cysteine know motif. Cell 73: 421-4 (1993); **Miyajima A et al** Cytokine receptors and signal transduction. ARI 10: 295-331 (1992); **Muegge K & Durum SK** Cytokines and transcription factors. Cytokine 2: 1-8 (1990); **Nathan C & Sporn M** Cytokines in context. JCB 113: 981-86 (1991); **Patterson PH** The emerging neuropoietic cytokine family: First CDF/LIF, CNTF and IL6; Next ONC, MGF, GCSF? Curr. Opin. Neurobiol. 2: 94-7 (1992); **Paul WE &**

Genes encoding cytokines, cytokine receptors, or cytokine-regulated genes.

Factor	Chromosomal location	Factor	Chromosomal location
ACT-2	17q21-23	IL7	8q12-q13
aFGF	5q31.3-q33.2	IL7 R p90	5p13
Amphiregulin	4q13-21	IL8	4q12-21
Angiogenin	14q11	IL8 R	2q35
BDNF	1p15.5	IL9	5q31-32
bek	10q25.3-q26	IL10	1
bFGF	4q25	IL11	19q13.3-13.4
BMP-1	8	IL12 p40	5q31-q33
BMP-2A	20p12	IL12 p35	3p12-q13.2
BMP-3	4p13-q21	IP10	4q12-21
BMP-5	6	IRF-1	5q31.1
BMP-6	6	JE	17q11. 2-12
BMP-7	20	KGF	15
CD40 ligand	Xq26.3-q27	*kit*	4q11-q12
CNTF	11q12.2	K-*sam*	10q26
CRIPTO	3	*lck*	1p35-p32
CTAP-III	4q12-q13	LD78	17
CTGF	6q23.1	LIF	22q12-q12.2
Defensin	8p23	lyn	8q13-qter
EGF	4q25-q27	MCAF	17
EGF/TGF-α-R	7p13-p12	MCP-1	17q11.2-q21.1
Endothelin 1	6p23-24	M-CSF	1p13-p21.2
Endothelin 1 R	4	M-CSF-R	5q33.3
EPA/TIMP	Xp11.1-p11.4	*met*	7p11-4
EPO	7q11-q22	MGSA	421
EPO-R	19p.13.3	Midkine	11p11 – p13
FGF-4	11q13	MIP1-α	17q11-q21
FGF-5	4q21	MIS	19p13.3-13.2
flg	8p12	MPR	6q25-q27
flt	13q12	MSP	3
fyn	6q21	Mx	21q22.3
GBP	22q12-q13.1	Myeloblastin	19
G-CSF	17q21-q22	NDF	8p21-p12
G-CSF R	1p32-34.3	NEGF2	11p13-p11
GIF-B	16	Neuroleukin	19cen-q12
GM-CSF	5q22-q31	Neurotrophin 3	12p13
gp130	5, 17	Neurotrophin 4	19q13.3
GRP	18q21	Neurotrophin 5	19
GRP R	Xp11-q11	Neurotrophin 6	19
Haptoglobin	16q22	NGF	1p13
HBNF	7q22 – 7qter	NGF-β-R	17q21-q22
HGF	7q11.2-21.1	Oncomodulin	7p1-p13
HGF R/*met*	7q21-q31	Oncostatin M	22q12-q12.2
HGF-like	3p21	PD-ECGF	22q13
HI30	9q22.3 – q33	PDGF-A	7p21-p22
I-309	17	PDGF-A-R	4q11
IFN-α	9p22	PDGF-B	22q12.3-q13.1
IFN-α-R	21q22.1	PDGF R B	5q33.3
IFN-β	9p22	PF4	4q12-21
IFN-β-R	21q21-qter	PlGF	14
IFN-γ	12q24.1	POMC	2p23
IFN-γ-R	6q15-q21, 21	RANTES	17q11. 2-q12
IGF-1	12q23	S100a	1q21
IGF-1-R	15q25-q26	S100b	21q22.2-q22.3
IGF-2	11p15.5	SCF	12q22-q24
IGF-2 R	6	TCA-3	11
IL1α	2q13	TGF-α	2p11-2p13
IL1β	2q13-q21	TGF-α/EGF-R	7p13-p12
IL1ra	2q13-14.1	TGF-β1	19q13
IL1 R type I	2q12	TGF-β2	1q41
IL2	4q26-q27	TGF-β3	14q24
IL2-R p55	10p15-p14	Thrombospondin	15q11-qter
IL2 R p75	22q11. 2-q12	TNF-α	6p23-6q12
IL2 R γ	Xq13	TNF-β	6p23-6q12
IL3	5q23-q31	Transferrin	3q21
IL4	5q23-31	Transthyretin	18 q12.1
IL5	5q23-q31	Uromodulin	16p 13.11
IL5 R α	3p26	Vitronectin	17q11
IL6	7p14-p21	Wnt-1	12q13
IL6 R	1q21		

Seder RA Lymphocyte responses and cytokines. Cell 76: 241-51 (1994); **Plata-Salamán CR** Immunoregulators in the nervous system. Neurosci. Biobehav. Rev. 15: 185-215 (1991); **Quesenberry P et al** Growth factors. Blood (Suppl. 1: 116-30 (1988); **Reeves R & Magnuson NS** Mechanisms regulating transient expression of mammalian cytokine genes and cellular oncogenes. Progr. Nucl. Acid Res. Mol. Biol. 38: 241-82 (1990); **Rozengurt E** Neuropeptides as cellular growth factors: Role of multiple signaling pathways. EJCI 21: 123-34 (1991); **Sarvetnick N** Transgenic mouse models for growth factor studies. MiE 198: 519-25 (1991); **Schultz RM** The potential role of cytokines in cancer therapy. Prog. Drug. Res. 39: 219-50 (1992); **Siegfried JM** Strategies for identification of peptide growth factors. Pharmacol. Ther. 56: 233-45 (1992); **Smith EM** Hormonal activities of cytokines. Chem. Immunol. 52: 154-69 (1992); **Sprang SR & Bazan JF** Cytokine structural taxonomy and mechanisms of receptor engagement. Curr. Biol. 3: 815-27 (1993); **Whicher JT & Evans SW** Cytokines in disease. Clin. Chem. 367: 1269-81 (1990); **Whitman M & Melton DA** Growth factors in early embryogenesis. ARC 5: 93-117 (1989)

Cytokine synthesis inhibitory factor: see CSIF.

Cytolysin: This factor is identical with Perforin 1 (see: Perforins).

Ishikawa H et al Molecular cloning of rat cytolysin. JI 143: 3069-73 (1989)

Cytolytic T lymphocyte differentiation factor: see: CDF.

Cytolytic factor: This factor is identical with » TNF-α.

Cytostatic factor: This factor is identical with » TNF-α. See also: CF (cytostatic factor).

Cytotaxin: Generic name given to substances that are directly chemotactic for cells. See: Chemotaxis.

Cytotoxic (killer) T cell activating factor: see: KAF.

Cytotoxic lymphocyte maturation factor: see: CLMF.

Cytotoxic cell-generating factor: see: CGF.

Cytotoxic differentiation factor: see: CDF.

Cytotoxicity-stimulating factor: see: CySF.

Cytotoxic T cell differentiation factor: This factor which is probably also identical with another factor designated » CDF (cytotoxic differentiation factor) is identical with » IL6.

Okada M et al BSF-2/IL6 functions as a killer helper factor in the *in vitro* induction of cytotoxic T cells. JI 141: 1543-9 (1988)

Cytotoxic tumor infiltrating lymphocytes: abbrev. CTIL. See: TILs (tumor-infiltrating lymphocytes).

Cytotoxin: see: CTX.

D

d2: This » ERG (early response gene) is identical with » Egr-1.

D10: (full name: D10G4.1) A murine T helper (TH2) cell line derived from AKR/J (H-2k) mice. D10 cells are MHC-restricted, antigen-specific helper T cells specific for the antigen egg white conalbumin. The cells secrete large amounts of » IL4, and also » GM-CSF, an IL1 inducer factor, and probably » IL10. The cells are used to assay » IL1 (see also: Bioassays). The growth of these cells and of a variant subline, designated ***MD10 cells***, depends on IL1 and they respond to IL1 isolated from almost all vertebrate species. The combined use of anti-IL1α and anti-IL1β mAb is needed to completely inhibit the proliferative response of D10 cells. Methotrexate inhibits the IL1-induced proliferation of D10 cells.
D10 cells have been reported to alter in tissue culture and to become insensitive to IL1 by beginning to produce IL1 itself.
The activity of IL1 is determined by measuring the incorporation of ^{3}H thymidine into the newly synthesized DNA of proliferating cells Cell proliferation can be determined also by employing the » MTT assay. The presence of IL6 significantly enhances D10 cell responses to IL1 in the presence of other costimulatory signals. Quantities of IL6 which alone were not costimulatory to D10 cells enhanced IL1-induced D10 proliferation significantly when Con A was used as a costimulus. No synergism between IL1 and » IL6 is observed with 3D3 (a clonotypic activating anti-TCR mAb) as a costimulus.
D10 cells also proliferate in response to lectins, » IL2, murine IL4, » IL7, and very high concentrations of murine » IL6. The assay is not affected by high concentrations of human IL6 or TNF-α and only minimally affected by high concentrations of murine TNF-α.The effects of IL2 and IL4 can be measured by using saturating concentrations of both factors. The assay is inhibited by plasma or serum components which can, however, be removed by precipitation with polyethylene glycol.
Depending upon the subclone of D10 cells, IL1 concentrations in the femto- to attogramme/mL region can be detected. Addition of saturating amounts of IL2, which neutralizes any effects of IL2 in the test sample, has been used to increase the specificity of the assay for IL1. ***D10S*** cells, for example, proliferate in response to attomolar concentrations of IL1α or IL1β in the absence of mitogens. Proliferation of this subclone is not observed in response to endotoxin, human or murine » IFN γ, » TNF-α, TNF-β, or » GM-CSF.
An alternative and entirely different method of detecting these factors is » Message amplification phenotyping.
D10 cells also proliferate in response to immobilized anti-CD3 antibody. Proliferation of D10 cells induced by anti-CD3 antibody is completely inhibited by » CsA (cyclosporine A), but CsA has little effect on proliferation induced by IL2 or IL4. The expression of IL4 and IL5 genes is differentially regulated in these cells: anti-CD3 antibodies and IL2 induce the expression of the cellular » oncogenes » *myc* and » *myb*. Anti-CD3 antibodies also induce the expression of » IL4, » IL5, and » GM-CSF. IL2 induces IL5 mRNA expression but not IL4 or GM-CSF mRNA. IL4 induces the expression of IL5 but not IL4 mRNA. The proliferation of D10 cells stimulated by IL1, antigens, and immobilized anti-CD3 antibody is blocked by » IFN-γ,which does not, however, block IL4 production.

Bohjanen PR et al Differential regulation of interleukin 4 and interleukin 5 gene expression: a comparison of T cell gene induction by anti-CD3 antibody or by exogenous lymphokines. PNAS 87: 5283-7 (1990); **Deibel MR Jr & Killar LM** One-step immunoadsorbent column purification of interleukin-4 from murine D10 supernatants. LR 7: 469-77 (1988); **Gajewski TF & Fitch FW** Anti-proliferative effect of IFN-γ in immune regulation. I. IFN-γ inhibits the proliferation of Th2 but not Th1 murine helper T lymphocyte clones. JI 140: 4245-52 (1988); **Hopkins SJ & Humphreys M** Simple, sensitive and specific bioassay of interleukin-1. JIM 120: 271-7 (1989); **Kaye J et al** Both a monoclonal antibody and antisera specific for determinants unique to individual cloned helper T cell lines can substitute for antigen and antigen-presenting cells in the activation of T cells. JEM 158: 836-56 (1983); **Kupper T et al** Growth of an interleukin 2/interleukin 4-dependent T cell line induced by granulocyte-macrophage colony-stimulating factor (GM-CSF). JI 138: 4288-

92 (1987); **Kupper T et al** Autocrine growth of T cells independent of interleukin 2: identification of interleukin 4 (IL 4, BSF-1) as an autocrine growth factor for a cloned antigen-specific helper T cell. JI 138: 4280-7 (1987); **Lacey DL et al** Interleukin 1 stimulates proliferation of a nontransformed T lymphocyte line in the absence of a co-mitogen. JI 139: 2649-55 (1987); **Mizutani H et al** Synergistic interactions of IL1 and IL6 in T cell activation. Mitogen but not antigen receptor-induced proliferation of a cloned T helper cell line is enhanced by exogenous IL6. JI 143: 896-901 (1989); **Orencole SF & Dinarello CA** Characterization of a subclone (D10S) of the D10.G4.1 helper T cell line which proliferates to attomolar concentrations of interleukin-1 in the absence of mitogens. Cytokine 1: 14-23 (1989); **Pecanha LM et al** Lymphokine control of type 2 antigen response. IL10 inhibits IL5- but not IL2-induced Ig secretion by T cell-independent antigens. JI 148: 3427-32 (1992); **Schreiber S et al** Cytokine pattern of Langerhans cells isolated from murine epidermal cell cultures. JI 149: 3524-34 (1992); **Segal R et al** The effects of methotrexate on the production and activity of interleukin-1. Arthritis Rheum. 32: 370-7 (1989); **Suda T et al** Biological activity of recombinant murine interleukin-6 in interleukin-1 T cell assays. JIM 120: 173-9 (1989); **Thor G & Brian AA** Glycosylation variants of murine interleukin-4: evidence for different functional properties. Immunology 75: 143-9 (1992); **Winstanley FP & Eckersall PD** Bioassay of bovine interleukin-1-like activity. Res. Vet Sci. 52: 273-6 (1992)

D10G4.1 cells: see: D10 cells.

D10S cells: see: D10 cells.

D36: A murine mast cell line that proliferates in response to human » IL10 in the presence of » IL4. The cells are used to assay human IL10 (see also: Bioassays). Cell proliferation is blocked completely by antibodies directed against human IL10. Human » IL1, IL2, IL3, IL4, IL5, IL6, IL9, IL12, » GM-CSF, and » TNF-a doe not appear to interfere with this assay.

Schlaak JF et al A sensitive and specific bioassay for the detection of human interleukin 10. JIM 168: 49-54 (1994)

Da: A transformed murine lymphoblastic cell line. This cell line has been used to identify and to isolate » LIF which was previously known as HILDA (human interleukin for Da). The cell line is therefore also used to detect this factor.

The cells are dependent upon » IL3, and » GM-CSF for survival and growth. DA cells are stimulated to proliferate also by murine and human » Epo. Antibody to Epo specifically blocks Epo-stimulated growth. Proliferation stimulated by Epo and GM-CSF is transient, decreasing within 24 to 48 hours of exposure. However, Epo acts cooperatively with GM-CSF to sustain proliferation. The cells also respond to » murine and human IL4. Da cells produce » IL6.

Da cell-based » bioassays are complicated and must be carried out in the presence of suitable cytokine-specific antibodies to inhibit unwanted factors. An alternative and entirely different method of detecting these factors is » Message amplification phenotyping.

Branch DR et al Identification of an erythropoietin-sensitive cell line. Blood 69: 1782-5 (1987); **Broudy VC et al** Dynamics of erythropoietin receptor expression on erythropoietin-responsive murine cell lines. Blood 75: 1622-6 (1990); **Gascan H et al** Characterization and NH_2-terminal amino acid sequence of natural human interleukin for DA cells: leukemia inhibitory factor. LR 8: 79-84 (1989); **Gascan H et al** Response of murine IL3-sensitive cell lines to cytokines of human and murine origin. LR 8: 79-84 (1989); **Hültner L et al** Interleukin 3 mediates interleukin 6 production in murine interleukin 3-dependent hemopoietic cells. GF 2: 43-51 (1989); **Moreau JF et al** Leukemia inhibitory factor is identical to the myeloid growth factor human interleukin for DA cells. N 326: 201-4 (1988); **Moreau JF et al** Characterization of a factor produced by human T cell clones exhibiting eosinophil-activating and burst-promoting activities. JI 138: 3844-9 (1987); **Sakaguchi M et al** The expression of functional erythropoietin receptors on an interleukin-3 dependent cell line. BBRC 146: 7-12 (1987); **Smith AG & Hooper ML** Inhibition of pluripotential embryonic stem cell differentiation by purified polypeptides. N 326: 197-200 (1989); **Tsao CJ et al** Expression of the functional erythropoietin receptors on interleukin 3-dependent murine cell lines. JI 140: 89-93 (1988); **Tsuda H et al** Mode of action of erythropoietin (Epo) in an Epo-dependent murine cell line. I. Involvement of adenosine 3',5'-cyclic monophosphate not as a second messenger but as a regulator of cell growth. EH 17: 211-7 (1989); **Tsuda H et al** Mode of action of erythropoietin (Epo) in an Epo-dependent murine cell line. II. Cell cycle dependency of Epo action. EH 17: 218-22 (1989)

DAUDI: A human B lymphoblastoid cell line established from a Burkitt lymphoma of a 16 year old male patient. The cells express the receptor for the IgG-Fc fragment and complement receptors on their cell surface but do not express HLA antigens as surface markers.

DAUDI cells are very sensitive to » IFN-α and their growth is inhibited by very low concentration of this interferon. Daudi cells arrest in the G1 phase of the » cell cycle in the presence of IFN-α. The cells express the leukocyte activation marker CD69 in response to IFN-α treatment. Some variant cells lines have been specifically selected for interferon-resistance.

DAUDI cells express CD54 (ICAM-1) which is a ligand for CD43 (sialophorin, leukosialin, large sialoglycoprotein or gp115); Daudi cells bind specifically to purified immobilized CD43. The cells also secrete an uncharacterized factor (see: THP-1-derived growth-promoting activity) that promotes the growth of a large variety of cell lines.

Among other things DAUDI cells are resistant against NK cells and are used to test the activity of

» LAK cells (lymphokine activated killer cells) (see also: Cell activation).
Daudi cells grow as disseminated tumors in mice with severe combined immune deficiency (SCID) after either i.v.. or i.p. injection (see also: Immunodeficient mice).

Deblandre GA et al CD69 is expressed on Daudi cells in response to interferon-α. Cytokine 4: 36-43 (1992); **Kumar R & Atlas I** Interferon α induces the expression of retinoblastoma gene product in human Burkitt lymphoma Daudi cells: role in growth regulation. PNAS 89: 6599-6603 (1992); **Hermann GG et al** LAK-cell-mediated cytotoxicity against tumor cell targets used to monitor the stimulatory effect of interleukin-2: cytotoxicity, target recognition and phenotype of effector cells lysing the Daudi, T24 and K562 tumor cell lines. Nat. Immun. 11: 7-16 (1992); **Nederman T et al** An *in vitro* bioassay for quantitation of human interferons by measurements of antiproliferative activity on a continuous human lymphoma cell line. Biologicals 18: 29-34 (1990); **Rosenstein V et al** CD43, a molecule defective in Wiskott-Aldrich syndrome, binds ICAM-1. N 354: 233-5 (1991); **Schomburg A et al** *In vivo* and *ex vivo* antitumor activity in patients receiving low-dose subcutaneous recombinant interleukin-2. Nat. Immun. 11: 133-43 (1992)

DBP: *DNA binding proteins* General term given to proteins with DNA-binding capacity, usually binding DNA in a sequence-specific manner. Some of these proteins are of particular interest in cytokine research because they act as transcription factors (see: Gene expression). They bind specifically to regions within the promoters of cytokine genes or genes the expression of which is inducible or can be repressed (see, for example, *fos*, NFIL6) in a cell type- or tissue-specific manner.

***Decapentaplegic*:** see: *dpp*.

Decidual suppressor factor: see: DSF.

DEF: Gene symbol for the human gene locus encoding » Defensins.

Defensins: Small basic unglycosylated proteins of 29-34 containing three intramolecular disulfide bonds, that cyclize and stabilize the proteins in a complexly folded triple-helical β-sheet configuration. These cysteine disulfide bonds are essential for the biological activities of defensins. The human genes have been designated HNP-1, HNP-2, and HNP-3 (H = human; NP = neutrophil peptides; abbrev. also: HP-1, HP-2, HP-3). HNP-1 and HNP-3 differ from each other by only a few nucleotide substitutions. Another human neutrophil defensin, designated HNP-4, is approximately 100-fold less abundant and may serve functions other than host defense.
Defensin mRNA is detected in normal bone marrow cells but not in peripheral blood leukocytes. The proteins are identical in sequence except at the N-terminal amino acid. The human gene locus encoding defensins, ***DEF***, maps to chromosome 8p23. Functional proteins arise by sequential posttranslational processing of a preproprotein of 93-95 amino acids, removing 64 aminoterminal amino acids.
Defensins were found initially in invertebrates as constituents of a primitive immune system. Defensins are also found in other species, including humans, rats, and rabbits. Rabbit microbicidal cationic protein s (see: MCP-1 - MCP-2) are identical with defensins.
At least nine mouse strains have been described that almost entirely lack defensins in their neutrophils, suggesting that they may be imperfect experimental models of infection in which neutrophil function is significant. Defensins are members of a family of related peptides which also include the so-called ***cryptdins*** (intestinal defensins; amino termini 3 to 6 amino acids longer than those of leukocyte-derived defensins) isolated from murine intestines and expressed in epithelial cells of intestinal crypts. and » ***corticostatins***.
Defensins are mainly found in segmented neutrophilic granulocytes and constitute approximately 30-50 % of all proteins of azurophilic granules. Defensins are released from these granules by transport into vacuoles generated after ingestion of microorganisms and destroy them in an oxygen-independent manner by permeabilization of the outer and inner membrane. In micromolar concentrations defensins are bactericidal for a variety of Gram-negative and -positive microorganism, mycobacteria, and yeasts, and some enveloped viruses. Defensins also possess a nonspecific cytotoxic activity against a wide range of normal and

```
Human HNP-1     ACYCRIPACIAGERRYGTCIYQGRLWAFCC
Rabbit NP-1    VVCACRRALCLPRERRAGFCRIRGRIHPLCCRR
Rat NP-1       VTCYCRRTRCGFRERLSGACGYRGRIYRLCCR
```

Amino acid sequences of some mammalian defensins.
Conserved cysteine residues participating in disulfide bond formation are shown in yellow.

malignant cells, including cells that are resistant to » TNF-α and » NKCF (natural killer cytotoxic factor). They kill cells by insertion into the cell membrane, permeabilizing the membranes by the creation of voltage-regulated channels.

The family of human defensins is diverse and is not restricted to expression in leukocytes. Human defensin -5 is highly expressed in Paneth cells of the small intestine. The existence of enteric defensins suggests that these peptides contribute to the antimicrobial barrier function of the small bowel mucosa, protecting the small intestine from bacterial overgrowth by autochthonous flora and from invasion by potential pathogens that cause infection via the peroral route, such as L. monocytogenes and Salmonella species.

HNP-1, the most abundant representative of human defensins, has been found to form very stable complexes with » Alpha 2-Macroglobulin which may function as a scavenger of defensins in inflamed tissues and may constitute an important mechanism for the regulation and containment of inflammation.

Apart from their activity on membranes defensins may also act on DNA, introducing single-strand breaks, and this may, at least in part, also account for some of their biological activities.

At nanomolar concentrations defensins (HNP-1 and HNP-2, but not HNP-3) are specific potent chemoattractants for monocytes (see also: Chemotaxis). The release of defensins by neutrophils may therefore facilitate the local accumulation of monocytes in areas of infection and/or » inflammation. Human neutrophil defensins have also been shown to be very potent inhibitors of purified protein kinase C. They block activation of PKC by the tumor promoter 12-O-tetradecanoyl-phorbol-13-acetate (see also: Phorbol esters) but they do not inhibit its binding to PKC.

Bateman A et al The effect of HP-1 and related neutrophil granule peptides on DNA synthesis in HL60 cells. Regul. Pept. 35: 135-43 (1991); **Bateman A et al** The isolation and identification of multiple forms of the neutrophil granule peptides from human leukemic cells. JBC 266: 7524-30 (1991); **Boman HG** Antibacterial peptides: key components needed in immunity. Cell 65: 205-7 (1991); **Cullor JS et al** Bactericidal potency and mechanistic specificity of neutrophil defensins against bovine mastitis pathogens. Vet. Microbiol. 29: 49-58 (1991); **Daher KA et al** Isolation and characterization of human defensin cDNA clones. PNAS 85: 7327-31 (1988); **Eisenhauer PB & Lehrer RI** Mouse neutrophils lack defensins. Infect. Immun. 60: 3446-7 (1992); **Eisenhauer PB et al** Cryptdins: antimicrobial defensins of the murine small intestine. Infect. Immunol. 60: 3556-65 (1992); **Ganz T et al** Defensins: microbicidal and cytotoxic peptides of mammalian host defense cells. Med. Microbiol. Immunol. 181: 99-105 (1992); **Gera JF & Lichtenstein A** Human neutrophil peptide defensins induce single strand DNA breaks in target cells. CI 138: 108-20 (1991); **Harwig SS et al** Characterization of defensin precursors in mature human neutrophils. Blood 79: 1532-7 (1992); **Hill CP et al** Crystal structure of defensin HNP-3, an amphiphilic dimer: mechanisms of membrane permeabilization. S 251: 1481-5 (1991); **Jones DE & Bevins CL** Paneth cells of the human small intestine express an antimicrobial peptide gene. JBC 267: 23216-25 (1992); **Lehrer et al** Defensins: antimicrobial and cytotoxic peptides of mammalian cells. ARI 11: 105-28 (1993); **Lichtenstein A** Mechanism of mammalian cell lysis mediated by peptide defensins. Evidence for an initial alteration of the plasma membrane. JCI 88: 93-100 (1991); **Lin MY et al** The defensin-related murine CRS1C gene: expression in Paneth cells and linkage to Defcr, the cryptdin locus. Genomics 14: 363-8 (1992); **Michaelson D et al** Cationic defensins arise from charge-neutralized propeptides: a mechanism for avoiding leukocyte autocytotoxicity? J. Leukoc. Biol. 51: 634-9 (1992); **Ouellette AJ & Lualdi JC** A novel mouse gene family coding for cationic, cysteine-rich peptides. Regulation in small intestine and cells of myeloid origin. JBC 265: 9831-7 (1990); **Ouelette AJ et al** Purification and primary structure of murine cryptdin-1, a Paneth cell Defensin. FL 304: 146-8 (1992); **Panyutich AV et al** An enzyme immunoassay for human defensins. JIM 141: 149-55 (1991); **Panyutich A & Ganz T** Activated α 2-macroglobulin is a principal defensin-binding protein. Am. J. Respir. Cell. Mol. Biol. 5: 101-6 (1991); **Ouellette AJ & Lualdi JC** A novel mouse gene family coding for cationic, cysteine-rich peptides: regulation in small intestine and cells of myeloid origin. JBC 265: 9831-7 (1990); **Selsted ME et al** Enteric Defensins: antibiotic peptide components of intestinal host defense. JCB 118: 929-36 (1992); **Sparkes RS et al** Assignment of defensin gene(s) to human chromosome 8p23. Genomics 5: 240-4 (1989); **Territo MC et al** Monocyte-chemotactic activity of defensins from human neutrophils. JCI 84: 2017-20 (1989); **Valore EV & Ganz T** Posttranslational processing of defensins in immature human myeloid cells. Blood 79: 1538-44 (1992)

delta: A transcription factor identical with » GATA-1.

Safrany G & Perry RP Characterization of the mouse gene that encodes the δ/YY1/NF-E1/UCRBP transcription factor. PNAS 90: 5559-63 (1993)

***Delta*:** A gene found in *Drosophila melanogaster* which encodes a 110 kDa transmembrane protein containing nine EGF-like domains in its extracellular sequence. The protein is an invertebrate homologue of » EGF involved in the control of embryonic development. *Delta* protein interacts with another protein, designated » *Notch*, and compensates the effects of certain lethal *Notch* mutations by down-regulating its synthesis. The mechanism responsible for this regulatory effect is thought to involve direct protein-protein interactions with the EGF-like sequence elements found in both proteins.

Kopczynski CC et al *Delta*, a *Drosophila* neurogenic gene, is transcriptionally complex and encodes a protein related to blood coagulation factors and epidermal growth factor of vertebrates. GD 2: 1723-35 (1988); **Kopczynski CC & Muskavitch MAT** Complex spatio-temporal accumulation of alternative transcripts from the neurogenic gene *Delta* during *Drosophila* embryogenesis. Development 107: 623-36 (1989); **Muskavitch MAT & Hoffman FM** Homologues of vertebrate growth factors in *Drosophila melanogaster* and other vertebrates. CTDB 24: 289-28 (1990); **Rebay I et al** Specific EGF repeats of *Notch* mediate interactions with *delta* and *serrate*: implications for *Notch* as a multifunctional receptor. Cell 67: 687-99 (1991); **Vässin H et al** The neurogenic gene *Delta* of *D. melanogaster* is expressed in neurogenic territories and encodes a putative transmembrane protein with EGF-like repeats. EJ 3431-40 (1987)

Delta assay: (Δ assay) A technique used to detect » hematopoietic stem cells (see also : Hematopoiesis). This assay is a short-term suspension culture assay in which potentially colony-forming cells (see: colony formation assay) are first grown in liquid culture for one week before they are replated in semisolid medium. Numbers of colonies obtained in this way are considered as an indicator of the number of hematopoietic progenitors of an early stage (earlier than those normally obtained after 14 days of growth in semisolid medium). Interpretation of the two-step liquid culture assay is based on the assumption that hematopoietic progenitor cells normally not detected in colony formation assays can divide during the 1 week period in suspension and give rise to progeny capable of generating colonies in secondary semisolid cultures. Increases in clonogenic cell numbers following the suspension culture period are taken as a manifestation of the self-renewal capacity of the cells under study (for example: » CD34-positive bone marrow cells). This assay is used to study the responses of early hematopoietic stem cells to hematopoietic growth factors (see: Hematopoietins) alone or in combination.

Lemoli RM et al Autologous bone marrow transplantation in acute myelogenous leukemia: *in vitro* treatment with myeloid-specific monoclonal antibodies and drugs in combination. Blood 77: 1829-36 (1991); **Lemoli RM et al** Autologous bone marrow transplantation in acute myelogenous leukemia: *in vitro* treatment with myeloid-specific monoclonal antibodies and drugs in combination. Blood 77: 1829-36 (1991); **Lemoli RM et al** Interleukin 11 stimulates the proliferation of human hematopoietic $CD34^+$ and $CD34^+CD33^-DR^-$ cells and synergises with stem cell factor, interleukin-3, and granulocyte-macrophage colony-stimulating factor. EH 21: 1668-72 (1993); **Moore MAS & Warren DJ** Synergy of interleukin-1 and granulocyte colony-stimulating factor: *In vivo* stimulation of stem cell recovery and hemopoietic regeneration following 5-fluoro-uracil treatment of mice. PNAS 84: 7134-7 (1987); Moore MAS Coordinate actions of hematopoietic growth factors in stimulation of bone marrow function. Chapter 29 in Sporn B & Roberts AB (eds) Handbook of Exp. Pharmacol. Vol. 95/II, Springer Verlag, Berlin; **Muench MO & Moore MA** Accelerated recovery of peripheral blood cell counts in mice transplanted with *in vitro* cytokine-expanded hematopoietic progenitors. EH 20: 611-8 (1992); **Muench MO et al** Interactions among colony-stimulating factors, IL1 β, IL6, and *kit*-ligand in the regulation of primitive murine hematopoietic cells EH 20: 339-49 (1992); **Nara N & McCulloch EA** The proliferation in suspension of the progenitors of the blast cells in acute myeloblastic leukemia. Blood 65: 1484-93 (1985); **Park CH et al** Divergent effect of growth factors on leukaemic progenitor cells: self-renewal (SR) vs differentiation (DF). Abstract. EH 16: 509 (1988)

DE-Proteins: *delayed early* General name given to proteins the synthesis of which is induced early after the binding of a growth factor to its receptor. See also: Gene expression.

DER6: A delayed early response gene (see also: ERG) identified in murine » 3T3 fibroblasts. It shows 90 % homology with human » MIF (macrophage inhibition factor).

Lanahan A et al Growth factor-induced delayed early response genes. MCB 12: 3919-29 (1992)

DER: *Drosophila EGF receptor* This is the *Drosophila* homologue of the human » EGF receptor gene. It is identical with two genes, designated ***torpedo*** and ***faint little ball***, respectively. Both genes are involved in the dorso-ventricular development of the embryo. DER has been shown to exert pleiotropic activities. Many DER mutations are embryonic lethal alleles. DER affects pattern formation, cell division, and cell death in eye imaginal discs. It has been shown also to determine neuroblast identity. See also: EGF.

Baker NE & Rubin GM Ellipse mutations in the *Drosophila* homologue of the EGF receptor affect pattern formation, cell division, and cell death in eye imaginal discs. Dev. Biol. 150: 381-396 (1992); **Goode S et al** The neurogenic locus brainiac cooperates with the *Drosophila* EGF receptor to establish the ovarian follicle and to determine its dorsal-ventral polarity. Development 116: 177-92 (1992); **Price JV et al** The maternal ventralising locus *torpedo* is allelic to *faint little ball*, an embryonic lethal, and encodes the *Drosophila* EGF receptor homologue. Cell 56: 1085-92 (1989); **Schejter ED & Shilo BZ** The *Drosophila* EGF receptor homologue (DER) gene is allelic to *faint little ball*, a locus essential for embryonic development. Cell 56: 1093-1104 (1989); **Raz E & Shilo BZ** Dissection of the faint little ball (flb) phenotype: determination of the development of the *Drosophila* central nervous system by early interactions in the ectoderm. Development 114: 113-123 (1992); **Raz E et al** Interallelic complementation among DER/flb alleles: implications for the mechanism of signal transduction by receptor-tyrosine kinases. Genetics 129: 191-201 (1991); **Shilo BZ & Raz E** Developmental control by the *Drosophila* EGF receptor homologue DER. TIG 7: 388-92 (1991)

Dermal fibroblast-derived cytokine: This factor is identical with » IL8.

Dexter-type long-term cultures: see: LTBMC (long-term bone marrow culture).

DF: *differentiation factor* This factor is identical with » G-CSF. See also: CSF for general information about colony stimulating factors.

Dexter TM et al The role of hemopoietic cell growth factor (interleukin 3) in the development of hemopoietic cells. Ciba Found. Symp. 116: 129-47 (1985)

DF 77: see: Transferrin.

D-Factor: *differentiation-stimulating factor* Several factors have been described under this name in the literature published before 1986. In most cases it is not clear whether they are identical or distinct factors. One activity described as D-factor which is produced by murine L929 and promotes the differentiation of the murine myeloid leukemia cell line » M1 into macrophages is identical with » LIF.
At least some of the activities ascribed to D-Factor in the older literature may also be due to the activity of » G-CSF (See also: CSF for general information about colony stimulating factors).

Gough NM & Williams RL The pleiotropic actions of leukemia inhibitory factor. CC 1: 77-80 (1989); **Hozumi M et al** Protein factors that regulate the growth and differentiation of mouse myeloid leukemia cells. Ciba Found. Symp. 148: 25-33; (1990); **Lowe DG et al** Genomic cloning and heterologous expression of human differentiation-stimulating factor. DNA 8: 351-9 (1989); **Tomida M et al** Characterization of a factor inducing differentiation of mouse myeloid leukaemic cells purified from conditioned medium of mouse Ehrlich ascites tumor cells. FL 178: 291-6 (1984); **Tomida M et al** Purification of a factor inducing differentiation of mouse myeloid leukemic M1 cells from conditioned medium of mouse fibroblast L929 cells. JBC 259: 10978-82 (1984)

DIA: *differentiation inhibiting activity* This factor was initially isolated from culture media of BRL cells (Buffalo rat liver). It inhibits the spontaneous occurring differentiation of embryonic stem cells *in vitro* at picomolar concentrations (see: ES cells) and is therefore also called ***ES cell differentiation inhibitory activity***. ES cells can be propagated in the presence of DIA without using a layer of » feeder cells normally required to maintain a homogeneous population of pluripotent stem cells. The factor is therefore also referred to as ***ES cell growth factor*** or ***ESCGF***. The factor is identical with » LIF.

Heath JK & Smith AG Growth factors in embryogenesis. BMB 45: 319-36 (1989); **Smith AG & Hooper ML** Buffalo rat liver cells produce a diffusible activity which inhibits the differentiation of murine embryonic carcinoma and embryonic stem cells. Dev. Biol. 121: 1-9 (1987); **Smith AG et al** Inhibition of pluripotential embryonic stem cell differentiation by purified polypeptides. N 336: 688-90 (1988); **Smith AG & Hooper ML** Inhibition of pluripotential embryonic stem cell differentiation by purified polypeptides. N 326: 197-200 (1989); **Williams RL et al** Myeloid leukemia inhibitory factor (LIF) maintains the developmental potential of embryonic stem cells. N 336: 684-7 (1988)

DIF: *differentiation inducing factor* Several factors have been described under this name in the literature published before 1986. In most cases it is not clear whether they are identical or distinct factors. One activity described as DIF which is produced by murine L929 cells, Ehrlich ascites cells and mitogen-stimulated spleen cells which promotes differentiation of the murine myeloid leukemia cell line » M1 and also induces osteolysis is identical with human » LIF, differing from LIF in one amino acid position.
One activity described as DIF which is produced by mitogen-stimulated lymphocytes and some T cell lines which promotes differentiation of the human promyelocytic leukemia cell line » HL-60 and also inhibits the growth of leukemic and normal progenitor cells is identical with » TNF-β.
One activity, derived from the human fibroblast cell line WI-26VA4, which induces the differentiation of the human monocytic leukemia » U937 cells into a monocyte/macrophage pathway, is identical with » IL6. The factor differs from genuine IL6 by the absence of the N-terminal proline.

Abe E et al Differentiation-inducing factor purified from conditioned medium of mitogen-stimulated spleen cell cultures stimulates bone resorption. PNAS 83: 5958-62 (1986); **Abe E et al** Macrophage differentiation-inducing factor from human monocytic cells is equivalent to murine leukemia inhibitory factor. JBC 264: 8941-5 (1989); **Abe E et al** Differentiation-inducing factor produced by the osteoblastic cell line MC3T3-E1 stimulates bone resorption by promoting osteoclast formation. J. Bone Miner. Res. 3: 635-45 (1988); **Noda M et al** Purification and characterization of human fibroblast derived differentiation inducing factor for human monoblastic leukemia cells identical to interleukin-6. Anticancer Res. 11: 961-8 (1991); **Olsson I et al** Myeloid cell differentiation: the differentiation inducing factors of myeloid leukemia cells. Leukemia 2: 16S-23S (1988); **Olsson IL et al** Priming of human myeloid leukaemic cell lines HL-60 and U-937 with retinoic acid for differentiation effects of cyclic adenosine 3': 5'-monophosphate-inducing agents and a T lymphocyte-derived differentiation factor. CR 42: 3928-33 (1982) **Shimizu Y et al** Purification and characterization of a T lymphocyte-derived differentiation inducing factor for human promyelocytic cell line (HL-60) and its relationship to lymphotoxin. Microbiol. Immunol. 33: 489-501 (1989); **Takeda K et al** Identity of differentiation inducing factor and tumor necrosis factor. N 323: 338-40 (1986)

DIF: *differentiation-inhibiting factor* This factor which promotes the differentiation of normal human bronchial epithelial cells into squamous cell types is identical with » TGF-β.

Masui T et al Type β transforming growth factor is the primary differentiation-inducing factor for normal human bronchial epithelial cells. PNAS 83: 2438-42 (1986)

Differential gene screening: see: Gene libraries.

Differential hybridization: see: Gene libraries.

Differential libraries: see: Gene libraries.

Differentiation antigens: General term used for cell surface proteins selectively expressed either at different stages of development of activation states (see also: Cell activation). See also: CD antigens.

Differentiation autoinducing activity: see: WEHI-3B.

Differentiation enhancing factor: This factor is produced by murine erythroleukemia (MEL) cells. It promotes erythroid differentiation. The factor possesses a unique sequence with no similarity to that of known proteins.

Sparatore B et al A vincristine-resistant murine erythroleukemia cell line secretes a differentiation enhancing factor. BBRC 173: 156-63 (1990); **Sparatore B et al** Differentiation of murine erythroleukemia cells by hexamethylenebisacetamide involves secretion and binding to membranes of a differentiation enhancing factor. BBRC 179: 153-60 (1991); **Sparatore B et al** Characterization of the biological role of murine erythroleukemia cells "differentiation enhancing factor" using antisense oligodeoxynucleotides. BBRC 193: 941-7 (1993)

Differentiation factor: abbrev. DF. This factor is identical with » G-CSF. See also: CSF for general information about colony stimulating factors.

Differentiation inducing factor: see: DIF.

Differentiation inducing factor for human monoblastic leukemia cells: This factor is secreted by a human fibroblast cell line. It induces the differentiation of the human monocytic leukemia cell line » U937 into monocytes and macrophages. The sequence of the first 13 aminoterminal amino acids is identical with that of » IL6 and the factor is most probably identical with IL6.

Noda M et al Purification and characterization of human fibroblast derived differentiation inducing factor for human monoblastic leukemia cells identical to interleukin 6. ACR 11: 961-8 (1991)

Differentiation inhibiting activity: see: DIA.

Differentiation-inhibiting factor: see: DIF.

Differentiation inhibiting factor for M1 cells: see: I-factor.

Differentiation-inhibiting protein: see: DIP.

Differentiation-retarding factor: see: DRF.

Differentiation-stimulating factor: see: D-Factor.

DIP: *differentiation-inhibiting protein* This poorly characterized factor is isolated from the blood plasma of a 60-year-old woman suffering from pure red cell aplasia. DIP activity is not detectable in the plasma of normal, healthy subjects. DIP functions as an erythropoiesis-inhibiting protein, selectively inhibiting the growth and differentiation of normal human and murine » BFU-E, but not » CFU-E cells or murine » CFU-GM. It also inhibits dimethyl sulfoxide-induced hemoglobin synthesis by Friend murine erythroleukemia cells without blocking their proliferation. DIP also blocks the differentiation of human erythroleukemia cells and this involves the inhibition of an early and critical activation of inactive membrane protein kinase C. The biological activities of DIP resemble those of » AGIF (autocrine differentiation-inhibiting factor).

Durkin JP et al Characterization of a novel erythropoiesis-inhibiting human protein. ANY 628: 233-40 (1991); **Durkin JP et al** Evidence that a novel human differentiation-inhibiting protein blocks the dimethyl sulfoxide-induced differentiation of erythroleukemia cells by inhibiting the activation of membrane protein kinase C. Cancer Res. 52: 6329-34 (1992); **Durkin JP et al** The identification and characterization of a novel human differentiation-inhibiting protein that selectively blocks erythroid differentiation. Blood 79: 1161-71 (1992)

Diphtheria toxin receptor: see: HB-EGF (heparin-binding EGF like factor).

(DL)BCGF: This factor isolated from the Dennert B cell lymphoma cell line (C. C3. 11. 75 = DL) is identical with » BCGF 2 and hence » IL5. For some aspects of nomenclature see also: BCGF (B cell growth factors).

Ambrus JL & Fauci AS Human B lymphoma cell line producing B cell growth factor. JCI 75: 732-9 (1985); **Dutton RW et al** Partial purification and characterization of a BCGF II from EL4 culture supernatants. JI 132: 2451-6 (1984); **Swain SL & Dutton RW** Production of a B cell growth-promoting activity (DL)BCGF, from a cloned T cell line and its assay on the BCL1 B cell tumor. JEM 156: 1821-34 (1982); **Swain SL et al** Evidence for two distinct classes of murine B cell growth factors with activities in different functional assays. JEM 158: 822-35 (1983)

(DL)TRF: This factor is most likely not identical with » (DL)BCGF (see also: TRF 1, T cell replacing factor 1) and may be identical with one of the factors showing » BCGF 2 (B cell growth factor 2 = IL4) activity.

Dutton RW et al Partial purification and characterization of a BCGF II from EL4 culture supernatants. JI 132: 2451-6 (1984)

DNA binding proteins: see: DBP.

Down regulatory protein of interleukin 2 receptor: see: RPT-1.

Downstream: see: Upstream.

dpp*: *decapentaplegic A gene isolated from the genome of *Drosophila melanogaster*. The *dpp* locus, which spans at least 55 kb, is involved in the dorsal-ventral development of the embryo and the morphogenesis of the imaginal disk (controlling the proliferation of epithelial cells). *dpp* is also involved in the organogenesis of the intestines.
A gene with similar biological activities is ***tolloid*** (tld) the absence of which leads to a partial conversion of dorsal into ventral ectoderm. *tld* and *dpp* are members of the » TGF-β family of proteins and is related to human bone morphogenetic proteins (see: BMP). *dpp* can functionally replace BMP-2 and BMP-4in an ectopic bone formation assay.
dpp protein appears to play a key role as a growth and differentiation factor in development, being responsible for the development of all dorsal structures. The expression of *dpp* is under positive control of the homeotic gene called *Ultrabithorax* and under negative control of the homeotic gene *Abdominal A*. In addition secreted *dpp* also influences gene expression in cells of the endoderm, inducing, for example, other homeotic genes (*labial*).
dpp protein also shows homology with a protein isolated from *Xenopus*- cells which also belongs to the » TGF-β family of proteins.
Gelbart WM The *decapentaplegic* gene: a TGF-β homologue controlling pattern formation in *Drosophila*. Development (Suppl.) 107: 65-74 (1989); **Immerglück K et al** Induction across germ layers in *Drosophila* mediated by a genetic cascade. Cell 62: 261-8 (1990); **Irish VF & Gelbart WM** The *decapentaplegic* gene is required for dorsal-ventral patterning of the *Drosophila* embryo. Genes Dev. 1: 868-79 (1987); **Kaplan FS et al** Fibrodysplasia ossificans progressiva: a clue from the fly? Calcif. Tissue Int. 47: 117-25 (1990); **Muskavitch MAT & Hoffman FM** Homologues of vertebrate growth factors in *Drosophila melanogaster* and other vertebrates. CTDB 24: 289-328 (1990); **Panganiban GEF et al** Biochemical characterization of the *Drosophila dpp* protein, a member of the transforming growth factor-β family of growth factors. MCB 10: 2669-77 (1990); **Panganiban GEF et al** A *Drosophila* growth factor homologue, *decapentaplegic*, regulates homeotic gene expression within and across germ layers during midgut morphogenesis. Development 110: 1041-50 (1990); **Panganiban GEF et al** Biochemical characterization of the *Drosophila dpp* protein, a member of the transforming growth factor β family of growth factors. MCB 10: 2669-77 (1990); **Posakony LG et al** Wing formation in *Drosophila melanogaster* requires *decapentaplegic* gene function along the anterior-posterior compartment boundary. Mech. Dev. 33: 69-82 (1990); **Raftery LA et al** The relationship of *decapentaplegic* and *engrailed* expression in *Drosophila* imaginal disks: Do these genes mark the anterior-posterior compartment boundary? Development 113: 27-33 (1991); **Ray RP et al** The control of cell fate along the dorsal-ventral axis of the *Drosophila* embryo. Development 113: 35-54 (1991); **Shimell MJ et al** The *Drosophila* dorsal-ventral patterning gene *tolloid* is related to human bone morphogenetic protein 1. Cell 67: 469-81 (1991); **St Johnson RD et al** The molecular organization of the *decapentaplegic* gene in *Drosophila melanogaster*. GD 4: 1114-27 (1990)

DRF: *differentiation retarding factor* A factor initially isolated from BRL cells (Buffalo rat liver) which inhibits the differentiation of embryonic stem cells (see also: ES cells) and embryonic carcinoma cells. The factor is identical with » LIF.
Koopman P & Cotton R A factor produced by feeder cells which inhibits embryonic carcinoma differentiation. ECR 154: 233-42 (1984); **Smith AG & Hooper ML** Buffalo rat liver cells produce a diffusible activity which inhibits the differentiation of murine embryonic carcinoma and embryonic stem cells. Dev. Biol. 121: 1-9 (1987)

***Drosophila* 60A:** A *Drosophila* gene encoding a secreted protein that belongs to the family of transforming growth factors (see: TGF-β). The *Drosophila* protein is more closely related to the human bone morphogenetic proteins, BMP-5, -6 and -7 (see: BMP) than to a TGF-like factor also isolated from *Drosophila*, designated » *dpp* (*decapentaplegic*).
Doctor JS et al Sequence, biochemical characterization, and developmental expression of a new member of the TGF-β superfamily in *Drosophila melanogaster*. Dev. Biol. 151: 491-505 (1992); **Wharton KA et al** *Drosophila* 60A gene, another transforming growth factor β family member, is closely related to human bone morphogenetic proteins. PNAS 88: 9214-8 (1991)

DSF: *decidual suppressor factor* Postimplantation murine decidual tissue from allopregnant C3H mice has been shown to release *in vitro* a potent immunosuppressive factor. It is probably involved in the local suppression of maternal immune reactions against the embryo by hormonally induced decidual suppressor cells and by molecules released from the placenta.
DSF is released by a population of small lymphocytic decidual suppressor cells. This factor is neutralized by antibodies directed against TGF-β2. DSF is closely related to TGF-β2 but as released from decidua, differs in size. DSF also has the ability to promote anchorage-independent growth of normal rat kidney fibroblasts similar to TGF-βs.
Clark DA et al Murine pregnancy decidua produces a unique immunosuppressive molecule related to transforming growth

factor β-2. JI 144: 3008-14 (1990); **Lea RG et al** Release of a transforming growth factor (TGF)-β 2-related suppressor factor from postimplantation murine decidual tissue can be correlated with the detection of a subpopulation of cells containing RNA for TGF-β 2. JI 148: 778-87 (1992)

Duffy blood group antigen: see: CKR-1.

DVR gene family: The DVR (*decapentaplegic*-Vg-related) gene family comprises a family of secreted proteins related to » TGF-β. It consists of at least 15 members, including » *dpp* (*decapentaplegic*) from *Drosophila*, » Vg-1 (vegetalising factor) from *Xenopus*, and the mammalian bone morphogenetic protein genes (see: BMP).
DVR proteins form part of a cascade of extracellular signaling molecules mediating inductive tissue interactions during development. The ectopic expression of DVR-4 causes » *Xenopus laevis* embryos to develop with an overall posterior and/or ventral character. DVR-4 induces ventral types of mesoderm in animal cap explants and overrides the dorsalizing effects of » activin A.
Jones CM et al DVR-4 (bone morphogenetic protein-4) as a posterior-ventralizing factor in *Xenopus* mesoderm induction. Development 115: 639-47 (1992); **Lyons KM et al** The DVR gene family in embryonic development. TIG 7: 408-12 (1991); **Lyons KM et al** The TGF-β-related DVR gene family in mammalian development. Ciba Found. Symp. 165: 219-30 (1992)

DW34: A murine cell line (see also: LTBMC, long-term bone marrow culture) that is completely dependent on » IL7 and is therefore used to assay this factor (see also: Bioassays). The activities of IL7 is determined by measuring the incorporation of ^{3}H thymidine into the newly synthesized DNA of the proliferating factor-dependent cells Cell proliferation can be determined also by employing the » MTT assay. An alternative and entirely different method of detecting IL7 is » Message amplification phenotyping.
Sudo T et al Interleukin 7 production and function in stromal cell-dependent B cell development. JEM 170: 333-8 (1989)

Dwarf mice: These animals are the result of homozygous autosomal recessive mutations at the *dw* (***Snell-Bagg mouse***) or the *df* (***Ames dwarf mouse***), respectively. Snell dwarf mice and exhibit a microcephalic cerebrum with hypomyelination, retarded neuronal growth, and underdevelopment of axons and dendrites.
The primary defect in these animals has been identified in the anterior pituitary and these mice are deficient in » Growth hormone and » Prolactin. They also show impaired production of other pituitary-derived hormones. The Snell-Bagg mouse has been shown to be deficient in thyroxine. Mutations in the gene encoding ***Pit-1***, a pituitary-specific transcription factor responsible for pituitary development and hormone expression in mammals, have been found to be responsible, at least in part, for the absence of somatotroph, lactotroph and thyrotroph cells and thus for the hormone deficiencies in Snell-Bagg dwarf mice. A point mutation in the Pit-1 has also been identified on one allele in a patient with combined pituitary hormone deficiency. Absence of Pit-1 also appears to be responsible for the absence of growth hormone releasing factor receptor expression in Snell-Bagg mice.
Dwarf mice show atrophy of peripheral lymphoid organs and the thymus and cellular depletion of the bone marrow (profound lack of B cell progenitors). Snell-Bagg mice have been shown to exhibit decreased peripheral blood cell counts affecting all lineages (erythrocytic, leukocytic, and platelets). This defect can be improved by treatment with human growth hormone.
Depressed primary IgM but not secondary IgG antibody responses, decreased graft-versus-host reactivity, decreases in the number of splenic T and B lymphocytes, and a prolonged survival of skin grafts are other indications of multiple immunological defects in dwarf mice. Immunodeficiencies observed in these mice provide evidence implicating anterior pituitary hormones in the regulation of immune responses as an important component of the » Neuroimmune network. Ectopic pituitary transplants, treatment with Prolactin, or administration of some » thymic hormones have been shown to restore immunocompetence in Ames dwarf mice but does not appear to correct B cell progenitor deficiencies in Snell-Bagg mice.
● **Reviews: Voss JW & Rosenfeld MG** Anterior pituitary development: short tales from dwarf mice. Cell 70: 527-30 (1992)
Barkley MS et al Prolactin status of hereditary dwarf mice. Endocrinol. 110: 2088-96 (1982); **Baroni C** Thymus, peripheral lymphoid tissues, and immunological responses of the pituitary dwarf mouse. Experientia 23: 282-3 (1967); **Bartke A** Histology of the anterior hypophysis, thyroid, and gonads of two types of dwarf mice. Anat. Rec. 149: 225-36 (1964); **Bouchon R et al** Prolactin similar to ectopic pituitary isograft restores responsiveness in Snell dwarf mice. Neuroreport 3: 210-2 (1992); **Buckwalter MS et al** Localization of the panhypopituitary dwarf mutation (df) on mouse chromosome 11 in an intersubspecific backcross. Genomics 10: 515-26 (1991); **Cross RJ et al** Immunologic disparity in the hypopituitary dwarf mouse. JI 148: 1347-52 (1992); **Duquesnoy RJ et al** Immunological studies on the Snell-Bagg pituitary dwarf mouse. PSEBM 133: 201-6 (1970); **Duquesnoy RJ** Immunodeficiency of the thymus-dependent system of the Ames dwarf mouse. JI 108: 1578-1590 (1972);

Esquifino AI et al Ectopic pituitary transplants restore immunocompetence in Ames dwarf mice. Acta Endocrinol. Copenh. 125: 67-72 (1991); **Fabris N et al** Hormones and the immunological capacity. III: the immunodeficiency disease of the hypopituitary Snell-Bagg dwarf mouse. Clin. Exp. Immunol. 9: 209-25 (1971); **Lewis UJ** Growth hormone in normal and dwarf mice. Mem. Soc. Endocrinol. 15: 179-91 (1967); **Lin C et al** Pit-1-dependent expression of the receptor for growth hormone releasing factor mediates pituitary cell growth. N 360: 765-8 (1992); **Murphy WJ et al** Immunologic and hematologic effects of neuroendocrine hormones. Studies on DW/J dwarf mice. JI 148: 3799-805 (1992); **Murphy WJ et al** Differential effects of growth hormone and prolactin on murine T cell development and function. JEM 178: 231-6 (1993); **Noguchi T** Retarded cerebral growth of hormone-deficient mice. Comp. Biochem. Physiol. C. 98: 239-48 (1991); **Phelps CJ & Hurley DL** Development of hypophysiotropic neuron abnormalities in GH- and PRL-deficient dwarf mice. Recent Prog. Horm. Res. 48: 489-96 (1993); **Radovick S et al** A mutation in the POU-homeodomain of Pit-1 responsible for combined pituitary hormone deficiency. S 257: 1115-8 (1992); **Rhodes SJ et al** A tissue-specific enhancer confers Pit-1-dependent morphogen inducibility and autoregulation on the pit-1 gene. Genes Dev. 7: 913-32 (1993); **Villanua MA et al** Changes in lymphoid organs of Ames dwarf mice after treatment with growth hormone, prolactin or ectopic pituitary transplants. J. Endocrinol. Invest. 15: 587-95 (1992)

E

EA: *enhancing activity* A general term often used to describe less well characterized factors which enhance a recognizable biological activity (e. g. » GM-EA = granulocyte-macrophage enhancing activity, a factor that promotes the generation of granulocytes and macrophages).

Ea3.17: A murine IL3-dependent pre-B cell line that is used to assay » murine IL3 (see also: Bioassays). The cells also respond to » IL2 and » murine or human » IL4. The constitutive expression of » *myc* » oncogene in these cells from a recombinant retrovirus increases growth rates in liquid cultures and a higher cloning efficiency in soft agar but does not eliminate their requirement for growth factors.
Gascan H et al Response of murine IL3-sensitive cell lines to cytokines of human and murine origin. LR 8: 79-84 (1989); **Hume CR et al** Constitutive c-*myc* expression enhances the response of murine mast cells to IL3, but does not eliminate their requirement for growth factors. O 2: 223-6 (1988)

EAF: *eosinophil-activating factor* This protein of ≈ 40 kDa is found the serum of individual with *Schistosoma mansoni* infections and in supernatants of cultured peripheral blood mononuclear cells from moderately eosinophilic individuals. EAF is also secreted upon stimulation by » TNF-α of a hybrid cell line produced by fusing human cervical keratinocytes with HeLa cells. High levels of EAF are produced by mononuclear cells of some but not by all asthma patients. Incubation of mononuclear cells with certain allergens enhances EAF production in sensitive individuals.
EAF enhances the capacity of human peripheral blood eosinophils to kill IgG antibody-coated schistosomula of *Schistosoma mansoni*. EAF also increases the ability of eosinophils to lyse antibody-coated, herpes simplex virus-infected Chang liver cells. It enhances the production of superoxide and hydrogen peroxide by eosinophils that occurs both spontaneously and in response to opsonized zymosan. EAF enhances eosinophil degranulation, both spontaneously and after incubation with opsonized zymosan. Enhanced degranulation is associated with release of eosinophil peroxidase and eosinophil cationic protein. EAF enhances IgG-dependent production of leukotriene C4 by human eosinophils in a dose- and time-dependent fashion. EAF increases the surface expression of the α chain of the complement receptor CR3 (CD11b).
EAF can be separated from » GM-CSF and » TNF both of which have eosinophil-activating activity when tested against schistosomula. Eosinophils are not activated by » IL1, » IL2, » IL3, » IFN-α, bacterial lipopolysaccharides or phorbol myristate acetate (see also: Phorbol esters). EAF is also not identical with » IL5 or » ECEF (eosinophil cytotoxicity-enhancing factor).
Fitzharris P et al The effects of eosinophil activating factor on IgG-dependent sulphidopeptide leukotriene generation by human eosinophils. Clin. Exp. Immunol. 66: 673-80 (1986); **Mazza G et al** The presence of eosinophil-activating mediators in sera from individuals with *Schistosoma mansoni* infections. EJI 21: 901-5 (1991); **Thorne KJ et al** Partial purification and biological properties of an eosinophil-activating factor. EJI 15: 1083-91 (1985); **Thorne KJ et al** A comparison of eosinophil-activating factor (EAF) with other monokines and lymphokines. EJI 16: 1143-9 (1986); **Thorne KJ et al** Eosinophil-activating factor (EAF) production by a human cell line (ESH 98) stimulated with tumor necrosis factor. Immunology 63: 545-50 (1988); **Thorne KJ et al** Production of eosinophil-activating factor (EAF) by peripheral blood mononuclear cells from asthma patients. Int. Arch. Allergy Appl. Immunol. 90: 345-51 (1989); **Thorne KJ et al** A new method for measuring eosinophil activating factors, based on the increased expression of CR3 α chain (CD11b) on the surface of activated eosinophils. JIM 133(1): 47-54 (1990)

Early growth response genes: see: Gene expression.

Early pregnancy factor: see: EPF.

EBI 1: *EBV-induced gene 1* This gene, among others, is abundantly expressed after infection of primary B lymphocytes with EBV. The cDNA sequence encodes a G protein-coupled peptide receptor. EBI 1 is expressed exclusively in B and T lymphocyte cell lines and in lymphoid tissues and is highly homologous to the » IL8 receptors.
Birkenbach M et al Epstein-Barr virus-induced genes: first

lymphocyte-specific G protein-coupled peptide receptors. J. Virol. 67: 2209-20 (1993)

EBVCS: *EBV cell surface antigen* see: CD23.

EC: (a) abbrev. for endothelial cells; (b) abbrev. for epidermal cells; (c) abbrev. for embryonic carcinoma); (d) as a prefix, occasionally used to denote "extracellular".

EC_{50}: *effective concentration* Used to indicate the concentration of a cytokine which elicits 50 % of the maximal biological response. See also: BRMP.

EC cells: Embryonal carcinoma cells. See ES cells (embryonic stem cells).

ECDGF: *embryonal carcinoma-derived growth factor* This heparin-binding factor (see also: HBGF) of 17.5 kDa is obtained from the culture supernatants of embryonal carcinoma cells (PC13 cell line) grown in serum-free medium (see also: SFM). The factor is a member of the heparin-binding growth factor family of proteins (see: HBGF). It induces the proliferation of PC13 cells, of its differentiated variants, and also of murine » 3T3 fibroblasts. Treatment of P13 cells with ECDGF leads to a decrease in the numbers of high-affinity receptors for » EGF. ECDGF also causes phosphorylation of the EGF receptor and activates protein kinase C. In combination with » TGF-β ECDGF induces the development of mesodermal tissues in frogs. ECDGF is assayed by its proliferation-inducing activity on » FBHE cells.

Heath JK & Isacke EM PC13 embryonal carcinoma-derived growth factor. EJ 3: 2957-62 (1984); **Heath JK et al** The role of EGF receptor transmodulation in embryonal carcinoma-derived growth factor-induced mitogenesis. EJ 5: 1809-14 (1986); **Heath JK & Rees AR** Growth factors in mammalian embryogenesis. Ciba Found. Symp. 116: 3-22 (1985); **Heath JK & Smith AG** Regulatory factors of embryonic stem cells. JCS Suppl. 10: 257-66 (1988); **Mahadevan LC et al** Embryonal carcinoma-derived growth factor activates protein kinase C *in vivo* and *in vitro*. EJ 6: 921-6 (1987); **Slack JM et al** The role of fibroblast growth factor in early *Xenopus* development. Development. 107 Suppl. 141-8 (1989); **van Veggel JH et al** PC13 embryonal carcinoma cells produce a heparin-binding growth factor ECR 169: 280-6 (1987) [erratum in ECR 171: 524]

ECDGF: *endothelial cell-derived growth factor* This poorly characterized factor of ≈ 10-30 kDa is found in endothelial cell-conditioned medium. ECDGF stimulates the growth of smooth muscle cells and fibroblasts but does not support bovine aortic endothelial (BAE) cell growth in limiting serum concentrations (0.2%). The factor is not identical with » PDGF. Heparin, dextran sulfate, and carageenan rapidly release non-PDGF-like ECDGF from BAE cell monolayers. The non-PDGF-like mitogen is associated with endothelial cell surfaces in contrast to the apparent constitutive release of PDGF-like mitogens. ECDGF induces an angiogenic response when implanted on the chick chorioallantoic membrane (see also: Angiogenesis factors).

Gajdusek CM & Schwartz SM Ability of endothelial cells to condition culture medium. JCP 110: 35-42 (1982); **Gajdusek CM** Release of endothelial cell-derived growth factor (ECDGF) by heparin. JCP 121: 13-21 (1984); **Gajdusek CM & Harris-Hooker SA** *In vitro* and *in vivo* effects of endothelial cell-derived growth factor. AEMB 172: 179-203 (1984); **Gajdusek CM & Schwartz SM** Comparison of intracellular and extracellular mitogenic activity. JCP 121: 316-22 (1984)

ECE: *endothelin-converting enzyme* see: ET (endothelin).

ECEF: *eosinophil cytotoxicity-enhancing factor* also abbrev. M-ECEF (monocyte-derived ECEF). This factor is produced by lipopolysaccharide-stimulated human monocytes. It is also secreted by » U937 cells. Multiple forms of this factor (10, 13, 14 kDa) have been described. Membrane-associated ECEF (mECEF), apparently an integral membrane component, is found in U937 cells, in 70% of monocytes and, at lower levels, on blood T lymphocytes. The factor is identical with » ADF (adult T cell leukemia-derived factor). Some M-ECEF activity has been attributed to » TNF.

M-ECEF enhances antibody-dependent cytotoxic activities of human eosinophils and increases the ability of human eosinophils to kill larvae of *Schistosoma mansoni*. The factor also increases the release of leukotriene C4 and other arachidonic acid metabolites by eosinophils.

Balcewicz-Sablinska MK et al Human eosinophil cytotoxicity-enhancing factor: II. Multiple forms synthesized by U937 cells and their relationship to thioredoxin/adult T cell leukemia-derived factor. JI 147: 2170-4 (1991); **Elsas PX et al** Eosinophil cytotoxicity enhancing factor: purification, characterization, and immunocytochemical localization on the monocyte surface. EJI 20: 1143-51 (1990); **Elsas PX et al** Selection of U937 histiocytic lymphoma cells highly responsive to phorbol ester-induced differentiation using monoclonal antibody to the eosinophil cytotoxicity enhancing factor. Blood 75: 2427-33 (1990); **Silberstein DS & David JR** Tumor necrosis factor enhances eosinophil toxicity to Schistosoma mansoni larvae. PNAS 83: 1055-9 (1986); **Silberstein DS et al** Characterization of a factor from the U937 cell line that enhances the toxicity of human eosinophils to Schistosoma mansoni larvae. JI 138: 3042-50 (1987); **Silberstein DS et al** Human eosinophil cytotoxicity-enhancing factor: purification, physical characteristics, and partial amino acid sequence of an active polypeptide. JI 143: 979-87 (1989)

EC-ETAF: *epidermal cell-derived thymocyte activating factor* see: ETAF.

ECGF: *embryonic carcinoma-derived growth factor* This factor is identical with » bFGF. The recommended name for *bFGF/ECGF* is ***FGF-2*** (see: Nomenclature Meeting Report and Recommendation, January 18, 1991, reprinted in: ANY Vol. 638, pp. xiii-xvi).
Heath JK & Isacke CM PC13 embryonal carcinoma-derived growth factor. EJ 3: 2957-62 (1984)

ECGF: *endothelial cell growth factor* This factor is released by activated macrophages and platelets (see also: Cell activation). It has a length of 155 amino acids. Its gene is on human chromosome 5q31.3-33. 2. The factor promotes the proliferation of vascular endothelial cells. It is identical with » aFGF.
Factors that also show ECGF activity include » PD-ECGF, » f-ECGF (fibroblast-derived endothelial cell growth factor) and » NB41-ECGF.
Jaye M et al Human endothelial cell growth factor: Cloning, nucleotide sequence, and chromosome localization. S 233: 541-5 (1986); **Lobb RR & Fett JW** Purification of two distinct growth factors from bovine neural tissue by heparin affinity chromatography. B 23: 6295-6299 (1984); **Maciag T et al** Heparin binds endothelial cell growth factor, the principal endothelial cell mitogen in bovine brain. S 225: 932-5 (1984); **Schreiber AB et al** A unique family of endothelial cell polypeptide mitogens: the antigenic and receptor cross-reactivity of bovine endothelial cell growth factor, brain-derived acidic fibroblast growth factor, and eye-derived growth factor-II. JCB 101: 1623-6 (1985); **Winkles JA et al** Human smooth vascular muscle cells express and respond to heparin-binding growth factor 1 (endothelial cell growth factor). PNAS 84: 7124-8 (1987)

ECGF-α: *endothelial cell growth factor α* This factor is identical with a truncated form of » ECGF-β.
Burgess WH et al Multiple forms of endothelial cell growth factor. JBC 260: 11389-92 (1986); **Burgess WH et al** Structural evidence that endothelial cell growth factor β is the precursor of both endothelial cell growth factor α and acidic fibroblast growth factor. PNAS 83: 7216-20 (1986); **Ohtaki T et al** Purification of acidic fibroblast growth factor from bovine omentum. BBRC 161: 169-75 (1989); **Schreiber AB et al** A unique family of endothelial cell polypeptide mitogens: the antigenic and receptor cross-reactivity of bovine endothelial cell growth factor, brain-derived acidic fibroblast growth factor, and eye-derived growth factor-II. JCB 101: 1623-6 (1985)

ECGF-β: *endothelial cell growth factor β* This factor is identical with » bFGF. Acidic fibroblast growth factor (see: aFGF) is derived from ECGF-β by posttranslational processing. Another factor, » ECGF-α, is also derived from ECGF-β in the same manner. ECGF-β represents a 20 amino acid aminoterminal extension of » ECGF-α and a 14 amino acid aminoterminal extension of » aFGF.
Burgess WH et al Structural evidence that endothelial cell growth factor β is the precursor of both endothelial cell growth factor α and acidic fibroblast growth factor. PNAS 83: 7216-20 (1986); **Ohtaki T et al** Purification of acidic fibroblast growth factor from bovine omentum. BBRC 161: 169-75 (1989); **Schreiber AB et al** A unique family of endothelial cell polypeptide mitogens: the antigenic and receptor cross-reactivity of bovine endothelial cell growth factor, brain-derived acidic fibroblast growth factor, and eye-derived growth factor-II. JCB 101: 1623-6 (1985)

ECGF: *(hypothalamus-derived) endothelial cell growth factor* This factor is identical with » aFGF.
Burgess WH et al Multiple forms of endothelial cell growth factor. JBC 260: 11389-92 (1986); **Schreiber AB et al** A unique family of endothelial cell polypeptide mitogens: the antigenic and receptor cross-reactivity of bovine endothelial cell growth factor, brain-derived acidic fibroblast growth factor, and eye-derived growth factor-II. JCB 101: 1623-6 (1985)

ECGF-1: *endothelial cell growth factor 1* This factor is the product of the human gene locus ECGF1. It is identical with » PD-ECGF (platelet-derived endothelial cell growth factor).
Stenman G et al Regional localization of the human platelet-derived endothelial cell growth factor (ECGF1) gene to chromosome 22q13. Cytogenet. Cell Genet. 59: 22-3 (1992)

ECGF-2a: *endothelial cell growth factor 2a* This factor is isolated from serum-free culture medium (see also: SFM, serum-free medium) of hepatoma cells. It stimulates the growth of endothelial cells. ECGF-2a is identical with ***HPSTI*** (human pancreatic secretory trypsin inhibitor), a specific serine proteinase inhibitor which is normally expressed and secreted by pancreatic cells. See also: ECGF-2b.
Hochstrasser K et al Proteinaseinhibitoren als tumorassoziierte Wachstumsfaktoren. Laryngo-Rhino-Otol. 68: 51-6 (1989); **McKeehan WL et al** Two apparent human endothelial cell growth factors from human hepatoma cells are tumor-associated proteinase inhibitors. JBC 261: 5378-83 (1986)

ECGF-2b: *endothelial cell growth factor 2b* This factor is isolated from the serum-free culture medium (see also: SFM, serum-free medium) of hepatoma cells. It stimulates the growth of endothelial cells.
ECGF-2b is identical with the proteins known as ***Bikunin***, ***urinary trypsin inhibitor*** and ***HI-30*** (human inhibitor 30 kDa). These proteins are specific inhibitors of serine proteases. They are the biologically active components of the inter-α-trypsin-inhibitor complex (see: ITI).
Examination of the mRNA reveals that the complex is a fusion protein of HI-30 and α1-Micro-

globulin. This fusion protein cannot be detected extracellularly and is probably a depot form of HI-30 from which the inhibitor is released by proteolytic cleavage. An increased secretion of HI-30 is observed during inflammatory processes (see also: Inflammation) and with some malignant tumors. The inhibitor has been shown to be overproduced by a factor of 10 during embryonic development and in the first view postnatal years. HI-30 can also be isolated from some tumor cell lines and is probably a growth factor for these cells.

Gebhard W et al Complementary DNA and derived amino acid sequence of the precursor of one of the three protein components of inter-α-trypsin inhibitor. FL 229: 63-7 (1988); **Hochstrasser K et al** Proteinaseinhibitoren als tumorassoziierte Wachstumsfaktoren. Laryngo-Rhino-Otol. Stuttg. 68: 51-6 (1989); **Hochstrasser K & Wachter E** Kunitz-type proteinase inhibitors derived by limited proteolysis of inter-α-trypsin inhibitor. 1. Determination of their amino acid sequence of the antitryptic domain by solid phase Edman degradation. Hoppe-Seyler's Z. Physiol. Chem. 360: 1285-96 (1979); **McKeehan WL et al** Two apparent human endothelial cell growth factors from human hepatoma cells are tumor-associated proteinase inhibitors. JBC 261: 5378-83 (1986); **Wachter E et al** Kunitz-type proteinase inhibitors derived by limited proteolysis of inter-α-trypsin inhibitor. 2. Characterization of a second inhibitory inactive domain by amino acid sequence determination. Hoppe-Seyler's Z. Physiol. Chem. 360: 1297-1303 (1979); **Wachter E & Hochstrasser K** Kunitz-type proteinase inhibitors derived by limited proteolysis of inter-α-trypsin inhibitor. 3. Sequence of the two Kunitz-type domains inside of the native inter-α-trypsin inhibitor and also of its cleavage products. Hoppe-Seyler's Z. Physiol. Chem. 360: 1305-11 (1979)

EC-GRAM: *epidermal cell-derived granulocyte-activating mediator* see: GRAM.

ECGS: *endothelial cell growth supplement* abbrev. also EnGS. Commercially available growth supplement prepared from bovine neural tissue. It is used as a relatively undefined source of growth factors (see also: ECGF, endothelial cell growth factor) and as a serum replacement (see also: SFM, serum-free medium) for establishing cell lines (see also: Cell culture). It is often used as a crude source of „ aFGF.

Bochsler PN et al Isolation and characterization of equine microvascular endothelial cells *in vitro*. Am. J. Vet. Res. 50: 1800-5 (1989); **Engelmann K & Friedl P** Optimization of culture conditions for human corneal endothelial cells. In Vitro Dell. Dev. Biol. 25: 1065-72 (1989); **Fawcett J et al** Isolation and properties in culture of human adrenal capillary endothelial cells. BBRC 174: 903-8 (1991); **Fickling SA et al** Characterization of human umbilical vein endothelial cell lines produced by transfection with the early region of SV40. ECR 201: 517-21 (1992); **Gordon EL et al** A comparison of primary cultures of rat cerebral microvascular endothelial cells to rat aortic endothelial cells. In Vitro Dell. Dev. Biol. 27: 312-26 (1991); **Jassal D et al** Growth of distal fetal rat lung epithelial cells in a defined serum-free medium. In Vitro Dell. Dev. Biol. 27: 625-32 (1991); **Nag AC et al** Factors controlling embryonic heart cell proliferation in serum-free synthetic media. In Vitro Dell. Dev. Biol. 21: 553-62 (1985); **Pintus C et al** Endothelial cell growth supplement: a cell cloning factor that promotes the growth of monoclonal antibody producing hybridoma cells JIM 61: 195-200 (1983); **Tontsch U & Bauer HC** Isolation, characterization, and long-term cultivation of porcine and murine cerebral capillary endothelial cells. Microvas. Res. 37: 148-61 (1989); **Van Scott MR et al** Effect of hormones on growth and function of cultured canine tracheal epithelial cells. Am. J. Physiol. 255: C237-45 (1988)

ECI: *eosinophil cytotoxicity inhibitor* This factor of 75 kDa is purified from serum of a human subject with severe allergic dermatitis whose eosinophils and neutrophils failed to respond to » TNF. The serum factor inhibits the cytotoxic function of TNF- and » GM-CSF-stimulated normal human eosinophils and also inhibits neutrophil adherence functions. ECI does not protect » L929 cells from the toxic effects of TNF. The aminoterminal amino acid sequence identifies ECI as the β chain of the C3 complement component (apparently free, but perhaps attached to very small fragments of the α chain). ECI is a component of a feedback mechanism that suppresses functions of cytokine-activated eosinophils in inflammation.

Minkoff MS et al Identification of C3 β chain as the human serum eosinophil cytotoxicity inhibitor. JEM 174: 1267-70 (1991); **Silberstein DS et al** A serum factor that suppresses the cytotoxic function of cytokine-stimulated human eosinophils. JEM 171: 681-93 (1990)

EC IL3: This factor is isolated from epidermal cells. It activates mast cells. The biological activity of this factor can be completely inhibited by anti-IL3 antibodies so that it can be assumed that the factor is identical with » IL3.

Danner M & Luger TA Human keratinocytes and epidermoid carcinoma cell lines produce a cytokine with interleukin-3 like activity. J. Invest. Dermatol. 88: 353-61 (1987); **Kock A & Luger TA** High-performance liquid chromatographic separation of distinct epidermal cell-derived cytokines. J. Chromatogr. 326: 129-36 (1985); **Luger TA et al** Epidermal cells synthesize a cytokine with interleukin 3-like properties. JI 134: 915-9 (1985); **Luger TA et al** Characterization of immunoregulatory cytokines produced by epidermal cells. Scand. J. Immunol. 21: 455-62 (1985)

ECM: *extracellular matrix* The extracellular matrix is a complex network of different combinations of collagens, proteoglycans (PG), hyaluronic acid, laminin, fibronectin, and many other glycoproteins including proteolytic enzymes involved in degradation and remodeling of the extracellular matrix (metalloproteinases; see: TIMP, tissue inhibitor of metalloproteinases). The extracellular matrix occupies the space between cells. Many of the components of the extracellular matrix are connected to proteins of the cytoskeleton by transmembrane proteins.

The extracellular matrix is more than a scaffold that fills extracellular spaces. Many of its components are engaged in processes mediating cell-to-cell interactions. In many instances the capacity of a cell to proliferate, differentiate, and to express specialized functions intimately depends on the presence and maintenance of an intact extracellular matrix (see, for example: Hematopoiesis). Some of its components, in particular the proteoglycans, function as modulators of the biological activities of growth factors. The organization composition, and physical properties the extracellular environment are also essential for the modulation of endothelial cell functions in angiogenic processes (see also: Angiogenesis factors).

Proteoglycans are proteins modified by glycosaminoglycans (GAG). Glycosaminoglycans are long-chain compounds of repeated disaccharide units. The four main types consist mainly of sulfated heparan sulfate/heparin, chondroitin sulfate/dermatan, keratan sulfate, and the non-sulfated glycosaminoglycan hyaluronic acid. Many proteoglycans contain a core protein which links them to the cellular membrane. Hyaluronic acid is the only extracellular oligosaccharide that is not covalently linked to a protein.

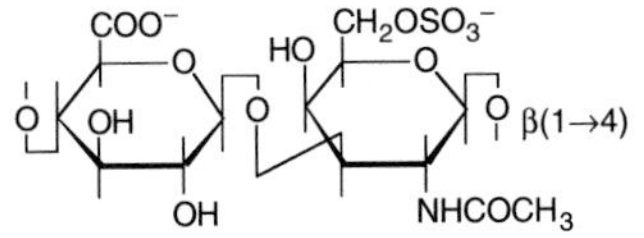

Glycosaminoglycans are negatively charged compounds and therefore bind unspecifically to many other substances, including growth factors. Some proteins are known to interact specifically with glycosaminoglycans. The interactions between glycosaminoglycans and cytokines are one of the important mechanisms underlying communication processes between cells that are mediated by secreted and locally acting factors.

The molecular mechanisms by which growth factors recognize glycosaminoglycans is still unknown. It probably requires the presence of positively charged a-helical protein domains functioning as heparin binding sites (see also: HBGF, heparin-binding growth factors).

Heparan sulfate proteoglycans (abbrev. HSPG), especially syndecan, bind to » bFGF with low affinity and protect this factor from degradation. At the same time this creates a matrix- and cell surface-associated reservoir from which the factor can be released either by heparin or by enzymes which specifically attack heparan sulfate (heparanases). Matrix-bound bFGF can also be released by plasmin as a non-covalent complex with heparan sulfate proteoglycan or glycosaminoglycan. A portion of the cell surface HSPG is anchored via a covalently linked glycosyl-phosphatidylinositol residue, which can be released by treatment with a specific phospholipase C. A fraction of the associated bFGF is also released by this treatment. Plasminogen present in the extracellular matrix is thought to participate in the activation of latent forms of » TGF-β.

Other factors, including » GM-CSF, » IL3 (multi-CSF), stem cell inhibitor (see: SCI), » LIF, » IFN-

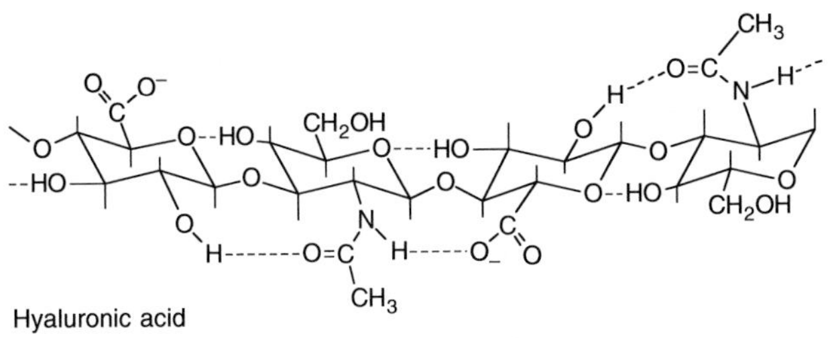

Chondroitin-6-sulfate
n = 20-60

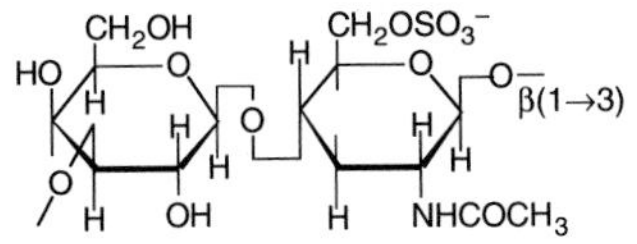

Keratan sulfate
n = 25

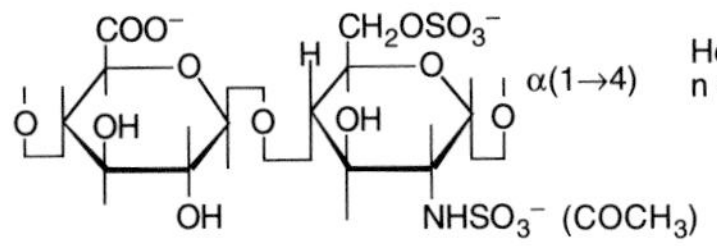

Heparan sulfate
n = 15-30

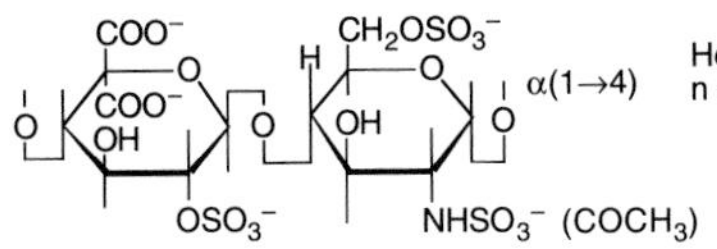

Heparin
n = 15-30

Hyaluronic acid
n = ca. 25000 β(1→4)-linked disaccharide units.

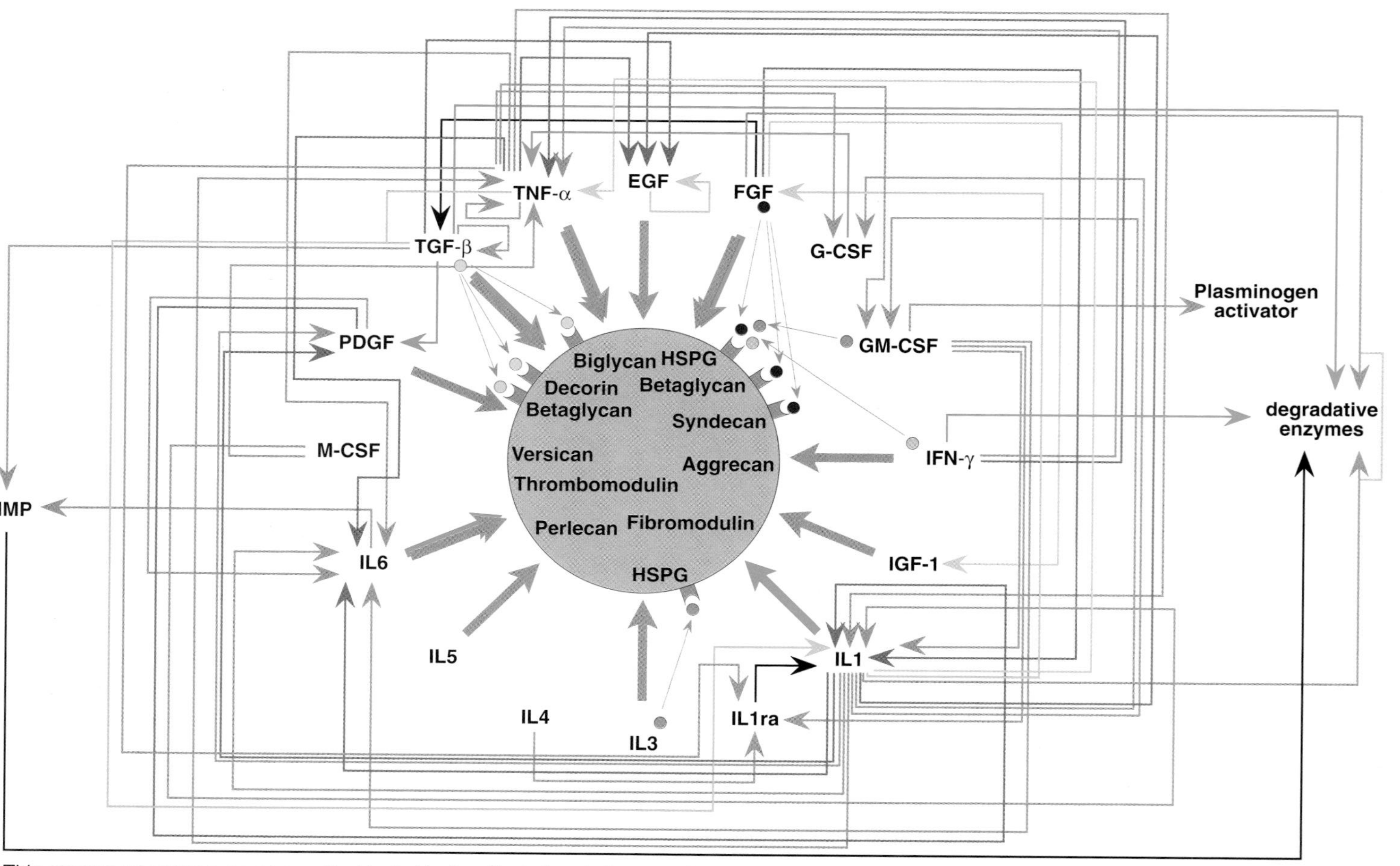

Thin arrows: magenta: synergizes with; black: blocks effect of; red: downregulates expression of factor; green: upregulates expression of factor; orange: downregulates expression of factor receptor; purple: upregulates expression of factor receptor .
Thick arrows: red: catabolic, i. e. inhibition of proteoglycan synthesis and/or promotion of proteoglycan degradation; green: anabolic, i. e. stimulation of proteoglycan synthesis and/or inhibition of proteoglycan degradation. Some cytokines have both activities, depending on cell type. Others have no direct effects but influence proteoglycan synthesis indirectly by modifying the activities of other cytokines.

γ, » PF4 (platelet factor 4) and » AR (amphiregulin) can be bound to the matrix in a similar manner. as complexes with heparan sulfate proteoglycans or heparin. » TGF-β has been shown to be bound to the proteoglycans betaglycan and decorin.

It has been shown that » bFGF and also » VEGF (vascular endothelial cell growth factor) only bind to their specific cell surface receptors if they are attached either to free of to membrane-bound heparan sulfate proteoglycans or other heparin-like components of the extracellular matrix. The presentation of matrix-bound factors may be one mechanism by which some factors interact with receptors on other nearby cells. This effectively establishes cell-to-cell contacts and may also induce biological activities in juxtaposed cells (see: Juxtacrine).

Kreis T & Vale R (eds) Guidebook to the extracellular matrix and adhesion proteins. Oxford University Press 1993)

● BIOCHEMISTRY & MOLECULAR BIOLOGY: **Cardin AD & Weintraub HJ** Molecular modeling of protein-glycosaminoglycan interactions. Arteriosclerosis 9: 21-32 (1989); **Hedman KM et al** Isolation of the pericellular matrix of human fibroblast cultures. JCB 81: 83-91 (1979); **Hook M et al** Cell surface glycosaminoglycan. ARB 53: 847-69 (1984); **Kallunki P et al** Cloning of human heparan sulfate proteoglycan core protein, assignment of the gene (HSPG2) to 1p36.1-p35, and identification of a Bam HI restriction fragment length polymorphism. Genomics 11: 389-96 (1991); **Kallunki P & Tryggvason K** Human basement membrane heparan sulfate proteoglycan core protein: A 467-kD protein containing multiple domains resembling elements of the low density lipoprotein receptor, laminin, neural cell adhesion molecules, and epidermal growth factor. JCB 116: 559-71 (1992); **McDonald JA** Extracellular matrix assembly. ARC 4: 183-208 (1988); **Yamada K & Weston JA** Isolation of a major cell surface glycoprotein from fibroblasts. PNAS 71: 3492-6 (1974)

● FUNCTIONS: **Akiyama SK et al** Cell surface receptors for extracellular matrix components. BBA Rev. Biomembranes 1031: 91-110 (1990); **Bashkin P et al** Release of cell surface-associated basic fibroblast growth factor by glycosylphosphatidylinositol-specific phospholipase C. JCP 151: 126-37 (1992); **Burgess AW** Cell surface heparan sulphate implicated in hemopoietic growth factor signaling. IT 9: 267-8 (1988); **Campbell AD & Wicha MS** Extracellular matrix and the hematopoietic microenvironment. J. Lab. Clin. Med. 112: 140-6 (1988); **Bissell MJ et al** How does the extracellular matrix direct gene expression? J. Theor. Biol. 99: 31-68 (1982); **Carey DJ** Control of growth and differentiation of vascular cells by extracellular matrix proteins. ARP 53: 161-77 (1991); **DiPersio CM et al** The extracellular matrix co-ordinately modulates liver transcription factors and hepatocyte morphology. MCB 11: 4405-14 (1991); **Flaumenhaft R & Rifkin DB** Extracellular matrix regulation of growth factor and protease activity. COCB 3: 817-23 (1991); **Gordon MY et al** Compartmentalization of a hemopoietic growth factor (GM-CSF) by glycosaminoglycans in the bone marrow environment. N 326: 403-5 (1987); **Gordon MY et al** Haemopoietic growth factors and receptors: bound and free. Cancer Cells 3: 127-33 (1991); **Ignotz RA** TGF-β and extracellular matrix-related influences on gene expression and phenotype. Crit. Rev. Eukaryotic Gene Expr. 1: 75-84 (1991); **Ingber D** Extracellular matrix and cell shape: Potential control points for inhibition of angiogenesis. JCBc 47: 236-41 (1991); **Klagsbrun M** The affinity of fibroblast growth factors (FGFs) for heparin; FGF-heparan sulfate interactions in cells and extracellular matrix. COCB 2: 857-63 (1990); **Kolset SO & Gallagher JT** Proteoglycans in hematopoietic cells. BBA 1032: 191-211 (1991); **Madri JA et al** Phenotypic modulation of endothelial cells by transforming growth factor-β depends upon the composition and organization of the extracellular matrix. JCB 106: 1375-84 (1988); **Massagué J** A helping hand from proteoglycans. Curr. Biol. 1: 117-9 (1991); **McDonald JA** Matrix regulation of cell shape and gene expression. COCB 1: 995-9 (1989); **Nietfeld JJ** Cytokines and proteoglycans Experientia 49: 456-69 (1993); **Rapraeger AC et al** Requirement of heparan sulfate for bFGF-mediated fibroblast growth and myoblast differentiation. S 252: 1705-8 (1991); **Roberts R et al** Heparan sulphate bound growth factors: a mechanism for stromal cell-mediated hematopoiesis. N 332: 376-8 (1988); **Ruoslahti E et al** Extracellular matrices and cell adhesion. Arteriosclerosis 5: 581-94 (1985); **Ruoslahti E & Yamaguchi Y** Proteoglycans as modulators of growth factor activities. Cell 64: 867-9 (1991); **Sakaguchi K et al** Identification of heparan sulfate proteoglycan as a high affinity receptor for acidic fibroblast growth factor (aFGF) in a parathyroid cell line. JBC 266: 7270-8 (1991); **Saksela O & Rifkind DB** Release of basic fibroblast growth factor-heparan sulfate complexes from endothelial cells by plasminogen activator-mediated proteolytic activity. JCB 110: 767-75 (1990); **Saksela O et al** Endothelial cell-derived heparan sulfate binds fibroblast growth factors and protects it from proteolytic degradation. JCB 107: 743-51 (1988); **Shimizu Y & Shaw S** Lymphocyte interactions with extracellular matrix. FJ 5: 2292-9 (1991); **Tan EM et al** Modulation of extracellular matrix gene expression by heparin and endothelial cell growth factor in human smooth muscle cells. Lab. Invest. 64: 474-82 (1991); **Vlodavsky I et al** Extracellular matrix-resident growth factors and enzymes: possible involvement in tumor metastasis and angiogenesis. Cancer Metastasis Rev. 9: 203-26 (1990); **Watt FM** The extracellular matrix and cell shape. TIBS11: 482-5 (1986); **Yayon A et al** Cell surface, heparin-like molecules are required for binding of basic fibroblast growth factor to its high affinity receptor. Cell 64: 841-8 (1991)

Some examples of the regulatory cytokine network involved in extracellular matrix metabolism.
A number of cytokines affect the extracellular matrix by either inhibiting proteoglycan synthesis and/or promoting proteoglycan degradation or by stimulating proteoglycan synthesis and/or inhibiting proteoglycan degradation. Proteoglycans have also been shown to bind cytokines: for example, fibroblast growth factors bind to betaglycan, extracellular matrix and plasma membrane HSPG (heparan sulfate proteoglycan), and syndecan; GM-CSF binds to extracellular matrix HSPG; IFN-γ binds to extracellular matrix HSPG; IL3 binds to extracellular matrix HSPG; TGF-β binds to betaglycan, biglycan, and decorin. In addition, many cytokines up- or down-regulate each other's expression and thus also indirectly influence processes within the extracellular matrix.

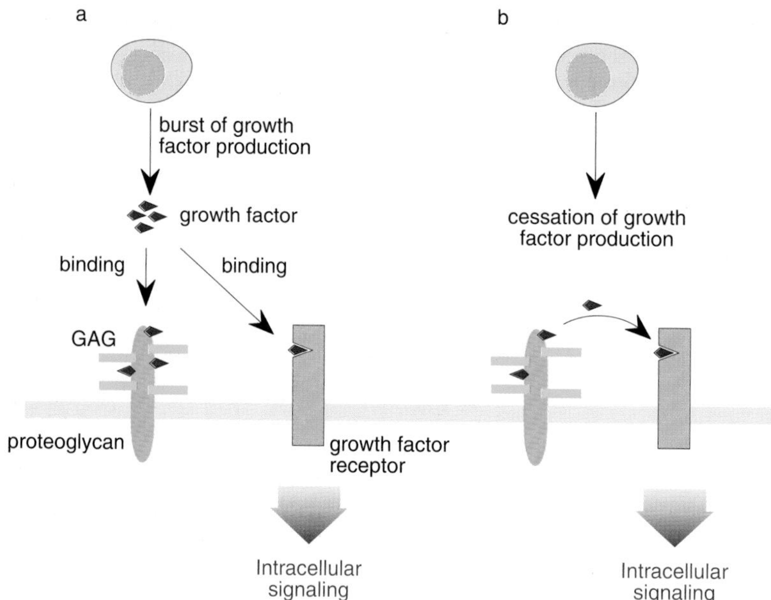

Proteoglycan as growth factor reservoirs.
Proteoglycans are believed to capture some growth factors and to prevent their rapid clearance after a local burst of growth factor production and release. Some growth factors bind to glycosaminoglycans and to their receptors, the latter inititating intracellular signaling responses (a).
When the external source stops production and release of the growth factor (b) the proteoglycans may act as cell surface reservoirs that go on releasing the growth factor. Binding of the growth factor to the proteoglycans can involve interaction with the glycosaminoglycan side chains (as with bFGF) or with the core protein of the glycosaminoglycan (as with TGF-β).

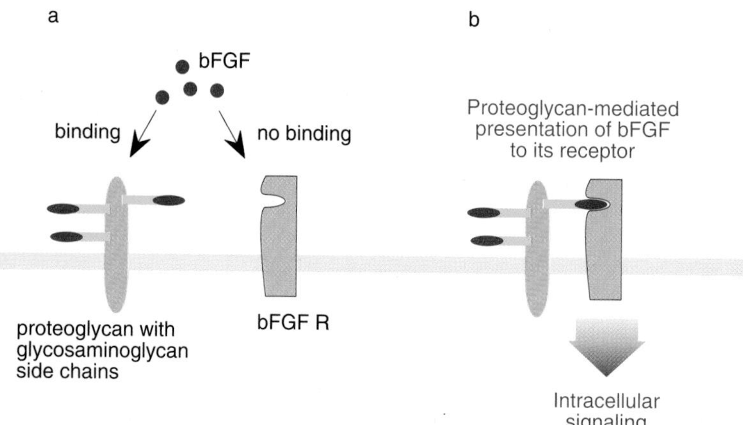

Proteoglycan-mediated presentation of bFGF to its receptor.
bFGF by itself does not bind to its high affinity cell surface receptor (a). bFGF receptors also do not seem to recognize the solution conformation of bFGF. Prior binding of bFGF to the glycosaminoglycan side chains of a proteogly can alter the conformation of bFGF in a way that enables it to bind to its receptor and to elicit intracellular signaling responses (b). High affinity bFGF receptors on CHO cells engineered to lack cell surface expression of heparan sulfate binding sites for bFGF have been shown to be unable to bind bFGF.

***E. coli*:** see: *Escherichia coli.*

ECRF3: An open reading frame encoded by Herpesvirus saimiri, a T-lymphotropic virus that causes fatal lymphoproliferative diseases in several non-human primates. ECRF3 encodes a putative seven-transmembrane-domain receptor that is remotely related (approximately 30% amino acid identity) to the known mammalian » chemokine receptors, namely » IL8 receptor types A and B and the receptors for » MIP-1α and » RANTES. ECRF3 encodes a functional receptor for » IL8, » MGSA (melanoma growth stimulatory activity), and NAP-2 (neutrophil activating peptide 2). The observation that the herpesvirus encodes a functional chemokine receptor suggests a novel role for these cytokines in the pathogenesis of viral infection by transmembrane signaling via the product of ECRF3. For other virus strategies utilizing receptor homologues see also: Viroceptor.

Ahuja SK & Murphy PM Molecular piracy of mammalian interleukin-8 receptor type B by herpesvirus saimiri. JBC 268: 20691-4 (1993)

ECSA: *erythroid colony-stimulating activity* This activity found in mouse plasma from normal, anemic, and hypoxic mice supports the growth of early erythroid progenitor cells » CFU-E. The bulk of ECSA activity is due to » EPO.

Sakata S et al Neutralization and immunoaffinity chromatography of erythroid colony-stimulating activity in mouse plasma by an anti-erythropoietin monoclonal antibody. Acta Haematol. 88: 175-84 (1992)

E-CSF: *erythrocyte colony stimulation factor* This factor is identical with » IL5. See: E-CSF (erythroid CSF).

E-CSF: *erythroid colony-stimulating factor* New name proposed for a factor originally called ***B-BPA*** (B lymphocyte-derived burst promoting activity) and identified as a component of serum-free normal human resting B lymphocyte-conditioned medium. Some transformed B lymphocyte cell lines have been found to produce an E-CSF-like activity. E-CSF activity has been found also in the splenic B lymphocyte fractions of mice, rats, rabbits, and calves.

E-CSF is an integral membrane protein of 28 kDa. It acts as a lineage-specific regulator of erythropoiesis (see also: Hematopoiesis), stimulating the proliferation of murine and human » BFU-E erythroid progenitors in the presence of » Epo, and increasing the sensitivity of early » CFU-E to Epo. Human E-CSF without Epo stimulates the proliferation of human and murine erythroleukemia cells and is synergistic with Epo for the hemoglobinization of Rauscher murine erythroleukemia cells. E-CSF is biochemically and immunochemically distinct from other known growth factors with erythroid burst-promoting activity (see: EPA, BPA), including » IL3, » GM-CSF.

Dainiak N & Cohen CM Surface membrane vesicles from mononuclear cells stimulate erythroid stem cells to proliferate in culture. Blood 60: 583-94 (1982); **Dainiak N & Cohen CM** Association of human MHC class II antigen and membrane bound erythroid burst promoting activity (mBPA) on vesicles exfoliated from the surfaces of B cells. Blood 80: 172 (abstract) (1992); **Feldman L et al** Purification of a membrane-derived human erythroid growth factor. PNAS 84: 6775-9 (1987); **Feldman L & Dainiak N** B lymphocyte derived erythroid burst-promoting activity is distinct from other known lymphokines. Blood 73: 1814-20 (1989); **Feldman L et al** B lymphocyte-derived burst-promoting activity is a pleiotropic erythroid colony-stimulating factor, E-CSF. EH 20: 1223-8 (1992); **Feldman L et al** A sensitive new bioassay for erythroid colony-stimulating factor. EH 21: 1657-62 (1993)

ECSF: *erythroid colony stimulating factor* This factor is identical with » IL3.

Johnson GR & Metcalf D Pure and mixed erythroid colony formation *in vitro* stimulated by spleen conditioned medium with no detectable erythropoietin. PNAS 74: 3879-82 (1977)

EDA: *eosinophil differentiation activity* see: EDF (eosinophil differentiation factor).

EDCF: *endothelium-derived contracting factor* This factor is isolated from supernatants of endothelial cell cultures. It is identical with » ET (endothelin).

Hickey KA et al Characterization of a coronary vasoconstrictor produced by endothelial cells in culture. Am. J. Physiol. 248: C550-C6 (1985); **Lüscher TF et al** Endothelium-derived contracting factors. Hypertension 19: 117-30 (1992)

EDF: *eosinophil differentiation factor* This is a generic name used to describe activities promoting the differentiation of eosinophils in cultivated bone marrow.

One particular factor which promotes the growth of early hematopoietic progenitor cells (see: BFU-E) in a » colony formation assay, leading to the development of eosinophilic colonies, is identical with » IL5 (correlating with the activity of a factor described as » BCGF-2). It is probably the same factor as » Eo-CSF (eosinophil colony-stimulating factor). Some of the activity ascribed to EDF in the older literature may also be due to IL4. While IL5 is a late-acting factor in the development of eosinophils, other factors, also possessing EDF activity, at least *in vitro*, are » IL3 and » GM-CSF. The most sensitive assay for EDF is the production of eosinophils in bulk bone marrow cultures, and this can be used to assay » IL5.

Campbell HD et al Molecular cloning, nucleotide sequence, and expression of the gene encoding human eosinophil differentiation factor. PNAS 84: 6629-33 (1987); **Campbell HD et al** Isolation, structure, and expression of cDNA and genomic clones for murine eosinophil differentiation factor. EJB 174: 345-52 (1988); **Clutterbuck E et al** Recombinant human interleukin 5 is an eosinophil differentiation factor but has no activity in standard human B cell growth factor assays. EJI 17: 1743-50 (1987); **Lopez AF et al** Murine eosinophil differentiation factor. An eosinophil-specific colony-stimulating factor with activity for human cells. JEM 163: 1085-99 (1986); **O'Garra A et al** Interleukin 4 (B cell growth factor II/eosinophil differentiation factor) is a mitogen and differentiation factor for preactivated murine B lymphocytes. PNAS 83: 5228-32 (1986); **Sanderson CJ et al** The production of lymphokines by primary alloreactive T-cell clones: a co-ordinate analysis of 233 clones in seven lymphokine assays. Immunology 56: 575-84 (1985); **Sanderson CJ et al** Identification of a lymphokine that stimulates eosinophil differentiation *in vitro*. Its relationship to interleukin 3, and functional properties of eosinophils produced in cultures. JEM 162: 60-74 (1985); **Sanderson CJ et al** Eosinophil differentiation factor also has B cell growth factor activity: proposed name interleukin 4. PNAS 83: 437-40 (1986); **Sanderson CJ et al** Control of eosinophilia. Int. Arch. Allergy. Appl. Immunol. 94: 122-6 (1991); **Strath M & Sanderson CJ** Detection of eosinophil differentiation factor and its relationship to eosinophilia in Mesocestoides corti-infected mice. EH 14: 16-20 (1986); **Swain SL & Dutton RW** Production of a B cell growth promoting activity, (DL) BCGF, from a cloned T cell line and its assay on the BCL 1 B cell tumor. JEM 156: 1821-34 (1982); **Warren DJ & Sanderson CJ** Production of a T cell hybrid producing a lymphokine stimulating eosinophil differentiation. Immunology 54: 615-23 (1985)

EDF: *erythroid differentiation factor* This homodimeric factor of 25 kDa is isolated from the culture medium of some human monocytic leukemia cell lines. It induces the differentiation of murine erythroleukemic cells and human K562 cells. This factor is a homodimeric protein composed of a subunit that is identical with the βA subunit of » Activin A. The factor is therefore frequently referred to as Activin A/EDF. The sequences of EDF derived from pigs, cattle, and rats are almost identical with the human factor.

Eto Y et al Purification and characterization of erythroid differentiation factor (EDF) from human leukemia cell line THP-1. BBRC 142: 1095-1103 (1987); **Murata M et al** Erythroid differentiation factor is encoded by the same mRNA as that of the inhibin β_A chain. PNAS 85: 2434-8 (1988); **Nishimura M et al** Effect of erythroid differentiation factor on megakaryocytic differentiation of L8057, a murine megakaryoblastic leukemia cell line. BBRC 181: 1042-47 (1991); **Schwall RH & Lai C** Erythroid differentiation bioassays for activin. MiE 198: 340-6 (1991)

EDGF: *endothelial derived growth factor* This poorly characterized factor is found in the conditioned medium of human umbilical vein endothelial cell cultures. The factor stimulates the proliferation of fibroblasts and induces quiescent fibroblasts to proliferate. EDGF also promotes the proliferation of both normal and pathological progenitors of bone marrow fibroblasts (CFU-F) by increasing both the number and the size of the colonies.

Bianchi-Scarra G et al Effect of endothelial cell conditioned medium on the growth of human bone marrow fibroblasts. JCP 123(3): 343-6 (1985)

EDGF: *eye-derived growth factor* Generic name given to various growth factors isolated from several ocular tissues. Specific factors that have been identified are two heparin-binding growth factors designated » EDGF 1 and » EDGF 2. See also: RDGF (retina-derived growth factor).

EDGF 1: *eye-derived growth factor* This factor isolated from bovine retina is the bovine homologue of human » bFGF. (see also: RDGF-α:, retina-derived growth factor).

Courty J et al [*In vitro* properties of various growth factors and *in vivo* effects] Biochimie 66: 419-28 (1984); **Courty J et al** Bovine retina contains three growth factor activities with different affinity to heparin: eye-derived growth factor I, II, III. Biochimie 67: 265-9 (1985); **Courty J et al** Evidence for FGF-like growth factor in adult bovine retina: analogies with EDGF I. BBRC 136: 102-8 (1986); **Moenner M et al** Evidence and characterization of the receptor to eye-derived growth factor I, the retinal form of basic fibroblast growth factor, on bovine epithelial lens cells. PNAS 83: 5024-8 (1986)

EDGF 2: *eye-derived growth factor* This factor isolated from bovine retina is the bovine homologue of human » aFGF. (see also: RDGF-β, retina-derived growth factor). The bovine EDGF 2 cDNA (4.1 kb) has a sequence coding for the 155 amino acids of bovine aFGF, and shows 87% identity with human aFGF.

Alterio J et al Characterization of a bovine acidic FGF cDNA clone and its expression in brain and retina. FL 242: 41-6 (1988); **Barritault D et al** Purification, characterization, and biological properties of the eye-derived growth factor from retina: analogies with brain-derived growth factor. J. Neurosci. Res. 8: 477-90 (1982) **Courty J et al** [*In vitro* properties of various growth factors and *in vivo* effects] Biochimie 66: 419-28 (1984); **Courty J et al** Bovine retina contains three growth factor activities with different affinity to heparin: eye-derived growth factor I, II, III. Biochimie 67: 265-9 (1985); **Courty J et al** Biochemical comparative studies between eye- and brain-derived growth factors. Biochimie 69: 511-6 (1987); **Gimenez-Gallego G et al** Brain-derived acidic fibroblast growth factor: complete amino acid sequence and homologies. S 230: 1385-8 (1985); **Schreiber AB et al** A unique family of endothelial cell polypeptide mitogens: the antigenic and receptor cross-reactivity of bovine endothelial cell growth factor, brain-derived acidic fibroblast growth factor, and eye-derived growth factor-II. JCB 101: 1623-6 (1985)

ED-LCA: *endothelial cell-derived lymphocyte chemotactic activity* This poorly characterized factor of ≈ 13-15 kDa was originally detected in culture supernatant of serotonin-stimulated bovine aortic endothelial cells. It is a chemoattractant for CD4/CD8-positive T lymphocytes and does not enhance the migration of neutrophils or monocytes. (see also: Chemotaxis).
A similar activity is also secreted by serotonin-stimulated human aortic endothelial cells but not by pulmonary arterial endothelial cells from the same individual. Secretion is not induced by other vasoactive amines such as histamine or » angiotensin II and is blocked by 5-HT2 receptor antagonists. A decrease in the uptake of serotonin and a reduction in serotonin-induced secretion of ED-LCA is observed in endothelial cells of atherosclerotic areas.
Farber HW & Beer DJ Restricted secretion of a T lymphocyte chemotactic cytokine by serotonin-stimulated cultured aortic endothelial cells. Circ. Res. 69: 257-65 (1991); **Farber HW** Cytokine secretion by human aortic endothelial cells is related to degree of atherosclerosis. Am. J. Physiol. 262: H1088-95 (1992)

EDMF: *endothelioma-derived motility factor* This poorly characterized factor of 40-65 kDa has been found to be secreted by polyoma virus middle T-transformed murine endothelioma cell lines. These tumors are used as models for studying vascular lesions such as hemangiomas, hemangiosarcomas, and Kaposi's sarcoma and tumor-associated angiogenesis because they produce fast-growing, hemorrhaging, cavernous blood-filled hemangiomas, mainly formed by recruited host endothelial cells. EDMF stimulates chemotaxis (motility induced by a gradient of soluble attractant), haptotaxis (motility in response to substrate-bound attractant), and chemoinvasion (migration through a layer of reconstituted basement membrane, Matrigel) of normal human, bovine, and murine endothelial cells. EDMF binds to heparin and its activity is inhibited by heparin. EDMF is not identical with » VEGF (vascular endothelial growth factor).
Taraboletti G et al Endothelial cell migration and invasiveness are induced by a soluble factor produced by murine endothelioma cells transformed by polyoma virus middle T oncogene. CR 53: 3812-6 (1993)

EDNAP: *endothelial-derived neutrophil-activating peptide* see: ENAP.

EF: *enhancing factor* A general term used to describe factors which enhance a recognizable biological activity as in IgA-EF (see: IL5), IgE-EF (see: IL4), and IgG1-EF (see: IL4). In these cases the biological activity is the enhanced synthesis of the immunoglobulins.

EEK: see addendum.

EF: *enhancing factor* This poorly characterized 14 kDa factor is isolated from the murine small intestines. It is produced predominantly from Paneth cells at the base of the crypts of Lieberkühn thought to be of functional importance as an unspecific defense system.
The factor appears to be a modulator of EGF activity, increasing the binding of » EGF to its receptor on » A431 cells.
Desai SJ et al Ontogeny of enhancing factor in mouse intestines and skin. Histochemistry 96: 371-4 (1991); **Mulherkar R et al** Expression of enhancing factor gene and its localization in mouse tissues. Histochemistry 96: 367-70 (1991)

EGF: *epidermal growth factor*
■ **Alternative Names:** HMGF (human milk growth factor); PGF (prostatic growth factor); urogastrone. See also: individual entries for further information.
■ **Sources:** EGF is found in varying concentrations in milk, saliva, urine (9-100 ng/mL), plasma (1 ng/mL) and also in most other body fluids. Cells in various organs, including brain, kidney, salivary gland, and stomach, produce this factor. The production of EGF has been shown to be stimulated by testosterone and to be inhibited by estrogens.
■ **Protein Characteristics:** EGF is a globular protein of 6.4 kDa consisting of 53 amino acids. It contains three intramolecular disulfide bonds essential for biological activity. EGF proteins are evolutionary closely conserved. Human EGF and murine EGF have 37 amino acids in common. Ca. 70 % homology is found between human EGF and EGFs isolated from other species. The relative positions of the cysteine residues is conserved (see also: EGF-like repeats).
Human EGF is synthesized as a very long prepro-protein of 1207 amino acids from which the factor (position 970-1023) is released by proteolytic cleavage. The EGF precursor is N-glycosylated and contains a hydrophobic domain allowing it to be anchored in the plasma membrane. In cells that do not cleave this precursor (kidney cells, for example, the membrane-bound form of the precursor may itself serve as a receptor for as yet unknown ligands. It may also be involved in » juxtacrine growth control mechanisms.
■ **Gene Structure:** The gene encoding the EGF precursor has a length of ≈ 110 kb. It contains 24

exons. 15 of these exons encode protein domains (see also: EGF-like repeats) which are homologous to domains found in other proteins. The human gene maps to chromosome 4q25-q27.

■ **Related Factors:** EGF is the prototype of a large family of EGF-like proteins (sometimes abbrev. EGF-F). The transforming growth factor » TGF-α shows a strong homology to EGF in the first 45 aminoterminal amino acids and also displays a similar arrangement of the cysteine residues. EGF-specific antibodies do not, however, neutralize TGF-α. Both factors are functionally analogous. They bind to the same receptor and show more or less the same biological activities.

The 19 kDa protein secreted by Vaccinia virus-infected cells (see: VVGF) also shows homology to EGF and displays the same mitogenic activities. The classification of many other proteins as members of the EGF family (see: Betacellulin, *Delta*, *Notch*, *lin*-12, AR (amphiregulin), uEGF-1) is based on some more or less pronounced sequence homologies and the presence of distinct » EGF-like repeats containing the conserved cysteine residues of EGF. Most of these proteins appear to be involved in protein-to-protein and also protein-to-cell interactions, possibly mediated by the EGF-like sequence domains. Most of these proteins have been shown to be involved in embryonic development.

■ **Receptor Structure, Gene(s), Expression:** EGF receptors are expressed in almost all types of tissues at receptor densities of 20000-50000 copies/cell (see also: A431). Parietal endoderm, mature skeletal muscles, and hematopoietic tissues do not express the receptor.

The EGF receptor is a 170 kDa transmembrane glycoprotein with a length of 1186 amino acids. It is identical with a previously described glycoprotein called » SA-7 (species antigen 7). The extracellular receptor domain has a length of 621 amino acids, including 11 glycosylated asparagine residues and 51 cysteine residues. This domain contains the EGF binding site and also binds mammalian » TGF-α.

The intracellular receptor domain has a length of 542 amino acids. It encodes an intrinsic tyrosine-specific protein kinase activity. This kinase catalyses the transfer of the γ-phosphate of ATP to a tyrosine residue of the receptor and also of some other intracellular proteins. The intracellular kinase domain of the EGF receptor is activated by binding of EGF or TGF-α to the extracellular receptor domain. The EGF receptor is also phosphorylated by protein kinase C at serine and threonine residues. For a specific transcription factor regulating EGF receptor gene expression see: ETF.

The human EGF receptor appears to exist in two conformational states binding EGF either with low or with high affinity. The human EGF receptor also binds EGF-like proteins such as » VGF (vaccinia growth factor) secreted by vaccinia virus-infected cells, and to a lesser extent » AR (amphiregulin). The ovine EGF receptor discriminated between EGF and TGF, binding TGF-α with an ≈ 200-fold higher affinity. The chicken homologue of the EGF receptor appears to be functionally different from the human receptor. Unlike the human receptor it is involved in regulating the self-renewal of erythroid precursor cells during » hematopoiesis (see also: *erb*). The *Drosophila* homologue of the EGF receptor is a gene called » DER.

The activity of the EGF receptor can be modulated by other growth factors, including » PDGF, » FGF and » Bombesin. Its synthesis is induced by » NGF. Gangliosides and sphingosine have been shown to modulate the kinase activity of the intracellular EGF receptor domain. Sphingosine activates protein kinase C which phosphorylates the receptor at several sites and thus reduces the affinity of the receptor for EGF (see also: Receptor transmodulation). IFN-α decreases the EGF receptor density.

The intracellular domain of the EGF receptor is almost identical (97% over 250 amino acids of the kinase domain) with an » oncogene called » *erb*B. This is a truncated form of the EGF receptor the aberrant expression of which leads to malignant transformation of cells. Another oncogene product, designated » *neu*, is also closely related to the EGF receptor. The protein designated » EGFBP (EGF binding protein) also binds EGF.

The EGF receptor gene has a length of over 50 kb and is on human chromosome 7p14-p12. The 5′ region of this gene does not contain a classical TATA box or a CAAT box (see: Gene expression). Six possible transcriptional start sites have been identified. Some of them encode soluble secreted forms of the receptor.

■ **Signal Transduction:** The earliest indication of EGF receptor binding are alterations of cellular morphology and an increase in glucose and ion transport. Some of these effects require the activation of the tyrosine kinase activity. Cells that have been exposed to EGF for at least 8 hours show elevated DNA, RNA, and protein synthesis and increased proliferation rates. These processes also require receptor-associated protein kinase activity because mutated receptors which still bind EGF but do not express the intrinsic kinase cannot be stimulated mitogenically.

Binding of EGF to its receptor leads to conformational changes involving receptor dimerisation and » autophosphorylation of receptor tyrosine residues. The receptor/ligand complexes form clusters which are internalized by the cells and subsequently degraded in endosomes. Dimerisation and phosphorylation of the receptor activates the receptor-associated phospholipase γ, leading to the formation of inositol-1,4,5-triphosphate and 1,2-diacylglycerol. Calcium (see also: Calcium ionophore) is released from intracellular depots. The high-affinity binding site for EGF is lost upon phosphorylation of the receptor by protein kinase C.

Mutations of the autophosphorylation site tyrosine-992 (altered to phenylalanine) prevent the synthesis of inositol phosphate. C-terminal truncated EGF receptors are also strongly impaired in their ability to activate inositol phosphate metabolism.

■ **BIOLOGICAL ACTIVITIES:** The biological activities of EGF are not species-specific. *In vivo* biological activities of EGF are best illustrated by its involvement in the embryonic development in mice. EGF causes a premature lid opening and also promotes the premature development of teeth. In premature lambs the injection of EGF induces the elevated maturation of lung alveolar membranes.

EGF has been shown to inhibit the secretion of gastric acids. It also modulates the synthesis of a number of hormones, including the secretion of » prolactin from pituitary tumors and of chorion gonadotropin from chorion carcinoma cells. In the central nervous system EGF influences the activity of some types of GABAergic and dopaminergic neurons. EGF may also have direct central actions as shown by its inhibitory activity on food intake in rats following intracerebroventricular injection.

EGF is a strong mitogen for many cells or ectodermal, mesodermal, and endodermal origin. EGF controls and stimulates the proliferation of epidermal and epithelial cells, including fibroblasts, kidney epithelial cells, human glial cells, ovary granulosa cells, and thyroid cells *in vitro*. EGF also stimulates the proliferation of embryonic cells. EGF-mediated proliferation of » 431 cells is inhibited by » EGFI (EGF inhibitor).The proliferation of some cell lines has been shown to be inhibited by EGF.

EGF acts as a differentiation factor for some cell types. It strongly influences the synthesis and turnover of proteins of the extracellular matrix (see: ECM), including fibronectin, collagens, laminin, and glycosaminoglycans. EGF increases the release of calcium from bone tissue and, like TGF-α, thus promotes bone resorption. EGF acts as a mitogen for the basal cells of the olfactory epithelium that give rise to olfactory neurons and » TGF-β promotes neurogenesis.

To a limited extent EGF also acts as an » angiogenesis factor because it is mitogenic for endothelial cells. The mitogenic activity of EGF for endothelial cells can be potentiated by thrombin.

EGF is a strong chemoattractant for fibroblasts and epithelial cells (see also: Chemotaxis). EGF alone and also in combination with other cytokines is an important factor mediating » wound healing processes.

In spleen cells murine and human EGF, but not TGF-α that binds to the same receptor, induce the synthesis of » IFN-γ.

In several *in vitro* systems murine and human TGF-α are functionally interchangeable with murine and human EGFs. Human TGF-α is as active as murine EGF in promoting eyelid opening in newborn mice. Human EGF is as potent as its murine homologue with respect to this biological activity.

In several cell types EGF down-regulates the expression of the receptor for » TGF-β so that it can be assumed that some of the EGF activities are indirect (see also: Autocrine).

■ **TRANSGENIC/KNOCK-OUT/ANTISENSE STUDIES:** The activity and tissue specificity of the human EGF receptor promoter have been studied in » transgenic mice expressing a transgene consisting of human 1EGF receptor promoter and enhancer sequences fused to the bacterial chloramphenicol acetyltransferase gene. Analysis for the presence of acetyltransferase activity in transgenic mice reveals that the human EGF receptor promoter construct is consistently active in the thymus and spleen. Thymocytes show no reporter enzyme activity suggesting that expression in the thymus is confined to the thymic stroma, which consists mainly of epithelial cells.

■ **ASSAYS:** EGF can be assayed by employing cell lines the proliferation of which is enhanced or inhibited by EGF (see: A431, NRK; see also: Bioassays). EGF can be assayed also with a sensitive ELISA test. EGF receptors are detected by ligand binding assays. For further information see also subentry "Assays" in the reference section.

■ **Clinical Use & Significance:** EGF may be a trophic substance for the gastrointestinal mucosa and may play a gastroprotective role due to its ability to stimulate the proliferation of mucosa cells. EGF has been shown to effectively promote healing of ulcers at concentrations that do not inhibit the synthesis of gastric acids. EGF, and also the EGF-like factors » TGF-α and » VGF (vaccinia growth factor) have been shown to promote regeneration of the epidermis and the generation of granulation tissues following severe burns (see also: Wound healing). Local treatment with EGF has also been described to improve healing of injured corneal tissues, for example after epikeratoplasty and in patients with persistent epithelial defects suffering from neurotrophic keratitis.

One study on the topical application of human recombinant EGF in patients with venous ulcerations of the lower extremities has revealed that EGF did not significantly enhance re-epithelialization of venous ulcers although the treatment led to a greater reduction in ulcer size and a larger number of healed ulcers.

In some human tumors the EGF receptor gene has been reported to be amplified, leading to an overexpression of EGF receptors on the cell surface (see also: *erb*B). In tissue culture and also in experimental animals several anti-EGF receptor antibodies have been shown to be capable of inhibiting the growth of these tumor cells.

The determination of the EGF receptor status in human mammary and ovarial carcinomas is of clinical importance. It has been shown that the disease-free interval and also total survival rates are much higher if tumors do not express the receptor. Tumor aggressiveness appears to be correlated with an increased expression of EGF receptors. It has been found that, from a prognostic point of view, the measurement of EGF receptor expression may be much more important than the detection of estrogen receptors in stomach, mammary, kidney, and salivary gland adenocarcinomas. When patients are divided into three groups on the basis of the expression of both c-*erb*B-2 oncoprotein and EGFR, those expressing both proteins show a worse prognosis than other groups.

The simultaneous expression of EGF receptors and » TGF-α in primary tumors of advanced gastric cancer appears to correlate significantly with a very poor prognosis. Several studies show that monoclonal antibodies directed against the EGF receptor may be useful therapeutic agents that can be used alone without conjugation to other cytotoxic moieties in the treatment of several tumor types expressing or overexpressing the EGF receptor.

Immunohistochemical staining of EGF (and also of TGF-β) in endoscopic biopsy specimens may be useful for the diagnosis of the penetrating type of early gastric cancer and also for the diagnosis of the initial lesion of linitis plastica-type gastric cancer.

It has been proposed that EGF may play a role in some cases of human male infertility, particularly those with unexplained oligospermia. Markedly decreased levels of circulating EGF and a decrease in spermatids in the testis and mature sperm in the epididymis have been observed after sialoadenectomy. The changes were corrected by administration of EGF.

● **Reviews: Carpenter G & Cohen S** Epidermal growth factor. ARB 48: 193-216 (1979); **Carpenter G** EGF: new tricks for an old growth factor. Curr. Opin. Cell. Biol. 5: 261-4 (1993); **Das M** Epidermal growth factor: mechanism of action. Int. Rev. Cytol. 78: 233-56 (1982); **Fisher DA & Lakshmanan J** Metabolism and effects of epidermal growth factor and related growth factors in mammals. Endocrinol. Rev. 11: 418-42 (1990); **Gill GN et al** Epidermal growth factor and its receptor. Mol. Cell. Endocrinol. 51: 169-86 (1987); **Mroczkowski B & Ball R** Epidermal growth factor: biology and properties of its gene and protein precursor. In: Habenicht A (edt) Growth factors, differentiation factors, and cytokines, pp. 18-30, Springer, Berlin 1990

● **Biochemistry & Molecular Biology: Bell GI et al** Human epidermal growth factor precursor: cDNA sequence, expressed *in vitro* and gene organization. NAR 14: 8427-46 (1986); **Brissenden JE et al** Human chromosomal mapping of genes for insulin-like growth factors I and II and epidermal growth factor. N 310: 781-4 (1984); **Davies RL et al** Genetic analysis of epidermal growth factor action: assignment of human epidermal growth factor receptor gene to chromosome 7. PNAS 77: 4188-92 (1980); **Dorow DS & Simpson RJ** Cloning and sequence analysis of a cDNA for rat epidermal growth factor. NAR 16: 9338 (1988); **Gray A et al** Nucleotide sequence of epidermal growth factor cDNA predicts a 128,000-molecular weight protein precursor. N 303: 722-5 (1983); **Heath WF & Merrifield RB** A synthetic approach to structure-function relationships in the murine epidermal growth factor molecule. PNAS 83: 6367-71 (1986); **Morton CC et al** Human genes for insulin-like growth factor I and II and epidermal growth factor are located on 12q22-q24.1, 11p15, and 4q25-q27, respectively. Cytogenet. Cell Genet. 41: 245-9 (1986); **Prestrelski SJ et al** Solution structure and dynamics of epidermal growth factor and transforming growth factor α. JBC 267: 319-22 (1992); **Savage CR jr et al** The primary structure of epidermal growth factor. JBC 247: 7612-21 (1972); **Schaudies RP et al** Proteolytic processing of epidermal growth factor within endosomes. BBRC 143: 710-5 (1987); **Simpson RJ et al** Rat epidermal growth factor: complete amino acid sequence. homology with the corresponding murine and human proteins; isolation of a form truncated at both ends with full *in vitro* biological activity. EJB 153: 629-37 (1985); **Urdea MS et al** Chemical synthesis of a gene for human epidermal growth factor urogastrone and its expression in yeast. PNAS 80: 7461-5 (1983)

● **Receptor Structure, Gene(s), Expression: Avivi A et al** Comparison of EGF receptor sequences as a guide to study the ligand binding site. O 6: 673-6 (1991); **Brown AB & Carpenter G** Acute regulation of the epidermal growth factor receptor in response to nerve growth factor. J. Neurochem. 57: 1740-9 (1991); **Cochet C et al** Demonstration of epidermal growth factor-induced receptor dimerisation in living cells using a chemical covalent cross-linking agent. JBC 263: 3290-5 (1988); **Cochet C et al** Interaction between the epidermal growth factor receptor and phosphoinositide kinases. JBC 266: 637-44 (1991); **Eisenkraft BL et al** α-Interferon down-regulates epidermal growth factor receptors on renal carcinoma cells: Relation to cellular responsiveness to the antiproliferative action of α-interferon. CR 51: 5881-7 (1991); **Greenfield C et al** EGF binding induces a conformational change in the external domain of its receptor. EJ 8: 4115-24 (1989); **Gunther N et al** The secreted form of the epidermal growth factor receptor: characterization and crystallization of the receptor-ligand complex. JBC 265: 22082-5 (1990); **Haley J et al** The human EGF receptor gene: structure of the 110 kb locus and identification of sequences regulating its transcription. OR 1: 375-96 (1987); **Hepler JR et al** Epidermal growth factor stimulates the rapid accumulation of inositol (1,4,5)-triphosphate and a rise in cytosolic calcium mobilized from intracellular stores in A431 cells. JBC 262: 2951-6 (1987); **Ishii S et al** Characterization and sequence of the promoter region of the human epidermal growth factor receptor gene. PNAS 82: 4920-4 (1985); **Kashles O et al** A dominant negative mutation suppresses the function of normal epidermal growth factor receptors by heterodimerisation. MCB 11: 1454-63 (1991); **Kawasaki K et al** Mega base map of the epidermal growth factor (EGF) receptor gene flanking regions and structure of the amplification unit in EGF receptor-hyperproducing squamous carcinoma cells. JJCR 79: 1174-83 (1988); **Lax I et al** Functional analysis of the ligand binding site of EGF-receptor utilizing chimaeric chicken/human receptor molecules. EJ 8: 421-7 (1989); **Oberg KC et al** Growth factor receptors: the epidermal growth factor receptor as a model. In: Habenicht A (edt) Growth factors, differentiation factors, and cytokines, pp. 3-17, Springer, Berlin 1990; **Petch LA et al** A truncated- secreted form of the epidermal growth factor receptor is encoded by an alternatively spliced transcript in normal rat tissues. MCB 10: 2973-82 (1990); **Simmen FA et al** Isolation of an evolutionarily conserved epidermal growth factor receptor cDNA from human A431 carcinoma cells. BBRC 124: 125-32 (1984); **Ullrich A et al** Human epidermal growth factor receptor cDNA sequence and aberrant expression of the amplified gene in A431 epidermoid carcinoma cells. N 309: 418-25 (1984)

● **Biological Activities: Cohen S** Isolation of a mouse submaxillary gland protein accelerating incisor eruption and eyelid opening in the new-born animal. JBC 237: 1555-62 (1962); **Ferrari G et al** Epidermal growth factor exerts neuronotrophic effects on dopaminergic and GABAergic CNS neurons: Comparison with basic fibroblast growth factor. J. Neurosci Res. 30: 493-7 (1991); **Mahanthappa NK & Schwarting GA** Peptide growth factor control of olfactory neurogenesis and neuron survival *in vitro*: roles of EGF and TGF-β s. Neuron 10: 293-305 (1993); **Plata-Salamán CR** Epidermal growth factor and the nervous system. Peptides 12: 653-63 (1991); **Plata-Salaman CR** Food intake suppression by growth factors and platelet peptides by direct action in the central nervous system. Neurosci. Lett. 94: 161-6 (1988); **Smith JM et al** Human transforming growth factor-α causes precocious eyelid opening in newborn mice. N 315: 515-6 (1985)

● **Transgenic/Knock-out/Antisense Studies: Jhappan C et al** An epidermal growth factor receptor promoter construct selectively expresses in the thymus and spleen of transgenic mice. CI 149: 99-106 (1993)

● **Assays: Ashizawa K & Cheng SY** A sensitive and rapid *in situ* immunoassay to quantitatively determine the cellular antigens in intact cultured cells. JBBM 24: 297-307 (1992); **Bowen-Pope DF & Ross R** Is epidermal growth factor present in human blood? Interference with the radioreceptor assay for epidermal growth factor. BBRC 114: 1036-41 (1983); **Carpenter G & Zendegui J** A biological assay for epidermal growth factor/urogastrone and related polypeptides. AB 153: 279-82 (1986); **Dagogo-Jack S et al** Homologous radioimmunoassay for epidermal growth factor in human saliva. J. Immunoassay 6: 125-36 (1985); **Dittadi R et al** Immunoprecipitation method for a radioligand binding assay of epidermal growth factor receptor. J. Nucl. Med. Allied Sci. 34: 225-30 (1990); **Dittadi R et al** Radioligand binding assay of epidermal growth factor receptor: causes of variability and standardization of the assay. Clin. Chem. 36: 849-54 (1990); **Falsette N et al** Measurement of occupied and non-occupied epidermal growth factor receptor sites in 216 human breast cancer biopsies. Breast Cancer Res. 20: 177-83 (1992); **Farley K et al** Development of solid-phase enzyme-linked immunosorbent assays for the determination of epidermal growth factor receptor and pp60c-*src* tyrosine protein kinase activity. AB 203: 151-7 (1992); **Formento JL et al** Epidermal growth factor receptor assay: validation of a single point method and application to breast cancer. Breast Cancer Res. Treat. 17: 211-9 (1991); **Grimaux M et al** A simplified immunoenzymetric assay of the epidermal growth factor receptor in breast tumors: evaluation in 282 cases. IJC 45: 255-62 (1990); **Gullick WJ et al** A radioimmunoassay for human epidermal growth factor receptor. AB 141: 253-61 (1984); **Hansen G et al** Gastric ulcer is accompanied by a decrease of epidermal growth factor in gastric juice and saliva. J. Clin. Chem. Clin. Biochem. 27: 539-45 (1989); **Hayashi T & Sakamoto S** Radioimmunoassay of human epidermal growth factor - hEGF levels in human body fluids. J. Pharmacobiodyn. 11: 146-51 (1988); **Iacopetta BJ et al** Epidermal growth factor in human and bovine milk. Acta Paediatr. 81: 287-91 (1992); **Iezzoni JC et al** Rapid colorimetric detection of epidermal growth factor receptor mRNA by in situ hybridization. J. Clin. Lab. Anal. 7: 247-51 (1993); **Jaspar JM & Franchimont P** Radioimmunoassay of human epidermal growth factor in human breast cyst fluid. Eur. J. Cancer Clin. Oncol. 21: 1343-8 (1985); **Kawamoto T et al** Quantitative assay of epidermal growth factor receptor in human squamous cell carcinomas of the oral region by an avidin-biotin method. Jpn. J. Cancer Res. 82: 403-10 (1991); **Kermode JC & Tritton TR** Receptor-purified, Bolton-Hunter radioiodinated, recombinant, human epidermal growth factor: an improved radioligand for receptor studies. J. Recept. Res. 9: 429-40 (1989-90); **Kienhuis CB et al** Scintillation proximity assay to study the interaction of epidermal growth factor with its receptor. J. Recept. Res. 12: 389-99 (1992); **Kishi H et al** Epidermal growth factor (EGF) in seminal plasma and prostatic gland: a radioreceptor assay. Arch. Androl. 20: 243-9 (1988); **Koenders PG et al** Epidermal growth factor receptors in human breast cancer: a plea for standardization of assay methodology. Eur. J. Cancer 28: 693-7 (1992); **Lin PH et al** Western blot detection of epidermal growth factor receptor from plasmalemma of culture cells using 125I-labelled epidermal growth factor. AB 167: 128-39 (1987); **Metcalfe RA et al** A simple bioassay for epidermal growth factor using pig thyrocytes. J. Endocrinol. 134: 449-57 (1992); **Orsini B et al** Radioimmunoassay of epidermal growth factor in human saliva and gastric juice. Clin. Biochem. 24: 135-41 (1991); **Radinsky R et al** A rapid colorimetric *in situ* messenger RNA hybridization technique for analysis of epidermal growth factor receptor in paraffin-embedded surgical specimens of human colon carcinomas. CR 53: 937-43 (1993); **Stoscheck CM** Enzyme-linked immunosorbent assay for the epidermal growth factor receptor. JCBc 43: 229-41 (1990); **Sunada H et al** A direct radioimmunoassay for human epidermal growth factor receptor using 32P-autophosphorylated receptor. AB 149: 438-47 (1985); **Tripathi RC et al** Radioimmunoassay of epidermal growth factor in human lenses at various stages of development of cataract. Exp. Eye Res. 53: 759-64 (1991); **van Zoelen EJ** The use of biological assays for detection of polypeptide growth factors. Prog. Growth Factor Res. 2: 131-52 (1990); for further information also see individual cell lines used in individual bioassays.

● **Clinical Use & Significance: Andersson A et al** Effects of EGF-dextran-tyrosine-131I conjugates on the clonogenic survival of cultured glioma cells. J. Neurooncol. 14: 213-23 (1992); **Berger MS et al** Evaluation of epidermal growth factor receptors in bladder tumors. Br. J. Cancer 56: 533-7 (1987); **Bilous M et al** Immunocytochemistry and *in situ* hybridization of epidermal growth factor receptor and relation to prognostic factors in breast cancer. Eur. J. Cancer 28A: 1033-7 (1992); **Brady LW** Malignant astrocytomas treated with iodine-125 labeled monoclonal antibody 425 against epidermal growth factor receptor: a phase II trial. Int. J. Radiat. Oncol. Biol. Phys. 22: 225-30 (1992); **Brown GL et al** Enhancement of wound healing by topical treatment with epidermal growth factor NEJM 321: 76-9 (1989); **Charpin C et al** Epidermal growth factor receptor in breast cancer: correlation of quantitative immunocytochemical assays to prognostic factors. Breast Cancer Res. Treat. 25: 203-10 (1993); **Daniele S et al** Treatment of persistent epithelial defects in neurotrophic keratitis with epidermal growth factor: a preliminary open study. Graefes Arch. Clin. Exp. Ophthalmol. 230: 314-7 (1992); **Falanga V et al** Topical use of human recombinant epidermal growth factor (h-EGF) in venous ulcers. J. Dermatol. Surg. Oncol. 18: 604-6 (1992); **Hawkins RA et al** Epidermal growth factor receptors in intracranial and breast tumors Their clinical significance. Brit. J. Cancer 63: 553-60 (1991); **Hirayama D et al** Immunohistochemical study of epidermal growth factor and transforming growth factor-β in the penetrating type of early gastric cancer. Hum. Pathol. 23: 681-5 (1992); **Kinoshita S** Clinical application of epidermal growth factor in ocular surface disorders. J. Dermatol. 19: 680-3 (1992); **Maurizi M et al** EGF receptor expression in primary laryngeal cancer: correlation with clinico-pathological features and prognostic significance. IJC 52: 862-6 (1992); **McKenzie SJ** Diagnostic utility of oncogenes and their products in human cancer. BBA 1072: 193-214 (1991); **Modjtahedi H et al** Immunotherapy of human tumor xenografts overexpressing the EGF receptor with rat antibodies that block growth factor-receptor interaction. Br. J. Cancer 67: 254-61 (1993); **Modjtahedi H et al** The human EGF receptor as a target for cancer therapy: six new rat mAbs against the receptor on the breast carcinoma MDA-MB 468. Br. J. Cancer 67: 247-53 (1993); **Nicholson S et al** Quantitative assays of epidermal growth factor receptor in human breast cancer: cut-off points of clinical relevance. IJC 42: 36-41 (1988); **Osaki A et al** Prognostic significance of co-expression of c-*erb*B-2 oncoprotein and epidermal growth factor receptor in breast cancer patients. Am. J. Surg. 164: 323-6 (1992); **Pastor JC & Calonge M** Epidermal growth factor and corneal wound healing. A multicenter study. Cornea 11: 311-4 (1992); **Read LC et al** Epidermal growth factor: physiological roles and therapeutic applications. Biotechnol. Ther. 1: 237-72 (1989-90); **Slamon DJ et al** Human breast cancer: correlation with relaps and survival with amplification of the HER-2/*neu* oncogene. S 235: 177-82 (1987); **Toi M et al** Epidermal growth factor receptor expression as a prognostic indicator in breast cancer. Eur. J. Cancer 27: 977-80 (1991); **Tsutsumi O et al** A physiological role of epidermal growth factor in male reproductive function. S 233: 975-7 (1986); **Yamamoto T et al** High incidence of amplification of the epidermal growth factor receptor gene in human squamous carcinoma cell lines. CR 46: 414-6 (1986); **Yonemura S et al** Interrelationship between transforming growth factor-α and epidermal growth factor receptor in advanced gastric cancer. Oncology 49: 157-61 (1992)

EGF1 – EGF3: see: PGF (prostatic growth factor).

EGFBP: *EGF binding protein* This protein is a member of the » kallikrein family of serine proteases structurally related to trypsin. It is identical with the prorenin converting enzyme (PRECE) found in the ICR mouse submandibular gland and the mGK-13 (mouse glandular kallikrein) gene product identified in Balb/c mice. EGFBP specifically processes » EGF. It is closely related by sequence and immunological cross-reactivity to γ-NGF and a β-NGF endopeptidase which specifically removes an N-terminal octapeptide from β-NGF (see: NGF).

EGF-BP forms sodium dodecyl sulfate-stable complexes with the protease » nexin II (PN II) secreted by normal human fibroblasts. These complexes then bind to the same cells and are rapidly internalized and degraded.

Anundi H et al Partial amino-acid sequence of the epidermal growth-factor-binding protein. EJI 129: 365-71 (1982); **Kim WS et al** The presence of two types of prorenin converting enzymes in the mouse submandibular gland. FL 293: 142-4 (1991)

EGF-F: abbrev. occasionally used for EGF-like factors see: EGF.

EGFI: *EGF inhibitor* This factor of 34 kDa has been isolated from rabbit kidney cells infected with malignant rabbit fibroma virus. EGFI inhibits the effects of » EGF on certain target cell lines. It also inhibits proliferation of EGF receptor-bearing » A431 cells *in vitro*. EGFI decreases EGF-induced phosphorylation of cellular proteins. The partial protein sequence of two fragments of EGFI shows striking similarity to two *ras*-like proteins.

Strayer DS & Mathew J A 34-kd protein with strong homology to ras-like proteins inhibits epidermal growth factor activity. Am. J. Pathol. 142: 1141-53 (1993)

EGF inhibitor: see: EGFI.

EGF-like repeats: EGF-(epidermal growth factor)-like sequence domains are variously referred to as EGF-like cysteine-rich repeated motif, EGF-like domains, or EGF-like modules. They have a length of ≈ 40 amino acids and are characterized by a conserved arrangement of six cysteine resi-

dues also found in EGF itself. These residues form three evenly spaced disulfide bonds. Such domains have been observed as structural elements in a number of cytokines and in many extracellular and cell surface proteins in which they frequently occur in multiple copies. Some of these repeats are recognized by antibodies directed against EGF.
The biological functions of these cysteine-rich sequence modules are largely unknown. Their existence suggests that at least some proteins containing them may also have some EGF-like functions. The examination of factor XII (Hageman factor) has revealed that it can act as a mitogen for cells such as human hepatoma cells (HepG2 cell line) which also respond to EGF.
A protein of *Drosophila*, designated » *Notch*, which is a member of the EGF family of proteins, contains 36 EGF repeats. Some of them have been proposed to function as discrete ligand-binding units, making this protein a multifunctional receptor which may potentially interact with several different proteins during development.

Apella E et al Structure and function of epidermal growth factor-like regions. FL 231: 1-4 (1988); **Rebay I et al** Specific EGF repeats of *Notch* mediate interactions with *Delta* and *Serrate*: implications for *Notch* as a multifunctional receptor. Cell 67: 687-99 (1991); **Schmeidler-Sapiro KT et al** Mitogenic effects of coagulation factor CII and factor XIIa on HepG2 cells. PNAS 88: 4382-5 (1991); **Stanescu V et al** Immunological detection of the EGF-like domain of the core proteins of large proteoglycans from human and baboon cartilage. Connect. Tissue Res. 26: 283-93 (1991)

EGFR-specific transcription factor: see: ETF.

Egr-1: *early response gene 1* An » ERG (early response gene) transcription of which is rapidly and transiently stimulated in fibroblasts, epithelial cells, and lymphocytes following mitogenic stimulation. In fibroblasts Egr-1 expression is induced, among other things, by » » bFGF, PDGF, » FGF, and » EGF. Expression of Egr-1 in mesangial cells is induced by » PDGF, » vasopressin, and » angiotensin and correlates with mitogenicity. In macrophages bacterial lipopolysaccharides (see: Endotoxins) are a potent inducer of Egr-1 expression. Expression of Egr-1 can be induced by treatment of cells with » Phorbol esters or » okadaic acid. Retinoic acid increases zif268 early gene expression in rat preosteoblastic cells. In neuronal cells expression of Egr-1 is regulated by synaptic activity. The 5' region of the Egr-1 gene includes four serum response elements (see: SRE), a cAMP-like response element (see: CRE), an AP1-like response element (see: jun), and an SP1 binding site.
EGR-1 has been shown to be identical with genes designated » ETR103, d2, » zif268, NGFI-A (see: NGFI), Krox-24 (see: Krox), and TIS8 (see: TIS). Egr-1 encodes a nuclear phosphoprotein with "zinc-binding finger" structure and is a transcription factor (see also: Gene expression), recognizing a short DNA sequence, GCGGGGGCG, in the promoters of a variety of genes. The gene is often, but not always, coregulated with » *fos*. The human EGR1 gene maps to chromosome 5q23-31. Egr-1 mRNA increases dramatically during cardiac and neural cell differentiation, and following membrane depolarization both *in vitro* and *in vivo*. It is rapidly induced in hippocampal neurons by synaptic NMDA receptor activation. It is expressed in response to renal ischemia and during compensatory renal hypertrophy in mice and has been found to be induced during the cellular response to radiation injury. Egr-1 is also expressed in murine B lymphocytes stimulated through their receptor for surface immunoglobulins. Egr-1 and the related Egr-2 (see below) have been shown to be involved in the induction of myeloid leukemia cell differentiation along the monocytic lineage and in the activation of human monocytes; its expression is rapidly and transiently induced by » GM-CSF. Egr-1 is essential for and restricts differentiation of myeloblasts along the macrophage lineage.
It has been shown recently that the Wilm´s tumor locus at human chromosome 11p13 encodes a zinc finger protein and that this protein also recognizes DNA binding sites of Egr-1, implying that the loss of DNA binding activity of the Wilm´s tumor gene may contribute to the tumorigenic process.
Egr transcription factors constitute an entire family of related proteins that has been highly conserved between species. These have in common a GC-rich target sequence (5´-GCGGGGGCG-3´), termed the ***EGR consensus box***, and binding to this sequence via a highly conserved zinc finger motif.
Egr-2, also known as krox-20 (see: krox), has been identified as a structurally related human gene encoding a protein of 406 amino acids which is coregulated with Egr-1 by fibroblast and lymphocyte mitogens, but not in other cell types. The human Egr-2 gene maps to chromosome 10q21-22. The human ***Egr-3*** gene encodes a protein of 387 amino acids. The Egr-3 gene has a single intron and maps to chromosome 8 at bands p21-23. Egr-3 expression is induced by mitogenic stimula-

tion of rodent and human fibroblasts and a monkey kidney epithelial cell line. ***Egr-4*** is identical with NGFI-C (see: NGFI). Another member of the Egr family is a tumor suppressor gene, ***WT1***, the Wilms' tumor susceptibility gene (see also: Oncogenes), that is implicated in Wilms' tumor. WT1 has been shown to regulate expression of » IGF (IGF2), » PDGF (PDGF-A) and some other genes showing expression in developing kidney.

● **Reviews: Madden SL & Rauscher FJ 3rd** Positive and negative regulation of transcription and cell growth mediated by the EGR family of zinc-finger gene products. ANY 684: 75-84 (1993); **Sukhatne VP et al** The Egr family of nuclear signal transducers. Am. J. Kidney Dis. 17: 615-8 (1991)

Alexandropoulos K et al v-*Fps*-responsiveness in the Egr-1 promoter is mediated by serum response elements. NAR 20: 2355-9 (1992); **Bernstein SH et al** Posttranscriptional regulation of the zinc finger-encoding EGR-1 gene by granulocyte-macrophage colony-stimulating factor in human U-937 monocytic leukemia cells: involvement of a pertussis toxin-sensitive G protein. Cell Growth Differ. 2: 273-8 (1991); **Cao XM et al** Identification and characterization of the Egr-1 gene product, a DNA-binding zinc finger protein induced by differentiation and growth signals. MCB 10: 1931-9 (1990); **Cole AJ et al** Rapid increase of an immediate early gene messenger RNA in hippocampal neurons by synaptic NMDA receptor activation. N 340: 474-6 (1989); **Coleman DL et al** Lipopolysaccharide induces Egr-1 mRNA and protein in murine peritoneal macrophages. JI 149: 3045-51 (1992); **Darland T et al** Regulation of Egr-1 (Zfp-6) and c-fos expression in differentiating embryonal carcinoma cells. O 6: 1367-76 (1990); **Gashler AL et al** A novel repression module, an extensive activation domain, and a bipartite nuclear localization signal defined in the immediate-early transcription factor Egr-1. MCB 13: 4556-71 (1993); **Gupta MP et al** Egr-1, a serum-inducible zinc finger protein, regulates transcription of the rat cardiac α-myosin heavy chain gene. JBC 266: 12813-6 (1991); **Hallahan DE et al** Protein kinase C mediates x-ray inducibility of nuclear signal transducers EGR1 and JUN. PNAS 88: 2156-60 (1991); **Huang RP et al** Characterization of the DNA-binding properties of the early growth response-1 (Egr-1) transcription factor: evidence for modulation by a redox mechanism. DNA Cell Biol. 12: 265-73 (1993); **Iwaki K et al** α- and β-adrenergic stimulation induces distinct patterns of immediate early gene expression in neonatal rat myocardial cells. fos/jun expression is associated with sarcomere assembly; Egr-1 induction is primarily an α 1-mediated response. JBC 265: 13809-17 (1990); **Joseph LJ et al** Molecular cloning, sequencing, and mapping of EGR2, a human early growth response gene encoding a protein with "zinc-binding finger" structure PNAS 85: 7164-8 (1988); **Kharbanda S et al** Expression of the early growth response 1 and 2 zinc finger genes during induction of monocytic differentiation. JCI 88: 571-7 (1991); **Liu JW et al** Granulocyte-macrophage colony-stimulating factor induces transcriptional activation of Egr-1 in murine peritoneal macrophages. JBC 266: 5929-33 (1991); **McMahon AP et al** Developmental expression of the putative transcription factor Egr-1 suggests that Egr-1 and c-fos are coregulated in some tissues. Development 108: 281-7 (1990); **Mundschau LJ & Faller DV** BALB/c-3T3 fibroblasts resistant to growth inhibition by β interferon exhibit aberrant platelet-derived growth factor, epidermal growth factor, and fibroblast growth factor signal transduction. MCB 11: 3148-54 (1991); **Nguyen HQ et al** The zinc finger transcription factor Egr-1 is essential for and restricts differentiation along the macrophage lineage. Cell 72: 197-209 (1993); **Patwardhan S et al** EGR3, a novel member of the Egr family of genes encoding immediate-early transcription factors. O 6: 917-28 (1991); **Quellette AJ et al** Expression of two "immediate early" genes, Egr-1 and c-fos, in response to renal ischemia and during compensatory renal hypertrophy in mice. JCI 85: 766-71 (1990); **Ragona G et al** The transcriptional factor Egr-1 is synthesized by baculovirus-infected insect cells in an active, DNA-binding form. DNA Cell Biol. 10: 61-6 (1991); **Rangnekar YM et al** The serum and TPA responsive promoter and intron-exon structure of EGR2, a human early growth response gene encoding a zinc finger protein. NAR 18: 2749-57 (1990); **Rauscher FJ 3rd et al** Binding of the Wilms' tumor locus zinc finger protein to the EGR-1 consensus sequence. S 250: 1259-62 (1990); **Rim M et al** Evidence that activation of the Egr-1 promoter by v-*Raf* involves serum response elements. O 7: 2065-8 (1992); **Rupprecht HD et al** Effect of vasoactive agents on induction of Egr-1 in rat mesangial cells: correlation with mitogenicity. Am. J. Physiol. 263: F623-36 (1992); **Sakamoto KM et al** 5' upstream sequence and genomic structure of the human primary response gene, EGR-1/TIS8. O 6: 867-71 (1991); **Seyfert VL et al** Differential expression of a zinc finger-encoding gene in response to positive versus negative signaling through receptor immunoglobulin in murine B lymphocytes. MCB 9: 2038-8 (1989); **Seyfert VL et al** Egr-1 expression in surface Ig-mediated B cell activation. Kinetics and association with protein kinase C activation. JI 145: 3647-53 (1990); **Suggs SV et al** cDNA sequence of the human cellular early growth response gene Egr-1. NAR 18: 4283 (1990); **Sukhatne VP et al** A zinc finger-encoding gene coregulated with c-fos during growth and differentiation, and after cellular depolarization. Cell 53: 37-43 (1988); **Suwa LJ et al** Retinoic acid increases zif268 early gene expression in rat preosteoblastic cells. MCB 11: 2503-10 (1991); **Tsai-Morris CH et al** 5' flanking sequence and genomic structure of Egr-1, a murine mitogen inducible zinc finger encoding gene. NAR 16: 8835-46 (1988); **Worley PF et al** Constitutive expression of zif268 in neocortex is regulated by synaptic activity. PNAS 88: 5106-10 (1991)

EGR consensus box: see: Egr-1.

EIF: *Epstein-Barr virus inducing factor* This factor is identical with » TGF-β.

Bauer G et al Epstein-Barr-virus induction by a serum factor. Characterization of the purified factor and the mechanism of its activation. JBC 257: 11411-5 (1982); **Bauer G et al** Tumor-promoting activity of Epstein-Barr-virus-inducing factor/transforming growth factor type β (EIF/TGF-β) is due to the induction of irreversible transformation. IJC 47: 881-8 (1991); **Fürstenberger G et al** Stimulatory role of transforming growth factors in multistage skin carcinogenesis: possible explanation for the tumor-inducing effect of wounding in initiated NMRI mouse skin. IJC 43: 915-21 (1989)

EIP: *epidermal inhibitory pentapeptide* This pentapeptide with the sequence pEEDSG has been isolated from mouse backskin. It inhibits proliferation of epithelial cells in mouse backskin and normal epidermal cell proliferation *in vitro* at a restricted and low dose level. EIP probably functions as an endogenous inhibitor of cell proliferation and is involved in the control of normal tissue homeostasis. A single application of EIP to hairless mouse skin is followed by a long-lasting period of reduced epidermal cell proliferation even under conditions of sustained rapid epidermal cell proliferation induced by an erythemic dose of ultraviolet light (UV-B). EIP has also been shown to affect hematopoietic tissues, effectively inhibiting » CFU-S cells at concentrations of 1-10 ng (but not at 1 μg).

It has been shown that the pentapeptide can be serine-phosphorylated and that the phosphorylated forms are protected from hydrolysis by a serum enzyme which normally yields a stable tripeptide (pEED). For other hematopoietic inhibitory peptides see also: AcSDKP, pEEDCK.

Bramucci M et al Epidermal inhibitory pentapeptide phosphorylated *in vitro* by calf thymus protein kinase NII is protected from serum enzyme hydrolysis. BBRC 183: 474-80 (1992); **Elgjo K et al** Inhibitory epidermal pentapeptide modulates proliferation and differentiation of transformed mouse epidermal cells *in vitro*. Virchows Arch. B. Cell. Pathol. 60: 161-4 (1991); **Elgjo K & Reichelt KL** Structure and function of growth inhibitory epidermal pentapeptide. ANY 548: 197-203 (1988); **Olson WM & Elgjo K** UVB-induced epidermal hyperproliferation is modified by a single, topical treatment with a mitosis inhibitory epidermal pentapeptide. J. Invest. Dermatol. 94: 101-6 (1990); **Robinson, P et al** Detection and characterization of growth inhibitory factors: epidermal inhibitory pentapeptide (EIP). In: McKay I & Leigh I (eds). Growth factors, a practical approach, IRL Press, pp. 133-55, Oxford (1992); **Whitehead PA et al** Identification and partial characterization of a serum enzyme which hydrolyses epidermal inhibitory pentapeptide. BBRC 175: 978-85 (1991)

EK-AF-1: *embryonic kidney-derived angiogenesis factor 1* This factor is isolated from embryonic kidney cells. It has the activity of an » Angiogenesis factor. The factor is identical with » aFGF.

Risau W & Ekblom P Growth factors and the embryonic kidney. PCBR 226: 147-56 (1986); **Risau W & Ekblom P** Production of a heparin-binding angiogenesis factor by the embryonic kidney. JCB 103: 1101-7 (1986)

EK-AF-2: *embryonic kidney-derived angiogenesis factor 2* This factor is isolated from embryonic kidney cells. It has the activity of an » Angiogenesis factor. The factor is identical with » bFGF.

see: EK-AF-1.

EK-derived basophil promoting activity: *epidermal keratinocyte-derived basophil promoting activity* This activity is secreted by human keratinocytes. It promotes the growth of early myeloid cells, in particular the development of basophils. Some of the activity is due to » IL3 and some to a soluble form of the cell surface antigen » CD23.

Jaffray P et al Epidermal keratinocyte-derived basophil promoting activity. Role of interleukin 3 and soluble CD23. JCI 90: 1242-7 (1992)

ELCF: *epidermal (cell-derived) lymphocyte chemotactic factor* This protein is isolated from epidermal cells. It is a strong chemoattractant for lymphocytes. The factor is probably identical with » MDNCF (monocyte-derived neutrophil chemotactic factor) and hence identical with» IL8. It therefore belongs to the » chemokine family of cytokines.

Zachariae C et al Epidermal lymphocyte chemotactic factor specifically attracts OKT4-positive lymphocytes. Arch. Dermatol. Res. 280: 354-7 (1988)

EL4: A murine T lymphoma (thymoma) cell line established from a chemically induced tumor in C57/BL/6N mouse. EL4 cells have been shown to secrete » IL2, » CSF, and MAF (macrophage activating factor). The cell line is used to assay » IL1 (see also: Bioassays). Many sublines of EL4 have been established but not all of them are responsive to IL1. The cells respond to IL1 isolated from almost all vertebrates. IL1 activity is determined indirectly by measuring the amounts of » IL2 induced by IL1 in another cell line, » CTLL-2. This so-called conversion assay can be performed also by cocultivation of EL4 cells with a CTLL variant that does not express thymidine kinase (EL4 6.1). EL4 also responds to murine » TNF-α. It is therefore necessary to use neutralizing antibodies to distinguish between these to factors.

A subclone of EL4, designated ***NOB-1***, is also used to assay IL1. It constitutively produces very little IL2 but in response to IL1 produces high concentrations of IL2.The range of detection is ≈ 500 fg/mL to 50 pg/mL. NOB-1 is not responsive to » TNF-α, » TNF-β, » IFN-γ, and bacterial lipopolysaccharides.

Both cell lines are also used to determine the levels an IL1 receptor antagonist (see: IL1ra) which competes with IL1 for receptor binding. Other cell lines used to assay IL1 are » LBRM 33 and » RPMI 1788. An alternative and entirely different method of detecting IL1 is » Message amplification phenotyping.

Recently, EL4 thymoma cells have been found to be a source of biologically active » CD40 ligand (see: TRAP, TNF-related activation protein). The cells also secrete an uncharacterized factor (see: THP-1-derived growth-promoting activity) that promotes the growth of a large variety of cell lines.

Chang AE et al A large scale method of separating multiple lymphokines secreted by the murine EL-4 thymoma. J. Immunopharmacol. 7: 17-31 (1985); **Gearing AJH et al** A simple sensitive bioassay for interleukin-1 which is unresponsive to 10(3) U/ml of interleukin-2. JIM 99: 7-11 (1987); **Le-Moal MA et al**

A sensitive IL2dependent assay for IL1. JIM 24: 23-30 (1988); **Nadeau RW et al** Quantification of recombinant human interleukin 1a by a specific two cell immunobioassay. JIM 168: 9-16 (1994); **Nakanishi K et al** Both interleukin 2 and a second T cell derived factor in EL4-supernatant have activity as differentiation factors in IgM synthesis. JEM 160: 1605-21 (1984); **Remvig L et al** Biological assays for interleukin 1 detection. Comparison of human T lymphocyte, murine thymocyte, and NOB-1 assays. Allergy 46: 59-67 (1991); **Simon PL et al** A modified assay for interleukin 1 (IL1). JIM 84: 85-94 (1985); **Tucci A et al** Effects of eleven cytokines and of IL1 and tumor necrosis factor inhibitors in a human B cell assay. JI 148: 2778-84 (1992)

EL4-BCGF: *EL4 B cell growth factor* This B cell growth factor that co-stimulates anti-IgM-stimulated splenic B cells is isolated from the culture supernatant of phorbol 12-myristate 13-acetate (PMA)-stimulated » EL4 cells (see also: Phorbol esters). It is identical with » BCGF-1. Both factors are identical with » IL4. See also: BCGF (B cell growth factor).

Dutton RW et al Partial purification and characterization of a BCGF II from EL4 culture supernatants. JI 132: 2451-6 (1984); **Farrar JJ et al** Biochemical and physicochemical characterization of mouse B cell growth factor: a lymphokine distinct from interleukin 2. JI 131: 1838-42 (1983); **Howard M et al** Identification of a T cell derived B cell growth factor distinct from interleukin 2. JEM 155: 914-23 (1982)

ELDF: *eosinophilic leukemia cell differentiation factor* This poorly characterized factor of ≈ 30-40 kDa (gel filtration) is constitutively secreted by a human adult T cell leukemia cell line (HIL-3). ELDF is a glycoprotein but glycosylation does not appear to be essential for biological activity. ELDF induces differentiation of the human eosinophilic leukemia cell line EoL-1, causing the expression of eosinophilic granules and segmented nuclei, » CD23 and an eosinophil differentiation antigen EO-1. The factor also induces EoL-1 cells to respond to an eosinophil chemotactic factor, platelet activating factor. The factor does not appear to be identical with » » IL2, IL3, » IL4, » IL5, » M-CSF, » G-CSF), TNF-α.

Morita M et al Differentiation of a human eosinophilic leukemia cell line (EoL-1) by a human T cell leukemia cell line (HIL-3)-derived factor. Blood 77: 1766-75 (1991); **Ohshima Y et al** Characterization of an eosinophilic leukemia cell differentiation factor (ELDF) produced by a human T cell leukemia cell line, HIL3. EH 21: 749-54 (1993)

ELDIF: *epidermal cell-derived lymphocyte differentiating factor* This poorly characterized factor is produced by cultured human keratinocytes. It inhibits *in vitro* the production of » IL2 and the proliferation of murine spleen cells.

Nicolas JF et al Epidermal cell-derived lymphocyte differentiating factor (ELDIF) inhibits *in vitro* lymphoproliferative responses and interleukin 2 production. J. Invest. Dermatol. 88: 161-6 (1987)

ELISA: *enzyme-linked immunosorbent assays* see: Cytokine assays.

ELMER: *Endogenous Ligand conferring MitogEnic Response* A poorly characterized factor isolated from sera obtained before delivery from women with preeclampsia. ELMER is an acid- and heat-labile protein of ≈ 160k Da. It is a potent mitogen for human fibroblasts but not for human endothelial cells.

ELMER may be a potential serum marker of preeclampsia; it may also play roles in the vasospasm and proliferative vascular lesion, termed atherosis, frequently associated with the preeclamptic syndrome.

Taylor RN et al Partial characterization of a novel growth factor from the blood of women with preeclampsia. J. Clin. Endocrinol. Metab. 70: 1285-91 (1990)

EL-TRF: *EL T cell replacing factor* This activity is produced by » EL4 cells. It promotes the synthesis of IgM by B cells. The factor is identical with » IL5. For some aspects of nomenclature see also: TRF (T cell replacing factors).

Nakanishi K et al Soluble factors involved in B cell differentiation: identification of two distinct T cell-replacing factors (TRF). JI 130: 2219-24 (1983); **Nakanishi K et al** Ig RNA expression in normal B cells stimulated with anti-IgM antibody and T cell-derived growth and differentiation factors. JEM 160: 1736-51 (1984); **Nakanishi K et al** Both interleukin 2 and a second T cell-derived factor in EL-4 supernatant have activity as differentiation factors in IgM synthesis. JEM 160: 1605-21 (1984)

EMAP: *endothelial-monocyte activating polypeptides* Two factors, designated EMAP I (≈ 44 kDa) and EMAP II (≈ 22 kDa) have been described to be secreted by murine meth A fibrosarcoma cells. EMAP I is identical with » Meth A factor. EMAP II has the ability to induce tissue factor procoagulant activity in endothelial cells. It induces monocyte migration and is also chemotactic for granulocytes. Injection of the polypeptide into mouse footpads elicits an inflammatory response with tissue swelling and polymorphonuclear leukocyte infiltration. The aminoterminal sequence of EMAP II appears to be unique and cloning of its cDNA shows that the factor is not related to any other known cytokine. The protein shows limited homology with von Willebrand antigen II over a stretch of 10 amino acids.

Kao J et al Endothelial monocyte-activating polypeptide II. A novel tumor-derived polypeptide that activates host-response mechanisms. JBC 267: 20239-47 (1992); **Kao J et al** Endothelial-monocyte activating polypeptides (EMAPs): tumor derived mediators which activate the host inflammatory response. Behring Inst. Mitt. 92: 92-106 (1993)

Embryonal carcinoma cells: see: ES cells (embryonic stem cells).

Embryonal carcinoma-derived growth factor: see: ECDGF.

Embryonic carcinoma-derived growth factor: see ECGF.

Embryonic growth/differentiation factor: see: GDF-1.

Embryonic kidney-derived angiogenesis factor 1: see: EK-AF-1.

Embryonic kidney-derived angiogenesis factor 2: see: EK-AF-2.

Embryonic stem cell growth factor: see: ESCGF.

ENA-78: This factor is produced by various human epithelial cell lines. It has a length of 78 amino acids. The factor activates neutrophils and is a chemoattractant for these cells. It increases intracellular calcium levels (see also: Calcium ionophore) and promotes exocytosis. The factor is produced together with other factors such as » IL8, » MGSA (melanoma growth stimulatory activity) following stimulation of the cells by » IL1 or » TNF-α. The primary sequence of ENA-78 shows that it belongs to the » chemokine family of cytokines.
The sequence of ENA-78 shows a 53 % sequence identity with » NAP-2 (neutrophil activating peptide 2) and GRO-α (see: MGSA). It shows 67 % identity with » AMCF (alveolar macrophage chemotactic factor). ENA-78 binds to the same receptors that also bind » IL8, » NAP-2 and GRO-α.
Strieter RM et al The detection of a novel neutrophil-activating peptide (ENA-78) using a sensitive ELISA. Immunol. Invest. 21: 589-96 (1992); **Walz A et al** Structure and neutrophil-activating properties of a novel inflammatory peptide (ENA-78) with homology to interleukin 8. JEM 174: 1355-62 (1991)

ENAP: *endothelial cell neutrophil activating peptide* This factor is isolated from activated human endothelial cells. It activates neutrophils (see also: Cell activation) and is a strong chemoattractant for these cells (see also: Chemotaxis). The factor occurs in two forms, designated ENAP-α (7,5 kDa) and ENAP-β (17 kDa), respectively. They are identical in their biochemical properties and biological activities with two other Neutrophil-activating proteins (see: MONAP, LYNAP). Both factors are identical with one of the molecular forms of » IL8.
Schroeder JM & Christophers E Secretion of novel and homologous neutrophil-activating peptides by LPS-stimulated human endothelial cells. JI 142: 244-51 (1989)

Endogenous ligand conferring mitogenic response: see: ELMER.

Endogenous pyrogens: see: EP.

Endorphins (α, β, γ): see: POMC (proopiomelanocortin).

Endothelial cell-derived growth factor: see: ECDGF.

Endothelial cell-derived lymphocyte chemotactic activity: see: ED-LCA.

Endothelial cell growth factor: see: ECGF.

Endothelial cell growth factor α: see: ECGF-α.

Endothelial cell growth factor 1: see: ECGF-1.

Endothelial cell growth factor 2a: see: ECGF-2a.

Endothelial cell growth factor 2b: see: ECGF-2b.

Endothelial cell growth supplement: see: ECGS.

Endothelial cell stimulating angiogenic factor: abbrev. ESAF. See: HUAF (human uterine angiogenesis factor).

Endothelial-derived neutrophil-activating peptide: see: ENAP.

Endothelial cell neutrophil activating peptide β: abbrev. β-ENAP. See: ENAP.

Endothelial derived growth factor: see: EDGF.

Endothelial-derived neutrophil chemotactic factor: see: NCF.

Embryonic receptor kinase: see addendum: EmRK2. **EmRK2:** see addendum.
EMT: see addendum.

Endothelial-monocyte activating polypeptides: see: EMAP.

Endothelins: see: ET.

Endothelioma-derived motility factor: see: EDMF.

Endothelium-derived contracting factor: see: EDCF.

Endotoxin: A term used to describe an essential lipopolysaccharide (LPS) component of the cell wall of Gram-negative bacteria. Together with phospholipids and membrane-bound proteins it is a constituent of the outer cell membrane. These lipopolysaccharides define many of the properties of host-parasite interactions. The diversity of the bacterial lipopolysaccharides is responsible for the characteristic antigenic properties of Gram-negative bacteria.

LPS consists of three structural elements. One is a hydrophobic component, called lipid A, which serves to anchor the molecule into the membrane. The second is a core oligosaccharide. The third component is a hydrophilic O-polysaccharide projecting into the extracellular space More than 150 different variants of the third component are known. The biological functions of the individual LPS components can be inferred, at least in part, from the properties of some known bacterial mutants expressing aberrant LPS variants. The O-polysaccharide appears to be involved in host-parasite interactions because its removal leads to a loss of virulence. If the proximal part of the core oligosaccharide is lost the bacteria become extremely sensitive to detergents, antibiotics, and bile salts. It can therefore be assumed that this region is in some way essential for the maintenance of outer membrane functions as a biological barrier. Mutations altering the lipid A component are mostly not viable, suggesting that it is important for the maintenance of outer membrane integrity as a whole.

Although bacterial endotoxins may differ considerably in their antigenicity they elicit the same physiological responses during infection of an organism with Gram-negative micro-organisms. They are mitogenic for B cells and function as polyclonal B cell activators, mediate the activation of macrophages (see also: Cell activation) and activate the complement cascade. They act as a physiological stimulus for the synthesis of » cytokines (proinflammatory cytokines such as » TNF-α, » IL1, » IL6, » IL8) and non-protein mediators, which in turn, are responsible for most pathophysiological consequences of a bacterial infection (see: Septic Shock, Shwartzman phenomenon, Inflammation).

The interaction of LPS with target cells and the subsequent initiation of endotoxin shock can be modulated by plasma proteins that bind LPS. The most important ***LPS-binding proteins*** known so far are high density lipoprotein (HDL), low den-

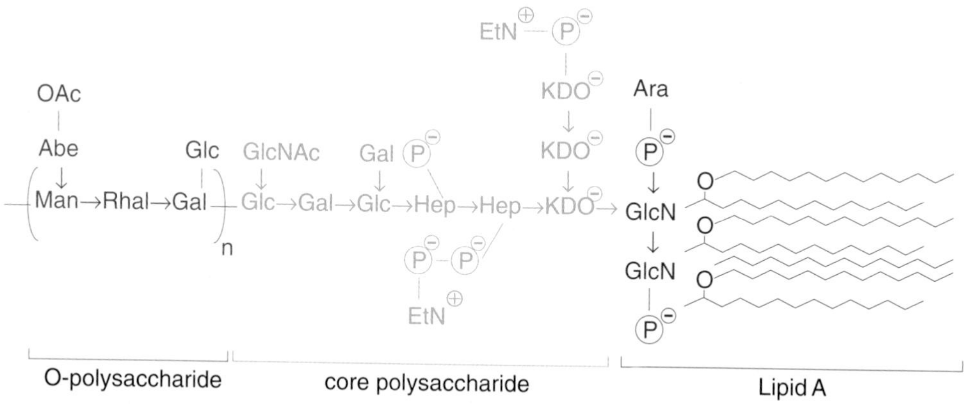

Structure of bacterial endotoxins.
As exemplified by Salmonella lipopolysaccharide the molecules has three distinct parts, designated O-polysacchar-rid (red), core-polysaccharid (green) and lipid A (blue).
Abe = Abequose; Ara = 4-Amino-Arabinose; EtN = Ethanolamine; Gal = D-Galactose; Glc = D-Glucose; GlcNAc = N-acetyl-D-Glucosamine; Hep = L-Glycero-Mannoheptose; KDO = 2-Keto-3-Deoxyoctonic acid; Man = D-Mannose; OAc = O-acetyl; P = phosphoric acid ester; Rha = L-Rhamnose.

sity lipoprotein (LDL), ***LPS-binding protein*** (***LBP***) and specific LPS antibodies. LPS-binding serum protein mediates binding of endotoxins to membrane-bound CD14 on monocytes. CD14 is also present as a soluble receptor in serum that can neutralize endotoxins so that they are unable to activate monocytes. LBP appears to be responsible for lethal endotoxemia (see also: Septic shock). LBP amplifies the production of » TNF at low concentrations of endotoxins and is detrimental in the presence of LPS at higher concentrations. Antibodies against LBP given at the same time as a lethal dose of LPS in mice have been shown to protect against the lethal effect of LPS.

The sensitivity of the organism to LPS has a genetic basis and varies considerably among different species. Sensitivity and tolerance can be altered/ induced reversibly both *in vivo* and *in vitro* by various experimental conditions, indicating a desensitization of the cytokine-producing cells. Hypersensitivity to LPS is observed in response to live or killed microorganisms, growing tumors, hepatotoxic agents or proteins to which the organism has been immunologically primed. The state of hypersensitivity is characterized by an increased ability to produce cytokines upon LPS challenge and an increased susceptibility to the toxic activity of » TNF-α. It has been shown that » IFN-γ is a central mediator in the development of LPS hypersensitivity. The mouse strain C3H/HeJ, for example, is LPS-resistant due to an impairment in the production of » IFN-γ.

A recently developed monoclonal anti-LPS antibody has been shown to cross-react with endotoxins of various types of bacteria and to cross-protect against endotoxin effects. The ***bactericidal/permeability-increasing protein*** (***BPI***) a 55 kDa cytotoxic cationic protein found in the azurophilic granules of polymorphonuclear leukocytes displays specific toxicity against Gram-negative bacteria by interacting with LPS from a broad range of species. A recombinant N-terminal fragment of BPI carries all the antibacterial activities of holo-BPI and is more potent than the holo-protein. It has been shown to protect animals against the lethal effects of administered LPS.

Baker PJ et al Bacterial polysaccharides, endotoxins, and immunomodulation. AEMB 319: 3-18 (1992); **Chen TV et al** Lipopolysaccharide receptors and signal transduction pathways in mononuclear phagocytes. CTMI 181: 169-88 (1992); **Derijk RH et al** The role of macrophages in the hypothalamic-pituitary-adrenal activation in response to endotoxin (LPS). Res. Immunol. 143: 224-9 (1992); **Doran JE** Biological effects of endotoxin. Curr. Stud. Hematol. Blood Transf. 59: 66-99 (1992); **Elsbach P & Weiss J** The bactericidal/permeability-increasing protein (BPI), a potent element in host-defense against Gram-negative bacteria and lipopolysaccharide. Immunobiology 187: 417-29 (1993); **Elsbach P & Weiss J** Bactericidal/permeability increasing protein and host defense against gram-negative bacteria and endotoxin. Curr. Opin. Immunol. 5: 103-7 (1993); **Fang KC** Monoclonal antibodies to endotoxin in the management of sepsis West. J. Med. 158: 393-9 (1993); **Fink MP** Adoptive immunotherapy of Gram-negative sepsis: use of monoclonal antibodies to lipopolysaccharide. Crit. Care Med. 21: S32-9 (1993); **Flad HD et al** Agonists and antagonists for lipopolysaccharide-induced cytokines. Immunobiology 187: 303-16 (1993); **Galanos C & Freudenberg MA** Mechanisms of endotoxin shock and endotoxin hypersensitivity. Immunobiology 187: 346-56 (1993); **Gallay P et al** Lipopolysaccharide-binding protein as a major plasma protein responsible for endotoxemic shock. PNAS 90: 9935-8 (1993); **Ghosh S et al** Endotoxin-induced organ injury. Crit. Care Med. 21: S19-24 (1993); **Hurley JC** Reappraisal of the role of endotoxin in the sepsis syndrome. The Lancet 341: 1133-1135 (1993); **Lynn WA & Golenbock DT** Lipopolysaccharide antagonists. IT 13: 271-6 (1992); **Martich GD et al** Response of man to endotoxin. Immunobiology 187: 403-16 (1993); **Mengozzi M & Ghezzi P** Cytokine down-regulation in endotoxin tolerance. ECN 4: 89-98 (1993); **Morrison DC et al** Identification and characterization of mammalian cell membrane receptors for LPS-endotoxin. AEMB 319: 23-9 (1992); **Morrison DC et al** Endotoxin receptors on mammalian cells Immunobiology 187: 212-26 (1993); **Raetz CRH** Biochemistry of endotoxins. ARB 59: 129-70 (1990); **Raetz CR et al** Gram-negative endotoxin: an extraordinary lipid with profound effects on eukaryotic signal transduction. FJ 5: 2652-60 (1991); **Schnaitman CA & Klena JD** Genetics of lipopolysaccharide biosynthesis in enteric bacteria. Microbiol. Rev. 57: 655-82 (1993); **Schumann RR** Function of lipopolysaccharide (LPS)-binding protein (LBP) and CD14, the receptor for LPS/LBP complexes: a short review. Res. Immunol. 143: 11-5 (1992); **Sultzer BM et al** Lipopolysaccharide nonresponder cells: the C3H/HeJ defect. Immunobiology 187: 257-71 (1993); **Tobias PS & Ulevitch RJ** Lipopolysaccharide binding protein and CD14 in LPS dependent macrophage activation. Immunobiology 187: 227-32 (1993); **Ulevitch RJ** Recognition of bacterial endotoxins by receptor-dependent mechanisms. AI 53: 267-89 (1993); **Young LS** Bacterial endotoxins: Cytokine mediators and new therapies for sepsis. Endotoxins and mediators - An introduction. PCBR 367: 1-7 (1991)

engrailed: see: Homeotic genes.

EnGS: *endothelial cell growth supplement* see: ECGS.

Enhancing activity: see: EA.

Enhancing factors: see: EF.

ENKAF: *epidermal (cell-derived) NK (cell activity) augmenting factor* This factor is found in the culture supernatants of squamous cell carcinomas and normal epidermal cells. It is a chemoattractant for NK cells (see also: Chemotaxis) and also stimulates NK cell activity. This poorly characterized factor i not identical with » IL1, » IL2 or interferons (see: IFN).

Luger TA et al Human epidermal cells and squamous carcinoma cells synthesize a cytokine that augments natural killer cell activity. JI 134: 2477-83 (1985); **Uchida A** The cytolytic and regulatory role of natural killer cells in human neoplasia. BBA 865: 329-40 (1986)

Enzyme-linked immunosorbent assays: abbrev. ELISA. See: Cytokine assays.

EoCP-1: *eosinophil chemotactic polypeptide* This protein is released by human platelets following their stimulation by thrombin. It is identical with » RANTES. A variant of EoCP, designated ***EoCP-2***, differs from EoCP-1 in two amino acid positions at the N-terminal end. Both proteins are strongly chemotactic for eosinophils (see also: Chemotaxis).

Kameyoshi Y et al Cytokine RANTES released by thrombin-stimulated platelets is a potent attractant for human eosinophils. JEM 176: 587-92 (1992)

EoCP-2: *eosinophil chemotactic polypeptide* see: EoCP-1.

Eo-CSF: *eosinophil colony-stimulating factor* This factor supports the development of eosinophilic colonies in a » colony formation assay. It is identical with » IL5 (see also: EDF, eosinophil differentiation factor). Some of the activity may be due to » GM-CSF or » IL3.

Lopez AF et al Murine eosinophil differentiation factor. An eosinophil-specific colony-stimulating factor with activity for human cells. JEM 163: 1085-99 (1986); **Nakamura Y et al** Factors that stimulate the proliferation and survival of eosinophils in eosinophilic pleural effusion: relationship to granulocyte/macrophage colony-stimulating factor, interleukin-5, and interleukin-3. Am. J. Respir. Cell. Mol. Biol. 8: 605-11 (1993); **Nicola NA et al** Separation of functionally distinct human granulocyte-macrophage colony-stimulating factors. Blood 54: 614-27 (1979); **Yokota T et al** Isolation and characterization of lymphokine cDNA clones encoding mouse and human IgA-enhancing factor and eosinophil colony-stimulating factor activities: relationship to interleukin 5. PNAS 84: 7388-92 (1987)

Eo-DF: *eosinophil differentiation factor* This factor is identical with » IL5. See also: EDF (eosinophil differentiation factor).

Eosinophil-activating factor: see: EAF.

Eosinophil chemotactic polypeptide: see: EoCP-1.

Eosinophil colony stimulating factor: see: Eo-CSF.

Eosinophil cytotoxicity-enhancing factor: see: ECEF.

Eosinophil cytotoxicity inhibitor: see: ECI.

Eosinophil differentiation factor: see: EDF; see also: Eo-DF.

Eosinophilic leukemia cell differentiation factor: see: ELDF.

Eosinophil stimulation promoter: see: ESP.

Eosinophil viability-enhancing activity: This factor is produced and released by human mononuclear cells. It enhances the viability of cultured eosinophils. The activity of the factor is completely neutralized by antibodies directed against » GM-CSF, but not by specific antibodies to » IL3 and IL5.

Burke LA et al Identification of the major activity derived from cultured human peripheral blood mononuclear cells, which enhances eosinophil viability, as granulocyte macrophage colony-stimulating factor (GM-CSF). J. Allergy. Clin. Immunol. 88: 226-35 (1991)

EP: *endogenous pyrogens* Endogenous pyrogens are substances that are produced by the host during » inflammation, trauma, or infection. They elevate the thermoregulatory set point in the hypothalamus and usually induce a rapid rise (within 12 minutes) in core temperature after intravenous injection.

One protein with EP activity was initially isolated from leukocytes and was also called leukocytic pyrogen (abbrev. LP). This protein is identical with » LEM (leukocytic endogenous mediator) and hence identical with » IL1.

Some of the EP activity described in the older literature may also be due to » TNF-α and » IFN-α which also possess fever-inducing activity. Other endogenous pyrogens are » IL6, » MIP (macrophage inflammatory protein), » CNTF (ciliary neurotrophic factor).

Atkins E The pathogenesis of fever. Physiol. Rev. 40: 580 (1960); **Baracos V et al** Stimulation of muscle protein degradation and prostaglandin E_2 release by leukocytic pyrogen. NEJM 308: 553-8 (1983); **Bodel P & Miller H** Pyrogen from mouse macrophages causes fever in mice. PSEBM 151: 93-96 (1976); **Cebula TA et al** Synthesis of four endogenous pyrogens by rabbit macrophages. J. Lab. Clin. Med. 94: 95 (1979); **Dinarello CA et al** Human leukocytic pyrogen: purification and development of a radioimmunoassay. PNAS 74: 4624-7 (1977); **Dinarello CA et al** Endogenous pyrogens. MiE 163: 495-510 (1988); **Duff GW & Durum SK** The pyrogenic and mitogenic actions of interleukin-1 are related. N 304: 449-51 (1983); **Gordon AH**

Eotaxin: see addendum.

& Parker ID A pyrogen derived from human white cells which is active in mice. Br. J. Exp. Pathol. 61: 534-9 (1980); **Matsushima K et al** Purification of human interleukin 1 from human monocyte culture supernatants and identity of thymocyte comitogenic factor, fibroblast proliferation factor, acute phase protein-inducing factor, and endogenous pyrogen. CI 92: 290-301 (1985); **Merriman CR et al** Comparison of leukocytic pyrogen and leukocytic endogenous mediator. PSEBM 154: 224-7 (1977); **Murphy PA et al** Endogenous pyrogens made by rabbit peritoneal exudate cells are identical with lymphocyte-activating factors made by rabbit alveolar macrophages. JI 124: 2498-2501 (1980); **Murphy PA et al** Further purification of rabbit leukocyte pyrogen. J. Lab. Clin. Med. 83: 310 (1974); **Taktak YS et al** Assay of pyrogens by interleukin-6 release from monocytic cell lines. J. Pharm. Pharmacol. 43: 578-82 (1991)

Ep: abbrev. for erythropoietin. See: Epo.

EPA: *erythroid potentiating activity* Generic name given to factors that support *in vitro* the growth of the normal early erythroid bone marrow progenitor cells » BFU-E and » CFU-E (see also: Hematopoiesis). Such factors, also called ***EPF*** (erythroid potentiating factor(s), were originally observed in culture supernatants of the human monocytic cell line » U937 and were found to differ from » CSF (colony stimulating factors) or » Epo. Such activities are also found to be secreted by other cells. It has been observed that hemin also possesses erythroid growth-potentiating activity.
EPA activity is also called » BPA (burst-promoting activity) in some of the old literature and then appears to refer to the activity on BFU-E rather than the more mature CFU-E. The term appears also to have been used for activities of human rather than murine origin. One of the factors showing EPA activity is probably identical with » EPA (erythroid promoting activity = TIMP). It has been shown that » IGF-I and IGF-II also influence the growth of human marrow erythroid progenitors via a direct mechanism involving the type I IGF receptor. For a membrane-bound form of a factor with BPA activity see: E-CSF (erythroid colony-stimulating factor).

Ascensao JL et al Production of erythroid potentiating factor(s) by a human monocytic cell line. Blood 57: 170-3 (1981); **Docherty AJ et al** Sequence of human tissue inhibitor of metalloproteinases and its identity to erythroid-potentiating activity. N 318: 66-9 (1985); **Gasson JC et al** Characterization of purified human erythroid potentiating activity. PCBR 184: 95-104 (1985); **Gauwerky CE et al** Human leukemia cell line K562 responds to erythroid-potentiating activity. Blood 59: 300-5 (1982); **Kaye FJ et al** The effect of hemin *in vitro* and *in vivo* on human erythroid progenitor cells Int. J. Cell Cloning 5: 74-88 (1987); **Merchav S et al** Comparative studies of the erythroid-potentiating effects of biosynthetic human insulin-like growth factors-I and -II. J. Clin. Endocrinol. Metab. 74: 447-52 (1992); **Niskanen E et al** Recombinant human erythroid potentiating activity enhances the effect of erythropoietin in mice. Eur. J. Haematol. 45: 267-70 (1990); **Tanno Y et al** Human interleukin-3-like activity, basophil and eosinophil growth promoting activities and colony stimulating factor derived from several cell lines. Int. Arch. Allergy. Appl. Immunol. 83: 1-5 (1987); **Westbrook CA et al** Purification and characterization of human T lymphocyte-derived erythroid-potentiating activity. JBC 259: 9992-6 (1984)

EPA: *erythroid promoting activity* This is a 28 kDa factor with a length of 207 amino acids inferred from the cDNA sequence (including 23 amino acids functioning as a secretory » signal sequence).
The activity of this protein is not species-specific. The factor stimulates the growth of the early hematopoietic progenitor cell types » BFU-E and » CFU-E (see also: Hematopoiesis).
The human gene encoding EPA has a length of ≈ 3 kb. It contains at least three exons and maps to chromosome Xp11.1-p11.4. The EPA gene is located within intron 6 of the SYN1 gene (synapsin 1), which has 13 exons.
The EPA gene is identical with a gene encoding a glycoprotein called » TIMP (tissue inhibitor of metalloproteinases). A cDNA sequence isolated from a mouse embryo fibroblast cDNA library, designated ***16C8*** probably is the murine counterpart of the human collagenase inhibitor with EPA activity. It has been suggested that abnormalities in the EPA/TIMP gene should be sought in disorders such as Menkes syndrome, which maps to approximately the same region.

Avalos BR et al K562 cells produce and respond to human erythroid-potentiating activity. Blood 71: 1720-5 (1988); **Docherty AJP et al** Sequence of human tissue inhibitor of metalloproteinases and its identity to erythroid-potentiating activity. N 318: 66-9 (1985); **Edwards DR et al** A growth-responsive gene (16C8) in normal mouse fibroblasts homologous to a human collagenase inhibitor with erythroid-potentiating activity: evidence for inducible and constitutive transcripts. NAR 14: 8863-78 (1986); **Gasson JC et al** Characterization of purified human erythroid-potentiating activity. PCBR 184: 95-104 (1985); **Gasson JC et al** Molecular characterization and expression of the gene encoding human erythroid potentiating activity. N 315: 768-71 (1985); **Gewert DR et al** Characterization and expression of a murine gene homologous to human EPA/TIMP: a virus-induced gene in the mouse. EJ 6: 651-7 (1987); **Huebner K et al** Localization of the gene encoding human erythroid-potentiating activity to chromosome region Xp11.1-Xp11.4. Am. J. Hum. Genet. 38: 819-26 (1986); **Niskanen E et al** *In vivo* effect of human erythroid-potentiating activity on hematopoiesis in mice. Blood 72: 806-10 (1988); **Strife A et al** Activities of four purified growth factors on highly enriched human hematopoietic progenitor cells. Blood 69: 1508-23 (1987); **Westbrook CA et al** Purification and characterization of human T lymphocyte-derived erythroid-potentiating activity. JBC 259: 9992-6 (1984)

EPF: *early pregnancy factor* EPF is a marker of the development of the embryo before and during

implantation. Its presence is revealed *in vitro* as a lymphocyte-modifying activity in maternal serum by a rosette inhibition assay. EPF is detected within hours of fertilization and is present for at least the first two-thirds of pregnancy, with continued detection dependent upon the presence of a viable embryo or fetus. EPF activity is a true indication of a viable embryo, since removal of the embryo results in a rapid removal of EPF response from serum. Passive immunization of mice against EPF leads to failure to maintain pregnancy.
EPF may not be a unique factor expressed as a novel, pregnancy-specific protein. It has been shown that » thioredoxin or thioredoxin-like molecules are mainly responsible for increased rosette inhibition titers.
Recent studies suggest a more general link of EPF with cell development and cell proliferation. EPF activity has been found in platelets, pregnancy urine, medium conditioned by estrous mouse ovaries (stimulated with prolactin and embryo-conditioned medium), medium conditioned by tumor cells, and serum from rats 24 h after partial hepatectomy. EPF is produced by proliferating tumor cells and by liver cells after partial hepatectomy. Passive immunization with anti-EPF monoclonal antibodies demonstrate that these cells need EPF for survival. Neutralization of EPF with monoclonal antibodies retards growth of some murine tumors both *in vitro* and *in vivo* and points to the presence of EPF-induced suppressor factor circulating in the serum of tumor-bearing mice. Active immunization with EPF suppresses the formation of immune ascites in BALB/c mice.
EPF does not appear to be a product of the pre-embryo, and addition of anti-EPF monoclonal antibodies to embryo cultures does not adversely affect development from the 2-cell to the blastocyst stage. However, neutralization of EPF *in vivo* by anti-EPF antibodies retards embryo development.
EPF has been shown to suppress the delayed-type hypersensitivity reaction and may act as an immunological response modifier of the maternal immune system, possibly reducing the expression of some cytokines.

Athanasas-Platsis S et al Antibodies to early pregnancy factor retard embryonic development in mice *in vivo*. J. Reprod. Fertil. 92: 443-51 (1991); **Cavanagh AC et al** Identification of a putative inhibitor of early pregnancy factor in mice. J. Reprod. Fertil. 91: 239-48 (1991); **Clarke FM** Identification of molecules and mechanisms involved in the 'early pregnancy factor' system. Reprod. Fertil. Dev. 4: 423-33 (1992); **Mesrogli M & Dieterle S** Embryonic losses after *in vitro* fertilization and embryo transfer. Acta Obstet. Gynecol. Scand. 72: 36-8 (1993); **Morton H et al** Early pregnancy factor has immunosuppressive and growth factor properties. Reprod. Fertil. Dev. 4: 411-22 (1992); **Quinn KA** Active immunization with EPF suppresses the formation of immune ascites in BALB/c mice. Immunol. Cell Biol. 69: 1-6 (1991); **Quinn KA & Morton H** Effect of monoclonal antibodies to early pregnancy factor (EPF) on the *in vivo* growth of transplantable murine tumours. Cancer-Immunol-Immunother. 34: 265-71 (1992); **Tonissen KF et al & Wells JRE** Isolation and characterization of human thioredoxin-endoding genes. Gene 102: 221-8 (1991)

EPF: *erythroid potentiating factor(s)* see: EPA (erythroid potentiating activity).

Epibolin: This factor is identical with » vitronectin.

Epidermal cell-derived granulocyte-activating mediator: abbrev. EC-GRAM. See: GRAM.

Epidermal cell-derived lymphocyte chemotactic factor: see: ELCF.

Epidermal cell-derived lymphocyte differentiating factor: see: ELDIF.

Epidermal cell-derived NK augmenting factor: see: ENKAF.

Epidermal cell-derived thymocyte activating factor: see: ETAF.

Epidermal inhibitory pentapeptide: see: EIP.

Epidermal keratinocyte-derived basophil promoting activity: see: EK-derived basophil promoting activity.

Epidermal lymphocyte chemotactic factor: see: ELCF.

Epithelial cell growth inhibiting factor: This factor is identical with one of the molecular forms of » TGF-β (TGF-β2).

Holley R et al Purification of kidney epithelial cell growth inhibitor. PNAS 77: 5989-92 (1980); **Tucker R et al** Growth inhibitor from BSC-1 cells closely related to platelet type β transforming growth factor. S 226: 705-7 (1984)

Epithelial transforming growth factor: see: TGFe.

Epithelins: Epithelins are small cysteine-rich ≈ 6 kDa proteins. These proteins have been isolated

eph- and elk-related kinase: see addendum: EEK. **EPH:** see addendum.

initially from rat kidney tissue and have now been isolated also as cDNAs from murine and human tissues. Epithelin mRNA is expressed in many epithelial cell types.
The two known epithelins are ***Epithelin 1*** (56 amino acids) and ***Epithelin 2*** (57 amino acids). They display 47 % sequence homology. The positions of all 12 cysteine residues is conserved in both proteins.
Both epithelins are synthesized from a common precursor (see: PCDGF, PC cell-derived growth factor). The two proteins act as agonist and antagonist on murine keratinocytes and fibroblasts: Epithelin 1 stimulates the proliferation of these cells and Epithelin 2 inhibits Epithelin 1-induced cell proliferation. Epithelins also inhibit the proliferation of human epidermal carcinoma cell lines such as » A431 and thus function as negative growth factors. Epithelins are also thought to be involved in inflammatory processes (see: Inflammation) and » wound healing.
The ***granulins***, initially isolated from rat leukocytes and bone marrow and later also from human bone marrow, are homologues of Epithelin 1 and also have growth modulatory activity, exhibiting proliferative and antiproliferative effects on epithelial cells *in vitro*. Granulins are expressed in myelogenous leukemic cell lines of promonocytic, promyelocytic, and proerythroid lineage, in fibroblasts, and very strongly in epithelial cell lines. The widespread occurrence of granulin mRNA in cells of the hematopoietic system and in epithelia implies important functions in these tissues.

Bateman A et al Granulins, a novel class of peptide from leukocytes. BBRC 173: 1161-8 (1990); **Bhandari V et al** Isolation and sequence of the granulin precursor cDNA from human bone marrow reveals tandem cysteine-rich granulin domains. PNAS 89: 1715-9 (1992); **Bhandari V & Bateman A** Structure and chromosomal location of the human granulin gene. BBRC188: 57-63 (1992); **Plowman GD et al** The epithelin precursor encodes two proteins with opposing activities on epithelial cell growth. JBC 267: 13073-8 (1992); **Shoyab M et al** Epithelins 1 and 2: isolation and characterization of two cysteine-rich growth-modulating proteins. PNAS 87: 7912-6 (1990)

Epo: *erythropoietin* also abbrev. Ep.
■ **ALTERNATIVE NAMES:** ECSA (erythroid colony-stimulating activity); ESF (erythropoiesis stimulating factor).
■ **SOURCES:** Epo is predominantly synthesized and secreted by tubular and juxtatubular capillary endothelial and interstitial cells of the kidney. Ca. 10-15 % of the total amount of Epo comes from extrarenal sources and is predominantly produced by hepatocytes and Kupffer cells of the liver.
In the fetus the liver is the main source of Epo. It appears that the switch in the synthesis from liver to kidney which takes place after birth is genetically determined. The synthesis of erythropoietin, in liver and kidney is inducible by anemia of

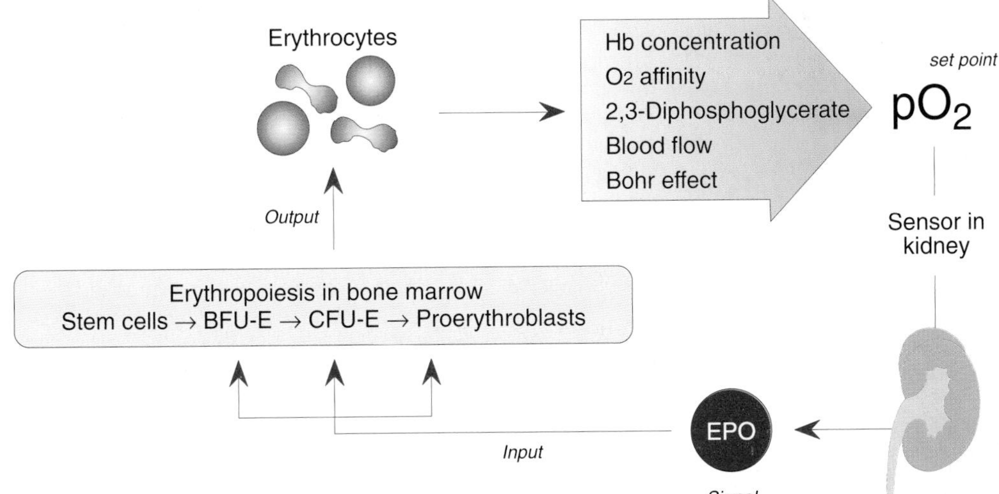

Physiological control of erythropoiesis.
The kidney functions as a sensory organ monitoring oxygen supply and demand. Depending upon the oxygen tension production of erythropoietin is either enhanced (under conditions of renal hypoxia) or reduced (under conditions of increased oxygen tension). Erythropoietin regulates generation of erythroid progenitor cells from bone marrow stem cells, their expansion, and differentiation into erythrocytes.

various origins, a fall of the arterial oxygen tension caused by either cardiopulmonary disorders or by a decrease of the oxygen tension in the inspiratory gas. Hepatic synthesis of Epo is enhanced after hepatic viral infections and exposure to hepatotoxic substances. Macrophages have also been described to produce Epo. The factor is also produced by some kidney and liver tumors, fibroleiomyomas of the uterus, cerebral hemangioblastomas, and dermoid cysts of the ovary. The synthesis of Epo *in vitro* can be inhibited in hepatoma cells and also in serum-free perfused rat kidneys by » IL1 and » TNF-α. This synthesis is not affected by » IL6, » TGF-β and » IFN-γ.

■ **PROTEIN CHARACTERISTICS:** Epo is a relative heat- and pH-stable acidic (pI = 4.5) protein of 34-37 kDa. It is N-glycosylated at Asn24, Asn36, Asn83 and O-glycosylated at Ser126. Several differently glycosylated variants are synthesized. In addition Epo is also sialylated. Epo contains two disulfide bonds (positions 7/161; 29/33). The protein is formed as a precursor of 193 amino acids which yields a mature protein of 166 amino acids. The sequences of simian, murine, and human Epo show a sequence identity of 95 % and 85 %, respectively.

Approximately 40 % of the molecular mass of Epo is due to its glycosylation. Glycosylation is an important factor determining the pharmacokinetic behavior of Epo *in vivo*. Non-glycosylated Epo has an extremely short biological half life. It still binds to its receptor and may even have a higher specific activity *in vitro*.

■ **GENE STRUCTURE:** The human Epo gene maps to chromosome 7q21-q22. It contains at least five exons. The DNA sequences of the human and the murine factor show 82 % homology. Transcriptional response of the Epo gene to hypoxia is mediated in part by promoter sequences and to a greater extent by a 24 bp sequence 3´ to the human Epo gene functioning as a hypoxia-responsive transcriptional enhancer.

■ **RECEPTOR STRUCTURE, GENE(S), EXPRESSION:** The biological activity of Epo is mediated by specific receptors expressed at 300-3000 copies/cell. The receptor is also expressed by cell types not responding to Epo. Pluripotent embryonic stem cells and early multipotent hematopoietic cells (see also: Hematopoiesis) express receptor transcripts. The commitment to non-erythroid lineages (e. g. macrophages and lymphocytes) is accompanied by the cessation of receptor expression.

The murine receptor is a protein of 507 amino acids with a single membrane-spanning domain. The cytoplasmic domain has a length of 236 amino acids. A point mutation at codon 129 of the murine Epo receptor gene results in constitutive activation. Mice expressing the aberrant receptor develop erythrocytosis and splenomegaly. Clonal growth factor-independent, proerythroblast cell lines that express Epo receptor have been isolated from the spleen of these animals.

Heterologous expression of the human cDNA in COS cells yields a protein of about 66 kD (508 amino acids). Both the cDNA and the protein sequence of the human receptor are 82% homologous to the murine receptor. The human Epo receptor gene consists of 8 exons and has a length of ≈ 6 kb. It maps to chromosome 19p13.3. It is a member of a cytokine receptor family including receptors for » Growth hormone, » IL6, and » IL2. Epo receptor signaling has recently been shown to involve the tyrosine kinase JAK2 (see: Janus kinases).

■ **SIGNAL TRANSDUCTION:** The Epo receptor is internalized after Epo has bound. The details of post receptor signal transduction processes are largely unknown. The activation of adenylate or guanylate cyclase does not seem to be involved although cAMP and cGMP may modulate receptor signals. Binding of Epo to its receptor activates phospholipases A2 and C. This leads to a release of membrane phospholipids (enhancement of lipoxygenase-mediated arachidonic acid metabolism), the synthesis of diacyl glycerol, an increase in intracellular calcium levels (see also: Calcium ionophore) and intracellular pH. Phospholipase C induction also leads to an increased expression of the » oncogenes » *fos* and » *myc*.

A point mutation in codon 129 of the murine Epo receptor has been shown to cause the constitutive activation of the receptor. Mice infected with recombinant viruses carrying the mutated receptor develop a pronounced erythrocytosis. Cells expressing the mutated receptor cause an erythroleukemia following their injection into experimental mice.

■ **BIOLOGICAL ACTIVITIES:** Human Epo is biologically active in rodents. Its synthesis is subject to a complex control circuit which links kidney and bone marrow in a feedback loop. Synthesis depends on venous oxygen partial pressure and is increased under hypoxic conditions. The oxygen sensor in the kidney is believed to be a heme protein. Epo production is also influenced by a variety of other humoral factors, including, among others, testosterone, thyroid hormone, » Growth hormone, and catecholamines. Several immunomodulatory » cytokines such as » IL1, » TNF-α and » IL6 have been shown to reduce the synthesis of Epo *in vitro*.

Epo is mainly a differentiation factor for late determined and differentiated progenitor cells of erythropoiesis. It determines their differentiation and maturation into erythrocytes. In addition Epo also regulates the proliferation of erythropoietic progenitor cells. The Epo sensitivity progressively increases from immature, but determined, » BFU-E) » CFU-E. Epo has not been shown to act on pluripotent stem cells (see also: Hematopoiesis).

The pathophysiological excess of Epo leads to erythrocytosis. This is accompanied by an increase in blood viscosity and cardiac output and may lead in some cases also to heart failure and pulmonary hypertension.

To a certain extent Epo is also a costimulator of megakaryocytopoiesis. The activity of Epo is synergised by » IL4. The suppression of erythropoiesis induced by » TNF-α can be abolished by exogenous Epo.

Epo has been shown to act as an » autocrine growth factor for certain human erythroleukemic cells. In addition, Epo is a mitogen and a chemoattractant for endothelial cells (see also: Chemotaxis). Epo also directly stimulates activated and differentiated B cells and enhances B cell immunoglobulin production and proliferation.

■ **TRANSGENIC/KNOCK-OUT/ANTISENSE STUDIES:** Studies of Epo transgenic mice have revealed that different DNA sequences flanking the Epo gene control liver versus kidney expression of the gene and that some of these sequences are located 3´ to the gene. Some 3´ flanking sequences of ≈ 50 nucleotides also function as an enhancer which can mediate transcriptional induction in response to hypoxia.

The consequences of a deregulated expression of Epo have been demonstrated in » transgenic animals (mice) expressing increased levels of human Epo in all transgenic tissues analyzed. Overexpression of Epo leads to a polycythemia and a general increase in erythrocyte, hematocrit and hemoglobin values. A significant reduction of platelets is also observed in these animals.

■ **ASSAYS:** Epo can be assayed by employing cell lines such as » HCD57, » NFS-60, » TF-1 and » UT-7, which respond to the factor (see also: Bioassays). Epo activity can also be assessed in a » colony formation assay by determining the number of » CFU-E from bone marrow cells. An alternative and entirely different detection method is » Message amplification phenotyping. For further information see also subentry "Assays" in the reference section.

■ **CLINICAL USE & SIGNIFICANCE:** Chronical kidney disease causes the destruction of Epo-producing cells in the kidney. The resulting lack of Epo frequently induces hyporegenerative normochrome normocytic anemias. The main clinical use of Epo is therefore the treatment of patients with severe kidney insufficiency (hematocrit below 0.3) who usually also receive transfusions. Renal anemia is frequently seen as a complication of terminal kidney insufficiency occurring in ≈ 50 % of dialysis patients. While dialysis is a means to overcome disturbances of water, electrolyte, and acid-base balance in uremic patients, it cannot balance the loss of endocrine kidney functions. Prolonged bleeding times have been shown to be improved by Epo treatment of uremic patients. In uremic patients on chronic maintenance hemodialysis treatment with recombinant Epo also improves platelet adhesion and aggregation in addition to and independent of its effect on the hematocrit. The most important complication in the treatment of renal anemia with Epo is hypertony. Increases in urea, potassium, and phosphate levels are also possible. An increase in blood viscosity must be considered. Iron deficiency is the main reason for insufficient response to recombinant Epo therapy. It can be overcome by concomitant intravenous iron supply.

Epo treatment has also been described to lead to an expansion of thrombopoietic progenitor cells and circulating platelets.

One important application of Epo may be the presurgical activation of erythropoiesis, allowing the collection of autologous donor blood.

The use of Epo has also been suggested for nonrenal forms of anemia induced, for example, by chronic infections, inflammatory processes, radiation therapy, and cytostatic drug treatment, and encouraging results in patients with non-renal anemia have been reported.

● **REVIEWS:** **Abels RI & Rudnick SA** Erythropoietin: evolving clinical applications. EH 19: 842-50 (1991); **Graber SE & Krantz SB** Erythropoietin: biology and clinical use. Hematol./Oncol. Clin. North Amer. 3: 369-400 (1989); **Jelkman W & Gross AJ** (eds) Erythropoietin. Springer, Berlin 1989; **Koury MJ & Bondurant MC** The molecular mechanism of erythropoietin action. EJB 210: 649-63 (1992); **Krantz SB** Erythropoietin. Blood 77: 419-34 (1991); **Tabbara IA** Erythropoietin. Biology and clinical applications. Arch. Intern. Med. 153: 298-304 (1993)

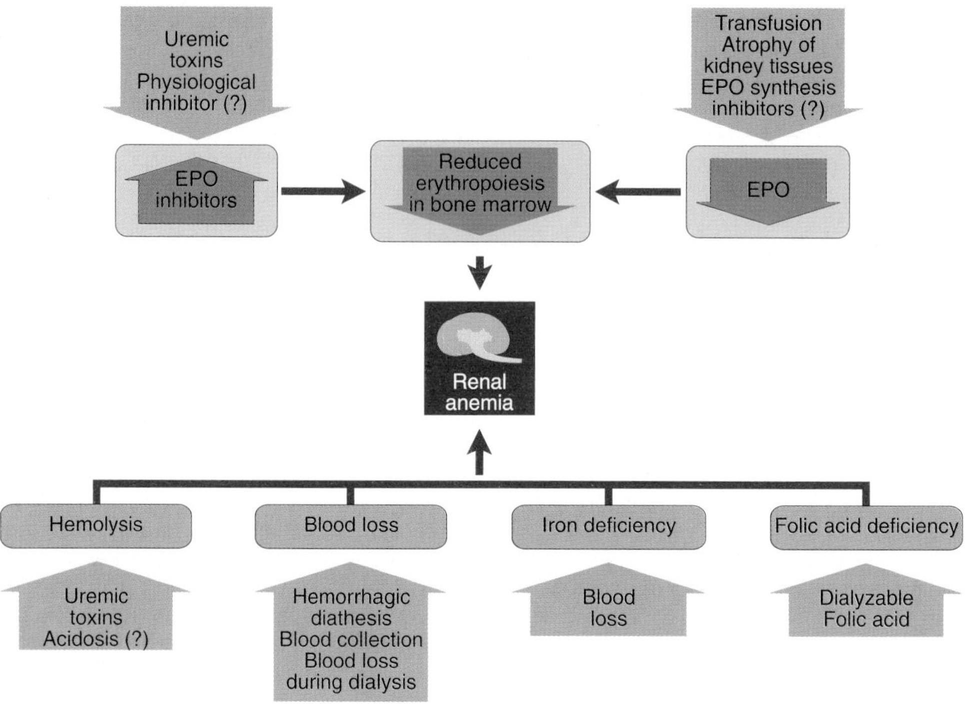

Pathogenesis of renal anemia.
Many factors, including hemolysis, blood loss, iron deficiency, and folic acid deficiency, contribute to renal anemia. Physiologically most important factor are those directly or indirectly causing a reduction in available erythropoietin.

● **BIOCHEMISTRY & MOLECULAR BIOLOGY: Browne JK et al** Erythropoietin: gene cloning, protein structure, and biological properties. CSHSQB 51: 693-702 (1986); **Goldberg MA et al** Regulation of the erythropoietin gene: evidence that the oxygen sensor is a heme protein. S 242: 1412-5 (1988); **Jacobs K et al** Isolation and characterization of genomic and cDNA clones of human erythropoietin. N 313: 806-10 (1985); **Koury ST et al** Quantitation of erythropoietin-producing cell in kidneys of mice by *in situ* hybridization: correlation with hematocrit, renal erythropoietin mRNA, and serum erythropoietin concentration. Blood 74: 645-51 (1989); **Law ML et al** Chromosomal assignment of the human erythropoietin gene and its DNA polymorphism. PNAS 83: 6920-4 (1986); **Lee-Huang S** Cloning and expression of human erythropoietin cDNA in *Escherichia coli*. PNAS 81: 2708-12 (1984); **Lin FH et al** Cloning and expression of the human erythropoietin gene. PNAS 82: 7580-4 (1985); **Lin FJ et al** Monkey erythropoietin gene: cloning, expression, and comparison with the human erythropoietin gene. Gene 44: 201-9 (1986); **Madan A & Curtin PT** A 24-base-pair sequence 3´ to the human erythropoietin gene contains a hypoxia-resp0onsive transcriptional enhancer. PNAS 90: 3928-32 (1993); **McDonald JD et al** Cloning, sequencing, and evolutionary analysis of the mouse erythropoietin gene. MCB 6: 842-8 (1986); **Recny M et al** Structural characteristics of natural human urinary and recombinant DNA-derived erythropoietin. JBC 262: 17156-63 (1987); **Sasaki H et al** Carbohydrate structure of erythropoietin expressed in Chinese hamster ovary cells by a human erythropoietin cDNA. JBC 262: 12059-76 (1987); **Semenza GL & Wang GL** A nuclear factor induced by hypoxia via *de novo* protein synthesis binds to the human erythropoietin gene enhancer at a site required for transcriptional activation. MCB 12: 5447-54 1992); **Shoemaker CB & Mitsock LD** Murine erythropoietin gene: cloning, expressing, and human gene homology. MCB 6: 849-59 (1986); **Watkins PC et al** Regional assignment of the erythropoietin gene to human chromosome region 7pter-q22. Cytogenet. Cell. Genet. 42: 214-8 (1986)

● **RECEPTORS: Bazan JF** A novel family of growth factor receptors: a common binding domain in the growth hormone, prolactin, erythropoietin, and IL6 receptors, and the p75 IL2 receptor β-chain. BBRC 164: 788-95 (1989); **D'Andrea AD et al** Expression cloning of the murine erythropoietin receptor. Cell 57: 277-85 (1989); **D'Andrea et al** The cytoplasmic region of the erythropoietin receptor contains nonoverlapping positive and negative growth- regulatory domains. MCB 11: 1980-7 (1991); **D'Andrea AD & Zon LI** Erythropoietin receptor: subunit structure and activation. JCI 86: 681-7 (1990); **D'Andrea AD et al** The cytoplasmic region of the erythropoietin receptor contains nonoverlapping positive and negative growth-regulatory domains. MCB 11: 1980-7 (1991); **Ersley AJ** Erythropoietin. NEJM 324: 1339-44 (1991); **Jones SS et al** Human erythropoietin receptor: cloning, expression, and biological characterization. Blood 76: 31-5 (1990); **Longmore GD et al** An activating mutation in the murine erythropoietin receptor induces erythroleukaemia in mice: a cytokine receptor superfamily oncogene. Cell 67: 1089-1102 (1991); **Maouche L et al** Cloning of the gene encoding the human erythropoietin receptor. Blood 78: 2557-63 (1991); **Mason-Garcia M & Beckman BS** Signal transduction in erythropoiesis. FJ 5: 2958-64 (1991); **Noguchi**

CT et al Cloning of the human erythropoietin receptor gene. Blood 78: 2548-56 (1991); **Penny LA & Forget BG** Genomic organization of the human erythropoietin receptor gene. Genomics 11: 974-80 (1991); **Winkelmann JC et al** The gene for the human erythropoietin receptor: analysis of the coding sequence and assignment on chromosome 19p. Blood 76: 24-30 (1990); **Youssoufian H et al** Genomic structure and transcription of the mouse erythropoietin receptor gene. MCB 10: 3675-82 (1990)

● **Biological Activities: Anagnostou A et al** Erythropoietin has a mitogenic and positive chemotactic effect on endothelial cells. PNAS 87: 5978-82 (1990); **Fandrey J & Jelkman WE** Interleukin 1 and tumor necrosis factor-α inhibit erythropoietin production *in vitro*. ANY 628: 250-5 (1991); **Geissler K et al** Recombinant human erythropoietin: A multipotential hemopoietic growth factor *in vivo* and *in vitro*. Contrib. Nephrol. 87: 1-10 (1990); **Gregory CJ** Erythropoietin sensitivity as a differentiation marker in the hemopoietic system. Studies of three erythropoietic colony responses in culture. JCP 89: 289-301 (1976); **Jelkman W et al** Monokines inhibiting erythropoietin production in human hepatoma cultures and in isolated perfused rat kidneys. Life Sci. 50: 301-8 (1992); **Kimata H et al** Human recombinant erythropoietin directly stimulates B cell immunoglobulin production and proliferation in serum-free medium. Clin. Exp. Immunol. 85: 151-6 (1991); **Kimata H et al** Erythropoietin enhances immunoglobulin production and proliferation by human plasma cells in a serum-free medium. Clin. Immunol. Immunopathol. 59: 495-501 (1991); **Kimata H et al** Effect of recombinant human erythropoietin on human IgE production *in vitro*. Clin. Exp. Immunol. 83: 483-7 (1991); **Koury MJ & Bondurant MC** Erythropoietin retards DNA breakdown and prevents programmed cell death in erythroid progenitor cells. S 248: 378-81 (1990); **Lim VS et al** Effect of recombinant human erythropoietin on renal function in humans. Kidney Int. 37: 131-6 (1990); **Mitjavila MT et al** Autocrine stimulation by erythropoietin and autonomous growth of human erythroid leukemic cells *in vitro*. JCI 88: 789-97 (1991)

● **Transgenic/Knock-out/Antisense Studies: Semenza GL et al** Polycythemia in transgenic mice expressing the human erythropoietin gene. PNAS 86: 2301-5 (1989); **Semenza GL et al** Human erythropoietin gene expression in transgenic mice: multiple transcription initiation sites and *cis*-acting regulatory elements. MCB 10: 930-8 (1990)

● **Assays: Andre M et al** Performance of an immunoradiometric assay of erythropoietin and results for specimens from anemic and polycythemic patients. Clin. Chem. 38: 758-63 (1992); **Hankins WD et al** Erythropoietin-dependent and erythropoietin-producing cell lines. Implications for research and for leukemia therapy. ANY 554: 21-8 (1989); **Kendall RGT et al** Storage and preparation of samples for erythropoietin radioimmunoassay. Clin. Lab. Haematol. 13: 189-96 (1991); **Krumvieh D et al** Comparison of relevant biological assays for the determination of biological active erythropoietin. Dev. Biol. Stand. 69: 15-22 (1988); **Ma DD et al** Assessment of an EIA for measuring human serum erythropoietin as compared with RIA and an in-vitro bioassay. Br. J. Haematol. 80: 431-6 (1992); **Noe G et al** A sensitive sandwich ELISA for measuring erythropoietin in human serum Br. J. Haematol. 80: 285-92 (1992); **Pauly JU et al** Highly specific and highly sensitive enzyme immunoassays for antibodies to human interleukin 3 (IL3) and human erythropoietin (EPO) in serum. Behring Inst. Mitt. 90: 112-25 (1991); **Sakata S & Enoki Y** Improved microbioassay for plasma erythropoietin based on CFU-E colony formation. Ann. Hematol. 64: 224-30 (1992); **Sanengen T et al** Immunoreactive erythropoietin and erythropoiesis stimulating factor(s) in plasma from hypertransfused neonatal and adult mice. Studies with a radioimmunoassay and a cell culture assay for erythropoietin. Acta Physiol. Scand. 135: 11-6 (1989); **Widness JA et al** A sensitive and specific erythropoietin immunoprecipitation assay: application to pharmacokinetic studies. J. Lab. Clin. Med. 119: 285-94 (1992); for further information see also individual cell lines used in individual bioassays.

● **Clinical Use & Significance: Beguin Y et al** Erythropoiesis in multiple myeloma: defective red cell production due to inappropriate erythropoietin production. Br. J. Haematol. 82: 648-53 (1992); **Blackburn ME et al** Anemia in children following cardiac transplantation: treatment with low dose human recombinant erythropoietin. Int. J. Cardiol. 36: 263-6 (1992); **Brockmöller J et al** The pharmacokinetics and pharmacodynamics of recombinant human erythropoietin in hemodialysis patients. Br. J. Clin. Pharmacol. 34: 499-508 (1992); **Brunati C et al** Circulating burst-forming-unit erythroid and the responsiveness to recombinant human erythropoietin in patients on regular hemodialytic treatment. Nephron 62: 150-4 (1992); **Cascinu S et al** Cisplatin-associated anemia treated with subcutaneous erythropoietin. A pilot study. Br. J. Cancer 67: 156-8 (1993); **Eschbach JW et al** Recombinant human erythropoietin in anemic patients with end-stage renal disease. Results of a phase III multicenter clinical trial. Ann. Intern. Med. 111: 992-1000 (1989); **Eschbach JW & Adamson JW** Guidelines for recombinant human erythropoietin therapy. Amer. J. Kidney Dis. 14 (Suppl.) 2-8 (1989); **Eschbach JW et al** The long-term effects of recombinant human erythropoietin on the cardiovascular system. Clin. Nephrol. 38: S98-103 (1992); **Foa P** Erythropoietin: Clinical applications. Acta Haematol. 86: 162-8 (1991); **Garnick MB** (edt) Erythropoietin in clinical applications. An international perspective. Marcel Dekker, New York 1990; **Goldraich I & Goldraich N** Once weekly subcutaneous administration of recombinant erythropoietin in children treated with CAPD. Adv. Perit. Dial. 8: 440-3 (1992); **Grutzmacher P et al** Effect of recombinant human erythropoietin on iron balance in maintenance hemodialysis: theoretical considerations, clinical experience and consequences. Clin-Nephrol. 38: S92-7 (1992); **Ludwig H et al** Erythropoietin treatment for chronic anemia of selected hematological malignancies and solid tumors. Ann. Oncol. 4: 161-7 (1993); **McMahon LP et al** Haemodynamic changes and physical performance at comparative levels of hemoglobin after long-term treatment with recombinant erythropoietin. Nephrol. Dial. Transplant. 7: 1199-206 (1992); **Miller CB et al** Impaired erythropoietin response to anemia after bone marrow transplantation. Blood 80: 2677-82 (1992); **Montini G et al** Pharmacokinetics and hematologic response to subcutaneous administration of recombinant human erythropoietin in children undergoing long-term peritoneal dialysis: a multicenter study. J. Pediatr. 122: 297-302 (1993); **Nagel RL et al** F reticulocyte response in sickle cell anemia treated with recombinant human erythropoietin: a double-blind study. Blood 81: 9-14 (1993); **Paganini EP & Miller T** Erythropoietin therapy in renal failure. Adv-Intern-Med. 38: 223-43 (1993); **Steegmann JL et al** Erythropoietin treatment in allogeneic BMT accelerates erythroid reconstitution: results of a prospective controlled randomized trial. Bone Marrow Transplant. 10: 541-6 (1992); **Takahashi K et al** Plasma concentrations of immunoreactive-endothelin in patients with chronic renal failure treated with recombinant human erythropoietin. Clin. Sci. 84: 47-50 (1993); **Tsukuda M et al** Clinical application of recombinant human erythropoietin for treatments in patients with head and neck cancer. Cancer Immunol. Immunother. 36: 52-6 (1993); **Zwaginga JJ et al** Treatment of uremic anemia with recombinant erythropoietin also reduces the defects in platelet adhesion and aggregation caused by uremic plasma. Thromb. Haemost. 66: 638-47 (1991)

Epstein-Barr virus inducing factor: see: EIF.

erb: *erb*A and *erb*B are » oncogenes found in the acutely transforming avian erythroblastosis virus (AEV). This virus induces lethal erythroleukemia and fibrosarcomas in chicken *in vivo* and transforms hematopoietic cells and fibroblasts *in vitro*. Cellular homologues of these genes have been isolated from the genomes of a number of vertebrates. Two distantly related classes of *erb*A genes exist in humans and mice, and multiple copies of one of the classes exist in the human genome. The cellular *erb*A gene maps to human chromosome 3p22-3p24.1. Related sequences are also found on chromosome 17q11.2-17q21. The cellular *erb*A/ TRα gene has been shown to give rise to multiple divergent receptor proteins that are identical for the first 370 amino acids and then diverge completely.

The viral *erb*B encodes a protein that is a truncated and mutated version of the cellular » EGF receptor which is a protein with intrinsic protein kinase activity that becomes activated after ligand binding. The viral gene lacks the extracellular ligand-binding domain and also some carboxy-terminal sequences. The intrinsic protein kinase activity of this aberrant receptor is expressed constitutively. Expression of this truncated receptor therefore mimics an activated form of the receptor. Since it is independent of the ligand binding to the receptor the consequence is malignant transformation of the cell. The cellular homologue of the viral *erb*B gene is now called c-*erb*B1.The *erb*B-related gene designated *erb*B-2 is identical with the » *neu* oncogene.

The cellular *erb*A gene has been shown to encode a mutated derivative of the intracellular thyroid hormone receptor, binding T3, T4, and TRIAC. Viral *erb*A is defective in binding thyroid hormones

The target cells for AEV are early erythroid burst forming units (see: BFU-E). In contrast to normal erythroid precursors (see also: Hematopoiesis), AEV-transformed erythroblasts in culture are characterized by extensive self-renewal in the absence of differentiation.

The two viral genes are examples of co-operating » oncogenes. The cellular and viral *erb*A genes have been shown to encode transcription factors that regulate the expression of thyroid hormone-responsive genes. Viral *erb*A has also been shown to selectively suppress the activity of three differentiation-specific erythrocyte genes and to inhibit the function of its normal endogenous cellular counterpart, resulting in a loss of hormone responsiveness and probably hormone-induced differentiation. Viral *erb*A enhances the transforming activity of the viral *erb*B gene and also restores the erythroblast-transforming potential of viral *erb*B mutants that have lost their capacity to transform cells. Viral *erb*A has been shown to enable a variety oncogenes encoding protein tyrosine kinases to transform erythroblasts (e. g. *src*, » *fes*, *sea*, » *fms*), all of which by themselves are unable to do so.

● Reviews: **Damm K** c-*erb*A: proto-oncogene or growth suppressor gene? ACT 59: 89-113 (1992); **Damm K** *Erb*A: tumor suppressor turned oncogene? FJ 7: 904-09 (1993); **De Nayer P** The thyroid hormone receptors: molecular basis of thyroid hormone resistance. Hormone Res. 38: 57-61 (1992); **Ghysdael J & Beug H** The leukemia oncogene v-*erb*A: a dominant negative version of ligand dependent transcription factors that regulates red cell differentiation? Cancer Surv. 14: 169-80 (1992); **Hayman M J***erb*B, growth factor receptor turned oncogene. TIG 2: 260-3 (1986); **Hayman MJ & Enrietto PJ** Cell transformation by the epidermal growth factor receptor and v-*erb*B. Cancer Cells 3: 301-7 (1991); **Maihle NJ et al** c-*erb*B and the epidermal growth-factor receptor: a molecule with dual identity. BBA 948: 287-304 (1989); **Nunez EA** The *erb*-A family receptors for thyroid hormones, steroids, vitamin D, and retinoic acid: characteristics and modulation. Curr. Opin. Cell Biol. 1: 177-85 (1989)

● Biochemistry & Function: **Debuire B et al** Sequencing the *erb*A gene of avian erythroblastosis virus reveals a new type of oncogene. S 224: 1456-9 (1984); **Downward J et al** Close similarity of epidermal growth factor receptor and v-*erb*B oncogene protein sequences. N 307: 605-8 (1983); **Flickinger TW et al** An alternatively processed mRNA from the Avian c-*erb*B gene encodes a soluble, truncated form of the receptor that can block ligand-dependent transformation. MCB 12: 883-93 (1992); **Hayman MJ & Enrietto PJ** Cell transformation by the epidermal growth factor receptor and v-*erb*B. CC 3: 302-7 (1991); **Jansson M et al** Isolation and characterization of multiple human genes homologous to the oncogenes of avian erythroblastosis virus. EJ 2: 561-5 (1983); **Maihle NJ et al** Native avian c-*erb*B gene expresses a secreted protein product corresponding to the ligand-binding domain of the receptor. PNAS 88: 1825-9 (1991); **Merlino GT et al** Structure and localization of genes encoding aberrant and normal epidermal growth factor receptor RNAs from A431 human carcinoma cells. MCB 5: 1722-34 (1985); **Pain B et al** EGF-R as a hemopoietic growth factor receptor: the c-*erb*B product is present in chicken erythrocytic progenitors and controls their self-renewal. Cell 65: 37-46 (1991); **Privalsky ML et al** The membrane glycoprotein encoded by the retroviral oncogene v-*erb*-B is structurally related to tyrosine-specific protein kinases. PNAS81: 704-7 (1984); **Sap J et al** The c-*erb*-A protein is a high-affinity receptor for thyroid hormone. N 324: 635-40 (1986); **Spurr NK et al** Chromosomal localization of the human homologues to the oncogenes *erb*A and B. EJ 3: 159-63 (1984); **Weinberger C et al** The c-*erb*-A gene encodes a thyroid hormone receptor. N 324: 641-6 (1986)

Erbstatin: see: Tyrphostins.

ERC: *erythropoietin-responsive cells* A collective term for committed » hematopoietic stem cells that are capable of producing cells of the erythroid lineage (see: Hematopoiesis). ERCs include » BFU-e and » CFU-e. Both cell types differentiate in response to » Epo.

Erf-1: This » ERG (early response gene) is the human homologue of TIS11 (see: TIS) and » cMG1. It is identical with the transcription factor NF-E1 or GF-1 expressed in megakaryocytic erythroid, and mast cell lineages and encodes a protein known as » GATA-1.

Barnard RC et al Coding sequence of ERF-1, the human homologue of Tis11b/cMG1, members of the Tis11 family of early response genes. NAR 21: 3580 (1993); **Ryo R et al** Megakaryocytic leukemia and platelet factor 4. Leuk. Lymphoma 8: 327-36 (1992)

ERG: *early response genes* General term used to describe genes (see: Gene expression) rapidly and transiently expressed by cells within 0.5 - 2 hours in response to environmental stimuli. ERG expression is low or absent in quiescent cells and transcription of these genes, unlike that of others, does not depend on *de novo* protein synthesis. Induction is unaffected by cycloheximide, suggesting that posttranslational modifications account for their activation. Expression of many ERGs is superinduced when cells are treated with an inducer plus protein synthesis inhibitors. Many of the characterized ERGs contain a sequence motif in their 3'-noncoding sequences that appears to make the message labile see: ARE, AU-rich element).
Examples of stimuli leading to ERG expression include binding of » cytokines or antigens to their receptors, treatment of cells with hormones, neurotransmitters, eicosanoids, arachidonic acid metabolites, prostaglandins, lipid ethers, » Phorbol esters » Okadaic acid, retinoic acid, cytostatic or cytotoxic drugs, protein synthesis inhibitors, cholera toxin, serum, or Ca^{2+} signals (see: calcium ionophore), cAMP signals, ionizing radiation, UV light, changes in osmotic pressure, oxygen deprivation, cellular adhesion or aggregation, membrane depolarization (synaptic activity in neuronal cells), surgery, stress, or tissue regenerative processes. ERG expression has been recorded to be induced also under such circumstances as phase-shifted circadian rhythms, exposure to brief visual or tactile experiences, or the presentation of tape-recorded songs to birds. Some of these findings suggest a role for genomic responses in neural pro-

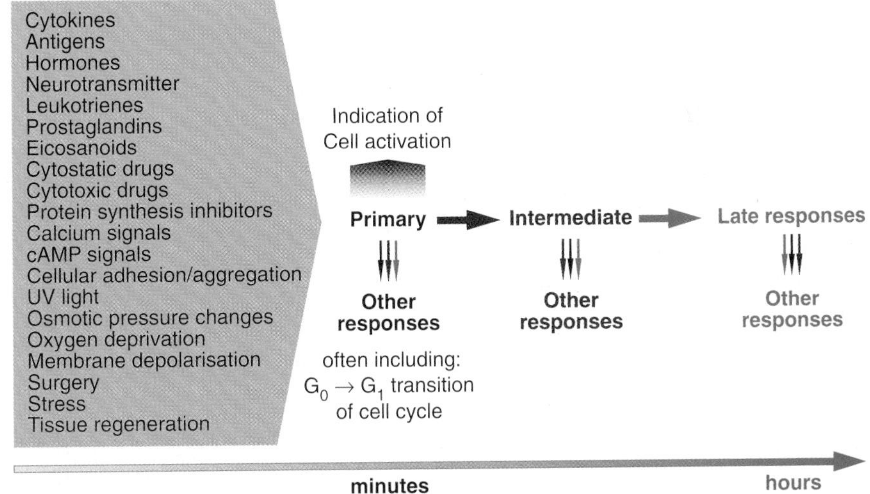

Early response gene expression and subsequent expression cascades.
Many different stimuli can lead to early gene activation processes within the cell. Examples of factors encoded by early response genes are transcription factors. Examples of intermediate genes include those encoding collagenase (metalloprotease), JE (cytokine), cathepsin L (protease), osteopontin and fibronectin (Extracellular matrix proteins), TIMP (metalloprotease inhibitor). Examples of late genes are those encoding dihydrofolate reductase and thymidine kinase (nucleotide metabolism), and histone H4 (chromatin structure). The three vertical colored arrows indicating other responses also indicate that there may not be a clear-cut distinction between early, intermediate, and late responses in that, for example, late response genes elicited by one of the triggers in one cell type may be an early response gene activated by another cytokine in the same or another cell type. See text for details.

cesses linked to pattern recognition, discrimination, or the formation of auditory/tactile/visual associations.

Early responses of cells to external stimuli are unusually complex and probably involve several hundred ERGs (for some examples of individual genes see: 3CH134, A20, B12, B61, Egr-1, *fos*, fra-1, GATA-1, *jun*, krox, *myb*, *myc*, pip92, *rel*, ST2, TSC-22; for groups of genes see: CRG, MyD, IRS, NGFI, TIS genes, TSG genes). This allows for great plasticity in cellular responses to the primary stimulus. However, a relatively small set of common primary response genes is frequently induced in a variety of cellular responses. Many ERGs are expressed in response to the same cytokines so that it must be assumed that cytokine-*specific* intracellular responses occur further downstream. Some ERGs appear to be expressed only in response to a more restricted sets of cytokines. In addition, some ERGs have been shown to be expressed in a cell-type or tissue-specific manner (for a technique allowing identification of ERGs, subtractive hybridization, see: Gene libraries).

ERGs often function to multiply and propagate the initial (gene activation) process. Many ERGs have been shown to be nuclear transcriptional activators or repressors. This subset of early response genes is called ***immediate early response genes*** or ***primary response genes*** and several of these genes are encoded by » oncogenes. ERGs encoding transcription factors are believed to act as intracellular "third messengers" that couple receptor-generated signals to activation-associated changes in gene expression, often forming large transcriptional complexes with a variety of other transcription factors and accessory proteins at so-called response elements within the promoters of the genes the transcription of which they modulate (as examples see:: ARRE, CLE, CRE, GRE, IRS, MRE, SRE, TRE). ERG transcription factors activate ***delayed-early response*** genes and ***late response genes***.

Other ERGs have been shown to encode protein kinases, other cytokines (see: JE, KC, MCP-1), receptors, cytoskeletal, and extracellular matrix genes, or specific enzymes (e. g. myeloperoxidase in human myeloid leukemic cells). Many ERGs encode so-called competence factors (see: Cell cycle) which allow the entry of resting cells into the cell cycle. Various stimuli lead to ERG expression without inducing cell proliferation. ERGs are believed to play a role in altering the physiologic status of cells in response to cytokines. Some ERGs are expressed during cellular differentiation. Detection of one or more expressed ERGs is an indication of » Cell activation and, in the absence of other recognizable features, indicates, for example, that a cytokine under study, has some effects on the cells being investigated because it elicits signal transduction. Aberrant expression of some oncogenes (e. g. » *abl*, or *ras*) can disrupt or deregulate normal ERG responses and thus can play an important role in cellular transformation processes. The proper temporal and/or spatial patterns of expression of ERGs and other genes thus regulates the phenotypes of cells.

Some viruses, such as HTLV are known to contain genes that induce constitutive expression of a high proportion of early response genes and this is one strategy by which they subvert the cellular machinery and exploit it to their own advantage (for other strategies see also: viroceptor).

The characterization of ERGs and the dissection of molecular pathways leading to differential and/or combinatorial expression of ERG subsets provide important insights into how cells respond to diverse extracellular signals. ERGs are critical parts of a complex signaling cascade with converging and/or diverging pathways that link environmental changes with intracellular responses. ERG expression creates regulatory hierarchies and provides independent and/or interrelated regulatory mechanisms that can confer distinct and interactive developmental functions. Clarification of how distinct sets of genes contribute to the problem and/or solution of pathological situations can be expected to guide efforts to diagnose and treat pathological conditions. The importance of ERGs is underlined by the observation that many of these genes are evolutionarily conserved (for a technique allowing establishment of ERG expression profiles see: Message amplification phenotyping).

Bravo R et al Growth factor-responsive genes in fibroblasts. Cell Growth Differ. 1: 305-9 (1990); **Cochran BH** Regulation of immediate early gene expression. NIDA Res. Monogr. 125: 3-24 (1993); **Farrar WL et al** Haemopoietic growth factor regulation of protein kinases and genes associated with cell proliferation. Ciba Found. Symp. 148: 127-37 (1990); **Herschman HR** Primary response genes induced by growth factors and tumor promoters. ARB 60: 281-319 (1991); **McMahon SB & Monroe JG** Role of primary response genes in generating cellular responses to growth factors. FL 6: 2707-15 (1992); **Mohn KL et al** The immediate-early growth response in regenerating liver and insulin-stimulated H-35 cells: comparison with serum-stimulated 3T3 cells and identification of 41 novel immediate-early genes. MCB 11: 381-90 (1991)

Mack KJ & Mack PA Induction of transcription factors in somatosensory cortex after tactile stimulation. Brain Res. Mol.

Brain Res. 12: 141-7 (1992); **Mello CV et al** Song presentation induces gene expression in the songbird forebrain. PNAS 89: 6818-22 (1992); **Rusak B et al** Circadian variation in photic regulation of immediate-early gene mRNAs in rat suprachiasmatic nucleus cells. Brain Res. Mol. Brain Res. 14: 124-30 (1992); **Rosen KM et al** Brief visual experience induces immediate early gene expression in the cat visual cortex. PNAS 89: 5437-41 (1992)

ERK: *extracellular signal-regulated kinases* A family of highly conserved cytosolic serine/threonine protein kinases also known as ***MAP kinases*** (mitogen-activated protein kinases). They are involved in second messenger mechanisms and are activated and tyrosine- or threonine-phosphorylated in response to a wide variety of extracellular signals transduced via ligand binding to receptors with tyrosine kinase activity (many of which are also cytokine receptors). ERKs are activated by MEKs (Map kinases or ERK kinases) and deactivated by specific phosphatases (see: CL100).

ERK protein kinases have been shown to be phosphorylated on tyrosine and threonine in response to mitogens, and apparently phosphorylate exogenous substrates on serine and/or threonine. This family of kinases probably serves as intermediates that depend on tyrosine phosphorylation to activate serine/threonine phosphorylation cascades. Individual family members participate in a protein kinase cascade leading to cell growth and differentiation and probably mediate specific responses in different developmental stages, in different cell types, or following exposure to different extracellular signals. ERKs control a broad spectrum of cellular events but are particularly known for their possible roles in » cell cycle progression and the control of cell division.

MAP kinases have been shown to phosphorylate and activate ribosomal protein S6 kinase while being themselves regulated by a cytoplasmic factor (MAP kinase activator).

Ahn NG et al The mitogen-activated protein kinase activator. Curr. Opin. Cell Biol. 4: 992-9 (1992); **Boulton TG & Cobb MH** Identification of multiple extracellular signal-regulated kinases (ERKs) with antipeptide antibodies. Cell. Regul. 2: 357-71 (1991); **Boulton TG et al** ERKs: a family of protein-serine/threonine kinases that are activated and tyrosine phosphorylated in response to insulin and NGF. Cell 65: 663-75 (1991); **Crews CM et al** Erks: their fifteen minutes has arrived. Cell Growth Differ. 3: 135-42 (1992); **Nishida E & Gotoh Y** Mitogen-activated protein kinase and cytoskeleton in mitogenic signal transduction. Int. Rev. Cytol. 138: 211-38 (1992); **Owaki H et al** Extracellular signal-regulated kinases in T cells: characterization of human ERK1 and ERK2 cDNAs. BBRC 182: 1416-22 (1992); **Pelech SL & Sanghera JS** Mitogen-activated protein kinases: versatile transducers for cell signaling. TIBS 17: 233-8 (1992); **Zheng CF & Guan KL** Cloning and characterization of two distinct human extracellular signal-regulated kinase activator kinases, MEK1 and MEK2. JBC 268: 11435-9 (1993)

Erythroid burst promoting activity: see: BPA (burst promoting activity).

Erythrocyte colony stimulation factor: see: E-CSF.

Erythroid colony-stimulating activity: see: ECSA.

Erythroid colony stimulating factor: see: ECSF.

Erythroid differentiation factor: see: EDF.

Erythroid potentiating activity: see: EPA.

Erythroid potentiating factor(s): abbrev. EPF. See: EPA (erythroid potentiating activity).

Erythroid promoting activity: see: EPA.

Erythropoiesis stimulating factor: see: ESF.

Erythropoietin: see: Epo.

Erythropoietin-responsive cells: see: ERC.

ESAF: *endothelial cell stimulating angiogenesis factor* see: HUAF (human uterine angiogenesis factor).

ESC-CEB derived endothelial cell growth factor: This factor, produced by embryonic stem cells (ESC) and cystic embryoid bodies (CEB) is identical with » aFGF. See also: ES cells.

Risau W et al Vasculogenesis and angiogenesis in embryonic-stem-cell-derived embryoid bodies. Development 102: 471-8 (1988)

ES cell differentiation inhibitory activity: see: DIA (differentiation inhibitory activity).

ES cell growth factor: see: DIA (differentiation inhibiting activity).

ES cells: *embryonic stem cells* These cells are non-transformed pluripotent cells derived from the inner cell masses of normal late blastocysts, which give rise to all organs of the growing fetus including the germ line. The establishment of permanent ES cell lines has been facilitated by coculture with » feeder cells (usually irradiated fibroblasts) or culture in medium conditioned by established teratocarcinoma stem cell lines and a number of other cells. These cells can also be propagated without a feeder cell layer in the presence of » DIA (differentiation inhibiting activity) shown to be identical with » LIF (leukemia inhibitory factor). LIF prevents spontaneous differentiation of the cells in culture and maintains their full developmental potential. ES cells retain their totipotent characteristics even after prolonged culture and can be induced to differentiate into practically all cell types, including germ line cells. The differentiation observed after removal of the factor and also treatment with inducing agents such as retinoic acid or 3-methoxybenzamide is frequently accompanied by the generation of complex cystoid embryoid bodies in which endodermal, mesodermal, and ectodermal formations can be detected.

Pluripotent stem cells can also be obtained from teratocarcinomas, tumors of ectodermal origin that originate from primordial germ cells (***embryonal carcinoma cells***, ***EC cells***). The pluripotency of these cells is proven by the ability of subclones derived from isolated single cells to differentiate into a wide variety of cell types and by the formation of teratocarcinomas when injected into mice.

ES and EC cells provide a population of normal pluripotent embryonic cells that can easily be manipulated. They are used as models for studies of cellular differentiation because it is assumed that the factors produced and secreted by these cells and also factors that induce biological activities in these cells are also important for the control of early embryonic development which, for various reasons, is intractable to direct conventional biochemical analysis. ES cells have been used, for example, to develop *in vitro* systems to study T and B cell development and to study the differentiation and generation in culture of multiple hematopoietic lineages (see also: Hematopoiesis).

The full developmental potential of the ES cells is also demonstrated by the development of chimaeric animals following injection of these cells into blastocysts, transfer of the manipulated blastocysts into pseudo-pregnant recipients, and development to term. ES cells originally introduced into blastocysts take part in the formation of all tissues, including the germ line. Such observations also imply that these cells remain responsive to all the regulatory cues which govern mammalian development.

Genetically manipulated ES cells harboring foreign DNA can be used to generate lines of true-breeding » transgenic animals. Simple coculture of pluripotent ES cells with morulae-stage embryos has also been shown to result in aggregates that also yield viable chimaeras at a similar frequency to that of blastocyst injection. Some recently developed ES cell lines allow derivation of completely cell culture-derived mice.

The use of ES cells in combination with the homologous recombination technology, which allows the targeted disruption of genes, offers the possibility of altering ES cells in a controlled manner and therefore of generating » transgenic animals with a predetermined genome. This approach has already been successful in generating loss-of-function mutations (null mutations) in » transgenic mice, lacking, e. g. expression of cytokine genes or genes involved in cytokine-mediated intracellular signal transduction pathways. While many of these studies have identified the essential role of the genes in question for cellular processes one of the results of such studies is the unexpected observation that in several instances these animals develop almost normally (see: subentry "Transgenic/Knock-out/Antisense studies" for individual factors).

Askew GR et al Site-directed point mutations in embryonic stem cells: a gene-targeting tag-and-exchange strategy. MCB 13: 4115-24 (1993); **Bradley A et al** Formation of germ-line chimaeras from embryo-derived teratocarcinoma cell lines. N 309: 255-6 (1984); **Bradley A & Robertson E** Embryo-derived stem cells: a tool for elucidating the developmental genetics of the mouse. Curr. Top. Dev. Biol. 20: 357-71 (1986); **Bradley A et al** Genetic manipulation of the mouse via gene targeting in embryonic stem cells. Ciba Found. Symp. 165: 256-69; discussion 269-76 (1992); **Brown DG et al** Criteria that optimize the potential of murine embryonic stem cells for *in vitro* and *in vivo* developmental studies. In Vitro Cell. Dev Biol. 28A: 773-8 (1992); **Capecchi MR** Altering the genome by homologous recombination. S 244: 1288-92 (1989); **Capecchi MR** The new mouse genetics: altering the genome by gene targeting. TIG 5: 70-6 (1989); **Chen U** Differentiation of mouse embryonic stem cells to lympho-hematopoietic lineages *in vitro*. Dev. Immunol. 2: 29-50 (1992); **Chen U** Careful maintenance of undifferentiated mouse embryonic stem cells is necessary for their capacity to differentiate to hematopoietic lineages *in vitro*. CTMI 177: 3-12 (1992); **Darrow AL et al** Maintenance and use of F9 teratocarcinoma cells. MiE 190: 110-7 (1990); **Doetschman T et al** Targeted correction of a mutant HPRT gene in mouse embryonic stem cells. N 330: 576-8 (1987); **Doetschman TC** Gene targeting in embryonic stem cells. BT 16: 89-101 (1991); **Evans MJ & Kaufman MH** Pluripotential cells grown directly from normal mouse embryos. Cancer Surveys 2: 185-206 (1983); **Evans MJ & Kaufman MH** Establishment in culture of pluripotential cells from mouse embryos. N 292: 154-6 (1981); **Friedrich G & Soriano P** Promoter traps in embryonic stem cells: a genetic screen to identify and mutate developmental genes in mice. Genes Dev. 5: 1513-23 (1991); **Fung-Leung WP & Mak TW** Embryonic stem cells and homologous recombination. Curr. Opin. Immunol. 4: 189-94 (1992); **Gossler A et al** Transgenesis by means of blastocyst-derived embryonic stem cell lines. PNAS 83: 9065-9 (1986); **Gutierrez-Ramos JC & Palacios R** *in vitro* differentiation of embryonic stem cells into lymphocyte precursors able to generate T and B lymphocytes *in vivo*. PNAS 89: 9171-5 (1992); **Hasty P et al** Target frequency and integration pattern for insertion and replacement vectors in embryonic stem cells. MCB 11: 4509-17 (1991); **Hasty P et al** The length of homology required for gene targeting in embryonic stem cells. MCB 11: 5586-91 (1991); **Joyner AL et al** The gene trap approach in embryonic stem cells: the potential for genetic screens in mice. Ciba Found. Symp. 165: 277-88; discussion 288-97 (1992); **Keller G et al** Haematopoietic commitment during embryonic stem cell differentiation in culture. MCB 13: 473-86 (1993); **Martin GR** Isolation of a pluripotent cell line from early mouse embryos cultured in medium conditioned by teratocarcinoma stem cells. PNAS 78: 7634-8 (1981); **Matsui Y et al** Derivation of pluripotential embryonic stem cells from murine primordial germ cells in culture. Cell 70: 841-7 (1992); **McClanahan T et al** Hematopoietic growth factor receptor genes as markers of lineage commitment during *in vitro* development of hematopoietic cells. Blood 81: 2903-15 (1993); **Mombaerts P et al** Creation of a large genomic deletion at the T cell antigen receptor β-subunit locus in mouse embryonic stem cells by gene targeting. PNAS 88: 3084-7 (1991); **Mortensen RM et al** Embryonic stem cells lacking a functional inhibitory G-protein subunit (α i2) produced by gene targeting of both alleles. PNAS 88: 7036-40 (1991); **Mortensen RM et al** Production of homozygous mutant ES cells with a single targeting construct. MCB 12: 2391-5 (1992); **Mummery CL et al** Characteristics of embryonic stem cell differentiation: a comparison with two embryonal carcinoma cell lines. Cell Differ. Dev. 30: 195-206 (1990); **Nagy A et al** Derivation of completely cell culture-derived mice from early passage embryonic cells. PNAS 90: 8424-8 (1993); **Nichols J et al** Establishment of germ-line-competent embryonic stem (ES) cells using differentiation inhibiting activity. Development 110: 1341-8 (1990); **Pascoe WS et al** Genes and functions: trapping and targeting in embryonic stem cells. BBA 1114: 209-21 (1992); **Pease S et al** Isolation of embryonic stem (ES) cells in media supplemented with recombinant leukemia inhibitory factor (LIF). Dev. Biol. 141: 344-52 (1990); **Pease S et al** Formation of germ-line chimeras from embryonic stem cells maintained with recombinant leukemia inhibitory factor. ECR 190: 209-11 (1990); **Reid LH et al** Cotransformation and gene targeting in mouse embryonic stem cells. MCB 11: 2769-77 (1991); **Robertson E et al** Germ-line transmission of genes introduced into cultured pluripotential cells by retroviral vector. N 323: 445-8 (1986); **Robertson EJ** (edt) Teratocarcinomas and embryonic stem cells: a practical approach. IRL Press, Oxford 1987; **Robertson EJ** Using embryonic stem cells to introduce mutations into the mouse germ line. Biol. Reprod. 44: 238-45 (1991); **Rossant J et al** Genome manipulation in embryonic stem cells. Phil. Trans. R. Soc. Lond. Biol. 339: 207-15 (1993); **Rubinstein M et al** Introduction of a point mutation into the mouse genome by homologous recombination in embryonic stem cells using a replacement type vector with a selectable marker. NAR 21: 2613-7 (1993); **Schmitt RM et al** Hematopoietic development of embryonic stem cells *in vitro*: cytokine and receptor gene expression. Genes Dev. 5: 728-40 (1991); **Smith AG et al** Inhibition of pluripotential embryonic stem cell differentiation by purified polypeptides. N 326: 197-200 (1989); **Smith AG** Culture and differentiation of embryonic stem cells. J. Tiss. Cult. Meth. 13: 89-94 (1991); **Smith AG** Mouse embryo stem cells: their identification, propagation, and manipulation. Semin. Cell Biol. 3: 385-99 (1992); **Snodgrass HR** Embryonic stem cells and *in vitro* hematopoiesis. JCBc 49: 225-30 (1992); **Stewart CL et al** Expression of foreign genes from retroviral vectors in mouse teratocarcinoma chimaeras. EJ 4: 3701-9 (1985); **te Riele H et al** Highly efficient gene targeting in embryonic stem cells through homologous recombination with isogenic DNA constructs. PNAS 89: 5128-32 (1992); **Thomas KR & Capecchi MR** Site-directed mutagenesis by gene targeting in mouse embryo-derived stem cells. Cell 51: 503-12 (1987); **Thomas KR et al** High-fidelity gene targeting in embryonic stem cells by using sequence replacement vectors. MCB 12: 2919-23 (1992); **Valancius V & Smithies O** Double-strand gap repair in a mammalian gene targeting reaction. MCB 11: 4389-97 (1991); **Valancius V & Smithies O** Testing an "in-out" targeting procedure for making subtle genomic modifications in mouse embryonic stem cells. MCB 11: 1402-8 (1991); **Watt FM** Cell culture models of differentiation. FJ 5: 287-94 (1991); **Wiles MV & Keller G** Multiple hematopoietic lineages develop from embryonic stem (ES) cells in culture. Development 111: 259-67 (1991); **Wood SA et al** Non-injection methods for the production of embryonic stem cell embryo chimaeras. N 365: 87-9 (1993); **Zheng H et al** Fidelity of targeted recombination in human fibroblasts and murine embryonic stem cells. PNAS 88: 8067-71 (1991); **Zimmer A** Manipulating the genome by homologous recombination in embryonic stem cells. ARN 15: 115-37 (1992)

ESCGF: *embryonic stem cell growth factor* see: DIA (differentiation inhibiting activity).

Escherichia coli: abbrev. *E. coli*. A small rod-shaped facultative anaerobic Gram-negative bacterium named after its discoverer, the German physician Theodor Escherich. *E. coli* has a length of ≈ 2-4 µm and a diameter of ≈1 µm. Most wild-type strains are motile although most laboratory strains do not possess flagellae. *E. coli* is a member of the larger family *Enterobacteriaceae* and is thus closely related to *Salmonella typhimurium*, *Shigella spp*, *Klebsiella*, and the plant pathogen *Erwinia spp*. *E. coli* can be identified by specific biochemical and physiological characteristics and by specific antigenic reactions (see also: Endotoxin).

E. coli is found in the human intestinal tract and constitutes ≈ 1 % of all microorganisms that can be cultivated from human faeces (≈10^7 bacteria/gram of faeces). It can survive for some time in the normal environment. *E. coli* is usually not pathogenic but it is a potential pathogen elsewhere in the body. Some strains of *E. coli* are cause urinary tract infections, bacteremia, peritonitis, and neonatal meningitis.

Work with *E. coli* has many advantages for molecular biologists. *E. coli* can be cultivated on simple media containing mineral salts, ammonium salts as a source of nitrogen, and simple sugars that serve as carbon sources. In rich media supplemented with undefined meat or yeast extracts and/or protein hydrolysates *E. coli* has a doubling time of

≈ 30 min. This yields so-called overnight cultures containing ≈ 3×10^9 bacteria per mL.

Discoveries such as the exchange of genetic material between bacterial cells by means of plasmids, and numerous experiments with bacterial viruses (bacteriophages) are milestones in the history of molecular biology and bacterial genetics. Work with bacteria, plasmids, and bacterial viruses has been of decisive importance in developing and refining the concept of genes (see also: Gene expression), and in elucidating the genetic code, mechanisms of DNA replication and DNA repair, and protein biosynthesis and numerous biochemical reaction pathways. This work has also been the foundation of modern techniques of genetic engineering.

Today *E. coli* is the best characterized microorganism. Many thousands of different and well-characterized mutants are known. Its genome consists of ≈ 4.6 kb of circular double-stranded DNA and ≈ 70 % of it has been sequenced. More than 1500 genes have been identified, and ≈ 80% of its biochemical reactions are known.

E. coli is still one of the standard organisms used to clone and express foreign genes as a result of its ease of manipulation and the quantity of protein obtainable. E. coli can be manipulated to synthesize more than 50 % of its total cell protein as a recombinant product. *E. coli* has been used to study heterologous expression of many cytokine genes and other proteins of interest although not all proteins are amenable to high-level expression. Several features of heterologous coding sequences have been shown to be responsible for limited translation efficiency, affecting both translation initiation and elongation. Moreover, high-level production of proteins often correlates with poor quality, leading, for example, to variable amounts of N-terminal heterogeneity. Unlike eukaryotic expression systems *E. coli* cannot be used to express foreign genes with introns (see also: Gene expression). In addition, *E. coli* does not possess the molecular machinery allowing correct post-translational processing of foreign proteins. Nevertheless, many of the bacteria-derived cytokines have been found to be biologically active in spite of the absence of, for example, glycosylation. For mammalian expression systems used in cytokine research see also: BHK cells, CHO cells, COS cells, Namalwa cells, Baculovirus expression system).

● **Basics: Neidhardt FC et al.** (edt) *Escherichia coli* and *Salmonella typhimurium*. Cellular and molecular biology. Vol. I, II. American Society for Microbiology. Washington, D. C.1987; **Miller JH** (edt) Experiments in molecular genetics. Cold Spring Harbor Laboratory, Cold Spring Harbor, New York, 1972; **Olins PO & Lee SC** Recent advances in heterologous gene expression in Escherichia coli. Curr. Opin. Biotechnol. 4: 520-525 (1993); **Stader JA & Silhavy TJ** Engineering *Escherichia coli* to secrete heterologous gene products. MiE 185: 166-87 (1990)

● **Expression of Cytokines and Other Proteins: Arakawa T et al** Acid unfolding and self-association of recombinant *Escherichia coli* derived human interferon γ. B 26: 5428-32 (1987); **Arcone R et al** Single-step purification and structural characterization of human interleukin-6 produced in *Escherichia coli* from a T7 RNA polymerase expression vector. EJB 198: 541-7 (1987); **Barone AD et al** The expression in *Escherichia coli* of recombinant human platelet factor 4, a protein with immunoregulatory activity. JBC 263: 8710-5 (1988); **Bergonzoni L et al** Characterization of a biologically active extracellular domain of fibroblast growth factor receptor 1 expressed in *Escherichia coli*. EJB 210: 823-29 (1988); **Brakenhoff JP et al** Analysis of human IL6 mutants expressed in *Escherichia coli*. Biologic activities are not affected by deletion of amino acids 1-28. JI 143: 1175-82 (1989); **Cantrell AS et al** Effects of second-codon mutations on expression of the insulin-like growth factor-II-encoding gene in *Escherichia coli*. Gene 98: 217-23 (1991); **Casagli MC et al** Different conformation of purified human recombinant interleukin 1 β from *Escherichia coli* and *Saccharomyces cerevisiae* is related to different level of biological activity. BBRC 162: 357-63 (1989); **Chanda PK et al** Expression of human interleukin-2 receptor cDNA in *E. coli*. BBRC 141: 804-11 (1986); **Chen AB et al** Quantitation of *E. coli* protein impurities in recombinant human interferon-γ. Appl. Biochem. Biotechnol. 36: 137-52 (1992); **Czarniecki CW et al** *In vitro* biological activities of *Escherichia coli*-derived bovine interferons-α, -β, and -γ. J. Interferon. Res. 6: 29-37 (1986); **Daumy GO et al** Isolation and characterization of biologically active murine interleukin-1 α derived from expression of a synthetic gene in *Escherichia coli*. BBA 998: 32-42 (1989); **DeLamarter JF et al** Recombinant murine GM-CSF from *E. coli* has biological activity and is neutralized by a specific antiserum. EJ 4: 2575-81 (1985); **Denefle P et al** Heterologous protein export in *Escherichia coli*: influence of bacterial signal peptides on the export of human interleukin 1 β. Gene 85: 499-510 (1989); **Devlin PE et al** Alteration of amino-terminal codons of human granulocyte-colony-stimulating factor increases expression levels and allows efficient processing by methionine aminopeptidase in *Escherichia coli*. Gene 65: 13-22 (1988); **Dicou E et al** Expression of recombinant human nerve growth factor in *Escherichia coli*. Neurochem. Int. 20: 129-34 (1992); **Dorr RT** Clinical properties of yeast-derived versus *Escherichia coli*-derived granulocyte-macrophage colony-stimulating factor. Clin. Ther. 15: 19-29 (1993); **Duffaud GD et al** Expression and secretion of foreign proteins in *Escherichia coli*. MiE 153: 492-507 (1987); **Dykes CW et al** Expression of atrial natriuretic factor as a cleavable fusion protein with chloramphenicol acetyltransferase in *Escherichia coli*. EJB 174: 411-6 (1988); **Fountoulakis M et al** Purification and biochemical characterization of a soluble human interferon γ receptor expressed in *Escherichia coli*. JBC 265: 13268-75 (1990); **Frigeri LG et al** Expression of biologically active recombinant rat IgE-binding protein in *Escherichia coli*. JBC 265: 20763-9 (1990); **Graber P et al** Purification, characterization and crystallization of selenomethionyl recombinant human interleukin-5 from *Escherichia coli*. EJB 212: 751-5 (1993); **Grenett HE et al** Isolation and characterization of biologically active murine interleukin-6 produced in *Escherichia coli*. Gene 101: 267-71 (1991); **Gronski P et al** *E. coli* derived human granulocyte-macrophage colony-stimulating factor (rh GM-CSF) available for clinical trials. Behring Inst. Mitt. 83:

246-9 (1988); **Guisez Y et al** High-level expression, purification, and renaturation of recombinant murine interleukin-2 from *Escherichia coli*. Protein Expr. Purif. 4: 240-6 (1993); **Honda S et al** Differential purification by immunoaffinity chromatography of two carboxy-terminal portion-deleted derivatives of recombinant human interferon-γ from *Escherichia coli*. J. Interferon Res. 7: 145-54 (1987); **Hou YT et al** Efficient expression of unfused human α D-interferon in *Escherichia coli* using overlapping termination and initiation codons (TGATG) in its signal sequence. J. Interferon Res. 6: 437-43 (1986); **Huang JJ et al** High-level expression in *Escherichia coli* of a soluble and fully active recombinant interleukin-1 β. Mol. Biol. Med. 4: 169-81 (1987); **Hummel M et al** Gene synthesis, expression in *Escherichia coli* and purification of immunoreactive human insulin-like growth factors I and II. Application of a modified HPLC separation technique for hydrophobic proteins. EJB 180: 555-61 (1989); **Ito H et al** Molecular cloning and expression in *Escherichia coli* of the cDNA coding for rabbit tumor necrosis factor. DNA 5: 149-56 (1986); **Iwane M et al** Expression of cDNA encoding human basic fibroblast growth factor in *E. coli*. BBRC 146: 470-7 (1987); **Jayaram B et al** Purification of human interleukin-4 produced in *Escherichia coli*. Gene 79: 345-54 (1989); **Jerala R et al** Cloning a synthetic gene for human stefin B and its expression in *E. coli*. FL 239: 41-4 (1988); **Joseph Liauzun E et al** Human recombinant interleukin-1 β isolated from *Escherichia coli* by simple osmotic shock. Gene 86: 291-5 (1990); **Jung G et al** High-cell density fermentation studies of recombinant *Escherichia coli* strains expressing human interleukin-1 β. Ann. Inst. Pasteur. Microbiol. 139: 129-46 (1988); **Kashima N et al** Expression of murine interleukin-2 cDNA in *E. coli* and biological activities of recombinant murine interleukin-2. J. Biochem. Tokyo 102: 715-24 (1987); **Ke Y et al** A rapid procedure for production of human basic fibroblast growth factor in *Escherichia coli* cells. BBA 1131: 307-10 (1992); **Ke YQ et al** High-level production of human acidic fibroblast growth factor in *E. coli* cells: inhibition of DNA synthesis in rat mammary fibroblasts at high concentrations of growth factor. BBRC 171: 963-71 (1990); **Kennett RH et al** Detection of *E. coli* colonies expressing the v-sis oncogene product with monoclonal antibodies made against synthetic peptides. JIM 85: 169-82 (1985); **Kikumoto Y et al** Purification and characterization of recombinant human interleukin-1 β produced in *Escherichia coli*. BBRC 147: 315-21 (1987); **Knoerzer W et al** Expression of synthetic genes encoding bovine and human basic fibroblast growth factors (bFGFs) in *Escherichia coli*. Gene 75: 21-30 (1989); **Knott JA et al** The isolation and characterization of human atrial natriuretic factor produced as a fusion protein in *Escherichia coli*. EJB 174: 405-10 (1988); **Langley KE et al** Purification and characterization of soluble forms of human and rat stem cell factor recombinantly expressed by *Escherichia coli* and by Chinese hamster ovary cells. ABB 295: 21-8 (1992); **Libby RT et al** Expression and purification of native human granulocyte-macrophage colony-stimulating factor from an *Escherichia coli* secretion vector. DNA 6: 221-9 (1987); **Lin LS et al** Purification of recombinant human interferon β expressed in *Escherichia coli*. MiE 119: 183-92 (1986); **Lindley I et al** Synthesis and expression in *Escherichia coli* of the gene encoding monocyte-derived neutrophil-activating factor: biological equivalence between natural and recombinant neutrophil-activating factor. PNAS 85: 9199-203 (1988); **Lorberboum Galski H et al** Cytotoxic activity of an interleukin 2-Pseudomonas exotoxin chimeric protein produced in *Escherichia coli*. PNAS 85: 1922-6 (1988); **Lu HS et al** Folding and oxidation of recombinant human granulocyte colony stimulating factor produced in *Escherichia coli*. Characterization of the disulfide-reduced intermediates and cysteine-serine analogs. JBC 267: 8770-7 (1992); **Lu HS et al** Identification of unusual replacement of methionine by norleucine in recombinant interleukin-2 produced by *E. coli*. BBRC 156: 807-13 (1988); **McDonald JR et al** Expression and characterization of recombinant human ciliary neurotrophic factor from *Escherichia coli*. BBA 1090: 70-80 (1991); **Metcalf D et al** *In vitro* actions on hemopoietic cells of recombinant murine GM-CSF purified after production in *Escherichia coli*: comparison with purified native GM-CSF. JCP 128: 421-31 (1986); **Miyamoto C et al** Molecular cloning and regulated expression of the human c-*myc* gene in *Escherichia coli* and *Saccharomyces cerevisiae*: comparison of the protein products. PNAS 82: 7232-6 (1985); **Moghaddam A & Bicknell R** Expression of platelet-derived endothelial cell growth factor in *Escherichia coli* and confirmation of its thymidine phosphorylase activity. B 31: 12141-6 (1992); **Moschera JA et al** Purification of recombinant human fibroblast interferon produced in *Escherichia coli*. MiE 119: 177-83 (1986); **Negro A et al** Synthesis and purification of biologically active rat brain-derived neurotrophic factor from *Escherichia coli*. BBRC 186: 1553-9 (1992); **Nilsson B et al** Expression and purification of recombinant insulin-like growth factors from *Escherichia coli*. MiE 198: 3-16 (1991); **Oka T et al** Synthesis and secretion of human epidermal growth factor by *Escherichia coli*. PNAS 82: 7212-6 (1985); **Pandey KN & Kanungo J** Expression of extracellular ligand-binding domain of murine guanylate cyclase/atrial natriuretic factor receptor cDNA in *Escherichia coli*. BBRC 190: 724-31 (1993); **Proudfoot AE et al** Preparation and characterization of human interleukin-5 expressed in recombinant *Escherichia coli*. BJ 270: 357-61 (1990); **Revel M et al** Biological activities of recombinant human IFN-β 2/IL6 (*E. coli*). ANY 557: 144-55 (1989); **Rose K et al** Human interleukin-5 expressed in *Escherichia coli* has N-terminal modifications. BJ 286: 825-8 (1992); **Rosenberg SA et al** Biological activity of recombinant human interleukin-2 produced in *Escherichia coli*. S 223: 1412-4 (1984); **Schoenfeld HJ et al** Efficient purification of recombinant human tumor necrosis factor β from *Escherichia coli* yields biologically active protein with a trimeric structure that binds to both tumor necrosis factor receptors. JBC 266: 3863-9 (1991); **Shapiro R et al** Expression of Met-(-1) angiogenin in *Escherichia coli*: conversion to the authentic less than Glu-1 protein. AB 175: 450-61 (1988); **Shatzman AR & Rosenberg M** Expression, identification, and characterization of recombinant gene products in *Escherichia coli*. MiE 152: 661-73 (1987); **Simons G et al** High-level expression of human interferon γ in *Escherichia coli* under control of the pL promoter of bacteriophage lambda. Gene 28: 55-64 (1984); **Squires CH et al** Production and characterization of human basic fibroblast growth factor from *Escherichia coli*. JBC 263: 16297-302 (1988); **Steinkasserer A et al** Human interleukin-1 receptor antagonist. High yield expression in *E. coli* and examination of cysteine residues. FL 310: 63-5 (1992); **Su X et al** Equine tumor necrosis factor α: cloning and expression in *Escherichia coli*, generation of monoclonal antibodies, and development of a sensitive enzyme-linked immunosorbent assay. Hybridoma 11: 715-27 (1992); **Su X et al** Production of recombinant porcine tumor necrosis factor α in a novel *E. coli* expression system. Biotechniques 13: 756-62 (1992); **Takamatsu H et al** Expression and purification of biologically active human OSF-1 in *Escherichia coli* BBRC 185: 224-30 (1992); **Tanabe Y et al** Purification of recombinant human tumor necrosis factor precursor from *Escherichia coli*. BBRC 179: 683-8 (1991); **Tanaka H & Kaneko T** Pharmacokinetic and pharmacodynamic comparisons between human granulocyte colony-stimulating factor purified from human bladder carcinoma cell line 5637 culture medium and recombinant human granulocyte colony-stimulating factor produced in *Escherichia coli*. J. Pharmacol. Exp. Ther. 262: 439-44 (1992); **Tocci MJ et al** Expression in *Escherichia coli* of fully active recombinant human IL 1 β: comparison with native human IL 1 β. JI 138: 1109-14 (1987); **Tsai LB et al** Control of misincorporation of de novo synthesized norleucine into recombinant interleukin-2 in *E. coli*. BBRC 156: 733-9 (1988); **Tsuji Y et al** Production in *Escherichia coli* of human thymosin β 4 as chimeric protein with human tumor necrosis factor. BI 18: 501-8 (1989); **Tsukui K et al** Two monoclonal antibodies distinguish between human recombinant interferon-α 5s produced by *Escherichia coli* and by mouse cells. Microbiol. Immunol. 30: 1271-9 (1986); **Upton FZ et al** Production and characterization of recombinant chicken insulin-like growth factor-I from *Escherichia coli*. J. Mol. Endocrinol. 9: 83-92 (1992); **Utsumi J et al** Characterization of *E. coli*-derived recombinant human interferon-β as compared with fibroblast human interferon-β. J. Biochem-Tokyo. 101: 1199-208 (1987); **van Kimmenade A et al** Expression, renaturation and purification of recombinant human interleukin 4 from *Escherichia coli*. EJB 173: 109-14 (1988); **Vandenbroeck K et al** Refolding and single-step purification of porcine interferon-γ from *Escherichia coli* inclusion bodies. Conditions for reconstitution of dimeric IFN-γ. EJB 215: 481-6 (1993); **Watanabe T et al** Molecular characterization of recombinant human acidic fibroblast growth factor produced in *E. coli*: comparative studies with human basic fibroblast growth factor. Mol. Endocrinol. 4: 869-79 (1990); **Windsor WT et al** Analysis of the conformation and stability of *Escherichia coli* derived recombinant human interleukin 4 by circular dichroism. B 30: 1259-64 (1991); **Wingfield P et al** Purification and characterization of human interleukin-1 α produced in *Escherichia coli*. EJB 165: 537-41 (1987); **Winkler ME et al** The purification of fully active recombinant transforming growth factor α produced in *Escherichia coli*. JBC 261: 13838-43 (1986); **Yamada T et al** Separation of recombinant human interleukin-2 and methionyl interleukin-2 produced in *Escherichia coli*. BBRC 135: 837-43 (1986); **Yamazaki S et al** Determination of the complete amino acid sequence of recombinant human γ-interferon produced in *Escherichia coli*. J. Interferon Res. 6: 331-6 (1986); **Yem AW et al** A two step purification of recombinant human interleukin-1 β expressed in *E. coli*. Immunol. Invest. 17: 551-9 (1988); **Zazo M et al** High-level synthesis in *Escherichia coli* of shortened and full-length human acidic fibroblast growth factor and purification in a form stable in aqueous solutions. Gene 113: 231-8 (1992); **Zucker K et al** Production and characterization of recombinant canine interferon-γ from *Escherichia coli*. J. Interferon Res. 13: 91-7 (1993)

ESF: *erythropoiesis stimulating factor* This factor is identical with » Epo. Sometimes the term refers to plasma erythropoietin. The term is also used frequently as a generic term for all factors that stimulate erythropoiesis measured in » bioassays, including Epo.

One particular factor with ESF activity has been found in the plasma of an anephric patient with nearly normal hematocrit. It is identical with IGF I (see: IGF) which may replace » Epo as a stimulator of erythropoiesis in some patients with anemia and renal failure.

Congote LF et al The N-terminal sequence of the major erythropoietic factor of an anephric patient is identical to insulin-like growth factor 1. J. Clin. Endocrinol. Metab. 72: 727-9 (1991); **Hellebostad M et al** Variations in erythropoiesis throughout a lifetime. Studies in a high-leukaemic mouse strain, the AKR/O strain, and a non-leukaemic strain, the WLO strain. Blut 61: 358-68 (1990); **Hellebostad M et al** Leukemia and anemia in AKR/O mice. I. Leukemia characteristics, hematological variables and erythropoiesis stimulating factor(s). APMIS 99: 869-78 (1991); **Meberg A et al** Plasma erythropoietin levels in mice during the growth period. Br. J. Haematol. 45: 569-74 (1980)

ESP: *eosinophil stimulation promoter* This activity is found in mitogen- or antigen-stimulated murine spleen cells infected with *Schistosoma mansoni*. It is also produced by granulomas induced by deposited eggs. The ESP activity is the result of two synergizing factors, namely » GM-CSF and » IL5.

Secor WE et al Eosinophils and immune mechanisms. VI. The synergistic combination of granulocyte-macrophage colony-stimulating factor and IL5 accounts for eosinophil-stimulation promoter activity in *Schistosoma mansoni*-infected mice. JI 144: 1484-9 (1990); **Potter K & Leid RW** A review of eosinophil chemotaxis and function in *Taenia taeniformis* infections in the laboratory rat. Vet. Parasitol. 20: 103-16 (1986)

Established cell lines see: Cell culture. See also individual entries.

ET: *endothelin*

■ **ALTERNATIVE NAMES:** EDCF (endothelium-derived contracting factor). See also: individual entry for further information.

■ **SOURCES:** Endothelins Are produced and secreted by endothelial cells and epithelial cells. The synthesis of ET-2 by human kidney carcinoma cells is decreased by » EGF. The synthesis of endothelins by vascular endothelial cells is increased by bacterial » endotoxins, » bombesin, thrombin, » IL1, » TGF-β, » PDGF, by vasoactive hormones such as » angiotensin and » vasopressin, by ET-1 itself, and even by hemodynamic shear forces or hypoxia. The synthesis of ET-1 is stimulated by » Activin A and inhibited by » Follistatin. A strong inducer of endothelin synthesis in mesangial cells is » TNF.

■ **PROTEIN CHARACTERISTICS:** Endothelins are a family of closely related peptides of 21 amino acids with two disulfide bonds. The four known species are isoforms encoded by four different genes. They are called ***ET-1***, ***ET-2***, ***ET-3*** and »***VIC*** (vasoactive intestinal contractor = Endothelin β = rat ET-2).

```
ET-1       CSCSSLMDKECVYFCHLDIIW
ET-2       CSCSSWLDKECVYFCHLDIIW
ET-3       CTCFTYKDKECVYYCHLDIIW
VIC        CSCNDWLDKECVYFCHLDIIW
SFTX S6b   CSCKDMTDKECLYFCHQDVIW
```

Sequences of some endothelins.
Conserved cystein residues are shown in yellow. Amino acid positions identical in at least four proteins are shown in blue.

Endothelins arise in several steps by processing of a preproprotein of ca 212 amino acids (preproET). These contain the sequences of the ET peptides and also highly conserved cysteine-rich ET-like sequences of 15 amino acids.

The immediate precursors (pro-endothelins; 38 amino acids) are also known as BIG (big endothelins). The enzyme responsible for processing the precursors is EC 3. 4. 24. 11. It is a membrane metalloendopeptidase also called ECE (endothelin-converting enzyme). This enzyme also processes other precursor proteins, for example those encoding enkephalins and » SP (substance P; see also: Tachykinins).

■ **Gene Structure:** The human ET-1 gene (EDN1) maps to chromosome 6p24-p23. It yields a 2,026-nucleotide mRNA, excluding the poly(A) tail, encoded in 5 exons distributed over 6,836 base pairs. At least 2 preproendothelin-1 mRNAs, differing in the 5´ untranslated region, are produced from a single gene by use of different promoters. The analysis of the tissue distribution of the two mRNAs shows a tissue-type specificity for one mRNA in brain and heart tissues.

The ET-2 gene (EDN2) is at 1p34, and the ET-3 (EDN2) gene maps to 20q13.2-q13.3.

■ **Related Factors:** At the protein level endothelins show a very high degree of homology to four peptide toxins found in snake venom (*Atractaspis engaddensis*) and to a-scorpion toxin which is known to activate voltage-gated sodium channels. The four snake venom proteins are called **sarafotoxins** (abbrev. SRTXa, b, c, d). ET-1 also displays some structural homology with Apamin from bee venom, which has also been shown to act on ion channels, and to conotoxin isolated from certain fish species.

■ **Receptor Structure, Gene(s), Expression:** The known endothelin receptors (≈ 48 kDa; 442 amino acids) are strongly glycosylated glycoproteins which show a high degree of sequence homology. They belong to the group of receptors coupled to G proteins.

The human ET(A) receptor gene spans more than 40 kb and contains 8 exons. It maps to chromosome 4. The ET(B) receptor gene is on chromosome 13. It spans 24 kb and comprises 7 exons. Every intron occurs near the border of the putative transmembrane domain in the coding region. Binding of ligands to the receptor protein induces the hydrolysis of phophoinositol. Binding of endothelins to their receptors also induces the transcription of some » oncogenes, including » *fos* and » *myc*.

Two distinct subclasses of ET receptors, namely, ET1-specific and ET-nonselective, are known. The ET1-specific type is called ET(A) and the nonselective type ET(B). One endothelin receptor isolated from bovine lung tissues shows 90 % sequence homology with a receptor isolated from rat vascular smooth muscle cells. A second type of receptor, isolated from rat lung tissue, shows only 55 % sequence homology with the rat vascular smooth muscle cell receptor.

Some of the endothelin receptors appear to be specific for a particular type of endothelin. Rat receptors isolated from cardiac muscle and uterus bind

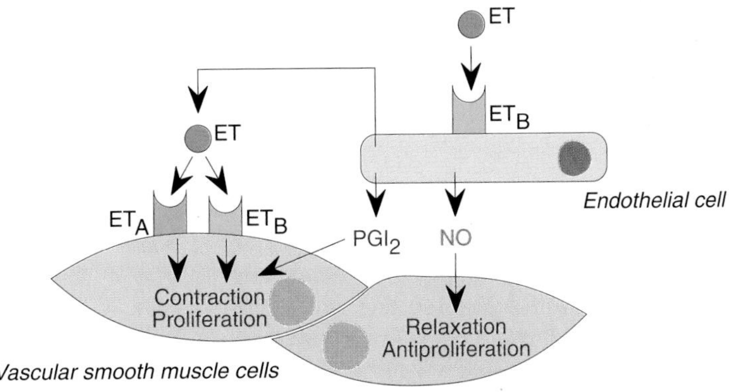

Vascular effects of endothelins.
Endothelin (ET), described previously as EDCF (endothelium-derived contracting factor) can activate receptors on vascular smooth muscle cells (ETA and ETB receptors) to elicit contraction and proliferation. By binding to its endothelial receptor (ETB) ET can cause synthesis of nitric oxide (NO) and prostacyclin (PGI2). NO can cause relaxation and possibly antiproliferative effects by a mechanism involving activation of cGMP. Prostacyclin stimulates the formation of cAMP and augments contraction/proliferation.

ET-1 and SRTX-b equally well with high affinity and ET-3 with low affinity. bET-1 (big ET-1) also binds to endothelin receptors, albeit with markedly reduced affinity.

■ **Biological Activities:** Endothelins are mainly known for their pharmacological activities. They are the most potent vasoconstrictors known. The vasoconstrictor activity of bET-1 (big ET-1) is ≈ 100-fold lower than that of ET-1 and it is not known whether this protein has a physiological role. Endothelins are potent positively inotropic and chronotropic agents for the myocardium. They increase the plasma levels of a number of vasoactive hormones.

The intravenous injection of ET or SRTX in mice is lethal. Endothelins increase the secretion of ANP (atrial natriuretic peptide), aldosterone and catecholamines. They reduce renal blood flow and glomerular filtration. ET-1 may be an important » autocrine factor in the regulation of inner medullary collecting duct (IMCD) physiology in the nephron. IMCD cells synthesize ET-1, express endothelin receptors, and respond to exogenous ET-1 by reducing $Na(^{+})$-$K(^{+})$-ATPase activity and water transport. The production of ET-1 by epithelia, smooth muscle, and fibroblasts of the urinary bladder suggests that it may act as an » autocrine hormone in the regulation of the bladder wall structure and smooth muscle tone.

In addition to their vasoconstrictor actions, endothelins have effects on the central nervous system and on neuronal excitability. They induce the depolarization of spinal neurons, the release of » vasopressin and oxytocin, and may be involved in the response of glial cells to tissue injuries.

Apart from their pharmacological activities endothelins also show some activities of » cytokines although they are usually not classified as cytokines due to their small sizes. Endothelins act as » paracrine modulators of smooth muscle cells and connective tissue cell functions. In human mammary carcinomas endothelins are » paracrine growth factors for mammary stromal cells. The synthesis of endothelins in these tumors is induced by » GRP (gastrin-releasing peptide) and glucocorticoids. Endothelin have also been shown to function as a growth-promoting factor for human thyroid epithelial cells (thyrocytes), presumably through intracellular calcium influx (see also: Calcium ionophore).

ET-1 induces the synthesis of » TGF-β, » PDGF, » thrombospondin, » ANF (atrial natriuretic factor). Endothelins stimulate the proliferation of smooth muscle cells, fibroblasts, mesangial cells, established glioma cells lines and primary astroglial cells. Endothelins are » autocrine growth factors for several human tumor cell lines. ET-1 secreted from human keratinocytes is a mitogen for human melanocytes and may play an essential role in the UV hyperpigmentation in the epidermis. The influences of ET-1 on DNA replication and » cell cycle are synergised by » EGF, » PDGF, » bFGF, and » TGF-α. ET-1 and ET-2 in combination with » EGF promote the growth of some cell types in soft agar (see also: colony formation assay) and this growth-promoting activity is almost as pronounced as that of » TGF-β. Endothelins are also involved in the regulation of » Hematopoiesis, with cells of the megakaryoblast/megakaryocyte lineage appearing to be the main targets.

■ **Clinical Use & Significance:** The detection of endothelins in the lung tissues of asthmatic patients suggests that these factors may play an important pathophysiological role as bronchoconstrictors. Elevated plasma concentrations of endothelins have been observed in hypertension, myocardial infarction, cardiogenic shock, Raynaud syndrome, and Crohn's disease.

● **Reviews: Leppaluoto J & Ruskoaho H** Endothelin peptides: biological activities, cellular signaling and clinical significance. Ann. Med. 24: 153-61 (1992); **Miller RC et al** Endothelins - from receptors to medicine. TIPS 14: 54-60 (1993); **O'Halloran DJ et al** Neuropeptides synthesized in the anterior pituitary: possible paracrine role. Mol. Cell. Endocrinol. 75: C7-C12 (1991); **Rubanyi GM & Parker Botelho LH** Endothelins. FJ 5: 2713-20 (1991); **Rubanyi GM** (edt) Endothelin. Oxford Univ. Press, New York, 1991; **Simonson MS & Dunn MJ** Endothelin peptides and the kidney. ARP 55: 249-65 (1993); **Sokolovsky M et al** Endothelins and sarafotoxins: physiological regulation, receptor subtypes and transmembrane signaling. TIBS 16: 261-4 (1991); **Sokolovsky M et al** Structure-function relationships of endothelins, sarafotoxins, and their receptor subtypes. J. Neurochem. 59: 809-21 (1992); **Opgenorth TJ et al** Endothelin-converting enzymes. FASEB J. 6: 2653-9 (1992)

● **Biochemistry & Molecular Biology: Arinami T et al** Chromosomal assignments of the human endothelin family genes: the endothelin-1 gene (EDN1) to 6p23- p24, the endothelin-2 gene (EDN2) to 1p34, and the endothelin-3 gene (EDN3) to 20q13.2-q13.3. Am. J. Hum. Genet. 48: 990-6 (1991); **Benatti L et al** Two preproendothelin 1 mRNAs transcribed by alternative promoters. JCI 91: 1149-56 (1993); **Bloch KD et al** Structural organization and chromosomal assignment of the gene encoding endothelin. JBC 264: 10851-7 (1989); **Bloch KD et al** cDNA cloning and chromosomal assignment of the gene encoding endothelin 3. JBC 264: 18156-61 (1989); **Brown MR et al** Transforming growth factor-β: Role in mediating serum-induced endothelin production by vascular endothelial cells. E 129: 2355-60 (1991); **Ehrenreich H et al** Selective autoregulation of endothelins in primary astrocyte cultures: endothelin receptor-mediated potentiation of endothelin-1 secretion. The New Biologist 3: 135-41 (1991); **Giaid A et al** Endothelin 1, an endothelium-derived peptide, is expressed in neurons of the human spinal cord and dorsal root ganglia. PNAS 86: 7634-8 (1989);

Inoue A et al The human preproendothelin-1 gene: complete nucleotide sequence and regulation of expression. JBC 264: 14954-9 (1989); **Inoue A et al** The human endothelin family: three structurally and pharmacologically distinct isopeptides predicted by three separate genes. PNAS 86: 2863-7 (1989); **Itoh Y et al** Cloning and sequence analysis of cDNA encoding the precursor of a human endothelium-derived vasoconstrictor peptide, endothelin. identity of human and porcine endothelin. FL 231: 440-4 (1988); **Lin HY et al** Cloning and functional expression of a vascular smooth muscle endothelin 1 receptor. PNAS 88: 3185-9 (1991); **Ohkubo S et al** Specific expression of human endothelin-2 (ET-2) gene in a renal adenocarcinoma cell line: molecular cloning of cDNA encoding the precursor of ET-2 and its characterization. FL 274: 136-40 (1990); **Saida K et al** A novel peptide, vasoactive intestinal contractor, of a new (endothelin) peptide family. JBC 264: 14613-6 (1989); **Springall DS et al** Endothelin immunoreactivity of airway epithelium in asthmatic patients. Lancet 337: 697-701 (1991); **Tokito F et al** Epidermal growth factor (EGF) decreases endothelin-2 (ET-2) production in human renal adenocarcinoma cells. FL 295: 17-21 (1991); **Williams DL et al** Sarafotoxin S6c: An agonist which distinguishes between endothelin receptor subtypes. BBRC 175: 556-61 (1991); **Yanagisawa M et al** A novel potent vasoconstrictor peptide produced by vascular endothelial cells. N 332: 411-5 (1988); **Yanagisawa M & Masaki T** Endothelin, a novel endothelium-derived peptide. Biochem. Pharmacol. 38: 1877-83 (1989)

● **Receptors: Arai H et al** Cloning and expression of a cDNA encoding an endothelin receptor. N 348: 730-2 (1990); **Arai H et al** The human endothelin-B receptor gene: structural organization and chromosomal assignment. JBC 268: 3463- 70 (1993); **Culouscou JM et al** Biochemical analysis of the epithelin receptor. JBC 268: 10458-62 (1993); **Cyr C et al** Cloning and chromosomal localization of a human endothelin ETA receptor. BBRC 181: 184-90 (1991); **Hosoda K et al** Organization, structure, chromosomal assignment, and expression of the gene encoding the human endothelin-A receptor. JBC 267: 18797-804 (1992); **Nakamuta M et al** Cloning and sequence analysis of a cDNA encoding human non-selective type of endothelin receptor. BBRC 177: 34- 9 (1991); **Ogawa Y et al** Molecular cloning of a non-isopeptide- selective human endothelin receptor. BBRC 178: 248-55 (1991); **Takayanagi R et al** Multiple subtypes of endothelin receptors in porcine tissues: characterization by ligand binding, affinity labeling and regional distribution. Regul. Pept. 32: 23-37 (1991)

● **Biological Activities: Battistini B et al** Growth regulatory properties of endothelins. Peptides 14: 385-99 (1993); **Brown MR et al** Transforming growth factor β: role in mediating serum-induced endothelin production by vascular endothelial cells. E 129: 2355-60 (1991); **Diochot S et al** Effects of endothelins on the human megakaryoblastic cell line MEG-01. Eur. J. Pharmacol. 227: 427-31 (1992); **Eguchi K et al** Stimulation of mitogenesis in human thyroid epithelial cells by endothelin. Acta Endocrinol. Copenh. 128: 215-20 (1993); **Ehrenreich H et al** Endothelins belong to the assortment of mast cell-derived and mast cell-bound cytokines. New Biol. 4: 147-56 (1992); **Fabregat I & Rozengurt E** Vasoactive intestinal contractor, a novel peptide, shares a common receptor with endothelin-1 and stimulates Ca^{2+} mobilization and DNA synthesis in Swiss 3T3 cells. BBRC 167: 161-7 (1990); **Fantoni G et al** Endothelin-1: a new autocrine/paracrine factor in rat testis. Am. J. Physiol. 265: E267-74 (1993); **Fujii Y et al** Endothelin as an autocrine factor in the regulation of parathyroid cells. PNAS 88: 4235-9 (1991); **Hahn AW et al** Stimulation of endothelin mRNA and secretion in rat vascular smooth muscle cells: a novel autocrine function. Cell Regul. 1: 649-59 (1990); **Imokawa G et al** Endothelins secreted from human keratinocytes are intrinsic mitogens for human melanocytes. JBC 267: 24675-80 (1992); **Ito H et al** Endothelin-1 is an autocrine/paracrine factor in the mechanism of angiotensin II-induced hypertrophy in cultured rat cardiomyocytes. JCI 92: 398-403 (1993); **Kohan DE & Padilla E** Endothelin-1 is an autocrine factor in rat inner medullary collecting ducts. Am. J. Physiol. 263: F607-12 (1992); **Kusuhara M et al** Stimulation of anchorage-independent cell growth by endothelin in NRK 49F cells. CR 52: 3011-4 (1992); **Peacock AJ et al** Endothelin-1 and endothelin-3 induce chemotaxis and replication of pulmonary artery fibroblasts. Am. J. Respir. Cell. Mol. Biol. 7: 492-9 (1992); **Saenz de Tejada I et al** Endothelin in the urinary bladder. I. Synthesis of endothelin-1 by epithelia, muscle and fibroblasts suggests autocrine and paracrine cellular regulation. J. Urol. 148: 1290-8 (1992); **Schrey MP et al** Bombesin and glucocorticoids stimulate human breast cancer cells to produce endothelin, a paracrine mitogen for breast stromal cells. CR 52: 1786-90 (1992); **Shichiri M et al** Endothelin-1 is an autocrine/paracrine growth factor for human cancer cell lines. JCI 87: 1867-71 (1991); **Simonson MS & Dunn MJ** Cellular signaling by peptides of the endothelin gene family. FJ 4: 2989-3000 (1990); **Yeh YC et al** Synergistic effects of endothelin-1 (ET-1) and transforming growth factor α (TGF-α) or epidermal growth factor (EGF) on DNA replication and G1 to S phase transition. Biosci. Rep. 11: 171-80 (1991); **Zachary I et al** Bombesin, vasopressin, and endothelin rapidly stimulate tyrosine phosphorylation in intact Swiss 3T3 cells. PNAS 88: 4577-81 (1991)

ETAF: *epidermal cell-derived thymocyte activating factor* ETAF appears to be a biological activity rather than a distinct factor. It is detected in a thymocyte costimulation assay and has been identified in, and isolated from, epidermal cells (***EC-ETAF***) and squamous cell carcinomas (***SCC-ETAF***).

ETAF has a multiplicity of divergent biologic effects. It induces a peripheral neutrophilia and *in vitro* is chemotactic for neutrophils and a potent T cell chemoattractant. It induces muscle proteolysis *in vitro*. ETAF-containing preparations have also been shown to stimulate hepatocytes *in vitro* to synthesize » acute phase proteins (see also: Acute phase reaction). ETAF has growth-promoting effects on keratinocytes and fibroblasts.

The major factors responsible for ETAF activity are the two molecular forms of » IL1 as shown by the fact that a monoclonal antibody directed against IL1α and IL1β blocks ETAF activity. The ETAF activity isolated from keratinocytes has occasionally been referred to as » IL1k. A costimulation assay reveals that » IL6 and » TNF-α also possess ETAF activity.

Goldminz D et al Keratinocyte membrane-associated epidermal cell-derived thymocyte activating factor (ETAF). J. Invest. Dermatol. 88: 97-100 (1987); **Granstein RD et al** *In vivo* inflammatory activity of epidermal cell-derived thymocyte activating factor and recombinant interleukin 1 in the mouse. JCI 77: 1020-7 (1986); **Horowitz MC et al** Human keratinocytes contain mRNA indistinguishable from monocyte interleukin-1 and mRNA keratinocyte epidermal cell-derived thymocyte activating factor is identical to interleukin-1. J. Cell. Med. 164: 2095-100 (1986); **Kupper TS et al** Human keratinocytes contain mRNA indistinguishable from monocyte interleukin 1 α and β mRNA. Keratinocyte epidermal cell-derived thymocyte-activating factor is identical to interleukin 1. JEM 164: 2095-2100 (1986); **Luger TA et al** Murine epidermal cell-derived thymocyte activating factor resembles murine interleukin 1. JI 128: 2147-52 (1982); **Luger TA et al** Chemotactic properties of partially purified human epidermal cell-derived thymocyte-activating factor (ETAF) for polymorphonuclear and mononuclear cells. JI 131: 816-20 (1983); **Luger TA et al** Characteristics of an epidermal cell thymocyte activating factor, ETAF, produced by human epidermal cells and a human squamous cell line. J. Invest. Dermatol. 81: 187-93 (1983); **Luger TA et al** Monoclonal anti-IL1 is directed against a common site of human IL1 α and IL1 β. Immunobiology 172: 346-56 (1986); **Sauder DN et al** Epidermal cell production of thymocyte activating factor (ETAF). J. Invest. Dermatol. 79: 34-9 (1982); **Sauder DN et al** Chemotactic cytokines: the role of leukocytic pyrogen and epidermal cell thymocyte-activating factor in neutrophil chemotaxis. JI 132: 828-32 (1984); **Sauder DN et al** Biologic properties of epidermal cell thymocyte-activating factor (ETAF). J. Invest. Dermatol. 85: 176s-9s (1985); **Sauder DN** Epidermal cell thymocyte-activating factor. Adv. Dermatol. 1: 237-46 (1986); **Sztein MB et al** An epidermal cell-derived cytokine triggers the *in vivo* synthesis of serum amyloid A by hepatocytes. JI 129: 87-90 (1982)

ETF: *EGFR-specific transcription factor* A transcription factor of 120 kDa originally identified by its ability to specifically stimulate expression of » EGF receptors. A binding site for ETF has also been identified in the promoter region of the gene encoding the receptor for » IGF-I and in the » ret gene promoter. ETF recognizes various GC-rich sequences in the promoter regions of various other genes and to appears to be a specific transcription factor for promoters which do not contain TATA box elements.

Itoh F et al Identification and analysis of the ret proto-oncogene promoter region in neuroblastoma cell lines and medullary thyroid carcinomas from MEN2A patients. O 7: 1201-6 (1992); **Kageyama R et al** A transcription factor active on the epidermal growth factor receptor gene. PNAS 85: 5016-20 (1988); **Kageyama R et al** Nuclear factor ETF specifically stimulates transcription from promoters without a TATA box. JBC 264: 15508-14 (1989); **Mamula PW & Goldfine ID** Cloning and characterization of the human insulin-like growth factor-I receptor gene 5'-flanking region. DNA Cell Biol. 11: 43-50 (1992)

ETR103: This » ERG (early response gene) is expressed in the course of macrophagic differentiation of » HL-60 cells stimulated by » Phorbol ester TPA. It is identical with » Egr-1.

Shimizu N et al This » ERG (early response gene) A gene coding for a zinc finger protein is induced during 12-O-tetradecanoylphorbol-13-acetate-stimulated HL-60 cell differentiation. J. Biochem. Tokyo 111: 272-7 (1992)

EUA: *extra unnecessary acronym* At least one EUA will be perpetuated forever in our memories by one of the finest acronyms ever coined within the biomedical community: see: UGF. For another EUA see: FOG.

Expression libraries: see: Gene libraries.

Extracellular matrix: see: ECM.

Extracellular signal-regulated kinases: see: ERK.

Eye-derived growth factor 1: see: EDGF 1.

Eye-derived growth factor 2: see: EDGF 2.

F

F1: *factor 1* This factor is produced by the human B cell line, Karpas 160. It is identical with » TNF-β.
Ni J et al Production and induction of a novel cytotoxin (factor 2) by a human B cell line. Cytokine 4: 305-12 (1992)

F2: *factor 2* This factor is produced by the human B cell line, Karpas 160. Its synthesis can be enhanced by treatment of the cells with PMA or sodium butyrate. This poorly characterized factor of 10-14 and 60-70 kDa is cytotoxic for a number of human tumor cell lines, including the human erythroleukemic cell line K562 which does not respond to » TNF. The biological activity cannot be neutralized by antibodies directed against » TNF, » TGF-β, » IFN-α and » IFN-γ. See also: F1.
Ni J et al Production and induction of a novel cytotoxin (factor 2) by a human B cell line. Cytokine 4: 305-12 (1992); **Ni J et al** Characterization and partial purification of a novel cytotoxic lymphokine (factor 2) produced by a human B cell line (Karpas 160). Int. Immunol. 4: 519-31 (1992); **Ni J & Karpas A** Relationship between a novel human cytotoxin (factor 2) produced by a B cell line (Karpas 160) and phorbol-myristate-acetate-associated cytotoxicity. Clin. Exp. Immunol. 88: 360-6 (1992); **Ni J & Karpas A** Isolation of a novel cytotoxic lymphokine (factor 2) from a human B cell line (Karpas 160b) by preparative isoelectric focusing in the rotofor cell and chromatofocusing. Cytokine 5: 31-7 (1993)

F5MIF-2: *macrophage migration inhibitory factor 2* This 45 kDa factor is secreted by some human T cell hybridomas. It inhibits the migration of macrophages. The factor is not identical with any other cytokine, including other migration inhibition factors (see also: MIF).
Hirose S et al Macrophage migration inhibitory factor (MIF) produced by a human T cell hybridoma clone. Microbiol. Immunol. 35: 235-45 (1991); **Oki S et al** Macrophage migration inhibitory factor (MIF) produced by a human T cell hybridoma clone. Lymphokine Cytokine-Res. 10: 273-80 (1991)

FABP: *fatty acid-binding protein* see: MDGI (mammary-derived growth inhibitor).

F-ACT1: This transcription factor, which binds to a sequence motif within the » DRE (serum response element), is identical with » GATA-1.
Lee TC et al Displacement of BrdUrd-induced YY1 by serum response factor activates skeletal α-actin transcription in embryonic myoblasts. PNAS 89: 9814-8 (1992)

Factor 1: see: F1.

Factor 2: see: F2.

Factor replacing macrophages: see: FRM.

FAF: *fibroblast activating factor* This is an operational definition of biological activities, secreted by various cells, which activate fibroblasts (see also: Cell activation). One distinct factor with FAF activity, described as fibroblast activating factor and fibroblast modulatory activity, is identical with » IL1. Some FAFs may be related to IL1. Another factor with FAF activity is produced by monocytes. This factor is called FAF-M. It is a poorly characterized 38 kDa protein which is not identical with IL1. Another factor with FAF activity is secreted by » U937 cells and called » U937-FAF. In addition, there are a number of poorly characterized factors that also possess FAF activity. Some of these factors have been designated » fibroblast proliferation factors.
Agelli M & Wahl SM Synthesis of biologically active fibroblast-activating factor (FAF) by *Xenopus* oocytes injected with T lymphocyte mRNA. CI 110: 183-90 (1987); **Austgulen R et al** *In vitro* cultured human monocytes release fibroblast proliferation factor(s) different from interleukin 1. J. Leukoc. Biol. 42: 1-8 (1987); **Delaporte F et al** [Keratinocyte-fibroblast interactions: I. Production by the keratinocytes of soluble factors stimulating the proliferation of normal human skin fibroblasts] Pathol. Biol. Paris 37: 875-80 (1989); **Dohlman JG et al** Generation of a unique fibroblast-activating factor by human monocytes. Immunology 52: 577-84 (1984); **Jordana M et al** Spontaneous *in vitro* release of alveolar-macrophage cytokines after the intratracheal instillation of bleomycin in rats. Characterization and kinetic studies. Am. Rev. Respir. Dis. 137: 1135-40 (1988); **Matsushima K et al** Purification of human interleukin 1 from human monocyte culture supernatants and identity of thymocyte comitogenic factor, fibroblast proliferation factor, acute phase protein-inducing factor, and endogenous pyrogen. CI 92: 290-301 (1985); **Ofosu-Appiah WA et al** Fibroblast-activating factor production by interleukin (IL2)-dependent T cell clones from rheumatoid arthritis patients and normal donors. Rheumatol. Int. 8: 219-24 (1988); **Salahuddin SZ et al** Lymphokine production by cultured human T cells transformed by human T cell leukemia lymphoma virus I. S 223: 703-7 (1984); **Shaked A et al** Isolation of fibroblast proliferation factor: distinction from

interleukin-1. Curr. Surg. 44: 487-90 (1987); **Turk CW et al** Immunological mediators of wound healing and fibrosis. JCP 5: s89-93 (1987); **Wahl SM & Gately CL** Modulation of fibroblast growth by a lymphokine of human T cell continuous T cell line origin. JI 130: 1226-30 (1983); **Wahl SM et al** Spontaneous production of fibroblast-activating factor(s) by synovial inflammatory cells. A potential mechanism for enhanced tissue destruction. JEM 161: 210-22 (1985); **Wahl SM et al** Bacterial cell wall-induced hepatic granulomas. An *in vivo* model of T cell-dependent fibrosis. JEM 163: 884-902 (1986)

faint little ball: see: DER (*Drosophila* EGF receptor).

Factor XII: see: EGF-like repeats.

Factor-dependent cell lines: Established cell lines that entirely depend for their survival and proliferation on the continuous presence of one or more of growth factors. Factor-dependent cell lines are capable of continuous growth in the presence of growth factors they require (see also: Colony formation assay; Delta assay; LTBMC, long-term bone marrow culture). Factor-dependent cell lines have been instrumental in analyzing in detail the processes leading to cell death following growth factor deprivation. Withdrawal of the growth factor usually leads to cell death by » apoptosis rather than to a cessation of growth. This is in marked contrast to normal cells which sooner or later undergo terminal differentiation and/or die or cells that have become growth factor-independent by producing, for example, » autocrine growth factors.

In spite of their unlimited growth capacities factor-dependent cell lines normally do not cause tumors *in vivo*. They are therefore regarded as immortalized counterparts of normal cells. The study of factor-dependent cell lines provides information about promotion of cell survival, proliferation, differentiation, and induction of functional activities by cytokines. They can be used for searching for factors or compounds that are capable to replace the cytokine which is required by the cells. Such cells have been used also to study intracellular signal processes taking place after binding of a cytokine to its receptor.

In addition, factor-dependent cell lines allow detailed analyses of processes leading to factor independence. Introduction of expression vectors directing the synthesis of the cytokine that the factor-dependent cell line requires usually creates an » autocrine loop which effectively abrogates factor dependence. These experiments frequently show that aberrant activation of a growth factor autocrine circuit can lead to malignant transformation. Hybrid cell lines constructed by fusing growth factor-dependent cell lines with their independent normal parental cells have been used to study the determinants governing factor (in)dependence. The observation that, in some cases, factor dependence is the dominant trait has been taken as evidence for the existence of suppressor activities negatively regulating growth factor expression. Some studies with factor-dependent cells and independently arising autonomous mutants derived from it have revealed the existence of distinct phenotype classes ranging from immortalization, loss of growth factor requirement for survival, loss of growth factor requirement for proliferation, and loss of growth factor-stimulated proliferation. Such studies indicate a divergence of the pathways of growth factor-regulated survival and proliferation and the various cell types may represent phenotypes occurring at intermediate stages in tumor-cell progression.

In many instances overexpression of an activated » oncogene in factor-dependent cell lines has been shown to abrogate factor dependence (see, for example: 32D, *abl*, FDC-P1; IC2) although in other cases constitutive expression of an oncogene only potentiates the response to the growth factor without eliminating the growth factor requirement.

Introduction of an expression vector directing the synthesis of a functional cytokine receptor into factor-dependent cell lines which itself does not express it can be used often to create factor dependence for the cytokine acting upon the newly introduced receptor. This technique is used frequently to study specific growth and differentiation signals initiated by the newly introduced receptor and its associated cytokine, in particular second messenger cascades, phosphorylation of proteins, and alterations of » gene expression. Such experiments demonstrate, among other things, that responding recipient cells express all of the intracellular components of the signaling pathway of the heterologous receptor necessary to evoke a mitogenic response and sustain continuous proliferation. In addition, this approach can be used to define structure-function relationships by investigating the influences of cytokine and/or receptor mutations and/or hybrids on mitogenic and other cellular responses (see also: Muteins).

Factor-dependent cell lines have been used also as models to study the functional expression of heterologous growth factor receptors or their reconstitution from individual components. Expression of a novel cytokine receptor can be used effectively to create new factor-dependent cell lines; it is even possible to use cloned receptor genes from one species and a recipient cell line of another species to create a cell line which then responds to the appropriate cytokine of another species to which it would normally not respond.

Factor-dependent cell lines have been instrumental in providing » bioassays for many cytokines. They can be used to screen hybridomas for the production of monoclonal antibodies that inhibit factor-dependent proliferation. For further information on factor-dependent cell lines and other cell lines employed for cytokine assays see individual cytokines (subentry Assays).

Ihle JN & Askew D Origins and properties of hematopoietic growth factor-dependent cell lines. Int. J. Cell Cloning 7: 68-91 (1989); **Oval J & Taetle R** Factor-dependent human leukemia cell lines: new models for regulation of acute non-lymphocytic leukemia cell growth and differentiation. Blood Rev. 4: 270-9 (1990); **Pollard JW et al** Independently arising macrophage mutants dissociate growth factor-regulated survival and proliferation. PNAS 88: 1474-8 (1991)

Fas antigen: see: Apo-1.

Fas ligand: see: Apo-1.

Fat-storing cell proliferation factor: see: Proliferation factor for fat-storing cells.

Fatty acid-binding protein: abbrev. FABP. see: MDGI (mammary-derived growth inhibitor).

FBHE: *fetal bovine heart endothelial* The FBHE cell line is used as a » bioassay to study the growth-promoting effects of heparin-binding growth factors (see also: HBGF). The cells do not proliferate even in serum-containing media unless a source of fibroblast growth factor is made available. These cells are therefore used to assay » aFGF and » bFGF. The cells have an absolute requirement for heparin. The cells also proliferate in response to » ECDGF (embryonal carcinoma-derived growth factor). The cells have been shown to react in a biphasic way to » TGF-β, which stimulates the growth of these cells at low concentrations and inhibits it at high concentrations. At concentrations above 1 ng/mL TGF-β also inhibits FGF-induced proliferation of FBHE cells. Different isoforms of TGF-β also differ in their individual activities.

Cheifetz S et al Distinct transforming growth factor-β (TGF-β) receptor subsets as determinants of cellular responsiveness to three TGF-β isoforms. JBC 265: 20533-8 (1990); **Müller G et al** Inhibitory action of transforming growth factor β on endothelial cells. PNAS 84: 5600-4 (1987); **Myoken Y et al** Bifunctional effects of transforming growth factor-β (TGF-β) on endothelial cell growth correlate with phenotypes of TGF-β binding sites. ECR 191: 299-304 (1990); **Van Zoelen EJJ et al** Identification and characterization of polypeptide growth factors secreted by murine embryonal carcinoma cells. Dev. Biol. 133: 272-83 (1989)

Fc receptors: Immunoglobulin-binding receptors expressed on almost all types of immune cells and a number of their cells (see: CD16, CD32, CD64). Some circulating forms function as » IBF (immunoglobulin binding factors).

FD: abbrev. for factor-dependent. See: Factor-dependent cell lines.

FDC-Pmix: (FDC = factor-dependent cells) A multipotent » hematopoietic stem cell line obtained by long-term murine bone marrow cultures (see: LTBMC). FDCP-Mix is able to differentiate along myeloid and erythroid lineages in response to specific cytokines, yielding mixed colonies in a » colony formation assay). These cells are thought to closely reflect the situation *in vivo* (see: Hematopoiesis). They are used as a model for studying the control of growth and differentiation processes of multipotent hematopoietic stem cells and have been used as target cells for the introduction of genes by retroviral transfer.

Morphologically these cells more or less resemble blast cells if kept in the presence of the colony-stimulating factor » IL3 (multi-CSF; see also: CSF) or horse serum. Replacement of horse serum by fetal calf serum or co-cultivation with stromal cells induces differentiation into all myeloid cell types. At low concentrations of IL3 the cells respond to » GM-CSF, » M-CSF, » G-CSF, or » Epo by developing into mature postmitotic cells. The types of mature cells developing from FDCP-Mix depend on the individual growth factors added to IL3. At high concentrations of IL3 the cells proliferate and self-renew without differentiating, irrespective of the presence of other growth factors. In the absence of IL3 the cells do not respond to other growth factors. Erythroid differentiation of FDCP-Mix in the presence of both IL3 and Epo is favored by » IL9 which by itself does not support proliferation of these cells.

Beck-Engeser G et al Retroviral vectors related to the myeloproliferative sarcoma virus allow efficient expression in hema-

Fas ligand: see addendum.

topoietic stem and precursor cell lines, but retroviral infection is reduced in more primitive cells. Hum. Gene Ther. 2: 61-70 (1991); **Bourette RP et al** Murine interleukin 9 stimulates the proliferation of mouse erythroid progenitor cells and favors the erythroid differentiation of multipotent FDCP-Mix cells. EH 20: 868-73 (1992); **Dexter TM et al** The role of growth factors in self-renewal and differentiation of hematopoietic stem cells. Philos. Trans. R. Soc. Lond. 327: 85-98 (1990); **Heberlein C et al** Retrotransposons as mutagens in the induction of growth autonomy in hematopoietic cells. O 5: 1799-807 (1990); **Heyworth CM et al** The biochemistry and biology of the myeloid hematopoietic cell growth factors. JCS 13: s57-s74 (1990); **Just U et al** Expression of the GM-CSF gene after retroviral transfer in hematopoietic stem cell lines induces synchronous granulocyte-macrophage differentiation. Cell 64: 1163-73 (1991); **Spooner E et al** Self-renewal and differentiation of IL3-dependent multipotential stem cells are modulated by stromal cells and serum factors. Differentiation 31: 111-8 (1986)

FDC-P1: *factor dependent cell P1* FDC-P1 are murine non-tumorigenic diploid granulocyte and monocyte/macrophage progenitor cells (see also: Hematopoiesis) originally obtained from long-term cultures of bone marrow cells supplemented with conditioned medium of » WEHI cells. The cells express Thy-1, Ly-5, H-11 and the macrophage-specific marker F4/80; they do not, however, respond to » M-CSF. The growth of these cells depends upon » IL3 and » GM-CSF (see also: FDC-Pmix) and are therefore also used for assaying these factors (see also: Bioassays). Active » phorbol esters (e.g., phorbol 12-myristate 13-acetate) can entirely replace IL3 to promote proliferation of FDC-P1 cells. The cells also respond to » IL4, » IGF-1, and » IFN-γ to a limited extent, with IL4 and IFN-γ mainly prolonging survival for several days and inducing some DNA synthesis. They secrete » IL6. An alternative and entirely different method of detecting these factors is » Message amplification phenotyping.

IL4 inhibits in a dose-dependent fashion either the IL3- or GM-CSF-induced growth of the FDC-P1 cells.

Withdrawal of IL3 arrests FDC-P1 cells in the G_0/G_1 phase of the » cell cycle which they re-enter synchronously after addition of the factor. The introduction of an » oncogene encoding a tyrosine-specific protein kinase, such as » *abl*, » *fms*, *src*, and » *trk* by means of a retroviral vector renders FDC-P1 cells independent of IL3. Expression of » *myc* has the same effect. The expression of c-*fms* allows FDC-P1 cells to grow in the absence of IL3 or GM-CSF provided that M-CSF is present. The expression of v-*fms* allows FDC-P1 to grow in the absence of any added hemopoietic growth factors, including M-CSF, although the addition of M-CSF enhances v-*fms* activity. V-*fms* cell lines grow to a higher cell density in suspension and are tumorigenic.

Leukemic transformation of FDC-P1 cells is observed if the cells are transplanted into irradiated mice. Many cell variants that are subsequently isolated from these animals have lost their growth factor requirements and express IL3 or GM-CSF constitutively. The activation and constitutive expression of these growth factor genes results from the transposition of a retroviral element (see: IAP, intracisternal A particle).

Injection of FDC-P1 cells into transgenic mice overexpressing » GM-CSF leads to death within three months with elevated blast cell numbers in the blood, massive organ infiltration by blast cells, and associated anemia and thrombocytopenia. All transgenic recipients contain transformed FDC-P1 cells able to produce rapidly-growing transplanted leukemias in syngeneic normal recipients. The leukemias appear to arise in the primary recipients by independent transformation events. All primary recipients at death contain fully transformed leukemic cells and a large population of untransformed FDC-P1 cells or cells with altered characteristics not yet representing full transformation. At present it is not known whether the FDC-P1 engrafted model has some validity for myelodysplasia. If so, sustained administration of GM-CSF to myelodysplastic patients possessing abnormal, potentially preleukemic, granulocyte-macrophage populations might increase the risk accumulating pretransformed or fully transformed leukemic cells.

FDC-P1 cell do not express a receptor for » FGF. Introduction of a gene encoding this receptor allows growth and differentiation of the cells in the presence of FGF and heparin, which is known to be required for the interaction of FGF (see also: bFGF) with its receptor (see also: HBGF, heparin-binding growth factors).

Certain combinations of colony-stimulating factors (see also: CSF), for example a combination of » GM-CSF and » M-CSF show the phenomenon of synergistic suppression. It is characterized by a partial suppression of the differentiation of macrophage colonies rather than an enhancement of macrophage-like colonies which should have been expected to be promoted by M-CSF. This observation may be clinically relevant in view of combination therapies employing different colony-stimulating factors.

Boswell HS et al A *ras* oncogene imparts growth factor independence to myeloid cells that abnormally regulate protein kinase C: a nonautocrine transformation pathway. EH 18: 452-60 (1990); **Cleveland JL et al** Tyrosine kinase oncogenes abrogate interleukin-3 dependence of murine myeloid cells through signaling pathways involving c-*myc* : conditional regulation of c-*myc* transcription by temperature-sensitive v-*abl*. MCB 9: 5685-95 (1989); **Dean M et al** Role of *myc* in the abrogation of IL3 dependence of myeloid FDC-P1 cells. OR 1: 279-96 (1987); **Dexter TM et al** Growth of factor-dependent hemopoietic precursor cell lines. JEM 152: 1036-47 (1980); **Dibb NJ et al** Expression of v-*fms* and c-*fms* in the hemopoietic cell line FDC-P1. Growth factors 2: 301-11 (1990); **Dunn AR & Wilks AF** Contributions of autocrine and non-autocrine mechanisms to tumorigenicity in a murine model for leukemia. Ciba Found. Symp. 148: 145-55 (1990); **Duhrsen U et al** *In vivo* transformation of factor-dependent hemopoietic cells: role of intracisternal A-particle transposition for growth factor gene activation. EJ 9: 1087-96 (1990); **Hapel AJ et al** Different colony-stimulating factors are detected by the "interleukin-3"-dependent cell lines FDC-Pl and 32D cl-23. Blood 64: 786-90 (1984); **Hapel AJ et al** Generation of an autocrine leukemia using a retroviral expression vector carrying the interleukin-3 gene. LR 5: 249-54 (1986); **Hültner L et al** Interleukin 3 mediates interleukin 6 production in murine interleukin 3-dependent hemopoietic cells. GF 2: 43-51 (1989); **Ihle JN et al** Phenotypic characteristics of cell lines requiring interleukin 3 for growth. JI 129: 1377-83 (1982); **Keith WN et al** Retrovirus-mediated transfer and expression of GM-CSF in hematopoietic cells. Br. J. Cancer 62: 388-94 (1990); **Kelso A** An assay for colony-stimulating factor (CSF) production by single T lymphocytes: estimation of the frequency of cells producing granulocyte-macrophage CSF and multilineage CSF within a T lymphocyte clone. JI 136: 2930-7 (1986); **Kelso A & Troutt AB** Survival of the myeloid progenitor cell line FDC-P1 is prolonged by interferon-γ or interleukin-4. GF 6: 233-42 (1992); **Kipreos ET & Wang JV** Reversible dependence on growth factor interleukin-3 in myeloid cells expressing temperature sensitive v-*abl* oncogene. OR 2: 277-84 (1988); **Li M & Bernard O** FDC-P1 myeloid cells engineered to express fibroblast growth factor receptor 1 proliferate and differentiate in the presence of fibroblast growth factor and heparin. PNAS 89: 3315-9 (1992); **London L & McKearn JP** The dual regulatory role of interleukin 4 is mediated through a direct effect on the target cell. EH 18: 1059-63 (1990); **Maraskovsky E et al** High-frequency activation of single $CD4^+$ and $CD8^+$ T cells to proliferate and secrete cytokines using anti-receptor antibodies and IL2. Int. Immunol. 3: 255-64 (1991); **Metcalf D et al** Synergistic suppression: anomalous inhibition of the proliferation of factor-dependent hemopoietic cells by combination of two colony-stimulating factors. PNAS 89: 2819-23 (1992); **Metcalf D & Rasko JE** Leukemic transformation of immortalized FDC-P1 cells engrafted in GM-CSF transgenic mice. Leukemia 7: 878-86 (1993); **McCubrey J et al** Abrogation of factor-dependence in two IL3-dependent cell lines can occur by two distinct mechanisms. OR 4: 97-109 (1989); **McCubrey J et al** Growth-promoting effects of insulin-like growth factor-1 (IGF-1) on hematopoietic cells: overexpression of introduced IGF-1 receptor abrogates interleukin-3 dependency of murine factor-dependent cells by a ligand-dependent mechanism. Blood 78: 921-9 (1991); **Redner RL et al** Variable pattern of *jun* and *fos* gene expression in different hematopoietic cell lines during interleukin 3-induced entry into the cell cycle. O 7: 43-50 (1992); **Rohrschneider LR & Metcalf D** Induction of macrophage colony-stimulating factor-dependent growth and differentiation after introduction of the murine c-*fms* gene into FDC-P1 cells. MCB 9: 5081-92 (1989); **Wang LM et al** Common elements in interleukin 4 and insulin signaling pathways in factor-dependent hematopoietic cells. PNAS 90: 4032-6 (1993); **Wheeler EF et al** The v-*fms* oncogene induces factor-independent growth and transformation of the interleukin-3-dependent myeloid cell line FDC-P1. MCB 7: 1673-80 (1987); **Whetton AD & Dexter TM** Myeloid hematopoietic growth factors. BBA 989: 111-32 (1989)

FDCp1 growth factor: A factor secreted by the rat/mouse T cell hybridoma cell line, PC60, following stimulation with » IL1 and » IL2. This factor is the rat homologue of » GM-CSF. A similar factor, ***FDC-P1 CSF*** (FDC-P1 colony-stimulating activity) is produced by murine irradiated bone marrow stomal cells. This factor is identical with GM-CSF.

Alberico TA et al Stromal growth factor production in irradiated lectin exposed long-term murine bone marrow cultures. Blood 69: 1120-7 (1987); **Quesenberry P et al** Bone marrow adherent cell hemopoietic growth factor production. PCBR 184: 247-56 (1985); **Vandenabeele P et al** Response of murine cell lines to an IL1/IL2-induced factor in a rat/mouse T hybridoma (PC60): different induction of cytokines by human IL1 α and IL1 β and partial amino acid sequence of rat GM-CSF. LR 9: 381-9 (1990)

FDC-P2: This cell line was derived from the mouse IL 3-dependent cell line » FDC-P1 after culture in » IL2-conditioned culture medium. IL2 growth responsiveness extends to human, rat, and mouse IL2. When cultured with murine IL3 the cells do not proliferate and appear to accumulate in the G1 phase of the » cell cycle. DNA synthesis is also induced efficiently in FDC-P2 cells by » IL4, insulin, and » IGF-1. Although these factors do not individually sustain long-term growth, a combination of IL4 with either insulin or IGF-I supports continuous growth. The cells also proliferate in response to » GM-CSF. The cells secrete » IL6.

Overexpression of the human activated *ras* oncogene introduced into FDC-P2 cells by means of an expression vector yields cells that show IL3-independent growth and tumorigenicity in nude mice (see also: Immunodeficient mice).

Dexter TM et al Growth of factor-dependent hemopoietic precursor cell lines. JEM 152: 1036-47 (1980); **Dexter TM et al** The role of hemopoietic cell growth factor (interleukin 3) in the development of hemopoietic cells. Ciba Found. Symp. 116: 129-47 (1985); **Gascan H et al** Response of murine IL3-sensitive cell lines to cytokines of human and murine origin. LR 8: 79-84 (1989); **Hültner L et al** Interleukin 3 mediates interleukin 6 production in murine interleukin 3-dependent hemopoietic cells. GF 2: 43-51 (1989); **Le Gros GS et al** Induction of IL 2 responsiveness in a murine IL 3-dependent cell line. JI 135: 4009-14 (1985); **Sakihama T et al** Monoclonal IgGs from an autoimmune MRL/Mp-lpr/lpr mouse induce an interleukin-3-dependent myeloid cell line to produce tumor necrosis factor α and interleukin-6. CI 132: 1-9 (1991); **Uemura N et al** Acquisition of interleukin-3 independence in FDC-P2 cells after transfection with the activated c-H-*ras* gene using a bovine papillomavirus-based plasmid vector. Blood 80: 3198-24 (1992); **Wang LM et al** IL4 activates a distinct signal transduction cascade from IL3 in factor-dependent myeloid cells. EJ 11: 4899-908 (1992); **Wang LM et al** Common elements in interleukin 4 and insulin signaling pathways in factor-dependent hematopoietic cells. PNAS 90: 4032-6 (1993); **Whetton AD & Dexter TM** Myeloid hematopoietic growth factors. BBA 989: 111-32 (1989)

FDGF: *fibroblast-derived growth factor* This factor is secreted by an SV40-transformed BHK (baby hamster kidney) cell line. It is a mitogen and a chemoattractant for endothelial cells. This factor belongs to the group of PDGF-like growth factors and is very like identical with one of the molecular forms of » PDGF.

Bürk RR Induction of cell proliferation by a migration factor released from a transformed cell line. ECR 101: 293-8 (1976); **Connolly DT et al** Human fibroblast-derived growth factor is a mitogen and chemoattractant for endothelial cells. BBRC 144: 705-12 (1987); **Dicker P et al** Similarities between fibroblast-derived growth factor and platelet-derived growth factor. ECR 135: 221-7 (1981); **Rozengurt E et al** Inhibition of epidermal growth factor binding to mouse cultured cells by fibroblast-derived growth factor. Evidence for an indirect mechanism. JBC 257: 3680-6 (1982); **Stroobant P et al** Highly purified fibroblast-derived growth factor, an SV40-transformed fibroblast-secreted mitogen, is closely related to platelet-derived growth factor. EJ 4: 1945-9 (1985); **Stroobant P et al** Purification of fibroblast-derived growth factor. MiE 147: 40-7 (1987)

FDGI: *fibroblast-derived growth inhibitor* This fibroblast-derived factor which behaves as an inhibitor for early-stage melanoma cells but as a mitogenic agent for more advanced stage melanoma cells is identical with » IL6.

Lu C et al Interleukin 6: a fibroblast-derived growth inhibitor of human melanoma cells from early but not advanced stages of tumor progression. PNAS 89: 9215-9 (1992)

FDNAP: *fibroblast-derived neutrophil activating peptide* This factor is identical with » IL8 and therefore belongs to the » chemokine family of cytokines. See: FINAP.

FDNCF: *fibroblast-derived neutrophil chemotactic factor* This factor is secreted by fibroblasts. It is a chemoattractant for human neutrophils (see also: Chemotaxis). FDNCF is identical with » IL8. However, its N-terminal end has been extended by 5 amino acids (77 instead of 72 amino acids).The factor is also referred to as ***AVLPR-IL8*** (A = alanine, V = valine, L = leucine, P = proline, R = arginine). FDNCP therefore belongs to the » chemokine family of cytokines.

Biologically active genuine IL8 is obtained from FDNCF by specific removal of the additional pentapeptide by plasmin.

Golds EE et al BJ 259: 585-8 (1989); **Nakagawa H et al** Generation of interleukin-8 by plasmin from AVLPR-interleukin-8, the human fibroblast-derived neutrophil chemotactic factor. FL 282: 412-4 (1991); **Schröder JM et al** IL1α or tumor necrosis factor α stimulate release of the NAP-1/IL8-related neutrophil chemotactic protein in human dermal fibroblasts. JI 144: 2223-32 (1990); **Van Damme J et al** The chemotactic activity for granulocytes produces by virally infected fibroblasts is identical to monocyte-derived interleukin 8. EJI 19: 1189-94 (1990)

f-ECGF: *fibroblast (derived) endothelial cell growth factor* This poorly characterized factor is isolated from the culture supernatants of human fibroblasts. It selectively supports the growth of endothelial cells *in vitro* and shows the activity of an » angiogenesis factor *in vivo*. The factor appears to be distinct from » aFGF and » bFGF.

Satoh T et al Characterization of a new endothelial cell growth factor (f-ECGF) partially purified from the supernatant of human fibroblast cells. Cell Struct. Funct. 14: 731-40 (1989)

Feeder cells: The cultivation *in vitro* of some fastidious cells, requiring, for example, the presence of a variety of growth factors of unknown identity, is an exacting task. One way to circumvent the problems usually associated with the establishment and maintenance of such cells is the use of a layer of feeder cells as a substrate on which other cells are grown. In some cases the use of feeder cells for certain types of cell cultures can be circumvented once the growth factors required by the cells are known (see for example: ES cells.).

The growth-promoting effect of feeder cells for other cells is mainly due to the fact that feeder cells secrete many soluble mediator substances into the ambient medium which, in turn, support the growth of other cells (see also: Paracrine). Purified growth factors and also conditioned media (see: CM) can be used in some instances to make fastidious cells independent of the presence of feeder cell layers.

Apart from the secreted factors feeder cells also provide a solid substrate which also supports the growth of other cells. This activity can be attributed to the provision of an intact extracellular matrix (see also: ECM) and matrix-associated factors (see also: Juxtacrine). For one specific application of feeder cells see also: LTBMC (long-term bone marrow cultures).

fes*:** *fes*, originally identified as an » oncogene in Gardner-Arnstein and Snyder-Theilen isolates of feline sarcoma virus (*fes*V), and ***fps, a transforming gene common to the Fujinami, PRC II, and UR 1 strains of avian sarcoma virus, correspond to a common cellular genetic locus which has remained highly conserved throughout vertebrate evolution. The human *fes/fps* gene extends over 11 kb and contains 17 introns. It maps to chromosome 15q26.1, distal to the breakpoint in chromosome 15 in the translocation commonly seen in acute promyelocytic leukemia, t(15;17) (q24;q22).

fes/fps encodes a non-receptor tyrosine-specific protein kinase of 92 kDa, also called ***NCP92***. It appears to be expressed predominantly in hematopoietic tissues (immature and differentiated cells of the myeloid lineage) and to be involved in hematopoietic growth factor signaling.

v-*fps* has been shown to activate promoters under the control of the » phorbol ester response element (see: TRE). It mediates induction of the 9E3 cytokine. *fes/fps* is involved in signaling through the » Epo, » GM-CSF, and » IL3 receptors. It rapidly becomes tyrosine-phosphorylated on treatment of human erythroleukemia » TF-1 cells with these growth factors and has been shown to become physically linked with the Epo and the GM-CSF receptor.

Human c-*fps/fes* when expressed at sufficiently high levels in murine » 3T3 cells can induce cellular transformation and this effect is reversible by treatment of the cells with » Herbimycin A. Overexpression of *fes/fps* in the growth factor-dependent myeloid cell line » FDC-P1has been shown to abrogate the requirement for exogenous growth factors. Expression of the mitogen-responsive transcription factor » Egr-1 is rapidly induced by v-*fps* in the absence of protein synthesis. The *fes* proto-oncogene p93c-*fes* and its associated tyrosine kinase activity is expressed in mature granulocytes, monocytes, differentiated » HL-60 leukemia cells, and leukemia cell lines » KG-1, THP-1, HEL, and » U-937, which can be induced to differentiate along the granulocyte/monocyte pathway. Cellular *fes* expression is absent in the » K562 cell line, which is resistant to myeloid differentiation; these cells acquire the ability to differentiate after introduction of a functionally expressed *fes* gene. Truncated c-*fps/fes* transcripts of ≈ 0.9 kb (normal length: 3 kb) have been detected in various human lymphoma and lymphoid leukemia cell lines, but not in normal untransformed hematopoietic cells.

■ **Transgenic/Knock-out/Antisense Studies:** The establishment of » transgenic mice that express v-*fps* shows that these animals have severe cardiac or neurologic abnormalities and a high incidence of lymphoid or mesenchymal tumors. Transgenic mice with v-*fps* under the transcriptional control of a 5' human β-globin promoter (GF) or with both 5' and 3' β-globin regulatory sequences (GEF) are viable and develop a spectrum of benign and malignant tumors, including lymphomas (mainly monoclonal and of T cell origin), thymomas, fibrosarcomas, angiosarcomas, hemangiomas, and neurofibrosarcomas.

Introduction of the cellular *fps/fes* coding sequence into the mouse germ line reveals transcription of the human c-*fps/fes* transgene to be highest in bone marrow, showing a tissue distribution identical to that of the endogenous mouse gene. Expression of the human proto-oncogene in bone marrow has been shown to be independent of the integration site, proportional to the transgene copy number, and to be of comparable efficiency to that of the endogenous mouse c-*fps/fes* gene. Elevated levels of normal human p92c-*fes* have no obvious effect on mouse development or hematopoiesis.

c-*fes* is expressed at high levels in the terminal stages of granulocytic differentiation. Inhibition of c-*fes* proto-oncogene expression by » antisense oligonucleotides has been studied myeloid cell lines and fresh leukemic promyelocytes of acute promyelocytic leukemia induced to differentiate with retinoic acid and dimethylsulfoxide. These cells do not differentiate but die by » apoptosis. This process can be prevented by » GM-CSF, but not by » IL3, » IL6, or » SCF. These observations suggest a possible role of the c-*fes* gene product in counteracting apoptosis during granulocytic differentiation.

Alexandropoulos K et al v-*fps*-responsiveness in the Egr-1 promoter is mediated by serum response elements. NAR 20: 2355-9 (1992); **Alexandropoulos K et al** Ha-Ras functions downstream from protein kinase C in v-*fps*-induced gene expression mediated by TPA response elements. O 8: 803-7 (1993); **Alcalay M et al** Characterization of human and mouse c-*fes* cDNA clones and identification of the 5' end of the gene. O 5: 267-75 (1990; **Borellini F et al** Increased DNA binding and transcriptional activity associated with transcription factor Sp1 in K562 cells transfected with the myeloid-specific c-*fes* tyrosine kinase gene. JBC 266: 15850-4 (1991); **Brach MA et al** Expression of the c-*fes* proto-oncogene in granulocyte-macrophage colony-stimulating factor-dependent acute myelogenous leukemia cells grown autonomously. Int. J. Cell Cloning. 9: 89-94 (1991); **Chen LH et al** Single amino acid substitution, from Glu1025 to Asp, of the *fps* oncogenic protein causes temperature sensitivity in transformation and kinase activity. Virology 155: 106-19 (1986); **Fang F et al** Effect of the mutation of tyrosine 713 in p93c-*fes* on its catalytic activity and ability to promote myeloid differentiation in K562 cells. B 32: 6995-7001 (1993); **Feldman RA et al** Selective potentiation of c-*fps/fes* transforming activity by a phosphatase inhibitor. OR 5: 187-97 (1990); **Groffen J et al** Transforming genes of avian (v-*fps*) and mammalian (v-*fes*) retroviruses correspond to a common cellular locus. Virology 125: 480-6 (1983); **Hanazono Y et al** Erythropoietin induces tyrosine phosphorylation and kinase activity of the c-*fps/fes* proto-oncogene product in human erythropoietin-responsive cells. Blood 81: 3193-6 (1993); **Hanazono Y et al** c-*fps/fes* protein-tyrosine kinase is implicated in a signaling pathway triggered by granulocyte-macrophage colony-stimulating factor and interleukin-3. EJ 12: 1641-6 (1993); **Harper ME et al** Chromosomal sublocalization of human c-*myb* and c-*fes* cellular onc genes. N 304: 169-71 (1983); **Hjermstad SJ et al** Regulation of the human c-*fes* protein tyrosine kinase (p93c-*fes*) by its *src* homology 2 domain and major autophosphorylation site (Tyr-713). O 8: 2283-92 (1993); **Huang CC et al** Nucleotide sequence of v-*fps* in the PRCII strain of avian sarcoma virus. J. Virol. 50: 125-31 (1984); **Jhanwar SC et al** Localization of the cellular oncogenes ABL, SIS, and *fes* on human germ-line chromosomes. Cytogenet. Cell. Genet. 38: 73-5 (1984); **Jucker M et al** Expression of truncated transcripts of the proto-oncogene c-*fps/fes* in human lymphoma and lymphoid leukemia cell lines. O 7: 943-52 (1992); **Kamps MP & Sefton BM** Identification of multiple novel polypeptide substrates of the v-*src*, v-*yes*, v-*fps*, v-ros, and v-*erb*-B oncogenic tyrosine protein kinases utilizing antisera against phosphotyrosine. O 2: 305-15 (1988); **Klinken SP** Erythroproliferation *in vitro* can be induced by *abl, fes, src, ras, bas, raf, raf/myc, erb* B and *cbl* oncogenes but not by *myc, myb* and *fos*. OR 3: 187-92 (1988); **Koch CA et al** The common src homology region 2 domain of cytoplasmic signaling proteins is a positive effector of v-*fps* tyrosine kinase function. MCB 9: 4131-40 (1989); **Meckling-Gill KA et al** A retrovirus encoding the v-*fps* protein-tyrosine kinase induces factor-independent growth and tumorigenicity in FDC-P1 cells. BBA 1137: 65-72 (1992); **Pawson T et al** Structure-function relationships in cellular and viral *fps/fes* cytoplasmic protein-tyrosine kinases. AEMB 234: 55-64 (1988); **Roebroek AJ et al** The structure of the human c-*fes/fps* proto-oncogene. EJ 4(11): 2897-903 (1985); **Roebroek AJ et al** Structure of the feline c-*fes/fps* proto-oncogene: genesis of a retroviral oncogene. J. Virol. 61: 2009-16 (1987); **Smithgall TE et al** Identification of the differentiation-associated p93 tyrosine protein kinase of HL-60 leukemia cells as the product of the human c-*fes* locus and its expression in myelomonocytic cells. JBC 263: 15050-5 (1988); **Smithgall TE et al** Construction of a cDNA for the human c-*fes* proto-oncogene protein-tyrosine kinase and its expression in a baculovirus system. B 31: 4828-33 (1992); **Sodroski JG et al** Transforming potential of a human proto-oncogene (c-*fps/fes*) locus. PNAS 81(10): 3039-43 (1984); **Yu G et al** K562 leukemia cells transfected with the human c-*fes* gene acquire the ability to undergo myeloid differentiation. JBC 264: 10276-81 (1989)

● **TRANSGENIC/KNOCK-OUT/ANTISENSE STUDIES: Chow LH et al** Progressive cardiac fibrosis and myocyte injury in v-*fps* transgenic mice. A model for primary disorders of connective tissue in the heart? Lab-Invest. 1991 Apr; 64(4): 457-62 (1991); **Ferrari S et al** Differential effects of c-*myb* and c-*fes* antisense oligodeoxynucleotides on granulocytic differentiation of human myeloid leukemia HL60 cells. Cell-Growth-Differ. 1990 Nov; 1: 543-8 (1990); **Greer P et al** Myeloid expression of the human c-*fps/fes* proto-oncogene in transgenic mice. MCB 10(6): 2521-7 (1990); **Manfredini R et al** Inhibition of c-*fes* expression by an antisense oligomer causes apoptosis of HL60 cells induced to granulocytic differentiation. JEM 178: 381-9 (1993); **Yee SP et al** Lymphoid and mesenchymal tumors in transgenic mice expressing the v-*fps* protein-tyrosine kinase. MCB 9: 5491-9 (1989); **Yee SP et al** Cardiac and neurological abnormalities in v-*fps* transgenic mice. PNAS 86: 5873-7 (1989)

Fetal liver kinase 1: abbrev. *flk*-1. see: KDR (kinase domain receptor).

FF-GF: *FF101-derived growth factor* This poorly characterized factor of ≈ 70 kDa is secreted by the rat hepatoma cell line, FF101, established for growth in serum-free, protein-free medium. FF-GF stimulates DNA synthesis of various cells from different origins. The growth-promoting activities of FF-GF are not inhibited by antibodies against » aFGF or » bFGF.

Matsuda H et al Purification and characterization of a novel growth factor (FF-GF) synthesized by a rat hepatoma cell line, FF101. BBRC 189: 654-61 (1992)

FGF: *fibroblast growth factor* FGFs are a family of related 16-18 kDa proteins controlling normal growth and differentiation of mesenchymal, epithelial, and neuroectodermal cell types.

Two main groups of FGFs are known. One type of FGF was initially isolated from brain tissue. It was identified by its proliferation-enhancing activities for murine fibroblasts (3T3 cells). Due to its basic pI the factor was named » bFGF (basic FGF; new name: FGF-2). This factor is the prototype of the FGF family.

Another factor, also isolated initially from brain tissues, is » aFGF (acidic FGF; new name: FGF-1). It was initially identified by its proliferation-enhancing activity for myoblasts.

Other homologous FGFs belonging to the same family are » *int*-2 (FGF-3), » FGF-5, » FGF-6, » K-FGF and » KGF (keratinocyte growth factor = FGF-7). All factors are products of different genes, some of which are » oncogenes (FGF-3, 4, 5). Some FGFs do not appear to be secreted (aFGF, bFGF) while others (FGF-3, 4, 5, 6) have a » signal sequence mediating secretion.

FGFs show a very high affinity to heparin and are therefore also sometimes referred to as heparin-binding growth factors (see: HBGF).

The nomenclature of these factors may sometimes be confusing. It has been proposed, therefore, to name them FGF-x, with x being a consecutive number It should be noted that some members of the FGF family are not mitogenic for fibroblasts either *in vitro* or *in vivo*.

Nomenclature Meeting Report and Recommendation, January 18, 1991, reprinted in: ANY Vol. 638, pp. xiii-xvi.

FGF-1: *fibroblast growth factor 1* The recommended name for » aFGF (see: FGF for references).

FGF-2: *fibroblast growth factor 2* The recommended name for » bFGF (see: FGF for references).

FGF-3: *fibroblast growth factor 3* The recommended name for the product of the » int2 oncogene (see: FGF for references).

FGF-4: *fibroblast growth factor 4* The recommended name for the product of the » *hst* oncogene (see: FGF for references).

FGF-5: *fibroblast growth factor 5* FGF-5 is a 29.5 kDa protein also called ***HBGF-5*** (heparin-binding growth factor 5), *hst*1 or HSTF1 (for some aspects of nomenclature see also: FGF). FGF-5 is the product of an » oncogene initially detected after transfection of DNA from a human bladder carcinoma cell line into murine NIH-3T3 cells (see: 3T3).
FGF-5 is a sialylated glycoprotein belonging to the fibroblast growth factor family of proteins (see: FGF). At the nucleotide level homology to human » bFGF is ≈ 44 %. The human FGF-5 gene consists of three exons and maps to chromosome 4q21. The biological activities of FGF-5 resemble those of » bFGF and » aFGF. The factor is mitogenic for fibroblasts and endothelial cells *in vitro*. Its biological functions *in vivo* are unknown. In the mouse FGF-5 is expressed during embryogenesis and also in some neurons of adult brain.

Bates B et al Biosynthesis of human fibroblast growth factor 5. MCB 11: 1840-5 (1991); **Burgess WH & Maciag T** The heparin-binding (fibroblast) growth factor family of proteins. ARB 58: 575-606 (1989); **Dionne CA et al** Chromosome assignment by polymerase chain reaction techniques: assignment of the oncogene FGF-5 to human chromosome 4. Biotechniques 8: 190-4 (1990); **Haub O et al** Expression of murine fibroblast growth factor-5 in the adult central nervous system. PNAS 87: 8022-6 (1990); **Hebert JM et al** Isolation of cDNAs encoding four mouse FGF family members and characterization of their expression during embryogenesis. Dev. Biol. 138: 454-63 (1990); **Lafage-Pochitaloff M et al** The human basic fibroblast growth factor gene is located on the long arm of chromosome 4 at bands q26-q27. OR 5: 241-4 (1990); genes. Mammalian Genome 2: 135-7 (1992); **Nguyen C et al** The FGF-related oncogenes *hst* and *int*.2, and the *bcl*.1 locus are contained within one megabase in band q13 of chromosome 11, while *fgf*.5 oncogene maps to 4q21. O 3: 703-8 (1988); **Zhan X et al** The human FGF-5 oncogene encodes a novel protein related to fibroblast growth factors. MCB 8: 3487-95 (1988)

FGF-6: *fibroblast growth factor 6* FGF-6 is a protein belonging to the fibroblast growth factor family of proteins (see: FGF). Its biological activities resemble those of » bFGF and » aFGF. The gene encoding FGF-6 is also called » *hst*-2 or *hst*-1-related (see: *hst*). The human gene maps to chromosome 12p13. FGF-6 is the product of an » oncogene initially isolated by screening a human genomic library with a *hst*-specific probe under relaxed hybridization conditions. The protein is synthesized as a precursor of 198 amino acids that includes a secretory » signal sequence. The amino acid sequence of FGF-6 shows 70% identity with the *hst* gene product over the C-terminal two-thirds of the protein.
The FGF-6 gene is expressed in human leukemia cell lines with platelet/megakaryocytic differentiation potential. FGF-6-transformed murine NIH-3T3 fibroblasts form a well-vascularized tumor in nude mice (see also: Immunodeficient mice), thus suggesting an angiogenic property similar to some other members of this gene family.

De Lapeyière D et al Structure, chromosome mapping, and expression of the murine FGF-6 gene. O 5: 823-31 (1990); **Iida S et al** Human *hst*-2 (FGF-6) oncogene: cDNA cloning and characterization. O 7: 303-9 (1992); **Marics I et al** Characterization of the HST-related FGF-6 gene, a new member of the fibroblast growth factor gene family. O 4: 335-40 (1989)

FGF-7: *fibroblast growth factor 7* The new name for the product of the gene encoding » KGF (keratinocyte growth factor). See also: FGF for some aspects of nomenclature.

Nomenclature Meeting Report and Recommendation, January 18, 1991, reprinted in: ANY Vol. 638, pp. xiii-xvi.

FGF-8: *fibroblast growth factor 8* see: AIGF (androgen-induced growth factor).

FGF-9: *fibroblast growth factor 9* The new name for a heparin-binding growth factor (see also: HBGF) originally termed GAF (glia activating factor). FGF-9 is secreted by a human glioma cell line and acts on cells of the central nervous system. It is a potent mitogen for glial cells. It activates O-2A progenitor cells, » PC12 cells, murine » 3T3 cells, and rat glial cells. The factor does not act on human umbilical vein endothelial cells but is active on fibroblasts.
Human FGF-9 is a protein of 208 amino acids with sequence similarity of ≈ 30 % to other members of the fibroblast growth factor family of proteins (see also: FGF). Cloning of rat FGF-9 reveals that FGF-9 sequences are strongly conserved. The introduction of FGF-9 cDNA into murine fibroblasts leads to malignant transformation of the cells, as has been found for other members of the FGF family.

Miyamoto M et al Molecular cloning of a novel cytokine cDNA encoding the ninth member of the fibroblast growth factor family, which has a unique secretion pattern. MCB 13: 4251-9 (1993); **Naruo K et al** Novel secretory heparin-binding factors from human glioma cells (glia-activating factors) involved in glial cell growth. JBC 268: 2857-64 (1993)

FGF-α: This factor is identical with » aFGF.
Bertolini J et al Rapid chromatographic isolation and immunoblot characterization of immunoreactive fibroblast growth factor-related polypeptides from various tissues. J. Chromatogr. 491: 49-60 (1989)

FGF-β: This factor is identical with » bFGF.
Bertolini J et al Rapid chromatographic isolation and immunoblot characterization of immunoreactive fibroblast growth factor-related polypeptides from various tissues. J. Chromatogr. 491: 49-60 (1989); **Guthridge M et al** Detection of FGF-β mRNA in chondrosarcoma cells by a new *in situ* hybridization technique with synthetic oligonucleotide probes. JBBM 22: 279-88 (1991); **Schmitt JF et al** A new quantitative polymerase chain reaction-high performance ion exchange liquid chromatographic method for the detection of fibroblast growth factor-β (FGF-β) gene amplification. JBBM 24: 119-33 (1992)

fgr*:** The cellular *fgr* gene, previously known as ***src2, on the short arm of human chromosome 1 (1p36.2-p36.1) is the cellular homologue of the oncogene of the acute transforming retrovirus (GR-FeSV, Gardner-Rasheed feline sarcoma virus). *fgr* encodes a tyrosine-specific protein kinase (59,4 kDa; 529 amino acids) belonging to the family of *src*-related protein kinases. Transforming versions of the cellular *fgr* proto-oncogene can arise by point mutations and removal of sequences from the 3´ end. Normal myelomonocytic cells and tissue macrophages are the major sites of *fgr* mRNA expression. The protein accumulates in human » HL60 cells when these are induced towards monocytic or granulocytic differentiation. Certain lymphomas (but not sarcomas or carcinomas) express *fgr*-related messenger RNA. It has been observed that the expression of the *fgr* (and » *hck*) protein-tyrosine kinases in acute myeloid leukemic blasts is associated with early commitment and differentiation events in the monocytic and granulocytic lineages. The levels of *fgr* RNA in Burkitt's lymphoma cells rise approximately 50-fold upon infection with EBV. The production of *fgr* in bone marrow-derived monocytic cells is induced by » M-CSF, » GM-CSF, and » IFN-γ.

fgr has been implicated in regulation of the behavior of activated phagocytes. Functional activation of human cultured macrophages with bacterial lipopolysaccharides decreases the level of *fgr* transcripts and at the same time augments the expression of » *hck*. *fgr* has been shown to be physically and functionally associated with » CD32 and thus appears to be involved in signal transduction pathways following interaction of Fc receptors with antibody-antigen complexes.

Overexpression of the viral *fgr* oncogene in murine epidermal keratinocytes which require the presence of » EGF, confers independence from insulin and » EGF.

Cheah MS et al *fgr* proto-oncogene mRNA induced in B lymphocytes by Epstein-Barr virus infection. N 319: 238-40 (1986); **Dracopoli NC et al** Localization of the *fgr* proto-oncogene on the genetic linkage map of human chromosome 1p. Genomics 3: 124-8 (1988); **Falco JP et al** Interactions of growth factors and retroviral oncogenes with mitogenic signal transduction pathways of Balb/MK keratinocytes. O 2: 573-8 (1988); **Faulkner L et al** Regulation of c-*fgr* messenger RNA levels in U937 cells treated with different modulating agents. Immunology 76: 65-71 (1992); Hamada F et al Association of immunoglobulin G Fc receptor II with *src*-like protein-tyrosine kinase *fgr* in neutrophils. PNAS 90: 6305-9 (1993); **Inoue K et al** Isolation and sequencing of cDNA clones homologous to the v-*fgr* oncogene from a human B lymphocyte cell line, IM-9. O 1: 301-4 (1987); **Katagiri K et al** Expression of *src* family genes during monocytic differentiation of HL-60 cells. JI 146: 701-7 (1991); **Katamine S et al** Primary structure of the human *fgr* proto-oncogene product p55c-*fgr*. MCB 8: 259-66 (1988); **King FJ & Cole MD** Molecular cloning and sequencing of the murine c-*fgr* gene. O 5: 337-44 (1990); **Le Beau MM et al** Evidence for two distinct c-src loci on human chromosomes 1 and 20. N 312: 70-1 (1984); **Link DC et al** Characterization of the 5' untranslated region of the human c-*fgr* gene and identification of the major myelomonocytic c-*fgr* promoter. O 7: 877-84 (1992); **Knutson JC** The level of c-*fgr* RNA is increased by EBNA-2, an Epstein-Barr virus gene required for B-cell immortalization. J. Virol. 64: 2530-6 (1990); **Ley TJ et al** Tissue-specific expression and developmental regulation of the human *fgr* proto-oncogene. MCB 9: 92-9 (1989); **Miyazaki Y et al** *fgr* proto-oncogene is expressed during terminal granulocytic differentiation of human promyelocytic HL60 cells. EH 21: 366-71 (1993); **Nishizawa M et al** Structure, expression, and chromosomal location of the human c-*fgr* gene. MCB 6: 511-7 (1986); **Notario V et al** Expression of the *fgr* proto-oncogene product as a function of myelomonocytic cell maturation. JCB 109: 3129-36 (1989); **Parker RC et al** Isolation of duplicated human c-src genes located on chromosomes 1 and 20. MCB 5: 831-8 (1985); **Patel M et al** Structure of the complete human c-*fgr* proto-oncogene and identification of multiple transcriptional start sites. O 5: 201-6 (1990); **Patel MS et al** A polymorphic microsatellite repeat is located close to the promoter region of the c-*fgr* proto-oncogene (*fgr*) at chromosome 1p36.2-p36.1. Hum. Mol. Genet. 1: 65 (1993); **Sator O et al** Diverse biologic properties imparted by the c-*fgr* proto-oncogene. JBC 267: 3460-5 (1992); **Tronick SR et al** Isolation and chromosomal localization of the human fgr proto-oncogene, a distinct member of the tyrosine kinase gene family. PNAS 82: 6595-9 (1985); **Willman CL et al** Differential expression and regulation of the c-*src* and c-*fgr* proto-oncogenes in myelomonocytic cells. PNAS 84: 4480-4 (1987); **Willman CL et al** Expression of the c-*fgr* and *hck* protein-tyrosine kinases in acute myeloid leukemic blasts is associated with early commitment and differentiation events in the monocytic and granulocytic lineages. Blood 77: 726-34 (1991); **Yi TL & Willman CL** Cloning of the murine c-*fgr* proto-oncogene cDNA and induction of c-*fgr* expression by proliferation and activation factors in normal bone marrow-derived monocytic cells. O 4: 1081-7 (1989); **Ziegler SF et al** Augmented expression of a myeloid-specific protein tyrosine kinase gene (*hck*) after macrophage activation. JEM 168: 1801-10 (1988)

***fgr*:** see addendum: *hck*.

Fibroblast activating factor: see: FAF.

Fibroblast derived differentiation inducing factor for human monoblastic leukemia cells: see: differentiation inducing factor for human monoblastic leukemia cells.

Fibroblast (derived) endothelial cell growth factor: see: f-ECGF.

Fibroblast-derived growth factor: see: FDGF.

Fibroblast-derived growth inhibitor: see: FDGI.

Fibroblast-derived macrophage activating factor: abbrev. FMAF. See MAF (macrophage activating factor).

Fibroblast-derived neutrophil activating peptide: see: FDNAP.

Fibroblast-derived neutrophil-activating protein: see: FINAP.

Fibroblast-derived neutrophil chemotactic factor: see: FDNCF.

Fibroblast-derived thymocyte-activating factor: see: FTAF.

Fibroblast interferon: This Interferon is identical with » IFN-β. See also: IFN for general information about interferons.

Fibroblast growth factors: see: FGF.

Fibroblast-inducible cytokine: see: FIC.

Fibroblast proliferation factor: One specific factor is identical with one of the molecular forms of » IL1. In addition there are other, less well characterized factors of the same name that are not identical with IL1. Some of these have also been referred to as » FAF (fibroblast activating factor).
Armendariz-Borunda- J et al Kupffer cells from CC1(4)-treated rat livers induce skin fibroblast and liver fat-storing cell proliferation in culture. Matrix 9: 150-8 (1989); **Austgulen R et al** *In vitro* cultured human monocytes release fibroblast proliferation factor(s) different from interleukin 1. J. Leukoc. Biol. 42: 1-8 (1987); **Matsushima K et al** Purification of human interleukin 1 from human monocyte culture supernatants and identity of thymocyte comitogenic factor, fibroblast proliferation factor, acute phase protein-inducing factor, and endogenous pyrogen. CI 92: 290-301 (1985); **Shaked A et al** Isolation of fibroblast proliferation factor: distinction from interleukin-1. Curr. Surg. 44: 487-90 (1987)

Fibroblast stimulating factor 1: see: FsF-1.

Fibroblast tumor cytotoxic factor: see: TCF.

Fibronectin-binding growth factor: This factor is isolated from bovine serum. It consists of two chains held together with disulphide bonds that are required for full activity. The biological activity of 25Kd is in a latent form that requires acidification, alkalinisation or exposure to urea for activation. At 50ng/ml the factor is mitogenic for murine fibroblasts. It is inhibitory at 10ng/ml for murine » 3T3 cells. Active 25Kd stimulates colony formation by fibroblasts on soft agar in the presence of 5ng/ml » EGF. The structural properties, chromatographic behavior, immunological and biological activities of 25Kd suggest a close relationship to » TGF-β.
Eskandarani HA & Ayad SR 25 kDa Isolation and characterization of 25Kd fibronectin-binding growth factor. Anticancer Res. 9: 695-707 (1989)

FIC: *fibroblast-inducible cytokine* The product of an immediate-early gene (see also: Gene expression), designated P16, isolated by differential screening of a cDNA library of serum-stimulated murine » 3T3 fibroblasts. FIC is a non-glycosylated monocytic cytokine of 97 amino acids. It displays 57 % homology to human and rabbit monocyte chemoattractant protein 1 (see: MCP-1). belongs to the » chemokine family of cytokines. It induces changes in intracellular calcium levels (see also: Calcium ionophore) in monocytes but not in platelets, neutrophils, or lymphocytes. Specific binding of FIC is observed in human monocytes, mouse monocytic cultured cells, and human endothelial cells, but not in lymphocytes, neutrophils, or primary mouse fibroblasts. The protein appears to bind to receptors on human monocytes that also bind murine and human MCP-1.
Almendral JM et al Complexity of the early genetic response to growth factors in mouse fibroblasts. MCB 8: 2140-8 (1988); **Heinrich JN et al** The product of a novel growth factor-activated gene, fic, is a biologically activity C-C-type cytokine. MCB 13: 2020-30 (1993)

Ficoll gradient centrifugation: Ficoll is a synthetic polymer made by copolymerisation of sucrose and epichlorhydrin. Among other things, it is a widely used medium for density centrifugation. Discontinuous Ficoll gradient centrifugation exploits physical properties such as density and size and is used to enrich and fractionate » hematopoietic stem cells and other hematopoietic progenitor cell types (see: Hematopoiesis) , i. e. to deplete from T lymphocytes.

Charbord P et al The separation of human cord blood by density gradient does not induce a major loss of progenitor cells. Bone Marrow Transplant. 9: s109-10 (1992); **Chau WK & Law P** Effect of L-phenylalanine methyl ester on the colony formation of hematopoietic progenitor cells from human bone marrow. Int. J. Cell Clon. 9: 211-9 (1991); **Cottler-Fox M et al** Mononuclear and $CD34^+$ stem cell recovery after automated Ficoll processing of marrow or peripheral blood stem cells for transplantation. PC>BR 377: 569-73 (1993); **Dy M et al** Concomitant histamine, interleukin 4, and interleukin 6 production by hematopoietic progenitor subsets in response to interleukin 3. EH 19: 934-40 (1991); **Iacone A et al** Density gradient separation of hematopoietic stem cells in autologous bone marrow transplantation. Haematologica 76: s18-21 (1991); **Jones RJ et al** Separation of pluripotent haematopoietic stem cells from spleen colony-forming cells. N 347: 188-9 (1990); **Law P et al** Density gradient isolation of peripheral blood mononuclear cells using a blood cell processor. Transfusion 28: 145-50 (1988); **Niskanen E et al** Separation by velocity sedimentation of human haemopoietic precursors forming colonies *in vivo* and *in vitro* cultures. Cell. Tissue Kinet. 18: 399-406 (1985); **Wagner JE jr** Isolation of primitive hematopoietic stem cells. Semin. Hematol. 29: s6-9 (1992)

F-IFN: *fibroblast interferon* This interferon is identical with » IFN-β. See also: IFN for general information about interferons.

Fi-IFN: *fibroblast interferon* This interferon is identical with » IFN-β. See also: IFN for general information about interferons.

Filgrastim: Generic name for non-glycosylated recombinant » G-CSF (= Neupogen, Amgen Inc. Thousand Oaks, California). See also: KW-2228 (Marograstim).

Anderson M Filgrastim. Conn. Med. 56: 298-9 (1992); **Glasser L et al** Measurement of serum granulocyte colony-stimulating factor in a patient with congenital agranulocytosis (Kostmann's syndrome). Am. J. Dis. Child. 145: 925-8 (1991); **Rogers KM** Topics in clinical pharmacology: filgrastim, a myeloid colony stimulating factor. Am. J. Med. Sci. 303: 429-31 (1992); **Scarim SK** G-CSF and GM-CSF. Pediatr. Nurs. 17: 501-2 (1991); **Sheridan WP et al** Effect of peripheral-blood progenitor cells mobilized by filgrastim (G-CSF) on platelet recovery after high-dose chemotherapy. Lancet 339: 640-4 (1992), Comment in: Lancet 339: 1410-11 (1992)

FINAP: *fibroblast-derived neutrophil-activating protein* This factor is produced by human skin fibroblasts after stimulation with IL1α or IL1β (see: IL1). Production of this factor by these cells is not induced by incubation with bacterial » endotoxins. Three forms of FINAP, designated ***FINAP-α*** (6,7 kDa), ***FINAP-β*** (3,6 kDa) and ***FINAP-γ*** (5,3 kDa), have been described. FINAP-α is recognized by antibodies directed against » NAP-1 (neutrophil activating protein 1)/IL8. It is the 77 amino acid form of IL8. FINAP-β is a truncated form of FINAP-α. Its amino acid sequence is identical with that of » MGSA (melanoma growth stimulatory activity).

FINAPs belong to the » chemokine family of cytokines.

Schröder JM et al Lipopolysaccharide-stimulated human monocytes secrete, apart from neutrophil-activating peptide 1/interleukin 8, a second neutrophil-activating protein. NH_2-terminal amino acid sequence identity with melanoma growth stimulatory activity. JEM 171: 1091-100 (1990); **Schröder JM et al** IL1-α or tumor necrosis factor α stimulate release of three (NAP-1/IL8-related neutrophil chemotactic proteins in human dermal fibroblasts. JI 144: 2223-32 (1990)

FINAP-α, β, γ: fibroblast-derived neutrophil-activating protein see: FINAP.

FK506: see » CsA (cyclosporin).

FKBP: *FK506 binding protein* see: CsA (cyclosporin).

FLEXFIT: see: Bioassays.

flg*:** *fms-like gene* This human gene has been isolated initially by sequence homology screening with sequences derived from the » *fms* gene. *flg* is expressed mainly in skin, brain, and lung tissues. It is not a member of the » *fms* family. *flg* encodes a receptor for » aFGF and » bFGF with tyrosine kinase activity and is also known as ***FGFR 1. It is identical with a gene also called N-*sam* (see: K-*sam*) and closely related to another protein called TK-14. The human *flg* gene maps to chromosome 8p12.

The *flg* receptor also binds Kaposi-FGF (see: K-FGF) albeit with a 15-fold reduced affinity in comparison to aFGF and bFGF. Binding of bFGF to the *flg* receptor requires interaction of the growth factor with heparan sulfate/proteoglycans of the extracellular matrix (see also: ECM). The expression of some mutated variants of the *flg* receptor leads to the malignant transformation of cells.

Avivi A et al *Flg*-2, a new member of the family of fibroblast growth factor receptors. O 6: 1089-92 (1991); **Fasel NJ et al** Isolation from mouse fibroblasts of a cDNA encoding a new form of the fibroblast growth factor receptor (*flg*). BBRC 178: 8-15 (1991); **Katoh O et al** Expression of the heparin-binding growth factor receptor genes in human megakaryocytic leukemia cells. BBRC 183: 83-92 (1992); **Mansukhani A et al** A murine fibroblast growth factor (FGF) receptor expressed in CHO cells is activated by basic FGF and Kaposi FGF. PNAS 87: 4378-82

(1990); **Mohammadi M et al** A tyrosine-phosphorylated carboxy-terminal peptide of the fibroblast growth factor receptor (*flg*) is a binding site for the SH2 domain of phospholipase C-γ1. MCB 11: 5068-78 (1991); **Orr-Urtreger A et al** Developmental expression of two murine fibroblast growth factor receptors, *flg* and *bek*. Development 113: 1419-34 (1991); **Ruta M et al** A novel protein tyrosine kinase gene whose expression is modulated during endothelial cell differentiation. O 3: 9-15 (1988); **Ruta M et al** Receptor for acidic fibroblast growth factor is related to the tyrosine kinase encoded by the *fms*-like gene (*flg*). PNAS 86: 8722-6 (1989); **Yayon A et al** Cell surface, heparin-like molecules are required for binding of basic fibroblast growth factor to its high affinity receptor. Cell 64: 841-8 (1991); **Yan G et al** Expression and transforming activity of a variant of the heparin-binding fibroblast growth factor receptor (*flg*) gene resulting from splicing of the α exon at an alternate 3'-acceptor site. BBRC183: 423-30 (1992)

***flk*-1:** *fetal liver kinase 1* see: KDR (kinase domain receptor).

flk*-2:** fetal liver kinase 2 This protein tyrosine kinase is found on murine hematopoietic stem cells. It appears to be identical with a *fms*-like kinase designated ***flt-3.

The ligand for *flk2/flt*-3 encodes a protein of 204 amino acids with an extracellular domain of 161 amino acids, a transmembrane domain of 22 amino acids, and a cytoplasmic domain of 21 amino acids. The protein also exists in a soluble form which stimulates the proliferation of defined subpopulations of murine bone marrow and fetal liver cells as well as human bone marrow cells highly enriched for » hematopoietic stem cells and primitive uncommitted progenitor cells.

Lyman SD et al Characterization of the protein encoded by the *flt*-3 (*flk*-2) receptor-like tyrosine kinase gene. O 8: 815-22 (1993); **Lyman SD et al** Molecular cloning of a ligand for the *flt3/flk*-2 tyrosine kinase receptor: a proliferative factor for primitive hematopoietic cells. Cell 75: 1157-67 (1993); **Matthews W et al** A receptor tyrosine kinase specific to hematopoietic stem cells and progenitor-enriched cell populations. Cell 651143-52 (1991); **Rosnet O et al** Murine *Flt*-3, a gene encoding a novel tyrosine kinase receptor of the PDGFR/CSF1R family. O 6:; 1641-50 (1991)

flt*:** *fms-like tyrosine kinase* This protein is related to » *fms* (the receptor for » M-CSF). *flt* is a transmembrane receptor protein that encodes a cytoplasmic domain functioning as a tyrosine-specific protein kinase. It is related to another receptor called » KDR and a *fms*-like tyrosine kinase 4 (FLT4***) cloned from a human HEL erythroleukemia cell library the gene of which maps to human chromosomal region 5q34-q35 slightly telomeric to the receptors for »M-CSF (see: *fms*) and » PDGR. Murine FLT4 maps to chromosome 11, region A5-B1.

On the basis of structural similarities *lck*, KDR, *flt*-4 constitute a subfamily of class III tyrosine kinases (see also: PTK, protein tyrosine kinase).

The human *flt* gene is identical with » K-*sam*. It maps to chromosome 13q12, proximal to the breakpoint of a balanced reciprocal chromosomal translocation [t(2;13)(q35;q14)] which has been identified in more than 50% of alveolar rhabdomyosarcomas. The *flt* gene, however, is not a target for disruption by this tumor-specific translocation.

The ligand for this receptor protein is » VEGF (vascular endothelial cell growth factor). It has been observed that the expression of the *flt* receptor is strongly reduced in many tumor cell lines. The *flt* receptor is expressed in developing and mature blood vessels and its ligand may therefore regulate endothelial differentiation, blood vessel growth, and vascular repair.

Aprelikova O et al *flt*4, a novel class III receptor tyrosine kinase in chromosome 5q33-qter. CR 52: 746-8 (1992); **de Vries C et al** The *fms*-like tyrosine kinase, a receptor for vascular endothelial growth factor. S 255: 989-91 (1992); **Galland F et al** Chromosomal localization of *flt*4, a novel receptor-type tyrosine kinase gene. Genomics 13: 475-8 (1992); **Galland F et al** The FLT4 gene encodes a transmembrane tyrosine kinase related to the vascular endothelial growth factor receptor. O 8: 1233-40 (1993); **Pajusola K et al** *flt*4 receptor tyrosine kinase contains seven immunoglobulin-like loops and is expressed in multiple human tissues and cell lines. CR 52: 5738-43 (1992); **Peters KG et al** Vascular endothelial growth factor receptor expression during embryogenesis and tissue repair suggests a role in endothelial differentiation and blood vessel growth. PNAS 90: 8915-9 (1993); **Rosnet O et al** Close physical linkage of the FLT1 and FLT3 genes on chromosome 13 in man and chromosome 5 in mouse. O 8: 173-9 (1993); **Satoh H et al** Regional localization of the human c-*ros*-1 on 6q22 and *flt* on 13q12. Jpn. J. Cancer Res. 78: 772-5 (1987); **Shibuya M et al** Nucleotide sequence and expression of a novel human receptor type tyrosine kinase gene (*flt*) closely related to the *fms* family. O 5: 519-24 (1990)

***flt*-3:** see: *flk*-2.

FLT4: see: *flt* (*fms*-like tyrosine kinase).

Fluorouracil: (5-Fluorouracil; abbrev. 5-FU) 5-FU is a substrate for thymidylate synthase, an enzyme which catalyses the conversion of deoxyuridine monophosphate (dUMP) to thymidine monophosphate (TMP) and thus plays a key role in the biosynthesis of pyrimidine bases. Thymidylate synthase converts 5-FU into to 5-fluorodeoxyuridylate *in vivo*. 5-FU is a so-called suicide inhibitor of thymidylate synthase, i. e. a substrate which the enzyme converts into a reactive irreversible inhibitor that immediately inactivates its catalytic activity. 5-FU is therefore an antimetabolite of nucleic acid synthesis. The compound can be converted also into 5-fluorouridine diphosphate or triphosphate and can be incorporated into RNA.

5-Fluorouracil

5-FU has been used extensively because, after systemic or local administration *in vivo* or treatment of cells *in vitro*, it has a temporary inhibitory effect on mouse bone marrow progenitors, killing most of them (see also: Hematopoiesis). 5-FU mainly affects multilineage colonies (see also: Colony formation assay) containing » CFU-GEMM. 5-FU has been shown to spare mainly non-cycling » hematopoietic stem cells (see also: Cell cycle). In addition, these surviving stem cells are more primitive than the average normal stem cells and also have a greater self-renewal potential and a high capacity to generate other hematopoietic cell types. Treatment with this compound, therefore, is one way to obtain cell populations with highly enriched primitive stem cell.
Treatment with 5-FU or other compounds (see: 4-HC, Hydroperoxycyclophosphamide) can be used also to study the types of progenitor cells arising from protected stem cells. Surviving cells can be assayed in long-term bone marrow cultures (see: LTBMC), by » limiting dilution analysis, » colony formation assays or by their ability to repopulate the bone marrow of lethally irradiated recipient mice (see: MRA, marrow repopulating ability, for an assay system). Such experiments demonstrate that » hematopoietic stem cell populations are heterogeneous with respect to self-renewal capacity, that the cell types differ in their expression of cell surface markers, and that they differ also in their requirements for the presence of certain cytokines. Cytoreductive treatment *in vivo* with 5-FU can be employed also to study the effects of cytokines and/or other compounds with respect to their protective and/or stimulatory effects on regenerating bone marrow cells and to study the differential toxicity for individual hematopoietic cell lineages.

Adam J & Rosendaal M Patterns of recovery of high proliferation potential colony-forming cells after stressing the hemopoietic system. I. Leuk. Res. 11: 421-7 (1987); **Baines P et al** Physical and kinetic properties of hemopoietic progenitor cell populations from mouse marrow detected in five different assay systems. Leuk. Res. 6: 81-8 (1982); **Baines P et al** Serum-free culture of primitive murine hemopoietic colony-forming cells. Int. J. Cell Cloning 4: 103-14 (1986); **Barker JE et al** Advantages of gradient vs. 5-fluorouracil enrichment of stem cells for retroviral-mediated gene transfer. EH 21: 47-54 (1993); **Botnik LE et al** Differential effects of cytotoxic agents on hematopoietic progenitors. CR 41: 2338-42 (1981); **Bradley TR et al** *In vivo* effects of interleukin-1 α on regenerating mouse bone marrow myeloid colony-forming cells after treatment with 5-fluorouracil. Leukemia 3: 893-6 (1989); **Donowitz GR & Quesenberry P** 5-Fluorouracil effect on cultured murine stem cell progeny and peripheral leukocytes. EH 14: 207-14 (1986); **Gordon MY et al** Plastic-adherent progenitor cells in human bone marrow. EH 15: 772-8 (1987); **Harrison DE & Lerner CP** Most primitive hematopoietic stem cells are stimulated to cycle rapidly after treatment with 5-fluorouracil. Blood 78: 1237-40 (1991); **Hodgson GS et al** The organization of hemopoietic tissue as inferred from the effects of 5-fluorouracil. EH 10: 26-35 (1982); **Hunt P et al** Evidence that stem cell factor is involved in the rebound thrombocytosis that follows 5-fluorouracil treatment. Blood 80: 904-11 (1992); **Jones BC et al** Enhanced megakaryocyte repopulating ability of stem cells surviving 5-fluorouracil treatment. EH 8: 61-4 (1980); **Kerk DK et al** Two classes of primitive pluripotent hemopoietic progenitor cells: separation by adherence. JCP 125: 127-34 (1985); **Koike K et al** Declining sensitivity to interleukin 3 of murine multipotential hemopoietic progenitors during their development. Application to a culture system that favors blast cell colony formation. JCI 77: 894-9 (1986); **Lerner C & Harrison DE** 5-Fluorouracil spares hemopoietic stem cells responsible for long-term repopulation. EH 18: 114-8 (1990); **Lohrmann HP & Schreml W** Cytotoxic drugs and the granulopoietic system. Recent Results Cancer Res. 81: 1-222 (1982); **Miyama-Inaba M et al** Isolation of murine pluripotent hemopoietic stem cells in the Go phase. BBRC 147: 687-94 (1987); **Moore MA & Warren DJ** Synergy of interleukin 1 and granulocyte colony-stimulating factor: *in vivo* stimulation of stem-cell recovery and hematopoietic regeneration following 5-fluorouracil treatment of mice. PNAS 84: 7134-8 (1987); **Mulder AH et al** Thymus regeneration by bone marrow cell suspensions differing in the potential to form early and late spleen colonies. EH 13: 768-75 (1985); **Nio Y et al** Comparative effects of a recombinant and a mutein type of granulocyte colony stimulating factor on the growth of Meth-A fibrosarcoma with 5-fluorouracil chemotherapy. Biotherapy 4: 81-6 (1992); **Ophir R et al** THF-γ 2, a thymic hormone, increases immunocompetence and survival in 5-fluorouracil-treated mice bearing MOPC-315 plasmacytoma. Cancer Immunol. Immunother. 30: 119-25 (1989); **O'Reilly M & Gamelli RL** Recombinant granulocyte-macrophage colony-stimulating factor improves hematopoietic recovery after 5-fluorouracil. J. Surg. Res. 45: 104-11 (1988); **Quesenberry P et al** Studies on the regulation of hemopoiesis. EH 13: s43-8 (1985); **Quesniaux VF et al** Use of 5-fluorouracil to analyze the effect of macrophage inflammatory protein-1 α on long-term reconstituting stem cells *in vivo*. Blood 81: 1497-504 (1993); **Rice A et al** 5-fluorouracil permits access to a primitive subpopulation of peripheral blood stem cells. Stem Cells Dayt. 11: s326-35 (1993); **Rich IN** The effect of 5-fluorouracil on erythropoiesis. Blood 77: 1164-70 (1991); **Rosendaal M & Adam J** Patterns of hemopoietic recovery after stress. II. Treatment with fluorouracil. Leuk. Res. 12: 479-85 (1988); **Santelli G & Valeriote F** Schedule-dependent cytotoxicity of 5-fluorouracil in mice. JNCI 76: 159-64 (1986); **Schütz JD et al** 5-Fluorouracil incorporation into DNA of CF-1 mouse bone marrow cells as a possible mechanism of toxicity. CR 44: 1358-63 (1984); **Schwartz GN et al** Countercurrent centrifugal elutriation (CCE) recovery profiles of hematopoietic stem cells in marrow from normal and 5-FU-treated mice. EH 14: 963-70 (1986); **Shibagaki T et al** Fraction of pluripotent hemopoietic stem cells in DNA synthesis varies with generation age. EH 14: 794-7 (1986); **Steward FM et al** Post-5-fluorouracil human marrow: stem cell characteristics and renewal properties after autologous marrow transplantation. Blood 81: 2283-9 (1993); **Suda T et al** Proliferative kinetics and differentiation of murine blast cell colonies in culture: evidence for variable G0 periods and constant doubling rates of early pluripotent hemopoietic progenitors. JCP 117: 308-18 (1983); **Suda J et al** Analysis of differentiation of mouse hemopoietic stem cells in culture by sequential replating of paired progenitors. Blood 64: 393-9 (1984); **Suda J et al** Purified interleukin-3 and erythropoietin support the terminal differentiation of hemopoietic progenitors in serum-free culture. Blood 67: 1002-6 (1986); **Szilvassy SJ & Cory S** Phenotypic and functional characterization of competitive long-term repopulating hematopoietic stem cells enriched from 5-fluorouracil-treated murine marrow. Blood 81: 2310-20 (1993); **Takatsuki F et al** Interleukin 6 perfusion stimulates reconstitution of the immune and hematopoietic systems after 5-fluorouracil treatment. CR 50: 2885-90 (1990); **Van Zant G** Studies of hematopoietic stem cells spared by 5-fluorouracil. JEM 159: 679-90 (1984); **Vetvicka V et al** Effects of 5-fluorouracil on B lymphocyte lineage cells. JI 137: 2405-10 (1986); **Von Wangenheim KH et al** 5-Fluorouracil treatment after irradiation impairs recovery of bone marrow functions. Radiat. Environ. Biophys. 26: 163-70 (1987); **Williams DE et al** Pluripotential hematopoietic stem cells in post-5-fluorouracil murine bone marrow express the Thy-1 antigen. JI 135: 1004-11 (1985); **Yeager AM et al** The effects of 5-fluorouracil on hematopoiesis: studies of murine megakaryocyte-CFC, granulocyte-macrophage-CFC, and peripheral blood cell levels. EH 11: 944-52 (1983)

FMAF: *fibroblast-derived macrophage activating factor* See MAF (macrophage activating factor).

fMLP: Formyl-Methionyl-Leucyl-phenylalanine (f-Met-Leu-Phe). N-formylated peptides are believed to derive from bacterial protein degradation or from mitochondrial proteins upon tissue damage. fMLP is a strong chemoattractant (see also: Chemotaxis) and, among other things, induces adherence, degranulation and production of tissue-destructive oxygen-derived free radicals in phagocytic cells. For experimental studies the receptor agonist fMLP is used as a general purpose agent to deliberately prime and activate granulocytes (see also: Cell activation). The stimulatory activity of fMLP is influenced negatively by » IL1 and positively by » TNF-α.
Receptors for bacterial N-formyl peptides are instrumental for neutrophil chemotactic locomotion and activation at sites of infection and » inflammation. fMLP receptor expression is upregulated by various cytokines.
The human fMLP receptor gene maps to chromosome 19. The receptor shows a pronounced sequence homology (69 %) to the receptor of » IL8. Two structural homologues of the fMLP receptor (FPR; formyl peptide receptor-like) that do not recognize fMLP as a ligand but probably function as chemotactic receptors have been identified on chromosome 19.
Stimulation of the fMLP receptor activates a phospholipase C which in turn results in production of the intracellular second messengers diacylglycerol and inositol triphosphate. These second messengers activate the protein kinase C and mobilize calcium from intracellular stores (see also: Calcium ionophore), respectively. The formyl peptide receptor is a G- protein coupled receptor. If this G protein is blocked by pertussis toxin, cells do not respond to fMLP.
Atkinson YH et al Recombinant human granulocyte-macrophage colony-stimulating factor (rH GM-CSF) regulates f Met-Leu-Phe receptors on human neutrophils. Immunology 64: 519-25 (1988); **Bao L et al** Mapping of genes for the human C5a receptor (C5AR), human FMLP receptor (FPR), and two FMLP receptor homologue orphan receptors (FPRH1, FPRH2) to chromosome 19. Genomics 13: 437-40 (1992); **Boulay F et al** The human N- formylpeptide receptor: characterization of two cDNA isolates and evidence for a new subfamily of G-protein-coupled receptors. B 29: 11123-33 (1990); **De Nardin E et al** Identification of a gene encoding for the human formyl peptide receptor. BI 26: 381-7 (1992); **Jesaitis AJ et al** Functional molecular complexes of human N-formyl chemoattractant receptors and actin. JI 154: 5653-65 (1993); **Perez HD et al** Cloning of the gene coding for a human receptor for formyl peptides. Characterization of a promoter region and evidence for polymorphic expression. B 31: 11595-9 (1992); **Prossnitz ER et al** Signal transducing properties of the N-formyl peptide receptor expressed in undifferentiated HL60 cells. JI 151: 5704-15 (1993); **Quehenberger O et al** Multiple domains of the N-formyl peptide receptor are required for high-affinity ligand binding. Construction and analysis of chimeric N-formyl peptide receptors. JBC 268: 18167-75 (1993)

***fms*:** c-*fms*-Oncogen. c-*fms* is the cellular homologue of a viral » oncogene called v-*fms* which is encoded by SM-FeSV (Susan McDonough strain of Feline sarcoma virus). *fms* encodes a transmembrane protein consisting of an extracellular ligand-binding domain of 512 amino acids, a transmembrane region of 25 amino acids, and a cytoplasmic domain of 435 amino acids. The cytoplasmic domain encoded a tyrosine-specific protein kinase activity. The human *fms* gene consists of 21 small exons interrupted by introns ranging in size from 6.3 kb to less than 0.1 kb.
fms is expressed on cells of the monocyte/macrophage lineage. It has been renamed ***CD115*** (see also: CD Antigens). The synthesis of this receptor can be enhanced by treatment of cells with dexamethasone. *fms* is a receptor for » M-CSF and is also called CSF1R (CSF-1 = M-CSF). Binding of the ligand activates the tyrosine kinase activity of the receptor. The viral oncogene encodes a protein

with a constitutive kinase. Mutations in the extracellular domain
(Leu301→Ser) or within the carboxyterminal domain (Tyr969→Phe) activate the functions of the cellular *fms* gene and enhance its transforming activity.

Mutations activating the M-CSF receptors have been observed in approximately 10 % of the patients with myelodysplastic syndromes. A deletion of both alleles of the CSF1R locus, which maps to human chromosome 5q33.2-3 in the vicinity of the receptor gene for » PDGF, is found in the bone marrow cells of some of these patients (see also: $5q^-$ syndrome). Breaks at this locus have been found in several cases of malignant histiocytosis, a neoplastic process characterized by fever, progressive wasting, lymphadenopathy, hepatosplenomegaly, and the proliferation of atypical histiocytes at all stages of maturation with frequent phagocytic activity.

Boultwood J et al Loss of both CSF1R (*fms*) alleles in patients with myelodysplasia and a chromosome 5 deletion. PNAS 88: 6176-80 (1991); **de Vries C et al** The *fms*-like tyrosine kinase, a receptor for vascular endothelial growth factor. S 255: 989-91 (1992); **Groffen J et al** Chromosomal location of the human c-*fms* oncogene. NAR 11: 6331-9 (1983); **Hampe A et al** Nucleotide sequence of the feline retroviral oncogene v-*fms* shows unexpected homology with oncogenes encoding tyrosine-specific protein kinases. PNAS 81: 85-9 (1984); **Hampe A et al** Nucleotide sequence and structural organization of the human *fms* proto-oncogene. OR 4: 9-17 (1989); **Kacinski BM et al** *fms* (CSF-1 receptor) and CSF-1 transcripts and protein are expressed by human breast carcinomas *in vivo* and *in vitro*. O 6: 941-52 (1991); **Le Beau MM et al** Assignment of the GM-CSF, CSF-1, and *fms* genes to human chromosome 5 provides evidence for linkage of a family of genes regulating hematopoiesis and for their involvement in the deletion (5q) in myeloid disorders. CSHSQB 51: 899-909 (1986); **Lyman SD & Rohrschneider LR** Analysis of functional domains of the v-*fms*-encoded protein of Susan McDonough strain feline sarcoma virus by linker insertion mutagenesis. MCB 7: 3287-96 (1987); **Manger R et al** Cell surface expression of the McDonough strain of feline sarcoma virus *fms* gene product (gp 140*fms*). Cell 39: 327-37 (1983); **Nienhuls AW et al** Expression of the human c-*fms* proto-oncogene in hematopoietic cells and its deletion in the syndrome. Cell 42: 421-8 (1985); **Reddy EP et al** (eds) The oncogene handbook. Elsevier Science Publishers, Amsterdam 1988; **Rettenmier CW et al** The colony-stimulating factor (CSF-1) (c-*fms* proto-oncogene product) and its ligand. JCS Suppl. 9: 27-44 (1988); **Ridge SA et al** *fms* mutations in myelodysplastic, leukaemic, and normal subjects. PNAS 87: 1377-80 (1990); **Roberts WM et al** Tandem linkage of human CSF-1 receptor (c-*fms*) and PDGF receptor genes. Cell 55: 655-61 (1988); **Rothwell VM & Rohrschneider LR** The murine c-*fms* cDNA: cloning, sequence analysis and retroviral expression. OR 1: 311-24 (1987); **Roussel MF et al** Molecular cloning of the c-*fms* locus and its assignment to human chromosome 5. J. Virol. 48: 770-3 (1983); **Roussel MF et al** A point mutation in the extracellular domain of the human CSF-1 receptor (c-*fms* proto-oncogene product) activates its transforming potential Cell 55: 979-88 (1988); **Sherr CJ et al** The c-*fms* proto-oncogene product is related to the receptor for the mononuclear phagocyte growth factor CSF-1. Cell 41: 665-76 (1985); **van Daalen T et al** Random mutagenesis of CSF-1 receptor (*fms*) reveals multiple sites for activating mutations within the extracellular domain. EJ 11: 551-7 (1992); **Wang Z et al** Identification of the ligand-binding regions in the macrophage colony-stimulating factor receptor extracellular domain. MCB 13: 5348-59 (1993)

***fms*-like tyrosine kinase:** see: *flt*.

FOG: The author's private extra unnecessary acronym (see: EUA) essentially meaning forgotten growth factor. *Mea culpa*.

Follicle stimulating hormone releasing protein: see: FRP.

Follicle-stimulating hormone (FSH) suppressing protein: This hormone is identical with » Follistatin.

Follistatin: This protein is also called ***follicle-stimulating hormone (FSH) suppressing protein*** (abbrev. FSP) and ***Activin-binding protein***. This protein is synthesized in the pituitary. It is also found in the ovaries, the decidual cells of the endometrium, and in some other tissues.

Follistatin is a monomeric glycoprotein of 315 amino acids. It is synthesized as a precursor of 344 amino acids. The protein is remarkably cysteine-rich (36 cysteine residues.

Follistatin suppresses the release of by FSH (follicle stimulating hormone) and » Luteinizing hormone) stimulated by gonadotropin-releasing hormone. It is therefore a natural antagonist of the biological activities of » Activin A. Follistatin binds to Activin A with high affinity; it does not bind to the Activin A receptor. Follistatin also binds » inhibins because of a common subunit with Activin A.

Follistatin can modulate activin action in a cell-type specific fashion, and thus may play an important role in regulating the bioavailability of activin. Apart from being an important physiological suppressor of hormone activities follistatin may also play a role in erythropoiesis (see also: Hematopoiesis). The simultaneous administration of Activin A (= erythroid differentiation factor, see: EDF) and Follistatin considerably reduces the number of erythroid progenitor cells in the bone marrow and the spleen of mice.

Follistatin binds to heparan sulfate chains of proteoglycans of the extracellular matrix (see also: ECM). This interaction is markedly reduced by heparin and heparan sulfate.

DePaolo LV et al Follistatin and activin: A potential intrinsic regulatory system within diverse tissues. PSEBM198: 500-12 (1991); **Esch FS et al** Structural characterization of follistatin: a novel follicle-stimulating hormone release-inhibiting polypeptide from the gonad. ME 1: 849-55 (1987); **Inouye S et al** Recombinant expression of human follistatin with 315 and 288 amino acids: chemical and biological comparison with native porcine follistatin. E 129: 815-22 (1991); **Inouye S et al** Site-specific mutagenesis of human follistatin. BBRC 179: 352-8 (1991); **Kaiser M et al** The rat follistatin gene is highly expressed in decidual tissue. E 126: 2768-70 (1990); **Klein R et al** The radioimmunoassay of follicle-stimulating hormone (FSH)-suppressing protein (FSP): stimulation of bovine granulosa cell FSP secretion by FSH. E 128: 1048-56 (1991); **Kogawa K et al** Activin binding protein is present in pituitary. E 128: 1434-40 (1991); **Ling N et al** Novel ovarian regulatory peptides: inhibin, activin, and follistatin. Clin. Obstet. Gynecol. 33: 690-702 (1990); **Mather JP et al** Follistatin modulates activin activity in a cell- and tissue-specific manner. Endocrinology 132: 2732-4 (1993); **Michel U et al** Rat follistatin: gonadal and extragonadal expression and evidence for alternative splicing. BBRC 173: 401-7 (1990); **Nakamura T et al** Activin-binding protein from rat ovary is follistatin. S 247: 836-8 (1990); **Nakamura T et al** Follistatin, an activin binding protein associates with heparan sulfate chains of proteoglycans on follicular granulosa cells. JBC 266: 19432-7 (1991); **Ogawa K et al** Expression of α, β A and β B subunits of inhibin or activin and follistatin in rat pancreatic islets FL 319: 217-20 (1993); **Saito S et al** Production of activin-binding protein by rat granulosa cells *in vitro*. BBRC 176: 413-22 (1991); **Shimonaka M et al** Follistatin binds to both activin and inhibin through the common subunit. E 128: 3313-5 (1991); **Shiozaki M et al** Evidence for the participation of endogenous activin A/erythroid differentiation factor in the regulation of erythropoiesis. PNAS 89: 1553-6 (1992); **Sugawara M et al** Radioimmunoassay of follistatin: application for *in vitro* fertilization procedures. J. Clin. Endocrinol. Metab. 71: 1672-4 (1990); **Tashiro K et al** Expression of mRNA for activin-binding protein (follistatin) during early embryonic development of *Xenopus laevis*. BBRC 174: 1022-7 (1991); **Ueno N et al** Isolation and partial characterization of follistatin: a single-chain M(r) 35,000 monomeric protein that inhibits the release of follicle-stimulating hormone. PNAS 84: 8282-6 (1987); **Van den Eijnden-van Raaij AJ et al** Differential expression of inhibin subunits and follistatin, but not of activin receptor type II, during early murine embryonic development. Dev. Biol. 154: 356-65 (1992)

Foreign cell-induced interferon: This interferon is identical with » IFN-α. See also: IFN for general information about interferons.

Formyl-Methionyl-Leucyl-Phenylalanine: see: fMLP.

fos*:** c-*fos* is the cellular homologue of the viral v-*fos* oncogene found in FBJ (Finkel-Biskis-Jinkins) and FBR murine osteosarcoma viruses (MSV) (see also: Oncogenes). The human *fos* gene maps to chromosome 14q21-q31. *fos* has been identified as ***TIS-28, a gene inducible in several cell types by » phorbol esters (see: TIS genes).

c-*fos* is thought to have an important role in signal transduction, cell proliferation, and differentiation. It is a nuclear protein which, in combination with other transcription factors (see, for example: *jun*) acts as a trans-activating regulator of gene expression.

fos is a so-called *immediate early response* gene which plays a key role in the early response of cells to growth factors. It is expressed within hours after binding of various growth factors to their corresponding receptors. *fos* is also involved in the control of cell growth and differentiation of embryonic hematopoietic and neuronal cells.

The expression of *fos* is itself subject to regulation by other transcription factors binding to some sequences in its promoter region (see: SRE, serum response element; see also: Figure showing the promoter structure of » IL6; see also: » NFIL6). Two elements in the *fos* promoter can mediate the induction of *fos* by » NGF.

One DNA region within the *fos* promoter binds a factor called ***SIF*** (*sis*-inducible factor) which mediates the specific induction of *fos* synthesis by » PDGF.

The aberrant expression of *fos* is responsible for the malignant transformation of cells. Resting T cells (G_0 phase of the » cell cycle) do not express *fos* protein. The activation of T cells through their antigen receptors or in response to many different mitogenic stimuli is accompanied by a rapid, but transient, increase in the synthesis of *fos* protein. This response can be abolished by treatment of the cells with » CsA (cyclosporin A) or antibodies directed against a cell surface receptor called CD11a which is the receptor for the cellular adhesion protein ICAM-1.

■ **TRANSGENIC/KNOCK-OUT/ANTISENSE STUDIES:** Chimaeric mouse strains generated by introduction of a functional c-*fos* gene into » ES cells and overexpressing *fos* protein show a normal pattern of embryonic development. However, young and also older animals show aberrations of bone development and subsequently develop bone tumors. c-*fos*/v-*jun* doubly transgenic animals also show perturbations of B cell development in bone marrow.

fos-transgenic mice (see also: transgenic animals) with deregulated expression of the *fos* gene also show an impaired growth of bones and cartilage and hematopoietic cell development. These animals, however, do not develop malignant tumors. Transgenic mice that express c-*fos* from the H2-Kb promoter in several organs have enlarged spleens and hyperplastic thymuses containing an increased number of thymic epithelial cells. The exogenous c-*fos* expression specifically affects thymic T cell development, leading to an increase of the fraction of mature thymocytes. B and T cell function is impaired, and H2-c-*fos* mice are immune deficient.

No gross phenotypic changes have been observed in transgenic mouse lines expressing low levels of *fos*, driven by the » Mx promoter, in skeletal muscle, brain and salivary glands and very high levels in spleen, liver, thymus, heart and kidney. Transgenic mice, having mouse proto-oncogene c-*fos* under the control of the murine MHC class I gene (H2-Kb) promoter, show accelerated rates chemical carcinogenesis.

The level of the c-*fos* gene transcripts is 100-fold greater in human term fetal membranes than in other normal human tissues and cells. The levels of c-*fos* expression in human amniotic and chorionic cells are close to that of v-*fos* expression leading to the induction of osteosarcomas in mice and transformation of fibroblasts *in vitro*.

Mouse embryonic stem cell lines (see: ES cells) have been generated in which both copies of the c-*fos* gene are specifically disrupted by homologous recombination. In one case the disruption of both copies of c-*fos* in these cells had no detectable effect on embryonic stem cell viability, growth rate, or differentiation potential. Embryonic stem cells lacking c-*fos* could differentiate into a wide range of cell types in tissue culture and also in chimeric mice. In another case mice lacking c-*fos* developed severe osteopetrosis, ossification of the marrow space, and absence of tooth eruption at ≈ 11 days. These mice showed delayed or absent gametogenesis, lymphopenia, and altered behavior, but some of these mice lived as long as their wild-type or heterozygous litter mates. It seems therefore, that despite the large body of information suggesting an important role for c-*fos* in growth and differentiation of most cell types, this gene is not essential for these processes in at least some cell types, but is involved in the development and function of several distinct tissues.

Angel P et al The role of *jun*, *fos*, and the AP-1 complex in cell proliferation and transformation. BBA Rev. Cancer 1072: 129-57 (1991); **Ekstrand AJ & Zech L** Human c-*fos* proto-oncogene mapped to chromosome 14, band q24.3-q31: possibilities for oncogene activation by chromosomal rearrangements in human neoplasms. ECR 169: 262-6 (1987); **Field SJ et al** Growth and differentiation of embryonic stem cells that lack an intact c-*fos* gene. PNAS 89: 9306-10 (1992); **Graham R & Gilman M** Distinct protein targets for signals acting at the c-*fos* serum-response element. S 251: 189-92 (1991); **Kovary K & Bravo R** The *Jun* and *Fos* families are both required for cell cycle progression in fibroblasts. MCB 11: 4466-72 (1991); **Metz R & Ziff E** cAMP stimulates the C/EPB-related transcription factor rNFIL6 to *trans*-locate to the nucleus and induce c-*fos* transcription. GD 5: 1764-6 (1991); **Perlmutter RM & Ziegler S** Proto-oncogene expression following lymphocyte activation. CTMT 35: 571-86 (1990); **Reddy EP et al** (eds) The oncogene handbook. Elsevier Science Publishers, Amsterdam 1988; **Rivera CM & Greenberg ME** Growth factor-induced gene expression: the ups and downs of c-*fos* regulation. New Biol. 2: 751-8 (1990); **Van Straaten F et al** Complete nucleotide sequence of a human c-*onc* gene: deduced amino acid sequence of the human c-*fos* gene protein. PNAS 80: 3183-7 (1983); **Verma IM & Sassone-Corsi P** Proto-oncogene *fos*: complex but versatile regulation. Cell 51: 513-4 (1987); **Visvader J et al** Two adjacent promoter elements mediate nerve growth factor activation of the c-*fos* gene and bind distinct nuclear complexes. PNAS 85: 9474-8 (1988); **Wagner BJ et al** The SIF binding element confers *sis*/PDGF inducibility onto the c-*fos* promoter. EJ 9: 4477-84 (1990); **Wang ZQ et al** A novel target cell for c-*fos*-induced oncogenesis: development of chondrogenic tumors in embryonic stem cell chimaeras. EJ 10: 2437-50 (1991); **Wang ZQ et al** Bone and hematopoietic defects in mice lacking c-*fos*. N 360: 741-5 (1992)

● **Transgenic/Knock-out/Antisense Studies: Bachiller D & Ruther U** Inducible expression of the proto-oncogene c-*fos* in transgenic mice. Arch. Geschwulstforsch. 60: 357-60 (1990); **Fujita K et al** B cell development is perturbed in bone marrow from c-*fos*/v-*jun* doubly transgenic mice. Int. Immunol. 5: 227-30 (1993); **Grigoriadis AE et al** Osteoblasts are target cells for transformation in c-*fos* transgenic mice. JCB 122: 685-701 (1993); **Johnson RS et al** Pleiotropic effects of a null mutation in the c-*fos* proto-oncogene Cell 71: 577-86 (1992); **Rüther U et al** Deregulated c-*fos* expression interferes with normal bone development in transgenic mice. N 325: 412-6 (1987); **Ruther U et al** c-*fos* expression interferes with thymus development in transgenic mice. Cell 53: 847-56 (1988); **Ruther U et al** c-*fos* expression induces bone tumors in transgenic mice. O 4: 861-5 (1989); **Ruther U & Wagner EF** The specific consequences of c-*fos* expression in transgenic mice. Prog. Nucleic Acid Res. Mol. Biol. 36: 235-45 (1989); **Sakai N** Chemical carcinogenesis is accelerated in c-*fos* transgenic mice. Kobe J. Med. Sci. 36: 37-53 (1990)

fps**:** see: *fes*.

fos**-related antigen 1:** see: fra-1.

fra-1: *fos-related antigen 1* An » ERG (early response gene) whose transcription is rapidly and transiently stimulated in diverse cell types following mitogenic stimulation of cells with serum. The protein shows several regions of extensive amino acid homology with » *fos*. Only a subset of the agents and conditions that activate *fos* expression also induce fra-1 expression. fra-1 has been shown to be one of the subunits of the NF-AT (nuclear factor for activated T cells) transcriptional complex which mediates the expression of » IL2.

Boise LH et al The NFAT-1 DNA binding complex in activated T cells contains Fra-1 and JunB. MCB 13: 1911-9 (1993); **Cohen DR & Curran T** fra-1: a serum-inducible, cellular immediate-early gene that encodes a *fos*-related antigen. MCB 8: 2063-9 (1988)

Friend spleen focus-forming virus: see: SFFV.

FRM: *factor replacing macrophages* This poorly characterized factor of 35 kDa has been isolated from the supernatant of murine resident peritoneal macrophage cultures. It can replace macrophages in the induction of the *in vitro* antibody response to sheep erythrocytes. When added at the outset of cultures it induces B cells to generate an antigen-specific antibody response. FRM also markedly reduces both proliferation and cell cycle progression of B cells stimulated with anti-μ plus » IL4. The addition of FRM to such cultures leads to a preferential accumulation of cells in the early G1 phase of the » Cell cycle and to a decreased frequency of cells in all other phases.
Souvannavong V et al Flow cytometric analysis of opposite effects of a monokine on proliferation and differentiation of murine B lymphocytes. Cytometry 13: 510-7 (1992)

FRP: *FSH-(follicle stimulating hormone) releasing protein* This protein is identical with the βA subunit of » Activin A.
Lumpkin MD et al Purification of FSH-releasing factor: its dissimilarity from LHRH of mammalian, avian, and piscian origin. Brain Res. Bull. 18: 175-8 (1987); **Vale W et al** Purification and characterization of an FSH releasing protein from porcine ovarian follicular fluid. N 321: 776-9 (1986)

FSdGF: *folliculo stellate cell-derived growth factor* This factor, isolated from bovine pituitary folliculo stellate cells, is identical with » VEGF (vascular endothelial cell growth factor).
Gospodarowicz D et al Isolation and characterization of a newly identified endothelial cell mitogen factor produced by pituitary derived folliculo stellate cells. PNAS 86: 7311-5 (1989)

FsF: *fibroblast stimulating factor* This poorly characterized factor is produced by cloned *Schistosoma mansoni* antigen-specific T cells. It served as a competence factor for murine fibroblasts (see also: Cell activation). FsF has an apparent molecular weight of 17 kDa and is not identical with » CSF, » IL3, and » IFN. A monoclonal anti-B cell-stimulating factor-1 (= IL4; see: BSF 1) antibody only partially blocks the fibroblast proliferation induced by T cell supernatants. It is not known whether FsF is identical with » FsF-1.
Lammie PJ et al Partial characterization of a fibroblast-stimulating factor produced by cloned murine T lymphocytes. Am. J. Pathol. 130: 289-95 (1988)

FsF-1: *fibroblast stimulating factor 1* This is an acidic factor isolated from chronic granulomas developing after infections with *Schistosoma mansoni* or *Schistosoma japonicum*. The factor is secreted by CD4-positive lymphocytes residing in these granulomas. The factor is highly mitogenic for fibroblasts and stimulates liver fat-storing cells *in vitro*. This poorly characterized factor of ≈ 60 kDa is not identical with » aFGF and its activity is not neutralized by aFGF-specific antibodies. It is a member of the heparin-binding growth factor family (see also: HBGF). The amino acid sequence derived from the cDNA clone demonstrates that FsF-1 is a new factor. It is not known whether FsF-1 is identical with » FsF.
Greenwel P et al Fibroblast-stimulating factor 1, a novel lymphokine produced in schistosomal egg granulomas, stimulates liver fat-storing cells *in vitro*. Infect. Immun. 61: 3985-7 (1993); **Prakash S et al** Fibroblast stimulation in schistosomiasis. XI. Purification to apparent homogeneity of fibroblast-stimulating factor 1, an acidic heparin-binding growth factor produced by schistosomal egg granulomas. JI 146: 1679-84 (1991); **Prakash S & Wyler DJ** Fibroblast stimulation in schistosomiasis. XII. Identification of $CD4^+$ lymphocytes within schistosomal egg granulomas as a source of an apparently novel fibroblast growth factor (FsF-1). JI 148: 3583-7 (1992); **Wyler DJ** Schistosomes, fibroblasts, and growth factors: how a worm causes liver scarring. New Biol. 3: 734-40 (1991)

FSH-releasing protein: This protein is identical with » Activin A. See also: FRP (FSH-(follicle stimulating hormone) releasing protein).

FSH release inhibitor: see: Inhibins.

FSH suppressing protein: This hormone is identical with » Follistatin.

FSP: *follicle-stimulating hormone (FSH) suppressing protein* This hormone is identical with » Follistatin.

FTAF: *fibroblast-derived thymocyte-activating factor* This poorly characterized factor is produced by fibroblasts. It activates thymocytes and also promotes their proliferation (see also: TAF, thymocyte activating factor). This factor inhibits the growth of those cells that produce it. This factor, or at least some of the biological activity ascribed to it, may be identical or closely related to one of the molecular forms of » IL1 because antisera directed against (IL1β reduce activity by ≈ 50 %. An FTAF activity found to be released by guinea pig fibroblasts has not been characterized.
Hanazawa S et al Bacteroides gingivalis fimbriae stimulate production of thymocyte-activating factor by human gingival fibroblasts. Infect. Immun. 56: 272-4 (1988); **Iribe H & Koga T** Partial characterization of thymocyte-activating factor derived from MDP-stimulated guinea pig fibroblasts. Microbiol. Immunol.

FRK: see addendum.

28: 1125-36 (1984); **Ohmori Y et al** Spontaneous production of thymocyte-activating factor by human gingival fibroblasts and its autoregulatory effect on their proliferation. Infect. Immun. 55: 947-54 (1987)

F-TCF: *fibroblast tumor cytotoxic factor* see: TCF (tumor cytotoxic factor).

Fumagillin: This antibiotic is obtained from *Aspergillus fumigatus Fresenius*, *In vitro* this compound inhibits the growth on endothelial cells. A derivative, O-(Chloroacetylcarbamoyl)-Fumagillol, known as ***AGM-1470*** or ***TNP-470***, shows the same activity at 50-fold reduced concentrations. In experimental animals fumagillin has been shown to be a potent ***angioinhibin***, i. e., a suppressor of blood vessel development (see also: Angiogenesis factors).

Most angioinhibins also display a pronounced anti-tumor activity. The growth of transplanted human solid tumors in mice can be inhibited effectively by systemic administration of fumagillin or its derivatives. These compounds prevent tumor growth if administered at the time of inoculation of the tumor cells and also stop the growth of massive established tumors. This activity is thought to be due to the ability of fumagillin to interfere with tumor angiogenesis. This view is supported by two observations. Fumagillins and derivatives are not cytotoxic for those cells *in vitro* that lead to massive tumor growth *in vivo*. The growth of tumors is inhibited if they form solid tumors. It is not inhibited if the same cells are grown in ascites fluid as a suspension of individual cells.

At present fumagillin and some related compounds are tested in clinical trials for their anti-tumor activities. A therapeutical intervention with these compounds may be of clinical significance also for non-tumor diseases such as diabetic retinopathy, hemangiomas, arthritis, and psoriasis which also involve neovascularisation processes. The permanent administration of fumagillin has been show to cause considerable weight losses in experimental animals. On the other hand, fumagillin derivatives appear to be promising because they do not show any toxic side effects which would prevent their long-term use.

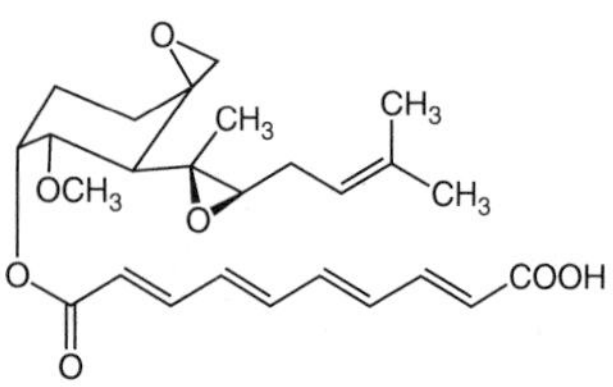

Corey EJ & Snider BB A total synthesis of fumagillin. J. Am. Chem. Soc. 94: 2549-50 (1972); **Ingber D et al** Synthetic analogs of fumagillin that inhibit angiogenesis and suppress tumor growth. N 348: 555-7 (1990); **Jaronski ST** Cytochemical evidence for RNA synthesis inhibition by fumagillin. J. Antibiot. (Tokyo) 25: 327-31 (1972); **Kusaka M et al** Potent anti-angiogenic action of AGM-1470: comparison to the fumagillin parent. BBRC 174: 1070-6 (1991); **Marui S et al** Chemical modification of fumagillin. I. 6-O-acyl, 6-O-sulfonyl, 6-O-alkyl, and 6-O-(N-substituted-carbamoyl)fumagillols. Chem. Pharm. Bull. Tokyo 40: 96-101 (1992); **Marui S & Kishimoto S** Chemical modification of fumagillin. II. 6-Amino-6-deoxyfumagillol and its derivatives. Chem. Pharm. Bull. Tokyo 40: 575-9 (1992); **Yamaoka M et al** Angiogenesis inhibitor TNP-470 (AGM-1470) potently inhibits tumor growth of hormone-independent human breast and prostate carcinoma cell lines. CR 53: 5233-6 (1993); **Yamaoka M et al** Inhibition of tumor growth and metastasis of rodent tumors by the angiogenesis inhibitor O-(chloroacetyl-carbamoyl)fumagillol (TNP-470; AGM-1470). CR 53: 4262-7 (1993)

fushi tarazu: see: Homeotic genes.

Fusion toxins: see: Cytokine fusion toxins.

Fx: A heat-stable protein of 5 kDa found in resting human platelets where it complexes the bulk of unpolymerised actin. Fx is identical with thymosin β 4 (see: Thymic hormones).

Authentic thymosin β 4 is functionally equivalent to Fx, forming a 1:1 complex with actin monomers and inhibiting polymerization. The widespread distribution and high intracellular concentration of thymosin β 4 suggests that it plays a significant role in regulating actin assembly in many cell types, for example, the morphological changes and actin redistribution occurring during the cytokinesis.

Thymosin β 10 has also been shown to function as an actin monomer sequestering protein.

Nachmias VT Small actin-binding proteins: the β-thymosin family. Curr. Opin. Cell Biol. 5: 56-62 (1993); **Otero A et al** Transcript levels of thymosin β 4, an actin-sequestering peptide, in cell proliferation. BBA 1176: 59-63 (1993); **Safer D et al** Thymosin β 4 and Fx, an actin-sequestering peptide, are indi-

Fumagillin

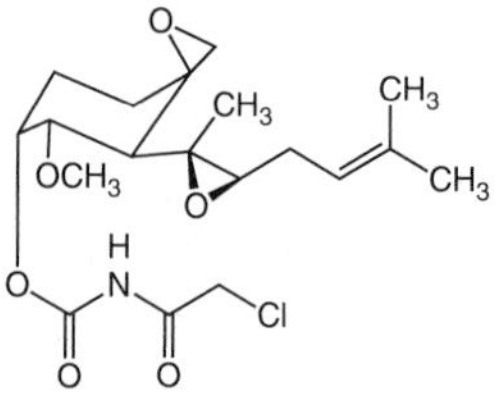

AGM-1470
(O-(Chloroacetylcarbamoyl)-Fumagillol)

stinguishable. JBC 266: 4029-32 (1991); **Sanders MC et al** Thymosin β 4 (Fx peptide) is a potent regulator of actin polymerization in living cells. PNAS 89: 4678-82 (1992); **Yu FX et al** Thymosin β 10 and thymosin β 4 are both actin monomer sequestering proteins. JBC 268: 502-9 (1993)

***fyn*:** *fyn* is a non-receptor cytoplasmic protein (59 kDa) of unknown function belonging to the *src* family of tyrosine kinases. It shows 74% amino acid identity to the tyrosine kinase domain of *src*. It was isolated originally on the basis of sequence homology to another *src* family member, » *yes*. *Fyn* contains an SH2 domain (see: *src* homology domain) which is necessary for its association with various target proteins. As a result of alternative splicing, *fyn* exists as two isoforms that differ exclusively within a short sequence spanning the end of the SH2 region and the beginning of the tyrosine protein kinase domain. While one p59f*yn* isoform (*fyn*B) is highly expressed in brain, the alternative product (*fyn*T) is principally found in T lymphocytes. The expression of *fyn* is regulated in both a developmental and cell type-specific manner. The human *fyn* gene maps to chromosome 6q21.

fyn plays a general role in growth control. It has been observed that mutant p59*fyn* with either a single amino acid substitution in the kinase domain or a deletion removing part of the SH2 domain produces a transforming protein, perhaps due to enhanced tyrosine kinase activity. *fyn* tyrosine kinase in the brain has been suggested to play a role in axonal growth and specialized functions of mature neurons and glia.

fyn has been shown to become associated after binding of » PDGF to its receptor. It also physically associates with the » IL7 receptor to mediate some of the intracellular signal processes. *fyn* has been shown to physically associate with the IL2 receptor. Elevated levels of *fyn* have been implicated in the development of tolerance in T lymphocyte anergy, a long-lived nonresponsiveness to normal antigenic stimulation for IL2 production. Only p59*fyn*T increases IL2 production in response to antigen stimulation of T lymphocytes.

fyn also becomes activated after binding of » M-CSF to its membrane receptor. *fyn* has also been shown to utilize several proteins as substrates that are associated with the T cell receptor (TCR)/CD3 complex. *fyn* has also been shown to associate with » CD23. Activation of platelets by thrombin and other physiological agonists leads to a dramatic increase in tyrosine phosphorylation of multiple cellular proteins, and one of the tyrosine kinases involved in this process has been shown to be *fyn*. *fyn* is associated with the major platelet membrane glycoprotein IV (GPIV, CD36), together with » *lyn* and » *yes*. *fyn* itself is a substrate of the CD45 tyrosine phosphatase.

fyn and other *src* family tyrosine kinases including» *blk*, » *lyn*, and perhaps » lck, are activated upon engagement of the B cell antigen receptor complex. These kinases then act directly or indirectly to phosphorylate and/or activate effector proteins including p42 (microtubule-associated protein kinase, MAPK), phospholipases C-γ 1 and C-γ 2, phosphatidylinositol 3-kinase (PI 3-K), and p21*ras*-GTPase-activating protein (GAP).

■ **TRANSGENIC/KNOCK-OUT/ANTISENSE STUDIES:** The biological activities of *fyn* have been analyzed in » transgenic mice that lack the thymic isoform of *fyn* but retain expression of the brain isoform of the protein. These *fyn*Tnull mice exhibit a specific lymphoid defect which is characterized by thymocytes being refractory to stimulation through the T cell receptor with mitogen or antigen, while peripheral T cells mature normally and reacquire significant signaling capabilities. These data confirm that the thymic isoform of *fyn* plays a pivotal role in T cell receptor signal transduction and demonstrate that additional developmentally regulated signaling components are required for T cell receptor-induced lymphocyte activation.

fyn mutant mice also show defects in long-term potentiation, spatial learning, and hippocampal development, suggesting that it may regulate the growth of neurons in the developing hippocampus and the strength of synaptic plasticity in the mature hippocampus.

Bare DJ et al p59*fyn* in rat brain is localized in developing axonal tracts and subpopulations of adult neurons and glia. O 8: 1429-36 (1993); **Burgess KE et al** CD5 acts as a tyrosine kinase substrate within a receptor complex comprising T-cell receptor zeta chain/CD3 and protein-tyrosine kinases p56lck and p59*fyn*. PNAS 89: 9311-5 (1992); **Cambier JC & Campbell KS** Membrane immunoglobulin and its accomplices: new lessons from an old receptor. FJ 6: 3207-17 (1992); **Cichowski K et al** p21*ras*-GAP association with *fyn*, *lyn*, and *yes* in thrombin-activated platelets. JBC 267: 5025-8 (1992); **Cooke MP et al** Regulation of T cell receptor signaling by a *src* family protein-tyrosine kinase (p59*fyn*). Cell 65: 281-91 (1991); **Courtneidge SA et al** Activation of *src* family kinases by colony stimulating factor-1, and their association with its receptor. EJ 12: 943-50 (1993); **Davidson D et al** Differential regulation of T cell antigen responsiveness by isoforms of the *src*-related tyrosine protein kinase p59*fyn*. JEM 175: 1483-92 (1992); **Gassmann M et al** Protein tyrosine kinase p59*fyn* is associated with the T cell receptor-CD3 complex in functional human lymphocytes. EJI 22: 283-6 (1992); **Gutkind JS et al** Thrombin-dependent association of phosphatidylinositol-3 kinase with p60c-*src* and p59*fyn* in human platelets. MCB 10: 3806-9 (1990); **Huang MM et al**

Membrane glycoprotein IV (CD36) is physically associated with the *fyn*, *lyn*, and *yes* protein-tyrosine kinases in human platelets. PNAS 88: 7844-8 (1991); **Ingraham CA et al** Cell type and developmental regulation of the *fyn* proto-oncogene in neural retina. O 7: 95-100 (1992); **Kawakami Y et al** Identification of *fyn*-encoded proteins in normal human blood cells. O 4: 389-91 (1989); **Kobayashi N et al** Functional coupling of the *src*-family protein tyrosine kinases p59*fyn* and p53/56*lyn* with the interleukin 2 receptor: implications for redundancy and pleiotropism in cytokine signal transduction. PNAS 90: 4201-5 (1993); **Lin J & Justement LB** The MB-1/B29 heterodimer couples the B cell antigen receptor to multiple *src* family protein tyrosine kinases. JI 149: 1548-55 (1992); **Maekawa N et al** Induction of Fc ε RII/CD23 on PHA-activated human peripheral blood T lymphocytes and the association of *fyn* tyrosine kinase with Fc ε RII/CD23. Res. Immunol. 143: 422-5 (1992); **Mustelin T et al** Regulation of the p59*fyn* protein tyrosine kinase by the CD45 phosphotyrosine phosphatase. EJI 22: 1173-8 (1992); **Noble ME et al** Crystal structure of the SH3 domain in human *fyn*; comparison of the three-dimensional structures of SH3 domains in tyrosine kinases and spectrin. EJ 12: 2617-24 (1993); **Pleiman CM et al** Mapping of sites on the *src* family protein tyrosine kinases p55*blk*, p59*fyn*, and p56*lyn* which interact with the effector molecules phospholipase C-γ 2, microtubule-associated protein kinase, GTPase-activating protein, and phosphatidylinositol 3-kinase. MCB 13: 5877-87 (1993); **Popescu NC et al** Chromosomal localization of the human FYN gene. O 1: 449-51 (1987); **Quill H et al** Anergic Th1 cells express altered levels of the protein tyrosine kinases p56*lck* and p59*fyn*. JI 149: 2887-93 (1992); **Semba K et al** Transformation of chicken embryo fibroblast cells by avian retroviruses containing the human *fyn* gene and its mutated genes. MCB 10: 3095-104 (1990); **Shiroo M et al** CD45 tyrosine phosphatase-activated p59*fyn* couples the T cell antigen receptor to pathways of diacylglycerol production, protein kinase C activation and calcium influx. EJ 11: 4887-97 (1992); **Sugi K et al** *fyn* tyrosine kinase associated with Fc ε RII/CD23: possible multiple roles in lymphocyte activation. PNAS 88: 9132-5 (1991); **Takayama T et al** A role for the *fyn* oncogene in metastasis of methylcholanthrene-induced fibrosarcoma A cells. IJC 54: 875-9 (1993); **Timson-Gauen LK et al** p59*fyn* tyrosine kinase associates with multiple T-cell receptor subunits through its unique amino-terminal domain. MCB 12: 5438-46 (1992); **Twamley GM et al** Association of *fyn* with the activated platelet-derived growth factor receptor: requirements for binding and phosphorylation. O 7: 1893-901 (1992); **Umemori H et al** Specific expressions of *fyn* and *lyn*, lymphocyte antigen receptor-associated tyrosine kinases, in the central nervous system. Brain Res. Mol. Brain Res. 16: 303-10 (1993); **Van Oers NS et al** Differential involvement of protein tyrosine kinases p56lck and p59*fyn* in T cell development. AEMB 323: 89-99 (1993); **Venkitaraman AR & Cowling RJ** Interleukin 7 receptor functions by recruiting the tyrosine kinase p59*fyn* through a segment of its cytoplasmic tail. PNAS 89: 12083-7 (1992)

● TRANSGENIC/KNOCK-OUT/ANTISENSE STUDIES: **Appleby MW et al** Defective T cell receptor signaling in mice lacking the thymic isoform of p59*fyn*. Cell 70: 751-62 (1992); **Grant SG et al** Impaired long-term potentiation, spatial learning, and hippocampal development in *fyn* mutant mice. S 258: 1903-10 (1993)

***Fyn*-related kinase:** see addendum: FRK.

G

G: Used as a prefix or suffix it denotes granulocytes such as in » G-CSF or » CFU-G (colony forming unit granulocytes).

G26: This factor is produced by activated human T lymphocytes (see also: Cell activation). It is identical with » ACT-2 and belongs to the » chemokine family of cytokines.
Miller MD et al A novel polypeptide secreted by activated human T lymphocytes. JI 143: 2907-16 (1989)

GAF: *gamma-interferon activation factor* GAF is a transcription factor (see: gene expression) specifically induced by treatment of cells with » IFN-γ. GAF exists in the cytoplasm in a latent form and is activated by phosphorylation immediately after binding of IFN-γ to its receptor. Phosphorylation of GAF is mediated by an IFN-regulated tyrosine kinase. The phosphorylated GAF protein translocates to the nucleus and specifically recognizes a nucleotide sequence, called ***GAS*** (γ-interferon activation site) which is found in the promoter region of genes the expression of which is induced by IFN-γ and overlaps with the interferon-stimulated response element, ISRE (see: IRS) also responsible for interferon-induced transcription.
The DNA-binding moiety of GAF is an IFN-responsive protein of 91 kDa which is also phosphorylated in response to » IFN-α and participates in the formation of the IFN-α-regulated transcriptional activator ISGF-3 (see also: IRS, interferon response sequence).
Decker T et al Cytoplasmic activation of GAF, an IFN-α-regulated DNA-binding factor. EJ 10: 927-32 (1991); **Eilers A et al** The response of γ interferon activation factor is under developmental control in cells of the macrophage lineage. MCB 13: 3245-54 (1993); **Shuai K et al** Interferon γ triggers transcription through cytoplasmic tyrosine phosphorylation of a 91 kDa DNA-binding protein. S 258: 1808-12 (1992)

GAF: *glia activating factor* see: FGF-9.

GAG: *glycosaminoglycan* see: ECM (extracellular matrix).

Galactoside-binding protein: see: GBP.

Gamma-BCDF: *B cell differentiation factor γ* see: BCDF γ.

Gamma-Endorphin: see: POMC (proopiomelanocortin).

Gamma-Interferon: see: IFN-γ. See also: IFN for general information about interferons.

Gamma-IP10: *gamma immune protein 10* see: IP-10.

Gamma-Lipotropin: see: POMC (proopiomelanocortin).

GAS: *gamma-interferon activation site* see: GAF (γ-interferon activation factor).

Gastrin: A hexadecapeptide hormone, also called ***cholecystokinin B***, synthesized from a larger precursor in the gastric antrum mucosa (G cells) in response to alkaline pH, mechanical stimulation, or vagus stimulation. Gastrin promotes the release of hydrochloric acid in the stomach and of digestive enzymes in the pancreas. Only the C-terminal tetrapeptide sequence of gastrin is responsible for all its physiologic actions. The synthesis is inhibited by acidic stomach juice. Cholecystokinin is a competitive inhibitor of gastrin.
Antral gastrin secretion and gene expression is inhibited by the » paracrine release of » Somatostatin from antral D cells. TGF-α and » EGF stimulate gastrin reporter gene constructs when transfected into pituitary GH4 cells. Somatostatin inhibits EGF stimulation of gastrin gene expression, which is in part mediated at the level of transcriptional regulation as somatostatin inhibits EGF stimulation of gastrin reporter gene constructs.
Apart from its pharmacological activities gastrin also has cytokine-like activities. It has been shown that gastrin and » TGF-α can act synergistically to stimulate pancreatic islet growth, although neither peptide alone is sufficient. In » transgenic animals (mice) pancreatic coexpression of both gastrin and

TGF-α significantly increases islet mass in mice expressing both transgenes. Gastrin has been shown to promote the growth of some colonic tumor cell lines and it has been suggested that aberrant expression of gastrin may contribute to deregulated proliferation of many colorectal carcinomas. Gastrin has also been shown to act as a growth factor for some pancreatic cancer cell lines. Gastrin may play a role also as a local growth factor in the developing colon. Gastrin also stimulates Ca^{2+} mobilization and clonal growth in small cell lung cancer cells. It has been suggested that gastrin receptor antagonists may be a therapeutic option for gastrin receptor-positive, gastro-intestinal tumors.

Bachwich D et al Identification of a cis-regulatory element mediating somatostatin inhibition of epidermal growth factor-stimulated gastrin gene transcription. Mol. Endocrinol. 6: 1175-84 (1992); **Blackmore M & Hirst BH** Autocrine stimulation of growth of AR4-2J rat pancreatic tumor cells by gastrin. Br. J. Cancer 66: 32-8 (1992); **Finley GG et al** Expression of the gastrin gene in the normal human colon and colorectal adenocarcinoma. CR 53: 2919-26 (1993); **Ishizuka J et al** The effect of gastrin on growth of human stomach cancer cells. Ann. Surg. 215: 528-34 (1992); **Kochman ML et al** Post-translational processing of gastrin in neoplastic human colonic tissues. BBRC 189: 1165-9 (1992); **Luttichau HR et al** Developmental expression of the gastrin and cholecystokinin genes in rat colon. Gastroenterology 104: 1092-8 (1993); **Rehfeld JW & Hilsted L** Gastrin and cancer. Adv. Clin. Chem. 29: 239-62 (1992); **Remy-Heinz N et al** Evidence for autocrine growth stimulation by a gastrin/CCK-like peptide of the gastric cancer HGT-1 cell line. Mol. Cell. Endocrinol. 93: 23-9 (1993); **Sethi T & Rozengurt E** Gastrin stimulates Ca^{2+} mobilization and clonal growth in small cell lung cancer cells. CR 52: 6031-5 (1992); **Van Solinge WW & Rehfeld JF** Co-transcription of the gastrin and cholecystokinin genes with selective translation of gastrin mRNA in a human gastric carcinoma cell line. FL 309: 47-50 (1992); **Wang TC et al** Pancreatic gastrin stimulates islet differentiation of transforming growth factor α-induced ductular precursor cells. JCI 92: 1349-56 (1993); **Watson SA et al** Inhibition of gastrin-stimulated growth of gastrointestinal tumor cells by octreotide and the gastrin/cholecystokinin receptor antagonists, proglumide and lorglumide. Eur. J. Cancer 28A: 1462-7 (1992); **Watson SA et al** Therapeutic effect of the gastrin receptor antagonist, CR2093 on gastrointestinal tumor cell growth. Br. J. Cancer 65: 879-83 (1992)

Gastrin-releasing peptide: see: GRP.

GATA-1: GATA-1, formerly known as » ***Eryf1***, » ***CSBP-1***, » ***F-ACT1***, » ***NF-E1***, » ***GF-1*** , » ***YY-1***, » ***UCRBP***, ***transcription factor δ***, is a DNA-binding protein of the 'zinc-finger' family that functions as a transcription factor (see also: Gene expression). GATA transcription factors form a » gene family comprising two related factors, designated » GATA-2 and GATA-3. The GATA factors share extensive homology in their DNA-binding domains and are coexpressed in erythroid cells.
The gene encoding GATA-1 has been identified as » ***cMG1***, ***CL100***, or » ***Erf-1***. GATA-1 recognizes the general consensus motif WGATAR and is expressed predominantly in erythroid cells. GATA-1 is highly expressed in the megakaryocytic and mast cell lineages, but not in other blood cell lineages or in non-hemopoietic cells. GATA-1 activity is required for normal erythroid development and is induced by treatment of cells with a variety of hematopoietic growth factors (see also: Hematopoietins). GATA-1 regulates erythroid-expressed genes in maturing erythroblasts, including the expression of the » Epo receptor. The extinction of erythroid genes after treatment of erythroleukemic cells with » Phorbol esters correlates with down-regulation of expression of NF-E1 and a related transcription factor, NF-E2.
δ thalassemia, a complex group of inherited disorders of globin genes characterized by impaired synthesis of the δ globin chain has been shown to be caused by a mutation at position -77 of the δ globin gene. This mutation lies within the GATA-1 binding site and abolishes GATA-1 binding, thus impairing the expression of the δ globin gene. The human GATA-1 gene has been mapped to chromosome Xp21-11.
GATA-1 has been implicated in retrovirus-induced induction of erythroleukaemia by specifically inducing Epo responsiveness in hematopoietic cells.. Two such viruses, murine ME26 virus, and avian E26 virus, encode a 135-kDa *gag-myb-ets* fusion protein which is localized in the nucleus and activates the GATA-1 gene, thus positively regulating the expression of the Epo receptor and leading to the proliferation of a unique population of Epo-responsive cells.
GATA-1 is autoregulatory, i.e. it has been shown to bind to its own promoter and up-regulate its own transcription. The activated glucocorticoid receptor has been shown to interfere with GATA-1 expression, thus explaining, at least in part, the inhibition of differentiation of some erythroleukemia cells by steroids. Elevated levels of GATA-1 have been observed in human leukemic K562 cells which differentiate in response to anthracycline antitumor drugs.

■ **Transgenic/Knock-out/Antisense Studies:** Disruption of the GATA-1 gene by homologous recombination in a murine embryonic stem cell line (see also: ES cells) results in their inability to generate mature red blood cells *in vivo* and *in vitro*, thus demonstrating that this gene plays an essential role in erythroid development and that other GATA-binding proteins cannot compensate for its absence.

● **Reviews: Simon MC** Transcription factor GATA-1 and erythroid development. PSEBM 202: 115-21 (1993)
Aurigemma RE et al Transactivation of erythroid transcription factor GATA-1 by a myb-ets-containing retrovirus. J. Virol. 66: 3056-61 (1992); **Caiulo A et al** Mapping the gene encoding the human erythroid transcriptional factor NFE1-GF1 to Xp11.23. Hum. Genet. 86: 388-90); **Chang TJ et al** Inhibition of mouse GATA-1 function by the glucocorticoid receptor: possible mechanism of steroid inhibition of erythroleukemia cell differentiation. Mol. Endocrinol. 7: 528-42 (1993); **Chiba T et al** GATA-1 transactivates erythropoietin receptor gene, and erythropoietin receptor-mediated signals enhance GATA-1 gene expression. NAR 19: 3843-8 (1991); **Matsuda M et al** δ thalassemia caused by disruption of the site for an erythroid-specific transcription factor, GATA-1, in the δ globin gene promoter. Blood 80: 1347-51 (1992); **Mignotte Y et al** The extinction of erythroid genes after tetradecanoylphorbol acetate treatment of erythroleukemic cells correlates with down-regulation of the tissue-specific factors NF-E1 and NF-E2. JBC 265: 22090-2 (1990); **Nicolis S et al** An erythroid specific enhancer upstream to the gene encoding the cell-type specific transcription factor GATA-1. NAR 19: 5285-91 (1991); **Sposi NM et al** Cell cycle-dependent initiation and lineage-dependent abrogation of GATA-1 expression in pure differentiating hematopoietic progenitors. PNAS 89: 6353-7 (1992); **Trenteseaux C et al** Increased expression of GATA-1 and NFE-2 erythroid-specific transcription factors during aclacinomycin-mediated differentiation of human erythroleukemic cells. Leukemia 7: 452-7 (1993); **Walters M & Martin DI** Functional erythroid promoters created by interaction of the transcription factor GATA-1 with CACCC and AP-1/NFE-2 elements. PNAS 89: 10444-8 (1992)
● **Transgenic/Knock-out/Antisense Studies: Pevny L et al** Erythroid differentiation in chimaeric mice blocked by a targeted mutation in the gene for transcription factor GATA-1. N 349: 257-60 (1991); **Simon MC et al** Rescue of the block to erythropoiesis in murine ES cells lacking a functional GATA-1 gene: analysis of a transcription factor in a developmental program. Nature Genet. 1: 92-8 (1992); **Simon MC et al** Rescue of erythroid development in gene targeted GATA-1- mouse embryonic stem cells. Nat. Genet. 1: 92-8 (1992)

GATA-2: A transcription factor belonging to the same » gene family as GATA-1 and » GATA-3. The GATA factors share extensive homology in their DNA-binding domains and are coexpressed in erythroid cells. GATA-2 has been shown to be required for preproendothelin (see: ET) gene expression in vascular endothelial cells.
Briegel K et al Ectopic expression of a conditional GATA-2/estrogen receptor chimera arrests erythroid differentiation in a hormone-dependent manner. Genes Dev. 7: 1097-109 (1993); **Dorfman DM et al** Human transcription factor GATA-2. Evidence for regulation of preproendothelin-1 gene expression in endothelial cells. JBC 267: 1279-85 (1992); **Lee ME et al** Cloning of the GATA-binding protein that regulates endothelin-1 gene expression in endothelial cells. JBC 266: 16188-92 (1991)

GATA-3: A transcription factor belonging to the same » gene family as GATA-1 and » GATA-2. The GATA factors share extensive homology in their DNA-binding domains and are coexpressed in erythroid cells. GATA-2 transcription factor has been implicated in regulating the self-renewal capacity of early erythroid progenitor cells (see also: Hematopoiesis). Human GATA-3 is most abundantly expressed in T cells and appears to play an important role in regulating the T cell-specific expression of T cell receptor subunit genes. Functional GATA-3 binding sites have also been identified within murine » CD8 α and » CD4 upstream regulatory sequences. Human GATA-3 has also been implicated in transcriptional activation of HIV-1 genes.
Briegel K et al Ectopic expression of a conditional GATA-2/estrogen receptor chimera arrests erythroid differentiation in a hormone-dependent manner. Genes Dev. 7: 1097-109 (1993); **Ho IC et al** Human GATA-3: a lineage-restricted transcription factor that regulates the expression of the T cell receptor α gene. EJ 10: 1187-92 (1991); **Ko LJ et al** Murine and human T-lymphocyte GATA-3 factors mediate transcription through a cis-regulatory element within the human T-cell receptor δ gene enhancer. MCB 11: 2778-84 (1991); **Landry DB et al** Functional GATA-3 binding sites within murine CD8 α upstream regulatory sequences. JEM 178: 941-9 (1993); **Yang Z & Engel JD** Human T cell transcription factor GATA-3 stimulates HIV-1 expression. NAR 21: 2831-6 (1993)

GATS: *growth stimulatory activity for TS1 cells* This protein of 25 kDa is found in conditioned medium of stimulated human connective tissue cells. It was initially isolated as a factor stimulating the growth of murine » TS1 cells that proliferate in response to murine » IL9. The factor is identical with » LIF (leukemia inhibitory factor).
Van Damme J et al Human growth factor for murine interleukin (IL)-9 responsive T cell lines: co-induction with IL6 in fibroblasts and identification as LIF/HILDA. EJI 22: 2801-8 (1992)

GBP: *β-galactoside-binding protein* GBP is constitutively produced by murine embryo fibroblasts. The cDNA sequence reveals a protein with a length of 134 amino acids (14,735 kDa). The corresponding gene contains four exons. The murine gene maps to the E region of chromosome 15 and the human gene maps to 22q12-q13.1.
The protein exists in two monomeric forms. One of these forms non-covalently binds to a glycan complex and this may facilitate » paracrine activities. Tetrameric forms of the protein retain biological activity but are not capable of binding to specific saccharide residues.
Murine GBP is a cytostatic protein. It is a negative growth factor arresting cells in the G_0 and the G_2 phase of the » cell cycle and reversibly inhibiting the growth of cells producing it (see also: Autocrine). The biological activity is mediated by a high-affinity membrane receptor.
HL14 is a cDNA clone overexpressed tenfold

GATA-2: see addendum.

during the induction of » apoptosis in the glucocorticoid-sensitive human leukemia cell line CEM C7. It encodes a β-galactoside binding protein which appears to be the human counterpart of mGBP.

Baldini A et al Mapping on human and mouse chromosomes of the gene for the β-galactoside-binding protein, an autocrine-negative growth factor. Genomics 15: 216-8 (1993); **Castronovo V et al** Identification of a 14-kDa laminin binding protein (HLBP14) in human melanoma cells that is identical to the 14-kDa galactoside binding lectin. ABB 297: 132-8 (1992); **Couraud PO et al** Molecular cloning characterization, and expression of a human 14 kDa lectin. JBC 264: 1310-6 (1989); **Goldstone SD & Lavin MF** Isolation of a cDNA clone, encoding a human β-galactoside binding protein, overexpressed during glucocorticoid-induced cell death. BBRC 178: 746-50 (1991); **Wells V & Mallucci L** Identification of an autocrine negative growth factor: mouse β-galactoside-binding protein is a cytostatic factor and cell growth regulator. Cell 64: 91-7 (1991); **Wells V & Mallucci L** Molecular expression of the negative growth factor murine β-galactoside binding protein (mGBP). BBA 1121: 239-44 (1992)

GCA: *granulocyte chemotactic activity* see: GCP (granulocyte chemotactic peptide).

G-CSA: *granulocytic neutrophil colony-stimulating activity* This activity is released by human peripheral blood leukocytes upon serum stimulation. G-CSA stimulates neutrophil colony formation from human bone marrow cells (see also: Colony formation assay). G-CSA has been identified as G-CSF.

Quesniaux VF et al Human serum stimulates the production of G-CSF, IL1, IL6 and IL8 by human peripheral blood leukocytes. Br. J. Haematol. 82: 6-12 (1992)

G-CSF: *granulocyte colony stimulating factor*

■ **Alternative Names:** 5637-derived factor; CSF-2 (colony-stimulating factor 2); CSF-β (colony stimulating factor β); DF (differentiation factor); D factor (leukemia cell differentiation inducing factor); G-CSA (granulocytic neutrophil colony-stimulating activity); GM-DF (granulocyte-macrophage and leukemia cell differentiation-inducing factor); LBGF (leukemic blast growth factor); M1 differentiation inducing activity; MGI-1G (macrophage-granulocyte inducer 1G); MGI-2 (macrophage-granulocyte inducer 2); NAP-IF (neutrophil alkaline phosphatase-inducing factor); pluripoietin, pluripoietin-β; pCSF (pluripotent colony stimulating factor); SCIF (suppressor cell-inducing factor); WEHI-3B differentiation inducing activity; see also: individual entries for further information.

A recombinant mutated form of human G-CSF is » KW-2228 (= Marograstim). Recombinant G-CSF is also marketed under the generic name » Filgrastim and Lenograstim and under the brand names Neupogen, Neutrogin, and Granocyte.

■ **Sources:** G-CSF was isolated initially as a factor supporting the growth of granulocyte-containing colonies in soft agar cultures (see also: CSF and Colony formation assay). The factor is secreted by activated monocytes and macrophages and neutrophils (see also: Cell activation). It is also produced by stromal cells, fibroblasts, and endothelial cells. Epithelial carcinomas, acute myeloid leukemia cells and various tumor cell lines (bladder carcinomas, medulloblastomas), also express this factor. The synthesis of G-CSF can be induced by bacterial » endotoxins, » TNF, » IL1 and » GM-CSF. Prostaglandin E_2 inhibits the synthesis of G-CSF.

■ **Protein Characteristics:** G-CSF is an O-glycosylated 19.6 kDa glycoprotein with a pI of 5.5. The biologically active form is a monomer. The analysis of its cDNA has revealed a protein of 207 amino acids containing a hydrophobic secretory » signal sequence of 30 amino acids. G-CSF contains 5 cysteine residues, four of which form disulfide bonds (positions 36-42; 64-74). The sugar moiety of this factor is not required for full biological activity.

Comparison of the primary sequence of this factor with those of the two other colony-stimulating factors, » GM-CSF and » M-CSF, shows that the three factors are not related to each other. Murine and human G-CSF show a sequence homology of ≈ 70 % at the DNA level and of 72 % at the protein level.

■ **Gene Structure:** The gene has a length of 2.5 kb and includes five exons. The human G-CSF gene maps to chromosome 17q21-q22. It is located in the vicinity of a translocation break point which frequently occurs in acute promyelocytic leukemias.

The nucleotide sequences of G-CSF and » IL6 reveal similarities which suggest that both genes are evolutionary related.

Differential splicing of the G-CSF mRNA can lead to the production of two variant forms of this protein. One of the resulting proteins is shortened by 3 amino acids.

The 5´region of the G-CSF gene contains several sequence domains called » CLE (conserved lymphokine element). They are binding sites for transcription factors (see also: Gene expression). One binding site, designated CK-1 (see: Cytokine 1) binds a transcription factor that mediates enhanced

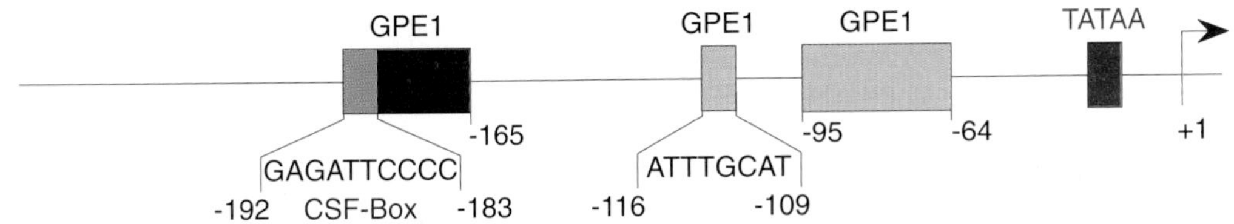

Structure of the G-CSF gene promoter.
The promoter contains a TATA box (red) typical for many eukaryotic genes. The promoter region contains also three sequence elements, designated GPE1-3 (G-CSF promoter elements), which can bind sequence-specific DNA binding proteins.
GPE1 consists of two regions. One (dark blue) is specific for the G-CSF gene, and another so-called CSF box (light blue) is found also in the promoter regions of the genes encoding murine and human GM-CSF and IL3. This region is important for the controlled expression of these genes. GPE2 contains a binding site for the transcription factor OTF (octamer transcription factor). Two proteins, designated OTF1 and OTF2 specifically bind to this region, and OTF2 is expressed specifically in B cells. Destruction of the CSF box by mutations causes an almost complete reduction of gene expression whereas mutations at the other sites only lead to a reduction of promoter activity. (J. Biol. Chem. 10: 5897-5902 (1990))

transcription of the G-CSF gene, inducible by several other cytokines.

■ **Receptor Structure, Gene(s), Expression:** The G-CSF receptor is expressed on all cells of the neutrophil/granulocyte lineage. It is also expressed in placenta cells, endothelial cells and various carcinoma cell lines. The human receptor has a length of 813 amino acids. It displays the classical elements of many other cytokine receptors, i. e. an extracellular ligand-binding domain, a transmembrane domain, and a cytoplasmic domain. The human receptor shows 62.5 % sequence homology to the murine receptor. The receptor binds G-CSF with high affinity (K_{dis} = 550 pM). The gene encoding the human G-CSF receptor (CSF3R) maps to chromosome 1p32-p34. 3.

At least four different forms of the human G-CSF receptor, resulting from alternative splicing of the mRNA, have been cloned from human placenta and » U937 myeloid leukemia cells. One variant contains a deletion of the transmembrane region and is probably a soluble form of the receptor. Another variant contains 27 additional amino acids in its cytoplasmic domain. This form has been described to be expressed in placenta.

■ **Biological Activities:** Human G-CSF is active in murine cells and vice versa. G-CSF stimulates the proliferation and differentiation of hematopoietic progenitor cells committed to the neutrophil/granulocyte lineage in a dose-dependent manner. At higher concentrations this factor induces the generation of colonies in soft agar cultures (see also: Colony formation assay) containing granulocytes and macrophages. The fully differentiated neutrophilic granulocytes are functionally activated by G-CSF (see also: Cell activation). G-CSF is a mitogen for some human myeloid leukemia cells and also for some carcinoma cell lines.

G-CSF synergises with some other cytokines, including » GM-CSF and » IL4. GM-CSF and G-CSF are required, for example, to develop neutrophilic colonies *in vitro*. The concerted action of G-CSF and » Epo is required to support the growth of mixed colonies of the early erythroid progenitors. A combination of » IL4 with G-CSF has been shown to lead to synergistic suppression of the growth of some human leukemic cell lines.

In vitro G-CSF enhances the antibody-dependent cellular cytotoxicity of granulocytes against tumor cells. G-CSF induces the synthesis of receptors for » fMLP (Formyl-Met-Leu-Phe) which, in turn, induces the prolonged production of oxygen radicals and the release of arachidonic acid. G-CSF also increases the synthesis of Fc-receptors for IgA. Unlike » GM-CSF, G-CSF does not induce the expression of cellular markers such as CD11A and CD11C.

■ **Assays:** G-CSF can be assayed in » colony formation assays employing murine bone marrow cells. Stimulation of proliferation of bone marrow cells induced by G-CSF is also used to assay this factor. G-CSF is also detected in specific » bioassays with cells lines that depend in their growth on the presence of G-CSF or that respond to this factor (see: AML 193; 32D; GNFS-50; M1; NFS-60; OCI/AML1a; WEHI). An alternative and entirely different detection method is » Message amplification phenotyping. For further information see also subentry "Assays" in the reference section.

■ **Clinical Use & Significance:** The main and most important clinical application of G-CSF is

G-CSF: see addendum.

probably the treatment of transient phases of leukopenia following chemo- and/or radiotherapy. G-CSF can be used to expand the myeloid cell lineage. It has been shown that the pretreatment with recombinant human G-CSF prior to marrow harvest can improve the graft by increasing the total number of myeloid lineage restricted progenitor cells, resulting in stable but not accelerated myeloid engraftment of autologous marrow.

Animal experiments and also the administration in humans have shown that the continuous administration of G-CSF for several days increases peripheral neutrophil blood counts by a factor of 10-15-fold without having any effects on erythrocytes and platelets. The daily administration is required to maintain elevated neutrophil levels. The duration of chemotherapy-induced neutropenia and neutropenia following autologous bone marrow transplantation is also markedly decreased by treatment with G-CSF.

Marked improvements of hematological parameters have also been observed in the treatment of patients with congenital neutropenia (Kostmann syndrome), patients with neutropenia caused by diffuse infiltration of the bone marrow by malignant Non-Hodgkin lymphomas, and in cases of chronic idiopathic neutropenia. One general effect of treatment with G-CSF appears to be a marked reduction of severe infections and episodes of fever which are normally observed to occur in these patients. G-CSF treatment also allows dose intensification with various antitumor drug regimes. The dosage of antibiotics can also be reduced.

It appears that the subcutaneous administration of G-CSF is more favorable at low doses than intravenous injection. Some animal experiments suggest that the effects of lethal irradiation can be abolished by a timely administration of G-CSF and that the complete reconstitution of hematopoiesis is possible. At present data from human irradiated patients are not yet available.

The transduction of murine tumor cells with a functional G-CSF gene has been shown to lead to the rejection of the genetically modified cells by syngeneic hosts (for cancer vaccines see also: Cytokine gene transfer).

● **Reviews: Demetri GD & Griffin JD** Granulocyte colony-stimulating factor and its receptor. Blood 78: 2791-2808 (1991); **Moore MAS** The clinical use of colony stimulating factors. ARI 9: 159-91 (1991); **Negrin RS & Greenberg PL** Therapy of hematopoietic disorders with recombinant colony-stimulating factors. Adv. Pharmacol. 23: 263-296 (1992)

● **Biochemistry & Molecular Biology: Asano M et al** Three individual regulatory elements of the promoter positively activate the transcription of the murine gene encoding granulocyte colony-stimulating factor. Gene 107: 241-6 (1991); **Hill CP et al** The structure of granulocyte-colony-stimulating factor and its relationship to other growth factors. PNAS 90: 5167-71 (1993); **Kanda N et al** Human gene coding for granulocyte colony-stimulating factor is assigned to the q21-q22 region of chromosome 17. Somat. Cell Mol. Genet. 13: 679-84 (1987); **Kuczek ES et al** A granulocyte-colony-stimulating factor gene promoter element responsive to inflammatory mediators is functionally distinct from an identical sequence in the granulocyte-macrophage colony-stimulating factor gene. JI 146: 2426-33 (1991); **Le Beau MM et al** Chromosomal localization of the human G-CSF gene to 17q11 proximal to the breakpoint of the t(15;17) in acute promyelocytic leukemia. Leukemia 1: 759-99 (1987); **Lu HS et al** Disulfide and secondary structures of recombinant human granulocyte colony-stimulating factor. ABB 268: 81-92 (1989); **Nagata S et al** The chromosomal gene structure and two mRNAs for human granulocyte colony-stimulating factor. EJ 5: 575-81 (1986); **Nagata S et al** Molecular cloning and expression of cDNA for human granulocyte colony-stimulating factor. N 319: 415-8 (1986); **Nagata S** Gene structure and function of granulocyte colony-stimulating factor. BioEssays 10: 113-7 (1989); **Simmers RN et al** Localization of the human G-CSF gene to the region of a breakpoint in the translocation typical of acute promyelocytic leukemia. Hum. Genet. 78: 134-6 (1988); **Simmers RN et al** Localization of the G-CSF gene on chromosome 17 proximal to the breakpoint in the t(15;17) in acute promyelocytic leukemia. Blood 70: 330-2 (1987); **Tani K et al** Implantation of human fibroblasts transfected with human granulocyte colony-stimulating factor cDNA into mice as a model of cytokine-supplement gene therapy. Blood 74: 1274-80 (1989); **Tsuchiya M et al** Isolation and characterization of the cDNA for murine granulocyte colony-stimulating factor. PNAS 83: 7633-7 (1986); **Tweardy DJ et al** Molecular cloning and characterization of a cDNA for human granulocyte colony-stimulating factor (G-CSF) from a glioblastoma multiforme cell line and localization of the B-CSF gene to chromosome band 17q21. OR 1: 209-20 (1987)

● **Receptors: Fukunaga R et al** Expression cloning of a receptor for murine granulocyte colony-stimulating factor. Cell 61: 341-50 (1990); **Fukunaga R et al** Purification and characterization of the receptor for murine granulocyte colony-stimulating factor. JBC 265: 14008-15 (1990); **Fukunaga R et al** Three different mRNAs encoding human granulocyte colony-stimulating factor receptor. PNAS 87: 8702-6 (1990); **Fukunaga R et al** Functional domains of the granulocyte colony-stimulating factor receptor. EMBO J. 10: 2855-65 (1991); **Inazawa J et al** Assignment of the human granulocyte colony-stimulating factor receptor gene (CSF3R) to chromosome 1 at region p35-p34. 3. Genomics 10: 1075-8 (1991); **Larsen A et al Expression cloning of a human granulocyte colony-stimulating factor receptor:** a structural mosaic of hematopoietin receptor, immunoglobulin, and fibronectin domains. JEM 172: 1559-70 (1990); **Tweardy DJ et al** Molecular cloning of cDNAs for the human granulocyte colony-stimulating factor receptor from HL-60 and mapping of the gene to chromosome region 1p32-34. Blood 79: 1148-54 (1992)

● **Biological Activities: Barber KE et al** Human granulocyte-macrophage progenitors and their sensitivity to cytotoxins: analysis by limiting dilution. Blood 70: 1773-6 (1987); **Begley CG et al** Purified colony-stimulating factors enhance the survival of human neutrophils and eosinophils *in vitro*: a rapid and sensitive microassay for colony-stimulating factors. Blood 68: 162-6 (1986); **Begley CG et al** Primary human myeloid leukemia cells: comparative responsiveness to proliferative stimulation by GM-CSF or G-CSF and membrane expression of CSF receptors. Leukemia 1: 1-8 (1987); **Chang JM et al** Long-term exposure to retrovirally expressed granulocyte-colony-stimulating factor induces a nonneoplastic granulocytic and progenitor cell hyperplasia without tissue damage in mice. JCI 84: 1488-96 (1989); **Khwaja A et al** Interactions of granulocyte-macrophage colony-stimulating factor (CSF), granulocyte CSF, and tumor necrosis factor α in the priming of the neutrophil respiratory burst. Blood 79: 745-53 (1992); **Lopez AF et al** Activation of granulocyte cytotoxic function by purified mouse colony-stimulating factor. JI 131: 2983-8 (1983); **Kitagawa S et al** Recombinant human granulocyte colony-stimulating factor enhances superoxide release in human granulocytes stimulated by the chemotactic peptide. BBRC 144: 1143-6 (1987); **Maekawa T et al** Synergistic suppression of the clonogenicity of U937 leukemic cells by combinations of recombinant human interleukin 4 and granulocyte colony-stimulating factor. EH 20: 1201-7 (1992); **Pebuske MJ et al** Growth response of human myeloid leukemia cells to colony-stimulating factors. EH 16: 360-6 (1988); **Pebuske MJ et al** Preferential response of acute myeloid leukemias with translocation involving chromosome 17 to human recombinant granulocyte colony-stimulating factor. Blood 72: 257-65 (1988); **Platzer E et al** Biological activities of a human pluripotent hematopoietic colony stimulating factor on normal and leukemic cells. JEM 162: 1788-1801 (1985); **Tanimura M et al** Neutrophil priming by granulocyte colony stimulating factor and its modulation by protein kinase inhibitors. Biochem. Pharmacol. 44: 1045-52 (1992); **Yuo A et al** Recombinant human granulocyte colony-stimulating factor repairs the abnormalities of neutrophils in patients with myelodysplastic syndromes and chronic myelogenous leukemia. Blood 70: 404-11 (1987)

● **Assays: Braman AM & Schwartz KA** A radioimmune microfilter plate assay for the detection of anti-granulocyte antibodies. Am. J. Hematol. 39: 194-201 (1992); **Clogston CL et al** Detection and quantitation of recombinant granulocyte colony-stimulating factor charge isoforms: comparative analysis by cationic-exchange chromatography, isoelectric focusing gel electrophoresis, and peptide mapping. AB 202: 375-83 (1992); **Hattori K et al** Quantitative *in vivo* assay of human granulocyte colony-stimulating factor using cyclophosphamide-induced neutropenic mice. Blood 75: 1228-33 (1990); **Kuwabara T et al** Highly sensitive enzyme-linked immunosorbent assay for marograstim (KW-2228), a mutant of human granulocyte colony-stimulating factor. J. Pharmacobiodyn. 15: 121-9 (1992); **Motojima H et al** Quantitative enzyme immunoassay for human granulocyte colony stimulating factor (G-CSF). JIM 118: 187-92 (1989); **Sallerfors B & Olofsson T** Granulocyte-macrophage colony-stimulating factor (GM-CSF) and granulocyte colony-stimulating factor (G-CSF) secretion by adherent monocytes measured by quantitative immunoassays. Eur. J. Haematol. 49: 199-207 (1992); **Shorter SC et al** Production of granulocyte colony-stimulating factor at the materno-foetal interface in human pregnancy. Immunology 75: 468-74 (1992); **Tanaka H & Kaneko T** Development of a competitive radioimmunoassay and a sandwich enzyme-linked immunosorbent assay for recombinant human granulocyte colony-stimulating factor. Application to a pharmacokinetic study in rats. J. Pharmacobiodyn. 15: 359-66 (1992); **Tie F et al** An improved ELISA with linear sweep voltammetry detection. JIM 149: 115-20 (1992); **Watanabe M et al** Mutant protein of recombinant human granulocyte colony-stimulating factor for receptor binding assay. AB 195: 38-44 (1991); for further information also see individual cell lines used in individual bioassays.

● **Clinical Use & Significance: Aso Y & Akaza H** Effect of recombinant human granulocyte colony-stimulating factor in patients receiving chemotherapy for urogenital cancer. Urological rhG-CSF Study Group. J. Urol. 147: 1060-4 (1992); **Boxer LA et al** Recombinant human granulocyte-colony-stimulating factor in the treatment of patients with neutropenia. Clin. Immunol. Immunopathol. 62: S39-46 (1992); **Bronchud MH et al** Phase I/II study of recombinant human granulocyte colony-stimulating factor in patients receiving intensive chemotherapy for small cell lung cancer. Br. J. Cancer 56: 809-13 (1989); **Chelstrom LM et al** Treatment of BCL-1 murine B cell leukemia with recombinant cytokines. Comparative analysis of the antileukemic potential of interleukin 1 β (IL1 β), interleukin 2 (IL2), interleukin-6 (IL6), tumor necrosis factor α (TNF α), granulocyte colony stimulating factor (G-CSF), granulocyte-macrophage colony stimulating factor (GM-CSF), and their combination. Leuk. Lymphoma 7: 79-86 (1992); **Colombo MP et al** Granulocyte colony-stimulating factor gene transfer suppresses tumorigenicity of a murine adenocarcinoma *in vivo*. JEM 173: 889-97 (1991); **DeLuca E et al** Prior chemotherapy does not prevent effective mobilisation by G-CSF of peripheral blood progenitor cells. Br. J. Cancer. 66: 893-9 (1992); **Donahue RE & Clark SC** Granulocyte colony-stimulating factors as therapeutic agents. Immunol. Ser. 57: 637-49 (1992); **Furukawa T et al** Successful treatment of chronic idiopathic neutropenia using recombinant granulocyte colony-stimulating factor. Ann. Hematol. 62: 22-4 (1991); **Gabrilove JL** Clinical applications of granulocyte colony stimulating factor (G-CSF). GF 6: 187-91 (1992); **Gianni AM et al** Granulocyte-macrophage colony-stimulating factor or granulocyte colony-stimulating factor infusion makes high-dose etoposide a safe outpatient regimen that is effective in lymphoma and myeloma patients. J. Clin. Oncol. 10: 1955-62 (1992); **Glaspy JA & Golde DW** Granulocyte colony-stimulating factor (G-CSF): preclinical and clinical studies. Semin. Oncol. 19: 386-94 (1992); **Glaspy JA et al** Treatment of hairy cell leukemia with granulocyte colony-stimulating factor and recombinant consensus interferon or recombinant interferon-α-2b. J. Immunother. 11: 198-208 (1992); **Glaspy JA et al** Therapy for neutropenia in hairy cell leukemia with recombinant human granulocyte colony-stimulating factor. Ann. Int. Med. 109: 789-95 (1988); **Hanazono Y et al** Treatment of acute nonlymphocytic leukemia by combination of recombinant human granulocyte colony-stimulating factor and cytotoxic agents: a report of six cases. Int. J. Hematol. 55: 243-8 (1992); **Hengge UR et al** Granulocyte colony-stimulating factor treatment in AIDS patients. Clin. Investig. 70: 922-6 (1992); **Herrmann F et al** G-CSF and M-CSF: preclinical and clinical results. Immunol. Ser. 57: 651-60 (1992); **Herrmann F** G-CSF: status quo and new indications. Infection 20: 183-8 (1992); **Jakubowski AA et al** Effects of human granulocyte colony-stimulating factor in patients with idiopathic neutropenia. NEJM 320: 38-42 (1989); **Johnsen HE et al** Increased yield of myeloid progenitor cells in bone marrow harvested for autologous transplantation by pretreatment with recombinant human granulocyte-colony stimulating factor. Bone Marrow Transplant. 10: 229-34 (1992); **Kawano Y et al** Effects of progenitor cell dose and preleukapheresis use of human recombinant granulocyte colony-stimulating factor on the recovery of hematopoiesis after blood stem cell autografting in children. EH 21: 103-8 (1993); **Krumwieh D et al** Preclinical studies on synergistic effects of IL1, IL3, G-CSF, and GM-CSF in cynomolgous monkeys. IJCC 8 (Suppl. 1) 229-48 (1990); **Lieschke GJ et al** Studies of oral neutrophil levels in patients receiving G-CSF after autologous marrow transplantation. Br. J. Haematol. 82: 589-95 (1992); **Marks LB et al** Reversal of radiation-induced neutropenia by granulocyte colony-stimulating factor. Med. Pediatr. Oncol. 20: 240-2 (1992); **Marlton PV et al** Granulocyte colony stimulating factor in the management of chronic neutropenia. Med. J. Aust. 156: 729-31 (1992); **Ohno R et al** No increase of leukemia relapse in newly diagnosed patients with acute myeloid leukemia who received granulocyte colony-stimulating factor for life-threatening infection during remission induction and consolidation therapy. Japan Adult Leukemia Study Group. Blood 81: 561-2 (1993); **Pettengell R et al** Granulocyte colony-stimulating factor to prevent dose-limiting neutropenia in non-Hodgkin's lymphoma: a randomized controlled trial. Blood 80: 1430-6 (1992); **Sarosy G et al** Phase I study of taxol and granulocyte colony-stimulating factor in patients with refractory ovarian cancer. J. Clin. Oncol. 10: 1165-70 (1992); **Seidman AD et al** Dose-intensification of MVAC with recombinant granulocyte colony-stimulating factor as initial therapy in advanced urothelial cancer. J. Clin. Oncol. 11: 408-14 (1993); **Sheridan WP et al** Granulocyte colony-stimulating factor and neutrophil recovery after high-dose chemotherapy and autologous bone marrow transplantation. Lancet 2: 891-5 (1989); **Sheridan WP et al** Effect of peripheral blood progenitor cells mobilized by filgastrim (G-CSF) on platelet recovery after high-dose chemotherapy. Lancet 339: 640-8 (1992); **Steward WP** Granulocyte and granulocyte-macrophage colony-stimulating factors. The Lancet 342: 153-7 (1993); **Stoppa AM et al** *In vivo* administration of recombinant granulocyte colony stimulating factor corrects acquired neutrophil function deficiency association with chronic graft-versus-host disease. Br. J. Haematol. 83: 169-70 (1993); **Tajiri J et al** Granulocyte colony-stimulating factor treatment of antithyroid drug-induced granulocytopenia. Arch. Intern. Med. 153: 509-14 (1993); **Takada M et al** The use of granulocyte colony-stimulating factor to shorten the interval between cycles of mitomycin C, vindesine, and cisplatin chemotherapy in non-small-cell lung cancer. Cancer Chemother. Pharmacol. 31: 182-6 (1992); **Tanaka J et al** Successful second allogeneic bone marrow transplantation in a relapsed acute myeloid leukemia patient with fungal liver abscess. Ann. Hematol. 65: 193-5 (1992); **Walls J et al** Case report: granulocyte colony-stimulating factor overcomes severe neutropenia of large granular lymphocytosis. Am. J. Med. Sci. 304: 363-5 (1992)

GCI: *granulocyte chemiluminescence inducer* This poorly characterized 60 kDa factor is released by mononuclear leukocytes following their stimulation by bacterial » endotoxins (see also: Cell activation). GCI activates the oxidative metabolism in segmented neutrophilic granulocytes.

Maly FE et al Induction of granulocyte chemiluminescence by a mediator derived from human monocytes. LR 5: 21-33 (1986)

GCF: *granulocyte chemotactic factor* see: GCP (granulocyte chemotactic peptide).

GCP: *granulocyte chemotactic peptide* This factor is also called ***GCA*** (granulocyte chemotactic activity), ***GCF*** (granulocyte chemotactic factor), or ***GCP-1***. The factor is produced by measles virus-infected fibroblasts or fibroblasts treated with poly(rI).poly(rC). It is also synthesized by an osteosarcoma cell line (MG-63) and a hepatoma cell line (Malavu). Its synthesis is inducible by » IL1. The factor is identical with » IL8 and therefore belongs to the » chemokine family of cytokines.

Peveri P et al A novel neutrophil-activating factor produced by human mononuclear phagocytes. JEM 167: 1547-59 (1988); **Rampart M et al** Granulocyte chemotactic protein/interleukin 8 induces plasma leakage and neutrophil accumulation in rabbit skin. Am. J. Pathol. 135: 21-5 (1989); **Van Damme J et al** Identification by sequence analysis of chemotactic factors for monocytes produced by normal and transformed cells stimulated with virus, double-stranded RNA or cytokine. EJI 19: 2367-73 (1989); **Van Damme J et al** The chemotactic activity for granulocytes produces by virally infected fibroblasts is identical to monocyte-derived interleukin 8. EJI 19: 1189-94 (1990); **Van Damme J et al** Characterization of granulocyte chemotactic activity from human cytokine-stimulated chondrocytes as interleukin 8. Cytokine 2: 106-11 (1990)

GCP-1: granulocyte chemotactic protein 1. see: GCP

GCP-2: *granulocyte chemotactic protein 2* This factor of 6 kDa is produced by stimulated human osteosarcoma cells (MG-63). It is co-produced with » IL8 and structurally related to the other members of the » IL8 family. Four different forms of GCP-2, showing differences in truncation at the amino terminus, have been described. Human and bovine GCP-2 are 67% identical at the amino acid level. GCP-2 sequences show only weak similarity with that of IL8, and human GCP-2 does not cross-react in a radioimmunoassay for IL8.

GCP-2 provokes a significant granulocyte infiltration in a rabbit skin model involving intradermal injection, albeit to a lesser extent than IL8 and related factors. GCP-2 does not attract monocytes *in vivo* and does not induce the cells *in vitro* to migrate or to produce enzyme.

Proost P et al Human and bovine granulocyte chemotactic protein-2: complete amino acid sequence and functional characterization as chemokines. B 32: 10170-10177 (1993); **Proost P et al** Identification of a novel granulocyte chemotactic protein (GCP-2) from human tumor cells. *In vitro* and *in vivo* comparison with natural forms of GRO, IP-10, and IL8. JI 150: 1000-10 (1993)

GDCF: *glioma (cell)-derived monocyte chemotactic factor* This factor is produced by human glioma cells. It is a chemoattractant for monocytes. An identical monocyte chemotactic activity (MCA) is found also in the culture fluid of PHA-stimulated human mononuclear leukocytes.

GDCF is encoded by the human homologue of the » JE gene and therefore belongs to the » chemokine family of cytokines.

GDCF is identical with » MCP-1 (monocyte chemoattractant protein 1), » MCP (monocyte chemotactic protein), » MCAF (monocyte chemotactic and activating factor), » HC11, » SMC-CF (smooth muscle cell chemotactic factor), » MCAF (monocyte chemotactic and activating factor), and » TDCF (tumor-derived chemotactic factor).

Kuratsu J et al Production and characterization of human glioma cell-derived monocyte chemotactic factor. JNCI 81: 347-51 (1989); **Yoshimura T et al** Purification and amino acid analysis of two human glioma-derived monocyte chemoattractants. JEM 169: 1449-59 (1989); **Yoshimura T et al** Purification and amino acid analysis of two monocyte chemoattractants produced by phytohemagglutinin-stimulated human blood mononuclear leukocytes. JI 142: 1956-62 (1289)

GDCF 2: *glioma-derived monocyte chemotactic factor 2* This factor is secreted by human glioma cells. It has a length of 76 amino acids, including 4 half-cysteines at positions 11, 12, 36, and 52; the amino acid composition of this protein is the same as that of » LDCF (lymphocyte-derived chemotactic factor). GDCF 2 is encoded by the human homologue of the » JE gene. The amino acid sequence is the same as the GDCF 2 therefore belongs to the » chemokine family of cytokines. GDCF 2 is chemotactic for human monocytes, but not neutrophils. See also: GDCF.

Robinson EA et al Complete amino acid sequence of a human monocyte chemoattractant, a putative mediator of cellular immune reactions. PNAS 86: 1850-4 (1989); **Yoshimura T et al** Purification and amino acid analysis of two human glioma-derived monocyte chemoattractants. JEM 169: 1449-59 (1989)

GDF: *granulocyte-derived factor* This poorly characterized factor of ≈ 12 kDa is secreted constitutively by normal polymorphonuclear neutrophils. It strongly enhances the uptake of 3H-thymidine by several leukemic cell lines in a dose-dependent manner and greatly increases the clonogenic efficiency of the responsive cells.

Ho CK et al Induction of thymidine kinase activity and clonal growth of certain leukemic cell lines by a granulocyte-derived factor. Blood 75: 2438-44 (1990)

GDF-1: *(embryonic) growth/differentiation factor 1* This factor is specifically expressed in embryonic brain and nerve tissues. It is a member of the family of transforming growth factors (see: TGF-β). GDF-1 is most homologous to *Xenopus* » Vg-1 (52%). At the DNA level the human and the murine factor show a sequence homology of 87 %. The GDF-1 cDNA predicts a protein of 357 amino acids (38.6 kDa). Genomic Southern analysis indicates that GDF-1 may be highly conserved across species. GDF-1, which is probably a secreted glycoprotein, functions as a growth and differentiation factor in neuronal tissues.

GDF-3, also known as ***Vgr-2*** (Vg related gene 2), and ***GDF-9*** are two other mammalian growth/differentiation factors belonging to the transforming growth family of factors although they do not show the same pattern of conserved cysteine residues characteristic of this family. Both factors, which are also related to » Vg-1 and human bone morphogenetic protein 4 (see: BMP), are probably also secreted. GDF-3 transcripts are detected primarily in adult bone marrow, spleen, thymus, and adipose tissue while GDF-9 transcripts are detected only in the ovary.

Lee SJ Identification of a novel member (GDF-1) of the transforming growth factor-β superfamily. ME 4: 1034-40 (1990); **Lee SJ** Expression of growth/differentiation factor 1 in the nervous system: Conservation of a 7bicistronic structure. PNAS 88: 4250-4 (1991); **McPherron AC & Lee SJ** GDF-3 and GDF-9: two new members of the transforming growth factor-β superfamily containing a novel pattern of cysteines. JBC 268: 3444-9 (1993)

GDF-3: *growth/differentiatio1n factor 3* see: GDF-1.

GDF-9: *growth/differentiation fac3tor 9* see: GDF-1.

GDGF: *glioma-derived growth factor* One specific factor, designated ***GDGF-1***, is a homodimer of a protein structurally closely related to the A chain of » PDGF. This PDGF-related factor also binds to the same receptors as PDGF-A. The factor known as ***GDGF-2*** is very likely to the heterodimeric PDGF consisting of the A and the B chain.

In contrast to PDGF GDGF-1 is characterized by a reduced mitogenic activity. It only stimulates weakly the » autophosphorylation of its receptor and does not possess chemotactic activities (see also: Chemotaxis). GDGF-1 modulates the expression of the receptor for » EGF (see also: Receptor transmodulation).

Betsholtz C et al Synthesis of a PDGF-like growth factor in human glioma and sarcoma cells suggests the expression of the cellular homologue of the transforming protein of simian sarcoma virus. BBRC 117: 176-82 (1983); **Hammacher A et al** A human glioma cell line secretes three structurally and functionally different dimeric forms of platelet-derived growth factor. EJB 176: 179-86 (1988); **Nistér M et al** A glioma-derived analog to platelet-derived growth factor. Demonstration of receptor competing activity and immunological crossreactivity. PNAS 81: 926-30 (1984); **Nistér M et al** A glioma-derived PDGF A chain homodimer has different functional activities from a PDGF AB heterodimer purified from human platelets. Cell 52: 791-9 (1988)

GDN: glia-derived nexin see: Nexin I.

GDNF: *glial cell line-derived neurotrophic factor* This factor was found initially in culture supernatants of rat B49 glial cell lines. It is a disulfide-bonded homodimeric glycosylated protein of 134

GDF-5–7: see addendum: GRB.

amino acids (18-22 kDa on reducing SDS gels). The amino acid sequences inferred from rat and human cDNAs are 93 % identical. The proteins contain seven conserved cysteine residues in the same relative spacing found in all members of the TGF-β superfamily of proteins.
In embryonic midbrain cultures GDNF promotes the survival and morphological differentiation of dopaminergic neurons and increases their high-affinity dopamine uptake. It does not increase total neuron or astrocyte numbers and does not influence transmitter uptake by serotoninergic neurons. It has been suggested that GDNF may be of clinical relevance in the treatment of Parkinson's disease which is marked by progressive degeneration of midbrain dopaminergic neurons.
Lin LFH et al GDNF: a glial cell line-derived neurotrophic factor for midbrain dopaminergic neurons. S 260: 1130-32 (1993)

GdNPF: glial-derived neurite promoting factor see: Nexin I.

GD-PDGF: *glioma-derived platelet-derived growth factor-related* This factor is isolated from human glioma cells and their supernatants. It is related to one of the molecular forms of » PDGF.
Nakamura T et al Glioma-derived PDGF-related protein presents as 17 kd intracellularly and assembled form induces actin reorganization. J. Neurooncol. 11: 215-24 (1991)

GD-VEGF: *glioma-derived vascular endothelial cell growth factor* This factor, a homodimer of 46 kDa (164 amino acids), was initially isolated from the culture supernatants of rat glioma cells. It is the rat homologue of human » VEGF (vascular endothelial cell growth factor). GD-VEGF shows 90 % identity with human VEGF and bovine VEGF at the protein level.
Conn G et al Purification of a glycoprotein vascular endothelial cell mitogen from a rat glioma-derived cell line. PNAS 87: 1323-7 (1990); **Conn G et al** Amino acid and cDNA sequences of a vascular endothelial cell mitogen that is homologous to platelet-derived growth factor. PNAS 87: 2628-32 (1990)

Geldanamycin: see: Herbimycin A.

GEMM: abbrev. for granulocyte-erythrocyte-monocyte-macrophage. See: CFU-GEMM.

Gene cloning: see: Gene libraries.

Gene-disrupted mice: see: Knock-out mice.

Gene expression: This general term describes the processes of transcription and translation, i. e. the formation of mRNA mediated by specific RNA polymerases, and the formation of proteins, guided by mRNA sequences at the ribosomes, respectively. The mRNA transcribed from a gene, and also the protein obtained by translation of this mRNA are generally referred to as the ***gene products***. The term ***heterologous gene expression*** usually refers to the production of recombinant gene products in cells of another species (see: Recombinant cytokines for expression systems and Gene libraries for a brief discussion of cloning strategies).
Many proteins, also including many » cytokines and their receptors, are synthesized as ***prepro-proteins*** or ***pro-proteins***. The presequence, also called ***leader sequence***, is found at the aminoterminal end of the sequence of the mature protein. It functions as a sequence that directs the unstable precursor protein to membranes (see: Signal sequences). The prosequence is thought to play a role in the correct folding and the correct formation of disulfide bonds of the mature protein.
It has been shown for some cytokines that the precursor forms may also have biological activity. The biological activity of these factors is therefore controlled at various levels.
The process of gene expression depends on many different factors. The proper temporal and/or spatial patterns of gene expression regulates the phenotypes of cells. In eukaryotic cells the extent of gene expression can be modulated at the level of the DNA, for example by various differences in the organization of the chromatin, by differences in DNA methylation, by DNA recombination processes, and by the number of gene copies.
At the level of the RNA processes such as initiation, elongation, and termination of transcription all interfere with gene expression levels. The activity of a gene may also be influenced by differential processing of the transcribed mRNA, mRNA stability (see, for example: ARE (AU-rich element)), differential codon usage, interaction with protein factors, differential splice phenomena, and varying translation efficiencies. Some mRNAs appear to be protected from rapid degradation by translation, whereas degradation is coupled to translation for other mRNAs. The molecular determinants of this selective effect of translation are unknown. At the protein level factors that influence the outcome of gene activity are protein stability, posttranslational modifications, correct three-dimensional folding, correct aggregation of subunits, and also the general instability, i. e. their biological half life, of the proteins themselves.

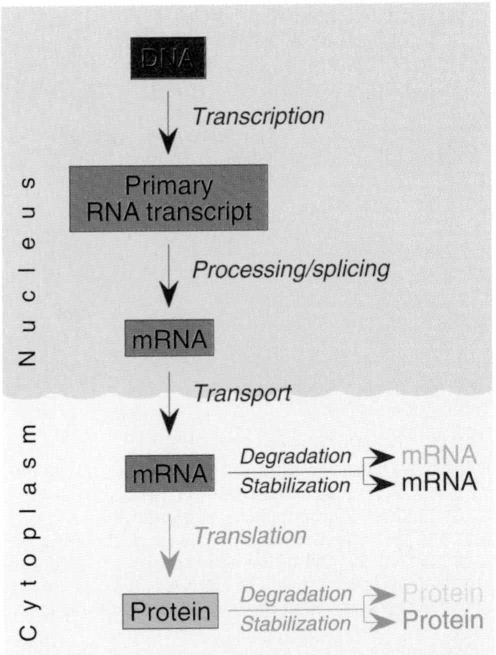

Flow of information and regulatory points during gene expression.
Gene expression involves transcription of DNA, which yields the primary RNA transcript, processing and splicing of the primary transcript to yield mature mRNA, and translation of the message into proteins. Gene expression can be controlled positively or negatively at all levels. Selective degradation or stabilization of mRNA or proteins are also utilized to control the availability of end products.

Many genes are always active (***constitutive gene expression***). Others are expressed in response to various physiological stimuli (***induced gene expression***), see also ERG (early response genes). The expression of individual genes can be restricted to certain cell types (***cell-specific gene expression***) or to certain tissues (***tissue-specific gene expression***).

These processes are mediated by a plethora of specific regulatory DNA regions found in the promoter regions of almost all genes. These regulatory sequences are frequently referred to as ***response elements*** (abbrev. RE) (see also: ARRE, CLE, CRE, GRE, IRS, MRE, SRE, TRE). They are binding sites for sequence-specific DNA binding proteins which are called ***transcription factors***. Some transcription factors are general purpose factors (***basal transcription factors***) required for transcription of all genes, while others act on specific genes or classes of genes by binding in a sequence-specific manner to response elements and other sequence motifs within the corresponding gene promoters. The expression of many of these transcription factors is developmentally and also tissue-specifically controlled and is itself subject to the action of other transcription factors and other accessory proteins such as nuclear receptors (see also: ERG, early response genes, Homeotic genes). Binding sites for transcription factors are often clustered and a variety of transcription factors have been found to form complexes with others or to compete with others for binding to overlapping DNA binding motifs. Several structural motifs have been found within those regions of transcription factor proteins recognizing and contacting DNA. Within each of these structural motifs, there are often families of related proteins (see also: Gene families) that recognize similar DNA sequences and are conserved throughout the eukaryotic kingdom.

Many cytokine genes are known to contain a number of different regulatory elements in their promoter regions. Their presence explains, at least in part, why the expression of some cytokines is modulated by other cytokines and different other physiological stimuli. Some cytokine and cytokine receptor genes contain cytokine-specific regulatory sequences binding cytokine-specific regulatory factors (see: IL6DBP, NF-IL6).

The distinct elements in the promoter regions of genes responsible for cell- and tissue-specific expression of genes can be used to construct chimaeric genes which can direct cell- and tissue-specific expression of a foreign gene if expressed in » transgenic animals.

Cytokines do not elicit biological responses directly. Instead, they stimulate or inhibit the production of specific DNA binding protein factors that, in turn, control the expression of other genes.

The signals generated by a cytokine binding to its receptor lead, within hours, to the transient activation of so-called ***immediate-early* genes** (abbrev. IEG), also known as ***primary response genes***, ***competence genes***, or ***early response* genes** (abbrev. ERG). A conservative estimation suggests the existence of several hundred different such genes (see: ERG for specific examples). Some of the early proteins have been shown to bind to promoter elements of so-called ***delayed early genes*** (DE genes) and to stimulate the synthesis of delayed early proteins. The simple interaction of a cytokine with its receptor therefore sets

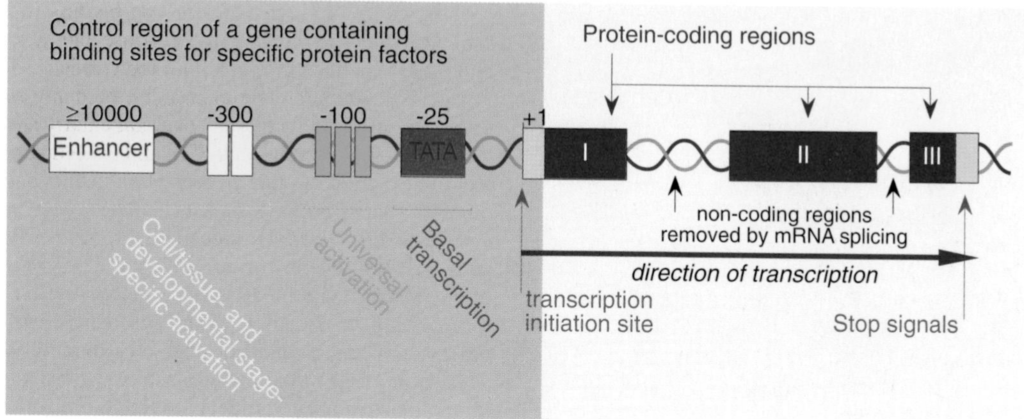

General structure of a eukaryotic gene.
Many eukaryotic genes contain specific protein-coding sequences, designated exons (boxes marked I, II, III) and non-coding regions, designated introns. Intron sequences are transcribed and are then removed from primary transcripts by splicing. Differential splicing processes also removing one or another exon sequence can yield mRNAs that are translated into proteins that then differ by the presence or absence of the sequences encoded by the particular exon. Transcription initiates at a nucleotide designated +1. Distinct signals at the end of the gene specify the end of the transcription unit and also specify whether the RNA becomes polyadenylated or not. The mRNA is thus longer than the particular part that is also translated. Transcribed but untranslated sequences at the beginning and at the end of the gene are shown as light red boxes.
The promoter region with a number of controlling sequences governing basal (red), inducible (green), and/or cell/tissue- or developmental stage-specific (yellow) expression of the gene are usually located in front of the gene (the 5´ side). Numbers above boxes indicate distance in nucleotides from the transcriptional start site at +1. These controlling sequences are binding sites for specific transcription factors and other DNA binding proteins. Some of these sites have been shown to overlap.

into motion an entire cascade of gene expression involving gene products which either act simultaneously or sequentially.

It is currently believed that transcription factors are the last link in the long chain of events beginning with binding of a cytokine to its specific receptor and ending with the cell- and tissue-specific expression or repression of genes and the associated alterations in the functional activities of a cell.

Almendral JM et al Complexity of the early genetic response to growth factors in mouse fibroblasts. MCB 8: 2140-8 (1988); **Bravo R** Growth factor inducible genes in fibroblasts. In: Habenicht A (edt) Growth factors, differentiation factors, and cytokines, pp. 324-43, Springer, Berlin 1990; **Clark AR & Docherty K** Negative regulation of transcription in eukaryotes. BJ 296: 521-41 (1993); **Cleveland DW** Gene regulation through messenger RNA stability. Curr. Opin. in Cell Biol. 1: 1148-53 (1989); **Cowell IG** Repression versus activation in the control of gene transcription. TIBS 19: 38-42 (1994); **Faist S & Meyer S** Compilation of vertebrate-encoded transcription factors. NAR 20: 3-36 (1992); **Falvey E & Schibler U** How are the regulators regulated. FJ 5: 309-14 (1991); **Hames BD & Glover DM** (eds) Transcription and splicing. IRL Press, Oxford 1988; **Harrison PR** Molecular mechanisms involved in the regulation of gene expression during cell differentiation and development. in: Dexter TM et al (eds) Colony-stimulating factors, Molecular and cellular biology, pp. 411-64, Marcel Dekker Inc. New York 1990); **Herschman HR** Primary response genes induced by growth factors and tumor promoters. ARB 60: 281-319 (1991); **Klug A** Protein designs for the specific recognition of DNA. Gene 135: 83-92 (1993); **Kozak M** An analysis of 5´ noncoding sequences from 699 vertebrate messenger RNAs. NAR 15: 8125-48 (1987); **Lanahan A et al** Growth factor-induced delayed early response genes. MCB 12: 3919-29 (1992); **Latchman DS** Eukaryotic transcription factors. Academic Press, London 1991; **Lau LF & Nathans D** Genes induced by serum growth factors. In: Cohen P & Foulkes JG (eds) The hormonal control regulation of gene transcription, pp. 165-201, Elsevier, New York, 1991; **Muegge K & Durum SK** Cytokines and transcription factors. Cytokine 2: 1-8 (1990); **Sachs AB** Messenger RNA degradation in eukaryotes. Cell 74: 413-21 (1993); **Schaffner W** How do different transcription factors binding the same DNA sequence sort out their jobs? TIG 5: 37-9 (1989); **Wingender E** Compilation of transcription regulating proteins. NAR 16: 1879-1902 (1988); **Woodgett JR** Early gene induction by growth factors. BMB 45: 529-40 (1989); **Zipfel PF et al** Mitogenic activation of human T cells induces two closely related genes which share structural similarity with a new family of secreted factors. JI 142: 1582-90 (1989)

Gene families: A term used to describe genes that display enough nucleotide sequence homologies to permit their classification as a member of a group of related genes. Such genes have usually arisen by gene duplication of amplification and have subsequently mutated. Examples of such families of genes are the genes encoding ribosomal

RNA, some histone and globin genes (globins α and β), genes encoding human » growth hormone, actin genes, genes encoding serine proteases vitellogenin genes, and major histocompatibility antigen genes.

Gene superfamilies or ***multigene families*** consists of groups of genes that do not show a marked homology among each other. Nevertheless, they are related to each other by the occurrence of subdomains within the proteins encoded by them. The genes belonging to a gene superfamily encode proteins that are, accordingly, classified as belonging to a particular protein family. Examples are genes encoding immunoglobulins, genes encoding serine proteases, many enzymes and receptors, and many genes encoding cytokine receptors (see: Cytokine receptor superfamily) or cytokines. Some designations such as » interleukins or lymphokines, for example, have no structural similarities as their common name might imply. Large superfamilies of cytokines include the » TGF-β superfamily (various TGF-β isoforms, » activin A, » inhibins, » BMP (bone morphogenetic factor), » *dpp* (*decapentaplegic*) and some others), the » PDGF superfamily (including » VEGF/VPF), the » EGF superfamily (EGF, TGF-α, » AR (amphiregulin), » HB-EGF, and some others), » chemokines, » FGF (fibroblast growth factor). For techniques allowing identification of gene families see: Gene libraries.

It has been proposed that the term family be used for proteins that have very closely related sequences with greater than 50 % amino acid identity. The term superfamily is used for proteins showing less than 50 % sequence identity but related in evolution and to have similar structures and types of function.

Barcley AN et al Leukocyte antigens factsbook. Academic Press, London 1993.

Gene libraries: A term used to describe a collection of DNA fragments derived from the genome of an organism and cloned randomly into suitable cloning vectors (plasmids, phages). If plasmid cloning vectors are used for the establishment of the library, this library essentially is a collection of host cells, each of which contains a plasmid with an inserted DNA fragment. If phages have been used to clone the DNA fragments, the library consists of a phage lysate with each phage containing an inserted fragment of DNA. Together the individual fragments in the collection of inserted molecules represent the entire genetic information of an organism and in this case the library is said to be a ***representative library***.

If the library has been established by using fragmented cellular DNA of an organism the library is said to be a ***genomic library***. The term ***genomic DNA clone*** or ***chromosomal DNA clone*** then refers to an individual cell carrying a cloning vector with one of the cellular DNA fragments or to a phage isolate with a specific DNA insert.

Many different techniques have been developed to isolate specific DNA clones from a library and if the process has been successful the specific DNA clone is said to have been cloned. Detection of an individual clone in a library can be achieved by employing strategies of ***nucleic acid hybridization*** in which short chemically synthesized labeled oligonucleotides are used to detect complementary sequences in individual cells or phages containing an insert. The sequences of such oligonucleotides used to identify the desired gene in the library can be derived, for example, from known protein sequences according to the rules of the genetic code. Related genes (see also: Gene families) can be identified by altering the hybridization conditions (e. g. ***low stringency hybridization***, i. e. hybridization under conditions that allows base mismatches) and/or by using so-called degenerated oligonucleotides (mixtures of oligonucleotides that differ from each other by base substitutions at identical and/or different positions). The desired gene can also be identified by the activities of the encoded gene product. In this case, one uses so-called ***expression libraries*** that have been established by cloning DNA fragments into special cloning vectors allowing the functional expression of cloned DNA fragments. Functional gene products and hence the desired clones can then be detected either by antibodies or other ligands that specifically recognize the encoded proteins or by exploiting a bioactivity of the gene product, if known.

In contrast to genomic libraries c***DNA libraries*** contain inserts of DNA fragments that correspond to the entire mRNAs of a cell. cDNA (or complementary DNA) is obtained by using the enzyme reverse transcriptase to copy mRNA sequences back into the corresponding DNA sequences. One of the advantages of a cDNA library is the improved frequency with which individual DNA fragments occur. This is due to the presence of multiple copies of a mRNA as opposed to DNA for any given gene. In addition, the representation of inserts derived from functionally expressed mRNAs

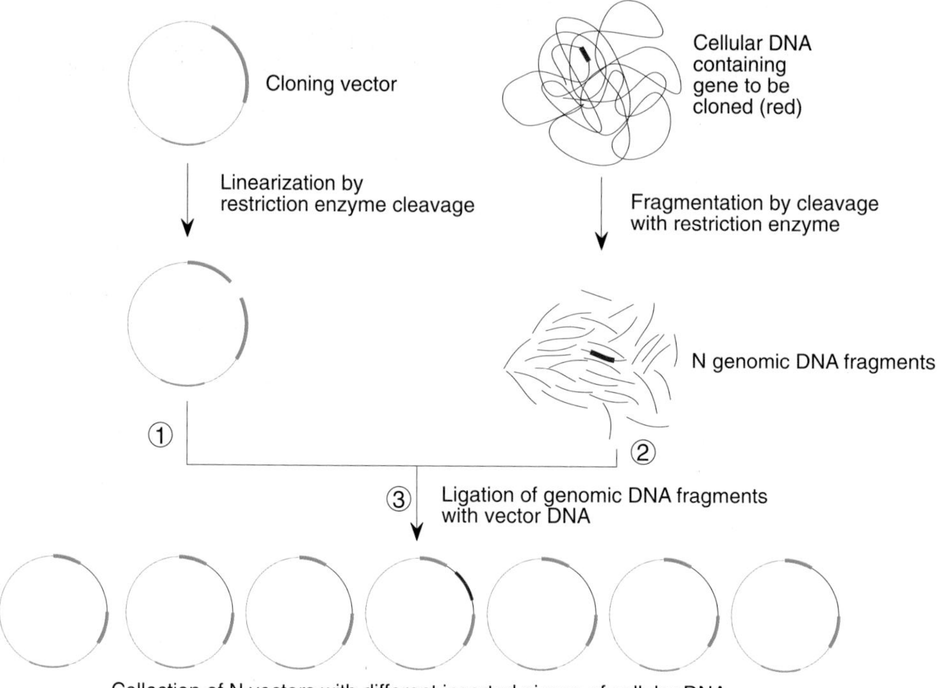

Collection of N vectors with different inserted pieces of cellular DNA

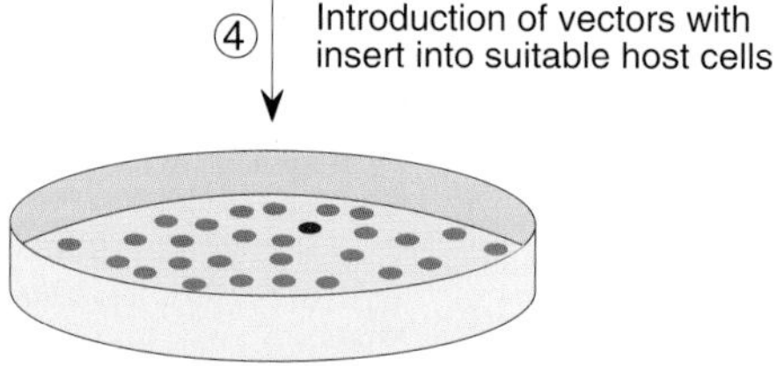

Collection of N cell colonies each containing a cloning vector with a different genomic insert.

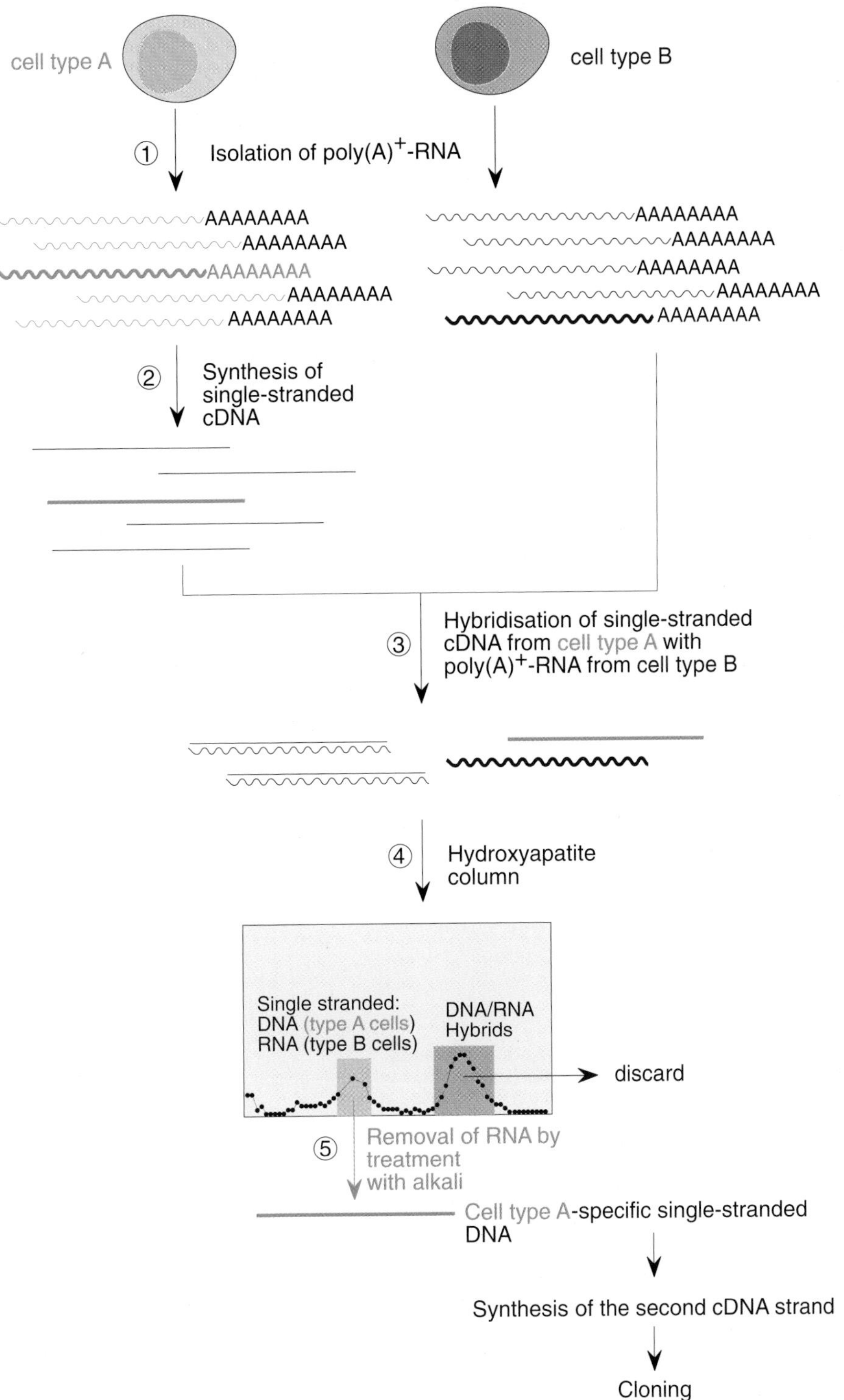
cell type A
cell type B
① Isolation of poly(A)+-RNA
AAAAAAAA
② Synthesis of single-stranded cDNA
③ Hybridisation of single-stranded cDNA from cell type A with poly(A)+-RNA from cell type B
④ Hydroxyapatite column
Single stranded: DNA (type A cells) RNA (type B cells)
DNA/RNA Hybrids
discard
⑤ Removal of RNA by treatment with alkali
Cell type A-specific single-stranded DNA
Synthesis of the second cDNA strand
Cloning

Legend to figure page 282

Principle of gene cloning and preparation of gene libraries.
A suitable cloning vector (in this example a circular plasmid) is linearized by cleavage with a suitable restriction enzyme (1). The cellular DNA containing the gene of interest (red) is also cleaved into fragments with this enzyme (2). For various reasons the cellular DNA is cleaved in a way yielding N overlapping fragments with a uniform length of ≈ 20 kb. Linearized vector DNA and genomic DNA fragments are then ligated in vitro, yielding a population of N recircularized cloning vector molecules, each of which contains an insert of cellular genomic DNA (3). These molecules are introduced into suitable host cells (4).
The collection of cloning vectors with inserts obtained at step 3 or the collection of cells obtained after step 4 are called a gene library. This library is called a representative library if the sum of the genomic DNA inserts found in the cloning vectors of all cells represents the entire cellular DNA. Only a representative library, i. e. only the presence of all DNA fragments generated originally, guarantees a reasonable chance of finding the desired gene. Statistical analysis shows that for human DNA more than 600000 cell clones must be screened in order to find a particular DNA fragment containing the desired gene with a probability of 99 %.
In principle libraries can also be made with cellular RNA as starting material. The resulting collection of host cells containing recombinant vectors is then called a cDNA library.
The art of cloning is to find the one particular cell (marked red) which contains the cloning vector with the gene of interest. This process is called library screening. If this goal has been achieved the gene in question is said to have been cloned.
Many different techniques are available for screening a library. The most important approaches involve screening by nucleic acid hybridization and screening by functional analysis. Nucleic acid hybridization requires some prior knowledge of the DNA sequence either of the gene to be cloned or of stretches of DNA in the vicinity of the gene to be cloned. Functional screening involves the use of expression vectors that allow cells containing the vector with the desired gene to express the corresponding protein. Under these circumstances, cells containing a vector with the desired gene can be identified by means of antibodies directed against the protein. Cells producing the desired gene product can also be identified in bioassays detecting protein activities, if these are available.

Legend to figure page 283

General strategy for subtractive library screening.
Consider two cell types, designated A and B. These cells may be completely different (e. g. brain and liver cells) or identical cells that differ from each other, for example, in their activation state or pretreatment with cytokines.
In the initial step (step 1) polyadenylated mRNA (jagged lines ending in AAAAA) is isolated from both types of cells. The RNA shown in green is expressed in cell type A only while that shown in blue is expressed in cell type B only. RNAs expressed in either cell type are shown in black.
RNA isolated from cell type A is used to generate single-stranded cDNA by reverse transcription (step 2). This cDNA is then hybridized with RNA from cell type B (step 3). Those RNAs (black) expressed in both cell types will form DNA/RNA hybrids while cDNA single strands derived from type A-specific mRNA not present in type B cells will remain single stranded. Likewise, RNA that is expressed in type B but not in type A cells will remain single stranded. Single and double stranded molecules are separated by hydroxyapatite chromatography (step 4). This yields fractions containing only single stranded DNA or RNA molecules. Removal of RNA by treatment with alkali yields a population of cDNA molecules derived from those RNAs specifically expressed in type A cells only. These can be converted into double stranded DNA which can then be cloned.
If the aim is to investigate those RNAs specifically expressed in type B cells only type B-specific RNA is converted into single stranded DNA and hybridized with RNA from type A cells. This scheme represents the “classical” way of generating subtractive libraries. The process has been simplified considerably by the application of PCR-based strategies.

found in a cDNA libraries also reflects the functional activities of the cells from which the library was established. Moreover, as cDNA clones are derived from mRNA, they do not contain any intervening sequences (introns). cDNA clones can therefore be used directly to express the proteins encoded by them (see also: Gene expression, Recombinant Cytokines). Many » cytokine genes have originally been isolated in the form of cDNA clones.
A special form a gene library that is gaining in importance is the ***subtractive library*** or ***differential library***. Screening of such libraries is often referred to as ***subtractive hybridization***, ***subtraction cloning***, ***differential gene screening***, ***differential hybridization***, or ***genomic difference cloning***. Use of subtractive libraries allows cloning of genes that are differentially expressed in two different cell types, i. e. genes expressed in a cell type-specific manner. The procedure can be used also to clone genes that are differentially expressed in response to environmental stimuli in the same type of cells. The major advantage of subtractive libraries over other libraries lies in the fact that it allows cloning of genes without any previous knowledge of its sequence or function (normally used as a criterion either for protein purification or for cDNA isolation in expression libraries), the only essential parameter being a difference in gene expression in the two types of cells being compared. Use of subtractive libraries has been instrumental in identifying new cytokine and cytokine receptor genes as well as a plethora of genes the expression of which is specifically induced by treatment of cells with cytokines or other stimuli (see also: ERG, early response genes).

● **Reviews: Schweinfest CW & Papas TS** Subtraction hybridization: an approach to the isolation of genes differentially expressed in cancer and other biological systems. Int. J. Oncology 1: 499-506 (1992); **Houge G** Simplified construction of a subtracted cDNA library using asymmetric PCR. PCR Methods Appl. 2: 204-9 (1993); for further references see any good text book on Genetic Engineering.

● **Examples: Beadling C et al** Isolation of interleukin 2-induced immediate-early genes. PNAS 90: 2719-23 (1993); **Bergsagel PL et al** Genes expressed selectively in plasmacytomas: markers of differentiation and transformation. CTMI 182: 223-8 (1992); **Boothman DA et al** Isolation of x-ray-inducible transcripts from radioresistant human melanoma cells. PNAS 90: 7200-4 (1993); **Dobner T et al** Differentiation-specific expression of a novel G protein-coupled receptor from Burkitt's lymphoma. EJI 22: 2795-9 (1992); **Doggett DL et al** Differential gene expression between young and senescent, quiescent WI-38 cells. Mech. Ageing Dev. 65: 239-55 (1992); **Hajnal A et al** Suppression of ras-mediated transformation. Differential expression of genes encoding extracellular matrix proteins in normal, transformed and revertant cells. Adv. Enzyme Regul. 33: 267-80 (1993); **Kumar S et al** Identification of a set of genes with developmentally down-regulated expression in the mouse brain. BBRC 185: 1155-61 (1992); **Marechal D et al** A subtractive hybridization method to isolate tissue-specific transcripts: application to the selection of new brain-specific products. AB 208: 330-3 (1993); **Orr SL et al** A new approach to understanding T cell development: the isolation and characterization of immature CD4⁻, CD8⁻, CD3⁻ T cell cDNAs by subtraction cloning. Mol. Biol. Cell. 3: 761-73 (1992); **Owens GP & Cohen JJ** Identification of genes involved in programmed cell death. Cancer Metastasis Res. 11: 149-56 (1992); **Wieland I et al** Isolation of DNA sequences deleted in lung cancer by genomic difference cloning. PNAS 89: 9705-9 (1992)

Gene therapy: see: Cytokine gene transfer.

Genetic ablation: Genetic ablation, also called ***genetic amputation*** or ***cell ablation***, is a technique that can be used to suppress selectively the growth of a certain cell population rather than the activity of an individual gene. This approach essentially involves construction and expression of a gene under the control of a gene promoter that is known to be active only in a certain cell type (see also: Gene expression). This promoter can be used to drive expression of cytotoxic protein (see also: Cytokine fusion toxin) or another gene product that renders the cells sensitive to drugs.
In principle genetic ablation can be used to suppress the growth of any desired cell line or cell type as long as cell or tissue type-specific regulatory sequences are available. The technique has been employed to create » transgenic animals lacking a particular cell type. It should be noted, however, that the expression of the toxin may already exert its effects in early phases of embryonal development, depending on the particular construct used, and that this may also suppress the further development of developing fetuses.
By coupling, for example, the A chain of diphtheria toxin the δ2 crystallin gene promoter it has been possible to suppress in part and also completely the growth of a particular cell population in the lens of transgenic mice and to produce microphthalmia in animals expressing the construct. Similar approaches have been used to suppress the growth of a cell population in the pancreas, of choroidal melanocytes, and of Schwann cells or Purkinje cells in the nervous system of transgenic animals. A line of transgenic mice has been identified with a recessive defect in lymphocyte or granulocyte function, presumably as a result of insertional mutagenesis by the integrated transgene. Transgenic mice homozygous for the transgene integrant show nearly complete absence

generalized lymphoproliferative disease: see addendum: Fas ligand.

of lymphocytes in peripheral lymph nodes and Peyer's patches, a severely diminished thymus medulla, and a greatly enlarged spleen.

By generating transgenic mice that express the herpes simplex virus 1-thymidine kinase (HSV-TK) gene under the transcriptional control of the murine » IL2 promoter it has been possible to create IL2-producing cells that are sensitive to the cytotoxic effects of the antiviral drug ganciclovir (GANC) and thus can be eliminated selectively from animals treated with this drug. A similar approach has been used to eliminate cells producing » IL4.

Genetic ablation may be of particular importance for » cytokine gene transfer into explanted tumor cells and appears to be a way to generate autologous or HLA-matched allogeneic tumor cell vaccines for the treatment of patients with various tumors.

Feddersen RM et al Disrupted cerebellar cortical development and progressive degeneration of Purkinje cells in SV40 T antigen transgenic mice. Neuron 9: 955-66 (1992); **Kamogawa Y et al** The relationship of IL4 and IFN-γ-producing T cells studied by lineage ablation of IL4 producing cells. Cell 75: 985-95 (1993); **Klein KL et al** Embryology and morphology of microphthalmia in transgenic mice expressing a γ F-crystallin/diphtheria toxin A hybrid gene. Lab. Invest. 67: 31-41 (1992); **Lo D et al** A recessive defect in lymphocyte or granulocyte function caused by an integrated transgene. Am. J. Pathol. 141: 1237-46 (1992); **Messing A et al** P0 promoter directs expression of reporter and toxin genes to Schwann cells of transgenic mice. Neuron 8: 507-20 (1992); **Minasi LE et al** The selective ablation of interleukin 2-producing cells isolated from transgenic mice. JEM 177: 1451-9 (1993); **Mintz B & Klein-Szanto AJ** Malignancy of eye melanomas originating in the retinal pigment epithelium of transgenic mice after genetic ablation of choroidal melanocytes. PNAS 89: 11421-5 (1992); **Moffat KG et al** Inducible cell ablation in *Drosophila* by cold-sensitive ricin A chain. Development 114: 681-7 (1992); **Neville DM jr et al** *In vivo* T cell ablation by a holo-immunotoxin directed at human CD3. PNAS 89: 2585-9 (1992); **O'Kane CJ & Moffat KG** Selective cell ablation and genetic surgery. Curr. Opin. Genet. Dev. 2: 602-7 (1992)

Genetically engineered cytokines: see: Muteins. See also: Recombinant Cytokines.

Genistein: (4´,5,7-Trihydroxyisoflavone) This isoflavone compound was initially detected in culture media of *Pseudomonas spp in search for **quercetin**-*like flavonoids that, like quercetin, inhibit protein tyrosine kinases . Unlike quercetin, which inhibits tyrosine and serine/threonine kinases, Genistein is a highly specific inhibitor of tyrosine-specific protein kinases (PTK) *in vitro* and *in vivo* with little activity on serine- and threonine-specific protein kinases.

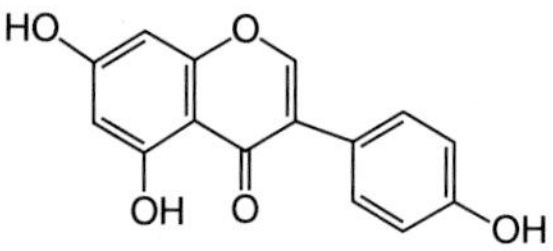

Genistein

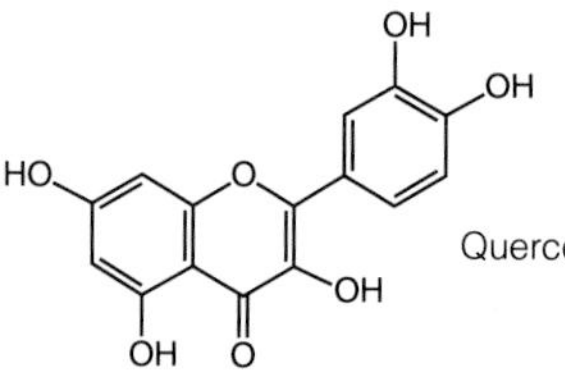

Quercetin

Genistein is not a competitive inhibitor of substrate binding. Instead it inhibits the binding of ATP to the kinases. Other ATP-dependent enzymes such as cAMP-dependent protein kinase, protein kinase C, phosphorylase kinase, 5´ nucleotidase, and phosphodiesterase are only inhibited weakly. The mechanism of action of genistein therefore differs from that of other specific inhibitors showing the same activity, for example the » tyrphostins. Genistein has also been shown to inhibit the activity of several enzymes other than tyrosine kinases, including β-galactosidase and DNA topoisomerases I and II. The activity on topoisomerases and its reported ability to induce strand breakage of DNA *in vivo* suggests an alternative mechanism whereby genistein can be cytotoxic.

Genistein can be used to test whether binding of a cytokine to its receptor or induction of the synthesis of a growth factor requires tyrosine-specific protein kinase activities for post receptor signaling processes. Genistein inhibits the receptors for » EGF, » PDGF and IGF-1 (see: IGF) which possess intracellular domains encoding intrinsic tyrosine-specific protein kinase activities. Genistein has been shown to induce differentiation of several cell lines, including human neuroblastoma and mouse erythroleukemia cells. It also inhibits keratinocyte differentiation.

In mesangial cells Genistein inhibits the synthesis of prostaglandins induced by » IL1.

Genistein has recently been identified also in the urine of healthy human subjects as one of several fractions able to inhibit the proliferation of cultured human neuroblastoma cells at micromolar concentrations. It is formed in the human body from precursors obtained by diet, suggesting that genistein is a natural antineoplastic agent present in diet and that it could be useful for the therapy of pediatric tumors. For other agents used to dissect cytokine-mediated signal transduction pathways see: Bryostatins, Calcium ionophore, Calphostin C, H8, Herbimycin A, K-252a, Lavendustin A, Phorbol esters, Okadaic acid, Staurosporine, Suramin, Tyrphostins, Vanadate.

● REVIEWS: **Casnellie JE** Protein kinase inhibitors: probes for the functions of protein phosphorylation. Adv. Pharmacol. 22: 167-205 (1991)
Abler A et al Genistein differentially inhibits postreceptor effects of insulin in rat adipocytes without inhibiting the insulin receptor kinase. JBC 267: 3946-51 (1992); **Akiyama T et al** Genistein, a specific inhibitor of tyrosine-specific protein kinases. JBC 262: 5592-5 (1987); **Akiyama T et al** Use and specificity of genistein as inhibitor of protein-tyrosine kinases. MiE 201: 362-70 (1991); **Atluru D et al** Genistein, a selective protein tyrosine kinase inhibitor, inhibits interleukin-2 and leukotriene B4 production from human mononuclear cells. Clin. Immunol. Immunopathol. 59: 379-87 (1991); **Constantinou A et al** Induction of differentiation and DNA strand breakage in human HL60 and K-562 leukemia cells by genistein. CR 50: 2618-24 (1990); **Coyne DW & Morrison AR** Effect of the tyrosine kinase inhibitor, genistein, on interleukin 1-stimulated PGE2 production in mesangial cells. BBRC 173: 718-24 (1990); **Honma Y et al** Induction by some protein kinase inhibitors of differentiation of a mouse megakaryoblastic cell line established by coinfection with Abelson murine leukemia virus and recombinant SV40 retrovirus. CR 51: 4649-55 (1991); **Satoh T et al** Inhibition of interleukin 3 and granulocyte-macrophage colony-stimulating factor stimulated increase of active *ras*.GTP by herbimycin A, a specific inhibitor of tyrosine kinases. JBC 267: 2537-41 (1992); **Schweigerer L et al** Identification in human urine of a natural growth inhibitor for cells derived from solid pediatric tumors. EJCI 22: 260-4 (1992); **Tomonaga T et al** Isoflavonoids, genistein, psi-tectorigenin, and orobol, increase cytoplasmic free calcium in isolated rat hepatocytes. BBRC 182: 894-9 (1992); **Yamashita Y et al** Induction of mammalian topoisomerase II dependent DNA cleavage by nonintercalative flavonoids, genistein and orobol. Biochem. Pharmacol. 39: 737-44 (1990)

Gene superfamily: see: Gene family.

Genomic libraries: see: Gene libraries.

Genomic difference cloning: see: Gene libraries.

GF: abbrev. for » growth factor.

GF-1: This transcription factor is identical with » GATA-1.

Zon LI et al The major human erythroid DNA-binding protein (GF-1): primary sequence and localization of the gene to the X chromosome. PNAS 87: 668-72 (1990)

GFD: *growth factor domain* see: EGF-like repeats.

GF-D8: A human myeloid leukemia cell line established from the peripheral blood of an 82-year-old man suffering from acute myeloblastic leukemia. GF-D8 cells are myeloblasts of immature progenitor origin. The long-term survival and proliferation of GF-D8 cells is dependent on the presence of either » GM-CSF or » IL3. Weak colony growth is also observed after exposure of GF-D8 cells to » SCF (stem cell factor) but not after exposure to » G-CSF, » M-CSF, IL1β (see: IL1), » IL2, » IL5, or » TNF-α. GM-CSF- and IL3-induced proliferation is dose dependent, with significant growth observed at concentrations as low as 0.1 ng/mL, but the combination of both factors has no synergistic effect.

Rambaldi A et al Establishment and characterization of a new granulocyte-macrophage colony-stimulating factor-dependent and interleukin-3-dependent human acute myeloid leukemia cell line (GF-D8). Blood 81: 1376-83 (1993)

GGF: *glial growth factor* This factor is isolated from the culture supernatant of glial cells, bovine pituitaries, and in large amounts from human schwannomas. GGF is a basic protein of 31 kDa that forms dimers of 56 kDa. This factor has been renamed GGF-I following the isolation of two other species, designated GGF-II (≈ 59 kDa) and GGF-III (≈ 45 kDa).

GGF was initially identified by its action on rat Schwann cells. GGF is also mitogenic for astrocytes and fibroblasts. Recent sequence data indicate that all three molecular forms of GGF are probably derived from a single copy gene which itself has been shown to encode *erb*B2 ligands (see: heregulin) and » ARIA (acetylcholine receptor inducing activity). These factors are collectively known as » neuregulins.

Brockes JP Studies on cultured Schwann cells: the induction of myelin synthesis, and the control of their proliferation by a new growth factor. J. Exp. Biol. 95: 215-30 (1981); **Brockes JP** Assay and isolation of glial growth factor from the bovine pituitary. MiE 147: 217-25 (1986); **Brockes JP et al** Glial growth factor-like activity in Schwann cell tumors. Ann. Neurol. 20: 317-22 (1986); **Eccleston PA et al** Control of peripheral glial cell proliferation: a comparison of the division rates of enteric glia and Schwann cells and their response to mitogens. Dev. Biol. 127: 409-18 (1987); **Goodearl ADJ et al** Purification of multiple forms of glial growth factor. JBC 268: 18095-102 (1993); **Lemke BE & Brockes JP** Identification and purification of glial growth factor. J. Neurosci. 4: 75-83 (1984); **Marchionni MA et al** Glial growth factors are alternatively spliced *erb*B2 ligands expressed in the nervous system. N 362: 312-8 (1993); **Ratner N et al** The neuronal cell-surface molecule mitogenic for Schwann cells is a heparin-binding protein. PNAS 85: 6992-6 (1988)

GGIF: *glial growth inhibitory factor* This poorly characterized factor is isolated from the culture medium of the murine neuroblastoma cell line NAs-1. It inhibits the stimulation of normal rat glioblasts induced by » GMF-β (glial maturation factor). See also: Glial growth inhibitor.

Kato T et al Functional dissociation of dual activities of glia maturation factor: inhibition of glial proliferation and preservation of differentiation by glial growth inhibitory factor. Brain Res. 430: 153-6 (1987)

GGPF: *glial growth promoting factor* GGPF is produced by antigen- or lectin-stimulated T lymphocytes and human T cell leukemia virus (HTLV)-infected cell lines. GGPF exists as a dimer of 30 kDa and a monomer of 18 kDa. Both forms have biological activity. The factor stimulates the proliferation of rat brain oligodendrocytes and astrocytes. The factor may be involved in remyelinization processes following traumatic nerve injuries.

Beneveniste EN et al Purification and characterization of a human T lymphocyte-derived glial growth-promoting factor. PNAS 82: 3930-4 (1985); **Merrill JE et al** T cell lines established from multiple sclerosis cerebrospinal fluid T cells using human retroviruses. J. Neuroimmunol. 21: 213-26 (1989)

GH: see: Growth hormone.

GI: abbrev. for growth inhibitor. Sometimes used as a suffix in abbreviations.

GIF: *glucocorticoid increasing factor* This poorly characterized factor is found in supernatants of mitogen-stimulated human and rat T lymphocytes. Upon injection into normal rats it causes a significant increase in corticosterone levels and also raises levels of ACTH (see: POMC, proopiomelanocortin). Intracerebroventricular injection results in increased glucocorticoid blood levels. Treatment with GIF does not raise corticosterone levels in rats treated with dexamethasone or subjected to hypophysectomy to block endogenous output of ACTH. GIF thus seems to be an important modulator within the » neuroimmune network, requiring a functional pituitary gland. The factor does not appear to be identical with » IL1, » IL2, endogenous pyrogens (see: EP), bacterial » endotoxins, or » IFN-γ.

Besedovsky H et al Lymphoid cells produce an immunoregulatory glucocorticoid increasing factor (GIF) acting through the pituitary gland. Clin. Exp. Immunol. 59: 622-8 (1985); **Besedovsky H et al** Host endocrine responses during tumor growth. PCBR 288: 203-13 (1989)

GIF: *glycosylation inhibiting factor* This protein is 98 % homologous with human » MIF (macrophage inhibition factor) over all 115 residues.

Tagaya Y et al Biochemical and biological characterization of a recombinant murine glycosylation inhibiting factor (GIF). JI 150: Abstract 47A (1993)

GIF: *growth inhibition factors* This generic term is occasionally used for cytokines inhibiting the growth of cells.

GIF: *growth inhibition factor* This factor, also called ***GIFB*** or ***GRIF*** (growth inhibitory factor, brain), is found in normal human brain tissue and is expressed exclusively in the nervous system (astrocytes in gray matter). GIF is not expressed in fetal brain.

GIF has a length of 68 amino acids and shows 63 % sequence identity with human metallothionein II at the protein level. All cystein residues are conserved between GIF and mammalian metallothioneins. Human GIF contains three zinc and four copper atoms per molecule.

The human GIFB gene is on chromosome 16, as are the genes encoding metallothionein which do not function as growth inhibitors. GIF inhibits survival and the outgrowth of neurites in cortical neurons. Drastically reduced concentrations of this factor are found in the brains of patients with Alzheimer's disease.

Kobayashi H et al Molecular cloning of rat growth inhibitory factor cDNA and the expression in the central nervous system. Brain. Res. Mol. Brain Res. 19: 188-94 (1993); **Tsuji S et al** Molecular cloning of human growth inhibitory factor cDNA and its down-regulation in Alzheimer's disease. EJ 11: 4843-50 (1992); **Uchida Y et al** The growth inhibitory factor that is deficient in the Alzheimer's disease brain is a 68 amino acid metallothionein-like protein. Neuron 7: 337-47 (1991)

GIF: *growth inhibiting factor* This poorly characterized factor has been identified in supernatants of the neoplastic V79 Chinese hamster cell line based on its ability to inhibit the proliferation of the same cell line. The partially purified activity reversibly inhibits the growth of a wide variety of human tumor cells, but not various normal human fibroblasts, in monolayer culture and in soft agar. GIF is not identical with » TGF-β, TNF-α, or » IL1.

Ichinose I et al Growth modulation of human tumor cells by a growth-inhibiting activity derived from tumorigenic V79 Chinese hamster cells. In Vitro Cell. Dev. Biol. 29A: 332-8 (1993)

GI-factor: *growth inhibitory factor* This poorly characterized 25 kDa factor is secreted by macrophages in murine lung tissues and bone marrow. This factor inhibits the growth of murine monocytic leukemia cell lines (Mm-A, J774.1). The human monocytic leukemia cell lines » U937 and THP-1 and the promyelocytic cell line » HL-60 do not respond to this factor. The factor also inhibits the growth of murine bone marrow cells stimulated by » M-CSF. The biological activity of this factor is not neutralized by antibodies directed against » IFN-α and » IFN-β.

Hozumi M et al Protein factors that regulate the growth and differentiation of mouse myeloid leukemia cells. Ciba Found. Symp. 148: 25-33; discussion 33-42 (1990); **Ikuta T et al** Normal mouse lung tissue produces a growth-inhibitory factor(s) preferential for mouse monocytic leukemia cells. Cancer Immunol. Immunother. 30: 139-44 (1989); **Kasukabe T et al** Purification of a novel growth inhibitory factor for partially differentiated myeloid leukemic cells. JBC 263: 5431-5 (1988)

GIFB: *growth inhibitory factor, brain* see: GIF (growth inhibition factor).

gIFN: abbrev. for Gamma-Interferon. See: IFN-γ. See also: IFN for general information about interferons.

GIP: *granulocyte inhibitory protein* This factor of 23 kDa (pI = 4. 0-4. 5) is isolated from the serum filtrate of a patient with chronic kidney failure. The known amino-terminal sequence of the GIP protein does not show any homology to known growth hormones and/or cytokines.
GIP inhibits glucose uptake, oxydative metabolism, chemotaxis, and cytotoxic activity of segmented neutrophil granulocytes. GIP is probably one of the major contributors of leukocyte dysfunction frequently observed in uremic patients. GIP has been shown to be a pluripotential modulator of inflammatory responses, specifically inducing the expression of » IL6 and » IL8 in human mesangial cells.

Hörl WH et al Physicochemical characterization of a polypeptide present in uremic serum that inhibits the biological activity of polymorphonuclear cells. PNAS 87: 6353-7 (1990); **Ziesche R et al** A granulocyte inhibitory protein overexpressed in chronic renal disease regulates expression of interleukin 6 and interleukin 8. PNAS 91: 301-5 (1994)

Gip17: see: nm23.

Glia activating factor: abbrev. GAF. See: FGF-9.

Glial cell line-derived neurotrophic factor: see: GDNF.

Glial-derived neurite promoting factor: abbrev. GdNPF. see: Nexin I.

Glia-derived nexin: abbrev. GDN. see: Nexin I.

Glial growth factor: see: GGF.

Glial growth inhibitor: This poorly characterized factor of ≈ 100 kDa is detected in von Recklinghausen neurofibroma extracts. It suppresses the growth of glioma, astrocytoma, glioblastoma, oligodendroglioma, and Schwannoma cell lines. See also: GGIF (glial growth inhibitory factor).

Asai K et al von Recklinghausen neurofibroma produces neuronal and glial growth-modulating factors. Brain Res. 556: 344-8 (1991)

Glial growth inhibitory factor: see: GGIF.

Glial growth promoting factor: see: GGPF.

Glial maturation factor β: see: GMF-β.

Glial promoting factor 1: GPF-1 see: OGF (oligodendroglia growth factor).

Glioblastoma-derived T cell suppressor factor: see: G-TsF.

Glioma cell-derived monocyte chemotactic factor: see: GDCF.

Glioma-derived growth factor: see: GDGF.

Glioma-derived monocyte chemotactic factor: see: GDCF.

Glioma-derived PDGF-related: see: GD-PDGF.

Glioma-derived suppressor factor: see: GSF.

Glioma-derived T cell suppressor factor: see: G-TsF.

Glioma-derived vascular endothelial cell growth factor: abbrev. GD-VEGF. see: VEGF.

Gliostatin: abbrev. GLS. This factor is isolated from extracts of a neurofibroma. It is a homodimer consisting of two 50 kDa subunits. This neurotrophic factor (see also: Neurotrophins) is a potent inhibitor of the proliferation of glioma cells and stimulated astroblasts. It is inactive on neuronal cells. Gliostatin probably plays an important role in the development and regeneration of the central nervous system. It is probably also involved in angiogenic processes connected with the development of the blood-brain barrier. The factor is identical with » PD-ECGF (platelet-derived endothelial cell growth factor.

Asai K et al A novel glial growth inhibitory factor, gliostatin, derived from neurofibroma. J. Neurochem. 59: 307-17 (1992); **Asai K et al** Neurotrophic action of gliostatin on cortical neu-

***gld*:** see addendum: Fas ligand.

rons. Identity of gliostatin and platelet-derived endothelial cell growth factor. JBC 267: 20311-6 (1992)

Glomerular permeability factor: see: GPF.

GLS: see: Gliostatin.

Glucocorticoid increasing factor: see: GIF.

Glucocorticoid response element: see: GRE.

Glucocorticoid-suppressible mitogenic (activity): see: GSM.

Glucosteroid response-modifying factor: see: GRMFT.

Glutathione S-transferase: abbrev. GST. See: MIF (migration inhibition factor).

Glycoprotein 2: abbrev. GP-2. see: uromodulin.

Glycosylation inhibiting factor: see: GIF.

Glucose-6-phosphate isomerase: abbrev. GPI. see: NLK (neuroleukin).

GM: Used as a prefix or suffix it denotes granulocytes such as in » GM-CSF or » CFU-GM (colony forming unit granulocytes/macrophages).

GMA: abbrev. for growth and maturation activity. Used in abbreviations as in eos.-GMA (a factor promoting the growth and maturation of eosinophils). This term is more or less an operational definition of a biological activity and does not refer to a distinct factor.
Salahuddin SZ et al Lymphokine production by cultured human T cells transformed by human T cell leukemia lymphoma virus I. S 223: 703-7 (1984)

GM-CSA: *granulocyte-macrophage colony-stimulating activity* This factor was originally isolated from an adherent murine marrow cell line (TC-1). It acts synergistically with » M-CSF to stimulate giant macrophage colonies. This factor, which also synergises IL 3-induced megakaryocyte colony formation, is identical with » GM-CSF.
Quesenberry P et al Multilineage synergistic activity produced by a murine adherent marrow cell line. Blood 69: 827-35 (1987); **Song ZX et al** Hematopoietic factor production by a cell line (TC-1) derived from adherent murine marrow cells. Blood 66: 273-81 (1985)

GM-CSF: *granulocyte-macrophage colony-stimulating factor*

■ **Alternative Names:** BPA (burst promoting activity); CSF-α (colony stimulating factor α); CSF-β (colony stimulating factor β); CSF 2 (colony stimulating factor 2); Eo-CSF (eosinophil colony-stimulating factor); Eosinophil viability-enhancing activity; ESP (eosinophil stimulation promoter); FDCp1-GF (FDCp1 growth factor = rat GM-CSF); G/M-CSA (granulocyte-macrophage colony-stimulating activity); HCGF (hematopoietic cell growth factor); HCSF (histamine-producing cell stimulating factor); KM-102-BPA (KM-102 burst promoting activity; KTGF (keratinocyte-derived T cell growth factor); LBGF (leukemic blast growth factor); MFF (macrophage fusion factor); MGI-1GM (macrophage-granulocyte inducer); NIF-T (neutrophil inhibition factor, T cell-derived neutrophil migration inhibition factor, neutrophil migration inhibition factor from T lymphocytes); Pluripoietin-α; TPO (thrombopoietin); WEHI-3B differentiation inducing activity. Recombinant GM-CSF is also marketed under the generic names Sargramostim and Molgramostim and under the brand names Leucomax, Leukine, and Prokine. See also: individual entries for further information.

■ **Sources:** This protein is secreted together with other factors by T cells and macrophages following their activation by antigens or mitogens (see also: Cell activation). Ca. 90 % of the secreted colony-stimulating activities are due GM-CSF (see also: CSF).

The synthesis of GM-CSF by various other cell types, e. g. endothelial cells and fibroblasts, is inducible by » TNF-α, » TNF-β, » IL1, » IL2 and » IFN. Some cell types express GM-CSF constitutively. Constitutive synthesis may also be the result of promoter insertion mutations (see: IAP, intracisternal A particle).

■ **Protein Characteristics:** GM-CSF is a monomeric protein of 127 amino acids with two glycosylation sites. The protein is synthesized as a precursor of 144 amino acids, which included a hydrophobic secretory » signal sequence at the aminoterminal end. The sugar moiety is not required for the full spectrum of biological activities. Non-glycosylated and glycosylated GM-CSF show the same activities *in vitro*. Fully glycosylated GM-CSF is biologically more active *in vivo* than the non-glycosylated protein. The different molecular weight forms of GM-CSF (14 kDa, 35 kDa) described in the literature are the result of

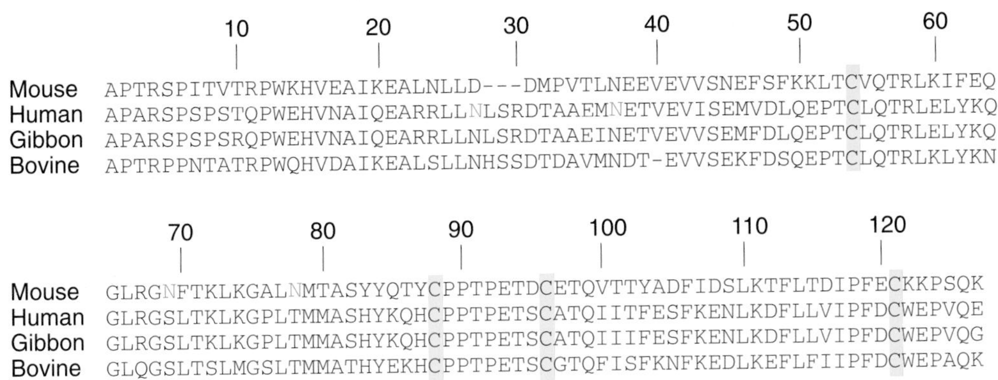

Sequences of GM-CSF from four different species.
Conserved cysteine residues are shown in yellow. The four molecules also show a high degree of sequence homology at other positions. The known N glycosylation sites (Asn = N) are shown in blue.
Point mutations at E21→P, L59→P, E63→P, L66→P, and L110→P) are known to destroy important α-helical regions within the molecule. These mutations reduce the biological activity by a factor of 100 to 10000. (J. Biol. Chem. 266: 5333-5341 (1991))

varying degrees of glycosylation. GM-CSF contains four cysteine residues (positions 54/96 and 88/121).

A comparison of the protein sequence of GM-CSF with those of the other colony-stimulating factors reveals that they are not related to each other. Human and murine GM-CSF display 60 % homology at the protein level and 70 % at the nucleotide level. The two factors do not, however, cross-react immunologically.

GM-CSF can be associated with the extracellular matrix of cells (see: ECM) as a complex with heparan sulfate proteoglycans. This allows storage of the factor in a biologically inactive form. The exact mechanism by which the factor is eventually released from these depots is not known.

GM-CSF can also be expressed as an integral membrane protein. The membrane-bound or matrix-associated forms of the factor can interact with receptors on other nearby cells. This effectively establishes cell-to-cell contacts and may also induce biological activities in juxtaposed cells (see: Juxtacrine).

■ **Gene Structure:** The human GM-CSF gene maps to chromosome 5q22-31 in the vicinity of other genes encoding hematopoietic growth factors (M-CSF, IL3, IL4, IL5) and the gene encoding the M-CSF receptor. The human gene has a length of ≈ 2. 5 kb and contains four exons. The distance between the GM-CSF gene and the IL3 gene is ≈ 9 kb.

It is noteworthy that deletions of the IL3/GM-CSF locus are frequently observed in patients » 5q$^-$syndrome. The biological significance of these deletions is, however, unclear.

The 5´ region of the GM-CSF gene contains several sequence elements known as » CLE (conserved lymphokine element). They function as binding sites for transcription factors (see also: Gene expression) modulating the expression of the GM-CSF gene.

■ **Receptor Structure, Gene(s), Expression:** GM-CSF receptors are expressed at densities of several 100 to several 1000 copies/cell on the cell surface of myeloid cells. The receptor is also expressed on non-hematopoietic cells such as endothelial cells and small cell lung carcinoma cells. Lymphoid cells and erythroid cells do not appear to express the receptor.

In receptor-positive cell lineages the receptor density decreases with increasing degrees of maturation.

The receptor shows significant homologies with other receptors for hematopoietic growth factors, including » IL2β, » IL3, » IL6, » IL7, Epo and the » prolactin receptors (see also: Cytokine receptor superfamily).

One cloned subunit of the GM-CSF receptor (GM-Rα, 45 kDa) binds GM-CSF with low affinity. The second subunit (GM-Rβ, 120 kDa) does not bind GM-CSF. GM-Rα is a protein of 400 amino acids that contains only a short cytoplasmic domain of 54 amino acids. The high affinity GM-CSF receptor is formed by the aggregation of the two receptor subunits. The GM-Rβ subunit of the receptor is also a constituent of other cytokine receptor systems. It is a component of the high affinity receptors for » IL3 and » IL5, both of which also contain a cytokine-specific subunit (see also: AIC2A). A soluble form of the GM-CSF receptor has also been described.

■ **Biological Activities:** Human GM-CSF is not active on murine cells and vice versa. GM-CSF was isolated initially as a factor stimulating the growth of macrophage/granulocyte-containing colonies in soft agar cultures (see also: Colony formation assay). GM-CSF is indispensable for the growth and development of granulocyte and macrophage progenitor cells. It stimulates myeloblasts and monoblasts and triggers irreversible differentiation of these cells. GM-CSF synergises with » Epo in the proliferation of erythroid and megakaryocytic progenitor cells. In combination with another colony-stimulating factor, » M-CSF, one observes the phenomenon of synergistic suppression, i. e., the combination of these two factors leads to a partial suppression of the generation of macrophage-containing cell colonies.

For some types of blast cells from patients with acute myeloid leukemia GM-CSF acts as an » autocrine growth factor. GM-CSF is a strong chemoattractant for neutrophils (see also: Chemotaxis). It enhances microbicidal activity, oxidative metabolism, and phagocytotic activity of neutrophils and macrophages. It also improves the cytotoxicity of these cells.

GM-CSF displays a less pronounced specificity than, for example, » G-CSF. It stimulates the proliferation and differentiation of neutrophilic, eosinophilic, and monocytic lineages. It also functionally activates the corresponding mature forms, enhancing, for example, to the expression of certain cell surface adhesion proteins (CD-11A, CD-11C). The overexpression of these proteins could be one explanation for the observed local accumulation of granulocytes at sites of » inflammation. In addition, GM-CSF also enhances expression of receptors for » fMLP (Formyl-Met-Leu-Phe) which is a stimulator of neutrophil activity.

At pico- to nanomolar concentrations GM-CSF is chemotactic for eosinophils and also influences the chemotactic behavior of these cells in response to other chemotactic factors.

In granulocytes GM-CSF stimulates the release of arachidonic acid metabolites and the increased generation of reactive oxygen species. The activation of the Na^+/H^+ antiport system leads to a rapid alkalization of the cytosol. Phagocytotic activities of neutrophil granulocytes and the cytotoxicity of eosinophils is also enhanced considerably by GM-CSF. Since GM-CSF is produced by cells present at sites of inflammatory responses (T lymphocytes, tissue macrophages, endothelial cells, mast cells) it can be assumed that it is an important mediator for inflammatory reactions.

The functional state of Langerhans cells of the skin is also influenced by GM-CSF. These cells are not capable of initiating primary immune responses, for example, contact sensibilization. They are converted to highly potent immunostimulatory dendritic cells by GM-CSF (and also » IL1). Langerhans cells therefore form an *in situ* reservoir for immunologically immature lymphoid dendritic cells. The maturation of these cells which is seen

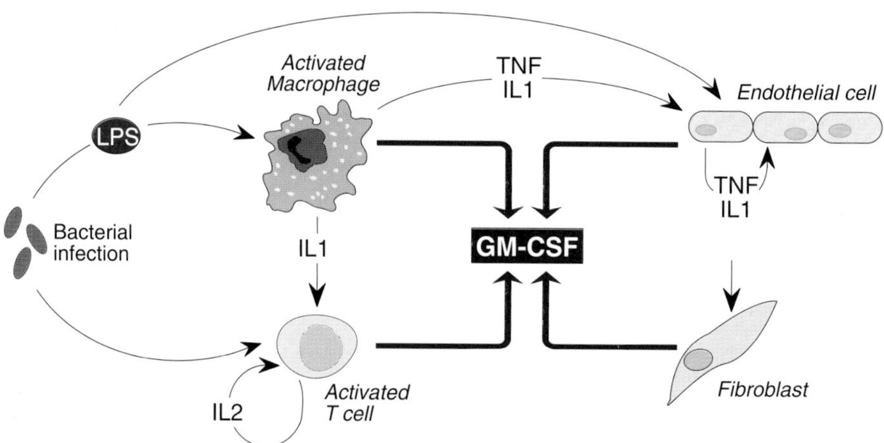

Some possible interactions leading to the synthesis and release of GM-CSF in response to infections.

as an increased ability to process antigens, can be down-regulated by » TNF-α.
At nanomolar concentrations GM-CSF induces the expression of Complement C3a receptors on basophils. Cells which normally do not respond to C3a and which have been activated by GM-CSF degranulate in response to the C3a stimulus. This is accompanied by the release of histamine and leukotriene C4. This process may be of significance in hypersensitivity reactions associated with inflammatory responses. GM-CSF has also been shown to be a potent inducer of trophoblast interferon (see: TP-1).
GM-CSF synergises with some other cytokines, including » IL1, » IL3 and » G-CSF. GM-CSF and G-CSF must act in concert to allow the development of neutrophil-containing colonies *in vitro*.
IL3 by itself only negligibly expands the number of circulating blood cells; a subsequent dose of GM-CSF, however, significantly increases cell numbers, probably because IL3 first leads to an expansion of those cells capable of responding to GM-CSF.
The observations that most IL3-dependent cell lines can also grow in the presence of GM-CSF and » IL4 and that several synergistic effects are observed between GM-CSF and IL4 suggest that these three factors perform similar functions in controlling the growth of cells. There are some indications that the mechanism of signal transduction contains at least some common factors.
Experiments with » oncogenes encoding a tyrosine-specific protein kinase have shown that the expression of the kinase activity in factor-dependent cells abolishes their dependence on GM-CSF, IL3 and IL4. The exact mechanism by which these factors regulate the proliferation and differentiation of cells is still unknown.

■ **Transgenic/Knock-out/Antisense Studies:** The consequences of a deregulated expression of GM-CSF have been studied in » transgenic animals (mice) harboring a constitutively expressed GM-CSF gene. The overexpression of the transgene encoding GM-CSF leads to pathological alterations in the retina and causes blindness and also causes muscle deterioration. These mice are characterized by a very pronounced increase in activated macrophages. In addition, the overexpression of GM-CSF leads to the activation of mature macrophages secreting large amounts of » IL1 and » TNF, suggesting that these cytokines may be responsible for some aspects of the transgenic mouse disease. Histopathological examination demonstrates a pronounced increase in the progenitor cell population of the monocytic lineage. GM-CSF-transgenic animals usually die within months from the massive tissue damages resulting from the overexpression of these factors. Similar results have been obtained with mice possessing a bone marrow manipulated to overexpress GM-CSF by transformation with suitable retrovirus vectors. These findings do not seem to be of clinical significance, though. The long-term treatment of primates and mice with GM-CSF has shown that life-threatening complications do not occur. For aspects of leukemic conversion of non-leukemic cells in GM-CSF-transgenic animals see: » FCP-P1.

GM-CSF knockouts: see addendum: GM-CSF.

■ **Assays:** GM-CSF can be assayed in a » colony formation assay by the development of colonies containing macrophages, neutrophils, eosinophils, and megakaryocytes. GM-CSF is also detected in specific » bioassays with cells lines that depend in their growth on the presence of GM-CSF or that respond to this factor (see: AML193; B6SUt-A; BAC1. 2F5; BCL_1; Da; FDC-P1; GF-D8; GM/SO; IC-2; MO7E; NFS-60; PT18; TALL-103; TF-1; UT-7). An alternative and entirely different detection method is » Message amplification phenotyping. For further information see also subentry "Assays" in the reference section.

■ **Clinical Use & Significance:** The usual dose, route and schedules for GM-CSF are 5-10 micrograms/kg/day either by 4-6 h intravenous infusion or by subcutaneous injection. At such doses, adverse effects are predominantly mild-to-moderate in nature, occur in 20-30% of patients and usually comprise fever, myalgia, malaise, rash and injection site reaction. Intravenous bolus or short infusions of GM-CSF are more likely to promote adverse effects. Some side effects are probably not due to direct actions of GM-CSF but are caused by the GM-CSF-induced secretion of other cytokines such as » TNF, » IL1 and » IL6 (see also: Acute phase reaction, Acute phase proteins).
GM-CSF can be employed for the physiological reconstitution of » hematopoiesis in all diseases characterized either by an aberrant maturation of blood cells or by a reduced production of leukocytes. The main and most important clinical application of GM-CSF is probably the treatment of life-threatening neutropenia following chemo- and/or radiotherapy, which is markedly reduced under GM-CSF treatment. GM-CSF can also be used to correct chemotherapy-induced cytopenias and to counteract cytopenia-related predisposition to infections and hemorrhages.

In order to avoid potential complications following the administration of GM-CSF careful clinical monitoring is required in certain patient groups, for example those with myelodysplastic syndrome, acute myeloid leukemia, inflammatory disease, autoimmune thrombocytopenia or malfunctional immunological responsiveness.

Several studies have demonstrated that the use of GM-CSF enhances tolerance to cytotoxic drug treatment and can be used to prevent dose reductions necessitated by the side effects of cytotoxic drug treatment. GM-CSF treatment frequently permits to increase the doses of cytotoxic drugs per course. These studies have also revealed a significantly reduced morbidity under GM-CSF treatment.

There are no indications that the treatment with hematopoietic growth factors induces a depletion of the stem cell population or the expansion of non-myeloid tumor cells. It has been observed, however, that GM-CSF treatment may lead to a transient expansion of clonogenic blast cells in patients with myelodysplastic syndromes and acute myeloid leukemia. The expansion of malignant clones is reversible if GM-CSF is withdrawn.

One of the advantages of GM-CSF treatment may be the induction of differentiation and the activation of resting leukemic cells to enter the cell cycle where they are susceptible to cell cycle-specific drugs.

At present, GM-CSF represents an important advance in bone marrow transplantation and has become a standard therapy. The drug can be given safely and does not appear to increase the risk of graft-versus-host disease or tumor relapse. GM-CSF enhances the reconstitution of the hematopoietic system in patients undergoing autologous or allogenic bone marrow transplantation and patients with delayed engraftment after bone marrow transplantation. The beneficial effects of GM-CSF following autologous bone marrow transplantation consist not only of a shorter period of absolute neutropenia, but also fewer significant infections, a diminished requirement for intravenous antibiotic administration, and a shorter overall duration of inpatient hospitalization. Since GM-CSF leads to a marked expansion of the peripheral pool of stem cells this may also be an alternative to traditional methods of collecting large amounts of these cells. Some animal experiments suggest that the effects of lethal irradiation can be abolished by a timely administration of GM-CSF and that the complete reconstitution of hematopoiesis is possible. At present data from human irradiated patients are not yet available.

Fusion proteins consisting of GM-CSF and tumor-derived tumor-specific variable regions of the immunoglobulin molecules expressed on malignant B cells have been constructed. These chimaeric tumor idiotypic proteins are strong immunogens that induce idiotype-specific antibodies without other carrier proteins or adjuvants. They have been shown to protect recipient animals from challenge with an otherwise lethal dose of B lymphoma tumor cells. It has been suggested that this approach may be applicable to the design of vaccines for a variety of other diseases. For other fusion proteins converting weakly immunogenic antigens into potent immunogens see: IL2 (subentry Clinical use & Significance).

The transduction of murine tumor cells with a functional GM-CSF gene has been shown to lead to the rejection of the genetically modified cells by syngeneic hosts (for cancer vaccines see also: Cytokine gene transfer).

● **Reviews: Demetri GD & Griffin JD** Granulocyte colony-stimulating factor and its receptor. Blood 78: 2791-2808 (1991); **Fan D et al** Granulocyte-macrophage colony-stimulating factor (GM-CSF) in the management of cancer. *In vivo* 5: 571-8 (1991); **Moore MAS** The clinical use of colony stimulating factors. ARI 9: 159-91 (1991); **Negrin RS & Greenberg PL** Therapy of hematopoietic disorders with recombinant colony-stimulating factors. Adv. Pharmacol. 23: 263-296 (1992); **Ruef C & Coleman DL** Granulocyte-macrophage colony-stimulating factor: pleiotropic cytokine with potential clinical usefulness. Rev. Infect. Dis. 12: 41-62 (1990)

● **Biochemistry & Molecular Biology: Barlow DP et al** Close genetic and physical linkage between the murine hemopoietic growth factor genes GM-CSF and multi-CSF (IL3). EJ 6: 617-23 (1987); **Cantrell MA et al** Cloning, sequence, and expression of human granulocyte/macrophage colony stimulating factor. PNAS 82: 6250-4 (1985); **Frolova EI et al** Linkage mapping of the human CSF2 and IL3 genes. PNAS 88: 4821-4 (1991); **Gough NM et al** Molecular cloning of cDNA encoding a murine hematopoietic growth regulator, granulocyte-macrophage colony stimulating factor. N 309: 763-7 (1984); **Gough NM et al** Structure and expression of the mRNA for murine granulocyte-macrophage colony stimulating factor. EJ 4: 645-53 (1985); **Gough NM et al** Localization of the human GM-CSF receptor gene to the X-Y pseudoautosomal region. N 345: 734-6 (1990); **Huebner K et al** The human gene encoding GM-CSF is at 5q21-32, the chromosome region deleted in the $5q^-$-anomaly. S 230: 1282-5 (1985); **Kaushansky K et al** Genomic cloning, characterization, and multilineage growth-promoting activity of human granulocyte-macrophage colony-stimulating factor. PNAS 83: 3101-5 (1986); **Kuczek ES et al** A granulocyte-colony-stimulating factor gene promoter element responsive to inflammatory mediators is functionally distinct from an identical sequence in the granulocyte-macrophage colony-stimulating factor gene. JI 146: 2426-33 (1991); **Lang RA et al** Transgenic mice expressing a hemopoietic growth factor gene (GM-CSF) develop accumulations of macrophages, blindness, and a fatal syndrome of tissue damage. Cell 51: 675-86 (1987);

Le Beau MM et al Assignment of the GM-CSF, CSF-1, and *fms* genes to human chromosome 5 provides evidence for linkage of a family of genes regulating hematopoiesis and for their involvement in the deletion (5q) in myeloid disorders. CSHSQB 51: 899-909 (1986); **Lee F et al** Isolation of cDNA for a human granulocyte-macrophage colony-stimulating factor by functional expression in mammalian cells. PNAS 82: 4360-4 (1985); **Metcalf D & Moore JG** Divergent disease patterns in granulocyte-macrophage colony-stimulating factor transgenic mice associated with different transgenic insertion sites. PNAS 85: 7767-71 (1988); **Miyatake S et al** Structure of the chromosomal gene for granulocyte-macrophage colony stimulating factor: comparison of the mouse and human genes. EJ 4: 2561-8 (1985); **Miyatake S et al** Characterization of the mouse granulocyte-macrophage colony-stimulating factor (GM-CSF) gene promoter: Nuclear factors that interact with an element shared by three lymphokine genes, those for GM-CSF, interleukin-4 (IL4), and IL5. MCB 11: 5894-901 (1991); **Nimer SD & Golde DW** The 5q$^-$ abnormality. Blood 70: 1705-12 (1987); **Price V et al** Expression, purification and characterization of recombinant murine granulocyte-macrophage colony-stimulating factor and bovine interleukin-2 from yeast. Gene 55: 287-93 (1987); **Reichert P et al** Crystallization and preliminary X-ray investigation of recombinant human granulocyte-macrophage colony-stimulating factor. JBC 265: 452-3 (1990); **Schanafelt AB et al** The amino-terminal helix of GM-CSF and IL5 governs high affinity binding to their receptors. EJ 10: 4105-12 (1991); **Stanley E et al** The structure and expression of the murine gene encoding granulocyte-macrophage colony stimulating factor: evidence for utilization of alternative promoters. EJ 4: 2569-73 (1985); **Tanaka H & Kaneko T** Pharmacokinetic and pharmacodynamic comparisons between human granulocyte colony-stimulating factor purified from human bladder carcinoma cell line 5637 culture medium and recombinant human granulocyte colony-stimulating factor produced in *Escherichia coli*. J. Pharmacol. Exp. Ther. 262: 439-44 (1992); **van Leeuwen BH et al** Molecular organization of the cytokine gene cluster involving the human IL3, IL4, IL5, and GM-CSF genes on human chromosome 5. Blood 73: 1142-8 (1989); **Wong GG et al** Human GM-CSF: molecular cloning of the complementary DNA and purification of the natural and recombinant proteins. S 228: 810-5 (1985); **Yang YC et al** The human genes for GM-CSF and IL3 are closely linked in tandem on chromosome 5. Blood 71: 958-61 (1988)

● **RECEPTORS: Chiba S et al** Characterization and molecular features of the cell surface receptor for human granulocyte-macrophage colony-stimulating factor. Leukemia 4: 29-36 (1990); **Fung MC et al** Distinguishing between mouse IL3 and IL3 receptor-like (IL5/GM-CSF receptor converter) mRNAs using the polymerase chain reaction method. JIM 149: 97-103 (1992); **Gearing DP et al** Expression cloning of a receptor for human granulocyte-macrophage colony-stimulating factor. EJ 8: 3667-76 (1989); **Hayashida K et al** Molecular cloning of a second subunit of the receptor for human granulocyte-macrophage colony-stimulating factor (GM-CSF): reconstitution of a high-affinity GM-CSF receptor. PNAS 87: 9655-9 (1990); **Park LS et al** Heterogeneity in human interleukin-3 receptors. A subclass that binds human granulocyte/macrophage colony stimulating factor. JBC 264: 5420-7 (1989); **Park LS et al** Cloning of the low-affinity murine granulocyte-macrophage colony-stimulating factor receptor and reconstitution of a high-affinity receptor complex. PNAS 89: 4295-9 (1992); **Raines MA et al** Identification and molecular cloning of a soluble human granulocyte-macrophage colony-stimulating factor receptor. PNAS 88: 8230-7 (1991)

● **BIOLOGICAL ACTIVITIES: Lemoli RM et al** Interleukin 11 stimulates the proliferation of human hematopoietic CD34$^+$ and CD34$^+$CD33$^-$DR$^-$ cells and synergises with stem cell factor, interleukin-3, and granulocyte-macrophage colony-stimulating factor. EH 21: 1668-72 (1993); **Lopez AF et al** Reciprocal inhibition of binding between interleukin-3 and granulocyte-macrophage colony-stimulating factor to human eosinophils. PNAS 86: 7022-6 (1989); **Metcalf D et al** Synergistic suppression: anomalous inhibition of the proliferation of factor-dependent hemopoietic cells by combination of two colony-stimulating factors. PNAS 89: 2819-23 (1992); **Monroy RL et al** Granulocyte-macrophage colony-stimulating factor: more than a hemopoietin. Clin. Immunol. Immunopathol. 54: 333-46 (1990); **Warringa RA et al** Modulation and induction of eosinophil chemotaxis by granulocyte-macrophage colony-stimulating factor and interleukin-3. Blood 77: 2694-700 (1991)

● **TRANSGENIC/KNOCK-OUT/ANTISENSE STUDIES: Cuthbertson RA & Lang RA** Developmental ocular disease in GM-CSF transgenic mice is mediated by autostimulated macrophages. Dev. Biol. 134: 119-29 (1989); **Cuthbertson RA et al** Macrophage products IL1 α, TNF α and bFGF may mediate multiple cytopathic effects in the developing eyes of GM-CSF transgenic mice. Exp. Eye Res. 51: 335-44 (1990); **Elliott MJ et al** Selective up-regulation of macrophage function in granulocyte-macrophage colony-stimulating factor transgenic mice. JI 147: 2957-63 (1991); **Gearing AJ et al** Elevated levels of GM-CSF and IL1 in the serum, peritoneal and pleural cavities of GM-CSF transgenic mice. Immunology 67: 216-20 (1989); **Lang RA et al** TNF α, IL1 α and bFGF are implicated in the complex disease of GM-CSF transgenic mice. GF 6: 131-8 (1992); **Metcalf D & Moore JG** Divergent disease patterns in granulocyte-macrophage colony-stimulating factor transgenic mice associated with different transgene insertion sites. PNAS 85: 7767-71 (1988); **Metcalf D** Mechanisms contributing to the sex difference in levels of granulocyte-macrophage colony-stimulating factor in the urine of GM-CSF transgenic mice. EH 16: 794-800 (1988); **Metcalf D et al** The excess numbers of peritoneal macrophages in granulocyte-macrophage colony-stimulating factor transgenic mice are generated by local proliferation. JEM 175: 877-84 (1992); **Metcalf D & Rasko JE** Leukemic transformation of immortalized FDC-P1 cells engrafted in GM-CSF transgenic mice. Leukemia 7: 878-86 (1993);**Tran HT et al** Anti-bacterial activity of peritoneal cells from transgenic mice producing high levels of GM-CSF. Immunology 71: 377-82 (1990)

● **ASSAYS: Cebon J et al** Pharmacokinetics of human granulocyte-macrophage colony-stimulating factor using a sensitive immunoassay Blood 72: 1340-7 (1988); **Katzen NA et al** Comparison of granulocyte-macrophage colony stimulating factor and interleukin-1 production from human peripheral blood mononuclear cells as measured by specific radioimmunoassays. ECN 3: 365-72 (1992); **Lewis CE et al** Measurement of cytokine release by human cells. A quantitative analysis at the single cell level using the reverse hemolytic plaque assay. JIM 127: 51-9 (1990); **Mortensen BT et al** Development and application of a sensitive radioimmunoassay for human granulocyte-macrophage colony-stimulating factor able to measure normal concentrations in blood. EH 21: 1366-70 (1993); **Oez S et al** A highly sensitive quantitative bioassay for human granulocyte-macrophage colony-stimulating factor. EH 18: 1108-11 (1990); **Roncaroli F et al** An immunoenzyme technique for the identification of granulocyte-macrophage colony-stimulating factor (GM-CSF) receptors using digoxigenated-GM-CSF. JIM 158: 191-6 (1993); **Sallerfors B & Olofsson T** Granulocyte-macrophage colony-stimulating factor (GM-CSF) and granulocyte colony-stimulating factor (G-CSF) secretion by adherent monocytes measured by quantitative immunoassays. Eur. J. Haematol. 49:

199-207 (1992); **Zenke G et al** A cocktail of three monoclonal antibodies significantly increases the sensitivity of an enzyme immunoassay for human granulocyte-macrophage colony-stimulating factor. J. Immunoassay 12: 185-206 (1991); for further information also see individual cell lines used in individual bioassays.

● **CLINICAL USE & SIGNIFICANCE:** **Armitage JO** The use of granulocyte-macrophage colony-stimulating factor in bone marrow transplantation. Semin. Hematol. 29: s14-8 (1992); **Cebon J et al** The dissociation of GM-CSF efficacy from toxicity according to route of administration: a pharmacodynamic study. Br. J. Haematol. 80: 144-50 (1992); **Chelstrom LM et al** Treatment of BCL-1 murine B cell leukemia with recombinant cytokines. Comparative analysis of the anti-leukemic potential of interleukin 1 β (IL1 β), interleukin 2 (IL2), interleukin-6 (IL6), tumor necrosis factor α (TNF α), granulocyte colony stimulating factor (G-CSF), granulocyte-macrophage colony stimulating factor (GM-CSF), and their combination. Leuk. Lymphoma 7: 79-86 (1992); **Demetri GD & Antman KH** Granulocyte-macrophage colony-stimulating factor (GM-CSF): preclinical and clinical investigations. Semin. Oncol. 19: 362-85 (1992); **De Witte T et al** Recombinant human granulocyte-macrophage colony-stimulating factor accelerates neutrophil and monocyte recovery after allogeneic T cell-depleted bone marrow transplantation. Blood 79: 1359-65 (1992); **Dranoff G et al** Vaccination with irradiated tumor cells engineered to secrete murine granulocyte-macrophage colony-stimulating factor stimulates potent, specific, and long-lasting anti-tumor immunity. PNAS 90: 3539-43 (1993); **Freund M & Kleine HD** The role of GM-CSF in infection. Infection 20: S84-92 (1992); **Gorin NC et al** Recombinant human granulocyte-macrophage colony-stimulating factor after high-dose chemotherapy and autologous bone marrow transplantation with unpurged and purged marrow in non-Hodgkin's lymphoma: a double-blind placebo-controlled trial. Blood 80: 1149-57 (1992); **Grant SM & Heel RC** Recombinant granulocyte-macrophage colony-stimulating factor (rGM-CSF). A review of its pharmacological properties and prospective role in the management of myelosuppression. Drugs 43: 516-60 (1992); **Gratwohl A et al** Granulocyte-macrophage colony stimulating factor (GM-CSF) in emergency treatment. Schweiz. Med. Wochenschr.121: 413-7 (1991); **Haas R et al** Recombinant human granulocyte-macrophage colony-stimulating factor (rhGM-CSF) subsequent to chemotherapy improves collection of blood stem cells for autografting in patients not eligible for bone marrow harvest. Bone Marrow Transplant. 9: 459-65 (1992); **Krumwieh D et al** Preclinical studies on synergistic effects of IL1, IL3, G-CSF, and GM-CSF in cynomolgous monkeys. IJCC 8 (Suppl. 1) 229-48 (1990); **Khwaja A et al** Recombinant human granulocyte-macrophage colony-stimulating factor after autologous bone marrow transplantation for malignant lymphoma: a British National Lymphoma Investigation double-blind, placebo-controlled trial. Br. J. Haematol. 82: 317-23 (1992); **Link H et al** A controlled trial of recombinant human granulocyte-macrophage colony-stimulating factor after total body irradiation, high-dose chemotherapy, and autologous bone marrow transplantation for acute lymphoblastic leukemia or malignant lymphoma. Blood 80: 2188-95 (1992); **Mitsuyashu RT & Golde DW** The clinical role of granulocyte-macrophage colony-stimulating factor. Hematol. Oncol. Clin. of North Amer. 3: 411-25 (1989); **Neidhart JA et al** Dosing regimen of granulocyte-macrophage colony-stimulating factor to support dose-intensive chemotherapy. J. Clin. Oncol. 10: 1460-9 (1992); **Nemunaitis J** Colony-stimulating factors: a new step in clinical practice. Part I. Ann. Med. 24: 439-44 (1992); **Nemunaitis J et al** Phase II trial of recombinant human granulocyte-macrophage colony-stimulating factor in patients undergoing allogeneic bone marrow transplantation from unrelated donors. Blood 79: 2572-7 (1992); **Neumanaitis J** Granulocyte-macrophage-colony-stimulating factor: a review from preclinical development to clinical application. Transfusion 33: 70-83 (1993); **Rosenfeld CS & Nemunaitis J** The role of granulocyte-macrophage colony-stimulating factor-stimulated progenitor cells in oncology. Semin. Hematol. 29: s19-26 (1992); **Schuster MW** Granulocyte-macrophage colony-stimulating factor (GM-CSF): what role in bone marrow transplantation? Infection 20: S95-9 (1992); **Shadduck RK** Granulocyte-macrophage colony-stimulating factor: present use and future directions. Semin. Hematol. 29: s38-42 (1992); **Shea TC et al** Sequential cycles of high-dose carboplatin administered with recombinant human granulocyte-macrophage colony-stimulating factor and repeated infusions of autologous peripheral-blood progenitor cells: a novel and effective method for delivering multiple courses of dose-intensive therapy. J. Clin. Oncol. 10: 464-73 (1992); **Singer JW** Role of colony-stimulating factors in bone marrow transplantation. Semin. Oncol. 19: s27-31 (1992); **Stern AC & Jones TC** The side-effect profile of GM-CSF. Infection 20: S124-7 (1992); **Sternberg CN et al** Escalated M-VAC chemotherapy and recombinant human granulocyte-macrophage colony stimulating factor (rhGM-CSF) in patients with advanced urothelial tract tumors. Ann. Oncol. 4: 403-7 (1993); **Steward WP et al** Granulocyte-macrophage colony-stimulating factor allows safe escalation of dose-intensity of chemotherapy in metastatic adult soft tissue sarcomas: a study of the European Organization for Research and Treatment of Cancer Soft Tissue and Bone Sarcoma Group. J. Clin. Oncol. 11: 15-21 (1993); **Steward WP** Granulocyte and granulocyte-macrophage colony-stimulating factors. The Lancet 342: 153-7 (1993); **Sureda A et al** GM-CSF administration enhances granulocytic recovery in purged autologous bone marrow transplantation for acute lymphoblastic leukemia. Prog. Clin. Biol. Res. 377: 315-20 (1992); **Tao MH & Levy R** Idiotype/granulocyte-macrophage colony-stimulating factor fusion protein as a vaccine for B cell lymphoma. N 362: 755-8 (1993); **Tapp H & Vowels M** Prophylactic use of GM-CSF in pediatric marrow transplantation. Transplant. Proc. 24: 2267-8 (1992); **Vadhan-Raj S et al** Use of granulocyte-macrophage colony-stimulating factor in hematopoietic disorders: biology and nature of response. Semin. Hematol. 29: s4-13 (1992); **Vadhan-Raj S** Clinical trials of granulocyte-macrophage colony-stimulating factor for the treatment of aplastic anemia. Immunol. Ser. 57: 661-70 (1992); **Valent P et al** Treatment of *de novo* acute myelogenous leukemia with recombinant granulocyte macrophage-colony-stimulating factor in combination with standard induction chemotherapy: effect of granulocyte macrophage-colony-stimulating factor on white blood cell counts. Med. Pediatr. Oncol. Suppl. 2: 18-22 (1992); **Weiss M & Belohradsky BH** Granulocyte-macrophage colony-stimulating factor (GM-CSF): a variety of possible applications in clinical medicine. Infection 20: S81-3 (1992)

GM-CSF receptor affinity converter: see: gp130 and subentry "receptor structure, gene(s), expression" for GM-CSF.

GM-DF: *granulocyte/macrophage and leukemia cell differentiation-inducing factor* This factor is identical with » G-CSF. See also: CSF for general information about colony-stimulating factors.

Dexter TM et al The role of hemopoietic cell growth factor (interleukin 3) in the development of hemopoietic cells. Ciba Found. Symp. 116: 129-47 (1985); **Gabrilove JL et al** Constitutive production of leukemia differentiation, colony-stimulating, erythroid burst-promoting, and pluripoietic factors by a human hepatoma cell line: characterization of the leukemia differentiation factor. Blood 66: 407-15 (1985); **Moore MA et al** Therapeutic implications of serum factors inhibiting proliferation and inducing differentiation of myeloid leukaemic cells. Blood Cells 9: 125-44 (1983)

GM-EA: *granulomonopoietic enhancing activity* see: GM-EF (granulomonopoietic enhancing factor).

GM-EF: *granulomonopoietic enhancing factor* This poorly characterized factor of 74 kDa is also called ***GM-EA*** (granulomonopoietic enhancing activity). It acts as an accessory protein that enhances the proliferation and/or maturation of the myeloid progenitor cells » CFU-GM and pre-CFU-GM. In combination with other colony-stimulating factors (see also: CSF) it promotes the development of granulocyte/macrophage-containing colonies. The factor has no effect on either erythroid (see: BFU-E) or mixed (granulocyte erythrocyte macrophage megakaryocyte colony-forming units (see: CFU-GEMM).
GM-EF is produced constitutively by mature macrophages (MDLM = monocyte-derived lipid-containing macrophage). The production of GM-EF is potentiated by bacterial lipopolysaccharides and is inhibited by treatment of the cells with » lactoferrin, prostaglandin E2, or » IFN-γ.
GM-EF is not identical with » IL1, » IL4 or » IL6 which are also known to enhance the growth of myelopoietic precursor cells in combination with other colony-stimulating factors. It has been reported to resemble a murine factor produced by adherent macrophages in long-term bone marrow cultures (see: CPA, colony-promoting activity).
GM-EF may be of clinical interest because it might allow reconstitution or improvement of myelopoiesis in patients with bone marrow dysplasia or myelosuppression if used in combination with other colony-stimulating factors such as » G-CSF or » GM-CSF.
Lieu CW et al Production of monoclonal antibody against granulomonopoietic enhancing activity (GM-EA). JIM 141: 133-8 (1991); **Wang SY et al** Long-term culture of human bone marrow macrophages: macrophage development is associated with the production of granulomonopoietic enhancing activity (GM-EA). JI 135: 1186-93 (1985); **Wang SY et al** Biological characterization of a granulomonopoietic enhancing activity derived from cultured human lipid-containing macrophages. Blood 65: 1181-90 (1985); **Wang SY et al** Down regulation of myelopoiesis by mediators inhibiting the production of macrophage-derived granulomonopoietic enhancing activity (GM-EA). Blood 72: 2001-6 (1988); **Wang SY et al** The effect of lipopolysaccharide on the production of GM-EA, GM-CSA, and PGE2 by human monocyte-derived lipid-containing macrophages. EH 16: 349-54 (1988); **Wang SY et al** Detection of the target progenitor cells of granulomonopoietic enhancing activity. Blood 76: 495-500 (1990); **Wang SY** Purification and characterization of human macrophage-derived granulomonopoietic enhancing factor (GM-EF). EH 20: 552-7 (1992)

GMF: *glial maturation factor(s)* A general term used to describe factors that modulate the functional maturation of glial cells. A distinct factor with GMF activity is » GMF-β.

GMF-β: *glia maturation factor β.* This factor is isolated from brain tissues and has been found in all vertebrate tissues analyzed so far. It is an acidic (pI = 5.2) protein of 17 kDa. The bovine and human factors have a length of 142 amino acids and contain three cysteine residues. Two of them (position 86/95) form a disulfide bond that is essential for bioactivity. The factor has a blocked aminoterminus (N-Acetyl-serine). It does not display any significant homology to other sequenced proteins. The factor is involved in the phenotypic differentiation of glial cells and neurons. The factor is expressed in the cytosol and on the cell surface of astrocytes. It is not secreted and presumably acts at short distances involving cell-to-cell interactions with target cells expressing a receptor (see also: Juxtacrine growth control).
The protein stimulates the proliferation of astroblasts and promotes the outgrowth of neurites. It inhibits the proliferation of tumors derived from this cell type. The unlimited growth of gliomas may be due to a defect in the transport of this protein from the cytosol to the cell surface. The reversible inhibition of a number of neuronal and non-neuronal neoplastic cells by GMF-β is also observed. The factor suppresses the mitogenic activity of » aFGF but aFGF does not synergise in its activity on astrocytes.
Harman K et al Glia maturation factor β stimulates axon regeneration in transsected rat sciatic nerve. Brain Res. 564: 332-5 (1991); **Kaplan R et al** Molecular cloning and expression of biologically active human glia maturation factor-β. J. Neurochem. 57: 483-90 (1991); **Lim R & Zaheer A** Structure and function of glia maturation factor β. AEMB 296: 161-4 (1991); **Lim R et al** Antiproliferative function of glia maturation factor β. Cell Regul. 1: 741-6 (1990); **Lim R et al** Cell-surface expression of glia maturation factor β in astrocytes FJ 4: 3360-3 (1990); **Lim R et al** Complete amino acid sequence of bovine glia maturation factor β. PNAS 87: 5233-7 (1990); **Lim R et al** Mitogenic activity of glia maturation factor. ECR 159: 335-43 (1985); **Lim R et al** Purification of bovine glia maturation factor and characterization with monoclonal antibody. B 24: 8070-4 (1985); **Lim R et al** Purification and characterization of glia maturation factor β. A growth regulator of neurons and glia. PNAS 86: 3901-5 (1989); **Lim R et al** Suppression of glioma growth *in vitro* and

in vivo by glia maturation factor. CR 46: 5241-7 (1986); **Zaheer A & Lim R** Disulfide isoforms of recombinant glia maturation factor β. BBRC 171: 746-51 (1990)

GM/SO: A human leukemia cell line used to assay (see also: Bioassays) the colony-stimulating factor » GM-CSF (see also: CSF). GM-CSF activity is determined by measuring the incorporation of ^{3}H thymidine into the newly synthesized DNA of proliferating cells. Cell proliferation can be determined also by employing the » MTT assay. An alternative and entirely different method of detecting GM-CSF is » Message amplification phenotyping.
Oez S et al A highly sensitive quantitative bioassay for human granulocyte-macrophage colony-stimulating factor. EH 18: 1108-11 (1990)

GNFS-60: A murine leukemia cell line obtained after retroviral infection. The cell line is used to assay the colony-stimulating factor » G-CSF (see also: CSF). The assay is not specific because this cell line also responds to » M-CSF and » IL6 (see also: Bioassays) thus requiring the use of suitable neutralizing factor-specific antibodies. See also: NFS-60 cell line.
Weinstein Y et al Truncation of the c-*myb* gene by a retroviral integration in an interleukin 3-dependent myeloid leukemia cell line. PNAS 83: 5010-4 (1986)

Gonadotropin II: see: Luteinizing hormone.

GOS: *G_0/G_1 switch genes* This is a general term for genes encoding mRNAs that are overexpressed in lymphocytes entering the G_1 phase of the » cell cycle.
Some of the genes are identical with known cytokine genes or genes known to be overexpressed following activation of cells by a cytokine. One human gene, designated ***GOS19-1***, was identified as an early G0/G1 switch gene in cultured blood mononuclear cells. It encodes a factor called » SCI (stem cell inhibitor) which inhibits stem cell proliferation. This factor is identical with » MIP (macrophage inflammatory protein; MIP1α). Other GOS genes are members of a family of genes known as » SIG (small inducible genes). Most of these genes encode proteins that belong to the » chemokine family of cytokines. ***GOS7*** is identical with the » oncogene » *fos*. GOS-30 encodes an » ERG (early response gene) and is identical with » Egr-1.
Blum S et al Three human homologues of a murine gene encoding an inhibitor of stem cell proliferation. DNA Cell Biol. 9: 589-602(1990); **Forsdyke DR** cDNA cloning of mRNAs which increase rapidly in human lymphocytes cultured with concanavalin A and cycloheximide. BBRC 129: 619-25 (1985); **Siderovski DP et al** A set of human putative lymphocyte G0/G1 switch genes includes genes homologous to rodent cytokine and zinc finger protein-encoding genes. DNA Cell Biol. 9: 579-87 (1990)

gp: abbrev. for glycoprotein. Suffixed numbers indicate the molecular mass, for example, » gp130, a glycoprotein with the apparent molecular mass of 130 kDa.

GP-2: *glycoprotein 2* see: uromodulin.

gp39: This cell surface glycoprotein of 39 kDa expressed on activated T cells has been identified as a ligand of » CD40. See: TRAP (TNF-related activation protein).
Hollenbaugh D et al The human T cell antigen gp39, a member of the TNF gene family, is a ligand for the CD40 receptor: expression of a soluble form of gp39 with B cell co-stimulatory activity. EJ 11: 4313-21 (1992)

gp130: This transmembrane glycoprotein with a length of 918 amino acids, including an intracellular domain of 277 amino acids, is a subunit constituent of several cytokine receptors, including those for » IL6, » IL11, » LIF, » Oncostatin M, and » CNTF (ciliary neurotrophic factor).
gp 130 participates in the formation of high-affinity receptors for these cytokines by binding to low affinity receptor chains. Accordingly, gp130 has also been called an ***affinity converter***. Ligand binding to a cytokine receptor leads to the dimerization of gp130 (shown for the IL6 receptor) or heterodimerization (shown for LIF, Oncostatin M, and CNTF receptors) with a gp130-related protein known as the LIFRβ subunit. Binding of the respective ligands is associated with the activation/association of a family of tyrosine kinases known as » Janus kinases, as the first step of intracellular signal transduction. Monoclonal antibodies directed against gp130 have been shown to inhibit IL6-mediated functions. Soluble forms of gp130 (sgp130) with molecular masses of 90 and 110 Kda have been found in human serum. They can inhibit biological functions of those cytokines utilizing receptor systems with gp130 as a component.
The human gp130 gene product appears to be homologous to two distinct chromosomal loci on chromosomes 5 and 17. The presence of two distinct gp130 gene sequences is restricted to primates and is not found in other vertebrates.
Davis S et al LIFRβ and gp130 as heterodimerizing signal transducers of the tripartite CNTF receptor. S 260: 1805-8 (1993); **Gearing GP et al** The IL6 signal transducer, gp130: an oncostatin M receptor and affinity converter for the LIF receptor. S 255: 1434-7 (1992); **Hibi M et al** Molecular cloning and expression

gonadotrope polypeptide: see addendum: Granins. **GP 87:** see addendum: Granins.

of an IL6 signal transducer, gp130. Cell 63: 1149-57 (1990); **Kidd VJ et al** Chromosomal localization of the IL6 receptor signal transducing subunit, gp130 (IL6ST). Somat. Cell. Mol. Genet. 18: 477-83 (1992); **Liu J et al** Interleukin-6 signal transducer gp130 mediates oncostatin M signaling. JBC 267: 16763-6 (1992); **Murakami M et al** Critical cytoplasmic region of the interleukin 6 signal transducer gp130 is conserved in the cytokine receptor family. PNAS 88: 11349-53 (1991); **Murakami M et al** IL6-induced homodimerization of gp130 and associated activation of a tyrosine kinase. S 260: 1808-10 (1993); **Narazaki M et al** Soluble forms of the interleukin-6 signal-transducing receptor component gp130 in human serum possessing a potential to inhibit signals through membrane-anchored gp130. Blood 82: 1120-6 (1993); **Saito T et al** Preparation of monoclonal antibodies against the IL6 signal transducer, gp130, that can inhibit IL6-mediated functions. JIM 163: 217-23 (1993); **Stahl N et al** Association and activation of Jak-Tyk kinases by CNTF-LIF-OSM-IL6b receptor components. S 263: 92 (95 (1994); **Taga T et al** Interleukin-6 triggers the association of its receptor with a possible signal transducer, gp130. Cell 58: 573-81 (1989); **Taga T et al** Functional inhibition of hematopoietic and neurotrophic cytokines by blocking the interleukin 6 signal transducer gp130. PNAS 89: 10998-1001 (1992); **Wang Y et al** Molecular cloning and characterization of the rat liver IL6 signal transducing molecule, gp130. Genomics 14: 666-72 (1992); **Yasukawa K et al** Association of recombinant soluble IL6-signal transducer, gp130, with a complex of IL 6 and soluble IL6 receptor, and establishment of an ELISA for soluble gp130. Immunol. Lett. 31: 123-30 (1992)

GPA: *growth-promoting activity* This factor is identical with » bFGF.

Lipton A et al Liver as a source of transformed-cell growth factor. In: Sato GH & Ross R (eds) Hormones and cell culture. Cold Spring Harb. Conf. Cell Prolif. Vol. 6: 461-75 (1979)

GPA: *growth-promoting activity* This factor is found in chick sciatic nerves. GPA mRNA is induced in embryonic chick eyes during the period of neuron cell death in ciliary ganglions. GPA mRNA is expressed specifically in the layer of the eye that contains the targets of the ciliary ganglions and in primary cultures of smooth muscle cells isolated from the choroid layer of the eye. In contrast to another protein, » CNTF (ciliary neurotrophic factor) biologically active GPA is released from cells transfected with a GPA cDNA.

Leung DW et al Cloning, expression during development, and evidence for release of a trophic factor for ciliary ganglion neurons. Neuron 8: 1045-53 (1992)

GPBP: granulocyte/pollen-binding protein This heat-stable protein has been found in normal human serum. It promotes the binding of granulocytes to timothy grass pollen. GPBP is identical with » transferrin. The pollen-binding property of transferrin is unrelated to iron transport.

Sass-Kuhn SP et al Human granulocyte/pollen-binding protein: recognition and identification as transferrin. JCI 73: 202-10 (1984)

GPF: *glomerular permeability factor* This poorly characterized factor of 60-160 kDa is produced by human T cell hybridomas derived from the T cells of a patient with mammal change nephrotic syndrome (MCNS). Intravenous injection of this factor into rats induces proteinuria. In normal human lymphocyte cultures GPF enhances Concanavalin-A-induced lymphocyte blastogenesis. GPF is cytotoxic to tumor cell lines of epithelial origin, but only cytostatic to tumor cells of hematopoietic origin.

Koyama A et al A glomerular permeability factor produced by human T cell hybridomas. Kidney Int. 40: 453-60 (1991)

GPF: *growth promoting factor* This poorly characterized 7-10 kDa factor is isolated from the serum-free culture supernatants (see also: SFM, serum-free medium) of the astrocytoma cell line GA-1. It promotes the proliferation and the differentiation of normal glioblasts.

Ito J et al Autocrine regulation of glial proliferation and differentiation: the induction of cytodifferentiation of postmitotic normal glioblasts by growth-promoting factor from astrocytoma cells. Brain Res. 374: 335-41 (1986)

GPF: *growth promoting factors* Also: GSA (growth-stimulating activities) General terms, sometimes also used in abbreviations, describing factors and more frequently, biological activities, that promote the growth of cells. These factors can be isolated, for example, from the conditioned medium of cells (see: CM). Some such factors can be isolated by exploiting some of their biochemical properties (see, for example: HBGF, heparin-binding growth factors).

GPF-1: *glial promoting factor 1* This factor is identical with » OGF (oligodendroglia growth factor).

GPF-2: *glial promoting factor 1* This poorly characterized 9 kDa factor is isolated from rat brain cortex. It stimulates the growth of astroglial cells.

Giulian D et al Brain peptides and glial growth. I. Glial promoting factors as regulators of gliogenesis in the developing and injured central nervous system. JCB 102: 803-11 (1986)

GPF-3: *glial promoting factor 1* This poorly characterized 6 kDa factor is isolated from rat brain. It stimulates the proliferation of oligodendrocytes.

Giulian D et al Brain peptides and glial growth. I. Glial promoting factors as regulators of gliogenesis in the developing and injured central nervous system. JCB 102: 803-11 (1986)

GPI: abbrev. for Glucose-6-phosphate isomerase see: NLK (neuroleukin).

GRAM: *granulocyte-activating mediators* This poorly characterized factor (17 and 44 kDa) is released by human epidermal cells and the epidermoid carcinoma cell line A431 (epidermal cell-derived granulocyte-activating mediator, abbrev. ***EC-GRAM***. Factor production is enhanced by bacterial lipopolysaccharides (see also: LPS), but not by silica particles and phytohemagglutinin. The activity stimulates human granulocytes to release significant levels of toxic oxygen radicals. Monocytes release ***M-GRAM*** (monocyte-derived GRAM) upon stimulation with bacterial lipopolysaccharides (LPS) which also induces a long-lasting chemiluminescence response in human granulocytes. M-GRAM is not identical with » IL1, » IL2, » IFN-α, » IFN-γ, » G-CSF, or » M-CSF, all of which do not induce a chemiluminescence response. Such a response is elicited by » GM-CSF and particularly by » TNF-α. TNF-α activity but not » TNF-β activity is found in M-GRAM samples, but antibodies neutralizing the GM-CSF or TNF activities in M-GRAM preparations do not substantially block the chemiluminescence signal.

Lit. **Kapp A et al** Granulocyte-activating mediators (GRAM). II. Generation by human epidermal cells – relation to GM-CSF. Arch. Dermatol. Res. 279: 470-7 (1987); **Kapp A et al** Granulocyte-activating mediators (GRAM): III. Further functional characterization of monocyte-derived GRAM. Arch. Dermatol. Res. 280: 346-53 (1988)

Gram-negative sepsis: See: Endotoxins, Septic shock.

Granocyte: Brand name for recombinant » G-CSF (Rhone-Poulenc Rorer, Cologne, FRG).

Granulin: see: Epithelins.

Granulocyte-activating mediators: see: GRAM.

Granulocyte chemiluminescence inducer: see: GCI.

Granulocyte chemotactic factor: see: GCF. See: GCP (granulocyte chemotactic peptide).

Granulocyte chemotactic peptide: see: GCP.

Granulocyte colony forming cells: abbrev. CFC-G or G-CFC. See: CFC.

Granulocyte colony stimulating activity: see: G-CSF.

Granulocyte colony stimulating factor: see: G-CSF.

Granulocyte-derived factor: see: GDF.

Granulocyte inhibitory protein: see: GIP.

Granulocyte/macrophage and leukemia cell differentiation-inducing factor: see: GM-DF.

Granulocyte-macrophage colony forming cells: abbrev. CFC-GM. See: CFC.

Granulocyte-macrophage colony stimulating factor: see: GM-CSF.

Granulocyte/pollen-binding protein: see: GPBP.

Granulomonopoietic enhancing activity: see: GM-EF (granulomonopoietic enhancing factor).

Granulomonopoietic enhancing factor: see: GM-EF.

GRE: *glucocorticoid response element* A short sequence region (TGTTCT) found in the promoter of some genes, including some cytokine genes such as » IL6, » IL8, the expression of which is subject to regulation by glucocorticoids. This sequence element is the binding site for regulatory proteins (glucocorticoid receptor and other regulatory proteins) that influence transcription (see also: Gene expression).

Dahlman-Wright K et al DNA-binding by the glucocorticoid receptor: a structural and functional analysis. J. Steroid Biochem. Mol. Biol. 41: 249-72 (1992); **Gehring U** The structure of glucocorticoid receptors. J. Steroid Biochem. Mol. Biol. 45: 183-90 (1993); **Gronemeyer H** Control of transcription activation by steroid hormone receptors. FJ 6: 2524-9 (1992); **Miesfeld RL** Molecular genetics of corticosteroid action. ARRD 141: S11-7 (1990); **Oakley RH & Cidlowski JA** Homologous down regulation of the glucocorticoid receptor: the molecular machinery. Crit. Rev. Eukaryot. Gene Exp. 3: 63-88 (1993)

GRF: *growth hormone releasing factor* see: Growth hormone.

GRIF: *growth inhibitory factor* see: GIF (growth inhibition factor).

GRMFT: *glucosteroid response-modifying factor* This poorly characterized factor is produced by mouse spleen cells and a murine T cell hybridoma, FS6 14.13.1 after stimulation with concanavalin

Granins: see addendum. **GRB:** see addendum.

A. GRMFT blocks glucosteroid suppression of helper T cell function and the growth of granulocyte/macrophage progenitor cells *in vitro*.
GRMFT does not appear to be identical with » IL1, » IL2, » IL3, » IFN-γ. The factor is probably not identical with various » TRFs (T cell replacing factors) or » CSFs (colony-stimulating factors) as FS6 14.13.1 have been reported not to secrete such activities.
Fairchild SS et al T cell-derived glucosteroid response-modifying factor (GRMFT): a unique lymphokine made by normal T lymphocytes and a T cell hybridoma. JI 132: 821-7 (1984)

GRO: *growth regulated oncogene* see: » MGSA (melanoma growth stimulatory activity).

Growth and maturation activity: see: GMA.

Growth/differentiation factor: see: GDF-1.

Growth factors: This collective term, originally referring to substances that promote cell growth, is used rather loosely now, comprising molecules that function as growth stimulators (mitogens) but also as growth inhibitors (sometimes referred to as negative growth factors), factors that stimulate cell migration (see: motogenic cytokines) and function as chemotactic agents (see also: Chemotaxis), factors that modulate differentiated functions of cells, and factors involved in » Apoptosis. Such factors can be diffusible factors and can also exist in membrane-anchored forms (see: Juxtacrine). In many instances the term is used as a synonym for » cytokines. See also: Recombinant cytokines.

Growth factor modules: Short distinct protein sequence domains that appear in a variety of proteins other than that in which it was first identified. For examples see: EGF-like repeats, src homology regions.

Growth factors: abbrev. GF. See: Cytokines.

Growth hormone: abbrev. ***GH,*** also known as ***somatotropic hormone*** or ***somatotropin***. GH is a single chain polypeptide hormone of 191 amino acids mainly produced by the adenohypophysis (anterior pituitary). It is released in response to the hypothalamus-derived Growth hormone releasing hormone (abbrev. ***GHRH***; also ***GRF***, growth hormone releasing factor) also known as ***somatotropin releasing factor/hormone*** (abbrev. ***SRF/SRH***), or ***somatoliberin***. Production of GH is inhibited by » Somatostatin and » Activin A and enhanced by » GRP (gastrin-releasing peptide), » IL1, » IL6, » VIP (vasoactive intestinal peptide), and some » thymic hormones.
The physiological activity for which GH is best known is the promotion of growth of bone, cartilage, and soft tissues. GH secretion is maximal at puberty and declines after 30 years of age. Overproduction of GH leads to acromegaly while lack of it causes dwarfism (see also: Laron-type dwarfism; Dwarf mice).
Detectable levels of GH are found throughout the remainder of adulthood, suggesting other functions in addition to promotion of growth. GH may be important for the maintenance of lean body mass; most growth-promoting effects of GH are mediated by » IGF-1 the synthesis of which is regulated by GH (see also IGF-BP, IGF binding proteins). The biological activities of GH are mediated by receptors belonging to the » Cytokine receptor superfamily. GH has been shown to bind to a receptor that also binds » Prolactin.
GH appears to control important immune functions (see also: Neuroimmune network). GH has been shown to be produced by T cells, B cells, and macrophages. In phytohemagglutinin-stimulated human lymphocytes GH appears to upregulate its own expression. Receptors for GH releasing hormone have also been found on cells of the immune system. GH appears to act as an enhancer of immune responses and is produced in considerable amounts by T helper cells. Hypothalamus-derived GH releasing hormone has been shown recently to elicit GH production by lymphocytes. GH augments the cytolytic activity of T cells, antibody synthesis, and granulocyte differentiation induced by » GM-CSF. GH also enhances production of » TNF-α, generation of superoxide anions from peritoneal macrophages, and natural killer activity. GH induces a chemotactic response in human monocytes which is inhibited by » Somatostatin. GH enhances the synthesis of some » thymic hormones. An active fragment of GH appears to account for some of the biological activity of » PM (pregnancy mitogen). GH has been shown to promote engraftment of murine or human T cells in severe combined immunodeficient mice (SCID) mice (see also: Immunodeficient mice).
Treatment of mice with recombinant human GH has been shown to partially counteract the myelosuppressive properties of azidothymidine, resulting in an increase in splenic hematopoietic progenitor cells (see also: Hematopoiesis). GH has been

Growth factor receptor-bound protein: see addendum: GRB.

shown to function as a » paracrine growth and differentiation factors in the hematopoietic system (see also: Hematopoietins).

It has been shown that hypophysectomy causes thymic atrophy and an impairment of immune functions. Immune deficits in hypophysectomised rats that are unable to mount an immune response are normalized by treatment with GH or » Prolactin.

■ TRANSGENIC/KNOCK-OUT/ANTISENSE STUDIES: The expression of » antisense oligonucleotides specific for growth hormone mRNA has recently been shown to inhibit proliferation of B and T cells, suggesting either » autocrine or » paracrine growth regulatory circuits. Although growth hormone itself does not appear to be mitogenic for lymphocytes it appears to be necessary for interleukin-induced lymphocyte proliferation.

● REVIEWS: **Gala RR** Prolactin and growth hormone in the regulation of the immune system. PSEBM 198: 513-27 (1991); **Kelley KW et al** Growth hormone, prolactin, and insulin-like growth factors: new jobs for old players. Brain Behav. Immunol. 6: 317-26 (1992); **Hooghe R et al** Growth hormone and prolactin are paracrine growth and differentiation factors in the haemopoietic system. IT 14: 212-4 (1993)

Baglia LA et al Production of immunoreactive forms of growth hormone by the Burkitt tumor serum-free cell line sfRamos. Endocrinology 130: 2446-54 (1992); **Fu YK et al** Growth hormone augments superoxide anion secretion of human neutrophils by binding to the prolactin receptor. JCI 89: 451-7 (1992); **Guarcello V et al** Growth hormone releasing hormone receptors on thymocytes and splenocytes from rats. CI 136: 291-302 (1991); **Edwards CK 3rd et al** *In vivo* administration of recombinant growth hormone or γ interferon activities macrophages: enhanced resistance to experimental *Salmonella typhimurium* infection is correlated with generation of reactive oxygen intermediates. Infect. Immunol. 60: 2514-21 (1992); **Edwards CK 3rd et al** The macrophage-activating properties of growth hormone. Cell. Mol. Neurobiol. 12: 499-510 (1992); **Goya RG et al** *In vivo* effects of growth hormone on thymus function in aging mice. Brain Behav. Immunol. 6: 341-54 (1992); **Hattori N et al** Immunoreactive growth hormone (GH) secretion by human lymphocytes: augmented release by exogenous GH. BBRC 168: 396-401 (1990); **Hattori N et al** Growth hormone (GH) secretion from human lymphocytes is up-regulated by GH, but not affected by insulin-like growth factor-I. J. Clin. Endocrinol. Metab. 76: 937-9 (1993); **Kelley KW** The role of growth hormone in modulation of the immune response. ANY 594: 95-103 (1990); **Kitaoka M et al** Inhibition of growth hormone secretion by activin A in human growth hormone-secreting tumor cells. Acta Endocrinol. (Copenh.) 124: 666-71 (1991); **Knyszynski A et al** Effects of growth hormone on thymocyte development from progenitor cells in the bone marrow. Brain Behav. Immunol. 6: 327-40 (1992); **Kooijman R et al** Effects of insulin-like growth factors and growth hormone on the *in vitro* proliferation of T lymphocytes. J. Neuroimmunol. 38: 95-104 (1992); **Murphy WJ et al** Growth hormone exerts hematopoietic growth-promoting effects *in vivo* and partially counteracts the myelosuppressive effects of azidothymidine. Blood 80: 1443-7 (1992); **Murphy WJ et al** Human growth hormone promotes engraftment of murine or human T cells in severe combined immunodeficient mice. PNAS 89: 4481-5 (1992); **Murphy WJ et al** Differential effects of growth hormone and prolactin on murine T cell development and function. JEM 178: 231-6 (1993); **Murphy WJ et al** Recombinant human growth hormone promotes human lymphocyte engraftment in immunodeficient mice and results in an increased incidence of human Epstein Barr virus-induced B-cell lymphoma. Brain Behav. Immunol. 6: 355-64 (1992); **Murphy WJ et al** Role of neuroendocrine hormones in murine T cell development. Growth hormone exerts thymopoietic effects *in vivo*. JI 149: 3851-7 (1992); **Nagy E et al** Regulation of immunity in rats by lactogenic and growth hormones. Acta Endocrinol. Copenh. 102: 351-7 (1983); **Rudman D et al** Effects of human growth hormone in men over 60 years old. NEJM 323: 1-6 (1990); **Rudman D** Growth hormone, body composition, and aging. J. Am. Geriatr. Soc. 33: 800-7 (1990); **Smith PE** The effect of hypophysectomy upon the involution of the thymus in the rat. Anat. Rec. 47: 119-29 (1930); **Spangelo BL et al** Thymosin fraction 5 stimulates prolactin and growth hormone release from anterior pituitary cells *in vitro*. Endocrinology 121: 2035-43 (1987); **Timsit J et al** Growth hormone and insulin-like growth factor-I stimulate hormonal function and proliferation of thymic epithelial cells. J. Clin. Endocrinol. Metab. 75: 183-8 (1992); **Varma S et al** Growth hormone secretion by human peripheral blood mononuclear cells detected by an enzyme-linked immunoplaque assay. J. Clin. Endocrinol. Metab. 76: 49-53 (1993); **Weigent DA et al** The production of growth hormone by subpopulations of rat mononuclear leukocytes. CI 135: 55-65 (1991); **Wiedermann CJ et al** Stimulation of monocyte chemotaxis by human growth hormone and its deactivation by somatostatin. Blood 82: 954-60 (1993); **Yoshida A et al** Recombinant human growth hormone stimulates B cell immunoglobulin synthesis and proliferation in serum-free medium. Acta Endocrinol. Copenh. 126: 524-9 (1992)

● TRANSGENIC/KNOCK-OUT/ANTISENSE STUDIES: **Weigent DA et al** An antisense oligodeoxynucleotide to growth hormone messenger ribonucleic acid inhibits lymphocyte proliferation. Endocrinology 128: 2053-7 (1991)

Growth hormone dependent binding protein: This protein is identical with one of the insulin-like growth factor binding proteins (see: IGF-BP).

Growth hormone independent binding protein: This protein is identical with one of the insulin-like growth factor binding proteins (see: IGF-BP).

Growth inhibition factor: see: GIF.

Growth inhibitory factor: see: GIF.

Growth inhibitor from BSC-1: see: BSC-1 cell growth inhibitor.

Growth regulated oncogene: abbrev. *gro*. see: MGSA (melanoma growth stimulatory activity).

Growth-promoting activity: see: GPA.

Growth-promoting factor: see: GPF.

Growth stimulatory activity for TS1 cells: see: GATS.

GRP: *gastrin-releasing peptide*

■ **Alternative Names:** Mammalian » bombesin
■ **Sources:** GRP is a neuropeptide found in the central and peripheral nervous system where it is secreted by neuronal and endocrine cells. The neurotransmitter acetylcholine stimulates the secretion of GRP by normal neuroendocrine cells in the lung. Some tumors derived from these cells (small cell neuroendocrine lung tumors) and also fibroblasts, smooth muscle cells, and pancreatic islet cells also produce GRP.
■ **Protein Characteristics:** GRP is a protein of 27 amino acids (sequence see: Bombesin). It is synthesized as a precursor of 148 amino acids that contains a hydrophobic secretory » signal sequence. The mature protein is modified by α-amidation. A biologically active shortened cleavage product of GRP consisting of the 10 carboxyterminal amino acids has also been described. The human factor differs from porcine GRP at five amino acid positions.
■ **Gene Structure:** The human gene encoding GRP maps to chromosome 18q21.
■ **Related Factors:** Human GRP is the homologue of peptides isolated from amphibians (see: Bombesin). It is also related to the bombesin-like factor neuromedin B (see: Tachykinins).
■ **Receptor Structure, Gene(s), Expression:** Two types of receptors have been described for GRP and related neuropeptides. One of these receptors is a protein of 384 amino acids. It contains seven transmembrane domains and is a member of family of G-protein-coupled receptors. This receptor binds GRP with high affinity. It also binds » bombesin and neuromedin B with low affinity. The GRP receptor gene maps to human chromosome Xp11-q11.
■ **Biological Activities:** GRP regulates various physiological processes, including hormone secretion (pancreas, pituitary, gastrointestinal tract), thermoregulation, circadian rhythms, blood glucose levels, and contraction of smooth muscle cells,

GRP stimulates the release of gastrointestinal peptide hormones and influences the release of many pituitary hormones such as » Growth hormone and » prolactin. Many of the biological activities of GRP resemble those of » bombesin.

Apart from the classical role of neurohumoral hormone GRP also displays some cytokine-like activities although it is not classified as a cytokine due to its small size. It is a potent mitogen for murine fibroblasts (3T3 cells) and human bronchial epithelial cells. GRP is an » autocrine growth factor for SCLC (small-cell lung cancer).
■ **Clinical Use & Significance:** Antibodies directed against GRP and also pharmacological antagonists are now being tested for their antitumor activities against small cell lung cancers. See also: Bombesin.

● **Reviews: Hildebrand P et al** Human gastrin-releasing peptide: biological potency in humans. Regulatory Peptides 26: 423-33 (1991); **McDonald TJ** The gastrin-releasing polypeptide (GRP) Adv. Metab. Dis. 11: 199-250 (1988)
● **Biochemistry & Molecular Biology: Battey JF et al** Molecular cloning of the bombesin/gastrin-releasing peptide receptor from Swiss 3T3 cells. PNAS 88: 395-9 (1991); **Conlon JM et al** Primary structures of the bombesin-like neuropeptides in frog brain show that bombesin is not the amphibian gastrin-releasing peptide. BBRC 178: 526-30 (1991); **Feldman RI et al** Purification and characterization of the bombesin/gastrin-releasing peptide receptor from Swiss 3T3 cells. JBC 265: 17364-72 (1990); **Hajri A et al** Gastrin-releasing peptide: *in vivo* and *in vitro* growth effects on an acinar pancreatic carcinoma. CR 52: 3726-32 (1992); **Sausville EA et al** Expression of the gastrin-releasing peptide gene in human small cell lung cancer: evidence for alternative processing resulting in three distinct mRNAs. JBC 261: 2451-7 (1986); **Spindel ER et al** Cloning and characterization of cDNAs encoding human gastrin-releasing peptide. PNAS 81: 5699-703 (1984)
● **Receptors: Battey J & Wada E** Two distinct receptor subtypes for mammalian bombesin-like peptides. TINS 14: 524-8 (1991); **Battey J et al** Molecular cloning of the bombesin/gastrin-releasing peptide precursor from Swiss 3T3 cells. PNAS 88: 395-9 (1991); **Kane MA et al** Isolation of a gastrin releasing peptide receptor from normal rat pancreas. Peptides 12: 207-13 (1991); **Kane MA et al** Isolation of the bombesin/gastrin-releasing peptide receptor from human small cell lung carcinoma NCI-H345 cells. JBC 266: 9486-93 (1991); **Schantz LJ et al** Assignment of the GRP receptor gene to the human X chromosome. Cytogenet. Cell Genet. 58: 2085-6 (1991); **Wada E et al** cDNA cloning, characterization, and brain region-specific expression of a neuromedin B-preferring bombesin receptor. Neuron 6: 421-30 (1991)
● **Biological Activities: Willey JC et al** Bombesin and the C-terminal tetradecapeptide of gastrin-releasing peptide are growth factors for human bronchial epithelial cells. ECR 153: 245-8 (1984)
● **Clinical Use & Significance: Avis IL et al** Preclinical evaluation of an anti-autocrine growth factor monoclonal antibody for treatment of patients with small-cell lung cancer. JNCI 83: 1470-6 (1991); **Mulshine J et al** Phase I evaluation of an anti-gastrin releasing peptide (GRP) monoclonal antibody (MoAB) in patients with advanced lung cancer. Proc. Am. Soc. Clin. Oncol. 7: 213 (1988); **Radulovic S et al** Inhibition of growth of HT-29 human colon cancer xenografts in nude mice by treatment with bombesin/gastrin releasing peptide antagonist (RC-3095). CR 51: 6006-9 (1991); **Sehti T et al** Growth of small cell lung cancer cells: stimulation by multiple neuropeptides and inhibition by broad spectrum antagonists *in vitro* and *in vivo*. CR 52: 2737s-42s (1992)

GSA: *growth-stimulating activities* see: GPF (growth promoting factors)

GSF: *glia cell stimulating factor* This poorly characterized factor of 10-30 kDa is produced by T lymphocytes and some T cell lines (MOLT-4F, RPMI 1788). It stimulates RNA and DNA synthesis in cultured murine glia cells and influences the functional activities of astrocyte progenitor cells.
Fontana A et al Glia cell stimulating factor (GSF): a new lymphokine. Part 1. Cellular sources and partial purification of murine GSF, role of cytoskeleton and protein synthesis in its production. J. Neuroimmunol. 2: 55-71 (1982); **Fontana A et al** Glia cell stimulating factor (GSF): a new lymphokine. Part 2. Cellular sources and partial purification of human GSF. J. Neuroimmunol. 2: 73-81 (1982)

GSF: *glioma-derived suppressor factor* This poorly characterized factor is secreted by human glial tumor cells. It suppresses the mitogen responsiveness of normal human peripheral blood lymphocytes. GSF also inhibits the production of » IL2 by mitogen-activated human T cells. GSF suppresses and in some cases completely inhibits the expression of functional high affinity IL2 receptors on activated T cells so that, in the presence of GSF, the addition of IL2 to mitogen activated human T cells does not restore their normal proliferative response. GSF may therefore be responsible for the abnormalities in cell-mediated and humoral immunity observed in T cells of patients with primary malignant intracranial tumors. For another factor interfering with IL2 receptor expression see: IL2R α chain inhibitory activity.
Elliott LH et al Inability of mitogen-activated lymphocytes obtained from patients with malignant primary intracranial tumors to express high affinity interleukin 2 receptors. JCI 86: 80-6 (1990); **Elliott LH et al** Suppression of high affinity IL2 receptors on mitogen activated lymphocytes by glioma-derived suppressor factor. J. Neurooncol. 14: 1-7 (1992)

GSM: *glucocorticoid-suppressible mitogenic (activity)* This factor is secreted by a rat hepatoma cell line (BDS.1) which are hypersensitive to the antiproliferative effects of glucocorticoids. GSM expression is suppressed by treatment of the cells with glucocorticoids. The factor is a 28 kDa homodimeric protein that functions as an » angiogenesis factor This biological activity of GSM is completely neutralized by antibodies directed against the A chain of » PDGF. GSM is identical with the AA-homodimer of PDGF. It has been suggested that glucocorticoids may regulate hepatoma growth *in vivo* by modulating PDGF-stimulated tumor vascularization.
Cook PW et al Partial characterization of a glucocorticoid suppressible mitogenic activity secreted from a rat hepatoma cell line hypersensitive to the antiproliferative effects of glucocorticoids. JBC 264: 14151-8 (1989); **Haraguchi T et al** Identification of the glucocorticoid-suppressible mitogen from rat hepatoma cells as an angiogenic platelet-derived growth factor A-chain homodimer. JBC 266: 18299-307 (1991)

GST: glutathione S-transferase see: MIF (migration inhibition factor).

G-TsF: *glioblastoma-derived T cell suppressor factor* This factor is secreted by the human glioblastoma cell line 308 which also constitutively secretes a soluble factor with biologic and biochemical characteristics of human monocyte-derived » IL1. G-TsF inhibits the effects of IL1 and » IL 2 on T lymphocytes. It inhibits the lectin-induced proliferation of thymocytes, the IL2-induced proliferation of T cells and the development of alloreactive cytotoxic T cells.
G-TsF has a length of 112 amino acids. At the protein level the factor shows a homology of 71 % to » TGF-β including the number and positions of nine cysteine residues. A comparison of the first 30 aminoterminal amino acids reveals complete sequence identity with CIF-B (cartilage-inducing factor, see: CIF-A) G-TsF is identical with TGF-β2 (see: TGF-β).
de Martin R et al Complementary DNA for human glioblastoma-derived T cell suppressor factor, a novel member of the transforming growth factor-β gene family. EJ 6: 3673-7 (1987); **Kuppner MC et al** Inhibition of lymphocyte function by glioblastoma-derived transforming growth factor β 2. J. Neurosurg. 71: 211-7 (1989); **Schwyzer M & Fontana A** Partial purification and biochemical characterization of a T cell suppressor factor produced by human glioblastoma cells. JI 134: 1003-9 (1985); **Siepl C et al** The glioblastoma-derived T cell suppressor factor/transforming growth factor-β 2 inhibits T cell growth without affecting the interaction of interleukin 2 with its receptor. EJI 18: 593-600 (1988); **Siepl C et al** Glioblastoma-cell-derived T cell suppressor factor (G-TsF). Sequence analysis and biologic mechanism of G-TsF. ANY 540: 437-9 (1988); **Wrann M et al** T cell suppressor factor from human glioblastoma cells is a 12.5 kd protein closely related to transforming growth factor β. EJ 6: 1633-6 (1987)

Guanine nucleotide exchange factor: see addendum: SOS1.

H

h: (also: hu) in abbreviations, such as in h-GM-CSF, abbrev. for human.

H-1: see: Hemopoietin 1.

H7: see: H8.

H8: N-(2-methylaminoethyl)-5-isochinoline sulfonamide. A compound derived from calmodulin antagonists that does no longer possess calmodulin-inhibitory activity. H8 selectively inhibits several cAMP-dependent protein kinases. H8 (affecting protein kinase A) and also some related derivatives, referred to as the H series, e. g. ***H7*** (non-selective, affecting protein kinase C), are used to test whether binding of a cytokine to its receptor involves cAMP-dependent protein kinases for post receptor signaling processes. A newly synthesized isoquinoline sulfonamide, designated ***H89*** is even more potent than H8 in inhibiting cAMP-dependent protein kinases.

$-SO_2-NH-(CH_2)_2-NH-CH_3 \times 2\,HCl$

H8 =

N-(2-methylaminoethyl)-5-isochinolinsulfonamid

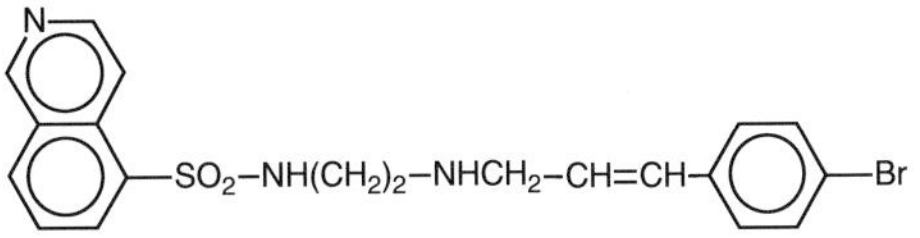

$-SO_2-NH(CH_2)_2-NHCH_2-CH=CH-$ ... $-Br$

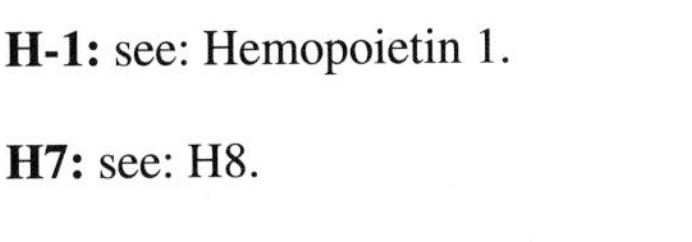

H89

For other agents used to dissect cytokine-mediated signal transduction pathways see: Bryostatins, Calcium ionophore, Calphostin C, Genistein, Herbimycin A, K-252a, Lavendustin A, Phorbol esters, Okadaic acid, Staurosporine, Suramin, Tyrphostins, Vanadate.

Hengel H et al Dissection of signals controlling T cell function and activation: H7, an inhibitor of protein kinase C, blocks induction of primary T cell proliferation by suppressing interleukin (IL)2 receptor expression without affecting IL2 production. EJI 21: 1575-82 (1991); **Hidaka H et al** Isoquinoline sulfonamides: novel and potent inhibitors of cyclic nucleotide-dependent protein kinase and protein kinase C. B 23: 5036-41(1984); **Hidaka H et al** Properties and use of H-series compounds as protein kinase inhibitors. MiE 201: 328-339 (1991); **Hidaka H & Kobayashi R** Pharmacology of protein kinase inhibitors. Annu. Rev. Pharmacol. Toxicol. 32: 377-97 (1992); **Nixon JS et al** Modulation of cellular processes by H7, a non-selective inhibitor of protein kinases. Agents Actions 32: 188-93 (1991)

H69-derived immunosuppressive factor: This poorly characterized factor of ≈ 192 kDa is secreted by human small cell lung cancer (SCLC) cell lines (H69, N857). It inhibits the blastogenic response of human peripheral blood lymphocytes to phytohemagglutinin or concanavalin A and also the cytotoxic activity of lymphokine-activated killer cells (see: LAK cells).

Ikeda T et al Characterization and purification of an immunosuppressive factor produced by a small cell lung cancer cell line. Jpn. J. Cancer Res. 82: 332-8 (1991)

H89: see: H8.

H400: This factor is identical with MIP-1β, one of the molecular forms of» MIP (macrophage inflammatory protein). It therefore belongs to the » chemokine family of cytokines.

Brown KD et al A family of small inducible proteins secreted by leukocytes are members of a new superfamily that includes leukocyte and fibroblast-derived inflammatory agents, growth factors, and indicators of various activation processes. JI 142: 679-87 (1989)

HAF: *human angiogenic factor* This poorly characterized factor of 67 kDa is found in the cell culture supernatants of the human melanoma cell line A-375/2 grown in serum-free media (see also: SFM). The factor is also produced by inflammatory macrophages regenerating tissues. It acts as an » Angiogenesis factor *in vivo* but is not mitogenic for endothelial cells.

Fruhbeis B et al Immunolocalization of an angiogenic factor (HAF) in normal, inflammatory and tumor tissues. IJC 42: 207-12 (1988); **Osthoff KS et al** Purification and characterization of a novel human angiogenic factor (h-AF). BBRC 146: 945-52 (1987)

Haptoglobin: A tetrameric plasma glycoprotein produced primarily by hepatocytes. Haptoglobulin is polymorphic due to the existence of two allelic forms of the α chains (α1, 9 kDa, 83 amino acids; and α2, 17.3 kDa, 142 amino acids). There are three different phenotypic haptoglobulins: type 1-1 (2α1/2β), 2-2(2α2/2β), 1-2(α1α2/2β). The glycosylated β chain (40 kDa, 245 amino acids unglycosylated 35 kDa) is the same in all Haptoglobin types. Type 1-1 haptoglobulin is monomeric but the other two types occur in several polymeric forms with molecular masses up to 1 million. Cloning of the human haptoglobin cDNA shows that haptoglobin is synthesized as a single polypeptide chain which is then cleaved at an arginine residue to generate its two subunits, α2 and β. The genes encoding the haptoglobin α and β chains have been mapped to human chromosome 16q22. A closely linked haptoglobin-related (Hpr) gene pair has been identified 2.2 kb downstream of the haptoglobin locus.

Haptoglobin is one of the » acute phase proteins produced during the » acute phase reaction. Its synthesis is induced by various cytokines, including » IL1 (previously identified as » haptoglobin inducing activity), » IL6, » CNTF (ciliary neurotrophic factor). The increased synthesis of haptoglobin mediated by IL6 is suppressed by » TNF-α. The gene encoding Haptoglobin contains an » IL6-responsive element (see: IL6RE).

The haptoglobin gene is actively transcribed in adult but not in fetal liver. Serum haptoglobulin levels are greatly increased when there is extensive tissue damage or necrosis and its principal function appears to prevent loss of body iron from intravascular hemolysis. Haptoglobulin forms complexes with free plasma oxyhemoglobulin that cannot be filtered by the kidneys. The heme porphyrin ring is enzymatically removed in the liver and the globin is subsequently degraded. Human haptoglobin probably also binds human myoglobin, albeit with a much lower affinity than that for hemoglobin.

Haptoglobin can modulate lymphocyte and macrophage functions, inhibiting polyclonal lymphocyte mitogenic responses to phytohemagglutinin or concanavalin and inhibiting or enhancing B cell mitogenesis in response to bacterial » endotoxins, depending on its concentration. Haptoglobin inhibits neutrophil respiratory burst activity in cells stimulated with » fMLP, arachidonic acid, and opsonized zymosan. Haptoglobin has been identified as one of the proteins in the synovial fluids of patients with rheumatoid arthritis that complexes hyaluronic acid and protects it from depolymerisation by activated phagocytes. Haptoglobin may have immunosuppressive activities (see: SER (suppressive E receptor). Haptoglobin-hamoglobin complexes in human plasma inhibit endothelium dependent relaxation.

Haptoglobin stimulates the formation of prostaglandin E2 in osteoblast-like cells isolated from neonatal mouse calvarial bones and potentiates the stimulatory effect of » bradykinin and thrombin on PGE2 formation. Enhanced production of haptoglobin seen in different inflammatory processes may therefore contribute to the destruction of bone by inducing the formation of prostanoids capable of stimulating bone resorption.

Haptoglobin has been shown recently to be an » angiogenesis factor. It is found in sera from patients with systemic vasculitis and serum haptoglobin levels in vasculitis patients correlate both with disease and angiogenic activity. It has been suggested that the increased levels of haptoglobin found in chronic inflammatory conditions may play an important role in tissue repair. In systemic vasculitis, haptoglobin might also compensate for ischemia by promoting development of collateral vessels.

Haptoglobin has been suggested to be a sensitive parameter for early detection of moderate hemolysis in pregnant women with HELLP syndrome, an extremely progressive form of gestosis characterized by hemolysis, elevated liver enzymes, and low platelet counts.

Baseler MW & Burrell R Purification of haptoglobin and its effects on lymphocyte and alveolar macrophage responses. Inflammation 7: 387-400 (1983); **Baumann H et al** Distinct regulation of the interleukin-1 and interleukin-6 response elements of the rat haptoglobin gene in rat and human hepatoma cells. MCB 10: 5967-76 (1990); **Bowman BH et al** Haptoglobin. MiE 163: 452-74 (1988); **Chow V et al** Biosynthesis of rabbit haptoglobin: chemical evidence for a single chain precursor. FL 153: 275-9 (1983); **Cid MC et al** Identification of haptoglobin as an angiogenic factor in sera from patients with systemic vasculitis. JCI 91: 977-85 (1993); **Edwards DH et al** Haptoglobin-hemoglobin complex in human plasma inhibits endothelium dependent relaxation: evidence that endothelium derived relaxing factor acts as a local autacoid. Cardiovasc. Res. 20: 549-56 (1986); **Frohlander N et al** Haptoglobin synergistically potentiates bradykinin and thrombin induced prostaglandin biosynthesis in isolated osteoblasts. BBRC 178: 343-51 (1991); **Heinderyckx M et al** Secretion of glycosylated human recombinant haptoglobin in baculovirus-infected insect cells. Mol. Biol. Rep. 13: 225-32 (1988-89); **Hutadilok N et al** Binding of haptoglobin, inter-α-trypsin inhibitor, and α 1 proteinase inhibitor to synovial fluid hyaluronate and the influence of these proteins on its degradation by oxygen derived free radicals. Ann. Rheum. Dis. 47: 377-85 (1988); **Kazim AL & Atassi MZ** Haemoglobin binding with haptoglobin. Localization of the haptoglobin-binding site on the α-chain of human hemoglobin. BJ 197: 507-10 (1981); **Kurosky A et al** Covalent structure of human haptoglobin: a serine protease homologue. PNAS 77: 3388-92 (1980); **Lerner UH & Frohlander N** Haptoglobin-stimulated bone resorption in neonatal mouse calvarial bones *in vitro*. Arthritis Rheum. 35: 587-91 (1992); **Maeda N** Nucleotide sequence of the haptoglobin and haptoglobin-related gene pair. The haptoglobin-related gene contains a retrovirus-like element. JBC 260: 6698-709 (1985); **Maeda N & Smithies O** The evolution of multigene families: human haptoglobin genes. ARG 20: 81-108 (1986); **Marinkovic S & Baumann H** Structure, hormonal regulation, and identification of the interleukin-6- and dexamethasone-responsive element of the rat haptoglobin gene. MCB 10: 1573-83 (1990); **McGill JR et al** Localization of the haptoglobin α and β genes (HPA and HPB) to human chromosome 16q22 by *in situ* hybridization. Cytogenet. Cell. Genet. 38: 155-7 (1984); **Murata H & Miyamoto T** Bovine haptoglobin as a possible immunomodulator in the sera of transported calves. Br. Vet. J. 149: 277-83 (1993); **Natsuka S et al** Augmentation of haptoglobin production in Hep3B cell line by a nuclear factor NF-IL6. FL 291: 58-62 (1991); **Oh SK et al** Specific binding of haptoglobin to human neutrophils and its functional consequences. J. Leukoc. Biol. 47: 142-8 (1990); **Oliviero S et al** The human haptoglobin gene: transcriptional regulation during development and acute phase induction. EJ 6: 1905-12 (1987); **Oliviero S & Cortese R** The human haptoglobin gene promoter: interleukin-6-responsive elements interact with a DNA-binding protein induced by interleukin-6 EJ 8: 1145-51 (1989); **Rademacher BE & Steele WJ** A general method for the isolation of haptoglobin 1-1, 2-1, and 2-2 from human plasma. AB 160: 119-26 (1987); **Raugei G et al** Sequence of human haptoglobin cDNA: evidence that the α and β subunits are coded by the same mRNA. NAR 11: 5811-9 (1983); **Raynes JG et al** Acute-phase protein synthesis in human hepatoma cells: differential regulation of serum amyloid A (SAA) and haptoglobin by interleukin-1 and interleukin-6. Clin. Exp. Immunol. 83: 488-91 (1991); **Sakata S et al** Human haptoglobin binds to human myoglobin. BBA 873: 312-5 (1986); **Valette I et al** Haptoglobin heavy and light chains. Structural properties, reassembly, and formation of minicomplex with hemoglobin. JBC 256: 672-9 (1981); **van der Straten A et al** Molecular cloning of human haptoglobin cDNA: evidence for a single mRNA coding for α 2 and β chains. EJ 2: 1003-7 (1983); **Wilke G et al** Haptoglobin as a sensitive marker of hemolysis in HELLP-syndrome. Int. J. Gynaecol. Obstet. 39: 29-34 (1992)

Haptoglobin inducing activity: This factor is identical with one of the molecular forms of » IL1. It induces the synthesis of » haptoglobin.

Matsushima K et al Purification of human interleukin 1 from human monocyte culture supernatants and identity of thymocyte comitogenic factor, fibroblast proliferation factor, acute phase protein-inducing factor, and endogenous pyrogen. CI 92: 290-301 (1985)

Haptotaxis: see: Chemotaxis.

HARP: *heparin affin regulatory peptide* This protein is identical with » HBNF (heparin-binding neurite-promoting factor).

HBBM: *heparin-binding brain mitogen* This protein was initially isolated as a heparin-binding protein (see also: HBGF, heparin-binding growth factors) from brain tissues of several species. It is identical with » HBNF (heparin-binding neurite-promoting factor).

Huber D et al Amino-terminal sequences of a novel heparin-binding protein with mitogenic activity for endothelial cells from human bovine, rat, and chick brain: high interspecies homology. Neurochem. Res. 15: 435-9 (1990)

HB-EGF: *heparin binding EGF-like factor* This factor is secreted by the cell line » U-937 and is also synthesized by macrophages and human vascular smooth muscle cells. It is a monomeric heparin-binding O-glycosylated protein of 86 amino acids (22 kDa) (see also: HBGF, heparin-binding growth factors). The existence of several N-terminally truncated forms has been described. The protein is synthesized from a precursor of 208 amino acids. It is a member of the » EGF family of proteins and shares ≈ 40 % sequence identity in its carboxyterminal domain with EGF and » TGF-α. HB-EGF has an aminoterminal extension absent in EGF and TGF-α. The factor competes with EGF for binding to the EGF receptor and also induces the » autophosphorylation of the EGF receptor.

The factor is a potent mitogen for murine » 3T3 cells, keratinocytes and smooth muscle cells, but not for endothelial cells. The mitogenic activity on smooth muscle cells is much stronger than that of » EGF and appears to involve interactions with cell surface heparan sulfate proteoglycans (see also: ECM, extracellular matrix). HB-EGF production in vascular smooth muscle cells is induced by » angiotensin. HB-EGF may have an important » autocrine role in the proliferation of these cells in vascular diseases such as atherosclerosis and hypertension.

The expression of the HB-EGF gene in human umbilical chord endothelial cells is induced immediately after treatment of the cells with » TNF-α or IL1β (see: IL1). The HB-EGFgene therefore encodes an immediate early protein (see also: Gene expression). HB-EGF is a major growth factor component of wound fluid and may play an important role in » wound healing.

The heparin binding domain of HB-EGF has been used to construct chimeric toxin proteins consisting of this domain and TGF-α. A membrane-anchored form of HB-EGF has been shown to serve as a receptor for diphtheria toxin entry into the cell.

Abraham JA et al Heparin-binding EGF-like growth factor: characterization of rat and mouse cDNA clones, protein domain conservation across species, and transcript expression in tissues. BBRC 190: 125-33 (1993); **Diuz Sm et al** Heparin-binding epi-

HARP: see addendum.

dermal growth factor-like growth factor expression in cultured fetal human vascular smooth muscle cells. JBC 268: 18330-4 (1993); **Higashiyama S et al** A heparin-binding growth factor secreted by macrophage like cells that is related to EGF. S 251: 936-9 (1991); **Higashiyama S et al** Structure of heparin-binding EGF-like growth factor. Multiple forms, primary structure, and glycosylation of the mature protein. JBC 267: 6205-12 (1992); **Higashiyama S et al** Heparin-binding EGF-like growth factor stimulation of smooth muscle cell migration: dependence on interactions with cell surface heparan sulfate. JCB 122: 933-40 (1993); **Marikovsky M et al** Appearance of heparin-binding EGF-like growth factor in wound fluid as a response to injury. PNAS 90: 3889-93 (1993); **Mesri EA et al** Heparin-binding transforming growth factor α-*Pseudomonas* exotoxin A. A heparan sulfate-modulated recombinant toxin cytotoxic to cancer cells and proliferating smooth muscle cells. JBC 268: 4853-62 (1993); **Naglich JG et al** Expression cloning of a diphtheria toxin receptor: identity with a heparin-binding EGF-like growth factor. Cell 69: 1051-61 (1992); **Temizer DH et al** Induction of heparin-binding epidermal growth factor-like growth factor mRNA by phorbol ester and angiotensin II in rat aortic smooth muscle cells. JBC 267: 24892-6 (1992); **Yoshizumi M et al** Tumor necrosis factor increases transcription of the heparin-binding epidermal growth factor-like growth factor gene in vascular endothelial cells. JBC 267: 9467-9 (1992)

HBGAF: *heparin binding cell growth associated factor* This neurotrophic and mitogenic factor is identical with » HBNF (heparin-binding neurite-promoting factor).

Kuo MD et al Characterization of heparin-binding growth-associated factor receptor in NIH 3T3 cells. BBRC 182: 188-94 (1992)

HB-GAM: *heparin binding growth-associated molecule* This protein of 15.3 kDa was initially isolated as a heparin-binding protein (see also: HBGF, heparin-binding growth factors) from brain tissues of several species. It was originally described to possess mitogenic activity for vascular endothelial cells and fibroblasts, but this does not seem to be the case. HB-GAM promotes growth of » SW-13 cells in soft agar. The factor is identical with » HBNF (heparin-binding neurite-promoting factor).

Courty J et al Mitogenic properties of a new endothelial cell growth factor related to pleiotrophin. BBRC 180: 145-51 (1991); **Hampton BS et al** Structural and functional characterization of full-length heparin-binding growth associated molecule. Mol. Biol. Cell. 3: 85-93 (1992); **Merenmies J & Rauvala H** Molecular cloning of the 18-kDa growth-associated protein of developing brain. JBC 265: 16721-4 (1990); **Raulo E et al** Secretion and biological activities of heparin-binding growth-associated molecule. Neurite outgrowth-promoting and mitogenic actions of the recombinant and tissue-derived protein. JBC 11408-16 (1992); **Wellstein A et al** A heparin-binding growth factor secreted from breast cancer cells homologous to a developmentally regulated cytokine. JBC 267: 2582-7 (1992)

HBGF: *heparin-binding growth factor* These factors, isolated from bovine neural tissue by their high affinity to heparin, are identical either with » aFGF or » bFGF.

Lobb RR & Fett JW Purification of two distinct growth factors from bovine neural tissue by heparin affinity chromatography. B 23: 6295-6299 (1984)

HBGF: *heparin-binding growth factors* This is a generic term to describe factors that can be isolated by affinity chromatography (***HSAC*** = heparin-sepharose affinity chromatography) due to their very high affinity for heparin.

The ability to bind to heparin appears to be a general property of many growth-regulating factors. HSAC is therefore a frequently used technique to isolate new growth factors from tissues or cell culture supernatants, allowing their purification almost in a single step.

Two classes of heparin-binding growth factors are generally distinguished. Class 1 (HBGF-1) comprises the more acidic proteins. Almost all of them are either identical with, or closely related to » aFGF. Class 2 (HBGF-2) comprises the more basic proteins. Almost all of them are either identical with, or closely related to » bFGF. In the meantime a number of other HBGFs have been isolated (see: HBGF-3 to HBGF-8).

It is assumed that binding to heparin stabilizes these proteins and protects them from degradation and inactivation. Since heparin and heparin derivatives are important components of the extracellular matrix (see: ECM) this structure may function as a depot for some of these growth factors. It has been observed that binding to heparin or related compounds is an essential step required for some growth factors to interact with their cell surface receptors. Heparin significantly enhances the biological activity of some heparin-binding growth factors (see for example: bFGF).

The biological activities of HBGFs can be assayed by their proliferation-inducing activities on the bovine cell line » FBHE.

Burgess WH & Maciag T The heparin-binding (fibroblast) growth factor family of proteins. ARB 58: 575-606 (1989); **Damon DH et al** Heparin potentiates the action of acidic fibroblast growth factor by prolonging its biological half-life. JCP 138: 221-6 (1989); **Górsky A et al** Immunomodulating activity of heparin. FJ 5: 2287-2291 (1991); **Hearn MT et al** High-performance liquid chromatography of amino acids, peptides and proteins. LXXIV. Separation of heparin-binding growth factors by reversed-phase high-performance liquid chromatography. J. Chromatogr. 26: 371-8 (1987); **Klagsbrun M** The affinity of fibroblast growth factors (FGFs) for heparin; FGF-heparan sulfate interactions in cells and extracellular matrix. COCB 2: 857-63 (1990); **Lobb RR et al** Purification of heparin-binding growth factors. AB 154: 1-14 (1986); **Lobb RR** Clinical applications of heparin-binding growth factors. EJCI 18: 321-36

(1988); **Lobb RR et al** Purification and characterization of heparin-binding endothelial cell growth factors. JBC 261: 1924-8 (1986); **Murphy PR et al** In-gel ligand blotting with 125I-heparin for detection of heparin-binding growth factors. AB 187: 197-201 (1990); **Shing Y et al** Heparin affinity: purification of a tumor-derived capillary endothelial cell growth factor. S 223: 1296-8 (1984); **van Zoelen EJ** The use of biological assays for detection of polypeptide growth factors. Prog. Growth Factor Res. 2: 131-52 (1990)

HBGF-1: *heparin-binding growth factor 1* This factor is also called » ECGF (endothelial cell growth factor) of which several forms have been described. It is usually taken to be identical with » aFGF but really is a processed form of ECGF. See also: HBGF, heparin-binding growth factors. The recommended name for HBGF-1/aFGF is ***FGF-1*** (see also: FGF).

Chiu IM et al Alternative splicing generates two forms of mRNA coding for human heparin-binding growth factor 1. O 5: 755-62 (1990); **Goodrich S et al** The nucleotide sequence of rat heparin binding growth factor I (HBGF-I). NAR 17: 2867(1989); **Shing Y** Bioaffinity chromatography of fibroblast growth factors. MiE 198: 91-5 (1991); **Wang WP et al** Cloning of the gene coding for human class 1 heparin-binding growth factor and its expression in fetal tissues. MCB 9: 2387-95 (1989); **Winkles JA et al** Human vascular smooth muscle cell both express and respond to heparin-binding growth factor I (endothelial cell growth factor). PNAS 84: 7124-8 (1987); **Winkles JA et al** Human smooth vascular muscle cells express and respond to heparin-binding growth factor 1 (endothelial cell growth factor). PNAS 84: 7124-8 (1987)

HBGF-1U: *heparin-binding growth factor 1* A deletion mutant of human » HBGF-1 (acidic fibroblast growth factor, aFGF), lacking the sequence Asn-Tyr-Lys-Lys-Pro-Lys-Leu. It is capable of initiating cellular » *fos* mRNA expression, one of the early responses observed after binding of HBGF-1 to its receptor, and polypeptide phosphorylation on tyrosine residues of its receptor at concentrations that do not induce either DNA synthesis or cell proliferation. HBGF-1U functions as an aFGF antagonist.

Imamura T et al Identification of a heparin-binding growth factor-1 nuclear translocation sequence by deletion mutation analysis. JBC 267: 5676-9 (1992)

HBGF-2: *heparin-binding growth factor 2* This factor is identical with » bFGF. The recommended name for HBGF-1/bFGF is ***FGF-2*** (see also: FGF). See also: HBGF, heparin-binding growth factors).

Schubert D et al Multiple influences of a heparin-binding growth factor on neuronal development. JCB 104: 635-43 (1987)

HBGF-3: *heparin-binding growth factor 3* HBGF-3 is the product of the murine » *int*-2 gene which is activated by the integration (*int*) of murine mammary tumor virus (see also: FGF). See also: HBGF, heparin-binding growth factors).

Burgess WH & Maciag T The heparin-binding (fibroblast) growth factor family of proteins. ARB 58: 575-606 (1989)

HBGF-4: *heparin-binding growth factor 4* This factor is identical with the » *hst* protein (see also: FGF). See also: HBGF, heparin-binding growth factors).

HBGF-5: *heparin-binding growth factor 5* This factor is identical with » FGF-5 (see also: FGF). See also: HBGF, heparin-binding growth factors).

HBGF-7: *heparin-binding growth factor 7* This factor (see also: HBGF, heparin-binding growth factors) is identical with » KGF (keratinocyte growth factor) (see also: FGF). See also: HBGF, heparin-binding growth factors).

Yan G et al Sequence of rat keratinocyte growth factor (heparin-binding growth factor type 7). In Vitro Cell. Dev. Biol. 27A: 437-8 (1991)

HBGF-8: *heparin-binding growth factor 8* This factor is isolated from bovine uterus. It is identical with » HBNF (heparin-binding neurite-promoting factor). See also: HBGF, heparin-binding growth factors).

Milner PD et al A novel 17 kD heparin-binding growth factor (HBGF-8) in bovine uterus: purification and N-terminal amino acid sequence. BBRC 165: 1096-1103 (1989)

HBNF: *heparin-binding neurite-promoting factor*
■ **ALTERNATIVE NAMES:** HARP (heparin affin regulatory peptide); HBBM (heparin-binding brain mitogen); HB-GAF (heparin-binding growth-associated factor); HB-GAM (heparin-binding growth-associated molecule); HBGF-8 (heparin-binding growth factor); HBNF (heparin-binding neurotrophic factor); OSF-1 (osteoblast-specific factor); p18; PTN (pleiotrophin). See also: individual entries for further information.
■ **SOURCES:** HBNF is found predominantly in brain tissues. HBNF mRNA appears to be expressed selectively in human meningothelial cells, but not in fibroblastic cells, blood vessels, and collagen bundles of benign melanomas. It is also expressed in bone tissues. Gradually increasing concentrations of HBNF-specific mRNA are observed during embryonic development. Concentrations are highest shortly before birth, and these levels are retained later in life. The expression of the HBNF gene is regulated by »

PDGF. In murine 3T3 fibroblasts the synthesis of HBNF is reduced by » bFGF.

■ **PROTEIN CHARACTERISTICS:** HBNF is a very basic non-glycosylated protein of 15 kDa with a length of 136 amino acids (***HBNF-a***). The protein contains ten cysteine residues which may be involved in the formation of disulfide bonds. The factor was initially isolated by its strong affinity to heparin (see also: HBGF, heparin-binding growth factors; hence its name: HBGF-8).
Differences in the molecular masses observed in SDS gels (18 kDa) and inferred from the sequence (15.5 kDa) are thought to be due to the altered electrophoretic mobility of this basic protein.
Two variants of this protein, called ***HBNF-b*** and ***HBNF-c***, have been observed during purification. They are very likely purification artifacts and are not observed if proteinase inhibitors are employed during the isolation steps.
HBNF is a secreted protein that is obtained by proteolytic processing of a precursor with a length of 168 amino acids.
HBNF is evolutionary strongly conserved. The sequences of human, murine, bovine, and rat HBNF show a sequence homology of ≈ 98 %. A sequence homology of 46 % is observed with a protein isolated from mouse, called » MK).

■ **GENE STRUCTURE:** The human HBNF gene, designated ***NEGF1*** (neurite growth promoting factor 1), contains seven exons and extends over ≈ 65 kb. The human gene maps to chromosome 7q22 – 7qter (for a related gene see: NEGF2). The mouse OSF-1 gene consists of at least 5 exons and 4 introns and spans > 32 kb. The overall organization of the mouse OSF-1 gene is similar and the locations of the three exon-intron junctions within the coding region are identical to the mouse gene encoding the differentiation-related factor midkine (see: MK). This similarity and the high degree of nucleotide sequence homology (≈ 55%) of mouse OSF-1 and mouse MK suggests that OSF-1 and MK are generated from a common ancestral gene and are members of a family of structurally and probably functionally related proteins.

■ **BIOLOGICAL ACTIVITIES:** This factor is expressed predominantly during postnatal development of the brain. It appears to be involved in the maturation and growth of the brain. HBNF induces the outgrowth of neurites in cultured neurons.
HBNF is a mitogen for capillary endothelial cells, fibroblasts, and some epithelial cell lines. It is also an » angiogenesis factor. The overexpression of bovine HBNF cDNA transforms murine » 3T3 fibroblasts into cells that form extensively metastasizing tumors in nude mice (see also: Immunodeficient mice). HBNF is also expressed in high levels in melanomas that are highly vascularised. A HB-GAM-like factor isolated from some human mammary carcinoma cells may be an » autocrine growth factor for these cells.

■ **ASSAY:** HBNF promotes growth of » SW-13 cells in soft agar which can therefore be used to assay this factor.

Böhlen P et al Isolation from bovine brain and structural characterization of HBNF, a heparin-binding neurotrophic factor. GF 4: 97-107 (1991); **Böhlen P & Kovesdi I** HBNF and MK, members of a novel gene family of heparin-binding proteins with potential roles in embryogenesis and brain function. PGFR 3: 143-57 (1991); **Bloch B et al** Expression of the HBNF (heparin-binding neurite-promoting factor) gene in the brain of fetal, neonatal and adult rat: an *in situ* hybridization study. Brain Res. Dev. Brain Res. 70: 267-78 (1992); **Chaucan AK et al** Pleiotrophin transforms NIH 3T3 cells and induces tumors in nude mice. PNAS 90: 679-682 (1993); **Hulmes JD et al** Comparison of the disulfide bond arrangements of human recombinant and bovine brain heparin binding neurite-promoting factors. BBRC 192: 738-46 (1993); **Katoh K et al** Genomic organization of the mouse OSF-1 gene. DNA Cell Biol. 11: 735-43 (1992); **Kovesdi I et al** Heparin-binding neurotrophic factor (HBNF) and MK, members of a new family of homologous, developmentally regulated proteins BBRC 172: 850-4 (1990); **Kretschmer PJ et al** Cloning, characterization, and developmental regulation of two members of a novel human gene family of neurite outgrowth-promoting proteins. GF 5: 99-114 (1991); **Lai S et al** Structure of the human heparin-binding growth factor gene pleiotrophin. BBRC 187: 1113-22 (1992); **Li YS et al** Cloning and expression of a developmentally regulated protein that induces mitogenic and neurite outgrowth activity. S 250: 1690-4 (1990); **Li Y et al** Pleiotrophin gene expression is highly restricted and is regulated by platelet-derived growth factor. BBRC 184: 427-32 (1992); **Mailleux P et al** The new growth factor pleiotrophin (HB-GAM) mRNA is selectively present in the meningothelial cells of human meningiomas. Neurosci. Lett. 142: 31-5 (1992); **Merenmies J** Cell density-dependent expression of heparin-binding growth-associated molecule (HB-GAM, p18) and its down-regulation by fibroblast growth factors. FL 307: 297-300 (1992); **Naito A et al** Similarity of the genomic structure between the two members in a new family of heparin-binding factors. BBRC 183: 701-7 (1992); **Raulo E et al** Secretion and biological activities of heparin-binding growth-associated molecule. Neurite outgrowth-promoting and mitogenic actions of the recombinant and tissue-derived protein. JBC 267: 11408-16 (1992); **Wanaka A et al** Developmentally regulated expression of pleiotrophin, a novel heparin binding growth factor, in the nervous system of the rat. Brain Res. Dev. Brain Res. 72: 133-44 (1993)

HBNF: *heparin-binding neurotrophic factor* This protein was initially isolated as a heparin-binding protein (see also: HBGF, heparin-binding growth factors) from brain tissues of several species. It is identical with » HBNF (heparin-binding neurite-promoting factor).

Kovesdi I et al Heparin-binding neurotrophic factor (HBNF) and MK, members of a new family of homologous, developmentally regulated proteins. BBRC 172: 850-4 (1990)

HBp17: This factor of 17 kDa is isolated from culture supernatants of the human cell line » A431. It is produced mainly by keratinocytes and squamous carcinoma cells. The factor binds non-covalently and reversibly to » aFGF and » bFGF and thus blocks their biological activities. This binding can be prevented by heparin. This binding protein may play a role as a physiological regulator and modulator of growth factor activities.

Wu DQ et al Characterization and molecular cloning of a putative binding protein for heparin-binding growth factors. JBC 266: 16778-85 (1991)

HC-11: HC-11 and ***HC-14*** are two genes the expression of which in macrophages is inducible by IFN-γ. HC-11 (76 amino acids) belongs to the » chemokine family of cytokines. The factor is identical with » MCP-1 (monocyte chemoattractant protein 1), » MCP (monocyte chemotactic protein), » MCAF (monocyte chemotactic and activating factor), » SMC-CF (smooth muscle cell chemotactic factor), » LDCF (lymphocyte-derived chemotactic factor), » GDCF (glioma-derived monocyte chemotactic factor), » TDCF (tumor-derived chemotactic factor). HC-14 is identical with » MCP-2 (

Chang H et al Cloning and expression of a γ-interferon inducible gene in monocytes: a new member of a cytokine gene family. Int. Immunol. 1: 388-97 (1989)

HC-14: see: HC-11.

HC-21: This gene was isolated as a cDNA from CD2-stimulated human lymphocytes. It is expressed in T lymphocytes and IL2-dependent T cell lines. HC-21 encodes a secreted protein of 69 amino acids derived from a precursor of 92 amino acids. The protein sequence encoded by HC-21 suggests that it belongs to the » chemokine family of cytokines. It is closely related to » MIP1β (macrophage inflammatory protein β, see: MIP).

Chang HC & Reinherz EL Isolation and characterization of a cDNA encoding a putative cytokine which is induced by stimulation via the CD2 structure on human lymphocytes. EJI 19: 1045-51 (1989)

HCD57: An erythroleukemia cell line derived from a mouse infected at birth with Friend murine leukemia virus (see also: SFFV). The cells proliferate in response to » Epo and require it for survival. The cells can therefore be used to assay this factor. The introduction of SFFV into these Epo-dependent cells efficiently and reproducibly gives rise to lines which express high levels of SFFV and are factor-independent. SFFV appears to be unique in its ability to abrogate the factor dependence of these cells, since infection of these cells with retroviruses carrying a variety of different » oncogenes has no effect. The induction of Epo independence by SFFV does not involve a classical » autocrine growth control mechanism, since there is no evidence that the factor-independent cells synthesize or secrete Epo or depend on it for their growth (see: SFFV).

Epo mediates the rapid phosphorylation of » *raf*-1 in these cells; c-*raf* » antisense oligodeoxyribonucleotides, which specifically decreases intracellular *raf*-1 levels, also substantially inhibit Epo-induced DNA synthesis.

Carroll MP et al Erythropoietin induces *raf*-1 activation and *raf*-1 is required for erythropoietin-mediated proliferation. JBC 266: 14964-9 (1991); **Ruscetti SK et al** Friend spleen focus-forming virus induces factor independence in an erythropoietin-dependent erythroleukemia cell line. J. Virol. 64: 1057-62 (1990); **Spivak JL et al** Erythropoietin is both a mitogen and a survival factor. Blood. 77: 1228-33 (1991)

HCGF: *hematopoietic cell growth factor* This factor is a lineage-indifferent hematopoietic regulatory factor originally found to be secreted by » WEHI-3 myelomonocytic leukemia cells. It promotes proliferation and development of all myeloid hematopoietic progenitor cells. In addition, the factor also facilitates self-renewal of primitive totipotent » hematopoietic stem cells (see also: CFU-S, CFU-Mix). HCGF can be used to generate continuously growing, non-leukemic, factor-dependent cell lines, *in vitro* (see also: FDC-P). In the absence of HCGF these cells die within hours. HCGF is identical with » IL3. Some HCGF activity may be ascribed to » GM-CSF.

Bazill GW et al Characterization and partial purification of a hematopoietic cell growth factor in WEHI-3 cell-conditioned medium. BJ 210: 747-59 (1983); **Dexter TM et al** The role of hemopoietic cell growth factor (interleukin 3) in the development of hemopoietic cells. Ciba Found. Symp. 116: 129-47 (1985)

HCGF: *Hodgkin's cell growth factor* This factor of 68 kDa is secreted by multinucleated Reed-Steinberg giant cells of Hodgkin lymphomas. HCGF may be an » autocrine growth factor for this cell type. HCGF is recognized by a monoclonal antibody directed against » IL4. Polyclonal anti-IL4 antibodies completely neutralize this factor. HCGF competes with recombinant IL4 for binding to the high-affinity IL4 receptor. Long-term culture of Reed-Steinberg cells in serum-free media (see also: SFM) is not inhibited in the presence of anti-IL4 antibodies so that some other factors must be involved in the growth of these cells.

Newcom SR et al Interleukin-4 is an autocrine growth factor secreted by the L-428 Reed-Steinberg cell. Blood 79: 191-7 (1992)

HCI: *human collagenase inhibitor* HCI has been identified as a secretory product of platelets and alveolar macrophages. It plays a major role in modulating the activity of interstitial collagenase and a number of connective tissue metalloendoproteases. HCI is identical with » TIMP (tissue inhibitor of metalloproteases).

Carmichael DF et al Primary structure and cDNA cloning of human fibroblast collagenase inhibitor. PNAS 83: 2407-11 (1986)

***hck*:** *hematopoietic cell kinase* (pronounced 'hick') This protein of 57 kDa (505 amino acids) has been identified as a product of a gene the expression of which in granulocytic and monocytic leukemia cells increases after the cells have been induced to differentiate. *hck* is primarily expressed in hematopoietic cells, particularly granulocytes. NK cells express very low levels (25-80 times less than monocytes) of mRNA encoding *hck. The* protein exists in at least two molecular forms arising by alternative splicing. *hck* is related to the *src* family of protein kinases and closely related to » *lck*, a lymphocyte-specific protein tyrosine kinase. Its gene maps to human chromosome 20q11-q12, a region that is affected by interstitial deletions in some acute myeloid leukemias and myeloproliferative disorders.

hck is an immediate early response gene (see: Gene expression) and has been implicated in regulation of the behavior of activated phagocytes. Functional activation of human cultured macrophages with bacterial lipopolysaccharides augments the expression of *hck* transcripts and of p59*hck*, but decreases the level of transcripts encoded by the closely related » *fgr*. Overexpression of an activated mutant of *hck* in the murine macrophage cell line BAC1.2F5 has been shown to augment production of » TNF-α in response to bacterial lipopolysaccharides, whereas inhibition of endogenous *hck* expression, by » antisense oligonucleotides, interferes with TNF synthesis.

It has been observed that the expression of the *hck* (and » *fgr*) protein-tyrosine kinases in acute myeloid leukemic blasts is associated with early commitment and differentiation events in the monocytic and granulocytic lineages.

Biondi A et al Expression of lineage-restricted protein tyrosine kinase genes in human natural killer cells. EJI 21: 843-6 (1991); **Boulet I et al** Lipopolysaccharide- and interferon-γ-induced expression of *hck* and *lyn* tyrosine kinases in murine bone marrow-derived macrophages. O 7: 703-10 (1992); **Cambier JC & Campbell KS** Membrane immunoglobulin and its accomplices: new lessons from an old receptor. FJ 6: 3207-17 (1992); **English BK et al** *hck* tyrosine kinase activity modulates tumor necrosis factor production by murine macrophages. JEM 178: 1017-22 (1993); **Hradetzky D et al** The genomic locus of the human hemopoietic-specific cell protein tyrosine kinase (PTK)-encoding gene (*hck*) confirms conservation of exon-intron structure among human PTKs of the *src* family. Gene 113: 275-80 (1992); **Klemsz MJ et al** Nucleotide sequence of the mouse *hck* gene. NAR 15: 9600 (1987); **Lichtenberg U et al** Human protein-tyrosine kinase gene *hck*: expression and structural analysis of the promoter region. O 7: 849-58 (1992); **Lock P et al** Functional analysis and nucleotide sequence of the promoter region of the murine *hck* gene. MCB 10: 4603-11 (1990); **Lock P et al** Two isoforms of murine *hck*, generated by utilization of alternative translational initiation codons, exhibit different patterns of subcellular localization. MCB 11: 4363-70 (1991); **Mufson RA** Induction of immediate early response genes by macrophage colony-stimulating factor in normal human monocytes. JI 145: 2333-9 (1990); **Quintrell N et al** Identification of a human gene (*hck*) that encodes a protein-tyrosine kinase and is expressed in hemopoietic cells. MCB 7: 2267-75 (1987); **Willman CL et al** Expression of the c-*fgr* and *hck* protein-tyrosine kinases in acute myeloid leukemic blasts is associated with early commitment and differentiation events in the monocytic and granulocytic lineages. Blood 77: 726-34 (1991); **Ziegler SF et al** Novel protein-tyrosine kinase gene (*hck*) preferentially expressed in cells of hematopoietic origin. MCB 7: 2276-85 (1987); **Ziegler SF et al** Augmented expression of a myeloid-specific protein tyrosine kinase gene (*hck*) after macrophage activation. JEM 168: 1801-10 (1988); **Ziegler SF et al** Structure and expression of the murine *hck* gene. O 6: 283-8 (1991)

HCP: *hematopoietic cell phosphatase* This protein tyrosine phosphatase, encoded by a gene on human chromosome 12p12-p13, is also known as ***PTP1C***, ***SHP***, or ***SHPTP1***. HCP is a cytoplasmic protein that contains a tyrosine phosphatase catalytic domain at the carboxyl terminus and two contiguous » *src* homology domains (SH2) at the amino terminus. It has been observed that » M-CSF induces tyrosine phosphorylation of HCP.

HCP has been shown to associate with » *kit*, which functions as a receptor for the hematopoietic growth factor » SCF, following ligand binding and » autophosphorylation of the *kit* receptor. Following ligand binding HCP also associates with the tyrosine-phosphorylated β chain of the » IL3 receptor and the » GM-CSF receptor. Overexpression of HCP in IL3-dependent cells suppresses growth and inhibition of HCP expression in these cells marginally increases the growth response to IL3. HCP may also dephosphorylate (and thereby inactivate Jak2 kinase (see: Janus kinases).

HCP probably plays an essential role for the down-regulation of growth signals in hematopoietic cells. A critical role for HCP in regulating » hematopoiesis is suggested also by the genetic defi-

hck: see addendum.

ciencies observed in mice carrying the mutation known as » motheaten.

Matthews RJ et al Characterization of hematopoietic intracellular protein tyrosine phosphatases: description of a phosphatase containing an SH2 domain and another enriched in proline-, glutamic acid-, serine-, and threonine-rich sequences. MCB 12: 2396-405 (1992); **Plutzky J et al** Isolation of a *src* homology 2-containing tyrosine phosphatase. PNAS 89: 1123-7 (1992); **Shen SH et al** A protein tyrosine phosphatase with sequence similarity to the SH2 domain of the protein tyrosine kinases. N 352: 736-9 (1991); **Yi T et al** Identification of novel protein tyrosine phosphatases of hematopoietic cells by PCR amplification. Blood 78: 2222-8 (1992); **Yi T et al** Protein tyrosine phosphatase containing SH2 domains: characterization, preferential expression in hematopoietic cells, and localization to human chromosome 12p12-p13. MCB 12: 836-46 (1992); **Yi T et al** Hematopoietic cell phosphatase associates with the interleukin-3 (IL3) receptor β chain and down-regulates IL3-induced tyrosine phosphorylation and mitogenesis. MCB 13: 7577-86 (1993)

HCSF: *histamine-producing cell stimulating factor* HCSF was initially observed in mixed lymphocyte cultures between donor and recipient of a mouse skin allograft. It strongly induces the synthesis of histamines in murine hepatocytes and causes an increased production of histamine from target cells present in bone marrow. HCSF activity was initially thought to be identical with » IL3 which also possesses HCSF activity. However, anti-IL3 antibodies do not inhibit the increase in histamine synthesis induced by HCSF while it strongly diminishes that induced by IL3. HCSF is identical with » GM-CSF.

Dy M et al Histamine-producing cell stimulating factor (HCSF) and interleukin 3 (IL3): evidence for two distinct molecular entities. JI 136: 208-12 (1986); **Dy M et al** Histamine-producing cell stimulating factor: a biological activity shared by interleukin 3 and granulocyte-macrophage colony-st8imulating factor. EJI 17: 1243-8 (1987); **Ihle JN et al** Biological properties of homogenous interleukin-3. I. Demonstration of WEHI-3 growth factor activity, mast cell growth factor activity, P cell-stimulating factor activity, colony-stimulating factor activity, and histamine-producing cell-stimulating factor activity. JI 131: 282-7 (1983)

HDGF: *heart-derived growth factor* This factor was purified from bovine myocardium. It stimulates the proliferation of vascular endothelial cells. The factor is identical with » aFGF.

Sasaki H et al Purification of acidic fibroblast growth factor from bovine heart and its localization in the cardiac myocytes. JBC 264: 17606-12 (1989)

HDGF: *hepatoma-derived growth factor* This hepatoma-derived growth factor acts as an angiogenic endothelial mitogen. It is identical with » bFGF. For another hepatoma-derived growth factor see: HuHGF.

Klagsbrun M et al Human tumor cells synthesize an endothelial cell growth factor that is structurally related to basic fibroblast growth factor. PNAS 83: 2448-52 (1986); **Lobb RR et al** Purification and characterization of heparin-binding endothelial cell growth factors. JBC 261: 1924-8 (1986)

HEK: see addendum.

HDGF: *hypothalamus-derived growth factor* Two types of this factor isolated from bovine hypothalamus have been described. The acidic aHDGF (anionic HDGF) is identical with » aFGF. The basic cHDGF (cationic HDGF) is identical with » bFGF. See also: BDGF (brain-derived growth factor).

Klagsbrun M & Shing Y Heparin affinity of anionic and cationic capillary endothelial cell growth factors: analysis of hypothalamus-derived growth factors and fibroblast growth factors. PNAS 82: 805-9 (1985)

HDGF 2: *hepatoma-derived growth factor 2* This factor is produced by a human hepatoma cell line. It is identical with » bFGF.

Klagsbrun M et al Human tumor cells synthesize an endothelial cell growth factor that is structurally related to basic fibroblast growth factor. PNAS 83: 2448-52 (1986)

HDLF: *Hodgkin-derived leukocyte factor* This poorly characterized factor of ≈ 70 kDa is produced by some Hodgkin's disease-derived cell lines. HDLF inhibits the chemotactic migration of neutrophils and stimulates the production of reactive oxygen species.

Schell-Frederick E et al Inhibition of human neutrophil migration by supernatants from Hodgkin's disease-derived cell lines. EJCI 18: 290-6 (1988)

HDNF: *hippocampus-derived neurotrophic factor* This factor is identical with » NT-3 (see also: Neurotrophins).

Ernfors P et al Identification of cells in rat brain and peripheral tissues expressing mRNA for members of the nerve growth factor family. Neuron 5: 511-26 (1990)

Heart-derived growth factor: see: HDGF.

Heavy chain class switching: see: Isotype switching.

Helper peak 1: see: HP1.

Hematopoiesis: (from Greek *haima* for blood and *poiein*, to make) Hematopoiesis is the dynamic and complex developmental process of the formation of new blood cells. The bone marrow with its intersinusoidal spaces is the site responsible for the generation of blood cells in healthy post-natal human beings. The nature of these structures is determined predominantly by a network of stromal cells and an amorphous substance in which the

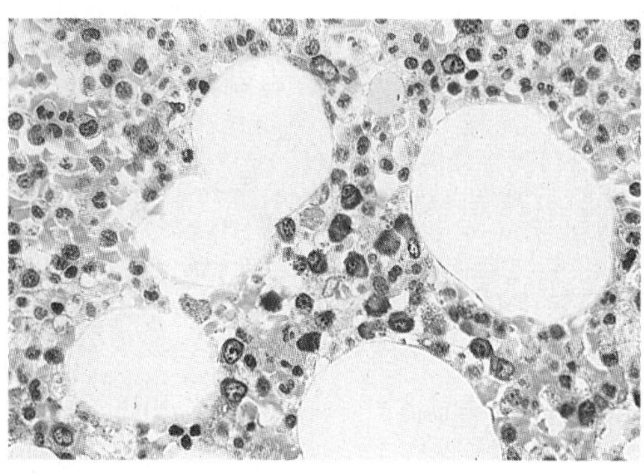
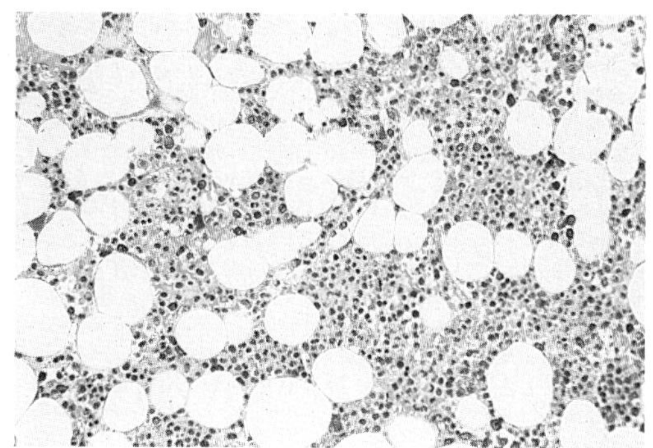

Hematopoietically active tissue in pelvic bone.
Courtesy of Dr. T. Friedrich; Institute of Pathology, University of Leipzig, FRG.

blood-forming cells are embedded. The morphologically discernible areas of active hematopoiesis are frequently referred to as ***cobblestone area***. The corresponding cells are known as » CAFC (cobblestone area-forming cells).
The microenvironment of a cell plays an important role in the differentiation of individual bone marrow cells. Further differentiation of cells into one of several lineages critically depends on the nature of factors acting on these cells at a particular time and at a particular concentration. The bone marrow stroma contains many different cell types, including macrophages, fibroblasts, endothelial cells, smooth muscle cells, T lymphocytes, monocytes etc. These cells, in combination with components of the extracellular matrix (see: ECM) and basement membranes, form the so-called ***hematopoietic (inductive) microenvironment*** (abbrev. ***HIM***) which maintains the functional integrity of this complex system of resident and circulating cells (see also: LTBMC (long-term bone marrow culture). Cells of the hematopoietic microenvironment show low or no detectable cell growth and are believed to be in the G0 phase of the » cell cycle.
All different types of blood cells are derived from a small common pool of totipotent cells, called » ***hematopoietic stem cells***, laid down in hematopoietic organs early during embryogenesis. These totipotent stem cells are also referred to as ***HSC*** (hematopoietic stems cells), **PHSC** (primitive (or pluripotent) hemopoietic stem cells), ***PLSC*** (pluripotent lymphoid stem cells), ***PPSC*** (pluripotent stem cells, or ***PSC***), and ***THSC*** (totipotent hematopoietic stem cells). They have the unique property of self-renewal (***self-renewal potential***, abbrev. SRP), i. e. they give rise to progeny identical in appearance and differentiation potential. These cells persist throughout adult life and are therefore responsible for the maintenance of hematopoiesis. This process is also called ***steady-state hematopoiesis*** or ***constitutive hematopoiesis***. The remarkable biological activities of the stem cells are illustrated by their ability to colonize the bone marrow of lethally irradiated animals (see: ***MRA***, marrow repopulating ability) and, by their lymphopoietic and myelopoietic potential, to reconstitute the entire hematopoietic system. Since no other cell types are capable of achieving this task long-term repopulating activity is used as a functional assay for pluripotent stem cells (see also: LTBMC; CFC; CFU-S). Pluripotent stem cells are quiescent (see also: SCI, stem cell inhibitor, CFU-S). This is shown by their resistance to treatment with » fluorouracil or » 4-HC (4-hydroperoxycyclophosphamide), which spare them and eliminates dividing cells without adversely affecting the long-term repopulating ability of bone marrow. These cells are of interest not only because of their developmental capacity but also because of their potential usefulness as a source of autologous bone marrow cells for the treatment of hematological disorders (see: 4-HC) and as vectors for gene therapy (see also: Cytokine gene transfer).
The totipotent » hematopoietic stem cells give rise to transit populations with restricted differentiation capacity. Stem cell differentiation *in vivo*

appears to be a stochastic process. The ***progenitor cells*** arising as the result of stem cell differentiation are called ***committed cells***, or, for historical reasons, colony-forming units (see: CFU; see also individual types: BFU-E, CFU-E, CFU-Eo, CFU-G, CFU-GEMM, CFU-GM, CFU-M, CFU-MEG). Committed cells, which can be identified by expression of specific lineage markers, comprise multipotent (MPSC), bipotent (BPPC), and unipotent (UPSC) cell types which are determined to differentiate into any of the hematopoietic lineages, i. e. ***lymphoid cells*** that ultimately give rise to B cells and T cells, and ***myeloid cells*** that eventually give rise to monocytes, platelets, granulocytes/monocytes (neutrophils, eosinophils, basophils), and erythrocytes. These lineages are defined by the nature of the fully differentiated cells eventually evolving from these precursor cells. The developmental potential of these cells is generally limited to only one or two of the hematopoietic lineages, and these cells progressively display the antigenic, biochemical, and morphological features characteristic of the mature cells of the appropriate lineages and lose their capacity for self-renewal. Their proliferation is normally tightly controlled and coupled to development. Cells leaving the bone marrow usually possess little or no proliferative potential; erythrocytes and platelets do not contain genetic material, and neutrophils have condensed DNA and cannot undergo replication. Some hematopoietic cells, including pre-T cells, mast cells, and monocytes can undergo further replication and development in various tissues.

Pluripotent stem cells are still difficult to characterize morphologically. The pool of determined stem cells can usually be differentiated in functional assays. Most of the distinct intermediate forms can be distinguished from each other by the stage-specific expression of cell surface markers (see also: CD antigens) and their dependence on the presence of one or several growth factors that are absolutely required for their survival and proliferation. All members of a particular lineage that are still capable of proliferation can give rise to malignant variants at all stages of differentiation.

Approximately 10^9 nucleated cells are produced per kilogram of body weight within the hematopoietic system per day. The continuity of vital functions is strictly dependent on the constant production of new cells and under normal conditions there is a quantitative and qualitative equilibrium for all blood cells, maintaining mature cell numbers within quite narrow limits. The life span of the fully differentiated mature forms of blood cells may vary considerably, being on the order of several hours for some cells (granulocytes), several weeks (erythrocytes) and several years (memory cells). Blood loss and cell losses, caused, for example, by a variety of pathological processes, infections, or treatment with cytotoxic drugs, can be compensated by an almost exponential increase in the generation of new cells. This process is called ***induced hematopoiesis***. Increased production of cells is largely restricted to the specific cell type that is required in the particular stress situation: hemolysis, for example induces erythroid hyperplasia, while granulocyte hyperplasia is observed in respose to bacterial infections. Alterations in the balance between self-renewal and differentiation can lead to the emergence of cells that survive and grow in situations unfavorable for the growth of normal cells and hence to the establishment of leukemias.

The self-renewal of the stem cell population in the bone marrow, the proliferation and differentiation of hematopoietic progenitor cells, their survival, and also all functional activities of the circulating mature forms are subject to regulation by a cascade of proteins that are generally known as ***hematopoietic growth factors*** or » ***hematopoietins***. These factors, which are products of stromal and other cells, and also many other low molecular weight factors (see also: AcSDKP; pEEDCK, thymic hormones) function as promoters or inhibitors (see also: Restrictins) of differentiation and positively or negatively influence hematopoietic processes. At the genetic level the molecular basis of commitment and differentiation of hematopoietic cells is still poorly understood. Recent progress has been made with the identification of » homeotic genes that appear to be of fundamental importance in these processes.

Allen TD & Dexter TM The essential cells of the hemopoietic microenvironment. EH 12: 517-21 (1984); **Bender JG et al** Regulation of myelopoiesis. Comp. Immunol. Microbiol. 10: 79-91 (1987); **Campbell AD & Wicha MS** Extracellular matrix and the hematopoietic microenvironment. J. Lab. Clin. Med. 112: 140-6 (1988); **Clapp DW** Somatic gene therapy into hematopoietic cells. Current status and future implications. Clin. Perinatol. 20: 155-68 (1993); **Cournoyer D & Caskey CT** Gene therapy of the immune system. ARI 11: 297-329 (1993); **Dexter TM** Biology of hematopoiesis. Oncology Rev. 4: 7-9 (1989); **Dexter TM & Spooncer E** Growth and differentiation in the hemopoietic system. ARC 3: 423-41 (1987); **Golde DW** (edt) Haematopoiesis, Curchill-Livingstone, Edinburgh 1984; **Gordon MG** Extracellular matrix of the marrow environment. Br. J. Haematol. 70: 1-4 (1988); **Gordon MY & Barrett AJ** Bone marrow disorders: the biological basis of clinical problems. Black-

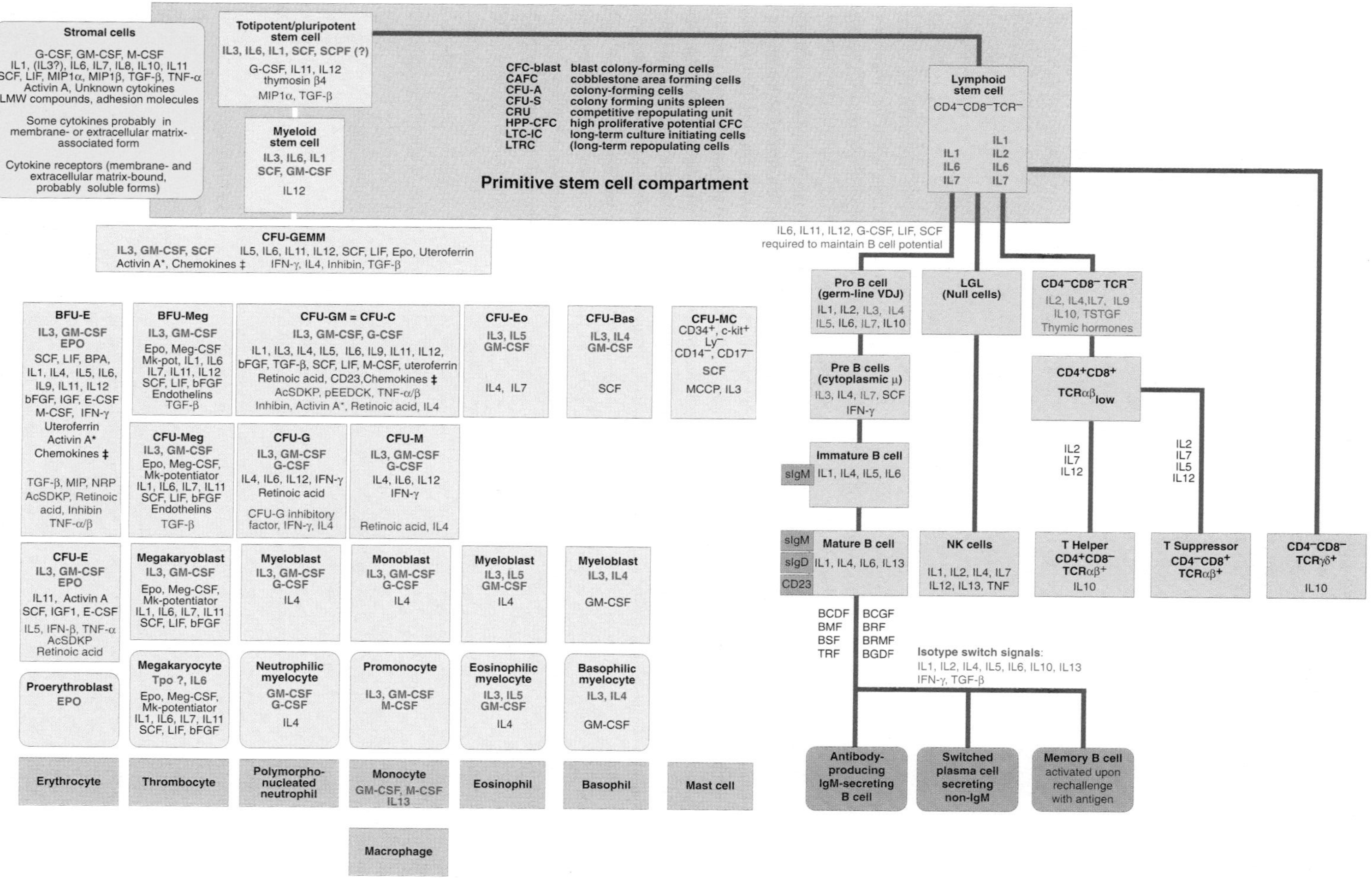
Stromal cells
G-CSF, GM-CSF, M-CSF
IL1, (IL3?), IL6, IL7, IL8, IL10, IL11
SCF, LIF, MIP1α, MIP1β, TGF-β, TNF-α
Activin A, Unknown cytokines
LMW compounds, adhesion molecules
Some cytokines probably in membrane- or extracellular matrix-associated form
Cytokine receptors (membrane- and extracellular matrix-bound, probably soluble forms)
Totipotent/pluripotent stem cell
IL3, IL6, IL1, SCF, SCPF (?)
G-CSF, IL11, IL12
thymosin β4
MIP1α, TGF-β
Myeloid stem cell
IL3, IL6, IL1
SCF, GM-CSF
IL12
CFC-blast blast colony-forming cells
CAFC cobblestone area forming cells
CFU-A colony-forming cells
CFU-S colony forming units spleen
CRU competitive repopulating unit
HPP-CFC high proliferative potential CFC
LTC-IC long-term culture initiating cells
LTRC (long-term repopulating cells
Primitive stem cell compartment
Lymphoid stem cell
CD4⁻CD8⁻TCR⁻
IL1 IL6 IL7
IL1 IL2 IL6 IL7
IL6, IL11, IL12, G-CSF, LIF, SCF
required to maintain B cell potential
CFU-GEMM
IL3, GM-CSF, SCF
Activin A*, Chemokines ‡
IL5, IL6, IL11, IL12, SCF, LIF, Epo, Uteroferrin
IFN-γ, IL4, Inhibin, TGF-β
BFU-E
IL3, GM-CSF
EPO
SCF, LIF, BPA, IL1, IL4, IL5, IL6, IL9, IL11, IL12
bFGF, IGF, E-CSF
M-CSF, IFN-γ
Uteroferrin
Activin A*
Chemokines ‡
TGF-β, MIP, NRP
AcSDKP, Retinoic acid, Inhibin
TNF-α/β
BFU-Meg
IL3, GM-CSF
Epo, Meg-CSF
Mk-pot, IL1, IL6
IL7, IL11, IL12
SCF, LIF, bFGF
Endothelins
TGF-β
CFU-GM = CFU-C
IL3, GM-CSF, G-CSF
IL1, IL3, IL4, IL5, IL6, IL9, IL11, IL12,
bFGF, TGF-β, SCF, LIF, M-CSF, uteroferrin
Retinoic acid, CD23,Chemokines ‡
AcSDKP, pEEDCK, TNF-α/β
Inhibin, Activin A*, Retinoic acid, IL4
CFU-Eo
IL3, IL5
GM-CSF
IL4, IL7
CFU-Bas
IL3, IL4
GM-CSF
SCF
CFU-MC
CD34⁺, c-kit⁺
Ly⁻
CD14⁻, CD17⁻
SCF
MCCP, IL3
CFU-Meg
IL3, GM-CSF
Epo, Meg-CSF,
Mk-potentiator
IL1, IL6, IL7, IL11
SCF, LIF, bFGF
Endothelins
TGF-β
CFU-G
IL3, GM-CSF
G-CSF
IL4, IL6, IL12, IFN-γ
Retinoic acid
CFU-G inhibitory factor, IFN-γ, IL4
CFU-M
IL3, GM-CSF
G-CSF
IL4, IL6, IL12
IFN-γ
Retinoic acid, IL4
CFU-E
IL3, GM-CSF
EPO
IL11, Activin A
SCF, IGF1, E-CSF
IL5, IFN-β, TNF-α
AcSDKP
Retinoic acid
Megakaryoblast
IL3, GM-CSF
Epo, Meg-CSF,
Mk-potentiator
IL1, IL6, IL7, IL11
SCF, LIF, bFGF
Myeloblast
IL3, GM-CSF
G-CSF
IL4
Monoblast
IL3, GM-CSF
G-CSF
IL4
Myeloblast
IL3, IL5
GM-CSF
IL4
Myeloblast
IL3, IL4
GM-CSF
Proerythroblast
EPO
Megakaryocyte
Tpo ?, IL6
Epo, Meg-CSF,
Mk-potentiator
IL1, IL6, IL7, IL11
SCF, LIF, bFGF
Neutrophilic myelocyte
GM-CSF
G-CSF
IL4
Promonocyte
IL3, GM-CSF
M-CSF
Eosinophilic myelocyte
IL3, IL5
GM-CSF
IL4
Basophilic myelocyte
IL3, IL4
GM-CSF
Erythrocyte
Thrombocyte
Polymorpho-nucleated neutrophil
Monocyte
GM-CSF, M-CSF
IL13
Eosinophil
Basophil
Mast cell
Macrophage
Pro B cell
(germ-line VDJ)
IL1, IL2, IL3, IL4
IL5, IL6, IL7, IL10
LGL
(Null cells)
CD4⁻CD8⁻ TCR⁻
IL2, IL4,IL7, IL9
IL10, TSTGF
Thymic hormones
Pre B cells
(cytoplasmic μ)
IL3, IL4, IL7, SCF
IFN-γ
CD4⁺CD8⁺
TCRαβlow
Immature B cell
sIgM
IL1, IL4, IL5, IL6
IL2
IL7
IL12
IL2
IL7
IL5
IL12
sIgM
sIgD
CD23
Mature B cell
IL1, IL4, IL6, IL13
NK cells
IL1, IL2, IL4, IL7
IL12, IL13, TNF
T Helper
CD4⁺CD8⁻
TCRαβ⁺
IL10
T Suppressor
CD4⁻CD8⁺
TCRαβ⁺
CD4⁻CD8⁻
TCRγδ⁺
IL10
BCDF
BMF
BSF
TRF
BCGF
BRF
BRMF
BGDF
Isotype switch signals:
IL1, IL2, IL4, IL5, IL6, IL10, IL13
IFN-γ, TGF-β
Antibody-producing IgM-secreting B cell
Switched plasma cell secreting non-IgM
Memory B cell
activated upon rechallenge with antigen

well Scientific Publications, Oxford 1985; **Greenberger JS** The hematopoietic microenvironment. Crit. Rev. Oncol. Hematol. 11: 65-84 (1991); **Greenberger JS** Toxic effects on the hematopoietic microenvironment. EH 19: 1101-9 (1991); **Mayani H et al** Biology of the hemopoietic microenvironment. Eur. J. Haematol. 49: 225-33 (1992); **Metcalf D** The clonal culture of hemopoietic cells. Elsevier Science Publishers, Amsterdam 1984; **Metcalf D** The molecular control of blood cells. Harvard University Press, Cambridge 1989; **Moore MAS et al** Molecular control of hemopoiesis. CIBA Found. Symp. 148: 43-61 (1990); **Ogawa M** Differentiation and proliferation of hematopoietic stem cells. Blood 81: 2844-53 (1993); **Pantel K & Nakeff A** The role of lymphoid cells in hematopoietic regulation. EH 21: 738-42 (1993); **Punt CJ** Regulation of hematopoietic cell function by protein tyrosine kinases. Leuk. Res. 16: 551-9 (1992); **Shin SS et al** Immunoarchitecture of normal human bone marrow: a study of frozen and fixed tissue sections. Hum. Pathol. 23: 686-94 (1992); **Shultz LD** Hematopoiesis and models of immunodeficiency. Semin. Immunol. 3: 397-408 (1992); **Singer JW et al** The human hematopoietic microenvironment. Rec. Adv. Haematol. 4: 1-24 (1985); **Tavassoli M & Friedenstein A** Haemopoietic stromal microenvironment. Am. J. Haematol. 15: 195-203 (1983); **Tavassoli M & Minguell JJ** Homing of hemopoietic progenitor cells to the marrow. PSEBM 196: 367-73 (1991); **Weiss L & Sakai H** The hemopoietic stroma. Am. J. Anatomy 170: 447-63 (1984); **Wilson FD** Studies of the hematopoietic microenvironment, hematopoietic stroma, and microenvironment. Bilb. Haematol. 48: 210-92 (1984)

Hematopoietic cell growth factor: see: HCGF.

Hematopoietic cell kinase: see: *hck*.

Hematopoietic cell phosphatase: see: HCP.

Hematopoietic CSF 309: see: CSF 309.

Hematopoietic cytokines: see: Hematopoietins.

Hematopoietic growth factors: see: Hematopoietins.

Hematopoietic inductive microenvironment: abbrev. HIM. see: Hematopoiesis.

Hematopoietic stem cells: These cells are very primitive totipotent cells with extensive self-maintenance and self-renewal capacity persisting throughout adult life. The term ***self-renewal capacity*** refers to the fact that these cells can give rise to progeny identical in appearance and differentiation potential. Hematopoietic stem cells are the only cells capable of colonizing lethally irradiated bone marrow to reconstitute the entire hematopoietic system. This activity is referred to as » ***MRA*** (marrow-repopulating ability) and has also

Simplified diagram of hematopoiesis and cytokines involved in its regulation.
Individual cell types are boxed. Early totipotent or pluripotent hematopoietic stem cells with self-renewing ability (green box) are not identifiable morphologically and their exact position within the hierarchy is not known. These cell types may themselves consist of a hierarchy of cells of increasing maturity. Hematopoietic stem cells and early progenitor cells give rise to a series of myeloid (red) or lymphoid (blue) cell types differing from each other by their degree of differentiation. The hematopoietic microenvironment (yellow box) plays an important role in the physiological regulation of these processes. Later stages derived from these stem cells are multipotent, bipotent, or unipotent progenitors that can be identified by means of sets of cell surface markers. Most likely these stages also consist of subpopulations of cell types differing in their responsiveness to various cytokines. The earliest morphologically identifiable cell types are marked by boxes with rounded corners; functionally mature end cells are marked by darker red or blue shading.
Factors shown in green are generally regarded as being more important than factors shown in black which, at least in vitro, appear to affect these cells types in a positive way. The latter can, for example, prime these cells and render them more responsive to other factors, enhance their survival, or promote act as maintenance or commitment factors, or augment their proliferation. Factors in red have been shown to affect growth, survival or differentiation in a negative way. In addition, all factors can influence each other, showing synergistic, antagonistic and/or additive effects. For abbreviations see individual entries in this dictionary.
The actual situation in vivo is more complicated than that depicted in this diagram. It has been shown that great importance must be attached to combinations of factors acting in synergy rather than individual factors acting on their own. The fact that some factors seem to disappear from the boxed cell types at some point of the hierarchy does not necessarily mean that these factors do not influence the particular cell type. It only means that data pertaining to their action on these particular cell types have not been reported. In addition, some factors appear to act on subsets of cell types. This is demonstrated by the activities of Activin A and some chemokines.
* Activin A suppresses the proliferation of IL3-responsive CFU-GM progenitors and stimulates the proliferation and differentiation of IL-3-responsive BFU-E progenitors. CFU-GM colony formation is inhibited by activin A only when the cells are stimulated with IL3, but not when stimulated with G-CSF, GM-CSF), or SCF plus G-CSF. BFU-E colony formation is enhanced by activin A only when the cells are exposed to IL3 plus Epo, but not when exposed to Epo or Epo plus SCF. [see: Blood 81(11): 2891-7 (1993)]
‡ MIP-1α, MIP-2α, PF4, IL8, and MCAF suppress in dose-response fashion colony formation of immature subsets of myeloid progenitor cells (CFU-GEMM, BFU-E, and CFU-GM) stimulated by GM-CSF plus SCF. MIP-1β blocks the suppressive effects of MIP-1α. MIP-2β or GRO-α block the suppressive effects of IL8 and PF4. [see: J. Immunol. 150(8 Pt 1): 3448-58].

Hematopoietic cell kinase: see addendum: hck.
Hematopoietic consensus tyrosine-lacking kinase: see addendum: HYL.

been used to develop a competitive repopulation assay to detect stem cells.

The hematopoietic stem cell pool consists of long-lived quiescent cells (see also: Cell cycle). Entrance of stem cells into the cell cycle is usually induced only in the presence of a variety of different known and unknown or poorly characterized cytokines. This is probably a safeguard mechanism preventing an exhaustion of the stem cell compartment through excess differentiation induced by individual late-acting differentiation-inducing cytokines which may be present, for example, under conditions of chronic infections.

Hematopoietic stem cells are quite rare and constitute ≈ 4-400/10^5 normal bone marrow cells, depending upon the assay used. Hematopoietic stem cells also circulate in the peripheral blood. Quantitative analyses of these primitive hematopoietic cell populations are important both for basic biology and for clinical applications. Long term bone marrow cultures (see: LTBMC) and growth in semisolid media (see also: Colony formation assays) have been used to develop assay systems for the purification, characterization, or detection of pluripotent stem cells or early precursors. Unfortunately, many conventional assays fail to measure the most important characteristics of primitive hemopoietic stem cells: long-term repopulating ability and maximal differentiating ability. Assays for clones or colonies arising from a single stem cell are called ***clonogenic assays***. Early hematopoietic stem cells have been defined as » Blast colony-forming cells, » CAFC (cobblestone area forming cells), » CFU-A; » CFC-Mix, » LTC-IC (long-term culture initiating cells), » LTRC (long-term repopulating cells), and » HPP-CFC (high proliferative potential colony-forming cells). *In vivo*, stem cells have been defined as » CFU-S (colony forming units spleen) and CRU (competitive repopulating unit); see: MRA, marrow repopulating ability). Enumeration of CRUs is now believed to yield one of the closest estimates of true hematopoietic stem cells.

The hematopoietic stem cell compartment is now believed to contain a continuum of cell types and some of them are unlikely to be readily detectable with *in vitro* clonogenic assays. Indeed, the individual cell types detected in stem cell assays are not identical and show differences with regard to support hematopoiesis, the length of time over which they function, and the differentiative potential.

Primitive cell types can be detected by » colony formation assays, » delta assay, » limiting dilution analysis, » LTBMC (long-term bone marrow culture), or by their ability to reconstitute hematopoiesis *in vivo* (see: MRA, marrow repopulating ability). They can be enriched selectively and studied further by treatment with » fluorouracil or » 4-HC (4-hydroperoxycyclophosphamide), the latter compound being particularly useful for tumor cell purging. They can be enriched by » Ficoll gradient centrifugation of total bone marrow. They can be isolated also by differential fluorescence-activated cell sorter (FACS) selection, employing various dyes such as Hoechst 33342 or rhodamine-123 (see: CFU-S, LTRC). Early hematopoietic stem cells are among those expressing the cell surface glycoprotein » CD34 or murine » Sca-1.

The molecular nature of the mechanisms underlying stem cell renewal and of mechanisms preventing exhaustion of stem cell pools throughout the entire life span of an organism are largely unknown. These processes critically depend on the maintenance of a functional hematopoietic microenvironment (see: Hematopoiesis) and the presence of a plethora of cytokines collectively known as » hematopoietins (hematopoietic growth factors). Recent progress has been made to define the molecular basis of commitment and differentiation of hematopoietic cells at the genetic level with the identification of » homeotic genes that appear to be of fundamental importance in these processes.

Ahmed T et al Peripheral blood stem cell mobilization by cytokines. J. Clin. Apheresis 7: 129-31 (1992); **Askew DS et al** Insertional mutagenesis and the transformation of hematopoietic stem cells. Hematol. Pathol. 7: 1-22 (1993); **Bernstein A** Molecular genetic approaches to the elucidation of hematopoietic stem cell function. Stem Cells Dayt. 11: s31-5 (1993); **Civin CI et al** Positive stem cell selection – basic science. PCBR 333: 387-401 (1990); **Civin CI** Identification and positive selection of human progenitor/stem cells for bone marrow transplantation. PCBR 377: 461-72 (1992); **Clapp DW** Somatic gene therapy into hematopoietic cells. Current status and future implications. Clin. Perinatol. 20: 155-68 (1993); **Cronkite EP et al** (eds) Haematopoietic stem cell physiology. Progress in Clinical and Biol. Res. 184, Alan R. Liss, New York, 1985; **Dexter TM et al** The role of growth factors in self-renewal and differentiation of hemopoietic stem cells. Phil. Trans. Roy. Soc. (London) B 327: 85-98 (1990); **Gee AP** Tumor cell purging and positive selection of hematopoietic stem cells. J. Clin. Apheresis 7: 135-7 (1992); **Golde DW & Takaku F** (eds) Haematopoietic stem cells. M. Dekker, New York, 1985; **Graham GJ & Pragnell IB** The hemopoietic stem cell: properties and control mechanisms. Semin. Cell Biol. 3: 423-34 (1992); **Hall PA & Watt FM** Stem cells: the generation and maintenance of cellular diversity. Development 106: 619-33 (1989); **Harrison DE et al** Primitive hemopoietic stem cells: direct assay of most productive populations by competitive repopulation with simple binomial, correlation and covariance calculations. EH 21: 206-19 (1993); **Harrison DE** Evaluating functional abilities of primitive hematopoietic stem cell populations. CTMI 177: 13-30 (1992); **Heath JK et al** Growth and differentiation factors of pluripotential stem cells. JCS 13: s75-s85 (1990); **Ikuta K et al** Lymphocyte development from stem cells. ARI 10: 759-83 (1992); **Janssen WE et al** Peripheral blood and bone marrow hematopoietic stem cells: are they the same? Semin. Oncol. 20: s19-27 (1993); **Keller G** Clonal analysis of hematopoietic stem cell development *in vivo* CTMI 177: 41-57 (1992); **Kessinger A** Utilization of peripheral blood stem cells in autotransplantation. Hematol. Oncol. Clin. North Am. 7: 535-45 (1993); **Lemischka IR** The hematopoietic stem cell and its clonal progeny: mechanisms regulating the hierarchy of primitive hematopoietic cells. Cancer Surv. 15: 3-18 (1992); **Meagher RC & Herzig RH** Techniques of harvesting and cryopreservation of stem cells. Hematol. Oncol. Clin. North Am. 7: 501-33 (1993); **Metcalf D** Hematopoietic regulators: redundancy or subtlety? Blood 82: 3515-23 (1993); **Moreb J & Zucali JR** The therapeutic potential of interleukin-1 and tumor necrosis factor on hematopoietic stem cells. Leuk. Lymphoma 8: 267-75 (1992); **Ogawa M** Differentiation and proliferation of hematopoietic stem cells. Blood 81: 2844-53 (1993); **Phillips RA** Comparison of different assays for multipotent hematopoietic stem cells. In: Ford RJ & Maizel AL (eds) Mediators in cell growth and differentiation. pp 135-45, Raven Press, New York 1985; **Quesenberry PJ** Stroma-dependent hematolymphopoietic stem cells. CTMI 177: 151-66 (1992); **Uchida N et al** Heterogeneity of hematopoietic stem cells. Curr. Opin. Immunol. 5: 177-84 (1993); **Wagner JE jr** Isolation of primitive hematopoietic stem cells. Semin. Hematol. 29: s6-9 (1992); **Wright EG & Pragnell IB** The stem cell compartment: assays and negative regulators. CTMI 177: 136-49 (1992); **Zipori D** The renewal and differentiation of hematopoietic stem cells. FJ 6: 2691-7 (1992); for further references see also: Hematopoiesis and individual cell types mentioned in this entry, Hematopoietins.

Hematopoietin 1: see: Hemopoietin 1.

Hematopoietin 2: This factor is found in the culture supernatants of » WEHI cells. It is identical with » IL3.

Bartelmez SH et al Lineage specific receptors used to identify a growth factor for developmentally early hemopoietic cells: assay of hemopoietin-2. JCP 122: 362-9 (1985)

Hematopoietin receptor superfamily: see: Cytokine receptor superfamily.

Hematopoietins: This is a generic name given to ***hematopoietic growth factors*** (abbrev. HGF) or ***hematopoietic cytokines***, which act on cells of the hematopoietic system (see: Hematopoiesis, Hematopoietic stem cells).

Hematopoietic growth factors are produced by many different cell types. They are either secreted or exist in membrane-bound or matrix-associated forms. and may therefore also have different modes of action such as » autocrine, » paracrine, or » juxtacrine growth control mechanisms. These factors are active at all stages of development. They are strictly regulated, i. e., they are synthesized by activated cells under certain conditions (see also: Gene expression, Cell activation) rather than being produced constitutively. Many observations point to the existence of an ordered hierarchy and a concerted action of factors involved in the development of the hematopoietic system. These factors are required for the maintenance of » hematopoietic stem cells, their proliferation, their differentiation into different hematopoietic lineages, and for the maintenance of a stable equilibrium between proliferation and differentiation. These factors allow an organism to shift this equilibrium to one or the other side, as required, for example, under stress conditions. Many of these factors overlap in their biological activities. Teleologically this guarantees a high efficiency and also allows substitution and/or complementation of individual components the functions of which may have been impaired, for example, under pathological conditions. In addition, responses elicited by these factors are usually contextual, i.e. they depend on the presence and concentration of other » cytokines and/or factors in the environment of the responding cells.

Hematopoietic growth factors in a wide sense include the various colony-stimulating factors (see: CSF), » Epo, » SCF (stem cell factor), » SCPF (stem cell proliferation factor), various » interleukins (» IL1, » IL3, » IL4, » IL5, » IL6, » IL11, » IL12), » LIF, TGF-β, » MIP1α, TNF-α, also many other low molecular weight factors (see also: AcSDKP; pEEDCK, thymic hormones), and several other proteins initially identified by some biological activities that have nothing to do with hematopoiesis (see: Cytokines). Many of these proteins are multifunctional. They act on very early differentiation stages or at later stages; they can also act in a lineage-specific manner or may influence more than one lineage. Proliferation and maturation of committed progenitors is controlled by late-acting lineage-specific factors such as » Epo, » M-CSF, » G-CSF, and » IL5. Multipotential progenitors beginning active cell proliferation are regulated by several overlapping cytokines, including » IL3, » GM-CSF, and » IL4. Triggering of cycling by dormant primitive progenitors and maintenance of B cell potential of the primitive progenitors appears to require interactions of early acting cytokines including » IL6, » G-CSF, » IL11, » IL12, » LIF, and » SCF.

The term ***type I factors*** is occasionally used to describe those factors involved in the regulation of hematopoiesis that act directly on some cell types.

This group includes » IL3 (multi-CSF) and » GM-CSF. Factors that synergise with colony-stimulating factors but that by themselves do not possess intrinsic colony-stimulating activity have occasionally been referred to as ***type 2 factors***. They include » IL1, » IL4, » IL5 and » IL6. ***type 3 factors*** are those modulating the hematopoietic growth by stimulating the release of colony-stimulating factors by their respective producer cells. They include » IL1, » IL2, »TNF-β and » IFN-γ.

Some factors negatively regulate processes of hematopoiesis (see: Restrictins). They may selectively inhibit the proliferation of some types of hematopoietic cells and may even induce cell death. The transforming growth factor » TGF-β, for example, predominantly acts on primitive hematopoietic cells and lymphoid cells. The factor called » MIP-α (macrophage inflammatory protein) shows similar activity on primitive myelopoietic cells. As a whole the pleiotropic action of many of these factors underscores the problems associated with nomenclatures more or less based on functional activities of mediator substances (see also: Cytokines, Lymphokines, Interleukins).

An important advancement in the studies of the biological activities of hematopoietic growth factors is the development of established cell lines that entirely depend for their survival and proliferation on the continuous presence of one or more of these factors, so-called » factor-dependent cell lines.

The detailed analysis of the biochemical and physiological properties of hematopoietic growth factors is also facilitated by some other developments. On the one hand, the availability of monoclonal antibodies directed against specific cell surface markers has allowed the identification, isolation, and characterization of individual well-defined clonogenic progenitor cell populations. On the other hand, the use of gene cloning techniques has allowed identification and characterization of many different factors and also their isolation in highly purified form. These factors are now also available in quantities allowing them to be tested and used under clinical conditions. One must not forget that most of these factors act at nano to picomolar concentrations within an organism. Their isolation from natural sources by classical biochemical separation techniques therefore would be an exacting task bringing biochemists to the limits of their skills.

Hematopoietic growth factors are now at the forefront of what can be achieved pharmacologically. They augment established forms of therapy and prophylaxis of primary and secondary immune defects exemplified, for example, by the substitution of immunoglobulins and bone marrow transplantation. Hematopoietic growth factors can be used for a physiological reconstitution of the hematopoietic system in all diseases associated either with an aberrant maturation of blood cells or with a reduced production of leukocytes. Use of these factors, either alone or in combination with each other is probably one of the most effective way to stimulate hematopoiesis *in vivo*. Pharmacological doses of these factors may even allow a further enhancement of hematopoiesis that has already been driven to its maximum by stress conditions.

The use of hematopoietic growth factors may also allow escalation of doses of chemotherapeutic agents, resulting, in turn, in increased tumor responses or cures. An unknown which has not been adequately evaluated in any clinical study to date and which may vary from tumor to tumor is the possibility that in some situations hematopoietic growth factors may also stimulate tumor growth. This possibility is suggested by observations of receptors for these growth factors on various tumor cell lines and of varying degrees of *in vitro* tumor cell proliferative responses to these factors. Increased chemotherapy doses and increases in tumor growth rates by 20-30% (probably not detected in clinical studies to date) would then cancel each other and the application of these factors would then have no significant beneficial effect.

The development of entirely new concepts of treatment (see also: Cytokine gene transfer) are also a direct consequence of our understanding of the biological activities of hematopoietic growth factors and their availability. Peripheral stem cell separation, for example, is a form of treatment which expands *in vivo* the fraction of stem cells in the peripheral blood by administration of colony-stimulating factors and/or other hematopoietic growth factors. These stem cells can be collected by various techniques and can be used for autologous re-infusion into patients undergoing intensive chemotherapy.

Axelrad AA Some hematopoietic negative regulators. Expl. Hematol. 18: 143-50 (1990); **Barge AJ** A review of the efficacy and tolerability of recombinant haematopoietic growth factors in bone marrow transplantation. Bone Marrow Transplant. 11: suppl. 2: 1-11 (1993); **Brach MA & Herrmann F** Hematopoietic growth factors: Interactions and regulation of production. Acta Haematol. 86: 128-37 (1991); **Bronchud M** Can hematopoietic growth factors be used to improve the success of cytotoxic chemotherapy? Anticancer Drugs 4: 127-39 (1993); **Dexter TM et al** The role of growth factors in self-renewal and differentiation of hemopoietic stem cells. Phil. Trans. Roy. Soc. (London) B 327: 85-98 (1990); **Dorshkind K** Regulation of hemopoiesis by bone marrow stromal cells and their products. ARI 8: 111-37 (1990); **Fleischman RA** Southwestern Internal Medicine Conference: clinical use of hematopoietic growth factors. Am. J. Med. Sci. 305: 248-73 (1993); **Graham GJ & Pragnell IB** Negative regulators of hemopoiesis – current advances. PGFR 2: 181-92 (1990); **Gordon MY et al** Haemopoietic growth factors and receptors: bound and free. Cancer Cells 3: 127-33 (1991); **Heath JK et al** Growth and differentiation factors of pluripotential stem cells. JCS 13: s75-s85 (1990); **Herrmann F & Mertelsmann R** Polypeptides controlling hematopoietic cell development and activation. I. *In vitro* results. Blut 58: 117-28 (1989); **Herrman F et al** Polypeptides controlling hematopoietic cell development and activation. II. Clinical results. Blut 58: 173-9 (1989); **Hooper WC** The role of transforming growth factor-β in hematopoiesis. A review. Leuk. Res. 15: 179-84 (1991); **Ihle JN & Askew D** Origins and properties of hematopoietic growth factor-dependent cell lines. IJCC 7: 68-91 (1989); **Kittler EL et al** Biologic significance of constitutive and subliminal growth factor production by bone marrow stroma. Blood 79: 3168-78 (1992); **Krystal G et al** Hematopoietic growth factor receptors. Hematol Pathol. 5: 141-62 (1991); **Lee F** Growth factors controlling the development of hemopoietic cells. PCBR 332: 385-90 (1990); **Lindemann A et al** Clinical evaluation of hematopoietic growth factors. AEMB 297: 93-102 (1991); **Nemunaitis J** Growth factors of the future. Leuk. Lymphoma 9: 329-36 (1993); **Nicola NA** Haemopoietic cell growth factors and their receptors. ARB 58: 45-77 (1989); **Ogawa M** Differentiation and proliferation of hematopoietic stem cells. Blood 81: 2844-53 (1993); **Roberts R et al** Heparan sulfate-bound growth factors: a mechanism for stromal cell mediated hemopoiesis. N 332: 376-8 (1988); **Robinson BE & Quesenberry PJ** Haematopoietic growth factors: overview and clinical applications. Am. J. Med. Sci. 300: Part I 163-70; Part II 237-44; Part III 311-21 (1990); **Williams ME & Quesenberry PJ** Hematopoietic growth factors. Hematol. Pathol. 6: 105-24 (1992)

Hemolymphopoietic growth factor-1: see: HLGF-1.

Hemopoietin 1: also: Hematopoietin 1; abbrev. H-1. This factor is produced by the human tumor cell line HBT5637. It promotes the growth of primitive hematopoietic precursor cells (see also: Hematopoiesis) *in vitro* in combination with » M-CSF, » GM-CSF and » IL3. This factor is identical with » IL1. See also: HP-1.

Bartelmez SH & Stanley ER Synergism between hemopoietic factors (HGFs) detected by their effects on cells leaving receptors for a lineage-specific HGF: assay for hemopoietin-1. JCP 122: 370-8 (1985); **Delwel R et al** Hemopoietin-1 activity of interleukin 1 (IL1) on acute myeloid leukemia colony-forming cells (AML-CFU) *in vitro*: IL1 induces production of tumor necrosis factor-α which synergizes with IL3 or granulocyte-macrophage colony-stimulating factor. Leukemia 4: 557-60 (1990); **Jubinsky PT & Stanley ER** Purification of hemopoietin 1: a multilineage hemopoietic growth factor. PNAS 82: 2764-8 (1985); **McNiece IK et al** Detection of murine hemopoietin-1 in media conditioned by EMT6 cells. EH 15: 854-8 (1987); **Mochizuki DY et al** Interleukin-1 regulates hematopoietic activity, a role previously associated to hemopoietin-1. PNAS 84: 5267-71 (1987); **Morrissey PJ & Mochizuki DY** Interleukin 1 is identical to hemopoietin 1: studies on its therapeutic effects on myelopoiesis and lymphopoiesis. Biotherapy 1: 281-91 (1989); **Stanley ER et al** Regulation of very primitive multipotential hemopoietic cells by hemopoietin-1. Cell 45: 667-74 (1986)

Hemoregulatory peptide: abbrev. HP. See: pEE-DCK.

Hemorrhagic factor: This factor is identical with » TNF-α.

Heparin affin regulatory peptide: see: HARP.

Heparin-binding brain mitogen: see: HBBM.

Heparin binding EGF-like factor: see: HB-EGF.

Heparin binding growth-associated molecule: see: HB-GAM.

Heparin-binding growth factor 1: see: HBGF-1.

Heparin-binding growth factor 2: see: HBGF-2.

Heparin-binding growth factor 3: see: HBGF-3.

Heparin-binding growth factor 4: abbrev. HBGF-4. see: *hst*.

Heparin-binding growth factor 5: abbrev. HBGF-5. see: FGF-5.

Heparin-binding growth factor 7: see: HBGF-7.

Heparin-binding growth factor 8: see: HBGF-8.

Heparin-binding growth factor alpha: see: HGF-α.

Heparin-binding growth factor beta: see: HGF-β.

Heparin-binding growth factors: see: HBGF.

Heparin-binding neurite-promoting factor: see: HBNF (heparin-binding neurite-promoting factor).

Heparin-binding neurotrophic factor: see: HBNF.

Heparin-binding secretory transforming factor 1: abbrev. HSTF1. see: *hst*.

Heparin neutralizing protein: This factor isolated from rabbit platelets is identical with » PF4 (platelet factor 4).

Ginsberg MH et al Purification of a heparin-neutralizing protein from rabbit platelets and its homology with human platelet factor 4. JBC 254: 12365-71 (1979)

Heparin-Sepharose affinity chromatography: abbrev. HSAC. See: HBGF (heparin-binding growth factors).

Hepatic sinusoidal endothelial cell-derived migration stimulating factor: see: HSE-MSF.

Hepatic stimulator substance: see: HSS.

Hepatocyte growth factor: see: HGF (two entries for two different factors).

Hepatocyte growth inhibitory factor: see: HGI.

Hepatocyte growth-stimulating factor: see: HGSF.

Hepatocyte proliferation factor: see: HPF.

Hepatocyte proliferation inhibitor: see: HPI.

Hepatocyte stimulating factor 1: see: HSF-1.

Hepatocyte stimulating factor 2: see: HSF-2.

Hepatocyte stimulating factor 3: see: HSF-3.

Hepatoma-derived growth factor: see: HDGF and HuHGF.

Hepatoma-derived growth factor 2: see: HDGF 2.

Hepatoma growth factor: see: HGF.

Hepatopoietin A: see: HPTA.

Hepatopoietin B: abbrev. HPTB. see: HPTA (hepatopoietin A).

Hepatotropin: This factor was initially isolated from rat serum and was found also at elevated concentrations after partial hepatectomy. It promotes hepatocyte growth in primary culture. It is identical with » HGF (hepatocyte growth factor).

Nakamura T et al Partial purification and characterization of hepatocyte growth factor from serum of hepatectomised rats. BBRC 122: 1450-9 (1984)

HER-2: abbrev. for human EGF receptor 2. See: *neu* oncogene.

Herbimycin A: A benzochinoid ansamycin antibiotic isolated from *Streptomyces* sp. MH237-CF8. This compound specifically inhibits the phosphorylation of tyrosine residues catalyzed by various protein kinases. ***Geldanamycin***, another benzochinoid antibiotic, isolated from *Streptomyces hygroscopicus*, is structurally related to Herbimycin A and shows the same biological activity. It is ≈ 10 times stronger than Herbimycin A but also more toxic.

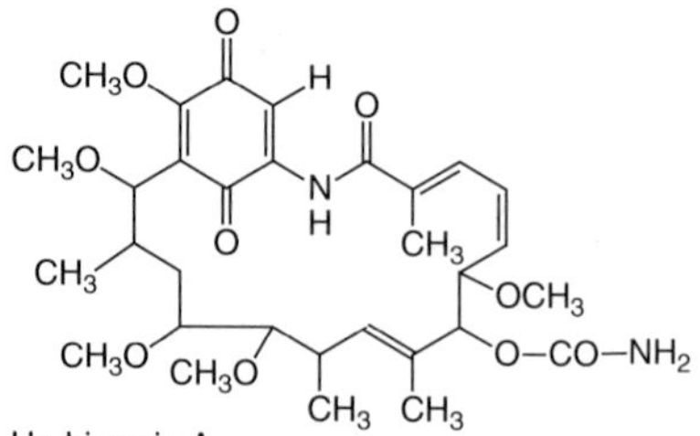

Herbimycin A

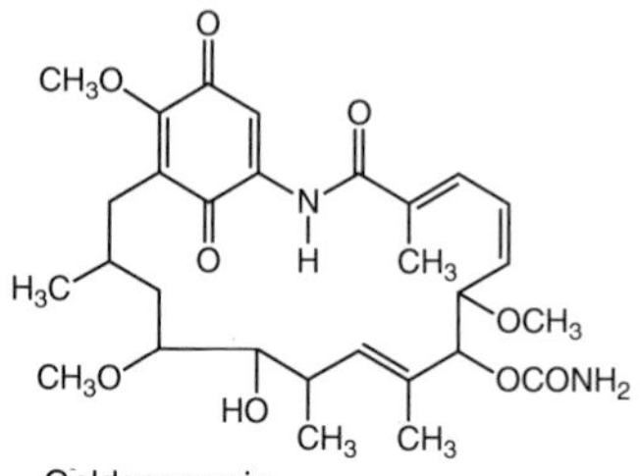

Geldanamycin

Several cell lines can be induced to differentiate by treatment with herbimycin A. In some instances the differentiation process has been shown to be the result of inhibition of tyrosine-specific protein kinase activities. Herbimycin A can be used to test whether binding of a cytokine to its receptor or induction of the synthesis of a growth factor requires tyrosine-specific protein kinase activities for post receptor signaling processes (for other agents used to dissect cytokine-mediated signal transduction pathways see: Bryostatins, Calcium ionophore, Calphostin C, Genistein, H8, K-252a, Lavendustin A, Phorbol esters, Okadaic acid, Staurosporine, Suramin, Tyrphostins, Vanadate).
In T cells Herbimycin A blocks the activity of some protein kinases such as the receptor-associated kinase » *lck* and another kinase called » *fyn*

both of which are involved in mediating responses of the T cell antigen receptor and also of » CD23. Herbimycin A thus inhibits » Cell activation and the proliferation of T cells by blocking the T cell antigen receptor-mediated production of » IL2 and also the synthesis of the IL2receptor. The growth of cells in the presence of » IL3 and » GM-CSF can also be blocked by Herbimycin A.

Transformed cells expressing activated » oncogenes encoding an intrinsic tyrosine-specific protein kinase including, for example, *src*, » *yes*, *fps* (see: *fes*). *ros*, » *abl*, and *erb*B (see: *neu*) revert to a normal phenotype following treatment with Herbimycin A. The transformed phenotype of cells induced by the expression of oncogenes *raf*, *ras* and » *myc* is not altered by Herbimycin A. In murine lymphoblastoma cells Herbimycin A inhibits the expression of the » *myc* oncogene. In addition, Herbimycin A is also a potent inhibitor of angiogenesis (see also: Angiogenesis factors).

In combination with retinoic acid Herbimycin A induces the neuronal differentiation of human neuroblastoma cells and may therefore be valuable in the drug treatment of these tumors.

● **CHEMISTRY: Omura S et al** Herbimycin, a new antibiotic produced by a strain of *streptomyces*. J. Antibiot. (Tokyo). 32: 255-61 (1979); **Shibata K et al** Chemical modification of herbimycin A. Synthesis and *in vivo* antitumor activities of halogenated and other related derivatives of herbimycin A. J. Antibiot. (Tokyo). 39: 415-23 (1986); **Uehara Y et al** Screening of agents which convert transformed morphology of Rous sarcoma virus-infected rat kidney cells to normal morphology: identification of an active agent as herbimycin and its inhibition of intracellular *src* kinase. JJCR 76: 672-5 (1985)

● **BIOLOGICAL ACTIVITIES: Fukazawa H et al** Specific inhibition of cytoplasmic protein tyrosine kinases by herbimycin A *in vitro*. Biochem. Pharmacol. 42: 1661-71 (1991); **Furuzaki A et al** Herbimycin A: an ansamycin antibiotic: X-ray crystal structure. J. Antibiot. (Tokyo). 33: 781-2 (1980); **Honma Y et al** Induction by some protein kinase inhibitors of differentiation of a mouse megakaryoblastic cell line established by coinfection with Abelson murine leukemia virus and recombinant SV40 retrovirus. CR 51: 4649-55 (1991); **Honma H et al** Herbimycin A, an inhibitor of tyrosine kinase, prolongs survival of mice inoculated with myeloid leukemia C1 cells with high expression of v-*abl* tyrosine kinase. CR 52: 4017-20 (1992); **Honma H et al** Effects of herbimycin A derivatives on growth and differentiation of K562 human leukemic cells. Anticancer Res. 12: 189-92 (1992); **Iwasaki T et al** Herbimycin A blocks IL1-induced NF-kappa B DNA-binding activity in lymphoid cell lines. FL 298: 240-4 (1992); **June CH et al** Inhibition of tyrosine phosphorylation prevents T cell receptor-mediated signal transduction. PNAS 87: 7722-6 (1990); **Murakami Y et al** Induction of hsp72/73 by herbimycin A, an inhibitor of transformation by tyrosine kinase oncogenes. ECR 195: 338-44 (1991); **Oikawa T et al** Powerful antiangiogenic activity of herbimycin A (named angiostatic antibiotic). J. Antibiot. (Tokyo) 42: 1202-4 (1989); **Preis PN et al** Neuronal cell differentiation of human neuroblastoma cells by retinoic acid plus herbimycin A. CR 48: 6530-4 (1988); **Satoh T et al** Inhibition of interleukin 3 and granulocyte-macrophage colony-stimulating factor-stimulated increase of active *ras*.GTP by herbimycin A, a specific inhibitor of tyrosine kinases. JBC 267: 2537-41 (1992); Shakarjian MP et al 3-Hydroxy-3-methylglutaryl-coenzyme A reductase inhibition in a rat mast cell line. Impairment of tyrosine kinase-dependent signal transduction and the subsequent degranulation response. JBC 268: 15252-9 (1993); **Uehara Y et al** Phenotypic change from transformed to normal induced by benzoquinoid ansamycins accompanies inactivation of p60*src* in rat kidney cells infected with Rous sarcoma virus. MCB 6: 2198-206 (1986); **Uehara Y et al** Use and selectivity of herbimycin A as inhibitor of protein-tyrosine kinases. ME 201: 370-9 (1991); **Uehara Y et al** Inhibition of transforming activity of tyrosine kinase oncogenes by herbimycin A. Virology 164: 294-8 (1988); **Yamaki H et al** Inhibition of c-myc gene expression in murine lymphoblastoma cells by geldanamycin and herbimycin, antibiotics of benzoquinoid ansamycin group. J. Antibiot. Tokyo 42: 604-10 (1989); **Yamashita T et al** A new activity of herbimycin A: inhibition of angiogenesis. J. Antibiot. (Tokyo) 42: 1015-7 (1989)

Heregulin: abbrev. HRG-α (gene symbol HGL). This glycoprotein of 45 kDa is secreted by several mammary carcinoma cell lines overexpressing the » *neu* oncogene. The factor is also secreted by some normal tissues. Rat, Rabbit, and human heregulin share high sequence homology and induce tyrosine phosphorylation in their target cells of human origin.

Heregulin is identical with » NDF (*neu* differentiation factor). It specifically induces phosphorylation of $p185^{neu}$, and is probably the natural ligand of the » *neu* oncogene receptor, activating the receptor at a specific developmental stage during embryogenesis. Its expression is confined predominantly to the central and peripheral nervous systems, including the neuroepithelium that lines the lateral ventricles of the brain, the ventral horn of the spinal cord, and the intestinal and dorsal root ganglia. Herregulin also binds to a related receptor, called *HER*4/$p180^{erbB4}$, in the absence of *neu*/*HER*-2. This receptor type is co-expressed with *neu*/*HER*-2 in most breast cancer cells responding to heregulin, whereas *HER*2-positive ovarian and fibroblast lines do not respond to heregulin.

At present it is not clear whether heregulin is identical with or related to » NAF (*neu* protein specific activating factor) which also binds to the *neu* receptor. Heregulin appears to be the human homologue of » ARIA (acetylcholine receptor inducing activity) and of bovine » GGF (glial growth factor).

p45 is a recently isolated protein related to heregulin which is a ligand for p180*erb*B2 (see also: *erb*).

Culouscou JM et al Characterization of a breast cancer cell differentiation factor that specifically activates the HER4/

p180*erb*B4 receptor. JBC 268: 18407-10 (1993); **Holmes WE et al** Identification of heregulin, a specific activator of p185*erb*B2. S 256: 1205-10 (1992); **Plowman GD et al** Heregulin induces tyrosine phosphorylation of HER4/p180^{erbB4}. N 366: 473-5 (1993)

Heterologous gene expression: see: Recombinant cytokines. See also: Gene expression.

H-FABP: *heart fatty acid-binding protein* see: MDGI (mammary-derived growth inhibitor).

HFB-1: This human myeloma cell line is used for » Bioassays of B cell differentiation factors (see also: BCDF). HFB-1 cells do not respond to human » IL4 or » IL6 and does not appear to proliferate in response to » IL1, » IL2, or » IFN-γ. They respond to » LMW-BCGF (low molecular weight B cell growth factor).

Callard RE The marmoset B lymphoblastoid cell line (B95-8) produces and responds to B cell growth and differentiation factors: role of shed CD23 (sCD23). Immunology 65: 379-84 (1988); **Hunter KW jr et al** Antibacterial activity of a human monoclonal antibody to *Haemophilus influenzae* type B capsular polysaccharide. Lancet 2(8302): 798-9 (1982); **Shields JG et al** The response of selected human B cell lines to B cell growth and differentiation factors. EJI 17: 535-40 (1987)

hFcRII: see: » CD32.

hFcRIII: see: » CD16.

hFDGI: *human fibroblast-derived growth inhibitor* see: FDGI.

HGF: *hemopoietic growth factors* see: Hematopoietins.

HGF: *hepatocyte growth factor* This factor is isolated from bovine liver. It is mitogenic for hepatocytes. This factor is identical with » bFGF and is not identical with HGF = scatter factor.

Ueno N et al Purification and partial characterization of a mitogenic factor from bovine liver: structural homology with basic fibroblast growth factor. Regul. Pept. 16: 135-45 (1986)

HGF: *hepatocyte growth factor*

■ **Alternative Names:** F-TCF (fibroblast tumor cytotoxic factor); HL-60-HGF; HPTA (Hepatopoietin A), SF (scatter factor), TCF (tumor cytotoxic factor). The factor is now frequently referred to as HGF/SF (for scatter factor). See also: individual entries for further information.

■ **Sources:** HGF was isolated initially from the serum of partially hepatectomised rats. It is produced predominantly in the liver and the pancreas. The liver cells mainly responsible for the synthesis of HGF are Kupffer cells and sinusoidal endothelial cells. The synthesis of HGF is stimulated in particular by injuries affecting liver tissues (hepatitis, ischemia, hepatectomy). In liver cells the expression of the HGF gene can be repressed strongly by » TGF-β. In human skin fibroblasts the synthesis of HGF is induced by both molecular forms of » IL1.

HGF can also be isolated from platelets, kidney, and human serum. HGF is also expressed in human placenta and is found predominantly in trophoblastic tumors. HGF is also synthesized by the » HL-60 cell line.

■ **Protein Characteristics:** HGF consists of two subunits held by a disulfide bond. The α subunit (69 kDa) has a length of 440 amino acids. The β subunit (34 kDa) has a length of 234 amino acids. The β chain of the factor shows ≈ 38 % homology at the protein level to the serine protease domain of plasminogen. HGF is a glycoprotein; the non-glycosylated factor is also bioactive. The β-chain of the murine and rat factor display ≈ 95 % sequence identity. The murine β-chain also shows ≈ 70 % sequence identity with rabbit » HPTA (hepatopoietin A).

A variant of HGF is obtained by alternative splicing. This variant, designated HGF/NK2, is no longer mitogenic and inhibits the HGF-induced proliferation of cells. This demonstrates that one and the same gene encodes a growth factor and a natural specific antagonist.

Both subunits of HGF are obtained by endoproteolytic cleavage of a common preproprotein of 728 amino acids (≈ 90 kDa) which itself is biologically inactive. The inactive precursor is mostly found in a matrix-associated form. Maturation of the precursor into the active heterodimer takes place in the extracellular environment and results from proteolytic cleavage by urokinase which thus acts as a HGF convertase.

■ **Gene Structure:** The HGF gene has a length of ≈ 70 kb and contains 18 exons. It encodes an mRNA of 6.3 kb. The human gene maps to chromosome 7q11.2-21.1. An IL6 responsive element (see also: Gene expression) and a binding site for an IL6-specific transcription factor, » NF-IL6 (see also: IL6) lies in close proximity to the transcription initiation site of the HGF gene.

■ **Receptor Structure, Gene(s), Expression:** The HGF receptor is encoded by the » *met* protooncogene on human chromosome 7p11-4. This gene encodes a protein with intrinsic tyrosine-specific protein kinase activity in its intracellular domain. The heavy chain is thought to play an important role in the interaction of HGF with its receptor and the light chain is further required for the tyrosine phosphorylation of the receptor.

The phosphorylated receptor binds phosphatidyl inositol-3-kinase which is responsible for post receptor signaling.

Alternative splicing leads to the formation of various isoforms of the receptor. It has been suggested that these isoforms are biologically active and may mediate different biologically responses. The existence of further non-*met* receptors for HGF has also been suggested.

■ **BIOLOGICAL ACTIVITIES:** The biological activities of HGF are not species-specific.

HGF is a hepatotrophic factor that stimulates the proliferation of hepatocytes *in vivo*. HGF is the most potent known mitogen for hepatocytes in primary culture. It is one of the major mitogens engaged in liver regeneration processes following partial hepatectomy and other liver injuries. HGF has also been shown to be involved in the regeneration of kidney tubules. It is a mitogen for kidney epithelial cells. *In vitro* HGF induces endothelial cells to proliferate and migrate (see also: Scatter factor).

HGF and » EGF activities are additive. The mitogenic effect of HGF is enhanced by norepinephine. The biological activity of HGF can be inhibited by suramin. HGF abolishes the inhibitory actions of » TGF-β on cellular proliferation. HGF has also been shown to be a priming factor for neutrophils (see also: Cell activation).

At present HGF is the only known factor influencing the three-dimensional growth of cells in three-dimensional collagen gels where it leads to the formation of tubular structures in growing epithelial cells. HGF is therefore a multifunctional cytokine that acts as a mitogen, a motogen (see: Scatter factor) and a morphogen, explaining its functions in organogenesis and tissue regeneration.

HGF also shows » paracrine activities in that it is produced by kidney mesangial cells and acts on renal epithelial cells. HGF also stimulates the growth of human melanoma cells and murine keratinocytes. The growth of some hepatocellular carcinomas and some human tumor cell lines (B6/F1 melanoma, KB cells) is inhibited by HGF. In the rabbit cornea, highly purified HGF promotes neovascularization at subnanomolar concentrations (see also: Angiogenesis factors).

HGF synergizes with » IL3 and » GM-CSF to stimulate colony formation of hematopoietic progenitor cells *in vitro* and may, therefore, also modulate » hematopoiesis.

■ **ASSAYS:** A sensitive ELISA assay is available. Other assay systems are based on the activity of HGF as » SF (scatter factor). For further information see also subentry "Assays" in the reference section.

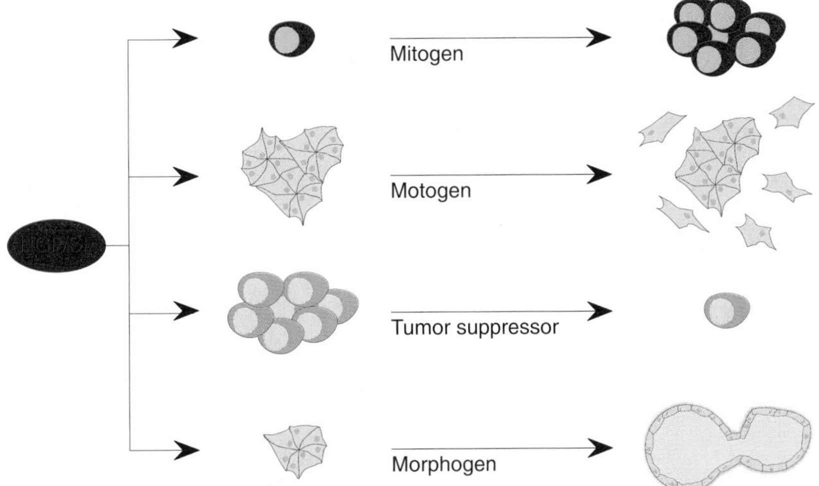

Schematic description of pleiotropic actions of HGF.
HGF stimulates cell growth and cell motility. It also induces epithelial-like structures and functions as a tumor suppressor for various tumor cell types. See text for details.

■ **Clinical Use & Significance:** This factor may be of clinical importance due to its involvement in tissue regeneration following hepatectomy.

● **Reviews: Gherardi E & Stoker M** Hepatocyte growth factor-scatter factor: Mitogen, motogen, and *met*. CC 3: 227-32 (1991); **Goldberg ID & Rosen EM** (eds) Hepatocyte growth factor-Scatter factor (HGF-SF) and the c-*met* receptor. Birkhäuser, Basel 1993; **Matsumoto K & Nakamura T** Hepatocyte growth factor: molecular structure, roles in liver regeneration, and other biological functions. Crit. Rev. Oncogen. 3: 27-54 (1992); **Michalopoulos GK & Zarnegar R** Hepatocyte growth factor. Hepatology 15: 149-55 (1992); **Rubin, JS et al** Hepatocyte growth factor/scatter factor and its receptor, the c-*met* proto-oncogene product. BBA 1155: 357-71 (1993)

● **Biochemistry & Molecular Biology: Aravamudan B et al** Characterization of the 5'-flanking region of the hepatocyte growth factor gene. BBRC 195: 346-53 (1993); **Asami O et al** Purification and characterization of hepatocyte growth factor from injured liver of carbon tetrachloride-treated rats. JB 109: 8-13 (1991); **Chan AM et al** Identification of a competitive HGF antagonist encoded by an alternative transcript. S 254: 1382-5 (1991); **Degen SJ et al** Characterization of the mouse cDNA and gene coding for a hepatocyte growth factor-like protein: expression during development. B 30: 9781-91 (1991); **Fukuyama R et al** Assignment of hepatocyte growth factor (HGF) gene to chromosome 7q21.1. Cytogen. Cell Genet. 58: 1921 (1991); **Gohda E et al** Purification and partial characterization of hepatocyte growth factor from plasma of a patient with fulminant hepatic failure. JCI 81: 414-9 (1988); **Higashio K et al** Identity of a tumor cytotoxic factor from human fibroblasts and hepatocyte growth factor. BBRC 170: 397-404 (1990); **Jiang W et al** Neutrophil priming by hepatocyte growth factor, a novel cytokine. Immunology 77: 147-9 (1992); **Lokker NA & Godowski PJ** Generation and characterization of a competitive antagonist of human hepatocyte growth factor, HGF/NK1. JBC 268: 17145-50 (1993); **Matsumoto K et al** Deletion of kringle domains or the N-terminal hairpin structure in hepatocyte growth factor results in marked decreases in related biological activities. BBRC 181: 691-9 (1991); **Miyazawa K et al** Molecular cloning and sequence analysis of cDNA for human hepatocyte growth factor BBRC 163: 967-73 (1989); **Montesano R et al** Identification of a fibroblast-derived epithelial morphogen as hepatocyte growth factor. Cell 87: 901-8 (1991); **Naka D et al** Activation of hepatocyte growth factor by proteolytic conversion of a single chain form to a heterodimer. JBC 267: 20114-9 (1992); **Nakamura T et al** Purification and characterization of a growth factor from rat platelets for mature parenchymal hepatocytes in primary culture. PNAS 83: 6489-93 (1986); **Nakamura T et al** Purification and subunit structure of hepatocyte growth factor from rat platelets. FL 224: 311-6 (1987); **Nakamura T et al** Molecular cloning and expression of human hepatocyte growth factor. N 342: 440-3 (1989); **Naldini L et al** Extracellular proteolytic cleavage by urokinase is required for activation of hepatocyte growth factor/scatter factor. EJ 13: 4825-33 (1992); **Okigaki M et al** Functional characterization of human hepatocyte growth factor mutants obtained by deletion of structural domains. B 31: 9555-61 (1992); **Rubin JS et al** A broad spectrum human lung fibroblast-derived mitogen is a variant of hepatocyte growth factor. PNAS 88: 415-9 (1991); **Saccone S et al** Regional mapping of the human hepatocyte growth factor (HGF)-scatter factor gene to chromosome 7q21.1. Genomics 13: 832-4 (1992); **Seki T et al** Isolation and expression of cDNA for different forms of hepatocyte growth factor from human leukocytes. BBRC 172: 321-7 (1990); **Seki T et al** Organization of the human hepatocyte growth factor-encoding gene. Gene 102: 213-9 (1991); **Tashiro K et al** Deduced primary structure of rat hepatocyte growth factor and expression of the mRNA in rat tissues. PNAS 87: 3200-4 (1990); **Yee CJ et al** Expression and characterization of biologically active human hepatocyte growth factor (HGF) by insect cells infected with HGF-recombinant baculovirus. B 32: 7922-31 (1993); **Zarnegar R et al** Localization of hepatocyte growth factor (HGF) gene on human chromosome 7. Genomics 12: 147-50 (1992)

● **Receptors: Bottaro DP et al** Identification of the hepatocyte growth factor receptor as the c-*met* proto-oncogene product. S 251: 802-4 (1991); **Chan AM et al** Identification of a competitive HGF antagonist encoded by an alternative transcript. S 254: 1382-5 (1991); **Graziani A et al** The tyrosine-phosphorylated hepatocyte growth factor/scatter factor receptor associates with phosphatidylinositol 3-kinase. JBC 266: 22087-90 (1991); **Higuchi O & Nakamura T** Identification and change in the receptor for hepatocyte growth factor in rat liver after partial hepatectomy or induced hepatitis. BBRC 176: 599-607 (1991); **Komada M et al** Characterization of hepatocyte growth factor receptors on Meth A cells. EJB 204: 857-64 (1992); **Naldini L et al** Hepatocyte growth factor stimulates the tyrosine kinase activity of the receptor encoded by the proto-oncogene C-*met*. O 6: 501-4 (1991); **Naldini L et al** Scatter factor and hepatocyte growth factor are indistinguishable ligands for the *met* receptor. EJ 10: 2867-78 (1991)

● **Biological Activities: Bussolino F et al** Hepatocyte growth factor is a potent angiogenic factor which stimulates endothelial cell motility and growth. JCB 119: 629-41 (1992); **Igawa T et al** Hepatocyte growth factor is a potent mitogen for cultured rabbit renal tubular epithelial cells. BBRC 174: 831-8 (1991); **Jiang W et al** Neutrophil priming by hepatocyte growth factor, a novel cytokine. Immunology 77: 147-9 (1992); **Kan M et al** Hepatocyte growth factor/hepatopoietin A stimulates the growth of rat kidney proximal tubule epithelial cells (RPTE), rat nonparenchymal liver cells, human melanoma cells, mouse keratinocytes and stimulates anchorage-independent growth of SV40-transformed RPTE. BBRC 174: 331-7 (1991); **Kmiecik TE et al** Hepatocyte growth factor is a synergistic factor for the growth of hematopoietic progenitor cells. Blood 80: 2454-7 (1992); **Matsumoto K et al** Hepatocyte growth factor is a potent stimulator of human melanocyte DNA synthesis and growth. BBRC 176: 45-51 (1991); **Matsumoto K et al** Marked stimulation of growth and motility of human keratinocytes by hepatocyte growth factor. ECR 196: 114-20 (1991); **Matsumoto K et al** Hepatocyte growth factor: molecular structure and implications for a central role in liver regeneration. J. Gastroenterol. Hepatol. 6: 509-19 (1991); **Matsumoto K et al** Up-regulation of hepatocyte growth factor gene expression by interleukin-1 in human skin fibroblasts. BBRC 188: 235-43 (1992); **Montesano R et al** Identification of a fibroblast-derived epithelial morphogen as hepatocyte growth factor. Cell 67: 901-8 (1991); **Shiota G et al** Hepatocyte growth factor inhibits growth of hepatocellular carcinoma cells. PNAS 89: 373-7 (1992); **Tajima H et al** Hepatocyte growth factor has potent anti-proliferative activity in various tumor cell lines. FL 291: 229-32 (1991); **Tajima H et al** Regulation of cell growth and motility by hepatocyte growth factor and receptor expression in various cell species. ECR 202: 423-31 (1992); **Wolf HK et al** Hepatocyte growth factor in human placenta and trophoblastic disease. Am. J. Pathol. 138: 1035-43 (1991)

● **Assays: Nishino T et al** Promyelocytic leukemia cell line, HL-60, produces human hepatocyte growth factor. BBRC 181: 323-30 (1991); **Rosen EM et al** Quantitation of cytokine-stimulated migration of endothelium and epithelium by a new assay using microcarrier beads. ECR 186: 22-31 (1990); **Shima N et al** ELISA for F-TCF (human hepatocyte growth factor/hHGF)/fibroblast-derived tumor cytotoxic factor antigen

HGF transgenics: see adendum: HGF.

employing monoclonal antibodies and its application to patients with liver diseases. Gastroenterol. Jpn. 26: 477-82 (1991); **Tsubouchi H et al** Levels of the human hepatocyte growth factor in serum of patients with various liver diseases determined by an enzyme-linked immunosorbent assay. Hepatology 13: 1-5 (1991)

● **Clinical Use & Significance: Tomiya T et al** Significance of serum human hepatocyte growth factor levels in patients with hepatic failure. Hepatology, 15: 1-4 (1992)

HGF: *hepatoma growth factor* This factor is identical with » bFGF.

Lobb RR & Fett JW Purification of two distinct growth factors from bovine neural tissue by heparin affinity chromatography. B 23: 6295-9 (1984); **Pettmann B et al** Purification of two astroglial growth factors from bovine brain. FL 189: 102-8 (1985)

HGF: *hybridoma growth factor* This factor was initially detected as a T cell-derived protein promoting the growth of hybridomas. It was found to be identical with previously described factors » 26 kDa protein, IFN-β2, and » BSF-2. The new name suggested for HGF is » IL6. ***L-HGF***, obtained from supernatants of virus-infected » L929 fibroblasts is identical with HGF and hence identical with IL6.

HGF activity is also used as an assay for IL6. In addition, HGF/IL6 can be used to increase the proportion of B cell hybridomas secreting monoclonal antibodies of desired specificity in culture.

Bazin R & Lemieux R Increased proportion of B cell hybridomas secreting monoclonal antibodies of desired specificity in cultures containing macrophage-derived hybridoma growth factor (IL6). JIM 116: 245-9 (1989); **Brakenhoff JP et al** Molecular cloning and expression of hybridoma growth factor in *Escherichia coli*. JI 139: 4116-21 (1987); **Nordan RD et al** Purification and aminoterminal amino acid sequence of a plasmacytoma growth factor derived from the murine macrophage cell line P388D1. JI 139: 813-7 (1987); **Van Damme J et al** Purification and characterization of human fibroblast-derived hybridoma growth factor identical to T cell-derived B cell stimulating factor-2 (interleukin-6). EJB 168: 543-50 (1987); **Van Damme J & Van Snick J** Induction of hybridoma growth factor (HGF) identical to IL6, in human fibroblasts by IL1: use of HGF activity in specific and sensitive biological assays for IL1 and IL6. Dev. Biol. Stand. 69: 31-8 (1988); **Van Snick J et al** Purification and NH2-terminal amino acid sequence of a new T cell-derived lymphokine with growth factor activity for B cell hybridomas. PNAS 83: 9679-83 (1986); **Van Damme J et al** Separation and comparison of two monokines with lymphocyte-activating factor activity: IL1 β and hybridoma growth factor (HGF). Identification of leukocyte-derived HGF as IL6. JI 140: 1534-41 (1988)

HGF: *hypothalamic growth factor* This factor is identical with » bFGF.

Hauschka PV et al Growth factors in bone matrix. JBC 261: 12665-74 (1986)

HGF-α: *Heparin-binding growth factor alpha* This factor, isolated from bovine brain and hypothalamus, is identical with » aFGF. See also: BDGF (brain-derived growth factor), HBGF (heparin-binding growth factors).

Lobb RR & Fett JW Purification of two distinct growth factors from bovine neural tissue by heparin affinity chromatography. B 23: 6295-9 (1984)

HGF-β: *Heparin-binding growth factor beta* This factor is identical with » bFGF. See also: BDGF (brain-derived growth factor), HBGF (heparin-binding growth factors).

see: HGF-α.

HGF-like protein: *hepatocyte growth factor-like protein* A cDNA encoding this protein has been isolated from a human liver cDNA library. It shows ≈ 50 % identity with » HGF (hepatocyte growth factor) and possesses an identical domain composition as hepatocyte growth factor. The factor/gene is also called ***macrophage stimulating 1*** (MST1) and is identical with » MSP (macrophage stimulating protein). The human gene maps to chromosome 3p21 (DNF15S2 locus) 444 bp upstream of the gene encoding acylpeptide hydrolase. The biological function of HGF-like protein has not yet been determined.

Degen SJF et al Assignment of the gene coding for hepatocyte growth factor-like protein to mouse chromosome 9. Genomics 13: 1368-9 (1992); **Han S et al** Characterization of the DNF15S2 locus on human chromosome 3: identification of a gene coding for four kringle domains with homology to hepatocyte growth factor. B 30: 9768-80 (1991)

HGF/SF: *hepatocyte growth factor-scatter factor* This is an alternative name for » HGF (hepatocyte growth factor) and » SF (scatter factor) both of which are identical.

HGI: *hepatocyte growth inhibitory factor* This factor of 15-40 kDa is produced and secreted by an HTLV-infected T cell line (ATL-2). It is not identical with » ADF (ATL [adult T cell leukemia]-derived factor) which is also secreted by this cell line. HGI appears to be secreted also by other cell lines that also secrete ADF while it is not detected in the conditioned media of cell lines producing little ADF. HGI inhibits the » EGF-dependent proliferation of primary rat hepatocytes.

Inamoto T et al Hepatocyte growth inhibitory factor derived from HTLV-I(+) T cell lines: effect on the epidermal growth factor-dependent proliferation of rat hepatocytes. Clin. Immunol. Immunopathol. 58: 366-76 (1991)

HGL: see: Heregulin.

HGSF: *hepatocyte growth-stimulating factor* This factor of 75 kDa is isolated from the livers of mice treated with carbon tetrachloride. This factor strongly stimulates DNA synthesis in cultured rat hepatocytes. This poorly characterized factor does not react with monoclonal or polyclonal antibodies directed against human » HGF (hepatocyte growth factor).

Another poorly characterized factor of the same name (90-110 kDa by gel filtration) has been isolated as a heparin-binding protein from normal calf serum. It stimulates DNA synthesis and proliferation in primary cultures of adult Balb/c mouse hepatocytes and in two liver-derived epithelial cell lines (C6 and C2.8) plated at low cell density in serum-free medium in the absence of » EGF.

Barone E et al Partial purification of a high molecular weight hepatocyte growth stimulating factor from normal calf serum. J. Biol. Regul. Homeost. Agents. 6: 121-31 (1992); **Gohda E et al** Purification and characterization of a mouse hepatocyte growth-stimulating factor from the liver of carbon tetrachloride-treated mice. J. Pharmacobiodyn. 15: 131-7 (1992); **Hayashi Y et al** Characterization of a mouse hepatocyte growth-stimulating factor in serum of mice treated with carbon tetrachloride. Chem. Pharm. Bull. Tokyo 40: _ 452-5 (1992)

HI-30: *human inhibitor 30 kDa* HI-30 is identical with » ECGF2b (endothelial cell growth factor 2b).

High molecular mass inhibitor factor: see: HMMIF.

High molecular weight B cell growth factor: see: HMW-BCGF.

High proliferative potential colony-forming cells: see: HPP-CFC. See also: LTBMC (long-term bone marrow culture).

HIL-3-derived factor: see: ELDF (eosinophilic leukemia cell differentiation factor).

HILDA: *human interleukin for Da cells* This factor of 41 kDa was originally isolated from several T lymphocyte clones obtained from rejected human kidney allografts and maintained for several months in recombinant IL2 and antigen-supplemented cultures.

It is also secreted by » 5637 cells and some other human tumor cell lines. This factor is a mitogen for murine » DA cells and a strong chemoattractant for eosinophils. It also activates murine and human eosinophils and displays burst-promoting activity (see also: BPA) on human marrow. The factor is identical with » LIF.

Gascan H et al Characterization and NH_2-terminal amino acid sequence of natural human interleukin for DA cells: leukemia inhibitory factor. JBC 264: 21509-15 (1989); **Godard A et al** Biochemical characterization and purification of HILDA, a human lymphokine active on eosinophils and bone marrow cells. Blood 71: 1618-23 (1988); **Moreau JF et al** Characterization of a factor produced by human T cell clones exhibiting eosinophil-activating and burst-promoting activities. JI 138: 3844-9 (1987); **Moreau JF et al** Capacity of alloreactive human T clones to produce factor(s) inducing proliferation of the IL3-dependent DA-1 murine cell line. I. Evidence that this production is under IL2 control. Ann. Inst. Pasteur Immunol. 137C: 25-37 (1986); **Moreau JF et al** Leukemia inhibitory factor is identical to the myeloid growth factor human interleukin for DA cells. N 336: 690-2 (1988); **Soulillou JP & Jacques Y** Constitutive production of human interleukin for DA cells/leukemia inhibitory factor by human tumor cell lines derived from various tissues. JI 144: 2592-8 (1990)

HIM: *hematopoietic inductive microenvironment* see: Hematopoiesis.

Hippocampus-derived neurotrophic factor: see: HDNF.

Histamine-induced suppressor factor: abbrev. HSF. See: Suppressor factors.

Histamine-producing cell stimulating factor: see: HCSF.

Histamine release inhibitory factor: see: HRIF.

Histamine-releasing factor(s): see: HRF.

HIV-1-inhibitory factor: This poorly characterized factor is secreted by activated » CD8-positive T cells. It blocks HIV-1 replication in naturally infected » CD4-positive T cells. The factor is not identical with » IFN-α2 which also inhibits HIV-1 replication. It is also not identical with » IFN-β, » IFN-γ, or » TNF-α, which do not possess blocking activity.

Brinchmann JE et al *In vitro* replication of HIV-1 in naturally infected $CD4^+$ T cells is inhibited by rIFN α 2 and by a soluble factor secreted by activated $CD8^+$ T cells, but not by rIFN β, rIFN γ, or recombinant tumor necrosis factor-α. J. Acquir. Immune Defic. Syndr. 4: 480-8 (1991)

HL14: see: GBP (β-galactoside-binding protein).

HL-60: This cell line was established from a 36 year old female patient with acute promyelocytic leukemia. The cells lack specific markers for lymphoid cells. They express receptors for Fc and complement. The cells grow into colonies in soft agar (see also: Colony formation assay), display phagocytotic activity, respond to a variety of chemotactic stimuli (see also: Chemotaxis) and can be induced to differentiate into monocytes and granulocytes by various cytokines. This is associated with the downregulation of the expression of » myeloblastin. HL60 cells have been used extensively as a model of neutrophil function.

Interferon » IFN-γ, but not other interferons, and also » LIF, » Activin A, » maturation inducer activity, and » DIF (differentiation inducing factor = TNF-β), but not colony-stimulating factors » G-CSF and » GM-CSF induce differentiation. Sodium butyrate and a poorly characterized T lymphocyte cell line-derived factor, which appears to be distinct from » IL1, » IL2, » IFN-γ, » IFN-α, » IL3, » EPA, and » GM-CSF, induces basophilic differentiation of HL-60 cells. Differentiation is also induced by » Phorbol esters. Differentiation requires the expression of the » *fes* oncogene.

The ability of HL-60 cells to form colonies in soft agar is inhibited by combinations of » IL6, » LIF, » GM-CSF and » G-CSF (see also: CSF, colony-stimulating factors). The growth of these cell is inhibited by » TGF-β at doses ranging from 0.025 to 2.5 ng/mL. HL-60 cells have been shown to secrete » HGF (hepatocyte growth factor) and » LDSF (leukemia cell-derived suppressor factor). The cells are also used as producers of » TNF-α. The cells also secrete an uncharacterized factor (see: THP-1-derived growth-promoting activity) that promotes the growth of a large variety of cell lines.

Breitman TR Growth and differentiation of human myeloid leukemia cell line HL60. MiE 190: 118-30 (1990); **Collins SJ et al** Continuous growth and differentiation of human myeloid leukaemic cells in suspension culture. N 270: 347-9 (1979); **Collins SJ** The HL-60 promyelocytic leukemia cell line: proliferation, differentiation, and cellular oncogene expression. Blood 70: 1233-44 (1987); **Fabian I et al** Differentiation and functional activity of human eosinophilic cells from an eosinophil HL-60 subline: response to recombinant hematopoietic growth factors. Blood 80: 788-94 (1992); **Harris P & Ralph P** Human leukaemic models of myelomonocytic development: a review of the HL60 and U937 cell lines. J. Leukocyte Biol. 37: 403-22 (1985); **Hutt-Taylor- SR et al** Sodium butyrate and a T lymphocyte cell line-derived differentiation factor induce basophilic differentiation of the human promyelocytic leukemia cell line HL-60. Blood 71: 209-15 (1988); **Leonard JP et al** Regulation of hematopoiesis IV: the role of interleukin 3 and bryostatin 1 in the growth of erythropoietic progenitors from normal and anemic W/W mice. Blood 72: 1492-6 (1988); **Maekawa T & Metcalf D** Clonal suppression of HL60 and U937 cells by recombinant human leukemia inhibitory factor in combination with GM-CSF or G-CSF. Leukemia 3: 270-6 (1989); **May WS et al** Antineoplastic bryostatins are multipotential stimulators of human hematopoietic progenitor cells. PNAS 84: 8483-7 (1987); **Nishino T et al** Promyelocytic leukemia cell line, HL-60, produces human hepatocyte growth factor. BBRC 181: 323-30 (1991); **Piacibello W et al** Differential effect of transforming growth factor β 1 on the proliferation of human lymphoid and myeloid leukemia cells. Haematologica 76: 460-6 (1991); **Treon SP et al** Growth restraint and differentiation by LPS/TNF-α/IFN-γ reorganization of the microtubule network in human leukemia cell lines. Leukemia 6(Suppl.3) 141S-45S (1992); **Wang SY et al** Induction of differentiation in HL-60 cells by retinoic acid and lymphocyte-derived differentiation-inducing factor but not by recombinant G-CSF and GM-CSF. Leuk. Res. 13: 1091-7 (1989); **Yamada R et al** Induction of differentiation of the human promyelocytic cell line HL-60 by activin/EDF. BBRC 187: 79-85 (1992)

HL-60-HGF: *HL-60 HGF-like immunoreactivity; HL-60 hepatocyte growth factor* This factor is secreted by PMA-stimulated » HL-60 cells. It is identical with HGF (hepatocyte growth factor).

Nishino T et al Promyelocytic leukemia cell line, HL-60, produces human hepatocyte growth factor. BBRC 181: 323-30 (1991)

HLGF-1: *hemolymphopoietic growth factor-1* This poorly characterized factor is a glycoprotein produced by the murine TC-1 cell line derived from adherent marrow cells. The factor acts synergistically » IL3 in inducing multilineage colony formation. It also synergises with IL3 in megakaryocyte colony formation. The factor may be identical with the ligand for the » *kit* receptor, » SCF (stem cell factor).

Quesenberry PJ et al Multilineage synergistic activity produced by a murine adherent marrow cell line. Blood 69: 827-35 (1987); **Quesenberry PJ et al** Long-term marrow cultures: human and murine systems. JCBc 45: 273-8 (1991)

HLI-α: *human leukocyte interferon* This interferon is identical with » IFN-α. See also: IFN for general information about interferons.

HMGF: *human milk growth factor* Human milk contains several species of growth factors. ***HMGF-III*** is the major fraction (≈ 75 %) of the mitogenic activity in human milk. It is very like identical with » EGF. Two other factors, designated ***HMG-I*** and ***HMG-II***, have not been characterized. See also: MDGF (milk-derived growth factor) and MGF (milk growth factor).

Lit. **Shing YW & Klagsbrun M** Human and bovine milk contain different sets of growth factors. E 115: 273-82 (1984); **Sullivan RC et al** Use of size-exclusion and ion-exchange high-performance liquid chromatography for the isolation of biologically active growth factors. J. Chromatogr. 266: 301-11 (1983)

HMGF III: see: HMGF human milk growth factor.

HMMIF: *high molecular mass inhibitor factor* This poorly characterized factor of 115 kDa is found in the supernatant of a $CD1^+/CD8^+$ T cell clone (GI-CO-T-9) established from a culture of an acute T lymphoblastic leukemia (T-ALL). This clone does not produce TNF-α, TNF-β, » IFN-α, IFN-τ, » IL1, » IL2 and has no natural killer-like activity. HMMIF inhibits the responsiveness of normal peripheral blood mononuclear cells to phytohemagglutinin. A factor of 80 kDa, designated ***LMMIF*** (low molecular mass inhibitor factor) has the same activity.
Montaldo PG et al A human acute leukemia-derived T cell line produces two inhibitor factors which suppress lymphocyte proliferation: characterization and purification of the molecules. LR 7: 413-27 (1988)

HMW-BCGF: *high molecular weight B cell growth factor* also abbrev. BCGF-H. This factor is isolated from T cells and some T cell and B cell lines following their stimulation with phytohemagglutinin. This factor of 50-60 kDa is also called 60 kDa-BCGF and Namalwa-BCGF.
HMW-BCGF is a mitogen for activated B cells (see also: BCGF, B cell growth factor; BCDF, B cell differentiation factor). The factor also inhibits immunoglobulin secretion and selectively expands certain B cell subpopulations. A receptor for HMW-BCGF is found on activated B cells. It is not expressed in resting B cells (see also: Cell activation).
Antigenic properties and also the functional activities of this factor show a pronounced homology to the complement factor B (Bb) activation fragment. Antibodies directed against HMW-BCGF recognize Bb and also inhibit the mitogenic activity for B cells. For some aspects of nomenclature see also: BCGF (B cell growth factors). It has been suggested that this factor be called IL14 (interleukin 14).
Ambrus JL & Fauci AS Human B lymphoma cell line producing B cell growth factor. JCI 75: 732-9 (1985); **Ambrus JL et al** Purification to homogeneity of a high molecular weight human B cell growth factor: demonstration of specific binding to activated B cells and development of a monoclonal antibody to the factor. JEM 162: 1319-35 (1985); **Ambrus JL et al** Intracellular signaling events associated with the induction of proliferation of normal human B lymphocytes by two different antigenically related human B cell growth factors (high molecular weight B cell growth factor (HMW-BCGF) and the complement factor Bb). JBC 266: 3702-8 (1991); **Ambrus JL et al** Identification of a cDNA for a human high-molecular-weight B cell growth factor. PNAS 90: 6330-4 (1993); **Ogawa N et al** Abnormal production of B cell growth factor in patients with systemic lupus erythematosus. Clin. Exp. Immunol. 89: 26-31 (1992)

HNP-1 to **-3:** see: Defensins.

Hodgkin-derived leukocyte factor: see: HDLF.

Hodgkin's cell growth factor: see: HCGF.

Hoechst 33342: A DNA binding dye that is used to identify » hematopoietic stem cells by fluorescence activated cell sorting. See: CFU-S (colony forming unit spleen) and LTRC (long-term repopulating cells).

Homeo box: see: Homeotic genes.

Homeo domain: see: Homeotic genes.

Homeotic genes: These genes, have been identified originally by mutations (homeotic mutations) causing aberrant segment development or one body structure to be replaced by a different one in the fruit fly *Drosophila melanogaster*. Examples of such genes are ***antennapaedia***, ***bithorax***, ***engrailed*** (see: *Wnt*), ***fushi tarazu***.
Homeotic genes comprise large » Gene families. The genes are organized in clusters expressed in the developing embryo in complex patterns and with a positional hierarchy. The proteins encoded by homeotic genes have been found to contain an evolutionarily highly conserved region, the so-called ***homeo box*** or ***homeo domain***. Homeobox-containing genes have been found in other organisms, including mouse, » *Xenopus laevis*, *Saccharomyces cerevisiae*, and man; some of them have been isolated on the basis of homologies with the fruit fly genes. Proteins containing a homeobox sequence domain are sequence-specific DNA-binding proteins that bind to multiple different genes and thereby determine the developmental fate of a cell. The proteins encoded by these genes control » Gene expression and modify expression patterns of genes both in developing as well as in adult tissues. They play an important role in mammalian embryonic pattern formation and are involved in oncogenic processes. Some of these genes play a crucial role in local pattern formation, while others are tissue-specific or ubiquitous transcription regulators.
Multiple homeobox genes are expressed in hematopoietic cell lineages. They are involved in commitment and differentiation of hematopoietic cells and their expression is lineage-specific so that it has been assumed that they function as regulators of lineage determination during hematopoiesis.
The expression of a gene, called HEX (hematopoietically expressed homeobox), is observed in

multipotential hematopoietic progenitors, (see: Hematopoiesis), and in cells of the B lymphocyte and myeloid lineages. HEX is not expressed in T lymphocytes or erythroid cells.

Two recently characterized human homeobox genes, HB24 and HB9, have found to be highly expressed in bone marrow cells enriched for » CD34-positive cells, present at low levels in unfractionated bone marrow cells, and essentially undetectable in bone marrow cells depleted of CD34 cells. Inhibition of HB24 expression in CD34-positive bone marrow cells via » antisense oligonucleotides impairs the proliferation of these cells in response to » IL3 and » GM-CSF. Conversely, transient expression of HB24 in CD34-positive cells inhibits their differentiation into mature hematopoietic cell types. HB24-specific RNA transcripts have been found to be elevated in bone marrow and peripheral blood mononuclear cells isolated from patients with acute myelogenous leukemia.

Human homeobox-containing genes of the HOX family have been shown to be coordinately regulated in blocks in myeloid cells whereas they appear to function as isolated genes in lymphoid cells. Aberrant expression of these genes has recently been related to leukemic phenotype. Six contiguous genes of the HOX2 locus are highly expressed in acute non-lymphocytic leukemia and are switched off in chronic myelogenous leukemia. It has been shown that treatment of normal murine bone marrow cells stimulated with appropriate colony stimulating factors with » antisense oligodeoxynucleotides of the mouse homeobox gene Hox 2.3 leads to a selective inhibition of myeloid colony formation without significant effect on erythroid and megakaryocytic hematopoiesis.

● **Reviews: Deschamps J & Meijlink F** Mammalian homeobox genes in normal development and neoplasia. Crit. Rev. Oncog. 3: 117-73 (1992)

Bedford FK et al HEX: a novel homeobox gene expressed during haematopoiesis and conserved between mouse and human. NAR 21(5): 1245-9 (1993); **Celetti A et al** Characteristic patterns of HOX gene expression in different types of human leukemia. IJC 53: 237-44 (1993); **Deguchi Y & Kehrl JH** Selective expression of two homeobox genes in CD34-positive cells from human bone marrow. Blood 78: 323-8 (1991); **Deguchi Y et al** A diverged homeobox gene is involved in the proliferation and lineage commitment of human hematopoietic progenitors and highly expressed in acute myelogenous leukemia. Blood 79: 2841-8 (1992); **Lawrence HJ & Largman C** Homeobox genes in normal hematopoiesis and leukemia. Blood 80: 2445-53 (1992); **Lowney P et al** A human Hox 1 homeobox gene exhibits myeloid-specific expression of alternative transcripts in human hematopoietic cells. NAR 19: 3443-9 (1991); **Magli MC et al** Coordinate regulation of HOX genes in human hematopoietic cells. PNAS 88: 6348-52 (1991); **Shen WF et al** Modulation of homeobox gene expression alters the phenotype of human hematopoietic cell lines. EJ 11: 983-9 (1992); **Vieille-Grosjean I et al** Identification of homeobox-containing genes expressed in hematopoietic blast cells. BBRC 185: 785-92 (1992); **Vieille-Grosjean I et al** Lineage and stage specific expression of HOX 1 genes in the human hematopoietic system. BBRC 183: 1124-30 (1992); **Wu J et al** Selective inhibition of normal murine myelopoiesis "*in vitro*" by a Hox 2.3 antisense oligodeoxynucleotide. Cell. Mol. Biol. 38: 367-76 (1992)

Homologous recombination: see: Knock-out mice.

HP: *hemoregulatory peptide* see: pEEDCK.

HP1: *helper peak 1* This activity was observed in supernatants of human mixed peripheral blood leukocyte cultures. It partially restored the antibody response of T cell-deficient adherent murine spleen cells and nude mouse spleen cells (see also: Immunodeficient mice) to the thymic dependent antigen SRC. The factor is identical with » TPF (thymocyte proliferative factor), » HP1 (helper peak 1), and » LAF (lymphocyte activating factor), and hence identical with » IL1.

Aarden LA et al Revised nomenclature for antigen-nonspecific T cell proliferation and helper factors. CI 48: 433-36 (1979); **Koopman WJ et al** Evidence for the identification of lymphocyte-activating factor as the adherent T cell-derived mediator responsible for enhanced antibody synthesis by nude mouse spleen cells. CI 35: 92-8 (1978)

HP1: *hematopoietin 1, hemopoietin 1* This factor is identical with » IL1. See also: Hemopoietin 1.

Aarden LA et al Revised nomenclature for antigen-nonspecific T cell proliferation and helper factors. CI 48: 433-36 (1979); **Mochizuki DY et al** Interleukin 1 regulates hematopoietic activity, a role previously ascribed to hemopoietin 1. PNAS 84: 5271-6 (1987) **Moore MA & Warren DJ** Synergy of interleukin-1 and granulocyte colony-stimulating factor: *in vivo* stimulation of stem-cell recovery and hematopoietic regeneration following 5-fluorouracil treatment of mice. PNAS 84: 7134-8 (1987)

HP1: *hybridoma/plasmacytoma factor 1* This factor was originally identified as a T cell-derived lymphokine with growth factor activity for B cell hybridomas and plasmacytomas (see: HPGF, PCT-GF). Although there is virtually no similarity between the NH2-terminal region of HP1 and its human biological counterpart (» 26-kDa protein = » IFN-β2 = » BSF-2 (B cell stimulatory factor-2) = » IL6), extensive amino acid similarities in the middle and COOH-terminal regions of these molecules suggest that HP1 is the murine homologue of human » IL6.

Simpson RJ et al Murine hybridoma/plasmacytoma growth factor: complete amino acid sequence and relation to human interleukin-6. EJB 176: 187-97 (1988); **Van Snick J et al** cDNA Cloning of Murine Interleukin HP1: Homology with Human Interleukin 6. EJI 18: 193-7 (1988)

HP-1 to **HP-3:** human neutrophil granule peptides. See: defensins.

HP1-SMP: see: SMP (somatomedin/insulin-like growth factor (IGF)-like polypeptide).

HP3-SMP: see: SMP (somatomedin/insulin-like growth factor (IGF)-like polypeptide).

HP2: *hematopoietin 2* This factor is identical with » IL3.

Bartelmez SH et al Lineage specific receptors used to identify a growth factor for developmentally early hemopoietic cells: assay of hemopoietin-2. JCP 122: 362-9 (1985)

HPA: *hypothalamic-pituitary-adrenocortical axis* see: Neuroimmune network.

HPF: *hepatocyte proliferation factor* This poorly characterized factor has been isolated from the cytosol of regenerating rat and porcine livers. It stimulates DNA synthesis in rat hepatocytes.

Schwarz LC et al The characterization and partial purification of hepatocyte proliferation factor. Ann. Surg. 202: 296-302 (1985)

HPGF: *hemopoietic cell growth factor* This factor was initially described as an activity found in culture supernatants of » WEHI cells that supported the growth of hematopoietic cells. It is identical with » IL3.

Bazill GEW et al Characterization and partial purification of a haemopoietic cell growth factor in WEHI-3 cell conditioned medium. BJ 210: 747-59 (1983); **Prestidge RL et al** Biochemical comparison of murine colony-stimulating factors secreted by a T cell lymphoma and a myelomonocytic leukemia. JI 133: 293-8 (1984)

HPGF: *human pituitary growth factor* This factor is identical with » bFGF.

Rowe JM et al Purification and characterization of a human pituitary growth factor. B 25: 6421-5 (1986)

HPGF: *hybridoma/plasmacytoma growth factor* This factor was initially described as an activity supporting the growth of hybridomas and plasmacytomas. It is produced by normal human fibroblasts and by a line of human osteosarcoma cells (MG-63). The aminoterminal sequence of the MG-63-derived protein was found to be identical with the » 26 kDa protein. HPGF is therefore identical with » IL6.

Nordan RD & Potter M A macrophage-derived factor required by plasmacytomas for survival and proliferation *in vitro*. S 233: 566-9 (1986); **Nordan RD et al** Purification and aminoterminal amino acid sequence of a plasmacytoma growth factor derived from the murine macrophage cell line P388D1. JI 139: 813-7 (1987); **Poupart P et al** B cell growth modulating and differentiating activity of recombinant human 26-kd protein (BSF-2, HuIFN-β 2, HPGF). EJ 6: 1219-24 (1987); **Van Damme J et al** Identification of the human 26-kD protein, interferon β2 (IFN-β2), as a B cell hybridoma/plasmacytoma growth factor induced by interleukin 1 and tumor necrosis factor. JEM 165: 914-9 (1987); **Van Damme J et al** Interleukin 1 and poly(rI)poly(rC) induce production of a hybridoma growth factor by human fibroblasts. EJI 17: 1-7 (1987)

HP-GSA: *hematopoietic progenitor cell growth-stimulating activity* This is a general term occasionally used to describe factors supporting the growth of early bone marrow progenitor cells. One distinct factor with this activity, isolated from a human melanoma cell line, is probably identical with » IL3. See also: Hematopoiesis.

Kraft AS & Zuckerman KS Production of multilineage hemopoietic growth-stimulating activities by a human melanoma cell line. EH 14: 867-72 (1986)

HPI: hepatocyte proliferation inhibitor This poorly characterized activity can be isolated from the cytosolic fractions of rat and human adult liver tissues. HPI is not found in human fetal liver > 20 weeks of age. HPI activity is found in proteins of 26 kDa, 17-19 kDa and 18-50 kDa and appears to consist of multiple proteins. HPI from rat and human sources have been reported to be identical. HPI appears to be an endogenous agent that inhibits DNA synthesis in rat and fetal chick hepatocytes, but not in malignant rat liver cells. Its precise role in liver growth and differentiation of liver cells is unclear. HPI activity is not due to » TGF-β.

Chen TS et al Fraction from human and rat liver which is inhibitory for proliferation of liver cells. Cytobios 59: 79-86 (1989); **Devi BG et al** Ontogeny of hepatocyte proliferation inhibitor activity during human liver development and its effect on cell proliferation in *in vivo* and *in vitro* studies. Biochem. Cell Biol. 71: 241-7 (1993); **Hugget AC et al** Characterization of a hepatic proliferation inhibitor (HPI): effect of HPI on the growth of normal liver cells - comparison with transforming growth factor β. JCBc 35: 305-14 (1987); **McMahon JB et al** Purification and properties of a rat liver protein that specifically inhibits the proliferation of non-malignant epithelial cells from rat liver. PNAS 79: 456-60 (1982)

HPP-CFC: *high proliferative potential colony-forming cells* These cells have been defined by their ability to form very large colonies (> 0.5 mm in diameter) in bone marrow cell cultures (see: LTBMC) containing approximately 50,000 cells, including progenitor and mature hematopoietic cells of the granulocyte, macrophage, and megakaryocyte lineages. HPP-CFCs have been divided into three different classes (HPP-CFC-1, -2, 3) in order of primitiveness. They are relatively resistant to treatment *in vivo* with the cytotoxic drug » fluorouracil or » 4-HC (4-hydroperoxycyclophosphamide) and have been shown to correlate with cells capable of repopulating the bone marrow of lethally irradiated mice (see also: MRA, marrow repopulating ability, for an assay system). HPP-CFC, unlike » LTC-IC (long-term culture-initiating cells), are not considered to be a measure of true pluripotential » hematopoietic stem cells but, rather, of their slightly more differentiated progeny (primitive hematopoietic progenitor cells, abbrev. PHPC). It has been suggested that » blast colony-forming cells resemble the most primitive HPP-CFC type. Numbers of HPP-CFCs can be determined by » limiting dilution analysis.

It has been shown that » TGF-β is a direct selective inhibitor of murine and human HPP-CFC. Various combination of cytokines have been shown to promote the expansion of HPP-CFC in liquid cultures (see: Delta assay) with » IL1, » IL6, and » SCF, alone or in combination, or SCF + » IL3, being the most effective. SCF in particular has been shown to synergistically interact with a number of cytokines at suboptimal concentrations to directly augment the proliferative capacity of HPP-CFC. The presence of lithium in Dexter-type long-term bone marrow cultures (see: LTBMC) has been shown to stimulate HPP-CFC, probably by induction of multiple growth factors from stromal cells.

Baines P et al Characterization of a developmentally early macrophage progenitor found in normal mouse marrow. Br. J. Haematol. 48: 147-53 (1981); **Baines P et al** Physical and kinetic properties of hemopoietic progenitor cell populations from mouse marrow detected in five different assay systems. Leuk. Res. 6: 81-8 (1982); **Bartelmez SH et al** Uncovering the heterogeneity of hematopoietic repopulating cells EH 19: 861-2 (1991); **Bertoncello I et al** The resolution, enrichment, and organization of normal bone marrow high proliferative potential colony-forming cell subsets on the basis of rhodamine-123 fluorescence. EH 19: 174-8 (1991); **Bertoncello I et al** An improved negative immunomagnetic selection strategy for the purification of primitive hemopoietic cells from normal bone marrow. EH 19: 95-100 (1991); **Boggs DR et al** Hematopoietic stem cells with high proliferative potential. Assay of their concentration in marrow by the frequency and duration of cure of W/Wv mice. JCI 70: 242-53 (1982); **Han M et al** *In vitro* expansion of murine hematopoietic progenitor cells in liquid cultures for bone marrow transplantation: effects of stem cell factor. Int. J. Hematol. 57: 113-20 (1993); **Lowry PA et al** Effects of rrSCF on multiple cytokine responsive HPP-CFC generated from SCA^+Lin^- murine hematopoietic progenitors. EH 19: 994-6 (1991); **Lowry PA et al** Stem cell factor induction of *in vitro* murine hematopoietic colony formation by "subliminal" cytokine combinations: the role of "anchor factors". Blood 80: 663-9 (1992); **McGrath HE et al** Lithium stimulation of HPP-CFC and stromal growth factor production in murine Dexter culture. JCP 151: 276-86 (1992); **McNiece IK et al** Subpopulations of mouse bone marrow high-proliferative potential colony-forming cells (HPP-CFC). EH 14: 856-60 (1986); **McNiece IK et al** Generation of murine hematopoietic precursor cells from macrophage high-proliferative potential colony-forming cells. EH 15: 972-7 (1987); **McNiece IK et al** Colony-forming cells with high proliferative potential (HPP-CFC). IJCC 8: 146-60 (1990); **McNiece IK et al** Stimulation of murine high proliferative potential colony forming cells by the combination of GM-CSF and CSF-1. Blood 72: 191-5 (1988); **Muench MO et al** The *in vitro* growth of murine high proliferative potential-colony forming cells is not enhanced by growth in a low oxygen atmosphere. Cytokine 4: 488-94 (1992); **Muench MO et al** Interactions among colony-stimulating factors, IL1 β, IL6, and *kit*-ligand in the regulation of primitive murine hematopoietic cells EH 20: 339-49 (1992); **Quesenberry PJ** The blueness of stem cells. EH 19: 725-8 (1991); **Robinson S & Riches A** Haematopoietic stem cell proliferation regulators investigated using an *in vitro* assay. J. Anat. 174: 153-62 (1991); **Spangrude GJ et al** Mouse hematopoietic stem cells. Blood 78: 1395-402 (1991); **Spangrude GJ** Hematopoietic stem-cell differentiation. Curr. Opin. Immunol. 3: 171-8 (1991); **Srour EF et al** Long-term generation and expansion of human primitive hematopoietic progenitor cells *in vitro*. Blood 81: 661-9 (1993); **Steward FM et al** Post-5-fluorouracil human marrow: stem cell characteristics and renewal properties after autologous marrow transplantation. Blood 81: 2283-9 (1993); **Wolf NS et al** *In vivo* and *in vitro* characterization of long-term repopulating primitive hematopoietic cells isolated by sequential Hoechst 33342-rhodamine 123 FACS selection. EH 21: 614-22 (1993)

HPSTI: *human pancreatic secretory trypsin inhibitor* see: » ECGF2a (endothelial cell growth factor 2a).

HPTA: *hepatopoietin A* This factor was initially isolated as a heparin-binding growth factor (see also: HBGF) from rabbit tissues and rat serum. It is a heterodimeric glycoprotein of 105 kDa that stimulates DNA synthesis in primary cultures of rat hepatocytes in serum-free media (see also: SFM). The factor is identical with » HGF (hepatocyte growth factor). The factor designated ***Hepatopoietin B*** (HPTB) is a glycolipid mitogenic for hepatocytes.

Fleig WE Liver-specific growth factors. Scand. J. Gastroenterol. Suppl. 151: 31-6 (1988); **Goldberg M** Purification and partial characterization of a liver cell proliferation factor called hepatopoietin. JCBc 27: 291-302 (1985); **Michalopoulos GK et al** Hepatopoietins A and B and hepatocyte growth factor. Dig. Dis. Sci. 36: 681-6 (1991); **Michalopopoulos GK** Liver regeneration: molecular mechanisms of growth control. FJ 4: 176-87 (1990); **Thaler FJ & Michalopoulos GK** Hepatopoietin A: partial characterization and trypsin activation of a hepatocyte growth factor. CR 45: 2545-9 (1985); **Zarnegar R et al** NH_2-terminal amino acid sequence of rabbit hepatopoietin A, a heparin-binding polypeptide growth factor for hepatocytes. BBRC 163: 1370-6 (1989); **Zarnegar R et al** Purification and biological characterization of human hepatopoietin A, a polypeptide growth factor for hepatocytes. CR 49: 3314-20 (1989); **Zarnegar R et al** Tissue distribution of hepatopoietin-A: heparin-binding polypeptide growth factor for hepatocytes. PNAS 87: 1252-6 (1990)

HPTB: *hepatopoietin B* see: HPTA (hepatopoietin A).

HRA-N: *neutrophil-derived histamine-releasing activity* see: HRF (histamine-releasing factors).

HRF: *histamine-releasing factors* This is a general term used for factors that induce the release of histamines from basophils and/or mast cells when stimulated with antigens or mitogens. HRF activities appear to be quite heterogeneous and to vary in physiochemical properties depending upon the cell source. Histamine release inhibitory activities have also been described (see: HRIF).

Apart from some poorly characterized factors with HRF activity » CTAP 3 (connective tissue activating peptide-3), » IL3, » MCAF (monocyte chemotactic and activating factor), » MCP-1 (monocyte chemoattractant protein-1), » NAF (neutrophil-activating factor), » NAP-1 (neutrophil activating protein-1), » NAP-2 (neutrophil activating protein-2), » SCF (stem cell factor) and » GM-CSF also possess HRF activity. A neutrophil-derived histamine-releasing activity of 1.4-2.4 kDa, designated ***HRA-N***, has been found to be released by human polymorphonuclear neutrophils. This activity causes dose-related histamine release from human basophils, isolated human cutaneous mast cells and rat basophil leukemia cells. It appears to be different from other factors with HRF activity released my mononuclear cells or platelets. HRA-N has also been shown to cause the release of serotonin but not of arachidonic acid metabolites from rat basophilic leukemia cells.

Alam R et al Interleukin-8 and RANTES inhibit basophil histamine release induced with monocyte chemotactic and activating factor/monocyte chemoattractant peptide-1 and histamine releasing factor. Am. J. Respir. Cell. Mol. Biol. 7: 427-33 (1992); **Alam R et al** Monocyte chemotactic and activating factor is a potent histamine-releasing factor for basophils. JCI 89: 723-8 (1992); **Haak-Frendscho M et al** Comparison of mononuclear cell and B lymphoblastoid histamine-releasing factor and their distinction from an IgE-binding factor. Clin. Immunol. Immunopathol.49: 72-82 (1988); **Grant JA et al** Histamine-releasing factors and inhibitory factors. Int. Arch. Allergy Appl. Immunol. 94: 141-3 (1991); **Igarashi Y et al** Human neutrophil-derived histamine-releasing activity (HRA-N) causes the release of serotonin but not arachidonic acid metabolites from rat basophilic leukemia cells. J. Allergy Clin. Immunol. 89: 1085-97 (1992); **Kaplan AP et al** Histamine-releasing factors. Int. Arch. Allergy Appl. Immunol. 94: 148-53 (1991); **Kuna P et al** IL8 inhibits histamine release from human basophils induced by histamine-releasing factors, connective tissue activating peptide III, and IL3. JI 147: 1920-4 (1991); **Liao TN & Hsieh KH** Characterization of histamine-releasing activity: role of cytokines and IgE heterogeneity. J. Clin. Immunol. 12: 248-58 (1992); **White MY & Kaliner MA** Neutrophils and mast cells. I. Human neutrophil-derived histamine-releasing activity. JI 139: 1624-30 (1987); **White MV et al** Neutrophils and mast cells. Comparison of neutrophil-derived histamine-releasing activity with other histamine-releasing factors. JI 141: 3575-83 (1988); **White MV et al** Neutrophils and mast cells: characterization of cells responsive to neutrophil-derived histamine-releasing activity (HRA-N). J. Allergy Clin. Immunol. 84: 773-80 (1989)

HRG-α: see: Heregulin.

HRIF: *histamine release inhibitory factor* This poorly characterized factor is a specific antagonist of histamine-releasing factors (see also: HRF). It is produced by peripheral blood mononuclear cells (B cells, T cells, monocytes) upon stimulation with histamine or mitogens such as Con A. It inhibits HRF-induced histamine release from basophils and mast cells. One particular factor with HRIF activity is » IL8.

Alam R et al Study of the cellular origin of histamine release inhibitory factor using highly purified subsets of mononuclear cells. JI 143: 2280-4 (1989); **Alam R et al** Detection of histamine release inhibitory factor- and histamine releasing factor-like activities in bronchoalveolar lavage fluids. Am. Rev. Respir. Dis. 141: 666-71 (1990); **Grant JA et al** Histamine-releasing factors and inhibitory factors. Int. Arch. Allergy Appl. Immunol. 94: 141-3 (1991); **Kuna P et al** IL8 inhibits histamine release from human basophils induced by histamine-releasing factors, connective tissue activating peptide III, and IL3. JI 147: 1920-4 (1991)

HS: *heparan sulfate* see: ECM (extracellular matrix).

HSA: *humoral stimulatory activity* This term describes an activity found in the serum of AML (acute myeloid leukemia) following treatment with cytotoxic antimetabolites (ara-C). This activity is responsible for the increased proliferation and an enhanced metabolization of the antimetabolite observed after an initial period of cytoreduction. Increased metabolization of araC then inhibits DNA synthesis in the growing cells. It is assumed that several different cytokines are responsible for this activity.

Karp JE et al Direct relationship of marrow cell growth and 1-β-D-arabinofuranosylcytosine metabolism. CR 44: 5046-50 (1984); **Karp JE et al** Effects of rhGM-CSF on intracellular ara-C pharmacology *in vitro* in acute myelocytic leukemia: comparability with drug-induced humoral stimulatory activity. Leukemia 4: 553-6 (1990)

HSAC: *heparin-sepharose affinity chromatography* see: HBGF (heparin-binding growth factors).

HSC: see: hematopoietic stem cells. See also: Hematopoiesis.

HSE-MSF: *hepatic sinusoidal endothelial cell-derived migration stimulating factor* This dimeric factor (disulfide-linked components of 110 and 67 kDa) is secreted by mouse hepatic sinusoidal endothelial cells. It differentially stimulates the migration of liver-metastatic lymphoma cells (RAW117-H10). The factor is a proteolytic fragment of murine complement component C3b.

Hamada JI et al A paracrine migration-stimulating factor for metastatic tumor cells secreted by mouse hepatic sinusoidal endothelial cells: identification as complement component C3b. CR 53: 4418-23 (1993)

HSF: *hepatocyte stimulating factor* This factor which is secreted by some hepatoma cell lines is identical with » IL6. Stimulation of the proliferation of hepatocytes is used as a » bioassay for this factor. See also: HSF-1.

Geiger T et al Cell-free-synthesized interleukin-6 (BSF-2/IFN-β 2) exhibits hepatocyte-stimulating activity. EJB 175: 181-6 (1988); **Jordana M et al** Spontaneous *in vitro* release of alveolar-macrophage cytokines after the intratracheal instillation of bleomycin in rats. Characterization and kinetic studies. Am. Rev. Respir. Dis. 137: 1135-40 (1988); **Northemann W et al** Production of interleukin 6 by hepatoma cells. Mol. Biol. Med. 7: 273-85 (1990)

HSF: *histamine-induced suppressor factor* see: Suppressor factors.

HSF-1: *hepatocyte stimulating factor 1* This factor was described initially as a monocyte/macrophage-derived regulatory protein stimulating the hepatic synthesis of several plasma proteins. It is identical with » IFN-β2 or » BSF-2 and hence is identical with » IL6.

Baumann H et al Interaction among hepatocyte-stimulating factors, interleukin 1, and glucocorticoids for regulation of acute phase plasma proteins in human hepatoma (HepG2) cells. JI 139: 4122-8 (1987); **Gauldie J et al** Interferon β2/B cell stimulatory factor 2 shares identity with monocyte derived hepatocyte stimulatory factor and regulates the major acute phase protein response in liver cells. PNAS 84: 7251-5 (1987)

HSF-2: *hepatocyte stimulating factor 2* This factor induces the production of » acute phase proteins (see also: acute phase reaction) in cultured hepatoma cells. It is identical with » LIF. At least some HSF-2 activity found in cell supernatants appears to be identical with » IL6 (= HSF-1).

Baumann H et al Interaction among hepatocyte-stimulating factors, interleukin 1, and glucocorticoids for regulation of acute phase plasma proteins in human hepatoma (HepG2) cells. JI 139: 4122-8 (1987); **Baumann H et al** Human hepatocyte-stimulating factor-III and interleukin-6 are structurally and immunologically distinct but regulate the production of the same acute phase plasma proteins. JBC 264: 8046-51 (1989); **Baumann H & Wong GG** Hepatocyte-stimulating factor 3 shares structural and functional identity with leukemia-inhibitory factor. JI 143: 1163-7 (1989)

HSF-3: *hepatocyte stimulating factor 3* This factor is produced constitutively by cultured keratinocytes and squamous carcinoma cell lines (COLO-16). It induces the same spectrum of » acute phase proteins (see also: acute phase reaction) in cultured hepatoma cells as » IL6. The factor is identical with » LIF, the activity of which is completely neutralized by antibodies directed against HSF-3.

Baumann H et al Human hepatocyte-stimulating factor-III and interleukin-6 are structurally and immunologically distinct but regulate the production of the same acute phase plasma proteins. JBC 264: 8046-51 (1989); **Baumann H & Wong GG** Hepatocyte-stimulating factor 3 shares structural and functional identity with leukemia-inhibitory factor. JI 143: 1163-7 (1989)

HSF: *hybridoma suppressor factor* This poorly characterized factor of ≈ 10-12 kDa is secreted by the human thymus cell hybridoma, 8E-24. It is a potent immunosuppressive factor which inhibits polyclonal immunoglobulin (Ig) production. The suppression is monocyte-dependent in that its suppressive activity for Ig production is not observed in monocyte-depleted lymphocyte cultures and is restored by addition of monocytes.

HSF significantly suppresses PHA induced » IL2 production of peripheral blood mononuclear cells in a dose dependent fashion without inhibiting the proliferative response of CTLL-20 target cells to IL2. HSF in the presence of monocytes modulates the function of CD4-positive cells by inhibiting IL2 production. The inhibition of IL2 production by HSF appears to be responsible for the suppression of antibody production. HSF inhibits PWM-induced IL2 production by CD4-positive cells but not by CD8-positive cells and its suppressive activity on Ig production is totally abrogated by preabsorption with CD4-positive cells, but not by CD8-positive cells. HSF suppresses IFN-γ activity produced by mitogen-stimulated peripheral blood monocytes without affecting the generation of lymphokines responsible for B cell growth and differentiation (see also: BCGF and BCDF). HSF does not inhibit the expression of either IL2R (p55) or transferrin receptors by activated T cells in spite of causing the suppression of IL2 production.

Haghighi AZ & Cathcart MK Subtractive antibody to a human immunosuppressive lymphokine affinity isolates a suppressive factor and blocks its function. Immunopharmacology 24: 65-76 (1992); **Murakami M et al** Human hybridoma suppressor factor (HSF) inhibits IL2 production in addition to suppressing immunoglobulin production. Hybridoma 7: 595-608 (1988); **Tomita Y et al** Human hybridoma suppressor factor acts selectively on CD4+ cells. Immunopharmacology 16: 199-205 (1988); **Tomita Y et al** Selective suppression of lymphokine production by human hybridoma suppressor factor (HSF). Immunobiology 181: 64-83 (1990)

HSPG: *heparan sulfate proteoglycan* see: ECM (extracellular matrix).

HSS: *hepatic stimulator substance* This poorly characterized substance of ≈ 12-18 kDa has been found in the regenerating liver tissues of several species including rat, rabbit, dog, pig, humans. The protein is negatively charged and contains disulfide bonds that are not required for its activity. HSS stimulates organ-specifically the growth of hepatocytes, hepatoma cells, and liver epithelial cells. It has also been shown to protect against acute liver failure induced by carbon tetrachloride poisoning in mice. In some cell lines HSS neutralizes the growth-inhibitory activities of » TGF-β. HSS displays no species specificity.

Fleig WE & Hoss G Partial purification of rat hepatic stimulator substance and characterization of its action on hepatoma cells and normal hepatocytes. Hepatology 9: 240-8 (1989); **Gupta S et al** Mitogenic effects of hepatic stimulator substance on cultured nonparenchymal liver epithelial cells. Hepatology 15: 485-91 (1992); **LaBrecque DR** Hepatic stimulator substance. Discovery, characteristics, and mechanism of action. Dig. Dis. Sci. 36: 669-73 (1991); **LaBrecque DR et al** Purification and physical-chemical characterization of hepatic stimulator substance. Hepatology 7: 100-6 (1987); **Mei MH et al** Hepatic stimulator substance protects against acute liver failure induced by carbon tetrachloride poisoning in mice. Hepatology 17: 638-44 (1993); **Yao ZQ et al** Human regenerative stimulator substance: partial purification and biological characterization of hepatic stimulator substance from human fetal liver cells. Hepatology 12: 1144-51 (1990)

hst*:** abbrev. from *h*uman *s*tomach *c*ancer. *hst* is also known as ***HBGF-4 (heparin-binding growth factor 4), ***FGFK*** and ***FGF-4*** (fibroblast growth factor 4). The official gene symbol is ***HSTF1*** (heparin-binding secretory transforming factor 1). The protein encoded by the *hst* gene has a molecular mass of 23 kDa. The human *hst*-1 gene maps to chromosome 11q13. 3 ≈ 35 kb downstream of the » *int*-2 gene. Both genes have the same orientation. In the mouse the two genes are less than 20 kb apart. An identical gene is the KS3 gene (see: K-FGF, Kaposi fibroblast growth factor).

In some human tumors the *hst*-1 and the *int*-2 genes are co-amplified. In tumors of the esophagus ≈ 50 % of the cells in primary tumors and all cells in metastases show this co-amplification. Both proteins show a pronounced sequence homology with » bFGF.

hst is produced and secreted by embryonic carcinoma and embryonic stem cells (see: ES cells) and functions as an » autocrine growth factor that also influences the differentiation of these cells. Their growth can be inhibited by antibodies directed against the *hst* protein. *hst*-transformed cells are very tumorigenic. The immunization of mice with *hst* protein significantly reduces the tumor takes following injection of syngeneic *hst*- or *ras*-transformed cells. The insertional activation of *hst* by MMTV (mouse mammary tumor virus) is one of the factors contributing to mammary tumorigenesis in » transgenic animals carrying the » *Wnt*-1 transgene.

FGF-4 has been implicated recently as one of the molecules that directs outgrowth and patterning of the limb during chick embryonic growth. The effects of FGF-4 on limb proliferation can be modulated by BMP-2 (see: BMP, bone morphogenetic proteins).

Adelaide J et al Chromosomal localization of the *hst* oncogene and its co-amplification with *int*.2 in a human melanoma. O 2: 413-6 (1988); **Brookes S et al** Linkage map of a region of human chromosome band 11q13 amplified in breast and squamous cell tumors. Genes Chromosomes Cancer 4: 290-301 (1992); **Delli-Bovi P et al** Processing, secretion, and biological properties of a novel growth factor of the fibroblast growth factor family with oncogenic potential. MCB 8: 2933-2941 (1988); **Hagemeijer A et al** Localization of the HST/FGFK gene with regard to 11q13 chromosomal breakpoint and fragile site. Genes Chromosome. Cancer 3: 210-4 (1991); **Nguyen C et al** The FGF-related oncogenes *hst* and *int*.2, and the *bcl*.1 locus are contained within one megabase in band q13 of chromosome 11, while *fgf*.5 oncogene maps to 4q21. O 3: 703-8 (1988); **Niswander L et al** FGF-4 replaces the apical ectodermal ridge and directs outgrowth and patterning of the limb. Cell 75: 579-87 (1993); **Reddy EP et al** (eds) The oncogene handbook. Elsevier Science Publishers, Amsterdam 1988; **Sakamoto H et al** Transforming gene from human stomach cancers and a non-cancerous portion of stomach mucosa. PNAS 83: 3997-4001 (1986); **Sugimura T et al** Molecular biology of the *hst*-1 gene. Ciba Found. Symp. 150: 79-89 (1990); **Tahara E** Growth factors and oncogenes in human gastrointestinal carcinomas. JCRCO 116: 121-31 (1990); **Taira M et al** cDNA sequence of human transforming gene *hst* and identification of the coding sequence required for transforming activity. PNAS 84: 2985-9 (1987); **Talarico D et al** The k-fgf/*hst* oncogene induces transformation through an autocrine mechanism that requires extracellular stimulation of the mitogenic pathway. MCB 11: 1138-45 (1991); **Talarico D et al** Protection of mice against tumor growth by immunization with an oncogene-related growth factor. PNAS 87: 4222-5 (1990); **Wada A et al** Two homologous oncogenes, HST1 and INT2, are closely located in human genome. BBRC 157: 828-35 (1988); **Yoshida T et al** Genomic sequence of *hst*, a transforming gene encoding a protein homologous to fibroblast growth factors and the *int*-2-encoded protein. PNAS 84: 7305-9 (1987); **Yoshida T et al** Identification and characterization of fibroblast growth factor-related transforming gene *hst*-1. MiE 198: 124-37 (1991); **Yoshida M et al** Human HST1 (HSTF-1) maps to chromosome band 11q13 and co-amplifies with the INT2 gene in human cancer. PNAS 85: 4861-4 (1988); **Yoshida T et al** Characterization of the *hst*-1 gene and its product. ANY 638: 27-37 (1991)

***hst*-1:** see: *hst*.

***hst*-2:** see: *hst*.

HSTF1: *heparin-binding secretory transforming factor 1* see: *hst*.

HT-2: This murine T helper cell-derived CD35-positive (C3b receptor) CD4- (Lyt-2) and CD8α-(L3T4)-negative cell line of BALB/c origin is absolutely dependent on exogenous IL2 for its growth. It is used to assay human, rat, and murine IL2 (see also: Bioassays). IL2 activity is determined by measuring the incorporation of ^{3}H thymidine into the newly synthesized DNA of the proliferating cells. Cell proliferation can be determined also by employing the » MTT assay. An alternative and entirely different method of detecting IL2 is » Message amplification phenotyping.

Proliferation of HT-2 cells in response to suboptimal doses of IL2 is significantly increased by dibutyryl guanosine cyclic monophosphate and inhibited by dibutyryl adenosine cyclic monophosphate. Complement C3 also has an enhancing effect on the IL2-dependent proliferation of HT-2 cells. High doses of isoniazid also increase the proliferation of HT-2 cells in the presence of suboptimal doses of interleukin 2. HT-2 proliferative responses to recombinant human IL2 are potentiated as much as fourfold by 10^{-10} M all-trans-retinoic acid.

Micellar gangliosides, in particular highly sialylated gangliosides, are potent inhibitors of the IL2-induced proliferation of HT-2 cells. They abolish both DNA and protein synthesis, and depressed cellular expansion, without affecting viability. These effects are reversible for at least 12 hr following ganglioside treatment. Ganglioside micelles and lipid vesicles containing gangliosides are able to bind IL2 and prevent binding of IL2 to high-affinity receptors on the lymphocyte surface. Glycophorin A, the major sialoglycoprotein of the human erythrocyte membrane, inhibits IL2-stimulated proliferation of HT-2 cells and the IL2-dependent cell line » CTLL-2 in a dose-dependent manner.

The proliferation of HT-2 cells engineered to express CD4 or CD8 in response to IL2 is inhibited by anti-CD4 and anti-CD8 antibodies without affecting the IL4 responses. Mutant CD8 lacking the cytoplasmic portion normally interacting with p56*lck* have no inhibitory effect, suggesting that CD4 and CD8 mediate negative regulation of T cell IL2 responses via cytoplasmically associated p56*lck*.

HT-2 cells also responds to » murine (but not human) IL4 and » GM-CSF, making it necessary to use either factor-specific antibodies or other cell lines (see: CT6 cells). IL2 and IL4 induce a synergistic proliferative response in HT-2 cells. IL4 stimulation of HT-2 cells appears to be independent of the action of IL2 since no IL2 mRNA is found in the IL4-stimulated cells by Northern blotting, the IL4 response is not blocked by anti-IL2R antibodies, and is not inhibited by cyclosporin A. IL4 plus PMA treatment of resting cells causes enhanced expression of IL2R and prepares cells to proliferate to IL2 alone.

Picomolar concentrations of TGF-β inhibit S-phase progression stimulated by IL2 or IL4. TGF-β pretreatment decreases the expression of high affinity IL2R and markedly reduces the IL2-stimulated transferrin receptor expression. It also inhibits the early increase in GM-CSF mRNA stimulated by IL2 or IL4. TGF-β also significantly reduces early (1 to 2 h) increases in cellular » *myc* mRNA levels stimulated by IL2 or IL4. HT-2 cells probably also produce » IL10.

Baker PE et al Monoclonal cytolytic T cell lines. JEM 149: 273 (1979); **Bartok I et al** Interaction between C3 and IL2; inhibition of C3b binding to CR1 by IL2. Immunol. Lett. 21: 131-7 (1989); **Brown M et al** IL4/B cell stimulatory factor 1 stimulates T cell growth by an IL2-independent mechanism. JI 141: 504-11 (1988); **Chu JW & Sharom FJ** Effect of micellar and bilayer gangliosides on proliferation of interleukin-2-dependent lymphocytes. CI 132: 319-38 (1991); **Chu JW & Sharom FJ** Glycophorin A interacts with interleukin-2 and inhibits interleukin-2-dependent T lymphocyte proliferation. CI 145: 223-39 (1992); **Fernandez-Botran R et al** Interactions between receptors for interleukin 2 and interleukin 4 on lines of helper T cells (HT-2) and B lymphoma cells (BCL1). JEM 169: 379-91 (1989); **Ho SN et al** Differential bioassay of interleukin 2 and interleukin 4. JIM 98: 99-104 (1987); **Jiang XL et al** Potentiation of IL2-induced T cell proliferation by retinoids. IJI 14: 195-204 (1992); **Karasuyama H et al** Autocrine growth and tumorigenicity of interleukin 2-dependent helper T cells transfected with IL2 gene. JEM 169: 13-25 (1989); **Kreft B et al** Detection of intracellular interleukin-10 by flow cytometry. JIM 156: 125-8 (1992); **Kucharz EJ & Goodwin JS** Dibutyryl guanosine cyclic monophosphate causes proliferation of interleukin-2-dependent T cells in the presence of suboptimal levels of interleukin-2. IJI 11: 687-90 (1989); **Kucharz EJ & Sierakowski SJ** Studies on immunomodulatory properties of isoniazid. III. Effect of isoniazid on proliferation of interleukin-dependent and interleukin-independent cell line. J. Hyg. Epidemiol. Microbiol. Immunol. 34: 305-8 (1990); **Kupper T et al** Autocrine growth of T cells independent of interleukin 2: identification of interleukin 4 (IL 4, BSF-1) as an autocrine growth factor for a cloned antigen-specific helper T cell. JI 138: 4280-7 (1987); **Kupper TS et al** Keratinocyte-derived T cell growth factor (KTGF) is identical to granulocyte-macrophage colony-stimulating factor (GM-CSF). J. Invest. Dermatol. 91: 185-8 (1988); **Kupper TS et al** Growth of an interleukin 2/interleukin 4-dependent T cell line induced by granulocyte-macrophage colony-stimulating factor (GM-CSF). JI 138: 4288-92 (1987); **Lichtman AH et al** B cell stimulatory factor 1 and not interleukin 2 is the autocrine growth factor for some helper T lymphocytes. PNAS 84: 824-7 (1987); **Mosman TR & Fong TAT** Specific assays for cytokine production by T cells JIM 116: 151-8 (1989); **Ruegemer JJ et al** Regulatory effects of transforming growth factor-β on IL2- and IL4-dependent T cell-cycle progression. JI 144: 1767-76 (1990); **Takahashi, K et al** CD4 and CD8 regulate interleukin 2 responses of T cells. PNAS 89: 5557-61 (1992); **Smith CA & Rennick DM** Characterization of a murine lymphokine distinct from interleukin 2 and interleukin 3 (IL3) possessing a T cell growth factor activity and a mast-cell growth factor activity that synergizes with IL3. PNAS 83: 1857-61 (1986); **Woods A et al** Granulocyte-macrophage colony stimulating factor produced by cloned L3T4a$^+$, class II-restricted T cells induces HT-2 cells to proliferate. JI 138: 4293-7 (1987)

HT29-IGFBP: *HT29 insulin-like growth factor binding protein* see: IGF-BP (insulin-like growth factor binding protein).

HTB 9: see: 5637 cell line.

HTCA: *human tumor clonogenic (cell) assay* This assay involves soft agar colony formation by human tumor cells (see: Colony formation assay). It is utilized generally to measure the sensitivity of human tumors to anticancer drugs. Results observed are taken to reflect the therapeutic efficacy of a drug. This test is also useful in assaying the cytotoxicity of lymphocytes (and factors secreted by them) against tumor cells. The test has also been used as an evaluation method for drug targeting.
Ali-Osman F & Beltz PA Optimization and characterization of the capillary human tumor clonogenic cell assay. CR 48: 715-24 (1988); **Berdel WE et al** Various human hematopoietic growth factors (interleukin 3, GM-CSF, G-CSF) stimulate clonal growth of nonhematopoietic tumor cells. Blood 73: 80-3 (1989); **Fujita J et al** Detection of cytotoxicity of freshly obtained lymphocytes and lymphocytes activated with recombinant interleukin 2 (IL2) against lung cancer cell lines by human tumor clonogenic assay (HTCA). Eur. J. Cancer Clin. Oncol. 22: 445-50 (1986); **Hanauske AR et al** The human tumor cloning assay in cancer research and therapy. a review with clinical correlations. Curr. Probl. Cancer 9: 1-66 (1985); **Hida T et al** Chemosensitivity and radiosensitivity of small cell lung cancer cell lines studied by a newly developed 3-(4,5-Dimethylthiazol-2-yl)-2,5-diphenyltetrazolium-bromide (MTT) hybrid assay. CR 49: 4785-90 (1989); **Matsuyama H et al** Direct and indirect effects of human interferon α on renal cell carcinoma: a new *in vitro* assay system for evaluating cytokine-mediated antitumor effects. Cancer Immunol. Immunother. 37: 84-8 (1993); **Nomori H et al** Detection of NK activity and antibody-dependent cellular cytotoxicity of lymphocytes by human tumor clonogenic assay – its correlation with the 51Cr release assay. IJC 35: 449-55 (1985); **Okadome M et al** Potential of human lymph node cells for antitumor activity mediated by interferon γ. Cancer 68: 2378-83 (1991); **Salmon SE** Human tumor clonogenic assays: growth conditions and applications. CGC 19: 21-8 (1986); **Tueni EA** Human tumor clonogenic assay: what is new? Eur. J. Cancer Clin. Oncol. 25: 1031-3 (1989)

hu: (also: h) in abbreviations, such as in h-GM-CSF and HuIFN-α abbrev. for human.

HUAF: *human uterine angiogenesis factor* HUAF is an extract derived from decidua tissues. The major component responsible for its activity as » Angiogenesis factor *in vivo* and *in vitro* is due to a mixture of basic and acidic fibroblast growth factors (see: aFGF, bFGF).
HUAF and also a number of embryonic tissues and human and bovine pineal glands contain a poorly characterized non-protein of low molecular mass, designated ***ESAF*** (endothelial cell stimulating angiogenesis factor) that appears to be a specific and potent mitogen for endothelial cells in microvessels. ESAF is released by chondrocytes during calcification and mediates cartilage breakdown. Raised levels of ESAF have been found in actively growing human intracranial tumors. ESAF has also been found in diseased human vitreous humor. ESAF activates latent matrix metalloproteinases by dissociating the complexes formed with tissue inhibitors of metalloproteinases (see: TIMP). ESAF and FGF have been shown to synergise *in vitro* and *in vivo*. ESAF has also been shown to be responsible for some of the activities described as » TAF (tumor angiogenesis factor).
Brown RA & McFarland CD Regulation of growth plate cartilage degradation *in vitro*: effects of calcification and a low molecular weight angiogenic factor (ESAF). Bone Miner. 17: 49-57 (1992); **McFarland CD et al** Production of endothelial cell stimulating angiogenesis factor (ESAF) by chondrocytes during *in vitro* cartilage calcification. Bone Miner. 11: 319-33 (1990); **McLaughlin B et al** Activation of the matrix metalloproteinase inhibitor complex by a low molecular weight angiogenic factor. BBA 1073: 295-8 (1991); **Odedra R & Weiss JB** A synergistic effect on microvessel endothelial cell proliferation between fibroblast growth factor and ESAF. BBRC 143: 947-53 (1987); **Taylor CM et al** Bovine and human pineal glands contain substantial quantities of endothelial cell stimulating angio-

genesis factor. J. Neural Transm. 71: 79-84 (1988); **Taylor CM et al** Endothelial cell-stimulating angiogenesis factor in vitreous from extraretinal neovascularizations. Invest. Ophthalmol. Vis. Sci. 30: 2174-8 (1989); **Taylor CM et al** Raised levels of latent collagenase activating angiogenesis factor (ESAF) are present in actively growing human intracranial tumors. Br. J. Cancer 64: 164-8 (1991); **Taylor CM et al** Concentrations of endothelial cell stimulating angiogenesis factor, a major component of human uterine angiogenesis factor, in human and bovine embryonic tissues and decidua. J. Reprod. Fert. 94: 445-9 (1992)

HuHGF: *hepatoma-derived growth factor* This poorly characterized factor of ≈ 64 kDa is secreted by a human hepatoma cell line established from hepatoma tissue of an adult male. The cells grow autonomously in a serum-free chemically defined medium. It stimulates DNA synthesis of Swiss mouse » 3T3 cells and of hepatoma cells producing the factor.
Nakamura H et al Partial purification and characterization of human hepatoma-derived growth factor. Clin. Chim. Acta 183: 273-84 (1989)

Human angiogenic factor: see: HAF.

Human collagenase inhibitor: see: HCI.

Human fibroblast-derived growth inhibitor: abbrev. hFDGI. See: FDGI.

Human glioma-derived T cell suppressor factor: see: G-TsF.

Human inhibitor 30 kDa: abbrev. HI-30. see: ITI (inter-α-trypsin inhibitor).

Human interleukin for Da cells: see: HILDA.

Human milk growth factor: see: HMGF.

Human pancreatic secretory trypsin inhibitor: abbrev. HPSTI. See: ECGF-2a (endothelial cell growth factor 2b).

Human pituitary growth factor: see: HPGF.

Human placenta-purified factor: This factor, isolated from placenta, is identical with » bFGF.
Story MT Cultured human foreskin fibroblasts produce a factor that stimulates their growth with properties similar to basic fibroblast growth factor. In Vitro Cell Dev. Biol. 25: 402-8 (1989)

Human seminal plasma factor: see: SP factor.

Humoral stimulatory activity: see: HSA.

Human tumor clonogenic (cell) assay: see: HTCA.

Human uterine angiogenesis factor: see: HUAF.

Hybridoma growth factor: see: HGF.

Hybridoma/plasmacytoma growth factor: see: HPGF.

Hybridoma suppressor factor: see: HSF.

Hydroperoxycyclophosphamide: see: 4-HC (Hydroperoxycyclophosphamide).

Hyper-IgM immunodeficiency syndrome: see: TRAP (TNF-related activation protein).

Hypothalamic growth factor: see: HGF.

Hypothalamic-pituitary-adrenocortical axis: abbrev. HPA. See: Neuroimmune network.

Hypothalamus-derived endothelial cell growth factor: see: ECGF. See also: HGF (hypothalamus growth factor).

Hypothalamus-derived growth factor A: abbrev. HDGF-A. See: HGF (hypothalamus growth factor).

Hypothalamus-derived growth factor C: abbrev. HDGF-C. See: HGF (hypothalamus growth factor).

HYL: see addendum.

I

I-309: I-309 is the name of a cDNA clone encoding a 15-16 kDa glycoprotein of 72 amino acids. This protein is secreted by activated T lymphocytes (see also: Cell activation). The human I-309 gene, designated ***SCYA1***, small inducible cytokine A1) maps to chromosome 17. I-309 belongs to the » chemokine family of cytokines. The murine homologue of I-309 is »***TCA-3***.
At concentrations of ≈ 100 nM I-309 is chemotactic for human monocytes (see also: Chemotaxis) and also activates them (see also: cell activation). It leads to a transient increase in cytoplasmic calcium levels (see also: Calcium ionophore) in monocytes. I-309 has no effects on neutrophils.
Miller MD & Krangel MS The human cytokine I-309 is a monocyte chemoattractant. PNAS 89: 2950-4 (1992); **Miller MD et al** Sequence and chromosomal location of the I-309 gene. Relationship to genes encoding a family of inflammatory cytokines. JI 145: 2737-44 (1990); **Miller MD et al** A novel polypeptide secreted by activated human T lymphocytes. JI 143: 2907-16 (1989); **Wilson SD et al** Expression and characterization of TCA3: a murine inflammatory protein. JI 145: 2745-50 (1990)

IAP: *immunosuppressive acidic protein* IAP is an α 1-acid glycoprotein that is produced mainly by macrophages. It exhibits various types of immunosuppressive activities. IAP lacks disease specificity but it has been shown to increase in the serum of patients with inflammatory neurological diseases and has been described as a tumor associated marker in some solid tumors and hematological diseases. Serum IAP levels can serve as a tumor marker in some tumor patients, with levels decreasing after surgery and increasing with recurrence and poor prognosis. IAP levels have been used for evaluating clinical course, therapeutic effectiveness, and prognosis of patients with various malignant tumors.
Castelli M et al Immunosuppressive acidic protein (IAP) and CA 125 assays in detection of human ovarian cancer: preliminary results. Int. J. Biol. Markers 2: 187-90 (1987); **Gluck S & Koster W** Immunosuppressive acidic protein in malignant diseases. Clinical relevance? Cancer 67: 610-2 (1991); **Igarashi T et al** Serum immunosuppressive acidic protein as a tumor marker for renal cell carcinoma. Eur. Urol. 19: 332-5 (1991); **Seki H et al** Intrathecal synthesis of immunosuppressive acidic protein (IAP) in patients with multiple sclerosis and other inflammatory neurological diseases. J. Neurol. Sci. 85: 259-66 (1988)

IAP: *intracisternal A particles* IAPs are retrovirus-like repetitive DNA elements, also known as retrotransposons, that are reiterated 2000-fold and widely dispersed in the mouse genome. They possess a retrovirus-like LTR region (long-terminal repeat) that functions as a promoter. Large variations in sequence and length have been observed among individually cloned IAP LTRs. They can be used to distinguish individual elements.
Many of these elements have a length of ≈ 7 kb but frequently one also observes truncated variants generated by deletions. These shorter IAPs resemble defective retroviruses. IAPs encode the genes *gag* and *pol* that are also found in retroviral genomes. These proteins can aggregate into virus-like particles. Truncated forms of the gag protein encoded by an IAP are secreted by various murine T cell hybridomas. They function as IgE binding proteins (see also: IgE-BF).
IAPs are found in many copies in the genomes of rodents. They are constitutively expressed at high levels in many mouse tumors (characteristically in plasmacytomas) and at lower levels in the thymus and activated splenic B cells. Expression in plasmacytomas appears to be selective for LTRs carrying a set of regulatory sequence variants that differ from those expressed in activated lymphocytes.
The integration of an IAP retrotransposon in the vicinity of a gene can lead to the transcriptional activation of this gene. This is exemplified by the constitute synthesis of » IL3 by » WEHI cells which contains an IL3-IAP element of ≈ 5 kb, and by the constitutive synthesis of » GM-CSF by some WEHI subclones and by » FDC-P1 cells. The insertion of an IAP element with concomitant constitutive expression of a cytokine has also been observed in » IL5-independent cell lines.
The activation of these and possibly also of other growth factor genes by the insertion of an IAP element is one of the mechanisms leading to growth factor independence in cell lines normally requiring a growth factor for survival and proliferation (see also: Autocrine).
Variants of the receptor for » IL6 have been descri-

bed in which the cytoplasmic domain of the receptor is replaced by IAP portions. These altered receptors are functionally active and are expressed at levels higher than the normal receptors. It has been suggested that these altered receptors may play a role in the establishment of some plasmacytomas that use IL6 as an » autocrine growth factor. The constitutive expression of IL6 in certain murine plasmacytomas has also been shown to be the result of IAP transposition.

Aota S et al Nucleotide sequence and molecular evolution of mouse retrovirus-like IAP elements. Gene 56: 1-12 (1987); **Ben-David L et al** A deletion and a rearrangement distinguish between the intracisternal A particle of Hox 2.4 and that of interleukin 3 in the same leukaemic cells. Virology 182: 382-7 (1991); **Blankenstein T et al** DNA rearrangement and constitutive expression of the interleukin 6 gene in a mouse plasmacytoma. JEM 171: 965-70 (1990); **Chang YA et al** Identification of a novel murine IAP-promoted placenta-expressed gene. NAR 19: 3667-72 (1991); **Christy R & Huang RC** Functional analysis of the long terminal repeats of intracisternal A particle genes: sequences within the U3 region determine both the efficiency and direction of promoter activity. MCB 8: 1093-1102 (1988); **Christy R et al** Nucleotide sequence of murine intracisternal A particle gene LTRs have extreme variability within the R region. NAR 13: 289-302 (1985); **Duhrsen U et al** *In vivo* transformation of factor-dependent hemopoietic cells: role of intracisternal A particle transposition for growth factor gene activation. EJ 9: 1087-96 (1990); **Galien R et al** *ras* oncogene activates the intracisternal A particle long terminal repeat promoter through a cAMP response element. O 6: 849-55 (1991); **Givol D** Activation of oncogenes by transposable elements. Biochem. Soc. Symp. 51: 183-96 (1986); **Gupta S et al** Detection of a human intracisternal retroviral particle associated with $CD4^+$ T cell deficiency. PNAS 89: 7831-5 (1992); **Hapel AJ & Young IG** Abnormal expression of interleukin 2 and leukemia. ACR 7: 661-7 (1987); **Heberlein C et al** Retrotransposons as mutagens in the induction of growth autonomy in hematopoietic cells. O 5: 1799-807 (1990); **Hirsch HH et al** Suppressible and nonsuppressible autocrine mast cell tumors are distinguished by insertion of an endogenous retroviral element (IAP) into the interleukin 3 gene. JEM 178: 403-11 (1993); **Kuff E & Lueders K** The intracisternal A particle gene family: structure and functional aspects. ACR 51: 183-276 (1988); **Kuff EL et al** cDNA clones encoding murine IgE-binding factors represent multiple structural variants of intracisternal A particle genes. PNAS 83: 6583-7 (1986); **Leslie KB et al** Intracisternal A-type particle-mediated activations of cytokine genes in a murine myelomonocytic leukemia: generation of functional cytokine mRNA by retroviral splicing events. MCB 11: 5562-70 (1991); **Lueders K & Kuff E** Transposition of intracisternal A particle genes. Prog. Nucl. Acid Res. Mol. Biol. 36: 173-86 (1989); **Lueders K et al** Selective expression of intracisternal A particle genes in established mouse plasmacytomas. MCB 13: 7439-46 (1993); **Mietz JA et al** Nucleotide sequence of a complete mouse intracisternal A particle genome: relationship to known aspects of particle assembly and function. J. Virol. 61: 3020-9 (1987); **Mietz JA et al** Selective activation of a discrete family of endogenous proviral elements in normal BALB/c lymphocytes, MCB 12: 220-8 (1992); **Ono M** Molecular biology of type A endogenous retrovirus. Kitasato Arch. Exp. Med. 63: 77-90 (1990); **Qin Z et al** The interleukin-6 gene locus seems to be a preferred target site for retrotransposon integration. Immunogenetics 33: 260-6 (1991); **Sugita T et al** Functional murine interleukin 6 receptor with the intracisternal A particle gene product at its cytoplasmic domain. Its possible role in plasmacytomagenesis. JEM 171: 2001-9 (1990); **Toh H et al** Retroviral gag and DNA endonuclease coding sequences in IgE-binding factor gene. N 318: 388-9 (1985); **Tohyama K et al** Establishment of an interleukin 5-dependent subclone from an interleukin 3-dependent murine hemopoietic progenitor cell line, LyD9, and its malignant transformation by autocrine secretion of interleukin 5. EJ 9: 1823-30 (1990); **Ymer S et al** Nucleotide sequence of the intracisternal A particle genome inserted 5´ to the interleukin 3 gene of the leukemia cell line WEHI-3B. NAR 14: 5901-18 (1986)

IBF: *immunoglobulin binding factors* IBFs are generated either by cleavage at the cell membrane or by splicing of a Fc receptor transmembrane exon (see also: CD16, CD32, CD64). In addition there are other proteins with IBF activity that are probably different from Fc receptors. One IBF found in human seminal plasma shows complete identity with ***prostatic secretory protein***, ***β-microseminoprotein*** and β-inhibin (see: Inhibins).
IBFs function as cytokines able to suppress the immunoglobulin production by normal and transformed B cells. IBFs appear to act as both immunoregulatory and growth regulatory factors and strongly inhibit the development of hybridoma B cell colonies

Fridman WH Fc receptors and immunoglobulin binding factors. FJ 5: 2684-90 (1991); **Fridman WH** Factors controlling immunoglobulin production and B cell proliferation. Bull. Cancer Paris. 1991; 78: 195-201 (1991); **Liang ZG et al** Immunoglobulin binding factor of seminal plasma: a secretory product of human prostate. Arch. Androl. 28: 159-64 (1992); **Liang ZG et al** Structural identity of immunoglobulin binding factor and prostatic secretory protein of human seminal plasma. BBRC 180: 356-9 (1991); **Sandilands GP et al** Immunohistochemical localization of a plasma protein (glycoprotein 60) which inhibits complement-mediated prevention of immune precipitation. Immunology 70: 303-8 (1990); **Teillaud JL et al** Regulation of hybridoma B cell proliferation by immunoglobulin-binding factor. Cancer Detect. Prev. 1988; 12: 115-23 (1988)

IC: sometimes used as a prefix in combination with growth factor acronyms to denote “intracellular” such as in IC-BCGF (intracellular B cell growth factor).

IC-2: A murine mast cell progenitor cell line. Growth and survival depend on » IL3 or » GM-CSF and the cells have been used to assay these factors. The cells are also stimulated by » Epo. The cells also respond to » IL4, although IL4-induced proliferation is transient. IL4 synergises with limiting concentrations of IL3 and GM-CSF and reduces proliferation induced by saturating concentrations of IL3 and GM-CSF. IL3-dependent growth is retarded by » TGF-β. The cells appear to express a low affinity receptor for » IL2 but do not respond to IL2.

IC2 cells have been found to survive and proliferate over a 48 h period in the absence of IL3 when incubated with micromolar concentrations of sodium orthovanadate. Vanadate also stimulated synthesis of nucleotides and protein in IC2 cells. Vanadate potentiates the effect of submaximal doses of IL3 but does not appear to act synergistically with IL3 to give a maximal proliferative response of the cells.

Upon introduction of a functional » EGF receptor EGF stimulates IC2/EGF cells for a short term growth response. In the presence of IL3 and EGF these cells differentiate into more mature mast cells and this differentiation is reversible when EGF is removed, thus providing a cellular system for *in vitro* study of mast cell differentiation.

Koyasu S et al Expression of interleukin 2 receptors on interleukin 3-dependent cell lines. JI 136: 984-7 (1986); **Koyasu S et al** Growth regulation of multi-factor-dependent myeloid cell lines: IL4, TGF-β and pertussis toxin modulate IL3- or GM-CSF-induced growth by controlling cell cycle length. Cell. Struct. Funct. 14: 459-71 (1989); **Ohno K et al** Production of granulocyte/macrophage colony-stimulating factor by cultured astrocytes. BBRC 169: 719-24 (1990); **Sakamoto T et al** Production of granulocyte-macrophage colony-stimulating factor by adult murine parenchymal liver cells (hepatocytes). Reg. Immunol. 3: 260-7 (1990-91) **Tojo A et al** Vanadate can replace interleukin 3 for transient growth of factor-dependent cells. ECR 171: 16-23 (1987); **Tsao CJ et al** Expression of the functional erythropoietin receptors on interleukin 3-dependent murine cell lines. JI 140: 89-93 (1988); **Wang HM et al** EGF induces differentiation of an IL3-dependent cell line expressing the EGF receptor. EJ 8: 3677-84 (1989); **Yonehara S et al** Identification of a cell surface 105 kd protein (Aic-2 antigen) which binds interleukin-3. Int. Immunol. 1990; 2: 143-50 (1990)

ICE: see: IL1β-Convertase.

ICIL1ra: *intracellular IL1ra* see: IL1ra (IL1 receptor antagonist).

ICSBP: *interferon consensus sequence binding protein* see: IRS (interferon response sequence).

ICSH: *interstitial cell-stimulating hormone* see: Luteinizing hormone.

IDS: *inhibitor of DNA synthesis* see: Suppressor factors.

IDF-45: *inhibitory diffusible factor 45* This monomeric factor of 45 kDa is isolated from the culture supernatants of confluent murine » 3T3 cells. The known aminoterminal sequences demonstrates that this factor is related to rat and murine IGFBP3 (insulin-like growth factor binding protein 3) and it has been suggested that IDF-45 be called murine IGFBP-3.

IDF-45 binds insulin-like growth factors (see: IGF and IGF-BP). It reversibly inhibits the growth of chicken fibroblasts and some other cell types. IDF-45 also inhibits the » bFGF-induced DNA synthesis in murine embryonic fibroblasts and chicken fibroblasts. It has little effect on » PDGF-, » EGF-, and insulin-induced DNA synthesis. The growth-inhibitory effects of IGF-45 are abolished if IGF-45 is complexed with IGF-I. Another IGF binding protein, IGF-BP1, shows a similar behavior (see: IGF-BP).

Blat C et al Isolation and aminoterminal sequence of a novel cellular growth inhibitor (inhibitory diffusible factor 45) secreted by 3T3 fibroblasts. JBC 264: 6021-4 (1989); **Blat C et al** Inhibitory diffusible factor 45 bifunctional activity. As a cell growth inhibitor and as an insulin-like growth factor I-binding protein. JBC 264: 12449-54 (1989); **Blat C et al** Purification from transformed mouse fibroblast of a cell growth inhibitor which is an IGF-binding protein. GF 6: 65-75 (1992); **Delbe J et al** Differences in inhibition by IDF45 (an inhibitory diffusible factor) of early RNA synthesis stimulation induced by pp60 v-*src* and various mitogens. JCP 142: 359-64 (1990); **Liu L et al** IGFBP-1, an insulin like growth factor binding protein, is a cell growth inhibitor. BBRC 174: 673-9 (1991); **Villaudy J et al** An IGF binding protein is an inhibitor of FGF stimulation. JCP 149: 492-6 (1991)

IE proteins: *Immediate early* General term for proteins usually expressed within minutes after binding of a growth factor to its receptor (see also: Gene expression). IE proteins are competence factors the expression of which coincides with the entry of resting cells into the » cell cycle (see also: Cell activation).

IF: A suffix sometimes used in acronyms of growth factor, such as in NAP-IF, to denote "inducing factor" or "inhibiting factor" or "inhibitory factor".

I factor: This factor is isolated from cell lysates and culture supernatants of the murine myeloid leukemia cell line » M1 (16 K I factor). It inhibits the differentiation of these cells into monocytes and macrophages. This factor is identical with nucleoside diphosphate kinase encoded by the » nm23 gene.

Hozumi M et al Protein factors that regulate the growth and differentiation of mouse myeloid leukemia cells. Ciba Found. Symp. 148: 25-33; (1990); **Okabe-Kado J et al** Purification of a factor inhibiting differentiation from conditioned medium of nondifferentiating mouse myeloid leukemia cells. JBC 263: 10994-9 (1988); **Okabe-Kado J et al** Identity of a differentiation inhibiting factor for mouse myeloid leukemia cells with NM23/nucleoside diphosphate kinase. BBRC 182: 987-94 (1992); **Okabe-Kado J** Factors inhibiting differentiation of myeloid leukemia cells. Crit. Rev. Oncog. 3: 293-319 (1992)

IFI-78k-Gen: *interferon-inducible 78 kDa protein* see: Mx protein.

IFN: abbrev. for interferons. By definition interferons are proteins that, at least in homologous cells, elicit a virus-unspecific antiviral activity. This activity requires new synthesis of RNA and proteins and is not observed in the presence of suitable RNA and protein synthesis inhibitors.
Apart from their antiviral activities interferons also possess antiproliferative and immunomodulating activities and influence the metabolism, growth and differentiation of cells in many different ways.
The three main human interferons are known as » IFN-α, » IFN-β and » IFN-γ. IFN-α and IFN-β are also called ***type 1 interferons***. Bovine » TP-1 (trophoblast protein 1) is also a type 1 interferon. IFN-γ has been designated ***type 2 interferon***. Another protein, called » IFNβ_2 is not an interferon but is identical with » IL6.
Some older names of interferons such as ***leukocyte interferon*** (IFN-α), ***fibroblast interferon*** (IFN-β) and ***immune interferon*** (IFN-γ) are still in use. These names are derived from the main producers and from the typical bioactivity and still reflect the concept of a typical producer cell, which has been superseded now because it is known that a plethora of different cell types are capable of producing interferons.
Interferons are a heterogeneous group of proteins with some similar biological activities that are distinguished from each other by many different physical and immunochemical properties. They are also encoded by different structural genes. Most interferons are multifunctional proteins with bioactivities that are strictly species-specific (see: TP-1, trophoblast protein 1, for an exception). These substances are synthesized following the activation of the immune system. The human interferons » IFN-β and » IFN-γ are encoded by two different single genes while human » IFN-α constitutes a family of at least 23 different genes.
Although there are some indications of constitutive interferon synthesis by some cell types interferons are generally inducible proteins. Their synthesis is induced by many different physiological and non-physiological inducers including, among others, nucleic acids, synthetic oligonucleotides (polyIC), pyran copolymers, lipopolysaccharides, and a number of low-molecular weight compounds.
The production of interferons in an organism may be localized or, depending on the strength of the inducing stimulus, systemic, eventually culminating in an interferonaemia. The detection of interferons in the circulation is usually an indication of some pathological state.
Interferons are mainly known for their antiviral activities against a wide spectrum of viruses. Interferons are synthesized, for example, by virus-infected cells and protect other, non-infected but virus-sensitive cells against infection for some time. In addition interferons are also known to have protective effects against some non-viral pathogens.
The antiviral activities of interferons are also used to detect these proteins in a » bioassay. Interferons possess very high specific activities. It is on the order of 10^9 units/mg of protein for IFN-α and IFN-β. One unit is defined as the amount of interferon that reduced virus multiplication by 50 % under standardized conditions.
Interferons are also potent immunomodulators. They can promote or inhibit the synthesis of antibodies by activated B cells and also activate macrophages, natural killer cells, and T cells. Interferons mainly influence early unspecific immune response processes mediated predominantly by monocytes/macrophages. Among other things interferons increase antigen and receptor expression in effector cells, induce the expression of new genes, inhibit the expression of some genes, and also prolong phases of the » cell cycle. Interferons also influence differentiation and developmental processes which is exemplified by their effects on the maturation of immature muscle cells, the induction of globin genes, the methylation of tRNA, and the expression of carcinoembryonic antigen on tumor cells.
Interferons also possess direct antiproliferative activities and are cytostatic or cytotoxic for a number of different tumor cell types. These activities are partly due to complex interactions with other growth factors and their receptors the expression of which may be stimulated or inhibited by interferons. Many growth factors are also capable of inducing the synthesis of interferons.
Hormone-like activities of interferons are observed in cells of the central nervous and the neuroendocrine system. Interferons modulate central

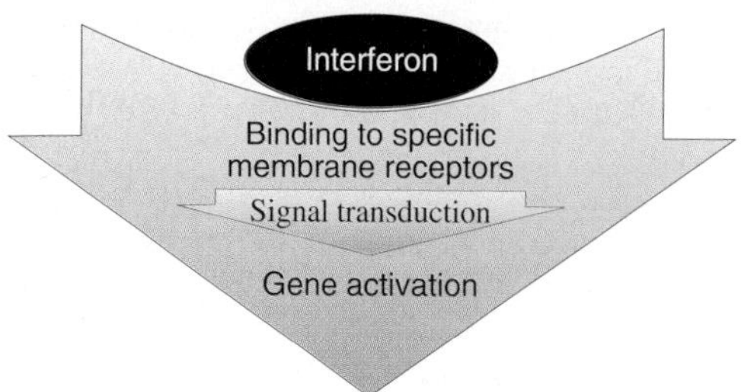

Antiviral activity
Inhibition of viral DNA replication

Antiproliferative activity
Alterations of cell membranes
Alterations of cytoskeleton
Stimulation of cell differentiation
Modulation of growth factor expression
Inhibition/induction of oncogene expression
Reversion of malignant cell phenotypes

Immunomodulatory activity
Induction of cytokine expression
Activation of macrophages
Activation of lymphocytes
Upregulation of HLA class I and II expression
Modulation of expression of tumor-associated antigens

opioid functions and can induce alterations in sleep and behavioral pattern s at higher doses administered systemically. IFN-α significantly increases the firing rate of spontaneous active cells in the hippocampus and the somatosensoric cortex.

IFN-α and IFN-β are thought to bind to the same receptor on the cell surface of target cells. This receptor differs from that of IFN-γ. Binding of an interferon to its receptor induces the expression of a number of new proteins. Some genes are specifically induced either by » IFN-α and » IFN-β or by » IFN-γ. Some of the newly formed proteins are nucleases, synthases and protein kinases that influence protein biosynthesis and are, at least in part, also responsible for some biological effects, for example, antiviral activities.

The induced protein kinase inactivates the eukaryotic initiation factor eIF-2 and thus inhibits the synthesis of new viral proteins. The expression of 2´,5´-polyadenylate synthetase which is also induced by interferons, leads to products that themselves activate endonucleases that in turn degrade viral mRNAs. Interferons do not show antiviral activities in undifferentiated embryonic carcinoma cells (EC cells) that do not express certain endonucleases.

Overview of IFN-induced proteins mediating the antiviral state.

Protein	Viruses and Mechanism of inhibition	Reference
2´,5´-oligoadenylate synthase/RNAase L	Picorna virus, HIV, Vaccinia virus RNA breakdown	Adv. Virus Res. 38: 147 (1990) Adv. Virus Res. 42: 57 (1993)
dsRNA-dependent protein kinase	Adenovirus, VSV, Reovirus VSV, Picorna virus inhibition of viral RNA translation	Adv. Virus Res. 38: 147 (1990) Adv. Virus Res. 42: 57 (1993)
Mx proteins	Influenza virus, Rhabdovirus Thogoto virus, VSV inhibition of primary transcription and/or post-transcriptional processes	Trends Cell Biol. 3: 268 (1993)
Soluble LDL receptor	VSV interference with virus assembly or budding	Science 262: 250 (1993)
9-27	HIV binding to viral RNAs	Science 259: 1314 (1993)
Nitric oxide synthase	Ectromelia virus, Vaccinia virus HSV-1 NO-mediated inhibition of rTNP reductase (?) deamination of DNA	Science 261: 1445 (1993)

Of major clinical interest are the antiviral and growth-inhibitory activities of interferons while their use as immunoregulators is still in its infancy. All interferons show a similar spectrum of side effects with adverse effects observable more or less in all organs. Some of the side effects may actually be dose-limiting although all symptoms appear to be readily reversible if treatment is discontinued.

Borden EC Interferons: Pleiotropic cellular modulators. Clin. Immunol. Immunopathol. 62: S18-S24 (1992); **Byrne GI & Turco J** Interferon and non-viral pathogens. Marcel Dekker, New York, 1988; **DeMaeyer E et al** Immuno-modulating properties of interferons. Phil. Trans. R. Soc. London. 299: 77-90 (1982); **DeMaeyer E & DeMaeyer-Guignard J** Interferons and other regulatory cytokines. J. Wiley, New York, 1988; **Fent K & Zbinden G** Toxicity of interferon and interleukin. TIPS 8: 100-5 (1987); **Garbe C & Krasagakis K** Effects of interferons and cytokines on melanoma cells. J. Invest. Dermatol. 100: 239s-44s (1993); **Kardamakis D** Interferons in the treatment of malignancies. *In vivo* 5: 589-98 (1991); **Kerr IM & Stark GR** The antiviral effects of the interferons and their inhibition. J. Interferon Res. 12: 237-40 (1992); **Murray HW et al** The interferons, macrophage activation, and host defense against nonviral pathogens. J. Interferon Res. 12: 319-22 (1992); **Pestka S et al** Interferons and their actions. ARB 56: 727-77 (1987); **Rubinstein M** Multiple interferon subtypes: The phenomenon and its relevance. J. Interferon Res. 11: (Suppl. Jan) 3-9 (1991); **Staeheli P** Interferon-induced proteins and the antiviral state. Adv. Virus Res. 38: 147-200 (1990); **Tanaka N & Taniguchi T** Cytokine gene regulation: regulatory cis elements and DNA binding factors involved in the interferon system. AI 52: 263-81 (1992); **Williams BRG** Transcriptional regulation of interferon-stimulated genes. EJB 200: 1-11 (1991)

IFN-α: interferon-alpha, abbrev. also aIFN.

■ **Alternative Names:** B cell interferon; buffy coat interferon; foreign cell-induced interferon; Leukocyte interferon (Le-IFN); lymphoblast interferon (LyIFN-α); lymphoblastoid interferon (LyIFN-α), Namalwa interferon, pH2-stable interferon; Type 1 Interferon. Some of the activity ascribed to » EP (endogenous pyrogen) can also be attributed to IFN-α. See also: individual entries for further information.

Preparations of leukocyte and lymphoblastoid interferon are mixtures of relatively undefined composition of various IFN-α subtypes. LeIFN-α may contain further factors that influence the proliferation and differentiation of cells (see also: Cytokines) while LyIFN-α may contain variable amounts of » IFN-β. For a synthetic IFN-α see also: Consensus interferon.

■ **Sources:** α interferons are produced by monocytes/macrophages, lymphoblastoid cells, fibroblasts, and a number of different cell types following induction by viruses, nucleic acids, glucocorticoid hormones, and low-molecular weight substances (n-butyrate, 5-bromodeoxy uridine).

■ **Protein Characteristics:** At least 23 different variants of α interferons are known. The individual proteins have molecular masses between 19-26 kDa and consist of proteins with lengths of 156-166 and 172 amino acids.

All α interferon subtypes possess a common conserved sequence region between amino acid positions 115-151 while the amino-terminal ends are variable. Many IFN-α subtypes differ in their sequences at only one or two positions. Naturally occurring variants also include proteins truncated by 10 amino acids at the carboxy-terminal end.

Disulfide bonds are formed between cysteines at positions 1/98 and 29/138. The disulfide bond 29/138 is essential for biological activity while the 1/98 bond can be reduces without affecting biological activity. All α interferons contain a potential glycosylation site but most subtypes are not glycosylated. In contrast to » IFN-γ IFN-α proteins are stable at pH2.

■ **Gene Structure:** There are at least 23 different IFN-α genes. They have a length of 1-2 kb and are clustered on human chromosome 9p22. It is not known whether all these genes are actually expressed following stimulation of the cells. In some cell systems expression of some subtypes (α1, α2, α4) is stronger than those of others. IFN-α genes do not contain intron sequences found in many other eukaryotic genes (see also: Gene expression).

Based upon the structures two types of IFN-α genes, designated class I and II, are distinguished. They encode proteins of 156-166 and172 amino acids, respectively.

Deletions covering 9p22 are frequently observed in cells of lymphoblastoid leukemias. It is not known to date whether this is of significance with respect to interferon expression.

■ **Receptor Structure, Gene(s), Expression:** The gene encoding the IFN-α receptor maps to human chromosome 21q22.1. IFN-α and » IFN-β are thought to bind to the same IFN binding subunit which is expressed in 100-5000 copies in IFN-α-sensitive and -resistant cells and is associated with other as yet unidentified proteins. The interferon » IFN-ω (Omega interferon) also binds to the IFN-α/IFN-β receptor. Another receptor expressed on B lymphocytes is identical with » CD21 (see also: CD antigens). This receptor also binds Epstein-Barr virus through its gp350/220 coat protein.

Signal transduction mechanisms elicited after binding of IFN-α to its receptors involves tyrosine phosphorylation of various non-receptor protein tyrosine kinases belonging to the » Janus kinases. Soluble forms of the IFN-α receptor, corresponding to truncated forms of the extracellular domain of the cell surface IFN-α receptor, have been found in human serum and in normal human urine.

■ **BIOLOGICAL ACTIVITIES:** All known subtypes of IFN-α show the same antiviral antiparasitic, antiproliferative activities in suitable bioassays although they may differ in relative activities.

Human IFN-α is also a potent antiviral substance in murine, porcine, and bovine cell systems. Human IFN-α is less active in rodent cells. Site-directed mutagenesis techniques have been used to create some variants of certain subtypes (α2) that display ≈ 100-fold enhanced antiviral activities in mouse cells.

IFN-α inhibits the expression of a number of cytokines in hematopoietic progenitor cells (see: Hematopoiesis) that in turn induce a state of competence in these cells allowing them to pass from the G_0 into the S phase of the » cell cycle.

The growth of some tumor cell types *in vitro* is inhibited by IFN-α which may also stimulate the synthesis of tumor-associated cell surface antigens. In renal carcinomas IFN-α reduces the expression of receptors for » EGF. IFN-α also inhibits the growth of fibroblasts and monocytes *in vitro*. IFN-α also inhibits the proliferation of B cell *in vitro* and blocks the synthesis of antibodies. IFN-α also selectively blocks the expression of some mitochondrial genes.

IFN-α specifically induces the expression of a number of genes (see also: Mx-Protein). These genes contain regulatory DNA sequences within their promoter regions (ISRE; interferon-α-stimulated response elements; see: IRS, interferon response sequence) that function as binding sites for a number of transcription factors. Some of these genes are also expressed in response to other interferons.

The occurrence of spontaneous antibodies directed against IFN-α has been observed in patients with certain types of autoimmune diseases, generalized virus infections, and a number of tumors. Some inbred strains of mice appear to produce constitutively antibodies directed against IFN-α or IFN-β.

■ **TRANSGENIC/KNOCK-OUT/ANTISENSE STUDIES:** Interferon expression in the testes of transgenic mice has been shown to cause sterility.

■ **ASSAYS:** IFN-α is assayed by a cytopathic effect reduction test employing human and bovine cell lines (see: MDBK, WISH). Minute amounts of IFN-α can be assayed also by detection of the » Mx protein specifically induced by this interferon. A sandwich ELISA employing bispecific monoclonal antibodies for rapid detection (10 units/mL = 0.1 ng/mL within 2-3 h) is available. For further information see also subentry "Assays" in the reference section.

■ **CLINICAL USE & SIGNIFICANCE:** IFN-α is mainly employed as a standard therapy for hairy cell leukemia, metastasizing renal carcinoma and AIDS-associated Kaposi sarcomas. It is also active against a number of other tumors and viral infections. IFN-α is approved by the Food and Drug Administration for the treatment of condyloma acuminata (genital or venereal warts).

Hairy cell leukemia constitutes ≈ 2 % of all leukemias. Treatment with IFN-α markedly improves blood and bone marrow parameters. The number of necessary blood transfusions is reduced and the frequency of life-threatening infections is also reduced.

Treatment of disseminated Kaposi sarcomas results in complete or partial remissions in ≈ 30-40 % of the patients. In patients with advanced malignant melanomas treatment with a combination of IFN-α and chemotherapy (Dacarbazin, DTIC) has been found to be particularly effective and to be superior to treatment with IFN-α alone. Complete remissions and also a significant increase in survival times have been observed in responders. Intralesional therapy with IFN-α has been found to cause almost complete disappearance of tumors in 80 % of patients with basaliomas.

Moderate and high doses of IFN-α are one of the most effective forms of treatment of metastasizing renal carcinomas. Response rates of combinations of vinblastin and IFN-α are ≈ 25 % higher than those with interferon alone. Response rates have been reported to be improved by combining IFN-α with antineoplastic agents or other cytokines. Combination therapy with systemically administered » IL2 and IFN-α has resulted in long-term remissions in 30% of patients with metastatic renal cell carcinoma.

Treatment of CML with IFN-α causes hematological remissions in most patients and has been shown to cause a complete elimination of the PH1-(Philadelphia chromosome)-positive cells in the bone marrow of some patients.

Prospective studies are now under way to evaluate the effectiveness of IFN-α in the treatment of non-Hodgkin lymphomas, cutaneous T cell lymphomas (Mycosis fungoides, Sézary syndrome), multiple myelomas, condylomata acuminata and chronic active hepatitis B.

Some patients treated with genetically engineered recombinant IFN-α2 have been shown to develop neutralizing antibodies against interferon. Increasing levels of antibodies correlate with increasing reoccurrence of the disease. Therefore, patients should be monitored for the presence and clinical relevance of IFN-α antibodies to determine those who could respond to alternative treatment. It has been assumed that the recombinant protein may possess an altered tertiary structure leading to the exposure of a novel immunoreactive epitope not normally recognized in natural IFN-α. Continuation of the treatment with natural purified IFN-α leads to the disappearance of antibodies and also causes remissions.

In many instances a combination of the various interferons has been found to cause synergistic effects. The antiviral/antiproliferative/antitumor properties of IFN-α is potentiated by febrile temperatures.

● **Reviews:** for further information see references cited for » IFN.

● **Biochemistry & Molecular Biology: De Maeyer-Guignard J & de Maeyer E** Natural antibodies to interferon-α and interferon-β are a common feature of inbred mouse strains. JI 136: 1708-11 (1986); **Derynck R et al** Expression of human interferon α cDNA in yeast. NAR 11: 1819-37 (1983); **Diaz MO et al** Homozygous deletion of the α- and β_1-interferon genes in human leukemia and derived cell lines. PNAS 85: 5259-63 (1988); **Hitzeman RA et al** Expression of human gene for interferon in yeast. N 293: 717-22 (1981); **John J et al** Isolation and characterization of a new mutant human cell line unresponsive to α and β interferons. MCB 11: 4189-95 (1991); **Maeda S et al** Production of human α-interferon in silkworm using a baculovirus vector. N 315: 592-4 (1985); **Ohlsson M et al** Close linkage of α and β interferons and infrequent duplication of β interferon in humans. PNAS 82: 4473-6 (1985); **Trent JM et al** Chromosomal localization of human leukocyte, fibroblast, and immune interferon genes by means of *in situ* hybridization. PNAS 79: 7809-13 (1982); **Weber H et al** Single amino acid changes that render human IFN-α 2 biologically active on mouse cells. EJ 6: 591-8 (1987)

● **Receptor Structure, Gene(s), Expression: Delcayre AX et al** Epstein Barr virus/complement C3d receptor is an interferon α receptor. EJ 10: 919-26 (1991); **Hannigan GE & Williams BRG** Signal transduction by interferon-α through arachidonic acid metabolism. S 251: 204-7 (1991); **Lutfalla G et al** The structure of the human interferon α/β receptor gene. JBC 267: 2802-9 (1992); **Novick D et al** Soluble interferon-α receptor molecules are present in body fluids. FL 314: 445-8 (1992); **Uzé G et al** Genetic transfer of a functional human interferon α receptor into mouse cells: cloning and expression of its cDNA. Cell 60: 225-34 (1990)

● **biological activities:** for further information see references cited for » IFN.

● **Transgenic/Knock-out/Antisense Studies: Hekman AC et al** Interferon expression in the testes of transgenic mice leads to sterility. JBC 263: 12151-5 (1988)

● **Assays: Andersson G et al** Application of four anti-human interferon-α monoclonal antibodies for immunoassay and comparative analysis of natural interferon-α mixtures. J. Interferon Res. 11: 53-60 (1991); **Grander D et al** Measurement of interferon sensitivity in tumor cells from fine-needle aspirations. Eur. J. Cancer 28A: 815-8 (1992); **Haines DS & Gillespie DH** RNA abundance measured by a lysate RNase protection assay. Biotechniques 12: 736-41 (1992); **Kagnoff MF et al** Detection by immunofluorescence of lymphokine-producing T cells. Immunol. Res. 10: 255-7 (1991); **Kita M et al** Production of human α- and γ-interferon is dependent on age and sex and is decreased in rheumatoid arthritis: a simple method for a large-scale assay. J. Clin. Lab. Anal. 5: 238-41 (1991); **Kontsekova E et al** Quadroma-secreted bi(interferon α 2-peroxidase)-specific antibody suitable for one-step immunoassay. Hybridoma 11: 461-8 (1992); **L'Haridon RM** Immunosorbent binding bioassay: a solid phase biological immunoassay for the titration of antisera to α interferons. J. Immunoassay 12: 99-112 (1991); **L'Haridon RM et al** Production of an hybridoma library to recombinant porcine α I interferon: a very sensitive assay (ISBBA) allows the detection of a large number of clones. Hybridoma 10: 35-47 (1991); **Loster K et al** Novel antibody coating of a magnetizable solid phase for use in enzyme immunoassays. JIM 148: 41-7 (1992); **Phillips TM** Measurement of recombinant interferon levels by high performance immunoaffinity chromatography in body fluids of cancer patients on interferon therapy. Biomed. Chromatogr. 6: 287-90 (1992); **Prummer O et al** Filter spot-ELISA for the enumeration of interferon-α antibody-secreting cells. JIM 130: 187-93 (1990); **Rosolen A et al** Detection of functional interferon α receptors in human neuroendocrine tumor cell lines using a new monoclonal antibody. ECN 3: 81-8 (1992); **Thavasu PW et al** Measuring cytokine levels in blood. Importance of anticoagulants, processing, and storage conditions. JIM 153: 115-24 (1992); for further information also see individual cell lines used in individual bioassays.

● **Clinical Use & Significance: Atzpodien J et al** Home therapy with recombinant interleukin-2 and interferon-α2b in advanced human malignancies. Lancet June 23, 1509-12 (1990); **Atzpodien J et al** α-Interferon and interleukin-2 in renal cell carcinoma: studies in nonhospitalized patients. Semin. Oncol. 18: s108-12 (1991); **Battezzati PM et al** Factors predicting early response to treatment with recombinant interferon α-2a in chronic non-A, non-B hepatitis. Preliminary report of a long-term trial. Ital. J. Gastroenterol. 24: 481-4 (1992); **Belldegrun A et al** Interferon-α primed tumor-infiltrating lymphocytes combined with interleukin-2 and interferon-α as therapy for metastatic renal cell carcinoma. J. Urol. 150: 1384-90 (1993); **Braken JB et al** Current status of interferon α in the treatment of chronic hepatitis B. Pharm. Weekbl. Sci. 14: 167-73 (1992); **Catani L et al** *In vitro* inhibition of interferon α-2a antiproliferative activity by antibodies developed during treatment for essential thrombocythemia. Haematologica 77: 318-21 (1992); **Christmas TI et al** Effect of interferon-α 2a on malignant mesothelioma. J. Interferon Res. 13: 9-12 (1993); **Diodati G et al** Interferon therapy of cryptogenic chronic active liver disease and its relationship to anti-HCV. Arch. Virol. Suppl. 4: 299-303 (1992); **Ebert T et al** The role of cytokines in the therapy of renal cell carcinoma. Rec. Results in Cancer Res. 126: 113-8 (1993); **Engelmann U et al** Interferon-α 2b instillation prophylaxis in superficial bladder cancer – a prospective, controlled three-armed trial. Project Group Bochum-Interferon and Superficial Bladder Cancer. Anticancer. Drugs. 3 Suppl 1: 33-7 (1992); **Fattovich G et al** A randomized controlled trial of lymphoblastoid interferon-α in patients with chronic hepatitis B lacking HBeAg. Hepatology 15: 584-9 (1992); **Ferrara F et al** Recombinant interferon-α 2A as maintenance treatment for patients with advanced stage chronic lymphocytic leukemia responding to chemotherapy. Am. J. Hematol. 41: 45-9 (1992); **Finter NB et al** The use of interferon-α in virus infections. Drugs 42: 749-65 (1991); **Frasci G et al**

Intraperitoneal (ip) cisplatin-mitoxantrone-interferon-α 2b in ovarian cancer patients with minimal residual disease. Gynecol. Oncol. 50: 60-7 (1993); **Gill PS et al** Phase I/II trials of α-interferon alone or in combination with zidovudine as maintenance therapy following induction chemotherapy in the treatment of acquired immunodeficiency syndrome-related Kaposi's sarcoma. Semin. Oncol. 18: s53-7 (1991); **Gutterman JU** The role of interferons in the treatment of hematological malignancies. Sem. Hematol. 25: 3-8 (1988); **Italian Cooperative Study Group on Chronic Myeloid Leukemia.** A prospective comparison of α-IFN and conventional chemotherapy in Ph+ chronic myeloid leukemia. Clinical and cytogenetic results at 2 years in 322 patients. Haematologica 77: 204-14 (1992); **Janson ET et al** Treatment with α-interferon versus α-interferon in combination with streptozocin and doxorubicin in patients with malignant carcinoid tumors: a randomized trial. Ann. Oncol. 3: 635-8 (1992); **Kardamakis D** Interferons in the treatment of malignancies. *In vivo* 5: 589-98 (1991); **Kaiser U** Generalized giant-cell tumor of bone: successful treatment of pulmonary metastases with interferon α, a case report. J. Cancer Res. Clin. Oncol. 119: 301-3 (1993); **Krown SE et al** Interferon-α, zidovudine, and granulocyte-macrophage colony-stimulating factor: a phase I AIDS Clinical Trials Group study in patients with Kaposi's sarcoma associated with AIDS. J. Clin. Oncol. 10: 1344-51 (1992); **Lind MJ et al** A phase II study of ifosfamide and α2b-interferon in advanced non-small-cell lung cancer. Cancer Chemother. Pharmacol. 28: 142-4 (1991); **Morabito F et al** α 2-interferon in B cell chronic lymphocytic leukemia: clinical response, serum cytokine levels, and immunophenotype modulation. Leukemia 7: 366-71 (1993); **Moreno M et al** Prospective, randomized controlled trial of interferon-α in children with chronic hepatitis B. Hepatology 13: 1035-9 (1991); **Mughal TI et al** Role of recombinant α-interferon in the treatment of advanced cutaneous malignant melanoma. Oncology 48: 365-8 (1991); **Nicoletto MO et al** Experience with intraperitoneal α-2a interferon. Oncology 49: 467-73 (1992); **Noel C et al** Acute and definitive renal failure in progressive multiple myeloma treated with recombinant interferon α-2a: report of two patients. Am. J. Hematol. 41: 298-9 (1992); **Orita K et al** Early phase II study of interferon-α and tumor necrosis factor-α combination in patients with advanced cancer. Acta Med. Okayama. 46: 103-12 (1992); **Pazdur R et al** Phase II evaluation of recombinant α-2a-interferon and continuous infusion of fluorouracil in previously untreated metastatic colorectal adenocarcinoma. Cancer 71: 1214-8 (1993); **Prummer O** Interferon-α antibodies in patients with renal cell carcinoma treated with recombinant interferon-α-2A in an adjuvant multicenter trial. The Delta-P Study Group. Cancer 71: 1828-34 (1993); **Ridolfi R et al** Evaluation of toxicity in 22 patients treated with subcutaneous interleukin-2, α-interferon with and without chemotherapy. J. Chemother. 4: 394-8 (1992); **Rosenberg SA** Combination therapy with interleukin-2 and α-interferon for the treatment of patients with advanced cancer. J. Clin. Oncol. 7: 1863-74 (1989); **Ruiz-Schuchter LM et al** Sequential chemotherapy and immunotherapy for the treatment of metastatic melanoma. J. Immunother. 12: 272-6 (1992); **Salmon SE et al** α-Interferon for remission maintenance: preliminary report on the Southwest Oncology Group Study. Semin. Oncol. 18: s33-6 (1991); **Sanchez A et al** Low-dose α-interferon treatment of chronic myeloid leukemia. Am. J. Hematol. 39: 61-2 (1992); **Sarosdy M** High-dose versus low-dose intravesical interferon-α 2b in the treatment of carcinoma *in situ*: a randomized, controlled study. Anticancer Drugs 3 Suppl 1: 13-7 (1992); **Sella A et al** Phase II study of interferon-α and chemotherapy (5-fluorouracil and mitomycin C) in metastatic renal cell cancer. J. Urol. 147: 573-7 (1992); **Smalley RV et al** A randomized comparison of two doses of human lymphoblastoid interferon-α in hairy cell leukemia. Wellcome HCL Study Group. Blood 78: 3133-41 (1991); **Stuart-Harris RC et al** The clinical application of the interferons: a review. NSW Therapeutic Assessment Group Med. J. Aust. 156: 869-72 (1992); **Veenhof CH et al** A dose-escalation study of recombinant interferon-α in patients with a metastatic carcinoid tumor. Eur. J. Cancer 28: 75-8 (1992); **Vogelzang NJ et al** Subcutaneous interleukin-2 plus interferon α-2a in metastatic renal cancer: an outpatient multicenter trial. J. Clin. Oncol. 11: 1809-16 (1993); **Yamamoto H et al** Interferon therapy for non-A, non-B hepatitis: a pilot study and review of the literature. Hepatogastroenterology 39: 377-80 (1992)

IFN-α II1: see: IFN-ω (omega interferon).

IFN-α Con1: see: Consensus interferon.

IFN-αLe: abbrev. for Leukocyte interferon. This interferon is identical with » IFN-α. See also: IFN for general information about interferons.

IFN-αLy: abbrev. for Lymphoblast interferon. This interferon is identical with » IFN-α. See also: IFN for general information about interferons.

IFN-β: Beta-Interferon, abbrev. bIFN.

■ **ALTERNATIVE NAMES:** Fibroblast interferon (Fi-IFN, F-IFN); Type 1 interferon; pH2-stable IFN; R1-GI factor (murine R1 growth inhibitory factor). See also: individual entries for further information.

■ **SOURCES:** IFN-β is produced mainly by fibroblasts and some epithelial cell types. The synthesis of IFN-β can be induced by common inducers of interferons, including viruses, double-stranded RNA, and micro-organisms. It is also induced by some cytokines such as » TNF and » IL1.

■ **PROTEIN CHARACTERISTICS:** IFN-β is a glycoprotein (≈ 20 % sugar moiety) of 20 kDa and has a length of 166 amino acids. Glycosylation is not required for biological activity *in vitro*.
The protein contains a disulfide bond Cys31/141) required for biological activity. At the DNA level IFN-β displays 34 % sequence homology with » IFN-β2 and ≈ 30 % homology with other » IFN-α subtypes. In contrast to IFN-γ IFN-β is stable at pH2.

■ **GENE STRUCTURE:** The human gene encoding IFN-β has a length of 777 bp and maps to chromosome 9p22 in the vicinity of the » IFN-α gene cluster. The IFN-β gene does not contain introns. At least three different genes encoding IFN-β have been found in the genomes of cattle.

■ **RECEPTOR STRUCTURE, GENE(S), EXPRESSION:** IFN-β binds to the same receptor as » IFN-α.

■ **BIOLOGICAL ACTIVITIES:** In contrast to » IFN-α IFN-β is strictly species-specific. IFN-β of other species is inactive in human cells.
IFN-β is involved in the regulation of unspecific humoral immune responses and immune responses against viral infections. IFN-β increases the expression of HLA class I antigens and blocks the expression of HLA class II antigens stimulated by IFN-γ. IFN-β stimulates the activity of NK cells and hence also antibody-dependent cytotoxicity (see: ADCC). The activity of T suppressor cells elicited by several stimuli is also stimulated by IFN-β. IFN-β enhances the synthesis of the low-affinity IgE receptor » CD23. In activated monocytes (see also: cell activation) IFN-β induces the synthesis of neopterin. It also enhances serum concentrations of » Beta-2-Microglobulin. IFN-β selectively inhibits the expression of some mitochondrial genes.
IFN-β shows antiproliferative activity against a number of cell lines established from solid tumors.

■ **TRANSGENIC/KNOCK-OUT/ANTISENSE STUDIES:** An enhanced viral resistance and abnormalities in fertility are seen in » transgenic animals (mice) expressing the human IFN-β gene.

■ **ASSAYS:** Minute amounts of IFN-β can be detected indirectly by measuring IFN-induced proteins such as » Mx protein. For further information see also subentry "Assays" in the reference section.

■ **CLINICAL USE & SIGNIFICANCE:** IFN-β can be used for topic treatment of condylomata acuminata. It is also suitable for the prophylactic use following surgical removal of large condylomas. Some studies suggest that IFN-β tends to prevent disease activity in patients with multiple sclerosis. IFN-β in combination with IFN-α has been used in the treatment of chronic active hepatitis B and appears to be most promising if the disease has not lasted longer than 5 years. The antiviral activity of IFN-β is demonstrated also in the treatment of severe childhood viral encephalitis.
A combination treatment in combination with acyclovir is more effective than treatment with acyclovir alone.
IFN-β is a lipophilic molecule that should be particularly useful for local tumor therapy due to its specific pharmacokinetics. It is hardly removed from the tumor tissues after intralesional administration and hence also shows little systemic side effects. Head and neck squamous carcinomas, mammary and cervical carcinomas, and also malignant melanomas respond well to treatment with IFN-β. IFN-β also appears to be very promising for the adjuvant therapy of malignant melanomas with a high potential for metastasis. Response rates have been reported to be improved by combining IFN-β with antineoplastic agents or other cytokines.
In many instances a combination of the various interferons has been found to cause synergistic effects. The antiviral/antiproliferative/antitumor properties of IFN-β is potentiated by febrile temperatures.

● **REVIEWS:** for further information see references cited for » IFN.

● **BIOCHEMISTRY & MOLECULAR BIOLOGY: Diaz MO et al** Homozygous deletion of the α- and β_1-interferon genes in human leukemia and derived cell lines. PNAS 85: 5259-63 (1988); **Hosoi K et al** Structural characterization of fibroblast human interferon-β. J. Interferon Res. 8: 375-84 (1988); **John J et al** Isolation and characterization of a new mutant human cell line unresponsive to α and β interferons. MCB 11: 4189-95 (1991); **Mark DF et al** Site-specific mutagenesis of the human fibroblast interferon gene. PNAS 81: 5662-6 (1984); **Ohlsson M et al** Close linkage of α and β interferons and infrequent duplication of β interferon in humans. PNAS 82: 4473-6 (1985); **Smith GE et al** Production of human β interferon in insect cells infected with a baculovirus expression vector. MCB 3: 2156-65 (1983); **Taniguchi T et al** The nucleotide sequence of human fibroblast interferon cDNA. Gene 10: 11-5 (1980); **Trent JM et al** Chromosomal localization of human leukocyte, fibroblast, and immune interferon genes by means of *in situ* hybridization. PNAS 79: 7809-13 (1982); **Utsumi J et al** Characterization of *E. coli*-derived recombinant human interferon-β as compared with fibroblast human interferon-β. JB 101: 1199-208 (1987); **Zilberstein A et al** Structure and expression of cDNA and genes for human interferon-β2, a distinct species inducible by growth-stimulatory cytokines. EJ 5: 2529-37 (1986)

● **RECEPTOR STRUCTURE, GENE(S), EXPRESSION: Lutfalla G et al** The structure of the human interferon α/β receptor gene. JBC 267: 2802-9 (1992)

● **BIOLOGICAL ACTIVITIES:** for further information see references cited for » IFN.

● **TRANSGENIC/KNOCK-OUT/ANTISENSE STUDIES: Chen XZ et al** Enhanced viral resistance in transgenic mice expressing the human β 1 interferon. J. Virol. 62: 3883-7 (1988); **Iwakura Y et al** Male sterility of transgenic mice carrying exogenous mouse interferon-β gene under the control of the metallothionein enhancer-promoter. EJ 7: 3757-62 (1988)

● **ASSAYS: Kagnoff MF et al** Detection by immunofluorescence of lymphokine-producing T cells. Immunol. Res. 10: 255-7 (1991); **Markaryan AN et al** Construction of expression vectors for gene fusions on the model of β-galactosidase-human fibroblast β-interferon for the purpose of immunoenzyme assay. ANY 646: 125-35 (1991)

● **CLINICAL USE & SIGNIFICANCE: Buzzi F et al** Combination of β-interferon and tamoxifen as a new way to overcome clinical resistance to tamoxifen in advanced breast cancer. Anticancer Res. 12: 869-71 (1992); **Capalbo M et al** Treatment of chronic hepatitis B with β interferon given intramuscularly: a pilot study. Ital. J. Gastroenterol. 24: 203-5 (1992); **Duggan DB et al** A phase II study of recombinant interleukin-2 with or without recombinant interferon-β in non-Hodgkin's lymphoma. A study of the Cancer and Leukemia Group B. J. Immunother. 12: 115-22 (1992); **Fierlbeck G et al** [Intralesional therapy of melanoma metastases with recombinant interferon-β] Hautarzt. 43: 16-21 (1992); **Kagawa T et al** A pilot study of long-term weekly interferon-β administration for chronic hepatitis B. Am. J.

Gastroenterol. 88: 212-6 (1993); **Panitch HS** Interferons in multiple sclerosis. A review of the evidence. Drugs 44: 946-62 (1992); **Pungetti D et al** [HPV infections in the lower genital tract in the female. Results of β-interferon treatment] Minerva Ginecol. 43: 469-74 (1991); **Resta L et al** Variations of lymphocyte sub-populations in vulvar condylomata during therapy with β-interferon. Eur. J. Gynaecol. Oncol. 13: 440-4 (1992); **Stuart-Harris RC et al** The clinical application of the interferons: a review. NSW Therapeutic Assessment Group Med. J. Aust. 156: 869-72 (1992); **Wintergerst U & Belohradsky BH** Acyclovir monotherapy versus acyclovir plus β-interferon in focal viral encephalitis in children. Infection 20: 207-12 (1992)

IFN-β2: Interferon β2. IFN-β2 was isolated initially as a factor synthesized by human fibroblasts in response to » IL1, » TNF, » PDGF or poly(I)poly(C) which also induce the synthesis of » IFN-β.

A weak antiviral activity was attributed to IFN-β2 but this was not confirmed in subsequent studies using pure recombinant factor. IFN-β2 is identical with the » 26 kDa protein and » HPGF (hybridoma-plasmacytoma growth factor)and hence is identical with » IL6.

Hirano T et al Absence of antiviral activity in recombinant B cell stimulatory factor 2 (BSF-2). Immunol. Lett. 17: 41-5 (1988); **Poupart P et al** B cell growth modulating and differentiating activity of recombinant human 26-Kd protein (BSF-2, HuIFN-b2, HPGF). EJ 6: 1219-24 (1987); **Reis LF et al** Antiviral action of tumor necrosis factor in human fibroblasts is not mediated by B cell stimulatory factor 2/IFN-β 2, and is inhibited by specific antibodies to IFN-β. JI 140: 1566-70 (1988); **Sehgal PB et al** Human β-2-interferon and B cell differentiation factor are identical. S 235: 731-2 (1987); **Van Damme J et al** Identification of the human 26-kDa protein, interferon β2 (IFN-β2), as a B cell hybridoma/plasmacytoma growth factor induced by interleukin 1 and tumor necrosis factor. JEM 165: 914-9 (1987); **Weissenbach J et al** Two interferon mRNAs in human fibroblasts: *in vitro* translation and *Escherichia coli* cloning studies. PNAS 77: 7152-6 (1980); **Zilberstein A et al** Structure and expression of cDNA and genes for human interferon-β2, a distinct species inducible by growth-stimulatory cytokines. EJ 5: 2529-37 (1986)

IFN-β-inducing factor: This factor stimulates the synthesis of » IFN-β in fibroblasts. It is identical with one of the two species of » IL1 (IL1β).

Bunning RA et al Homogeneous interferon-β-inducing 22 k factor (IL1β) has connective tissue cell stimulating activities. BBRC 139: 1150-7 (1986); **Cameron P et al** Amino acid sequence analysis of human interleukin 1 (IL1). Evidence for biochemically distinct forms of IL1. JEM 162: 790-801 (1985)

IFN-γ: Gamma-Interferon, abbrev. gIFN.

■ **ALTERNATIVE NAMES:** Immune interferon (IIF), Type 2 interferon, T interferon, antigen-induced interferon, mitogen-induced interferon, pH2-labile IFN.

Many of the activities described in the older publications as » MAF (macrophage activating factor) or » TRF (T cell replacing factor) are due to IFN-γ.

■ **SOURCES:** IFN-γ is produced mainly by T cells and natural killer cells activated by antigens, mitogens, or alloantigens. It is produced by » CD4-positive and » CD8-positive lymphocytes. The synthesis of IFN-γ is induced, among other things, by » IL2, » bFGF, and » EGF. The synthesis of IFN-γ is inhibited by 1α,25-Dihydroxy vitamin D3, dexamethasone and » CsA (Ciclosporin).

B cells also produce IFN-γ, and constitutive synthesis has been observed in many established human B cell lines.

■ **PROTEIN CHARACTERISTICS:** IFN-γ is a dimeric protein with subunits of 146 amino acids. The protein is glycosylated at two sites. The pI is 8.3-8.5. IFN-γ is synthesized as a precursor protein of 166 amino acids including a secretory » signal sequence of 23 amino acids. Two molecular forms of the biologically active protein of 20 and 25 kDa have been described. Both of them are glycosylated at position 25. The 25 kDa form is also glycosylated at position 97. The observed differences of natural IFN-γ with respect to molecular mass and charge are due to variable glycosylation patterns. 40-60 kDa forms observed under non-denaturing conditions are dimers and tetramers of IFN-γ. Recombinant IFN-γ isolated from *Escherichia coli* is also biologically active and glycosylation therefore is not required for biological activity. IFN-γ contains two cysteine residues which are not involved in disulfide bonding.

At least six different variants of naturally occurring IFN-γ have been described. They differ from each other by variable lengths of the carboxyterminal ends. The biological activities of these variants do not differ from recombinant *Escherichia coli*-derived IFN-γ. It has been proposed that at least some of these variants are the result of proteolytic cleavage by exopeptidases and hence constitute purification artifacts. In contrast to » IFN-α and » IFN-β » IFN-γ is labile at pH 2.

IFN-γ can exist in a form associated with the extracellular matrix (see: ECM) and may therefore exert » juxtacrine growth control.

IFN-γ does not display significant homology with the other two interferons, » IFN-α and » IFN-β. Murine and human IFN-γ show ≈ 40 % sequence homology at the protein level.

■ **GENE STRUCTURE:** The human gene has a length of ≈ 6 kb. It contains four exons and maps to chromosome 12q24.1.

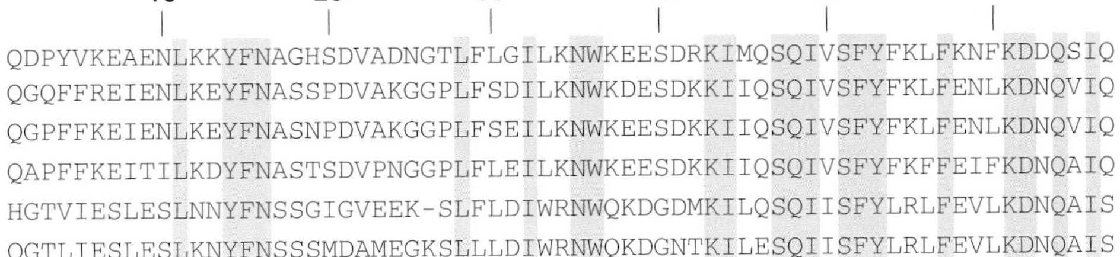

```
         10        20        30        40        50        60
QDPYVKEAENLKKYFNAGHSDVADNGTLFLGILKNWKEESDRKIMQSQIVSFYFKLFKNFKDDQSIQ
QGQFFREIENLKEYFNASSPDVAKGGPLFSDILKNWKDESDKKIIQSQIVSFYFKLFENLKDNQVIQ
QGPFFKEIENLKEYFNASNPDVAKGGPLFSEILKNWKEESDKKIIQSQIVSFYFKLFENLKDNQVIQ
QAPFFKEITILKDYFNASTSDVPNGGPLFLEILKNWKEESDKKIIQSQIVSFYFKFFEIFKDNQAIQ
HGTVIESLESLNNYFNSSGIGVEEK-SLFLDIWRNWQKDGDMKILQSQIISFYLRLFEVLKDNQAIS
QGTLIESLESLKNYFNSSSMDAMEGKSLLLDIWRNWQKDGNTKILESQIISFYLRLFEVLKDNQAIS
```

```
70        80        90        100       110       120       130
KSVETIKEDMNVKFFNSNKKKRDDFEKLTNYSVTGLNVQRKAIHELIQVMAELSPAAKTGKRKRSQMLFRG
RSMDIIKQDMFQKFLNGSSEKLEDFKKLIQIPVDDLQIQRKAINELIQVMNDLSPKSNLRKRKRSQNLFQG
RSMDIIKQDMFQKKLNGSSEKLEDFKRLIQIPVDDLQIQRKAINELIQVMNDLSPKSNLRKRKRSQNLFRG
RSMDVIKQDMFQRFLNGSSGKLNDFEKLIKIPVDNLQIQRKAISELIKVMNDLSPRSNLRKRKRSQTMFQG
NNISVIESHLITTFFSNSKAKKDAFMSIAKFEVNNPQVQRQAFNELIRVVHQLLPESSLRKRKRSRC
NNISVIESHLITNFFSNSKAKKDAFMSIAKFEVNNPQIQHKAVNELIRVIHQLSPESSLRKRKRSRC
```

Comparison of IFN-γ cDNA sequences from different species.
Sequences from bottom to top are human, bovine, sheep, porcine, mouse, and rat.
Amino acid positions conserved in all six species are shown in blue. The homologies become even more pronounced if only two or three species are considered or if conservative amino acid exchanges are also taken into account.

■ **Receptor Structure, Gene(s), Expression:** A number of binding proteins with molecular masses between 70 and 160 kDa have been described for IFN-γ. They are expressed on all types of human cells with the exception of mature erythrocytes. The receptor expressed in monocytes and other hematopoietic cells has a molecular mass of 140 kDa. A 54 kDa protein is observed in other cell types. The expression of the IFN-γ receptor on human peripheral blood monocytes is significantly increased by » GM-CSF.

IFN-γ receptors are N- and O-glycosylated and bind IFN-γ with an affinity of $10^{-10} - 10^{-11}$ M. The extracellular domain of the human 54 kDa receptor has a length of 229 amino acids and a transmembrane domain of 20 amino acids. The intracellular domain has a length of 222 amino acids and is probably involved in signal transduction.

The gene encoding this receptor maps to human chromosome 6. This receptor is not related to the receptors for » IFN-α and » IFN-β. At the protein level the murine and the human receptor show a sequence homology of 52 %.

IFN-γ/ receptor complexes are rapidly internalized by endocytosis. In mouse/human cellular hybrids carrying human chromosome 6 and therefore expressing the receptor IFN-γ binds to this receptor without eliciting any biological responses. The cells respond to IFN-γ if human chromosome 21 is also introduced into these cells. These observations suggest that another still unknown protein encoded by a gene on human chromosome 21 is required to generate a functional receptor molecule. This protein interacts with the extracellular domain of the receptor and, like the 54 kDa molecule, is also species-specific.

A soluble form of the IFN-γ receptor has also been described. Soluble receptors have also been found for » IL1 (see: IL1ra, IL1 receptor antagonist), » IL2, » IL4, » IL6, » IL7, » IGF and » TNF-α. They probably function as physiological regulators of cytokine activities (see also: Cytokine inhibitors) by inhibiting receptor binding or act as transport proteins.

■ **Biological Activities:** IFN-γ has antiviral and antiparasitic activities and also inhibits the proliferation of a number of normal and transformed cells. IFN-γ synergises with TNF-α and » TNF-β in inhibiting the proliferation of various cell types. The growth inhibitory activities of IFN-γ are more pronounced than those of the other interferons. However, the main biological activity of IFN-γ appears to be immunomodulatory in contrast to the other interferons which are mainly antiviral.

In T helper cells » IL2 induces the synthesis of IFN-γ and other cytokines. IFN-γ acts synergistically with » IL1 and » IL2 and appears to be requir-

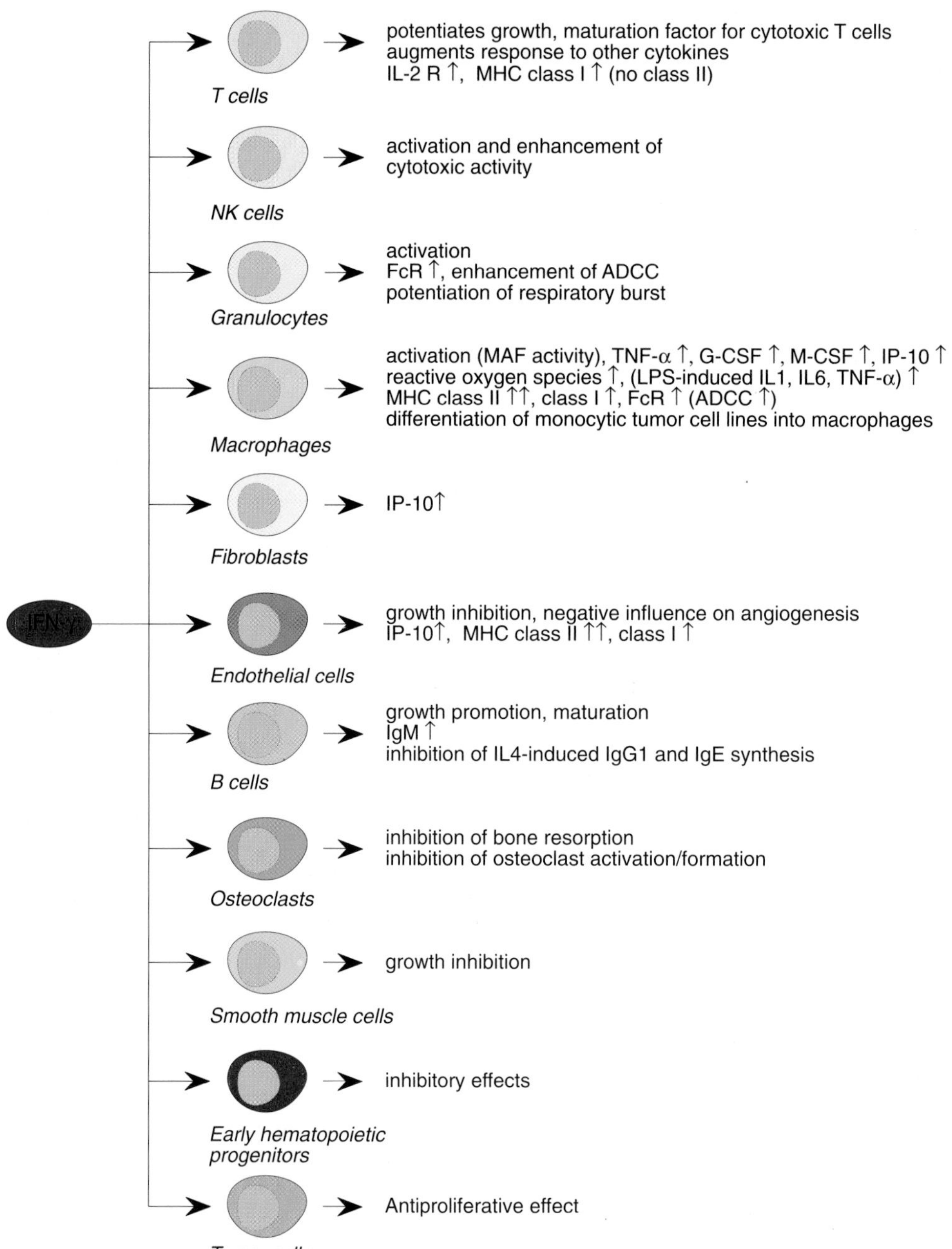

Schematic representation of the multiple activities of IFN-γ at the cellular level.

ed for the expression of IL2 receptors on the cell surface of T lymphocytes. Blocking of the IL2 receptor by specific antibodies also inhibits the synthesis of IFN-γ. IFN-γ thus influences cell-mediated mechanisms of cytotoxicity. IFN-γ is a modulator of T cell growth and functional differentiation. It is a growth-promoting factor for T lymphocytes and potentiates the response of these cells to mitogens or growth factors.
The human promyelocytic leukemia cell line » HL60 can be induced to differentiate by a number of stimuli. IFN-γ, but not other interferons, specifically induces differentiation of these cells into monocytes.
IFN-γ inhibits the growth of B cells induced by » IL4. IFN-γ and Anti-Ig costimulate the proliferation of human B cells but not of murine B cells. IFN-γ also inhibits the production of IgG1 and IgE elicited by IL4 in LPS-stimulated B cells. IFN-γ regulates the expression of MHC class 2 genes and is the only interferon that stimulates the expression of these proteins. IFN-γ also stimulates the expression of Ia antigens on the cell surface, the expression of » CD4 (see also: CD antigens) in T helper cells, and the expression of high-affinity receptors for IgG (see also: CD16, CD32, CD64) in myeloid cell lines, neutrophils, and monocytes. In monocytes and macrophages IFN-γ induces the secretion of » TNF-α and the transcription of genes encoding » G-CSF and » M-CSF. In macrophages IFN-γ stimulates the release of reactive oxygen species. IFN-γ is also involved in processes of bone growth and inhibits bone resorption probably by partial inhibition of the formation of osteoclasts.
IFN-γ inhibits the proliferation of smooth muscle cells of the arterial intima *in vitro* and *in vivo* and therefore probably functions as an endogenous inhibitor for vascular overreactions such as stenosis following injuries of arteries. IFN-γ inhibits the proliferation of endothelial cells and the synthesis of collagens by myofibroblasts. It thus functions as an inhibitor of capillary growth mediated by myofibroblasts and fibroblast growth factors (see: FGF) and » PDGF.
IFN-γ specifically induces the transcription of a number of genes. These genes contain regulatory DNA sequences within their promoter regions (ISRE; interferon-α-stimulated response elements; see: IRS, interferon response sequence) that function as binding sites for a number of transcription factors (see: GAF; γ-interferon activation factor). Some of these genes are also expressed in response to other interferons.

■ **TRANSGENIC/KNOCK-OUT/ANTISENSE STUDIES:** The expression of IFN-γ in the pancreas of » transgenic animals (mice) has been shown to precipitate autoimmune diabetes. IFN-γ produced locally in the pancreas of transgenic animals causes lymphocyte infiltration and islet cell destruction. However, new islet cells are formed continuously from duct cells. The IFN-γ induced islet neogenesis is similar to embryonic islet morphogenesis and offers a model system for studying factors modulating islet development.
Mice with a targeted disruption of the IFN-γ gene have been developed by homologous recombination in » ES (embryonic stem) cells. These animals develop normally and are healthy in the absence of pathogens. They are characterized, however, by an impaired production of macrophage antimicrobial products and reduced expression of macrophage major histocompatibility complex class II antigens. IFN-γ-deficient mice are killed by normally sublethal doses of the intracellular pathogen *Mycobacterium bovis*. Splenocytes show uncontrolled proliferation in response to mitogen and alloantigen. After a mixed lymphocyte reaction, T cell cytolytic activity is found to be enhanced against allogeneic target cells. Resting splenic natural killer cell activity is reduced in these mice.
Mice lacking the IFN-γ receptor have also been created by homologous recombination. These animals have no overt anomalies and show normal cytotoxic and T helper cell responses. However, mutant mice show an increased susceptibility to infection by *Listeria monocytogenes* and vaccinia virus. Immunoglobulin isotype analysis reveals that IFN-γ is required for a normal antigen-specific immunoglobulin G2a response.

■ **ASSAYS:** IFN-γ can be detected by sensitive immunoassays. A specific ELISA test allows detection of individual cells producing IFN-γ. Minute amounts of IFN-γ can be detected indirectly by measuring IFN-induced proteins such as » Mx protein. The induction of the synthesis of » IP-10 (immune protein 10) has also been used to measure IFN-γ. A new bioassay employs induction of indoleamine 2,3-dioxygenase activity in » 2D9 cells. A sensitive radioreceptor assay is also available. For further information see also subentry "Assays" in the reference section.

■ **CLINICAL USE & SIGNIFICANCE:** Like the other interferons IFN-γ can be used as an antiviral and antiparasitic agent. IFN-γ has been shown to be effective in the treatment of chronic polyarthritis.

This treatment, which probably involves a modulation of macrophage activities, significantly reduces joint aches and improves various clinical parameters and allows reduction of corticosteroid doses.

In spite of the antiproliferative activities of IFN-γ hopes that this interferon may be of use in the treatment of various malignancies have been disappointed. In many instances a combination treatment with other interferons and cytokines, and also with chemotherapy has only been moderately effective.

IFN-γ may be of value in the treatment of opportunistic infections in AIDS patients. It has also been shown to reduce inflammation, clinical symptoms, and eosinophilia in severe atopic dermatitis.

Experiments are in progress to insert IFN-γ genes into human tumor cells by genetic engineering as a means for generating autologous or HLA-matched allogeneic tumor cell vaccines for the treatment of patients with renal cell carcinoma and other cancers. The transduction of murine tumor cells with a functional IFN-γ gene has been shown to lead to the rejection of the genetically modified cells by syngeneic hosts (for cancer vaccines see also: Cytokine gene transfer).

● **Reviews: De Maeyer E & De Maeyer-Guignard J** Interferon-γ. Curr. Opin. Immunol. 4: 321-6 (1992); **Farrar MA & Schreiber RD** The molecular cell biology of interferon-γ and its receptor. ARI 11: 571-611 (1993); **Gray PW & Goeddel DV** Molecular biology of interferon-γ. Lymphokines 13: 151-62 (1987); **Möller G** (edt) γ interferon. IR Vol. 97, 1987; **Vilcek J et al** Interferon-γ: a lymphokine for all sea sons. Lymphokines 11: 1-32 (1985); for further information see also references cited for » IFN.

● **Biochemistry & Molecular Biology: Balick SE et al** Three-dimensional structure of recombinant human interferon-γ. S 252: 698-702 (1991); **Basu M et al** Purification and partial characterization of a receptor protein for mouse interferon γ. PNAS 85: 6282-6 (1988); **Dayton MA et al** Human B cell lines express the interferon γ gene. Cytokine 4: 454-60 (1992); **Decker T et al** Cytoplasmic activation of GAF, an IFN-γ-regulated DNA-binding factor. EJ 10: 927-32 (1991); **Devos R et al** Molecular cloning of human immune interferon cDNA and its expression in eukaryotic cells. NAR 10: 2487-2502, (1982); **Dijkema R et al** Cloning and expression of the chromosomal immune interferon gene of the rat. EJ 4: 761-7 (1985); **Fan XD et al** Molecular cloning of a gene selectively induced by γ interferon from human macrophage cell line U937. MCB 9: 1922-8 (1989); **Gray PW & Goeddel DV** Structure of the human immune interferon gene. N 298: 859-63 (1982); **Gray PW & Goeddel DV** Cloning and expression of murine interferon cDNA. PNAS 80: 5842-6 (1983); **McInnes CJ et al** The molecular cloning of the ovine γ-interferon cDNA using the polymerase chain reaction. NAR 18: 4012 (1990); **Naylor S L et al** Human immune interferon gene is located on chromosome 12. JEM 157: 1020-7 (1983); **Samudzi CT et al** Crystal structure of recombinant rabbit interferon-γ at 2 7-Å resolution. JBC 266: 21791-7 (1991); **Simons G et al** High level expression of human interferon-γ in *E. coli* under control of the P_L promoter of bacteriophage λ. Gene 28: 55-64 (1984); **Trent JM et al** Chromosomal localization of human leukocyte, fibroblast, and immune interferon genes by means of *in situ* hybridization. PNAS 79: 7809-13 (1982)

● **Receptor Structure, Gene(s), Expression: Aguet M et al** Molecular cloning and expression of the interferon γ receptor. Cell 55: 273-80 (1988); **Calderon J et al** Purification and characterization of the human interferon-γ receptor from placenta. PNAS 85: 4837-41 (1988); **Finbloom DS et al** Culture of human monocytes with granulocyte-macrophage colony-stimulating factor results in enhancement of IFN-γ receptors but suppression of IFN-γ-induced expression of the gene IP-10. JI 150: 2383-90 (1993); **Gibbs VC et al** The extracellular domain of the human interferon γ receptor interacts with a species-specific signal transducer. MCB 11: 5860-6 (1991); **Gray PW et al** Cloning and expression of the cDNA for the murine interferon-γ receptor. PNAS 86: 8497-501 (1989); **Kalina U et al** The human γ interferon receptor accessory factor encoded by chromosome 21 transduces the signal for the induction of 2',5'-oligoadenylate-synthetase, resistance to virus cytopathic effect, and major histocompatibility complex class I antigens. J. Virol. 67: 1702-6 (1993); **Munro S & Maniatis T** Expression cloning of the murine interferon-γ receptor cDNA. PNAS 86: 9248-52 (1989); **Rashidbaigi A et al** The gene for the human immune interferon receptor is located on chromosome 6. PNAS 83: 384-8 (1986)

● **Biological Activities:** for further information see references cited for » IFN.

● **Transgenic/Knock-out/Antisense Studies: Cooper AM et al** Disseminated tuberculosis in interferon g gene-disrupted mice. JEM 178: 2243-7 (1993); **Dalton DK et al** Multiple defects of immune cell function in mice with disrupted interferon-γ genes S 259: 1739-42 (1993); **Gu D & Sarvetnick N** Epithelial cell proliferation and islet neogenesis in IFN-γ transgenic mice. Development 118: 33-46 (1993); **Huang S et al** Immune response in mice that lack the interferon-γ receptor. S 259: 1742-5 (1993); **Kamijo R et al** Generation of nitric oxide and induction of major histocompatibility complex class II antigen in macrophages from mice lacking the interferon γ receptor. PNAS 90: 6626-30 (1993); **Sarvetnick N et al** Insulin-dependent diabetes mellitus induced in transgenic mice by ectopic expression of class II MHC and interferon-γ. Cell 52: 773-82 (1988); **Sarvetnick N et al** Inflammatory destruction of pancreatic β cells in γ-interferon transgenic mice. CSHSQB 54: 837-42 (1989); **Sarvetnick N et al** Regeneration of pancreatic endocrine cells in interferon-γ transgenic mice. AEMB 321: 85-9 (1992)

● **Assays: Agrewala JN et al** Measurement and computation of murine interleukin-4 and interferon-γ by exploiting the unique abilities of these lymphokines to induce the secretion of IgG1 and IgG2a. J. Immunoassay 14: 83-97 (1993); **Baigrie RJ et al** Effect of major surgery on the release of interferon γ by peripheral blood mononuclear cells: an investigation at the single cell level using the reverse hemolytic plaque assay. Cytokine 4: 63-5 (1992); **Barker JN et al** Detection of interferon-γ mRNA in psoriatic epidermis by polymerase chain reaction. J. Dermatol. Sci. 2: 106-11 (1991); **Billman-Jacobe H et al** A comparison of the interferon γ assay with the absorbed ELISA for the diagnosis of Johne's disease in cattle. Aust. Vet. J. 69: 25-8 (1992); **Chen AB et al** Quantitation of *E. coli* protein impurities in recombinant human interferon-γ. Appl. Biochem. Biotechnol. 36: 137-52 (1992); **Czerkinsky C et al** Reverse ELISPOT assay for clonal analysis of cytokine production. I. Enumeration of γ-interferon-secreting cells. JIM 110: 29-36 (1988); **De Groote D et al** Novel method for the measurement of cytokine production by a one-stage procedure. JIM 163: 259-67 (1993);

Ellaurie M et al A simplified human whole blood assay for measurement of dust mite-specific γ interferon production *in vitro*. Ann Allergy 66: 143-7 (1991); **Elsasser-Beile U et al** Cytokine levels in whole blood cell cultures as parameters of the cellular immunologic activity in patients with malignant melanoma and basal cell carcinoma. Cancer 71: 231-6 (1993); **Jung T et al** Detection of intracellular cytokines by flow cytometry. JIM 159: 197-207 (1993); **Kabilan L et al** Detection of intracellular expression and secretion of interferon-γ at the single-cell level after activation of human T cells with tetanus toxoid *in vitro*. EJI 20: 1085-9 (1990); **Kita M et al** Production of human α- and γ-interferon is dependent on age and sex and is decreased in rheumatoid arthritis: a simple method for a large-scale assay. J. Clin. Lab. Anal. 5: 238-41 (1991); **Kominami G et al** Preparation and characterization of labelled interferon-γ and the development of radioreceptor assay for interferon-γ. J. Pharm. Biomed. Anal. 9: 387-91 (1991); **Lebedin YS et al** Serum levels of interleukin 4, interleukin 6 and interferon-γ following *in vivo* isotype-specific activation of IgE synthesis in humans. Int. Arch. Allergy Appl. Immunol. 96: 92-4 (1991); **Le Moal MA et al** Improvement of an ELISA bioassay for the routine titration of murine interferon-γ. Res. Immunol. 140: 613-24 (1989); **Lewis CE et al** Measurement of cytokine release by human cells. A quantitative analysis at the single cell level using the reverse hemolytic plaque assay. JIM 127: 51-9 (1990); **Merville P et al** Detection of single cells secreting IFN-γ, IL6, and IL10 in irreversibly rejected human kidney allografts, and their modulation by IL2 and IL4. Transplantation 55: 639-46 (1993); **Migliorini P et al** Macrophage NO2- production as a sensitive and rapid assay for the quantitation of murine IFN-γ. JIM 139: 107-14 (1991); **Prescott S et al** Radio-immunoassay detection of interferon-γ in urine after intravesical Evans BCG therapy. J. Urol. 144: 1248-51 (1990); **Ruzgas TA et al** Ellipsometric immunosensors for the determination of γ-interferon and human serum albumin. Biosens. Bioelecton. 7: 305-8 (1992); and interferon-γ. JIM 137: 47-54 (1991); **Skidmore BJ et al** Enumeration of cytokine-secreting cells at the single-cell level. EJI 19: 1591-7 (1989); **Stefanos S & Pestka S** A convenient assay for detection of soluble interferon receptors. J. Biol. Regul. Homeost. Agents 4: 57-9 (1990); **Taguchi T et al** Detection of individual mouse splenic T cells producing IFN-γ and IL5 using the enzyme-linked immunospot (ELISPOT) assay. JIM 128: 65-73 (1990); **Thavasu PW et al** Measuring cytokine levels in blood. Importance of anticoagulants, processing, and storage conditions. JIM 153: 115-24 (1992); **van der Meide PH et al** Assessment of the inhibitory effect of immunosuppressive agents on rat T cell interferon-γ production using an ELISPOT assay. JIM 144: 203-13 (1991); **van der Meide PH et al** Complement-mediated inactivation of interferon-γ in ELISA systems. J Immunoassay 12: 65-82 (1991); **Wilson AB et al** A competitive inhibition ELISA for the quantification of human interferon-γ. JIM 162: 247-55 (1993)

● Clinical Use & Significance: **Digel W et al** Pharmacokinetics and biological activity in subcutaneous long-term administration of recombinant interferon-γ in cancer patients. Cancer Immunol. Immunother. 34: 169-74 (1991); **Ernstoff MS et al** Immunological effects of treatment with sequential administration of recombinant interferon γ and α in patients with metastatic renal cell carcinoma during a phase I trial. CR 52: 851-6 (1992); **Freundlich B et al** Treatment of systemic sclerosis with recombinant interferon-γ. A phase I/II clinical trial. Arthritis Rheum. 35: 1134-42 (1992); **Gansbacher B et al** Retroviral vector-mediated γ-interferon gene transfer into tumor cells generates potent and long lasting antitumor immunity. CR 50: 7820-5 (1990); **Gastl G et al** Retroviral vector-mediated lymphokine gene transfer into human renal cancer cells. CR 52: 6229-36 (1992); **Gill PJ et al** Chronic granulomatous disease presenting with osteomyelitis: favorable response to treatment with interferon-γ. J. Pediatr. Orthop. 12: 398-400 (1992); **Hanifin JM et al** Recombinant interferon γ therapy for atopic dermatitis. (1993); J. Am. Acad. Dermatol. 28: 189-97 (1993); **Kardamakis D** Interferons in the treatment of malignancies. *In vivo* 5: 589-98 (1991); **Landmann R et al** Prolonged interferon-γ application by subcutaneous infusion in cancer patients: differential response of serum CD14, neopterin, and monocyte HLA class I and II antigens. J. Interferon. Res. 12: 103-11 (1992); **Lemmel EM et al** Multicenter double-blind trial of interferon-γ versus placebo in the treatment of rheumatoid arthritis. Arthritis Rheum. 34: 1621-2 (1991); **Lumsden AJ et al** Improved efficacy of doxorubicin by simultaneous treatment with interferon-γ and interleukin-2. In Vivo 6: 553-8 (1992); **Machold KP et al** Recombinant human interferon γ in the treatment of rheumatoid arthritis: double blind placebo controlled study. Ann. Rheum. Dis. 51: 1039-43 (1992); **Mahmoud HH et al** Phase I study of recombinant human interferon γ in children with relapsed acute leukemia. Leukemia 6: 1181-4 (1992); **Margolin KA et al** Phase I trial of interleukin-2 plus γ-interferon. J. Immunother. 11: 50-5 (1992); **Polk HC Jr et al** A randomized prospective clinical trial to determine the efficacy of interferon-γ in severely injured patients. Am. J. Surg. 163: 191-6 (1992); **Reichel RP et al** Clinical study with recombinant interferon γ versus interferon α-2c in patients with condylomata acuminata. Int. J. STD. AIDS. 3: 350-4 (1992); **Schiller JH et al** Clinical and biologic effects of combination therapy with γ-interferon and tumor necrosis factor. Cancer 69: 562-71 (1992); **Stuart-Harris RC et al** The clinical application of the interferons: a review. NSW Therapeutic Assessment Group Med. J. Aust. 156: 869-72 (1992); **Taylor CE et al** A Southwest Oncology Group Phase I study of the sequential combination of recombinant interferon-γ and recombinant interleukin-2 in patients with cancer. J. Immunother. 11: 176-83 (1992); **Viens P et al** Interleukin-2 in association with increasing doses of interferon-γ in patients with advanced cancer. J. Immunother. 11: 218-24 (1992); **Wandl UB et al** Combination therapy with interferon α-2b plus low-dose interferon γ in pretreated patients with Ph-positive chronic myelogenous leukemia. Br. J. Haematol. 81: 516-9 (1992); **Watanabe Y et al** Exogenous expression of mouse interferon γ cDNA in mouse neuroblastoma C1300 cells results in reduced tumorigenicity by augmented anti-tumor immunity. PNAS 86: 9456-60 (1989); **Weiner LM** Applications of γ-interferon in cancer therapy. Mol. Biother. 3: 186-91 (1991); **Weiner LM** Applications of γ-interferon in cancer therapy. Mol. Biother. 3: 186-91 (1991); **Yoneda K et al** Induction of lymphokine-activated killer cells with low-dose interleukin 2 and interferon-γ in oral cancer patients. J. Clin. Immunol. 12: 289-99 (1992)

IFN-ω: *Omega-interferon* IFN-ω is a natural component of human leukocyte interferon (see also: LeIFN). This interferon is also called ***IFN-α II1***. It displays a high degree of homology with various » IFN-α species including positions of the cysteine residues involved in disulfide bonds. However, sequence divergence allows classification as a unique protein family.

IFN-ω binds to the same receptors as » IFN-α and » IFN-β. To date the exact biological activities and the physiological role of this interferon are unknown. A related protein is bovine trophoblast protein » TP-1.

Adolf GR et al Human interferon omega 1: isolation of the gene, expression in Chinese hamster ovary cells and characterization of the recombinant protein. BBA 1089: 167-74 (1991); **Adolf GR et al** Purification and characterization of natural human interferon omega 1. Two alternative cleavage sites for the signal peptidase. JBC 265: 9290-5 (1990); **Adolf GR et al** Monoclonal antibodies and enzyme immunoassays specific for human interferon (IFN) omega 1: evidence that IFN-omega 1 is a component of human leukocyte IFN. Virology 175: 410-7 (1990); **Adolf GR et al** Antigenic structure of human interferon omega 1 (interferon α II1): comparison with other human interferons. J. Gen. Virol. 68: 1669-76 (1987); **Flores I et al** Human interferon-ω (omega) binds to the α/β receptor. JBC 266: 19875-7 (1991); **Hauptmann R & Swetly P** A novel class of human type 1 interferons. NAR 13: 4739-49 (1985); **Kontsek P et al** Are the acid-labile interferon α and interferon omega 1 identical? Virology 181: 416-8 (1991); **Mege D et al** The porcine family of interferon-omega: Cloning, structural analysis, and functional studies of five related genes. J. Interferon Res. 11: 341-50 (1991)

IgA-EF: *IgA enhancing factor* This factor is found in supernatants of a subset of helper T cell clones. It can enhance IgA synthesis in cultures of lipopolysaccharide-stimulated T cell-depleted spleen cells (see also: Isotype switching). IL4 can potentiate the effect of the IgA-enhancing factor. IgA-enhancing factor activity is mediated by the same polypeptide that has been characterized as » BCGF 2 (B cell growth factor 2), » TRF (T cell-replacing factor), and » EDF (eosinophil-differentiation factor). IgA-EF is identical with » IL5.

Bond MW et al A mouse T cell product that preferentially enhances IgA production. II: physico-chemical characterization. JI 139: 3691-6 (1987); **Coffman RL et al** A mouse T cell product that preferentially enhances IgA production. I. Biological characterization. JI 139: 3685-90 (1987); **Murray PD et al** Interleukin 5 and interleukin 4 produced by Peyer's patch T cells selectively enhance immunoglobulin A expression. JI 139: 2669-74 (1987); **Yokota T et al** Isolation and characterization of lymphokine cDNA clones encoding mouse and human IgA-enhancing factor and eosinophil colony-stimulating factor activities: relationship to interleukin 5. PNAS 84: 7388-92 (1987)

IGBP: *insulin-like growth factor binding protein(s)* see: IGF-BP.

IgE-BF: *IgE binding factor* The synthesis and function of IgE is dependent on IgE-binding proteins, which include cell surface IgE receptors and IgE-binding lymphokines. One of the known factors with affinity for IgE is the secreted form of the cell surface antigen » CD23 (see also: CD antigens). For another factor with IgE-binding activity see: IAP (intracisternal A particle).

Chen PO et al A rapid screening assay for detection of IgE-binding factors in humans. JIM 58: 59-71 (1983); **Delespesse G et al** Human IgE-binding factors. IT 10: 159-64 (1989); **Delespesse G & Sarfati M** IgE-binding factors (soluble CD23) and IgE regulation. Res. Immunol. 141: 75-7 (1990); **Liu FT et al** Identification of an IgE-binding protein by molecular cloning. PNAS 82: 4100-4 (1985); **Yanagihara Y et al** Establishment of a sensitive radio immunoassay for the detection of human IgE-binding factor (soluble CD23). Int. Arch. Allergy Immunol. 98: 189-99 (1992)

IgE-EF: *IgE enhancing factor* This factor is found in supernatants of a subset of helper T cell clones. It can enhance synthesis of IgE and IgG1 in cultures of lipopolysaccharide-stimulated T cell-depleted spleen cells. The factor is identical with the ***IgE switch factor*** that causes » (isotype switching. Both factors are identical with » IL4.

Coffman RL & Carty J A T cell activity than enhances polyclonal IgE production and its inhibition by interferon-γ. JI 136: 949-54 (1986); **Snapper CM et al** Differential regulation of IgG1 and IgE synthesis by interleukin 4. JEM 167: 183-96 (1988)

IgE switch factor: see: IgE-EF.

IGF: *insulin-like growth factor*

■ **Alternative Names:** MSA (multiplication-stimulating activity); NSILA (non-suppressible insulin-like activity); somatomedins; SF (sulfation factor); SFA (sulfation factor activity); SGF (skeletal growth factor); SMP (somatomedin/insulin-like growth factor (IGF)-like polypeptides). See also: individual entries for further information.

■ **Sources:** Both types of IGF are synthesized in many fetal and adult tissues. IGF-1 is produced constitutively in large amounts in the liver (≈ 10 mg/day). It is also produced locally in many other tissues including kidney, heart, lung, fat tissues, and various glandular tissues. IGF-1 is also produced by chondroblasts, fibroblasts, and osteoclasts. Normal serum levels in healthy human subjects are 150-250 μg/L for IGF-1 and 400-900 μg/L for IGF-2. These levels may vary with age, sex, hormonal status, and nutritional status. Lowered levels of IGF-1 are observed after removal of the pituitary. Human milk also contains these factors (see: MGF, milk growth factor).

■ **Protein Characteristics:** Two different IGFs have been described. They are called IGF-1 (= somatomedin-C) and IGF-2. IGF-1 has a length of 70 amino acids (7.6 kDa). IGF-2 has a length of 67 amino acids (7.4 kDa). A variant of IGF-2, probably encoded by a separate gene, has been described. This protein (10 kDa) has a substitution of cys-gly-asp for ser-33 and a carboxyterminal extension of 21 residues. The structure of both IGF proteins is homologous to human pro-insulin. Ca.

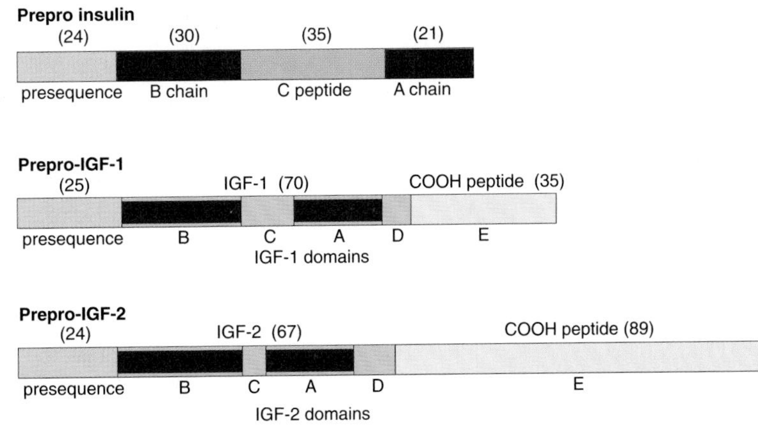

Schematic representation of protein domains (B,C,A,D,E) in the prepro forms of IGF-1, IGF-2 and insulin. Regions showing homologies with insulin are indicated by dark blue rectangles. The lengths of individual fragments (in amino acids) are indicated in brackets.

50 % sequence homology are observed in individual protein domains.

Both IGFs contain three intramolecular disulfide bonds. They display Ca. 62 % sequence homology with each other and 47 % identity with insulin. IGF-1 and IGF-2 do not cross-react immunologically with each other. Rat and mouse IGF-1 differ from each other by a single amino acid and differ from human IGF-1 in three and four amino acids. respectively. Rat MSA differs from IGF-2 by 5 amino acids.

IGF-1 and IGF-2 are obtained by processing of longer precursor molecules at the amino- and carboxyterminal ends. Several shortened variants have been described which presumably arise by differential processing of the precursor protein.

■ **Gene Structure:** IGF-1 and IGF-2 are encoded by two different genes which are expressed differentially in different tissues and at different times of development. The human IGF-1 gene contains five exons and has a length of ≈ 90 kb. It maps to human chromosome 12q23 in close vicinity of the gene encoding phenylalanine hydroxylase. The murine IGF-1 gene maps to chromosome 10. The human IGF-2 gene contains five exons and has a length of ≈ 30 kb. It maps to chromosome 11p15 and is flanked at the 5´ end by the gene encoding tyrosine hydroxylase and at the 3´ end by the insulin gene. The IGF-2 and insulin genes have the same polarity and are separated by 12.6 kb of intergenic DNA that includes a dispersed middle repetitive Alu sequence.

The IGF-2 gene has been shown to be subject to parental imprinting with the paternal IGF-2 gene being expressed while the maternal allele being silent in most tissues.

■ **Related Factors:** IGF-1 and IGF-2 are members of a protein family of related proteins with hormonal activities including nerve growth factor (see: NGF) and the female sex hormone relaxin.

■ **Receptor Structure, Gene(s), Expression:** The IGF-1 receptor is a transmembrane glycoprotein of 350 kDa which is generated from a precursor of 1367 amino acids. The receptor is a heterotetrameric disulfide-linked protein. It consists of two α (135 kDa) subunits, which bind IGF-1, and two β (90 kDa) subunits. The β subunit possesses intrinsic tyrosine-specific protein kinase activity. This kinase domain shows ≈ 84 % homology with the insulin receptor. IGF-1 also binds to the insulin receptor and vice versa although the heterologous ligand always binds ≈ 100-fold less well. Binding of IGF-1 to its receptor leads to the » autophosphorylation of the β subunit and also to the phosphorylation of some cytoplasmic substrate proteins. The gene encoding the IGF-1 receptor maps to human chromosome 15q25-q26 in the vicinity of the *fes* oncogene.

The IGF-2 receptor is a single-chain protein of 250 kDa. It is a high affinity binding protein for IGF-2 and shows very low affinity for IGF-1 and none for insulin. The cytoplasmic domain of the receptor does not possess intrinsic tyrosine kinase activity. This receptor identical with cation-independent mannose-6-phosphate receptor (see: MPR) involved in lysosomal enzyme targeting. The IGF-2R gene is also subject to parental imprinting with the maternal allele being expressed. A soluble truncat-

ed form of the IGF-2 receptor has also been described. The gene maps to human chromosome 6.
A number of IGF binding proteins lacking receptor signal transduction functions have been found in serum (see: IGFBP, IGF binding proteins). These factors modulate IGF activities. Some of them also possess stimulating effects *in vitro* or are growth inhibitory (see: IDF-45).
Soluble receptors and binding proteins have also been found for » IL1 (see: IL1ra, IL1 receptor antagonist), » IL2, » IL4, » IL6, » IL7, » TNF-α and » IFN-γ. They probably function as physiological regulators of cytokine activities (see also: Cytokine inhibitors) by inhibiting receptor binding or act as transport proteins.

■ **BIOLOGICAL ACTIVITIES:** The two known IGFs were isolated initially as serum factors with insulin-like activities that could not be neutralized by antibodies directed against insulin.
In healthy subjects exogenous IGF-1, like insulin, causes hypoglycemia and a transient decrease in free serum fatty acids. Binding of IGF-1 to carrier proteins prevents the establishment of a permanent hypoglycemia in spite of high serum IGF concentrations. These carrier proteins also increase plasma half lives of IGF and prevent the release of IGF from the blood stream into interstitial spaces.
Cell types responding to IGFs include adipocytes, chondrocytes, epithelial cells, fibroblasts, glial cells, hepatocytes, muscle cells, and osteoblasts. IGFs are mitogenic *in vitro* for a number of mesodermal cell types. IGF-1 is usually a stronger mitogen than IGF-2. IGF-1 is an » autocrine growth factor for astrocytes and a differentiation factor for oligodendrocytes and neurons. IGF-2 is an » autocrine growth factor for human rhabdomyosarcomas and also influences the motility of these cells. It is also an autocrine growth factor for neuroblastoma cells.
IGF influences the cellular differentiation of ovary granulosa cells. In the ovary IGF-1 enhances the activities of estrogens and androgens and upregulates the expression of receptors for » luteinising hormone. IGF-1 and IGF-2 also promote oocyte maturation. In the male reproductive tract IGF-1 (and also » EGF) stimulate androgen production in Leydig cells.
IGF-1 stimulates collagen and matrix synthesis by bone cells *in vitro*. The synthesis of IGF-1 in bone cells is inhibited by cortisol. IGF-1 is considered to be one of the major anabolic factors regulating the metabolism of joint cartilage *in vivo*.
IGF-2 has long been considered to be the fetal form of IGF-1. Although this strict separation can no longer be made IGF-2 still plays an important role as embryonic and fetal growth factor during embryogenesis.
The main factor regulating plasma concentrations of IGF is » growth hormone which regulates the expression of the IGF-1 gene and also controls the expression of some IGF binding proteins. Growth hormone stimulates the synthesis of IGF-1 in hepatocytes, fibroblasts, chondrocytes. In fibroblasts the synthesis of IGF-1 is also stimulated by » PDGF and » FGF. IGF-1 is more dependent on growth hormone than IGF-2. IGF-1 promotes the growth of hypophysectomized or diabetic rats and also of Snell dwarf mice with congenital growth hormone deficiency. While growth hormone is ineffective in diabetic animals their growth is enhanced by IGF-1.
IGF-1 establishes a negative feedback loop in inhibiting the secretion of growth hormone. This, in turn, reduces the secretion of GHRH (growth hormone releasing hormone) and stimulates the release of » Somatostatin in the hypothalamus. Serum levels of IGF-1 are also modulated by » prolactin, placental lactogen, thyroid hormones and steroids which act, at least in part, by influencing growth hormone levels. IGF-1 is therefore responsible for many, if not all, growth-promoting activities ascribed to » growth hormone.
IGF-1 induces skeletal growth and also differentiation of myoblasts during fetal development.
IGF, which is probably also produced by stromal cells of the bone marrow also influences » hematopoiesis, in particular the development of the red blood cell lineage (see: BFU-E, CFU-E). In serum-free cultures (see also: SFM, serum-free medium) IGF-1 functions as a colony-stimulating factors for erythroid cells (see also: CSF). The ***erythropoiesis-stimulating activity*** found in the serum of anephric patients has been identified as IGF-1. In neonatal or hypophysectomized experimental animals the injection of IGF-1 also enhances erythropoiesis.

■ **TRANSGENIC/KNOCK-OUT/ANTISENSE STUDIES:** The analysis of brain growth and myelination in a transgenic mouse line that overexpresses IGF-I indicate that IGF-I is a potent inducer of brain growth and myelination *in vivo*. At postnatal day 55, when brain growth and myelination are essentially complete in normal mice, the brains of transgenic mice are 55% larger than those of controls owing to an increase in cell size and apparently in cell number. Total myelin content of the transgenic

mice is also enhanced, and this is primarily the result of an increase in myelin production per oligodendrocyte.
Mice that carry a targeted disruption (see also: ES cells) of the gene encoding IGF-I have been constructed. Heterozygous animals are ≈ 10-20 % smaller than wild-type litter mates and have lower than normal levels of IGF-I. These mice are healthy and fertile and all tissues appear histologically normal. Homozygous mutant mice are < 60 % body weight of wild type animals. More than 95 % of these animals die perinatally. Histophathology is characterized by a gross underdevelopment of muscle tissues. Lung tissues of late embryonic and neonates are less organized with ill-defined alveolae. IGF-I thus appears to be essential for correct embryonic development in mice.
Mice that carry a targeted disruption (see also: ES cells) of the gene encoding IGF-II have been constructed. Transmission of this mutation through the male germline results in heterozygous progeny that are growth deficient. In contrast, when the disrupted gene is transmitted maternally, the heterozygous offspring are phenotypically normal. Homozygous mutants are indistinguishable in appearance from growth-deficient heterozygous siblings. The analysis of transcripts from the wild-type and mutated alleles indicates that only the paternal allele is expressed in embryos, while the maternal allele is silent. An exception is the choroid plexus and leptomeninges, where both alleles are transcriptionally active. These results demonstrate that IGF-II is indispensable for normal embryonic growth and that the IGF-II gene is subject to tissue-specific parental imprinting. Mice carrying a targeted disruption of the IGF-1 gene also exhibit growth deficiencies similar in severity to that observed in IGF-2 null mutants. Some of the IGF-1-deficient dwarf mice die shortly after birth, but some survive and reach adulthood.
Null mutants of the IGF-1 receptor gene die invariably at birth and exhibit a more severe growth deficiency (45 % normal size. These animals have a general organ hypoplasia. Double mutants deficient in IGF-1 and IGF-1 receptor expression do not differ in phenotype from single mutants of the IGF-1 receptor. Double mutants of IGF-2 and IGF-1 receptor and of IGF-1/IGF-2 are phenotypically identical, showing a dwarfism with 30 % of the normal size.
The use of » antisense oligonucleotide inhibiting production and secretion of IGF-I by a human embryonic lung fibroblast cell line (WI-38) has shown that IGF-I acts as an » autocrine growth factor for these cells.

■ **Assays:** IGF can be detected by sensitive immunoassays and radioreceptor assays. IGF is also detected by measurement of sulfate incorporation into porcine cartilage cells. A » bioassay involving the use of the » MCF-7 breast cancer cell line is also available. For further information see also subentry "Assays" in the reference section.

■ **Clinical Use & Significance:** IGF-1 levels are used in the diagnosis of pathological states affecting levels of » growth hormone since normal levels of IGF-1 practically exclude a growth hormone deficiency. In patients with acromegaly (enhanced growth hormone levels) IGF-1 levels are also increased and these levels correlate with the severity of the disease. The importance of IGF-1 as a laboratory parameter is based on the fact that the expression of IGF-1, in contrast to growth hormone, does not show daily fluctuations. IGF-1 levels can be measured without a provocation test, in contrast to growth hormone.
IGF-1 is not found in patients with » Laron-type dwarfism, a congenital endocrinological disease characterized by lacking growth hormone receptors.
IGF-1 also binds to the insulin receptor and therefore induces hypoglycemia. Growth hormone induces an increase in blood glucose levels and stimulates the secretion of insulin and may therefore precipitate diabetes and insulin resistance. IGF-1, on the other hand, promotes insulin actions by inhibiting growth hormone. IGF-1 may therefore be of interest in improving relative insulin sensitivity in metabolic diseases associated with insulin resistance such as adiposity, diabetes mellitus and hypertriglyceridemia.
IGF-1 may also be of therapeutical interest since it lowers total cholesterol/HDL ratios. There are some indications that IGF-1, unlike insulin, may not affect lipogenetic activities of insulin and the down-r3egulation of insulin receptors and also atherogenetic processes to the same extent as insulin. IGF-1 may be of interest in the treatment of renal complications associated with diabetes mellitus since IGF-1 stimulates kidney functions. The infusion of IGF-1 increases creatinine clearance and renal plasma flow by ≈ 30 % without causing proteinura and without altering glomerular filtration pressures.
In degenerative joint diseases local application of IGF-1 may be of interest due to its ability to stimu-

late osteoblast activity and the production of collagen. The factor may also be valuable in fracture healing and in the treatment of postmenopausal osteoporosis.

It has also been suggested that centrally administered IGF-1 may have therapeutic potential for brain injury since it reduces neuronal loss after unilateral hypoxic-ischemic injury in experimental animals.

Rat hippocampal and human cortical neurons have been shown to be protected by IGFs against induced damage induced by iron, which is believed to contribute to the process of cell damage and death resulting from ischemic and traumatic insults by catalyzing the oxidation of protein and lipids.

It has been demonstrated that the continuous subcutaneous infusion of IGF-1 reduces gut atrophy and bacterial translocation in severely burned experimental animals. Recombinant human IGF-1 may be useful, therefore, to improve gut mucosal function and reduce infectious morbidity in severely traumatized or septic patients.

Due to its activity as an » angiogenesis factor IGF-1 has been implicated as one of the stimuli for retinal neovascularization in proliferative diabetic retinopathy.

● **Reviews: Cohick WS & Clemmons DR** The insulin-like growth factors. ARP 55: 131-53 (1993); **Daughaday WH & Rotwein P** Insulin-like growth factors I and II. Peptide, messenger ribonucleic acid and gene structures, serum, and tissue concentrations. Endocrinol. Rev. 10: 69-91 (1989); **Florini JR et al** Hormones, growth factors, and myogenic differentiation. ARP 53: 201-16 (1991); **Humbel RE** Insulin-like growth factors 1 and 2. EJB 190: 445-62 (1990); **LeRoith D et al** Insulin-like growth factors and their receptors as growth regulators in normal physiology and pathologic states. Trends Endocrinol. Metabolism 2: 134-9 (1991); **Oh Y et al** New concepts in insulin-like growth factor receptor physiology. Growth Regul. 3: 113-23 (1993); **Schofield PM** (ed.) The Insulin-like growth factors: Structure and Biological Functions. Oxford University Press 1992; **Spencer EM** (edt) Modern concepts of insulin-like growth factors. Elsevier Sci. Pub., New York, 1991

● **Biochemistry & Molecular Biology: Bayne S et al** Primary sequences of insulin-like growth factors 1 and 2 isolated from porcine plasma. J. Chromat: Biomedical Applications 562: 391-402 (1991); **Bell GI et al** Sequence of a cDNA clone encoding human preproinsulin-like growth factor II. N 310: 775-7 (1984); **Bell GI et al** Isolation of the human insulin-like growth factor genes: Insulin-like growth factor II and insulin genes are contiguous. PNAS 82: 6450-4 (1985); **Bowcock A & Sartorelli V** Polymorphism and mapping of the IGF-1 gene, and absence of association with stature among African Pygmies. Hum. Genet. 85: 349-54 (1990); **Brissenden JE et al** Human chromosomal mapping of genes for insulin-like growth factors I and II and epidermal growth factor. N 310: 781-4 (1984); **Chairotti L et al** Structure of the rat insulin-like growth factor 2 transcriptional unit. ME 2: 1115-22 (1988); **de Pagter-Holthuizen P et al** Chromosomal localization and preliminary characterization of the human gene encoding insulin-like growth factor II. Hum. Genet. 69: 170-3 (1985); **de Pagter-Holthuizen P et al** Organization of the human genes for insulin-like growth factors 1 and 2. FL 195: 179-84 (1986); **Fotsis T et al** Nucleotide sequence of the bovine insulin-like growth factor 1 (IGF-1) and its IGF-1A precursor. NAR 18: 676 (1990); **Jansen M et al** Sequence of cDNA encoding human insulin-like growth factor I precursor. N 306: 609-11 (1983); **Morton CC et al** Human genes for insulin-like growth factor I and II and epidermal growth factor are located on 12q22-q24.1, 11p15, and 4q25-q27, respectively. Cytogenet. Cell Genet. 41: 245-9 (1986); **Nilsson B et al** Efficient secretion and purification of human insulin-like growth factor I with a gene fusion vector in *Staphylococci*. NAR 13: 1151-6 (1985); **Nilsson B et al** Expression and purification of recombinant insulin-like growth factors from *Escherichia coli*. MiE 198: 3-16 (1991); **Rall LB et al** Human insulin-like growth factor I and II messenger RNA: isolation of complementary DNA and analysis of expression. MiE 146: 239-48 (1987); **Rinderknecht E & Humbel R** The amino acid sequence of human insulin-like growth factor-1 and its structural homology with insulin. JBC 253: 2769-73 (1978); **Rinderknecht E & Humbel R** Primary structure of human insulin-like growth factor-2. FL 89: 283-6 (1978); **Rotwein P et al** Organization and sequence of the human insulin-like growth factor 1 gene. JBC 261: 4828-32 (1986); **Rotwein P** Structure, evolution, expression and regulation of insulin-like growth factors I and II. GF 5: 3-18 (1991); **Schofield PN** Molecular biology of the insulin-like growth factors: Gene structure and expression. Acta Paed. Scand. 80, Suppl. 372: 83-90 (1991); **Tamura K et al** Primary structure of rat insulin-like growth factor I and its biological activities. JBC 264: 5616-21 (1989); **Taylor BA & Grieco D** Localization of the gene encoding insulin-like growth factor I on mouse chromosome 10. Cytogenet. Cell Genet. 56: 57-8 (1991); **Tricoli JV et al** Localization of insulin-like growth factor genes to human chromosomes 11 and 12. N 310: 784-6 (1984); **Ullrich A et al** Isolation of the human insulin-like growth factor I gene using a single synthetic DNA probe. EJ 3: 361-4 (1984); **Whitfield HJ et al** Isolation of a cDNA clone encoding rat insulin-like growth factor-II precursor. N 312: 277-80 (1984); **Zumstein PP et al** Amino acid sequence of a variant pro-form of insulin-like growth factor II. PNAS 82: 3169-72 (1985)

● **Receptor Structure, Gene(s), Expression: Barlow DP et al** The mouse insulin-like growth factor type 2 receptor is imprinted and closely linked to the Tme locus. N 349: 84-7 (1991); **Czech MP** Signal transmission by the insulin-like growth factors. Cell 59: 235-8 (1989); **Haig D & Graham C** Genomic imprinting and the strange case of the insulin-like growth factor 2 receptor. Cell 64: 1045-6 (1991); **MacDonald R et al** Serum form of the rat insulin-like growth factor II/mannose-6-phosphate receptor is truncated in the carboxy-terminal domain. JBC 264: 3256-61 (1989); **Neely EK et al** Insulin-like growth factor receptors. Acta Paed. Scand. 80, Suppl. 372: 116-23 (1991); **Nissley P & Lopaczynski W** Insulin-like growth factor receptors. GF 5: 29-43 (1991); **Sundaresan S & Francke U** Insulin-like growth factor I receptor gene is concordant with c-fes proto-oncogene and mouse chromosome 7 in somatic cell hybrids. Somat. Cell Mol. Genet. 15: 373-6 (1989); **Tong PK et al** The cation-independent mannose-6-phosphate receptor binds insulin-like growth factor 2. JBC 263: 2585-8 (1987); **Ullrich A et al** Insulin-like growth factor 1 receptor primary structure: comparison with insulin receptor suggests structural determinants that define functional specificity. EJ 5: 2503-12 (1986); **Ullrich A** Insulin-like growth factor I receptor cDNA cloning. MiE 198: 17-25 (1991)

● **Biological Activities: Aron DC** Insulin-like growth factor I and erythropoiesis. BioFactors 3: 211-6 (1992); **Chatelain P et al** Paracrine and autocrine regulation of insulin-like growth factor I. Acta Paed. Scand. 80, Suppl. 372: 92-5 (1991); **Congote LF et al** The N-terminal sequence of the major erythropoietic factor of an anephric patient is identical to insulin-like growth factor 1. J. Clin. Endocrinol. Metab. 72: 727-9 (1991); **El-Brady OM et al** Insulin-like growth factor II acts as an autocrine growth and motility factor in human rhabdomyosarcoma tumors. Cell Growth Differ. 1: 325-31 (1990); **Gluckman PD et al** The endocrine role of insulin-like growth factor I. Acta Paed. Scand. 80, Suppl. 372: 97-105 (1991); **Gluckman PD et al** The role of the insulin-like growth factor system in neuronal rescue. ANY 692: 138-48 (1993); **McCarthy TL et al** Cortisol inhibits the synthesis of insulin-like growth factor I in bone cell cultures. E 126: 1569-75 (1990); **Shoyab M et al** Purification of a colon cancer cell growth inhibitor and its identification as an insulin-like growth factor binding protein. CR 51: 2813-9 (1991)

● **Transgenic/Knock-out/Antisense Studies: Baker J et al** Role of insulin-like growth factors in embryonic and postnatal growth. Cell 75: 73-82 (1993); **Behringer RR et al** Expression of insulin-like growth factor I stimulates normal somatic growth in growth hormone-deficient transgenic mice. Endocrinology 127: 1033-40 (1990); **Carson MJ et al** Insulin-like growth factor I increases brain growth and central nervous system myelination in transgenic mice. Neuron 10: 729-40 (1993); **De Chiara TM et al** Parental imprinting of the mouse insulin-like growth factor II gene. Cell 64: 849-59 (1991); **D'Ercole AJ** Expression of insulin-like growth factor-I in transgenic mice. ANY 692: 149-60 (1993); **Doi T et al** Glomerular lesions in mice transgenic for growth hormone and insulin-like growth factor-I. I. Relationship between increased glomerular size and mesangial sclerosis. Am. J. Pathol. 137: 541-52 (1990); **Liu JP et al** Mice carrying null mutations of the genes encoding insulin-like growth factor I (IGF-1) and type 1 IGF receptor (Igf1r). Cell 75: 59-72 (1993); **Mathews LS et al** Expression of insulin-like growth factor I in transgenic mice with elevated levels of growth hormone is correlated with growth. Endocrinology 123: 433-7 (1988); **Mathews LS et al** Growth enhancement of transgenic mice expressing human insulin-like growth factor I. Endocrinology 123: 2827-33 (1988); **Moats-Straats BM et al** Insulin-like growth factor-I (IGF-I) antisense oligodeoxynucleotide mediated inhibition of DNA synthesis by WI-38 cells: evidence for autocrine actions of IGF-I. Mol. Endocrinol. 7: 171-80 (1993); **Powell-Braxton L et al** IGF-1 is required for normal embryonic growth in mice. Genes Dev. 7: 2609-17 (1993); **Quaife CJ et al** Histopathology associated with elevated levels of growth hormone and insulin-like growth factor I in transgenic mice. Endocrinology 124: 40-8 (1989)

● **Assays: Asakawa K et al** Radioimmunoassay for insulin-like growth factor II (IGF-II). Endocrinol. Jpn. 37: 607-14 (1990); **Bang P et al** Comparison of acid ethanol extraction and acid gel filtration prior to IGF-I and IGF-II radioimmunoassays: improvement of determinations in acid ethanol extracts by the use of truncated IGF-I as radioligand. Acta Endocrinol. Copenh. 124: 620-9 (1991); **Baxter RC** Radioimmunoassay for insulin-like growth factor (IGF) II: interference by pure IGF-binding proteins. J. Immunoassay 11: 445-58 (1990); **Baxter RC et al** Natural and recombinant DNA-derived human insulin-like growth factor-I compared for use in radioligand assays. Clin. Chem. 33: 544-8 (1987); **Bicsak TA et al** Insulin-like growth factor binding protein measurement: sodium dodecyl sulfate-stable complexes with insulin-like growth factor in serum prevent accurate assessment of total binding protein content by ligand blotting. AB 191: 75-9 (1990); **Binoux M et al** Specific assay for insulin-like growth factor (IGF) II using the IGF binding proteins extracted from human cerebrospinal fluid. J. Clin. Endocrinol. Metab. 63: 1151-5 (1986); **Blum WF et al** Growth hormone resistance and inhibition of somatomedin activity by excess of insulin-like growth factor binding protein in uremia. Pediatr. Nephrol. 5: 539-44 (1991); **Blum WF et al** A specific radioimmunoassay for insulin-like growth factor II: the interference of IGF binding proteins can be blocked by excess IGF-I. Acta Endocrinol. Copenh. 118: 374-80 (1988); **Bobek G et al** Radioimmunoassay of soluble insulin-like growth factor-II/mannose 6-phosphate receptor: developmental regulation of receptor release by rat tissues in culture. Endocrinology 130: 3387-94 (1992); **Bowsher RR et al** Measurement of insulin-like growth factor-II in physiological fluids and tissues. I. An improved extraction procedure and radioimmunoassay for human and rat fluids. Endocrinology 128: 805-4 (1991); **Breier BH et al** Radioimmunoassay for insulin-like growth factor-I: solutions to some potential problems and pitfalls. J. Endocrinol. 128: 347-57 (1991); **Daughaday WH** Radioligand assays for insulin-like growth factor II. MiE 146: 248-59 (1987); **Daughaday WH & Trivedi B** Measurement of derivatives of proinsulin-like growth factor-II in serum by a radioimmunoassay directed against the E-domain in normal subjects and patients with nonislet cell tumor hypoglycemia. J. Clin. Endocrinol. Metab. 75: 110-5 (1992); **Daughaday WH & Trivedi B** Heterogeneity of serum peptides with immunoactivity detected by a radioimmunoassay for proinsulin-like growth factor-II E domain: description of a free E domain peptide in serum. J. Clin. Endocrinol. Metab. 75: 641-5 (1992); **Daughaday WH et al** Serum somatomedin binding proteins: physiologic significance and interference in radioligand assay. J. Lab. Clin. Med. 109: 355-63 (1987); **Delta A & Hitchcock E** Rapid detection of gene expression. Mol. Cell. Probes. 5: 437-43 (1991); **Furlanetto RW & Marino JM** Radioimmunoassay of somatomedin C/insulin-like growth factor I. MiE 146: 216-26 (1987); **Giannella-Neto D et al** Evaluation of a radioimmunoassay for somatomedin-C/insulin-like growth factor I (Sm-C/IGF-I) in human plasma. Clin. Chim. Acta. 188: 253-60 (1990); **Glick RP et al** Radioimmunoassay of insulin-like growth factors in cyst fluid of central nervous system tumors. J. Neurosurg. 74: 972-8 (1991); **Hintz RL et al** A sensitive radioimmunoassay for somatomedin-C/insulin-like growth-factor I based on synthetic insulin-like growth factor 57-70. Horm. Metab. Res. 20: 344-7 (1988); **Johnson TR et al** Newly synthesized RNA: simultaneous measurement in intact cells of transcription rates and RNA stability of insulin-like growth factor I, actin, and albumin in growth hormone-stimulated hepatocytes. PNAS 88: 5287-91 (1991); **Kao PC et al** Assay of somatomedin C by cartridge extraction prior to radioimmunoassay with antiserum developed against synthetic somatomedin C. Ann. Clin. Lab. Sci. 18: 120-30 (1988); **Karey KP et al** Human recombinant insulin-like growth factor I. II. Binding characterization and radioreceptor assay development using Balb/c 3T3 mouse embryo fibroblasts. In Vitro Cell Dev. Biol. 24: 1107-13 (1988); **Lee WH et al** Measurement of insulin-like growth factor-II in physiological fluids and tissues. II. Extraction quantification in rat tissues. Endocrinology 128: 815-22 (1991); **McMorris FA et al** Insulin-like growth factor I/somatomedin C: a potent inducer of oligodendrocyte development. PNAS 83: 822-6 (1986); **Mesiano S et al** Failure of acid-ethanol treatment to prevent interference by binding proteins in radioligand assays for the insulin-like growth factors. J. Endocrinol. 119: 453-60 (1988); **Mitchell ML et al** Radioimmunoassay of somatomedin-C in filter paper discs containing dried blood. Clin. Chem. 33: 536-8 (1987); **Miyakawa M et al** Radioimmunoassay for insulin-like growth factor I (IGF-I) using biosynthetic IGF-I. Endocrinol. Jpn. 33: 795-801 (1986); **Moller C et al** Quantitative comparison of insulin-like growth factor mRNA levels in human and rat tissues analyzed by a solution hybridization assay. J. Mol. Endocrinol. 7: 13-22 (1991); **Morrell DJ et al** A monoclonal antibody to human insulin-like growth factor-I: characterization, use in radioimmunoassay and effect on the biological activities of the growth factor. J. Mol. Endocrinol. 2: 201-6 (1989); **Pekonen F et al** A monoclonal antibody-based immunoradiometric assay for low molecular weight insulin-like growth factor binding protein/placental protein 12. J. Immunoassay. 10: 325-37 (1989); **Perdue JF et al** Development of a specific radioimmuno assay for E domain containing forms of insulin-like growth factor II. AEMB 293: 45-56 (1991); **Pezzino V et al** Radioimmunoassay for human insulin-like growth factor-I receptor: applicability to breast carcinoma specimens and cell lines. Metabolism 40: 861-5 (1991); **Pfeifle B et al** Radioimmunoassay for the measurement of insulin-like growth factor I in patients with pituitary disease in comparison with commercially available somatomedin-C radioimmunoassays. J. Clin. Chem. Clin. Biochem. 2: 393-8 (1986); **Rutanen EM et al** Measurement of insulin-like growth factor binding protein-1 in cervical/vaginal secretions: comparison with the ROM-check Membrane Immunoassay in the diagnosis of ruptured fetal membranes. Clin. Chim. Acta 214: 73-81 (1993); **Silbergeld A et al** A comparison of IGF-I levels measured by two commercially available radioimmunoassays. Acta Endocrinol. Copenh. 119: 333-8 (1988); **Spadoni GL et al** Determination of IGF-II levels in human serum using the erythroleukemia cell line K562. J. Endocrinol. Invest. 13: 97-102 (1990); **Tamura K et al** Enzyme-linked Immunosorbent assay for human insulin-like growth factor I using monoclonal and polyclonal antibodies with defined epitope recognition. J. Endocrinol. 125: 327-35 (1990); **van Buul-Offers S et al** The bovine placenta: a specific radioreceptor assay for both insulin-like growth factor I and both insulin-like growth factor II. Acta Endocrinol. Copenh. 118: 306-13 (1988); **van Zoelen EJ** The use of biological assays for detection of polypeptide growth factors. Prog. Growth Factor Res. 2: 131-52 (1990); **Wang DY et al** Radioimmunoassayable insulin-like growth factor-I in human breast cyst fluid. Eur. J. Cancer Clin. Oncol. 25: 867-72 (1989); **Zangger I et al** Insulin-like growth factor I and II in 14 animal species and man as determined by three radioligand assays and two bioassays. Acta Endocrinol. Copenh. 114: 107-12 (1987)

● **Clinical Use & Significance:** **Becker W et al** A comparison of ePTFE membranes alone or in combination with platelet-derived growth factors and insulin-like growth factor-I or demineralized freeze-dried bone in promoting bone formation around immediate extraction socket implants. J. Periodontol. 63: 929-40 (1992); **Clemmons DR & Underwood LE** Role of insulin-like growth factors and growth hormone in reversing catabolic states. Horm-Res. 38: s37-40 (1992); **Cotterill AM et al** The therapeutic potential of recombinant human insulin-like growth factor-I. Clin. Endocrinol. Oxf. 37: 11-5 (1992); **Froesch ER et al** Therapeutic potential of insulin-like growth factor I. Teratog. Carcinog. Mutagen. 1: 254-60 (1990); **Gluckman PD & Ambler GR** Therapeutic use of insulin-like growth factor I: lessons from *in vivo* animal studies. Acta Paediatr. 81 Suppl 383: 134-6 (1992); **Gluckman P et al** A role for IGF-1 in the rescue of CNS neurons following hypoxic-ischemic injury. BBRC 182: 593-9 (1992); **Grant MB et al** Insulin-like growth factor I acts as an angiogenic agent in rabbit cornea and retina: comparative studies with basic fibroblast growth factor. Diabetologia 36: 282-91 (1993); **Huang KF et al** Insulin-like growth factor 1 (IGF-1) reduces gut atrophy and bacterial translocation after severe burn injury. Arch. Surg. 128: 47-53 (1993); **Kopple JD** The rationale for the use of growth hormone or insulin-like growth factor I in adult patients with renal failure. Miner. Electrolyte Metab. 18: 269-75 (1992); **Laron Z** Somatomedin-1 (insulin-like growth factor-I) in clinical use. Facts and potential. Drugs 45: 1-8 (1993); **O'Shea M et al** Insulin-like growth factor I and the kidney. Semin. Nephrol. 13: 96-108 (1993); **Rutherford RB et al** Platelet-derived and insulin-like growth factors stimulate regeneration of periodontal attachment in monkeys. J. Periodontal. Res. 27: 285-90 (1992); **Takano K et al** Repeated sc administration of recombinant human insulin-like growth factor I (IGF-I) to human subjects for 7 days. Growth Regul. 1: 23-8 (1991); **Usala AL et al** Brief report: treatment of insulin-resistant diabetic ketoacidosis with insulin-like growth factor I in an adolescent with insulin-dependent diabetes. NEJM 327: 853-7 (1992); **Zhang Y et al** Basic FGF, NGF, and IGFs protect hippocampal and cortical neurons against iron-induced degeneration. J. Cereb. Blood Flow Metab. 13: 378-88 (1993); **Zenobi PD et al** Insulin-like growth factor-I improves glucose and lipid metabolism in type 2 diabetes mellitus. JCI 90: 2234-41 (1992)

IGF-1: *insulin-like growth factor1* see: IGF.

IGF-2: *insulin-like growth factor2* see: IGF.

IGF-BP: *insulin-like growth factor binding protein(s)* also abbrev. IGBP or IBP. IGF-BPs are found in various body fluids such as blood serum, amniotic fluid, and liquor. They are synthesized in the liver and are also produced by various tumor cell lines and cell types. Some cells produce and secrete some IGF-BPs constitutively and the synthesis of some IGF-BPs is regulated, among other things, by IGFs themselves. IGFs also appear to promote directly the proteolytic degradation of some IGF-BPs (IGFBP-4) into fragments that do not bind IGFs, thus providing a mechanism by which IGFs may increase their own availability and/or activity in biological fluids.

IGF-BPs are high affinity binding proteins for » IGF. The major fraction of IGFs circulating in the blood are bound non-covalently to these carrier proteins, forming complexes of 140-150 kDa (large complex) and 40 kDa (small complex) which may be monomeric and oligomeric $(IGF\text{-}BP)_n$ complexes. Some IGFBP/IGF complexes may contain additional complexed proteins. The formation of these complexes may be inhibited by glycosaminoglycans. IGFs can be released from these complexes by treatment with acids, heparin, proteases, and plasmin.

IGF binding proteins modulate IGF activities by increasing their plasma half lives and by inhibiting or promoting the interactions of IGFs with their receptors on certain target cells. In addition these binding proteins provide a reservoir for IGF in pericellular spaces. Some IGFBPs also have stimulating effects *in vitro* and some may inhibit the growth of cells (see also: IDF-45). Some granulosa cell-derived IGF-BPs appear to function as antigonadotropins at the level of the ovary. Small-cell lung cancer cells have been found to produce and release IGF binding proteins which differ from those found in liver and placenta. It has been suggested that they may function as mediators in the » autocrine and/or » paracrine growth regulation of IGFs in these tumors.

IGF-BPs are cysteine-rich proteins of which various molecular forms are known that may also differ in the extent of glycosylation. They have no sequence homology with the IGF receptors. The proteins display strong sequence homologies, suggesting that they are encoded by a closely related family of genes. The number (18) and position of the cysteine residues is conserved in almost all IGF-BPs. There are some indications that IGF-BPs can be phosphorylated and that phosphorylation also alters their biological activities. At present at least six different IGF binding proteins are known. They differ in their binding efficiencies.

BP-1 is a protein of 34 kDa (= ***BP-34***) 25 and 28 kDa forms (25 k and 27 k IGF-BP) of this protein have also been described and probably arise by differences in glycosylation. The gene encoding this protein has a length of 5.2 kb and contains four exons. It maps to human chromosome 7p14-p12 at a distance of ≈ 20 kb from the IGF-BP3 gene. The transcription of the BP-1 gene is repressed by insulin while inhibitors of glucose uptake such as cytochalasin B enhance the synthesis of BP-1. Cortisol also increases plasma levels of BP-1 in humans. The synthesis of BP-1 by human hepatoma cells is enhanced by » EGF. BP-1 has equal affinity for IGF-1 and IGF-2. Serine phosphorylation of BP-1 has been shown to alter its affinity for IGF-I and IGF-II and to modify its capacity to modulate cellular responses to the IGFs.

The growth-inhibitory activities of IGF-BP1 resemble those of another IGF-binding protein called » IDF-45. BP-28 inhibits serum IGF bioactivity on cartilage *in vitro*. BP-1 has also been shown to be an inhibitor if IGF mitogenic activities for human breast cancer cells. BP-28 may also be one of the IGF-I inhibitors observed in diabetic serum and may play a role in retarded growth and delayed puberty often seen in the adolescent diabetic. BP-1 enhances the mitogenic effect of IGF-I, but not that of IGF-II, in cultured human keratinocytes and fibroblasts.

In the human osteosarcoma cell line MG-63, BP-1 can form an IGF reservoir in the pericellular space surrounding the cells by forming complexes that are incapable of binding to the IGF receptors. These complexes can be dissolved by plasmin. The secretion of plasminogen activators by osteosarcoma cells and the availability of plasminogen in the extravascular tissues may provide a regulatory system in osteosarcoma cells in which pericellular plasmin affects the availability of IGFs to their membrane receptors.

34 k IGF-BP is identical with » PP12 (placental protein 12). The protein is also known as ***AFBP*** (amniotic fluid binding protein), ***α-pregnancy-associated binding protein***, ***growth hormone independent binding protein***, ***binding protein 28***, ***binding protein 26***, and ***binding protein 25***.

BP-1 is found predominantly in the placenta and the amniotic fluid. The predominant sites of BP-1 transcription in the human fetal kidney are those with most active differentiation. Elevated serum levels have been observed in patients with » Laron type dwarfism and » growth hormone deficiency. High serum levels of BP-1 are found in newborns and it has been suggested that this could be important in protecting them from hypoglycemia.

BP-2 (27 kDa, 24 kDa; 289 amino acids) is observed mainly in brain and liquor, showing complex patterns of gene expression during postnatal brain development. Elevated levels BP-2 have been observed in the serum of prostate cancer patients. BP-2 expression does not depend on » growth hormone. The gene maps to human chromosome 2q33-q34. It contains four exons and has a length of 32 kb. BP-2 has been shown to be secreted by intestinal epithelial cells and to capable of limiting the mitogenic activity of both exogenous and endogenous IGFs by blocking the association of the growth factors with cell surface binding sites. BP-2 has also been implicated in myeloblast differentiation. It shows preferential affinity for IGF-2.

BP-3 (264 amino acids, 53 kDa; = ***BP-53***) is the major IGF binding protein present in serum of humans and animals. It is also present in the α granules of platelets. BP-3 shows a similar affinity for IGF-I and IGF-II. The mature protein is cysteine-rich and has a length of 264 amino acids. Its amino acid sequence is 33 % identical with that of BP-2. IGFBP-3 is also known as ***growth hormone dependent binding protein***, acid stable subunit of the 140 K IGF complex, and ***binding protein 29***. The human gene has a length of 8.9 kb and contains 5 exons. It maps to chromosome 7p14-p12 in the vicinity of the IGF-BP1 gene. Smaller fragments of BP-3 consisting of various C- or N-terminally truncated forms have also been described. A proteolytic enzyme specific for BP3 has been isolated from serum of pregnant women.

The 140 kDa IGF binding protein complex in human serum consists of three subunits: an acid-labile, non-IGF-binding glycoprotein (α-subunit), BP-3 (β-subunit), and IGF-I or IGF-II (γ-subunit). Glycosaminoglycans have been shown to inhibit complex formation without affecting the binary complex. Since the ternary IGF-binding protein complex cannot cross the capillary barrier, a decrease in the affinity of the complex, mediated by circulating or cell-associated glycosaminoglycans, may be important in the passage of IGFs and IGFBP-3 to the tissues.

BP-3 inhibits follicle-stimulating hormone. Markedly decreased levels of BP-3 are observed in patients with » growth hormone deficiencies (see also: Laron dwarfism), while markedly elevated levels are observed in patients with high levels of » growth hormone (acromegaly). In murine fibroblasts the synthesis of IGF-BP-3 is stimulated by mitogenic growth factors such as » Bombesin, » Vasopressin, » PDGF, and » EGF. In human skin fibroblasts the synthesis of IGF-BP3 is stimulated by » TGF-β.

Stimulation by serum of dense cultures of murine » 3T3 fibroblast cells rapidly induces increased synthesis of a growth inhibitor identified as murine IGFBP-3. Secretion of the factor is also induced by » bFGF, » PDGF, and insulin. bFGF- stimulated DNA synthesis is arrested when accumulation of mIGFBP-3 is maximal, suggesting that the accumulation of mIGFBP-3 may induce a feedback regulation of cell growth.

BP-4 (237 amino acids) is the predominant IGF binding proteins expressed by human osteoblast-like cells. It is identical with a protein known as ***colon cancer cell growth inhibitor***. The human gene is on chromosome 17q12-q21 in the vicinity of the BRCA1 (hereditary breast-ovarian cancer gene) gene. This protein, which is also known as ***HT29-IGF-BP*** inhibits the growth of the human adenocarcinoma cell line HT29. BP-4 has been found to inhibit both basal and IGF-mediated chick pelvic cartilage growth *in vitro*. BP-4 is subject to proteolysis by a BP-4-specific protease induced by IGF-II that modifies BP-4 structure and function. This posttranslational regulation of BP-4 may provide a means for cooperative control of local cell growth by IGF-I and IGF-II.

BP-5 is a protein of 23 kDa (252 amino acids) encoded by a gene on human chromosome 5. It is highly expressed in the kidney but is also found in other tissues. Human osteoblast-like bone cells in culture secrete several IGF-binding proteins, including BP-5. It displays similar and relatively low affinities for IGF-I and IGF-II. It enhances mitogenesis in the presence of IGF-I or IGF-II and also stimulates mitogenesis in the absence of exogenous or endogenous IGF.

Human ***BP-6*** has a length of 216 amino acids (22.8 kDa) and is an O-glycosylated protein. It is abundant in cerebrospinal fluid and has a marked preferential binding affinity for IGF-II over IGF-I. The gene maps to human chromosome 12. Levels of BP-6 (and also of BP-5) have been found to be increased in human breast cancer cells treated with estradiol and IGF-I and may thus contribute to mitogenesis.

A 29 kDa IGF-BP has been found in human bone. This protein has a much higher affinity for IGF-II than IGF-I and potentiates the proliferative actions of IGF-II on bone cells.

Report on the nomenclature of the IGF binding proteins. Endocrinology 130: 1736-7 (1992)

● Reviews: **Baxter RC** The insulin-like growth factors and their binding proteins. Comp. Biochem. Physiol. B 91: 229-35 (1988); **Baxter RC** Insulin-like growth factor (IGF) binding proteins: The role of serum IGFBPs in regulating IGF availability. Acta Paed. Scand. 80, Suppl. 372: 107-14 (1991); **Cohen P et al** Clinical aspects of insulin-like growth factor binding proteins. Acta Endocrinol. 124, Suppl. 2: 74-85 (1991); **Lamson G** Insulin-like growth factor binding proteins: Structural and molecular relationships. GF 5: 19-28 (1991); **Minuto F et al** Paracrine actions of IGF binding proteins. Acta Endocrinol. 124, Suppl. 2: 63-9 (1991); **Ooi GT** Insulin-like growth factor-binding proteins (IGFBPs): more than just 1, 2, 3. Mol. Cell. Endocrinol. C39-C43 (1990); **Sara VR & Hall K** Insulin-like growth factors and their binding proteins. Physiol. Rev. 70: 591-615 (1990); **Shimasaki S & Ling N** Identification and molecular characterization of insulin-like growth factor binding proteins (IGFBP-1, -2, -3, -4, -5, and -6). PGFR 3: 243-66 (1991)

● General & Uncharacterized: **Bautista CM et al** Isolation of a novel insulin-like growth factor (IGF) binding protein from human bone: A potential candidate for fixing IGF-II in human bone. BBRC 176: 756-63 (1991); **Busby WH et al** Purified preparations of the amniotic fluid-derived insulin-like growth factor binding protein contain multimeric forms that are biologically active. Endocrinol. 125: 773-7 (1989); **Jaques G et al** Production of insulin-like growth factor binding proteins by small-cell lung cancer cell lines. ECR 184: 396-406 (1989)

● **BP1: Angervo M et al** Epidermal growth factor enhances insulin-like growth factor binding protein-1 synthesis in human hepatoma cells. BBRC 189: 1177-83 (1992); **Bell SC et al** Monoclonal antibodies to human secretory "pregnancy-associated endometrial α 1-globulin," an insulin-like growth factor binding protein: characterization and use in radioimmunoassay, Western blots, and immunohistochemistry. AM. J. Reprod. Immunol. 20: 87-96 (1989); **Bernardini S et al** Plasma levels of insulin-like growth factor binding protein-1, and growth hormone binding protein activity from birth to the third month of life. Acta Endocrinol. Copenh. 127: 313-8 (1992); **Brinkman A et al** Organization of the gene encoding the insulin-like. growth factor binding protein IBP-1. BBRC 157: 898-907 (1988); **Campbell PG et al** Involvement of the plasmin system in dissociation of the insulin-like growth factor binding protein complex. Endocrinology 130: 1401-12 (1992); **Conover CA et al** Cortisol increases plasma insulin-like growth factor binding protein-1 in humans. Acta Endocrinol. Copenh. 128: 140-3 (1993); **Ehren-**

borg E et al Contiguous localization of the genes encoding human insulin-like growth factor binding proteins 1 (IGBP1) and 3 (IGBP3) on chromosome 7. Genomics 12: 497-502 (1992); **Ekstrand J et al** The gene for insulin-like growth factor binding protein 1 is localized to human chromosomal region 7p14-p12. Genomics 6: 413-8 (1990); **Frost RA & Tseng L** Insulin-like growth factor binding protein 1 is phosphorylated by cultured human endometrial stromal cells and multiple protein kinases *in vitro*. JBC 266: 18082-8 (1991); **Frost RA et al** Insulin-like growth factor binding protein-1 inhibits the mitogenic effect of insulin-like growth factors and progestins in human endometrial stromal cells. Biol. Reprod 49: 104-11 (1993); **Jones JI et al** Identification of the sites of phosphorylation in insulin-like growth factor binding protein-1. Regulation of its affinity by phosphorylation of serine 101. JBC 268: 1125-31 (1993); **Kratz G et al** Effect of recombinant IGF binding protein-1 on primary cultures of human keratinocytes and fibroblasts: selective enhancement of IGF-1 but not IGF-2-induced cell proliferation. ECR 202: 381-5 (1992); **Lewitt MS et al** Insulin-like growth factor binding protein 1 modulates blood glucose levels. Endocrinology 129: 2254-6 (1991); **Liu F et al** Characterization of insulin-like growth factor binding proteins in human serum from patients with chronic renal failure. J. Clin. Endocrinol. Metab. 70: 620-8 (1990); **McGuire WL jr et al** Regulation of insulin-like growth factor-binding protein (IGFBP) expression by breast cancer cells: use of IGFBP-1 as an inhibitor of insulin-like growth factor action. JNCI 84: 1336-41 (1992); **Oh Y et al** Characterization of the affinities of insulin-like growth factor (IGF)-binding proteins 1-4 for IGF-I, IGF-II, IGF-I/insulin hybrid, and IGF-I analogs. Endocrinology 132: 1337-44 (1993); **Perkel VS et al** An inhibitory insulin-like growth factor binding protein (In-IGFBP) from human prostatic cell conditioned medium reveals N-terminal sequence identity with bone derived In-IGFBP. J. Clin. Endocrinol. Metab. 71: 533-5 (1990); **Powell DR et al** Insulin inhibits transcription of the human gene for insulin-like growth factor-binding protein-1. JCB 266: 18868-76 (1991); **Roghani M et al** Isolation from human cerebrospinal fluid of a new insulin-like growth factor binding protein with a selective affinity for IGF-II. FL 255: 253-8 (1989); **Rosenfeld RG et al** Structural and immunological comparison of insulin-like growth factor binding proteins of cerebrospinal and amniotic fluids. J. Clin. Endocrinol. Metabol. 68: 638-46 (1989); **Rutanen EM & Seppala M** Insulin-like growth factor binding protein-1 in female reproductive functions. Int. J. Gynaecol. Obstet. 39: 3-89 (1992); **Suikkari AM et al** Expression of insulin-like growth factor binding protein-1 mRNA in human fetal kidney. Kidney Int. 42: 749-54 (1992); **Taylor AM et al** The growth hormone independent insulin-like growth factor-I binding protein BP-28 is associated with serum insulin-like growth factor-I inhibitory bioactivity in adolescent insulin-dependent diabetics. Clin. Endocrinol. Oxf. 32: 229-39 (1990); **Waites GT et al** Human pregnancy-associated endometrial α 1-globulin, an insulin-like growth factor binding protein: immunohistological localization in the decidua and placenta during pregnancy employing monoclonal antibodies. J. Endocrinol. 120: 351-7 (1989); **Wang HS et al** Purification and assay of insulin-like growth factor-binding protein-1: measurement of circulating levels throughout pregnancy. J. Endocrinol. 128: 161-8 (1991); **Wang HS & Chard T** The role of insulin-like growth factor-I and insulin-like growth factor-binding protein-1 in the control of human fetal growth. J. Endocrinol. 132: 11-9 (1992); **Yang YW et al** Identification of rat cell lines that preferentially express insulin-like growth factor binding proteins rIGFBP-1, 2, or 3. ME 4: 29-38 (1990)

● **BP2: Agarwal N et al** Sequence analysis, expression and chromosomal localizationof a gene, isolated from a subtracted human retina cDNA library, that encodes an insulin-like growth factor binding protein (IGFBP2). Exp. Eye Res. 52: 549-61 (1991); **Alitalio T et al** The gene encoding human low molecular weight insulin-like growth factor binding protein (IGF-BP25): regional localization to 7p12-p13 and description of a DNA polymorphism. Hum. Genet. 83: 335-8 (1989); **Binkert C et al** Structure of the human insulin-like growth factor binding protein-2 gene. Mol. Endocrinol. 6: 826-36 (1992); **Binkert C et al** Cloning, sequence analysis and expression of a cDNA encoding a novel insulin-like growth factor binding protein (IGFBP-2). EJ 8: 2497-502 (1989); **Bourner MJ et al** Cloning and sequence determination of bovine insulin-like growth factor binding protein-2 (IGFBP-2): Comparison of its structural and functional properties with IGFBP-1. JCBc 48: 215-26 (1992); **Cohen P et al** Elevated levels of insulin-like growth factor-binding protein-2 in the serum of prostate cancer patients. J. Clin. Endocrinol. Metabol. 76: 1031-5 (1993); **Ehrenborg E et al** Structure and localizationof the human insulin-like growth factor-binding protein 2 gene. BBRC 176: 1250-5 (1991); **Ernst CW et al** Gene expression and secretion of insulin-like growth factor-binding proteins during myoblast differentiation. Endocrinology 130: 607-15 (1992); **Feyen JH et al** Recombinant human (Cys281) insulin-like growth factor binding protein 2 inhibits both basal and insulin-like growth factor I-stimulated proliferation and collagen synthesis in fetal rat calvariae. JBC 266: 19469-74 (1991); **Landwehr J et al** Cloning and characterization of the gene encoding murine insulin-like growth factor-binding protein-2, mIGFBP-2. Gene 124: 281-6 (1993); **Lee WH et al** Localization of insulin-like growth factor binding protein-2 messenger RNA during postnatal brain development: correlation with insulin-like growth factors I and II. Neuroscience 53: 251-65 (1993); **Oh Y et al** Characterization of the affinities of insulin-like growth factor (IGF)-binding proteins 1-4 for IGF-I, IGF-II, IGF-I/insulin hybrid, and IGF-I analogs. Endocrinology 132: 1337-44 (1993); **Park JH et al** Secretion of insulin-like growth factor II (IGF-II) and IGF-binding protein-2 by intestinal epithelial (IEC-6) cells: implications for autocrine growth regulation. Endocrinology 131: 1359-68 (1992); **Streck RD et al** Insulin-like growth factor I and II and insulin-like growth factor binding protein-2 RNAs are expressed in adjacent tissues within rat embryonic and fetal limbs. Dev. Biol. 151: 586-96 (1992); **Wood WI et al** Cloning and expression of the growth hormone-dependent insulin-like growth factor-binding protein. Molec. Endocr. 2: 1176-85, (1988); **Yang YW et al** Identification of rat cell lines that preferentially express insulin-like growth factor binding proteins rIGFBP-1, 2, or 3. ME 4: 29-38 (1990)

● **BP3: Baxter RC** Glycosaminoglycans inhibit formation of the 140 kDa insulin-like growth factor binding protein complex. BJ 271: 773-7 (1990); **Conover CA et al** Structural and biological characterization of bovine insulin-like growth factor binding protein-3. E 127: 2795-803 (1990); **Corps AN & Brown KD** Mitogens regulate the production of insulin-like growth factor binding protein by Swiss 3T3 cells. Endocrinology 128: 1057-64 (1991); **deMellow JSM & Baxter RC** Growth hormone-dependent insulin-like growth factor (IGF) binding protein both inhibits and potentiates IGF1-stimulated DNA synthesis in human skin fibroblasts. BBRC 156: 199-204 (1988); **Giudice LC et al** Identification of insulin-like growth factor binding protein 3 (IGFBP-3) and IGFBP-2 in human follicular fluid. J. Clin. Endocrinol. Metab. 71: 1330-8 (1990); **Hasegawa Y et al** Western ligand blot assay for human growth hormone-dependent insulin-like growth factor binding protein (IGFBP-3): the serum levels in patients with classical growth hormone deficiency. Endocrinol. Jpn. 39: 121-7 (1992); **Lamson G et al** A simple assay for proteolysis of IGFBP-3. J. Clin. Endocrinol. Metabol. 72: 1391-3 (1991); **Liu L et al** Synthesis and secretion by mouse

3T3 cells of an inhibitor of cell growth (mIGFBP-3): correlation with cell proliferation. Biol. Cell. 76: 125-30 (1992); **Martin JL & Baxter RC** Transforming growth factor β stimulates production of insulin-like growth factor binding protein 3 by human skin fibroblasts. Endocrinology 128: 1325-33 (1991); **Oh Y et al** Characterization of the affinities of insulin-like growth factor (IGF)-binding proteins 1-4 for IGF-I, IGF-II, IGF-I/insulin hybrid, and IGF-I analogs. Endocrinology 132: 1337-44 (1993); **Schmid C et al** Intact but not truncated insulin-like growth factor binding protein-3 (IGFBP-3) blocks IGF I-induced stimulation of osteoblasts: Control of IGF signaling to bone cells by IGFBP-3-specific proteolysis? BBRC 179: 579-85 (1991); **Shimasaki S et al** Complementary DNA structure of the high molecular weight rat insulin-like growth factor binding protein (IGF-BP3) and tissue distribution of its mRNA. BBRC 165: 907-12 (1989); **Shimasaki S et al** Structural characterization of a follicle-stimulating hormone action inhibitor in porcine ovarian follicular fluid. Its identification as the insulin-like growth factor binding protein. JBC 265: 2198-202 (1990); **Spencer EM et al** Insulin-like growth factor binding protein-3 is present in the α-granules of platelets. Endocrinology 132: 996-1001 (1993); **Wood WI et al** Cloning and expression of the growth hormone-dependent insulin-like growth factor binding protein. Mol. Endocrinol. 2: 1176-85 (1988); **Yang YW et al** Identification of rat cell lines that preferentially express insulin-like growth factor binding proteins rIGFBP-1, 2, or 3. ME 4: 29-38 (1990)

● **BP4: Bajalica S et al** Localization of the human insulin-like growth-factor-binding protein 4 gene to chromosomal region 17q12- 21.1. Hum. Genet. 89: 234-6 (1992); **Conover CA et al** Posttranslational regulation of insulin-like growth factor binding protein-4 in normal and transformed human fibroblasts. Insulin-like growth factor dependence and biological studies. JCI 91: 1129-37 (1993); **Culouscou JM & Shoyab M** Purification of a colon cancer cell growth inhibitor and its identification as an insulin-like growth factor binding protein. CR 51: 2813-9 (1991); **Fowlkes J & Freemark M** Evidence for a novel insulin-like growth factor (IGF)-dependent protease regulating IGF-binding protein-4 in dermal fibroblasts. Endocrinology 131: 2071-6 (1992); **Gao L et al** Structure of the rat insulin-like growth factor binding protein-4 gene. BBRC 190: 1053-9 (1993); **Kiefer MC et al** Identification and molecular cloning of two new 30-kDa insulin-like growth factor binding proteins isolated from adult human serum. JBC 266: 9043-9 (1991); **Kiefer MC et al** Characterization of recombinant human insulin-like growth factor binding proteins 4, 5, and 6 produced in yeast. JBC 267: 12692-9 (1992); LaTour D et al Inhibitory insulin-like growth factor binding protein: cloning, complete sequence, and physiological regulation. Mol. Endocrinol. 4: 1806-14 (1990); **Mohan S & Baylink DJ** Evidence that the inhibition of TE85 human bone cell proliferation by agents which stimulate cAMP production may in part be mediated by changes in the IGF-II regulatory system. Growth Regul. 1: 110-8 (1991); **Neely EK & Rosenfeld RG** Insulin-like growth factors (IGFs) reduce IGF-binding protein-4 (IGFBP-4) concentration and stimulate IGFBP-3 independently of IGF receptors in human fibroblasts and epidermal cells. Endocrinology 130: 985-93 (1992); **Oh Y et al** Characterization of the affinities of insulin-like growth factor (IGF)-binding proteins 1-4 for IGF-I, IGF-II, IGF-I/insulin hybrid, and IGF-I analogs. Endocrinology 132: 1337-44 (1993); **Shimasaki S et al** Characterization of an insulin-like growth factor binding protein (IGFBP-4) produced by the B104 rat neuronal cell line: chemical and biological properties and differential synthesis by sublines. Endocrinology 129: 1006-15 (1991); **Shimasaki S et al** Molecular cloning of the cDNAs encoding a novel insulin-like growth factor binding protein from rat and human. Mol. Endocrinol. 4: 1451-8 (1990); **Schiltz PM et al** Insulin-like growth factor binding protein-4 inhibits both basal and IGF-mediated chick pelvic cartilage growth *in vitro*. J. Bone Miner. Res. 8: 391-6 (1993); **Tonin P et al** The human insulin-like growth factor-binding protein 4 gene maps to chromosome region 17q12-q21.1 and is close to the gene for hereditary breast-ovarian cancer. Genomics 18: 414-417 (1993)

● **BP5: Andress DL & Birnbaum RS** Human osteoblast-derived insulin-like growth factor (IGF) binding protein-5 stimulates osteoblast mitogenesis and potentiates IGF action. JBC 267: 22467-72 (1992); **Backeljauw PF et al** Synthesis and regulation of insulin-like growth factor binding protein-5 in FRTL-5 cells. Endocrinology 132: 1677-81 (1993); **Kiefer MC et al** Identification and molecular cloning of two new 30-kDa insulin-like growth factor binding proteins isolated from adult human serum. JBC 266: 9043-9 (1991); **Kiefer MC et al** Characterization of recombinant human insulin-like growth factor binding proteins 4, 5, and 6 produced in yeast. JBC 267: 12692-9 (1992); **Sheikh MS et al** Identification of the insulin-like growth factor binding proteins 5 and 6 (IGFBP-5 and 6) in human breast cancer cells. BBRC 183: 1003-10 (1992); Shimasaki S et al Identification of five different insulin-like growth factor binding proteins (IGFBPs) from adult rat serum and molecular cloning of a novel IGFBP-5 in rat and human. JBC 266: 10646-53 (1991); **Zhu X et al** Cloning of the rat insulin- like growth factor binding protein-5 gene and DNA sequence analysis of its promoter region. BBRC 190: 1045-52 (1993)

● **BP6: Bach LA et al** Human insulin-like growth factor binding protein-6 is O-glycosylated. BBRC 186: 301-7 (1992); **Baxter RC & Saunders H** Radioimmunoassay of insulin-like growth factor-binding protein-6 in human serum and other body fluids. J. Endocrinol. 134: 133-9 (1992); **Kiefer MC et al** Characterization of recombinant human insulin-like growth factor binding proteins 4, 5, and 6 produced in yeast. JBC 267: 12692-9 (1992); **Sheikh MS et al** Identification of the insulin-like growth factor binding proteins 5 and 6 (IGFBP-5 and 6) in human breast cancer cells. BBRC 183: 1003-10 (1992); **Shimasaki S et al** Isolation and molecular cloning of insulin-like growth factor-binding protein 6. Mol. Endocrinol. 5: 938-48 (1991); **Zhu X et al** Structural characterization of the rat insulin-like growth factor binding protein-6 gene. BBRC 191: 1237-43 (1993)

● **TRANSGENIC/KNOCK-OUT/ANTISENSE STUDIES: Camacho-Hubner C et al** Regulation of insulin-like growth factor (IGF) binding proteins in transgenic mice with altered expression of growth hormone and IGF-I. Endocrinology 129: 1201-6 (1991)

IgG1-enhancing factor: abbrev. IgG1-EF. This factor induces the synthesis of IgG1 (see also: Isotype switching). It is identical with » IL4.

Snapper CM et al Differential regulation of IgG1 and IgE synthesis by interleukin 4. JEM 167: 183-96 (1988); **Vitetta ES et al** Serological, biochemical, and functional identity of B cell-stimulatory factor 1 and B cell differentiation factor for IgG1. JEM 162: 1726-31 (1985)

IgG1 induction factor: abbrev. IgG1-IF. This factor is produced by activated B cells. It elevates the synthesis of IgG1 and suppresses IgG3 and IgG2b responses in lipopolysaccharide-stimulated murine spleen cell cultures (see also: Isotype switching).

The factor is identical with » BSF p1 (B cell stimulating factor p1) and shows the same biological

activities as » BSF 1 (B cell stimulating factor), » BCGF 2 (B cell growth factor) and » TCGF (T cell growth factor). The factor is identical with » IL4.
Isakson PC et al T cell-derived B cell differentiation factor(s). Effect on the isotype switch of murine B cells. JEM 155: 734-48 (1982); **Noma Y et al** Cloning of cDNA encoding the murine IgG1 induction factor by a novel strategy using SP6 promoter. N 319: 640-6 (1986); **Severinson E et al** IgG1 induction factor. Int. Rev. Immunol. 2: 143-56 (1987); **Severinson E et al** Interleukin 4 (IgG1 induction factor): a multifunctional lymphokine acting also on T cells. EJI 17: 67-72 (1987); **Sideras P et al** Secretion of IgG1 induction factor by T cell clones and hybridomas. EJI 15: 586-93 (1985); **Sideras P et al** Partial biochemical characterization of IgG1-inducing factor. EJI 15: 593-8 (1985); **Sideras P et al** IgG1 induction factor: a single molecular entity with multiple biological functions. AEMB 213: 227-36 (1987); **Snapper CM et al** Differential regulation of IgG1 and IgE synthesis by interleukin 4. JEM 167: 183-96 (1988); **Vitetta E et al** Serological, biochemical, and functional identity of B cell-stimulatory factor 1 and B cell differentiation factor for IgG1. JEM 162: 1726-31 (1985)

IgG1 secretion factor: abbrev. IgG1-SF. This factor induces the synthesis of IgG1 (see also: Isotype switching). It is identical with » IL4.
Sideras P et al Partial biochemical characterization of IgG1-inducing factor. EJI 15: 593-8 (1985); **Snapper CM et al** Differential regulation of IgG1 and IgE synthesis by interleukin 4. JEM 167: 183-96 (1988); **Vitetta ES et al** T cell-derived lymphokines that induce IgM and IgG secretion in activated murine B cells. IR 78: 137-57 (1984)

IgG2b-inducing factor: This poorly characterized factor of 50-70 kDa is present in the synovial fluid of rheumatoid arthritis patients. It functions as a T cell-replacing factor (see also: TRF) which selectively induces IgG2b antibody formation in lipopolysaccharide-activated mouse spleen cells *in vitro* and *in vivo*. IgG2b induction is not caused by » IL6, » IL1, » TGF-β, » PDGF, » NGF, » FGF, » EGF, elastase, collagenase, or phospholipase A2 (see also: Isotype switching).
The B cell activity of this factor has been studied in CBA/N mice which have an X-linked B cell immunodeficiency which manifests itself as a defective humoral response to certain thymus-independent antigens (TI-2). The factor appears to be capable of reconstituting partly the B cell deficiency in CBA/N splenic B cells *in vitro*. Cells from CBA/N mice cannot respond to » IL4 after preactivation with bacterial lipopolysaccharide with production of IgG1 antibodies *in vitro*. IgG2b-inducing factor completely restores a normal IL4-induced IgG1 response.
Abedi-Valugerdi M et al Synovial fluid from rheumatoid arthritis patients induces polyclonal antibody formation *in vivo*. Scand. J. Immunol. 30: 587-96 (1989) **Abedi-Valugerdi M et al** Relationship between IgG2b-inducing activity in rheumatoid arthritis synovial fluid and other well-known cytokines and inflammatory mediators. Arthritis Rheum. 34: 1461-5 (1991); **Abedi-Valugerdi M et al** Partial biochemical characterization and purification of IgG2b inducing factor as a new cytokine from synovial fluid of patients with rheumatoid arthritis. Scand. J. Immunol. 37: 430-6 (1993); **Ridderstad A et al** Rheumatoid synovial fluid reconstitutes the B cell defect in CBA/N mice. Scand. J. Immunol. 30: 749-53 (1989)

IIF: *immune interferon* This interferon is identical with » IFN-γ. See also: IFN for general information about interferons.

IIF-2: *(tumor) invasion-inhibiting factor 2* This factor is isolated from bovine liver. It has a length of 20 amino acids and is identical with the carboxyterminal region (positions 69-89) of a DNA-binding non-histone protein belonging to the group of HMG 17 proteins (high mobility group). The factor suppresses the chemotactic migration of highly metastasizing B16 melanoma cells and inhibits the generation of lung metastases if it is injected at the same time as the tumor cells into mice.
Isoai A et al Tumor invasion inhibiting factor 2: primary structure and inhibitory effect on invasion *in vitro* and pulmonary metastasis of tumor cells. CR 52: 1422-6 (1992); **Isoai A et al** Purification and characterization of tumor invasion-inhibiting factors. JJCR 81: 909-14 (1990)

IL: abbrev. for » Interleukin. Individual interleukins are indicated by numbers; see: IL1 to IL13.

IL1: Interleukin 1.
■ **Alternative Names:** Adherence-promoting factor; APPIF (acute phase protein inducing factor); BAF (B cell activating factor) = IL1α; BCAF (B cell activating factor) = IL1α; BCDF (B cell differentiation factor); BDF (B cell differentiation factor); Catabolin = IL1β; EP (endogenous pyrogen) = IL1α; CETAF (corneal epithelial cell-derived thymocyte activating factor); ETAF (epidermal cell-derived thymocyte activating factor); FAF (fibroblast activating factor); fibroblast proliferation factor; H1 (hematopoietin-1) = IL1β; haptoglobin inducing activity; HP1 (helper peak 1); HP1 (hematopoietin 1); IFN-β-inducing factor = IL1β; Interleukin-β (= IL1β); IL1k (k = keratinocyte); LAF (lymphocyte activating factor) = IL1-α; LEM (leukocyte endogenous mediator) = IL1α; LP (leukocytic pyrogen); MCF (mononuclear cell factor) = IL1α; MNCF (mononuclear cell factor) = IL1α; MP (mitogenic protein) = IL1α; MPIF (muscle proteolysis-inducing factor); NRA (neutrophil releasing activity); OAF (osteoclast activating fac-

tor) = IL1β; ODC factor (ornithine decarboxylase-inducing factor); PIF (proteolysis inducing factor); PMN factor (polymorphonuclear leukocyte-derived lymphocyte proliferation-potentiating factor); SAA (serum amyloid A) inducer; TAF (thymocyte activating factor); TCF (thymocyte comitogenic factor); TMP (thymocyte mitogenic protein); TPF (thymocyte proliferation factor); TR (T cell replacing factor); TRF-M (macrophage-derived T cell replacing factor); TRF-3 (T cell replacing factor 3) = IL1α; tumor inhibitory factor 2 (= IL1α); TSF (thymocyte stimulating factor). See also: individual entries for further information.

■ **SOURCES:** Monocytes are the main source of secreted IL1. They express predominantly IL1β while human keratinocytes express large amounts of IL1α. Murine macrophages display a transition from IL1β to IL1α production during maturation of monocytes into inflammatory macrophages.

IL1 is also produced by activated macrophages from different sources (alveolar macrophages, Kupffer cells, adherent spleen and peritoneal macrophages) and also by peripheral neutrophil granulocytes. Endothelial cells, fibroblasts, smooth muscle cells, keratinocytes, Langerhans cells of the skin, osteoclasts, astrocytes, epithelial cells of the thymus and the cornea, T cells, B cells, NK cells and many melanoma cell lines also produce IL1. The constitutive production of IL1 by human umbilical vein endothelial cells is inhibited by » transthyretin.

The concentrations of IL1 observed in the cerebrospinal fluid are due to local synthesis and also due to the direct transport of IL1 through the blood-brain barrier by means of a saturable carrier system and the ability of activated T lymphocytes to pass this barrier.

■ **PROTEIN CHARACTERISTICS:** There are two functionally almost equivalent forms of IL1, IL1α (17 kDa, 159 amino acids; pI = 5,0) and IL1β (17 kDa, 153 amino acids; pI = 7,0) that are encoded by two different genes. IL1-β is the predominant form in humans while it is IL1-α in mice.

At the protein level IL1α and IL1β display ≈ 27 % homology mainly restricted to the carboxy-terminal region; the names therefore suggest a relationship that does not really exist. On the other hand the three-dimensional of the two IL1 forms are almost identical. Both forms are spherical proteins devoid of α-helical regions, and both forms also bind to the same receptor.

Rabbit, mouse, and human IL1-α show ≈ 61-65 % sequence homology at the protein level. Only 27-33 % sequence homology is observed with IL1β isolated from these three species. Rat IL1 and human-IL1α show 65 % homology.

IL1-α and IL1-β are synthesized as precursors of ≈ 35 kDa (271 amino acid precursor for IL1α and 269 amino acids for IL1β). The mature proteins are generated by proteolytic cleavage by a number of proteases. Cells of the myelomonocytic lineage also express a IL1β-specific protease (see: IL1-Convertase) that can be inhibited by protease inhibitors such as » pentamidine.

Low molecular weight forms of IL1 (11, 4, 2 kDa) have also been found in serum and are secreted in the urine. The IL1α precursor is biologically active but not the IL1β precursor.

IL1 is unusual in that the intracellular precursors do not contain a recognizable hydrophobic secretory » signal sequence that would allow secretion of the protein by classical secretory pathways involving the endoplasmic reticulum/Golgi system. Mature forms of ILα and ILβ and also their precursors are found to be secreted by murine macrophages after stimulation with bacterial lipopolysaccharides. At least IL1-β has been shown to be cleaved by an LPS-inducible protease after externalization. IL1β precursors can be cleaved by proteases including elastase, cathepsin G, and collagenase, which are the major proteases released at sites of inflammation.

Apart from secreted forms of IL1, a biologically active 22 kDa form associated with the cell surface membrane has also been described. It consists mainly of IL1α and may be involved in » juxtacrine growth control involving the interaction with receptors in adjacent cells. A nuclear targeting sequence (see: Signal sequences) has been identified in the IL1-a precursor, suggesting that this factor may also act in the cell nucleus (see also: Intracrine).

■ **GENE STRUCTURE:** IL1α and IL1β are encoded by two different genes of different lengths (IL1α = 12 kb; IL1β = 9.7 kb) and similar organization comprising seven exons. The IL1α mRNA has a length of 2-2.3 kb that for IL1β a length of 1.6-1.7 kb. At the DNA level both genes show a homology of ≈ 45 %. The sequences of rat and human IL1α show ≈ 73 % sequence homology.

The IL1β promoter is ≈ 10-50-fold stronger than that of the IL1α gene. The human IL1 genes map to chromosome 2 (IL1α = 2q13; IL1β = 2q13-q21) in the vicinity of the gene encoding the IL1 receptor antagonist » IL1ra. In the mouse the two genes map to chromosome 2 and are ≈ 50 kb apart. They have the same orientation and the IL1β gene is 5´ to the IL1α gene.

■ **RECEPTOR STRUCTURE, GENE(S), EXPRESSION:** Both forms of IL1 bind to the same receptor and therefore also show similar if not identical biological activities. The IL1β but not the IL1α precursor must be processed before it can bind to the receptor.

Two kinds of IL1 receptors that bind with different affinities (10^{-9} – 10^{-10} and 10^{-11} M) have been described. IL1α and IL1β block the binding of each other to the receptor.

The receptor isolated from T cells (80 kDa; p80) is expressed predominantly on T cells and cells of mesenchymal origin. It binds both types of IL1 with equal affinity. This type is also called ***type I receptor***. It has been designated ***CDw121a*** (see also: CD Antigens). It is encoded by a gene mapping to human chromosome 2q12. The receptor contains three extracellular immunoglobulin-like domains and therefore is a member of the immunoglobulin superfamily of proteins. The entire protein has a length of 576 amino acids consisting of an extracellular ligand-binding domain and a cytoplasmic domain of 213 amino acids. This receptor contains seven N-glycosylation sites. The extracellular domain and also the transmembrane region of 21 amino acids do not show any relationship to other proteins.

The ***type II receptor*** of 60 kDa (p60) consists of 398 amino acids with a short cytoplasmic domain of 29 amino acids. It has been designated ***CDw121b*** (see also: CD Antigens). It is isolated from B cells, granulocytes, and macrophages. It is expressed predominantly on B cells and cells of the myelomonocytic lineage and is encoded by a separate gene. Specific IL1 receptors have been found in the central nervous system that bind IL1β much stronger than IL1α. Both types of IL1 receptors also bind the IL1 receptor antagonist » IL1ra.

Biological activities of IL1 have been observed also at concentration ranges of 10^{-13} – 10^{-15} M which may suggest the existence of high affinity receptor conformations or the existence of other high affinity receptors.

IL1 receptors are expressed at various densities from several thousand up to ≈ 10000/cell in many different cell types. Many cell types also constitutively express IL1 receptors at low densities. Receptor densities generally increases following the stimulation of cells by IL1. The incubation with glucocorticoids increases IL1 receptor densities ≈ 10-fold.

IL1 receptor-mediated signal transduction involves adenylate cyclase which transiently increases intracellular cAMP levels. A cAMP-dependent protein kinase (PKA) and a pertussis toxin-sensitive GTP binding protein of 46 kDa are also involved. Binding of IL1 to its receptor activates transcription factor NF-κB (see also: Gene expression). Protein kinase C (PKC), calcium (see also: Calcium ionophore), phosphatidyl inositol and related metabolites do not appear to be involved in signal transduction.

Vaccinia and cowpox virus encode an IL1β-specific binding protein that is capable of modulating the biological activities of IL1 (see: B15R).

■ **BIOLOGICAL ACTIVITIES:** IL1α and IL1β are biologically more or less equivalent pleiotropic factors that act locally and also systemically. Only a few functional differences between the factors have been described; only IL1β appears to be expressed constitutively in the brain.

The plethora of biological activities is exemplified by the many different acronyms under which IL1 has been described. IL1s do not show species specificity with the exception of human IL1β that is inactive in porcine cells. Some of the biological activities of IL1 are mediated indirectly by the induction of the synthesis of other mediators including ACTH (Corticotropin; see: POMC, proopiomelanocortin), PGE_2, » PF4 (platelet factor 4), » CSF (colony stimulating factor), » IL6, and » IL8.

The synthesis of IL1 can be induced by other cytokines including » TNF-α, » IFN-α, » IFN-β and » IFN-γ and also by bacterial » endotoxins, viruses, mitogens, and antigens. In human skin fibroblasts In IL1α and TNF-α induce the synthesis of IL1β. In pheochromocytoma cells (see also: PC12) » NGF induces the synthesis of IL1. Human mononuclear cells are very sensitive to bacterial endotoxins and synthesize IL1 in response to picogramm amounts of endotoxin per mL. In human monocytes LPS induces ≈ tenfold more IL1-β than IL1-α-specific mRNA and the respective proteins.

The synthesis of IL1 is controlled by a complex feedback loop since IL1 is also capable of inhibiting or promoting its own synthesis, depending on conditions and cell types.

In vitro PGE_2, the synthesis of which is increased by IL1, and glucocorticoids inhibit the synthesis of IL1. The PGE_2-mediated inhibition of IL1 synthesis, like the inhibition of IL1 synthesis caused by » IFN-α and » IFN-γ, is mediated by an increase of

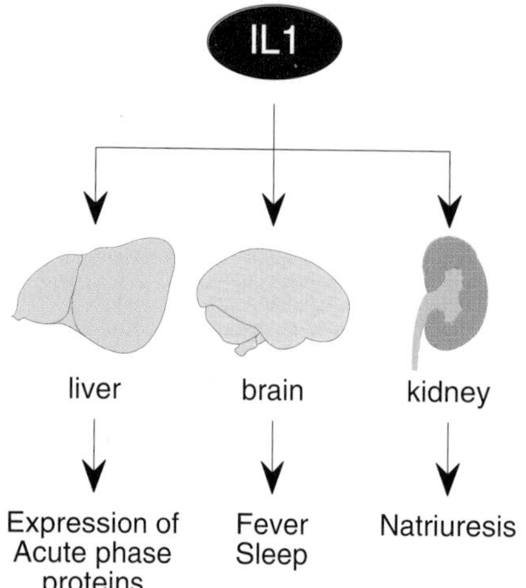

Actions of IL1 on liver, brain, and kidney.
IL1 induces the synthesis of acute phase proteins in liver hepatocytes. In the brain IL1 influences those centers involved in regulation of slow wave sleep. In the kidney IL1 promotes natriuresis.

intracellular cAMP levels. IL6 can also inhibit the synthesis of IL1. The synthesis of IL1 is also controlled by natural inhibitors (see: IL1ra, IL1 receptor antagonist), themselves induced by IL1, and is also inhibited by lipoproteins, lipids, and α_2-Macroglobulin. IL1 induces the expression of the » oncogenes » *fos* and » *myc*.

The main biological activity of IL1 is the stimulation of T helper cells which are induced to secrete » IL2 and to express IL2 receptors. Virus-infected macrophages produce large amounts of an IL1 inhibitor that may support opportunistic infections and transformation of cells in patients with T cell maturation defects (see: IL1ra).

IL1 acts directly on B cells, promoting their proliferation and the synthesis of immunoglobulins. IL1 also functions as a priming-factor that makes B cells responsive to » IL5. IL1 stimulates the proliferation and activation of NK cells and fibroblasts, thymocytes, glioblastoma cells. It also promotes the proliferation of astroglia and microglia and may be involved in pathological processes such as astrogliosis and demyelinisation. The IL1-mediated proliferation of lymphocytes is inhibited by TGF-β1 and TGF-β2 (see: TGF-β).

A mechanism of » autocrine growth control by IL1 has been suggested for leukemic blast cells in which the uncontrolled synthesis of IL1 is thought to lead to the production of colony-stimulating factors (see also: CSF) that in turn promote the proliferation of these cells. In combination with other cytokines IL1 appears to be an autocrine growth factor for human gastric and thyroid carcinoma cells. The growth-promoting activities of IL1 are mediated indirectly in some systems by regulating the expression of high affinity receptors for another cytokine, » FGF.

IL1 has also been shown to be radioprotective.

IL1 also has antiproliferative and cytocidal activities on certain tumor cell types. It supports the monocyte-mediated tumor cytotoxicity and induces tumor regression. IL1 is cytotoxic for insulin-producing β cells of the Langerhans islets of the pancreas.

IL1 inhibits the growth of endothelial cells *in vivo* and *in vitro*. It also inhibits the growth of rat hepatocytes in culture.

IL1 causes many alterations of endothelial functions *in vivo*. It promotes thrombotic processes and attenuates anticoagulatory mechanisms. IL1 therefore plays an important role in pathological processes such as venous thrombosis, arteriosclerosis, vasculitis, and disseminated intravasal coagulation.

IL1 promotes the adhesion of neutrophils, monocytes, T cells, and B cells by enhancing the expression of adhesion molecules such as CAM-1 (intercellular adhesion molecule) and ELAM (endothelial leukocyte adhesion molecule). The expression of membrane-associated thrombomodulin is decreased by IL1.

IL1 also influences the functional activities of Langerhans cells of the skin. These cells are not capable of eliciting primary immune responses (e. g. contact sensibilisation). IL1 (and also » GM-CSF) convert these cells into potent immunostimulatory dendritic cells. The Langerhans cells therefore constitute an *in situ* reservoir for immunologically immature lymphoid dendritic cells. The increased ability of maturated Langerhans cell to process antigens is decreased by » TNF-α.

IL1 in combination with other cytokines is an important mediator of inflammatory reactions (see also: Inflammation). IL1 markedly enhances the metabolism of arachidonic acid (in particular of prostacyclin and PGE_2) in inflammatory cells such as fibroblasts, synovial cells, chondrocytes, endothelial cells, hepatocytes, and osteoclasts. In addition one observes an increased secretion of inflammatory proteins such as neutral proteases (collagenase, elastase and plasminogen activator). This activity of IL1 antagonizes the effects of »

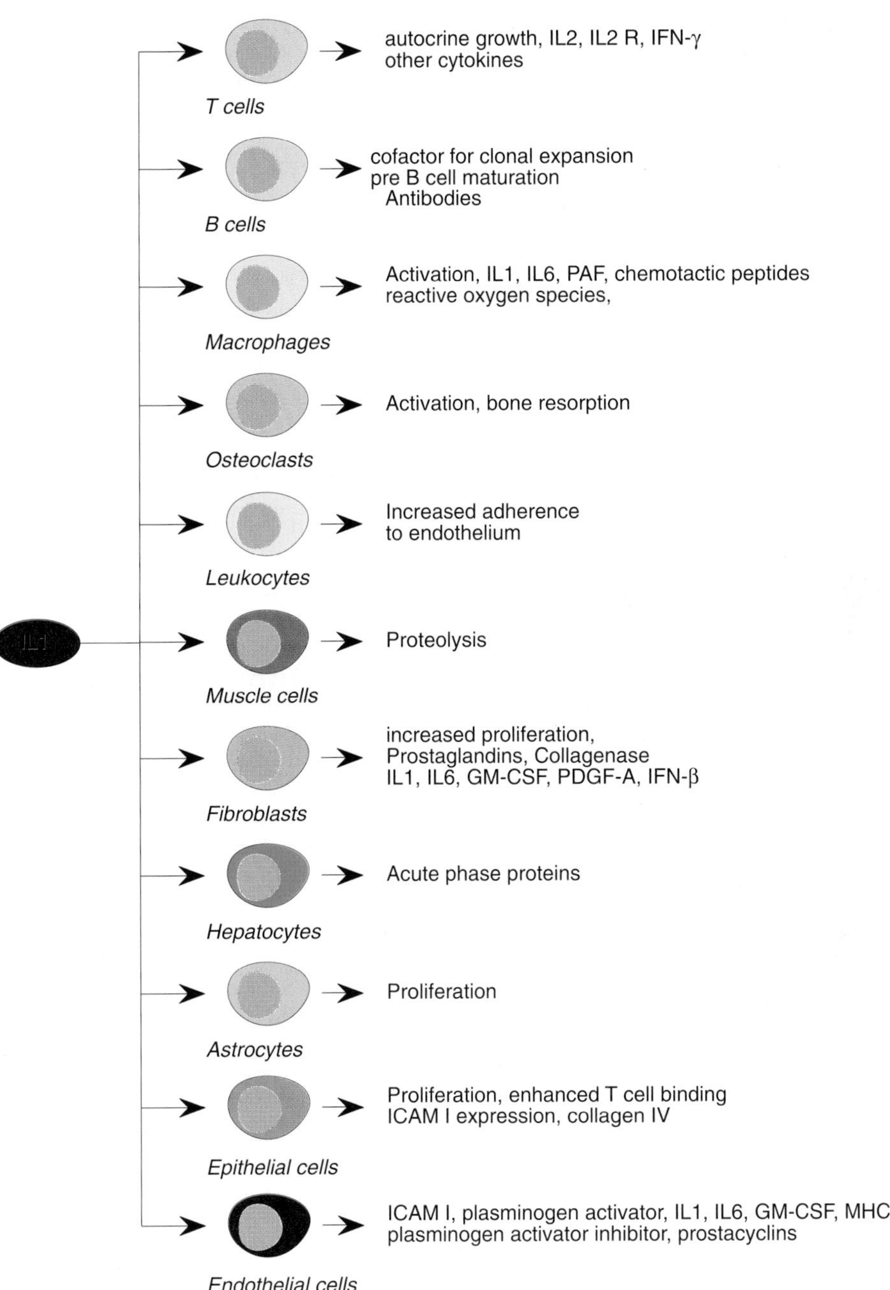

Schematic representation of the multiple activities of IL1 at the cellular level.

TGF-β on the extracellular matrix (see also: ECM).

IL1 is also a strong chemoattractant for leukocytes (see also: Chemotaxis); *in vivo* the injection of IL1 leads to the local accumulation of neutrophils at the site of injection. IL1 also activates oxidative metabolism in neutrophils.

In combination with » TNF IL1 appears to be involved in the generation of lytic bone lesions. IL1 activates osteoclasts and therefore suppresses the formation of new bone. Low concentrations of IL1, however, promote new bone growth. IL1 inhibits the enzyme lipoprotein lipase in adipocytes. In vascular smooth muscle cells and skin fibroblasts IL1 induces the synthesis of » bFGF which is a mitogen for these cells.

Like » IL2 IL1 also modulates the electrophysiological behavior of neurons. IL1 also directly affects the central nervous system as an afferent signal modulating the release of a number of hormones and activates the hypothalamic-pituitary-adrenocortical axis (see: Neuroimmune network). In the hypothalamus IL1 activates the serotoninergic system. IL1 also functions as an endogenous pyrogen (see: EP) and induces a significant elevation of body temperature by causing the release of prostaglandins in the thermoregulatory center of the hypothalamus. This activity is inhibited by α-Melanocyte-stimulating hormone (α-MSH; see: POMC, proopiomelanocortin). IL1, like IL6, stimulates the synthesis of ACTH (Corticotropin) in the pituitary and also induces the release of » luteinising hormone, » growth hormone, » thyrotropin, and CRF (corticotropin releasing factor; see: POMC). IL1 also enhances expression of the » POMC gene (Pro-opiomelanocortin) which yields ACTH, β-Endorphin, α-, β-, and c-MSH. Glucocorticoids synthesized in response to ACTH inhibit the production of IL6, IL1 and TNF *in vivo*, thus establishing a sort of negative feedback loop between the immune system and neuroendocrine functions. α-MSH inhibits the IL1 activities by an as yet unknown mechanism. IL1 induces the synthesis of corticostatin by granulocytes which in turn inhibits the ACTH-induces synthesis of glucocorticosteroids. Glucocorticosteroids on the other hand increase the expression of IL1 receptors. In blood serum IL1 has been shown to decrease plasma concentrations of iron and zinc. In Schwann cells and fibroblast-like cells of nervous tissue IL1 induces the synthesis of » NGF.

IL1β immunoreactive nerve fibers in human hypothalamus innervate those endocrine and autonomous nuclei controlling central aspects of the » acute phase reaction. Astrocytes proliferate in the presence of IL1 and release » IL3 which is a growth factor for microglia and peritoneal macrophages. In Astrocytes IL1 also promotes the synthesis of » GM-CSF, » IL6 and » TNF. In the central nervous system IL1 is also involved in the induction of the so-called *slow-wave* sleep. IL1β can prevent the depression of antibody responses observed during sleep deprivation. The important role of IL1 for the connections between the brain and the immune system is also exemplified by the observation that IL1 after intraperitoneal injection into Wistar rats stimulates the metabolism of the neurotransmitter noradrenalin in the hypothalamus and the dorsal posterior brain stem. Some neuropeptides such as neurotensin, » SP (substance P), and neurokinin A and B (see: Tachykinins) are themselves capable of enhancing the synthesis of IL1. In certain neurons of the rat hypothalamus IL1β induces the synthesis of » Somatostatin.

IL1 synergises with some other factors. The effect of IL1 on the proliferation of thymocytes involves the stimulation of IL2 synthesis and the expression of IL2 receptors on T cells. For certain T helper cell populations and also for B cells IL1 functions as an additional growth factor. IL1 potentiates the effects of colony-stimulating factors (see also: CSF) and promotes the generation of myeloid progenitor cells from stem cells (see also: Hematopoiesis). IL1 synergises with » GM-CSF in the induction of macrophage colony growth. IL1 can also induce the synthesis of » G-CSF and » M-CSF by bone marrow stromal cells and stimulates the synthesis of GM-CSF and G-CSF by human skin cells and of GM-CSF by peripheral blood lymphocytes. By enhancing the expression of receptors for colony-stimulating factors IL1 is involved in various processes of » hematopoiesis. IL1 also induces the proliferation of pluripotent bone marrow progenitor cells.

■ **ASSAYS:** IL1 can be detected in a number of » Bioassays employing several cell lines (see: A375, D10; EL4; LBRM 33; Mono Mac 6; RPMI 1788; T1165). IL1 can also be detected indirectly by determining concentrations of IL1-induced » HGF (hybridoma growth factor = IL6), IL1-induced » LIF activity or IL1-induced production of prostaglandin E2 by fibroblasts. IL1 can be detected in a thymocyte costimulation assay. It can also be assayed by sensitive ELISAs and RIAs. Intracellular IL1 is detected by two color immunofluorescence flow cytometry. An alternative and entirely different detection method is » Message amplification phenotyping. A radioligand assay is available for the detection of IL1 receptors. For further information see also subentry "Assays" in the reference section.

■ **Clinical Use & Significance:** IL1 may be of clinical significance due to its activities as a stimulator of T cells. It may be useful after massive immunosuppression and/or therapy with cytostatic drugs. IL1-α has been shown to accelerate the recovery of platelets after high-dose carboplatin therapy and may be clinically useful in preventing or treating thrombocytopenia induced by chemotherapy.

Since IL1 is also involved in » hematopoiesis, acting directly on primitive hematopoietic cells and inducing the synthesis of colony-stimulating factors (see: CSF), it may be useful as a stimulator of hematopoiesis either alone or in combination with other substances. Several studies suggest that combinations of IL1 and lineage specific growth factors may produce synergistic hematological effects and that IL1 may have a possible protective effect on chemotherapy-induced myelosuppression.

Peripheral blood mononuclear cells of patients with aplastic anemia show markedly decreased IL1 production. Aplastic patients with low IL 1 production are distinguished by the severity of their disease and the degree of neutropenia while myelodysplastic patients with comparable degrees of pancytopenia have normal IL 1 production. This deficient hematopoietic growth factor production in a bone marrow failure syndrome may contribute to the pathogenesis of some cases of aplastic anemia and to susceptibility to infection.

Animal experiments have shown that IL1 has radioprotective activities. IL1 enhances the capacity of allogeneic bone marrow (BM) cells to promote survival of mice given doses of radiation (1,200 to 1,350 cGy) that are significantly higher than those generally used for bone marrow ablation (850 to 950 cGy). IL1 may therefore prove clinically useful in promoting BM engraftment.

It has been shown that IL1 also promotes » wound healing. This activity is thought to involve effects on angiogenesis (see also: Angiogenesis factors), the promotion of fibroblast proliferation, and the chemotactic activity on neutrophils.

The clinical application of IL1 is hampered by the very high toxicity of this compound since minute amounts of IL1 can induce » septic shock. The use of IL1 receptor antagonists appears to be more promising that the use of IL1. Since elevated amounts of IL1 are found in the synovial fluid of patients with rheumatoid arthritis and osteoarthritis receptor antagonists may be of advantage (see also: IL1ra). These antagonists have also been used to prevent IL1-induced » septic shock in experimental animals. They have also been used to prevent colitis in rabbits induced by immune complexes.

IL1α may be a pathogenetic factor in the complex processes leading to vascular occlusion and an important *in situ* indicator of and a potential participant in vascular injury. It is consistently present in all vessels with sclerotic histopathologic changes following aortocoronary bypass grafting of saphenous veins and internal mammary arteries.

● **Reviews: Bomford R & Henderson B** (eds) Interleukin-1, inflammation and disease. Elsevier, New York 1989; **Brazel D et al** Interleukin-1, characterization of the molecule, functional activity, and clinical implications. Biotechnol. Ther. 2: 241-67 (1991); **di Giovine FS & Duff GW** Interleukin 1: the first interleukin. IT 11: 13-20 (1990); **Dinarello CA** Interleukin-1 and interleukin-1 antagonism. Blood 77: 1627-52 (1991); **Dinarello CA** Interleukin-1 and its receptor. CRC Crit. Rev. Immunol. 9: 1-20 (1989); **Dinarello CA & Wolff SM** The role of interleukin-1 in disease. NEJM 328: 106-13 (1993); **Dinarello CA** Modalities for reducing interleukin 1 activity in disease. TIPS 14: 155-9 (1993); **Fibbe WE et al** The biological activities of interleukin-1. Blut 59: 147-56 (1989); **Krakauer T** Human Interleukin-1. CRC Crit. Rev. Immunol. 6: 213-44 (1986); **Sauder DN** Interleukin 1. Arch. Dermatol. 125: 679-82 (1989); **Zurawski G** Analyzing lymphokine-receptor interactions of IL1 and IL2 by recombinant-DNA technology. TibTech 9: 250-7 (1991)

● **Biochemistry & Molecular Biology: Auron PE et al** Nucleotide sequence of human monocyte interleukin-1 precursor cDNA. PNAS 81: 7907-11 (1984); **Auron PE et al** Human and murine interleukin 1 possess sequence and structural similarities. J. Mol. Cell. Immunol. 2: 169-77 (1985); **Auron PE & Webb AC** The structure and regulation of the human prointerleukin 1 β gene. Ann. Inst. Pasteur Immunol. 138: 462-9 (1987); **Banks WA** Human interleukin (IL) 1 α, murine IL1 α, and murine IL1 β are transported from blood to brain in the mouse by a shared saturable mechanism. J. Pharmacol. Exp. Ther. 259: 988-96 (1991); **Bensi G et al** Human interleukin-1 β gene. Gene 52: 95-101 (1987); **Beuscher HU et al** Transition from interleukin 1 β (IL1 β) to IL1 α production during maturation of inflammatory macrophages *in vivo*. JEM 175: 1793-7 (1992); **Beuscher HU et al** IL1β is secreted by activated murine macrophages as biologically inactive precursor. JI 144: 2179-83 (1990); **Black RA et al** Activation of IL1β by a co-induced protease. FL 247: 386-90 (1989); **Cameron PM et al** Purification to homogeneity and amino acid sequence analysis of two anionic species of human interleukin-1. JEM 164: 237-50 (1986); **Clark BD et al** Genomic sequence for prointerleukin-1 β: possible evolution from a reverse transcribed prointerleukin-1 α gene. NAR 14: 7897-914 (1985); **Daumy GO et al** Isolation and characterization of biologically active murine interleukin-1 α derived from expression of a synthetic gene in *Escherichia coli*. BBA 998: 32-42 (1989); **DeChiara TM et al** Structure-function analysis of murine interleukin 1: biologically active polypeptides are at least 127 amino acids long and are derived from the carboxyl terminus of a 270-amino acid precursor. PNAS 83:

8303-7 (1985); **Furutani Y et al** Cloning and characterization of the cDNAs for human and rabbit interleukin-1 precursor. NAR 13: 5869-82 (1985); **Furutani Y et al** Complete nucleotide sequence of the gene for human interleukin-1 α. NAR 14: 3167-79 (1986); **Gray PW et al** Two interleukin 1 genes in the mouse: Cloning and expression of the cDNA for murine interleukin 1-β. JI 137: 3644-8 (1987); **Hazuda D et al** Purification and characterization of human recombinant precursor interleukin 1β. JBC 264: 1689-93 (1989); **Huang JJ et al** Characterization of murine IL1 β. Isolation, expression, and purification. JI 140: 3838-43 (1988); **Hume M et al** Regulation of interleukin-1β production by glucocorticoids in human monocytes: The mechanism of action depends on the activation signal. BBRC 180: 1383-9 (1991); **Kawashima H et al** Structure-activity relationships in human interleukin-1 α: identification of key residues for expression of biological activities. Protein Eng. 5: 171-6 (1992); **Krakauer T** Purification to homogeneity of biologically active human interleukin 1. Prep. Biochem. 14: 449-70 (1984-85; **Kronheim SR et al** Human interleukin 1. purification to homogeneity. JEM 161: 490-502 (1985); **Kurt-Jones EA et al** Identification of a membrane-associated interleukin 1 in macrophages. PNAS 82: 1204-8 (1985); **Lafage M et al** The human interleukin 1 α gene is located on the long arm of chromosome 2 at band q13. Blood 73: 104-7 (1989); **Leong SR et al** The nucleotide sequence for the cDNA of bovine interleukin-1 α. NAR 16: 9053 (1988); **Lomedico PT et al** Cloning and expression of murine interleukin-1 in *Escherichia coli.* N 312: 458-92 (1984); **Lomedico PT et al** Cloning and expression of murine, human, and rabbit interleukin 1 genes. Lymphokines 13: 139-50 (1987); **Luger TA et al** Monoclonal anti-IL1 is directed against a common site of human IL1 α and IL1 β. Immunobiology 172: 346-56 (1986); **Maliszewski CR et al** Porcine ILI α cDNA nucleotide sequence. NAR 18: 4282 (1990); **Maliszewski CR et al** Cloning, sequence and expression of bovine interleukin 1-α and interleukin 1-β complementary DNAs. Mol. Immunol. 25: 429-37 (1988); **March CJ et al** Cloning, sequence, and expression of two distinct human interleukin-1 complementary cDNAs. N 315: 641-5 (1985); **Modi WS et al** Chromosomal localization of the human interleukin 1 α (IL1 α) gene. Genomics 2: 310-4 (1988); **Mosley B et al** Determination of the minimum polypeptide lengths of the functionally active sites of human interleukins 1α and 1β. PNAS 84: 4572-6 (1987); **Nishida T et al** Molecular cloning and expression of rat interleukin-1 α cDNA. JB 105: 351-7 (1989); **Priestle JP et al** Crystal structure of the cytokine interleukin-1β. EJ 7: 339-45 (1988); **Rubartelli A et al** A novel secretory pathway for interleukin 1-β, a protein lacking a signal sequence. EJ 9: 1503-10 (1990); **Shabon U et al** Human melanoma cells transcribe interleukin 1 genes identical to those of monocytes. CR 51: 3334-5 (1991); **Silver AR et al** The IL1 α and β genes are closely linked (less than 70 kb) on mouse chromosome 2. Somat. Cell. Mol. Genet. 16: 549-56 (1990); **Vandenabeele P et al** Development of a simple, sensitive and specific bioassay for interleukin-1 based on the proliferation of RPMI 1788 cells: comparison with other bioassays for IL1. JIM 135: 25-32 (1990); **Van Oostrum J et al** The structure of murine interleukin-1β at 2. 8 Å resolution. J. Struct Biol. 107: 189-95 (1991); **Warner SJC et al** Interleukin 1 induces IL1 gene expression in smooth muscle cells. JEM 165: 1316-31 (1987); **Webb AC et al** Interleukin 1 gene (IL1) assigned to long arm of human chromosome 2. LR 5: 77-85 (1986); **Wessendorf JH et al** Identification of a nuclear localization sequence within the structure of the human interleukin-1 α precursor. JBC 268: 22100-4 (1993); **Yem AW et al** Resolution and biological properties of three N-terminal analogs of recombinant human interleukin-1 β. LR 7: 85-92 (1988); **Zurawski SM et al** Expression in *Escherichia coli* of synthetic human interleukin-1 α genes encoding the processed active protein, mutant proteins, and β-galactosidase fusion proteins. Gene 49: 61-8 (1986)

● **RECEPTOR STRUCTURE, GENE(S), EXPRESSION: Bomsztyk K et al** Evidence for different interleukin 1 receptors in murine B and T cells. PNAS 86: 8034-8 (1989); **Chizzonite R et al** Two high affinity interleukin 1 receptor represent separate gene products. PNAS 86: 8029-33 (1989); **Copeland NG et al** Chromosomal location of murine and human IL1 receptor genes. Genomics 9: 44-50 (1991); **Cunningham ET & De Souza EB** Interleukin 1 receptors in the brain and endocrine tissues. IT 14: 171-6 (1993); **Dower SK et al** The cell surface receptors for interleukin 1 α and interleukin 1 β are identical. N 324: 266-8 (1986); **Dower SK et al** Biology of the IL1 receptor. J. Invest. Dermatol. 94: 68s-75s (1990); **Larrick JW** Native interleukin 1 inhibitors. IT 10: 61-6 (1989), Comment in: IT10: 222 (1989); **McMahan CJ et al** A novel IL1 receptor, cloned from B cells by mammalian expression, is expressed in many cell types. EJ 10: 2821-32 (1991); **Mosley B et al** The IL1 receptor binds the human interleukin-1α precursor but not the human interleukin-1β precursor. JBC 262: 2941-4 (1987); **Sims JE et al** cDNA expression cloning of the IL1 receptor, a member of the immunoglobulin superfamily. S 241: 585-9 (1988); **Symons JA et al** Purification and characterization of a novel soluble receptor for interleukin 1. JEM 174: 1251-4 (1991)

● **BIOLOGICAL ACTIVITIES: Alheim K et al** Interleukin 1 expression is inducible by nerve growth factor in PC12 pheochromocytoma cells. PNAS 88: 9302-6 (1991); **Beck G et al** Interleukin 1: a common endogenous mediator of inflammation and the local Shwartzman reaction. JI 136: 3025-31 (1986); **Bernton EW et al** Release of multiple hormones by a direct action of interleukin 1 on pituitary cells. S 238: 511-2 (1987); **Besedovsky H et al** Immunoregulatory feedback between interleukin 1 and glucocorticoid hormones. S 233: 652-4 (1986); **Breder CD et al** Interleukin 1 immunoreactive innervation of the human hypothalamus. S 240: 321-4 (1988); **Brown R et al** Interleukin 1 and muramyl-dipeptide can prevent decreased antibody response associated with sleep deprivation. Brain Behav. Immunol. 3: 320-30 (1989); **Comens PG et al** Interleukin 1 is a potent modulator of insulin-secretion from isolated rat islets of Langerhans. Diabetes 36: 963-70 (1987); **Endres S et al** Interleukin-1 in the pathogenesis of fever. EJCI 17: 469-74 (1987); **Gemma C et al** Activation of the hypothalamic serotoninergic system by central interleukin-1. Eur. J. Pharmacol. 209: 139-40 (1991); **Giulian D et al** Interleukin 1 is an astroglial growth factor in the developing brain. J. Neurosci. 8: 709-14 (1988); **Helle M et al** Interleukin-6 is involved in interleukin 1-induced activities. EJI 18: 957-9 (1988); **Hermus ARMM & Sweep CGJ** Cytokines and the hypothalamic-pituitary-adrenal axis. JSBMB 37: 867-71 (1990); **Hestdal K et al** *In vivo* effect of interleukin-1 α on hematopoiesis: role of colony-stimulating factor receptor modulation. Blood 80: 2486-94 (1992); **Ito R et al** Interleukin 1α acts as an autocrine growth stimulator for human gastric carcinoma cells. CR 53: 4102-6 (1993); **Kimura T et al** Central effects of interleukin-1 on blood pressure, thermogenesis, and the release of vasopressin, ACTH, and atrial natriuretic peptide. ANY 689: 330-45 (1993); **Kirkham B** Interleukin 1, immune activation pathways, and different mechanisms in osteoarthritis and rheumatoid arthritis. Ann. Rheum. Dis. 50: 395-400 (1991); **Koyama N et al** Role of recombinant interleukin 1 compared to recombinant T cell replacing factor/interleukin 5 in B cell differentiation. Immunology 63: 277-83 (1988); **Lachman LB et al** Natural and recombinant human interleukin 1β is cytotoxic for human melanoma cells. JI 136: 3098-3102 (1986); **Last-Barney K et al** Synergistic and overlapping activities of tumor necrosis factor-α and IL1. JI 141: 527-30 (1988); **Lindholm D et al** Interleukin 1 regulates synthesis of nerve growth factor in non-neuronal cells of rat sciatic nerve. N 330: 658-9 (1987); **Lovett D et al** Macrophage cytotoxicity: interleukin 1 as a mediator of tumor cytostasis. JI 136: 340-7 (1986); **Marley S et al** Interleukin 1 has positive and negative regulatory effects in human long-term bone marrow culture. EH 20: 75-9 (1992); **McMahan CJ et al** A novel IL1 receptor cloned from B cells by mammalian expression is expressed in many cell types. EJ 10: 2821-32 (1991); **Merrill JE** Effects of Interleukin 1 and tumor necrosis factor α on astrocytes, microglia, oligodendrocytes, and glial precursors *in vitro*. Dev. Neurosci. 13: 130-7 (1991); **Muller K et al** 1,α,25-dihydroxyvitamin D3 and a novel vitamin D analog MC903 are potent inhibitors of human interleukin 1 *in vitro*. Immunol. Lett. 17: 361-5 (1988); **Nakamura T et al** Interleukin-1 β is a potent growth inhibitor of adult rat hepatocytes in primary culture. ECR 179: 488-97 (1988); **Neta R et al** Interdependence of the radioprotective effects of human recombinant interleukin 1 α, tumor necrosis factor α, granulocyte colony-stimulating factor, and murine recombinant granulocyte-macrophage colony-stimulating factor. JI 140: 108-11 (1988); **Onozaki K et al** Human interleukin 1 is cytocidal factor for several tumor cell lines. JI 135: 3962-8 (1985); **Rettori V et al** Central action of interleukin-1 in altering the release of TSH, growth hormone, and prolactin in the male rat. J. Neurosci. Res. 18: 179-83 (1987); **Roh MS et al** Direct stimulation of the adrenal cortex by interleukin 1. Surgery 102: 140-6 (1987); **Rothwell NJ** Functions and mechanisms of interleukin 1 in the brain. TIPS 12: 430-6 (1991); **Rupp EA et al** Specific bioactivities on monocyte-derived interleukin 1 and interleukin 1 β are similar to each other on cultured murine thymocytes and on cultured human connective tissue cells. JCI 78: 836-9 (1986); **Suda T et al** Interleukin-1 stimulates corticotropin-releasing factor gene expression in rat hypothalamus. E 126: 1223-8 (1990); **Whitnall MH et al** Effects of interleukin-1 on the stress-responsive and -nonresponsive subtypes of corticotropin-releasing hormone neurosecretory axons. Endocrinology 131: 37-44 (1992)

● Assays: **Anderson J et al** Effects of acetate dialysate on transforming growth factor β 1, interleukin, and β 2-microglobulin plasma levels. Kidney Int. 40: 1110-7 (1991); **Bendtzen K et al** Measurement of human IL1 by LIF induction, pancreatic islet cell cytotoxicity, and bone resorption. LR 5: S93-S98 (1986); **Bristow AF et al** Interleukin-1 β production *in vivo* and *in vitro* in rats and mice measured using specific immunoradiometric assays. J. Mol. Endocrinol. 7: 1-7 (1991); **Cannon JG et al** Interleukin-1 β in human plasma: optimization of blood collection, plasma extraction, and radioimmunoassay methods. LR 7: 457-67 (1988); **Capper SJ et al** Specific radioimmunoassays for IL 1 α and IL 1 β in plasma at physiological and acidic pH: determination of immunoreactive forms by gel filtration and radioligand binding studies. Cytokine. 2: 182-9 (1990); **Carroll GJ & Bell MC** IgM class immunoglobulin with high rheumatoid factor activity interferes with the measurement of interleukin 1 β J. Rheumatol. 18: 1266-9 (1991); **Conlon PJ** A rapid biologic assay for the detection of interleukin 1. JI 131: 1280-2 (1983); **de Caestecker MP et al** The detection of intracytoplasmic interleukin-1 α, interleukin-1 β and tumor necrosis factor α expression in human monocytes using two color immunofluorescence flow cytometry. JIM 154: 11-20 (1992); **Dinarello CA** ELISA kits based on monoclonal antibodies do not measure total IL1 β synthesis [comment for: Herzyk DJ et al] JIM148: 255-9 (1992); **Dularay B et al** IL1 secreting cell assay and its application to cells from patients with rheumatoid arthritis. Br. J. Rheumatol. 31: 19-24 (1992); **Elsasser-Beile U et al** Cytokine levels in whole blood cell cultures as parameters of the cellular immunologic activity in patients with malignant melanoma and basal cell carcinoma. Cancer 71: 231-6 (1993); **Falk W et al** A new assay for interleukin 1 in the presence of interleukin 2. JIM 99: 47-52 (1987); **Gaffney EV et al** Antibodies and their role in interleukin-1 research. IR 119: 181-201 (1991); **Gallay P et al** Characterization and detection of naturally occurring antibodies against IL1 α and IL1 β in normal human plasma. ECN 2: 329-38 (1991); **Gnocchi P et al** Development and applications of a radioimmunoassay (RIA) for the *in vitro* and *in vivo* quantification of murine IL1 β. Lymphokine Cytokine Res. 11: 257-63 (1992); **Grassi J et al** Determination of IL1 α, IL1 β and IL2 in biological media using specific enzyme immunometric assays. IR 119: 125-45 (1991); **Hansen MB et al** Human anti-interleukin 1 α antibodies. Immunol. Lett. 30: 133-9 (1991); **Herzyk DJ et al** Sandwich ELISA formats designed to detect 17 kDa IL1 β significantly underestimate 35 kDa IL1 β [see comments = Dinarello, C. A.] JIM 148: 243-54 (1992); **Herzyk DJ & Wewers MD** ELISA detection of IL1 β in human sera needs independent confirmation. False positives in hospitalized patients. Am. Rev. Respir Dis. 147: 139-42 (1993); **Hopkins SJ & Humphreys M** Bioassay of interleukin-1 in serum and plasma following removal of inhibitory activity with polyethylene glycol. JIM 133: 127-31 (1990); **Horuk R et al** A rapid and direct method for the detection and quantification of interleukin 1 receptors using 96 well filtration plates. JIM 119: 255-8 (1989); **Kenney JS et al** Monoclonal antibodies to human recombinant interleukin 1 (IL1)β: quantitation of IL1 β and inhibition of biological activity. JI 138: 4236-42 (1987); **Kinane DF et al** Bioassay of interleukin 1 (IL1) in human gingival crevicular fluid during experimental gingivitis. Arch. Oral. Biol. 37: 153-6 (1992); **Le-Moal MA et al** A sensitive IL2dependent assay for IL1. JIM 24: 23-30 (1988); **Lewis CE et al** Measurement of cytokine release by human cells. A quantitative analysis at the single cell level using the reverse hemolytic plaque assay. JIM 127: 51-9 (1990); **Lisi PJ et al** Development and use of a radioimmunoassay for human interleukin-1 β. LR 6: 229-44 (1987); **Lonnemann G et al** A radioimmunoassay for human interleukin-1 α: measurement of IL1 α produced *in vitro* by human blood mononuclear cells stimulated with endotoxin. LR 7: 75-84 (1988); **Mae N et al** Identification of high-affinity anti-IL1 α autoantibodies in normal human serum as an interfering substance in a sensitive enzyme-linked immunosorbent assay for IL1 α. Lymphokine Cytokine Res. 10: 61-8 (1991); **Mukaida N et al** Establishment of a highly sensitive enzyme-linked immunosorbent assay for interleukin 1 α employing a fluorogenic substrate. JIM 107: 41-6 (1988); **Newton R & Covington M** The activation of human fibroblast prostaglandin E production by interleukin 1. CI 110: 338-49 (1987); **Nordstrom I & Ferrua B** Reverse ELISPOT assay for clonal analysis of cytokine production. II. Enumeration of interleukin-1-secreting cells by amplified (avidin-biotin anti-peroxidase) assay. JIM 150: 199-206 (1992); **Pennline KJ et al** Detection of *in vivo*-induced IL1 mRNA in murine cells by flow cytometry (FC) and fluorescent *in situ* hybridization(FISH). Lymphokine Cytokine Res. 11: 65-71 (1992); **Poole S et al** Development and application of radioimmunoassays for interleukin-1 α and interleukin-1 β. JIM 116: 259-64 (1989); **Rafferty B et al** Measurement of cytokine production by the monocytic cell line Mono Mac 6 using novel immunoradiometric assays for interleukin-1 β and interleukin-6. JIM 144: 69-76 (1991); **Remvig L et al** Biological assays for interleukin 1 detection. Comparison of human T lymphocyte, murine thymocyte, and NOB-1 assays. Allergy 46: 59-67 (1991); **Remvig L et al** A standardized human T lymphocyte proliferation assay for detecting soluble accessory factors from monocytes. II. Interleukin 1 production and detection. Acta Pathol. Microbiol. Immunol. Scand. 93: 91-6 (1985); **Riske F et al** Measurement of biologically active interleukin-1 by a soluble receptor binding assay. AB 185: 206-12 (1990); **Scapigliati G et al** Quantitation of biologically active IL1 by a sensitive assay based on immobilized human IL1 receptor type II (IL1RII). JIM 138: 31-8 (1991); **Slack J et al** Application of the MultiScreen system to cytokine radioreceptor assays. Biotechniques 7: 1132-8 (1989); **Stya M et al** Development and characterization of two neutralizing monoclonal antibodies to human interleukin 1 α. J. Biol. Response Mod. 7: 162-72 (1988); **Sunahara N et al** Differential determination of recombinant human interleukin-1 α and its deamidated derivative by two sandwich enzyme immunoassays using monoclonal antibodies. JIM 119: 75-82 (1989); **Thavasu PW et al** Measuring cytokine levels in blood. Importance of anticoagulants, processing, and storage conditions. JIM 153: 115-24 (1992); **Thorpe R et al** Sensitive and specific immunoradiometric assays for human interleukin-1 α. LR 7: 119-27 (1988); **Van Damme J & Van Snick J** Induction of hybridoma growth factor (HGF) identical to IL6 in human fibroblasts by IL1: use of HGF activity in specific and sensitive biological assays for IL1 and IL6. Dev. Biol. Stand. 69: 31-8 (1988); **Zheng RQ et al** Detection of interleukin-6 and interleukin-1 production in human thyroid epithelial cells by non-radioactive *in situ* hybridization and immunohistochemical methods. Clin. Exp. Immunol. 83: 314-9 (1991); **Zimecki M et al** A comparison of the conventional and the rosette method of determination of interleukin 1 (IL1) activity. The advantages of the rosette method. Arch. Immunol. Ther. Exp. Warsz. 36: 655-60 (1988); **Zimecki M & Wieczorek Z** Interleukin 1 decreases the level of thymocytes forming rosettes with autologous erythrocytes: a new method of determination of interleukin 1 activity. Arch. Immunol. Ther. Exp. Warsz. 35: 3719 (1987); for further information also see individual cell lines used in individual bioassays.

● **Clinical Use & Significance: Brody JI et al** Interleukin-1 α as a factor in occlusive vascular disease Am. J. Clin. Pathol. 97: 8-13 (1992), comments in Am. J. Clin. Pathol. 97: 1-3 (1992); **Chelstrom LM et al** Treatment of BCL-1 murine B cell leukemia with recombinant cytokines. Comparative analysis of the anti-leukemic potential of interleukin 1 β (IL1 β), interleukin 2 (IL2), interleukin-6 (IL6), tumor necrosis factor α (TNF α), granulocyte colony stimulating factor (G-CSF), granulocyte-macrophage colony stimulating factor (GM-CSF), and their combination. Leuk. Lymphoma 7: 79-86 (1992); **Crown J et al** Interleukin-1: biological effects in human hematopoiesis. Leuk. Lymphoma 9: 433-40 (1993); **Dubois CM et al** *In vivo* interleukin-1 (IL1) administration indirectly promotes type II IL1 receptor expression on hematopoietic bone marrow cells: Novel mechanism for the hematopoietic effects of IL1. Blood 78: 2841-7 (1991); **Dullens HFJ et al** Cancer treatment with interleukins 1, 4, and 6 and combinations of cytokines: a review. *In vivo* 5: 567-70 (1991); **Fibbe WE & Willemze R** The role of interleukin-1 in hematopoiesis. Acta Haematol. 86: 148-54 (1991); **Johnson CS** Modulation of chemotherapy antineoplastic agents with biologic agents: enhancement of antitumor activities by interleukin-1. Curr. Opin. Oncol. 4: 1108-15 (1992); **Krumwieh D et al** Preclinical studies on synergistic effects of IL1, IL3, G-CSF, and GM-CSF in cynomolgous monkeys. IJCC 8 (Suppl. 1) 229-48 (1990); **Marumo K et al** Enhancement of lymphokine-activated killer activity induction *in vitro* by interleukin-1 administered in patients with urological malignancies. J. Immunother. 11: 191-7 (1992); **Monroy RL et al** Therapeutic evaluation of interleukin-1 for stimulation of hematopoiesis in primates after autologous bone marrow transplantation. Biotherapy 4: 97-108 (1992); **Moreb J & Zucali JR** The therapeutic potential of interleukin-1 and tumor necrosis factor on hematopoietic stem cells. Leuk. Lymphoma 8: 267-75 (1992); **Nakao S et al** Decreased interleukin 1 production in aplastic anemia. Br. J. Haematol. 71: 431-6 (1989); **Oppenheim JJ et al** Interleukin-1 enhances survival of lethally irradiated mice treated with allogeneic bone marrow cells. Blood 74: 2257-63 (1989); **Smith-JW 2d** The effects of treatment with interleukin-1 α on platelet recovery after high-dose carboplatin. NEJM 328: 756-61 (1993); **Smith JW 2d** The toxic and hematological effects of interleukin-1 α administered in a phase I trial to patients with advanced malignancies. J. Clin. Oncol. 10: 1141-52 (1992); **Starnes HF Jr** Biological effects and possible clinical applications of interleukin 1. Semin. Hematol. 28, Suppl. 2: 34-41 (1991); **Tanaka S et al** Preventive effects of interleukin 1 β for ACNU-induced myelosuppression in malignant brain tumors: the experimental and preliminary clinical studies. J. Neurooncol. 14: 159-68 (1992); **Walsh CE et al** A trial of recombinant human interleukin-1 in patients with severe refractory aplastic anemia. Br. J. Haematol. 80: 106-10 (1992)

IL1α: see: IL1.

IL1β: see: IL1.

IL1β-Convertase: *IL1β converting enzyme* (abbrev. ICE) This enzyme is a cysteine protease found in monocytes, lymphocytes, neutrophils, resting and activated T lymphocytes, placenta tissue, and several B lymphoblastoid cell lines. It is a heterodimeric protein composed of a 10 kDa and a 20 or 22 kDa subunit encoded by a common precursor of 45 kDa. The human gene maps to chromosome 11q13-q23, a site frequently involved in chromosomal rearrangements in human cancers, including a number of leukemias and lymphomas. The two subunits are derived from a single proenzyme, possibly by autoproteolysis. The murine gene maps to chromosome 9. At the protein level the homology between the murine and the human protein is 68 % (71 % at the nucleotide level).

The murine convertase shows broad constitutive expression and is detected in mononuclear phagocyte and T lymphocyte cell lines and in spleen, heart, brain, and adrenal glands. Its expression in mononuclear phagocytes is up-regulated by treatment with bacterial lipopolysaccharide or » IFN-γ. The enzyme cleaves the 31 kDa precursor of human IL1β (see: IL1) at two sequence-related sites: Asp27-Gly28 (site 1) and Asp116-Ala117 (site 2). Cleavage at site 2 yields the biologically active cytokine of 17 kDa. The enzyme does not cleave the IL1-α precursor. Since the selective inhibition of the enzyme in human blood monocytes blocks the production of mature IL1β, the enzyme may be a potential therapeutic target. The design of potent peptide aldehyde inhibitors have already been described and the similarities of the human and murine enzymes will make it possible to assess the therapeutic potential of human enzyme inhibitors in murine models of disease.

Human keratinocytes (and other non-bone marrow-derived cells) do not contain an IL1 convertase activity or any activity capable of processing 31-kD IL1β. These cells produce IL1β in an inactive form that can be processed only after leaving the cell.
Cowpox virus encodes a specific inhibitor of this protease. This protein, encoded by the ***crmA*** (cytokine response modifier A) or ***38K*** gene allows the virus to suppress inflammatory and antiviral reactions elicited by IL1. Microinjection of the *crm*A gene into chicken dorsal root ganglion neurons has been found to prevent cell death induced by withdrawal of » NGF, suggesting that ICE may play a role in neuronal death in vertebrates. For other anti-inflammatory viral strategies see also: viroceptor.
Murine IL1β-Convertase has been found to be the mammalian homologue of the cysteine protease encoded by the nematode*C. elegans* cell death gene *ced*-3. Overexpression of the murine convertase in Rat-1 cells leads to cell death by » apoptosis. This cell death can be inhibited by simultaneous overexpression of the *crm*A gene or by » *bcl*-2. Murine IL1β-Convertase may thus function during mammalian development to cause programmed cell death. The substrate that this enzyme acts upon to cause cell death is presently unknown.

Cerretti DP et al Molecular cloning of the interleukin 1β converting enzyme. S 256: 97-100 (1992); **Gagliardini V et al** Prevention of vertebrate neuronal death by the crmA gene. S 263: 826-8 (1994); **Howard AD et al** IL1-converting enzyme requires aspartic acid residues for processing of the IL1β precursor at two distinct sites and does not cleave 31 kDa IL1α. JI 147: 2964-9 (1991); **Kostura MJ et al** Identification of a monocyte-specific pre-interleukin 1β convertase activity. PNAS 86: 5227-31 (1989); **Nett MA et al** Molecular cloning of the murine IL1 β converting enzyme cDNA. JI 149: 3254-9 (1992); **Miller DK et al** Purification and characterization of active human interleukin 1β converting enzyme from THP.1 monocytic cells. JBC 268: 18062-9 (1993); **Miura M et al** Induction of apoptosis in fibroblasts by IL1β-converting enzyme, a mammalian homologue of the *C. elegans* cell death gene *ced*-3. Cell 75: 653-660 (1993); **Mizutani H et al** Human keratinocytes produce but do not process pro-interleukin-1 (IL1) β. Different strategies of IL1 production and processing in monocytes and keratinocytes. JCI 87: 1066-71 (1991); **Molineaux SM et al** Interleukin 1 β (IL1 β) processing in murine macrophages requires a structurally conserved homologue of human IL1 β converting enzyme. PNAS 90: 1809-13 (1993); **Palumbo GJ et al** Inhibition of an inflammatory response is mediated by a 38 kDa protein of cowpox virus. Virology 171: 262-73 (1989); **Ray CA et al** Viral inhibition of inflammation: cowpox virus encodes an inhibitor of interleukin 1β converting enzyme. Cell 69: 597-604 (1992); **Thornberry NA et al** A novel heterodimeric cysteine protease is required for interleukin-1 β processing in monocytes. N 356: 768-74 (1992); **Yuan J et al** The *C. elegans* cell death gene ced-3 encodes a protein similar to mammalian interleukin-1b-converting enzyme. Cell 75: 641-52 (1993)

IL1γ: This name is occasionally used for the IL1 receptor antagonists (see: L-1ra) since their amino acid sequences are related to both forms of IL1.

IL1-inducible 26k factor: see: 26 kDa protein.

IL1 inhibitory protein: see: IL1ra (IL1 receptor antagonist).

IL1k: "k" signifies "keratinocytes". This factor is identical with » ETAF (epidermal cell-derived thymocyte activating factor) which is identical with » IL1.

Sauder DN et al Chemotactic cytokines: the role of leukocytic pyrogen and epidermal cell thymocyte-activating factor in neutrophil chemotaxis. JI 132: 828-32 (1984); **Sauder DN et al** Biology and molecular biology of epidermal cell-derived thymocyte activating factor. ANY 548: 241-52 (1988)

IL1ra: IL*1 receptor antagonist* also abbrev. IL1RN (gene symbol). This factor is also called IL1 Inhibitor, **IRAP** (IL1 receptor antagonist protein), or **IL1γ**. Three other factors functioning as IL1 receptor antagonists isolated from the urine of a patient with monocytic leukemia, from IgG-stimulated monocytes, and from » U937 cells are identical with IL1ra. An intracellular form of IL1ra is called ***ICIL1ra*** (intracellular IL1ra). The intracellular form of IL1ra has been found to be produced by keratinocytes and other epithelial cells, macrophages and fibroblasts, while the secretory molecules is produced by monocytes, macrophages, neutrophils, fibroblasts and other cells.
IL1ra is isolated from human monocytes and other cells also expressing IL1 as a glycoprotein of 23-25 kDa. The human gene contains four exons and maps to chromosome 2q13-14.1 in the vicinity of the two genes encoding the two molecular forms of IL1. IL1ra shows ≈ 30 % homology with IL1-β at the protein level. IL1ra does not possess IL1-like activities.
Another protein that inhibits IL1 activities is the poorly characterized protein of 70-80 kDa called ***contra-IL1***. For another IL1 inhibitor that is not identical with IL1ra see also: M20 interleukin-1 inhibitor.
Another poorly characterized factor, ***IL1 inhibitory protein***, is released together with IL1 from human alveolar macrophages. It also functions as an IL1 receptor antagonist, blocking binding of IL1 to its receptor. In addition there are a number of poorly characterized factors also inhibiting the bioactivities of IL1.

Another IL1 inhibitor or 97.4 kDa, designated ***ILS***, has been observed in supernatants of human gingival organ cultures. ILS is produced mainly by keratinocytes. It inhibits the effects of both murine » IL1 and IL2 on thymocyte proliferation. ILS blocks the IL1-induced expression of the CD1 (T6) antigen, a highly specific marker for human Langerhans cells.

Urine from patients with monocytic leukemia have been found to contain high levels of IL1 inhibitor(s) of ≈ 25-35 kDa that may be distinct from other inhibitors already described.

IL1ra (152 amino acids) is synthesized as a precursor protein containing a classical secretory » signal sequence of 25 amino acids that allows the secretion of the factor via the endoplasmic reticulum/Golgi pathway. Apart from the secreted forms of IL1ra intracellular forms of this factor have also been described. Mouse and rat IL1ra display 77 % and 75 % amino acid sequence homology, respectively, to human IL1ra.

Glycosylated and also non-glycosylated IL1ra bind to the IL1 receptor with almost the same affinity as IL1. Binding of IL1ra to the IL1 receptor does not initiate intracellular signal transduction processes normally occurring after engagement of the receptor with IL1. IL1ra functions as a competitive inhibitor of IL1 receptor binding *in vitro* and *in vivo* and antagonizes the activities of both molecular forms of IL1 without having agonist effects itself.

This naturally occurring IL1 inhibitor is not species-specific and appears to be a physiologically important regulator of IL1 expression. The differentiation of monocytes into macrophages *in vitro* is associated with the constitutive synthesis of this receptor antagonist. This synthesis is enhanced in the presence of » GM-CSF. In human monocytes the expression of IL1ra is enhanced by » IL4, which at the same time also down-regulates the production of » IL1 and TNF-α. Human polymorphonuclear cells, after appropriate stimulation (bacterial lipopolysaccharides, » IL4, » G-CSF, » GM-CSF, » TNF-α) also synthesize IL1ra. In synovial fibroblasts the synthesis of IL1ra is markedly enhanced by » IL1, TNF-α, or » PDGF. The synthesis of IL1ra has been shown to be inducible by IL1β administration in humans.

IL1ra inhibits the release of both forms of » IL1, the secretion of » IL2, and the expression of IL2 receptors on the cell surface. It also blocks the stimulation of prostaglandin E2 synthesis in synovial cells and the proliferation of thymocytes. IL1ra also inhibits the release of leukotriene B4 by LPS-stimulated human monocytes. In isolated pancreatic islet cells IL1ra blocks the release of insulin. IL1ra has also been shown to block IL1-mediated induction of expression of human immunodeficiency virus in infected cells.

■ **Clinical Use & Significance:** IL1ra can prevent the » septic shock usually observed during Gram-negative sepsis in animal experiments. IL1ra acts as a non-steroidal anti-inflammatory substance and may be useful in the treatment of chronic inflammatory diseases such as autoimmune diabetes and arthritis. Graft-versus-host reactions have also been shown to be inhibited by administration of IL1ra.

IL1ra may also be of clinical relevance since it is known that IL1 synthesized by myeloma cells, Hodgkin lymphoma cells, B cells, T cells, and a variety of leukemic cell types in turn promotes the synthesis of colony-stimulating factors (see: CSF). It has been shown that spontaneous as well as » IL1-stimulated proliferation of acute myelogenous leukemia cells *in vitro* can be inhibited significantly by recombinant IL1ra in a dose dependent manner. This effect correlates with a reduction or disappearance in culture supernatants of » GM-CSF, which is normally secreted by these cells. Northern blot analysis performed on freshly isolated, uncultured leukemic blasts demonstrates constitutive expression of the IL1β gene in 19 of 23 AML cases analyzed and simultaneous expression of IL1ra in 3 of 23 cases. An imbalanced secretion of IL1 and its natural receptor antagonist may, therefore, contribute to the unrestricted growth of AML cells.

It has been suggested that IL1ra may improve hemorrhagic shock survival by preventing ATP depletion in vital organs.

Virus-infected macrophages produce large amounts of IL1 inhibitors and this may support opportunistic infections or malignant transformation of cells in patients with T cell maturation defects.

IL1ra has also been shown to inhibit augmentation of metastasis induced by » IL1 or bacterial lipopolysaccharide in a human melanoma/nude mouse system (see also: Immunodeficient mice).

● **Reviews: Arend WP** Interleukin-1 receptor antagonist. AI 54: 167-227 (1993); **Dinarello CA & Thompson RC** Blocking IL1: interleukin 1 receptor antagonist *in vivo* and *in vitro*. IT 12: 404-10 (1991); **Larrick JW** Native interleukin 1 inhibitors. IT 10: 61 (1989)

● **Biochemistry & Molecular Biology: Arend WP et al** An IL1 inhibitor from human monocytes. Production and characterization of biological properties. JI 143: 1851-8 (1989);

Application of IL1ra in various disease models.

Disease model	Reference
Rabbits	
Death from endotoxin, LPS, or *E. coli*	Nature 348: 550 (1990)
	FASEB J. 5: 338 (1991)
Hemodynamic shock and tissue damage after *E. coli*	FASEB J. 5: 338 (1991)
Hemodynamic shock from *Staphylococcus epidermidis*	Infect. Immun. 61: 3342 (1993)
Inflammatory bowel disease	J. Clin. Invest. 86: 972 (1990)
Muramyl dipeptide/IL1-induced sleep and fever	Am. J. Physiol. 265: R907 (1993)
Mice	
Death from LPS	J. Exp. Med. 173: 1029 (1991)
Cerebral malaria	Infect. Immun. 59: 1188 (1991)
Collagen-induced arthritis	Arthritis Rheum. 33: S20 (abstract)
Hypoglycemia and CSF production after endotoxin	Infect. Immun. 59: 1188 (1991)
	Infect. Immun. 59: 2494 (1991)
Neutrophil accumulation during inflammatory peritonitis	J. Exp. Med. 173: 931 (1991)
Sciatic nerve regeneration	J. Neuroimmunol. 39: 75 (1992)
Graft-versus-host disease	Blood 78: 1915 (1992)
Shock and death after *Bacillus anthracis*	Proc. Natl. Acad. Sci. 90: 10198 (1993)
Bleomycin-induced pulmonary fibrosis	Cytokine 5: 57 (1993)
Hemorrhagic shock	Surgery 114: 278 (1993)
IL1-mediated hypercalcemia	J. Bone Miner. Res. 8: 583 (1993)
Rats	
Death from *Klebsiella pneumoniae*	Infect. Immun. 61: 926 (1993)
Experimental enterocolitis	Agents Actions 34: 187 (1991)
LPS-induced pulmonary inflammation	Am. J. Pathol. 138: 521 (1991)
Acute phase response	Am. J. Physiol. 265: R739 (1993)
Ischaemic and excitotoxic brain damage	Cerebrovasc. Brain Metab. Rev. 5: 178
(1993)	
Carrageenan-induced pleurisy	Inflammation 17: 121 (1993)
Experimental crescentic glomerulonephritis	Kidney Int. 43: 479 (1993)
Acid secretion	Life Sci. 52: 785 (1993)
Muscle proteolysis	Ann. Surg. 216: 381 (1992)
LPS-induced fever	Am. J. Physiol. 263: R653 (1992)
Gram-negative sepsis	Surgery 112: 188 (1992)
IL1β-induced food and water intake suppression	Physiol. Behav. 51: 1277 (1992)
Baboons	
Hemodynamic shock and tissue damage after *E. coli*	FASEB J. 5: 338 (1991)
IL1-induced monocytopenia	Ann. Surg. 218: 79 (1993)
Guinea pigs	
Airway dysfunction	Am. J. Respir. Cell. Mol. Biol. 8: 365
(1993)	
Humans	
Neutrophilia after *E. coli* endotoxin	Blood 82: 2985 (1993)
Experimental metastasis of melanoma cells in nude mice	Cancer Res. 53: 5051 (1993)
Suppression of clonogenic growth of AML/CML progenitors *in vitro*	Leuk. Lymphoma 10: 407 (1993)
	Leukemia 6: 898 (1992)
	Blood 79: 1938 (1992)

Arend WP Interleukin 1 receptor antagonist. A new member of the interleukin 1 family. JCI 88: 1445-51 (1991); **Balavoine JF et al** Prostaglandin E2 and collagenase production by fibroblasts and synovial cells is regulated by urine-derived human interleukin 1 and inhibitor(s). JCI 78: 1120-4 (1986); **Bargetzi MJ et al** Interleukin-1β induces interleukin-1 receptor antagonist and tumor necrosis factor binding protein in humans. CR 53: 4010-3 (1993); **Bienkowski MJ et al** Purification and characterization of interleukin 1 receptor level antagonist proteins from THP-1 cells. JBC 265: 14505-11 (1990); **Carter DB et al** Purification, cloning, expression, and biological characterization of an interleukin-1 receptor antagonist protein. N 344: 633-7 (1990); **Eisenberg SP et al** Primary structure and functional expression from complementary DNA of a human interleukin 1 receptor antagonist. N 343: 341-6 (1990); **Haskill S et al** cDNA cloning of an intracellular form of the human interleukin 1 receptor antagonist associated with epithelium. PNAS 88: 3681-5 (1991); **Lennard A et al** Cloning and chromosome mapping of the human interleukin-1 receptor antagonist gene. Cytokine 4: 83-9 (1992); **Matsushime H et al** Cloning and expression of murine interleukin-1 receptor antagonist in macrophages stimulated by colony-stimulating factor 1. Blood 78: 616-23 (1991); **Patterson D et al** The human interleukin-1 receptor antagonist (IL1RN) gene is located in the chromosome 2q14 region. Genomics 15: 173-6 (1993); **Seckinger P et al** A urine inhibitory interleukin 1 activity that blocks ligand binding. JI 139: 1546-9 (1987); **Shuck ME et al** Cloning, heterologous expression and characterization of murine interleukin 1 receptor antagonist protein. EJI 21: 2775-80 (1991); **Steinkasserer A et al** Length variation within intron 2 of the human IL1 receptor antagonist protein gene (IL1RN). NAR 19: 5095 (1991); **Steinkasserer A et al** The human IL1 receptor antagonist gene (IL1RN) maps to chromosome 2q14-q21, in the region of the IL1 α and IL1 β loci. Genomics 13: 654-7 (1992); **Steinkasserer A et al** Chromosomal mapping of the human interleukin-1 receptor antagonist gene (IL1RN) and isolation of specific YAC clones. Agents Actions 38 Spec No: C59-60 (1993); **Takeuchi M et al** Characterization of IL1 inhibitory factor released from human alveolar macrophages as IL1 receptor antagonist. Clin. Exp. Immunol. 88: 181-7 (1992); **Walsh LJ et al** Isolation and purification of ILS, an interleukin 1 inhibitor produced by human gingival epithelial cells. Clin. Exp. Immunol. 68: 366-74 (1987)

● **BIOLOGICAL ACTIVITIES: Balavoine JF et al** Prostaglandin E2 and collagenase production by fibroblasts and synovial cells is regulated by urine-derived human interleukin 1 and inhibitor(s). JCI 78: 1120-4 (1986); **Bargetzki MJ et al** Interleukin 1β induces interleukin-1 receptor antagonist and tumor necrosis factor binding proteins in humans. CR 53: 4010-13 (1993); **Conti P et al** Inhibition of interleukin-1 (α and β), interleukin-2 secretion and surface expression of interleukin-2 receptor (IL2R) by a novel cytokine interleukin-1 receptor antagonist (IL1ra). Scand. J. Immunol. 36: 27-33 (1992); **Conti P et al** Human recombinant IL1 receptor antagonist (IL1Ra) inhibits leukotriene B4 generation from human monocyte suspensions stimulated by lipopolysaccharide (LPS). Clin. Exp. Immunol. 91: 526-31 (1993); **Dayer-Metroz MD et al** A natural interleukin 1 (IL1) inhibitor counteracts the inhibitory effect of IL1 on insulin production in cultured rat pancreatic islets. J. Autoimmun. 2: 163-71 (1989); **Dripps DJ et al** Interleukin 1 (IL1); receptor antagonist binds to the 80-kDa IL1 receptor but does not initiate IL1 signal transduction. JBC 266: 10331-6 (1991); **Eisenberg SP et al** Primary structure and functional expression from complementary DNA of a human interleukin-1 receptor antagonist. N 343: 341-6 (1990); **Firestein GS et al** IL1 receptor antagonist protein production and gene expression in rheumatoid arthritis and osteoarthritis synovium. JI 149: 1054-62 (1992); **Hannum CH et al** Interleukin-1 receptor antagonist activity of a human interleukin-1 inhibitor. N 343: 336-40 (1990); **Hawes AS et al** Comparison of peripheral blood leukocyte kinetics after live *Escherichia coli*, endotoxin, or interleukin-1 α administration. Studies using a novel interleukin-1 receptor antagonist. Ann. Surg. 218: 79-90 (1993); **Janson RW et al** Production of IL1 receptor antagonist by human *in vitro*-derived macrophages. Effects of lipopolysaccharides and granulocyte-macrophage colony-stimulating factor. JI 147: 4218-23 (1991); **Koch AE et al** Expression of interleukin-1 and interleukin-1 receptor antagonist by human rheumatoid synovial tissue macrophages. Clin. Immunol. Immunopathol. 65: 23-9 (1992); **Martel-Pelletier J et al** The synthesis of IL1 receptor antagonist (IL1ra) by synovial fibroblasts is markedly increased by the cytokines TNF-α and IL1. BBA 1175: 302-5 (1993); **Numerof RP et al** Suppression of IL2-induced SAA gene expression in mice by the administration of an IL1 receptor antagonist. Cytokine 4: 555-60 (1992); **Orino E et al** IL4 up-regulates IL1 receptor antagonist gene expression and its production in human blood monocytes. JI 149: 925-31 (1992); **Poli G et al** Interleukin 1 induces expression of the human immunodeficiency virus alone and in synergy with interleukin 6 in chronically infected U1 cells: inhibition of inductive effects by the interleukin 1 receptor antagonist. PNAS 91: 108-12 (1994); **Rambaldi A et al** Modulation of cell proliferation and cytokine production in acute myeloblastic leukemia by interleukin-1 receptor antagonist and lack of its expression by leukemic cells. Blood 78: 3248-53 (1991); **Re F et al** Expression of interleukin-1 receptor antagonist (IL1ra) by human circulating polymorphonuclear cells. EJI 23: 570-3 (1993); **Scala G et al** Accessory cell function of human B cells. I. Production of both interleukin 1-like activity and an interleukin 1 inhibitory factor by an EBV-transformed human B cell line. JEM 159: 1637-52 (1984); **Walsh LJ & Seymour GJ** Interleukin 1 induces CD1 antigen expression on human gingival epithelial cells. J. Invest. Dermatol. 90: 13-6 (1988)

● **ASSAYS: Malyak M et al** IL1ra ELISA: reduction and alkylation of synovial fluid eliminates interference by IgM rheumatoid factors. JIM 140: 281-8 (1991)

● **CLINICAL USE & SIGNIFICANCE: Arendt WP & Coll BP** Interaction of recombinant monocyte-dericed interleukin 1 receptor antagonist with rheumatoid synovial cells. Cytokine 3: 407-13 (1991); **Chirivi RGS et al** Interleukin 1 receptor antagonist inhibits augmentation of metastasis induced by interleukin 1 or lipopolysaccharide in a human melanoma/nude mouse system. CR 53: 5051-4 (1993); **McCarthy PL Jr et al** Inhibition of interleukin-1 by an interleukin-1 receptor antagonist prevents graft-versus-host disease. Blood 78: 1915-8 (1991); **Ohlsson K et al** Interleukin 1 receptor antagonist reduces mortality from endotoxin shock. N 348: 550-4 (1990); **Pellicane JV et al** Interleukin-1 receptor antagonist improves survival and preserves organ adenosine-5'-triphosphate after hemorrhagic shock. Surgery 114: 278-83; discussion 283-4 (1993); **Rambaldi A et al** Modulation of cell proliferation and cytokine production in AML by recombinant interleukin-1 receptor antagonist. Leukemia 7: S10-2 (1993); **Wakabayashi G et al** A specific receptor antagonist for interleukin 1 prevents *Escherichia coli*-induced shock in rabbits. FJ 5: 338-43 (1991)

IL1 receptor antagonist protein: abbrev. IRAP. see: IL1ra (IL1 receptor antagonist).

IL1 inhibitor: see: IL1ra (IL1 receptor antagonist). See also: M20 interleukin-1 inhibitor.

IL1RN: IL1 receptor antagonist. See: IL1ra.

IL2: Interleukin 2.

■ **Alternative Names:** BF (blastogenic factor); Co-stimulator; EDF (eosinophil differentiation factor); KHF (killer cell helper factor); LMF (lymphocyte mitogenic factor); LCM factor (lymphocyte-conditioned medium factor); LPF (lymphocyte proliferation factor); MAF-C I (macrophage-activating factor for cytotoxicity I); PFC-EA (plaque forming cell enhancing factor); SCIF (secondary cytotoxic T cell-inducing factor); TCGF (T cell growth factor); TCPA (T colony-promoting activity); TDF (thymocyte differentiation factor); TGP-3; TMF (thymocyte mitogenic factor); TMF (T cell maturation factor); TMF (T cell mitogenic factor); TRF 3 (T cell replacing factor 3); TSF (thymocyte stimulating factor). See also: individual entries for further information.

■ **Sources:** Under physiological conditions IL2 is produced mainly by CD4-positive T cells following their activation by mitogens or allogens (see also: cell activation). Several secondary signals are required for maximal expression of IL2. Resting cells do not produce IL2. *In vitro* the synthesis of IL2 is inhibited by dexamethasone or » CsA (ciclosporin A). Transformed T and B cells, leukemia cells, » LAK cells (lymphokine-activated killer cells) and NK cells also secrete IL2. Vitamin E can enhance the production of IL2. With increasing age the antigen- and mitogen-stimulated synthesis of IL2 and hence also T cell-mediated immune responses decrease.

■ **Protein Characteristics:** IL2 is a protein of 133 amino acids (15.4 kDa) with a slightly basic pI. It does not display sequence homology to any other factors. Murine and human IL2 display a homology of ≈ 65 %. IL2 is synthesized as a precursor protein of 153 amino acids with the first 20 aminoterminal amino acids functioning as a hydrophobic secretory » signal sequence. The protein contains a single disulfide bond (positions Cys58/105) essential for biological activity.

IL2 is O-glycosylated at threonine at position 3. Variants with different molecular masses and charges are due to variable glycosylation. Non-glycosylated IL2 is also biologically active. Glycosylation appears to promote elimination of the factor by hepatocytes.

A dimeric form of human IL2, produced by the action of a transglutaminase isolated from regenerating fish optic nerves, has been shown to be a cytotoxic factor for rat brain oligodendrocytes in culture.

■ **Gene Structure:** The human IL2 gene contains four exons. The IL2 gene maps to human chromosome 4q26-28 (murine chromosome 3). Translocations, deletions, and/or chromosomal gene amplifications of the IL2 gene have been observed neither under physiological nor under pathological conditions (see: 5q⁻ syndrome for other cytokines).

The homology of murine and human IL2 is 72% at the nucleotide level in the coding region.

The synthesis of IL2 is regulated at the level of transcription (see also: ARRE). T cells contain a labile repressor that modulates the post-transcriptional processing of IL2 mRNA precursors. This repressor prevents post-transcriptional processing so that ≈ 98 % of the IL2 mRNA remain unprocessed.

The promoter region of the IL2 gene contains several binding sites for specific transcription factors (see also: Gene expression) that allows the regulated activation or repression of IL2 gene activity. The transcription factor » NF-AT regulates the expression of the IL2 gene in T cells. A factor designated » TCF8, prevents IL2 gene expression.

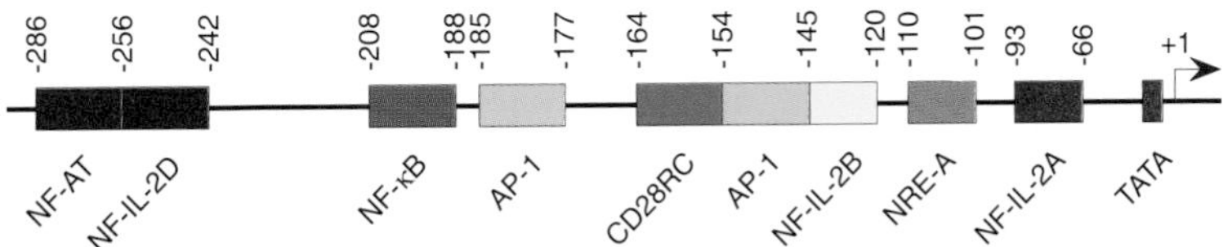

The promoter/enhancer region of the IL2 gene.
The IL2 promoter contains a classical TATA box sequence. +1 indicates the transcriptional start site. The region located in front of this site (nucleotide positions –66 - –286) contains a variety of distinct short sequence elements which function as binding sites for transcription factors involved in the regulation of gene expression. NRE-A (negative response element) is a sequence region involved in the suppression of IL2 expression. The CD28RC region (CD28 response complex) possesses enhancer activity.

■ **Receptor Structure, Gene(s), Expression:** The biological activities of IL2 are mediated by a membrane receptor that is expressed almost exclusively on activated, but not on resting, T cells (see also: cell activation) at densities of 4–12 × 10^3 receptors/cell. Activated B cells and resting mononuclear leukocytes rarely express this receptor. The expression of the IL2 receptor is modulated by » IL5 and » IL6. For other factors specifically influencing IL2 receptor expression see: » IL2R/p55 (Tac) inducing factor, » IL2 receptor inducing factor, » IL2R α chain inhibitory activity.
Three different types of IL2 receptors are distinguished that are expressed differentially and independently. The high affinity IL2 receptor (Kdis ~ 10 pM) constitutes approximately 10 % of all IL2 receptors expressed by a cells. This receptor is a membrane receptor complex consisting of the two subunits IL2Rα (***TAC antigen*** =T cell activation antigen; p55) and IL2Rβ (p75; new designation: ***CD122***, see also: CD Antigens) as the ligand binding domains and a γ chain as a signaling component (see below). p75 is expressed constitutively on resting T lymphocytes, NK cells, and a number of other cell types while the expression of p55 is usually observed only after cell activation. p55 is, however, synthesized constitutively by a number of tumor cells and by HTLV-1-infected cells.
IL2 receptor expression of monocytes is induced by » IFN-γ, so that these cells become tumor-cytotoxic. In T cells the expression of p75 can be reduced by » IL3.
An intermediate affinity IL2 receptor (Kdis = 100 pM) consists of the p75 subunit and a γ chain (see below) while a low affinity receptor (Kdis = 10 nM) is formed by p55 alone.
p55 has a length of 251 amino acids with an extracellular domain of 219 amino acids an a very short cytoplasmic domain of 13 amino acids. The p55 gene maps to human chromosome 10p14-p15. The expression of p55 is regulated by a nuclear protein called » RPT-1.
p75 has a length of 525 amino acids with an extracellular domain of 214 amino acids an a cytoplasmic domain of 286 amino acids. The p75 gene contains 10 exons and has a length of ≈ 24 kb. It maps to human chromosome 22q11. 2-q12 and to murine chromosome 15 (band E).
A third 64 kDa subunit of the IL2 receptor, designated γ, has been described recently. Murine and human γ subunits of the receptor have ≈ 70 % sequence identity at the nucleotide and amino acid levels. This subunit is required for the generation of high and intermediate affinity IL2 receptors but does not bind IL2 by itself. These two receptor types consist of a αβγ heterotrimer and a βγ heterodimer, respectively. The gene encoding the γ subunit of the IL2 receptor maps to human chromosome Xq13, spans ≈ 4.2 kb and contains eight exons. Relationships to markers in linkage studies suggest that this gene and SCIDX1, the gene for X-linked severe combined immunodeficiency, have the same location. Moreover, in each of 3 unrelated patients with X-linked SCID, a different mutation in the IL2Rγ gene has been observed. The γ subunit of the IL2 receptor has been shown recently to be a component of the receptors for » IL4 and » IL7. It is probably also a component of the » IL13

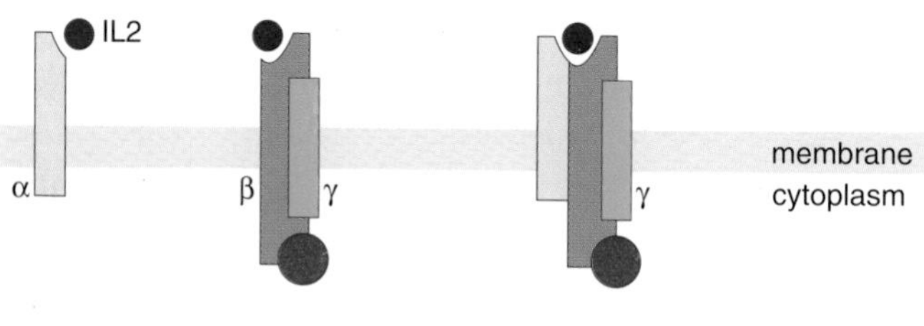

Schematic model of the known subunits of the IL2 receptor.
The low affinity receptor contains the α chain of 55 kDa which is identical with the CD25/TAC antigen. The α chain has 219 amino acids in its extracellular domain, 19 amino acids in the transmembrane region, and a short cytoplasmic tail of 13 amino acids.
Intermediate affinity receptors are dimers of the 75 kDa β chain (CD122) and a γ chain. The β chain has an extracellular domain of 214 amino acids, a transmembrane region of 25 amino acids, and a cytoplasmic domain of 286 amino acids. High affinity receptors contain all three chains. The γ chain has been shown recently to be a component of other cytokine receptor systems, including those for » IL4, IL7 and possible also IL9 and IL13.
The β chain of the IL2 receptor has been shown to be associated with *src* family protein tyrosine kinases (red circle) such as *lck* or *fyn*. Signaling pathways associated with the γ subunit have not yet been elucidated.

receptor. These findings can explain the severity of the immune defect in X-linked immunodeficiency. The amino acids at positions 267-317 lying directly adjacent to the transmembrane region of p75 are involved in IL2-mediated signal transduction. In addition the IL2 receptor is associated with a number of other proteins (p22, p40, p100) which are thought to be involved in mediating conformational changes in the receptor chains, receptor-mediated endocytosis, and further signal transduction processes. One of the identified proteins is the 95 kDa cell adhesion molecule ICAM-1 which probably focuses IL2 receptors at regions of cell-to-cell contacts and thus may mediate » paracrine activities, e. g. during IL2-mediated stimulation of T cells. Another protein associated with p75 is a tyrosine-specific protein kinase called » *lck*. The observation that IL2-induced proliferation of cells is inhibited by specific inhibitors of protein tyrosine kinases in an *lck*-negative cell line suggests that other kinases may also be associated with IL2 receptors. Two such kinases, called » *fyn* and » *lyn*, have been identified. In addition, IL2 receptor signaling may also be mediated by » *vav*.

Activated lymphocytes continuously secrete a 42 kDa fragment of the TAC antigen. This fragment circulates in the serum and plasma and functions as a soluble IL2 receptor (***sIL2R***). The concentrations of this soluble receptor vary markedly in different pathological situations, e. g. infections, autoimmune diseases, leukemias, or after organ transplantation. Levels may increase up to 100-fold. The levels of sIL2R appear to correlate with the severity of HIV-induces diseases and may be of diagnostic value also in other settings.

■ **BIOLOGICAL ACTIVITIES:** Mouse and human IL2 both cause proliferation of T cells of the homologous species at high efficiency. Human IL 2 also stimulates proliferation of mouse T cells at similar concentrations, whereas mouse IL 2 stimulates human T cells at a lower (sixfold to 170-fold) efficiency.

IL2 is a growth factor for all subpopulations of T lymphocytes. It is an antigen-unspecific prolifera-

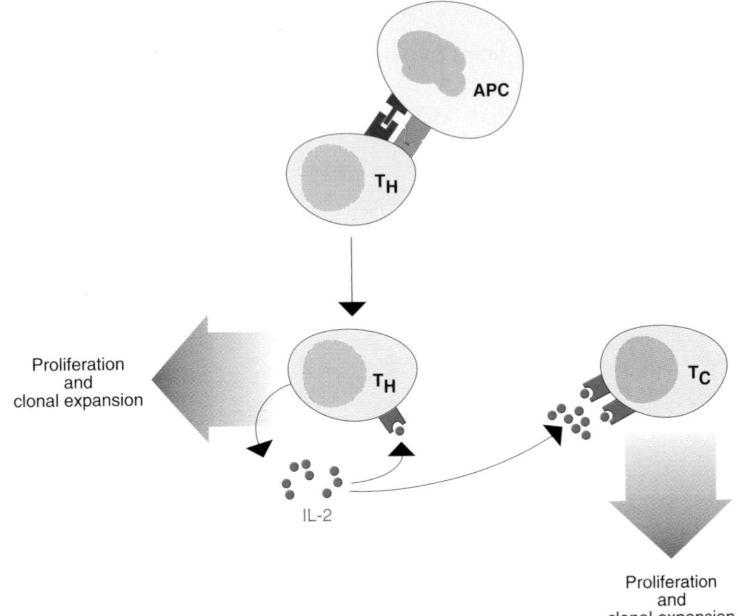

Simplified view of the role of IL2 in the development of cytotoxic T lymphocytes.
CD8-positive cytotoxic T lymphocytes (TC) that have recognized MHC class I associated antigens on cells are stimulated to express the IL2 receptor and thus can respond to IL2 produced by T helper cells (TH). This leads to the proliferation of cytotoxic T lymphocytes and to clonal expansion.
The helper cells themselves are triggered to produce IL2 after having been activated through interaction with an MHC class II associated antigen on an antigen-presenting cell (APC). This process leads to the expression of IL2 receptors on TH cells which then also start to proliferate. In addition these interactions lead to the secretion of a number of other cytokines.

tion factor for T cells that induces » cell cycle progression in resting cells and thus allows clonal expansion of activated T lymphocytes. This effect is modulated by hormones such as » prolactin.
The p55 receptor subunit is expressed in adult T cell leukemia (ATL). Since freshly isolated leukemic cells also secrete IL2 and respond to it IL2 may function as an » autocrine growth factor for these cells capable of worsening ATL.
IL2 also promotes the proliferation of activated B cells also this requires the presence of additional factors, for example, » IL4. *In vitro* IL2 also stimulates the growth of oligodendroglial cells.
Due to its effects on T and B cells IL2 is a central regulator of immune responses. It also plays a role in anti-inflammatory reactions (see also: Inflammation), in » Hematopoiesis and in tumor surveillance. IL2 stimulates the synthesis of » IFN-γ in peripheral leukocytes and also induces the secretion of » IL1, » TNF-α and » TNF-β.
The induction of the secretion of tumoricidal cytokines apart from the activity in the expansion of » LAK cells (lymphokine-activated killer cells) are probably the main factors responsible for the antitumor activity of IL2.
IL2 damages the blood-brain barrier and the integrity of the endothelium of brain vessel. These effects may be the underlying causes of neuropsychiatric side effects such as fatigue, disorientation, and depression, frequently observed under IL2 therapy. IL2 also alters the electrophysiological behavior of neurons.
A dimeric form of human IL2, produced by the action of a transglutaminase isolated from regenerating fish optic nerves, has been shown to be a cytotoxic factor for rat brain oligodendrocytes in culture. It has been suggested that dimerization of IL2 may provide a mechanism to permit nerve growth under conditions in which oligodendrocytes inhibit neuronal regeneration.

■ **TRANSGENIC/KNOCK-OUT/ANTISENSE STUDIES:** The consequences of a deregulated expression of IL2 have been studied in » transgenic animals (mice) expressing either the human IL2 gene or the TAC IL2 receptor. Transgenic mice harboring the human gene for IL2 or the TAC-IL2 receptor constitutively express IL2 in the thymus, spleen, bone marrow, lungs, muscle, and skin. Among other things these animals show pronounced growth retardation and die prematurely. Histologically one observes a selective loss of Purkinje cells in the cerebellum and a focal infiltration of lymphocytes, leading to pneumonia. The spleen shows a massive increase of cells with the marker spectrum thy1$^+$/CD3$^-$4$^-$8$^-$. The expression of the human IL2 gene which has been transferred into murine T cells by means of a retrovirus vector leads to growth autonomy and malignant transformation of the cells.
Transgenic mice expressing murine interleukin IL2 constitutively in islet β cells die at an early age because of a predominant macrophage inflammatory response that destroys the exocrine pancreas. Animals with lower levels of IL2 survive and have islets that become increasingly infiltrated with lymphocytes over time without, however, developing autoimmunity to islet antigens.
The requirement for IL2 has been tested *in vivo* using IL2-deficient mice generated by targeted recombination » ES cells. Mice homozygous for the IL2 gene mutation are normal with regard to thymocyte and peripheral T cell subset composition during the first 3-4 weeks of age. However, a dysregulation of the immune system is manifested by reduced polyclonal *in vitro* T cell responses and by dramatic changes in the isotype levels of serum immunoglobulins. IL2-deficient mice have been shown to develop an unlimited inflammatory bowel disease later in life (past 6 weeks of age) with striking clinical and histological similarity to ulcerative colitis in humans.
IL2-deficient mice have also been analyzed with respect to immune responses *in vivo*. Primary and secondary cytotoxic T cell responses against vaccinia and lymphocytic choriomeningitis virus have been found to be within normal ranges. T helper cell responses are delayed but biologically functional. Natural killer cell activities are markedly reduced but inducible. B cells reactivity to vesicular stomatitis virus is not impaired. These observations suggest that other lymphokines may compensate for the IL2 defects *in vivo*. IL2-deficient mice have high levels of IgG1 and IgE isotypes in their sera. This was taken as evidence that » IL4 was probably involved in immunoglobulin class switching and compensated, at least in part, for the IL2 deficiency. The recent generation of double-negative mice not expressing IL2 and IL4 shows, however, that these also have normal major T cell subsets and B cells, indicating that both interleukins are not essential for development of the immune system. It was even observed that IL2/IL4 double negative mice displayed higher levels of T cell proliferation.
The functions of IL2-producing cells have also been studied by creating transgenic mice in which

IL2-producing cells, rendered sensitive to the cytotoxic effects of the antiviral drug gancyclovir, can be eliminated selectively (cell ablation) at different stages of development. These studies show that some proliferative responses of T cells to antigens occur via an alternative pathway not requiring IL2 synthesis and release.

■ **Assays:** IL2 can be assayed in » Bioassays employing cell lines that respond to the factor (see: ATH8; CT6; CTLL; FDCP-Mix; HT-2; NKC3; TALL-103). Specific ELISA assays for IL2 and enzyme immunoassays for the soluble receptor are also available. The soluble receptor can also be detected by employing biotinylated IL2 and flow-through cytometry or ELISA assays. An alternative and entirely different detection method is » Message amplification phenotyping. For IL2 units see also: BRMP unit. For further information see also subentry "Assays" in the reference section.

■ **Clinical Use & Significance:** IL2 displays significant anti-tumor activity for a variety of tumor cell types since it supports the proliferation and clonal expansion of T cells that specifically attack certain tumor types (see: Adoptive immunotherapy, LAK cells (lymphokine-activated killer cells) and » TILs (tumor-infiltrating lymphocytes) for clinical applications). IL2 is increasingly used to treat patients with cancers refractory to conventional treatment. Combination therapy with systemically administered IL2 has resulted in long-term remissions in 30% of patients with metastatic renal cell carcinoma, for which there is no standard treatment. Objective and long-lived clinical responses have also been documented in a proportion of patients with melanoma or acute myeloid leukemia.

In spite of the stimulation of a range of host anti-tumor defense mechanisms, only 30-40 per cent of patients who are treated respond to IL2 therapy as assessed by a reduction in tumor volume. More randomized clinical trials are required to evaluate precisely the role of IL2 in various therapeutic combinations and to ascertain the optimum therapeutic regimens for individual tumor types. The therapeutic results obtained in humans have been less beneficial and more selective than those in experimental animals and more basic research is required to ascertain how IL2 produces its antitumor effects.

A study with 15 patients with metastatic colorectal carcinoma prior to and during treatment with a continuous intravenous infusion of IL2 suggests that measurement of serum levels of C-reactive protein (CRP) may permit to distinguish those cancer patients who are most likely to respond to and benefit from IL2 therapy.

One of the reasons may be that IL2 has been used more or less as a chemotherapeutic compound in the highest tolerable dose, which mainly stimulates nonspecific » LAK cell activity through low-affinity IL2 receptors without leading to systemic immunity. High dose IL2 therapy, accordingly, is also associated with a great number of unwanted toxic side-effects. IL2 has additional effects on other components of the cellular immune system, including B cells and macrophages, and induces the secretion of other soluble mediators, including » TNF-α, » TNF-β, and » IFN-γ. These effects may contribute to the antitumor activity of IL2 as well as to its dose-related toxicity.

Several studies have demonstrated that the application of intratumoral low doses of IL2 can be highly effective against cancer and without toxic side-effects. IL2-toxin protein chimeras are currently under investigation in the treatment of cancer (see: Cytokine fusion toxins).

Protocols are now under intense study to elevate the immunogenicity of tumors by the introduction of a functional IL2 gene (see also: Cytokine gene transfer). It is hoped that the enhanced expression and secretion of IL2 by tumor cells engineered by means of this sophisticated technology will lead to the activation of T cells and thus provide a modality for the treatment of these tumors and their metastases. The transduction of murine tumor cells with a functional IL2 gene has been shown to lead to the rejection of the genetically modified cells by syngeneic hosts. Altered tumor cells expressing IL2 also increase systemic immunity.

Cellular immunodeficiency diseases, especially those with impaired IL2 production are successfully treated by daily injections of hIL2. IL2 is also effective on some patients with antibody deficiency, probably caused by the lack of T cell help for B cells.

A number of diseases have been described to be associated with the aberrant expression of IL2 or IL2 receptors, including Hodgkin's disease, Graft-versus-Host reaction, multiple sclerosis, rheumatoid arthritis, systemic lupus erythematosis, type I diabetes, lepromatous leprosy, AIDS, immunodeficiency syndrome, severe burn traumas, and allogenic bone marrow transplantation.

Fusion proteins consisting of IL2 and the weakly immunogenic Herpes simplex virus (HSV-1) glycoprotein D have been constructed. They have been reported to enhance both humoral and cellular immunity against HSV-1 and to protect mice immunized with the fusion protein against HSC-1 infections. For other fusion proteins converting weakly immunogenic antigens into potent immunogens see: GM-CSF (subentry Clinical use & Significance).

● **Reviews: Nistico G** Communications among central nervous system, neuroendocrine and immune systems: Interleukin 2. Prog. Neurobiol. 40: 463-75 (1993); **Smith K** Interleukin 2: inception, impact, and implications. S 240: 1169-76 (1988); **Taniguchi T & Minami Y** The IL2/IL2 receptor system: a current overview. Cell 73: 5-8 (1993);**Tartour E et al** Current status of interleukin-2 therapy in cancer. Biomed. Pharmacother. 46: 473-84 (1992); **von Rohr A & Thatcher N** Clinical applications of interleukin 2. PGFR 4: 229-46 (1992); **Waldmann TA et al** The interleukin-2 receptor: a target for immunotherapy. ANY 685: 603-10 (1993); **Waxman J & Balkwill F** (eds) Interleukin 2. Blackwell Scientific Publ. Oxford 1992; **Williams TM et al** Interleukin 2: Basic biology and therapeutic use. Hematol. Pathol. 5: 45-55 (1991); **Zurawski G** Analyzing lymphokine-receptor interactions of IL1 and IL2 by recombinant-DNA technology. TibTech 9: 250-7 (1991)

● **Biochemistry & Molecular Biology: Brandhuber BJ et al** Three-dimensional structure of the interleukin-2. S 232: 1707-9 (1987); **Cerretti DP et al** Cloning, sequence, and expression of bovine interleukin 2. PNAS 83: 3223-7 (1986); **Chen SJ et al** A viral long terminal repeat in the interleukin 2 gene of a cell line that constitutively produces interleukin 2. PNAS 82: 7284-8 (1985); **Conradt HS et al** Secretion of glycosylated human interleukin-2 by recombinant mammalian cell lines. Carbohydrate Res. 149: 443-50 (1986); **Cosic I et al** Prediction of "hot spots" in interleukin 2 based on informational spectrum characteristics of growth-regulating factors. Comparison with experimental data. Biochimie 71: 333-42 (1989); **Degrave W et al** Cloning and structure of the human interleukin 2 chromosomal gene. EJ 2: 2349-53 (1983); **Degrave W et al** Cloning and structure of a mouse interleukin-2 chromosomal gene. Mol. Biol. Rep. 11: 57-61 (1986); **Devos R et al** Molecular cloning of human interleukin 2 cDNA and its expression in *E. coli.* NAR 11: 4370-3 (1983); **Durand DB et al** Characterization of antigen receptor response elements within the interleukin-2 enhancer. MCB 8: 1715-24 (1988); **Fiorentino L et al** Assignment of the interleukin-2 locus to mouse chromosome 3. Genomics 5: 6513 (1989); **Fujita T et al** Structure of the human interleukin 2 gene. PNAS 80: 7437-41 (1983); **Fujita T** Molecular cloning of human and murine interleukin-2 genes and their expression in various host cells. J. Invest. Dermatol. 85: 180s-2 (1985); **Fuse A et al** Organization and structure of the mouse interleukin-2 gene. NAR 12: 9323-31 (1984); **Goodall JC et al** cDNA Cloning of ovine interleukin-2 by PCR. NAR 18: 5883 (1990); **Goodall JC et al** cDNA cloning of porcine interleukin 2 by polymerase chain reaction. BBA Gene Structure and Expression 1089: 257-8 (1991); **Holbrook NJ et al** T cell growth factor: complete nucleotide sequence and organization of the gene in normal and malignant cells. PNAS 81: 1634-8 (1984); **Hoyos B et al** Kappa B specific DNA binding proteins: role in the regulation of human interleukin-2 gene expression. S 244: 457-60 (1989); **Kaempfer R et al** Regulation of human interleukin 2 gene expression. Lymphokines 13: 59-72 (1987); **Kato K et al** Purification and characterization of recombinant human interleukin-2 produced in *Escherichia coli.* BBRC 130: 692-9 (1985); **Landgraf BE et al** Conformational perturbation of interleukin-2: A strategy for the design of cytokine analogs. Proteins 9: 207-16 (1991); **Maeda S et al** Cloning of interleukin 2 mRNAs from human tonsils. BBRC 115: 1040-7 (1983); **Matsui H et al** Molecular cloning and expression of the human interleukin 2 gene. Lymphokines 12: 1-12 (1985); **Mita S et al** Isolation and characterization of a human interleukin 2 gene. BBRC 117: 114-21 (1983); **Mevissen ML et al** Identification of structural differences between different forms of interleukin-2 (IL2) using anti-(human recombinant) IL2 monoclonal antibodies. Cytokine 3: 54-9 (1991); **Muegge K et al** Interleukin-1 costimulatory activity on the interleukin-2 promoter via AP-1. S 246: 249-51 (1989); **Munoz A et al** Cloning and expression of human interleukin 2 in *Streptomyces lividans* using the *Escherichia coli* consensus promoter. BBRC 133: 511-9 (1985); **Price V et al** Expression, purification and characterization of recombinant murine granulocyte-macrophage colony-stimulating factor and bovine interleukin-2 from yeast. Gene 55: 287-93 (1987); **Robb R et al** Amino acid sequence and posttranslational modification of human interleukin-2. PNAS 81: 6486-90 (1984); **Sato T et al** New approaches for the high-level expression of human interleukin-2 cDNA in *Escherichia coli.* J. Biochem. Tokyo 101: 525-34 (1987); **Seigal LJ et al** Gene for T cell growth factor: location on human chromosome 4q and feline chromosome B1. S 223: 175-78 (1983); **Sharon M et al** Novel interleukin-2 receptor subunit detected by cross-linking under high-affinity conditions. S 234: 859-63 (1986); **Shows T et al** Interleukin-2 (IL2) is assigned to human chromosome 4. Som. Cell. Molec. Genet. 10: 315-8 (1984); **Taniguchi T et al** Structure and expression of a cloned cDNA for human interleukin-2. N 302: 305-10 (1983); **Yamada G et al** Retroviral expression of the human interleukin 2 gene in a murine T cell line results in cell growth autonomy and tumorigenicity. EJ 6: 2705-9 (1987); **Yokota T et al** Use of a cDNA expression vector for isolation of mouse interleukin 2 cDNA clones: expression of T cell growth-factor activity after transfection of monkey cells. PNAS 82: 68-72 (1985); **Zurawski SM et al** Definition and spatial location of mouse interleukin 2 residues that interact with its heterotrimeric receptor. EJ 12: 5113-9 (1993)

● **Receptor Structure, Gene(s), Expression: Arima N et al** Pseudo-high affinity interleukin 2 (IL2) receptor lacks the third component that is essential for functional IL2 binding and signaling. JEM 176: 1265-72 (1992); **Asao H et al** Reconstitution of functional interleukin 2 receptor complexes on fibroblastoid cells: involvement of the cytoplasmic domain of the g chain in two distinct signaling pathways. PNAS 90: 4127-31 (1993); **Bazan JF** A novel family of growth factor receptors: a common binding domain in the growth hormone, prolactin, erythropoietin, and IL6 receptors, and the p75 IL2 receptor β-chain. BBRC 164: 788-95 (1989); **Campbell HD et al** Assignment of the interleukin-2 receptor-β chain gene (IL2RB) to band E on mouse chromosome 15. Genomics 12: 179-80 (1992); **Cao X et al** Characterization of cDNAs encoding the murine interleukin 2 receptor (IL2R) γ chain: chromosomal mapping and tissue specificity of IL2R γ-chain expression. PNAS 90: 8464-8 (1993); **Cosman D et al** Cloning, sequence, and expression of human interleukin-2 receptor. N 312: 768-71 (1984); **Gnarra JR et al** Human interleukin 2 receptor β-chain gene: chromosomal localization and identification of 5´ regulatory sequences. PNAS 87: 3440-4 (1990); **Greene WC et al** Isolation and expression of complementary DNAs encoding the human interleukin 2 receptor. CR 45: 4563s-7s (1985); **Hatakeyama M et al** Interleukin 2 receptor β chain gene: generation of the three receptor forms by cloned human α and β chain cDNAs. S 244: 551-6 (1989); **Ishida N et al** Molecular cloning and structure of the human interleukin 2 receptor gene. NAR 13: 7579-89 (1985); **Kobayashi N et al** Functional coupling of the *src*-family protein tyrosine kinases p59*fyn* and p53/56*lyn* with the interleukin 2 receptor: implications for redundancy and pleiotropism in cytokine signal transduction. PNAS 90: 4201-5 (1993); **Kobayashi N et al** Cloning and sequencing of the cDNA encoding a mouse IL2 receptor γ. Gene 130: 303-4 (1993); **Kondo M et al** Sharing of the interleukin-2 (IL2) receptor γ chain between receptors for IL2 and IL4. S 262: 1874-7 (1993); **Kucharz EJ et al** Serum inhibitors of interleukin 2. Life Sci. 42: 1485-91 (1988); **Kumaki S et al** Cloning of the mouse interleukin 2 receptor γ chain: demonstration of functional differences between the mouse and human receptors. BBRC 193: 356-63 (1993); **Leonard WJ et al** Molecular cloning and expression of cDNAs for the human interleukin-2 receptor. N 311: 626-35 (1984); **Leonard WJ et al** Structure of the human interleukin-2 receptor gene. S 230: 633-9 (1985); **Leonard WJ et al** Localizationof the gene encoding the human interleukin-2 receptor on chromosome 10. S 228: 1547-9 (1985); **Mills GB et al** Transmembrane signaling by interleukin 2. Biochem. Soc. Trans. 19: 277-87 (1991); **Miller J et al** Nucleotide sequence and expression of a mouse interleukin 2 receptor cDNA. JI 134: 4212-7 (1985); **Minami Y et al** The interleukin-2 receptors: insights into a complex signaling mechanism. BBA 1114: 163-77 (1992); **Mishi M et al** Expression of functional Interleukin 2 receptors in human light chain (Tac) transgenic mice. N 331: 267-9 (1988); **Miyazaki T et al** The integrity of the conserved `WS motif' common to IL2 and other cytokine receptors is essential for ligand binding and signal transduction. EMBO J. 10: 3191-7 (1991); **Nikaido T et al** Molecular cloning of cDNA encoding human interleukin-2 receptor. N 311: 631-5 (1984); **Noguchi M et al** Interleukin-2 receptor γ chain mutation results in X-linked severe combined immunodeficiency in humans. Cell 73: 147-57 1993); **Noguchi M et al** Characterization of the human interleukin-2 receptor γ chain gene. JBC 268: 13601-8 (1993); **Onishi R et al** Interleukin-3-induced downregulation of the expression of interleukin-2 receptor β chain in human T cells. Blood 78: 2908-17 (1991); **Puck JM et al** The γ chain of I. L-2 receptor (IL2R-γ) maps to Xq13 and is mutated in human X-linked severe combined immunodeficiency, SCIDX1. (Abstract) X Chromosome Workshop, St. Louis, Mo., May 10 (1993); **Robb RJ et al** Structure-function relationships for the interleukin 2 receptor: location of ligand and antibody binding sites on the Tac receptor chain by mutational analysis. PNAS 85: 5654-8 (1988); **Rubin LA & Nelson DL** The soluble interleukin-2 receptor: biology, function. Ann. Int. Med. 113: 619-27 (1990); **Russell SM et al** Interleukin 2 receptor g chain: a functional component of the interleukin-4 receptor. S 262: 1880-3 (1993); **Saito Y et al** Biochemical evidence for a third chain of the interleukin-2 receptor. JCB 266: 22186-91 (1991); **Sauvé K et al** Localization in human interleukin 2 of the binding site to the a chain (p55) of the interleukin 2 receptor. PNAS 88: 4636-40 (1991); **Shibuya H et al** The human interleukin-2 receptor β-chain: genomic organization, promoter analysis, and chromosomal assignment. NAR 18: 3697-703 (1990); **Shimuzu A et al** Nucleotide sequence of mouse IL2 receptor cDNA and its comparison with the human IL2 receptor gene. NAR 13: 1505-16 (1985); **Smith KA** The interleukin 2 receptor. ARC 5: 397-425 (1989); **Strom TB et al** Interleukin 2 receptor-directed therapies: antibody-based or cytokine-based targeting molecules. ARM 44: 343-53 (1993); **Sugamura K et al** The IL2/IL2 receptor system: involvement of a novel receptor subunit, γ chain, in growth signal transduction. Tohoku J. Exp. Med. 168: 231-7 (1992); **Takeshita T et al** Cloning of the γ chain of the human IL2 receptor. S 257: 379-82 (1992); **Voss SD et al** Identification of a direct interaction between interleukin 2 and the p64 interleukin 2 receptor γ chain. PNAS 90: 2428-32 (1993); **Waldmann TA** The interleukin-2 receptor. JBC 266: 2681-4 (1991); **Waldmann TA et al** The multichain interleukin-2 receptor: A target for immunotherapy. Ann. Intern. Med. 116: 148-60 (1992)

● **Biological Activities:** for information see references under reviews.

● **Transgenic/Knock-out/Antisense Studies: Allison J et al** Inflammation but not autoimmunity occurs in transgenic mice expressing constitutive levels of interleukin-2 in islet β cells. EJI 22: 1115-21 (1992); **Gutierrez-Ramos JC et al** Analysis of T cell subpopulations in human IL2R α transgenic mice: expansion of Thy1.2- thymocytes and depletion of double-positive T cell precursors. Res. Immunol. 140: 661-74 (1989); **Ishida Y et al** Effects of the deregulated expression of human Interleukin 2 in transgenic mice. Int. Immunol. 1: 113-20 (1989); **Ishida Y et al** Immunological abnormalities in human interleukin-2 or interleukin-2/interleukin-2 receptor L chain transgenic mice. Int. Symp. Princess Takamatsu Cancer Res. Fund. 19: 61-71 (1988); **Ishida Y et al** Expansion of natural killer cells but not T cells in human interleukin 2/Interleukin 2 receptor (Tac) transgenic mice. JEM 170: 1103-15 (1989); **Ishida Y et al** Immune dysfunctions and activation of natural killer cells in human IL2 and IL2/IL2 receptor L-chain transgenic mice. CSHSQB 54: 695-704 (1989); **Katsuki M et al** Lymphocyte infiltration into cerebellum in transgenic mice carrying human IL2 gene. Int. Immunol. 1: 214-8 (1989); **Kündig TM et al** Immune responses in interleukin-2-deficient mice. S 262: 1059-61 (1993); **Loriann E et al** The selective ablation of interleukin 2-producing cells isolated from transgenic mice. JEM 177: 1451-9 (1993); **Minasi LE et al** The selective ablation of interleukin 2-producing cells isolated from transgenic mice. JEM 177: 1451-9 (1993); **Nishi M et al** Expression of functional interleukin-2 receptors in human light chain/Tac transgenic mice. N 331: 267-9 (1988); **Ohta M et al** Anomalies in transgenic mice carrying the human interleukin-2 gene. Tokai J. Exp. Clin. Med. 15: 307-15 (1990); **Sadlack B et al** Ulcerative colitis-like disease in mice with a disrupted interleukin-2 gene. Cell 75: 253-61 (1993); **Sadlack B et al** Development and proliferation of lymphocytes in mice deficient for both interleukins-2 and 4. EJI 24: 281-4 (1994); **Schorle H et al** Development and function of T cells in mice rendered interleukin-2 deficient by gene targeting. N 352: 621-4 (1991)

● **Assays: Baran D et al** New radioimmunoassay for the detection of free IL2 receptor in supernatants of murine helper and cytotoxic T cell lines. Ann. Inst. Pasteur Immunol. 137D: 23-31 (1986); **Brandt E et al** Binding characteristics of a monoclonal antibody against human IL2 and its application for IL2 measurement. LR 5: S35-S42 (1986); **Burger CJ et al** An improved data analysis method for interleukin 2 microassay. Comput. Biol. Med. 16: 377-90 (1986); **Cardenas JM et al** Human interleukin 2. Quantitation by a sensitive radioimmunoassay. JIM 89: 181-9 (1986); **Cohen N et al** Comparison of ELISA and a flow cytometric method in the daily monitoring of soluble interleukin 2 receptor in liver transplant recipients. Transpl. Proc. 23: 1426-7 (1991); **Cohen N et al** An improved method for the detection of soluble interleukin 2 receptors in liver transplant recipients by flow cytometry. Transplantation 51: 417-21 (1991); **Deacock S et al** A rapid limiting dilution assay for measuring frequencies of alloreactive, interleukin-2-producing T cells in humans. JIM 147: 83-92 (1992); **Dusch A et al** A comparison of assays for the response of primary human T cells upon stimulation with interleukin-2, interleukin-4 and interleukin-7. ECN 3: 97-102 (1992); **Elsasser-Beile U et al** Cytokine levels in whole blood cell cultures as parameters of the cellular immunologic activity in patients with malignant melanoma and basal cell carcinoma. Cancer 71: 231-6 (1993); **Ferrua B et al** Human interleukin 2. Detection at the picomolar level by sandwich enzyme immunoassay. JIM 97: 215-20 (1987); **Goldstein AM et al** A competitive enzyme-linked immunoassay (ELISA) for the measurement of soluble human interleukin 2 receptor (IL2R, Tac protein). JIM 107: 103-9 (1988); **Grassi J et al** Determination of IL1 α, IL1 β and IL2 in biological media using specific enzyme immunometric assays. IR 119: 125-45 (1991); **Heeg K et al** A rapid colorimetric assay for the determination of IL2-producing helper T cell frequencies. JIM 77: 237-46 (1985); **Heslan JM et al** Differentiation between vascular permeability factor and IL2 in lymphocyte supernatants from patients with minimal-change nephrotic syndrome. Clin. Exp. Immunol. 86: 157-62 (1991); **Hirooka Y et al** Impaired production of interleukin-2(IL2) in patients with Graves' disease by newly developed IL2 radioimmunoassay. Exp. Clin. Endocrinol. 96: 64-72 (1990); **Honda M et al** Fluorescence sandwich enzyme-linked immunosorbent assay for detecting human interleukin 2 receptors. JIM 110: 129-36 (1988); **Howell WM et al** Detection of IL2 at mRNA and protein levels in synovial infiltrates from inflammatory arthropathies using biotinylated oligonucleotide probes *in situ*. Clin. Exp. Immunol. 86: 393-8 (1991); **Ide M et al** Neutralizing monoclonal antibodies against recombinant human interleukin 2. JIM 101: 57-62 (1987); **Igietseme JU & Herscowitz HB** Quantitative measurement of T lymphocyte activation by an enzyme-linked immunosorbent assay (ELISA) detecting interleukin 2 receptor expression. JIM 97: 123-31 (1987); **Igietseme JU & Herscowitz HB** A modified *in situ* enzyme-linked immunosorbent assay for quantitating interleukin-2 activity employing monoclonal anti-IL2 receptor antibody. JIM 108: 145-52 (1988); **Jung T et al** Detection of intracellular cytokines by flow cytometry. JIM 159: 197-207 (1993); **Labalette-Houache M et al** Improved permeabilization procedure for flow cytometric detection of internal antigens. Analysis of interleukin-2 production. JIM 138: 143-53 (1991); **Lee TP et al** A simple radioimmunoassay of human interleukin-2. J. Immunoassay 9: 193-206 (1988); **Lewis CE et al** Measurement of cytokine release by human cells. A quantitative analysis at the single cell level using the reverse hemolytic plaque assay. JIM 127: 51-9 (1990); **Lindqvist C et al** Improved method for multisample evaluation of ligand binding to cells. Application to IL2 receptor measurement. JIM 113: 231-5 (1988); **Moriya N et al** An enzyme-immunoassay for human interleukin-2. J. Immunoassay 8: 131-43 (1987); **Mosmann TR et al** Species-specificity of T cell stimulating activities of IL 2 and BSF-1 (IL 4): comparison of normal and recombinant, mouse and human IL 2 and BSF-1 (IL 4). JI 138: 1813-6 (1987); **Nadeau RW et al** Quantification of recombinant interleukin-2 in human serum by a specific immunobioassay. Anal. Chem. 61: 1732-6 (1989); **Ohike Y et al** A radioimmunoassay that sandwiches human interleukin-2 between radiolabeled monoclonal antibody and the receptor on a hematopoietic cell line. JIM 87: 245-9 (1986); **Osawa H et al** Enzyme-linked immunosorbent assay of mouse interleukin 2 receptors. JIM 92: 109-15 (1986); **Pruett SB et al** A quantitative non-isotopic bioassay for interleukin 2. Immunol. Invest. 14: 541-8 (1985); **Redmond S et al** Monoclonal antibodies for purification and assay of IL2. LR 5: S29-S34 (1986); **Santos JL et al** Evaluation of lymphocyte activation by flow cytometric determination of interleukin-2 (CD25) receptor. J. Clin. Lab. Immunol. 34: 145-9 (1991); **Sette A et al** A microcomputer program for probit analysis of interleukin 2 (IL2) titration data. JIM 86: 265-77 (1986); **Sharief MK et al** Determination of interleukin-2 in cerebrospinal fluid by a sensitive enzyme-linked immunosorbent assay. JIM 147: 51-6 (1992); **Tada H et al** An improved colorimetric assay for interleukin 2. JIM 93: 157-65 (1986); **Takaishi M et al** Assay for the soluble interleukin-2 receptor by sandwich enzyme linked immunosorbent assay. Jpn. J. Med. Sci. Biol. 43: 151-61 (1990); **Taki S et al** Biotinylation of human interleukin 2 for flow cytometry analysis of interleukin 2 receptors. JIM 122: 33-41 (1989); **Wang H & Beckner SK** A colorimetric method for detection of specific ligand binding. AB 204: 59-64 (1992); **Yussim A et al** The assay of soluble IL2R receptor levels in graft microenvironment can be helpful in differential diagnosis of transplant dysfunction. Transpl. Proc. 23: 260-2 (1991); for further information also see individual cell lines used in individual bioassays.

● **Clinical Use & Significance:** **Blaise D et al** Prevention of acute GVHD by *in vivo* use of anti-interleukin-2 receptor monoclonal antibody (33B3 1): A feasibility trial in 15 patients. Bone Marrow Transplant. 8: 105-11 (1991); **Broom J et al** Interleukin 2 therapy in cancer: identification of responders. Br. J. Cancer. 66: 1185-7 (1992); **Bubeník J et al** IL2 gene transfer in immunotherapy of cancer: Local administration of IL2-activated lymphocytes and X63-m-IL2 cells constitutively producing IL2 inhibits growth of plasmacytomas in syngeneic mice. Nat. Immun. Cell Growth Regul. 10: 247-55 (1991); **Chelstrom LM et al** Treatment of BCL-1 murine B cell leukemia with recombinant cytokines. Comparative analysis of the anti-leukemic potential of interleukin 1 β (IL1 β), interleukin 2 (IL2), interleukin-6 (IL6), tumor necrosis factor α (TNF α), granulocyte colony stimulating factor (G-CSF), granulocyte-macrophage colony stimulating factor (GM-CSF), and their combination. Leuk. Lymphoma 7: 79-86 (1992); **Ebert T et al** The role of cytokines in the therapy of renal cell carcinoma. Rec. Results in Cancer Res. 126: 113-8 (1993); **Fearon ER et al** Interleukin-2 production by tumor cells bypasses T helper function in the generation of an antitumor response. Cell 60: 397-403 (1990); **Gambacorti-Passerini C et al** A pilot phase II trial of continuous-infusion interleukin-2 followed by lymphokine-activated killer cell therapy and bolus-infusion interleukin-2 in renal cancer. J. Immunother. 13: 43-8 (1993); **Gansbacher B et al** Interleukin 2 gene transfer into tumor cells abrogates tumorigenicity and induces protective immunity. JEM 172: 1217-24 (1990); **Gastl G et al** Retroviral vector-mediated lymphokine gene transfer into human renal cancer cells. CR 52: 6229-36 (1992); **Heys SD et al** Interleukin 2 therapy: current role in surgical oncological practice. Br. J. Surg. 80: 155-62 (1993); **Hinuma S et al** A novel strategy for converting recombinant viral protein into high immunogenic antigen. FL 288: 138-42 (1991); **Huland E et al** Interleukin-2 by inhalation: local therapy for metastatic renal cell carcinoma. J. Urol. 147: 344-8 (1992); **Kaplan G et al** Rational immunotherapy with interleukin 2. BT 10: 157-62 (1992); **Kolitz JE & Mertelsmann R** The immunotherapy of human cancer with interleukin 2: Present status and future directions. Cancer Invest. 9: 529-42 (1991); **Maas RA et al** Interleukin-2 in cancer treatment: disappointing or (still) promising? A review. Cancer Immunol. Immunother. 36: 141-8 (1993); **Palmer PA et al** Continuous infusion of recombinant interleukin-2 with or without autologous lymphokine activated killer cells for the treatment of advanced renal cell carcinoma. Eur. J. Cancer 28A: 1038-44 (1992); **Porgador A et al** Anti-metastatic vaccination of tumor-bearing mice with IL2-gene-inserted tumor cells. IJC 53: 471-7 (1993); **Prentice HG et al** The role of immunotherapy in the treatment of acute myeloblastic leukemia: from allogeneic bone marrow transplantation to the application of interleukin 2. Cancer. Treat. Res. 64: 121-34 (1993); **Pui CH et al** Serum interleukin-2 receptor: clinical and biological implications. Leukemia 3: 323-7 (1989); **Ridolfi R et al** Evaluation of toxicity in 22 patients treated with subcutaneous interleukin-2, α-interferon with and without chemotherapy. J. Chemother. 4: 394-8 (1992); **Rosenberg SA et al** Experience with the use of high-dose interleukin-2 in the treatment of 652 cancer patients. Ann. Surg. 210: 474-85 (1989); **Schendel DJ & Gansbacher B** Tumor-specific lysis of human renal cell carcinomas by tumor-infiltrating lymphocytes: modulation of recognition through retroviral transduction of tumor cells with interleukin 2 complementary DNA and exogenous α interferon treatment. CR 53:

4020-5 (1993); **Schomburg A et al** Hematotoxicity of interleukin-2 in man: clinical effects and comparison of various treatment regimens. Acta Haematol. 89: 119-31 (1993); **Schoof DD et al** Survival characteristics of metastatic renal cell carcinoma patients treated with lymphokine-activated killer cells plus interleukin-2. Urology 41: 534-9 (1993); **Schröder TJ et al** A multicenter study to evaluate a novel assay for quantitation of soluble interleukin 2 receptor in renal transplant recipients. Transplantation 53: 34-40 (1992); **Smith KA** Lowest dose interleukin-2 immunotherapy. Blood 81: 1414-23 (1993); **Stoter G et al** Interleukin 2. The experience of the Rotterdam Cancer Institute, Daniel den Hoed Kliniek. Biotherapy 2: 261-5 (1990); **Strom TB et al** Interleukin 2 receptor-directed immunosuppressive therapies: antibody-based or cytokine-based targeting molecules. IR 129: 131-63 (1992); **Vial T & Descotes J** Clinical toxicity of interleukin-2. Drug. Saf. 7: 417-33 (1992); **Vogelzang NJ et al** Subcutaneous interleukin-2 plus interferon alfa-2a in metastatic renal cancer: an outpatient multicenter trial. J. Clin. Oncol. 11: 1809-16 (1993); **Von der Maase H et al** Recombinant interleukin-2 in metastatic renal cell carcinoma – A European multicenter phase II study. Eur. J. Cancer 27: 1583-9 (1991)

IL2R alpha chain inhibitory activity: *tac inhibitory activity*, p29. This factor is produced by adherent cells of HIV-infected patients. It is also secreted by HIV-infected » U937 cells.

This poorly characterized factor of 29 kDa (p29) is not cytotoxic. It is not identical with » TNF-α, » IFN-α, or » IFN-γ. It inhibits the expression of the α subunit of the IL2 receptor but not the expression of the β subunit in normal activated T cells. It also does not inhibit the production of IL2 by these cells. The factor also inhibits mitogen-induced proliferation of normal T cells. For another factor interfering with IL2 receptor expression see: GSF (glioma-derived suppressor factor).

Ammar A et al Production of a tac inhibitory activity by adherent cells of HIV-infected subjects at different clinical stages. J. Acquir. Immune Defic. Syndr. 4: 1208-17 (1991); **Ammar A et al** Biological and biochemical characterization of a factor produced spontaneously by adherent cells of human immunodeficiency virus-infected patients inhibiting interleukin-2 receptor α chain (Tac) expression on normal T cells. JCI 87: 2048-55 (1991); **Ammar A et al** Human immunodeficiency virus-infected adherent cell-derived inhibitory factor (p29) inhibits normal T cell proliferation through decreased expression of high affinity interleukin-2 receptors and production of interleukin-2. JCI 90: 8-14 (1992)

IL2 enhancing factor: see: BGEF-2 (B cell derived growth enhancing factor 2).

IL2R/p55 (Tac) inducing factor: abbrev. TIA. This factor is secreted by HTLV-I-transformed T cells isolated from patients with adult T cell leukemia. The factor is identical with » ADF (adult T cell leukemia-derived factor). See also: IL2 receptor inducing factor (abbrev. IL2RIF).

Okamoto T et al Human thioredoxin/adult T cell leukemia-derived factor activates the enhancer binding protein of human immunodeficiency virus type 1 by thiol redox control mechanism. Int. Immunol. 4: 811-9 (1992); **Tagaya Y et al** IL2R/p55 (Tac) inducing factor. Purification and characterization of adult T cell leukemia-derived factor. JI 140: 2614-20 (1988); **Tagaya Y et al** Transcription of IL2 receptor gene is stimulated by ATL-derived factor produced by HTLV-I(+) T cell lines. Immunol. Lett. 15: 221-8 (1987); **Yodoi J et al** Interleukin 2 receptor-inducing factor(s) in adult T cell leukemia. Acta Haematol. 78: 56-63 (1987)

IL2 receptor inducing factor: abbrev. IL2RIF. This factor is secreted by HTLV-I-transformed T cells isolated from patients with adult T cell leukemia. The factor is identical with » ADF (adult T cell leukemia-derived factor). See also: IL2R/p55 (Tac) inducing factor (abbrev. TIA).

Another T cell IL2 receptor-inducing activity is found in the supernatants of a stimulated TH1 cell clone. This factor induces IL2 receptor expression and proliferation of TH1 clones (see also: TH!/TH2 cytokines). It is at present unclear whether this is a unique factor. Another activity which induces IL2 receptor expression and IL2 responsiveness in murine T cells after cross-linking of T cell receptors is RIF. At least one of the IL2-RIF factors appears to be identical with » IL12 (see: TSF, T cell stimulating factor).

Kato T et al Induction of IL2 receptor expression and proliferation of T cell clones by a novel cytokine(s). CI 142: 79-93 (1992); **Takakura K et al** Interleukin-2 receptor/p55/Tac)-inducing activity in porcine follicular fluids. E 125: 618-623 (1989); **Wagner H et al** Cross-linking of T cell receptors is insufficient to induce IL2 responsiveness (activation) in resting Lyt-2$^+$ T cells. IL4 or RIF are essential as second signal. ANY 532: 128-35 (1988); **Wagner H & Heeg K** Two distinct signals regulate induction of IL2 responsiveness in CD8$^+$ murine T cells. Immunology 64: 433-8 (1988); **Yodoi J et al** Interleukin 2 receptor-inducing factor(s) in adult T cell leukemia. Acta Haematol. 78: 56-63 (1987)

IL2RIF: see: IL2 receptor inducing factor.

IL3: Interleukin 3.

■ **ALTERNATIVE NAMES:** 20α-Steroid-Dehydrogenase-inducing factor, BP (burst promoting activity); BPA (blood progenitor activator); BPA (burst promoting activity); CFU-s (colony forming unit spleen); CFU-SA (colony forming unit stimulating activity); CSF-2α (colony stimulating factor); CSF-2β (colony stimulating factor); EC IL3 (epidermal cell IL3); ECSF (erythroid colony stimulating factor); EK-derived basophil promoting activity (epidermal keratinocyte-derived basophil promoting activity); Eo-CSF (eosinophil colony stimulating factor); HCGF (hematopoietic cell growth factor); HCSF (histamine-producing cell stimulating factor); HP2 (hemopoietin 2); HPGF (hemopoietic cell growth factor); maturation inducer activity; MCGF (mast cell growth factor); MCSA (multi-colony stimulating activity); MEG-CSA (megakaryocyte colony stimulating activity); MEG-CSF (megakaryocyte colony stimulating factor); MGF (mast cell growth factor); mixed colony stimulating factor; Multi-CSF (multi colony stimulating factor); multi HGF (multilineage hemopoietic growth factor); Multipoietin; NC cell growth factor (natural cytotoxic cell growth factor); Neutrophil-granulocyte colony stimulating factor; PCSA (P cell stimulating activity); PCSF (P cell stimulating factor); P factor; Pluripotential stem cell-supporting factor; PSF (persisting cell stimulating factor); PSF (P cell stimulating factor); PSF (progenitor stimulating factor); PSH (panspecific hemopoietin); SA (synergistic activity); SAF (stem cell activating factor); Thy1-inducing factor; WEHI-3 factor; WEHI 3 hematopoietic growth factor; WGF (WEHI 3 growth factor). See also: individual entries for further information.

■ **Sources:** IL3 is produced mainly by antigen- and mitogen-activated T cells (see also: cell activation), but also by keratinocytes, NK cells, mast cells, endothelial cells, and monocytes. Some cell lines produce IL3 constitutively. Substances that inhibit the activation of T lymphocytes such as glucocorticoids or » CsA (Ciclosporin A) also inhibit the production of IL3. For the constitutive synthesis of IL3 see also: IAP (intracisternal A particle).

IL3 may be associated with the extracellular matrix (see: ECM) in the form of complexes with heparan sulfate/proteoglycan. It can thus be stored in a biologically inactive form but it may also exert » juxtacrine activities. The molecular mechanisms underlying the release from extracellular matrix stores is still unknown.

■ **Protein Characteristics:** Human IL3 is a protein of 15-17 kDa (133 amino acids). It is synthesized as a precursor containing a hydrophobic secretory » signal sequence of 19 amino acids. IL3 contains two putative glycosylation sites at positions 15 and 70 and contains a single disulfide bond (Cys16/84). The analysis of bacterial-derived recombinant IL3 shows that glycosylation is not required for the activity of IL3.

IL3 sequences are evolutionarily less well conserved. Human and murine IL3 show ≈ 29 % homology at the protein level while murine and rat IL3 show ≈ 54 % homology.

■ **Gene Structure:** The human IL3 has a length of ≈ 2.2 kb and contains five exons. The gene maps to human chromosome 5q23-31. Murine and human IL3 genes structurally resemble each other but at the nucleotide level the homology is ≈ 49 %. The murine IL3 gene maps to chromosome 17.

The human IL3 gene is located in close vicinity to other cytokine genes, including those encoding » GM-CSF, » M-CSF, » IL4 and IL5 The distance between the IL3 and the GM-CSF gene is ≈ 9 kb with the IL3 gene on the 5´side of the GM-CSF gene. Deletions of the IL3/GM-CSF locus are frequently observed in patients with myelodysplastic syndrome (see also: $5q^-$-Syndrome).

■ **Receptor Structure, Gene(s), Expression:** IL3 receptors are expressed on macrophages, mast cells, eosinophils, megakaryocytes, basophils,

```
Gibbon   QAPMTQTTSLKTSW-VNCSNMIDEIITHLKQPPLPLLDFNNLNGEDQDILMENNLRRPNL
Human    QAPMTQTTSLKTSW-VNCSNMIDEIITHLKQPPLPLLDFNNLNGEDQDILMENNLRRPNL
Mouse    QASISGRDTHRLTRTINCSSIVKEIIGKL---PEPEL----KTDDEGPSLRNKSFRRVNL
Rat      QISDRGSDAHHLLRTIDCRTIALEILVKL---PVSGL----NNSDDKANLRNSTLRRVNL

Gibbon   EAFNKAVKSL--QNASAIESILKNLPPCLPMATAAPTRHPIRIKDGDWNEFRRKLKFYLK
Human    EAFNRAVKSL--QNASAIESILKNLLPCLPLATAAPTRHPIHIKDGDWNEFRRKLTFYLK
Mouse    SKFVESQGEVDPEDRYVIKSNLQKLNCCLPTSANDSALPGVFIR--DLDDFRKKLRFYMV
Rat      DEFLKSQEEFDSQDTTDIKSKLQKLKCCIPAAASDSVLPGVYNK--DLDDFKKKLRFYVI

Gibbon   TLENEQAQQMTLSLEIS
Human    TLENAQAQQTTLSLAIF
Mouse    HLNDLETVLTSRPPQPASGSVSPNRGTVEC
Rat      HLKDLQPVSVSRPPQPTSSSDNFRPMTVEC
```

Comparison of IL3 sequences from different species.
Identical amino acid positions in Gibbon and human IL3 are shown in blue; those identical in rat and mouse IL3 are shown in green. The two conserved cysteine residues are shown in yellow.

bone marrow progenitor cells, and various myeloid leukemia cells.

IL3/receptor complexes have a Kdis of 10^{-9} – 10^{-10} M. Binding of IL3 to its receptor causes specific phosphorylation of a 150 kDa membrane glycoprotein.

Biological activities are also detected at the concentration range of 10^{-13} – 10^{-15} M which suggests the existence either of other high affinity receptor conformations or other receptors.

The high affinity IL3 receptor is formed by aggregation of the IL3-specific subunit of the receptor with another subunit called GM-Rβ which constitutes the larger subunit of the receptor for GM-CSF and is also involved as a subunit in the formation of » IL5 receptors (see also: AIC2A). This subunit which is common to all three receptors is also called Beta c.

Binding of IL3 to its receptor causes the phosphorylation of the receptor and also of some other cytoplasmic proteins (p70, p56, p38) and also activates protein kinase C in IL3-dependent cell lines. The non-receptor tyrosine kinase » *lyn* has been implicated in IL3-mediated signal transduction processes.

■ **BIOLOGICAL ACTIVITIES:** The biological activities of IL3 are species-specific. The plethora of biological activities is exemplified by the many different acronyms under which IL3 has been described. IL3 is a growth factor that establishes the link between the immune system and the hematopoietic system. It supports the proliferation and development of almost all types of hematopoietic progenitor cells (see also: Hematopoiesis and CFU). In rhesus monkeys IL3 causes the expansion of all types of circulating hematopoietic progenitor cells. IL3 also supports the differentiation of early non-lineage-committed hematopoietic progenitor cells into granulocytes, macrophages, erythroid cells, megakaryocytes, and mast cell colonies. IL3 also stimulates clonal growth of non-hematopoietic stromal cells in bone marrow cultures (see also: LTBMC, long-term bone marrow culture). IL3 is a priming factor for » hematopoietic stem cells *in vitro* and *in vivo* (see also: cell activation) that makes the cells responsive to later-acting factors such as » Epo, » GM-CSF and » IL6. IL3 also induces the increased expression of receptors for colony-stimulating factors (see: CSF). At pico- to nanomolar concentrations IL3 is a chemoattractant for eosinophils and also influences the chemotactic behavior of these cells in response to other chemotactically active factors.

IL3 does not appear to be an obligatory factor in normal » Hematopoiesis. The athymic nude mouse (see also: Immunodeficient mice) which lacks T cells responsible for IL3 production displays a normal pattern of hematopoietic development. In the lymphoid lineage IL3 supports the growth of preB cells.

Experiments with » oncogenes encoding a tyrosine-specific protein kinase show that the expression of the oncogene-encoded tyrosine kinase abolishes IL3 requirements (and also GM-CSF and IL4 requirements) in factor-dependent cells. The exact mechanism by which these factors regulate proliferation and differentiation is still unknown.

IL3 induces the proliferation of mast cells and macrophages and causes the synthesis of histamines by mast cells and phagocytosis in macrophages. In mast cells IL3 prevents the expression of class II MHC molecules normally induced by » IFN-γ.

IL3 induces the expression of complement C3a receptors in basophils at nanomolar concentrations. A transient stimulation by IL3 is required to make basophils responsive for further activation (see also: cell activation) by » IL8 which by itself does not possess histamine releasing activity. Tissue mast cells do not respond to IL3. In LPS-stimulated macrophages IL3 significantly enhances the secretion of other cytokines including » IL1, » IL6 and » TNF. *In vitro* IL3 also stimulates the proliferation of keratinocytes. This observation has also been made in rhesus monkeys treated with recombinant IL3 but at present it is unclear whether IL3 is involved in skin reactions associated with keratinocyte growth and acanthosis.

IL3 also specifically induces the production of enzymes involved in cellular metabolism, differentiation and DNA/RNA metabolism. Among other things IL3 induces the expression of 20-α-steroid dehydrogenase and of histidine and Ornithine decarboxylase.

■ **ASSAYS:** IL3 can be detected in » Bioassays employing responsive cell lines (see: AML 193; 32D; B6SUt-A; B13; Da; Ea3.17; FDC-P1; GF-D8; IC-2; KMT-2; L138.8A; LyD9; MO7E; NFS-60; PT18; TALL-103; TF-1; TMD2; UT-7). A sensitive ELISA test for IL3 is also available. An alternative and entirely different detection method is » Message amplification phenotyping. For further information see also subentry "Assays" in the reference section.

■ **CLINICAL USE & SIGNIFICANCE:** IL3 alone or in combination with other colony-stimulating factors (see also: CSF) or » Epo is probably useful in the reconstitution of bone marrow and in stimulating erythropoiesis (see also: Hematopoiesis). IL3 also appears a valuable substance in view of its ability to stimulate the proliferation of very early hematopoietic progenitor cells such as » CFU-GEMM, » BFU-E, and » CFU-GM, which can be enhanced even more by simultaneous administration of other factors such as » GM-CSF. These effects can be exploited in the generation of stem cells, allowing stem cell separation and re-infusion into patients undergoing, for example, intensive chemotherapy. It may also be useful in the treatment of anemic patients with congenital defects such as BFU-E deficiency. IL3 appears to be of limited benefit in patients who are severely aplastic after autologous bone marrow transplantation and have very low levels of bone marrow progenitors.

In patients with myelodysplastic syndromes, who frequently present with anemia, leukopenia and thrombocytopenia due to defective maturation of bone marrow cells initial phase I/II trials with IL3 have demonstrated an increase of neutrophil counts in 59%, of platelet counts in 34%, and in reticulocyte counts in 25% of the patients.

It has been demonstrated that IL3 administered subcutaneously appears to be an interesting hematopoietic growth factor for reduction of chemotherapy-induced myelotoxicity in patients with advanced ovarian cancer.

IL3 is probably of major importance in the treatment of thrombopenias, secondary hematopoietic insufficiencies and certain types of refractory anemia. The administration of IL3 followed by subsequent course of IL6 greatly improves platelet counts. This combination may therefore allow an effective control of thrombocytopenias and severe bleeding frequently associated with this condition. Some monocytic and myeloid leukemias the growth of which is inhibited almost completely by antibodies directed against IL3 have been described in the mouse These leukemias are characterized by the overexpression of the IL3 gene. In humans, however, such cases have not been described to date. A number of malignant myeloid and lymphoid leukemias respond to IL3. *In vitro* ALL blast cells can be made more responsive to cytotoxic drugs such as cytosine arabinoside by treatment with IL3. At present it is unclear whether this also occurs *in vivo*.

● **REVIEWS: Frendl G** Interleukin 3: from colony-stimulating factor to pluripotent immunoregulatory cytokine. IJI 14: 421-30 (1992); **Ihle JN** Interleukin-3 and hematopoiesis. Chem. Immunol. 51: 65-106 (1992); **Schrader JW** The panspecific hemopoietin of activated T lymphocytes. ARI 4: 205-30 (1986); **Schrader JW** The panspecific hemopoietin interleukin 3: physiology and pathology. Lymphokines 15: 281-311 (1988); **Valent P et al** Why clinicians should be interested in Interleukin 3. Blut 61: 338-45 (1990); **Wagemaker G et al** Interleukin-3. Biotherapy 2: 337-45 (1990)

● **BIOCHEMISTRY & MOLECULAR BIOLOGY: Andrews NC et al** Regulation of the human interleukin 3 gene. Trans. Assoc. Am. Physicians 102: 240-51 (1989); **Barlow DP et al** Close genetic and physical linkage between the murine hemopoietic growth factor genes GM-CSF and multi-CSF (IL3). EJ 6: 617-23 (1987); **Burger H et al** Nucleotide sequence of the gene encoding Rhesus monkey (*Macaca mulatta*) interleukin-3. NAR 18: 6718 (1990); **Campbell HD et al** Cloning and nucleotide sequence of the murine interleukin-3 gene. EJB 150: 297-304 (1985); **Cohen DR et al** Cloning and expression of the rat interleukin 3 gene. NAR 14: 3641-58 (1986); **Clark-Lewis I et al** Automated chemical synthesis of a protein growth factor for hemopoietic cells, interleukin 3. S 231: 134-9 (1986); **Dorssers L et al** Characterization of a human multi-lineage-colony-stimulating factor cDNA clone identified by a conserved non-coding sequence in mouse interleukin-3. Gene 55: 115-24 (1987); **Dunbar CE et al** COOH-terminal-modified interleukin 3 is retained intracellularly and stimulates autocrine growth. S 245: 1493-6 (1989); **Fung MC et al** Molecular cloning of cDNA for murine interleukin-3. N 307: 233-7 (1984); **Frolova EI et al** Linkage mapping of the human CSF2 and IL3 genes. PNAS 88: 4821-4 (1991); **Fung MC et al** Molecular cloning of cDNA for murine interleukin-3. N 307: 233-6 (1984); **Le Beau MM et al** The interleukin 3 gene is located on human chromosome 5 and is deleted in myeloid leukaemias with a deletion of 5q. PNAS 84: 5913-7 (1987); **Lokker N et al** Structure-activity relationship study of human interleukin 3: role of the C-terminal region for biological activity. EJ 10: 2125-31 (1991); **Miyajima A et al** Expression of murine and human granulocyte-macrophage colony stimulating factors in *S. cerevisiae*: mutagenesis of the potential glycosylation sites. EJ 5: 1193-7 (1986); **Miyajima A et al** Use of the silkworm, *Bombyx mori*, and an insect baculovirus vector for high-level expression and secretion of biologically active mouse interleukin-3. Gene 58: 273-81 (1987); **Miyatake S et al** Structure of the chromosomal gene for murine interleukin 3. PNAS 82: 316-20 (1985); **Otsuka T et al** Isolation and characterization of an expressible cDNA encoding human IL3. Induction of IL3 mRNA in human T cell clones. JI 140: 2288-95 (1988); **Todokoro K et al** Isolation and characterization of a genomic DDD mouse interleukin-3 gene. Gene 39: 103-7 (1985); **van Leeuwen BH et al** Molecular organization of the cytokine gene cluster involving the human IL3, IL4, IL5, and GM-CSF genes on human chromosome 5. Blood 73: 1142-8 (1989); **Yang YC & Clark SC** Structure of the gene for human interleukin-3 or multi-CSF. PCBR 251: 3-11 (1987); **Yang YC et al** Human IL3 (multi-CSF): identification by expression cloning of a novel hematopoietic growth factor related to murine IL3. Cell 47: 3-10 (1986); **Yang YC & Clark SC** Cloning of the human interleukin-3 gene. Lymphokines 15: 375-91 (1988); **Yang YC et al** The human genes for GM-CSF and IL3 are closely linked in tandem on chromosome 5. Blood 71: 958-61 (1988); **Yokata T et al** Isolation and characterization of a mouse cDNA clone that expresses mast cell growth-factor activity in monkey cells. PNAS 81: 1070-3 (1984)

● **RECEPTOR STRUCTURE, GENE(S), EXPRESSION: Duronio V et al** Two polypeptides identified by interleukin 3 cross-linking represent distinct components of the interleukin 3 receptor. EH

20: 505-11 (1992); **Fung MC et al** Distinguishing between mouse IL3 and IL3 receptor-like (IL5/GM-CSF receptor converter) mRNAs using the polymerase chain reaction method. JIM 149: 97-103 (1992); **Hara T & Miyajima A** Two distinct functional high affinity receptors for mouse interleukin-3 (IL3). EJ 11: 1875-84 (1992); **Kitamura T et al** Expression cloning of the human IL3 receptor cDNA reveals shared β subunit for the human IL3 and GM-CSF receptors. Cell 66: 1165-74 (1991); **Lopez AF et al** Reciprocal inhibition of binding between interleukin-3 and granulocyte-macrophage colony-stimulating factor to human eosinophils. PNAS 86: 7022-6 (1989); **Mui AL et al** Purification of the murine interleukin 3 receptor. JBC 267: 16523-30 (1992); **Mui AL et al** Ligand-induced phosphorylation of the murine interleukin 3 receptor signals its cleavage. PNAS 89: 10812-6 (1992); **Park LS et al** Heterogeneity in human interleukin-3 receptors. A subclass that binds human granulocyte/macrophage colony stimulating factor. JBC 264: 5420-7 (1989); **Urdal DL et al** Molecular characterization of colony-stimulating factors and their receptors: human interleukin-3. ANY 554: 167-76 (1989)

● **Biological Activities: Aglietta M et al** Interleukin-3 *in vivo*: kinetic of response of target cells. Blood 82: 2054-61 (1993); **Cohen L et al** Interleukin 3 enhances cytokine production by LPS-stimulated macrophages. Immunol. Lett. 28: 121-6 (1991); **Kobayashi M et al** Interleukin-3 is significantly more effective than other colony-stimulating factors in long-term maintenance of human bone marrow-derived colony-forming cells *in vitro*. Blood 73: 1836-41 (1989); **Lemoli RM et al** Interleukin 11 stimulates the proliferation of human hematopoietic $CD34^+$ and $CD34^+CD33^-DR^-$ cells and synergises with stem cell factor, interleukin-3, and granulocyte-macrophage colony-stimulating factor. EH 21: 1668-72 (1993); **Warringa RA et al** Modulation and induction of eosinophil chemotaxis by granulocyte-macrophage colony-stimulating factor and interleukin-3. Blood 77: 2694-700 (1991); **Whetton AD & Dexter TM** The mode of action of interleukin 3 in promoting survival, proliferation, and differentiation of hemopoietic progenitor cells. Lymphokines 15: 355-74 (1988)

● **Assays: Knopf HP & Papoian R** Preparation of europium-streptavidin in a time-resolved fluoroimmunoassay for interleukin-3. JIM 138: 233-6 (1991); **Knopf HP et al** A time-resolved fluoroimmunoassay for recombinant human interleukin-3. Ann. Clin. Biochem. 30: 69-71 (1993); **Papoian R et al** A sensitive ELISA for measuring recombinant human interleukin-3 in human plasma or serum. JIM 145: 161-5 (1991); **Pauly JU et al** Highly specific and highly sensitive enzyme immunoassays for antibodies to human interleukin 3 (IL3) and human erythropoietin (EPO) in serum. Behring Inst. Mitt. 90: 112-25 (1991); **Pauly JU et al** An enzyme linked immunosorbent assay for the detection of human interleukin 3 in serum. Behring Inst. Mitt. 90: 104-11 (1991); **Wang H & Beckner SK** A colorimetric method for detection of specific ligand binding. AB 204: 59-64 (1992); **Ziltener HJ et al** Sandwich enzyme immunoassay for murine IL3. Cytokine 1: 56-61 (1989); for further information also see individual cell lines used in individual bioassays.

● **Clinical Use & Significance: Biesma B et al** Effects of interleukin-3 after chemotherapy for advanced ovarian cancer. Blood 80: 1141-8 (1992); **Brugger W et al** Mobilization of peripheral blood progenitor cells by sequential administration of interleukin-3 and granulocyte-macrophage colony-stimulating factor following polychemotherapy with etoposide, ifosfamide, and cisplatin. Blood 79: 1193-200 (1992); **Crump M et al** Interleukin-3 followed by GM-CSF for delayed engraftment after autologous bone marrow transplantation. EH 21: 405-410 (1993); **Ganser A et al** Effects of recombinant human Interleukin 3 in patients with normal hematopoiesis and in patients with bone marrow failure. Blood 76: 666-76 (1990); **Ganser A et al** Preclinical and clinical evaluation of interleukin 3. Behring Inst. Mitt. 90: 50-61 (1991); **Ganser A et al** Interleukin-3 in the treatment of myelodysplastic syndromes Int. J. Clin. Lab. Res. 1992; 22: 125-8 (1992); **Ganser A et al** Sequential *in vivo* treatment with two recombinant human hematopoietic growth factors (interleukin-3 and granulocyte-macrophage colony-stimulating factor) as a new therapeutic modality to stimulate hematopoiesis: results of a phase I study. Blood 79: 2583-91 (1992); **Geissler K et al** Effect of interleukin-3 on responsiveness to granulocyte-colony-stimulating factor in severe aplastic anemia. Ann. Intern. Med. 117: 223-5 (1992); **Gillio AP et al** Treatment of Diamond-Blackfan anemia with recombinant human interleukin-3. Blood 82: 744-51 (1993); **Kindler V et al** Stimulation of hematopoiesis *in vivo* by recombinant bacterial murine interleukin 3. PNAS 83: 1001-5 (1985); **Krumwieh D et al** Preclinical studies on synergistic effects of IL1, IL3, G-CSF, and GM-CSF in cynomolgous monkeys. IJCC 8 (Suppl. 1) 229-48 (1990); **Kurzrock R et al** Phase I study of recombinant human interleukin-3 in patients with bone marrow failure. J. Clin. Oncol. 9: 1241-50 (1991); **Lindemann A et al** Biologic effects of recombinant human interleukin-3 *in vivo*. J. Clin. Oncol. 9: 2120-7 (1991); **Naparstek E et al** Enhanced marrow recovery by short preincubation of marrow allografts with human recombinant interleukin-3 and granulocyte-macrophage colony-stimulating factor. Blood 80: 1673-8 (1992); **Orazi A et al** Recombinant human interleukin-3 and recombinant human granulocyte-macrophage colony-stimulating factor administered *in vivo* after high-dose cyclophosphamide cancer chemotherapy: effect on hematopoiesis and microenvironment in human bone marrow. Blood 79: 2610-9 (1992)

IL3-IAP: see: IAP (intracisternal A particle).

IL3 inhibitor p16: see: SOD (superoxide dismutase).

IL3-LA: *IL3-like activity* This poorly characterized factor of 25-30 kDa is released spontaneously by cultured human peripheral-blood mononuclear cells. It stimulates the proliferative activity of several IL3-dependent cell lines and promotes the development of basophils in cultured human cord blood cells.

Fishman P et al Spontaneous release of a factor with interleukin-3-like activity by human lymphocytes and monocytes. Nat. Immun. Cell Growth Regul. 9: 334-41 (1990); **Fishman P et al** The relationship between interleukin-3-like activity and basophil production in chronic myeloid leukemia patients. Immunol. Lett. 28: 73-7 (1991); **Fishman P et al** Cultured human cord blood cells spontaneously produce a factor with basophil-promoting activity. Immunol. Lett. 34: 189-93 (1992)

IL3 receptor affinity converter: see: gp130 and subentry "receptor structure, gene(s), expression" for IL3.

IL4: Interleukin 4.

■ **Alternative Names:** IaIF (MHC class II (Ia) inducing factor); BCDF-ε: (B cell differentiation factor ε); BCDF-γ. (B cell differentiation factor γ); BCGFγ (B cell growth factor γ); BCGF1 (B cell growth factor 1); BSF-1 (B cell stimulatory factor 1); BSFp1 (B cell stimulatory factor p1); EL4-BCGF (EL4 B cell growth factor); HCGF (Hodgkin's cell growth factor); IgE-enhancing factor; IgG1-enhancing factor; IgG1-induction factor; MCGF-2 (mast cell growth factor 2); MFF (macrophage fusion factor); TCGF-2 (T cell growth factor 2); THCGF (thymocyte growth factor). See also: individual entries for further information.

■ **SOURCES:** IL4 is produced mainly by a subpopulation of activated T cells (TH2, CD4-positive helper cells; see also: TH1/TH2 cytokines) which are the biologically most active helper cells for B cells and which also secrete » IL5 and » IL6. Another subpopulation (TH1) also produces IL4 albeit to a lesser extent. Non-T/Non-B cells of the mast cell lineage also produce IL4.

■ **PROTEIN CHARACTERISTICS:** IL4 is a protein of 129 amino acids (20 kDa) that is synthesized as a precursor containing a hydrophobic secretory » signal sequence of 24 amino acids. IL4 is glycosylated at two arginine residues (positions 38 and 105) and contains six cysteine residues involved in disulfide bond formation. The disulfide bonds are essential for biological activity. Some glycosylation variants of IL4 have been described that differ in their biological activities. A comparison of murine and human IL4 shows that both proteins only diverge at positions 91-128.

An IL4 variant, Y124D, in which Tyr124 of the recombinant human protein is substituted by an aspartic acid residue, binds with high affinity to the IL4 receptor (KD = 310 pM). This variant is a powerful antagonist for the IL4 receptor system. It retains no detectable proliferative activity for T cells and competitively inhibits IL4-dependent T cell proliferation (K(i) = 620 pM).

■ **GENE STRUCTURE:** The human IL4 gene contains four exons and has a length of ≈ 10 kb. It maps to chromosome 5q23-31. The murine gene maps to chromosome 11. The IL4 gene is in close proximity to other genes encoding hematopoietic growth factors (see: GM-CSF, M-CSF, IL3, IL5). The distance between the IL4 and the IL5 gene is ≈ 90-240 kb.

At the nucleotide level the human and the murine IL4 gene display ≈ 70 % homology. The 5′ region of the IL4 contains several sequence elements, designated » CLE (conserved lymphokine element), that are binding sites for transcription factors controlling the expression of this and other genes (see also: Gene expression). A sequence motif, called P sequence (CGAAAATTTCC) in the 5′ region of the human IL4 gene (positions –79 - –69) is the binding site for a nuclear factor, called NF(P), mediating the response to T cell activation signals (see also: cell activation).

■ **RECEPTOR STRUCTURE, GENE(S), EXPRESSION:** The biological activities of IL4 are mediated by a specific receptor (Kdis = 20-100 pM) which is expressed at densities of 100-5000 copies/cell. The extracellular domain of the IL4 receptor is related to the receptors for » Epo, » IL6, and the β chain of the » IL2 receptor. It has been given the name ***CDw124*** (see also: CD Antigens).

The cDNA for the murine IL4 receptor encodes a transmembrane protein of 810 amino acids (including a secretory » signal sequence). This receptor has a large intracellular domain of 553 amino acids. The human receptor has an extracellular domain of 207 amino acids, a transmembrane domain of 24 residues, and a large intracellular domain of 569 amino acids.

The IL4 receptor has been shown recently to contain the γ subunit of the » IL2 receptor as a signaling component. This γ subunit is also associated with the receptors for » IL4 and » IL7 and probably also of » IL13. These findings can explain the severity of the immune defect in X-linked immunodeficiency.

Two forms of the receptor have been described, one of which is secreted. The secreted receptor only contains the extracellular IL4 binding domain and is capable of blocking IL4 activities. An IL4 binding protein (***IL4BP***) that binds IL4 with the same affinity as the IL4 receptor has also been shown to be a soluble IL4 receptor variant. These soluble receptors probably function as physiological regulators of cytokine activities (see also: Cytokine inhibitors) by inhibiting receptor binding or act as transport proteins. Indeed, studies with » transgenic animals constitutively expressing elevated levels of a soluble IL4 receptor have shown that, despite an almost normal development, these animals show a markedly delayed rejection of allografts, suggesting a critical role of IL4 in alloreactivity.

In experimental animals the soluble IL4 receptor can block IgE secretion by neutralizing endogenous IL4. It can also enhance, in a dose-dependent manner, the biologic effects of exogenously administered IL4, presumably by altering the biodistribution of the cytokine.

Soluble receptors or binding proteins have also been described for » IL1 (see: IL1ra, IL1 receptor antagonist), » IL2, » IL6, » IL7, » TNF-α, » IGF, and » IFN-γ.

The mechanisms of IL4-mediated intracellular signal transduction are largely unknown. IL4 influences intracellular calcium levels (see also: Calcium ionophore) and also the metabolism of inositol phospholipids and protein kinase C.

■ **BIOLOGICAL ACTIVITIES:** The biological activities of IL4 are species-specific; mouse IL4 is inactive on human cells and human IL4 is inactive on murine cells.

IL4 promotes the proliferation and differentiation of activated B cells (see also: CD40), the expression of class II MHC antigens, and of low-affinity IgE receptors (see: CD23) in resting B cells. IL4 is probably an » autocrine growth factor for Hodgkin's lymphomas (see also: HCGF, Hodgkin's cell growth factor).

IL4 enhances expression of class II MHC antigens on B cells. It can promote their capacity to respond to other B cell stimuli and to present antigens for T cells. This may be one way to promote the clonal expansion of specific B cells and the immune system may thus be able to respond to very low concentrations of antigens. The production of IL4 by Non-B Non-T cells is stimulated if these cells interact with other cells via their Fc receptors for IgE or IgG. This effect can be enhanced by » IL3. IL2 and PAF (platelet activating factor) induce the synthesis of IL4 while » TGF-β inhibits it.

IL4 inhibits the IL2-induced activation of NK cells. IL4 stimulates the proliferation of thymocytes with the marker spectrum $CD4^-8^-$, $CD4^+8^-$and $CD4^-8^+$. In CD4-positive cells IL4 induces the expression of » CD8 (see also: CD antigens).

IL3 antagonizes the IL2-induced effects in B cells and causes a slow decrease of the expression of IL2 receptors, thus inhibiting the IL2-stimulated proliferation of human B cells. In activated B cells

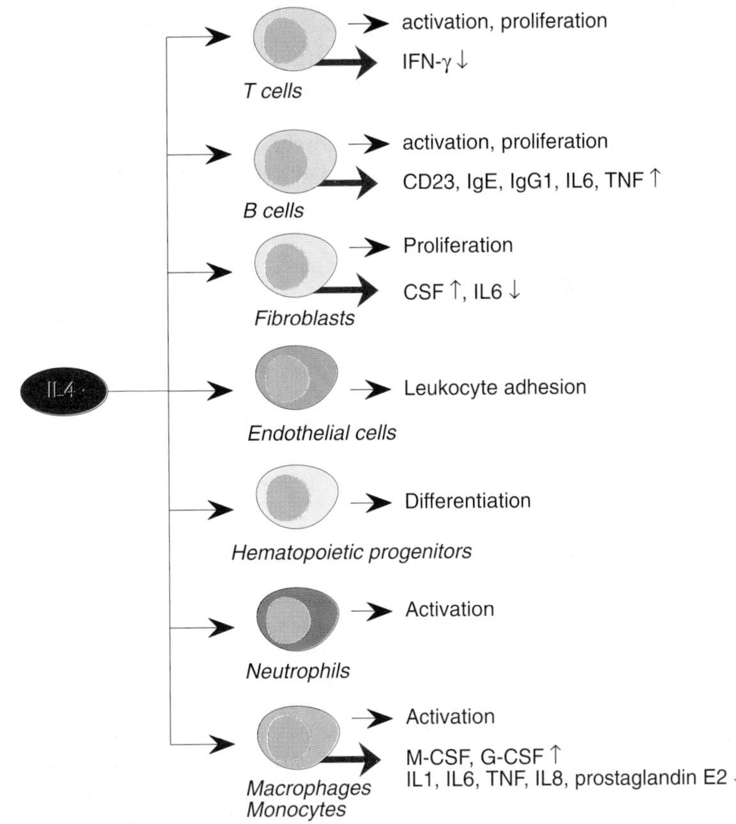

Schematic representation of the multiple activities of IL4 at the cellular level.

IL4 stimulates the synthesis of IgG_1 and IgE and inhibits the synthesis of IgM, IgG_3, IgG_{2a} and IgG_{2b}. This » isotype switching induced by IL4 in B cells is antagonized by » IFN-γ. The growth of multiple myelomas can be suppressed by IL4 which inhibits the synthesis of » IL6, a myeloma growth factor. IL4 also inhibits the synthesis of IL6 in human alveolar macrophages.

Pretreatment of macrophages with IL4 prevents the production of » IL1, » TNF-α and prostaglandins in response to activation of the cells by bacterial » endotoxins or » IFN-γ.

IL4 synergises with » Epo and G-CSF/Epo in the generation of colonies containing granulocytes or erythroid progenitor cells in a » colony formation assay.

Experiments with » oncogenes encoding a tyrosine-specific protein kinase show that the expression of the oncogene-encoded tyrosine kinase abolishes IL4 requirements (and also GM-CSF and IL3 requirements) in factor-dependent cells. The exact mechanism by which these factors regulate proliferation and differentiation is still unknown.

■ **TRANSGENIC/KNOCK-OUT/ANTISENSE STUDIES:** The consequences of the deregulated expression of IL4 has been studied in » transgenic animals (mice) overexpressing human IL4 in all tissues. In these animals one observes the excessive production of serum IgE levels. The expression of the IL4 transgene in the thymus interferes with T cell maturation and strongly reduces the population of immature $CD4^+8^+$ thymocytes and peripheral T cells while the numbers of mature $CD8^+$ thymocytes increases excessively. In addition these animals suffer from severe inflammatory blepharitis. The levels of IL4 cannot be increased over a certain level because the uncontrolled overexpression leads to the death of these animals. The introduction of a recombinant IL4 gene and the resulting overproduction of IL4 leads to factor-independence in IL4-dependent cells although it does not cause cell transformation.

Several cytokine mRNA such as those coding for » IL5, » IL6, IFN-γ and also the IL4 receptor are found to be up-regulated in IL4-transgenic mice, whereas » IL1, » IL2, » IL3, » TNF-α and » TNF-β mRNA levels do not seem to be influenced by IL4.

The expression of IL4 in transgenic mice has profound effects on B cell function. B cells from these animals have an increased size and show elevated expression of class II MHC. The animals show elevated serum levels of IgG1 and IgE and decreased levels of IgG2a, IgG2b, and IgG3. Antigen-specific antibody responses to immunization with hapten-carrier conjugates are also affected by IL4. Treatment of IL4-transgenic mice with IL4-specific monoclonal antibodies reverses many of the isotype alterations, indicating that they arise as a direct consequence of IL4 secretion.

IL4-transgenic mice established from 129/Sc mice that are genetically resistant to infection with *Leishmania major* have been shown to be susceptible to *Leishmania* infections, thus demonstrating that Th2 cell-derived IL4 promotes disease development in cutaneous Leishmaniasis.

Transgenic mice overexpressing IL4 under the control of lymphocyte-specific regulatory sequences derived from the » *lck* gene have been shown to develop osteopetrosis as the result of markedly decreased bone formation by osteoblasts.

The biological activities of IL4 have also been studied in IL4-deficient mice generated by targeted disruption of the IL4 gene in embryonic stem cells (see: ES cells). Thymus and T cell subsets develop normally. IL4 appears to play an important role in the establishment of a functional TH2 immune response (see also: TH1/TH2 cytokines). CD4-positive T cells from naive IL4-deficient mice fail to produce TH2-derived cytokines after *in vitro* stimulation. The levels of TH2 cytokines IL5, » IL9, and IL10 from CD4-positive T cells obtained after infection of these mice with the nematode *Nippostrongylus brasiliensis* are significantly reduced. IL4-deficient mice have also been shown to be resistant to retrovirus-induced immunodeficiency syndrome. Double mutant mice, deficient in IL4 and » IL2 have been shown to be viable and to have normal major T cell subsets and B cells, indicating that neither IL2 nor IL4 are essential for development of the immune system. IL4-deficient mice have been shown also to develop resistance against retrovirus-induced immunodeficiency (MAIDS), displaying no lethalities and delayed development of T cell abnormalities usually associated with MAIDS.

■ **ASSAYS:** The classical detection method for IL4 is a B cell costimulation assay measuring the enhanced proliferation of stimulated purified B cells. IL4 can be detected also in » Bioassays, employing IL4-responsive cells (see: BALM-4; BCL_1; CT.4S; CTL44; CTLL; Da; FDCP-Mix; HT-2; L4; L138.8A; MO7E; MC/9; NFS-60; Ramos, Sez 627, TF-1; TS-1). A specific detection method for human IL4 is the induction of » CD23 in a number of B cell lines with CD23 detected either by flow-through cytometry or by a fluorescence immunoassay. An alternative and entirely different detection method is » Message amplification phenotyping. For further information see also subentry "Assays" in the reference section.

■ **Clinical Use & Significance:** IL4 may be of clinical importance in the treatment of inflammatory diseases and autoimmune diseases since it inhibits the production of inflammatory cytokines such as » IL1, » IL6 and » TNF-α by monocytes and of » TNF by T cells (see also: Inflammation). IL4 may also be useful in the treatment of solid tumors, of hematopoietic systemic diseases, and of immune defects. IL4 inhibits the growth of colon and mammary carcinomas. It has been shown to augment the development of lymphokine-activated killer cells (see: LAK cells).

IL4 may play an essential role in the pathogenesis of chronic lymphocytic leukemia disease, which is characterized by the accumulation of slow-dividing and long-lived monoclonal B cells arrested at the intermediate stage of their differentiation. by preventing both the death and the proliferation of the malignant B cells. It protects chronic lymphocytic leukemic B cells from death by » apoptosis and upregulates the expression of a protective gene, » *bcl*-2.

The transduction of murine tumor cells with a functional IL4 gene has been shown to lead to the rejection of the genetically modified cells by syngeneic hosts. Altered tumor cells expressing IL4 also increase systemic immunity. Mice vaccinated with transduced cells reject a subsequent challenge of non-transduced cells, and, in some cases, a pre-existing tumor (for cancer vaccines see also: Cytokine gene transfer).

● **Reviews: Boulay JL & Paul WE** The interleukin-4 family of lymphokines. Curr. Opin. Immunol. 4: 294-8 (1992); **Dullens HFJ et al** Cancer treatment with interleukins 1, 4, and 6 and combinations of cytokines: a review. *In vivo* 5: 567-70 (1991); **Jansen JH et al** Interleukin-4. A regulatory protein. Blut 60: 269-74 (1990); **Paul WE** Interleukin 4/B cell stimulatory factor 1: one lymphokine, many functions. FJ 1: 456-61 (1987); **Paul WE & Ohara J** B cell stimulatory factor-1/Interleukin 4. ARI 5: 429-59 (1987); **Paul WE** Interleukin 4: A prototypic immunoregulatory lymphokine. Blood 77: 1859-70 (1991); **Yokota T et al** Molecular biology of interleukin 4 and interleukin 5 genes and biology of their products that stimulate B cell, T cells, and hemopoietic cells. IR 102: 137-87 (1988); **Callard RE** Immunoregulation by interleukin-4 in man. Brit. J. Haematol. 78: 293-9 (1991)

● **Biochemistry & Molecular Biology: Abe E et al** An 11-base-pair DNA sequence motif apparently unique to the human interleukin 4 gene confers responsiveness to T cell activation signals. PNAS 89: 2864-8 (1992); **Arai N et al** Complete nucleotide sequence of the chromosomal gene for human IL4 and its expression. JI 142: 274-82 (1989); **Blankenstein T et al** Retroviral interleukin 4 gene transfer into an interleukin 4-dependent cell line results in autocrine growth but not in tumorigenicity. EJI 20: 935-8 (1990); **Boothby M et al** A DNA binding protein regulated by IL4 and by differentiation in B cells. S 242: 1559-62 (1988); **Chandrasekharappa SC et al** A long-range restriction map of the interleukin-4 and interleukin-5 linkage group on chromosome 5. Genomics 6: 94-9 (1990); **Eder A et al** The 5′ region of the human interleukin 4 gene: structure and potential regulatory elements. NAR16: 772 (1988); **Kruse N et al** Conversion of human interleukin-4 into a high affinity antagonist by a single amino acid replacement. EJ 11: 3237-44 (1992); **Kruse N et al** Two distinct functional sites of human interleukin 4 are identified by variants impaired in either receptor binding or receptor activation. EJ 12: 5121-9 (1993); **Le HV et al** Isolation and characterization of multiple variants of recombinant human interleukin 4 expressed in mammalian cells. JBC 263: 10817-23 (1988); **Le Beau MM et al** Interleukin 4 and interleukin 5 map to human chromosome 5 in a region encoding growth factors and receptors and are deleted in myeloid leukaemias with a del(5q). Blood 73: 647-50 (1989); **Lee F et al** Isolation and characterization of a mouse interleukin cDNA clone that expresses B cell stimulatory factor 1 activities and T cell and mast cell-stimulating activities. PNAS 83: 2061-5 (1986); **McKnight AJ et al** Molecular cloning of rat interleukin 4 cDNA and analysis of the cytokine repertoire of subsets of $CD4^+$ T cells. EJI 21: 1187-94 (1991); **Noma Y et al** Cloning of cDNA encoding the murine IgG1 induction factor by a novel strategy using SP6 promoter. N 319: 640-6 (1986); **Otsuka T et al** Structural analysis of the mouse chromosomal gene encoding interleukin 4 which expresses B cell, T cell and mast cell stimulating activities. NAR 15: 333-44 (1987); **Redfield C et al** Secondary structure and topology of human interleukin 4 in solution. B 30: 11029-35 (1991); **Sutherland GR et al** Interleukin 4 is at 5q31 and interleukin 6 is at 7p15. Hum. Genet. 79: 335-7 (1988); **Szabo SJ et al** Identification of cis-acting regulatory elements controlling interleukin 4 gene expression in T cells: roles for NF-Y and NF-ATc. MCB 13: 4793-805 (1993); **Takahashi M et al** Chromosomal mapping of the mouse IL4 and human IL5 genes. Genomics 4: 47-52 (1989); **van Leeuwen BH et al** Molecular organization of the cytokine gene cluster involving the human IL3, IL4, IL5, and GM-CSF genes on human chromosome 5. Blood 73: 1142-8 (1989); **Yokota T et al** Isolation and characterization of a human interleukin cDNA clone, homologous to mouse B cell stimulatory factor 1, that expresses B cell and T cell-stimulating activities. PNAS 83: 5894-8 (1986)

● **Receptor Structure, Gene(s), Expression: Cabrillat H et al** High affinity binding of human interleukin 4 to cell lines. BBRC 149: 995-1001 (1987); **Finney M et al** Regulation of the interleukin 4 signal in human B lymphocytes. Biochem. Soc. Trans. 19: 287-91 (1991); **Harada N et al** Expression cloning of a cDNA encoding the murine interleukin 4 receptor based on ligand binding. PNAS 87: 857-61 (1990); **Idzerda RL et al** Human IL4 receptor confers biological responsiveness and defines a novel receptor superfamily. JEM 171: 861-73 (1990); **Jacobs CA et al** Characterization and pharmacokinetic parameters of recombinant soluble interleukin-4 receptor. Blood 77: 2396-403 (1991); **Kondo M et al** Sharing of the interleukin-2 (IL2) receptor g chain between receptors for IL2 and IL4. S 262: 1874-7 (1993); **Mosley B et al** The murine interleukin-4 receptor: molecular cloning and characterization of secreted and membrane-bound forms. Cell 59: 335-48 (1989); **Russell SM et al** Interleukin 2 receptor g chain: a functional component of the interleukin-4 receptor. S 262: 1880-3 (1993); **Sato TA et al** Recombinant soluble murine IL4 receptor can inhibit or enhance IgE responses *in vivo*. JI 150: 2717-23 (1993); **Suzuki H et al**

Gene mapping of mouse IL4 receptor: The loci of interleukin 4 (IL4) receptor gene and lymphocyte function associated antigen 1 (LFA-1) gene are closely linked on chromosome 7. Immunogenetics 34: 252-6 (1991); **Wrighton N et al** The murine interleukin-4 receptor gene: genomic structure, expression and potential for alternative splicing. GF 6: 103-18 (1992)

● **Biological Activities: Hermann F et al** Interleukin-4 inhibits growth of multiple myelomas by suppressing interleukin-6 expression. Blood 78: 2070-4 (1991); **Karray S et al** Interleukin 4 counteracts the interleukin 2-induced proliferation of monoclonal B cells. JEM 168: 85-94 (1988); **Mosmann TR et al** Species-specificity of T cell stimulating activities of IL 2 and BSF-1 (IL 4): comparison of normal and recombinant, mouse and human IL 2 and BSF-1 (IL 4). JI 138: 1813-6 (1987); **Ogo T et al** Effects of interleukin 4 on stromal cell-associated bone marrow culture. EH 19: 899-904 (1991); **Pritchard MA et al** The interleukin-4 receptor gene (IL4R) maps to 16p11. 2-16p12. 1 in human and to the distal region of mouse chromosome 7. Genomics 10: 801-6 (1991); **Spits H et al** IL4 inhibits IL2-mediated induction of human lymphokine activated killer cells, but not the generation of antigen specific cytotoxic T lymphocytes in mixed leukocyte cultures. JI 141: 29-36 (1988); **Swain SL et al** The role of IL4 and IL5: characterization of a distinct helper T cell subset that makes IL4 and IL5 (Th2) and requires priming before induction of lymphokine secretion. IR 102: 77-105 (1988); **Thor G & Brian AA** Glycosylation variants of murine interleukin 4: evidence for different functional properties. Immunology 75: 143-9 (1992)

● **Transgenic/Knock-out/Antisense Studies: Burstein HJ et al** Humoral immune functions in IL4 transgenic mice. JI 147: 2950-6 (1991); **Kanagawa O et al** Resistance of mice deficient in IL4 to retrovirus-induced immunodeficiency syndrome (MAIDS). S 268: 240-2 (1993); **Kopf M et al** Disruption of the murine IL4 gene blocks Th2 cytokine responses. N 362: 245-8 (1993); **Kühn R et al** Generation and analysis of interleukin-4 deficient mice. S 254: 707-10 (1991); **Leal LM et al** Interleukin-4 transgenic mice of resistant background are susceptible to *Leishmania* major infection. EJI 23: 566-9 (1993); **Lewis DB et al** Osteoporosis induced in mice by overproduction of interleukin 4. PNAS 90: 11618-22 (1993); **Maliszewski CR et al** Delayed allograft rejection in mice transgenic for a soluble form of the IL4 receptor. CI 143: 434-48 (1992); **Müller W et al** Major histocompatibility complex class II hyperexpression on B cells in interleukin 4-transgenic mice does not lead to B cell proliferation and hypergammaglobulinemia. EJI 21: 921-5 (1991); **Sadlack B et al** Development and proliferation of lymphocytes in mice deficient for both interleukins-2 and 4. EJI 24: 281-4 (1994); **Tepper RI et al** IL4 induces allergic-like inflammatory disease and alters T cell development in transgenic mice. Cell 62: 457-67 (1990)

● **Assays: Agrewala JN et al** Measurement and computation of murine interleukin-4 and interferon-γ by exploiting the unique abilities of these lymphokines to induce the secretion of IgG1 and IgG2a. J. Immunoassay 14: 83-97 (1993); **Callard RE et al** Assay for human B cell growth and differentiation factors. in: Clemens MJ et al (eds) Lymphokines and Interferons. A practical Approach, pp. 345-64, IRL Press, Oxford 1987; **Custer MC & Lotze MT** A biologic assay for IL4. Rapid fluorescence assay for IL4 detection in supernatants and serum. JIM 128: 109-17 (1990); **Dusch A et al** A comparison of assays for the response of primary human T cells upon stimulation with interleukin-2, interleukin-4 and interleukin-7. ECN 3: 97-102 (1992); **Ishizuka T et al** An ultrasensitive system to detect IL4: enzyme-linked immunosorbent assay (ELISA) combined with an avidin-biotin and enzyme amplification system. JIM 153: 213-22 (1992); **Jung T et al** Detection of intracellular cytokines by flow cytometry. JIM 159: 197-207 (1993); **Kanagawa O et al** Resistance of mice deficient in IL4 to retrovirus-induced immunodeficiency syndrome (MAIDS). S 262: 240-2 (1993); **Lebedin YS et al** Serum levels of interleukin 4, interleukin 6 and interferon-γ following *in vivo* isotype-specific activation of IgE synthesis in humans. Int. Arch. Allergy Appl. Immunol. 96: 92-4 (1991); **Maeda M et al** Application of a human T cell line derived from a Sezary syndrome patient for human interleukin 4 assay. Immunol. Lett. 18: 247-53 (1988); **Noe G et al** A sensitive sandwich ELISA for measuring erythropoietin in human serum. Br. J. Haematol. 80: 285-92 (1992); **Platzer C et al** Analysis of cytokine mRNA levels in interleukin-4-transgenic mice by quantitative polymerase chain reaction. EJI 22: 1179-84 (1992); **Seder RA et al** Increased frequency of interleukin 4-producing T cells as a result of polyclonal priming. Use of a single-cell assay to detect interleukin 4-producing cells. EJI 21: 1241-7 (1991); for further information also see individual cell lines used in individual bioassays.

● **Clinical Use & Significance: Cornacoff JB et al** Preclinical evaluation of recombinant human interleukin-4. Toxicol. Lett. 64-65 Spec No: 299-310 (1992); **Dancescu M et al** Interleukin 4 protects chronic lymphocytic leukemic B cells from death by apoptosis and upregulates *bcl*-2 expression. JEM 176: 1319-26 (1992); **Dullens HFJ et al** Cancer treatment with interleukins 1, 4, and 6 and combinations of cytokines: a review. *In vivo* 5: 567-70 (1991); **Gallagher G & Zaloom Y** Peritumoral IL4 treatment induces systemic inhibition of tumor growth in experimental melanoma. Anticancer Res. 12: 1019-24 (1992); **Gilleece MH et al** Recombinant human interleukin 4 (IL4) given as daily subcutaneous injections – a phase I dose toxicity trial. Br. J. Cancer. 66: 204-10 (1992); **Golumbek PT et al** Treatment of established renal cancer by cells engineered to secrete interleukin 4. S 254: 713-6 (1991); **Li WQ et al** Lack of tumorigenicity of interleukin 4 autocrine growing cells seems related to the anti-tumor function of interleukin 4. Mol. Immunol. 27: 1331-7 (1990); **Okabe M et al** Inhibitory anti-tumor effects of interleukin-4 on Philadelphia chromosome-positive acute lymphocytic leukemia and other hematopoietic malignancies. Leuk. Lymphoma 8: 57-63 (1992); **Rubin JT & Lotze MT** Acute gastric mucosal injury associated with the systemic administration of interleukin-4. Surgery 111: 274-80 (1992); **Tepper RI** Murine interleukin-4 displays potent anti-tumor activity *in vivo*. Cell 57: 503-12 (1989); **Tepper RI** The tumor-cytokine transplantation assay and the antitumor activity of interleukin-4. Bone Marrow Transplant. 9: s177-81 (1992); **Tepper RI et al** An eosinophil-dependent mechanism for the antitumor effect of interleukin-4. S 257: 548-51 (1992); **Toi M et al** Inhibition of colon and breast carcinoma cell growth by interleukin-4. CR 52: 275-9 (1992)

IL4BP: abbrev. for IL4 binding protein see: IL4.

IL5: Interleukin 5.

■ **Alternative Names:** B151-TRF (B151 T cell replacing factor); B151-TRF1 (B151 T cell replacing factor; see: B151-TRF); BCDF (B cell differentiation factor); BCDFα (B cell differentiation factor α); BCDFμ (B cell differentiation factor μ); BCGF-2 (B cell growth factor 2); (DL)BCGF (Dennert line B cell growth factor); BGDF (B cell growth and differentiation factor); CFU-Eo GSF (colony forming unit eosinophil growth stimulating factor); EDF (eosinophil differentiation fac-

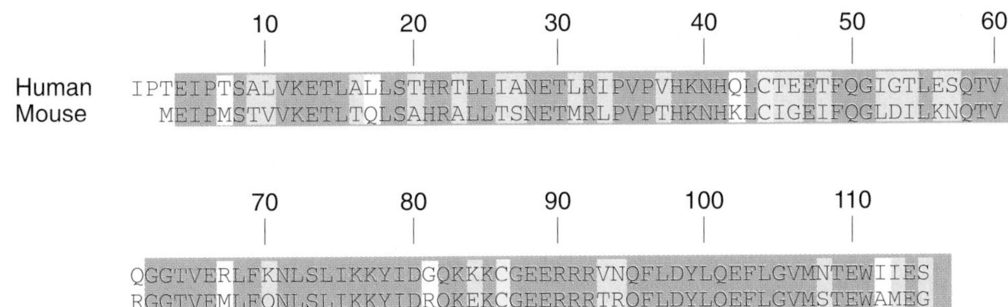

Sequence comparison of human and murine IL5.
Cysteine residues involved in disulfide bond formation are shown in yellow. Identical amino acid positions are shown in blue. Positions in green indicate conserved amino acid exchanges. Positions that differ between the two sequences are shown in white.

tor); Eo-CSF (eosinophil colony-stimulating factor); Eo-DF (eosinophil differentiation factor); ESP (eosinophil stimulation promoter); IgA-EF (IgA-enhancing factor); KHF (killer helper factor); TCRF-2 (T cell replacing factor 2); TRF-1 (T cell replacing factor). See also: individual entries for further information.

■ **SOURCES:** IL5 is produced by T cells. For the constitutive synthesis see also: IAP (intracisternal A particle).

■ **PROTEIN CHARACTERISTICS:** Murine IL5 cDNA encodes a protein of 113 amino acids: the human protein has a length of 115 amino acids. The biologically active form of IL5 is an N-glycosylated antiparallel homodimer linked by disulfide bonds. Monomeric forms are biologically inactive. Variable molecular masses of the native protein are caused by heterogeneous glycosylation. Non-glycosylated IL5 is also biologically active.

Murine and human IL5 are ≈ 70 % identical. While murine and human IL5 have the same specific activity on human cells murine IL5 is ≈ 50-100-fold more active on murine cells than human IL5. The generation of human/murine chimaeric proteins has revealed that the carboxyterminal end of the protein is re4sponsible for this species specificity. This region probably interacts with the receptor.

■ **GENE STRUCTURE:** The human IL5 gene has a length of ≈ 4 kb and contains four exons. It maps to human chromosome 5q23-31 in the vicinity of genes encoding » GM-CSF, » M-CSF, » IL3 and » IL4 which all have the same orientation and are located within a 500 kb fragment of DNA. The distance between the IL5 and IL4 gene is ≈ 90-240 kb. The 5´ region of the IL5-gene contains several elements, known as » CLE (conserved lymphokine element), which are binding sites for transcription factors controlling the expression of this gene (see also: Gene expression).

■ **RECEPTOR STRUCTURE, GENE(S), EXPRESSION:** High affinity (≈ 30-100/cell) and low affinity (≈ 200-400/cell) receptors for IL5 are expressed in all hematopoietic and lymphoid cells. The gene for the human IL5 receptor α subunit maps to chromosome 3p26.

The low affinity receptor (Kdis = 30 nM) has a molecular mass of 50 kDa (p60). The cDNA of a cloned murine IL5 receptor encodes a glycoprotein of 415 amino acids containing a secretory » signal sequence of 17 amino acids. The extracellular domain of this protein has a length of 322 amino acids. The transmembrane domain has a length of 22 amino acids and the cytoplasmic domain has a length of 54 amino acids. The receptor contains another protein that does not bind IL5. This second subunit is expressed in some » IL3-dependent cell lines that do not bind IL5. The introduction of p60 into these cells generates a high affinity IL5 receptor. The second subunit of the IL5 receptor is identical with » GM-Rβ which is the larger subunit of the receptor for the colony-stimulating factor » GM-CSF. GM-Rβ is also involved in the generation of a high affinity receptor for » IL3. Soluble forms of the IL5 receptor have also been described.

■ **BIOLOGICAL ACTIVITIES:** Human IL5 is active in murine cells albeit with greatly reduced biological activity. Murine IL5 is active in human cells.

IL5 is a specific » hematopoietin (hematopoietic growth factor) that is responsible for the growth and differentiation of eosinophils. IL5 promotes the growth of early hematopoietic progenitor cells

» BFU-E while it causes differentiation of » CFU-E the proliferation of which is inhibited by IL5. IL5 strongly stimulates the proliferation, activation, and differentiation of eosinophilic granulocytes (see also: cell activation). B cells can be made responsive to IL5 by treatment with suboptimal doses of » IL1.

IL5 also promotes the generation of cytotoxic T cells from thymocytes. In thymocytes IL5 induces the expression of high affinity IL2 receptors.

In contrast to human IL5 murine IL5 also acts on B cells (see also: BCDF, B cell differentiation factor). It induces the proliferation of pre-activated B cells and their differentiation. Murine IL5 also stimulates the production and secretion of IgM and IgA by B cells stimulated by bacterial » endotoxins (see also: Isotype switching).

■ **TRANSGENIC/KNOCK-OUT/ANTISENSE STUDIES:** The consequences of the deregulated expression of IL5 have been studied in » transgenic animals (mice) expressing the human factor. Mice overexpressing IL5 have a characteristic lifelong continuous blood and spleen eosinophilia and eosinophil infiltration in lungs and gut but they remain normal and do not show any signs of tissue damage. *In vivo* administration of antibodies to murine IL5 receptor inhibits eosinophilia in IL5 transgenic mice. Transgenic mice expressing the IL5 gene under the control of a metallothionein promoter have elevated levels of IL5 in the serum which can be increased further by administration of cadmium. These animals show an increase in the levels of serum IgM and IgA. Treatment with cadmium results in the expansion of a population of Ly-1-positive B cell in the spleen, which have been considered to produce autoantibodies, and which are also selectively developed by IL5 in long-term bone marrow culture (see: LTBMC). The serum of these animals also contains elevated concentrations of polyreactive anti-DNA antibodies of IgM class. Studies of immunoglobulin gene usage in IL5 transgenic mice show that IL5 maintains CD5-positive B cells that have a fetus-type of immunoglobulin gene usage. IL5 may thus be responsible for prolonging the life span of immature CD5-positive B cells, which subsequently mature to CD5-negative B cells that secrete polyreactive natural antibodies.

■ **ASSAYS:** IL5 can be detected in » Bioassays employing cells that respond to this factor (see: B13; BCL_1; T88-M; TALL-103; TF-1). IL5 can also be detected by sensitive immunoassays. Another assay involves the detection of eosinophilic colonies in a colony formation assay (see: EDF, eosinophil differentiation factor) or the detection of eosinophil peroxidase. An alternative and entirely different detection method is » Message amplification phenotyping. For further information see also subentry "Assays" in the reference section.

■ **CLINICAL USE & SIGNIFICANCE:** A possible clinical application is suggested by the activity of IL5 on eosinophils. Animal experiments have shown that eosinophilia elicited by nematode infections in mice and the concomitant infiltration of the lung with eosinophils can be prevented by administration of monoclonal animals directed against IL5.

● **REVIEWS: Sanderson CJ et al** Molecular and cellular biology of eosinophil differentiation factor (interleukin-5) and its effect on human and mouse B cells. IR 102: 29-50 (1988); **Takatsu K et al** T cell-replacing factor (TRF)/interleukin 5 (IL5) molecular and functional properties. IR 102: 107-35 (1988); **Takatsu K** Interleukin-5. Curr. Opin. Immunol. 4: 299-306 (1992); **Yokota T et al** Molecular biology of interleukin 4 and interleukin 5 genes and biology of their products that stimulate B cell, T cells, and hemopoietic cells. IR 102: 137-87 (1988); **Takatsu K** Interleukin 5 (IL5) and its receptor. Microbiol Immunol. 35: 593-606 (1991)

● **BIOCHEMISTRY & MOLECULAR BIOLOGY: Azuma C et al** Cloning of cDNA for human T cell replacing factor (interleukin-5) and comparison with the murine homologue. NAR 14: 9149-58 (1986); **Chandrasekharappa SC et al** A long-range restriction map of the interleukin-4 and interleukin-5 linkage group on chromosome 5. Genomics 6: 94-9 (1990); **Ingley E et al** Production and purification of recombinant human interleukin 5 from yeast and baculovirus expression systems. EJB 196: 623-9 (1991); **Ingley E et al** Production and purification of recombinant human interleukin-5 from yeast and baculovirus expression systems. EJB 196: 623-9 (1991); **Kinashi T et al** Cloning of complementary DNA encoding T cell-replacing factor and identity with B cell growth factor 2. N 324: 70-3 (1986); **Le Beau MM et al** Interleukin 4 and interleukin 5 map to human chromosome 5 in a region encoding growth factors and receptors and are deleted in myeloid leukaemias with a del(5q). Blood 73: 647-50 (1989); **McKenzie ANJ et al** Structure-function analysis of interleukin-5 utilizing mouse/human chimeric molecules. EJ 10: 1191-3 (1991); **Milburn MV et al** A novel dimer configuration revealed by the crystal structure at 2.4 Å resolution of human interleukin-5. N 363: 172-6 (1993); **Sutherland GR et al** Interleukin 5 is at 5q31 and is deleted in the $5q^-$ syndrome. Blood 71: 1150-2 (1988); **Takahashi M et al** Chromosomal mapping of the mouse IL4 and human IL5 genes. Genomics 4: 47-52 (1989); **Tanabe T et al** Molecular cloning and structure of the human interleukin-5 gene. JBC 262: 16580-4 (1987); **Tavernier J et al** Expression of human and murine interleukin-5 in eukaryotic systems. DNA 8: 491-501 (1989); **van Leeuwen BH et al** Molecular organization of the cytokine gene cluster involving the human IL3, IL4, IL5, and GM-CSF genes on human chromosome 5. Blood 73: 1142-8 (1989); **Yokota T et al** Isolation and characterization of lymphokine cDNA clones encoding mouse and human IgA-enhancing factor and eosinophil colony-stimulating factor activities: relationship to interleukin 5. PNAS 84: 7388-92 (1987)

● **RECEPTOR STRUCTURE, GENE(S), EXPRESSION: Devos R et al** Amino acid sequence analysis of a mouse interleukin 5 receptor protein reveals homology with a mouse interleukin 3 receptor protein. EJI 21: 1315-7 (1991); **Devos R et al** Molecular basis of a high affinity murine interleukin-5 receptor. EJ 10: 2133-7 (1991); **Fung MC et al** Distinguishing between mouse IL3 and IL3 receptor-like (IL5/GM-CSF receptor converter) mRNAs using the polymerase chain reaction method. JIM 149: 97-103 (1992); **Jacob CO et al** Mapping of the interleukin 5 receptor gene to human chromosome 3 p25-p26 and to mouse chromosome 6 close to the *raf*-1 locus with polymorphic tandem repeat sequences. Mamm. Genome. 4: 435-9 (1993); **Murata Y et al** Molecular cloning and expression of the human interleukin 5 receptor. JEM 175: 341-51 (1992); **Takaki S et al** Molecular cloning and expression of the murine interleukin-5 receptor. EJ 9: 4367-74 (1990); **Takaki S et al** Identification of the second subunit of the murine interleukin-5 receptor: interleukin-3 receptor-like protein AIC2B is a component of the high affinity interleukin-5 receptor. EJ 10: 2833-8 (1991); **Tavernier J et al** A human high affinity interleukin-5 receptor (IL5R) is composed of an IL5-specific α chain and a β chain shared with the receptor for GM-CSF. Cell 66: 1175-84 (1991); **Tuypens T et al** Organization and chromosomal localization of the human interleukin 5 receptor α-chain gene. ECN 3: 451-9 (1992)

● **Biological Activities:** **Clutterbuck E et al** Recombinant human interleukin 5 is an eosinophil differentiation factor but has no activity in standard human B cell growth factor assays. EJI 17: 1743-50 (1987); **Koyama N et al** Role of recombinant interleukin 1 compared to recombinant T cell replacing factor/interleukin 5 in B cell differentiation. Immunology 63: 277-83 (1988); **Nakao K et al** Effects of erythroid differentiation factor (EDF) on proliferation and differentiation of human hematopoietic progenitors. EH 19: 1090-5 (1991); **Ramos T** Interleukin 5 is a differentiation factor for cytotoxic T lymphocytes. Immunol. Lett. 21: 277-84 (1989); **Swain SL et al** The role of IL4 and IL5: characterization of a distinct helper T cell subset that makes IL4 and IL5 (Th2) and requires priming before induction of lymphokine secretion. IR 102: 77-105 (1988)

● **Transgenic/Knock-out/Antisense Studies:** **Dent LA** Eosinophilia in transgenic mice expressing interleukin 5. JEM 172: 1425-31 (1990); **Hitoshi Y et al** *In vivo* administration of antibody to murine IL5 receptor inhibits eosinophilia of IL5 transgenic mice. Int. Immunol. 3: 135-9 (1991); **Katoh S et al** Maintenance of $CD5^+$ B cells at an early developmental stage by interleukin-5: evidence from immunoglobulin gene usage in interleukin-5 transgenic mice. DNA Cell Biol. 12: 481-91 (1993); **Strath M et al** Infection of IL5 transgenic mice with *Mesocestoides corti* induces very high levels of IL5 but depressed production of eosinophils. EH 20: 229-34 (1992); **Tominaga A et al** Transgenic mice expressing a B cell growth and differentiation factor gene develop eosinophilia and autoantibody production. JEM 173: 429-37 (1991)

● **Assays:** **Abrams JS et al** Strategies of anti-cytokine monoclonal antibody development: immunoassay of IL10 and IL5 in clinical samples. IR 127: 5-24 (1992); **Fukuda Y et al** A sandwich enzyme-linked immunosorbent assay for human interleukin 5. JIM 143: 89-94 (1991); **Guiffre A et al** A quantitative polymerase chain reaction assay for interleukin 5 messenger RNA. AB 212: 50-7 (1993); **Kagnoff MF et al** Detection by immunofluorescence of lymphokine-producing T cells. Immunol. Res. 10: 255-7 (1991); **Kikuchi Y et al** Biochemical and functional characterization of soluble form of IL5 receptor α (sIL5Rα). JIM 167: 289-98 (1994); **McNamee LA et al** Production, characterization, and use of monoclonal antibodies to human interleukin 5 in an enzyme-linked immunosorbent assay. JIM 141: 81-8 (1991); **O'Garra A & Sanderson CJ** Eosinophilic differentiation factor and its associated B cell growth factor activities. in: Clemens MJ et al (eds) Lymphokines and Interferons. A practical Approach, pp. 323-43, IRL Press, Oxford 1987; **Schoenbeck S et al** Detection of individual Peyer's patch T cells that produce interleukin-5 and interferon-γ. JIM 137: 47-54 (1991); **Taguchi T et al** Detection of individual mouse splenic T cells producing IFN-γ and IL5 using the enzyme-linked immunospot (ELISPOT) assay. JIM 128: 65-73 (1990); for further information also see individual cell lines used in individual bioassays.

● **Clinical Use & Significance:** **Baumann MA et al** Effects of interleukin-5 on acute myeloid leukemias. Am. J. Hematol. 39: 269-74 (1992); **Coffman RL et al** Antibody to interleukin-5 inhibits helminth-induced eosinophilia in mice. S 245: 308-10 (1989)

IL6: Interleukin 6.

■ **Alternative Names:** 26 kDa Protein; BCDF (B cell differentiation factor); BCSF (B cell stimulatory factor); BSF-2 (B cell stimulation factor 2); BSF-p2 (B cell stimulation factor p2); CDF (CAT (or ChAT) development factor, choline acetyltransferase development factor); CDF (cytolytic differentiation factor for T lymphocytes); CDF (cytolytic T cell differentiation factor); CDF (cytotoxic T cell differentiation factor); CPA (colony promoting activity); CSF 309 (hematopoietic colony-stimulating factor 309); DIF (differentiation inducing factor); differentiation inducing factor for human monoblastic leukemia cells; FDGI (fibroblast-derived growth inhibitor); fibroblast derived differentiation inducing factor for human monoblastic leukemia cells; HGF (hybridoma growth factor); HP1; HPGF (hybridoma/plasmacytoma growth factor); HSF (hepatocyte stimulatory factor); HSF-1 (hepatocyte stimulatory factor 1); $IFN\beta_2$ (Interferon β_2); ILHP1 (interleukin-hemopoietin-1); ILHP1 (interleukin-hybridoma/plasmacytoma-1); L-GI factor (murine lung-derived growth inhibitory factor); L-HGF (L929-derived hybridoma growth factor; macrophage-granulocyte inducing protein 2; MGI-2A (macrophage-granulocyte inducer 2A); Mk potentiator; Myeloma GF (myeloma growth factor); NKAF (natural killer (cell) activating factor); PCT-GF (plasmacytoma growth factor); TAF (T cell activating factor); THCGF (thymocyte growth factor); TPO (thrombopoietin); TSF (thymocyte stimulating factor); WI-26-VA4 factor. See also: individual entries for further information.

■ **Sources:** IL6 is produced by many different cell types. The main sources *in vivo* are stimulated monocytes, fibroblasts, and endothelial cells. Macrophages, T and B lymphocytes, granulocytes, smooth muscle cells, eosinophils, chondrocytes, osteoblasts, mast cells, glial cells, and keratinocytes also produce IL6 after stimulation.

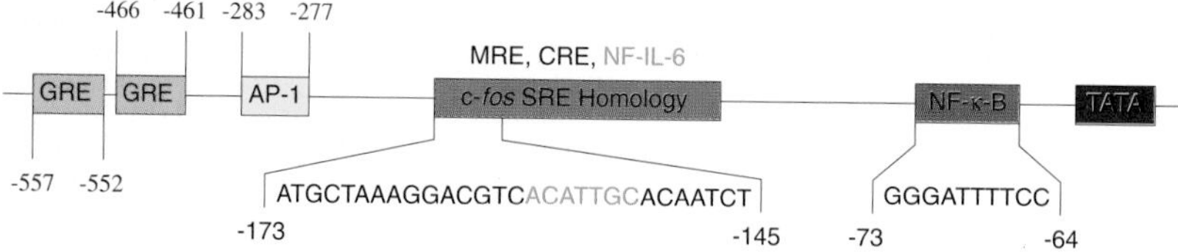

Structure of the IL6 promoter.
The promoter region contains several distinct binding sites for transcription factors:
AP-1 (transcription factor AP-1), CRE (cyclic AMP responsive element), GRE (glucocorticoid response element), MRE (multi-response element) NF-κ-B (transcription factor NF-κ-B), NF-IL-6 (IL-6-specific nuclear factor), SRE (serum responsive element).

Glioblastoma cells constitutively produce IL6 and the factor can be detected also in the cerebrospinal fluid. Human milk also contains IL6 (see: MGF, milk growth factor).

Cardiac myxomas and cervical and bladder carcinomas secrete large amounts of IL6. Very high levels of IL6 are also detected in the synovial fluid of patients with rheumatoid arthritis but not with osteoarthritis.

Physiological stimuli for the synthesis of IL6 are » IL1, bacterial » endotoxins, » TNF, » PDGF, and » Oncostatin M. In fibroblasts the synthesis of IL6 is stimulated by » IFN-β, » TNF-α, » PDGF, and viral infections. In thymic stromal cells the synthesis of IL6 can be induced by » NGF. IL6 can also stimulate or inhibits its own synthesis, depending upon the cell type. For the constitutive production of IL6 see also: IAP (intracisternal A particle).

IL6 is also produced in the anterior lobe of the pituitary and can be induced by bacterial endotoxins and all substances elevating intracellular levels of cAMP (prostaglandin E2, forskolin, cholera toxin). VIP (vasoactive intestinal peptide) also stimulates the synthesis of IL6 in the anterior lobe of the pituitary while » growth hormone releasing hormone is inactive. Macrolide antibiotics spiramycin and erythromycin stimulate the synthesis of IL6 in LPS-activated human monocytes but not the synthesis of IL1 or TNF-α.

Glucocorticoids inhibit the synthesis of IL6 and » IL4 or TGF-β reduce its synthesis. The synthesis of IL6 by human alveolar macrophages is inhibited by » IL4.

In serum IL6 is complexed with » Alpha-2-Macroglobulin (α_2M) which protects IL6 from cleavage by proteases and possibly functions as a transport protein.

■ **Protein Characteristics:** IL6 is a protein of 184 amino acids glycosylated at positions 73 and 172. It is synthesized as a precursor protein of 212 amino acids. Monocytes express at least five different molecular forms of IL6 with molecular masses of 21.5-28 kDa. They mainly differ by post-translational alterations such as glycosylation and phosphorylation.

IL6 isolated from various cell types shows some microheterogeneity in its N terminus. A 42-45 kDa form has been observed in plasma that is probably complexed with a carrier protein » Alpha-2-Macroglobulin (α_2M) Murine and human IL6 show 65 % sequence homology at the DNA level and 42 % homology at the protein level.

■ **Gene Structure:** The human IL6 gene has a length of ≈ 5 kb and contains five exons. It maps to human chromosome 7p21-p14 between the markers D7S135 and D7S370. The murine gene maps to chromosome 5. The nucleotide sequences of IL6 and » G-CSF genes resemble each other in a way suggesting a possible evolutionary relationship.

The IL6 gene promoter contains many different regulatory elements allowing the induction of expression by various stimuli, including glucocorticoids and cAMP (see also: Gene expression). The NF-κB binding site is responsible in non-lymphoid cells for the induction of the IL6 gene expression by » IL1 or » TNF. In lymphoid cells a factor related to the *rel* oncogene (IL6*k*B binding factor II) functions as a repressor that prevents the interaction of transcription factors with the IL6-κB binding site.

■ **Receptor Structure, Gene(s), Expression:** The IL6 receptor is expressed on T cells, mitogen-activated B cells, peripheral monocytes and some macrophage- and B cell-derived tumor cell types. It is not expressed in resting B cells but in resting T cells. In hepatocytes the IL6 receptor expression is enhanced after treatment with IL6 or » IL1. In several cell types the expression of the IL6 receptor is also enhanced by glucocorticoids. For the over-expression of the IL6 receptor see also: IAP

(intracisternal A particle). The IL6 receptor gene maps to human chromosome 1q21.
The IL6 receptor is a strongly glycosylated protein of 80 kDa and a length of 449 amino acids. It has been designated ***CD126*** (see also: CD Antigens). It is synthesized as a precursor of 468 amino acids. The molecular structure resembles that of receptors for » M-CSF, » PDGF and » IL1 in that the receptor contains an immunoglobulin-like sequence domain in the aminoterminal region of the extracellular receptor domain.
The intracellular domain of the IL6 receptor has a length of ≈ 82 amino acids and does not show any homology to other proteins involved in intracellular signal transduction.
Two different forms of the receptor have been described that bind IL6 with different affinities (Kdis = 10^{-9} and 10^{-11} M) and most likely arise by post-translational modification of the same receptor protein.
Biological activities of IL6 have also been found at concentrations of 10^{-13} – 10^{-15} M suggesting either the existence of other high-affinity receptor conformations or the existence of further receptor molecules with higher affinities.
IL6 receptor-mediated signal transduction involves protein kinase C and also adenylate cyclase.
The complex formed between IL6 and its receptor associates with a transmembrane glycoprotein, gp130 (918 amino acids; cytoplasmic domain of 277 amino acids), that is involved in signal transduction. Binding of IL6 to its receptor leads to disulfide-linked homodimerization of gp130 and the associated activation of a tyrosine kinase as the first step of signal transduction. gp130 is also expressed in cells that do not express IL6 receptors. It has been found to be a component of other receptors, including those for » IL11, » LIF, » Oncostatin M, and » CNTF (ciliary neurotrophic factor). This explains why » LIF, » CNTF (ciliary neurotrophic factor) and IL6 share many biological activities although the factors themselves are not related to each other.
A soluble form of the IL6 receptor (see: IL6R-SUP (soluble urinary protein)) has also been described that also interacts with gp130. These soluble receptors probably function as physiological regulators of cytokine activities (see also: Cytokine inhibitors) by inhibiting receptor binding or act as transport proteins. Similar soluble receptors or binding proteins have also been described for » IL1 (see: IL1ra, IL1 receptor antagonist), » IL2, » IL4, » IL7, » TNF-α, » IGF, and » IFN-γ.

■ **Biological Activities:** Human IL6 is biologically active in monkeys, rats, and mice. Murine IL3 is not active in human cells.
The plethora of biological activities is exemplified by the many different acronyms under which IL6 has been described. IL6 is a pleiotropic cytokine influencing antigen-specific immune responses and inflammatory reactions (see also: Inflammation). It is one of the major physiological mediators of the » acute phase reaction.
In hepatocytes IL6 in combination with glucocorticoids induces the synthesis of metallothioneins and increases intracellular zinc levels, thus preventing CCL_4-induced hepatotoxicity.
IL6 is a neurotrophic factor for cholinergic neurons that promotes their survival in culture (see also: Neurotrophins). Some neuronal cell lines can be induced to differentiate by IL6.
IL6, like » IL1, stimulates the synthesis of ACTH (Corticotropin) in the pituitary. Glucocorticoids synthesized in response to ACTH inhibit the production of IL6, IL1 and TNF *in vivo*, thus establishing a sort of negative feedback loop between the immune system and neuroendocrine functions.
In astrocytes IL6 induces the synthesis of » NGF.
IL6 is a B cell differentiation factor *in vivo* and *in vitro* (see also: BCDF) and an activation factor for T cells (see also: cell activation). In the presence of » IL2 IL6 induces the differentiation of mature and immature T cells into cytotoxic T cells. IL6 also induces the proliferation of thymocytes and probably plays a role in the development of thymic T cells.

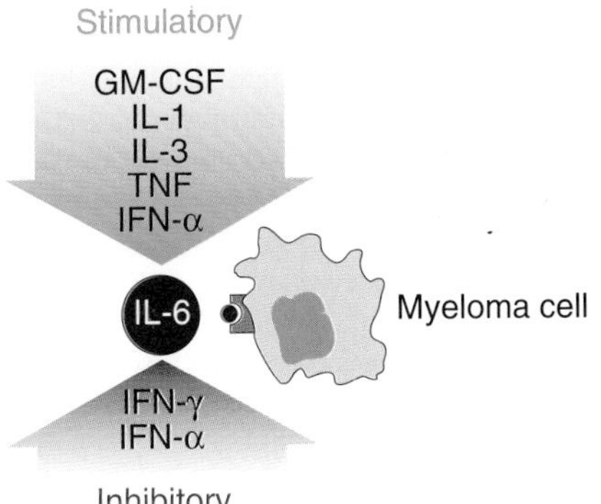

IL6 action on myeloma cells.
IL6 appears to be a major regulator of myeloma cell growth. Growth is stimulated or inhibited, however, by a number of other cytokines.
Low concentrations (< 60 U/mL)of IFN-α are stimulatory whereas higher doses (> 100 U/mL) are inhibitory.

IL6 is capable of inducing the final maturation of B cells into immunoglobulin-secreting plasma cells if the cells have been pre-activated by » IL4. In B cells IL6 stimulates the secretion of antibodies to such a degree that serum IgG1 levels can rise 120-400-fold.

IL6 at concentrations of only 0.002 ng/mL is one of the major » autocrine growth factor for many human myelomas. The growth of these cells can be inhibited by monoclonal antibodies directed against IL6. It can also be inhibited by the introduction of » antisense oligonucleotides against IL6 or by » IL4. The growth-inhibitory effects of corticosteroids on myeloma cells is probably due to the steroid-induced reduction in the expression of IL5. The growth of human IL6-dependent myeloma cells can also be inhibited by » IFN-γ. IL6 may also function as an » autocrine growth factor for other tumor types, some of which have been found to secrete IL6 constitutively. IL6 has been shown to be an autocrine growth factor for *in vitro* cervical tumor cell growth. On the other hand IL6 blocks the growth of some solid tumors such as mammary carcinomas, cervical carcinomas, human lung cancer cell lines, histiocytic lymphomas, and melanomas. The growth-inhibitory effects are probably compensated by an IL6-induced reduction in cell-to-cell interactions which also promotes increased metastasizing potential.

IL6 has been found to be expressed in murine blastocysts before hematopoietic differentiation. This suggests that IL6 also regulates the growth and development of trophoblasts or embryonic stem cells.

IL6 and » IL3 synergise *in vitro* in promoting the proliferation of multipotent hematopoietic progenitor cells. IL6 is also a thrombopoietin that induces the maturation of megakaryocytes *in vitro* and increases platelet counts *in vivo*. In murine, but not in human bone marrow cultures (see also: LTBMC, long-term bone marrow culture) IL6 shows activities resembling those of » GM-CSF.

Plasmacytoma cells produce IL6 and also the IL6 receptor. It has been suggested that these cells are stimulated in an » autocrine fashion. A » paracrine mechanism involving the presence of two different cell populations, one producing the factor and the other expressing the receptor, has also been described

■ **Transgenic/Knock-out/Antisense Studies:** The consequences of a deregulated expression of IL6 have been evaluated in » transgenic animals (mice) expressing the human IL6 gene. These animals display dramatically elevated serum levels of IL6 and polyclonal IgG1. In addition one observes a massive plasmacytosis and all typical indications of a proliferative mesangial glomerulonephritis which is probably due to the excessive overproduction of the cytokine.

Transgenic mice expressing the IL6 gene under the control of the human keratin gene promoter express IL6 in squamous epithelial cells. These mice are much smaller than normal mice and they also show a retarded development of the coat. An increased proliferation of epidermal cells and an infiltration of leukocytes are not observed. Other transgenic mice constitutively expressing murine IL6 at a level sufficient to induce IL6-responsive genes show less immature B cells in bone marrow and expression of IL6-inducible liver genes, but the mice appear healthy and can easily be used for breeding. A high incidence of lymphomas associated with different tissues is observed only in mice older than 18 months.

Intracerebral overexpression of IL6 in transgenic mice has been shown to be associated with neurological syndromes the severity of which correlate with the levels of IL6 expression. These mice are characterized by runting, tremor, neurodegeneration, astrocytosis, angiogenesis, and induction of acute-phase proteins production, suggesting a direct pathogenic role of IL6 in inflammatory, infectious, and neurodegenerative CNS diseases.

The biological activities of IL6 have also been studied in IL6-deficient mice generated by targeted disruption of the IL6 gene in embryonic stem cells (see: ES cells). IL6-deficient mice of both sexes are viable and fertile and do not present any evident phenotypic abnormality. However, analysis of bone metabolism in females demonstrates that these animals have higher rates of bone turnover than control litter mates. In these mice, ovariectomy, which is known to result in increased bone turnover with bone resorption exceeding bone formation in normal animals, does not induce any change in either bone mass or bone remodeling rates. These results suggest that IL6 plays an important role in the local regulation of bone turnover and appears to be essential for bone loss caused by estrogen deficiency.

■ **Assays:** IL6 can be detected in » Bioassays employing IL6-responsive cell lines (see: 7TD1; B9; CESS, KT-3; M1, MH60-BSF2, MO7E; Mono Mac 6; NFS-60; PIL-6; SKW6-Cl4; T1165; XG-1). IL6 can also be assayed by its activity as a

IL6 knockouts: see addendum: IL6.

hybridoma growth factor (see: HGF) Sensitive immunoassays and colorimetric tests are also available. An alternative and entirely different detection method is » Message amplification phenotyping. An ELISA assay exists for detecting the receptor-associated gp130 protein. For further information see also subentry "Assays" in the reference section.

■ **Clinical Use & Significance:** The determination of IL6 serum levels may be useful to monitor the activity of myelomas and to calculate tumor cell masses. Low IL6 serum levels are observed in monoclonal gammopathies and in smouldering myelomas while IL6 serum levels are markedly increased in patients with progressive disease and also in patients with plasma cell leukemia. A blockade of the IL6 receptor or the inhibition of IL6 by monoclonal antibodies may be a way to delay or prevent the maturation of B cells into plasma cells, suggesting also a new therapeutical concept involving cytostatic drugs coupled to monoclonal antibodies.

The deregulated expression of IL6 is probably one of the major factor involved in the pathogenesis of a number of diseases. The excessive overproduction of IL6 (and other B cell differentiation factors; see also: BCDF) has been observed in various pathological conditions such as rheumatoid arthritis, multiple myeloma, Lennert syndrome (histiocytic lymphoma), Castleman's disease (lymphadenopathy with massive infiltration of plasma cells, hyper γ-globulinemia, anemia, and enhanced concentrations of » acute phase proteins), cardiac myxomas and liver cirrhosis. The constitutive synthesis of IL6 by glioblastomas and the secretion of IL6 into the cerebrospinal fluid may explain the elevated levels of » acute phase proteins and immune complexes in the serum.

IL6 probably also plays a role in the pathogenesis of chronic polyarthritis (together with » IL1 and » IL8) since excessive concentrations of IL6 are found in the synovial fluid.

It has been suggested that IL6, due to its effects on hematopoietic cells, may be suitable for the treatment of certain types of anemia and thrombocytopenia. Pretreatment with IL3 and subsequent administration of IL6 has been shown to increase platelet counts considerable. A combination of these two factors could therefore be valuable for the effective control of thromocytopenias and the very severe complications (excessive bleeding) usually associated with this condition. In combination with other cytokines (e. g. » IL2) IL6 may be useful in the treatment of some tumor types.

Very high levels of IL6 in the cerebrospinal fluid are frequently observed in bacterial (up to 500 ng/mL) and viral meningitis (10 – 1000 ng/mL). The detection of elevated concentrations of IL6 in the urine of transplanted patients may be an early indicator of a graft-versus-host reaction. The detection of IL6 in the amniotic fluid may be an indication of intra-amniotic infections. In inflammatory intestinal diseases elevated plasma levels of IL6 may be an indicator of disease status. In patients with mesangioproliferative glomerulonephritis elevated urine levels of IL6 are also an indicator of disease status.

Monitoring of postoperative serum IL6 levels may be more helpful than monitoring of C-reactive protein levels for estimation of inflammatory status and early detection of an » acute-phase response.

The transduction of murine tumor cells with a functional IL6 gene has been shown to lead to the rejection of the genetically modified cells by syngeneic hosts. Altered tumor cells expressing IL4 also increase systemic immunity. Mice vaccinated with transduced cells reject a subsequent challenge of non-transduced cells, and, in some cases, a preexisting tumor (for cancer vaccines see also: Cytokine gene transfer).

Serum and urinary IL6 levels have been shown to be predicting factors of Kawasaki disease activity.

● **Reviews: Akira S et al** Biology of multifunctional cytokines: IL 6 and related molecules (IL 1 and TNF) FJ 4: 2860-7 (1990); **Bauer J & Herrmann F** Interleukin-6 in clinical medicine. Ann. Hematol. 62: 203-10 (1991); **Bowcock AM et al** The molecular biology of β-2-interferon/interleukin 6 (IFN β 2). ANY 557: 345-52 (1989); **Brach MA & Herrmann F** Interleukin 6: presence and future. Int. J. Clin. Lab. Res. 22: 143-51 (1992); **Dullens HFJ et al** Cancer treatment with interleukins 1, 4, and 6 and combinations of cytokines: a review. *In vivo* 5: 567-70 (1991); **Van Snick J** Interleukin 6; an overview. ARI 8: 253-78 (1990); **Kishimoto T** The biology of interleukin-6. Blood 74: 1-10 (1990); **Lee F et al** Interleukin-6: a multifunctional regulator of growth and differentiation. ANY 557: 215-29 (1989); **Patterson PH** The emerging neuropoietic cytokine family: first CDF/LIF, CNTF and IL6; next ONC, MGF, GCSF? Curr. Opin. Neurobiol. 2: 94-7 (1992); **Van Oers MH et al** Interleukin-6, a new target for therapy in multiple myeloma? Ann. Hematol. 66: 219-23 (1993); **Van Snick J** From hybridoma and plasmacytoma growth factors to interleukin 6. in: Del Guerico P & Cruse JM (eds): B lymphocytes: Function and Regulation. Contrib. Microbiol. Immunol. Karger, Basel, Vol. 11, pp. 73-95 (1989); **Wolvekamp MC & Marquet RL** Interleukin-6: historical background, genetics and biological significance. Immunol. Lett. 24: 1-9 (1990)

● **Biochemistry & Molecular Biology: Akira S et al** A nuclear factor for IL6 expression (NF-IL6) is a member of a c/EBP family. EJ 9: 1897-906 (1990); **Arcone R et al** Single-step purification and structural characterization of human interleukin 6 produced in *Escherichia coli* from a T7 RNA polymerase expression vector. EJB 198: 541-7 (1991); **Blankenstein T et al** DNA rearrangement and constitutive expression of the interleukin 6 gene in a mouse plasmacytoma. JEM 171: 965-70 (1990); **Bowcock AM et al** The human "interferon-β 2/hepatocyte stimulating factor/interleukin-6" gene: DNA polymorphism studies and localization to chromosome 7p21. Genomics 3: 8-16 (1988); **Brakenhoff JPJ et al** Molecular cloning and expression of hybridoma growth factor in *Escherichia coli.* JI 139: 4116-21 (1987); **Ferguson-Smith AC et al** Regional localizationof the interferon-β 2/B cell stimulatory factor 2/hepatocyte stimulating factor gene to human chromosome 7p15-p21. Genomics 2: 203-8 (1988); **Grenett HE et al** Cloning and sequence analysis of the cDNA for murine interleukin-6. NAR 18: 6455 (1990); **Grenett HE et al** Isolation and characterization of biologically active murine interleukin-6 produced in *Escherichia coli.* Gene 101: 267-71 (1991); **Guisez Y et al** Production and purification of recombinant human interleukin-6 secreted by the yeast *Saccharomyces cerevisiae.* EJB 198: 217-22 (1991); **Hattori M et al** acute phase reaction induces a specific complex between hepatic nuclear proteins and the interleukin 6 response element of the rat α-2-macroglobulin gene. PNAS 87: 2364-8 (1990); **Hirano T et al** Complementary DNA for novel human interleukin (BSF-2) that induces B lymphocytes to produce immunoglobulin. N 324: 73-6 (1986); **Isshiki H et al** Constitutive and IL1 inducible factors interact with the IL1 responsive element in the IL6 gene. MCB 10: 2757-64 (1990); **Levy Y et al** Interleukin-6 antisense oligonucleotides inhibit the growth of human myeloma cell lines. JCI 88: 696-9 (1991); **Mock BA et al** The murine IL6 gene maps to the proximal region of chromosome 5. JI 142: 1372-6 (1989); **Nakayama K et al** A lymphoid cell-specific nuclear factor containing c-*Rel*-like proteins preferentially interacts with Interleukin 6 kB-related motifs whose activities are repressed in lymphoid cells. MCB 12: 1736-46 (1992); **Natsuka S et al** Macrophage differentiation-specific expression of NF-IL6, a transcription factor for interleukin-6. Blood 79: 460-6 (1992); **Nishimura C et al** Site-specific mutagenesis of human interleukin-6 and its biological activity. FL 281: 167-9 (1991); **Qin Z et al** The interleukin-6 gene locus seems to be a preferred target site for retrotransposon integration. Immunogenetics 33: 260-6 (1991); **Ray A et al** A multiple cytokine-responsive and 2nd messenger-responsive element in the enhancer of the human interleukin-6 gene: similarities with v-*fos* gene regulation. MCB 9: 5537-47 (1989); **Ray A et al** On the mechanism for efficient repression of the interleukin-6 promoter by glucocorticoids: enhancer, TATA box, and RNA start site (lnr motif) occlusion. MCB 10: 5736-46 (1990); **Savino R et al** Saturation mutagenesis of the human interleukin 6 receptor binding site: implications for its three-dimensional structure. PNAS 90: 4067-71 (1993); **Schooltink H et al** Structural and functional studies on the human hepatic interleukin-6 receptor. Molecular cloning and overexpression in HepG2 cells. BJ 277: 659-64 (1991); **Seghal PB et al** Human chromosome 7 carries the β2-interferon gene. PNAS 83: 5219-22 (1986); **Spirer J et al** The interleukin-6-dependent DNA-binding protein gene (transcription factor 5 TCF5) maps to human chromosome 20 and rat chromosome 3, the IL6 receptor locus (IL6R) to human chromosome 1 and rat chromosome 2, and the rat IL6 gene to rat chromosome 4. Genomics 10: 539-46 (1991); **Sutherland GR et al** Interleukin 4 is at 5q31 and interleukin 6 is at 7p15. Hum. Genet. 79: 335-7 (1988); **Tanabe O et al** Genomic structure of the murine IL6 gene. High degree of conservation of potential regulatory sequences between mouse and human. JI 141: 3875-81 (1988); **Tsunasawa S et al** Complementary DNA for a novel human interleukin (BSF-2) that induces B lymphocytes to produce immunoglobulin. N 324: 73-6 (1986); **Van Snick J et al** Purification and NH_2-terminal amino acid sequence of a T cell-derived lymphokine with growth factor activity for B cell hybridomas. PNAS 83: 9679-83 (1986); **Yasukawa K et al** Structure and expression of human B cell stimulatory factor-2 (BSF-2/IL6) gene. EJ 6: 2939-45 (1987); **Zilberstein A et al** Structure and expression of cDNA and genes for human interferon-β2, a distinct species inducible by growth-stimulatory cytokines. EJ 5: 2529-37 (1986); **Szpirer J et al** The interleukin 6-dependent DNA binding protein gene (transcription factor 5: TCF5) maps to human chromosome 20 and rat chromosome 3, the IL 6 receptor locus (IL6R) to human chromosome 1 and rat chromosome 2, and the rat IL6 gene to rat chromosome 4. Genomics 10: 539-46 (1991)

● **Receptor Structure, Gene(s), Expression: Bazan JF** A novel family of growth factor receptors: a common binding domain in the growth hormone, prolactin, erythropoietin, and IL6 receptors, and the p75 IL2 receptor β-chain. BBRC 164: 788-95 (1989); **Chen-Kiang S et al** Nuclear signaling by interleukin-6. Curr. Opin. Immunol. 5: 124-8 (1993); **Kluck PM et al** The human interleukin-6 receptor α chain gene is localized on chromosome 1 band q21. Hum. Genet. 90: 542-4 (1993); **Yamasaki K et al** Cloning and expression of the human interleukin-6 (BSF-2/IFNβ2) receptor. S 241: 825-8 (1988)

● **Biological Activities: Bergui L et al** Interleukin 3 and interleukin 6 synergistically promote the proliferation and differentiation of malignant plasma cell precursors in multiple myeloma. JEM 170: 613-9 (1989); **Brandt SJ et al** Dysregulated interleukin 6 expression produces a syndrome resembling Castleman's disease in mice. JCI 86: 592-9 (1990); **Eustace D et al** Interleukin-6 (IL6) functions as an autocrine growth factor in cervical carcinomas *in vitro*. Gynecol. Oncol. 50: 15-9 (1993); **Hata H et al** Interleukin-6 gene expression in multiple myeloma: a characteristic of immature tumor cells. Blood 81: 3357-64 (1993); **Hawley RG et al** Transplantable myeloproliferative disease induced in mice by an interleukin 6 retrovirus. JEM 176: 1149-63 (1992); **Hermann F et al** Interleukin-4 inhibits growth of multiple myelomas by suppressing interleukin-6 expression. Blood 78: 2070-4 (1991); **Hermus ARMM & Sweep CGJ** Cytokines and the hypothalamic-pituitary-adrenal axis. JSBMB 37: 867-71 (1990); **Heinrich PC et al** Interleukin-6 and the acute phase response. BJ 269, Suppl. 51-66 (1990); **Helle M et al** Interleukin-6 is involved in interleukin 1-induced activities. EJI 18: 957-9 (1988); **Helle M et al** Differential induction of interleukin-6 production by monocytes, endothelial cells and smooth muscle cells. PCBR 367: 61-71 (1991); **Hirano T** Interleukin-6 and its relation to inflammation and disease. Clin. Immunol. Immunopathol. 62: S60-S65 (1992); **Horii Y et al** Involvement of interleukin-6 in mesangial proliferative glomerulonephritis. JI 143: 3949-55 (1989); **Ishibashi T et al** Human interleukin 6 is a direct promoter for maturation of megakaryocytes *in vitro*. PNAS 86: 5953-7 (1989); **Jumming L & Vilcek J** Interleukin 6: a multifunctional cytokine regulating immune reactions and the acute phase protein response. Lab. Invest. 61: 588-602 (1988); **Mullberg J et al** The soluble interleukin-6 receptor is generated by shedding. EJI 23: 473-80 (1993); **Nachbaur DM et al** Serum levels of interleukin-6 in multiple myeloma and other hematological disorders: correlation with disease activity and other prognostic parameters. Ann. Hematol. 62: 54-8 (1991); **Oldenburg HS et al** Cachexia and the acute-phase protein response in inflammation are regulated by interleukin-6. EJI 23: 1889-94 (1993); **Roodman GD et al** Interleukin 6. A potential autocrine/paracrine factor in Paget's disease of bone. JCI 89: 46-52 (1992); **Ryffel B et al** Pathology induced by interleukin-6. Int. Rev. Exp. Pathol. 1993; 34: 79-89 (1993); **Satoh T et al** Induction of neuronal differentiation in PC12 cells by B cell stimulatory factor 2/interleukin 6. MCB 8:

3546-9 (1988); **Screpanti I et al** Neuromodulatory loop mediated by nerve growth factor and interleukin 6 in thymic stromal cell cultures. PNAS 89: 3209-12 (1992); **Sehgal PB** Interleukin 6: molecular pathophysiology. J. Invest. Dermatol. 94: 2S-6S (1990); **Straneva JE et al** Is interleukin 6 the physiological regulator of thrombopoiesis? EH 20: 47-50 (1992); **Takizawa H et al** Growth inhibition of human lung cancer cell lines by interleukin 6 *in vitro*: a possible role in tumor growth via an autocrine mechanism. CR 15; 53: 4175-81 (1993); **Van Meir E et al** Human glioblastoma cells release interleukin 6 *in vivo* and *in vitro*. CR 50: 6683-8 (1990); **Zhang XG et al** Interleukin 6 is a potent myeloma cell growth factor in patients with aggressive multiple myeloma. Blood 74: 11-3 (1989)

● TRANSGENIC/KNOCK-OUT/ANTISENSE STUDIES: **Campbell IL et al** Neurologic disease induced in transgenic mice by cerebral overexpression of interleukin 6. PNAS 90: 10061-5 (1993); **Poli V et al** Interleukin-6 deficient mice are protected from bone loss caused by estrogen depletion. EJ 13: 1189-96 (1994); **Suematsu S et al** IgG1 plasmacytosis in IL6 transgenic mice. PNAS 86: 7547-51 (1989); **Suematsu S et al** Generation of plasmacytomas with the chromosomal translocation t(12;15) in interleukin 6 transgenic mice. PNAS 89: 232-5 (1992); **Turksen K et al** Interleukin 6: insights to its function in skin by overexpression in transgenic mice. PNAS 89: 5069-72 (1992); **Woodroofe C et al** Long-term consequences of interleukin-6 overexpression in transgenic mice. DNA Cell Biol. 11: 587-92 (1992)

● ASSAYS: **Anderson J et al** Effects of acetate dialysate on transforming growth factor β 1, interleukin, and β 2-microglobulin plasma levels. Kidney Int. 40: 1110-7 (1991); **De Groote D et al** Novel method for the measurement of cytokine production by a one-stage procedure. JIM 163: 259-67 (1993); **Guba SC et al** Bone marrow stromal fibroblasts secrete interleukin-6 and granulocyte-macrophage colony-stimulating factor in the absence of inflammatory stimulation: demonstration by serum-free bioassay, enzyme-linked immunosorbent assay, and reverse transcriptase polymerase chain reaction. Blood 80: 1190-8 (1992); **Helle M et al** Sensitive ELISA for interleukin-6. Detection of IL6 in biological fluids: synovial fluids and sera. JIM 138: 47-56 (1991); **Ida N et al** An enzyme-linked immunosorbent assay for the measurement of human interleukin 6. JIM 133: 279-84 (1990); **Jones KP et al** Measurement of interleukin-6 in bronchoalveolar lavage fluid by radioimmunoassay: differences between patients with interstitial lung disease and control subjects. Clin. Exp. Immunol. 83: 30-4 (1991); **Lebedin YS et al** Serum levels of interleukin 4, interleukin 6 and interferon-γ following *in vivo* isotype-specific activation of IgE synthesis in humans. Int. Arch. Allergy Appl. Immunol. 96: 92-4 (1991); **Lewis CE et al** Measurement of cytokine release by human cells. A quantitative analysis at the single cell level using the reverse hemolytic plaque assay. JIM 127: 51-9 (1990); **Ling ZD et al** Particle concentration fluorescence immunoassay for measuring interleukin-6 receptor numbers. Cytokine 3: 17-20 (1991); **Lu ZY et al** High amounts of circulating interleukin (IL)-6 in the form of monomeric immune complexes during anti-IL6 therapy. Towards a new methodology for measuring overall cytokine production in human *in vivo*. EJI 22: 2819-24 (1992); **Merville P et al** Detection of single cells secreting IFN-γ, IL6, and IL10 in irreversibly rejected human kidney allografts, and their modulation by IL2 and IL4. Transplantation 55: 639-46 (1993); **Novick D et al** Monoclonal antibodies to the soluble human IL6 receptor: affinity purification, ELISA, and inhibition of ligand binding. Hybridoma 10: 137-46 (1991); **Ogata A et al** A new highly sensitive immunoassay for cytokines by dissociation-enhanced lanthanide fluoroimmunoassay (DELFIA). JIM 148: 15-22 (1992); **Ohno M & Abe T** Rapid colorimetric assay for the quantification of leukemia inhibitory factor (LIF) and interleukin 6 (IL6). JIM 145: 199-203 (1991); **Peppard JY et al** A simple and rapid radioimmunoassay for human interleukin-6. JIM 148: 23-8 (1992); **Rafferty B et al** Measurement of cytokine production by the monocytic cell line Mono Mac 6 using novel immunoradiometric assays for interleukin-1 β and interleukin-6. JIM 144: 69-76 (1991); **Sawamura M et al** Characterization of 5B12.1, a monoclonal antibody specific for IL6. GF 3: 181-90 (1990); **Shirai A et al** Detection and quantitation of cells secreting IL6 under physiologic conditions in BALB/c mice. JI 150: 793-9 (1993); **Solary E et al** Radioimmunoassay for the measurement of serum IL6 and its correlation with tumor cell mass parameters in multiple myeloma. Am. J. Hematol. 39: 163-7 (1992); **Suzuki A et al** Estimation of interleukin 6 production by reverse transcriptase-polymerase chain reaction in four human myeloma cell lines. Leuk. Res. 15: 1043-50 (1991); **Taktak YS et al** Assay of pyrogens by interleukin-6 release from monocytic cell lines. J. Pharm. Pharmacol. 43: 578-82 (1991); **Teppo AM et al** Radioimmunoassay of interleukin-6 in plasma. Clin. Chem. 37: 1691-5 (1991); **Thavasu PW et al** Measuring cytokine levels in blood. Importance of anticoagulants, processing, and storage conditions. JIM 153: 115-24 (1992); **Van Damme J & Van Snick J** Induction of hybridoma growth factor (HGF) identical to IL6 in human fibroblasts by IL1: use of HGF activity in specific and sensitive biological assays for IL1 and IL6. Dev. Biol. Stand. 69: 31-8 (1988); **Walz G et al** The role of interleukin-6 in mitogenic T cell activation: detection of interleukin-2 heteronuclear RNA by polymerase chain reaction. CI 134: 511-9 (1991); **Yanagawa H et al** Interleukin-4 downregulates interleukin-6 production by human alveolar macrophages at protein and mRNA levels. Microbiol. Immunol. 35: 879-93 (1991); **Zheng RQ et al** Detection of interleukin-6 and interleukin-1 production in human thyroid epithelial cells by non-radioactive *in situ* hybridization and immunohistochemical methods. Clin. Exp. Immunol. 83: 314-9 (1991); for further information also see individual cell lines used in individual bioassays.

● CLINICAL USE & SIGNIFICANCE: **Asano S et al** *In vivo* effects of recombinant human interleukin-6 in primates: stimulated production of platelets. Blood 75: 1602-5 (1990); **Chelstrom LM et al** Treatment of BCL-1 murine B cell leukemia with recombinant cytokines. Comparative analysis of the anti-leukemic potential of interleukin 1 β (IL1 β), interleukin 2 (IL2), interleukin-6 (IL6), tumor necrosis factor α (TNF α), granulocyte colony stimulating factor (G-CSF), granulocyte-macrophage colony stimulating factor (GM-CSF), and their combination. Leuk. Lymphoma 7: 79-86 (1992); **Chow YM et al** Serum and urinary interleukin-6 (IL6) levels as predicting factors of Kawasaki disease activity. Acta Paediatr. Sin. 34: 77-83 (1993); **Dullens HFJ et al** Cancer treatment with interleukins 1, 4, and 6 and combinations of cytokines: a review. *In vivo* 5: 567-70 (1991); **Eisenthal A et al** Antitumor effects of recombinant interleukin-6 expressed in eukaryotic cells. Cancer. Immunol. Immunother. 36: 101-7 (1993); **Furukawa S et al** Kawasaki disease differs from anaphylactoid purpurea and measles with regard to tumor necrosis factor-α and interleukin 6 in serum. Eur. J. Pediatr. 151: 44-7 (1992); **Givon T et al** Antitumor effects of human recombinant interleukin-6 on acute myeloid leukemia in mice and in cell cultures. Blood 79: 2392-8 (1992); **Hsu SM et al** Expression of interleukin-6 in Castleman's disease. Hum. Pathol. 24: 833-9 (1993); **Jablons DM et al** IL6/IFN-β-2 as a circulating hormone: induction by cytokine administration in humans. JI 142: 1542-7 (1989); **Leger-Ravet MB et al** Interleukin-6 gene expression in Castleman's disease. Blood 78: 2923-30 (1991); **Martinez-Maza O & Berek J** Interleukin 6 and cancer treatment. *In vivo* 5: 583-8 (1991); **Mule JJ et al** Clinical application of IL6 in cancer therapy. Res. Immunol. 143: 777-9 (1992);

Ohzato H et al Interleukin-6 as a new indicator of inflammatory status: detection of serum levels of interleukin-6 and C-reactive protein after surgery. Surgery 111: 201-9 (1992); **Porgador A et al** Interleukin 6 gene transfection into Lewis lung carcinoma tumor cells suppresses the malignant phenotype and confers immunotherapeutic competence against parental metastatic cells. CR 35: 218-34 (1992); **Weber J et al** Phase I trial of subcutaneous interleukin-6 in patients with advanced malignancies. J. Clin. Oncol. 11: 499-506 (1993); **Yoshizaki K et al** Pathological significance of interleukin 6 (IL6/BSF-2) in Castleman's disease. Blood 74: 1360-7 (1989)

IL6DBP: *Interleukin 6-dependent binding protein* This transcriptions factor is identical with » NF-IL6. See also: IL6RE.

IL6RE: *IL6 responsive element* A short nucleotide sequence (CTGGGA) found in the promoter region of genes the expression of which is modulated by » IL6. It is the binding protein for nuclear proteins (see: NF-IL6 = IL6DBP) that function as transcription factors and control the expression of these genes in response to an IL6 stimulus. Such IL6-specific elements have been found in a number of genes encoding » acute phase proteins (e. g. » haptoglobin, hemopexin, C-reactive protein; see also: acute phase reaction).

Akira S et al A nuclear factor for IL6 expression (NF-IL6) is a member of a c/EBP family. EJ 9: 1897-906 (1990); **Descombes P et al** LAP, a novel member of the C/EBP gene family, encodes a liver-enriched transcriptional activator protein. Genes Dev. 4: 1541-51 (1990); **Hattori M et al** acute phase reaction induces a specific complex between hepatic nuclear proteins and the interleukin 6 response element of the rat α-2-macroglobulin gene. PNAS 87: 2364-8 (1990); **Natsuka S et al** Macrophage differentiation-specific expression of NF-IL6, a transcription factor for interleukin-6. Blood 79: 460-6 (1992); **Poli V et al** IL6DBP, a nuclear protein involved in interleukin-6 signal transduction, defines a new family of leucine zipper proteins related to C/EBP. Cell 63: 643-653 (1990)

IL6R-SUP: *Interleukin 6 receptor soluble urinary protein* A urinary protein isolated by affinity chromatography of crude human urinary proteins on recombinant IL6 columns. This protein is identical with as a truncated 50 kDa soluble form of the 80 kDa » IL6 cellular receptor.

IL6-R-SUP enhances the growth stimulation of mouse plasmacytoma » T1165 by suboptimal concentrations of human recombinant IL6. The growth-inhibitory effect of IL6 on human breast carcinoma cells is enhanced by addition of IL6-R-SUP although these cells possess abundant IL6 receptors. With IL6-R-SUP, complete growth inhibition by IL6 is achieved and the cells become more sensitive to low levels of IL6. These effects are prevented by a monoclonal antibody against IL6-R-SUP which blocks IL6 binding to cells. The naturally occurring IL6-R-SUP may help to increase the growth-regulatory action of IL6.

Novick D et al Soluble cytokine receptors are present in normal human urine. JEM 170: 1409-14 (1989); **Novick D et al** Purification of soluble cytokine receptors from normal human urine by ligand-affinity and immunoaffinity chromatography. J. Chromatogr. 510: 331-7 (1990); **Novick D et al** Enhancement of interleukin 6 cytostatic effect on human breast carcinoma cells by soluble IL6 receptor from urine and reversion by monoclonal antibody. Cytokine 4: 6-11 (1992)

IL7: Interleukin 7.

■ **ALTERNATIVE NAMES:** B cell precursor growth-promoting activity; Lpo (lymphopoietin-1); LP-1 (lymphopoietin 1); PBGF (pre B cell growth factor); THCGF (thymocyte growth factor). See also: individual entries for further information.

■ **SOURCES:** IL7 is secreted constitutively by adherent bone marrow stromal cells and thymic cells. Murine and human keratinocytes have also been shown to express and secrete IL7.

■ **PROTEIN CHARACTERISTICS:** Murine IL7 is a glycoprotein of 25 kDa that contains six cysteine residues. It is derived from a precursor protein containing a classical secretory » signal sequence of 25 amino acids. The disulfide bonds are essential for the biological activity of the protein.

Human (152 amino acids; 17.4 kDa) and murine IL7 (129 amino acids) show 60 % sequence homology at the protein level. The human protein is 17 amino acids longer than the murine protein since the human gene contains an additional exon.

■ **GENE STRUCTURE:** The human IL7 gene has a length of ≈ 33 kb and contains six exons. The gene maps to chromosome 8q12-q13. The murine IL7 gene maps to chromosome 3. It has a length of ≈ 56 kb and contains five exons.

■ **RECEPTOR STRUCTURE, GENE(S), EXPRESSION:** The human IL7 receptor is an integral strongly glycosylated membrane proteins of 76 kDa expressed on activated T cells. This receptor has been designated as ***CDw127*** (see also: CD Antigens). In addition there is a second (unrelated) receptor form of 90 kDa that is also expressed on unstimulated human T cells. The receptor gene encoding the p90 receptor molecules maps to chromosome 5p13. The ability of peripheral blood mononuclear cells to proliferate in response to IL7 correlates with the expression of the 76 kDa receptor, and not with the expression of the 90 kDa form. The IL7 receptor may contain other components in addition to the p90 and p70 IL7 binding species.

IL7 receptors are expressed on pre-B cells and their progenitors. They are not expressed on mature B cells. IL7 receptors are also expressed on bone marrow macrophages. Functional IL7 receptors are found on the cell surface of multiphenotypic, biphenotypic, and immature lymphoid progenitors of B cells with the gene arrangement of the heavy immunoglobulin chain such as those observed in the germ line. Cells expressing the heavy chain of immunoglobulin or cells that have undergone gene arrangements do not express the receptor.
A soluble IL7 receptor form has also been described. Such soluble receptors and binding proteins have also been found for » IL1 (see: IL1ra, IL1 receptor antagonist), » IL2, » IL4, » IL6, » IGF, » TNF-α and » IFN-γ. They probably function as physiological regulators of cytokine activities (see also: Cytokine inhibitors) by inhibiting receptor binding or act as transport proteins.
After binding of IL7 to its receptor the receptor associates with a non-receptor protein kinase called » *fyn*, which appears to mediate some of the intracellular signals. One signaling component of the IL7 receptor has been shown recently to be identical with the γ subunit of the » IL2 receptor. This subunit is also a component of the » IL4 receptor and is probably also shared with the » IL13 receptor.
■ **BIOLOGICAL ACTIVITIES:** The activity of human IL7 is not species-specific. IL7 stimulates the proliferation of pre-B and pro-B cells without affecting their differentiation. IL7 can replace murine bone marrow stromal cells in supporting the extended growth of both pre-B cells and pro-B cells (see also: LTBMC, long-term bone marrow culture). The protein does not act on mature B cells. The IL7-dependent proliferation of B cell precursors is inhibited by » TGF-β. IL7 also selectively supports the maturation of megakaryocytes. In normal pre-B cells IL7 induces the expression of the » *myc* oncogene.
Some of the effects of IL7 on natural killer cells appear to be mediated by » TNF-α since the effects of IL7 are inhibited by an antibody directed against TNF-α.
IL7 stimulates the proliferation of early and mature activated T cells and this activity is synergised by suboptimal doses of » IL1. Unstimulated human T cells, which also express high affinity IL7 receptors, do not proliferate in response to IL7. IL7 stimulates the proliferation of thymocytes with the markers $CD4^-8^-$, $CD4^+8^-$, $CD4^-8^+$ and is therefore an important differentiation factor for functionally different subpopulation of T lymphocytes. Thymocytes proliferate in response to IL7 independently of three other known T cell growth factors, » IL2, » IL4, and » IL7.
In human peripheral monocytes IL7 induces the synthesis of some inflammatory mediators such as » IL1, » IL6 and » MIP (macrophage inflammatory protein) (see also: Inflammation). IL7 also enhances the expression and secretion of » IL3 and » GM-CSF in activated human T cells. IL7 downregulates expression of TGF-β in macrophages which has been suggested as an inhibitor of the antitumor immune response.
IL7 has recently been found to sustain the expression of genes known to control rearrangement of the T cell receptor β gene and to be a cofactor for such rearrangements during early T cell development in mice.
■ **TRANSGENIC/KNOCK-OUT/ANTISENSE STUDIES:** The role of IL7 in the development of the lymphoid system has been studied in transgenic animals (mice) carrying an IL7 cDNA fused to an immunoglobulin heavy chain promoter and enhancer. The transgene is expressed in the bone marrow, lymph nodes, spleen, thymus, and skin. The expression of the transgene leads to a perturbation of T cell development characterized by a marked reduction of $CD4^+$ $CD8^+$ (double-positive) thymocytes. The animals also develop a progressive cutaneous disorder involving a dermal lymphoid infiltrate that results in progressive alopecia, hyperkeratosis, and exfoliation. The expression of the transgene also provokes the development of a lymphoproliferative disorder that induces B and T cell lymphomas within the first 4 months of life, demonstrating that IL7 can act as an » oncogene in the living organism.
A high incidence of severe lymphoproliferative disease has been observed also in a newly generated strain of transgenic mice expressing murine IL7 under the control of the E α (MHC class II) promoter. These animals are characterized by the selective expansion of cells at an early stage of B cell development and expansion of cells phenotypically identical to bipotent (B/macrophage) stem cell populations found in midgestation embryonic liver. These cells are oligoclonal, or in rare cases monoclonal, and include clones of cells with unrearranged Ig heavy chain loci.
■ **ASSAYS:** IL7 can be assayed by its growth-promoting activity on pre-B cells in long-term Whitlock-Witte bone marrow cultures (see also:

LTBMC, long-term bone marrow culture). It can also be assayed in » Bioassays employing cell lines that respond to the factor (see: 1xN/2b; 2E8; CT6; DW34; MH11; Nb2). An alternative and entirely different detection method is » Message amplification phenotyping. For further information see also subentry "Assays" in the reference section.

■ **CLINICAL USE & SIGNIFICANCE:** IL7 may be of clinical significance for » adoptive immunotherapy since it is capable *in vivo* to cause the $CD4^+$-T cell-dependent destruction of tumor cells (see also: LAK cells, lymphokine activated killer cells; see also: Adoptive immunotherapy). A comparison of the IL7- and IL2-induced activities of cytotoxic T lymphocytes shows similar activities on lung metastases of murine sarcomas. IL7 has also been shown to induce LAK cell activity comparable quantitatively to that induced by IL2 in cells obtained from patients early after autologous or syngeneic bone marrow transplantation. It induces an even greater LAK cell activity *in vitro* in peripheral blood mononuclear cells obtained after autologous bone marrow transplantation and preactivated *in vivo* by IL2 therapy. It has therefore been suggested that IL7, alone or in combination with IL2, may be used as a consolidative immunotherapy for malignancies in patients after autologous bone marrow transplantation.

The administration of human recombinant IL7 to normal and irradiated mice increases the numbers of lymphocytes and some immature cells of the myeloid lineage.

The transduction of murine tumor cells with a functional IL7 gene has been shown to lead to the rejection of the genetically modified cells by syngeneic hosts (for cancer vaccines see also: Cytokine gene transfer).

A participation of IL7 in the pathogenesis of inflammatory skin diseases and cutaneous T cell lymphomas is suggested by the growth-promoting effects of IL7 and its synthesis by keratinocytes.

● **REVIEWS: Park LS et al** The role of IL7 and its receptor in B cell ontogeny. AEMB 323: 125-9 (1992); **Melchers F et al** B lymphocyte lineage-committed, IL7 and stroma cell-reactive progenitors and precursors, and their differentiation to B cells. AEMB 323: 111-7 (1992)

● **BIOCHEMISTRY & MOLECULAR BIOLOGY: Brunton LL et al** An STS in the human IL7 gene located at 18q12-13. NAR 18: 1315 (1990); **Goodwin RG et al** Cloning of the human and murine interleukin-7 receptors: demonstration of a soluble form and homology to a new receptor superfamily. Cell 60: 941-51 (1990); **Goodwin RG et al** Human interleukin 7: molecular cloning and growth factor activity on human and murine B lineage cells. PNAS 86: 302-6 (1989); **Lupton SD et al** Characterization of the human and murine IL7 genes. JI 144: 3592-601 (1990); **Sutherland GR et al** The gene for human interleukin 7 (IL7 is at 8q12-13. Hum. Genet. 82: 371-2 (1989)

● **RECEPTOR STRUCTURE, GENE(S), EXPRESSION: Armitage RJ et al** Identification of a novel low-affinity receptor for human interleukin 7. Blood 79: 1738-45 (1992); **Goodwin RG et al** Cloning of the human and murine interleukin-7 receptors: demonstration of a soluble form and homology to a new receptor superfamily. Cell 60: 941-51 (1990); **Lynch M et al** The interleukin-7 receptor gene is at 5p13 Hum. Genet. 89: 566-8 (1992); **Noguchi M et al** Interleukin-2 receptor γ chain: a functional component of the interleukin-7 receptor. S 262: 1877-80 (1993); **Page TH et al** Characterization of a novel high affinity human IL7 receptor. JI 151: 4753-63 (1993); **Park LS et al** Murine interleukin 7 (IL7) receptor. Characterization of an IL7-dependent cell line. JEM 171: 1073-89 (1990); **Pleiman CM et al** Organization of the murine and human interleukin-7 receptor genes: two mRNAs generated by differential splicing and presence of a type I-interferon-inducible promoter. MCB 11: 3052-9 (1991); **Sutherland GR et al** The gene for human interleukin 7 (IL7) is at 8q12-13. Hum. Genet. 82: 371-2 (1989); **Welch PA et al** Human IL7: a novel T cell growth factor. JI 143: 3562-7 (1989)

● **BIOLOGICAL ACTIVITIES: Alderson MR et al** Interleukin-7 enhances cytolytic T lymphocyte generation and induces lymphokine activated killer cells from human peripheral blood. JEM 172: 577-87 (1990); **Armitage RJ et al** Regulation of human T cell proliferation by IL7. JI 144: 938-41 (1990); **Chazen CD et al** Interleukin 7 is a T cell growth factor. PNAS 86: 5923-7 (1989); **Conlon PJ et al** Murine thymocytes proliferate in direct response to interleukin-7. Blood 74: 1368-73 (1989); **Dokter WH et al** IL7 enhances the expression of IL3 and granulocyte-macrophage-CSF mRNA in activated human T cells by post-transcriptional mechanisms. JI 150: 2584-90 (1993); **Dubinett SM et al** Down-regulation of macrophage transforming growth factor-b messenger RNA expression by IL7. JI 151: 6670-80 (1993); **Grabstein KH et al** Regulation of T cell proliferation by IL7. JI 140: 3015-20 (1990); **Henney CS** Interleukin 7: effects on early events in lymphopoiesis. IT 10: 170-3 (1989); **Heufler C et al** Interleukin 7 is produced by murine and human keratinocytes. JEM 178: 1109-14 (1993); **Joshi PC & Choi YS** Human interleukin 7 is a B cell growth factor for activated B cells. EJI 21: 681-6 (1991); **Lee G et al** Normal B cell precursors responsive to recombinant murine IL7 and inhibition of IL7 activity by transforming growth factor-β. JI 142: 3875-83 (1989); **Morrissey PJ et al** Recombinant interleukin 7, pre-B cell growth factor, has costimulatory activity on purified mature T cells. JEM 169: 707-16 (1989); **Morrow MA et al** Interleukin-7 induces N-*myc* and c-*myc* expression in normal precursor B lymphocytes. GD 6: 61-70 (1992); **Muegge K et al** Interleukin 7: a cofactor for V(D)J rearrangement of the T cell receptor β gene. S 261: 93-5 (1993); **Namen AE et al** Stimulation of B cell progenitors by cloned murine interleukin-7. N 333: 571-3 (1988); **Namen AE et al** B cell precursor growth-promoting activity. JEM 167: 988-1002 (1988); **Naume B et al** A comparative study of IL12 (cytotoxic lymphocyte maturation factor), IL2, and IL7-induced effects on immunomagnetically purified $CD56^+$ NK cells. JI 148: 2429-36 (1992); **Naume B et al** Gene expression and secretion of cytokines and cytokine receptors from highly purified $CD56^+$ natural killer cells stimulated with interleukin-2, interleukin-7 and interleukin-12. EJI 23: 1831-8 (1993); **Saeland S et al** Interleukin-7 induces the proliferation of normal human B cell precursors. Blood 78: 2229-38 (1991); **Sakata T et al** Constitutive expression of interleukin-7 mRNA and production of IL7 by a cloned murine thymic stromal cell line. J. Leukocyte Biol. 48: 205-12 (1990); **Skjonsberg C et al** Interleukin 7 differentiates a subgroup of acute lymphoblastic leukaemias. Blood 77: 2445-50 (1991); **Tushinski RJ et al** The effects of interleukin 7 (IL7) on human bone marrow *in vitro*. EH 19: 749-54 (1991); **Watson JD et al** Effect of IL7 on the growth of fetal thymocytes in culture. JI 143: 1215-22 (1989); **Welch PA et al** Human IL7: a novel T cell growth factor. JI 143: 3562-7 (1989)

● **Transgenic/Knock-out/Antisense Studies: Fisher AG et al** Lymphoproliferative disorders in an IL7 transgenic mouse line. Leukemia 2: S66-8 (1993); **Rich BE et al** Cutaneous lymphoproliferation and lymphomas in interleukin 7 transgenic mice. JEM 177: 305-16 (1993); **Samaridis J et al** Development of lymphocytes in interleukin 7-transgenic mice. EJI 21: 453-60 (1991)

● **Assays: Dusch A et al** A comparison of assays for the response of primary human T cells upon stimulation with interleukin-2, interleukin-4 and interleukin-7. ECN 3: 97-102 (1992); **Namen AE et al** B cell precursor growth-promoting activity: purification and characterization of a growth factor active on lymphocyte precursors. JEM 167: 988-1002 (1988); for further information also see individual cell lines used in individual bioassays.

● **Clinical Use & Significance: Aoki T et al** Expression of murine interleukin 7 in a murine glioma cell line results in reduced tumorigenicity *in vivo*. PNAS 89: 3850-54 (1992); **Faltynek CR et al** Administration of human recombinant IL7 to normal and irradiated mice increases the numbers of lymphocytes and some immature cells of the myeloid lineage. JI 149: 1276-82 (1992); **Hock H et al** Interleukin 7 induces $CD4^+$ T cell-dependent tumor rejection. JEM 174: 1291-8 (1991); **Jicha DL et al** Interleukin 7 generates antitumor cytotoxic T lymphocytes against murine sarcomas with efficacy in cellular adoptive immunotherapy. JEM 174: 1511-5 (1991); **McBride WH et al** Genetic modification of a murine fibrosarcoma to produce IL7 stimulates host cell infiltration and tumor immunity. CR 52: 3931-7 (1992); **Pavletic Z et al** Induction by interleukin 7 of lymphokine-activated killer activity in lymphocytes from autologous and syngeneic marrow transplant recipients before and after systemic interleukin 2 therapy. EH 21: 1371-8 (1993); **Stotter H et al** IL7 induces human lymphokine activated killer cell activity and is regulated by IL4. JI 146: 150-5 (1991)

IL8: Interleukin 8.

■ **Alternative Names:** 3-10C; ANAP (anionic neutrophil-activating peptide); Chemotaxin; EDNAP (endothelial-derived neutrophil-activating peptide); ENAP (endothelial cell neutrophil activating peptide); FDNAP (fibroblast-derived neutrophil-activating peptide); FINAP (fibroblast-derived neutrophil-activating protein); GCF (granulocyte chemotactic factor); GCP (granulocyte chemotactic peptide); LAI (leukocyte adhesion inhibitor); LCF (lymphocyte chemotactic factor); LDNAP (leukocyte-derived neutrophil-activating peptide); LIF (leukocyte inhibitory factor); LYNAP (lymphocyte-derived neutrophil-activating peptide); MDNAP (monocyte-derived neutrophil-activating peptide); MDNCF (monocyte-derived neutrophil chemotactic factor); MOC (monocyte-derived chemotaxin); MONAP (monocyte-derived neutrophil-activating peptide); NAF (neutrophil-activating factor); NAP-1 (neutrophil activating peptide-1); NCF (neutrophil chemotactic factor); NCP (neutrophil chemotactic protein); PLF (psoriatic leukotactic factor); TCF (T cell chemotactic factor); TSG-1 (tumor necrosis factor stimulated gene; see: TSG genes). See also: individual entries for further information.

■ **Sources:** IL8 is produced by stimulated monocytes but not by tissue macrophages and T lymphocytes. IL8 is also produced by macrophages, fibroblasts, endothelial cells, keratinocytes, melanocytes, hepatocytes, chondrocytes, and a number of tumor cell lines.

In many cell types the synthesis of IL8 is strongly stimulated by » IL1 and » TNF-α. In human skin fibroblasts the expression of IL8 is enhanced by » leukoregulin. The interferon » IFN-γ can function as a costimulator. The synthesis of IL8 is also induced by phytohemagglutinins, concanavalin A, double-stranded RNA, » phorbol ester, sodium urate crystals, viruses, and bacterial lipopolysaccharides. The expression of IL8 from resting and stimulated human blood monocytes is up-regulated by » IL7. In chondrocytes the synthesis of IL8 is stimulated by IL1β (see: IL1), » TNF-α and lipopolysaccharides. In human astrocytes the synthesis and secretion of IL8 is induced by » IL1 and » TNF-α. Glucocorticoids, » IL4, » TGF-β, inhibitors of 5´ lipoxygenase, and 1,25$(OH)_2$ vitamin D_3 inhibit the synthesis of IL8. IL8 is constitutively and commonly produced by various carcinoma cell lines and this synthesis may be related to the elevation of serum IL8 in patients with hepatocellular carcinoma.

■ **Protein Characteristics:** IL8 is a non-glycosylated protein of 8 kDa (72 amino acids). It is produced by processing of a precursor protein of 99 amino acids. Processing of this precursor by specific proteases yields N-terminal variants of IL8. ***AVLPR-IL8*** is a truncated IL8 variant lacking the first five amino acids. This shortened protein is identical with » FDNCF (fibroblast-derived neutrophil chemotactic factor). Longer forms of IL8 (79 and 77 amino acids) and shorter forms (69 amino acids) have also been isolated from culture supernatants of LPS-stimulated lymphocytes (see: MDNCF, monocyte-derived neutrophil chemotactic factor and FINAP, fibroblast-derived neutrophil activating proteins), IL1- or TNF-stimulated fibroblasts (see: NAP-1, neutrophil activating protein), and polyI: C-stimulated endothelial cells (see: LAI, leukocyte adhesion inhibitor). The predominant form of IL8 produced by endothelial cells (and also

by anchorage-dependent cells and human glioblastoma cells) is the 77 amino acid variant.
The IL8 protein contains four cysteine residues participating in disulfide bridges (Cys-7/Cys-34; Cys-9/Cys-50).

■ **GENE STRUCTURE:** The human IL8 gene has a length of 5.1 kb and contains four exons. It maps to human chromosome 4q12-q21. The mRNA consists of a 101 base 5´ untranslated region, an open reading frame of 297 bases, and a long 3´ untranslated region of 1.2 kb. The 5' flanking region of the IL8 gene contains potential binding sites for several nuclear factors including activation factor-1, activation factor-2, IFN regulatory factor-1, hepatocyte nuclear factor-1, a glucocorticoid responsive element, and a heat shock element.

■ **RELATED FACTORS:** The proteins » *mig* (monokine induced by γ interferon), » PF4 (platelet factor 4), » MGSA (melanoma growth stimulatory activity), and a number of several other factors belonging to the » chemokine family of cytokines are related to IL8.

■ **RECEPTOR STRUCTURE, GENE(S), EXPRESSION:** The IL8 receptor is a dimeric glycoprotein consisting of a 59 kDa and a 67 kDa subunit. It has been given the name ***CDw128*** (see also: CD Antigens). It is expressed in many different cell types including those not responding to IL8. The receptor density is ≈ 20000/cell in neutrophils and ≈ 300/cell in T lymphocytes.
The IL8 receptor is a member of a family of G protein-coupled receptors. There are at least two different IL8 receptor types. The type I receptor specifically binds IL8 (Kd = 0.8-4 nM). The type II receptor (Kd for IL8 = 0.3-2 nM) also binds the IL8-related factors » MGSA (melanoma growth stimulating activity), GRO (see: MGSA), MIP-2 (see: MIP, macrophage inflammatory protein), and » NAP-2 (neutrophil activating peptide 2). Both receptor genes map to human chromosome 2q35. For a protein highly homologous to the IL8 receptors see: EBI 1 (EBV-induced gene 1) and ECRF3. The MGSA/Gro/IL8/MIP receptor has been identified as CKR-1, which is identical with the Duffy blood group antigen that appears to be a receptor for the malarial parasite *Plasmodium vivax*..

■ **BIOLOGICAL ACTIVITIES:** The activities of IL8 are not species-specific. Human IL8 is also active in rodent and rabbit cells. The biological activities of IL8 resemble those of a related protein, » NAP-2 (neutrophil-activating peptide 2).
IL8 differs from all other cytokines in its ability to specifically activate neutrophil granulocytes. In neutrophils IL8 causes a transient increase in cytosolic calcium levels (see also: Calcium ionophore) and the release of enzymes from granules. IL8 also enhances the metabolism of reactive oxygen species and increases chemotaxis and the enhanced expression of adhesion molecules. A pre-activation by » IL3 is required to render basophils and neutrophils susceptible to further activation by IL8. IL8 alone does not release histamines. IL8 actually inhibits histamine release from human basophils induced by histamine-releasing factors (see also: HRF), » CTAP 3 (connective tissue activating peptide III), and IL3. IL8 is also involved in mediating pain.
IL8 antagonizes IL4-induced IgE production by human B cells without affecting IgM, IgG1, IgG2, IgG3, IgG4, or IgA production. IL8 directly affects B cells through a specific mechanism that is different from » IFN-γ, IFN-α, or prostaglandin E2.
The intravenous administration of IL8 in baboons causes a severe, albeit transient, granulocytopenia, followed by a granulocytosis which persists as long as sufficient IL8 levels are maintained.
IL8 is chemotactic for all known types of migratory immune cells (see also: Chemotaxis). IL8 inhibits the adhesion of leukocytes to activated endothelial cells (see also: cell activation) and therefore possesses anti-inflammatory activities (see also: LAI, leukocyte adhesion inhibitor) (see also: Inflammation). The 72 amino acid form of IL8 is ≈ ten-fold more potent in inhibiting neutrophil adhesion that the 77 amino acid variant. IL8 is a mitogen for epidermal cells. *In vivo* IL8 strongly binds to erythrocytes. This absorption may be of physiological importance in the regulation of inflammatory reactions since IL8 bound to erythrocytes no longer activates neutrophils. Macrophage-derived IL8 functions as an » angiogenesis factor and may play a role in angiogenesis-dependent disorders such as rheumatoid arthritis, tumor growth, and » wound repair.

■ **TRANSGENIC/KNOCK-OUT/ANTISENSE STUDIES:** Blockade of IL8 expression in some human melanoma cell lines by » Antisense oligonucleotides has shown that IL8 functions as an » Autocrine growth factor for these cells. A participation of IL8 in macrophage-mediated angiogenesis (see also: Angiogenesis factors) is suggested by the observation that IL8 antisense oligonucleotides to block monocyte-induced angiogenic activity.

■ **ASSAYS:** IL8 can be detected in assays measuring the migration of buffy coat leukocytes from agarose blocks. Sensitive immunoassays are also available. For further information see also subentry "Assays" in the reference section.

■ **Clinical Use & Significance:** IL8 may be of clinical relevance in psoriasis and rheumatoid arthritis. Elevated concentrations are observed in psoriatic scales and this may explain the high proliferation rate observed in these cells. IL8 may be also a marker of different inflammatory processes.

IL8 (and also » IL1 and » IL6) probably plays a role in the pathogenesis of chronic polyarthritis since excessive amounts of this factor are found in synovial fluids. The activation of neutrophils may enhance the migration of cells into the capillaries of the joints. These cells are thought to pass through the capillaries and enter the surrounding tissues thus causing a constant stream of inflammatory cells through the joints.

Human recombinant IL8 has been shown that the lesion responsible for defective neutrophil functions in patients with myelodysplastic syndrome can be restored without stimulating myeloid progenitor cells. IL8 may therefore be able to reduce the risks of lethal infections in these patients without the potential risk of stimulating leukemic clones.

● **Reviews: Baggiolini M et al** Interleukin 8 and related chemotactic cytokines – CXC and CC chemokines. Adv. Immunol. 55: 97-179 (1994); **Matsushima K & Oppenheim JJ** Interleukin 8 and MCAF: novel inflammatory cytokines inducible by IL1 and TNF. Cytokine 1: 2-13 (1989); **Mukaida N et al** Regulation of human interleukin 8 gene expression and binding of several other members of the intercrine family to receptors for interleukin-8. AEMB 305: 31-8 (1991); **Schröder JM & Christophers E** The biology of NAP-1/IL8, a neutrophil-activating cytokine. Immunol. Ser. 57: 387-416 (1992); **Zwahlen R et al** *In vitro* and *in vivo* activity and pathophysiology of human interleukin-8 and related peptides. Int. Rev. Exp. Pathol. 34: 27-42 (1993)

● **Biochemistry & Molecular Biology: Aloisi F et al** Production of hemolymphopoietic cytokines (IL6, IL8, colony-stimulating factors) by normal human astrocytes in response to IL1 β and tumor necrosis factor-α. JI 149: 2358-66 (1992); **Apella E et al** Determination of the primary structure of NAP-1/IL8 and a monocyte chemoattractant protein, MCP-1/MCAF. PCBR 349: 405-17 (1990); **Auer M et al** Crystallization and preliminary X-ray crystallographic study of interleukin-8. FL 265: 30-2 (1990); **Clore GM & Gronenborn AM** NMR and X-ray analysis of the three-dimensional structure of interleukin-8. Cytokines 4: 18-40 (1992); **Clore GM et al** Three-dimensional structure of Interleukin-8 in solution. B 29: 1689-96 (1990); **Farber JM** A macrophage mRNA selectively induced by γ-interferon encodes a member of the platelet factor 4 family of cytokines. PNAS 87: 5238-42 (1990); **Furuta R et al** Production and characterization of recombinant human neutrophil chemotactic factor. JB 106: 436-41 (1989); **Hebert CA et al** Endothelial and leukocyte forms of IL8: conversion by thrombin and interactions with neutrophils. JI 145: 3033-40 (1990); **Hébert CA et al** Scanning mutagenesis of interleukin-8 identifies a cluster of residues required for receptor binding. JCB 266: 18989-94 (1991); **Kang XQ et al** Production, purification, and characterization of human recombinant IL8 from the eukaryotic vector expression system baculovirus. Protein Expr. Purif. 3: 313-21 (1992); **Lee J et al** Characterization of complementary DNA clones encoding the rabbit IL8 receptor. JI 148: 1261-4 (1992); **Modi WS et al** Monocyte-derived neutrophil chemotactic factor (MDNCF/IL8) resides in a gene cluster along with several other members of the platelet factor 4 gene superfamily. Hum. Genet. 84: 185-7 (1990); **Mukaida N et al** Genomic structure of the human monocyte-derived neutrophil chemotactic factor IL8. JI 143: 1366-71 (1989); **Mukaida N et al** Cooperative interaction of nuclear factor-kappa B- and cis-regulatory enhancer binding protein-like factor binding elements in activating the interleukin-8 gene by pro-inflammatory cytokines. JBC 265: 21128-33 (1990); **Schröder JM** Generation of NAP-1 and related peptides in psoriasis and other inflammatory skin diseases. Cytokines 4: 54-76 (1992); **Sakamoto K et al** Interleukin-8 is constitutively and commonly produced by various human carcinoma cell lines. Int. J. Clin. Lab. Res. 22: 216-9 (1992); **Sica A et al** IL 1 transcriptionally activates the monocyte-derived neutrophil chemotactic factor/IL 8 gene in endothelial cells. Immunology 69: 548-53 (1990); **Standiford TJ et al** IL7 up-regulates the expression of IL8 from resting and stimulated human blood monocytes. JI 149: 2035-9 (1992); **Strieter RM et al** Induction and regulation of interleukin 8 gene expression. AEMB 305: 23-30 (1991); **Tada M et al** Human glioblastoma cells produce 77 amino acid interleukin-8 (IL8(77)). J. Neurooncol. 16: 25-34 (1993); **Yoshimura TK et al** Neutrophil chemotactic factor produced by lipopolysaccharide (LPS)-stimulated human blood mononuclear leukocytes: partial characterization and separation from interleukin 1 (IL 1). JI 139: 788-93 (1987)

● **Receptor Structure, Gene(s), Expression: Barnett ML et al** Characterization of interleukin-8 receptors in human neutrophil membranes: regulation by guanine nucleotides. BBA 1177: 275-82 (1993); **Cerretti DP et al** Molecular characterization of receptors for human interleukin-8, GRO/melanoma growth-stimulatory activity and neutrophil activating peptide-2. Mol. Immunol. 30: 359-67 (1993); **Darbonne WC et al** Red blood cells are a sink for interleukin 8, a leukocyte chemotaxin. JCI 88: 1362-9 (1991); **Hebert CA et al** Partial functional mapping of the human interleukin-8 type A receptor. Identification of a major ligand binding domain. JBC 268: 18549-53 (1993); **Holmes WE et al** Structure and functional expression of a human Interleukin-8 receptor. S 253: 1278-80 (1991); **LaRosa GJ et al** Amino terminus of the interleukin-8 receptor is a major determinant of receptor subtype specificity. JBC 267: 25402-6 (1992); **Morris SW et al** Assignment of the genes encoding human interleukin-8 receptor types 1 and 2 and an interleukin-8 receptor pseudogene to chromosome 2q35. Genomics 14: 685-91 (1992); **Murphy PM & Tiffany HL** Cloning of complementary DNA encoding a functional human Interleukin-8 receptor. S 253: 1280-3 (1991); **Petersen F et al** Lymphokine Cytokine Res. Neutrophil-activating polypeptides IL8 and NAP-2 induce identical signal transduction pathways in the regulation of lysosomal enzyme release. 10: 35-41 (1991); **Schumacher C et al** High- and low-affinity binding of GROα and neutrophil-activating peptide 2 to interleukin 8 receptors on human neutrophils. PNAS 89: 10542-6 (1992); **Thomas KM et al** The interleukin-8 receptor is encoded by a neutrophil-specific cDNA clone, F3R. JBC 266: 14839-41 (1991); **Wu DQ et al G** protein-coupled signal transduction pathways for interleukin 8. S 261: 101-3 (1993)

● **Biological Activities: Baggilioni M et al** Neutrophil-activating peptide-1/interleukin 8, a novel cytokine that activates neutrophils. JCI 84: 1045-9 (1989); **Cunha FQ et al** Interleukin-8 as a mediator of sympathetic pain. Br. J. Pharmacol. 104: 765-7 (1991); **Gimbrone MA jr et al** Endothelial interleukin 8: a novel inhibitor of leukocyte-endothelial interactions. S 246: 1601-3 (1989); **Hechtman DH et al** Intravascular IL8. Inhibitor of polymorphonuclear leukocyte accumulation at sites of acute inflammation. JI 147: 883-92 (1991); **Kimata H et al** Interleukin 8 (IL8) selectively inhibits immunoglobulin E production induced by IL4 in human B cells. JEM 176: 1227-31 (1992); **Koch AE et al** Interleukin-8 as a macrophage-derived mediator of angiogenesis. S 258: 1798-801 (1992); **Krieger M et al** Activation of human basophils through the IL8 receptor. JI 149: 2662-7 (1992); **Kuijpers TW et al** Neutrophil migration across monolayers of cytokine-prestimulated endothelial cells: a role for platelet-activating factor and IL8. JCB 117: 565-72 (1992); **Kuna P et al** IL8 inhibits histamine release from human basophils induced by histamine-releasing factors, connective tissue activating peptide III, and IL3. JI 147: 1920-4 (1991); **Kunkel SL et al** Interleukin-8 (IL8): the major neutrophil chemotactic factor in the lung. Exp. Lung Res. 17: 17-23 (1991); **Larsen CG et al** Production of interleukin 8 by human dermal fibroblasts and keratinocytes in response to interleukin IL1 or tumor necrosis factor. Immunology 68: 31-6 (1989); **Leonard EJ & Yoshimura T** Neutrophil attractant protein-1 (NAP-2) (Interleukin-8). Am. J. Respir. Cell. Mol. Biol. 2: 479-86 (1990); **Lotz M et al** Cartilage and joint inflammation. Regulation of IL8 expression by human articular chondrocytes. JI 138: 466-73 (1992); **Moser B et al** Neutrophil-activating peptide 2 and gro/melanoma growth-stimulatory activity interact with neutrophil-activating peptide 1/interleukin 8 receptors on human neutrophils. JBC 266: 10666-71 (1991); **Nourshargh S et al** A comparative study of the neutrophil stimulatory activity *in vitro* and pro-inflammatory properties *in vivo* of 72 amino acid and 77 amino acid IL8. JI 148: 106-11 (1992); **Rathanaswami P et al** Expression of the cytokine RANTES in human rheumatoid synovial fibroblasts. Differential regulation of RANTES and interleukin-8 genes by inflammatory cytokines. JBC 268: 5834-9 (1993); **Ribeiro RA et al** IL8 causes *in vivo* neutrophil migration by a cell-dependent mechanism. Immunology 73: 472-7 (1991); **Smith WB et al** Interleukin 8 induces neutrophil transendothelial migration. Immunology 72: 65-72 (1991); **Sticherling M et al** Immunohistochemical studies on NAP-1/IL8 in contact eczema and atopic dermatitis. Arch. Dermatol. Res. 284: 82-5 (1992); **Terkeltaub R et al** Monocyte-derived neutrophil chemotactic factor/interleukin-8 is a potential mediator of crystal-induced inflammation. Arthritis and Rheumatism 34: 894-903 (1991); **Thornton AJ et al** Cytokine-induced gene expression of a neutrophil chemotactic factor/IL 8 in human hepatocytes. JI 144: 2609-13 (1990); **Van Damme J et al** The neutrophil-activating proteins interleukin 8 and β-thromboglobulin: *in vitro* and *in vivo* comparison of NH2-terminally processed forms. EJI 20: 2113-8 (1990); **Van Zee KJ et al** Effects of intravenous IL8 administration in nonhuman primates. JI 148: 1746-52 (1992)

● **Transgenic/Knock-out/Antisense Studies: Koch AE et al** Interleukin-8 as a macrophage-derived mediator of angiogenesis. S 258: 1798-801 (1992); **Schadendorf D et al** IL8 produced by human malignant melanoma cells *in vitro* is an essential autocrine growth factor. JI 151: 2667-75 (1993)

● **Assays: Bignold LP** Measurement of chemotaxis of polymorphonuclear leukocytes *in vitro*. The problems of the control of gradients of chemotactic factors, of the control of the cells and of the separation of chemotaxis from chemokinesis. JIM 108: 1-18 (1988); **DeForge LE & Remick DG** Sandwich ELISA for detection of picogram quantities of interleukin 8. Immunol. Invest. 20: 89-97 (1991); **Elner VM et al** Human corneal interleukin-8. IL1 and TNF-induced gene expression and secretion. Am. J. Pathol. 139: 977-88 (1991); **Ida N et al** A highly sensitive enzyme-linked immunosorbent assay for the measurement of interleukin-8 in biological fluids. JIM 156: 27-38 (1992); **Ko V et al** A sensitive enzyme-linked immunosorbent assay for human interleukin-8. JIM 149: 227-35 (1992); **Kristensen MS et al** Quantitative determination of IL1 α-induced IL8 mRNA levels in cultured human keratinocytes, dermal fibroblasts, endothelial cells, and monocytes. J. Invest. Dermatol. 97: 506-10 (1991); **Paludan K & Thestrup-Pedersen K** Use of the polymerase chain reaction in quantification of interleukin 8 mRNA in minute epidermal samples. J. Invest. Dermatol. 99: 830-5 (1992); **Rampart M et al** Development and application of a radioimmunoassay for interleukin-8: detection of interleukin-8 in synovial fluids from patients with inflammatory joint disease. Lab. Invest. 66: 512-8 (1992); **Romero R et al** Neutrophil attractant/activating peptide-1/interleukin 8 in term and preterm parturition. Am. J. Obstet. Gynecol. 165: 813-20 (1991); **Sticherling M et al** Production and characterization of monoclonal antibodies against the novel neutrophil activating peptide NAP/IL8. JI 143: 1628-34 (1989); **Tilg H et al** A method for the detection of erythrocyte-bound interleukin-8 in humans during interleukin-1 immunotherapy. JIM 163: 253-8 (1993)

● **Clinical Use & Significance: Gillitzer R et al** Upper keratinocytes of psoriatic skin lesions express high levels of NAP-1/IL8 mRNA *in situ*. J. Invest. Dermatol. 97: 73-9 (1991); **Koch JE et al** Synovial tissue macrophage as a source of the chemotactic cytokine IL8. JI 147: 2187-95 (1991); **Peichl P et al** Presence of NAP-1/IL8 in synovial fluids indicates a possible pathogenic role in rheumatoid arthritis. Scand. J. Immunol. 34: 333-9 (1991); **Van Zee KJ et al** IL8 in septic shock, endotoxemia, and after IL1 administration. JI 146: 3478-82 (1991); **Zwierzina H et al** Recombinant human interleukin-8 restores function in neutrophils from patients with myelodysplastic syndrome without stimulating myeloid progenitor cells. Scand. J. Immunol. 37: 322-8 (1993)

IL9: Interleukin 9.

■ **Alternative Names:** MCGF (mast cell growth factor); MEA (mast cell growth enhancing activity); megakaryoblast growth factor; P40; TCGF 3 (T cell growth factor 3). See also: individual entries for further information.

■ **Sources:** IL9 can be isolated from culture supernatants of mitogen- or antigen-stimulated T helper cells. In primary lymphocyte cultures it is produced predominantly by » CD4-positive cells. The synthesis of IL9 can be induced by » phorbol ester and » calcium ionophore.

■ **Protein Characteristics:** IL9 is a protein of 14 kDa. The cDNA encodes a protein of 144 amino acids with a secretory » signal sequence of 18 amino acids. IL9 is glycosylated extensively, has an isoelectric point of pI = 10 and contains ten cysteine residues involved in disulfide bonding. Human and murine IL9 show 69 & sequence homology at the nucleotide level and 55 % homology at the protein level.

■ **Gene Structure:** The human and murine IL9 genes have a length of ≈ 4 kb and contain five

exons. The murine and the human IL9 genes share ≈ 69 % Homology. The human gene maps to chromosome 5q31-32 which is a chromosomal region sometimes deleted in patients with myelodysplastic syndrome (see also: 5q⁻ syndrome). The murine IL9 gene maps to chromosome 13 distal to Tcrg and proximal to Dhfr.

■ **RECEPTOR STRUCTURE, GENE(S), EXPRESSION:** The murine receptor is a glycoprotein of 64 kDa (54 kDa unglycosylated; 468 amino acids) that binds IL9 with a dissociation constant of 100 pM. The human receptor gene encodes a protein of 522 amino acids with 53 % homology to the mouse IL9 receptor. Homologies are particularly pronounced in the extracellular regions (67 %). Variants of the receptor arising by alternative splicing and lacking the sequences encoding the transmembrane and cytoplasmic domains have also been described.

The receptor is expressed in IL9 responsive cells at a density of ≈ 3000/cell. Like many other cytokines binding of IL9 to its receptor leads to the formation of complexes that are rapidly internalized and degraded by the cells. The receptor is expressed in membrane-bound and soluble forms.

■ **BIOLOGICAL ACTIVITIES:** Murine IL9 stimulates human cells but human IL9 is inactive on murine cells. IL9 stimulates the proliferation of a number of T helper cell clones in the absence of antigens or antigen-presenting cells. It does not promote the proliferation of freshly isolated T cells or cytolytic T cells. The activity as a growth factor therefore appears to be restricted to a distinct subpopulation of cells or may be discernible only in cells of a particular activation state (see also: cell activation).

After activation of the T cell receptor » IL1 is a very potent costimulator for the synthesis of IL9. Antibodies directed against » IL2 block the synthesis of IL9 in human T lymphocytes. IL9 potentiates the » IL4-induced production of IgG, IgM, and IgE by normal human B lymphocytes. IL9 also induces functional changes such as secretion of » IL6 by mast cell lines. IL9 is also a potent regulator of the expression of protease genes like those belonging to the granzyme family.

In the presence of » IL3 IL9 enhances the proliferation of bone marrow mast cells. IL9 synergises with » Epo and selectively supports a subpopulation of an early class of » BFU-E that respond to » IL3 (see also: FDCP-Mix). IL9 also stimulations erythroid progenitors in normal human bone marrow in the absence of serum. IL9 also stimulates colony formation of » CFU-E. IL9 induces the production of » IL6 by mast cells and some other cell types. This effect is synergised by » IL4.

IL9 also stimulates the proliferation of fetal thymocytes and murine thymic lymphomas in response to » IL2. IL9 also promotes the proliferation of some leukemia cell lines.

The introduction of an expressed IL9 gene in factor-dependent cells abolishes the growth factor requirement. If these cells are injected intraperitoneally or subcutaneously into syngeneic animals they develop T cell tumors. It can therefore be assumed that the uncontrolled expression of IL9 is involved in the development of certain types of T cell tumors.

■ **ASSAYS:** IL9 can be detected in a » Bioassay employing cells that respond to the factor (see: L138.8A; TS-1). An alternative and entirely different detection method is » Message amplification phenotyping. For further information see also subentry "Assays" in the reference section.

■ **CLINICAL USE & SIGNIFICANCE:** IL9 is expressed by Reed-Sternberg cells and Hodgkin lymphoma cells and some large aplastic lymphoma cells, while non-Hodgkin lymphomas and peripheral T cell lymphomas do not express it. IL9 may therefore play a role in the pathogenesis of Hodgkin's disease and large cell anaplastic lymphoma and has been shown to function as an » autocrine growth factor for Hodgkin's cell lines. Constitutive expression of IL9 has also been demonstrated in HTLV-1-transformed T cells.

● **REVIEWS: Quesniaux VF** Interleukins 9, 10, 11 and 12 and *kit* ligand: a brief overview. Res. Immunol. 143: 385-400 (1992); **Renauld JC et al** Interleukin 9. AI 54: 79-97 (1993); **Yang YC** Human interleukin-9: a new cytokine in hematopoiesis. Leuk. Lymphoma 8: 441-7 (1992)

● **BIOCHEMISTRY & MOLECULAR BIOLOGY: Kelleher K et al** Human interleukin-9: Genomic sequence, chromosomal location, and sequences essential for its expression in human T cell leukemia virus (HTLV)-I-transformed human T cells. Blood 77: 1436-41 (1991); **Mock BA et al** IL9 maps to mouse chromosome 13 and human chromosome 5. Immunogenetics 31: 265-70 (1990); **Modi WS et al** Regional localizationof the human glutaminase (GLS) and interleukin 9 (IL9) genes by *in situ* hybridization. Cytogenet. Cell Genet. 57: 114-6 (1991); **Yang YC et al** Expression cloning of a cDNA encoding a novel human hematopoietic growth factor: human homologue of murine T cell growth factor P40. Blood 74: 1880-4 (1989)

● **RECEPTOR STRUCTURE, GENE(S), EXPRESSION: Renauld JC et al** Expression cloning of the murine and human interleukin 9 receptor cDNAs. PNAS 89: 5690-4 (1992)

● **BIOLOGICAL ACTIVITIES: Birner A et al** Recombinant murine interleukin 9 enhances the erythropoietin-dependent colony formation of human BFU-E. EH 20: 541-5 (1992); **Donahue RE et al** Human P40 T cell growth factor supports erythroid colony formation. Blood 75: 2271-5 (1990); **Dugas B et al** Interleukin-9 potentiates the interleukin 4-induced immunoglobulin (IgG, IgM, and IgE) production by normal human B lymphocytes. EJI

23: 1687-92 (1993); **Dugas B et al** Functional interaction between interleukin-9/P40 and interleukin-4 in the induction of IgE production by normal human B lymphocytes. Biotechnol. Ther. 4: 31-42 (1993); **Holbrook ST et al** Effect of interleukin-9 on clonogenic maturation and cell-cycle status of fetal and adult hematopoietic progenitors Blood 77: 2129-34 (1991); **Houssiau FA et al** IL2 dependence of IL9 expression in human T lymphocytes. JI 148: 3147-51 (1992); **Hültner L & Möller J** Mast cell growth-enhancing activity (MEA) stimulates interleukin 6 production in a mouse bone marrow-derived mast cell line and a malignant subline. EH 18: 873-7 (1990); **Lu L et al** Human interleukin (IL)-9 specifically stimulates proliferation of $CD34^+DR^+CD33^-$ erythroid progenitors in normal human bone marrow in the absence of serum. EH 20: 418-24 (1992); **Schaafsma MR et al** Interleukin-9 stimulates the proliferation of enriched human erythroid progenitor cells: additive effect with GM-CSF. Ann. Hematol. 66: 45-9 (1993); **Schmitt E et al** IL1 serves as a secondary signal for IL9 expression. JI 147: 3848-54 (1991); **Sonoda Y et al** Human interleukin-9 supports formation of a subpopulation of erythroid bursts that are responsive to interleukin-3. Am. J. Hematol. 41: 84-91 (1992); **Uyttenhove C et al** Autonomous growth and tumorigenicity induced by P40/interleukin 9 cDNA transfection of a mouse P40-dependent T cell line. JEM 173: 519-22 (1991); **Vink A et al** Interleukin-9 stimulates *in vitro* growth of mouse thymic lymphomas. EJI 23: 1134-8 (1993); **Williams DE et al** T cell growth factor P40 promotes the proliferation of myeloid cell lines and enhances erythroid burst formation by normal murine bone marrow cells *in vitro*. Blood 76: 906-11 (1990)

● **ASSAYS:** **Renauld JC et al** Cloning and expression of a cDNA for the human homologue of mouse T cell and mast cell growth factor P40. Cytokine 2: 9-12 (1990); for further information also see individual cell lines used in individual bioassays.

● **CLINICAL USE & SIGNIFICANCE:** **Gruss HJ et al** Interleukin 9 is expressed by primary and cultured Hodgkin and Reed-Sternberg cells. CR 52: 1026-31 (1992); **Kelleher K et al** Human interleukin-9: genomic sequence, chromosomal location, and sequences essential for its expression in human T cell leukemia virus (HTLV)-I-transformed human T cells. Blood 77: 1436-41 (1991); **Merz H et al** Interleukin 9 expression in human malignant lymphomas: unique association with Hodgkin's disease and large cell anaplastic lymphoma. Blood 78: 1311-7 (1991)

IL10: Interleukin 10.

■ **ALTERNATIVE NAMES:** BTCGF (B cell-derived T cell growth factor); CSIF (cytokine synthesis inhibitory factor); TGIF (T cell growth inhibitory factor). See also: individual entries for further information.

■ **SOURCES:** IL10 is produced by murine T cells (Th2 cells but not Th1 cells) following their stimulation by lectins (see also: TH1/TH2 cytokines). The main source for B cell-derived IL10 in mice are Ly-1 (B-1) B cells that express CD5 (Ly-1) and CD11 (Mac-1). Murine keratinocytes also produce IL10.

In humans IL10 is produced by activated CD8-positive peripheral blood T cells, by Th0, Th1-, and Th2-like $CD4^+$ T cell clones after both antigen-specific and polyclonal activation, by B cell lymphomas, and by LPS-activated monocytes and mast cells. B cell lines derived from patients with acquired immunodeficiency syndrome and Burkitt's lymphoma constitutively secrete large quantities of IL10. The synthesis of IL10 by monocytes is inhibited by » IL4 and IL10.

■ **PROTEIN CHARACTERISTICS:** IL10 is a homodimeric protein with subunits having a length of 160 amino acids. Human IL10 shows 73 % amino acid homology with murine IL10.

■ **GENE STRUCTURE:** The human IL10 contains four exons. It is closely related to the product of the BCRF-1 gene (*Bam* HI C fragment rightward reading frame) of Epstein-Barr virus (84% homology at the protein level); These two proteins are more closely related to each other than human and murine IL10. BCRF-1 has therefore also been called viral IL10 (vIL10). The human IL10 gene maps to chromosome 1. The human IL10 shows 81 % homology with murine IL10 at the nucleotide level.

■ **RECEPTOR STRUCTURE, GENE(S), EXPRESSION:** A receptor has been identified on murine and human cells by using radiolabeled IL10. Mouse IL10 is capable of blocking binding of human IL10 to mouse but not human cells. The murine IL10 receptor has been cloned. This receptor is a protein of ≈ 110 kDa that binds murine IL10 specifically. This receptor is structurally related to receptors for » IFN.

■ **BIOLOGICAL ACTIVITIES:** IL10 inhibits the synthesis of a number of cytokines such as » IFN-γ, » IL2 and TNF-β in Th1 subpopulations of T cells but not of Th2 cells (see also: TH1/TH2 cytokines). This activity is antagonized by » IL4. The inhibitory effect on IFN-γ production is indirect and appears to be the result of a suppression of » IL12 synthesis by accessory cells. In the human system, IL10 is produced by, and down-regulates the function of, Th1 and Th2 cells. In macrophages stimulated by bacterial lipopolysaccharides (see also: cell activation) IL10 inhibits the synthesis of » IL1, » IL6 and » TNF-α by promoting, among other things, the degradation of cytokine mRNA. It also leads to an inhibition of antigen presentation. In human monocytes IFN-γ and IL10 antagonize each other's production and function. IL10 has also been shown to be a physiologic antagonist of » IL12.

IL10 also inhibits mitogen- or anti-CD3-induced proliferation of T cells in the presence of accessory cells and reduces the production of IFN-γ and IL2. Exogenous IL2 and IL4 inhibit the proliferation-inhibitory effect but do not influence the produc-

tion of IFN-γ. In LPS-stimulated macrophages IFN-γ increases the synthesis of » IL6 by inhibiting the production of IL10. IL10 (CSIF) appears to be responsible for most or all of the ability of Th2 supernatants to inhibit cytokine synthesis by Th1 cells.
IL10 inhibits *in vivo* the secretion of » TNF-α and protects against the lethality of endotoxin in a murine model of » septic shock if administered before challenging the mice with bacterial lipopolysaccharides.
IL10 inhibits » IL5- but not IL2-induced Ig secretion by T cell-independent antigens.
Murine Ly-1 cells are the principal source of IL10, cells. In contrast to normal B cells, Ly-1 B cells express greatly elevated constitutive and inducible levels of IL10. These cells also have the distinctive property of continuous self-replenishment. The continuous treatment of newborn mice with anti-IL10 antibodies leads to a depletion of the Ly-1 B cells while maintaining a normal population of splenic B cells. These mice also contain greatly reduced serum immunoglobulin M levels and are also impaired in their antibody responses to specific antigens. IL10 is therefore a regulator of Ly-1 B cell development. The mechanism of Ly-1 B cell depletion appears to involve the increased production of » IFN-γ since coadministration of neutralizing anti-IFN-γ antibodies substantially restores the number of peritoneal-resident Ly-1 B cells in these mice.
IL10 acts as a costimulator of the proliferation of mast cells (in the presence of IL3 and/or IL4) and peripheral lymphocytes. While IL10 alone has not effects on the proliferation of mast cells and their progenitors a combination of IL10 and » IL4 has the same growth-promoting effects as » IL3 alone. Optimal growth of mast cells is achieved by a combination of IL3, IL4 and IL10. Due to its effects on mast cells IL10 probably plays a role in the development of mastocytosis frequently observed after parasitic infections by potentiating the effects of IL3 and IL4.
IL10 is also a costimulator for the growth of mature and immature thymocytes (together with » IL2, » IL4 and » IL7) and functions as a cytotoxic T cell differentiation factor, promoting a higher number of IL2-activated cytotoxic T lymphocyte precursors to proliferate and differentiate into cytotoxic effector cells. IL10 sustains viability of B cells *in vitro* and also stimulates B cells and promotes their differentiation (see also: BCDF, B cell differentiation factor). It enhances the expression of MHC class II antigens on B cells whereas it inhibits MHC class II expression on monocytes. In B cells activated via their antigen receptors or via » CD40 IL10 induces the secretion of IgG, IgA and IgM. This effect is synergised by » IL4 while the synthesis of immunoglobulins induced by IL10 is antagonized by » TGF-β. The activation of macrophages can be prevented by IL10.
It has been shown that human IL10 is a potent and specific chemoattractant for human T lymphocytes. The chemotactic activity is directed towards » CD8-positive cells and not towards » CD4-positive cells. IL10 also inhibits the chemotactic response of $CD4^+$ cells, but not of $CD8^+$ cells, towards » IL8.
Viral ***BCRF-1*** protein does not support the proliferation of mast cells although it blocks the synthesis of cytokines like IL10. The viral protein is expressed during the productive phase of EBV proliferation in B lymphocytes. It has been assumed that this protein suppresses antiviral immune responses by specifically inhibiting the synthesis of » IFN-γ and other cytokines which would kill the host cell expressing viral antigens at the cell surface. By employing this strategy EBV therefore has a powerful mechanism providing some selective growth advantages over other viruses (see also: Viroceptor). BCRF1 also enhances the reactivation of virus-specific cytotoxic T cell and HLA-unrestricted killer cell responses. Since BCRF1 is expressed during the lytic cycle of EBV, its may also enhance immune responses to EBV-infected cells during periods of virus replication *in vivo*, thus effectively limiting its own replication to maintain the apathogenic virus carrier state that is characteristic of EBV. BCRF1 has also been shown to inhibit superoxide anion production by human monocytes. For another virus-encoded protein increasing survival of virus-infected cells see: BHRF1.
IL10 has also been found to be produced by a number of AIDS lymphomas and it has been suggested that this factor may stimulate the proliferation of malignant cells in an » autocrine manner with EBV and HIV synergistically triggering its production.

■ **TRANSGENIC/KNOCK-OUT/ANTISENSE STUDIES:** Transgenic expression of IL10 in the islets of Langerhans of » transgenic mice has been shown to lead to a pronounced pancreatic inflammation, without inflammation of the islets of Langerhans and without diabetes. This recruitment of inflammatory cells to the pancreas is in marked contrast to the biological activities of IL10 *in vitro* which indicate that IL10 is a powerful immunosuppressive cytokine. IL10, therefore, provides a potent recruitment signal for leukocyte migration *in vivo* and these effects are relevant for *in vivo* therapeutic applications of IL10.

Transgenic mice in which the IL10 gene has been inactivated by targeted mutation in » ES cells have been generated to study the biological activities of IL10. These mice show normal development of lymphocytes and antibody responses. However, most animals are growth retarded and anemic and suffer from chronic enterocolitis. Extensive mucosal hyperplasia, inflammatory reactions, and aberrant expression of major histocompatibility complex class II molecules on epithelia are noted in the intestines. The observation that these animals only develop a local inflammation limited to the proximal colon if kept under specific pathogen-free conditions suggests that IL10 is an essential immunoregulator in the intestinal tract and that the generalized bowel inflammation in IL10-deficient animals is due to uncontrolled immune responses stimulated by enteric antigens.

■ **ASSAYS:** IL10 can be detected with a sensitive ELISA assay. The murine mast cell line » D36 can be used to bioassay human IL10. The intracellular factor can also be detected by flow cytometry. For further information see also subentry "Assays" in the reference section.

■ **CLINICAL USE & SIGNIFICANCE:** The introduction of an IL10 expression vector into » CHO cells has been used to analyze the consequences of local IL10 production *in vivo*. These altered cells were no longer tumorigenic in nude mice or severe combined immunodeficient SCID mice (see also: Immunodeficient mice) and also suppressed the growth of equal numbers of co-injected normal CHO cells. While normal CHO tumors are usually substantially infiltrated by macrophages were virtually absent within CHO-IL10 tumor tissues, suggesting that IL10 indirectly suppresses tumor growth of certain tumors by inhibiting infiltration of macrophages which may provide tumor growth promoting activity.

IL10 has been detected in the sera of a subgroup of patients with active non-Hodgkin's lymphoma. IL10 levels appear to correlate with a poor survival in patients with intermediate or high-grade non-Hodgkin's lymphoma.

● **REVIEWS: de Waal-Malefyt R et al** Interleukin 10. Curr. Opin. Immunol. 4: 314-20 (1992); **Howard M & O'Garra A** Biological properties of interleukin 10. IT 13: 198-200 (1992); **Howard M et al** Biological properties of interleukin 10. J. Clin. Immunol. 12: 239-47 (1992); **Quesniaux VF** Interleukins 9, 10, 11 and 12 and *kit* ligand: a brief overview. Res. Immunol. 143: 385-400 (1992); **Spits H & de Waal-Malefyt R** Functional characterization of human IL10. Int. Arch. Allergy Immunol. 99: 8-15 (1992); **Zlotnik A & Moore KW** Interleukin 10. Cytokines 3: 366-71 (1991)

● **BIOCHEMISTRY & MOLECULAR BIOLOGY: Hsu DH et al** Expression of IL10 activity by Epstein-Barr virus protein BCRFI. S 250: 830-2 (1990); **Kim JM et al** Structure of the mouse IL10 gene and chromosomal localizationof the mouse and human genes. JI 148: 3618-23 (1992); **Moore KW et al** Evolving principles in immunopathology: Interleukin 10 and its relationship to Epstein-Barr virus protein BCRF1. Springer Semin. Immunopathol. 13: 157-66 (1991); **Mosmann TR** Isolation of MAbs specific for IL4, IL5, and IL6, and a new TH2-specific cytokine, cytokine synthesis inhibitory factor (CSIF, IL10), using a solid phase radioimmunoadsorbent assay: blocking and non-blocking anti-CSIF MAbs. JI 145: 2938-45 (1990); **Vieira P et al** Isolation and expression of human cytokine synthesis inhibitory factor cDNA clones: homology to Epstein-Barr virus open reading frame BCRFI. PNAS 88: 1172-6 (1991); **Windsor WT et al** Disulfide bond assignments and secondary structure analysis of human and murine interleukin 10. B 32: 8807-15 (1993); **Yssel H et al** IL10 is produced by subsets of human $CD4^+$ T cell clones and peripheral blood T cells. JI 149: 2378-84 (1992)

● **RECEPTOR STRUCTURE, GENE(S), EXPRESSION: Suk Yue Ho A et al** A receptor for interleukin 10 is related to interferon receptors. PNAS 90: 11267-71 (1993); **Tan JC et al** Characterization of interleukin-10 receptors on human and mouse cells. JBC 268: 21053-9 (1993)

● **BIOLOGICAL ACTIVITIES: Benjamin D et al** Human B cell interleukin-10: B cell lines derived from patients with acquired immunodeficiency syndrome and Burkitt's lymphoma constitutively secrete large quantities of interleukin-10. Blood 80: 1289-98 (1992); **Bogdan C et al** Macrophage deactivation by interleukin 10. JEM 174: 1549-55 (1991); **Bogdan C et al** Contrasting mechanisms for suppression of macrophage cytokine release by transforming growth factor-β and interleukin-10. JBC 267: 23301-8 (1992); **Chen WF & Zlotnik A** IL10: a novel cytotoxic T cell differentiation factor. JI 147: 528-34 (1991); **Chomarat P et al** Interferon γ inhibits interleukin 10 production by monocytes. JEM 177: 523-7 (1993); **D'Andrea A et al** Interleukin 10 (IL10) inhibits human lymphocyte interferon g production by suppressing natural killer cell stimulatory factor/IL12 synthesis in accessory cells. JEM 178: 1041-8 (1993); **Defrance T et al** Interleukin 10 and transforming growth factor β cooperate to induce anti-CD40-activated naive human B cells to secrete immunoglobulin A. JEM 175: 671-82 (1992); **Del-Prete G et al** Human IL10 is produced by both type 1 helper (Th1) and type 2 helper (Th2) T cell clones and inhibits their antigen-specific proliferation and cytokine production. JI 150: 353-60 (1993); **De Waal Malefyt R et al** Interleukin 10 (IL10) inhibits cytokine synthesis by human monocytes: An autoregulatory role of IL10 produced by monocytes. JEM 174: 1209-20 (1991); **De Waal Malefyt R et al** Interleukin 10 (IL10) and viral IL10 strongly reduce antigen-specific human T cell proliferation by diminishing the antigen-presenting capacity of monocytes via downregulation of class II major histocompatibility complex expression. JEM 174: 915-24 (1991); **Emelie D et al** *In vivo* production of interleukin-10 by malignant cells in AIDS lymphomas. EJI 22: 2937-42 (1992); **Enk AH & Katz SI** Identification and induction of keratinocyte-derived interleukin 10. JI 149: 92-95 (1992); **Fiorentino DF et al** A. IL10 acts on the antigen-presenting cell to inhibit cytokine production by Th1 cells. JI 146: 3444-51 (1991);

Gerard C et al Interleukin 10 reduces the release of tumor necrosis factor and prevents lethality in experimental endotoxemia. JEM 177: 547-50 (1993); **Go N et al** Interleukin 10, a novel B cell stimulatory factor: unresponsiveness of X chromosome-linked immunodeficiency B cells. JEM 172: 1625-31 (1990); **Hsu DH et al** Differential effects of IL4 and IL10 on IL2-induced IFN-γ synthesis and lymphokine-activated killer activity. Int. Immunol. 4: 563-9 (1992); **Ishida H et al** Continuous anti-interleukin 10 antibody administration depletes mice of Ly-1 B cells but not conventional B cells. JEM 175: 1213-20 (1992); **Jinquan T et al** Human IL10 is a chemoattractant for $CD8^+$ T lymphocytes and an inhibitor of IL9-induced $CD4^+$ T lymphocyte migration. JI 151: 4545-51 (1993); **Miyazaki I et al** Viral interleukin 10 is critical for the induction of B cell growth transformation by Epstein-Barr virus. JEM 178: 439-47 (1993); **Mosmann TR** Role of a new cytokine, interleukin-10, in the cross-regulation of T helper cells. ANY 628: 337-44 (1991); **Niiro H et al** Epstein-Barr virus BCRF1 gene product (viral interleukin 10) inhibits superoxide anion production by human monocytes. Lymphokine Cytokine Res. 11: 209-14 (1992); **O'Garra A et al** Production of cytokines by mouse B cells: B lymphomas and normal B cells produce interleukin 10. Int. Immunol. 2: 821-32 (1990); **O'Garra A et al** Ly-1 B (B-1) cells are the main source of B cell-derived interleukin 10. EJI 22: 711-7 (1992); **Pecanha LM et al** Lymphokine control of type 2 antigen response. IL10 inhibits IL5- but not IL2-induced Ig secretion by T cell-independent antigens. JI 148: 3427-32 (1992); **Pecanha LM et al** IL10 inhibits T cell-independent but not T cell-dependent responses *in vitro*. JI 150: 3215-23 (1993); **Ralph P et al** IL10, T lymphocyte inhibitor of human blood cell production of IL1 and tumor necrosis factor. JI 148: 808-14 (1992); **Rousset F et al** Interleukin 10 is a potent growth and differentiation factor for activated human B lymphocytes. PNAS 89: 1890-3 (1992); **Schlaak JF et al** Differential effects of IL10 on proliferation and cytokine production of human γ/δ and α/β T cells. Scand. J. Immunol. 39: 209-15 (1994); **Stewart JP et al** The interleukin-10 homologue encoded by Epstein-Barr virus enhances the reactivation of virus-specific cytotoxic T cell and HLA-unrestricted killer cell responses. Virology 191: 773-82 (1992); **Street NE & Mosmann TR** Functional diversity of T lymphocytes due to secretion of different cytokine patterns. FJ 5: 171-7 (1991); **Taga K & Tosato G** IL10 inhibits human T cell proliferation and IL2 production. JI 148: 1143-8 (1992); **Thompson-Snipes LA et al** Interleukin 10: a novel stimulatory factor for mast cells and their progenitors. JEM 173: 507-10 (1991)

● **Assays: Abrams JS et al** Strategies of anti-cytokine monoclonal antibody development: immunoassay of IL10 and IL5 in clinical samples. IR 127: 5-24 (1992); **Fiorentino DF et al** IL10 inhibits cytokine production by activated macrophages. JI 147: 3815-22 (1991); **Kreft B et al** Detection of intracellular interleukin-10 by flow cytometry. JIM 156: 125-8 (1992); **Merville P et al** Detection of single cells secreting IFN-γ, IL6, and IL10 in irreversibly rejected human kidney allografts, and their modulation by IL2 and IL4. Transplantation 55: 639-46 (1993); **Mosmann TR et al** Isolation of monoclonal antibodies specific for IL4, IL5, IL6, and a new Th2-specific cytokine (IL10), cytokine synthesis inhibitory factor, by using a solid phase radioimmunoadsorbent assay. JI 145: 2938-45 (1990)

● **Transgenic/Knock-out/Antisense Studies: Kühn R et al** Interleukin-10-deficient mice develop chronic enterocolitis. Cell 75: 263-74 (1993); **Wogensen L et al** Leukocyte extravasation into the pancreatic tissue in transgenic mice expressing interleukin 10 in the islets of Langerhans. JEM 178: 175-85 (1993)

● **Clinical Use & Significance: Blay JY et al** Serum interleukin-10 in non-Hodgkin's lymphoma: a prognostic factor. Blood 82: 2169-74 (1993); **Richter G et al** Interleukin 10 transfected into Chinese hamster ovary cells prevents tumor growth and macrophage infiltration. CR 15; 53: 4134-7 (1993)

IL11: Interleukin 11.

■ **Alternative Names:** AGIF (adipogenesis inhibitory factor); megakaryocyte colony-stimulating factor, plasmacytoma stimulator activity; T1154 mitogenic activity. See also: individual entries for further information.

■ **Sources:** IL11 is secreted by bone marrow stromal cells (fibroblasts) and is also produced by a number of mesenchymal cells. The human thyroid carcinoma cell line NIM-1 constitutively secretes IL11. The synthesis of IL11 is induced by » IL1 and » phorbol esters.

■ **Protein Characteristics:** IL11 is a non-glycosylated protein of ≈ 23 kDa (179 amino acids as inferred from the cDNA sequence). The isoelectric point is pH 6.3. The protein does not contain cysteines. IL11 is not related to any other proteins.

■ **Gene Structure:** The human IL11 genes contains five exons and has a length of 7 kb. It maps to chromosome 19q13. 3-13. 4. Primate and human IL11 sequences show homology of 97 %. The human IL11 gene contains an IL1 response element in its 5´ regulatory region (see also: Gene expression).

■ **Receptor Structure, Gene(s), Expression:** Murine » 3T3 fibroblasts express high affinity IL11 receptors at a density of ≈ 5000/cell. The receptor is a receptor of 151 kDa. The IL11 receptor utilizes » gp130 as its signal transducer, which is also a component of other cytokine receptors.

■ **Biological Activities:** IL11 promotes primary and secondary immune responses *in vitro* and *in vivo* and modulates antigen-specific antibody reactions. IL11 promotes the proliferation of » IL6-dependent plasmacytoma cell lines in the presence of neutralizing IL6 antibodies. IL11 also stimulates the T cell-dependent development of IgG-secreting B cells in spleen cell cultures. Blast cells cultured in a medium containing IL11 and » Epo differentiate specifically into macrophages.

Like » IL6 and » G-CSF IL11 synergises with » IL3 in stimulating the generation of human and murine megakaryocyte colonies *in vitro* (see also: Colony formation assay). IL11 may therefore be an important regulator of megakaryocytopoiesis (see also: Meg-CSF, TPO). IL11 may regulate malignant cells of the megakaryocytic lineage in part by an » autocrine growth control mechanism. In combination with » IL4 IL11 also enhances the colony formation by primitive hematopoietic pre-

cursor cells. IL11 specifically shortens the G_0 period of the » cell cycle of stem cells.
In the presence of IL3 or » SCF (stem cell factor) IL11 has profound stimulatory effects on primitive multilineage hematopoietic progenitors, pre-CFC(multi), as well as on precursors representing various stages of erythroid differentiation observable *in vitro*, including CFC(multi), » BFU-E, and » CFU-E. The combination of SCF with IL11 also stimulates highly proliferative erythroid progenitors.
IL11 inhibits the differentiation of preadipocytes (see: AGIF, adipogenesis inhibitory factor). It induces the synthesis of some » acute phase proteins by hepatocytes (see also: acute phase reaction); the qualitative and quantitative effects resemble those observed with » IL6 or » LIF. IL11 synergises with » IL1.
The physiological consequences of the overexpression of human IL11 have been studied in lethally irradiated mice having received transplanted murine bone marrow cells infected with a recombinant retrovirus bearing a human IL11 gene. Long-term reconstituted primary and secondary recipients express high levels of IL11 and show loss of body fat, thymus atrophy, some alterations in plasma protein levels, frequent inflammation of the eyelids, and often a hyperactive state. A sustained rise in peripheral platelet levels but no changes in the total number of circulating leukocytes is seen in the majority of the transplanted animals despite a > 20-fold elevation in myeloid progenitor cell content in the spleen.

■ **ASSAYS:** IL11 can be detected in a » Bioassay employing cells that respond to this factor (see: T1165). An alternative and entirely different detection method is » Message amplification phenotyping. For further information see also subentry "Assays" in the reference section.

● **REVIEWS: Kawashima I & Takiguchi Y** Interleukin-11: a novel stroma-derived cytokine. PGFR 4: 191-206 (1992); **Quesniaux VF** Interleukins 9, 10, 11 and 12 and *kit* ligand: a brief overview. Res. Immunol. 143: 385-400 (1992); **Quesniaux VF et al** Review of a novel hematopoietic cytokine, interleukin-11. Int. Rev. Exp. Pathol. 34: 205-14 (1993)

● **BIOCHEMISTRY & MOLECULAR BIOLOGY: Kelleher K et al** Human interleukin-9: genomic sequence, chromosomal location, and sequences essential for its expression in human T cell leukemia virus (HTLV)-I-transformed human T cells. Blood 77: 1436-41 (1991); **McKinley D et al** Genomic sequence and chromosomal location of human interleukin-11 gene (IL11). Genomics 13: 814-9 (1992); **Paul SR et al** Molecular cloning of a cDNA encoding interleukin 11, a stromal cell-derived lymphopoietic and hematopoietic cytokine PNAS 87: 7512-6 (1990); **Paul SR & Schendel P** The cloning and biological characterization of recombinant human interleukin 11. IJCC 10: 135-43 (1992); **Yang-Feng TL et al** Assignment of the gene encoding human interleukin-11 to chromosome 19q13.3-q13.4. Cytogenet. Cell Genet. 58: 2027 (1991)

● **RECEPTOR STRUCTURE, GENE(S), EXPRESSION: Yin T et al** Characterization of interleukin 11 receptor and protein tyrosine phosphorylation induced by interleukin 11 in mouse 3T3-L1 cells. JBC 267: 8347-51 (1992); **Yin T et al** Involvement of IL6 signal transducer gp130 in IL11-mediated signal transduction. JI 151: 2555-61 (1993)

● **BIOLOGICAL ACTIVITIES: Anderson KC et al** Interleukin-11 promotes accessory cell-dependent B cell differentiation in humans. Blood 80: 2797-804 (1992); **Bauman H et al** Interleukin-11 regulates the hepatic expression of the same plasma protein genes as interleukin 6. JBC 266: 20424-7 (1991); **Bruno E et al** Effects of recombinant interleukin 11 on human megakaryocyte progenitor cells. EH 19: 378-81 (1991); **Burstein SA et al** Leukemia inhibitory factor and interleukin-11 promote maturation of murine and human megakaryocytes *in vitro*. JCP 153: 305-12 (1992); **Hangoc G et al** *In vivo* effects of recombinant interleukin-11 on myelopoiesis in mice. Blood 81: 965-72 (1993); **Hawley RG et al** Progenitor cell hyperplasia with rare development of myeloid leukemia in interleukin 11 bone marrow chimeras. JEM 178: 1175-88 (1993); **Hirayama F et al** Clonal proliferation of murine lymphohemopoietic progenitors in culture. PNAS 89: 5907-11 (1992); **Keller DC et al** Interleukin-11 inhibits adipogenesis and stimulates myelopoiesis in human long-term marrow cultures. Blood 82: 1428-35 (1993); **Kobayashi S et al** Interleukin-11 acts as an autocrine growth factor for human megakaryoblastic cell lines. Blood 81: 889-93 (1993); **Lemoli RM et al** Interleukin 11 stimulates the proliferation of human hematopoietic $CD34^+$ and $CD34^+CD33^-DR^-$ cells and synergises with stem cell factor, interleukin-3, and granulocyte-macrophage colony-stimulating factor. EH 21: 1668-72 (1993); **Musashi M et al** Direct and synergistic effects of interleukin 11 on murine hemopoiesis in culture. PNAS 88: 765-9 (1991); **Musashi M et al** Synergistic interactions between interleukin-11 and interleukin-4 in support of proliferation of primitive hematopoietic progenitors of mice. Blood 78: 1448-51 (1991); **Neben TY et al** Recombinant human interleukin-11 stimulates megakaryocytopoiesis and increases peripheral platelets in normal and splenectomized mice. Blood 81: 901-8 (1993); **Paul SR et al** Lack of a role of interleukin 11 in the growth of multiple myeloma. Leuk. Res. 16: 247-52 (1992); **Quesniaux VF et al** Interleukin-11 stimulates multiple phases of erythropoiesis *in vitro*. 80: 1218-23 (1992); **Schibler KR et al** Effect of interleukin-11 on cycling status and clonogenic maturation of fetal and adult hematopoietic progenitors. Blood 80: 900-3 (1992); **Teramura M et al** Interleukin-11 enhances human megakaryocytopoiesis *in vitro*. Blood 79: 327-31 (1992); **Tsuji K et al** Enhancement of murine hematopoiesis by synergistic interactions between steel factor (ligand for c-*kit*), interleukin-11, and other early acting factors in culture. Blood 79: 2855-60 (1992); **Yin T et al** Enhancement of *in vitro* and *in vivo* antigen-specific antibody responses by interleukin 11. JEM 175: 211-6 (1992); **Yonemura Y et al** Synergistic effects of interleukin 3 and interleukin 11 on murine megakaryopoiesis in serum-free culture. EH 20: 1011-6 (1992)

● **ASSAYS: Tohyama K et al** Production of multiple growth factors by a newly established human thyroid carcinoma cell line. Jpn. J. Cancer Res. 83: 153-8 (1992); for further information also see individual cell lines used in individual bioassays.

IL12: Interleukin 12.

■ **ALTERNATIVE NAMES:** CLMF (cytotoxic lymphocyte maturation factor); NKSF (natural killer cell stimulatory factor); TcMF (CTL maturation factor; see: CLMF); TSF (T cell stimulating factor). See also: individual entries for further information.

■ **SOURCES:** IL12 is secreted by peripheral lymphocytes after induction. It is produced mainly by B cells and to a lesser extent by T cells. The most powerful inducers of IL12 are bacteria, bacterial products, and parasites. IL12 is produced after stimulation with » phorbol esters or » calcium ionophore by human B lymphoblastoid cells. The human lymphoblastoid cell line NC37 secretes large amounts of the uncomplexed 40 kDa subunit of IL12.

■ **PROTEIN CHARACTERISTICS:** IL12 is a heterodimeric 70 kDa glycoprotein consisting of a 40 kDa subunit (p40, 306 amino acids; 10 % carbohydrate) and a 35 kDa subunit (p35, 197 amino acids; 20 % carbohydrate) linked by disulfide bonds that are essential for the biological activity of IL12. p40 contains 10 cysteines and a binding site for heparin (see also: HBGF, heparin-binding growth factors); p35 contains 7 cysteines.

The two subunits of IL12 are not related to any other known proteins. p40 shows some homology with the extracellular domain of the receptor for » IL6, and p35 appears to be a homologue of IL6.

■ **GENE STRUCTURE:** The gene encoding the p40 subunit of IL12 (IL12B) maps to human chromosome 5q31-q33 in the same region that also harbors other cytokine genes (see: $5q^-$ syndrome). The gene encoding the p35 subunit of IL12 (IL12A) maps to human chromosome 3p12-q13.2. The expression of the two genes is regulated independently of each other

■ **RECEPTOR STRUCTURE, GENE(S), EXPRESSION:** The IL12 receptor appears to be a single protein of ≈ 110 kDa. Up to 1000-9000 high affinity IL12 receptors/cell are expressed on peripheral blood mononuclear cells activated by various T cell mitogens or by » IL2. IL12 receptors are present on activated $CD4^+$ and $CD8^+$ T cells and on activated $CD56^+$ natural killer cells. Resting peripheral blood mononuclear cells, tonsillar B cells, or tonsillar B cells activated by anti-IgM/Dx, anti-IgM/Dx + IL2, or SAC + IL2 do not express the receptor. High affinity IL12 receptors are expressed constitutively on a transformed marmoset NK-like cell line, HVS.SILVA 40.

Binding of IL12 to its receptor can be prevented by monoclonal antibodies directed against the p40 subunit which therefore contains the binding site. The p40 subunit of IL12 shows homology with the extracellular domain of the » IL6 receptor.

■ **BIOLOGICAL ACTIVITIES:** Human IL12 is not active in murine lymphocytes. Hybrid heterodimers consisting of murine p35 and human p40 subunits retain bioactivity on murine cells; however, the combination of human p35 and murine p40 is completely inactive on murine cells. Murine IL12 is active on both murine and human lymphocytes.

The p40 subunit of murine IL12 subunit p40 (IL12p40) has been shown to specifically antagonize the effects of the IL12 heterodimer in different assay systems and to function as an endogenous specific inhibitor for the IL12 heterodimer.

IL12 stimulates the proliferation of human lymphoblasts activated by phytohemagglutinin (see also: cell activation). IL12 activates CD56-positive NK cells, and this activity is blocked by antibodies specific for » TNF-α. IL12 promotes specific allogenic CTL reactions. IL12 synergizes also with anti-CD3 antibodies and with allogeneic stimulation in mixed lymphocyte cultures in inducing T cell proliferation.

In peripheral lymphocytes of the TH1 type IL12 induces the synthesis of » IFN-γ and » IL2, and » TNF. TNF-α also appears to be involved in mediating the effects of IL12 on natural killer cells since the effects of IL12 are inhibited by an antibody directed against TNF-α. IL12 and TNF-α are co-stimulators for IFN-γ production with IL12 maximizing the IFN-γ response; the production of IL12, TNF, and IFN-γ is inhibited by » IL10. In TH2 helper cells IL12 reduces the synthesis of » IL4, » IL5, and » IL10 (see also: TH1/TH2 cytokines).

IL12 synergises with suboptimal amounts of » IL2 in promoting the proliferation of mononuclear cells in the peripheral blood and in promoting the generation of » LAK cells (lymphokine activated killer cells). Picomolar concentrations of IL12 are as effective as nanomolar concentrations of IL2 in augmenting the cytolytic activity of natural killer cells expanded *in vivo* by IL2. IL12 also acts as a co-mitogen and potentiates the IL2-induced proliferation of resting peripheral cells.

IL12 is probably involved also in the selection of immunoglobulin isotypes. At picomolar concentrations IL12 markedly inhibits the synthesis of IgE by IL4-stimulated peripheral blood mononuclear cells also in the presence of antibodies directed against » IFN-γ.

Murine IL12 has been shown to alter $CD4^+$ subset differentiation and to be involved in the induction of protective immunity against intracellular parasitic infections in mice. It cures mice infected with *Leishmania major*.

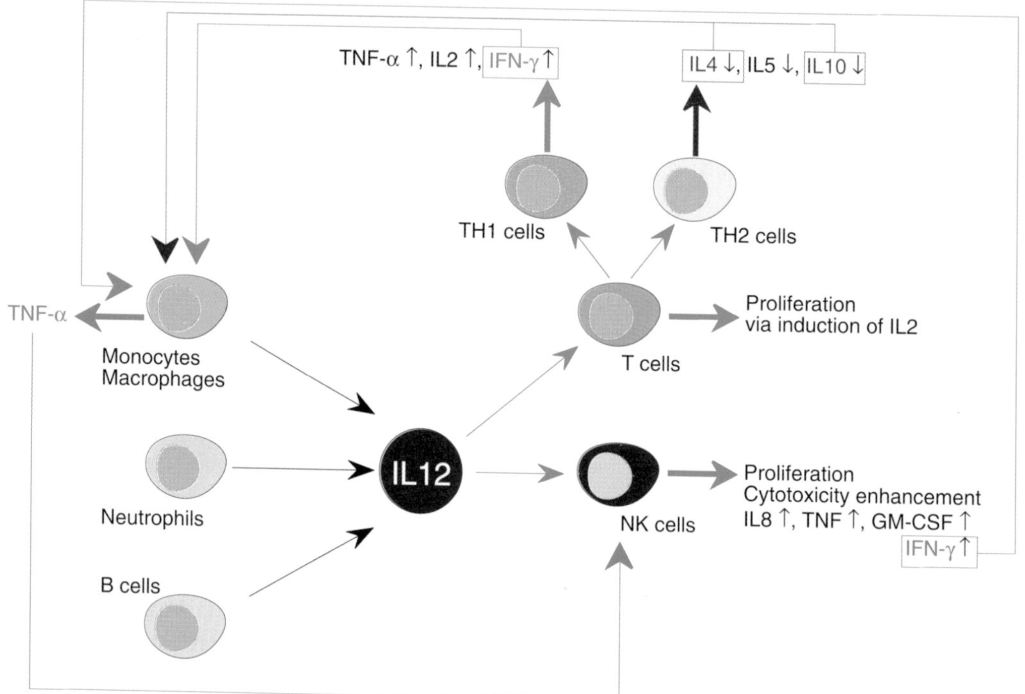

Production, major biological effects, and immunoregulatory function of IL12.
B cells, neutrophils and macrophages/monocytes produce IL12 in response to bacterial products or parasites. The major targets of IL12 are NK cells and T cells. In NK cells IL12 induces proliferation, enhances cell-mediated cytotoxicity, and induces cytokine secretion. T cells proliferate in response to IL12. IL12 also favors development of TH1 subsets of cells and prevents differentiation of TH2 cells.
Several possible positive and negative feedback loops can be envisaged.
TH1 cell and NK cell-derived IFN-γ further augment production of IL12 by macrophages. IL10 and IL4, both products of TH2 cells, suppress macrophage production of IL12 and are thus part of a negative feedback loop. IL10 also inhibits the production of TNF-α (stimulatory for NK cells) and IFN-γ. Green lines/arrow heads: stimulation/production; red lines/arrow heads: inhibition.

IL12 enhances » SCF (stem cell factor)-induced myelopoiesis of primitive bone marrow progenitor cells and synergizes with colony-stimulating factors (see: CSF) to induce proliferation. IL12 also has synergistic effects on more committed bone marrow progenitors, synergising with IL3, IL11, or IL3 plus SCF.

■ **Clinical Use & Significance:** IL12 is of potential clinical interest since it allows to reduce doses of IL2 required for the generation of » LAK cells (lymphokine-activated killer cells). This may reduce the toxicity of higher doses of IL2 used for » Adoptive immunotherapy. IL12 may be useful in expanding an antigen-specific T cell population, as the culture of cytotoxic T lymphocytes with IL12 and low-dose IL2 leads to proliferation only in response to an antigen co-signal. IL12 has been shown to augment natural killer-cell mediated cytotoxicity in a number of conditions, including patients with hairy cell leukemia.
The ability of IL12 to induce the synthesis of IFN-γ and to stimulate the proliferation of resting peripheral cells may also be of interest.

■ **Assays:** IL12 can be assayed by its activity as » NKSF (natural killer cell stimulatory factor) or CLMF (cytotoxic lymphocyte maturation factor).

● **Reviews: Quesniaux VF** Interleukins 9, 10, 11 and 12 and *kit* ligand: a brief overview. Res. Immunol. 143: 385-400 (1992); **Trinchieri G et al** Natural killer cell stimulatory factor (NKSF) or interleukin 12 is a key regulator of immune response in inflammation. PGFR 4: 355-68 (1992); **Trinchieri G et al** Interleukin-12 and its role in the generation of TH1 cells. IT 14: 335-8 (1993)

● **Biochemistry & Molecular Biology: Gearing DP & Cosman D** Homology of the p40 subunit of natural killer cell stimulatory factor (NKSF) with the extracellular domain of the interleukin-6 receptor. Cell 66: 9-10 (1991); **Merberg DM et al** Sequence similarity between NKSF and the IL6/G-CSF family.

IT 13: 77-8 (1992); **Podlaski FJ et al** Molecular characterization of interleukin 12. ABB 294: 230-7 (1992); **Schoenhaut DS et al** Cloning and expression of murine IL12. JI 148: 3433-40 (1992); **Sieburth D et al** Assignment of genes encoding a unique cytokine (IL12) composed of two unrelated subunits to chromosomes 3 and 5. Genomics 14: 59-62 (1992)

● **Receptor Structure, Gene(s), Expression: Desai BB et al** IL12 receptor. II. Distribution and regulation of receptor expression. JI 148: 3125-32 (1992); **Chizzonite R et al** IL12 receptor. I. Characterization of the receptor on phytohemagglutinin-activated human lymphoblasts. JI 148: 3117-24 (1992); **Gearing DP & Cosman D** Homology of the p40 subunit of natural killer cell stimulatory factor (NKSF) with the extracellular domain of the interleukin-6 receptor. Cell 66: 9-10 (1991)

● **Biological Activities: Bertagnolli MM et al** IL12 augments antigen-dependent proliferation of activated T lymphocytes. JI 149: 3778-83 (1992); **Chan SH et al** Induction of interferon γ production by natural killer cell stimulatory factor: characterization of the responder cells and synergy with other inducers. JEM 173: 869-79 (1991); **Chan SH et al** Mechanisms of IFN-γ induction by natural killer cell stimulatory factor (NKSF/IL12): Role of transcription and mRNA stability in the synergistic interaction between NKSF and IL2. JI 148: 92-8 (1992); **Chizzonite R et al** IL12: monoclonal antibodies specific for the 40 kDa subunit block receptor binding and biological activity on activated human lymphoblasts. JI 147: 1548-56 (1991); **D'Andrea A et al** Production of natural killer cell stimulatory factor (interleukin 12) by peripheral blood mononuclear cells. JEM 176: 1387-98 (1992); **D'Andrea A et al** Interleukin 10 (IL10) inhibits human lymphocyte interferon γ production by suppressing natural killer cell stimulatory factor/IL12 synthesis in accessory cells. JEM 178: 1041-8 (1993); **Gately MK et al** Regulation of human lymphocyte proliferation by a heterodimeric cytokine, IL12 (cytotoxic lymphocyte maturation factor). JI 147: 874-82 (1991); **Germann T et al** Interleukin-12/T cell stimulating factor, a cytokine with multiple effects on T helper type 1 (Th1) but not on Th2 cells. EJI 23: 1762-70 (1993); **Heinzel FP et al** Recombinant interleukin 12 cures mice infected with *Leishmania major*. JEM 177: 1505-9 (1993); **Jacobsen SE et al** Cytotoxic lymphocyte maturation factor (interleukin 12) is a synergistic growth factor for hematopoietic stem cells. JEM 178: 413-8 (1993); **Kiniwa M et al** Recombinant interleukin-12 suppresses the synthesis of immunoglobulin E by interleukin-4 stimulated human lymphocytes. JCI 90: 262-6 (1992); **Mattner F et al** The interleukin-12 subunit p40 specifically inhibits effects of the interleukin-12 heterodimer. EJI 23: 2202-8 (1993); **Naume B et al** A comparative study of IL12 (cytotoxic lymphocyte maturation factor), IL2, and IL7-induced effects on immunomagnetically purified CD56$^+$ NK cells. JI 148: 2429-36 (1992); **Naume B et al** Gene expression and secretion of cytokines and cytokine receptors from highly purified CD56$^+$ natural killer cells stimulated with interleukin-2, interleukin-7 and interleukin-12. EJI 23: 1831-8 (1993); **Perussia B et al** Natural killer (NK) cell stimulatory factor or IL12 has differential effects on the proliferation of TCR-α β$^+$, TCR-γ δ$^+$ T lymphocytes, and NK cells. JI 149: 3495-502 (1992); **Ploemacher RE et al** Interleukin-12 synergizes with interleukin-3 and steel factor to enhance recovery of murine hemopoietic stem cells in liquid culture. Leukemia 7: 1381-8 (1993); **Ploemacher RE et al** Interleukin-12 enhances interleukin-3 dependent multilineage hematopoietic colony formation stimulated by interleukin-11 or steel factor. Leukemia 7: 1374-80 (1993); **Robertson MJ et al** Response of human natural killer (NK) cells to NK cell stimulatory factor (NKSF): cytolytic activity and proliferation of NK cells are differentially regulated by NKSF. JEM 175: 779-88 (1992); **Tripp CS et al** Interleukin 12 and tumor necrosis factor α are costimulators of interferon γ production by natural killer cells in severe combined immunodeficiency mice with listeriosis, and interleukin 10 is a physiologic antagonist. PNAS 90: 3725-9 (1993); **Valiante NM et al** Role of the production of natural killer cell stimulatory factor (NKSF/IL12) in the ability of B cell lines to stimulate T and NK cell proliferation. CI 145: 187-98 (1992); **Wu CY et al** IL12 induces the production of IFN-γ by neonatal human CD4 T cells. JI 151: 1938-49 (1993)

● **Clinical Use & Significance: Bigda J et al** Interleukin 12 augments natural killer-cell mediated cytotoxicity in hairy cell leukemia. Leuk. Lymphoma 10: 121-5 (1993); **Chehimi J et al** Natural killer (NK) cell stimulatory factor increases the cytotoxic activity of NK cells from both healthy donors and human immunodeficiency virus-infected patients. JEM 175: 789-96 (1992); **Chehimi J et al** Enhancing effect of natural killer cell stimulatory factor (NKSF/interleukin-12) on cell-mediated cytotoxicity against tumor-derived and virus-infected cells. EJI 23: 1826-30 (1993); **Lieberman MD et al** Natural killer cell stimulatory factor (NKSF) augments natural killer cell and antibody-dependent tumoricidal response against colon carcinoma cell lines. J. Surg. Res. 50: 410-5 (1991)

IL13: Interleukin 13. A protein originally detected by differential screening of a subtracted cDNA library prepared from activated peripheral blood mononuclear cells treated with anti-CD28. This cytokine with a relative molecular mass of ≈ 10 kDa, designated ***NC30***, is probably the human homologue of murine » P600, produced by TH2 helper T cell clones following activation (see also: TH1/TH2 cytokines). IL13 is expressed in activated human T cells but not in human heart, brain, placenta, lung, liver, and skeletal muscle tissues. The human IL13 cDNA is 66 % identical with that of the mouse. IL13 is a non-glycosylated protein of 132 amino acids that contains five cysteine residues. A variant containing an additional glutamine residue at position 98 has been described. It arises from alternative splicing at an intron-exon boundary and appears to have the same specific biological activity as IL13 lacking this additional amino acid.

The structure of the gene for IL13 is closely similar to that of » IL3, » IL5, » IL4, and » GM-CSF, and all five genes are clustered on 5q (see also: 5q$^-$ syndrome). The IL13 gene is located between the GM-CSF and the IL4 gene, with the IL13 gene in close vicinity of the IL4 gene [cen-- IL3--GM-CSF--IL13--IL4--IL5--tel]. The human IL13 gene is a single copy gene and contains four exons. It shows a high degree of sequence identity with the murine IL13 gene.

IL13 down-modulates macrophage activity, reducing the production of inflammatory cytokines in response to IFN-γ or bacterial lypopolysaccharides. IL13 also decreases the production of nitric oxide by activated macrophages, leading to a decrease in parasiticidal activity. IL13 induces human monocyte differentiation, enhances survival time in culture, and also induces B cell differentiation and proliferation and » Isotype switching (see also: BGDF, B cell growth and differentiation factors). It induces » IL4-independent IgG4 and IgE synthesis in human B cells and germ-line IgE heavy chain gene transcription. IL13 induces considerable levels of IgM and IgG, but no IgA, in cultures of highly purified surface IgD-positive or total B cells in the presence of an activated CD4-positive T cell clone. IL13 induces proliferation and differentiation of human B cells activated by the » CD40 ligand (see: TRAP). IL13 synergizes with » IL2 in regulating » IFN-γ synthesis in large granular lymphocytes. IL13 has been shown to inhibit strongly LPS-induced tissue factor expression and to reduce the pyrogenic effects of » IL1 or » TNF, thus protecting endothelial and monocyte surfaces against inflammatory mediator-induced procoagulant changes. IL13 inhibits human immunodeficiency virus type 1 production in primary blood-derived human macrophages *in vitro*.

IL13 has been observed to increase the IL2-induced killer activity of » LAK cells (lymphokine-activated killer cells). Systemic immunity to parental tumor cells can be induced by IL13-secreting tumor cells *in vivo*.

IL13, like IL4, induces » CD23 expression on B cells and enhances CD72, and class II major histocompatibility complex antigen expression.

Human and murine IL13 activities are assayed by employing human » TF-1 erythroleukemia cells which proliferate in response to this factor. The activity of IL13 is independent of IL4, but these two cytokines have common signaling pathways. IL13 competitively inhibits binding of human IL4 to functional IL4 receptors on the human premyeloid cell line » TF-1 that responds to IL4 and IL13.

● REVIEWS: **Zurawski G & de Vries JE** Interleukin 13, an interleukin 4-like cytokine that acts on monocytes and B cells, but not on T cells. IT 15: 19-26 (1994)

Briere F et al IL10 and IL13 as B cell growth and differentiation factors. Nouv. Rev. Fr. Hematol. 35: 233-5 (1993); **Cocks BG et al** IL13 induces proliferation and differentiation of human B cells activated by the CD40 ligand. Int. Immunol. 5: 657-63 (1993); **Doherty TM et al** Modulation of murine macrophage function by IL13. JI 151: 7151-60 (1993); **Herbert JM et al** IL4 and IL13 exhibit comparable abilities to reduce pyrogen-induced expression of procoagulant activity in endothelial cells and monocytes. FL 328: 268-70 (1993); **McKenzie ANJ et al** Interleukin 13, a T cell-derived cytokine that regulates human monocyte and B cell functions. PNAS 90: 3735-9 (1993); **Minty A et al** Molecular cloning of the MCP-3 chemokine gene and regulation of its expression. Eur. Cytokine. Netw. 4: 99-110 (1993); **McKenzie AN et al** Structural comparison and chromosomal localization of the human and mouse IL13 genes. JI 150: 5436-44 (1993); **Minty E et al** Interleukin-13 is a new human lymphokine regulating inflammatory and immune responses. N 352: 248-50 (1993); **Montaner LJ et al** Interleukin 13 inhibits human immunodeficiency virus type 1 production in primary blood-

Comparison of some biological activities of human IL13 and IL4.

Activity	Human IL13	Human IL4
B cells		
Proliferation	stimulatory	stimulatory
Isotype switching	stimulatory	stimulatory
Expression of CD23	up regulation	up regulation
Expression of MHC class II	up regulation	up regulation
T cells		
Proliferation	stimulatory	negative effect
Expression of CD8α	up regulation	down regulation
Macrophages/Monocytes		
Expression of MHC class II	up regulation	up regulation
Expression of CD23	up regulation	up regulation
Expression of Fcg receptors	down regulation	down regulation
Expression of proinflammatory cytokines	down regulation	down regulation
Expression of IL1ra	up regulation	up regulation
Expression of IL8, MIP-1a	down regulation	down regulation
Function as antigen presenting cell	stimulatory	stimulatory
Activity in ADCC	down regulation	down regulation

derived human macrophages *in vitro*. JEM 178: 743-7 (1993); **Morgan JG et al** The selective isolation of novel cDNAs encoded by the regions surrounding the human interleukin 4 and 5 genes. NAR 20: 5173-9 (1992); **Punnonen J et al** Interleukin 13 induces interleukin 4-dependent IgG4 and IgE synthesis and CD23 expression by human B cells. PNAS 90: 3730-4 (1993); **Zurawsky SM et al** Receptors for interleukin 13 and interleukin 4 are complex and share a novel component that functions in signal transduction. EJ 12: 2663-70 (1993)

IL14: Interleukin 14. This name has been proposed for » HMW-BCGF (high molecular weight BCGF).

IL-B4: see: Interleukin B.

ILF: *interleukin enhancer binding factor* ILF is a DNA binding protein that binds to purine-rich sequence motifs in the promoter of the » IL2 gene. ILF also binds to purine-rich regulatory motifs in the HIV-1 long terminal repeat. The protein is expressed constitutively in lymphoid and non lymphoid tissues. The corresponding gene maps to human chromosome 17q25.

Li C et al Cloning of a cellular factor, interleukin binding factor, that binds to NFAT-like motifs in the human immunodeficiency virus long terminal repeat. PNAS 88: 7739-7743 (1991); **Li C et al** Characterizationand chromosomal mapping of the gene encoding the cellular DNA binding protein ILF. Genomics 13: 665-71 (1992)

ILGF: *insulin-like growth factors* see: IGF.

ILHP1: *interleukin-hemopoietin-1* This T cell-derived factor was identified by its ability to support the growth of a subset of B cell hybridomas (H = Hybridoma; P = plasmacytoma). It is identical with » IL6.

Caypas S et al Identification of an interleukin HP1-like plasmacytoma growth factor produced by L cells in response to viral infection. JI 139: 2965-9 (1987); **Coulie PG et al** Interleukin-HP1-related hybridoma and plasmacytoma growth factors induced by lipopolysaccharide *in vivo*. EJI 17: 1217-80 (1987); **Simpson RJ et al** Murine hybridoma/plasmacytoma growth factor. Complete amino-acid sequence and relation to human interleukin-6. EJB 176: 187-97 (1988); **Van Snick J et al** Purification and NH2-terminal amino acid sequence of a new T cell-derived lymphokine with growth factor activity for B cell hybridomas. PNAS 83: 9679-83 (1986); **Van Snick J et al** Interleukin-HP1, a T cell-derived hybridoma growth factor that supports the *in vitro* growth of murine plasmacytomas. JEM 165: 641-9 (1987); **Van Snick J et al** cDNA cloning of murine interleukin-HP1: homology with human interleukin 6. EJI 18: 193-7 (1988)

Ilotropin: This poorly characterized factor of ≈ 29-44 kDa specifically promotes the growth of pancreatic islet cells.

Pittenger GL et al The partial isolation and characterization of ilotropin, a novel islet-specific growth factor. AEMB 321: 123-30; discussion 131-2 (1992)

IL15: see addendum.

ILS: An IL1 inhibitor expressed by epithelial cells. See: IL1ra (IL1 receptor antagonist).

Immediate early genes: see: Gene expression.

Immune-neuroendocrine axis: see: Neuroimmune network.

Immune interferon: This interferon is identical with » IFN-γ. See also: IFN for general information about interferons.

Immune protein 10 gamma: see: IP-10.

Immunoassays: see: Cytokine assays.

Immunocytokines: A term used occasionally for proteins modulating the growth and functional activities of immune cells. It is more or less a synonym for » cytokines.

Immunodeficient mice: A variety of genetic mutations are known that impair immune functions in mice. Some of these mutant strains are widely used in cytokine research to study a variety of physiologic and disease processes.

Immunodeficient mouse models are useful for understanding mechanisms supporting allogeneic and xenogeneic tumor growth and progression. These mice are used as models of normal » Hematopoiesis, to study the ontogeny of immune system development, to investigate the growth of lymphomas and leukemic cell types, and to study the influences of cytokines and new therapeutic strategies. Acceptance of multiple tissue xenografts permits these mice to be used also as intermediate models for host-specific, fastidious organisms for which a small animal model has not been available previously.

Genetic loci affecting immune responses include ***nu*** (nude), ***scid*** (severe combined immunodeficiency), ***beige***, and ***xid*** (X-linked immunodeficiency). The various mouse mutants have differing immunological properties and therefore permit complementary studies. The extent to which some of these mutations interfere with immune functions can vary with the genetic background. It has been shown, for example, that the immune defects in CH3 scid mice are more severe than those in the original CB17 mice.

■ **Beige:** These mice have a deficiency in cytotoxic T cells and natural killer (NK) cells. The beige mutation has been shown to affect lysosomal storage and to cause cells to accumulate abnormally large cytoplasmic granules. The effect of the beige mutation on the immune system is indirect.

■ **Bnx:** The ***bnx*** mouse has defects in three loci and has been obtained by crossing of beige, nude and xid mice. These mice have natural killer cells and intact non-lymphoid resistance systems (macrophages). They still mount immune responses and animals with even mild subclinical infections are extremely resistant to xenografting. They have been used to study » Hematopoiesis after engraftment of normal human hematopoietic tissues.

■ **Nude:** The murine recessive nude mutation on chromosome 11 arose spontaneously. Viability and fertility are severely reduced but can be improved under specific pathogen-free conditions.

Homozygotes are hairless from birth and completely lack a thymus due to a failure of development of the thymic anlage. The thymic rudiment of homozygous nude mice fails to attract lymphoid cells but their lymphocytes are normal in their ability to populate implanted normal thymuses. The lack of the thymus leads to many defects of the immune system, including depletion of lymphocytes in thymus-dependent regions of lymph nodes and spleen, a greatly reduced lymphocyte population composed almost entirely of B cells, very poor response to thymus-dependent antigens but relatively normal IgM responses to thymus-independent antigens. Nude mice can produce low levels of mature T cells through extrathymic tissues, particularly when they are exposed to foreign antigens. Nude mice are naturally thymectomized animals and are, therefore, used for investigations of the role of the thymus in immune reactions. They easily accept xenografts.

■ **Scid:** The murine recessive scid mutation on chromosome 16 arose in the C.B-17 inbred strain (BALB/c.C57BL/Ka-Igh-1b). Homozygous scid mice are fertile and can survive a year or more under specific pathogen-free conditions.

The scid mutation affects differentiation of stem cells into mature lymphocytes but does not impair myeloid differentiation (see also: Hematopoiesis). The scid mouse has a thymus, lymph nodes, and splenic follicles virtually devoid of lymphocytes. Homozygotes are deficient in B and T cell functions and lack detectable B cells and pre-B cells. Their spleen cells do not respond to B or T cell mitogens. Scid mice appear to have a generalized radiation repair defect which renders them more sensitive than wild-type mice to the effects of ionizing irradiation.

Scid mice have normal numbers of radiation-sensitive stem cells with defective proliferative capacity. The rearrangement of immunoglobulin and T cell receptor genes normally occurring in B and T lymphocytes is not observed and is probably due to defects in VDJ recombinases. Most homozygous scid animals have no detectable levels of IgM, IgG1, IgG2a, IgG2b, IgG3, or IgA although some young and adult scid mice are leaky and express low levels of one to three of these immunoglobulins.

Scid mice have normal numbers of natural killer cells, the main effectors of non-MHC-restricted immunity which also includes recognition of hemopoietic histocompatibility antigens on bone marrow allografts. These animals thus provide a lymphoid-free system in which to study NK effector functions.

Scid mice have been used also to study the role of cytokines on the differentiation and function of hematopoietic cell types *in vivo*, to evaluate possible lineage relationships between T and B lymphocytes and lymphoid-like cells, and to study interactions of selected normal and abnormal (e. g. autoimmune) lymphoid subsets.

Scid mice can be used for long-term engraftment of human hematolymphoid organs fetal liver, bone marrow, thymus, lymph node, skin, and mucosal tissues. Scid mice have been used to create mouse/human hematolymphoid chimeras, designated ***SCID-hu mice***. Engraftment of certain combinations of human organs (e. g. human thymus and intact fetal liver) into scid mice results in long-term multilineage human » Hematopoiesis *in vivo*. These animals thus serve as models for human hematolymphoid differentiation and function. They have been used also for as preclinical models to assess the *in vivo* effects of hematopoietic cytokines on human hematopoiesis.

■ **Xid:** The murine xid defect is X-linked and involves dysfunction of a cytoplasmic tyrosine kinase, designated ***Btk*** (Bruton's tyrosine kinase; also called ***bpk*** = B cell progenitor kinase, or ***atk*** = agammaglobulinemia tyrosine kinase) which is also defective in human X-linked agammaglobulinemia, a human immunodeficiency caused by failure of pre-B cells in the bone marrow to develop into circulating mature cells. The mutation in the xid locus affects » LAK cell activity (lymphokine-aktivated killer cells) and also B cell responses to certain thymus-independent antigens. Xid mice show an expansion of a functionally abnormal B cell population that phenotypically resembles immature B cells.

● BEIGE: **Raffa RB et al** The combined immunological and antinociceptive defects of beige-J mice: the possible existence of a 'µ-repressin'. Life Sci. 52: 1-8 (1993); **Roder J** The beige mutation in the mouse. I. A stem cell predetermined impairment in natural killer cell function. JI 123: 2168-723 (1979); **Talmadge J et al** Role of NK cells in tumor growth and metastasis in beige mice. N 284: 622-4 (1980)

● BNX: **Biondi A et al** Human T-cell lymphoblastic lymphoma expressing the T-cell receptor γ/δ established in immune-deficient (bg/nu/xid) mice. Leukemia 7: 281-9 (1993); **Dick JE** Establishment of assays for human hematopoietic cells in immune deficient mice. CTMI 152: 219-24 (1989); **Ferrari G et al** Transfer of the ADA gene into human ADA-deficient T lymphocytes reconstitutes specific immune functions. Blood 80: 1120-4 (1992); **Jicha DL et al** The persistence of human peripheral lymphocytes, tumor infiltrating lymphocytes, and colon adenocarcinomas in immunodeficient mice. J. Immunother. 11: 19-29 (1992); **Kamel-Reid S et al** A model of human acute lymphoblastic leukemia in immune-deficient scid mice. S 246: 1597-600 (1989); **Kollmann TR et al** The concurrent maturation of mouse and human thymocytes in human fetal thymus implanted in NIH-beige-nude-xid mice is associated with the reconstitution of the murine immune system. JEM 177: 821-32 (1993); **von Kalle C et al** Growth of Hodgkin cell lines in severely combined immunodeficient mice. IJC 52: 887-91 (1992)

● NUDE: **Flanagan SP** "Nude", a new hairless gene with pleiotropic effects in the mouse. Genet. Res. 8: 295-309 (1966); **Fogh J & Giovanella BC** (eds) The nude mouse in experiment and clinical research Academic Press 1982; **Fogh J et al** One hundred and twenty-seven cultured human tumor cell lines producing tumors in nude mice. JNCI 59: 221-6 (1977); **Nilsson K et al** Tumorigenicity of human hematopoietic cell lines in athymic nude mice. IJC 19: 337-44 (1977); **Pantelouris EM** Absence of thymus in a mouse mutant. N 217: 370-1 (1968)

● SCID: **Bosma GC et al** A severe combined immunodeficiency mutation in the mouse N 301: 527-30 (1983); **Bosma GC et al** The mouse mutation severe combined immune deficiency (scid) is on chromosome 16. Immunogenetics 29: 54-7 (1989); **Bosma MJ & Carroll AM** The SCID mouse mutant: Definition, characterization, and potential uses. ARI 9: 323-50 (1991); **Bosma MJ** B and T cell leakiness in the scid mouse mutant. Immunodefic. Rev. 3: 261-76 (1992); **Custer RP et al** Severe combined immune deficiency (SCID) in the mouse. Pathology, reconstitution, neoplasms. Am. J. Pathol. 120: 464-77 (1985); **Dick JE** Immune-deficient mice as models of normal and leukemic human hematopoiesis. Cancer Cells 3: 39-48 (1991); **Dorshkind K et al** Functional status of cells from lymphoid and myeloid tissues in mice with severe combined immunodeficiency disease. JI 132: 1804-8 (1984); **Dorshkind K et al** Natural killer (NK) cells are present in mice with severe combined immunodeficiency (scid). JI 134: 3798-801 (1985); **Duchosal MA** SCID mice in the study of human autoimmune diseases. Springer Semin. Immunopathol. 14: 159-77 (1992); **Fulop GM & Phillips RA** Full reconstitution of the immune deficiency in scid mice with normal stem cells requires low-dose irradiation of the recipients. JI 136: 4438-43 (1986); **Fulop GM & Phillips RA** Use of scid mice to identify and quantitate lymphoid-restricted stem cell in long-term bone marrow cultures. Blood 74: 1537-44 (1989); **Fulop G & Phillips R** The scid mutation in mice causes a general defect in radiation repair. N 134: 479-82 (1990); **Greenwood JD** Xenogeneic PBL-scid mice: their potential and current limitations. Lab. Anim. Sci. 43: 151-5 (1993); **Hardy RR et al** Repopulation of SCID mice with fetal-derived B lineage cells. CTMI 182: 73-80 (1992); **Koyoizumi S et al** Implantation and maintenance of functional human bone marrow in SCID-hu mice. Blood 79: 1704-11 (1992); **Koyoizumi S et al** Preclinical analysis of cytokine therapy in the SCID-hu mouse. Blood 81: 1479-88 (1993); **Lapidot T et al** Cytokine stimulation of multilineage hematopoiesis from immature human cells engrafted in SCID mice. S 255: 1137-41 (1992); **Martino G et al** The fate of human peripheral blood lymphocytes after transplantation into SCID mice. EJI 23: 1023-8 (1993); **McCune JM et al** The SCID-hu mouse: a small animal model for HIV infection and pathogenesis. ARI 9: 399-429 (1991); **McCune JM** The SCID-hu mouse: a small animal model for the analysis of human hematolymphoid differentiation and function. Bone Marrow Transplant. 9: s74-6 (1992); **Melchers F et al** Progenitors and precursor B lymphocytes of mice. Proliferation and differentiation *in vitro* and population, differentiation and turnover in SCID mice *in vivo* of normal and abnormal cells. CTMI 182: 3-12 (1992); **Namikawa R et al** Long-term hematopoiesis in the SCID-hu mouse. JEM 172: 1055-63 (1990); **Nonoyama S et al** Strain-dependent leakiness of mice with severe combined immune deficiency. JI 150: 3817-24 (1993); **Peault B et al** Thy-1-expressing $CD34^+$ human cells express multiple hematopoietic potentialities *in vitro* and in SCID-hu mice. Nouv. Rev. Fr. Hematol. 35: 91-3 (1993); **Phillips RA et al** Growth of human tumors in immune-deficient scid mice and nude mice. CTMI 152: 259-63 (1989); **Ratajczak MZ et al** *In vivo* treatment of human leukemia in a scid mouse model with c-*myb* antisense oligodeoxynucleotides. PNAS 89: 11823-7 (1992); **Schuler W et al** Rearrangement of antigen receptor genes is defective in mice with severe combined immune deficiency. Cell 46: 963-72 (1986); **Taniguchi S et al** Hemopoietic stem-cell compartment of the SCID mouse: double-exponential survival curve after γ irradiation. PNAS 90: 4354-8 (1993); **Taylor PC** Current status review: the severe combined immunodeficient (SCID) mouse: xenogeneic-SCID chimeras in the investigation of human autoimmune disease. Int. J. Exp. Pathol. 73: 251-9 (1992); **Vandekerckhove BA et al** Human hematopoietic cells and thymic epithelial cells induce tolerance via different mechanisms in the SCID-hu mouse thymus. JEM 175: 1033-43 (1992); **Vladutiu AO** The severe combined immunodeficient (SCID) mouse as a model for the study of autoimmune diseases. Clin. Exp. Immunol. 93: 1-8 (1993); **Waller EK et al** Human T-cell development in SCID-hu mice: staphylococcal enterotoxins induce specific clonal deletions, proliferation, and anergy. Blood 80: 3144-56 (1992)

● XID: **Karagogeos D et al** Early arrest of B cell development in nude, X-linked immune-deficient mice. EJI 16: 1125-30 (1986); **Rawlings DJ et al** Mutation of unique region of Bruton's tyrosine kinase in immunodeficient XID mice. S 261: 358-61 (1993); **Scher I et al** X-linked B lymphocyte immune defect in CBA/N mice. II. Studies of the mechanisms underlying the immune defect. JEM 142: 142: 637-50 (1975); **Sprent J et al** Physiology of B cells in mice with X-linked immunodeficiency. I. Size, migratory properties, and turnover of the B cell population. JI 134: 1442-8 (1985); **Thomas JD et al** Colocalization of X-linked agammaglobulinemia and X-linked immunodeficiency genes. S 261: 355-8 (1993); **Tsukada S et al** Deficient expression of a B cell cytoplasmic tyrosine kinase in human X-linked agammaglobulinemia. 72: 279-90 (1993)

Immunoglobulin binding factors: see: IBF.

Immunoglobulin heavy chain class switching: see: Isotype switching.

Immunoglobulin (Ig) production stimulating factor: see: IPSF.

Immunomodulation: A general term used to describe the specific and unspecific alterations of immune system functions by a very heterogeneous group of microbial or synthetic substances and products of activated cells (see also: cell activation) such as » cytokines or » thymic hormones. Immunomodulators are also called biological response modifiers (abbrev. BRM). Their mechanism of action usually involves a complicated interplay of various regulator and effector systems. Immunomodulators may enhance (immunoprophylaxis, immunostimulation), restore (immunosubstitution, immunorestauration) or suppress (immunosuppression, immunodeviation) immunological reactivity.
Baseler MW & Urba WJ Immunological monitoring and clinical trials of biological response modifiers. Cancer Chemother. Biol. Response Modif. 12: 231-50 (1991); **Clark JW** Biological response modifiers. Cancer Chemother. Biol. Response Modif. 12: 193-212 (1991); **Rees RC** Cytokines as biological response modifiers. J. Clin. Pathol. 45: 93-8 (1992)

Immunophilins: A general term for a family of cytoplasmic proteins with immunoregulatory activities also involved in fundamental cellular processes such as protein folding and protein transport. Some prominent members of immunophilins are binding proteins for the immunosuppressants » CsA (Ciclosporin A) and FK506 (see: CsA).

Immunoradiometric assay: abbrev. IRMA. See: Cytokine assays.

Immunosuppressive acidic protein: see: IAP.

Immunotoxins: General term for cytotoxic proteins generated by covalent linkage of an antibody to a cytocidal toxin protein. Specific binding to target cells is effected by the antibody moiety of this hybrid protein while the toxin portion is used to specifically inactivate the target cells. For special constructs see also: mitotoxins, saporin, Cytokine fusion toxins.
Frankel AE Immunotoxin therapy of cancer. Oncology Huntingt. 7: 69-78 (1993); **Wawrzynczak EJ & Derbyshire EJ** Immunotoxins: the power and the glory. IT 13: 381-3 (1992)

Immunotransmitter: A term used occasionally for proteins modulating the growth and functional activities of immune cells. It is a synonym for » cytokines.

Induced Hematopoiesis**:** see: Hematopoiesis.

Inflammation: A general name for reactions occurring after most kinds of tissue injuries or infections or immunologic stimulation as a defense against foreign or altered endogenous substances. Inflammatory reactions involve a number of biochemical and cellular alterations the extent of which correlates with the extent of the initial trauma.
The most prominent systemic manifestation of inflammation is an elevation of body temperature and a variety of biochemical alterations known as the » acute phase reaction which leads to the synthesis of » acute phase proteins in the liver. The local inflammatory reaction is characterized by an initial increase in blood flow to the site of injury, enhanced vascular permeability, and the ordered and directional influx and selective accumulation of different effector cells from the peripheral blood at the site of injury. Influx of antigen non-specific but highly destructive cells (neutrophils) is one of the earliest stages of the inflammatory response. These cells mount a rapid, non-specific phagocytic response. At a later stage monocyte-macrophages and cells of other lymphocyte lineages (specific subsets of T and B cells) appear at the site of injury. These cell types are associated with antigen-specific and more tightly regulated immune responses and once activated also produce protective and inflammatory molecules.. An exudation of plasma into the lesion in the early stage is also observed.
Inflammatory cells express increasing numbers of cell-surface proteins and glycoproteins known as cell adhesion molecules. Endothelial cells are also activated during the initial phase of the inflammatory response and then express, among other things, adhesion molecule counterreceptors. The regulated expression of these molecules allows for the precise trafficking of circulating leukocytes to inflammatory sites. Cellular attachment of immune cells to endothelial cells lining blood vessels surrounding the inflammatory site prevents them from being swept past the site of infection or tissue damage and is a crucial step required for the subsequent emigration of these cells into the surrounding inflammatory tissues (extravasation).

Some cytokines involved in inflammatory reactions.

Cytokine	Major activity
Chemokines	Lymphocyte chemotaxis
G-CSF	Expansion of granulocyte populations
GM-CSF	Expansion of granulocyte/macrophage populations
IFN-α/β	Antiviral activity
IFN-γ	Macrophage activation Cytokine synthesis co-stimulator Stimulation of antigen presentation
IL1α	B cell activation
IL1β	Mediation of pain and fever
IL2	T cell proliferation Macrophage activation LAK cell activity
IL3	Differentiation of hematopoietic stem cells
IL4	IgE induction Inhibition of macrophage migration
IL5	Eosinophil differentiation
IL6	Induction of acute phase protein synthesis Terminal maturation of B cells
IL8	Neutrophil chemotaxis
IL9	B cell activation Mast cell growth factor
IL10	Repression of cytokine synthesis in T cells
IL11	Stimulation of megakaryocytopoiesis Induction of acute phase protein synthesis Stimulation of IgG synthesis
IL12	Cytotoxic lymphocyte maturation Stimulation of natural killer cells
M-CSF	Expansion of macrophage populations
PDGF	Fibroblast chemotaxis Monocyte/neutrophil activation Connective tissue stroma formation
TGF-α	Regulation of cell growth
TGF-β	Regulation of cell growth
TNF-α	T cell/B cell proliferation and function Macrophage activation Induction of procoagulant activity by endothelial cells Fibroblast activation/proliferation
TNF-β	Similar to TNF-α

The highly efficient process of cellular influx to inflammatory sites is mediated by a plethora of mediator substances supporting and dispersing inflammation. These mediators are found in the serum or tissue fluids, are released by degranulating cells, and are also secreted by inflammatory cells upon activation, or activated endothelial cells in blood vessels at the site of inflammation. They serve as muscle-active and edema-promoting substances, chemotaxins (see: Chemokines; see also: Chemotaxis), and cellular activators and inducers of all kinds of effector cells engaged in the inflammatory response.

Inflammatory mediators include some well studied compounds such as anaphylatoxins of the complement cascade, kinins of the coagulation system, leukotriens, prostaglandins, and many other lipid mediators. Another group of mediators are neuropeptides such as » tachykinins, » VIP (vasoactive intestinal peptide), and » VPF (vascular permeability factor). These substances enhance capillary permeability and have vasodilatatory and bronchoconstrictory activity and also increase the production of mucus. A number of » cytokines, known collectively as pro-inflammatory cytokines because they accelerate inflammation, also regulate inflammatory reactions either directly or by their ability to induce the synthesis of cellular adhesion molecules or other cytokines in certain cell types. These cytokines have overlapping biological functions and form a complicated » cytokine network (see also: Neuroimmune network). They include, among other things, various » chemokines, a number of » interleukins, » TNF-α, » CSF (colony-stimulating factors), and » IFN-γ and various neuromodulatory factors (see: Neuroimmune network). The observed redundancy among the different cytokines and other mediators of inflammation generally guarantees a substitution or complementation of individual components that may have been inactivated under pathological conditions. Under normal circumstances these cascades of inflammatory reactions induced by the mediators are strictly regulated. Failure to do so can lead to multiple organ failure (see: Septic shock). Inflammatory mediators and suitable inhibitors are, therefore, of key interest for modulating and ameliorating the effects of inflammatory reactions and their sequelae.

Arai KI et al Cytokines: co-ordinators of immune and inflammatory responses. ARB 59: 783-836 (1990); **Beutler BI & Cerami A** Tumor necrosis, cachexia, shock, and inflammation: a common mediator. ARB 57: 505-19 (1988); **Billiau A & Vandekerckhove F** Cytokines and their interactions with other inflammatory mediators in the pathogenesis of sepsis and septic shock. EJCI 21: 559-73 (1991); **Bomford R & Henderson B** (eds) Interleukin 1, inflammation and disease. Elsevier Science Publish. New York 1989; **Butcher EC** Leukocyte-endothelial cell recognition: three (or more) steps to specificity and diversity. Cell 67: 1033-6 (1991); **Cerami A** Inflammatory cytokines. Clin. Immunol. Immunopathol. 62: S3-S10 (1992); **Colten HR** Tissue-specific regulation of inflammation. J. Appl. Physiol. 72: 1-7 (1992); **Cronstein BN & Weissmann G** The adhesion molecules of inflammation. Arthritis Rheum. 36: 147-57 (1993); **Emery P et al** Systemic mediators of inflammation. Brit. J. Hosp. Med. 45: 164-8 (1991); **Faist E et al** (eds) Immune consequences of trauma, shock, and sepsis, Springer, Berlin 1989; **Gallin JI et al** (eds) Inflammation: Basic principles and clinical correlates. Raven Press, New York 1988; **Hirano T** Interleukin-6 and its relation to inflammation and disease. Clin. Immunol. Immunopathol. 62: S60-S65 (1992); **Lasky LA** Combinatorial mediators of inflammation? Curr. Biol. 3: 366-8 (1993); **Movat HZ et al** Acute inflammation in gram-negative infection: endotoxin, interleukin 1, tumor necrosis factor, and neutrophils. Fed. Proc. 46: 97-104 (1987); **Oppenheim JJ et al** Properties of the novel pro-inflammatory supergene "intercrine" cytokine family. ARI 9: 617-48 (1991); **Smith CW** Endothelial adhesion molecules and their role in inflammation. Can. J. Physiol. Pharmacol. 71: 76-87 (1993); **Wahl SM et al** Role of growth factors in inflammation and repair. JCB 40: 193-9 (1989); **Whicher JT & Evans SW** Cytokines in disease. Clin. Chem. 367: 1269-81 (1990)

Inhibins: Proteins that inhibit the secretion of FSH (follicle stimulating hormone) in the pituitary (FSH release inhibitor), influence the development of granulosa cells and inhibit the FSH-stimulated synthesis of estrogens. At least one form of Inhibins is identical with » SCF (Sertoli cell factor). The FSH-stimulated release of inhibin can be potentiated by » IGF-1 and is blocked by » EGF and » TGF-α.

Inhibins are related to » TGF-β (≈ 35 % sequence homology in the carboxyterminal region). A similar activity as that of inhibin is shown by » Follistatin, a binding protein for » Activin A. Follistatin binds inhibin and activin A by binding to the subunit common to both factors.

It has been suggested that the 2 forms of the inhibin β subunit be referred to as β-A and β-B. The β dimer of inhibin, which stimulates FSH secretion, should be called » activin; the homodimer of the β-A subunit is to be termed » activin A and the heterodimer consisting of 1 β-A and 1 β-B subunit termed activin A-B.

The β chain homodimer of inhibin is identical with » Activin A. It is also called FSH-(follicle stimulating hormone) releasing protein (FRP). A protein called ***Alpha-N-Peptide*** is derived by posttranslational processing of the α subunit of inhibin. It acts locally in follicles and influences ovulation. The genes encoding Inhibin-α, βA and βB map to chromosomes 2q33-qter, 2cen-q13, and 7p15-p13, respectively.

β inhibin also shows some sequence relationship with TGF-like proteins of *Drosophila* (see: *dpp*) and *Xenopus laevis* (see: *Vg*-1). It is identical with the erythroid differentiation factor » EDF and FRP (follicle stimulating hormone releasing protein) and has also been found to function as an immunoglobulin binding protein (see: IBF).

Inhibins have been shown to regulate the proliferation of multipotential and erythroid progenitor cells in human bone marrow cultures and to modulate hemoglobin accumulation in purified human erythroid progenitor cells (see also: Activin A). *in vivo* to reduce significantly absolute numbers of marrow progenitors (» CFU-GEMM), spleen nucleated cellularity, and also absolute numbers of early progenitors (» CFU-GM, BFU-E, CFU-GEMM) in the spleen.

■ **Clinical Use & Significance:** Inhibins may have clinical significance as a growth inhibitory and an erythroid differentiation factor in the treatment of erythroleukemia and as a general myeloprotective agent.

It has been found that inhibin is produced also by granulosa cell tumors and that measurement of serum inhibin levels is a useful early marker for primary, recurrent, and residual granulosa-cell tumors. Another clinical application that has been explored is male contraception through FSH suppression.

■ **Transgenic/Knock-out/Antisense Studies:** A targeted deletion of the α-inhibin gene has been generated by homologous recombination in mouse embryonic stem cells (see: ES cells). Mice homozygous for the null allele (inhibin-deficient) initially develop normally but ultimately develop

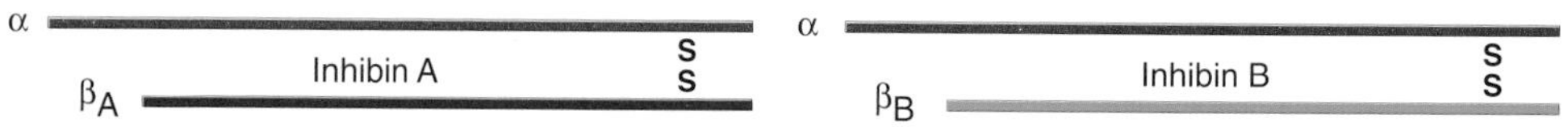

Molecular structure of inhibins.
There are two different inhibins, designated A and B. Inhibin A is a heterodimer of a unique βA subunit and another subunit called α. Activin B is a heterodimer of the unique βB subunit and the α subunit found in inhibin A. Homo- and heterodimeric combinations of β-A and β-B form activin A, activin B, and activin A-B.

mixed or incompletely differentiated gonadal stromal tumors either unilaterally or bilaterally. Inhibin thus possesses tumor-suppressor activity and is an important negative regulator of gonadal stromal cell proliferation.

Bremner WJ Inhibin: from hypothesis to clinical application. NEJM 321: 826-827 (1989); **Burger HG et al** Inhibin: definition and nomenclature, including related substances. J. Clin. Endocr. Metab. 66: 885-6 (1988); **de Jong FH et al** Inhibin and related proteins: localization, regulation, and effects. AEMB 274: 271-93 (1990); **Findlay JK et al** Role of inhibin-related peptides as intragonadal regulators. Reprod. Fertil. Dev. 2: 205-18 (1990); **Hangoc G et al** Effects *in vivo* of recombinant human inhibin on myelopoiesis in mice. EH 20: 243-6 (1992); **Healy D et al** Inhibin and related peptides in pregnancy. J. Clin. Endocrinol. Metabol. 4: 233-47 (1990); **Hillier SG et al** Control of immunoactive inhibin production by human granulosa cells. Clin. Endocrinol. (Oxf.) 35: 71-8 (1991); **Lappohn RE et al** Inhibin as a marker for granulosa-cell tumors. NEJM 321: 790-793 (1989); **Ling N et al** Pituitary FSH is released by a heterodimer of the β-subunits from the two forms of inhibin. N 321: 779-82 (1986); **Ling N et al** Novel ovarian regulatory peptides: inhibin, activin, and follistatin. Clin. Obstet. Gynecol. 33: 690-702 (1990); **Mason AJ et al** Structure of two human ovarian inhibins. BBRC 135: 957-964 (1986); **Mason A et al** Complementary DNA sequences of ovarian follicular fluid inhibin show precursor structure and homology with transforming growth factor β. N 318: 659-63 (1985); **Murata M et al** Erythroid differentiation factor is encoded by the same mRNA as that of the inhibin β_A chain. PNAS 85: 2434-8 (1988); **Rivier C et al** Studies of inhibin family hormones: a review. Hormone Res. 28: 104-18 (1987); **Sawchenko PE et al** Inhibin β in central neural pathways involved in the control of oxytocin secretion. N 334: 615-7 (1988); **Seideh NG et al** Complete amino acid sequence of human seminal plasma β-inhibin. FL 175: 349-55 (1984); **Sheth AR** Inhibins. Boca Raton, Florida, CRC Press 1987; **Vale W et al** The inhibin/activin family of hormones and growth factors. In: Sporn MA & Roberts AB (eds) Peptide growth factors and their receptors. Handbook of Experimental Pharmacology, Vol. 95: 211-48, Springer, Berlin 1990; **Yu J et al** Importance of FSH-releasing protein and inhibin in erythrodifferentiation. N 330: 765-7 (1987)

● **Transgenic/Knock-out/Antisense Studies: Matzuk MM et al** α-inhibin is a tumor-suppressor gene with gonadal specificity in mice N 360: 313-9 (1992)

Inhibitor of DNA synthesis: abbrev. IDS. See: Suppressor factors.

Inhibitor of IL2 activity: abbrev. INH-IL2. See: Suppressor factors.

Inhibitory diffusible factor 45: see: IDF-45.

Inhibitory factor for cellular DNA: see: STIF.

INH-IL2: *inhibitor of IL2 activity* see: Suppressor factors.

Insulin-like growth factor I: see: IGF.

Insulin-like growth factor II: see: IGF.

Insulin-related factor: see: IRF.

***int*-1**: This gene has been renamed *Wnt*-1. see: *Wnt*.

int*-2**: *int*-2 was initially identified as a cellular proto-oncogene (see: Oncogene) the expression of which was found to be activated by the integration of viral DNA sequences into neighboring chromosomal DNA. The protein encoded by the *int*-2 gene is also called ***FGF-3 (fibroblast growth factor 3; see also: FGF). FGF-3 is involved in the differentiation of mesodermal structures during embryonic development and functions as a growth factor for epithelial cells. *int*-2 is a homologue of » bFGF. Unlike bFGF *int*-2 possesses a secretory » signal sequence that allows secretion of the protein by classical secretory pathways via the endoplasmic reticulum and Golgi system. Some of the secreted protein has been shown to be associated with the extracellular matrix (see: ECM). The factor can be released from the extracellular matrix by soluble glucosaminoglycans such as heparin (see also: HBGF, heparin-binding growth factors), heparan sulfate, or dermatan sulfate.

The introduction of a functional *int*-2 gene into the murine mammary epithelial cell line HC11 demonstrates that *int*-2 protein can functionally replace » bFGF as a growth and differentiation factor.

The introduction of the *int*-2 gene into murine » 3T3 fibroblasts causes morphological transformation of the cells. High concentrations of heparin inhibit the growth of the cells and revert the transformed phenotype back to normal.

In the mouse embryo, *int*-2 transcripts are detected in the rhombencephalon at a time when the induction of the inner ear occurs. *int*-2 constitutes a signal for the induction of the primordium of the inner ear, the otic vesicle. Its formation can be inhibited by *int*-2-specific » antisense oligonucleotides and by antibodies against the *int*-2 protein. In the absence of the rhombencephalon the inductive signal can be mimicked by » bFGF.

The *int*-2 gene maps to human chromosome 11q13. 3 in the immediate vicinity of the » *hst* gene which also encodes a growth factor. Both genes have been found to be co-amplified in ≈ 50 % of all primary and in all metastasizing carcinomas of the esophagus.

■ **Transgenic/Knock-out/Antisense Studies:** The overexpression of *int*-2 under the control of the mouse mammary tumor virus (MMTV) promoter leads to hyperplasia in mammary and prostatic tissues of » transgenic animals (mouse). The insertional activation of *int*-2 by MMTV is one of the factors contributing to mammary tumorigenesis in » transgenic animals carrying the » *Wnt*-1 transgene.

● **Reviews: Dickson C et al** The structure and function of the *int*-2 oncogene. PGFR 1: 123-32 (1989); **Dickson C et al** Expression, processing, and properties of *int*-2. ANY 638: 18-26 (1991); **Dickson C et al** Characterization of *int*-2: a member of the fibroblast growth factor family. JCS 13: s87-s96 (1990); **Nusse R** The *int* genes in mammary tumorigenesis and in normal development. TIG 4: 291-5 (1988)

● **Biochemistry & Molecular Biology: Brookes S et al** Sequence organization of the human *int*-2 gene and its expression in teratocarcinoma cells. O 4: 429-36 (1989); **Brookes S et al** Linkage map of a region of human chromosome band 11q13 amplified in breast and squamous cell tumors. Genes Chromosomes Cancer 4: 290-301 (1992); **Casey G et al** Characterizationand chromosome assignment of the human homologue of *int*-2, a potential proto-oncogene. MCB 6: 502-10 (1986); **Dickson C et al** Characterization of *int*-2: a member of the fibroblast growth factor family. JCS Suppl. 13: 87-96 (1990); **Dickson C et al** Int-2: a member of the fibroblast growth factor family has different subcellular fates depending on the choice of initiation codon. Enzyme 44: 225-34 (1992); **Kiefer P et al** The *Int*-2/Fgf-3 oncogene product is secreted and associates with extracellular matrix: implications for cell transformation. MCB 11: 5929-36 (1991); **Moore R et al** Sequence, topography, and protein coding potential of mouse *int*-2: a putative oncogene activated by mouse mammary tumor virus. EJ 5: 919-24 (1986); **Nguyen C et al** The FGF-related oncogenes *hst* and *int*.2, and the *bcl*.1 locus are contained within one megabase in band q13 of chromosome 11, while *fgf*.5 oncogene maps to 4q21. O 3: 703-8 (1988); **Smith R et al** Multiple RNAs expressed from the *int*-2 gene in mouse embryonal carcinoma cell lines encode a protein with homology to fibroblast growth factor. EJ 4: 1013-22 (1988); **Wada A et al** Two homologous oncogenes, HST1 and INT2, are closely located in human genome. BBRC 157: 828-35 (1988); **Wilkinson DG et al** Expression pattern of the FGF-related proto-oncogene *int*-2 suggests multiple roles in fetal development. Development 105: 131-8 (1989); **Yoshida T et al** Genomic sequence of *hst*, a transforming gene encoding a protein homologous to fibroblast growth factors and the *int*-2-encoded protein. PNAS 84: 7305-9 (1987)

● **Biological Activities: Goldfarb M et al** Cell transformation by *Int*-2, a member of the fibroblast growth factor family. O 6: 65-71 (1991); **Paterno GD et al** Mesoderm-inducing properties of INT-2 and kFGF: two oncogene-encoded growth factors related to FGF. Development 106: 79-83 (1990); **Represa J et al** The *int*-2 proto- oncogene is responsible for induction of the inner ear. N 353: 561-3 (1991); **Tahara E** Growth factors and oncogenes in human gastrointestinal carcinomas. JCRCO 116: 121-31 (1990); **Venesio T et al** The *int*-2 gene product acts as a growth factor and substitutes for basic fibroblast growth factor in promoting the differentiation of a normal mouse mammary epithelial cell line. Cell Growth Differ. 3: 63-71 (1992)

● **Transgenic/Knock-out/Antisense Studies: Kwan H et al** Transgenes expressing the *Wnt*-1 and *int*-2 proto-oncogenes cooperate during mammary carcinogenesis in doubly transgenic mice. MCB 12: 147-54 (1992); **Muller WJ et al** The *int*-2 gene product acts as an epithelial growth factor in transgenic mice. EJ 9: 907-13 (1990); **Ornitz DM et al** Int-2, an autocrine and/or ultra-short-range effector in transgenic mammary tissue transplants. JNCI 84: 887-92 (1992); **Stamp G et al** Nonuniform expression of a mouse mammary tumor virus-driven *int*-2/Fgf-3 transgene in pregnancy-responsive breast tumors. Cell Growth Diff. 3: 929-38 (1992); **Tutrone RF jr et al** Benign prostatic hyperplasia in a transgenic mouse: a new hormonally sensitive investigatory model. J. Urol. 149: 633-9 (1993)

Inter-alpha-trypsin inhibitor: see: ITI.

Intercrines: A general term for a family of small, activation-inducible, mainly chemotactic (see also: Chemotaxis) and pro-inflammatory (see also: Inflammation) proteins. This family of structurally related proteins has been renamed as » chemokines.

Oppenheim JJ et al Properties of the novel proinflammatory supergene "intercrine" cytokine family. ARI 9: 617-48 (1991)

Interferon consensus sequence binding protein: abbrev. ICSBP. see: IRS (interferon response sequence).

Interferons: see: IFN for general information about interferons and individual entries for » IFN-α, » IFN-β, and » IFN-γ.

Interferon-inducible 78 kDa protein: abbrev. IFI-78k gene. see: Mx protein.

Interferon regulatory factor: abbrev. IRF. see: IRS (interferon response sequence).

Interferon regulatory factor 1: abbrev. IRF1 This factor is identical with ISGF3 (interferon-stimulated gene factor 3). see: IRS (interferon response sequence).

Interferon regulatory factor 2: abbrev. IRF2. see: IRS (interferon response sequence).

Interferon response sequence: see: IRS.

Interferon-stimulated gene factor 2: abbrev. ISGF2. see: IRS (interferon response sequence).

Interferon-stimulated gene factor 3: abbrev. ISGF3. The factor is identical with IRF1 (interferon regulatory factor 1). see: IRS (interferon response sequence).

Interferon-stimulated response element: abbrev. ISRE. see: IRS (interferon response sequence).

Interleukin 1 to **Interleukin 14:** see: IL1 to IL14.

Interleukin 1 receptor antagonist: see: IL1ra.

Interleukin 1 Inhibitor: see: IL1ra (IL1 receptor antagonist) and M20 Interleukin-1 inhibitor.

Interleukin-beta: IL1β see: IL1.

Interleukin B: abbrev. IL-B. New name for a factor formerly called ***BEF*** (B cell-derived enhancing factor) or ***IL-B4***. IL-B from both normal and transformed B cells consists of two subunits of similar size and amino acid composition.
This factor was originally described as a non-immunoglobulin regulatory factor spontaneously produced by murine B lymphocytes and B cell lines that enhances the *in vitro* antigen-driven antibody response of unfractionated spleen cells stimulated by thymus-dependent antigens. BEF is also biologically active *in vivo* and significantly enhances primary IgM and IgG antibody responses in mice. IL-B selectively prevents the differentiation of suppressor T lymphocytes from precursors into effectors. IL-B prevents expression of isotype-specific Fc receptors and the release of soluble factors suppressing production of the corresponding isotype in the T cell hybrid T2D4.
del Guercio P et al Regulatory function of Thy-1-negative cells: V. A lymphokine of B cell origin (BEF) induces *in vitro* high antibody response in genetically selected low responder mice. J. Immunogenet. 12: 45-53 (1985); **del Guercio P et al** B lymphocyte regulation of the immune system. I. *In vivo* biologic activity of the novel lymphokine, B cell-derived enhancing factor (BEF). JI 134: 996-1002 (1985);**del Guercio P et al** B lymphocyte regulation of the immune system. II. Inhibition of Fc receptor expression of lymphocytes by BEF, a lymphokine of B cell origin. JI 134: 3926-33 (1985); **del Guercio P et al** B lymphocyte regulation of the immune system. III. Preparation and characterization of rabbit antibodies that inhibit the biological activity of the lymphokine, B cell-derived enhancing factor (BEF). CI 98: 333-40 (1986); **del Guercio P et al** Characterization of murine interleukin B by a monoclonal antibody. N 329: 445-7 (1987); **del Guercio P et al** Inhibitory activity of interleukin B on the suppressor T cell hybrid T2D4. JI 138(10): 3295-9 (1987)

Interleukins: Collective term for a group of structurally and functionally distinct soluble proteins secreted by different types of leukocytes that are involved in cell-to-cell communication.
A factor is classified as an interleukin if its primary structure is known and the factor has been characterized biochemically.
Interleukins are involved in processes of » cell activation, cell differentiation, proliferation, and cell-to-cell interactions. A definition of interleukins on the basis of their functions is problematic since these factors are produced by many different cells, probably all cell types, and since they also possess a wide spectrum of biological activities.
The term interleukin (between leukocytes) has never been meant to imply that these factors can act only on lymphocytes. The same interleukins can be produced by many different cell types and an individual interleukin may act on different cell types, eliciting different biological responses, depending on the particular type of cell.
The expression of interleukins is usually strictly regulated, i. e. the factors are often not secreted constitutively. They are synthesized after cell activation, following a physiological or non-physiological stimulus. Some interleukins are autoregulatory and regulate their own synthesis or the expression of their own receptors. For detailed information see individual entries: IL1 to IL14. See also: Recombinant cytokines.
Aarden LA et al Revised nomenclature for antigen-nonspecific T cell proliferation and helper factors. CI 48: 433-36 (1979); **Arai KI et al** Cytokines: co-ordinators of immune and inflammatory responses. ARB 59: 783-836 (1990); **Callard RE** (edt) Cytokines and B lymphocytes. Academic Press, London, 1991; **Dawson MM** Lymphokines and interleukins. Open University Press, Milton Keynes, 1991; **Fent K & Zbinden G** Toxicity of interferon and interleukin. TIPS 8: 100-5 (1987); **Mizel SB** The interleukins. FJ 3: 2379-88 (1989); **O'Garra A** Interleukins and the immune system 1. Lancet 1: 943-47 (1989); **Paul WE** Nomenclature of lymphokines which regulate B lymphocytes. Mol. Immunol. 21: 343-5 (1984); **Paul WE** Pleiotropy and redundancy: T cell-derived lymphokines in the immune response. Cell 57: 521-4 (1989); **Parry DAD et al** Conformation homologies among the cytokines: interleukins and colony-stimulating factors. J. Mol. Recog. 1: 107-11 (1988); **Smith KA** Draft proposals for Interleukin nomenclature. IT 7: 321-2 (1986)

Interleukin enhancer binding factor: see: ILF.

Interleukin for Da cells: see: HILDA.

Interleukin Hemopoietin-1: see: ILHP1.

Interleukin HP1: abbrev. ILHP1. A murine T cell-derived lymphokine with growth factor activity for B cell hybridomas and plasmacytomas (see also: PCT-GF, plasmacytoma growth factor). This factor was later shown to be produced also by macrophages and fibroblasts. It is identical with » IL6.
Cayphas S et al Identification of an interleukin HP1-like plasmacytoma growth factor produced by L cells in response to viral infection. JI 139: 2965-9 (1987); **Van Snick J et al** Interleukin-HP1, a T cell-derived hybridoma growth factor that supports the

Interleukin 15: see addendum: IL15.

in vitro growth of murine plasmacytomas. JEM 165: 641-9 (1987); **Van Snick J et al** cDNA cloning of murine interleukin-HP1: homology with human interleukin 6. EJI 18: 193-7 (1988); **Vink A et al** B cell growth and differentiation activity of interleukin-HP1 and related murine plasmacytoma growth factors. Synergy with interleukin 1. EJI 18: 607-12 (1988)

Interstitial cell-stimulating hormone: abbrev. ICSH. See: Luteinizing hormone.

Intracellular IL1ra: abbrev. ICIL1ra. see: IL1ra (IL1 receptor antagonist).

Intracrine: A mechanism of growth control involving the direct action of a cytokine within a cell. It has been observed that some factors, including » bFGF, » EGF, » NGF and » PDGF, produce factor/receptor complexes that are rapidly internalized by the cells and translocated to the nucleus without being degraded. In addition, the activity of some growth factors cannot be inhibited completely by antibodies directed against these factors.

An intracrine mechanism of growth control would create an internal » autocrine loop requiring the presence of suitable intracellular biologically active receptors. The mechanisms underlying intracrine growth control are largely unknown. They might involve direct modulation of DNA replication and/or transcription (see also: Gene expression). Since some cytokines, including » aFGF, » bFGF, » CNTF and » PD-ECGF do not possess secretory signal sequences allowing their release by classical secretory pathways via the endoplasmic reticulum/Golgi system an intracrine activity could be a further mechanism explaining at least some of their biological actions.

It has been demonstrated that » bFGF contains a nuclear transport sequence (abbrev. *NTS*) in its aminoterminal sequence (see also: Signal sequence) that mediates the transport of this factor into the nucleus. Deletion of this targeting sequence generates a variant of bFGF that is no longer mitogenic but retains all other activities normally elicited after binding of this factor to its receptor, including tyrosine phosphorylation and induction of early gene synthesis.

For other mechanisms of growth control see also: Autocrine, Juxtacrine, Paracrine, Retrocrine.

Baldin V et al Translocation of bFGF to the nucleus is G_1 phase cell cycle specific in bovine aortic endothelial cells. EJ 9: 1511-7 (1990); **Imamura T et al** Identification of a heparin-binding growth factor-1 nuclear translocation sequence by deletion mutation analysis. JBC 267: 5676-9 (1992); **Imamura T et al** Recovery of mitogenic activity of a growth factor mutant with a nuclear translocation sequence. S 249: 1567-70 (1990); **Logan A** Intracrine regulation at the nucleus – a further mechanism of growth factor activity? J. Endocrinol. 125: 339-43 (1990); **Mignatti P & Rifkin DB** Release of basic fibroblast growth factor, an angiogenic factor devoid of secretory signal sequence: A trivial phenomenon or a novel secretion mechanism? JCB 47: 201-7 (1991); **Yan G et al** Exon switching and activation of stromal and embryonic fibroblast growth factor (FGF)-FGF receptor genes in prostate epithelial cells accompany stromal independence and malignancy. MCB 13: 4513-22 (1993)

Intracisternal A particles: see: IAP.

INTRON-A: Recombinant IFN-α2b (Schering-Plough Corp, Kenilworth, NJ; Essex pharma). See: IFN-α.

Invasion-inhibiting factor 2: see IIF-2.

IP-10: *immune protein 10* also: γ-IP10. IP-10 is also called » C7. The expression of IP-10 from a variety of cells, including monocytes, endothelial cells, keratinocytes, and fibroblasts, is induced by » IFN-γ and » TNF-α. The induction of the synthesis of IP-10 is used as an assay of » IFN-γ activities.

The protein has a length of 98 amino acids. It shows homology to » PF4 (platelet factor 4) and belongs to the » chemokine family of cytokines. IP-10 is also related to a gene called CRG-2 (see: CRG, cytokine responsive gene) which encodes a product that also belongs to the chemokine family of proteins. The human IP-10 genes contains four exons and maps to chromosome 4q12-21 in the vicinity of the cytokine genes belonging to the chemokine family of proteins.

γ-IP10 has been detected in keratinocytes, lymphocytes, monocytes, and endothelial cells in immunologically mediated processes, such as positive tuberculin skin tests, and in growth-activated keratinocytes, such as in psoriasis. Keratinocytes in normal epidermis do not produce γ-IP10. *In vivo* (mouse) IP-10 is synthesized predominantly in the liver and the kidney after intravenous injection of inflammatory agents, and in particular after injection of » IFN-γ, and may play an important role in the response of liver and kidney to systemic inflammation. IP-10 mRNA expression is more sensitive to suppression by » IL4 when stimulated by bacterial lipopolysaccharides than by IFN-γ/IL2.

It has been suggested that IP-10 may play an important role in hypersensitivity reactions of the delayed type. Increased levels of IP-10 are found in psoriatic plaques characterized by the infiltration of

Interleukin T: see addendum. **Invasion stimulating factor:** see addendum: ISF.

neutrophils. IP-10, however, does not activate neutrophils.

IP-10 probably also plays a role in regulation of the growth of early hematopoietic progenitor cells. It has been shown to suppress *in vitro* colony formation of highly enriched » CD34-positive cells in the presence of » SCF, » GM-CSF, or SCF + » Epo, but not in their absence with the exception of SCF. Tumor cells genetically engineered to express high levels of murine IP-10 have been shown to elicit a powerful host-mediated antitumor effect *in vivo* which appears to be mediated by the recruitment of inflammatory infiltrates composed of lymphocytes, neutrophils, and monocytes.

Dewald B et al IP-10, a γ-interferon-inducible protein related to interleukin-8, lacks neutrophil activating properties. Immunol. Lett. 32: 81-4 (1992); **Finbloom DS et al** Culture of human monocytes with granulocyte-macrophage colony-stimulating factor results in enhancement of IFN-γ receptors but suppression of IFN-γ-induced expression of the gene IP-10. JI 150: 2383-90 (1993); **Gautam S et al** IL4 suppresses cytokine gene expression induced by IFN-γ and/or IL2 in murine peritoneal macrophages. JI 148: 1725-30 (1992); **Gottlieb AB et al** Detection of a γ interferon-induced protein, IP-10, in psoriatic plaques. JEM 168: 941-8 (1988); **Kaplan G et al** The expression of a γ interferon-induced protein (IP-10) in delayed immune responses in human skin. JEM 166: 1098-108 (1987); **Luster AD et al** γ-interferon transcriptionally regulates an early-response gene containing homology to platelet proteins. N. 315: 672-6 (1985); **Luster AD et al** Genomic characterization of a γ-interferon-inducible gene (IP-10) and identification of an interferon-inducible hypersensitive site. MCB 7: 3723-31 (1987); **Luster AD et al** Biochemical characterization of a γ-interferon-inducible gene (IP-10). JEM 166: 1084-97 (1987); **Luster AD et al** Interferon-inducible gene maps to a chromosomal band associated with a (4;11) translocation in acute leukaemic cells. PNAS 84: 2868-71 (1987); **Luster AD & Leder P** IP-10, a C-X-C-chemokine, elicits a potent thymus-dependent antitumor response *in vivo*. JEM 178: 1057-65 (1993); **Narumi S et al** Tissue-specific expression of murine IP-10 mRNA following systemic treatment with interferon γ. Leukoc. Biol. 52: 27-33 (1992); **Ohmori V & Hamilton TA** Ca^{2+} and calmodulin selectively regulate lipopolysaccharide-inducible cytokine mRNA expression in murine peritoneal macrophages. JI 148: 538-45 (1992); **Ohmori Y et al** Tumor necrosis factor-α induces cell type and tissue-specific expression of chemoattractant cytokines *in vivo*. Am. J. Pathol. 142: 861-70 (1993); **Sarris AH et al** Human interferon-inducible protein 10: expression and purification of recombinant protein demonstrate inhibition of early human hematopoietic progenitors. JEM 178: 1127-32 (1993); **Smoller BR & Krueger J** Detection of cytokine-induced protein γ-immune protein-10 (γ-IP10) in atypical melanocytic proliferations. J. Am. Acad. Dermatol. 25: 627-31 (1991); **Wathelet MG et al** Cloning and chromosomal location of human genes inducible by type I interferon. Somat. Cell Molec. Genet. 14: 415-26 (1988)

IPSF: *immunoglobulin (Ig) production stimulating factor* This poorly characterized factor is found culture supernatant and lysate of the human lymphoblastoid » Namalwa cell line. Partially purified IPSF is a macromolecule of about 500 kDa containing a 72 kDa protein as a major component. The factor stimulates proliferation and Ig production of human-to-human hybridoma HB4C5 cells but not that of mouse-to-mouse hybridomas. IPSF also stimulates Ig production of human-to-human hybridomas derived from NAT-30 cells, a human fusion partner derived from Namalwa cells, but not that of other human-to-human or mouse-to-mouse hybridomas.

Yamada K et al Partial purification and characterization of immunoglobulin production stimulating factor derived from Namalwa cells. *In vitro* Cell. Dev. Biol. 25: 243-7 (1989)

ir: as a prefix such as in irIL3, irACTH, irIGF-1, this abbrev. stands for "immunoreactive", i. e. a substance reacting with antibodies directed against the respective protein (IL3, ACTH, IGF-1). The observation that a protein reacts with a specific antibody essentially means that the protein in question contains epitopes also occurring in the protein against which the antibody was developed in the first place. Immunoreactivity is not a biochemical proof that the two proteins are identical. Since many cytokines contain sequence domains and characteristic protein sequence modules allowing them to be grouped into larger families of proteins the use of a known specific antibody may allow related factors belonging to the same protein family to be detected and isolated (see also: Bioassays, Cytokine assays).

IR: *immunoregulator* see also: Immunomodulation.

IRAP: *IL1 receptor antagonist protein* see: IL1ra (Interleukin 1 receptor antagonist).

ir-FGF: *immunoreactive fibroblast growth factor* see: MDGF (macrophage-derived growth factor).

IRF: *insulin-related factor* This factor is isolated from serum-free culture supernatants of the teratoma cell line 1246-3A. The factor is immunologically and biochemically related to insulin and differs from insulin-like growth factors (see: IGF). IRF binds to the insulin receptor and functions as a stimulatory » autocrine growth factor for some insulin-dependent cell lines.

Yamada Y & Serrero G Autocrine growth induced by the insulin-related factor in the insulin-independent teratoma cell line 1246-3A. PNAS 85: 5936-40 (1988)

IRF: *interferon regulatory factor* see: IRS (interferon response sequence).

IRF1: *interferon regulatory factor* This factor is identical with ISGF3 (interferon-stimulated gene factor 3). see: IRS (interferon response sequence).

IRF2: *interferon regulatory factor* see: IRS (interferon response sequence).

IRMA: *immunoradiometric assays* see: Cytokine assays.

IRS: *interferon response sequence* A short DNA sequence found in the promoter regions of several genes the expression of which is modulated by interferons (see: IFN). This region, also called interferon-stimulated response element (abbrev. ***ISRE***) is the binding site for regulatory proteins influencing the transcription of genes (see also: Gene expression). The different strengths of IFN-γ-induced promoter activities mediated by ISRE is modulated, at least in part, by control sequences either overlapping with ISRE or lying in the vicinity of the ISRE sequence. One of these sequences is called, for example, GAS and binds » GAF (γ-interferon activation factor).
One protein, called ***ICSBP*** (interferon consensus sequence binding protein) binds to homologous interferon response elements found in a variety of IFN-inducible genes and represses transcription of several genes induced by interferons.
Another protein factor binding to ISRE is ***ISGF3*** (interferon-stimulated gene factor 3). The expression of this factor is specifically induced by binding of » IFN-α to its receptor. It consists of a DNA-binding subunit of 48 kDa (γ subunit) that is associated with three other non-DNA-binding proteins of 84, 91 (***p91***) and 113 kDa, respectively. These three proteins are also called ***ISGF3 Alpha proteins***. They are phosphorylated by a specific phosphotyrosine kinase that can be inhibited by » Genistein. These proteins are phosphorylated in the cytoplasm and then migrate into the nucleus. ISGF3 shows significant homology to another interferon-reactive factor called IRF1, and the 91 kDa protein has been described as a component of another transcriptional activator, » GAF (γ-interferon activation factor). p91 is also a component of the signaling pathway for » CNTF (ciliary neurotrophic factor).
The expression of ISGF3 is not induced by » IFN-γ. Its synthesis is enhanced, however if cells treated with IFN-γ are subsequently treated with IFN-α.

IRF1 (interferon-responsive factor, interferon regulatory factor) is a transcription factor which is identical with CRG-1 (see: CRG, cytokine responsive gene). It is also called ***ISGF2*** (interferon-stimulated gene factor 2). The human IRF1 gene maps to chromosome 5q31.1. The IRF1 gene is approximately 200 kb telomeric to the » IL5 gene. The genes encoding IL3 and GM-CSF are located at least 200 kb, but not more than 1600 kb, telomeric to this region.
IRF1 is synthesized *de novo* after treatment of cells with type I interferons. IRF1 has multiple binding sites in the promoter region of the gene encoding » IFN-β (IRS = AAGTGA, AAATGA; ***PRD*** = positive regulatory domain). It regulates the induced expression of this interferon and also the expression of some IFN-β-inducible genes (e. g. class I Histocompatibility antigens, 2´,5´-oligoadenylate synthetase), and of some cytokine genes. The constitutive expression of the murine ISGF2 gene causes a strong resistance against various RNA viruses. Deletions of the IRF-1 gene have been found in many cases of leukemia and myelodysplasia (see also: 5q$^-$ syndrome).
PRD itself contains some subdomains. PRD is the binding site for proteins called ***PRDI-BF*** (positive regulatory domain binding factors) which repress the expression of the IFN-β interferon gene. Some PRD domains overlap with domains responsible for the regulation of gene expression by cAMP (see also: CRE, cyclic AMP-responsive element) thus demonstrating that this gene is subject to a very complex regulatory circuit.
The factors » TNF, » IL1, » IL6, and » LIF are potent inducers of IRF-1 and IRF-2. IRF-1 has also been shown to be induced specifically by » prolactin. The terminal differentiation of myelomonocytic » M1 cells induced by IL6 or LIF can be blocked by treatment with anti-IFN-β antisera or IRF-1-specific » Antisense oligonucleotides. The constitutive expression of IRF1 in » transgenic animals (mice) leads to a dramatic reduction in the number of B lymphocytes. It has been shown that a tumor suppressor gene (see: Oncogenes) hypothesized to lie in the 5q31 region is the gene encoding IRF1. This gene is commonly deleted in human leukemia and preleukemic myelodysplasia. The activation of genes induced by IRF-1 can be prevented by another inhibitory factor called ***IRF 2*** (also called ***PRDI-BFc***; c = constitutive or ***ISGF-1***). An imbalance in the IRF-1/IRF-2 ratio may lead to the dysregulation of cell growth as a critical step for oncogenesis.

 IRF-1 knockouts: see addendum: IRF-1. **IRS:** see addendum: IRF-1.

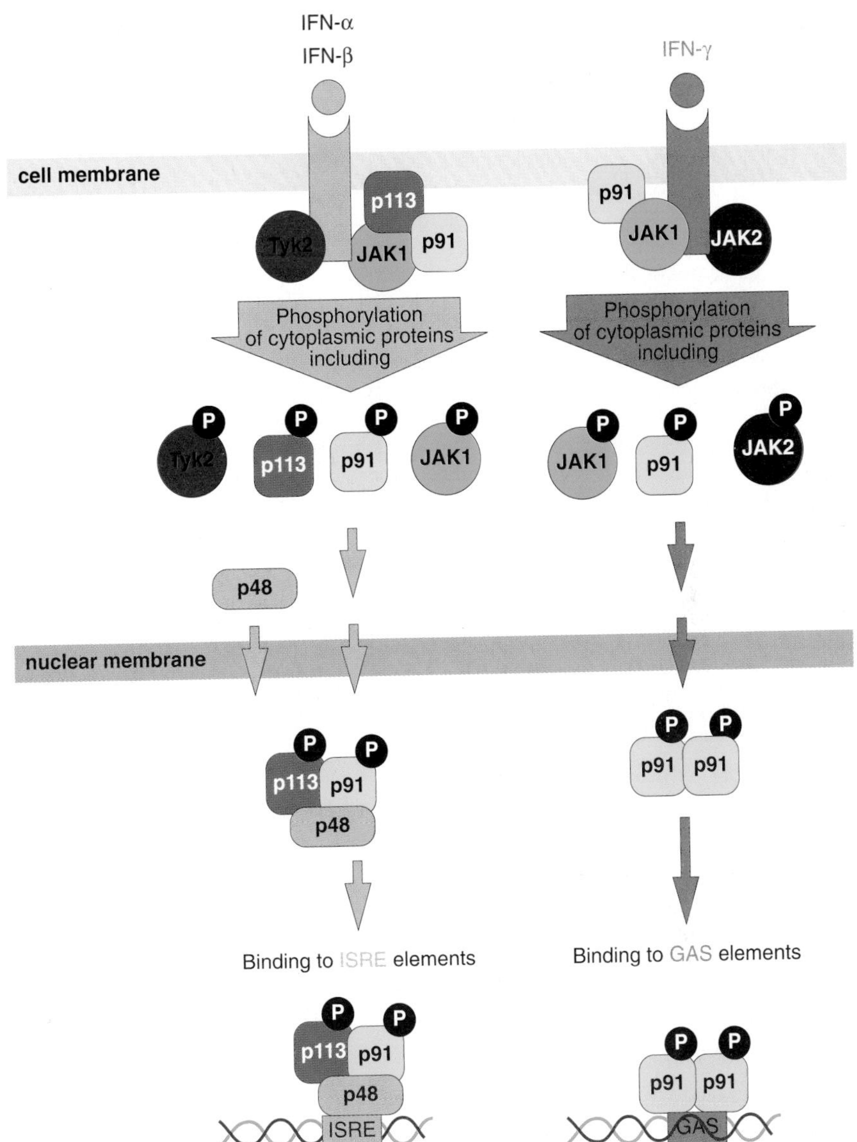

Signaling cascades initiated by interferons.
Intracellular signaling events initiated in response to binding of interferons to their membrane receptors involves several cytoplasmic proteins directly associated with IFN-α/IFN-β or IFN-γ receptors or bound as complexes, possibly also involving membrane anchoring. Some of the proteins are shared by the two receptor types. Binding of interferons to their receptors leads to the phosphorylation of these proteins and other cytoplasmic proteins. Following phosphorylation these proteins form complexes (p113/p91/p48 in the case of the IFN-α/IFN-β receptor, and p91/p91 in the case of the IFN-γ receptor). p113/p91/p48 complexes bind to ISRE elements in the promoter regions of some genes whereas p91/p91 complexes bind to GAS elements.

■ **Transgenic/Knock-out/Antisense Studies:** The biological consequences of IRF-1 and IRF-2 expression have been studied in » transgenic animals carrying targeted disruptions of either gene, generated by homologous recombination in » ES cells. IRF-1 deficiency is associated with a lack of the normally observed type I interferon induction by poly(I).poly(C), while type I interferon inducibility following infection with Newcastle disease virus remains unaffected. IRF-1-deficient mice also show a profound reduction in TCRαβ^{+} CD4^{-}CD8^{+} T cells, suggesting that IRF-1 plays a critical role in T cell development. IRF-2-deficient mice exhibit bone marrow suppression of hematopoiesis and B lymphopoiesis. These animals die after infection with lymphocytic choriomeningitis virus. IRF-2-deficient cells show up-regulated type I interferon induction by infection with Newcastle disease virus.

Abdollahi A et al Interferon regulatory factor 1 is a myeloid differentiation primary response gene induced by interleukin 6 and leukemia inhibitory factor: role in growth inhibition. Cell Growth Differ. 2: 401-7 (1991); Driggers PH et al An interferon g-regulated protein that binds the interferon-inducible enhancer element of major histocompatibility complex class I genes. PNAS 87: 3743-7 (1990); **Cohen L & Hiscott J** Characterization of TH3, an induction-specific protein interacting with the interferon β promoter. Virology 191: 589-99 (1992); **Du W & Maniatis T** An ATF/CREB binding site protein is required for virus induction of the human interferon β gene. PNAS 89: 2150-4 (1992); **Fu XY et al** ISGF3, the transcriptional activator induced by interferon α, consists of multiple interacting polypeptide chains. PNAS 87: 8555-9 (1990); **Fu XY et al** The proteins of ISGF-3, the interferon α-induced transcriptional activator, define a gene family involved in signal transduction. PNAS 89: 7840-3 (1992); **Fu XY et al** A transcription factor with SH2 and SH3 domains is directly activated by an interferon α-induced cytoplasmic protein tyrosine kinase(s). Cell 70: 323-35 (1992); **Fujita T et al** Induction of endogenous IFN-α and IFN-β genes by a regulatory transcription factor, IRF-1. N 337: 270-2 (1989); **Harada H et al** Structurally similar but functionally distinct factors, IRF-1 and IRF-2, bind to the same regulatory elements of IFN and IFN-inducible genes. Cell 58: 729-39 (1989); **Harada H et al** Absence of the type I IFN system in EC cells: transcriptional activator (IRF-1) and repressor (IRF-2) genes are developmentally regulated. Cell 63: 303-12 (1990); **Improta T et al** Interferon-γ potentiates the antiviral activity and the expression of interferon-stimulated genes induced by interferon-α in U937 cells. J. Interferon Res. 12: 87-94 (1992); **Kanno Y et al** The genomic structure of the murine ICSBP gene reveals the presence of the γ interferon-responsive element, to which an ISGF3a subunit (or similar) molecule binds. MCB 13: 3951-63 (1993); **Keller AD & Maniatis T** Only two of the five zinc fingers of the eukaryotic transcriptional repressor PRDI-BF1 are required for sequence-specific DNA binding. MCB 12: 1940-9 (1992); **Keller AD & Maniatis T** Identification and characterization of a novel repressor of β-interferon gene expression. Genes Dev. 5: 868-79 (1991); **Kirchhoff S et al** Interferon regulatory factor 1 (IRF-1) mediates cell growth inhibition by transactivation of downstream target genes. NAR 21: 2881-9 (1993); **Miyamoto M et al** Regulated expression of a gene encoding a nuclear factor, IRF-1, that specifically binds to IFN-β gene regulatory elements. Cell 54: 903-13 (1988); **Nelson NJ et al** ICSBP, a new member of the IRF family suppresses IFN-induced gene transcription. MCB 13: 588-99 (1993); **Palombella VJ & Maniatis T** Inducible processing of interferon regulatory factor-2. MCB 12: 3325-36 (1992); **Pine R** Constitutive expression of an ISGF2/IRF1 transgene leads to interferon-independent activation of interferon-inducible genes and resistance to virus infection. J. Virol. 66: 4470-8 (1992); **Politis AD et al** Modulation of interferon consensus sequence binding protein mRNA in murine peritoneal macrophages. Induction by IFN-γ and down-regulation by IFN-α, dexamethasone, and protein kinase inhibitors. JI 148: 801-7 (1992); **Reis LFL et al** Critical role of a common transcription factor, IRF-1, in the regulation of IFN-β and IFN-inducible genes. EJ 11: 185-93 (1992); **Schindler C et al** Proteins of transcription factor ISGF-3: one gene encodes the 91-and 84-kDa ISGF-3 proteins that are activated by interferon α. PNAS 89: 7836-9 (1992); **Schwarz LA et al** Interferon regulatory factor-1 is inducible by prolactin, interleukin-2 and concanavalin A in T cells. Mol. Cell. Endocrinol. 86: 103-10 (1992); **Strehlow I & Decker T** Transcriptional induction of IFN-γ-responsive genes is modulated by DNA surrounding the interferon stimulation response element. NAR 20: 3865-72 (1992); **Tanaka N et al** Recognition DNA sequences of interferon regulatory factor 1 (IRF-1) and IRF-2, regulators of cell growth and the interferon system. MCB 13: 4531-8 (1993); **Veals SA** Subunit of an α-interferon-responsive transcription factor is related to interferon regulatory factor and *myb* families of DNA-binding proteins. MCB 12: 3315-24 (1992); **Weisz AP et al** Cloning and characterization of the human ICSBP, a possible negative regulator that binds enhancer elements common to interferon inducible genes. JBC 267: 25589-96 (1992); **Whiteside ST et al** Identification of novel factors that bind to the PRD I region of the human β-interferon promoter. NAR 20: 1531-8 (1992); **Williams BRG** Transcriptional regulation of interferon-stimulated genes. EJB 200: 1-11 (1991); **Willman CL et al** Deletion of IRF-1, mapping to chromosome 5q31.1, in human leukemia and preleukemic myelodysplasia. S 259: 968-711(1993); **Yamada G et al** Specific depletion of the B cell population induced by aberrant expression of human interferon regulatory factor 1 gene in transgenic mice. PNAS 88: 532-6 (1991); **Yu-Lee LY et al** Interferon-regulatory factor 1 is an immediate-early gene under transcriptional regulation by prolactin in Nb2 T cells. MCB 10: 3087-94 (1990)

● **Transgenic/Knock-out/Antisense Studies: Lim SP & Hui KM** Characterization of a novel IRF-1-deficient mutant cell line. Immunogenetics 39: 168-77 (1994); **Matsuyama T et al** Targeted disruption of IRF-1 or IRF-2 results in abnormal type I IFN gene induction and aberrant lymphocyte development. Cell 75: 83-97 (1993)

ISC: *immunoglobulin secreting cells*

ISG: *interferon-stimulated genes* General term for genes that are activated specifically in response to interferons (see: IFN).

ISGF2: *interferon-stimulated gene factor 2* The factor is identical with IRF1 (interferon-responsive factor, interferon regulatory factor). See: IRS (interferon response sequence).

ISF: see addendum.

ISGF3: *interferon-stimulated gene factor 3* see: IRS (interferon response sequence).

Isotype switching: Mature IgM- and IgD-expressing B cells undergo the process of immunoglobulin heavy chain class switching (isotype switching) after antigenic stimulation. This allows their progeny to produce antibodies with heavy chains of different classes, such as γ, α, and ε. Isotype switching from IgM to other Ig isotypes typically occurs subsequent to a first exposure to antigen. At the molecular level this process involves the orderly somatic rearrangement (deletion) of immunoglobulin heavy chain constant region (C_H) genes and/or alternative RNA splicing of germline transcripts.

Isotype switching in appropriately activated cells is regulated by T helper cells through physical contact with B cells (e. g. through » CD40) and some cytokines. These factors are also collectively referred to as ***switch factors***. The mechanism of action of these factors seems to be related to their

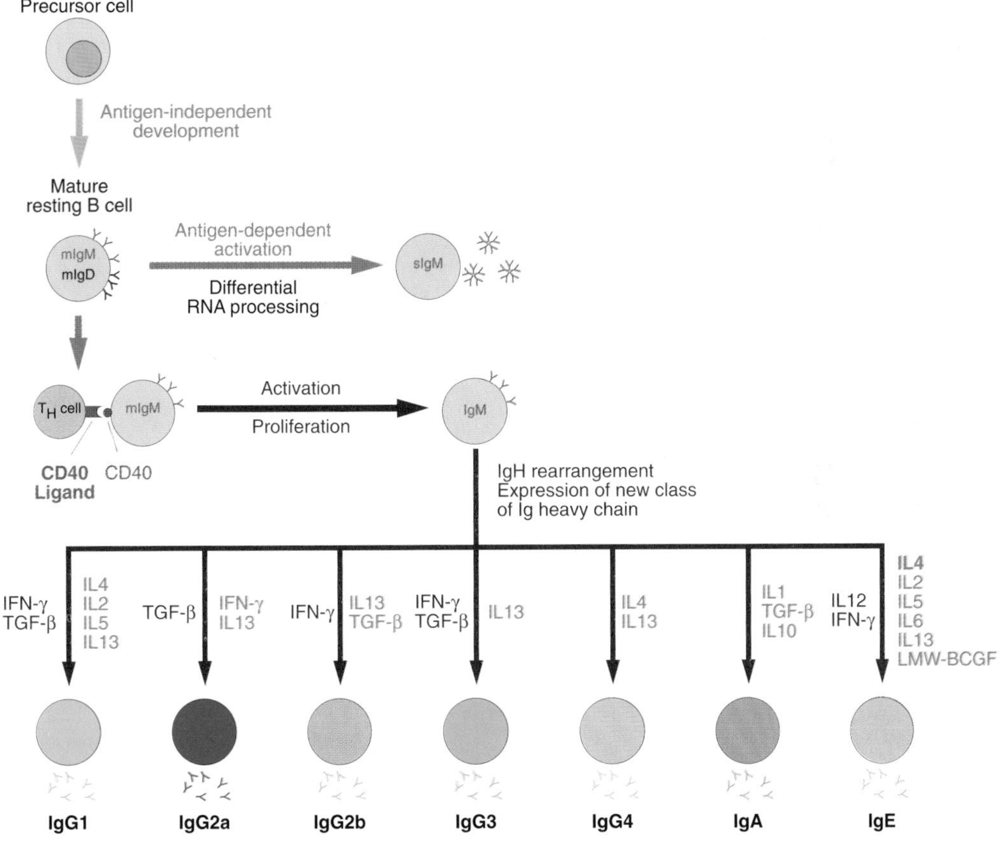

Role of cytokines in isotype switching.
Precursor cell-derived resting mature B cells expressing membrane-bound immunoglobulins can interact via their CD40 cell surface molecule with CD40 ligands expressed on T helper cells. This leads to activation and proliferation of B cells that undergo class switching. Some known cytokines promoting class switching (green color) are shown alongside the arrows while those inhibiting expression of immunoglobulin isotypes are shown in red.
Mutations in the CD40 ligand have been shown to be the cause of the X-linked hyper-immunoglobulin M syndrome (HIM) which is characterized by the production of higher than normal amounts of IgM and the inability to carry out class switching. HIM patients are susceptible to opportunistic infections due to the failure to produce the full set of antibodies.

ability to increase or decrease the transcriptional activation of particular switch regions and to enhance recombination events. Some of the cytokines known to be involved in isotype switching, either alone or in combination with each other or additional trigger signals, are » IL4 (identified previously as » IgE-EF, » IgG1-enhancing factor, » IgG1 induction factor, » IgG1 secretion factor), » IL5 (identified previously as » IgA-EF), » TGF-β, » IL1, » IL2, » IL6, » IL13, and some poorly characterized factors (see: IgG2b-inducing factor). IL4 and IL5 synergize to enhance the IgG1 response. Optimal IgG1 responses also require IL2. IL1 can enhance IgA production in the presence of IL5. IL6 rapidly enhances IgG1 class switches. TGF-β has been shown to inhibit the production of many immunoglobulin isotypes (IgM, IgG1, IgG2a, IgG3) but does not appear to affect IgG2b. TGF-β appears to specifically induce surface IgA-negative cells to switch to IgA production. The combination of IL10 and TGF-β induces naive B cells to secrete IgA1 and IgA2 as a consequence of isotype switching.

Some cytokines such as » IFN-γ and » IFN-α can act both as inducers or inhibitors of isotype switching. IFN-γ is known to inhibit the ability of IL4 to enhance IgG1 and IgE synthesis. IFN-γ also inhibits IgG1, IgG2b, IgG3 secretion but can stimulate IgG2a secretion.

● REVIEWS: **Finkelman FD et al** Lymphokine control of *in vivo* immunoglobulin isotype selection. ARI 8: 303-33 (1990); **Purkerson J & Isakson P** A two-signal model for regulation of immunoglobulin isotype switching. FJ 6: 3245-52 (1992); **Schultz CL & Coffman RL** Control of isotype switching by T cells and cytokines. Curr. Opin. Immunol. 3: 350-4 (1991); **Snapper CM & Mond JJ** Towards a comprehensive view of immunoglobulin class switching. IT 14: 15-7 (1993); **Teale JM & Estes DM** Immunoglobulin isotype regulation. In: Callard RE (Edt) Cytokines and B lymphocytes, pp. 173-93, Academic Press, London (1990); **Vercelli D & Geha RS** Regulation of isotype switching. Curr. Opin. Immunol. 4: 794-7 (1992)

Briere F et al IL10 and IL13 as B cell growth and differentiation factors. Nouv. Rev. Fr. Hematol. 35: 233-5 (1993); **Cocks BG et al** IL13 induces proliferation and differentiation of human B cells activated by the CD40 ligand. Int-Immunol. 1993 Jun; 5: 657-63 (1993); **Ehrhardt RO et al** Effect of transforming growth factor (TGF)-β 1 on IgA isotype expression. TGF-β 1 induces a small increase in sIgA+ B cells regardless of the method of B cell activation. JI148: 3830-6 (1992); **Gerondakis S et al** Structure and expression of murine germ-line immunoglobulin epsilon heavy chain transcripts induced by interleukin 4. PNAS 87: 1581-5 (1990); **Iwasato T et al** Biased distribution of recombination sites within S regions upon immunoglobulin class switch recombination induced by transforming growth factor β and lipopolysaccharide. JEM 175: 1539-46 (1992); **Jabara HH et al** Hydrocortisone and IL4 induce IgE isotype switching in human B cells. JI 147: 1557-60 (1991); **Kim PH & Kagnoff MF** Transforming growth factor β 1 increases IgA isotype switching at the clonal level. JI 145: 3773-8 (1990); **Kunimoto DY et al** The roles of IL4, TGF-β and LPS in IgA switching. ECN 3: 407-15 (1992); **Lebman DA et al** Molecular characterization of germ-line immunoglobulin A transcripts produced during transforming growth factor type β-induced isotype switching. PNAS 87: 3962-6 (1990); **Mandler R et al** IL4 induction of IgE class switching by lipopolysaccharide-activated murine B cells occurs predominantly through sequential switching. JI 150: 407-18 (1993); **Matsuoka M et al** Switch circular DNA formed in cytokine-treated mouse splenocytes: evidence for intramolecular DNA deletion in immunoglobulin class switching. Cell 62: 135-42 (1990); **McIntyre TM et al** Transforming growth factor β 1 selectivity stimulates immunoglobulin G2b secretion by lipopolysaccharide-activated murine B cells. JEM 177: 1031-7 (1993); **Purkerson JM & Isakson PC** Isotype switching in anti-immunoglobulin-activated B lymphoblasts: differential requirements for interleukin 4 and other lymphokines to elicit membrane vs. secreted IgG1. EJI 21: 707-14 (1991); **Purkerson JM & Isakson PC** Interleukin 5 (IL5) provides a signal that is required in addition to IL4 for isotype switching to immunoglobulin (Ig) G1 and IgE. JEM 175: 973-82 (1992); **Purkerson J & Isakson P** A two-signal model for regulation of immunoglobulin isotype switching. FJ 6: 3245-52 (1992); **Roper RL et al** Prostaglandin E2 promotes IL4-induced IgE and IgG1 synthesis. JI 145: 2644-51 (1990); **Shapira SK et al** Deletional switch recombination occurs in interleukin-4-induced isotype switching to IgE expression by human B cells. PNAS 88: 7528-32 (1991); **Siebenkotten G et al** The murine IgG1/IgE class switch program. EJI 22: 1827-34 (1992); **Sonoda E et al** Differential regulation of IgA production by TGF-β and IL5: TGF-β induces surface IgA-positive cells bearing IL5 receptor, whereas IL5 promotes their survival and maturation into IgA-secreting cells. CI 140: 158-72 (1992); **Splawski JB et al** Immunoregulatory role of CD40 in human B cell differentiation. JI 150: 1276-85 (1993); **Whitley MZ et al** Distinct IL4 response mechanisms of the MHC gene A α in different mouse B cell lines. Mol. Immunol. 30: 821-32 (1993); **Whitmore AC et al** Ig isotype switching in B lymphocytes. The effect of T cell-derived interleukins, cytokines, cholera toxin, and antigen on isotype switch frequency of a cloned B cell lymphoma. Int. Immunol. 3: 95-103 (1991); **Xu L et al** Replacement of germ-line epsilon promoter by gene targeting alters control of immunoglobulin heavy chain class switching. PNAS 90: 3705-9 (1993)

ISRE: *interferon-stimulated response element* see: IRS.

IT: Immunotherapy. see: Immunomodulation.

ITI: *Inter-alpha-trypsin inhibitor* ITI is a serine protease inhibitor of ≈ 220 kDa mainly synthesized in the liver and found in human serum at concentrations of 0.4-0.5 g/L. Substrates of ITI include trypsin, chymotrypsin, and a neutrophil elastase.

The inhibitor consists of three evolutionarily related polypeptide chains called heavy chains (H1, 92 kDa; H2, 98 kDa; H3, 107 kDa) that are cross-linked by chondroitin sulfate which itself is linked by an O-glycosidic bond with a smaller subunit (L chain), variably called ***Bikunin***, ***urinary trypsin inhibitor*** (UTI) or ***HI-30*** (human inhibitor 30 kDa). Bikunin is linked via chondroitin sulfate with a fourth protein. The entire complex is called ***Pre-Alpha-Trypsin-Inhibitor***. The cleavage of the bikunin precursor protein yields ***α-1-Microglobulin*** (= ***protein HC***), which appears to be involved in regulation of the inflammatory process. Protein HC is normally present as a free form and as a complex with IgA in the blood, spinal fluid and urine in relatively low concentration but is present in high concentration in the urine of patients with tubular proteinuria and in blood and urine of patients on renal dialysis.

Bikunin is identical with » ECGF-2b (endothelial cell growth factor 2b). Bikunin and α-1-Microglobulin have been shown to be » acute phase proteins (see: acute phase reaction) in rats.

The gene encoding HI-30 (which gives also rise to protein HC) has a length of 17.8 kb, contains 10 exons, and maps to human chromosome 9q22.3 – q33; its genetic designation is ***AMBP*** (α-1-microglobulin/bikunin precursor). The genes encoding H1 and H3 map to human chromosome 3p21.2-p21.1. The gene encoding H2 maps to human chromosome 10p15.

Balduyck M & Mizon J Inter-α-trypsin inhibitor and its plasma and urine derivatives. Ann. Biol. Clin. Paris 49: 273-81 (1991); **Bourguignon J et al** Human inter-α-trypsin inhibitor. Synthesis and maturation in hepatoma HepG2 cells. BJ 261: 305-8 (1989); **Diarra-Mehrpour M et al** Human plasma inter- α-trypsin inhibitor is encoded by four genes on three chromosomes. EJB 179: 147-54 (1989); **Diarra-Mehrpour M et al** Structural analysis of the human inter-α-trypsin inhibitor light chain gene. EJB 191: 131-8 (1990); **Diarra-Mehrpour M et al** Human inter-α-trypsin inhibitor: full-length cDNA sequence of the heavy chain H1. BBA 1132: 114-8 (1992); **Gressier B et al** Crossed immunoelectrophoresis does not allow accurate determination of inter-α-trypsin inhibitor and its derivatives in plasma. Biol. Chem. Hoppe-Seyler 371: 865-70 (1990); **Lindqvist A et al** Rat α 1 microglobulin: co-expression in liver with the light chain of inter-α trypsin inhibitor. BBA 1130: 63-7 (1992); **Lionny C et al** Inter-α-trypsin-inhibitor and its derivatives in inflammatory syndromes. Presse Med. 20: 203-6 (1991); **Luckenbach C et al** Genetic polymorphism of inter-α-trypsin inhibitor (ITI): formal genetic and linkage analyses. Hum. Genet. 87: 89-90 (1991); **Mendez E et al** Human protein HC and its IgA complex are inhibitors of neutrophil chemotaxis. PNAS 83: 1472-5 (1986); **Odum L** Inter-α-trypsin inhibitor and pre-α trypsin inhibitor in health and disease. Determination by immunoelectrophoresis and immunoblotting. Biol. Chem. Hoppe-Seyler 371: 1153-8 (1990); **Salier JP et al** Isolation and characterization of cDNAs encoding the heavy chain of human inter-α-trypsin inhibitor (IATI): unambiguous evidence for multipolypeptide chain structure of IATI. PNAS 84: 8272-6 (1987); **Salier JP et al** The genes for the inter-α inhibitor family share a homologous organization in human and mouse. Mamm. Genome 2: 233-9 (1992); **Salier JP et al** Homologous chromosomal locations of the four genes for inter-α-inhibitor and pre-α-inhibitor family in human and mouse: assignment of the ancestral gene for the lipocalin superfamily. Genomics 14: 83-8 (1992); **Sjoberg EM & Fries E** Biosynthesis of bikunin (urinary trypsin inhibitor) in rat hepatocytes. ABB 295: 217-22 (1992); **Tavakkol A** Molecular cloning of porcine α 1 microglobulin/HI-30 reveals developmental and tissue-specific expression of two variant messenger ribonucleic acids. BBA 1088: 47-56 (1991); **Traboni C et al** The gene coding for proteins HC and HI-30 of inter-α-trypsin inhibitor maps to 9q22.3 – q33. Cytogenet. Cell. Genet. 50: 46-8 (1989) **Trefz G et al** Establishment of an enzyme-linked immuno-sorbent assay for urinary trypsin inhibitor by using a monoclonal antibody. J. Immunoassay 12: 347-69 (1991); **Vetr H & Gebhard W** Structure of the human α-1 microglobulin-bikunin gene. Biol. Chem. Hoppe-Seyler 371: 1185-96 (1990); **Vogt U & Cleve H** A "new" genetic polymorphism of a human serum protein: inter-α-trypsin inhibitor. Hum. Genet. 84: 151-4 (1990)

IX 207-887: A thiofuran derivative initially used for the treatment of arthritic diseases. At therapeutic concentrations this compound blocks the release of » IL1 by human monocytes and murine peritoneal macrophages. The compound may therefore be a valuable agent allowing to ameliorate or to inhibit the pro-inflammatory actions of IL1.

Schnyder J et al Inhibition of interleukin 1 release by IX 207-887 (1990). Agents Actions. 30: 350-62 (1990)

IxN/2b: see: 1xN/2b.

J

J82 cytokine: see: BCDC (bladder carcinoma-derived cytokine).

J774: A murine macrophage cell line established from a tumor that arose in a female BALB/c mouse. Its growth is inhibited by dextran sulfate, purified protein derivative, and bacterial lipopolysaccharides. This cell line synthesizes large amounts of lysozyme and exhibits minor cytolysis but predominantly antibody-dependent phagocytosis. The cells have been shown to express cell-bound receptors for immunoglobulin and complement. J774 is used in » bioassays to detect » M-CSF since its growth is dependent on this factor. The cells constitutively secrete » IL1.

Ralph P & Nakoinz I Direct toxic effects of immunopotentiators on monocytic, myelomonocytic and histiocytic or macrophage tumor cells in culture. CR 37: 546-50 (1977); **Ralph P & Nakoinz I** Antibody-dependent killing of erythrocyte and tumor targets by macrophage-related cell lines: enhancement by PPD and LPS. JI 119: 950-4 (1977); **Ralph P & Nakoinz I** Phagocytosis and cytolysis by a macrophage tumor and its cloned cell line. N 257: 393-4 (1975); **Ralph P & Nakoinz I** Lysozyme synthesis by established human and murine histiocytic lymphoma cell line. JEM 143: 1528-33 (1976); **Snyderman R et al** Biologic and biochemical activities of continuous macrophage cell lines P388D1 and J744.I. JI 119: 2060-6 (1977); **Whetton AD & Dexter TM** Myeloid hematopoietic growth factors. BBA 989: 111-32 (1989)

J774-derived cytotoxic factor: A cytotoxic factor produced by the murine J774 cell line after stimulation with bacterial lipopolysaccharides and 12-o-tetradecanoyl phorbol-13-acetate (see also: Phorbol esters). The factor is most likely murine » TNF. (see also: CTX, cytotoxin).

Sakurai A et al Macrophage cell line, J774, producing a tumor necrosis factor. Jpn. J. Exp. Med. 56: 195-9 (1986)

JAK1: see: Janus kinases.

JAK2: see: Janus kinases.

Janus kinases: or abbrev. *j*ust *a*nother *k*inase. Name of a recently described family of non-receptor protein tyrosine kinases of ≈ 130 kDa, comprising ***JAK1***, ***JAK2,*** and ***Tyk2***. These proteins have a second kinase-like domain, do not possess so-called » *src* homology domains (SH2, and SH3), and have no membrane-spanning domains.

A cell line lacking JAK1 has been shown to be completely defective in interferon responses. JAK1 has been shown to be essential for signal transduction processes mediated by the receptors for » IFN-α and » IFN-γ.

JAK2 has been shown to be associated with the receptors for » growth hormone and » Epo and is involved in signal transduction processes mediated by these factors, coupling, for example, Epo binding to its receptor to tyrosine phosphorylation and mitogenesis. JAK2 is also activated by » IL3, » GM-CSF, » G-CSF, and IFN-γ. JAK2 has been shown to be essential for signal transduction processes of the » IFN-γ receptor.

Tyk2 has recently been shown to be essential for signal transduction processes of the » IFN-α receptor.

All members of the Janus kinase family have been shown to be essential components of receptor-mediated signal transduction processes involving a receptor subunit common for » CNTF (ciliary neurotrophic factor), » LIF, » Oncostatin M, and » IL6 (see also: gp130).

Argetsinger LS et al Identification of JAK2 as a growth hormone receptor-associated tyrosine kinase. Cell 74: 237-44 (1993); **Harpur AG et al** JAK2, a third member of the JAK family of protein tyrosine kinases. O 7: 1347-53 (1992); **Müller M et al** The protein tyrosine kinase JAK1 complements defects in interferon-α/β and -γ signal transduction. N 366: 129-35 (1993); **Silvennoinen O et al** Structure of the JAK 2 protein tyrosine kinase and its role in IL3 signal transduction. PNAS 90: 8429-33 (1993); **Stahl N et al** Association and activation of Jak-Tyk kinases by CNTF-LIF-OSM-IL6b receptor components. S 263: 92 (95 (1994); **Velasquez L et al** A protein tyrosine kinase in the interferon α/β signaling pathway. Cell 70: 313-22 (1992); **Witthuhn BA et al** JAK 2 associates with the erythropoietin receptor and is tyrosine phosphorylated and activated following stimulation with erythropoietin. Cell 74: 227-36 (1993)

JE: This gene is also called » ***SCY A2*** (small inducible cytokine A2) or » TSG-8. It was initially identified in murine » 3T3 fibroblasts as a gene strongly expressed after treatment of the cells with growth factors and mitogens. The murine gene maps to chromosome 11.

JAK3: see addendum.

Janus kinases: see addendum: Janus kinases; see also addendum: JAK3.

A potent inducer of JE synthesis in 3T3 cells is » PDGF. In murine macrophages JE is also induced by » M-CSF, » IL1, » EGF, bacterial » endotoxins, » IFN-γ, and synthetic dsRNA. In Osteoblasts and epithelial cell lines JE is induced by » TGF-β. Its synthesis is also induced by » TNF-α. Fibroblasts and endothelial cells can also produce JE.

The expression of JE is a marker of cells responding to mitogenic stimuli since JE is not expressed in resting cells. JE is a so-called competence gene (see also: Cell activation), i. e. genes the expression of which coincides with the entry of resting cells into the » cell cycle. JE is also overexpressed during renal ischemia.

The human JE gene contains three exons and maps to chromosome 17q11. 2-12. Its coding sequence shows 59 % homology with that of the gene encoding » M-CSF and 61 % homology with that encoding » IFN-α. Murine and rat JE gene sequence display 82 % sequence homology. Exon-intron junctions are conserved in the murine and human gene. Murine and human JE proteins differ in that the murine protein contains a carboxyterminal extension of 48 amino acids.

Human JE is a protein of 12 kDa (148 amino acids). The secreted protein is a glycoprotein of 25 kDa that does not display any sequence homology with M-CSF, IFN-α and any other cytokine at the protein level.

The JE protein is also a cytokine. It is identical with the chemotactic protein » MCP-1 (monocyte chemoattractant protein 1) and belongs to the » chemokine family of cytokines. JE is also identical with » LDCF (lymphocyte-derived chemotactic factor), » TDCF (tumor-derived chemotactic factor), » HC11, » SMC-CF (smooth muscle cell chemotactic factor), » MCAF (monocyte chemotactic and activating factor), » GDCF (glioma-derived monocyte chemotactic factor).

■ **Clinical Use & Significance:** JE/MCP-1 appears to mediate the antitumor activity of monocytes: malignant cells engineered to express a JE/MCP-1 gene do not form tumors in nude mice (see also: Immunodeficient mice) and also suppress the growth of other tumor cells transplanted simultaneously. JE is therefore of potential interest in the therapy of malignant tumors (for cancer vaccines see also: Cytokine gene transfer).

Cochran B et al Molecular cloning of gene sequences regulated by platelet-derived growth factor. Cell 33: 939-47 (1983); **Hanazawa S et al** Transforming growth factor-β-induced gene expression of monocyte chemoattractant JE in mouse osteoblastic cells, MC3T3-E1. BBRC 180: 1130-6 (1991); **Kallin B et al** Cloning of a growth arrest-specific and transforming growth factor β-regulated gene, TI 1, from an epithelial cell line. MCB 11: 5338-45 (1991); **Kawahara RS & Deuel TF** Platelet-derived growth factor-inducible gene JE is a member of a family of small inducible genes related to platelet factor 4. JBC 264: 679-82 (1989); **Kawahara RS et al** Glucocorticoids inhibit the transcriptional induction of JE, a platelet-derived growth factor-inducible gene. JBC 266: 13261-6 (1991); **Kohase M et al** A cytokine network in human diploid fibroblasts: interactions of β-interferons, tumor necrosis factor, platelet-derived growth factor, and interleukin 1. MCB 7: 273-80 (1987); **Medici I et al** Improved method for purification of human platelet factor 4 by affinity and ion-exchange chromatography. Thromb. Res. 54: 277-87 (1989); **Ohmori Y et al** Tumor necrosis factor-α induces cell type and tissue-specific expression of chemoattractant cytokines *in vivo*. Am. J. Pathol. 142: 861-70 (1993); **Rollins BJ et al** Cloning and expression of JE, a PDGF-inducible gene with cytokine-like properties. PNAS 85: 3738-42 (1988); **Rollins BJ et al** The human homologue of the JE gene encodes a monocyte secretory protein. MCB 9: 4687-95 (1989); **Rollins BJ & Sunday ME** Suppression of tumor formation *in vivo* by expression of the JE gene in malignant cells. MCB 11: 3125-31 (1991); **Rollins BJ** JE/MCP-1: An early-response gene encodes a monocyte-specific cytokine. Cancer Cells 3: 517-24 (1991); **Rollins BJ et al** Assignment of the human small inducible cytokine A2 gene, SCYA2 (encoding JE or MCP-1) to 17q11.2-12 – evolutionary relatedness of cytokines clustered at the same locus. Genomics 10: 489-92 (1991); **Safirstein R et al** Expression of cytokine-like genes JE and KC is increased during renal ischemia. Am. J. Physiol. Renal, Fluid Electrolyte Physiol. 261: F1095-F101 (1991); **Smith A et al** Sigje, a member of the small inducible gene family that includes platelet factor 4 and melanoma growth stimulatory activity, is on mouse chromosome 11. Cytogenet. Cell. Genet. 52: 194-6 (1989); **Taubman MB et al** JE mRNA accumulates rapidly in aortic injury and in platelet-derived growth factor-stimulated vascular smooth muscle cells. Circ. Res. 70: 314-25 (1992); **Timmer HTM et al** Analysis of the rat JE gene promoter identifies an AP-1 binding site essential for basal expression but not for TPA induction. NAR 18: 23-34 (1990); **Van Damme J et al** Production and identification of natural monocyte chemotactic protein from virally infected murine fibroblasts. Relationship with the product of the mouse competence (JE) gene. EJB 199: 223-9 (1991); **Yoshimura T & Leonard EJ** Secretion by human fibroblasts of monocyte chemoattractant protein-1, the product of gene JE. JI 144: 2377-83 (1990)

JEG-3 cell factor: This factor is secreted by the chorion carcinoma cell line JEG-3. It is also the product of normal T lymphocytes. It inhibits the antigen- and mitogen-stimulated proliferation of mononuclear cells. The factor contains a noncovalently linked subcomponent, lipid suppressor substance. The protein is probably identical with or closely related to » SIF (suppressor cell induction factor).

Wolf RL et al Characterization of an immune suppressor from transformed human trophoblastic JEG-3 cells. CI 78: 356-7 (1983); **Wolf RL** Human placental cells that regulate lymphocyte function. Pediatr. Res. 23: 212-8 (1988)

JR-2(82): A human EBV-transformed B cell line. Two subclones, designated D3 and B10 have been developed. These cells are used to assay B cell differentiation factors. The cells proliferate spontaneously at high cell densities but die in the absence of these factors.

D3 expresses surface IgG and differentiates in the presence of » LMW-BCGF, to secrete IgG. B10 lacks surface and cytoplasmic Ig and fails to differentiate in response to LMW-BCGF. » CD23 cannot be induced on B10 by incubation with either LMW-BCGF or IL4. B10 does not shed CD23 and shed CD23 is not a growth factor for either cloned line. Expression of CD23 on D3 cells is not affected by preincubation with LMW-BCGF. Neither B10 or D3 cells respond to recombinant factors » IL1, » IL2, » IL4, » IL6, » TNF-α, TNF-β, » IFN-γ, or to » HMW-BCGF (Namalwa), alone or in combination.

Warrington RJ The characterization of a human B cell line utilisable for the assay of B cell growth factors. JIM 100: 117-22 (1987); **Warrington RJ** Low molecular weight B cell growth factor-responsive cloned human B cell lines. I. Phenotypic differences and lack of requirement for CD23 (Fc epsilon RII). JI 143: 2546-52 (1989)

jTCGF: Jurkat *T cell growth factor* See: TCGF.

***jun*:** (abbrev. from Japanese ju-nana for 17) The » oncogene *jun* is the putative transforming gene of avian sarcoma virus 17 which induces fibrosarcomas in chickens and can transform a number of avian cell types *in vitro* by the action of v-*jun*. The human *jun* gene maps to chromosome 1p31-p32. The human gene does not contain introns and displays ≈ 80 % sequence identity with the viral gene. *jun* belongs to a family of related genes that includes junD and junB.

jun encodes the transcription factor ***AP-1*** (activator protein 1) which recognizes the AP-1 consensus sequence TGACTCA, a response element that confers sensitivity to the tumor-promoting » phorbol ester TPA (see also: TRE, TPA response element). *jun* itself forms homodimers or heterodimers with junD and junB and also interacts with the » oncogene product » *fos*, forming *jun-fos* heterodimers.

The AP-1 transcription factors are considered immediate-early response genes (see also: Gene expression) and are thought to be involved in a wide range of transcriptional regulatory processes linked to cellular proliferation and differentiation.

The stimulation of » IL2 gene expression by » IL1 is due to the interaction of AP-1 with a distinct regulatory region within the promoter of the IL2 gene. The expression of the *jun* gene is controlled by its own gene product. In addition, *jun* expression is also activated by a number of growth factors that stimulate cell proliferation. AP-1, together with some other proteins that bind to it, stimulates the transcription of many other genes encoding products involved in » cell activation and the regulation of cell growth. Recent studies demonstrate that *jun* is essential for » cell cycle progression and DNA replication.

■ **Transgenic/Knock-out/Antisense Studies:** Transgenic mice carrying the v-*jun* oncogene, driven by the promoter of the widely expressed H-2KK MHC class I antigen gene are initially phenotypically normal, but after full-thickness wounding they show abnormal wound repair, characterized by hyperplastic granulation tissue. Wounding appears to be a prerequisite for tumorigenesis, suggesting collaboration between the expressed transgene and a wound-related event. Many of the lesions observed in these mice are slowly progressive because of continuing fibroblast proliferation, and over 2-5 months some give rise to dermal fibrosarcomas.

c-*fos*/v-*jun* doubly transgenic animals also show perturbations of B cell development in bone marrow and it is thought that the deregulated expression of AP-1 interferes with IL7-mediated proliferation and differentiation of immature B cells. *jun*-transgneic mice also develop heterogeneous malignant sarcomas by a multistage mechanism following wounding.

Mouse embryonic stem cell lines (see: ES cells) in which both copies of the c-*jun* gene have been inactivated by homologous recombination have also been generated. The disruption of both copies of the c-*jun* gene has no apparent effect on ES cell viability, growth rate and *in vitro* differentiation potential. The subcutaneous injection of ES cells lacking c-*jun* into syngeneic mice leads to a drastic reduction in the formation of teratocarcinomas, suggesting that functional c-*jun* protein is essential for efficient tumor growth *in vivo*. Heterozygous transgenic mice generated by these ES cells appear normal, but homozygous mice that do not express c-*jun* are not viable. The embryos die at mid- to late-gestation and exhibit impaired hepatogenesis characterized by a pronounced hypoplasia of the epithelial compartment. These mice also have altered fetal liver erythropoiesis and develop generalized tissue edema which is most prominent in the brain. The embryonic lethal phenotype of these c-*jun* null mutations indicates that there are some tissues in which overlapping expression or redundant functions of the other *jun* family members cannot compensate for the loss of c-*jun* functions although *jun* family members can form biologically active heterodimeric complexes and can substitute for the absence of c-*jun* in some *in vitro* assays.

Angel P et al Oncogene *jun* encodes a sequence-specific *trans*-activator similar to AP-1. N 332: 166-71 (1988); **Angel P et al** The *jun* proto-oncogene is positively autoregulated by its product, *jun*/AP-1. Cell 55: 875-85 (1988); **Angel P et al** The role of *jun*, *fos*, and the AP-1 complex in cell proliferation and transformation. BBA Rev. Cancer 1072: 129-57 (1991); **Bohmann D et al** Human proto-oncogene c-*jun* encodes a DNA binding protein with structural and functional properties of transcription factor AP-1. S 238: 1386-92 (1987); **Bos TJ et al** v- *jun* encodes a nuclear protein with enhancer binding properties of AP-1. Cell 52: 705-12 (1988); **Brach MA et al** The mitogenic response to tumor necrosis factor α requires c-*jun*/AP-1. MCB 13: 4284-90 (1993); **Brach MA et al** Identification of NF-*jun*, a novel inducible transcription factor that regulates c-*jun* gene transcription. EJ 11: 1479-86 (1992); **Haluska FG et al** Localization of the human *jun* proto-oncogene to chromosome region 1p31-32. PNAS 85: 2215-8 (1988); **Hattori FG et al** Structure and chromosomal localizationof the functional intronless human *jun* proto-oncogene. PNAS 85: 9148-52 (1988); **Jonat C et al** Antitumor promotion and anti-inflammation: down-modulation of AP-1 (*fos*/*jun*) activity by glucocorticoid hormone. Cell 62: 1189-204 (1990); **Kovary K & Bravo R** The *Jun* and *Fos* families are both required for cell cycle progression in fibroblasts. MCB 11: 4466-72 (1991); **Lamph WH et al** Induction of proto-oncogene *jun*/AP-1 by serum and TPA. N 334: 629-31 (1988); **Lee W et al** Purified transcription factor AP-1 interacts with TPA-inducible enhancer elements. Cell 49: 741-52 (1987); **Maki Y et al** Avian sarcoma virus 17 carries the *jun* oncogene. PNAS 84: 2848-52 (1987); **Muegge K et al** Interleukin-1 costimulatory activity on the interleukin-2 promoter via AP-1. S 246: 249-51 (1989); **Murakami Y et al** The nuclear proto-oncogenes c-*jun* and c-*fos* as regulators of DNA replication. PNAS 88: 3947-51 (1991); **Vogt PK et al** Homology between the DNA-binding domain of the GCN4 regulatory protein of yeast and the carboxy-terminal region of a protein coded for by the oncogene *jun*. PNAS 84: 3316-9 (1987); **Vogt PK & Bos TJ** *Jun*: oncogene and transcription factor. ACR 55: 2-36 (1990); **Xanthoudakis S & Curran T** Identification and characterization of *Ref*-1, a nuclear protein that facilitates AP-1 DNA-binding activity. EJ 11: 653-65 (1992)

● **TRANSGENIC/KNOCK-OUT/ANTISENSE STUDIES: Fujita K et al** B cell development is perturbed in bone marrow from c-*fos*/v-*jun* doubly transgenic mice. Int. Immunol. 5: 227-30 (1993); **Johnson RS et al** A null mutation at the c-*jun* locus causes embryonic lethality and retarded cell growth in culture. Genes Dev. 7: 1309-17 (1993); **Hilberg F & Wagner EF** Embryonic stem (ES) cells lacking functional c-*jun*: consequences for growth and differentiation, AP-1 activity and tumorigenicity. O 7: 2371-80 (1992); **Hilberg F et al** c-*jun* is essential for normal mouse development and hepatogenesis. N 365: 179-81 (1993); **Marshall GM & Vanhamme L** Characterization of sarcoma cell lines from v-*jun* transgenic mice. CR 53: 622-6 (1993); **Schuh AC et al** Obligatory wounding requirement for tumorigenesis in v-*jun* transgenic mice. N 346: 756-60 (1990); **Schuh AC et al** Skeletal muscle arises as a late event during development of wound sarcomas in v-*jun* transgenic mice. O 7: 667-76 (1992); **Schuh AC et al** Altered growth and spontaneous transformation of cells cultured from v-*jun* transgenic mice recapitulate wound-induced multistage tumorigenesis. Cell Growth Differ. 4: 177-84 (1993); **Vanhamme L et al** Tumor necrosis factor α and interleukin 1 α induce anchorage independence in v-*jun* transgenic murine cells. CR 53: 615-21 (1993)

JURKAT: A CD4-positive T cell leukemia cell line. It is used in cytokine research as a producer of » IL2 and also served as a source for the isolation of the IL2 gene (see also: TCGF, T cell growth factor). Large amounts of IL2 are produced after stimulation of the cells with » phorbol esters and either lectins or monoclonal antibodies directed against the T cell antigen receptor. IL2 is not produced in the absence of stimuli. The synthesis of IL2 by JURKAT cells is inhibited by sodium butyrate.

Jurkat cells and variants derived from it are also used to study the physiology of lymphoblastoid T cells and T cell activation. The cells appear to secrete an antiproliferative protein (see: 160 SF). Proliferation of Jurkat cells is inhibited by monoclonal antibodies directed against » IGF receptors (IGF-I and IGF-II), suggesting that both factors are involved in the proliferation of this cell line.

Baier TG et al Influence of antibodies against IGF-I, insulin or their receptors on proliferation of human acute lymphoblastic leukemia cell lines. Leuk. Res. 16: 807-14 (1992); **Koretzky GA et al** Restoration of T cell receptor-mediated signal transduction by transfection of CD45 cDNA into a CD45-deficient variant of the JURKAT T cell line. JI 149: 1138-42 (1992); **Le Gros GS et al** The effects of sodium butyrate on lymphokine production. LR 4: 221-7 (1985); **Pawelec G et al** Constitutive interleukin 2 production by the JURKAT human leukemic T cell line. EJI 12: 387-92 (1982); **Peyron JF et al** The CD45 protein tyrosine phosphatase is required for the completion of the activation program leading to lymphokine production in the JURKAT human T cell line. Int. Immunol. 3: 1357-66 (1991); **Weiss A et al** The role of T3 surface molecules in the activation of human T cells: a two-stimulus requirement for IL 2 production reflects events occurring at a pre-translational level. JI 133: 123-8 (1984)

Just another kinase: see: Janus kinases.

Juxtacrine: A mechanism of growth control and intercellular communication involving specific cell-to-cell contacts. These are established by the interaction of a membrane-bound form of a growth factor that is normally secreted with a receptor on an adjacent cell. The membrane-bound form of the growth factor then elicits the same spectrum of responses as the soluble factor.

The membrane bound form of the factor is frequently an incompletely processed biologically active precursor of the secreted form of the factor. It can also be generated by alternative splicing of the corresponding mRNA. Juxtacrine growth control has been described to be elicited, among others, by membrane-anchored forms of » AR (amphiregulin), » Betacellulin, » EGF, » GM-CSF, » HB-EGF (heparin-binding EGF-like growth factor), » M-CSF, » SCF (stem cell factor), » SDGF

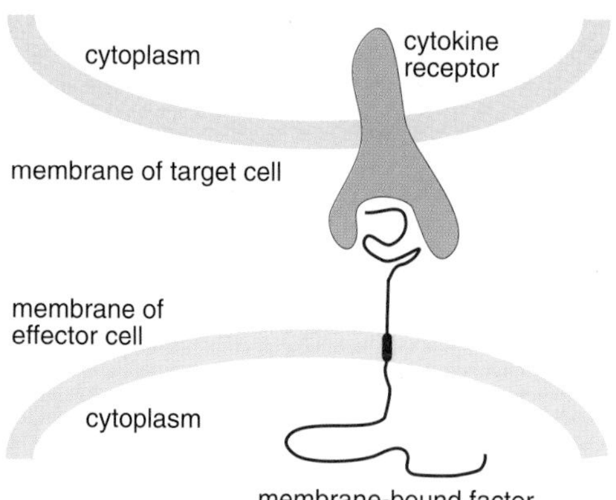

Juxtacrine growth control.
A membrane-bound cytokine expressed by an effector cell can interact with its receptor expressed by a neighboring target cell. This membrane-bound factor, which is as biologically active as a freely diffusible factor, thus can mediate specific cell-to-cell contacts.

(schwannoma-derived growth factor), » VGF, » CD23 ligand, » CD 27 ligand, » CD40 ligand, » IL1, » TGF-α, and » TNF-α.

Another mechanism of juxtacrine interaction involving extracellular matrix-associated (see: ECM) rather than membrane-anchored forms of growth factors has been described for » GM-CSF, » IL3, » IFN-γ, »SCI (stem cell inhibitor), and » LIF.

Since these specific interactions constitute a controlled form of cell adhesion juxtacrine interactions are probably involved in many developmental processes requiring directed and specific cell-to-cell interactions. This form of interaction allows developmentally important signals to be transmitted specifically between neighboring cells. Juxtacrine interactions are probably also involved in the control of » hematopoiesis, allowing specific interactions between hematopoietic cells and the surrounding stromal cells. Experimentally this form of specific cell interactions is accessible by introduction of genes encoding specific receptors into cells that normally do not express this receptor and the subsequent detection of engineered cells adhering to the stroma.

For other mechanisms of growth control involving diffusible growth factors rather than membrane-bound forms see also: Autocrine, Intracrine, Paracrine, Retrocrine.

Anklesaria P et al Cell-cell adhesion mediated by binding of membrane-anchored transforming growth factor α to epidermal growth factor receptor promotes cell proliferation. PNAS 87: 3289-93 (1990); **Bosenberg MW & Massagué J** Juxtacrine cell signaling molecules. Curr. Opin. Cell Biol. 5: 832-838 (1993); **Dainiak N** Cell membrane family of growth regulatory factors. PCBR 352: 49-61 (1990); **Gordon MY et al** Haemopoietic growth factors and receptors: bound and free. Cancer Cells 3: 127-33 (1991); **Greenberger JS** The hematopoietic microenvironment. Crit. Rev. Oncol. Hematol. 11: 65-84 (1991); **Lorant DE et al** Coexpression of GMP-140 and PAF by endothelium stimulated by histamine or thrombin: a juxtracrine system for adhesion and activation of neutrophils. JCB 115: 223-34 (1991); **Pandiella A & Massagué J** Cleavage of the membrane precursor for transforming growth factor α is a regulated process. PNAS 88: 1726-30 (1991); **Massagué J & Pandiella A** Membrane-anchored growth factors ARB 62: 515-41 (1993); **Perez C et al** A non-secretable cell surface mutant of tumor necrosis factor (TNF) kills by cell-to-cell contact. Cell 63: 251-8 (1990); **Stein J et al** Direct stimulation of cells expressing receptors for macrophage colony stimulating factor (CSF-1) by a plasma membrane-bound precursor of human CSF-1. Blood 76: 1308-14 (1990); **Wong ST et al** The TGF-α precursor expressed on the cell surface binds to the EGF receptor on adjacent cells, leading to signal transduction. Cell 56: 495-506 (1989)

K

K-252a: This fungal alkaloid kinase inhibitor is a derivative of » Staurosporine. It is an efficient inhibitor of protein kinases (IC50s of 10 to 30 nM), including protein kinase C and serine/threonine protein kinases and is, therefore, used to study the involvement of these kinases in post receptor signaling processes (for other agents used to dissect cytokine-mediated signal transduction pathways see: Bryostatins, Calcium ionophore, Calphostin C, Genistein, H8, Herbimycin A, Lavendustin A, Phorbol esters, Okadaic acid, Staurosporine, Suramin, Tyrphostins, Vanadate). It inhibits some of the activities of » NGF in » PC12 cells and has no effect on biological activities mediated by » EGF and » bFGF. Gangliosides prevent the inhibition by K-252a of NGF responses in PC12 cells.

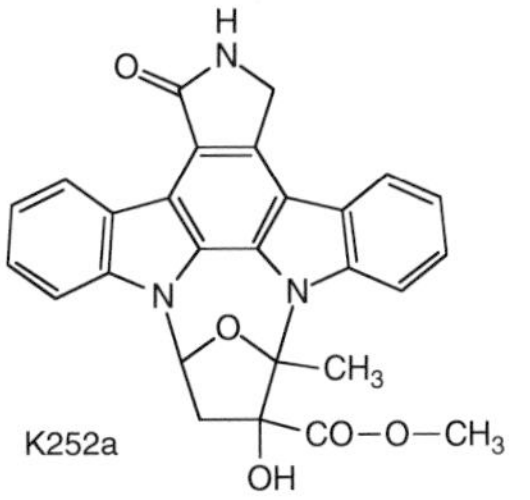

K252a

K-252a selectively inhibits the » autophosphorylation of NGF and » neurotrophin receptors (see: *trk*) and thus blocks subsequent signal transduction mechanisms, including the activation of p21*ras* that is normally observed after binding of NGF to *trk*. Some derivatives of K-252a are potent antiproliferative compounds for several human tumor cell lines not responding to the parent compound.

Akinaga S Antitumor effect of KT6124, a novel derivative of protein kinase inhibitor K-252a, and its mechanism of action. Cancer Chemother. Pharmacol. 29: 266-72 (1992); **Berg MM et al** K-252a inhibits nerve growth factor-induced *trk* proto-oncogene tyrosine phosphorylation and kinase activity. JBC 267: 13-6 (1992); **Ferrari G et al** Gangliosides prevent the inhibition by K-252a of NGF responses in PC12 cells. Brain. Res. Dev. Brain. Res. 65: 35-42 (1992); **Knusel B & Hefti F** K-252 compounds: modulators of neurotrophin signal transduction. J. Neurochem. 59: 1987-96 (1992); **Muroya K et al** Specific inhibition of NGF receptor tyrosine kinase activity by K-252a. BBA 1135: 353-6 (1992); **Nagashima K et al** Inhibition of nerve growth factor-induced neurite outgrowth of PC12 cells by a protein kinase inhibitor which does not permeate the cell membrane. FL 293: 119-23 (1991); **Nye SH et al** K-252a and staurosporine selectively block autophosphorylation of neurotrophin receptors and neurotrophin-mediated responses. Mol. Biol. Cell. 3: 677-86 (1992); **Ohmichi M et al** Inhibition of the cellular actions of nerve growth factor by staurosporine and K252A results from the attenuation of the activity of the *trk* tyrosine kinase. B 31: 4034-9 (1992); **Tapley P et al** K252a is a selective inhibitor of the tyrosine protein kinase activity of the *trk* family of oncogenes and neurotrophin receptors. O 7: 371-81 (1992)

K562-derived T cell inhibitory factor: see: K-TIF.

KAF: *cytotoxic (killer) T cell activating factor* This poorly characterized factor is an acidic protein of 70-100 kDa produced by some macrophage-like cell lines and the human cell line THP-1 (monocytic leukemia). It activates human cytotoxic killer T lymphocytes *in vitro* against autologous tumors.

Maeda N et al Augmentation of human cytotoxic T lymphocytes against autologous tumor by a factor released from human monocytic leukemia cell line. Jpn. J. Cancer Res. 80: 537-45 (1989)

KAF: *keratinocyte-derived autocrine factor* The known 20 aminoterminal amino acids of KAF are completely identical with » AR (amphiregulin) and the factor is therefore also called ***keratinocyte-derived amphiregulin***. Antibodies directed against KAF also recognise amphiregulin.
KAF is produced by keratinocytes and mammary epithelial cells. It stimulates the growth of human keratinocytes and some murine cell lines. Its mitogenic activity is inhibited by heparin and dextran sulfate (see also: HBGF, heparin-binding growth factors). Heparin sulfate also inhibits the competition of KAF for the » EGF receptor.

Cook PW et al A heparin sulfate-regulated human keratinocyte autocrine factor is similar or identical to amphiregulin. MCB 11: 2547-57 (1991); **Cook PW et al** Inhibition of autonomous human keratinocyte proliferation and amphiregulin mitogenic activity by sulfated polysaccharides. *In vitro* Cell. Dev. Biol. 28A: 218-22 (1992)

KAF: *kidney angiogenic factor* This factor was initially isolated from bovine kidney as a mitogenic for endothelial cells. It is identical with » bFGF.

Baird A et al Isolation and partial characterization of an endothelial cell growth factor from bovine kidney: homology with basic fibroblast growth factor. Regul. Pept. 12: 201-3 (1985)

Kallidin: This factor is identical with » Bradykinin.

Kallikreins: Kallikreins are present in many tissues, such as the pancreas, salivary glands, pituitary, prostate, and testes. Humans possess 3 fully characterized kallikrein-like genes. The gene expressed in kidney, pancreas and salivary gland (KLK), and the gene encoding prostate-specific antigen (APS) have been localized to chromosome 19q13.2-qter. The third gene, hGK-1, which has highest homology to and a similar tissue specificity of expression as the APS gene, maps to human chromosome 19q13.3-q13.4.

Contact of human plasma with a negatively charged surface such as dextran sulfate activates prekallikrein to kallikrein, which releases the vasoactive peptide bradykinin from high-molecular-weight kininogen. In addition to its classical role as an intermediate enzyme of the kallikrein kinin system glandular (tissue) kallikreins are involved in the posttranslational modification of polypeptide hormones precursors and growth factors. Kallikreins belong to a family of serine proteases structurally related to trypsin. Kallikrein (kininogen, kininogenase = EC3.4.21.8) cleaves proteins between arginine and lysine residues.

One member of the kallikrein family is a subunit of » NGF. This γ-NGF endopeptidase shows the arginylesterase activity characteristic of kallikreins and specifically removes an N-terminal octapeptide (NGF^{1-8}) from β-NGF. A related endopeptidase, immunologically cross-reacting with γ-NGF, is known as » EGFBP (EGF binding protein). The other subunit of NGF, α-NGF, is also a member of the glandular kallikrein gene family but does not possess protease activity.

Apart from its role as a bradykinin-releasing enzyme kallikrein may also be involved in the regulation of immune functions. *In vitro* glandular kallikrein strongly enhances the spontaneous and mitogen-induced proliferation of lymphocytes and this effect can be abolished by serine proteinase inhibitors. Kallikrein is also chemotactic for polymorphonuclear leukocytes.

Baker AR & Shine J Human kidney kallikrein: cDNA cloning and sequence analysis. DNA 4: 445-450 (1985); **Evans BA et al** Structure and chromosomal localization of the human renal kallikrein gene. B 27: 3124-9 (1988); **Fahnestock M et al** β-NGF-endopeptidase: structure and activity of a kallikrein encoded by the gene mGK-22. B 30: 3443-50 (1991); **Hu ZQ et al** Enhancement of lymphocyte proliferation by mouse glandular kallikrein. Immunol. Lett. 32: 85-9 (1992); **Kaplan AP et al** A prealbumin activator of prekallikrein. III. Appearance of chemotactic activity for neutrophils. JEM 135: 81-97 (1972); **Mason AJ et al** Structure of mouse kallikrein gene family suggests a role in specific processing of biologically active peptides. N 303: 300-7 (1983); **Nishikawa K et al** Generation of vasoactive peptide bradykinin from human umbilical vein endothelium-bound high molecular weight kininogen by plasma kallikrein. Blood 80: 1980-8 (1992); **Qin H et al** Localization of human glandular kallikrein-1 gene to chromosome 19q13.3-13.4 by *in situ* hybridization. Hum. Hered. 41: 222-6 (1991); **Richards RI et al** Human glandular Kallikrein genes: genetic and physical mapping of the KLK1 locus using a highly polymorphic microsatellite PCR marker. Genomics 11: 77-82 (1991); **Riegman PHJ et al** The prostate-specific antigen gene and the human glandular kallikrein-1 gene are tandemly located on chromosome 19. FL 247: 123-6 (1989); **Riegman PHJ et al** Characterization of the human kallikrein locus. Genomics 14: 6-11 (1992); **Schapira M et al** Purified human plasma kallikrein aggregates human blood neutrophils. JCI 69: 1199-1201 (1982); **Schedlich LJ et al** Primary structure of a human glandular kallikrein gene. DNA 6: 429-37 (1987); **Shimamura T et al** A cytotoxic serine proteinase isolated from mouse submandibular gland. Immunol-Lett. 22: 155-9 (1989); **Qin H et al** Localization of human glandular kallikrein-1 gene to chromosome 19q13.3-13.4 by *in situ* hybridization. Hum. Hered. 41: 222-6 (1991)

Kallistatin: This human plasma protein is a specific inhibitor of human tissue » kallikrein. It represents a major inhibitor of human tissue kallikrein in the circulation.

Zhou GX et al Kallistatin: a novel human tissue kallikrein inhibitor; purification, characterization, and reactive center sequence. JBC 267: 25873-80 (1992)

β-NGF endopeptidase	ILGGFKCEKNSQPWQVAVYYLDEYLCGBVL
EGF-BP type A	ILGGFKCEKNSQPWQVAVYYLDEYLCGBVL
mGK-22	ILGGFKCEKNSQPWQVAVYYLDEYLCGBVL
EGF-BP type B	VVGGFNCEKNSQPWQVAVYYQKEHICGBVL
EGF-BP type C	IVGGFKCEKNSQPWHVAVYRYNEYICGBVL
γ-NGF	IVGGFKCEKNSQPWHVAVYRYTQYLCGBVL

Comparison of aminoterminal sequences of murine kallikrein mGK-22, EGF binding proteins and γ-NGF.

Kaposi-FGF: Kaposi fibroblast growth factor. see: KFGF.

Kaposi fibroblast growth factor: see: KFGF.

KATO-III cell-derived stomach cancer amplified gene: see: K-*sam*.

KC: A protein initially identified by its overexpression in murine monocytes and macrophages following in response to » PDGF and » M-CSF. KC is involved in neutrophil chemotaxis and activation. The synthesis of KC in vascular endothelial cells is induced by thrombin. Its synthesis is also induced by » TNF-α.
KC is the murine homologue of the human GRO gene (see: MGSA, melanoma growth stimulating activity) and therefore belongs to the » chemokine family of cytokines.

Cochran B et al Molecular cloning of gene sequences regulated by platelet-derived growth factor. Cell 33: 939-47 (1983); **Introna M et al** The effect of LPS on expression of the early competence gene JE and gene KC in murine peripheral macrophages. JI 138: 3891-6 (1987); **Oquendo P et al** The platelet-derived growth factor-inducible *KC* gene encodes a secretory protein related to platelet α-granule proteins. JBC 264: 4133-7 (1989); **Ohmori Y et al** Tumor necrosis factor-α induces cell type and tissue-specific expression of chemoattractant cytokines *in vivo*. Am. J. Pathol. 142: 861-70 (1993); **Safirstein R et al** Expression of cytokine-like genes JE and KC is increased during renal ischemia. Am. J. Physiol. Renal, Fluid Electrolyte Physiol. 261: F1095-F1101 (1991); **Shen XY et al** Thrombin-induced expression of the KC gene in cultured aortic endothelial cells. Involvement of proteolytic activity and protein kinase C. BBA 1049: 145-50 (1990)

KDR: *kinase insert domain receptor* A gene isolated from a human endothelial cell cDNA library. It encodes a receptor tyrosine kinase. The human gene maps to chromosome 4q31.2-q32 and encodes a transcript of ≈ 7 kb. The KDR protein is one of the receptors for » VEGF (vascular endothelial cell growth factor). The KDR receptor tyrosine kinase shares structural similarities with a recently reported receptor for VEGF, » *flt*, in a manner reminiscent of the similarities between the α and β forms of the » PDGF receptors.
The KDR gene is the human homologue of *flk*-1 (fetal liver kinase 1), a receptor tyrosine kinase closely related to » *flt*, cloned from mouse cell populations enriched for » hematopoietic stem and progenitor cells. Human KDR and mouse *flk*-1 show 85 % amino acid identity. Mouse *flk*-1 is selectively expressed in vascular endothelium and has also been shown to function as a VEGF receptor.
On the basis of structural similarities KDR, *flt*, *flk*-1, and some other related receptors constitute a subfamily of class III tyrosine kinases (see also: PTK, protein tyrosine kinase).
It has been shown that *flk*-1 is expressed abundantly in proliferating endothelial cells of the vascular sprouts (see also: Angiogenesis factors) and branching vessels of embryonic and early postnatal brain and that its expression is reduced drastically in adult brain where proliferation has ceased. *flk*-1 is also expressed in the blood islands in the yolk sac of embryos. The expression of this receptor therefore correlates with the development of the vascular system and with endothelial cell proliferation.

Matthews W et al A receptor tyrosine kinase cDNA isolated from a population of enriched primitive hematopoietic cells and exhibiting close genetic linkage to c-*kit*. PNAS 88: 9026-30 (1991); **Millauer B et al** High affinity VEGF binding and developmental expression suggest *flk*-1 as a major regulator of vasculogenesis and angiogenesis. Cell 72: 835-46 (1993); **Terman BI et al** Identification of a new endothelial cell growth factor receptor tyrosine kinase. O 6: 1677-83 (1991); **Terman BI et al** The KDR gene maps to human chromosome 4q31.2-q32, a locus which is distinct from locations for other type III growth factor receptor tyrosine kinases. Cytogenet. Cell. Genet. 60: 247-9 (1992); **Terman BI et al** Identification of the KDR tyrosine kinase as a receptor for vascular endothelial cell growth factor. BBRC 187: 1579-86 (1992); **Quinn TP et al** Fetal liver kinase 1 is a receptor for vascular endothelial growth factor and is selectively expressed in vascular endothelium. PNAS 90: 7533-7 (1993)

Keratinocyte-derived amphiregulin: see: KAF (keratinocyte-derived autocrine factor).

Keratinocyte-derived autocrine factor: see: KAF.

Keratinocyte epidermal cell-derived thymocyte-activating factor: This factor is identical with » IL1. See also: ETAF (epidermal thymocyte-activating factor).

Kupper TS et al Human keratinocytes contain mRNA indistinguishable from monocyte interleukin 1 α and β mRNA. Keratinocyte epidermal cell-derived thymocyte-activating factor is identical to interleukin 1. JEM 164: 2095-100 (1986)

Keratinocyte growth factor: see: KGF.

K-FGF: *Kaposi fibroblast growth factor* This factor of 22 kDa is secreted by Kaposi sarcoma cells. It is also known as ***KS-FGF*** (Kaposi sarcoma FGF), ***KS protein*** and ***KS3 protein***. The recommended name for this factor is FGF-4 (see also: FGF, fibroblast growth factors)
K-FGF is the product of the KS oncogene initially

detected after transfection of murine 3T3 cells with DNA from human Kaposi sarcomas. Expression of the activated oncogene results in cell transformation, which can be inhibited by » suramin. This oncogene is identical with the » *hst* oncogene isolated from human stomach cancers. It has been proposed that the therapeutic effect of » IFN-α in AIDS-KS may be based on antiangiogenesis activity (see also: Angiogenesis factors) by suppressing proto-oncogenes-oncogenes of the FGF family including K-FGF.

K-FGF is closely related to » bFGF produced by many normal cell types. K-FGF acts as a mitogen for fibroblasts, endothelial cells and melanocytes. In contrast to bFGF K-FGF contains a classical hydrophobic » signal sequence allowing release of this factor by classical secretory pathways (secretion via endoplasmic reticulum/Golgi system). K-FGF is also produced and secreted by embryonic carcinoma and embryonic stem cells (see: ES cells) and functions as an » autocrine growth factor that also influences the differentiation of these cells. K-FGF also plays a role in the differentiation of the mesoderm during embryonic development. K-FGF/hst has also been shown to have angiogenic activity in neural transplants (see also: Angiogenesis factors).

Binding of K-FGF to the FGF receptor requires the simultaneous interaction with heparan sulfate proteoglycans of the extracellular matrix (see: ECM) as demonstrated by the ability of heparitinase to inhibit receptor binding and biological activity.

Basilico C et al Expression and activation of the K-*fgf* oncogene. ANY 567: 95-103 (1989); **Brustle O et al** Angiogenic activity of the K-fgf/hst oncogene in neural transplants. O 7: 1177-83 (1992); **Damen JE et al** Transformation and amplification of the K-fgf proto-oncogene in NIH-3T3 cells, and induction of metastatic potential. BBA 1097: 103-10 (1991); **Delli Bovi P et al** An oncogene isolated by transfection of Kaposi's sarcoma DNA encodes a growth factor that is a member of the FGF family. Cell 50: 729-37 (1987); **Fuller-Pace F et al** Cell transformation by kFGF requires secretion but not glycosylation. JCB 115: 547-55 (1991); **Hebert JM et al** Isolation of cDNAs encoding four mouse FGF family members and characterization of their expression during embryogenesis. Dev. Biol. 138: 454-63 (1990); **Huebner K et al** The FGF-related oncogene K-FGF maps to human chromosome region 11q13 possible near *int*-2. OR 3: 263-70 (1988); **Olwin BB & Rapraeger A** Repression of myogenic differentiation by aFGF, bFGF, and K-FGF is dependent on cellular heparan sulfate. JCB 118: 631-9 (1992); **Paterno GD et al** Mesoderm-inducing properties of INT-2 and kFGF: two oncogene-encoded growth factors related to FGF. Development 106: 79-83 (1990); **Sinkovics JG** Kaposi's sarcoma: Its oncogenes and growth factors. CRC Crit. Rev. Oncology/Hematology 11: 87-107 (1991); **Wellstein A et al** Autocrine growth stimulation by secreted Kaposi fibroblast growth factor but not by endogenous basic fibroblast growth factor. Cell Growth Differ. 1: 63-71 (1990)

KG-1: A human acute myeloid leukemia cell line established from a patient with acute myelogenous leukemia that had evolved from an erythroleukemia. It represents an early stage of hematopoietic differentiation (see also: Hematopoiesis). A subline of KG-1, designated KG1a, has lost myeloid features, acquired new karyotypic markers, and has several characteristics associated with immature T cells. Both KG-1 and KG1a transcribe unrearranged IgH genes.

The cells are used in » colony formation assays instead of human bone marrow cells (see also: LTBMC, long-term bone marrow culture) to assay the colony-stimulating activity of cytokines (see also: CSF). The growth of these cell is inhibited by » TGF-β at doses ranging from 0.025 to 2.5 ng/mL.

K562 cells do not express the cellular » *fes* oncogene. cell line and are resistant to myeloid differentiation; these cells acquire the ability to differentiate after introduction of a functionally expressed *fes* gene. Differentiation of KG-1 cells into macrophage-like cell types can be induced by treatment of the cells with » phorbol esters.

The synthesis of the enzyme γ-glutamyl transferase, a multifunctional enzyme mediating, among other things the conversion of leukotriene C4 to leukotriene D4 and known to be required for normal myeloid proliferation and differentiation, is induced in KG-1 cells by » IL1, » IL3, » GM-CSF, and » TNF-α, but not by » IL6, » G-CSF, or » M-CSF.

Aota F et al Monoclonal antibody against myeloid leukaemia cell line (KG-1). CR 43: 1093-6 (1983); **Furley AJ et al** Divergent molecular phenotypes of KG1 and KG1a myeloid cell lines. Blood 68: 1101-7 (1986); **Kiss Z et al** Differential effects of various protein kinase C activators on protein phosphorylation in human acute myeloblastic leukaemia cell line KG-1 and its phorbol ester-resistant subline KG-1a. CR 47: 1302-7 (1987); **Koeffler HP et al** An undifferentiated variant derived from the human acute myelogenous leukaemia cell line KG-1. Blood 56: 265-73 (1980); **Koeffler HP & Golde DW** Human myeloid leukemia cell lines: a review. Blood 56: 344-50 (1980); **Koeffler HP & Golde DW** Humoral modulation of human acute myelogenous leukemia cell growth *in vitro*. CR 40: 1858-62 (1980); **Koeffler HP et al** An undifferentiated variant derived from the human acute myelogenous leukemia cell line (KG-1). Blood 56: 265-73 (1980); **Miller AM et al** Haematopoietic growth factor induction of γ-glutamyl transferase in the KG-1 myeloid cell line. EH 21: 9-15 (1993); **Niskanen E et al** Responsiveness of a human myelogenous leukaemia cell line (KG-1) to humoral factors *in vivo*. Leukaemia Res. 4: 203-8 (1980); **Piacibello W et al** Differential effect of transforming growth factor β 1 on the proliferation of human lymphoid and myeloid leukemia cells. Haematologica 76: 460-6 (1991); **Treon SP et al** Growth restraint and differentiation by LPS/TNF-α/IFN-γ reorganization of the microtubule network in human leukemia cell lines. Leukemia 6(Suppl.3) 141S-45S (1992); **Yu G et al** K562 leukemia cells transfected with the human c-*fes* gene acquire the ability to undergo myeloid differentiation. JBC 264: 10276-81 (1989)

KGF: *keratinocyte growth factor* KGF was found in conditioned medium of a human embryonic lung fibroblast cell line. It is secreted in large amounts by fibroblast-like stromal cells in epithelial tissues. The factor is also called HBGF-7 (heparin-binding growth factor 7) and the recommended new name is FGF-7 (fibroblast growth factor 7); see also: FGF. KGF is a member of the fibroblast growth factor family (see: bFGF as prototype). The bovine counterpart of KGF is » SDGF-3 (spleen-derived growth factor).

KGF is a protein of 22.5 kDa with a length of 194 amino acids inferred from the cDNA sequence. The human gene contains three exons and maps to human chromosome 15. The gene is unusual in that a portion of it, comprising exon 2, exon 3, the intron between them, and a 3′ noncoding segment of the KGF transcript, is amplified in the human genome and dispersed to multiple human chromosomes. Unlike the majority of described pseudogenes these copies are transcriptionally active and differentially regulated in various tissues. These multiple copies of KGF-related genes are also observed in the genomic DNAs of chimpanzee and gorilla, but not in gibbon, African green monkey, macaques, mice, or chickens.

The KGF receptor encodes a tyrosine kinase and is a member of the family of receptors binding » aFGF and » bFGF. Binding of KGF to its receptor is competed by » aFGF and ≈ 20-fold less well by » bFGF. The KGF receptor differs from the FGFR-2 receptor (see: *bek*) by a divergent stretch of 49 amino acids in its extracellular domain. The two receptors arise by differential splicing and the KGF receptor transcript is specifically found in epithelial cells. Thus, two growth factor receptors with different ligand-binding specificities and expression patterns are encoded by alternative transcripts of the same gene. A synthetic peptide (His199 - Tyr223) corresponding to part of the predicted sequence of the KGF receptor alternative exon blocks KGF mitogenic activity, the interaction between KGF and its receptor, and the interaction between KGF and a neutralizing KGF-specific monoclonal antibody.

KGF is a potent specific mitogen for many epithelial cells but not for fibroblasts and endothelial cells. KGF is thought to play an important role in the » paracrine growth control of normal epithelial cells. KGF stimulates the proliferation of primary and secondary human keratinocytes to the same extent than » EGF.

■ **Clinical Use & Significance:** KGF is probably one of the major factors playing a role in tissue repair following skin injuries (see also: Wound healing). Its mRNA is induced ≈ 160-fold in basal keratinocytes while mRNAs levels of » aFGF, » bFGF, and » FGF-5 are only slightly induced (2- to 10-fold) during wound healing, and no expression of FGF-3 (see: *int*), FGF-4 (see: *hst*), and » FGF-6 is detected in normal and wounded skin. Topical application of KGF has recently been shown to improve reepithelialisation in a porcine wound healing model.

■ **Transgenic/Knock-out/Antisense Studies:** The introduction of a KGF fusion gene into » transgenic animals (mice) and the directed expression of the gene in keratinocytes leads to many phenotypic changes including a frail and weak appearance, grossly wrinkled skin, gross increase in epidermal thickness accompanied by alterations in epidermal growth and differentiation, marked suppression of hair follicle morphogenesis, suppression of adipogenesis, elevated salivation, gross transformations in the epidermis and tongue epithelium, and altered differentiation of salivary glands.

Aaronson SA et al Keratinocyte growth factor. A fibroblast growth factor family member with unusual target cell specificity. ANY 638: 62-77 (1991); **Bottaro DP et al** Characterization of the receptor for keratinocyte growth factor. Evidence for multiple fibroblast growth factor receptors. JBC 265: 12767-70 (1990); **Bottaro DP et al** A keratinocyte growth factor receptor-derived peptide antagonist identifies part of the ligand binding site. JBC 268: 9180-3 (1993); **Chiu ML & O'Keefe EJ** Placental keratinocyte growth factor: partial purification and comparison with epidermal growth factor. ABB 269: 75-85 (1989); **Finch PW et al** Human KGF is FGF-related with properties of a paracrine effector of epithelial cell growth. S 245: 752-5 (1989); **Gilchrist BA et al** Characterization and partial purification of keratinocyte growth factor from the hypothalamus. JCP 120: 377-83 (1984); **Kelley MJ et al** Emergence of the keratinocyte growth factor multigene family during the great ape radiation. PNAS 89: 9287-91 (1992); **Marchese C et al** Human keratinocyte growth factor activity on proliferation and differentiation of human keratinocytes: differentiation response distinguishes KGF from EGF family. JCP 144: 326-32 (1990); **Miki T et al** Expression of cDNA cloning of the KGF receptor by creation of a transforming autocrine loop. S 251: 72-5 (1991); **Miki T et al** Determination of ligand-binding specificity by alternative splicing: two distinct growth factor receptors encoded by a single gene. PNAS 89: 246-50 (1992); **Ron D et al** Expression of biologically active recombinant keratinocyte growth factor. Structure/function analysis of amino-terminal truncation mutants. JBC 268: 2984-8 (1993); **Rubin JS et al** Purification and characterization of a newly identified growth factor specific for epithelial cells. PNAS 86: 802-6 (1989); **Staiano-Coichi L et al** Human keratinocyte growth factor effects in a porcine model of epidermal wound healing. JEM 178: 865-78

(1993); **Werner S et al** Large induction of keratinocyte growth factor expression in the dermis during wound healing. PNAS 89: 6896-900 (1992); **Yan GC et al** Sequence of rat keratinocyte growth factor (heparin-binding growth factor type 7). In Vitro Cell Dev. Biol. 27A 437-8 (1991); **Yan GC et al** Exon switching and activation of stromal and embryonic fibroblast growth factor (FGF)-FGF receptor genes in prostate epithelial cells accompany stromal independence and malignancy. MCB 13: 4513-22 (1993); Heparin-binding keratinocyte growth factor (FGF-7) is a candidate stromal-to-epithelial cell prostate andromedin. Mol. Endocrinol. 6: 2123-8 (1992)

● **Transgenic/Knock-out/Antisense Studies: Guo L et al** Targeting expression of keratinocyte growth factor to keratinocytes elicits striking changes in epithelial differentiation in transgenic mice. EJ 12: 973-86 (1993)

KGF: *kidney growth factor* This poorly characterized factor was isolated from the blood plasma of heminephrectomised rats. The protein is heat- and trypsin-resistant. It stimulates the DNA synthesis in murine kidneys and kidney cell lines and also stimulates the uptake of sodium by these cells. KGF may be a renotropic protein which can play a key role in the renal compensatory growth after uninephrectomy.

Esbrit P et al Biological properties of a renotropic protein present in plasma of uninephrectomised rats. Renal Physiol. Biochem. 14: 224-35 (1991)

KHF: *killer cell helper factor* KHF activities of 15-20 kDa and 45-50 kDa were initially isolated as proteins that promote the development of activated thymocytes (see also: Cell activation) into cytotoxic T cells. The 15-20 kDa factor has been shown to be identical with » IL2. The other KHF activity may be due to » IL5 and » IFN-γ.

Aarden LA et al Revised nomenclature for antigen-nonspecific T cell proliferation and helper factors. CI 48: 433-36 (1979); **Kaieda T et al** A human helper T cell clone secreting both killer helper factor(s) and T cell-replacing factor(s). JI 129: 46-51 (1982); **Takatsu K et al** Interleukin 5, a T cell-derived B cell differentiation factor also induces cytotoxic T lymphocytes. PNAS 84: 4234-40 (1987)

Ki-1: see: CD30.

Kidney angiogenic factor: see: KAF.

Kidney growth factor: see: KGF.

Killer cell helper factor: see: KHF.

Killer factor: see: NKCF (natural killer cytotoxic factor)

Killer helper factor: see: KHF.

Killer T cell activating factor: see: KAF.

Kinase insert: see: PTK (protein tyrosine kinase).

Kinin 9: alternative name for Bradykinin.

***kit*:** The *kit* proto-oncogene (see: oncogenes) encodes a transmembrane receptor with intrinsic tyrosine-specific protein kinase activity in its intracellular domain. It is the cellular homologue of the viral *kit* oncogene of HZ4-FSV (Hardy-Zuckerman 4 feline sarcoma virus). The kit receptor has been renamed CD117 (see also: CD Antigens). The ligand for the *kit* receptor is » SCF (stem cell factor).

This gene was initially identified by many deletions and point mutations of the murine dominant White spotting (W) locus on chromosome 5. Mutations at this locus produces deficiencies in three migratory cell populations, namely, the pluripotent » hematopoietic stem cell, the migrating melanoblast during early embryonic development, and the primordial germ cell during this same period of development.

The plethora of W mutations available for molecular analysis offers a unique opportunity to dissect the role of a tyrosine kinase receptor and its cognate ligand during development in a fashion not possible for most other mammalian genes. Mutations at the murine white locus lead to alterations in coat colors, impair the development of gonads, and lead to a reduced production of the erythroid precursor cells » BFU-E (see also: Hematopoiesis). The c-*kit* protein is probably also involved in the proliferation and differentiation of placenta tissue. The *kit* receptor is structurally closely related to the receptors for » PDGF and » M-CSF (see: *fms* oncogene). It possesses a bipartite protein kinase domain characterized by the insertion of a so-called kinase insert (see: PTK, protein tyrosine kinase). This insert contains a binding site for phosphatidylinositol 3′ kinase. The extracellular portion of the *kit*-encoded receptor comprises five immunoglobulin (Ig)-like domains.

The *kit* receptor density is markedly reduced by binding of its cognate ligand. This also leads to the immediate dimerisation and » autophosphorylation of the receptor that, in turn, activates its intrinsic protein kinase activity. In mast cells the receptor density is also decreased by treatment of the cells with » IL3, » GM-CSF and » Epo, but not by » IL4. Engagement of the receptor leads to the rapid phosphorylation of the oncogene product » *vav* which appears to play an important role in *kit*-mediated signal transduction.

The human c-*kit* proto-oncogene maps to chromosome 4q11-q12 in the same region also encoding one of the » PDGF receptors (PDGFRA). The two genes have been located on a DNA fragment of ≈ 700 kb. The human *kit* gene has a length of more than 70 kb and contains 21 exons. The longest transcript is 5230 bp and is alternatively spliced (see also: Gene expression).

A deletion of the *kit* gene also encompassing the PDGFRA locus (one receptor for » PDGF) is the underlying cause of human piebaldism, a rare autosomal dominant disorder of pigmentation, characterized by congenital patches of white skin and hair from which melanocytes are absent. Missense (Gly664→Arg) and other types of mutations in this gene locus have also been shown to occur in such patients; they account for a continuous range of phenotypes in human piebaldism.

c-*kit* has been found to be aberrantly expressed almost exclusively in small-cell lung cancer (SCLC) among various types of solid tumors. Binding of the ligand, SCF, to this receptor mediates chemotaxis and moderate *in vitro* cell growth.

Blume-Jensen P et al Activation of the human c-*kit* product by ligand-induced dimerisation mediates circular actin reorganization and chemotaxis. EJ 10: 4121-8 (1991); **Catlett JP et al** c-*kit* Expression by CD34⁺ bone marrow progenitors and inhibition of response to recombinant human interleukin-3 following exposure to c-*kit* antisense oligonucleotides. Blood 78: 3186-91 (1991); **Chabot B et al** The proto-oncogene c-*kit* encoding a transmembrane tyrosine kinase receptor maps to the mouse W locus. N 335: 88-89 (1988); **d'Auriol L et al** Localization of the human c-*kit* proto-oncogene on the q11-q12 region of chromosome 4. Hum. Genet. 78: 374-6 (1988); **Dubreuil P et al** The c-*fms* gene complements the mitogenic defect in mast cells derived from mutant W mice but not mi (microphthalmia) mice. PNAS 88: 2341-5 (1991); **Flanagan JG & Leder P** The *kit* ligand: a cell surface molecule altered in steel mutant fibroblasts. Cell 63: 185-94 (1990); **Fleischman RA et al** Deletion of the c-*kit* proto-oncogene in the human developmental defect piebald trait. PNAS 88: 10885-9 (1991); **Fleischman RA** Human piebald trait resulting from a dominant negative mutant allele of the c-*kit* membrane receptor gene. JCI 89: 1713-7 (1992); **Giebel LB & Spritz RA** Mutation of the *kit* (mast/stem cell growth factor receptor) proto-oncogene in human piebaldism. PNAS 88: 8696-9 (1991); **Giebel LB et al** Organization and nucleotide sequence of the human *kit* (mast/stem cell growth factor receptor) proto-oncogene. O 7: 2207-7 (1992); **Handel MA et al** Developmental abnormalities in Steel17H mice result from a splicing defect in the steel factor cytoplasmic tail. Genes Dev. 6: 1832-42 (1992); **Huang E et al** The hematopopietic growth factor KL is encoded by the SL locus and is the ligand of the c-*kit* receptor, the gene product of the *W* locus. Cell 63: 225-33 (1990); **Lev S et al** Interkinase domain of *kit* contains the binding site for phosphatidylinositol 3' kinase. PNAS 89: 678-82 (1992); **Lev S et al** Dimerization and activation of the *kit* receptor by monovalent and bivalent binding of the stem cell factor. JBC 267: 15970-7 (1992); **Lev S et al** Interspecies molecular chimeras of *kit* help define the binding site of the stem cell factor. MCB 13: 2224-34 (1993); **Nocka K et al** Molecular bases of dominant negative and loss of function mutations at the murine c-*kit*/white spotting locus: W-37, W-v, W-41 and W. EJ 9: 1805-13 (1990); **Qiu F et al** Primary structure of c-*kit*: relationship with the CSF-1/PDGF receptor kinase family – oncogenic activation of v-*kit* involves deletion of extracellular domain and C-terminus. EJ 7: 1003-11 (1988); **Ratajczak MZ et al** Role of the *kit* proto-oncogene in normal and malignant human hematopoiesis. PNAS 89: 1710-4 (1992); **Reith AD** ω mutant mice with mild or severe developmental defects contain distinct point mutations in the kinase domain of the c-*kit* receptor. GD 4: 390-400 (1990); **Sekido Y et al** Recombinant human stem cell factor mediates chemotaxis of small-cell lung cancer cell lines aberrantly expressing the c-*kit* proto-oncogene. CR 53: 1709-14 (1993); **Spritz RA et al** Dominant negative and loss of function mutations of the c-*kit* (mast/stem cell growth factor receptor) proto-oncogene in human piebaldism. Am. J. Hum. Genet. 50: 261-9 (1992); **Spritz RA et al** Deletion of the *kit* and PDGFRA genes in a patient with piebaldism. Am. J. Med. Genet. 44: 492-5 (1992); **Spritz RA et al** Mutations of the *kit* (mast/stem cell growth factor receptor) proto-oncogene account for a continuous range of phenotypes in human piebaldism. Am. J. Hum. Genet. 51: 1058-65 (1992); **Spritz RA** Lack of apparent hematologic abnormalities in human patients with c-*kit* (stem cell factor receptor) gene mutations. Blood 79: 2497-9 (1992); **Spritz RA et al** Novel mutations of the *kit* (mast/stem cell growth factor receptor) proto-oncogene in human piebaldism. J. Invest. Dermatol. 101: 22-5 (1993); **Toyota M et al** Expression of two types of *kit* ligand mRNAs in human tumor cells. Int. J. Hematol. 55: 301-4 (1992); **Vandenbark GR et al** Cloning and structural analysis of the human c-*kit* gene. O 7: 1259-66 (1992); **Welham MJ & Schrader JW** Modulation of c-*kit* mRNA and protein by hemopoietic growth factors. MCB 11: 2901-4 (1991)

Kit225: A human T cell line established from a patient with T cell chronic lymphocytic leukemia. The cells express the markers CD3, » CD4 and are » CD8-negative. The cells are not infected with HTLV I or II. Kit 225 cells express a large amount of » IL2 receptors constitutively and their growth is absolutely dependent on IL2. No other stimuli, such as lectins or antigens, are required for maintaining the responsiveness to IL2.

Arima N et al Pseudo-high affinity interleukin 2 (IL2) receptor lacks the third component that is essential for functional IL2 binding and signaling. JEM 176: 1265-72 (1992); **Hori T et al** Establishment of an interleukin 2-dependent human T cell line from a patient with T cell chronic lymphocytic leukemia who is not infected with human T cell leukemia/lymphoma virus. Blood 70: 1069-72 (1987); **Sawami H et al** Signal transduction by interleukin 2 in human T cells: activation of tyrosine and ribosomal S6 kinases and cell-cycle regulatory genes. JCP 151: 367-77 (1992)

***Kit* ligand:** This factor is identical with » SCF (stem cell factor). The receptor for SCF is » *kit*.

KL: *kit ligand* This factor is identical with SCF (stem cell factor) which binds to the receptor named » *kit*.

KM102: This human stromal cell line (see also: Hematopoiesis) has been obtained from adherent bone marrow cell layers of Dexter-type long-term bone marrow cultures (see: LTBMC) after immortalisation with SV40. The cells have been shown to express » GM-CSF and » IL11 mRNA. They support the growth of early hematopoietic progenitor types » CFU-GM, » CFU-G, » CFU-M, and myeloid leukemic cells. KM102 cells do not support B leukemic cells.

Harigaya K & Handa H Generation of functional clonal cell lines from human bone marrow stroma. PNAS 82: 3477-80 (1985); **Kawashima I et al** Molecular cloning of cDNA encoding adipogenesis inhibitory factor and identity with interleukin-11. FL 283: 199-202 (1991); **Kohama T et al** A burst-promoting activity derived from the human bone marrow stromal cell line KM-102 is identical to the granulocyte-macrophage colony-stimulating factor. EH 16: 603-8 (1988); **Manabe A et al** Bone marrow-derived stromal cells prevent apoptotic cell death in B lineage acute lymphoblastic leukemia. Blood 79: 2370-7 (1992); **Ohkawa H & Harigaya K** Effect of direct cell-to-cell interaction between the KM-102 clonal human marrow stromal cell line and the HL-60 myeloid leukemic cell line on the differentiation and proliferation of the HL-60 line. CR 47: 2879-82 (1987)

KM-102-BPA: *KM-102 burst promoting activity* This factor is isolated from culture supernatant of the human bone marrow stromal cell line KM-102. It is identical with » GM-CSF. See also: BPA (burst promoting activity). KM-102 cells are also a source of » IL11 (see: AGIF (adipogenesis inhibitory factor).

Kohama T et al A burst-promoting activity derived from the human bone marrow stromal cell line KM-102 is identical to the granulocyte-macrophage colony-stimulating factor. EH 16: 603-8 (1988)

KMT-2: A human cell line established from umbilical cord blood cells. Morphologic and cytochemical studies (peroxidase-negative, Sudan-black-negative, chloroacetate esterase-negative, PAS-positive, nonspecific esterase-positive) and phenotyping (HLA-DR, My7 = CD13, My9 = CD33, My10 = CD34, MCS-2, LeuM1-positive, glycophorin A-negative, and P2-negative) suggest that KMT-2 cells are myelomonocytic cells, probably of immature progenitor origin.
The cell line is used to assay » IL3 (see also: Bioassays). IL3 activity is determined by measuring the incorporation of ^{3}H thymidine into the newly synthesized DNA of the proliferating factor-dependent cells. Cell proliferation can be determined also by employing the » MTT assay. An alternative and entirely different method of detecting IL3 is » Message amplification phenotyping. The cells also respond to » GM-CSF and slightly to » IL6. Their growth is not supported by » IL1, » IL2, » IL4, » IL5, or » Epo.
KMT-2 cells form colonies in soft agar (see: Colony formation assay). Colony formation is suppressed by » LD78. The conditioned medium of KMT-2 cells cultured with LD78 suppresses colony formation by » CD34-positive cells, suggesting that LD78 induces factors that are inhibitory for hematopoiesis. LD78 also inhibits the » Epo- and» IL3-induced differentiation of KMT-2 cells into neutrophils, macrophages and megakaryocytes.

Shiozaki H et al Suppressive effect of LD78 on the proliferation of human hemopoietic progenitors. Jpn. J. Cancer Res. 83: 499-504 (1992); **Tamura S et al** A new hematopoietic cell line, KMT-2, having human interleukin 3 receptors. Blood 76: 501-7 (1990)

Knock-out mice: A general term (sometimes abbrev. GKO, gene knock-out) relating to artificially generated null mutations of a gene, including those encoding cytokines or their receptors. Such mouse mutants, which are devoid of a particular cytokine or receptor function, are created by a process called ***homologous recombination*** or ***targeted disruption***. It involves the inactivation of an endogenous gene by insertion of cloned sequences. This takes place in » ES (embryonic stem) cells which are then used to produce » Transgenic animals. Another way to inactivate specifically a given cytokine gene is the use of » Antisense oligonucleotides (see also: Genetic ablation).
Since most cytokines have ubiquitous biological activities, their physiologic significance as normal regulators of physiology is often difficult to assess. Studies of gene functions in suitable knock-out mice are of particular importance in cytokine research because, unlike *in vitro* studies, they provide information about the true *in vivo* functions of a given cytokine by studying the effects of their absence. In some instances studies with cytokine knock-out mice have already revealed either that these mice develop normally and/or do not show the pronounced effects on the immune system and other organs one would have expected from the known *in vitro* activities of the cytokines (for a good example see: IL2, IL4).
If information about knock-out studies is available for a given cytokine or its receptor it can be found in this dictionary as a special subentry (Transgenic/Knock-out/Antisense studies) for each particular cytokine.

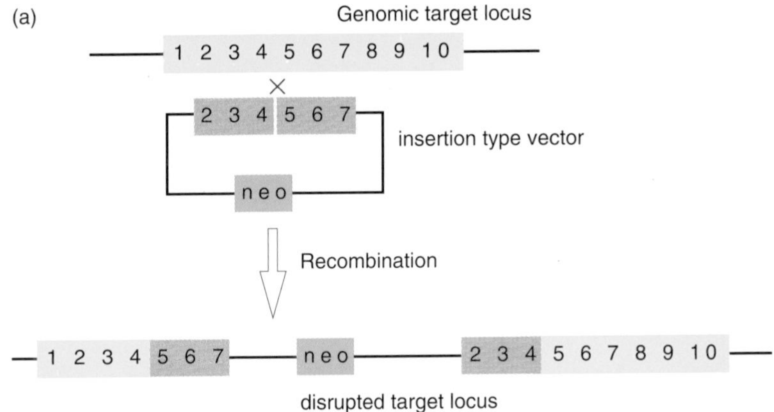

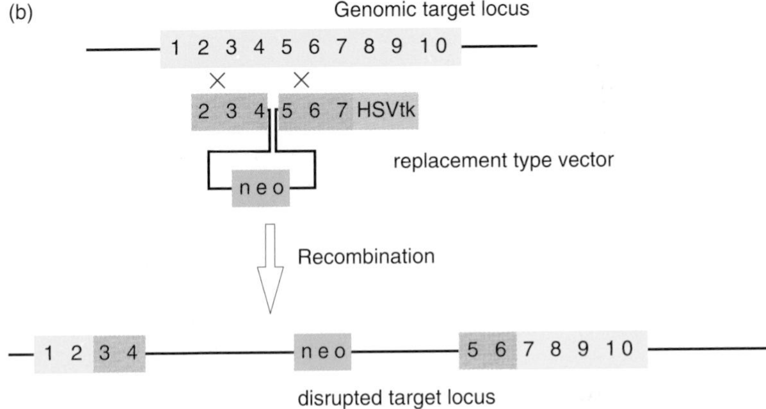

Principle of gene targeting involving the use of insertion or replacement type vectors.
(a) Use of insertion type vectors involves a single cross-over between genomic target sequences and homologous sequences at either end of the targeting vector. The neomycin resistance gene contained within the vector serves as a positive selectable marker.
(b) Gene targeting using replacement type vectors requires two cross-over events. The positive selection marker (neo) is retained while the negative selectable marker (HSV thymidine kinase) is lost. The advantage of this system is the fact that cells harboring randomly and unspecifically integrated gene constructs still carry the thymidine kinase gene. These cells can be eliminated selectively by using thymidine kinase as a selective marker. One disadvantage of the system may be the time required to handle ES cells *in vitro.* These cells have a certain tendency to differentiate and this can favor the subsequent generation of genetic mosaics rather than the desired germ line integration when the cells are used to create transgenic animals.

Selected phenotypes of some knock-out mice.

Disrupted gene	Major effect
abl	increased perinatal mortality; abnormal spleen, head, and eye development; major reductions in adult bone marrow B cell progenitors
bcl-2	growth retardation, early death, massive apoptotic involution of thymus and spleen, disappearance of T and B lymphocytes from bone marrow, thymus and periphery, increased sensitivity of T cells to glucocorticoids and γ-irradiation
BDNF	severe deficiencies in coordination and balance; excessive degeneration in several sensory ganglia; survival of sympathetic, midbrain dopaminergic and motor neurons unaffected
β2-microglobulin	mice fertile and apparently healthy, little if any functional MHC class I antigen expression, no mature $CD4^-CD8^+$ cells; defects in $CD4^-CD8^+$ T cell-mediated cytotoxicity
CNTF	no functional or morphological changes during first postnatal weeks, progressive atrophy and loss of motor neurons with increasing age
CREB	dwarf phenotype with atrophied pituitary glands markedly deficient in somatotroph but not other cell types
fos	severe osteopetrosis, ossification of the marrow space, delayed or absent gametogenesis, lymphopenia; altered behavior; some animals without discernible phenotype
fyn	thymocytes refractory to stimulation through T cell receptor; normal maturation of peripheral T cells mature; long-term potentiation, spatial learning; defects in hippocampal development
GATA-1	inability to generate mature erythrocytes
HCP	(spontaneous mutation: motheaten) multiple T and B cells defects; multiple hematopoietic defects
IFN-γ	normal development in the absence of pathogens; impaired production of macrophage antimicrobial products; reduced expression of macrophage MHCclass II antigens; high susceptibility to sublethal doses of *Mycobacterium bovis*; uncontrolled proliferation of splenocytes in response to mitogen and alloantigen; enhanced T cell cytolytic activity against allogeneic target cells; reduced activity of resting splenic natural killer cells
IFN-γ R	no overt anomalies; normal cytotoxic and T helper cell responses; increased susceptibility to *Listeria monocytogenes* and vaccinia virus; no antigen-specific Ig G2a response
IGF-I	< 60 % body weight of wild type animals; gross underdevelopment of muscle tissues; ill-defined alveolae in lung tissues of late embryos and neonates; perinatal death of more than 95 % of animals
IGF-I R	< 45 % normal size; general organ hypoplasia; death at birth
IGF-II	transmission through male germline: heterozygous growth-deficient progeny ; transmission through maternal germline: phenotypically normal heterozygotes; homozygous mutants indistinguishable from growth-deficient heterozygotes
IGF-I/IGF-IR	double mutant similar to IGF-I R mutants
IGF-II/IGF-IR	double mutant; dwarfism; 30 % of normal size
IGF-I/IGF-II	double mutant; dwarfism; 30 % of normal size
IL2	normal thymocyte/peripheral T cell subset composition during 3-4 weeks of age; reduced polyclonal *in vitro* T cell responses; dramatic changes in the isotype levels of serum immunoglobulins; unlimited inflammatory bowel disease past 6 weeks of age; T helper cell responses delayed but biologically functional; natural killer cell activities markedly reduced but inducible; high levels of serum IgG1 and IgE isotypes
IL2 /IL4	double mutant; viable; normal major T cell subsets and B cells; higher levels of T cell proliferation; resistance against retrovirus-induced immunodeficiency (MAIDS)
IL4	thymus and T cell subsets develop normally; failure of naive $CD4^+$ cells to produce TH2-derived cytokines after *in vitro* stimulation; resistance to retrovirus-induced immunodeficiency syndrome
IL6	viable and fertile; no osteoporosis after estrogen deficiency

Disrupted gene	Major effect
IL10	normal development of lymphocytes and antibody responses; growth retardation and anemia; development of chronic enterocolitis; extensive mucosal hyperplasia, inflammatory reactions; aberrant expression of MHC class II molecules on intestinal epithelia
α-inhibin	development of mixed or incompletely differentiated gonadal stromal tumors
IRF-1	lack of the normally observed type I IFN induction by poly(I).poly(C); profound reduction in TCRαβ$^+$ CD4$^-$CD8$^+$ T cells
IRF-2	bone marrow suppression of hematopoiesis and B lymphopoiesis; animals die after infection with lymphocytic choriomeningitis virus; up-regulated type I IFN induction by infection with Newcastle disease virus
jun	death at mid-gestation; drastic reduction in the formation of teratocarcinomas by injected ES cells lacking *jun*; impaired hepatogenesis with pronounced hypoplasia of epithelial compartment; altered fetal liver erythropoiesis; generalized tissue edema
lck	pronounced thymic atrophy; dramatic reduction in the double-positive (CD4$^+$CD8$^+$) thymocyte population; mature, single-positive thymocytes not detectable; only very few peripheral T cells
LIF	viable; failure to implant in the uterus; rescue by transfer to the uterus of normal mice; dramatically decreased numbers of stem cells in spleen and bone marrow; numbers of committed progenitors reduced in spleen but not in bone marrow
M-CSF	spontaneous *op/op* mouse (osteopetrosis); skeletal abnormalities; almost complete absence of osteoclasts; severe deficiency of mononuclear phagocytes; elevated serum levels of GM-CSF
myb	all hematopoietic lineages affected with the exception of megakaryocytes; inability to switch from yolk sac to liver erythropoiesis; death at approximately 15.5 days of gestation due to severe anemia
myc	embryonic lethal at day 9.5
NGF p75 R	animals viable and fertile; markedly decreased sensory innervation by calcitonin gene-related peptide- and substance P-immunoreactive fibers; no dysfunctions of sympathetic neurons; embryonic cutaneous sensory trigeminal neurons require more NGF for survival
pim-1	bone marrow-derived mast cells with distinct growth disadvantage in the presence of IL3, but not when stimulated by IL4, IL9, or SCF
pit-1	Snell-Bagg dwarf mice; microcephalic cerebrum with hypomyelination, retarded neuronal growth, and underdevelopment of axons and dendrites; growth hormone and prolactin deficiency
SCF	embryonic lethal (*Steel* mutation); pronounced anemia and severe impairment of development of the hematopoietic system characterized by a very pronounced deficiency of mast cells
SCF R (kit)	(*white* locus); alterations in coat colors; impairment of gonad development; reduction of early erythroid precursor cells
TGF-α	mice are viable and fertile; aberrant whiskers, fur and hair follicles; eye abnormalities of variable incidence and severity
TGF-β1	no gross developmental abnormalities; death at about 20 days after birth due to wasting syndrome; multifocal, mixed inflammatory cell response and tissue necrosis; massive infiltration of lymphocytes and macrophages in many organs; elevated levels of both the class I and class II MHC
TNF p55 R	thymocyte development and lymphocyte populations unaltered; clonal deletion of potentially self-reactive T cells not impaired; mice resistant to lethal dosage of bacterial lipopolysaccharides or enterotoxins; mice readily succumb to infections with *L. monocytogenes*
*trk*B	mice develop to birth; death soon after birth from multiple neuronal deficiencies
wnt-1	death often before birth; sometimes survival with severe ataxia; severe abnormalities in development of mesencephalon and metencephalon

krox: ***Krox 20***, also known as Egr-2 (see: Egr-1), and ***Krox 24*** are » ERGs (early response gene) that are transiently activated during G0/G1 transition of the » Cell cycle in murine cells. The genes encode zinc finger proteins that function as transcription factors (see also: Gene Expression). The genes are regulated through a serum response element (see: SRE) and a cAMP responsive element (see: CRE) and are members of the Egr family of transcription factors (see: Egr-1). Their expression can be induced by » Phorbol esters and activation of the signal transduction pathways mediated by protein kinase A (PKA) or protein kinase C (PKC). Both genes have been found to be activated in response to the tax gene encoded by HTLV I. The human Krox20 gene maps to chromosome 10q21.1 to q22.1. Krox 24 has been shown to be identical with » Egr-1.

■ **TRANSGENIC/KNOCK-OUT/ANTISENSE STUDIES:** Krox-20 has been implicated in the control of segmentation during hindbrain development in mice. Mice in which Krox-20 has been inactivated by targeted deletion in » ES cells die during the first two weeks after birth and show a marked reduction or elimination of hindbrain segments r3 and r5.

Alexandre C et al Transactivation of Krox-20 and Krox-24 promoters by the HTLV-1 Tax protein through common regulatory elements. O 6: 1851-7 (1991); **Chavrier P et al** Structure, chromosome location, and expression of the mouse zinc finger gene Krox-20: multiple gene products and coregulation with the proto-oncogene c-fos. MCB 9: 787-97 (1989); **Janssen-Timmen U et al** Structure, chromosome mapping and regulation of the mouse zinc-finger gene Krox-24; evidence for a common regulatory pathway for immediate-early serum-response genes. Gene 80: 325-36 (1989); **Mack KJ et al** krox 20 messenger RNA and protein expression in the adult central nervous system. Brain Res. Mol. Brain Res. 14: 117-23 (1992); **Mechta F et al** Stimulation of protein kinase C or protein kinase A mediated signal transduction pathways shows three modes of response among serum inducible genes. New Biol. 1: 297-304 (1989)

● **TRANSGENIC/KNOCK-OUT/ANTISENSE STUDIES: Schneider-Maunoury S et al** Disruption of Krox-20 results in alteration of rhombomeres 3 and 5 in the developing hindbrain. Cell 75: 1199-1214 (1993)

K-*sam*: *KATO-III cell-derived stomach cancer amplified gene* This gene was initially identified as a gene amplified and overexpressed in poorly differentiated human stomach cancers. It shows significant homology with the chicken and human receptor for » bFGF

K-*sam* is also called FGFR-3, arising from differential splicing of the FGF-R2 receptor gene (see: bFGF). It encodes a protein of 682 amino acids with an intrinsic tyrosine-specific protein kinase activity in its intracellular domain. This receptor is also synthesized in a secreted form which may also lack the kinase domain.

K-*sam* shows close homology to a receptor for the murine keratinocyte growth factor (see: KGF) » *bek*, and the receptors called » *flg* and » TK-14. The human K-*sam* gene maps to chromosome 10q26. Another gene related to K-*sam* is N-*sam*. N-*sam* is a homologue of the chicken bFGF Cek1 receptor. It is identical with » *flg*.

Sam3, isolated from a mouse brain complementary DNA library, is a protein of 800 amino acids with high homology to mouse K-sam/*bek* (67%) and N-*sam*/*flg* (63%). The sam3 protein also has high homology to human FGFR3 (92%) and chicken cek2 (80%) proteins and most likely is a mouse counterpart of human FGFR3 and chicken cek2 proteins. The relative amounts of K-*sam*/*bek*, N-*sam*/*flg*, and sam3/FGFR3 receptor mRNAs are different depending on the organs examined. These gene products may, therefore, have different biological functions in organ development.

Gilbert E et al Control of *bek* and K-*sam* sites in alternative splicing of the fibroblast growth factor receptor 2 pre-mRNA. MCB 13: 5461-8 (1993); **Hattori Y et al** An amplified gene in stomach cancer, K-*sam*, a member of heparin-binding growth factor receptor gene. PNAS 87: 5983-7 (1990); **Hattori Y et al** K-*sam*-related gene, N-*sam*, encodes fibroblast growth factor receptor and is expressed in T lymphocytic tumors. CR 52: 3367-71 (1992); **Katoh M et al** K-*sam* gene encodes secreted as well as transmembrane receptor tyrosine kinase. PNAS 89: 2960-4 (1992); **Katoh O et al** Expression of the heparin-binding growth factor receptor genes in human megakaryocytic leukemia cells. BBRC 183: 83-92 (1992); **Katoh O et al** Isolation of the complementary DNA encoding a mouse heparin-binding growth factor receptor with the use of a unique kinase insert sequence. CR 53: 1136-41 (1993); **Keegan K et al** Isolation of an additional member of the fibroblast growth factor receptor family, FGFR-3. PNAS 88: 1095-9 (1991); **Mattei MG et al** Assignment by *in situ* hybridization of a fibroblast growth factor receptor gene to human chromosome band 10q26. Hum. Genet. 87: 84-6 (1991)

KS3: see: K-FGF (Kaposi fibroblast growth factor).

KS-FGF: see: K-FGF (Kaposi fibroblast growth factor).

KS protein: *Kaposi sarcoma protein* see: K-FGF (Kaposi fibroblast growth factor).

KT-3: A T lymphoma cell line established from a patient with Lennert's lymphoma. The cells show macrophage-dependent growth and macrophage-derived factors are able to replace macrophage functions. KT-3 cells proliferate in response to » IL2, » IL4, or » IL6 (not murine IL6) but not in re-

Kostman syndrome: see addendum: G-CSF.

sponse to » IL1α or β, » IL3, » G-CSF, » GM-CSF, » IFN-α, » IFN-β, » IFN-γ, or » TNF-α. Polyclonal rabbit anti-IL6 antibody almost completely neutralize the activities of macrophage-derived factors or IL6 but not IL2 or IL4. The cell line has been used in » bioassays to detect human IL6.

Shimizu S et al Interleukin-6 (B-cell stimulatory factor 2)-dependent growth of a Lennert's lymphoma-derived T-cell line (KT-3). Blood 72: 1826-8 (1988)

KTGF: *keratinocyte-derived T cell growth factor* This factor is isolated from culture supernatants of neonatal keratinocytes and established keratinocyte cell lines. It supports the growth of the IL2-dependent helper T cell line » HT-2 and does not support the growth of the » CTLL-2 cell line derived from cytotoxic T cells. KTGF was initially believed to be identical with » IL2 or BSF 1 (B cell stimulating factor 1 = Interleukin 4). It is identical with » GM-CSF.

Coleman DL et al Characterization of a keratinocyte-derived T cell growth factor distinct from interleukin 2 and B cell stimulatory factor 1. JI 138: 3314-8 (1987); **Chodakewitz J et al** Keratinocyte-derived granulocyte/macrophage colony-stimulating factor induces DNA synthesis by peritoneal macrophages. JI 140: 832-6 (1988); **Kupper TS et al** Keratinocyte-derived T cell growth factor: a T cell growth factor functionally distinct from interleukin 2. PNAS 83: 4451-5 (1986); **Kupper TS et al** Keratinocyte-derived T cell growth factor (KTGF) is identical to granulocyte-macrophage colony-stimulating factor (GM-CSF). J. Invest. Dermatol. 91: 185-8 (1988)

K-TIF: *K562-derived T cell inhibitory factor* This poorly characterized factor or 30-35 kDa (pI = 6.0-6.4) is secreted by the human erythroleukemic cell line K562. It specifically binds to receptors of mononuclear cells and inhibits the growth of T lymphocytes activated via the CD3 pathway.

De Felice M et al Identification and characterization of a T cell growth inhibitory factor produced by K562 erythromyeloid cells. CI 138: 55-63 (1991)

KW-2228: Syn. Marograstim. A recombinant, N-terminally modified form of human » G-CSF in which Thr1, Leu3, Gly4, Pro5, and Cys17 are respectively substituted with Ala, Thr, Tyr, Arg, and Ser.

KW-2228 is more stable than the genuine recombinant factor. It displays a high specific activity, has an increased plasma half-life and improved pharmacokinetic behavior. The factor supports granulopoiesis in the mouse much better than the parent factor. It stimulates the growth of » CFU-G in soft agar (see also: Colony formation assay) and of leukemic blast cell progenitors much better than genuine recombinant G-CSF. See also: Filgrastim.

Kato Y et al *In vivo* effect of human granulocyte colony-stimulating factor derivatives on hematopoiesis in primates. Acta Haematol. 86: 70-8 (1991); **Kuga T et al** Mutagenesis of human granulocyte colony stimulating factor. BBRC 159: 103-11 (1989); **Kuwabara T et al** Highly sensitive enzyme-linked immunosorbent assay for marograstim (KW-2228), a mutant of human granulocyte colony-stimulating factor. J. Pharmacobiodyn. 15: 121-9 (1992); **Nio Y et al** Comparative effects of a recombinant and a mutein type of granulocyte colony stimulating factor on the growth of Meth-A fibrosarcoma with 5-fluorouracil chemotherapy. Biotherapy 4: 81-6 (1992); **Suzuki T et al** Enhanced effect of mutant granulocyte-colony-stimulating factor (KW-2228) on the growth of normal and leukemic hemopoietic progenitor cells in comparison with recombinant human granulocyte-colony-stimulating factor (G-CSF). Acta Haematol. 87: 181-9 (1992)

KWN: *key word networking* A term pertaining to the extensive use of unavoidable cross-references in this cytokine dictionary. Partly caused by synonyms, acronyms, and » EUAs. Possibly one elaborate way to annoy readers and certainly also a desperate attempt by the author to account for the intricacies of the » cytokine network and the » Neuroimmune network. *Ultra posse nemo obligatur* (we speak broken Latin).

L

L1: see: Calgranulins.

L2G25B: This protein was initially identified as a human T lymphocyte-specific cDNA clone the expression of which was inducible by a number of stimuli. L2G25B is identical with human MIP1-α (see: MIP, macrophage inflammatory protein).
Kwon BS et al Expression characteristics of two potential T cell mediator genes. CI 121: 414-22 (1989); **Kwon BS & Weissman SM** cDNA sequences of two inducible T cell genes. PNAS 86: 1963-7 (1989); **Oh KO et al** Identification of cell surface receptors for murine macrophage inflammatory protein-1 α. JI 147: 2978-83 (1991)

L3T4: This cell surface protein is identical with » CD4 (see also: CD antigens).

L4: This human B lymphoma cell line is used to assay B cell growth and differentiation factors. The cells proliferate in response to » IL4 and are used to assay this factor (see also: Bioassays). L4 cells do not respond to human »IL5.
Callard RE The marmoset B lymphoblastoid cell line (B95-8) produces and responds to B cell growth and differentiation factors: role of shed CD23 (sCD23). Immunology 65: 379-84 (1988); **Shields JG et al** The response of selected human B cell lines to B cell growth and differentiation factors. EJI 17: 535-40 (1987)

L87/4: This human stromal cell line (see also: Hematopoiesis) has been obtained from adherent bone marrow cell layers of long-term bone marrow cultures (see: LTBMC) of a hematologically normal patient after immortalisation with SV40. The cells have been shown by » message amplification phenotyping to express » IL1β, » IL6 and IL6 receptors, » IL7, » IL8, » IL10, » IL11, » LIF, » SCF and its receptor » *kit*, » TGF-β, » G-CSF, » GM-CSF, » M-CSF, » MIP-1α, and » TNF-α. By the same method they are negative for the expression of » MIP-1β, IFN-γ, » IL2, » IL3, » IL4, » IL5, » IL9.
The cells support the long-term culture of enriched human » CD34-positive cells and of the BL70 Burkitt lymphoma cell line. The most remarkable feature of this cell line is its radioresistance. The cells persist as growth-arrested adherent layers following irradiation exceeding 20 Gy. Another radioresistant cell line, designated ***L88/5*** has similar features.
Thalmeier K et al Establishment of two permanent human bone marrow stromal cell lines with long-term post-irradiation feeder capacity. Blood 83(7): 1799-807 (1994)

L88/5: see: L87/4.

L138.8A: A mouse bone marrow-derived » factor-dependent mast cell line. This cell line responds to » IL3, » IL4, or » IL9 and can be used in » bioassays for these factors. The cells secrete » IL6 in response to all three growth factors with IL9 being the most potent inducer.
Hültner L et al Interleukin 3 mediates interleukin 6 production in murine interleukin 3-dependent hemopoietic cells. GF 2: 43-51 (1989); **Hültner L & Moeller J** Mast cell growth-enhancing activity (MEA) stimulates interleukin 6 production in a mouse bone marrow-derived mast cell line and a malignant subline. EH 18: 873-7 (1990)

L929: This murine aneuploid fibrosarcoma cell line is used to assay » TNF-α and » TNF-β (see also: Bioassays) which lyse cells sensitized by pretreatment with actinomycin. Treatment with TNF initiates » apoptosis and subsequent cell death. The level of detection is on the order of 50 pg/mL. Another cell line which is ≈ 5-10-fold more sensitive is WEHI-164 (see: WEHI-3B). Some L929 subclones have been developed that allow detection of TNF at concentrations of ≈ 30 pmol/L (≈ 500 fg/mL). TNF-resistant sublines have also been established. An alternative and entirely different detection method is » Message amplification phenotyping.
L929 cells are greater than 10 times more sensitive to TNF-mediated cytotoxicity in the presence of lithium chloride which induces TNF-α secretion. In nude mice (see also: Immunodeficient mice), a combination of TNF and LiCl leads to hemorrhagic necrosis and growth inhibition of L929 tumors, whereas little effect is observed when TNF is administered alone.
TNF-α and » TNF-β can be distinguished from each other by using suitable neutralizing anti-

bodies. The L929 cytotoxicity assay can be influenced considerably by heat-inactivated human serum (56 ° C, 30 min) which is also cytotoxic and may lead to a detachment of cells.
Extracellular ATP and adenosine, but not AMP or GTP in concentrations of 0.5 to 2.5 mM inhibit TNF-induced cytolysis of L929 cells in the presence of actinomycin D when present throughout the entire assay period.
Treatment of the cells with TNF, » IL1, poly(I)-poly(C), » IFN-β, » calcium ionophore A23187, or phorbol 12-myristate 13-acetate (see also: Phorbol esters), but not with » EGF, dibutyryl-cAMP, or the adenylate cyclase activator forskolin induces interferon regulatory factor 1 (see: IRS, interferon response sequence). Treatment of the cells with TNF or » IL1 also induces the synthesis of IFN-β. TNF-α stimulates the release of arachidonic acid from these cells. The addition of hydrocortisone or nordihydroguaiaretic acid (NDGA) decreases the cytotoxic effect of TNF-α but exogenously added arachidonate or linoleate, indomethacin and eicosatetraynoic acid (ETYA) are without effect. Suppression of arachidonate metabolism by steroids effectively inhibits TNF-mediated cytolysis of L929 cells. An increase in intracellular cAMP levels (obtained by adding cell-permeable dibutyryl cAMP to the culture medium)or culture of L929 cells in the presence of reagents elevating intracellular cAMP concentrations.
Recombinant human TNF-α alone has no effect on L929 tumor cells at 100 units/ml for 20 h of continuous exposure. However, under the same conditions, rHTNF markedly enhances the cytotoxicity of Adriamycin, actinomycin D, 4'-(9-acridinylamino)-methanesulfon-m-anisidide, teniposide (VM 26), and etoposide (VP 16), all targeted at DNA topoisomerase II. VM-26 can lower the TNF LD_{50} to femtomolar levels. The topoisomerase II inhibitors novobiocin and coumermycin, which bind to the enzyme ATPase site, protect L929 cells from TNF cytotoxicity but enhance TNF cytotoxicity in human ME-180 cells.
L929 cells are severely deficient in gap junctional communication and known cell-cell adhesion molecules. L929 cells secrete a neurotrophic factor that is closely related to the β subunit of » NGF. Other NGF subunits are not produced. L929 also secrete a factor that promotes the differentiation of porcine peripheral blood mononuclear cells into cells of macrophage/monocyte lineage. This factor is identical with murine » M-CSF. L929 cell-conditioned medium is therefore frequently used as a crude source of murine M-CSF. L929 cells also secrete a chemotactic factor specific for monocytes which is identical to murine » JE. The cells also secrete an uncharacterized factor (see: THP-1-derived growth-promoting activity) that promotes the growth of a large variety of cell lines.

Aggarwal BB et al Human tumor necrosis factor: production, purification, and characterization. JBC 260: 2345-2354 (1985); **Alexander RB et al** Synergistic enhancement by tumor necrosis factor of *in vitro* cytotoxicity from chemotherapeutic drugs targeted at DNA topoisomerase II. CR 47: 2403-6 (1987); **Baarsch MJ et al** Detection of tumor necrosis factor α from porcine alveolar macrophages using an L929 fibroblast bioassay. JIM 140: 15-22 (1991); **Baloch Z et al** Synergistic interactions between tumor necrosis factor and inhibitors of DNA topoisomerase I and II. JI 145: 2908-13 (1990); **Belizario JE & Dinarello CA** Interleukin 1, interleukin 6, tumor necrosis factor, and transforming growth factor β increase cell resistance to tumor necrosis factor cytotoxicity by growth arrest in the G1 phase of the cell cycle. CR 51: 2379-85 (1991); **Beyaert R et al** Lithium chloride potentiates tumor necrosis factor-mediated cytotoxicity *in vitro* and *in vivo*. PNAS 86: 9494-8 (1989); **Beyaert R et al** Lithium chloride potentiates tumor necrosis factor-induced and interleukin 1-induced cytokine and cytokine receptor expression. Cytokine 3: 284-91 (1991); **Branch DR et al** A specific and reliable bioassay for the detection of femtomolar levels of human and murine tumor necrosis factors. JIM 143: 251-61 (1991); **Chun M & Hoffmann MK** Intracellular cAMP regulates the cytotoxicity of recombinant tumor necrosis factor for L cells *in vitro*. LR 6: 161-7 (1987); **Dicou E et al** The neurotrophic factor nerve growth factor (NGF) synthesized by murine L-929 cells is not associated with α and γ subunits CR Seances Acad. Sci. III 297: 523-5 (1983); **Fujita T et al** Induction of the transcription factor IRF-1 and interferon-β mRNAs by cytokines and activators of second-messenger pathways. PNAS 86: 9936-40 (1989); **Genovesi EV et al** *In vitro* induction of swine peripheral blood monocyte proliferation by the fibroblast-derived murine hematopoietic growth factor CSF-1. Vet. Immunol. Immunopathol. 23: 223-44 (1989); **Hay H & Cohen J** Studies on the specificity of the L929 cell bioassay for the measurement of tumor necrosis factor. J. Clin. Lab. Immunol. 29: 151-5 (1989); **Hepburn A et al** Modulation of tumor necrosis factor-α cytotoxicity in L929 cells by bacterial toxins, hydrocortisone and inhibitors of arachidonic acid metabolism. BBRC 149: 815-822 (1987); **Hwu P et al** Use of soluble recombinant TNF receptor to improve detection of TNF secretion in cultures of tumor infiltrating lymphocytes. JIM 151: 139-47 (1992); **Jeffes EW et al** A simple nonisotopic *in vitro* bioassay for LT and TNF employing sodium fluoride-treated L-929 target cells that detects picogram quantities of LT and TNF and is as sensitive as TNF assays done with ELISA methodology. Lymphokine Cytokine Res. 10: 147-51 (1991); **Keisari V et al** A colorimetric microtiter assay for the quantitation of cytokine activity on adherent cells in tissue culture. JIM 146: 155-61 (1992); **Kinzer D & Lehmann V** Extracellular ATP and adenosine modulate tumor necrosis factor-induced lysis of L929 cells in the presence of actinomycin D. JI 146: 2708-11 (1991); **Kleinerman ES et al** Lithium chloride stimulates human monocytes to secrete tumor necrosis factor/cachectin. J. Leukoc. Biol. 46: 484-92 (1989); **Matthews N et al** Tumor cell killing by tumor necrosis factor: inhibition by anaerobic conditions, free-radical scavengers and inhibitors of arachidonate metabolism. Immunology 62: 153-5 (1987); **Meager A et al** Assays for tumor necrosis factor and related cytokines. JIM 116: 1-17 (1989); **Musil LS et al** Differential phosphorylation of the gap junction protein connexin43 in junctional communication-competent and -deficient cell lines. JCB 111: 2077-88 (1990); **Neale ML & Matthews N** Development of tumor cell resistance to tumor necrosis factor does not confer resistance to cytotoxic drugs. Eur. J. Cancer Clin. Oncol. 25: 133-7 (1989); **Scuderi P et al** Human sera and culture supernatants from human tumors and diploid fetal fibroblasts suppress tumor necrosis factor secretion *in vitro*. J. Leukoc. Biol. 46: 34-40 (1989); **Spofford B et al** Cell-mediated immunity *in vitro*: a highly sensitive assay for human lymphotoxin. JI 112: 2111-6 (1974); **Tomkins P et al** The L929 cell bioassay for murine tumor necrosis factor is not influenced by other murine cytokines JIM 151: 313-5 (1992); **Utsugi T et al** Synergistic antitumor effects of topoisomerase inhibitors and natural cell-mediated cytotoxicity. CR 49: 1429-33 (1989); **Van Damme J et al** Production and identification of natural monocyte chemotactic protein from virally infected murine fibroblasts. Relationship with the product of the mouse competence (JE) gene. EJB 199: 223-9 (1991)

L929-derived hybridoma growth factor: abbrev. L-HGF. See: HGF.

LA: (a) Occasionally used as a suffix (e. g. EGF-LA) to denote "like-activity". This "like" activity may later turn out to be identical with the genuine factor. It may, however, also be a distinct factor; (b) occasionally used as a prefix to denote "low-affinity" (receptor) such as in low affinity interleukin 2 receptor (LA-IL 2R).

Lactoferrin: Lactoferrin is a glycoprotein of ≈ 77 kDa. It is one of the so-called siderophilins, non-heme iron (III)-binding proteins with two binding sites for iron (see also: Transferrin). Lactoferrin is found in milk and also in the granules of polymorphonuclear leukocytes, from which it is released when the cells are activated. Lactoferrin may contribute to antimicrobial activity by making iron unavailable to microorganisms.

Apart from its function as an iron carrier lactoferrin, like transferrin, may act as a » cytokine. Specific receptors for lactoferrin are detected on the surface of different hematopoietic cell types. Lactoferrin can inhibit the growth of » CFU-GM and » BFU-E hematopoietic precursor cells. This myelosuppression can be blocked by » IL6, but not by » IL1. Lactoferrin has been shown to decrease production and release of cytokines from monocytes and macrophages, including » IL1, which is known to trigger other cells to release colony-stimulating factors (see: CSF). Lactoferrin has been found to be identical with » CIF (colony inhibitory factor); major abnormalities of granulopoiesis have not been observed, however, in the rare genetic disorder of lactoferrin deficiency.

Lactoferrin released from neutrophils phagocytosing opsonized particles inhibits proliferation of mixed lymphocyte cultures, suppresses antibody production, and regulates natural killer cell activity. Lactoferrin interacts with bacterial lipopolysaccharide (see: LPS) in such a fashion as to affect its binding to myeloid cells. It inhibits LPS priming of neutrophils. Lactoferrin also influences the release of monokines by endothelial cells, and functions as a growth factor for lymphocytic cell lines cultured in serum-free medium.

Bagby GF jr et al A monokine regulates colony-stimulating activity production by vascular endothelial cells. Blood 62: 663-8 (1983); **Broxmeyer HE et al** The opposing actions *in vivo* on murine myelopoiesis of purified preparations of lactoferrin and the colony stimulating factors. Blood Cells 13: 31-48 (1987); **Cohen MS et al** Interaction of lactoferrin and lipopolysaccharide (LPS): effects on the antioxidant property of lactoferrin and the ability of LPS to prime human neutrophils for enhanced superoxide formation. J. Infect. Dis. 166: 1375-8 (1992); **Crouch SP et al** Regulation of cytokine release from mononuclear cells by the iron-binding protein lactoferrin. Blood 80: 235-40 (1992); **Delforge A et al** Lactoferrin: its role as a regulator of human granulopoiesis? ANY 459: 85-96 (1985); **Djeha A & Brock JH** Effect of transferrin, lactoferrin and chelated iron on human T lymphocytes. Br. J. Haematol. 80: 235-41 (1992); **Gahr M et al** Influence of lactoferrin on the function of human polymorphonuclear leukocytes and monocytes. J. Leukoc. Biol. 49: 427-33 (1991); **Gentile P & Broxmeyer HE** Interleukin-6 ablates the accessory cell-mediated suppressive effects of lactoferrin on human hematopoietic progenitor cell proliferation *in vitro*. ANY 628: 74-83 (1991); **Hangoc G et al** Influence of T lymphocytes and lactoferrin on the survival-promoting effects of IL1 and IL6 on human bone marrow granulocyte-macrophage and erythroid progenitor cells. EH 19: 697-703 (1991); **Hashizume S et al** Identification of lactoferrin as an essential growth factor for human lymphocytic cell lines in serum-free medium. BBA 763: 377-82 (1983); **McCormick JA et al** Lactoferrin-inducible monocyte cytotoxicity for K562 cells and decay of natural killer lymphocyte cytotoxicity. Clin. Exp. Immunol. 83: 154-6 (1991); **Miyazawa K et al** Lactoferrin-lipopolysaccharide interactions. Effect on lactoferrin binding to monocyte/macrophage-differentiated HL-60 cells. JI 146: 723-9 (1991); **Shau H et al** Modulation of natural killer and lymphokine-activated killer cell cytotoxicity by lactoferrin. J. Leukoc. Biol. 51: 343-9 (1992); **Srivastava CH et al** Regulation of human bone marrow lactoferrin and myeloperoxidase gene expression by tumor necrosis factor-α. JI 146: 1^014-9 (1991)

Ladsin: *large adhesive scatter factor* This factor has been isolated from cell culture supernatants of the human gastric carcinoma cell line STKM-1. It is a large protein consisting of disulfide-linked subunits of 140, 150, and 160 kDa. The factor binds to heparin (see also: HBGF). Ladsin disrupts intercellular connections and inhibits cell-to-cell binding, resulting in marked dispersion or scattering of sensitive cells into disconnected individual cells. Ladsin has been shown to promote scattering of a wide variety of cells, including epithelial, endothelial, and fibroblastic cells. Unlike other » motogenic cytokines Ladsin appears to be a general cell motility factor.

Miyazaki K et al A large cell-adhesive scatter factor secreted by human gastric carcinoma cells. PNAS 90: 11767-71 (1993)

LAF: *lymphocyte activating factor* The term LAF (lymphocyte activating factor) activity refers to the ability of the supernatants of human monocytes incubated with endotoxin to stimulate the proliferation of murine thymocytes in the presence of phytohemagglutinin. This activity described in the older literature is identical with » TPF (thymocyte proliferative factor) and » HP1 (helper peak 1) and hence identical with » IL1. Some of the LAF activity reported in older references may also be due to » IL6. IL1-like factors and many other poorly characterized factors also possess LAF activity.

Aarden LA et al Revised nomenclature for antigen-nonspecific T cell proliferation and helper factors. CI 48: 433-36 (1979); **Blyden G & Handschumacher RE** Purification and properties of human lymphocyte-activating factor (LAF). JI 118: 1631-8 (1977); **Calderon J et al** The modulation of lymphocyte functions by molecules secreted by macrophages. JEM 142: 151-64 (1975); **Gery I & Handschumacher RE** Potentiation of T lymphocyte response to mitogens. III. Properties of the mediator(s) from adherent cells. CI 11: 162-9 (1974); **Gery I et al** Potentiation of the T lymphocyte response to mitogens: 1. the responding cell. JEM 136: 128 (1972); **Gery I & Waksman BH** Potentiation of the T lymphocyte response to mitogens. II. The cellular source of potentiating mediator(s). JEM 136: 143 (1972); **Koopman WJ et al** Evidence for the identification of lymphocyte-activating factor as the adherent cell-derived mediator responsible for enhanced antibody synthesis by nude mouse spleen cells. CI 35: 92-8 (1978); **Lachman LB et al** Partial purification of human lymphocyte-activating factor (LAF) by ultrafiltration and electrophoresis techniques. JI 119: 2019-23 (1977); **Lachman LB et al** Preparation and characterization of lymphocyte-activating factor (LAF) from acute monocytic and myelomonocytic leukaemia cells. CI 41: 199-206 (1978); **Mizel SB** Physicochemical characterization of lymphocyte activating factor (LAF). JI 122: 2167-72 (1979); **Paetkau V et al** Cellular origins and targets of Costimulator (IL2). IR 51: 157-75 (1980); **Smith KA et al** Lymphocyte activating factor promotes T cell growth factor production by cloned murine lymphoma cells. N 287: 853-5 (1980); **Togawa A et al** Characterization of lymphocyte-activating factor (LAF) produced by human mononuclear cells: biochemical relationship of high and low molecular weight forms of LAF. JI 122: 2112-7 (1979); **Van Damme J et al** Separation and comparison of two monokines with lymphocyte-activating factor activity: IL1 β and hybridoma growth factor (HGF). Identification of leukocyte-derived HGF as IL6. JI 140: 1534-41 (1988) **Wood DD et al** The four biochemically distinct species of human interleukin 1 all exhibit similar biologic activities. JI 134: 895-903 (1985)

LAG-1: *lymphocyte activation gene 1* This gene was isolated from human lymphocytes. It encodes a 14 kDa protein with a length of 69 amino acids. Based on structural homologies this protein belongs to the » chemokine family of cytokines. It is identical with » MIP1β (see: MIP, macrophage inflammatory protein).

The LAG-1 gene maps to chromosome 17q21. Its promoter contains a » CSF-Box (colony stimulating factor box).

Baixeras E et al Cloning and expression of a lymphocyte activation gene (LAG-1). Mol. Immunol. 27: 1091-1102 (1990)

LAI: *leukocyte adhesion inhibitor* This factor is produced by endothelial cells activated by » IL1, » TNF, and bacterial » endotoxins (see also: Cell activation). It prevents the adhesion of neutrophils to the endothelium and thus protects it from damage caused by neutrophils. LAI therefore has anti-inflammatory activities (see also: Inflammation).

LAI is identical with one of the molecular forms of » IL8. It has an extension of 5 N-terminal amino acids. This 77 amino acid form of IL8 is quickly converted into the chemotactically more active 72 amino acid form of IL8 by enzymes such as thrombin.

Gimbrone MA jr et al Endothelial interleukin 8: a novel inhibitor of leukocyte-endothelial interactions. S 246: 1601-3 (1989); **Hebert CA et al** Endothelial and leukocyte forms of IL8: conversion by thrombin and interactions with neutrophils. JI 145: 3033-40 (1990)

LAK-1: This poorly characterized cell surface antigen of 120 kDa is found on the cell surface of » LAK cells (lymphokine-activated killer cells). Antibodies directed against this protein also inhibit LAK cell activity.

Zocchi MR et al A novel 120 kDa surface antigen expressed by a subset of human lymphocytes. JEM 166: 319-26 (1987)

LAK cell induction helper factor: see: LHF.

LAK helper factor: see: LHF.

LAK cells: *lymphokine-activated killer cells* These cells constitute a subpopulation of peripheral blood mononuclear blood cells that are collected from tumor patients and expanded *in vitro* by treatment with » IL2. Higher yields of LAK cells may be obtained by treatment of peripheral blood lymphocytes with lectin and then with recombinant IL2 (***LILAK cells***).

LAK cells develop a broad range of lytic activity against fresh tumor cells and cultured tumor cell lines. Ca. 10^{10} autologous *ex vivo*-activated LAK cells are re-infused into tumor patients for the

treatment of various tumors (see: Adoptive immunotherapy).
Ca. 90 % of the LAK cells are derived from precursor cells with the immunological marker spectrum $CD3^-$, $CD11^+$, $CD14^-$, $CD16^+$, $CD56^+$ (see also: CD antigens). Following their activation they carry the markers $CD2^+$, $CD3^-$, $CD56^+$ and hence are activated natural killer (NK) cells. IL2 activation also induces a significant increase in the expression of the CD25 antigen which functions as an IL2 receptor. LAK cells also express a poorly characterized cell surface antigen called » LAK-1. LAK cells display an MHC-unrestricted cytolysis against a wide variety of fresh and cultured tumor cells.
The molecular basis of the interactions between LAK cells and their target cell has not yet been elucidated in detail. Binding to target cells involves Fc receptors on the LAK cell and several other receptors including the complement 3 receptor and several intercellular adhesion molecules and their ligands (see also: ADCC, antibody-dependent cellular cytotoxicity). CD58 (LFA-3), CD54 (ICAM-1) and CD11a/CD18 (LFA-1) adhesion molecules are some of the molecules regulating non-specific major histocompatibility complex-unrestricted and CD3/T cell receptor (TcR)-triggered cytotoxicity.
Some factors effectively mediating the lysis of tumor cells by LAK cells are » perforins (induced by IL2), and » leukalexins, both of which permeabilise cellular membranes. Cell lysis also involves proteolytic enzymes (granzymes) and cytotoxic NK cell factors such as » NKCF (NK cytotoxic factor). Some observations suggest in addition to rapid cytolysis, antitumor activities of LAK cells also involve slow-acting cytokine production. They appear to secrete, for example, a cytotoxic factor of ≈ 57 kDa. A cytotoxic factor found in the membrane fractions of adherent LAK cells is » M-CTX. Membrane-bound forms of » TNF-α may also contribute to cell lysis.
Several protocols have been developed to generate LAK cells and to enhance their cytolytic activities. It appears that many different cytokines and other substances are involved in the initial generation of LAK cell activities either as » autocrine or » paracrine factors.
Adherent LAK cells (abbrev. ALAK) are isolated from peripheral blood-derived non-MHC-restricted LAK cells by *in vitro* cultivation of LAK cells on nylon fibres or plastic surfaces. These cells are ≈ up to 10-fold more cytotoxic than non-adherent cells. Some of this increased cytotoxicity may be the result of » cell activation, which is known to be initiated by mere contact of cells with inert surfaces, and the resulting enhanced production of cytokines.
The aggressiveness of LAK cells can also be enhanced by *in vitro* treatment with » TNF-α or » IFN-γ. TNF-α appears to promote the migration of lymphocytes into the tumor (see also TILs, tumor-infiltrating lymphocytes) while IFN-γ increases the expression of the major histocompatibility antigens. An increased cytotoxic activity of LAK cells may also be obtained by treatment with » IL4. The addition of » GM-CSF in the presence of IL2 during the induction of LAK cells has also been described to enhance the number of effector cells with substantial cytotoxic activity. The IL2-induced LAK activity is also significantly up-regulated in the presence of cisplatin/FK-565 (see also: CsA, cyclosporin). Purified protein A from *Staphylococcus aureus* Cowan I strain in combination with IL2 significantly enhances the cytotoxicity and lytic ability of LAK cells against NK sensitive and NK-resistant tumor cells. Cultivation of LAK cells at higher oxygen concentrations also appears to promote their generation and to enhance their functional activities. L-arginine has been shown to be an essential substrate for optimal generation of LAK cells. Supplemental dietary L-arginine also appears to enhance lymphocyte cytotoxic mechanisms and to potentiate IL2 immunotherapy.
The use of anti-CD3 antibodies, which are also capable of activating T cells, in combination with

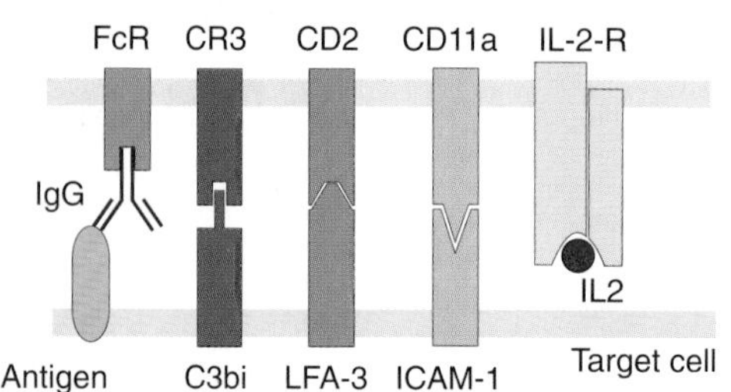

Interaction of LAK cells with tumor cells.
LAK cells that have been activated by binding of IL2 to its receptor can interact with target cells by a variety of adhesive interactions, many of which have not been elucidated sufficiently. The process involves binding of CD2 and CD11a with their counterreceptors, LFA-3 (lymphocyte function antigen 3) and ICAM-1 (intercellular adhesion molecule 1), respectively. In addition, Fc receptors or Complement receptor 3 on LAK cells can interact with antigens or C3bi, respectively, on the target cell.

IL2 increases the yields of LAK cells by a factor of 10-1000 over that achieved by IL2 alone. It has been observed that » CIK (cytokine-induced killer cells) can be generated by use of anti-CD3 antibodies and various cytokines, which are even more cytotoxic than LAK cells.

Factors such as » TNF-α, » TNF-β and also » IL1 can be employed with lower doses of IL2 to generate LAK cells which are as cytolytic as those obtained by high doses of IL2. The treatment of peripheral lymphocytes and monocytes with » IL7 has been shown to stimulate endogenous production of IL2. In addition, IL7 also promotes the $CD4^+$-T cell-dependent destruction of tumor cells and appears to induce significant LAK cell activity, either alone or in combination with IL2. The interleukin » IL12 has also been observed to be of importance. It facilitates both non-specific LAK cell activities and specific cytotoxic T lymphocyte responses and is capable of reducing the amounts of IL2 required for LAK cell generation *in vitro*. The combination of IL4 and IL12 results in a synergistic proliferative activity (8-fold) in $CD56^+$ natural killer cells and a marked increase in the cell yields.

LAK cells may play an important role in regulation of monocytes, which are important accessory cells in the activation of T cells for specific antigen recognition. IL2 can induce LAK cells to down-regulate antigen presentation function in monocytes activated by » GM-CSF and » IL3) but not by » IFN-γ.

Several inhibitors of LAK cell activity may also play an important role in limiting the therapeutic efficacy of IL2 and/or LAK cells. *in vitro*, » IL4, » TGF-β, steroids, and also prostaglandin E_2, known to stimulate adenylate cyclase, interfere with the generation of LAK cells if added during the activation period. *In vivo* the activity of LAK cells is inhibited by steroids and suppressor cells. Peripheral blood neutrophil polymorphonuclear leukocytes inhibit the cytolytic efficiency of IL2)-activated LAK cells in a dose-dependent manner although the supernatants of cultured neutrophils do not contain soluble factors eliciting this response. Tumors may also produce factors which inhibit cytokine secretion from LAK cells or may develop other means of escaping the cytolytic activities of LAK cells. It has been observed, for example, that for some unknown reasons natural killer cells of many patients with advanced solid tumors are defective in their ability to respond by proliferation to IL2 even in the presence of exogenously supplied growth factors. In addition, fibrin deposition around tumor and/or effector cells can also protect tumor cells from immune destruction and diminish the efficiency of the cytotoxic LAK/NK cells. The resistance of some tumor cell types against adoptive immunotherapy appears to be due, at least in part, by the ability of » IL2 itself to protect tumor cells expressing the IL2 receptor against LAK activity by an as yet unknown mechanism.

The generation of LAK cells for re-infusion into tumor patients is an exacting and cumbersome task. It has been shown recently that LAK cell activity can also be generated *in vivo* following, for example, the subcutaneous administration of low doses of IL2 as an alternative to conventional systemic treatment involving continuous intravenous infusion of (highly toxic) high doses of IL2. NK cell activity appears to be maintained for ≈ 6 months after discontinuation of IL2 administration.

For some types of tumors, e. g. advanced renal cell carcinoma, a review of cases shows that the use of LAK cells does not lead to higher response rates or to prolonged response duration, progression-free survival or survival. Adoptive transfer of *in vitro*-generated LAK cells therefore is not necessarily required for the antineoplastic activities of IL2 and an increased efficacy of the IL2 treatment. For a novel approach in cancer therapy involving the use of genetically engineered cells see: Cytokine gene transfer.

● Reviews: **Grimm EA et al** Lymphokine-activated killer cells. Induction and function. ANY 532: 380-6 (1988); **Hiserodt JC** Lymphokine-activated killer cells: biology and relevance to disease. Cancer Invest. 11: 420-39 (1993); **Lindemann A et al** Lymphokine activated killer cells. Blut 59: 375-84 (1989); **Melief CJ** Tumor eradication by adoptive transfer of cytotoxic T lymphocytes. Adv. Cancer Res. 58: 143-75 (1992); **Palmer PA et al** Continuous infusion of recombinant interleukin-2 with or without autologous lymphokine activated killer cells for the treatment of advanced renal cell carcinoma. Eur. J. Cancer 28A: 1038-44 (1992); **Young JDE & Liu CC** Multiple mechanisms of lymphocyte-mediated killing. IT 9: 140-4 (1988)

● Basic Aspects and Trials: **Ballas ZK & Rasmussen W** Lymphokine-activated killer cells. VII. IL4 induces an $NK1.1^+CD8\ \alpha^+\beta^-$ TCR-α β $B220^+$ lymphokine-activated killer subset. JI 150: 17-30 (1993); **Basse P et al** Tissue distribution of adoptively transferred adherent lymphokine-activated killer cells assessed by different cell labels. Cancer Immunol. Immunother. 34: 221-7 (1992); **Caligiuri MA et al** Extended continuous infusion low-dose recombinant interleukin-2 in advanced cancer: prolonged immunomodulation without significant toxicity. J. Clin. Oncol. 9: 2110-9 (1991); **Clayman GL et al** Immunomodulation of the induction phase of lymphokine-activated killer activity by acute phase proteins. Otolaryngol. Head Neck Surg. 105: 26-34 (1991); **Colborn D et al** Expansion of lymphokine-activated killer cells for clinical use utilizing a novel culture device. JIM 119: 247-54 (1989); **Crump III WL et al** Synergy of human recombinant interleukin 1 with interleukin 2 in the generation of lymphokine-activated killer cells. CR 49: 149-53 (1989); **Echarti C & Maurer HR** Lymphokine-activated killer cells: determination of their tumor cytolytic capacity by a clonogenic microassay using agar capillaries. JIM 143: 41-7 (1991); **Ettinghausen SE et al** Haematologic effects of immunotherapy with lymphokine-activated killer cells and recombinant interleukin-2 in cancer patients. Blood 69: 1654-60 (1987); **Finkelstein DM & Miller RG** Effect of culture media on lymphokine-activated killer effector phenotype and lytic capacity. Cancer Immunol. Immunother. 33: 103-8 (1991); **Fujimoto T et al** Evaluation of basic procedures for adoptive immunotherapy for gastric cancer. Biotherapy 5: 153-63 (1992); **Galandrini R et al** Adhesion molecule-mediated signals regulate major histocompatibility complex-unrestricted and CD3/T cell receptor-triggered cytotoxicity. EJI 22: 2047-53 (1992); **Gambacorti-Passerini C et al** A pilot phase II trial of continuous-infusion interleukin-2 followed by lymphokine-activated killer cell therapy and bolus-infusion interleukin-2 in renal cancer. J. Immunother. 13: 43-8 (1993); **Geller RL et al** Generation of lymphokine-activated killer activity in T cells: Possible regulatory circuits. JI 146: 3280-8 (1991); **Grimm EA et al** TGF-β inhibits the *in vitro* induction of lymphokine-activated killing activity. Cancer Immunol. Immunother. 27: 53-8 (1988); **Gunji Y et al** Fibrin formation inhibits the *in vitro* cytotoxic activity of human natural and lymphokine-activated killer cells. Blood Coagul. Fibrinolysis. 1: 663-72 (1990); **Hank JA et al** *In vivo* induction of the lymphokine-activated killer phenomenon: interleukin-2 dependent human non-major histocompatibility complex-restricted cytotoxicity generated *in vivo* during administration of human recombinant interleukin-2. CR 48: 1965-71 (1988); **Herrmann GG et al** LAK-cell-mediated cytotoxicity against tumor cell targets used to monitor the stimulatory effect of interleukin-2: cytotoxicity, target recognition and phenotype of effector cells lysing the Daudi, T24 and K562 tumor cell lines. Nat. Immun. 11: 7-16 (1992); **Iho S et al** Characteristics of interleukin-6-enhanced lymphokine-activated killer cell function. CI 135: 66-77 (1991); **Kasid A et al** Effects of transforming growth factor-β on human lymphokine-activated killer cell precursors. Autocrine inhibition of cellular proliferation and differentiation to immune killer cells. JI 141: 690-8 (1988); **Kato K et al** Augmentation by tumor necrosis factor α of the systemic therapeutic effect of lymphokine-activated killer cells in adoptive immunotherapy of murine tumor. Jpn. J. Cancer Res. 82: 464-9 (1991); **Lazenby AW et al** IL1 synergy with IL2 in the generation of lymphokine activated killer cells is mediated by TNF-α and β (lymphotoxin). Cytokine 4: 479-87 (1992); **Lieberman MD et al** Enhancement of interleukin-2 immunotherapy with L-arginine. Ann. Surg. 215: 157-65 (1992); **Lotze MT et al** Lysis of fresh and cultured autologous tumor by human lymphocytes cultured in T cell growth factor. CR 41: 4420-5 (1981); **Lotze MT et al** *In vivo* administration of purified human interleukin-2. II. half-life, immunological effects, and expansion of peripheral lymphoid cells *in vivo* with recombinant IL2. JI 135: 2865-75 (1985); **Maas RA et al** Mechanisms of tumor regression induced by low doses of interleukin-2. *In vivo* 5: 637-42 (1991); **Marumo K et al** Enhancement of lymphokine-activated killer activity induction *in vitro* by interleukin-1 administered in patients with urological malignancies. J. Immunother. 11: 191-7 (1992); **Matossian-Rogers A et al** Tumor necrosis factor-α enhances the cytolytic and cytostatic capacity of interleukin-2 activated killer cells. Brit. J. Cancer 59: 573-7 (1989); **McKenzie RS et al** Identification of a novel CD56⁻ lymphokine-activated killer cell precursor in cancer patients receiving recombinant interleukin 2. CR 52: 6318-22 (1992); **Mehta S et al** Lymphokine Cytokine Res. Lymphokine-activated effector cells: modulation of activity by cytokines. 11: 73-7 (1992); **Melder RJ et al** A new approach to generating antitumor effectors for adoptive immunotherapy using human adherent lymphokine-activated killer cells. CR 48: 3461-9 (1988); **Miller JS et al** Adherent lymphokine-activated killer cells suppress autologous human normal bone marrow progenitors. Blood 77: 2389-95 (1991); **Mule JJ et al** Transforming growth factor-β inhibits the *in vitro* generation of lymphokine-activated killer cells and cytotoxic T cells. Cancer Immunol. Immunother. 26: 95-100 (1988); **Murata M et al** Development of a new culture system for human lymphokine-activated killer cells: Comparison with a conventional static culture method. Cytotechnol. 7: 75-83 (1991); **Naume B et al** Immunomagnetic isolation of NK and LAK cells. JIM 136: 1-9 (1991); **Naume B et al** Synergistic effects of interleukin 4 and interleukin 12 on NK cell proliferation. Cytokine 5: 38-46 (1993); **Nishimura T et al** Combination tumor-immunotherapy with recombinant tumor necrosis factor and recombinant interleukin-2 in mice. IJC 40: 255-61 (1987); **Ottonello L et al** Suppression of lymphokine-activated killer (LAK) cell function by neutrophil polymorphonuclear leukocytes. J. Clin. Lab. Immunol. 34: 37-40 (1991); **Owen-Schaub LB et al** Synergy of tumor necrosis factor and interleukin 2 in the activation of human cytotoxic lymphocytes: effect of tumor necrosis factor α and interleukin 2 in the generation of human lymphokine-activated killer cell cytotoxicity. CR 48: 788-92 (1988); **Papamichail M & Baxevanis CN** γ-interferon enhances the cytotoxic activity of interleukin-2-induced peripheral blood lymphocyte (LAK) cells, tumor infiltrating lymphocytes (TIL), and effusion associated lymphocytes. J. Chemother. 4: 387-93 (1992); **Philips JH & Lanier LL** Dissection of the lymphokine-activated killer phenomenon. Relative contribution of peripheral blood natural killer cells and T lymphocytes to cytolysis. JEM 164: 814-25 (1986); **Reiter Z & Taylor MW** Interleukin 2 protects hairy leukemic cells from lymphokine-activated killer cell-mediated cytotoxicity. CR 53: 3555-60 (1993); **Schoof DD et al** Survival characteristics of metastatic renal cell carcinoma patients treated with lymphokine-activated killer cells plus interleukin-2. Urology 41: 534-9 (1993); **Shimizu K et al** A high density cell culture system for generation of human lymphokine-activated killer (LAK) cells for clinical use in adoptive immunotherapy. J. Clin. Lab. Immunol. 32: 41-7 (1990); **Shimizu Y et al** Proliferation and cytotoxicity of lectin-induced lymphokine-activated killer (LILAK) cells. J. Gastroenterol. Hepatol. 6: 485-90 (1991); **Singh KP et al** Protein A potentiates lymphokine-activated killer cell induction in normal and melanoma patient lymphocytes. Immunopharmacol. Immunotoxicol. 14: 73-103 (1992); **Sondel PM et al** Destruction of autologous human lymphocytes by interleukin 2-activated cytotoxic cells. JI 137: 502-11 (1986); **Tachibana I et al** Generation of a small cell lung cancer variant resistant to lymphokine-activated killer (LAK) cells: association with resistance to a LAK cell-derived, cytostatic factor. CR 52: 3310-6 (1992); **Wersall P et al** Simplified long term large scale production of highly active human LAK cells for therapy. Med. Oncol. Tumor Pharmacother. 7: 257-63 (1990); **Yannelli JR** The preparation of effector cells for use in the adoptive cellular immunotherapy of human cancer. JIM 139: 1-16 (1991)

LANR: *low-affinity neurotrophin receptor* see: LNGFR (low affinity nerve growth factor receptor).

LAP: *liver activator protein* see: NF-IL6.

LA-PF4: *low affinity platelet factor 4* This factor is identical with » CTAP-3 (connective tissue-activating protein). See also: Beta-Thromboglobulin.

Large adhesive scatter factor: see: Ladsin.

Large granular lymphocyte-derived cytotoxic factor: see: LGL-CF.

Laron-type dwarfism: abbrev. LTD. A congenital autosomal recessive disorder characterized by a general retardation of growth, obesity, and a number of other disorders. Associated endocrinological disorders include hypoglycemia and abnormal concentrations of free fatty acids. Affected patients also show enhanced serum levels of » growth hormone and markedly decreased levels of IGF-1 (see: IGF) the synthesis of which is normally regulated by growth hormone.
The underlying genetic defect of this "growth hormone resistance" are partial deletions, stop codons, and missense mutations in the gene encoding the growth hormone receptor on human chromosome 17. Most of these patients also lack a 150 kDa IGF-1 carrier protein (IGF-BP3; see: IGF-BP, insulin-like growth factor binding proteins) that complexes IGF normally circulating in the serum. Treatment with IGF-1 induces the synthesis of this carrier protein.
Hormonal and biochemical responses observed after administration of IGF to these patients resemble those observed normally after treatment with growth hormone.
Amselem S et al Laron dwarfism and mutations of the growth hormone receptor gene. NEJM 321: 989-95 (1989); **Baumann G et al** Absence of the plasma growth hormone binding protein in Laron-type dwarfism. J. Clin. Endocrinol. Metab. 65: 814-6 (1987); **Buchanan CR et al** Laron-type dwarfism with apparently normal high affinity serum growth hormone-binding protein. Clin. Endocrinol. Oxf. 35: 179-85 (1991); **Cotterill AM et al** The insulin-like growth factor (IGF-)-binding proteins and IGF bioactivity in Laron-type dwarfism. J. Clin. Endocrinol. Metab. 74: 56-63 (1992); **Daughaday WH et al** Defective sulfation factor generation: a possible etiological link in dwarfism. Trans. Assoc. Am. Phys. 82: 129-40 (1969); **Daughaday WH & Trivedi B** Absence of serum growth hormone binding protein in patients with growth hormone receptor deficiency (Laron dwarfism). PNAS 84: 4636-40 (1987); **Fielder PJ et al** Expression of serum insulin-like growth factors, insulin-like growth factor-binding proteins, and the growth hormone-binding protein in heterozygote relatives of Ecuadorian growth hormone receptor deficient patients. J. Clin. Endocrinol. Metab. 74: 743-50 (1992); **Gourmelen M et al** Effects of exogenous insulin-like growth factor I on insulin-like growth factor binding proteins in a case of growth hormone insensitivity (Laron-type). Acta Paediatr. Scand. 377: 115-7 (1991); **Guevara-Aguirre J et al** Growth hormone receptor deficiency (Laron syndrome): clinical and genetic characteristics. Acta Paediatr. Scand. Suppl. 377: 96-103 (1991); **Kanety H et al** Long-term treatment of Laron type dwarfs with insulin-like growth factor-1 increases serum insulin-like growth factor-binding protein-3 in the absence of growth hormone activity. Acta Endocrinol. Copenh. 128: 144-9 (1993); **Laron Z et al** Biochemical and hormonal changes induced by one-week of administration of rIGF-1 to patients with Laron type dwarfism. Clin. Endocrinol. Oxf. 35: 145-50 (1991); **Laron Z et al** Effects of insulin-like growth factor on linear growth, head circumference, and body fat in patients with Laron-type dwarfism. Lancet 339: 1258-61 (1992); **Laron Z et al** IGF binding protein 3 in patients with Laron type dwarfism: effect of exogenous rIGF-I. Clin. Endocrinol. Oxf. 36: 301-4 (1992); **Laron Z et al** Growth hormone and insulin-like growth factor regulate insulin-like growth factor-binding protein-1 in Laron type dwarfism, growth hormone deficiency and constitutional short stature. Acta Endocrinol. Copenh. 127: 351-8 (1992); **Laron Z** Somatomedin-1 (insulin-like growth factor-I) in clinical use. Facts and potential. Drugs 45: 1-8 (1993); **Rosenbloom AL** The chronicle of growth hormone receptor deficiency (Laron syndrome). Acta Paediatr. 81 Suppl 383: 117-20 (1992); **Walker JL et al** Effects of the infusion of insulin-like growth factor I in a child with growth hormone insensitivity syndrome (Laron dwarfism). NEJM 324: 1483-8 (1991); **Walker JL et al** Stimulation of statural growth by recombinant insulin-like growth factor I in a child with growth hormone insensitivity syndrome (Laron type). J. Pediatr. 121: 641-6 (1992); **Wilton P** Treatment with recombinant human insulin-like growth factor I of children with growth hormone receptor deficiency (Laron syndrome). Kabi Pharmacia Study Group on Insulin-like Growth Factor I Treatment in Growth Hormone Insensitivity Syndromes. Acta Paediatr. 81 Suppl 383: 137-42 (1992)

Late-acting T cell replacing factor: see: TRF (T cell replacing factors).

Latent cytokines: Biologically inactive forms of cytokines. Some factors, such as » TGF-β exist as high molecular weight complexes from which biologically active molecules can be released by dissociating the complex. The biological activities of other factors are modulated by binding proteins (see, for example: IGF-BP; TNF-BF; Alpha 2-Macroglobulin). Other factors require proteolytic processing to yield biologically active factors (see, for example: IL1 convertase).

LATI: *lymphokine-activated tumor inhibition* This is an experimental variant of the lymphokine-mediated immunotherapy of tumors (see also: Adoptive immunotherapy) involving the use of » LAK cells (lymphokine-activated killer cells. This approach essentially involves local injections of low doses of » IL2 into the tumor mass to induce non-reactive lymphocytes accumulated in the tumor mass to develop into reactive cells. A similar activation of cells has also been observed with the local injection of » IL4 and immunoreactive peptide fragments of » IL1.

Experimentally LATI has been successful in the treatment of poorly immunoreactive tumors.

Bosco MC et al Ability of interleukin-4 to elicit an immunoreaction against a murine tumor. G. Batteriol. Virol. Immunol. 81: 3-9 (1988); **Forni G et al** Tumor inhibition by interleukin-2 at the tumor/host interface. BBA 865: 307-27 (1986); **Forni G et al** Interleukin 2 activated tumor inhibition *in vivo* depends on the systemic involvement of host immunoreactivity. JI 138: 4033-41 (1987); **Forni G et al** Lymphokine-activated tumor inhibition in mice. Ability of a nonapeptide of the human IL1 β to recruit antitumor reactivity in recipient mice. JI 142: 712-8 (1989)

Lavendustin A: This compound has been isolated from strains of *Streptomyces griseolavendus.* It is a highly specific inhibitor of tyrosine-specific protein kinases with almost no activity for serine- and threonine-specific protein kinases. Lavendustin A is not a competitive inhibitor of substrate binding. Instead it inhibits the binding of ATP to the kinases.

Lavendustin A

Lavendustin A can be used to test whether binding of a cytokine to its receptor or induction of the synthesis of a growth factor requires tyrosine-specific protein kinase activities for post receptor signaling processes. Lavendustin A inhibits the receptors for » EGF, » PDGF and IGF-1 (see: IGF). For other agents used to dissect cytokine-mediated signal transduction pathways see: Bryostatins, Calcium ionophore, Calphostin C, Genistein, H8, Herbimycin A, K-252a, Phorbol esters, Okadaic acid, Staurosporine, Suramin, Tyrphostins, Vanadate.

Hsu CYJ et al Kinetic analysis of the inhibition of the epidermal growth factor receptor tyrosine kinase by lavendustin-A and its analog. JCB 266: 21105-12 (1991); **Onoda T et al** Isolation of a novel tyrosine kinase inhibitor, lavendustin A, from *Streptomyces griseolavendus*. J. Natl. Prod. 52: 1252-7 (1989); **Onoda T et al** Inhibition of tyrosine kinase and epidermal growth factor receptor internalization by lavendustin A methyl ester in cultured A431 cells. Drugs Exp. Clin. Res. 16: 249-53 (1990); **Umezawa K** Inhibitors of oncogene product functions. Jap. J. Cancer Chemother. 17: 315-21 (1990)

LBGF: *leukemic blast growth factor* This factor is found in culture supernatants of the human bladder carcinoma cell line HTB9. It supports the proliferation of leukemic blast cells. The factor is probably identical with » GM-CSF which is also secreted in high amounts by these cells.

Another factors with LBGF activity has been found in phytohemagglutinin-stimulated leukocyte-conditioned medium. It supports colony formation (see: Colony formation assay) of a subpopulation of circulating blast cells from the peripheral blood of acute myeloblastic leukemia patients and is probably identical with » G-CSF.

Asano Y et al Effect of human G-CSF on clonogenic cells in acute myeloblastic leukemia. Eur. J. Cancer Clin Oncol. 24: 1285-7 (1988); **Hayashi S et al** Human bladder carcinoma cell line HTB9, which secretes a factor to stimulate clonogenic leukemic blast growth, expresses the granulocyte-macrophage colony-stimulating factor gene. Jpn. J. Cancer Res. 78: 1224-8 (1987); **Hoang T et al** Production of leukemic blast growth factor by a human bladder carcinoma cell line. Blood 66: 748-51 (1985)

LBIF: *lymphocyte blastogenesis inhibitory factor* This factor of 57 kDa is isolated from the culture supernatants of the human lymphoma cell line » U937. It inhibits the synthesis of various immunoglobulin species by LPS-stimulated spleen cells and also influences the proliferation of T lymphocytes. The factor strongly inhibits the growth of a number of different tumor cell lines and is also cytotoxic or cytostatic for some of these cells.
This factor is probably synthesized by human cells infected with mycoplasms and may in fact be a product of *Mycoplasma arginini.* LBIF displays the activity of an arginine deiminase and may play a role in the immunosuppression caused by some arginine-metabolising mycoplasms observed *in vivo.*

Sugimura K et al Fast protein liquid chromatography of lymphocyte blastogenesis inhibitory factor produced by the human macrophage-like cell line U937. J. Chromatogr. 440: 131-40 (1988); **Sugimura K et al** A cytokine, lymphocyte blastogenesis inhibitory factor (LBIF), arrests mitogen-stimulated T lymphocytes at early G1 phase with no influence on interleukin 2 production and interleukin 2 receptor light chain expression. EJI 19: 1357-64 (1989); **Sugimura K et al** Abnormal behavior of γ-committed B lymphocytes probed by a lymphocyte blastogenesis inhibitory factor in autoimmune MRL mice. EJI 20: 1899-904 (1990); **Sugimura K et al** Tumor growth inhibitory activity of a lymphocyte blastogenesis inhibitory factor. CR 50: 345-9 (1990); **Sugimura K et al** A lymphocyte blastogenesis inhibitory factor (LBIF) reversibly arrests a human melanoma cell line, A375, at G1 and G2 phases of cell cycle. ECR 188: 272-8 (1990); **Sugimura K et al** Identification and purification of arginine deiminase that originated from Mycoplasma arginini. Infect. Immun. 58: 2510-5 (1990)

LBP: *LPS binding protein* see: Endotoxin.

LBP: LIF binding protein see: LIF.

LBRM 33: This murine T cell lymphoma line (B10.BR mouse) is used to assay » IL1 (see also:

Bioassays). The activity of IL1 is determined indirectly by measuring the amounts of IL1-induced » IL2 in another cell line (see: CTLL). An alternative and entirely different detection method is » Message amplification phenotyping. LBRM 33 cells also secrete colony-stimulating factor » IL3. The ***LBRM TG6*** cell line is a thioguanine-resistant derivative of the LBRM 33 murine lymphoma cell line. When incubated in the presence of PHA and » IL1, the cells also produce » IL2. LBRM TG cells are equally sensitive to both human and mouse IL1. The synthesis of IL2 and IL3 by activated LBRM-33 cells is coordinately inhibited by sodium butyrate.

Alheim K et al Interleukin 1 expression is inducible by nerve growth factor in PC12 pheochromocytoma cells. PNAS 88: 9302-6 (1991); **Larrick JW et al** An improved assay for the detection of interleukin 1. JIM 79: 39-45 (1985); **Le Gros GS et al** The effects of sodium butyrate on lymphokine production. LR 4: 221-7 (1985); **Palacios R** Monoclonal antibodies against human Ia antigens stimulate monocytes to secrete interleukin 1. PNAS 82: 6652-6 (1985); **Segal R et al** The effects of methotrexate on the production and activity of interleukin-1. Arthritis Rheum. 32: 370-7 (1989); **Suda T et al** Biological activity of recombinant murine interleukin-6 in interleukin-1 T cell assays. JIM 120: 173-8 (1989)

L-cell CSF: *L cell colony stimulating activity* This factor is identical with » M-CSF. Antibodies directed against L-cell CSF neutralize the biological activity of » Urinary colony-stimulating factor.

Dexter TM & Shadduck RK The regulation of hemopoiesis in long-term bone marrow cultures: I. role of L-cell CSF. JCP 102: 279-86 (1980); **Guilbert LJ & Stanley ER** Specific interaction of murine colony-stimulating factor with mononuclear phagocytic cells. JCB 85: 153-9 (1980); **Stanley ER et al** Factors regulating macrophage production and growth: identity of colony-stimulating factor and macrophage growth factor. JEM 143: 631-47 (1976); **Stanley ER & Heard PM** Factors regulating macrophage production and growth. JBC 252: 4305 (1977); **Waheed A & Shadduck RK** Purification of colony-stimulating factor by affinity chromatography. Blood 60: 238-44 (1982)

LCF: *lymphocyte chemokinetic factor(s)* These poorly characterized factors are chemotactic for human lymphocytes. Some such proteins of 5 and 35 kDa are produced by human tonsillar lymphocytes.

McFadden RG et al Lymphocyte chemokinetic factors derived from human tonsils: modulation by 1,25-dihydroxyvitamin D3 (calcitriol). Am. J. Respir. Cell. Mol. Biol. 4: 42-9 (1991)

LCF: *lymphocyte chemoattractant factor* See: LCF (lymphocyte chemotactic factors) LCF (lymphocyte chemoattractant factor) is a tetrameric glycoprotein of 56 kDa produced by activated T lymphocytes. LCF binds to » CD4 and stimulates migration of CD4-positive lymphocytes and monocytes. The factor also stimulations the migration of eosinophils. It does not stimulate degranulation, leukotriene C4 release, or respiratory burst activity of human eosinophils.

Cruikshank WW et al Lymphocyte chemoattractant factor induces CD4-dependent intracytoplasmic signaling in lymphocytes. JI 146: 2928-34 (1991); **Rand TH et al** CD4-mediated stimulation of human eosinophils: lymphocyte chemoattractant factor and other CD4-binding ligands elicit eosinophil migration. JEM 173: 1521-8 (1991); **Weller PF** Cytokine regulation of eosinophil function. Clin. Immunol. Immunopathol. 62: S55-9 (1992)

LCF: *lymphocyte chemotactic factors* These poorly characterized factors of 14-17 and 40-50 kDa are synthesized by antigen- or mitogen-stimulated suppressor T cells and cytotoxic T cells. The factors are chemotactic for CD4-positive helper T cells (see also: Chemotaxis). They are not identical with either » IL1 and » IL2. The synthesis of these factors is inhibited by β-endorphin and met-enkephalin. Some of the LCF activity is probably due to » IL8.

Brown SL & Van Epps DE Suppression of T lymphocyte chemotactic factor production by the opioid peptides β-endorphin and met-enkephalin. JI 134: 3384-90 (1985); **Cole D et al** Defective T lymphocyte chemotactic factor production in patients with established malignancy. Clin. Immunol. Immunopathol. 38: 209-21 (1986); **Potter JW & Van Epps DE** Separation and purification of lymphocyte chemotactic factor (LCF) and interleukin 2 produced by human peripheral blood mononuclear cells. CI 105: 9-22 (1987); **Shimokawa Y et al** Lymphocyte chemotaxis in inflammation. VII. Isolation and purification of chemotactic factors for T lymphocytes from PPD-induced delayed hypersensitivity skin reaction site in the guinea pig. Immunology 51: 275-85 (1984); **Thornton AJ et al** Kupffer cell-derived cytokines induce the synthesis of a leukocyte chemotactic peptide, interleukin 8, in human hepatoma and primary hepatocyte cultures. Hepatology 14: 1112-22 (1991); **Wilkinson PC** Lymphocyte locomotion *in vitro*: the role of growth activators and chemoattractants. Biomed. Pharmacother. 41: 329-36 (1987)

LC-GF: *leukemia cell-growth factor* This poorly characterized factor is produced by a murine cultured cell line (MKM-O) of endothelial origin that was established from a tumor of a BALB/C (nu/nu) mouse inoculated with human hepatoma tissue fragments. The cells also produce an unidentified colony-stimulating factor (see: CSF) which acts on both human and murine bone marrow-derived granulocytes and macrophage colony-forming units (see: CFU-GM). LC-GF activity is also produced by rat vascular endothelial cells. LC-GF promotes the growth of human leukemia cell lines (» HL-60, HSB-2, CEM, » DAUDI and » K562). For other leukemia-associated growth factors see also: LDA, LGPF, LDGF, LGF.

LCF: see addendum.

Yoshimura T et al Colony-stimulating factor and leukemia cell-growth factor produced by a murine endothelial cell line, MKM-O. Eur. J. Haematol. 39: 136-43 (1987)

LCIA: *leukemia cell inhibitory activity* This poorly characterized cell-surface-associated glycoprotein is expressed by an endothelial-like clone derived from mouse bone marrow stroma (MBA-2.1) and also by some thymic stroma cell lines. LCIA is not detected in anchorage-dependent cells of non-hemopoietic origin. LCIA selectively inhibits the growth of plasmacytomas. Tumors of the lymphoid lineage and plasmacytomas in particular are the most sensitive to LCIA. Myeloid, macrophage, and erythroleukemia tumors are resistant to LCIA, as are normal hemopoietic target cells including pluripotent stem cells, myeloid progenitor cells, and mitogen-stimulated spleen cells.

Tamir M et al Thymus-derived stromal cell lines. Int. J. Cell Cloning 5: 289-301 (1987); **Zipori D et al** Differentiation stage and lineage-specific inhibitor from the stroma of mouse bone marrow that restricts lymphoma cell growth. PNAS 83: 4547-51 (1986)

***lck*:** The *lck* gene is a proto-oncogene (see: oncogenes) encoding a lymphocyte-specific tyrosine-specific protein kinase of 56 kDa ($p56^{lck}$) belonging to the *src* family of tyrosine kinases. This gene, which maps to human chromosome 1p35-p32, is involved in the control of lymphocyte proliferation. *lck* is expressed exclusively in lymphoid cells, predominantly in thymocytes and peripheral T cells.

In T lymphocytes the *lck* kinase is associated with the cytoplasmic domains of the cell surface proteins » CD4 and » CD8 (see also: CD antigens). Its intrinsic protein kinase activity is activated following binding of ligands to CD4 and CD8. *lck* thus mediates post receptor signal transduction pathways. *lck* is also associated with the » CD16 receptor.

Increased activity of the *lck* tyrosine kinase is also observed after activation of T cells by » IL2. *lck* has recently been found to be associated also with the β chain of the IL2 receptor. Results of studies involving the use of an *lck*-deficient subline of the IL2-dependent cytotoxic T cell line » CTLL-2 indicate that p56*lck* expression is not obligatory for IL2-mediated T cell growth stimulation; however, this *lck* plays a central role in the generation T cell-mediated cytotoxic responses.

Virus-induced murine lymphomas have been found to have rearrangement of the *lck* gene. The overexpression of *lck* gene leads to the development of lymphoid tumors. Joining of the *lck* gene with the T cell antigen receptor β subunit gene has been observed in a patient with T lymphoblastic leukemia and a translocation t(1;7)(p34;q34) as the sole karyotypic abnormality.

■ **TRANSGENIC/KNOCK-OUT/ANTISENSE STUDIES:** An *lck* null mutation created by homologous recombination in embryonic stem cells (see: ES cells) has been created to evaluate the role of p56*lck* in T cell development and activation. *lck*-deficient mice show a pronounced thymic atrophy, with a dramatic reduction in the double-positive ($CD4^+CD8^+$) thymocyte population. Mature, single-positive thymocytes are not detectable in these mice and there are only very few peripheral T cells. These results demonstrate the crucial role of this T cell-specific tyrosine kinase in thymocyte development. Similar findings suggesting a critical function early in T cell development have been reported with » transgenic mice expressing a dominant negative *lck* transgene.

Burnett RC et al The *lck* gene is involved in the t(1;7)(p34;q34) in the T cell acute lymphoblastic. leukemia derived cell line, HSB-2. Genes Chromosomes Cancer 3: 461-7 (1991); **Garvin AM et al** Structure of the murine *lck* gene and its rearrangement in a murine lymphoma cell line. MCB 8: 3058-64 (1988); **Hatakeyama M et al** Interaction of the IL2 receptor with the *src*-family kinase $p56^{lck}$: identification of novel intermolecular association. S 252: 1523-8 (1991); **Horak ID et al** T lymphocyte interleukin 2-dependent tyrosine protein kinase signal transduction involves the activation of $p56^{lck}$. PNAS 88: 1996-2000 (1991); **Karnitz L et al** Effects of p56*lck* deficiency on the growth and cytolytic effector function of an interleukin-2-dependent cytotoxic T cell line. MCB 12: 4521-30 (1992); **Kobayashi N et al** Functional coupling of the *src* family protein tyrosine kinases p59*fyn* and p53/56*lyn* with the interleukin 2 receptor: implications for redundancy and pleiotropism in cytokine signal transduction. PNAS 90: 4201-5 (1993); **Marth JD et al** Localization of a lymphocyte-specific protein tyrosine kinase gene (*lck*) at a site of frequent chromosomal abnormalities in human lymphomas. PNAS 83: 7400-4 (1986); **Marth JD et al** Translational activation of the *lck* proto-oncogene. N 332: 171-3 (1988); **Minami Y et al** Association of p56*lck* with IL2 receptor β chain is critical for the IL2 induced activation of p56lck. EJ 12: 759-68 (1993); **Perlmutter RM & Ziegler S** Proto-oncogene expression following lymphocyte activation. CTMT 35: 571-86 (1990); **Rouer E et al** Structure of the human *lck* gene: differences in genomic organization within src-related genes affect only N-terminal exons. Gene 84: 105-13 (1989); **Shaw AS et al** The *lck* protein tyrosine kinase interacts with the cytoplasmic tail of the CD4 glycoprotein through its unique aminoterminal domain. Cell 59: 627-36 (1989); **Shaw AS et al** Short related sequences in the cytoplasmic domains of CD4 and CD8 mediate binding to the aminoterminal domain of the $p56^{lck}$ tyrosine protein kinase. MCB 10: 1853-62 (1990); **Tycko B et al** Chromosomal translocations joining *lck* and TCRB loci in human T cell leukemia. JEM 174: 867-73 (1991); **Van Oers NS et al** Differential involvement of protein tyrosine kinases p56lck and p59*fyn* in T cell development. AEMB 323: 89-99 (1993); **Xu H & Littman DR** A kinase-independent function of *lck* in potentiating antigen-specific T cell activation. Cell 74: 633-43 (1993)

● TRANSGENIC/KNOCK-OUT/ANTISENSE STUDIES: **Levin SD et al** A dominant-negative transgene defines a role for p56lck in thymopoiesis. EJ 12: 1671-80 (1993); **Molina TJ et al** Profound block in thymocyte development in mice lacking p56lck. N 357: 161-4 (1992)

LCM: *leukocyte* (or *lymphocyte*)*-conditioned medium* see: CM.

LCM factor: *(factor from) lymphocyte-conditioned medium* This factor found in the growth medium of phytohemagglutinin-stimulated T cells is identical with » IL2.

Klein B et al The role of interleukin 1 and interleukin 2 in human T colony formation. CI 77: 348-56 (1983)

LD78: This factor was initially isolated from stimulated human tonsillar lymphocytes as a cDNA clone. The gene is also called GOS19 (see: GOS) and » ***pAT 464***. LD78 is produced by T cells, B cells, and various tumor cell lines. Glucocorticoids, » CsA (cyclosporin) and » TGF-β inhibit the synthesis of this factors. The human gene maps to chromosome 17. It contains three exons and encodes a secreted protein with a length of 92 amino acids. At physiological ionic strength, the 8 kDa LD78 molecule exists as soluble, heterogeneous, multimeric complexes of 100 to > 250 kDa. Based on sequence similarities LD78 belongs to the » chemokine family of cytokines. LD78 is the human homologue of murine » MIP (macrophage inflammatory protein).

LD78 probably plays an inhibitory role in hematopoiesis. Its action is thought to be indirect through induction of the synthesis of other inhibitory factors (see also: KMT-2). LD78 promotes migration of human peripheral T lymphocytes at concentrations from 10^{-11} M to 10^{-7} M. LD78 enhances the adherence of monocytes to endothelial cells.

LD78 receptors have been found on several cell lines, including » HL-60, and » U937 cells. High and low affinity receptors are expressed on U937 cells at a density of 30000-90000/cell. The receptor has a molecular mass of ≈ 52 kDa and is distinct from receptors for » IL8 or » IFN-γ.

Forsdyke DR cDNA cloning of mRNAs which increase rapidly in human lymphocytes cultured with concanavalin A and cycloheximide. BBRC 129: 619-25 (1985); **Graves DT et al** Identification of monocyte chemotactic activity produced by malignant cells. S 245: 1490-3 (1989); **Irving SG et al** Two inflammatory mediator cytokine genes are closely linked and variably amplified on chromosome 17. NAR 18: 3261-70 (1990); **Nakao M et al** Structures of human genes coding for cytokine LD78 and their expression. MCB 10: 3646-58 (1990); **Nomiyama H et al** Characterization of cytokine LD78 gene promoters: positive and negative transcriptional factors bind to a negative regulatory element common to LD78, interleukin-3, and granulocyte-macrophage colony-stimulating factor gene promoters. MCB 13: 2787-801 (1993); **Obaru K et al** A cDNA clone used to study mRNA inducible in human tonsillar lymphocytes by a tumor promoter. JB 99: 885-94 (1986); **Patel SR et al** Characterization of the quaternary structure and conformational properties of the human stem cell inhibitor protein LD78 in solution. B 32: 5466-71 (1993); **Shiozaki H et al** Suppressive effect of LD78 on the proliferation of human hemopoietic progenitors. Jpn. J. Cancer Res. 83: 499-504 (1992); **Takatsuki Y et al** Synthesis of a novel cytokine and its gene (LD78) expression in hematopoietic fresh tumor cells and cell lines. JCI 84: 1707-12 (1989); **Tanaka J et al** T cell chemotactic activity of cytokine LD78: a comparative study with interleukin-8, a chemotactic factor for the T cell CD45RA^{+} phenotype. Int. Arch. Allergy Immunol. 100: 201-8 (1993); **Widmer U et al** Genomic structure of murine macrophage inflammatory protein-1 α and conservation of potential regulatory sequences with a human homologue, LD78. JI 146: 4031-40 (1991); **Yamamura Y et al** Synthesis of a novel cytokine and its gene (LD78) expressions in hematopoietic fresh tumor cells and cell lines. JCI 84: 1707-12 (1989); **Yamamura Y et al** Identification and characterization of specific receptors for the LD78 cytokine. Int. J. Hematol. 55: 131-7 (1992)

LDA: *leukemia-differentiating activity* This poorly characterized factor of ≈ 40 kDa is found in medium conditioned by the LD-1 melanoma, a » G-CSF secreting human tumor line. Partially-purified LDA induces » HL-60 cells to produce superoxide, become phagocytic, and to develop macrophage-like morphology and surface markers. The LDA markedly suppresses clonal growth in agar of HL-60 cells, and cells of the human myeloid leukemia lines PBL 985 and » K562, but does not suppress clonal growth of the B lymphoblast lines Raji and Daudi. LDA activity is not neutralized by antibodies to » G-CSF, » GM-CSF, » IFN-α, » IFN-γ, » TNF, urokinase, and tissue plasminogen activator. For other leukemia-associated growth factors see also: LC-GF, LGPF, LDGF, LGF.

Lilly MB & Kraft AS Leukemia-differentiating activity expressed by the human melanoma cell line LD-1. Leuk. Res. 12: 217-25 (1988)

LDA: abbrev. for » Limiting dilution analysis.

LDCF: *lymphocyte-derived chemotactic factor* This protein is produced by antigen- and mitogen-stimulated CD4-positive lymphocytes. It is encoded by the » JE gene and has a length of 76 amino acids (12.5 kDa). LDCF therefore belongs to the » chemokine family of cytokines.

LDCF is identical with » MCP-1 (monocyte chemoattractant protein 1), » MCP (monocyte chemotactic protein), » MCAF (monocyte chemotactic and activating factor), » HC11, » SMC-CF (smooth muscle cell chemotactic factor), » GDCF-2 (glioma-derived monocyte chemotactic factor),

» MCAF (monocyte chemotactic and activating factor), » TDCF (tumor-derived chemotactic factor).
The protein is chemotactic for monocytes (see also: Chemotaxis), but not for neutrophils. It also stimulates monocyte-mediated antitumor activity. LDCF probably also plays a role in delayed hypersensitivity reactions, mediating the migration of inflammatory cells (see also: Inflammation). It may also be indirectly involved in the rejection of allotransplants.

Altman LC et al A human mononuclear leukocyte chemotactic factor: characterization, specificity, and kinetics of production by homologous leukocytes. JI 110: 801-10 (1983); **Antonaci S et al** Monocyte chemotactic responsiveness triggered by lymphocyte-derived chemotactic factor in aged donors. Definitive report. LR 4: 359-65 (1985); **Leonard EJ & Meltzer MS** Characterization of mouse lymphocyte derived chemotactic factor. CI 26: 200-10 (1976); **Yoshimura T et al** Purification and amino acid analysis of two monocyte chemoattractants produced by phytohemagglutinin-stimulated human blood mononuclear leukocytes. JI 142: 1956-62 (1289)

LDECF: *lymphocyte-derived eosinophil chemotactic factors* These poorly characterized factors of ≈ 45-60 kDa were isolated from human CD4-positive T lymphocytes. LDECF-PD (pI 7-8) is derived from patients with parasite disease. Its production depends on antigen or mitogen stimulation. LDECF-HES (pI 6) is derived from patients with hypereosinophilic syndrome (HES). Its synthesis does not require stimulation of cells.
Both factors are chemotactic for eosinophils from healthy individuals. Eosinophils from patients with HES are attracted by LDECF-HES but not LDECF-PD. LDECF-HES enhances expression of » CD32 and » CD16 but not that of CD35 on eosinophils. LDECF-PD enhances expression of CD35 and » CD16 but not of » CD32. LDECF-PD but not LDECF-HES suppresses the release of eosinophilic cationic protein (ECP) from eosinophils. Monoclonal antibodies against » GM-CSM, » IL3, and » IL5 do not inhibit LDECF activity.

Hirashima M et al Functional heterogeneity of human eosinophil chemotactic lymphokines. Lymphokine Cytokine Res. 10: 481-6 (1991)

LDF: *lymphocyte-derived fibroblast activating factor* see: FAF (fibroblast activating factors).

LDGF: *leukemia-derived growth factor* This factor is produced by the human acute lymphoblastoid T cell leukemia cell line MOLT-4 and some other malignant T lymphoid cells. This factor of 5-15 kDa is not identical with IL2. It stimulates the proliferation of the producer cells and other T cell lines not responding to IL2. LDGF does not stimulate the proliferation of B lymphoblastoid and myeloid cell lines. A factor with an identical biological behavior is also secreted by murine T cell lymphomas. LDGF is not identical with another factor of the same name abbrev. » LGF. For other leukemia-associated growth factors see also: LDA, LGPF, LC-GF.

Hays EF et al Autocrine growth of murine lymphoma cells. JNCI 80: 116-21 (1988); **Hays EF et al** Leukaemia-derived growth factor (non-interleukin 2) produced by murine lymphoma T cell lines. PNAS 81: 7807-11 (1984); **Uittenbogaart CH et al** Leukaemia-derived growth factors produced by human malignant T lymphoid cell lines. CR 46: 1219-23 (1986); **Uittenbogaart CH et al** Leukaemia-derived growth factor (non-interleukin 2) produced by a human malignant T lymphoid cell line. PNAS 79: 7004-8 (1982)

LDGF-1: *lung-derived growth factor 1* see: lung-derived growth factor.

LDGF-M4: see: LDGF (leukemia-derived growth factor).

LDNAP: *leukocyte-derived neutrophil-activating peptide* see: LYNAP.

LDSF: *leukemia cell-derived suppressor factor* This poorly characterized factor of 66 kDa is produced constitutively by » HL60 leukemia cells. It suppresses *in vitro* proliferation and activation of normal human lymphocytes. LDSF is not cytolytic to lymphocytes and does not affect IL2 or transferrin receptor expression.

Abolhassani M & Chiao JW Purification and characterization of a human leukemia cell-derived immunosuppressive factor. Prep. Biochem. 21: 25-33 (1991)

Lectin-induced lymphokine-activated killer cells: abbrev. LILAK cells. See: LAK cells.

LeIFN: *leukocyte interferon* This interferon is identical with » IFN-α.
Natural leukocyte interferon is usually a mixture of undefined composition, containing mainly different subtypes of » IFN-α (see also: IFN-ω). The natural product may also contain further factors influencing cell proliferation and differentiation.

LEM: *leukocyte endogenous mediator* This factor was initially isolated from leukocytes as a protein inducing the synthesis of » acute phase proteins in the liver (see: acute phase reaction). It was also found to reduce plasma concentrations of iron and zinc and to induce neutrophilia. It is identical with LP (leukocytic pyrogen; see: EP, endogenous pyrogens) and hence identical with » IL1.

Merriman CR et al Comparison of leukocytic pyrogen and leukocytic endogenous mediator. PSEBM 154: 224-7 (1977); **Whicher JT et al** Defective production of leucocytic endogenous mediator by peripheral blood leukocytes of patients with systemic sclerosis, systemic lupus erythematosus, rheumatoid arthritis, and mixed connective tissue disease. Clin. Exp. Immunol. 65: 80-9 (1986)

Lenograstim: Generic name for glycosylated recombinant » G-CSF (Chugai) marketed under the brand name Neutrogin.

USAN Council. List No. 340. New names. Lenograstim. Clin. Pharmacol. Ther. 52(1): 113 (1992)

Lens 10K protein: see: 10 K protein.

Leu-2: Alternative name for » CD8 after the monoclonal antibody recognizing this cell surface protein (see also: CD antigens).

Leu-3: Alternative name for » CD4 after the monoclonal antibody recognizing this cell surface protein (see also: CD antigens).

Leu-5: Alternative name for » CD2 after the monoclonal antibody recognizing this cell surface protein (see also: CD antigens).

Leu-11: Alternative name for » CD16 after the monoclonal antibody recognizing this cell surface protein (see also: CD antigens).

Leukemia cell-derived suppressor factor: see: LDSF.

Leukemia cell-growth factor: see: LC-GF.

Leukemia cell inhibitory activity: see: LCIA.

Leukemia-derived growth factor: see: LDGF.

Leukemia-derived growth factor 1: abbrev. LGF-1. see: LGF.

Leukemia-derived growth factor 2: abbrev. LGF-2. see: LGF.

Leukemia-derived transforming growth factor: This poorly characterized factor has been isolated from the serum-free culture medium of SMS-SB cells, a human pre-B acute lymphoblastic leukemia cell line. The factor is biochemically distinct from known fibroblast mitogens such as » EGF, » TGF-α, » TGF-β, and the endothelial cell growth factor/fibroblast growth factor family of mitogens (see: FGF). The factor is not mitogenic for normal or leukemic lymphocytes or granulocyte/macrophage progenitors.

Zack J et al Characterization of a leukemia-derived transforming growth factor. Leukemia 1: 737-45 (1987)

Leukemia-differentiating activity: see: LDA.

Leukemia-growth-promoting factor: see: LGPF.

Leukemia inhibitory factor: see: LIF. This factor is not identical with LIF= leukocyte inhibitory factor.

Leukemic blast growth factor: see: LBGF.

Leukemic cell growth-promoting factor: see: LGF.

Leukalexins: This is the generic name for a group of poorly characterized proteins with cytolytic activity. They are synthesized by cytotoxic T lymphocytes together with other cytolytic proteins such as » NKCF (natural killer cytotoxic factor) and » perforins. These proteins are involved in cell killing by » LAK cells (lymphokine-activated killer cells). It has been suggested that these proteins may directly damage DNA.

Joag S et al Mechanisms of lymphocyte-mediated lysis. JCBc 39: 239-52 (1989); **Konigsberg PJ & Podack ER** A cytolytic protein isolated from murine cytolytic CTL granules. Fed. Proc. 46: 1225, abstract 5295 (1987); **Liu CC et al** Identification, isolation, and characterization of a novel cytotoxin in murine cytolytic lymphocytes. Cell 51: 393-403 (1987)

Leukine: Brand name for recombinant glycosylated » GM-CSF marketed under the generic name Sargramostim (Immunex).

Leukocyte adhesion inhibitor: see: LAI.

Leukocyte-derived neutrophil-activating peptide: abbrev. LDNAP. See: LYNAP.

Leukocyte endogenous mediator: see: LEM.

Leukocyte infiltration inducing factor: This poorly characterized factor of 45-60 kDa is also called ***PRF*** (PMNL recruiting factor). It is secret-

LESTR: see addendum.
Leukocyte-derived 7-transmembrane domain receptor: see addendum.

ed by rabbit peritoneal macrophages and human macrophages following stimulation by bacterial » endotoxins. In rabbits this factor induces the local infiltration of segmented neutrophil granulocytes at the site of intradermal injection The factor is not identical with » IL1, » TNF-α, » GM-CSF and » IL6.

A similar factor, called ***PRA*** (PMNL recruiting activity), is also secreted by the human macrophage line THP-1. Some of the "recruiting activity" can be attributed to » IFN-γ.

Issekutz AC et al Rabbit alveolar macrophages stimulated with endotoxin and lung fragments from endotoxemic rabbits produce a leukocyte infiltration-inducing factor. Exp. Lung Res. 17: 803-19 (1991); **Issekutz TB et al** The recruitment of lymphocytes into the skin by T cell lymphokines: the role of γ interferon. Clin. Exp. Immunol. 73: 70-5 (1988); **Megyeri P et al** An endotoxin-induced factor distinct from interleukin 1 and tumor necrosis factor α produced by the THP-1 human macrophage line stimulates polymorphonuclear leukocyte infiltration *in vivo*. J. Leukoc. Biol. 47: 70-8 (1990)

Leukocyte inhibitory factor: see: LIF.

Leukocyte interferon: abbrev. HLI-α, human leukocyte interferon). This interferon is identical with » IFN-α. See also: IFN for general information about interferons.

Leukocyte migration enhancing factor: see: MEF.

Leukocyte migration inhibitory factor: see: LIF.

Leukocyte migration stimulation factor: see: LMSF.

Leukocytic pyrogen: see: LP. See also: EP (endogenous pyrogens).

Leukomax: Brand name for recombinant nonglycosylated » GM-CSF (Shering Plough/Sandoz) marketed under the generic name Molgramostim.

Leukoregulin: abbrev. LR, LRG. This glycoprotein of 50 kDa is secreted by activated natural killer lymphocytes. Leukoregulin shows tumor growth inhibitory activity and enhances membrane permeability and the uptake of cytotoxic drugs by tumors cells without affecting these processes in normal cells.

In human skin fibroblasts leukoregulin selectively induces the synthesis of collagenase, upregulates stromelysin-1 gene expression, and induces the expression and secretion of » IL8. This protein is therefore probably involved in the physiological regulation of extracellular matrix degradation (see also: ECM). It may be involved in the pathogenesis of diseases characterized by an increased collagen metabolism.

At concentrations of 1-100 mM leukoregulin in combination with acyclovir (acycloguanosine) significantly enhances the replication of Herpes Simplex Type 1 DNA and the release of infectious virus particles by infected cells.

Barnett SC & Evans CH Influence of extracellular calcium on cell permeabilisation and growth regulation by the lymphokine leukoregulin. JCB 43: 89-101 (1990); **Evans CH** Leukoregulin. PCBR 288: 259-70 (1989); **Evans CH** Preparative isoelectric focusing in ampholine electrofocusing columns versus immobiline polyacrylamide gel for the purification of biologically active leukoregulin. AB 177: 358-63 (1989); **Evans CH & Baker PD** Tumor-inhibitory antibiotic uptake facilitated by leukoregulin: a new approach to drug delivery. JNCI 80: 861-4 (1988); **Fishelson Z et al** The human lymphokine leukoregulin induces cell resistance to complement-mediated lysis. Immunol. Lett. 32: 35-42 (1992); **Furbert-Harris PM & Evans CH** Leukoregulin upregulation of tumor cell sensitivity to natural killer and lymphokine-activated killer cell cytotoxicity. Cancer Immunol. Immunother. 30: 86-91 (1989); **Hooks JJ et al** Leukoregulin, a novel cytokine, enhances the anti-herpes virus actions of acyclovir. Clin. Immunol. Immunopathol. 60: 244-53 (1991); **Mauviel A et al** Transcriptional activation of fibroblast collagenase gene expression by a novel lymphokine. leukoregulin. JBC 267: 5644-8 (1992); **Mauviel A et al** Modulation of human dermal fibroblast extracellular matrix metabolism by the lymphokine leukoregulin. JCB 113: 1455-62 (1991); **Mauviel A et al** Leukoregulin, a T cell-derived cytokine, induces IL8 gene expression and secretion in human skin fibroblasts. Demonstration and secretion in human skin fibroblasts. Demonstration of enhanced NF-kappa B binding and NF-kappa B-driven promoter activity. JI 149: 2969-76 (1992); **Mauviel A et al** Leukoregulin, a T cell derived cytokine, upregulates stromelysin-1 gene expression in human dermal fibroblasts: evidence for the role of AP-1 in transcriptional activation. JCBc 50: 53-61 (1992); **Merchant RE et al** Leukoregulin inhibits the growth of human glioblastoma *in vitro*. J. Neuroimmunol. 13: 41-5 (1986); **Ortaldo JR et al** Analysis of cytostatic/cytotoxic lymphokines: relationship of natural killer cytotoxic factor to recombinant lymphotoxin, recombinant tumor necrosis factor, and leukoregulin. JI 137: 2857-63 (1986); **Ransom JH et al** Leukoregulin, a direct-acting anticancer immunological hormone that is distinct from lymphotoxin and interferon. Cancer Res. 45: 851-62 (1985); **Sheehy PA et al** Activation of ion channels in tumor cells by leukoregulin, a cytostatic lymphokine. JNCI 80: 868-71 (1988)

LFA-2: This cell surface protein is identical with » CD2 (see also: CD antigens).

LFA-3: This cell surface protein is also called CD58. It is a ligand for » CD2 (see also: CD antigens).

LGEF: *lymphocyte growth enhancing factor* This poorly characterized factor is secreted by the human myeloid cell line » HL60. It is a costimulating

Leukocyte janus kinase: see addendum: JAK3.
Leukocyte tyrosine kinase: see addendum: LTK.

factor that enhances the growth of mitogen-stimulated T cells. The activity cannot be neutralized by antibodies directed against » IL1, » IL2, » IL3 and » IL6.
Abolhassani M & Chiao JW Identification of a co-stimulatory factor for human T cell proliferation. Cancer Lett. 56: 71-6 (1991)

LGF: *leukemia-derived growth factor* This activity is isolated from the serum-free culture supernatants (see also: SFM, serum-free medium) of the human erythroleukemic cell line K-562 T1. It consists of two factors called LGF-1 (20 kDa) and LGF-2. LGF-1 promotes the proliferation of a number of human leukemia cell lines. The first 30 aminoterminal amino acids of LGF-1 are identical with ubiquitin, a protein of 8.6 kDa found in eukaryotic cells. In the nucleus ubiquitin is conjugated to histone 2A to form the nuclear protein A24 which may play a role in regulation of chromatin structure, and in the cytoplasm is part of an ATP-dependent non-lysosomal proteolytic pathway. The physiological significance of ubiquitin has not yet been fully resolved. Ubiquitin purified from bovine thymus does not show cell proliferating activity for any cells tested.
LGF-2 supports the growth of fibroblasts. LGF is not identical with Leukemia-derived growth factor abbrev. » LDGF. For other leukemia-associated growth factors see also: LC-GF, LDA, LGPF.
Mihara A et al N-terminal amino acid sequence of leukaemia-derived growth factor (LGF) from human erythroleukemia cell culture. *In vitro* Cell. Dev. Biol. 23: 317-22 (1987)

LGF: *leukemic cell-growth-promoting factor* This poorly characterized factor of 18 kDa is produced by murine and human fibroblasts upon treatment with » TGF-β. The factor supports the growth of leukemic cells.
Komatsu K et al Transforming growth factor(TGF)-β 1 induces leukemic cell-growth-promoting activity in fibroblast cells. Cell. Biol. Int. 17: 433-40 (1993)

LGF-1: *leukemia-derived growth factor* see: LGF.

LGF-2: *leukemia-derived growth factor* see: LGF.

LGF: *liver growth factor* This factor isolated from plasma of partially hepatectomized rats induces DNA synthesis in hepatocytes *in vivo* and *in vitro*. It has been identified as an albumin-bilirubin complex. A similar factor has been identified in the serum of human patients with hepatitis.
Diaz-Gil JJ et al Identification of a liver growth factor as an albumin-bilirubin complex. BJ 243: 443-8 (1987); **Diaz-Gil JJ et al** Liver growth factor purified from human plasma is an albumin-bilirubin complex. Mol. Biol. Med. 6: 197-207 (1989)

LGF: *lymphoma growth factor* This poorly characterized factor has been found to be secreted by many different murine primary X-ray-induced thymic lymphomas (PXTL). It serves as an » autocrine growth factor for these cells. LGF-dependent cells are non-tumorigenic or poorly tumorigenic and do not clone in soft agar. Upon progression PXTL cells become growth factor-independent, are highly tumorigenic *in vivo* and clone in soft agar.
LGF has no » IL1, » IL2, or » IL3 activity and LGF-secreting cells do not synthesize detectable IL1, -2, or -3 mRNA. LGF contains no detectable » IFN-α, » IFN-β, » IFN-γ or » GM-CSF activity and purified » EGF, » TGF-β, and interleukin preparations are inactive on LGF-dependent PXTL cells.
Haas M et al Autocrine growth and progression of murine X-ray-induced T cell lymphomas. EJ 5: 1775-82 (1986)

L-GI factor: *lung-derived growth inhibitory factor* This factor is secreted by murine lung cells. It inhibits the growth of murine monocytic leukemia cells and supports the proliferation of IL6-dependent » MH60.BSF2 cells. The factor is probably identical with murine » IL6.
Kasukabe T et al Characterization of growth inhibitory factors for mouse monocytic leukaemia cells. Leuk. Res. 16: 139-44 (1992)

LGL-CF: *large granular lymphocyte-derived cytotoxic factor* This poorly characterized factor is produced by large granular lymphocytes of cancer patients during interaction with autologous tumor cells. Supernatants produced by culture of large granular lymphocytes alone lysates of these cells do not contain detectable cytolytic activity. LGL-CF mediates lysing of autologous and allogeneic fresh human tumor cells. Treatment with monoclonal and polyclonal antibodies directed against » TNF-α, » TNF-β, » IFN-α, IFN-γ, or » IL1α, alone or in combination, do not inhibit the cytotoxic activity of this factor against fresh human tumor cells.
Uchida A & Klein E Generation of cytotoxic factor by human large granular lymphocytes during interaction with autologous tumor cells: lysis of fresh human tumor cells. JNCI 80: 1398-403 (1988); **Uchida A et al** Lysing of fresh human tumor by a cytotoxic factor derived from autologous large granular lymphocytes independently of other known cytokines. Cancer Immunol. Immunother. 31: 60-4 (1990)

LGPF: *leukemia-growth-promoting factor* This poorly characterized factor of ≈ 25 kDa is found in the conditioned medium of the thymic reticulo-epithelial-like cell line B6TE. It stimulates the growth of a murine leukemia subline (L17R). The factor is not identical with » IL1, » IL2, » IL3, and » GM-CSF. The growth of L17R leukemia cells is not only stimulated by LGPF, but also by pituitary and brain fibroblast growth factor (see: FGF). It has been suggested that LGPF may be a member of the FGF family of proteins. For other leukemia-associated growth factors see also: LC-GF, LDA, LDGF, LGF.

Miyazawa T et al Growth of a cultured leukemia subline was promoted by conditioned medium of thymic reticuloepithelial-like cells (B6TE). Leuk. Res. 7: 637-46 (1983); **Miyazawa T et al** Isolation and characteristics of a leukemia-growth-promoting factor from calf thymus. Leuk. Res. 9: 1315-21 (1985)

LH: see: Luteinizing hormone.

LHF: *LAK helper factor* also: LAK cell induction helper factor. This poorly characterized factor is produced by activated murine spleen cells. It enhances the generation of » LAK cells (lymphokine-activated killer cells) by suboptimal concentrations of » IL2. This factor is probably not identical with » IL1, » IL2, » TNF and » IFN.

Kawase I et al Augmentation of murine lymphokine-activated killer cell induction by a factor produced by Nocardia rubra cell wall skeleton-stimulated T cells. JJCR 80: 1098-1105 (1989)

L-HGF: *L929-derived hybridoma growth factor* see: HGF (hybridoma growth factor).

LIF: *leukocyte (migration) inhibitory factor* Leukocyte inhibitory factor is a lymphokine that inhibits the random and directed migration of polymorphonuclear (PMN) leukocytes. LIF also induces specific granule secretion by PMNs and potentiates many responses mediated by the chemotactic compound fMLP (formyl-methionyl-leucyl-phenylalanine). It activates human neutrophils and macrophages to release leukotriene B4 and thromboxanes.

LIF was purified initially from a human non-T, non-B leukemia cell line (Reh). One factor with LIF activity that influences the adherence of neutrophils to endothelial cells and thus facilitates their migration into the interstitial space is probably identical with » LAI (leukocyte adhesion inhibitor).

The two factors, designated ***B-LIF*** and ***T-LIF*** (B/T cell-derived leukocyte (migration) inhibitory factor) also inhibit the migration of leukocytes. They have not yet been properly characterized.

The term LIF is frequently used as an operational definition of an activity rather than for a distinct factor. See also: LMIF (lymphocyte migration inhibitory factor).

Borish L & Rocklin R Physiological studies with human leukocyte inhibitory factor. Immunol. Ser. 57: 373-85 (1992); **Conti P** Leukocyte inhibitory factor activates human neutrophils and macrophages to release leukotriene B4 and thromboxanes. Cytokine 2: 142-8 (1990); **Meshulam DH et al** Purification of a lymphoid cell line product with leukocyte inhibitory factor activity. PNAS 79: 601-5 (1982); **Rocklin RE et al** Partial characterization of a lymphoid cell line (Reh) product with leukocyte inhibitory factor (LIF) activity. JI 127: 534-9 (1981); **Rosen A et al** A T helper cell x Molt4 human hybridoma constitutively producing B cell stimulatory and inhibitory factors. LR 5: 185-204 (1986); **Salahuddin SZ et al** Lymphokine production by cultured human T cells transformed by human T cell leukaemia lymphoma virus I. S 223: 703-7 (1984); **Schainberg H et al** Leukocyte inhibitory factor stimulates neutrophil-endothelial cell adhesion. JI 141: 3055-60 (1988); **Szigeti R & Rosen A** Studies on leukocyte migration inhibitory factor (LIF) produced by activated T and B cells. LR 7: 11-20 (1988)

LIF: *leukemia inhibitory factor; myeloid leukemia inhibitory factor*

■ **ALTERNATIVE NAMES:** adult bovine aortic endothelial (ABAE) cell growth-inhibitory activity; CDF (cholinergic differentiation factor); CNDF (cholinergic neuronal differentiation factor); D factor (differentiation-stimulating factor); DIA (differentiation inhibiting activity); DIF (differentiation inducing factor); DRF (differentiation retarding factor); ES cell growth factor; ESCGF (embryonic stem cell growth factor); GATS (growth stimulatory activity for TS1 cells; HILDA (human interleukin for Da cells); HSF-2 (hepatocyte stimulating factor 2); HSF-3 (hepatocyte stimulating factor 3); Lipoprotein-Lipase-Inhibitor; M1 differentiation inducing activity; MLPLI (melanoma-derived lipoprotein lipase inhibitor); OAF (osteoclast activating factor). See also: individual entries for further information.

■ **SOURCES:** This factor is produced by various fibroblast cell lines, antigen-stimulated alloreactive T lymphocytes, mitogen-activated spleen cells, and Krebs and Ehrlich ascites cells. The factor is also produced by activated monocytes (see also: Cell activation). Human lung fibroblasts and umbilical chord endothelial cells produce LIF constitutively. The synthesis in mesenchymal cells of LIF can be induced by IL1α (see: IL1), » TGF-β, » EGF, and » bFGF.

■ **PROTEIN CHARACTERISTICS:** LIF is a heavily and variably glycosylated 58 kDa protein. with a

length of 179 amino acids. Glycosylation does not appear to be essential for bioactivity. Two different glycosylation variants have been designated as » LIF-A and » LIF-B.
The murine and human factors show a homology of 79 % at the amino acid level. Both factors show a high degree of conservative amino acid exchanges. The factor is produced as a soluble protein and also as a variant bound to the extracellular matrix (see also: ECM). Both variants arise from alternatively spliced transcripts.

■ **GENE STRUCTURE:** The human gene has a length of 6.2 kb. It contains three exons and an unusually long 3´ untranslated region of ≈ 3.2 kb. The transcript has a length of 4.1 kb.
The human gene maps to chromosome 22q12-q12.2 ≈ 15 kb away from a chromosomal breakpoint for a translocation (t(11;22) (q24;q12). This translocation is found in practically all Ewing sarcomas and peripheral neuroepitheliomas and serves as a cytogenetic marker for these two tumor types. The human gene encoding » Oncostatin M lies at a distance of ≈ 16 kb.
The murine LIF gene maps to chromosome 11 (subbands A1-A2) which is more or less syntenic with human chromosome 22q. The murine gene maps to a chromosomal region with a high incidence of chromosomal alterations in embryonic carcinoma cells.
The murine and human LIF genes show a homology of ≈ 75 % mainly restricted to the coding sequences but also including some parts of the 5´and 3´ ends of the transcription unit. An exon corresponding to the 5' end of a variant LIF transcript in the mouse that encodes a potentially matrix-associated form of LIF is not conserved in the human, ovine and porcine genes.

■ **RECEPTOR STRUCTURE, GENE(S), EXPRESSION:** Macrophages, monocytes, and their precursors express ≈ 300-500 high-affinity receptors (Kdis = 10-200 pM) per cell.
Another receptor expressed with a density of 2000-6000 copies per cell (Kdis = 1-3 nM) has also been described.
One component of the LIF receptor is a 130 kDa glycoprotein, » gp130. gp130 is also a component of the receptor for » IL6, » IL11, » CNTF (ciliary neurotrophic factor), and » oncostatin M. It functions as the signal transducer subunit for the receptor. This finding also explains that these factors share several common biological activities although the factors themselves are not related to each other.
A glycoprotein of ≈ 90 kDa, designated ***LBP*** (LIF binding protein, ***LIFRβ***) that specifically binds LIF has been isolated from normal mouse serum. It is a soluble truncated form of the α chain of the cellular receptor and probably serves as an inhibitor of the systemic effects of locally produced LIF because it acts as a competitive inhibitor of LIF binding to its cellular receptor. This protein has been found also to be involved in the formation of functional receptors for » CNTF (ciliary neurotrophic factor) and » Oncostatin M and heterodimerises with gp130.

■ **BIOLOGICAL ACTIVITIES:** The biological activities of human LIF are not species-specific; human LIF is fully active in murine cells and even binds with higher affinity to the murine LIF receptor. Murine LIF, however, binds only to the mouse receptor.
Some of the bioactivities of LIF can be mediated by the factor bound to the extracellular matrix (see: ECM) acting on nearby cells (see also: Juxtacrine). Some observations point to the fact that the diffusible form of LIF and the membrane-bound form differ in their biological activities.
LIF was initially isolated as a factor that inhibits many, but not all, myeloid leukemia cells and induces their differentiation into macrophages. For some myeloid leukemia cell lines such as » M1 a short pulse of LIF is sufficient to initiate irreversible differentiation. LIF also prolongs the » cell cycle of stem cells in some AML cell lines probably by increasing the time spent in the G_2-M-G_1 phase of the cell cycle.
LIF by itself does not possess colony-stimulating activity (see also: CSF) for early human bone marrow cells expressing the cell surface marker » CD34 (see also: CD antigens). It also does not influence the number, size, and differentiation of normal colonies induced by » GM-CSF, » G-CSF, » M-CSF and » IL3 in a » colony formation assay. The colony formation of IL3-dependent primitive blast cells, however, is promoted by LIF almost as well as by » G-CSF or » IL6. This suggests that LIF plays a role in the regulation of very early hematopoietic cells (see also: Hematopoiesis). Cultured normal human bone marrow stromal cells constitutively express LIF message. Exposure of these cells to » IL1, » TGF-β, and »TNF-α (but not » IFN α) increases the level of LIF RNA. Cultured stromal cells derived from patients with chronic myelogenous leukemia show enhanced LIF expression. LIF may participate, either alone or through interaction with other cytokines, in the bone marrow

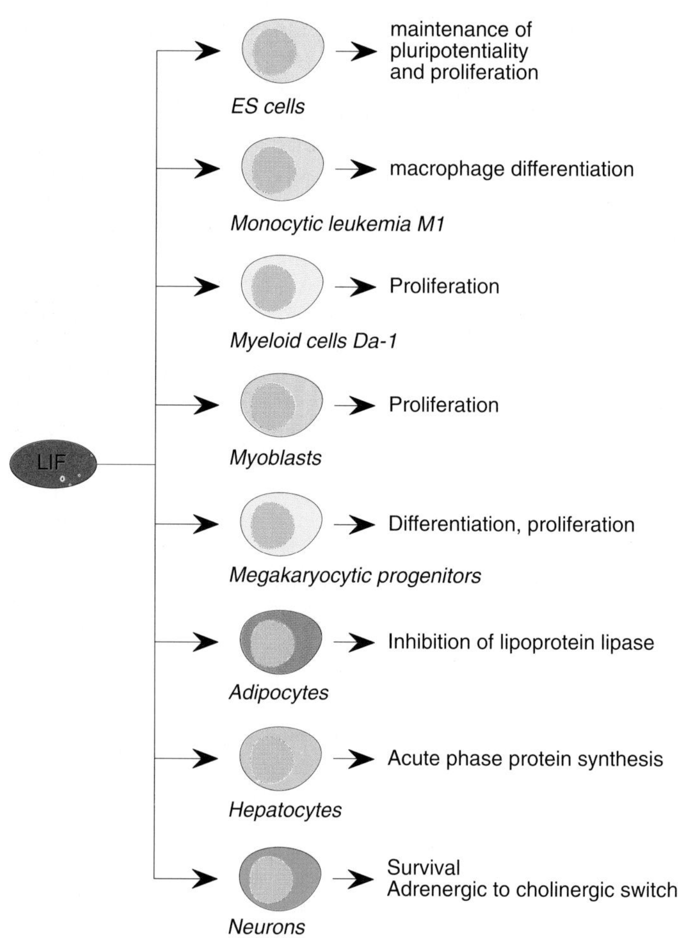

Schematic representation of the multiple activities of LIF at the cellular level.

microenvironment-mediated influence on both normal and malignant hematopoietic processes.

LIF has been shown to be a potent inhibitor of endothelial cell proliferation. LIF is a factor that inhibits adipogenesis by inhibiting the lipoprotein lipase in adipocytes. The inhibition of this enzyme probably reduces uptake of fatty acids by adipocytes and leads to catabolism of lipids in fat tissue. Mice with a high blood level of LIF suffer from weight loss and are also cachectic. In addition one also observes an accumulation of osteoblasts in the bone marrow and formation of new bone tissues and calcifications in the heart and skeletal muscles. A marked increase in the number of hematopoietic cells is found in spleen and liver. In mice injected with LIF megakaryocyte and platelet counts rise.

In cultured neuronal cells LIF influences the type of neurotransmitter expressed and induces differentiation into cell types expressing cholinergic transmitter. LIF induces the expression of choline acetyltransferase and represses, among other things, the expression of tyrosine hydroxylase and dopamine β-hydroxylase. LIF also influences the development of sensory neurons. LIF is probably involved in the regulation of growth and development of the nervous system. *In vitro* it stimulates the differentiation of embryonic neural crest cells into cell types resembling sensory neurons. It also supports the development and maintenance of such cells. LIF has been shown by » message amplification phenotyping to induce synthesis of mRNAs for choline acetyltransferase, » Somatostatin, » SP (substance P), » VIP (vasoactive

intestinal polypeptide), cholecystokinin, and enkephalin in cultured sympathetic neurons.
The expression of LIF in murine blastocysts before onset of hematopoietic differentiation suggests that LIF regulates the growth and development of trophoblasts and/or embryonic stem cells. LIF is expressed very early during the differentiation of » ES cells (embryonic stem cells) and inhibits their differentiation as long as it is present. This factor is therefore important for culturing these cells and for generating » transgenic animals.
By a mechanism requiring the synthesis of prostaglandins LIF also stimulates bone resorption and bone formation and may therefore be involved in regulating osteoblast and osteoclast functions.
LIF induces the synthesis of » acute phase proteins in hepatocytes (see also: acute phase reaction).
The expression of LIF in uterine tissues is initiated at a time corresponding to the nidation of the blastocyst. It may therefore also be involved in the growth regulation of the blastocyst and the implantation of the embryo.

■ **TRANSGENIC/KNOCK-OUT/ANTISENSE STUDIES:** The overexpression of LIF in transgenic mice has been shown to be lethal, leading to the absence of mesodermal cells and the inhibition of gastrulation in the embryos. Chimaeric mouse embryos overexpressing the diffusible form of LIF cDNA look essentially normal. Chimerae expressing LIF associated with the extracellular matrix show an abnormal proliferation of tissues and the absence of differentiated mesoderm. They have not undertaken the normal pathway of gastrulation.
LIF-deficient mice have been generated by targeted disruption of the gene in » ES cells and subsequent generation of » transgenic animals. In spite of its important functions *in vitro* for the development of » ES cells LIF does not appear to be important for the development of viable adults. Embryos of homozygous LIF-/- mice fail to implant in the uterus but can be rescued by transfer to the uterus of normal LIF+/+ or +/- mice.
The analysis of these animals shows that LIF appears to be required for the survival of the normal pool of stem cells, but not for their terminal differentiation. LIF-negative animals have dramatically decreased numbers of stem cells in spleen and bone marrow (see also: Hematopoiesis). The remaining LIF-negative stem cells are pluripotent as shown by their ability to promote long-term survival of lethally irradiated wild-type animals. LIF-negative stem cells appear to interact differently with the splenic and medullary microenvironment since the numbers of committed progenitors are reduced in the spleen but not the bone marrow. Heterozygous LIF-negative animals have an intermediate phenotype, implying that LIF has a dosage effect. The defects in stem cell number can be compensated by administration of exogenous LIF.

■ **ASSAYS:** LIF can be assayed in a » bioassay employing cell lines that depend on, or respond to, the factor (see: Da; M1; NBFL). A colorimetric assay and an ELISA assay are also available. An alternative and entirely different detection method is » Message amplification phenotyping. For further information see also subentry "Assays" in the reference section.

■ **CLINICAL USE & SIGNIFICANCE:** The enhancement of platelet counts observed in experimental animals suggest that LIF, either alone or in combination with other factors, may be valuable in the treatment of thrombocytopenias. It may also be possible to use this factor in combination with colony-stimulating factors » G-CSF and » GM-CSF to induce differentiation of myeloid leukemia cells.
LIF has been found in the urine, but not in the serum, of patients showing an acute graft-versus-host reaction following kidney transplantation. The detection of LIF may therefore be of diagnostic value.

● **REVIEWS: Gearing DP** Leukaemia inhibitory factor: does the cap fit? ANY 628: 9-18 (1991); **Gough NM & Williams RL** The pleiotropic actions of leukaemia inhibitory factor. CC 1: 77-80 (1989); **Hilton DJ & Gough NM** Leukemia inhibitory factor: A biological perspective. JCBc 46: 21-6 (1991); **Hilton DJ** LIF: lots of interesting functions. TIBS 17: 72-6 (1992); **Kurzrock R et al** LIF: not just a leukaemia inhibitory factor. Endocrinol. Rev. 12: 208-17 (1991); **Metcalf D** The leukaemia inhibitory factor (LIF). IJCC 9: 95-108 (1991); **Metcalf D** Leukemia inhibitory factor - a puzzling polyfunctional regulator. GF 7: 169-73 (1992); **Patterson PH** The emerging neuropoietic cytokine family: first CDF/LIF, CNTF and IL6; next ONC, MGF, GCSF? Curr. Opin. Neurobiol. 2: 94-7 (1992); **Rao MS & Landis SC** Cell interactions that determine sympathetic neuron transmitter phenotype and the neurokines that mediate them. J. Neurobiol. 24: 215-32 (1993)

● **BIOCHEMISTRY & MOLECULAR BIOLOGY: Budarf M et al** Human differentiation-stimulating factor (leukaemia inhibitory factor, human interleukin DA) gene maps distal to the Ewing sarcoma breakpoint on 22q. Cytogenet. Cell. Genet. 52: 19-22 (1989); **Doolittle D et al** Tandem linkage of genes coding for leukemia inhibitory factor (LIF) and oncostatin M (OSM) on human chromosome 22. Cytogenet. Cell Genet. 64: 240-244 (1993); **Gearing DP et al** Molecular cloning and expression of cDNA encoding a murine myeloid leukaemia inhibitory factor (LIF). EJ 6: 3995-4002 (1987); **Gearing DP et al** Production of leukaemia inhibitory factor in *Escherichia coli* by a novel procedure and its use in maintaining embryonic stem cells in cul-

LIF transgenics: see addendum: LIF.

ture. BT 1: 1157-61 (1989); **Gough NM et al** Molecular cloning and expression of the human homologue of the murine gene encoding myeloid leukaemia inhibitory factor. PNAS 85: 2623-7 (1988); **Gough NM et al** Molecular biology of the leukaemia inhibitory factor gene. Ciba Found. Symp. 167: 24-38 (1992); **Jeffery E et al** Close proximity of the genes for leukemia inhibitory factor and oncostatin M. Cytokine 5: 107-11 (1993); **Kola I et al** Localization of the murine leukaemia inhibitory factor gene near the centromere on chromosome 11. GF 2: 235-40 (1990); **Lowe DG et al** Genomic cloning and heterologous expression of human differentiation-stimulating factor. DNA 8: 351-9 (1989); **Lubbert M et al** Expression of leukaemia inhibitory factor is regulated in human mesenchymal cells. Leukemia 5: 361-5 (1991); **Moreau JF et al** Leukaemia inhibitory factor is identical to the myeloid growth factor human interleukin for DA cells. N 326: 201-4 (1988); **Owczarek CM et al** Inter-species chimeras of leukaemia inhibitory factor define a major human receptor-binding determinant. EJ 12: 3487-95 (1993); **Rathjen PD et al** Differentiation inhibiting activity is produced in matrix-associated and diffusible forms that are generated by alternate promoter usage. Cell 62: 1105-14 (1990); **Selleri L et al** Molecular localisation of the translocation of Ewing sarcoma by chromosomal *in situ* suppression hybridization. PNAS 88: 887-91 (1991); **Stahl J et al** Structural organization of the genes for murine and human leukaemia inhibitory factor. Evolutionary conservation of coding and non-coding regions. JBC 265: 8833-41 (1990); **Sutherland GR et al** The gene for human leukaemia inhibitory factor (LIF) maps to 22q12. Leukaemia 3: 9-13 (1989); **Wilson TA et al** Cross-species comparison of the sequence of the leukaemia inhibitory factor gene and its protein. EJB 204: 21-30 (1992)

● **Receptors: Gearing DP et al** Leukaemia inhibitory factor receptor is structurally related to the IL6 signal transducer, gp130. EJ 10: 2839-48 (1991); **Gearing DP et al** Reconstitution of high affinity leukaemia inhibitory factor (LIF) receptors in hemopoietic cells transfected with the cloned human LIF receptor. Ciba Found. Symp. 167: 245-55 (1992); **Gearing GP et al** The IL6 signal transducer, gp130: an oncostatin M receptor and affinity converter for the LIF receptor. S 255: 1434-7 (1992); **Godard A et al** High and low affinity receptors for human interleukin for DA cells/leukemia inhibitory factor on human cells. Molecular characterization and cellular distribution. JBC 267: 3214-22 (1992); **Hilton DJ et al** Distribution and binding properties of receptors for leukaemia inhibitory factor. Ciba Found. Symp. 167: 227-39 (1992); **Layton MJ et al** A major binding protein for leukemia inhibitory factor in normal mouse serum: identification as a soluble form of the cellular receptor. PNAS 89: 8616-20 (1992)

● **Biological Activities: Bernard C et al** Regulation of neurotransmitter synthesis: from neurone to gene. J. Physiol. Paris 85: 97-104 (1991); **Bhatt H et al** Uterine expression of leukemia inhibitory factor coincides with the onset of blastocyst implantation. PNAS 88: 11408-12 (1991); **Burstein SA et al** Leukemia inhibitory factor and interleukin-11 promote maturation of murine and human megakaryocytes *in vitro*. JCP 153: 305-12 (1992); **Fann MJ Patterson PH** A novel approach to screen for cytokine effects on neuronal gene expression. J. Neurochem. 61: 1349-55 (1993); **Ferrara N et al** Pituitary follicular cells secrete an inhibitor of aortic endothelial cell growth: identification as leukemia inhibitory factor. PNAS 89: 698-702 (1992); **Hirayoshi K et al** Both D factor/LIF and IL6 inhibit the differentiation of mouse teratocarcinoma F9 cells. FL 282: 401-4 (1991); **Ishimi Y et al** Leukemia inhibitory factor/differentiation-stimulating factor (LIF/D-factor): regulation of its production and possible roles in bone metabolism. JCP 152: 71-8 (1992); **Leary AG et al** Leukemia inhibitory factor differentiation-inhibiting activity/human interleukin for DA cells augments proliferation of human hematopoietic stem cells. Blood 75: 1960-4 (1990); **Lorenzo JA et al** Leukemia inhibitory factor (LIF) inhibits basal bone resorption in fetal rat long bone cultures. Cytokine 2: 266-71 (1990); **Lowe C et al** Regulation of osteoblast proliferation by leukaemia inhibitory factor. J. Bone Miner. Res. 6: 1277-83 (1991); **Maekawa T & Metcalf D** Clonal suppression of HL60 and U937 cells by recombinant human leukaemia inhibitory factor in combination with GM-CSF or G-CSF. Leukemia 3: 270-6 (1989); **Martin TJ et al** Leukaemia inhibitory factor and bone cell function. Ciba Found. Symp. 167: 141-50 (1992); **Metcalf D & Gearing DP** A fatal syndrome in mice engrafted with cells producing high levels of leukaemia inhibitory factor (LIF). PNAS 86: 5948-52 (1989); **Metcalf D et al** Effects of injected leukaemia inhibitory factor on hematopoietic and other tissue in mice. Blood 76: 50-6 (1990); **Metcalf D et al** Leukaemia inhibitory factor can potentiate murine megakaryocyte production *in vitro*. Blood 77: 2150-3 (1991); **Murphy M et al** Generation of sensory neurons is stimulated by leukemia inhibitory factor. PNAS 88: 3498-501 (1991); **Reid LR et al** Leukemia inhibitory factor: a novel bone-active cytokine. E 126: 1416-20 (1990); **Verfaillie C & McGlave P** Leukemia inhibitory factor/human interleukin for DA cells: a growth factor that stimulates the *in vitro* development of multipotential human hematopoietic progenitors. Blood 77: 263-70 (1991); **Wetzler M et al** Constitutive expression of leukemia inhibitory factor RNA by human bone marrow stromal cells and modulation by IL1, TNF-α, and TGF-β. EH 19: 347-51 (1991); **Williams RL et al** Myeloid leukaemia inhibitory factor (LIF) maintains the developmental potential of embryonic stem cells. N 336: 684-7 (1988)

● **Transgenic/Knock-out/Antisense Studies: Conquet F et al** Inhibited gastrulation in mouse embryos overexpressing the leukemia inhibitory factor. PNAS 89: 8195-9 (1992); **Escary JL et al** Leukaemia inhibitory factor is necessary for maintenance of hematopoietic stem cells and thymocyte stimulation. N 363: 361-4 (1993); **Stewart CL et al** Blastocyst implantation depends on maternal expression of leukaemia inhibitory factor N 359: 76-9 (1992)

● **Assays: Bendtzen K et al** Measurement of human IL1 by LIF induction, pancreatic islet cell cytotoxicity, and bone resorption. LR 5: S93-8 (1986); **De Groote D et al** An ELISA for the measurement of human leukemia inhibitory factor in biological fluids and culture supernatants. JIM 167: 253-61 (1994); **Kim KJ et al** Detection of human leukemia inhibitory factor by monoclonal antibody based ELISA. JIM 156: 9-17 (1992); **Ohno M & Abe T** Rapid colorimetric assay for the quantification of leukaemia inhibitory factor (LIF) and interleukin 6 (IL6). JIM 145: 199-203 (1991); **Waring P et al** Leukemia inhibitory factor levels are elevated in septic shock and various inflammatory body fluids. JCI 90: 2031-7 (1992)

● **Clinical Use & Significance: Metcalf D et al** Actions of leukaemia inhibitory factor on megakaryocyte and platelet formation. Ciba Found. Symp. 167: 174-82 (1992); **Taupin JL et al** HILDA/LIF urinary excretion during acute kidney rejection. Transplantation 53: 655-8 (1992); **Waring P et al** Leukemia inhibitory factor levels are elevated in septic shock and various inflammatory body fluids. JCI 90: 2031-7 (1992)

LIF-A: LIF-A and LIF-B are found in conditioned culture media of Krebs II ascites cells. Both factors are probably different glycosylation variants of the same protein, » LIF (leukemia inhibitory factor). LIF-A is identical with » M1 differentiation inducing activity.

Hilton DJ et al Resolution and purification of three distinct factors produced by Krebs ascites cells which have differentiation-inducing activity on murine myeloid leukemic cell lines. JBC 263: 9238-43 (1988)

LIF-B: see: LIF-a.

LIF binding protein: abbrev. LBP. See: LIF.

LILAK cells: *lectin-induced lymphokine-activated killer cells* see: LAK cells.

Limiting dilution analysis: abbrev. LDA. A type of analysis initially used in immunology to determine in a population of lymphocytic cells the unknown frequency of discrete clones of lymphocytes that respond to a specific antigen or with a particular effector function. It is the only way to assess, in a quantitative manner, immune responses in humans at the level of individual cells and the technique therefore is a valuable non-invasive prognostic and diagnostic tool that can be carried out as often as required. Limiting dilution analysis methods have subsequently been adapted for other purposes, allowing, for example, detection, quantification, and functional analysis of other individual (rare) cell types by investigating their clonal growth.

If, for example, the number of » hematopoietic stem cells in a particular preparation of bone marrow (see also: Hematopoiesis, LTBMC) is considered, the true number of stem cells in the marrow cannot be determined by simply employing a » colony formation assay to measure plating efficiencies because stem cells replicate and give rise to multiple progenitors cell types which, in turn, can give rise to multiple colonies. The observed number of colonies, therefore, is not a true measure of the number of stem cells initially present in the marrow (this is quite unlike the general procedures used in » Cell culture for "ordinary" adherent cells to determine their clone-forming efficiencies: each viable cell will give rise to a single colony and the ratio obtained by dividing the number of observed colonies by the number of inoculated cells will yield the plating efficiency as a measure of cell viability in, for example, a freshly thawed stock of cells).

To obtain a more appropriate estimate of the true number of stem cells present in the hematopoietic cell population under study various dilutions of cell suspensions are plated into microwell plates for an appropriate culture period and under conditions that are optimal for the development of the effector cell types that are being measured. As shown in the example some wells will then be found to contain one (or more) colonies while others will not. If the number of cells per microwell for each dilution is plotted against the negative logarithm of the fraction of wells *without* colonies for each dilution (i. e. those that did not contain a stem cell with colony-forming capacity) one will ideally be able to draw a straight line originating from the zero point of the coordinate axes. Poisson statistics shows that exactly one stem cell is present among the (known) number of inoculated cells if 37 % of the wells are empty (do not contain a colony). It is therefore possible to infer from the graph the number of bone marrow cells required to ensure the presence of at least one viable and replicating stem cell (for an *in vivo* assay based on the same principles of statistical analysis see: MRA (marrow repopulating ability).

The analysis of limiting dilution assays is often complicated by inter-assay variabilities and difficulties in using linear regression in comparing the fit of each analysis to a single-hit Poisson model. Gross and reproducible deviations from single-hit kinetics in a particular experiment would, for example, indicate the presence of accessory cells or agents with suppressor activities for clonal cell growth.

This type of analysis is valuable, for example, in comparing the "qualities" of different bone marrow preparations to be used for bone marrow transplantation (see also: 4-HC), in determining the effects of individual » cytokines on the ability to promote or inhibit proliferation of stem cell or individual progenitor types derived thereof, in determining the ability of a clonal » stromal cell line to replace the heterogeneous microenvironment of fresh stroma for *in vitro* stem cell support, or in determining pool sizes of particular (cytokine-producing) cells or their depletion/persistence in response to drug treatment. Limiting dilution analysis is also a relatively simple means of separating particular cell functions from the potential effects of accessory cells or suppressor factors and adsorption.

Counting of empty wells in limiting dilution assays guarantees that negative responses are scored. If positive wells were scored to estimate frequencies one would never be sure whether one or more precursors in a well were responsible for any positive responses. This type of analysis, however, can provide specific information concerning, for example, the proliferative capacities of individual cells types. Identical numbers of "empty" wells but gross differences in the total number of colonies observed in "positive" wells in a comparison of two different starting populations would suggest differences in the functional capacities of the colony-forming cell types contained in the two cell populations.

● **THEORY: Bonnefoix T & Sotto JJ** The standard $\chi 2$ test used in limiting dilution assays is insufficient for estimating the goodness-of-fit to the single-hit Poisson model. JIM 167: 21-33 (1994); **Burleson JA et al** Use of logistic regression in comparing multiple limiting dilution analyses of antigen-reactive T cells. JIM 159: 47-52 (1993); **Cobb L et al** Comparison of statistical methods for the analysis of limiting dilution assays. In Vitro Cell Dev. Biol. 25: 76-81 (1989); **Coller HA & Coller BS** Poisson statistical analysis of repetitive subcloning by the limiting dilution technique as a way of assessing hybridoma monoclonality. MiE 121: 412-7 (1986); **Lefkovits J & Weidman H** Limiting dilution analysis of cells in the immune system. Cambridge Univ. Press, London 1979; **Sette A et al** A BASIC microcomputer program for data analysis of limiting dilution assays. Comput. Appl. Biosci. Apr; 5(2): 161 (1989); **Schmehl MK et al** Power analysis of statistical methods for comparing treatment differences from limiting dilution assays. In Vitro Cell Dev. Biol. 25: 69-75 (1989); **Strijbosch LW et al** Limiting dilution assays. Experimental design and statistical analysis. JIM 97: 133-40 (1987); Waldmann H et al Limiting dilution analysis In: Klaus GGB (edt) Lymphocytes, a practical approach, chapter 8. IRL Press, Oxford 1987;

● **APPLICATIONS: Albertini MR et al** Limiting dilution analysis of lymphokine-activated killer cell precursor frequencies in peripheral blood lymphocytes of cancer patients receiving interleukin-2 therapy. J. Biol. Response. Mod. 9: 456-62 (1990); **Alp NJ et al** Automation of limiting dilution cytotoxicity assays. JIM 129: 269-76 (1990); **Barber KE et al** Human granulocyte-macrophage progenitors and their sensitivity to cytotoxins: analysis by limiting dilution. Blood 70: 1773-6 (1987); **Cerrone MC & Kuhn RE** Description of a urease-based microELISA for the analysis of limiting dilution microcultures. JIM 138: 65-75 (1991); **Chen Y et al** Studies on human CFU-Mix microvolume culture by use of limiting dilution assay. J. Tongji Med. Univ. 10: 19-22 (1990); **Deacock S et al** A rapid limiting dilution assay for measuring frequencies of alloreactive, interleukin-2-producing T cells in humans. JIM 147: 83-92 (1992); **Hibi T & Dosch HM** Limiting dilution analysis of the B cell compartment in human bone marrow. EJI 16: 139-45 (1986); **Katsura Y et al** Limiting dilution analysis of the stem cells for T cell lineage. JI 137: 2434-9 (1986); **Kelso A** An assay for colony-stimulating factor (CSF) production by single T lymphocytes: estimation of the frequency of cells producing granulocyte-macrophage CSF and multilineage CSF within a T lymphocyte clone. JI 136: 2930-7 (1986); **Lantz O et al** Persistence of donor-specific IL2-secreting cells and cytotoxic T lymphocyte precursors in human kidney transplant recipients evidenced by limiting dilution analysis. JI 144:

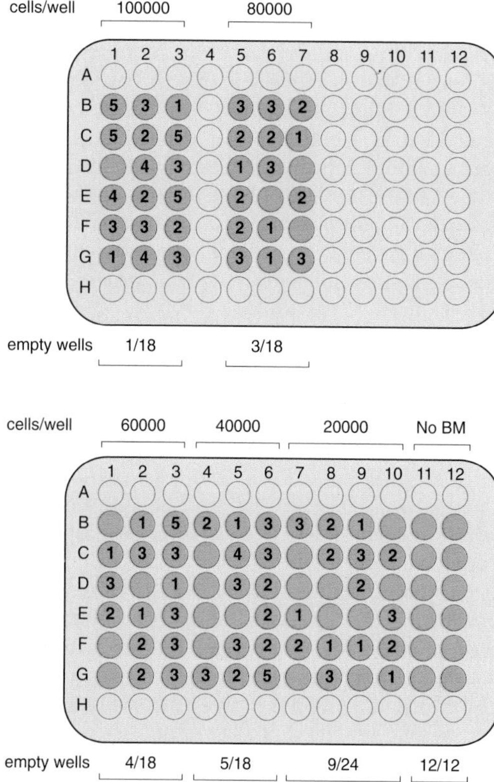

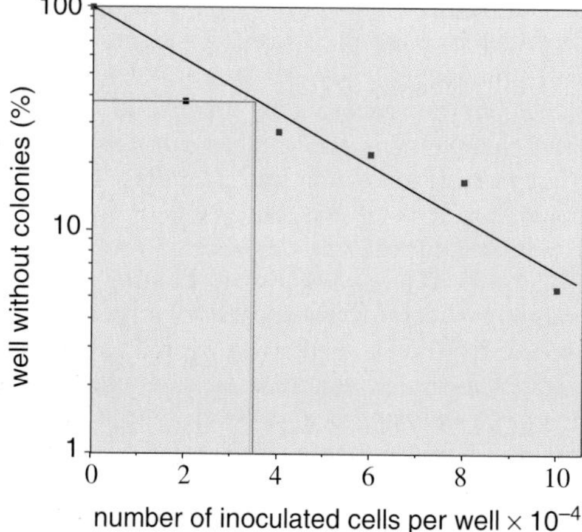

Example of a limiting dilution analysis to determine hematopoietic stem cell numbers.
An indicated number of hematopoietic cells were inoculated. The number of colonies arising after incubation is indicated in each well. Plotting of the number of empty wells for each dilution against the number of inoculated cells per well in a semilog fashion yields a curve from which the number of stem cells present in the inoculated hematopoietic cell population can be inferred.

3748-55 (1990); **Lattime EC et al** Limiting dilution analysis of TNF producing cells in C3H/HeJ mice. JI 141: 3422-8 (1988); **Mergenthaler HG & Dörmer P** Hematopoiesis studied by limiting dilution of either hematopoietic stem or stromal cells in two-step human micro long-term bone marrow cultures. ANY 628: 172-4 (1991); **Moretti L et al** Limiting dilution analysis of IL2 producing cells: I. Studies of normal human peripheral blood. Haematologica 77: 463-9 (1992); **Moretti L et al** Quantitation of T cell depletion by limiting dilution analysis. Haematologica 76: 188-92 (1991); **Neben S et al** Quantitation of murine hematopoietic stem cells *in vitro* by limiting dilution analysis of cobblestone area formation on a clonal stromal cell line. EH 21: 438-43 (1993); **Ploemacher RE et al** An *in vitro* limiting-dilution assay of long-term repopulating hematopoietic stem cells in the mouse. Blood 74: 2755-63 (1989); **Rocha B & Bandeira A** Limiting dilution analysis of interleukin-2-producing mature T cells. Interleukin 2 secretion is an exclusive property of L3T4+ lymphocytes. Scand. J. Immunol. 27: 47-57 (1988); **Sharrock CE et al** Limiting dilution analysis of human T cells: a useful clinical tool. IT 11: 281-6 (1990); **Takaue Y et al** A simplified and rapid limiting dilution assay of T lymphocytes. CI 133: 526-31 (1991); **Takaue Y et al** Cell-mediated suppression of human hematopoiesis: evaluation by limiting-dilution analysis of hematopoietic progenitors. Am. J. Hematol. 32: 205-11 (1989); **Ventura GJ et al** Analysis of peripheral blood granulocyte-macrophage colony growth by limiting dilution assay. EH 17: 125-9 (1989); **Vie H et al** Limiting dilution analysis (LDA) of cells responding to recombinant interleukin-2 without previous stimulation: evidence that all responding cells are lymphokine-activated potent effectors. Immunology. 57: 351-7 (1986); **Watanabe T et al** Quantitative analysis of the effect of colony-stimulating factors on human marrow progenitor growth in liquid-suspension cultures: application of limiting dilution assay. EH 17: 120-4 (1989)

***lin*-12:** (abbrev. for lineage-defective) This is the name of a homeotic gene found in the nematode *Caenorhabditis elegans*. This gene encodes a transmembrane protein involved in embryonic development. This protein belongs to the » EGF family of proteins and contains 11 EGF-like repeat elements.

Greenwald I *lin*-12, a nematode homeotic gene, is homologous to a set of mammalian proteins that includes epidermal growth factor. Cell 43: 583-90 (1985); **Lambie EJ & Kimble J** Two homologous regulatory genes, lin-12 and glp-1, having overlapping functions. Development 112: 231-40 (1991); **Muskavitch MAT & Hoffman FM** Homologues of vertebrate growth factors in *Drosophila melanogaster* and other vertebrates. CTDB 24: 289-328 (1990)

Lipoprotein lipase inhibitor: see: MLPLI (melanoma-derived lipoprotein lipase inhibitor)

Lipotropin: see: POMC (proopiomelanocortin).

Litorin: see: Bombesin.

Liver activator protein: abbrev. LAP. This transcription factor is identical with » NF-IL6.

L-JAK: see addendum: JAK3.

Liver-derived immunoinhibitory factor: This poorly characterized factor of 28 kDa is isolated from perfused murine livers and murine hepatocytes. The monomeric factor inhibits *in vitro* the IL2-induced proliferation of spleen cells and also the growth of rat hepatoma cells (H35 cells) and murine B16- melanoma cells.

Tzung SP et al Isolation and characterization of a novel liver-derived immunoinhibitory factor. Hepatology 14: 888-94 (1991)

Liver growth factor: see: LGF.

LK: abbrev. for » lymphokines.

L-M: This highly TNF-sensitive tumorigenic murine fibroblast cell line is used to assay » TNF-α and » TNF-β (see also: Bioassays) by measuring the degree of cell lysis induced by these factors in the presence of actinomycin D under serum-free conditions. Both factors can be distinguished from each other by using appropriate neutralizing antibodies. Recombinant TNF-β expressed in *Escherichia coli* is also cytolytic for L-M cells. Oligomeric TNF is cytotoxic towards L-M cells, whereas monomeric TNF is virtually inactive. L-M cells are 3 to 4 times more sensitive to cytolysis by TNF-β than TNF-α under serum-free assay conditions.

The cytotoxicity of recombinant human TNF against L-M cells in combination with Mitomycin C, Adriamycin, Cytosine arabinoside, Actinomycin D, Daunomycin, Cisplatin, Vincristine, and 5-Fluorouracil, based on the concentration necessary for 50% inhibition of cell growth (IC50), is 4 to 347 times as high as that of TNF alone. Micromolar concentrations of glucocorticoids render L-M cells less sensitive to the cytotoxic activity of murine TNF. The cytotoxicity of TNF on L-M cells is also reduced by lysosomotropic agents, DMSO (hydroxyl radical scavenger), NDGA (lipoxygenase inhibitor), and sodium azide (mitochondrial respiration inhibitor). The cytotoxicity of recombinant human TNF against L-M cells in incubation for 12 h at 38.5 and 40 °C based on the concentration necessary for 50% cytotoxicity is, respectively, 125 and more than 500 times as high as in similar incubation at 37 °C.

Catecholamines, 4-alkylcatechols and their diacetylated derivatives are potent inducers of the synthesis of » NGF in L-M cells.

An alternative and entirely different method of detecting IL6 is » Message amplification phenotyping.

Corti A et al Oligomeric tumor necrosis factor α slowly converts into inactive forms at bioactive levels. BJ 284: 905-10 (1992); **Furakawa Y et al** Stimulatory effect of 4-alkylcatechols and their diacetylated derivatives on the synthesis of nerve growth factor. Biochem. Pharmacol. 40: 2337-42 (1990); **Hahn CJ et al** Effect of human recombinant tumor necrosis factor on the growth of different human and mouse long-term hematopoietic cell lines. J. Leukoc. Biol. 40: 21-8 (1986); **Kramer SM et al** Comparison of TNF-α and TNF-β cytolytic biological activities in a serum-free bioassay. LR 5: S139-43 (1986); **Kramer SM & Carver M** E. Serum-free *in vitro* bioassay for the detection of tumor necrosis factor. JIM 93: 201-6 (1986); **Kull FC Jr** Reduction in tumor necrosis factor receptor affinity and cytotoxicity by glucocorticoids. BBRC 153: 402-9 (1988); **Fomsgaard A et al** Detection of tumor necrosis factor from lipopolysaccharide-stimulated human mononuclear cells by enzyme-linked immunosorbent assay and cytotoxicity bioassay. Scand. J. Immunol. 27: 143-7 (1988); **Meager A et al** Assays for tumor necrosis factor and related cytokines. JIM 116: 1-17 (1989); **Nargi FE & Yang TJ** Optimization of the L-M cell bioassay for quantitating tumor necrosis factor α in serum and plasma. JIM 159: 81-91 (1993); **Watanabe N et al** Synergistic cytotoxicity of recombinant human TNF and various anti-cancer drugs. Immunopharmacol. Immunotoxicol. 10: 117-27 (1988); **Watanabe N et al** Cytocidal mechanism of TNF: effects of lysosomal enzyme and hydroxyl radical inhibitors on cytotoxicity. Immunopharmacol. Immunotoxicol. 10: 109-16 (1988); **Watanabe N et al** Synergistic cytotoxic and antitumor effects of recombinant human tumor necrosis factor and hyperthermia. CR 48: 650-3 (1988)

LMF: *lymphocyte migration factor(s)* These poorly characterized factors of 30-70 kDa are produced by tumor-infiltrating neutrophils. They mediate the infiltration of cytotoxic and suppressor T lymphocytes into the tumor mass.

Shijubo N et al Two distinct mechanisms involved in the infiltration of lymphocytes into tumors. JJCR 79: 1111-8 (1988); **Shijubo N et al** Functional analysis of mononuclear cells infiltrating into tumors. V. A soluble factor involved in the regulation of cytotoxic/suppressor T cell infiltration into tumors. JI 142: 2961-7 (1989); **Yamaki T et al** Functional analysis of mononuclear cells infiltrating into tumors. III. Soluble factors involved in the regulation of T lymphocyte infiltration into tumors. JI 140: 4388-96 (1988)

LMF: *lymphocyte mitogenic factor* This factor is identical with » IL2.

In addition there are other poorly characterized factors with LMF activity identified in human cell supernatants that are not identical with » IL1 or » IL2 and that function as a B cell differentiation factor (see: BCDF).

Duncan MR et al Concanavalin A-induced human lymphocyte mitogenic factor: activity distinct from interleukin 1 and 2. JI 129: 56-62 (1982); **Postlethwaite AE** Isolation and characterization of lymphocyte mitogenic factor released *in vivo* during cell-mediated immune reaction in the guinea pig. JI 125: 1955-8 (1981)

LMIF: *lymphocyte migration inhibitory factor* This poorly characterized factor of ≈14 kDa exists as a monomer and a d. It is produced by macrophages and inhibits the migration of lymphocytes. The term is also used as an operational definition for a biological activity rather than for a distinct factor. See also: LIF (leukocyte (migration) inhibitory factor).

Ye S & Cheung HT Regulation of lymphocyte motility by macrophages: characterization of a lymphocyte migration inhibitory factor derived from a macrophage-like cell line. CI 122: 231-43 (1989)

LMMIF: *low molecular mass inhibitor factor* see: HMMIF.

LMSF: *leukocyte migration stimulation factor* This term is used as an operational definition for a biological activity promoting the migration of leukocytes rather than for a distinct factor.

LMW-BCGF: *low molecular weight B cell growth factors* This activity is isolated from activated T cells (see also: Cell activation). It is also referred to as ***BCGF-L*** and ***BCGF-12KD***. Most of the activities described as LMW-BCGF are poorly characterized factors that act as B cell growth factors (see also: BCGF for some problems of nomenclature, and » JR-2(82)). The results of most early studies employing LMW-BCGF must be interpreted with caution since it was shown that the commercially available preparation frequently used in these studies contained » IL2, » IFN-γ, » TNF-α and » GM-CSF.

One particular LMW-BCGF produced by T cells is BCGF-12KD (12 kDa) the sequence of which is not identical to any other known cytokine acting on B cells. The human gene encoding BCGF-12KD maps to chromosome 1q23-25, a region which has been reported to be subject to translocations in B cell leukemias. In humans LMW-BCGF also acts as a T cell replacing factor (see: TRF) that supports the long-term growth of anti-IgM-stimulated B cells (BCDF, B cell differentiation factor; for another system allowing long-term growth of normal B cells see also CD40). The factor is also an » autocrine growth factor *in vivo* and *in vitro* for B-CLL (chronic lymphocytic B cell leukemia). The growth of these cells is inhibited by using a specific antibody directed against BCGF-12KD. Antibodies directed against BCGF-12kD also inhibit the *in vitro* growth of several EBV-transformed B cell lines, suggesting that it functions as an » autocrine growth factor for these cells. BCGF-12kD also appears to alter a regulatory step which may be

involved in the EBV transformation of B cells since culture of EBV-transformed cell lines results in a time-dependent reduction of episomal EBV sequences. LMW-BCDF has been shown to regulate » CD23 gene expression. LMW-BCGF is also capable of inducing poke weed mitogen-activated IgM-positive/A- B cells to switch to IgA synthesis.
Bertolini JN et al Human low molecular weight B cell growth factor induces surface IgM^+/A- B cells to express and secrete IgA. JI 149: 1771-8 (1992); **Butler JL et al** Development of a T cell hybridoma secreting B cell growth factor. JEM 157: 60-8 (1983); **Butler JL et al** Development of a human T cell-hybridoma secreting separate B cell growth and differentiation factors. PNAS 81: 2475-8 (1984); **Defrance T & Banchereau J** Role of cytokines in the ontogeny, activation, and proliferation of B lymphocytes. In: Callard RE (edt) Cytokines and B lymphocytes, pp. 64-114, Academic Press, London 1990; **Ford RJ et al** Soluble factor activation of human B lymphocytes. N 294: 261-3 (1981); **Fournier S et al** Low-molecular weight B cell growth factor (BCGF-12KD) as an autocrine growth factor in B cell chronic lymphocytic leukemia. EJI 22: 1927-30 (1992); **Kolb JP et al** Intracellular signaling events associated with the induction of DNA synthesis in human B lymphocytes. I. Stimulation of PKC-dependent and -independent pathways by LMW-BCGF. CI 146: 117-30 (1993); **Kumar A et al** Human BCGF-12kD functions as an autocrine growth factor in transformed B cells. ECN 1: 109-13 (1990); **Mehta SR et al** Purification of human B cell growth factor. JI 135: 3298-302 (1985); **Mehta SR et al** Characterization of the cell surface receptors for human B cell growth factor of 12,000 molecular weight. JI 137: 2210-4 (1986); **Morgan J et al** Negative modulation of Epstein-Barr virus episomes by a human B cell growth factor. J. Virol. 63: 3190-4 (1989); **Ogawa N et al** Abnormal production of B cell growth factor in patients with systemic lupus erythematosus. Clin. Exp. Immunol. 89: 26-31 (1992); **Sharma S et al** Molecular cloning and expression of a human B cell growth factor gene in *Escherichia coli*. S 235: 1489-92 (1987); **Uckun FM et al** B cell growth factor receptor expression and B cell growth factor response of leukemic B cell precursors and B lineage lymphoid progenitor cells. Blood 70: 1020-4 (1987); **Wörmann B et al** Low molecular weight B cell growth factor induces proliferation of human B cell precursor acute lymphoblastic leukemias. Blood 70: 132-8 (1987); **Yang MP et al** Proliferative responses of a bovine leukemia virus-infected lymphoblastoid B cell line by its culture supernatant and cytokines. J. Vet. Med. Sci. 54: 255-9 (1992); **Yoshizaki K et al** Characterization of human B cell growth factor (BCGF) from a cloned T cell or mitogen-stimulated T cells. JI 130: 1241-6 (1983)

LNGFR: *low affinity nerve growth factor receptor* A low-affinity receptor for » NGF. This receptor also binds some » neurotrophins and is therefore sometimes referred to as ***LANR*** (low-affinity neurotrophin receptor).
Rodriguez-Tebar A et al Binding of neurotrophin-3 to its neuronal receptors and interactions with nerve growth factor and brain-derived neurotrophic factor. EJ 11: 917-22 (1992)

Local hormones: see: autacoids.

Long-term bone marrow culture: see: LTBMC.

Long-term culture-initiating cells: see: LTC-IC. See also: LTBMC (long-term bone marrow culture).

Long-term repopulating cells: see: LTRC. See also: MRA (marrow repopulating ability) for an assay system).

Low affinity nerve growth factor receptor: see: LNGFR.

Low-affinity neurotrophin receptor: abbrev. LANR. see: LNGFR (low affinity nerve growth factor receptor).

Low affinity platelet factor 4: abbrev. LA-PF4. see: Beta-Thromboglobulin.

Low molecular mass inhibitor factor: see: HMMIF.

Low molecular weight B cell growth factor: see: LMW-BCGF.

LP: *leukocytic pyrogen* Leukocytic pyrogen is an endogenous mediator of fever. It is synthesized and released from mononuclear phagocytes following activation by several microbial and immunologically-derived substances. LP is identical with » IL1. See also: EP (endogenous pyrogens).
Baracos V et al Stimulation of muscle protein degradation and prostaglandin E2 release by leukocytic pyrogen (interleukin-1). A mechanism for the increased degradation of muscle proteins during fever. NEJM 308: 553-8 (1983); **Sauder DN et al** Chemotactic cytokines: the role of leukocytic pyrogen and epidermal cell thymocyte-activating factor in neutrophil chemotaxis. JI 132: 828-32 (1984)

LP-1: *lymphopoietin 1* abbrev. also Lpo. This factor is identical with » IL7.
Chantry D et al Interleukin 7 (murine pre-B cell growth factor(lymphopoietin 1) stimulates thymocyte growth: regulation by transforming growth factor β. EJI 19: 783-6 (1989); **Namen A et al** B cell precursor growth-promoting activity. Purification and characterization of a growth factor active on lymphocyte precursors. JEM 167: 988-1002 (1988)

LPF: *lymphocyte proliferation factor* This factor is identical with » IL2.

Lpo: *lymphopoietin* see: LP-1.

lpr: *lymphoproliferative* A recessive mutation observed during inbreeding in a stock derived

from mouse strains LG, AKR, C3H, and C57BL/6. The mutation gld (generalized lymphoproliferative disease) produces an abnormal T cell population indistinguishable from that of lpr.
Homozygotes of inbred MRL/Mp-lpr/lpr mice develop an autoimmune syndrome with massive lymphoproliferation and immune complex glomerulonephrosis, serologically and pathologically resembling human systemic lupus erythematosus. There is enhanced T helper cell activity and early-onset generalized polyclonal B cell activation. One observes an excess of abnormal non-malignant T cells (TCRα/β$^+$, Thy-1$^+$, Lyt-1$^+$ (= CD5), Ly-5$^+$ (= CD45 = B220), CD2$^-$, CD4$^-$, Lyt-2$^-$ (= CD8)) coexisting with normal T cells. The vast majority of peripheral T cells of lpr/lpr mice express the cell surface glycoprotein CD44 (pgp-1), which has been shown to be a marker of murine memory T lymphocytes. These cells produce increased concentrations of » IFN-γ and » IL4. There is a progressive defect in the ability of lymphocytes to produce and respond to » IL2, a general increase in the production of » IL3, increased production of » IL4 by hepatic lymphocytes. A gradual age-related increase in the production of » IL6 has also been described.
One of the underlying causes of the immunological abnormalities is a defect in the Fas antigen (see: Apo-1). This cell surface protein mediates » apoptosis and may play an important role in the negative selection of autoreactive T cells in the thymus.
Persistently high levels of serum » M-CSF in MRL-lpr mice, not present in normal mice, have been reported. M-CSF transcripts in MRL-lpr renal cortex appear to increase with an increase in the severity of lupus nephritis. Cultured MRL-lpr glomerular mesangial cells, however, have lower basal secreted and membrane bound M-CSF.
Polymorphonuclear leukocytes from MRL/lpr mice exhibit a marked defect in the amplification of FcR-mediated phagocytosis stimulated by various inflammatory mediators. This defect is acquired and correlates with the onset of the autoimmune disease observed in this strain. The cells also exhibit a defect in extravasation into the thioglycollate-inflamed peritoneum, and these effects can be attributed to increased levels of » TGF-β.
Patients with a progressive lymphoproliferative disorder associated with autoimmunity resembling the lymphoproliferative/autoimmune disease seen in lpr and gld mice have recently been described.

Brennan DC et al Cultured mesangial cells from autoimmune MRL-lpr mice have decreased secreted and surface M-CSF. Kidney Int. 42: 279-84 (1992); **Budd RC et al** Elevated production of interferon-γ and interleukin 4 by mature T cells from autoimmune lpr mice correlates with Pgp-1 (CD44) expression. EJI 21: 1081-4 (1991); **Cohen PL & Eisenberg RA** Lpr and gld: single gene models of systemic autoimmunity and lymphoproliferative disease. ARI 9: 243-69 (1991); **Cohen PL & Eisenberg RA** The lpr and gld genes in systemic autoimmunity: life and death in the Fas lane. IT 13: 427-8 (1992); **Davidson WF et al** Dissociation of severe lupus-like disease from polyclonal B cell activation and IL 2 deficiency in C3H-lpr/lpr mice. JI 133: 1048-56 (1984); **Davidson WF et al** Phenotypic, functional, and molecular genetic comparisons of the abnormal lymphoid cells of C3H-lpr/lpr and C3H-gld/gld mice. JI 126: 4075-84 (1986); **Davidson WF et al** Cytokine secretion by C3H-lpr and -gld T cells. Hypersecretion of IFN-γ and tumor necrosis factor-α by stimulated CD4$^+$ T cells. JI 146: 4138-48 (1991); **Gresham HD et al** Defective neutrophil function in the autoimmune mouse strain MRL/lpr. Potential role of transforming growth factor-β. JI 146: 3911-21 (1991); **Magilavy DB et al** Liver of MRL/lpr mice contain interleukin-4-producing lymphocytes and accessory cells that support the proliferation of Th2 helper T lymphocyte clones. EJI 22: 2359-65 (1992); **Matsuzaki Y et al** Evidence for the existence of two parallel differentiation pathways in the thymus of MRL lpr/lpr mice. JI 149: 1069-74 (1991); **Morse HC III et al** Abnormalities induced by the mutant gene lpr: expansion of a unique lymphocyte subset. JI 129: 2612-5 (1982); **Palacios R** Spontaneous production of interleukin 3 by T lymphocytes from autoimmune MRL/Mp-lpr/lpr mice. EJI 14: 599-605 (1984); **Sneller MC et al** A novel lymphoproliferative/autoimmune syndrome resembling murine lpr/gld disease. JCI 90: 334-41 (1992); **Tang B et al** Age-associated increase in interleukin 6 in MRL/lpr mice. Int. Immunol. 3: 273-8 (1991); **Van Houten N & Budd RC** Accelerated programmed cell death of MRL-lpr/lpr T lymphocytes. JI 149: 2513-7 (1992); **Wofsy D et al** Deficient interleukin 2 activity in MRL/Mp and C57BL/6J mice bearing the lpr gene. JEM 154: 1671-80 (1981); **Yui MA et al** Increased macrophage colony-stimulating factor in neonatal and adult autoimmune MRL-lpr mice. Am. J. Pathol. 139: 255-61 (1991)

LPS: lipopolysaccharide (a) any compound consisting of covalently linked lipids and polysaccharides. (b) more frequently used to denote a component of bacterial cell walls. See: Endotoxin.

LPS binding proteins: see: Endotoxin.

LR: abbrev. for » Leukoregulin.

LRG: abbrev. for » Leukoregulin.

LRIF: (natural killer) lysis resistance inducing factor see: NK-RIF (natural killer resistance inducing factor).

LSC: *lymphoid stem cell* Term used to describe a precursor cell type in the lymphoid lineage that gives rise to pro-B and pre-B cells which, in turn, develop into B cells and plasma cells (see also:

Hematopoiesis). The cells are characterized by expressing the cell surface proteins CD10, CD38 and HLA-DR as markers.

LSCF: *lung squamous cell carcinoma-derived immunosuppressive factor* This poorly characterized factor of ≈ 22 kDa is produced by the lung squamous carcinoma cell line QG56. LSCF is a tumor-derived immunosuppressive factor that inhibits proliferation, cytotoxicity, and expression of cytokine mRNA of T cells in a dose-dependent manner. The activity of the LSCF is not blocked by anti-TGF-β sera, and LSCF does not suppress the proliferation of TGF-β-sensitive mink lung cells (Mv1Lu).

Yoshino I et al Characterization of lung squamous cell carcinoma-derived T-cell suppressive factor. Cancer 72: 2347-57 (1993)

LT: *lymphotoxin* This factor is identical with » TNF-β. See also: TNF. It has been suggested that LT be renamed LT-α since a novel protein, » LT-β, forms heteromeric complexes with the original lymphotoxin.

Aggarwal BB et al Primary structure of human lymphotoxin derived from 1788 lymphoblastoid cell line. JBC 260: 2334-44 (1985)

LT-α: *lymphotoxin* α see: LT.

LT-β: *lymphotoxin* β see » TNF-β. This glycoprotein of 33 kDa is found on the surface of activated T cells, B cells, and LAK cells as a complex with LT-α (= TNF-β) thus providing a means of anchoring TNF in the membrane.

LT-β displays significant homology with » TNF-α, » TNF-β, and the ligand for » CD 40 (see: TRAP, TNF-related activation protein). The LT-β gene maps next to the TNF-α/TNF-β locus in the major histocompatibility complex.

Androlewicz MJ et al Lymphotoxin is expressed as a heteromeric complex with a distinct 33 kDa glycoprotein on the surface of an activated human T cell hybridoma. JBC 267: 2542-47 (1992); **Browning JL et al** Lymphotoxin and an associated 33 kDa glycoprotein are expressed on the surface of an activated human T cell hybridoma. JI 147: 1230-37 (1991); **Browning JL et al** Lymphotoxin β, a novel member of the TNF family that forms a heterodimeric complex with lymphotoxin on the cell surface. Cell 72: 847-56 (1993)

LTBMC: *long-term bone marrow culture* The long-term culture of bone marrow cells employs primary adherent layers of stromal cells as competent regulatory » feeder cells. Maintenance of » hematopoiesis *in vivo* and *in vitro* critically depends on the functional integrity of a heterogeneous population of stromal cells. These cells form the so-called hematopoietic inductive microenvironment that maintains the functional integrity of this complex system of resident and circulating hematopoietic cells. Stromal cells are an important source of » cytokines and low molecular weight substances required for the controlled differentiation and proliferation of hematopoietic progenitor cells and provide a complex functional » ECM (extracellular matrix) allowing direct cell-to-cell contacts between different cell types. In addition, at least some cytokines have been found to be expressed on the cell membranes of stromal cells or to bind to their extracellular matrix, suggesting that these cytokines could function as » paracrine mediators. Some experiments in which bone marrow cells in culture are separated from the stromal layer by a microporous membrane indicate that, in addition to known early-acting cytokines (IL3, SCF, LIF, or G-CSF), stroma cells also produce other soluble factors. These can either alone or in synergy with defined cytokines conserve primitive stem cells, induce early differentiation of a fraction of the primitive progenitors, and prevent their terminal differentiation.

The term "***stromal cells***" is used rather loosely and the true histogeneic origin of these mesenchyme-derived cells is still uncertain. Stromal cells include, among others, macrophages, fibroblasts, adipocytes, and endothelial cells and are frequently defined as non-hematopoietic cells. They generally do not express T lymphocyte markers (CD2, CD3, CD4, CD8), B lineage markers (B220 = CD45, CD19, CD20), and myeloid markers (CD14, CD15, Gr-1, CD11b = Mac-1, Mac-2). Instead, stromal cell lines usually express many antigens commonly associated with non-hematopoietic cells.

Adherent stromal cell layers in LTMC are established by growing bone marrow cells over a period of several weeks. These are then irradiated to permanently eliminate resident hematopoietic cells, before they are seeded with new bone marrow or progenitor cells enriched by various techniques (see, for example: fluorouracil, 4-HC, Ficoll gradient centrifugation). This process of reinitiation of LTBMC is called ***recharging***. The heterogeneous composition of the stroma makes it extremely difficult to analyze the role of individual cell types in hematopoietic development. To study the precise role of stromal cells in hematopoiesis numerous ***stromal cell lines*** have therefore been established from bone marrow and a variety of other tissues,

LT-β: see addendum.

including spleen, thymus, and non-hematopoietic organs as kidney, lung, skin, or mammary tumors. It has not yet been established conclusively that cultured stromal cells encompass all types of stromal cells identified *in vivo* and that they retain all of their functional properties *in vitro*. It also remains to be clarified whether different morphologies represent different cell types differing in function and/or lineage. Cloned lines have been observed to spontaneously generate sublines that can differ dramatically in their functional capacities. The predominant function of the particular organ from which these different stromal cell lines were derived is not necessarily reflected in their *in vitro* capacities. It has been shown that stromal cell lines, independent of their morphological features, like inhomogeneous primary stromal cell layers established from bone marrow, can be used to maintain hematopoietic cell development *ex vivo* for prolonged periods of time. Stromal cell lines constitutively produce a number of cytokines, including » GM-CSF, » G-CSF, » M-CSF, » SCF, » LIF, » IL1, » IL6, » IL7, » IL8, » IL11, » IGF, » TGF-β, and usually respond to external stimuli by changing their pattern of cytokine secretion (for an analysis technique see: Message amplification phenotyping). Apart from being useful in maintaining early progenitor cells without the loss of reconstituting capacity immortalized bone marrow stromal cell line are also potential sources of novel hematopoietic growth factors (see for example: AGIF, adipogenesis inhibitory factor).

While a number of spontaneously immortalized murine stromal cell lines supporting long-term development of hematopoietic cells *in vitro* are available there are comparatively few reports of corresponding human lines. Most human stromal cell lines have been established by immortalisation with human adenoviruses, Simian virus SV40, or by the introduction of the SV40-derived T antigen (see: KM102, SC-MSC, PK-2, SCL1-24, ST-1). Although they have been used to analyze stromal cell-progenitor cell interactions these cell lines still have major drawbacks in that they cannot be passaged indefinitely without undergoing a characteristic crisis leading to cell death beyond a certain number of cell generations and in that they cannot be inhibited by irradiation without detachment from culture flasks. In addition, many of the early human stromal cell lines show a decreased contact inhibition (see also: Cell culture) and have not been used widely to analyze hematopoietic proliferation and differentiation.

The recent development of new human stromal cell lines that persist as growth-arrested adherent feeder cell layers following ionizing irradiation (≥ 20 Gy) and support long-term bone marrow cultures of enriched » CD34-positive cells will probably facilitate studies of feeder dependence of hematopoietic cell development in long-term cultures (see: L87/4).

Two types of culture techniques allowing establishment long-term cultures of hematopoietic cells are generally distinguished. The ***Whitlock-Witte system*** was initially developed for murine bone marrow to obtain stromal layers devoid of hematopoietic cells. This culture system uses a "poor" culture medium containing 5 % fetal calf serum without cortisone. It is a lymphoid culture system and permits the growth of freshly isolated bone marrow cells that form confluent adherent stromal cell layers within 2-3 weeks, which support the growth of B lymphocytes. Lymphoid cell growth becomes apparent by 3-5 weeks. Some modifications in cell culture parameters also allow the selective proliferation and differentiation of all developmental stages of pre-B cells and B lymphocytes. Cultured cells can reconstitute the B lymphocyte compartment in immune-compromised mice. Whitlock-Witte cultures do not maintain primitive multilineage hematopoietic precursors such as » CFU-S.

Another culture technique, known as ***Dexter-type long-term culture***, is a myelopoietic culture system that allows maintenance of murine and human hematopoiesis *in vitro* over a period of several months. This system is particularly useful to maintain myeloid progenitor cells. Dexter-type culture systems differ from Whitlock-Witte cultures by making use of a rich culture medium containing high concentrations of horse serum and hydrocortisone and lower incubation temperatures. While Dexter-type cultures do not allow differentiation or proliferation of lymphoid cells, their stroma supports the maintenance of B and T lymphocyte precursors and pluripotent stem cells such as » CFU-S.

It has been shown that mouse Dexter cultures can be switched to Whitlock-Witte conditions by exchanging the composition of the medium. Such ***switch cultures*** are characterized by a cessation of myelopoiesis and the appearance of B lymphoid cells. Purified stromal layers initiated under Whitlock-Witte conditions can be switched to Dexter conditions and then can sustain myelopoiesis, demonstrating that all stromal cell components required to support myelopoiesis are maintained in both types of cultures. Several long-term culture systems employing various cytokines and perfusion bioreactor systems have also been described. They appear to yield enough human bone marrow progenitor cells to be used either for marrow transplantation therapies or gene therapy. LTBMC is regarded as an experimentally accessible *in vitro* model of » hematopoiesis *in vivo*. Reconstitution of long-term multilineage hematopoiesis in irradiated long-term bone marrow cultures has been proposed to be analogous to reconstitution of irradiated whole animals (see also: MRA, marrow repopulating ability). The continued retrieval of hematopoietic progenitor cells such as » CFU-GEMM, » BFU-E, » CFU-E, or »CFU-GM from long-term marrow cultures (LTMC) is usually taken as evidence for the presence of more primitive hematopoietic stem cells that proliferate and differentiate in these cultures. Sophisticated analyses such as » limiting dilution analysis of particular importance to rule out that the continued presence of progenitors in such cultures merely results from survival and/or limited self-renewal of progenitor cells present when the culture was initiated. Long term bone marrow cultures have been used to develop assay systems for the purification and characterization of pluripotent stem cells and early precursors defined as » CFU-S (colony forming units spleen), » CAFC (cobblestone area forming cells), » LTC-IC (long-term culture initiating cells), and » HPP-CFC (high proliferative potential colony-forming cells) thought to be *in vitro* equivalents of stem cells with self-renewal capacities and the ability to repopulate irradiated bone marrow (see: MRA, marrow-repopulating ability). Long-term culture systems reproducing all aspects of marrow function are also used to characterize the various cell populations of bone marrow responsible for the orderly sequence of proliferation and differentiation events maintaining blood cell numbers and to elucidate the mechanisms underlying gene expression changes and molecular signaling events associated with hematopoiesis.

● Reviews: **Dexter TM et al** Long-term marrow culture: an overview of techniques and experiences. In: Wright DG & Greenberger J (eds) Long-term bone marrow culture, Alan R. Liss, 1984; **Dexter TM & Spooncer E** Growth and differentiation in the hemopoietic system. ARC 3: 423-41 (1987); **Dorshkind K** Regulation of hemopoiesis by bone marrow stromal cells and their products. ARI 8: 111-37 (1990); **Eaves CJ et al** Molecular analysis of primitive hematopoietic cell proliferation control mechanisms. ANY 628: 298-306 (1991); **Eaves CJ et al** Regulation of primitive human hematopoietic cells in long-term marrow culture. Semin. Hematol. 28: 126-31 (1991); **Eaves CJ et al** Methodology of long-term culture of human hemopoietic cells. J. Tiss. Cult. Meth. 13: 55-62 (1991); **Golde DW & Takaku F** (eds) Haematopoietic stem cells. M. Dekker, New York, 1985; **Metcalf D** Hemopoietic colonies: *in vitro* cloning of normal and leukaemic cells. Recent Res. Cancer Res. 61: 1-227 (1977); **Metcalf D** Clonal culture of hemopoietic cells: techniques and applications. Elsevier, Amsterdam 1984; **Wright DG & Greenberger JS** (eds) Long term bone marrow culture. Alan Liss, New York 1984

● Stromal Cell Lines: **Aizawa S et al** Establishment of a variety of human bone marrow stromal cell lines by the recombinant SV40-adenovirus vector. JCP 148: 245-51 (1991); **Deryugina EI & Müller-Sieburg CE** Stromal cells in long-term cultures: keys to the elucidation of hematopoietic development? Crit. Rev. Immunol. 13: 115-50 (1993); **Kagami Y et al** Genetically modified human bone marrow stromal cells using a retroviral vector carrying the human GM-CSF gene. PCBR 377: 309-14 (1992); **Novotny JR et al** Cloned stromal cell lines derived from human Whitlock/Witte-type long-term bone marrow cultures. EH 18: 775-84 (1990); **Zipori D & Tamir M** Stromal cells of hemopoietic origin. IJCC 7: 281-91 (1989); for a list of murine stromal cells see review by Deryugina EI & Müller-Sieburg CE. Further references for cell lines mentioned here see individual cell lines.

● Culture Techniques: **Chang J et al** Long-term bone marrow cultures: their use in autologous marrow transplantation. Cancer Cells 1: 17-24 (1989); **Coutinho LH et al** The use of cultured bone marrow cells in autologous transplantation. PCBR 333: 415-32 (1990); **Dexter TM et al** Conditions controlling the proliferation of hemopoietic stem cells *in vitro*. JCP 91: 335-44 (1977); **Dexter TM et al** Maintenance of hemopoietic stem cells and production of differentiated progeny in allogeneic and semi-allogeneic bone marrow chimeras *in vitro*. JEM 145: 1612-6 (1977); **Fibach E & Rachmilewitz EA** The two-step liquid culture: a novel procedure for studying maturation of human normal and pathological erythroid precursors. Stem Cells Dayt. 11: s36-41 (1993); **Gartner S & Kaplan H** Long-term culture of human bone marrow cells. PNAS 77: 4756-9 (1980); **Hocking WG & Golde DW** Long-term human bone marrow cultures. Blood 56: 118-24 (1980); **Kittler EL** Biologic significance of constitutive and subliminal growth factor production by bone marrow stroma. Blood 15: 79: 3168-78 (1992); **Koller MR et al** Expansion of primitive human hematopoietic progenitors in a perfusion bioreactor system with IL3, IL6, and stem cell factor. BT 11: 358-63 (1993); **Mergenthaler HG & Dörmer P** Hematopoiesis studied by limiting dilution of either hematopoietic stem or stromal cells in two-step human micro long-term bone marrow cultures. ANY 628: 172-4 (1991); **Palsson BO et al** Expansion of human bone marrow progenitor cells in a high cell density continuous perfusion system. BT 11: 368-72 (1993); **Quesenberry P et al** Long-term marrow cultures: human and murine systems. JCBc 45: 273-8 (1991); **Schultz JS et al** Enhancement of cell production in long-term bone marrow culture. IJCC 10: 161-5 (1992); **Varma A et al** Can Dexter cultures support stem cell proliferation? EH 20: 87-91 (1992); **Verfaillie CM** Soluble factor(s) produced by human bone marrow stroma increase cytokine-induced proliferation and maturation of primitive hematopoietic progenitors while preventing their terminal differentiation. Blood 82: 2045-53 (1993); **Weilbaecher K et al** Culture of phenotypically defined hematopoietic stem cells and other progenitors at limiting dilution on Dexter monolayers. Blood 78: 945-52 (1991); **Whitlock CA & Witte ON** Long-term culture of B lymphocytes and their precursors for murine bone marrow. PNAS 79: 3608-12 (1982); **Whitlock CA & Witte ON** Long term culture of murine bone marrow precursors of B lymphocytes. MiE 150: 275-86 (1987); **Whitlock C et al** *In vitro* analysis of murine B cell development. ARI 3: 213-35 (1985); **Whitlock CA et al** Murine B cell lymphopoiesis in long term culture. JIM 67: 353-69 (1984)

LTBP: *latent TGF binding protein* see: TGF-β.

LTC: *long-term culture* see: Cell culture. See also: LTBMC (long-term bone marrow culture).

LTC-IC: *long-term culture-initiating cells* Most, if not all, colony-forming cells, which give rise to colonies of mature progeny within one to three weeks in » colony formation assays, represent an intermediate stage of hematopoietic progenitor that rapidly disappears as they differentiate or die. LTC-ICs allow reconstitution of long-term multilineage hematopoiesis in irradiated long-term bone marrow cultures (see: LTBMC) when cultured on supportive fibroblast monolayers and can be studied further by treatment with agents such as » fluorouracil or » 4-HC (4-hydroperoxycyclophosphamide).

LTC-ICs are a distinct, rare primitive hematopoietic cell type found in bone marrow and also circulating in normal blood, which generate clonogenic cell progeny detectable after a minimum of 5 weeks incubation on suitable feeder cell layers. As long-term culture initiating ability is considered to be analogous to reconstitution of hematopoiesis in irradiated whole animals determination of LTC-ICs (see: limiting dilution analysis) is thought to be a good measure of true pluripotent » hematopoietic stem cells.

Bartelmez SH et al Uncovering the heterogeneity of hematopoietic repopulating cells EH 19: 861-2 (1991); **Bertoncello I** Status of high proliferative potential colony-forming cells in the hematopoietic stem cell hierarchy. CTMI 177: 83-94 (1992); **Dooley DC & Law P** Detection and quantitation of long-term culture-initiating cells in normal human peripheral blood. EH 20: 156-60 (1992); **Eaves CJ et al** Molecular analysis of primitive hematopoietic cell proliferation control mechanisms. ANY 628: 298-306 (1991); **Eaves CJ et al** Methodology of long-term culture of human hemopoietic cells. J. Tiss. Cult. Meth. 13: 55-62 (1991); **Eaves CJ et al** The human hematopoietic stem cell *in vitro* and *in vivo* Blood Cells 18: 301-7 (1992); **Hughes PF et al** Retroviral gene transfer to primitive normal and leukemic hematopoietic cells using clinically applicable procedures. JCI 89: 1817-24 (1992); **Quesenberry PJ** The blueness of stem cells. EH 19: 725-8 (1991); **Rogers JA & Berman JW** A tumor necrosis factor-responsive long-term culture-initiating cell is associated with the stromal layer of mouse long-term bone marrow cultures. PNAS 90: 5777-80 (1993); **Spangrude GJ et al** Mouse hematopoietic stem cells. Blood 78: 1395-402 (1991); **Spangrude GJ** Hematopoietic stem-cell differentiation. Curr. Opin. Immunol. 3: 171-8 (1991); **Sutherland HJ et al** Characterization and partial purification of human marrow cells capable of initiating long-term hematopoiesis *in vitro*. Blood 74: 1563-70 (1989); **Sutherland HJ et al** Alternative mechanisms with and without steel factor support primitive human hematopoiesis. Blood 81: 1465-70 (1993); **Udomsakdi C et al** Separation of functionally distinct subpopulations of primitive human hematopoietic cells using rhodamine-123. EH 19: 338-42 (1991); **Udomsakdi C et al** Characterization of primitive hematopoietic cells in normal human peripheral blood. Blood 80: 2513-21 (1992)

LTD: abbrev. for Laron type dwarfism.

L-TGF-β: *latent TGF* see: TGF-β. See also: Latent cytokines.

LTIF: *lymphocyte transformation inhibition factor* This poorly characterized activity has been detected in the serum and plasma of patients with active multiple sclerosis and patients with some lymphoproliferative disorders. Partial purification separates LTIF from serum immunoglobulins, but this leads to an accelerated loss of biological activity. LTIF inhibits blast transformation in mitogen-stimulated T lymphocytes.

Wernicke JF et al A humoral lymphocyte transformation inhibition factor in multiple sclerosis. J. Clin. Lab. Immunol. 12: 187-95 (1983)

LTRC: *long-term repopulating cells* This term refers to subpopulations of very primitive » hematopoietic stem cells (***PHSC***, primitive hemopoietic stem cells) that are capable of repopulating the myeloid and the lymphoid B and T cell compartments in irradiated recipients *in vivo* (see: MRA, marrow repopulating ability) and/or that possess high proliferative potential *in vitro* (see: HPP-CFC). Such cells can be isolated by fluorescence-activated cell sorter (FACS) selection of density gradient-enriched, lineage-depleted marrow cells as cells binding the DNA binding dye, ***Hoechst 33342*** and the mitochondrial binding dye, ***rhodamine 123*** (Rh-123). Staining with these dyes also allows differentiation of LTRC from ***short-term repopulating cells*** which provide sufficient offspring to protect irradiated mice until pluripotent » hematopoietic stem cells can reconstitute hematopoiesis. LTRCs can also be selected as survivors after treatment of bone marrow with » fluorouracil or » 4-HC (4-hydroperoxycyclophosphamide). They have also been identified as cells differentially adhering to plastic surfaces and/or stromal cell layers.

Bartelmez SH et al Uncovering the heterogeneity of hematopoietic repopulating cells EH 19: 861-2 (1991); **Bearpark AD & Gordon MY** Adhesive properties distinguish sub-populations of hemopoietic stem cells with different spleen colony-forming and marrow repopulating capacities. Bone Marrow Transplant.

LTK: see addendum.

4: 625-8 (1989); **Brecher G et al** Self-renewal of the long-term repopulating stem cell. PNAS 90: 6028-31 (1993); **Harrison DE et al** Primitive hemopoietic stem cells: direct assay of most productive populations by competitive repopulation with simple binomial, correlation and covariance calculations. EH 21: 206-19 (1993); **Okada S et al** Sequential analysis of hematopoietic reconstitution achieved by transplantation of hematopoietic stem cells. Blood 81: 1720-5 (1993); **Szilvassy SJ & Corry S** Phenotypic and functional characterization of competitive long-term repopulating hematopoietic stem cells enriched from 5-fluorouracil-treated murine marrow. Blood 81: 2310-20 (1993); **van der Sluijs JP et al** Marrow repopulating cells, but not CFU-S, establish long-term *in vitro* hemopoiesis on a marrow-derived stromal layer. EH 18: 893-6 (1990); **van der Sluijs JP et al** Loss of long-term repopulating ability in long-term bone marrow culture. Leukemia 7: 725-32 (1993); **Vecchini F et al** Purified murine hematopoietic stem cells function longer on nonirradiated W41/Wv than on +/+ irradiated stroma. Blood 81: 1489-96 (1993); **Visser JW et al** Culture of hematopoietic stem cells purified from murine bone marrow. Semin. Hematol. 28: 117-25 (1991); **Wolf NS et al** *In vivo* and *in vitro* characterization of long-term repopulating primitive hematopoietic cells isolated by sequential Hoechst 33342-rhodamine 123 FACS selection. EH 21: 614-22 (1993)

LUCT: *lung carcinoma-derived chemotaxin* This factor of 10 kDa is found in the culture supernatants of the human lung giant cell carcinoma cell line LU65C. LUCT is strongly related to » 3-10C. It identical with » IL8 and therefore belongs to the » chemokine family of cytokines.

Hotta K et al Coding region structure of interleukin-8 gene of human lung giant cell carcinoma LU65C cells that produce LUCT/interleukin-8: homogeneity in interleukin-8 genes. Immunol. Lett. 24: 165-70 (1990), Erratum in Immunol. Lett. 27: 258 (1991); **Suzuki K et al** Purification and partial primary sequence of a chemotactic protein for polymorphonuclear leukocytes derived from human lung giant cell carcinoma LU65C cells. JEM 169: 1895-901 (1989)

Lung carcinoma-derived chemotaxin: see: LUCT.

Lung-derived growth factor: This factor is found in culture supernatants of lung tissue fragments. It is a mitogen for certain tumor cells that preferentially metastasize into lung tissues (rat mammary adenocarcinoma, murine B6 melanoma). It is not secreted by non-metastasizing cell variants derived from the parent lines. Lung-derived growth factor is identical with the iron transport protein » transferrin. It may be identical with another growth factor called ***LDGF-1*** (lung-derived growth factor 1).

Cavanaugh PG et al Purification and characterization of a Mr approximately 66,000 lung-derived (paracrine) growth factor that preferentially stimulates the *in vitro* proliferation of lung-metastasizing tumor cells. JCBc 43: 127-38 (1990) **Cavanaugh PG & Nicolson GL** Lung-derived growth factor that stimulates the growth of lung-metastasizing tumor cells: identification as transferrin. JCBc 47: 261-71 (1991); **Cavanaugh PG & Nicolson GL** Purification and some properties of a lung-derived growth factor that differentially stimulates the growth of tumor cells metastatic to the lung. CR 49: 3928-33 (1989); **Cerra RF & Nathanson SD** Organ-specific chemotactic factors present in lung extracellular matrix. J. Surg. Res. 46: 422-6 (1989); **Nicolson GL et al** Adhesive, invasive, and growth properties of selected metastatic variants of a murine large-cell lymphoma. Invasion Metastasis. 9: 102-16 (1989); **Nicolson GL et al** Differential stimulation of the growth of lung-metastasizing tumor cells by lung (paracrine) growth factors: identification of transferrin-like mitogens in lung tissue-conditioned medium. Monogr. Natl. Cancer Inst. 1992: 153-61 (1992)

Lung-derived growth inhibitory factor: see: L-GI factor.

Lung squamous cell carcinoma-derived immunosuppressive factor: see: LSCF.

Luteinizing hormone: abbrev. LH. This hormone is also known as ***gonadotropin II***, or ***interstitial cell-stimulating hormone*** (abbrev. ICSH). The hormone consists of an α chain that is identical with ***follicle stimulating hormone*** (abbrev. FSH), » thyrotropin, or ***chorionic gonadotropin***, and a β chain that is hormone-specific.

LH is produced in the anterior pituitary together with FHS in response to luteinising hormone releasing factor (abbrev. LHRH, also: GRF = gonadotropin releasing factor or gonadoliberin). Its synthesis is also influenced by » IL1, and » IL6. Production of LH is suppressed by » Follistatin. Induction of LH receptors mediated by follicle stimulating hormone is inhibited by » bFGF. LH receptor expression is upregulated by » IGF-1.

Apart from its documented physiological role LH appears to be involved also in immune functions thus establishing it as one component of the » neuroimmune network.

LH has been shown to be produced also by lymphocytes. LH is secreted by human thymocytes and appears to act as a co-mitogen in lymphoproliferation in an » autocrine fashion. Immunoactivation enhances the concentration of LHRH and its gene expression in human peripheral T lymphocytes. Marked elevation of luteinising hormone production by lymphocytes is seen in response to treatment with LHRH, and thymocyte-derived LHRH has been shown to be identical with the hypothalamic releasing factor.

Azad N et al Immunoactivation enhances the concentration of luteinizing hormone-releasing hormone peptide and its gene expression in human peripheral T-lymphocytes. Endocrinology 133: 215-23 (1993); **Maier CC et al** Thymocytes express a mRNA that is identical to hypothalamic luteinizing hormone-releasing hormone mRNA. Cell. Mol. Neurobiol. 12: 447-54 (1992); **Sabharwal P et al** Human thymocytes secrete luteinizing hormone: an autocrine regulator of T-cell proliferation. BBRC 187: 1187-92 (1992)

Ly-2: see: TAP (T cell activating protein).

LyD9: A murine Ly-1$^+$ Mac-1$^+$ B-220$^+$ CC11$^+$ precursor B cell line that depends on » IL3 for its growth and is used in a » bioassay for this factor. The cells die by » apoptosis in response to growth factor deprivation. The cells have been shown to grow in response to a combination of » TNF-α and » IL2, although neither factor alone can stimulate proliferation.

DNA synthesis but not actual growth of LyD9 cells is stimulated to some extent by » IL7. Antibodies against » IL4 and » IL6 or against receptors for » IL2, » IL3, and » IL5 do not inhibit this effect. IL7 alone or in various combinations with other cytokines (from » IL1 α to IL6) do not induce differentiation into IgM$^+$ B cells regardless of the presence of bacterial lipopolysaccharide.

Upon coculture with bone marrow accessory (or stroma) cells or with dendritic cells and T cells LyD9 cells differentiate *in vitro* into , macrophages, neutrophils, and mature B cells producing IgM and IgG. The induction of differentiation by coculture with bone marrow stroma cells is blocked by antibodies directed against lymphocyte function-associated antigen 1 or » IL4.

A tumorigenic » IL5-dependent subline of LyD9 cells, designated K-5, has been established and found to express IL5 constitutively due to transcriptional activation caused by the insertion of an » IAP (intracisternal A particle) element to the 5´ flanking region of the IL5 gene. LyD9 cells have been used also to develop sublines dependent on » IL4 (line K-4) or » GM-CSF (lines K-GM and LS-1).

Kinashi T et al Differentiation of an interleukin 3-dependent precursor B-cell clone into immunoglobulin-producing cells *in vitro*. PNAS 85: 4473-7 (1988); **Kinashi T et al** An interleukin-4-dependent precursor clone is an intermediate of the differentiation pathway from an interleukin-3-dependent precursor clone into myeloid cells as well as B lymphocytes. Int. Immunol. 1: 11-9 (1989); **Lee KH et al** Different stromal cell lines support lineage-selective differentiation of the multipotential bone marrow stem cell clone LyD9. JEM 173: 1257-66 (1991); **Saito Y et al** Interleukin 2 and tumor necrosis factor α are complementary for proliferation of the hematopoietic stem cell line LyD9. GF 7: 297-303 (1992); **Takeda S et al** *In vitro* effects of recombinant interleukin 7 on growth and differentiation of bone marrow pro-B- and pro-T-lymphocyte clones and fetal thymocyte clones. PNAS 86: 1634-8 (1989); **Tashiro K et al** Germline transcripts of the immunoglobulin heavy-chain and T cell receptor genes in a murine hematopoietic stem cell line LyD9 and its derivative cell lines. Immunol. Lett. 28: 147-54 (1991); **Tohyama K et al** Establishment of an interleukin 5-dependent subclone from an interleukin 3-dependent murine hemopoietic progenitor cell line, LyD9, and its malignant transformation by autocrine secretion of interleukin 5. EJ 9: 1823-30 (1990)

LyIFN-α: abbrev. for lymphoblast(oid) interferon. This interferon is identical with » IFN-α. See also: IFN for general information about interferons. Due to difficulties in purification the natural product may contain variable amounts of » IFN-β (≈ 10 %).

Lymphoblast interferon: see: LyIFN-α.

Lymphoblastoid interferon: see: IFN-α.

Lymphocyte activating factor: see: LAF.

Lymphocyte activation gene 1: see: LAG-1.

Lymphocyte blastogenesis inhibitory factor: see: LBIF.

Lymphocyte chemoattractant factor: see: LCF.

Lymphocyte chemokinetic factor(s): see: LCF.

Lymphocyte chemotactic factor: see: LCF.

Lymphocyte-derived chemotactic factor: see: LDCF.

Lymphocyte-derived fibroblast activating factor: see: LDF.

Lymphocyte-derived neutrophil-activating peptide: see: LYNAP.

Lymphocyte growth enhancing factor: see: LGEF.

Lymphocyte migration factor(s): see: LMF.

Lymphocyte migration inhibitory factor: see: LMIF.

Lymphocyte mitogenic factor: see: LMF.

Lymphocyte proliferation factor: see: LPF.

Lymphocyte proliferation inhibitory factors: This term has been used in some data banks and often refers to different (often poorly characterized) inhibitors of » IL1 activities (see: IL1ra, IL1 receptor antagonist). Many factors that negatively affect lymphocyte functions have also been indexed under this term (see: Suppressor factors, SIF).

Lymphocyte proliferation-potentiating factor: see: PMN factor.

Lymphocyte transformation inhibition factor: see: LTIF.

Lymphoid cells: see: Hematopoiesis.

Lymphokines: This is a generic name for a number of unrelated soluble proteins of 12-30 kDa produced by various populations of lymphocytes following their stimulation by an antigen or other mode of » cell activation. These factors act antigen-unspecifically at nano to picomolar concentrations. They may, however, preferentially stimulate some reaction in the presence of antigens.
This term was coined in the 60ies and initially referred to soluble non-immunoglobulin proteins produced by antigen-stimulated lymphocytes. Later this term was also used for other factors produced by mitogen-stimulated lymphocytes, cultured cell lines, non-lymphoid cells, and factors found in body fluids and urine.
The nomenclature is problematic since it was shown that these factors are not secreted exclusively by cells of the immune system. These factors also do not act specifically on these cells. The term is still used, however, as a general term denoting factors produced by cells of the immune system and acting upon them. For further information see: » Cytokines. See also: Recombinant cytokines.

Lymphokine-activated killer cells: see: LAK cells.

Lymphokine-activated tumor inhibition: see: LATI.

Lymphokine gene transfer: see: Cytokine gene transfer.

Lymphoma growth factor: see: LGF.

Lymphopoietin-1: abbrev. Lpo-1. This factor is identical with » IL7.

Lymphotoxin: see: LT.

Lymphotoxin α: The α is frequently omitted (see: TNF). This factor is identical with » TNF-β.

Lymphotoxin β: see: TNF-β.

***lyn*:** *lyn* is a non-receptor cytoplasmic tyrosine kinase belonging to the *src* family of tyrosine kinases. It was isolated originally on the basis of sequence homology to another *src* family member, » *yes*. The human gene maps to chromosome 8 in the region q13-qter.
lyn is expressed predominantly in hemopoietic cells. Two different forms of *lyn* of 56 and 53 kDa arise by differential splicing. *lyn* is expressed preferentially in B lymphocytes but very little in normal T lymphocytes. *lyn* transcripts are found also in the granular layer of the adult mouse cerebellum.
lyn has been shown to be a signal transducing molecule for membrane-bound immunoglobulin M. *lyn* and other *src* family tyrosine kinases including» *blk*, » *fyn*, and perhaps » *lck*, are activated upon engagement of the B cell antigen receptor complex. These kinases then act directly or indirectly to phosphorylate and/or activate effector proteins including p42 (microtubule-associated protein kinase, MAPK), phospholipases C-γ 1 and C-γ 2, phosphatidylinositol 3-kinase (PI 3-K), and p21*ras*-GTPase-activating protein (GAP).
lyn has also been implicated in signaling through the » IL3 receptor. It has been shown to be physically associated with the » IL2 receptor.
lyn may be involved in a signaling pathway of neuroblasts committed to neuronal differentiation. *lyn* transcripts have been found preferentially at early stages in human neuroblastomas whereas they were barely detectable in highly malignant tumors.
Activation of platelets by thrombin and other physiological agonists leads to a dramatic increase in tyrosine phosphorylation of multiple cellular proteins, and one of the tyrosine kinases involved in this process has been shown to be *lyn*. *lyn* is associated with the major platelet membrane glycoprotein IV (GPIV, CD36), together with » *fyn* and » *yes*.

Bielke W et al Expression of the B cell-associated tyrosine kinase gene *lyn* in primary neuroblastoma tumors and its modulation during the differentiation of neuroblastoma cell lines. BBRC 186: 1403-9 (1992); **Cambier JC & Campbell KS** Membrane immunoglobulin and its accomplices: new lessons from an old receptor. FJ 6: 3207-17 (1992); **Cichowski K et al** p21*ras*GAP association with *fyn*, *lyn*, and *yes* in thrombin-activated platelets. JBC 267: 5025-8 (1992); **Huang MM et al** Membrane glycoprotein IV (CD36) is physically associated with the *fyn*, *lyn*, and *yes* protein-tyrosine kinases in human platelets.

PNAS 88: 7844-8 (1991); **Kobayashi N et al** Functional coupling of the *src*-family protein tyrosine kinases p59*fyn* and p53/56*lyn* with the interleukin 2 receptor: implications for redundancy and pleiotropism in cytokine signal transduction. PNAS 90: 4201-5 (1993); **Law DA et al** Examination of B lymphoid cell lines for membrane immunoglobulin-stimulated tyrosine phosphorylation and *src*-family tyrosine kinase mRNA expression. Mol. Immunol. 29: 917-26 (1992); **Lin J & Justement LB** The MB-1/B29 heterodimer couples the B cell antigen receptor to multiple *src* family protein tyrosine kinases. JI 149: 1548-55 (1992); **Meier RW et al** *lyn*, a *src*-like tyrosine-specific protein kinase, is expressed in HL60 cells induced to monocyte-like or granulocyte-like cells. BBRC 185: 91-5 (1992); **O'Connor R et al** Phenotypic changes induced by interleukin-2 (IL2) and IL3 in an immature T-lymphocytic leukemia are associated with regulated expression of IL2 receptor β chain and of protein tyrosine kinases LCK and *lyn*. Blood 80: 1017-25 (1992); **Pleiman CM et al** Mapping of sites on the *src* family protein tyrosine kinases p55*blk*, p59*fyn*, and p56*lyn* which interact with the effector molecules phospholipase C-γ 2, microtubule-associated protein kinase, GTPase-activating protein, and phosphatidylinositol 3-kinase. MCB 13: 5877-87 (1993); **Stanley E et al** Alternatively spliced murine *lyn* mRNAs encode distinct proteins. MCB 11: 3399-406 (1991); **Torigoe T et al** Interleukin-3 regulates the activity of the *lyn* protein-tyrosine kinase in myeloid-committed leukemic cell lines. Blood 80: 617-24 (1992); **Torigoe T et al** Interleukin 2 regulates the activity of the *lyn* protein-tyrosine kinase in a B-cell line. PNAS 89: 2674-8 (1992); **Torigoe T et al** Regulation of *src*-family protein tyrosine kinases by interleukins, IL2, and IL3. Leukemia 6: s94-s97 (1992); **Umemori H et al** Specific expressions of *fyn* and *lyn*, lymphocyte antigen receptor-associated tyrosine kinases, in the central nervous system. Brain Res. Mol. Brain Res. 16: 303-10 (1992); **Yamanashi Y et al** The *yes*-related cellular gene *lyn* encodes a possible tyrosine kinase similar to p56lck. MCB 7: 237-43 (1987); **Yamanashi Y et al** Selective expression of a protein-tyrosine kinase, p56*lyn*, in hematopoietic cells and association with production of human T-cell lymphotropic virus type I. PNAS 86: 6538-42 (1989); **Yamanashi Y et al** Association of B cell antigen receptor with protein tyrosine kinase *lyn*. S 251: 192-4 (1991); **Yamanashi Y et al** Activation of *src*-like protein-tyrosine kinase *lyn* and its association with phosphatidylinositol 3-kinase upon B-cell antigen receptor-mediated signaling. PNAS 89: 1118-22 (1992)

LYNAP: *lymphocyte-derived neutrophil-activating peptide* This factor is produced by lymphocytes. It activates neutrophils. It is identical with » IL8 and therefore belongs to the » chemokine family of cytokines.

Gregory H et al Structure determination of a human lymphocyte derived neutrophil activating peptide (LYNAP). BBRC 151: 883-90 (1988)

Ly-PLP: *lymphocyte prolactin-like polypeptide* This protein is found in the culture supernatants of ConA-stimulated murine thymus cells. It is related to » prolactin.

Sha GN et al Identification and characterization of a prolactin-like polypeptide synthesized by mitogen-stimulated murine lymphocytes. Int. Immunol. 3: 297-304 (1991)

Lysis resistance inducing factor: abbrev. LRIF. see: NK-RIF (natural killer resistance inducing factor).

Lyt-2: This cell surface protein is identical with » CD8 (see also: CD antigens).

M

m: (a) a prefixed m in abbrev. such as mEGF (epidermal growth factor), denotes murine; (b) a prefixed m in abbrev. such as mIL1, may denote membrane-bound, membrane-associated (in contrast to the soluble secreted and circulating factor).

M1: This murine myelomonocytic cell line is used to assay » LIF (see also: Bioassays) by its ability to induce the generation of colonies of differentiated cells in a » colony formation assay. The assay is not specific since » IL6 shows the same biological activity and both factors in combination show enhanced actions. However, LIF is 16-25-fold more active than IL6. Induction of differentiation in M1 leukemic colonies by both LIF and IL6 is enhanced by the addition of » G-CSF or » M-CSF but not by » GM-CSF or » IL3. Differentiation of M1 cells into macrophage-like cells is also observed with » Oncostatin M. The cells also secrete an uncharacterized factor (see: THP-1-derived growth-promoting activity) that promotes the growth of a large variety of cell lines.
A more specific assay for LIF is the inhibition of differentiation in colonies of normal murine embryonic stem cells. M1 cells are also used to assay » G-CSF since this factor is required for growth of the cells. An alternative and entirely different method of detecting these factors is » Message amplification phenotyping.
The M1 cell line is frequently used as a model for studying the mechanisms of cell differentiation. The cells can be induced to terminally differentiate by myelopoietic differentiation factors ($M1D^+$ cells) and were used to develop various sublines that show differential responses to a number of differentiation-inducing proteins. M1 cells constitutively produce low levels of IL6 and production is enhanced by LIF. M1 cells respond to IL6 with activation of a terminal differentiation program which includes activation of genes for » IL6, » IL1, » GM-CSF, M-CSF,» TNF, and » TGF-β1, and for receptors for some of these proteins, thus establishing a network of positive and negative regulatory cytokines. There are clones of M1 cells that differentiate with IL6 but not with LIF and another M1 clone that differentiates with either IL6 or LIF. Differentiation induced by IL6 or » LIF is inhibited by » TGF-β. Cells induced to differentiate with IL6 undergo » apoptosis on withdrawal of IL6, and can be rescued from apoptosis by IL6, » IL3, M-CSF, G-CSF or IL1, but not by GM-CSF. The constitutive overexpression of the *myc* oncogene in M1 cells blocks the differentiation of the cells induced by IL6 and LIF. The cells remain in a stage between immature blast cells and mature macrophages. The same effect, i. e. inhibition of terminal differentiation, is also achieved by treatment with anti-anti-IFN-β antisera or » antisense oligonucleotides of IRF-1 (see: IRS, interferon response sequence). The introduction of the p53 tumor suppressor gene (see: oncogenes) into M1 cells which normally do not express this protein, induces cell death by » apoptosis. Apoptotic cell death in turn can be prevented in these cells by treatment with IL6. For other factors inhibiting the differentiation of M1 cells see also: I factor.

Bruce AG et al Oncostatin M is a differentiation factor for myeloid leukemia cells. JI 149: 1271-5 (1992); **Hoffmann-Liebermann B & Liebermann DA** Interleukin-6 and leukemia inhibitory factor-induced terminal differentiation of myeloid leukemia cells is blocked at an intermediate stage by constitutive c-*myc*. MCB 11: 2375-81 (1991); **Hozumi M et al** Protein factors that regulate the growth and differentiation of mouse myeloid leukemia cells. Ciba Found. Symp. 148: 25-33; (1990); **Ichikawa Y** Differentiation of a cell line of myeloid leukemia. JCP 74: 223-34 (1969); **Lotem J et al** Clonal variation in susceptibility to differentiation by different protein inducers in the myeloid leukemia cell line M1. Leukemia 3: 804-7 (1989); **Lotem J & Sachs L** Regulation of leukemic cells by interleukin 6 and leukemia inhibitory factor. Ciba Found. Symp. 167: 80-8 (1992); **Metcalf D** Actions and interactions of G-CSF, LIF, and IL6 on normal and leukemic murine cells. Leukemia 3: 349-55 (1989); **Miyaura C et al** Recombinant human interleukin 6 (B cell stimulatory factor 2) is a potent inducer of differentiation of mouse myeloid leukemia cells (M1). FL 234: 17-21 (1988); **Okabe-Kado J** Factors inhibiting differentiation of myeloid leukemia cells. Crit. Rev. Oncog. 3: 293-319 (1992); **Ruhl S & Pluznik DH** Dissociation of early and late markers of myeloid differentiation by interferon-γ and interleukin-6. JCP 155: 130-8 (1993); **Shabo Y et al** The myeloid blood cell differentiation-inducing protein MGI-2A is interleukin 6. Blood 72: 2070-3 (1988); **Whetton AD & Dexter TM** Myeloid hematopoietic growth factors. BBA 989: 111-32 (1989); **Williams RL et al** Myeloid leukemia inhibitory factor (LIF) maintains the developmental potential of embryonic stem cells. N 336: 684-7 (1988); **Yonish-Rouack E et al** Wild-type p53 induces apoptosis of myeloid leukemic cells that is inhibited by interleukin-6. N 352: 345-7 (1991)

M1 differentiation inducing activity: This factor is found in conditioned culture media of Krebs II ascites cells. Some of the differentiation-inducing activity is due to » G-CSF and » » LIF (leukemia inhibitory factor).
Hilton DJ et al Resolution and purification of three distinct factors produced by Krebs ascites cells which have differentiation-inducing activity on murine myeloid leukemic cell lines. JBC 263: 9238-43 (1988)

M6P-R: Mannose-6-phosphate receptor. see: MPR.

M20 interleukin-1 inhibitor: This poorly characterized factor of ≈ 52 kDa is secreted by the human myelomonocytic cell line M20. It specifically blocks IL1-induced processes *in vitro* and *in vivo*. Among other things the factor ameliorates acute inflammatory responses such as fever, leukocytosis, enlargement of lymph nodes, and the expression of » acute phase proteins in experimental animals.
The factor reversibly inhibits the growth of myeloid cells in the presence of IL1 and the growth of normal hematopoietic progenitor cells and freshly isolated leukemia cells stimulated by » GM-CSF. For another IL1 inhibitor that is not identical with the M20 factor see also: IL1ra (IL1 receptor antagonist).
Barak V et al Interleukin 1 inhibitory activity secreted by a human myelomonocytic cell line (M20). EJI 16: 1449-52 (1986); **Barak V et al** The specific IL1 inhibitor from the human M20 cell line is distinct from the IL1 receptor antagonist. Lymphokine Cytokine Res. 10: 437-42 (1991); **Peled T et al** Effect of M20 interleukin-1 inhibitor on normal and leukemic human myeloid progenitors. Blood 79: 1172-7 (1992); **Peritt D et al** The M20 IL1 inhibitor. I. Purification by preparative isoelectric focusing in free solution. JIM 155: 159-65 (1992); **Peritt D et al** The M20 IL1 inhibitor. II. Biological characterization. JIM 155: 167-74 (1992); **Vivian B et al** The M20 IL1 inhibitor prevents onset of adjuvant arthritis. Biotherapy 4: 317-23 (1992)

M119: see: *mig* (monokine induced by γ interferon).

Ma: abbrev. for macrophage.

Macroglobulin α2: see: Alpha-2-Macroglobulin.

Macrophage activating factor: see: MAF.

Macrophage-activating factor for cytotoxicity I: see: MAF-C I.

Macrophage aggregating factor: see: MIF.

Macrophage arming factor: see: SMAF (specific macrophage arming factor).

Macrophage chemotactic factor: see: MCF.

Macrophage chemotaxis inhibitor: see: MCI.

Macrophage colony-forming cells: abbrev. CFC-M. See: CFC.

Macrophage cytotoxic factor: see: MCF.

Macrophage cytotoxicity-inducing factor: see: MCF.

Macrophage cytotoxicity-inducing factor 2: see: MCIF 2.

Macrophage cytotoxin: see: MCT.

Macrophage deactivation factor: see: MDF.

Macrophage-dependent fibroblast stimulating activity: see: MFSA.

Macrophage-derived angiogenic activity: see: MDAA.

Macrophage-derived blastogenic factor: see: MBF.

Macrophage-derived chemoattractant: see: MDC.

Macrophage-derived cytotoxic factor: This factor is secreted by murine peritoneal macrophages following their stimulation by » calcium ionophore and bacterial lipopolysaccharides. It is probably identical with » TNF-α.
Dong Z Studies on characteristics of macrophage-derived cytotoxic factor. Chung Kuo I Hsueh Ko Hsueh Yuan Hsueh Pao 12: 375-9 (1990)

Macrophage-derived deactivating factor: see: MDF.

Macrophage-derived endothelial cell inhibitor: see: MD-ECI.

Macrophage-derived fibroblast growth factor: see: MDGF (macrophage-derived growth factor).

Macrophage-derived growth factor: see: MDGF.

Macrophage-derived T cell replacing factor: see: TRF-M.

Macrophage fusion factor 2: see: MFF.

Macrophage-granulocyte inducer: see: MGI.

Macrophage-granulocyte inducing protein 2A: This factor is identical with » IL6.
Shabo Y et al The myeloid blood cell differentiation-inducing protein MGI-2A is interleukin 6. Blood 72: 2070-3 (1988)

Macrophage growth factor: see: MGF (two entries).

Macrophage inflammatory protein: see: MIP.

Macrophage migration enhancement factor: see: MEF.

Macrophage migration inhibitory factor: see: MIF (migration inhibition factor).

Macrophage migration inhibitory factor: abbrev. MMIF. See: MIF (migration inhibition factor).

Macrophage migration inhibitory factor 1: abbrev. MIF-1. see: MIF (migration inhibition factor).

Macrophage migration inhibitory factor 2: abbrev. MIF-1. see: MIF (migration inhibition factor)

Macrophage procoagulant activity inducing factor: abbrev. MPCA inducing factor. see: MPIF (macrophage procoagulant inducing factor).

Macrophage procoagulant inducing factor: see: MPIF (macrophage procoagulant inducing factor).

Macrophage-released neutrophil chemotactic factor: see: MNCF.

Macrophage slowing factor: see: MSF.

Macrophage stimulating 1: abbrev. MST1. See: HGF-like protein (hepatocyte growth factor-like protein)

Macrophage stimulating protein: see: MSP.

Madin-Darby bovine kidney cell line: see: MDBK.

MAF: *macrophage activating factor* This activity was initially observed in culture supernatants of mitogen- and antigen-activated lymphocytes (see also: Cell activation). MAF activates macrophages and allows them to act as cytotoxic cells that non-specifically kill tumor cells. MAF is more or less an operational definition for a particular biological activity rather than the name of a distinct factor. The main factor responsible for MAF activity is » IFN-γ. Some of the MAF activity is due to » GM-CSF or » M-CSF, » IL3, or other factors.
In addition numerous biochemically poorly characterized factors have been shown to possess MAF activity. Some factors that normally are not generally classified as » cytokines also possess MAF activity including the » acute phase protein C-reactive protein (see also: acute phase reaction).
Two factors, designated ***FMAF*** I and II, fibroblast-derived MAF (5 and 10 kDa), have been found in the extracellular matrix of a human embryo fibroblast cell strain. They stimulate the attachment ability of macrophages and their production of superoxide anions [3H]UTP incorporation The factors do not show significant » CSF (colony stimulating factor) and » IFN activities.
Cameron DJ Characterization of human macrophage activation factor (MAF) prepared from antigen-stimulated lymphocytes. J. Clin. Lab. Immunol. 13: 47-50 (1984); **Chantry D et al** Granulocyte-macrophage colony stimulating factor induces both HLA-DR expression and cytokine production by human monocytes. Cytokine 2: 60-7 (1990); **Clark-Lewis I et al** Preparation of T cell growth factor free from interferon and factors stimulating hemopoietic cells and mast cells. JIM 51: 311-22 (1982); **De Groot JW et al** Differences in the induction of macrophage cytotoxicity by the specific T lymphocyte factor, specific macrophage arming factor (SMAF), and the lymphokine macrophage activating factor (MAF). Immunobiology 179: 131-44 (1989); **Frendl G & Beller DI** Regulation of macrophage activation by IL3. I. IL3 functions as a macrophage-activating factor with unique properties, inducing Ia and lymphocyte function-associated antigen-1 but not cytotoxicity. JI 144: 3392-9 (1990); **Fukuzawa Y et al** Biological and biochemical characterization of macrophage activating factor (MAF) in murine lymphocytes: physicochemical similarity of MAF to γ interferon (IFN-γ). Microbiol. Immunol. 28: 691-702 (1984); **Kleinschmidt WJ & Schultz RM** Similarities of murine γ interferon and the lymphokine that renders macrophages cytotoxic. J. Interferon Res. 2: 291-9 (1982); **Meltzer MS et al** Macrophage

activation for tumor cytotoxicity: induction of macrophage tumoricidal activity by lymphokines from EL-4, a continuous T cell line. JI 129: 2802-7 (1982); **Miyagawa N et al** Effect of C-reactive protein on peritoneal macrophages. I. Human C-reactive protein inhibits migration of guinea pig peritoneal macrophages. Microbiol. Immunol. 32: 709-19 (1988); **Miyagawa N et al** Effect of C-reactive protein on peritoneal macrophages. II. Human C-reactive protein activates peritoneal macrophages of guinea pigs to release superoxide anion *in vitro*. Microbiol. Immunol. 32: 721-31 (1988); **Nathan CF et al** Identification of IFN-γ as the lymphokine that activates human macrophage oxidative metabolism and antimicrobial activity. JEM 158: 670-89 (1983); **Nathan CF et al** Activation of human macrophages: comparison of other cytokines with interferon-γ. JEM 160: 600-5 (1984); **Okai Y** Heterogeneous macrophage-activating factors from extracellular matrix of human embryo fibroblasts. Immunol. Lett. 17: 145-9 (1988); **Remold HG & Mednis AD** Migration inhibitory factor. MiE 116: 379-94 (1985); **Sadlik JR et al** Lymphocyte supernatant-induced human monocyte tumoricidal activity: dependence on the presence of γ-interferon. CR 45: 1940-5 (1985); **Salahuddin SZ et al** Lymphokine production by cultured human T cells transformed by human T cell leukemia lymphoma virus I. S 223: 703-7 (1984); **Schreiber RD et al** Macrophage-activating factor produced by a T cell hybridoma: physicochemical and biosynthetic resemblance to γ-interferon. JI 131: 826-33 (1983); **Schultz RM & Kleinschmidt WJ** Functional identity between murine γ interferon and macrophage activating factor. N 305: 239-40 (1983); **Svedersky LP et al** Biological and antigenic similarities of murine interferon and macrophage activation factor. JEM 159: 812-27 (1984); **Taniyama T et al** Constitutive production of novel macrophage-activating factor(s) by human T cell hybridomas. Clin. Invest. Med. 13: 305-12 (1990); **Warren MK & Ralph P** Macrophage growth factor (CSF-1) stimulates human monocyte production of interferon, tumor necrosis factor, and colony stimulating activity. JI 137: 2180-5 (1986)

MAF: *macrophage arming factor* see: SMAF (specific macrophage arming factor).

MAF-C I: *macrophage-activating factor for cytotoxicity I* A priming and a triggering factor are required to convert macrophages into cells that are cytotoxic to various tumor cells. IMAF-C I is a priming macrophage activating factor (see also: MAF) found in the culture supernatant of the human T cell hybridoma, H3-E9-6. This factor is identical with » IL2.

Higashi N et al Identification of human T cell hybridoma-derived macrophage activating factor as interleukin-2. J. Biochem. Tokyo 113: 715-20 (1993)

Mafosfamide: see: 4-HC (Hydroperoxycyclophosphamide).

Maintenance factors: see: Survival factors.

Mammalian bombesin: see: GRP (gastrin-releasing peptide).

Mammary-derived growth factor 1: see: MDGF-1.

Mammary-derived growth inhibition factor: see: MDGI (mammary-derived growth inhibitor).

Mammary-derived growth inhibitor: see: MDGI.

Mammary tumor-derived growth factor: see: MTGF.

Mammastatin: abbrev. MS. This factor (47 and 65 kDa) is isolated from normal human and murine mammary epithelial cells. It appears to be specifically expressed in these tissues and its expression decreased in transformed mammary cells. This poorly characterized factor selectively inhibits the growth of transformed mammary epithelial cell lines. It is inactive on non-mammary transformed cells lines. The factor is not identical with » TGF-β.

Ervin PR Jr et al Production of mammastatin, a tissue-specific growth inhibitor, by normal human mammary cells. S 244: 1585-7 (1989)

Manchester inhibitor of stem cell proliferation: see: SCI (stem cell inhibitor).

Graham GJ & Pragnell IB Negative regulators of hemopoiesis – current advances. Progr. Growth Factor Res. 2: 181-92 (1990)

Mannose-6-phosphate receptor: see: MPR.

MANS: *membrane-associated neurotransmitter-stimulating factor* MANS modulates sympathetic neurotransmitter expression and promotes ciliary neuron survival in cell culture. MANS induces the production of » SP (substance P) in ganglion cells *in vitro* and *in vivo*. Its biological effects and biochemical properties are similar to those of » CNTF (ciliary neurotrophic factor). MANS preparations have been shown to contain CNTF or a CNTF-like protein and has recently been described to be identical with CNTF.

Adler JE et al Partial purification and characterization of a membrane-derived factor regulating neurotransmitter phenotypic expression. PNAS 86: 1080-3 (1989); **AR MS et al** Membrane-associated neurotransmitter stimulating factor is very similar to ciliary neurotrophic factor. Dev. Biol. 153: 411-6(1992); **AR MS & Landis SC** Cell interactions that determine sympathetic neuron transmitter phenotype and the neurokines that mediate them. J. Neurobiol. 24: 215-32 (1993); **Wong V & Kessler JA** Solubilization of a membrane factor that stimulates levels of substance P and choline acetyltransferase in sympathetic neurons. PNAS 84: 8726-9 (1987)

MAP kinases: see: ERK (extracellular signal-regulated kinases).

MAPP: *message amplification phenotyping* see: Message amplification phenotyping.

MARC: MARC has been identified in a cDNA library of the stimulated mouse mast cell line, CPII. It is strongly upregulated at the mRNA level after the physiological challenge of the cells with immunoglobulin IgE plus antigen. Unstimulated cells and a number of other, different cell lines (uninduced and induced) and mouse tissues do not express the protein. MARC has been suggested to be involved in certain acute and chronic pathological mast cell-driven diseases.
MARC protein has a length of 97 amino acids and is a member of the » chemokine family of cytokines. It displays a high degree of homology at the protein level to several other cytokines of the chemokine family, including human » JE (58.2 %), human » MCP-1 (56 %), » RANTES (36 %), MIP (34 %), and » murine TCA3 (= human I-309; 39.5 %). Murine MARC protein may be the homologue of human MCP-3 (see: MCP-2).
Kulmburg PA et al Immunoglobulin E plus antigen challenge induces a novel intercrine/chemokine in mouse mast cells. JEM 176: 1773-8 (1992)

Marograstim: see: KW-2228.

Marrow repopulating ability: see: MRA.

Mast cell arming T cell factor: abbrev. MTCF. see: SMAF (specific macrophage arming factor).

Mast cell-committed progenitor proliferation factor: see: MCCP proliferation factor.

Mast cell costimulatory activity: see: MCA.

Mast cell growth enhancing activity: see: MEA.

Mast cell growth factor: see: MCGF, MGF.

Mast cell growth factor 2: abbrev. also MGF-2. see: MCGF-2.

Mast cell growth factor 3: MCGF-3.

MAT: *monocyto-angiotropin* see: angiotropin.

Maturation inducer activity: This activity was isolated from culture supernatants of mitogen-activated lymphocytes. It promotes the differentiation of the promyelocytic leukemia cell line » HL-60 into monocytes and macrophages. Some of the activity is due to » IL3. Another poorly characterized protein of 52-56 kDa has also been described to possess the same activity.
Abolhassani M et al Two separate differentiation inducing proteins for human myeloid leukemia cells and their isolation from normal lymphocytes. PSEBM 195: 288-91 (1990)

MBF: *macrophage-derived blastogenic factor* This poorly characterized factor of 29-35 kDa is secreted from human or murine macrophages stimulated with galactose oxidase, a well-characterized enzyme able to induce marked polyclonal activation of lymphocytes. It stimulates resting T lymphocytes to proliferate and to produce » IFN-γ. The activity of MBF is not neutralized by antibodies to » IL1.
Antonelli G et al A macrophage-derived factor different from interleukin 1 and able to induce interferon-γ and lymphoproliferation in resting T lymphocytes. CI 113: 376-86 (1988)

MC/9: A murine mast cell line that depends on » IL3 for its growth and this is blocked by anti-IL3 antibodies. The cells also express receptors for » IL4 and also proliferate in response to this cytokine. The cell line shows only weak responses to either cytokine and strongly proliferates in the presence of a combination of the two cytokines. Saturating amounts of one of the cytokines can therefore be used to measure an additional growth response caused by the other cytokine. The sensitivity limit for IL3 is ≈ 5 pg/mL.
Cytokine-induced proliferation of MC/9 cells is inhibited by rapamycin. This compound, however, does not inhibit cytokine production by these cells (in contrast to the related FK506 and » CsA (cyclosporine A).
MC9 cells also express phospholipase and lipoxygenase activity when stimulated with IgE and hapten, and a 5-lipoxygenase activity when stimulated with the ionophore A23187.
Abrams JS & Pearce MK Development of rat anti-mouse interleukin 3 monoclonal antibodies which neutralize bioactivity *in vitro*. JI 140: 131-7 (1988); **Bryant RW et al** Modulation of the 5-lipoxygenase activity of MC-9 mast cells: activation by hydroperoxides. Prostaglandins 32: 615-27 (1986); **Hatfield SM et al** Rapamycin and FK506 differentially inhibit mast cell cytokine production and cytokine-induced proliferation and act as reciprocal antagonists. J. Pharmacol. Exp. Ther. 261: 970-6 (1992); **Hatfield SM & Roehm NW** Cyclosporine and FK506 inhibition of murine mast cell cytokine production. J. Pharmacol. Exp. Ther. 260: 680-8 (1992); **Ishida H et al** Evaluation of murine interleukin 4 (IL4) receptor expression using anti-receptor

monoclonal antibodies and S1 nuclease protection analyses. CI 136: 142-54 ((1991); **Miyajima A et al** Use of the silkworm, *Bombyx mori*, and an insect baculovirus vector for high-level expression and secretion of biologically active mouse interleukin-3. Gene 58: 273-81 (1987); **Mosman TR & Fong TAT** Specific assays for cytokine production by T cells JIM 116: 151-8 (1989); **Musch MW et al** Ionophore-stimulated lipoxygenase activity and histamine release in a cloned murine mast cell, MC9. Prostaglandins 29: _405-30 (1985); **Musch MW & Siegel MI** Antigenic stimulated release of arachidonic acid, lipoxygenase activity and histamine release in a cloned murine mast cell MC9. BBRC 126: 517-25 (1985); **Nabel G et al** Inducer T lymphocytes synthesize a factor that stimulates proliferation of cloned mast cells. N 291: 332-4 (1981); **Smith CA & Rennick DM** Characterization of a murine lymphokine distinct from interleukin 2 and interleukin 3 (IL3) possessing a T cell growth factor activity and a mast-cell growth factor activity that synergizes with IL3. PNAS 83: 1857-61 (1986)

MCA: *mast cell costimulatory activity* This poorly characterized factor of 35-40 kDa is found in culture supernatants of a TH2 helper cell clone (ST2/K.9) stimulated by concanavalin A. It is produced preferentially by TH2 helper cells(see also: TH1/TH2 cytokines). MCA is a glycoprotein that enhances the proliferation of mucosal mast cells promoted by a combination of » IL3 and » IL4. MCA does not stimulate mast cell proliferation in the absence of a combination of IL3 and IL4 or in the presence of either IL3 or IL4 alone. MCA is not identical with » IL2 to » IL7, » IL9, » IFN-γ, » TNF-α, » TNF-β, » GM-CSF, » G-CSF, » M-CSF, » Epo, » LIF, or » EGF.

Schmitt E et al Characterization of a T cell-derived mast cell costimulatory activity (MCA) that acts synergistically with interleukin 3 and interleukin 4 on the growth of murine mast cells. Cytokine 2: 407-15 (1990)

MCA: *monocyte chemotactic activity* This is a general term referring to factors affecting monocyte chemotaxis. Many of these activities belong to the » chemokine family of cytokines.

MCAF: *monocyte chemotactic and activating factor* This factor of 15 kDa (76 amino acids) is also called MCF (macrophage chemotactic factor) or » TSG-8 (see: TSG genes). It is produced by monocytes and macrophages, fibroblasts, endothelial cells, keratinocytes, smooth muscle cells, astrocytes. and various tumor cells lines. The factor was initially detected during the purification of » IL8. Inducers for the synthesis of MCAF are » IL1, which also prolongs the biological half life of » M-CSF mRNA, » TNF, » PDGF, phytohemagglutinin, bacterial lipopolysaccharides, hydroxy urea, double-stranded RNA and viruses. Glucocorticoids inhibit the synthesis of MCAF.

MCAF belongs to the » chemokine family of cytokines. It is encoded by the » JE gene on human chromosome 17. This factor is identical with » MCP-1 (monocyte chemoattractant protein 1), » MCP (monocyte chemotactic protein), » HC11, » SMC-CF (smooth muscle cell chemotactic factor), » LDCF (lymphocyte-derived chemotactic factor), » GDCF (glioma-derived monocyte chemotactic factor), » TDCF (tumor-derived chemotactic factor).

MCAF is a strong chemoattractant for monocytes (see also: Chemotaxis), it increases the production of reactive oxygen species and enhances the release of N-acetyl-β-glucuronaminidase. MCAF promotes the cytostatic activities of cultured human monocytes against a number of tumor cells lines. In basophils MCAF is one of the most potent cytokines inducing the release of histamines. This specific activity is antagonized by another cytokine, » RANTES.

Keratinocyte-derived MCAF may be important in the regulation of cutaneous monocyte trafficking and may also be responsible for the recruitment of Langerhans cells and dermal dendrocytes, which share many phenotypic features with monocytes/macrophages, to their anatomic locations in skin. MCAF (and » IL8) may play a role also in growth regulation and spreading of melanomas.

The biological activities of MCAF are mediated by a specific receptor. A second species of receptors also appears to bind MCAF and MIP-1α and MIP-1β (see: MIP, macrophage inflammatory protein).

Alam R et al Monocyte chemotactic and activating factor is a potent histamine-releasing factor for basophils. JCI 89: 723-8 (1992); **Apella E et al** Determination of the primary structure of NAP-1/IL8 and a monocyte chemoattractant protein, MCP-1/MCAF. PCBR 349: 405-17 (1990); **Barker JN et al** Monocyte chemotaxis and activating factor production by keratinocytes in response to IFN-γ. JI 146: 1192-7 (1991); **Furutani Y et al** Cloning and sequencing of the cDNA for human monocyte chemotactic and activating factor (MCAF). BBRC 159: 249-55 (1989); **Gronenborn AM & Clore GM** Modeling of the three-dimensional structure of the monocyte chemoattractant and activating protein MCAF/MCP-1 on the basis of the solution structure of interleukin 8. Prot. Engin. 4: 263-9 (1991); **Kasahara T et al** IL1 and TNF-α induction of IL8 and monocyte chemotactic and activating factor (MCAF) mRNA expression in a human astrocytoma cell line. Immunology 74: 60-7 (1991); **Kuna P et al** Monocyte chemotactic and activating factor is a potent histamine-releasing factor for human basophils. JEM 175: 489-93 (1992); **Larsen CG et al** Production of monocyte chemotactic and activating factor (MCAF) by human dermal fibroblasts in response to interleukin 1 or tumor necrosis factor. BBRC 160: 1403-8 (1989); **Matsushima K & Oppenheim JJ** Interleukin 8 and MCAF: novel inflammatory cytokines inducible by IL1 and TNF. Cytokine 1: 2-13 (1989); **Matsushima K et al** Purification and characterization of a novel monocyte chemotactic and activating factor produced by a human myelomonocytic cell line. JEM 169: 1485-90 (1989); **Wang JM et al** Human recombinant macrophage inflammatory protein-1 α and -β and monocyte chemotactic and activating factor utilize common and unique receptors on human monocytes. JI 150: 3022-9 (1993); **Yoshimura T et al** Purification and amino acid analysis of two human monocyte chemoattractants produced by phytohemagglutinin-stimulated human blood mononuclear leukocytes. JI 142: 1956-62 (1989); **Yoshimura T et al** Identification of high affinity receptors for human monocyte chemoattractant protein-1 (MCP-1) on human monocytes. JI 145: 292-7 (1990); **Yoshimura T & Leonard EJ** Secretion by human fibroblasts of monocyte chemoattractant protein-1, the product of gene JE. JI 144: 2377-83 (1990); **Zachariae COC et al** Properties of monocyte chemotactic and activating factor (MCAF) purified from a human fibrosarcoma cell line JEM 171: 2177-82 (1990); **Zachariae CO et al** Expression and secretion of leukocyte chemotactic cytokines by normal human melanocytes and melanoma cells. J. Invest. Dermatol. 97: 593-9 (1991)

MCCF: *mast cell-committed progenitor* A type of tissue cells developing into mast cells.

MCCP proliferation factor: *mast cell-committed progenitor proliferation factor* This poorly characterized factor (> 5 kDa, < 50 kDa) is found in the conditioned medium of fibroblasts. It allows late-stage murine mast cell-committed progenitors to form colonies in methylcellulose (see also: Colony formation assays). The activity of the factor is not absorbed by immobilized antibodies to » NGF and its activity cannot be mimicked by » IL1, » IL2, » IL4, » GM-CSF, » G-CSF, » M-CSF, » IFN-α, » IFN-β, » IFN-γ, » NGF, » EGF, serum fibronectin, heparin, or a number of glycosaminoglycans (see also: ECM, extracellular matrix). When grown on fibroblast monolayers in the presence of this factor mast cell-committed progenitors populations can be obtained that are virtually uncontaminated with other hematopoietic progenitors (see also: Hematopoiesis).

Jarboe DL et al The mast cell-committed progenitor. I. Description of a cell capable of IL3-independent proliferation and differentiation without contact with fibroblasts. JI 142: 2405-17 (1989)

MCF: *macrophage chemotactic factor* This factor is isolated from a T cell hybridoma. It is identical with » MCAF (monocyte chemotactic and activating factor).

Sasaki Y et al Macrophage chemotactic factor partially purified from granulomatous inflammation. CI 134: 171-9 (1991); **Yoshizuka N et al** Macrophage chemotactic factor (MCF) produced by a human T cell hybridoma clone. CI 123: 212-25 (1989)

MCF: *macrophage cytotoxic factor* At least one factor with MCF activity, derived from astrocytes and cytotoxic for oligodendrocytes, is identical with » TNF-α. It is not clear whether other factors with MCF activity, derived from macrophages or macrophage hybridomas, are identical with TNF-α or distinct factors.

Robbins DS et al Production of cytotoxic factor for oligodendrocytes by stimulated astrocytes. JI 139: 2593-7 (1987); **Sherwood ER et al** Glucan stimulates production of antitumor cytolytic/cytostatic factor(s) by macrophages. J. Biol. Response Mod. 5: 504-26 (1986); **Takeda Y et al** Purification and characterization of a cytotoxic factor produced by a mouse macrophage hybridoma. CI 96: 277-89 (1985)

MCF: *macrophage cytotoxicity-inducing factors* These factors of 29 kDa (P29) and 14,7 kDa (P14.7) have been isolated from the culture supernatants of a human T cell hybridoma. The two factors are not related to each other. They activate monocytes and increase their tumor cytotoxicity. They induce the synthesis of » IL1 in macrophages. The analysis of the N-terminal sequence of P29 shows that it is not related to any other known cytokine (for other macrophage-activating factors see: MAF).

Jones CM et al Purification and amino acid analysis of a human macrophage cytotoxicity-inducing factor (MCF). EH 19: 704-9 (1991); **Jones CM** Characterization of a human monocyte cytotoxicity-inducing factor (MCF). Immunobiology 178: 229-49 (1988); **Jones CM** Identification of a human monocyte cytotoxicity-inducing factor from T cell hybridomas produced from Sezary's cells. JI 137: 571-7 (1986)

MCF: *mononuclear cell factor* also abbrev. MNCF. This factor was originally described as an activity that activates chondrocytes, induces the synthesis of prostaglandin E2 in fibroblasts, promotes the degradation of cartilage tissues by inducing the synthesis of collagenases in fibroblasts and rheumatoid synovial cells, and generally influences the metabolism of the extracelluar matrix (see also: ECM). MCF is identical with » IL1.

Balavoine JF et al Prostaglandin E2 and collagenase production by fibroblasts and synovial cells is regulated by urine-derived human interleukin 1 and inhibitor(s). JCI 78: 1120-4 (1986); **Dayer JM et al** Collagens act as ligands to stimulate human monocytes to produce mononuclear cell factor and prostaglandins. Coll. Relat. Res. 2: 523-40 (1982); **Dayer JM et al** Collagenase production by rheumatoid synovial cells: stimulation by a human lymphocyte factor. S 195: 181-83 (1977); **Dayer JM et al** Participation of monocyte-macrophage, and lymphocytes in the production of a factor which stimulates collagenase and prostaglandin release by rheumatoid synovial cells. JCI 64: 1386-92 (1979); **Dayer JM et al** Prostaglandin production by rheumatoid synovial cells: stimulation by a factor from human mononuclear cells. JEM 145: 1399-1404 (1977); **Krane SM et al** Mononuclear cell-conditioned medium containing mononuclear cell factor (MCF) homologous with interleukin-1, stimulates collagen and fibronectin synthesis by adherent rheumatoid synovial cells: effects of prostaglandin E2 and indomethacin. Collagen Res. 5: 99-117 (1985); **McCroskery PA et al** Stimulation of procollagenase synthesis in human rheumatoid synovial fibroblasts by mononuclear cell factor/interleukin 1. FL 191: 7-12 (1985); **Nar-**

della FA et al Self associating IgG rheumatoid factors stimulate monocytes to release prostaglandin and mononuclear cell factor that stimulates collagenase and prostaglandin production by synovial cells. Theumatol. Int. 3: 183-6 (1983)

MCF: *monocyte cytotoxic factor* This factor is secreted by activated human monocytes. It is cytotoxic for tumor cells. The factor is probably identical with » TNF-α or » TNF-β.

Bersani L et al Involvement of tumor necrosis factor in monocyte-mediated rapid killing of actinomycin D-pretreated WEHI 164 sarcoma cells. Immunology 59: 323-5 (1986); **Espevik T & Nissen-Meyer J** A highly sensitive cell line, WEHI 164 clone 13, for measuring cytotoxic factor/tumor necrosis factor from human monocytes. JIM 95: 99-105 (1986); **Nissen-Meyer J & Kildahl-Andersen O** Purification of cytostatic protein factors released from human monocytes. Scand. J. Immunol. 20: 317-25 (1984); **Uchida A & Yanagawa E** Natural cytotoxicity of human blood monocytes: production of monocyte cytotoxic factors (MCF) during interaction with tumor cells. Immunol. Lett. 8: 311-6 (1984); **Uchida A** The cytolytic and regulatory role of natural killer cells in human neoplasia. BBA 865: 329-40 (1986); **Ziegler-Heitbrock HW et al** Tumor necrosis factor as effector molecule in monocyte mediated cytotoxicity. CR 46: 5947-52 (1986)

MCF-7: A human breast cancer cell line that has retained several characteristics of differentiated mammary epithelium, including the ability to process estradiol via cytoplasmic estrogen receptors. The cell line is used, among other things, to measure » IGFs (see also: bioassays). These cells are absolutely dependent on exogenous growth factors and become quiescent if cultured in growth factor-inactivated serum in the absence of estrogen or phenol red. Resting cells can be restimulated to proliferate by insulin or IGFs, particularly in synergism with estradiol. The cells respond much stronger to IGF-I than to IGF-II. The growth-promoting effects of insulin are slightly reduced by » TGF-β. Other growth factors such as » EGF, » PDGF, or » TGF-β do not stimulate proliferation of resting MCF-7 cells and do not influence IGF assays.

In vitro, picomolar concentrations of » IL2 directly inhibit MCF-7 cell proliferation after 12 days of culture, while nanomolar doses of IL2 significantly stimulate MCF-7 cell growth over the same time period. In addition, micromolar concentrations of IL2 have virtually no effect on the *in vitro* proliferation of MCF-7 cells.

Paciotti GF & Tamarkin L Interleukin-2 differentially affects the proliferation of a hormone-dependent and a hormone-independent human breast cancer cell line *in vitro* and *in vivo*. Anticancer Res. 8: 1233-8 (1988); **Soule HD et al** JNCI 51: 1409-16 (1973); **Van der Burg B et al** Mitogenic stimulation of human breast cancer cells in a growth-factor-defined medium: synergistic action of insulin and estrogen. JCP 134: 101-8 (1988); **Van der Burg B et al** Direct effects of estrogen on c-*fos* and c-*myc* proto-oncogene expression and cellular proliferation in human breast cancer cells. Mol. Cell. Endocrinol. 64: 223-8 (1989); **Van Zoelen EJJ et al** Production of insulin-like growth factors, platelet-derived growth factor, and transforming growth factors and their role in the density-dependent growth regulation of a differentiated embryonal carcinoma cell line. Endocrinology 124: 2029-41 (1989)

MCGF: *mast cell growth factor* This factor is identical with » IL3. The mouse IL3 gene was isolated from a mouse sperm DNA library based on homology with the mouse mast cell growth-factor (MCGF) cDNA sequence. Some of the MCGF activity may be due to » IL4 (see: MCGF-2) or » IL9.

Ihle JN et al Biological properties of homogenous interleukin-3. I. Demonstration of WEHI-3 growth factor activity, mast cell growth factor activity, P cell-stimulating factor activity, colony-stimulating factor activity, and histamine-producing cell-stimulating factor activity. JI 131: 282-7 (1983); **Miyatake S et al** Structure of the chromosomal gene for murine interleukin 3. PNAS 82: 316-20 (1985); **Nabel G et al** Inducer T lymphocytes synthesize a factor that stimulates proliferation of cloned mast cells. N 291: 332-4 (1981); **Rennick DM** A cloned MCGF cDNA encodes a multilineage hematopoietic growth factor: multiple activities of interleukin 3. JI 134: 910-4 (1985); **Yokota T et al** Isolation and characterization of a mouse cDNA clone that expresses mast cell growth-factor activity in monkey cells. PNAS 81: 1070-4 (1984); **Yung YP et al** Long-term *in vitro* culture of murine mast cells. II. Purification of a mast cell growth factor and its dissociation from TCGF. JI 127: 794-9 (1981)

MCGF-2: *mast cell growth factor 2* This factor is identical with » IL4.

Brown MA et al B cell stimulatory factor-1/interleukin-4 mRNA is expressed by normal and transformed mast cells. Cell 50: 809-18 (1987); **Smith CA & Rennick DM** Characterization of a murine lymphokine distinct from interleukin 2 and interleukin 3 (IL3) possessing a T cell growth factor activity and a mast cell growth factor activity that synergizes with IL3. PNAS 83: 1857-61 (1986)

MCGF-3: *mast cell growth factor 2* This poorly characterized factor is produced by a murine bone marrow stromal cell line. It synergises with » IL3 and » IL4 in promoting the growth of the IL3-dependent mast cell line NFS/N1.

Boswell HS et al A novel mast cell growth factor (MCGF-3) produced by marrow-adherent cells that synergises with interleukin 3 and interleukin 4. EH 18: 794-800 (1990)

MCI: *macrophage chemotaxis inhibitor* This factor is isolated from murine tumor cells. It prevents the migration and accumulation of macrophages in tumor tissues and also inhibits inflammatory reactions (see also: Inflammation).

This protein is also found in the serum of human cancer patients. The protein is related to a coat protein called p15E found in several retroviruses (MMTV, FLV, MoLV, RLV).

Cianciolo GJ Anti-inflammatory proteins associated with human and murine neoplasms. BBA 865: 69-82 (1986)

MCIF 2: *macrophage cytotoxicity-inducing factor* A poorly characterized factor found in the supernatant of murine T cells in limiting dilution cultures and a long-term T cell clone. MCIF2 synergizes with » IFN-γ in inducing macrophage tumor cytotoxicity; IFN-γ and MCIF2 alone are ineffective. It appears that IFN-γ serves as the priming and MCIF2 as the triggering signal for macrophage activation (see also: Cell activation).

Hamann U & Krammer PH Activation of macrophage tumor cytotoxicity by the synergism of two T cell-derived lymphokines: immune interferon (IFN-γ) and macrophage cytotoxicity-inducing factor 2 (MCIF2). EJI 15: 18-24 (1985); **Krammer PH et al** Priming and triggering of tumoricidal and schistosomulicidal macrophages by two sequential lymphokine signals: interferon-γ and macrophage cytotoxicity inducing factor 2. JI 135: 3258-63 (1985)

MCP: *monocyte chemotactic protein* This factor is encoded by the human » JE gene. It is chemotactic for monocytes (see also: Chemotaxis). This factor is identical with » MCAF (monocyte chemotactic and activating factor), » MCP-1 (monocyte chemoattractant protein 1), » HC11, » SMC-CF (smooth muscle cell chemotactic factor), » LDCF (lymphocyte-derived chemotactic factor), » GDCF (glioma-derived monocyte chemotactic factor), » TDCF (tumor-derived chemotactic factor).

MCP belongs to the » chemokine family of cytokines. It has been shown to bind to a receptor called » CKR-1 (chemokine receptor 1).

Decock B et al Identification of the monocyte chemotactic protein from human osteosarcoma cells and monocytes: detection of a novel N-terminally processed form. BBRC 167: 904-9 (1990); **Shyy YJ et al** Structure of human monocyte chemotactic protein gene and its regulation by TPA. BBRC 169: 346-51 (1990)

MCP-1 - MCP2: *microbicidal cationic protein* These microbicidal proteins have been isolated from rabbit lung macrophages. They are identical with the »defensins NP-1 and NP-2, respectively.

Patterson-Delafield J et al Microbicidal cationic proteins in rabbits alveolar macrophages: a potential host defense mechanism. Infect. Immun. 30: 180-92 (1980)

MCP-1: *monocyte chemoattractant protein-1* MCP-1 is encoded by the human » JE gene at chromosome 17q11.2-q21.1, also known as »***SCY A2*** (small inducible cytokine A2). Antibodies directed against murine MCP-1 cross-react with the human factor. The human factor is 49 amino acids shorter at the aminoterminus than the rodent factor.

MCP-1 belongs to the » chemokine family of cytokines. It is identical with » MCAF (monocyte chemotactic and activating factor), » MCP (monocyte chemotactic protein), » HC11, » SMC-CF (smooth muscle cell chemotactic factor), » LDCF (lymphocyte-derived chemotactic factor), » GDCF (glioma-derived monocyte chemotactic factor), » TDCF (tumor-derived chemotactic factor) and » P6. MCP-1 shows a high degree of homology with » MARC protein).

MCP-1 is expressed by monocytes, vascular endothelial cells, smooth muscle cells, glomerular mesangial cell, osteoblastic cells, and human pulmonary type II-like epithelial cells in culture. It is constitutively produced by the human glioma U-105MG cell line. MCP1 mRNA is induced in human peripheral blood mononuclear leukocytes by phytohemagglutinin (PHA), lipopolysaccharide, and » IL1, but not by » IL2, » TNF, or » IFN-γ. In mesangial cells the synthesis and release of MCP-1 is rapidly induced by IgG complexes, but not monomeric IgG or F(ab')2 fragments of IgG.

MCP-1 is chemotactic for monocytes but not neutrophils. Maximal induction of migration is observed at a concentration of 10 ng/ml (see also: Chemotaxis). Point mutations have been described at two amino acid positions which alter the factor so that it is then also chemotactic for neutrophils. Elevated levels of MCP-1 are observed in macrophage-rich atherosclerotic plaques. The factor activates the tumoricidal activity of monocytes and macrophages *in vivo*. It regulates the expression of cell surface antigens (CD11c, CD11b) and the expression of cytokines (IL1, IL6). MCP-1 is a potent activator of human basophils, inducing the degranulation and the release of histamines. In basophils activated by » IL3, » IL5 or » GM-CSF MCP-1 enhances the synthesis of leukotriene C4. IL1, TNA-α, PDGF, TGF-β, and LIF induce the synthesis of MCP-1 in human articular chondrocytes, which may thus play an active role in the initiation and progression of degenerative and inflammatory arthropathies by promoting monocyte influx and activation in synovial joints.

Beall CJ et al Conversion of monocyte chemoattractant protein-1 into a neutrophil attractant by substitution of two amino acids. JBC 267: 3455-9 (1992); **Bischoff SC et al** Monocyte chemotactic protein 1 is a potent activator of human basophils. JEM 175: 1271-5 (1992); **Bischoff SC et al** RANTES and related chemokines activate human basophil granulocytes through different G protein-coupled receptors. EJI 23: 761-7 (1993);

MCP-1: see addendum.

Evanoff HL et al A sensitive ELISA for the detection of human monocyte chemoattractant protein-1 (MCP-1). Immunol. Invest. 21: 39-45 (1992); **Hora K et al** Receptors for IgG complexes activate synthesis of monocyte chemoattractant peptide 1 and colony-stimulating factor 1. PNAS 89: 1745-9 (1992); **Jiang Y et al** Monocyte chemoattractant protein-1 regulates adhesion molecule expression and cytokine production in human monocytes. JI 148: 2423-8 (1992); **Kuna P et al** Characterization of the human basophil response to cytokines, growth factors, and histamine releasing factors of the intercrine/chemokine family. JI 150: 1932-43 (1993); **Leonard EJ et al** Human monocyte chemoattractant protein-1 (MCP-1). IT 11: 97-101 (1990); **Leonard EJ et al** Biological aspects of monocyte chemoattractant protein-1 (MCP-1). AEMB 305: 57-64 (1991); **Matsushima K et al** Purification and characterization of a novel monocyte chemotactic and activating factor produced by a human myelomonocytic cell line. JEM 169: 1485-90 (1989); **Mehrabian M et al** Localization of monocyte chemotactic protein-1 gene (SCYA2) to human chromosome 17q11.2-q21.1. Genomics 9: 200-1 (991); **Robinson EA et al** Complete amino acid sequence of a human monocyte chemoattractant, a putative mediator of cellular immune reactions. PNAS 86: 1850-4 (1989); **Rollins BJ** JE/MCP-1: An early-response gene encodes a monocyte-specific cytokine. Cancer Cells 3: 517-24 (1991); **Rovin BH et al** Cytokine-induced production of monocyte chemoattractant protein-1 by cultured human mesangial cells. JI 148: 2148-53 (1992); **Standiford TJ et al** Alveolar macrophage-derived cytokines induce monocyte chemoattractant protein-1 expression from human pulmonary type II-like epithelial cells. JBC 266: 9912-8 (1991); **Valente AJ et al** Characterization of monocyte chemotactic protein-1 binding to human monocytes. BBRC 176: 309-14 (1991); **Van Damme J et al** Identification by sequence analysis of chemotactic factor for monocytes produced by normal and transformed cells stimulated with virus, double-stranded RNA, or cytokine. EJI 19: 2367-73 (1989); **Villiger PM et al** Monocyte chemoattractant protein-1 (MCP-1) expression in human articular cartilage. Induction by peptide regulatory factors and differential effects of dexamethasone and retinoic acid. JCI 1 90: 488-96 (1992); **Wempe F et al** Gene expression and cDNA cloning identified a major basic protein constituent of bovine seminal plasma as bovine monocyte-chemoattractant protein-1 (MCP-1). DNA Cell Biol. 10: 671-9 (1991); **Williams SR et al** Regulated expression of monocyte chemoattractant protein-1 in normal human osteoblastic cells. Am. J. Physiol. 263: C194-9 (1992); **Ylä-Herttuala S et al** Expression of monocyte chemoattractant protein 1 in macrophage-rich areas of human and rabbit atherosclerotic lesions. PNAS 88: 5252-6 (1991); **Yoshimura T et al** Purification and amino acid analysis of two human glioma-derived monocyte chemoattractants. JEM 169: 1449-59 (1989); **Yoshimura T et al** Human monocyte chemoattractant protein (MCP-1): full-length cDNA cloning, expression in mitogen-stimulated blood mononuclear leukocytes, and sequence similarity to mouse competence gene JE. FL 244: 487-93 (1989); **Yoshimura T et al** Identification of high affinity receptors for human monocyte chemoattractant protein-1 (MCP-1) on human monocytes. JI 145: 292-7 (1990); **Yoshimura T et al** Production and characterization of mouse monoclonal antibodies against human monocyte chemoattractant protein-1. JI 147: 2229-33 (1991); **Yoshimura T et al** Molecular cloning of rat monocyte chemoattractant protein-1 (MCP-1) and its expression in rat spleen cells and tumor cell lines. BBRC 174: 504-9 (1991); **Yoshimura T & Leonard EJ** Human monocyte chemoattractant protein-1 (MCP-1). AEMB 305: 47-56 (1991); **Yoshimura T & Leonard EJ** Human monocyte chemoattractant protein-1: Structure and function. Cytokines 4: 131-52 (1992); **Yoshimura T & Yuhki N** Neutrophil attractant/activation protein-1 and monocyte chemoattractant protein-1 in rabbit: cDNA cloning and their expression in spleen cells. JI 146: 3483-8 (1991)

MCP-2: *monocyte chemoattractant protein-2* This factor is secreted by the human osteosarcoma cell line MG-63 which also secretes some other chemotactic factors including » IL8 and » MCP-1. MCP-2 and also another factor, ***MCP-3*** (designated NC28 cDNA) are closely related to » MCP-1. The MCP-3 protein sequence shows 74% identity with MCP-1 and 58 % homology with MCP-2. Secreted MCP-3 differs from MCP-1 in being N-glycosylated. MCP-2 appears to be identical with » HC14. Both factors specifically attract monocytes, but not neutrophils, *in vitro*. Intradermal injection of these two proteins into rabbits causes the selective recruitment of monocytes at the site of injection. Levels of MCP-3 mRNAs in peripheral blood mononuclear cells are increased by » IFN-γ and decreased by » IL13.

MCP-2 and MCP-3 show a high degree of homology (≈ 50 %) at the protein level with » MARC protein and human MCP-3 may be the human homologue of the mouse MARC gene. MCP-3 is often produced by tumor cell lines and regulates protease secretion by macrophages; its production may therefore contribute to invasion and metastasis of cancer cells.

Minty A et al Molecular cloning of the MCP-3 chemokine gene and regulation of its expression. ECN 4: 99-110 (1993); **Opdenakker G et al** Human monocyte chemotactic protein-3 (MCP-3): molecular cloning of the cDNA and comparison with other chemokines. BBRC 191: 535-42 (1993); **Van Damme J et al** Structural and functional identification of two human, tumor-derived monocyte chemotactic proteins (MCP-2 and MCP-3) belonging to the chemokine family. JEM 176: 59-65 (1992)

MCP-3: *monocyte chemoattractant protein-3* see: MCP-2.

MCSA: *multi-colony stimulating activity* This factor is identical with » IL3. See: Multi-CSF.

M-CSF: *macrophage colony-stimulating factor*
■ **Alternative Names:** CSA (colony stimulating activity); CSF (colony stimulating factor; especially in the older literature); CSF-1 (colony stimulating factor 1); L-cell CSF (L cell colony stimulating factor); MGF (macrophage growth factor); MGI-1M (macrophage-granulocyte inducer); Urinary colony -stimulating factor (abbrev. CSF-HU). See also: individual entries for further information.
■ **Sources:** M-CSF is produced by monocytes, granulocytes, endothelial cells, and fibroblasts.

Activated B cells and T cells (see also: Cell activation) and also a number of tumor cell lines are also capable of synthesizing this factor. M-CSF has been found to be synthesized by uterine epithelial cells *in vivo*. The factor is also found in human urine.
The synthesis of M-CSF can be induced by » IL1, » TNF-α, » IFN-γ, » GM-CSF and » PDGF. Prostaglandins, glucocorticoids, » TGF-β and substances that raise intracellular levels of cAMP inhibit the synthesis of M-CSF.
The murine *op* allele (see: *op/op* mice) is caused by a mutation of the M-CSF gene. These mice are characterized by a complete absence of this factor.
■ **PROTEIN CHARACTERISTICS:** M-CSF is a homodimeric glycoprotein the subunits of which are linked by disulfide bonds. The sugar moiety is not required for the full spectrum of biological activities. Different molecular forms of M-CSF with lengths of 256 (***M-CSF-α***), 554 (***M-CSF-β***) and 438 (***M-CSF-γ***) amino acids have been described. They arise by translation of alternatively spliced mRNAs. Another splice variant is called ***CSF-4***.
M-CSF-β is a secreted protein that does not occur in a membrane-bound form. M-CSF-α is expressed as an integral membrane protein that is slowly released by proteolytic cleavage. The membrane-bound form of M-CSF can interact with receptors on near-by cells and therefore mediates specific cell-to-cell contacts (see: Juxtacrine). Some high-molecular weight forms of murine M-CSF have been described. These forms are complexed with proteoglycan of the intracellular matrix (see: ECM) and interact with collagen type V. A comparison of the primary sequence of M-CSF with those of the other human colony-stimulating factors (GM-CSF, G-CSF) reveals that the three factors are not related to each other.
■ **GENE STRUCTURE:** The human M-CSF has a length of ≈ 20 kb and contains ten exons. The gene was originally located to human chromosome 5q33 in close proximity to genes encoding other hematopoietic growth factors » GM-CSF, » IL3, » IL4 and » IL5. The gene has now been reassigned to chromosome 1p13-p212 in the vicinity to the amylase genes. The murine M-CSF gene has been shown to be the site of the mutation in a form of osteopetrosis (see: op/op).
■ **RECEPTOR STRUCTURE, GENE(S), EXPRESSION:** The biological activities of M-CSF are mediated by a receptor of165 kDa encoded by a gene mapping to human chromosome 5q33. 3. The M-CSF receptor is identical with the proto-oncogene » *fms*. It has been renamed ***CD115*** (see also: CD Antigens). The gene is ≈ 800 bp from the » PDGF receptor gene. The expression of both receptor genes is probably controlled by common regulatory sequences since the M-CSF receptor gene does not possess regulatory regions of its own.
The receptor is a transmembrane protein with an extracellular ligand-binding domain of 512 amino acids, an intramembrane domain of 25 amino acids, and a cytoplasmic domain of 435 amino acids encoding a bipartite tyrosine kinase interrupted by a so-called kinase insert (see also: PTK, protein tyrosine kinase).
Binding of the ligand leads to the » autophosphorylation of the receptor. Ligand/receptor complexes are internalized by the cell and degraded. Effects associated with ligand binding include the stimulation of protein biosynthesis, glucose transport, and Na^+/K^+-ATPase activity. Intracellular signaling also involves a G-protein. After binding of M-CSF to its receptor the receptor associates with non-receptor protein kinases called » *fyn* and » *yes*, which appear to mediate some of the intracellular signals.
M-CSF circulating in the serum is mainly removed by binding to macrophage receptors and subsequent internalization and degradation.
Experiments with a functional M-CSF receptor gene have been used to study the plasticity of B lymphoid progenitor cells. Murine cells develop into which a cloned receptor gene was introduced can differentiate into macrophages in the presence of M-CSF. Similar results are obtained by introduction of » *myc* and *raf* » oncogenes into normal bone marrow cells using a retrovirus vector. This phenomenon is known as lineage switching. The switching process can be inhibited by » IL7 in these cells. Ca. 5-10 % of all acute leukemias are biphenotypic and express lymphoid and myeloid markers. These leukemias cannot be assigned to a distinct lineage and may be the result of transformation of biphenotypic progenitor cells.
■ **BIOLOGICAL ACTIVITIES:** Human M-CSF is active in mouse and rat cells. The murine factor is active in rat cells but inactive in human cells.
M-CSF was isolated initially as a factor stimulating the growth of macrophage/granulocyte-containing colonies in soft agar cultures (see also: Colony formation assay). M-CSF influences the proliferation and differentiation of » hematopoietic stem cells into macrophages but mainly the growth survival and differentiation of monocytes. In combination with another colony-stimulating

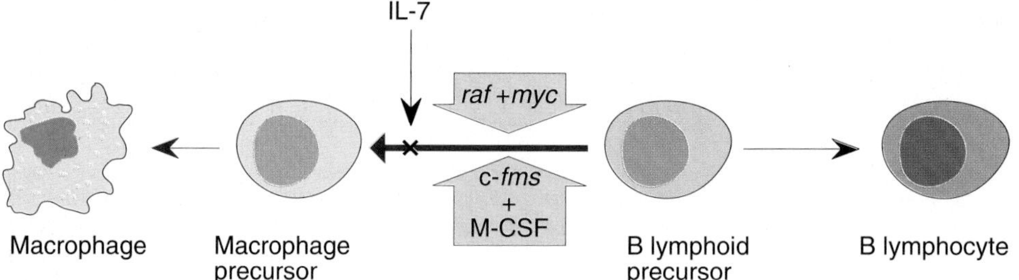

Plasticity of B lymphoid progenitor cells: lineage switching.
B lymphoid progenitor cells display a certain plasticity and can differentiate into macrophages in response to suitable stimuli. This is shown in experiments involving the introduction of a cloned M-CSF receptor gene (c-fms) into murine B lymphoid progenitors. In the presence of M-CSF these cells differentiate into macrophages. The switch is blocked by IL7. Similar results are obtained by introduction of the oncogenes v-raf and v-myc into normal bone marrow cells. Approx. 5-10 % of acute leukemias have been shown to express lymphoid and myeloid cell markers and thus cannot be assigned to either of the two lineages. It has been suggested that these cells arise after transformation of normal biphenotypic progenitor cells (Cell 64: 337-350 (1991)).

factor, » GM-CSF, one observes the phenomenon of synergistic suppression, i. e., the combination of these two factors leads to a partial suppression of the generation of macrophage-containing cell colonies

M-CSF is a specific factor in that the proliferation-inducing activity is more or less restricted to the macrophage lineage. M-CSF also is a potent stimulator of functional activities of monocytes.

In normal human macrophages M-CSF induces antibody-dependent cellular cytotoxicity (see: ADCC).

In monocytes and macrophages M-CSF induces the synthesis of » IL1, » G-CSF, » IFN, » TNF, plasminogen activator, thromboplastin, prostaglandins and thromboxanes and also oxidative metabolism. M-CSF synergises with » IL1, » IL3 and » IL6 in the stimulation of proliferation and the differentiation of primitive hematopoietic cells into macrophages.

Since M-CSF is also synthesized by uterine epithelial cells it may be involved in the maintenance of normal placental functions, acting on the decidua and trophoblasts.

In experimental animals M-CSF reduces plasma levels of cholesterol.

In humans M-CSF enhances monocyte counts and leads to an expansion of neutrophilic granulocytes.

■ **Assays:** M-CSF can be assayed in a » colony formation assay by the development of colonies containing macrophages. M-CSF is also detected in specific » bioassays with cells lines that depend in their growth on the presence of M-CSF or that respond to this factor (see: BAC1. 2F5; GNFS-50; J774). An alternative and entirely different detection method is » Message amplification phenotyping. For further information see also subentry "Assays" in the reference section.

■ **Clinical Use & Significance:** M-CSF may be clinically relevant in its capacity to reconstitute the hematopoietic system in combination with other hematopoietic factors.

M-CSF accelerates the recovery of the leukocyte pool following bone marrow transplantation. In some cases of acute myeloid leukemia M-CSF induces the terminal differentiation of these cells into cell types that do no longer proliferate.

M-CSF may be a tumor marker for ovarian tumors and tumors of the endometrium for which it may act as an » autocrine growth factor. Since M-CSF induces the synthesis of some inflammatory proteins (see also: Inflammation) it may be involved in inflammatory reactions which should then be amenable to manipulation with suitable inhibitors.

● **Reviews: Ralph P et al** Macrophage growth and stimulating factor, M-CSF. PCBR 338: 43-63 (1990); **Roth P & Stanley ER** The biology of CSF-1 and its receptor. CTMI 181: 141-67 (1992)

● **Biochemistry & Molecular Biology: Ben-Avram CM et al** Isolation and expression of complementary DNAs encoding the human interleukin 2 receptor. PNAS 82: 4486-9 (1985); **Cerretti TR et al** Human macrophage colony stimulating factor: alternative RNA and protein processing from a single gene. Mol. Immunol. 25: 761-70 (1988); **DeLamarter JF et al** Nucleotide sequence of a cDNA encoding murine CSF-1 (macrophage CSF). NAR 15: 2389-90 (1987); **Kawasaki ES et al** Molecular cloning of a complementary DNA encoding human macrophage-specific colony stimulating factor (CSF-1) S 230: 291-6 (1985); **Ladner MB et al** Human CSF-1: gene structure and alternative splicing of mRNA precursors. EJ 6: 2693-8 (1987); **Ladner MB et al** cDNA cloning and expression of murine CSF-1 from L929 cells. PNAS 85: 6706-10 (1988); **Le Beau MM et al** Assignment of the GM-CSF, CSF-1, and *fms* genes to human chromosome 5 provides evidence for linkage of a family of genes regulating hematopoiesis and for their involvement in the deletion (5q) in myeloid disorders. CSHSQB 51: 899-909 (1986); **Morris SW et al** Reassignment of the human CSF1 gene to chromosome 1p13-p21. Blood 78: 2013-20 (1991); **Pampfer S et al** Expression of colony-stimulating factor 1 (CSF-1) messenger RNA in human endometrial glands during the menstrual cycle: molecular cloning of a novel transcript that predicts a cell surface form of CSF-1. Mol. Endocrinol. 5: 1931-8 (1991); **Price LKH et al** The predominant form of secreted colony stimulating factor-1 is a proteoglycan. JBC 267: 2190-9 (1992); **Rajavashisth TB et al** Cloning and tissue specific expression of mouse macrophage colony-stimulating factor mRNA. PNAS 84: 1157-61 (1987); **Ryseck RP et al** The macrophage-colony stimulating factor gene is a growth factor-inducible immediate early gene in fibroblasts. The New Biologist 3: 151-7 (1991); **Saltman DL et al** Reassignment of the human macrophage colony stimulating factor gene to chromosome 1p13-21. BBRC 182: 1139-43 (1992); **Suzu S et al** Biological activity of a proteoglycan form of macrophage colony-stimulating factor and its binding to type V collagen. JBC 267: 16812-5 (1992); **Wong GG et al** Human CSF-1: molecular cloning and expression of 4-kb cDNA encoding the human urinary protein. S 235: 1504-8 (1987)

● **Receptors: Eccles MR** Genes encoding the platelet-derived growth factor (PDGF) receptor and colony-stimulating factor 1 (CSF-1) receptor are physically associated in mice as in humans. Gene, 108: 285-8 (1991); **Rettenmier CW et al** The colony-stimulating factor (CSF-1) receptor (c-*fms* proto-oncogene product) and its ligand. JCS Suppl. 9: 27-44 (1988); **Ridge SA et al** *fms* mutations in myelodysplastic, leukemic, and normal subjects. PNAS 87: 1377-80 (1990); **Roberts WM et al** Tandem linkage of human CSF-1 receptor (c-*fms*) and PDGF receptor genes. Cell 55: 655-61 (1988); **Roberts WM et al** Transcription of the human colony-stimulating factor-1 receptor gene is regulated by separate tissue-specific promoters. Blood 79: 586-93 (1992); **Rothwell VM & Rohrschneider LR** Murine c-*fms* cDNA: cloning, sequence analysis, and retroviral expression. OR 1: 311-24 (1987); **Roussel MF et al** A point mutation in the extracellular domain of the human CSF-1 receptor (c-*fms* proto-oncogene product) activates its transforming potential. Cell 55: 973-88 (1988); **Sherr CJ** Colony-stimulating factor-1-receptor. Blood 75: 1-12 (1990); **Suzu S et al** Identification of a high molecular weight macrophage colony-stimulating factor as a glycosaminoglycan-containing species. JBC 267: 4345-8 (1992)

● **Biological Activities: Arceci RJ et al** The temporal expression and location of colony stimulating factor-1 (CSF-1) and its receptor in the female reproductive tract are consistent with CSF-1 regulated placental development. PNAS 86: 8818-22 (1989); **Hattersley G et al** Macrophage colony stimulating factor (M-CSF); is essential for osteoclast formation *in vitro*. BBRC 177: 526-31 (1991); **Metcalf D et al** Synergistic suppression: anomalous inhibition of the proliferation of factor-dependent hemopoietic cells by combination of two colony-stimulating factors. PNAS 89: 2819-23 (1992); **Sherr CJ** Mitogenic response to colony-stimulating factor 1. TIG 7: 398-402 (1991)

● **Assays: Das SK et al** Human colony-stimulating factor (CSF-1) radioimmunoassay: resolution of three subclasses of human colony-stimulating factors. Blood 58: 630-41 (1981); **Hanamura T et al** Quantitation and identification of human monocytic colony-stimulating factor in human serum by enzyme-linked immunosorbent assay. Blood 72: 886-92 (1988); **Kawano Y et al** Measurement of serum levels of macrophage colony-stimulating factor (M-CSF) in patients with uremia. EH 21: 220-3 (1993); **Suzu S et al** Characterization of macrophage colony-stimulating factor in body fluids by immunoblot analysis Blood 77: 2160-5 (1991); for further information also see individual cell lines used in individual bioassays.

● **Clinical Use & Significance: Garnick MB & Stoudemire JB** Preclinical and clinical evaluation of recombinant human macrophage colony-stimulating factor (rhM-CSF). IJCC 8: 356-73 (1990); **Herrmann F et al** G-CSF and M-CSF: preclinical and clinical results. Immunol. Ser. 57: 651-60 (1992); **Munn DH & Cheung NK** Preclinical and clinical studies of macrophage colony-stimulating factor. Semin. Oncol. 19: 395-407 (1992); **Redman BG et al** Phase I trial of recombinant macrophage colony-stimulating factor by rapid intravenous infusion in patients with cancer. J Immunother. 12: 50-4 (1992); **Van de Pol CJ & Garnick MB** Clinical applications of recombinant macrophage-colony-stimulating factor (rhM-CSF). Biotechnol. Ther. 2: 231-9 (1991); **Zamkoff KW et al** A phase I trial of recombinant human macrophage colony-stimulating factor by rapid intravenous infusion in patients with refractory malignancy. J. Immunother. 11: 103-10 (1992)

M-CSF-α to **M-CSF-γ:** see: M-CSF.

MCT: *macrophage cytotoxin* This factor is identical with » TNF-α.

MC-TAF: *mesangial cell thymocyte activating factor* This factor, produced by mesangial cells, is most likely identical with one of the molecular forms of » IL1.

Lovett DH et al Thymocyte activating factor produced by glomerular mesangial cells. In: Interleukins, Lymphokines, and Cytokines. Academic Press, pp. 415-8 , New York 1983

M-CTX: This poorly characterized factor is isolated from membrane fractions obtained from highly purified » LAK cells (lymphokine-activated killer cells). The factor is cytolytic for a number of target cells and induces their death by » apoptosis. The factor is not identical with » perforins which also possess cytolytic activity. The activity of M-CTX cannot be inhibited by antibodies directed against » TNF-α, TNF-β and IFN-γ.

Felgar RE & Hiserodt JC Identification and partial characterization of a novel plasma membrane-associated lytic factor isolated from highly purified adherent lymphokine-activated killer cells. CI 141: 32-46 (1992)

MD10 cells: see: D10 cells.

MDAA: *macrophage-derived angiogenic activity* This poorly characterized factor is secreted by activated macrophages (see also: Cell activation). It was observed during the purification of » MDGF (macrophage-derived growth factor) and may be identical with MDGF.

Shimokado K et al A significant part of macrophage-derived growth factor consists of at least two forms of PDGF. Cell 43: 277-86 (1985)

MDBK: Madin-Darby bovine kidney (ATCC CCL 22). This spontaneously transformed continuous epithelial cell line is used to assay human » IFN-α which is quantified by performing a cytopathic effect inhibition assay of vesicular stomatitis virus. The cells secrete Insulin-like growth factor binding protein IGFBP-2 (see: IGFBP).

Treatment of MDBK cells with human IFN)-α2 results in a dose-dependent inhibition of cell growth and decreases the number of cell-surface » EGF receptors and a reduction in the affinity of EGF for its receptor. Human IFN-β and -γ, which exhibit little antiviral and antiproliferative activities on MDBK cells, have little effect on cell growth or the binding of EGF to these cells.

Chloroquine, which raises the pH in lysosomes and endosomes, counteracts the induction by interferon of the antiviral state but has no effects on the induction of 2',5'-oligoadenylate synthetase. Ethanol induces 2',5'-oligoadenylate synthetase and antiviral activities through IFN-β production.

Binder S et al [Adaptation of permanent bovine kidney cells to serum free hormone supplemented cell culture media] Zentralbl. Veterinärmed. B. 37: 696-700 (1990); **Bridgman R et al** The sensitivity of domestic animal cell lines to eight recombinant human interferons. J. Interferon Res. 8: 1-4 (1988); **Capellier M et al** [Critical study of an assay of human leukocyte and lymphoblastoid interferon on MDBK cells, in microplaque, by the "dye uptake" method. I. Biological factors] Ann. Pharm. Fr. 44: 217-27 (1986); **Capellier M et al** [Critical study of the titration of human leukocytic and lymphoblastoid interferon on MDBK cells, in microplaque by the so-called dye-uptake method. II. Statistical analysis] Ann. Pharm. Fr. 44: 285-91 (1986); **Chelbi-Alix MK & Thang MN** Multiple molecular forms of interferon display different specific activities in the induction of the antiviral state and 2'5' oligoadenylate synthetase. BBRC 141: 1042-50 (1986); **Chelbi-Alix MK & Thang MN** Chloroquine impairs the interferon-induced antiviral state without affecting the 2',5'-oligoadenylate synthetase. JBC 260: 7960-4 (1985); **Chelbi-Alix MK & Chousterman S** Ethanol induces 2',5'-oligoadenylate synthetase and antiviral activities through interferon-β production. JBC 267: 1741-5 (1992); **Feinstein S et al** Studies on cell binding and internalization of human lymphoblastoid (Namalva) interferon. J. Interferon Res. 5: 65-76 (1985); **Kramer MJ et al** Cell and virus sensitivity studies with recombinant human α interferons. J. Interferon Res. 3: 425-35 (1983); **Ransohoff RM et al** Effect of human α A interferon on influenza virus replication in MDBK cells. J. Virol. 56: 1049-52 (1985); **Tonew M & Gluck B** [Interferon sensitivity of various cell lines] J. Basic Microbiol. 26: 173-9 (1986); **Zoon KC et al** Modulation of epidermal growth factor receptors by human α interferon. PNAS 83: 8226-30 (1986)

MDC: *macrophage-derived chemoattractant* A heterogeneous poorly characterized group of protein factors of 12-15 kDa which are secreted by murine macrophages and monocytes.

These factors are potent chemoattractants for neutrophilic granulocytes. They also enhance their bactericidal activity and stimulate the release of lysozyme. Some of these proteins may be members of the » chemokine family of cytokines.

Yoshimura T et al Neutrophil chemotactic factor produced by lipopolysaccharide (LPS)-stimulated human blood mononuclear leukocytes. Partial characterization and separation from interleukin 1 (IL 1). JI 139: 788-93 (1987)

MD-ECI: *macrophage-derived endothelial cell inhibitor* This poorly characterized factor is secreted by human mononuclear cells and isolated by heparin affinity chromatography (see also: HBGF). MD-ECI inhibits basal and FGF-stimulated endothelial cell growth. Its effect on endothelial cells is dose-dependent, nontoxic, and reversible. MD-ECI is distinct from the known endothelial cell growth inhibitors such as » TGF-β and TNF-α.

Besner GE & Klagsbrun M Macrophages secrete a heparin-binding inhibitor of endothelial cell growth. Microvasc. Res. 42: 187-97 (1991)

MDF: *macrophage deactivation factor* Two species of MDF (13 kDa and 66 kDa) have been described. MDF activity is secreted by a number of murine normal and tumor cell lines. It reversibly blocks the » IFN-γ-induced enhanced oxygen metabolism, the production of nitrogen oxides, the capacity to kill intracellular parasites, and the expression of Ia antigens in macrophages and monocytes. Activated macrophages become deactivated (see also: Cell activation).

The 66 kDa protein appears to be identical with albumin according to the known N-terminal sequences. The other protein is not identical with » TGF-β which also possesses MDF activity. In addition, » IL4, » IL10 and » TGF-β have been shown also to have strong macrophage-deactivating effects although they cannot be categorized as purely macrophage deactivating since they also exert macrophage-activating effects.

Bogdan C & Nathan C Modulation of macrophage function by transforming growth factor β, interleukin-4, and interleukin-10. ANY 685: 713-39 (1993); **Ding A et al** Macrophage deactivating factor and transforming growth factor-β 1, -β-2, and β-3 inhibit induction of macrophage nitrogen oxide synthesis by IFN-γ. JI 145: 940-4 (1990); **Ellermann-Eriksen S et al** Differential sensitivity of macrophages from herpes simplex virus-resistant and -susceptible mice to respiratory burst priming by interferon-α/β. J. Gen. Virol. 70: 2139-47 (1989); **Nathan CF & Tsunawaki S** Secretion of toxic oxygen products by macrophages: regulatory cytokines and their effects on the oxidase. Ciba Found. Symp. 118: 211-30 (1986); **Nathan C** Mechanism and modulation of macrophage activation. Behring Inst. Mitt. Feb 88: 200-7 (1991); **Srimal S & Nathan C** Purification of macrophage deactivating factor. JEM 171: 1347-61 (1990); **Tsunawaki S & Nathan CF** Macrophage deactivation. Altered kinetic properties of superoxide-producing enzyme after exposure to tumor cell-conditioned medium. JEM 164: 1319-31 (1986); **Tsunawaki S et al** Comparison of transforming growth factor β and a macrophage-deactivating polypeptide from tumor cells. Differences in antigenicity and mechanism of action. JI 142: 3462-8 (1989)

MDF: *macrophage-derived deactivating factor* This poorly characterized factor (3 and 11 kDa) is released by granuloma-like lesions of mice (giant and epithelioid macrophages). The factor strongly inhibits production of superoxide anions in mouse, guinea pig, human, macrophages and mouse neutrophils activated with » phorbol esters, all-trans-retinal, or fMet-Leu-Phe.

Camarero VC et al Leukocyte-deactivating factor from macrophages: partial purification and biochemical characterization. A novel cytokine. JCP 157: 84-9 (1993)

MDF: *mesangial cell proliferating factors* MDF activities (100-70 and 8-12 kDa) are found in the conditioned media from human peripheral blood leukocytes treated with bacterial lipopolysaccharides. Both MDFs induce the proliferation of cultured mesangial cells and do not induce the proliferation of thymocytes. MDF is different from » IL1 or » IL6. The activity of the low molecular weight factor is partially inhibited by antibodies directed against PDGF, suggesting that some of the activity may be due to PDGF. The high molecular weight factor is not affected by anti-PDGF antibodies.

Morioka T et al Production by cultured human monocytes of mesangial cell proliferation factor(s) differing from interleukin-1 and interleukin-6. Clin. Exp. Immunol. 83: 182-6 (1991)

MDGF: *macrophage-derived growth factor* These factors are also called macrophage-derived fibroblast growth factor. MDGF is a collective term for some factors secreted by activated macrophages (see also: Cell activation) which stimulate the proliferation of fibroblasts, smooth muscle cells and endothelial cells. Some MDGF activities are probably identical with those called ***AMDGF*** (alveolar macrophage-derived growth factor).
The major fraction of MDGF activity is caused by » PDGF. Some MDGF activity may also be due to fibroblast growth factors (see: FGF), probably » aFGF and sometimes referred to as ***ir-FGF*** (immunoreactive fibroblast growth factor). See also: MDGF (monocyte-derived growth factor).

Baird A et al Immunoreactive fibroblast growth factor in cells of peritoneal exudate suggests its identity with macrophage-derived growth factor. BBRC 126: 358-64 (1985); **Baird A et al** Immunoreactive fibroblast growth factor in cells of peritoneal exudate suggests its identity with macrophage-derived growth factor. BBRC 126: 358-64 (1985); **Denholm EM & Phan SH** The effects of bleomycin on alveolar macrophage growth factor secretion. Am. J. Pathol. 134: 355-63 (1989); **Dvonch YM et al** Changes in growth factor levels in human wound fluid. Surgery 112: 18-23 (1992); **Kovacs EJ & Kelley J** Lymphokine regulation of macrophage-derived growth factor secretion following pulmonary injury. Am. J. Pathol. 121: 261-8 (1985); **Shimokado K et al** A significant part of macrophage-derived growth factor consists of at least two forms of PDGF. Cell 43: 277-86 (1985)

MDGF: *melanoma-derived growth factor* This factor is identical with » bFGF. See also: MeGF (melanocyte growth factor).

Lobb RR et al Purification and characterization of heparin-binding endothelial cell growth factors. JBC 261: 1924-8 (1986)

MDGF: *monocyte-derived growth factor* This factor of 16 kDa is produced by human monocytes. It is a monomeric form of » PDGF. MDGF may have a different primary structure since it exhibits different sensitivities to either formic acid or CNBr cleavage compared to PDGF A or B chain molecules. See also: MDGF (macrophage-derived growth factor).

Pencev D & Grotendorst GR Human peripheral blood monocytes secrete a unique form of PDGF. OR 3: 333-42 (1988)

MDGF: *monocyte-derived growth factor* This poorly characterized factor (pI 5.0; ≈ 33-42 kDa) is found in conditioned medium of mitogen-stimulated human peripheral blood lymphocytes. It stimulates the proliferation of fibroblast and smooth muscle cells. This activity is enhanced by » IL1. MDGF does not effectively compete with » PDGF for receptor binding and is not inhibited by PDGF antibodies. See also: MDGF (macrophage-derived growth factor).

Bonin PD et al Interleukin-1 promotes proliferation of vascular smooth muscle cells in coordination with PDGF or a monocyte derived growth factor. ECR 181: 475-82 (1989); **Singh JP & Bonin PD** Purification and biochemical properties of a human monocyte-derived growth factor. PNAS 85: 6374-8 (1988)

MDGF-1: *mammary-derived growth factor* This is an acidic (pI = 4,8) N-glycosylated protein of 62 kDa which is isolated from rodent mammary tumors, primary human mammary tumors and human milk. The protein does not appear to be related to any other cytokine according to the known N-terminal sequence of 18 amino acids.
MDGF-1 selectively amplifies the production of type IV collagen in cultures of mammary cells. It also stimulates the proliferation of normal and malignant rodent and human mammary epithelial cells and may function as an » autocrine or » paracrine factor produced by and acting on normal and malignant human breast epithelial cells possessing MDGF1 receptors.

The biological activity of this protein is mediated by a poorly characterized receptor of 120-140 kDa. Binding of MDGF-1 to its receptor leads to the tyrosine-specific phosphorylation of an unknown membrane-associated protein of 180-185 kDa specifically expressed in receptor-positive cells. Binding of MDGF-1 to its receptor is specific and is not competed by an excess of » EGF, » bFGF, and » IGF-1.

Bano M et al Purification of a mammary-derived growth factor from human milk and human mammary tumors. JBC 260: 5745-52 (1985); **Bano M et al** Characterization of mammary-derived growth factor 1 receptors and response in human mammary epithelial cell lines. JBC 265: 1874-80 (1990); **Bano M et al** Production and characterization of mammary-derived growth factor 1 in mammary epithelial cell lines. B 31: 610-6 (1992); **Bano M et al** Receptor-induced phosphorylation by mammary-derived growth factor 1 in mammary epithelial cell lines. JBC 267: 10389-92 (1992); **Kurebayashi J et al** MDGF-1: a new tyrosine kinase-associated growth factor. In: Moody TW (edt), Growth Factors, Peptides and Receptors, pp 311-22, Plenum Press New York (1993)

MDGF-2: *milk-derived growth factor 2* This factor is isolated from defatted milk and acid/ethanol extracts of primary human mammary tumors. The factor is identical with » TGF-α. This factor may be identical with » MGF (milk growth factor) also isolated from human milk. Some of the mitogenic activity of milk reported in the literature may also be due to the presence of » IGF.

Baxter RC et al Immunoreactive somatomedin-C/insulin-like growth factor I and its binding protein in human milk. J. Clin. Endocrinol. Metab. 58: 955-9 (1984); **Zwiebel JA et al** Partial purification of transforming growth factors from human milk. CR 46: 933-9 (1986)

MDGI: *mammary-derived growth inhibitor* This protein of 14.5 kDa (133 amino acids) is also called ***mammary-derived growth inhibition factor***. The protein is isolated from bovine mammary glands. It does not appear to be related to any other known growth inhibitors.

The protein belongs to a family of structurally related proteins and shows 95 % sequence homology with FABP (fatty acid-binding protein). It is also closely related to CRBP (cellular retinoid-binding protein), P422 (adipocyte differentiation associated protein) and myelin P-2.

MDGI partially inhibits the growth of a number of normal and transformed epithelial cell lines established from mammary glands. The expression of MDGI mRNA and of the protein coincides with the differentiation of mammary glandular tissues. The protein may therefore be involved in the development of the mammary gland. MDGI has also been shown to influence the differentiation of murine pluripotent embryonic stem cell lines. MDGI also influences the β-adrenergic activities of heart myocytes.

Billich S et al Cloning of a full-length complementary DNA for fatty acid binding protein from bovine heart. EJB 175: 549-56 (1988); **Böhmer FD et al** A polypeptide growth inhibitor isolated from lactating bovine mammary gland (MDGI) is a lipid-carrying protein. JCBc 38: 199-204 (1988); **Böhmer FD et al** Antibodies against mammary-derived growth inhibitor (MDGI) react with a fibroblast growth inhibitor and with heart fatty acid binding protein. BBRC 148: 1425-31 (1987); **Brandt R et al** A 13 kilodalton protein purified from milk fat globule membranes is closely related to a mammary-derived growth inhibitor. B 27: 1420-5 (1988); **Grosse R et al** Purification, biological assay, and immunoassay of mammary-derived growth inhibitor. MiE 198: 425-40 (1991); **Kurtz A et al** Developmental regulation of mammary-derived growth inhibitor expression in bovine mammary tissue. JCB 110: 1179-89 (1990); **Lehmann W et al** Response of different mammary epithelial cell lines to a mammary-derived growth inhibitor (MDGI). Biomed. Biochim. Acta 48: 143-51 (1989); **Lehmann W et al** Effect of a mammary-derived growth inhibitor on the expression of the oncogenes c-*fos*, c-*myc* and c-*ras*. FL 244: 185-7 (1989); **Muller T et al** A mammary-derived growth inhibitor (MDGI) related 70 kDa antigen identified in nuclei of mammary epithelial cells. JCP 138: 415-23 (1989); **Schoentgen F et al** Fatty acid binding protein from bovine brain. Amino acid sequence and some properties. EJB 185: 35-40 (1989); **Spener F et al** Characteristics of fatty acid-binding proteins and their relationship to mammary-derived growth inhibitor. Mol. Cell. Biochem. 98: 57-68 (1990); **Vogel F et al** Characterization and function dependent localization of mammary-derived growth inhibitor (MDGI) in mammary glands of bovine and mice. Acta Histochem. Suppl. 40: 77-80 (1990); **Wallukat G et al** Modulation of the β-adrenergic response in cultured rat heart cells. II. Mammary-derived growth inhibitor (MDGI) blocks induction of β-adrenergic supersensitivity. Dissociation from lipid-binding activity of MDGI. Mol. Cell. Biochem. 102: 49-60 (1991); **Wobus AM et al** Differentiation-promoting effects of mammary-derived growth inhibitor (MDGI) on pluripotent mouse embryonic stem cells. Virchows Arch. B. Cell. Pathol. 59: 339-42 (1990)

Mdk: abbrev. also: MK. See: Midkine.

MDNAP: *monocyte-derived neutrophil activating peptide* This factor is identical with » IL8. It belongs to the » chemokine family of cytokines.

Lindley I et al Synthesis and expression in *Escherichia coli* the gene encoding monocyte-derived neutrophil-activating factor: biological equivalence between natural and recombinant neutrophil-activating factor. PNSA 85: 9199-203 (1988)

MDNCF: *monocyte-derived neutrophil chemotactic factor* MDNCF is isolated from monocytes. It is chemotactic for neutrophils (see also: Chemotaxis). The protein is not identical with » IL1 and » TNF initially thought to be responsible for this activity. Several forms of MDNCF (a, b, c) have been described. MDNCF-a and MDNCF-b differ from MDNCF-a by seven and five amino acids, respectively, at the N-terminus.

MDNCF is identical with an anionic neutrophil-activating protein, called » ANAP, isolated from psoriasis scales. This factor is identical with » IL8 and therefore belongs to the » chemokine family of cytokines.

Farina PR et al Monocyte-derived neutrophil chemotactic factor (MDNCF): a stimulator of neutrophil function. FJ 3: A1333 (1989); **Hotta K et al** Coding region structure of interleukin-8 gene of human lung giant cell carcinoma LU65C cells that produce LUCT/interleukin-8: homogeneity in interleukin-8 genes. Immunol. Lett. 24: 165-70 (1990), Erratum in Immunol. Lett. 27: 258 (1991); **Matsushima K et al** Molecular cloning of a human monocyte-derived neutrophil chemotactic factor (MDNCF) and the induction of MDNCF mRNA by interleukin 1 and tumor necrosis factor. JEM 167: 1883-93 (1988); **Matsushima K et al** Purification and characterization of a novel monocyte chemotactic and activating factor produced by a human myelomonocytic cell line. JEM 169: 1485-90 (1989); **Modi WS et al** Monocyte-derived neutrophil chemotactic factor (MDNCF/IL8) resides in a gene cluster along with several other members of the platelet factor 4 gene superfamily. Hum. Genet. 84: 185-7 (1990); **Samanta AK et al** Identification and characterization of a specific receptor for monocyte-derived neutrophil chemotactic factor (MDNCF) on human neutrophils. JEM 169: 1185-9 (1989); **Schröder JM et al** Purification and partial biochemical characterization of a human monocyte-derived, neutrophil-activating peptide that lacks interleukin 1 activity. JI 139: 3474-83 (1987); **Tanaka S et al** Synthesis and biological characterization of monocyte-derived neutrophil chemotactic factor FL 236: 467-70 (1987); **Yoshimura T et al** Three forms of monocyte-derived neutrophil chemotactic factor (MDNCF) distinguished by different length of the amino-terminal sequence. Mol. Immunol. 84: 87-93 (1989); **Zipfel PF et al** Mitogenic activation of human T cells induces two closely related genes which share structural similarities with a new family of secreted factors. JI 142: 1582-90 (1989)

MEA: *mast cell growth enhancing activity* This factor is isolated from the growth medium of spleen cells. It synergises with » IL3 in promoting the growth of bone marrow mast cells and in inducing the production of » IL6 by these cells. The factor is identical with » IL9.

Hültner L et al Thiol-sensitive mast cell lines derived from mouse bone marrow respond to a mast cell growth-enhancing activity different from both IL3 and IL4 JI 142: 3440-6 (1989); **Hültner L et al** Mast cell growth-enhancing activity (MEA) is structurally related and functionally identical to the novel mouse T cell growth factor P40/TCGFIII EJI 20: 1413-6 (1990); **Hültner L & Moeller J et al** Mast cell growth-enhancing activity (MEA) stimulates interleukin 6 production in a mouse bone marrow-derived mast cell line and a malignant subline. EH 18: 873-7 (1990); **Moeller J et al** Purification of MEA, a mast cell growth-enhancing activity to apparent homogeneity and its partial amino acid sequencing. JI 144: 4231-4 (1990)

M-ECEF: *monocyte-derived eosinophil cytotoxicity-enhancing factor* see: ECEF.

MECIF: *monocyte-derived endothelial cell inhibitory factor* This poorly characterized factor is isolated from non-activated monocytes (see also: Cell activation). It inhibits the growth of human vascular endothelial cells and induces the expression of ELAM-1 (E-Selectin, endothelium-leukocyte adhesion receptor 1) and significantly increases prostacyclin secretion. MECIF activity appears to be immunologically distinct from » IL1b, » TNF-α, and » TNF-β.

Vilette D et al Identification of an endothelial cell growth inhibitory activity produced by human monocytes. ECR 188: 219-55 (1990)

MEF: *(leukocyte) migration enhancing factor* This is an operational definition of factors enhancing the migration of leukocytes. The MEF activity of factors can be inhibited by other factors that inhibit migration of these cells (see: MIF, migration inhibition factor). See also: Motogenic cytokines.

Salahuddin SZ et al Lymphokine production by cultured human T cells transformed by human T cell leukemia lymphoma virus I. S 223: 703-7 (1984); **Yamaki T et al** Functional analysis of mononuclear cells infiltrating into tumors. III. Soluble factors involved in the regulation of T lymphocyte infiltration into tumors. JI 140: 4388-96 (1988)

MEF: *macrophage migration enhancement factor* This poorly characterized factor of ≈ 28 kDa is isolated from suppressor cell-like lymphoid spleen cells. It displays positive chemokinetic but not chemotactic activity for macrophages (see also: Chemokinesis).

Gordon MR et al Induction of a macrophage migration enhancement factor after desensitization of tuberculin-positive rabbits with purified protein derivative. Infect. Immun. 51: 134-40 (1986); **Gordon MR & Myrvik QN** Biophysical characterization of macrophage migration enhancement factor. Mol. Immunol. 24: 1227-36 (1987)

Megakaryoblast growth factor: This factor is identical with » IL9.

Megakaryocyte colony stimulating activity: see: MEG-CSA.

Megakaryocyte colony stimulating factor: see: MEG-CSF.

Megakaryocyte potentiator: see: Mk-potentiator.

Megakaryocyte-stimulatory-factor: This factor isolated from rat bone marrow megakaryocytes is identical with » PF4 (platelet factor 4).

Greenberg SM et al *In vitro* stimulation of megakaryocyte maturation by megakaryocyte stimulatory factor. JBC 262: 3269-77 (1987)

Megakaryocyte colony stimulating factor: see addendum: thrombopoietin.
Megakaryocyte growth and development factor: see addendum: MGDF.
Megakaryocyte maturation factor: see addendum: thrombopoietin.

MEG-CSA: *megakaryocyte colony stimulating activity* This factor stimulates the growth of megakaryocyte-containing colonies in a » colony formation assay. It is identical with » IL3. See: Meg-CSF.

MEG-CSF: *megakaryocyte colony stimulating factor* abbrev. also: ***Mk-CSF***. Several factors stimulating the growth of megakaryocyte-containing colonies in a » colony formation assay have been described (see also: Hematopoiesis). One of the Meg-CSF activities is identical with » IL3. Other factors that may also regulate the survival, proliferation, and differentiation of progenitor cells into immature megakaryocytes in a colony formation assay include » GM-CSF and » Epo. The factors » aFGF, » bFGF, » IL1, » IL2 and » IL4, » IL11 play a stimulatory or co-stimulatory role. In addition there are some other factors with this activity or potentiating it (see: Mk potentiator) that have not yet been characterized (TPO, thrombopoietin).

A unique Meg-CSF has been isolated from the urine of patients with aplastic anemia. This factor stimulates CFU-Meg in the absence of adherent accessory cells and in serum-free cultures. It is not identical with » GM-CSF, » IL3, » M-CSF, » G-CSF, » IL1α, » IL6, » IL9 or » IL11.

Briddell RA & Hoffman R Cytokine regulation of the human burst-forming unit megakaryocyte. Blood 76: 516-22 (1990); **de Alarcon PA** Megakaryocyte colony-stimulating factor (Mk-CSF): its physiologic significance. Blood Cells 15: 173-85 (1989); **Erickson-Miller CL et al** Megakaryocyte colony-stimulating factor (Meg-CSF) is a unique cytokine specific for the megakaryocyte lineage. Br. J. Haematol. 84: 197-203 (1993); **Hamaguchi H et al** Interaction of monocytes and T cells in the regulation of normal human megakaryocytopoiesis *in vitro*: role of IL1 and IL2. Br. J. Haematol. 76: 12-20 (1990); **Han ZC et al** Megakaryocytopoiesis: characterization and regulation in normal and pathologic states. Int. J. Hematol. 54: 3-14 (1991); **Han ZC et al** New insights into the regulation of megakaryocytopoiesis by hematopoietic and fibroblastic growth factors and transforming growth factor β 1. Br. J. Haematol. 81: 1-5 (1992); **Hill RJ et al** Correlation of *in vitro* and *in vivo* biological activities during the partial purification of thrombopoietin. EH 20: 354-60 (1992); **Hoffman R et al** New insights into the regulation of human megakaryocytopoiesis. Blood Cells 13: 75-86 (1987); **Mazur EM et al** Modest stimulatory effect of recombinant human GM-CSF on colony growth from peripheral blood human megakaryocyte progenitor cells. EH 15: 1128-33 (1987); **Murphy MJ Jr** Megakaryocyte colony-stimulating factor and thrombopoiesis. Hematol. Oncol. Clin. North Am. 3: 465-78 (1989); **Ogata K et al** Partial purification and characterization of human megakaryocyte colony-stimulating factor (Meg-CSF). IJCC 8: 103-20 (1990); **Peschel C et al** Effects of B cell stimulatory factor 1/interleukin 4 on hematopoietic progenitor cells. Blood 70: 254-63 (1987); **Robinson BE et al** Recombinant murine granulocyte macrophage colony-stimulating factor has megakaryocyte colony-stimulating activity and augments megakaryocyte colony stimulation by interleukin 3. JCI 79: 1648-52 (1987); **Satoh K et al** Simultaneous assay for megakaryocyte colony-stimulating factor and megakaryocyte potentiator and its application. J. Lab. Clin. Med. 116: 162-71 (1990); **Sparrow RL & Williams N** Megakaryocyte colony-stimulating factor: its identity to interleukin 3. PCBR 215: 123-8 (1986)

MeGF: *melanocyte growth factor* This factor is a heparin-binding growth factor (see also: HBGF) isolated from human epidermis. It promotes the growth of melanocytes. The activity of this factor can be neutralized completely by antibodies directed against » bFGF. MeGF is probably identical with bFGF. Another factor, called ***melanocyte growth stimulating factor***, is isolated from bovine melanocytes of the iris. This factor is also closely related to bFGF. See also: MDGF (melanoma-derived growth factor).

Horikawa T et al Melanocyte growth factor in normal human skin. Pigment Cell Res. 4: 48-51 (1991); **Plouet J & Gospodarowicz D** Iris-derived melanocytes contain a growth factor that resembles basic fibroblast growth factor. Exp. Eye Res. 51: 519-29 (1990)

Meg-Pot: *megakaryocyte potentiator* see: Mk potentiator.

Melanocyte growth factor: see: MeGF.

Melanocyte growth stimulating factor: see: MeGF.

Melanocyte-stimulating hormone: abbrev. MSH. see: POMC (proopiomelanocortin).

Melanoma-derived growth factor: see: MDGF.

Melanoma-derived growth factor: This factor is identical with » PDGF.

Westermark B et al Human melanoma cell lines of primary and metastatic origin express the genes encoding the chains of platelet-derived growth factor (PDGF) and produce a PDGF-like growth factor. PNAS 83: 7197-200 (1986)

Melanoma-derived lipoprotein lipase inhibitor: See: MLPLI.

Melanoma growth stimulatory activity: see: MGSA.

Melanotropin: see: POMC (proopiomelanocortin).

Membrane-anchored growth factors: see: Juxtacrine.

Membrane-associated neurotransmitter-stimulating factor: see: MANS.

Membrane attack complex inhibitor: This inhibitor is identical with » Vitronectin.

Mesangial cell proliferating factor: see: MDF.

Mesangial cell thymocyte activating factor: see: MC-TAF.

Mesoderm inducing factor: abbrev. MIF. see: XTC-MIF.

Message amplification phenotyping: abbrev. MAPP.
This technique is also called ***mRNA phenotyping***. It is essentially an application of ***RT-PCR*** (reverse transcriptase polymerase chain reaction) or ***cDNA-PCR***. PCR is an *in vitro* technique allowing the selective enrichment of minute amounts of DNA regions of defined lengths and defined sequences from a heterogeneous population of other DNA molecules.
This technique exploits some of the properties of DNA polymerases which require a short double-stranded region (primer) to produce a double-stranded molecule from single-stranded starting material.
In principle the DNA containing the region to be amplified is made single-stranded by heating. It is then mixed with two chemically synthesized strand-specific oligonucleotides (PCR primers) complementary to regions flanking the sequence to be amplified. Under suitable conditions the oligonucleotides bind to each of the single strands containing the region to be amplified. The short double-stranded stretch of DNA formed between an oligonucleotide primer and a single-stranded DNA molecule is the starting point for DNA polymerase that can now copy the long DNA into a double-stranded molecule. The newly formed double strands are then heated to convert them again into single strands. These again provide templates to which primer oligonucleotides can bind to provide new substrates for the polymerase. A repetition of DNA denaturation, formation of hybrid molecules consisting of long DNA strands and PCR primers, polymerization by DNA polymerase is a chain reaction that leads to the selective exponential 10^6-10^7-fold amplification of those DNA sequences flanked by the primer oligonucleotides.
The reaction mixture is subjected to gel electrophoresis and amplified bands can then be visualized by appropriate staining of the DNA.
One of the most important prerequisites of this technique is an exact knowledge of the sequence of the DNA to be amplified because this sequence information is required to design suitable primers.
This versatile technique which is now common routine in many molecular biology laboratories is also used to detect transcripts for » cytokines or early response genes (see: ERG) expressed by cells in response to cytokines and other environmental stimuli. As a rule the synthesis of cytokines/ERGs is induced and transient so that a specific cytokine/ERG mRNA is present only if the cells under study also produce the cytokine/ERG. For many cytokines it has been demonstrated by now that the levels of cytokine-specific mRNA correlate well with levels of secreted factors. Message amplification phenotyping of cytokine-specific mRNAs therefore is an alternative to conventional » bioassays employing either » factor-dependent cell lines or cells that respond to cytokines. Detection of cytokines by detection their mRNAs has the additional advantage over conventional bioassays that species-specific PCR primers can be used, thus eliminating some pitfalls of bioassays that usually respond to more than one cytokine and mainly are not species-specific. Determination of ERGs by message amplification phenotyping yields a profile of genes or subsets thereof expressed in response to various conditions and allows dissection of molecular pathways of signal transduction by revealing individual components (for inhibitors used to dissect cytokine-mediated signal transduction pathways see: Bryostatins, Calcium ionophore, Calphostin C, Genistein, H8, Herbimycin A, K-252a, Lavendustin A, Phorbol esters, Okadaic acid, Suramin, Tyrphostins, Vanadate).
An additional advantage of transcript detection by the PCR technique is that it usually only requires $\approx$ 30 reaction cycles, which can be performed within a few hours. The system has been mechanized by the introduction of programmable thermal cyclers.
In view of the fact that cytokines are essential pathophysiological mediators the aberrant expression of which is responsible for, or associated with, many different disease processes PCR measurement of cytokine levels are of substantial clinical value. This very sensitive methodology is a valuable tool in the detection and quantitation of

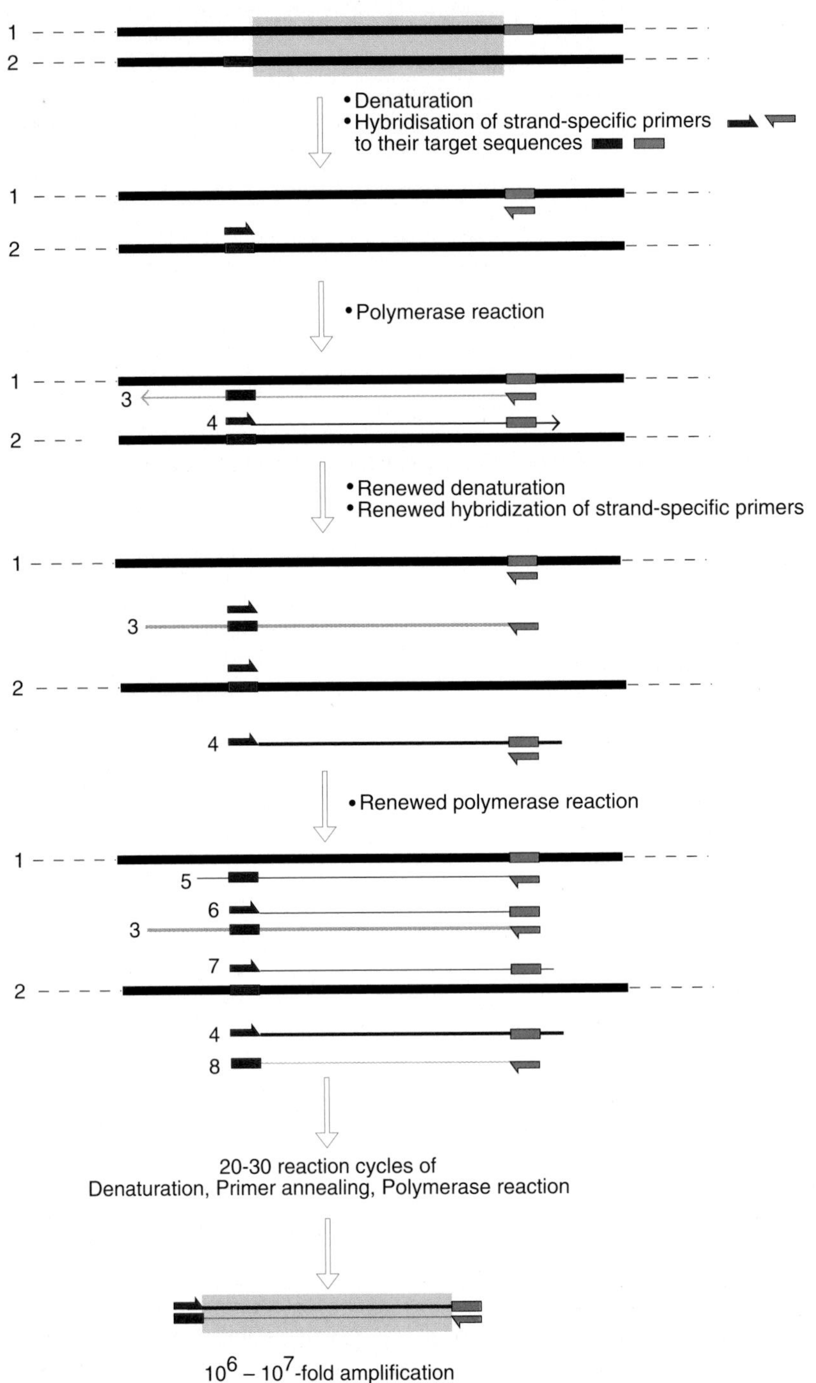
1
2
• Denaturation
• Hybridisation of strand-specific primers
to their target sequences
1
2
• Polymerase reaction
1
3
4
2
• Renewed denaturation
• Renewed hybridization of strand-specific primers
1
3
2
4
• Renewed polymerase reaction
1
5
6
3
7
2
4
8
20-30 reaction cycles of
Denaturation, Primer annealing, Polymerase reaction
10^6 – 10^7-fold amplification
of fragment flanked by primer binding sites

cytokine gene expression when only small amounts of tissue or cells are available. The amount of required material is on the order of 500 ng of nucleic acids. This amount is easily obtained from ≈ 10^6 cells. PCR amplification, however, can also be carried out with DNA from single cells and the lower limit of linearity on standard curves of some quantitative PCR amplification reactions is as little as 0.01 fg of input DNA, which corresponded to approximately 20 cells. 10-100 pg of cytoplasmic RNA are usually sufficient to allow detection of gene transcription.

PCR amplification can be performed with a variety of different primers and therefore allows to establish ***cytokine expression profiles*** with an accuracy and ease that is rarely seen with conventional » bioassays. The sensitivity of this approach is demonstrated by the capability to detect a twofold difference in mRNA levels between hemizygous and heterozygous animals (see also: Transgenic animals).

Adolff CH et al Polymerase chain reaction amplification of messages for growth factors in cells from human bronchoalveolar lavage fluids. Inflammation 15: 259-68 (1991); **Albino AP et al** Induction of growth factor RNA expression in human malignant melanoma: markers of transformation. CR 51: 4815-20 (1991); **Avraham H et al** Cytokine gene expression and synthesis by human megakaryocytic cells. IJCC 10: 70-9 (1992); **Ballagi-Pordany A et al** Quantitative determination of mRNA phenotypes by the polymerase chain reaction. AB 196: 89-94 (1991); **Bickler SW et al** A method for rapid analysis of peripheral blood mononuclear leukocyte cytokine mRNA. EH 20: 980-5 (1992); **Brenner CA et al** Cytokine MAPPing: observation and quantification of cytokine mRNA in small numbers of cells using the polymerase chain reaction. In: Balkwill FR (edt) Cytokines – A practical approach, pp. 51-9, IRL Press, Oxford 1991; **Brenner CA et al** Message amplification phenotyping, a technique to simultaneously measure multiple messenger RNAs from small numbers of cells. BioTechniques 7: 1096-103 (1989); **Carding SR et al** A polymerase chain reaction assay for the detection and quantitation of cytokine gene expression in small numbers of cells. JIM 151: 277-87 (1992); **Cherwinski HM et al** Two types of mouse helper T cell clone. III. Further differences in lymphokine synthesis between Th1 and Th2 clones revealed by RNA hybridization, functionally monospecific bioassays, and monoclonal antibodies. JEM 166: 1229-44 (1987); **Erlich HA et al** Recent advances in the polymerase chain reaction. S 252: 1643-51 (1991); **Erlich HA** (edt) PCR Technology; Principles and applications for DNA amplification. Stockton Press, New York 1989; **Fann MJ Patterson PH** A novel approach to screen for cytokine effects on neuronal gene expression. J. Neurochem. 61: 1349-55 (1993); **Frazer A et al** Detection of mRNA for the transforming growth factor β family in human articular chondrocytes by the polymerase chain reaction. BBRC 180: 602-8 (1991); **Gilliland G et al** Analysis of cytokine mRNA and DNA detection and quantitation by competitive polymerase chain reaction. PNAS 87: 2725-9 (1990); **Goodall JC et al** cDNA cloning of porcine interleukin 2 by polymerase chain reaction. BBA Gene Structure and Expression 1089: 257-8 (1991); **Innis MA et al** (eds) PCR Protocols, a guide to methods and applications. Academic Press, London 1990; **Ishida N et al** Message amplification phenotyping of an inherited δ aminolevulinate dehydratase deficiency in a family with acute hepatic porphyria. BBRC 172: 237-42 (1990); **Kaashoek JGJ et al** Cytokine production by the bladder carcinoma cell line 5637: Rapid analysis of mRNA expression levels using a cDNA-PCR procedure. Lymphokine and Cytokine Research 10: 231-5 (1991); **Kawasaki ES et al** Diagnosis of chronic myeloid and acute lymphocytic leukemias by detection of leukemia-specific mRNA sequences amplified *in vitro*. PNAS 85: 5698-702 (1988); **Larrick JW et al** A novel technique to study immunoinflammatory cell function: message amplification phenotyping, MAPP. Int. J. Immunother. 5: 95-102 (1989); **McInnes CJ et al** The molecular cloning of the ovine γ-interferon cDNA using the polymerase chain reaction. NAR 18: 4012-(1990); **Melby PC et al** Quantitative measurement of human cytokine gene expression by polymerase chain reaction. JIM 159: 235-44 (1993); **Mohler**

Principle of the polymerase chain reaction (PCR).
This technique allows the enzymatic amplification of a double-stranded DNA region (shaded blue on top strands) flanked by two known sequences (green rectangle on strand #1 top right and red rectangle on strand #2 bottom left). These two sequences can bind complementary short oligonucleotides (green and red half arrows) called primers. Primers are synthesized chemically and are designed in a way that allows strand-specific binding, i.e. the green primer binds to its green target site on one strand and the red primer binds to its target site on the other strand.
The first step of the PCR reaction separates the double strands into single strands (strands 1 and 2) by heat denaturation. The second step is a process called primer annealing, i. e. the formation of partially double-stranded regions by interaction with the primer binding sites on each strand with the corresponding primer. DNA polymerase is then used to extend the primer region and to synthesize full-length complementary strands (light and dark blue lines) of strands 1 and 2, yielding two new double strands 1–3 and 2–4. This reaction requires the presence of deoxynucleoside triphosphates which serve as substrates for the polymerase. The process of denaturation, primer annealing, and primer extension can be repeated with the new double strands 1–3 and 2–4 since they also contain the primer binding sites. The reaction products of this second reaction cycle are four double strands: 1–5, 2–7, 3–6, 4–8.
Sequence regions synthesized and extended beyond the primer binding sites are obtained only from the original DNA strands. After a few reaction cycles the newly synthesized strands are terminated by the original primer binding sites. By repeating the process of denaturation, primer annealing and primer extension some 20 to 30 times one obtains an exponential amplification of the DNA sequence located between the two primer binding sites.
Many different variants and strategies have been described for enzymatic amplification of DNA sequences. With some modifications the process can be used also if only one of the two flanking primer binding sequences are known. The process can be used also with RNA, e. g. cytokine-specific mRNA. In this case the RNA is first converted into DNA by reverse transcription (RT-PCR), utilizing the enzyme reverse transcriptase.The cDNA obtained in this way is then subjected to a normal PCR reaction.

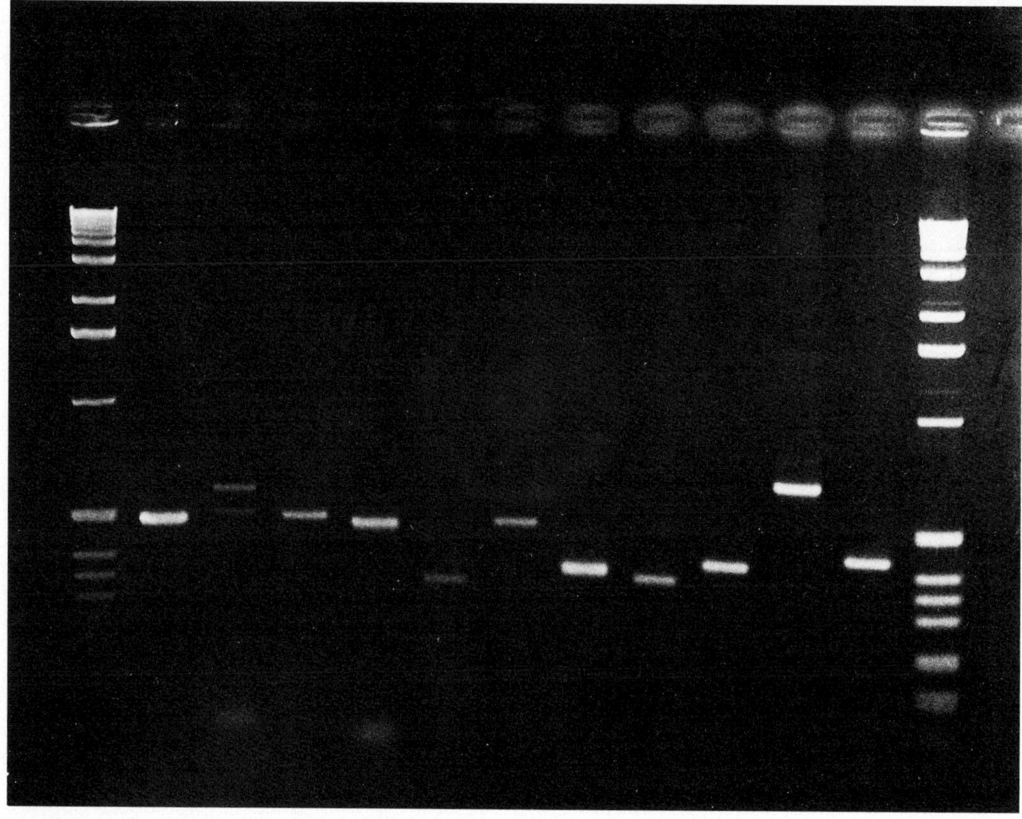

Example of the analysis of cells by PCR analysis of cytokine-specific mRNAs.
Amplified DNA fragments obtained after PCR were separated by gel electrophoresis and visualized by staining with ethidium bromide. Lanes 1 and 13 contain a 1 kb ladder as marker.

Lane 2:	486 bp fragment;	detection of c-*kit* in a melanoma cell line.
Lane3:	591 bp fragment;	detection of c-*kit* ligand in a melanoma cell line.
Lane4:	503 bp fragment;	detection of LIF in 5637 cells.
Lane 5:	481 bp fragment;	detection of IL3 in stimulated lymphocytes.
Lane 6:	331 bp fragment;	detection of IL4 in a stimulated B cell line (B2).
Lane 7:	495 bp fragment;	detection of IL6 in unstimulated B2 cells.
Lane 8:	365 bp fragment;	detection of IL8 in 5637 cells.
Lane 9:	351 bp fragment;	detection of IL9 in stimulated B2 cells.
Lane 10:	396 bp fragment;	detection of GM-CSF in 5637 cells.
Lane 11:	671 bp fragment;	detection of M-CSF in 5637cells.
Lane 12:	431 bp fragment;	detection of IL1β in 5637 cells.

Courtesy of K. Thalmeier, GSF Hämatologikum, München.

KM & Butler LD Quantitation of cytokine mRNA levels utilizing the reverse transcriptase-polymerase chain reaction following primary antigen-specific sensitization *in vivo*. I. Verification of linearity, reproducibility and specificity. Mol. Immunol. 28: 437-47 (1991); **Nelson PA et al** Effect of immunosuppressive drugs on cytokine gene transcription studied by message amplification phenotyping (MAPPing) polymerase chain reaction. Transplant. Proc. 23: 2867-9 (1991); **O'Garra A & Vieira P** Polymerase chain reaction for detection of cytokine gene expression. Curr. Opin. Immunol. 4: 211-5 (1992); **Paludan K & Thestrup-Pedersen K** Use of the polymerase chain reaction in quantification of interleukin 8 mRNA in minute epidermal samples. J. Invest. Dermatol. 99: 830-5 (1992); **Platzer C et al** Analysis of cytokine mRNA levels in interleukin-4-transgenic mice by quantitative polymerase chain reaction. EJI 22: 1179-84 (1992); **Rappolee DA et al** Wound macrophages express TGF-α and other growth factors *in vivo*: analysis by mRNA phenotyping. S 241: 708-12 (1988); **Rappolee DA et al** Novel method for studying mRNA phenotypes in single or small numbers of cells. JCBc 39: 1-11 (1989); **Saito S et al** Cytokine production by CD16⁻CD56bright natural killer cells in the human early pregnancy decidua. Int. Immunol. 5: 559-63 (1993); **Scheuermann RH & Bauer SR** Polymerase chain reaction-based mRNA quantification using an internal standard: analysis of oncogene expression. MiE 218: 446-73 (1993); **Shaffer AL et al** Amplification, detection, and automated sequencing of gibbon interleukin-2 mRNA by Thermus aquaticus DNA polymerase reverse transcription and polymerase chain reaction. AB 190: 292-6 (1990); **Sorg R et al** Rapid and sensitive mRNA phenotyping for interleukins (IL1 to IL6) and colony-stimulating factors (G-CSF, M-CSF, and GM-CSF) by reverse transcription and subsequent polymerase chain reaction. EH 19: 882-7 (1991); **Timblin C et al** Application for PCR technology to subtractive cDNA cloning: identification of genes expressed specifically in murine plasmacytoma cells. NAR 18: 1587-93 (1990); **Walker KB et al** Role of Th-1 lymphocytes in the development of protective immunity against *Mycobacterium leprae*. Analysis of lymphocyte function by polymerase chain reaction detection of cytokine messenger RNA. JI 148: 1885-9 (1992); **Wang AM et al** Quantitation of mRNA by the polymerase chain reaction. PNAS 86: 9717-21 (1989); **Wesselingh SL et al** Detection of cytokine mRNA in astrocyte cultures using the polymerase chain reaction. LR 9: 177-85 (1990); **Wolf SS & Cohen A** Expression of cytokines and their receptors by human thymocytes and thymic stromal cells. Immunology 77: 362-8 (1992); **Yamamura M et al** Defining protective responses to pathogens: cytokine profiles in leprosy lesions. S 254: 277-9 (1991)

***met*:** The activated *met* gene was originally detected in a human cell line that had been transformed by exposure to N-methyl-N'-nitro-N-nitrosoguanidine. c-*met* has been found to be overexpressed in some primary human colorectal carcinomas. c-*met* mRNA is predominantly expressed in epithelial cells, but has been detected in several murine hematopoietic progenitor cell lines, suggesting that *met* might function during » hematopoiesis.

The gene maps to human chromosome 7q21-q31. This is also a region associated with nonrandom chromosomal deletions observed in a portion of patients with acute nonlymphocytic leukemia. The murine *met* gene maps to chromosome 6 closely linked with the CFTR (cystic fibrosis transmembrane conductance regulator) gene.

The *met* gene encodes a 190 kDa heterodimeric transmembrane protein (p190$^{\alpha\beta}$) consisting of a heavily glycosylated 45-50 kDa (α) and a 145 kDa (β) subunit linked by two disulfide bonds. The α subunit is extracellular and the N-terminal end of the β subunit is also on the extracellular side. The C-terminal cytoplasmic domain of the β-chain encodes a tyrosine-specific protein kinase. The β-chain also contains the extracellular ligand binding site. Both subunits originate from glycosylation and proteolytic cleavage of a common precursor of 170 kDa (unglycosylated: 150 kDa). Truncated forms of the *met* protein (130 kDa, 140 kDa, lacking the intracellular tyrosine kinase domain, have also been described.

Under physiological conditions the intrinsic protein kinase activity of *met* is activated only by binding of the ligand to the receptor, leading to the » autophosphorylation of the receptor. The activated oncogene encodes a receptor that is constitutively autophosphorylated in some human tumor cell lines due to the amplification or overexpression of the *met* gene. The tyrosine kinase activity of the *met* receptor is negatively modulated by protein kinase C (PKC). An increase of intracellular Ca^{2+} has a similar inhibitory effect *in vivo* and this effect is independent of PKC. The tyrosine-phosphorylated receptor also phosphorylates various other cytoplasmic signal transducers, including the 85-kDa subunit of phosphatidylinositol 3-kinase (Pl 3-kinase), *ras*GAP, phospholipase-C γ (PLC-γ), and » *fyn*, a tyrosine kinase of the *src* family.

The *met* protein is a receptor for » HGF (hepatocyte growth factor).

Bardelli A et al Autophosphorylation promotes complex formation of recombinant hepatocyte growth factor receptor with cytoplasmic effectors containing SH2 domains. O 7: 1973-8 (1992); **Bottaro DP et al** Identification of the hepatocyte growth factor receptor as the c-*met* proto-oncogene product. S 251: 802-4 (1991); **Chan AML et al** Characterization of the mouse *met* proto-oncogene. O 2: 593-600 (1988); **Cooper CS et al** The *met* oncogene: from detection by transfection to transmembrane receptor for hepatocyte growth factor. O 7: 3-7 (1992); **Dean M et al** The human *met* oncogene is related to the tyrosine kinase oncogenes. N 318: 385-8 (1985); **Ferracini R et al** Identification of the major autophosphorylation site of the *met*/hepatocyte growth factor receptor tyrosine kinase. JBC 266: 19558-64 (1991); **Gandino L et al** Protein kinase-c activation inhibits tyrosine phosphorylation of the c-*met* protein. O 5: 721-5 (1990); **Gandino L et al** Intracellular calcium regulates the tyrosine kinase receptor encoded by the *met* oncogene. JBC 266: 16098-104 (1991); **Gherardi E & Stoker M** Hepatocyte growth factor-scatter factor: Mitogen, motogen, and *met*. CC 3: 227-32 (1991); **Giordano S et al** Biosynthesis of the protein encoded by the c-*met* proto-oncogene. O 4: 1383-8 (1989); **Giordano S et al**

transfer of motogenic and invasive response to scatter factor/hepatocyte growth factor by transfection of human *met* oncogene. PNAS 90: 649-53 (1993); **Gonzatti-Haces M et al** Characterization of the TPR-*met* oncogene p65 and the *met* proto-oncogene p140 protein-tyrosine kinases. PNAS 85: 21-5 (1988); **Higuchi O & Nakamura T** Identification and change in the receptor for hepatocyte growth factor in rat liver after partial hepatectomy or induced hepatitis. BBRC 176: 599-607 (1991); **Higuchi O et al** Expression of c-*met* proto-oncogene in COS cells induces the signal transducing high-affinity receptor for hepatocyte growth factor. FL 301: 282-6 (1992); **Kelley KA et al** Expression and chromosome localization of the murine cystic fibrosis transmembrane conductance regulator. Genomics 13: 381-8 (1992); **Liu C et al** Overexpression of c-*met* proto-oncogene but not epidermal growth factor receptor or c-*erb*B-2 in primary human colorectal carcinomas. O 7: 181-5 (1992); **Mondino A et al** Defective posttranslational processing activates the tyrosine kinase encoded by the *met* proto-oncogene (Hepatocyte growth factor receptor). MCB 11: 6084-92 (1991); **Naldini L et al** Hepatocyte growth factor (HGF) stimulates the tyrosine kinase activity of the receptor encoded by the proto-oncogene C-*met*. O 6: 501-4 (1991); **Naldini L et al** Scatter factor and hepatocyte growth factor are indistinguishable ligands for the *met* receptor. EJ 10: 2867-78 (1991); **Naldini L et al** The tyrosine kinase encoded by the *met* proto-oncogene is activated by autophosphorylation. MCB 11: 1793-803 (1991); **Park M et al** Mechanisms of *met* oncogene activation Cell 45: 895-904 (1986); **Park M et al** Sequence of *met* proto-oncogene cDNAs features characteristic of the tyrosine kinase family of growth-factor receptors. PNAS 84: 6379-6383 (1987); **Ponzetto C et al** c-*met* is amplified but not mutated in a cell line with an activated *met* tyrosine kinase. O 6: 553-9 (1991); **Prat M et al** The receptor encoded by the human c-*met* oncogene is expressed in hepatocytes, in epithelial cells, and in solid tumors. IJC 49: 323-8 (1991); **Prat M et al** C-terminal-truncated forms of the hepatocyte growth factor receptor. MCB 11: 5954-62 (1991); **Rodriguez GA et al** Alternative splicing generates isoforms of the *met* receptor tyrosine kinase which undergo differential processing. MCB 11: 2962-70 (1991); **Rong S et al** Tumorigenicity of the *met* proto-oncogene and the gene for hepatocyte growth factor. MCB 12: 5152-8 (1992); **Tempest PR et al** The activated human *met* gene encodes a protein tyrosine kinase. FL 209: 357-361 (1986); **Tempest PR et al** Structure of the *met* protein and variation of *met* protein kinase activity among human tumor cell lines. Br. J. Cancer 58: 3-7 (1988)

Metastasis inhibition factor: see: nm23.

Met-Enkephalin: see: POMC (proopiomelanocortin).

Meth A factor: This factor of ≈ 44 kDa is isolated from Meth A fibrosarcoma cells. It is also produced by a number of human and rodent tumors. The factor induces the synthesis of tissue thromboplastin (factor III) by endothelial cells. It enhances the procoagulant activity of the endothelium mediated by » TNF. Meth A factor has been renamed » EMAP I (endothelial-monocyte activating polypeptide).

Clauss M et al A polypeptide factor produced by fibrosarcoma cells that induces endothelial tissue factor and enhances the procoagulant response to tumor necrosis factor/cachectin. JBC 265: 7078-83 (1990)

MF: *maintenance factor* sometimes used in abbrev. such as CFU-s MF (colony forming unit spleen maintenance factor = IL3).

MFF: *macrophage fusion factor* This factor is isolated from the culture supernatants of murine concanavalin A-stimulated thymocytes. It induces the fusion of bone marrow cells and alveolar macrophages to large multinucleated cells that produce high levels of » IL1.

One factor responsible for this activity is identical with » IL4. The activity of IL4 is stimulated considerably by » IL6, » IFN-γ and » GM-CSF. The MFF activity of the thymocyte supernatant is only partially blocked by anti-IL4 antibodies. The residual activity is mainly attributed to GM-CSF.

Abe E Granulocyte-macrophage colony-stimulating factor is a major macrophage fusion factor present in conditioned medium of concanavalin A-stimulated spleen cell cultures. JI 147: 1810-5 (1991); **Galindo B et al** Fusion of normal rabbit alveolar macrophages induced by supernatant fluids from BCG-sensitised lymph node cells after elicitation by antigens. Infect. Immun. 9: 212-6 (1974); **Sone S** Functions of multinucleated giant cells formed by fusing rat alveolar macrophages with lymphokines containing macrophage fusion factor. LR 3: 163-73 (1984)

MFSA: *macrophage-dependent fibroblast stimulating activity* This factor is probably identical with » IL1. See: FAF (fibroblast activating factors).

MGF: *macrophage growth factor* This factor is identical with » bFGF.

Baird A et al Immunoreactive fibroblast growth factor in cells of peritoneal exudate suggests its identity with macrophage-derived growth factor. BBRC 126: 358-64 (1985)

MGF: *macrophage growth factor* This factor was isolated initially as an activity stimulating cell division in adherent peritoneal macrophages. It is identical with » M-CSF. See also: CSF.

Cifone M & Defendi V Cyclic expression of a growth conditioning factor (MGF) on the cell surface. N 252: 151-3 (1974); **Mauel J & Defendi V** Regulation of DNA synthesis in mouse macrophages. I. Sources and purification of a the macrophage growth factor (MGF). ECR 65: 33-43 (1971); **Stanley ER et al** Factors regulating macrophage production and growth: identity of colony-stimulating factor and macrophage growth factor. JEM 143: 631-45 (1976)

MGF: *mast cell growth factor* This factor supports the growth of mast cells. It is identical with » SCF (stem cell factor). Direct evidence for the identity of MGF and SCF comes from an analysis of the steel-Dickie mutation in the mouse, symbolized Sl(d), that was shown to consist of a 4 kb intragenic deletion in MGF genomic sequences, resulting in the formation of a soluble truncated growth factor lacking both transmembrane and cytoplasmic domains.

Anderson DM et al Molecular cloning of mast cell growth factor, a hemopoietin that is active in both membrane bound and soluble forms. Cell 63: 235-43 (1990), Erratum in Cell 63: 1112 (1990); **Brannan CI et al** Steel-Dickie mutation encodes a c-*kit* ligand lacking transmembrane and cytoplasmic domains. PNAS 88: 4671-4674 (1991); **Broxmeyer HE et al** Effect of murine mast cell growth factor (c-*kit* proto-oncogene ligand) on colony formation by human marrow hematopoietic progenitor cells. Blood 77: 2142-9 (1991); **Copeland NG et al** Mast cell growth factor maps near the *Sl* locus and is structurally altered in a number of *Steel* alleles. Cell 63: 175-83 (1990); **Dolci S et al** Requirement for mast cell growth factor for primordial germ cell survival in culture. N 352: 809-11 (1991); **Williams DE & Lyman SD** Molecular and cellular biology of mast cell growth factor: the gene product of the murine steel locus. Behring Inst. Mitt. 90: 23-7 (1991)

MGF: *milk growth factor* This factor is isolated from bovine milk. Two forms of this factor have been described, ***MGF-a*** and ***MGF-b***. MGF-a is identical with TGF-β2 (see: TGF-β) and is the predominant form (85 %). The first 29 amino acids of MGF-b are identical with those of TGF-β1. MGF may be identical with » MDGF (milk-derived growth factor).

Some of the mitogenic activity of human milk or whey reported in the literature is due to the presence of » IGF, IGF binding proteins (see: IGF-BP), TNF-α, » IL6, and other as yet unidentified growth factors.

Milk-borne EGF has been shown to survive in the gastrointestinal tract and to be delivered in active form to peripheral organs. These factors may therefore play an important role in developing mammals in general, for example, acceleration of the maturation of the intestinal portion of the digestive system, and also for the development of neonatal host defenses. IL6 contained in milk is closely associated with the local production of IgA in the breast.

Baxter RC et al Immunoreactive somatomedin-C/insulin-like growth factor I and its binding protein in human milk. J. Clin. Endocrinol. Metab. 58: 955-9 (1984); **Bocci V et al** What is the role of cytokines in human colostrum? J. Biol. Regul. Homeost. Agents 5: 121-4 (1991); **Cox DA & Muerk RR** Isolation and characterization of milk growth factor, a transforming growth factor β 2-related polypeptide from bovine milk. EJB 197: 353-8 (1991); **Iacopetta BJ et al** Epidermal growth factor in human and bovine milk. Acta Paediatr. 81: 287-91 (1992); **Ichiba H et al** Measurement of growth promoting activity in human milk using a fetal small intestinal cell line. Biol. Neonate 61: 47-53 (1992); **Jin Y et al** Separation, purification, and sequence identification of TGF-β 1 and TGF-β 2 from bovine milk. J. Protein Chem. 10: 565-75 (1991); **Kohno Y et al** The effect of human milk on DNA synthesis of neonatal rat hepatocytes in primary culture. Pediatr. Res. 29: 251-5 (1991); **Koldovsky O et al** The developing gastrointestinal tract and milk-borne epidermal growth factor. AEMB 310: 99-105 (1991); **McCleary MJ** Epidermal growth factor: an important constituent of human milk. J. Hum. Lact. 7: 123-8 (1991); **Rudloff HE et al** Tumor necrosis factor-α in human milk. Pediatr. Res. 31: 29-33 (1992); **Saito S et al** Detection of IL6 in human milk and its involvement in IgA production. J. Reprod. Immunol. 20: 267-76 (1991); **Stoeck M et al** Comparison of the immunosuppressive properties of milk growth factor and transforming growth factors β 1 and β 2. JI 143: 3258-65 (1989); **Stoeck M et al** Transforming growth factors β 1 and β 2 as well as milk growth factor decrease anti-CD3-induced proliferation of human lymphocytes without inhibiting the anti-CD3-mediated increase of Ca^{2+} and the activation of protein kinase C. FL 249: 289-92 (1989); **Okada M et al** Transforming growth factor (TGF)-α in human milk. Life Sci. 48: 1151-6 (1991) **Polk DH** Do breast milk derived hormones play a role in neonatal development? Early Hum. Dev. 29: 329-31 (1992)

MGF: *multipotent growth factor* This factor is identical with » SCF (stem cell factor).

MGF: *myeloid growth factor* This factor is identical with » LIF.

Moreau JF et al Leukemia inhibitory factor is identical to the myeloid growth factor human interleukin for DA cells. N 326: 201-4 (1988); **Williams RL et al** Myeloid leukemia inhibitory factor (LIF) maintains the developmental potential of embryonic stem cells. N 336: 684-7 (1988)

MGF: *myelomonocytic growth factor* see: cMGF (chicken myelomonocytic growth factor).

MGF: *myogenic growth factor* This factor is identical with » bFGF.

Kardami E et al Myogenic growth factor present in skeletal muscle is purified by heparin affinity chromatography. PNAS 82: 8044-7 (1985)

MGF: *myxoma growth factor* This factor is encoded by a gene from the genome of myxoviruses. Sequence homologies reveal that this protein is a member of the EGF/TGF family of proteins (see also: EGF, TGF-α).

Myxoma virus is a leporipoxvirus of rabbits that induces a lethal syndrome characterized by disseminated tumor-like lesions, generalized immunosuppression, and secondary Gram-negative bacterial infection. Viral deletion mutants lacking MGF sequences do not cause the characteristic disease symptoms of myxomatosis. Injected rabbits develop only a benign, localized nodule at the site of injection that later regresses completely, leaving the animals resistant to challenge with

MGF: see addendum.

wild-type myxoma virus. MGF may therefore be involved in subverting the immune responses of the host (see also: Viroceptor).

Recombinant myxoma virus variants in which MGF was disrupted and replaced with either » VVGF (vaccinia virus growth factor), » SFGF (Shope fibroma growth factor), or rat » TGF-α produce infections which are both clinically and histopathologically indistinguishable from wild-type myxomatosis. This suggests that these EGF-like growth factors, which are diverse with respect to primary structure and origin, have similar biological functions in the context of myxoma virus pathogenesis and are mitogenic for the same target cells.

Opgenorth A et al Deletion analysis of two tandemly arranged virulence genes in myxoma virus, M11L and myxoma growth factor. J. Virol. 66: 4720-31 (1992); **Opgenorth A et al** Transforming growth factor α, Shope fibroma growth factor, and vaccinia growth factor can replace myxoma growth factor in the induction of myxomatosis in rabbits. Virology 192: 701-9 (1993); **Upton C et al** Mapping and sequencing of a gene from myxoma virus that is related to those encoding epidermal growth factor and transforming growth factor α. J. Virol. 61: 1271-5 (1987)

MGI: *macrophage-granulocyte inducer* This is an older acronym denoting several colony-stimulating factors (see also: CSF). They were initially identified as activities supporting the growth of hematopoietic colonies in a » Colony formation assay. The specificity, i. e. the predominant cell type within colonies supported by MGIs, is indicated by a suffix:

MGI-1G = granulocyte-specific (see: G-CSF); MGI-1GM = granulocyte/macrophage-specific (see: GM-CSF); MGI-1M = macrophage-specific (see: M-CSF).

Landau T & Sachs L Characterization of the inducer required for the development of macrophage and granulocytic colonies. PNAS 68: 2540-4 (1971)

MGI-1G: *macrophage-granulocyte inducer* see: MGI.

MGI-1GM: *macrophage-granulocyte inducer* see: MGI.

MGI-1M: *macrophage-granulocyte inducer* see: MGI.

MGI-2: *macrophage-granulocyte inducer* This factor is identical with » G-CSF. See also: MGI-2A.

Nicola NA et al Identification of the human analog of a regulator that induces differentiation in murine leukemic cells. N 314: 625-8 (1985)

MGI-2A: *macrophage-granulocyte inducer* This protein, sometimes also referred to as MGI-2, was initially detected by its ability to induce the differentiation of myeloid leukemic cell line » M1. It is identical with » IL6. It is not identical with a factor of the same name (= G-CSF).

Lotem J et al Clonal variation in susceptibility to differentiation by different protein inducers in the myeloid leukemia cell line M1. Leukemia 3: 804-7 (1989); **Shabo Y et al** The myeloid blood cell differentiation-inducing protein MGI-2A is interleukin 6. Blood 72: 2070-3 (1988)

M-GRAM: *monocyte-derived granulocyte-activating mediator* see: GRAM.

MGSA: *melanoma growth stimulatory activity* Human MGSA (73 amino acids) is encoded by a gene mapping to chromosome 4q21. This gene is called ***GRO*** (growth regulated oncogene) and contains four exons. The *GRO* gene was initially identified by its constitutive overexpression in spontaneously transformed Chinese hamster fibroblasts. Three different gro genes have been described. At the nucleotide level they show a homology of ≈ 90 %. GROα encodes MGSA. Several variants are known (see also: FINAP, fibroblast-derived neutrophil activating peptide).The proteins encoded by GROβ and GROγ are identical with murine MIPα and MIPβ (see: MIP, macrophage inflammatory protein). 53 % sequence identity is observed between GROα and a cytokine called » ENA-78. The human cytokine » NAP-3 (neutrophil-activating peptide 3) is identical with MGSA.

MGSA belongs to the » chemokine family of cytokines. Some members of the chemokine family are encoded by genes lying in the vicinity of the GRO gene at 4q21.

A related protein is encoded by the murine » *mig* gene. Its synthesis in macrophages is specifically induced by » IFN-γ. *mig* in turn is related to a protein called » KC the synthesis of which in murine cells is induced by » PDGF. KC is the murine homologue of the human GRO genes. A rat homologue of MGSA is » CINC. The avian homologue of the GRO gene is » 9E3.

The three GRO genes are expressed in a tissue-specific manner. GRO proteins are found predominantly in activated monocytes (see also: Cell activation) but are expressed also in fibroblasts, endothelial cells, synovial cells, and several tumor cell lines. The expression of the GRO genes is specifically induced by » IL1 in cells the proliferation of which is also stimulated by » IL1. Other inducers of GRO expression are » TNF and » IL6. All three factors are capable of induced the synthesis of GRO proteins ≈ 100-fold under suitable conditions.

MGSA has inflammatory (see also: Inflammation) and growth-regulating properties. It promotes the growth of fibroblasts. It is an » autocrine growth factor for melanoma cells. MGSA is a potent chemoattractant for neutrophils (see also: Chemotaxis) and enhances their degranulation and the release of lysosomal enzymes. Increased concentrations of MGSA have been observed in psoriatic scales which are also characterized by an increased infiltration of neutrophils.

MGSA and GRO proteins are functionally related to » IL8 and also bind to the same receptor. For another function of the MGSA/Gro/IL8/MIP receptor see also: CKR-1. For a viral NAP-2 receptor homologue see also: ECRF3.

● Reviews: **Sager R** GRO as a cytokine. In: Oppenheim JJ et al (eds) Molecular and cellular biology of cytokines. pp. 327-32, Wiley-Liss, New York, 1990); **Sager R et al** GRO: A novel chemotactic cytokine. AEMB 305: 73-7 (1991)

● Biochemistry & Molecular Biology: **Anisowicz A et al** Constitutive overexpression of a growth-regulated gene in transformed Chinese hamster and human cells. PNAS 84: 7188-92 (1987); **Anisowicz A et al** An NF-kappa B-like transcription factor mediates IL1/TNF-α induction of *gro* in human fibroblasts. JI 147: 520-7 (1991); **Anisowicz A et al** Functional diversity of *gro* gene expression in human fibroblasts and mammary epithelial cells. PNAS 85: 9645-9 (1988); **Baker NE et al** Nucleotide sequence of the human melanoma growth stimulatory activity (MGSA) gene. NAR 18: 6453 (1990); **Balentien E et al** Recombinant expression, biochemical characterization, and biological activities of the human MGSA/gro protein. B 29: 10225-33 (1990), [erratum in B 30: 594 (1991]; **Farber JM** A macrophage mRNA selectively induced by γ-interferon encodes a member of the platelet factor 4 family of cytokines. PNAS 87: 5238-42 (1990); **Golds EE et al** Inflammatory cytokines induce synthesis and secretion of gro protein and a neutrophil chemotactic factor but not β 2-microglobulin in human synovial cells and fibroblasts. BJ 259: 585-8 (1989); **Haskill S et al** Identification of three related human GRO genes encoding cytokine functions. PNAS 87: 7732-6 (1990); **Horuk R et al** Purification, receptor binding analysis, and biological characterization of human melanoma growth stimulating activity (MGSA): evidence for a novel MGSA receptor. JBC 268: 541-6 (1993); **Iida N & Grotendorst GR** Cloning and sequencing of a new gro transcript from activated human monocytes: Expression in leukocytes and wound tissue. MCB 10: 5596-9 (1990); **Jose PJ et al** Identification of a second neutrophil-chemoattractant cytokine generated during an inflammatory reaction in the rabbit peritoneal cavity *in vivo*. Purification, partial amino acid sequence, and structural relationship to melanoma growth stimulatory activity. BJ 278: 493-7 (1991); **Richmond A et al** Purification of melanoma growth stimulatory activity. JCP 129: 375-84 (1986); **Richmond A et al** Molecular characterization and chromosomal mapping of melanoma growth stimulatory activity, a growth factor structurally related to β-thromboglobulin. EJ 7: 2025-33 (1988); **Richmond A et al** Melanoma growth stimulatory activity: isolation from human melanoma tumors and characterization of tissue distribution. JCB 36: 185-98 (1988); **Schroeder JM et al** Lipopolysaccharide-stimulated human monocytes secrete, apart from neutrophil-activating peptide 1/interleukin 8, a second neutrophil-activating protein. JEM 171: 1091-100 (1990); **Stoeckle MY** Removal of a 3′ non-coding sequence is an initial step in degradation of gro α mRNA and is regulated by interleukin 1. NAR 20: 1123-7 (1992); **Thomas HG et al** Purification and characterization of recombinant melanoma growth stimulating activity. MiE 198: 373-83 (1991); **Watanabe K et al** The neutrophil chemoattractant produced by the rat kidney epithelial cell line NRK-52E is a protein related to the KC/gro protein. JBC 264: 19559-63 (1989); **Zipfel PF et al** Mitogenic activation of human T cells induces two closely related genes which share structural similarity with a new family of secreted factors. JI 142: 1582-90 (1989)

● Receptors: **Cheng QC et al** The melanoma growth stimulatory activity receptor consists of two proteins. Ligand binding results in enhanced tyrosine phosphorylation. JI 148: 451-6 (1992); **Cerretti DP et al** Molecular characterization of receptors for human interleukin-8, GRO/melanoma growth-stimulatory activity and neutrophil activating peptide-2. Mol. Immunol. 30: 359-67 (1993)

● Biological Activities: **Golds EE et al** Inflammatory cytokines induce synthesis and secretion of gro protein and a neutrophil chemotactic factor but not β2-microglobulin in human synovial cells and fibroblasts. BJ 259: 585-8 (1989); **Iida M et al** Level of neutrophil chemotactic factor/gro, a member of the interleukin 8 family, associated with lipopolysaccharide induced inflammation in rats. Infect. Immun. 60: 1268-72 (1992); **Magazin M et al** The biological activities of gro β and IL8 on human neutrophils are overlapping but not identical. ECN 3: 461-7 (1992); **Moser B et al** Neutrophil-activating peptide 2 and gro/melanoma growth-stimulatory activity interact with neutrophil-activating peptide 1/interleukin 8 receptors on human neutrophils. JBC 266: 10666-71 (1991); **Wen DZ et al** Expression and secretion of gro/MGSA by stimulated human endothelial cells. EJ 8: 1761-6 (1989)

MH11: A murine pre-B cell clone ($B220^+$ $MB-1^+$ sIgM-) that requires a stromal cell line, ST2, from day 13 fetal liver for its growth. The proliferation of MH11 is stimulated by » IL7, but IL7 cannot support MH11 growth without ST2 feeder cells. IL7 in combination with » SCF (stem cell factor) abolishes the requirement for stromal cells. The growth of MH11 on ST2 is inhibited almost completely by antibodies directed against SCF.

Inui S & Sakaguchi N Establishment of a murine pre-B cell clone dependent on interleukin-7 and stem cell factor. Immunol. Lett. 34: 279-88 (1992)

MH60-BSF2: A murine hybridoma cell line producing monoclonal antibodies (mAb) specific to human » IL6. MH60.BSF2 cells produce IgM chi type mAb (α BSF2-60) unable to neutralize IL 6 activity. The cell line depend on IL6 for its *in vitro* growth and are used to assay this factor (see also: Bioassays). The activity of IL6 is determined by measuring the incorporation of ^3H thymidine into the newly synthesized DNA of the proliferating cells. Cell proliferation can be determined also by employing the » MTT assay.

Recombinant human » IL 1 (α and β), » IL 2, » IL

3, » IL 4, » IFN-γ, » IFN-β, »G-CSF, or recombinant murine IL 3, IL 4, IL 5 do not induce the *in vitro* growth of MH60.BSF2 cells. An alternative and entirely different detection method is » Message amplification phenotyping.

Matsuda T et al Establishment of an interleukin 6 (IL 6)/B cell stimulatory factor 2-dependent cell line and preparation of anti-IL 6 monoclonal antibodies. EJI 18: 951-6 (1988); **Muraguchi A et al** The essential role of B cell stimulatory factor 2 (BSF-2/IL6) for the terminal differentiation of B cells. JEM 167: 332-44 (1988); **Nakajima K & Wall R** Interleukin-6 signals activating junB and TIS11 gene transcription in a B cell hybridoma. MCB 11: 1409-18 (1991)

MHC class II (Ia) inducing factor: see: Ia-inducing factor.

Microcystin-LR: see: Okadaic acid.

Microglobulin α_1: see: ITI (Inter-alpha-trypsin inhibitor).

Microglobulin β_2: abbrev. β_2M. see: Beta-2-Microglobulin.

Microseminoprotein β: see: IBF (immunoglobulin binding factors).

Midkine: see: MK.

MIF: *mesoderm inducing factor* see: XTC-MIF.

MIF: *mesoderm inducing factors* General term for factors inducing the formation of the mesoderm of amphibian embryos such as » *Xenopus laevis*. Mesoderm formation involves the interaction of cells of the vegetal hemisphere of the embryo with overlying equatorial and animal pole cells. Three classes of MIFs have been discovered. These are members of the »TGF-β, »FGF, and » *Wnt* families of proteins.

Smith JC Purifying and assaying mesoderm-inducing factors from vertebrate embryos. In: Hartley DA (ed.) Cellular interactions in development. Practical Approach Series, IRL Press, Oxford 1993, pp181-204

MIF: *migration inhibition factor,* ***macrophage migration inhibitory factor (MMIF)*** An operational definition for factors that inhibit *in vitro* the migration of macrophages out of capillary tubes or small pieces of tissue. Many proteins have been described to possess MIF activity, including » TNF, » IFN-γ, » IL1, » IL2, »GM-CSF (see also: LIF (leukocyte migration inhibitory factor). The proteins known as ***MRP 8*** and ***MRP 14*** (MIF-related protein) are MIF-related proteins that are recognized by some monoclonal antibodies directed against MIF activities. They are related to » Calgranulins.

Another protein, called ***MIF-1***, with MIF activity is secreted by a T cell hybridoma. Another poorly characterized migration inhibitory factor is » ***F5MIF-2***. Another factor influencing the chemotactic behavior of neutrophilic macrophages is » HDLF (Hodgkin-derived leukocyte factor). In addition there are many other poorly characterized factors with MIF activity that cannot be blocked by antibodies against known cytokines and that are therefore probably distinct factors.

MMIF has been cloned and has been shown to be a major secreted protein released by anterior pituitary cells in response to bacterial lipopolysaccharides. It is identical with » ***Sarcolectin binding protein***. Pituitary-derived MIF contributes to circulating MIF present in the post-acute phase of endotoxinemia (see also: Septic shock). Recombinant murine MIF greatly enhances lethality when co-injected with lipopolysaccharides and anti-MIF antibodies confer protection against lethal endotoxinemia. One particular factor with MIF activity is also found in chicken murine and human embryonic eye lenses (see: 10K protein). A 12 kDa factor of 115 amino acids that also inhibits the migration of macrophages is produced by antigen-stimulated T cells. These proteins show extensive sequence homology with each other and with a rat liver protein, designated » ***TRANSMIF***, a murine protein designated ***GIF*** (glycosylation inhibiting factor), and a mouse delayed early response gene, designated » ***DER6***. These highly homologous proteins, sharing sequence homologies of up to 98 %, are structurally related to the theta-class of ***glutathione S transferase*** (GST), a subclass of GSTs thought to be the most ancient evolutionary GST class. All proteins have retained a glutathione binding domain and demonstrate transferase activity. GSTs protect against xenobiotic chemicals by catalyzing the conjugation of glutathione, which renders them non-toxic.

MIF is known as a mediator of cellular immunity with specific effects on the differentiation of mononuclear phagocytes. The expression of MIF activity correlates well with delayed hypersensitivity and cellular immunity in humans and MIF is now recognized as a principal cytokine modulating T cell/macrophage interactions in the expression of delayed hypersensitivity and acquired cellular immunity. Recombinant human MIF is responsible for the specific activation of macrophages to kill intracellular parasites such as *Mycobacterium avium* and *Leishmania donovani*.

Recombinant human MIF has been shown to activate macrophage expression of NO, a potent antimicrobial agent that is produced by the NO synthase-mediated oxidation of terminal L-arginine guanidine nitrogen atoms.

Recombinant human MIF up-regulates expression of genes encoding HLA-DR and IL1β (see: IL1) and elaboration of IL1β by human monocyte-derived macrophages. Human recombinant MIF can also activate cultured human peripheral blood monocytes and monocyte-derived macrophages to become cytotoxic for tumor cells *in vitro*. MIF also induces macrophages to produce » TNF-α and » IL1β.

Administration of soluble bovine serum albumin or human immunodeficiency virus 120 kDa glycoprotein to mice in the presence of recombinant MIF together with incomplete Freund's adjuvant induces a strong T cell proliferative response comparable to that of complete Freund's adjuvant. Recombinant MIF also increases antibody production, especially of IgG1 and IgM, in mice. MIF may be useful, therefore, as an adjuvant in the development of vaccines.

MIF activity can be detected in the synovia of patients with rheumatoid arthritis. The expression of MIF at sites of inflammation also suggests a role for the mediator in regulating the function of macrophages in host defense.

Bernhagen J et al MIF is a pituitary-derived cytokine that potentiates lethal endotoxinemia. N 365: 756-9 (1993); **Bodeker BG et al** Culture conditions for selective and enhanced release of various murine lymphokines (CSF, IL2, MCF, MIF and TRF). Dev. Biol. Stand. 55: 241-5 (1983); **Blocki FA et al** MIF proteins are theta-class glutathione S-transferase homologues. Protein Sci. 2: 2095-102 (1993); **Bloom BR & Bennett B** Mechanisms of a reaction *in vitro* associated with delayed-type hypersensitivity. S 153: 80-2 (1966); **Cunha FQ et al** Recombinant migration inhibitory factor induces nitric oxide synthase in murine macrophages. JI 150: 1908-12 (1993); **David JR** MIF as a glutathione S-transferase. Parasitology Today 9: 315-6 (1993); **Gomez RS et al** Detection of migration inhibitory factor (MIF) by a monoclonal antibody in the microvasculature of inflamed skin. Arch. Dermatol. Res. 282: 374-8 (1990); **Herriott MJ et al** Mechanistic differences between migration inhibitory factor (MIF) and IFN-γ for macrophage activation. MIF and IFN-γ synergize with lipid A to mediate migration inhibition but only IFN-γ induces production of TNF-α and nitric oxide. JI 150: 4524-31 (1993); **Hirose S et al** Macrophage migration inhibitory factor (MIF) produced by a human T cell hybridoma clone. Microbiol. Immunol. 35: 235-45 (1991); **Hoon DS et al** Inhibition of lymphocyte motility by interleukin 2. Clin. Exp. Immunol. 66: 566-73 (1986); **Kawaguchi T et al** T cell-derived migration inhibitory factor and colony-stimulating factor share common structural elements. Blood 67: 1619-23 (1986); **Miyagawa N et al** Effect of C-reactive protein on peritoneal macrophages. I. Human C-reactive protein inhibits migration of guinea pig peritoneal macrophages. Microbiol. Immunol. 32: 709-19 (1988); **Miyagawa N et al** Effect of C-reactive protein on peritoneal macrophages. II. Human C-reactive protein activates peritoneal macrophages of guinea pigs to release superoxide anion *in vitro*. Microbiol. Immunol. 32: 721-31 (1988); **Murao S et al** A protein complex expressed during terminal differentiation of monomyelocytic cells is an inhibitor of cell growth. Cell Growth Diff. 1: 447-54 (1990); **Oki S et al** Macrophage migration inhibitory factor (MIF) produced by a human T cell hybridoma clone. Lymphokine Cytokine Res. 10: 273-80 (1991); **Pozzi LA & Weiser WY** Human recombinant migration inhibitory factor activates human macrophages to kill tumor cells. CI 145: 372-9 (1992); **Remold HG & Mednis AD** Migration inhibitory factor. MiE 116: 379-94 (1985); **Rocklin RE et al** *In vitro* lymphocyte response in patients with immunologic disorders: correlation of prediction of macrophage inhibitory factor with delayed hypersensitivity. NEJM 282: 1349-3 (1970); **Salahuddin SZ et al** Lymphokine production by cultured human T cells transformed by human T cell leukemia lymphoma virus I. S 223: 703-7 (1984); **Sorg C et al** Migration inhibitory factor. Springer Semin. Immunopathol. 7: 311 (1984); **Steinbeck MJ et al** Activation of bovine neutrophils by recombinant interferon-γ. CI 98: 137-44 (1986); **Orme IM et al** Inhibition of growth of Mycobacterium avium in murine and human mononuclear phagocytes by migration inhibitory factor. Infect. Immun. 61: 338-42 (1993); **Weiser WY et al** Molecular cloning of a cDNA encoding human macrophage migration inhibitory factor. PNAS 86: 7522-6 (1989); **Weiser WY et al** Dissociation of human interferon-γ-like activity from migration inhibition factor. CI 88: 109-22 (1984); **Weiser WY et al** Human recombinant migration inhibitory factor activates human macrophages to kill *Leishmania donovani*. JI 147: 2006-11 (1991); **Weiser WY et al** Recombinant human migration inhibitory factor has adjuvant activity. PNAS 89: 8049-52 (1992); Wistow GJ et al A macrophage migration inhibitory factor is expressed in the differentiating cells of the eye lens. PNAS 90: 1272-5 (1993)

MIF: *Müllerian inhibiting factor* see: MIS (Müllerian inhibiting substance).

MIF-1: *macrophage migration inhibitory factor 1* see: MIF (migration inhibition factor).

MIF-2: *macrophage migration inhibitory factor 2* see: F5MIF-2. See also: MIF (migration inhibition factor).

MIF-related protein 8: abbrev. MRP 8. see: Calgranulins.

MIF-related protein 14: abbrev. MRP 14. see: Calgranulins.

mig: *monokine induced by gamma interferon* This factor of 14.4 kDa is isolated from murine macrophages. This protein is also called ***M119*** (name of the corresponding cDNA clone). It is identical with CRG-10 (see: CRG, cytokine responsive

gene) and closely related to » PF4, » MGSA (melanoma growth stimulating activity) and » IL8. Human *mig* has a length of 103 amino acids (11.7 kDa). The protein belongs to the » chemokine family of cytokines.
The synthesis of *mig* is specifically induced by » IFN-γ, but not by IFN-α or bacterial lipopolysaccharides. It is thought that this protein is involved in modulating the growth, motility, and activation state of cells participating in inflammatory reactions.

Farber JM A macrophage mRNA selectively induced by γ-interferon encodes a member of the platelet factor 4 family of cytokines. PNAS 87: 5238-5242 (1990); **Farber JM** Hu*mig*: a new human member of the chemokine family of cytokines. BBRC 192: 223-30 (1993)

Migration inhibition factor: see: MIF.

Migration inhibition factor-related protein 8: abbrev. MRP 8. see: Calgranulins.

Migration inhibition factor-related protein 14: abbrev. MRP 14. see: Calgranulins.

Migration stimulating factor: see: MSF.

Migration stimulation factor: see: MStF.

***mil*:** see: *raf*.

Milk-derived growth factor 2: see: MDGF-2.

Milk growth factor: see: HMGF (human milk growth factor) and MGF (milk growth factor).

Minicytokines: A term occasionally used to denote factors that are not normally classified as » cytokines due to their small molecular weights. Minicytokines include regulatory peptide factors such as » bombesin, » bradykinin, » ET (endothelin), » GRP (gastrin-releasing peptide), » SP (substance P), » tachykinins, » vasopressin and » VIP (vasoactive intestinal peptide), many » thymic hormones, and very low molecular weight peptides such as » AcSDKP, » pEEDCK, and the pentapeptide » EIP (epidermal inhibitory pentapeptide). The biological activities of minicytokines overlap considerably with those of other cytokines and these activities are also markedly influenced by other cytokines.

MIP: *macrophage inflammatory protein* Two murine MIP proteins, designated ***MIP-1*** and ***MIP-2*** have been described. MIP-1 is an acidic protein. The two variants, designated ***MIP-1α*** and ***MIP-1β*** have a length of 69 amino acids (7.8 kDa). At the protein level they show a homology of 60 %. Their cDNAs show a homology of 57 %. The cDNA sequences of MIPα and MIPβ demonstrate that they are the murine homologues of human GROβ and GROγ (see: MGSA, melanoma growth stimulatory activity).
MIP1α is also called » EP (endogenous pyrogen), SISα (small inducible secreted), » L2G25B, SCYA3, and TY5. MIP1β is also called » EP (endogenous pyrogen), SISγ (small inducible secreted) and » H400. The human homologues have also been isolated. Murine MIP-1α is identical with human » 464.1 and is also called » LD78 or GOS-19-1 (see: GOS). Murine MIP-1β is identical with human » 744.1 and » Act-2.
The human genes map to chromosome 17q11-q21 with the human MIP-1α and MIP-1β genes arranged in tandem head-to-head orientation. The two murine genes map to chromosome 11.
Both MIP proteins belong to the » chemokine family of cytokines. MIP-1α has been shown to bind to a receptor called » CKR-1 (chemokine receptor 1).
Chromosomal aberrations in region 17q11 are frequently observed in Recklinghausen neurofibromatosis. Reciprocal translocations involving this regions are also observed in 90-100 % of acute promyelocytic leukemia. At present it is not known whether there is an involvement of these genes in these diseases.
The two MIP proteins are the major factors produced by macrophages following their stimulation with bacterial » endotoxin. Both proteins activate human neutrophilic, eosinophilic, and basophilic granulocytes (see also: Cell activation) and appear to be involved in acute neutrophilic inflammation. Both forms of MIP-1stimulate the production of reactive oxygen species in neutrophils and the release of lysosomal enzymes. They also induce the synthesis of other proinflammatory cytokines such as » IL1, » IL6 and » TNF in fibroblasts and macrophages (see also: Inflammation). MIP-1α is a potent basophil agonist, inducing a rapid change of cytosolic free calcium (see also: Calcium ionophore), the release of histamine and sulfido-leukotrienes, and chemotaxis. Murine MIP-1α is the primary stimulator of » TNF secretion by macrophages, whereas MIP-1β antagonizes the inductive effects of MIP-1α. In human monocytes the production of MIP-1β can be induced by bacterial lipopolysaccharides and » IL7.

MIP-1β is most effective at augmenting adhesion of CD8-positive T cells to the vascular cell adhesion molecule VCAM-1 and it does so by being present on the surface of endothelial cells complexed with endothelial proteoglycans.

The two factors also synergise with hematopoietic growth factors (see: Hematopoietins). The two forms of MIP-1 enhance the activities of » GM-CSF and promote the growth of more mature hematopoietic progenitor cells (see also: Hematopoiesis). MIP-1α (but not MIP-1β) also acts as an inhibitor of the proliferation of early hematopoietic stem cells and has therefore been called » SCI (stem cell inhibitor). MIP-1α has been shown to be capable of protecting hematopoietic progenitor cells from the cytotoxic effects of the cell cycle-specific drug hydroxyurea *in vivo* and may therefore also be of clinical significance.

The biological activities of MIP-1α and MIP-1β are mediated by receptors that bind both factors. A second species of receptors for these two factors also appears to bind » MCAF (monocyte chemotactic and activating factor).

MIP-2 is a basic protein of ≈ 6 kDa. It is extremely chemotactic for segmented neutrophilic granulocytes (see also: Chemotaxis) and also synergises with GM-CSF and M-CSF. In contrast to MIP-1 MIP-2 also induces the degranulation of human neutrophils but it does not enhance oxidative metabolism. MIP-2 binds to the same receptor as » IL8. MIP-1α and MIP-2 are also expressed by epidermal Langerhans cells and their expression is downregulated upon culture and maturation into dendritic cells.

■ **Clinical Use & Significance:** Based on the effects *in vivo* in mice it can be expected that MIP proteins may soon be tested for clinical efficacy for myeloprotection or dampening blood cell production in general.

● **Biochemistry & Molecular Biology: Blum S et al** Three human homologues of a murine gene encoding an inhibitor of stem cell proliferation. DNA Cell Biol. 9: 589-602 (1990); **Clements JM et al** Biological and structural properties of MIP-1 α expressed in yeast. Cytokine 4: 76-82 (1992); **Devatelis G et al** Cloning and characterization of a cDNA for murine macrophage inflammatory protein (MIP), a novel monokine with inflammatory and chemokinetic properties. JEM 167: 1939-44 (1988) (erratum in JEM 170: 2189 (1989)); **Farber JM** A macrophage mRNA selectively induced by γ-interferon encodes a member of the platelet factor 4 family of cytokines. PNAS 87: 5238-42 (1990); **Haskill S et al** Identification of three related human GRO genes encoding cytokine functions. PNAS 87: 7732-6 (1990); **Irving SG et al** Two inflammatory mediator cytokine genes are closely linked and variably amplified on chromosome 17q. NAR 18: 3261-70 (1990); **Sherry B et al** Resolution of the two components of macrophage inflammatory protein 1, and cloning and characterization of one of those components, macrophage inflammatory protein 1β. JEM 168: 2251-9 (1988); **Sherry B et al** Macrophage inflammatory proteins 1 and 2: An overview. Cytokines 4: 117-30 (1992); **Tekamp-Olson P et al** Cloning and characterization of cDNAs for murine macrophage inflammatory protein 2 and its human homologues. JEM 172: 911-9, (1990); **Wang JM et al** Human recombinant macrophage inflammatory protein-1 α and -β and monocyte chemotactic and activating factor utilize common and unique receptors on human monocytes. JI 150: 3022-9 (1993); **Widmer U et al** Genomic cloning and promoter analysis of macrophage inflammatory protein (MIP)-2, MIP-1α, and MIP-1β, members of the chemokine superfamily of proinflammatory cytokines. JI 150: 4996-5012 (1993); **Wolpe SD et al** Macrophages secrete a novel heparin-binding protein with inflammatory and neutrophil chemotactic properties. JEM 167: 570-81 (1988); **Wolpe SD et al** Identification and characterization of macrophage inflammatory protein 2. PNAS 86: 612-6 (1989); **Wolpe SD et al** Macrophage inflammatory protein 1 and 2: members of a novel superfamily of cytokines. FJ 3: 2565-73 (1989); **Ziegler S F et al** Induction of macrophage inflammatory protein-1β gene expression in human monocytes by lipopolysaccharide and IL7. JI 147: 2234-9 (1991)

● **Biological Activities: Appelberg R** Macrophage inflammatory proteins MIP-1 and MIP-2 are involved in T cell-mediated neutrophil recruitment. J. Leukoc. Biol. 52: 303-6 (1992); **Appelberg R** Interferon-γ (IFN-γ) and macrophage inflammatory proteins (MIP)-1 and -2 are involved in the regulation of the T cell-dependent chronic peritoneal neutrophilia of mice infected with mycobacteria. Clin. Exp. Immunol. 89: 269-73 (1992); **Bischoff SC et al** RANTES and related chemokines activate human basophil granulocytes through different G protein-coupled receptors. EJI 23: 761-7 (1993); **Broxmeyer HE et al** Myelopoietic enhancing effects of murine macrophage inflammatory protein 1 and 2 on colony formation *in vitro* by murine and human bone marrow granulocyte/macrophage progenitor cells. JEM 170: 1583-94 (1989); **Broxmeyer HE et al** Comparative analysis of the human macrophage inflammatory protein family of cytokines (chemokines) on proliferation of human myeloid progenitor cells. Interacting effects involving suppression, synergistic suppression, and blocking of suppression. JI 150: 3448-58 (1993); **Devatelis G et al** Macrophage inflammatory protein-1: a prostaglandin-independent endogenous pyrogen. S 243: 1066-8 (1989); **Fahey III TJ et al** Macrophage inflammatory protein 1 modulates macrophage function. JI 148: 2764-9 (1992); **Graham GJ et al** Identification and characterization of an inhibitor of hemopoietic stem cell proliferation. N 344: 442-4 (1990); **Heufler C et al** Cytokine gene expression in murine epidermal cell suspensions: interleukin 1 β and macrophage inflammatory protein 1 α are selectively expressed in Langerhans cells but are differentially regulated in culture. JEM 176: 1221-6 (1992); **Kuna P et al** Characterization of the human basophil response to cytokines, growth factors, and histamine releasing factors of the intercrine/chemokine family. JI 150: 1932-43 (1993); **Miñano FJ et al** Fever induced by macrophage inflammatory protein-1 (MIP-1) in rats: Hypothalamic sites of action. Brain Res. Bull. 27: 701-6 (1991); **Tanaka Y et al** T cell adhesion induced by proteoglycan-immobilized cytokine MIP-1 β. N 361: 79-82 (1993)

● **Clinical Use & Significance: Dunlop DJ et al** Demonstration of stem cell inhibition and myeloprotective effects of SCI/rhMIP-1α *in vivo*. Blood 79: 2221-5 (1992); **Lord BI et al** Macrophage-inflammatory protein protects multipotent hematopoietic cells from the cytotoxic effects of hydroxyurea *in vivo*. Blood 79: 2605-9 (1992); **Maze R et al** Myelosuppressive effects *in vivo* of purified recombinant murine macrophage inflammatory protein-1α. JI 149: 1004-9 (1992)

MIP-1: see: MIP (macrophage inflammatory protein).

MIP-2: see: MIP (macrophage inflammatory protein).

MIS: *Müllerian inhibiting substance.*

■ **ALTERNATIVE NAMES:** ***Müllerian inhibiting substance*** (abbrev. MIF), ***Müllerian inhibiting hormone***, ***anti-Müllerian hormone*** (abbrev. AMH).

■ **PROTEIN CHARACTERISTICS:** MIS (145 kDa) has a length of 560 amino acids and contains 12 cysteine residues. It is a homodimeric glycoprotein of 145 kDa with subunits that are linked by disulfide bonds. MIS is a member of the transforming growth factor family of proteins (see: TGF-β).

■ **GENE STRUCTURE:** The human MIS gene has five exons and maps to chromosome 19p13.3-13.2; the murine gene is on chromosome 10.

■ **RECEPTOR STRUCTURE, GENE(S), EXPRESSION:** A low abundance high affinity membrane receptor of 88 kDa has been identified on human » A431 cells.

■ **BIOLOGICAL ACTIVITIES:** MIS is expressed during the development of mammalian embryos. It is secreted by Sertoli cells in undifferentiated gonads and also by granulosa cells of the ovaries. It causes the regression of the Müllerian duct in male embryos which develops into tubes, uterus and upper vagina of female embryos. MIS has also been implicated in fetal lung development.

It has been reported that MIS also inhibits the meiosis in oocytes, » autophosphorylation of the » EGF receptor, and the growth of tumor cell lines. These findings were not reproducible with highly purified preparations of this factor.

■ **TRANSGENIC/KNOCK-OUT/ANTISENSE STUDIES:** The overexpression of MIS in » transgenic animals (mice) causes the aberrant sexual development characterized by a complete lack of uterus and oviduct and a blind vagina. A femininization is observed in male animals.

Transgenic mice carrying a fusion gene composed of human MIS transcriptional regulatory sequences linked to the SV40 T antigen gene specifically develop testicular tumors composed of a cell type histologically resembling the Sertoli cell. The lack of pathology at other sites suggests tissue-restricted expression of the transgene and the MIS regulatory sequence may therefore be useful to study the biology of Sertoli cells, which cannot be propagated *in vitro*.

■ **CLINICAL USE & SIGNIFICANCE:** The male sexual differentiation defect in humans known as persistent Müllerian duct syndrome (male internal pseudohermaphroditism, hernia uteri inguinale) is characterized by the failure of Müllerian duct regression in otherwise normal males caused by mutations in the MIS gene. Markedly elevated serum level of MIS have been observed in a woman with an ovarian sex-cord tumor with annular tubules, a rare tumor with the characteristics of both granulosa and Sertoli cells. An infant with Hirschsprung's disease and clinical features of persistent Müllerian duct syndrome has been shown to have a Müllerian inhibiting substance deficiency.

● **BIOCHEMISTRY & MOLECULAR BIOLOGY: Cate R et al** Isolation of the bovine and human genes for Müllerian inhibiting substance and expression of the human gene in animal cells. Cell 45: 685-98 (1986); **Cohen-Haguenauer O et al** Mapping of the gene for anti-Müllerian hormone to the short arm of human chromosome 19. Cytogenet. Cell Genet. 44: 2-6 (1987); **Donahue PK et al** A graded organ culture assay for the detection of Müllerian inhibiting substance. J. Surg. Res. 23: 141-8 (1977); **Josso N** Anti-Müllerian hormone: hormone or growth factor? PGFR 2: 169-79 (1990); **King TR et al** Mapping anti-müllerian hormone (Amh) and related sequences in the mouse: identification of a new region of homology between MMU10 and HSA19p. Genomics 11: 273-83 (1991); **Hudson PL et al** An immunoassay to detect human müllerian inhibiting substance in males and females during normal development. J. Clin. Endocrinol. Metab. 70: 16-22 (1990); **MacLaughlin DT et al** Bioassay, purification, cloning, and expression of Müllerian inhibiting substance. MiE 198: 358-69 (1991); **MacLaughlin DT et al** Müllerian duct regression and antiproliferative bioactivities of Müllerian inhibiting substance reside in its carboxy-terminal domain. Endocrinology 131: 291-6 (1992); **Miller WL** Immunoassays for human müllerian inhibitory factor (MIF): new insights into the physiology of MIF. J. Clin. Endocrinol. Metab. 70: 8-10 (1990); **Pepinsky RB et al** Proteolytic processing of Müllerian inhibiting substance produces a transforming growth factor-β-like fragment. JBC 263: 18961-65 (1988); **Picard JY & Josso N** Purification of testicular anti-Müllerian hormone allowing direct visualization of the pure glycoprotein and determination of yield and purification factor. Mol. Cell Endocrinol. 34: 23-9 (1984); **Picard JY et al** Cloning and expression of cDNA for anti-Müllerian hormone. PNAS 83: 5464-8 (1986); **Ragin RC et al** Human Müllerian inhibiting substance: enhanced purification imparts biochemical stability and restores antiproliferative effects. Prot. Express. Purif. 3: 236-45 (1992); **Wilson CA et al** Müllerian inhibiting substance requires its N-terminal domain for maintenance of biological activity, a novel finding within the transforming growth factor-β superfamily. Mol. Endocrinol. 7: 247-57 (1993)

● **RECEPTORS: Catlin E et al** Identification of a receptor for Müllerian Inhibiting substance. Endocrinology 133: 3007-13 (1993)

● **BIOLOGICAL ACTIVITIES: Catlin EA et al** Müllerian inhibiting substance blocks epidermal growth factor receptor phosphorylation in fetal rat lung membranes. Metabolism 40: 1178-84 (1991); **Catlin EA et al** Sex-specific fetal lung development and Müllerian inhibiting substance. Am. Rev. Respir. Dis. 141: 466-70 (1990); **Coughlin JP et al** Müllerian inhibiting substance blocks autophosphorylation of the EGF receptor by inhibiting tyrosine kinase. Mol. Cell. Endocrinol. 49: 75-86 (1987); **Wallen JW et al** Minimal antiproliferative effect of recombinant Müllerian inhibiting substance on gynecological tumor cell lines and tumor explants. CR 49: 2005-11 (1989)

● **TRANSGENIC/KNOCK-OUT/ANTISENSE STUDIES: Behringer RR et al** Abnormal sexual development in transgenic mice chronically expressing Müllerian inhibitory substance. N 345: 167-70 (1990); **Peschon JJ et al** Directed expression of an oncogene to Sertoli cells in transgenic mice using müllerian inhibiting substance regulatory sequences. Mol. Endocrinol 6: 1403-11 (1992)

● **CLINICAL USE & SIGNIFICANCE: Beheshti M et al** Familial persistent müllerian duct syndrome. J. Urol. 131: 968-9 (1984); **Brook CGD et al** Familial occurrence of persistent müllerian structures in otherwise normal males. Brit. Med. J. 1: 771-3 (1973); **Carre-Eusebe D et al** Variants of the anti-Müllerian hormone gene in a compound heterozygote with the persistent Müllerian duct syndrome and his family. Hum. Genet. 90: 389-94 (1992); **Cass DT & Hutson J** Association of Hirschsprung's disease and Müllerian inhibiting substance deficiency. J. Pediatr. Surg. 27: 1596-9 (1992); **Guerrier D et al** The persistent müllerian duct syndrome: a molecular approach. J. Clin. Endocr. Metab. 68: 46-52 (1989); **Gustafson ML et al** Müllerian inhibiting substance as a marker for ovarian sex-cord tumor. NEJM 326: 466-71 (1992); **Knebelmann B et al** Anti-Müllerian hormone Bruxelles: a nonsense mutation associated with the persistent Müllerian duct syndrome. PNAS 88: 3767-71 (1991)

Mitogen activated kinases: abbrev. MAP kinases. See: ERK (extracellular signal-regulated kinases).

Mitogenic protein: see: MP.

Mitogen-induced interferon: This interferon is identical with » IFN-γ. See also: IFN for general information about interferons.

Mitogen-regulated proteins: abbrev. MRP. see: Cell cycle.

Mitotoxins: This is a general term for fusion proteins obtained by translation of mRNAs derived from hybrid genes in which a gene encoding a growth factor or a cytokine has been fused to another gene encoding a cytotoxic protein. The resulting fusion proteins specifically bind to the growth factor receptor *in vitro* and *in vivo*. The internalization of receptor/hybrid protein complexes by cells releases the toxic component of the hybrid protein and thus allows these cells to be specifically killed. See also: Cytokine fusion toxins.

Lappi DA & Baird A Mitotoxins: growth factor-targeted cytotoxic molecules. PGFR 2: 223-36 (1990)

Mixed colony stimulating factor: This factor is identical with » IL3.

Hara H & Ogawa M Murine hemopoietic colonies in culture containing normoblasts, macrophages, and megakaryocytes. Am. J. Haematol. 4: 23 (1978)

MK: *midkine* abbrev. also: Mdk. A murine gene the expression of which is observed in mid-gestation of embryonic development. MK is also expressed at high levels in both embryonal carcinoma and pluripotential embryonic stem cells and their differentiated derivatives. The gene was originally identified as a gene overexpressed in these cells in response to treatment with retinoic acid. Human midkine is identical with » NEGF2 (neurite growth promoting factor 2).

The murine and human MK genes have a length of 3-4 kb and comprise five exons. The human MK gene maps to chromosome 11p11 – p13. MK is expressed as a precursor of 143 amino acids. Cleavage of a 22 amino acid secretory » signal sequence yields the mature protein of 14 kDa.

MK is not glycosylated. Its amino acid sequence displays ≈ 50 % homology to » HBNF (heparin-binding neurite-promoting factor). MK proteins are evolutionary conserved (human/mouse 87 %; human/chicken 65%; mouse/chicken 64%).

A gene isolated in the genome of chicken, » RIHB (retinoic acid-inducible heparin-binding protein) is probably the chicken homologue of the murine MK gene (see also: HBGF, heparin-binding growth factors).

The biological functions of MK are unknown. Its activity on neurons resembles that of » HBNF. MK is a mitogen for PC12 pheochromocytoma cells. MK is a potent mitogen for neurectodermal precursor cell types generated by treatment of 1009 embryonic carcinoma cells with retinoic acid but has no mitogenic or neurotrophic effects on more mature 1009-derived neuronal cell types. MK is active as an *in vitro* neurotrophic factor for E12 chick sympathetic neurons and its activity is markedly potentiated by binding the factor to tissue-culture plastic in the presence of heparin. MK is a multifunctional neuroregulatory molecule the biological activity of which depends upon association with components of the extracellular matrix (see also: ECM).

Böhlen P & Kovesdi I HBNF and MK, members of a novel gene family of heparin-binding proteins with potential roles in embryogenesis and brain function. PGFR 3: 143-57 (1991); **Fabri L et al** Structural characterization of native and recombinant forms of the neurotrophic cytokine MK. J. Chromatogr. 646: 213-25 (1993); **Kadomatsu K et al** cDNA cloning and sequencing of a new gene intensely expressed in early differentiation stages of embryonal carcinoma cells and in mid-gestation period of mouse embryogenesis. BBRC 151: 1312-8 (1988); **Kaname, T et al** Midkine gene (MDK), a gene for prenatal. differentiation and neuroregulation, maps to band 11p11.2 by fluorescence in situ hybridization. Genomics 17: 514-5 (1993); **Katoh K et al** Genomic organization of the mouse OSF-1 gene.

DNA Cell Biol. 11: 735-43 (1992); **Kretschmer PJ et al** Cloning, characterization, and developmental regulation of two members of a novel human gene family of neurite outgrowth-promoting proteins. GF 5: 99-114 (1991); **Muramatsu H & Muramatsu T** Purification of recombinant midkine and examination of its biological activities: functional comparison of new heparin binding factors. BBRC 177: 652-8 (1991); **Muramatsu T** Midkine (MK), the product of a retinoic acid responsive gene, and pleiotrophin constitute a new protein family regulating growth and differentiation. Int. J. Dev. Biol. 37: 183-8 (1993); **Nurcombe V et al** MK: a pluripotential embryonic stem-cell-derived neuroregulatory factor. Development 116: 1175-83 (1992); **Raulais D et al** A new heparin-binding protein regulated by retinoic acid from chick embryos. BBRC 175: 708-15 (1991); **Simon-Chazottes D et al** Chromosomal localization of two cell surface-associated molecules of potential importance in development: midkine (Mdk) and basigin (Bsg). Mamm. Genome 2: 269-71 (1992); **Tomomura M et al** A retinoic acid responsive gene, *MK*, found in the teratocarcinoma system. Heterogeneity of the transcript and the nature of the translation product. JBC 265: 10765-70 (1990); **Tomomura M et al** A retinoic acid responsive gene, *MK*, produces a secreted protein with heparin binding activity. BBRC 171: 603-9 (1990); **Tsutsui JI et al** A new family of heparin-binding factors: strong conservation of midkine (MK) sequences between the human and the mouse. BBRC 176: 792-7 (1991); **Uehara K et al** Genomic structure of human midkine (MK), a retinoic acid-responsive growth/differentiation factor. J. Biochem. Tokyo 111: 563-7 (1992)

Mk-CSF: *megakaryocyte colony-stimulating factor* see: MEG-CSF.

Mk potentiator: *megakaryocyte potentiator* abbrev. also MEG-Pot. This poorly characterized murine factor is produced by bone marrow macrophages. It does not support the growth of megakaryocytic colonies by its own but potentiates megakaryocyte colony growth induced by suboptimal amounts of » Meg-CSF. Meg-Pot activity can be provided by purified recombinant » IL6 but is most likely that there are other factors possessing Meg-Pot activity (see also: thrombopoietin).

Banu N et al Tissue sources of murine megakaryocyte potentiator: biochemical and immunological studies. Br. J. Haematol. 75: 313-8 (1990); **Hill RJ et al** Correlation of *in vitro* and *in vivo* biological activities during the partial purification of thrombopoietin. EH 20: 354-60 (1992); **Jackson H et al** The nature of an accessory cell in bone marrow stimulating murine megakaryopoiesis. EH 20: 241-4 (1992); **Oon SH & Williams N** Biochemical characterization of an *in vitro* murine megakaryocyte growth activity: megakaryocyte potentiator. Leuk. Res. 10: 403-11 (1986); **Satoh K et al** Simultaneous assay for megakaryocyte colony-stimulating factor and megakaryocyte potentiator and its application. J. Lab. Clin. Med. 116: 162-71 (1990); **Williams N et al** Studies on paracrine regulation of murine megakaryocytopoiesis. PCBR 356: 167-79 (1990); **Williams N et al** Recombinant interleukin 6 stimulates immature murine megakaryocytes. EH 18: 69-72 (1990)

MLA-144: This is a lymphoblastoid CD4-negative T helper cell line established from a spontaneous lymphosarcoma of the gibbon (*Hylobates lar*). This line releases gibbon ape leukemia virus and should be handled as potentially biohazardous material. Conditioned media from MLA 144 cells support growth and DNA synthesis of T cells from humans, several other species of primates, and also from mice and rabbits. This cell line is used because it constitutively produces » IL2 and » IL3. The cells are IL2-dependent and express low numbers of the 55 kDa receptor for IL2. The trace elements lithium and selenium enhance IL2 production in these cells.

Aspinall R & O'Gorman A Increased yields of IL2 in media conditioned by MLA 144 cells. JIM 101: 79-84 (1987); **Brown RL et al** Development of a serum-free medium which supports the long-term growth of human and nonhuman primate lymphoid cells. JCP 115: 191-8 (1983); **Brown RL et al** Modulation of interleukin 2 release from a primate lymphoid cell line in serum-free and serum-containing media. CI 92: 14-21 (1985); **Kaplan DR et al** Membrane-associated interleukin 2 epitopes on the surface of human T lymphocytes. JI 140: 819-26 (1988); **Linch DC & Donahue RE** Production of human active BPA from TCGF independent T cell lines that do not excrete HTLV: proof of direct action of MLA-144 derived BPA using purified BFU-E. Br. J. Haematol. 61: 71-82 (1985); **McGroarty RJ et al** Immunogold labeling of the low-affinity (55 kd) IL2 receptor on the surface of IL2 receptor-bearing cultured cells and mitogen-activated peripheral blood lymphocytes. J. Leukoc. Biol. 48: 213-9 (1990); **Rabin H et al** Spontaneous release of a factor with properties of T cell growth factor from a continuous line of primate tumor T cells. JI 127: 1852-1856 (1981); **Wu YY & Yang XH** Enhancement of interleukin 2 production in human and Gibbon T cells after *in vitro* treatment with lithium. PSEBM 198: 620-4 (1991); **Yuan Lin & Robb RJ** Induction of interleukin 2 production in the gibbon ape T cell line MLA 144. LR 4: 1-4 (1985)

ML-CCL64: see: CCL-64.

MLPLI: *melanoma-derived lipoprotein lipase inhibitor* This factor of 40 kDa is isolated from melanoma cells. The factor can also be isolated from a number of tumor cell lines that induce a metabolic cachexia syndrome if grown as transplanted tumors in experimental animals. MLPLI inhibits the lipoprotein lipase in adipocytes. Inhibition of this enzyme probably reduces the uptake of fatty acids and leads to the catabolism of lipids in fat tissues. The amino-terminal sequence of this factor is identical with that of » LIF.

A lipoprotein lipase inhibitor that inhibits the activity of lipoprotein lipase in fully differentiated 3T3-L1 adipocytes is produced also by SEKI, a cachexia-inducing human melanoma cell line.

Kawakami, M et al Suppression of lipoprotein lipase in 3T3-L1 cells by a mediator produced by SEKI melanoma, a cachexia-inducing human melanoma cell line. J. Biochem. Tokyo 109: 78-82 (1991); **Mori M et al** Purification of a lipoprotein lipase-inhi-

Mk-CSF: see addendum: thrombopoietin. **ML:** see addendum: thrombopoietin.

biting protein produced by a melanoma cell line associated with cancer cachexia. BBRC 160: 1085-92 (1989); **Mori M et al** Cancer cachexia syndrome developed in nude mice bearing melanoma cells producing leukemia inhibitory factor. CR 51: 6656-9 (1991)

MMIF: *macrophage migration inhibitory factor* see: MIF (migration inhibitory factor).

MNCF: *macrophage-released neutrophil chemotactic factor* This poorly characterized factor of 54 kDa is produced by rat peritoneal macrophages stimulated with bacterial lipopolysaccharides. It induces neutrophil migration that is not blocked by glucocorticoids. MNCF is probably not identical with » IL1β, » IL6, » IL8, or » TNF-α.

Lit. **Dias-Baruffi M et al** Macrophage-released neutrophil chemotactic factor (MNCF) induces PMN-neutrophil migration through lectin-like activity. Agents Actions 38 Spec No: C54-6 (1993)

MNCF: *mononuclear cell factor* see: MCF.

M-NFS-60: see: NFS-60.

MNSF: *monoclonal nonspecific suppressor factor* This factor of 12 kDa is produced by a murine T cell hybridoma. It suppresses the antibody response by cultured mononuclear cells to bacterial lipopolysaccharides. MNSF also inhibits the proliferative responses of B cells to T cell and B cell mitogens. The activity of MNSF is markedly inhibited by » IL2. MNSF also inhibits the growth of the plasmacytoma line MOPC-31C and of » EL4 cells, but not of murine » L929 fibroblasts. MNSF synergises with » TNF-α and increases the cytotoxic activity of TNF-α for L929 fibroblasts.

The N-terminal amino acid sequence of the protein shows significant (60%) homology to transforming growth factor β2 (see: TGF-β). A similar human factor, reacting with a monoclonal antibody, is also produced by concanavalin A-stimulated T cells.

Nakamura M et al Characterization of monoclonal nonspecific suppressor factor (MNSF) with the use of a monoclonal antibody. JI 138: 1799-803 (1987); **Nakamura M et al** Mode of action of monoclonal-nonspecific suppressor factor (MNSF) produced by murine hybridoma. CI 116: 230-9 (1988); **Nakamura M et al** Characterization of N-terminal amino acid sequence of monoclonal nonspecific suppressor factor. CI 139: 139-44 (1992); **Nakamura M et al** IFN-γ enhances the expression of cell surface receptors for monoclonal nonspecific suppressor factor. CI 139: 131-8 (1992)

Mo: abbrev. for monocytes.

MO7E: A human megakaryoblast-like leukemia cell line expressing CD13, » CD33, » CD34, CD36, CD45 (T200), and platelet glycoprotein 2b-3a (CD41(CD61). It is strictly dependent on either » IL3 or » GM-CSF and is therefore used to assay these factors (see also: Bioassays). Activities are determined by measuring the incorporation of ^{3}H thymidine into the newly synthesized DNA of proliferating cells. Cell proliferation can be determined also by employing the » MTT assay. An alternative and entirely different method of detecting IL6 is » Message amplification phenotyping. The bioassay is not specific because the cells also respond to » IL2, » IL6, » IL4, and » TNF-α. The cells also respond to » SCF (stem cell factor) which is a mitogen for these cells and can completely replace GM-CSF or IL3, mainly by inducing the synthesis of GM-CSF. SCF does not induce transcripts for any other cytokine to which the cells are responsive. SCF also synergises with » IL9 in stimulating MO7e cell proliferation. Proliferation of M07E cells induced by » GM-CSF and » SCF has been found to be stimulated in a dose-dependent manner by » IL2. IL3-supported colony formation of MO7E cells is significantly enhanced by » IFN-γ.

Avanzi GC et al Selective growth response to IL3 of a human leukemic cell line with megakaryoblastic features. Br. J. Haematol. 69: 359-66 (1988); **Avanzi GC et al** M-O7e human leukemic factor-dependent cell line provides a rapid and sensitive bioassay for the human cytokines GM-CSF and IL3. JCP 145: 458-64 (1990); **Bonsi L et al** M-07e cell bioassay detects stromal cell production of granulocyte-macrophage colony stimulating factor and stem cell factor in normal and in Diamond-Blackfan anemia bone marrow. Stem Cells Dayt. 11: s131-4 (1993); **Hendrie PC et al** Mast cell growth factor (c-*kit* ligand) enhances cytokine stimulation of proliferation of the human factor-dependent cell line, MO7e. EH 19: 1031-7 (1991); **Horie M & Broxmeyer HE** Involvement of immediate-early gene expression in the synergistic effects of steel factor in combination with granulocyte-macrophage colony-stimulating factor or interleukin-3 on proliferation of a human factor-dependent cell line. JBC 268: 968-73 (1993); **Kanakura Y et al** Functional expression of interleukin 2 receptor in a human factor-dependent megakaryoblastic leukemia cell line: evidence that granulocyte-macrophage colony-stimulating factor inhibits interleukin 2 binding to its receptor. CR 53: 675-80 (1993); **Kiss C et al** Human stem cell factor (c-*kit* ligand) induces an autocrine loop of growth in a GM-CSF-dependent megakaryocytic leukemia cell line. Leukemia 7: 235-40 (1993); **Kuriu A et al** Proliferation of human myeloid leukemia cell line associated with the tyrosine-phosphorylation and activation of the proto-oncogene c-*kit* product. Blood 78: 2834-40 (1991); **Miyazawa K et al** Comparative analysis of signaling pathways between mast cell growth factor (c-*kit* ligand) and granulocyte-macrophage colony-stimulating factor in a human factor-dependent myeloid cell line involves phosphorylation of *raf*-1, GTPase-activating protein and mitogen-activated protein kinase. EH 19: 1110-23 (1991); **Miyazawa K et al** Recombinant human interleukin-9 induces protein tyrosine phosphorylation and synergizes with steel factor to stimulate proliferation of the human factor-dependent cell line, M07e. Blood 80: 1685-92 (1992); **Murohashi I & Hoang T** Interferon-γ enhances growth factor-dependent proliferation of clonogenic cells in acute myeloblastic leukemia. Blood 78: 1085-95 (1991)

MOC: *monocyte-derived chemotaxin* This factor is identical with » IL8 and therefore belongs to the » chemokine family of cytokines.

Kownatzki E et al Stimulation of human neutrophilic granulocytes by two monocyte-derived cytokines. Agents Actions 26: 180-2 (1989)

Molgramostim: Generic name for recombinant non-glycosylated » GM-CSF (Shering Plough/ Sandoz) marketed under the brand name Leucomax.

MONAP: *monocyte-derived neutrophil-activating peptide* This factor is produced by monocytes and activates neutrophils. It is identical with » IL8 and therefore belongs to the » chemokine family of cytokines.

Kowalski J & Denhardt DT Regulation of the mRNA for monocyte-derived neutrophil-activating peptide in differentiating HL60 promyelocytes. MCB 9: 1946-57 (1989); **Schröder JM et al** Purification and partial biologic characterization of a human lymphocyte-derived peptide with potent neutrophil-stimulating activity. JI 140: 3524-32 (1987); **Schröder JM et al** Identification of different charged species of a human monocyte-derived neutrophil activating peptide (MONAP). BBRC 152: 277-84 (1988)

Monoclonal nonspecific suppressor factor: see: MNSF.

Monocyte chemoattractant protein 1: see: MCP-1.

Monocyte chemotactic and activating factor: see: MCAF.

Monocyte chemotactic protein: see: MCP.

Monocyte cytotoxic factor: see: MCF.

Monocyte cytotoxin: This factor is identical with » TNF-α.

Monocyte-derived chemotaxin: see: MOC.

Monocyte-derived endothelial cell inhibitory factor: see: MECIF.

Monocyte-derived granulocyte-activating mediator: abbrev. M-GRAM. See: GRAM.

Monocyte-derived growth factor: see: MDGF.

Monocyte-derived human B cell growth factor: This factor is identical with » IL6. See also: BCGF (B cell growth factor).

Monocyte-derived neutrophil-activating peptide see: MONAP.

Monocyte-derived neutrophil-chemotactic factor: see: MDNCF.

Monocyte-derived thymocyte proliferation factor: see: TPF (thymocyte proliferation factor).

Monocyte suppressor factor: see: MSF.

Monocyto-angiotropin: abbrev. MAT. see: angiotropin.

Monokines: This is a generic name for a number of unrelated soluble proteins produced by monocytes following their activation of damage. The nomenclature is problematic since it was shown that these factors are not secreted exclusively by monocytes. See: Cytokines for further information.

Monokine induced by gamma interferon: see: *mig*.

Mono Mac 6: A monocytic cell line that is used to assay IL1β (see: IL1) and » IL6. The cells secrete large amounts of IL6 and also » TNF if stimulated by bacterial » endotoxins. IL1 is also produced but is not secreted. The synthesis of TNF can be suppressed by various gangliosides. The cell line is also used to study monocyte-specific transcription factors (see also: Gene expression).

Haas JG et al Constitutive monocyte-restricted activity of NF-M, a nuclear factor that binds to a C/EBP motif. JI 149: 237-43 (1992); **Rafferty B et al** Measurement of cytokine production by the monocytic cell line Mono Mac 6 using novel immunoradiometric assays for interleukin 1 and interleukin 6. JIM 144: 69-74 (1991); **Ziegler-Heitbrock HW et al** *In vitro* desensitization to lipopolysaccharide suppresses tumor necrosis factor, interleukin-1 and interleukin-6 gene expression in a similar fashion. Immunology 75: 264-8 (1992); **Ziegler-Heitbrock HW et al** Gangliosides suppress tumor necrosis factor production in human monocytes. JI 148: 1753-8 (1992)

Mononuclear cell factor: also abbrev. MNCF. See: MCF.

motheaten: Name of a mutation that arose spontaneously in C57BL/6J mice. The mutation maps to the distal end of chromosome 6. Two alleles of this locus are known, designated *me* and *mev* (***viable me***). The disease in homozygotes of mev/mev is very similar to that of me/me mice but progresses more slowly.

A few days after birth animals homozygous for the recessive *me* locus develop subepidermal lesions which disrupt hair follicles and pigmentation and give the mice their characteristic motheaten appearance. The animals suffer from an early onset autoimmunity and severe immunodeficiency. Even under specific pathogen-free conditions animals die after 6 to 8 weeks from an unusual pneumonia which begins as early as 3 days and leads to a massive accumulation of granulocytes, macrophages, and lymphocytes in the alveoli.

Serum immunoglobulins, particularly IgM, are markedly elevated. The number of B cells bearing surface immunoglobulins is greatly reduced, and the B cells present are of the immature rather that adult type. The thymus appears normal for the first 3-4 weeks, after which there is a depletion of cortical thymocytes and a diminution in the size of the organ until it is atrophic. The number of T cells is normal, but several T cell functions are defective. Prothymocytes in bone marrow of *me/me* and mev/mev mice are present in essentially normal numbers, as determined by intrathymic injection, but apparently lack the ability to home effectively to the thymus, as determined by intravenous transfer. Natural killer cell activity is virtually absent, and » TGF-β is thought to be responsible, at least in part, for the defective NK differentiation. The mice also show abnormal immunoglobulin deposits in the kidney, thymus, skin, and lung. The serum contains factors cytotoxic to thymus and skin.

Bone marrow contains normal » hematopoietic stem cells with a normal number of colony-forming units (see also: CFU, Hematopoiesis) but shows aberrant growth during » LTBMC (long-term bone marrow culture). In viable motheaten mice, there is a major shift in hematopoiesis from bone marrow to spleen. The numbers of erythroid precursor cells » CFU-E are dramatically increased in spleen and show increased sensitivity to » Epo. The bone marrow CFU-E population is significantly diminished. The population of » BFU-E (erythroid burst-forming units) is diminished both in bone marrow and in spleen, and *mev/mev* spleen cell-conditioned medium shows a 40-fold reduction in burst-promoting activity (see: BPA).

The B cells of *mev/mev* mice produce high levels of a unique ***B cell maturation factor*** (see: BMF). Their serum also contains a factor that increases the activity of this BMF by up to three orders of magnitude. The action of these two factors probably causes the observed great increase in the number of plasma cells, leaving few or no resting B cells able to respond to specific antigens and thus precipitating the severe immunodeficiency of motheaten mice.

It has been shown recently that the genetic defects in motheaten mice are caused by a mutation in the gene encoding » HCP (hematopoietic cell phosphatase) which modulates the functions of several cytokines involved in hematopoiesis. The *me* allele is characterized by a complete absence of HCP while the *mev* allele shows reduced levels of the phosphatase.

Clark EA et al Mutations in mice that influence natural killer (NK) cell activity. Immunogenetics 12: 601-13 (1981); **Davidson WF et al** Phenotypic and functional effects of the motheaten gene on murine T and B lymphocytes. JI 122: 884-91 (1979); **Green MC & Shultz LD** Motheaten, an immunodeficient mutant of the mouse. I. Genetics and pathology. J. Hered. 66: 250-8 (1975); **Greiner DL et al** Defective lymphopoiesis in bone marrow of motheaten (me/me) and viable motheaten (mev/mev) mutant mice. I. Analysis of development of prothymocytes, early B lineage cells, and terminal deoxynucleotidyl transferase-positive cells. JEM 164: 1129-44 (1986); **Haar JL et al** Defective *in vitro* migratory capacity of bone marrow cells from viable motheaten mice in response to normal thymus culture supernatants. EH 17: 21-4 (1989); **Hayashi S et al** Lymphohemopoiesis in culture is prevented by interaction with adherent bone marrow cells from mutant viable Motheaten mice. JI 140: 2139-47 (1988); **Koo GC et al** Suppressive effects of monocytic cells and transforming growth factor-β on natural killer cell differentiation in autoimmune viable motheaten mutant mice. JI 147: 1194-200 (1991); **McCoy KL et al** Spontaneous production of colony-stimulating activity by splenic Mac-1 antigen-positive cells from autoimmune motheaten mice. JI 132: 272-6 (1984); **McCoy KL et al** Effects of the motheaten gene on murine B cell production. EH 13: 554-9 (1985); **Medlock ES et al** Defective lymphopoiesis in the bone marrow of Motheaten (me/me) and viable Motheaten (mev/mev) mutant mice. II. Description of a microenvironmental defect for the generation of terminal deoxynucleotidyltransferase-positive bone marrow cells *in vitro*. JI 138: 3590-7 (1987); **Shultz LD & Green MC** Motheaten, an immunodeficient mutant of the mouse. II. Depressed immune response and elevated serum immunoglobulins. JI 116: 936-43 (1976); **Shultz LD et al** Hematopoietic stem cell function in motheaten mice. EH 11: 667-80 (1983); **Shultz LD et al** "Viable motheaten", a new allele at the motheaten locus. I. Pathology. Am. J. Pathol. 116: 179-92 (1984); **Shultz LD** Pleiotropic effects of deleterious alleles at the "motheaten" locus. Curr. Top. Microbiol. Immunol. 137: 216-22 (1988); **Shultz LD et al** Mutations at the murine motheaten locus are within the hematopoietic cell protein tyrosine phosphatase (Hcph) gene. Cell 73: 1445-54 (1993); **Sidman CL et al** The mouse mutant "motheaten". II. Functional studies of the immune system. JI 121: 2399-404 (1978); **Sidman CL et al** Novel B cell maturation factor from spontaneously autoimmune viable motheaten mice. PNAS 81: 7199-202 (1984); **Sidman CL et al** A serum-derived molecule from autoimmune viable motheaten mice potentiates the action of a B cell maturation factor. JI 135: 870-2 (1985); **Sidman CL et al** Murine "viable motheaten" mutation reveals a gene critical to the development of both B and T lymphocytes. PNAS 86: 6279-82 (1989); **Tsui HW et al** Motheaten and viable motheaten mice have mutations in the hematopoietic cell phosphatase gene. Nature Genetics 4: 124-9 (1993); **Van Zant G & Shultz L** Hematologic abnormalities of the immunodeficient mouse mutant, viable Motheaten (mev). EH 17: 81-7 (1989)

Motogenic cytokines: A general term for cytokines that influence to motility and migration of cells. These factors are sometimes also called ***cell scattering factors*** (abbrev. CSF). They probably play an important role in embryonic development, » wound healing, » inflammation, and the invasive growth and metastatic dissemination of tumor cells (see also: Chemotaxis). Some motogenic cytokines act as » autocrine factors on the cells that produce them. Motogenic cytokines include » AMF (autocrine motility factor), ATX (autotaxin), » Ladsin, » MSF (migration stimulating factor), » SF (scatter factor) and » vitronectin. The action of some of these motogenic cytokines is blocked by inhibitory factors (see also: MIF, migration inhibition factor).

Liotta LA et al Tumor cell motility. Semin. Cancer Biol. 2: 111-4 (1991); **Rosen EM & Goldberg ID** Protein factors which regulate cell motility. In Vitro Cell. Dev. Biol. 25: 1079-87 (1989); **Seiki M et al** Comparison of autocrine mechanisms promoting motility in two metastatic cell lines: human melanoma and *ras*-transfected NIH3T3 cells. IJC 49: 717-20 (1991); **Stoker M & Gherardi E** Regulation of cell movement: the motogenic cytokines. BBA 1072: 81-102 (1991); **Stracke ML et al** Cell motility: a principal requirement for metastasis. Experientia Suppl. 59: 147-62 (1991)

Mouse sarcoma 180-derived growth factor: This heparin-binding factor (see also: HBGF) is found in medium conditioned by mouse sarcoma 180 cells. It stimulates the growth of capillary endothelial cells. The factor appears to be identical with » VEGF.

Rosenthal RA et al Conditioned medium from mouse sarcoma 180 cells contains vascular endothelial growth factor. GF 4: 53-9 (1990)

MP: *mitogenic protein* This macrophage-derived factor was initially described as a protein stimulating thymocyte proliferation. It is identical with IL1α (see: IL1).

Aarden LA et al Revised nomenclature for antigen-nonspecific T cell proliferation and helper factors. CI 48: 433-36 (1979); **Unanue ER et al** Synthesis and secretion of a mitogenic protein by macrophages: description of a superinduction phenomenon. JI 119: 925-31 (1977)

MP6-derived B cell stimulatory factor: see: BSF MP6.

MPC: abbrev. for myelopoietic progenitor cells. See: Hematopoiesis.

MPCA inducing factor: *macrophage procoagulant activity inducing factor* see: MPIF (macrophage procoagulant inducing factor).

MPG: abbrev. for membrane proteoglycans. see: ECM (extracellular matrix).

MPIF: *macrophage procoagulant inducing factor* This poorly characterized factor of ≈ 55 kDa is also called ***MPCA inducing factor*** (macrophage procoagulant activity inducing factor). It is produced by Th0 and Th1 helper T cells but not by Th2 cells (see also: TH1/TH2 cytokines). The factor promotes the procoagulant properties of macrophages (deposition of fibrin) and is a chemoattractant for phagocytosing cells. It is probably involved in delayed type hypersensitivity reactions. The factor is not identical with » IFN-α, » IFN-β, » IL1, IL2, » GM-CSF, M-CSF, TNF-α and TNF-β.

Bentel J & Atkinson K Cytokine activity after human bone marrow transplantation. II. Production of macrophage procoagulant activity and the cytokine regulating its production, macrophage procoagulant inducing factor. Br. J. Haematol. 69: 181-7 (1988); **Chung SW et al** The comparative effects of cyclosporin A and 16,16 dimethyl prostaglandin E2 on the allogeneic induction of monocyte/macrophage procoagulant activity and the cytokines macrophage procoagulant inducing factor and interleukin 2. Immunology 74: 670-6 (1991); **Fan ST & Edgington TS** Clonal analysis of mechanisms of murine T helper cell collaboration with effector cells of macrophage lineage. JI 141: 1819-27 (1988); **Fan ST et al** Clonal analysis of CD4$^+$ T helper cell subsets that induce the monocyte procoagulant response. CI 128: 52-62 (1990); **Gregory SA et al** Monocyte procoagulant inducing factor: a lymphokine involved in the T cell-instructed monocyte procoagulant response to antigen. JI 137: 3231-9 (1986); **Ryan J & Geczy CL** Macrophage procoagulant inducing factor. *In vivo* properties and chemotactic activity for phagocytic cells. JI 141: 2110-7 (1988); **Ryan J & Geczy C** Coagulation and the expression of cell-mediated immunity. Immunol. Cell. Biol. 65: 127-39 (1987); **Ryan J & Geczy CL** Characterization and purification of mouse macrophage procoagulant inducing factor. JI 137: 2864-70 (1986)

MPIF: *muscle proteolysis-inducing factor* see: PIF (proteolysis inducing factor).

***mpl*:** see: MPLV (myeloproliferative leukemia virus).

MPLV: *myeloproliferative leukemia virus* MPLV is a murine replication-defective acute leukemo-

 c-*mpl* ligand: see addendum: thrombopoietin. ***mpl* ligand:** see addendum: thrombopoietin.

genic retrovirus. It harbors a viral » oncogene called ***mpl***. The corresponding cellular counterpart (c-*mpl*) is predominantly expressed in hematopoietic tissues, in particular in immature hematopoietic precursor cells, but also in spleen, and fetal liver. *mpl* is not expressed in the thymus. The human oncogene maps to chromosome 1p34. Its sequence is strongly conserved among mammalians (81 % amino acid identity between humans and mouse).
The infection of adult mice with MPLV leads to hematopoietic disorders characterized by the extreme proliferation and differentiation of a number of hematopoietic lineages (see also: Hematopoiesis). The hematopoietic progenitor cells isolated from infected mice are immortalized and do not require exogenous growth factors in contrast to their normal factor-dependent counterparts.
Cells obtained from thymus and bone marrow grow in soft agar and develop into terminally differentiated erythroid, megakaryocytic, granulocytic cells or macrophages (see also: Colony formation assay).
mpl is expressed as a fusion protein with part of the viral coat protein of Friend murine leukemia virus. *mpl* is a member of the hematopoietic growth factor receptor superfamily and is closely related to receptors for » IL2 to » IL7, » GM-CSF, » G-CSF and » Epo (see: Cytokine receptor superfamily). The viral *mpl* gene is a truncated form of the cellular oncogene and probably functions as a receptor for an as yet unknown hematopoietic growth factor. A soluble form of this putative receptor is obtained by alternative splicing of the corresponding mRNA.
Le Coniat M et al The human homologue of the myeloproliferative virus maps to chromosome band 1p34. Hum. Genet. 83: 194-6 (1989); **Penciolelli JF et al** Genetic analysis of myeloproliferative leukemia virus, a novel acute leukemogenic replication-defective retrovirus. J. Virol. 61: 579-83 (1987); **Skoda RC et al** Murine c-*mpl*: a member of the hematopoietic growth factor receptor superfamily that transduces a proliferative signal. EJ 12: 2645-53 (1993); **Souyri M et al** A putative truncated cytokine receptor gene transduced by the myeloproliferative leukemia virus immortalizes hematopoietic progenitors. Cell 63: 1137-47 (1990); **Vigon I et al** Molecular cloning and characterization of *mpl*, the human homologue of the v-*mpl* oncogene: identification of a member of the hematopoietic growth factor receptor superfamily. PNAS 89: 5640-4 (1992); **Vigon I et al** Characterization of the murine Mpl proto-oncogene, a member of the hematopoietic cytokine receptor family: molecular cloning, chromosomal location and evidence for a function in cell growth. O 8: 2607-15 (1993)

MPR: *mannose-6-phosphate receptor* abbrev. also M6P-R. Two mannose-6-phosphate receptors have been described. They are called MPR46 (46 kDa) and MPR300. Both receptors mediate endocytosis of mannose-containing ligands and the intracellular transport of newly synthesized proteins containing mannose 6 phosphate recognition markers into lysosomes. No preferential binding of lysosomal enzymes to one of the two MPRs has been observed.
MRP300 is a cation-independent mannose-6-phosphate receptor (***CIM6PR***). It is encoded by a gene mapping to human chromosome 6q25-q27. This receptor also binds insulin-like growth factor IGF-2 (see: IGF). Binding of mannose and IGF-2 is independent of each other and utilizes two different binding sites. The biological consequences of this receptor having binding sites for unrelated molecules are unclear. On the one hand some species such as chicken and frogs possess MPR300 that does not contain a binding site for IGF-2. On the other hand the receptor is coupled to an inhibitory G-protein that is activated by binding of IGF-2, i. e. IGF-2 induces a signal that is transmitted to the cytoplasm. The mannose-6-phosphate receptor has also been shown to bind » TGF-β and » proliferin (see also: MRP, mitogen regulated protein).
Both types of receptors are expressed during the embryonic development of the mouse at different times and in different tissues. Tissues expressing MPR46 usually do not express MPR300 and vice versa.
■ **TRANSGENIC/KNOCK-OUT/ANTISENSE STUDIES:** The biological activities of both types of mannose 6 phosphate receptors have been investigated by creation of » transgenic animals in which each receptor gene has been inactivated by targeted disruption (see also: ES cells. Mice carrying homozygous mutations of either of the two receptors appear phenotypically normal, indicating mechanisms that compensate the receptor deficiencies *in vivo*. Both types of receptor deficiencies result in a misrouting of multiple lysosomal enzymes. Increased levels of phosphorylated lysosomal enzymes are present in body fluids of homozygous animals and lysosomal enzymes are also found to be secreted into the extracellular medium by several cell types *in vitro*.
Canfield WM & Kornfeld S The chicken liver cation-independent mannose 6-phosphate receptor lacks the high affinity binding site for insulin-like growth factor II. JBC 264: 7100-3 (1989); **Clairmont KB & Czech MP** Chicken and *Xenopus* mannose 6-phosphate receptors fail to bind insulin-like growth factor II. JBC 264: 16390-2 (1989); **Dennis PA & Rifkin DB** Cellular activation of latent transforming growth factor β requires binding to the cation-independent mannose 6-phos-

phate/insulin-like growth factor type II receptor. PNAS 88: 580-4 (1991); **Kiess W et al** Biochemical evidence that the type II insulin-like growth factor receptor is identical to the cation-independent mannose 6-phosphate receptor. JBC 263: 9339-44 (1988); **Lee SJ & Nathans D** Proliferin secreted by cultured cells binds to mannose 6-phosphate receptors. JBC 263: 3521-7 (1988); **Laureys G et al** Chromosomal mapping of the gene for the type II insulin-like growth factor receptor/cation-independent mannose 6-phosphate receptor in man and mouse. Genomics 3: 224-9 (1988); **Ma ZM et al** Cloning, sequencing, and functional characterization of the murine 46 kDa mannose 6-phosphate receptor. JBC 266: 10589-95 (1991); **Matzner U et al** Expression of the two mannose 6-phosphate receptors is spatially and temporally different during mouse embryogenesis. Development 114: 965-72 (1992); **Oshima A et al** The human cation-independent mannose 6-phosphate receptor. Cloning and sequence of the full-length cDNA and expression of functional receptor in COS cells. JBC 263: 2553-62 (1988); **Rosorius O** Characterization of phosphorylation sites in the cytoplasmic domain of the 300 kDa mannose-6-phosphate receptor. BJ 292: 833-8 (1993); **Vignon F & Rochefort H** Interactions of pro-cathepsin D and IGF-II on the mannose-6-phosphate/IGF-II receptor. Breast Cancer Res. Treat. 22: 47-57 (1992)

● **Transgenic/Knock-out/Antisense Studies: Köster A et al** Targeted disruption of the Mr 46000 mannose 6 phosphate receptor gene in mice results in misrouting of lysosomal proteins. EJ 12: 5219-23 (1993); **Ludwig T et al** Targeted disruption of the mouse cation-dependent mannose 6 phosphate receptor results in partial missorting of multiple lysosomal enzymes. EJ 12: 5225-35 (1993)

MPSC: *multipotent stem cells* see: Hematopoiesis.

MRA: *marrow repopulating ability* This term refers to primitive totipotent » hematopoietic stem cells with self-renewal capacity that are capable of repopulating the bone marrow of lethally irradiated mice (see also: LTRC, long-term repopulating cells). It is not known whether every repopulating cell contributes equally well to the various differentiated hematopoietic lineages (see: Hematopoiesis).

Two different types of cells with marrow repopulating ability have been distinguished in the mouse. Initial engraftment (short-term repopulation) is due to cells that also produce spleen colonies (see: CFU-S). Long-term engraftment is attributed to a different cell type but is only possible if the animals also receive short-term repopulating cells. A cell type which is more primitive than » CFU-S (pre-CFU-S) is considered to be responsible for long-term marrow repopulating ability.

For obvious reasons no equivalent experimental assays allowing detection of early hematopoietic cells with long-term marrow repopulating activity. Such assays would be highly desirable in view of bone marrow transplantation. Cells can be assayed in long-term bone marrow cultures (see: LTBMC), by » limiting dilution analysis, and/or » colony formation assays, but these available alternatives are poorly standardized and the results can only be evaluated in retrospect. Assays of hematopoietic progenitor cell types such as » BFU-E, » CFU-GM, or » CFU-GEMM do not predict marrow reconstituting ability of cells obtained after incubation of marrow cells with cytotoxic drugs such as » 4-HC (4-hydroperoxycyclophosphamide) or » fluorouracil to enrich for early progenitors.

Unfortunately, many conventional assays (see: Hematopoietic stem cells) fail to measure the most important characteristics of primitive hemopoietic stem cells: long-term repopulating ability and maximal differentiating ability . The ***competitive repopulation assay*** examines the precursors from which most differentiated cells of myeloid and lymphoid lineages (see: Hematopoiesis) are descended. Enumeration of the so-called ***CRU*** (***competitive repopulating unit***) detectable in these assays is now believed to yield one of the closest estimates of true hematopoietic stem cells.

The assay employs bone marrow cells from which stem cells have been enriched, for example, by treatment with » fluorouracil or 4-HC and/or subsequent culture *in vitro* (see: LTBMC, long-term bone marrow culture). These test cells can be marked, for example, by infection with helper-free neomycin-resistant recombinant retroviruses, to distinguish them, by virtue of their uniquely marked clonal progeny, from endogenous bone marrow cells of syngeneic recipient animals. If the recipient animals are female, male test cells can be used and identified by the presence of a Y chromosome by means of a Y chromosome-specific gene probe.

Harvested test cells are then injected into lethally irradiated recipients either alone or together with recipient-derived marrow cells with selectively compromised long-term repopulating potential. Compromised bone marrow cells are obtained by serial transplantation of marrow cells from syngeneic recipient mice. These cell populations contain a near normal frequency of » CFU-S (spleen colony-forming units) and various *in vitro* clonogenic cells (see: Colony formation assay) and can alone rescue recipient animals from the lethal effects of radiation in short-term assays. However, these compromised cells exhibit a markedly reduced competitive long-term hematopoietic repopulating ability. Coinjection of compromised cells with test cells ensures the survival of the recipient

animals and also provides a selective pressure to identify a class of stem cells that possess a high capacity for competitive repopulation in a long-term assay (≥ 5 weeks). The relative frequencies of limited numbers of hematopoietic stem cells in various test cell suspension contributing to the long-term repopulation of recipient animals can then be calculated by Poisson statistics (see also: Limiting dilution analysis). The term ***competitive repopulating units*** (abbrev. ***CRU***) is used to distinguish between the true number of competitive repopulating cells in a suspension of test cells and the presumably minimum number detectable by the *in vivo* assay. As stem cell differentiation has been shown to be a stochastic process it must be assumed that some totipotent cells will not express their full differentiation potential *in vivo* because they happen to become lineage-restricted within their first cell division.
The competitive repopulation assay can provide critical information about stem cell numbers during ontogeny and after genetic, biologic, or pharmacological manipulation of test cells *in vitro*. The *in vivo* model of competitive repopulation should provide the opportunity to explore genetic loci potentially interfering with the process of test cell engraftment. The analysis of test cells propagated *in vitro* has revealed, for example, that some totipotent hematopoietic stem cells can be maintained and amplified over extensive time periods *in vitro* without diminution of their long-term *in vivo* repopulating potential. It has been found also that *in vivo* administration of » SCF significantly increases the absolute number of primitive hemopoietic stem cells detectable in these assays. Assays of repopulating stem cell function by the competitive repopulation assay have shown that IL3 or IL4 alone preserve stem cell function *in vitro*. The combinations of IL3 and IL6, and IL3 and G-CSF increase stem cell function approximately twofold. The combinations of IL3 + G-CSF + IL6, and IL4 and IL6 have been found to decrease stem cell function approximately fourfold.

Bartelmez SH et al Uncovering the heterogeneity of hematopoietic repopulating cells. EH 19: 861-2 (1991); **Bodine DM et al** Effects of hematopoietic growth factors on the survival of primitive stem cells in liquid suspension culture. Blood 78: 914-20 (1991); **Bodine DM et al** *In vivo* administration of stem cell factor to mice increases the absolute number of pluripotent hematopoietic stem cells. Blood 82: 445-55 (1993); **Brecher G et al** Self-renewal of the long-term repopulating stem cell. PNAS 90: 6028-31 (1993); **Fraser CC et al** Proliferation of totipotent hematopoietic stem cells *in vitro* with retention of long-term competitive *in vivo* reconstituting ability. PNAS 89: 1968-72 (1992); **Harrison DE** Competitive repopulation in unirradiated normal recipients Blood 81: 2473-4 (1993); **Harrison DE et al** Primitive hemopoietic stem cells: direct assay of most productive populations by competitive repopulation with simple binomial, correlation and covariance calculations. EH 21: 206-19 (1993); **Neben S et al** Mobilization of hematopoietic stem and progenitor cell subpopulations from the marrow to the blood of mice following cyclophosphamide and/or granulocyte colony-stimulating factor. Blood 81: 1960-7 (1993); **Orlic D & Bodine DM** Pluripotent hematopoietic stem cells of low and high density can repopulate W/Wv mice. EH 20: 1291-5 (1992); **Spangrude GJ & Brooks DM** Phenotypic analysis of mouse hematopoietic stem cells shows a Thy-1-negative subset. Blood 80: 1957-64 (1992); **Szilvassy SJ et al** Quantitative assay for totipotent reconstituting hematopoietic stem cells by a competitive repopulation strategy. PNAS 87: 8736-40 (1990); **Szilvassy SJ & Cory S** Phenotypic and functional characterization of competitive long-term repopulating hematopoietic stem cells enriched from 5-fluorouracil-treated murine marrow. Blood 81: 2310-20 (1993); **Quesniaux YF et al** Use of 5-fluorouracil to analyze the effect of macrophage inflammatory protein-1 α on long-term reconstituting stem cells *in vivo*. Blood 81: 1497-504 (1993); **Van Zant G et al** Genotype-restricted growth and aging patterns in hematopoietic stem cell populations of allophenic mice. JEM 171: 1547-65 (1990); **Van Zant G et al** Differentiation of chimeric bone marrow *in vivo* reveals genotype-restricted contributions to hematopoiesis. EH 19: 941-9 (1991); **Wineman JP et al** CD4 is expressed on murine pluripotent hematopoietic stem cells. Blood 80: 1717-24 (1992)

MRE: *multiresponse element* A general term describing short DNA sequences found in the promoter region of some cytokine genes and other genes. These sequences usually consist of multiple and sometimes overlapping binding sites for DNA binding factors that function as regulatory proteins for » Gene expression (see also: ERG, early response genes).

mRNA phenotyping: see: Message amplification phenotyping.

MRP: *mitogen-regulated proteins* see: Cell cycle.

MRP: *mitogen-regulated protein* MRP is also called » PLF (proliferin). It is a secreted glycoprotein of ≈ 34 kDa. Its synthesis by some immortalized murine cells is stimulated by » EGF, » FGF, » TGF-α, tumor promoters, and serum. The protein is not produced by primary mouse embryo fibroblasts. The transformation of these cells by SV40 or Moloney sarcoma virus leads to the inhibition of MRP synthesis.
MRP is predominantly expressed by trophoblastic giant cells in the developing placenta. It is also released into the blood circulation. The synthesis of MRP ceases when the placenta stops growing. TGF-α stimulates the synthesis of MRP in tropho-

blasts while » TGF-β reduces the effect of TGF-α. MRP is thought to act as an endocrine or » paracrine cytokine regulating the proliferation of placenta cells. MRP binds to the mannose-6-phosphate receptor (see: MPR).

Chiang CP & Nilsen-Hamilton M Opposite and selective effects of epidermal growth factor and human platelet transforming growth factor-β on the production of secreted proteins by murine 3T3 cells and human fibroblasts. JBC 261: 10478-81 (1986); **Edwards DR et al** Evidence that post-transcriptional changes in the expression of mitogen regulated protein accompany immortalization of mouse cells. BBRC 147: 467-73 (1987); **Hamilton RT & Millis AJ** Developmental roles for growth factor-regulated secreted proteins. CTDB 24: 193-218 (1990); **Lee SJ et al** Trophoblastic giant cells of the mouse placenta as the site of proliferin synthesis. Endocrinology 122: 1761-8 (1988); **Parfett CLJ et al** Characterization of a cDNA clone encoding murine mitogen-regulated protein: regulation of mRNA levels in mortal and immortal cell lines. MCB 5: 3289-92 (1985); **Wilder EL & Linzer DIH** Expression of multiple proliferin genes in mouse cells. MCB 6: 3283-6 (1986)

MRP 8: *MIF (migration inhibition factor)-related protein 8* see: Calgranulins.

MRP 14: *MIF (migration inhibition factor)-related protein 14* see: Calgranulins.

MS: abbrev. for » mammastatin.

MSA: *multiplication-stimulating activity* This factor is isolated from rat tissues and secreted by some rat cell lines, e. g. BRL-3A cells (Buffalo rat liver cells). It is identical with human » IGF (IGF-2, insulin-like growth factor).

Dull TJ et al Insulin-like growth factor II precursor gene organization in relation to insulin gene family. N 310: 777-81 (1984); **Kato Y et al** Selective stimulation of sulfated glycosaminoglycan synthesis by multiplication-stimulating activity, cartilage-derived factor, and bone-derived growth factor. Comparison of their actions on cultured chondrocytes with those of fibroblast growth factor and Rhodamine fibrosarcoma-derived growth factor. BBA 716: 232-9 (1982); **Marquardt H et al** Purification and primary structure of a polypeptide with multiplication-stimulating activity from rat liver cell cultures. Homology with human insulin-like growth factor II. JBC 256: 6859-65 (1981); **Meyer PW & Schalch DS** Somatomedin synthesis by a subclone of Buffalo rat liver cells: characterization and evidence for immediate secretion of de novo synthesized hormone. Endocrinology 113: 588-95 (1983); **Moses AC et al** Increased levels of multiplication stimulating activity, an insulin-like growth factor, in foetal rat serum. PNAS 77: 3649-53 (1980)

MSF: *macrophage slowing factor* This is an operational definition for factors that reduce the migration of macrophages. See also: MIF (macrophage migration inhibition factor). The detection of MSF activity was an indication of the successful sensibilization of lymphocytes for antigens and also for » cell activation.

Brown KA et al A reappraisal of the macrophage electrophoretic mobility (MEM) test for the measurement of lymphokine activity. JIM 82: 189-98 (1985); **Bubenik J et al** A comparison of the macrophage electrophoretic mobility test and the leukocyte migration inhibition test in the detection of histocompatibility antigens on tumor cells. Neoplasma 28: 185-93 (1981)

MSF: *migration stimulating factor* This factor is secreted by human fetal but not adult fibroblasts and mammary carcinoma cells. MSF is a proline-rich protein of ≈ 70 kDa. It binds to heparin and this may be important for its *in vivo* activities (see also: HBGF, heparin-binding growth factors). MSF is not related to other motility factors such as » AMF (autocrine motility factor) and » SF (scatter factor).

MSF stimulates the migration of confluent fibroblasts through a collagen gel. It may be an » autocrine motility factor acting on its producer cells. The factor is thought to play a role in early embryonic development. It may also be involved in angiogenesis (see also: Angiogenesis factors). The factor may also promote the migration of embryonic fibroblasts into epithelial tumors.

The term MSF is also used as an operational definition for a biological activity rather than for a distinct factor.

Chen WY et al Differences between adult and foetal fibroblasts in the regulation of hyaluronate synthesis: correlation with migratory activity. JCS 94: 577-84 (1989); **Grey AM et al** Purification of the migration stimulating factor produced by fetal and breast cancer patient fibroblasts. PNAS 86: 2438-42 (1989); **Schor SL et al** Fetal and cancer patient fibroblasts produce an autocrine migration-stimulating factor not made by normal adult cells. JCS 90: 391-9 (1988); **Schor SL et al** Mechanism of action of the migration stimulating factor produced by fetal and cancer patient fibroblasts: effect on hyaluronic acid synthesis. In Vitro Cell. Dev. Biol. 25: 737-46 (1989); **Schor SL et al** Heterogeneity amongst fibroblasts in the production of migration stimulating factor (MSF): implications for cancer pathogenesis. Experientia Suppl. 59: 127-46 (1991); **Stoker M & Gherardi E** Regulation of cell movement: the motogenic cytokines. BBA 1072: 81-102 (1991)

MSF: *monocyte suppressor factor* This factor of 50-60 kDa is secreted by human monocytes. It inhibits the antigen-induced proliferation of T cells. In a thymocyte costimulation assay this factor inhibits the activity of soluble » IL1but not that of membrane-bound IL1 (IL1α). MSF has also been suggested to inhibit the release of IL1 from its membrane-bound form. MSF is identical with PAI (plasminogen activator inhibitor).

Goeken NE et al Monocyte suppressor factor is plasminogen activator inhibitor: inhibition of membrane-bound but not soluble IL1. JI 143: 603-8 (1989)

MSH: melanocyte-stimulating hormone. See: POMC (proopiomelanocortin).

MSP: *macrophage stimulating protein* This factor of 70 kDa is a disulfide-linked heterodimer (47 and 22 kDa subunits) isolated from human blood plasma. High levels of MSP RNA are found in liver tissues. The human MSP gene maps to chromosome 3p21 but related sequences are also found on chromosome 1. MSP protein shows 45 % identity with » HGF (hepatocyte growth factor). MSP is identical with » HGF-like protein (hepatocyte growth factor-like protein) and is also called ***MST1*** (macrophage stimulating 1).
MSP makes mouse resident peritoneal macrophages responsive to chemoattractants and causes the appearance of long cytoplasmic processes and pinocytic vesicles in freshly plated macrophages. MSP also induces phagocytosis via the C3b receptor, CR1.
Skeel et al Macrophage stimulating protein: purification, partial amino acid sequence, and cellular activity. JEM 173: 1227-34 (1991); **Yoshimura T et al** Cloning, sequencing, and expression of human macrophage stimulating protein (MSP, MST1) confirms MSP as a member of the family of kringle proteins and locates the MSP gene on chromosome 3. JBC 268: 15461-8 (1993)

MST1: *macrophage stimulating 1* See: HGF-like protein (hepatocyte growth factor-like protein)

MStF: *migration stimulation factor* This poorly characterized factor is produced by murine T cells. It is chemotactic for lymphocytes. Another factor of the same name is also produced by CD8-positive cells. It may be a distinct factor.
Gauthier-Rahman S et al Differential expression of migration inhibitory and migration stimulatory factors in two lines of mice genetically selected for high or low responsiveness to phytohemagglutinin. 1. Migration stimulatory factor(s) from T and B cells. Exp. Clin. Immunogenet. 8: 140-58 (1991); **Gauthier-Rahman S et al** Differential expression of migration inhibitory and migration stimulatory factors in two lines of mice genetically selected for high or low responsiveness to phytohemagglutinin. 2. Effects of mitogenic or allogeneic stimulation. Exp. Clin. Immunogenet. 8: 159-76 (1991); **Mendez-Samperio P et al** Production of leukocyte migration inhibition factor (LIF) and migration stimulation factor (MStF) by $CD4^+$ and $CD8^+$ human lymphocytes subsets and T cell clones. Rev. Invest. Clin. 41: 107-15 (1989)

M-T7 protein: This protein of 37 kDa is the major secreted protein from cells infected with myxoma virus. It is encoded by the M-T7 open reading frame of the virus genome. The protein has significant sequence similarity to the human and mouse receptors for » IFN-γ and specifically binds rabbit IFN-γ. It also inhibits the biological activity of extracellular IFN-γ. Since this interferon is one of the key regulatory cytokines in the host immune response against viral infections the biological function of this viral protein may therefore be the suppression of IFN-mediated functions. For similar viral strategies of subverting host immune responses see also: Viroceptor.
Upton C et al Encoding of a homologue of the IFN-γ receptor by myxoma virus. S 258: 1369-72 (1992)

MTCF: *mast cell arming T cell factor* see: SMAF (specific macrophage arming factor).

MTD: abbrev. for maximal tolerated dose.

MTGF: *mammary tumor-derived growth factor* This factor is identical with » bFGF.
Rowe JM et al Purification and characterization of a newly identified growth factor specific for epithelial cells. CR 46: 1408-12 (1986)

MTT assay: This test is a quantitative colorimetric method to determine cell proliferation. It utilizes the yellow tetrazolium salt [3-(4,5-Dimethylthiazol-2-yl)-2,5-diphenyltetrazolium-bromide] which is metabolized by mitochondrial succinic dehydrogenase activity of proliferating cells to yield a purple formazan reaction product. Cells do not require functional mitochondria as demonstrated by the observation that no differences are observed in formazan production by normal cells or respiratory-defective cells in which mitochondria have been poisoned by the nucleic acid toxin, ethidium bromide.

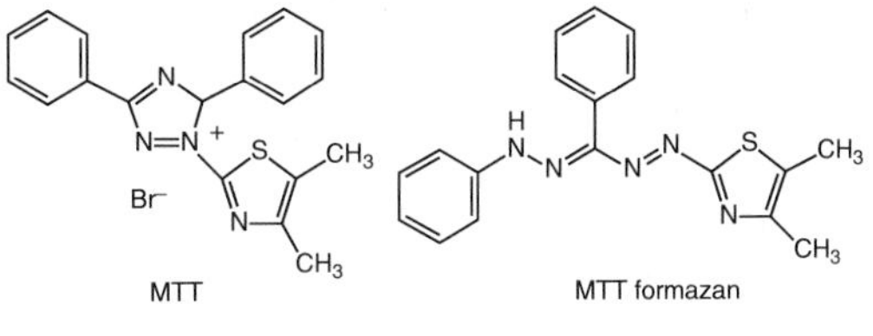

MTT

MTT formazan

Both serum and plasma, from a variety of species, non-specifically reduce both MTT and XTT tetrazolium salts to a colored formazan product. This effect appears to be particularly pronounced with fetal bovine serum.
A variety of refinements and modifications of the MTT assay have been described. The use of the tetrazolium derivative XTT (3'-[1-[(phenylamino)-carbonyl]-3,4-tetrazolium]-bis(4-methoxy-6-nitro)benzene-sulfonic acid hydrate) is of advantage because XTT is cleaved to a water soluble formazan product and thus allows elimina-

tion of the solubilization step required with MTT. The MTT assay is an alternative to the thymidine incorporation test which measures cell proliferation by determining the amounts of incorporated ^{3}H thymidine into freshly synthesized DNA. It can be used also to determine proliferation of mycoplasm-infected cells which would interfere with thymidine incorporation measurements. The proliferative profiles of cells as determined by the colorimetric assay, which essentially measures energy metabolism, or radioisotope assay usually do not show large differences.

The MTT assay has also been adapted to quantitate cytotoxic activities and growth inhibitory activities of cytokines.

Buttke TM et al Use of an aqueous tetrazolium/formazan assay to measure viability and proliferation of lymphokine dependent cell lines. JIM 157: 233-8 (1993); **Dusch A et al** A comparison of assays for the response of primary human T cells upon stimulation with interleukin-2, interleukin-4 and interleukin-7. ECN 3: 97-102 (1992); **Ferrari M et al** MTT colorimetric assay for testing macrophage cytotoxic activity *in vitro*. JIM 131: 165-72 (1990); **Garn H et al** An improved MTT assay using the electron-coupling agent manadione. JIM 168: 253-6 (1994); **Green LM et al** Rapid colorimetric assay for cell viability: application to the quantitation of cytotoxic and growth inhibitory lymphokines. JIM 70: 257-68 (1984); **Hansen MB et al** Re-examination and further development of a precise and rapid dye method for measuring cell growth/cell kill. JIM 119: 203-10 (1989); **Heeg K et al** A rapid colorimetric assay for the determination of IL2-producing helper T cell frequencies. JIM 77: 237-46 (1985); **Hida T et al** Chemosensitivity and radiosensitivity of small cell lung cancer cell lines studied by a newly developed 3-(4,5-Dimethylthiazol-2-yl)-2,5-diphenyltetrazolium-bromide (MTT) hybrid assay. CR 49: 4785-90 (1989); **Hussain RF et al** A new approach for measurement of cytotoxicity using colorimetric assay. JIM 160: 89-96 (1993); **Jiao H et al** A new 3-(4,5-dimethylthiazol-2-yl)-2,5-diphenyltetrazolium bromide (MTT) assay for testing macrophage cytotoxicity to L1210 and its drug-resistant cell lines *in vitro*. Cancer Immunol. Immunother. 35: 412-6 (1992); **Kajio T et al** Quantitative colorimetric assay for basic fibroblast growth factor using bovine endothelial cells and heparin. J. Pharmacol. Toxicol. Methods 28: 9-14 (1992); **Keisari Y** A colorimetric microtiter assay for the quantitation of cytokine activity on adherent cells in tissue culture. JIM 146: 155-61 (1992); **Loveland BE et al** Validation of the MTT dye assay for enumeration of cells in proliferative and antiproliferative assays. BI 27: 501-10 (1992); **Mosmann T et al** Rapid colorimetric assay for cellular growth and survival: application to proliferation and cytotoxicity assays. JIM 65: 55-63 (1983); **Mosman TR & Fong TAT** Specific assays for cytokine production by T cells JIM 116: 151-8 (1989); **Nargi FE & Yang TJ** Optimization of the L-M cell bioassay for quantitating tumor necrosis factor α in serum and plasma. JIM 159: 81-91 (1993); **Ohno M & Abe T** Rapid colorimetric assay for the quantification of leukemia inhibitory factor (LIF) and interleukin-6 (IL6). JIM 145: 199-203 (1991); **Roehm NW et al** An improved colorimetric assay for cell proliferation and viability utilizing the tetrazolium salt XTT. JIM 142: 257-65 (1991); **Sladowski D et al** An improved MTT assay. JIM 157: 203-7 (1993); **van de Loosdrecht AA et al** Cell mediated cytotoxicity against U 937 cells by human monocytes and macrophages in a modified colorimetric MTT assay. A methodological study. JIM 141: 15-22 (1991)

mu: *murine* Used in abbrev. such as mu-GM-CSFto indicate the murine origin of a factor.

Müllerian inhibiting factor: see: MIS.

Müllerian inhibiting hormone: see: MIS.

Müllerian inhibiting substance: see: MIS.

Multi-colony stimulating activity: see: MCSA.

Multi-CSF: This factor is a colony-stimulating factor (see also: CSF). It was termed Multi-CSF or ***MCSA*** (colony-stimulating activity) because of its ability to stimulate *in vitro* the development of colonies containing a variety of hematopoietic cell lineages. The factor is identical with » IL3.

The term Multi-CSF has been used also as a generic name for other factors, for example: » IL4, supporting the growth of various hematopoietic lineages.

Cutler RL et al Purification of multipotential colony stimulating factor from pokeweed mitogen stimulated mouse spleen cell conditioned medium. JBC 260: 6579-87 (1985); **Ihle JN et al** Biological properties of homogenous interleukin-3. I. Demonstration of WEHI-3 growth factor activity, mast cell growth factor activity, P cell-stimulating factor activity, colony-stimulating factor activity, and histamine-producing cell-stimulating factor activity. JI 131: 282-7 (1983); **Kishi K et al** Murine B cell stimulatory factor-1 (BSF-1)/interleukin-4 (IL4) is a multilineage colony-stimulating factor that acts directly on primitive hemopoietic progenitors. JCP 139: 463-8 (1989); **Ymer S et al** Constitutive synthesis of interleukin-3 by leukemia cell line WEHI-3B is due to retroviral insertion near the gene. N 317: 255-8 (1985)

Multigen families: see: gene family.

multi-HGF: *multilineage hemopoietic growth factor* This factor is identical with » IL3.

Iscove NN et al The multilineage hemopoietic growth factor (multi HGF) from murine WEHI-3D cells. LR 3: 67-74 (1984)

Multilineage colony stimulating activity: This factor is identical with » IL3.

Multilineage hemopoietic growth factor: see: multi-HGF.

Multiplication-stimulating activity: see: MSA.

Multipoietin: This factor is a colony-stimulating factor (see also: CSF). The (old) name derives from its ability to stimulate *in vitro* the development of colonies containing a variety of hematopoietic cell lineages. The factor is identical with » IL3.

Multipotent growth factor: see: MGF.

Multipotent stem cells: abbrev. MPSC. See: Hematopoiesis.

Multiresponse element: see: MRE.

Muscle proteolysis-inducing factor: see: MPIF.

Muteins: *mutant proteins* A general term describing mutated recombinant proteins with single or multiple amino acid substitutions. These proteins are frequently derived from cloned genes that have been subjected to site-directed or random mutagenesis, or from completely synthetic genes. High-resolution mutational and structural analyses have revealed a great deal about the molecular basis for cytokine actions. From these studies it has been possible to engineer homologues of cytokines with altered biological properties. Cytokines with greater specific activity in cell proliferation assays, enhanced antitumor activities, or improved pharmacokinetic properties are superior to the parent proteins and are therefore particularly useful in clinical applications. For specific examples of cytokine muteins see: KW-2228, pIXY321, Consensus interferon.

Mv1Lu: see: CCL-64.

MV-3D9: This cell line was established from mink lung fibroblasts. It is used to assay » TGF-β (see also: Bioassays) which inhibits the proliferation of these cells. Inhibition of proliferation is detected by measuring the decrease of ^{3}H thymidine incorporation into the DNA. An alternative and entirely different detection method is » Message amplification phenotyping.

Like B & Massague J The antiproliferative effect of type β transforming growth factor occurs at a level distal from receptors for growth-activating factors. JBC 261: 13426-9 (1986); **Su HC et al** A role for transforming growth factor-β 1 in regulating natural killer cell and T lymphocyte proliferative responses during acute infection with lymphocytic choriomeningitis virus. JI 147: 2717-27 (1991)

Mx protein: The existence of the Mx gene (mx symbol derived from myxovirus = influenza resistance) was initially inferred from genetic analysis of certain strains of mice that display an inborn resistance against infection with influenza A and B viruses. Mx-homologous genes have also been found in all other eukaryotes investigated so far. Some Mx proteins display a broader antiviral profile than the one observed for Mx1 in mice. Others, however, are seemingly devoid of antiviral activities, including human MxB, rat Mx3, and duck Mx. The proteins are also expressed in species that are not resistant to these viruses.

The expression of the murine Mx protein (Mx1) mediates resistance against influenza and VSV (vesicular stomatitis virus). The transfection of the Mx1$^+$ allele into cells that are susceptible to influenza virus demonstrates that the expression of this gene is required and sufficient to protect against influenza virus infections. The protein accumulates in the cell nucleus and probably blocks the synthesis of viral mRNAs presumably by interaction with viral polymerases. The expression of the autosomal dominant Mx1$^+$ allele induces the expression of » IFN-α and » IFN-β.

The analysis of » transgenic animals (mice) expressing high levels of the Mx protein demonstrates that these animals are protected against the consequences of a lethal infection with influenza virus while low-level expression animals are not protected or protected only if they are infected with high doses of viruses. These observations suggest that influenza viral pathogenesis is determined by a subtle balance between the dose of the infecting virus and the levels of the antiviral host factor Mx1. Transgenic studies also demonstrate that mice can be rendered resistant to a virulent infection by "intracellular immunization" achieved through germline transformation.

The human Mx protein (MxA = Mx1 = p78; ***IFI-78k*** = interferon-induced 78 kDa protein) has a length of 662 amino acids. It is encoded by the IFI-78k gene (interferon-inducible 78 kDa protein) on chromosome 21q22.3. The protein is produced in large amounts in the cytoplasm of cells treated with interferons (see: IFN for general information). The detection of Mx protein is used to detect minute amounts of biologically active interferons. Mx proteins have recently been shown to function as GTPases that can perform multiple rounds of GTP hydrolysis in the absence of accessory factors. Mx proteins show significant homology to other GTPases such as the VPS1 gene of *Saccharomyces cerevisiae* required for the sorting of soluble vacuolar proteins, and mammalian dynamin which is involved in the initial stages of endocytosis. Mx and the other two proteins have unique functions and cannot complement each other. Based on functional and sequence similarities it has been suggested that Mx may also be involved in protein trafficking.

Arnheiter H & Meier E Mx proteins: antiviral proteins by chance or by necessity? New Biol. 2: 851-7 (1990); **Horisberger MA et al** Cloning and sequence analysis of cDNAs for interferon- and virus-induced human Mx proteins reveal that they contain putative guanine nucleotide-binding sites: functional study of the corresponding gene promoter. J. Virol. 64: 1171-81 (1990); **Horisberger MA et al** cDNA cloning and assignment to chromosome 21 of IFI-78K gene, the human equivalent of murine Mx gene. Somat. Cell. Molec. Genet. 14: 123-31 (1988); **Horisberger MA & Gunst MC** Interferon-induced proteins: identification of Mx proteins in various mammalian species. Virology 180: 185-90 (1991); **Horisberger MA & Hochkeppel HK** IFN-α induced human 78 kD protein: purification and homologies with the mouse Mx protein, production of monoclonal antibodies, and potentiation effect of IFN-γ. J. Interferon Res. 7: 331-43 (1987); **Meier E et al** A family of interferon-induced Mx-related mRNAs encodes cytoplasmic and nuclear proteins in rat cells. J. Virol. 62: 2386-93 (1988); **Meier E et al** Activity of rat Mx proteins against a rhabdovirus. J. Virol. 64: 6263-9 (1990); **Müller M & Brem G** Disease resistance in farm animals. Experientia 47: 923-34 (1991); **Noteborn M et al** Transport of the murine Mx protein into the nucleus is dependent on a basic carboxy-terminal sequence. J. Interferon Res. 7: 657-69 (1987); **Nakayama M et al** Interferon-inducible mouse Mx1 protein that confers resistance to influenza virus is GTPase. JCB 266: 21404-8 (1991); **Simon A et al** Interferon-regulated Mx genes are not responsive to interleukin 1, tumor necrosis factor, and other cytokines. J. Virol. 65: 968-71 (1991); **Staeheli P et al** Transcriptional activation of the mouse Mx gene by type I interferon. MCB 6: 4770-4 (1986); **Staeheli P et al** Mx proteins: GTPases with antiviral activity. Trends Cell Biol. 3: 268-72 (1993); **Towbin H et al** A whole blood immunoassay for the interferon-inducible human Mx protein. J. Interferon Res. 12: 67-74 (1992); **Zürcher T et al** Mechanism of human MxA protein action: variants with changed antiviral properties. EJ 11: 1657-61 (1992)

● **Transgenic/Knock-out/Antisense Studies:** **Arnheiter H et al** Transgenic mice with intracellular immunity to influenza virus. Cell 62: 51-61 (1990); **Kolb E et al** Resistance to influenza virus infection of Mx transgenic mice expressing Mx protein under the control of two constitutive promoters. J. Virol. 66: 1709-16 (1992)

MY10: see: CD34.

***myb*:** c-*myb* is the cellular counterpart of the transforming » oncogene v-*myb* of avian myeloblastosis virus and the E26 leukemia virus which cause myeloid leukemia in chicken. The human gene maps to 6q22-6q23. This is the point of break in translocations involved in T cell acute lymphatic leukemia and in some ovarian cancers and melanomas. The *myb* gene encodes a nuclear protein involved in transcriptional regulation (see also: Gene expression). The DNA-binding domain of *myb* recognizes a specific *myb*-responsive element with the consensus sequence (C/T)G(A/G((A/C/G)GTT (A/G).

myb is predominantly expressed at high levels in hematopoietic cells and neural cells and rarely in fibroblasts or epithelial cells. The expression of *myb* appears to be essential for cell proliferation. It is required for normal hematopoiesis in fetal liver and for proliferation of hematopoietic progenitor cells *in vitro*. Down-regulation of c-*myb* gene expression is a prerequisite for erythroid differentiation induced by » Epo.

Suppression of *myb* and » *myc* has been shown to occur upon induction of terminal differentiation but not upon induction of growth inhibition in myeloid leukemia cells. During differentiation of hematopoietic and neuroblastoma cell lines, expression of c-*myb* is down-regulated, and constitutive expression of this gene blocks differentiation. In some » factor-dependent cells lines the constitutive overexpression of *myb* releases the cells from their requirement of insulin-like growth factors (see: IGF). Murine fibroblasts constitutively overexpressing *myb* and another oncogene, » *myc*, can be cultivated in serum-free media (see also: SFM) in the absence of additional growth factors.

Changes in endogenous *myb* genes have been implicated in both human and murine hematopoietic tumors. Oncogenic activation of c-*myb* in chicken hematopoietic cells and a number of murine myeloid tumor cell lines has been shown to be a consequence of 5′ and 3′ truncations.

■ **Transgenic/Knock-out/Antisense Studies:** Since *myb* appears to be critical for hematopoietic cell proliferation and development the disruption of its functions by » antisense oligonucleotides has been tried as an experimental therapeutic strategy for controlling leukemic cell growth *in vivo*. Animals transplanted with a human chronic myelogenous leukemia cell hybrid and treated with the antisense *myb* survived at least 3.5 times longer than control animals and displayed significantly less disease at the two sites most frequently involved by leukemic cell infiltration, the CNS and the ovary. The targeted disruption of the *myb* gene in » transgenic animals generated from engineered » ES cells demonstrates that all hematopoietic lineages are affected by the absence of c-*myb* expression with the exception of megakaryocytes. Homozygous mutant mice are unable to switch from yolk sac to liver erythropoiesis and die at approximately 15.5 days of gestation due to severe anemia.

Barletta C et al Relationship between the c-*myb* locus and the 6q- chromosomal aberration in leukemias and lymphomas. S 235: 1064-7 (1987); **Bender TP & Kuehl W** Murine *myb* proto-oncogene mRNA: cDNA sequence and evidence for 5´ heterogeneity. PNAS 83: 3204-8 (1986); **Franchini G et al** Structural organization and expression of human DNA sequences related to the transforming gene of avian myeloblastosis virus. PNAS 80: 7385-9 (1983); **Gonda TJ & Metcalf D** Expression of *myb*, *myc*, and *fos* proto-oncogenes during the differentiation of a murine myeloid leukemia. N 310: 249-51 (1984); **Grasser FA et al** Protein truncation is required for the activation of the c-*myb* proto-oncogene. MCB 11: 3987-96 (1991); **Harper ME et al** Chromosomal sublocalization of human c-*myb* and c-*fes* cellular onc genes. N 304: 169-71 (1983); **Howe KM & Watson RJ** Nucleotide preferences in sequence-specific recognition of DNA by c-*myb*. NAR 19: 3913-9 (1991); **Kuehl WM et al** Expression and function of the c-*myb* oncogene during hematopoietic differentiation. CTMI 141: 318-23 (1988); **Majello B et al** Human c-*myb* proto-oncogene: nucleotide sequence of cDNA and organization of the genomic locus. PNAS 83: 9636-40 (1986); **Meese E et al** Molecular mapping of the oncogene *myb* and rearrangements in malignant melanoma. Genes Chromosomes Cancer 1: 88-94 (1989); **Park JG & Reddy E P** Large-scale molecular mapping of human c-*myb* locus: c-*myb* proto-oncogene is not involved in 6q- abnormalities of lymphoid tumors. O 7: 1603-9 (1992); **Pelicci PG et al** Amplification of the c-*myb* oncogene in a case of human acute myelogenous leukemia. S 224: 1117-21 (1984); **Perlmutter RM & Ziegler S** Proto-oncogene expression following lymphocyte activation. CTMT 35: 571-86 (1990); **Ratajczak MZ et al** *In vivo* treatment of human leukemia in a scid mouse model with c-*myb* antisense oligodeoxynucleotides. PNAS 89: 11823-7 (1992); **Reiss K et al** The proto-oncogene c-*myb* increases the expression of insulin-like growth factor 1 and insulin-like growth factor 1 receptor messenger RNAs by a transcriptional mechanism. CR 51: 5997-6000 (1991); **Rushlow KE et al** Nucleotide sequence of the transforming gene of avian myeloblastosis virus. S 216: 1421-3 (1982); **Selvakumaran M et al** Deregulated c-*myb* disrupts interleukin-6- or leukemia inhibitory factor-induced myeloid differentiation prior to c-*myc:* role in leukemogenesis. MCB 12: 2493-500 (1992); **Shen-Ong GLC** The *myb* oncogene. BBA 1032: 39-52 (1991); **Szczylik C et al** Regulation of proliferation and cytokine expression of bone marrow fibroblasts: role of c-*myb*. JEM 178: 997-1005 (1993); **Todokoro K et al** Down-regulation of c-*myb* gene expression is a prerequisite for erythropoietin-induced erythroid differentiation. PNAS 85: 8900-4 (1988); **Travali S et al** Constitutively expressed c-*myb* abrogates the requirement for insulin-like growth factor 1 in 3T3 fibroblasts. MCB 11: 731-6 (1991); **Yokota J et al** Alterations of *myc*, *myb*, and *ras*(Ha) proto-oncogenes in cancers are frequent and show clinical correlation. S 231: 261-5 (1986); **Zabel BU et al** Regional assignment of human proto-oncogene c-*myb* to 6q21 – qter. Som. Cell Mol. Genet. 10: 105-8 (1984)

● **TRANSGENIC/KNOCK-OUT/ANTISENSE STUDIES: Gewirtz AM & Calabretta B** A c-*myb* antisense oligodeoxynucleotide inhibits normal human hematopoiesis *in vitro*. S 242: 1303-6 (1988); **McClinton D et al** Differentiation of mouse erythroleukaemia cells is blocked by late up-regulation of a c-*myb* transgene. MCB 10: 705-10 (1990); **Mucenski ML et al** A functional c-*myb* gene is required for normal murine fetal hepatic hematopoiesis. Cell 65: 677-89 (1991)

myc: The cellular c-*myc* oncogene is a member of a family of related genes including ***N-myc*** and **L-*myc***. The genes encode nuclear DNA-binding phosphoproteins that are involved in the regulation of gene expression and/or DNA replication during cell growth and differentiation. Sequences of the *myc* oncogene have been highly conserved througout evolution, from *Drosophila* to vertebrates.

c-*myc* encodes a protein of 65 kDa which is expressed in almost all normal and transformed cells. The expression correlates with the proliferation state of the cells. Transcription is repressed in quiescent or terminally differentiated cells. Expression of myc is generally induced after mitogenic stimulation of cells or serum induction (see also: ERG, early response genes). *myc* therefore is an important positive regulator of cell growth and proliferation. *myc* has also been demonstrated to be a potent inducer of » apoptosis when expressed in the absence of serum or growth factors. Apoptosis may also serve as a protective mechanism to prevent tumorigenicity elicited by deregulated *myc* expression. One of the mechanisms to counteract *myc* -mediated apoptosis is the expression of the » *bcl*-2 oncogene.

The expression of N-*myc* (human chromosome 2p24) and L-*myc* is restricted to certain types of undifferentiated cells. c-*myc* is expressed in lymphoid cells of all types of development. Pre-B cells express N-*myc* while more differentiated B cells do not express it. In normal pre-B cells the expression of all three genes is greatly enhanced by » IL7. In resting mouse fibroblasts (3T3 cells) the stimulation of cellular DNA synthesis induced by » PDGF can also be achieved by overexpression of *myc* instead of the PDGF stimulus. Overexpression of *myc*, however, is insufficient to drive resting cells into the S phase of the » cell cycle. This requires further growth factors such as » IGF or other less well characterized factors found in platelet-poor plasma. The simultaneous overexpression of *myc* and » *myb* is sufficient to allow growth of cells in serum-free media (see: SFM) in the absence of other growth factors.

Chromosomal translocations of the myc gene are a feature of all Burkitt's lymphoma. A translocation of the *myc* gene at 8q24 to a locus on chromosome 14 encoding the heavy chain of immunoglobulins is observed in ≈ 80-85 % of Burkitt's lymphomas. Translocations to chromosomes 2 (≈ 10 %) and 22 (≈5 %) are also observed. An aberrant expression of the *myc* gene and *myc* translocations are observed in a number of tumors. The involvement of the *myc* oncogene in translocations is the prototype in the relationship between chromosomal abnormalities and oncogenes.

■ **TRANSGENIC/KNOCK-OUT/ANTISENSE STUDIES:** Transgenic animals aberrantly expressing the *myc* oncogene develop lymphoproliferative B cell defects. This demonstrates that this gene is somehow involved in the regulation of lymphocyte proliferation. The constitutive expression of *myc* blocks further differentiation in murine erythroleukemia cells. The study of » transgenic animals (mice) expressing the *myc* gene reveals that the deregulated expression of the *myc* gene may be a heritable predisposing factor responsible for the development of mammary adenocarcinomas. Other additional tissue-specific factors may be required for tumorigenesis in other tissues, including pancreas, lung, brain, and the salivary glands. Crosses between *myc* - and *ras*-transgenic mice yielding doubly-transgenic animals develop tumors much faster not only in the mammary gland but also in other organs.

High level expression of c-*myc* driven by the mouse Thy-1 transcriptional unit results in the development of thymic tumors containing proliferating thymocytes and expanded populations of epithelial cells. Eμ(immunoglobulin heavy chain enhancer)*bcl*-2/*myc* mice show hyperproliferation of pre-B and B cells and develop tumors much faster than Eμ-*myc* mice. These tumors are derived from primitive hematopoietic cells, perhaps a lymphoid-committed stem cell. In contrast to Eμ-*myc*, Eμ-pim-1 transgenic mice are predisposed to T cell lymphomas. Double transgenic Eμ-*myc* /Eμ-pim-1 mice develop pre-B cell leukemia before birth.

Homologous recombination in embryonic stem cells (see: ES cells) has been used to generate both heterozygous and homozygous c-*myc* mutant ES cell lines that do not express any *myc* protein. The analysis of mice generated from these ES cells shows that the *myc* null mutation is lethal in homozygotes between 9.5 and 10.5 days of gestation. Pathologic abnormalities include the heart, pericardium, neural tube, and delay or failure in turning of the embryo. Heterozygous females have reduced fertility due to embryonic resorption before 9.5 days of gestation. c-*myc* protein therefore is necessary for embryonic survival beyond 10.5 days of gestation and appears to be dispensable for cell division both in ES cell lines and in the embryo before that time.

Disruption of the ***N-myc*** gene by targeted deletion in » ES cells and subsequent generation of » transgenic animals show that mice deficient in N-*myc* expression die at ≈ 9.5 days of gestation. N-myc thus plays an important role in development.

Babiss LE & Freidman JM Regulation of N-*myc* expression: use of an adenovirus vector to demonstrate post-transcriptional control. MCB 10: 6700-8 (1990); **Bernard O et al** Sequence of the murine and human cellular *myc* oncogenes and two modes of *myc* transcription resulting from chromosome translocation in B lymphoid tumors. EJ 2: 2373-85 (1983); **Biro S et al** Inhibitory effects of antisense oligodeoxynucleotides targeting c-*myc* mRNA on smooth muscle cell proliferation and migration. PNAS 90: 654-8 (1993); **Bissonnette RP et al** Apoptotic cell death induced by c-*myc* is inhibited by *bcl*-2. N 359: 552-4 (1992); **Blackwell TK et al** Sequence-specific DNA binding by the c-*myc* protein. S 250: 1149-51 (1990); **Cole MD** The *myc* oncogene: its role in transformation and differentiation. ARG 20: 361-84 (1986); **Collum RG & Alt FW** Are *myc* proteins transcription factors? CC 2: 69-75 (1990); **Dalla Favera et al** Human c-*myc* oncogene is located on the region of chromosome 8 that is translocated in Burkitt lymphoma cells. PNAS 79: 7824-7 (1982); **DePinho RA et al** Structure and expression of the murine N-*myc* gene. PNAS 83: 1827-31 (1986); **DePinho RA et al** *myc* family oncogenes in the development of normal and neoplastic cells. ACR 57: 1-46 (1991); **Evan GI & Littlewood TD** The role of c-myc in cell growth. Curr. Opin. Genet. Dev. 3: 44-9 (1993); **Fanidi A et al** Cooperative interaction between c-*myc* and *bcl*-2 proto-oncogenes N 359: 554-6 (1992); **Garson JA et al** Novel non-isotopic *in situ* hybridization technique detects small (1kb) unique sequences in routinely G-banded human chromosomes: fine mapping of N-*myc* and β-NGF genes. NAR 15: 4761-70 (1987); **Kelly K & Siebenlist U** The regulation and expression of c-*myc* in normal and malignant cells. ARI 4: 317-38 (1986); **Lüscher B & Eisenman RN** New light on *Myc* and *Myb*. Part I. *Myc*. GD 4: 2025-35 (1990); **Perlmutter RM & Ziegler S** Proto-oncogene expression following lymphocyte activation. CTMT 35: 571-86 (1990); **Spangler R & Sytkowski AJ** c-*myc* is an erythropoietin early response gene in normal erythroid cells: evidence for a protein kinase C-mediated signal. Blood 79: 52-7 (1992); **Taub R et al** Translocation of the c-*myc* gene into the immunoglobulin heavy chain locus in human Burkitt lymphoma and murine plasmacytoma cells. PNAS 79: 7837-41 (1982); **Wagner AJ et al** *myc* -mediated apoptosis is blocked by ectopic expression of *bcl*-2. MCB 13: 2432-40 (1993); **Watt R et al** Nucleotide sequence of cloned cDNA of human c-*myc* oncogene. N 303: 725-8 (1983); **Yokota J et al** Alterations of *myc*, *myb*, and *ras*(Ha) proto-oncogenes in cancers are frequent and show clinical correlation. Science 231: 261-5 (1986); **Zimmerman K & Alt FW** Expression and function of *myc* family genes. Crit. Rev. Oncogenesis 2: 75-95 (1990)

● **TRANSGENIC/KNOCK-OUT/ANTISENSE STUDIES: Davis AC et al** A null c-*myc* mutation causes lethality before 10.5 days of gestation in homozygotes and reduced fertility in heterozygous female mice. Genes Dev. 7: 671-82 (1993); **Dildrop R et al** Differential expression of *myc* -family genes during development: normal and deregulated N-*myc* expression in transgenic mice. CTMI 141: 100-9 (1988); **Harris AW et al** The E μ-*myc* transgenic mouse. A model for high-incidence spontaneous lymphoma and leukemia of early B cells. JEM 167: 353-71 (1988); **Haupt Y et al** Retroviral infection accelerates T lymphomagenesis in E μ-N-*ras* transgenic mice by activating c-*myc* or N-*myc*. O 7: 981-6 (1992); **Iwamoto T et al** Preferential development of pre-B lymphomas with drastically down-regulated N-*myc* in the E μ-ret transgenic mice. EJI 21: 1809-14 (1991); **Sawai S et al** Embryonic lethality resulting from disruption of both N-*myc* alleles in mouse zygotes. The New Biologist. 3: 861-9 (1991); **Scheuermann RH & Bauer SR** Tumorigenesis in transgenic mice expressing the c-*myc* oncogene with various lymphoid enhancer elements. CTMI 166: 221-31 (1990); **Schmidt EV et al** Transgenic mice bearing the human c-*myc* gene activated by an immunoglobulin enhancer: a pre-B cell lymphoma model. PNAS 85: 6047-51 (1988); **Schoenenberger CA et al** Targeted c-*myc* gene expression in mammary glands of transgenic mice induces mammary tumors with constitutive milk protein gene transcription. EJ 7: 169-75 (1988); **Sidman CL et al** Multiple mechanisms of tumorigenesis in E μ-*myc* transgenic mice. CR 53: 1665-9 (1993); **Sinn E et al** Coexpression of MMTV/v-Ha-*ras* and MMTV/c-*myc* genes in transgenic mice: synergistic action of oncogenes *in vivo*. Cell 49: 465-75 (1987); **Strasser A et al** Novel primitive lymphoid tumors induced in transgenic mice by cooperation between *myc* and *bcl*-2. N 348: 331-3 (1990); **Verbeek S et al** Mice bearing the E μ-*myc* and E μ-pim-1 transgenes develop pre-B cell leukemia prenatally. MCB 11: 1176-9 (1991); **Wang Y et al** Functional homology between N-*myc* and c-*myc* in murine plasmacytomagenesis: plasmacytoma development in N-*myc* transgenic mice. O 7: 1241-7 (1992); **Zimmerman K et al** Differential regulation of the N-*myc* gene in transfected cells and transgenic mice. MCB 10: 2096-103 (1990)

MyD: *myeloid differentiation primary response (gene)* These genes, of which at least 12 have been identified, are » ERGs (early response gene) whose transcription is rapidly and transiently stimulated in myeloid cells in response to various differentiation and growth inhibitory stimuli. It has been shown that » IL6 and » LIF trigger the same immediate early response upon induction of myeloid leukemia differentiation.

MyD88, ***MyD116*** and ***MyD118*** are rapidly and transiently induced in M1 cells following induction of terminal differentiation and growth arrest by » IL6 or treatment of the cells with » IL1 or bacterial lipopolysaccharides.

The open reading frame (LMW23-NL) in the African swine fever virus genome and the neurovirulence-associated gene (ICP34.5) of herpes simplex virus possesses striking similarity to MyD116, suggesting a role for these viral genes in determining viral host range.

Abdollahi A et al Sequence and expression of a cDNA encoding MyD118: a novel myeloid differentiation primary response gene induced by multiple cytokines. O 6: 165-7 (1991); **Lord KA et al** Dissection of the immediate early response of myeloid leukemia cells to terminal differentiation and growth inhibitory stimuli. Cell Growth Differ. 1: 637-45 (1990); **Lord KA et al** Nucleotide sequence and expression of a cDNA encoding MyD88, a novel myeloid differentiation primary response gene induced by IL6. O 5: 1095-7 (1990); **Lord KA et al** Sequence of MyD116 cDNA: a novel myeloid differentiation primary response gene induced by IL6. NAR 18: 2823 (1990); **Lord KA et al** Complexity of the immediate early response of myeloid cells to terminal differentiation and growth arrest includes ICAM-1, Jun-B and histone variants. O 5: 387-96 (1990); **Lord KA et al** Leukemia inhibitory factor and interleukin-6 trigger the same immediate early response, including tyrosine phosphorylation, upon induction of myeloid leukemia differentiation. MCB 11: 4371-9 (1991); **Sussman MD et al** Identification of an African swine fever virus gene with similarity to a myeloid differentiation primary response gene and a neurovirulence-associated gene of herpes simplex virus. J. Virol. 66: 5586-9 (1992)

MyD88: see: MyD.

MyD116: see: MyD.

MyD118: see: MyD.

Myelin P2: see: MDGI (mammary-derived growth inhibitor).

Myeloblastin: A serine protease originally identified in the human leukemia cell line » HL-60 cells. It is present in the azurophil granules of human polymorphonuclear leukocytes from which it is translocated to the cell surface during activation.

Myeloblastin is identical with ***proteinase 3***, a neutral serine protease of human neutrophils. It is also called ***AGP7*** and has been identified as the ***p29 autoantigen*** (c-ANCA, antineutrophil cytoplasmic autoantibody) of Wegener's granulomatosis which, in its classic form, is manifested by necrotizing granulomas of the upper and lower respiratory tract, focal necrotizing vasculitis involving both arteries and veins, and focal necrotizing glomerulitis.

Myeloblastin shows 42 % homology with the cationic antimicrobial protein » CAP 37. The myeloblastin gene spans approximately 6.5 kb and consists of five exons. It maps to human chromosome 19 in the vicinity of other genes encoding serine proteases found in azurophil granules of granulocytes.

Myeloblastin degrades a variety of extracellular matrix proteins (see: ECM), including elastin *in vitro* and causes emphysema when administered by tracheal insufflation to hamsters.

Myeloblastin is involved in the control of growth and differentiation of leukemic cells. HL-60 cells undergo terminal differentiation when exposed to monocytic and granulocytic inducers. Myeloblastin mRNA is undetectable in fully differentiated HL-60 cells as well as in human peripheral blood monocytes. Inhibition of myeloblastin expression by suitable » antisense oligodeoxynucleotides inhibits proliferation and induces differentiation of promyelocyte-like leukemia cells. Evidence from Northern analysis suggests that PR-3 expression is primarily confined to the promyelocytic/myelocytic stage of bone marrow development.

Bories D et al Down-regulation of a serine protease, myeloblastin, causes growth arrest and differentiation of promyelocytic leukemia cells. Cell 59: 959-68 (1989); **Campanelli D et al** Cloning of cDNA for proteinase 3: a serine protease, antibiotic, and autoantigen from human neutrophils. JEM 172: 1709-15

(1990); **Gross WL et al** Antineutrophil cytoplasmic autoantibody-associated diseases: a rheumatologist's perspective. Am. J. Kidney Dis. 18: 175-9 (1991); **Gupta SK et al** Identity of Wegener's autoantigen (p29) with proteinase 3 and myeloblastin Blood 76: 2162 (1990); **Kao RC et al** Proteinase 3: a distinct human polymorphonuclear leukocyte proteinase that produces emphysema in hamsters. JCI 82: 1963-73 (1988); **Labbaye C et al** Wegener autoantigen and myeloblastin are encoded by a single mRNA. PNAS 88: 9253-6 (1991); **Lesavre P** Antineutrophil cytoplasmic autoantibodies antigen specificity. Am. J. Kidney Dis. 18: 159-63 (1991); **Niles JL et al** Wegener's granulomatosis autoantigen is a novel neutrophil serine proteinase. Blood 74: 1888-93 (1989); **Sturrock AB et al** Structure, chromosomal assignment, and expression of the gene for proteinase-3. The Wegener's granulomatosis autoantigen. JBC 267: 21193-9 (1992); **Zimmer M et al** Three human elastase-like genes coordinately expressed in the myelomonocyte lineage are organized as a single genetic locus on 19pter. PNAS 89: 8215-9 (1992)

Myeloid growth factor: see: MGF.

Myeloid leukemia inhibitory factor: see: LIF.

Myeloid-associated proteins: see: Calgranulins.

Myeloid cells: see: Hematopoiesis.

Myeloma GF: *myeloma growth factor* This factor is identical with » IL6.

Myelomonocytic growth factor: abbrev. MGF. see: cMGF (chicken myelomonocytic growth factor).

Myeloproliferative leukemia virus: see: MPLV.

Myogenic growth factor: see: MGF.

Myxoma growth factor: see: MGF.

N

n: sometimes used as a prefix in abbrev. such as in n-IL2 to denote the natural or authentic factor and to distinguish it from recombinant factors.

N51: A secretory protein produced by mouse fibroblasts after stimulation with serum, » PDGF, or » bombesin. N51 expression is not induced by » EGF, » FGF, or the tumor promoter tetradecanoyl phorbol acetate (see also: Phorbol esters). The protein belongs to the » chemokine family of cytokines.

Ryseck RP et al Cloning and sequence of a secretory protein induced by growth factors in mouse fibroblasts. ECR 180: 266-75 (1989)

Na1 cells: A cell line established by transformation of human peripheral T cells with HTLV. These cells were used originally as sources of B cell growth and differentiation factors (see: BCDF, B cell differentiation factor; and BGDF, B cell growth and differentiation factor), in particular of » BSF-p2 (B cell stimulating factor p2).

Hirano T et al Purification to homogeneity and characterization of human B cell differentiation factor (BCDF or BSFp-2). PNAS 82: 5490-94 (1985); **Shimizu K et al** Immortalization of BGDF (BCGF II)- and BCDF-producing T cells by human T cell leukemia virus (HTLV) and characterization of human BGDF (BCGF II). JI 134: 1728-33 (1985)

NAF: *neu protein specific activating factor* This poorly characterized factor is isolated from transformed human T cells. It stimulates the intrinsic tyrosine-specific protein kinase of the » *neu* oncogene, induces receptor dimerisation and internalization of ligand/receptor complexes (see also: Autophosphorylation). The factor also enhances the proliferation of cells expressing *neu*. NAF does not interact with the » EGF receptor which is closely related to *neu*. At present it is not clear whether NAF is identical with or related to other proteins also interacting with *neu*, » NDF (*neu* differentiation factor), and » heregulin.

Davis JG et al Isolation and characterization of a *neu* protein-specific activating factor from human ATL-2 cell conditioned medium. BBRC 179: 1536-42 (1991); **Dobashi K et al** Characterization of a *neu*/c-*erb* B-2 protein-specific activating factor. PNAS 88: 8582-6 (1992)

NAF: *neutrophil-activating factor* NAF is also called » NAP-1 (neutrophil activating protein). NAF has a length of 72 amino acids. Variants of 77 (+AVLPR), 70 and 69 amino acids have also been described. NAF belongs to the » chemokine family of cytokines. The factor is identical with » IL8.

NAF was initially identified as a factor inducing the release of enzymes from neutrophils. NAF promotes the release of histamines, leukotrienes in IL3-activated basophils, stimulates the production of reactive oxygen species, increases the intracellular calcium levels (see also: Calcium ionophore), and also increases the adherence of neutrophilic granulocytes to endothelial cells.

Carveth HJ et al Neutrophil activating factor (NAF) induces polymorphonuclear leukocyte adherence to endothelial cells and to subendothelial matrix proteins. BBRC 162: 387-93 (1989); **Colditz I et al** *In vivo* inflammatory activity of neutrophil-activating factor, a novel chemotactic peptide derived from human monocytes. Am. J. Pathol. 134: 755-60 (1989); **Dahinden CA et al** The neutrophil-activating peptide NAF/NAP-1 induces histamine and leukotriene release by interleukin-3-primed basophils. JEM 170: 1787-92 (1989); **Lindley I et al** Synthesis and expression in *Escherichia coli* of the gene encoding monocyte-derived neutrophil-activating factor: biological equivalence between natural and recombinant neutrophil-activating factor. PNAS 85: 9199-203 (1988); **Peveri P et al** A novel neutrophil-activating factor produced by human mononuclear phagocytes. JEM 167: 1547-59 (1988); **Thelen M et al** Mechanism of neutrophil activation by NAF, a novel monocyte-derived peptide agonist. FJ 2: 2702-6 (1988); **Walz A et al** Purification and amino acid sequencing of NAF, a novel neutrophil-activating factor produced by monocytes. BBRC 149: 755-61 (1987); **Walz A et al** Structure and properties of a novel neutrophil-activating factor (NAF) produced by human monocytes. Agents Actions 26: 148-50 (1989); **Zipfel PF et al** Mitogenic activation of human T cells induces two closely related genes which share structural similarities with a new family of secreted factors. JI 142: 1582-90 (1989)

NAK1: An » ERG (early response gene) whose transcription is rapidly and transiently stimulated by growth-stimulating agents, such as adenosine diphosphate, in monkey kidney cells (BSC-1), by phytohemagglutinin in human lymphocytes, and by serum stimulation of arrested fibroblasts. It is expressed in human fetal muscle and adult liver, brain, and thyroid. NAK1 is induced in human melanoma cells exposed to » IL1. NAK1 probably is a nuclear receptor belonging to the human thyroid hormone receptor class. NAK is identical with mouse » nur77 and rat » NGFI-B.

Nakai A et al A human early response gene homologous to murine nur77 and rat NGFI-B, and related to the nuclear receptor superfamily. Mol. Endocrinol. 4: 1438-43 (1990); **Rangnekar YY et al** Interleukin-1-inducible tumor growth arrest is characterized by activation of cell type-specific "early" gene expression programs. JBC 267: 6240-8 (1992)

Namalwa: A human lymphoblastoid EBV-positive (integration sites at 1p32, 1q31, 5q21, 13q21, and 16p13) EBV non-producer cell line established from a Burkitt lymphoma. The cells grow as tumors in nude mice (see also: Immunodeficient mice). The long-term cultivation of Namalwa cells has led to the presence of complex chromosomal rearrangements, including a large number of unidentified marker chromosomes.

The cells secrete low amounts of a monoclonal antibody (IgM, λ light chain) of unknown specificity. There are many different phenotypic sublines (e. g. CSN/70, IPN/45, PNT and KN2) that differ markedly in the levels of expression of MHC class II DP, DQ, and DE antigens and appear to be arrested at different stages of differentiation. Treatment of the cells with » IFN-γ enhances the expression of HLA class II and » *myc* without inducing an antiviral state. Namalwa and various sublines have acquired mutations of the cellular » oncogene *myc* during *in vitro* culture, caused by insertion of type D retroviruses. Namalwa cells express the common acute lymphoblastic leukemia antigen (CALLA).

Namalwa cells have been used as a source of lymphoblastoid interferon (see: IFN-α) and are used as an alternative to other mammalian expression systems including » CHO (Chinese hamster ovary) cells, » BHK cells, » COS cells or the » Baculovirus expression system to express recombinant cytokines. One of the reasons is the ease with which Namalwa cells can be grown in chemically defined serum-free media (see also: SFM) and on a large scale allowing industrial mass production.

Caporossi D et al Specific sites for EBV association in the Namalwa Burkitt lymphoma cell line and in a lymphoblastoid line transformed *in vitro* with EBV. Cytogenet. Cell. Genet. 48: 220-3 (1988); **der Stepani I et al** Interferon γ is active on human lymphoblastoid Namalva cells without inducing an antiviral state. Biochem. Pharmacol. 37: 3271-6 (1988); **Finter NB et al** Interferon production from human cell cultures. CSHSQB 51: 571-5 (1986); **Guy K et al** MHC class II antigen and immunoglobulin expression in spontaneous phenotypic variants of the Burkitt's lymphoma cell line Namalwa. Immunology 59: 603-10 (1986); **Guy K et al** Variant sublines of the human B lymphoma cells Namalwa are at different stages of differentiation. Immunology 61: 383-6 (1987); **Guy K et al** Recurrent mutation of immunoglobulin and c-*myc* genes and differential expression of cell surface antigens occur in variant cell lines derived from a Burkitt lymphoma. IJC 45: 109-18 (1990); **Hosoi S** [Production of useful recombinant proteins using Namalwa KJM-1 cells adapted to serum-free medium] Hum. Cell. 4: 321-8 (1991); **Lazar A et al** Human lymphoblastoid interferon for clinical trials: large scale purification and safety tests. Dev. Biol. Stand. 55: 231-8 (1983); **Leonard JE et al** Establishment of a human B cell tumor in athymic mice. CR 47: 2899-902 (1987); **Lawrence JB et al** Sensitive, high-resolution chromatin and chromosome mapping *in situ*: presence and orientation of two closely integrated copies of EBV in a lymphoma line. Cell 52: 51-61 (1988); **May LT et al** Expression of the native α and β interferon genes in human cells. Virology 129: 116-26 (1983); **Middleton PG et al** Insertion of SMRV-H viral DNA at the c-*myc* gene locus of a BL cell line and presence in established cell lines. IJC 52: 451-4 (1992); **Okamoto M et al** Amplification and high-level expression of a cDNA for human granulocyte-macrophage colony-stimulating factor in human lymphoblastoid Namalwa cells. BT June 1990, pp. 550-3; **Okamoto M et al** Purification and characterization of three forms of differently glycosylated recombinant human granulocyte-macrophage colony-stimulating factor. ABB 286: 562-8 (1991); **Ruppersberger P et al** Characterization of marker chromosomes in Namalwa cells by chromosomal *in situ* suppression (CISS) hybridization and R-banding. Genes Chromosome Cancer 3: 394-9 (1991); **Shuttleworth J et al** Expression of interferon-α and interferon-β genes in human lymphoblastoid (Namalwa) cells. EJB 133: 399-404 (1983); **Vaillant F et al** Some biochemical properties of human lymphoblastoid Namalwa cells grown anaerobically. BI 23: 571-80 (1991); **Wang F et al** A bicistronic Epstein-Barr virus mRNA encodes two nuclear proteins in latently infected, growth-transformed lymphocytes. J. Virol. 61: 945-54 (1987); **Wurm F et al** Long term cultivation of Namalwa cells for interferon production: stable cytogenetic markers for identification of cells in spite of drastic chromosomal variation. Dev. Biol. Stand. 60: 393-403 (1985); **Yanagi H et al** High-level expression of human erythropoietin cDNA in stably transfected Namalwa cells. J. Ferment. Bioeng. 68: 257-63 (1989); **Yanagi H et al** Recombinant human erythropoietin produced by Namalwa cells. DNA 8: 419-27 (1989); **Yanagi H et al** Expression of human erythropoietin cDNA in human lymphoblastoid Namalwa cells: the inconsistency of a stable expression level with transient expression efficiency. Gene 76: 19-26 (1989); **Zoon KC et al** Purification and characterization of multiple components of human lymphoblastoid interferon-α. JBC 15210-6 (1992)

Namalwa BCGF: *Namalwa B cell growth factor* see: HMW-BCGF (high molecular weight B cell growth factor).

Namalwa interferon: This interferon originally isolated from » Namalwa cells, is identical with » IFN-α. See also: IFN for general information about interferons.

NAP: see: neural antiproliferative protein.

NAP-1: *neutrophil activating protein-1* This protein was initially isolated from stimulated mononuclear phagocytes. The protein is also called » NAF (neutrophil activating factor). NAP-1 is identical with » TCF (T lymphocyte chemotactic factor). NAP-1 is the N-terminal processed form of » LAI (leukocyte adhesion inhibitor). NAP-1 is identical with » IL8 and therefore belongs to the » chemokine family of cytokines.

The protein is chemotactic for T lymphocytes and basophils. It induces the release of histamines from basophils. Elevated levels of antibodies directed against NAP-1 are observed in patients with rheumatoid arthritis

The expression of the NAP-1 gene in fibroblasts can be downregulated by glucocorticoids. Its expression in human T lymphocytes is inhibited by » CsA (Ciclosporin).

Apella E et al Determination of the primary structure of NAP-1/IL8 and a monocyte chemoattractant protein, MCP-1/MCAF. PCBR 349: 405-17 (1990); **Baggiolini M et al** Neutrophil-activating peptide-1/interleukin 8, a novel cytokine that activates neutrophils. JCI 84: 1045-9 (1989); **Baggiolini M et al** Neutrophil activation and the effects of interleukin-8/neutrophil-activating peptide 1 (IL8/NAP-1). Cytokines 4: 1-17 (1992); **Clark-Lewis I et al** Chemical synthesis, purification, and characterization of two inflammatory proteins, neutrophil activating peptide 1 and neutrophil activating peptide 2. B 30: 3128-35 (1991); **Dahinden CA et al** The neutrophil-activating peptide NAF/NAP-1 induces histamine and leukotriene release by interleukin-3-primed basophils. JEM 170: 1787-92 (1989); **Detmers PA et al** Differential effects of neutrophil-activating peptide 1/IL8 and its homologues on leukocyte adhesion and phagocytosis. JI 147: 4211-7 (1991); **Leonard EJ et al** Leukocyte specificity and binding of human neutrophil attractant/activation protein-1. JI 144: 1323-30 (1990); **Leonard EJ et al** Chemotactic activity and receptor binding of neutrophil attractant/activation protein-1 (NAP-1) and structurally related host defense cytokines: interaction of NAP-2 with the NAP-1 receptor. J. Leukoc. Biol. 49: 258-65 (1991); **Peichl P et al** Human neutrophil activating peptide/interleukin 8 acts as an autoantigen in rheumatoid arthritis. Ann. Rheum. Dis. 51: 19-22 (1992); **Schröder JM** Generation of NAP-1 and related peptides in psoriasis and other inflammatory skin diseases. Cytokines 4: 54-76 (1992); **Sticherling M et al** Production and characterization of monoclonal antibodies against the novel neutrophil activating peptide NAP/IL8. JI 143: 1628-34 (1989); **Tobler A et al** Glucocorticoids downregulate gene expression of GM-CSF, NAP-1/IL8, and IL6, but not of M-CSF in human fibroblasts. Blood 79: 45-51 (1992); **White MV et al** Neutrophil attractant/activation protein (NAP-1) causes human basophil histamine release. Immunol. Lett. 22: 151-4 (1989); **Yoshimura T & Yuhki N** Neutrophil attractant/activation protein-1 and monocyte chemoattractant protein-1 in rabbit: cDNA cloning and their expression in spleen cells. JI 146: 3483-8 (1991); **Zipfel PF et al** Induction of members of the IL8/NAP-1 gene family in human T lymphocytes is suppressed by cyclosporin A. BBRC 181: 179-83 (1991)

NAP-2: *neutrophil activating protein-2* The expression of NAP-2 is induced in human monocytes by treatment with bacterial » endotoxins. NAP-2 is a protein with a length of 70 amino acids. The sequence is identical with the 70 carboxyterminal amino acids of PBP (platelet basic protein) from which it is released by proteolytic cleavage. NAP-2 is generated by cathepsin G. PBP is also the precursor of some other protein fragments with the activity of cytokines (see: Beta-Thromboglobulin). At the protein level NAP-2 displays 53 % identity with another cytokine, » ENA-78. NAP-2 belongs to the » chemokine family of cytokines. C-terminally truncated forms of NAP-2 showing a four-fold higher activity than authentic NAP-2 have also been described.

NAP-2 stimulates the release of elastase from neutrophils. It shows some of the activities of » IL8. It is ≈ 100-fold less chemotactic for neutrophils than IL8. In neutrophils NAP-2 induces the uptake of calcium (see also: Calcium ionophore), induces their degranulation, and the production of reactive oxygen species. Pretreatment of neutrophils with » MGSA (melanoma growth stimulating activity) or IL8 suppresses the formation of reactive oxygen. Basophils activated by » IL3 (see also: cell activation) produce histamines and IL8 in response to NAP-2.

NAP-2 interacts with the » NAP-1 receptor and binds to one type of receptors that also binds » IL8 and related factors. For a viral NAP-2 receptor homologue see also: ECRF3.

Brandt E et al A novel molecular variant of the neutrophil-activating peptide NAP-2 with enhanced biological activity is truncated at the C-terminus: identification by antibodies with defined epitope specificity. Mol. Immunol. 30: 30: 979-91 (1993); **Car BD et al** Formation of neutrophil-activating peptide 2 from platelet-derived connective-tissue-activating peptide III by different tissue proteinases. BJ 275: 581-4 (1991); **Clark-Lewis I et al** Chemical synthesis, purification, and characterization of two inflammatory proteins, neutrophil activating peptide 1 and neutrophil activating peptide 2. B 30: 3128-35 (1991); **Cohen AB et al** Generation of the neutrophil-activating peptide-2 by cathepsin G and cathepsin G-treated human platelets. Am. J. Physiol. 263: L249-56 (1992); **Holt JC et al** Isolation, characterization, and immunological detection of neutrophil- activating peptide 2: A proteolytic degradation product of platelet basic protein. PSEBM 199: 171-7 (1992); **Leonard EJ et al** Chemotactic activity and receptor binding of neutrophil attractant/activation protein (NAP-1) and structurally related host defense cytokines: interaction of NAP-2 with the NAP-1 receptor. J. Leuk. Biol. 49: 258-65 (1991); **Moser B et al** Neutrophil-activating peptide 2 and *gro*/melanoma growth-stimulatory activity interact with neutrophil-activating peptide 1/interleukin 8 receptors on human neutrophils. JBC 266: 10666-71 (1991); **Reddigari SR et al** Connective tissue-activating peptide-III and its derivative, neutrophil-activating peptide-2, release histamine from human basophils. J. Allergy Clin. Immunol. 89: 666-72 (1992); **Van Damme J et al** The neutrophil-activating proteins interleukin 8 and β-thromboglobulin: *in vitro* and *in vivo* comparison of NH2-terminally processed forms. EJI 20: 2113-8 (1990); **Waltz A & Baggiolini M** A novel cleavage product of β-thromboglobulin formed in cultures of stimulated mononuclear cells activates human neutrophils. BBRC 159: 969-75 (1986); **Walz A et al** Effects of neutrophil-activating peptide NAP-2, platelet basic protein, connective tissue-activating peptide III, and platelet factor 4 on human neutrophils. JEM 170: 1745-50 (1989); **Waltz A & Baggiolini M** Generation of the neutrophil-activating peptide NAP-2 from platelet basic protein or connective tissue-activating peptide III through monocyte proteases. JEM 171: 449-54 (1990); **Walz A** Generation and properties of neutrophil-activating peptide 2. Cytokines 4: 77-95 (1992)

NAP-3: *neutrophil activating protein-3* This protein of 5,3 kDa was initially observed in purifying » NAP-1. It is produced by human monocytes stimulated with bacterial » endotoxins. NAP-3 displays ≈ 40 % homology in its N-terminal sequence to » NAP-1. The aminoterminal sequence is identical with » MGSA (melanoma growth stimulating activity).

Schröder JM et al Lipopolysaccharide-stimulated human monocytes secrete, apart from neutrophil activating peptide 1/interleukin 8, a second neutrophil-activating protein. NH_2-terminal amino acid sequence identity with melanoma growth stimulatory activity. JEM 171: 1091-100 (1991)

NAP-4: *neutrophil activating protein-4* This protein of ≈ 8 kDa is isolated from a lysate of human platelets. It can be purified on affinity columns with antibodies directed against » NAP-1. The N-terminal sequence demonstrates that NAP-4 is a distinct factor with a strong homology to » PF4 (platelet factor 4). NAP-4 lacks two cysteine residues that are normally conserved in members of the PF4 family of proteins. NAP-4 induces chemotactic migration of human neutrophils.

Schröder JM et al Identification of a novel platelet-derived neutrophil chemotactic polypeptide with structural homology to platelet factor 4. BBRC 172: 898-904 (1990)

NAP-IF: *neutrophil alkaline phosphatase-inducing factor* This factor is isolated from the cystic fluid of a human squamous carcinoma. It induces the expression of alkaline phosphatase in human neutrophils which is an indication of cell maturation. The factor is identical with » G-CSF.

Sato N et al Identification of neutrophil alkaline phosphatase-inducing factor in cystic fluid of a human squamous cell carcinoma as granulocyte colony-stimulating factor. JCP 137: 272-6 (1988); **Sato N et al** Characterization of neutrophil alkaline phosphatase-inducing factor (NAP-IF). JCP 124: 255-60 (1985)

National Institute of Allergy & Infectious Diseases: see: NIAID.

National Institute for Biological Standards & Control: see: NIBSC.

Natural cytotoxic cell growth factor: see: NC cell growth factor.

Natural cytotoxic factor: abbrev. NCF. See: NKCF (natural killer cytotoxic factor).

Natural killer (cell) activating factor: see: NKAF.

Natural killer cell stimulatory factor: abbrev. NKSF. see: CLMF.

Natural killer colony-inhibiting activity: abbrev. NK-CIA. see: NKCF.

Natural killer cytotoxic factor: see: NKCF.

Natural killer enhancing factor: see: NKEF.

Natural-killer leukocyte chemotactic factor: see: NK-LCF.

Natural killer lysis resistance inducing factor: see: NK-LRIF.

Natural killer resistance inducing factor: see: NK-RIF.

Natural suppressor cell-derived suppressor factor: see: NS suppressor factor.

Nb2: A lactogen-dependent rat pre-T cell lymphoma cell line that proliferates in response to human and mouse » IL7 in culture medium containing 10% horse serum. Under serum-free culture condition human IL7 is much less effective. The effectiveness of IL7 on Nb2 cells is completely abolished by antibody to IL7, but not by antibody to » IL2. Proliferation of Nb2 cells is inhibited by high concentrations of human » growth hormone (≈ 70 μM) and dexamethasone.

Nb2 cells also proliferate in response to a factor isolated from the serum of pregnant women (see: PM, pregnancy mitogen), in response to » prolactin or an analog of human growth hormone.

Fletcher-Chiappini SE et al Glucocorticoid-prolactin interactions in Nb2 lymphoma cells: antiproliferative versus anticytolytic effects. PSEBM 202: 345-52 (1993); **Fuh G et al** Mechanism-based design of prolactin receptor antagonists. JBC 268: 5373-81 (1993); **Gala RR & Shevack EM** Evidence for the release of a prolactin-like substance by mouse lymphocytes and macrophages. PSEBM 205: 12-19 (1994); **Gout PW et al** Prolactin-stimulated growth of cell cultures established from malignant Nb rat lymphoma. CR 40: 2433-6 (1980); **Kornberg LJ & Liberti JP** Stimulation of Nb2 mitogenesis by an analog of human growth hormone (110-127). BI 28: 873-9 (1992); **Lawson DM et al** Rat lymphoma cell bioassay for prolactin: observations on its use and comparisons with radioimmunoassay. Life

Sci. 31: 3063-70 (1982); **Moy JA & Lawson DM** Serum specifically potentiates the mitogenic response of Nb2 lymphoma cells to rat prolactin. Endocrinology 123: 1314-9 (1988); **Nader S et al** Comparison of Nb2 lymphoma cell bioassay with immunoassay for human prolactin: the role of estrogen. J. Endocrinol. Invest. 15: 303-5 (1992); **Ohmae Y et al** Biological activities of synthesized 20K and 22K hGH in Nb2 bioassay and IM-9 radioreceptor assay. Endocrinol. Jpn. 36: 9-13 (1989); **Peabody CA et al** Prolactin bioassay and hyperprolactinemia. J. Endocrinol. Invest. 15: 497-9 (1992); **Sadeghi H & Wang BS** Proliferation of Nb2 lymphoma cells *in vitro* in response to interleukin-7. Immunol. Lett. 34: 105-8 (1992)

NB41-ECGF: *NB41 endothelial cell growth factor* This poorly characterized dimeric glycoprotein of ≈ 43-51 kDa is secreted by the murine neuroblastoma cell line NB41. It stimulates the proliferation of human umbilical chord endothelial cells. It is not related to other proteins with the same activity (see also: ECGF).

Levy AP et al An endothelial cell growth factor from the mouse neuroblastoma cell line NB41. GF 2: 9-19 (1989)

NBCF: *neonatal brain-derived carcinostatic factor* This poorly characterized factor or 62 kDa (pI = 9.1) is secreted by neonatal mouse brains but not by adult or fetal brain tissues and other organs. NBCF treated with neuraminidase loses part of its activity, suggesting glycan moieties required for the activity. NBCF secretion is also inhibited by tunicamycin at non-cytotoxic doses.

NBCF inhibits growth and DNA synthesis of malignant cells preferentially over those of normal cells. NBCF is more cytocidal for undifferentiated neuroblastoma cells than for other tumor cell types.

Miwa N et al Tumor growth-inhibitory glycoprotein secreted from the mouse brain at terminal stage of the ontogeny: molecular homogeneity and requirement of the retained protein conformation for exhibition of the cytotoxic action. BBA 1034: 309-17 (1990); **Miwa N** [Neonatal brain-derived carcinostatic factor (NBCF) – cytocidal action to neuroblastoma cells and molecular characters as a glycoprotein] Hum. Cell. 3: 137-45 (1990)

NBFL: A human neuroblastoma cell line. The cells are induced to produce » VIP (vasoactive intestinal peptide) in response to » CNTF (ciliary neurotrophic factor), » LIF (leukemia inhibitory factor), and » Oncostatin M and have been used to assay these cytokines.

Rao MS et al Oncostatin M regulates VIP expression in a human neuroblastoma cell line. Neuroreport 3: 865-8 (1992)

NC30: see: IL13.

NC cell growth factor: *natural cytotoxic cell growth factor* This factor supports the *in vitro* growth of natural cytotoxic cells. It is identical with » IL3.

Djeu JY et al Selective growth of natural cytotoxic but not natural killer effector cells in interleukin-3. N 306: 788-91 (1983); **Ihle JN et al** Interleukin 3: regulation of a lineage of lymphocytes expressing 20 α hydroxysteroid dehydrogenase. In: Interleukins, Lymphokines, and Cytokines. Academic Press, pp. 113-21 , New York 1983

NCF: *natural cytotoxic factor* see: NKCF (natural killer cytotoxic factor).

NCF: *neutrophil chemotactic factor* This protein is produced by monocytes, endothelial cells, and fibroblasts. It is identical with » IL8 and therefore belongs to the » chemokine family of cytokines. Its synthesis can be induced by » TNF-α, bacterial » endotoxins, and IL1β (see: IL1).

In addition, there are other, as yet unidentified factors having the same activity. Some of these factors have molecular weights below 1 kDa (see: fMLP) and are not proteins (such as leukotrienes). Others appear to be derivatives of complement components. The term NCF is also used as a generic name for factors that are chemotactic for neutrophils.

Dixit VM et al Molecular cloning of an endothelial derived neutrophil chemotactic factor: identity with monocyte derived factor. FJ 3: A305; Abstract 456 (1989); **Ehlers WH et al** Neutrophil chemotactic factors derived from conjunctival epithelial cells: preliminary biochemical characterization. CLAO J. 17: 65-8 (1991); **Furuta R et al** Production and characterization of recombinant human neutrophil chemotactic factor. J. Biochem. Tokyo 106: 436-41 (1989); **Ozaki T et al** Neutrophil chemotactic factors in the respiratory tract of patients with chronic airway diseases or idiopathic pulmonary fibrosis. Am. Rev. Respir. Dis. 145: 85-91 (1992); **Strieter RM et al** Endothelial cell gene expression of a neutrophil chemotactic factor by TNFα, LPS, and IL1β. S 243: 1467-9 (1989); **Strieter RM et al** Monokine-induced neutrophil chemotactic factor gene expression in human fibroblasts. JBC 264: 10621-6 (1989)

NDCF: *neutrophil-derived monocyte chemotactic factor* This poorly characterized factor of 12-13 kDa is produced by polymorphonuclear leukocytes following their stimulation by *Bacillus* Calmette-Guerin. The protein is chemotactic for monocytes (see also: Chemotaxis).

Antony VB et al Bacillus Calmette-Guérin-stimulated neutrophils release chemotaxins for monocytes in rabbit pleural spaces and *in vitro*. JCI 76: 1514-21 (1985)

NDF: *neu differentiation factor* NDF is isolated from *ras*-transformed fibroblasts. It is identical with » heregulin (human homologue). NDF is the ligand for the » *neu* oncogene and probably also the ligand of an as yet unidentified receptor. At present it is not clear whether NDF is identical

with or related to » NAF (*neu* protein specific activating factor) which also binds to the *neu* receptor. NDF appears to be the rat homologue of » ARIA (acetylcholine receptor inducing activity) and bovine » GGF (glial growth factor).
The interaction of NDF with *neu* has been found to be cell-specific: *neu* proteins of ovarian and fibroblastic cells are not stimulated by NDF.
NDF is a heparin-binding glycoprotein (see also: HBGF) of 44 kDa with intramolecular disulfide bonds. NDF stimulates tyrosine phosphorylation of the *neu* receptor. NDF induces the phenotypic differentiation of human mammary carcinoma cell lines into milk-producing growth-arrested cells. The human NDF gene maps to chromosome 8p21-p12.
NDF has been shown to induce expression of intercellular adhesion molecule 1 (ICAM-1) in cultured AU-565 human adenocarcinoma cells.
In mouse embryos NDF expression is confined predominantly to the central and peripheral nervous systems, including the neuroepithelium that lines the lateral ventricles of the brain, the ventral horn of the spinal cord, and the intestinal as well as dorsal root ganglia.

Bacus SS et al *neu* differentiation factor (heregulin) induces expression of intercellular adhesion molecule 1: implications for mammary tumors. CR 53: 5251-61 (1993); **Lee J & Wood, WI** Assignment of heregulin (HGL) to human chromosome 8p22-p11 by PCR analysis of somatic cell hybrid DNA. Genomics 16: 790-1 (1993); **Orr-Urtreger A et al** Neural expression and chromosomal mapping of *neu* differentiation factor to 8p12-p21. PNAS 90: 1867-71 (1993); **Peles E et al** Isolation of the *neu*/HER2 stimulatory ligand: a 44 kd glycoprotein that induces differentiation of mammary tumor cells. Cell 69: 205-16 (1992); **Peles E et al** Cell-type specific interaction of *neu* differentiation factor (NDF/heregulin) with *neu/HER*-2 suggests complex ligand-receptor relationships. EJ 12: 961-71 (1993); **Wen D et al** *neu* differentiation factor: a transmembrane glycoprotein containing an EGF domain and an immunoglobulin homology unit. Cell 69: 559-72 (1992)

NDLCF: *neutrophil-derived lymphocyte chemotactic factor* This poorly characterized factor is found in the ascites fluid of tumor patients following intraperitoneal administration of a preparation of Streptococcus (OK-432). The factor stimulates the migration of cytotoxic lymphocytes. See also: LCF.

Hayashi Y & Torisu M New approach to management of malignant ascites with streptococcal preparation OK-432. III. OK-432 attracts natural killer cells through a chemotactic factor released from activated neutrophils. Surgery 107: 74-84 (1990)

NDP kinase: Nucleoside diphosphate kinase. See: nm23.

Necrosin: This factor was initially found in the serum-free supernatants of the murine macrophage-like cell line » J774.1 that had been treated with bacterial » endotoxin. It exists as a multimer of 70 kDa consisting of 15 kDa subunits. The factor is cytotoxic for normal and malignant bovine and human cells lines. It is identical with » TNF-β.

Kull FC & Cuatrecasas P Necrosin: purification and properties of a cytotoxin derived from a murine macrophage-like cell line. PNAS 81: 7932-6 (1984)

NEF: *neurite extension factor* NEF is a disulfide-bonded dimer of a protein closely related to » S100. NEF induces process outgrowth from chick embryo cerebral cortical neurons. It also induces rapid morphological differentiation of Neuro-2A murine neuroblastoma cells, leading to the formation of elaborate multipolar neurites.

Kligman D & Shieh LS Neurite extension factor induces rapid morphological differentiation of mouse neuroblastoma cells in defined medium. Brain Res. 430: 296-300 (1987)

Negative growth factors: see: Growth factors.

NEGF1: *neurite growth promoting factor 1* This human gene encodes » HBNF (heparin-binding neurite promoting factor).

NEGF2: *neurite growth promoting factor 2* The human NEGF2 gene maps to chromosome 11p13-p11. The mature protein has a length of 121 amino acids and is 46% homologous with another heparin-binding neurite outgrowth-promoting factor, NEGF1 (see: HBNF, heparin-binding neurite promoting factor). NEGF-2 is also known as » MK (midkine).
The protein also possesses neurite outgrowth-promoting activity and the gene is a member of a highly conserved, developmentally regulated human gene family. NEGF2 probably plays a role in nervous system development and/or maintenance and its expression is developmentally regulated. NEGF2 protein is barely detectable before and after approximately one half to two thirds of the gestation period.

Eddy RL et al A human gene family of neurite outgrowth-promoting proteins: heparin-binding neurite outgrowth promoting factor maps to 11p11-11p13. Cytogenet. Cell Genet. 58: 1958 (1991)

NEL-GF: *neu/erb B2 ligand growth factor* see: *neu*.

Neonatal brain-derived carcinostatic factor: see: NBCF.

Nerve growth factor: see: NGF.

Nerve growth factor inducible genes: see: NGFI.

***neu*:** This is an activated cellular » oncogene initially isolated from rat neuro/glioblastomas obtained after transplacental induction of tumors with ethylnitroso urea. Lymphoid tissues do not appear to express *neu*.

neu encodes a transmembrane glycoprotein of 185 kDa (p185). Some carcinoma cells secrete a soluble form of the extracellular domain of p185^{neu}. *neu* is closely related to the receptors for » EGF and » TGF-α found in human cells. The human homologue of *neu* is also called ***HER-2*** (= human EGF receptor), ***NGL*** ((to avoid confusion with the neuraminidase locus, which is also symbolized *neu*), and ***c-erbB2*** (see also: *erb*). The gene maps to human chromosome 17q21. The corresponding rat gene maps to chromosome 10 which is syntenic with human chromosome 17.

A single point mutation (Val664→Glu664) in the transmembrane region of the *neu* protein is responsible for the transforming activity of the oncogene. This mutation activates the intracellular intrinsic tyrosine-specific kinase domain of the receptor protein, causing the » autophosphorylation of the receptor at position Tyr1248 and also the phosphorylation of some other cellular proteins. Mutations which replace Val664 by lysine, glycine, histidine, or tyrosine do not activate the oncogene. The activating mutation appears to be specific for experimental tumors obtained by chemical carcinogenesis since a similar activating mutation is not observed in naturally occurring human tumors (mammary, brain, stomach, pancreas). The mechanism underlying the activation of this oncogene is still unclear since NMR studies reveal that the three-dimensional structure of the receptor is not altered by the activating mutation in the transmembrane region. Binding of anti-*neu* antibodies to the *neu* receptor decreases the intrinsic tyrosine kinase activity of the receptor and causes a reversion of transformed cells to a non-transformed phenotype. In spite of the similarity between p185^{neu} and the EGF receptor the oncogene product does not bind EGF or TGF-α which also binds to the EGF receptor. Some interaction between *neu* and EGF receptors take place, however, since binding of EGF to the EGF receptor leads to the phosphorylation of *neu*. A natural ligand of p185^{neu} is » heregulin, also known as » NDF (*neu* differentiation factor). At present it is not clear whether heregulin/NDF is identical with » NAF (*neu* protein specific activating factor). A 30 kDa glycoprotein (gp30), secreted by estrogen receptor-negative breast cancer cell lines, has also been described to be a specific ligand for the *neu* receptor. gp30 is a heparin-binding protein (see also: HBGF) and also appears to induce the phosphorylation of the EGF receptor. » ARIA (acetylcholine receptor inducing activity), and » GGF (glial growth factor) have also been shown to be ligands for the *neu* oncogene and these factors are collectively known as » neuregulins. A *neu/erb* B2 ligand growth factor (***NEL-GF***; 25 kDa), stimulating the tyrosine-specific autophosphorylation of the *neu/erb* B2, has been purified from bovine kidney. It is a mitogen for several cell lines, including » 3T3 mouse fibroblasts, human epidermoid carcinoma cell line A431, and human breast cancer cells. NEL-GF appears to be distinct from the other *neu* ligands described so far.

The expression of the *neu* gene and its protein by Schwann cells both *in vivo* and *in vitro* suggests

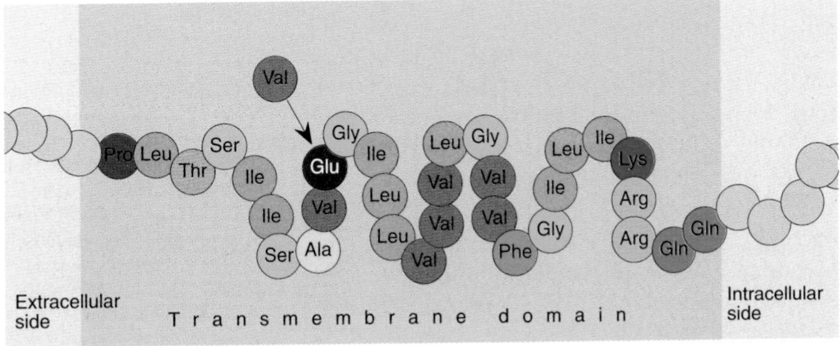

Transmembrane domain of the protein encoded by the *neu* proto-oncogene.
A single mutation within the transmembrane region of the protein leading to a glutamine residue in place of a valine residue results in an activated oncogene.

that p185[neu] may play a role in the regulation of Schwann cell proliferation or differentiation.

■ **CLINICAL USE & SIGNIFICANCE:** THE amplification and overexpression of *neu* is found in many adenocarcinomas. This is of prognostic value for evaluation of relapse and survival of tumor patients. A comparison of clinical data demonstrates that the detection of amplification of *neu* in mammary carcinomas is particularly useful and allows better prognoses than other clinical parameters used so far (age, tumor size, hormone status). The strong overexpression of *neu* is observed in ≈ 25 % of all primary mammary carcinomas and correlates with a poor prognosis. Overexpression of *neu* is an indicator of increased risk of developing recurrent disease in women with node-negative breast cancer.

The majority of patients co-expressing *neu* and the intercellular adhesion molecule ICAM-1(which is induced by » NDF, *neu* differentiation factor) have no lymph node involvement, unlike most *neu*-positive but ICAM-1-negative tumors, which metastasize to the lymphatic system.

■ **TRANSGENIC/KNOCK-OUT/ANTISENSE STUDIES:** The expression of the activated c-*neu* oncogene in » transgenic mice has been shown to be the sole factor responsible for causing the induction of mammary adenocarcinomas. These transgenic mice develop mammary carcinomas extending over the entire epithelium of each gland. In other tissues such as the parotid gland or the epididymis the expression of the c-*neu* oncogene leads to hypertrophy and hypoplasia without accompanying malignant transformation. The activated oncogene therefore only causes malignant transformation in a limited spectrum of tissues. The effect of mammary gland-specific overexpression of the unactivated *neu* proto-oncogene has also been assessed. Transgenic mice carrying unactivated *neu* under the transcriptional control of the mouse mammary tumor virus promoter/enhancer develop focal mammary tumors after long latency. Many of the tumor-bearing transgenic mice develop secondary metastatic tumors in the lung.

The pathogenicity of the human *neu*/c-*erb*B-2 oncogene has been evaluated in transgenic mice harboring a construct comprising the promoter-enhancer region of the MMTV LTR and a constitutively activated allele of the human c-*erb*B-2 gene. Expression of the transgene is observed in kidney, lung, mammary gland, salivary gland, Harderian gland, and in epithelial cells of the male reproductive tract. All transgenic mice die within four months of birth most likely from preneoplastic lesions in kidney and lung causing organ failure.

● **BIOCHEMISTRY & MOLECULAR BIOLOGY: Akiyama T et al** The product of the human c-*erb*B-2 gene: a 185-kilodalton glycoprotein with tyrosine kinase activity. Science 232: 1644-6 (1986); **Bargmann CJ et al** The *neu* oncogene encodes an epidermal factor receptor-related protein. N 319: 226-30 (1986); **Di Fiore PP et al** Cloning, expression, and biological effects of *erb*B-2/*neu* gene in mammalian cells. MiE 198: 272-7 (1991); **Fendly BM et al** Characterization of murine monoclonal antibodies reactive to either the human epidermal growth factor receptor or HER2/*neu* oncogene. CR 50: 1550-8 (1990); **Fukushige SI et al** Localization of a novel v-*erb*B-related gene, c-*erb*B-2, on human chromosome 17 and its amplification in a gastric cancer cell line. MCB 6: 955-8 (1986); **Gullick WJ et al** Three dimensional structure of the transmembrane region of the proto-oncogenic and oncogenic forms of the *neu* protein. EJ 11: 43-8 (1992); **Kokai Y et al** Stage- and tissue-specific expression of the *neu* oncogene in rat development. PNAS 84: 8498-501 (1987); **Schechter AL et al** The *neu* oncogene: an *erb*-B-related gene encoding a 185,000-Mr tumor antigen. N 312: 513-6 (1984); **Schechter AL et al** The *neu* gene: an *erb*B-homologous gene distinct from and unlinked to the gene encoding the EGF receptor. S 229: 976-8 (1985); **Tal M et al** Human HER2 (*neu*) promoter: evidence for multiple mechanisms for transcriptional initiation. MCB 7: 2597-601 (1987); **Yamamoto T et al** Similarity of protein encoded by the human c-*erb*-B-2 gene to epidermal growth factor receptor. N 319: 230-4 (1986); **Yan DH & Hung MC** Identification and characterization of a novel enhancer for the rat *neu* promoter. MCB 11: 1875-82 (1991)

● **RECEPTOR FUNCTION: Cohen JA et al** Expression of the *neu* proto-oncogene by Schwann cells during peripheral nerve development and Wallerian degeneration. J. Neurosci. Res. 31: 622-34 (1992); **Huang SS et al** Differential processing and turnover of the oncogenically activated *neu*/*erb* B2 gene product and its normal cellular counterpart. JBC 165: 3340-6 (1990); **Huang SS & Huang JS** Purification and characterization of the *neu*/*erb* B2 ligand-growth factor from bovine kidney. JBC 267: 11508-12 (1992); **Lupu R et al** Characterization of a growth factor that binds exclusively to the *erb*B-2 receptor and induces cellular responses. PNAS 89: 2287-91 (1992); **Lupu R et al** Purification and characterization of a novel growth factor from human breast cancer cells. B 31: 7330-40 (1992); **Wada T et al** Intermolecular association of the protein p185[neu] and EGF receptor modulates EGF receptor function. Cell 61: 1339-47 (1990); **Yarden Y & Peles E** Biochemical analysis of the ligand for the *neu* oncogenic receptor. B 30: 3543-550 (1991); **Peles E et al** Oncogenic forms of the *neu*/HER2 tyrosine kinase are permanently coupled to phospholipase Cγ. EJ 10: 2077-86 (1991)

● **ONCOGENIC POTENTIAL: Akiyama T et al** The transforming potential of the c-*erb*B-2 protein is regulated by its autophosphorylation at the carboxyl-terminal domain. MCB 11: 833-42 (1991); **Bargmann CI & Weinberg A** Oncogenic activation of the *neu*-encoded receptor protein by point mutation and deletion. EJ 7: 2043-52 (1988); **Di Fiore PP et al** *erb*B-2 is a potent oncogene when overexpressed in NIH/3T3 cells. Science 237: 178-82 (1987); **Drebin JA et al** Down-modulation of an oncogene protein product and reversion of the transformed phenotype by monoclonal antibodies. Cell 41: 695-706 (1985); **Lehtola L et al** Constitutively activated *neu* oncoprotein tyrosine kinase interferes with growth factor-induced signals for gene activation. JCBc 35: 69-81 (1991)

● **TRANSGENIC/KNOCK-OUT/ANTISENSE STUDIES: Bouchard L et al** Stochastic appearance of mammary tumors in transgenic mice carrying the MMTV/c-*neu* oncogene. Cell 57: 931-6 (1989); **Guy CT et al** Expression of the *neu* proto-oncogene in the mammary epithelium of transgenic mice induces metastatic disease. PNAS 89: 10578-82 (1992); **Hayes C et al** Expression of the *neu* oncogene under the transcriptional control of the myelin basic protein gene in transgenic mice: generation of transformed glial cells. J. Neurosci. Res. 31: 175-87 (1992); **Lucchini F et al** Early and multifocal tumors in breast, salivary, harderian and epididymal tissues developed in MMTY-*neu* transgenic mice. Cancer Lett. 64: 203-9 (1992); **Muller WJ et al** Single-step induction of mammary adenocarcinoma in transgenic mice bearing the activated c-*neu* oncogene. Cell 54: 105-15 (1988); **Stocklin E et al** An activated allele of the c-*erb*B-2 oncogene impairs kidney and lung function and causes early death of transgenic mice. JCB 122: 199-208 (1993)

● CLINICAL USE & SIGNIFICANCE: **Bacus SS et al** *neu* differentiation factor (heregulin) induces expression of intercellular adhesion molecule 1: implications for mammary tumors. CR 53: 5251-61 (1993); **Hurtt MR et al** Amplification of epidermal growth factor receptor gene in gliomas: Histopathology and prognosis. J. Neuropathol. Exp. Neurol. 51: 84-90 (1992); **Hynes NE** Amplification and overexpression of the *erb*B-2 gene in human tumors: its involvement in tumor development, significance as a prognostic factor, and potential as a target for cancer therapy. Semin. Cancer Biol. 4: 19-26 (1993); **Kern JA et al** p185^{neu} expression in human lung adenocarcinomas predicts shortened survival. CR 50: 5184-91 (1990); **King BL et al** *neu* proto-oncogene amplification and expression in ovarian adenocarcinoma cell lines. Am. J. Pathol. 140: 23-31 (1992); **Lundy J et al** Expression of *neu* protein, epidermal growth factor receptor, and transforming growth factor α in breast cancer: Correlation with clinicopathologic parameters. Am. J. Pathol. 138: 1527-34 (1991); **McKenzie SJ** Diagnostic utility of oncogenes and their products in human cancer. BBA 1072: 193-214 (1991); **Paik S et al** Clinical significance of *erb*B2 protein overexpression. Cancer Treat. Res. 61: 181-91 (1992); **Press MF et al** *Her-2/neu* expression in node-negative breast cancer: direct tissue quantitation by computerized image analysis and association of overexpression with increased risk of recurrent disease. CR 53: 4960-70 (1993); **Slamon DJ et al** Human breast cancer: correlation of relapse and survival with amplification of the *HER-2/neu* oncogene. S 235: 177-82 (1987); **Slamon DJ et al** Studies of the *HER-2/neu* proto-oncogene in human breast and ovarian cancer. Science 244: 707-12 (1989); **Torp SH et al** Amplification of the epidermal growth factor receptor gene in human gliomas. ACR 11: 2095-8 (1991); **van de Vijver MJ et al** *neu*-protein overexpression in breast cancer: association with comedo-type ductal carcinoma *in situ* and limited prognostic value in stage II breast cancer. NEJM 319: 1239-45 (1988); **Zabrescky JR et al** The extracellular domain of p185/*neu* is released from the surface of human breast carcinoma cells, SK-BR-3. JBC 266: 1716-20 (1991)

***neu* differentiation factor:** see: NDF.

***neu/HER*-2 stimulatory ligand:** see: NDF (*neu* differentiation factor.

Neupogen: Trademark for recombinant » G-CSF (Amgen Inc. Thousand Oaks, California). See: Filgrastim.

***neu* protein specific activating factor:** see: NAF.

Neural antiproliferative protein: abbrev. NAP. This protein of 55 kDa is found in the conditioned medium of Schwann cell and Schwannoma cultures. It completely inhibits proliferation of Schwann cells but not proliferation of immortalized and transformed Schwann cell types and thus functions as an » autocrine proliferation inhibitor. The action of the antiproliferative protein is probably indirect. The 55 kDa protein has been found to possess metalloprotease activity and stromelysin immunoreactivity, sharing many properties with stromelysin isolated from other sources, including the ability to cleave fibronectin. Limited proteolysis of fibronectin by the Schwann cell-derived protease generates an aminoterminal 29 kDa fibronectin fragment which itself expresses a potent antiproliferative activity for cultured Schwann cells. Normal and transformed Schwann cell types have been shown to secrete the proform of stromelysin, but transformed cells do not produce activated stromelysin and thus cannot generate the antiproliferative fragment of fibronectin. Once activated, a Schwann cell-derived protease similar to stromelysin may therefore cleave fibronectin and generate an antiproliferative activity which can maintain normal Schwann cell quiescence *in vitro*.

Muir D et al Schwann cell proliferation *in vitro* is under negative autocrine control. JCB 111: 2663-71 (1990); **Muir D & Manthorpe M** Stromelysin generates a fibronectin fragment that inhibits Schwann cell proliferation. JCB 116: 177-85 (1992)

Neuregulins: Collective term proposed for a group of factors encoded by alternatively spliced transcripts derived from a common gene which function as ligands for the » *neu* oncogene. These factors include » NDF (*neu* differentiation factor = » Heregulin), » ARIA (acetylcholine receptor inducing activity), and » GGF (glial growth factor).

Marchionni MA et al Glial growth factors are alternatively spliced *erb*B2 ligands expressed in the nervous system. N 362: 312-8 (1993)

Neurite extension factor: see: NEF.

Neurite-inducing factor: see: NIF.

Neurite outgrowth-promoting proteins: see: HBNF (heparin-binding neurite-promoting factor).

Neurite-promoting factor: see: NPF.

Neurite retraction factor: see: NRF.

Neuroblastoma growth factor: This poorly characterized factor of less than 5 kDa is detected in von Recklinghausen neurofibroma extracts. It promotes the proliferation of human neuroblastoma cells and the survival and neurite-extension of rat cortical neurons.
Distinct growth factors for human neuroblastoma cells are also » VIP (vasoactive intestinal peptide), » IGF-2, and » TNF-α.

Asai K et al von Recklinghausen neurofibroma produces neuronal and glial growth-modulating factors. Brain Res. 556: 344-8 (1991)

Neuroblastoma growth inhibitory factor: see: NGIF.

Neuroendocrine stress axis: see: Neuroimmune network.

Neuroimmune network: A substantial body of information now indicates that the nervous and immune systems are integrated and form an interdependent neuroimmune network (see also: Cytokine network). The field concentrating on studies of interactions between the immune system and neuroendocrine organs (including, for example, brain, pituitary, thyroid, parathyroid, pancreas, adrenal glands, testes, ovary) has a number of names, including ***psychoneuroimmunology*** and ***neuroimmunoendocrinology***.
Communication between the two systems and a reciprocal flow of information is suggested by several observations. Like many other physiological responses immune reactions can be conditioned in a classical Pavlovian fashion. Physical or emotional stress and psychiatric illness activate the endocrine system and can compromise immunological functions. Their profound effects on immune responses can, in turn, elicit marked physiological and chemical changes in the brain.
Direct modulation of the immune system by neuroendocrine influences can also be inferred from the innervation of immune organs such as spleen, thymus, and lymph nodes by sympathetic and sensory neurons. For example, about two thirds of mast cells in the rat intestinal mucosa have contacts with subepithelial peptidergic neurons.
A direct participation of the sympathetic nervous system in immune functions is also evidenced by the observation that sympathectomy and lesioning of specific regions of the brain can both enhance and/or suppress immune responses. Defined hypothalamic ablation has been shown, for example, to reduce the number of large granular lymphocytes, the activity of natural killer cells, the ratio of T helper/T suppressor cells, and the level of circulating B cells. Firing rates of hypothalamic neurons have been shown to be altered during immune responses. Anterior hypothalamic ablation has been shown to be associated with diminished immunoreactivity. Hypophysectomy is also known to cause an impairment of humoral and cell-mediated immunity, which can be corrected by treatment with neuromodulatory mediators such as » Prolactin, or » Growth hormone (see also: Dwarf mice). Hypophysectomy in rats and mice has been shown to decrease antibody responses, to prolong survival of grafted tissues, to decrease lymphocyte proliferation, to reduce spleenic natural killer cell activity, and to cause an inability to develop adjuvant arthritis.
A neural supply of immune organs allows for local delivery of neuropeptides and other neuromediators at high concentrations, which can then act on receptors expressed on immunocytes. The reciprocal entry of immune cells into the nervous system has also been observed. Monocytes, macrophages and T cells are able to cross the blood brain barrier. Macrophages can persist for very long intervals as resident microglia of the brain and constitute ≈ 10 % of the total glial cell population. Activated T cells are retained for days if they react specifically with central nervous system antigens. A variety of stimuli have been shown to induce expression of MHC molecules on astrocytes, microglia, and oligodendrocytes, which then can function to present antigens and to become targets for cytotoxic T cells. Functionally significant concentrations of some neuropeptides are also found at sites of immune and inflammatory reactions (see also: Inflammation).
Many different classes of molecules, including » cytokines, neurohormones, neurotransmitters, and many non-peptide mediators are involved in the amplification, coordination, and regulation of communication pathways within the neuroimmune system. It appears that these substances act on their classical neuroendocrine target cells and also serve as endogenous regulators of the immune system, acting on receptors expressed by the immune cells and thus possessing » autocrine and/or » paracrine functions. Circulating peripheral immune cells thus may establish a mobile source of neuromodulators, allowing these molecules to reach virtually all types of cells within an organism. It should be remembered that most, if

not all, endocrine glands contain a high number of lymphoid cells. For example, these cells can make up to 20 % of the total cell number in the case of the adrenal gland. Moreover, many classical » cytokines of the immune system have been shown to be produced by a variety of brain cells, including neurons and glial cells.

Neurohormonal involvement in immune reactions has been known for some time, in particular through the immunosuppressive effects of glucocorticoid hormones. Production from the adrenal gland of corticosterone in rodents and of cortisol in humans is stimulated by ACTH (see: POMC, proopiomelanocortin). Hypophysectomy causes adrenal atrophy because of adrenal dependence on pituitary-derived ACTH. Steroid hormones inhibit secretion of ACTH by the pituitary and also turn off the production of specific hypothalamic release factors, which positively regulate the synthesis of pituitary hormones. Pituitary and/or hypothalamic hormones in turn are usually controlled negatively by end products of the particular neuroendocrine cascade; glucocorticoid hormones, for example, suppress ACTH production (see also: GIF, glucocorticoid increasing factor). These interactions form the basis of the physiologically important regulatory entity known as the ***hypothalamic-pituitary-adrenocortical axis*** (abbrev. HPA) or ***stress axis***, which thus integrates functions of the hypothalamus, the pituitary, and the adrenal glands (see also: Acute Phase reaction). Recent evidence indicates that a similar situation exists in the immune system; for example, ACTH/endorphins and » growth hormone production by lymphocytes

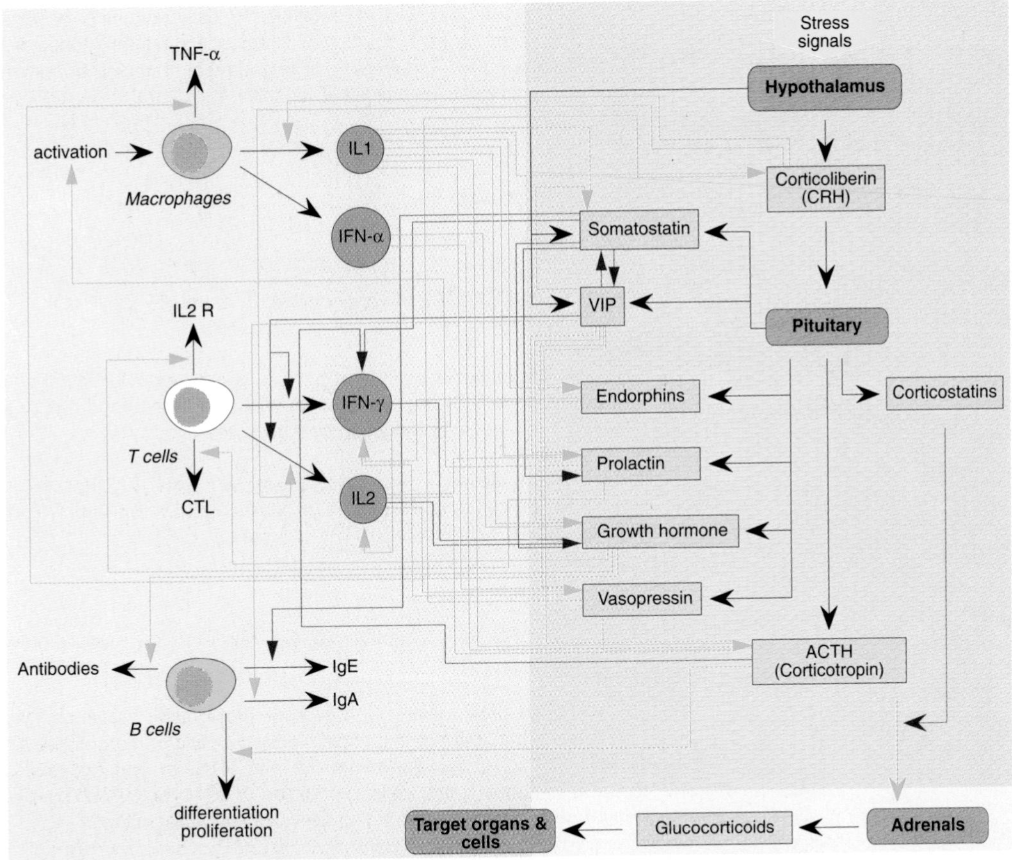

Simplified scheme of integrated regulation of the hypothalamic-pituitary-immune axis.
Some major hypothalamic-pituitary effects on immune functions and cross-regulation by immune cell-derived cytokines are shown. Neuromediation of immune functions by neural cells and neural cell activities of immune cytokines are not shown.
Black arrows: release of; green arrows: stimulation or enhancement of expression or release; red arrows: inhibition.

Neurotransmitters and hormones with immunomodulatory properties.

Acetylcholine	Bone marrow lymphocyte and macrophage numbers ↑
ACTH	Glucocorticoids ↑, costimulator for B cell growth/differentiation, modulation of antibody production, inhibition of IFN-γ synthesis, NK cell activity ↑
α-MSH	IL1, TNF-α, C5a-mediated neutrophil chemotaxis ↓
β-Endorphin	Antibody synthesis ↑↓, macrophage activation ↑↓, T cell activation ↑↓
Catecholamines	Mitogen-stimulated lymphocyte proliferation ↓
CRH	Lymphocyte production of ACTH ↑, IL2 expression ↑, NK cell activity ↑
Dynorphin	T cell stimulation
Glucocorticoids	Antibody production ↓, NK activity ↓, cytokine production ↓
Growth hormone	Enhancer of immune responses , cytolytic activity of T cells, ↑, antibody synthesis↑, GM-CSF-induced granulocyte differentiation↑, TNF-α production in macrophages ↑, generation of superoxide anions in macrophages↑, chemotactic response in human monocytes ↑, synthesis of some thymic hormones ↑, promotion of engraftment of murine or human T cells in severe combined immunodeficient mice (SCID) mice
Melatonin	Antibody synthesis ↑
Met-Enkephalin	T cell activation (low dose ↑; high dose ↓)
Oxytocin	T cell proliferation ↑
Prolactin	differentiation, proliferation, and function of lymphocytes ↑, astrocyte proliferation ↑, macrophage activation, IL2 production ↑
Somatostatin	Growth hormone-induced chemotactic response of human monocytes ↓, prolactin and growth hormone-induced immunoenhancing effects ↓, lymphocyte proliferation ↓, IgE production ↓, IFN-γ secretion ↓, superoxide production by monocytes ↓,
Substance P	Lymphocyte proliferation/differentiation ↑, immunoglobulin synthesis ↑, chemotactic migration of neutrophils ↓, mast cell degranulation ↑, production of IL1, IL6 and TNF-α, IFN-γ by monocytes/lymphocytes ↑, production of IL2 by T cells ↑,
Tyroxine	T cell activation ↑
Vasopressin	Synthesis of NGF ↑, release of ACTH ↑, modulation of T cell proliferation, modulation of central actions of IL1
VIP	Mitogen-stimulated lymphocyte proliferation ↑, thymocyte proliferation ↓, lymphocyte migration ↓, NK cell activity ↓, macrophage tumoricidal activity ↓, B cell immunoglobulin production ↓↑ B cell proliferation ↑, production of IL2 and IFN-γ by lymphocytes ↓, autocrine growth factor for neuroblastoma cells, production of IFN-α and IFN-β by glial cells ↑, secretagogue for neuron survival-promoting activities, astroglial mitogen

is regulated by corticotropin releasing factor and growth hormone releasing hormone, respectively. Other neuromodulators released from the neuronal systems and influencing cells of the immune system include, for example, α-MSH and β-Endorphin (see: POMC, proopiomelanocortin), follicle stimulating hormone, » Angiotensin, » Growth hormone, » Luteinising hormone, » IGF, neuronal differentiation factors (see below), » Prolactin, » Somatostatin, » SP (substance P) and related » Tachykinins, » Suppressin, » Thymic hormones, Thyrotropin, » Vasopressin, » VIP (vasoactive intestinal peptide), and many other small oligopeptides, neurotransmitters, and non-protein molecules (see also: Neuropoietins, Neurotrophins). Many of these factors have been shown to be released also by activated cells of the immune system. Immune cells have been shown to produce authentic neuromodulatory molecules identical with those also produced by brain and nerve cells as well as genetically determined variants arising both by alternative splicing of mRNA and by unique proteolytic cleavages of the preproproteins (see, for example: POMC). Immune cells also appear to utilize authentic receptors for these factors as well as receptors that differ in specificity, affinity, and signal transduction mechanisms from those in the nervous system. Precisely how these factors can modulate immunity and/or neuronal processes (cell growth, survival, and differentiation) remains to be determined.

In addition to classical neuromodulatory factors, classical cytokine mediators including, for example, » IL1, » IL2, » IL6, » TNF-α, » LIF, » IFNs (interferons), and » bFGF, also have potent neuroendocrine activities.

IL1 is produced by pituitary cells. It has been shown to be as potent as corticotropin-releasing hormone in some systems to induce the production of ACTH (see: POMC). Primary isolated pituitary cells can respond to IL1 by releasing ACTH, » Growth hormone, » Luteinizing hormone, and » thyrotropin. IL1 inhibits the production of » Prolactin by these cells. Antibodies to corticotropin releasing factor can block corticosteroid induction when IL1 is injected intravenously. The enhanced synthesis of ACTH causes a rise in corticosteroids which can, in turn, decrease IL1 production by macrophages. IL1 has also be shown to be a potent inducer of neuronal differentiation factors and instructive factors that can alter, for example, utilization of neurotransmitter types in specific neurons (see below).

Enhancement of » POMC gene expression and a differential release of different pituitary hormones *in vitro* has also been described for » IL2, which stimulates release of ACTH, » Prolactin, and » Thyrotropin while inhibiting the basal release of » Luteinizing hormone, follicle stimulating hormone, and » Growth hormone. A possible feedback mechanism for processes mediated by » IL1 is suggested by the fact that » IL1 induces T cells to produce IL2 although IL2 has not yet been shown to act directly at the level of the hypothalamus.

It has been shown that » IL6, which is also produced by pituitary cells, is also a more potent secretagogue for ACTH than corticotropin-releasing hormone in some systems. Intravenous injection of IL6 causes a dose-dependent increase in plasma ACTH levels. As seen with IL1 this rise of ACTH can be blocked by administration of antibodies directed against corticotropin-releasing hormone. Again, regulatory feedback loops are suggested by the inhibition of IL6 production by dexamethasone.

TNF-α is produced by astrocytes and probably also by pituitary cells. It stimulates adrenal and inhibits thyroid functions. All interferons have been shown to act within the hypothalamus-pituitary-adrenal axis. IFN-α has been observed to cause a rise in » Growth hormone, ACTH, and cortisol in some systems while lowering levels of » Growth hormone and » Thyrotropin in others. IFN-γ has been shown to inhibit corticostatin releasing hormone-induced ACTH secretion by cultured pituitary cells.

A protein that inhibits the proliferation of a myeloid cell line and induces differentiation of macrophage characteristics, » LIF (leukemia inhibitory factor), has recently been shown to be identical with the cholinergic differentiation factor » CDF, a protein that influences a fundamental aspect of neuronal identity, i. e. neurotransmitter and neuropeptide phenotype. These findings suggest analogies with the control of phenotypic decisions made in the hematopoietic system (see: Hematopoiesis) in which a plethora of chemically diverse » Hematopoietins induce distinct but partially overlapping differentiation responses. Similarly, » ARIA (acetylcholine receptor inducing activity)/Heregulin and its receptor, the » *neu* oncogene, have been implicated in Schwann cell functions and adenocarcinoma pathology and may thus be components of regulatory circuits linking neuroimmune processes with tumor physiology.

One of the interferons, » IFN-β, has been found to

Action of cytokines on cells of the nervous system.

Cytokine	Source	Activity
Activin A	monocytes brain cells bone marrow cells	neuronal survival factor inhibition of neuronal differentiation suppression of ACTH release inhibition of growth hormone release
aFGF	brain cells	promotion of glial precursor proliferation stimulation of neuronal proliferation and survival inhibition of astrocyte/oligodendrocyte precursor differentiation neuronal survival factor
AIM	serum	astrocyte progenitor differentiation
α2 macroglobulin	astroglial cells	neuronal differentiation
ANF	CNS endothelial cells	antiproliferative effects on astrocytes
AST-CF	astrocytes	cytotoxic activity for oligodendrocytes
Astrostatin	neurons	inhibition of type I astrocyte growth
bFGF	brain cells	promotion of glial precursor proliferation stimulation of neuronal proliferation and survival inhibition of astrocyte/oligodendrocyte precursor differentiation neuronal survival factor
BDNF	neurons Schwann cells	neuronal survival factor potentiation of synaptic activity
Bombesin	neurons	induction and secretion of neuropeptides and hormones
Bradykinin		induction of pain
CNTF	astrocytes in CNS Schwann cells in PNS	neuronal survival factor promotion of astrocyte/oligodendrocyte precursor differentiation induction of somatostatin, VIP, substance P production induction of fever
EGF	various cell types	modulation of function of GABAergic and dopaminergic neurons neuronal survival factor stimulation of proliferation of some neuronal precursor cells stimulation of glial cell growth modulation of higher functions (food intake reduction)
Endothelins	astrocytes neurons mast cells	depolarization of spinal neurons induction of vasopressin and oxytocin release glial cell functions following injury astrocyte proliferation
FGF-9	glial cells	promotion of glial cell proliferation activation of oligodendrocyte/astrocyte precursors
GDF-1	glial cells	promotion of astrocyte proliferation growth and differentiation factor
GDNF	glial cells	growth and survival factor for dopaminergic neurons
GGIF	brain cells	glial cell growth inhibitor
GGPF	T cells	promotion of astrocyte/oligodendrocyte proliferation promotion of remyelinization in response to injuries
GIF	astrocytes of gray matter	inhibition of survival and neurite outgrowth in cortical neurons
GMF-β	brain cells	promotion of neuronal and glial cell differentiation promotion of astroblast proliferation inhibition of growth of astroblast-derived tumors suppression of mitogenic activity of aFGF
GPA	sciatic neurons	promotion of differentiation of ciliary ganglion neurons
GPF	astrocytes	promotion of glial cell proliferation
GRP	neurons fibroblasts	induction of growth hormone and prolactin secretion thermoregulation circadian rhythms
GSF	T cells	promotion of glial cell growth modulation of astrocyte progenitor cell activity
GSF	glial cells	suppression of lymphocyte functions
HBNF	meningothelial cells	promotion of neuronal differentiation
IGF	astrocytes various cell types	promotion of astrocyte/microglia proliferation promotion of myelinization promotion of differentiation of oligodendrocytes and neuronal cells neuronal survival factor
IL1	astrocytes T cells B cells NK cells	promotion of astrocyte proliferation induction of release of IL3, GM-CSF, IL6, TNF by astrocytes activation of hypothalamic-pituitary-adrenal gland axis thermoregulation induction of NGF release by non-neuronal cells

Cytokine	Source	Activity
IL2	T cells	cytotoxic factor for oligodendrocytes modulation of electrobehaviour of neurons destruction of blood brain barrier functions
IL6	various cell types	cholinergic neuron survival factor neuronal differentiation factor stimulation of ACTH release induction of NGF release by astrocytes
LIF		neuronal survival factor differentiation of sensory neurons induction of cholinergic differentiation induction of somatostatin, VIP, substance P production
Midkine		activities resembling those of HBNF
NAP	Schwann cells	inhibition of Schwann cell proliferation
NEF		differentiation of neurons
Neuregulins	glial cells Schwann cells	glial growth factors formation and maintenance of chemical synapses (?) acetylcholine receptor expression myelin synthesis
Neuroleukin	T cells glial cells	differentiation factor for some NGF-unresponsive sensory neurons
Nexin I	fibroblasts glial cells	neuronal differentiation factor promotion of axonal regeneration
NGF	brain various tissues/cells	survival and functional differentiation of neurons
NGIF	glial cells	neuronal differentiation factor
NPF	muscle tissue	neuronal differentiation and survival factor
NSE	neurons neuroendocrine cells	neuronal survival factor
NT-3	neurons muscle cells	neuronal survival factor
OGF	neurons	stimulation of astroglial cells and oligodendrocytes promotion of myelinization
OP-1	kidney cells nervous system	induction of expression of neuronal adhesion molecules
PD-ECGF	platelets brain cells	stimulation of astroblast growth inhibition of glioma cell growth
PDGF	glial cells platelets other cell types	proliferation factor for oligodendrocytes/type I astrocytes inhibition of glial precursor cell growth promotion of nerve regeneration
PEDF	retinal epithelium	neuronal differentiation factor
S100	astroglia neurons Schwann cells	neuronal differentiation factor promotion of glial cell proliferation modulation of glial cell functions
SDGF	Schwann cells	growth factor for Schwann cells and astrocytes neuronal survival factor
TGF-α	various cell types	neuronal cell growth inhibitor modulation of serotonin metabolism
TGF-β	various cell types	neuronal cell growth inhibitor inhibition of neuronal survival suppression of microglia activities promotion of Schwann cell proliferation promotion of neurogenesis in olfactory epithelium
Thrombospondin	platelets	promotion of axonal growth
TNF-α	astrocytes microglia various cell types	cytotoxic for oligodendrocytes induction of proliferation of astroglia and microglia
VEGF		vascularization of glial tumors
VIP	hypothalamus pituitary	astroglial mitogen induction of neuron survival factor activities

be an unconditioned stimulus signal responsible for the bidirectional communication which links the central nervous system with the immune system (augmentation of natural killer cell activity).

The brain and other neuronal tissues have been shown to be the most abundant sources of » bFGF. bFGF has important functions in the development and maintenance of the nervous system, stimulating glial cells to proliferate and having a variety of trophic effects on neurons. Its synthesis is upregulated, among other things, by » IL1, » TNF-α, and » TGF-β. In the pituitary bFGF regulates the secretion of » thyrotropin and » prolactin.

Neuroimmunoactive factors thus are components of a complex circuitry of positive and negative signals and (auto)regulatory feedback and feedforward loops controlling neuroimmune functions and maintaining homeostasis by allowing crosstalk between the two systems. The complexity of the neuroimmune network is illustrated by the number and variety of hormonal and neural actions of individual modulatory factors and the observations that endocrine target cells for these factors themselves are able to produce them. These mediators can act coordinately on the same cells or sequentially on different cells, they can induce or repress the synthesis of cytokines and/or their receptors or receptor subunits, and generally alter patterns of » Gene expression following interaction with their receptors.

It has been suggested that the immune system possesses sensory functions with leukocytes identifying stimuli/stressor signals that are not recognizable by the central and peripheral nervous system (see also: Acute phase reaction). Cytokine receptors on cells of the immune and nervous system seem to play a sensory and regulatory role enabling the brain to monitor the progress of immune responses. The brain may be able also to modulate immune responses, for example, by using its neuroimmunomodulatory factors to alter the functional capacities of immune cells. Some of these factors may also be able to induce subtle shifts in immune cell populations that secrete immunomodulating cytokines (see also: TH1/TH2 cytokines). Many of these observations now help to clarify phenomena that have long been described but have had no physiological explanation. The complexity of the system suggests that there will be no single unifying role of one individual factor in nervous and immune system interactions. Eventually, however, elucidation of the mechanisms underlying communication systems within the body will help to integrate such diverse and seemingly incompatible fields as psychology/psychopathology, Ayurveda (a comprehensive Indian system of medicine based on Hindu philosophy), and physiology.

Berczi I (edt) Pituitary function and immunity. CRC Press, Boca Raton 1986; **Blalock JE** The immune system as a sensory organ. JI 132: 1067-70 (1984); **Blalock JE et al** A complete regulatory loop between the immune and neuroendocrine systems. Fed. Proc. 44: 108-11 (1985); **Blalock JE et al** The pituitary-adrenocortical axis and the immune system. Clin. Endocrinol. Metab. 14: 1021-38 (1985); **Blalock, JE** (edt) Neuroimmunoendocrinology, 2nd edition; VOl 43 of Chemical Immunology, Karger, Basel, 1992; **Carr DJJ** The role of endogenous opioids and their receptors in the immune system. PSEBM 198: 710-20 (1991); **Cooper JR et al** (eds) The biochemical basis of neuropharmacology. 6th edition. Oxford University Press 1991; **Dube KC et al** Personality types in Ayurveda. Am. J. Chinese Medicine 11: 25-34 (1983); **Dunn AJ** Nervous system-immune system interactions: an overview. J. Recept. Res. 8: 589-607 (1988); **Gala RR** Prolactin and growth hormone in the regulation of the immune system. PSEBM 198: 513-27 (1991); **Goetzl EJ & Spector NH** (eds) Neuroimmune networks: physiology and diseases. AR Liss, New York 1989; **Goetzl EJ & Sreedharan SP** Mediators of communication and adaptation in the neuroendocrine and immune system. FJ 6: 2646-52 (1992); **Hall NR et al** Thymic regulation of the hypothalamic-pituitary-gonadal axis. Int. J. Immunopharmacol. 14: 353-9 (1992); **Hermus ARMM & Sweep CGJ** Cytokines and the hypothalamic-pituitary-adrenal axis. JSBMB 37: 867-71 (1990); **Hiramoto R et al** Identification of specific pathways of communication between the CNS and NK cell system. Life Sci. 53: 527-40 (1993); **Johnson HM et al** Neuroendocrine peptide hormone regulation of immunity. Chem. Immunol. 52: 49-83 (1992); **Jonakait GM** Neural-immune interactions in sympathetic ganglia. TINS 16: 419-23 (1993); **Khansari DN et al** Effects of stress on the immune system. IT 11: 170-75 (1990); **Martignoni E et al** The brain as a target for adrenocortical steroids: cognitive implications. Psychoneuroendocrinology 17: 343-54 (1992); **Nistico G** Communications among central nervous system, neuroendocrine and immune systems: Interleukin 2. Prog. Neurobiol. 40: 463-75 (1993); **Payne LC et al** Hypothalamic releasing hormones mediating the effects of interleukin 1 on sleep. JCBc 53: 309-13 (1993); Patterson PH & Nawa H Neuronal differentiation factors/cytokines and synaptic plasticity. Cell 72(Suppl.): 123-37 (1993); **Pepin MC et al** Impaired type II glucocorticoid-receptor function in mice bearing antisense RNA transgene. N 355: 725-8 (1992); **Roszman TL & Brooks WH** Signaling pathways of the neuroendocrine-immune network. Chem. Immunol. 52: 170-90 (1992); **Singh VK** Immunoregulatory role of neuropeptides. Prog. Drug. Res. 38: 149-69 (1992); **Smith EM & Blalock JE** A molecular basis for interactions between the immune and neuroendocrine systems. Int. J. Neurosci. 38: 455-64 (1988); **Solvason HB et al** The identity of the unconditioned stimulus to the central nervous system is interferon-β. J. Neuroimmunol. 45: 75-81 (1993); **Swanson LW & Sawchenko PE** Hypothalamic integration: organization of the paraventricular and supraoptic nuclei. Annu. Rev. Neurosci. 6: 269-424 (1983)

Neurokines: see: Neuropoietins.

Neurokinin α: This factor is identical with neurokinin A (substance K, neuromedin L). see: Tachykinins.

Neurokinin 1: abbrev. NK1. This factor is identical with » SP (substance P) see: Tachykinins.

Neurokinin B: This factor is identical with neurokinin β and neuromedin K. see: Tachykinins.

Neurokinin β: This factor is identical with neurokininin B and neuromedin K. see: Tachykinins.

Neuroleukin: see: NLK.

Neuromedins: see: Tachykinins.

Neuromedin K: This factor is identical with NKB (Neurokinin B = Neurokinin β). see: Tachykinins.

Neuromedin L: This factor is identical with neurokinin A (substance K, neurokinin α). see: Tachykinins.

Neuronal differentiation factors: see: Neurotrophins.

Neuronotrophins: see: Neurotrophins.

Neuron-specific enolase: see: NSE.

Neuropeptide K: abbrev. NPK. see: Tachykinins.

Neuropeptide γ: abbrev. NPγ: see: Tachykinins.

Neuropoietins: A generic name for factors also known as ***neuropoietic cytokines*** or ***neurokines*** that regulate cell numbers in the nervous system and influence functional activities of neurons, e. g. by determining the type of neurotransmitter used by certain neurons. Neuropoietins thus are important mediators within the » Neuroimmune network. Factors with this activity include » CNTF (ciliary neuronotrophic factor), » CDF (cholinergic differentiation factor = LIF (leukemia inhibitory factor), and » Oncostatin M. These factors share some common tertiary structures and also address receptors sharing a common subunit.
The term neuropoietins is used rather loosely and may also refer to » Neurotrophins and factors produced in, or acting upon, the nervous system such as » bFGF, » interleukins, or » TGF-β.

Bazan JF Neuropoietic cytokines in the hematopoietic fold. Neuron 7: 197-208 (1991); **Fann MJ & Patterson PH** Neuropoietic cytokines and activin A differentially regulate the phenotype of cultured sympathetic neurons. PNAS 91: 43-7 (1994); **Hall AK & Rao MS** Cytokines and neurokines: related ligands and related receptors. TINS 15: 35-7 (1992); **Manthorpe M et al** An automated colorimetric microassay for neuronotrophic factors. Dev. Brain Res. 25: 191-8 (1986); **Meakin SO & Shooter EM** The nerve growth factor family of receptors. TINS 15: 323-31 (1992); **Patterson PH** The emerging neuropoietic cytokine family: First CDF/LIF, CNTF and IL6; Next ONC, MGF, GCSF? Curr. Opin. Neurobiol. 2: 94-7 (1992); **Rao MS & Landis SC** Cell interactions that determine sympathetic neuron transmitter phenotype and the neurokines that mediate them. J. Neurobiol. 24: 215-32 (1993); **Schinstine M et al** Intracerebral delivery of growth factors: potential application of genetically modified fibroblasts. PGFR 3: 57-66 (1991)

Neuropoietic cytokines: see: Neuropoietins.

Neurotrophic factors: see: Neurotrophins.

Neurotrophins: also: Neuronotrophins. This is a generic term to denote a number of neurotrophic factors enhancing neuronal differentiation and promoting the survival of neurons that are normally destined to die during different phases of the development of the nervous system (see also: neuropoietins). Most of these factors are active at extremely low concentrations (NGF: 10^{-12} M).
Neurotrophic factors promote neuronal survival both *in vivo* and *in vitro* and also support the general growth of neurons. Known neurotrophins include » BDNF (brain-derived neurotrophic factor), » NGF, » NT-3 (neurotrophin 3), » NT-4, » NT-5, » CNTF (ciliary neuronotrophic factor), and » Purpurin. Their biological activities are mediated by receptors encoded by the » *trk* gene. It has been possible, by combination of structural elements from » NGF, » BDNF and » NT-3, to engineer a multifunctional pan-neurotrophin that efficiently activates all » *trk* receptors and displays multiple neurotrophic specificities. Another neuronal survival factor is » NSE (neuron-specific enolase).
BDNF, NGF and NT-3 are sometimes also referred to collectively as the ***NGF family*** of proteins. Other factors with neurotrophic activities normally not classified as neurotrophins and often possessing a broader spectrum of functions are » » EGF, » HBNF (heparin-binding neurite-promoting factor), IGF-2 (see: IGF), » aFGF and » bFGF, » PDGF, » NSE (neuron-specific enolase), and » Activin A. For antiproliferative growth factors affecting neural cells see also: Neural antiproliferative protein, Astrostatine, GGIF (glial growth inhibitory factor).
Neurite-promoting factors (abbr. NPF) do not promote neuronal survival or general growth by themselves but are required in addition to a neurotrophic factor to induce outgrowth of axonal or

Neurotrophins: see addendum.

dendritic processes. Factors with NPF activity include » NGF, » S100, » GMF-β (glia maturation factor), proteoglycans (see also: ECM, extracellular matrix), merosin, fibronectin, collagens, cell adhesion molecules, and laminin.

Neuronal differentiation factors influence transmitter phenotypes without affecting neuronal survival. *In vitro* assays using recombinant neurotrophic factors and distributions of their mRNAs and proteins have indicated that members of a neurotrophic gene family may play sequential and complementary roles during development and in the adult nervous system.

Up to 80% of the neurons in diverse cell populations within the forming nervous system die during normal vertebral development. This is thought to ensure that adequate numbers of neurons establish appropriate innervation densities with effector organs or other neuronal populations. The neurotrophic theory of nervous system development postulates that limited amounts of neurotrophic factors are available during the ontogenesis of the vertebrate nervous system. Early nerve cell progenitors compete with each other for these factors that are produced in the developing target organs. Only those neurons that encounter these factors in a critical phase of their development will survive. The target organs therefore can regulate the density of innervation with the help of these factors, thus providing an effective means of matching neuron and target cell populations. The availability of limited amounts of survival factors and the competition for these factors regulates the number of cells in the nervous system and also in other organs (see also: Apoptosis). The upregulated expression of genes such as » *bcl*-2 has been shown to rescue factor-dependent neurons from apoptotic cell death.

These factors are of potential clinical interest since they influence the functional activities and survival of distinct neuronal populations within the peripheral and central nervous system. It has also been suggested that, apart from local injection, grafting of genetically modified factor-producing fibroblasts within the nervous system could be used as a delivery method for these factors.

Bailey K et al Neurotrophins: A group of structurally related proteins supporting the survival of neurons. Life Sciences Res. Rep. 51: 145-56 (1991); **Barde YA** What, if anything, is a neurotrophic factor? TINS 11: 343-6 (1988); **Barde YA** The nerve growth factor family. PGFR 2: 237-48 (1990); **Bothwell M** Keeping track of neurotrophin receptors. Cell 65: 915-8 (1991); **Early CJ & Higgins GA** Neurotrophin receptor gene expression. Methods Neurosci. 9: 166-78 (1992); **Ebendal T** Function and evolution in the NGF family and its receptors. J. Neurosci. Res. 32: 461-70 (1992); **Fallon J & Loughlin S** (eds) Neurotrophic factors. Academic Press, San Diego 1993); **Glass DJ & Yancopoulos GD** The neurotrophins and their receptors. Trends Cell Biol. 3: 262-8 (1993); **Ibanez CF et al** An extended surface of binding to *Trk* tyrosine kinase receptors in NGF and BDNF allows the engineering of a multifunctional pan-neurotrophin. EJ 12: 2281-93 (1993); **Johnson EMR jr. & Deckwerth TL** Molecular mechanism of developmental neuronal death. ARN 16: 31-46 (1993); **Korsching S** The neurotrophic factor concept: a reexamination. J. Neurosci. 13: 2739-48 (1993); **Manthorpe M et al** Detection and analysis of growth factors affecting neural cells. Neuromethods 23: 87-137, Boulton A et al (edt) The Humana Press Inc. (1992); **Meakin SO & Shooter EM** The nerve growth factor family of receptors. TINS 15: 323-31 (1992); **Mudge AW** Motor neurons find their factors. N 363: 213-4 (1992); **Purves D & Lichtman JW** Principles of neural development. Sunderland MA Sinauer Associates, pp. 131-49 (1985); **Richardson PM** Neurotrophic factors in regeneration. Curr.

Survival-promoting or phenotype-altering actions of some neurotrophic factors on overlapping and distinct populations of peripheral and central neurons.

Type of neuron	NGF	BDNF	NT3	NT4/5	CNTF
Cerebellar	–	+	++	+	+
Dorsal root ganglion	++	++	++	+	+
Hippocampal	–	+	+	+	+
Motor	–	++	+	+	++
Nigral (dopaminergic)	–	++	++	++	–
Nodose sensory	–	+	++	–	++
Parasympathetic	–	–	–	–	++
Retinal ganglion	–	+	+	?	++
Serptal (cholinergic)	+	+	+	+	+
Striatal (Cholinergic/other)	+	+	+	+	–
Sympathetic	++	±	+	+	+

Arbitrary scores: high activity: ++; low activity: +; no activity: –
Modified from Ann. Rev. Medicinal Chem. 28: 11-17 (1993)

Opin. Neurobiol. 1: 401-6 (1991); **Rodriguez-Tebar A et al** Neurotrophins: structural relatedness and receptor interactions. Phil. Trans. R. Soc. Lond (Biol. 331: 255-8 (1991); **Schinstine M et al** Intracerebral delivery of growth factors: potential application of genetically modified fibroblasts. PGFR 3: 57-66 (1991); **Snider WD & Johnson EM** Neurotrophic molecules. Ann. Neurol. 26: 489-506 (1989); **Strand FL et al** Neuropeptide hormones as neurotrophic factors. Physiol. Rev. 71: 1017-46 (1991); **Thoenen H** The changing scene of neurotrophic factors. TINS 14: 165-70 (1991); **Vantini G** The pharmacological potential of neurotrophins: a perspective. Psychoneuroendocrinology 17: 401-10 (1992); **Vantini G & Skaper SD** Neurotrophic factors: from physiology to pharmacology? Pharmacol. Res. 26: 1-15 (1992); **Yamamori T et al** Molecular mechanisms for generation of neural diversity and specificity: roles of polypeptide factors in development of postmitotic neurons. Neurosci. Res. 12: 545-82 (1992); **Yancopoulos GD et al** Neurotrophic factors, their receptors, and the signal transduction pathways they activate. CSHSQB 55: 371-9 (1990)

Neurotrophin-3 to **neurotrophin-5:** see: NT-3, NT-4, and NT-5, respectively.

Neutrogin: Brand name for glycosylated recombinant » G-CSF (Chugai) marketed under the generic name Lenograstim.

Neutrophil-activating factor: see: NAF.

Neutrophil-activating peptide 1 to **Neutrophil-activating peptide 4:** see: NAP-1, NAP-2, NAP-3, and NAP-4, respectively.

Neutrophil alkaline phosphatase-inducing factor: see: NAP-IF.

Neutrophil antigen NA: see: CD16.

Neutrophil chemotactic factor: see: NCF.

Neutrophil colony-stimulating activity: see: G-CSA.

Neutrophil-derived colony inhibitory factor: see: CIF.

Neutrophil-derived histamine-releasing activity: abbrev. HRA-N. See: HRF (histamine-releasing factors).

Neutrophil-derived lymphocyte chemotactic factor: see: NDLCF.

Neutrophil-derived monocyte chemotactic factor: see: NDCF.

Neutrophil-granulocyte colony stimulating factor: This factor which stimulates exclusively the formation of neutrophil granulocyte colonies is identical with » IL3.

Moore MA G-CSF: its relationship to leukemia differentiation-inducing activity and other hemopoietic regulators. JCP 1: s53-64 (1982)

Neutrophil migration inhibition factor from T lymphocytes: see: NIF-T.

Neutrophil (migration) inhibitory factor: abbrev. NIF. see: NIF-T.

Neutrophil recruitment inhibitory factor: see: NRIF.

Neutrophil releasing activity: see: NRA.

New Zealand black mice serum factor: see: NZB-SF.

Nexin I: (protease nexin I; abbrev. PN-1) PN-1 is a heparin binding glycoprotein (see also: HBGF) of 44 kDa released by human foreskin fibroblasts and cultured glial cells. The protein is also present on the surface of fibroblasts, bound to the extracellular matrix (see: ECM). PN-1 represents about 1% of their secreted protein. It is a broad specificity inhibitor of trypsin-like serine proteases, preferentially binding to thrombin, urokinase, trypsin, and plasmid. Heparin accelerates the rate of thrombin inhibition. PN-1 slowly inhibits Factor Xa and the γ subunit of » NGF but does not inhibit chymotrypsin-like proteases or leukocyte elastase. PN-1 forms covalent complexes with its target proteases, subsequently binding to cells via specific receptors. It forms sodium dodecyl sulfate-stable complexes with a protein named EGF-BP (epidermal growth factor binding protein).

Two forms of PN-1, designated α and β, are expressed in mammalian cells. They differ by the insertion of 3 nucleotides into the coding sequence, and apparently arise by utilization of an alternative splice acceptor site in the PN-1 gene. α-PN-1 contains an arginine residue at position 310, whereas β- PN-1 contains threonine-lysine residues. The amino acid sequence of β- PN-1 is identical with that of ***glial-derived neurite promoting factor*** (abbrev. ***GdNPF***), also known as ***glia-derived nexin*** (abbrev. ***GDN***) which promote neurite extension in cultures of neuroblastoma cells, chick sympathetic neurons, and hippocampal neurons. The neurite outgrowth activity of PN-1 depends on inhibition of thrombin, presumably because thrombin brings about neurite retraction.

PN-1 is the human homologue of ***NPF*** (neurite promoting factor) which is produced by rat C6 glioma cells in culture and has neurite promoting activity on a neuroblastoma cell line.

After selective delayed neuronal death glia-derived nexin is produced by astrocytes in the vicinity of degenerating and preserved neuronal structures and may play a role in structural rearrangements of the central nervous system. PN-1 probably also plays a protective role against extravasated serine proteases.

PN-1 is synthesized by muscle and localized at neuromuscular synapses. It inhibits the destruction of myotube matrix by inactivating the plasmin/plasminogen activation system and may be the physiologic regulator of this system during muscle development *in vivo*. *In vivo*, PN-1 is constitutively expressed in all parts of the olfactory system, where axonal regeneration and neurogenesis occur continuously throughout life. The factor may therefore be important for axonal regeneration *in vivo* and could have a role in Wallerian degeneration and peripheral nerve regeneration.

PN-1 may also function as a major extravascular inhibitor of T cell serine proteinase-1 (TSP-1) released from cytotoxic T lymphocytes, which most probably is involved in cell killing by cytotoxic T cells. A form of PN-1 is expressed on the platelet surface during platelet activation.

Choi BH et al Protease nexin-1. Localization in the human brain suggests a protective role against extravasated serine proteases. Am. J. Pathol. 137: 741-7 (1990); **Cunningham DD & Gurwitz D** Proteolytic regulation of neurite outgrowth from neuroblastoma cells by thrombin and protease nexin-1. JCBc 39: 55-64 (1989); **Farmer L et al** Glia-derived nexin potentiates neurite extension in hippocampal pyramidal cells *in vitro*. Dev. Neurosci. 12: 73-80 (1990); **Festoff BW et al** Plasminogen activators and inhibitors in the neuromuscular system: III. The serpin protease nexin I is synthesized by muscle and localized at neuromuscular synapses. JCP 147: 76-86 (1991); **Gloor S et al** A glia-derived neurite promoting factor with protease inhibitory activity belongs to the protease nexins. Cell 47: 687-93 (1986); **Gronke RS et al** A form of protease nexin I is expressed on the platelet surface during platelet activation. Blood 73: 472-8 (1989); **Gurwitz D & Cunningham DD** Thrombin modulates and reverses neuroblastoma neurite outgrowth. PNAS 85: 3440-4 (1988); **Gurwitz D & Cunningham DD** Neurite outgrowth activity of protease nexin-1 on neuroblastoma cells requires thrombin inhibition. JCP 142: 155-62 (1990); **Gurwitz D et al** Protease nexin-1 complexes and inhibits T cell serine proteinase-1. BBRC161: 300-4 (1989); **Knauer DJ et al** The glioma cell-derived neurite promoting activity protein is functionally and immunologically related to human protease nexin-I. JCP 132: 318-24 (1987); **McGrogan M et al** Molecular cloning and expression of two forms of human protease nexin I. BT 6: 172-7 (1988); **Meier R et al** Induction of glia-derived nexin after lesion of a peripheral nerve. N 342: 548-50 (1989); **Rao JS et al** Protease nexin I, a serpin, inhibits plasminogen-dependent degradation of muscle extracellular matrix. Muscle Nerve. 12: 640-6 (1989); **Rosenblatt DE et al** Identification of a protease inhibitor produced by astrocytes that is structurally and functionally homologous to human protease nexin-I. Brain Res. 415: 40-8 (1987); **Rovelli G et al** Characterization of the heparin-binding site of glia-derived nexin/protease nexin-1. B 31: 3542-9 (1992); **Scott RW et al** Protease nexin. Properties and a modified purification procedure. JBC 260: 7029-34 (1985); **Seilheimer B & Schachner M** Regulation of neural cell adhesion molecule expression on cultured mouse Schwann cells by nerve growth factor. EJ 6: 1611-6 (1987); **Sommer J et al** cDNA sequence coding for a rat glia-derived nexin and its homology to members of the serpin family. B 26: 6407-10 (1987); **Sommer J et al** Synthesis of glia-derived nexin in yeast. Gene 85: 453-9 (1989); **Stone SR et al** Glial-derived neurite-promoting factor is a slow-binding inhibitor of trypsin, thrombin, and urokinase. ABB 252: 237-44 (1987); **Van Nostrand WE; & Cunningham DD** Purification of protease nexin II from human fibroblasts. JBC 262: 8508-14 (1987); **Zum AD et al** A glia-derived nexin promotes neurite outgrowth in cultured chick sympathetic neurons. Dev. Neurosci. 1988; 10: 17-24 (1988)

NF: *nuclear factor* A general term for proteins localized in the nuclei of cells. Many of these proteins play a role in » gene expression as DNA binding proteins specifically recognizing distinct sequences in the promoter regions of genes (see also: figures showing the promoter structures of » G-CSF, » IL2, » IL6, » TNF-β genes). See also: NF-IL6, IRS (interferon response sequence), NF-AT, CLE (conserved lymphokine element).

NF-AT: *nuclear factor for activated T cells* This transcription factor (see also: Gene expression) is specifically expressed in activated T cells (see also: Cell activation). It binds to a specific DNA region in the promoter of the » IL2 gene (positions –285 - –254; see also: graphic showing the promoter structure of IL2). NF-AT is a multisubunit transcription factor consisting of at least three DNA-binding polypeptides: a pre-existing phosphoprotein subunit called NFATp and homodimers or heterodimers of » *fos* and » *jun* family proteins (see also: fra-1, *fos*-related antigen 1). NF-AT complexes regulate transcription of the IL2 gene. NF-AT is thought to be largely responsible for the T cell-specific inducibility of IL2 expression. NFATp has been implicated also in the transcriptional regulation of other cytokine genes.

The expression of NF-AT is inhibited by » CsA (Ciclosporin). Binding of NF-AT to its recognition site within the IL2 promoter is itself positively regulated by a modulatory protein of 17 kDa which is found in non-stimulated T cells and many other cell types.

An abnormality in the expression of NF-AT has been found to be involved in the multiple lymphokine deficiency of a patient with severe combined immunodeficiency. This patient had a normal number of circulating T cells but the T lymphocytes responded poorly to mitogens. The defect could be corrected *in vitro* and *in vivo* by recombinant IL2. The patient's T lymphocytes showed decreased levels in the expression of » IL2, » IL3, » IL4, and » IL5 mRNAs. The expression by T lymphocytes of other cytokines such as » GM-CSF and » IL6, both of which are not T lymphocyte-restricted, was not affected.

Brunvand MW et al Nuclear factors interacting with the mitogen-responsive regulatory region of the interleukin 2 gene. JBC 263: 18904-10 (1988); **Castigli E et al** Molecular basis of a multiple lymphokine deficiency in a patient with severe combined immunodeficiency. PNAS 90: 4728-32 (1993); **Chatila T et al** Primary combined immunodeficiency resulting from defective transcription of multiple T cell lymphokine genes. PNAS 87: 10033-10037 (1990); **Granelli-Piperno A & McHugh P** Characterization of a protein that regulates DNA-binding activity of NF-AT, the nuclear factor of activated T cells. PNAS 88: 11431-4 (1992); **Karttunen J & Shastri N** Measurement of ligand-induced activation in single viable T cells using the lacZ reporter gene. PNAS 88: 3972-6 (1991); **McCaffrey PG et al** A T cell nuclear factor resembling NF-AT binds to an NF-kappaB site and to the conserved lymphokine promoter sequence "cytokine-1". JBC 267: 1864-71 (1992); **McCaffrey PG et al** Isolation of the cyclosporin-sensitive T cell transcription factor NFATp. Science 262: 750-754 (1993); **Pahwa R et al** Recombinant interleukin 2 therapy in severe combined immunodeficiency disease. PNAS 86: 5069-5073 (1989); **Paliogianni F et al** Negative transcriptional regulation of human interleukin 2 (IL2) gene by glucocorticoids through interference with nuclear transcription factors AP-1 and NF-AT. JCI 91: 1481-9 (1993); **Verweij CL et al** Cell type specificity and activation requirements for NFAT-1 (nuclear factor of activated T cells) transcriptional activity determined by a new method using transgenic mice to assay transcriptional activity of an individual nuclear factor. JBC 265: 15788-95 (1990); **Yaseen NR et al** Comparative analysis of NFAT (nuclear factor of activated T cells) complex in human T and B lymphocytes. JBC 268: 14285-93 (1993)

NF-E1: A transcription factor that is identical with » GATA-1.

Safrany G & Perry RP Characterization of the mouse gene that encodes the δ/YY1/NF-E1/UCRBP transcription factor. PNAS 90: 5559-63 (1993); **Yamamoto M et al** Activity and tissue-specific expression of the transcription factor NF-E1 multigene family. Genes Dev. 4: 1650-62 (1990)

NF-IL6: *nuclear factor for IL6* NFIL6 is a transcription factor of 32 kDa (see also: Gene expression), also known as CCAAT/enhancer binding protein β (C/EBP β) that is identical with ***TCF5*** (transcription factor 5), ***LAP*** (liver activator protein), ***SF-B*** (silencer factor B) and ***IL6-DBP*** (IL6-dependent DNA binding protein). It has been proposed that NF-IL6 be called C/EBPβ (CCAAT/enhancer binding protein). NF-IL6 is a transcription factor belonging to a family of DNA-binding proteins predominantly characterized by a leucine zipper motif that mediates dimerization and a basic DNA binding domain. The human NF-IL6 gene maps to chromosome 20q13.

The expression of NF-IL6 is induced by » IL6. The NF-IL6 binding site, ATTAGGACAT, is found in the promoters of various cytokine genes and genes the expression of which is regulated by cytokines e. g. in some genes encoding » acute phase proteins (including the genes encoding » haptoglobin and hemopexin)and in genes such as GST-P (glutathione transferase P) the expression of which is strongly induced during the development of rat liver tumors.

Akira S et al A nuclear factor for IL6 expression (NF-IL6) is a member of a C/EBP family. EJ 9: 1897-906 (1990); **Akira S et al** A nuclear factor for the IL6 gene (NF-IL6). Chem. Immunol. 51: 299-322 (1992); **Cao Z et al** Regulated expression of three C/EBP isoforms during adipose conversion of 3T3-L1 cells. Genes Dev. 5: 1538-1552 (1991); **Chang CJ et al** Molecular cloning of a transcription factor, AGP/EBP, that belongs to members of the C/EBP family. MCB 10: 6642-53 (1990); **Descombes P et al** LAP, a novel member of the C/EBP gene family, encodes a liver-enriched transcriptional activator protein. GD 3: 1541-51 (1990); **Hendricks-Taylor LR et al** The CCAAT/enhancer binding protein (C/EBP α) gene (CEBPA) maps to human chromosome 19q13.1 and the related nuclear factor NF-IL6 (C/EBP β) gene (CEBPB) maps to human chromosome 20q13.1. Genomics 14: 12-7 (1992); **Imagawa M et al** SF-B binds to a negative element in glutathione transferase P gene is similar or identical to trans-activator LAP/IL6DBP. BBRC 179: 293-300 (1991); **Kinoshita S et al** A member of the C/EBP family, NF-IL6β, forms a heterodimer and transcriptionally synergizes with NF-IL6. PNAS 89: 1473-6 (1992); **Metz R & Ziff E** cAMP stimulates the C/EPB-related transcription factor rNFIL6 to *trans*-locate to the nucleus and induce c-*fos* transcription. GD 5: 1764-6 (1991); **Natsuka S et al** Augmentation of haptoglobin production in Hep3B cell line by a nuclear factor NF-IL6. FL 291: 58-62 (1991); **Natsuka S et al** Macrophage differentiation-specific expression of NF-IL6, a transcription factor for interleukin-6. Blood 79: 460-6 (1992); **Nishio Y et al** A nuclear factor for interleukin-6 expression (NF-IL6) and the glucocorticoid receptor synergistically activate transcription of the rat α 1-acid glycoprotein gene via direct protein-protein interaction. MCB 13: 1854-62 (1993); **Poli V et al** IL6DBP, a nuclear protein involved in interleukin-6 signal transduction, defines a new family of leucine zipper proteins related to C/EBP. Cell 63: 643-53 (1990); **Szpirer J et al** The interleukin 6-dependent DNA binding protein gene (transcription factor 5: TCF5) maps to human chromosome 20 and rat chromosome 3, the IL 6 receptor locus (IL6R) to human chromosome 1 and rat chromosome 2, and the rat IL6 gene to rat chromosome 4. Genomics 10: 539-46 (1991)

NFS-60: A murine myeloblastic cell line established from leukemic cells obtained after infection of (NFS X DBA/2) F1 adult mice with Cas Br-M murine leukemia virus. NFS-60 cells are used to

assay murine and human » G-CSF (see also: Bioassays) and are ≈ 10-30 times more sensitive than » 32D cells. The NFS-60 cells are totally dependent on » IL3 for growth and maintenance of viability *in vitro*. Neutralizing antibodies must therefore be used to discriminate between these factors. An alternative and entirely different detection method is » Message amplification phenotyping.
The cells also respond to murine IL3, » murine or human IL4, » human IL6, » murine GM-CSF and » Epo. Growth of NFS-60 cells in response to IL3, GM-CSF, G-CSF, » M-CSF, IL4, or IL6 is inhibited in a dose-dependent manner by » TGF-β1. The cells also secrete IL6.
The promyelocytic state of NFS-60 cells is maintained in the presence of » G-CSF. The cells differentiate into neutrophils and macrophages in the presence of IL3 and GM-CSF. They yield hemoglobinised colonies in a » colony formation assay in the presence of Epo.
NFS-60 cells have a rearrangement of the cellular » *myb* locus, involving the integration of a retrovirus into the region of the gene corresponding to the sixth exon of the avian c-*myb* locus. The cells produce *myb* protein with a truncated C-terminus and a normal N-terminus.
Several sublines of NFS-60 cells have been established. ***M-NFS-60*** grows rapidly and continuously in human or murine » M-CSF as round, nonadherent cells. The cells respond well to M-CSF and IL3, weakly to G-CSF and not at all to murine GM-CSF, IL4, or human IL1. A monoclonal antibody to human M-CSF specifically inhibits only M-NFS-60 proliferation induced by the human growth factor.

Gascan H et al Response of murine IL3-sensitive cell lines to cytokines of human and murine origin. LR 8: 79-84 (1989); **Hara K et al** Bipotential murine hemopoietic cell line (NFS-60) that is responsive to IL3, GM-CSF, G-CSF, and erythropoietin. EH 16: 256-61 (1988); **Hültner L et al** Interleukin 3 mediates interleukin 6 production in murine interleukin 3-dependent hemopoietic cells. GF 2: 43-51 (1989); **Keller JR et al** Transforming growth factor β: possible roles in the regulation of normal and leukemic hematopoietic cell growth. JCBc 39: 175-84 (1989); **Matsuda S et al** Human granulocyte colony-stimulating factor specifically binds to murine myeloblastic NFS-60 cells and activates their guanosine triphosphate binding proteins/adenylate cyclase system. Blood 74: 2343-8 (1989); **Metcalf D** Actions and interactions of G-CSF, LIF, and IL6 on normal and leukemic murine cells. Leukemia 3: 349-55 (1989); **Nakoinz I et al** Differentiation of the IL3-dependent NFS-60 cell line and adaptation to growth in macrophage colony-stimulating factor. JI 145: 860-4 (1990); **Saito K et al** Bioassays of recombinant human granulocyte colony stimulating factor – evaluations of the measurement of biological activity and correlations among *in vitro* colony-forming, NFS-60, AML-193, and *in vivo* CPA-mouse methods. Jap. Pharmacol. Therapeutics. 18 (Suppl.) 9: 31-41 (1990); **Shimane M et al** Sensitive assay *in vitro* of biological activity for recombinant human G-CSF (rG-CSF) using NFS-60 cells. Jap. Pharmacol. Therapeutics. 18 (Suppl.) 9: 23-9 (1990); **Shirafuji N et al** A new bioassay for human granulocyte colony-stimulating factor (hG-CSF) using murine myeloblastic NFS-60 cells as targets and estimation of its levels in sera from normal healthy persons and patients with infectious and hematological disorders. EH 17: 116-9 (1989); **Tohyama K et al** Detection of granulocyte colony-stimulating factor produced by a newly established human hepatoma cell line using a simple bioassay system. Jpn. J. Cancer Res. 80: 335-40 (1989); **Weinstein Y et al** Truncation of the c-*myb* gene by a retroviral integration in an interleukin 3-dependent myeloid leukemia cell line. PNAS 83: 5010-4 (1986); **Whetton AD & Dexter TM** Myeloid hematopoietic growth factors. BBA 989: 111-32 (1989)

NGF: *nerve growth factor*

■ **Sources:** NGF was initially isolated from the submaxillary glands of mice. NGF is synthesized in the hypothalamus, pituitary, thyroid gland, testes, and the epididymis. It is also produced by various cell types including vascular smooth muscle cells and fibroblasts.
The expression of NGF in specific neurons of the central nervous system (cortex, hippocampus) can be influenced positively by glutamate-mediated or negatively by GABAergic neuronal activity.
Serum, phorbol-12-myristate-13-acetate (see also: Phorbol esters), and vitamin D3 are potent inducers of NGF synthesis. Glucocorticoids inhibit the synthesis of NGF.
The synthesis of NGF in astrocytes is enhanced by » IL1, TNF-α, » PDGF and » TGF-β. These factors do not influence the synthesis of NGF in Schwann cells; NGF synthesis in Schwann cells is decreased by TGF-β.

■ **Protein Characteristics:** The 7S form of NGF initially isolated from the submaxillary gland of mice is a complex of three proteins (α, β, and γ) stabilized by zinc ions. The 26 kDa β subunit is a homodimer of two disulfide-bonded proteins with a length of 118 amino acids and displays the biological activity of NGF.
NGF is synthesized as a preproprotein of 305 amino acids including a secretory signals sequence of 18 amino acids and a prosequence of 103 amino acids. The mature factor is obtained by proteolytic processing at the N- and C-terminus of the precursor protein. NGF sequences are evolutionary conserved and display little sequence deviations in various species.
The α and γ subunits of the NGF complex are members of the » kallikrein family of proteins that are structurally related to trypsin-like serine proteases. The γ subunit of NGF is related to an EGF-binding protein (EGFBP; see kallikreins). It is

```
                   10        20        30        40        50        60
                    |         |         |         |         |         |
mouse       RSSTHPVFHMGEFSVCDSVSVWVGDKTTATDIKGKEVTVLAEVNINNSVFRQYFFETKCR
human       RSSSHPIFHRGEFSVCDSVSVWVGDKTTATDIKGKEVMVLGEVNINNSVFKQYFFETKCR
chicken     R-TAHPVLHRGEFSVCDSVSMWVGDKTTATDIKGKEVTVLGEVNINNNVFKQYFFETKCR
bovine      RSSSHPVFHRGEFSVCDSISVWVGDKTTATDIKGKEVMVLGEVNINNSVFKQYFFETKCR
guinea pig  RSSTHPVFHMGEFSVCDSVSVWVADKTTATDIKGKEVTVLAEVNANNNVFKQYFFETKCR

                   70        80        90       100       110       120
                    |         |         |         |         |         |
mouse       ASNPVESG RGIDSKHWNSYCTTTHTFVKALTTDEKQAAWRFIRIDTACVCVLSRKATRRG
human       DPNPVDSGCRGIDSKHWNSYCTTTHTFVKALTMDGKQAAWRFIRIDTACVCVLSRKAVRRA
chicken     DPRPVSSGCRGIDAKHWNSYCTTTHTFVKALTMEGKQAAWRFIRIDTACVCVLSRKSGR-P
bovine      DPNPVDSGCRGIDAKHWNSYCTTTHTFVKALTMDGKQAAWRFIRIDTACVCVLSRKZGQRA
guinea pig  DPSPVESGCRGIDSKHWNSYCTTTHTFVKALTTDNKQAAWRFIRIDTACVCVLNRKAARRG
```

Comparison of NGF sequences from different species.
This sequence comparison shows the strong evolutionary conservation of NGF sequences among species. Identical amino acid positions are shown in blue. Conserved cysteine residues are shown in yellow.

probably involved in the processing of the β subunit of NGF while the functions of the NGF α subunit are still unclear.

■ **Gene Structure:** The murine gene encoding β NGF has a length of at least 45 kb and contains four exons. The 3´ end of the fourth exons contains the coding sequence for β NGF. The mouse genes for α and γ NGF, located on mouse chromosome 7, are contiguous, transcribed from the same DNA strand, and separated by 5.3 kb of intergenic DNA. The human NGF gene maps to chromosome 1p13 and is organized in a similar way.

Differential splicing of the primary transcripts yields two different mRNAs that are also differentially expressed in different tissues. The α and γ subunits of NGF are encoded by separate genes. They show a marked sequence homology and are probably derived from a common ancestor.

The first exon of the NGF gene contains a so-called phorbol ester-responsive element (see also: Phorbol esters) which functions as a binding site for transcription factors (see also: Gene expression) that enhance the expression of the NGF gene.

■ **Related Factors:** Two proteins, » BDNF (brain-derived neurotrophic factor) and » NT-3 (neurotrophin-3) are related to NGF. They show some overlapping but not identical biological activities. The three proteins form the NGF family of proteins. They are also collectively referred to as » neurotrophins.

■ **Receptor Structure, Gene(s), Expression:** One receptor that is responsible for mediating most of the activities of NGF is preferentially expressed in neuronal tissues. This glycoprotein of 140 kDa ($gp140^{trk}$) is the product of the » *trk* gene. It possesses an intrinsic tyrosine-specific protein kinase in its intracellular domain. The expression of *trk* in monocytes as well as other findings suggest that » NGF, in addition to its neurotrophic function, is an immunoregulatory cytokine acting on monocytes.

A low-affinity receptor (kdis = 10^{-9} – 10^{-8}) of unknown physiological role is encoded by the LNGFR gene (low affinity nerve growth factor receptor). LNGFR is a transmembrane cysteine-rich glycoprotein of 399 amino acids (p75; $gp80^{LNGFR}$) with a very short intracellular domain.

The low-affinity receptor is expressed by many neurons that do not respond to NGF. It is also expressed in non-neuronal tissues such as testes, muscles, lymphoid tissues, and lymphocytes. LNGFR is also a receptor for » BDNF (brain-derived neurotrophic factor) and » NT-3.

It has been suggested that the combination of *trk* and LNGFR yields a high-affinity NGF receptor (kdis = 10^{-11} M) but this issue is still unclear as *trk* alone has been shown to allow cell growth in the absence of LNGFR in heterologous systems.

The intracellular signal transduction mechanisms initiated by binding of NGF to its receptor are largely unknown. The participation of cAMP as second messenger has been ruled out. Signal transduction may involve a G-protein. The expression of the *ras* oncogene also appears to be necessary for signaling since an inhibition of *ras* expression also inhibits NGF-mediated differentiation of certain cells (see: PC12). For NGF-inducible genes: see: NGFI.

■ **BIOLOGICAL ACTIVITIES:** NGF is not a mitogen so that its name is actually a misnomer. It is mainly responsible for the survival and the differentiation and the functional activities of sensory and sympathetic neurons in the peripheral nervous system. It also plays an important role in the development and functional activities of cholinergic neurons in the central nervous system.

The continuous infusion of NGF prevents the death of neurons in rat, following experimental transsection of the fimbria hippocampi in the fore brain. The treatment of newborn rats and mice with anti-NGF antibodies leads to the almost complete degeneration of neurons of the sympathetic nervous system and causes a variety of neuroendocrine deficits.

Immortalized rat fibroblasts, genetically altered to secrete NGF, have been shown to decrease excitotoxic lesion size by 80 % after implantation in rat brain near the striatum 7 days before striatal infusion of excitotoxic quantities of an NMDA-receptor agonist.

NGF induces the synthesis of a number of specific transmitter-like peptides in sensory neurons, including » SP (substance P; see also: tachykinins), » Somatostatin, and» VIP (vasoactive intestinal polypeptide). NGF inhibits the release of noradrenaline in sympathetic nerve ends and also acts as an inhibitory neuromodulator for adrenergic processes, possibly building an inhibitory feedback loop for catecholamine-stimulated synthesis of NGF.

Since NGF is also synthesized in non-neuronal tissues it may have a much wider spectrum of biological activities than thought previously. NGF stimulates chemotactic migration of human polymorphonuclear leukocytes *in vitro* (see also: Chemotaxis). Subdermal injection of NGF in mice also stimulates rapid and marked chemotactic recruitment of leukocytes at nanomolar concentrations. NGF also promotes the proliferation of mast cells. NGF enhances histamine release and strongly modulates the formation of lipid mediators by basophils in response to various stimuli.

NGF stimulates the growth and differentiation of B cells and the growth of T cells and of some tumor cell types. NGF inhibits immunoglobulin production by various human plasma cell. NGF-induced inhibition of Ig production is restored by » IL6. NGF also influences the differentiation of eosinophils and basophils. Since these cells are involved in immune and inflammatory reactions (see also: Inflammation) NGF may play an accessory role in the regulation of these processes.

The cytokines » IL1, IL6, and » bFGF are potent inducers of NGF. NGF induces the synthesis of IL1 in pheochromocytoma cells (see: PC12) which in turn acts as a growth factor for glial cells and induces the synthesis of NGF following nerve injuries.

In thymic stromal cells NGF induces the synthesis of » IL6. NGF induces the synthesis of the » oncogenes » *fos* and » *myc*. and also influences the expression of » EGF.

■ **TRANSGENIC/KNOCK-OUT/ANTISENSE STUDIES:** Experiments with » transgenic animals (mice) reveal that NGF may act as a growth factor under certain circumstances. The ectopic expression in the pituitary of NGF driven by a » prolactin promoter causes a pronounced hyperplasia of this gland with size enlargements of up to 10-100-fold. The overexpression of NGF in the pancreas of transgenic mice induces the enhanced innervation of this tissue.

In transgenic mice expressing the β subunit of NGF in sympathetic neurons using the human dopamine β-hydroxylase promoter the sympathetic trunk and nerves growing to peripheral tissues are enlarged and contain an increased number of sympathetic fibers. Although sympathetic axons reach peripheral tissues, terminal sympathetic innervation within tissues is decreased. This effect can be reversed in the pancreas by overexpression of NGF in pancreatic islets. It has been suggested that NGF gradients are not required to guide sympathetic axons to their targets, but are required for the establishment of the normal density and pattern of sympathetic innervation within target tissues.

A targeted mutation of the gene encoding the low affinity NGF receptor (LNGFR, p75) has been generated by homologous recombination in » ES cells. Mice homozygous for the mutation are viable and fertile. Mutant mice show markedly decreased sensory innervation by calcitonin gene-related peptide- and substance P-immunoreactive fibers and this is associated with loss of heat sensitivity and the development of ulcers in the distal extremities. Dysfunctions of sympathetic neurons have not been observed in these mice. Embryonic cranial sensory and sympathetic neurons from p75-deficient embryos respond normally to NGF, » BDNF, » NT-

NGF knockouts: see addendum: NGF.

3, and »NT-4/5 at saturating concentrations. Dose responses of sympathetic and visceral sensory neurons from mutant embryos are also normal. However, embryonic cutaneous sensory trigeminal neurons isolated from mutant embryos require more NGF for half-maximal survival, indicating that p75 enhances the sensitivity of NGF-dependent cutaneous sensory neurons to NGF.

■ **ASSAYS:** NGF and NGF-like activities are assayed in a » bioassay, measuring the outgrowth of neurites in the pheochromocytoma cell line » PC12 or its derivatives. NGF is also assayed by its activity on embryonic sensory neurons. A sensitive enzyme immunoassay is also available. NGF can also be detected by a modification of the » cell blot assay. An alternative and entirely different detection method is » Message amplification phenotyping. For further information see also subentry "Assays" in the reference section.

■ **CLINICAL USE & SIGNIFICANCE:** Vitamin D3 and some metabolically active precursors are potent inducers of the synthesis of NGF at concentrations of 100 pM. This may have implications for the treatment of neurodegenerative diseases with NGF. The infusion of human NGF into brains of primates has been shown to prevent the degeneration of cholinergic neurons. It may therefore be possible to use NGF to protect cholinergic neurons in Alzheimer's disease which is characterized by a selective degeneration of these cells. Another possible application might be the treatment of diabetes-associated polyneuropathies. In experimental animals NGF prevents chemotherapy-induced neuropathies.

Rat hippocampal and human cortical neurons have been shown to be protected by NGF against induced damage induced by iron, which is believed to contribute to the process of cell damage and death resulting from ischemic and traumatic insults by catalyzing the oxidation of protein and lipids.

Increased levels of anti-NGF antibodies have been found in the sera of HSV-infected patients. Since NGF promotes the latency of HSV *in vitro* it is thought that these antibodies may play a role in the course of these viral infections.

● **REVIEWS: Barde YA** The nerve growth factor family. PGFR 2: 237-48 (1990); **Gage FH et al** Nerve growth factor function in the central nervous system. CTMI 165: 71-94 (1991); **Levi A & Alemà S** The mechanism of action of nerve growth factor. ARPT 31: 205-28 (1991); **Rush RA** (edt) Nerve growth factors. Chichester, John Wiley & Sons, 1989; **Varon S et al** Nerve growth factor in CNS repair and regeneration. AEMB 296: 267-76 (1991); see also literature cited for » neurotrophins.

● **BIOCHEMISTRY & MOLECULAR BIOLOGY: Barnett J et al** Physicochemical characterization of recombinant human nerve growth factor produced in insect cells with a baculovirus vector. J. Neurochem. 57: 1052-61 (1991); **Borsani G et al** Human fetal brain β-nerve growth factor cDNA: Molecular cloning of 5' and 3' untranslated regions. Neurosci. Lett. 127: 117-20 (1991); **Buxser S et al** Single-step purification and biological activity of human nerve growth factor produced from insect cells. J. Neurochem. 56: 1012-8 (1991); **Carriero F et al** Structure and expression of the nerve growth factor gene in *Xenopus* oocytes and embryos. Mol. Reprod. Dev. 29: 313-22 (1991); **Dicou E** Expression of recombinant human nerve growth factor in *Escherichia coli*. Neurochem. Int. 20: 129-34 (1992); **Dracopoli NC et al** Two thyroid hormone regulated genes, the β-subunits of nerve growth factor (NGFB) and thyroid stimulating hormone (TSHB), are located less than 310 kb apart in both human and mouse genomes. Genomics 3: 161-7 (1988); **Ebendal T et al** Structure and expression of the chicken β nerve growth factor gene. EJ 5: 1483-7 (1986); **Edwards RH et al** Processing of the native nerve growth factor precursor to form biologically active nerve growth factor. JBC 263: 6810-5 (1988); **Evans BA & Richards RI** Genes for the α and γ subunits of mouse nerve growth factor are contiguous. EJ 4: 133-8 (1985); **Fahnestock M et al** β-NGF-endopeptidase: structure and activity of a kallikrein encoded by the gene mGK-22. B 30: 3443-50 (1991); **Fahnestock M** Structure and biosynthesis of nerve growth factor. CTMI 165: 1-26 (1991); **Franke U et al** The human gene for the β subunit of nerve growth factor is located on the proximal short arm of chromosome 1. S 222: 1251-8 (1983); **Garson JA et al** Novel non-isotopic *in situ* hybridization technique detects small (1kb) unique sequences in routinely G-banded human chromosomes: fine mapping of N-*myc* and β-NGF genes. NAR 15: 4761-70 (1987); **Ibáñez CF et al** Biological and immunological properties of recombinant human, rat, and chicken nerve growth factors: a comparative study. J. Neurochem. 57: 1033-41 (1991); **McDonald NQ et al** New protein fold revealed by a 2. 3-Å resolution crystal structure of nerve growth factor. N 354: 411-4 (1991); **Meier R et al** Molecular cloning of bovine and chick nerve growth factor (NGF): delineation of conserved and unconserved domains and their relationship to the biological activity and antigenicity of NGF. EJ 5: 1489-93 (1986); **Narhi LO et al** Comparison of the biophysical characteristics of human brain-derived neurotrophic factor, neurotrophin-3, and nerve growth factor. JBC 268: 13309-17 (1993); **Radeke MJ et al** Gene transfer and molecular cloning of the rat nerve growth factor receptor. N 325: 593-7 (1987); **Scott J et al** Isolation and nucleotide sequence of a cDNA encoding the precursor of mouse nerve growth factor. N 302: 538-40 (1983); **Selby MJ** The mouse nerve growth factor gene: structure and expression. MCB 7: 3057-64 (1987); **Suter U et al** Two conserved domains in the NGF propeptide are necessary and sufficient for the biosynthesis of correctly processed and biologically active NGF. EJ 10: 2395-400 (1991); **Ullrich A et al** Human β nerve growth factor sequences highly homologous to that of mouse. N 303: 821-5 (1983); **Ullrich A et al** Isolation of a cDNA clone coding for the γ-subunit of mouse nerve growth factor using a high-stringency selection procedure. DNA 3: 387-92 (1984); **Vigé X et al** Mechanism of nerve growth factor mRNA regulation by interleukin-1 and basic fibroblast growth factor in primary cultures of rat astrocytes. Mol. Pharmacol. 40: 186-92 (1991); **Wion D et al** Molecular cloning of the avian β-nerve growth factor gene: transcription in brain. FL 203: 82-6 (1986)

● **RECEPTORS: Barker PA & Murphy RA** The nerve growth factor receptor: a multicomponent system that mediates the actions of the neurotrophin family of proteins. Mol. Cell. Biochem. 110: 1-15 (1992); **Bothwell M** Tissue localization of nerve growth factor and nerve growth factor receptors. CTMI 165: 55-70 (1991); **Chao MV et al** Gene transfer and molecular cloning of the human NGF receptor. S 232: 518-21 (1986); **Chao MV** Distinctive features of nerve growth factor: structure and function. In: Habenicht A (edt) Growth factors, differentiation factors, and cytokines, pp. 65-81, Springer, Berlin 1990; **Chao MV** The membrane receptor for nerve growth factor. CTMI 165: 39-54 (1991); **DiStefano PS & Johnston EM** Identification of a truncated form of the nerve growth factor receptor. PNAS 85: 270-4 (1988); **Ebendal T** Function and evolution in the NGF family and its receptors. J. Neurosci. Res. 32: 461-70 (1992); **Hempstead BL et al** High-affinity NGF binding requires coexpression of the *trk* proto-oncogene and the low-affinity NGF receptor. N 350: 678-83 (1991); **Kaplan DR et al** The *trk* proto-oncogene product: a signal transducing receptor for nerve growth factor. S 252: 554-8 (1991); **Klein R et al** The *trk* proto-oncogene encodes a receptor for nerve growth factor. Cell 65: 189-97 (1991); **Meakin SO & Shooter EM** The nerve growth factor family of receptors. TINS 15: 323-31 (1992); **Ohmichi M et al** Nerve growth factor binds to the 140 kd *trk* proto-oncogene product and stimulates its association with the *src* homology domain of phospholipase C γ1. BBRC 179: 217-23 (1991); **Rodriguez-Tebar A et al** Binding of neurotrophin-3 to its neuronal receptors and interactions with nerve growth factor and brain-derived neurotrophic factor. EJ 11: 917-22 (1992)

● **Biological Activities: Alheim K et al** Interleukin 1 expression is inducible by nerve growth factor in PC12 pheochromocytoma cells. PNAS 88: 9302-6 (1991); **Bischoff SC & Dahinden CA** Effect of nerve growth factor on the release of inflammatory mediators by mature human basophils. Blood 79: 2662-9 (1992); **Borrelli E et al** Pituitary hyperplasia induced by ectopic expression of nerve growth factor. PNAS 89: 2764-8 (1992); **Boyle MD et al** Nerve growth factor: a chemotactic factor for polymorphonuclear leukocytes *in vivo*. JI 134: 564-8 (1985); **Brodie D & Gelfand EW** Functional nerve growth factor receptors on human B lymphocytes. Interaction with IL2. JI 148: 3492-7 (1992); **Brown AB & Carpenter G** Acute regulation of the epidermal growth factor receptor in response to nerve growth factor. J. Neurochem. 57: 1740-9 (1991); **Bruni A et al** Interaction between nerve growth factor and lysophosphatidylserine on rat peritoneal mast cells. FL 138: 190-2 (1982); **Creedon D & Tuttle JB** Nerve growth factor synthesis in vascular smooth muscle. Hypertension 18: 730-41 (1991); **Dicou E et al** Naturally occurring antibodies against nerve growth factor in human and rabbit sera: comparison between control and herpes simplex virus-infected patients. J. Neuroimmunol. 34: 153-8 (1991); **Frim DM et al** Effects of biologically delivered NGF, BDNF and bFGF on striatal excitotoxic lesions. Neuroreport 4: 367-70 (1993); **Kimata H et al** Nerve growth factor specifically induces human IgG4 production. EJI 21: 137-41 (1991); **Kimata H et al** Nerve growth factor inhibits immunoglobulin production by but not proliferation of human plasma cell lines. Clin. Immunol. Immunopathol. 60: 145-51 (1991); **Kimata H et al** Stimulation of Ig production and growth of human lymphoblastoid B-cell lines by nerve growth factor. Immunology 72: 451-2 (1991); **Lindholm D et al** Interleukin 1 regulates synthesis of nerve growth factor in non-neuronal cells of rat sciatic nerve. N 330: 658-9 (1987); **Manning PT et al** Protection from guanethidine-induced neuronal destruction by nerve growth factor: effect of NGF on immune function. Brain Res. 340: 61-9 (1985); **Matsuda H et al** Nerve growth factor promotes human hemopoietic colony growth and differentiation. PNAS 85: 6508-12 (1988); **Matsuoka I et al** Differential regulation of nerve growth factor and brain-derived neurotrophic factor expression in the peripheral nervous system. ANY 633: 550-2 (1991); **Mazurek N et al** Nerve growth factor induces mast cell degranulation without changing intracellular calcium levels. FL 198: 315-20 (1986); **Otten U et al** Nerve growth factor induces growth and differentiation of human B lymphocytes. PNAS 86: 10059-63 (1989); **Pearce FL & Thompson HL** Some characteristics of histamine secretion from rat peritoneal mast cells stimulated with nerve growth factor. J. Physiol. (London) 372: 379-93 (1986); **Screpanti I et al** Neuromodulatory loop mediated by nerve growth factor and interleukin 6 in thymic stromal cell cultures. PNAS 89: 3209-12 (1992); **Sieber-Blum M** Role of the neurotrophic factors BDNF and NGF in the commitment of pluripotent neural crest cells. Neuron 6: 949-55 (1991); **Suter U** NGF/BDNF chimaeric proteins: analysis of neurotrophin specificity by homologue-scanning mutagenesis. J. Neurosci. 12: 306-18 (1992); **Thoenen H et al** The synthesis of nerve growth factor and brain-derived neurotrophic factor in hippocampal and cortical neurons is regulated by specific transmitter systems. ANY 640: 86-90 (1991); **Thoenen H et al** The physiological function of nerve growth factor in the central nervous system: comparison with the periphery. Rev. Physiol. Biochem. Pharmacol. 109: 145-78 (1987); **Thorpe LW & Perez-Polo JR** The influence of nerve growth factor on the *in vitro* proliferative response of rat spleen lymphocytes. J. Neurosci. Res. 18: 134-9 (1987); **Ueyama T et al** New role of nerve growth factor – an inhibitory neuromodulator of adrenergic transmission. Brain Res. 559: 293-6 (1991); **Wion D et al** 1,25-Dihydroxy vitamin D3 is a potent inducer of nerve growth factor synthesis. J. Neurosci. Res. 28: 110-4 (1991); **Yan H et al** Chimaeric NGF-EGF receptors define domains responsible for neuronal differentiation. S 252: 561-3 (1991); **Yoshida K & Gage FH** Cooperative regulation of nerve growth factor synthesis and secretion in fibroblasts and astrocytes by fibroblast growth factor and other cytokines. Brain Res. 569: 14-25 (1992)

● **Transgenic/Knock-out/Antisense Studies: Alexander JM et al** Cell-specific and developmental regulation of a nerve growth factor-human growth hormone fusion gene in transgenic mice. Neuron 3: 133-9 (1989); **Davies AM et al** p75-deficient trigeminal sensory neurons have an altered response to NGF but not to other neurotrophins. Neuron 11: 565-74 (1993); **Edwards RH et al** Directed expression of NGF to pancreatic β cells in transgenic mice leads to selective hyperinnervation of the islets. Cell 58: 161-70 (1989); **Hoyle GW et al** Expression of NGF in sympathetic neurons leads to excessive axon outgrowth from ganglia but decreased terminal innervation within tissues. Neuron 10: 1019-34 (1993); **Lee KF et al** Targeted mutation of the gene encoding the low affinity NGF receptor p75 leads to deficits in the peripheral sensory nervous system. Cell 69: 737-49 (1992)

● **Assays: Fahnestock M** Detection and assay of nerve growth factor mRNA. MiE 198: 48-61 (1991); **Kenigsberg RL et al** Two distinct monoclonal antibodies raised against mouse β nerve growth factor. Generation of bi-specific anti-nerve growth factor anti-horseradish peroxidase antibodies for use in a homogeneous enzyme immunoassay. JIM 136: 247-57 (1991); **Lakshmanan J et al** β nerve growth factor in developing mouse cerebral cortical synaptosomes: measurement by competitive radioimmunoassay and bioassay. Pediatr. Res. 20: 391-5 (1986); **Lakshmanan J et al** β nerve growth factor measurements in mouse serum. J. Neurochem. 46: 882-91 (1986); **Larkfors L et al** Methylmercury-induced alterations in the nerve growth factor level in the developing brain. Brain Res. Dev. Brain Res. 62: 287-91 (1991); **Lorigados L et al** Two-site enzyme immunoassay for β NGF applied to human patient sera. J. Neurosci. Res. 32: 329-39 (1992); **Murase K et al** Highly sensitive enzyme immunoassay for β-nerve growth factor (NGF): a tool for measurement of NGF level in rat serum. BI 22: 807-13 (1990); **Murase K et al** Development of sensitive enzyme immunoassay for human nerve growth factor. BI 25: 29-34 (1991); **Naher-Noe M et al** Determination of nerve growth factor concentrations in human samples by two-site immunoenzymometric assay and bioassay. Eur. J. Clin. Chem. Clin. Biochem. 31: 375-80 (1993); **Nishio T et al** Detailed distribution of nerve growth factor in rat brain determined by a highly sensitive enzyme immunoassay. Exp. Neurol. 116: 76-84 (1992); **Nishizuka M et al** Age- and sex related differences in the nerve growth factor distribution in the rat brain. Brain Res. Bull. 27: 685-8 (1991); **Soderstrom S et al** Recombinant human β-nerve growth factor (NGF): biological activity and properties in an enzyme immunoassay. J. Neurosci. Res. 27: 665-77 (1990); **Stephani U et al** Nerve growth factor (NGF) in serum: evaluation of serum NGF levels with a sensitive bioassay employing embryonic sensory neurons. J. Neurosci. Res. 17: 25-35 (1987); **Tuszynski MH & Gage FH** *In vivo* assay of neuron-specific effects of nerve growth factor. MiE 198: 35-48 (1991); for further information also see individual cell lines used in individual bioassays.

● Clinical Use & Significance: **Apfel SC et al** Nerve growth factor prevents experimental cisplatin neuropathy. Ann. Neurol. 31: 76-80 (1992); **Tuszynski MH et al** Recombinant human nerve growth factor infusions prevent cholinergic neuronal degeneration in the adult primate brain. Ann. Neurol. 30: 625-36 (1991); **Zhang Y et al** Basic FGF, NGF, and IGFs protect hippocampal and cortical neurons against iron-induced degeneration. J. Cereb. Blood Flow Metab. 13: 378-88 (1993)

NGF family: This family comprises » NGF, » BDNF (brain-derived neurotrophic factor) and » NT-3 (neurotrophin 3). See also: Neurotrophins.

Barde YA The nerve growth factor family. PGFR 2: 237-48 (1990)

NGFI: *nerve growth factor inducible* These genes are » ERGs (early response gene) whose transcription is rapidly and transiently stimulated by » NGF and » SDGF in » PC12 rat pheochromocytoma cells, and other signals that initiate growth and differentiation of cells. NGFI constitutes a family of related proteins that function as transcription factors. All NGFI proteins recognize a short DNA sequence, GCGGGGGCG, in the promoters of genes.

NGFI-A has been shown to be identical with » Egr-1, TIS-8 (see: TIS), and » zif268. Thyroid hormone up-regulates NGFI-A gene expression in rat brain during development. ***NGFI-B*** (identical with TIS1, see: TIS; identical with » NAK1 and » nur-77) and **NGFI-C** (also known as Egr-4; see: Egr-1) also encode zinc finger proteins that function as a transcription factor (see also: Gene expression). NGFI-B has been shown to regulate in adrenocortical cells an essential steroidogenic enzyme, steroid 21-hydroxylase, that is subject to regulation by ACTH.

The human NGFI-C has been localized to chromosome 2p13. This region contains a constitutive fragile site that is associated with chromosomal breakpoints and translocations characteristic of some chronic lymphocytic leukemias.

In vivo, a rapid, dramatic increase in NGFI-B mRNA has been observed in the cerebral cortex, midbrain, and cerebellum of animals that experienced a convulsant-induced seizure.

Changelian PS et al Structure of the NGFI-A gene and detection of upstream sequences responsible for its transcriptional induction by nerve growth factor. PNAS 86: 377-81 (1989); **Crosby SD et al** The early response gene NGFI-C encodes a zinc finger transcriptional activator and is a member of the GCGGGGGCG (GSG) element-binding protein family. MCB 11: 3835-41 (1991); **Crosby SD et al** Neural-specific expression, genomic structure, and chromosomal localization of the gene encoding the zinc-finger transcription factor NGFI-C PNAS 89: 4739-43 (1992); **Day LM et al** The zinc finger protein NGFI-A exists in both nuclear and cytoplasmic forms in nerve growth factor-stimulated PC12 cells. JBC 265: 15253-60 (1990); **De Franco C et al** Nerve growth factor induces transcription of NGFIA through complex regulatory elements that are also sensitive to serum and phorbol 12-myristate 13-acetate. Mol. Endocrinol. 7: 365-79 (1993); **Fahrner TJ et al** The NGFI-B protein, an inducible member of the thyroid/steroid receptor family, is rapidly modified posttranslationally. MCB 10: 6454-9 (1990); **Kimura H & Schubert D** Schwannoma-derived growth factor promotes the neuronal differentiation and survival of PC12 cells. JCB 116: 777-83 (1992); **Lee HJ et al** The use of a DNA-binding domain replacement method for the detection of a potential TR3 orphan receptor response element in the mouse mammary tumor virus long terminal repeat. BBRC 193: 97-103 (1993); **Matheny C et al** Differential activation of NGF receptor and early response genes in neural crest-derived cells. Brain Res. Mol. Brain Res. 13: 75-81 (1992); **Paulsen RE et al** Domains regulating transcriptional activity of the inducible orphan receptor NGFI-B. JBC 267: 16491-6 (1992); **Pipaon C et al** Thyroid hormone up-regulates NGFI-A gene expression in rat brain during development. JBC 267: 21-3 (1992); **Watson MA & Milbrandt J** The NGFI-B gene, a transcriptionally inducible member of the steroid receptor gene superfamily: genomic structure and expression in rat brain after seizure induction. MCB 9: 4213-9 (1989); **Wilson TE et al** Identification of the DNA binding site for NGFI-B by genetic selection in yeast. S 252: 1296-300 (1991); **Wilson TE et al** The orphan nuclear receptor NGFI-B regulates expression of the gene encoding steroid 21-hydroxylase. MCB 13: 861-8 (1993)

NGFI-A: see: NGFI.

NGFI-B: see: NGFI.

NGFI-C: see: NGFI.

NGIF: *neuroblastoma growth inhibitory factor* This poorly characterized factor (75 kDa by gel filtration) has been observed in the conditioned medium of normal rat glioblasts. Cell growth rates and DNA synthesis in neuroblastoma cells (Neuro2a, NS-20Y, and N1E-115) cultured in the presence of NGIF are markedly inhibited. The factor induces morphological differentiation includ-

ing neural process formation. NGIF does not alter the growth rate or the morphology of non-neuronal cells such as glial cell lines (C6 and 354A) or fibroblasts (3T3). The ability of NGIF to suppress preferentially the neural growth suggests its regulatory role in normal brain development.
A factor of ≈ 97 kDa with NGIF activities that also inhibits » IL2-dependent T cell growth has been observed in media conditioned by human glioblastoma cells.

Fontana A et al Glioblastoma cells release interleukin 1 and factors inhibiting interleukin 2-mediated effects. JI 132: 1837-44 (1984); **Sakazaki Y et al** Characterization and partial purification of neuroblastoma growth inhibitory factor from the culture medium of glioblasts. Brain Res. 262: 125-35 (1983)

NGL: *neuro/glioblastoma-derived* Human gene symbol for the » *neu* oncogene.

NIAID: *National Institute of Allergy & Infectious Diseases* Bethesda, Maryland, USA. This institution provides reference standards for most interleukins and other cytokines with immunomodulatory activities.

NIBSC: *National Institute for Biological Standards & Control* Blanche Lane, South Mimms, Potters Bar, Herts, UK. This institution provides reference standards for most inerleukins and other cytokines with immunomodulatory activities.

NIF: *neurite-inducing factor* This factor of ≈ 20 kDa is isolated from murine submaxillary glands. It induces the outgrowth of neurites in the pheochromocytoma cell line » PC12. The factor is not identical with » NGF since its activity is not blocked by NGF-specific antibodies.

Ludecke G & Unsicker K Mitogenic effect of neurotrophic factors on human IMR 32 neuroblastoma cells. Cancer 65: 2270-8 (1990); **Wagner JA** NIF (neurite-inducing factor): a novel peptide inducing neurite formation in PC12 cells. J. Neurosci. 6: 61-7 (1986)

NIF: *neutrophil (migration) inhibitory factor* This factor is secreted by T lymphocytes. It is identical with » NIF-T.

NIF-T: *neutrophil inhibition factor from T lymphocytes* This factor of 22-26 kDa is secreted by T lymphocytes. It is chemotactic for neutrophils and stimulates the growth of granulocyte- and macrophage-containing colonies from bone marrow in a » colony formation assay. It also stimulates colony formation of » KG-1 cells. The factor is identical with » GM-CSF.

Gasson JC et al Purified human granulocyte-macrophage colony-stimulating factor: direct action on neutrophils. S 226: 1339-42 (1984); **Koeffler HP et al** Characterization of a novel HTLV-infected cell line. Blood 64: 482-90 (1984); **Weisbart RH et al** Further purification of neutrophil migration inhibition factor from T lymphocytes (NIF-T): evidence that NIF-T and leukocyte inhibitory factor (LIF) are immunologically distinct. Clin. Immunol. Immunopathol. 32: 269-74 (1984); **Wong LG et al** Effect of methylprednisolone on the production of neutrophil migration inhibition factor by T lymphocytes (NIF-T). Immunopharmacology 3: 179-85 (1981)

NIH3T3: see: 3T3 cells.

Nil-2-a: see: TCF8.

NK: *neurokinin* see: Tachykinins.

NK1: *neurokinin 1* This factor is identical with » SP (substance P).

NK2: *neurokinin 2* This factor is identical with neurokinin A. See: » Tachykinins.

NK3: *neurokinin 3* This factor is identical with neurokinin B. See: » Tachykinins.

NKA: *neurokinin A* This factor is identical with substance K. See: » Tachykinins.

NKB: *neurokinin B* This factor is identical with neuromedin K. See: » Tachykinins.

NKAF: *natural killer (cell) activating factor* This poorly characterized factor is secreted by activated murine lymphocytes (see also: Cell activation). It is a positive regulator of immune responses and increases the activity of natural killer cells. The factor is not identical with » IL2 or » IFNs. Some NKAF activity may be due to » IL6.

Kobayashi Y et al Natural killer (NK) cell activating factor produced by a human T cell hybridoma. Microbiol. Immunol. 35: 981-93 (1991); **Ichimura O et al** Characterization of mouse natural killer cell activating factor (NKAF), induced by OK-432. Evidence for interferon- and interleukin 2-independent NK cell activation. Brit. J. Cancer 50: 97-108 (1984); **Luger TA et al** IFN-β 2/IL6 augments the activity of human natural killer cells. JI 143: 1206-9 (1989); **Shitara K et al** Natural killer (NK) cell-activated factor released from murine thymocytes stimulated with an antitumor streptococcal preparation, OK-432. JI 134: 1039-47 (1985); **Uchida A** The cytolytic and regulatory role of natural killer cells in human neoplasia. BBA 865: 329-40 (1986)

NKC3: A murine natural killer cell line that depends on » IL2 for its growth. The cells are used as indicator cells for IL2 » bioassays.

Moriya N et al An enzyme-immunoassay for human interleukin-2. J. Immunoassay 8: 131-43 (1987); **Tada H et al** An improved colorimetric assay for interleukin 2. JIM 93: 157-65 (1986); **Yamada T et al** Importance of disulfide linkage for constructing the biologically active human interleukin-2. ABB 257: 194-9 (1987)

NKCF: *natural killer cytotoxic factor* NKCF is also known as ***NK-CIA*** (NK colony-inhibiting activity) or simply killer factor. This term is an operational definition for factors of ≈ 20-40 kDa released by NK cells bound to NK-sensitive target cells. NKCF shows selective cytotoxicity for NK-sensitive target cells (see also: LAK cells, lymphokine-activated killer cells). This biological activity is due to the formation of transmembrane channels that make the cellular membrane permeable.

The release of NKCF can also occur as a response to mitogenic cytokines. The production of NKCF and of its lytic activity can be stimulated by interferons (see: IFN). On the other hand, interferons may also inhibit the lysis of target cells by NK cells and this is due to the inhibition of NKCF release.

At least some of the NKCF activity is caused by » TNF-β.

Bialas T et al Distinction of partially purified human natural killer cytotoxic factor from recombinant human tumor necrosis factor and recombinant human lymphotoxin. CR 48: 891-8 (1988); **Degliantoni G et al** Natural killer (NK) cell-derived hematopoietic colony inhibiting activity and NK cytotoxic factor. Relationship with tumor necrosis factor and synergisms with immune interferon. JEM 162: 1512-30 (1985); **Graves SS et al** Studies on the lethal stage of natural killer cell-mediated cytotoxicity. I. Both phorbol ester and ionophore are required for release of natural killer cytotoxic factor (NKCF), suggesting a role for protein kinase C activity. JI 137: 1977-84 (1984); **Ortaldo JR et al** Analysis of cytostatic/cytotoxic lymphokines. Relationship of natural killer cytotoxic factor to recombinant lymphotoxin, recombinant tumor necrosis factor, and leukoregulin. JI 137: 2857-63 (1986); **Peters PM et al** Natural killer-sensitive targets stimulate production of TNF-α but not TNF-β (lymphotoxin) by highly purified human peripheral blood large granular lymphocytes. JI 137: 2592-8 (1986); **Ramirez R et al** Mechanisms involved in NK resistance induced by interferon-γ. CI 140: 248-56 (1992); **Roozemond RC et al** Liposomes can function as targets for natural killer cytotoxic factor but not for tumor necrosis factor. JI 142: 1209-16 (1989); **Sashchenko LP et al** Separation of the pore-forming and cytotoxic activities from natural killer cell cytotoxic factor. FL 226: 261-4 (1988); **Winkler-Pickett RT et al** Analysis of rat natural killer cytotoxic factor (NKCF): mechanism of action and relationship to other cytotoxic/cytostatic factors. CI 135: 42-54 (1991); **Wright SC & Bonavida B** Selective lysis of NK-sensitive target cells by a soluble mediator released from murine spleen cells and human peripheral blood lymphocytes. JI 126: 1516-21 (1981); **Wright SC & Bonavida B** Studies on the mechanism of natural killer (NK) cell-mediated cytotoxicity (CMC). I. Release of cytotoxic factors specific for NK-sensitive target cells (NKCF) during coculture of NK effector cells and NK target cells JI 129: 433-9 (1982); **Wright SC & Bonavida B** Studies on the mechanism of natural killer cell-mediated cytotoxicity. VII. Functional comparison of human natural killer cytotoxic factor with recombinant lymphotoxin and tumor necrosis factor. JI 138: 1791-8 (1987); **Young JDE & Liu CC** Multiple mechanisms of lymphocyte-mediated killing. IT 9: 140-4 (1988)

NK-CIA: *natural killer colony-inhibiting activity* see: NKCF.

NKEF: *natural killer enhancing factor* This poorly characterized factor is isolated from cytosolic extracts of human red blood cells. NKEF is probably a dimeric protein consisting of two 24 kDa subunits that may also aggregate noncovalently to form polymers of over 300 kDa in aqueous solution. Partial sequences of two NKEF peptides indicate that NKEF does not share significant homology with any known sequences. Some sequence homology is observed with a murine erythroleukemia-related gene encoding a protein of 257 amino acids, MER5, preferentially synthesized in murine erythroleukemia cells and presumably involved in the early period of differentiation of murine erythroid cells.

NKEF augments natural killer cytotoxicity *in vitro* against tumor cells if precoated on plastic surfaces.

Nemoto Y et al Antisense RNA of the latent period gene (MER5) inhibits the differentiation of murine erythroleukaemia cells. Gene 91: 261-5 (1990); **Shau H et al** Identification of a natural killer enhancing factor (NKEF). CI 147: 1-11 (1993); **Yamamoto T et al** Cloning of a housekeeping-type gene (MER5) preferentially expressed in murine erythroleukemia cells. Gene 80: 337-43 (1989)

NK-LCF: *natural killer leukocyte chemotactic factor* This poorly characterized factor of > 6.5 kDa is released by degranulating NK cells. It is chemotactic for mononuclear lymphocytes and neutrophil granulocytes and NK cells (see also: Chemotaxis). This factor is probably also involved in tumor cell killing since NK-sensitive tumor cells can cause the degranulation of NK cells and hence promote the further accumulation of NK cells and phagocytosing cells attracted by NK-LCF.

Greenberg AH et al NK-leukocyte chemotactic factor (NK-LCF): a large granular lymphocyte (LGL) granule-associated chemotactic factor. JI 137: 3224-30 (1986)

NK-LRIF: *natural killer lysis resistance inducing factor* This poorly characterized factor is secreted by some tumor cells. It blocks the cytotoxic activity of NK cells.

Serrano R et al Identification of a tumor factor inducing resistance to NK cell lysis. Immunol. Lett. 20: 311-6 (1989)

NK-RIF: *natural killer resistance inducing factor* This poorly characterized factor of ≈ 12.6 kDa is secreted by concanavalin A-activated rat spleen cells. It induces resistance against NK cell-mediated lysis in some lymphoma cell lines. The factor also induces the expression of MHC class I antigens and reduces the growth rates of lymphoma cells.

Saxena RK et al Modulation of major histocompatibility complex antigens and inhibition of proliferative activity of YAC lymphoma cells by a natural killer lysis resistance-inducing factor (NK-LRIF). Nat. Immun. Cell Growth Regul. 8: 197-208 (1989); **Saxena RK et al** Properties and characterization of a rat spleen cell-derived factor that induces resistance to natural killer cell lysis in YAC lymphoma cells. JI 141: 1782-7 (1988); **Saxena RK** A spleen cell derived factor imparts resistance to NK cell mediated lysis in a mouse lymphoma cell line. Immunol. Lett. 15: 105-8 (1987)

NKSF: *natural killer cell stimulatory factor* This factor was initially isolated from a lymphoblastoid B cell line and found to stimulate the activity of natural killer cells. The factor was renamed » IL12.

Kobayashi M et al Identification and purification of natural killer cell stimulatory factor (NKSF). JEM 170: 827-45 (1989); **Stern AS et al** Purification to homogeneity and partial characterization of cytotoxic lymphocyte maturation factor from human B lymphoid cells. PNAS 87: 6808-12 (1990); **Wolf SF et al** Cloning of cDNA for natural killer cell stimulatory factor, a heterodimeric cytokine with multiple biologic effects on T and natural killer cells. JI 146: 3074-81 (1991)

NLK: *neuroleukin* This factor of 56 kDa was initially detected in murine skeletal, muscle, heart, and kidney tissues. It is also produced by lectin-stimulated T cells. Some glioma cell lines produce NLK constitutively.

NLK promotes the survival of embryonic spinal neurons and sensory neurons not responding to » NGF. It does not appear to act on sympathetic and parasympathetic neurons. NLK induces the maturation of polyclonal B cells into Ig-G secreting cells in the presence of monocytes and T cells. Neuroleukin acts early in the *in vitro* response that leads to formation of antibody-secreting cells. Continued production of immunoglobulin by differentiated antibody- secreting cells is neuroleukin-independent. Neuroleukin does not behave as a B cell differentiation factor (see: BCDF) or B cell growth factor (see: BCGF) in several bioassays.

The analysis of the cDNA encoding phosphohexose isomerase (Glucose-6-phosphate isomerase, GPI; EC 5. 3. 1. 9) derived from various species including yeast has revealed that NLK and this enzyme are 90 % homologous at the amino acid level. Mouse and human neuroleukin cDNAs express GPI enzyme activity when transfected into monkey COS cells. The two genes are probably identical; they map to the long arm of human chromosome 19 (19cen-q12). NLK expressed in a heterologous cell system from the recombinant cDNA clone also possesses enzymatic activity. NLK may be generated by proteolytic processing of the enzyme.

One particular domain of NLK shows a marked homology with a conserved region of the 120 kDa coat protein of HIV-1. A peptide fragment of 19 amino acids obtained from this coat protein and also gp120 block the biological activities of NLK. The biological and possibly also pathological significance of these findings are unknown.

Baumann M & Brand K Purification and characterization of phosphohexose isomerase from human gastrointestinal carcinoma and its potential relationship to neuroleukin. CR 48: 7018-21 (1988); **Chaput M et al** The neurotrophic factor neuroleukin is 90 % homologous with phosphohexose isomerase. N 332: 454-7 (1988); **Faik P et al** Mouse glucose-6-phosphate isomerase and neuroleukin have identical 3´ sequences. N 332: 445-7 (1988); **Gurney ME et al** Molecular cloning and expression of neuroleukin, a neurotrophic factor for spinal and sensory neurons. S 234: 566-74 (1986); **Gurney ME et al** Neuroleukin: a lymphokine product of lectin-stimulated T cells. S 234: 574-81 (1986); **Hallbook F et al** Development and regional expression of chicken neuroleukin (glucose-6-phosphate isomerase) messenger RNA. J. Neurosci. Res. 23: 142-51 (1989); **Spear GT et al** Quantitative immunoassay of recombinant murine neuroleukin. Neuroscience 27: 41-8 (1988); **Spear GT & Gurney ME** Neuroleukin secretion is highly regulated in T cells but constitutive in C6 glioma cells. ANY 1988; 540: 407-8 (1988); **Sun AQ et al** Isolation and characterization of human glucose-6-phosphate isomerase isoforms containing two different size subunits. ABB 283: 120-9 (1990); **Tekamp-Olson P et al** The isolation, characterization and nucleotide sequence of the phosphoglucoisomerase gene of *Saccharomyces cerevisiae*. Gene. 73: 153-61 (1988)

NLS: *nuclear localization sequence* See: Signal sequence.

nm23: The nm23 gene encodes a nucleoside diphosphate kinase (p19/nm23; EC.2.7.4.6; NDP kinase). This enzyme is a hexamer of two different subunits (A, B, 152 amino acids each, displaying 88 % homology with each other). These subunits form all possible isoenzymes (A_6, A_5B... AB_5, B_6) which differ in their isoelectrical points.

NDP kinase catalyses the phosphorylation of nucleoside diphosphates into triphosphates required for the biosynthesis of nucleic acids. NDP kinase can also phosphorylate GDP in GTP-binding proteins and therefore acts as an activator for such proteins.

The A chain of the enzyme is identical with human nm23 which is encoded by a gene mapping to chromosome 17q22. nm23 is also called ***metastasis inhibition factor***. The B chain of the enzyme (gene on chromosome 17q21.3) is called ***nm23-H1***. nm23 protein is identical with » ***I factor***, a cytokine inhibiting the differentiation of some murine cell lines. The human gene encoding the cellular » *myc* purine-binding transcription factor PuF has been identified by screening of a cervical carcinoma cell complementary DNA library with a DNA fragment containing PuF binding sites and was found to show perfect identity with the human nm23-H2 nucleoside diphosphate kinase gene. nm23/NDP kinase has also been found on the cell surface of most human hematopoietic and some non-hematopoietic cell lines, indicating an extracellular role in addition to the reported intracellular functions.

nm23 shows 78 % homology with a *Drosophila* gene product called ***awd*** (abnormal wing disk) the disruption of which arrests cells in the metaphase. *awd* protein is associated with the microtubules and plays a role in the polymerization of spindle fibres. The ***Gip17*** protein of *Dictyostelium* shows ≈ 75 % homology to human nm23 and *Drosophila awd*.

An NDP kinase isolated from rat cells displays ≈ 88 % homology with human nm23 at the protein level as inferred from the rat kinase cDNA sequence. This rat protein is a binding protein for chromoglycate, an antiasthmatic compound that inhibits secretory reactions in mast cells mediated by $F_c\varepsilon$ receptors.

nm23 RNA levels are highest in cell lines with low metastatic potential but do not correlate with cell sensitivity to host immune responses. A markedly reduced expression of nm23 caused, among other things, by deletion mutations, correlates with a high metastasizing potential of some tumors including mammary and colorectal carcinomas. This is however not a general phenomenon and is not observed in all tumors. The expression of nm23 is enhanced in many solid tumors. In human neuroblastomas the enhanced expression of nm23 correlates well with the progressive development of an aggressively growing tumor. Low nm23 RNA levels are associated with histopathologic indications of high metastatic potential of human breast tumors. Somatic allelic deletion of the nm23 gene has been observed in DNA from human breast, renal, colorectal, and lung carcinomas. Mutations have also been observed in aggressive childhood neuroblastomas.

Barnes R et al Low nm23 protein expression in infiltrating ductal breast carcinomas correlates with reduced patient survival. Am. J. Pathol. 139: 245-50 (1991); **Bevilacqua G et al** Association of low nm23 RNA levels in human primary infiltrating ductal breast carcinomas with lymph node involvement and other histopathological indicators of high metastatic potential. CR 49: 5185-90 (1989); **Biggs J et al** A *Drosophila* gene that is homologous to a mammalian gene associated with tumor metastasis codes for a nucleoside diphosphate kinase. Cell 63: 933-40 (1990); **Caligo MA et al** Decreasing expression of NM23 gene in metastatic murine mammary tumors of viral etiology (MMTV). Anticancer Res. 12: 969-73 (1992); **Cohn KH et al** Association of nm23-H1 allelic deletions with distant metastases in colorectal carcinoma. Lancet 338: 722-4 (1991); **Dumas C et al** X-ray structure of nucleoside diphosphate kinase EJ 11: 3202-8 (1992); **Gilles AM et al** Nucleoside diphosphate kinase from human erythrocytes. Structural characterization of the two polypeptide chains responsible for heterogeneity of the hexameric enzyme. JBC 266: 8784-9 (1991); **Golden A et al** Nucleoside diphosphate kinases, nm23, and tumor metastasis: possible biochemical mechanisms. Cancer Treat. Res. 63: 345-58 (1992); **Hailat N et al** High levels of p19/nm23 protein in neuroblastoma are associated with advanced stage disease and with N-*myc* gene amplification. JCI 88: 341-5 (1991); **Hemmerich S et al** A cromoglycate binding protein from rat mast cells of a leukemia line is a nucleoside diphosphate kinase. B 31: 4574-9 (1992); **Hennessy C et al** Expression of the antimetastatic gene nm23 in human breast cancer: an association with good prognosis. JNCI 83: 281-5 (1991); **Hirayama R et al** Positive relationship between expression of antimetastatic factor (nm23 gene product or nucleoside diphosphate kinase) and good prognosis in human breast cancer. JNCI 83: 1249-50 (1991); **Keim D et al** Proliferation-related expression of p19/nm23 nucleoside diphosphate kinase. JCI 89: 919-24 (1992); **Lacombe ML et al** Overexpression of nucleoside diphosphate kinase (Nm23) in solid tumors. Eur. J. Cancer 27: 1302-07 (1991); **Leone A et al** Somatic allelic deletion of nm23 in human cancer. CR 51: 2490-3 (1991); **Leone A et al** Reduced tumor incidence, metastatic potential, and cytokine responsiveness of nm23-transfected melanoma cells. Cell 65: 25-35 (1991); **Leone A et al** Evidence for nm23 RNA overexpression, DNA amplification and mutation in aggressive childhood neuroblastomas. O 8: 855-65 (1993); **Nakamori S et al** Expression of nucleoside diphosphate kinase/nm23 gene product in human pancreatic cancer: an association with lymph node metastasis and tumor invasion. Clin. Exp. Metastasis 11: 151-8 (1993); **Nakayama T et al** Expression in human hepatocellular carcinoma of nucleoside diphosphate kinase, a homologue of the nm23 gene product. JNCI 84: 1349-54 (1992); **Nakayama H et al** Reduced expression of nm23 is associated with metastasis of human gastric carcinomas. Jpn. J. Cancer Res. 84: 184-90 (1993); **Postel EH et al** Human c-*myc* transcription factor PuF identified as nm23-H2 nucleoside diphosphate kinase, a candidate suppressor of tumor metastasis S 261: 478-80 (1993); **Randazzo PA et al** Activation of a small GTP-binding protein by nucleoside diphosphate kinase. S 254: 850-3 (1991); **Rosengard AM et al** Reduced nm23/Awd protein in tumor metastasis and aberrant *Drosophila* development. N 342: 177-80 (1989); **Sastre-Garau X et al** Nucleoside diphosphate kinase/NM23 expression in breast cancer: lack of correlation with lymph-node metastasis. IJC 50: 533-8 (1992); **Stahl JA et al** Identification of a second human nm23 gene, nm23-H2. CR 51: 445-9 (1991); **Steeg PS et al** Tumor metastasis and nm23: current concepts. Cancer Cells 3: 257-62 (1991); **Steeg PS et al** Evidence for a novel gene associated with low tumor metastatic potential. JNCI 80: 200-4 (1988); **Urano T et al** Expression of nm23/NDP kinase proteins on the cell surface. O 8: 1371-6 (1993); **Varesco L et al** The NM23 gene maps to human chromosome band 17q22 and shows a restriction fragment length polymorphism with BglII. Genes Chromosomes Cancer 4: 84-8 (1992); **Wallet V et al** *Dictyostelium* nucleoside diphosphate kinase highly homologous to nm23 and *awd* proteins involved in mammalian tumor metastasis and *Drosophila* development. JNCI 82: 1199-202 (1990); **Wang L et al** Mutation in the nm23 gene is associated with metastasis in colorectal cancer. CR 53: 717-20 (1993)

nm23-H1: see: nm23.

NMDGF: *nonmyocyte-derived growth factor* This poorly characterized factor of 45-50 kDa is isolated from the culture supernatant of cardiac non-myocytes (mainly fibroblasts surrounding myocytes). NMDGF is a heparin-binding protein (see also: HBGF). It stimulates the growth of myocytes *in vitro*. The biological activity of NMDGF is not blocked by antibodies directed against » PDGF, » TNF, » aFGF, » bFGF and » TGF-β which are also found in cardiac tissues.

Long CS et al A growth factor for cardiac myocytes is produced by cardiac nonmyocytes. Cell. Regul. 2: 1081-96 (1991)

NOB-1: A murine thymoma cell line. see: EL4.

nodal: A murine gene identified by its inactivation through retroviral insertion. nodal encodes a secreted protein belonging to the TGF-β superfamily of proteins. The protein is about equally related to » activin A and » BMPs (bone morphogenetic proteins). Disruption of the nodal gene leads to a failure to form the primitive streak in early embryogensis, a lack of axial mesoderm tissue, and an overproduction of ectoderm and extraembryonic ectoderm.

Conlon FL et al A novel retrovirally induced embryonic lethal mutation in the mouse: Assessment of the developmental fate of embryonic stem cells homozygous for the 413.d proviral insertion. Development 111: 969-81 (1991); **Iannaccone PM et al** Insertional mutation of a gene involved in growth regulation of the early mouse embryo. Dev. Dynamics 194: 198-208 (1992); **Zhou X et al** Nodal is a novel TGF-β like gene expressed in the mouse node during gastrulation. N 361: 543-7 (1993

Non-Interleukin 2: see: LDGF (leukemia-derived growth factor).

Non-myocyte-derived growth factor: see: NMDGF

Non-suppressible insulin-like activity: see: NSILA.

Non-suppressible insulin-like protein: see: NSILP.

***Notch*:** The name of a gene in the genome of *Drosophila melanogaster.* The gene encodes a transcript of 10,148 bp. Notch is an integral membrane protein of 2703 amino acids. Notch is the *Drosophila* homologue of human » EGF receptor. It is involved in early embryonic development and the differentiation of the neuroectoderm. Null mutations of the notch gene cause the transformation of most epidermal cells into neuroblasts. The *Ax* (*Abruptex*) allele of notch maps to the extracellular domain of the protein and causes the lost of adult sensory organs.

The intracellular domain of notch has a length of ≈ 1000 amino acids and is composed of a number of different sequence domains. The extracellular domain of notch contains 36 » EGF-like repeats that differ slightly in sequence. Some of these repeats are involved in the dimerisation and multimerisation of the notch protein. Other repeats function as receptor domains for proteins involved in the differentiation of cells into neural and epidermal precursors. Two of the 36 EGF-like repeats in the extracellular domain of notch interact with another protein, called » delta and possibly also with other proteins containing such repeat elements.

Notch-1, a mouse homologue of *Drosophila* Notch is a protein of 2531 amino acids protein containing a signal peptide, » 36 EGF-like repeats, 3 Notch/lin-12 repeats, a transmembrane domain, and 6 cdc10/ankyrin repeats.

A homologue of *Notch*, ***Xotch***, isolated from » *Xenopus laevis,* can functionally replace Notch. It also interacts with delta protein. The human homologue of *Drosophila* notch is called ***TAN-1***.

Artavanis-Tsakonas S et al The *Notch* locus and the cell biology of neuroblast segregation. ARC 7: 427-52 (1991); **Del Amo FF et al** Cloning, analysis, and chromosomal localization of *Notch*-1, a mouse homologue of *Drosophila Notch*. Genomics 15: 259-64 (1993); **Ellisen L et al** *TAN-1*, the human homologue of the *Drosophila Notch* gene, is broken by chromosomal translocations in T lymphoblastic neoplasms. Cell 66: 649-61 (1991); **Kelley MR et al** Mutations altering the structure of epidermal growth-factor like coding sequences at the *Drosophila Notch* locus. Cell 51: 539-48 (1987); **Kidd S et al** Sequence of the *notch* locus: relationship of the encoded protein to mammalian clotting and growth factors. MCB 6: 3094-108 (1986); **Muskavitch MAT & Hoffman FM** Homologues of vertebrate growth factors in *Drosophila melanogaster* and other vertebrates. CTDB 24: 289-328 (1990); **Rebay I et al** Specific EGF repeats of *Notch* mediate interactions with *delta* and *serrate*: implications for *Notch* as a multifunctional receptor. Cell 67: 687-99 (1991); **Weinmaster G et al** A homologue of *Drosophila Notch* expressed during mammalian development. Development 113: 199-205 (1991); **Wharton KA et al** Nucleotide sequence from the neurogenic locus *notch* implies a gene product that shares homology with proteins containing EGF-like repeats. Cell 43: 567-81 (1985)

NPF: *neurite-promoting factor* This poorly characterized factor is isolated from muscle tissues. It induces the outgrowth of neurites in embryonic neurons. The factor is also a survival factor for cultured motor neurons and sensory and sympathetic neurons.

Jeong SJ et al A neurite-promoting factor from muscle supports the survival of cultured chicken spinal motor neurons. J. Neurobiol. 22: 462-74 (1991)

NPK: *neuropeptide K* see: Tachykinins.

NPγ: *neuropeptide γ* see: Tachykinins.

NRA: *neutrophil releasing activity* This factor which leads to an enhanced release of bone marrow neutrophils and increases levels of circulating neutrophils is identical with » IL1.

Dinarello CA The biology of interleukin 1 and comparison to tumor necrosis factor. Immunol. Lett. 16: 227-31 (1987); **Kaushansky K et al** Interleukin 1 stimulates fibroblasts to synthesize granulocyte-macrophage and granulocyte colony-stimulating factors. Mechanism for the hematopoietic response to inflammation. JCI 81: 92-7 (1988); **Ulich TR et al** Kinetics and mechanisms of recombinant human interleukin 1 and tumor necrosis factor-α-induced changes in circulating numbers of neutrophils and lymphocytes. JI 139: 3406-15 (1987)

NRF: *neurite-retraction factor* This poorly characterized factor of ≈ 70 kDa is secreted by the human neuroblastoma/glioma hybrid cell line NG108-15. It reduces the retraction of neurites in morphologically differentiated cells grown in the presence of dibutyryl-cAMP in serum-free media (see also: SFM).

Ghahary A et al A serum factor inducing neurite retraction of morphologically differentiated neuroblastoma × glioma NG108-15 cells. JCS 92: 251-6 (1989)

NRIF: *neutrophil recruitment inhibitory factor* This poorly characterized activity isolated from lipopolysaccharide-stimulated macrophage monolayers specifically blocks carrageenin-induced neutrophil migration into the peritoneal cavities of rats, following intravenous injection.

Tavares BM et al Macrophages stimulated with lipopolysaccharide release a selective neutrophil recruitment inhibitory factor: an *in vivo* demonstration Braz. J. Med. Biol. Res. 22: 733-6 (1989)

NRK-49F: (NRK = normal rat kidney) A cell line established from a mixed population of rat kidney and fibroblasts cells. The cultivation of these cells in serum-free media requires » EGF, insulin, and retinoic acid.

The cells are used to assay transforming growth factors (see: TGF). The » TGF-β assay is based on the stimulation of colony formation of NRK cells grown in soft agar(see also: Colony formation assay) in the presence of » EGF, which overcomes contact inhibition, and 10 % serum (ED_{50} = 100pg/mL). TGF-α alone is not sufficient to support growth of these cells in soft agar but requires the presence of TGF-β or EGF. A more sensitive assay for this factor is provided by » CCL-64 cells. An alternative and entirely different detection method is » Message amplification phenotyping.
NRK cells growth to confluency in serum-containing medium become quiescent when they are transferred to serum-free medium. They can be restimulated to proliferate by » EGF, » PDGF, and heparin-binding growth factors (see: HBGF), but not by » IGF, TGF-β, or retinoic acid.

Assoian RK et al Transforming growth factor β in human platelets. JBC 258: 7155-9 (1983); **Bradshaw GL & Dubes GR** Polyoma virus transformation of rat kidney fibroblasts results in loss of requirement for insulin and retinoic acid. J. Gen. Virol. 64: 2311-5 (1983); **Marquardt H & Todaro GJ** Human transforming growth factor. Production by a melanoma cell line, purification, and initial characterization. JBC 257: 5220-5 (1982); **Rizzino A** Behavior of transforming growth factors in serum-free media: an improved assay for transforming growth factors. In Vitro 20: 815-22 (1984); **Rizzino A et al** Induction and modulation of anchorage-independent growth by platelet-derived growth factor, fibroblast growth factor, and transforming growth factor-β. CR 46: 2816-20 (1986); **Rizzino A** Soft agar growth assays for transforming growth factors and mitogenic peptides. MiE 146: 341-53 (1987); **Van Zoelen EJJ et al** Transforming growth factor-β and retinoic acid modulate phenotypic transformation of normal rat kidney cells induced by epidermal growth factor and platelet-derived growth factor. JBC 261: 5003-9 (1986); **Van Zoelen EJJ et al** PDGF-like growth factor induces EGF-potentiated phenotypic transformation of normal rat kidney cells in the absence of TGFβ. BBRC 141: 1229-35 (1986); **Van Zoelen EJJ et al** The role of polypeptide growth factors in phenotypic transformation of normal rat kidney cells. JBC 263: 64-68 (1988)

NRP: *negative regulatory protein* This poorly characterized factor of 79 kDa has been found in bone marrow supernatants from C57BL/6 (B6) mice. It rapidly, specifically, and reversibly inhibits DNA synthesis of » BFU-E (erythroid burst-forming unit) *in vitro*. The action of » TGF-β on DNA synthesis of BFU-E was identical to that of NRP in time scale, reversibility, and opposition by IL3, but NRP is not identical with TGF-β.

Axelrad AA et al A protein (NRP) that negatively regulates erythroid stem cell proliferation: antagonism to IL3 stimulation. PCBR 352: 79-86 (1990); **Del Rizzo DF et al** Negative regulation of DNA synthesis in early erythropoietic progenitor cells (BFU-E) by a protein purified from the medium of C57BL/6 mouse marrow cells. PNAS 85: 4320-4 (1988); **Del Rizzo DF et al** Interleukin 3 opposes the action of negative regulatory protein

(NRP) and of transforming growth factor-β (TGF-β) in their inhibition of DNA synthesis of the erythroid stem cell BFU-E. EH 18: 138-42 (1990)

N-*sam*: see: K-*sam*.

NSE: *neuron-specific enolase* At least three genes encode the different isoforms of the glycolytic enzyme enolase. NSE, (γ enolase, EC 4.2.1.11) a unique isoform specifically expressed in neurons and neural-related cells (neuroendocrine cells, pituicytes, and many tumor cells, but not in glia), is encoded by a gene containing 12 exons distributed over 9213 nucleotides. Immunostaining or detection of serum NSE are used as a tumor marker of neuroendocrine cancers. NSE is of special diagnostic value as a panendocrine marker of neuroendocrine tumors that do not produce hormones and peptides.

NSE has been found to be identical with a 44 kDa neuronal survival factor isolated from bovine brain that neuronal survival activity for the cultured neocortical neurons (see also: neurotrophins).

Oliva D et al Complete structure of the human gene encoding neuron-specific enolase. Genomics 10: 157-65 (1991); **Takei N et al** Neuronal survival factor from bovine brain is identical to neuron-specific enolase. J. Neurochem. 57: 1178-84 (1991)

NSILA: *non-suppressible insulin-like activity* This factor is identical with » IGF and IGF-related factors.

Froesch ER et al Insulin-like growth factor (IGF-NSILA): structure, function, and physiology. In: Sato GH & Ross R (eds) Hormones and cell culture. Cold Spring Harb. Conf. Cell Prolif. Vol. 6: 61-77 (1979)

NSILP: *non-suppressible insulin-like protein* This factor is identical with » somatomedins, and hence identical with » IGF and IGF-related factors in combination with corresponding binding proteins (see: IGF-BP).

nsINH: *antigen non-specific inhibitor* This poorly characterized factor (30-35 and 69-65 kDa) is produced by human T lymphocytes cultured *in vitro* for 5 days with *Candida albicans* purified polysaccharide (MPPS). nsINH blocks antigen-driven cell proliferation and the development of natural killer cells when added at the beginning of peripheral blood mononuclear cell culture. nsINH inhibits the production of » IL1, » IL2, the expression of IL2 receptor, and the synthesis of » IFN-γ. Addition of IL2 to the culture fully reverses the suppressive effect of nsINH.

Gilardini-Montani MS et al Regulation of self-major histocompatibility complex reactive human T cell clones. Int. J. Immunopharmacol. 12: 255-60 (1990); **Lombardi G et al** A non-specific inhibitor produced by *Candida albicans* activated T cells impairs cell proliferation by inhibiting interleukin-1 production. Clin. Exp. Immunol. 60: 303-10 (1985); **Lombardi G et al** Monocyte subsets in the production of inhibitory factor by *Candida albicans*-activated human T cells. Immunology 56: 373-6 (1985); **Lombardi E et al** Mechanism of action of an antigen nonspecific inhibitory factor produced by human T cells stimulated by MPPS and PPD. CI 98: 434-43 (1986)

NS suppressor factor: *natural suppressor cell-derived suppressor factor* This factor of 10-13 kDa is secreted by murine bone marrow natural suppressor cells activated by recombinant » IL3 or recombinant » GM-CSF. These cells, which non-specifically suppress immune responses, are generally found at sites of hemopoietic generation or regeneration. NS cells, are non-adherent, radioresistant non-T cells resident in the bone marrow, and have natural cytotoxic (NC), but not natural killer (NK) activity.

The factor is a potent inhibitor of myeloid colony formation at concentrations below those required for immunosuppression. It inhibits the growth of granulocyte-macrophage colony-forming units (see: CFU-GM), granulocyte erythrocyte macrophage megakaryocyte colony-forming units (see: CFU-GEMM), and erythroid colony-forming units (see: CFU-E). The NS suppressor factor strongly inhibits proliferation of the TGF-β-sensitive tumor cell line, A549. The activity of the factor is inhibited by neutralizing antibodies directed against » TGF-β.

Moore SC et al Transforming growth factor-β is the major mediator of natural suppressor cells derived from normal bone marrow. J. Leukoc. Biol. 52: 596-601 (1992); **Moore SC et al** Bone marrow natural suppressor cells inhibit the growth of myeloid progenitor cells and the synthesis of colony-stimulating factors. EH 20: 1178-83 (1992)

NT-3: *neurotrophin-3* also: neuronotrophin 3; abbrev. also: NTF-3. This factor is also called **HDNF** (hippocampus-derived neurotrophic factor). It is found in neurons of the central nervous system. NT-3 is also expressed in muscles and its expression is downregulated in denervated muscles. Many human gliomas express and secrete NT-3.

NT-3 is a basic protein (pI = 9,3) of 119 amino acids. It is synthesized as a precursor with a secretory signal sequence of 18 amino acids and a prosequence of 121 amino acids. The proteins obtained from various mammals are also identical and also display a conserved tissue distribution. Chicken NT-3 differs from its mammalian counterpart by a single conservative amino acid substitution. The human NT-3 gene maps to chromosome 12p13; the murine gene is located on chromosome 6.

Some protein domains of NT-3 are identical with those of » NGF and » BDNF (brain-derived neurotrophic factor). Total homology between these proteins is ≈ 50 % (see also: neurotrophins). The variable domains of NT-3 are probably involved in determining the specificity of expression in certain types of neurons. NT-3 is related to » NT-4 (60 % homology at the protein level) and » NT-5.

NT-3 selectively supports the survival of neuronal cell populations. NT-3 has been shown recently to prevent death of cultured embryonic rat spinal motor neurons at picomolar concentrations. NT-3 has been shown to enhance sprouting of corticospinal tract during development and after adult spinal cord lesion. The activities of NT-3 and » BDNF are additive in some systems. NT-3 acts as a mitogen for cells of the neural crest in serum-free media.

NT-3 has been shown to rapidly potentiate the spontaneous and impulse-evoked synaptic activity of developing neuromuscular synapses in culture and thus appears to be involved in the regulation of functions of developing synapses.

The biological activities of NT-3 are mediated by a receptor belonging to the » *trk* family of receptors with intrinsic tyrosine-specific protein kinase activity. NT-3 only binds weakly to the *trk* receptor which is a high-affinity receptor for » NGF. It has been possible, by combination of structural elements from » NGF, » BDNF and NT-3, to engineer a multifunctional pan-neurotrophin that efficiently activates all *trk* receptors and displays multiple neurotrophic specificities.

● REVIEWS: **Barde YA** The nerve growth factor family. PGFR 2: 237-48 (1990); **Thoenen H** The changing scene of neurotrophic factors. TINS 14: 165-70 (1991); **Vantini G** The pharmacological potential of neurotrophins: a perspective. Psychoneuroendocrinology 17: 401-10 (1992); see also literature cited for » neurotrophins.

● BIOCHEMISTRY & MOLECULAR BIOLOGY: **Ernfors P et al** Molecular cloning and neurotrophic activities of a protein with structural similarities to nerve growth factor. PNAS 87: 5454-8 (1990); **Friedman WJ et al** Transient and persistent expression of NT-3/HDNF mRNA in the rat brain during postnatal development. J. Neurosci. 11: 1577-84 (1991); **Gotz R et al** Production and characterization of recombinant mouse neurotrophin-3. EJB 204: 745-9 (1992); **Hamel W et al** Neurotrophin gene expression by cell lines derived from human gliomas. J. Neurosci. Res. 34: 147-57 (1993); **Hohn A et al** Identification and characterization of a novel member of the nerve growth factor/brain-derived neutrotrophic factor family. N 344: 339-41 (1990); **Jones KR & Reichardt LF** Molecular cloning of a human gene that is a member of the nerve growth factor family. PNAS 87: 8060-4 (1990); **Kaisho Y et al** Cloning and expression of a cDNA encoding a novel human neurotrophic factor. FL 266: 187-91 (1990); **Maisonpierre PC et al** Neurotrophin-3: a neurotrophic factor related to NGF and BDNF. S 247: 1446-51 (1990); **Maisonpierre PC et al** Human and rat brain-derived neurotrophic factor and neurotrophin-3: gene structures, distributions, and chromosomal localisations. Genomics 10: 558-68 (1991); **Maisonpierre PC et al** Gene sequences of chicken BDNF and NT-3. DNA Seq. 3: 49-54 (1992); **Narhi LO et al** Comparison of the biophysical characteristics of human brain-derived neurotrophic factor, neurotrophin-3, and nerve growth factor. JBC 268: 13309-17 (1993); **Özcelik T et al** Chromosomal mapping of brain-derived neurotrophic factor and neurotrophin 3. Genomics 10: 569-75 (1991); **Radziejewski C et al** Dimeric structure and conformational stability of brain-derived neurotrophic factor and neurotrophin-3. B 31: 4431-6 (1992); **Rosenthal A et al** Primary structure and biological activity of a novel human neurotrophic factor. Neuron 4: 767-73 (1990)

● RECEPTORS: see: *trk*.

● BIOLOGICAL ACTIVITIES: **Hamel W et al** Neurotrophin gene expression by cell lines derived from human gliomas. J. Neurosci. Res. 34: 147-57 (1993); **Henderson CE et al** Neurotrophins promote motor neuron survival and are present in embryonic limb bud. N 363: 266-270 (1993); **Kalcheim C et al** Neurotrophin 3 is a mitogen for cultured neural crest cells. PNAS 89: 1661-5 (1992); **Koliatsos VE et al** Evidence that brain-derived neurotrophic factor is a trophic factor for motor neurons *in vivo*. Neuron 10: 359-67 (1993); **Lohof AM et al** Potentiation of developing neuromuscular synapses by the neurotrophins NT-3 and BDNF. N 363: 350-3 (1993); **Schecterson LC & Bothwell M** Novel roles for neurotrophins are suggested by BDNF and NT-3 mRNA expression in developing neurons. Neuron 9: 449-63 (1992); **Schnell L et al** Neurotrophin-3 enhances sprouting of corticospinal tract during development and after adult spinal cord lesion. N 367: 170-3 (1994)

NT-3 knockouts: see addendum: NT-3.

NT-4: *neurotrophin-4* This factor of 123 amino acids is synthesized as a precursor of 236 amino acids. The amino acid sequence of NT-4 is less conserved than those of other » neurotrophins. It displays 50-60 % homology with » NT-3, » NGF and » BDNF (brain-derived neurotrophic factor). NT-4 has a similar activity as » NT-3. NT-4 has been shown to be a target-derived neurotrophic factor for neurons of the trigeminal ganglion. It binds to the low-affinity receptor for » NGF and to the *trk*B receptor (see: *trk*). The human NT-4 gene maps to chromosome19q13.3.

Ibanez CF et al Neurotrophin-4 is a target-derived neurotrophic factor for neurons of the trigeminal ganglion. Development 117: 1345-53 (1993); **Ip NY et al** Mammalian neurotrophin-4: structure, chromosomal localization, tissue distribution, and receptor specificity. PNAS 89: 3060-4 (1992); **Hallböök F et al** Evolutionary studies of the nerve growth factor family reveal a novel member abundantly expressed in *Xenopus* ovary. Neuron 6: 845-58 (1991); **Klein R et al** The *trk*B tyrosine protein kinase is a receptor for neurotrophin-4. Neuron 8: 947-56 (1992); see also literature cited for » neurotrophins.

NT-5: *neurotrophin-5* NT-5 is a homologue of » NT-4 which was initially isolated from » *Xenopus* *laevis*. NT-4 has more or less the same activities as » NT-3. The human NT-5 gene maps to chromosome 19. NT-5 has been shown recently to prevent death of cultured embryonic rat spinal motor neurons at picomolar concentrations. See also: Neurotrophins.

Berkemeier LR et al Neurotrophin-5: A novel neurotrophic factor that activates *trk* and *trk*B. Neuron 7: 857-66 (1991); **Berkemeier LR et al** Human chromosome 19 contains the neurotrophin-5 gene locus and three related genes that may encode novel acidic neurotrophins. Somat. Cell Molec. Genet. 18: 233-45 (1992); **Henderson CE et al** Neurotrophins promote motor neuron survival and are present in embryonic limb bud. N 363: 266-270 (1993); see also literature cited for » neurotrophins.

NT-6: *neurotrophin-6* NT-6 has been found to comprise three acidic proteins (α, β, γ) that are 95 % identical to each other and 75 % identical with » NT-5.

Berkemeier LR et al Human chromosome 19 contains the neurotrophin-5 gene locus and three related genes that may encode novel acidic neurotrophins. Somat. Cell Molec. Genet. 18: 233-45 (1992)

NTF-3: Alternative abbrev. for » NT-3 (neurotrophin-3).

NTS: *nuclear transport sequence* or *nuclear targeting sequence*. see: Signal sequences.

Nuclear factors: see: NF.

Nuclear factor for activated T cells: see: NF-AT.

Nuclear factor for IL6: see: NF-IL6.

Nuclear localization sequence: abbrev. NLS. See: signal sequences.

Nuclear polyhedrosis viruses: See: Baculovirus expression system.

Nuclear targeting sequence: abbrev. NTS. See: signal sequences.

Nuclear transport sequence: See: signal sequences.

Nucleoside diphosphate kinase: NDP kinase. see: nm23.

Nude mouse: see: Immunodeficient mice.

Null mutations of cytokine genes: see: Knockout mice.

nup475: *nuclear protein 475* An » ERG (early response gene) whose transcription is rapidly and transiently stimulated by cross-linking of membrane immunoglobulin on B lymphocytes. Nup475 encodes a transcription factor (see also: Gene expression). The gene is identical with ***GOS24*** (see: GOS), a member of a set of putative G0/G1 switch regulatory genes (see also: Cell cycle) that are expressed transiently within 1-2 hr of the addition of lectin or cycloheximide to human blood mononuclear cells. Nup475 is also identical with ***TTP***, ***TIS11*** (see: TIS), and ***Zfp36***. Expression of nup475 is rapidly induced by » GM-CSF in a factor-dependent cell line, » 32D.

Du Bois RN et al A growth factor-inducible nuclear protein with a novel cysteine/histidine repetitive sequence. JBC 265: 19185-91 (1990); **Ma Q & Herschman HR** A corrected sequence for the predicted protein from the mitogen-inducible TIS11 primary response gene. O 6: 1277-8 (1991); **Mittelstadt PR & De Franco AL** Induction of early response genes by cross-linking membrane Ig on B lymphocytes. JI 150: 4822-32 (1993); **Taylor GA et al** The human TTP protein: sequence, alignment with related proteins, and chromosomal localization of the mouse and human genes. NAR 19: 3454 (1991); **Varnum BC et al** Granulocyte-macrophage colony-stimulating factor and tetradecanoyl phorbol acetate induce a distinct, restricted subset of primary-response TIS genes in both proliferating and terminally differentiated myeloid cells. MCB 9: 3580-3 (1989)

Nur-77: *nuclear receptor 77* An » ERG (early response gene) the transcription of which is rapidly and transiently stimulated by cross-linking of membrane immunoglobulin on B lymphocytes. Nur-77 is inducible by » NGF or membrane depolarization in the rat pheochromocytoma cell line » PC12, by serum growth factors in fibroblasts, and during liver regeneration. nur77 is identical with » NAK1 and NGFI-B (see: NGFI). The protein has extensive sequence homology to members of the steroid/thyroid hormone receptor superfamily. By analogy to steroid receptors, the nur77 protein is thought to act as a ligand-dependent transcription factor (see also: Gene expression) that regulates the genomic response to growth factors. nur-77 is the mouse homologue of the human TR3 orphan receptor and identical with ***RNR-1*** (regenerating liver nuclear receptor).

Nur-77 has been implicated in the regulation of negative selection during T cell development. Nur-77 is a member of a family of proteins encoding nuclear receptors. One of these, designated ***Nurr1*** appears to be expressed specifically in the brain.

■ **Transgenic/Knock-out/Antisense Studies:** Blocking of nur-77 expression by » Antisense oligonucleotides directed against nur-77 has been shown to inhibit » apoptosis in cells stimulated via the T cell receptor.

Davis IJ et al Transcriptional activation by Nur77, a growth factor-inducible member of the steroid hormone receptor superfamily. Mol. Endocrinol. 5: 854-9 (1991); **Law SW et al** Identification of a new brain-specific transcription factor, NURR1. Mol. Endocrinol. 6: 2129-35 (1992); **Lee HJ et al** The use of a DNA-binding domain replacement method for the detection of a potential TR3 orphan receptor response element in the mouse mammary tumor virus long terminal repeat. BBRC 193: 97-103 (1993); **Mittelstadt PR & De Franco AL** Induction of early response genes by cross-linking membrane Ig on B lymphocytes. JI 150: 4822-32 (1993); **Von JK & Lau LF** Transcriptional activation of the inducible nuclear receptor gene nur77 by nerve growth factor and membrane depolarization in PC12 cells. JBC 268: 9148-55 (1993); **Woronicz JD et al** Requirement for the orphan steroid receptor Nur77 in apoptosis of T cell hybridomas. N 367: 277-81 (1994)

● **Transgenic/Knock-out/Antisense Studies: Liu ZG et al** Apoptotic signals delivered through the T cell receptor of a T cell hybrid require the immediate early gene nur 77. N 367: 281-4 (1994)

Nurr-1: see: Nur-77.

NZB-SF: *New Zealand black serum factor* NZB is an inbred strain of black mice with a high incidence of autoimmune defects characterized, among other things, by pre-T, T and B cell defects, defects in DNA repair, and abnormal levels of thymic hormones. NZB mice also suffer from glomerular and tubular kidney dysfunctions and develop a severe autoimmune hemolytic anemia.

NZB-SF is a glycoprotein of 60 kDa isolated from the serum of young animals. NZB-SF is distinguishable from » IL1α, » IL2, » IL3, » IL4, » IL5, » IL6, » GM-CSF, » IFN-γ, and » TNF-α. The factor enhances the maturation and proliferation of pre-B cells in bone marrow. It also influences the composition of B cell populations in the bone marrow and the spleen *in vivo*.

Jyonouchi H et al NZB serum factor (NZB-SF): B precursor cell maturation factor identified in murine lupus. I. Identification of 60 kDa glycoprotein as the major component from both spleen cell supernatant and serum. CI 132: 223-35 (1991); **Jyonouchi H et al** NZB serum factor (NZB-SF) – B precursor cell maturation factor. II. *In vivo* effects of NZB-SF or mAb against NZB-SF on B lineage cell population. Clin. Immunol. Immunopathol. 59: 388-97 (1991)

O

o: sometimes uses as a prefix to denote "of ovine origin".

OAF: *osteoclast activating factor* This activity is secreted by macrophages obtained from osteoporosis patients. It enhances bone resorption. OAF is probably a mixture of IL1β (see: IL1), » TNF-α, » LIF, and some other locally produced factors with IL1β being the major factor.
Abe E et al Differentiation-inducing factor purified from conditioned medium of mitogen-treated spleen cell cultures stimulates bone resorption. PNAS 83: 5958-62 (1986); **Dewhirst FE et al** Purification and partial sequence of human osteoclast activating factor: identity with interleukin 1β. JI 135: 2562-8 (1985); **Gowen M et al** An interleukin-1-like factor stimulates bone resorption *in vitro* N 306: 378-80 (1983)

OCI/AML1a: This lymphoblastoid cell line has been established from leukemic blast progenitors of a patient with acute myeloblastic leukemia after continuous culture in the presence of » G-CSF. OCI/AML1a cells do not respond to » GM-CSF, » IL3, IL1 or » SCF. The stimulatory effect of G-CSF on OCI/AML1a cells is almost completely blocked by monoclonal anti-G-CSF antibody.
Nara N Colony-stimulating factor (CSF)-dependent growth of two leukemia cell lines. Leuk. Lymphoma 7: 331-5 (1992)

O-CSF: *osteoclast colony-stimulating factor* This poorly characterized factor of 17 kDa is secreted by murine mammary carcinoma cells. It causes marked bone resorption in animals and promotes the differentiation of osteoclasts in murine bone marrow cultures.
Lee MY et al Isolation of a murine osteoclast colony-stimulating factor. PNAS 88: 8500-4 (1991)

ODC factor: *ornithine decarboxylase-inducing factor* This poorly characterized factor of ≈ 79 kDa factor is isolated from the ascites fluid of Ehrlich ascites cells. It influences the differentiation of liver cells, reduces catalase activity and influences the metabolism of thyroid hormones.
Another, obtained from the culture medium of a murine macrophage cell line (P388D1), also induces ornithine decarboxylase in the liver and spleen upon injection into mice. It is identical with » IL1 or a molecule closely related to it.
Endo Y et al Induction of ornithine decarboxylase in the liver and spleen of mice by interleukin 1-like factors produced from a macrophage cell line. BBA 838: 343-50 (1985); **Imamura K et al** Purification of ornithine decarboxylase-inducing factor from cell-free ascites fluid of Ehrlich ascites tumor and its characteristics. JJCR 82: 315-24 (1991); **Sasaki K et al** Mechanism of hepatic ornithine decarboxylase induction by the ornithine decarboxylase-inducing factor isolated from tumor ascites fluid: determination of target cells for the factor in the liver. J. Biochem. Tokyo 94: 949-59 (1983)

ODGF: *osteosarcoma-derived growth factor* This factor is isolated from an osteosarcoma cell line. It is identical with the A chain of » PDGF.
Heldin CH et al A human osteosarcoma cell line secretes a growth factor structurally related to a homodimer of PDGF A chains. N 319: 511-4 (1986); **Heldin CH et al** Chemical and biological properties of a growth factor from human cultured osteosarcoma cells: resemblance with platelet-derived growth factor. JCP 105: 235-46 (1980)

OGF: *oligodendroglia growth factor* This poorly characterized factor of 16 kDa is also called ***GPF-1*** (glial promoting factor 1). It is produced by neuronal cells. The factor selectively stimulates the growth of astroglial cells and oligodendrocytes. In induces the synthesis of myelin proteins *in vivo*.
Giulian D et al Brain peptides and glial growth. I. Glial promoting factors as regulators of gliogenesis in the developing and injured central nervous system. JCB 102: 803-11 (1986); **Giulian D et al** A growth factor from neuronal cell lines stimulates myelin protein synthesis in mammalian brain. J. Neurosci. 11: 327-36 (1991)

OGF: *osteoclast growth-inducing factor* This poorly characterized factor of > 19 kDa is produced in chicken and mouse bone cultures. It induces the growth of osteoclasts.
Dickson IR & Scheven BA Regulation of new osteoclast formation by a bone cell-derived macromolecular factor. BBRC 159: 1383-90 (1989)

OGF: *ovarian growth factor* This factor has been isolated from bovine pituitaries. It is mitogenic for ovatian cells and has the activity of an » angiogenesis factor. OGF is either identical with » aFGF or » bFGF or very closely related.

Gospodarowicz D et al Purification of a growth factor for ovarian cells from bovine pituitary glands. PNAS 71: 2295-9 (1974); **Makris A et al .**
The nonluteal porcine ovary as a source of angiogenic activity. Endocrinol. 115: 1672-7 (1984)

OIF: *osteoinductive factor* This factor is isolated from demineralised bovine bone matrix. It has been suggested that it be renamed » osteoglycin.
OIF is a glycoprotein. Two variants of 22 and 28 kDa are known. The non-glycosylated factor has a molecular mass of ≈ 12 kDa and contains a disulfide bond between cysteines 62 and 95.
The mature factor of 105 amino acids is obtained by cleavage of the C-terminal end of a precursor protein of 299 amino acids (298 amino acids in humans). Human and bovine OIF are 94 % identical at the protein level.
In vivo OIF supports the growth of bones and induces the differentiation of bone marrow cells into cells with osteoblast functions. OIF inhibits osteoclasts. Subcutaneous administration of the factor induces the ectopic generation of bones.
The bone-inducing activity of OIF probably involves interactions of the factor with the extracellular matrix (see also: ECM). Its activity is synergised strongly by » CIF-A (cartilage-inducing factor) and » TGF-β but not by » PDGF, » FGF and IGF-1 (see: IGF).
Chondrocytes which dedifferentiate in culture and lose their ability to synthesize collagen type II and proteoglycans can retain these functions if treated with OIF. This activity of OIF is strongly enhanced by a combination of » EGF, » PDGF, » aFGF and insulin.
Bentz H Purification and characterization of a unique osteoinductive factor from bovine bone. JBC 264: 20805-10 (1989); **Bentz H** Amino acid sequence of bovine osteoinductive factor. JBC 265: 5024-9 (1990); **Bentz H** Transforming growth factor β 2 enhances the osteoinductive activity of a bovine bone-inducing fraction containing bone morphogenetic protein 2 and 3. Matrix 11: 269-75 (1991); **Kukita A et al** Osteoinductive factor inhibits formation of human osteoclast-like cells. PNAS 87: 3023-6 (1990); **Madisen L et al** Molecular cloning of a novel bone-forming compound: osteoinductive factor. DNA and Cell Biol. 9: 303-9 (1990); **Oreffo ROC et al** Inhibitory effects of the bone-derived growth factors osteoinductive factor and transforming growth factor β on isolated osteoclasts. E 126: 3069-75 (1990)

OIP: *osteogenesis inhibitory protein* This poorly characterized factor is obtained from bone material. It antagonizes the biological activities of » BMP (bone morphogenetic protein) *in vivo* and *in vitro*. OIP reduces the BMP-induced differentiation of mesenchymal cells and inhibits the synthesis of glycosaminoglycans (see also: ECM).
Brownell AG Osteogenesis inhibitory protein: A (P)review. Connect. Tissue Res. 24: 13-6 (1990)

Okadaic acid: This is a tumor promoter on mouse skin and a potent and specific cell-permeating inhibitor of protein phosphatases (serine phosphatases 2A and 1) produced by the marine black sponge *Halichondria okadai*. Three other main types of okadaic acid class compounds, ***calyculin A***, isolated from the marine sponge *Discodermia*

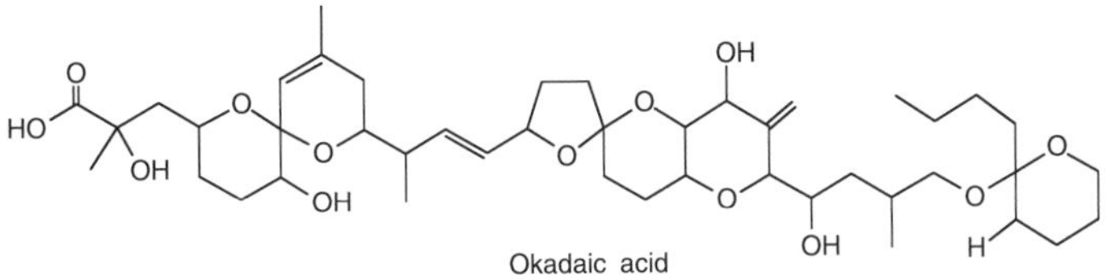

Okadaic acid

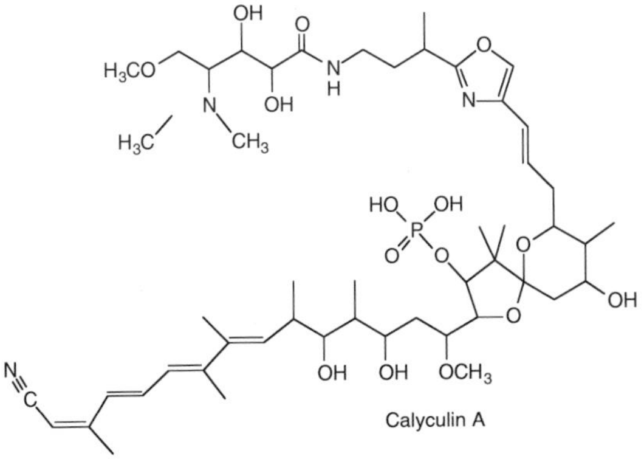

Calyculin A

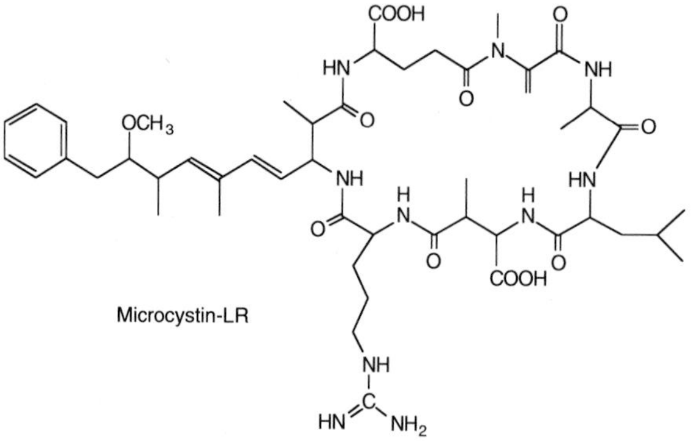

Microcystin-LR

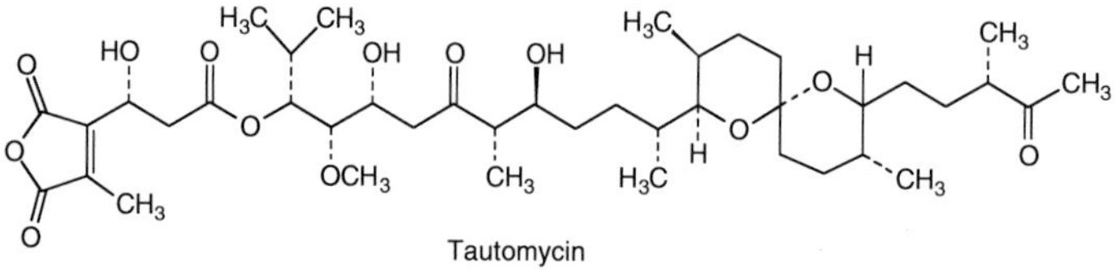

Tautomycin

calyx, ***microcystin-LR***, a cyclic heptapeptide product of the blue-green alga *Microcystis aeruginosa*, and ***tautomycin***, found in *Streptomyces spirover ticillatus* also selectively inhibit protein serine/threonine but not tyrosine phosphatase activity.

Okadaic acid completely inhibits phosphatase 2A at nanomolar concentrations and of type 1 phosphatases at 1 μM. Phosphatase 2B is significantly inhibited only at concentrations > 1 μM. Microcystin appears to be equally potent inhibitor for both types of phosphatase. Calyculin A inhibits protein phosphatase 2A.

The compounds cause a significant increase in total cellular protein phosphorylation by blocking dephosphorylation of phosphorylated proteins. Examination of inhibitor-sensitive processes therefore reveals the importance of protein serine/threonine phosphorylation states in the regulation of cellular processes and provides evidence for physiological functions of protein phosphatases. Okadaic acid and the other inhibitors can be used to mimic some of the effects normally elicited by » cytokines such as transcriptional activation of » ERGs (early response genes), induction of » Cell activation, prevention of cellular adhesion and cellular aggregation, production and release of growth factors, differentiation, inhibition of » apoptosis, and proliferation. Its use allows the identification of target molecules of intracellular signal transduction pathways that become phosphorylated in response to environmental stimuli. For other agents used to dissect cytokine-mediated signal transduction pathways see: Bryostatins, Calcium ionophore, Calphostin C, Genistein, H8, Herbimycin A, K-252a, Lavendustin A, Phorbol esters, Staurosporine, Suramin, Tyrphostins, Vanadate.

● **Reviews: Fujiki H & Suganuma M** Tumor promotion by inhibitors of protein phosphatases 1 and 2A: the okadaic acid class of compounds. ACR 61: 143-94 (1993); **Schonthal A** Okadaic acid – a valuable new tool for the study of signal transduction and cell cycle regulation? New Biol. 4: 16-21 (1992)

Adunyah SE et al Induction of differentiation and c-*jun* expression in human leukemic cells by okadaic acid, an inhibitor of protein phosphatases. JCP 151: 415-26 (1992); **Candeo P et al** Pathological phosphorylation causes neuronal death: effect of okadaic acid in primary culture of cerebellar granule cells. J. Neurochem. 59: 1558-61 (1992); **Chiou JY & Westhead EW** Okadaic acid, a protein phosphatase inhibitor, inhibits nerve growth factor-directed neurite outgrowth in PC12 cells. J. Neurochem. 59: 1963-6 (1992); **Cohen P et al** An improved procedure for identifying and quantitating protein phosphatases in mammalian tissues. FL 250: 596-600 (1989); **Ganapathi MK et al** Okadaic acid, an inhibitor of protein phosphatases 1 and 2A, inhibits induction of acute-phase proteins by interleukin-6 alone or in combination with interleukin-1 in human hepatoma cell lines. BJ 284: 645-8 (1992); **Guy GR et al** Okadaic acid mimics multiple changes in early protein phosphorylation and gene expression induced by tumor necrosis factor or interleukin-1. JBC 267: 1846-52 (1992); **Higuchi M & Aggarwal BB** Okadaic acid induces down-modulation and shedding of tumor necrosis factor receptors. Comparison with another tumor promoter, phorbol ester. JBC 268: 5624-31 (1993); **Holladay K et al** Okadaic acid induces the expression of both early and secondary response genes in mouse keratinocytes. Mol. Carcinog. 5: 16-24 (1992); **Ishida Y et al** Treatment of myeloid leukemic cells with the phosphatase inhibitor okadaic acid induces cell cycle arrest at either G1/S or G2/M depending on dose. JCP 150: 484-92 (1992); **Ishihara H et al** Calyculin A and okadaic acid: inhibitors of protein phosphatase activity. BBRC 159: 871-7 (1989); **Kharbanda S et al** Regulation of c-*jun* expression during induction of monocytic differentiation by okadaic acid. Cell Growth Differ. 3: 391-9 (1992); **Kharbanda S et al** Transcriptional regulation of the early growth response 1 gene in human myeloid leukemia cells by okadaic acid. Cell Growth Differ. 4: 17-23 (1993); **Lu DJ et al** Modulation of neutrophil activation by okadaic acid, a protein phosphatase inhibitor. Am-J-Physiol. 1992 Jan; 262(1 Pt 1): C39-49 (1992); **Luukkainen R et al** Isolation and identification of eight microcystins from thirteen Oscillatoria agardhii strains and structure of a new microcystin. Appl. Environ. Microbiol. 59: 2204-9 (1993); **MacKintosh C et al** Cyanobacterial microcystin-LR is a potent and specific inhibitor of protein phosphatases 1 and 2A from both mammals and higher plants. FL 264: 187-92 (1990); **Oikawa T et al** Okadaic acid is a potent angiogenesis inducer. Jpn. J. Cancer Res. 83: 6-9 (1992); **Park K et al** Inhibition of myogenesis by okadaic acid, an inhibitor of protein phosphatases, 1 and 2A, correlates with the induction of AP1. JBC 267(15): 10810-5 (1992); **Rieckmann P et al** Okadaic acid is a potent inducer of AP-1, NF-kappa B, and tumor necrosis factor-α in human B lymphocytes. BBRC 187: 51-7 (1992); **Sakurada K et al** Comparative effects of protein phosphatase inhibitors (okadaic acid and calyculin A) on human leukemia HL60, HL60/ADR and K562 cells. BBRC 187: 488-92 (1992); **Song Q et al** Inhibition of apoptosis in human tumor cells by okadaic acid. JCP 153: 550-6 (1992); **Song Q & Lavin MF** Calyculin A, a potent inhibitor of phosphatases-1 and -2A, prevents apoptosis. BBRC 190: 47-55 (1993); **Stotts RR et al** Structural modifications imparting reduced toxicity in microcystins from Microcystis spp. Toxicon 31: 783-9 (1993); **Suganuma M et al** Structurally different members of the okadaic acid class also selectively inhibit protein serine/threonine but not tyrosine phosphatase activity. Toxicon 30: 873-8 (1992); **Takai A et al** Estimation of the rate constants associated with the inhibitory effect of okadaic acid on type 2A protein phosphatase by time-course analysis. BJ 287: 101-6 (1992); **Takai A et al** Inhibitory effect of okadaic acid derivatives on protein phosphatases. A study on structure-affinity relationship. BJ 284: 539-44 (1992); **Walker TR & Watson SP** Okadaic acid inhibits activation of phospholipase C in human platelets by mimicking the actions of protein kinases A and C. Br. J. Pharmacol. 105: 627-31 (1992)

OKT: *Ortho-Kung-T cell* A general abbrev. for a panel of monoclonal antibodies recognizing a variety of antigens on the cell surface of leukocytes. Kung is the name of the scientist who developed these antibodies. The OKT series of monoclonal antibodies are widely used for the analysis of human peripheral blood T lymphocytes. OKT designations are still frequently used as synonyms for some cell surface markers recognized by these antibodies (see » CD antigens).

OKT3-induced cytokine-related syndrome: A clinical syndrome observed after treatment with OKT3, an antibody directed against CD3, which is part of the T cell receptor complex. It is characterized by the excessive release of cytokines such as » TNF, » IFN-γ and » IL2 into the circulation. Pretreatment with corticosteroids or administration of anti-TNF antibodies can markedly ameliorate the symptoms caused by the release of these cytokines. In mice methylprednisolon almost completely inhibits the systemic release of these cytokines if given before the administration of anti-TNF antibodies. Anti-IFN-γ antibodies have been found to protect mice against pathological changes induced by injection of anti-CD3 antibodies and to reduce dramatically the incidence of diarrhea and severity of hypothermia as well as mortality rates.

Alegre ML et al Evidence that pentoxifylline reduces anti-CD3 monoclonal antibody-induced cytokine release syndrome. Transplantation 52: 674-9 (1991); **Alegre ML et al** Cytokine release syndrome induced by the 145-2C11 anti-CD3 monoclonal antibody in mice: prevention by high doses of methylprednisolone. JI 146: 1184-91 (1991); **Chatenoud L et al** Corticosteroid inhibition of the OKT3-induced cytokine related syndrome – dosage and kinetics prerequisites. Transplantation 51: 334-48 (1991); **Ferran C et al** Anti-TNF abrogates the cytokine-related anti-CD3-induced syndrome. Transplant. Proc. 23: 849-50 (1991); **Matthys O et al** Modification of the anti-CD3-induced cytokine release syndrome by anti-interferon-γ or anti-interleukin-6 antibody treatment: protective effects and biphasic changes in blood cytokine levels. EJI 23: 2209-16 (1993); **Peces R et al** High-dose methylprednisolone inhibits the OKT3-induced cytokine-related syndrome Nephron 63: 118 (1993)

OKT4: Alternative name for » CD4 after the monoclonal antibody recognizing this cell surface protein (see also: CD antigens).

OKT8: Alternative name for » CD8 after the monoclonal antibody recognizing this cell surface protein (see also: CD antigens).

OKT10: Alternative name for » CD23 after the monoclonal antibody recognizing this cell surface protein (see also: CD antigens).

OKT11: Alternative name for » CD2 after the monoclonal antibody recognizing this cell surface protein (see also: CD antigens).

Oligodendroglia growth factor: see: OGF.

OM: abbrev. for » Oncostatin M.

Omega-interferon: see: IFN-ω. See also: IFN for general information about interferons.

Oncogenes: Some retroviruses rapidly induce tumors when inoculated into animals and can transform cells *in vitro*. Oncogenes were initially detected as retroviral genes directly responsible for the immortalisation of eukaryotic cells and

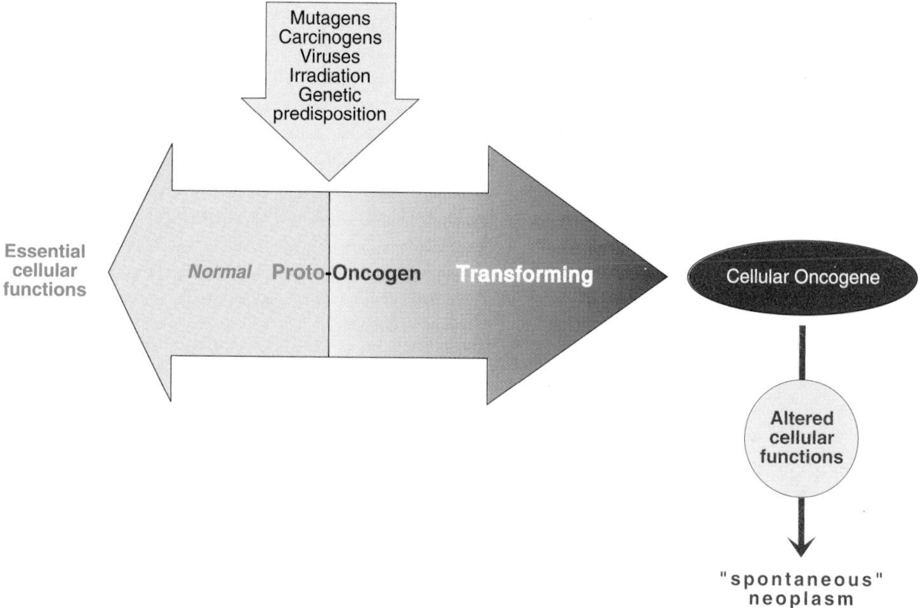

The dualistic nature of cellular proto-oncogenes.
The controlled expression of cellular proto-oncogenes plays an important role for normal cellular processes, including activation, proliferation, and differentiation. "Activation" of a proto-oncogenes disrupts normal cellular processes and causes malignant transformation of cells. Oncogene "activation" can be the result of gene amplification (e. g. c-myc, c-erbB), point mutations (e. g. c-ras, c-src, c-neu), insertions (e. g. c-myc, c-myb, c-mos, c-erbB, c-int-1, c-raf), or translocation (c-myc, c-abl). These alterations per se can cause malignant transformation. Often, however, it is the accumulation of several genetic alterations, including the loss of tumor suppressor genes, that causes the development of malignant tumors.

malignant transformation both *in vitro* and *in vivo*. Many of the viral oncogenes (general abbrev. ***v-onc***) have cellular homologues (general abbrev. ***c-onc***) known as proto-oncogenes. Evidence indicates that viral *onc* genes originated from cellular genes by recombination between a parent non-transforming virus and host cellular DNA. Their gene products are generally referred to as oncoproteins. Proto-oncogenes are evolutionary highly conserved genes required for normal cell functions including proliferation and differentiation.

In many instances the study of retroviral oncogene functions has provided insights into the role of cellular proto-oncogenes. The central biological functions of proto-oncogenes is underscored by the finding that homologous genes have been maintained in the genomes of lower eukaryotes (yeasts), insects, and humans.

At present ≈ 80 oncogenes are known but it is expected that their number is on the order of 200-300. Cellular oncogenes have been taken up into the genomes of retroviruses in the cause of evolution. Retroviral oncogenes have lost the intron-exon structures of their cellular counterparts (see also: Gene expression).

Some cellular proto-oncogenes are communication genes encoding products displaying homologies to growth factors (see: *hst*, *int*-2, *sis*). Other cellular proto-oncogenes encode growth factor receptors (see: *fms*, *neu*, *erb*). Yet other cellular proto-oncogenes (see: *fos*, *jun*, *vav*) are involved in different ways in signal transduction mechanisms associated with binding of growth factors to their receptors. Many of these genes have been shown to play an important role as so-called early response genes (see: ERG) that are rapidly and transiently activated in response to environmental stimuli. Aberrant expression of an oncogene in » factor-dependent cell lines frequently short-circuits factor dependencies and creates cells that can survive and grow in the absence of the growth factor. Under normal circumstances the activity of proto-oncogenes does not cause cell transformation. Genetic alterations including deletions, insertions, point mutations and translocations can activate cellular proto-oncogenes into oncogenes with

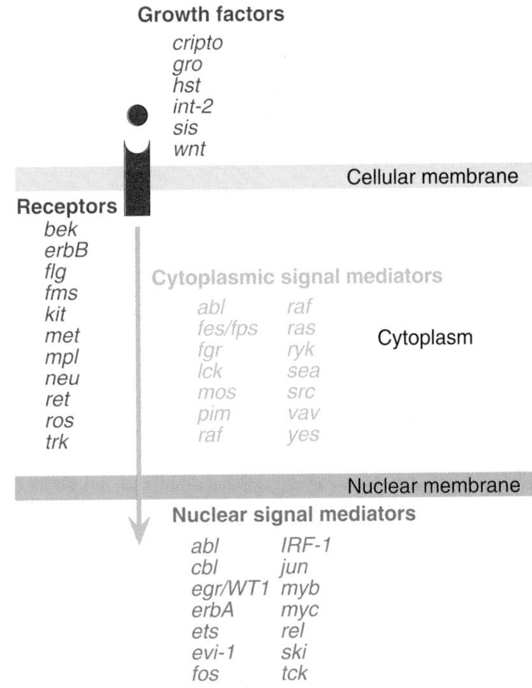

An overview of some oncogenes.
Many oncogene products are involved in growth factor-mediated processes. Some act as cytokines or function as membrane-bound cytokine receptors. Others are receptor-associated and/or localised within the cytoplasm or the nucleus. They are involved in intracellular signal transduction processes initiated by binding of a cytokine to its receptor.

transforming potential. Gene dose effects such as elevated expression caused, for example, by gene amplification or promoter insertion may also be responsible for the activated phenotype of a proto-oncogene. In many instances the detection of aberrantly expressed oncogenes is of clinical relevance and may have prognostic value.

The activities of activated proto-oncogenes are one of the key mechanisms involved in the malignant transformation of cells and the development of malignant cells that have escaped the constraints of normal regulated cell growth and differentiation. Activated oncogenes are involved in different phases of the initiation, progression, and/or metastasis of tumor cells. Most oncogenes transform cells by taking over the roles of growth factors, hormone receptors, G proteins, intracellular effector systems, and transcription factors and expressing these functions in an uncontrolled manner. The activated forms of proto-oncogenes encoding growth factor receptors are frequently altered in a way that allows them to initiate intracellular signal transduction pathways which would normally be expressed transiently following ligand binding to the receptor. Malignant cells then frequently show an ability to regulate their own growth by the release of » autocrine growth modulatory substances. Most important, the growth of certain malignant cell types has been shown to depend on such autocrine growth circuits. Information is rapidly emerging concerning autocrine growth factors in selected human malignancies and strategies allowing such circuits to be broken. Several oncogenes are competence genes involved in » cell activation that are expressed in target cells encountering a cytokine.

The so-called ***anti-oncogenes*** or tumor suppressor genes are oncogenes involved in negative growth control. These genes may encode growth factors that inhibit the growth of tumor cells or they may interact in a number of different ways with the regulation of growth factor expression and function.

The loss of growth inhibitory genes, caused for example, by deletion of alleles, is one of the key features in the genesis of tumors. One particular

Chromosomal localization of some oncogenes on human chromosomes.

Oncogene	Chromosomal localization
abl	9q34.1
bcl-2	18q21
cbl	11q23.3
erb1	7p13-q11.2
erb2	17
erbA1	3p22-p24
ets-1	11q23-q24
ets-2	21q22.3
fes/fps	15q26.1
FGF-5	4q21
FGF-6	12p13
fgr	1p36.1-36.2
fms	5q33.3
fos	14q21-q31
Ha-ras1	11p15
hst/FGF-4	11q13
int-2	11q13. 3
jun	1p31-p32
Ki-ras2	12p12-ter
kit	4q11-q21
K-*sam*	10q26
L-myc	1p32
met	7q21-q31
mos	8q22
mpl	1p34
myb	6q22-q24
myc	8q24
neu	17q21
N-myc	2p24
N-ras	1cen-1p21
pim-1	6p21.2
raf1	3p25
raf2	3p25
rel	10q11.2
ros	6q22
ryk	3q11-25
sis	22q12-q13
ski	1q22-qter
src-1	20q12-q13
src-2	1p36-p34
trk	1q23-1q24
vav	19p12-19p13.2
yes	18p11.32

tumor suppressor gene, called p53 or TP53, for example, is a nucleophosphoprotein of 375 amino acids. It is a sequence-specific DNA binding protein that functions as a transcription factor (see also: Gene expression) which can down-modulate the activity of several promoters. Mutations in the coding sequences of p53 have been shown to be one of the major aberrations observed in human tumors. Mutations in the p53 tumor suppressor gene have also been implicated in the pathogenesis of a dominantly inherited familial cancer syndrome (Li-Fraumeni syndrome). Transgenic mice overproducing a mutated p53 protein are predisposed to the development of lung carcinomas, bone and soft-tissue sarcomas, as well as lymphoid malignancies. Complete absence of p53 in mice shows that these mice are developmentally normal but are prone to develop a variety of neoplasms by 6 months of age.

p53 is not expressed in certain myeloleukemic cell lines (see also: M1). The introduction of a functional p53 gene into these cells initiates cell death by » apoptosis. Cell death can be prevented by treatment of the cells with » IL6 which causes the differentiation of M1 cells into monocytes. Reconstitution of wild-type p53 functions in transformed cells which have lost p53 has been shown to cause growth arrest.

It may be that the malignant transformation of cells is caused by aberrant expression of oncogenes and also by the activation of genes encoding products supporting the survival of cells that are normally destined to die (see: *bcl*-2). The lack of expression of a tumor suppressor gene product normally causing apoptotic cell death could then guarantee the survival of a tumor cell.

Experiments with mice harboring the null mutation of p53 demonstrate that p53 is not required for the normal development of the embryo. These animals appear to be prone to the generation of a number of different spontaneous tumors.

One particularly valuable tool for studying the effects and consequences of oncogene expression are » transgenic animals. In spite of the potentially lethal effects of untimely and/or unregulated oncogene expression such genes can be studied, for example, if the expression is restricted to a specific tissue, which can be achieved by choosing suitable regulatory elements. (see also: Gene expression).

The use of oncogene-transgenic animals allows the spectrum of tissues susceptible to the transforming activity of an oncogene, the relation between multistep oncogenesis and the co-operativity of oncogenes, and the effect of oncogenes on growth and differentiation to be studied in great detail during normal and malignant development. Viral oncogenes such as the large T antigen gene of SV40 virus, polyoma virus large and middle T genes, bovine papilloma virus, human JC and BK viruses, human T cell leukemia virus tat oncogene, the LTR sequence of Rous sarcoma virus, and

Overview of some known tumor suppressor genes in man.

Gene	Tumor	Protein function	Reference
α6 integrin	Mammary tumors	Adhesion molecule	FASEB J 7: 964 (1993)
APC	Adenomatous polyposis	?	Cell 66: 589 (1991)
DCC	Colorectal carcinoma	NCAM-like adhesion molecule	Blood 81: 2696 (1993)
HCC	Hepatocellular carcinoma	?	Cancer Detect. Prev. 17: 405 (1993)
K-*rev*1	?	GTP-binding protein	Cell 56: 77 (1989)
MCC	Colorectal carcinoma	?	Science 251: 1366 (1991)
NF1	Neurofibromatosis	GTPase-activatin protein	Cell 62: 599 (1990)
NF2	Neurofibromatosis, Schwannomas	cytoskeletal component link protein	Cell 72: 791 (1993)
nm23	Breast and colorectal carcinoma	Nucleoside diphosphate kinase	Cancer Cells 3: 257 (1991)
p53	lung, brain, breast tumors	transcription factor	Biochem. Biophys. Acta 1155: 181 (1993)
prohibitin	Breast cancer	?	Cancer Res. 52: 1643 (1992)
PTPγ	Renal cell, lung carcinoma	protein tyrosine phosphatase	Proc. Natl. Acad. Sci. 88: 5036 (1991)
RB-1	Retinoblastoma, osteosarcoma	nuclear transcriptional repressor	Cell 58: 1085 (1989)
VHL	Von Hippel-Lindau disease	?	Science 260: 1317 (1993)
WT1	Wilm's kidney tumor	Zinc finger transcription factor	FASEB J 7: 896 (1993)

For reviews see also: **Bryant PJ** Towards the cellular functions of tumor suppressors. Trends Cell Biol. 3: 31-5 (1993); **Levine AJ** The tumor suppressor genes. Annu. Rev. Biochem. 62: 623-51 (1993); **Weinberg R** Tumor suppressor genes. Neuron 11: 191-6 (1993); Yamamoto T Molecular basis of cancer: oncogenes and tumor suppressor genes. Microbiol. Immunol. 37: 11-22 (1993) human hepatitis B virus have been used for generating transgenic mice. Cellular oncogenes used to establish transgenic mice include » *fos*, » *fms*, » *myc*, *ras*, » *neu*, and *mos* oncogenes and various others.

Adams JM & Cory S Oncogene co-operation in leukemogenesis. Cancer Surv. 15: 119-41 (1992); **Adamson ED** Oncogenes in development. Development 99: 449-71 (1987); **Alitalo K & Schwab M** Oncogene amplification in tumor cells. ACR 47: 235-82 (1986); **Balmain A & Brown K** Oncogene activation in chemical carcinogenesis. ACR 51: 147-82 (1988); **Berns A** Tumorigenesis in transgenic mice: identification and characterization of synergizing oncogenes. JCBc 47: 130-5 (1991); **Bishop JM** Molecular themes in oncogenesis. Cell 64: 235-248 (1991); **Bouck N** Tumor angiogenesis: the role of oncogenes and tumor suppressor genes. CC 2: 179-85 (1990); **Cantley LC et al** Oncogenes and signal transduction. Cell 64: 281-302 (1991); **Compere SJ et al** Oncogenes in transgenic mice. BBA 948: 129-49 (1988); **Donehower LA et al** Mice deficient for p53 are developmentally normal but susceptible to spontaneous tumors. N 356: 215-21 (1992); **Ernberg IT** Oncogenes and growth factors - An updating. Eur J. Surg. 157, Suppl. 561: 15-20 (1991); **Forrester LM et al** Proto-oncogenes in mammalian development. Curr. Opin. Genetics Dev. 2: 38-44 (1992); **Fowlis DJ & Balmain A** Oncogenes and tumor suppressor genes in transgenic mouse models of neoplasia. Eur. J. Cancer 29A: 638-45 (1993); **Freeman RS & Donoghue DJ** Protein kinases and proto-oncogenes: Biochemical regulators of the eukaryotic cell cycle. B 30: 2293-302 (1991); **Friend SH et al** Oncogenes and tumor suppressing genes. NEJM 318: 618-22 (1988); **Gossen J & Vijg J** Transgenic mice as model systems for studying gene mutations *in vivo*. TIG 9: 27-31 (1993); **Haluska FG et al** Oncogene activation by chromosome translocation in human malignancy. ARG 21: 321-47 (1987); **Hunter T** Cooperation between oncogenes. Cell 64: 249-70 (1991); **Iversen OH** The hunt for endogenous growth-inhibitory and/or tumor suppression factors: their role in physiological and pathological growth regulation. ACR 57: 413-57 (1991); **Klinken SP** Erythroproliferation *in vitro* can be induced by *abl*, *fes*, *src*, *ras*, *bas*, *raf*, *raf/myc*, *erb* B and *cbl* oncogenes but not by *myc*, *myb* and *fos*. OR 3: 187-92 (1988); **Lavigueur A et al** High incidence of lung, bone, and lymphoid tumors in transgenic mice overexpressing mutant alleles of the p53 oncogene. MCB 9: 3982-91 (1989); **Lavigueur A & Bernstein A** p53 transgenic mice: accelerated erythroleukemia induction by Friend virus. O 6: 2197-201 (1991); **Lee SW et al** Positive selection of candidate tumor-suppressor genes by subtractive hybridization. PNAS 88: 2825-9 (1991); **Levine AJ & Momand J** Tumor suppressor genes: the p53 and retinoblastoma genes and gene products. BBA 1032: 119-36 (1990); **Lotem J & Sachs L** Hematopoietic cells from mice deficient in wild-type p53 are more resistant to induction of apoptosis by some agents. Blood 82: 1092-6 (1993); **Macara IG** Oncogenes and cellular signal transduction. Physiol. Rev. 69: 797-820 (1989); **Masuda H et al** Specificity of proto-oncogene amplification in human malignant diseases. Mol. Biol. Med. 4: 213-27 (1987); **McKenzie SJ** Diagnostic utility of oncogenes and their products in human cancer. BBA Rev. Cancer 1072: 193-214 (1991); **Mitchell CD** Recessive oncogenes, anti-oncogenes and tumor suppression. Br. Med. Bull. 47: 136-56 (1991); **Niman HL** Use of monoclonal antibodies as probes for oncogene products. Immunol. Ser. 53: 189-204 (1990); **Ong G & Sikora K** Oncogenes and their importance in clinical oncology. Diagn. Oncol. 1: 69-74 (1991); **Oren M** p53: the ultimate tumor suppressor gene. FJ 6: 3169-76 (1992); **Perlmutter RM & Ziegler S.** Proto-oncogene expression following lymphocyte activation. CTMT 35: 571-86 (1990); **Reddy EP et al** (eds) The oncogene handbook. Elsevier Science Publishers, Amsterdam 1988; **Sager R** Tumor suppressor genes: the puzzle and the promise. S 246: 1406-12 (1989); **Sarvetnick N** Transgenic mouse models for growth factor studies. MiE 198: 519-25 (1991); **Schwab M & Amier LC** Amplification of cellular oncogenes: a predictor of clinical outcome in human cancer. Genes, Chromosomes & Cancer 1: 181-94 (1990); **Sibbitt WL Jr** Oncogenes, growth factors, and autoimmune diseases. Anticancer Research 11: 97-114 (1991); **Travali S et al** Oncogenes in growth and development. FJ 4: 3209-14 (1990); **Van Beveren C & Verma IM** Homology among oncogenes. CTMI 123: 73-98 (1986); **Weinberg RA** (edt) Oncogenes and the molecular origins of cancer. Cold Spring Harbor Laboratory Press, Cold Spring Harbor, N. Y. 1989; **Yonish-Rouack E et al** Wild-type p53 induces apoptosis of myeloid leukaemic cells that is inhibited by interleukin-6. N 352: 345-7 (1991)

Onco M: see: Oncostatin M.

Oncomodulin: A small calcium-binding protein of 108 amino acids originally isolated from rat tumors (Morris hepatoma 5123). It is structurally and evolutionarily closely related to parvalbumin. The human oncomodulin gene maps to chromosome 7p1-p13. The rat oncomodulin gene (≈ 9 kb) has been shown to be under the control of a solitary long terminal repeat (LTR) element derived from an endogenous intracisternal A particle (see also: IAP) in normal cells. The integration of the IAP particle genome within the rat oncomodulin gene occurred after the rat and the mouse became distinct species.

Oncomodulin is an oncodevelopmental protein expressed in placental and extraembryonic tissue and re-expressed in a wide variety of tumors. Although oncomodulin has been shown to be expressed at high levels in some chemically transformed rat fibroblast cell lines it has neither an immortalizing nor transforming activities *in vitro*. Oncomodulin has been shown to inhibit glutathione reductase *in vivo* and *in vitro*.

Ahmed FR et al Structure of oncomodulin refined at 1.85 A resolution. An example of extensive molecular aggregation via Ca^{2+}. J. Mol. Biol. 216: 127-40 (1990); **Banville D et al** The intracisternal A particle derived solo LTR promoter of the rat oncomodulin gene is not present in the mouse gene. Genetica 86: 85-97 (1992); **Berchtold MW** Structure and expression of genes encoding the three-domain Ca^{2+}-binding proteins parvalbumin and oncomodulin. BBA 1009: 201-15 (1989); **MacManus JP et al** Oncomodulin in normal and transformed cells. AEMB 269: 107-10 (1990); **Mes-Masson AM et al** Expression of oncomodulin does not lead to the transformation or immortalization of mammalian cells *in vitro*. JCS 94: 517-25 (1989); **Palmer EJ et al** Inhibition of glutathione reductase by oncomodulin. ABB 277: 149-54 (1990)

Oncoproteins: A general term referring to the gene products of » oncogenes.

Oncostatin M: abbrev. OSM, OM, Onco M.

■ **Sources:** Oncostatin M is produced by activated monocytes, adherent macrophages, activated T cells (see also: Cell activation) and various T cell lines. The factor is also produced by » U937 cells treated with » phorbol esters.

■ **Protein Characteristics:** Oncostatin M is a monomeric glycoprotein of 28 kDa with a length of 196 amino acids. The factor is synthesized as a precursor of 252 amino acids by cleavage of a secretory » signal sequence of 25 amino acids from the N-terminal end and cleavage of 31 amino acids from the C-terminal end. Oncostatin M contains five cysteine residues. Four of these residues form disulfide bonds (C6–C127; C49–C167). Disulfide bond C49–C167 is essential for the biological activity of this factor.

■ **Gene Structure:** The Oncostatin M gene has a length of 4 kb and contains three exons. It encodes a transcript of ≈ 2 kb. The human gene maps to chromosome 22q12-q12.2 at a distance of 16 kb from the gene encoding » LIF (leukemia inhibitory factor), with the OSM gene located upstream of the LIF gene..

■ **Related Factors:** Oncostatin M resembles » LIF, » G-CSF, » IL6 and CNTF (ciliary neurotrophic factor) with respect to genome organization, amino acid sequence, and primary structure. These factors are members of a gene family.

■ **Receptor Structure, Gene(s), Expression:** Oncostatin M binds to a high-affinity receptor of ≈ 150-160 kDa. The Oncostatin M/receptor complex is rapidly internalized and degraded. The factor also binds to the high-affinity receptor for » LIF but not to the low affinity receptor of LIF.

The receptor component responsible for high-affinity binding is a glycoprotein, » gp130, that serves as a low-affinity receptor for oncostatin M. This component is also found in receptors for » IL6, » IL11, » LIF (leukemia inhibitory factor), and » CNTF (ciliary neurotrophic factor). The non-receptor tyrosine kinase » *yes* has been found to be involved in intracellular signal transduction processes induced by binding of oncostatin M to its receptor.

■ **Biological Activities:** Oncostatin M inhibits the growth of several tumor cell lines (A375 melanoma, lung carcinomas). The antiproliferative activity of oncostatin M for some cell lines is synergised by TGF-β and IFN-γ. It promotes the growth of human fibroblasts, vascular smooth muscle cells, and some normal cell lines. Oncostatin can inhibit the proliferation of murine » M1 myeloid leukemic cells and induces their differentiation into macrophage-like cells, a function shared by » LIF, G-CSF, and » IL6. Oncostatin M is a potent growth factor for HIV-related Kaposi sarcomas, promoting a change in morphology and promoting the growth of these cells in soft agar (see also: Colony formation assay). Oncostatin M also induces the synthesis of increased amounts of » IL6 in Kaposi sarcoma cells.

Oncostatin M also influences the growth of cultured endothelial cells and regulates the production of plasminogen activator by these cells. It has been shown to induce the expression of » G-CSF and » GM-CSF by human endothelial cells and may thus play a role in the regulation of myelopoiesis (see also: Hematopoiesis). In hepatocytes oncostatin M induces the synthesis of » acute phase proteins. In human endothelial cells oncostatin M induces the synthesis of » IL6, and this activity is synergised with » TGF-β but not with interferons. Oncostatin M enhances the uptake of LDL (low density lipoprotein) and induces the expression of LDL receptors in the human hepatocyte cell line HepG2. In human neuroblastoma cells oncostatin M regulates the expression of » VIP (vasoactive intestinal peptide) (see: NBFL).

■ **Clinical Use & Significance:** This factor may be of clinical relevance due to its activity as a growth inhibitor for some tumor cell lines. It may also be important as a factor involved in the growth control of hemopoietic cells as demonstrated by its ability to promote the differentiation of M1 cells and its ability to induce the expression of hematopoietic growth factors.

Brown TJ et al Purification and characterization of cytostatic lymphokines produced by activated human T lymphocytes: synergistic antiproliferative activity of transforming growth factor β1, interferon γ and oncostatin M for human melanoma cells. JI 139: 2977-83 (1987); **Brown TJ et al** Regulation of IL6 expression by oncostatin M. JI 147: 2175-80 (1991); **Brown TJ et al** Regulation of granulocyte colony-stimulating factor and granulocyte-macrophage colony-stimulating factor expression by oncostatin M. Blood 82: 33-7 (1993); **Bruce AG et al** Oncostatin M is a differentiation factor for myeloid leukemia cells. JI 149: 1271-5 (1992); **Doolittle D et al** Tandem linkage of genes coding for leukemia inhibitory factor (LIF) and oncostatin M (OSM) on human chromosome 22. Cytogenet. Cell Genet. 64: 240-244 (1993); **Gearing DP & Bruce AG** Oncostatin M binds the high-affinity leukemia inhibitory factor receptor. New Biol. 4: 61-5 (1992); **Gearing DP et al** The IL6 signal transducer, gp130: an oncostatin M receptor and affinity converter for the LIF receptor. S 255: 1434-7 (1992); **Grove RI et al** Oncostatin M up-regulates low density lipoprotein receptors in HepG2 cells by a novel mechanism. JBC 266: 18194-9 (1991); **Grove RI et al** Oncostatin M is a mitogen for rabbit vascular smooth muscle cells. PNAS 90: 823-7 (1993); **Hamilton JA et al** Oncostatin M stimulates urokinase-type plasminogen activator activity in human synovial fibroblasts. BBRC 180: 652-9 (1991); **Horn D et al** Regulation of cell growth by recombinant oncostatin M. GF 2: 157-65 (1990); **Jeffery E et al** Close proximity of the genes for leukemia inhibitory factor and oncostatin M. Cytokine 5: 107-11 (1993); **Kallestad JC et al** Disulfide bond assignment and identification of regions required for functional activity of oncostatin M. JBC 266: 8940-5 (1991); **Linsley PS et al** Cleavage of a hydrophilic C-terminal domain increases growth-inhibitory activity of oncostatin M. MCB 10: 1882-90 (1990); **Linsley PS et al** Identification and characterization of cellular receptors for the growth regulator, oncostatin M. JBC 264: 4282-9 (1989); **Liu J et al** Interleukin-6 signal transducer gp130 mediates oncostatin M signaling. JBC 267: 16763-6 (1992); **Malik N et al** Molecular cloning, sequence analysis, and functional expression of a novel growth regulator, oncostatin M. MCB 9: 2847-53 (1989); **McDonald VL et al** Selection and characterization of a variant of human melanoma cell line, A375, resistant to growth inhibitory effects of oncostatin M (OM): coresistant to interleukin 6 (IL6). GF 9: 167-75 (1993); **Miles SA** Oncostatin M as a potent mitogen for AIDS-Kaposi's sarcoma-derived cells. S 255: 1432-4 (1992); **Nair BC et al** Identification of a major growth factor for AIDS-Kaposi's sarcoma cells as Oncostatin M. S 255: 1430-2 (1992); **Radka SF et al** Abrogation of the antiproliferative activity of oncostatin M by a monoclonal antibody. Cytokine 4: 221-6 (1992); **Rao MS et al** Oncostatin M regulates VIP expression in a human neuroblastoma cell line. Neuroreport 3: 865-8 (1992); **Richards CD et al** Recombinant oncostatin M stimulates the production of acute phase proteins in HepG2 cells and rat primary hepatocytes *in vitro*. JI 148: 1731-6 (1992); **Rose TM & Bruce AG** Oncostatin M is a member of a cytokine family that includes leukemia-inhibitory factor, granulocyte colony-stimulating factor, and interleukin 6. PNAS 88: 8641-5 (1991); **Rose TM et al** The genes for oncostatin M (OSM) and leukemia inhibitory factor (LIF) are tightly linked on human chromosome 22. Genomics 17: 136-140 (1993); **Zarling JM et al** Oncostatin M: a growth regulator produced by differentiated histiocytic lymphoma cells. PNAS 83: 9739-43 (1986)

OP-1: *osteogenic protein 1* This factor is produced in the kidney and in the nervous system. It induces new bone growth. The factor is a member of the » TGF family of proteins. It is identical with BMP-7 (bone morphogenetic protein) and forms heterodimers with BMP-2A. See: BMP (bone morphogenetic protein).

Treatment of the neuroblastoma-glioma hybrid cell line NG108-15 with recombinant human OP-1 induces alterations in cell shape, formation of epithelioid sheets, and aggregation of cells into multilayered clusters. This is accompanied with the induction of three isoforms of neural cell adhesion molecules (N-CAMs). These observations suggest that OP-1 or a homologue may participate in the regulation of N-CAM during nervous system development and regeneration.

Kuber ST et al Bovine osteogenic protein is composed of dimers of OP-1 and BMP-2A, two members of the transforming growth factor-β superfamily. JBC 265: 13198-205 (1990); **Özkaynak E et al** OP-1 cDNA encodes an osteogenic protein in the TGF-β family. EJ 9: 2085-93 (1990); **Özkaynak E et al** Murine osteogenic protein (OP-1): high levels of mRNA in kidney. BBRC 179: 116-23 (1991); **Perides G et al** Induction of the neural cell adhesion molecule and neuronal aggregation by osteogenic protein 1. PNAS 89: 10326-30 (1992); **Sampath TK et al** Bovine osteogenic protein is composed of dimers of OP-1 and BMP-2A, two members of the transforming growth factor β superfamily. JBC 265: 13198-205 (1990)

OP-2: *osteogenic protein 2* OP-2 is discovered in mouse embryo and human hippocampus cDNA libraries. The TGF-β domain of OP-2 shows 74% identity to » OP-1, 75% to Vgr-1 (see: Vg-1), and 76% to BMP-5 (see: BMP, bone morphogenetic proteins). Mouse embryos express relatively high levels of OP-2 mRNA. OP-2 is also known as ***BMP-8*** (see: BMP, bone morphogenetic protein).

Özkaynak E et al Osteogenic protein-2. A new member of the transforming growth factor-β superfamily expressed early in embryogenesis. JBC 267: 25220-7 (1992)

Opiate peptides: see: POMC (proopiomelanocortin).

op/op **mouse:** The recessive *op* (osteopetrotic) mutation arose spontaneously in the C57BL/6J-dw stock of mice. The mutation maps to chromosome 3. It is a nonsense mutation in codon 277 of the gene encoding » M-CSF.

Animals with this mutation are toothless. They also show skeletal abnormalities, possess a lower body weight and display impaired fertility. The animals are also characterized by an almost complete absence of osteoclasts and have a severe deficiency of mononuclear phagocytes. Elevated concentrations of the colony-stimulating factor » GM-CSF are found in the serum.

Cell lines established from *op/op* mice secrete large amounts of GM-CSF and also » IL2, » IL3 and » IL4. The genetic defect can be corrected partially by treatment of osteopetrotic mice with M-CSF at doses maintaining normal circulating M-CSF concentrations. Early restoration of circulating M-CSF is required for rescue of the toothless phenotype, but only partially restores body weight. The deficiencies of pleural and peritoneal cavity macrophages and the reduced female fertility are not corrected by restoration of circulating M-CSF. At present it is not known whether a similar lethal disease of bone resorption in humans (osteopetrosis, marble bones, osteochondrodysplasia, Albers-Schönberg disease) is also caused by mutations in the M-CSF gene. Some data indicate normal levels of biologically active M-CSF in osteopetrosis patients.

Felix R et al Macrophage colony-stimulating factor restores *in vivo* bone resorption in the op/op osteopetrotic mouse. E 127:

2592-4 (1990); **Felix R et al** Impairment of macrophage colony-stimulating factor production and lack of resident bone marrow macrophages in the osteopetrotic op/op mouse. J. Bone Miner. Res. 5: 781-9 (1990); **Franz T** Immortalised fibroblastoid cells of osteopetrotic mutant mice (op/op) do not secrete CSF-1 and do not inhibit CSF-1 activity released by normal cells. EH 19: 170-3 (1991); **Kodama H et al** Congenital osteoclast deficiency in osteopetrotic (Op/op) mice is cured by injections of macrophage colony-stimulating factor. JEM 173: 269-72 (1991); **Kodama H et al** Essential role of macrophage colony-stimulating factor in the osteoclast differentiation supported by stromal cells. JEM 173: 1291-4 (1991); **Marks SC & Lane PW** Osteopetrosis, a new recessive skeletal mutation on chromosome 12 of the mouse. J. Hered. 67: 11-8 (1976); **Marks SC Jr et al** Congenitally osteopetrotic (op/op) mice are not cured by transplants of spleen or bone marrow cells from normal litter mates. Metab. Bone Dis. Relat. Res. 5: 183-6 (1984); **Orchard PJ et al** Circulating macrophage colony-stimulating factor is not reduced in malignant osteopetrosis. EH 20: 103-5 (1992); **Pollard JW et al** A pregnancy defect in the osteopetrotic (op/op) mouse demonstrates the requirement for CSF-1 in female fertility. Dev. Biol. 148: 273-83 (1991); **Shultz LD** Hematopoiesis and models of immunodeficiency. Semin. Immunol. 3: 397-408 (1992); **Takahashi N et al** Deficiency of osteoclasts in osteopetrotic mice is due to a defect in the local microenvironment provided by osteoblastic cells. E 128: 1792-6 (1991); **Wiktor-Jedrzejczak W et al** Total absence of colony-stimulating factor 1 in the macrophage-deficient osteopetrotic (op/op) mouse. PNAS 87: 4828-32 (1990), Erratum in PNAS 88: 5937; **Wiktor-Jedrzejczak W et al** Correction by CSF-1 of defects in the osteopetrotic op/op mouse suggests local, developmental, and humoral requirements for this growth factor. EH 19: 1049-54 (1991); **Yoshida H et al** The murine mutation osteopetrosis is in the coding region of the macrophage colony stimulating factor gene. N 345: 442-4 (1990)

Ornithine decarboxylase-inducing factor: see: ODC factor.

Orphan receptor: A general term for receptors the ligand of which has not been identified.

Orthovanadate: see: Vanadate.

OSF-1: *osteoblast-specific factor* This factor was initially isolated as a heparin-binding growth factor (see also: HBGF). It is identical with » HBNF (heparin-binding neurite-promoting factor).

Katoh K et al Genomic organization of the mouse OSF-1 gene. DNA Cell Biol. 11: 735-43 (1992); **Takamatsu H et al** Expression and purification of biologically active human OSF-1 in *Escherichia coli*. BBRC 185: 224-30 (1992); **Tezuka K et al** Isolation of mouse and human cDNA clones encoding a protein expressed specifically in osteoblasts and brain tissues. BBRC 173: 246-51 (1990)

OSM: abbrev. for » Oncostatin M.

Osteoblast-specific factor: see: OSF-1.

Osteoclast activating factor: see: OAF.

Osteoclast colony-stimulating factor: see: O-CSF.

Osteoclast growth-inducing factor: see: OGF.

Osteogenic protein 1: see: OP-1.

Osteogenic protein 2: see: OP-2.

Osteogenin: This factor with bone growth-promoting activity is identical with BMP 3. See: BMP (bone morphogenetic protein).

Luyten FP et al Purification and partial amino acid sequence of osteogenin, a protein initiating bone differentiation JBC 264: 13377-80 (1989); **Sampath TK et al** Isolation of osteogenin, an extracellular matrix-associated, bone-inductive protein, by heparin affinity chromatography. PNAS 84: 7109-13 (1987)

Osteoglycin: A new name proposed for » OIF (osteoinductive factor).

Bentz H Transforming growth factor β 2 enhances the osteoinductive activity of a bovine bone-inducing fraction containing bone morphogenetic protein 2 and 3. Matrix 11: 269-75 (1991)

Osteoinductive factor: see: OIF.

Osteosarcoma-derived growth factor: see: ODGF.

Ovarian growth factor: see: OGF.

P

p: (a) abbrev. for protein. suffixed numbers indicate molecular mass such as in p53 (53 kDa); an additional g indicates a glycoprotein such as is gp185. (b) depending upon context the prefixed p may indicate the porcine origin of a factor (c) in some abbrev. a prefixed p such as in pLD78 (see: LD78) and » pAT464 a cDNA or DNA clone (p = plasmid). Some cytokine names are derived from such clone names by omitting the p (LD78).

P2 myelin: see: MDGI (mammary-derived growth inhibitor).

P6: A basic protein of 76 amino acids that is the major basic protein constituent of bovine seminal plasma. The protein is identical with bovine » MCP-1 (monocyte chemoattractant protein 1). It displays 80 % sequence homology at the cDNA and 72 % at the protein level with human MCP-1.
Wempe F et al Gene expression and cDNA cloning identified a major basic protein constituent of bovine seminal plasma as bovine monocyte chemoattractant protein 1 (MCP-1). DNA Cell Biol. 10: 671-9 (1991)

p16 IL3 inhibitor: see: SOD (superoxide dismutase).

p18: A heparin-binding growth factor (see also: HBGF) of 18 kDa isolated from brain tissues of several species and also secreted by a human mammary carcinoma cell line. It is identical with » HBNF (heparin-binding neurite-promoting factor).
Kuo MD et al Amino acid sequence and characterization of a heparin-binding neurite-promoting factor (p18) from bovine brain. JBC 265: 18749-52 (1990); **Rauvala H** An 18-kd heparin-binding protein of developing brain that is distinct from fibroblast growth factors. EJ 8: 2933-41 (1989); **Wellstein A et al** A heparin-binding growth factor secreted from breast cancer cells homologous to a developmentally regulated cytokine. JBC 267: 2582-7 (1992)

p19/nm23: see: nm23.

p29: see: IL2R alpha chain inhibitory activity.

p29 autoantigen: see: myeloblastin.

P40: This glycoprotein of 126 amino acids supports the growth of a number of T helper cell lines in the absence of » IL2, » IL4 and antigens. It is identical with » IL9.
Renauld JC et al Human P40/Il9: expression in activated $CD4^+$ cells, genomic organization, and comparison with the mouse gene. JI 144: 4235-41 (1990); **Renauld JC et al** Cloning and expression of a cDNA for the human homologue of mouse T cell and mast cell growth factor P40. Cytokine 2: 9-12 (1990); **Simpson RJ et al** Complete amino acid sequence of a new murine T cell growth factor P40. EJB 183: 715-22 (1989); **Uyttenhove C et al** Functional and structural characterization of P40, a mouse glycoprotein with T cell growth factor activity. PNAS 85: 6934-8 (1988); **Van Snick J et al** Cloning and characterization of a cDNA for a new mouse T cell growth factor (P40). JEM 169: 363-8 (1989)

p53: Also called TP53. This is a tumor suppressor gene. See: Oncogene.

p55: This protein of 55 kDa that is also called TAC antigen or » CD25 is identical with the α chain of the receptor for » IL2.

p71/73: see: TIS genes.

p75: This protein of 75 kDa is identical with the β chain of the receptor for » IL2.

p91: see: IRS (interferon response sequence).

P422: *adipocyte differentiation associated protein* see: MDGI (mammary-derived growth inhibitor).

P500: This factor is also called SIS-ε (see also: SIS, small inducible secreted). It is identical with » TCA-3 and belongs to the » chemokine family of cytokines.
Brown KD et al A family of small inducible proteins secreted by leukocytes are members of a new superfamily that includes leukocyte and fibroblast-derived inflammatory agents, growth factors, and indicators of various activation processes. JI 142: 679-87 (1989)

P600: A murine gene identified as a cDNA sequence that is produced by murine TH2 helper T cell clones following activation, but not by TH1 helper cells (see also: TH1/TH2 cytokines). The

human cDNA sequence is 66 % identical with the mouse cDNA and has been named » IL13.
Brown KD et al A family of small inducible proteins secreted by leukocytes are members of a new superfamily that includes leukocyte and fibroblast-derived inflammatory agents, growth factors, and indicators of various activation processes. JI 142: 679-87 (1989); **Cherwinski HM et al** Two types of mouse helper T cell clone. III. Further differences in lymphokine synthesis between Th1 and Th2 clones revealed by RNA hybridization, functionally monospecific bioassays, and monoclonal antibodies. JEM 166: 1229-44 (1987); **Morgan JG et al** The selective isolation of novel cDNAs encoded by the regions surrounding the human interleukin 4 and 5 genes. NAR 20: 5173-9 (1992)

PAF: *placental angiogenic factor* This factor is isolated from human placenta. It is a form of » bFGF with an aminoterminal extension.
Sommer A et al A form of human basic fibroblast growth factor with an extended amino terminus. BBRC 144: 543-50 (1987)

PAF: *proliferation amplifying factor* This poorly characterized factor of ≈ 31 kDa is found in culture supernatants of the lymphoma cell line M12.4.1. It modulates the proliferation of hematopoietic cells.
Yao ZJ et al Modulating activity of B lymphocytes on hemopoiesis (II) – the purification of proliferation amplifying factor (PAF). Sci. Sin. B. 31: 702-9 (1988)

PAI-BP: *plasminogen activator inhibitor 1 binding protein* This factor is identical with » Vitronectin.

Panspecific hemopoietin: see: PSH.

Paracrine growth control: A mechanism of growth control involving the synthesis of a soluble factor by one cell that influences the growth and functional activities of near-by cells expressing the receptor for this factor. This mechanism of local growth control therefore differs from that mediated by endocrine factors involving the transport of a factor to the site of action by the circulation (endocrine actions of hormones).
Many paracrine factors are higher molecular weight peptide factors. Paracrine actions have also been described for low molecular weight substances; the neurotransmitter acetylcholine, for example is a paracrine growth factor for small cell lung carcinoma cells.
For other mechanisms of factor-mediated growth control see also: Autocrine, Intracrine, Juxtacrine Retrocrine.
Brandi ML Cellular models for the analysis of paracrine communications in parathyroid tissue. J. Endocrinol. Invest. 16: 303-14 (1993); **Garbers DL** Guanylyl cyclase receptors and their endocrine, paracrine, and autocrine ligands. Cell 71: 1-4 (1992); **Gol-Winkler R** Paracrine action of transforming growth factors. Clin. Endocrinol. Metab. 15: 99-115 (1986); **Nicolson GL** Paracrine/autocrine growth mechanisms in tumor metastasis. Oncol. Res. 4: 389-99 (1992); **Nicolson GL** Cancer progression and growth: relationship of paracrine and autocrine growth mechanisms to organ preference of metastasis. ECR 204: 171-80 (1993)

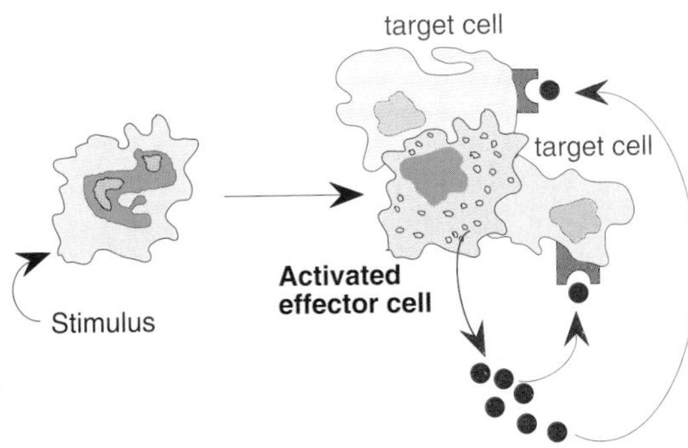

Paracrine growth control.
Stimulation of a cell leads to the expression and release of one or more mediators. The mediator molecule released by the activated effector cell can interact with membrane receptors expressed on neighboring cells and thus can elicit further responses in these cells.

Paracrine growth-stimulatory factor: see: PGSF.

Parathymosin alpha: see: Thymic hormones.

pAT 464: also: T464. Name of a cDNA clone encoding the cytokine » LD78.
Zipfel PF et al Mitogenic activation of human T cells induces two closely related genes which share structural similarity with a new family of secreted factors. JI 142: 1582-90 (1989); **Zipfel PF et al** Employment of antipeptide antisera to distinguish two closely related cytokines induced in human T cells. Lymphokine. Cytokine. Res. 11: 141-8 (1992)

pAT 744: also: AT744. Name of a cDNA clone encoding the cytokine » ACT-2.
see: pAT 464

***Patch*:** A mutation found in certain strains of mice. The mutation is lethal and causes severe anatomical malformations in homozygous animals. Heterozygous animals show hair coloration anomalies. The molecular defect is caused by a deletion of the gene encoding the α subunit of the receptor for » PDGF.
Morrison-Graham K et al A PDGF receptor mutation in the mouse (Patch) perturbs the development of a non-neuronal subset of neural crest-derived cells. Development 115: 133-42 (1992); **Orr-Urtreger A et al** Developmental expression of the α receptor for platelet-derived growth factor, which is deleted in the embryonic lethal Patch mutation. Development 115: 289-303 (1992); **Stephenson DA et al** Platelet-derived growth factor

Pancreastatin: see addendum: Granins. **Parastatin:** see addendum: Granins.
Parathyroid secretory protein I: see addendum: Granins.

receptor α-subunit gene (Pdgfra) is deleted in the mouse patch (Ph) mutation. PNAS 88: 6-10 (1991)

PBEF: *pre B cell colony enhancing factor* The cDNA for this factor has been isolated from a human peripheral blood lymphocyte cDNA. PBEF is a secreted protein of 52 kDa that is not related to any other known cytokine. Its expression is induced by pokeweed mitogen and superinduced by cycloheximide. The PBEF gene is transcribed mainly in human bone marrow, liver tissue, and muscle cells. PBEF by itself is not active. It acts on early B lineage precursor cells and synergizes with » SCF and » IL7 in pre B cell colony formation. PBEF is inactive with cells of the myeloid or erythroid lineage.

Samal B et al Cloning and characterization of the cDNA encoding a novel human pre-B cell colony-enhancing factor MCB 14: 1431-7 (1994)

PBGF: *pre B cell growth factor* This factor is identical with » IL7.

Humphries RK et al Activation of multiple hemopoietic growth factor genes in Abelson virus-transformed myeloid cells. EH 16: 774-81 (1988); **Landreth KS & Dorshkind K** Pre-B cell generation potentiated by soluble factors from a bone marrow stromal cell line. JI 140: 845-52 (1988); **Namen A et al** B cell precursor growth-promoting activity. Purification and characterization of a growth factor active on lymphocyte precursors. JEM 167: 988-1002 (1988)

PBP: *platelet basic protein* PBP is the biochemical precursor for a number of protein factors with cytokine activities. See: Beta-Thromboglobulin.

PC-4: An » ERG (early response gene) whose transcription is rapidly and transiently stimulated by treatment of cells with mitogens or agents that induce differentiation. PC-4 expression is induced by treatment of cells with » Phorbol esters such as TPA and PC-4 is identical with » TIS7 (see: TIS).

Jahner D & Hunter T The stimulation of quiescent rat fibroblasts by v-src and v-fps oncogenic protein-tyrosine kinases leads to the induction of a subset of immediate early genes. O 6: 1259-68 (1991); **Kreider BL & Rovera G** The immediate early gene response to a differentiative stimulus is disrupted by the v-*abl* and v-*ras* oncogenes. O 135-40 (1992)

PCD: abbrev. for programmed cell death. See: Apoptosis

P cell stimulating activity: see: PCSA.

P cell stimulating factor: see: PSF.

PC12: A cell line established from a spontaneous rat pheochromocytoma derived from chromaffin cells of the suprarenal medulla. These chromaffin cells are closely related to sympathetic neurons and can differentiate into sympathetic neurons in the presence of » NGF. PC12 cells are used for studies of the biological effects of NGF and » neurotrophins. They have been used extensively to investigate the molecular biology of » ERGs (early response genes), which has led to the identification of many different » TIS genes.

PC12 cell can be maintained indefinitely in culture in the absence of NGF. Addition of this growth factor induces differentiation into cell types resembling sympathetic neurons, and neurite outgrowth is used as a » bioassay for NGF. The activities of NGF can be inhibited specifically by » K-252-a. The differentiation can be recognized by the outgrowth of long neurites, the synthesis of neuronal enzymes such as choline acetyltransferase, and the generation of an excitable membrane potential. NGF also induces the synthesis of » IL1 in PC12 cells.

The outgrowth of neurites in PC12 cells can also be induced by » bFGF which is as potent as NGF in this respect. The injection of anti-*ras* antibodies into PC12 cells blocks the NGF- and bFGF-induced cell differentiation, suggesting that the » oncogene *ras* is an essential component of receptor-mediated signal transduction processes.

Alheim K et al Interleukin 1 expression is inducible by nerve growth factor in PC12 pheochromocytoma cells. PNAS 88: 9302-6 (1991); **Altin JG et al** Microinjection of a p21*ras* antibody into PC12 cells inhibits neurite outgrowth induced by nerve growth factor and basic fibroblast growth factor. GF 4: 145-55 (1991); **Greene LA & Tischler AS** Establishment of a noradrenergic clonal line of rat adrenal pheocromocytoma cells which respond to nerve growth factor. PNAS 73: 2424-8 (1976); **Greene LA & Tischler AS** PC12 pheocromocytoma cultures in neurological research. Adv. Cell. Neurobiol. 3: 373-414 (1982); **Sano M et al** A convenient bioassay for NGF using a new subline of PC12 pheochromocytoma cells (PC12D). Brain Res. 459: 404-6 (1988); **Wagner JA** The fibroblast growth factors: an emerging family of neural growth factors. CTMI 165: 95-118 (1991)

PC13 embryonal carcinoma-derived growth factor: see: ECDGF.

PCDGF: *PC cell-derived growth factor* This factor of 88 kDa (68 kDa deglycosylated) is secreted by the PC cell line, a highly tumorigenic insulin-independent variant of the teratoma-derived adipogenic cell line 1246. PCDGF is a growth promoting activity for murine » 3T3 fibroblasts and an » autocrine growth factor for the producer cells. Based on partial sequence analysis it has been suggested that PCDGF may be a potential precursor for » epithelins and/or granulin.

PBSF: see addendum.

Zhou J et al Purification of an autocrine growth factor homologous with mouse epithelin precursor from a highly tumorigenic cell line. JBC 268: 10863-9 (1993)

PCF: *peritoneal cytotoxic factor* A cytotoxic factor released by macrophages into the peritoneal fluid following treatment of tumor-bearing mice with the streptococcal preparation OK-432. The factor is most likely identical with » TNF.

Nagamuta M et al [Induction of a cytotoxic factor into the peritoneal fluid (PCF) by OK-432. II. Physicochemical and immunological characteristics of PCF] Gan To Kagaku Ryoho 13: 2358-62 (1986); **Watanabe N et al** Therapeutic effect of endogenous tumor necrosis factor on ascites Meth A sarcoma. J. Immunopharmacol. 8: 271-83 (1986); **Yamamoto A et al** Release of tumor necrosis factor (TNF) into mouse peritoneal fluids by OK-432, a streptococcal preparation. Immunopharmacology 11: 79-86 (1986)

PCR: abbrev. for polymerase chain reaction. For detection of cytokine gene expression by PCR see: Message amplification phenotyping.

PCSA: *P cell stimulating activity* This factor was initially described as a protein promoting the growth of P cells, a homogenous population of mast cell-like cells. It is identical with » IL3.

Ihle JN et al Biological properties of homogenous interleukin-3. I. Demonstration of WEHI-3 growth factor activity, mast cell growth factor activity, P cell-stimulating factor activity, colony-stimulating factor activity, and histamine-producing cell-stimulating factor activity. JI 131: 282-7 (1983); **Schrader JW et al** The persisting (P) cell: histamine content, regulation by a T cell derived factor, origin from bone marrow precursor, and relationship to mast cells. PNAS 78: 323-7 (1981); **Schrader JW et al** P-cell stimulating factor: biochemistry, biology, and role in oncogenesis. Contemp. Top. Mol. Immunol. 10: 121-46 (1985)

PCSF: *P cell stimulating factor* see: PSF (P cell stimulating factor).

pCSF: *pluripotent colony stimulating factor* This factor is identical with » G-CSF.

Welte K et al Purification and biochemical characterization of human pluripotent hematopoietic colony-stimulating factor. PNAS 82: 1526-30 (1985)

PCT-GF: *plasmacytoma growth factor* also abbrev. PGF. This factor, purified from conditioned medium of the murine macrophage cell line P388D1, was initially described as a protein promoting the growth of plasmacytoma cell lines (see for example: T1165). Withdrawal of the factor led to an arrest of cells in the G1 phase of the » cell cycle. The factor is identical with murine » IL6.

Nordan RP & Potter MA A macrophage-derived factor required by plasmacytomas for survival and proliferation *in vitro*. S 233: 566-9 (1986); **Nordan RP et al** Purification and NH_2-terminal sequence of a plasmacytoma growth factor derived from the murine macrophage cell line P388D1. JI 139: 813-7 (1987); **Nordan RP et al** The role of plasmacytoma growth factor in the *in vitro* responses of murine plasmacytoma cells. ANY 557: 200-5 (1989); **Sawamura M et al** Characterization of 5B12.1, a monoclonal antibody specific for IL6. GF 3: 181-90 (1990); **Van Damme J et al** Purification and characterization of human fibroblast-derived hybridoma growth factor identical to T cell-derived B cell stimulating factor-2 (interleukin-6). EJB 168: 543-50 (1987); **Vink A et al** Mouse plasmacytoma growth *in vivo*: enhancement by interleukin 6 (IL6) and inhibition by antibodies directed against IL6 or its receptor. JEM 172: 997-1000 (1990)

PD-ECGF: *platelet-derived endothelial cell growth factor*

■ **ALTERNATIVE NAMES:** ECGF1 (endothelial cell growth factor 1); gliostatin; Platelet-derived endothelial cell mitogen. See also: individual entries for further information.

■ **SOURCES:** This factor was initially isolated from platelets (Platelets from ≈ 800-1000 L of blood yield ≈ 30-40 μg of the factor). In contrast to other platelet-derived factors that are isolated from α granules, PDECGF is a cytoplasmic protein.
PD-ECGF is also found in connective tissue of the placenta (≈ 40 μg/placenta). The factor is also synthesized by fibroblasts, vascular smooth muscle cells, and a number of transformed cell lines. PD-ECGF is not produced by endothelial cells.

■ **PROTEIN CHARACTERISTICS:** PD-ECGF is an acidic non-glycosylated protein of 45 kDa. It is synthesized as a precursor of 482 amino acids from which it is derived by N-terminal processing. The protein isolated from placenta contains five additional amino acids at the N-terminus. PD-ECGF can be phosphorylated *in vivo* at serine residues but the biological significance of this phosphorylation step is unknown.
A comparison of the cDNA sequences encoding human ***thymidine phosphorylase*** (dThdPase) that catalyses the reversible phosphorolysis of thymidine and deoxyuridine into the corresponding bases and 2-deoxyribose-1-phosphate and PD-ECGF sequences reveals that 120 amino acids are identical in both proteins (positions 125-244 in PD-ECGF). It has been shown recently that PD-ECGF and thymidine phosphorylase are the products of the same gene. Purified *Escherichia coli* thymidine phosphorylase is also mitogenic for endothelial cells.
PD-ECGF, like » aFGF, » bFGF, and » CNTF (ciliary neuronotrophic factor), does not possess a » signal sequence that would allow secretion of the factor by classical secretion pathways (endoplas-

matic reticulum/Golgi system). The mechanism underlying the release of PD-ECGF is unknown. The factor may be released by lysis of platelets. PD-ECGF does not bind to heparin and its biological activity is not potentiated by heparin (see also: HBGF, heparin-binding growth factors).

■ **GENE STRUCTURE:** The gene encoding PD-ECGF has a length of more than 4.3 kb and contains ten exons. The promoter region does not contain classical TATA and CAAT boxes. Southern blot analysis of genomic DNA of various vertebrate species demonstrates a high sequence conservation. The human PD-ECGF gene maps to chromosome 22q13 distal to the locus for » PDGF-B (platelet-derived growth factor B; 22q12.3-q13.1). At present it is not clear whether these genes are involved in the many chromosome 22 aberrations documented, for example, for malignant gliomas, schwannomas, meningiomas, and hemangiopericytomas.

■ **RECEPTOR STRUCTURE, GENE(S), EXPRESSION:** No information available.

■ **BIOLOGICAL ACTIVITIES:** PD-ECGF is chemotactic for vascular endothelial cells and monocytes (see also: Chemotaxis). It also selectively supports the growth of these cells in contrast to many other » angiogenesis factors that are also mitogenic for fibroblasts and muscle cells.

PD-ECGF is not chemotactic for vascular smooth muscle cells. Angiogenesis can be inhibited by monoclonal antibodies directed against PD-ECGF. Transfection of PD-ECGF cDNA into cells harboring an activated Ha-*ras* » oncogene results in the generation of cells that produce strongly vascularised tumors in nude mice (see also: Immunodeficient mice) while control cells that do not produce PD-ECGF lead to the growth of poorly vascularised tumors.

Since PD-ECGF exclusively stimulates endothelial cells and is also produced by platelets it is probably involved in the maintenance of intact blood vessels. It may be that the factor is responsible for the repair of injured endothelial layers before other factors, e. g. platelet-derived » PDGF causes the increased deposition of cells to the lesion and thus promotes vessel obstructions.

PD-ECGF plays an important role in the generation of new blood vessels, for example, in the embryo and in the placenta which is particularly rich in blood vessels. PD-ECGF is probably involved as a local factor in » wound healing and the associated processes of new blood vessel formation.

PD-ECGF has recently been shown to possess thymidine phosphorylase activity, suggesting that the factor may also have effects on DNA synthesis. It has been suggested that human thymidine phosphorylase promotes endothelial cell proliferation by reducing thymidine levels that would otherwise be inhibitory to endothelial cell growth.

■ **ASSAYS:** A two-site enzyme immunoassay is available.

■ **CLINICAL USE & SIGNIFICANCE:** Several pharmacokinetic parameters and the specificity for endothelial cells suggest that PD-ECGF may be useful as a therapeutic agent to stimulate re-endothelialization *in vivo*, or, in view of its thymidine phosphorylase activity, in chemotherapy, by decreasing the pool of available thymidine.

PF-ECGF has been implicated in the development of multidrug resistance frequently evolving during treatment of various malignant tumors. Thymidine phosphorylase is one of the activating enzymes for fluorinated pyrimidines. The sensitivity of human KB cells transfected with PD-ECGF cDNA to doxifluridine has been shown to be considerably higher than that of non-transfected KB cells.

● **REVIEWS: Miyazono K et al** Platelet-derived endothelial cell growth factor. PGFR 3: 207-17 (1991)

● **BIOCHEMISTRY & MOLECULAR BIOLOGY: Barton GJ et al** Human platelet-derived endothelial cell growth factor is homologous to *Escherichia coli* thymidine phosphorylase. Protein Sci. 1: 688-90 (1992); **Finnis C et al** Expression of recombinant platelet-derived endothelial cell growth factor in the yeast *Saccharomyces cerevisiae*. Yeast, 8: 57-60 (1992); **Finnis C et al** Thymidine phosphorylase activity of platelet-derived endothelial cell growth factor is responsible for endothelial cell mitogenicity. EJB 212: 201-10 (1993); **Hagiwara K et al** Organization and chromosomal localization of the human platelet-derived endothelial cell growth factor gene. MCB 11: 2125-32 (1991); **Heldin CH et al** Purification, cloning, and expression of platelet-derived endothelial cell growth factor. MiE 198: 383-90 (1991); **Heldin CH et al** Platelet-derived endothelial growth factor. JCBc 47: 208-10 (1991); **Ishikawa F et al** Identification of angiogenic activity and the cloning and expression of platelet-derived endothelial cell growth factor. N 338: 557-61 (1989); **Miyazono K et al** Purification and properties of an endothelial cell growth factor from human platelets. JBC 262: 4098-103 (1987); **Miyazono K et al** High-yield purification of platelet-derived endothelial cell growth factor: structural characterization and establishment of a specific antiserum. B 28: 1704-10 (1989); **Moghaddam A & Bicknell R** Expression of platelet-derived endothelial cell growth factor in *Escherichia coli* and confirmation of its thymidine phosphorylase activity. B 31: 12141-6 (1992); **Stenman G et al** Regional localization of the human platelet-derived endothelial cell growth factor (ECGF1) gene to chromosome 22q13. Cytogenet. Cell Genet. 59: 22-3 (1992); **Usuki K et al** Production of platelet-derived endothelial cell growth factor by normal and transformed human cells in culture. PNAS 86: 7427-31 (1989); **Usuki K et al** Localization of platelet-derived endothelial cell growth factor in human placenta and purification of an alternatively processed form. Cell Regul. 1: 577-96 (1990); **Usuki K et al** Covalent linkage between nucleotides and platelet-derived endothelial cell growth factor. JBC 266: 20525-31 (1991); **Usuki K et al** Platelet-derived endothelial cell growth factor has thymidine phosphorylase activity. BBRC 184: 1311-6 (1992)

● **Assays: Hirano T et al** Establishment of an enzyme immunoassay system for gliostatin/platelet-derived endothelial cell growth factor (PD-ECGF). BBA 1176: 299-304 (1993)

● **Clinical Use & Significance: Akiyama S** [Molecular basis for resistance to anticancer agents and reversal of the resistance] Hum. Cell. 6: 1-6 (1993); **Waltenberger J et al** Platelet-derived endothelial cell growth factor. Pharmacokinetics, organ distribution and degradation after intravenous administration in rats. FL 313: 129-32 (1992)

PDGF: *platelet-derived growth factor*

■ **Alternative Names:** FDGF (fibroblast-derived growth factor); GDGF (glioma-derived growth factor); GDGF-1 (glioma-derived growth factor = PDGF-AA homodimer); GDGF-2 (glioma-derived growth factor = PDGF-AB heterodimer); GSM (glucocorticoid-suppressible mitogenic activity = PDGF-AA homodimer); MDF (mesangial cell proliferating factors); MDGF (monocyte-derived growth factor); ODGF (osteosarcoma-derived growth factor); T47D factor.
PDGF is the major component of » MDF and » MDGF. Some of the MDGF activity may be due to fibroblast growth factors (see: FGF). See also: individual entries for further information.

■ **Sources:** PDGF is synthesized mainly by megakaryocytes. It is stored in the α granules of platelets from which it is released after activation of platelets for example by thrombin (see also: Cell activation).
A plethora of other cell types also synthesize PDGF including macrophages, endothelial cells, fibroblasts, glial cells, astrocytes, myoblasts, smooth muscle cells, and a number of tumor cell lines. *In vivo* PDGF is synthesized by man cells as a protein consisting of one of the two PDGF chains. Platelets synthesize a mixture of the three possible isoforms (70 % AB, 20 % BB, 10 % AA) while EGF-stimulated fibroblasts synthesize AA homodimers. Activated macrophages and placental cytotrophoblasts produce the BB homodimer.
The synthesis of PDGF can be induced by » IL1, » IL6, » TNF-α, » TGF-β and » EGF.

■ **Protein Characteristics:** PDGF is a heat-stable positively charged (pI = 9.8-10) hydrophilic protein of 30 kDa. It consists of two related peptide chains, ***PDGF-A*** or ***PDGF-I*** (16 kDa, 124 amino acids), and ***PDGF-B*** or ***PDGF-II*** (14 kDa, 140 amino acids) that contain intramolecular disulfide bonds. The subunits are linked by disulfide bonds. All possible isoforms, i. e. AA, BB, and AB are biologically active.
Several variant forms of PDGF-A have been described. She display ≈ 60 % homology with PDGF-B. PDGF-B is identical with the gene product of the cellular » oncogene » *sis*. The amino-terminal end of PDGF-B is almost identical with $p28^{sis}$, the transforming oncoprotein of simian sarcoma virus (see: *sis*).
The three isoforms of PDGF bind to PDGF with different affinities and also differ in their biological activities.

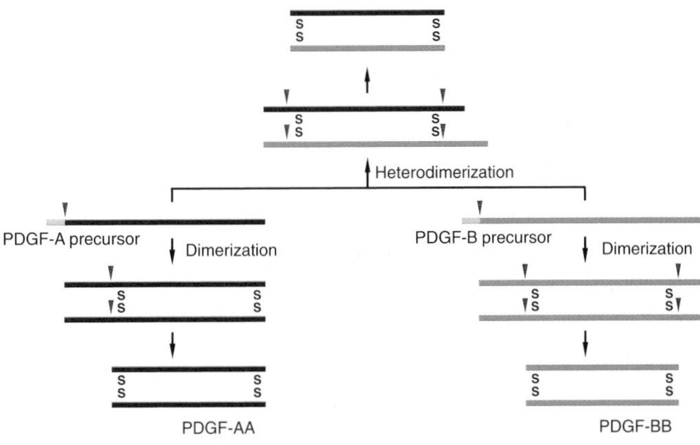

Biosynthesis of PDGF forms.
PDGF-A and PDGF-B are derived from two different precursors containing an N-terminal signal sequence (pink). This is cleaved off. The precursors of the A and the B chain are assembled into disulfide-bonded homo- and heterodimers that can undergo proteolytic cleavage to form mature PDGF-AA, PDGF-BB, and PDGF-AB. Cleavage sites are marked by red arrow heads.

Mutant forms of PDGF have been created. They create wild-type/mutant heterodimers deficient in assembly or processing. Such mutants behave as dominant negative mutants and are useful in investigation of the role of PDGF in normal and pathological conditions.

■ **Gene Structure:** The gene encoding PDGF-A has a length of ≈ 24 kb and contains at least seven exons. It maps to human chromosome 7p21-p22. The PDGF-A variants arise by differential splicing of the mRNA.

The PDGF-B gene has a length of ≈ 24 kb and contains at least seven exons. It maps to human chromosome 22q12.3-q13. 1 in the vicinity of the gene encoding » PD-ECGF (platelet-derived endothelial cell growth factor) at 22q13.

The PDGF genes are evolutionary conserved from mammals to amphibians. Both genes display a similar exon-intron structures and contain long (> 1 kb) 5′ and 3′ untranslated regions. The activity of both genes is regulated independently of each other in some cell types and co-ordinately in some other cell types.

■ **Receptor Structure, Gene(s), Expression:** PDGF receptors are expressed at a density of 40000-300000 copies per cell in fibroblasts, osteoblasts, chondroblasts, smooth muscle cells, glial cells, and endothelial cells. Epithelial cells do not appear to express receptors. The synthesis of PDGF receptors is subject to autoregulation by PDGF.

Two related receptors, called ***PDGFRα*** or ***PDGFRβ*** and ***PDGFRA*** or ***PDGFRB*** have been described. Binding of the ligand leads to the formation of receptor/ligand aggregates that are internalized by the cell. Binding of PDGF to its receptors can be inhibited by » Suramin.

The intracellular domain of the PDGF receptor possesses protein kinase activity. The kinase domain is interrupted by the insertion of a so-called kinase insert (see also: PTK, protein tyrosine kinase). Following ligand binding the carboxyterminal cytoplasmic domain of the receptor becomes phosphorylated at least at two positions (Y751 and Y857) (see also: Autophosphorylation).

The cytoplasmic domain of the PDGF receptor is complexed with several other proteins including phosphatidyl inositol-3-kinase (PI3), phospholipase C (PLC-γ), a GTPase-activating enzyme (GAP), and » *raf*-1. Complexes are formed after tyrosine » autophosphorylation of the receptor.

Phosphorylation at Y751 is essential for the association of the receptor with PI3. Both autophosphorylation sites must be phosphorylated to allow complex formation with GAP.

Some of the signal transduction processes mediated by PDGFRα and PDGFRβ are unique and some are also common.

Truncated forms of the receptors that contain only the extracellular domain have also been described. They are secreted and behave as antagonists of PDGF-stimulated processes.

The monomeric and dimeric forms of the receptors have different affinities for the different forms of PDGF. ***PDGFRα*** binds PDGF-A and PDGF-B while ***PDGFRβ*** only binds PDGF-B. PDGF-BB to the dimeric receptors αα, αβ, and ββ; PDGF-AA only binds to the αα receptor, and PDGF-AB binds to αα and αβ receptor types.

The gene encoding PDGFRB maps to human chromosome 5q33.3, flanked proximally by the GM-CSF gene and only 500 bp away from the » *fms* oncogene that functions as a receptor for » M-CSF. These two receptor genes are probably regulated together since the M-CSF receptor gene does not contain a regulatory region of its own. The gene encoding PDGFRA maps to human chromosome 4q11-q12 in the same region (≈ 700 kb) as the » *kit* oncogene that is the receptor for » SCF (stem cell factor).

A PDGF-B-derived peptide comprising the sequences between positions 116-121 and 157-163 (ANFLVWEIVRKK) functions as a PDGF antagonist. It competes with PDGF for binding to the α and β receptors and inhibits the dimerisation and autophosphorylation of the receptor.

Neomycin is a PDGF isoform- and receptor-specific antagonist, specifically inhibiting the autophosphorylation of the α-receptor by PDGF-BB, with less effect on the phosphorylation induced by PDGF-AA,and no effect on the phosphorylation of the β-receptor by PDGF-BB. After binding of PDGF to its receptor the receptor associates with non-receptor protein kinases called » *fyn* and » *yes*, which appear to mediate some of the intracellular signals.

■ **Biological Activities:** In contrast to many other cytokines PDGF is not released into the circulation. The biological half-life is less than two minutes after intravenous administration.

PDGF binds to several plasma proteins and also to proteins of the extracellular matrix (see: ECM) which facilitates local concentration of the factor. The factor functions as a local » autocrine and » paracrine growth factor. In the adult organism PDGF is involved in » wound healing processes.

The aberrant expression of PDGF is observed with vascular proliferative diseases such as atherosclerosis. PDGF and PDGF-like factors are autocrine growth factor for meningiomas.
The induction of various forms of PDGF by » TNF-β is responsible for its inhibitory activities on proliferation. At concentrations of 1-2 fg/cell TGF-β is a growth inhibitor for smooth muscle cells, fibroblasts, and chondrocytes while higher concentrations promote proliferation. This bimodal activity is mediated by PDGF-AA the synthesis and secretion of which is stimulated by low concentrations of TGF-β while higher concentrations of TGF-β down-regulate the expression of the PDGF receptor and hence the biological activities of PDGF.
PDGF-AA, in contrast to PDGF-AB and PDGF-BB, is a poor mitogen for vascular smooth muscle cells. However, together with » bFGF, which up-regulates PDGF-α receptors, it acts synergistically on DNA synthesis of these cells. PDGF does not act on epithelial cells and endothelial cells because these cell types do not express PDGF receptors. It has been shown that PDGF-BB is a potent mediator of connective tissue stroma formation in xenotransplanted human melanoma cells genetically engineered to secrete PDGF-BB, thereby facilitating the formation of a functional vascular system in the tumor (see also: Angiogenesis factors).
PDGF regulates the synthesis of its own receptor and also influences the expression of membrane receptors for » IL1, » EGF, 5-Hydroxytryptamine, LDL, transferrin, and muscarinergic receptors (see also: receptor transmodulation). PDGF induces the expression of many new genes in a plethora of cell types. Many of these genes also function as cytokines (see: Chemokines).
The functional consequences of the heterogeneity of PDGF are largely unknown. The dimeric form of PDGF is mainly mitogenic for cells of mesenchymal origin while monomeric forms of PDGF are mainly chemotactic (see also: Chemotaxis). Depending upon the cell type, the various isoforms of PDGF may stimulate or inhibit chemotactic responses.
PDGF supports the proliferation and mobility of oligodendrocytes/type I astrocytes and inhibits the premature differentiation of their progenitor cells. The neurotrophic and neuroregulatory activities (see also: Neurotrophins) of PDGF are also demonstrated by the fact that PDGF is involved in the development of the nervous system since PDGF-B receptors are expressed in almost all areas of the central nervous system.
At low concentrations PDGF is a chemoattractant for fibroblasts. PDGF is also chemotactic and activating for monocytes and neutrophils. (see also: cell activation) and activates the arachidonic acid cascade in these cells. PDGF is also a potent modulator of T cell functions. PDGF is also a potent chemoattractant for several hematopoietic cell types and cells of mesenchymal origin. PDGF is a potent vasoconstrictor.
PDGF-A and the α receptor are selectively and transiently expressed in early embryonic development of vertebrates and also in embryonic carcinoma cells. The analysis of the murine dominant » *Patch* (Ph) mutation has revealed that deletions of the α receptor causes severe anatomical alterations. These mutations are lethal. It can therefore be assumed that the function of PDGF during embryonic development is different from those in adult organisms.

■ **TRANSGENIC/KNOCK-OUT/ANTISENSE STUDIES:** A transgenic model in which the chloramphenicol acetyltransferase gene was placed under transcriptional control of the PDGF B-chain promoter demonstrates that the transgene is preferentially expressed within neural cell bodies in the cortex, hippocampus, and cerebellum. PDGF may act as a neuronal regulatory agent with neuronal release of PDGF contributing to nerve regeneration and to glial proliferation that leads to gliosis and scarring.

■ **ASSAYS:** PDGF can be assayed by means of a competitive enzymoimmunoassay or a radioreceptor assay or in a » bioassay using » 3T3 fibroblasts. For further information see also subentry "Assays" in the reference section.

■ **CLINICAL USE & SIGNIFICANCE:** Some initial studies suggest that PDGF (alone and in combination) may be useful in promoting bone formation and in soft tissue repair (see also: Wound healing). Local application of recombinant PDGF to bed ulcers has been shown to shorten the time required for healing.
PDGF has been implicated in the pathogenesis of fibroproliferative disorders and is closely associated with the presence of established disease. Markedly elevated levels of PDGF have been observed in obliterative bronchiolitis, a frequently fatal fibroproliferative disorder leading to the occlusion of small airways and affecting up to 30 % of lung transplant recipients.
It has been suggested that » autocrine loops established by co-expression of PDGF and its receptor may contribute to the growth of malignant astrocytomas in humans.

● **Reviews: Guha A** Platelet-derived growth factor: A general review with emphasis on astrocytomas. Pediatr. Neurosurg. 17: 14-20 (1992); **Habenicht AJR et al** Platelet-derived growth factor – a growth factor with an expanding role in health and disease. Klin. Wochenschr. 68: 53-9 (1990); **Khachigian LM & Chesterman CN** Platelet-derived growth factor and alternative splicing: a review. Pathology 24: 280-90 (1992); **Ross R et al** Platelet-derived growth factor and its role in health and disease. Phil. Trans. Roy. Soc. (Lond.) B 327: 155-69 (1990); **Westermark B & Sorg C** (eds) Biology of platelet-derived growth factor. Karger, Basel 1993

● **Biochemistry & Molecular Biology: Betsholtz C et al** cDNA sequence and chromosomal localization of human platelet-derived growth factor A-chain and its expression in tumor cell lines. N 320: 695-9 (1986); **Bontron DT et al** Platelet-derived growth factor A chain: gene structure, chromosomal location, and basis for alternative mRNA splicing. PNAS 85: 1492-6 (1988); **Collins T et al** Cultured human endothelial cells express platelet-derived growth factor B chains: cDNA cloning and structural analysis. N 316: 748-50 (1985); **Cook AL et al** Purification and analysis of proteinase-resistant mutants of recombinant platelet-derived growth factor-BB exhibiting improved biological activity. BJ 281: 57-65 (1992); **Craig S et al** Characterization of the structure and conformation of platelet-derived growth factor-BB (PDGF-BB) and proteinase-resistant mutants of PDGF-BB expressed in *Saccharomyces cerevisiae*. BJ 281: 67-72 (1992); **Dalla Favera R et al** Chromosomal localization of the human homologue (c-*sis*) of the Simian sarcoma virus *onc* gene. S 218: 686-8 (1982); **Kaetzel DM et al** Transcriptional control of the platelet-derived growth factor subunit genes. BioFactors 4: 71-81 (1993); **LaRochelle WJ et al** A novel mechanism regulating growth factor association with the cell surface: Identification of a PDGF retention domain. GD 5: 1191-9 (1991); **Mercola M et al** Dominant-negative mutants of a platelet-derived growth factor gene. Genes Dev. 4: 2333-41 (1990); **Rao CH et al** Structure and sequence of the human c-*sis*/platelet-derived growth factor 2 (*sis*/PDGF2) transcriptional unit. PNAS 83: 2392-6 (1986); **Rorsman F et al** Structural characterization of the human platelet-derived growth factor A-chain gene: alternative exon usage predicts two different precursor proteins. MCB 8: 571-7 (1988); **Stenman G et al** Sublocalization of the PDGF-A chain gene to chromosome 7 band q11.23 by *in situ* hybridization. ECR 178: 180-4 (1988); **Stenman G et al** The human platelet-derived growth factor α chain (PDGFA) gene maps to chromosome 7p22. Cytogenet. Cell. Genet. 60: 206-7 (1992); **Swan DC et al** Chromosomal mapping of the Simian sarcoma *onc* gene analog in human cells. PNAS 79: 4691-5 (1982); **Vassbotn FS et al** Reversion of autocrine transformation by a dominant negative platelet-derived growth factor mutant. MCB 13: 4066-76 (1993)

● **Receptors: Auger KR et al** PDGF-dependent tyrosine phosphorylation stimulates production of novel polyphosphoinositides in intact cells. Cell 57: 167-75 (1989); **Claesson-Welsh L et al** cDNA cloning and expression of a human platelet-derived growth factor (PDGF) receptor specific for B-chain-containing PDGF molecules. MCB 8: 3476-86 (1988); **Claesson-Welsh L et al** Cloning and expression of human platelet-derived growth factor α and β receptors. MiE 198: 72-7 (1991); **Daniel TO & Kumjian DA** Platelet-derived growth factor receptors and phospholipase C activation. Kidney Int. 41: 575-80 (1992); **Duan DS et al** A functional soluble extracellular region of the platelet-derived growth factor (PDGF) receptor antagonizes PDGF-stimulated responses. JBC 266: 413-8 (1991); **Eccles MR** Genes encoding the platelet-derived growth factor (PDGF) receptor and colony-stimulating factor 1 (CSF-1) receptor are physically associated in mice as in humans. Gene 108: 285-8 (1991); **Engström U et al** Identification of a peptide antagonist for platelet-derived growth factor. JBC 267: 16581-7 (1992); **Eriksson A et al** PDGFα- and β-receptors activate unique and common signal transduction pathways. EJ 11: 543-50 (1992); **Gronwald RG et al** Cloning and expression of a cDNA coding for the human platelet-derived growth factor receptor: evidence for more than one receptor class. PNAS 85: 3435-9 (1989); **Gronwald RG et al** The human PDGF receptor α-subunit gene maps to chromosome 4 in close proximity to c-*kit*. Hum. Genet. 85: 383-5 (1990); **Heldin CH & Westermark B** Signal transduction by the receptors for platelet-derived growth factor. JCS 96: 193-6 (1990); **Heldin CH et al** Platelet-derived growth factor: Isoform-specific signaling via heterodimeric or homodimeric receptor complexes. Kidney Int. 41: 571-4 (1992); **Hsieh C et al** Chromosomal localization of the gene for AA-type platelet-derived growth factor receptor (PDGFRA); in humans and mice. Cytogen. Cell Genet. 56: 160-3 (1991); **Matsui T et al** Isolation of a novel receptor cDNA establishes the existence of two PDGF receptor genes. S 243: 800-4 (1989); **Roberts WM et al** Tandem linkage of human CSF-1 receptor (c-*fms*) and PDGF receptor genes. Cell 55: 655-61 (1988); **Stenman G et al** Human PDGFA receptor gene maps to the same region on chromosome 4 as the *kit* oncogene. Genes Chromosom. Cancer 1: 155-8 (1989); **Yarden Y et al** Structure of the receptor for platelet-derived growth factor helps define a family of closely related growth factor receptors. N 323: 226-32 (1986)

● **Biological Activities: Adams EF** Autocrine control of human meningioma proliferation: secretion of platelet-derived growth factor-like molecules. IJC 49: 398-402 (1991); **Beckmann MP et al** Comparison of biological properties and transforming potential of human PDGF-A and PDGF-B chains. S 241: 1346-9 (1988); **Cochran BH et al** Molecular cloning of gene sequences regulated by platelet-derived growth factor. Cell 33: 939-47 (1983); **Daynes RA et al** Platelet-derived growth factor is a potent biologic response modifier of T cells. JEM 174: 1323-33 (1991); **Escobedo JA & Williams LT** A PDGF domain essential for mitogenesis but not for many other responses to PDGF. N 335: 85-7 (1988); **Forsberg K et al** Platelet-derived growth factor (PDGF) in oncogenesis: Development of a vascular connective tissue stroma in xenotransplanted human melanoma producing PDGF-BB. PNAS 90: 393-7 (1993); **Graves DT et al** The potential role of platelet-derived growth factor as an autocrine or paracrine factor for human bone cells. Connect. Tissue Res. 23: 209-18 (1989); **Janat MF & Liau G** Transforming growth factor β1 is a powerful modulator of platelet- derived growth factor action in vascular smooth muscle cells. JCP 150: 232-42 (1992); **Kalthoff H et al** Modulation of platelet-derived growth factor A- and B-chain/c-*sis* mRNA by tumor necrosis factor and other agents in adenocarcinoma cells. O 6: 1015-21 (1991); **Pringle N et al** Platelet-derived growth factor in central nervous system gliogenesis. ANY 633: 160-8 (1991); **Risau W et al** Platelet-derived growth factor is angiogenic *in vivo*. GF 7: 261-6 (1992); **Siegbahn A et al** Differential effects of the various isoforms of platelet-derived growth factor on chemotaxis of fibroblasts, monocytes, and granulocytes. JCI 85: 916-20 (1990); **Silver BJ** Platelet-derived growth factor in human malignancy. BioFactors 3: 217-27 (1992); **Smits A et al** Neurotrophic activity of platelet-derived growth factor (PDGF): Rat neuronal cells possess functional PDGF β-type receptors and respond to PDGF. PNAS 88: 8159-63 (1991); **Vassbotn FS et al** Neomycin is a platelet-derived growth factor (PDGF) antagonist that allows discrimination of PDGF α- and β-receptor signals in cells expressing both receptor types. JBC 267: 15635-41 (1992)

● **Transgenic/Knock-out/Antisense Studies: Fries JW & Collins T.** Platelet-derived growth factor expression in a transgenic model. Kidney Int. 41: 584-9 (1992); **Sasahara M et al**

PDGF B-chain in neurons of the central nervous system, posterior pituitary, and in a transgenic model. Cell 64: 217-27 (1991)
● Assays: **Czyrski JA & Gawlikowski W** Activity of human, bovine and porcine platelet-derived growth factor in a radioreceptor assay with human placental membrane protein. FL 219: 331-4 (1987); **Gersuk GM et al** Platelet-derived growth factor concentrations in platelet-poor plasma and urine from patients with myeloproliferative disorders. Blood 74: 2330-4 (1989); **Singh JP** A radioreceptor assay for platelet-derived growth factor. MiE 147: 13-22 (1987); **Rydziel S et al** Determination and expression of platelet-derived growth factor-AA in bone cell cultures. Endocrinol. 130: 1916-22 (1992); **Shiraishi T et al** Radioimmunoassay of human platelet-derived growth factor using monoclonal antibody toward a synthetic 73-97 fragment of its B-chain. Clin. Chim Acta 184: 65-74 (1989); **van Zoelen EJ** The use of biological assays for detection of polypeptide growth factors. PGFR 2: 131-52 (1990); **Walker C et al** Characterization of platelet-derived growth factor and platelet-derived growth factor receptor expression in asbestos-induced rat mesothelioma. CR 52: 301-6 (1992)
● Clinical Use & Significance: **Antoniades HN et al** Malignant epithelial cells in primary human lung carcinomas coexpress *in vivo* platelet-derived growth factor (PDGF) and PDGF receptor mRNA and their protein products. PNAS 89: 3942-6 (1991); **Becker W et al** A comparison of ePTFE membranes alone or in combination with platelet-derived growth factors and insulin-like growth factor-I or demineralized freeze-dried bone in promoting bone formation around immediate extraction socket implants. J. Periodontol. 63: 929-40 (1992); **Dosick SM** A prospective randomized trial of autologous platelet-derived wound healing factors for treatment of chronic nonhealing wounds: a preliminary report J. Vasc. Surg. 16: 125; discussion 126-8 (1992); **Glover JL** A prospective randomized trial of autologous platelet-derived wound healing factors for treatment of chronic nonhealing wounds: a preliminary report. J. Vasc. Surg. 16: 124-5; discussion 126-8 (1992); **Greenhalgh DG et al** PDGF and FGF stimulate wound healing in the genetically diabetic mouse. AM. J. Pathol. 136: 1235-46 (1990); **Hermanson M et al** Platelet-derived growth factor and its receptors in human glioma tissue: expression of messenger RNA and protein suggests the presence of autocrine and paracrine loops. CR 52: 3213-9 (1992); **Hertz MI et al** Obliterative bronchiolitis after lung transplantation: a fibroproliferative disorder associated with platelet-derived growth factor. PNAS 89: 10385-9 (1992); **Loss LE et al** A prospective randomized trial of autologous platelet-derived wound healing factors for treatment of chronic nonhealing wounds: a preliminary report. J. Vasc. Surg. 16: 125-6; discussion 126-8 (1992); **Pierce GF et al** Role of platelet-derived growth factor in wound healing. JCBc 45: 319-26 (1991); **Pierce GF et al** Platelet-derived growth factor-BB and transforming growth factor β 1 selectively modulate glycosaminoglycans, collagen, and myofibroblasts in excisional wounds. Am. J. Pathol. 138: 629-46 (1991); **Robson MC et al** Recombinant human platelet-derived growth factor-BB for the treatment of chronic pressure ulcers. Ann. Plast. Surg. 29: 193-201 (1992); **Robson MC et al** Platelet-derived growth factor BB for the treatment of chronic pressure ulcers. Lancet 339: 23-5 (1992); **Rutherford RB et al** Platelet-derived and insulin-like growth factors stimulate regeneration of periodontal attachment in monkeys. J. Periodontal. Res. 27: 285-90 (1992); **Shama SM et al** Dominant-negative mutants of platelet-derived growth factor revert the transformed phenotype of human astocytoma cells. MCB 13: 7203-12 (1993); **Smits A et al** Expression of platelet-derived growth factor and its receptors in proliferative disorders of fibroblastic origin. Am. J. Pathol. 140: 639-48 (1992); **Sprugel KH et al** Platelet-derived growth factor and impaired wound healing. PCBR 365: 327-40 (1991)

PDGF-1: This factor is identical with the A chain of » PDGF.
Collins T et al Cultured human endothelial cells express platelet-derived growth factor B chain: cDNA cloning and structural analysis. N 316: 748-50 (1985)

PDGF-2: This factor is identical with the B chain » PDGF.
see: PDGF-1.

PDGF-A: This factor is identical with the A chain » PDGF.

PDGF-B: This factor is identical with the B chain » PDGF.

PDGI: *platelet-derived (endothelial cell) growth inhibitor* This factor was initially isolated from platelets. It inhibits the growth of endothelial cells. PDGI is most probably identical with » TGF-β.
Brown MT & Clemmons DR Platelets contain a peptide inhibitor of endothelial cell replication and growth. PNAS 83: 3321-5 (1986); **Magyar-Lehmann S & Bohlen P** Purification of platelet-derived endothelial cell growth inhibitor and its characterization as transforming growth factor-β type 1. Experientia 48: 374-9 (1992)

PDMF: *placental-derived mitogenic factor* This poorly characterized factor is detected in tissue homogenates of 14 to 22 week human placentas. It stimulates the growth of fetal adrenocortical cells. The factor does not appear to be identical with » PDGF or » IGF, which are not mitogenic for these cells.
Simonian MH et al Placental-derived mitogenic factor for human fetal adrenocortical cell cultures. In Vitro Cell. Dev. Biol. 23: 57-62 (1987)

PDS: *plasma-derived serum* Whole blood serum (abbrev. WBS) has been shown to contain a variety of growth factors, including »EGF and » PDGF. Cell proliferation studies are therefore performed frequently by using plasma-derived serum. PDS is formed from blood plasma by removal of fibrin, fibrinogen, and some other (protein) components through precipitation and/or dialysis. Media supplemented with PDS are used in proliferation studies because they are thought to be growth factor restricted. PDS-supplemented media are also used to identify new cytokines.

PDWHF: *platelet-derived wound healing formula* A relatively undefined preparation of wound healing factors prepared from platelet α-granules. Topically applied (autologous) PDWHF accelerates wound healing and healing of chronic (nonhealing) cutaneous ulcers by stimulating angiogenesis, fibroblast proliferation, and collagen synthesis. It also appears to play a role in the healing processes of spinal cord injuries.

Atri SC et al Use of homologous platelet factors in achieving total healing of recalcitrant skin ulcers. Surgery 108: 508-12 (1990); **Doucette MM et al** Amputation prevention in a high-risk population through comprehensive wound-healing protocol. Arch. Phys. Med. Rehabil. 70: 780-5 (1989); **Hiraizumi Y et al** The effect of growth factor formula (platelet derived wound healing formula) in experimental spinal cord injuries. J. Am. Paraplegia Soc. 15: 7-13 (1992); **Hiraizumi Y et al** *In vivo* angiogenesis by platelet-derived wound-healing formula in injured spinal cord. Brain Res. Bull. 30: 353-7 (1993); **Knighton DR et al** Stimulation of repair in chronic, nonhealing, cutaneous ulcers using platelet-derived wound healing formula. Surg. Gynecol. Obstet. 170: 56-60 (1990); **Krupski WC et al** A prospective randomized trial of autologous platelet-derived wound healing factors for treatment of chronic nonhealing wounds: a preliminary report. J. Vasc. Surg. 14: 526-32 (1991); **Loss LE et al** A prospective randomized trial of autologous platelet-derived wound healing factors for treatment of chronic nonhealing wounds: a preliminary report J. Vasc. Surg. 16: 125-6 (1992); **Steed D et al** Clinical trials with purified platelet releasate. PCBR 365: 103-13 (1991)

PEDF: *pigment epithelium-derived factor* A protein of 50 kDa secreted by pigment epithelial cells of the fetal human retina. PEDF exhibits ≈ 27 % amino acid identity with human α1-antitrypsin, human α2-plasmin, and human α1-antichymotrypsin and ≈ 42 % homology if conservative substitutions are taken into account. These proteins are members of a large family of serine antiproteases (serpin family). PEDF lacks significant homology to a proposed consensus sequence for the serpin reactive center region and thus is unlikely to function as an antiprotease.

PEDF induces a neuronal phenotype in cultured human retinoblastoma cells (neoplastic counterpart of the human neural retina) and functions as a neurite-promoting factor. Differentiated cells also show increased expression of neuron-specific enolase and the 200 kDa neurofilament subunit.

Steele FR et al Pigment epithelium-derived factor: neurotrophic activity and identification as a member of the serine protease inhibitor gene family. PNAS 90: 1526-30 (1992); **Tombran-Tink J et al** Neurotrophic activity of interphotoreceptor matrix on human Y79 retinoblastoma cells. J. Comp. Neurol. 317: 175-86 (1992)

pEEDCK: This pentapeptide with the sequence PyroGlu-Glu-Asp-Cys-Lys is also called ***hemoregulatory peptide*** (abbrev. HP or HP5). It was initially identified in extracts of rodent bone marrow and human leukocytes. At present it is not known whether pEEDCK is the cleavage product of a larger precursor. The sequence of pEEDCK has been shown to be present in Giα proteins.

This peptide is a negative regulator of » hematopoiesis, inhibiting the formation of » CFU-GM colonies *in vitro*, and significantly and reversibly reducing the numbers of primitive » CFU-S and » CFU-GM *in vivo* in the mouse at concentrations as low as 10^{-13} M (see also: Hematopoietic stem cells). The peptide has no effects on the growth of erythroid progenitors » BFU-E or » CFU-E.

pEEDCK has been shown to dimerise through the formation of intermolecular disulfide bonds. The dimer, called HP5b, has no inhibitory activity and is strongly stimulatory for murine and human » CFU-GM.

It has been suggested that this peptide may be a useful myeloprotective agent in cancer chemotherapy by keeping hemopoietic stem cells in their normal nonproliferative state and thus preventing damage by certain cell cycle-specific cytostatic drugs. It has been shown that pEEDCK inhibits cytotoxic drug-induced recruitment of murine primitive CFU-S cells into active cell cycle. For other hematopoietic inhibitory peptides see also: AcSDKP, EIP (epidermal inhibitory pentapeptide).

King AG et al Regulation of colony-stimulating activity production from bone marrow stromal cells by the hematoregulatory peptide, HP-5 EH 20: 223-8 (1992); **Laerum OD et al** Inhibitory effects of a synthetic pentapeptide on hematopoietic stem cells *in vivo* and *in vitro*. EH 12: 7-17 (1984); **Laerum OD & Paukovits WR** Modification of murine hemopoiesis *in vivo* by a synthetic hemoregulatory peptide (HP5b). Differentiation 27: 106-112 (1984); **Laerum OD et al** Selectivity of hemoregulatory peptide (HP5b) action in culture. IJCC 8: 431-44 (1990); **Laerum OD et al** The dimer of hemoregulatory peptide HP5b) stimulates mouse and human myelopoiesis *in vitro*. EH 16: 274-80 (1988); **Laerum OD et al** The sequence of the hemoregulatory peptide is present in Gi α proteins. FL 269: 11-4 (1990); **Paukovits WR** Stem cell (CFU-S) inhibitory peptides: biological properties of application of pEEDCK as hemoprotector in cytostatic tumor therapy. Exp. Hematol. Today 72-80 (1989); **Paukovits WR et al** the use of hemoregulatory peptides (pEEDCK monomer and dimer) for reduction of cytostatic drug induced hemopoietic damage. Cancer Treat. Rev. 17: 347-54 (1990); **Paukovits WR et al** Hemoregulatory peptide pGlu-Glu-Asp-Cys-Lys: a new synthetic derivative for avoiding dimerization and loss of inhibitory activity. Mol. Pharmacol. 38: 401-9 (1990); **Paukovits WR et al** Prevention of hematotoxic side effects of cytostatic drugs in mice by a synthetic hemoregulatory peptide. CR 50: 328-32 (1990); **Paukovits WR et al** Inhibition of hematopoietic stem cell proliferation by hemoregulatory peptide pyroGlu-Glu-Asp-Cys-Lys (pEEDCK) provides protection against short-term neutropenia and long-term damage. ANY 628: 92-104 (1991); **Paukovits WR et al** Protection from arabinofuranosylcytosine and n-mustard-induced myelotoxicity using hemoregulatory peptide pGlu-Glu-Asp-Cys-Lys monomer and dimer. Blood 77: 1313-9 (1991); **Paukovits WR et al** Pre-CFU-S quiescence and stem cell exhaustion after cytostatic drug treatment: protective effects of the inhibitory peptide pGlu-Glu-Asp-Cys-Lys (pEEDCK). Blood 81: 1755-61 (1993); **Veiby OP**

et al Indirect stimulation of hemopoiesis by hemoregulatory peptide (HP5b) dimer in murine long-term bone marrow cultures. EH 20: 192-5 (1992)

Pentamidine: An aromatic diamidine that is used to treat *Pneumocystis carinii* infections. Pentamidine is also a specific and very potent inhibitor of the release of » IL1 by macrophages. It acts by inhibiting the proteolytic cleavage of the IL1 precursor and thus prevents the generation of the secreted 17 kDa form of IL1. The release of » TNF by LPS-stimulated macrophages is also inhibited. Pentamidine, like » pentoxifylline can therefore be used to ameliorate the pathophysiological consequences of » septic shock.

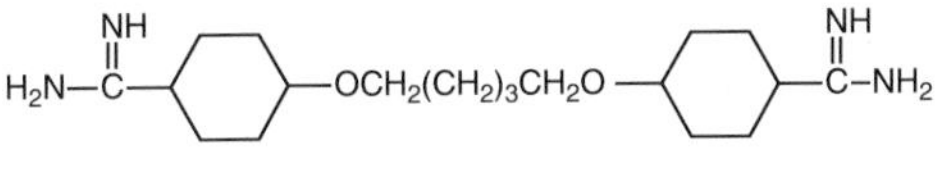

NH NH
$H_2N{-}C{-}$ $-OCH_2(CH_2)_3CH_2O-$ $-C{-}NH_2$

Pentamidine

Corsini E et al Modulation of tumor necrosis factor release from alveolar macrophages treated with pentamidine isothionate. IJI 14: 121-30 (1992); **Geratz JD et al** New aromatic diamidines with central α-oxyalkane or α,ω-dioxyalkane chains. Structure activity relationships for the inhibitor of trypsin, pancreatic kallikrein, and thrombin for the inhibition of the overall coagulation process. J. Med. Chem. 18: 477-81 (1975); **Rosenthal GJ et al** Pentamidine: an inhibitor of Interleukin-1 that acts via a post-translational event. Toxicol. Appl. Pharmacol. 107: 555-61 (1991); **Rosenthal GJ et al** Pentamidine blocks the pathophysiologic effects of endotoxemia through inhibition of cytokine release. Toxicol. Appl. Pharmacol. 112: 222-8 (1992)

Pentoxifylline: (Trental) abbrev. PTXF, PTX, Px, PTF. A methylxanthine derivative that improves red cell deformability, decreases platelet and red cell aggregation, and decreases fibrinogen and plasma viscosity. It has been used therefore in various types of peripheral and cerebrovascular diseases.

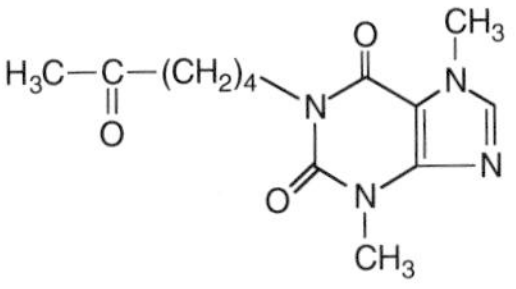

$H_3C{-}C(=O){-}(CH_2)_4{-}$ O CH_3 N N N N O CH_3

Pentoxifylline

Pentoxifylline has no effect on normal leukocytes but affects functions of activated neutrophils, decreasing adherence to endothelial cells or other surfaces, superoxide and lysozyme release, and increasing chemotaxis. It inhibits neutrophil adhesion to cultured endothelium cells and thus modulates leukocyte-endothelium and leukocyte-platelets interactions which are important factors in the development of » inflammation and thrombosis. Pentoxifylline at least 50 µg/mL suppresses NK cell function by inducing PGE2 synthesis from effector peripheral-blood mononuclear cells. Pentoxifylline inhibits the synthesis of » TNF-α *in vivo* and *in vitro* (see also: OKT3-induced cytokine-related syndrome). Pentoxifylline prevents the TNF-induced suppression of endothelial cell surface thrombomodulin. It decreases cellular and tissue damages mediated by » IL1- or » TNF-activated neutrophils and protects experimental mice against lethal endotoxin shocks (see: septic shock). Treatment with this compound also reduces the levels of circulating TNF and » IL6 and thus the compound acts as an antipyretic drug. It also prevents TNF-induced lung injury and cytokine-induced multiorgan failure. Simultaneous administration of pentoxifyllin and » IL2 appears to reduce the toxicity of IL2 treatment, at least in experimental animals, without altering the antitumor efficacy. Pentoxifylline and other methyl xanthines inhibit IL2 receptor expression in human lymphocytes. In humans pentoxifyllin reduces the amounts of circulating TNF without influencing the endogenous synthesis of other interleukins. Pentoxifylline has been shown to prevent biochemical and histological changes associated with an animal model of liver disease and to inhibit the proliferation of fibroblasts mediated by » PDGF.

Baumgartner JD Management of septic shock: new approaches. Curr. Clin. Top. Infect. Dis. 12: 165-87 (1992); **Berman B et al** Pentoxifylline inhibits certain constitutive and tumor necrosis factor-α-induced activities of human normal dermal fibroblasts. J. Invest. Dermatol. 98: 706-12 (1992); **Bianco JA et al** Phase I-II trial of pentoxifylline for the prevention of transplant-related toxicities following bone marrow transplantation. Blood 78: 1205-11 (1991); **Chao CC et al** Cytokine release from microglia: differential inhibition by pentoxifylline and dexamethasone. J. Infect. Dis. 166: 847-53 (1992); **Doherty GM et al** Pentoxifylline suppression of tumor necrosis factor gene transcription. Surgery 110: 192-8 (1991); **Edwards MJ et al** Pentoxifylline inhibits interleukin-2-induced toxicity in C57BL/6 mice but preserves antitumor efficacy. JCI 90: 637-641 (1992); **Han J et al** Dexamethasone and pentoxifylline inhibit endotoxin-induced cachectin/tumor necrosis factor synthesis at separate points in the signaling pathway. JEM 172: 391-4 (1990); **Ishizaka A et al** Prevention of interleukin 2-induced acute lung injury in guinea pigs by pentoxifylline. J. Appl. Physiol. 67: 2432-7 (1989); **LeMay LG et al** The effects of pentoxifylline on lipopolysaccharide (LPS) fever, plasma interleukin 6 (IL 6), and tumor necrosis factor (TNF) in the rat. Cytokine 2: 300-6 (1990); **Lilly CM et al** Pentoxifylline prevents tumor necrosis factor-induced lung injury. Am. Dev. Respir. Dis. 139: 1361-8 (1989); **Novick WJ jr et al** New pharmacological studies with pentoxifylline. BT 27: 449-54 (1990); **Ohdama S et al** Pentoxifylline prevents tumor necrosis factor-induced suppression of endothelial cell surface thrombomodulin. Thromb. Res. 62: 745-55 (1991); **Peterson TC** Pentoxifylline prevents fibrosis in an animal model and inhibits platelet-derived growth factor-driven proliferation of fibroblasts. Hepatology 17: 486-93 (1993); **Rao KM et al** Pentoxifylline and other methyl xanthines inhibit interleukin-2 receptor expression in human lymphocytes. CI 135: 314-25 (1991); **Reed WR & DeGowin RL** Suppressive effects of pentoxifylline on natural killer cell activity. J. Lab. Clin. Med. 119: 763-71 (1992); **Schade UF** Pentoxifylline increases survival in murine endotoxin shock and decreases formation of tumor necrosis factor. Circ. Shock. 31: 171-81 (1990); **Schandene L et al** Differential effects of pentoxifylline on the production of tumor necrosis factor-α (TNF-α) and interleukin-6 (IL6) by monocytes and T cells. Immunology 76: 30-4 (1992); **Sullivan GW et al** Inhibition of the inflammatory action of interleukin-1 and tumor necrosis factor (α) on neutrophil function by pentoxifylline. Infect. Immun. 56: 1722-9 (1988); **van Leenen D et al** Pentoxifylline attenuates neutrophil activation in experimental endotoxemia in chimpanzees. JI 151: 2318-25 (1993); **Wang P et al** Mechanism of the beneficial effects of pentoxifylline on hepatocellular function after trauma hemorrhage and resuscitation. Surgery 112: 451-7 (1992); **Weinberg JB et al** Inhibition of tumor necrosis factor-α (TNF-α) and interleukin-1 β (IL1 β) messenger RNA (mRNA) expression in HL-60 leukemia cells by pentoxifylline and dexamethasone: dissociation of acivicin-induced TNF-α and IL1 β mRNA expression from acivicin-induced monocytoid differentiation. Blood 79: 3337-43 (1992); **Saez-Llorens X et al** Pentoxifylline modulates meningeal inflammation in experimental bacterial meningitis. Antimicrob. Agents Chemother. 34: 837-43 (1990); **Zabel P et al** Effects of pentoxifylline in endotoxinemia in human volunteers. PCBR 367: 207-13 (1991)

Perforins: A class of cytolytic proteins that make membranes permeable by polymerizing in the presence of calcium and by forming transmembrane channels. These proteins, together with » leukalexins, and » NKCF (natural killer cytotoxic factor) are the major factors responsible for the cytotoxic activities of lymphocytes. They may therefore be important in » adoptive immunotherapy involving the use of lymphokine-activated killer cells (see: LAK cells).

Perforin 1 (pore-forming protein, abbrev. ***PFP***) is a glycoprotein of 70 kDa (534 amino acids) localized in cytoplasmic granules. It is identical with » Cytolysin and » C9-related protein. It is synthesized by cytotoxic T lymphocytes and natural killer cells. The P-1 gene maps to human chromosome 10q22 and to murine chromosome 10. The human and the murine gene consists of three exons. The two proteins display 68 % sequence identity (human/rat 69 %; mouse/rat: 85 %). The positions of the 20 cysteine residues are conserved in all species.

The biological activity of perforins is inhibited if the protein is complexed with » Vitronectin. In peripheral leukocytes, $CD3^-$, $CD4^-$ and CD8-positive T cells the synthesis of perforin is induced by » IL2. The IL2-induced synthesis of perforins can be enhanced by » IL6 which by itself is not an inducer. Activation of cells by » IL1 also induces synthesis of perforin. In LAK cells the synthesis of perforin is decreased by » IL4.

Acha-Orbea H et al Inhibition of lymphocyte mediated cytotoxicity by perforin antisense oligonucleotides. EJ 12: 3815-9 (1990); **Fink TM et al** Human perforin (PRF1) maps to 10q22, a region that is syntenic with mouse chromosome 10. Genomics 13: 1300-2 (1992); **Hameed A et al** Immunohistochemical identification of cytotoxic lymphocytes using human perforin monoclonal antibody. Am. J. Pathol. 140: 1025-30 (1992); **Henkart PA** Mechanisms of lymphocyte-mediated cytotoxicity. ARI 3: 31-58 (1985); **Lichtenheld MG** Structure and function of human perforin. N 335: 448-51 (1988); **Lichtenheld MG et al** Structure of the human perforin gene: a simple gene organization with interesting potential regulatory sequences. JI 143: 4267-74 (1989); **Liu CC** Identification and characterization of a pore-forming protein of human peripheral blood natural killer cells. JEM 164: 2061-76 (1986); **Lowrey DM et al** Isolation and characterization of cytotoxic granules from human lymphokine (interleukin 2) activated killer cells. Cancer Res. 48: 4681-8 (1988); **Lu P et al** Perforin expression in human peripheral blood mononuclear cells. Definition of an IL2-independent pathway of perforin induction in $CD8^+$ T cells. JI 148: 3354-60 (1992); **Masson D & Tschopp J** Isolation of a lytic, pore-forming protein (perforin) from cytolytic T lymphocytes. JBC 260: 9069-72 (1985); **Ojcius DM & Young JDE** Cytolytic pore-forming proteins and peptides: is there a common structural motif? TIBS 16: 225-9 (1991); **Podack ER** Molecular mechanisms of lymphocyte-mediated tumor cell lysis. IT 6: 21-7 (1985); **Podack ER et al** Structure, function and expression of murine and human perforin 1 (P1). IR 103: 203-11 (1988); **Podack ER et al** Isolation and biochemical and functional characterization of perforin 1 from cytolytic T cell granules. PNAS 82: 8629-33 (1985); **Podack ER et al** A central role of perforin in cytolysis? ARI 9: 129-57 (1991); **Podack ER & Kupfer A** T cell effector functions: mechanisms for delivery of cytotoxicity and help. ARC 7: 479-504 (1991); **Shinkai Y et al** Molecular cloning and chromosomal assignment of the human perforin (PFP) gene. Immunogenetics. 30: 452-7 (1989); **Smyth MJ et al** IL2 and IL6 synergise to augment the pore-forming protein gene expression and cytotoxic potential of human peripheral blood T cells. JI 145: 1156-66 (1990); **Trapani JA et al** Genomic organization of the mouse pore- forming protein (perforin) gene and localization to chromosome 10: similarities to and differences from C9. JEM 171: 545-57 (1990); **Young JDE & Cohn ZA** Cellular and humoral mechanisms of cytotoxicity: structural and functional analogies. AI 41: 269-332 (1987); **Young JDE et al** Properties of a purified pore-forming protein (perforin 1) isolated from H2-restricted cytotoxic T cell granules. JEM 164: 144-55 (1986); **Young JDE et al** Purification and characterization of a cytolytic pore-forming protein from granules of cloned lymphocytes with natural killer activity. Cell 44: 849-59 (1986); **Zalman LS et al** The cytolytic protein of human lymphocytes related to the ninth component (C9) of human complement: isolation from anti-CD3-activated peripheral blood mononuclear cells. PNAS 84: 2426-9 (1987)

Peripheral stem cell separation: see: hematopoiesis.

Peritoneal cytotoxic factor: see: PCF.

Persisting cell stimulating activity: see: PCSA.

Persisting cell stimulating factor: see: PSF.

Peptidyl-prolyl *cis-trans*-isomerase: see: CsA (Ciclosporin A).

P-Factor: This factor is identical with » IL3.

Park LS Characterization of the cell surface receptor for a multilineage colony-stimulating factor (CSF-2α). JBC 261: 205-10 (1986)

PF4: *platelet factor 4*

■ **ALTERNATIVE NAMES:** Heparin neutralizing protein; Megakaryocyte-stimulatory-factor; PF4 is identical with » CTAP-3 (connective tissue-activating protein = LA-PF4, low affinity PF4) which, in turn, is the precursor for » Beta-Thromboglobulin.

■ **SOURCES:** PF4 is synthesized in megakaryocytes and platelets. It is synthesized as a complex with a proteoglycan carrier of 350 kDa. This complex is released from the α granules of activated platelets. The synthesis of PF4 is enhanced by » IL1. All agents that activate platelets also induce the synthesis of PF4. PF4 is also found in mast cell granules and on the endothelium of human umbilical veins, but not arteries.

A monocyte chemotactic activity released by cultures of unfractionated peripheral blood cells is also identical with platelet factor 4.

■ **PROTEIN CHARACTERISTICS:** PF4 is a protein of 7.8 kDa (70 amino acids). It is formed by proteolytic cleavage of a 15 kDa precursor that is known as PBP (platelet basic protein) (see: Beta-Thromboglobulin). PF4 monomers form dimers and tetramers in varying molar ratios under certain solution conditions. The biologically active state for PF4 at physiological pH and ionic strength is the tetramer.

Non-human PF4 have additional amino acids at the N terminus while the C terminus is highly conserved.

■ **GENE STRUCTURE:** The human PF4 gene (≈ 1 kb) contains three exons and maps to chromosome 4q12-21 in the vicinity of the »Beta-Thromboglobulin gene (distance ≈ 7 kb) and in the vicinity of genes belonging to the » Chemokine family of cytokines within a region of ≈ 700 kb. PF4 is encoded by a » SIG (small inducible gene), so called because of its small size and its stimulation with platelet activation.

A variant of the PF4 gene, designated PF4V1, PF4-ALT, or PF4A, comprises 3 exons. Exon 1 codes for a hydrophobic leader sequence of 34 amino acids that has 70% sequence homology with the leader sequence of PF4. Exon 2 codes for a segment of 42 amino acids that is 100% identical with the corresponding segment of the mature PF4 sequence. Exon 3 codes for the 28 carboxy-terminal residues corresponding to a domain specifying heparin-binding and cellular chemotaxis. The PF4V1 gene also maps to 4q12-q21. The PF4V1 and the gene encoding » beta-thromboglobulin are located in a region of less than 190 kb. The 2 PF4 genes lie within about 6 kb of each other and are oriented in the same direction. The beta- thromboglobulin gene lies upstream of the PF4 gene.

■ **RELATED FACTORS:** PF-4 belongs to the » chemokine family of cytokines. The term platelet factor 4 superfamily is used for a subfamily of these factors.

■ **RECEPTOR STRUCTURE, GENE(S), EXPRESSION:** As yet no cell surface receptor has been described. It has been shown, however, that slight modifications of the N terminus of PF4 yield a variant that binds to the » IL8 receptor and then functions as a potent neutrophil activator and attractant.

```
        MKSAVLFLLGIIFLEQCGVRGTLVIRNARCSCISTSRGTIHYKSLKDLKQFAPSPNCNKTEIIATL-KNGDQTCLDPDSANVKKLNKEWEKK-
        MNQSAAVIFCLILLGLSGTQGIPLARTVRCNCIHIDDGPVRMRAIGKLEIIPASLSCPRVEIIATMKKNDEQRCLNPESKTIKNLNKAFSQK-
        MNQTAILICCLIFLTLSGIQGVPLSRTVRCTCISISNQPVNPRSLEKLEIIPASQFCPRVEIIATMKKKGEKRCLNPESKAIKNLLKAVSKE-
   MIPATRSLLCAALLLLATSRLATGAPIANELRCQCLQTMAG-IHLKNIQSLKVLPSGPHCTQTEVIATL-KNGREACLDPEAPLVQKIVQKMLKG-
 -PSNPRLLRVALLLLLLVAAGRRAAGASVATELRCQCLQTLQG-IHPKNIQSVNVKSPGPHCAQTEVIATL-KNGRKACLNPASPIVKKIIEKMLNS-
  MTSKLAVALLAAFLISAALCEGAVLPRSAKELRCQCIKTYSKPFHPKFIKELRVIESGPHCANTEIIVKL-SDGRELCLDPKENWVQRVVEKFLKR-
                      NLAKGKEESLDSDLYAELRCMCIKTTSG-IHPKNIQSLEVIGKGTHCNQVEVIATL-KDGRKICLDPDAPRIKKIVQKKLAG-
                                     EAEEDGDLQCLCVKT-TSQVRPRHITSLEVIKAGPHCPTAQLIATLTKNGRKICLDLQAPLYKKIIKKLLES-
   MNGKLGAVLALLLVSAALSQGRTLVKMGNELRCQCISTHSKFIHPKSIQDVKLTPSGPHCKNVEIIATL-KDGREVCLDPTAPWVQLIVKALMAK-
```

Sequence comparison of some members of the PF4 family.
From bottom to top the sequences are: m M119, mCRG-2, hIP-10, mKC/GRO, hGRO/MGSA, hIL8, hCTAP-III, hPF4, and c9E3 (m = murine; h = human; c = chicken). M119 is the sequence derived from the cDNA of the murine mig gene. mKC is the murine homolog of human GRO/MGSA. CTAP-III is the precursor of β-thromboglobulin. 9E3 is the chicken homolog of human GRO/MGSA. Conserved cysteine residues are shown in yellow.

■ **Biological Activities:** PF4 is chemotactic (see also: Chemotaxis) for inflammatory cells such as neutrophils and monocytes (see also: Inflammation). It activates neutrophils (see also: cell activation) and induces their degranulation. It does not induce the production of reactive oxygen species in these cells.

PF4 has a half-life in plasma of less than 3 minutes, and its rapid clearance appears to be a function of binding to the vascular endothelium. Once bound to the endothelium, PF4 can be released by heparin in a time-dependent manner. PF4 neutralizes the anticoagulatory activity of heparin sulfate in the extracellular matrix (see also: ECM) of endothelial cells. It inhibits local antithrombin III activity and thus promotes coagulation. Human PF4 has been found to stimulate the activity of leukocyte elastase and to inhibit collagenases. PF4 induces the expression of intercellular adhesion molecule I (ICAM-1) in cultured endothelial cells. It causes adherence of neutrophils to endothelial cells.

PF4 accelerates the generation of blood clots at the sites of injuries and initiates many cellular processes of » wound healing. PF4 also has immunomodulatory activities and can enhance the primary immune response in immunosuppressed mice. PF4 blocks the binding of » bFGF to its receptor and inhibits the migration of vascular endothelial cells. PF4 is a chemoattractant for fibroblasts and also increases their proliferation rates. In certain human hepatoma cell lines PF4 inhibits the binding of » TGF-β to its low-affinity receptor.

PF4 also has central actions and suppresses the food intake for at least 12 hours following intracerebroventricular injection in rats.

PF4 has been shown to be a potent inhibitor of angiogenesis in mice (see also: Angiogenesis factors). Daily administration of PF4 inhibits tumor growth of transplanted human colon carcinomas and melanomas. The angiostatic activity is due to the inhibition of local migration and proliferation of endothelial cells that are required for the formation of new blood vessels within the tumor mass. PF4-241, an analog of human PF4 lacking affinity for heparin, retains the ability to suppress the growth of various murine and human tumors transplanted in mice.

In vitro PF4 is a potent inhibitor of megakaroycytopoiesis (see also: Hematopoiesis). PF4 is a marker for the earliest members of the megakaryocytic lineage. PF4 has been shown to inhibit the growth of the human erythroleukemia cell line HEL, an effect that is abrogated by the addition of heparin.

■ **Assays:** Several immunoassays are available to determine PF4 concentrations.

■ **Transgenic/Knock-out/Antisense Studies:** Experiments with transgenic animals (mice) expressing a marker gene under the control of the PF4 promoter (see also: Gene expression) demonstrate that megakaryocytes are the only hematopoietic progenitor cells expressing the marker gene. The PF4 promoter region therefore contains sequence elements responsible for this tissue-specific gene expression.

■ **Clinical Use & Significance:** PF4, like » Beta-Thromboglobulin, is an important marker for an enhanced platelet activation (see also: Cell activation). Plasma PF4 is an index of platelet aggregation and thromboembolic risk.

PF4 may be useful in the treatment of tumors due to its marked anti-angiogenic activity. The finding that this activity is independent of heparin binding may allow the development of PF4-based angiostatic agents with reduced toxicity and improved bioavailability. Intralesional injection of PF4 has been shown to inhibit the growth of the B16 melanoma in syngeneic mice and the HCT-16 human colon carcinoma in nude mice (see also: Immunodeficient mice). PF4 may be implicated in pathological and physiological processes of bone and has been found to inhibit the growth of human osteoblast-like osteosarcoma cells.

Reversal of heparin anticoagulation in humans by protamine sulfate has been shown to result in various adverse reactions including leukopenia, thrombocytopenia, activation of complement, increased vascular permeability, systemic hypotension, pulmonary vasoconstriction, and pulmonary edema. Recombinant PF4 efficiently reverses heparin anticoagulation without adverse effects of heparin-protamine complexes and may, therefore, be an appropriate substitute for protamine sulfate in patients undergoing cardiovascular surgery and other procedures that require heparin anticoagulation.

It has been suggested that PF4 mRNA expression should be a marker of mature megakaryoblasts and that its expression in megakaryoblastic leukemia may indicate that a patient will have long survival and a good response to chemotherapy.

● **Reviews:** **Poncz M** Molecular biology of the megakaryocyte-specific gene platelet factor 4. PCBR 356: 65-76 (1990); **Stoeckle MY & Barker KA** Two burgeoning families of platelet factor 4-related proteins: mediators of the inflammatory response. New Biol. 2: 313-23 (1990); **Zucker MB & Katz IR**

Platelet factor 4: production, structure, and physiologic and immunologic action. PSEBM 198: 693-702 (1991)

● **Biochemistry & Molecular Biology: Barone AD et al** The expression in *Escherichia coli* of recombinant human platelet factor 4, a protein with immunoregulatory activity. JBC 263: 8710-5 (1988); **Bock PE et al** The multiple complexes formed by the interaction of platelet factor 4 with heparin. BJ 191: 769-76 (1980); **Chen MJ & Mayo KH** Human platelet factor 4 subunit association/dissociation thermodynamics and kinetics. B 30: 6402-11 (1991); **Ciaglowski RE & Walz DA** Isolation and amino acid sequence of bovine platelet factor 4. Comp. Biochem. Physiol. 82B: 715-9 (1985); **Deuel TF et al** Amino acid sequence of human platelet factor 4. PNAS 74: 2256-8 (1977); **Doi T et al** Structure of the rat platelet factor 4 gene: a marker for megakaryocyte differentiation. MCB 7: 898-904 (1987); **Eisman R et al** Structural and functional comparison of the genes for human platelet factor 4 and PF4alt. Blood 76: 336-44 (1990); **Green CJ et al** Identification and characterization of PF4varl, a human gene variant of platelet factor 4. MCB 9: 1445-51 (1989); **Griffin CA et al** Human platelet factor 4 gene is mapped to 4q12-q21. Cytogenet. Cell. Genet. 45: 67-9 (1987); **Holt JC et al** Characterization of human platelet basic protein, a precursor form of low-affinity platelet factor 4 and β-thromboglobulin. B 25: 1988-96 (1986); **Holt JC & Niewiarowski S** Platelet basic protein, low-affinity platelet factor 4, and β-thromboglobulin: purification and identification. MiE 169: 224-33 (1989); **Kaplan KL & Niewiarowski S** Nomenclature of secreted platelet proteins – report of the working party on secreted platelet proteins of the subcommittee on platelets. Thromb. Haemost. 53: 282-4 (1985); **Levine SP et al** Platelet factor 4 and the platelet secreted proteoglycan: immunologic characterization by crossed immunoelectrophoresis. Blood 75: 902-10 (1990); **Majumdar S et al** Characterization of the human β-thromboglobulin gene. Comparison with the gene for platelet factor 4. JBC 266: 5785-9 (1991); **Marshall SE et al** The interaction of platelet factor 4 with heparins of different chain length. BBA 797: 34-9 (1984); **Mayo KH et al** Molten globule monomer to condensed dimer: role of disulfide bonds in platelet factor-4 folding and subunit association. B 31: 12255-65 (1992); **Medici I et al** Improved method for purification of human platelet factor 4 by affinity and ion-exchange chromatography. Thromb. Res. 54: 277-87 (1989); **Park KS et al** Biological and biochemical properties of recombinant platelet factor 4 demonstrate identity with the native protein. Blood 75: 1290-5 (1990); **Poncz M et al** Cloning and characterization of platelet factor 4 cDNA derived from a human erythroleukemic cell line. Blood 69: 219-23 (1987); **Ravid K et al** Transcriptional regulation of the rat platelet factor 4 gene: interaction between an enhancer/silencer domain and the GATA site. MCB 11: 6116-27 (1991); **Rucinski B et al** Human platelet factor 4 and its C-terminal peptides: heparin binding and clearance from the circulation. Thromb. Haemost. 63: 493-8 (1990); **Shigeta O et al** Ovine platelet factor 4: purification, amino acid sequence, radioimmunoassay and comparison with platelet factor 4 of other species. Thromb. Res.64: 509-20 (1991); **St Charles R et al** The three-dimensional structure of bovine platelet factor 4 at 3. 0 Å resolution. JBC 264: 2092-99 (1989); **Stoeckle MY & Barker KA** Two burgeoning families of platelet factor 4-related proteins: mediators of the inflammatory response. New Biol. 2: 313-23 (1990); **Tunnacliffe A et al** Genes for β-thromboglobulin and platelet factor 4 are closely linked and form part of a cluster of related genes on chromosome 4. Blood 79: 2896-900 (1992); **Van Damme J et al** Identification by sequence analysis of chemotactic factors for monocytes produced by normal and transformed cells stimulated with virus, double-stranded RNA or cytokine. EJI 19: 2367-73 (1989); **Villanueva GB et al** Circular dichroism of platelet factor 4. ABB 170: 1745-50 (1989); **Walz DA et al** Primary structure of human platelet factor 4. Thromb. Res. 11: 833-98 (1977)

● **Receptor Structure, Gene(s), Expression: Clark-Lewis I et al** Platelet factor 4 binds to interleukin 8 receptors and activates neutrophils when its N terminus is modified with Glu-Leu-Arg. PNAS 90: 3574-7 (1993)

● **Biological Activities: Abgrall JF et al** Spontaneous *in vitro* megakaryocyte colony formation in primary thrombocythemia: relation to platelet factor 4 plasma level and β-thromboglobulin/platelet factor 4 ratio. Acta Haematol. 87: 118-21 (1992); **Bebawy ST et al** *In vitro* effects of platelet factor 4 on normal human neutrophil functions. J. Leuk. Biol. 39: 423-34 (1986); **Brindley LL et al** Stimulation of histamine release from human basophils by human platelet factor 4. JCI 72: 1218-23 (1983); **Carr ME et al** Platelet factor 4 enhances fibrin fiber polymerization. Thromb. Res. 45: 539-43 (1987); **Deuel TF et al** Platelet factor 4 is chemotactic for neutrophils and monocytes. PNAS 78: 4548-87 (1981); **Dumenco LL et al** Inhibition of the activation of Hageman factor (factor XII) by platelet factor 4. J. Lab. Clin. Med. 112: 394-400 (1988); **Gewirtz AM et al** Inhibition of human megakaryocytopoiesis *in vitro* by platelet factor 4 (PF4) and a synthetic COOH-terminal PF4 peptide. JCI 83: 1477-86 (1989); **Gregg EO et al** Immunomodulatory properties of platelet factor 4: prevention of concanavalin A suppressor-induction *in vitro* and augmentation of an antigen-specific delayed-type hypersensitivity response *in vivo*. Immunology 70: 230-4 (1990); **Han ZC et al** Negative regulation of human megakaryocytopoiesis by human platelet factor 4 (PF4) and connective tissue-activating peptide (CTAP-III). IJCC 8: 253-9 (1990); **Han ZC et al** Platelet factor 4 inhibits human megakaryocytopoiesis *in vitro*. Blood 75: 1234-9 (1990); **Han ZC et al** Negative regulation of human megakaryocytopoiesis by human platelet factor 4 and β thromboglobulin: comparative analysis in bone marrow cultures from normal individuals and patients with essential thrombocythemia and immune thrombocytopenic purpura. Br. J. Haematol. 74: 395-401 (1990); **Han ZC et al** Inhibitory effect of platelet factor 4 (PF4) on the growth of human erythroleukemia cells: proposed mechanism of action of PF4. J. Lab. Clin. Med. 120: 645-60 (1992); **Hiti-Harper J et al** Platelet factor 4: an inhibitor of collagenase. S 199: 991-2 (1978); **Katz IR et al** Protease-induced immunoregulatory activity for platelet factor 4. PNAS 83: 3491-5 (1986); **Katz IR et al** Alleviation of immunosuppression *in vitro* by recombinant platelet factor 4. Int. Immunol. 4: 183-90 (1992); **Lonsky SA et al** Stimulation of human granulocyte elastase by platelet factor 4 and heparin. BBRC 85: 1113-8 (1978); **Maione TE et al** Inhibition of angiogenesis by recombinant human platelet factor 4 and related peptides. S 247: 77-9 (1990); **Maione TE et al** Inhibition of tumor growth in mice by an analog of platelet factor 4 that lacks affinity for heparin and retains potent angiostatic activity. CR 51: 2077-83 (1991); **Osterman DG et al** The carboxy-terminal tridecapeptide of platelet factor 4 is a potent chemotactic agent for monocytes. BBRC 107: 130-5 (1982); **Plata-Salaman CR** Food intake suppression by growth factors and platelet peptides by direct action in the central nervous system. Neurosci. Lett. 94: 161-6 (1988); **Sato Y et al** Platelet factor 4 blocks the binding of basic fibroblast growth factor to the receptor and inhibits the spontaneous migration of vascular endothelial cells. BBRC 172: 595-600 (1990); **Sato Y et al** Carboxyl-terminal heparin-binding fragments of platelet factor 4 retain the blocking effect on the receptor binding of basic fibroblast growth factor. Jpn. J. Cancer Res. 84: 485-8 (1993); **Sharpe RJ et al** Induction of intercellular adhesion molecule-1 (ICAM-1) in cultured endothelium by recombinant platelet factor 4. J. Invest. Dermatol. 94: 578 (1990); **Sharpe RJ et al** Induction of local inflammation by recombinant human platelet factor 4 in the mouse. CI 137: 72-80 (1991); **Senior RM et al** Chemotactic activity of platelet α granule proteins for fibroblasts. JCB 96: 382-5 (1983); **Whitson RH et al** Platelet factor 4 selectively inhibits binding of TGF-β to the type I TGF-β 1 receptor. JCBc 47: 31-42 (1991); **Zucker MB et al** Immunoregulatory activity of peptides related to platelet factor 4. PNAS 86: 7571-4 (1989)

● Assays: **Bauch HJ et al** A double antibody sandwich microELISA for measuring human platelet factor 4. Thromb. Res. 37: 573-82 (1985); **Hegyi E & Nakeff A** Ultrastructural localization of platelet factor 4 in rat megakaryocytes and platelets by gold-labeled antibody detection. EH 17: 223-8 (1989); **Hegyi E et al** Immunogold probing of platelet factor 4 in different ploidy classes of rat megakaryocytes sorted by flow cytometry. EH 18: 789-93 (1990); **Lane DA et al** Detection of enhanced *in vivo* platelet α-granule release in different patient groups—comparison of β-thromboglobulin, platelet factor 4 and thrombospondin assays. Thromb. Haemost. 52: 183-7 (1984); **Levine SP et al** Platelet factor 4 and the platelet secreted proteoglycan: immunologic characterization by crossed immunoelectrophoresis. Blood 75: 902-10 (1990); **McLaren KM & Pepper DS** The immunoelectronmicroscopic localization of human platelet factor 4 in tissue mast cells. Histochem. J. 15: 795-800 (1983); **Osei-Bonsu A et al** A new medium improves sampling for determination of platelet factor 4 in human plasma. Wien. Klin. Wochenschr. 99: 595-600 (1987); **Ramachandran B et al** Polymerase chain reaction (PCR) for detection of PstI polymorphism in the human PF4 gene. NAR 18: 5919 (1990); **Sander HJ et al** Immunocytochemical localization of fibrinogen, platelet factor 4, and β thromboglobulin in thin frozen sections of human blood platelets. JCI 72: 1277-87 (1983); **Shimizu T et al** Determination of low affinity platelet factor 4 in frozen and thawed human platelets by the newly developed enzyme immunoassay system. Tohoku J. Exp. Med. 142: 453-60 (1984); **Shimizu T et** Sensitive enzyme immunoassay for the measurement of platelet factor 4 in blood plasma. Clin. Chim. Acta 138: 151-61 (1984); **Stenberg PE et al** Optimal techniques for the immunocytochemical demonstration of β-thromboglobulin, platelet factor 4, and fibrinogen in the α granules of unstimulated platelets. Histochem. J. 16: 983-1001 (1984); **Strauss W et al** Serial determinations of PF4 and β TG: comparisons between multiple venipunctures vs a catheter infusion system. Thromb. Haemost. 59: 491-4 (1988); **Takahashi H et al** Measurement of platelet factor 4 and β-thromboglobulin by an enzyme-linked immunosorbent assay Clin. Chim. Acta 175: 113-4 (1988)

● Transgenic/Knock-out/Antisense Studies: **Ravid K et al** Selective targeting of gene products with the megakaryocyte platelet factor 4 promoter. PNAS 88: 1521-5 (1991)

● Clinical Use & Significance: **Cook JJ et al** Platelet factor 4 efficiently reverses heparin anticoagulation in the rat without adverse effects of heparin-protamine complexes. Circulation 85: 1102-9 (1992); **De Caterina R et al** Platelet activation in angina at rest. Evidence by paired measurement of plasma β-thromboglobulin and platelet factor 4. Eur. Heart J. 9: 913-22 (1988); **Ffrench P et al** Comparative evaluation of plasma thrombospondin β-thromboglobulin and platelet factor 4 in acute myocardial infarction. Thromb. Res. 39: 619-24 (1985); **Gomi T et al** Plasma β-thromboglobulin to platelet factor 4 ratios as indices of vascular complications in essential hypertension. J. Hypertens. 6: 389-92 (1988); **Kaplan KL & Owen J** Plasma levels of β-thromboglobulin and platelet factor 4 as indices of platelet activation *in vivo*. Blood 57: 199-202 (1981); **Lorenz R & Brauer M** Platelet factor 4 (PF4) in septicaemia. Infection 16: 273-6 (1988); **Maione TE et al** Inhibition of angiogenesis by recombinant human platelet factor 4 and related peptides. S 247: 77-9 (1990); **McRoyan DK et al** β-Thromboglobulin and platelet factor 4 levels in a case of acute thrombotic thrombocytopenic purpura. South. Med. J. 78: 745-8 (1985); **Ravid K et al** Selective targeting of gene products with the megakaryocyte platelet factor 4 promoter. PNAS 88: 1521-5 (1991); **Ryo R et al** Platelet factor 4 mRNA expression in cells from a patient with megakaryoblastic crisis of chronic myelogenous leukemia. Cancer 67: 960-4 (1991); **Ryo R et al** Megakaryocytic leukemia and platelet factor 4. Leuk. Lymphoma 8: 327-36 (1992); **Sharpe RJ et al** Growth inhibition of murine melanoma and human colon carcinoma by recombinant human platelet factor 4. JNCI 82: 848-53 (1990); **Tatakis DN** Human platelet factor 4 is a direct inhibitor of human osteoblast-like osteosarcoma cell growth. BBRC 187: 287-93 (1992); **Yamauchi K et al** Plasma β-thromboglobulin and platelet factor 4 concentrations in patients with atrial fibrillation. Jpn. Heart J. 27: 481-7 (1986)

PF4 superfamily: see: Chemokines.

PFC-EA: *plaque forming cell enhancing factor* This factor is identical with » IL2.

Paetkau V et al Cellular origins and targets of Costimulator (IL2). IR 51: 157-75 (1980); **Watson J et al** The purification and quantitation of helper T cell-replacing factors secreted by murine spleen cells activated by Concanavalin A. JI 122: 209-15 (1979)

PFP: *pore-forming protein* see: Perforins.

PGF: *pituitary growth factor* This factor is identical with » bFGF.

Esch F et al Primary structure of bovine pituitary basic fibroblast growth factor (FGF) and comparison with the amino-terminal sequence of bovine brain acidic FGF. PNAS 82: 6507-11 (1985); **Gospodarowicz D et al** Isolation of brain fibroblast growth factor by heparin-sepharose affinity chromatography: identity with pituitary fibroblast growth factor. PNAS 81: 6963-7 (1984); **Rowe JM et al** Purification and characterisation of a human pituitary growth factor. B 25: 6421-5 (1986); **Smith JA et al** Brain and pituitary fibroblast growth factor activities behave identically on three independent high performance liquid chromatography systems. BBRC 119: 311-8 (1984)

PGF: *plasmacytoma growth factor* see: PCT-GF.

PGF: general abbrev. for polypeptide growth factor(s). See also: Growth factors.

PGF: *prostatic growth factor* abbrev. also PrGF. Three principal families of growth factors appear to be potentially involved in the development of benign prostatic hypertrophy: »EGF, » FGF, and » TGF-β. PGF is also called » BGF (bone growth factor) and is identical with » bFGF. Other PGFs described in the literature compete with » EGF for binding to the EGF receptor. These proteins are different processed forms of EGF or identical with EGF.

Desgrancchaps F et al [Prostatic growth factors and benign hypertrophy of the prostate. Current knowledge and perspectives] Prog. Urol. 2: 1031-44 (1992); **Muraki J et al** Human prostatic cancer cell derived growth factor stimulating fibroblasts and its mechanisms. Urology 38: 489-92 (1991); **Nishi N et al** Rat prostatic growth factors: purification and characterisation of high and low molecular weight epidermal growth factors from rat dorsolateral prostate. BBA 1095: 268-75 (1991); **Story MT et al** Preliminary characterisation and evaluation of techniques for the isolation of prostate-derived growth factor. PCBR 145: 197-216 (1984); **Story MT et al** Prostatic growth factor: purification and structural relationship to basic fibroblast growth factor. B 26: 3843-9 (1987); **Story MT** Polypeptide modulators of prostatic growth and development. Cancer Surv. 11: 123-46 (1991)

PGSF: *paracrine growth-stimulatory factor* This poorly characterized factor of ≈ 30 kDa was found in the culture supernatants of a spontaneously transformed rat liver epithelial cell line grown in serum-free medium. PGSF stimulates the proliferation of » NRK cells and induces anchorage-independent growth in soft agar culture without » EGF or TGF-β. PGSF slightly inhibits the growth of the spontaneously transformed rat liver epithelial cells which also appear to elaborate an unidentified low molecular weight autocrine growth-stimulatory factor (AGSF).

Narita T Spontaneously transformed rat-liver epithelial-cell line producing autocrine and paracrine growth factors. IJC 52: 455-60 (1992)

pH2-labile interferon: This interferon is identical with » IFN-γ. See also: IFN for general information about interferons.

pH2-stable interferon: This interferon is identical with » IFN-α or » IFN-β. See also: IFN for general information about interferons.

Phagocytosis inducing factor: see: PIF.

Phorbol esters: Phorbol esters are tumor-promoting compounds originally detected in oil prepared from seeds of *Croton tiglium*. Some esters, such as phorbol 12-myristate 13-acetate (abbrev. PMA) or 12-O-tetradecanoyl-phorbol-13-acetate (abbrev. TPA) have highly pleiotropic effects on cells in culture and on tissues *in vivo* and are frequently used in immunology as activators of T cells and polyclonal B cell activators *in vitro*. PMA acts as a direct agonist of protein kinase C, which plays a pivotal role in the cascade of events leading to B cell activation (see also: Cell activation; BCGF, B cell growth factor). Treatment of a variety of cells with phorbol esters either induces or inhibits differentiation.

Phorbol esters activate the expression of a variety of » ERGs (early-response genes) through protein kinase C-dependent pathways. Many such genes have been identified as TPA-inducible genes in » PC12 cells (see: TIS genes). The rapid and coordinated expression of cytokines following treatment of cells with phorbol esters and other agents is due, at least in part, to the presence of regulatory sequence elements (see, for example: TRE, CLE) in the corresponding gene promoters (see also: Gene expression).

Treatment of cells with phorbol esters frequently mimicks the effects of various cytokines and induces their synthesis and the synthesis of other gene products. Cultured cells stimulated with phorbol esters have been employed frequently in the search for new secreted cytokines. Active phorbol esters (e.g., phorbol 12-myristate 13-acetate) can entirely replace » IL3 to promote proliferation of » FDC-P1 cells and have been shown also to be able to replace » IL2 in IL2-dependent human cell lines. For other agents used to dissect cytokine-mediated signal transduction pathways see: Bryostatins, Calcium ionophore, Calphostin C, Genistein, H8, Herbimycin A, K-252a, Lavendustin A, Okadaic acid, Staurosporine, Suramin, Tyrphostins, Vanadate.

Blumberg PM Protein kinase C as the receptor for the phorbol ester tumor promoters: sixth Rhoads memorial award lecture. CR 48: 1-8 (1988); **Castagna M** Phorbol esters as signal transducers and tumor promoters. Biol. Cell. 59: 3-13 (1987); **Gschwendt M et al** Protein kinase C activation by phorbol esters: do cysteine-rich regions and pseudosubstrate motifs play a role? TIBS 16: 167-9 (1991); **Higuchi M & Aggarwal BB** Okadaic acid induces down-modulation and shedding of tumor necrosis factor receptors. Comparison with another tumor promoter, phorbol ester. JBC 268: 5624-31 (1993); **Johnson MD et al** Molecular cloning of gene sequences regulated by tumor promoters and mitogens through protein kinase C. MCB 7: 2821-9 (1987); **Murphy JJ & Norton JD** Phorbol ester induction of early response gene expression in lymphocytic leukemia and normal human B cells. Leuk. Res. 17: 657-62 (1993); **Petricoin EF 3rd et al** Modulation of interferon signaling in human fibroblasts by phorbol esters. MCB 12: 4486-95 (1992); **Pluznik DH & Bickel M** Phorbol esters, lipopolysaccharides and colony stimulating factor production. PCBR 352: 205-13 (1990); **Takeshita T et al** Phorbol esters can persistently replace interleukin-2 (IL2) for the growth of a human IL2-dependent T-cell line. JCP 136(2): 319-25 (1988)

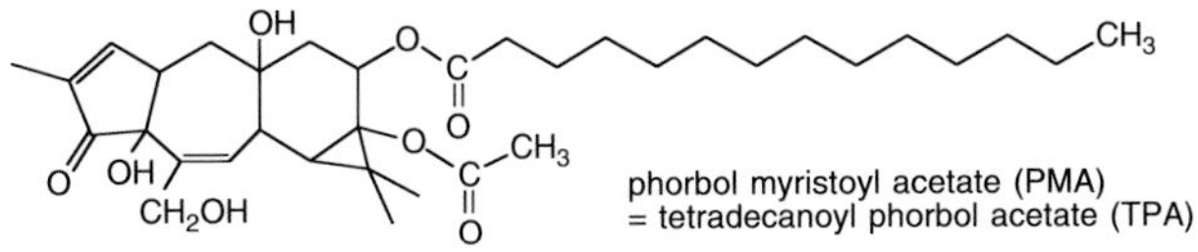

phorbol myristoyl acetate (PMA)
= tetradecanoyl phorbol acetate (TPA)

Phosphohexose isomerase: see: NLK (neuroleukin).

PHPC: *primitive hematopoietic progenitor cells* A general term referring to the more differentiated offspring of » Hematopoietic stem cells. See also: Hematopoiesis.

PHSC: *primitive (or pluripotent) hemopoietic stem cells* see: Hematopoiesis, Hematopoietic stem cells.

Physalaemin: see: » Tachykinins.

PIA: *proliferation inducing activity* A general term for factors that promote the growth of cells. Such factors can be isolated from conditioned media (see: CM). They can also be isolated on the basis of some of their biochemical properties (see: HBGF, heparin-binding growth factors).

PIF: *phagocytosis inducing factor* This poorly characterized factor is produced by T cells. It enhances the phagocytic activities in culture of monocytic cell lines. The factor is not identical with » IFN-α and » IFN-γ.

Hinuma S et al Constitutive production of phagocytosis inducing factor(s) in a monocyte-macrophage lineage cell line (THP-1) by retrovirus-transformed human T cell lines. Microbiol. Immunol. 28: 935-47 (1984); **Margolick JB et al** Production of phagocytosis-inducing factor and expression of 4B4 antigen by cloned human T cells before and after transformation with HTLV-I. CI 111: 196-203 (1988); **Margolick JB et al** Human $T4^+$ lymphocytes produce a phagocytosis-inducing factor (PIF) distinct from interferon-α and interferon-γ. JI 136: 546-54 (1986)

PIF: *proteolysis inducing factor* (also abbrev. ***MPIF***, muscle proteolysis inducing factor)This factor was initially identified in the serum of febrile patients. It is also produced by monocytes. The factor of ≈ 4 kDa induces the accelerated release of amino acids by muscle cells, the uptake of amino acids by liver cells, and the synthesis of » acute phase proteins (see: acute phase reaction). PIF is identical with a cleavage product of » IL1 that can function as an active mediator of IL 1 effects.

Belizario JE et al Bioactivity of skeletal muscle proteolysis-inducing factors in the plasma proteins from cancer patients with weight loss. Br. J. Cancer 63: 705-10 (1991); **Clowes GH et al** Muscle proteolysis induced by a circulating peptide in patients with sepsis or trauma. NEJM 308: 545-52 (1983); **Dinarello CA et al** Cleavage of human interleukin 1: isolation of a peptide fragment from plasma of febrile humans and activated monocytes. JI 133: 1332-8 (1984); **Loda M et al** Induction of hepatic protein synthesis by a peptide in blood plasma of patients with sepsis and trauma. Surgery 96: 204-13 (1984); **Romette J et al** Inflammatory syndrome and changes in plasma proteins. Pathol. Biol. Paris 34: 1006-12 (1986)

Pigment epithelium-derived factor: see: PEDF.

PIL-6: A cell line established from the murine plasmacytoma MOPC-104E. This cell line responds to murine and to human » IL6, but not to any other cytokines and has been used as an indicator cell line for the detection of IL6.

Takai Y et al Enhanced production of interleukin-6 in mice with type II collagen-induced arthritis. Arthritis Rheum. 32: 594-600 (1989)

***Pim*:** The *pim* gene is an » oncogene encoding a cytoplasmic non-receptor serine/threonine kinase (34 kDa, 313 amino acids) that is expressed primarily in B lymphoid and myeloid cell lines. It has been shown that viral leukemogenesis in mice is often initiated by proviral activation of the highly conserved cellular gene *pim*-1, and the human *pim*-1 gene is a homologue of the murine retroviral insertion site m*pim*-1. Human *pim*-1 protein has 94% identity with murine *pim*-1 whereas the nucleotide sequences of the two genes are 88% identical.
The gene is on human chromosome 6p21.2, contains six exons, and has a length of ≈ 5 kb. Two different proteins are encoded by this gene, and these arise by differential use of two alternative initiation sites.
The 33 kDa product of the *pim* gene is highly expressed in the liver and spleen during fetal hematopoiesis. It is only slightly expressed in circulating granulocytes in adults. It is overexpressed in hematopoietic malignancies, particularly in myeloid and lymphoid acute leukemias.
Expression of *pim*-1 can be induced in various myeloid leukemia cell lines by » IL3, » GM-CSF, » G-CSF, and » IL6, suggesting that *pim*-1 is utilised as part of the intracellular signaling cascade initiated by these myeloid growth factors binding to their respective receptors.
■ **TRANSGENIC/KNOCK-OUT/ANTISENSE STUDIES:** Transgenic mice that overexpress *pim*-1 have been shown to have a low incidence of spontaneous T cell lymphomas and an increased susceptibility to Moloney murine leukemia virus and N-ethyl-N-nitrosourea-induced lymphomas. Apart from a slight enlargement of the spleen, no abnormalities are found in prelymphomatous transgenic animals. Inactivation of the *pim*-1 gene in the germline of mice results in mice with a surprisingly subtle phenotype. Bone marrow-derived mast cells lacking

pim-1 expression have a distinct growth disadvantage when grown on » IL3, but not when stimulated by » IL4, » IL9, or » SCF.

Amson R et al The human proto-oncogene product p33*pim* is expressed during fetal hematopoiesis and in diverse leukemias. PNAS 86: 8857-61 (1989); **Domen J et al** Comparison of the human and mouse *pim*-1 cDNAs: nucleotide sequence and immunological identification of the *in vitro* synthesized *pim*-1 protein. OR 1: 103-12 (1987); **Friedmann M et al** Characterization of the proto-oncogene *pim*-1: kinase activity and substrate recognition sequence. ABB 298: 594-601 (1992); **Hoover D et al** Recombinant human *pim*-1 protein exhibits serine/threonine kinase activity. JBC 266: 14018-23 (1991); **Lilly M et al** Sustained expression of the *pim*-1 kinase is specifically induced in myeloid cells by cytokines whose receptors are structurally related. O 7: 727-32 (1992); **Meeker TC et al** The human *pim*-1 gene is selectively transcribed in different hemato-lymphoid cell lines in spite of a G + C-rich housekeeping promoter. MCB 10: 1680-8 (1990); **Meeker TC et al** Characterization of the human *pim*-1 gene: a putative proto-oncogene coding for a tissue specific member of the protein kinase family. OR 1: 87-101 (1987); **Nagarajan L & Narayana L** Transcriptional attenuation of *pim*-1 gene. BBRC 190: 435-9 (1993); **Reeves R et al** Primary structure of the putative human oncogene, *pim*-1. Gene 90: 303-7 (1990); **Saris CJM et al** The *pim*-1 oncogene encodes two related protein-serine/threonine kinases by alternative initiation at AUG and CUG. EJ 10: 655-664 (1991); **Selten G et al** The primary structure of the putative oncogene *pim*-1 shows extensive homology with protein kinases. Cell 46: 603-11 (1986)

● **TRANSGENIC/KNOCK-OUT/ANTISENSE STUDIES: Breuer M et al** Very high frequency of lymphoma induction by a chemical carcinogen in *pim*-1 transgenic mice. N 340: 61-3 (1989); **Breuer M et al** Carcinogen-induced lymphomagenesis in *pim*-1 transgenic mice: dose dependence and involvement of myc and *ras*. CR 51: 958-63 (1991); **Domen J et al** Impaired interleukin-3 response in *pim*-1-deficient bone marrow-derived mast cells. Blood 82: 1445-52 (1993); **te Riele H et al** Consecutive inactivation of both alleles of the *pim*-1 proto-oncogene by homologous recombination in embryonic stem cells. N 348: 649-51 (1990); **Van Lohuizen M et al** Predisposition to lymphomagenesis in *pim*-1 transgenic mice: cooperation with c-*myc* and N-*myc* in murine leukemia virus-induced tumors. Cell 56: 673-82 (1989)

pip92: An » ERG (early response gene) inducible by serum growth factors in fibroblasts. It is also induced in the rat pheochromocytoma cell line » PC12 by agents that cause proliferation, neuronal differentiation, and membrane depolarization. It is also expressed after cross-linking of membrane immunoglobulin on B lymphocytes.

Charles CH et al Pip92: a short-lived, growth factor-inducible protein in BALB/c 3T3 and PC12 cells. MCB 10: 6769-74 (1990); **Mittelstadt PR & De Franco AL** Induction of early response genes by cross-linking membrane Ig on B lymphocytes. JI 150: 4822-32 (1993)

Pit-1: see: Dwarf mice.

Pitressin: Alternative name for » Vasopressin.

Pituitary-adrenocortical axis: see: Neuroimmune network.

Pituitary-derived chondrocyte growth factor: This factor is identical with » bFGF.

Pituitary FGF: This factor is isolated from bovine brain tissues. It is identical with » aFGF:

Böhlen P et al Acidic fibroblast growth factor (FGF) from bovine brain: amino-terminal sequence and comparison with basic FGF. EJ 4: 1951-6 (1985)

Pituitary growth factor: abbrev. also HPGF (H = human). see: PGF.

Pituitrin P: This factor is identical with » Vasopressin.

pIXY321: A fusion protein derived from the coding sequences of » GM-CSF and » IL3, obtained from a genetically engineered plasmid expressed in yeast cells. The two parent molecules are connected by a short, flexible linker sequence in the fusion protein. pIXY321 exhibits greater receptor binding affinity and colony-stimulating activity than either GM-CSF or IL3. It has also beem shown to be more potent on a molar basis than the combination of the two parent factors. pIXY321 allows the stimulation and expansion of multilineage hematopoiesis from immature bone marrow progenitor cells. Clinical trials employing pIXY321 are currently underway in humans.

Bruno E et al Recombinant GM-CSF/IL3 fusion protein: its effect on *in vitro* human megakaryocytopoiesis. EH 20: 494-99 (1992); **Buescher ES et al** The effects of treatment with PIXY321 (GM-CSF/IL3 fusion protein) on human polymorphonuclear leukocyte function. EH 21: 1467-72 (1993); **Cairo MS et al** The *in vitro* effects of stem cell factor and PIXY321 on myeloid progenitor formation (CFU-GM) from immunomagnetic separated $CD34^+$ cord blood. Pediatr. Res. 32: 277-81 (1992); **Curtis BM et al** Enhanced hematopoietic activity of a human granulocyte/macrophage colony-stimulating factor-interleukin 3 fusion protein. PNAS 88: 5809-13 (1991); **Hamilton RF jr et al** The *in vivo* effects of PIXY321 therapy on human monocyte activity. J. Leukoc. Biol. 53: 640-50 (1993); **Lapidot T et al** Cytokine stimulation of multilineage hematopoiesis from immature human cells engrafted in SCID mice. S 255: 1137-41 (1992); **McAlister EB et al** Ex vivo expansion of peripheral blood progenitor cells with recombinant cytokines. EH 20: 626-8 (1992); **McCrady CW et al** Modulation of the activity of a human granulocyte-macrophage colony-stimulating factor/interleukin-3 fusion protein (pIXY321) by the macrocyclic lactone proteine kinase C activator bryostatin 1. EH 21: 893-900 (1993); **Williams DE & Park LS** Hematopoietic effects of a granulocyte-macrophage colony-stimulating factor/interleukin-3 fusion protein. Cancer 67: s2705-7 (1991)

PK-2: This human stromal cell line (see also: Hematopoiesis) has been obtained from Dexter-type long-term bone marrow cultures (see: LTBMC) after immortalisation with SV40. The cells have been shown to express mRNA for » M-CSF, » GM-CSF, » G-CSF, » IL1, and » TNF-α (see also: Message amplification phenotyping). Their functional capacities in supporting hematopoiesis in culture have not yet been determined.

Nemunaitis J et al Human marrow stromal cells: response to interleukin-6 (IL6) and control of IL6 expression. Blood 74: 1929-35 (1989)

Placenta growth factor: see: PlGF.

Placental angiogenic factor: see: PAF.

Placental-derived mitogenic factor: see: PDMF.

Placental protein 12: see: PP12.

Plaque forming cell enhancing factor: see: PFC-EA.

Plasmacytoma growth factor: see: PCT-GF

Plasmacytoma stimulatory activity: This factor is identical with » IL11.

Paul SR et al Molecular cloning of a cDNA encoding interleukin 11, a stromal cell-derived lymphopoietic and hematopoietic cytokine PNAS 87: 7512-6 (1990)

Plasmacytoma-derived suppressor factor: see: Suppressor factors.

Plasma-derived serum: see: PDS.

Plasminogen activator inhibitor 1 binding protein: abbrev. PAI-BP. This factor is identical with » Vitronectin.

Platelet basic protein: abbrev. PBP. see: Beta-Thromboglobulin.

Platelet factor 4: see: PF4.

Platelet factor 4 superfamily: see: Chemokine.

Platelet-derived endothelial cell growth factor: see: PD-ECGF.

Platelet-derived (endothelial cell) growth inhibitor: see: PDGI.

Platelet-derived endothelial cell mitogen: see: PD-ECGF (platelet-derived endothelial cell growth factor).

Platelet-derived growth factor: see: PDGF.

Platelet-derived wound healing formula: see: PDWHF.

Platelet-poor plasma: see: PPP.

Pleiotrophin: see: PTN.

PLF: *proliferin* This protein was initially characterized as a mRNA expressed a few hours after stimulation of resting murine » 3T3 cells with serum or » PDGF. Expression of this mRNA reaches a high level during the transition from the G_1 to the S phase of the cell » cycle. The mRNA, which is expressed at high levels in murine placenta, encodes a protein of 244 amino acids that is secreted as a glycoprotein of 34 kDa (21,5 kDa unglycosylated). Proliferin binds to the mannose-6-phosphate receptor (see: MPR). PLF is identical with » MRP (mitogen-regulated protein) and, by sequence homologies, is a member of the prolactin/growth-hormone family of proteins. The murine PLF gene maps to chromosome 13.

Connor AM et al Characterization of a mouse mitogen-regulated protein/proliferin gene and its promoter: a member of the growth hormone/prolactin gene superfamily. BBA 1009: 75-82 (1989); **Jackson-Grusby LL et al** Chromosomal mapping of the prolactin/growth hormone gene family in the mouse. Endocrinology 122: 2462-6 (1988); **Lee SJ & Nathans D** Proliferin secreted by cultured cells binds to mannose 6-phosphate receptors. JBC 263: 3521-7 (1988); **Linzer DI & Nathans D** Nucleotide sequence of a growth-related mRNA encoding a member of the prolactin-growth hormone family. PNAS 81: 4255-9 (1984); **Linzer DI et al** Identification of proliferin mRNA and protein in mouse placenta. PNAS 82: 4356-9 (1985); **Linzer DI & Nathans D** A new member of the prolactin-growth hormone gene family expressed in mouse placenta. EJ 4: 1419-23 (1985); **Linzer DI & Mordacq JC** Transcriptional regulation of proliferin gene expression in response to serum in transfected mouse cells. EJ 6: 2281-8 (1987); **Nilsen-Hamilton M et al** Relationship between mitogen-regulated protein (MRP) and proliferin (PLF), a member of the prolactin/growth hormone family. Gene 51: 163-70 (1987); **Wilder EL & Linzer DI** Expression of multiple proliferin genes in mouse cells. MCB 6: 3283-6 (1986)

PLF: *psoriatic leukotactic factor* This factor is found in large amounts in psoriatic scales. It is chemotactic for leukocytes. The factor is probably identical with » IL8 and therefore belongs to the » chemokine family of cytokines. See also: LCF (lymphocyte chemotactic factor) and GCP (granulocyte chemotactic peptide).

Tagami J et al Psoriatic leukotactic factor. Further physicochemical characterization and effect on the epidermal cells. Arch. Dermatol. Res. 272: 201-13 (1982)

PlGF: *placenta growth factor* This factor was isolated initially as a cDNA from a human placenta cDNA library. PlGF is also expressed in human umbilical vein endothelial cells. The cDNA encodes a protein of 149 amino acids which displays 51 % sequence identity with a sequence domain of » VPF (vascular permeability factor). The protein shows an overall amino acid identity with » PDGF of less than 30 %, but the eight cysteine residues of the mature portions of the PDGF chains are perfectly conserved in PlGF (and VEGF).
The biologically active form of this protein is a disulfide-linked dimer. The N-glycosylated dimeric protein is secreted and, like VPF, stimulates the proliferation of endothelial cell lines and functions as an » angiogenesis factor. The human gene encoding PlGF maps to chromosome 14. At least two different mRNAs are produced from this single-copy gene in different cell lines and tissues. The two different isolated cDNAs are identical except for the insertion of a highly basic 21 amino acid stretch at the carboxyl end of the protein. PlGF appears to be highly conserved in evolution, hybridizing to sequences present in the genomic DNA of *Drosophila*, *Xenopus*, chicken and mouse.
PlGF has been renamed ***PlGF-1*** after the discovery of ***PlGF-2***, which has a 21 amino acid insertion not present in PlGF. PlGF-1 and PlGF-2 compete in a dose-dependent way with the 165 amino acid form of » VEGF for receptor binding on endothelial cells.
Hauser S & Weich HA A heparin-binding form of placenta growth factor (PlGF-2) is expressed in human umbilical vein endothelial cells and in placenta. GF 9: 259-68 (1993); **Maglione D et al** Isolation of a human placenta cDNA coding for a protein related to the vascular permeability factor. PNAS 88: 9267-71 (1991); **Maglione D et al** Two alternative mRNAs coding for the angiogenic factor, placenta growth factor (PlGF), are transcribed from a single gene of chromosome 14. O 8: 925-31 (1993); **Maglione D et al** Translation of the placenta growth factor mRNA is severely affected by a small open reading frame localized in the 5' untranslated region. GF 8: 141-52 (1993)

PlGF-1: *placenta growth factor* see: PlGF.

PlGF-2: *placenta growth factor* see: PlGF.

PLSC: *pluripotent lymphoid stem cells* These cells give rise to various lymphoid cell types. See: Hematopoiesis.

Pluripoietin: abbrev. Ppo. *(human) pluripotent colony-stimulating factor* abbrev. pCSF. This factor was described initially as a myeloid colony-stimulating activity (see also: CSF, colony-stimulating factors) and a myeloid leukemia cell line differentiation factor. It is identical with » G-CSF. It is also called pluripoietin-β. Pluripoietin-α is identical with » GM-CSF.
Gabrilove JL et al Pluripoietin a: a second human hematopoietic colony-stimulating factor produced by the human bladder carcinoma cell line 5637. PNAS 83: 2478-2482 (1986); **Harris G et al** Distinct differentiation-inducing activities of γ-interferon and cytokine factors acting on the human promyelocytic leukemia cell line HL-60. CR 45: 3090-5 (1985); **Strife A et al** Activities of four purified growth factors on highly enriched human hematopoietic progenitor cells. Blood 69: 1508-23 (1987); **Welte K et al** Purification and biochemical characterization of human pluripotent hematopoietic colony-stimulating factor. PNAS 82: 1526-30 (1985)

Pluripoietin-α: *(human) pluripotent colony-stimulating factor* abbrev. pCSF. This factor is identical with » GM-CSF.
see: pluripoietin.

Pluripoietin-β: *(human) pluripotent colony-stimulating factor* abbrev. pCSF. This factor is identical with » G-CSF.
see: pluripoietin.

Pluripotent colony stimulating factor: abbrev. pCSF. see: Pluripoietin, Pluripoietin-α, and -β.

Pluripotential stem cell-supporting factor: This factor is most likely identical with » IL3.
Clark-Lewis I et al Preparation of T cell growth factor free from interferon and factors stimulating hemopoietic cells and mast cells. JIM 51: 311-22 (1982)

PM: *pregnancy mitogen* This factor of ≈ 10 kDa is isolated from the serum of pregnant women. It is present in serum samples from second trimester as well as term-pregnant women, but not in those of adult men or cycling females. PM induces rapid proliferation of the rat lymphoma cell line » Nb2. PM activity in the Nb2 cell bioassay is not affected by the presence of » prolactin antiserum. Its activity is immunoneutralized by coincubation with anti-placental lactogen serum and, to a lesser extent, by anti-growth hormone serum (see also: Growth hormone). Examination of the preliminary amino acid composition of PM reveals differences from that of a bioactive fragment of growth hormone and a corresponding portion of placental lactogen.
Kineman RD et al Purification from human pregnancy serum of a low molecular weight mitogen similar to placental lactogen and growth hormone. PSEBM 197: 441-9 (1991)

PMA: *phorbol 12-myristate 13-acetate* see: Phorbol esters.

PMN-CCF: *polymorphonuclear cell crystal-induced chemotactic factor* see: CCF.

PMN-derived lymphocyte proliferation-potentiating factor: see: PMN factor.

PMN factor: *polymorphonuclear leukocytes (PMN)-derived lymphocyte proliferation-potentiating factor* This factor of 18,5 kDa is released by rabbit early inflammatory peritoneal exudate cells (98% of polymorphonuclear leukocytes, PMN) stimulated with kaolin. It augments proliferation in a thymocyte comitogenic assay. PMN factor is considered to be the rabbit homologue of human »IL1 (IL1β), based on its physicochemical properties and the known aminoterminal amino acid sequence.

Goto F et al Purification and partial sequence of rabbit polymorphonuclear leukocyte-derived lymphocyte proliferation potentiating factor resembling IL1 β. JI 140: 1153-8 (1988)

PMNL (polymorphonuclear leukocyte) recruiting activity: abbrev. PRA. see: Leukocyte infiltration inducing factor.

PMNL (polymorphonuclear leukocyte) recruiting factor: abbrev. PRF. see: Leukocyte infiltration inducing factor.

POF: *prostatic osteoblastic factor* This factor is identical with » bFGF.

Jacobs SC et al Prostatic osteoblastic factor. Invest. Urol. 17: 195-8 (1979); **Story MT et al** Prostatic growth factor: purification and structural relationship to basic fibroblast growth factor. B 26: 3843-9 (1987)

Polyergin: This name was proposed as an alternative name for » TGF-β to point to its many different biological activities. Polyergin is identical with » BSC-1 cell growth inhibitor.

Hanks SK et al Amino acid sequence of the BCS-1 cell growth inhibitor (polyergin) deduced from the nucleotide sequence of the cDNA. PNAS 85: 79-82 (1988)

Polymerase chain reaction: abbrev. PCR. See: Message amplification phenotyping.

Polymorphonuclear leukocyte recruiting factor: abbrev. PRF. see: Leukocyte infiltration inducing factor.

POMC: *proopiomelanocortin* POMC (***pro-ACTH-Endorphin***) is a glycosylated 31 kDa protein precursor posttranslational processing of which yields several neuroactive peptides upon specific cleavage and possibly a great number of as yet unidentified small peptides that may be pharmacologically active. The POMC gene (human chromosome 2p23) contains two large introns: one, of about 3.5 kb, interrupts the N- terminal fragment of the common precursor; the other contains the sequence for a portion of the 5′ untranslated portion of the mRNA, all of the signal peptide, and 8 amino acids of the N-terminal fragment. The overall arrangement of introns and exons in the POMC gene is virtually identical in all mammalian species.

Expression and release of POMC-derived peptides from the pituitary is triggered by hypothalamus-derived ***corticotropin releasing factor*** (abbrev. ***CRF***), a protein of 41 amino acids derived by enzymatic cleavage from a 191 amino acid pre-prohormone and is under negative control of glucocorticoids. IL1 and IL6 enhance the expression of the POMC gene. Hormonal control of POMC gene transcription and release of peptide products derived from the POMC precursor is tissue-specific; for example, glucocorticoids specifically inhibit anterior but not intermediate pituitary POMC transcription.

POMC cleavage products include a large N-terminal fragment, which yields ***γ-MSH*** and possibly other unidentified cleavage products, ***ACTH*** (corticotropin, 39 amino acids), ***lipotropin***, ***α-MSH*** (melanocyte stimulating hormone; melanotropin; acetylated and amidated ACTH1-13), ***β-MSH***, ***β-endorphin***, and one other. ACTH and β-lipotropin (β-LPH) are derived from the proopiomelanocortin precursor peptide. α-MSH and ***CLIP*** (corticotropin-like intermediate peptide; ACTH18-39) are formed from ACTH. γ-LPH is the aminoterminal part of β-LPH, and β-endorphin is derived from carboxyterminal sequences of β-LPH. β-MSH is contained within γ-LPH. β-Endorphin can yield ***γ-Endorphin***, ***α-Endorphin***, and ***Met-Enkephalin*** by progressive shortening of the molecule at the carboxyterminal end (Met-Enkephalin is, however, encoded by a distinct gene).The endorphins and Met-Enkephalin are also known as ***opioid peptides***.

Individual products of the POMC protein have been shown to act on immune cells and to be produced by them, thus establishing close links between immune cells and the nervous system (see also: Neuroimmune network). On the other hand, this view has been challenged recently in the light of recent detailed analysis of POMC expression in human peripheral blood mononuclear cells. It has

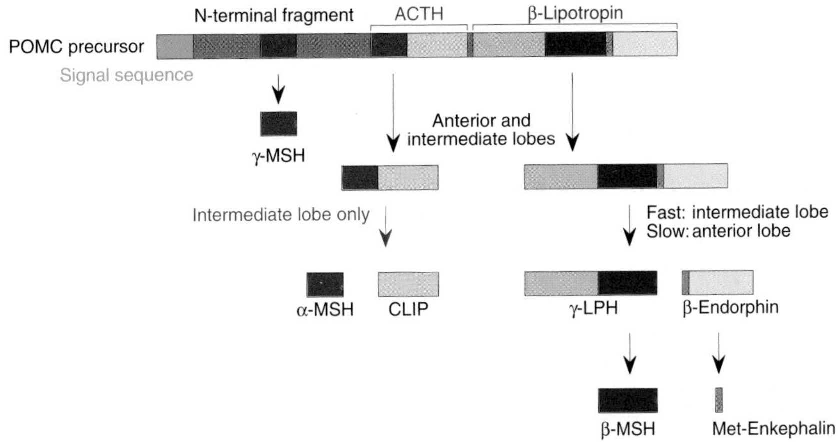

POMC processing.
Processing of the POMC precursor in the pituitary yields various bioactive fragments, some of which are produced only in the intermediate lobe. See text for details.

been suggested, however, that these studies, showing the non-expression of POMC in these cells, do not preclude the possibility of rare subsets of lymphoid cells producing POMC and peptides derived thereof.

■ **ACTH:** Stimuli for the production of ACTH, among others, are viral infections, bacterial lipopolysaccharides, Corticotropin releasing factor, » vasopressin, » IL1, » IL6, and thymopentin (see: Thymic hormones). ACTH activities are also modulated by » Corticostatins.

Apart from its production in the pituitary authentic ACTH has been found to be produced also by macrophages and lymphocytes. The releasing hormone CRP has been shown also to elicit lymphocyte production of ACTH. The mechanism of action is probably indirect and appears to involve CRF-induced macrophage production of » IL1 which then elicits expression of POMC in B lymphocytes. In hypothalamic neurons IL1 has been shown to cause release of CRF. Differential regulation of POMC is suggested by observations that the final set of proteins obtained by cleavage of the POMC precursor in the anterior lobe of the pituitary gland differs from that in the intermediate lobe. The processing pattern of POMC in leukocytes also differs from that observed in the pituitary.

The classical role of ACTH (39 amino acids) is to elicit a glucocorticoid response from the adrenal glands in situations of stress (see also: Neuroimmune network). At therapeutic concentrations (10^{-4} M) » suramin inhibits the ACTH-stimulated release of corticosteroids probably by direct interaction with ACTH. The synthesis of cortisol, which is mediated by ACTH, is inhibited also by TGF-β.

ACTH has been shown to promote growth and differentiation of human tonsillar B cells, enhancing the proliferation of activated B cells in the presence (but not in the absence) of B cell growth factors (see: BCGF) or » IL2 at nanomolar concentrations. In combination with » IL5 ACTH functions as a late-acting B cell growth factor (see also: BCGF). ACTH also increases IL2-induced secretion of IgM and IgG at low concentrations (10^{-9} – 10^{-13} M) whereas it has the opposite effect at high concentrations (10^{-6} M). ACTH has also been shown to inhibit production of » IFN-γ by interfering with helper cell functions. Corticotropin releasing factor has also been shown to augment IL2 expression on T lymphocytes, to enhance natural killer activity, and to stimulate B cell proliferation. ACTH has been shown to counteract the negative effects of glucocorticoids on NK cell activity.

■ **ENDORPHINS:** β-Endorphin (and also α-endorphin, γ-endorphin derived from it, and Met-enkephalin, the sequence of which is contained within the endorphin sequence) have been found to be produced also by macrophages and lymphocytes. The releasing hormone CRP has been shown also to elicit lymphocyte production of endorphins. Endorphins have been shown to enhance T cell proliferation and production of » IL2 and » IFN-γ. β-Endorphin also inhibits expression of the IL2 receptor in ConA-stimulated splenic lymphocytes. Moreover, endorphins enhance the natural cytotoxicity of lymphocytes and macrophages towards tumor cells, stimulate human peripheral blood mononuclear cell chemotaxis, inhibit production of T cell chemotactic factor(s), inhibit production of » IFN-γ by cultured human peripheral blood mononuclear cells, and inhibit expression of MHC class II antigens. β-Endorphin appears to act differentially: its C-terminal moiety enhances T cell proliferation, whereas this stimulatory effect can be prevented by peptides that possess the N-terminal enkephalin sequence. β-Endorphin has been shown to counteract the negative effects of glucocorticoids on NK cell activity.

■ **MSH:** IL1 is an endogenous pyrogen (see: EP) that induces the release of prostaglandins in the thermoregulatory center of the hypothalamus. This activity is inhibited by α-MSH. α-MSH has been shown to antagonize » IL1, » TNF-α, and C5a-mediated neutrophil chemotaxis. It acts as an antipyretic anti-inflammatory mediator. α-MSH has been shown to inhibit the production of IFN-γ by antigen-stimulated lymph node cells.

■ **TRANSGENIC/KNOCK-OUT/ANTISENSE STUDIES:** Transgenic animals overexpressing corticotropin releasing factor exhibit endocrine abnormalities involving the hypothalamic-pituitary-adrenal axis, such as elevated plasma levels of ACTH and glucocorticoids. These animals display physical changes similar to those of patients with Cushing's syndrome. Similar observations have been made with transgenic mice developing adrenocorticotropic hormone-secreting pituitary tumors due to the expression of a viral tumor antigen (Polyoma large T antigen).

● **POMC: Benjannet S et al** PC1 and PC2 are proprotein convertases capable of cleaving proopiomelanocortin at distinct pairs of basic residues. PNAS 88: 3564-8 (1991); **Chang ACY et al** Structural organization of human genomic DNA encoding the pro-opiomelanocortin peptide. PNAS 77: 4890-4894 (1980); **Cochet M et al** Characterization of the structural gene and putative 5-prime-regulatory sequences for human proopiomelanocortin. N 297: 335-339 (1982); **Chretien M et al** From β-lipotropin to β-endorphin and 'pro-opio-melanocortin.' Canad. J. Biochem. 57: 1111-1121 (1979); **Eipper BA & Mains, RE** Structure and biosynthesis of pro- adrenocorticotropin-endorphin and related peptides. Endocrine Rev. 1: 1-27 (1980); **Jones MT & Gilham B** Factors involved in the regulation of adrenocorticotropic hormone/b lipotropic hormone. Physiol. Rev. 68: 743-818 (1988); **Kraus J et al** Regulatory elements of the human proopiomelanocortin gene promoter. DNA Cell Biol. 12: 527-36 (1993); **Loh VP et al** Intracellular trafficking and processing of pro-opiomelanocortin. Cell Biophys. 19: 73-83 (1991); **Mains RE & Eipper BA** The tissue-specific processing of pro-ACTH/endorphin: Recent advances and unsolved problems. Trends Endocrinol. Metab. 1: 388-94 (1990); **Monig H et al** Structure of the POMC promoter region in pituitary and extrapituitary ACTH producing tumors. Exp. Clin. Endocrinol. 101: 36-8 (1993); **Nakai Y et al** Molecular mechanisms of glucocorticoid inhibition of human proopiomelanocortin gene transcription. JSBMB 40: 301-6 (1991); **Riegel AT et al** Proopiomelanocortin gene promoter elements required for constitutive and glucocorticoid-repressed transcription. ME 5: 1973-82 (1991); **Roberts JL et al** Regulation of pituitary proopiomelanocortin gene expression. In: Herz A (edt) Handbook of Exp. Pharmacol. 104/I: 347-77, Springer, Berlin (1993); **Sacerdote P et al** Pharmacological modulation of neuropeptides in peripheral mononuclear cells. J. Neuroimmunol. 32: 35-41 (1991); **Sharp B & Linner K** What do we know about the expression of proopiomelanocortin transcripts and related peptides in lymphoid tissue? Endocrinology 133: 1921A-21B (1993); **Therrien M & Drouin J** Molecular determinants for cell specificity and glucocorticoid repression of the proopiomelanocortin gene. ANY 680: 663-71 (1993); **Van Woudenberg AD et al** Analysis of proopiomelanocortin (POMC) messenger ribonucleic acid and POMC-derived peptides in human peripheral blood mononuclear cells: no evidence for a lymphocyte-derived POMC system. Endocrinology 133: 1922-33 (1993)

● **ACTH: Alvarez-Mon M et al** A potential role for adrenocorticotropin in regulating human B lymphocyte functions. JI 135: 3823-6 (1985); **Bennett HPJ et al** Confirmation of the 1-20 amino acid sequence of human adrenocorticotropin. BJ 133: 11-13 (1973); **Brooks KH** Adrenocorticotropin (ACTH) functions as a late-acting B cell growth factor and synergizes with interleukin 5. J. Mol. Cell. Immunol. 4: 327-35 (1990); **Carr DJJ et al** Corticotropin-releasing hormone augments natural killer cell activity through a naloxone-sensitive pathway. J. Neuroimmunol. 28: 53-61 (1990); **Gatti G et al** Interplay *in vitro* between ACTH, β-endorphin, and glucocorticoids in the modulation of spontaneous and lymphokine-inducible human natural killer (NK) cell activity. Brain Behav. Immunol. 7: 16-28 (1993); **Johnson HM et al** Regulation of lymphokine γ-interferon production by corticotropin. JI 132: 246-50 (1984); **Karanth S & McCann SM** Anterior pituitary hormone control by interleukin 2. PNAS 88: 2961-5 (1991); **McGillis JP et al** Stimulation of rat B lymphocyte proliferation by corticotropin-releasing factor. J. Neurosci. Res. 23: 346-52 (1989); **Lee TH et al** On the structure of human corticotropin (adrenocorticotropic hormone). JBC 236: 2970-2974 (1961); **McGlone JJ et al** Adrenocorticotropin stimulates natural killer cell activity. E 129: 1653-8 (1991); **Mesiano S et al** Basic fibroblast growth factor expression is regulated by corticotropin in the human fetal adrenal: A model for adrenal growth regulation. PNAS 88: 5428-32 (1991); **Ohgo S et al** Interleukin-1 (IL1) stimulates the release of corticotropin-releasing factor (CRF) from superfused rat hypothalamo-neurohypophyseal complexes (HNC) independently of the histaminergic mechanism. Brain Res. 558: 217-23 (1991); **Owens MJ & Nemeroff CB** Physiology and pharmacology of corticotropin-releasing factor. Pharmacol. Rev. 43: 425-73 (1991); **Plotsky PM et al** Central activin administration modulates corticotropin-releasing hormone and adrenocorticotropin secretion. E 128: 2520-5 (1991); **Smith EM et al** Corticotropin releasing factor induction of leukocyte-derived immunoreactive ACTH and endorphins. N 322: 881-2 (1986); **Singh VK** Stimulatory effect of corticotropin-releasing neurohormone on human lymphocyte proliferation and interleukin 2 receptor expression. J. Neurimmunol. 23: 257-62 (1989); **Suda T et al** Interleukin-1 stimulates corticotropin-releasing factor gene expression in rat hypothalamus. E 126: 1223-8 (1990); **Webster EL et al** Upregulation of interleukin-1 receptors in mouse AtT-20 pituitary tumor cells following treatment with corticotropin-releasing factor. E 129: 2796-8 (1991); **Yamashiro D & Li CH** Adrenocorticotropins.

44. Total synthesis of the human hormone by the solid-phase method. J. Am. Chem. Soc. 95: 1310-1315 (1973)

● **MSH: Brown SL et al** Suppression of T lymphocyte chemotactic factor production by the opioid peptides β-endorphin and Met-enkephalin. JI 134: 3384-90 (1985); **Harris, JI** Structure of a melanocyte-stimulating hormone from the human pituitary gland. N 184: 167-169 (1959); **Jegou S et al** Regulation of α-melanocyte-stimulating hormone release from hypothalamic neurons. ANY 680: 260-78 (1993); **Leu SJC & Singh VK** Modulation of natural killer cell-mediated lysis by corticotropin-releasing neurohormone. J. Neuroimmunol. 33: 253-60 (1991); **Martin LW et al** Neuropeptide α-MSH antagonizes IL6- and TNF-induced fever. Peptides 12: 297-9 (1991); **Mason MJ & Van Epps DE** Modulation of interleukin 1, tumor necrosis factor, and C5a-mediated neutrophil migration by α-melanocyte stimulating hormone (MSH). JI 142: 1646-51 (1989); **Nordlund JJ** α-Melanocyte-stimulating hormone: A ubiquitous cytokine with pigmenting effects. JAMA 266: 2753-4 (1991); **Smith EM et al** Immunosuppressive effects of corticotropin and melanotropin and their possible significance in human immunodeficiency virus infection. PNAS 89: 782-6 (1992); **Taylor AW et al** Identification of α-melanocyte stimulating hormone as a potential immunosuppressive factor in aqueous humor. Curr. Eye Res. 11: 1199-206 (1992); **Uehara Y et al** Carboxyl-terminal tripeptide of α-melanocyte-stimulating hormone antagonizes interleukin-1-induced anorexia. Eur. J. Pharmacol. 220: 119-22 (1992); **Villar M et al** Central and peripheral actions of α-MSH in the thermoregulation of rats. Peptides 12: 1441-3 (1991)

● **ENDORPHINS: Chiappelli F et al** Differential effect of β-endorphin on three human cytotoxic cell populations. IJI 13: 291-7 (1991); **Davis TP & Crowell SL** β-Endorphin is metabolized *in vitro* by human small lung cancer to γ-Endorphin which stimulates clonal growth. In: Moody TW (edt), Growth Factors, Peptides and Receptors, pp 389-400, Plenum Press New York (1993); **Faith RE et al** Neuroimmunomodulation with enkephalins: enhancement of human natural killer (NK) cell activity *in vitro*. Clin. Immunol. Immunopharmacol. 31: 412-8 (1984); **Garcia I et al** β-endorphin inhibits interleukin-2 release and expression of interleukin-2 receptors in concanavalin A-stimulated splenic lymphocytes. Lymphokine Cytokine Res. 11: 339-45 (1992); **Gilman SC et al** β-Endorphin enhances lymphocyte proliferative responses. PNAS 79: 4226-30 (1982); **Gilmore W & Weiner LP** β-endorphin enhances interleukin (IL2) production in murine lymphocytes. J. Neuroimmunol. 18: 125-8 (1988); **Gilmore W et al** The enhancement of polyclonal T cell proliferation by β-endorphin. Brain Res. Bull. 24: 687-92 (1990); **Hemmick LM & Bidlack JM** Endorphin peptides enhance mitogen-induced T cell proliferation which has been suppressed by prostaglandins. AEMB 288: 211-4 (1991); **Hemmick L & Bidlack JM** β-Endorphin stimulates rat T lymphocyte proliferation. J. Neuroimmunol. 29: 239-48 (1990); **Kay N et al** Interaction between endogenous opioids and IL2 on PHA-stimulated human lymphocytes. Immunology 70: 485-91 (1990); **Mathews PM et al** Enhancement of natural cytotoxicity by β-endorphin. JI 13: 1658-62 (1983); **McCain HW et al** β-Endorphin modulates human immunity via non-opiate receptor mechanisms. Life Sci. 31: 1619-24 (1982); **Morgano A et al** Expression of HLA class II antigens and proliferative capacity in autologous mixed lymphocyte reactions of human T lymphocytes exposed *in vitro* to α-endorphin. Brain Behav. Immun. 3: 214-22 (1989); **Oleson DR & Johnson DR** Regulation of human natural cytotoxicity by enkephalins and selective opiate agonists. Brain Behav. Immunol. 2: 171-86 (1988); **Peterson PK et al** Opioid-mediated suppression of interferon-γ production by cultured peripheral blood mononuclear cells. JCI 80: 824-31 (1987); **Van den Bergh P et al** Two opposing modes of action of β-endorphin on lymphocyte function. Immunology 72: 537-43 (1991); **Van den Bergh P et al** Identification of two moieties of β-endorphin with opposing effects on rat T-cell proliferation. Immunology 79: 18-23 (1993); **Van Epps DE & Saland L** β-Endorphin and Met-enkephalin stimulate human peripheral blood mononuclear cell chemotaxis. JI 132: 3046-53 (1984)

● **TRANSGENIC/KNOCK-OUT/ANTISENSE STUDIES: Helseth A et al** Transgenic mice that develop pituitary tumors. A model for Cushing's disease. Am. J. Pathol. 140: 1071-80 (1992); **Stenzel-Poore MP** Development of Cushing's syndrome in corticotropin-releasing factor transgenic mice. Endocrinology 130: 3378-86 (1992); **Tremblay Y et al** Pituitary-specific expression and glucocorticoid regulation of a proopiomelanocortin fusion gene in transgenic mice. PNAS 85: 8890-4 (1988)

Poxvirus growth factor: This virus-encoded protein belongs to the family of epidermal growth factor (ee: EGF). See VVGF (vaccinia virus growth factor).

Blomquist MC et al Vaccinia virus 19-kilodalton protein: relationship to several mammalian proteins including two growth factors. PNAS 81: 7363-7 (1984)

pp: *partially purified* Sometimes used as a suffix, such as in ppIL1.

PP12: *placental protein 12* This protein of 34 kDa is produced by the decidua under the influence of progesterone. It is also found in amniotic fluid and human serum. PP12 is also produced in the liver and by human hepatoma cell lines. The protein is synthesized as a precursor of 259 amino acids including a secretory » signal sequence of 25 amino acids.

PP12 is identical with IGF-BP1, a protein that binds » IGF (see: IGF-BP). Since the decidua and also the placenta express IGF receptors PP12 probably functions as a local modulator of IGF bioactivity and bioavailability. Alterations in the IGF-BP serum levels have been shown to correlate with endometrial dysfunctions.

Bischof P Three pregnancy proteins (PP12, PP14, and PAPP-A): their biological and clinical relevance. Am. J. Perinatol. 6: 110-6 (1989); **Fazleabas AT et al** Characterization of an insulin-like growth factor binding protein, analogous to human pregnancy associated endometrial α 1 globulin in decidua of the baboon (*Papio anubis*) placenta. Biol. Reprod. 40: 873-85 (1989); **Frauman AG et al** The binding characteristics and biological effects in FRTL5 cells of placental protein 12, an insulin-like growth factor binding protein purified from human amniotic fluid. E 124: 2289-96 (1989); **Halperin R et al** Identification, immunoaffinity purification, and partial characterization of a human decidua-associated protein. J. Reprod. Fertil. 88: 159-65 (1990); **Julkunen M et al** Primary structure of human insulin-like growth factor binding protein/placental protein 12 and tissue-specific expression of its mRNA. FL 236: 295-302 (1988); **Murphy LJ et al** Identification and characterization of a rat decidual insulin-like growth factor-binding protein complementary DNA. ME 4: 329-36 (1990); **Pekonen F et al** A monoclonal antibody-based immunoradiometric assay for low molecular weight insulin-like growth factor binding protein/placental protein 12. J. Immunoassay 10: 325-37 (1989); **Waites GT et al** Human "pregnancy-associated endometrial α 1-globulin", a 32 kDa insulin-like growth factor-binding protein: immunohistological distribution and localization in the adult and fetus. J. Endocrinol. 124: 333-9 (1990)

PPBP: *pro-platelet basic protein* see: Beta-Thromboglobulin.

PPIase: abbrev. for peptidyl-prolyl *cis-trans* isomerase. see: CsA (Ciclosporin A).

PPP: *platelet-poor plasma* This is a special preparation of plasma that is almost free of platelets and lacks » PDGF. The optimal growth of some fibroblasts and some fibroblast-like cell lines (see: 3T3) can be effected in media supplemented with » PDGF and PPP. Under certain growth conditions PPP can be replaced by » IGF in short-term cultures.
Cochran BH et al Post-transcriptional control of protein synthesis in Balb/c-3T3 cells by platelet-derived growth factor and platelet-poor plasma. JCP 109: 429-38 (1981); **Gaffney EV & Graves DT et al** Evidence that a human osteosarcoma cell line which secretes a mitogen similar to platelet-derived growth factor requires growth factors present in platelet-poor plasma. CR 43: 83-7 (1983); **Grimaldi MA** Regulation of human amnion cell growth and morphology by sera, plasma, and growth factors. Cell Tissue Res. 220: 611-21 (1981); **Gotlieb AI & Wong MK** The effect of platelet poor plasma serum on endothelial cell spreading. Artery 9: 59-68 (1981); **Wharton W et al** Inhibition of BALB/c-3T3 cells in late G1: commitment to DNA synthesis controlled by somatomedin C. JCP 107: 31-9 (1981)

PPPS: *platelet-poor plasma serum* see: PPP (platelet-poor plasma).

PPSC: *pluripotent stem cells* abbrev. also PSC. Pluripotent bone marrow stem cells which give rise to hematopoietic cell types (see also: Hematopoiesis). These cells express the cell surface markers CD7, » CD34, CD19 (see also: CD antigens) and HLA-DR. See also: LTBMC (long-term bone marrow culture).

PPT: *preprotachykinin* see: Tachykinins.

PRA: *PMNL (polymorphonuclear leukocyte) recruiting activity* see: Leukocyte infiltration inducing factor.

PRD: *positive regulatory domain* see: IRS (interferon response sequence).

PrDGF: *prostate-derived growth factor* This factor is probably identical with » aFGF. See also: Prostatropin. Another protein, called *prostatic growth factor*, is identical with » bFGF (see: PrGF).
Maehama S et al Purification and partial characterization of prostate-derived growth factor. PNAS 83: 8162-6 (1986)

Preadipocyte stimulating factor: see: PSF.

Prealbumin: see: transthyretin.

Pre-alpha-trypsin inhibitor: see: ITI (Inter-α-trypsin inhibitor).

Pre B cell colony enhancing factor: see: PBEF.

Pre-B cell growth factor: see: PBGF.

Pregnancy mitogen: see: PM.

Pregnancy recognition hormone: see: TP-1.

Prepro-proteins: see: Gene expression.

PRF: *peptide regulatory factors* see: Cytokines.

PRF: *PMNL (polymorphonuclear leukocyte) recruiting factor* see: Leukocyte infiltration inducing factor.

PrGF: *prostatic growth factor* abbrev. also PGF. This protein was initially isolated as a heparin-binding growth factor (see: HBGF) displaying the activity of an angiogenesis factor. It is identical with » bFGF.
Another protein, called *prostatic growth factor*, is identical with » aFGF (see also: Prostatropin). See also: PrDGF (prostate-derived growth factor).
Grunz H et al Induction of mesodermal tissues by acidic and basic heparin-binding growth factors. Cell. Differ. 22: 183-9 (1988); **Mydlo JH et al** Heparin-binding growth factor isolated from human prostatic extracts. Prostate 12: 343-55 (1988); **Mydlo JH et al** Expression of basic fibroblast growth factor mRNA in benign prostatic hyperplasia and prostatic carcinomas. Prostate 13: 241-7 (1988); **Story MT et al** Prostatic growth factor: purification and structural relationship to basic fibroblast growth factor. B 26: 3843-9 (1987)

Primary cell cultures: see: Cell culture.

Primary response genes: see: Gene expression.

Priming signals: see: Cell activation.

Primitive hematopoietic progenitor cells: abbrev. PHPC. See. HPP-CFC (high proliferative potential colony-forming cells).

Pre-B cell growth stimulating factor: see addendum: PBSF.

Progenitor stimulating factor: see: PSF.

Progenitor cells: see: Hematopoiesis.

Programmed cell death: see: Apoptosis.

Progression factors: see: Cell activation.

Pro-insulin: see: IGF.

Prokine: Brand name for recombinant glycosylated » GM-CSF marketed under the generic name Sargramostim (Immunex).

Prolactin: Prolactin, also known as ***lactotropin***, ***mammotropin***, ***luteotropic hormone*** (LTH), or ***luteotropin*** is a neuroendocrine pituitary hormone of 23 kDa. It is produced in increasing amounts during pregnancy and during suckling and acts primarily on the mammary gland by initiating and maintaining lactation in the postpartal phase. A potent inhibitor of basal secretion of prolactin by the pituitary is » TGF-β. Its secretion from the pituitary is stimulated by » VIP (vasoactive intestinal peptide).

Prolactin has also been shown to have cytokine-like activities and to have important immunoregulatory activities. It contributes to the development of lymphoid tissues and the maintenance of physiological immune function and also modulates a variety of T cell immune responses (see also: Dwarf mice). Hypophysectomy has been shown to lead to a regression of the thymus and a loss of immune competence which can be restored by treatment with prolactin.

In vitro studies suggest that lymphocytes are an important target tissue for circulating prolactin. Prolactin stimulates ornithine decarboxylase and activates protein kinase C, which are pivotal enzymes in the differentiation, proliferation, and function of lymphocytes. Prolactin antibodies inhibit lymphocyte proliferation. Prolactin induces » IL2 receptors on the surface of lymphocytes and acts as a progression factor during lymphocyte proliferation stimulated by » IL2, probably functioning in the nucleus without binding to its cell surface receptor. Some human B lymphoblastoid cells lines have been reported to produce prolactin constitutively.

A prolactin-like molecule is synthesized and secreted by concanavalin A- or phytohemagglutinin-stimulated human peripheral blood mononuclear cells and functions in an » autocrine manner as a growth factor for lymphoproliferation. A 29 kDa prolactin-like protein is also secreted as an » autocrine growth factor by Ramos Burkitt lymphoma cells during continuous serum-free growth.

In addition to triggering resting lymphocytes to cell division, the hormone can also control the magnitude of their response to polyclonal stimuli. Prolactin promotes the proliferation of » Nb2 pre-T cell lymphoma cells. In these cells prolactin induces the biphasic expression of a transcription factor, IRF-1 (interferon regulatory factor; see: IRS, interferon response sequence) and may be involved in » cell cycle activation and S phase progression. Prolactin can also protect Nb2 cells against glucocorticoid-receptor-mediated induction of » apoptosis.

Elevated levels of prolactin have been observed in patients during acute cardiac allograft rejection. It has been assumed that » CsA (ciclosporin A), which was used to suppress graft rejection, can act as an antagonist to prolactin binding to the prolactin receptor on lymphocytes.

Prolactin has been reported to activate cellular proliferation in nonreproductive tissue, such as liver, spleen, and thymus. It induces significant proliferation in aortic smooth muscle cells and also enhances proliferation of these cells induced by » PDGF. Prolactin also appears to be directly mitogenic for pancreatic β cells. Prolactin is also mitogenic for cultured astrocytes.

Prolactin has been found to enhance superoxide anion generation and hydrogen peroxide release from murine macrophages. » Growth hormone augments superoxide anion secretion of human neutrophils by binding to the prolactin receptor.

A 16 kDa aminoterminal fragment of prolactin, formed by enzymatic cleavage of intact 23 kDa prolactin in the pituitary gland and in target tissues for prolactin has been shown to inhibit the growth of bovine brain capillary endothelial cells, while intact prolactin is inactive. This activity is mediated by specific, high affinity, saturable binding sites (52 kDa, 32 kDa) for the 16 kDa protein which is distinct from the normal prolactin receptor. This cleavage product is a potent inhibitor of angiogenesis (see also: Angiogenesis factor). 16 kDa prolactin has been shown to be a potent mitogen for mammary epithelial cells via prolactin receptors.

● REVIEWS: **Gala RR** Prolactin and growth hormone in the regulation of the immune system. PSEBM 198: 513-27 (1991); **Hartmann D et al** Inhibition of lymphocyte proliferation to PRL. FJ 3: 2194-9 (1989); **Jara LJ et al** Prolactin, immunoregulation, and autoimmune diseases. Semin. Arthritis Rheum. 20:

273-84 (1991); **Kelley KW et al** Growth hormone, prolactin, and insulin-like growth factors: new jobs for old players. Brain Behav. Immunol. 6: 317-26 (1992); **Hooghe R et al** Growth hormone and prolactin are paracrine growth and differentiation factors in the haemopoietic system. IT 14: 212-4 (1993); **Wallis M** The expanding growth hormone/prolactin family. J. Mol. Endocrinol. 9: 185-8 (1992)

Baglia LA et al An Epstein-Barr virus-negative Burkitt lymphoma cell line (sfRamos) secretes a prolactin-like protein during continuous growth in serum-free medium. Endocrinology 128: 2266-72 (1991); **Berczi I & Nagy E** A possible role of prolactin in adjuvant arthritis. Arthritis Rheum. 25: 591-4 (1982); **Bernton EW et al** Suppression of macrophage activation and T lymphocyte function in hypoprolactinemic mice. S 239: 401-4 (1988); **Billestrup N & Nielsen JH** The stimulatory effect of growth hormone, prolactin, and placental lactogen on β-cell proliferation is not mediated by insulin-like growth factor-I. Endocrinology 129: 883-8 (1991); **Carrier M et al** Prolactin as a marker of rejection in human heart transplantation. J. Heart Transplant. 6: 290-2 (1987); **Clapp C & Weiner RI** A specific, high affinity, saturable binding site for the 16-kilodalton fragment of prolactin on capillary endothelial cells. Endocrinology 130: 1380-6 (1992); **Clapp C et al** The 16-kilodalton N-terminal fragment of human prolactin is a potent inhibitor of angiogenesis. Endocrinology 133: 1292-9 (1993); **Clevenger CV et al** Regulation of interleukin-2driven T lymphocyte proliferation by prolactin. PNAS 87: 6460-4 (1990); **Clevenger CV et al** Requirement of nuclear prolactin for interleukin-2-stimulated proliferation of T lymphocytes. S 253: 77-9 (1991); **Clevenger CV et al** Requirement for prolactin during cell cycle regulated gene expression in cloned T lymphocytes. Endocrinology 130: 3216-22 (1992); **Cooke NE et al** Human prolactin. cDNA structural analysis and evolutionary comparisons. J. Cell Biol. Chem. 256: 4007-16 (1981); **DeVito WJ et al** Prolactin-stimulated mitogenesis of cultured astrocytes is mediated by a protein kinase C-dependent mechanism. J. Neurochem. 60: 832-42 (1993); **Di Mattia GE et al** A human B lymphoblastoid cell line produces prolactin. Endocrinology 122: 2508-17 (1988); **Ferrara N et al** The 16K fragment of prolactin specifically inhibits basal or fibroblast growth factor stimulated growth of capillary endothelial cells. Endocrinology 129: 896-900 (1991); **Fletcher-Chiappini SE et al** Glucocorticoid-prolactin interactions in Nb2 lymphoma cells: antiproliferative versus anticytolytic effects. PSEBM 202: 345-52 (1993); **Frawley LS et al** Stimulation of prolactin secretion in rhesus monkeys by vasoactive intestinal polypeptide. Neuroendocrinol. 33: 79-83 (1981); **Fu YK et al** Growth hormone augments superoxide anion secretion of human neutrophils by binding to the prolactin receptor. JCI 89: 451-7 (1992); **Goffin V et al** Alanine-scanning mutagenesis of human prolactin: importance of the 58-74 region for bioactivity. Mol. Endocrinol. 6: 1381-92 (1992); **Krown KA et al** Prolactin isoform 2 as an autocrine growth factor for GH3 cells. Endocrinology 131: 595-602 (1992); **Matera L et al** Modulatory effect of prolactin on the resting and mitogen-induced activity of T, B, and NK lymphocytes. Brain Behav. Immun. 6: 409-17 (1992); **Meyer N et al** Prolactin-induced proliferation of the Nb2 T lymphoma is associated with protein kinase-C-independent phosphorylation of stathmin. Endocrinology 131: 1977-84 (1992); **Murata T & Ying SY** Transforming growth factor-β and activin inhibit basal secretion of prolactin in a pituitary monolayer culture system. PSEBM 198: 599-605 (1991); **Pahnke VG et al** Radioreceptor assay for lactogenic hormones based on membranes from rat mammary tumor. Horm. Metab. Res. 18: 680-5 (1986); **Russell DH et al** Prolactin receptors on human T and B lymphocytes. Antagonism of prolactin binding by cyclosporine. JI 134: 3027-31 (1985); **Sabharwal P et al** Prolactin synthesized and secreted by human peripheral blood mononuclear cells: an autocrine growth factor for lymphoproliferation. PNAS 89: 7713-6 (1992); **Sarkar DK et al** Transforming growth factor-β 1 messenger RNA and protein expression in the pituitary gland: its action on prolactin secretion and lactotropic growth. Mol. Endocrinol. 6: 1825-33 (1992); **Sauro MD & Zorn NE** Prolactin induces proliferation of vascular smooth muscle cells through a protein kinase C-dependent mechanism. JCP 148: 133-8 (1991); **Schwarz LA et al** Interferon regulatory factor-1 is inducible by prolactin, interleukin-2 and concanavalin A in T cells. Mol. Cell. Endocrinol. 86: 103-10 (1992); **Shah GN et al** Identification and characterization of a prolactin-like polypeptide synthesized by mitogen-stimulated murine lymphocytes. Int. Immunol. 3: 297-304 (1991); **Simionescu L et al** Radioreceptor assay of human prolactin using rabbit mammary receptors. Endocrinologie 25: 199-208 (1987); **Suzuki M et al** Radioreceptor assay of serum prolactin using nitrocellulose membrane-immobilized mammary prolactin receptor. AB 200: 42-6 (1992); **Wilner ML et al** The effect of hypoprolactinemia alone and in combination with cyclosporine on allograft rejection. Transplantation 49: 264-7 (1990)

Proleukin: Tradename of recombinant human » IL2 (Cetus, Emeryville, CA, USA).

Proliferation amplifying factor: see: PAF.

Proliferation factor for fat-storing cells: This poorly characterized factor of ≈ 60 kDa is found in the conditioned medium of early serum-free monolayer cultures of hepatocytes isolated from normal rat liver. It stimulates the proliferation of nonconfluent fat-storing cells (Ito cells or parasinusoidal lipocytes) maintained under serum-reduced conditions.

Gressner AM et al Identification and partial characterization of a hepatocyte-derived factor promoting proliferation of cultured fat-storing cells (parasinusoidal lipocytes). Hepatology 16: 1250-66 (1992)

Proliferation inducing activity: see: PIA.

Proliferin: see: PLF.

Proopiomelanocortin: see: POMC.

Pro-Platelet basic protein: abbrev. PPBP. See: Beta-Thromboglobulin.

Proproteins: see: Gene expression.

Prorenin converting enzyme: see: EGFBP (EGF binding protein)

Prostate carcinoma cell proliferation-inhibiting factor: see: SP factor (seminal plasma factor).

Prostate-derived growth factor: see: PrDGF.

Prostate epithelial cell growth factor: see: Prostatropin.

Prostatic growth factor: see: PrGF.

Prostatic osteoblastic factor: see: POF.

Prostatic secretory protein: see: IBF (immunoglobulin binding factors).

Prostatropin: *prostate epithelial cell growth factor* This factor is identical with » aFGF. It contains, however, an N-terminal extension of 15 amino acids and is blocked at the N-terminus by acetylation (see also: PrDGF, prostate-derived growth factor). Another protein, called *prostatic growth factor*, is identical with » bFGF (see: PrGF).
Crabb JW et al Complete primary structure of prostatropin, a prostate epithelial cell growth factor. B 25: 4988-93 (1986); **Crabb JW et al** Characterization of multiple forms of prostatropin (prostate epithelial cell growth factor) from bovine brain. BBRC 136: 1155-61 (1986); **Grunz H et al** Induction of mesodermal tissues by acidic and basic heparin-binding growth factors. Cell. Differ. 22: 183-9 (1988)

Protease Nexin I: see: Nexin I.

Proteinase 3: see: myeloblastin.

Protein family: see: Gene family.

Protein X: This factor is identical with » Vitronectin.

Proteoglycans: see: ECM (extracellular matrix).

Proteolysis inducing factor: see: PIF.

Prothymosin alpha: see: Thymic hormones.

Proto-oncogenes: see: Oncogenes.

PSC: see: PPSC (pluripotent stem cells).

PSF: *persisting (P) cell stimulating factor* Persisting (P) cells resembling mast cells appear in cultures of murine lymphoid or bone marrow cells and are capable of long-term growth *in vitro* in the presence of a T cell-derived growth factor. This factor (also abbrev. PCSF) stimulates the proliferation of many types of hemopoietic progenitor cells including the pluripotential hemopoietic stem cells. PSF is identical with » IL3.
Clark-Lewis I et al Purification to apparent homogeneity of a factor stimulating the growth of multiple lineages of hemopoietic cells. JBC 259: 7488-94 (1984); **Crapper RM et al** Stimulation of bone marrow-derived and peritoneal macrophages by a T lymphocyte-derived hemopoietic growth factor, persisting cell-stimulating factor. Blood 66: 859-65 (1985); **Ihle JN et al** Biological properties of homogenous interleukin-3. I. Demonstration of WEHI-3 growth factor activity, mast cell growth factor activity, P cell-stimulating factor activity, colony-stimulating factor activity, and histamine-producing cell-stimulating factor activity. JI 131: 282-7 (1983); **Schrader JW et al** The persisting (P) cell: histamine content, regulation by a T cell-derived factor, origin from a bone marrow precursor, and relationship to mast cells. PNAS 78: 323-7 (1981); **Schrader JW & Clark-Lewis I** A T cell-derived factor stimulating multipotential hemopoietic stem cells: molecular weight and distinction from T cell growth factor and T cell-derived granulocyte-macrophage colony-stimulating factor. JI 129: 30-5 (1982); **Schrader JW et al** P cell stimulating factor: characterization, action on multiple lineages of bone marrow-derived cells, and role in oncogenesis. IR 76: 79-104 (1983); **Schrader JW** Role of a single hemopoietic growth factor in multiple proliferative disorders of hemopoietic and related cells. Lancet 2(8395) 137-44 (1983); **Schrader JW & Crapper RM** Autogenous production of a hemopoietic growth factor, persisting-cell-stimulating factor, as a mechanism for transformation of bone marrow-derived cells. PNAS 80: 6892-6 (1983)

PSF: *preadipocyte stimulating factor* This poorly characterized factor of 63 kDa is found in rat serum and induces the adipogenic conversion, biochemical differentiation, and mitogenesis in primary cultures of rat preadipocytes. A similar activity is found in mouse serum. PSF activity is not due to insulin, » IGF-1, » growth hormone, glucocorticoids, combinations of these hormones, or » M-CSF, » GM-CSF, » IL1, » IL2, » IL3, » neuroleukin, or » TNF.
Li ZH et al Preadipocyte stimulating factor in rat serum: evidence for a discrete 63 kDa protein that promotes cell differentiation of rat preadipocytes in primary cultures. JCP 141: 543-57 (1989)

PSF: *progenitor stimulating factor* This factor which promotes the growth of early hematopoietic progenitor cells (see also: Hematopoiesis), is identical with » IL3.

PSH: *panspecific hemopoietin* This factor is identical with » IL3.
Schrader JW The panspecific hemopoietin of activated T lymphocytes (Interleukin-3). ARI 4: 205-30 (1986)

Psoriatic leukotactic factor: see: PLF.

Psychoneuroimmunology: see: Neuroimmune axis.

PT-18: A murine mast/basophil cell line functionally and cytochemically similar to mucosal mast cells that depends on » IL3 or » GM-CSF for its growth. They have been used in a » bioassay for these factors. The cells appear to secrete » TNF-α. Rapamycin inhibits Proliferation of PT18 cells, is inhibited by rapamycin, and this inhibition is prevented in a competitive manner by FK506, a structural analog of rapamycin (see: CsA, cyclosporin A).

Akahane K et al IL1 α induces granulocyte-macrophage colony-stimulating factor gene expression in murine B lymphocyte cell lines via mRNA stabilization. JI 146: 4190-6 (1991); **Bickel M et al** Granulocyte-macrophage colony-stimulating factor regulation in murine T cells and its relation to cyclosporin A. EH 16: 691-5 (1988); **Bressler RB et al** Inhibition of the growth of IL3-dependent mast cells from murine bone marrow by recombinant granulocyte macrophage-colony-stimulating factor. JI 143: 135-9 (1989); **Hultsch T et al** The effect of the immunophilin ligands rapamycin and FK506 on proliferation of mast cells and other hematopoietic cell lines. Mol. Biol. Cell. 3: 981-7 (1992); **Pluznik DH et al** Colony-stimulating factor (CSF) controls proliferation of CSF-dependent cells by acting during the G1 phase of the cell cycle. PNAS 81: 7451-5 (1984); **Richards AL et al** Natural cytotoxic cell-specific cytotoxic factor produced by IL3-dependent basophilic/mast cells. Relationship to TNF. JI 141: 3061-6 (1988)

PTC: *papillary thyroid carcinoma* see: ret.

PTF: see: Pentoxifyllin.

PTK: *protein tyrosine kinase* Protein kinases are enzymes that attach phosphate residues to protein substrates.

Many cytokine receptors (see also: Cytokine receptor superfamily) are transmembrane proteins with an extracellular domain that contains the ligand binding site and an intracellular domain encoding a tyrosine-specific protein kinase. The protein kinase domain of some cytokine receptors, including those for » M-CSF, » PDGF and » SCF (stem cell factor) is split by the insertion of the so-called ***kinase insert***.

Kinase domains catalyze the transfer of the γ-phosphates of ATP to specific tyrosine residues of other proteins following the binding of a growth factor ligand to the receptor (for some examples see: *flg*, *flt*, *fms*, KDR, *kit*, K-*sam*, *met*, *trk*, receptors for » Activin A, » aFGF, » bFGF, » BDNF, » EGF, » HGF (hepatocyte growth factor), » IGF, » KGF (keratinocyte growth factor), » NGF, » PDGF, » VEGF). Receptor phosphorylation can be reversed by specific phosphatases (for specific examples see: HCP, 3CH134, CL100). In many cases these have not yet been studied in detail.

The activity of the receptors is modulated by » autophosphorylation of the receptor protein. Protein-tyrosine kinases play a pivotal role in the response to growth factor signals. The phosphorylation of the receptor protein generates a cascade of signals, including association with non-receptor proteins (see also: *src* homology domains), that eventually culminates in the new expression of genes (see also: ERG, early response genes). Several inhibitors are available to study the importance of tyrosine phosphorylation of receptor and other proteins (for examples of general protein kinase inhibitors see: Bryostatins, Defensins, Genistein, H8, Herbimycin A; K-252a; Lavendustin A; Phorbol esters, Staurosporine, Suramin, tyrphostins).

Mutations affecting the receptor kinase activity or the ability of the receptor to become phosphorylated at cytoplasmic tyrosine residues have been shown to cause cellular transformation, implicating PTKs both in normal growth control and malignant growth.

Many cytokine receptors do not encode intrinsic protein tyrosine kinases. They have been shown to associate with other intracellular proteins. Some of these possess protein kinase activity and thus play an important role in receptor-mediated signal transduction (for examples see: *fyn*, *hck*, Janus kinases, *lck*, *lyn*, *pim*, TKF , *yes*).

PTN: *pleiotrophin* This factor was isolated initially as a heparin-binding growth factor (see also: HBGF) from the brain tissues of several species. PTN is identical with » HBNF (heparin-binding neurite-promoting factor).

Li YS et al Cloning and expression of a developmentally regulated protein that induces mitogenic and neurite outgrowth activity. S 250: 1690-4 (1990); **Li YS et al** Characterization of the human pleiotrophin gene: promoter region and chromosomal localization. JBC 267: 26011-6 (1992); **Milner et al** Cloning, nucleotide sequence, and chromosome localization of the human pleiotrophin gene. B 31: 12023-8 (1992)

PTP1C: see: HCP (hematopoietic cell phosphatase).

PTX: see: Pentoxifyllin.

PTXF: see: Pentoxifyllin.

PU-34: An SV40-transformed immortalized macaque bone marrow stromal cell line. These stromal cell lines are used as » feeder cells in the establishment of long-term bone marrow cultures (see: LTBMC). The cells express type I collagen and laminin in the extracellular matrix (see: ECM). The cells also express » IL6, » IL7, » GM-CSF, » M-CSF, » G-CSF, » LIF, and » IL11.

Paul SR et al Stromal cell-associated hematopoiesis: immortalisation and characterization of a primate bone marrow-derived stromal cell line. Blood 77: 1723-33 (1991)

Purpurin: Retinal purpurin is a secreted protein of 20 kDa (196 amino acids) found in neural retina, but not in other tissues such as brain, heart, liver, or muscle. The protein is highly concentrated in the neural retina between the pigmented epithelium and the outer segments of the photoreceptor cells and is synthesized by photoreceptor cells. Purpurin has approximately 50% sequence homology with liver-derived ***RBP*** (serum retinol binding protein), and is a member of the $\alpha 2$ microglobulin superfamily. Purpurin binds retinol and may play a major role in retinol transport across the interphotoreceptor cell matrix. The protein interacts with heparin and heparan sulfate proteoglycan (see also: ECM, extracellular matrix), but not with other glycosaminoglycans. Purpurin stimulates neural retina cell-substratum adhesion and prolongs the survival of neural retina cells in culture. Purpurin also supports the survival of dissociated ciliary ganglion cells and thus acts as a ciliary neurotrophic factors (see also: Neurotrophins).

Berman P et al Sequence analysis, cellular localization, and expression of a neuroretina adhesion and cell survival molecule. Cell 51: 135-42 (1987); **Melhus H et al** Retinol-binding protein and transthyretin expressed in HeLa cells form a complex in the endoplasmic reticulum in both the absence and the presence of retinol. ECR 197: 119-24 (1991); **Schubert D & LaCorbiere M** Isolation of an adhesion-mediating protein from chick neural retina adherons. JCB 101: 1071-7 (1985); **Schubert D & LaCorbiere M** Role of purpurin in neural retina histogenesis. PCBR 217B: 3-16 (1986); **Schubert D et al** A chick neural retina adhesion and survival molecule is a retinol-binding protein. JCB 102: 2295-301 (1986)

Px: see: Pentoxifyllin.

PY: *phosphotyrosine* see: PTK.

PYP: abbrev. for protein containing phosphotyrosine (P = phosphate; Y = tyrosine). see: PTK.

Pyrogens: see: EP (endogenous pyrogens).

Q

Quality control of cytokines: see: NIAID (National Institute of Allergy & Infectious Diseases), NIBSC (National Institute for Biological Standards & Control).

Quantification of cytokine receptor data: see: Scatchard plots (see also: Cytokine receptor superfamilies).

Quantification of cytokines: see: Bioassays, Colony stimulation assay, Cytokine assays, Message amplification phenotyping. See also individual cytokines (subentry "Assays", which lists cell lines used to assay cytokines) and references cited therein.

Quantitative analyses of » hematopoietic stem cells: see: Hematopoietic stem cells. See also: colony formation assays, delta assay, limiting dilution analysis, LTBMC (long-term bone marrow culture), MRA (marrow repopulating ability; for competitive repopulation assays).

Quercetin: see: Genistein.

R

R: A suffixed R with or without hyphen denotes receptor such as in GM-CSF-R, IL6R, EpoR, EGFR.

r: (a) recombinant; also sometimes re; see: rec. see: Recombinant cytokines. (b) depending upon context a prefixed r may also denote "of rat origin", rEGF (epidermal growth factor) for example.

R1-GI factor: This factor is secreted by a murine myeloblastic leukemia cell line. It inhibits the growth of murine monocytic leukemia cells. The factor is very likely identical with murine » IFN-β.
Kasukabe T et al Characterization of growth inhibitory factors for mouse monocytic leukemia cells. Leuk. Res. 16: 139-44 (1992)

Rabbit peritoneal factor: see: RPF.

Rabbit serum-derived growth inhibitor: see: RSGI.

Radioreceptor assays: abbrev. RRA. See: Cytokine assays.

raf: *raf* genes are proto-oncogenes (see: Oncogene). They encode three serine and threonine-specific protein kinases, called *raf*-1 (c-*raf*), A-*raf* and B-*raf* that are localized in the cytoplasm.
c-*raf*-1 is the cellular homologue of v-*raf* an oncogene found in the acute transforming replication-defective type C murine sarcoma virus 3661. The ***mil*** oncogene, a second oncogene in the avian retrovirus MH2, which contains the » *myc* oncogene, is the avian equivalent of the murine *raf* oncogene. The *raf*-1 gene maps to human chromosome 3p25; *raf*-2 is a processed pseudogene on human chromosome 4pter-p15.
raf-1 is a protein of 70-75 kDa that is expressed in all organs and cell lines. A- and B-*raf* are expressed in urogenital tissues and brain, respectively.
Activated *raf*-1 expresses a constitutive protein kinase and functions as an intracellular activator of cell growth. A revertant cell line, generated from v-*raf*-transformed 3T3 fibroblasts has been found to be deficient in the induction of » ERGs (early response genes) by serum and » phorbol esters. This oncogene therefore is important for the regulation of some responses mediated by these stimuli. Several cytokines induce the synthesis of *raf*-1 kinase, including » PDGF, » M-CSF, » GM-CSF, » EGF, » IL2 and » IL3. The activation of *raf* kinase, for example by the protein kinase activity of the intracellular domain of the » EGF receptor eventually activates the transcription of genes, among them also cytokine genes with promoters that contain the binding site for transcription factor AP-1/PEA3 (see: *jun*).
An activated RAF gene has been identified in the stomach cancer of a Japanese patient and there is some evidence for a relationship of RAF1 to renal cell carcinoma. A radiation-resistant laryngeal carcinoma cell line has also been shown to contain altered RAF1 sequences. The transforming DNA in a human glioblastoma line has been found to be identical with the RAF gene.
■ **TRANSGENIC/KNOCK-OUT/ANTISENSE STUDIES:** *Raf*-1 functions have been inhibited by expressing c-*raf*-1 antisense RNA or kinase-defective c-*raf*-1 mutants. Antisense RNA for c-*raf*-1 interferes with proliferation of normal » 3T3 fibroblast cells and reverts *raf*-transformed cells. Inhibition of *raf* blocks proliferation and transformation by Ki- and Ha-*ras* oncogenes.
Beck TW et al A complete coding sequence of the human a-*raf*-1 oncogene and transforming activity of a human a-*raf* carrying retrovirus. NAR 15: 595-609 (1987); **Bonner T et al** The human homologues of the *raf* (*mil*) oncogene are located on human chromosomes 3 and 4. S 223: 71-4 (1984); **Bonner T et al** Structure and biological activity of human homologues of the *raf/mil* oncogene. MCB 5: 1400-7 (1985); **Bonner TI et al** The complete coding sequence of the human *raf* oncogene and the corresponding structure of the c-*raf*-1 gene. NAR 14: 1009-15 (1986); **Fukui M et al** Detection of a *raf*-related and two other transforming DNA sequences in human tumors maintained in nude mice. PNAS 82: 5954-8 (1985); **Heidecker G et al** Mutational activation of c-*raf*-1 and definition of the minimal transforming sequence. MCB 10: 2503-12 (1990); **Ikawa S et al** B-*raf*, a new member of the *raf* family, is activated by DNA rearrangement. MCB 8: 2651-4 (1988); **Kanakura Y et al** Granulocyte-macrophage colony-stimulating factor and interleukin-3 induce rapid phosphorylation and activation of the proto-oncogene *Raf*-1 in a

human factor-dependent myeloid cell line. Blood 77: 243-8 (1991); **Kasid U et al** The *raf* oncogene is associated with a radiation-resistant human laryngeal cancer. S 237: 1039-41 (1987); **Klinken SP** Erythroproliferation *in vitro* can be induced by *abl*, *fes*, *src*, *ras*, *bas*, *raf*, *raf/myc*, *erb* B and *cbl* oncogenes but not by *myc*, *myb* and *fos*. OR 3: 187-92 (1988); **Klinken SP et al** Hemopoietic lineage switch: v-*raf* oncogene converts Emu-*myc* transgenic B cells into macrophages. Cell 53: 857-67 (1988); **Kolch W et al** *Raf* revertant cells resist transformation by non-nuclear oncogenes and are deficient in the induction of early response genes by TPA and serum. O 8: 361-70 (1993); **Li P et al** *raf*-1: a kinase currently without a cause but not lacking in effects. Cell 64: 479-82 (1991); **Rapp UR et al** Structure and biological activity of v-*raf*, a unique oncogene transduced by a retrovirus. PNAS 80: 4218-22 (1983); **Rapp UR et al** *raf* family serine/threonine protein kinases in mitogen signal transduction. CSHSQB 53: 173-84 (1988); **Shimizu K et al** Molecular cloning of an activated human oncogene, homologous to v-*raf*, from primary stomach cancer. PNAS 82: 5641-5 (1985); **Teyssier JR et al** Recurrent deletion of the short arm of chromosome 3 in human renal cell carcinoma: shift of the c-*raf* 1 locus. JNCI 77: 1187-91 (1986)

● TRANSGENIC/KNOCK-OUT/ANTISENSE STUDIES: **Kolch W et al** *Raf*-1 protein kinase is required for growth of induced NIH/3T3 cells. N 349: 426-8 (1991)

Radioimmunoassay: abbrev. RIA. See: Cytokine assays.

RAI: *ribonuclease/angiogenin inhibitor* see: Angiogenin.

Ramos: An EBV-negative, HLA class I-positive B lymphoblastoid cell line derived from a Burkitt lymphoma. Ramos cells, or a subline, designated Ramos.G6.C10, are used in a » bioassay of human » IL4. The assay is based on the induction of » CD23 expression induced in these cells by the cytokine. The transformation of Ramos cells with EBV yields cell lines that proliferate in response to » IL1. A subline of Ramos, designated sfRamos (for serum-free) has been shown to depend on » Prolactin or » Growth hormone for continuous growth in serum-free media.

Baglia LA et al Production of immunoreactive forms of growth hormone by the Burkitt tumor serum-free cell line sfRamos. Endocrinology 130: 2446-54 (1992); **Blazar BA & Murphy AM** Induction of B cell responsiveness to growth factors by Epstein-Barr virus conversion: comparison of endogenous factors and interleukin-1. Clin. Exp. Immunol. 80: 62-8 (1990); **Custer MC & Lotze MT** A biologic assay for IL4. Rapid fluorescence assay for IL4 detection in supernatants and serum. JIM 128: 109-17 (1990); **Siegel JP & Mostowski HS** A bioassay for the measurement of human interleukin-4. JIM 132: 287-95 (1990)

RANTES: *regulated upon activation, normal T expressed, and presumably secreted* This factor of 8 kDa is also called ***SISδ*** (small inducible secreted), SCY A5 (see: SCY family of cytokines), and » EoCP-1 (eosinophil chemotactic polypeptide). At the protein level murine RANTES shows 90 % homology with human RANTES. RANTES belongs to the » chemokine family of cytokines.

RANTES is expressed by an early response gene (see: ERG). The human gene maps to chromosome 17q11. 2-q12 in the vicinity of other cytokine genes belonging to the » chemokine family. The synthesis of RANTES is induced by » TNF-α and IL1-α (see: IL1)but not by » TGF-β, » IFN-γ and » IL6.

RANTES is produced by circulating T cells and T cell clones in culture but not by any T cell lines tested so far. The expression of RANTES is inhibited following stimulation of T lymphocytes.

RANTES is chemotactic for T cells, human eosinophils and basophils (see also: Chemotaxis) and plays an active role in recruiting leukocytes into inflammatory sites. RANTES also activates eosinophils to release, for example, eosinophilic cationic protein. It changes the density of eosinophils and makes them hypodense, which is thought to represent a state of generalized activation and is most often associated with diseases such as asthma and allergic rhinitis.

RANTES increases the adherence of monocytes to endothelial cells. It selectively supports the migration of monocytes and T lymphocytes expressing the cell surface markers » CD4 and UCHL1. These cells are thought to be pre-stimulated helper T cells with memory T cell functions. RANTES activates human basophils from some select basophil donors and causes the release of histamines (see also: HRF, histamine-releasing factors). On the other hand RANTES can also inhibit the release of histamines from basophils induced by several cytokines, including one of the most potent histamine inducers, » MCAF ().

RANTES is expressed by human synovial fibroblasts and may therefore participate in the ongoing inflammatory process in rheumatoid arthritis.

High affinity receptors for RANTES (≈ 700 binding sites/cell; Kd = 700 pM) have been identified on the human monocytic leukemia cell line THP-1, which responds to RANTES in chemotaxis and calcium mobilization assays. The chemotactic response of THP-1 cells to RANTES is markedly inhibited by preincubation with » MCAF (monocyte chemotactic and activating factor) or » MIP-1α (macrophage inflammatory protein). Binding of RANTES to monocytic cells is competed for by MCAF and MIP-1α.

Another RANTES receptor is a member of the G protein-coupled receptor superfamily. It shows ≈

33 % identity with receptors for » IL8 and may be the human homologue of the » US28 open reading frame of human cytomegalovirus. RANTES has also been shown to bind to a receptor called » CKR-1 which functions as a receptor for *Plasmodium vivax*. For another viral RANTES receptor homologue see also: ECRF3.

■ **Clinical Use & Significance:** It has been observed that antibodies to RANTES can dramatically inhibit the cellular infiltration associated with experimental mesangioproliferative nephritis. In addition, RANTES appears to be highly expressed in human renal allografts undergoing cellular rejection.

Alam R et al Interleukin-8 and RANTES inhibit basophil histamine release induced with monocyte chemotactic and activating factor/monocyte chemoattractant peptide-1 and histamine releasing factor. Am. J. Respir. Cell. Mol. Biol. 7: 427-33 (1992); **Alam R et al** RANTES is a chemotactic and activating factor for human eosinophils. JI 150: 3442-8 (1993); **Bischoff SC et al** RANTES and related chemokines activate human basophil granulocytes through different G protein-coupled receptors. EJI 23: 761-7 (1993); **Brown KD et al** A family of small inducible proteins secreted by leukocytes are members of a new superfamily that includes leukocyte and fibroblast-derived inflammatory agents, growth factors, and indicators of various activation processes. JI 142: 679-87 (1989); **Donlon TA et al** Localization of a human T cell-specific gene, RANTES (D17S136E), to chromosome 17q11. 2-q12. Genomics 6: 548-53 (1990); **Gao JL et al** Structure and functional expression of the human macrophage inflammatory protein 1 α/RANTES receptor. JEM 177: 1421-7 (1993); **Heeger P et al** Isolation and characterization of cDNA from renal tubular epithelium encoding murine RANTES. Kidney Int. 41: 220-5 (1992); **Kameyoshi Y et al** Cytokine RANTES released by thrombin-stimulated platelets is a potent attractant for human eosinophils. JEM 176: 587-92 (1992); **Kuna P et al** RANTES, a monocyte and T lymphocyte chemotactic cytokine releases histamine from human basophils. JI 149: 636-42 (1992); **Kuna P et al** Characterization of the human basophil response to cytokines, growth factors, and histamine releasing factors of the intercrine/chemokine family. JI 150: 1932-43 (1993); **Neote K et al** Molecular cloning, functional expression, and signaling characteristics of a C-C chemokine receptor. Cell 72: 415-25 (1993); **Rathanaswami P et al** Expression of the cytokine RANTES in human rheumatoid synovial fibroblasts. Differential regulation of RANTES and interleukin-8 genes by inflammatory cytokines. JBC 268: 5834-9 (1993); **Rot A et al** RANTES and macrophage inflammatory protein 1 α induce the migration and activation of normal human eosinophil granulocytes. JEM 176: 1489-95 (1992); **Schall T et al** The isolation and sequence of a novel gene from a human functional T cell line. JI 141: 1018-25 (1988); **Schall TJ** A human T cell-specific molecules is a member of a new gene family. JI 141: 1018-25 (1988); **Schall TJ** Selective attraction of monocytes and T lymphocytes of the memory phenotype by cytokine RANTES. N 334: 769-71 (1990); **Schall TJ** Biology of the RANTES/SIS cytokine family. Cytokine 3: 165-83 (1991); **Schall TJ et al** Molecular cloning and expression of the murine RANTES cytokine: structural and functional conservation between mouse and man. EJI 22: 1477-81 (1992); **Wang JM et al** Identification of RANTES receptors on human monocytic cells: competition for binding and desensitization by homologous chemotactic cytokines. JEM 177: 699-705 (1993); **Wiedermann CJ et al** Monocyte haptotaxis induces by the RANTES chemokine. Curr. Biol. 3: 735-9 (1993)

● **Clinical Use & Significance: Pattison J et al** RANTES chemokine expression in cell mediated transplant rejection of the kidney. Lancet #### in press (1993)

Rapamycin: see: CsA.

RA-SF: *rheumatoid arthritis synovial fluid* see: TRF (T cell replacing factors).

RBP: *serum retinol binding protein* see: Purpurin.

RDGF: *retina-derived growth factor* maybe also: *retinoblastoma-derived growth factor*. Generic name given to various growth factors isolated from retina and tumors of the retina on the basis of their ability to stimulate the proliferation of capillary endothelial cells *in vitro*. Specific factors that have been identified are » RDGF-α and » RDGF-β. Also see: EDGF (eye-derived growth factor).

RDGF-α: *retina-derived growth factor* This factor was isolated initially from bovine retina. It was also called » EDGF (eye-derived growth factor). The factor is a mitogen for endothelial cells and murine » 3T3 fibroblasts. The factor is also an » angiogenesis factor. It promotes the outgrowth of neurites in rat pheochromocytoma cells (see: PC12). This activity is potentiated by heparin. The factor is identical with » aFGF.

Baird A et al Retina-derived endothelial cell growth factors: partial molecular characterization and identity with acidic and basic fibroblast growth factor. B 24: 7855-9 (1985); **Barritault D et al** Purification, characterization, and biological properties of the eye-derived growth factor from retina: analogies with brain-derived growth factor. J. Neurosci. Res. 8: 477-90 (1982); **D'Amore PA & Klagsbrun M** Endothelial cell mitogens derived from retina and hypothalamus: biochemical and biological similarities. JCB 99: 1545-9 (1984); **Sullivan RC et al** Use of size-exclusion and ion-exchange high-performance liquid chromatography for the isolation of biologically active growth factors. J. Chromatogr. 266: 301-11 (1983); **Wagner JA & D'Amore PA** Neurite outgrowth induced by an endothelial cell mitogen isolated from retina. JCB 103: 1363-7 (1986)

RDGF-β: *retina-derived growth factor* This factor was isolated initially from bovine retina. The factor is a mitogen for endothelial cells and murine » 3T3 fibroblasts. The factor is also an » angiogenesis factor. The factor is identical with » bFGF. see: RDGF-α.

RE: *response element* see: Gene expression.

re: see: rec.

Reaferon: Recombinant α 2 interferon (in Russian publications). See: IFN-α.

rec: (also r or re) as a prefix such as in rEGF (epidermal growth factor), recEGF, or rhu GM-CSF (hu = human) or remIL6 (recombinant, murine) this indicates a recombinant factor obtained by expressing the corresponding gene in a heterologous cell system (see: Recombinant cytokines).

Receptor down-regulation: A term describing the loss of surface receptors for growth factors and cytokines observed after binding of a growth factor to its receptor. Receptor down-regulation is the result of ligand-induced internalization of receptor/ligand complexes and their degradation by lysosomes. In addition, internalized receptor molecules can also be recycled
Sorkin A & Waters CM Endocytosis of growth factor receptors BioEssays 15: 375-82 (1993)

Receptor inducing factor: see: RIF.

Receptor protein tyrosine kinases: abbrev. RPTK. See: PTK (protein tyrosine kinases).

Receptor shedding: This term relates to the ability of cells to cleave off the extracellular domain of a cytokine receptor and to release it into the circulation as a soluble product. In many instances these cleavage products are still capable of binding their ligands. They can therefore neutralize freely circulating cytokine molecules. The remaining portion of the membrane-bound receptor does not bind ligands any longer and hence the cells become desensibilised to the action of a particular cytokine. Such soluble receptors have been described for a variety of cytokines, including, for example, » IL4, » IL6, » IL7, » IGF, » TNF-α (see also: TNF-BF, TNF blocking factor) and » IFN-γ. Receptor shedding may constitute a further level of cytokine control mechanisms allowing their activities to be fine tuned *in vivo*. It may be of physiological importance because elevated levels of circulating cytokines as are observed, for example, during » septic shock and » acute phase reactions could be neutralized at least in part by this process. Administration of such receptor fragments might be a novel way to counteract the life-threatening complications of an endotoxemia (see also: Cytokine inhibitors).

Receptor transmodulation: This term relates to the regulation of the expression of cytokine receptors. It essentially means that the expression of a particular cytokine receptor on the cell surface is either stimulated or inhibited by another cytokine,

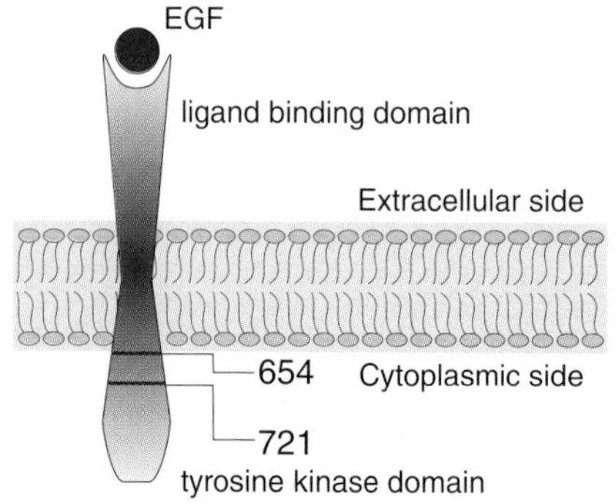

654	721	Ligand binding	Kinase activity	Trans-modulation
T	K	+	+	+
T	A	+	-	-
T	M	+	-	-
A	K	+	+	-
Y	K	+	+	-

Functional analysis of the EGF receptor by analysis of point mutations.
Residues 654 (threonine, T) and 721 (lysine, K) appear to be of particular importance for the biological activity of the EGF receptor. Introduction of other amino acids at these positions, for example, alanine (A), methionine (M), tyrosine (Y) lead to the loss of intrinsic tyrosine kinase activity within the cytoplasmic domain of the receptor. The ability of the receptor to bind EGF remains unaltered. Cells expressing mutated receptors do no longer show a mitogenic response to EGF.
Such experiments show that the tyrosine kinase activity of the EGF receptor plays a crucial role in subsequent intracellular signaling. All amino acid replacements also abolish receptor transmodulation, i. e. expression of the EGF receptor can no longer be modulated by cytokines other than EGF. For example, PDGF (platelet-derived growth factor) has been shown to transmodulate the EGF receptor. Following binding to its own receptor PDGF activates an intracellular calcium-dependent protein kinase that phosphorylates the EGF receptor at position 654. This reduces the affinity of EGF to its receptor as well as the intrinsic tyrosine kinase activity of the EGF receptor and hence adversely affects mitogenicity of EGF.

Receptor-mediated label-transfer assay: see addendum: RELAY.

Some examples of the regulation of cytokine receptor expression by other cytokines.

Factor	Cell type	Receptor	Effect	Reference
bFGF	rat granulosa cells	EGF R	↓	Biol. Reprod. 47: 202-12 (1992)
bFGF	smooth muscle cells	PDGF-α R	↑	J. Biol. Chem. 267: 18032-9 (1992)
bFGF	mouse muscle cells	IGF-1 R	↑	Mol. Endocrinol. 5: 678-84 (1991)
ET-1	rat osteosarcoma	ET-B R		Biochem. Biophys. Res. Commun. 186: 342-7 (1992)
GM-CSF	mouse macrophages	GM-CSF R	↑	J. Immunol. 149: 96-102 (1992)
IFN-α	human neuroendocrine tumors	IFN-α R	↓	Eur. Cytokine. Netw. 3: 81-8 (1992)
IFN-α	cultured human HIV-infected cells	TNF-α R	↑	J. Exp. Pathol. 5: 111-22 (1990)
IFN-γ	human HL-60 cells	M-CSF R	↑	Exp. Hematol. 19: 250-6 (1991)
IFN-γ	human breast carcinoma	EGF R	↓	Anticancer. Res. 11: 347-51 (1991)
IFN-γ	psoriatic epithelium cells			J. Invest. Dermatol. 94: 135S-140S (1990)
IFN-γ	bovine lymphocytes	TNF-α R	↓	Lymphokine. Res. 9: 43-58 (1990)
IFN-γ	human eosinophilic leukemia cells	IL2 R (Tac)	↑	J. Immunol. 143: 147-52 (1989)
IGF-1	human keratinocytes	EGF R	↑	J. Invest. Dermatol. 96: 419-24 (1991)
IGF-2	murine myoblasts	IGF-1 R	↓	J. Clin. Invest. 87: 1212-9 (1991)
IGF-BP3	bovine fibroblasts	IGF-1 R	↑*	Endocrinology 129: 710-6 (1991)
IL1	murine T helper type 2 cells	IL1 type 1 R	↓	Immunology 75: 427-34 (1992)
IL1	human synovial cells	Bradykinin R	↑	J. Pharmacol. Exp. Ther. 260: 384-92 (1992)
IL1	murine osteoblasts	PDGF-α R	↑	J. Biol. Chem. 266: 10143-7 (1991)
IL1	mouse macrophages, mouse bone marrow cells	M-CSF R	↓	Blood 77: 1923-8 (1991)
IL1	human gingival fibroblasts	EGF R	↓	J. Immunol. 142: 126-33 (1989)
IL2	murine cytotoxic T lymphocytes	IL4 R	↑	Immunology 70: 492-7 (1990)
IL2	human leukocytes	IL2-β R	↓	Cell. Immunol. 141: 409-21 (1992)
IL2	murine T helper type 2 cells	IL1 R type 1	↑	Immunology 75: 427-34 (1992)
IL2	human chronic lymphocytic leukemia cells, bovine lymphocytes	TNF-α R	↑	Blood 76: 1607-13 (1990) Lymphokine. Res. 9: 43-58 (1990)
IL3	mouse macrophages	GM-CSF R	↑	J. Immunol. 149: 96-102 (1992)
IL3	murine and human bone marrow cells	IL1 R	↑	J. Exp. Med. 172: 737-44 (1990)
IL4	murine T helper type 2 cells	IL1 type 1 R	↑	Eur. J. Immunol. 22: 153-7 (1992)
IL4	large granular lymphocytes, LAK cells NK cell	IL2β R	↓	Cancer. Lett. 64: 43-9 (1992) Pathobiology 60: 72-5 (1992) Int. Immunol. 3: 517-25 (1991) J. Immunol. 144: 2211-5 (1990)
IL6	human monocytes	IFN-γ R	↑	J. Immunol. 149: 1671-5 (1992)
IL6	human lymphoma (U937 cells) human hepatoma cells	TNF-α R	↑	Immunology 75: 669-73 (1992) Cytokine 3: 149-54 (1991)
IL7	human T cells	IL4 R	↑***	Int. Immunol. 2: 1039-45 (1990)
NGF	rat forebrain neurons	trk R	↑	Neuron 9: 465-78 (1992)
PDGF	murine and human embryonic mesenchymal cells	TGF-β R	↑	Exp. Cell. Res. 192: 1-9 (1991)
PDGF	3T3 fibroblasts human glioma cell line	EGF R	↓	J. Biol. Chem. 266: 2746-52 (1991) J. Biol. Chem. 265: 1847-51 (1990) Biochem. J. 264: 15-20 (1989)
PDGF-AA	3T3 fibroblasts	PDGF-β R	↓	J. Biol. Chem. 267: 11888-97 (1992)
TGF-α	human pancreatic cancer cells	EGF R	↓	Int. J. Pancreatol. 7: 71-81 (1990)
TGF-β	rabbit chondrocytes murine lymphoid and myeloid progenitor cell lines	IL1 R	↓	Exp. Cell. Res. 195: 376-85 (1991) J. Exp. Med. 172: 737-44 (1990)
TGF-β	murine B cells	IL5 R	↓	Cell. Immunol. 140: 158-72 (1992)
TGF-β	fetal rat hepatocytes	EGF R	↓↑**	J. Biol. Chem. 266: 13238-42 (1991)
TGF-β	murine EL4 thymona	IL2 R	↑	J. Cell. Physiol. 147: 460-9 (1991)
TGF-β	murine hematopoietic progenitor cells	GM-CSF R IL3 R G-CSF R	↓	Blood 77: 1706-16 (1991)
TGF-β	human smooth muscle cells, fibroblasts, chondrocytes	PDGF-α R	↓	Cell 63: 515-24 (1990)
TNF-α	human promyelocytic HL60 cell line human monoblastic cell line U937	TGF-β R		Blood 75: 626-32 (1990)
TNF-α	human AML blast cells	c-kit R	↑	Blood 80: 1224-30 (1992)
TNF-α	murine hematopoietic progenitor cells	CSF R	↓	J. Exp. Med. 175: 1759-72 (1992)
TNF-α	human monocytes human pancreatic carcinoma cells	IFN-γ R	↑	J. Immunol. 149: 1671-5 (1992) J. Interferon. Res. 11: 61-7 (1991)
TNF-α	human cytotoxic T lymphocytes	IL2 R	↑	Cancer. Immunol. Immunother. 35: 83-91 (1992)
TNF-α	human acute myeloid leukemia cells, granulocytes	G-CSF R	↓	J. Clin. Invest. 87: 838-41 (1991)
TNF-α	human acute myeloid leukemia cells	GM-CSF R IL3 R	↑	Blood 77: 989-95 (1991)
TNF-α	human gingival fibroblasts	EGF R	↓	J. Immunol. 142: 126-33 (1989)
TNF-α	murine macrophages	M-CSF R	↓	Blood 73: 307-11 (1989)
TNF-α	human neutrophils	GM-CSF R	↓	Proc. Natl. Acad. Sci. 87: 93-7 (1990)

* IGF-BP3 blocks IGF-I-induced IGF receptor down-regulation
** biphasic response; first receptor expression reduction, then progressive increase in receptor expression
*** almost exclusively on cells expressing the p55 IL2Rα subunit

thus allowing the effective control of biological activities of cytokines by other cytokines. These activities on receptor expression and/or function may be concentration-dependent with cytokines exerting differential effects on other cytokine receptors, depending on receptor occupancy.
Some cytokines are capable of lowering the number of receptors for another cytokine. This process is also called down-regulation of a receptor. It effectively blocks the action of a cytokine acting on a receptor the density of which has been lowered. The reverse, i. e. the new expression of a receptor or the increase in receptor density is called up-regulation of a receptor. This process makes cells sensitive for another cytokine.
This interplay between individual cytokine receptors and the resulting hierarchical action of the corresponding cytokines allows complicated feedback loops to be established, which are required for the ordered control of growth and differentiation of cells (see also: Autocrine growth control).
In addition to gross alterations in receptor densities one may also observe cytokine-induced alterations in receptor activities (e. g. » autophosphorylation, receptor-associated reactions, alterations in receptor affinities for particular ligands). At the molecular level, receptor transmodulation may be associated, among other things, with altered rates of receptor internalization, altered rates of degradation of internalized receptor pools, and alterations in receptor biosynthesis controlled at multiple levels, i. e. transcriptional, pre-translational and post-translational pathways.
The clinical relevance of the receptor transmodulation phenomenon is indicated by the fact that diminished receptor expression is usually associated with functional impairments in cellular activities, that the altered expression of cytokine receptors has been found in a number of different pathological settings, and that a variety of drugs (including, for example » CsA (cyclosporine A), » Pentoxifylline, or » Suramin) can be used effectively to interfere with receptor expression.

Recombinant cytokines: Sources for human cytokines are extremely poor and the expression of

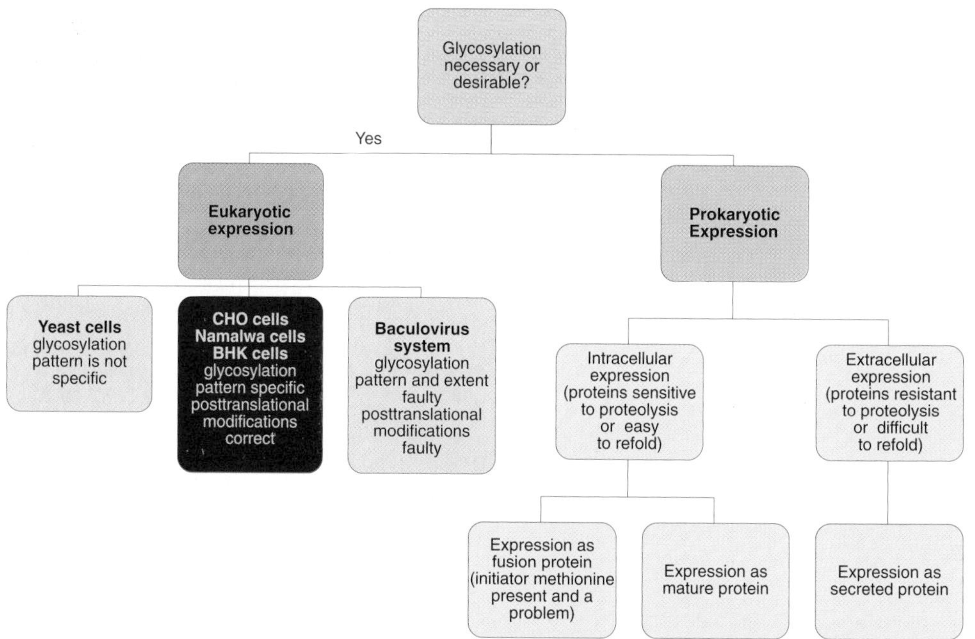

Simple decision tree for the selection of expression systems for recombinant cytokines.
Cloning and expression of cytokine genes in prokaryotic and eukaryotic systems differ from each other in many aspects. The most important decision to be taken is that of posttranslational processing. If this causes no problem then the cytokine can be produced in prokaryotic cells. If posttranslational processing poses a problem the cytokine can be expressed in yeasts, mammalian cells or insect cells. The best results with respect to posttranslational processing, including glycosylation, are usually obtained with mammalian cells. Expression in yeasts or insect cells can yield proteins with unspecific glycosylation patterns and other faulty posttranslational modifications that may interfere with the bioactivity of the expressed protein.

these factors may vary considerably, depending on the physiological state of the body. Generally, levels are too low to be considered as sources of factors for clinical use. As an example, 8.5 mg of » M-CSF have been isolated from 10,000 L of human urine.
Recombinant cytokines, i. e. cytokines produced by expression from suitable cloning vectors containing the desired cytokine gene, can be expressed in yeast (see: *Saccharomyces cerevisiae* expression system), bacteria (see: *Escherichia coli* expression system), mammalian cells (see: BHK, CHO, COS, Namalwa), or insect cell systems (see: Baculovirus expression system). Expression in each system results in a protein that differs, to a varying extent, from native molecules. Alterations can include absence of glycosylation (*E. coli*), alterations in glycosylation pattern (yeast, mammalian and insect cells), slight alterations in amino acid sequence (all systems). Proteins expressed in mature form in different host cells can differ also in their specific activities for several reasons. A review of laboratory studies shows that differences in physiochemical properties result in variations in the pharmacokinetics, biologic activity, and immunogenicity of cytokines expressed in different host cells. The expression system can influence the pharmacokinetic properties, biologic activity, and clinical toxicity of recombinant proteins. Protein variations may also lead to an increased clinical toxicity. On the other hand, expression vectors are useful for the construction of recombinant forms of cytokines to investigate structure/function relationships. Heterologous expression systems have been employed to express streamlined cytokines engineered for better clinical efficacy (see, for example: Muteins).

Thatcher DR Large-scale production of hematopoietic growth factors. in: Dexter TM et al (eds) Colony-stimulating factors, Molecular and cellular biology, pp. 329-57, Marcel Dekker Inc. New York 1990)

Recharging of long-term bone marrow cultures: see: LTBMC.

Regenerating liver nuclear receptor: abbrev. RNR-1. See: nur-77.

Response elements: abbrev. RE. A term denoting short sequence regions in the regulatory regions of genes (see also: Gene expression). These sequences are binding sites for transcription factors that regulate gene expression. Many genes including many cytokine genes contain ***cytokine response elements*** which are binding sites for cytokine-specific transcription factors (see, for example: IL6RE (IL6 response element)).

Restrictins: Name given to putative hematopoietic factors produced by bone marrow stromal cells (see: Hematopoiesis) that function as differentiation antagonists and promote stem cell renewal. The murine stromal cell line MBA-2.1 produces an activity, designated restrictin-P, which is specifically inhibitory to the growth of plasmacytomas and mature B cell lymphomas and arrests them in the G_0/G_1 phase of the » cell cycle.
Restrictin is also the name of a chick neural extracellular matrix protein implicated in neural cell attachment and found to be associated with the cell surface recognition protein F11.

Honigwachs-Shaanani J et al Restrictin-P: the first member of a putative family of novel inhibitors. ANY 628: 287-97 (1991); **Kadouri A et al** Dynamic changes in cytokine secretion by stromal cells during prolonged maintenance under protein-free conditions. IJCC 10: 299-308 (1992); **Norenberg U et al** The chicken neural extracellular matrix molecule restrictin: similarity with EGF-, fibronectin type III-, and fibrinogen-like motifs. Neuron 8: 849-63 (1992); **Zipori D** Role of stromal cell factors (restrictins) in microorganization of hemopoietic tissues. PCBR 352: 115-22 (1990); **Zipori D** The renewal and differentiation of hematopoietic stem cells. FJ 6: 2691-7 (1992)

***ret*:** The *ret* » oncogene encodes a cell surface receptor for an as yet unknown ligand. The protein possesses a tyrosine kinase domain. The human *ret* gene maps to chromosome 10q11.2. Alternative polyadenylation and splicing of mRNA yields four major mRNA species of different lengths in human malignant cell lines and rat tissues.
The *ret* oncogene has been found to be activated in many human papillary thyroid carcinomas but not in other thyroid tumors. This gene is referred to as ***PTC*** (papillary thyroid carcinoma) and is a rearranged form of the *ret* proto-oncogene. Mutations in the *ret* oncogene appear to be responsible for multiple endocrine neoplasia type 2A (MEN 2A), a dominantly inherited cancer syndrome that affects tissues derived from neural ectoderm and is characterized by medullary thyroid carcinoma and phaeochromocytoma.
■ **TRANSGENIC/KNOCK-OUT/ANTISENSE STUDIES:** A possible role for the *ret* oncogene in the proliferation of neural crest cells is suggested by the development of neuroblastoma in » transgenic animals overexpressing ret. Introduction of the *ret* oncogene into mice that are deficient for the » *kit* gene (receptor for » SCF) shows that *ret* can com-

RELAY: see addendum. **Renotropin:** see addendum.

pensate for the defect of *kit* during embryogenesis and postnatal life.

● REVIEWS: **Schneider R et al** The human proto-oncogene *ret*: a communicative cadherin? TIBS 17: 468-9 (1992)
Grieco M et al PTC is a novel rearranged form of the *ret* proto-oncogene and is frequently detected *in vivo* in human thyroid papillary carcinomas. Cell 60: 557-63 (1990); **Ishizaka Y et al** Human *ret* proto-oncogene mapped to chromosome 10q11.2. O 4: 1519-21 (1989); **Mulligan LM et al** Germ-line mutations of the *ret* proto-oncogene in multiple endocrine neoplasia type 2A. N 363: 458-60 (1993); **Nagao M et al** Expression of *ret* proto-oncogene in human neuroblastomas. Jpn. J. Cancer Res. 81: 309-12 (1990); **Santoro M et al** *ret* oncogene activation in human thyroid neoplasms is restricted to the papillary cancer subtype. JCI 89: 1517-22 (1992); **Tahira T et al** Characterization of *ret* proto-oncogene mRNAs encoding two isoforms of the protein product in a human neuroblastoma cell line. O 5: 97-102 (1990)
● TRANSGENIC/KNOCK-OUT/ANTISENSE STUDIES: **Iwamoto T et al** Oncogenicity of the *ret* transforming gene in MMTV/*ret* transgenic mice. O 5: 535-42 (1990); **Iwamoto T et al** The *ret* oncogene can induce melanogenesis and melanocyte development in Wv/Wv mice. ECR 200: 410-5 (1992); **Iwamoto T et al** Neuroblastoma in a transgenic mouse carrying a metallothionein/*ret* fusion gene. Br. J. Cancer 67: 504-7 (1993)

Retina-derived growth factor α: see: RDGF-α.

Retina-derived growth factor β: see: RDGF-β.

Retinoic acid-inducible heparin-binding protein: see: RIHB.

Retinoid-binding protein: abbrev. CRBP (cellular retinoid binding protein). see: MDGI (mammary-derived growth inhibitor).

Retrocrine: This term has been suggested to describe a type of growth control by which soluble forms of receptors that are normally a component of the cell surface membrane interact with distant target cells by binding to membrane-bound forms of cytokines that are normally secreted. For other types of growth control see also: Autocrine, Intracrine, Juxtacrine, Paracrine.
Butera ST Cytokine involvement in viral permissiveness and the progression of HIV disease. JCBc 53: 336-42 (1993)

Reverse hemolytic plaque assay: abbrev. RHPA. See: Cytokine assays.

Reverse transcriptase polymerase chain reaction: abbrev. RT-PCR. see: Message amplification phenotyping.

RG 50862: see: Tyrphostins.

RG 50864: see: Tyrphostins.

rh: see: rhu

Rheumatoid arthritis synovial fluid: abbrev. RA-SF. See: TRF (T cell replacing factors).

Rhodamine-123: A mitochondrial binding dye that is used to identify » hematopoietic stem cells by fluorescence activated cell sorting. See: CFU-S (colony forming unit spleen) and LTRC (long-term repopulating cells).

RHPA: *reverse hemolytic plaque assay* see: Cytokine assays.

rhu: also abbrev. rh. Abbrev. for recombinant human. Used in abbrev. for factors such as in rhu-GM-CSF to denote that the human factor has been obtained by heterologous expression of the cloned gene.

RIA: *radioimmunoassay* see: Cytokine assays.

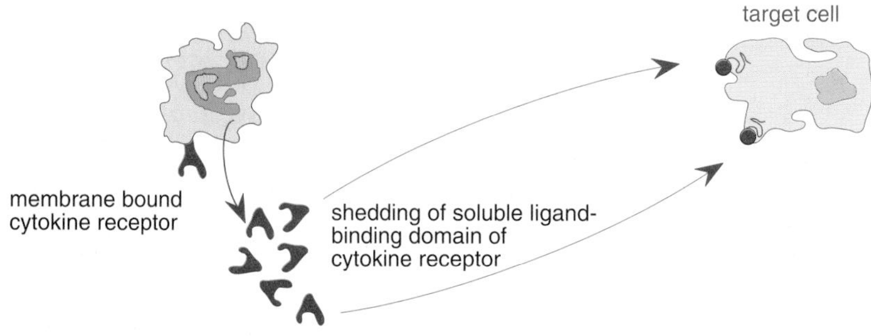

Retrocrine growth control mechanism.
See text for details.

RIF: *receptor inducing factor* This poorly characterized glycoprotein of ≈ 44 kDa is produced by the murine macrophage cell lineP388-D1. Glycosylation is not required for the biological activity of this factor. RIF functions as a costimulator of » IL2 receptor expression in resting CD8-positive T cells activated via the CD3 T cell receptor. RIF is a competence factor that bestows upon cells the ability to respond to IL2. See also: IL2 receptor inducing factor.

Hardt C et al Activation of murine $CD8^+$ lymphocytes: two distinct signals regulate c-*myc* and interleukin 2 receptor RNA expression. EJI 17: 1711-7 (1987); **Hardt C et al** Functional and biochemical characteristics of a murine interleukin 2 receptor-inducing factor. EJI 17: 209-16 (1987); **Schmidberger R et al** Primary activation of murine CD8 T cells via cross-linking of T3 cell surface structures: two signals regulate induction of interleukin 2 responsiveness. EJI 18: 277-82 (1988); **Wagner H & Heeg K** Two distinct signals regulate induction of IL2 responsiveness in $CD8^+$ murine T cells. Immunology 64: 433-8 (1988)

RIHB: *retinoic acid-inducible heparin-binding protein* see: This highly basic protein of 121 amino acids is isolated from chick embryos where it is expressed during early chick embryogenesis. The expression of RIHB is strongly induced by retinoic acid.

RIHB is derived from a larger precursor protein by removal of a » signal sequence of 21 amino acids. The protein contains 31 lysine and arginine residues and 10 cysteine residues. The sequence has 64 % homology at the protein level with murine » MK (midkine) and also considerable homology with » HBNF (heparin-binding neurite-promoting factor). RIHB is considered to be the avian homologue of MK. It has been suggested to be involved in the differentiation of chondrocytes but it is not mitogenic for this cell type.

Cockshutt AM et al Retinoic acid-induced heparin-binding factor (RIHB) mRNA and protein are strongly induced in chick embryo chondrocytes treated with retinoic acid. ECR 207: 430-438 (1993); **Urios P et al** Molecular cloning of RI-HB, a heparin-binding protein regulated by retinoic acid. BBRC 175: 617-24 (1991); **Vigny M et al** Identification of a new heparin-binding protein localized within chick basement membranes. EJB 186: 733-740 (1989)

RISBASES: *Ribonucleases with special biological action* A general term for ribonucleases with non-catalytic activities such as in » Angiogenin. Other members are selectively neurotoxic RNases, seminal RNAases, a lectin, and the self-incompatibility factors from a flowering plant.

D'Alessio G. et al Seminal RNase: a unique member of the ribonuclease superfamily. TIBS 16: 104-6 (1991)

RNR-1: *regenerating liver nuclear receptor* see: nur-77.

Roferon A: Recombinant IFN-α2a (Hoffman-La Roche). See: IFN-α.

Romurtide: A synthetic muramyl dipeptide (MDP) derivative that is a potent inducer of cytokine synthesis. An MDP derivative, MDP-Lys(L18) = N α-(N-acetylmuramyl-L-alanyl-D-isoglutaminyl)-N ε-stearoyl-L-lysine), also called Muroctasin, is used as a hematopoietic agent for restoration of leukopenia in cancer patients treated with radiotherapy and chemotherapy,

Azuma I Development of the cytokine inducer romurtide: experimental studies and clinical applications. TIPS 13: 425-7 (1992); **Azuma I** Synthetic immunoadjuvants: application to non-specific host stimulation and potentiation of vaccine immunogenicity. Vaccine 10: 1000-6 (1992)

RPF: *rabbit peritoneal factor* This factor of ≈ 6.8 kDa is isolated from rabbit peritoneal cavity fluid following injection of zymosan. RPF is a strong chemoattractant for neutrophils (see also: Chemotaxis) and induces edemas *in vivo*. The known aminoterminal amino acid sequence suggests that RPF is the rabbit homologue of human » IL8.

Beaubien B et al A novel neutrophil chemoattractant generated during an inflammatory reaction in the rabbit peritoneal cavity *in vivo*. BJ 271: 797-801 (1990)

RPMI 1788: (RPMI = Roswell Park Memorial Institute). An EBV-transformed hematopoietic cell line established from peripheral blood leukocytes of a normal 33 year old male patient. The cells secrete IgM light chains and express HLA 2,7 antigens. The cell line is used as a producer of » TNF-β.

The cells are also used in a » Bioassay to detect » IL1 which stimulates the proliferation of these cells. The assay appears to be specific for IL1 and is not influenced by the presence of high concentrations of murine or human » TNF and human » IL6. It has been reported that RPMI 1788 cells are constitutive producers of IL6.

Hutchins D et al Production and regulation of interleukin 6 in human B lymphoid cells. EJI 20: 961-8 (1990); **Vandenabeele P et al** Development of a simple, sensitive and specific bioassay for interleukin-1 based on the proliferation of RPMI 1788 cells: comparison with other bioassays for IL1. JIM 135: 25-32 (1990)

RPT-1: This murine protein of 41 kDa (353 amino acids) is also called ***Down regulatory protein of interleukin 2 receptor***. The protein is found in the cell nucleus and inhibits the expression of the α chain of the receptor for » IL2.

Patarca R et al *rpt*-1, an intracellular protein from helper/inducer T cells that regulates gene expression of interleukin 2 receptor and human immunodeficiency virus type 1 PNAS 85: 2733-7 (1988) [erratum in PNAS 85: 5224]

RPTK: *receptor protein tyrosine kinase* see: PTK (protein tyrosine kinases).

RRA: *radioreceptor assays* see: Cytokine assays.

RSGI: *rabbit serum-derived growth inhibitor* This poorly characterized factor is isolated from rabbit serum. It inhibits the growth of the tumorigenic rat cell line RSV-BRL and some human tumor cell lines. The factor is probably not identical with » TGF and » TNF. A variant of this factor, designated GI-I (18 kDa), specifically inhibits the growth of rodent cell lines while another variant, GI-II (36 kDa) inhibits all tested tumor cell lines.

Kimura T et al Purification and properties of growth inhibitor from normal rabbit serum. J. Biochem. Tokyo 110: 423-8 (1991); **Kimura T et al** Purification and partial characterization of two types of growth-inhibitory protein latently present in rabbit serum. BBA 1118: 239-48 (1992)

RT-PCR: *reverse transcriptase polymerase chain reaction* see: Message amplification phenotyping.

RTK: *receptor tyrosine kinase* see: PTK (protein tyrosine kinase).

ryk: An » oncogene originally identified in the genome of the RPL30 virus, an acute oncogenic avian retrovirus isolated from chicken tumors. The corresponding human *ryk* tyrosine kinase (607 amino acids) has been found to represent a ubiquitously expressed gene. Comparison of the human and mouse *ryk* sequences shows a 92% conservation at the nucleotide level and 97% at the amino acid level. The human *ryk* gene maps to chromosome 3q11-25.

Hovens CM et al RYK, a receptor tyrosine kinase-related molecule with unusual kinase domain motifs. PNAS 89: 11818-22 (1992); **Jia R et al** A novel oncogene, v-ryk, encoding a truncated receptor tyrosine kinase is transduced into the RPL30 virus without loss of viral sequences. J. Virol. 66: 5975-87 (1992); **Stacker SA et al** Molecular cloning and chromosomal localization of the human homologue of a receptor related to tyrosine kinases (RYK). O 8: 1347-56 (1993); **Tamagnone L et al** The human ryk cDNA sequence predicts a protein containing two putative transmembrane segments and a tyrosine kinase catalytic domain. O 8: 2009-14 (1993)

S

s: (a) sometimes used as a prefix in acronyms to denote the soluble form of a factor, i. e. a factor found in the circulation as in sIL2-R (soluble IL2receptor) and sCD23 (soluble CD23); (b) a prefixed s can also mean "surface" such as in sIgM (i. e. IgM associated with the cell surface membrane).

S100:

■ **Sources:** S100 is found at high concentrations in astroglial cells of the central nervous system and in discrete neurons in the mesencephalic and motor trigeminal, facial, and lemniscus nuclei. The protein is also expressed in Schwann cells of the peripheral nervous system. The S100 protein subunits, S100α and S100β, are expressed selectively by specific cell types. The S100a(0) protein (α α form of S100) is found predominantly in the cardiac and skeletal muscle of various mammals at much higher levels than in the nervous system. S100β is expressed in glial cells at levels ≈ tenfold higher than in most other cell types. S100 is also found in epidermal Langerhans cells, myoepithelial cells of the salivary and mammary glands, and in many other cells of various organs.

■ **Protein Characteristics:** S100 is a small acidic disulfide-linked, dimeric calcium-binding protein (2 sites per subunit) of 21 kDa. S100 protein exists in three forms, designated S100a(0) (α/α form), S100a (α/β form), and S100b (β/β form). The β-subunit of S100 protein (S100β) is highly conserved in the mammalian brain.

■ **Gene Structure:** The gene coding for human S100β maps to chromosome 21q22.2-q22.3, a chromosomal locus the duplication of which has been implicated in the generation of Down Syndrome. The murine S100β gene maps to chromosome 10. The human S100α subunit gene maps to chromosome 1q21. The human S100β gene is composed of 3 exons, the first of which specifies the 5'-untranslated region, while the second and third each encode a single calcium-binding domain. The promoter region contains a cAMP-responsive elements (see: CRE) and several other potential regulatory transcription elements.

■ **Related Factors:** S100 proteins form a family of related calcium-binding proteins. S100 is biochemically related to » Calgranulins. S100 shows 55% sequence homology with the chemotactic protein » CP-10. A closely related protein is » NEF (neurite extension factor)

■ **Biological Activities:** S100 is secreted by cultured astroglial cells and functions as a trophic factor for a number of neuronal cell types (see also: Neurotrophins), stimulating the differentiation of immature neurons. It promotes the survival of these cells *in vitro* and induces the outgrowth of neurites. S100 also enhances the proliferation of glial cells at nanomolar concentrations. S100β may be involved in the coordinate development and maintenance of the central nervous system by synchronously stimulating the differentiation of neurons and the proliferation of astroglia. A disulfide-bonded dimer of S100β appears to be the biologically active secreted form of S100.
S100 also inhibits the activities of some tyrosine-specific protein kinases involved in receptor-mediated processes of intracellular signal transduction.
The production of S100ao is markedly increased during the differentiation of murine 3T3-L1 cells into adipocytes. The protein may also participate in the function of adipocytes. S100ao protein in 3T3-L1 adipocytes is released by incubation with the lipolytic hormone, adrenocorticotropic hormone or catecholamines, in a cyclic-AMP-dependent manner.
It has been suggested that the natural proteolysis of S100 in brain tissue may lead to the formation of biologically active oligopeptide products that are involved, in particular, in the modulation of the functional activity of central benzodiazepine receptors.

■ **Transgenic/Knock-out/Antisense Studies:** Transgenic animals (mice) overexpressing the human S100β gene tolerate 10-100-fold higher than normal levels of S100β gene expression in the brain without any gross physical or behavioral abnormalities.

The inhibition of S100β expression by » antisense oligonucleotides reveals that this protein is involved in the regulation of glial cell morphology, cytoskeletal organization, and cell proliferation. Rat C6 glioma cells containing an S100β antisense gene and C6 cells treated with S100β antisense oligodeoxynucleotides show a flattened cell morphology, a more organized microfilament network, and a decrease in cell growth rate. S100b is involved in the regulation of microtubule assembly in brain. It has been demonstrated that S100β binds to some kinds of the microtubule-associated tau protein and mediates phosphorylation of tau protein by protein kinase II. The disruption of the microtubular cytoskeleton causes a specific reduction in the level of S100 protein mRNA in C6 cells.

■ **CLINICAL USE & SIGNIFICANCE:** The levels of S100β protein, mRNA, and specific neurotrophic activity are elevated 10-20-fold in extracts of temporal lobe from Alzheimer's disease patients. The cells containing the increased S100β are reactive astrocytes; the neuritic plaques are surrounded by S100β-containing astrocytes. The elevated levels of S100β provides a link between the prominent reactive gliosis and neuritic plaque formation in this disease and raises the possibility that S100β contributes to Alzheimer's disease neuropathology. These patients also appear to express autoantibodies directed against S100.

In spite of the lack of specificity to any tissue antibodies against protein S100 are widely used in the diagnosis of human tumors, mainly for the differentiation between neurogenic and soft tissue tumors. A positive response to S100 protein is regularly observed in benign Schwannomas, neurofibromas, and granular cell tumors, while fibroblastic tumors are S100-negative). S100 antibodies are also employed for the differentiation between nonpigmented melanomas (positive) and anaplastic carcinomas (negative).

● **REVIEWS: Isobe T et al** Chemistry and cell biology of neuron- and glia-specific proteins. Arch. Histol. Cytol. 52: 24-32 (1989); **Kligman D & Hilt DC** The S100 protein family. TIBS 13: 437-43 (1988); **Marshak DR et al** S100 β as a neurotrophic factor. Prog. Brain Res. 86: 169-81 (1990); **Van Eldik LJ & Zimmer DB** Approaches to study the role of S100 proteins in calcium-dependent cellular responses. J. Dairy Sci. 71: 2028-34 (1988)

● **BIOCHEMISTRY & MOLECULAR BIOLOGY: Allore RJ et al** Cloning and expression of the human S100 β gene. JBC 265: 15537-43 (1990); **Barger SW et al** Disulfide-linked S100 β dimers and signal transduction. BBA 1160: 105-12 (1992); **Duncan AM et al** Refined sublocalization of the human gene encoding the β subunit of the S100 protein (S100B) and confirmation of a subtle t(9;21) translocation using *in situ* hybridization. Cytogenet. Cell. Genet. 50: 234-5 (1989); **Engelkamp D et al** S100 α, CAPL, and CACY: molecular cloning and expression analysis of three calcium-binding proteins from human heart. B 31: 10258-64 (1992); **Engelkamp D et al** Six S100 genes are clustered on human chromosome 1q21: identification of two genes coding for the two previously unreported calcium-binding proteins S100D and S100E. PNAS 90: 6547-51 (1993); **Fleminger G et al** Calcium-modulated conformational affinity chromatography. Application to the purification of calmodulin and S100 proteins. J. Chromatogr. 597: 263-70 (1992); **Glenney JR jr et al** Isolation of a new member of the S100 protein family: amino acid sequence, tissue, and subcellular distribution. JCB 108: 569-78 (1989); **Haimoto H & Kato K** S100a0 (α α) protein in cardiac muscle. Isolation from human cardiac muscle and ultrastructural localization. EJB 171: 409-15 (1988); **Haimoto H et al** Differential distribution of immunoreactive S100-α and S100-β proteins in normal nonnervous human tissues. Lab. Invest. 57: 489-98 (1987); **Herrera, GA et al** S-100 protein expression by primary and metastatic adenocarcinomas. Am. J. Clin. Path. 89: 168-76 (1988); **Kato K et al** S100a0 (α α) protein: distribution in muscle tissues of various animals and purification from human pectoral muscle. J. Neurochem. 46: 1555-60 (1986); **Le Cam A** Natural tyrosine kinase inhibitors. Pathol. Biol. Paris 39: 796-800 (1991); **MacDonald G et al** Fine structure physical mapping of the region of mouse chromosome 10 homologous to human chromosome 21. Genomics 11: 317-23 (1991); **Mely V & Gerard D** Intra- and interchain disulfide bond generation in S100b protein. J. Neurochem. 55: 1100-6 (1990); **Mely V & Gerard D** Structural and ion-binding properties of an S100b protein mixed disulfide: comparison with the reappraised native S100b protein properties. ABB 279: 174-82 (1990); **Mely Y & Gerard D** Large-scale, one-step purification of oxidized and reduced forms of bovine brain S100b protein by HPLC. J. Neurochem. 50: 739-44 (1988); **Moore BW** A soluble protein characteristic of the nervous system. BBRC 19: 739-44 (1965); **Morii K et al** Structure and chromosome assignment of human S100 α and β subunit genes BBRC 175: 185-91 (1991) [erratum in BBRC 177: 894]; **Shimizu A et al** A molecular genetic linkage map of mouse chromosome 10, including the *myb*, S100b, Pah, Sl, and Ifg genes. Biochem. Genet. 30: 529-35 (1992); **Van Eldik LJ et al** Synthesis and expression of a gene coding for the calcium-modulated protein S100 β and designed for cassette-based, site-directed mutagenesis. JBC 263: 7830-7 (1988); **Zimmer DB et al** Isolation of a rat S100α cDNA and distribution of its mRNA in rat tissues. Brain Res. Bull. 27: 157-62 (1991)

● **BIOLOGICAL ACTIVITIES: Barger SW & Van Eldik LJ** S100 β stimulates calcium fluxes in glial and neuronal cells. JBC 267: 9689-94 (1992); **Baudier J & Cole RD** Interactions between the microtubule-associated tau proteins and S100β regulate tau phosphorylation by the Ca^{2+}/calmodulin-dependent protein kinase II. JBC 263: 5876-83 (1988); **Bhattacharyya A et al** S100 is present in developing chicken neurons and Schwann cells and promotes motor neuron survival *in vivo*. J. Neurobiol. 23: 451-66 (1992); **Deloulme JC et al** Interactions of S100 proteins with protein kinase substrates. Biological implication. AEMB 269: 153-7 (1990); **Dunn R et al** Reduction in S100 protein β subunit mRNA in C6 rat glioma cells following treatment with anti-microtubular drugs. JBC 262: 3562-6 (1987); **Hesketh J & Baudier J** Evidence that S100 proteins regulate microtubule assembly and stability in rat brain extracts. Int. J. Biochem. 18: 691-5 (1986); **Kato K et al** Induction of S100 protein in 3T3-L1 cells during differentiation to adipocytes and its liberating by lipolytic hormones. EJB 177: 461-6 (1988); **Poletaev AB et al** [Limited proteolysis of brain-specific protein S100. Isolation, physico-chemical and immunochemical characteristics of the neuropeptide AT-1-1] Biokhimiia 53: 985-90 (1988); **Selinfreund RH et al** Neurotrophic protein S100β stimulates glial cell proliferation. PNAS 88: 3554-8 (1991); **Winningham-**

Najor F et al Neurite extension and neuronal survival activities of recombinant S100 β proteins that differ in the content and position of cysteine residues. JCB 109: 3063-71 (1989)

● Transgenic/Knock-out/Antisense Studies: **Friend WC et al** Cell-specific expression of high levels of human S100 β in transgenic mouse brain is dependent on gene dosage. J. Neurosci. 12: 4337-46 (1992); **Selinfreund RH et al** Antisense inhibition of glial S100 β production results in alterations in cell morphology, cytoskeletal organization, and cell proliferation. JCB 111: 2021-8 (1990)

● Clinical Use & Significance: **Allore R et al** Gene encoding the β subunit of S100 protein is on chromosome 21: implications for Down syndrome. S 239: 1311-3 (1988); **Baumal R et al** Role of antibody to S100 protein in diagnostic pathology. Lab. Invest. 59: 152-4 (1988); **Jankovic BD & Djordjijevic D** Differential appearance of autoantibodies to human brain S100 protein, neuron specific enolase and myelin basic protein in psychiatric patients. Int. J. Neurosci. 60: 119-27 (1991); **Marks A & Allore R** S100 protein and Down syndrome. Bioessays 12: 381-3 (1990); **Marshak DR et al** Increased S100β neurotrophic activity in Alzheimer's disease temporal lobe. Neurobiol. Aging 13: 1-7 (1992)

SA: *synergistic activity* This factor was initially described as a protein that stimulates the generation of macrophage-like progenitor cells in combination with » M-CSF. It is identical with » IL3.

Kriegler AB et al Partial purification and characterization of a growth factor for macrophage progenitor cells with high proliferative potential in mouse bone marrow. Blood 60: 503-8 (1982)

SA-7: *species antigen 7* This cell surface glycoprotein of 165 kDa identified as a species antigen of human cells is identical with the receptor for » EGF.

Aden DP & Knowles BB Cell surface antigens coded for by the human chromosome 7. Immunogenetics 3: 209-11 (1976); **Carlin CR & Knowles BB** Identity of human epidermal growth factor (EGF) receptor with glycoprotein SA-7: evidence for differential phosphorylation of the two components of the EGF receptor from A431 cells. PNAS 79: 5026-30 (1982)

SAA inducer: *serum amyloid A inducer* This protein was initially isolated as a factor stimulating the production of the » acute phase protein serum Amyloid A. It is identical with » IL1.

McAdam KPWJ & Dinarello CA Induction of serum amyloid A synthesis by human leukocytic pyrogen. In: Agarwal MD (edt) Bacterial endotoxins and host response. Elsevier/North Holland, Amsterdam, pp. 167ff. (1980); **Sipe JD et al** Detection of a mediator derived from endotoxin stimulated macrophages that induces the acute phase SAA response in mice. JEM 150: 597-606 (1979); **Sztein MB et al** The role of macrophages in the acute phase response: SAA inducer is closely related to lymphocyte-activating factor and endogenous pyrogen. CI 63: 164-76 (1981); **Sztein MB et al** An epidermal cell-derived cytokine triggers the *in vivo* synthesis of serum amyloid A by hepatocytes. JI 129: 87-90 (1982)

***Saccharomyces cerevisiae* expression system:** [baker's yeast] *Saccharomyces cerevisiae* is a unicellular organism. The cells have a diameter of ≈ 3-5 μm and resemble plant and animals cells with respect to complexity in spite of their small size. *Saccharomyces cerevisiae* is one of the best characterized organisms at the molecular biological level. Many different genetic, biochemical, and other techniques are available to study this organism. Yeast cells and the many available mutants are easily propagated *in vitro* on chemically defined media. A particular advantage is the small size of the yeast genome ($1,4 \times 10^7$ bp) which facilitates molecular biological studies, for example the establishment of genome libraries.

Saccharomyces cerevisiae and also some other species such as *Schizosaccharomyces pombe* and *Kluyveromyces lactis* are frequently used as simple eukaryotic host organisms for the expression of heterologous proteins. Unlike prokaryotic expression systems (see: *Escherichia coli*) yeast cells express other eukaryotic genes even if they contain introns (see also: Gene expression) and most proteins are also processed and modified correctly. Accordingly, this organism is often used to produce recombinant growth factors (see: Recombinant cytokines). A disadvantage of this system is that glycosylation of recombinant proteins in yeast cells involves addition of large mannose-rich branched carbohydrate core structures at each N-glycosylation site not normally found on authentic mammalian proteins. If glycosylation is of importance, the only alternatives are mammalian expression systems (see: BHK, CHO, COS, Namalwa), or insect cell systems (see: Baculovirus expression system).

● Basics: **Barnett JA et al** Yeasts: characteristics and identification. Cambridge University Press, Cambridge 1983; **Bitter GA et al** Expression and secretion vectors for yeast. MiE 153: 516-544 (1987); **Broach JR et al** (edt) The molecular biology of the yeast *Saccharomyces*. Cold Spring Harbor Laboratories, Cold Spring Harbor, New York, 1982; **Buckholz RG** Yeast systems for the expression of heterologous gene products. Curr. Opin. Biotechnol. 4: 538-42 (1993); **Carter BL et al** Expression and secretion of foreign genes in yeast. In: Glover, D. M. (edt) DNA cloning, a practical approach, Vol. 3, pp. 141-161, IRL Press, Oxford 1987; **Cross F et al** Conjugation in *Saccharomyces cerevisiae*. Ann. Rev. Cell Biol. 4: 429-457 (1988); **Gunge N** Yeast DNA plasmids. Ann. Rev. Microbiol. 37: 253-276 (1983); **Hartwell LH** Biochemical genetics of yeast. Ann. Rev. Genet. 4: 373 (1970); **Herskowitz I** Life cycle of the budding yeast *Saccharomyces cerevisiae*. Microbiol. Rev. 52: 536-553 (1988); **Hicks J & Fox CF** Yeast cell biology. A. R. Liss, New York, 1986; **Innis MA** Glycosylation of heterologous proteins in *Saccharomyces cerevisiae*. Biotechnology 13: 233-46 (1989); **Nasim A et al** (edt) Molecular biology of the fission yeast. Academic Press, London 1989; **Petes TD** Molecular genetics of yeast. Ann. Rev. Biochem. 49: 845-876 (1980); **Romanos MA et al** Foreign gene expression in yeast: a review. Yeast 8: 423-88

(1992); **Rose A H & Harrison JS** The yeasts, Vol. 1 and 2. Academic Press, New York 1969 (Vol 1), 1971 (Vol 2); **Sherman F et al** Laboratory course manual for methods in yeast genetics. Cold Spring Harbor Laboratory, Cold Spring Harbor, New York 1986; **Sherman F et al** Methods in yeast genetics. Cold Spring Harbor Laboratory, Cold spring Harbor, New York, 1982; **Spencer J** Yeast genetics, a manual of methods. Springer Verlag, Berlin, 1989; **Strathern JN et al** (edt) The molecular biology of the yeast *Saccharomyces cerevisiae*. Vol. I. Life cycle and inheritance. Cold Spring Harbor Laboratory, Cold Spring Harbor, New York 1981; **Strathern JN et al** (edt) The molecular biology of the yeast *Saccharomyces cerevisiae*. Vol. II. Metabolism and gene expression. Cold Spring Harbor Laboratory, Cold Spring Harbor, New York 1982

● **Expression of Cytokines and Other Proteins:** **Axelsson K et al** Disulfide arrangement of human insulin-like growth factor I derived from yeast and plasma. EJB 206: 987-94 (1992); **Baldari C et al** A novel leader peptide which allows efficient secretion of a fragment of human interleukin 1 β in *Saccharomyces cerevisiae*. EJ 6: 229-34 (1987); **Barr PJ et al** Expression and processing of biologically active fibroblast growth factors in the yeast *Saccharomyces cerevisiae*. JBC 263: 16471-8 (1988); **Bayne ML et al** Expression, purification and characterization of recombinant human insulin-like growth factor I in yeast. Gene 66: 235-44 (1988); **Calderon-Cacia M et al** Incomplete process of recombinant human platelet-derived growth factor produced in yeast and its effect on the biological activity. BBRC 187: 1193-9 (1992); **Casagli MC et al** Different conformation of purified human recombinant interleukin 1 β from *Escherichia coli* and *Saccharomyces cerevisiae* is related to different level of biological activity. BBRC 162: 357-63 (1989); **Cosman D et al** Human interleukin-3 and granulocyte-macrophage colony stimulating factor: site-specific mutagenesis and expression in yeast. Dev. Biol. Stand. 69: 9-13 (1988); **Craig S et al** Characterization of the structure and conformation of platelet-derived growth factor-BB (PDGF-BB) and proteinase-resistant mutants of PDGF-BB expressed in *Saccharomyces cerevisiae*. BJ 281: 67-72 (1992); **Devenish RJ et al** Construction of expression vectors for the production of interferons in yeast. Dev. Biol. Stand. 67: 185-99 (1987); **Dorr RT** Clinical properties of yeast-derived versus *Escherichia coli*-derived granulocyte-macrophage colony-stimulating factor. Clin. Ther. 15: 19-29 (1993); **Elliott-S et al** Yeast-derived recombinant human insulin-like growth factor I: production, purification, and structural characterization. J. Protein Chem. 9: 95-104 (1990); **Finnis C et al** Expression of recombinant platelet-derived endothelial cell growth factor in the yeast *Saccharomyces cerevisiae*. Yeast 8: 57-60 (1992); **George-Nascimento C et al** Characterization of recombinant human epidermal growth factor produced in yeast. B 27: 797-802 (1988); **Gillis S et al** Production of recombinant human colony stimulating factors in yeast. Behring Inst. Mitt. 83: 1-7 (1988); **Gough NM et al** Biochemical characterization of murine leukemia inhibitory factor produced by Krebs ascites and by yeast cells. Blood Cells 14: 431-42 (1988); **Guisez-Y et al** Production and purification of recombinant human interleukin-6 secreted by the yeast *Saccharomyces cerevisiae*. EJB 198: 217-22 (1991); **Hard K et al** O-mannosylation of recombinant human insulin-like growth factor I (IGF-I) produced in *Saccharomyces cerevisiae*. FL 248: 111-4 (1989); **Herrmann F et al** Yeast-expressed granulocyte-macrophage colony-stimulating factor in cancer patients: a phase 2 clinical study. Behring Inst. Mitt. 83: 107-18 (1988); **Hitzeman RA et al** Construction of expression vectors for secretion of human interferons by yeast. MiE 119: 424-33 (1986); **Ingley E et al** Production and purification of recombinant human interleukin-5 from yeast and baculovirus expression systems. EJB 196: 623-9 (1991); **Kalsner I et al** Comparison of the carbohydrate moieties of recombinant soluble Fc ε receptor (sFc ε RII/sCD23) expressed in *Saccharomyces cerevisiae* and Chinese hamster ovary cells. Different O-glycosylation sites are used by yeast and mammalian cells. Glycoconj. J. 9: 209-16 (1992); **Kanaya E et al** Synthesis and secretion of human nerve growth factor by *Saccharomyces cerevisiae*. Gene 83: 65-74 (1989); **Kiefer MC et al** Characterization of recombinant human insulin-like growth factor binding proteins 4, 5, and 6 produced in yeast. JBC 267: 12692-9 (1992); **Miyajima A et al** Secretion of mature mouse interleukin-2 by *Saccharomyces cerevisiae*: use of a general secretion vector containing promoter and leader sequences of the mating pheromone α-factor. Gene 37: 155-61 (1985); **Miyamoto C et al** Molecular cloning and regulated expression of the human c-*myc* gene in *Escherichia coli* and *Saccharomyces cerevisiae*: comparison of the protein products. PNAS 82: 7232-6 (1985); **Moonen P et al** Increased biological activity of deglycosylated recombinant human granulocyte/macrophage colony-stimulating factor produced by yeast or animal cells. PNAS 84: 4428-31 (1987); **Ostman A et al** Expression of three recombinant homodimeric isoforms of PDGF in *Saccharomyces cerevisiae*: evidence for difference in receptor binding and functional activities. GF 1: 271-81 (1989); **Price V et al** Expression, purification and characterization of recombinant murine granulocyte-macrophage colony-stimulating factor and bovine interleukin-2 from yeast. Gene 55: 287-93 (1987); **Sedmak JJ et al** High levels of circulating neutralizing antibody in normal animals to recombinant mouse interferon-β produced in yeast. J. Interferon Res. 9: S61-5 (1989); **Settineri CA et al** Characterization of O-glycosylation sites in recombinant B-chain of platelet-derived growth factor expressed in yeast using liquid secondary ion mass spectrometry, tandem mass spectrometry and Edman sequence analysis. Biomed. Environ. Mass Spectrom. 19: 665-76 (1990);**Sommer J et al** Synthesis of glia-derived nexin in yeast. Gene 85: 453-9 (1989);**Wang J et al** Purification and characterization of recombinant tissue kallikrein from *Escherichia coli* and yeast. BJ 276: 63-71 (1991)

SAF: *Sézary T cell activating factor* This factor of 28 kDa has been found to be produced by mitogen-activated peripheral blood mononuclear cells from four of five patients with Sézary syndrome, a leukemic form of cutaneous T cell lymphoma characterized by circulating neoplastic CD4-positive T cells. T cells derived from these patients normally proliferate poorly in response to conventional T cell mitogens. SAF renders non-proliferating resting T cells from leukemic patients and healthy donors responsive to » IL2 in the absence of a costimulator. Several cytokines and combinations thereof, have been shown to be devoid of SAF activity, including » IL1 to » IL7, » GM-CSF, » TNF-α, » TNF-β, » ADF, » EGF, and » IFN-γ.

Abrams JT et al Sézary T cell activating factor induces functional interleukin 2 receptors on T cells derived from patients with Sézary syndrome. CR 53: 5501-6 (1993)

SAF: *stem cell activating factor* This factor enhances the formation of » hematopoietic stem cells in bone marrow (see also: Hematopoiesis). It is identical with » IL3.

Migliaccio AR & Visser JW Proliferation of purified murine hemopoietic stem cells in serum-free cultures stimulated with purified stem cell activating factor. EH 14: 1043-8 (1986); **Mulder AH et al** Thymus regeneration by bone marrow cell suspensions differing in the potential to form early and late spleen colonies. EH 13: 768-75 (1985)

Sam3**:** see: K-*sam* (KATO-III cell-derived stomach cancer amplified gene).

Sanarelli-Shwartzman phenomenon: see: Shwartzman phenomenon.

SAP: abbrev. for » Saporin.

SAP-1: *SRF accessory protein* see: SRF (serum response factor).

Saporin: abbrev. SAP. A toxic protein isolated from *Saponaria officinalis*. Saporin inhibits protein biosynthesis by binding to a ribosomal protein. Saporin functions as an RNA-N-glycosidase specifically depurinising the 28 S RNA of ribosomes.
Saporin is used to construct fusion proteins with growth factors. These fusion proteins are bifunctional reagents that specifically bind to receptors by means of their growth factor moiety. The fusion protein/receptor complexes are then internalized by the cells which are then killed by the toxic moiety of the fusion protein (see also: Cytokine fusion toxins). Cells not expressing growth factor receptors corresponding to the growth factor moiety of the fusion protein survive.

● Biochemistry & Molecular Biology: **Benatti L et al** A Saporin-6 cDNA containing a precursor sequence coding for a carboxyl-terminal extension. FL 291: 285-8 (1991); **Benatti L et al** Nucleotide sequence of cDNA coding for saporin 6, a type 1 ribosome-inactivating protein from *Saponaria officinalis*. EJB 183: 465-70 (1989); **Fordham-Skelton AP et al** Synthesis of saporin gene probes from partial protein sequence data: use of inosine oligonucleotides, genomic DNA, and the polymerase chain reaction. Mol. Gen. Genet. 221: 134-8 (1990); **Fordham-Skelton AP et al** Characterization of saporin genes: *In vitro* expression and ribosome inactivation. Mol. Gen. Genet. 229: 460-6 (1991); **Ippoliti R et al** A ribosomal protein is specifically recognized by saporin, a plant toxin which inhibits protein synthesis. FL 298: 145-8 (1992)

● Chimaeric Constructs: **Beitz JG et al** Antitumor activity of basic fibroblast growth factor-saporin mitotoxin *in vitro* and *in vivo*. CR 52: 227-30 (1992); **Flavell DJ et al** Characteristics and performance of a bispecific F (ab'γ)2 antibody for delivering saporin to a CD7+ human acute T cell leukemia cell line. Br. J. Cancer 64: 274-80 (1991); **Gasperi-Campani A et al** Inhibition of growth of breast cancer cells *in vitro* by the ribosome-inactivating protein Saporin 6. ACR 11: 1007-12 (1991); **Lappi DA et al** Biological and chemical characterization of basic FGF-saporin mitotoxin. BBRC 160: 917-23 (1989); **Lappi DA et al** The basic fibroblast growth factor-saporin mitotoxin acts through the basic fibroblast growth factor receptor. JCP 147: 17-26 (1991); **Lappi DA et al** Characterization of a saporin mitotoxin specifically cytotoxic to cells bearing the granulocyte-macrophage colony-stimulating factor receptor. GF 9: 31-9 (1993); **Martineau D et al** Basic fibroblast growth factor-saporin mitotoxin: *In vitro* studies of its cell-killing activity and of substances that alter that activity. ANY 638: 438-41 (1991); **Prieto I et al** Expression and characterization of a basic fibroblast growth factor-saporin fusion protein in *Escherichia coli*. ANY 638: 434-7 (1991); **Siena S et al** Synthesis and characterization of an antihuman T lymphocyte saporin immunotoxin (OKT1-SAP) with *in vivo* stability into nonhuman primates. Blood 72: 756-65 (1988)

Sarcolectin: see addendum.

Sarafotoxins: abbrev. SRTX. see: ET (Endothelins).

Sarcolectin binding protein: Sarcolectin is an interferon antagonist and growth promoter that has affinity for negatively charged carbohydrates. A sarcolectin binding protein of ≈ 12 kDa isolated from human placenta has been shown to be identical with human » MIF (macrophage migration inhibition factor).

Zeng FY et al The major binding protein of the interferon antagonist sarcolectin in human placenta is a macrophage migration inhibitory factor. ABB 303: 74-80 (1993)

Sargramostim: Generic name for recombinant » GM-CSF marketed under the brand names Leukine or Prokine (Immunex).

Sarcoma growth factor: abbrev. SGF. This factor was initially isolated from the culture supernatants of virus-transformed cells. It is a mixture of » TGF-α and » TGF-β.

SASP: *surface-associated sulfhydryl protein* see: ADF (ATL (adult T cell leukemia)-derived factor).

SBF: *suppressive B cell factor* This poorly characterized factor of ≈ 34 kDa is produced by B lymphocytes expressing Fc receptors for IgG, stimulated with IgG immune complexes. It is an autoregulatory B cell factor which suppresses polyclonal immunoglobulin production. Hyper-Ia expression, plasma membrane depolarization, and activation of phosphatidylinositol hydrolysis of resting B cells, all of which are induced by the stimulation with anti-μ antibody, are significantly suppressed by the pretreatment of cells with SBF. It has been observed that SBF production by rheumatoid arthritis (RA) patients' peripheral blood B lymphocytes correlates inversely with disease activity and *in vitro* rheumatoid factor production.

Ohno T et al Inhibitory mechanism of the proliferative responses of resting B cells: feedback regulation by a lymphokine (suppressive B cell factor) produced by Fc receptor-stimulated B cells. Immunology 61: 35-41 (1987); **Ohno T et al** Biochemical analysis of inhibitory effects of a lymphokine suppressive B cell factor on the activation process of resting B cells. CI 112: 27-39 (1988); **Pisko EJ et al** Decreased production of suppressive-B cell factor by synovial membrane B lymphocytes in rheumatoid arthritis. Clin. Exp. Rheumatol. 6: 239-45 (1988); **Tsujimura K et al** Comparative studies on FcR (FcRII, FcRIII, and FcR α) functions of murine B cells. JI 144: 4571-8 (1990)

Sca-1: This murine cell surface antigen is also called ***Ly-6A/E***. It is the murine homologue of human » CD34 expressed on early hematopoietic progenitor cells (see also: Hematopoiesis). Antibodies directed against Sca-1 are used to fractionate early hematopoietic progenitor cells that have marrow-repopulating ability *in vivo* (see also: MRA, marrow repopulating ability for an assay system) and are thus considered to represent very early forms of murine » hematopoietic stem cells.

Spangrude GJ & Brooks DM Phenotypic analysis of mouse hematopoietic stem cells shows a Thy-1-negative subset. Blood 80: 1957-64 (1992); **Szilvassy SJ & Cory S** Phenotypic and functional characterization of competitive long-term repopulating hematopoietic stem cells enriched from 5-fluorouracil-treated murine marrow. Blood 81: 2310-20 (1993); **Uchida N & Weissman IL** Searching for hematopoietic stem cells: Evidence that Thy-1.1lo Lin^- $Sca\text{-}1^+$ cells are the only stem cells in C57BL/Ka-Thy-1.1 bone marrow. JEM 175: 175-84 (1992); **Williams N et al** Recombinant rat stem cell factor stimulates the amplification and differentiation of fractionated mouse stem cell populations. Blood 79: 58-64 (1992)

Scatchard plots: Graphical representation of equations describing protein-ligand binding and analyzing equilibrium binding data. Raw data (dpm) are obtained from ligand binding experiments such as radioreceptor assays or radioimmunoassays. Characteristic to both methods is the saturable, specific, competitive and reversible interaction between ligands and their receptors or suitable antibodies. Such assays are widely used to determine cytokine concentrations (see also: Cytokine assays).

Scatchard analyses have provided valuable information about cytokine receptor affinities and often have revealed the existence of low and high affinity binding sites. If concentrations of a radiolabeled bound ligand for a cytokine receptor are plotted versus the ratio of bound/free ligand one obtains curves that are either linear or curvilinear. The slope of a straight line in a Scatchard diagram is related to the dissociation constant of ligand-receptor complexes while the intercept on the abscissa reveals the number of available receptors per cell. Straight lines indicate that a single affinity class of binding sites is present. Deviations from straight lines are more difficult to interpret and cannot be resolved simply into two or more linear components; nonlinear curves essentially show the existence of more than one classes of binding sites.

Abramson SN et al Evaluation of models for analysis of radioligand binding data. Mol. Pharmacol. 31: 103-11 (1987); **Carman-Krzan M** Radioligand-receptor binding in membrane receptor research. Prog. Med. Chem. 23: 41-89 (1986); **Cooke RR & McIntosh JE** The analysis of data from breast cancer estrogen and progesterone receptor assays: Scatchard plots are inferior to direct fitting by computer. Clin. Chim. Acta 154: 171-9 (1986); **Crabbe J** Correct use of Scatchard plots. TIBS 15: 12-3 (1990); **Cusack B & Richelson E** A method for radioligand binding assays using a robotic workstation. J. Recept. Res. 13: 123-34 (1993); **Dahlquist FW** The meaning of scatchard and Hill plots. MiE 48: 270-99 (1978); **DeBlasi A et al** Calculating receptor number from binding experiments using same compound as radioligand and competitor. TIPS 10: 227-9 (1989); **Ehlert FJ et al** Estimation of the affinities of allosteric ligands using radioligand binding and pharmacological null methods. Mol. Pharmacol. 33: 187-94 (1988); **Ensing K et al** Centrifugation or filtration in quantitative radioreceptor assays. JBBM 13: 85-96 (1986); **Faguet GB** Number of receptor sites from Scatchard and Klotz graphs: complementary approaches. JCBc 31: 243-50 (1986); **Ferkany JW** The radioreceptor assay: a simple, sensitive and rapid analytical procedure. Life Sci. 41: 881-4 (1987); **Galley WC et al** A simplified analysis of Scatchard plots for systems with two interacting binding sites. Biopolymers 27: 79-86 (1988); **Gray HE & Luttge WG** Linearization of two ligand-one binding site scatchard plot and the "IC50" competitive inhibition plot: application to the simplified graphical determination of equilibrium constants. Life Sci. 42: 231-7 (1988); **Hogg PJ & Winzor DJ** Effects of ligand multivalency in binding studies: a general counterpart of the Scatchard analysis. BBA 843: 159-63 (1985); **Ishida T et al** Interaction of protein with a self-associating ligand. Deviation from a hyperbolic binding curve and the appearance of apparent co-operativity in the Scatchard plot. J. Theor. Biol. 130: 49-66 (1988); **Kermode JC** The curvilinear Scatchard plot. Experimental artifact or receptor heterogeneity? Biochem. Pharmacol. 38: 2053-60 (1989); **Ketelslegers JM et al** The choice of erroneous models of hormone-receptor interactions: a consequence of illegitimate utilization of Scatchard graphs. Biochem. Pharmacol. 33: 707-10 (1984); **Kind G et al** A simple, low-cost computer program for Scatchard plot analysis of binding data in steroid hormone receptor assays. Exp. Clin. Endocrinol. 85: 263-8 (1985); **Liebl B et al** A partially automated radioligand binding assay system for use in clinical and pharmaceutical research. J. Recept. Res. 1993; 13: 369-78 (1993); **Light KE** Analyzing nonlinear scatchard plots. S 223: 76-8 (1984); **Martin RL et al** A simple method for calculating the dissociation constant of a receptor (or enzyme).unlabeled ligand complex from radioligand displacement measurements. ABB 284: 26-9 (1991); **McPherson GA** Analysis of radioligand binding experiments. A collection of computer programs for the IBM PC. J. Pharmacol. Methods 14: 213-28 (1985); **Mendel CM et al** The effect of ligand heterogeneity on the Scatchard plot. Particular relevance to lipoprotein binding analysis. JBC 260: 3451-5 (1985); **Motulsky HJ & Mahan LC** The kinetics of competitive radioligand binding predicted by the law of mass action. Mol. Pharmacol. 25: 1-9 (1984); **Nicosia S** More about the misuse of Scatchard plots in binding studies. Pharmacol. Res. Commun. 20: 733-7 (1988); **Ratkowsky DA & Reedy TJ** Choosing near-linear parameters in the four-parameter logistic model for radioligand and related assays. Biometrics 42: 575-82 (1986); **Rodbard D et al** Computer analysis of radioligand data: advantages, problems, and pitfalls. NIDA Res. Monogr. 70: 209-22 (1986); **Sagripanti JL et al** A simple computer program for Scatchard plot analysis of hormone receptors including statistical analysis on a low cost desk top calculator. Acta Physiol. Pharmacol. Latinoam. 34: 45-53 (1984); **Scatchard G** The attractions of protein for small molecules and ions. ANY 51: 660-72 (1949); **Schliebs R & Bigl V** A simple non-parametric procedure for rapid estimation of Scatchard parameters. Gen. Physiol. Biophys. 3: 271-5 (1984); **Tomlinson G** Inhibition of radioligand binding to receptors: a competitive business. TIPS 9: 159-62 (1988); **Tomlinson G & Hnatowich MR** Apparent competitive inhibition of radioligand binding to receptors: experimental and theoretical considerations in the analysis of equilibrium binding data. J. Recept. Res. 8: 809-30 (1988); **Venturino A et al** A simplified competition data analysis for radioligand specific activity determination. Int. J. Rad. Appl. Instrum. B 17: 233-7 (1990); **Whitcomb DC et al** Theoretical basis for a new *in vivo* radioreceptor assay for polypeptide hormones. Am. J. Physiol. 249: E555-60 (1985); **Wofsy C & Goldstein B** Interpretation of Scatchard plots for aggregating receptor systems. Math. Biosci. 112: 115-54 (1992); **Zierler K** Misuse of nonlinear Scatchard plots. TIBS 14: 314-7 (1989)

Scatter-Factor: see: SF.

SCC-ETAF: *squamous cell carcinoma epidermal cell-derived thymocyte activating factor* see: ETAF.

SCF: *Sertoli cell factor* This factor is secreted by rat Sertoli cells. It is identical with » Inhibin.
Another poorly characterized factor of the same name (> 10 kDa) found in Sertoli cell-conditioned media influences steroidogenesis in Leydig cells. Its production by Sertoli cell cultures is stimulated by follicle stimulating hormone, dibutyryl-cAMP, L-isoproterenol and glucagon but is not affected by androgens. It stimulates an early step in the steroidogenic pathway but at the same time hampers the conversion of C21-precursors into androgens. The activity of the Sertoli cell factor is not affected by an » Luteinizing hormone releasing hormone antagonist, and maximally effective concentrations of the factor and releasing hormone have additive effects. Several permanent cell lines (B16, Bowes, BHK, Ratec, RK13, Vero) and also H-540 Leydig cell tumor cells produce a factor with comparable biological effects on Leydig cells. These activities are not due to » IGF-I known to maintain steroidogenesis in Leydig tumor monocultures and in cocultures with Sertoli cells.
Sertoli cells have also been found to secrete » bFGF-like factors, » transferrin, and a variety of other poorly characterized factors that act on Leydig cells. For other Sertoli cell-derived factors see also: SCSGF (Sertoli cell secreted growth factor).

Seethalakshmi L et al Pituitary binding of 3H-labeled Sertoli cell factor *in vitro*: a potential radioreceptor assay for inhibin. Endocrinology 115: 1289-94 (1984); **Wagle JR et al** Effect of hypotonic treatment on Sertoli cell purity and function in culture. In Vitro Cell Dev. Biol. 22: 325-31 (1986)
Verhoeven G & Cailleau J A factor in spent media from Sertoli-cell-enriched cultures that stimulates steroidogenesis in Leydig cells. Mol. Cell. Endocrinol. 40: 57-68 (1985); **Verhoeven G & Cailleau J A** Specificity and partial purification of a factor in spent media from Sertoli cell-enriched cultures that stimulates steroidogenesis in Leydig cells. J. Steroid Biochem. 25: 393-402 (1986); **Verhoeven G & Cailleau J A** Rat tumor Leydig cells as a test system for the study of Sertoli cell factors that stimulate steroidogenesis. J. Androl. 12: 9-17 (1991)
Smith EP et al A rat Sertoli cell factor similar to basic fibroblast growth factor increases c-*fos* messenger ribonucleic acid in cultured Sertoli cells. Mol. Endocrinol. 3: 954-61 (1989)
Carreau S et al Stimulation of adult rat Leydig cell aromatase activity by a Sertoli cell factor. Endocrinology 122: 1103-9 (1988); **Gilmont RR et al** Synthesis of transferrin and transferrin mRNA in bovine Sertoli cells in culture and *in vivo*: sequence of partial cDNA clone for bovine transferrin. Biol. Reprod. 43: 139-50 (1990); **McKinnell C & Sharpe RM** The role of specific germ cell types in modulation of the secretion of androgen-regulated proteins (ARPs) by stage VI-VIII seminiferous tubules from the adult rat. Mol. Cell. Endocrinol. 1992 Feb; 83: 219-31 (1992); **Murai T et al** A partial characterization of a Sertoli cell-secreted protein stimulating Leydig cell testosterone production. Endocrinol. Jpn. 39: 209-15 (1992); **Ojeifo JO et al** Sertoli cell-secreted protein(s) stimulates DNA synthesis in purified rat Leydig cells *in vitro*. J. Reprod. Fertil. 90: 93-108 (1990); **Papadopoulos V** Identification and purification of a human Sertoli cell-secreted protein (hSCSP-80) stimulating Leydig cell steroid biosynthesis. J. Clin. Endocrinol. Metab. 72: 1332-9 (1991); **Papadopoulos V** Spent media from immature seminiferous tubules and Sertoli cells inhibit adult rat Leydig cell aromatase activity. Horm. Metab. Res. 19: 62-4 (1987)

SCF: *stem cell factor.*
■ **Alternative Names:** HLGF-1 (hemolymphopoietic growth factor-1); KL (*kit* ligand), MGF (multipotent growth factor), MGF (mast cell growth factor), SCGF (stem cell growth factor); SLF (steel factor). See also: individual entries for further information.
■ **Sources:** SCF is a stromal cell-derived cytokine synthesized by fibroblasts and other cell types.
■ **Protein Characteristics:** SCF is a protein of 164 amino acids (BRL cells, rat) or 248 amino acids (human). The protein is extensively N- and O-glycosylated. Glycosylation is not required for biological activity since the non-glycosylated recombinant protein produced in *Escherichia coli* is biologically fully active. The domain structure and some of the biochemical properties of SCF resemble those of » M-CSF.
The murine protein forms dimers. It displays ≈ 80 % sequence homology with the human factor. Pro-

Scatter factor-like: see addendum: SFL.

teolytic cleavage of a longer membrane-bound precursor protein of 220 and 248 amino acids, respectively, yields the active murine and human factor. The precursor also possesses biological activity. Differential expression of soluble and membrane-bound forms of the factor (designated KL-1 and KL-2) is achieved by alternative splicing of the corresponding mRNAs. The membrane-bound form of SCF can mediate specific cell-to-cell contacts by interaction with receptors on nearby cells (see: Juxtacrine).

■ **Gene Structure:** The gene encoding SCF maps to human chromosome 12q22-q24. It has a length of ≈ 70 kb and contains 21 exons.

The murine KL gene is identical with the *SL*-(*Steel*) locus identified by many lethal mutations. It maps to murine chromosome 10.

■ **Receptor Structure, Gene(s), Expression:** The receptor for SCF is the » oncogene called » *kit*. It has been renamed ***CD117*** (see also: CD Antigens).

■ **Biological Activities:** Rat SCF is as active as human SCF with respect to the induction of proliferation of bone marrow stem cells. Human SCF is ≈ 800-fold less active on murine cells than rat SCF. Murine SCF is active on human cells and can be employed in » LTBMC (long-term bone marrow cultures).

Mutated alleles of SCF with deletions in the coding region (Sl^{j}, Sl^{gb}, Sl^{8H}, Sl^{10H}, Sl^{18H}) and also mutations in the receptor gene (see: *kit*) are lethal and cause the early death of homozygous embryos in mice. These animals display a pronounced anemia and severe impairment of development of the hematopoietic system characterized by a very pronounced deficiency of mast cells. The anemia and mast cell deficiency observed in animals with mutations in the Sl gene can be abolished by administration of SCF.

In vitro SCF is a growth factor for primitive lymphoid and myeloid hematopoietic bone marrow progenitor cells expressing the early cell surface marker » CD34 (see also: CD antigens) or murine » Sca-1 (see also: Hematopoiesis); see also: BFU-e (burst-forming unit erythroid). The action of SCF on mouse and human hematopoietic progenitors is inhibited by » TGF-β.

To date it is unclear whether SCF really is *the* stem cell factor because there are some indications that cells exist which are even more primitive than those responding to SCF and other factors with SCF activity have been reported (see: SCPF, stem cell proliferation factor). At least SCF appears to be one factor that is responsible for the maintenance of basal constitutive and steady-state » hematopoiesis

The biological activities of SCF are synergised considerably by colony-stimulating factors » GM-CSF and » G-CSF, and also by » IL7, » Epo and some other growth and differentiation factors.

In combination with IL7 SCF stimulates the proliferation of pre-B cells. The combination of SCF and » IL7 does not stimulate the expansion or differentiation of B220-negative (= CD45-negative) lymphoid precursors but can act synergistically in the clonal proliferation of B220-positive cells. In combination with Epo SCF stimulates the proliferation of early erythroid cells (see: BFU-E), and in combination with G-CSF SCF stimulates the proliferation of granulocytes.

In vivo administration of SCF significantly raises the numbers of megakaryocytes and circulating platelet counts. SCF may be an important regulator of platelet production under both normal and physiologically disturbed situations.

The subcutaneous administration of recombinant human SCF to baboons (*Papio cynocephalus*) or cynomolgus monkeys (*Macaca fascicularis*) leads to a pronounced expansion of mast cell populations in many anatomical sites, with numbers of mast cells in some organs by more than 100-fold. There is no clinical evidence of mast cell activation, and discontinuation of treatment with SCF results in a rapid decline of mast cell numbers nearly to baseline levels.

SCF is also a potent chemoattractant for cells, e. g. mast cells, expressing the *kit* receptor. One response to SCF in these cells is a characteristic rearrangement of the actin filaments of the cytoskeleton. At concentrations ≈ 10-100 fold lower than those eliciting cell proliferation SCF induces the synthesis of histamines and leukotriene C4 in mast cells. It has been shown that IL3-dependent mast cells undergo apoptosis on removal of IL3 and that this can be prevented by the addition of SCF, thus demonstrating how these principle mast cell growth factors may act in concert to regulate mast cell number under physiologic conditions.

Some human solid tumor cell lines display high-affinity c-*kit* receptors and produce SCF, suggesting the possibility that » autocrine production of SCF by c-*kit* receptor-bearing tumor cells may enhance cell growth in tumor cell lines.

■ **Transgenic/Knock-out/Antisense Studies:** A transgene encoding the SV40 large T antigen that is specifically expressed in a subset of thymic epithelial cells around birth causes an increase in the number of immortal thymic epithelial cells, increases thymic mass, and expands thymopoiesis. One of the immortalized cell lines derived from thymic epithelium expresses, among other things, SCF, the genes for major histocompatibility complex class I and II, Thy-1, » IL6, » IL7, » M-CSF, » IL4 receptors, and » TGF-β. The expression of SCF is independent of IL4 and is enhanced dramatically in response to IL4. Since SCF, in concert with commitment factors, channels progenitors into hemopoietic lineages. these experiments reveal a specific interaction between thymocytes and a specialized subset of thymic epithelial cells. Transgenic mouse lines carrying the mouse metallothionein/ret fusion gene develop severe melanosis and melanocytic tumors. Crosses of these mice with mice carrying mutants in the SCF receptor demonstrate that the introduction of a functional ret oncogene can compensate for the defect of c-*kit* during both embryogenesis and postnatal life.

■ **Assays:** SCF can be assayed by its growth-promoting activity on pre-B cells in long-term Whitlock-Witte bone marrow cultures (see also: LTBMC, long-term bone marrow culture). It can also be assayed in » Bioassays employing cell lines that respond to the factor (see: MH11; MO7E). An alternative and entirely different detection method is » Message amplification phenotyping.

■ **Clinical Use & Significance:** SCF is of clinical significance because of its ability to induce differentiation in lymphoid and erythroid progenitor cells and mast cells. Since this factor acts on very early hematopoietic progenitor cells its use in combination with other cytokines might be of considerable interest in the treatment of myelodysplastic syndromes and after bone marrow transplantation. Improved *in vitro* erythropoiesis with SCF has been demonstrated to occur in several types of inherited marrow failure syndromes, including Diamond-Blackfan anemia, Fanconi's anemia, dyskeratosis congenita, amegakaryocytic thrombocytopenia, and transient erythroblastopenia of childhood. *In vivo* administration of recombinant SCF has been shown to lead to an expansion of human marrow » hematopoietic stem cells and progenitor cells. Some data suggest that the combination of SCF and » IL11 may be useful in humans undergoing myeloablative therapies. Treatment with SCF stimulates the circulation of cells that engraft and rescue lethally irradiated baboons.

● **Reviews: Galli MC et al** The kit ligand, stem cell factor. Adv. Immunol. 55: 1-96 (1994); **Galli MC et al** The biology of stem cell factor, a new hematopoietic growth factor involved in stem cell regulation. Int. J. Clin. Lab. Res. 23: 70-7 (1993); **Lyman SD & Williams DE** Biological activities and potential therapeutic uses of *steel* factor. A new growth factor active on multiple hematopoietic lineages. Am. J. Pediatr. Hematol. Oncol. 14: 1-7 (1992); **Quesniaux VF** Interleukins 9, 10, 11 and 12 and *kit* ligand: a brief overview. Res. Immunol. 143: 385-400 (1992); **Williams DE & Lyman SD** Characterization of the gene product of the *Steel* locus. PGFR 3: 235-42 (1991); **Williams DE & Lyman SD** Molecular and cellular biology of mast cell growth factor: the gene product of the murine *steel* locus. Behring Inst. Mitt. 90: 23-7 (1991)

● **Biochemistry & Molecular Biology: Anderson DM et al** Alternate splicing of mRNAs encoding human mast cell growth factor and localization of the gene to chromosome 12q22-24. Cell Growth Diff. 2: 373-8 (1991); **Arakawa T et al** Glycosylated and unglycosylated recombinant-derived human stem cell factors are dimeric and have extensive regular secondary structure. JBC 266: 18942-8 (1991); **Bazan JF** Genetic and structural homology of stem cell factor and macrophage colony-stimulating factor. Cell 65: 9-10 (1991); **Brannan CI et al** Steel-Dickie mutation encodes a c-*kit* ligand lacking transmembrane and cytoplasmic domains. PNAS 88: 4671-4 (1991); **Geissler EN et al** Stem cell factor (SCF), a novel hematopoietic growth factor and ligand for c-*kit* tyrosine kinase receptor, maps on human chromosome 12 between 12q14. 3 and 12qter. Somatic Cell Mol. Genet. 17: 207-14 (1991); **Giebel LB et al** Organization and nucleotide sequence of the human *kit* (mast/stem cell growth factor receptor) proto-oncogene. O 7: 2207-17 (1992); **Huang E et al** The hematopoietic growth factor KL is encoded by the *Sl* locus and is the ligand of the c-*kit* receptor, the gene product of the *W* locus. Cell 63: 225-33 (1990); **Huang EJ et al** Differential expression and processing of two cell associated forms of the *kit*-ligand: KL-1 and KL-2. Mol. Biol. Cell. 3: 349-62 (1992); **Langley KE et al** Purification and characterization of soluble forms of human and rat stem cell factor recombinantly expressed by *Escherichia coli* and by Chinese hamster ovary cells. ABB 295: 21-8 (1992); **Langley KE et al** Soluble stem cell factor in human serum. Blood 81: 656-60 (1993); **Lu HS et al** Amino acid sequence and post-translational modification of stem cell factor isolated from buffalo rat liver cell-conditioned medium. JBC 266: 8102-7 (1991); **Lu HS et al** Post-translational processing of membrane-associated recombinant human stem cell factor expressed in Chinese hamster ovary cells. ABB 298: 150-8 (1992); **Martin FH et al** Primary structure and functional expression of rat and human stem cell factor DNAs. Cell 63: 203-11 (1990); **Nishikawa M et al** Deletion mutagenesis of stem cell factor defines the C-terminal sequences essential for its biological activity. BBRC 188: 292-7 (1992); **Nocka K et al** Molecular bases of dominant negative and loss of function mutations at the murine c-*kit*/white spotting locus: W37, Wv, W41 and W. EJ 9: 1805-13 (1990); **Vandenbark GR et al** Cloning and structural analysis of the human c-*kit* gene. O 7: 1259-66 (1992); **Williams DE et al** Identification of a ligand for the c-*kit* proto-oncogene. Cell 63: 167-74 (1990); **Witte ON** Steel locus defines new multipotent growth factor Cell 63: 5-6 (1990); **Szebo KM et al** Stem cell factor is encoded at the *SL* locus of the mouse and is the ligand for the c-*kit* tyrosine kinase receptor. Cell 63: 213-24 (1990); **Szebo KM et al** Identification, purification, and biological characterization of hemopoietic stem cell factor from buffalo rat liver-conditioned medium. Cell 63: 195-201 (1990)

● **Biological Activities: Andrews RG et al** A c-*kit* ligand, recombinant human stem cell factor, mediates reversible expansion of multiple CD34$^+$ colony-forming cell types in blood and marrow of baboons. Blood 80: 920-7 (1992); **Avraham H et al** Interaction of human bone marrow fibroblasts with megakaryocytes: role of the c-*kit* ligand. Blood 80: 1679-84 (1992); **Avraham H et al** Effects of the stem cell factor, c-*kit* ligand, on human megakaryocytic cells. Blood 79: 365-71 (1992); **Berdel WE et al** Recombinant human stem cell factor stimulates growth of a human glioblastoma cell line expressing c-*kit* protooncogene. CR 52: 3498-502 (1992); **Billips LG et al** Differential roles of stromal cells, interleukin-7, and *kit*-ligand in the regulation of B lymphopoiesis. Blood 79: 1185-92 (1992); **Bischoff SC & Dahinden CA** c-*kit* ligand: a unique potentiator of mediator release by human lung mast cells. JEM 175: 237-44 (1992); **Bodine DM et al** Stem cell factor increases colony-forming unit-spleen number *in vitro* in synergy with interleukin-6, and *in vivo* in Sl/Sld mice as a single factor. Blood 79: 913-9 (1992); **Bodine DM et al** *In vivo* administration of stem cell factor to mice increases the absolute number of pluripotent hematopoietic stem cells. Blood 82: 445-55 (1993); **Brandt J et al** Role of c-*kit* ligand in the expansion of human hematopoietic progenitor cells. Blood 79: 634-41 (1992); **Broudy VC et al** Blasts from patients with acute myelogenous leukemia express functional receptors for stem cell factor. Blood 80: 60-7 (1992); **Cairo MS et al** The *in vitro* effects of stem cell factor and PIXY321 on myeloid progenitor formation (CFU-GM) from immunomagnetic separated CD34$^+$ cord blood. Pediatr. Res. 32: 277-81 (1992); **Carlesso N et al** Human recombinant stem cell factor stimulates *in vitro* proliferation of acute myeloid leukemia cells and expands the clonogenic cell pool. Leukemia 6: 642-8 (1992); **Carow CE et al** Mast cell growth factor (c-*kit* ligand) supports the growth of human multipotential progenitor cells with a high replating potential. Blood 78: 2216-21 (1991); **Chow FP et al** Effects of *in vivo* administration of stem cell factor on thrombopoiesis in normal and immunodeficient mice. EH 21: 1255-62 (1993); **Dai CH et al** Human burst-forming units-erythroid need direct interaction with stem cell factor for further development. Blood 78: 2493-7 (1991); **de Vos S et al** Transforming growth factor-β 1 interferes with the proliferation-inducing activity of stem cell factor in myelogenous leukemia blasts through functional down-regulation of the c-*kit* proto-oncogene product. CR 53: 3638-42 (1993); **Funk PE et al** Activity of stem cell factor and IL7 in combination on normal bone marrow B lineage cells. JI 150: 748-52 (1993); **Galli SJ et al** Reversible expansion of primate mast cell populations *in vivo* by stem cell factor. JCI 91: 148-52 (1993); **Heyworth CM et al** Stem cell factor directly stimulates the development of enriched granulocyte-macrophage colony-forming cells and promotes the effects of other colony-stimulating factors. Blood 80: 2230-6 (1992); **Hoffman R et al** The *in vitro* and *in vivo* effects of stem cell factor on human hematopoiesis. Stem Cells Dayt. 11: s76-82 (1993); **Hunt P et al** Evidence that stem cell factor is involved in the rebound thrombocytosis that follows 5-fluorouracil treatment. Blood 80: 904-11 (1992); **Irani AM et al** Recombinant human stem cell factor stimulates differentiation of mast cells from dispersed human fetal liver cells. Blood 80: 3009-21 (1992); **Kirshenbaum AS et al** Effect of IL3 and stem cell factor on the appearance of human basophils and mast cells from CD34$^+$ pluripotent progenitor cells. JI 148: 772-7 (1991); **Lemoli RM et al** Interleukin 11 stimulates the proliferation of human hematopoietic CD34$^+$ and CD34$^+$CD33$^-$DR$^-$ cells and synergises with stem cell factor, interleukin-3, and granulocyte-macrophage colony-stimulating factor. EH 21: 1668-72 (1993); **Lowry PA et al** Stem cell factor induction of *in vitro* murine hematopoietic colony formation by "subliminal" cytokine combinations: the role of "anchor factors". Blood 80: 663-9 (1992); **Matsui Y et al** Embryonic expression of a hematopoietic growth factor encoded by the *Sl* locus and a ligand for c-*kit*. N 347: 667-9 (1990); **McNiece IK et al** Transforming growth factor β inhibits the action of stem cell factor on mouse and human hematopoietic progenitors. IJCC 10: 80-6 (1992); **McNiece IK et al** Stem cell factor enhances *in vivo* effects of granulocyte colony stimulating factor for stimulating mobilization of peripheral blood progenitor cells. Stem Cells Dayt. 11: s36-41 (1993); **Meininger CJ et al** The c-*kit* receptor ligand functions as a mast cell chemoattractant. Blood 79: 958-63 (1992); **Mekori YA et al** IL3-dependent murine mast cells undergo apoptosis on removal of IL3. Prevention of apoptosis by c-*kit* ligand. JI 151: 3775-84 (1993); **Metcalf D & Nicola NA** Direct proliferative actions of stem cell factor on murine bone marrow cells *in vitro*: effects of combination with colony-stimulating factors. PNAS 88: 6239-43 (1991); **Migliaccio G et al** Stem cell factor induces proliferation and differentiation of highly enriched murine hemopoietic cells. PNAS 88: 7420-4 (1991); **Migliaccio G et al** Long-term generation of colony-forming cells in liquid culture of CD34$^+$ cord blood cells in the presence of recombinant human stem cell factor. Blood 79: 2620-7 (1992); **Muench MO et al** Interactions among colony-stimulating factors, IL1 β, IL6, and *kit*-ligand in the regulation of primitive murine hematopoietic cells. EH 20: 339-49 (1992); **Neta R et al** Inhibition of c-*kit* ligand/steel factor by antibodies reduces survival of lethally irradiated mice. Blood 81: 324-7 (1993); **Pietsch T et al** Effects of human stem cell factor (c-*kit* ligand) on proliferation of myeloid leukemia cells: heterogeneity in response and synergy with other hematopoietic growth factors. Blood 80: 1199-206 (1992); **Pietsch T et al** Human stem cell factor is a growth factor for myeloid leukemia cells. Recent Results Cancer Res. 131: 329-38 (1993); **Toksoz D et al** Support of human hematopoiesis in long-term bone marrow cultures by murine stromal cells selectively expressing the membrane-bound and secreted forms of the human homologue of the steel gene product, stem cell factor. PNAS 89: 7350-4 (1992); **Turner AM et al** Nonhematopoietic tumor cell lines express stem cell factor and display c-*kit* receptors. Blood 80: 374-81 (1992); **Valent P et al** Induction of differentiation of human mast cells from bone marrow and peripheral blood mononuclear cells by recombinant human stem cell factor/*kit*-ligand in long-term culture. Blood 80: 2237-45 (1992); **Zsebo KM et al** Radioprotection of mice by recombinant rat stem cell factor. PNAS 89: 9464-8 (1992)

● **Transgenic/Knock-out/Antisense Studies: Iwamoto T et al** The ret oncogene can induce melanogenesis and melanocyte development in Wv/Wv mice. ECR 200: 410-5 (1992); **Moll J et al** Thymic hyperplasia in transgenic mice caused by immortal epithelial cells expressing c-*kit* ligand. EJI 22: 1587-94 (1992); **Ray P et al** Ectopic expression of a c-kitW42 minigene in transgenic mice: recapitulation of W phenotypes and evidence for c-*kit* function in melanoblast progenitors. Genes Dev. 5: 2265-73 (1991)

● **Assays: Langley KE et al** Soluble stem cell factor in human serum. Blood 81: 656-60 (1993)

● **Clinical Use & Significance: Alter BP et al** Effect of stem cell factor on *in vitro* erythropoiesis in patients with bone marrow failure syndromes. Blood 80: 3000-8 (1992); **Andrews RG et al** Recombinant human stem cell factor, a c-*kit* ligand, stimulates hematopoiesis in primates. Blood 78: 1975-80 (1991); **Andrews RG et al** The ligand for c-*kit*, stem cell factor, stimulates the circulation of cells that engraft lethally irradiated baboons. Blood 80: 2715-20 (1992); **Backx B et al** *kit* ligand improves *in vitro* erythropoiesis in myelodysplastic syndrome. Blood 80: 1213-7 (1992); **Bagnara GP et al** Effect of stem cell factor on colony growth from acquired and constitutional (Fanconi) aplastic anemia. Blood 80: 382-7 (1992); **Bernstein ID et al** Recombinant human stem cell factor enhances the formation of colonies by CD34$^+$ and CD34$^+$lin$^-$ cells, and the generation of colony-forming cell progeny from CD34$^+$lin$^-$ cells cultured with interleukin-3, granulocyte colony-stimulating factor, or granulocyte-macrophage colony-stimulating factor. Blood 77: 2316-21 (1991); **Brugger W et al** *Ex vivo* expansion of enriched peripheral blood CD34$^+$ progenitor cells by stem cell factor, interleukin-1 β (IL1 β), IL6, IL3, interferon-γ, and erythropoietin. Blood 81: 2579-84 (1993); **Du XX et al** Comparative effects of *in vivo* treatment using interleukin-11 and stem cell factor on reconstitution in mice after bone marrow transplantation. Blood 82: 1016-22 (1993); **Firkin F et al** Expansion of hemopoietic activity in long-term culture of human bone marrow by c-*kit* ligand (stem cell factor). GF 8: 135-40 (1993); **Goselink HM et al** Effect of mast cell growth factor (c-*kit* ligand) on clonogenic leukemic precursor cells. Blood 80: 750-7 (1992); **Hintz-Obertreis P et al** Studies on the efficacy of mast cell growth factor (c-*kit* ligand) *in vitro* as well as *in vivo*. Behring Inst. Mitt. 90: 14-22 (1991); **Lynch DH et al** Pharmacokinetic parameters of recombinant mast cell growth factor (rMGF). Lymphokine Cytokine Res. 11: 233-43 (1993); **McNiece IK et al** Recombinant human stem cell factor synergises with GM-CSF, G-CSF, IL3 and Epo to stimulate human progenitor cells of the myeloid and erythroid lineages. EH 19: 226-31 (1991); **Migliaccio G et al** Effects of recombinant human stem cell factor (SCF) on the growth of human progenitor cells *in vitro*. JCP 148: 503-9 (1991); **Sekhsaria S & Malech HL** Recombinant human stem cell factor enhances myeloid colony growth from human peripheral blood progenitors. Blood 81: 2125-30 (1993); **Sieff CA et al** The production of steel factor mRNA in Diamond-Blackfan anemia long-term cultures and interactions of steel factor with erythropoietin and interleukin-3. Br. J. Haematol. 82: 640-7 (1992); **Tong J et al** *In vivo* administration of recombinant methionyl human stem cell factor expands the number of human marrow hematopoietic stem cells. Blood 82: 784-91 (1993); **Ulich TR et al** Stem cell factor in combination with granulocyte colony-stimulating factor (CSF) or granulocyte-macrophage CSF synergistically increases granulopoiesis *in vivo*. Blood 78: 1954-62 (1991); **Wodnar-Filipowicz A et al** Stem cell factor stimulates the *in vitro* growth of bone marrow cells from aplastic anemia patients. Blood 79: 3196-202 (1992)

Scid mouse: see: Immunodeficient mice.

Scid-Hu mouse: see: Immunodeficient mice.

SCGF: *stem cell growth factor* see: SCF (stem cell factor).

SCGF: *stem cell growth factor* This factor of 20 kDa is secreted by the human KPB-M15 myeloid cell line but not by a variety of T or myeloid-monocytoid cell lines tested. The factor is also found in human placental conditioned medium. SCGF has erythroid burst-promoting activity (see: BPA) and potentiates the formation of granulocyte-macrophage-colonies from bone marrow cells *in vitro*.

Hiraoka A et al Production of human hematopoietic survival and growth factor by a myeloid leukemia cell line (KPB-M15) and placenta as detected by a monoclonal antibody. CR 47: 5025-30 (1987); **Hiraoka A et al** Monoclonal antibodies against human hematopoietic survival and growth factor. Biomed. Biochim. Acta. 46: 419-27 (1987); **Hiraoka A et al** Further characterization of the biological properties of human hematopoietic survival and growth factor. Exp. Cell Biol. 57: 27-34 (1989)

Schwannoma-derived growth factor: see: SDGF.

SCI: *stem cell inhibition (factor), stem cell inhibitor* This factor is functionally, antigenically, and probably structurally identical with the bone marrow-derived inhibitor of stem cell proliferation described by Lord and Wright (Manchester) and referred to as the » Manchester inhibitor. The factor has been purified by monitoring its activity in the » CFU-A stem cell assay (see also: Hematopoietic stem cells).

SCI is encoded by the human GOS19-1 gene (see: GOS). It is identical with one of the » MIP (macrophage inflammatory protein) proteins (MIP-1α) and » LD78 and belongs to the » chemokine family of cytokines. The inhibitor is a product of bone marrow macrophages, activated lymphocytes, and monocytes. SCI mRNA can be induced in monocytes by bacterial lipopolysaccharide and» IL1, » IL2, » IL6 while IFN-γ decreases the expression of SCI.

SCI is an inhibitor of stem cell proliferation *in vitro* and *in vivo*. It specifically and reversibly inhibits the growth of early hematopoietic progenitor cells (see also: Hematopoiesis). It inhibits the growth of » CFU-S and » BFU-E stimulated by a combination of » IL3 and » Epo. In the presence of » TGF-β SCI is only a weak inhibitor. The factor also exists in a form associated with the extracellular matrix, thus enabling it to establish specific cell-to-cell contacts by interacting with receptors on near-by cells (see: Juxtacrine). SCI is also active on clonogenic epidermal cells and may thus be a more general stem cell inhibitor than was previously believed.

While only a low level of SCI expression is detectable in normal human bone marrow nucleated cells, very significant increases in the levels of SCI transcripts are observed in the same cells from patients with aplastic anemia and myelodysplastic syndrome.

SCI also inhibits the proliferation of epidermal keratinocytes *in vitro*. SCI mRNA is present in epidermal Langerhans cells but not in keratinocytes, suggesting an important growth regulatory function for SCI in keratopoiesis.

Blum S et al Three human homologues of a murine gene encoding an inhibitor of stem cell proliferation. DNA Cell Biol. 9: 589-602 (1990); **Graham GJ et al** Identification and characterization of an inhibitor of hemopoietic stem cell proliferation. N 344: 442-4 (1990); **Graham GJ & Pragnell IB** SCI/MIP-1 α: a potent stem cell inhibitor with potential roles in development. Dev. Biol. 151: 377-81 (1992); **Graham GJ et al** Purification and biochemical characterization of human and murine stem cell inhibitors (SCI). GF 7: 151-60 (1992); **Lord BI et al** Inhibitor of stem cell proliferation in normal bone marrow. Brit. J. Haematol. 34: 441-5 (1976); **Maciejewski JP et al** Expression of stem cell inhibitor (SCI) gene in patients with bone marrow failure. EH 20: 1112-7 (1992); **Maltman J et al** Transforming growth factor β: is it a downregulator of stem cell inhibition by macrophage inflammatory protein 1α? JEM 178: 925-32 (1993); **Parkinson EK et al** Hemopoietic stem cell inhibitor (SCI/MIP-1 α) also inhibits clonogenic epidermal keratinocyte proliferation. J. Invest. Dermatol. 101: 113-7 (1993); **Quesniaux VF et al** Use of 5-fluorouracil to analyze the effect of macrophage inflammatory protein-1 α on long-term reconstituting stem cells *in vivo*. Blood 81: 1497-504 (1993)

SCIF: *secondary cytotoxic T cell inducing factor* This factor is identical with » IL2.

Aarden LA et al Revised nomenclature for antigen-nonspecific T cell proliferation and helper factors. CI 48: 433-36 (1979); **Wagner H & Röllinghoff** MT-T cell interactions during *in vitro* cytotoxic allograft responses. I. Soluble products from activated Ly1+ T cells trigger autonomously antigen-primed Ly23+ T cells to cell proliferation and cytolytic activity. JEM 148: 1523-38 (1978)

SCIF: *suppressor cell-inducing factor* This poorly characterized factor is secreted by blood monocytes from patients with active pulmonary tuberculous. It induces suppressor cells which depress lymphocyte blastogenesis.

A factor with a similar activity has also been observed to be secreted by tumor cells of a patient with a squamous cell carcinoma of the maxilla transplanted into nude mice (see also: Immunodeficient mice). The patient manifested marked leukocytosis and this was also observed in transplanted mice together with splenomegaly. Surgical excision of the tumor resulted in a dramatic reduction of leukocyte count and spleen weight. Coculture of splenocytes from tumor-bearing mice with normal spleen cells inhibited blastogenesis in response to mitogen. The biological activity of the immune suppressive activity is neutralized by antibodies directed against » G-CSF, suggesting that it is G-CSF or a closely related factor.

See also: SIF (suppressor cell induction factor).

Fujiwara H et al Release of a suppressor cell-inducing factor by monocytes from patients with pulmonary tuberculosis. Immunology 72: 194-8 (1991); **Nishimura R** [Studies on the pathophysiology of paraneoplastic syndromes: both cancer cells and host immune cells are responsible for the pathophysiology of leukocytosis associated with oral cancer] Osaka Daigaku Shigaku Zasshi. 35: 147-79 (1990)

SCL (1-24): Several human stromal cell lines (see also: Hematopoiesis) have been described under this name. They have been obtained from Whitlock-Witte-type long-term bone marrow cultures (see: LTBMC) of fetal and adult bone marrow after immortalisation with SV40. The cells have been shown to express » IL6. Some lines also express » » G-CSF and » GM-CSF, » PDGF mRNA, » M-CSF, » SCF. The cells do not produce » IL2, » IL7, or » TNF. The cells do not express GM-CSF or G-CSF mRNA. The cells maintain myelopoiesis for up to seven weeks in cultures initiated with » CD34-positive cells.

Cicuttini FM et al Support of human cord blood progenitor cells on human stromal cell lines transformed by SV40 large T antigen under the influence of an inducible (metallothionein) promoter. Blood 80: 102-12 (1992)

SCM: abbrev. for serum-containing medium. See also: SFM (serum-free medium).

SCPF: *stem cell proliferation factor* This poorly characterized factor of 32 kDa (secreted form) and 37 kDa (membrane-bound form) is isolated from a human mixed germ cell tumor. Purified SCPF induces the appearance of blast-like cells with a high replating efficiency in clonogenic assays These cells express the cell surface marker » CD34. SCPF either alone or in combination with » IL3 maintains primitive » hematopoietic stem cells in short-term and long-term cultures (see also: LTBMC). For another "stem cell factor" see: SCF.

Lawman MJP et al Stem cell proliferation factor (SCPF): a novel cytokine capable of proliferating CD34+ cells in long-term culture. EH p1022, abstract 48 (1993)

SCSGF: *Sertoli cell secreted growth factor* This poorly characterized factor of < 9.5 kDa is found in conditioned medium from Sertoli cells. Addition of follicle-stimulating hormone, testosterone, retinol, and insulin to the Sertoli cells increased the secretion of the mitogenic activity.

The factor is a potent mitogen that can markedly stimulate the proliferation of several different cell lines of endoderm or mesoderm origin in the presence or absence of serum. The mitogen blocks epidermal growth factor (EGF) binding to its receptor but appears to be distinct from » EGF and TGF-α. For other Sertoli cell-derived factors see also: SCF (Sertoli cell factor).

Buch JP et al Partial characterization of a unique growth factor secreted by human Sertoli cells. Fertil. Steril. 49: 658-65 (1988); **Holmes SD et al** Rat Sertoli cells secrete a growth factor that blocks epidermal growth factor (EGF) binding to its receptor. JBC 261: 4076-80 (1986); **Lamb DJ et al** Partial characterization of a unique mitogenic activity secreted by rat Sertoli cells. Mol. Cell. Endocrinol. 79: 1-12 (1991)

SCY A1: *small inducible cytokine A1* The SCY A1 gene encodes the human inflammatory cytokine » I-309.
see also references for » SCY family of cytokines.

SCY A2: *small inducible cytokine A2* This gene is identical with the » JE gene encoding » MCP-1 (Monocyte chemotactic protein 1). It belongs to a family of cytokines which can be grouped into two subfamilies based on structure and chromosomal location, namely 17q and 4q (see: Chemokines).
Rollins BJ et al Assignment of the human small inducible cytokine A2 gene, SCYA2 (encoding JE or MCP-1) to 17q11.2-12 – evolutionary relatedness of cytokines clustered at the same locus. Genomics 10: 489-92 (1991); **Mehrabian M et al** Localization of monocyte chemotactic protein-1 gene (SCYA2) to human chromosome 17q11.2-q21.1. Genomics 9: 200-3 (1991); see also references for » SCY family of cytokines.

SCY A3: *small inducible cytokine A3* This factor is identical with MIP-1α (macrophage inflammatory protein). See: MIP.
see also references for » SCY family of cytokines.

SCY A4: *small inducible cytokine A4* The small inducible cytokine A4 was originally identified as an activation cDNA in differential screening of a cDNA library prepared from mRNA from stimulated peripheral blood lymphocytes. It was referred to as » Act-2. This factor is identical with MIP-1β (macrophage inflammatory protein). See: MIP.
see also references for » SCY family of cytokines.

SCY A5: *small inducible cytokine A5* The SCY A5 gene encodes the cytokine now known as » RANTES.
see also references for » SCY family of cytokines.

SCY family of cytokines: *small cytokine* This family of structurally related, activation-inducible cytokines is now more commonly referred to as the » chemokine superfamily. See also: SCY A1, A2, A3, A4, A5.
Sherry B & Cerami A Small cytokine superfamily. Curr. Opin. Immunol. 3: 56-60 (1991); **Widmer U et al** Genomic structure of murine macrophage inflammatory protein-1 α and conservation of potential regulatory sequences with a human homologue, LD78. JI 146: 4031-40 (1991)

SDGF: *schwannoma-derived growth factor* This factor of 17.7 kDa is isolated from serum-free culture supernatants of a human Schwannoma cell line (JS1) derived from the sciatic nerve. Its amino acid sequence reveals that SDGF is a member of the epidermal growth factor family of proteins (see also: EGF). It displays 76 % homology with » AR (Amphiregulin).
SDGF promotes the growth of Schwann cells, astrocytes, and fibroblasts. It induces the differentiation of rat pheochromocytoma cells (see: PC12) and also supports the survival of these cells in serum-free media (see also: SFM). SDGF must be transported into the nucleus to exert its mitogenic activity and probably interacts with a nuclear receptor (see: NGFI, nerve growth factor inducible). SDGF has been found to act as an » autocrine growth factor for the androgen-dependent SC2G cell line derived from Shinogi mouse mammary carcinoma SC115.
Kimura H et al Structure, expression, and function of a schwannoma-derived growth factor. N 348: 257-60 (1990); **Kimura H & Schubert D** Schwannoma-derived growth factor promotes the neuronal differentiation and survival of PC12 cells. JCB 116: 777-83 (1992); **Kimura H et al** Schwannoma-derived growth factor must be transported into the nucleus to exert its mitogenic activity. PNAS 90: 2165-9 (1993); **Sonoda H et al** Androgen-responsive expression and mitogenic activity of schwannoma-derived growth factor on an androgen-dependent Shionogi mouse mammary carcinoma cell line. BBRC 185: 103-9 (1992)

SDGF: *smooth (muscle cell)-derived growth factor* This poorly characterized factor is produced by the smooth muscle cells of arterial walls and stimulates their proliferation. The factor is not identical with » PDGF.
Morisaki N et al Effects of smooth muscle cell derived growth factor (SDGF) in combination with other growth factors on smooth muscle cells. Arteriosclerosis. 78: 61-7 (1989); **Morisaki N et al** Secretion of a new growth factor, smooth muscle cell derived growth factor, distinct from platelet-derived growth factor by cultured rabbit aortic smooth muscle cells. FL 230: 186-90 (1988)

SDGF: *spleen-derived growth factor* This poorly characterized glycoprotein of 8-12 kDa is isolated from spleen extracts. The factor specifically promotes the growth of normal fibroblasts. This proliferation-enhancing effect is inhibited by » Suramin. The factor also promotes the proliferation of hepatocytes in primary culture in a hormonally defined medium containing insulin, » EGFr, glucagon, » growth hormone and » prolactin. SDGF may be a mixture of fragments of proteins derived from the extracellular matrix (see also: ECM).
Dittrich W et al Biological properties and partial purification of a growth factor from porcine spleen. ECR 188: 172-4 (1990); **Suzuki T et al** A novel growth factor in rat spleen which promotes proliferation of hepatocytes in primary culture. BBRC 153: 1123-8 (1988)

SDGF-3: *spleen-derived growth factor 3* This factor has been isolated as a heparin-binding mitogen for rat hepatocytes from bovine spleen. SDGF-3 is also mitogenic for mouse epidermal keratinocytes but not for mouse fibroblasts. The mitogenic activity of SDGF-3 is abolished by a neutralizing monoclonal antibody specific for » KGF (keratinocyte growth factor) and it is believed that this bovine spleen-derived hepatocyte mitogen is the bovine homologue of KGF.

Suzuki M et al Spleen-derived growth factor, SDGF-3, is identified as keratinocyte growth factor (KGF). FL 328: 17-20 (1993)

SDIP: *spleen-derived immunosuppressive peptide* see: Thymic hormones.

SDKP: Acetyl-N-Ser-Asp-Lys-Pro. See: AcSDKP.

SDMF: *Smooth muscle cell-derived migration factor* This poorly characterized factor of 58 kDa is isolated from culture supernatants of rat and rabbit aorta. The factor stimulates the chemotactic migration of smooth muscle cells but is not mitogenic for these cells. SDMF does not enhance the migration of endothelial cells from either human umbilical cord vein or rabbit retinal tissue. The factor may play a role in the migration of medial smooth muscle cells into the intima during intimal thickening in atherosclerotic tissues. The factor is not identical with » PDGF, fibronectin, and other known » motogenic cytokines.

Koyama N et al Secretion of a potent new migration factor for smooth muscle cells (SMC) by cultured SMC. Atherosclerosis 86: 219-26 (1991); **Koyama N et al** Purification and characterization of an autocrine migration factor for vascular smooth muscle cells (SMC), SMC-derived migration factor. JBC 268: 13301-8 (1993)

Secondary cytotoxic T cell-inducing factor: see: SCIF.

Secretory proteins: see: Signal sequence.

Self-acting growth factors: see: Autocrine growth control.

Seminal plasma factor: see: SP factor.

Seminiferous growth factor: see: SGF.

Senescense: see: Cell culture.

Septic Shock: abbrev. ***SSS*** (septic shock syndrome). A severe life-threatening and frequently lethal hemodynamic break-down observed after Gram-negative septicemia and mainly caused by bacterial » endotoxins.

Management of the shock-specific symptoms is still one of the most challenging problems faced by microbiologists and clinicians. These symptoms are characterized by hypotension, insufficient tissue perfusion, uncontrollable bleeding, and multisystem organ failure caused mainly by hypoxia, tissue acidosis, and severe local alterations of metabolism. The development of a septicemia is frequently recognized only in a relatively late stage by the drop in blood pressure.

The massive deterioration of homeostasis, also known as disseminated intravasal coagulation involves blood vessels, platelets, blood coagulation and fibrinolytic processes, the presence or absence of inhibitors, the kallikrein-kininogen system, and complement.

At the cellular level the shock syndrome is elicited by endogenous mediators. Although the list of shock mediators currently comprises more than 150 candidates a careful analysis reveals that only a few are causally associated with shock symptoms, including histamine, complement C5a, β-endorphin, thromboxane B2, platelet activating factor, and oxygen free radicals. Plasma levels of » alpha-2-macroglobulin, an inhibitor of different proteinases, have been described to be reduced in patients with sepsis and to be associated with fatal outcome in some studies. The major proinflammatory cytokines involved in septic shock are » IL1, » IL6 and » TNF-α which are released by macrophages (see also: Cell activation) following their activation by bacterial » endotoxins.

The so-called toxic shock syndrome (abbrev. TSS) observed mainly in younger women is caused by tampons contaminated with *Staphylococcus aureus*. These bacteria produce an exotoxin of 23.1 kDa, ***TSST-1*** (toxic shock syndrome toxin), which induces the synthesis of » IL1 and » TNF. IL1 causes tachycardia and hypotension. It synergises with TNF the activity of which is also potentiated by IFN-γ. TNF mainly acts on endothelial cells and increases their procoagulatory activity. Activated endothelial cells also express a number of adhesion molecules that facilitate the adhesion of leukocytes to the endothelium. The accumulation of inflammatory cells further contributes to the tissue destruction (see also: Inflammation).

Rabbits challenged with a lethal dose of endotoxin

Secretogranin I: see addendum: Granins. **Secretogranin II:** see addendum: Granins. **Secretoneurin:** see addendum: Granins. **Secretory protein I:** see addendum: Granins. **Sem-5:** see addendum: ASH.

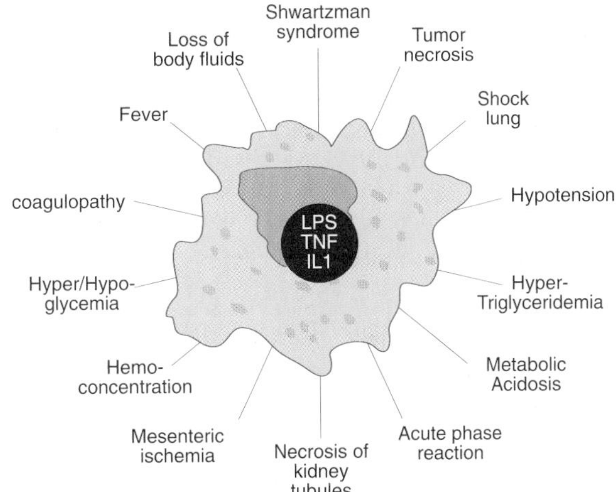

Central role macrophage-derived cytokines released in response to bacterial LPS.
The many different metabolic, physiological, and histopathological alterations observed during septic shock are a mainly a consequence of IL1 and TNF released by activated macrophages.

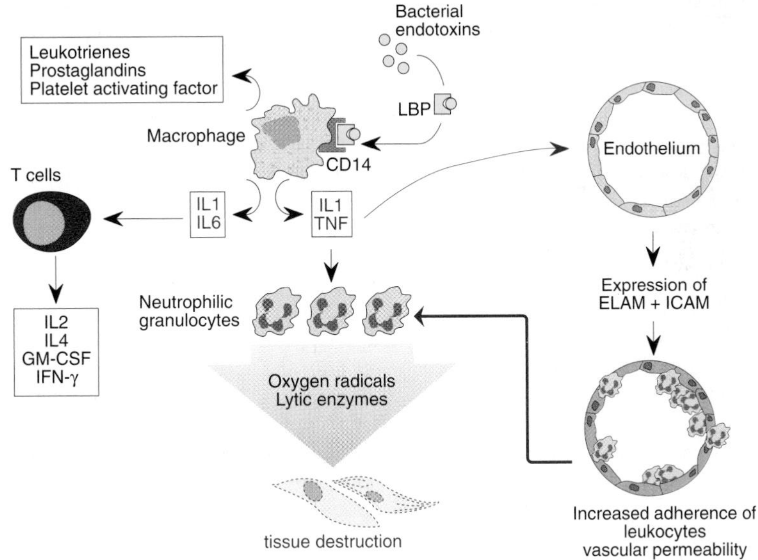

Simplified scheme showing the mechanism of inflammatory responses during septic shock.
Macrophages activated by bacterial endotoxins through binding of lipopolysaccharide binding protein (LBP-Endotoxin complexes to the CD14 receptor express several cytokines including IL1, IL6, IL8, TNF-α. The activation of the arachidonic acid pathway yields leukotrienes and prostaglandins, both of which have a range of physiological effects, including increasing vascular permeability and altering blood flow.
Some of the cytokines produced by activated macrophages activate T cells to produce more cytokines. These mediators trigger neutrophilic granulocytes to produce oxygen radicals and lytic enzymes. Endothelial cells react to these cytokines with elevated expression of ELAMs (endothelial-leukocyte adhesion molecules) and ICAMs (intercellular leukocyte adhesion molecules) which enhances leukocyte adherence and also triggers further production of cytokines. Activated adherent leukocytes show an increased cytotoxicity for invading microorganisms and can also attack body tissues.

produce several milligrams of TNF/kg of body weight which quickly reaches all tissues by the blood circulation. The symptoms observed after administration of pure TNF are almost identical with those observed after an endotoxin shock (see also: Shwartzman phenomenon). The severe effects of an endotoxinemia can be abolished almost completely by administration of antibodies directed against TNF. The importance of TNF in septic shock is illustrated by studies of mice in which one of the TNF receptors has been deleted by homologous recombination (see TNF-α, subentry Transgenic/Knock-out/Antisense studies). TNF receptor-deficient mice are resistant to endotoxic shock although they still succumb to infections with some other pathogens.

In humans TNF serum concentrations in excess of 1 ng/mL are frequently an indication of the lethal outcome of bacterial sepsis. However, absolute serum concentrations of cytokines involved in the pathophysiology of septic shock are normally no reliable indicators of the severity of the shock state and also do not allow prediction of the clinical outcome.

A markedly enhanced activity of TNF is observed in hepatectomised mice and also in mice treated with hepatotoxins such as galactosamin. This may be due to a block of an endogenous feedback mechanism controlling toxic concentrations of TNF since the liver is the major site of IL6 synthesis which in turn is responsible for the induction of the synthesis of » acute phase proteins (see: acute phase reaction).

Monoclonal antibodies directed against bacterial » endotoxins inactivate the bacterial toxin and therefore act the beginning of the causal chain of events. Such antibodies are of prophylactic value only and cannot be used to treat acute cases. Animal experiments with genetically engineered IL1 receptor antagonists (see: IL1ra) have shown that this substance positively influences blood pressure, initial leukopenia, and later leukocytosis in septic shock. This antagonist can be used also at least 2 hours after administration of endotoxins. At this stage untreated animals already display the first symptoms of shock. Administration of endotoxins at doses that normally lead to the death of 8 out of 10 animals within 48 hours after administration nine out of ten animals survive if the receptor antagonist is given at the same time as the endotoxin.

While untreated animals show severe destruction of lung architecture characterized by edemas of the alveolar walls and massive infiltration of blood cells, animals treated with the receptor antagonist and surviving the administration of endotoxins only display small focal bleeding. They usually recover completely within a week.

Unfortunately the IL1 receptor antagonist has a very short plasma half-life and therefore optimal effects are obtained only at relatively high doses on the order of $\approx$ 100 mg/kg body weight. Work is currently in progress to genetically engineer this factor to obtain variants with higher specific activity and improved biological half-lives. For other drug treatments involving substances that inhibit cytokine synthesis see, for example: Pentoxifyllin.

Ashkenazi A et al Protection against endotoxic shock by a tumor necrosis factor receptor immunoadhesin. PNAS 88: 10535-9 (1991); **Baumgartner JD** Management of septic shock: new approaches. Curr. Clin. Top. Infect. Dis. 12: 165-87 (1992); **Beck G et al** Interleukin 1: a common endogenous mediator of inflammation and the local Shwartzman reaction. JI 136: 3025-31 (1986); **Bernard C & Tedgui A** Cytokine network and the vessel wall. Insights into septic shock pathogenesis. ECN 3: 19-33 (1992); **Beutler BI & Cerami A** Tumor necrosis, cachexia, shock, and inflammation: a common mediator. ARB 57: 505-19 (1988); **Billiau A & Vandekerckhove F** Cytokines and their interactions with other inflammatory mediators in the pathogenesis of sepsis and septic shock. EJCI 21: 559-73 (1991); **Bodmer M et al** Preclinical review of anti-tumor necrosis factor monoclonal antibodies. Crit. Care Med. 21: S441-6 (1993); **Bone RC** The pathogenesis of sepsis. Ann. Intern. Med. 115: 457-69 (1991); **Brandtzaeg P et al** Severe human septic shock involves more than tumor necrosis factor. PCBR 367: 25-42 (1991); **Calandra T & Glauser MP** Cytokines and septic shock. Diagn. Microbiol. Infect. Dis. 13: 377-81 (1990); **Calandra T et al** High circulating levels of interleukin-6 in patients with septic shock: Evolution during sepsis, prognostic value, and interplay with other cytokines, Am. J. Med. 91: 23-9 (1991); **Calandra T et al** Anti-lipopolysaccharide and anti-tumor necrosis factor/cachectin antibodies for the treatment of Gram-negative bacteremia and septic shock. PCBR 367: 141-59 (1991); **Cohen J** Antibodies to tumor necrosis factor in the treatment of severe sepsis: Rationale and early clinical experience. PCBR 367: 187-96 (1991); **Colman RW** The role of plasma proteases in septic shock. NEJM 320: 1207-9 (1989); **de Boer JP et al** Interplay of complement and cytokines in the pathogenesis of septic shock. Immunopharmacology 24: 135-48 (1992); **Dinarello CA** The proinflammatory cytokines interleukin-1 and tumor necrosis factor and treatment of the septic shock syndrome. J. Infect. Dis. 163: 1177-84 (1991); **Dofferhoff ASM et al** Patterns of cytokines, plasma endotoxin, plasminogen activator inhibitor, and acute phase proteins during the treatment of severe sepsis in humans. Crit. Care Med. 20: 185-92 (1992); **Dunn DL** Role of endotoxin and host cytokines in septic shock. Chest 100: 164-8 (Suppl.) (1991); **Endo S et al** Two types of septic shock classified by the plasma levels of cytokines and endotoxin. Circ. Shock 38: 264-74 (1992); **Ertel W et al** The complex pattern of cytokines in sepsis: Association between prostaglandins, cachectin, and interleukins. Ann. Surg. 214: 141-8 (1991); **Faist E et al** (eds) Immune consequences of trauma, shock, and sepsis, Springer, Berlin 1989; **Glauser MP et al** Septic shock: pathogenesis. Lancet 338: 732-6 (1991); **Grau GE et al** Prevention of human TNF-induced cutaneous Shwartzman reaction and acute mortality in mice treated with anti-human TNF monoclonal antibodies. Clin. Exp. Immunol. 84: 411-4 (1991); **Harbour DV et al** Role of leukocyte-derived pro-opiomelanocortin peptides in endotoxic shock. Circ. Shock, 35: 181-91 (1991); **Hesselvik JF et al** Coagulation, fibrinolysis, and kallikrein systems in sepsis: relation to outcome. Crit. Care Med. 17: 724-33 (1989); **Hurley JC** Reappraisal of the role of endotoxin in the sepsis syndrome. The Lancet 341: 1133-1135 (1993); **Jafari HS & McCracken GH Jr** Sepsis and septic shock: a review for clinicians. Pediatr. Infect. Dis. J. 11: 739-48 (1992); **Kunkel SL et al** The role of TNF in diverse pathologic processes. Biotherapy 3: 135-41 (1991); **Meek M et al** The Baltimore Sepsis Scale: measurement of sepsis in patients with burns using a new scoring system. J. Burn Care Rehabil. 12: 564-8 (1991); **Ohlsson K et al** Interleukin-1 receptor antagonist reduces mortality from endotoxin shock. N 348: 550-4 (1990); **Parsonnet J et al** Induction of human interleukin 1 by toxic shock syndrome toxin 1. J. Infect. Dis. 151: 514-22 (1985); **Pinsky MR et al** Serum cytokine levels in human septic shock. Relation to multiple-system organ failure and mortality. Chest 103: 565-75 (1993); **St John RC & Dorinsky PM** Immunologic therapy for ARDS, septic shock, and multiple-organ failure. Chest 103: 932-43 (1993); **Strieter RM et al** Role of tumor necrosis factor-α in disease states and inflammation. Crit. Care Med. 21: S447-63 (1993); **Tracey KJ et al** Anti-cachectin/TNF monoclonal antibodies prevent septic shock during lethal bacteremia. N 330: 662-4 (1987); **Tracey KJ et al** TNF and other cytokines in the metabolism of septic shock and cachexia. Clin. Nutrition 11: 1-11 (1992); **Van Deventer SJH** Immunotherapy of Gram-negative septicemia. PCBR 367: 135-9 (1991); **Wortel CH et al** Therapy of Gram-negative sepsis in man with anti-endotoxin antibodies: A review. PCBR 367: 161-78 (1991); **Young LS** Bacterial endotoxins: Cytokine mediators and new therapies for sepsis. Endotoxins and mediators – An introduction. PCBR 367: 1-7 (1991); **Zentella A et al** The role of cachectin/TNF and other cytokines in sepsis. PCBR 367: 9-24 (1991); for further information see also: acute phase reaction, acute phase proteins,» inflammation.

SER: *suppressor E receptor factor* see: Suppressor factors.

SER: *suppressive E receptor* This factor was identified in the malignant effusions derived from ovarian, lung or head and neck cancer patients. It inhibits T cell-dependent immune responses *in vivo* and *in vitro*, including the inhibition of E-rosetting and may be a negative feed-back regulator of immune response produced by activated macrophages. Amino acid sequences of the two polymeric components of SER are identical with the β and α 2 subunits of normal adult » haptoglobin. Tumor-associated SER is 100-1,000 times more potent an immunosuppressor than normal plasma haptoglobin and is immunochemically analogous to the neonatal form.

Oh SK et al An analogy between fetal haptoglobin and a potent immunosuppressant in cancer. CR 47: 5120-6 (1987); **Oh SK et al** Monoclonal antibody to SER immune suppressor detects polymeric forms of haptoglobin. Hybridoma 8: 449-66 (1989); **Oh SK et al** Interference with immune response at the level of generating effector cells by tumor-associated haptoglobin. JNCI 82: 934-40 (1990); **Oh SK et al** Quantitative differentiation of the haptoglobin-related gene product from haptoglobin in human plasma: a possible test for tumor-associated antigen. Hybridoma 11: 1-12 (1992)

Sertoli cell secreted growth factor: see: SCSGF.

Sertoli cell factor: see: SCF.

Serum amyloid A inducer: see: SAA inducer.

Serum response element: see: SRE.

Serum response factor: see: SRF.

Serum response factor accessory protein: abbrev. SAP-1. see: SRF (serum response factor).

Serum retinol binding protein: abbrev. RBP. See: Purpurin.

Serum spreading factor: abbrev. SF. This factor is identical with » Vitronectin.

Serum thymic factor: now called thymulin. see: Thymic hormones.

Severe combined immunodeficiency mouse: see: Immunodeficient mice.

Sézary T cell activating factor: see: SAF.

Sez627: An » IL2-dependent T cell line derived from a patient with Sézary syndrome (leukemic form of cutaneous T cell lymphoma). Sez 627 respond only to human » IL4 and human IL2 but not to other available lymphokines. In combination with the murine IL2-dependent » CTLL-2 cell line, which responds to both human and murine IL2 but not to human IL4, human IL4 activity can be discriminated from human IL2 activity.

Maeda M et al Application of a human T cell line derived from a Sézary syndrome patient for human interleukin 4 assay. Immunol. Lett. 18: 247-53 (1988)

SF: *scatter factor*

This factor is identical with » HGF (hepatocyte growth factor) and is usually now referred to as HGF or HGF/SF (see: HGF for further information).

Scatter factor is produced by fibroblasts, smooth muscle cells, epithelial cells, keratinocytes and various other cell lines. It was initially isolated and characterized as a factor that influences the motility of cells at picomolar concentrations.

Severe congenital neutropenia: see addendum: G-CSF.

Scatter factor disrupts desmosomal junctions between epithelial cells and induces a motile fibroblast-like phenotype in individual cells. The factor therefore also influences the invasive growth of tumor cells derived from epithelial cells and may also be involved in processes of » wound healing and early embryonic development.
For some cell types including keratinocytes and mammary epithelial cells scatter factor is merely a motility factor. It is also an » autocrine factor that influences the motility of the cells that produce it. It is a potent mitogen for hepatocytes and also a morphogen (see: HGF, hepatocyte growth factor). Scatter factor binds to heparin and this may be important for its activities *in vivo* (see also: HBGF, heparin-binding growth factor. The actions of scatter factor are inhibited by » suramin.
Scatter factor has been shown to be an » Angiogenesis factor *in vivo*. It induces cultured microvascular endothelial cells to accumulate and secrete significantly increased quantities of urokinase, an enzyme associated with development of an invasive endothelial phenotype during angiogenesis. Scatter factor is present surrounding sites of blood vessel formation in psoriatic skin and may act as a » paracrine mediator in pathologic angiogenesis associated with human inflammatory disease.
Adams JC et al Production of scatter factor by ndk, a strain of epithelial cells, and inhibition of scatter factor activity by suramin. JCS 98: 385-94 (1991); **Bhargava MM et al** Purification, characterization and mechanism of action of scatter factor from human placenta. Experientia Suppl. 59: 63-75 (1991); **Coffer A et al** Purification and characterization of biologically active scatter factor from *ras*-transformed NIH-3T3 conditioned medium. BJ 278: 35-41 (1991); **Dowrick PG et al** Scatter factor affects major changes in the cytoskeletal organization of epithelial cells. Cytokine 3: 299-310 (1991); **Furlong RA et al** Comparison of biological and immunochemical properties indicates that scatter factor and hepatocyte growth factor are indistinguishable. JCS 100: 173-7 (1991); **Gherardi E et al** Purification of scatter factor, a fibroblast-derived basic protein that modulates epithelial interactions and movement. PNAS 86: 5844-8 (1989); **Gherardi E et al** Purification and characterization of scatter factor. Experientia Suppl. 59: 53-62 (1991); **Hofmann R et al** Scatter factor is a glycoprotein but glycosylation is not required for its activity. BBA 1120: 343-50 (1992); **Konishi T et al** Scatter factor from human embryonic lung fibroblasts is probably identical to hepatocyte growth factor. BBRC 180: 763-5 (1991); **Naldini L et al** Scatter factor and hepatocyte growth factor are indistinguishable ligands for the *met* receptor. EJ 10: 2867-78 (1991); **Nickoloff BJ et al** Scatter factor induces blood vessel formation *in vivo*. PNAS 90: 1937-41 (1993); **Rosen EM et al** Purified scatter factor stimulates epithelial and vascular endothelial cell migration. PSEBM195: 34-43 (1990); **Stoker M et al** Scatter factor is a fibroblast-derived modulator of epithelial cell mobility. N 327: 239-42 (1987); **Stoker M & Gherardi E** Regulation of cell movement: the motogenic cytokines. BBA 1072: 81-102 (1991); **Stoker M** Effect of scatter factor on motility of epithelial cells and fibroblasts. JCP 139: 565-9 (1989); **Weidner KM et al** Scatter factor: molecular characteristics and effect on the invasiveness of epithelial cells. JCB 1112097-108 (1990); **Weidner KM et al** Evidence for the identity of human scatter factor and human hepatocyte growth factor. PNAS 88: 7001-5 (1991)

SF: as a suffix such as in IgG1-SF this abbrev. may denote a "secretion factor", i. e. a factor that induces the secretion of another factor.

SF: *serum spreading factor* This factor is identical with » Vitronectin.

SF: *sulfation factor* This factor is identical with a factor earlier known as » somatomedin and now known as » IGF.
Daughaday WH et al Somatomedin. Proposed designation for sulphation factor. N 235: 107 (1972); **Salmon WD & Daughaday WH** A hormonally controlled serum factor which stimulates sulfate incorporation by cartilage *in vitro*. J. Lab. Clin. Med. 49: 825-36 (1957)

SF-1: *synergistic factor 1* see: 5637.

SFA: *sulfation factor activity* This factor is identical with a factor earlier known as somatomedin and now known as » IGF.

SF-B: *silencer factor B* This transcription factor is identical with » NF-IL6.

SFCM: *serum-free conditioned medium* See: CM.

SFFV: *spleen focus-forming virus* SFFV (Friend SFFV) is a replication-defective retrovirus that does not contain an » oncogene. It causes acute erythroleukemia in mice. The virus also induces the » Epo-independent proliferation of cells and makes cells independent of Epo (see also: HCD57).
The viral gene product responsible for these effects is an envelope glycoprotein of 55 kDa (gp55) that functions as an analog of Epo in binding to and activating the Epo receptor.
Cell lines that require » IL3 for their growth can be made IL3-independent by introduction of a biologically active Epo receptor and gp55 (see also: BaF3). In some virus-induced erythroleukemic cells rearrangements and amplification of the Epo gene lead to the constitutive synthesis of Epo. Alternatively such cells can also express an activated form of the Epo receptor which is caused by the insertion of viral promoter regions into the 5′ region of the Epo receptor gene.

The expression of gp55 and the activation of the Epo receptor per se are not sufficient for the malignant transformation of the cells. The establishment of a malignant leukemia cell also requires the inactivation of a tumor suppressor gene called p53 (see also: Apoptosis, Oncogenes) and the activation of certain oncogenes of the *ets* family (*fli*-1, *spi*-1).

Aizawa S et al *Env*-derived gp55 gene of Friend spleen focus forming virus specifically induces neoplastic proliferation of erythroid progenitor cells. EJ 9: 2107-16 (1990); **Ben-David Y & Bernstein A** Friend virus-induced erythroleukaemia and the multistage nature of cancer. Cell 66: 831-4 (1991); **Hino M et al** Unregulated expression of the erythropoietin receptor gene caused by insertion of spleen focus-forming virus long terminal repeat in a murine erythroleukaemia cell line. MCB 11: 5527-33 (1991); **Li JP et al** Activation of cell growth by binding of Friend spleen focus-forming virus gp55 glycoprotein to the erythropoietin receptor. N 343: 762-4 (1990); **Ruscetti SK et al** Friend spleen focus-forming virus induces factor independence in an erythropoietin-dependent erythroleukaemia cell line. J. Virol. 63: 1057-62 (1990); **Zon LI et al** The erythropoietin receptor transmembrane region is necessary for activation by the Friend spleen focus-forming virus gp55 glycoprotein. MCB 12: 2949-57 (1992)

SFGF: *Shope fibroma growth factor* This protein of 80 amino acids is encoded by a gene of Shope fibroma virus. The protein is expressed in infected cells at an early stage of viral replication. The factor, like » VGF (vaccinia growth factor) and » MGF (myxoma growth factor) belongs to the EGF family of proteins (see: EGF). SFGF can functionally replace MGF.

SFGF is a major virulence factor in viral infection with MRV (malignant rabbit fibroma virus), a recombinant virus with a myxoma virus background possessing some terminal sequences derived from Shope fibroma virus. SFGF is responsible for at least some of the cellular proliferation observed at tumor sites. Viruses lacking the growth factor gene are attenuated. Animals infected with wild-type MRV uniformly develop a fatal syndrome involving disseminated tumors accompanied by purulent conjunctivitis and rhinitis. SFGF deletion mutants of MRV cause similar initial signs of the MRV disease syndrome. However, ≈ 75% of infected animals completely recover from the viral and secondary bacterial infections and become immune to subsequent MRV challenge.

Chang W et al The genome of Shope Fibroma virus, a tumorigenic poxvirus, contains a growth factor gene with sequence similarities to those encoding epidermal growth factor. MCB 7: 535-40 (1987); **Opgenorth A et al** Deletion of the growth factor gene related to EGF and TGF α reduces virulence of malignant rabbit fibroma virus. Virology 186: 175-91 (1992)

SFM: *serum-free medium* Such media are completely chemically defined culture media that allow higher eukaryotic cells to be grown in the absence of serum (see also: Cell culture). Several such media are commercially available. Cell lines to be grown in such media usually have to be adapted to it to obtain sublines that show optimal growth and can be cultured indefinitely. In many instances long-term growth of cells in such media still pose some problems.

The use of chemically defined media facilitates analysis of factors secreted by cells (see also: CM, conditioned medium) and also analysis of cells in response to growth factors or mitogens. It has been shown that formulation of such media critically depends on the purity of the ingredients since cytokines have very high specific biological activities in the picomolar range; some lots of albumin have been observed to contain trace amounts of cytokines (e. g. » BPA; burst promoting activity) that may interfere with » bioassays.

Barnes D et al (eds) Methods for serum-free culture of neuronal and lymphoid cells. Alan R. Liss, New York, 1984.

SGF: *sarcoma growth factors* This factor is secreted by mouse cells transformed by Moloney sarcoma virus. Expression of this factor in normal cells leads to a transformed phenotype which is reversible when the factor is removed. The biological activity of this factor can be abolished by antibodies directed against the receptor for » EGF. SGF is a mixture of the transforming growth factors (see: TGF) » TGF-α and » TGF-β which also bind to the EGF receptor.

When it was found that SGF-like activities were also present in extracts of tumor cells other than sarcomas the name was changed into "transforming growth factor" (see: TGF).

Anzano MA et al Sarcoma growth factor from conditioned medium of virally transformed cells is composed of both type α and type β transforming growth factor. PNAS 80: 6264-8 (1983); **Carpenter G et al** Antibodies to the epidermal growth factor receptor block the biological activities of sarcoma growth factor. PNAS 80: 5627-30 (1983); **De Larco JE & Todaro GJ** Growth factors from murine sarcoma virus-transformed cells. PNAS 75: 4001-5 (1978); **De Larco JE et al** Properties of a sarcoma-growth-factor-like peptide from cells transformed by a temperature-sensitive sarcoma virus. JCP 109: 143-52 (1981); **Massague J et al** Affinity labeling of a transforming growth factor receptor that does not interact with epidermal growth factor. PNAS 79: 6822-6 (1982); **Rizzino A et al** Embryonal carcinoma cell growth and differentiation. Production of and response to molecules with transforming growth factor activity. ECR 143: 143-52 (1983); **Roberts AB et al** Transforming growth factors: isolation of polypeptides from virally and chemically transformed cells by ethanol extraction. PNAS 77: 3494-8 (1980)

SFGF transgenics: see addendum: SFGF. **SFL:** see addendum. **SGF:** see addendum.

SGF: *seminiferous growth factor* This poorly characterized factor is isolated from mammalian testes. It enhances the proliferation of murine » 3T3 fibroblasts and is also a mitogen for Leydig cells, Sertoli cells, and capillary endothelial cells. The biological activities of SGF are potentiated by heparin (10 μg/mL).
Antibodies directed against » aFGF, » FGF-5 and » bFGF do not inhibit the biological activity of SGF. Some of the activity attributed to SGF may be due to » IL1.
Bellve AR & Zheng WX Different pleiotropic actions of seminiferous growth factor (SGF), acidic fibroblast growth factor (aFGF), and basic fibroblast growth factor (bFGF). ANY 564: 116-31 (1989); **Braunhut SJ et al** The seminiferous growth factor induces proliferation of TM4 cells in serum-free medium. Biol. Reprod. 42: 639-48 (1990); **Gustafsson K et al** Isolation and partial characterization of an interleukin 1-like factor from rat testis interstitial fluid. J. Reprod. Immunol. 14: 139-50 (1988); **Zheng WX et al** Pleiotropic actions of the seminiferous growth factor on two testicular cell lines: comparison with acidic and basic fibroblast growth factor. GF 3: 73-82 (1990)

SGF: *skeletal growth factor* This factor is obtained from demineralised human bones. It is also produced by osteoblast-like cell lines (MC3T3-E1). The factor induces the synthesis of DNA in cultured bone cells and stimulates the synthesis of collagen in bone organ cultures. The factor is identical with IGF-2 (see: IGF).
Farley JR & Baylink DJ Purification of a skeletal growth factor from human bone. B 21: 3502-7 (1982); **Jennings JC et al** Purification of bovine skeletal growth factor. MiE 146: 281-94 (1987); **Linkhart S et al** Human skeletal growth factor stimulates collagen synthesis and inhibits proliferation in a clonal osteoblast cell line (MC3T3-E1) JCP 128: 307-12 (1986); **Mohan S et al** Bone-derived factors active on bone cells. Calcif. Tissue Int. 36: 139-45 (1984); **Mohan S et al** Primary structure of human skeletal growth factor: homology with human insulin-like growth factor II. BBA 966: 44-55 (1988); **Wergedal JE et al** Skeletal growth factor is produced by human osteoblast-like cells in culture. BBA 889: 163-70 (1986)

SH2 domains: see: *src* homology domains.

SH3 domains: see: *src* homology domains.

Shedding of receptors: see: Receptor shedding.

Shock syndrome: see: Septic shock.

Shope fibroma growth factor: see: SFGF.

Short ear: A mouse mutant in the gene encoding BMP-5 (see: BMP, bone morphogenetic proteins).

SHP: see: HCP (hematopoietic cell phosphatase).

SHPTP1: see: HCP (hematopoietic cell phosphatase).

Shwartzman phenomenon: (Sanarelli-Shwartzman phenomenon) This phenomenon describes a shock syndrome (see also: Septic shock) initially observed in rabbits receiving repeated injections of bacterial » endotoxins. Animals are presensibilised after the first injection of the toxin. A second course leads to local and/or disseminated coagulation, a blocking of peripheral blood vessels by fibrin-rich precipitates, and local inflammation and tissue necrosis in kidney, lung, and heart.
These symptoms resemble those also found in » septic shock and are caused by the endotoxin-induced release of a variety of inflammatory cytokines (see also: Inflammation).
Beck G et al Interleukin 1: a common endogenous mediator of inflammation and the local Shwartzman reaction. JI 136: 3025-31 (1986); **Goto Y et al** The effects of endotoxin on microcirculatory and lymphatic dynamics in the rat (Shwartzman phenomenon) Circulatory Shock 8: 533-42 (1981); **Grau GE et al** Prevention of human TNF-induced cutaneous Shwartzman reaction and acute mortality in mice treated with anti-human TNF monoclonal antibodies. Clin. Exp. Immunol. 84: 411-4 (1991); **Movat HZ et al** Acute inflammation and a Shwartzman-like reaction induced by interleukin 1 and tumor necrosis factor: synergistic action of the cytokines in the induction of inflammation and microvascular injury. Am. J. Pathol. 129: 463-76 (1987)

SICRI: *substance immunologically cross-reactive with insulin* This protein of 150 kDa immunologically cross-reacts with insulin. The protein is found in murine myeloid leukemia cells and human and murine melanoma cells. It functions as an » autocrine growth factor supporting the growth of these cells in soft agar (see also: Colony formation assay). The factor is also a mitogen for a number of human and murine transformed and normal cells. SICRI serum levels mirror the advancing development of the tumors.
Kruslin B et al Growth factors in human tumors. Res. Exp. Med. Berl. 189: 91-9 (1989); **Levanat S & Pavelic K** Substance immunologically cross-reactive with insulin from murine myeloid leukemia: purification and characterization. Biol. Chem. Hoppe-Seyler 371: 249-54 (1990); **Levanat S & Pavelic K** Isolation and purification of a substance immunologically cross-reactive with insulin (SICRI) from tumor tissue. Int. J. Biochem. 21: 509-15 (1989); **Osmak M** Substance immunologically cross-reactive with insulin (SICRI) stimulates cell division. Oncology 46: 54-7 (1989)

SIF: *soluble inhibitory factor* see: JEG-3 cell factor.

SIF: *suppressor cell induction factor* These poorly characterized factors of 43 kDa (SIF-α) and ≈ 6

kDa (SIF-β) are isolated from culture supernatants of the IL3-dependent suppressor cell line M1-A5 and also from supernatants of spleen cells of mice with transplanted M1 tumors. Both factors activate unspecific suppressor cells in unprimed populations of spleen cells that suppress humoral and cell-mediated immune responses.
A closely related or even identical factor is secreted by the chorion carcinoma cell line JEG-3 and can be isolated also from human placenta. SIF from placental and lymphocyte sources are probably identical and both factors inhibit the antigen- and mitogen-stimulated proliferation and antibody-producing responses of mononuclear cells.
See also: SCIF (suppressor cell-inducing factor).
Almawi WY et al Suppressor cell-inducing factor: a new lymphokine secreted by a natural suppressor cell line with natural cytotoxic activity. I. Purification to apparent homogeneity and initial characterization of suppressor cell-inducing factor. JI 141: 2529-35 (1988); **Almawi WY et al** Induction of suppression by a murine nonspecific suppressor inducer cell line (M1-A5). III. Partial purification of the suppressor cell-inducing factors. J. Mol. Cell. Immunol. 3: 157-66 (1987); **Almawi WY et al** Cyclic AMP as the second messenger for prostaglandin E in modulating suppressor cell-activation by natural suppressor/cytotoxic cells. IJI 9: 697-704 (1987); **Almawi WY & Pope BL** Induction of suppression by a murine nonspecific suppressor-inducer cell line (M1-A5). CI 96: 199-209 (1985); **Pope BL et al** Secretion of a suppressor cell inducing factor by an interleukin 3-dependent cell line with natural cytotoxic activity. III. Comparison with other interleukin 3-dependent cell lines. CI 116: 1-11 (1988); **Pope BL** Secretion of a suppressor cell-inducing factor by an interleukin-3-dependent cell line with natural cytotoxic activity. II. Range, potency, and kinetics of suppressive activity. Immunobiology 174: 107-18 (1987); **Pope BL** Secretion of a suppressor cell-inducing factor by an interleukin-3-dependent cell line with natural cytotoxic activity. CI 100: 97-111 (1986); **Wolf RL** Human placental cells that regulate lymphocyte function. Pediatr. Res. 23: 212-8 (1988); **Wolf RL et al** Characterization of an immune suppressor from transformed human trophoblastic JEG-3 cells. CI 78: 356-67 (1983)

SIG: *small inducible gene(s)* General term for a number of genes that constitute the » chemokine family of cytokines. These genes are normally not expressed in normal cells. Their expression is induced after » cell activation by a number of cytokines involved in inflammatory processes (see also: Inflammation) or » wound healing.
Members of the SIG family of factors include » JE, » KC, a variety of proteins found in the platelet α granules, » MGSA (melanoma growth stimulatory activity), » PF4 platelet factor 4), PBP (platelet basic protein; see: Beta-Thromboglobulin), » LDCF (lymphocyte-derived chemotactic factor), » RANTES, » SMC-CF (smooth muscle cell chemotactic factor).
Brown KD et al A family of small inducible proteins secreted by leukocytes are members of a new superfamily that includes leukocyte and fibroblast-derived inflammatory agents, growth factors, and indicators of various activation processes. JI 142: 679-87 (1989)

Sigje: *small inducible gene JE* see: SIG, see: JE.

Signal sequence: Proteins, including most cytokines, that are secreted by cells into the ambient medium are called secretory proteins. Their passage through intracellular membranes and also their export through the outer cellular membrane is mediated by signal sequences or leader sequences of ≈ 15-20 amino acids that are part of the preproteins. This leader sequence is cleaved off to yield the mature protein.
Leader sequences usually start with some polar amino acids while the internal amino acids are usually strongly hydrophobic. The signal sequence is recognized by soluble signal recognition particles (abbrev. SRP). The complex of nascent polypeptide chains and ribosomes is directed to the membrane of the endoplasmic reticulum where it reacts with a docking protein. Further synthesis of the growing polypeptide chain then proceeds at membrane-bound ribosomes. A special signal peptidase is responsible for the cleavage of the signal sequence and the remaining mature protein is then translocated through the membrane.
Signal peptidases have been isolated from many different species. They consist of one or more subunits. These enzymes do not recognize linear sequences of amino acids within the signal sequence but rather certain structural motifs. Inhibitors of signal peptides have not yet been described.
Some cytokines including » TNF, IL1α and IL1β (see: IL1) that are secreted by macrophages possess secretory signal sequences that are much longer than those found in the same factors secreted by lymphocytes (≥ 70 amino acids). This may be an indication of important differences in the mechanisms underlying the intracellular transport of these proteins.
Other cytokines including » aFGF, » bFGF, » CNTF (ciliary neuronotrophic factor) and » PD-ECGF (platelet-derived endothelial cell growth factor) do not possess signal sequences allowing their secretion via the classical endoplasmic reticulum/Golgi pathway. The mechanism by which these factors are released by cells is still unknown. They may be released after cell lysis, by processing of factors in the membrane or via lysosomal

vesicles. If the normal bFGF gene is furnished with a suitable hydrophobic secretory signal sequence the expression of the altered factor leads to malignant cell transformation.

A mechanism of » intracrine growth control has been suggested for some cytokines lacking a typical secretory signal sequence. Some of these factors, for example » bFGF and » AR (amphiregulin), contain positively charged nuclear transport or nuclear targeting sequences (abbrev. ***NTS***), also called ***NLS*** (nuclear localization sequence), which direct the protein into the cell nucleus after receptor-mediated endocytosis of the growth factor. No strict consensus sequences have been found for nuclear targeting sequences. Passage through the nuclear pore complex, a complex 125000 kDa array of proteins, is an energy-dependent process that requires recognition by the NLS by import receptors and interaction with import factors. It has been observed that deletion of the nuclear localization sequence in bFGF results in a factor that is no longer mitogenic while it retains all other activities usually elicited by binding to its receptor, including tyrosine phosphorylation and induction of early gene synthesis (see: ERG, early response genes). The re-introduction of a heterologous NLS also restores nuclear targeting.

Signal sequences play an important role in genetic engineering since their introduction into suitable cloning vectors, secretion vectors, specifically allows heterologous proteins, for example cytokines, to be secreted into the ambient medium which facilitates their subsequent purification.

Andrews DW et al The role of the N region in signal sequence and signal-anchor function. JBC 267: 7761-9 (1992); **Boyd D & Beckwith J** The role of charged amino acids in the localization of secreted and membrane proteins. Cell 62: 1031-3 (1990); **Bugler B et al** Alternative initiation of translation determines cytoplasmic and nuclear localization of basic fibroblast growth factor. MCB 11: 573-7 (1991); **Cao Y et al** Characterization of the nuclear translocation of acidic fibroblast growth factor. JCS 104: 77-87 (1993); **Imamura T et al** Identification of a heparin-binding growth factor-1 nuclear translocation sequence by deletion mutation analysis. JBC 267: 5676-9 (1992); **Imamura T et al** Recovery of mitogenic activity of a growth factor mutant with a nuclear translocation sequence. S 249: 1567-70 (1990); **Migliaccio G et al** The signal sequence receptor, unlike the signal recognition particle receptor, is not essential for protein translocation. JCB 117: 15-25 (1992); **Müller M** Proteolysis in protein import and export: Signal peptide processing in eu- and prokaryotes. Experientia 48: 118-29 (1992); **Newmeyer DD** The nuclear pore complex and nucleocytoplasmic transport. Curr. Opin. Cell Biol. 5: 395-407 (1993); **Quarto N et al** The NH_2-terminal extension of high molecular weight bFGF is a nuclear targeting signal. JCP 147: 311-8 (1991); **Randall LL et al** Export of protein: a biochemical view. Ann. Rev. Microbiol. 41: 507-41 (1987); **Silver PA** How proteins enter the nucleus. Cell 64: 489-97 (1991); **Stewart M** Nuclear pore structure and function. Semin. Cell Biol. 3: 267-77 (1992); **Verner K & Schatz G** Protein translocation across membranes. S 241: 1307-13 (1988); **Von Heijne G** Signal sequences. The limits of variation. J. Mol. Biol. 184: 99-105 (1985); **Von Heijne G** A new method for predicting signal sequence cleavage sites. NAR 14: 4683-90 (1986); **Watson MEE** Compilation of published signal sequences. NAR 12: 5144-64 (1984); **Yamasaki L & Lanford RE** Nuclear transport: a guide to import receptors. Trends Cell Biol. 2: 123-7 (1992)

SIH: *somatotropin release inhibiting hormone* See: Somatostatin.

Silencer factor B: abbrev. SF-B. This transcription factor is identical with » NF-IL6.

Simian BSC-1 cell growth inhibitor: see: BSC-1 cell growth inhibitor.

SIRS: *soluble immune response suppressor* A factor of ≈ 10 kDa that is secreted by mitogen, antigen- and interferon-stimulated cells. SIRS is also found in the serum of patients with steroid-responsive nephrotic syndrome which frequently have suppressed immune responses. Measurement of SIRS levels in these patients, however, is not a sensitive enough assay for predicting the response to treatment.

SIRS requires activation by oxidation and is probably a metalloprotein. The N-terminal amino acid sequence of one molecular form of SIRS (SIRS-α7) displays significant homology with certain neurotoxins (short neurotoxin 1) found in sea snake, adder, and cobra species.

SIRS is detected by its capacity to suppress the induction of plaque forming cells in pokeweed mitogen-stimulated *in vitro* lymphocyte culture in a dose dependent manner. Specificity is determined by the ability of monoclonal anti-SIRS coated beads to specifically absorb the suppressor activity.

SIRS is a nonspecific inhibitor of cell-mediated and humoral immune responses and cellular proliferation *in vivo* and *in vitro*. SIRS acts primarily during the induction of cytotoxic T lymphocytes and not at the effector phase. SIRS antiserum blocks the suppressive activity of Con A- or IFN-activated suppressor cells. SIRS may be a selective inhibitor of IL1 activity with respect to T and B cells, rendering them unresponsive to IL1 activation and/or maturation signals. In mice the intravenous administration of SIRS inhibits leukocytosis normally induced by » IL1. SIRS also acts as an antipyrogen and in rabbits prevents fever normally induced by bacterial lipopolysaccharides much more effec-

Signal transducers and activators of transcription: see addendum: STAT proteins.

tively than acetyl salicylic acid (see also: EP, endogenous pyrogens). For other factors with similar activities see also: suppressor factors.

Aune TM et al Purification and initial characterization of the lymphokine soluble immune response suppressor. JI 131: 2848-52 (1983); **Aune TM et al** Soluble immune response suppressor. MiE 116: 395-402 (1985); **Cheng IK et al** The role of soluble immune response suppressor lymphokine in the prediction of steroid responsiveness in idiopathic nephrotic syndrome. Clin. Nephrol. 32: 168-72 (1989); **Devens BH et al** Antipeptide antibody specific for the N-terminal of soluble immune response suppressor neutralizes concanavalin A and IFN-induced suppressor cell activity in an *in vitro* cytotoxic T lymphocyte response. JI 141: 3148-55 (1988); **Nowowiejski-Wieder I et al** Cell-free translation of the lymphokine soluble immune response suppressor (SIRS) and characterization of its mRNA. JI 132: 556-68 (1984); **Schnaper HW et al** Identification and initial characterization of concanavalin A- and interferon-induced human suppressor factors: evidence for a human equivalent of murine soluble immune response suppressor (SIRS). JI 132: 2429-35 (1984); **Schnaper HW** A regulatory system for soluble immune response suppressor production in steroid-responsive nephrotic syndrome. Kidney Int. 38: 151-9 (1990); **Schnaper HW** Divalent metal requirement for soluble immune response suppressor (SIRS) activity. CI 118: 157-65 (1989); **Webb DR et al** Putative N-terminal sequence of murine soluble immune response suppressor (SIRS): significant homology with short neurotoxin I. Int. Immunol. 2: 765-74 (1990); **Zimecki M et al** Inhibition of interleukin 1 (IL1)-elicited leukocytosis and LPS-induced fever by soluble immune response suppressor (SIRS). Immunopharmacology 19: 39-46 (1990); **Zimecki M et al** Control of interleukin 1 (IL1) activity. I. Inhibition of IL1 activity by soluble immune response suppressor (SIRS) *in vitro*. Arch. Immunol. Ther. Exp. Warsz. 39: 213-26 (1991)

sis: c-*sis* is a cellular » oncogene that is the cellular homologue of the viral v-*sis* oncogene of simian sarcoma virus (SSV) derived from the wooly monkey. The human gene maps to chromosome 22q12.3- 22q13.1. It extends over a region of ≈ 12 kb, including 1.2 kb of v-*sis*-related sequences interrupted by four intervening sequences. The c-*sis* gene closely resembles the gene encoding » PDGF (platelet-derived growth factor, B chain). The *sis* protein is a homologue of PDGF. Anti-PDGF antibodies have a direct effect on acute transformation by SSV in that they inhibit both proliferation and SSV-induced morphological changes in human diploid fibroblasts. This finding was one of the first observations suggesting a direct link between the molecular biology of normal mitogenesis and virus-induced oncogenesis caused by a growth factor.

The expression of the c-*sis* oncogene is regulated, among other things, by » TGF-β and some of the bioactivities of TGF-β are therefore indirectly caused by PDGF/*sis*.

In human meningiomas and astrocytomas the specific receptors for *sis* and PDGF-B are co-expressed while normal astrocytes and normal cells of the meninges do not express the *sis* receptor. *sis* may therefore be involved in the manifestation of central nervous system neoplasms by establishing » autocrine growth control loops. An » autocrine growth stimulation by PDGF-B/*sis* has been implicated in the pathologic proliferation of endothelial cells characteristically found in glioblastomas.

Bartram CR et al Localization of the human c-*sis* oncogene in Ph-1-positive and Ph-1-negative chronic myelocytic leukemia by in situ hybridization. Blood 63: 223-25 (1984); **Bejcek BE et al** Transformation by v-*sis* occurs by an internal autocrine mechanism. S 245: 1496-9 (1989); **Bonthron DT et al** Structure of the murine c-*sis* proto-oncogene (*sis*, PDGFB) encoding the B chain of platelet-derived growth factor. Genomics 10: 287-92 (1991); **Chiu IM et al** Nucleotide sequence analysis identifies the human c-*sis* proto-oncogene as a structural gene for platelet-derived growth factor. Cell 37: 123-9 (1984); **Dalla Favera R et al** Chromosomal localization of the human homologue (c-*sis*) of the Simian sarcoma virus *onc* gene. S 218: 686-8 (1982); **Deuel TF et al** Expression of a platelet-derived growth factor-like protein in simian sarcoma virus transformed cells. S 221: 1348-50 (1983); **Doolittle RF et al** Simian sarcoma virus *onc* gene v-*sis*, is derived from the gene (or genes) encoding a platelet-derived growth factor. S 221: 275-7 (1983); **Espino PC et al** The aminoterminal region of pp60*src* has a modulatory role and contains multiple sites of tyrosine phosphorylation. O 5: 283-93 (1990); **Hermansson M et al** Endothelial cell hyperplasia in human glioblastoma: coexpression of mRNA for platelet-derived growth factor (PDGF) B chain and PDGF receptor suggests autocrine growth stimulation. PNAS 85: 7748-52 (1988); **Huang JS et al** Transforming protein of simian sarcoma virus stimulates autocrine growth of SSV-transformed cells through PDGF cell surface receptors. Cell 39: 79-87 (1984); **Jhanwar SC et al** Localization of the cellular oncogenes ABL, SIS, and FES on human germ-line chromosomes. Cytogenet. Cell. Genet. 38: 73-5 (1984); **Jin HM et al** Identification and characterization of an essential, activating regulatory element of the human *sis*/PDGFB promoter in human megakaryocytes. PNAS 90: 7563-7 (1993); **Johnsson A et al** Antibodies against platelet-derived growth factor inhibit acute transformation by simian sarcoma virus. N 317: 438-40 (1985); **Josephs SF et al** Transforming potential of human c-*sis* nucleotide sequences encoding platelet-derived growth factor. S 225: 636-9 (1984); **Leal F et al** Evidence that the v-*sis* gene product transforms by interaction with the receptor for platelet-derived growth factor. S 230: 327-30 (1985); **Maxwell M et al** Human meningiomas co-express platelet-derived growth factor (PDGF) and PDGF-receptor genes and their protein products. IJC 46: 16-21 (1990); **Maxwell M et al** Coexpression of platelet-derived growth factor (PDGF) and PDGF-receptor genes by primary human astrocytomas may contribute to their development and maintenance. JCI 86: 131-40 (1990); **McClintok JT et al** Detection of c-*sis* proto-oncogene transcripts by direct enzyme-labeled cDNA probes and *in situ* hybridization. In Vitro Cell Dev. Biol. 28A: 102-8 (1992); **Owen AJ et al** Simian sarcoma virus-transformed cells secrete a mitogen identical to platelet-derived growth factor. S 225: 54-6 (1984); **Ratner L et al** Nucleotide sequence of transforming human c-*sis* cDNA clones with homology to platelet-derived growth factor. NAR 13: 5007-18 (1985); **Silver BJ** Platelet-derived growth factor in human malignancy. BioFactors 3: 217-27 (1992); **Sonobe MH et al** Imbalanced expression of cellular nuclear oncogenes caused by v-*sis*/PDGF-2. O 6: 1531-7 (1991); **Swan DC et al** Chromosomal mapping of the simian sarcoma virus *onc* gene analog in human cells. PNAS 79: 4691-5 (1982); **Waterfield MD et al** Platelet-derived growth factor is structurally related to the putative transforming protein p28*sis* of simian sarcoma virus. N 304: 35-9 (1983); **Yeh HJ et al** PDGF A-chain gene is expressed by mammalian neurons during development and in maturity. Cell 64: 209-16 (1991)

SIS: *small inducible secreted* General name given to a number of small secreted proteins with chemotactic (see also: Chemotaxis) and pro-inflammatory activities (see also: Inflammation). SIS-α is identical with MIP-1α (see: MIP, macrophage inflammatory protein). SIS-γ is identical with MIP-1β (see: MIP, macrophage inflammatory protein). SIS-ε is identical with » P500 and » TCA-3. These proteins are members of the » chemokine family of cytokines.
Brown KD et al A family of small inducible proteins secreted by leukocytes are members of a new superfamily that includes leukocyte and fibroblast-derived inflammatory agents, growth factors, and indicators of various activation processes. JI 142: 679-87 (1989); **Schall TJ** Biology of the RANTES/SIS cytokine family. Cytokine 3: 165-83 (1991)

***SIS*-δ:** This protein is identical with » RANTES. See also: *SIS* (small inducible secreted).

***SIS*-ε:** This protein is identical with » P500. See also: *SIS* (small inducible secreted).

SISS-B: *human soluble immune suppressor of B cell responses* see: suppressor factors.

SISS-T: *human soluble immune suppressor of T cell responses* see: suppressor factors.

SK: *substance K* see: Tachykinins.

Skeletal growth factor: see: SGF.

SKW6-Cl4: (SKW6.4) A human B lymphoblastoid EBV-transformed cell line. It is used, among other things, for detecting B cell differentiation factors (see: BCDF), » BIF (B cell inducing factor), or » TRF (T cell replacing factors) in » bioassays by measuring the induction of the synthesis of immunoglobulins (IgM). SKW6-C14 cells do not respond to B cell growth factors (see: BCGF).
The cells appear to secrete » IL1 and differentiate into IgM-secreting cells in response to exogenous IL1. IgM secretion is stimulated by » IL2, » IL4, and » IL6 alone without the need of a prior activation signal (in contrast to normal B cells which require it before secreting immunoglobulin in response to cytokine stimulation). Immunoglobulin secretion can be inhibited by the serine/threonine kinase inhibitor H7 (see: H8), and the tyrosine kinase inhibitor » Genistein, demonstrating the participation of protein phosphorylation steps required for immunoglobulin synthesis to begin. Neither » CsA (cyclosporin A) nor 6-mercaptopurine effectively inhibit IL6-induced IgM production by SKW6.4 cells. SKW6.4, which does not produce IL6, can be induced to IL6 production by phorbol 12-myristate 13-acetate (see also: Phorbol esters), » IL1 (α and β) and » IL4. Proliferation of these cells is inhibited by treatment with antibodies directed against insulin or insulin receptor.
Baier TG et al Influence of antibodies against IGF-I, insulin or their receptors on proliferation of human acute lymphoblastic leukemia cell lines. Leuk. Res. 16: 807-14 (1992); **Goldstein D & Kim A** Immunoglobulin secretion and phosphorylation of common proteins are induced by IL2, IL4, and IL6 in the factor responsive human B cell line, SKW6.4. JI 151: 6701-11 (1993); **Harigai M et al** Rheumatoid adherent synovial cells produce B cell differentiation factor activity neutralizable by antibody to B cell stimulatory factor-2/interleukin 6. J. Rheumatol. 15: 1616-22 (1988); **Hutchins D et al** Production and regulation of interleukin 6 in human B lymphoid cells. EJI 20: 961-8 (1990); **Jandl RC et al** Interleukin-1 stimulation of human B lymphoblast differentiation. Clin. Immunol. Immunopathol. 46: 115-21 (1988); **Ralph P et al** IgM and IgG secretion in human B cell lines regulated by B cell-inducing factors (BIF) and phorbol ester. Immunol. Lett. 7: 17-23 (1983); **Rock F et al** Overexpression and structure-function analysis of a bioengineered IL2/IL6 chimeric lymphokine. Protein Eng. 5: 583-91 (1992); **Saiki O & Ralph P** Clonal differences in response to T cell replacing factors in a human lymphoblast cell line. EJI 13: 31-4 (1983); **Stevens C et al** The effects of immunosuppressive agents on *in vitro* production of human immunoglobulins. Transplantation 51: 1240-4 (1992)

SL: *suppressor lymphokine* This poorly characterized factor of 66 kDa is secreted by the human T leukemia cell line 81-66-45. The factor inhibits phytohemagglutinin-stimulated proliferation of normal peripheral blood T cells. SL is cytostatic and its addition to the peripheral blood lymphocytes does not change the cell number or cell viability.
Abolhassani M et al Purification of a suppressor lymphokine (SL) from a human T cell line. Immunol. Invest. 18: 741-51 (1989)

SLF: *steel factor* This factor is identical with » SCF (stem cell factor).

***SL* locus:** This murine gene locus encodes » SCF (stem cell factor).

SmA: abbrev. for Somatomedin A. see: Somatomedins.

SMAF: *specific macrophage arming factor* This poorly characterized factor of 65-85 kDa is produced by T lymphocytes in the presence of tumor cells. Its synthesis does not depend on the presence of antigen-presenting macrophages. The factor activates peritoneal macrophages and enables them to kill tumor cells. This activity differs in several respects from the generation of unspecific cytotoxic macrophages induced by » MAF (macrophage activating factor). SMAF is closely related to and may be identical with ***MTCF*** (mast cell arming T cell factor) that influences mast cell-dependent hypersensitivity reactions.
One particular MAF activity was described in the earlier literature as a factor that rendered macrophages cytotoxic following treatment with supernatant of a mixed lymphocyte culture between skin graft recipients and donor spleen cells.
De Groot JW et al Differences in the induction of macrophage cytotoxicity by the specific T lymphocyte factor, specific macrophage arming factor (SMAF), and the lymphokine macrophage activating factor (MAF). Immunobiology 179: 131-44 (1989); **De Weger RA et al** Initial immunochemical characterization of specific macrophage arming factor. Cancer Immunol. Immunother. 30: 21-7 (1989); **Dimitriu A** Suppression of macrophage arming by corticosteroids. CI 21: 79-87 (1976); **Dullens HF et al** Production of specific macrophage arming factor precedes cytotoxic T lymphocyte activity *in vivo* during tumor rejection. Cancer Immunol. Immunother. 30: 28-33 (1989); **Vandebriel RJ et al** Two specific T cell factors that initiate immune responses in murine allograft systems. A comparison of biological functions. JI 143: 66-73 (1989)

Small cytokine gene family: abbrev. scy. see: Chemokines.

SmC: abbrev. for Somatomedin C. see: Somatomedins.

SMCF: *smooth muscle cell chemotactic factor* see: SMC-CF.

SMC-CF: *smooth muscle cell chemotactic factor* also abbrev. SMCF. This factor was purified from the serum-free medium conditioned by cultured baboon aortic medial smooth muscle cells as a monomeric protein of 14.5 kDa that was chemotactic for monocytes. This protein is encoded by the human » JE gene.
SMC-CF belongs to the » chemokine family of cytokines. It is identical with » MCP-1 (monocyte chemoattractant protein 1), » MCP (monocyte chemotactic protein), » HC11, » MCAF (monocyte chemotactic and activating factor), » LDCF (lymphocyte-derived chemotactic factor), » GDCF (glioma-derived monocyte chemotactic factor), » TDCF (tumor-derived chemotactic factor).
Cushing SD et al Minimally modified low density lipoprotein induces monocyte chemotactic protein 1 in human endothelial cells and smooth muscle cells. PNAS 87: 5134-8 (1990); **Graves DT et al** Identification of monocyte chemotactic activity produced by malignant cells. S 2435: 1490-3 (1989); **Jiang Y et al** Post-translational modification of a monocyte-specific chemoattractant synthesized by glioma, osteosarcoma, and vascular smooth muscle cells. JBC 265: 18318-21 (1990); **Valente AJ et al** Purification of a monocyte chemotactic factor secreted by nonhuman primate vascular cells in culture. B 27: 4162-8 (1988)

SMC-derived migration factor: see: SDMF.

Smooth muscle cell chemotactic factor: see: SMC-CF.

Smooth muscle cell derived growth factor: see: SDGF.

Smooth muscle cell-derived migration factor: see: SDMF.

SMP: *somatomedin/insulin-like growth factor (IGF)-like polypeptides* Activities purified from the serum-free conditioned medium of cultured rat epithelial-like cells. The major components of SMP (at least five polypeptides), designated HP1-SMP and HP3-SMP, stimulate DNA synthesis in cultured human fibroblasts and sulfation in cultured chick embryonic cartilage. The partial amino acid sequences of HP1- and HP3-SMP identify them as rat IGF-II (see: IGF).
Tanaka H et al Identification of a family of insulin-like growth factor II secreted by cultured rat epithelial-like cell line 18,54-SF: application of a monoclonal antibody. Endocrinology 124: 870-7 (1989)

SN: abbrev. for supernatant. This abbrev. usually refers to conditioned medium (see: CM).

Snell-Bagg mice: see: Dwarf mice.

SOD: *superoxide dismutase* A monomeric protein of 16 kDa (p16) has been isolated from media conditioned by a murine marrow-derived cell line (PB6) and from mouse marrow supernatants. It antagonizes IL3-dependent proliferation of cells in culture and reversibly inhibits DNA synthesis of erythroid progenitor cells (see: BFU-E) *in vitro*. This inhibitor is identical with murine cytosolic Cu, Zn-containing superoxide dismutase (EC1.15.1.1), an enzyme that has been highly conserved through evolution which causes the disproportionation of superoxide and converts superoxide anions to H_2O_2. Superoxide dismutase thus affects the cycling and growth factor responses of hematopoietic cells.

Pluthero FG et al Purification of an inhibitor of erythroid progenitor cell cycling and antagonist to interleukin 3 from mouse marrow cell supernatants and its identification as cytosolic superoxide dismutase. JCB 111: 1217-23 (1990)

Sodium vanadate: see: Vanadate.

Soft agar test: see: Colony formation assay.

Soluble immune response suppressor: see: SIRS. See also: suppressor factors.

Soluble immune suppressor of B cell responses: abbrev. SISS-B. see: suppressor factors.

Soluble immune suppressor of T cell responses: abbrev. SISS-T. see: suppressor factors.

Soluble inhibitory factor: abbrev. SIF. see: JEG-3 cell factor.

Soluble suppressor factor: abbrev. SSF. see: suppressor factors.

Somatic cytokine gene transfer: see: Cytokine gene transfer.

Somatoliberin: see: Growth hormone.

Somatomedins: Collective term for a family of peptides found in human plasma that stimulate cartilage growth by inducing the synthesis of collagen and other proteins. Somatomedin A is a mixture of IGF-1 and IGF-2. Somatomedin C, the major member of this group of proteins, is identical with » IGF-1. The somatomedins are bound to larger carrier proteins in the circulation. Somatomedins are synthesized in mesenchymal cells of multiple organs, especially in the liver and kidneys. The synthesis of somatomedins is stimulated by » growth hormone.

Somatomedins stimulate cartilage growth and mitosis and growth of several extraskeletal cell types. Somatomedins also display insulin-like activity in adipose tissue. Somatomedins are of clinical relevance for the diagnosis of growth disturbances due to pituitary disorders.

It has been suggested that a defect in IGF-I production may be one factor controlling growth in African pygmies, who fail to respond to exogenous hormone in the presence of normal serum levels of growth hormone and of somatomedin. For another IGF-1 defect see: Laron-type dwarfism.

Daughaday WH et al Somatomedin: proposed designation for sulphation factor. N 235: 107 only (1972); **Enberg G et al** The characterization of somatomedin A, isolated by microcomputer-controlled chromatography, reveals an apparent identity to insulin-like growth factor 1. EJB 143: 117-24 (1984); **Klapper DG et al** Sequence of somatomedin-C: confirmation of identity with insulin-like growth factor I. E 112: 2215-7 (1983); **Merimee TJ et al** Dwarfism in the pygmy: an isolated deficiency of insulin-like growth factor I. NEJM 305: 965-968 (1981); **Rimoin DL et al** Peripheral subresponsiveness to human growth hormone in the African pygmies. NEJM 281: 1383-8 (1969); **Rubin JS et al** Isolation and partial sequence analysis of rat basic somatomedin. Endocrinology 110: 734-40 (1982)

Somatostatin: Somatostatin, also known as ***somatotropin release inhibiting hormone***, abbrev. ***SIH***, is a peptide of 14 amino acids found in the hypothalamus and central and peripheral nervous system. ***Angiopeptin*** is a stable analog of somatostatin. Release of somatostatin from the hypothalamus has been shown to be inhibited by » VIP (vasoactive intestinal peptide). Cholinergic differentiation factor » CDF, » CNTF (ciliary neuronotrophic factor), and » NGF have been shown to induce synthesis of Somatostatin. Release of Somatostatin is also enhanced by » IGF-1 and » IL1β. Somatostatin inhibits the secretion of » Growth hormone, » VIP (vasoactive intestinal peptide), » Prolactin, » Thyrotropin, glucagon, insulin, secretin, » Gastrin, and cholecystokinin.

Immune cells have been found to express somatostatin receptors, and Somatostatin has been detected also in platelets, mononuclear leukocytes, mast cells, and polymorphonuclear leukocytes. The immunoenhancing activities of » Growth hormone and » Prolactin are thus inhibited, and somatostatin may be involved in a regulatory circuit of positive and negative signals controlling immune functions (see also: Neuroimmune network).

Somatostatin has been shown to inhibit proliferation of lymphoblastic cell lines and stimulation of human T lymphocytes by phytohemagglutinin at very low concentrations. At nanomolar concentrations Somatostatin also inhibits the proliferation of spleen and Peyer's patches-derived lymphocytes. Somatostatin also inhibits other immune responses, including » IFN-γ secretion, endotoxin-induced leukocytosis, and release of colony stimulating factors (see: CSF). Somatostatin inhibits IgE production by human mononuclear cells or purified B cells in coculture with monocytes or T cells. Somatostatin counteracts » chemotaxis of human monocytes induced by » Growth hormone and inhibits release of superoxides in human monocytes. Somatostatin has been shown to inhibit endothelial adhesiveness for mononuclear cells induced by » IL1. Somatostatin receptors have been identified in a high proportion of human malignant lymphomas.

Blum AM et al Substance P and somatostatin can modulate the amount of IgG2a secreted in response to schistosome egg antigens in murine Schistosomiasis mansoni. JI 151: 6994-7004 (1993); **Eglezos A et al** *In vivo* inhibition of the rat primary antibody response to antigenic stimulation by somatostatin. Immunol. Cell. Biol. 71: 125-9 (1993); **Epelbaum J et al** Vasoactive intestinal peptide inhibits release of somatostatin from hypothalamus *in vitro*. Eur. J. Pharmacol. 58: 493-5 (1979); **Goetzl EJ et al** Endogenous somatostatin-like peptides of rat basophilic leukemic cells. JI 35: 2707-12 (1985); **Kimata H et al** Differential effect of vasoactive intestinal peptide, somatostatin, and substance P on human IgE and IgG subclass production. CI 144: 429-42 (1992); **Kimata H et al** Effect of vasoactive intestinal peptide, somatostatin, and substance P on spontaneous IgE and IgG4 production in atopic patients. JI 150: 4630-40 (1993); **Lygren I et al** Vasoactive intestinal peptide and somatostatin in leukocytes. Scand. J. Clin. Lab. Invest. 44: 347-51 (1984); **Niedermühlbichler M & Wiedermann CJ** Suppression of superoxide release from human monocytes by somatostatin-related peptides. Regul. Pept. 41: 39-47 (1992); **Reubi JC et al** *In vitro* and *in vivo* detection of somatostatin receptors in human malignant lymphomas. ICJ 50: 895-900 (1992); **Stanisz AM et al** Differential effects of vasoactive intestinal peptide, substance P, and somatostatin on immunoglobulin synthesis and proliferation by lymphocytes from Peyer's patches, mesenteric lymph nodes, and spleen. JI 136: 152-6 (1986); **Tang SC et al** Regulation of human T lymphoblast growth by sensory neuropeptides: augmentation of cholecystokinin-induced inhibition of Molt-4 proliferation by somatostatin and vasoactive intestinal peptide *in vitro*. Immunol. Lett. 34: 237-42 (1992); **Wiedermann CJ et al** Stimulation of monocyte chemotaxis by human growth hormone and its deactivation by somatostatin. Blood 82: 954-60 (1993)

Somatotropic hormone: see: Growth hormone.

Somatotropin: see: Growth hormone.

Somatotropin release inhibiting hormone: abbrev. SIH. See: Somatostatin.

Somatotropin releasing factor: abbrev. SRF. See: Growth hormone.

SP: *substance P* (P = powder, the designation originating from early studies using powdered equine brain and intestine extracts) This small undecapeptide (RPKPQQFFGLM-NH_2) is also known as ***Neurokinin 1*** (abbrev. NK1). It is a member of the »Tachykinins.
Substance P is generated from a precursor protein which is processed by the membrane metalloendopeptidase EC 3. 4. 24. 11 which is also knows as ECE (endothelin-converting enzyme; see: ET, endothelins).
SP is found in the brain and the intestines and is also secreted by eosinophils, endothelial cells, and macrophages. In the central nervous system SP functions as a neurotransmitter in sensory neurons of the spinal cord. It is a neurogenic mediator of neurogenic » inflammation and hyperreactivity. Its biological activities are mediated by a specific receptor (NK1 receptor; see: Tachykinins) the expression of which in human astrocytoma cells is reduced by » IL1 and » TNF.
Apart from its classical role as a neurotransmitter SP also possesses activities of » cytokines although it is not classified as a cytokine due to its small size.
SP stimulates the proliferation and/or differentiation of activated lymphocytes and B cell lymphomas *in vivo* and *in vitro*. It induces the synthesis of immunoglobulins. SP is a late-acting B cell differentiation factor (see: BCDF). Its activities are, however, depend on other signals initiating B cell differentiation. SP also causes mast cell degranulation and the release of histamine. It has been shown to cause mast cell-dependent granulocyte infiltration and to influence lymphocyte traffic. Substance P has been suggested to be involved in rheumatoid arthritis by its ability to stimulate macrophage-like synoviocytes in the joints.
SP inhibits the chemotactic migration of neutrophils induced by » IL8 and » fMLP. SP stimulates the synthesis of » IL1, » IL6 and » TNF-α and » IFN-γ in monocytes and cultured lymphocytes and could therefore indirectly influence » acute phase reactions and inflammatory processes (see: Inflammation). SP also stimulates the production of » IL1 by astrocytes.
IL1, » LIF, » CNTF (ciliary neuronotrophic factor), » MANS (membrane-associated neurotransmitter-stimulating molecule), » NGF, » CNDF (cholinergic neuronal differentiation factor) and TGF-β induce the synthesis of SP in ganglion cells *in vitro* and *in vivo*. IL1 and TNF-α also influence the synthesis of SP in chromaffin sympathetic cells (adrenal medulla, paraganglia). SP may function as a maturation factor for neurons in culture.
SP has been shown to have potent stimulatory effects on » hematopoiesis, most probably mediated through the induction of » IL3 and » GM-CSF release from human bone marrow mononuclear cells.

SOS1: see addendum.

IL1 and SP synergise in promoting the proliferation of murine » 3T3 fibroblast cells. It acts as a competence factor for human skin fibroblasts and induces the release of growth-regulatory arachidonic acid metabolites (see also: Cell activation, Cell cycle). SP strongly enhances the expression and release of » IL2 in activated human T cells and murine lymphocytes. Several analogs of SP inhibit the growth of a number of tumor cells including small cell lung carcinoma.

Neutrophils express the cell surface marker CD10, which is a neutral endopeptidase (EC24.11). This peptidase also cleaves SP and produces several peptide fragments that do not bind to the SP receptor but influence the functional activities of the cells.

Bar-Shavit Z et al Enhancement of phagocytosis – a newly found activity of substance P in its N-terminal tetrapeptide sequence. BBRC 94: 1445-51 (1980); **Blum AM et al** Substance P and somatostatin can modulate the amount of IgG2a secreted in response to schistosome egg antigens in murine Schistosomiasis mansoni. JI 151: 6994-7004 (1993); **Calvo CF et al** Substance P enhances IL2 expression in activated human T cells. JI 148: 3498-504 (1992); **Eskay RL et al** Interleukin-1α and tumor necrosis factor-α differentially regulate enkephalin, vasoactive intestinal polypeptide, neurotensin, and substance P biosynthesis in chromaffin cells. E 130: 2252-8 (1992); **Everard MJ et al** *In vitro* effects of substance P analog [D-Arg1, D-Phe5, D-Trp7,9, Leu11] substance P on human tumor and normal cell growth. Br. J. Cancer 65: 388-92 (1992); **Freidin M & Kessler JA** Cytokine regulation of substance P expression in sympathetic neurons. PNAS 88: 3200-3 (1991); **Gether U et al** Stable expression of high affinity NK1 (substance P) and NK2 (neurokinin A) receptors but low affinity NK3 (neurokinin B) receptors in transfected CHO cells. FL 296: 241-4 (1992); **Gilchrist CA et al** Identification of nerve growth factor-responsive sequences within the 5' region of the bovine preprotachykinin gene. DNA Cell Biol. 10: 743-9 (1991); **Harmar AJ et al** Identification and cDNA sequence of δ-preprotachykinin, a fourth splicing variant of the rat substance P precursor. FL 275: 22-4 (1990); **Hershey AD & Krause JE** Molecular characterization of a functional cDNA encoding the rat substance P receptor. S 247: 958-62 (1990); **Jacquin TD et al** Substance P immunoreactivity of rat brain stem neurons in primary culture. J. Neurosci. Res. 31: 131-5 (1992); **Jansco N et al** Direct evidence for neurogenic inflammation and its prevention by denervation and by pretreatment with capsaicin. Br. J. Pharmacol. 31: 138-51 (1967); **Johnson CL & Johnson CG** Tumor necrosis factor and interleukin-1 down-regulate receptors for substance P in human astrocytoma cells. Brain Res. 564: 79-85 (1991); **Kähler CM et al** Substance P: a competence factor from human fibroblast proliferation that induces the release of growth regulatory arachidonic acid metabolites. JCP 156: 579-87 (1993); **Kawaguchi Y et al** Sequence analysis of cloned cDNA for rat substance P precursor: existence of a third substance P precursor. BBRC 139: 1040-6 (1986); **Lembeck F et al** Increase of substance P in primary afferent nerves during chronic pain. Neuropeptides 1: 175-80 (1981); **Lotz M et al** Substance P activation of rheumatoid synoviocytes: neural pathway in pathogenesis of arthritis. S 235: 893-5 (1987); **Mantyh PW** Substance P and the inflammatory and immune response. ANY 632: 263-71 (1991); **Marasco WA et al** Substance P binds to the formylpeptide chemotaxis receptor on the rabbit neutrophil. BBRC 99: 1065-72 (1982); **Martin FC et al** Substance P stimulates IL1 production by astrocytes via intracellular calcium. Brain-Res. 599: 13-8 (1992); **Nawa H et al** Nucleotide sequence of cloned cDNA for two types of bovine brain substance P precursors. N 306: 32-6 (1984); **Nilsson J et al** Stimulation of connective tissue cell growth by substance P and substance K. N 315: 61-4 (1985); **Pascual DW et al** The cytokine-like action of substance P upon B cell differentiation. Reg. Immunol. 4: 100-4 (1992); **Payan GD** Neuropeptides and inflammation: the role of substance P. ARM 40: 341-52 (1989); **Rameshwar P et al** Immunoregulatory effects of neuropeptides. **Rameshwar P et al** Stimulation of interleukin-2 production by substance p. J. Neuroimmunol. 37: 65-74 (1992); Stimulation of IL2 production in murine lymphocytes by substance P and related tachykinins. JI 151: 2484-96 (1993); **Rameshwar P et al** *In vitro* stimulatory effect of substance P on hematopoiesis. Blood 81: 391-8 (1993); **Ruff MR et al** Substance P receptor-mediated chemotaxis of human monocytes. Peptides 2: 107-11 (1985); **Sethi T & Rozengurt E** Multiple neuropeptides stimulate clonal growth of small cell lung cancer: Effects of bradykinin, vasopressin, cholecystokinin, galanin, and neurotensin. CR 51: 3621-3 (1991); **Skidgel RA et al** Metabolism of substance P and bradykinin by human neutrophils. Biochem. Pharmacol. 41: 1335-44 (1991); **Soder O & Hellström PM** Neuropeptide regulation of human thymocyte, guinea pig T lymphocyte and rat B lymphocyte mitogenesis. Int. Arch. Allergy Appl. Immunol. 84: 205-11 (1987); **Stanisz AM et al** Differential effects of vasoactive intestinal peptide, substance P, and somatostatin on immunoglobulin synthesis and proliferation by lymphocytes from Peyer's patches, mesenteric lymph nodes, and spleen. JI 136: 152-6 (1986); **Takeda Y et al** Molecular cloning, structural characterization and functional expression of the human substance P receptor. BBRC 179: 1232-40 (1991); **Wang L et al** Differential processing of substance P and neurokinin A by plasma dipeptidyl(amino)peptidase IV, aminopeptidase M and angiotensin converting enzyme. Peptides 12: 1357-64 (1991); **Wiedermann CJ et al** *In vitro* human polymorphonuclear leukocyte chemokinesis and human monocyte chemotaxis are different activities of aminoterminal and carboxyterminal substance P. Naunyn-Schmiedeberg's Arch. Pharmacol. 340: 665-8 (1986)

SP5: *splenopentin* see: splenin.

sp18: *secreted protein 18 kDa* see: CsA (ciclosporin).

Specific macrophage arming factor: see: SMAF.

SP factor *(human) seminal plasma factor* (prostate carcinoma cell proliferation-inhibiting factor) This factor of 25 kDa is isolated from human seminal plasma. It strongly inhibits DNA synthesis in human metastasizing androgen-dependent prostatic carcinomas. The biological activity of this factor is completely inhibited by antibodies directed against » TGF-β. The factor is probably identical with TGF-β or is a member of the same protein family.

Lokeshwar BL & Block NL Isolation of a prostate carcinoma cell proliferation-inhibiting factor from human seminal plasma and its similarity to transforming growth factor β. CR 52: 5821-5 (1992)

Spleen colony assay: The earliest clonal assay permitting the study of the kinetics of multipotential » hematopoietic stem cells (see: CFU-S).

Spleen-derived growth factor: see: SDGF.

Spleen-derived growth factor 3: see: SDGF-3.

Spleen-derived immunosuppressive peptide: abbrev. SDIP. See: Thymic hormones.

Spleen focus-forming virus: see: SFFV.

Splenin: This protein of 48 amino acids is found in human and bovine spleen and lymph nodes but not other tissues. It was originally described as thymopoietin III. Human splenin differs from human thymopoietin, one of the » thymic hormones, at four amino acid positions. A synthetic pentapeptide corresponding to residues 32-36 (Arg-Lys-Glu-Val-Tyr), called ***splenopentin*** (***SP5***), has the same biological activities as splenin.
Splenin and splenopentin do not affect neuromuscular transmission (like thymopoietin and thymopentin), and they induce both T and B cell precursors.
SP5 stimulates the differentiation of virgin B and T lymphocytes and significantly enhances the reconstitution of immune reactions after immune suppression. In mice SP5 induces bone marrow cell maturation and promotes the reconstitution of the immune response after total body irradiation. SP5 and derivatives enhance the phagocytic capabilities of human granulocytes in a dose-dependent manner.
The human acute lymphoblastoid T cell leukemia cell line MOLT-4 responds to splenin with elevations of intracellular cGMP but not cAMP. In long-term human bone marrow cell cultures (see also: LTBMC) Sp-5 alone has only a marginal activity, but it potentiates the effects of » GM-CSF. Splenin, splenopentin and synthetic derivatives have been used as an immunomodulator of humoral and cellular immunity in the complex treatment of patients with cancer, hay fever, and immunological defects.

Audhya T et al Contrasting biological activities of thymopoietin and splenin, two closely related polypeptide products of thymus and spleen. PNAS 81: 2847-9 (1984); **Audhya T et al** Isolation and complete amino acid sequence of human thymopoietin and splenin. PNAS 84: 3545-9 (1987); **Baker B et al** Selected human T cell lines respond to thymopoietin with intracellular cyclic GMP elevations. Immunopharmacology 16: 115-22 (1988); **Eckert R et al** Splenopentin-induced reconstitution of the immune response after total body irradiation: optimization of treatment regime. Exp. Clin. Endocrinol. 94: 223-5 (1989); **Eckert R et al** Splenopentin (DAc SP-5)-accelerated reconstitution of antibody formation after syngeneic bone marrow transplantation. Exp. Clin. Endocrinol. 94: 219-22 (1989); **Gruner S et al** Stimulation of the recruitment of epidermal Langerhans cells by splenopentin. Arch. Dermatol. Res. 281: 526-9 (1990); **Maciejewski J et al** Cytofluorometric and cytomorphologic analysis of human bone marrow cells derived from stromal cultures stimulated by granulocyte-macrophage colony-stimulating factor, interferon-γ and splenopentin pentapeptide. EJI 20: 1209-13 (1990); **Simon HU et al** Clinical, biochemical and immunological effectiveness of diacetyl-splenopentin (BCH 069) in hay fever. Allergol. Immunopathol. Madr. 18: 155-60 (1990); **Weber HA et al** Splenopentin (DAc-SP-5) accelerates the restoration of myelopoietic and immune systems after sublethal radiation in mice. IJI 12: 761-8 (1990)

Splenopentin: see: splenin.

Spodoptera frugiperda: see: Baculovirus expression system.

S protein: This factor is identical with » Vitronectin.

SRBC receptor: *sheep red blood cell receptor* This cell surface protein is identical with » CD2 (see also: CD antigens).

src* homology domains:** Name given to highly conserved noncatalytical structural sequence domains initially detected in the *src* » oncogene, which encodes a 60 kDa protein tyrosine kinase. The ***SH2 domain (≈ 100 amino acids) has complex regulatory properties and modulates the protein-tyrosine kinase activity present in the carboxy-terminal half of the *src* protein. A single point mutation in the SH2 region of the cellular *src* gene can activate its transforming potential. The ***SH3 domain*** (≈ 50-60 amino acids) is a distinct motif the biological function of which has not been elucidated completely. SH3 domains bind to specific proline-rich protein sequences and are involved in protein-protein interactions. It has been suggested that, together with SH2, SH3 domains may modulate interactions with the cytoskeleton and membrane.
Two classes of SH2-containing proteins are known. A conserved noncatalytic SH2 domain is located immediately N-terminal to the kinase domains of all cytoplasmic protein-tyrosine kinases. Several proteins implicated in the regulation of cellular responses to mitogenic stimuli and in the control of normal and abnormal cell growth and differentiation contain such non-catalytic domains, e. g. phos-

phatidylinositol 3-kinase p85, phospholipase C-γ, GTPase activating protein, and several proteins encoded by » oncogenes. Conserved SH2 and SH3 domains have also been found in interferon-stimulated gene factor 3 (ISGF3; see: IRS, interferon response sequence), the primary transcription factor induced by » IFN-α that becomes tyrosine phosphorylated both *in vitro* and *in vivo* in response to IFN-α.

SH domains have been shown to mediate the protein-protein interaction with activated, i. e. tyrosine phosphorylated, receptor tyrosine protein kinases (see also: Autophosphorylation) and also to confer high-affinity phosphotyrosine-dependent binding to other proteins, thereby mediating the formation of heteromeric signaling protein complexes at or near the plasma membrane. Different SH2 domains also exhibit distinct binding specificities for phosphoserine/phosphothreonine-containing proteins. Although there are numerous phosphorylated sites on activated proteins, each of the SH2-containing proteins bind only a subset of phosphorylated sites with high affinity. The specificity appears to be mediated by the residues immediately carbocyterminal to the phosphotyrosine residues in the peptide target and probably also by the regions of variability in the SH2 domains.

Protein tyrosine kinases are often encoded in the cytoplasmic domains of cytokine receptors and are activated by binding of a cytokine to its receptor. Alternatively, cytokine receptors lacking intrinsic kinase activities are often linked with cytoplasmic protein tyrosine kinases. SH2 domain-containing proteins associate with both types of proteins, once they have become phosphorylated, and thus play a general role in cellular signaling pathways. Such associations have been observed, for example, with » *fms* (the receptor for » M-CSF), » *kit* (the receptor for » SCF), » PDGF, » EGF, » Epo, » IL4, and » IL3 receptors.

● Reviews: **Carpenter G** Receptor tyrosine kinase substrates: *src* homology domains and signal transduction. FJ 6: 3283-9 (1992); **Koch CA et al** SH2 and SH3 domains: elements that control interactions of cytoplasmic signaling proteins. S 252: 668-74 (1991); **Margolis B** Proteins with SH2 domains: transducers in the tyrosine kinase signaling pathway. Cell Growth Differ. 3: 73-80 (1992); **Mayer BJ & Baltimore D** Signaling through SH2 and SH3 domains. Trends Cell Biol. 3: 8-13 (1993); **Montminy M** Trying on a new pair of SH2s S 261: 1694-5 (1993); **Musacchio A et al** SH3 - an abundant protein domain in search of a function. FL 307: 55-61 (1992); **Pawson T** SH2 and SH3 domains. Curr. Opin. Struct. Biol. 2: 432-7 (1992); **Pawson T & Schlessinger J** SH2 and SH3 domains. Curr. Biol. 3: 434-42 (1992); **Pawson T & Gish GD** SH2 and SH3 domains: from structure to function. Cell 71: 359-62 (1992)

● Specific Examples: **Anderson D et al** Binding of SH2 domains of phospholipase C γ 1, GAP, and *src* to activated growth factor receptors. S 250: 979-82 (1990); **Bardelli A et al** Autophosphorylation promotes complex formation of recombinant hepatocyte growth factor receptor with cytoplasmic effectors containing SH2 domains. O 7: 1973-8 (1992); **Bibbins KB et al** Binding of the *src* SH2 domain to phosphopeptides is determined by residues in both the SH2 domain and the phosphopeptides. MCB 13: 7278-87 (1993); **Booker GW et al** Structure of an SH2 domain of the p85 α subunit of phosphatidylinositol-3-OH kinase. N 358: 684-7 (1992); **Cooper JA & Kashishian A** *In vivo* binding properties of SH2 domains from GTPase-activating protein and phosphatidylinositol 3-kinase. MCB 13: 1737-45 (1993); **Eck MJ et al** Recognition of a high-affinity phosphotyrosyl peptide by the *src* homology-2 domain of p56lck. N 362: 87-91 (1993); **Escobedo JA et al** cDNA cloning of a novel 85 kd protein that has SH2 domains and regulates binding of PI3-kinase to the PDGF β-receptor. Cell 65: 75-82 (1991); **Felder S et al** SH2 domains exhibit high-affinity binding to tyrosine-phosphorylated peptides yet also exhibit rapid dissociation and exchange. MCB 13: 1449-55 (1993); **Fu XY** A transcription factor with SH2 and SH3 domains is directly activated by an interferon α-induced cytoplasmic protein tyrosine kinase(s). Cell 70: 323-35 (1992); **Klippel A et al** The C-terminal SH2 domain of p85 accounts for the high affinity and specificity of the binding of phosphatidylinositol 3-kinase to phosphorylated platelet-derived growth factor β receptor. MCB 12: 1451-9 (1992); **Liu X et al** Regulation of c-*src* tyrosine kinase activity by the *src* SH2 domain. O 8: 1119-26 (1993);**Margolis B** The tyrosine phosphorylated carboxyterminus of the EGF receptor is a binding site for GAP and PLC-γ. EJ 9: 4375-80 (1990); **Margolis B et al** High-efficiency expression/cloning of epidermal growth factor-receptor-binding proteins with *src* homology 2 domains. PNAS 89: 8894-8 (192); **Mayer BJ et al** The noncatalytic *src* homology region 2 segment of *abl* tyrosine kinase binds to tyrosine-phosphorylated cellular proteins with high affinity. PNAS 88: 627-31 (1991); **McGlade CJ et al** SH2 domains of the p85 α subunit of phosphatidylinositol 3-kinase regulate binding to growth factor receptors. MCB 12: 991-7 (1992); **Mohammadi M et al** A tyrosine-phosphorylated carboxy-terminal peptide of the fibroblast growth factor receptor (*flg*) is a binding site for the SH2 domain of phospholipase C-γ 1. MCB 11: 5068-78 (191); **Moran MF et al** *src* homology region 2 domains direct protein-protein interactions in signal transduction. PNAS 87: 8622-6 (1990); **Muller AJ et al** En bloc substitution of the *src* homology region 2 domain activates the transforming potential of the c-*abl* protein tyrosine kinase. PNAS 90: 3457-61 (1993); **Myers MG jr et al** IRS-1 activates phosphatidylinositol 3'-kinase by associating with *src* homology 2 domains of p85. PNAS 89: 10350-4 (1992); **O'Brien MC et al** Activation of the proto-oncogene p60c-*src* by point mutations in the SH2 domain. MCB 10: 2855-62 (1990); **Ohmichi M et al** Activation of phosphatidylinositol-3 kinase by nerve growth factor involves indirect coupling of the *trk* proto-oncogene with *src* homology 2 domains. Neuron 9: 769-77 (1992); **Overduin M et al** Three-dimensional solution structure of the *src* homology 2 domain of c-*abl*. Cell 70: 697-704 (1992); **Pendergast AM et al** BCR sequences essential for transformation by the BCR-*abl* oncogene bind to the *abl* SH2 regulatory domain in a non-phosphotyrosine-dependent manner. Cell 66: 161-71 (1991); **Piccione E et al** Phosphatidylinositol 3-kinase p85 SH2 domain specificity defined by direct phosphopeptide/SH2 domain binding. B 32: 3197-202 (1993); **Reedijk M et al** Tyr721 regulates specific binding of the CSF-1 receptor kinase insert to PI 3'-kinase SH2 domains: a model for SH2-mediated receptor-target interactions. EJ 11: 1365-72

(1992); **Ren R et al** Identification of a ten-amino acid proline-rich SH3 binding site. S 259: 1157-61 (1993); **Reynolds PJ et al** Functional analysis of the SH2 and SH3 domains of the *lck* tyrosine protein kinase. O 7: 1949-55 (1992); **Rotin D et al** Presence of SH2 domains of phospholipase C γ 1 enhances substrate phosphorylation by increasing the affinity toward the epidermal growth factor receptor. JBC 267: 9678-83 (1992); **Rotin D et al** SH2 domains prevent tyrosine dephosphorylation of the EGF receptor: identification of Tyr992 as the high-affinity binding site for SH2 domains of phospholipase C γ. EJ 11: 559-67 (1992); **Sadowski HB & Gilman MZ** Cell-free activation of a DNA-binding protein by epidermal growth factor. N 362: 79-83 (1993); **Sierke SL et al** Structural basis of interactions between epidermal growth factor receptor and SH2 domain proteins. BBRC 191: 45-54 (1993); **Veillette A et al** Regulation of the enzymatic function of the lymphocyte-specific tyrosine protein kinase p56lck by the non-catalytic SH2 and SH3 domains. O 7: 971-80 (1992); **Waksman G et al** Binding of a high affinity phosphotyrosyl peptide to the *src* SH2 domain: crystal structure of the complexed and peptide-free forms. Cell 72: 779-90 (1993); **Wood ER et al** Quantitative analysis of SH2 domain binding. Evidence for specificity and competition. JBC 267: 14138-44 (1992); **Yamamoto K et al** Association of phosphorylated insulin-like growth factor-I receptor with the SH2 domains of phosphatidylinositol 3-kinase p85. JBC 267: 11337-43 (1992); **Zhou S et al** SH2 domains recognize specific phosphopeptide sequences. Cell 72: 767-78 (1993); **Zhu G et al** Direct analysis of the binding of *src*-homology 2 domains of phospholipase C to the activated epidermal growth factor receptor. PNAS 89: 9559-63 (1992); **Zhu G et al** Direct analysis of the binding of the *abl src* homology 2 domain to the activated epidermal growth factor receptor. JBC 268: 1775-9 (1993)

SRE: *serum response element* A short sequence found in the promoter regions of a number of cytokine genes and genes the expression of which is stimulated by cytokines and growth factors (see also: Gene expression). This sequence element mediates the inducible expression of these genes by several external and intracellular signaling molecules including serum (and factors contained therein), 12-O-tetradecanoylphorbol-13-acetate (see also: Phorbol esters), cAMP, and membrane-associated tyrosine kinases such as *src*, *fps*, » *raf*, and *ras*.

SRE contains the conserved sequence $CC(A/T)_6GG$, also called ***CArG box***, which forms a core binding site for » SRF (serum response factor) and for other DNA-binding proteins that are either closely related to or distinct from SRF (see: SRF) and that bind to the CArG box or to related sites extending beyond and encompassing the CArg box.

de Belle I et al Identification of a multiprotein complex interacting with the c-*fos* serum response element. MCB 11: 2752-9 (1991); **Greenberg ME et al** Mutation of the c-*fos* gene dyad symmetry element inhibits serum inducibility of transcription *in vivo* and the nuclear regulatory factor binding *in vitro*. MCB 7: 1217-25 (1987); **Treisman R** The serum response element. TIBS 17: 423-6 (1992)

c-*src* tyrosine kinase: see addendum: CSK.

SRF: *serum response factor* A dimeric phosphorylated transcription factor of 67 kDa (see also: Gene expression) that specifically binds to serum response elements (see: SRE) in the promoters of some cytokine genes and other genes the expression of which is inducible by cytokines and growth factors.

SRF modulates the binding of another protein ($p62^{TCF}$ = ternary complex factor) to the serum response element. $p62^{TCF}$ interacts with SRF and also binds to sequences in the vicinity of CArG box (see: SRE). Another accessory protein binding to the response element and interacting with other DNA-binding proteins at this site is ***SAP-1*** (SRF accessory protein). Binding of these proteins to the response element results in the formation of a multiprotein complex. A recently identified »„ zinc finger protein, designated SRE-ZBP also binds to the » SRE.

Attar RM & Gilman MZ Expression cloning of a novel zinc finger protein that binds to the c-fos serum response element. MCB 12: 2432-43 (1992); **Dalton S & Treisman R** Characterization of SAP-1, a protein recruited by serum response factor to the c-*fos* serum response element. Cell 68: 597-612 (1992); **Dalton S et al** Isolation and characterization of SRF accessory proteins. Philos. Trans. R. Soc. Lond. Biol. 340: 325-32 (1993); **Janknecht R et al** Identification of multiple SRF N-terminal phosphorylation sites affecting DNA binding properties. EJ 11: 1045-54 (1992); **Marais R et al** Casein kinase II phosphorylation increases the rate of serum response factor-binding site exchange. EJ 11: 97-105 (1992); **Misra et al** The serum response factor is extensively modified by phosphorylation following its synthesis in serum-stimulated fibroblasts. MCB 11: 4545-54 (1991); **Norman C et al** Isolation and properties of cDNA clones encoding SRF, a transcription factor that binds to the c-*fos* serum response element. Cell 55: 989-1003 (1988); **Pollock R & Treisman R** Human SRF-related proteins: DNA-binding properties and potential regulatory targets. GD 5: 2327-41 (1991); **Prywes R & Roeder RG** Purification of the c-*fos* enhancer-binding protein. MCB 7: 3482-9 (1987); **Sharrocks AD et al** Identification of amino acids essential for DNA binding and dimerisation in p67SRF: implications for a novel DNA-binding motif. MCB 13: 123-32 (1993); **Treisman R** Identification and purification of a polypeptide that binds to the c-*fos* serum response element. EJ 6: 2711-7 (1987); **Treisman R** The serum response element. TIBS 17: 423-6 (1992)

SRF: *somatotropin releasing factor* see: Growth hormone.

SRF accessory protein: abbrev. SAP-1. See: SRF (serum response factor).

SRP: abbrev. for self-renewal potential. see: Hematopoiesis.

SRTX: Sarafotoxin. see: ET (Endothelins).

SSF: *soluble suppressor factor* see: suppressor factors.

SSS: abbrev. for septic shock syndrome. See: Septic Shock.

ST-1: This human stromal cell line (see also: Hematopoiesis) has been obtained from adherent fetal liver cell layers of Dexter-type long-term bone marrow cultures (see: LTBMC). The cells have been established by spontaneous immortalisation without using SV40. The cells have been shown to express » M-CSF, » IL6 mRNA, and undefined factors stimulating early hematopoietic progenitors » BFU-E, » CFU-GM, » CFU-G, and » CFU-M. The cells do not produce » IL3. Upon treatment with » IL1 the cells produce » GM-CSF and » G-CSF mRNA. They support the growth of myeloid progenitor cells from fetal liver for up to four weeks.

Tsai S et al Isolation of a human stromal cell strain secreting hemopoietic growth factors. JCP 127: 137-45 (1986); **Yang-YC et al** Interleukin-1 regulation of hematopoietic growth factor production by human stromal fibroblasts. JCP 134: 292-6 (1988)

ST2: This murine ST2 gene is expressed in growth-stimulated BALB/c-3T3 cells as an » ERG (early response gene) following mitogenic stimuli. Expression is superinduced in the presence of cycloheximide by murine » IFN-β. This gene and a related gene, designated ***ST2L***, encode proteins of unknown function that are similar to the extracellular portions of the » IL1 receptors. The human homologue of murine ST2 has been found to be expressed in lymphocytes and also shows remarkable similarity to the extracellular portions of the IL1 receptors.

Takagi T et al Identification of the product of the murine ST2 gene. BBRC 1178: 194-200 (1993); **Tominaga S et al** Nucleotide sequence of a complementary DNA for human ST2. BBA 1171: 215-8 (1992); **Yanagisawa K et al** Murine ST2 gene is a member of the primary response gene family induced by growth factors. FL 302: 51-3 (1992); **Yanagisawa K et al** Presence of a novel primary response gene ST2L, encoding a product highly similar to the interleukin 1 receptor type 1. FL 318: 83-7 (1993)

ST2L: see: ST2.

Staurosporine: Staurosporine is a microbial alkaloid isolated from *Streptomyces* species. It has antifungal and hypotensive activities and is one of the most potent general inhibitors of protein kinases including protein kinase C, protein kinase A, p60v-*src* protein kinase, and » EGF receptor-encoded protein kinase.

Staurosporine

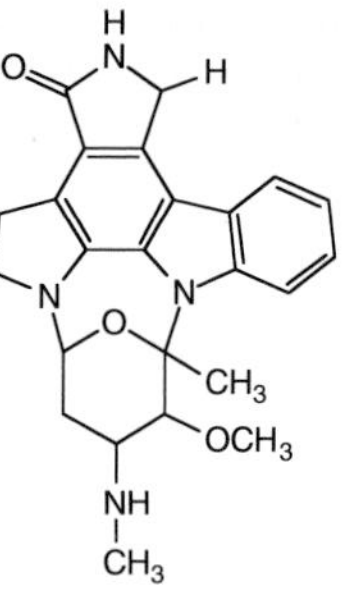

Staurosporine shows a strong cytotoxic effect on the growth of various mammalian cells. It also induces differentiation and affects various functions of platelets and smooth muscle cells. Staurosporine has been shown to inhibit tumor formation from grafted murine papilloma cells by inducing terminal squamous differentiation. It induces the production of » IL8 by human neutrophils and also reverses the negative effects of » IFN-γ on IL8 production. IL2-driven mitogenesis in » IL2-dependent cells is also inhibited by staurosporine. Staurosporine also appears to be a potent inhibitor of angiogenesis (see also: Angiogenesis factors). ***CGP41251*** is a less toxic derivative of staurosporine that is a potent reversible inhibitor of protein kinase C and » PDGF-mediated signal transduction. The staurosporine-related compound ***UCN-01***, isolated from *Streptomyces* sp. No. 126 which also produces Staurosporine, selectively inhibits protein kinase C. UCN-01 possesses antitumor activity *in vivo* against a variety of tumor cells. For another staurosporine-related kinase inhibitor see: K252a.

For other agents used to dissect cytokine-mediated signal transduction pathways see: Bryostatins, Calcium ionophore, Calphostin C, Genistein, H8, Herbimycin A, K-252a, Lavendustin A, Phorbol esters, Okadaic acid, Suramin, Tyrphostins, Vanadate.

● **Reviews: Casnellie JE** Protein kinase inhibitors: probes for the functions of protein phosphorylation. Adv. Pharmacol. 22: 167-205 (1991); **Hidaka H & Kobayashi R** Pharmacology of protein kinase inhibitors. Annu. Rev. Pharmacol. Toxicol. 32: 377-97 (1992); **Tamaoki T** Use and specificity of staurosporine, UCN-01, and calphostin C as protein kinase inhibitors. MiE 201: 340-47 (1991)

Andrejauska-Buchdunger E & Regenass U Differential inhibition of the epidermal growth factor-, platelet-derived growth factor-, and protein kinase C-mediated signal transduction pathways by the staurosporine derivative CGP 41251. CR 52: 5353-8 (1992); **Cassatella MA et al** Studies on the regulatory mechanisms of interleukin-8 gene expression in resting and IFN-γ-treated neutrophils: evidence on the capability of staurosporine of inducing the production of interleukin-8 by human neutrophils. BBRC 190: 660-7 (1993); **Oikawa T et al** Inhibition of

STAT proteins: see addendum.

angiogenesis by staurosporine, a potent protein kinase inhibitor. J. Antibiot. Tokyo 45: 1155-60 (1992); **Strickland JE et al** Inhibition of tumor formation from grafted murine papilloma cells by treatment of grafts with staurosporine, an inducer of squamous differentiation. Carcinogenesis 14: 205-9 (1993); **Zilberman Y & Gutman Y** Multiple effects of staurosporine, a kinase inhibitor, on thymocyte functions. Comparison with the effect of tyrosine kinase inhibitors. Biochem. Pharmacol. 44: 1563-8 (1992)

Steady-state hematopoiesis**:** see: Hematopoiesis.

Steel factor: This factor is identical with » SCF (stem cell factor).

Lyman SD & Williams DE Biological activities and potential therapeutic uses of steel factor. A new growth factor active on multiple hematopoietic lineages. Am. J. Pediatr. Hematol. Oncol. 14: 1-7 (1992)

Stefin A: see: Calgranulins.

Stem cell activating factor: see: SAF.

Stem cell factor: see: SCF.

Stem cell growth factor: see: SCGF.

Stem cell inhibition factor: see: SCI.

Stem cell proliferation factor: see: SCPF.

Stem cells: see: Hematopoietic stem cells.

Stem cell separation: see: Hematopoiesis.

STIF: *stimulated rat T cell-derived inhibitory factor for cellular DNA* This poorly characterized unglycosylated factor of ≈ 45-50 kDa is found in culture supernatants of concanavalin A-stimulated SD rat suppressor T cells. STIF is not identical with » IL2 or IFN-γ. It inhibits the DNA synthesis of mouse bone marrow and leukemia cells. STIF also inhibits the DNA synthesis of a variety of normal and neoplastic cells from rats and humans in a dose-dependent fashion. Inhibition of cellular DNA synthesis induced by STIF is reversible by culturing the cells in STIF-free medium. STIF inhibits proliferative responses of rat lymphocytes to T cell mitogens, Con A and phytohemagglutinin, and a B cell mitogen, lipopolysaccharide. It also inhibits » IL2-dependent growth of cloned T572 cells. STIF also inhibits blastogenesis and cytotoxic T cell generation in allogeneic mixed lymphocyte reaction. The release of IL2 from concanavalin A-stimulated T cells is also inhibited.

Stem cell tyrosine kinase 1: see addendum: STK1.

Chiba K et al Stimulated rat T cell-derived inhibitory factor for cellular DNA synthesis (STIF). I. Isolation and characterization. JI 134: 1019-25 (1985); **Chiba K et al** Stimulated rat T cell-derived inhibitory factor for cellular DNA synthesis (STIF). II. Kinetics of the production and characterization of the producer. JI 134: 1026-31 (1985); **Chiba K et al** Stimulated rat T cell-derived inhibitory factor for cellular DNA synthesis (STIF). III. Effect on cell proliferation and immune responses. JI 134: 3172-8 (1985)

Stimulated rat T cell-derived inhibitory factor for cellular DNA: see: STIF.

Stress axis: see: Neuroimmune network.

Stromal cell lines: see: LTBMC (long-term bone marrow culture).

Subcultures: see: Cell culture.

Substance immunologically cross-reactive with insulin: see: SICRI.

Substance K: This factor is identical with NKA (Neurokinin A). See: Tachykinins.

Substance P: see: SP.

Subtraction cloning: see: Gene libraries.

Subtractive hybridization: see: Gene libraries.

Subtractive libraries: see: Gene libraries.

Sulfation factor: see: SF.

Sup: supernatant. This suffix usually denotes conditioned medium. See: CM.

Supergene families: see: Gene families.

Superoxide dismutase: see: SOD.

Suppressin: abbrev. SPN. This factor has been isolated from bovine pituitary extracts. It is also produced and secreted by primary mouse splenocytes, human peripheral blood lymphocytes, and rat primary pituitary cell cultures. The analysis of rat pituitary cell types capable of producing suppressin (somatotrophs, lactotrophs, corticotrophs, thyrotrophs, and mammosomatotrophs) has revealed that suppressin is mainly produced in the adenohypophysis and that these cell types account for more than 70 % of adenomas in human pituitaries

STK1: see addendum.

Suppressin is a monomeric protein of 63 kDa. It is a powerful inhibitor of cell proliferation in mitogen-stimulated murine splenocytes, neoplastic cells of neuroendocrine or immune origin, human peripheral blood lymphocytes, and primary rat pituitary cells at picomole and nanomole concentrations.

Preincubation of murine leukocytes with suppressin enhances their natural killer cell activity and induces the production of » IFN-α and » IFN-β. Suppressin also potentiates the activities of » IFN-γ in augmenting natural killer activities. *In vivo*, suppressin increases the time of survival of C57BL/6 mice injected with » EL-4 lymphoma cells. Suppressin has also been shown to inhibit production of IgA, IgG, and IgM in response to concanavalin A in a dose-dependent manner.

Ban EM et al Identification of suppressin-producing cells in the rat pituitary. Endocrinology 133: 241-7 (1993); **Carr DJ et al** Immunomodulatory characteristics of a novel antiproliferative protein, suppressin. J. Neuroimmunol. 30: 179-87 (1990); **LeBoef RD et al** Isolation, purification, and partial characterization of suppressin, a novel inhibitor of cell proliferation. JBC 265: 158-65 (1990); **Le Boef RD et al** Cellular effects of suppressin: a biological response modifier of cells of the immune system. Prog. Neuroimmunol. 3: 176-65 (1990); **Le Boef RD & Blalock JE** Biochemical and cellular characteristics of suppressin - a novel pituitary-derived inhibitor of neuroendocrine and immune cell proliferation. ANY 594: 393-5 (1990)

Suppressive B cell factor: see: SBF.

Suppressor cell induction factor: see: SIF; SCIF

Suppressive E receptor: see: SER.

Suppressor E receptor factor: abbrev. SER. See: Suppressor factors.

Suppressor factors: A generic name given to some biochemically poorly characterized soluble factors secreted either 6-48 hours or later (3-4 days) by mitogen- and antigen-stimulated cells. These factors suppress immune responses usually antigen-unspecifically. Some factors also inhibit the proliferation of normal and neoplastic cell types.

Suppressive human lymphokines have been grouped into four categories: primarily stimulatory lymphokines that also mediate certain suppressive activities, suppressive lymphokines produced during altered states of immunity, suppressive lymphokines produced by exogenously stimulated lymphocytes, and suppressive lymphokines produced by unstimulated lymphocytes.

Factors possessing suppressor activity include » IFN-γ and » IFN-α, and a number of poorly characterized factors such as ***HSF*** (histamine-induced suppressor factor) released by histamine-stimulated cells, ***IDS*** (inhibitor of DNA synthesis) released by mitogen-stimulated lymphocytes, ***INH-IL2*** (inhibitor of IL2 activity) released by Concanavalin A-stimulated spleen cells, ***plasmacytoma-derived suppressor factor*** released by plasmacytomas, ***SER*** (suppressor E receptor factor) released by malignant ascites tumors, » ***SIRS*** (soluble immune response suppressor) released by mitogen-stimulated peripheral lymphocytes, ***SISS-B*** (human soluble immune suppressor of B cell responses) and ***SISS-T*** (human soluble immune suppressor of T cell responses) released by mitogen-stimulated peripheral lymphocytes, 57 kDa ***SSF*** (soluble suppressor factor) released by HTLV-III-infected CD4-positive lymphocytes capable of inhibiting T cell-dependent immune reactivity, 67 kDa ***SSF-H*** (papillomavirus-induced soluble suppressor factor) obtained from patients with extensive papillomavirus infections such as condyloma acuminatum (CA) and epidermodysplasia verruciformis (EV), depressing the proliferative responses of T cells to phytohemagglutinin-P (PHA-P), the production of »IL2, and inhibiting the mitogenic activity of IL2 on » CTLL-2 cells.

Aune TM Role and function of antigen nonspecific suppressor factors. CRC Crit. Rev. Immunol. 7: 93-130 (1987); **Beer DJ et al** Cellular interactions in the generation and expression of histamine-induced suppressor activity. CI 69: 101-12 (1982); **Chopra V et al** Purification and partial characterization of a soluble suppressor factor from papillomavirus-infected patients. Viral. Immunol. 4: 249-58 (1991); **Chopra V & Tyring SK** Suppression of interleukin-2 production and activity by factor(s) released by peripheral blood mononuclear cells during papillomavirus infections. Viral. Immunol. 4: 237-48 (1991); **Fairchild RL & Moorhead JW** Soluble factors in tolerance and contact sensitivity to DNFB in mice. X. IL2 is the activation signal mediating release of synthesized suppressor factor. CI 133: 147-60 (1991); **Fleisher TA et al** Soluble suppressor supernatants elaborated by concanavalin A-activated human mononuclear cells. II. Characterization of a soluble suppressor of B cell immunoglobulin production. JI 126: 1192-7 (1981); **Greene WC et al** Soluble suppressor supernatants elaborated by concanavalin A-activated human mononuclear cells. I. Characterization of a soluble suppressor of T cell proliferation. JI 126: 1185-91 (1981); **Halpern MT & Schwartz SA** Modulation of a human immunosuppressive lymphokine by monosaccharides. CI 136: 29-40 (1991); **Halpern MT** Human nonspecific suppressive lymphokines. J. Clin. Immunol. 11: 1-12 (1991); **Honda M et al** Characterization and partial purification of a specific interleukin 2 inhibitor. JI 135: 1834-9 (1985); **Jegasothy BV et al** Regulatory substances produced by lymphocytes. IV. IDS (inhibitor of DNA synthesis) inhibits stimulated lymphocyte proliferation by activation of membrane adenylate cyclase at a restriction point in late G. Immunochemistry

15: 551 (1978); **Laurence J et al** Soluble suppressor factors in patients with acquired immune deficiency syndrome and its prodrome. Elaboration *in vitro* by T lymphocyte-adherent cell interactions. JCI 72: 2072-81 (1983); **Laurence J et al** A soluble inhibitor of T lymphocyte function induced by HIV-1 infection of CD4⁺ T cells: characterization of a cellular protein and its relationship to p15E. CI 128: 337-52 (1990); **Oh SK & Lapenson DP** E receptor-related immunosuppressive factor in malignant pleural fluid and plasma: molecular mechanism of action on DNA polymerase α. JI 135: 355-61 (1985); **Rich RR & Pierce CW** Biological expression of lymphocyte activation. II. Generation of a population of thymus-derived suppressor lymphocytes. JEM 137: 649 (1973); **Ullrich S & Zolla-Pazner S** Immunoregulatory circuits in myeloma. Clin. Haematol. 11: 87-111 (1982); **Wilkins JA et al** The production of immunoregulatory factors by a human macrophage-like cell line. I. Characterization of an inhibitor of lymphocyte DNA synthesis. CI 75: 328-36 (1983)

Suppressor lymphokine: see: SL.

Suramin: A hexasulfonated naphthylurea compound of 1429 Da that was used initially as a drug for the treatment of trypanosomiasis.

Suramin blocks the biological activities of a number of growth factors including interactions with their corresponding receptors and receptor-mediated signal transduction processes. In some cells suramin inhibits processes involving protein kinase C. Phosphatidylinositol kinase and Diacylglycerol kinase are also inhibited by suramin (for other agents used to dissect cytokine-mediated signal transduction pathways see: Bryostatins, Calcium ionophore, Calphostin C, Genistein, H8, Herbimycin A, K-252a, Lavendustin A, Phorbol esters, Okadaic acid, Staurosporine, Tyrphostins, Vanadate).

Suramin appears to inhibit the activities of » IL2 by competition for binding to the IL2 receptor.

At therapeutical concentrations (10^{-4} M) Suramin inhibits the ACTH-stimulated release (see: POMC, proopiomelanocortin) of corticosteroids which in turn influence the synthesis and secretion of many other cytokines. This process is thought to involve the direct interaction between suramin and ACTH.

Suramin also inhibits the » autocrine stimulation of transformed cells expressing an aberrant » bFGF and reverses the transformed phenotype of these cells. Kaposi sarcomas, which produce FGF as an autocrine growth factor have been shown to be able to escape growth inhibition by suramin. This effect is probably due to the ability of suramin to induce increased expression of » bFGF, » FGF5, and FGF receptors.

Suramin inhibits the proliferation of osteosarcoma cells induced by IGF-1 (see: IGF) and blocks the proliferation of human rhabdomyosarcomas which produce IGF-2 (see: IGF) as an » autocrine growth factor.

Suramin binds to » PDGF and thus reduces binding of PDGF to its receptor. This in turn prevents autocrine stimulation of PDGF-dependent cell proliferation.

Suramin also has direct influences on the extracellular matrix (see also: ECM). It modulates the expression of laminin and » thrombospondin and thus alters the cell adhesion properties of tumor cells. It also inhibits bone resorbtion induced by several cytokines.

Suramin also inhibits some enzymes localized in the cell nucleus, including DNA and RNA polymerase, terminal deoxynucleotidyl transferase, reverse transcriptase, and DNA topoisomerase II.

Suramin induces cell differentiation in some cell types. It inhibits angiogenesis (see: angiogenesis factors) and also inhibits the invasive growth of some tumor cells.

Suramin has been used in the treatment of metastasizing kidney tumors, adenocarcinomas, lymphomas, and prostatic carcinomas.

Adams JC et al Production of scatter factor by ndk, a strain of epithelial cells, and inhibition of scatter factor activity by suramin. JCS 98: 385-94 (1991); **Alzani R et al** Suramin induces deoligomerization of human tumor necrosis factor α. JBC 268: 12526-9 (1993); **Betsholtz C et al** Efficient reversion of simian sarcoma virus-transformation and inhibition of growth factor-induced mitogenesis by suramin. PNAS 83: 6440-4 (1986); **Bojanowski K et al** Suramin is an inhibitor of DNA topoisomerase II *in vitro* and in Chinese hamster fibrosarcoma cells.

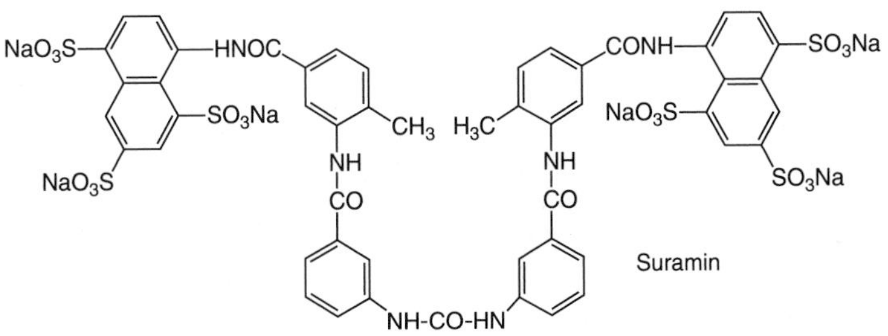

NaO_3S — HNOC — CONH — SO_3Na
SO_3Na CH_3 H_3C NaO_3S
NaO_3S NH NH SO_3Na
CO CO
NH-CO-HN

Suramin

PNAS 89: 3025-9 (1992); **Fantini J et al** Suramin inhibits cell growth and glycolytic activity and triggers differentiation of human colic adenocarcinoma cell clone HT29-D4. JBC 264: 10282-6 (1989); **Garrett JS et al** Blockade of autocrine stimulation in simian sarcoma virus-transformed cells reverses down-regulation of platelet-derived growth factor receptors. PNAS 81: 7466-70 (1984); **Hannink M & Donoghue DJ** Autocrine stimulation by the v-*sis* gene product requires a ligand-receptor interaction at the cell surface. JCB 107: 287-98 (1988); **Hosang M** Suramin binds to platelet-derived growth factor and inhibits its biological activity. JCBc 29: 265-73 (1985); **Huang YQ et al** Increased expression of fibroblast growth factors (FGFs) and their receptor by protamine and suramin on Kaposi's sarcoma-derived cells. Anticancer Res. 13: 887-90 (1993); **La Rocca RV et al** Suramin: prototype of a new generation of antitumor compounds. CC 2: 106-15 (1990); **La Rocca RV et al** Use of suramin in treatment of prostatic carcinoma refractory to conventional hormonal manipulation. Urol. Clin. North Am. 18: 123-9 (1991); **Larsen AK** Suramin: an anticancer drug with unique biological effects. Cancer Chemother. Pharmacol. 32: 96-8 (1993); **Minniti CP et al** Suramin inhibits the growth of human rhabdomyosarcoma by interrupting the insulin-like growth factor II autocrine growth loop. CR 52: 1830-5 (1992); **Myers C et al** Suramin: a novel growth factor antagonist with activity in hormone-refractory metastatic prostate cancer. J. Clin. Oncol. 10: 881-9 (1992); **Scher HI et al** Use of adaptive control with feedback to individualize suramin dosing. CR 52: 64-70 (1992); **Sjolund M & Thyberg J** Suramin inhibits binding and degradation of platelet-derived growth factor in arterial smooth muscle cells but does not interfere with autocrine stimulation of DNA synthesis. Cell Tissue Res. 256: 35-43 (1989); Takano S et al Angiosuppressive and antiproliferative actions of suramin: a growth factor antagonist. In: Moody TW (edt), Growth Factors, Peptides and Receptors, pp 255-64, Plenum Press New York (1993); **Tsiquaye K & Zuckerman A** Suramin inhibits duck hepatitis B virus DNA polymerase activity. J. Hepatol. 1: 663-9 (1985); **Voogd TE et al** Recent research on the biological activity of suramin. Pharmacol. Rev. 45: 177-203 (1993); **Walther MM et al** Suramin inhibits bone resorption and reduces osteoblast number in a neonatal mouse calvarial bone resorption assay. Endocrinology 131: 2263-70 (1992); **Zaniboni A** Suramin: the discovery of an old anticancer drug. Med. Oncol. Tumor Pharmacother. 7: 287-90 (1990)

Surface-associated sulfhydryl protein: abbrev. SASP. See: ADF (adult T cell leukemia-derived factor).

Survival factors: A general term sometimes used to indicate that a given cytokine or "growth factor" can maintain cells in culture for extended periods of time without essentially providing a growth stimulatory signal (see, for example: neurotrophins). Such activities are also referred to as ***Maintenance factor***. Factor deprivation often results in cell death by » apoptosis (see also: Factor-dependent cell lines). Some mitogens also function as survival factors for certain cell types. One function of survival factors *in vivo* probably is to allow cells to undergo precommitted differentiation without inducing their differentiation. Maintenance/survival factors have been suggested to be responsible for the maintenance of the functional activities of » hematopoietic stem cells.

Suspension cultures: see: Cell culture.

SC-MSC: Several human stromal cell lines (see also: Hematopoiesis) have been described under this name. They have been obtained from long-term bone marrow cultures (see: LTBMC) of hematologically normal patients and patients with hematological diseases after immortalisation with SV40. The cells have been shown to express » M-CSF mRNA and » TGF-β mRNA. Some lines also express » IL6 mRNA and » IL1β mRNA (see also: Message amplification phenotyping). The cells do not express GM-CSF or G-CSF mRNA. Their functional capacities in supporting hematopoiesis in culture have not yet been determined.

Andrews DF 3rd et al Sodium vanadate, a tyrosine phosphatase inhibitor, affects expression of hematopoietic growth factors and extracellular matrix RNAs in SV40-transformed human marrow stromal cells. EH 20: 449-53 (1992); **Singer JW et al** Simian virus 40-transformed adherent cells from human long-term marrow cultures: cloned cell lines produce cells with stromal and hematopoietic characteristics. Blood 70: 464-74 (1987); **Slack JL et al** Regulation of cytokine and growth factor gene expression in human bone marrow stromal cells transformed with simian virus 40. Blood 75: 2319-27 (1990)

SW-13: These cells are derived from a human adenocarcinoma of the adrenal cortex. They form only a few small colonies when suspended in soft agar at low cell densities. The numbers and sizes of colonies increase dramatically following stimulation with serum-free medium conditioned by SW-13 cells, indicating the possibility of » autocrine growth control. Some of this colony-stimulating activity is due to » IL1. The cells also secrete a factor biologically related to, or identical with » PDGF and » bFGF. They also secrete » TGFe (transforming growth factor type e).
The cells grow in soft agar in response to » aFGF, » bFGF, » IL1, and » TGFe, and this effect is inhibited by heparin. bFGF activity can be functionally replaced by » K-FGF or » *int*-2 (= FGF-3).

Corin SJ et al Enhancement of anchorage-independent growth of a human adrenal carcinoma cell line by endogenously produced basic fibroblast growth factor. IJC 46: 516-21 (1990); **Halper J & Carter BJ** Modulation of growth of human carcinoma SW-13 cells by heparin and growth factors. JCP 141: 16-23 (1989); **Hamburger AW & White CP** Autocrine growth factors for human tumor clonogenic cells. IJCC 3: 399-406 (1985); **Hamburger AW et al** Stimulation of anchorage-independent growth of human tumor cells by interleukin 1. CR 47: 5612-5 (1987); **Merlo GR et al** The mouse *int*-2 gene exhibits basic fibroblast growth factor activity in a basic fibroblast growth factor-responsive cell line. Cell Growth Diff. 1: 463-72 (1990); **O'Donnell KA et al** Production of platelet-derived growth factor-like protein(s) by a human carcinoma cell line. In Vitro Cell Dev. Biol. 25: 381-4 (1989); **Venesio T et al** The *int*-2 gene product acts as a growth factor and substitutes for basic fibroblast growth factor in promoting the differentiation of a normal mouse mammary epithelial cell line. Cell Growth Diff. 3: 63-71 (1992); **Wellstein A et al** Autocrine growth stimulation by secreted Kaposi fibroblast growth factor but not by endogenous basic fibroblast growth factor. Cell Growth Diff. 1: 63-71 (1990)

Switch factors: see: Isotype switching.

Synergistic activity: see: SA.

Synergistic factor 1: abbrev. SF-1. See: 5637.

Sweat gland factor: see addendum: SGF.

T

T2: A transcriptionally active open reading frame found within the terminal inverted repeats of Shope fibroma virus. It encodes a protein related to the cellular » TNF receptor. T2 is secreted by virus-infected cells. It specifically binds both TNF-α and TNF-β and inhibits binding of these cytokines to native TNF receptors on cells.
The terminal regions of two other Leporipoxviruses (myxoma virus and malignant rabbit fibroma virus) also encode a closely related T2 homologue with all the structural motifs predicted for a secreted TNF binding protein. These viruses are extremely invasive and capable of inducing extensive immunosuppression in rabbits. T2 and related proteins may therefore be involved in determining the virulence of the virus by modulating the host immune response to the infection. T2 represents a soluble form of the type I TNF receptor secreted from virally infected cells. The function of this soluble receptor form is to immunosuppress the host by abrogating the potentially destructive effects of TNF. For similar viral strategies of subverting host immune responses see also: Viroceptor.
Myxoma viruses with a deleted T2 gene grow normally in tissue culture. However, upon infection of susceptible rabbits the viral disease is significantly attenuated. The majority of infected rabbits are able to mount an effective immune response to the infection and completely recover, becoming immune to subsequent challenge with wild type myxoma virus.
Howard ST et al Vaccinia virus homologues of the Shope fibroma virus inverted terminal repeat proteins and a discontinuous ORF related to the tumor necrosis factor receptor family. Virology 180: 633-47 (1991); **Smith CA et al** T2 open reading frame from the Shope fibroma virus encodes a soluble form of the TNF receptor. BBRC 176: 335-42 (1991); **Upton C et al** Myxoma virus expresses a secreted protein with homology to the tumor necrosis factor receptor gene family that contributes to viral virulence. Virology 184: 370-82 (1991)

T4: Alternative name for » CD4 (see also: CD antigens).

T8: Alternative name for » CD8 (see also: CD antigens).

T11: Alternative name for » CD2 (see also: CD antigens).

T24 BCDF: *T-24 B cell differentiation factor* T24 is a human transitional cell bladder carcinoma cell line. T24 cells constitutively produce a B cell differentiation factor (see: BCDF) that induces immunoglobulin secretion from lymphoblastoid » CESS cells, other B lymphocytes from B cell chronic lymphocytic leukemia, and normal human B cells without causing clonal expansion. This factor is identical with » IL6.
Bubenick J et al Established cell line of urinary bladder carcinoma (T24) containing specific tumor specific antigen. IJC 11: 765 (1973); **Christie JF et al** The human bladder carcinoma line T-24 secretes a human B cell differentiation factor. Immunology 60: 467-9 (1987); **Gallagher G et al** T-24.B cell differentiation factor induces immunoglobulin secretion in human B cells without prior cell replication. Immunology 60: 523-9 (1987); **Hirano T et al** Complementary DNA for novel human interleukin (BSF-2) that induces B lymphocytes to produce immunoglobulin. N 324: 73-6 (1986); **Rawle FC et al** B cell growth and differentiation induced by supernatants of transformed epithelial cell lines. EJI 16: 1017-9 (1986)

T47D factor: This factor is secreted by the human mammary carcinoma cell line T47D. It is a mitogen for murine » 3T3 fibroblasts. The factor is probably identical with one of the molecular forms of » PDGF. Its biological activity is completely blocked by anti-PDGF antibodies.
Rozengurt E et al Production of PDGF-like growth factor by breast cancer cell lines. IJC 36: 247-52 (1985)

T88-M: A murine early B-lineage cell line that depends on » IL5 for its growth. It is used in » bioassays for this factor. The cells also proliferate in response to » IL3. The cells have been used extensively to characterize the murine IL5 receptor.
Introduction of a functional » IL5 gene by means of a retroviral expression vector into T88-M cells creates an » autocrine loop, generating IL5-independent cells that are tumorigenic *in vivo*.
Blankenstein T et al Retroviral interleukin 5 gene transfer into interleukin 5-dependent growing cell lines results in autocrine growth and tumorigenicity. EJI 20: 2699-705 (1990); **Murata Y et al** Interleukin 5 and interleukin 3 induce serine and tyrosine phosphorylations of several cellular proteins in an interleukin 5-

dependent cell line. BBRC 173: 1102-8 (1990); **Yamaguchi N et al** Characterization of the murine interleukin 5 receptor by using a monoclonal antibody. Int. Immunol. 2: 181-7 (1990)

T90/44 antigen: see: CD28.

T1165: A murine plasmacytoma cell line established from ascites tumors of BALB/cAnPt mice. Growth of the cells depends on » IL6 which was originally detected as » PCT-GF (plasmacytoma growth factor) secreted by the murine macrophage cell line P388D1. The cells are therefore used to assay this factor in a » bioassay. The proliferation of T1165 can also be induced by » GM-CSF and to a lesser extent by » IL1 and » TNF-α. The assay is therefore only specific in the presence of suitable factor-specific antibodies.

T1165 cells can be grown in the presence of IL6-specific antibodies if » IL11 is added to the growth medium. The cells can be used therefore to assay IL11. An alternative and entirely different method of detecting these factors is » Message amplification phenotyping.

Le J et al Tumor necrosis factor and interleukin 1 can act as essential growth factors in a murine plasmacytoma line. LR 7: 99-106 (1988); **Neckers LM & Nordan RP** Regulation of murine plasmacytoma transferrin receptor expression and G1 traversal by plasmacytoma cell growth factor. JCP 135: 495-501 (1988); **Nordan RP & Potter MA** A macrophage-derived factor required by plasmacytomas for survival and proliferation *in vitro*. S 233: 566-9 (1986); **Paul SR et al** Molecular cloning of a cDNA encoding interleukin 11, a stromal cell-derived lymphopoietic and hematopoietic cytokine PNAS 87: 7512-6 (1990)

T1154 mitogenic activity: This factor is isolated from the supernatants of » T1165 cells. It identical with » IL11.

Paul SR et al Molecular cloning of a cDNA encoding interleukin 11, a stromal cell-derived lymphopoietic and hematopoietic cytokine PNAS 87: 7512-6 (1990)

TAC antigen: TAC = T cell activation. This T cell activation marker which is also called » CD25, was initially characterized as the T cell growth factor (TCGF) binding protein. CD25 is identical with the α-chain of the receptor for » IL2.

Leonard WJ et al A monoclonal antibody that appears to recognize the receptor for human T cell growth factor; partial characterization of the receptor. N 300: 267-9 (1982); **Robb RJ & Greene WC** Direct demonstration of the identity of T cell growth factor binding protein and the Tac antigen. JEM 158: 1332-7 (1983); **Uchiyama T et al** A monoclonal antibody (anti-Tac) reactive with activated and functionally mature human T cells. Production of anti-Tac monoclonal antibody and distribution of Tac(+) cells. JI 126: 1393-7 (1981)

Tac inducing factor: see: TIA.

Tac inhibitory activity: see: IL2R alpha chain inhibitory activity.

Tachykinins: A generic name for a family of closely related short neuropeptides initially identified by their activities as neurotransmitters. Many of these compounds were initially identified in mollusks and amphibians (frog ***Physalaemin***) and were later also found in vertebrates and invertebrates (*Drosophila*, locusts). These peptides play an important role as neuromodulators and are involved in such different processes as stress regulation (see also: Neuroimmune network), pain, and control of vascular tonus.

The tachykinin family includes » SP (substance P = NK1, Neurokinin 1), ***Neurokinin A*** (= NK2, Neurokinin 2, ***substance K***, abbrev. SK; = Neurokinin α, ***Neuromedin L***), ***Neurokinin B*** (= NK3, Neurokinin 3, Neurokinin β, ***Neuromedin K***).

Some tachykinins arise by differential splicing of the RNA transcript of the prepro-tachykinin (PTT) gene and by differential post-translation processing of the prepro-protein. Three forms of message (α, β, and γ) arise by alternative splicing events with the β and γ forms of preprotachykinins encoding both substance P and neurokinin A, and the α form containing only the substance P sequence. The expression of the PTT gene is subject to regulation by » NGF. Neuropeptide γ is an N-terminal extended form of neurokinin A (γ-preprotachykinin 72-92).

Substance P (SP)	NH_2 - MLGFFQQPKPR
Neurokinin A (NKA), Substance K	NH_2 - MLGFFSDTKH
Neuropeptide K (NPK)	NH_2 - MLGVFSDTKHRKHSIQGHGYLAKLLAVQKEISSDAD
Neuropeptide γ (NPγ)	NH_2 - MLGVFSDTKHRKHSIQGHGYD
Neurokinin B (Neuromedin B)	NH_2 - MLGVFSDHMD

Amino acid sequences of some tachykinins.
This sequence comparison shows the conserved aminoterminal sequences. All factors with the exception of Neurokinin B (NKB) are encoded by the PPTI gene (Prepro-Tachykinin-I) codiert. NKB is encoded by the PPT II gene.

The three known tachykinin receptors belong to the group of receptors coupled to G-proteins. These receptors are characterized by a highly conserved transmembrane domain which passes seven times through the membrane. The receptors (neurokinin receptors) are called ***NK1*** (substance P-specific), ***NK2*** (Neurokinin A-specific; binds neuropeptide γ) and ***NK3*** (Neurokinin B-specific). These receptors also bind » Bombesin and Bombesin-like peptides such as » GRP (gastrin-releasing hormone).

Apart from the classical role as neurotransmitters tachykinins also show cytokine-like activities although it is not classified as a cytokine due to its small size. Tachykinins stimulate the proliferation of T cells, enhance the mitogen-induced release of cytokines such as » IFN-γ, » TNF-α, » IL1 and » IL6 by granulocytes and macrophages, enhance the secretion of immunoglobulins, and also mediate chemotactic and phagocytic processes (see also: Chemotaxis).

Tachykinins activate neutrophilic granulocytes (see also: Cell activation), they induce the synthesis of adhesion molecules in endothelial cells, and they are involved in the regulation of non-infectious inflammatory processes. The neuromedins B and C stimulate the growth of fibroblasts and smooth muscle cells. They are also mitogenic for small cell lung carcinomas and act as » autocrine growth factors. Neurokinin A and » SP (substance P) inhibit chemotactic migration of neutrophils induced by » IL8. Neurokinin A and B and also substance P promote the synthesis of » IL1.

Bevis CL & Zasloff M Peptides from frog skin. ARB 59: 395-414 (1990); **Buck SH et al** Tachykinins and their receptors: Pharmacology, biochemistry and molecular biology advance a neuropeptide story to the forefront of science. Neurochem. Int. 18: 167-70 (1991); **Burcher E et al** Neuropeptide γ, the most potent contractile tachykinin in human isolated bronchus, acts via a non-classical NK2 receptor. Neuropeptides 20: 79-82 (1991); **Castiglione R de & Gozzini L** Non-mammalian peptides: structure determination synthesis, and biological activity. Chimica*oggi* April 1991, pp. 9-15; **Chang MM & Leeman SE** Isolation of a sialogogic peptide from bovine hypothalamus tissue and its characterization as substance P. JBC 245: 4784-90 (1970); **Dam TV et al** γ-preprotachykinin-(72-92)-peptide amide: an endogenous preprotachykinin I gene-derived peptide that preferentially binds to neurokinin 2 receptors. PNAS 87: 246-250 (1990); **Eglezos A et al** Modulation of the immune response by tachykinins. Immunol. Cell. Biol. 69: 285-94 (1991); **Gilchrist CA et al** Identification of nerve growth factor-responsive sequences within the 5' region of the bovine preprotachykinin gene. DNA Cell Biol. 10: 743-9 (1991); **Gilchrist CA et al** Regulation of preprotachykinin gene expression by nerve growth factor. ANY 632: 391-3 (1991); **Harmar AJ et al** Identification and cDNA sequence of δ-preprotachykinin, a fourth splicing variant of the rat substance P precursor. FL 275: 22-4 (1990); **Helke CJ et al** Diversity in mammalian tachykinin peptidergic neurons: multiple peptides, receptors, and regulatory mechanisms. FJ 4: 1606-15 (1990); **Hershey AD & Krause JE** Molecular characterization of a functional cDNA encoding the rat substance P receptor. S 247: 958-62 (1990); **Kage R et al** Neuropeptide γ: a peptide isolated from rabbit intestine that is derived from γ-preprotachykinin. J. Neurochem. 50: 1412-7 (1988); **Kanagawa K et al** Neuromedin K: a novel mammalian tachykinin identified in porcine spinal cord. BBRC 114: 533-540 (1983); **Kimura S et al** Pharmacological characterization of novel mammalian tachykinins, neurokinin α and neurokinin β. Neurosci. Res. 2: 97-104 (1984); **Krause JE et al** Three rat preprotachykinin mRNAs encode the neuropeptides substance P and neurokinin A. PNAS 84: 881-5 (1987); **Lemaire I** Bombesin-related peptides modulate interleukin-1 production by alveolar macrophages. Neuropeptides, 20: 217-23 (1991); **Li XJ et al** Cloning, heterologous expression and developmental regulation of a *Drosophila* receptor for tachykinin-like peptides. EJ 10: 3221-9 (1991); **Lotz M et al** Effect of neuropeptides on production of inflammatory cytokines by human monocytes. S 241: 1218-21 (1988); **Maggio JE** Tachykinins. ARN 11: 13-28 (1990); **Masu Y et al** cDNA cloning of bovine substance K receptor through oocyte expression system. N 329: 836-8 (1987); **Munekata E** Neurokinin A and B. Comp. Biochem. Physiol. [C] 98C: 171-9 (1991); **Nakanishi S** Structure and regulation of the preprotachykinin gene. TINS 9: 41-4 (1986); **Nawa H et al** Tissue-specific generation of two preprotachykinin mRNAs from one gene by alternative RNA splicing. N 312: 729-34 (1984); **O'Halloran DJ et al** Neuropeptides synthesized in the anterior pituitary: possible paracrine role. Mol. Cell. Endocrinol. 75: C7-C12 (1991); **Ohkubo H & Nakanishi S** Molecular characterization of the three tachykinin receptors. ANY 632: 53-62 (1991); **Regoli D et al** Pharmacological receptors for substance P and neurokinins. Life Sciences 40: 109-117 (1987); **Rozengurt E** Neuropeptides as cellular growth factors: Role of multiple signaling pathways. EJCI 21: 123-34 (1991); **Schoofs L et al** Locustatachykinin I and II, two novel insect neuropeptides with homology to peptides of the vertebrate tachykinin family. FL 261: 397-401 (1990); **Schüller HM** Receptor-mediated mitogenic signals and lung cancer. Cancer Cells 3: 496-503 (1991); **Sethi T & Rozengurt E** Multiple neuropeptides stimulate clonal growth of small cell lung cancer: effects of bradykinin, vasopressin, cholecystokinin, galanin, and neurotensin. CR 51: 3621-3 (1991); **Shigemoto R et al** Cloning and expression of a rat neuromedin K receptor cDNA. JBC 265: 623-8 (1990); **Takeda Y et al** Analysis of tachykinin peptide family gene expression patterns by combined high performance liquid chromatography, radioimmunoassay. Methods Neurosci. 6: 119-30 (1991); **Weihe E et al** The tachykinin neuroimmune connection in inflammatory pain. ANY 632: 283-95 (1991); **Wiedermann CJ et al** Priming of normal human neutrophils by tachykinins: Tuftsin-like inhibition of *in vitro* chemotaxis stimulated by formylpeptide or interleukin-8. Regul. Pept. 36: 359-68 (1991); **Yokota Y et al** Molecular characterization of a functional cDNA for rat substance P receptor. JBC 264: 17649-52 (1989)

TAF: *T cell activating factor* This factor was isolated initially by its ability to activate purified resting peripheral lymph node T cells to produce » IL2. This murine protein is identical with » IL6.

Garman RD & Raulet DH Characterization of a novel murine T cell-activating factor. JI 138: 1121-9 (1987); **Garman RD et al** B cell-stimulatory factor 2 (β 2 interferon) functions as a second signal for interleukin 2 production by mature murine T cells. PNAS 84: 7628-33 (1987); **Van Snick J** From hybridoma and plasmacytoma growth factors to interleukin 6. in: Del Guerico P & Cruse JM (eds): B lymphocytes: Function and Regulation. Contrib. Microbiol. Immunol. Karger, Basel, Vol. 11, pp. 73-95 (1989)

TAF: *thymocyte activating factor* This factor is identical with one of the molecular forms of » IL1. See also: ETAF (epidermal cell-derived thymocyte activating factor) and MC-TAF (mesangial cell-derived TAF). In addition there are other poorly characterized factors also possessing TAF activity.

Okai Y & Yamashita U Different inducing activities for cytotoxic T cells by heterogeneous thymocyte-activating factors from SV40-transformed human embryo fibroblasts. Immunol. Lett. 14: 1-7 (1986); **Sauder ND et al** Epidermal cell production of thymocyte activating factor. JI 130: 34-9 (1982); **Takada H et al** Induction of interleukin-1 and -6 in human gingival fibroblast cultures stimulated with Bacteroides lipopolysaccharides. Infect. Immun. 59: 295-301 (1991)

TAF: *tumor angiogenesis factor* A generic name for activities promoting the growth of new blood vessels by stimulating endothelial cell growth and cell migration. Such activities have been found in various human tumor cell lines and cell lines of other species as well as in fetal tissues and human or bovine decidua. Some of the activity is identical with » bFGF and several TAFs have been characterized as ribonucleoproteins. A major component in preparations with TAF activity appears to be a poorly characterized non-protein (see: HUAF, human uterine angiogenesis factor). See also: angiogenesis factors.

Bard RH et al Detection of tumor angiogenesis factor in adenocarcinoma of kidney. Urology 27: 447-50 (1986); **Battiwalla ZF et al** An antiserum to tumor angiogenesis factor: therapeutic approach to solid tumors. Anticancer Res. 9: 1809-13 (1989); **Dastur ZH** Regression of S-180 sarcoma using a xenogeneic antiserum. Indian J. Exp. Biol. 29: 393-5 (1991); **Deshpande RG & Shethna YI** Isolation & characterization of tumor angiogenesis factor from solid tumors & body fluids from cancer patients. Indian J. Med. Res. 90: 241-7 (1989); **Folkman J** Tumor angiogenesis. ACR 43: 175-203 (1985); **Ishiwata I et al** Establishment of HUOCA-II, a human ovarian clear cell adenocarcinoma cell line, and its angiogenic activity. JNCI 78: 667-73 (1987); **Ishiwata I et al** Tumor angiogenic activity of gynecological tumor cell lines on the chorioallantoic membrane. Gynecol. Oncol. 29: 87-93 (1988); **Ishiwata I et al** Effect of tumor angiogenesis factor on proliferation of endothelial cell and tube formation. Virchows Arch. A Pathol. Anat. Histopathol. 417: 473-6 (1990); **Olander JV et al** An assay measuring the stimulation of several types of bovine endothelial cells by growth factor(s) derived from cultured human tumor cells. In Vitro 18: 99-107 (1982); **Sekiya S et al** Tumor angiogenesis activity of human choriocarcinoma cells grown *in vitro*. Gynecol. Oncol. 25: 271-80 (1986)

TAL: *tumor-associated lymphocytes* See: TILs (tumor-infiltrating lymphocytes).

TALL-103: A permanent cell line established after culture in the presence of » IL3 of bone marrow cells from a child with an immature ($CD2^+$, $CD5^+$, $CD7^+$) acute T lymphocytic leukemia (T-ALL).
The strictly IL3-dependent subline, ***TALL-103/3***, has lost the T cell-specific markers and expresses a myeloid phenotype ($CD15^+$, $CD33^+$). These cells grow in synthetic (serum-free) medium also in the presence of either » GM-CSF or » IL5, in which they retain a myeloid phenotype. TALL-103/3 cells can still be phenotypically converted to the lymphoid lineage upon addition of » IL2, thus maintaining their bipotentiality.
TALL-103/2 is a subline obtained by switching of the IL3-dependent cells at an early passage to medium containing only human IL2. These cells have a T lymphoid phenotype ($CD2^+$, $CD3^+$, $CD8^+$, TCR-$\gamma\delta^+$, $CD7^+$) and still express the CD33 myeloid antigen originally present on the IL3 expanded population. The TALL-103/2 cells strictly require IL2 for growth, are irreversibly committed to the lymphoid lineage, and cannot survive in the presence of any other hemopoietic growth factor. IL1α synergizes with IL2 in supporting the short and long term growth of this cell line, whereas » IL4 abrogates its growth. Functionally, the TALL-103/2 cells display MHC-nonrestricted cytotoxic activity that is significantly enhanced by addition of either » IL4, » IL6, or IFN-γ.

O'Connor R et al Growth factor-dependent differentiation along the myeloid and lymphoid lineages in an immature acute T lymphocytic leukemia. JI 145: 3779-87 (1990); **O'Connor R et al** Growth factor requirements of childhood acute T lymphoblastic leukemia: correlation between presence of chromosomal abnormalities and ability to grow permanently *in vitro*. Blood 77: 1534-45 (1991); **O'Connor R et al** Phenotypic changes induced by interleukin-2 (IL2) and IL3 in an immature T-lymphocytic leukemia are associated with regulated expression of IL2 receptor β chain and of protein tyrosine kinases *lck* and *lyn*. Blood 80: 1017-25 (1992); **Santoli D et al** Synergistic and antagonistic effects of IL1 α and IL4, respectively, on the IL2-dependent growth of a T cell receptor-$\gamma\delta^+$ human T leukemia cell line. JI 144: 4703-11 (1990); **Valtieri M et al** Establishment and characterization of an undifferentiated human T leukemia cell line which requires granulocyte-macrophage colony stimulatory factor for growth. JI 138: 4042-50 (1987)

TAM: *tumor-associated macrophages* See: TILs (tumor-infiltrating lymphocytes).

TAMF: *tumor autocrine motility factor* see: AMF (autocrine motility factor).

Tamm-Horsfall glycoprotein: abbrev. THP. See: Uromodulin.

TAN-1: This human gene locus encodes the human homologue of the *Drosophila* gene » *Notch.*

TAP: *T cell activation protein* This protein is encoded by the murine Ly2 locus. It is expressed in various organs. TAP is a cysteine-rich protein which is anchored in the membrane by phosphatidylinositol. It influences the synthesis of » IL2 and is responsible for the maximal expression of IL2 together with other secondary signals.
Reiser H et al Cloning and expression of a cDNA for the T cell activating protein TAP. PNAS 85: 2255-9 (1988); **Yeh ET et al** The expression, function, and ontogeny of a novel T cell-activating protein, TAP, in the thymus. JI 137: 1232-8 (1986)

Targeted disruption of cytokine functions: see: Knock-out mice.

Tautomycin: see: Okadaic acid.

T-Bam: see: 5c8.

TBPI: *tumor necrosis factor binding protein* see: TNF-BF.

TCA: abbrev. for T cell activation. See also: Cell activation.

TCA-3: *T cell activation* This factor is produced by T cells. It is selectively expressed during activation via the antigen-receptor pathway or stimulation of the cells by concanavalin A. Expression of TCA-3 in response to Con A is blocked by cyclosporin A treatment.
TCA-3 belongs to the » chemokine family of cytokines. It is the murine homologue of human » I-309. The gene which contains at least three introns and has a length less than 4.7 kb maps to murine chromosome 11. The cDNA of TCA-3 is 512 bases in length excluding poly(A) and encodes a predicted 92-amino acid protein having the characteristics of a secreted polypeptide of approximately 69 amino acids. Another factor, designated ***P500***, is probably a splice variant of the TCA-3 mRNA.
Brown KD et al A family of small inducible proteins secreted by leukocytes are members of a new superfamily that includes leukocyte and fibroblast-derived inflammatory agents, growth factors, and indicators of various activation processes. JI 142: 679-87 (1989); **Burd PA et al** Cloning and characterization of a novel T cell activation gene. JI 138: 3126-31 (1987); **Luo Y et al** Serologic analysis of a murine chemokine, TCA3. JI 150: 971-9 (1993); **Wilson SD et al** Expression and characterization of TCA3: a murine inflammatory protein. JI 145: 2745-50 (1990); **Wilson SD et al** Clustering of cytokine genes on mouse chromosome 11. JEM 171: 1301-14 (1990)

T cell activating factor: see: TAF.

T cell activation protein: see: TAP.

T cell chemotactic factor: see: TCF.

T cell-derived inhibitory factor for cellular DNA: see: STIF.

T cell-derived leukocyte (migration) inhibitory factor: abbrev. T-LIF. See: LIF (leukocyte (migration) inhibitory factor).

T cell-derived neutrophil migration inhibition factor: see: NIF-T.

T cell growth factor: see: TCGF.

T cell growth factor 2: see: TCGF-2.

T cell growth factor 3: see: TCGF-3.

T cell growth inhibitory factor: see: TGIF.

T cell maturation factor: see: TMF.

T cell replacing factor: see: TRF.

T cell replacing factor of macrophages: see: TRFM.

T cell replacing factor 1: see: TRF-1.

T cell replacing factor 2: see: TRF-2.

T cell replacing factor 3: see: TRF-3.

T cell stimulating factor: see: TSF (two entries).

T cell suppressor factor: abbrev. TsF. see: G-TsF (glioblastoma-derived T cell suppressor factor).

TCF: *T lymphocyte chemotactic factor* This factor, which is chemotactic for neutrophils and T lymphocytes, is found in culture supernatants of phytohemagglutinin-stimulated human blood mononuclear leukocytes. The amino-terminal amino acid sequence of TCF is identical with that of » NAP-1 (neutrophil-activating protein).
Larsen CG et al The neutrophil-activating protein (NAP-1) is also chemotactic for T lymphocytes. S 243: 1464-6 (1989)

TCF: *thymocyte comitogenic factor* This factor is identical with » IL1.

Matsushima K et al Purification of human interleukin 1 from human monocyte culture supernatants and identity of thymocyte comitogenic factor, fibroblast proliferation factor, acute phase protein-inducing factor, and endogenous pyrogen. CI 92: 290-301 (1985)

TCF: *tumor cytotoxic factor* This factor is also called ***F-TCF*** (fibroblast tumor cytotoxic factor). It is detected in culture supernatants of human embryonic lung fibroblast cells. TCF is cytotoxic for a number of tumor cell lines (e. g. Sarcoma 180) and cytostatic for human KB cells. TCF is a potent mitogen for hepatocytes, human endothelial cells, and melanocytes.

One of the factors with TCF activity is completely identical with » HGF (hepatocyte growth factor) isolated from placenta. Another form with a deletion of 5 amino acids has slightly altered heparin binding and hepatocyte growth stimulating activities. However, both factors show the same specific activities in tumor cytotoxicity assays.

Coffer A et al Purification and characterization of biologically active scatter factor from *ras*-transformed NIH 3T3 conditioned medium. Biochem. J. 278: 35-41 (1991); **Higashio K et al** Identity of a tumor cytotoxic factor from human fibroblasts and hepatocyte growth factor. BBRC 170: 397-404 (1990); **Shima N et al** Tumor cytotoxic factor/hepatocyte growth factor from human fibroblasts: cloning of its cDNA, purification and characterization of recombinant protein. BBRC 180: 1151-8 (1991); **Shima N et al** A fibroblast-derived tumor cytotoxic factor/F-TCF (hepatocyte growth factor/HGF) has multiple functions *in vitro*. Cell. Biol. Int. Rep. 15: 397-408 (1991); **Shima N et al** ELISA for F-TCF (human hepatocyte growth factor/hHGF)/fibroblast-derived tumor cytotoxic factor antigen employing monoclonal antibodies and its application to patients with liver diseases. Gastroenterol. Jpn. 26: 477-82 (1991); **Sone S et al** Effector mechanism of human monocyte-mediated cytotoxicity: role of a new tumor cytotoxic factor distinct from interleukin 1 and tumor necrosis factor α. Biotherapy 1: 233-43 (1989)

TCF5: *transcription factor 5* This transcription factor is identical with » NF-IL6.

TCF8: This human gene on chromosome 10p12 encodes a » zinc finger protein, designated, ***Nil-2-a***, which inhibits T lymphocyte-specific » IL2 gene expression by binding to a negative regulatory domain 100 nucleotides 5' of the IL2 transcription start site.

Williams TM et al The TCF8 gene encoding a zinc finger protein (Nil-2-a) resides on human chromosome 10p11.2. Genomics 14: 194-6 (1992)

TCGF: *T cell growth factor* A generic name for a number of soluble factors found in the culture supernatants of mitogen-stimulated leukocytes that support the proliferation of activated T cells (see also: Cell activation) and T cell lines. This name was coined in the beginning of cytokine research. It is still employed to characterize the functional activities of new factors (TCGF activity).

The "classical" TCGF was initially isolated from normal peripheral blood lymphocytes and from » Jurkat cells (jTCGF). This TCGF is identical with » IL2. Other factors with TCGF activity are » IL4, » IL7 and » IL10.

Aarden LA et al Revised nomenclature for antigen-nonspecific T cell proliferation and helper factors. CI 48: 433-36 (1979); **Bohlen P et al** Isolation and partial characterization of human T cell growth factor. BBRC 117: 623-30 (1983); **Brown M et al** IL4/B cell stimulatory factor 1 stimulates T cell growth by an IL2-independent mechanism. JI 141: 504-11 (1988); **Clark SC et al** Human T cell growth factor: partial amino acid sequence, cDNA cloning, and organization and expression in normal and leukemic cells. PNAS 81: 2543-7 (1984); **Clark-Lewis I et al** Preparation of T cell growth factor free from interferon and factors stimulating hemopoietic cells and mast cells. JIM 51: 311-22 (1982); **Gillis S et al** T cell growth factor: parameters of production and a quantitative microassay for activity. JI 120: 2027-31 (1978); **Kupper T et al** Autocrine growth of T cells independent of interleukin 2: identification of interleukin 4 (IL 4, BSF-1) as an autocrine growth factor for a cloned antigen-specific helper T cell. JI 138: 4280-7 (1987); **Mier JW & Gallo RC** Purification and some characteristics of human T cell growth factor from phytohemagglutinin-stimulated lymphocyte-conditioned media. PNAS 77: 6134-8 (1980); **Mier JW & Gallo RC** The purification and properties of human T cell growth factor. JI 128: 1122-7 (1982); **Milstone DS & Parker CW** Purification of primate T cell growth factor to apparent homogeneity by reverse phase high pressure liquid chromatography: evidence for two highly active molecularly distinct species. BBRC 115: 762-8 (1983); **Robb RJ et al** Purification and partial sequence analysis of human T cell growth factor. PNAS 80: 5990-4 (1983); **Smith KA** T cell growth factor. IR 51: 337-57 (1980)

TCGF: *transformed cell growth factor* This factor is identical with » TGF-α or » TGF-β.

Lipton A et al Liver as a source of transformed-cell growth factor. In: Sato GH & Ross R (eds) Hormones and cell culture. Cold Spring Harb. Conf. Cell Prolif. Vol. 6: 461-75 (1979)

TCGF 2: *T cell growth factor 2* This factor is identical with » IL4.

Leanderson T et al B cell growth factors: distinction from T cell growth factor and B cell maturation factor. PNAS 79: 7455-9 (1982); **Mosmann TR et al** T cell and mast cell lines respond to B cell stimulatory factor-1. PNAS 83: 5654-8 (1986); **Sideras P et al** IgG1 induction factor: a single molecular entity with multiple biological functions. AEMB 213: 227-36 (1987)

TCGF 3: *T cell growth factor 3* This factor is secreted by various T cell lines following their activation by antigens or lectins. It is identical with » IL9.

Schmitt E et al TCGF III/P40 is produced by naive murine $CD4^+$ T cells but is not a general T cell growth factor. EJI 19: 2167-70 (1989)

TCGF-β: *T cell growth factor β* This factor is identical with » IL7.
Suda T & Zlotnik A *In vitro* induction of CD8 expression on thymic pre-T cells. II. Characterization of $CD3^-CD4^-CD8\ \alpha^+$ cells generated *in vitro* by culturing $CD25^+CD3^-CD4^-CD8^-$ thymocytes with T cell growth factor-β and tumor necrosis factor-α. JI 149: 71-6 (1992)

TCGF(IL2)-receptor inducing factor: This factor is identical with » ADF (adult T cell leukemia-derived factor).
Okada M et al TCGF(IL2)-receptor inducing factor(s). II. Possible role of ATL-derived factor (ADF) on constitutive IL2 receptor expression of HTLV-I(+) T cell lines. JI 135: 3995-4003 (1985); **Yodoi J et al** TCGF(IL2)-receptor inducing factor(s). I. Regulation of IL 2 receptor on a natural killer-like cell line (YT cells). JI 134: 1623-30 (1985)

TcMF: *CTL maturation factor* see: CLMF (cytotoxic lymphocyte maturation factor).

T colony-promoting activity: see: TCPA.

TCPA: *T colony-promoting activity* This factor is found in culture supernatants with phytohemagglutinin-stimulated peripheral blood lymphocytes. It promotes the growth of T lymphocyte colonies in agar. The factor is identical with » IL2.
Claesson MH et al Mouse B- and T-cell colony formation *in vitro*. I. Separation of colony-promoting and -inhibiting activities in concanavalin A rat spleen conditioned medium. Scand. J. Immunol. 19: 205-10 (1984); **Jourdan M et al** Control of human T-colony formation by interleukin-2. Immunology 54: 249-53 (1985); **Klein B et al** The role of interleukin 1 and interleukin 2 in human T colony formation. CI 77: 348-56 (1983); **Rey A et al** The role of interleukin-2 in T colony formation by human pre-T cells (pTCFC). Clin. Exp. Immunol. 58: 154-60 (1984)

TCSF: *tumor cell-derived collagenase stimulatory factor* This factor of 58 kDa is produced by the human lung carcinoma cell line LX-1. It stimulates the production of interstitial collagenase in fibroblasts. TCSF also stimulates expression in human fibroblasts of mRNA for stromelysin 1 and 72-kDa gelatinase/type IV collagenase. It has been suggested that tumor cell interaction with fibroblasts via TCSF may lead to increased degradation of interstitial or basement membrane matrix components and thus may enhance tumor cell invasion. The known partial sequence shows that this factor is not related to any other known cytokine.
Ellis SM et al Monoclonal antibody preparation and purification of a tumor cell collagenase-stimulatory factor. CR 49: 3385-91 (1989); **Javadpour N & Guirguis R** Tumor collagenase stimulating factor (TCSF) and tumor autocrine motility factor (TAMF) in bladder cancer. PCBR 370: 393-8 (1991); **Kataoka H et al** Tumor cell-derived collagenase-stimulatory factor increases expression of interstitial collagenase, stromelysin, and 72-kDa gelatinase. CR 53: 3154-8 (1993); **Nabeshima K et al** Partial sequence and characterization of the tumor cell-derived collagenase stimulatory factor. ABB 285: 90-6 (1991)

TDCF: *tumor-derived chemotactic factors* A general term for factors isolated from the culture supernatants of a number of human and murine tumor cell lines. These factors are chemotactic for mononuclear phagocytes (see also: Chemotaxis). One distinct TDCF is encoded by the human » JE gene. It belongs to the » chemokine family of cytokines and is one important determinant of macrophage infiltration in tumors. This TDCF is identical with » MCP-1 (monocyte chemoattractant protein 1), » MCP (monocyte chemotactic protein), » MCAF (monocyte chemotactic and activating factor), » HC11, » SMC-CF (smooth muscle cell chemotactic factor), » GDCF (glioma-derived monocyte chemotactic factor).
Bottazzi B et al Regulation of the macrophage content of neoplasms by chemoattractants. S 220: 210-2 (1983); **Bottazzi B et al** A chemoattractant expressed in human sarcoma cells (tumor-derived chemotactic factor, TDCF) is identical to monocyte chemoattractant protein-1/monocyte chemotactic and activating factor (MCP-1/MCAF). IJC 45: 795-7 (1990); **Walter S et al** Macrophage infiltration and growth of sarcoma clones expressing different amounts of monocyte chemotactic protein/JE. IJC 49: 431-5 (1991)

TDF: *thymocyte differentiating factor* This factor is identical with » IL2.

TDF: *tumor degenerating factor* This factor of 30 kDa, which is probably glycosylated, is secreted by human embryonic fibroblasts. Its synthesis is increased by treatment with human » IFN-α. The coculture of human fibroblasts with KB cells also augments TDF production. TDF induces degenerative changes in human tumor cells *in vitro*, including the formation of cell-free areas, and is effective on human KB, HeLa, FL and of hepatoma cells, but not on murine » L929, » 3T3, SV-3T3 cells, bovine » MDBK cells, and rabbit cells, or monkey cells, suggesting that TDF has species specificity. TDF is not effective on human non-transformed cells, namely various human fibroblasts. The coculture of human fibroblasts with KB cells augmented TDF production. The activity of this factor is inhibited by fibronectin which also complexes TDF.
TDF may be involved in the regulation of the interaction between tumor cells and interstitial cells.
Imanishi J et al [Production and characterization of human tumor degenerating factor (TDF)] CR. Soc. Seances. Soc. Biol. Fil. 177: 570-3 (1983); **Imanishi J et al** [Trial of purifying tumor degenerating factor of human fibroblasts] CR. Soc. Seances.

Soc. Biol. Fil. 178: 313-6 (1984); **Imanishi J et al** [Effect of interferons on the activity of human tumor degeneration factor (TDF)] CR. Soc. Seances. Soc. Biol. Fil. 179: 549-53 (1985); **Imanishi J et al** [Inhibition of colony formation by a human tumor degenerating factor (TDF); enhancement of this effect by interferons] CR. Soc. Seances. Soc. Biol. Fil. 180: 596-600 (1986); **Tanaka A et al** Production and characterization of tumor-degenerating factor. JNCI 74: 575-81 (1985); **Tanaka A et al** Mutual inhibitory activities of tumor-degenerating factor (TDF) and fibronectin. Cell Struct. Funct. 13: 459-70 (1988); **Yamamura Y et al** [Tumor-degenerating factor (TDF)] Hum. Cell. 3: 113-7 (1990)

TDGF-1: *teratocacinoma-derived growth factor* see: CRIPTO.

TDSF: *tumor-derived suppressor factor* This poorly characterized factor of 69-77 kDa is found in serum-free supernatants of the human melanoma cell line G361. TDSF strongly suppresses the generation of tumoricidal lymphokine-activated killer cells in response to » IL2 (see: LAK cells). TDSF is not identical with TGF-β and may be a novel immunoregulatory cytokine.

A tumor-derived suppressor factor has also been found in patients with acute myeloid leukemia. At present it is not known whether the two TDSFs are identical.

Lim SH et al Production of tumor-derived suppressor factor in patients with acute myeloid leukemia. Leuk. Res. 15: 263-8 (1991); **Somers SS et al** Comparison of transforming growth factor β and a human tumor-derived suppressor factor. Cancer Immunol. Immunother. 33: 217-22 (1991)

Teratocacinoma-derived growth factor: abbrev. TDGF-1. See: CRIPTO.

Tetradecanoyl phorbol acetate: see: Phorbol esters.

Tetradecanoyl phorbol acetate-inducible sequences: see: TIS genes.

Tetrazolium salt reduction test: see: MTT assay.

TF: *transcription factor* see: Gene expression.

TF-1: A human premyeloid cell line established from a patient with erythroleukemia. The survival of these cells is supported by » IL1, » IL4, » IL6 and » M-CSF. TF-1 cells are employed for » bioassays of various cytokines and proliferates in response to » IL3, » IL4, » IL5, » GM-CSF, » Epo, and » SCF (stem cell factor). Factor activities are determined by measuring the incorporation of ^{3}H thymidine into the newly synthesized DNA of proliferating cells. Cell proliferation can be determined also by employing the » MTT assay. The specificity of theses assays must be controlled by employing suitable neutralizing antibodies. An alternative and entirely different method of detecting these factors is » Message amplification phenotyping.

TF-1 cells have recently been shown to respond to human and murine » IL13, which also competitively inhibits the binding of IL4 to its receptor. TH-1 cell proliferation is used to assay this factor. The inhibition of IL5-induced proliferation by » TGF-β has recently been described to be a sensitive bioassay for TGF-β. The assay is made specific by using specific neutralizing antibodies against TGF-β. Cytokines such as » IFN-β, IFN-γ, and » TNF-α have been shown to be ≈ 100-fold less potent, and » IFN-α or TNF-β to be ≈ 1000-fold less potent than TGF-β in their ability to inhibit IL5-stimulated proliferation of TF-1 cells.

Hintz-Obertreis P et al Development of a rapid, highly sensitive, non-radioactive assay system for hematopoietic growth factors. Behring Inst. Mitt. 90: 99-103 (1991); **Kitamura T et al** Establishment and characterization of a unique human cell line that proliferates dependently on GM-CSF, IL3, or erythropoietin. JCP 140: 323-34 (1989); **Kitamura T et al** Identification and analysis of human erythropoietin receptors on a factor-dependent cell line, TF-1. Blood 73: 375-80 (1989); **Kitamura T et al** IL1 up-regulates the expression of cytokine receptors on a factor-dependent human hemopoietic cell line, TF-1. Int. Immunol. 3: 571-7 (1991); **Nishikawa M et al** Deletion mutagenesis of stem cell factor defines the C-terminal sequences essential for its biological activity. BBRC 188: 292-7 (1992); **Randall LA et al** A novel sensitive bioassay for transforming growth factor β. JIM 164: 61-7 (1993)

TF5: *thymosin fraction 5* See: Thymic hormones.

TG6: see: LBRM 33.

TGF: *transforming growth factor* Transforming growth factors were initially identified as proteins secreted by virus-transformed mouse cells (see: SGF, sarcoma growth factors). These factors were found to induce a reversible transformed phenotype in suitable non-neoplastic fibroblastic indicator cell lines *in vitro*. This phenotype was operationally characterized by a loss of density-dependent growth, i. e. overgrowth of cells, in monolayer cultures (see also: Cell culture), characteristic changes in cellular morphology, and the ability of the cells to grow as non-adherent cell colonies in soft agar (see also: Colony formation assay). The name TGF was chosen to indicate the fact that the original growth factor activities of sarcomas were also present in tumor cells other than sarcomas.

Peptides representing two distinct classes of TGFs, designated » TGF-α and » TGF-β, have been purified to homogeneity. They are distinguished both chemically by their unique amino acid sequences and biologically by their different activities on cells.

Type α TGFs are single chain peptides of 50-53 amino acids cross-linked by three disulfide bonds. They have strong homology to » EGF with which they compete for receptor binding. Type β TGFs have a homodimeric structure comprised of two chains of 112 amino acids, each containing nine cysteine residues; TGF-β binds to a unique cell surface receptor. Type α TGFs are usually mitogenic for fibroblasts, whereas type β TGFs have bifunctional effects on cell growth and can either stimulate or inhibit growth of the same cells, depending on conditions. The interactions of type α and β TGFs can be either synergistic or antagonistic.

Roberts AB et al Transforming growth factors: isolation of polypeptides from virally and chemically transformed cells by ethanol extraction. PNAS 77: 3494-8 (1980)

TGF-α: *transforming growth factor* α

■ **ALTERNATIVE NAMES:** MDGF-2 (milk-derived growth factor 2); TGF-1 (transforming growth factor 1); TCGF (transformed cell growth factor). See also: individual entries for further information.

■ **SOURCES:** TGF-α is produced by a number of human carcinomas and cell lines transformed by viral and cellular » oncogenes. The factor is also found in urine and plasma. It is also produced by some non-transformed cells during the development of mammalian embryos. It is synthesized in pituitary cells and in some regions of the adult brain. The production of TGF-α in pituitary cells in inhibited by » TGF-β. TGF-α is produced by keratinocytes, macrophages, hepatocytes, platelets. Its synthesis is stimulated by infection with viruses. In mammary tissue its synthesis is induced by estrogens.

■ **PROTEIN CHARACTERISTICS:** The protein has a length of 50 amino acids. It is not related to » TGF-β. The factor is obtained by proteolytic processing of a transmembrane precursor protein of 160 amino acids. Membrane-associated elastase-like serine proteases have been implicated in the processing of the pro-TGF-α form. Rat and murine TGF-α are almost identical with the human factor. TGF-α is closely related to » EGF and TGF-α is therefore sometimes called EGF-like transforming growth factor. The biological activities of these two growth factors are very similar and both bind to the same receptor. The transmembrane precursor of TGF-α can bind to EGF receptors on nearby cells. This enables it to establish specific cell-to-cell contacts and to elicit developmentally important responses in neighboring cells (see also: Juxtacrine).

■ **GENE STRUCTURE:** The gene encoding TGF-α has a length of ≈ 100 kb and contains six exons. It maps to human chromosome 2p11-2p13. Breakpoints in the variant Burkitt lymphoma translocation t(2;8) occur within these bands but it is not known whether this is relevant. The TGF-α gene has been shown to be allelic with the recessive mutation *waved*-1 in mice (see subentry Transgenic/Knock-out/Antisense studies).

■ **RECEPTOR STRUCTURE, GENE(S), EXPRESSION:** The TGF-α receptor is identical with the » EGF receptor. It is encoded by the cellular » *erb* » oncogene. Binding of TGF-α to this receptor is not as strong as that of EGF. In contrast to the mammalian receptor the avian receptor distinguishes between EGF and TGF-α.

■ **BIOLOGICAL ACTIVITIES:** In several *in vitro* systems murine and human TGF-α are functionally interchangeable with murine and human EGFs. Human TGF-α is as active as murine EGF in promoting eyelid opening in newborn mice. Human EGF is as potent as its murine homologue with respect to this biological property.

The biological activities of TGF-α resemble those of EGF since both factors bind to the same receptor. Some biological activities of TGF-α are, however, stronger than those of EGF. TGF-α is thought to be the fetal form of EGF. The physiological role of TGF-α is probably the control of epidermal development during development and differentiation of the cells. TGF-α is probably also involved in the regeneration of liver tissues.

TGF-α was initially detected as a protein that synergises with transforming growth factor TGF-β in supporting the growth of anchorage-independent cells in soft agar (see also: Colony formation assay). TGF-α induces the long-term proliferation of murine and chicken early hematopoietic progenitor cells such as » BFU-E without causing differentiation. In these species TGF-α is therefore a hematopoietic growth factor (see also: » Hematopoietins). In the presence of insulin and » Epo TGF-α induces the terminal differentiation of BFU-E cells into erythrocytes. TGF-α is an » autocrine growth factor for ovary carcinomas.

TGF-α stimulates the proliferation of cultured

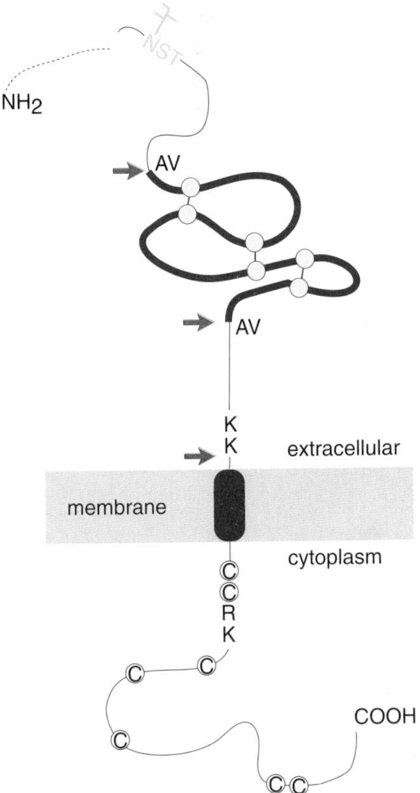

Structure of the membrane-bound TGF-α precursor. The 50 amino acids of mature TGF-α are shown as a thick blue line containing three disulfide bonds (yellow circles). This sequence is flanked by two sites that can be cleaved proteolytically (red arrows). Cleavage of mature TGF-α proceeds in two steps. The first step is cleavage at the aminoterminal AV (Ala-Val) site. Another cleavage site in the vicinity of the transmembrane region (dark blue rounded rectangle) is located at Lys96-97 (KK). The intracellular domain of the factor is cysteine rich and remains anchored in the membrane. The dotted line marks the cleaved signal sequence at the N terminus. NST is Asn-Ser-Thr. The protein is glycosylated at the Asn (N) within this sequence.

endothelial cells and is a more potent » angiogenesis factor than EGF in several bioassays. TGF-α may play a role in the vascularisation of tumor tissues since it is produced by tumor cells and macrophages residing in this tissue.

TGF-α also affects bone formation and remodeling by inhibition of the synthesis of collagen and release of calcium (see also: Calcium ionophore). These effects are more pronounced than those of EGF. TGF-α also promotes the generation of osteoblast-like cells in long-term bone marrow cultures.

TGF-α-transgenics: see addendum.

Keratinocytes harboring a constitutively expressed TGF-α gene develop benign skin papillomas in nude mice (see also: Immunodeficient mice).

An increased production of TGF-α is usually associated with neoplastic transformation of various cell types. The expression of a membrane-bound TGF-α precursor leads to the transformation of » NRK (normal rat kidney) cells.

■ **Transgenic/Knock-out/Antisense Studies:** The role of TGF-α is exemplified in » transgenic animals expressing the factor in all tissues. These animals are characterized by a marked hyperproliferative state in some tissues. Overexpression of TGF-α mediated by the mMT-1 promoter induces epithelial hyperplasia, liver neoplasia, and the abnormal development of the mammary gland (carcinomas) and pancreas (metaplasia). The overexpression of a TGF-α transgene in mice has been described to lead to a high incidence of liver tumors. TGF-α may function as a promoter of liver carcinogenesis through its effect as an » autocrine inducer of hepatocyte proliferation with TGF-α overexpression also favoring tumor progression. The transgenic TGF-α mice also show reduced NK cell activity and increased plasma estradiol concentrations. TGF-α expression in transgenic mice also induces dramatic structural and functional lesions of the glandular stomach that are similar to Menetrier's disease in humans. These animals develop severe adenomatous hyperplasia resulting in nodular thickening or hypertrophy of the gastric mucosa. Secretions obtained from affected stomachs do not contain detectable gastric acid, suggesting a pronounced impairment of parietal cell functions.

Transgenic mice overexpressing TGF-α under control of the metallothionein promoter have, on average, 20% reductions in body and carcass weights compared to nontransgenic litter mates. This loss results from significant decreases in the comparative weights of bone, muscle, and especially fat. Despite its effects on adipose development, introduction of the metallothionein-TGF-α transgene into the ob/ob genetic background of obese mice does not suppress the marked obesity characteristic of this mutation.

Behavioral alterations (lengthened immobility in the swim test and elevated levels of aggression in the resident-intruder test) have been observed in some transgenic mice overexpressing TGF-α. These animals also have a reduced ratio between the metabolite of serotonin (5-HT), and 5-HT in the brain. It has been suggested that TGF α may influence behavior by affecting the uptake of 5-HT in neurons.

Transgenic animals co-expressing TGF-α and the » *myc* gene develop pancreas and liver tumors with a more malignant appearance than transgenic animals expressing *myc* alone. TGF-α transgenic mice show a dramatic enhancement in the incidence of liver tumors induced by diverse agents, demonstrating that TGF-α overexpression complements both initiation and promotion of hepatocarcinogenesis.

The physiological role of TGF-α has also been explored by disruption of the mouse gene by homologous recombination in embryonic stem cells (see: ES cells). Homozygous mutant mice are viable and fertile, but display pronounced waviness of the whiskers and fur, accompanied by abnormal curvature, disorientation, and misalignment of the hair follicles. Homozygous and, to a lesser extent, heterozygous mice also display eye abnormalities of variable incidence and severity, including open eyelids at birth, reduced eyeball size, and superficial opacity. Histological examination reveals eyelid and anterior segment dysgenesis, corneal inflammation and scarring, and lens and retinal defects. Wound healing in these tissues was not impaired. Similar hair and eye defects have been previously associated with the recessive mutation *waved*-1 (wa-1), and crosses between wa-1 homozygotes and TGF-α-deficient mice confirm that wa-1 and TGF-α are allelic.

■ **ASSAYS:** TGF-α can be assayed in a » colony formation assay by measuring the factor-induced substrate-independent growth of fibroblasts (see: NRK-49F). It can also be detected by means of a specific radioligand assay. For further information see also subentry "Assays" in the reference section.

■ **CLINICAL USE & SIGNIFICANCE:** Inhibitors of TGF-α may be of importance because of its activity as an » angiogenesis factor for tumor cells. A genetically engineered fusion protein consisting of TNF-α and *Pseudomonas* toxin (see also: Cytokine fusion toxins) is capable of killing tumor cells expressing increased amounts of the EGF receptor. An enhanced production of TGF-α has been observed in the development of psoriatic lesions. The factor may therefore also be involved in the pathogenesis of psoriasis. The simultaneous expression of TGF-α and » EGF receptors in primary tumors of advanced gastric cancer appears to correlate significantly with a very poor prognosis.

● **REVIEWS: Derynck R** The physiology of transforming growth factor-α. ACR 58: 27-52 (1992); **Pandiella A & Massague J** Transforming growth factor-α. Biochem. Soc. Trans. 19 259-62 (1991)

● **BIOCHEMISTRY & MOLECULAR BIOLOGY: Bosenberg MW et al** The cytoplasmic carboxy-terminal amino acid specifies cleavage of membrane TGFα into soluble growth factor. Cell 71: 1157-65 (1992); **Brissenden JE et al** Mapping of transforming growth factor α gene on human chromosome 2 close to the breakpoint of the Burkitt's lymphoma t(2;8) variant translocation. CR 45: 5593-7 (1985); **Cappelluti E et al** Potential role of two novel elastase-like enzymes in processing pro-transforming growth factor-α. B 32: 551-60 (1993); **Chiang T et al** Molecular cloning of the complementary DNA encoding for the hamster TGF-α mature peptide. Carcinogenesis 12: 529-32 (1991); **Jakobovits EB et al** The human transforming growth factor α promoter directs transcription initiation from a single site in the absence of a TATA-sequence. MCB 8: 5549-54 (1988); **Lee DC et al** Cloning and sequence analysis of a cDNA for rat transforming growth factor-α. N 313: 489-91 (1985); **Luetteke NC et al** TGFα deficiency results in hair follicle and eye abnormalities in targeted and waved-1 mice. **Massagué J** Transforming growth α, a model for membrane-anchored growth factors. JBC 265: 21393-6 (1990); **Prestrelski SJ et al** Solution structure and dynamics of epidermal growth factor and transforming growth factor α. JBC 267: 319-22 (1992); **Raymond VW et al** Regulation of transforming growth factor a messenger RNA expression in a chemically transformed rat hepatic epithelial cell line by phorbol ester and hormones CR 49: 3608-12 (1989); **Tricoli JV et al** The gene for human transforming growth factor α is on the short arm of chromosome 2. Cytogenet. Cell. Genet. 42: 94-8 (1986)

● **RECEPTORS: Wong ST et al** The TGF-α precursor expressed on the cell surface binds to the EGF receptor on adjacent cells, leading to signal transduction. Cell 56: 495-506 (1989)

● **BIOLOGICAL ACTIVITIES: Anklesaria P et al** Cell-cell adhesion mediated by binding of membrane-anchored transforming growth factor α to epidermal growth factor receptors promotes cell proliferation. PNAS 87: 3289-93 (1990); **Ju WD et al** Tumorigenic transformation of NIH 3T3 cells by the autocrine synthesis of transforming growth factor α. The New Biologist 3: 380-8 (1991); **Mead JE & Fausto N** Transforming growth factor α may be a physiological regulator of liver regeneration by means of an autocrine mechanism. PNAS 86: 1558-62 (1989); **Mueller SG & Kudlow JE** Transforming growth factor β (TGF-β) inhibits TGF α expression in bovine anterior pituitary-derived cells. ME 5: 1439-46 (1991); **Schreiber AB et al** Transforming growth factor α: a more potent angiogenic mediator than epidermal growth factor. S 232: 1250-3 (1986); **Smith JM et al** Human transforming growth factor-α causes precocious eyelid opening in newborn mice. N 315: 515-6 (1985); **Stromberg K et al** Transforming growth factor-α acts as an autocrine growth factor in ovarian carcinoma cell lines. CR 52: 341-7 (1992)

● **TRANSGENIC/KNOCK-OUT/ANTISENSE STUDIES: Bockman DE & Merlino G** Cytological changes in the pancreas of transgenic mice overexpressing transforming growth factor α. Gastroenterology 103: 1883-92 (1992); **Dempsey PJ et al** Possible role of transforming growth factor α in the pathogenesis of Menetrier's disease: supportive evidence form humans and transgenic mice. Gastroenterology 103: 1950-63 (1992); **Halter SA et al** Distinctive patterns of hyperplasia in transgenic mice with mouse mammary tumor virus transforming growth factor-α. Characterization of mammary gland and skin proliferations. Am. J. Pathol. 140: 1131-46 (1992); of the TGF-α integral membrane precursors induces transformation in nrk cells. O 5: 1213-21 (1990); **Hilakivi-Clarke LA et al** Alterations in behavior, steroid hormones and natural killer cell activity in male transgenic TGF α mice. Brain Res. 588: 97-103 (1992); **Hilakivi-Clarke LA et al** Opposing behavioral alterations in male and female transgenic TGF α mice: association with tumor susceptibility. Br. J. Cancer 67: 1026-30 (1993); **Hilakivi-Clarke LA & Goldberg R** Effects of tryptophan and serotonin uptake inhibitors on behavior in male transgenic transforming growth factor α mice. Eur. J. Pharmacol. 237: 101-8 (1993); **Jhappan C et al** TGF α overexpression in transgenic mice induces liver neoplasia and abnormal development of the mammary gland and pancreas. Cell 61: 1137-46 (1990); **Lee GH et al** Development of liver tumors in transforming growth factor α transgenic mice. CR 52: 5162-70 (1992); **Luetteke NC et al** TGF α deficiency results in hair follicle and eye abnormalities in targeted and waved-1 mice. Cell 73: 263-78 (1993); **Luetteke NC et al** Regulation of fat and muscle development by transforming growth factor α in transgenic mice and in cultured cells. Cell Growth Differ. 4: 203-13 (1993); **Mann GB et al** Mice with a null mutation of the TGF α gene have abnormal skin architecture, wavy hair, and curly whiskers and often develop corneal inflammation. Cell 73: 249-61 (1993); **Matsui Y et al** Development of mammary hyperplasia and neoplasia in MMTV-TGF α transgenic mice. Cell 61: 1147-55 (1990); **Sandgren EP et al** Overexpression of TGF α in transgenic mice: induction of epithelial hyperplasia, pancreatic metaplasia and carcinoma of the breast. Cell 61: 1121-35 (1990); **Sandgren EP et al** Transforming growth factor α dramatically enhances oncogene-induced carcinogenesis in transgenic mouse pancreas and liver. MCB 13: 320-30 (1993); **Sandgren EP et al** Transforming growth factor α dramatically enhances oncogene-induced carcinogenesis in transgenic mouse pancreas and liver. MCB 13: 320-330 (1993); **Takagi H et al** Molecular and genetic analysis of liver oncogenesis in transforming growth factor α transgenic mice. CR 52: 5171-7 (1992); **Takagi H et al** Hypertrophic gastropathy resembling Menetrier's disease in transgenic mice overexpressing transforming growth factor α in the stomach. JCI 90: 1161-7 (1992); **Takagi H et al** Collaboration between growth factors and diverse chemical carcinogens in hepatocarcinogenesis of transforming growth factor α transgenic mice. CR 53: 4329-36 (1993); **Vassar R & Fuchs E** Transgenic mice provide new insights into the role of TGF-α during epidermal development and differentiation. GD 5: 714-27 (1991)

● **Assays: Baldwin GS & Zhang QX** Measurement of gastrin and transforming growth factor α messenger RNA levels in colonic carcinoma cell lines by quantitative polymerase chain reaction. CR 52: 2251-7 (1992); **Hoffmann R & Cameron LA** scintillation proximity assay for transforming growth factor α (TGF α). Biochem. Soc. Trans. 19: 238S (1991); **Frolik CA & De-Larco JE** Radioreceptor assays for transforming growth factors. MiE 146: 95-102 (1987); **Lucas C et al** Generation of antibodies and assays for transforming growth factor α. MiE 198: 185-91 (1991); **Hoffman R & Cameron L** Characterization of a scintillation proximity assay to detect modulators of transforming growth factor α (TGF α) binding. AB 203: 70-5 (1992); **Inagaki H et al** A new sandwich enzyme-linked immunosorbent assay (ELISA) for transforming growth factor α (TGF α) based upon conformational modification by antibody binding. JIM 128: 27-37 (1990); **Kamiya Y et al** Transforming growth factor-α activity in effusions: comparison of radioimmunoassay and radioreceptor assay. Life Sci. 52: 1381-6 (1993); **Okada M et al** Transforming growth factor (TGF)-α in human milk. Life Sci. 48: 1151-6 (1991); **Smith J & McLachlan JC** Identification of a novel growth factor with transforming activity secreted by individual chick embryos. Development 109: 905-10 (1990)

● **Clinical Use & Significance: Chaudhary VK et al** Activity of a recombinant fusion protein between transforming growth factor type α and *Pseudomonas* toxin. PNAS 84: 4538-42 (1987); **Debinski W et al** Substitution of foreign protein sequences into a chimeric toxin composed of transforming growth factor α and *Pseudomonas* exotoxin. MCB 11: 1751-3 (1991); ; **Yonemura S et al** Interrelationship between transforming growth factor-α and epidermal growth factor receptor in advanced gastric cancer. Oncology 49: 157-61 (1992)

TGF-β: *transforming growth factor β* also abbr.: TGF-B or B-TGF.

■ **Alternative Names:** Aqueous humor lymphocyte inhibitory activity; CIF-A (cartilage-inducing factor A = TGF-β1); CIF-B (cartilage-inducing factor B = TGF-β2); DIF (differentiation-inhibiting factor); DSF (decidual suppressor factor = TGF-β2); EIF (Epstein-Barr virus inducing factor); epithelial cell growth inhibiting factor; G-TsF (glioma-derived T cell suppressor factor); MDGF (milk-derived growth factor); MGF (milk growth factor); MGF-a (milk growth factor = TGF-β2); MGF-b (milk growth factor = TGF-β1); PDGI (platelet-derived (endothelial cell) growth inhibitor = TGF-β1); polyergin; Simian BSC-1 cell growth inhibitor; SP factor; TCGF (transformed cell growth factor); TGF-β2; TGF-2; TGI (tissue-derived growth inhibitor); TIF-1 (tumor-inducing factor 1). See also: individual entries for further information.

■ **Sources:** Platelets yield milligram amounts of TGF-β/ kilogram. The factor and its isoforms (see below) can also be isolated from other tissues (μg TGF/kg) and is found predominantly in spleen and bone tissues. Human milk also contains this factor (see: MGF, milk growth factor). It is also synthesized for example by macrophages (TGF-β1), lymphocytes (TGF-β1), endothelial cells (TGF-β1), keratinocytes (TGF-β2), granulosa cells (TGF-β2), chondrocytes (TGF-β1), glioblastoma cells (TGF-β2), leukemia cells (TGF-β1).

Depending upon cell type or conditions, the secretion of TGF-β can be induced by a number of different stimuli including steroids, retinoids, » EGF, » NGF, activators of lymphocytes (see also: Cell activation), vitamin D3, and » IL1. The synthesis of TGF-β can be inhibited by EGF, » FGF, dexamethasone, calcium (see also: Calcium ionophore), retinoids and follicle-stimulating hormone. TGF-β also influences the expression of its own gene and this may be important in » Wound healing.

TGF-β can exist associated with the extracellular matrix (see: ECM) as a complex with betaglycan and decorin. This allows the factor to be stored in a biologically inactive form. The exact molecular mechanisms underlying its release from these reservoirs is unknown.

■ **PROTEIN CHARACTERISTICS:** TGF-β exists in at least five isoforms, known as β1 to β5 that are not related to TGF-α. Their amino acid sequences display homologies on the order of 70-80 %. TGF-β1 is the prevalent form and is found almost ubiquitously while the other isoforms are expressed in a more limited spectrum of cells and tissues.

The biologically active forms of all isoforms are disulfide-linked homodimers. Disulfide-linked heterodimers of TGF isoforms have also been reported. The heat- and acid-stable monomeric subunits have a length of 112 amino acids. TGF-β4 contains two additional amino acids in the vicinity of the aminoterminal end.

The isoforms of TGF-β arise by proteolytic cleavage of longer precursors (TGF-β1: 390 amino acids, TGF-β2: 412 amino acids, TGF-β3: 412 amino acids, TGF-β4: 304 amino acids, TGF-β5: 382 amino acids). The isoforms are derived from the carboxyterminal ends of these precursors.

Isoforms isolated from different species are evolutionarily closely conserved and have sequence identities on the order of 98 %. Mature human, porcine, simian and bovine TGF-β1 are identical and differ from murine TGF-β1 in a single amino acid position. Human and chicken TGF-β1 are also identical.

Almost all forms of TGF-β are released as biologically inactive forms that are also known as ***L-TGF*** (latent TGF). Latent forms are complexes of TGF-β, an aminoterminal portion of the TGF-β precursor, designated ***TGF-LAP*** (TGF-latency associated peptide), and a specific binding protein (205 kDa; 125-160 kDa in platelets), known as ***LTBP*** (latent TGF binding protein, 1394 amino acids).

L-TGF has been shown to be localized at the cell surface by binding to the mannose-6-phosphate/insulin-like growth factor II receptor (see: MPR) through mannose-6-phosphate-containing carbohydrates of TGF-LAP and through interaction with the L-TGF binding protein. LTBP may be processed in a cell-type specific manner. Biologically active TGF-β results after dissociation from the LAP complex. The nature of the activation mechanism of L-TGF *in vivo* is unclear. It may involve direct cell-to-cell contacts, proteases, specifically plasmin, and has recently also been shown to be mediated by transglutaminases (EC 2.3.2.13). Another potent physiologic regulator of TGF-β activation is » thrombospondin. TGF-β2 is unique among various isoforms in that it lacks an RGD integrin-binding sequence in its precursor.

The mechanisms of TGF-β activation are not known in detail. The main fraction of the factor in the serum is covalently attached to the » acute phase protein » Alpha-2-macroglobulin (α_2M) the synthesis of which is known to be induced several hundred-fold by » IL6. α_2M/TGF-β complexes are believed to represent TGF-β molecules released by platelets after tissue injuries and destined to degradation.

Mutant forms of TGF-β have been created. They form wild-type/mutant heterodimers deficient in assembly or processing. Such mutants behave as dominant negative mutants and are useful in in-

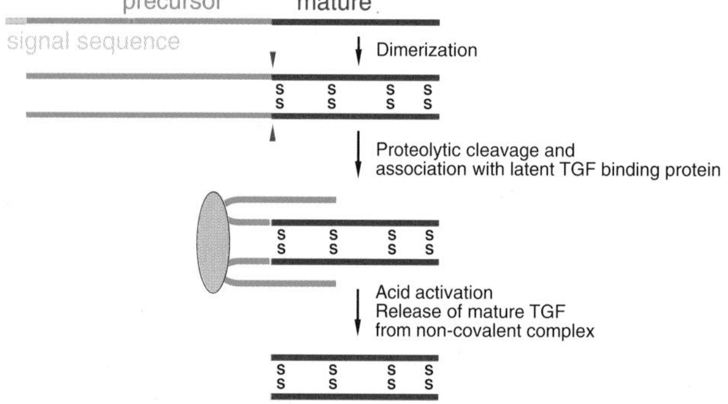

Biosynthesis and processing of mature TGF-β.
TGF-β (dark blue) is synthesized initially as a long precursor molecule containing a signal sequence. When this is removed the remaining part of the molecule forms a disulfide-bonded dimer. Proteolytic cleavage (red arrow heads) yields a complex in which the mature TGF-β protein remains associated with the rest of the precursor molecule (green). The complex is held together by means of a binding protein. TGF-β in this complex is said to be in a latent form. Acid activation releases mature biologically active TGF-β from this non-covalent complex.

vestigation of the role of TGF-β in normal and pathological conditions.

■ **GENE STRUCTURE:** The different isoforms of TGF-β are encoded by different genes. All genes have a length of more than 100 kb and contain seven exons. The genes map to different chromosomes. The TGF-β1 gene maps to human chromosome 19q13; TGF-β2 maps to human chromosome 1q41; TGF-β3 maps to human chromosome 14q24. These genes are differentially expressed. The TGF-β3 gene is strongly expressed in embryonic heart and lung tissues but only marginally in liver, spleen, and kidney tissues. TGF-β1 is strongly expressed in spleen tissues.

■ **RELATED FACTORS:** TGF-β is the prototype of a family of proteins known as the TGF-β superfamily. This family includes » inhibins, » Activin A, » MIS (Müllerian inhibiting substance), » BMP (bone morphogenetic protein), » *dpp* (*decapentaplegic*) and » *Vg*-1. MNSF (monoclonal nonspecific suppressor factor) shows 60 % sequence identity with TGF-β2.

■ **RECEPTOR STRUCTURE, GENE(S), EXPRESSION:** An entire family of glycoprotein receptors for TGF-β has emerged. Some of these proteins do not bind TGFβ-related factors belonging to the TGF-β family. Some receptors bind TGF-β and the related factors » Activin A and » Inhibin. A type V receptor (proteoglycan, 400 kDa) has also recently been described.

Type 1 receptors and ***type 2 receptors*** have a molecular mass of 53 and 75 kDa, respectively. Signaling through the receptor requires heterodimeric complexes between both types of receptors. The type II receptor has been shown to bind the ligand, but it is incapable of mediating TGF-β responses in the absence of the type I receptor. The type 1 receptor is expressed predominantly in hematopoietic progenitor cells. The type 2 receptor encodes a protein with an intracellular domain that functions as a serine/threonine kinase.

Individual TGF-b isotypes bind to the receptors with varying affinities. For example, TGF-β1 binds ≈ tenfold better than TGF-β2. In several cell types the expression of the TGF-β receptors is decreased by » EGF (see also: receptor transmodulation). In endothelial cells the expression of the TGF-β receptor is decreased by » bFGF.

Almost all types of cells express a 560 kDa high-affinity receptor for TGF-β, called ***type 3 receptor***. The receptor density can vary between 600 and 80000 per cell. This receptor type is not expressed in primary epithelial, endothelial, and lymphoid cells.

The type 3 receptor is a proteoglycan (Betaglycan; see also: ECM) with a length of 853 amino acids. This receptor contains a transmembrane domain and a short cytoplasmic domain of 41 amino acids. The type 3 receptor binds TGF-β1 and TGF-β2 equally well. Membrane betaglycan appears to be a regulator of TGF-β access to the signaling receptors; it increases binding of TGF-β to the signaling receptor, enhances cell responsiveness to TGF-β, and eliminates some of the marked biological differences between different TGF-β isoforms.

The major TGF-β1-binding protein co-existing with TGF-β receptors I and II on human umbilical vein endothelial cells appears to be ***endoglin***, a dimeric membrane glycoprotein of 95 kDa that shares regions of sequence identity with betaglycan and that is expressed at high levels on human vascular endothelial cells.

```
Vg-1        RPRRKRSYSKLPFTASNICKKRHLYVEFK-DVGWQNWVIAPQGYMANYCYGECPYPLTEILN
TGF-β       RHRRALDTNYCFSSTEKNCCVRQLYIDFRKDLGWK-WIHEPKGYHANFCLGPCPYIWS-LL-
β-Inhibin   HPHRRRRRGLECDGKVNICCKKQFFVSF-KDIGWNDWIIAPSGYHANYCEGECPSHIAGTSG
dpp         RNKRHARRPTRRKNHDDTCRRHSLYVDFS-DVGWDDWIVAPLGYDAYYCHGKCPFPLADHFN

Vg-1        GSNHAILQTLVHSIEPEDIP----LPCCVPTKMSPISMLFYDNNDNVVLRHYENMAVDECGCR
TGF-β       DTQYSKVLALYQQHNPGASA----APCCVPQALEPLPIVYYV-GRKPKVEQLSNMIVRSCKCS
β-Inhibin   SSLSFHSTVINYYRMRGHSPFANLKSCCVPTKLRSMSMLYYDDGQNIIKKDIQNMIVEECGCS
dpp         STNHAVVQTLVNNNNPGKVP----KACCVPTQLDSVAMLYLNDQSTVVLKNYQEMTVVGCGCR
```

Comparison of some sequences of factors belonging to the TGF-β family.
Conserved cysteine residues are shown in yellow. Identical amino acid positions in at least two proteins are shown in blue.

Two human receptors, designated ***TSR-I*** and ***AcrR-I***, have been cloned and found to encode transmembrane serine/threonine kinases distantly related to known mammalian TGF-β and » activin A receptors. These new receptor types form heterodimeric complexes with type II TGF-β receptors which possess signaling capacities differing from those of the other receptor types.
Retinoblastoma cells, undifferentiated embryonic stem cells and several leukemic cell types do not express TGF receptors. In endothelial cells the expression of TGF receptors is down-regulated by » FGF.
TGF receptor-associated signal transduction mechanisms are largely unknown. None of the receptors appears to encode a tyrosine-specific protein kinase found in the intracellular domain of many other cytokine receptors (see also: PTK, protein tyrosine kinase). Signal transduction may involve a G-protein.
■ **Biological Activities:** The biological activities of TGF-β are not species-specific. The various TGF-β isotypes share many biological activities and their actions on cells are qualitatively similar in most cases although there are a few examples of distinct activities. In some systems TGF-β3 appears to be more active than the other isotypes. TGF-β2 is the only variant that does not inhibit the growth of endothelial cells. TGF-β2 and 3, but not TGF-β1, inhibit the survival of cultured embryonic chick ciliary ganglionic neurons. The most pronounced differences in the TGF-β isoforms is their spatially and temporally distinct expression of both the mRNAs and proteins in developing tissues, regenerating tissues, and in pathologic responses.
TGF-β is the most potent known growth inhibitor for normal and transformed epithelial cells, endothelial cells, fibroblasts, neuronal cells, lymphoid cells and other hematopoietic cell types (see also: CFU-S), hepatocytes, and keratinocytes.
TGF-β inhibits the proliferation of T lymphocytes by down-regulating predominantly IL2-mediated proliferative signals. It also inhibits the growth of natural killer cells *in vivo* and deactivates macrophages. TGF-b blocks the antitumor activity mediated *in vivo* by » IL2 and transferred lymphokine-activated or tumor-infiltrating lymphocytes (see: LAK cells, TIL). TGF-β inhibits the synthesis of » GM-CSF, » IL3, and the expression of the receptor for » G-CSF. It also inhibits the » IL3-, » GM-CSF-, and » M-CSF-induced growth of immature hematopoietic progenitor cells, in particular of » CFU-GEMM. TGF-β1 inhibits CFU-GEMM ≈ 100-fold stronger than TGF-β2. TGF-β also inhibits megakaryocytopoiesis (see: CFU-GM). In many cell types TGF-β antagonizes the biological activities of » EGF, » PDGF, » aFGF and » bFGF. TGF-β1 and TGF-β2 inhibit the IL1-induced proliferation of lymphocytes. The latent form of TGF-β is a strong inhibitor of erythroleukemia cell lines.
The extent of growth inhibition induced by TGF-β depends on the cell type, on the concentration of TGF-β, and on the presence of other factors. Epithelial cells of the lung and keratinocytes are practically arrested in their growth. The growth of some cells is inhibited by TGF-β although these cells secrete TGF-β complexes and can growth in the presence of these biologically inactive complexes. The growth-inhibitory activities of TGF-β can be abolished by » HGF (hepatocyte growth factor). Some mutations occurring in the TGF precursor protein have been described that act as dominant negative mutations. These altered proteins can inhibit the secretion of normal TGF isotypes if they are produced simultaneously.
At concentrations of 1-2 fg/cell TGF-β is a growth inhibitor for smooth muscle cells, fibroblasts, and chondrocytes. At higher concentrations TGF-β stimulates the growth of these cells. This bimodal activity is mediated in part by » PDGF (A chain homodimer) the synthesis and secretion of which is stimulated by low concentrations of TGF-β, while higher concentrations lower the expression of the PDGF receptors and hence diminish the biological effects of PDGF.
Although TGF-β inhibits the growth of endothelial cells it is an » angiogenesis factor in several bioassays. Under certain circumstances TGF-β may, however, also inhibit angiogenesis. It has been suggested that overproduction of TGF-β1 by tumor cells can contribute to neovascularization and may help promote tumor development *in vivo*. TGF-β is an » autocrine growth factor for malignant gliomas. TGF-β stimulates the growth of some cell types of mesenchymal origin including fibroblasts and osteoblasts *in vivo* and *in vitro*. TGF-β1 and TGF-β2 promote the proliferation of Schwann cells. TGF-β induces the synthesis of bone matrix proteins in osteoblasts. The anabolic activity on bone cells is antagonized by glucocorticoids. Factors that promote bone resorption (IL1, 1,25-Dihydroxy-vitamin D3, parathormone) induce the synthesis of TGF-β in bone cells while calcitonin, an inhibitor of bone resorption, reduces

the synthesis of TGF-β. TGF-β regulates the expression of the » PDGF β chain (see also: *sis* oncogene).
TGF-β appears to be one of the major factors promoting neurogenesis in the olfactory epithelium that gives rise to olfactory neurons. In pituitary cells TGF-β inhibits the synthesis of » TGF-α. TGF-β also suppresses activation and » GM-CSF- or » M-CSF-induced proliferation of microglia; it suppresses the IFN-γ-induced expression of class II MHC antigens and the production of » IL1, » IL6, and TNF-α by these cells and may thus play an important role in the development of various diseases in the central nervous system by inhibiting the functions of microglia in inflammation or in immunoregulation.
TGF-β stimulates the synthesis of the major matrix proteins (see also: ECM) including collagen, proteoglycans, glycosaminoglycans, fibronectin, integrins, » thrombospondin, osteonectin, osteopontin. It inhibits their degradation mainly by inhibiting the synthesis of neutral metalloproteinases (see: » TIMP, tissue inhibitor of metalloproteases; see also: EPA = erythroid promoting activity) and by increasing the synthesis of proteinase inhibitors.
TGF-β also regulates the expression of plasminogen activator and plasminogen activator inhibitor. The gene encoding plasminogen activator I inhibitor contains a specific TGF-β1 responsive element in its promoter region which mediates the binding of specific transcription factors (see also: Gene expression). TGF-β also inhibits the secretion of proteases (plasminogen activator, cathepsin L, stromelysin, collagenase). These activities of TGF-β are antagonized by » IL1.
TGF-β1 modulates the interaction of tumor cells with the cellular matrix. It is therefore probably also involved in metastatic processes together with other protein factors. TGFβ3 is involved in the early development of the heart. It is responsible for the transformation of epithelial cells within the membranes into mesenchymal cells.
TGF-β has mainly suppressive effects on the immune system since it inhibits the IL2-dependent proliferation of T and B lymphocytes. TGF-β inhibits the proliferation of B lymphocytes, IL2-induces T lymphocytes, and IL1-induced thymocytes. It also inhibits the maturation of B cells. It also suppresses the interferon-induced cytotoxic activity of natural killer cells, the activity of cytotoxic T lymphocytes, and the proliferation of lymphokine-activated killer cell precursors.
TGF-β also influences the secretion of immunoglobulins by B lymphocytes. The synthesis of IgG and IgM by B lymphocytes is inhibited while the synthesis of IgA is stimulated (see also: Isotype switching).
TGF-β modulates the synthesis of » acute phase proteins by hepatocytes and inhibits the ACTH-induced synthesis of cortisol.
TGF-β1 is the most potent known chemoattractant for neutrophils (see also: Chemotaxis). It specifically attracts neutrophils at concentrations of 1 fM and is therefore ≈ 150000-fold more potent than » fMLP (Formyl-Met-Leu-Phe).
TGF-β stimulates the expression of » IL1, » PDGF, » bFGF in monocytes and inhibits the synthesis of » TNF-α, » TNF-β and » IFN-γ in adherent human monocytes and murine macrophages.
TGF-β stimulates the synthesis of estrogens, mediated by FSH (follicle-stimulating hormone), the synthesis of progesterone in ovarian granulosa cells. It inhibits the synthesis of testosterone in Leydig cells.

■ **TRANSGENIC/KNOCK-OUT/ANTISENSE STUDIES:** The biological consequences of the overexpression of TGF-β have been studied in » transgenic animals (mice) overexpressing the factor in the epidermis. These animals are characterized by a marked reduction in the number of replicating cells in the epidermis and hair follicles. They show a compact orthohyperkeratosis and a reduction in the number of hair follicles. The skin is shiny and tautly stretched, and the animals are rigid and appear to be restricted in their ability to move and breathe- These animals usually die within 24 hrs after birth.
Overexpression of TGF-β has been investigated also in transgenic mice harboring a fusion gene consisting of a porcine TGF-β1 cDNA placed under the control of regulatory elements of the pregnancy-responsive mouse whey-acidic protein gene. Females from two of four transgenic lines have been found to be unable to lactate due to inhibition of the formation of lobuloalveolar structures and suppression of production of endogenous milk protein. TGF-β1 thus may play an important *in vivo* role in regulating the development and function of the mammary gland.
Disruption of the TGF-β1 gene by homologous recombination in murine embryonic stem cells (see: ES cells) enables mice to be generated that carry the disrupted allele. Animals homozygous for the mutated TGF allele do not produce detec-

table amounts of either TGF-β1 RNA or protein. They show no gross developmental abnormalities, but about 20 days after birth they succumb to a wasting syndrome accompanied by a multifocal, mixed inflammatory cell response and tissue necrosis, leading to organ failure and death. Pathological examination reveals an excessive inflammatory response with massive infiltration of lymphocytes and macrophages in many organs, but primarily in heart and lungs. Many lesions resemble those found in autoimmune disorders, graft-versus-host disease, or certain viral diseases. This phenotype suggests a prominent role for TGF-β1 in homeostatic regulation of immune cell proliferation and extravasation into tissues. It has been shown that TGF-b1-negative mice express elevated levels of both the class I and class II major histocompatibility antigens.

■ **Assays:** TGF-β can be detected in a » Bioassay involving the use of cell lines that respond to the factor (see: AKR-2B; CCL-64; NRK; MV-3D9; NRK-49F; TF-1). TGF-β can also be assayed in sensitive ELISA test or by a radioreceptor assay or with colorimetric methods. An alternative and entirely different detection method is » Message amplification phenotyping. For further information see also subentry "Assays" in the reference section.

■ **Clinical Use & Significance:** Many of the biological activities of TGF-β point to the fact that it may be a potent regulator of » wound healing and of bone fracture healing. Local application of TGF-β has been shown to accelerate wound healing. In combination with bone morphogenetic protein 2 (see: BMP) TGF-β plays an important role in the development of ossification of the posterior longitudinal ligament of the cervical spine. The factor may be helpful in the treatment of traumatic tissue injuries and in the treatment of osteoporosis. TGF-β2 has been used for the treatment of full-thickness macular holes and has been shown to improve visual acuity. In one incisional wound healing rat model systemic administration of TGF-β has been shown to enhance wound healing and to reverse age- or glucocorticoid-impaired wound healing even if given 24 hours *before* wounding. Although the exact mechanism of action is unclear it has been suggested that this may be due to the ability of TGF-β to induce its own synthesis. TGF-β may therefore prime macrophages and fibroblasts throughout the body to respond more effectively to future injury and if this could be verified systemic treatment with one priming does of TGF-β would represent a novel approach to endocrine replacement therapy.

Antagonists of TGF-β may be valuable in the treatment of fibrotic disorders which are often associated with increases in TGF-β activities. TGF-β has been implicated as a mediator of a range of inflammatory disorders such as rheumatoid arthritis and nephritis, myelofibrosis, scleroderma, and pulmonary fibrosis.

Animal experiments demonstrate that TGF-β has cardioprotective activities following ischemic injuries. This activity is probably due to the inhibition of processes mediated by » TNF-α. In addition TGF-β minimizes the loss of the protective factor EDRF (endothelium-derived relaxing factors). EDRF is identical with nitrogen monoxide that mediates acetylcholine- and thrombin-induced vasodilatation and that also inhibits platelet aggregation.

Immunohistochemical staining of TGF-β (and also of EGF) in endoscopic biopsy specimens may be useful for the diagnosis of the penetrating type of early gastric cancer and also for the diagnosis of the initial lesion of linitis plastica-type gastric cancer.

Expression of TGF-β1 does not appear to correlate with histological grade, estrogen receptor status, EGF receptor status, and Ki-67 labeling in mammary carcinomas. However, prominent reactivity appears to correlate with node status and metastasizing potential, suggesting that TGF-β1 may be a determining factor for invasion and metastasis.

TGF-β produced by neonatal rat cardiac myocytes stabilizes the beating rate and sustains their spontaneous rhythmic beating in serum-free medium. This effect is inhibited by I IL1β (see: IL1), an inflammatory mediator secreted by immune cells during myocardial injury, and TGF-β can overcome this inhibition. TGF-β may therefore be an important regulator of contractile function of the heart and have significant implications for understanding cardiac physiology in health and disease.

● **Reviews: Barnard JA et al** The cell biology of transforming growth factor β. BBA 1032: 79-87 (1990); **Bonewald LF & Mundy GR** Role of transforming growth factor β in bone remodeling: A review. Connect. Tissue Res. 23: 201-8 (1989); **Border WA & Ruoslahti E** Transforming growth factor-β in disease: the dark side of tissue repair. JCI 90: 1-7 (1992); **Burt DW** Evolutionary grouping of the transforming growth factor-β superfamily. BBRC 184: 590-5 (1992); **Gentry LE et al** Type 1 transforming growth factor β: amplified expression and secretion of mature and precursor polypeptides in Chinese hamster ovary cells. MCB 7: 3418-27 (1987); **Hooper WC** The role of transforming growth factor-β in hematopoiesis. A review. Leukemia Res. 15: 179-84 (1991); **Ignotz RA** TGF-β and extracellular matrix-related influences on gene expression and phenotype. Crit. Rev. Eukaryotic Gene Expr. 1: 75-84 (1991); **Joyce ME et al** Transforming growth factor-β in the regulation of fracture repair. Orthop. Clin. North Am. 21: 199-209 (1990); **Kingsley DM** The TGF-β superfamily: new members, new receptors, and new genetic tests of function in different organisms. Genes Dev. 8: 133-46 (1994); **Logan A & Berry M** Transforming growth factor β1 and basic fibroblast growth factor in the injured CNS. TIPS 14: 337-43 (1993); **Miyazono K et al** Receptors for transforming growth factor β. Adv. Immunol. 55: 181-220 (1994); **Miyazono K & Heldin CH** Structure, function and possible clinical application of transforming growth factor-β. J. Dermatol. 19: 644-7 (1992); **Pfeilschifter J** Transforming growth factor-β. In: Habenicht A (edt) Growth factors, differentiation factors, and cytokines, pp. 56-64, Springer, Berlin 1990); **Roberts AB et al** Transforming growth factor β: biochemistry and roles in embryogenesis, tissue repair and remodeling, and carcinogenesis. Recent Progr. Horm. Res. 44: 157-97 (1988); **Roberts AB & Sporn MB** The transforming growth factor-βs. In: Sporn MB & Roberts AB (eds) Handbook of experimental pharmacology, Vol. 95, I. Peptide growth factors and their receptors, pp. 419-72, Springer, Berlin 1990; **Roberts AB & Sporn MB** Physiological actions and clinical applications of transforming growth factor-β (TGF-β). GF 8: 1-9 (1993); **Sporn MB & Roberts AB** Transforming growth factor-β: recent progress and new challenges. JCB 119: 1017-21 (1992)

● BIOCHEMISTRY & MOLECULAR BIOLOGY: **Barton DE et al** Chromosomal mapping of genes for transforming factors β2 and β3 in man and mouse: dispersion of the TGF-β gene family. OR 3: 323-31 (1988); **Burmester JK et al** Characterization of distinct functional domains of transforming growth factor β. PNAS 90: 8628-32 (1993); **Burt DW et al** Comparative analysis of human and chicken transforming growth factor-β2 and-β3 promoters. J. Mol. Endocrinol. 7: 175-83 (1991); **Cheifetz S et al** The transforming growth factor-β receptor type III is a membrane proteoglycan. JBC 263: 16984-91 (1988); **Denhez F et al** Cloning by polymerase chain reaction of a new TGF-β, TGF-β3. GF 3: 139-46 (1990); **Derynck R et al** Human transforming growth factor-β cDNA sequence and expression in tumor cell lines. N 316: 701-5 (1985); **Derynck R et al** A new type of transforming growth factor-β, TGF-β 3. EJ 7: 3737-43 (1988); **Derynck R et al** Intron-exon structure of human transforming growth factor-β precursor gene. NAR 15: 3188-9 (1987); **Dickinson ME et al** Chromosomal localization of seven members of the murine TGF-β superfamily suggests close linkage to several morphogenetic mutant loci. Genomics 6: 505-20 (1990); **Flaumenhaft R et al** Role of the latent TGF-β binding protein in the activation of latent TGF-β by co-cultures of endothelial and smooth muscle cells. JCB 122: 995-1002 (1993); **Hoffmann FM** Transforming growth factor-β-related genes in *Drosophila* and vertebrate development. COCB 3: 947-52 (1991); **Jakowlew SB et al** Complementary deoxyribonucleic acid cloning of a messenger ribonucleic acid encoding transforming growth factor β 4 from chicken embryo chondrocytes. ME 2: 1186-95 (1988); **Kojima S et al** Requirement for transglutaminase in the activation of latent transforming growth factor-β in bovine endothelial cells. JCB 121: 439-48 (1993); **Kondaiah P et al** Identification of a novel transforming growth factor-β (TGF-β5) in *Xenopus laevis*. JBC 265: 1089-93 (1990); **Kramer IM et al** TGF-β1 induces phosphorylation of the cyclic AMP responsive element binding protein in ML-CCI64 cells. EJ 10: 1083-9 (1991); **Lopez AR et al** Dominant negative mutants of transforming growth factor-β inhibit the secretion of different transforming growth factor β isoforms. MCB 12: 1674-9 (1992); **Lyons RM et al** Transforming growth factors and the regulation of cell proliferation. EJB 187: 467-73 (1990); **Miller DA et al** Murine transforming growth factor-β2 cDNA sequence and expression in adult tissues and embryos. ME 3: 1108-14 (1989); **Noma T et al** Molecular cloning and structure of the human transforming growth factor β 2 gene promoter. GF 4: 247-55 (1991); **Ogawa Y et al** Purification and characterization of transforming growth factor-β 2.3 and -β 1.2 heterodimers from bovine bone. JBC 267: 2325-8 (1992); **Olofsson A et al** Transforming growth factor-β 1, -β 2, and -β 3 secreted by a human glioblastoma cell line. Identification of small and different forms of large latent complexes. JBC 267: 19482-8 (1992); **Riccio A et al** Transforming growth factor β1-responsive element: closely associated binding sites for USF and CCAAT-binding transcription factor-nuclear factor I in the type 1 plasminogen activator inhibitor gene. MCB 12: 1846-55 (1992); **Schultz-Cherry S & Murphy-Ullrich JE** Thrombospondin causes activation of latent transforming growth factor-β secreted by endothelial cells by a novel mechanism. JCB 122: 923-32 (1993); **Sharples K et al** Cloning and sequence analysis of simian transforming growth factor-β cDNA. DNA 6: 239-44 (1987); **ten Dijke P et al** Identification of a new member of the transforming growth factor-β gene family. PNAS 85: 4715-9 (1988)

● RECEPTORS: **Attisano L et al** Identification of human activin and TGFβ type I receptors that form heterodimeric kinase complexes with type II receptors. Cell 75: 671-80 (1993); **Bassing CH et al** A transforming growth factor b type I receptor that signals to activate gene expression. S 263: 87-9 (1994); **Cheifetz S et al** Distinct transforming growth factor-β (TGF-β) receptor subsets as determinants of cellular responsiveness to three TGF-β isoforms. JBC 265: 20533-8 (1990); **Cheifetz S et al** Endoglin is a component of the transforming growth factor-β receptor system in human endothelial cells. JBC 267: 19027-30 (1992); **Chen RH et al** Inactivation of the type II receptor reveals two receptor pathways for the diverse TGF-β activities S 260: 1335-8 (1993 (; **Fafeur V et al** Basic FGF treatment of endothelial cells down-regulates the 85-KDa TGF β receptor subtype and decreases the growth inhibitory response to TGF-β 1. GF 3: 237-45 (1990); **Franzén P et al** Cloning of TGFβ type I receptor that forms a heterodimeric complex with TGFβ type II receptor. Cell 75: 681-92 (1993); **Hirai R & Kaji K** Transforming growth factor β 1-specific binding proteins on human vascular endothelial cells. Exp. Cell. Res. 201: 119-25 (1992); **Lin HY et al** Expression cloning of the TGF-β type II receptor, a functional transmembrane serine/threonine kinase. Cell 68: 775-85 (1992); **López-Casillas F et al** Betaglycan presents ligand to the TGFβ signaling receptor. Cell 73: 1435-44 (1993); **Massagué J et al** Mediators of TGF-β action: TGF-β receptors and TGF-β binding proteoglycans. ANY 593: 59-72 (1990); **Massagué J** Receptors for the TGF-β family. Cell 69: 1067-70 (1992); **Miyazono K et al** A role of the latent TGF-β1-binding protein in the assembly and secretion of TGF-β1. EJ 10: 1091-101 (1991); **O'Grady P et al** Expression of a new type high molecular weight receptor (type V receptor) of transforming growth factor β in normal and transformed cells. BBRC 179: 378-85 (1991); **O'Grady P et al** Transforming growth factor β (TGF-β) type V receptor has a TGF-β-stimulated serine/threonine-specific autophosphorylation activity. JBC 267: 21033-7 (1992); **Wang XF et al** Expression cloning and characterization of the TGF-β type III receptor. Cell 67: 797-805 (1991); **Wieser R et al** Signaling activity of transforming growth factor b type II receptors lacking specific domains in the cytoplasmic region. MCB 13: 7239-47 (1993)

● BIOLOGICAL ACTIVITIES: **Ahuja SS et al** Effect of transforming growth factor-β on early and late activation events in human T cells. JI 150: 3109-18 (1993); **Battegay EJ et al** TGF-β induces bimodal proliferation of connective tissue cells via complex control of an autocrine PDGF loop. Cell 63: 515-24 (1990); **Centrella M et al** Transforming growth factor-β and remodeling of bone. J. Bone Joint Surg. [Am.] 73A: 1418-28 (1991); **Feige JJ et al** Transforming growth factor β1: An autocrine regulator of adrenocortical steroidogenesis. Endocrine Res. 17: 267-79 (1991); **Filmus J & Kerbel RS** Development of resistance mechanisms to the growth-inhibitory effects of transforming growth factor-β during tumor progression. Curr. Opin. Oncol. 5: 123-9 (1993); **Iruela-Arispe ML & Sage EH** Endothelial cells exhibiting angiogenesis *in vitro* proliferate in response to TGF-b1. JCBc 52: 414-430 (1993); **Jacobsen SE et al** Transforming growth factor-β trans-modulates the expression of colony stimulating factor receptors on murine hematopoietic progenitor cell lines. Blood 77: 1706-16 (1991); **Janat MF & Liau G** Transforming growth factor β1 is a powerful modulator of platelet-derived growth factor action in vascular smooth muscle cells. JCP 150: 232-42 (1992); **Jennings MT et al** TGFβ1 and TGFβ2 are potential growth regulators for low-grade and malignant gliomas *in vitro*: Evidence in support of an autocrine hypothesis. IJC 49: 129-39 (1991); **Lebman DA et al** Molecular characterization of germ-line immunoglobulin A transcripts produced during transforming growth factor type β-induced isotype switching. PNAS 87: 3962-6 (1990); **Lefer AM et al** Mediation of cardioprotection by transforming growth factor-β. S 249: 61-4 (1990); **Lopez AR** Dominant negative mutants of transforming growth factor-β1 inhibit the secretion of different transforming growth factor-β isoforms. MCB 12: 1674-9 (1992); **Madri JA et al** Phenotypic modulation of endothelial cells by transforming growth factor-β depends upon the composition and organization of the extracellular matrix. JCB 106: 1375-84 (1988); **Marcelli C et al** *In vivo* effects of human recombinant transforming growth factor β on bone turnover in normal mice. J. Bone Miner. Res. 5: 1087-96 (1990); **McCartney-Francis N et al** TGF-β regulates production of growth factors and TGF-β by human peripheral blood monocytes. GF 4: 27-35 (1990); **Mahanthappa NK & Schwarting GA** Peptide growth factor control of olfactory neurogenesis and neuron survival *in vitro*: roles of EGF and TGF-β s. Neuron 10: 293-305 (1993); **Meager A** Assays for transforming growth factor β. JIM 141: 1-14 (1991); **Moses HL et al** TGF-β stimulation and inhibition of cell proliferation: new mechanistic insights. Cell 63: 245-7 (1990); **Mueller SG & Kudlow JE** Transforming growth factor β (TGF-β) inhibits TGF α expression in bovine anterior pituitary-derived cells. ME 5: 1439-46 (1991); **Myoken Y et al** Bifunctional effects of transforming growth factor-β (TGF-β) on endothelial cell growth correlate with phenotypes of TGF-β binding sites. ECR 191: 299-304 (1990); **Piao YF et al** Latent form of transforming growth factor β 1 acts as a potent growth inhibitor on a human erythroleukaemia cell line. BBRC 167: 27-32 (1990); **Reibmann J et al** Transforming growth factor β1, a potent chemoattractant for human neutrophils, bypasses classic signal transduction pathways. PNAS 88: 6805-9 (1991); **Roberts AB & Sporn MB** Differential expression of the TGF-β isoforms in embryogenesis suggests specific roles in developing and adult tissues. Mol. Reprod. Dev. 32: 91-8 (1992); **Schraufnagel DE et al** The effect of transforming growth factor-α on airway angiogenesis. J. Thorac. Cardiovasc. Surg. 104: 1582-8 (1992); **Su HC et al** A role for transforming growth factor-β 1 in regulating natural killer cell and T lymphocyte proliferative responses during acute infection with lymphocytic choriomeningitis virus. JI 147: 2717-27 (1991); **Suzumura A et al** Transforming growth factor-β suppresses activation and proliferation of microglia *in vitro*. JI 151: 2150-8 (1993); **Ueki N et al** Excessive production of transforming growth-factor β 1 can play an important role in the development of tumorigenesis by its action for angiogenesis: validity of neutralizing antibodies to block tumor growth. BBA 1137: 189-96 (1992); **Wahl SM et al** Transforming growth factor-β is a potent immunosuppressive agent that inhibits IL1 dependent lymphocyte proliferation. JI 140: 3026-32 (1988)

● **Transgenic Studies:** **Erickson HP** Gene knockouts of c-*src*, transforming growth factor β1, and tenascin suggest superfluous, nonfunctional expression of proteins. JCB 5: 194-203 (1993); **Geiser AG et al** Transforming growth factor β1 (TGF-β1) controls expression of major histocompatibility genes in the postnatal mouse: aberrant histocompatibility antigen expression in the pathogenesis of the TGF-β1 null mouse phenotype. PNAS 90: 9944-8 (1993); **Jhappan C et al** Targeting expression of a transforming growth factor β 1 transgene to the pregnant mammary gland inhibits alveolar development and lactation. EJ 12: 1835-45 (1993); **Kulkarni AB et al** Transforming growth factor β1 null mutation in mice causes excessive inflammatory response and early death. PNAS 90: 770-4 (1993); **Sellheyer K et al** Inhibition of skin development by overexpression of transforming growth factor β1 in the epidermis of transgenic mice. PNAS 90: 5237-41 (1993); **Shull MM et al** Targeted disruption of the mouse transforming growth factor-β 1 gene results in multifocal inflammatory disease. N 359: 693-9 (1992)

● **Assays:** **Absher M et al** A rapid colorimetric bioassay for transforming growth factor β (TGF-β) using a microwell plate reader JIM 138: 301-3 (1991); **Anderson J et al** Effects of acetate dialysate on transforming growth factor β 1, interleukin, and β 2-microglobulin plasma levels. Kidney Int. 40: 1110-7 (1991); **Flanders KC et al** Antibodies to transforming growth factor-β 2 peptides: specific detection of TGF-β 2 in immunoassays. GF 3: 45-52 (1990); **Flaumenhaft R & Rifkin DB** Cell density dependent effects of TGF-β demonstrated by a plasminogen activator-based assay for TGF-β. JCP 152: 48-55 (1992); **Frazer A et al** Detection of mRNA for the transforming growth factor β family in human articular chondrocytes by the polymerase chain reaction. BBRC 180: 602-8 (1991); **Frolik CA & DeLarco JE** Radioreceptor assays for transforming growth factors. MiE 146: 95-102 (1987); **Lucas C et al** The autocrine production of transforming growth factor-β 1 during lymphocyte activation. A study with a monoclonal antibody-based ELISA. JI 145: 1415-22 (1990); **Lucas C et al** Generation of antibodies and assays for transforming growth factor β. MiE 198: 303-16 (1991); **Merino J et al** The measurement of transforming growth factor type β (TGF β) levels produced by peripheral blood mononuclear cells requires the efficient elimination of contaminating platelets. JIM 153: 151-9 (1992); **Ogawa Y & Seyedin SM** Purification of transforming growth factor β 1 and β 2 from bovine bone and cell culture assays. MiE 198: 317-27 (1991); **Smith J & McLachlan JC** Identification of a novel growth factor with transforming activity secreted by individual chick embryos. Development 109: 905-10 (1990); **van Zoelen EJ** The use of biological assays for detection of polypeptide growth factors. PGFR 2: 131-52 (1990); **Wakefield LM** An assay for type-β transforming growth factor receptor. MiE 146: 167-73 (1987); for further information also see individual cell lines used in individual bioassays.

● **Clinical Use & Significance:** Symposium on clinical applications of TGF-β. London, 12-14 June 1990. Proceedings. Ciba Found. Symp. 157: 1-241, Chichester-Wiley, 1991; **Beck LS et al** One systemic administration of transforming growth factor β1 reverses age- or glucocorticoid-impaired wound healing. JCI 92: 2841-9 (1993); **Border WA et al** Suppression of experimental glomerulonephritis by antiserum against transforming growth factor β1. N 346: 371-4 (1990); **Glaser BM et al** Transforming growth factor-β 2 for the treatment of full-thickness macular holes. A prospective randomized study. Ophthalmology 99: 1162-72 (1992); **Hirayama D et al** Immunohistochemical study of epidermal growth factor and transforming growth factor-β in the penetrating type of early gastric cancer. Hum. Pathol. 23: 681-5 (1992); **Kawaguchi H et al** Immunohistochemical demonstration of bone morphogenetic protein-2 and transforming growth factor-β in the ossification of the posterior longitudinal ligament of the cervical spine. Spine 17: S33-6 (1992); **Migdalska A et al** Growth inhibitory effects of transforming growth factor-β 1 *in vivo*. GF 4: 239-45 (1991); **Mustoe TA et al** Growth factor-induced acceleration of tissue repair through direct and inductive activities in a rabbit dermal ulcer model. JCI 87: 694-703 (1991); **Nasim MM et al** Transforming growth factor α expression in normal gastric mucosa, intestinal metaplasia, dysplasia and gastric carcinoma - an immunohistochemical study. Histopathology 20: 339-43 (1992); **Roberts AB et al** Role of transforming growth factor-β in maintenance of function of cultured neonatal cardiac myocytes. Autocrine action and reversal of damaging effects of interleukin-1. JCI 90: 2056-62 (1992); **Wakefield LM et al** Roles for transforming growth factors-β in the genesis, prevention, and treatment of breast cancer. Cancer Treat. Res. 61: 97-136 (1992); **Walker RA & Dearing SJ** Transforming growth factor β 1 in ductal carcinoma *in situ* and invasive carcinomas of the breast. Eur. J. Cancer 28: 641-4 (1992)

TGF-γ: *transforming growth factor* γ This factor is identical with » TGF-β.

Moses HL et al Transforming growth factor production by chemically transformed cells. CR 41: 2842-8 (1981); **Moses HL et al** Role of transforming growth factors in neoplastic transformation. In: Veneziale CM (edt), pp. 147-67, Control of cell growth and proliferation. Van Nostrand Reinhold, New York, 1984

TGFe: *epithelial transforming growth factor* This factor of 22 kDa was initially isolated from epithelial tissues. It is also present in normal tissues, in particular in the pituitary, plasma, and platelets. It supports the growth of various carcinoma cell lines (see: SW-13) in soft agar cultures (see also: Colony formation assay) and monolayer cultures. TGFe stimulates both anchorage-dependent and -independent growth of epithelial and fibroblastic cells of nonneoplastic origin. The mitogenic activity of TGFe in monolayer is slightly less than that of » bFGF, equipotent to that of » EGF, and greater than that of » IGF-1. TGFe is also a potent mitogen for normal human epidermal keratinocytes.

Human TGFe does not bind to heparin and fails to stimulate growth of endothelial cells in monolayer culture. Biologically active truncated forms of TGFe have also been described.

TGFe does not bind to the receptors for TGF-β. Amino acid analysis and partial sequence analysis demonstrate that TGFe is not related to any known growth factor.

Brown CA & Halper J Mitogenic effects of transforming growth factor type 3 on epithelial and fibroblastic cells – comparison with other growth factors. ECR 190: 233-42 (1990); **Dunnington DJ et al** Characterization and partial purification of human epithelial transforming growth factor. JCBc 44: 229-39 (1990); **Dunnington DJ et al** Identification of a low molecular weight form of epithelial transforming growth factor. Life Sci. 47: 2059-63 (1990); **Halper J & Moses HL** Purification and characterization of a novel transforming growth factor. CR 47: 4552-9 (1987); **Halper J & Carter BJ** Modulation of growth of human carcinoma SW-13 cells by heparin and growth factors. JCP 141: 16-23 (1989); **Halper J et al** Presence of growth factors in human pituitary. Lab. Invest. 66: 639-45 (1992); **Parnell PG et al** Purification of transforming growth factor type e. JCBc 42: 111-6 (1990); **Parnell PG et al** Transforming growth factor e: amino acid analysis and partial amino acid sequence. GF 7: 65-72 (1992)

TGF-LAP: *TGF-latency associated peptide* see: TGF-β.

TGI: *tissue-derived growth inhibitor* A general term for inhibitory growth factors. The known factors of this kind are either identical with » TGF-β or are members of the same protein family.

TGIF: *T cell growth inhibitory factor* This factor is secreted by the CD8-positive suppressor T cell line 13G2. It inhibits the proliferation of type 1 helper T cell lines. The biological activity of this factor is completely inhibited by antibodies directed against » IL1o. The factor is identical with IL10.

Hisatsune T et al A suppressive lymphokine derived from Ts clone 13G2 is IL10. Lymphokine Cytokine Res. 11: 87-93 (1992)

TGIF: *thank God its Friday* (and the book is finished).

TGIF: *tumor growth inhibitory factor* This poorly characterized factor of ≈ 43 kDa is secreted by peripheral blood mononuclear cells activated by a preparation of streptococci (OK-432). The factor inhibits the growth of the human stomach carcinoma cell line MK-1 and is cytostatic for seven out of ten human tumor cell lines tested. A combination of suboptimal concentrations of this factor and » IFN-γ also inhibit the growth of these cells. The activity of TGIF cannot be blocked by antibodies directed against human » IFN or » TNF. This factor may be of interest in the treatment of patients undergoing OK-432 cancer therapy.

Katano M et al Induction of tumor growth inhibitory factor (TGIF) in human mononuclear cells by OK-432, a streptococcal preparation. Cancer Immunol. Immunother. 27: 198-204 (1988); **Katano M et al** A possible clinical application of multicytokine-producing cytotoxic mononuclear cell (MCCM) therapy. Biotherapy 3: 373-9 (1991); **Katano M et al** Synergistic effects of TGIF and interferon-γ (IFN-γ) on tumor cell growth. TGIF and IFN-γ. Biotherapy 2: 33-40 (1990); **Katano M et al** The antitumor effect of locally injecting human peripheral blood mononuclear cells treated with OK-432 into the tumor site: the possible role of a tumor growth inhibitory factor (TGIF). Jpn. J.

Surg. 20: 76-82 (1990); **Kubota E et al** Tumoricidal effect of human PBMC following stimulation with OK-432 and its application for locoregional immunotherapy in head and neck cancer patients. J. Craniomaxillofac. Surg. 21: 30-7 (1993)

TGP: *thymocyte growth peptide* see: Thymic hormones.

TGP-3: recombinant » IL2 (Takeda Chemical Industries, Ltd, Osaka).

TH1/TH2 cytokines: The term TH1 and TH2 cytokines refers to the patterns of cytokines secreted by two different subpopulations of murine » CD4-positive T cells that determine the outcome of an antigenic response toward humoral or cell-mediated immunity.

Type 1 helper cells (TH1), but not type 2 helper cells (TH2), secrete » IL2, » IFN-γ, and » TNF-β whereas TH2 cells, but not TH1 cells, express » IL4, » IL5, » IL6, and » IL10. Murine TH2 cells, but not TH1 cells, also express » P600, the human counterpart of which has been identified as » IL13. The molecular mechanisms underlying the evolution of these two different cell types from common precursors are still unknown. Studies with mice carrying null mutations of the » IL4 gene have shown that IL4 plays an important role in the establishment of a functional TH2 immune response. The different patterns of cytokine secretion correspond with different functions as immune effectors. TH1 cells promote cell-mediated effector responses. TH2 cells are mainly helper cells that influence B cell development and augment humoral responses such as the secretion of antibodies, predominantly of IgE, by B cells. Both types of TH cells influence each other by the cytokines they secrete; IFN-γ, for example, can downregulate TH2 clones while TH2 cytokines, such as » IL10 (see also: CSIF, cytokine synthesis inhibitory factor), can suppress TH1 functions. IFN-γ has also been shown to inhibit the proliferation of murine TH2 cells but not that of TH1 helper T lymphocyte clones. It thus appears that these functional subsets are mutually antagonistic such that the decision of which subset predominates within an infection may also determine its outcome.

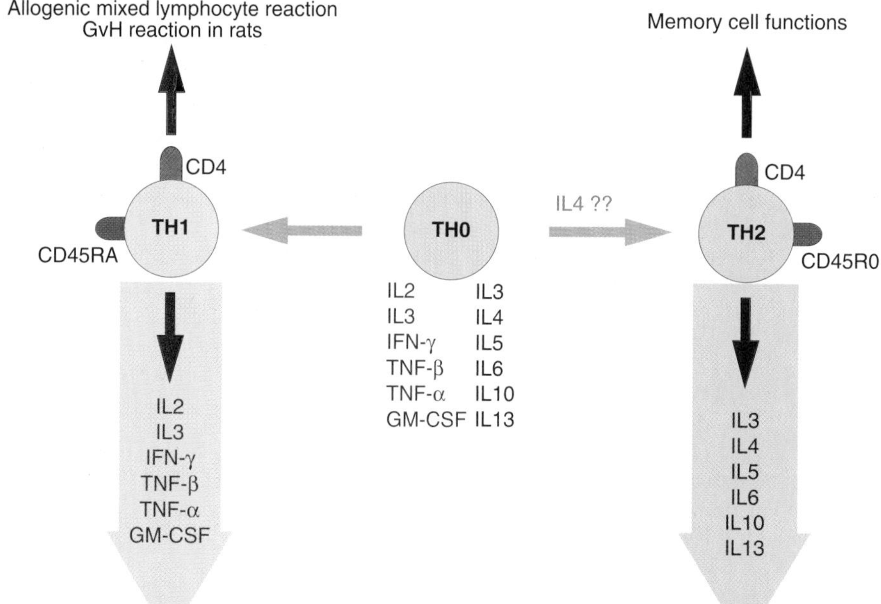

Comparison of TH1 and TH2 helper cells.
Both cell types probably arise from a common precursor. They differ in function and also in their pattern of secreted cytokines. See text for details.

Murine » IL12 has been shown recently to alter CD4+ subset differentiation and to be involved in the induction of protective immunity against intracellular parasitic infections in mice. IL12 appears to prevent deleterious TH2 T cell responses and to promote curative TH1 responses in an IFN-γ-dependent fashion during murine leishmaniasis. Another exogenous factor also influencing the development of undifferentiated CD4+ T cells towards either the TH1 or TH2 phenotype is » TGF-β.

A third subset of memory cells, designated TH0 cells and believed to be precursor cells that develop into either TH1 or TH2 cells, can produce all cytokines found to be secreted either by TH1 or TH2 cells at low levels.

It appears that different types of helper cell populations resembling those observed in mice are also found in humans. However, the differences in cytokine expression seems to be quantitative rather than qualitative.

Bloom BR et al Revisiting and revising T suppressor cells. IT 13: 131-6 (1992); **Bottomly K** A functional dichotomy in CD4+ T lymphocytes. IT 9: 268-74 (1988); **Cherwinski HM et al** Two types of mouse helper T cell clone. III. Further differences in lymphokine synthesis between Th1 and Th2 clones revealed by RNA hybridization, functionally monospecific bioassays, and monoclonal antibodies. JEM 166: 1229-44 (1987); **Clerici M & Shearer GM** A TH1 to TH2 switch is a critical step in the etiology of HIV infections. IT 14: 107-111 (1993); **del Prete GF et al** Purified protein derivative of *Mycobacterium tuberculosis* and excretory-secretory antigen(s) of *Toxocara canis* expand *in vitro* human T cells with stable and opposite (type 1 T helper or type 2 T helper) profile of cytokine production. JCI 88: 346-50 (1991); **de Vries JE et al** Do human TH1 and TH2 CD4+ clones exist? Res. Immunol. 142: 59-63 (1991); **Fiorentino DF et al** IL10 acts on the antigen-presenting cell to inhibit cytokine production by Th1 cells. JI 146: 3444-51 (1991); **Firestein GS et al** A new murine CD4+ T cell subset with an unrestricted cytokine profile. JI 143: 518-25 (1989); **Gajewski TF & Fitch FW** Antiproliferative effect of IFN-γ in immune regulation. I. IFN-γ inhibits the proliferation of Th2 but not Th1 murine helper T lymphocyte clones. JI 140: 4245-52 (1988); **Heinzel FP et al** Recombinant interleukin 12 cures mice infected with *Leishmania major*. JEM 177: 1505-9 (1993); **Hikida M et al** Suppression of interleukin 4 production from type 2 helper T cell clone by antisense oligodeoxynucleotide. Immunol. Lett. 34: 297-302 (1992); **Kamogawa Y et al** The relationship of IL4 and IFN-γ-producing T cells studied by lineage ablation of IL4 producing cells. Cell 75: 985-95 (1993); **Kapsenberg ML et al** Functional subsets of allergen-reactive human CD4+ T cells. IT 12: 392-5 (1992); **Kawakami K & Parker DC** Differences between T helper cell type I (Th1) and Th2 cell lines in signaling pathways for induction of contact-dependent T cell help. EJI 22: 85-93 (1992); **Kelso A et al** Heterogeneity in lymphokine profiles of CD4+ and CD8+ T cells and clones activated *in vivo* and *in vitro*. IR 123: 85-114 (1991); **Lederer JA et al** Regulation of cytokine gene expression in T helper cell subsets. JI 152: 77-86 (1994); **Maggi E et al** Reduced production of interleukin 2 and interferon-γ and enhanced helper activity for IgG synthesis by cloned CD4+ T cells from patients with AIDS. EJI 17: 1685-90 (1987); **Mosmann TR et al** Two types of murine helper T cell clone. I. Definition according to profiles of lymphokine activities and secreted proteins. JI 136: 2348-57 (1986); **Mosmann TR & Coffman RL** Heterogeneity of cytokine secretion patterns and functions of helper T cells. AI 111-47 (1989); **Mosmann TR & Coffman RL** TH1 and TH2 cells: different patterns of lymphokine secretion lead to different functional properties. ARI 7: 145-73 (1989); **Paul WE & Seder RA** Lymphocyte responses and cytokines. Cell 76: 241-51 (1994); **Rocken M et al** A common precursor for CD4+ T cells producing IL2 or IL4. JI 148: 1031-6 (1992); **Romagnani S** Type 1 T helper and type 2 T helper cells: functions, regulation and role in protection and disease. Int. J. Clin. Lab. Res. 21: 152-8 (1991); **Romagnani S** Human TH1 and TH2 subsets: doubt no more. IT 12: 256-7 (1991); **Rooney JW et al** A common factor regulates both Th1- and Th2-specific cytokine gene expression. EJ 13: 625-33 (1994); **Trinchieri G et al** Interleukin-12 and its role in the generation of TH1 cells. IT 14: 335-8 (1993); **Vowels BR et al** Aberrant cytokine production by Sezary syndrome patients: cytokine secretion pattern resembles murine Th2 cells. J. Invest. Dermatol. 99: 90-4 (1992); **Williams ME et al** Activation of functionally distinct subsets of CD4+ T lymphocytes. Res. Immunol. 142: 23-8 (1991)

THF: *thymic humoral factor* see: Thymic hormones.

THGF: *thymocyte growth factor* This poorly characterized factor of 20 kDa is purified from the supernatant of a thymic T lymphocyte-precursor cell line (TC.SC-1/2.0). It stimulates the growth of unstimulated splenocytes and thymocytes. While suboptimal doses of mitogen or phorbol myristate acetate (see also: Phorbol esters) do not enhance the response THGF stimulates the growth of thymocytes synergistically with » IL2. THGF also stimulates the formation of hematopoietic spleen colonies in irradiated mice.

Talaev VYu et al Purification and characterization of a thymocyte growth factor. 1. Purification. Biomed Sci. 2: 511-4 (1991); **Talaev VYu et al** Purification and characterization of a thymocyte growth factor. 2. Biological activity of the thymocyte growth factor. Biomed Sci. 2: 515-9 (1991)

Thioredoxin: see: BSF MP6 (B cell stimulating factor MP-6). See also: » ADF (adult T cell leukemia-derived factor).

THP: *Tamm-Horsfall Glycoprotein* see: Uromodulin.

THP-1-derived growth-promoting activity: This poorly characterized factor of ≈ 50-70 kDa is secreted by the human myelomonocytic cell line THP-1. The factor promotes the growth of human and mouse myeloid cells (» M1, K562, » U937, » HL60), mouse T cells (» EL4), human B cells (» Daudi, Raji), mouse mastocytoma cells (» IC2),

human melanoma cells, mouse transformed fibroblast cells (» L929), human lung fibroblast cells, and mouse bone marrow fibroblast/stromal-like cells.
The factor does not appear to be identical with » IL1, » IL6, » IL8, » TNF, » IFN, » CSF, and » LIF, known to be secreted by THP-1 cells. The factor also appears to be different from » IL3, » HGF (hepatocyte growth factor), or » IL12.
Takeuchi A et al Human myelomonocytic cell line THP-1 produces a novel growth-promoting factor with a wide target cell spectrum. CR 53: 1871-6 (1993)

Thromboglobulin β: see: Beta-Thromboglobulin.

Thrombopoiesis-stimulating factor: abbrev. TSF. See: TPO (thrombopoietin).

Thrombopoietin: see: TPO.

Thrombospondin: abbrev. ***TSP***. A glycoprotein of 420 kDa with three identical disulfide-bonded subunits. TSP is present in high concentrations in the platelet α-granule and can readily be secreted following platelet activation. It is also synthesized by a variety of normal and transformed cells and secreted into the extracellular matrix (see: ECM). Some cells appear to express also monomeric forms of TSP. To date, four genes have been identified that encode members of the TSP » gene family. The human TSP1 gene maps to chromosome 15q11-qter.
TSP expression is highly regulated by different hormones and » cytokines and is developmentally controlled. High levels are expressed in developing heart, muscle, bone and brain of the embryo. TSP has a limited distribution in normal adult tissues and is highly expressed in response to injury and » inflammation.
TSP is an adhesive protein that can bind to many different cell types, including hematopoietic progenitor cells. It possesses a number of functional domains that serve as receptor recognizing regions. The amino-terminal heparin binding domain interacts with heparin, other glycosaminoglycans and glycolipids and likely recognizes specific cell surface proteoglycans. The central disulfide cross-linked region is responsible for high affinity binding to the » CD antigen » CD36 (glycoprotein IV). TSP has been demonstrated to bind to integrins of the α v β 3 and α IIb β 3 class. The carboxy-terminal region of TSP also contains at least one binding epitope for a cell receptor. TSP can interact with fibrinogen and fibronectin.
TSP modulates a number of processes, including aggregation of platelets and adhesion/migration of cells during embryogenesis and » wound healing. Interactions of TSP and CD36 have been implicated in phagocytic clearance of senescent neutrophils, which may limit inflammatory tissue injury and promote resolution. TSP functions as an inhibitor of the fibrinolytic enzyme plasmin and of neutrophil elastase, which is primarily responsible for degrading and solubilizing connective tissue during inflammatory processes.
TSP appears to be intimately involved in the regulation of cellular proliferation. It is expressed as an » ERG (early response gene) following treatment of cells with various mitogens and » cytokines. It modulates the progression of cells through the » cell cycle. It augments the proliferative response to » EGF. TSP inhibits endothelial cell chemotactic response to » bFGF and reduces their proliferative response to serum and bFGF. Overexpression of thrombospondin in murine » 3T3 cells has been shown to result in serum and anchorage independent growth. TSP inhibits the *in vitro* proliferation of endothelial cells from a variety of sources, including human umbilical vein. TSP stimulates the growth of vascular smooth muscle cells and human foreskin fibroblasts. Biosynthesis, secretion, and utilization as an attachment factor in keratinocytes has been shown to be decreased by » IFN-γ. A combination of IFN-γ and » TNF-α inhibits TSP production in these cells. It has been shown that TSP can bind » TGF-β under physiological conditions. TSP may protect TGF-β from extracellular inactivators. TSP-associated TGF-β is biologically active and that at least some growth modulating activities of TSP are due to TGF-β. TSP and several proteolytic fragments have been shown to possess neurite-promoting factor activity that can influence neuronal development by selectively enhancing axonal growth.
TSP has anti-angiogenic properties (see also: Angiogenesis factors) and these may be due to inhibition of the pericellular proteolysis required for endothelial cell migration and endothelial cell proliferation.
TSP levels correlate directly with the growth and invasive phenotype and inversely with the degree of differentiation of squamous carcinoma cells.
■ **TRANSGENIC/KNOCK-OUT/ANTISENSE STUDIES:** It has been shown that inhibition of TSP production by introduction of » antisense oligonucleoti-

Thrombopoietin: see addendum.

des into a human squamous carcinoma cell line with high TSP production and an invasive phenotype leads to reduced growth rates *in vitro*. After subcutaneous inoculation into nude mice (see also: Immunodeficient mice), the modified tumor cells form either no tumors or tumors that are slow growing and highly differentiated.

● **Reviews: Asch AS & Nachman RL** Thrombospondin: phenomenology to function. Prog. Hemost. Thromb. 9: 157-76 (1989); **Lahav J** The functions of thrombospondin and its involvement in physiology and pathophysiology. BBA 1182: 1-14 (1993); **Mosher DF** Physiology of thrombospondin. ARM 41: 85-97 (1990); **Walz DA** Thrombospondin as a mediator of cancer cell adhesion in metastasis. Cancer Metastasis Rev. 11: 313-24 (1992)

Bagavandoss P & Wilks JW Specific inhibition of endothelial cell proliferation by thrombospondin. BBRC 170: 867-72 (1990); **Bagavandoss P et al** Recombinant truncated thrombospondin-1 monomer modulates endothelial cell plasminogen activator inhibitor 1 accumulation and proliferation *in vitro*. BBRC 192: 325-32 (1993); **Castle VP et al** High level thrombospondin 1 expression in two NIH 3T3 cloned lines confers serum- and anchorage-independent growth. JBC 268: 2899-903 (1993); **Donoviel DB et al** Thrombospondin gene expression is associated with mitogenesis in 3T3 cells: induction by basic fibroblast growth factor. JCP 145: 16-23 (1990); **Framson P & Bornstein P** A serum response element and a binding site for NF-Y mediate the serum response of the human thrombospondin 1 gene. JBC 268: 4989-96 (1993); **Good DJ et al** A tumor suppressor-dependent inhibitor of angiogenesis is immunologically and functionally indistinguishable from a fragment of thrombospondin. PNAS 87: 6624-8 (1990); **Hennessy SW et al** Complete thrombospondin mRNA sequence includes potential regulatory sites in the 3' untranslated region. JCB 108: 729-36 (1989); **Hogg PJ et al** Thrombospondin is a slow tight-binding inhibitor of plasmin. B 31: 265-9 (1992); **Hogg PJ et al** Thrombospondin is a tight-binding competitive inhibitor of neutrophil elastase. JBC 268: 7139-46 (1993); **Huber AR et al** Monocyte diapedesis through an *in vitro* vessel wall construct: inhibition with monoclonal antibodies to thrombospondin. J. Leukoc. Biol. 52: 524-8 (1992); **Iruela-Arispe ML et al** Thrombospondin exerts an antiangiogenic effect on cord formation by endothelial cells *in vitro*. PNAS 88: 5026-30 (1991); **LaBell TL et al** Thrombospondin II: partial cDNA sequence, chromosome location, and expression of a second member of the thrombospondin gene family in humans. Genomics 12: 421-9 (1992); **Laherty CD et al** Characterization of the promoter region of the human thrombospondin gene. DNA sequences within the first intron increase transcription. JBC 264: 11222-7 (1989); **Lawler J et al** The evolution of the thrombospondin gene family. J. Mol. Evol. 36: 509-16 (1993); **Long MW & Dixit VM** Thrombospondin functions as a cytoadhesion molecule for human hematopoietic progenitor cells. Blood 75: 2311-8 (1990); **Murphy-Ullrich JE et al** Transforming growth factor-β complexes with thrombospondin. Mol. Biol. Cell. 3: 181-8 (1992); **Nickoloff BJ et al** Inhibitory effect of γ interferon on cultured human keratinocyte thrombospondin production, distribution, and biologic activities. J. Invest. Dermatol. 91: 213-8 (1988); **Nicosia RF & Tuszynski GP** Matrix-bound thrombospondin promotes angiogenesis in vitro. JCB 124: 183-93 (1994); **O'Rourke KM et al** Thrombospondin 1 and thrombospondin 2 are expressed as both homo- and heterotrimers. JBC 267: 24921-4 (1992); **O'Shea KS et al** Thrombospondin and a 140 kd fragment promote adhesion and neurite outgrowth from embryonic central and peripheral neurons and from PC12 cells. Neuron 7: 231-7 (1991); **Osterhout DJ et al** Thrombospondin promotes process outgrowth in neurons from the peripheral and central nervous systems. Dev. Biol. 150: 256-65 (1992); **Phan SH et al** Stimulation of fibroblast proliferation by thrombospondin. BBRC 163: 56-63 (1989); **Savill J et al** Thrombospondin cooperates with CD36 and the vitronectin receptor in macrophage recognition of neutrophils undergoing apoptosis. JCI 90: 1513-22 (1992); **Schultz-Cherry S & Murphy-Ullrich JE** Thrombospondin causes activation of latent transforming growth factor-β secreted by endothelial cells by a novel mechanism. JCB 122: 923-32 (1993); **Scott-Burden T et al** Induction of thrombospondin expression in vascular smooth muscle cells by angiotensin II. J. Cardiovasc. Pharmacol. 16: S17-20 (1990); **Taraboletti G et al** Platelet thrombospondin modulates endothelial cell adhesion, motility, and growth: a potential angiogenesis regulatory factor. JCB 111: 765-72 (1990); **Taraboletti G et al** Thrombospondin modulates basic fibroblast growth factor activities on endothelial cells. EXS. 1992; 61: 210-3 (1992); **Tuszynski GP et al** Thrombospondin levels in patients with malignancy. Thromb. Haemost. 67: 607-11 (1992); **Wolf FW et al** Structure and chromosomal localization of the human thrombospondin gene. Genomics 6: 685-91 (1990)

● **Transgenic/Knock-out/Antisense Studies: Castle V et al** Antisense-mediated reduction in thrombospondin reverses the malignant phenotype of a human squamous carcinoma. JCI 87: 1883-8 (1991); **Silverstein RL et al** Sense and antisense cDNA transfection of CD36 (glycoprotein IV) in melanoma cells. Role of CD36 as a thrombospondin receptor. JBC 267: 16607-12 (1992)

THSC: *totipotent hematopoietic stem cells* see: Hematopoietic stem cells. See also: Hematopoiesis.

Thy1-inducing factor: see: TIF.

Thymic hormones: The thymic epithelium is involved in the differentiation of T cells through several mechanisms, involving the direct contact with stem cells and the secretion of a variety of biochemically characterized proteins, called thymic hormones. The intrathymic environment is characterized by a complex network of » paracrine, » autocrine, and endocrine signals, involving both cytokines and thymic peptides. Human recombinant » IL1 induces thymic epithelial cell proliferation and » IFN-γ and » TNF-α enhance the effects of IL1. Thymic epithelial cells respond to IL1 with proliferation and secretion of » IL6 and » GM-CSF.

The endocrine influences of thymic hormones decline with age and are associated with "thymic menopause" and cellular immune senescence, contributing to the development of diseases in the aged. The exact physiological roles of thymic hormones are largely unknown. These biologically active proteins and peptides are present in various thymus fractions and participate in the regulation, differentiation, and function of thymus-derived lymphocytes; they may also act directly or indirectly on other cells and may have more general functions than those in cellular immunity and differentiation of lymphoid cells. Thymic preparations and some purified factors have been used as immunomodulators of humoral and cellular immunity in patients with primary immunodeficiencies, autoimmune, and neoplastic disorders.

Parathymosin alpha, a peptide isolated from rat thymus, contains ≈ 101 amino acids. It shows 43% structural identity with thymosin α 1 (see below) and prothymosin α in the first 30 aminoterminal amino acids. Parathymosin α appears to modulate the action of prothymosin α in protecting sensitive strains of mice against opportunistic infection with *Candida albicans* and has been shown to block the *in vivo* immunoenhancing effects of prothymosin α.

Prothymosin alpha from rat thymus contains 111 amino acid residues. The sequence of the first 28 amino acids at the aminoterminus is identical to that of calf thymosin α 1 (see below). Bovine prothymosin α (109 amino acids) closely resembles human prothymosin α, with only two substitutions, glutamic acid for aspartic acid at position 31 and alanine for serine at position 83. Human prothymosin (111 amino acids) also contains the thymosin α 1 sequence at its aminoterminus. It contains deletions corresponding to positions Gln39 and Lys108 of the rat polypeptide and also differs from rat prothymosin α at positions corresponding to residues 87, 92, and 102 of the latter, with substitutions of alanine for proline, alanine for valine, and aspartic acid for glutamic acid, respectively. The human prothymosin α gene maps to chromosome 2.***Thymosin alpha 1m*** (3.1 kDa) is the biologically active N-terminal fragment of prothymosin α (12 kDa).

Prothymosin α mRNA is expressed in mouse and human kidney, liver, spleen, normal lymphocytes (predominantly T cells), human T cell leukemia virus-infected T cells, and myeloma cells (B cell lineage). The amount of this protein seems to be associated with the relative size of the nucleus and is inducible during cell growth. An RNA-linked protein from the cytoplasm of mouse cells has been found to be identical with prothymosin α and very similar RNA-protein complexes have also been identified in human, bovine and yeast cells. Prothymosin α appears to be a rather abundant nuclear protein in rat thymus.

Prothymosin α enhances allo- and auto-antigen-induced human T cell proliferation. It has been shown to enhance MHC class II antigen expression in antigen-presenting cells and may therefore be important in the regulation of immune responses. Prothymosin α has been shown to potentiate or fully restore the deficient cytotoxic effector function of peripheral mononuclear cells in patients with advanced malignancies. Human prothymosin is significantly less active than rat prothymosin in protecting mice against infection with *Candida albicans* and in stimulating release *in vivo* of migration inhibitory factor.

Serum thymic factor (abbrev. FTS, facteur thymique serique) is a nonapeptide. It is now called ***thymulin***. ***SDIP*** (spleen-derived immunosuppressive peptide) is closely related to FTS.

Thymulin needs a zinc ion to express its immunoregulatory properties (***Zn-FTS***). The secretion of zinc-thymulin by human thymic epithelial cells and its action on T lymphocyte proliferation are regulated by IL1. IL1 also induces the synthesis of metallothioneins thymic epithelial cells, which are thought to be involved in the transfer of zinc to thymulin. The production of thymulin by human and murine thymic epithelial cells is also stimulated by » prolactin and » growth hormone *in vivo* and *in vitro*; the effect of growth hormone being mediated probably by » IGF. Thymulin influences a variety of T cell differentiation markers and functions *in vivo* and enhances several functions of the various T cell subsets in normal or partially thymus-deficient recipients. It stimulates DNA synthesis in short-term cultures of thymocytes. Thymulin has been shown to protect mice from death after whole-body X-irradiation and this radioprotective effect is probably due to the enhancement of hematologic recovery in the animals. It has been demonstrated that a progressive reduction of the plasma level of thymulin occurs with advancing age. Cancer patients also show much lower thymulin levels. The administration of an amino acid combination (lysine-arginine) to elderly individuals and cancer patients can increase the synthesis and/or release of thymulin to normal levels and also increases the number of peripheral T cell subsets.

The production of thymulin by human and murine thymic epithelial cells is stimulated by » prolactin *in vivo* and *in vitro*.

Thymic humoral factor (abbrev. ***THF***) is a polypeptide of 3.2 kDa. It promotes » IL2 production in intact and thymus-deprived mice and augments lymphocyte proliferation *in vitro* and *in vivo*. The octapeptide ***THF-gamma 2*** has the amino acid sequence Leu-Glu-Asp-Gly-Pro-Lys-Phe-Leu. It has been shown to increase the effectiveness of combined chemotherapy and immunotherapy in plasmacytoma-bearing mice and to restore immunocompetence after chemotherapy.

THF leads to clonal expansion, differentiation and maturation of T cell lymphocytes. THF γ2 has also been shown to play a stimulatory role on human myeloid and erythroid hematopoietic progenitor cells. THF augments mixed lymphocyte reactions, graft-versus-host reactivity, cytotoxic responses, and production of IL2. Clinical data based on the use of THF indicate that it can normalize the ratio between helper and suppressor/cytotoxic subsets. THF reconstitutes the defective cell-mediated immunity in patients with various types of neoplasms and secondary immune deficiencies as a result of chemo and/or radiotherapy.

Neutral endopeptidase (NEP; EC 3.4.24.11 = common acute lymphoblastic leukemia antigen, CALLA) has been shown to hydrolyze THF γ 2 with high efficiency. In view of the presence of NEP in bone-marrow cells and other cells of the immune system this enzyme may play a role in modulating the activity of peptides such as THF.

Thymocyte growth peptide (abbrev. TGP) is an N-terminally blocked ribofolate nonapeptide (Glu-Ala-Lys-Ser-Gln-Gly-Gly-Ser-Asn) that can be purified from sheep, human and calf thymus. The blocking moiety consists of a formylpteroyl group bound to the N-terminal glutamyl residue of the nonapeptide. Native TGP contains zinc, and Zn^{2+} counteracts degradation of the molecule and is required for full biological activity. TGP initiates DNA synthesis in immature thymocytes.

Thymomodulin (Ellem Industria Farmaceutica S.P.A, Milan, Italy) is a calf thymus acid lysate derivative, composed of several peptides with a molecular weight range of 1-10 kDa. It seems to directly stimulate lymphocytes to secrete » GM-CSF and to modulate macrophage-lymphocyte interactions, resulting in the release of » TNF and » GM-CSF. Thymomodulin also enhances macrophage HLA-DR expression and increases mitogen-induced T cell proliferation. It can improve the clinical symptoms associated with infections, allergies and malignancies, and can improve immunological functions during aging.

Thymopoietin (abbrev. ***Tpo***) is a protein of 49 amino acids. Three variants having largely identical sequences have been described. Thymopoietin III (TP-III) is a 49 amino acid monomeric peptide isolated from bovine spleen and later renamed » Splenin. ***TP-5*** (***thymopentin*** or ***thymopoietin 32-36***) is a pentapeptide (Arg-Lys-Asp-Val-Tyr) corresponding to amino acids 32-36 of thymopoietin. It appears to represent the active site of thymopoietin in that it has all the biological activities of the native hormone. Some analogs of TP-5, for example ***TP-3*** (Arg-Lys-Asp), and ***TP-4*** (Arg-Lys-Asp-Val) exhibit significant immuno-stimulating potencies *in vitro* and *in vivo* exceeding those of thymopentin. Thymopoietin and thymopentin affects neuromuscular transmission. Thymopoietin specifically interacts with the neuronal nicotinic α-bungarotoxin receptor population and can regulate the toxin binding sites in chromaffin cells in culture. It can cause a complete block of contractile responses evoked by stimulation of the phrenic nerve diaphragm junction of the rat *in vitro* and is only slightly less potent than α-bungarotoxin at concentrations of $\approx 10^{-8}$ M.

Thymopoietin induces the phenotypic differentiation of T precursor cells *in vitro* while inhibiting phenotypic differentiation of B cells. Thymopentin is able to increase IL2 production and IL2 receptor expression. The increased IL2 synthesis/IL2 receptor expression observed after treatment with thymopentin leads to an increased intrinsic T cell responsiveness and probably also to an increase in the size of the responsive T cell pool and thus may be the crucial mechanism of the immunopotentiating activity of the drug. TP-5 enhances bone marrow natural killer cells, probably by permitting the maturation of their precursors, and also natural killer activities in peripheral blood mononuclear cells of patients having received the drug three times per week for one month (50 mg sc). TP-5 has also been found to be useful in the treatment of a subgroup of patients with Sezary syndrome (a cutaneous T cell lymphoma). Thymopentin also induces ACTH-like immunoreactivity release by human lymphocytes.

Thymosin fraction 5 (***TF5***), is a partially purified preparation obtained from calf thymus and contains 40-60 peptides. It can correct many of the immunological deficiencies resulting from the lack of thymosin function in animal models and in humans. TF5 is able to enhance the spontaneous natural killer activity of normal human large granular lymphocytes. TF5 exhibits additive effects with recombinant » IFN-α in enhancing natural killer activity *in vitro*. TF5 also enhances » IL2 production, » IL2 receptor expression, and » IFN-γ production in mitogen-stimulated large granular lymphocytes. These effects are probably due to thymosin-α 1 as the active species in TF5. Thymic extracts have also been shown to enhance » TNF-α production by cord blood lymphocytes and of » IL1 by human peripheral blood monocytes. TF5 has also been reported to modulate the secretion of some hypothalamic peptides and pituitary hormones, including » prolactin and » growth hormone. Thymosin fraction 5 has been shown to prolong the survival of some lung cancer patients.

Thymostimulin (***TP1***) is also a partially purified preparation of thymic peptides and has been shown to enhance the production of IL2 and IFN-γ in cord blood lymphocytes. A study of immunocompromised surgical patients has suggested that treatment with thymostimulin is effective in avoiding or reducing postoperative infection rates. Thymostimulin has been shown to significantly enhance the number of early erythroid progenitor cells (see: BFU-E) in elderly subjects, probably by its effects on T lymphocyte functions. The imbalance of phagocyte functions in chronic obstructive pulmonary disease (COPD) has been shown to be significantly improved by treatment with thymostimulin. TP1 also appears to restore depressed monocyte and dendritic cell functions in patients with head and neck cancer.

Thymosin alpha 1 is a protein of 28 amino acids which is derived from a 113 amino acid precursor protein, prothymosin α. A variant lacking four carboxyterminal amino acids [***des-(25-28)-thymosin alpha 1***] and another variant possessing seven additional carboxyterminal amino acids [***thymosin alpha 11***] have been isolated from calf thymus fractions. Thymosin α 11, in doses of less than 300 ng per mouse, protects susceptible inbred murine strains against opportunistic infections with *Candida albicans* and shows approximately the same potency as thymosin α 1.

Thymosin α 1 can induce *in vitro* differentiation of murine T cell precursors and human thymocytes. It also induces terminal differentiation of functionally immature cord blood lymphocytes and has been shown to induce the production of » IL2, high affinity IL2 receptors, and B cell growth factors by peripheral blood mononuclear cells. Helper/inducer (Th, $CD4^+$) and cytotoxic/suppressor (Tc, $CD8^+$) T cell populations are targets of thymosin activity. In nude mice (see also: Immunodeficient mice) thymosin-α 1 appears to exert its effect at an early stage of T cell differentiation and induces a T cell subpopulation capable of producing » IL3 (but not yet IL2). Thymosin α 1 (and also β 4) have been shown to increase the efficiency of antigen presentation by macrophages and it has been suggested that the activation of the macrophages at the time of antigen presentation may be an initial step in the regulation of the immune function by these two thymic hormones. Thymosin α 1 has also been found to be an endogenous modulator of α-thrombin activity. Thymosin α 1 has been suggested to play a role as an » autocrine growth factor for the human breast cancer cell line MCF-7 in culture. Thymosin α 1 has been shown to down-regulate the growth of human non-small cell lung cancer cells *in vitro* and *in vivo*.

β-thymosins constitute a family of related proteins showing a high degree of sequence homology and microheterogeneity.

Thymosin beta 4 is a protein of 43 amino acids with an acetylated aminoterminus originally isolated from calf thymus. It is also produced by some cell lines, including myoblasts and fibroblasts and also appears to be produced by some human medullary thyroid carcinomas. Thymosin β 4 has been shown to be induced during » GM-CSF-induced differentiation of bone marrow cells. Human thymosin β 4 has been found to be identical with the human interferon-inducible gene » ***6-26***. Thymosin β 4 has also been found to be identical with » ***Fx***, a protein involved in actin polymerization.

Thymosin β 4 induces the expression of terminal deoxynucleotidyl transferase (EC 2.7.7.31) in transferase-negative murine thymocytes *in vivo* and *in vitro* and thus appears to act on lymphoid stem cells, probably controlling the early stages of the maturation process of thymus-dependent lymphocytes. Thymosin β 4 has also been found to be an activator of calmodulin-dependent enzymes in the hypothalamus. It also improves and normalizes suppressor cell activities in diabetic patients. Thymosin β 4 can undergo enzymatic cleavage *in vitro* and *in vivo* and yields a tetrapeptide that functions as a regulator of the hematopoietic system (see: AcSDKP).

A fusion protein constructed from » TNF with human thymosin β 4 has been shown to have significant antitumor effects in experimental tumor-bearing animals after systemic injection. The fusion protein had a significantly higher half-life in serum than the original TNF and caused regressions of tumors at concentrations at which TNF was inactive.

Thymosin beta 8 contains 39 amino acid residues, of which 31 are identical to the corresponding amino acid residues in thymosin β 4. The NH2 terminus of thymosin β 8 is acetylalanine, compared with acetylserine in thymosin β 4.

Thymosin beta 9 (41 amino acids) is identical with thymosin β 8 except for the presence of an additional dipeptide, -Ala-LysOH, at the carboxyterminal terminus. 32 of its 41 amino acids are identical with those of thymosin β 4.

Thymosin beta 10 is composed of 43 amino acid residues and shows 75% sequence homology with thymosin β 4. A comparison of Human and rat thymosin β 10 cDNA sequences reveal 100% identity for the deduced amino acid sequence and 95% nucleotide identity for the coding region. Thymosin β 10 has been found to be abundantly expressed in embryonic/fetal human brain but absent in adult tissues and probably plays an important role in early neuroembryogenesis and neural maturation. Thymosin β-10 mRNA appears to be specifically expressed in highly metastatic human melanoma cells lines and may be progression marker for human cutaneous melanoma. Thymosin β 10 has also been shown to function as an actin monomer sequestering protein (see: Fx).

Thymosin beta 11 contains 41 amino acids and is 78% homologous to thymosin β 4.

Thymosin therapy has been reported to promotes disease remission and cessation of hepatitis B virus replication in patients with chronic viral infection.

● **Reviews: Dardenne M & Savino W** Neuroendocrine circuits controlling the physiology of the thymic epithelium. ANY 650: 85-90 (1992); **Hadden JW** Thymic endocrinology. IJI 14: 345-52 (1992); **Low TL & Goldstein AL** Thymic hormones: an overview. MiE 116: 213-9 (1985); **Oates KK et al** Mechanism of action of the thymosins: modulation of lymphokines, receptors, and T cell differentiation antigens. Immunol. Ser. 45: 273-88 (1989); **Schulof RS et al** Thymic peptide hormones: basic properties and clinical applications in cancer. Crit. Rev. Oncol. Hematol. 3: 309-76 (1985); **Sztein MB & Goldstein AL** Thymic hormones - a clinical update. Springer Semin. Immunopathol. 9: 1-18 (1986)

● **Thymic Epithelium: Galy AH et al** Effects of cytokines on human thymic epithelial cells in culture: IL1 induces thymic epithelial cell proliferation and change in morphology. CI 124: 13-27 (1989)

● **Parathymosin Alpha: Clinton M et al** The sequence of human parathymosin deduced from a cloned human kidney cDNA. BBRC 158: 855-62 (1989); **Haritos AA et al** Parathymosin α: a peptide from rat tissues with structural homology to prothymosin α. PNAS 82: 1050-3 (1985); **Frangou-Lazaridis M et al** Prothymosin α and parathymosin: amino acid sequences deduced from the cloned rat spleen cDNAs. ABB 263: 305-10 (1988)

● **Prothymosin Alpha: Baxevanis CN et al** Enhancement of human T lymphocyte function by prothymosin α: increased production of interleukin-2 and expression of interleukin-2 receptors in normal human peripheral blood T lymphocytes. Immunopharmacol. Immunotoxicol. 12: 595-617 (1990); **Baxevanis CN et al** Prothymosin α enhances human and murine MHC class II surface antigen expression and messenger RNA accumulation. JI 148: 1979-84 (1992); **Baxevanis CN et al** Prothymosin α restores depressed allogeneic cell-mediated lympholysis and natural-killer-cell activity in patients with cancer. IJC 53: 264-8 (1993); **Eschenfeldt WH & Berger SL** The human prothymosin α gene is polymorphic and induced upon growth stimulation: evidence using a cloned cDNA. PNAS 83: 9403-7 (1986); **Eschenfeldt WH et al** Isolation and partial sequencing of the human prothymosin α gene family. Evidence against export of the gene products. JBC 264: 7546-55 (1989); **Grangou-Laxaridis M et al** Prothymosin α and parathymosin: amino acid sequences deduced from the cloned rat spleen cDNAs. Arch. Biochem. Biophys. 263: 305-10 (1988); **Gomez-Marquez J & Segade F** Prothymosin α is a nuclear protein. FL 226: 217-9 (1988); **Haritos AA et al** Primary structure of rat thymus prothymosin α. PNAS 82: 343-6 (1985); **Haritos AA et al** Prothymosin α and α 1-like peptides. MiE 116: 255-65 (1985); **Makarova T et al** Prothymosin α is an evolutionary conserved protein covalently linked to a small RNA. FL 257: 247-50 (1989); **Palvimo J et al** Identification of a low-Mr acidic nuclear protein as prothymosin α. FL 277: 257-60 (1990); **Pan LX et al** Human prothymosin α: amino acid sequence and immunologic properties. ABB 250: 197-201 (1986); **Panneerselvam C et al** The amino acid sequence of bovine thymus prothymosin α. ABB 265: 454-7 (1988); **Szabo P et al** Prothymosin α gene in humans: organization of its promoter region and localization to chromosome 2. Hum. Genet. 90: 629-34 (1993)

● **Serum Thymic Factor: Auger G et al** Synthesis and biological activity of eight thymulin analogs. Biol. Chem. Hoppe-Seyler. 368: 463-70 (1987); **Bach JF & Dardenne M** Thymulin, a zinc-dependent hormone. Med. Oncol. Tumor Pharmacother. 6: 25-9 (1989); **Coto JA et al** Interleukin 1 regulates secretion of zinc-thymulin by human thymic epithelial cells and its action on T lymphocyte proliferation and nuclear protein kinase C. PNAS 89: 7752-6 (1992); **Dardenne M et al** Neuroendocrine control of thymic hormonal production. I. Prolactin stimulates *in vivo* and *in vitro* the production of thymulin by human and murine thymic epithelial cells. Endocrinology 125: 3-12 (1989); **Kobayashi H et al** Serum thymic factor as a radioprotective agent promoting survival after X-irradiation. Experientia 46: 484-6 (1990); **Lenfant M et al** Relationship between a spleen-derived immunosuppressive peptide 'SDIP' and the 'Facteur thymique serique' (FTS): biochemical and biological comparison of the two factors. Immunology 48: 635-45 (1983); **Mocchegiani E et al** Recovery of low thymic hormone levels in cancer patients by lysine-arginine combination. IJI 12: 365-71 (1990); **Safieh-Garabedian B et al** Thymulin and its role in immunomodulation. J. Autoimmun. 5: 547-55 (1992); **Timsit J et al** Growth hormone and insulin-like growth factor-I stimulate hormonal function and proliferation of thymic epithelial cells. J. Clin. Endocrinol. Metab. 75: 183-8 (1992)

● **Thymic Humoral Factor: Barak Y et al** Thymic humoral factor-γ 2, an immunoregulatory peptide, enhances human hematopoietic progenitor cell growth. EH 20: 173-7 (1992); **Burstein Y et al** Thymic humoral factor γ 2: purification and amino acid sequence of an immunoregulatory peptide from calf thymus. B 27: 4066-71 (1988); **Handzel ZT et al** Immunomodulation of T cell deficiency in humans by thymic humoral factor: from crude extract to synthetic thymic humoral factor-γ 2. J. Biol. Response Mod. 9: 269-78 (1990); **Indig FE et al** Hydrolysis of thymic humoral factor γ 2 by neutral endopeptidase (EC 3.4.24.11). BJ 278: 891-4 (1991); **Ophir R et al** THF-γ 2, a thymic hormone, increases immunocompetence and survival in 5-fluorouracil-treated mice bearing MOPC-315 plasmacytoma.

Cancer Immunol. Immunother. 30: 119-25 (1989); **Ophir R et al** A synthetic thymic hormone, THF-γ 2, repairs immunodeficiency of mice cured of plasmacytoma by melphalan. IJC 45: 1190-4 (1990); **Umiel T et al** THF, a thymic hormone, promotes interleukin-2 production in intact and thymus-deprived mice. J. Biol. Response. Mod. 3: 423-34 (1984)

● **THYMOCYTE GROWTH PEPTIDE: Ernstrom U et al** Purification of thymocyte growth peptide (TGP) from sheep thymus. Relationship to FTS/thymulin. Biosci. Rep. 10: 403-12 (1990); **Ernstrom U** Identification of a mammalian growth factor as a ribofolate peptide. Biosci. Rep. 11: 119-30 (1991)

● **THYMOMODULIN: Balbi B et al** Thymomodulin increases release of granulocyte-macrophage colony stimulating factor and of tumor necrosis factor *in vitro*. Eur. Respir. J. 5: 1097-103 (1992); **Balbi B et al** Thymomodulin increases HLA-DR expression by macrophages but not T lymphocyte proliferation in autologous mixed leukocyte reaction. Eur. Respir. J. 6: 102-9 (1993); **Cavagni G et al** "Food allergy in children: an attempt to improve the effects of the elimination diet with an immunomodulating agent (thymomodulin). A double-blind clinical trial". Immunopharmacol. Immunotoxicol. 11: 131-42 (1989); **Galli L et al** [Preventive effect of thymomodulin in recurrent respiratory infections in children] Pediatr. Med. Chir. 12: 229-32 (1990); **Kouttab NM et al** Thymomodulin: biological properties and clinical applications. Med. Oncol. Tumor Pharmacother. 6: 5-9 (1989); **Maiorano V et al** Thymomodulin increases the depressed production of superoxide anion by alveolar macrophages in patients with chronic bronchitis. Int. J. Tissue React. 11: 21-5 (1989)

● **THYMOPOIETIN: Abiko T & Sekino H** Synthesis of the revised amino acid sequence of thymopoietin II and examination of its immunological effect on the impaired T lymphocyte transformation of a uremic patient with pneumonia. Chem. Pharm. Bull. Tokyo. 35: 2016-24 (1987); **Abiko T** Syntheses and structure-activity relationships of thymopoietin. AEMB 223: 153-5 (1987); **Audhya T et al** Complete amino acid sequences of bovine thymopoietins I, II, and III: closely homologous polypeptides. B 20: 6195-200 (1981); **Audhya T et al** Isolation and complete amino acid sequence of human thymopoietin and splenin. PNAS 84: 3545-9 (1987); **Barcellini W et al** *In vivo* immunopotentiating activity of thymopentin in aging humans: modulation of IL2 receptor expression. Clin. Immunol. Immunopathol. 48: 140-9 (1988); **Bernengo MG et al** Thymopentin in Sezary syndrome. JNCI 84: 1341-6 (1992); **Buzzetti R et al** Thymopentin induces release of ACTH-like immunoreactivity by human lymphocytes. J. Clin. Lab. Immunol. 29: 157-9 (1989); **Denes L et al** Selective restoration of immunosuppressive effect of cytotoxic agents by thymopoietin fragments. Cancer Immunol. Immunother. 32: 51-4 (1990); **Faist E et al** Immunomodulatory therapy with thymopentin and indomethacin. Successful restoration of interleukin-2 synthesis in patients undergoing major surgery. Ann. Surg. 214: 264-73 (1991); **Fiorilli M et al** *In vitro* enhancement of bone marrow natural killer cells after incubation with thymopoietin32-36 (TP-5). Thymus 5: 375-82 (1989); **Goldstein G & Audhya TK** Thymopoietin to thymopentin: experimental studies. Surv. Immunol. Res. 4: s1-10 (1985); **Hahn GS & Hamburger RN** Evolutionary relationship of thymopoietin to immunoglobulins and cellular recognition molecules. JI 126: 459-62 (1981); **Heavner GA et al** Structural requirements for the biological activity of thymopentin analogs. ABB 242: 248-55 (1985); **Hu C et al** *In vivo* enhancement of NK-cell activity by thymopentin. IJI 12: 193-7 (1990); **Kisfaludy L et al** Immuno-regulating peptides, I. Synthesis and structure-activity relationships of thymopentin analogs. Hoppe-Seylers Z. Physiol. Chem. 364: 933-40 (1983); **Lin CY & Low TL** A comparative study on the immunological effects of bovine and porcine thymic extracts: induction of lymphoproliferative response and enhancement of interleukin-2, γ-interferon and tumor necrotic factor production *in vitro* on cord blood lymphocytes. Immunopharmacology 18: 1-10 (1989); **Quik M et al** Thymopoietin, a thymic polypeptide, regulates nicotinic α-bungarotoxin sites in chromaffin cells in culture. Mol. Pharmacol. 37: 90-7 (1990); **Rajnavolgyi E et al** The influence of new thymopoietin derivatives on the immune response of inbred mice. IJI 8: 167-77 (1986); **Zevin-Sonkin D et al** Molecular cloning of the bovine thymopoietin gene and its expression in different calf tissues: evidence for a predominant expression in thymocytes. Immunol. Lett. 31: 301-9 (1992)

● **THYMOSTIMULIN: Balleari E et al** *In vivo* hemopoietic activity of thymic extract 'Thymustimulin' in aged healthy humans. Thymus 19: 59-63 (1992); **Ciconi E et al** [Perioperative treatment with thymostimulin in patients with stomach and colorectal neoplasms. Our experience with 114 cases] Minerva Chir. 47: 939-40 (1992); **Lai N et al** [Postoperative infections: the use of thymostimulin (TP1) in patients at risk] G. Chir. 13: 377-8 (1992); **Lin CY et al** Enhancement of interleukin-2 and γ-interferon production *in vitro* on cord blood lymphocytes and *in vivo* on primary cellular immunodeficiency patients with thymic extract (thymostimulin). J. Clin. Immunol. 8: 103-7 (1988); **Mantovani G et al** [Controlled trial of thymostimulin treatment of patients with primary carcinoma of the larynx resected surgically. Immunological and clinical evaluation and therapeutic prospects] Recenti Prog. Med. 83: 303-6 (1992); **Surico N & Tavassoli K** Effect of immunostimulating therapy on the immunocompetent system in breast carcinoma. Panminerva Med. 34: 172-80 (1992); **Tas MP et al** Depressed monocyte polarization and clustering of dendritic cells in patients with head and neck cancer: *in vitro* restoration of this immunosuppression by thymic hormones. Cancer Immunol. Immunother. 36: 108-14 (1993); **Tortorella C et al** Thymostimulin administration modulates polymorph metabolic pathway in patients with chronic obstructive pulmonary disease. Immunopharmacol. Immunotoxicol. 14: 421-37 (1992)

● **THYMOSINS: Caldarella J et al** Thymosin α 11: a peptide related to thymosin α 1 isolated from calf thymosin fraction 5. PNAS 80: 7424-7 (1983); **Cohen MH et al** Thymosin fraction 5 and intensive combination chemotherapy prolonging the survival of patients with small cell lung cancer. JAMA 241: 1813-21 (1979); **Conlon JM et al** Isolation and structural characterization of thymosin-β 4 from a human medullary thyroid carcinoma. J. Endocrinol. 118: 155-9 (1988); **Condon MR & Hall AK** Expression of thymosin β-4 and related genes in developing human brain. J. Mol. Neurosci. 3: 165-70 (1992); **Dugina TN et al** [Thymosin α(1) - an endogenous modulator of α-thrombin recognition site] Biull. Eksp. Biol. Med. 114: 260-2 (1992); **Erickson-Viitanen S et al** Thymosin β 10, a new analog of thymosin β 4 in mammalian tissues. ABB 225: 407-13 (1983); **Erickson-Viitanen S & Horecker BL** Thymosin β 11: a peptide from trout liver homologous to thymosin β 4. ABB 233: 815-20 (1984); **Galoyan AA et al** A hypothalamic activator of calmodulin-dependent enzymes is thymosin β 4 (1-39). Neurochem. Res. 17: 773-7 (1992); **Gomez-Marquez J et al** Thymosin-β 4 gene. Preliminary characterization and expression in tissues, thymic cells, and lymphocytes. JI 143: 2740-4 (1989); **Gondo H et al** Differential expression of the human thymosin-β 4 gene in lymphocytes, macrophages, and granulocytes. JI 139: 3840-8 (1987); **Goodall GJ et al** Thymosin β 4 in cultured mammalian cell lines. ABB 221: 598-601 (1983); **Goodall GJ & Horecker BL** Molecular cloning of the cDNA for rat spleen thymosin β 10 and the deduced amino acid sequence. ABB 256: 402-5 (1987); **Hall AK** Developmental regulation of thymosin β 10 mRNA in the human brain. Brain Res. Mol. Brain Res. 9: 175-7 (1991);

Hannappel E et al Thymosins β 8 and β 9: two new peptides isolated from calf thymus homologous to thymosin β 4. 79: 1708-11 (1982); **Haritos AA** α-thymosins: relationships in structure, distribution, and function. Isozymes Curr. Top. Biol. Med. Res. 14: 123-52 (1987); **Ho AD et al** Terminal differentiation of cord blood lymphocytes induced by thymosin fraction 5 and thymosin α 1. Scand. J. Immunol. 21: 221-5 (1985); **Horecker BL et al** Thymosin β 4-like peptides. MiE 116: 265-9 (1985); **Hu SK et al** Thymosin enhances the production of IL1 α by human peripheral blood monocytes. Lymphokine Res. 8: 203-14 (1989); **Kouttab NM et al** Production of human B and T cell growth factors is enhanced by thymic hormones. Immunopharmacology 16: 97-105 (1988); **Leichtling KD et al** Thymosin α 1 modulates the expression of high affinity interleukin-2 receptors on normal human lymphocytes. IJI 12: 19-29 (1990); **Low TL et al** Complete amino acid sequence of bovine thymosin β 4: a thymic hormone that induces terminal deoxynucleotidyl transferase activity in thymocyte populations. PNAS 78: 1162-6 (1981); **Low TL & Goldstein AL** Thymosin β 4. MiE 116: 248-55 (1985); **Low TL & Goldstein AL** Thymosin α 1 and polypeptide β 1. MiE 116: 233-48 (1985); **Low TL & Goldstein AL** Thymosin fraction 5 and 5A. MiE 116: 219-339 (1985); **Low TL & Goldstein AL** Thymosins: structure, function and therapeutic applications. Thymus 6: 27-42 (1984); **McCreary V et al** Sequence of a human kidney cDNA clone encoding thymosin β 10. BBRC 152: 862-6 (1988); **Moody TW et al** Thymosin α 1 down-regulates the growth of human non-small cell lung cancer cells *in vitro* and *in vivo*. CR 53: 5214-8 (1993); **Moscinski LC et al** Identification of a series of differentiation-associated gene sequences from GM-CSF-stimulated bone marrow. O 5: 31-7 (1990); **Mutchnick MG et al** Thymosin treatment of chronic hepatitis B: a placebo-controlled pilot trial. Hepatology 14: 409-15 and 567-9 (1991); **Noguchi K et al** Antitumor activity of a novel chimera tumor necrosis factor (TNF-STH) constructed by connecting rTNF-S with thymosin β 4 against murine syngeneic tumors. J. Immunother. 10: 105-11 (1991); **Oates KK & Coss MC** Biochemical and immunohistological identification of thymosin α-1 in MCF-7 breast cancer cells. Thymus 17: 147-54 (1991); **Ohta Y et al** Thymosin-α 1 increases the capability to produce interleukin-3 but not interleukin-2 in nu/nu mice. J. Biol. Response Mod. 6: 181-93 (1987); **Serrate SA et al** Modulation of human natural killer cell cytotoxic activity, lymphokine production, and interleukin 2 receptor expression by thymic hormones. JI 139: 2338-43 (1987); **Shimamura R et al** Expression of the thymosin β 4 gene during differentiation of hematopoietic cells. Blood 76: 977-84 (1990); **Spangelo BL et al** Biology and chemistry of thymosin peptides. Modulators of immunity and neuroendocrine circuits. ANY 496: 196-204 (1987); **Spangelo BL et al** Thymosin fraction 5 stimulates prolactin and growth hormone release from anterior pituitary cells *in vitro*. Endocrinology 121: 2035-43 (1987); **Sztein MB & Serrate SA** Characterization of the immunoregulatory properties of thymosin α 1 on interleukin-2 production and interleukin-2 receptor expression in normal human lymphocytes. IJI 11: 789-800 (1989); **Talmadge JE et al** Thymosin: immunomodulatory and therapeutic characteristics. PCBR 161: 457-65 (1984); **Tzehoval E et al** Thymosins α 1 and β 4 potentiate the antigen-presenting capacity of macrophages. Immunopharmacology 18: 107-13 (1989); **Wang SH et al** Effects of thymosin and insulin on suppressor T cell in type 1 diabetes. Diabetes Res. 19: 21-9 (1992); **Watts JD et al** Thymosins: both nuclear and cytoplasmic proteins. EJB 192: 643-51 (1990); **Weterman MA et al** Thymosin β-10 expression in melanoma cell lines and melanocytic lesions: a new progression marker for human cutaneous melanoma. IJC 53: 278-84 (1993); **Wetzel R et al** Production of biologically active N α-desacetylthymosin α 1 in *Escherichia coli* through expression of a chemically synthesized gene. B 19: 6096-104 (1980)

Thymic humoral factor: abbrev. THF. See: Thymic hormones.

Thymic stroma-derived T cell growth factor: see: TSTGF.

Thymidine phosphorylase: see: PD-ECGF (platelet-derived endothelial cell growth factor).

Thymocyte activating factor: see: TAF.

Thymocyte comitogenic factor: see: TCF.

Thymocyte differentiation factor: see: TDF.

Thymocyte growth factor: see: THGF.

Thymocyte growth peptide: abbrev. TGP. See: Thymic hormones.

Thymocyte mitogenic factor: see: TMF.

Thymocyte mitogenic protein: see: TMP.

Thymocyte proliferation factor: see: TPF.

Thymocyte stimulating factor: see: TSF.

Thymomodulin: see: Thymic hormones.

Thymopentin: abbrev. TP5. see: Thymic hormones.

Thymopoietin: see: Thymic hormones.

Thymosin alpha: see: Thymic hormones.

Thymosin beta: see: Thymic hormones.

Thymosin fraction 5: see: Thymic hormones.

Thymostimulin: abbrev. TP1. See: Thymic hormones.

Thymotaxin: This protein is secreted by a rat thymic epithelial cell line. It is chemotactic for rat bone marrow hematopoietic precursors ($Thy1^+$ immature lymphoid cells devoid of T cell, B cell, and myeloid cell differentiation markers). Thymotaxin is identical with » Beta2-microglobulin.

Thymic stroma-derived T-cell inhibitory factor: see addendum: TSTIF.

Dargemont C et al Thymotaxin, a chemotactic protein, is identical to β2-microglobulin. EJ 2: 1061-5 (1989); **Deugnier MA et al** Characterization of rat T cell precursors sorted by chemotactic migration toward thymotaxin. Cell 56: 1073-83 (1989)

Thymulin: (= serum thymic factor) see: Thymic hormones.

Thyroid stimulating hormone: see: Thyrotropin.

Thyrotropin: abbrev. ***TSH*** (thyroid stimulating hormone). A glycoprotein hormone consisting of two protein chains, one of which is identical with a subunit of » Luteinizing hormone. Thyrotropin is produced in the anterior pituitary in response to thyrotropin releasing hormone (thyroliberin; thyrotropic hormone releasing factor, abbrev. TRF). Secretion of TRF is promoted by noradrenalin and inhibited by serotonin. Its synthesis is also influenced by » IL1, and » IL6 and enhanced by » bFGF. Thyrotropin stimulates the thyroid gland to secrete thyroid hormones such as thyroxin and triiodothyronin. These two hormones inhibit the secretion of TRF and thyrotropin. Thyrotropin has been shown to stimulate secretion of » prolactin and acts as a neurotransmitter in the central nervous system. Apart from its well-known physiological role thyrotropin appears to be involved in the modulation of immune responses within the » neuroimmune network.
Thyrotropin has been shown to be produced also by human peripheral blood cells and to be produced constitutively by human T leukemia cell lines. Marked elevation of thyrotropin production by lymphocytes is seen in response of thyrotropin releasing hormone. TSH has been shown to enhance *in vitro* T cell-dependent and T cell-independent antibody responses. It enhances proliferation of lymphocytes stimulated by suboptimal concentrations of » IL2 and enhances IL2-induced NK cell activity. TSH also enhances production of superoxide anions by stimulated macrophages.
Blalock JE et al Enhancement of the *in vitro* antibody response by thyrotropin. BBRC 125: 30-4 (1985); **Coutelier JP et al** Binding and functional effects of thyroid stimulating hormone on human immune cells. J. Clin. Immunol. 10: 204-10 (1990); **Harbour DV et al** Differential expression and regulation of thyrotropin (TSH) in T cell lines. Mol. Cell. Endocrinol. 64: 229-41 (1989); **Koshida H & Kotake Y** Thyrotropin-releasing hormone enhances the superoxide anion production of rabbit peritoneal macrophages stimulated with N-formyl-methionyl-leucyl-phenylalanine and opsonized zymosan. Life Sci. 53: 725-31 (1993); **Kruger TE et al** Thyrotropin: an endogenous regulator of the *in vitro* immune response. JI 142: 744-7 (1989); **Provinciali M et al** Improvement in the proliferative capacity and natural killer cell activity of murine spleen lymphocytes by thyrotropin. Int. J. Immunopharmacol. 14: 865-70 (1992); **Smith EM et al** Human lymphocyte production of immunoreactive thyrotropin. PNAS 80: 6010-3 (1983)

Thyroxine-binding prealbumin: see: transthyretin.

TIA: see: IL2R/p55 (Tac) inducing factor.

TIF: *Thy1-inducing factor* This factor induces the expression of the Thy-1 antigen in murine lymphocytes residing in the thymus and in peripheral lymphocytes. It is identical with » IL3.
Schrader JW et al Expression of the Thy-1 antigen is not limited to T cells in cultures of mouse hemopoietic cells. PNAS 79: 4161-5 (1983)

TIF-1: *tumor-inducing factor 1* This factor is secreted by the human rhabdomyosarcoma cell line A673 together with » TIF-2. This factor is identical with » TGF-β.
Iwata KK et al Isolation of tumor cell growth-inhibiting factors from a human rhabdomyosarcoma cell line. CR 45: 2689-94 (1985)

TIF-2: *tumor inhibitory factor 2* This factor is secreted by the human rhabdomyosarcoma cell line A673. It inhibits the growth of these cells, suggesting that it may act as an » autocrine growth inhibitor, and also of lung and mammary carcinoma cell lines. It stimulates the growth of normal human fibroblasts. The known N-terminal sequence (19 amino acids) suggest that this factor is identical with IL1α (see: IL1).
Fryling C et al Two distinct tumor cell growth-inhibiting factors from a human rhabdomyosarcoma cell line. CR 45: 2695-9 (1985); **Fryling C et al** Purification and characterization of tumor inhibitory factor 2: its identity to interleukin 1. CR 49: 3333-7 (1989)

TIL: *tumor infiltrating lymphocytes* Lymphocyte infiltration into a tumor can be regarded as an expression of host immunity against the tumor, but tumor-infiltrating lymphocytes have little (***CTIL***; c = cytotoxic) or no cytotoxicity. It has been observed that cells resembling TILs, » CIK (cytokine-induced killer cells), can be generated which are even more cytotoxic.
TILs, which in humans include $CD4^+$ and $CD8^+$ cells, are isolated directly from tumor tissues and can be expanded *in vitro* by culture in the presence of » IL2. Lymphocyte expansion and proliferation is significantly enhanced in the presence of » IL4. These cells which already recognized the tumor *in vivo* are capable of accumulating at the site of the

tumor following their re-introduction into the tumor patient. These cells are activated by the treatment with IL2 and are frequently more aggressive than normal lymphokine activated cells (see also: LAK cells); in the mouse TILs have been shown to be 50 to 100 times more potent than » LAK (lymphokine activated killer) cells. These cells may be suitable to reach and destroy distant metastases. The cytotoxic activities of TILs are enhanced by » IFN-γ. The antitumor activity of TILs *in vivo* is blocked by » TGF-β.

TILs also are suitable targets for the introduction of cytokine genes as a novel form of cancer therapy (see: Cytokine gene transfer). The tumor targeting nature of TILs creates the possibility of using them as a vehicle to deliver gene products specifically to tumor tissues. Genetically modified TILs carrying an expressed gene encoding » TNF have been used in the treatment of patients with advanced melanoma. The anti-tumor activity of murine TILs against modified and unmodified tumor cells has been shown to be increased after introduction of MHC class I antigens into tumor cells, and this approach may also be of potential significance as an immunomodulating strategy. For another variant of treatment of tumors with TILs see: LATI (lymphokine-activated tumor inhibition).

Alexander RB & Rosenberg SA Adoptively transferred tumor-infiltrating lymphocytes can cure established metastatic tumor in mice and persist long-term *in vivo* as functional memory T lymphocytes. J. Immunother. 10: 389-97 (1991); **Bukowski RM et al** Clinical results and characterization of tumor-infiltrating lymphocytes with or without recombinant interleukin 2 in human metastatic renal cell carcinoma. CR 51: 4199-205 (1991); **Chin Y et al** Large scale expansion of human tumor infiltrating lymphocytes with surface-modified stimulator cells for adoptive immunotherapy. Anticancer Res. 12: 733-6 (1992); **Dillman RO et al** Continuous interleukin-2 and tumor-infiltrating lymphocytes as treatment of advanced melanoma: A National Biotherapy Study Group trial. Cancer 68: 1-8 (1991); **Figlin RA** Cancer immunotherapy using tumor-infiltrating lymphocytes. Semin. Hematol. 1992 Apr; 29: s33-5 (1992); **Favrot MC & Philip T** Treatment of patients with advanced cancer using tumor infiltrating lymphocytes transduced with the gene of resistance to neomycin. Hum. Gene Ther. 3: 533-42 (1992); **Gotoh K et al** Augmentation of cytotoxicity of tumor-infiltrating lymphocytes by biological response modifiers. IJI 13: 485-92 (1991); **Hwu P et al** Use of soluble recombinant TNF receptor to improve detection of TNF secretion in cultures of tumor infiltrating lymphocytes. JIM 151: 139-47 (1992); **Jemma C et al** *In vitro* and *in vivo* comparison of the activity of human lymphokine-activated killer (LAK) cells and adherent LAK cells. J. Immunother. 10: 189-99 (1991); **Kasid A et al** Human gene transfer: characterization of human tumor-infiltrating lymphocytes as vehicles for retroviral-mediated gene transfer in man. PNAS 87: 473-7 (1990); **Nishimura T et al** Recombinant interleukin-2-expanded tumor infiltrating lymphocytes from human renal cell cancer do not exhibit autologous tumor cell-specific cytotoxicity. Urologia Internationalis 47, Suppl 1: 83-5 (1991); **Papamichail M & Baxevanis CN** γ-interferon enhances the cytotoxic activity of interleukin-2-induced peripheral blood lymphocyte (LAK) cells, tumor infiltrating lymphocytes (TIL), and effusion associated lymphocytes. J. Chemother. 4: 387-93 (1992); **Peyret C et al** Regulatory effects of interleukin-4 on tumor-infiltrating lymphocytes derived from human renal cell carcinoma. J. Surg. Res. 53: 602-9 (1992); **Rosenberg SA et al** Gene transfer into humans: immunotherapy of patients with advanced melanoma using tumor infiltrating lymphocytes modified by retroviral gene transduction. NEJM 323: 570-8 (1990); **Rosenberg SA** Immunotherapy and gene therapy of cancer. CR 51: s5074-s9 (1991); **Schendel DJ & Gansbacher B** Tumor-specific lysis of human renal cell carcinomas by tumor-infiltrating lymphocytes: modulation of recognition through retroviral transduction of tumor cells with interleukin 2 complementary DNA and exogenous α interferon treatment. CR 53: 4020-5 (1993); **Sivanandham M et al** Prospects for gene therapy and lymphokine therapy for metastatic melanoma. Ann. Plast. Surg. 28: 114-8 (1992); **Whiteside TL** Cancer therapy with tumor-infiltrating lymphocytes: evaluation of potential and limitations. *In vivo* 5: 553-60 (1991); **Whiteside TL et al** Tumor-infiltrating lymphocytes. Potential and limitations to their use for cancer therapy. Crit. Rev. Oncol. Hematol. 12: 25-47 (1992)

TIMP: *tissue inhibitor of metalloproteases* Metalloproteinases of the extracellular matrix (see: ECM) are a family of secreted proteolytic enzymes that are involved in the biosynthesis of connective tissue. The synthesis and secretion of matrix metalloproteinases (MMPs) is induced in various cell types by a number of cytokines including » EGF, » PDGF, » IL1, » IL6, » IL8, IL1β, » TNF-α, and » bFGF. Metalloproteinases degrade constituents of the basal membrane and the extracellular matrix, including collagens, proteoglycans, gelatin, fibronectin, laminin, and elastin, under physiological and pathological conditions. The activities of these enzymes also facilitate the invasive migration of cells.

The biological activities of the proteases is subject to a complex regulation also involving specific inhibitors, called TIMP (tissue inhibitor of metalloproteinases). Many proforms of these metalloproteinases form complexes with these inhibitors. TIMP is a major regulator of extracellular matrix synthesis and degradation.

The two inhibitors are called TIMP-1 and TIMP-2. TIMP-1 is a protein of 28 kDa, TIMP-2 of 21 kDa. TIMP-1 and TIMP-2 exhibit an overal similarity of 71 % at the protein level. TIMP-1 is also the same as » HCI (human collagenase inhibitor). The murine ***3/10*** gene is the homologue of human TIMP-2. Murine and human TIMP-2 display 96 % homology at the protein level. Homology between murine TIMP-1 and murine TIMP-2 is 42 %. The TIMP-1 gene lies within intron 6 (out of 13) of the

X-linked synapsin I (SYNI) gene in both man (Xp11.1-p11.4) and mouse. It is transcribed in the opposite direction to the SYNI gene in the mouse. The disruption of the TIMP-1 gene in pluripotent embryonic stem cells increases the invasive properties of these cells in an *in vitro* assay. The introduction of a TIMP-1 gene into invasively growing melanoma cells (B16-F10) leads to the overproduction of this inhibitor and suppresses invasive growth. Apart from their biological activities as inhibitors of metalloproteinases TIMP-1 and TIMP-2 also functions as growth factors for many human and murine cell types.

TIMP-2 inhibits the activities of transin, matrin (pump-1), 72 kDa gelatinase/type IV collagenase (MMP-2; matrix metalloproteinase), and interstitial collagenase. TIMP-2 regulates not only the activity of the mature enzyme but also the autolytic processing of the proenzyme. TIMP 2 has recently been found to be secreted by mouse folliculo-stellate cell. It acts as a cell survival factor for endocrine cells in the anterior pituitary gland.

Both inhibitors also influence the capacity of erythroid progenitor cells to grow as colonies in soft agar (see also: Colony formation assay). TIMPs are identical with » EPA (erythroid promoting activity).

Alexander CM & Werb Z Targeted disruption of the tissue inhibitor of metalloproteinases gene increases the invasive behavior of primitive mesenchymal cells derived from embryonic stem cells *in vitro*. JCB 118: 727-39 (1992); **Alitalo R et al** metalloproteinases and *jun/fos* transcription factor complex characterize tumor promoter-induced megakaryoblastic differentiation of K562 leukemia cells. Blood 75: 1974-82 (1990); **Bertaux B et al** Growth stimulation of human keratinocytes by tissue inhibitor of metalloproteinases. J. Invest. Dermatol. 97: 679-85 (1991); **Campbell CE et al** Identification of a serum- and phorbol ester-responsive element in the murine tissue inhibitor of metalloproteinase gene. JBC 266: 7199-206 (1991); **Clark IM et al** Polyclonal and monoclonal antibodies against human tissue inhibitor of metalloproteinases (TIMP) and the design of an enzyme-linked immunosorbent assay to measure TIMP. Matrix 11: 76-85 (1991); **Cocuzzi ET et al** Expression and purification of mouse TIMP-1 from *Escherichia coli*. FL 307: 375-8 (1992); **DeClerck VA et al** Inhibition of invasion and metastasis in cells transfected with an inhibitor of metalloproteinases. CR 52: 701-8 (1992); **Derry JMJ & Barnard PJ** Physical linkage of the A-*raf*-1, properdin, synapsin I, and TIMP genes on the human and mouse X chromosomes. Genomics 12: 632-8 (1992); **Docherty AJ et al** Sequence of human tissue inhibitor of metalloproteinases and its identity to erythroid-potentiating activity. N 318: 66-9 (1985); **Fridman R et al** Domain structure of human 72-kDa gelatinase/type IV collagenase. Characterization of proteolytic activity and identification of the tissue inhibitor of metalloproteinase-2 (TIMP-2) binding regions. JBC 267: 15398-405 (1992); **Hayakawa T et al** Tissue inhibitor of metalloproteinases from human bone marrow stromal cell line KM 102 has erythroid-potentiating activity, suggesting its possibly bifunctional role in the hematopoietic microenvironment. FL 268: 125-8 (1990); **Hayakawa T et al** Growth-promoting activity of tissue inhibitor of metalloproteinases-1 (TIMP-1) for a wide range of cells: A possible new growth factor in serum. FL 298: 29-32 (1992); **Huebner K et al** Localization of the gene encoding human erythroid-potentiating activity to chromosome region Xp11.1-Xp11.4. Am. J. Hum. Genet. 38: 819-26 (1986); **Ito A et al** Calmodulin differentially modulates the interleukin 1-induced biosynthesis of tissue inhibitor of metalloproteinases and matrix metalloproteinases in human uterine cervical fibroblasts. JBC 266: 13598-601 (1991); **Khokha R et al** Suppression of invasion by inducible expression of tissue inhibitor of metalloproteinase-1 (TIMP-1) in B16-F10 melanoma cells. JNCI 84: 1017-22 (1992); **Kishi JI et al** Purification and characterization of a new tissue inhibitor of metalloproteinases (TIMP-2) from mouse colon 26 tumor cells. Matrix 11: 10-6 (1991); **Kolkenbrock H et al** The complex between a tissue inhibitor of metalloproteinases (TIMP-2) and 72-kDa progelatinase is a metalloproteinase inhibitor. EJB 198: 775-81 (1991); **Kubota S et al** Transforming growth factor-β suppresses the invasiveness of human fibrosarcoma cells *in vitro* by increasing expression of tissue inhibitor of metalloprotease. BBRC176: 129-36 (1991); **Leco KJ et al** Differential regulation of TIMP-1 and TIMP-2 mRNA expression in normal and Ha-*ras*-transformed murine fibroblasts. Gene 117: 209-17 (1992); **Lotz M & Guerne PA** Interleukin-6 induces the synthesis of tissue inhibitor of metalloproteinases-1/erythroid potentiating activity (TIMP-1/EPA). JBC 266: 2017-20 (1991); **Matsumoto H et al** Newly established murine pituitary folliculo-stellate-like cell line (TtT/GF) secretes potent pituitary glandular cell survival factors, one of which corresponds to metalloproteinase inhibitor. BBRC 194: 909-15 (1993); **Murate T et al** Erythroid potentiating activity of tissue inhibitor of metalloproteinases on the differentiation of erythropoietin-responsive mouse erythroleukemia cell line, ELM-I-1-3, is closely related to its cell growth potentiating activity. EH 21: 169-76 (1993); **Murphy G et al** The N-terminal domain of tissue inhibitor of metalloproteinases retains metalloproteinase inhibitory activity. B 30: 8097-102 (1991); **Muscat GE et al** Proliferin, a prolactin/growth hormone-like peptide represses myogenic-specific transcription by the suppression of an essential serum response factor-like DNA-binding activity. ME 5: 802-14 (1991); **Nemeth JA & Goolsby CL** TIMP-2, a growth-stimulatory protein from SV40-transformed human fibroblasts. ECR 207: 376-82 (1993); **Ogata V et al** Matrix metalloproteinase 3 (stromelysin) activates the precursor for the human matrix metalloproteinase 9. JBC 267: 3581-4 (1992); **Osthues A et al** Isolation and characterization of tissue inhibitors of metalloproteinases (TIMP-1 and TIMP-2) from human rheumatoid synovial fluid. FL 296: 16-20 (1992); **Roswit WT et al** Purification and sequence analysis of two rat tissue inhibitors of metalloproteinases. ABB 292: 402-10 (1992); **Sato H et al** Expression of genes encoding type IV collagen-degrading metalloproteinases and tissue inhibitors of metalloproteinases in various human tumor cells. O 7: 77-83 (1992); **Schumacher A et al** Murine cyclophilin-S1: A variant peptidyl-prolyl isomerase with a putative signal sequence expressed in differentiating F9 cells. BBA Gene Struct. Expression 1129: 13-22 (1991); **Shimizu S et al** Cloning and sequencing of the cDNA encoding a mouse tissue inhibitor of metalloproteinase-2. Gene 114: 291-2 (1992); **Spurr NK et al** Chromosomal assignment of the gene encoding the human tissue inhibitor of metalloproteinases to Xp11.1-p11.4. Ann. Hum. Genet. 51: 189-94 (1987); **Stetler-Stevenson WG et al** Tissue inhibitor of metalloproteinase-2 (TIMP-2) has erythroid-potentiating activity. FL 296: 231-4 (1992); **Ward RV et al** The purification of tissue inhibitor of metalloproteinases-2 from its 72 kDa progelatinase complex. Demonstration of the biochemical similarities of tissue in-

TIMP-3: see addendum.

hibitor of metalloproteinases-2 and tissue inhibitor of metalloproteinases-1. BJ 278: 179-87 (1991); **Willard HF et al** Regional localization of the TIMP gene on the human X chromosome: extension of a conserved synteny and linkage group on proximal Xp. Hum. Genet. 81: 234-8 (1989); **Wright JK et al** Transforming growth factor β stimulates the production of the tissue inhibitor of metalloproteinases (TIMP) by human synovial and skin fibroblasts. BBA Mol. Cell Res. 1094: 207-10 (1991)

T interferon: An alternative older name for » IFN-γ.

TIS genes: *tetradecanoyl phorbol acetate-inducible sequences* TIS genes represent a family of immediate early response genes (see: ERG) originally identified as transcripts induced rapidly by the » Phorbol ester TPA in murine » 3T3 fibroblasts. More than 50 of these genes have been described. They are inducible in response of a variety of » cytokines in a variety of cell types and some of them have also been reported to be expressed in response to treatment of cells with heavy metals. Each TIS gene has a distinct tissue specificity and/or developmental profile.

TIS1 and ***TIS8*** are identical with NGFIB, and NGFIA, respectively, two genes induced by » NGF in pheochromocytoma cells (see: NGFI). ***TIS7*** is identical with » PC-4. ***TIS-10*** encodes a prostaglandin synthase/cyclooxygenase distinct from prostaglandin synthase/cyclooxygenase. (EC 1.14.99.1), previously cloned from mouse, man, and sheep. Its synthesis in inhibitible by dexamethasone. TIS-10 is identical with the inducible protein ***p71/73*** of mouse macrophages which is expressed in response to one of several stimuli that trigger the expression of cytolytic activity when these cells have previously been primed for tumor cell killing by » IFN-γ.***Tis 11*** is identical with » » ERF-1, » cMG1 or Nup475 and encodes the erythroid-specific transcription factor » GATA-1. ***TIS-21*** encodes a protein of 170 amino acids that is inducible by » NGF. It is homologous with the human ***BTG1 gene*** and the murine ***PC3 gene***. The human and murine proteins are 100 % identical and the corresponding chicken gene product shows a homology of 91 %. BTG-1 has antiproliferative functions. ***TIS-28*** has been identified as the » *fos* oncogene.

Expression of TIS-8, TIS-11, and some other TIS genes is rapidly induced by » GM-CSF in a factor-dependent cell line, » 32D. TIS-1, TIS-8, and TIS-21 are also inducible by » bFGF in PC12 cells. TIS-11 is also inducible by » EGF in PC12 cells.

Altin JG et al Differential induction of primary-response (TIS) genes in PC12 pheochromocytoma cells and the unresponsive variant PC12nnr5. JBC 266: 5401-6 (1991); **Barnard RC et al** Coding sequence of ERF-1, the human homologue of Tis11b/cMG1, members of the Tis11 family of early response genes. NAR 21: 3580 (1993); Ryo R et al Megakaryocytic leukemia and platelet factor 4. Leuk. Lymphoma 8: 327-36 (1992); **Batistatou A et al** Nerve growth factor employs multiple pathways to induce primary response genes in PC12 cells. Mol. Biol. Cell. 3: 363-71 (1992); **Cicatiello L et al** Identification of a specific pattern of "immediate-early" gene activation induced by estrogen during mitogenic stimulation of rat uterine cells. Receptor 3: 17-30 (1993); **Epner DE & Herschman HR** Heavy metals induce expression of the TPA-inducible sequence (TIS) genes. JCP 148: 68-74 (1991); **Fletcher BS et al** Structure and expression of TIS21, a primary response gene induced by growth factors and tumor promoters. JBC 266: 14511-8 (1991); **Fletcher BS et al** Structure of the mitogen-inducible TIS10 gene and demonstration that the TIS10-encoded protein is a functional prostaglandin G/H synthase. JBC 267: 4338-44 (1992); **Gubits RM et al** Immediate early gene induction after neonatal hypoxia-ischemia. Brain Res. Mol. Brain Res. 18: 228-38 (1993); **Kaneda N et al** Sequence of a rat TIS11 cDNA, an immediate early gene induced by growth factors and phorbol esters. Gene 118: 289-91 (1992); **Kujubu DA et al** Induction of transiently expressed genes in PC-12 pheochromocytoma cells. O 1: 257-62 (1987); **Kujubu DA et al** TIS10, a phorbol ester tumor promoter-inducible mRNA from Swiss 3T3 cells, encodes a novel prostaglandin synthase/cyclooxygenase homologue. JBC 266: 12866-72 (1991); **Lim RW et al** Cloning of tetradecanoyl phorbol ester-induced 'primary response' sequences and their expression in density-arrested Swiss 3T3 cells and a TPA non-proliferative variant. O 1: 263-70 (1987); **Phillips TA et al** The mouse macrophage activation-associated marker protein, p71/73, is an inducible prostaglandin endoperoxide synthase (cyclooxygenase). J. Leukoc. >Biol. 53: 411-9 (1993); **Rouault JP et al** Sequence analysis reveals that the BTG1 anti-proliferative gene is conserved throughout evolution in its coding and 3' non-coding regions. Gene 129: 303-6 (1993); **Tippetts MT et al** Tumor promoter-inducible genes are differentially expressed in the developing mouse. MCB 8: 4570-2 (1988); **Varnum BC et al** Granulocyte-macrophage colony-stimulating factor and tetradecanoyl phorbol acetate induce a distinct, restricted subset of primary-response TIS genes in both proliferating and terminally differentiated myeloid cells. MCB 9: 3580-3 (1989); Nucleotide sequence of a cDNA encoding TIS11, a message induced in Swiss 3T3 cells by the tumor promoter tetradecanoyl phorbol acetate. O 4: 119-20 (1989)

TIS1: see: TIS genes.

TIS7: see: TIS genes.

TIS8: see: TIS genes.

TIS10: see: TIS genes.

TIS11: see: TIS genes.

TIS21: see: TIS genes.

TIS28: see: TIS genes.

Tissue-derived growth inhibitor: see: TGI.

Tissue inhibitor of metalloproteases: see: TIMP. See also: EPA (erythroid promoting activity).

TK-14: This human gene, isolated from a tumor cDNA library, encodes a receptor for » aFGF and » bFGF. The TK-14 gene is a homologue of the » K-*sam* gene and is also closely related to murine » *bek* and human » *flg* which also function as receptors for fibroblast growth factors.

Houssaint E et al Related fibroblast growth factor receptor genes exist in the human genome. PNAS 87: 8180-4 (1990)

TKF: *tumor-killing factor* This poorly characterized factor, which appears to be distinct from » TNF-α or TNF-β, is produced by human monocytic cells. It is detected in the supernatants of some human macrophage-monocyte hybridomas after stimulation with PMA, polypeptone, and retinoic acid.

Taniyama T et al Demonstration of a novel tumor-killing factor secreted from human macrophage-monocyte hybridomas. JI 141: 4061-6 (1988)

TKF: *tyrosine kinase-related to fibroblast growth factor receptor* A tyrosine-specific protein kinase exclusively expressed in lung, tissues. Sequence analysis of the cDNA demonstrates that TKF is a member of the fibroblast growth factor receptor family (see: aFGF and bFGF). In contrast to other members of this receptor family TKF is found in the cytosol.

Holtrich U et al Two additional protein-tyrosine kinases expressed in human lung: fourth member of the fibroblast growth factor receptor family and an intracellular protein-tyrosine kinase. PNAS 88: 10411-5 (1991)

T-LIF: *T cell-derived leukocyte (migration) inhibitory factor* see: LIF (leukocyte (migration) inhibitory factor).

T lymphocyte chemotactic factor: see: TCF.

T lymphocyte-derived neutrophil migration inhibition factor: see: NIF-T.

TMD2: This lymphoblastoid cell line has been established from the peripheral blood of a patient with chronic lymphocytic leukemia in the acute phase after continuous culture in the presence of » IL3. TMD2 cells depend on the presence of IL3. Other interleukins, including » IL1, » IL2, » IL4, » IL5, » IL6, » GM-CSF, » G-CSF, or » SCF do not support the growth or survival of TMD2 cells. Anti-IL3-antibody blocks the stimulatory effect of IL3 on these cells.

Nara N Colony-stimulating factor (CSF)-dependent growth of two leukemia cell lines. Leuk. Lymphoma 7: 331-5 (1992); **Tohda S et al** Establishment of an interleukin-3-dependent leukemic cell line from a patient with chronic lymphocytic leukemia in the acute phase. Blood 78: 1789-94 (1991)

TMF: *thymocyte mitogenic factor* This factor was initially detected in cell culture supernatants as an activity promoting the proliferation of lectin-stimulated thymocytes. It is identical with » IL2.

Aarden LA et al Revised nomenclature for antigen-nonspecific T cell proliferation and helper factors. CI 48: 433-36 (1979); **Farrar JJ et al** Biochemical relationship of thymocyte mitogenic factor and factors enhancing humoral and cell-mediated immune responses. JI 121: 1353-60 (1978); **Paetkau V et al** Cellular origins and targets of Costimulator (IL2). IR 51: 157-75 (1980)

TMF: *T cell maturation factor* See: TMF (thymocyte mitogenic factor).

TMF: *T cell mitogenic factor* See: TMF (thymocyte mitogenic factor).

TMIF: *tumor migration inhibition factor* This poorly characterized factor of ≈ 6 kDa reversibly inhibits the migration of a number of tumor cell types. Human and murine TMIF influences the adherence of tumor cells to endothelial cell monolayers *in vitro* as seen by diminished attachment and enhanced dissociation. Partially purified TMIF is also cytostatic for a number of tumor cell lines.

Antonia SJ et al The elaboration of a small molecular weight cytostatic factor by lymphoblastoid lines and activated lymphocytes. LR 5: 301-12 (1986); **Cohen MC & Cohen S** The role of lymphokines in neoplastic disease. Hum. Pathol. 17: 264-70 (1986); **Cohen MC et al** Adherence of tumor cells to endothelial monolayers: inhibition by lymphokines. CI 95: 247-57 (1985); **Cohen MC et al** *In vitro* migration of tumor cells from human neoplasms: inhibition by lymphokines. Clin. Immunol. Immunopathol. 34: 94-9 (1985); **Donskoy M et al** Lymphokine-induced migration inhibition of murine tumor cells derived from solid neoplasms. CR 44: 3870-2 (1984); **D'Silva H et al** The tumor disappearance reaction: an *in vivo* effect of a noncytotoxic lymphokine active against tumor cells. Clin. Immunol. Immunopathol. 34: 326-32 (1985)

TMP: *thymocyte mitogenic protein* This factor is secreted by macrophages following their stimulation with bacterial lipopolysaccharides and other inducers. It is identical with » LAF (lymphocyte activating factor) and hence identical with » IL1.

Lu CY & Unanue ER Ontogeny of murine macrophages: functions related to antigen presentation. Infect. Immun. 36: 169-75 (1982); **Tenu JP et al** Stimulation of thymocyte mitogenic protein secretion and of cytostatic activity of mouse peritoneal macrophages by trehalose dimycolate and muramyldipeptide. EJI 10: 647-53 (1980)

TNBSA-F: This factor of 35-55 kDa is induced by intravenous injection of TNBSA (trinitrobenzene sulfonic acid) in mice and prevents delayed-type hypersensitivity reactions (picryl chloride contact sensitivity test). The factor is produced by pooled spleen and lymph node cells *in vitro*. The inhibitory activity is not present in supernatants from lymphoid cells of sham-treated mice.
TNBSA-F induces a dose-dependent unresponsiveness of » HT-2 cells to » IL2. The inhibitory effect of TNBSA-F is not due to the presence of » TGF-β, soluble immune-response suppressor (see: SIRS), » IFN-γ, or » JE.

Ferreri NR et al Inhibition of IL2-dependent proliferation by a prostaglandin-dependent suppressor factor. JI 150: 2102-11 (1993)

TNF: *tumor necrosis factor* TNF was found originally in mouse serum after intravenous injection of bacterial endotoxin into mice primed with viable Mycobacterium bovis, strain Bacillus Calmette-Guerin (BCG). TNF was then shown to be present also in sera of rats, rabbits and guinea pigs. TNF-containing serum from mice is cytotoxic or cytostatic to a number of mouse and human transformed cell lines, but less or not toxic to normal cells *in vitro*. It causes necrosis of transplantable tumors in mice.
The two molecular species of TNF are known as » TNF-α (cachectin) and » TNF-β (Lymphotoxin). In many articles no distinction is made between TNF-α and TNF-β. In most cases the term TNF alone refers to TNF-α = cachectin, while TNF-β is usually referred to as lymphotoxin. An alternative name may be TNF/lymphotoxin.

TNF-α: tumor necrosis factor alpha.
■ **Alternative Names:** Cachectin; CF (cytotoxic factor); CTX (cytotoxin); DIF (differentiation inducing factor); EP (endogenous pyrogen), some of the EP activity may be due to » IL1 and » IFN-α); hemorrhagic factor; macrophage-derived cytotoxic factor; J774-derived cytotoxic factor; MCF (macrophage cytotoxic factor); MCT (macrophage cytotoxin); PCF (peritoneal cytotoxic factor). See also: individual entries for further information.
■ **Sources:** TNF is secreted by macrophages, monocytes, neutrophils, T cells, NK cells following their stimulation by lipopolysaccharides. CD4-positive cells secrete TNF-α while CD8-positive cells secrete little or no TNF-α. Stimulated peripheral neutrophilic granulocytes but also unstimulated cells and also a number of transformed cell lines, astrocytes, microglial, smooth muscle cells, and fibroblasts also secrete TNF. Human milk also contains this factor (see: MGF, milk growth factor).
The synthesis of TNF-α is induced by many different stimuli including interferons (see: IFN), » IL2, » GM-CSF, » SP (substance P; see also: tachykinins), » Bradykinin, Immune complexes, inhibitors of cyclooxygenase and PAF (platelet activating factor).
The production of TNF is inhibited by » IL6, TGF-β, vitamin D3, prostaglandin E_2, dexamethasone, » CsA (ciclosporin A), and antagonists of PAF (platelet activating factor).
■ **Protein Characteristics:** Human TNF-α is a non-glycosylated protein of 17 kDa and a length of 157 amino acids. Murine TNF-α is N-glycosylated. Homology with » TNF-β is ≈ 30 %.
TNF-α forms dimers and trimers.
The 17 kDa form of the factor is produced by processing of a precursor protein of 233 amino acids. A transmembrane form of 26 kDa has also been described.
TNF-α contains a single disulfide bond that can be destroyed without altering the biological activity of the factor. Mutations Ala84→Val and Val91→Ala reduce the cytotoxic activity of the factor almost completely. These sites are involved in receptor binding. The deletion of 7 N-terminal amino acids and the replacement of Pro8Ser9Asp10 by ArgLysArg yields a mutated factor with an ≈ 10-fold enhanced antitumor activity and increased receptor binding, as demonstrated by the » L-M cell assay, while at the same time reducing the toxicity.
■ **Gene Structure:** The gene has a length of ≈ 3,6 kb and contains four exons. The primary transcript has a length of 2762 nucleotides and encodes a precursor protein of 233 amino acids. The aminoterminal 78 amino acids function as a presequence (see also: Gene expression).
The human gene maps to chromosome 6p23-6q12. It is located between class I HLA region for HLA-B and the gene encoding complement factor C. The gene encoding TNF-β is ≈ 1.2 kb downstream of the TNF-α gene. However, both genes are regulated independently. The two genes also lie close to each other on murine chromosome 17.
■ **Receptor Structure, Gene(s), Expression:** Ca. 500-10000 high-affinity receptors (K_a = 2,5 × 10^{-9} M) for TNF-α are expressed on all somatic cell types with the exception of erythrocytes.

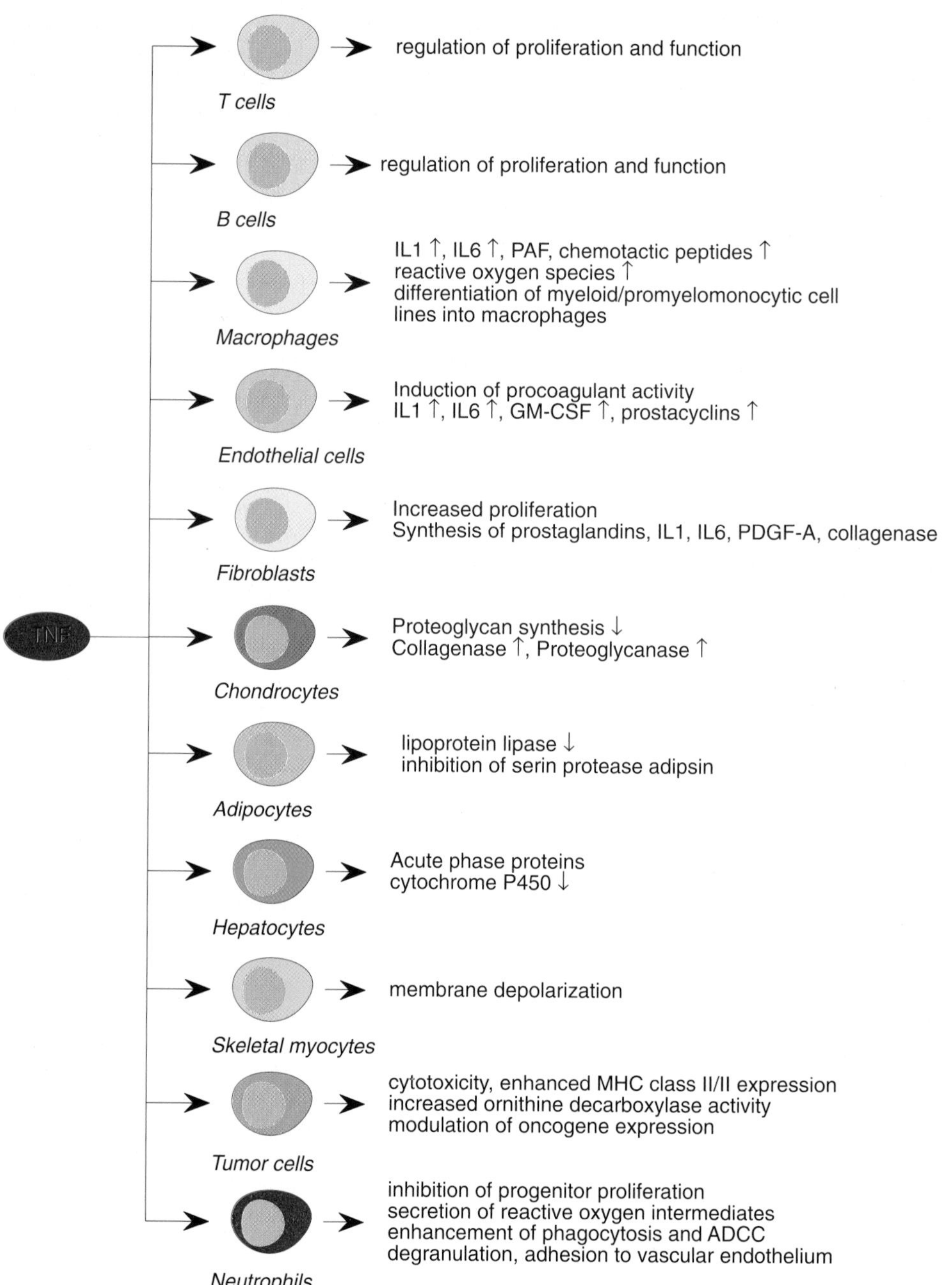

Schematic representation of the multiple activities of TNFs at the cellular level.

Two receptors of 55 kDa (TNF-R1; new designation: ***CD120a***, see also: CD Antigens) and 75 kDa (TNF-R2; ; new designation: ***CD120b***, see also: CD Antigens) have been described. One receptor is a glycosylated protein of 455 amino acids that contains an extracellular domain of 171 and a cytoplasmic domain of 221 amino acids. Sequence homologies in the cysteine-rich domains of the extracellular portion reveal that the receptor is related to the low-affinity receptor of » NGF and to human cell surface antigen » CD40.
The two known receptors bind both TNF-α and TNF-β. p55 is expressed particularly on cells susceptible to the cytotoxic action of TNF. p75 is also present on many cell types, especially those of myeloid origin. It is strongly expressed on stimulated T and B lymphocytes. The differential activities of TNF on various cell types, i. e. growth-promoting and growth-inhibiting activities, are probably mediated by the differential expression and/or regulation of multiple receptors in combination with other distinct receptor-associated proteins. p55 appears to play a critical role in host defenses against microorganisms and their pathogenic factors (see subentry Transgenic/Knock-out/Antisense studies).
A third receptor subtype is expressed in normal human liver. It binds TNF-α but not » TNF-β. Some viruses contain genes encoding secreted proteins with TNF binding properties that are closely homologous to the p55 and p75 TNF receptors (see: viroceptor).
Differential effects of the two receptor subtypes have also been found in TNF-mediated adhesion of leukocytes to the endothelium. It appears that engagement of the p55 receptor specifically leads to the induction of the cellular adhesion molecules ICAM-1, E-selectin, V-CAM-1, and CD44, while engagement of both the p55 and the p75 receptor induces expression of α2 integrin.
Truncated soluble forms of the receptor have also been found. The soluble forms, in particular the soluble extracellular domain of the p60 receptor, block the antiproliferative effects of TNF and may therefore modulate the harmful effects of TNF.
Apart from the membrane-bound receptors several soluble proteins that bind TNF have been described. These proteins of ≈ 30 kDa, called ***TBPI*** and ***TBPII*** (tumor necrosis factor binding proteins; see: TNF-BP), are derived from the TNF-binding domain of the membrane receptor. They can be isolated from urine and serum and probably function as physiological regulators of TNF activities by inhibiting binding of TNF to its receptor (see: TNF-BF, TNF blocking factor).
Receptor densities are reduced by » IL1 and tumor promoters such as » phorbol esters. The expression of TNF-α receptor density is induced by » IFN-α, IFN-β, and » IFN-γ.

■ **Biological Activities:** Human TNF-α is active on murine cells with a slightly reduced specific activity. In general, TNF-α and » TNF-β display similar spectra of biological activities in *in vitro* systems, although TNF-β is often less potent or displays apparent partial agonist activity.
TNF-α shows a wide spectrum of biological activities. It causes cytolysis and cytostasis of many tumor cell lines *in vitro*. Sensitive cells die within hours after exposure to picomolar concentrations of the factor and this involves, at least in part, mitochondria-derived second messenger molecules serving as common mediators of TNF cytotoxic and gene-regulatory signaling pathways. The factor induces hemorrhagic necrosis of transplanted tumors. Within hours after injection TNF-α leads to the destruction of small blood vessels within malignant tumors. The factor also enhances phagocytosis and cytotoxicity in neutrophilic granulocytes and also modulates the expression of many other proteins, including » *fos*, » *myc*, » IL1 and » IL6 (see: TIS genes).
The 26 kDa form of TNF is found predominantly on activated monocytes and T cells (see also: cell activation). It is also biologically active and mediates cell destruction by direct cell-to-cell contacts (see: Juxtacrine).
In vivo TNF-α in combination with » IL1 is responsible for many alterations of the endothelium. It inhibits anticoagulatory mechanisms and promotes thrombotic processes and therefore plays an important role in pathological processes such as venous thromboses, arteriosclerosis, vasculitis, and disseminated intravasal coagulation (see also: Septic shock). The expression of membrane thrombomodulin is decreased by TNF-α. TNF-α is a potent chemoattractant for neutrophils (see also: Chemotaxis) and also increases their adherence to the endothelium (see also subentry: receptors). The chemotactic properties of » fMLP (Formyl-Met-Leu-Phe) for neutrophils are enhanced by TNF-α. TNF-α induces the synthesis of a number of chemoattractant cytokines, including » IP-10, » JE, » KC, in a cell-type and tissue-specific manner.
Although TNF inhibits the growth of endothelial cells *in vitro* it is a potent » angiogenesis factor *in*

vivo. The angiogenic activity of TNF is significantly inhibited by IFN-γ.

TNF-α is a growth factor for normal human diploid fibroblasts. It promotes the synthesis of collagenase and prostaglandin E_2 in fibroblasts. It may also function as an » autocrine growth factor for human chronic lymphocytic leukemia cells *in vivo* and has been described to be an autocrine growth factor for neuroblastoma cells. The autocrine growth-promoting activity is inhibited by » IL4.

In resting macrophages TNF induces the synthesis of » IL1 and prostaglandin E_2. It also stimulates phagocytosis and the synthesis of superoxide dismutase in macrophages. TNF activates osteoclasts and thus induces bone resorption.

TNF-α inhibits the synthesis of lipoprotein lipase and thus suppresses lipogenetic metabolism in adipocytes.

In leukocyte and lymphocyte progenitors TNF stimulates the expression of class I and II HLA and differentiation antigens, and the production of IL1, colony-stimulating factors (see: CSF), » IFN-γ, arachidonic acid metabolism. It also stimulates the biosynthesis of collagenases in endothelial cells and synovial cells.

The interleukin » IL6 suppresses the synthesis of » IL1 induced by bacterial » endotoxins and TNF, and the endotoxin-induced synthesis of TNF.

The neurotransmitter » SP (substance P; see also: tachykinins) induces the synthesis of TNF and IL1 in macrophages. IL1, like IL6, stimulates the synthesis of ACTH (corticotropin) in the pituitary. Glucocorticoids synthesized in response to ACTH in turn inhibit the synthesis of IL6, IL1 and TNF *in vivo*, thus establishing a negative feedback loop between the immune system and neuroendocrine functions.

TNF promotes the proliferation of astroglia and microglia and therefore may be involved in pathological processes such as astrogliosis and demyelinisation.

TNF-α enhances the proliferation of T cells induced by various stimuli in the absence of » IL2. Some subpopulations of T cells only respond to IL2 in the presence of TNF-α. In The presence of IL2 TNF-α promotes the proliferation and differentiation of B cells.

The functional capacities of skin Langerhans cells are also influenced by TNF-α. These cells are not capable of initiating primary immune responses such as contact sensibilisation. They are converted into immunostimulatory dendritic cells by » GM-CSF and also » IL1. These cells therefore are a reservoir for immunologically immature lymphoid dendritic cells. The enhanced ability of maturated Langerhans cells to process antigens is significantly reduced by TNF-α.

Although TNF-α is also required for normal immune responses the overexpression has severe pathological consequences. TNF-α is the major mediator of cachexia observed in tumor patients (hence its name: cachectin). TNF is also responsible for some of the severe effects during Gram-negative sepsis (see: Septic shock).

TNF mediates part of the cell-mediated immunity against obligate and facultative bacteria and parasites. It confers protection against Listeria monocytogenes infections, and anti-TNF antibodies weaken the ability of mice to cope with these infections.

■ **TRANSGENIC/KNOCK-OUT/ANTISENSE STUDIES:** The consequences of a deregulated expression of TNF-α have been studied in » transgenic animals (mice) expressing the human factor. Overexpression of TNF-α precipitates chronic inflammatory polyarthritis which can be prevented completely by treatment of the animals with monoclonal antibodies directed against the human factor. Transgenic mice constitutively overexpressing human TNF-α in their T cell compartment under the control of human CD2 gene regulatory signals have been shown to develop marked histologic and cellular changes locally in their lymphoid organ and a lethal wasting syndrome associated with widespread vascular thrombosis and tissue necrosis. These changes could be neutralized also by the administration of monoclonal antibodies specific for human TNF-α. In another TNF-α transgenic mouse model the expression of the transgene in murine pancreatic β cells results in severe and permanent insulitis without evolution towards diabetes.

Transgenic mouse line bearing a reporter gene construct have been constructed in which chloramphenicol acetyl transferase coding sequences are expressed under the control of TNF regulatory sequences. In these animals expression of the transferase within tissues reflects TNF production. Analysis of these animals shows that the transferase is constitutively expressed in both the fetal and maternal thymuses, and in the placenta, but in no other tissues. Crosses between transgenic and non-transgenic mice indicate that the trophoblast, rather than the decidua or uterus, is the source of transferase activity. A soluble TNF receptor/IgG heavy chain chimeric protein, which strongly inhibits TNF activity *in vitro* and *in vivo*, can cross the placenta, but does not interrupt pregnancy, and has no obvious effect on fetal development. These findings suggest that TNF may not be required for completion of a normal gestation.

Recent studies with mice deficient in p55 receptor expression (generated by targeted homologous recombination in embryonic stem cells (see: ES cells)) demonstrate that thymocyte development and lymphocyte populations are unaltered and that clonal deletion of potentially self-reactive T cells is not impaired. The loss of p55 functions in these mice, which still express the p75 TNF receptor, renders mice resistant to lethal dosages of either bacterial lipopolysaccharides or enterotoxins (see also: Septic shock). These mice, however, readily succumb to infections with L. monocytogenes.

■ **Assays:** TNF-α can be detected in » bioassays involving cell lines that respond to it (see: BT-20, CT6, EL4; L929; L-M; MO7E; T1165; WEHI-3B). TNF-α can also be detected by a sensitive sandwich enzyme immunoassay, ELISA and an immunoradiometric assay (IRMA). Intracellular factor is detected by two color immunofluorescence flow cytometry. An alternative and entirely different detection method is » Message amplification phenotyping. For further information see also subentry "Assays" in the reference section.

■ **Clinical Use & Significance:** In contrast to chemotherapeutic drugs TNF specifically attacks malignant cells. Extensive preclinical studies have documented a direct cytostatic and cytotoxic effect of TNF-α against subcutaneous human xenografts and lymph node metastases in nude mice (see also: Immunodeficient mice), as well as a variety of immunomodulatory effects on various immune effector cells, including neutrophils, macrophages, and T cells. Single- and multiple-dose phase I studies have confirmed that TNF can be safely administered to patients with advanced malignancies in a dose range associated with anticancer effect without concomitant serious toxicities such as shock and cachexia. However, clinical trials on the whole have unfortunately so far failed to demonstrate significant improvements in cancer treatment, with TNF resistance and TNF-induced systemic toxicity being two major limitations for the use of TNF as an antineoplastic agent in most cases. The combined use of TNF and cytotoxic or immune modulatory agents, particularly » IFN-γ and possibly » IL2, may be of advantage in the treatment of some tumors. In some cases intratumoral application of TNF has been found to be of advantage in tumor control.

Some mutant forms of TNF-β with selective activity on the p55 receptor have been described recently. It has been shown that activation of the p55 receptor is sufficient to trigger cytotoxic activity towards transformed cells. Some of these mutants have been described to retain their antitumor activity in nude mice (see also: Immunodeficient mice) carrying transplanted human tumors. It is hoped that such mutant forms may induce less systemic toxicity in man.

TNF can be used to increase the aggressiveness of lymphokine-activated killer cells (see: LAK cells). There are some indications that inhibitors of TNF may be of advantage. Since TNF is found in the synovial fluid of patients suffering from arthritis, these inhibitors may be helpful in ameliorating the disease and this has been shown to be the case in animal models of severe collagen-induced arthritis. Inhibitors may also ameliorate the severe consequences of » septic shock.

TNF-α appears to be an important » autocrine factor promoting the survival of hairy cell leukemia cells. It may therefore be important in the pathogenesis of this disease.

Studies with an experimental fibrosarcoma metastasis model have shown that TNF induces significant enhancement of the number of metastases in the lung. It has been suggested that low doses of endogenous TNF or administration of TNF during cytokine therapy may enhance the metastatic potential of circulating tumor cells. The transduction of murine tumor cells with a functional TNF-α gene has been shown to lead to the rejection of the genetically modified cells by syngeneic hosts (for cancer vaccines see also: Cytokine gene transfer).

TNF-α has also been shown to protect hematopoietic progenitors against irradiation and cytotoxic agents, suggesting that it may have some potential therapeutic applications in aplasia induced by chemotherapy or bone marrow transplantation.

One case of severe therapy-resistant Morbus Crohn has been treated with monoclonal antibodies directed against TNF-α. Treatment has been reported to have resulted in a complete remission lasting for three months.

● **Reviews: Aggarwal BB & Vilcek J** (eds) Tumor necrosis factor: structure, function, and mechanism of action. Marcel Dekker Inc. 1992; **Beutler B & Cerami A** Tumor necrosis, cachexia, shock and inflammation: a common mediator. ARB 57: 505-18 (1988); **Beutler B & Cerami A** The biology of

Efficacy of anti-TNF therapies in disease.

Type	Benefits	Reference
Baboons		
E. coli bacteremia	Hemodynamic improvement	Proc. Natl. Acad. Sci 89: 4845 (1992)
Staphylococcus aureus sepsis	Prevention of death	J. Trauma 33: 568 (1992)
Mice		
E. coli bacteremia	Survival	Infect. Immun. 60: 4133 (1992)
Endotoxemia	None	J. Clin. Invest. 91: 1459 (1993) J. Immunol. 148: 2724 (1992)
Peritoneal sepsis	None	J. Clin. Invest. 91: 1459 (1993) J. Immunol. 148: 2724 (1992)
Salmonella infection	Protection	Microb. Pathog. 14: 473 (1993)
Streptococcus pyrogenes sepsis	None	FEMS Microbiol. Lett. 110: 175-8 (1993)
E. coli/Bacteroides fragilis peritonitis	Reduction of mortality	Arch. Surg. 128: 77 (1993)
TCDD toxicity	Reduction of mortality	Toxicol. Appl. Pharmacol. 117: 126 (1992)
Collagen-induced arthritis	Amelioration of disease	Proc. Natl. Acad. Sci. 89: 9784 (1992)
Non-obese diabetic mice	Hypersensitivity to irradiation Development of insulitis	Int. Immunol. 4: 611 (1992)
Rats		
Neonatal streptococcal sepsis	Prolonged survival time	Infect. Immun. 61: 227 (1993)
Tumor-induced anorexia	Normalization of body temperature Enhanced food intake	Am. J. Physiol. 265: R615 (1993)
Liver allograft	Suppression of rejection Enhancement of infectious complications	Transplant. Proc. 25: 128 (1993
Cardiac allografts	Prologation of alograft survival	Transplantation 53: 283 (1992)
Human		
Rheumatoid arthritis	Improvement of clinical parameters	Clin. Exp. Rheumatol. 11: S173 (1993)
Cerebral malaria	Reduction of fever	Q. J. Med. 86: 91 (1993)
OKT3-induced cytokine release syndrome	Prevention of effect	Transplant. Proc. 25: 47 (1993)
Severe sepsis or septic shock	Tolerated/beneficial no survival benefit	Crit. Care Med. 21: 318 (1993)
Acute refractory GVHD	Beneficial in some patients Largely ineffective to prevent GVHD recurrence	Blood 79: 3362 (1992)
Septic shock	Improvement of left ventricular function	Chest 101: 810 (1992)

cachectin/TNF – a primary mediator of the host response. ARI 7: 625-55 (1989); **Bock G** (edt) Tumor necrosis factor and related cytotoxins. John Wiley and Sons, Chichester, Ciba Foundation Symp. 131, 1987; **Bodmer M et al** Preclinical review of anti-tumor necrosis factor monoclonal antibodies. Crit. Care Med. 21: S441-6 (1993); **Bonavida B & Granger G** (eds) Tumor necrosis factor: structure, mechanisms of action, role in disease and therapy. Karger, Basel 1990; **Gifford GE & Duckworth DH** Introduction to TNF and related lymphokines. Biotherapy 3: 103-11 (1991); **Strieter RM et al** Role of tumor necrosis factor-α in disease states and inflammation. Crit. Care Med. 21: S447-63 (1993); **Tracey KJ & Cerami A** Tumor necrosis factor: an updated review of its biology. Crit. Care Med. 21: S415-22 (1993)

● **Biochemistry & Molecular Biology: Aggarwal BB et al** Human tumor necrosis factor. Production, purification, and characterization. JBC 260: 2345-54 (1985); **Carroll MC et al** Linkage map of the human major histocompatibility complex including the tumor necrosis factor genes. PNAS 84: 8535-9 (1987); **Corti A et al** Antigenic regions of tumor necrosis factor α and their topographic relationships with structural/functional domains. Mol. Immunol. 471-9 (1992); **Davis JM et al** Structure of human tumor necrosis factor α derived from recombinant DNA. B 26: 1322-6 (1987); **Eck MJ & Sprang SR** The structure of tumor necrosius factor α at 2.6 Å resolution: implications for receptor binding. JBC 264: 17595-605 (1989); **Fransen L et al** Cloning of mouse tumor necrosis factor cDNA and its eukaryotic expression. NAR 13: 4417-29 (1985); **Goeddel DV et al** Tumor necrosis factors: gene structure and biological activities. CSHSQB 51: 597-609 (1986); **Jones EY et al** Structure of tumor necrosis factor. N 338: 225-8 (1989); **Li CB et al** Cloning and expression of murine lymphotoxin cDNA. JI 138: 4496-501 (1987); **Marmenout A et al** Molecular cloning and expression of human tumor necrosis factor and comparison with mouse tumor necrosis factor. EJB 152: 515-22 (1985); **Müller U et al** Tumor necrosis factor and lymphotoxin genes map close to H-2D in the mouse major histocompatibility complex. N 325: 265-7 (1987); **Nakamura S et al** A novel recombinant tumor necrosis factor-α mutant with increased anti-tumor activity and lower toxicity. IJC 48: 744-8 (1991); **Nedospasov SA et al** The genes for tumor necrosis factor (TNF-α) and lymphotoxin (TNF-β) are tandemly arranged on chromosome 17 of the mouse. NAR 14: 7713-25 (1986); **Nedwin GE et al** Human lymphotoxin and tumor necrosis factor genes. Structure, homology, and chromosomal localization. NAR 13: 6361-73 (1985); **Nedwin GE et al** Effect of interleukin 2, interferon-γ, and mitogens on the production of tumor necrosis factors α and β. JI 135: 2492-7 (1985); **Pennica D et al** Cloning and expression in *Escherichia coli* of the cDNA for murine tumor necrosis factor. PNAS 82: 6060-4 (1985); **Pennica D et al** Human tumor necrosis factor: precursor structure, expression, and homology to lymphotoxin. N 312: 724-9 (1984); **Pennica D et al** Cloning and expression in *Escherichia coli* of the cDNA for murine tumor necrosis factor. PNAS 82: 6060-4 (1985); **Ragoussis J et al** Localization of the genes for tumor necrosis factor and lymphotoxin between the HLA class I and III regions by field inversion gel electrophoresis. Immunogenetics 27: 66-9 (1988); **Schall TJ et al** Molecular cloning and expression of a receptor for human tumor necrosis factor. Cell 61: 361-70 (1990); **Shirai T et al** Cloning and expression in *Escherichia coli* of the gene for human tumor necrosis factor. N 313: 803-6 (1985); **Spies T et al** Genes for the tumor necrosis factors α and β are linked to the human major histocompatibility complex. PNAS 83: 8699-702 (1986); **Spriggs DR et al** Genomic structure, induction, and production of TNF-α. Immunol. Ser. 56: 3-34 (1992); **Van Ostade X et al** Localization of the active site of human tumor necrosis factor (hTNF) by mutational analysis. EJ 10: 827-36 (1991); **Yamagishi J et al** Mutational analysis of structure-activity relationships in human tumor necrosis factor-α. Protein Eng. 3: 713-9 (1990); **Young AJ et al** Primary structure of ovine tumor necrosis factor α cDNA. NAR 18: 6723 (1990); **Wang AM et al** Molecular cloning of the complementary DNA for human tumor necrosis factor. S 228: 149-54 (1985)

● **Receptors: Banner DW et al** Crystal structure of the soluble human 55 kd TNF receptor-human TNFβ complex: implications for TNF receptor activation. Cell 73: 431-45 (1993); **Brakebusch C et al** Cytoplasmic truncation of the p55 tumor necrosis factor receptor abolishes signaling, but not induced shedding of the receptor. EJ 11: 943-50 (1992); **Dembic Z et al** Two human TNF receptors have similar extracellular but distinct intracellular domain sequences. Cytokine 2: 231-7 (1990); **Gatanaga T et al** Purification and characterization of an inhibitor (soluble tumor necrosis factor receptor) for tumor necrosis factor and lymphotoxin obtained from the serum ultrafiltrates of human cancer patients. PNAS 87: 8781-4 (1990); **Goodwin RG et al** Molecular cloning and expression of the type 1 and type 2 murine receptors for tumor necrosis factor. MCB 11: 3020-6 (1991); **Gray PW et al** Cloning of human tumor necrosis factor (TNF) receptor cDNA and expression of recombinant soluble TNF-binding protein. PNAS 87: 7380-4 (1990); **Heller RA et al** Complementary DNA cloning of a receptor for tumor necrosis factor and demonstration of a shed form of the receptor. PNAS 87: 6151-5 (1990); **Heller RA et al 7**The p70 tumor necrosis factor receptor mediates cytotoxicity. Cell 70: 47-56 (1992); **Kohno T et al** A second tumor necrosis factor receptor gene product can be shed as a naturally occurring tumor necrosis factor inhibitor. PNAS 87: 8331-5 (1990); **Loetscher H et al** Molecular cloning and expression of the human 55 kd tumor necrosis factor receptor. Cell 61: 351-9 (1990); **Loetscher H et al** Recombinant 55 kDa TNF receptor stoichometry of binding to TNFα and TNFβ and inhibition of TNF activity. JBC 266: 18324-9 (1991); **Mackey F et al** Tumor necrosis factor α (TNF-α)-induced cell adhesion to human endothelial cells is under dominant control of one TNF receptor type, TNF-R55. JEM 177: 1277-86 (1993); **Marsters SA et al I**dentification of cysteine-rich domains of the type 1 tumor necrosis factor receptor involved in ligand binding. JBC 267: 5747-50 (1992); **Nophar Y et al** Soluble forms of tumor necrosis factor receptors (TNF-Rs). The cDNA for the type I TNF-R cloned using amino acid sequence data of its soluble form, encodes both the cell surface and a soluble form of the receptor. EJ 9: 3269-78 (1990); **Pandita R et al** Interferon-γ induces cell surface expression for both types of tumor necrosis factor receptors. FL 312: 87-90 (1992); **Schwalb DM et al** Identification of a new receptor subtype for tumor necrosis factor-α. JBC 268: 9949-52 (1993); **Smith C et al** A receptor for tumor necrosis factor defines an unusual family of cellular and viral proteins. S 248: 1019-23 (1990); **Tartaglia LA et al** The two different receptors for tumor necrosis factor mediate distinct cellular responses. PNAS 88: 9292-6 (1991); **Tartaglia LA et al** Tumor necrosis factor signaling: a dominant negative mutation suppresses the activation of the 55 kDa tumor necrosis factor receptor. JBC 267: 4304-7 (1992); **Tartaglia LA et al** A novel domain within the 55 kd TNF receptor signals cell death. Cell 74: 845-53 (1993)

● **Biological Activities: Bonavida B** Immunomodulatory effect of tumor necrosis factor. Biotherapy 3: 127-33 (1991); **Brouckaert P et al** Tumor necrosis factor, its receptors and the connection with interleukin 1 and interleukin 6. Immunobiology 187: 317-29 (1993); **Camussi G et al** The molecular action of tumor necrosis factor-α. EJB 202: 3-14 (1991); **Cordingley FT et al** Tumor necrosis factor as an autocrine tumor growth factor for chronic B cell malignancies. Lancet I: 969-71 (1988); **Frater-Schröder M et al** Tumor necrosis factor type a, a potent inhibitor of endothelial cell growth *in vitro* is angiogenic *in vivo*. PNAS 84: 5277-81 (1987); **Goillot E et al** Tumor necrosis factor as an autocrine growth factor for neuroblastoma. CR 52: 3194-200 (1992); **Grunfeld C & Feingold KR** The metabolic effects of tumor necrosis factor and other cytokines. Biotherapy 3: 143-58 (1991); **Larrick JW & Wright SC** Cytotoxic mechanism of tumor necrosis factor-α. FJ 4: 3215-32 (1990); **Larrick JW & Kunchel SL** The role of tumor necrosis factor and interleukin in the immunoinflammatory response. Pharm. Res. 5: 129-39 (1988); **Last-Barney K et al** Synergistic and overlapping activities of tumor necrosis factor-α and IL1. JI 141: 527-30 (1988); **Merrill JE** Effects of Interleukin 1 and tumor necrosis factor α on astrocytes, microglia, oligodendrocytes, and glial precursors *in vitro*. Dev. Neurosci. 13: 130-7 (1991); **Nacy CA et al** Tumor necrosis factor-α: Central regulatory cytokine in the induction of macrophage antimicrobial activities. Pathobiology 59: 182-4 (1991); **Perez C et al** A nonsecretable cell surface mutant of tumor necrosis factor (TNF) kills by cell-to-cell contact. Cell 63: 251-8 (1990); **Sato N et al** Actions of TNF and IFN-γ on angiogenesis *in vitro*. J. Invest. Dermatol. 95: 85S-9S (1990); **Schiller JH et al** Tumor necrosis factor, but not other hematopoietic growth factors, prolongs the survival of hairy cell leukemia cells. Leuk. Res. 16: 337-46 (1992); **Schulze-Osthoff K et al** Depletion of the mitochondrial electron transport abrogates the cytotoxic and gene-inductive effects of TNF. EJ 12: 3095-3104 (1993); **Tartaglia LA et al** The two different receptors for tumor necrosis factor mediate distinct cellular responses. PNAS 88: 9292-6 (1991); **Trinchieri G** Effects of TNF and lymphotoxin on the hematopoietic system. Immunol. Ser. 56: 289-313 (1992)

● **Transgenic/Knock-out/Antisense Studies: Cheng J et al** Cachexia and graft-vs.-host-disease-type skin changes in keratin promoter-driven TNF α transgenic mice. Genes Dev. 6: 1444-56 (1992); **Giroir BP et al** The biosynthesis of tumor necrosis factor during pregnancy: studies with a CAT reporter transgene and TNF inhibitors. ECN 3: 533-8 (1992); **Higuchi Y et al** Expression of a tumor necrosis factor α transgene in murine pancreatic β cells results in severe and permanent insulitis without evolution towards diabetes. JEM 176: 1719-31 (1992); **Keffer J et al** Transgenic mice expressing human tumor necrosis factor: a predictive genetic model of arthritis. EJ 10: 4025-31 (1991); **Pfeffer K et al** Mice deficient for the 55 kd tumor necrosis factor receptor are resistant to endotoxic shock, yet succumb to *L. monocytogenes* infection. Cell 73: 457-67 (1993); **Picarella DE et al** Insulitis in transgenic mice expressing tumor necrosis factor β (lymphotoxin) in the pancreas. PNAS 89: 10036-40

(1992); **Picarella DE et al** Transgenic tumor necrosis factor (TNF)-α production in pancreatic islets leads to insulitis, not diabetes. Distinct patterns of inflammation in TNF-α and TNF-β transgenic mice. JI 150: 4136-50 (1993); **Probert L et al** Wasting, ischemia, and lymphoid abnormalities in mice expressing T cell-targeted human tumor necrosis factor transgenes. JI 151: 1894-1906 (1993)

● **Assays: Corti-A et al** Oligomeric tumor necrosis factor α slowly converts into inactive forms at bioactive levels. BJ 284: 905-10 (1992); **de Caestecker MP et al** The detection of intracytoplasmic interleukin-1 α, interleukin-1 β and tumor necrosis factor α expression in human monocytes using two color immunofluorescence flow cytometry. JIM 154: 11-20 (1992); **De Groote D et al** Novel method for the measurement of cytokine production by a one-stage procedure. JIM 163: 259-67 (1993); **Elsasser-Beile U et al** Cytokine levels in whole blood cell cultures as parameters of the cellular immunologic activity in patients with malignant melanoma and basal cell carcinoma. Cancer 71: 231-6 (1993); **Engelberts I et al** Evaluation of measurement of human TNF in plasma by ELISA. Lymphokine Cytokine Res. 10: 69-76 (1991); **Engelberts I et al** Evidence for different effect of soluble TNF receptors on various TNF measurements in human biological fluids. Lancet 338: 515-6 (1991); **Erikaku T et al** Bioluminescent immunoassay using a monomeric Fab'-photoprotein aequorin conjugate. BBRC 174: 1331-6 (1991); **Exley AR & Cohen J** Optimal collection of blood samples for the measurement of tumor necrosis factor α. Cytokine 2: 353-6 (1990); **Fomsgaard A et al** Detection of tumor necrosis factor from lipopolysaccharide-stimulated human mononuclear cells by enzyme-linked immunosorbent assay and cytotoxicity bioassay. Scand. J. Immunol. 27: 143-7 (1988); **Furukawa S et al** Kawasaki disease differs from anaphylactoid purpurea and measles with regard to tumor necrosis factor-α and interleukin 6 in serum. Eur. J. Pediatr. 151: 44-7 (1992); **Higuchi M & Aggarwal BB** Microtiter plate radioreceptor assay for tumor necrosis factor and its receptors in large numbers of samples. AB 204: 53-8 (1992); **Holobaugh PA & McChesney DC** Effect of anticoagulants and heat on the detection of tumor necrosis factor in murine blood. JIM 135: 95-9 (1990); **Hwu P et al** Use of soluble recombinant TNF receptor to improve detection of TNF secretion in cultures of tumor infiltrating lymphocytes. JIM 151: 139-47 (1992); **Lamb WR et al** A peroxidase-linked enzyme immunoassay for tumor necrosis factor α utilizing alternative colorimetric or chemilumimetric substrates. JIM 155: 215-23 (1992); **Liabakk NB et al** A rapid and sensitive immunoassay for tumor necrosis factor using magnetic monodisperse polymer particles. JIM 134: 253-9 (1990); **Liabakk NB et al** Development of immunoassays for the detection of soluble tumor necrosis factor receptors. JIM 141: 237-43 (1991); **McLaughlin PJ et al** Improvement in sensitivity of enzyme-linked immunosorbent assay for tumor necrosis factor. Immunol. Cell. Biol. 68: 51-5 (1990); **Meager A et al** Assays for tumor necrosis factor and related cytokines. JIM 116: 1-17 (1989); **Ogata A et al** A new highly sensitive immunoassay for cytokines by dissociation-enhanced lanthanide fluoroimmunoassay (DELFIA). JIM 148: 15-22 (1992); **Oremek G et al** [Enzyme immunoassay stability of α tumor necrosis factor in plasma and serum] Med. Klin. 87: 626-30; 636 (1992); **Rossomando EF & White L** A novel method for the detection of TNF-α in gingival crevicular fluid. J. Periodontol. 64: 445-9 (1993); **Skidmore BJ et al** Enumeration of cytokine-secreting cells at the single-cell level. EJI 19: 1591-7 (1989); **Teppo AM & Maury CP** Radioimmunoassay of tumor necrosis factor in serum. Clin. Chem. 33: 2024-7 (1987); **Thavasu PW et al** Measuring cytokine levels in blood. Importance of anticoagulants, processing, and storage conditions. JIM 153: 115-24 (1992); **van Kooten C et al** Interleukin-4 inhibits both paracrine and autocrine tumor necrosis factor-α-induced proliferation of B chronic lymphocytic leukemia cells. Blood 80: 1299-306 (1992); **Voigt HJ & Steib L** Tumor necrosis factor and pregnancy – a contribution to the immunology of reproduction. Arch. Gynecol. Obstet. 246: 223-6 (1989); **Zheng RQ et al** Detection of *in vivo* production of tumor necrosis factor-α by human thyroid epithelial cells. Immunology 75: 456-62 (1992); for further information also see individual cell lines used in individual bioassays.

● **Clinical Use & Significance: Asher AL et al** Murine tumor cells transduced with the gene for tumor necrosis factor-α. Evidence for paracrine immune effects of tumor necrosis factor against tumors. JI 146: 3227-34 (1991); **Blankenstein T et al** Tumor suppression after tumor cell-targeted tumor necrosis factor α gene transfer. JEM 173: 1047-52 (1991); **Brown TD et al** A phase II trial of recombinant tumor necrosis factor in patients with adenocarcinoma of the pancreas: A Southwest Oncology Group Study. J. Immunother. 10: 376-8 (1991); **Chelstrom LM et al** Treatment of BCL-1 murine B cell leukemia with recombinant cytokines. Comparative analysis of the anti-leukemic potential of interleukin 1 β (IL1 β), interleukin 2 (IL2), interleukin-6 (IL6), tumor necrosis factor α (TNF α), granulocyte colony stimulating factor (G-CSF), granulocyte-macrophage colony stimulating factor (GM-CSF), and their combination. Leuk. Lymphoma 7: 79-86 (1992); **Derkx B et al** Tumor necrosis factor antibody treatment in Crohn's disease. Lancet 342: 173-4 (1993); **Feldman ER et al** Phase II trial of recombinant tumor necrosis factor in disseminated malignant melanoma. Am. J. Clin. Oncol. 15: 256-9 (1992); **Gerain J et al** High serum levels of TNF-α after its administration for isolation perfusion of the limb. Cytokine 4: 585-91 (1992); **Hersh EM et al** Phase II studies of recombinant human tumor necrosis factor α in patients with malignant disease: A summary of the Southwest Oncology Group experience. J. Immunother. 10: 426-31 (1991); **Kaplan LD et al** A phase I/II study of recombinant tumor necrosis factor and recombinant interferon γ in patients with AIDS-related complex. Biotechnol. Ther. 1: 229-36 (1989-90); **Lienard D et al** In transit metastases of malignant melanoma treated by high dose rTNF α in combination with interferon-γ and melphalan in isolation perfusion. World. J. Surg. 16: 234-40 (1992); **Lindemann A et al** High level secretion of tumor necrosis factor-α contributes to hematopoietic failure in hairy cell leukemia. Blood 73: 880-4 (1989); **Mavligit GM et al** Regional biologic therapy. Hepatic arterial infusion of recombinant human tumor necrosis factor in patients with liver metastases. Cancer 69: 557-61 (1992); **Mohler KM et al** Soluble tumor necrosis factor (TNF) receptors are effective therapeutic agents in lethal endotoxemia and function simultaneously as both TNF carriers and TNF antagonists. JI 151: 1548-61 (1993); **Moreb J & Zucali JR** The therapeutic potential of interleukin-1 and tumor necrosis factor on hematopoietic stem cells. Leuk. Lymphoma 8: 267-75 (1992); **Moritz T et al** Tumor necrosis factor α modifies resistance to interferon α *in vivo*: first clinical data. Cancer Immunol. Immunother. 35: 342-6 (1992); **Muggia FM et al** High incidence of coagulopathy in phase II studies of recombinant tumor necrosis factor in advanced pancreatic and gastric cancers. Anticancer Drugs 3: 211-7 (1992); **Negrier MS et al** Phase I trial of recombinant interleukin-2 followed by recombinant tumor necrosis factor in patients with metastatic cancer. J. Immunother. 11: 93-102 (1992); **Ohnuma T et al** Effects of natural interferon α, natural tumor necrosis factor α and their combination on human mesothelioma xenografts in nude mice. Cancer Immunol. Immunother. 36: 31-6 (1993); **Orosz P et al** Enhancement of experimental metastasis by tumor necrosis factor. JEM 177: 1391-8 (1993); **Saks S & Rosenblum M** Recombinant human TNF-α: preclinical studies and results from early clinical trials.

Immunol. Ser. 56: 567-87 (1992); **Serretta V et al** Intravesical therapy of superficial bladder transitional cell carcinoma with tumor necrosis factor-α: preliminary report of a phase I-II study. Eur. Urol. 22: 112-4 (1992); **Sternberg CN et al** Recombinant tumor necrosis factor for superficial bladder tumors. Ann. Oncol. 3: 741-5 (1992); **Taguchi T & Sohmura Y** Clinical studies with TNF. Biotherapy 3: 177-86 (1991); **Teng MN et al** Long-term inhibition of tumor growth by tumor necrosis factor in the absence of cachexia or T cell immunity. PNAS 88: 3535-9 (1991); **van der Schelling GP et al** A phase I study of local treatment of liver metastases with recombinant tumor necrosis factor. Eur. J. Cancer 28A: 1073-8 (1992); **Van Ostade X et al** Human TNF mutants with selective activity on the p55 receptor. N 361: 266-9 (1993); **Warren DJ et al** Tumor necrosis factor induces cell cycle arrest in multipotential hematopoietic stem cells: a possible radioprotective mechanism. Eur. J. Haematol. 45: 158-63 (1990); **Williams RO et al** Anti-tumor necrosis factor ameliorates joint disease in murine collagen-induced arthritis. PNAS 89: 9784-8 (1992); **Yoshida J et al** Clinical effect of intra-arterial tumor necrosis factor-α for malignant glioma. J. Neurosurg. 77: 78-83 (1992)

TNF-α INH: *tumor necrosis factor α inhibitor* This factor of 33 kDa is isolated from the urine of febrile patients. It inhibits TNF-α-induced cytotoxicity in » L929 cells. It also blocks prostaglandin E2 production and expression of cell-associated interleukin-1 by human dermal fibroblasts, providing evidence for antiinflammatory activity. TNF-β-induced cytotoxicity is only slightly affected by the inhibitor. TNF-α INH blocks of TNF-α binding to the promonocytic cell line » U937. Preincubation of TNF-α with the inhibitor increases binding inhibition, suggesting an interaction between TNF-α and the inhibitor. The inhibitor blocks TNF-α-induced respiratory burst in human neutrophils; inhibitor-treated TNF-α is unable to stimulate a neutrophil lucigenin-dependent chemiluminescence response and superoxide formation. Treatment of TNF with the inhibitor also significantly reduces the priming ability of TNF-α for a response to the chemotactic peptide » f-MLP. TNF-α INH also blocks class I antigen expression in the human Colo 205 tumor cell line. TNF-α INH affects TNF-α synergism with IFN-γ-induced HLA-DR antigen expression but had no effect on IFN-γ activity.

The NH2-terminal amino acid sequence of the inhibitor appears to be unique. TNF-α INH is also expressed as a membrane protein and is probably a soluble form of the TNF receptor itself (see also: TNF-BF (TNF blocking factor)).

Another inhibitor of TNF has been isolated from the human histiocytic lymphoma cell line » U937 that is capable of inhibiting both TNF-α and TNF-β. It is also a soluble fragment of a TNF receptor. Another inhibitor of TNF-α, designated ***TNF-α I***, has been observed in the serum AND pleural effusions of cancer patients. It is probably also a receptor fragment.

Engelmann H et al Two tumor necrosis factor-binding proteins purified from human urine. Evidence for immunological cross-reactivity with cell surface tumor necrosis factor receptors. JBC 265: 1531-6 (1990); **Ferrante A et al** Inhibition of tumor necrosis factor α (TNF-α)-induced neutrophil respiratory burst by a TNF inhibitor. Immunology 72: 440-2 (1991); **Kohno T et al** A second tumor necrosis factor receptor gene product can shed a naturally occurring tumor necrosis factor inhibitor. PNAS 87: 8331-5 (1990); **Martinet N et al** Characterization of a tumor necrosis factor-α inhibitor activity in cancer patients. Am. J. Respir. Cell. Mol. Biol. 6: 510-5 (1992); **Seckinger P et al** Purification and biologic characterization of a specific tumor necrosis factor α inhibitor. JBC 264: 11966-73 (1989); **Seckinger P et al** Tumor necrosis factor inhibitor: purification, NH2-terminal amino acid sequence and evidence for anti-inflammatory and immunomodulatory activities. EJI 20: 1167-74 (1990); **Seckinger P et al** Characterization of a tumor necrosis factor α (TNF-α) inhibitor: evidence of immunological cross-reactivity with the TNF receptor. PNAS 87: 5188-92 (1990)

TNF-β: tumor necrosis factor beta.

■ **ALTERNATIVE NAMES:** Coley's toxin; CTX (cytotoxin); DIF (differentiation inducing factor); F1 (factor 1); hemorrhagic factor; LT (lymphotoxin = LT-α; see also: LT-β); necrosin; NKCF (natural killer cytotoxic factor); NKCIA (natural killer colony-inhibitory activity). See also: individual entries for further information.

■ **SOURCES:** This factor is produced predominantly by mitogen-stimulated T lymphocytes and leukocytes. The factor is also secreted by fibroblasts, astrocytes, myeloma cells, endothelial cells, epithelial cells and a number of transformed cell lines. The synthesis of TNF-β is stimulated by interferons (see: IFN) and » IL2. Some pre-B cell lines and Abelson murine leukemia virus-transformed pre-B cell lines constitutively produce TNF-β.

■ **PROTEIN CHARACTERISTICS:** TNF-β is a protein of 171 amino acids N-glycosylated at position 62. Some cell lines secrete different glycosylated forms of the factor that may also differ in their biological activities. The protein does not contain disulfide bonds and forms heteromers with » LT-β that anchors the complexes in the membrane. TNF-β and TNF-α show ≈ 30 % sequence homology. Murine and human TNF-β are highly homologous (74%). Recombinant human proteins with deletions of 27 amino acids from the N terminus appear to be biologically active in several bioassays.

■ **GENE STRUCTURE:** The gene has a length of ≈3 kb and contains four exons. It encodes a primary

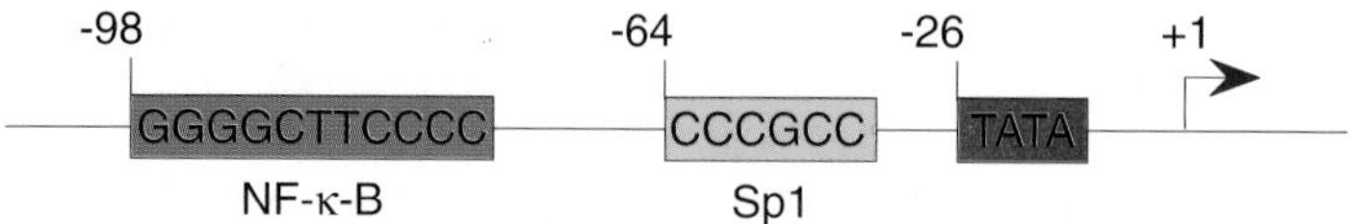

Structure of the TNF-β promoter.
+1 marks the transcriptional start site. Negative numbers indicate nucleotide positions 5´ to the transcriptional start site.

transcript of 2038 nucleotides yielding a mRNA of 1.4 kb. The gene maps to human chromosome 6p23-6q12 ≈ 1.2 kb apart from the » TNF-α gene. However, both genes are regulated independently. The 5´region of the TNF-β promoter contains a poly(dA-dT)-rich sequence that binds the non-histone protein HMG-I which is involved in the regulation of the constitutive expression of the gene.

■ **Receptor Structure, Gene(s), Expression:** TNF-β binds to the same receptor as » TNF-α.

■ **Biological Activities:** TNF-β acts on a plethora of different cells. This activity is not species-specific. Human TNF-β acts on murine cells but shows a slightly reduced specific activity. In general, TNF-β and » TNF-α display similar spectra of biological activities in *in vitro* systems, although TNF-β is often less potent or displays apparent partial agonist activity.

TNF-β induces the synthesis of » GM-CSF, » G-CSF, » IL1, collagenase, and prostaglandin E_2 in fibroblasts.

TNF-β is cytolytic or cytostatic for many tumor cells. In monocytes TNF-β induces the terminal differentiation and the synthesis of » G-CSF. TNF-β is a mitogen for B lymphocytes.

In neutrophils TNF-β induces the production of reactive oxygen species. It is also a chemoattractant for these cells (see also: Chemotaxis), increases phagocytosis, and also increases adhesion to the endothelium.

TNF-β inhibits the growth of osteoclasts and keratinocytes. Although TNF-β binds to the same receptor as TNF-α it is not involved in the establishment of an endotoxin shock (see also: Septic shock).

TNF-β promotes the proliferation of fibroblasts and is probably involved in processes of » wound healing *in vivo*. Hemorrhagic necrosis of tumors induced by TNF-β *in vivo* is probably the result of an inhibition of the growth of endothelial cells and the activity of TNF-β as an anti-angiogenesis factor (see: Angiogenesis factors).

Administration of TNF induces metabolic acidosis, decreases the partial pressure of CO_2, induces the synthesis of stress hormones such as epinephrine, norepinephrine, and glucagon, and also alters glucose metabolism.

■ **Transgenic/Knock-out/Antisense Studies:** The biological activities of TNF-β have been studied in » transgenic animals (mice) expressing the murine TNF-β gene under the control of the rat insulin II promoter in the pancreas, kidney, and skin. The expression of TNF-β in the pancreas of transgenic mice results in a leukocytic inflammatory infiltrate consisting primarily of B220+ IgM+ B cells and CD4+ and CD8+ T cells. The insulitis is reminiscent of the early stages of diabetes, though the mice do not progress to diabetes.

■ **Assays:** TNF-β can be detected in » bioassays involving cell lines that respond to it (see: BT-20, L929; L-M; WEHI-3B). TNF-β can also be detected by an ELISA test and an immunoradiometric assay (IRMA). An alternative and entirely different detection method is » Message amplification phenotyping. For further information see also subentry "Assays" in the reference section.

■ **Clinical Use & Significance:** The clinical application of this factor is only in its initial stages. The intrapleural administration of TNF-β may significantly reduce liquid volumes in some metastasizing ascites tumors. TNF-β levels in the sera of patients with meningococcal septicemia have been shown to correlate with morbidity and mortality.

● **Reviews: Aggarwal B & Vilcek J** (eds) Tumor necrosis factor: structure, function, and mechanism of action. Marcel Dekker Inc. 1992; **Bonavida B & Granger G** (eds) Tumor necrosis factor: structure, mechanisms of action, role in disease and therapy. Karger, Basel 1990; **Paul NL & Ruddle NH** Lymphotoxin. ARI 6: 407-38 (1987)

● **Biochemistry & Molecular Biology: Aggarwal BB et al** Primary structure of human lymphotoxin derived from 1788 lymphoblastoid cell line. JBC 260: 2334-44 (1985); **Aggarwal BB** Structure of tumor necrosis factor and its receptor. Biotherapy 3: 113-20 (1991); **Benjamin D et al** Human B cell TNF-β microheterogeneity. Lymphokine Cytokine Res. 11: 45-54 (1992); **Carroll MC et al** Linkage map of the human major histocompatibility complex including the tumor necrosis factor genes. PNAS 84: 8535-9 (1987); **Eck MJ et al** The structure of human lymphotoxin (tumor necrosius factor β) at 1.9 Å resolution:. JBC 267: 2119-22 (1992); **Evans AM et al** Mapping of prolactin and tumor necrosis factor β genes on human chromosome 6p using lymphoblastoid cell deletion mutants. Somat. Cell

TNF-β knockouts: see addendum: TNF-β.

Mol. Genet. 15: 203-13 (1989); **Fashena SJ et al** A poly(dA-dT) upstream activating sequence binds high-mobility group I protein and contributes to lymphotoxin (tumor necrosis factor-β) gene regulation. MCB 12: 894-903 (1992); **Gardner SM et al** Mouse lymphotoxin and tumor necrosis factor: Structural analysis of the cloned genes, physical linkage, and chromosomal position. JI 139: 476-83 (1987); **Goeddel DV et al** Tumor necrosis factors: gene structure and biological activities. CSHSQB 51: 597-609 (1986); **Gray PW et al** Cloning and expression of cDNA for human lymphotoxin, a lymphotoxin with tumor necrosis activity. N 312: 721-4 (1984); **Kobayashi Y et al** Cloning and expression of human lymphotoxin mRNA derived from a human T cell hybridoma. JB 100: 727-33 (1986); **Lantz M et al** Lymphotoxin produced by human B and T cell lines appears in two distinct forms. Mol. Immunol. 28: 9-16 (1991); **Matsuyama N et al** Nucleotide sequence of a cDNA encoding human tumor necrosis factor β from B lymphoblastoid cell RPMI 1788. FL 302: 141-4 (1992); **Müller U et al** Tumor necrosis factor and lymphotoxin genes map close to H-2D in the mouse major histocompatibility complex. N 325: 265-7 (1987); **Nedospasov SA et al** The genes for tumor necrosis factor (TNF-α) and lymphotoxin (TNF-β) are tandemly arranged on chromosome 17 of the mouse. NAR 14: 7713-25 (1986); **Nedospasov SA et al** DNA sequence polymorphism at the human tumor necrosis factor (TNF) locus. Numerous TNF/lymphotoxin alleles tagged by two closely linked microsatellites in the upstream region of the lymphotoxin (TNF-β) gene. JI 147: 1053-9 (1991); **Nedwin GE et al** Human lymphotoxin and tumor necrosis factor genes: structure homology and chromosomal location. NAR 13: 6361-7 (1985); **Nishikawa S et al** 27 amino acid residues can be deleted from the N-terminus of human lymphotoxin without impairment of its cytotoxic activity. J. Mol. Recognit. 3: 94-9 (1990); **Pennica D et al** Human tumor necrosis factor: precursor structure, expression, and homology to lymphotoxin. N 312: 724-9 (1984); **Ragoussis J et al** Localization of the genes for tumor necrosis factor and lymphotoxin between the HLA class I and III regions by field inversion gel electrophoresis. Immunogenetics. 27: 66-9 (1988); **Ruddle NH et al** Lymphotoxin: cloning, regulation and mechanism of killing. Ciba Found. Symp. 131: 64-82 (1987); **Schoenfeld HJ et al** Efficient purification of recombinant human tumor necrosis factor β from *Escherichia coli* yields biologically active protein with a trimeric structure that binds to both tumor necrosis factor receptors. JBC 266: 3863-9 (1991)

● **Receptors:** see: TNF-α.

● **Biological Activities:** **Kunkel SL et al** The role of TNF in diverse pathologic processes. Biotherapy 3: 135-41 (1991); **Sato N et al** Actions of TNF and IFN-γ on angiogenesis *in vitro*. J. Invest. Dermatol. 95: 85S-9S (1990); **Trinchieri G** Effects of TNF and lymphotoxin on the hematopoietic system. Immunol. Ser. 56: 289-313 (1992)

● **Transgenic/Knock-out/Antisense Studies:** **Picarella DE et al** Insulitis in transgenic mice expressing tumor necrosis factor β (lymphotoxin) in the pancreas. PNAS 89: 10036-40 (1992)

● **Assays:** **Adolf GR & Lamche HR** Highly sensitive enzyme immunoassay for human lymphotoxin (tumor necrosis factor β) in serum. JIM 130: 177-85 (1990); **Engelberts I et al** Evaluation of measurement of human TNF in plasma by ELISA. Lymphokine Cytokine Res. 10: 69-76 (1991); **Meager A et al** A two-site sandwich immunoradiometric assay of human lymphotoxin with monoclonal antibodies and its applications. JIM 104: 31-42 (1987); **Skidmore BJ et al** Enumeration of cytokine-secreting cells at the single-cell level. EJI 19: 1591-7 (1989); **Tada H et al** An enzyme immunoassay for human lymphotoxin. J. Immunoassay 10: 93-105 (1989); **Tada H et al** Bispecific antibody-producing hybrid hybridoma and its use in one-step immunoassays for human lymphotoxin. Hybridoma 8: 73-83 (1989); for further information also see individual cell lines used in individual bioassays.

● **Clinical Use & Significance:** **Haranaka K** Tumor necrosis factor: how to improve the anti-tumor activity and decrease accompanying side effects for therapeutic application. J. Biol. Response Modif. 7: 525-34 (1988); **Waage A et al** Association between tumor necrosis factor in serum and fatal outcome in patients with meningococcal disease. Lancet 1: 355-7 (1987)

TNF-BF: *TNF blocking factor* This factor of 28 kDa is isolated from serum ultrafiltrates of tumor patients. It is obtained by processing of the amino-terminal end of the extracellular domain of the » TNF-α membrane receptor. A similar factor is known as TNF binding protein (***TNFBP I***). It is a soluble 60 kDa fragment of the extracellular » TNF-α receptor domain.

TNF-BP I is detected in serum, urine, and cell culture supernatants of several human tumor cell lines. Serum TNF-BP I is significantly elevated in patients with burns and markedly increased in patients with renal failure.

TNF-BF inhibits the cytolytic activities of TNF-α and TNF-β *in vitro*. In mice administration of TNF-BF prevents the necrosis of tumor tissues induced by injection of TNF-α.

Soluble TNF blocking factors probably function as physiological inhibitors of TNF activities by binding to circulating TNF and preventing the interactions of these factors with their membrane receptors. These factors may also be a strategy by which tumors protect themselves against the toxic effects of TNF. At present data correlating serum concentrations of these inhibitors and tumor progression or metastasis are not available.

TNF-BF is probably related to another inhibitor of TNF-α activity, TNF-α I, found in the serum of cancer patients and in pleural effusions. It is at present unclear whether TNF-BF is identical with the TNF inhibitor » TNF-α INH (tumor necrosis factor α inhibitor).

Aderka D et al Increased serum levels of soluble receptors for tumor necrosis factor in cancer patients. CR 51: 5602-07 (1991); **Adolf GR & Apfler I** A monoclonal antibody-based enzyme immunoassay for quantitation of human tumor necrosis factor binding protein I, a soluble fragment of the 60 kDa TNF receptor, in biological fluids. JIM 143: 127-36 (1991); **Adolf GR & Fühbeis B** Monoclonal antibodies to soluble human TNF receptor (TNF binding protein) enhance its ability to block TNF toxicity. Cytokine 4: 180-4 (1992); **Bargetzki MJ et al** Interleukin 1β induces interleukin-1 receptor antagonist and tumor necrosis factor binding proteins in humans. CR 53: 4010-13 (1993); **Engelmann H et al** A tumor necrosis factor-binding protein purified to homogeneity from human urine protects cells from tumor necrosis factor toxicity. JBC 264: 11974-80 (1989); **Gatanaga T et al** Purification and characterization of an inhibitor (soluble tumor necrosis factor receptor) for tumor necrosis factor and lymphotoxin obtained from the serum ultrafiltrates of human cancer patients. PNAS 87: 8781-4 (1990); **Lantz M et al** Characterization *in vitro* of a human tumor necrosis factor binding protein. A soluble form of a tumor necrosis factor receptor. JCI 86: 1396-1402 (1990); **Lantz M et al** On the binding of tumor necrosis factor (TNF) to heparin and the release *in vivo* of the TNF-binding protein I by heparin. JCI 88: 2026-31 (1991); **Martinet N et al** Characterization of a tumor necrosis factor-α inhibitor activity in cancer patients. Am. J. Respir. Cell. Mol. Biol. 6: 510-5 (1992); **Nophar Y et al** Soluble forms of tumor necrosis factor receptors (TNF-Rs). The cDNA for the type 1 TNF-R, cloned using amino acid sequence data of its soluble form, encodes both the cell surface and a soluble form of the receptor. EJ 10: 3269-78 (1990) **Ohlsson I et al** A tumor necrosis factor binding protein (TNFBP) – physiological antagonist of TNF. Biotherapy 3: 159-65 (1991)

TNFBP: *TNF binding protein* see: TNF-BF (TNF blocking factor).

TNF-related activation protein: see: TRAP.

TNF-stimulated genes: see TSG genes

TNP-470: A derivative of » Fumagillin.

***tolloid*:** see: *dpp* (*decapentaplegic*).

***torpedo*:** see: DER (*Drosophila* EGF receptor).

Toxic cytokine fusion toxins: see: Cytokine fusion toxins.

Toxic shock syndrome: abbrev. TSS. see: Septic shock.

Toxic shock syndrome toxin: abbrev. TSST. see: Septic shock.

TP-1: *trophoblast protein 1* This protein, also known as ***pregnancy recognition hormone*** or ***tropoblastin***, can be isolated from bovine and ovine sources. It plays an important role in the maternal recognition and establishment of pregnancy by providing an antiluteolytic signal. The massive amounts of TP-1 produced during the period of pregnancy establishment is stimulated at least in part by maternal » GM-CSF.

TP-1 is a member of the type I interferons (collective term for » IFN-α and » IFN-β) and is also related to human » IFN-ω (Omega-interferon). The sequences of the bovine (22 and 24 kDa) and the ovine protein (17 kDa) are 85% identical and about 79% identical to that for bovine IFN-α II. The protein is as potent an antiviral agent as any known interferon. TP-1 also has other functional characteristics of IFNs such as antiproliferative and immunosuppressive activities. It inhibits cell proliferation across species and does not appear to exhibit toxicity at high concentrations. Ovine TP-1 is a potent activator of natural killer cells *in vitro*. TP-1 may have therapeutic potential as an antitumor agent without the toxic effects generally associated with IFNs.

Charlier M et al Cloning and expression of cDNA encoding ovine trophoblastin: its identity with a class-II α interferon. Gene 77: 341-8 (1989); **Cross JC et al** Constitutive and trophoblast-specific expression of a class of bovine interferon genes. PNAS 88: 3817-21 (1991); **Helmer SD et al** Identification of bovine trophoblast protein-1, a secretory protein immunologically related to ovine trophoblast protein-1. J. Reprod. Fertil. 79: 83-91 (1987); **Imakawa K et al** Interferon-like sequence of ovine trophoblast protein secreted by embryonic trophoectoderm. N 330: 377-9 (1987); **Imakawa K et al** Molecular cloning and characterization of complementary deoxyribonucleic acids corresponding to bovine trophoblast protein-1: a comparison with ovine trophoblast protein-1 and bovine interferon-α II. Mol. Endocrinol. 3: 127-39 (1989); **Imakawa K et al** A novel role for GM-CSF: enhancement of pregnancy specific interferon production, ovine trophoblast protein-1. Endocrinology 132: 1869-71 (1993); **Klemann SW et al** The production, purification, and bioactivity of recombinant bovine trophoblast protein-1 (bovine trophoblast interferon). Mol. Endocrinol. 4: 1506-14 (1990); **Leaman DW et al** Genes for the trophoblast interferons and their distribution among mammals. Reprod. Fertil. Dev. 4: 349-53 (1992); **Nephew KP et al** Differential expression of distinct mRNAs for ovine trophoblast protein-1 and related sheep type I interferons. Biol. Reprod. 48: 768-78 (1993); **Plante C et al** Purification of bovine trophoblast protein-1 complex and quantification of its microheterogeneous variants as affected by culture conditions. J. Reprod. Immunol. 18: 271-91 (1990); **Pontzer CH et al** Antiproliferative activity of a pregnancy recognition hormone, ovine trophoblast protein-1. CR 51: 5304-7 (1991); **Roberts RM et al** The polypeptides and genes for ovine and bovine trophoblast protein-1. J. Reprod. Fertil. 43: s3-12 (1991); **Roberts RM et al** Interferons at the placental interface. J. Reprod. Fertil. 1990; 41: s63-74 (1990); **Skopets B et al** Inhibition of lymphocyte proliferation by bovine trophoblast protein-1 (type I trophoblast interferon) and bovine interferon-α I1. Vet. Immunol. Immunopathol. 34: 81-96 (1992); **Tuo W et al** Natural killer cell activity of lymphocytes exposed to ovine, type I, trophoblast interferon. Am. J. Reprod. Immunol. 29: 26-34 (1993)

TP1: *thymostimulin* See: Thymic hormones.

TP5: *thymopentin* see: Thymic hormones.

TP53: Alternative name of the tumor suppressor gene p53. see: Oncogene.

TPA: *12-O-tetradecanoyl-phorbol-13-acetate* see: Phorbol esters.

TPA inducible sequences: see: TIS genes.

TPA responsive element: see: TRE.

TPF: *thymocyte proliferation factor* This factor was found initially in the supernatants of human mixed peripheral blood leukocyte cultures. It is identical with » LAF (lymphocyte activating factor) and » HP1 (helper peak 1) and hence identical with » IL1.

Another factor with TPF activity is released by human monocyte-macrophages treated with fibrogenic silica dust. It also causes fibroblast activation as measured by an increase in fibroblast proliferation. The factor is most likely identical with » IL1.

Koopman WJ et al Evidence for the identification of lymphocyte-activating factor as the adherent T cell-derived mediator responsible for enhanced antibody synthesis by nude mouse spleen cells. CI 35: 92-8 (1978); **Schmidt JA et al** Silica-stimulated monocytes release fibroblast proliferation factors identical to interleukin 1. A potential role for interleukin 1 in the pathogenesis of silicosis. JCI 73: 1462-72 (1984)

TPO: *thrombopoietin* A general term, also known as ***TSF*** (thrombopoiesis-stimulating factor) for mediator substances that function as natural stimulators of megakaryocytopoiesis and, thus, promote the differentiation of platelets (see also: Hematopoiesis). TPO stimulates an increase in megakaryocyte size and number, DNA content, endomitosis, and maturation. It also increases the number of small acetyl-cholinesterase-positive cells that are early precursor cells of the megakaryocytic lineage. TSF has been described to be identical with » GM-CSF and » IL6 in various assays. Another TSF species, as yet uncharacterized, is a protein of 40-47 kDa.

Although IL6 possesses thrombopoietin activity that stimulates *in vitro* the production of megakaryocytes and increases platelet counts *in vivo* it does not appear to be the only factor with Tpo activity. There is some evidence that IL6 and also GM-CSF are not identical (*in vivo*) with TSF activities described by various researchers so that the true nature of TPO remains elusive. It has been postulated that the activity described as » Meg-Pot (megakaryocyte potentiator) is a thrombopoietin (see also: Meg-CSF). Enhancement of human megakaryocytopoiesis *in vitro* has recently been shown to be enhanced by » IL11. A recombinant fusion protein consisting of » GM-CSF and » IL3 (see: pIXY321) has been shown to support various stages of human megakaryocytopoiesis.

Briddell RA et al Role of cytokines in sustaining long-term human megakaryocytopoiesis *in vitro*. Blood 79: 332-7 (1992); **Bruno E et al** Interacting cytokines regulate *in vitro* human megakaryocytopoiesis. Blood 73: 671-7 (1989); **Bruno E & Hoffman R** Effect of interleukin 6 on *in vitro* human megakaryocytopoiesis: its interaction with other cytokines. EH 17: 1038-43 (1989); **Bruno E et al** Recombinant GM-CSF/IL3 fusion protein: its effect on *in vitro* human megakaryocytopoiesis. EH 20: 494-9 (1992); **Carter CD & McDonald TP** Thrombopoietin from human embryonic kidney cells causes increased thrombocytopoiesis in sublethally irradiated mice. Radiat. Res. 132: 74-81 (1992); **Dessypris EN et al** Thrombopoiesis-stimulating factor: its effects on megakaryocyte colony formation *in vitro* and its relation to human granulocyte-macrophage colony-stimulating factor. EH 18: 754-7 (1990); **Gastl G et al** High IL6 levels in ascitic fluid correlate with reactive thrombocytosis in patients with epithelial ovarian cancer. Br. J. Haematol. 83: 433-41 (1993); **Hill RJ et al** Stimulation of thrombopoiesis in mice by human recombinant interleukin 6. JCI 85: 1242-7 (1990); **Hill RJ et al** Correlation of *in vitro* and *in vivo* biological activities during the partial purification of thrombopoietin. EH 20: 354-60 (1992); **Hill RJ et al** Evidence that interleukin-6 does not play a role in the stimulation of platelet production after induction of acute thrombocytopenia. Blood 80: 346-51 (1992); **Hoffman R** Regulation of megakaryocytopoiesis. Blood 74: 1196-212 (1989); **Ishibashi T et al** Interleukin-6 is a potent thrombopoietic factor *in vivo* in mice. Blood 74: 1241-4 (1989); **Ishibashi T et al** Effect of recombinant granulocyte-macrophage colony-stimulating factor on murine thrombocytopoiesis *in vitro* and *in vivo*. Blood 75: 1433-8 (1990); **Long MW et al** Synergistic regulation of human megakaryocyte development. JCI 82: 1779-86 (1988); **McDonald TP** The regulation of megakaryocyte and platelet production. IJCC 7: 139-55 (1989); **McDonald TP** A four-step procedure for the purification of thrombopoietin. EH 17: 865-71 (1989); **McDonald TP & Jackson CW** Thrombopoietin derived from human embryonic kidney cells stimulates an increase in DNA content of murine megakaryocytes *in vivo*. EH 18: 758-63 (1990); **McDonald TP et al** Comparative effects of thrombopoietin and interleukin-6 on murine megakaryocytopoiesis and platelet production. Blood 77: 735-40 (1991); **McDonald TP** Thrombopoietin. Its biology, clinical aspects, and possibilities. Am. J. Pediatr. Hematol. Oncol. 14: 8-21 (1992); **Nagasawa T et al** Thrombopoietic activity of human interleukin-6. FL 260: 176-8 (1990); **Straneva JE et al** Effects of thrombocytopoiesis-stimulating factor on terminal cytoplasmic maturation of human megakaryocytes. EH 17: 1122-7 (1989); **Straneva JE et al** Is interleukin 6 the physiological regulator of thrombopoiesis? EH 20: 47-50 (1992); **Williams N et al** Two-factor requirement for murine megakaryocyte colony formation. JCP 110: 101-4 (1982); **Williams N** Megakaryocyte growth factors. Immunol. Ser. 49: 215-29 (1990); **Williams N** Stimulators of megakaryocyte development and platelet production. PGFR 2: 81-95 (1990); **Williams N** Is thrombopoietin interleukin 6? EH 19: 714-8 (1991); **Yannucchi AM et al** *In vivo* stimulation of megakaryocytopoiesis by recombinant murine granulocyte-macrophage colony-stimulating factor. Blood 76: 1473-80 (1990); **Zeidler C et al** *In vivo* effects of interleukin-6 on thrombopoiesis in healthy and irradiated primates. Blood 80: 2740-5 (1992)

Thrombopoietin: see addendum.

Tpo: *thymopoietin* see: Thymic hormones.

Transcription factor 5: abbrev. TCF5. This transcription factor is identical with » NF-IL6.

Transcription factors: see: Gene expression.

Transferrin: Transferrin (often abbrev. TRF) is a glycoprotein of ≈ 77 kDa (679 amino acids). Transferrin is identical with » ***GPBP*** (granulocyte/pollen-binding protein). The human transferrin gene maps to chromosome 3q21. Transferrin is one of the so-called siderophilins, non-heme iron (III)-binding proteins with two binding sites for iron (see also: Lactoferrin).

Transferrin is the iron-transport protein of vertebrate serum and donates iron to cells, e. g. reticulocytes and their precursors) through interaction with a specific membrane receptor, CD71. It is synthesized in the liver, but lower amounts are also produced in other organs, such as testis and brain. A number of genetic transferrin variants are recognized, as is a disease characterized by a congenital absence of transferrin. Iron is dissociated from transferrin in a nonlysosomal acidic compartment of the cell. After dissociation of iron, transferrin and its receptor return undegraded to the extracellular environment and the cell membrane, respectively.

Transferrin appears to be indispensable for most cells growing in tissue culture. It is frequently referred to as a growth factor because, in analogy to other growth factor-receptor interactions, proliferating cells express high numbers of transferrin receptors, and the binding of transferrin to their receptors is needed for cells to initiate and maintain their DNA synthesis.

Apart from its role as an iron transport protein transferrin acts as a cytokine and has functions that may not be related to its iron-carrying capacity. It, or a closely related factor, is an » autocrine autostimulatory growth factor for the acute promyelocytic leukemia cell line » HL-60. The addition of transferrin receptor antibodies inhibits the stimulatory action of the endogenous transferrin related activity. Transferrin has also been identified as one of the factors promoting the clonal growth of murine granulocyte and macrophage precursors cultured *in vitro* under serum-free conditions.

ML-1 human myeloblastic leukemia cells differentiate to monocyte/macrophage-like cells by the sequential action of competence and progression factors (see also: Cell cycle). Competence is induced by » TNF-α, » TGF-β, and the » phorbol ester tetradecanoylphorbol acetate. A progression factor, designated ***DF77***, isolated as a 77 kDa glycoprotein from mitogen-stimulated human leukocyte-conditioned medium, is an isoform of human transferrin. The expression of the transferrin receptor has been found to be up-regulated by iron in human monocytes-macrophages.

Transferrin is a pituitary-derived growth factor for MTW9/PL2 rat mammary tumor cells cultivated under serum-free conditions and has been shown to be a growth factor for human prostatic carcinomas. A neurotrophic factor (NTF; see also Neurotrophins) isolated from chicken nerves that stimulates the incorporation of tritiated thymidine and supports myotube formation is also transferrin or a transferrin-like factor.

Transferrin has also be described as a » Lung-derived growth factor.

Beach RL et al The identification of neurotrophic factor as a transferrin. FL 156: 151-6 (1983); **Bowman BH et al** Transferrin: evolution and genetic regulation of expression. Adv. Genet. 25: 1-38 (1988); **Briere N et al** Synergistic influence of epidermal growth factor, insulin and transferrin on human fetal kidney in culture. Biofactors 3: 113-20 (1991); **Chan RY et al** Transferrin-receptor-independent but iron-dependent proliferation of variant Chinese hamster ovary cells. ECR 202: 326-36 (1992); **Denstman S et al** Identification of transferrin as a progression factor for ML-1 human myeloblastic leukemia cell differentiation. JBC 266: 14873-6 (1991); **Dittmann KH & Petrides PE** A 41 kDa transferrin related molecule acts as an autocrine growth factor for HL-60 cells. BBRC 176: 473-8 (1991); **Djeha A et al** Transferrin synthesis by macrophages: up-regulation by γ-interferon and effect on lymphocyte proliferation. FEMS Microbiol. Immunol. 5: 279-82 (1992); **Djeha A & Brock JH** Effect of transferrin, lactoferrin and chelated iron on human T lymphocytes. Br. J. Haematol. 80: 235-41 (1992); **Huerre C et al** The structural gene for transferrin (TF) maps to 3q21-3qter. Ann. Genet. 27: 5-10 (1984); **Iizuka Y et al** Colony formation of granulocyte (CFU-g) and macrophage (CFU-m) precursors in serum- and albumin-free culture: effect of transferrin on clonal growth. Exp. Cell Biol. 54: 275-80 (1986); **MacGillivray RTA et al** The complete amino acid sequence of human serum transferrin. PNAS 79: 2504-8 (1982); **Nicolson GL et al** Differential stimulation of the growth of lung-metastasizing tumor cells by lung (paracrine) growth factors: identification of transferrin-like mitogens in lung tissue-conditioned medium. Monogr. Natl. Cancer Inst. 1992: 153-61 (1992); **Park I et al** Organization of the human transferrin gene: direct evidence that it originated by gene duplication. PNAS 82: 3149-53 (1985); **Pflüger KH et al** Transferrin derivatives with growth factor activities in acute myeloblastic leukemia: an autocrine/paracrine pathway. Hämatol. Bluttransfus. 33: 87-94 (1990); **Riss TL & Sirbasku DA** Purification and identification of transferrin as a major pituitary-derived mitogen for MTW9/PL2 rat mammary tumor cells. In Vitro Cell. Dev. Biol. 23: 841-9 (1987); **Ross MC & Zetter BR** Selective stimulation of prostatic carcinoma cell proliferation by transferrin. PNAS 89: 6197-201 (1992); **Sutherland R et al** Ubiquitous cell-surface glycoprotein on tumor cells is proliferation-associated receptor for transferrin. PNAS 78: 4515-9 (1981); **Testa U et al** Iron up-modulates the expression of transferrin receptors during monocyte-macrophage maturation. JBC 264: 13181-7 (1989); **Uzan G et al** Molecular cloning and sequence analysis of cDNA for human transferrin. BBRC 119: 273-81 (1984); **Van der Pouw-Kraan T et al** Human transferrin allows efficient IgE production by anti-CD3-stimulated human lymphocytes at low cell densities. EJI 21: 385-90 (1991); **Yang F et al** Human transferrin: cDNA characterization and chromosomal localization. PNAS 81: 2752-6 (1984); **Yoshinari K et al**

A growth-promoting factor for human myeloid leukemia cells from horse serum identified as horse serum ferritin. BBA 1010: 28-34 (1989)

Transformed cell growth factor: see: TCGF.

Transforming growth factor: see: TGF.

Transforming growth factor α: see: TGF-α.

Transforming growth factor β: see: TGF-β.

Transgenic animals: The term "transgenic" describes organisms that contain in their chromosomes stably integrated copies of genes or gene constructs derived from other species or not normally found in the host animals. These so-called transgenes may or may not be expressed. Transgenic animals can be generated by introducing cloned DNA of the foreign genes into fertilized oocytes by micro-injection into pronuclei. These oocytes are subsequently transferred into the uterus of a pseudo-pregnant recipient animal and develop to term. Another strategy involves the use of embryonic stem cells (see: ES cells). These cells have the advantage that they can be manipulated easily *in vitro* before they are used to generate transgenics.

Animals developing from manipulated cells contain the foreign gene in all somatic cells and also in germ-line cells if the foreign gene was integrated into the genome of the recipient cell before the first cell division. Integration at a later stage yields genetic mosaics consisting of normal cells with a normal genome and cells with a transgenome. Transgenes integrated into the genome of germ line cells of the transgenic animals are transmitted to their offspring according to the Mendel's Law.

Primary animals developing after gene transfer are always hemizygous because the transgene is not integrated into both copies of a homologous pair of chromosomes. Pure homozygous lines are obtained by crossing of two suitable hemizygous (founder) animals.

The importance of gene transfer techniques for biomedical and molecular biological basic research cannot be overemphasized. Transgenic animals have been used to determine the cis-acting DNA elements responsible for the tissue-specific expression of genes, and current technology now permits the targeted expression of a protein to a particular organ or cell type within an organ. It is a common feature that the transgene is expressed properly both spatially and temporally. So far, a large number of transgenic mice have been generated expressing, e.g. » oncogenes, viral genes, immunoglobulins, cytokines, and cell surface and MHC antigens. A wide variety of oncogenes and proto-oncogenes from viral and cellular sources have been inserted into the germline of mice with subsequent development of neoplasia. Many of the published reports describe similarities between morphologic features of the transgenic mice tumors and those occurring naturally in humans. Transgenic mice bearing an oncogene targeted for the expression in a specific tissue can reveal how that oncogene influences differentiation and help to delineate the pathways to malignancy. Transgenic animals have also been powerful models for the investigation of various features of the immune system, particularly for studies of lymphocyte differentiation and tolerance.

In many instances transgenic animal models indirectly also provide an enormous increase in our knowledge of human diseases, providing opportunities to dissect the various molecular pathways involved in such processes. Transgenic animals are also studied to design gene constructs that function in a reproducible, predictable manner and that can be inserted at sites predetermined by the researchers as an important prerequisite for the ultimate goal: gene therapy (≈ 5% of established transgenic lines carry insertional mutations, caused by the integration of the transgene per se, leading to inherited disorders and developmental abnormalities). In addition, the characterization of the insertion sites of exogenous sequences in transgenic mice can identify loci that are potentially useful for the genetic analysis of the mammalian genome.

The analyses of transgenic animal models can be carried out with a degree of sophistication unattainable by any other technique. The analysis of transgenic animals allows, for example to elucidate the control mechanisms underlying embryonic development and the development and differentiation of embryonic cells, cancer cells (see: Oncogenes), T cells, and B cells. Transgenic animals also facilitate the analysis of the complex network of interactions between different genes including tissue-specific gene expression and additive polygenic actions of genes. Gene transfer techniques also allow studies of the effects and consequences of the expression of developmentally regulated genes the uncontrolled expression

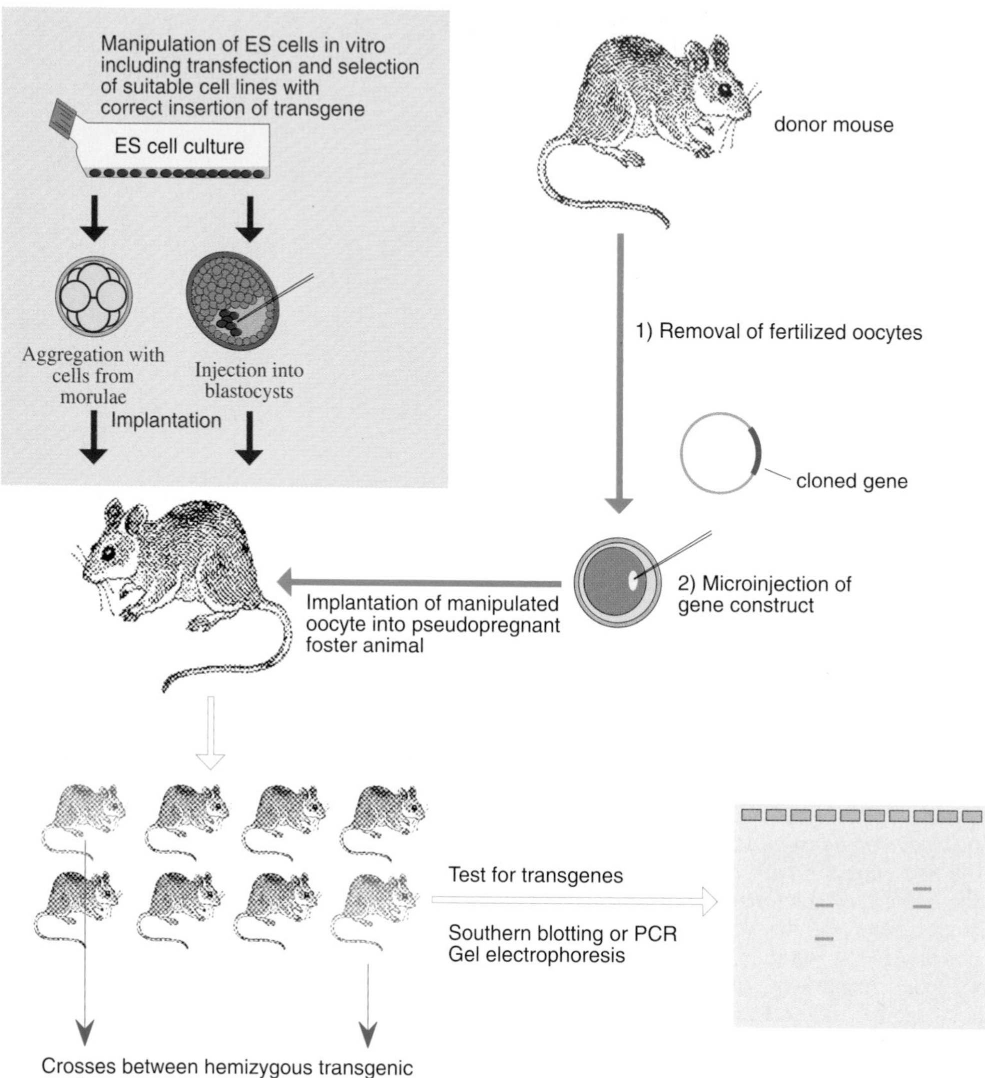

Strategies to generate transgenic animals.
In principle two strategies are available. The classical technique involves microinjection of gene constructs into fertilized oozytes obtained from a donor animal. These are then implanted into a pseudopregnant foster animal and carried out to term. Other strategies rely on in vitro manipulation of embryonic stem cells (ES cells) which can be grown as populations of pure totipotent cells in vitro in the presence of suitable feeder cell layers or LIF (leukemia inhibitory factor). Manipulated ES cells are either aggregated with morula stage cells or injected into blastocysts. Aggregated cells or manipulated blastocysts are again implanted into pseudopregnant animals where they can develop to term. Offspring obtained is tested by Southern blotting or Polymerase chain reaction for the presence of the transgene. Founder animals, which are hemizygous for the transgene, can then be crossed to obtain pure homozygous lines. ES cell manipulation is particularly useful for gene targeting experiments designed to create null mutations of an endogenous gene since all manipulations and also the successful disruption of a gene can be confirmed in vitro before the manipulated cells are used to generate live transgenic animals.

of which may be lethal. Such genes can be investigated if the expression is restricted to a specific tissue which can be achieved by choosing suitable regulatory elements (see also: Gene expression).

Transgenic animal models have now been established for many different cytokines (see subentry: Transgenic/Knock-out/Antisense studies for individual cytokines). These animal models either over-expressing or constitutively expressing a particular cytokine gene are of particular interest in cytokine research because the analysis of these animals reveals at least some of the complex molecular biological effects of single cytokines in a living organism that cannot be elucidated by studying cell cultures. These models can be used, for example, to study the cell-specific, tissue-specific, and developmentally regulated expression of individual cytokines and their pathophysiological consequences. Information from such experiments is frequently directly applicable to the understanding of pathogenetic mechanisms and will help to design new therapeutic approaches. It should be noted, however, there are also many examples of transgenic mouse models in which the particular consequences of transgene expression or non-expression are very different from the phenotypes associated with the human genetic diseases upon which these models were bases (for instance, Lesh-Nyhan syndrome, Duchenne muscular dystrophy, phenylketonuria).

Essentially the techniques used currently for the generation of transgenic animals expand the normal genome by the *addition* of a gene. Gene targeting technologies, i. e. the homologous recombination between specific chromosomal DNA sequences and exogenously introduced DNA sequences) have been developed and applied to pluripotent, mouse embryonic stem cells. Gene targeting provides the means to create mice of specifically altered genotype in which endogenous genes in their natural position in a cell genome are disrupted by the insertion of exogenous DNA. These transgenic animals carry Null mutations (loss-of-function mutations) of a particular gene and are frequently referred to as ***Knock-out mice***. Many of the phenotypes expressed by these animals are as expected from the *in vitro* assay data, but some are unexpected in that some animals with disrupted cytokine genes show an almost normal behavior. For techniques allowing studies of the effects of disrupted cytokine gene expression see: Antisense oligonucleotides, Genetic ablation.

Adams JM & Cory S Transgenic models for hemopoietic malignancies. BBA 1072: 9-31 (1991); **Bluethmann H** Analysis of the immune system with transgenic mice: T cell development. Experientia 47: 884-90 (1991); **Compere SJ et al** Oncogenes in transgenic mice. BBA 948: 129-49 (1988); **Fowlis DJ & Balmain A** Oncogenes and tumor suppressor genes in transgenic mouse models of neoplasia. Eur. J. Cancer 29A: 638-45 (1993); **Hanahan D** Transgenic mice as probes into complex systems. S 246: 1265-75 (1989); **Harris AW et al** Transgenic mouse models for hematopoietic tumorigenesis. CTMI 141: 82-93 (1988); **Iglesias A** Analysis of the immune system with transgenic mice: B cell development and lymphokines. Experientia 47: 878-84 (1991); **Ishida Y et al** Expansion of natural killer cells but not T cells in human interleukin 2/Interleukin 2 receptor (Tac) transgenic mice. JEM 170: 1103-15 (1989); **Ishida Y et al** Effects of the deregulated expression of human Interleukin-2 in transgenic mice. Int. Immunol. 1: 113-20 (1989); **Isola LM & Gordon JW** Transgenic animals: a new era in developmental biology and medicine. BT 16: 3-20 (1991); **Jhappan C et al** TGF α overexpression in transgenic mice induces liver neoplasia and abnormal development of the mammary gland and pancreas. Cell 61: 1137-46 (1990); **Keffer J et al** Transgenic mice expressing human tumor necrosis factor: A predictive genetic model of arthritis. EJ 10: 4025-31 (1991); **Koretsky AP** Investigation of cell physiology in the animal using transgenic technology. Am. J. Physiol. 262: C261-75 (1992); **Lang RA et al** Transgenic mice expressing a hemopoietic growth factor gene (GM-CSF) develop accumulations of macrophages, blindness, and a fatal syndrome of tissue damage. Cell 51: 675-86 (1987); **Lübbert M et al** Animal models for the biological effects of continuous high cytokine levels. Blut 61: 253-7 (1990); **Matsui Y et al** Development of mammary hyperplasia and neoplasia in MMTV-TGF α transgenic mice. Cell 61: 1147-55 (1990); **Metcalf D & Moore JG** Divergent disease patterns in granulocyte-macrophage colony-stimulating factor transgenic mice associated with different transgenic insertion sites. PNAS 85: 7767-71 (1988); **Mishi M et al** Expression of functional Interleukin 2 receptors in human light chain (Tac) transgenic mice. N 331: 267-9 (1988); **Morahan G** Transgenic mice as immune system models. Curr. Opin. Immunol. 3: 219-23 (1991); **Ravid K et al** Selective targeting of gene products with the megakaryocyte platelet factor 4 promoter. PNAS 88: 1521-5 (1991); **Roemer K et al** Knock-in and knock-out. Transgenes, Development and Disease: A Keystone Symposium sponsored by Genentech and Immunex, Tamarron, CO, USA, January 12-18, 1991. New Biol. 3: 331-5 (1991); **Samaridis J et al** Development of lymphocytes in interleukin 7-transgenic mice. EJI 21: 453-60 (1991); **Sandgren EP et al** Overexpression of TGF α in transgenic mice: induction of epithelial hyperplasia, pancreatic metaplasia and carcinoma of the breast. Cell 61: 1121-35 (1990); **Sarvetnick N** Transgenic mouse models for growth factor studies. MiE 198: 519-25 (1991); **Semenza GL et al** Human erythropoietin gene expression in transgenic mice: multiple transcription initiation sites and *cis*-acting regulatory elements. MCB 10: 930-8 (1990); **Sigmund CD** Major approaches for generating and analyzing transgenic mice. An overview. Hypertension 22: 599-607 (1993); **Suematsu S et al** IgG1 plasmacytosis in IL6 transgenic mice. PNAS 86: 7547-51 (1989); **Tepper RI et al** IL4 induces allergic-like inflammatory disease and alters T cell development in transgenic mice. Cell 62: 457-67 (1990); **Tominaga A et al** Transgenic mice expressing a B cell growth and differentiation factor gene develop eosinophilia and autoantibody production. JEM 173: 429-37 (1991); **Vassar R & Fuchs E** Transgenic mice provide new insights into the role of TGF-α during epidermal development and differentiation. GD 5: 714-27 (1991)

TRANSMIF: A macrophage migration inhibition factor isolated from rat liver. The protein shows glutathione S-transferase activity and induces the expression of the enzyme NO synthase in macrophages. The protein is highly homologous with human » MIF (migration inhibition factor).
Blocki FA et al Rat liver protein linking chemical and immunological detoxification systems. N 360: 269-70 (1992)

Transmodulation: see: Receptor transmodulation.

Transthyretin: abbrev. TTR. This protein of 55 kDa is composed of four identical subunits. It is also known as ***thyroxin-binding prealbumin*** or ***prealbumin***. The gene maps to human chromosome 18 q12.1 and consists of four exons.
Transthyretin plays an important role in the plasma transport of thyroxin and retinol, interacting with T4 and retinol (vitamin A) binding protein (RBP). TTR is secreted by hepatocytes into the serum and by the choroid plexus epithelium into the cerebral spinal fluid of most vertebrates. It is expressed also in small amounts by several other tissues. By synthesizing an important hormone carrier protein, the choroid plexus may be an important link in the chemical communication between the central nervous system and the bloodstream. Variant types of TTR are known to be closely associated with familial amyloidotic polyneuropathy (FAP), an autosomal dominant genetic disorder, and have been identified as the major components of the amyloid fibril deposits.
Transthyretin has been identified recently as an inhibitor of constitutive interleukin-1 (IL1) production by human umbilical vein endothelial cells found in human serum. TTR also inhibits LPS-stimulated IL1 production by cells of the human monocytic leukemia line THP-1 and monocyte-derived IL1β production. TTR probably functions by inhibiting processing of newly synthesized peptides for secretion. TTR may therefore act as an endogenous antiinflammatory mediator.
Borish L et al Transthyretin is an inhibitor of monocyte and endothelial cell interleukin-1 production. Inflammation 16: 471-84 (1992); **Furuya H et al** Production of recombinant human transthyretin with biological activities toward the understanding of the molecular basis of familial amyloidotic polyneuropathy (FAP). B 30: 2415-21 (1991); **Mita S et al** Cloning and sequence analysis of cDNA for human prealbumin. BBRC 124: 558-64 (1984); **Sakaki Y et al** Generation of transgenic mice producing a human transthyretin variant: a possible mouse model for familial amyloidotic polyneuropathy. BBRC 139: 794-9 (1986); **Sakaki Y et al** Human transthyretin (prealbumin) gene and molecular genetics of familial amyloidotic polyneuropathy. Mol. Biol. Med. 6: 161-8 (1989); **Whitehead AS et al** Cloning of human prealbumin complementary DNA. Localization of the gene to chromosome 18 and detection of a variant prealbumin allele in a family with familial amyloid polyneuropathy. Mol. Biol. Med. 2: 411-23 (1984); **Yi S et al** Systemic amyloidosis in transgenic mice carrying the human mutant transthyretin (Met30) gene. Pathologic similarity to human familial amyloidotic polyneuropathy, type I. Am. J. Pathol. 138: 403-12 (1991)

TRAP: *TNF-related activation protein* This human protein of 29.3 kDa (261 amino acids; glycosylated: ***gp39***) is a member of the » TNF family of proteins. It was isolated as a cDNA clone from a collection of T cell activation genes (see also: Cell activation). The protein shows 82.8 % and 77.4 % identity at the cDNA and protein level, respectively, with a similar protein isolated from murine » EL4 thymoma cells. Both proteins are the ligands for the » CD40 cell surface antigen expressed on resting B cells. The expression of this T cell membrane protein is responsible for inducing T cell-dependent B cell activation. The CD40 ligand stimulates B cell proliferation and secretion of all immunoglobulin isotypes in the presence of cytokines.
CD40 ligand has recently been shown to induce cytokine production and tumoricidal activity in peripheral blood monocytes. It also costimulates proliferation of activated T cells and this is accompanied by the production of » IFN-γ, » TNF-α, and » IL2. Activation of T cells in the absence of a costimulus induces » CD25 expression (p55 IL2 receptor) and the synthesis of more CD40 ligand. Expression of gp39 on murine T helper cells and CD4-positive T cells is inhibited by » IFN-γ. It is inhibited by TGF-β only on T helper type 2 cells. The human gene encoding TRAP maps to chromosome Xq26.3-q27. Deletions, point mutations and frameshift mutations clustering within a limited region of the CD40 ligand extracellular domain have been found to be the basis of a rare X-linked immunodeficiency syndrome (Hyper-IgM immunodeficiency syndrome, HIGM1) characterized by recurrent bacterial infections, very low or absent IgG, IgA, and IgE, and normal to increased IgM and IgD serum levels.
Human CD4-positive T cells infected by human immune deficiency virus-1, which causes severe dysfunction of cellular immunity, but paradoxically results in intense polyclonal activation of B cells, do not express the CD40 ligand.
Alderson MR et al CD40 expression by human monocytes: regulation by cytokines and activation of monocytes by the ligand for CD40. JEM 178: 669-74 (1993); **Armitage RJ et al** Molecular and biological characterization of a murine ligand for CD40. N 357: 80-2 (1992); **Armitage RJ et al** Identification of

a source of biologically active CD40 ligand. EJI 22: 2071-6 (1992); **Armitage RJ et al** CD40 ligand is a T cell growth factor. EJI 23: 2326-31 (1993); **Aruffo A et al** The CD40 ligand, gp39, is defective in activated T cells from patients with X-linked hyper-IgM syndrome. Cell 72: 291-300 (1993); **Callard RE et al** CD40 ligand and its role in X-linked hyper IgM syndrome. IT 14: 559-64 (1993; **Castle BE et al** Regulation of expression of the ligand for CD40 on T helper lymphocytes. JI 151: 1777-88 (1993); **DiSanto JP et al** CD40 ligand mutations in X-linked immunodeficiency with hyper-IgM. N 361: 541-543 (1993); **Graf D et al** Cloning of TRAP, a ligand for CD40 on human T cells. EJI 22: 3191-3194 (1992); **Hollenbaugh D et al** The human T cell antigen gp39, a member of the TNF gene family, is a ligand for the CD40 receptor: expression of a soluble form of gp39 with B cell co-stimulatory activity. EJ 11: 4313-21 (1992); **Jabara HH et al** CD40 and IgE: synergism between anti-CD40 monoclonal antibody and interleukin 4 in the induction of IgE synthesis by highly purified human B cells. JEM 172: 1861-4 (1990); **Korthäuser U et al** Defective expression of T cell CD40 ligand causes X-linked immunodeficiency with hyper-IgM. N 361: 539-541 (1993); **Lane P et al** Activated human T cells express a ligand for the human B cell-associated antigen CD40 which participates in T cell-dependent activation of B lymphocytes. EJI 22: 2573-8 (1992); **Lederman S et al** Identification of a novel surface protein on activated $CD4^+$ T cells that induces contact-dependent B cell differentiation (help). JEM 175: 1091-101 (1992); **Macchia D et al** Membrane tumor necrosis factor-α is involved in the polyclonal B cell activation induced by HIV-infected human T cells. N 363: 464-6 (1993); **Marshall LS et al** The molecular basis for T cell help in humoral immunity: CD40 and its ligand, gp39. J. Clin. Immunol. 13: 165-74 (1993); **Noelle RJ et al** A 39-kDa protein on activated helper T cells binds CD40 and transduces the signal for cognate activation of B cells. PNAS 89: 6550-4 (1992); **Noelle RJ et al** CD40 and its ligand, an essential ligand-receptor pair for thymus-dependent B cell activation. IT 13: 431-3 (1992); **Roy M et al** The regulation of the expression of gp39, the CD40 ligand, on normal and cloned $CD4^+$ T cells. JI 151: 2497-510 (1993); **Spriggs MK et al** Recombinant human CD40 ligand stimulates B cell proliferation and immunoglobulin E secretion. JEM 176: 1543-50 (1992)

TRE: *TPA responsive element* A short sequence (TGACTCA) in the promoter region of some cytokine genes and other genes. It mediates the expression of these genes in response to some external stimuli, for example » phorbol esters such as TPA. TRE is the binding site for a transcription factor (see: Gene expression) called AP-1. This factor is encoded by the » *jun* gene. In many different cell types treatment with phorbol esters (e.g. 4 β-phorbol 12-myristate 13-acetate, PMA) leads to the activation of protein-kinase C (PKC) and subsequently to the activation of the AP-1-responsive gene expression. The TRE sequence also binds other transcription factors, for example CREB (see: CRE, cAMP responsive element).

Angel P et al Phorbol-ester-inducible genes contain a common *cis* element recognized by a TPA-modulated *trans*-acting factor. Cell 49: 729-739 (1987); **de Groot RP & Kruijer W** Transcriptional activation by TGF β 1 mediated by the dyad symmetry element (DSE) and the TPA responsive element (TRE). BBRC 168: 1074-81 (1990); **Hirai S & Yaniv M** *jun* DNA-binding is modulated by mutations between the leucines or by direct interaction of *fos* with the TGACTCA sequence. New Biol. 1: 181-91 (1989); **Hoeffler JP et al** Distinct adenosine 3',5'-monophosphate and phorbol ester-responsive signal transduction pathways converge at the level of transcriptional activation by the interactions of DNA-binding proteins. Mol. Endocrinol. 3: 868-80 (1989); **Kim SJ et al** Activation of the second promoter of the transforming growth factor-β 1 gene by transforming growth factor-β 1 and phorbol ester occurs through the same target sequences. JBC 264: 19373-8 (1989); **Ray A et al** A multiple cytokine- and second messenger-responsive element in the enhancer of the human interleukin-6 gene: similarities with c-*fos* gene regulation. MCB 9: 5537-47 (1989); **Scotto L et al** Type β 1 transforming growth factor gene expression. A corrected mRNA structure reveals a downstream phorbol ester responsive element in human cells. JBC 265: 2203-8 (1990); **Serkkola E & Hurme M** Synergism between protein-kinase C and cAMP-dependent pathways in the expression of the interleukin-1 β gene is mediated via the activator-protein-1 (AP-1) enhancer activity. EJB 212: 243-9 (1993); **Yelcich A & Ziff EB** Functional analysis of an isolated *fos* promoter element with AP-1 site homology reveals cell type-specific transcriptional properties. MCB 10: 6273-82 (1990)

Trental: see: Pentoxifyllin.

TRF: *T cell replacing factors* A normal immune response requires the co-ordinated interaction and also the specific contact between T and B cells. Some of the helper functions of T cells are due to the continuous production and secretion of specific growth factors. The term TRF was coined at the beginning of cytokine research to describe soluble factors found in the culture supernatants of stimulated cells that render the synthesis of antibodies by B cells independent of the simultaneous presence of T cells. In many instances TRF is an operational definition of biological activities and does not signify biochemically identified distinct factors, at least not in the older literature. In addition, combinations of cytokines can also have TRF activities (e. g. a combination of » IL2, » IFN-γ, and » TNF-α). Factors present in culture supernatants of human T cell hybrid clones or mitogen-stimulated T cells that induce antibody secretion by B cell lines have also been called » BCDF (B cell differentiation factor) or » BCDF 2.

The pleiotropic actions of many growth factors (see also: Cytokines) and also some observed species-specific activities demonstrate that the nomenclature based on function is not suitable. Nevertheless the term is still used to point to one particular activity of a factor, i. e. the ability to replace T cells. The interleukin » IL5 has TRF activity for antigen-specific antibody reactions in mice but not in humans. In humans » IL2 possesses TRF activity. Likewise, » IFN-γ functions as a TRF in mice but not in humans. Even simple statements about TRF activity must be specified further. IL2, for example has TRF activity for sparse and semiconfluent B cell cultures; the production of antibodies is not restored by IL2 in confluent cultures of resting B cells. An activity described as ***late-acting T cell replacing factor*** (TRF) appears to be identical with » IL5.

An unknown factor with TRF activity, designated ***RA-SF*** (rheumatoid arthritis synovial fluid) has been observed in the synovial fluid of patients with rheumatoid arthritis. This factor has the ability to replace T cells for the induction of antibody secretion by human blood lymphoid cells stimulated by pokeweed mitogen (PWM) *in vitro*. RA-SF also has the capacity to act as a B cell-stimulatory factor of mouse splenic lymphocytes in the presence of lipopolysaccharide (LPS). RA-SF has properties similar to » IL4 in that it induces differentiation of antibody secretion in the LPS-pretreated mouse cell, but unlike IL4, which gives IgG1 and IgE, it selectively induces IgG2b synthesis. RA-SF is biologically distinct from » IL2, IL4, » IL5 or » (IFN-γ.

It has been observed that concanavalin A, frequently used to activate B cells also possesses TRF activity in the presence of IL2

Aarden LA et al Revised nomenclature for antigen-nonspecific T cell proliferation and helper factors. CI 48: 433-36 (1979); **Bodeker BG et al** Culture conditions for selective and enhanced release of various murine lymphokines (CSF, IL2, MCF, MIF and TRF). Dev. Biol. Stand. 55: 241-5 (1983); **Callard RE et al** T cell help in human antigen-specific antibody responses can be replaced by interleukin 2. EJI 16: 1037-42 (1986); **Callard RE & Smith SH** Interleukin 2 and low molecular weight B cell growth factor are T cell-replacing factors for different subpopulations of human B cells. EJI 18: 1635-8 (1988); **Clark-Lewis I et al** Preparation of T cell growth factor free from interferon and factors stimulating hemopoietic cells and mast cells. JIM 51: 311-22 (1982); **Hoffmann MK & Watson J** Helper T cell-replacing factors secreted by thymus-derived cells and macrophages: cellular requirements for B cell activation and synergistic properties. JI 122: 1371-5 (1979); **Möller E & Strom H** Biological characterization of T cell-replacing factor in the synovial fluid of rheumatoid arthritis patients. Scand. J. Immunol. 27: 717-24 (1988); **O'Garra A et al** B cell factors are pleiotropic. IT 9: 45-54 (1988); **Sauerwein RW et al** Analysis of T cell-replacing factor-like activity: potent induction of T helper activity for human B cells by residual concanavalin A and interleukin 2. EJI 17: 1145-50 (1987); **Smith SH et al** Human T cell-replacing factor(s): a comparison of recombinant and purified human B cell growth and differentiation factors. EJI 19: 2045-9 (1989); **Swain SL et al** T cell replacing factors in the B cell response to antigen. IR 63: 111-28 (1982); **Vazquez A et al** Differentiation factors for human specific B cell response. EJI 16: 803-8 (1986); **Watson J et al** Purification of murine helper T cell-replacing factors. PCBR 66: 513-22 (1981)

TRF: *T cell replacing factor* This factor has the activity of » BCGF 2 (B cell growth factor 2). A subsequent comparison of the cloned cDNA sequences reveals that this specific TRF is identical with » IL5. See also: TRF (T cell replacing factors).

Harada N et al Production of a monoclonal antibody useful in the molecular characterization of murine T cell-replacing factor/B cell growth factor II. PNAS 84: 4581-5 (1987); **Kinashi T et al** Cloning of a complementary DNA encoding T cell replacing factor and identity with B cell growth factor II. N 324: 70-3 (1986); **Schimpl A & Wecker E** Replacement of T cell function by a T cell product. N 237: 15-7 (1972); **Takatsu K et al** Antigen-induced T cell-replacing factor (TRF). III. establishment of T cell hybrid clone continuously producing TRF and functional analysis of released TRF. JI 125: 2646-53 (1980); **Tominaga A et al** Molecular properties and regulation of mRNA expression for murine T cell-replacing factor/IL5. JI 140: 1175-81 (1988)

TRF: *T cell replacing factor* This factor may be identical with murine (but not human) » IFN-γ. For some aspects of nomenclature see also: TRF (T cell replacing factors).

Brunswick M & Lake P Obligatory role of γ interferon in T cell-replacing factor-dependent, antigen-specific murine B cell responses. JEM 161: 953-71 (1985)

TRF: abbrev. for » Transferrin.

TRF 1: *T cell replacing factor 1* This factor is identical with » IL5. For some aspects of nomenclature see also: TRF (T cell replacing factors).

A factor found in culture supernatants of an antigen-stimulated long-term alloreactive T cell line, C.C3.11.75, also contains a T cell replacing factor activity for the B cell response to antigen. This factor, designated ***((DL)TRF)*** is probably identical with TRF 1 and the TRF showing BCGF-2 activity.

Gilbert KM & Hoffmann MK Lymphokine-induced suppressor B cells. Immunology 46: 545-53 (1982); **Harada N et al** BCGFII activity on activated B cells of a purified murine T cell replacing factor (TRF) from a T cell hybridoma (B151k12). JI 134: 3944-51 (1985); **Kinashi T et al** Cloning of a complementary DNA encoding T cell replacing factor and identity with B cell growth factor II. N 324: 70-3 (1986); **Schimpfl A & Wecker E** Replacement of T cell function by a T cell product. N (New Biology) 237: 15-7 (1972); **Swain SL & Dutton RW** Production of a B cell growth promoting activity (DL) BCGF, from a cloned T cell line and its assay on the BCL 1 B cell tumor. JEM 156: 1821-34 (1982); **Swain SL et al** Culture supernatants of a stimulated T cell line have helper activity that acts synergistically with interleukin 2 in the response of B cells to antigen. PNAS 78: 2517-21 (1981); **Wetzel GD et al** A monoclonal T cell-replacing activity can act directly on B cells to enhance clonal expansion. JEM 156: 306-11 (1982)

TRF 2: *T cell replacing factor 2* This poorly characterized factor functions as a differentiation factor for resting B cells and supports their proliferation. For some aspects of nomenclature see also: TRF (T cell replacing factors).

O'Garra A et al B cell factors are pleiotropic. IT 9: 45-54 (1988)

TRF 3: *T cell replacing factor 3* This factor is identical with IL1α (see: IL1). For some aspects of nomenclature see also: TRF (T cell replacing factors).

Aarden LA et al Revised nomenclature for antigen-nonspecific T cell proliferation and helper factors. CI 48: 433-36 (1979); **Watson J** The role of humoral factors in the initiation of *in vitro* primary immune responses. III. Characterization of factors that replace thymus-derived cells. JI 111: 1301-13 (1973)

TRFM: *macrophage-derived T cell replacing factor* This factor is identical with » IL1. For some aspects of nomenclature see also: TRF (T cell replacing factors).

Aarden LA et al Revised nomenclature for antigen-nonspecific T cell proliferation and helper factors. CI 48: 433-36 (1979)

Trichoplusia ni: see: Baculovirus expression system.

Trigger signals: see: Cell activation.

***trk*:** The *trk* gene encodes a transmembrane protein of 140 kDa with an intracellular domain functioning as a tyrosine-specific protein kinase. The human gene maps to chromosome 1q23-1q24. *trk* is preferentially expressed in neuronal tissues. Transforming alleles of the *trk* gene were originally isolated as » oncogenes from human colon and thyroid carcinomas.
The transforming *trk* oncogene is generated by fusion of the gene encoding non-muscle tropomyosin and other DNA sequences located 5´to the *trk* locus with the receptor-like kinase domain of *trk* (hence its name: *t*ropomyosin *r*eceptor *k*inase). A homologue of *trk* from *Drosophila*, designated D*trk* (160 kDa), encodes a protein closely related to *trk*. It functions as a neural adhesion protein and is expressed during embryonic development.
trk is expressed during embryonic development and is involved in morphogenetic signaling. *trk* is the receptor for » NGF (*trk*A). Functional *trk* receptors have also been found on human monocytes. Binding of NGF leads to the activation of the intrinsic protein kinase activity and to the » autophosphorylation of the *trk* protein. Association of *trk* with another protein, LNGFR, which is a low affinity receptor for » NGF, yields the high-affinity NGF receptor. *trk* also functions as the receptor of the » neurotrophin » NT-3.
Other members of the *trk* family (*trk*B, *trk*C) have different affinities for neurotrophins. *trk*B (145 kDa) which is expressed in regions of the nervous system not expressing *trk*A binds » BDNF (brain-derived neurotrophic factor) and the neurotrophins » NT-3, » NT-4, and » NT-5. *trk*C (145 kDa) is relatively specific for NT-3. *trk*C has been shown to encode multiple NT-3 receptors that differ from each other by the presence of 14 or 25 amino acids between two kinase subdomains.
It has been possible, by combination of structural elements from NGF, BDNF and NT-3, to engineer a pan-neurotrophin that efficiently activates all *trk* receptors and displays multiple neurotrophic specificities.
The expression of *trk* in monocytes and activated CD4-positive T cell clones as well as other findings suggest that » NGF and also other neurotrophins, in addition to their neurotrophic function, are immunoregulatory cytokines acting on monocytes.

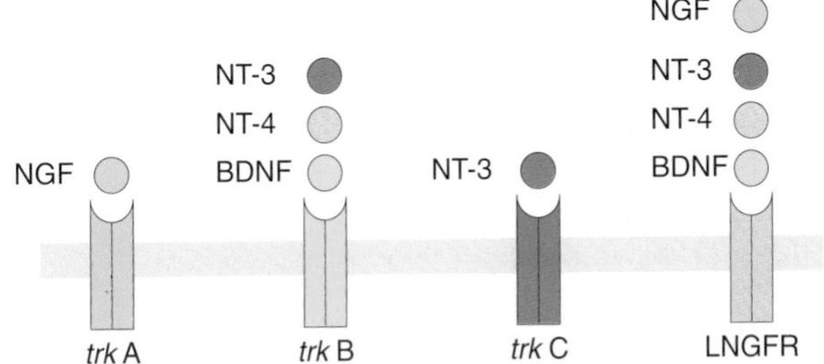

Interaction of neurotrophins with their receptors.
While NGF and NT-3 specifically bind to *trk* A and *trk* C, respectively, *trk* B can be activated by both BDNF and NT4 and also by NT-3 in some cases. All of the neurotrophins bind to the low affinity NGF receptor LNGFR.

***trk*:** see addendum.

■ **Transgenic/Knock-out/Antisense Studies:** The biological consequences of *trk*B expression have been studied in » transgenic animals carrying targeted disruptions of the gene, generated by homologous recombination in » ES cells. *trk*B-deficient mice develop to birth. However, they die soon after birth and show multiple central and peripheral nervous system neuronal deficiencies.

■ **Clinical Use & Significance:** In human neuroblastomas, a childhood tumor of neural crest origin, *trk* expression strongly correlates with favorable tumor stage (I, II, and IVS vs. III and IV), younger age (< 1 year vs. > or = 1 year), normal N-*myc* copy number, and low level of N-*myc* expression (see also: *myc*). A high level of *trk* expression is strongly predictive of a favorable outcome. A tumor with a functional receptor may be dependent on the neurotrophin NGF for survival and may regress in its absence, allowing a new approach to the treatment of certain patients with neuroblastoma.

Barbacid M et al The *trk* family of tyrosine protein kinase receptors. BBA Rev. Cancer 1072: 103-27 (1991); **Bongarzone I et al** High frequency of activation of tyrosine kinase oncogenes in human papillary thyroid carcinoma. O 4: 1457-62 (1989); **Bothwell M** Keeping track of neurotrophin receptors. Cell 65: 915-8 (1991); **Cordon-Cardo C et al** The *trk* tyrosine protein kinase mediates the mitogenic properties of nerve growth factor and neurotrophin-3. Cell 66: 173-83 (1991); **Coulier F et al** Mechanism of activation of the human *trk* oncogene. MCB 9: 15-23 (1989); **Ebendal T** Function and evolution in the NGF family and its receptors. J. Neurosci. Res. 32: 461-70 (1992); **Ehrhard PB et al** Expression of functional *trk* proto-oncogene in human monocytes. PNAS 90: 5423-5427 (1993); **Ehrhard PB et al** Expression of nerve growth factor and nerve growth factor receptor tyrosine kinase *trk* in activated CD4-positive T cell clones. PNAS 90: 10984-8 (1993); **Glass DJ et al** *trk*B mediates BDNF/NT-3-dependent survival and proliferation in fibroblasts lacking the low affinity NGF receptor. Cell 68: 405-13 (1991); **Glass DJ & Yancopoulos GD** The neurotrophins and their receptors. Trends Cell Biol. 3: 262-8 (1993); **Hempstead BL et al** High-affinity NGF binding requires coexpression of the *trk* proto-oncogene and the low-affinity NGF receptor. N 350: 678-83 (1991); **Ibanez CF et al** An extended surface of binding to *Trk* tyrosine kinase receptors in NGF and BDNF allows the engineering of a multifunctional pan-neurotrophin. EJ 12: 2281-93 (1993); **Kaplan DR et al** The *trk* proto-oncogene product: a signal transducing receptor for nerve growth factor. S 252: 554-8 (1991); **Kaplan DR et al** Tyrosine phosphorylation and tyrosine kinase activity of *trk* proto-oncogene product induced by NGF. N 350: 158-60 (1991); **Klein R et al** The *trk* proto-oncogene encodes a receptor for nerve growth factor. Cell 65: 189-97 (1991); **Klein R et al** The *trk*B tyrosine protein kinase is a receptor for brain-derived neurotrophic factor and Neurotrophin-3. Cell 66: 395-403 (1991); **Klein R et al** *trk*B, a novel tyrosine protein kinase receptor expressed during mouse neural development. EJ 8: 3701-9 (1989); **Klein R et al** The *trk*B tyrosine protein kinase is a receptor for neurotrophin-4. Neuron 8: 947-56 (1992); **Lamballe F et al** *trk*C, a new member of the *trk* family of tyrosine protein kinases, is a receptor for neurotrophin-3. Cell 66: 967-79 (1991); **Lamballe F et al** *trk*C encodes multiple neurotrophin-3 receptors with distinct biological properties and substrate specificities. EJ 12: 3083-94 (1993); **Loeb DM et al** The *trk* proto-oncogene rescues NGF responsiveness in mutant NGF-nonresponsive PC12 cell lines. Cell 66: 961-6 (1991); **Martin-Zanca D et al** A human oncogene formed by the fusion of truncated tropomyosin and protein tyrosine kinase sequences. N 319: 743-8 (1986); **Martin-Zanca D et al** Molecular and biochemical characterization of the human *trk* proto-oncogene. MCB 9: 24-33 (1991); **Martin-Zanca D et al** Expression of the *trk* proto-oncogene is restricted to the sensory cranial and spinal ganglia of neural crest origin in mouse development. GD 4: 683-94 (1990); **Meakin SO & Shooter EM** The nerve growth factor family of receptors. TINS 15: 323-31 (1992); **Middlemas DS et al** *trk*B, a neural receptor protein-tyrosine kinase: evidence for a full-length and two truncated receptors. MCB 11: 143-53 (1991); **Miozzo M et al** Human *trk* proto-oncogene maps to chromosome 1q32-q41. O 5: 1411-4 (1990); **Mitra G et al** Identification and biochemical characterization of p70(*trk*), product of the human *trk* oncogene. PNAS 84: 6707-11 (1987); **Mitra G** Mutational analysis of conserved residues in the tyrosine kinase domain of the human *trk* oncogene. O 6: 2237-41 (1991); **Morris CM et al** Localization of the *trk* proto-oncogene to human chromosome bands 1q23-1q24. O 6: 1093-5 (1991); **Mulherkar R & Coulier F** Overexpression of human *trk* proto-oncogene into mouse cells using an inducible vector system. BBRC 177: 90-6 (1991); **Pulido D et al** D*trk*, a *Drosophila* gene related to the *trk* family of neurotrophin receptors, encodes a novel class of neural cell adhesion molecule. EJ 11: 391-404 (1992); **Rodriguez-Tebar A et al** Binding of neurotrophin-3 to its neuronal receptors and interactions with nerve growth factor and brain-derived neurotrophic factor. EJ 11: 917-22 (1992); **Soppet D et al** The neurotrophic factors brain-derived neurotrophic factor and neurotrophin-3 are ligands for the *trk*B tyrosine kinase receptor. Cell 65: 895-903 (1991); **Squinto SP et al** *trk*B encodes a functional receptor for brain-derived neurotrophic factor and neurotrophin-3, but not nerve growth factor. Cell 65: 885-93 (1991)

● **Transgenic/Knock-out/Antisense Studies: Klein R et al** Targeted disruption of the *trk*B neurotrophin receptor gene results in nervous system lesions and neonatal death. Cell 75: 113-22 (1993)

● **Clinical Use & Significance: Matsushima H & Bogenmann E** Expression of *trk*A cDNA in neuroblastomas mediates differentiation *in vitro* and *in vivo*. MCB 13: 7447-56 (1993); **Nakagawara A et al** Association between high levels of expression of the *trk* gene and favorable outcome in human neuroblastoma. NEJM 328: 847-54 (1993); **Suzuki T et al** Lack of high-affinity nerve growth factor receptors in aggressive neuroblastomas. JNCI 85: 377-84 (1993)

Trophoblastin: see: TP-1.

Trophoblast protein 1: see: TP-1.

Trp-Ser-X-Trp-Ser motif: see: Cytokine receptor superfamily.

TRX: abbrev. for thioredoxin. see: » ADF (adult T cell leukemia-derived factor).

Trypsin inhibitor: see: ITI (Inter-α-trypsin inhibitor).

TS1: This murine helper T cell line is used in a » bioassay to detect » IL9. Growth and survival of these cells depend on IL9 and » IL4. An alternative and entirely different detection method is » Message amplification phenotyping. The cells also respond to » LIF (see: GATS, growth stimulatory activity for TS1 cells).
Uyttenhove C et al Functional and structural characterization of P40, a mouse glycoprotein with T cell growth factor activity. PNAS 85: 6934-8 (1988)

TSC-22: An » ERG (early response gene) expressed rapidly and transiently in murine osteoblastic cells after stimulation with » TGF-β for 1-2 hours. TSC-22 gene expression is also induced by » phorbol esters , serum, cholera toxin, or dexamethasone, but not appreciably by » EGF. The gene encodes a protein of 18 kDa (143 amino acids). Its function is unknown.
Shibanuma M et al Isolation of a gene encoding a putative leucine zipper structure that is induced by transforming growth factor β 1 and other growth factors. JBC 267: 10219-24 (1992)

TSF: *T cell stimulating factor* This is identical with » IL2.
Aarden LA et al Revised nomenclature for antigen-nonspecific T cell proliferation and helper factors. CI 48: 433-36 (1979)

TSF: *T cell stimulating factor* This factor is produced by accessory cells of the immune system. It induces the synthesis by TH1 and natural killer cells of » IFN-γ in combination with » IL2. Production of IFN-γ is insensitive to the immunosuppressive drug » CsA (cyclosporin A). TSF enhances IL2-induced proliferation of mouse TH1 cells. TSF and IL2 also induce homotypic cell aggregation of TH1 cells.
TSF is identical with at least one of the known » IL2 receptor inducing factors. TSF has been shown recently to be identical with murine » IL12.
Germann T et al An antigen-independent physiological activation pathway for L3T4+ T lymphocytes. EJI 17: 775-81 (1987); **Germann T et al** Components of an antigen-/T cell receptor-independent pathway of lymphokine production. EJI 21: 1857-61 (1992); **Germann T et al** Interleukin 12/T cell stimulating factor, a cytokine with multiple effects on Z helper type 1 (TH1) but not on TH2 cells. EJI 23: 1762-70 (1993); **Germann T et al** Requirements for the growth of TH1 lymphocyte clones. EJI 20: 2035-40 (1990); **Kato T et al** Induction of IL2 receptor expression and proliferation of T cell clones by a novel cytokine(s). CI 142: 79-93 (1992); **Matsui M et al** Dissociation between the expression of IL 2 receptor and IL 2 receptor mRNA in the antigen-specific T cell clone stimulated by the specific antigen with B cell APC. JI 139: 373-9 (1987)

TSF: *thrombopoiesis-stimulating factor* see: TPO (thrombopoietin).

TSF: *thymocyte stimulating factor* This factor was isolated initially as a protein that induces proliferation of lectin-stimulated thymocytes without being itself mitogenic. The factor is identical with » IL2. Some TSF activity is also due to » IL1 or » IL6 or a combination of the two factors.
Aarden LA et al Revised nomenclature for antigen-nonspecific T cell proliferation and helper factors. CI 48: 433-36 (1979); **Chen DM & DiSabato G** Further studies on the thymocyte-stimulating factor. CI 22: 211-24 (1976); **Chen DM & DiSabato G** Some effects of thymocyte-stimulating factor. CI 30: 195-203 (1977); **DiSabato G et al** Production by murine spleen cells of an activity stimulating the PHA-responsivenes of thymus lymphocytes. CI 17: 495-504 (1975); **Elias JA et al** A synergistic interaction of IL6 and IL1 mediates the thymocyte-stimulating activity produced by recombinant IL1-stimulated fibroblasts. JI 142: 509-14 (1989); **Paetkau V et al** Cellular origins and targets of Costimulator (IL2). IR 51: 157-75 (1980)

TsF: *T cell suppression factor* see: G-TsF.

TSG: *tumor suppressor gene* see: Oncogenes.

TSG genes: *tumor necrosis factor-stimulated gene sequences* Transcripts of TSG genes were identified originally as genes induced rapidly and transiently after treatment of human fibroblasts with » TNF-α for three hours. They represent a family of immediate early response genes (see: ERG), some of which function as transcription factors (see also: Gene expression). ***TSG-1*** has been shown to be identical with the gene for » IL8. ***TSG-6*** is a secretory protein that been identified as a member of the hyaluronate binding protein family. It is also induced by treatment of cells with » IL1.***TSG-8*** corresponds to the gene for » MCAF (monocyte chemotactic and activating factor). ***TSG-21*** and ***TSG-27*** are identical to the genes for collagenase and stromelysin, respectively.
Lee TH et al Isolation and characterization of eight tumor necrosis factor-induced gene sequences from human fibroblasts. MCB 1982-8 (1990); **Lee TH et al** A novel secretory tumor necrosis factor-inducible protein (TSG-6) is a member of the family of hyaluronate binding proteins, closely related to the adhesion receptor CD44. JCB 116: 545-57 (1992)

TSG1: see: TSG genes.

TSG6: see: TSG genes.

TSG8: see: TSG genes.

TSG21: see: TSG genes.

TSG27: see: TSG genes.

TSH: abbrev. for » Thyrotropin.

TSP: see: Thrombospondin.

TSS: *toxic shock syndrome* see: Septic shock.

TSST: *toxic shock syndrome toxin* see: Septic shock.

TSTGF: *thymic stroma-derived T cell growth factor* This factor poorly characterized factor of 25 kDa is isolated from the culture supernatants of the fibroblast-like cell line MRL104.8a established from thymic stroma. This factor functionally replaces these cells which can be used to support the growth of » IL2-dependent T helper cell lines in the absence of IL2 and antigens. TSTGF also supports the growth of immature thymocytes in the presence of » IL1, » IL2 or » IL4 and influences the differentiation of thymocytes. The biological activity of TSTGF is inhibited by heparin and glycosaminoglycans of the extracellular matrix (see: ECM).

Fujiwara H et al Thymic stroma-derived T cell growth factor: its role in the growth of immature thymocytes and T cell repertoire selection. Int. Symp. Princess Takamatsu Cancer Res. Fund 19: 115-26 (1988); **Fujiwara H et al** Proliferation and differentiation of immature thymocytes induced by a thymic stromal cell clone. Thymus 16: 159-72 (1990); **Kimura K et al** Role of glycosaminoglycans in the regulation of T cell proliferation induced by thymic stroma-derived T cell growth factor. JI 146: 2618-24 (1991); **Kosaka H et al** Thymic stroma-derived T cell growth factor (TSTGF). III. Its ability to promote T cell proliferation without stimulating interleukin 2 or 4-dependent autocrine mechanism. J. Leukoc. Biol. 47: 121-8 (1990); **Kosaka H et al** Model for clonal elimination in the thymus. PNAS 86: 3773-7 (1989); **Mizushima Y et al** Thymic stroma-derived T cell growth factor. IV. Capacity of TSTGF to promote the growth of L3T4-Lyt-2-thymocytes by synergy with phorbol myristate acetate or various IL. JI 142: 1195-202 (1989); **Ogata M et al** Thymic stroma-derived T cell growth factor (TSTGF). I. Functional distinction of TSTGF from interleukin 2 and 4 and its preferential growth-promoting effect on helper T cell clones. JI 139: 2675-82 (1987); **Sato S et al** Thymic stroma-derived T cell growth factor (TSTGF) II. Biochemical and functional characterization. J. Leukoc. Biol. 44: 149-57 (1988); **Tatsumi Y et al** Differentiation of thymocytes from $CD3^-CD4^-CD8^-$ through $CD3^-CD4^-CD8^+$ into more mature stages induced by a thymic stromal cell clone. PNAS 87: 2750-4 (1990)

TTP: This » ERG (early response gene) is identical with » Nup475.

Tumor angiogenesis: see: Angiogenesis factors.

Tumor angiogenesis factor: see: TAF.

Tumor autocrine motility factor: abbrev. TAMF. see: AMF.

Tumor cell cytotoxin: This factor is identical with » TNF-α.

Tumor cell-derived collagenase stimulatory factor: see: TCSF.

Tumor cell-targeted cytokine gene transfer: see: Cytokine gene transfer.

Tumor cytotoxic factor: see: TCF.

Tumor-derived chemotactic factors: see: TDCF.

Tumor degenerating factor: see: TDF.

Tumor-derived suppressor factor: see: TDSF.

Tumor growth inhibitory factor: see: TGIF.

Tumor-inducing factor 1: see: TIF-1.

Tumor infiltrating lymphocytes: See: TILs.

Tumor inhibitory factor 2: see: TIF-2.

Tumor invasion-inhibiting factor 2: see: IIF-2.

Tumor-killing factor: see: TKF.

Tumor migration inhibition factor: see: TMIF.

Tumor necrosis factor: see: TNF. see also: TNF-α, TNF-β.

Tumor necrosis factor α: see: TNF-α.

Tumor necrosis factor β: see: TNF-β.

Tumor necrosis factor binding proteins: see: TNF-BF (TNF blocking factor).

Tumor necrosis factor α inhibitor: see: TNF-α INH.

Tumor necrosis factor-stimulated genes: see: TSG genes.

Tumor vaccination: see: Cytokine gene transfer.

Tumor vascular permeability factor: see: VPF (vascular permeability factor).

TSTIF: see addendum.

TY5: This factor is identical with MIP-1α (see: MIP, macrophage inflammatory protein) and therefore belongs to the » chemokine family of cytokines.

Brown KD et al A family of small inducible proteins secreted by leukocytes are members of a new superfamily that includes leukocyte and fibroblast-derived inflammatory agents, growth factors, and indicators of various activation processes. JI 142: 679-87 (1989)

Tyk2: see: Janus kinases.

Type 1 hematopoietins: see: Hematopoiesis.

Type 1 helper cell cytokine: see: TH1/TH2 Cytokines.

Type 2 helper cell cytokine: (TH2 cytokines). see: TH1/TH2 Cytokines.

Type 1 interferons: Collective term for the acid-stable interferons » IFN-α and » IFN-β which differ immunologically and with regard to some biologic and physiochemical properties. See also: IFN for general information about interferons.

Type-2-astrocyte-inducing factor: This poorly characterized factor of 20 kDa is produced by cultured rat brain cells and type 1 astrocytes. It induces the development of type 2 astrocytes. This activity resembles that of » CNTF (ciliary neurotrophic factor).

Lillien LE et al Type-2-astrocyte development in rat brain cultures is initiated by a CNTF-like protein produced by type-1 astrocytes. Neuron 1: 485-94 (1988)

Type 2 hematopoietins: see: Hematopoiesis.

Type 2 interferon: This interferon is identical with » IFN-γ. See also: IFN for general information about interferons.

Tyrosine kinase-related to fibroblast growth factor receptor: see: TKF.

Tyrphostins: A collective term for a large number of synthetic *tyr*osine and *phos*phorylation *in*hibitors such as ***AG126***, ***AG213***, ***AG370***, ***AG490***, ***RG50862***, and ***RG50864*** that specifically inhibit tyrosine-specific protein kinases (see: PTK). They are derived from ***Erbstatin*** a compound initially isolated from *Streptomyces* MH435-hF3 that inhibits the » autophosphorylation of the receptor for » EGF.

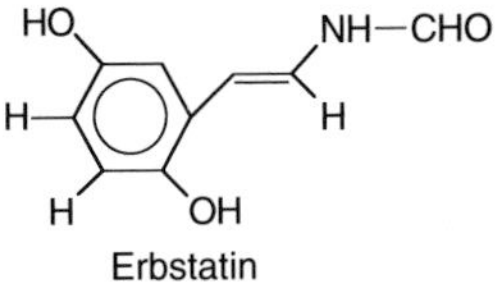

HO, NH–CHO, H, H, H, OH

Erbstatin

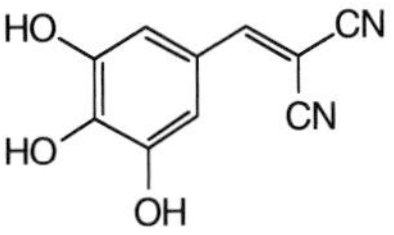

HO, CN, CN, HO, OH

Tyrphostin

Tyrphostins inhibit for example receptors for » EGF, » PDGF and IGF-1 (see: IGF) the intracellular domains of which encode intrinsic tyrosine-specific protein kinase activities. They act as competitive inhibitors of substrate binding without affecting the binding of ATP. The mechanism of action of tyrphostins therefore differs from that of » Genistein, another inhibitor of tyrosine-specific protein kinases. Erbstatin and many tyrphostins have been shown to inhibit the growth factor-induced proliferation of various tumors and thus have antineoplastic activities. Tyrphostins also inhibit the thrombin-induced activation of platelets. Some tyrphostins discriminate between the highly homologous EGF receptor kinase (HER1) and *Erb*B2/*neu* kinase (HER2; see: *neu*). These findings may lead to selective tyrosine kinase blockers for the treatment of diseases in which *Erb*B2/*neu* is involved.

Tyrphostins can be used to test whether binding of a cytokine to its receptor or induction of the synthesis of a growth factor requires tyrosine-specific protein kinase activities for post receptor signaling processes (for other agents used to dissect cytokine-mediated signal transduction pathways see: Bryostatins, Calcium ionophore, Calphostin C, Genistein, H8, Herbimycin A, K-252a, Lavendustin A, Phorbol esters, Okadaic acid, Staurosporine, Suramin, Vanadate).

Bishop WR et al Inhibition of protein kinase C by the tyrosine kinase inhibitor erbstatin. Biochem. Pharmacol. 40: 2129-35 (1990); **Daya-Makin M et al** Erbstatin and tyrphostin block protein serine kinase activation and meiotic maturation of sea star oocytes. BBA 1093: 87-94 (1991); **Gazit A et al** Tyrphostins I: synthesis and biological activity of protein tyrosine kinase inhibitors. J. Med. Chem. 32: 2344-52 (1989); **Gazit A et al** Tyrphostins. 2. Heterocyclic and α-substituted benzylidenemalononitrile tyrphostins as potent inhibitors of EGF receptor and ErbB2/neu tyrosine kinases. J. Med. Chem. 34: 1896-907 (1991); **Honma Y et al** Induction by some protein kinase inhibitors of differentiation of a mouse megakaryoblastic cell line established by coinfection with Abelson murine leukemia virus and recombinant SV40 retrovirus. CR 51: 4649-55 (1991); **Hori T et al** Inhibition of tyrosine kinase and *src* oncogene functions

tyk1: see addendum: LTK.

by stable erbstatin analogs. J. Antibiot. Tokyo 45: 280-2 (1992); **Isshiki K et al** Inhibition of tyrosine protein kinase by synthetic erbstatin analogs. J. Antibiot. (Tokyo) 40: 1209-10 (1987); **Imoto M et al** Antitumor activity of erbstatin, a tyrosine protein kinase inhibitor. JJCR 78: 329-32 (1987); **Imoto M et al** *In situ* inhibition of tyrosine protein kinase by erbstatin. BI 15: 989-95 (1987); **Isshiki K et al** Effective synthesis of erbstatin and its analogs. J. Antibiot. (Tokyo) 40: 1207-8 (1987); **Levitzki A et al** Inhibition of protein-tyrosine kinases by tyrphostins. MiE 201: 347-61 (1991); **Lyall RM et al** Tyrphostins inhibit epidermal growth factor (EGF)-receptor tyrosine kinase activity in living cells and EGF-stimulated cell proliferation. JBC 264: 14503-9 (1989); **Naccache PH et al** Selective inhibition of human neutrophil functional responsiveness by erbstatin, an inhibitor of tyrosine protein kinase. Blood 76: 2098-104 (1990); **Nakamura H et al** The structure of an epidermal growth factor-receptor kinase inhibitor, erbstatin. J. Antibiot. (Tokyo) 39: 314-5 (1986), [erratum in J. Antibiot. (Tokyo) 39: 1191]; **Rendu F et al** Tyrosine kinase blockers: new platelet activation inhibitors. Blood Coagul. Fibrinolysis 1: 713-6 (1990); **Salari H et al** Erbstatin blocks platelet activating factor-induced protein-tyrosine phosphorylation, polyphosphoinositide hydrolysis, protein kinase C activation, serotonin secretion, and aggregation of rabbit platelets. FL 263: 104-8 (1990); **Takekura N et al** Effects of tyrosine kinase inhibitor, erbstatin, on cell growth and growth-factor/receptor gene expression in human gastric carcinoma cells. IJC 47: 938-42 (1991); **Toi M et al** Antineoplastic effect of erbstatin on human mammary and oesophageal tumors in nude mice. Eur. J. Cancer 26: 722-4 (1990); **Umezawa H et al** Studies on a new epidermal growth factor-receptor kinase inhibitor, erbstatin, produced by MH435-hF3. J. Antibiot. Tokyo. 39: 170-3 (1986); **Umezawa K & Imoto M** Use of erbstatin as a protein-tyrosine kinase inhibitor. MiE 201: 379-85 (1991); **Umezawa H et al** Inhibition of epidermal growth factor-induced DNA synthesis by tyrosine kinase inhibitors. FL 29: 198-200 (1990); **Yaish P et al** Blocking of EGF-dependent cell proliferation by EGF receptor kinase inhibitors. S 242: 933-5 (1988)

U

U937: (ATCC: CRL 1593) A human cell line established from a diffuse histiocytic lymphoma of a 37 year old male patient. This cell line is one of the few cell lines displaying many monocytic characteristics and has thus served as a model for monocyte/macrophage differentiation *in vitro*. The cells are committed to the macrophage branch of the myeloid lineage and can be induced by a variety of agents to mature from a promonocytic into a monocytic stage of development. This process is accompanied by the acquisition of a number of morphological and functional attributes normally associated with mature macrophages and requires the expression of the » *fes* » oncogene.

The cell line constitutively produces » IL1 and also produces » GM-CSF, » ECEF (eosinophil cytotoxicity-enhancing factor), » VPF (vascular permeability factor), and » LBIF (lymphocyte blastogenesis inhibitory factor). The cells also secrete an uncharacterized factor (see: THP-1-derived growth-promoting activity) that promotes the growth of a large variety of cell lines.

U937 can be activated by treatment with supernatants of lymphocyte cultures (see also: Cell activation). They display a pronounced antibody-dependent cellular cytotoxicity (see: ADCC) against a number of erythroid and neoplastic target cells. » Phorbol esters, retinoic acid, 1-25 dihydroxyvitamin D3, » IL6 (see: DIF, differentiation inducing factor), » IFN-γ, and TNF-α induce the differentiation of U937 cells into monocyte/macrophage-like cells secreting » M-CSF. GM-CSF inhibits the colony growth of U937 cells in agar culture. This effect is, at least in part, due to the induction of secretion of TNF-α. Incubation of U937 cells » GM-CSF makes them responsive to induction of » TNF by bacterial lipopolysaccharides. Treatment with » IL6 inhibits TNF production. In the presence of GM-CSF differentiation is induced by » Oncostatin M.

A marked suppression of clonogenicity is observed using combinations of » LIF and » G-CSF. U937 cells respond by diminished spontaneous migration when confronted with affinity-purified soluble fragments of » CD23.

The cytotoxic activity of » TNF-α for U937 cells is markedly reduced by pretreatment with » ADF (adult T cell leukemia-derived factor) and enhanced by low concentrations of IL6. The growth of U937 cell is inhibited by » TGF-β at doses ranging from 0.025 to 2.5 ng/mL. It is also inhibited by a monocyte differentiation factor of 45-55 kDa, distinct from » IFN-γ, TNF-α, or » GM-CSF, produced by activated $CD4^+$ and $CD8^+$ T cells.

Variants of U937 cells capable of long-term growth in serum-free culture have also been described. Monoclonal antibodies to the eosinophil cytotoxicity-enhancing factor (see: ECEF) have been used to select U937 variants that differentiate in response to suboptimal doses of the » phorbol ester PMA.

Aderka D et al IL6 inhibits lipopolysaccharide-induced tumor necrosis factor production in cultured human monocytes, U937 cells, and in mice. JI 143: 3517-23 (1989); **Bohbot A et al** U937 cell line: impact of CSFs, IL6, and IFN-γ on the differentiation and the Leu-CAM proteins expression. EH 21: 564-72 (1993); **Bruce AG et al** Oncostatin M is a differentiation factor for myeloid leukemia cells. JI 149: 1271-5 (1992); **Cairns JA et al** Interleukin-6 regulates the cytotoxic effect of tumor necrosis factor on U937 cells. Immunology 75: 669-73 (1992); **Cannistra SA et al** Human granulocyte-macrophage colony-stimulating factor induces expression of the tumor necrosis factor gene by the U937 cell line and by normal human monocytes. JCI 79: 1720-8 (1987); **Chorvath B et al** Human monoblastoid cell line U937 cultured in protein-free medium: immunophenotype, cytochemical, and biochemical markers. Neoplasma 38: 483-92 (1991); **Clement LT et al** Characterization of a monocyte differentiation factor distinct from γ-interferon, tumor necrosis factor, or GM-colony-stimulating factor that regulates the growth and functional capabilities of the U937 monocytic cell line. J. Leukoc. Biol. 44: 101-10 (1988); **Connolly DT et al** Human vascular permeability factor. Isolation from U937 cells. JBC 264: 20017-24 (1989); **Dolecki GJ & Connolly DT** Effects of a variety of cytokines and inducing agents on vascular permeability factor mRNA levels in U937 cells. BBRC 180: 572-8 (1991); **Duits AJ et al** Selective enhancement of Leu-CAM expression by interleukin-6 during differentiation of human promonocytic U937 cells. Scand. J. Immunol. 33: 151-9 (1991); **Flores-Romo L et al** Soluble fragments of the low-affinity IgE receptor (CD23) inhibit the spontaneous migration of U937 monocytic cells: neutralization of MIF-activity by a CD23 antibody. Immunology. 67: 547-9 (1989); **Harris P & Ralph P** Human leukaemic models of myelomonocytic development: a review of the HL60 and U937 cell lines. J. Leukocyte Biol. 37:

403-22 (1985); **Koren HS et al** *In vitro* activation of a human macrophage-like cell line. N 279: 328-31 (1979); **Liu MY & Wu MC** Induction of human monocyte cell line U937 differentiation and CSF-1 production by phorbol ester. EH 20: 974-9 (1992); **Maekawa T & Metcalf D** Clonal suppression of HL60 and U937 cells by recombinant human leukemia inhibitory factor in combination with GM-CSF or G-CSF. Leukemia 3: 270-6 (1989); **Matsuda M et al** Protective activity of adult T cell leukemia-derived factor (ADF) against tumor necrosis factor-dependent cytotoxicity on U937 cells. JI 147: 3837-41 (1991); **Minta JO et al** *In vitro* induction of cytolytic and functional differentiation of monocyte-like cell line U937 with phorbol myristate acetate. Am. J. Pathol. 119: 111-26 (1985); **Oberg F et al** Characterization of a U-937 subline which can be induced to differentiate in serum-free medium. IJC 50: 153-60 (1992); **Piacibello W et al** Differential effect of transforming growth factor β 1 on the proliferation of human lymphoid and myeloid leukemia cells. Haematologica 76: 460-6 (1991); **Peck R & Bollag W** Potentiation of retinoid-induced differentiation of HL-60 and U937 cell lines by cytokines. Eur. J. Cancer 27: 53-7 (1991); **Shipley JM et al** Karyotypic analysis of the human monoblastic cell line U937. CGC 30: 277-84 (1988); **Sundström C & Nilsson K** Establishment and characterization of a human histiocytic cell line (U-937). IJC 17: 565-77 (1976); **Testa U et al** Effects of endogenous and exogenous interferons on the differentiation of human monocyte cell line U937. CR 48: 82-8 (1988); **Treon SP et al** Growth restraint and differentiation by LPS/TNF-α/IFN-γ reorganization of the microtubule network in human leukemia cell lines. Leukemia 6(Suppl.3) 141S-45S (1992); **Zuckerman SH et al** Synergistc effect of granulocyte-macrophage colony-stimulating factor and on the differentiation of the human monocytic cell line U937. Blood 71: 619-24 (1988)

U937 differentiation-inducing factor: This fibroblast-derived factor which induces the differentiation of » U937 cells into monocytes/macrophages is identical with » IL6.

Noda M et al Purification and characterization of human fibroblast derived differentiation inducing factor for human monoblastic leukemia cells identical to interleukin-6. Anticancer Res. 11: 961-8 (1991)

U937-FAF: *U937-derived fibroblast-activating factor* This poorly characterized factor of 16-18 kDa is secreted by » U937 cells. This factor promotes the proliferation of fibroblasts. It is not identical with » IL1, » TNF and » bFGF and appears to be a unique factor as demonstrated by the known N-terminal amino acid sequence (see also: FAF, fibroblast activating factor). U937-FAF also stimulates the synthesis and release of prostaglandin E2 and proteoglycans by fibroblasts. See also: FAF (fibroblast activating factors).

Demeter J et al Isolation and partial characterization of the structures of fibroblast activating factor-related proteins from U937 cells. Immunology 72: 350-4 (1991); **Turck CW et al** Isolation and partial characterization of a fibroblast-activating factor generated by U937 human monocytic leukocytes. JI 141: 1225-30 (1988); **Turk CW et al** Diverse responses of human fibroblasts to a highly purified fibroblast-activating factor from the U937 line of human monocytes. Immunology 68: 410-5 (1989)

U937SF: *U937 suppressor factor* This poorly characterized factor of 69 kDa is secreted by » U937 cells. U937SF inhibits antigen/mitogen-induced proliferation of human peripheral blood mononuclear cells. It also depresses » IL2 production and IL2 receptor (CD25) expression in peripheral blood mononuclear cells stimulated with an antigen but not with a mitogen. U937SF does not affect monocyte functions such as antigen processing and IL1 production or the expression of T cell receptor (TCR) or CD3 molecules on the surface of lymphocytes.

Ohnishi K et al Pattern of the action of a suppressor factor produced by a human macrophage-like cell line, U937. Immunol. Cell. Biol. 70: 89-96 (1992)

uAMF: *urinary autocrine motility factor* see: AMF (autocrine motility factor)

Ubiquitin: see: LGF (leukemia-derived growth factor).

UCRBP: *UCR binding protein* A transcription factor initially detected by its ability to bind to the upstream conserved region (UCR) of Moloney murine leukemia virus and to regulate promoter activity in a negative way. It is identical with » GATA-1.

Flanagan JR et al Cloning of a negative transcription factor that binds to the upstream conserved region of Moloney murine leukemia virus. MCB 12: 38-44 (1992); **Safrany G & Perry RP** Characterization of the mouse gene that encodes the δ/YY1/NF-E1/UCRBP transcription factor. PNAS 90: 5559-63 (1993)

UCRM: *universal cytokine response mediator* UCRM is a large protein complex consisting of various transcription factors (see: Gene expression) and accessory proteins that mediates universal primary and delayed signaling responses (see: ERG, early response gene) initiated by binding of almost all cytokines to their receptors.

Ajoke IS et al Universal cytokine receptor signaling: complexes for all seasons. Macromol. Cell. Signal. 1: 261-88 (1994); **Jodan DES** A universal acceptor for cytokine receptor mediated signals. Jpn. J. Cytokine Sig. Transd. 6: 111-116 (1994)

UCN-01: see Staurosporine.

UDGF: *uterine-derived growth factor* UDGF-α isolated from porcine uterus tissue is identical with » aFGF while UDGF-β is identical with » bFGF.

Brigstock DR et al Purification and characterization of heparin-binding growth factors from porcine uterus. BJ 266: 273-82 (1990); **Ikeda T & Sirbasku DA** Purification and properties of a mammary uterine pituitary tumor cell growth factor from pregnant sheep uterus. JBC 259: 4049-64 (1984)

uEGF-1: This factor is isolated from the purple sea urchin *Strongylocentrotus purpuratus*. The factor is involved in the regulation of embryonic development. It is a homologue of » EGF.
Muskavitch MAT & Hoffman FM Homologues of vertebrate growth factors in *Drosophila melanogaster* and other vertebrates. CTDB 24: 289-28 (1990)

UGF: *unidentified growth factor* This extra unnecessary acronym (EUA) was accidentally unterred from the depths of electronic data banks by searching for UGF instead of » UDGF. Long live our remarkable ability to create such fine acronyms.
Alenier JC & Combs GF Jr Effects on feed palatability of ingredients believed to contain unidentified growth factors for poultry. Poult. Sci. 60: 215-24 (1981)

Under-agarose leukocyte migration inhibition technique: see: Chemotaxis.

Unidentified growth factor: see: UGF.

Universal cytokine response mediator: see: UCRM.

UPSC: *unipotent stem cells* see: Hematopoiesis.

Upstream: This term has two meanings. In molecular biology it refers to sites or directions when moving along a nucleic acid sequence; upstream of where you are then means further into the direction of the 5´ end of the nucleic acid molecule (promoter sequences, for example, lie upstream of coding sequences). The reverse is ***downstream***, meaning into the direction of the 3´ end of a molecule.
In cytokine research the term upstream is often used within the context of signal transduction pathways. If these are seen as a hierarchy of events beginning by binding of a ligand to a cell surface receptor which then elicits a cascade of second messenger pathways and culminates in the modulation of » Gene expression by transcription factors upstream of where you are refers to those events further up the hierarchy. ***Downstream signals*** are those further down the hierarchy of events. For some agents used to dissect cytokine-mediated signal transduction pathways and to determine signal hierarchies see: Bryostatins, Calcium ionophore, Calphostin C, Genistein, H8, Herbimycin A, K-252a, Lavendustin A, Phorbol esters, Okadaic acid, Staurosporine, Suramin, Tyrphostins, Vanadate.

Urinary autocrine motility factor: abbrev. uAMF. see: AMF (autocrine motility factor)

Urinary colony-stimulating factor: This activity, isolated from human urine, is neutralized by antibodies directed against » L-cell CSF and is identical with » M-CSF.
Gao G et al Characterization of deglycosylated human urinary colony-stimulating factor (CSF-1). Comp. Biochem. Physiol. B. 89: 551-5 (1988); **Hanamura T et al** Quantitation and identification of human monocytic colony-stimulating factor in human serum by enzyme-linked immunosorbent assay. Blood 72: 886-92 (1988); **Motoyoshi K et al** Granulocyte-macrophage colony-stimulating and binding activities of purified human urinary colony-stimulating factor to murine and human bone marrow cells. Blood 60: 1378-86 (1982); **Motoyoshi K et al** Recombinant and native human urinary colony-stimulating factor directly augments granulocytic and granulocyte-macrophage colony-stimulating factor production of human peripheral blood monocytes. EH 17: 68-71 (1989); **Sakai N et al** Occurrence of a monocyte/macrophage colony-stimulating factor in the continuous ambulatory peritoneal dialysis fluids and its chromatographic behaviors and antigenicity compared with human urinary colony-stimulating factor. JCP 118: 1-5 (1984); **Tao X et al** Isolation and characterization of human urinary colony-stimulating factor. Biol. Chem. Hoppe-Seyler 368: 187-94 (1987); **Wang FF & Goldwasser E** Purification of a human urinary colony-stimulating factor. JCBc 21: 263-75 (1983); **Zhu FD et al** Induction of tumor necrosis factor by macrophage colony-stimulating factor *in vivo*. J. Biol. Response Mod. 9: 339-42 (1990)

Urinary trypsin inhibitor: abbrev. UTI. see: ITI (inter-α-trypsin inhibitor).

URO: abbrev. for » urogastrone.

Urodilatin: A protein of 32 amino acids isolated from human urine and produced in the kidney. It is identical in sequence to » ANF (atrial natriuretic factor) but contains an N-terminal tetrapeptide extension [ANF (95-126)]. Urodilatin binds to ANF receptors. Urodilatin is not affected by proximal tubule neutral endopeptidase 24:11 (NEP) which destroys most of intrarenal luminal ANP and kinins.
Phosphorylation of urodilatin decreases its vasorelaxant potency, while dephosphorylation of phosphorylated urodilatin by acidic phosphatase completely restores bioactivity. Urodilatin seems to exert more potent renal effects than ANF [ANF (99-126)] and has been used to prevent kidney failure after heart transplantation.
Feller SM et al Urodilatin: a newly described member of the ANP family. TIPS 10: 93-4 (1989); **Hummel M et al** Urodilatin: a new peptide with beneficial effects in the postoperative therapy of cardiac transplant recipients. Clin. Investig. 70: 674-82 (1992); **Hummel M et al** Urodilatin, a new therapy to prevent kidney failure after heart transplantation. J. Heart Lung Transpl. 12: 209-17 (1993); **Valentin JP & Humphreys MH** Urodilatin: a paracrine renal natriuretic peptide. Semin. Nephrol. 13: 61-70 (1993)

Urogastrone: abbrev. URO; also: URG. This factor of 53 amino acids was initially isolated from human urine. It is found predominantly in the duodenum and in the salivary glands. γ-urogastrone (52 amino acids) lacks the carboxyterminal arginine of β-urogastrone. Urogastrone is a potent inhibitor of gastric acid secretion and also promotes epithelial cell proliferation. Urogastrone and » EGF have similar biological activities and a high degree of homology of amino acid sequence (70 %).

Gregory H Isolation and structure of urogastrone and its relationship to epidermal growth factor. N 257: 325-27 (1975); **Smith J et al** Chemical synthesis and cloning of a gene for human β-urogastrone. NAR 10: 4467-82 (1982)

Uromodulin: This glycoprotein of 85 kDa (639 amino acids including a » signal sequence of 24 amino acids), also called ***uromucoid***. was initially found in the urine of pregnant women. It is produced by kidney cells of Henle´s loop. Uromodulin is also secreted by HLA-DR monocytes and the neo-fibroblasts which derive from them.

The protein is identical with ***THP*** (***Tamm-Horsfall glycoprotein***) found in large amounts in normal urine. It is closely related (88 % homology at the protein level) to glycoprotein ***GP-2*** found in the zymogen granules of the exocrine pancreas. The gene encoding human uromodulin maps to chromosome 16p13.11.

The role of Tamm-Horsfall protein in physiological and pathological states, and its regulation, are still unknown. Tamm-Horsfall protein is anchored to the cytoplasmic membrane via a phosphatidyl-inositol group, from which it can be released upon cleavage by a specific phospholipase.

Uromodulin is a protein with immunosuppressive activities that are mainly due to the covalently attached oligosaccharides. Tamm-Horsfall protein probably has a different glycosylation pattern since it does not have the same immunosuppressive properties as uromodulin. Uromodulin inhibits antigen-specific T cell reactions. It binds » IL1 and » TNF-α with high affinity and thus blocks the activity of this interleukin (see also: Cytokine inhibitors). Uromodulin appears to act as an endogenous lectin and also binds » IL2 and » TNF through its sugar side chains although is does not appear to influence their biological activities.

Bringuier AF et al T lymphocyte control of HLA-DR blood monocyte differentiation into neo-fibroblasts. Further evidence of pluripotential secreting functions of HLA-DR monocytes, involving not only collagen but also uromodulin, amyloid-β peptide, α-fetoprotein and carcinoembryonic antigen. Biomed. Pharmacother. 46: 91-108 (1992); **Brown K et al** Uromodulin, an immunosuppressive protein derived from pregnancy urine, is an inhibitor of interleukin 1. PNAS 83: 9119-23 (1987); **Hession C et al** Uromodulin (Tamm-Horsfall glycoprotein): a renal ligand for lymphokines. S 237: 1479-84 (1987); **Hoops TC & Rindler MJ** Isolation of the cDNA encoding glycoprotein 2 (GP-2), the major zymogen granule membrane protein. Homology to uromodulin/Tamm-Horsfall protein. JBC 266: 4257-63 (1991); **Jeanpierre C et al** Chromosomal assignment of the uromodulin gene (UMOD) to 16p13.11. Cytogenet. Cell. Genet. 62: 185-7 (1993); **Kumar S & Muchmore AV** Tamm-Horsfall protein – Uromodulin (1950-1990). Kidney Int. 37: 1395-401 (1990); **Moonen P et al** Native cytokines do not bind to uromodulin (Tamm-Horsfall glycoprotein). FL 226: 314-8 (1988); **Muchmore AV & Decker JM** Uromodulin. An immunosuppressive 85-kilodalton glycoprotein isolated from human pregnancy urine is a high affinity ligand for recombinant interleukin 1-α. JBC 261: 13404-7 (1986); **Muchmore AV et al** Evidence that specific high-mannose oligosaccharides can directly inhibit antigen-driven T cell responses. J. Leukoc. Biol. 48: 457-64 (1990); **Muchmore AV et al** *In vitro* evidence that carbohydrate moieties derived from uromodulin, an 85000 dalton immunosuppressive glycoprotein isolated from human pregnancy urine, are immunosuppressive in the absence of intact protein. JI 138: 2547-53 (1987); **Pennica D et al** Identification of human uromodulin as the Tamm-Horsfall urinary glycoprotein. S 236: 83-8 (1987); **Rindler MJ et al** Uromodulin (Tamm-Horsfall glycoprotein/uromucoid) is a phosphatidylinositol-linked membrane protein. JBC 265: 20784-9 (1990); **Sathyamoorthy N et al** Evidence that specific high mannose structure directly regulate multiple cellular activities. Mol. Cell. Biochem. 102: 139-47 (1991); **Serafini-Cessi F et al** Rapid isolation of Tamm-Horsfall glycoprotein (uromodulin) from human urine. JIM 120: 185-9 (1989); **Sherblom AP et al** The lectin-like interaction between recombinant tumor necrosis factor and uromodulin. JBC 263: 5418-24 (1988); **Winkelstein A et al** Uromodulin: a specific inhibitor of IL1-initiated human T cell colony formation. Immunopharmacology 20: 201-5 (1990)

Uromucoid: see: Uromodulin.

US28: An open reading frame found in the genome of human cytomegalovirus. It encodes a protein containing seven putative membrane-spanning domains, and a series of well-defined motifs characteristic of the rhodopsin-like G protein-coupled receptors. This protein is similar in sequence (33 % identity) to the recently cloned receptor for the human macrophage inflammatory protein » MIP-1α and » RANTES which has been suggested to be the human homologue of US28 (see also: CKE-1).

Chee MS et al Human cytomegalovirus encodes three G protein-coupled receptor homologues N 344: 774-7 (1990); **Gao JL et al** Structure and functional expression of the human macrophage inflammatory protein 1 α/RANTES receptor. JEM 177: 1421-7 (1993); **Neote K et al** Molecular cloning, functional expression, and signaling characteristics of a C-C chemokine receptor. Cell 72: 415-25 (1993)

UT-7: A cell line established from the bone marrow of a patient with acute megakaryoblastic leu-

kemia. The cells express GPIIb/IIIa (CD41a), GPIb (CD42b), MY 7 (CD13), MY 9 (CD33), glycophorin A and PPO (platelet peroxidase activity). Cytogenetic analysis shows a human male near-tetraploid karyotype with a modal chromosome number of 92-96. Treatment of the cells with phorbol myristate acetate (see also: Phorbol esters) dramatically increases the synthesis of » PF4 (platelet factor 4) and » β-thromboglobulin, which are specifically synthesized in the process of megakaryocyte maturation. PMA treatment suppresses the growth of UT-7 cells and enhances their differentiation and maturation.
The growth of UT-7 cells strictly depends on » GM-CSF (1 ng/ml), » IL3 (10 units/ml) or » Epo (1 unit/ml). Therefore UT-7 cells are useful to assay these factors. The proliferation of UT-7 cells is stimulated also by » IL6 which is, however, unable to maintain the line in long-term culture.
Dusanter-Fourt I et al Erythropoietin induces the tyrosine phosphorylation of its own receptor in human erythropoietin-responsive cells. JBC 267: 10670-5 (1992); **Komatsu N et al** Establishment and characterization of a human leukemic cell line with megakaryocytic features: dependency on granulocyte-macrophage colony-stimulating factor, interleukin 3, or erythropoietin for growth and survival. CR 51: 341-8 (1991); **Komatsu N et al** Erythropoietin rapidly induces tyrosine phosphorylation in the human erythropoietin-dependent cell line, UT-7. Blood 80: 53-9 (1992); **Miura Y et al** Growth and differentiation of two human megakaryoblastic cell lines; CMK and UT-7. PCBR 356: 259-70 (1990)

Uterine angiogenesis factor: see: HUAF (human uterine angiogenesis factor).

Uterine-derived growth factor: see: UDGF.

Uteroferrin: Uteroferrin is a purple-colored progesterone-induced glycoprotein of 35 kDa containing two molecules of iron. It is the major secretory product of the porcine uterus under the influence of progesterone and supplies iron to the developing fetuses during pregnancy.
Uteroferrin has been shown to be an intracellular tartrate-resistant acid phosphatase that shares many properties with the type 5 tartrate-resistant acid phosphatase in human placenta, the chondrocytes of osteoclastic bone tumors, and spleens of hairy cell leukemia. Uteroferrin also exists in a heterodimeric form associated with one of three uteroferrin-associated proteins that have high amino acid sequence homology with serine protease inhibitors.
Human and porcine uteroferrin have been shown to function as a hematopoietic growth factor that act on » BFU-E, CFU-GM, and CFU-GEMM hematopoietic progenitors. One of the functions of uteroferrin may be the initiation of hematopoiesis and colonization of the hematopoietic organs during the first and second trimesters of gestation.
Bazer FW et al Uteroferrin: a progesterone-induced hematopoietic growth factor of uterine origin. EH 19: 910-5 (1991); **Ketcham CM et al** The type 5 acid phosphatase from spleen of humans with hairy cell leukemia. JBC 260: 5768-76 (1985); **Ling P & Roberts RM** Uteroferrin and intracellular tartrate-resistant acid phosphatases are the products of the same gene. JBC 268: 6896-902 (1993)

UTI: *urinary trypsin inhibitor* see: ITI (inter-α-trypsin inhibitor).

V

v: In combinations a prefixed c signifies "viral", for example v-*sis* (the viral » *sis* » oncogene) or vIL10 (viral » IL10).

Vaccination against cancer: see: Cytokine gene transfer.

Vaccinia 19 kDa protein: This protein is identical with » VVGF (vaccinia virus growth factor).

Vaccinia growth factor: see: » VVGF (vaccinia virus growth factor).

Vaccinia virus growth factor: see: VVGF.

Vanadate: Sodium orthovanadate (Na_3VO_4) at micromolar concentrations is a potent general inhibitor of phosphatases. Vanadate also inhibits phosphotyrosine phosphatases in cultured cells. Vanadate is not a complete inhibitor as some phosphatases are not inhibited. Most published reports show a residual (5-10 %) phosphotyrosine phosphatase activity in inhibitor concentrations of up to 1mM vanadate.

Treatment of cells with this inhibitor results in an increase of proteins phosphorylated at tyrosine residues. Since transient phosphorylation of proteins on tyrosine is an important step in receptor-mediated signal transduction of many cytokines this agent can be used to mimic cytokine actions and to study the involvement of tyrosine-phosphorylation and/or specific phosphoprotein phosphatases in cytokine-mediated receptor signaling and to dissect molecular signal transduction pathways (for other agents used to dissect cytokine-mediated signal transduction pathways see: Bryostatins, Calcium ionophore, Calphostin C, Genistein, H8, Herbimycin A, K-252a, Lavendustin A, Phorbol esters, Okadaic acid, Staurosporine, Suramin, Tyrphostins).

Andrews DF 3rd et al Sodium vanadate, a tyrosine phosphatase inhibitor, affects expression of hematopoietic growth factors and extracellular matrix RNAs in SV40-transformed human marrow stromal cells. EH 20: 449-53 (1992); **Chao W et al** Protein tyrosine phosphorylation and regulation of the receptor for platelet-activating factor in rat Kupffer cells. Effect of sodium vanadate. BJ 288: 777-84 (1992); **Dai CH & Krantz S** Vanadate mimics the effect of stem cell factor on highly purified human erythroid burst-forming units *in vitro*, but not the effect of erythropoietin. EH 20: 1055-60 (1992); **Feldman RA et al** Selective potentiation of c-*fps/fes* transforming activity by a phosphatase inhibitor. OR 5: 187-97 (1990); **Gordon JA** Use of vanadate as protein-phosphotyrosine phosphatase inhibitor. MiE 201: 477-82 (1991); **Hecht D & Zick Y** Selective inhibition of protein tyrosine phosphatase activities by H2O2 and vanadate *in vitro*. BBRC 188: 773-9 (1992); **Igarashi K et al** *In vitro* activation of a transcription factor by γ interferon requires a membrane-associated tyrosine kinase and is mimicked by vanadate. MCB 13: 3984-9 (1993); **Kanakura Y et al** Signal transduction of the human granulocyte-macrophage colony-stimulating factor and interleukin-3 receptors involves tyrosine phosphorylation of a common set of cytoplasmic proteins. Blood 76: 706-15 (1990); **Kanakura Y et al** Phorbol 12-myristate 13-acetate inhibits granulocyte-macrophage colony stimulating factor-induced protein tyrosine phosphorylation in a human factor-dependent hematopoietic cell line. JBC 266: 490-5 (1991); **Sorensen PH et al** Interleukin-3, GM-CSF, and TPA induce distinct phosphorylation events in an interleukin 3-dependent multipotential cell line. Blood 73: 406-18 (1989); **Trudel S et al** Activation of permeabilized HL60 cells by vanadate. Evidence for divergent signaling pathways. BJ 269: 127-31 (1990); **Trudel S et al** Mechanism of vanadate-induced activation of tyrosine phosphorylation and of the respiratory burst in HL60 cells. Role of reduced oxygen metabolites. BJ 276: 611-9 (1991); **Zippel R et al** Inhibition of phosphotyrosine phosphatases reveals candidate substrates of the PDGF receptor kinase. Eur. J. Cell Biol. 50: 428-34 (1989)

VAS: abbrev. for » vasculotropin.

Vascular endothelial cell growth factor: see: VEGF.

Vascular endothelial cell proliferation factor: see: VEGF (vascular endothelial cell growth factor).

Vascular expansion: see: Angiogenesis factors.

Vascular permeability factor: see: VGF. See also: VEGF (vascular endothelial cell growth factor).

Vasculogenesis: see: Angiogenesis factors.

Vasculotropin: abbrev. VAS. This factor was initially isolated as an angiogenic protein (see also:

Angiogenesis factors) mitogenic for endothelial cells. It is identical with » VEGF (vascular endothelial cell growth factor).

Favard C et al Purification and biological properties of vasculotropin, a new angiogenic cytokine. Biol. Cell 73: 1-6 (1991); **Plouet J et al** Isolation and characterization of a newly identified endothelial cell mitogen produced by pituitary derived AtT-20 cells. EJ 8: 3801-6 (1989); **Moukadiri H et al** [Vasculotropin: a new angiogenic growth factor] Pathol. Biol. Paris 39: 153-6 (1991)

Vasoactive intestinal contractor: see: VIC.

Vasoactive intestinal peptide: see: VIP.

Vasopressin: abbrev. VP. This protein is also called ***antidiuretic hormon**e* (ADH), ***adiuretin***, ***vasotocin***, ***pituitrin P*** and ***pitressin***. It is a cyclic nonapeptide (1,6 disulfide bridge) synthesized in the hypothalamus and stored in the posterior lobe of the pituitary from which it is released into the circulation as necessary. The term ***arginine vasopressin*** (abbrev. ***AVP***) refers to the presence of an arginine residue at position 8 (which can also be occupied by lysine); residues 3 and 4 have also been shown to be variable.

```
Phe——————Gln
 |        |
Tyr      Asn
 |        |
Cys—S-S—Cys—Pro—Arg—Gly—NH2
```

AVP

AVP has also been found in other tissues, including testis, ovary, uterus, adrenal gland, superior cervical ganglion, and thymus. AVP is derived from a larger precursor, pre-pro-arginine-vasopressin-neurophysin II.

Vasopressin regulates osmotic pressure in body fluids via a specific vasopressor receptor (V1). It has direct antidiuretic activity in the kidney, mediated by the antidiuretic receptor V2, and promotes re-adsorption of water in the distal convoluted tubules of the kidney. It also causes vasoconstriction in peripheral small blood vessels by stimulating smooth muscle cells in the cell walls to contract. Binding sites for AVP have been found also on human peripheral blood mononuclear cells.

Apart from the classical role as a hormone vasopressin also acts as» cytokine although it is not classified as such due to its small size. In serum-free media (see also: SFM) vasopressin at nanomolar concentrations stimulates DNA synthesis and proliferation of murine 3T3 fibroblasts in the absence of other growth factors. This activity is potentiated by insulin. The mitogenic activity of vasopressin is associated with the activation of Ca^{2+}-mobilising G proteins. Vasopressin is an » autocrine growth factor for small cell lung cancers. The synthesis and secretion of vasopressin can be induced by » IL1, and IL2, but not by » IL6. In resting murine fibroblasts (see also: 3T3) pretreatment of the cells with vasopressin modulates the activities of » PDGF by heterologous desensibilisation of the receptor (see also: Receptor transmodulation). In vascular smooth muscle cells vasopressin enhances the synthesis of » NGF mRNA.

AVP has been shown to play an important role in the modulation of the stress response (see: Neuroimmune axis). It directly stimulates the release of ACTH and affects ACTH levels indirectly by enhancing production of corticotropin releasing factor (see: POMC, proopiomelanocortin). AVP also plays an important role in the regulation of » IFN-γ production by providing a helper signal (induction of » IL2 by T cells). IL2 itself has been shown to induce production of AVP in some systems. Chronic intracerebral infusion of AVP has also been shown to modulate behavioral effects of » IL1.

■ **TRANSGENIC/KNOCK-OUT/ANTISENSE STUDIES:** The bioactivities of vasopressin have been investigated in » transgenic animals expressing the arginine-vasopressin transgene in a tissue-specific manner. Animals homozygous for the vasopressin transgene have increased basal plasma levels of vasopressin peptide but have no apparent change in basal water metabolism.

Bell J et al Identification and characterization of [125I]arginine vasopressin binding sites on human peripheral blood mononuclear cells. Life Sci. 52: 95-105 (1993); **Bernard-Weil E** Role played by vasopressin (and of an adrenal postpituitary imbalance) in the development of cancerous diseases. Med. Hypotheses 37: 127-36 (1992); **Bluthe RM & Danzer R** Chronic intracerebral infusions of vasopressin and vasopressin antagonist modulate behavioral effects of interleukin-1 in rat. Brain Res. Bull. 29: 897-900 (1992); **Casting NW** Criteria for establishing a physiological role for brain peptides. A case in point: the role of vasopressin in thermoregulation during fever and antipyresis. Brain Res. Rev. 14: 143-53 (1989); **Clements JA & Funder JW** Arginine vasopressin (AVP) and AVP-like immune reactivity in peripheral tissues. Endocrinol. Rev. 7: 449-60 (1986); **Domin J & Rozengurt E** Heterologous desensitization of platelet-derived growth factor-mediated arachidonic acid release and prostaglandin synthesis. JBC 267: 15217-23 (1992); **Doris PA** Vasopressin and central integrative processes. Neuroendocrinology 8: 75-85 (1984); **Gibbs DM** Vasopressin and oxytocin: hypothalamic modulations of the stress response: a review. Psychoneuroendocrinology 11: 131-40 (1986); **Hunt NH et al** Role of vasopressin in the mitogenic response of rat bone marrow cells to hemorrhage. J. Endocrinol. 72: 5-16 (1977); **Ivell R & Burbach JPH** The molecular biology of vasopressin and oxytocin genes. J. Neuroendocrinol. 3: 583-5 (1991); **Johnson HM et al** Vasopressin replacement of interleukin 2 requirement in γ interferon production. Lymphokine activity of a neuroendocrine hormone. JI 129: 963-86 (1982); **Johnson HM & Torres BA** Regulation of lymphokine production by arginine vasopressin and oxytocin: modulation of lymphocyte function by neurohypohyseal hormone. JI 135: 773s-5s (1985); **Lutz W et al** Vasopressin receptor-mediated endocytosis: Current view. Am. J. Physiol. Renal, Fluid Electrolyte Physiol. 261: F1-F13 (1991); **Manning M et al** Carboxy terminus of vasopressin required for activity but not binding. N 308: 652-53 (1984); **Naito Y et al** Effects of interleukins on plasma arginine vasopressin and oxytocin levels in conscious freely moving rats. BBRC 174: 1189-95 (1991); **Pardy K et al** The influence of interleukin-2 on vasopressin and oxytocin gene expression in the rodent hypothalamus. J. Neuroimmunol. 42: 131-8 (1993); **Rozengurt E et al** Vasopressin stimulation of mouse 3T3 cell growth. PNAS 76: 1284-7 (1979); **Russell WE & Bucher NLR** Vasopressin modulates liver regeneration in the Brattleboro rat. Am. J. Physiol. 245: G321-4 (1983); **Sethi T & Rozengurt E** Multiple neuropeptides stimulate clonal growth of small cell lung cancer: effects of bradykinin, vasopressin, cholecystokinin, galanin, and neurotensin. CR 51: 3621-3 (1991); **Vallotton MB** The multiple faces of the vasopressin receptors. Mol. Cell. Endocrinol. 78: C73-6 (1991); **Verbeeck MA et al** Expression of the vasopressin and gastrin-releasing peptide genes in small cell lung carcinoma cell lines. Pathobiology 60: 136-42 (1992); **Woll P & Rozengurt E** Neuropeptides as growth regulators. BMB 45: 492-505 (1989); **Woll PJ & Rozengurt E** Two classes of antagonist interact with receptors for the mitogenic neuropeptides bombesin, bradykinin, and vasopressin. GF 1: 75-83 (1988); **Zachary I et al** Bombesin, vasopressin, and endothelin rapidly stimulate tyrosine phosphorylation in intact Swiss 3T3 cells. PNAS 88: 4577-81 (1991)

● TRANSGENIC/KNOCK-OUT/ANTISENSE STUDIES: **Grant FD et al** Transgenic mouse models of vasopressin expression. Hypertension 22: 640-5 (1993)

Vasotocin: Alternative name for » Vasopressin.

***vav*:** The *vav* » oncogene was found to be generated by a genetic rearrangement during gene transfer assays. The oncogene is activated by genetic rearrangements leading to the truncation of aminoterminal sequences. *vav* maps to human chromosome 19p12-19p13.2 and is closely linked to the insulin receptor. Chromosome region 19p13 is involved in different karyotypic abnormalities in a variety of malignancies including melanomas and leukemias.

The *vav* cDNA encodes an open reading frame of 2391 capable of directing the synthesis of a protein of 797 amino acids (95 kDa). The *vav* proto-oncogene shares homology with the *dbl* oncogene encoding a GDP-GTP exchange factor for the *ras*-like polypeptide CDC42Hs (and itself possesses GDP-GTP exchange activity), the *bcr* gene which recombines with the » *abl* oncogene in certain forms of leukemia, and a yeast gene (CDC24) involved in cytoskeletal organization. *vav* contains sequence motifs commonly found in transcription factors (see also: Gene expression), such as helix-loop-helix, leucine-zipper and zinc-finger motifs and nuclear localization signals (see: Signal sequences), as well as a single SH2 and two SH3 domains (see: *src* homology domains) which can be tyrosine-phosphorylated.

The normal *vav* allele, the *vav* proto-oncogene, is transcribed exclusively in cells of hematopoietic origin, including those of erythroid, lymphoid and myeloid lineages. *vav* mRNA has been found in 49 of 50 murine hematopoietic cell lines representing diverse hematopoietic lineages, and *in situ* hybridization in embryos shows *vav* expression to be confined to the only hematopoietically active tissue, fetal liver.

The function of *vav* is unknown. *vav* is rapidly phosphorylated on tyrosine residues in response to various ligands. It has been shown to participate in the signaling processes that mediate the antigen-induced activation of B lymphocytes. It plays a role in receptor-mediated signaling of Å IL2. Stimulation of the T cell antigen receptor on normal human peripheral blood lymphocytes or on human leukemic T cells, and the crosslinking of IgE receptors on rat basophilic leukemia cells, both promote the phosphorylation of tyrosine residues in *vav*. The activation of the receptor for » EGF leads to marked tyrosine phosphorylation of *vav* in cells transiently expressing *vav*, and *vav* associates with the receptor through its SH2 domain. Tyrosine phosphorylation of *vav* is greatly enhanced in the two human cell lines » M07E and »TF-1 in response to » SCF (stem cell factor), but not in response to » IL3 or » GM-CSF.

■ TRANSGENIC/KNOCK-OUT/ANTISENSE STUDIES: Disruption of the *vav* gene by homologous recombination in murine embryonic stem cells (see: ES cells) by the expression of » antisense oligonucleotides has been shown to disrupt the ability of these cells to differentiate into hematopoietic cell types. *vav* thus plays a critical role in the development of hematopoietic cells from totipotent cells.

Adams JM et al The hematopoietically expressed *vav* proto-oncogene shares homology with the dbl GDP-GTP exchange factor, the bcr gene and a yeast gene (CDC24) involved in cytoskeletal organization. O 7: 611-8 (1992); **Alai M et al** Steel factor stimulates the tyrosine phosphorylation of the proto-oncogene product, p95vav, in human hemopoietic cells. JBC 267: 18021-5 (1992); **Bustelo XR & Barbacid M** Tyrosine phosphorylation of the *vav* proto-oncogene product in activated B

Vasostatin: see addendum: Granins.

cells. S 256: 1196-9 (1992); **Bustelo XR et al** Product of *vav* proto-oncogene defines a new class of tyrosine protein kinase substrates N 356: 68-71 (1992); **Coppola J et al** Mechanism of activation of the *vav* proto-oncogene. Cell Growth Differ. 2: 95-105 (1991); **Evans GA et al** Interleukin-2 induces tyrosine phosphorylation of the *vav* proto-oncogene product in human T cells: lack of requirement for the tyrosine kinase *lck*. BJ 294: 339-42 (1993); **Katzav S et al** *vav*, a novel human oncogene derived from a locus ubiquitously expressed in hematopoietic cells. EJ 8: 2283-90 (1989); **Katzav S et al** Loss of the amino-terminal helix-loop-helix domain of the *vav* proto-oncogene activates its transforming potential. MCB 11: 1912-20 (1991); **Katzav S** *vav*: a molecule for all hemopoiesis? Br. J. Haematol. 81: 141-4 (1992); **Margolis B et al** Tyrosine phosphorylation of *vav* proto-oncogene product containing SH2 domain and transcription factor motifs. N356: 71-4 (1992); **Martinerie C et al** The human *vav* proto-oncogene maps to chromosome region 19p12-19p13.2. Hum. Genet. 86: 65-8 (1990); **Puil L & Pawson T** Cell regulation: Vagaries of *vav*. *vav*, a protein that contains an intriguing array of motifs, participates in the signaling pathways of distinct hematopoietic receptors and may represent a new class of signal transducing molecules. Curr. Biol. 2: 275-7 (1992)
● **TRANSGENIC/KNOCK-OUT/ANTISENSE STUDIES: Wulf GM et al** Inhibition of hematopoietic development from embryonic stem cells by antisense *vav* RNA. EJ 12: 5065-74 (1993)

vegetalising factor: see: *Vg*-1.

***vegetal-specific-related*-1:** This factor is identical with BMP-6 (= DVR-6; see: DVR). See: BMP (bone morphogenetic protein). See also: Bg-1 (vegetalising factor).

VEGF: *vascular endothelial cell growth factor* also: VEG/PF (vascular endothelial growth factor/vascular permeability factor).

■ **ALTERNATIVE NAMES:** GD-VEGF (glioma-derived vascular endothelial cell growth factor), Mouse sarcoma 180-derived growth factor; VAS (Vasculotropin), Vascular endothelial cell proliferation factor; VPF (vascular permeability factor). See also: individual entries for further information.

■ **SOURCES:** VEGF has been isolated from bovine pituitaries. The protein is also produced by murine neuroblastoma cell lines and a plethora of other tumor cells, tumors, and normal cell types, including macrophages, lung epithelial cells, kidney epithelial cells, follicular cell in the pituitary, corpus luteum cells, aortic smooth muscle cells.

■ **PROTEIN CHARACTERISTICS:** VEGF is a homodimeric heavily glycosylated protein of 46-48 kDa (24 kDa subunits). Glycosylation is not required, however, for biological activity. The subunits are linked by disulphide bonds. The human factor occurs in several molecular variants of 121, 165 and 189, 206 amino acids, arising by alternative splicing of the mRNA. The 165 amino acid form of the factor is the most common form in most tissues. Kaposi sarcomas express VEGF-121 and VEGF-165. The 189 amino acid variant of VEGF is identical with » VPF (vascular permeability factor).

VEGF-121 and VEGF-165 are soluble secreted forms of the factor while VEGF-189 and VEGF-206 are mostly bound to heparin-containing proteoglycans in the cell surface or in the basement membrane (see also: ECM, extracellular matrix). The bioavailability of VEGF is probably regulated at the genetic level by alternative splicing that determines whether VEGF will be soluble or incorporated into a biological reservoir and also through proteolysis following plasminogen activation.

Rat and bovine VEGF are one amino acid shorter than the human factor, and the bovine and human sequences show a homology of 95 %.

VEGF is not related with fibroblast growth factors (see: FGF) and only displays limited homology (18%) to the β chain of » PDGF. However, the positions of all eight cysteine residues are conserved in VEGF and PDGF.

In contrast to other factors mitogenic for endothelial cells such as » aFGF, » bFGF and » PDGF VEGF is synthesized as a precursor containing a typical hydrophobic secretory » signal sequence of 26 amino acids. Glycosylation is not required for efficient secretion of VEGF.

■ **GENE STRUCTURE:** The human gene has a length of ≈ 12 kb and contains eight exons. Four species of mRNA encoding VEGFs have been identified and found to be expressed in a tissue-specific manner. They arise from differential splicing with the 165 amino acid form of VEGF lacking sequences encoded by exon 6 and the 121 amino acid form lacking exon 6 and 7 sequences.

■ **RECEPTOR STRUCTURE, GENE(S), EXPRESSION:** A high-affinity glycoprotein receptor of 170-235 kDa is expressed on vascular endothelial cells. The interaction of VEGF with heparin-like molecules of the extracellular matrix (see also: ECM) is required for efficient receptor binding. Protamine sulfate and suramin are capable of replacing the receptor-bound factor. The high-affinity receptor for VEGF has been identified as the gene product of the » *flt* (*fms*-like) gene. Another receptor for VEGF is » KDR. A factor that competes with the 165 amino acid form of VEGF for receptor binding is » PlGF (placenta growth factor).

The binding of VEGF to » alpha 2 macroglobulin inhibits its receptor binding ability, indicating that α2M may function as a VEGF removal and inactivation factor. Heparin and heparan sulfate, but not other glycosaminoglycans such as chondroitin sulfate, efficiently inhibit the binding of VEGF to α2M.

■ **Biological Activities:** The biological activities of VEGF are not species-specific. The different isoforms of VEGF have different properties *in vitro* and this may also apply to their *in vivo* functions.

VEGF is a highly specific mitogen for vascular endothelial cells. *In vitro* the two shorter forms of VEGF stimulates the proliferation of macrovascular endothelial cells. VEGF does not appear to enhance the proliferation of other cell types. VEGF significantly influence vascular permeability and is a strong angiogenic protein in several bioassays (see also: Angiogenesis factors) and probably also plays a role in neovascularisation under physiological conditions. A potent synergism between VEGF and » bFGF in the induction of angiogenesis has been observed. It has been suggested that VEGF released from smooth muscle cells and macrophages may play a role in the development of arteriosclerotic diseases.

In endothelial cells VEGF induces the synthesis of von Willebrand factor. It is also a potent chemoattractant for monocytes (see also: Chemotaxis) and thus has procoagulatory activities. In microvascular endothelial cells VEGF induces the synthesis of plasminogen activator and plasminogen activator inhibitor type 1. VEGF also induces the synthesis of the metalloproteinase, interstitial collagenase, which degrades interstitial collagen types I-III under normal physiological conditions. In several organs the expression of VEGF appears to be regulated during development. VEGF plays a role in the development and function of primate follicles and the ovarian corpus luteum, supporting the proliferation of blood vessels. The differentiation of adipocytes, of pheochromocytomas, and myocytes is accompanied by the controlled expression of VEGF.

■ **Assays:** VEGF can be assayed by an immunofluorometric test. An alternative and entirely different detection method is » Message amplification phenotyping. For further information see also subentry "Assays" in the reference section.

■ **Clinical Use & Significance:** VEGF is probably important in the pathophysiology of neuronal and other tumors, probably functioning as a potent angiogenesis factor for human gliomas. Its synthesis is also induced by hypoxia. The extravasation of cells observed as a response to VEGF may be an important factor determining the colonization of distant sites. Due to its influences on vascular permeability VEGF may also be involved in altering blood-brain-barrier functions under normal and pathological conditions. The production of VPF in human malignant glioma cells expressing EGF receptors is significantly increased by » EGF. VEGF released by glioma cells *in situ* most likely accounts for the clinical features of glioblastoma multiforme tumors in patients, including striking tumor angiogenesis, increased cerebral edema and hypercoagulability manifesting as focal tumor necrosis, deep vein thrombosis, or pulmonary embolism.

VEGF secreted from the stromal cells may be responsible for the endothelial cell proliferation in capillary hemangioblastomas which are composed of abundant microvasculature and primitive angiogenic elements represented by stromal cells. The production and secretion of VEGF by human retinal pigment epithelial cells may be important in the pathogenesis of ocular neovascularization.

The treatment of nude mice (see also: Immunodeficient mice)carrying transplanted human rhabdomyosarcoma, glioblastoma or leuomyosarcoma cells with antibodies directed against VEGF inhibits tumor growth. The observation that the growth of these tumors *in vitro* remains unaffected by the antibody demonstrates that the inhibition of angiogenesis in the transplanted tumors is one of the major causes of tumor growth suppression. The expression of VEGF-121 or VEGF-165 in » CHO cells confers the ability to form tumors in nude mice.

● **Reviews: Connolly D T** Vascular permeability factor: a unique regulator of blood vessel function. JCBc 47: 219-23 (1991); **Ferrara N et al** The vascular endothelial growth factor family of polypeptides. JCBc 47: 211-8 (1991); **Ferrara N et al** Molecular and biological properties of the vascular endothelial growth factor family of proteins. Endocrin. Rev. 13: 18-32 (1992); **Kim K J et al** The vascular endothelial growth factor proteins: identification of biologically relevant regions by neutralizing monoclonal antibodies. GF 7: 53-64 (1992); **Klagsbrun M & Soker S** VEGF/VPF: the angiogenesis factor found? Current Biol. 3: 699-702 (1993)

● **Biochemistry & Molecular Biology: Berse B et al** Vascular permeability factor (vascular endothelial growth factor) gene is expressed differentially in normal tissues, macrophages, and tumors. Mol. Biol. Cell. 12: 211-20 (1992); **Cohen T et al** High levels of biologically active vascular endothelial growth factor (VEGF) are produced by the baculovirus expression system. GF 7: 131-8 (1992); **Favard C et al** Purification and biological properties of vasculotropin, a new angiogenic cytokine. Biol. Cell 73: 1-6 (1991); **Ferrara N et al** Purification and cloning of vascular endothelial growth factor secreted by pituitary folliculo-stellate cells. MiE 198: 391-404 (1991); **Ferrara N &**

Henzel WJ Pituitary follicular cells secrete a novel heparin-binding growth factor specific for vascular endothelial cells. BBRC 161: 851-8 (1989); **Fiebich, BL et al** Synthesis and assembly of functionally active human vascular endothelial growth factor homodimers in insect cells. EJB 211: 19-26 (1993); **Houck KA et al** The vascular endothelial growth factor family: Identification of a fourth molecular species and characterization of alternative splicing of RNA. ME 5: 1806-14 (1991); **Houck KA et al** Dual regulation of vascular endothelial growth factor bioavailability by genetic and proteolytic mechanisms. JBC 267: 26031-7 (1992); **Keck PJ et al** Vascular permeability factor, an endothelial cell mitogen related to PDGF. S 246: 1309-12 (1989); **Kim KJ et al** The vascular endothelial growth factor proteins: identification of biologically relevant regions by neutralizing monoclonal antibodies. GF 7: 53-64 (1992); **Moghaddam A & Bicknell R** Expression of platelet-derived endothelial cell growth factor in *Escherichia coli* and confirmation of its thymidine phosphorylase activity. B 31: 12141-6 (1992); **Myoken Y et al** Vascular endothelial cell growth factor (VEGF) produced by A431 human epidermoid carcinoma cells and identification of VEGF membrane binding sites. PNAS 88: 5819-23 (1991); **Rosenthal RA et al** Conditioned medium from mouse sarcoma 180 cells contains vascular endothelial growth factor. GF 4: 53-9 (1990); **Senger DR et al** Purification and NH_2-terminal amino acid sequence of guinea pig tumor-secreted vascular permeability factor. CR 50: 1774-8 (1990); **Tischer E et al** The human gene for vascular endothelial growth factor. Multiple protein forms are encoded through alternative exon splicing. JBC 266: 11947-54 (1991); **Yeo TK et al** Glycosylation is essential for efficient secretion but not for permeability-enhancing activity of vascular permeability factor (vascular endothelial growth factor). BBRC 179: 1568-75 (1991)

● **Receptors: de Vries C et al** The *fms*-like tyrosine kinase, a receptor for vascular endothelial growth factor. S 255: 989-91 (1992); **Galland F et al** The FLT4 gene encodes a transmembrane tyrosine kinase related to the vascular endothelial growth factor receptor. O 8: 1233-40 (1993); **Gitay-Goren H et al** The binding of vascular endothelial growth factor to its receptors is dependent on cell surface-associated heparin-like molecules. JBC 267: 6093-8 (1992); **Millauer B et al** High affinity VEGF binding and developmental expression suggest Flk-1 as a major regulator of vasculogenesis and angiogenesis. Cell 72: 835-46 (1993); **Myoken Y et al** Vascular endothelial cell growth factor (VEGF) produced by A-431 human epidermoid carcinoma cells and identification of VEGF membrane binding sites. PNAS 88: 5819-23 (1991); **Shen H et al** Characterization of vascular permeability factor/vascular endothelial growth factor receptors on mononuclear phagocytes. Blood 81: 2767-73 (1993); **Terman BI et al** Identification of the KDR tyrosine kinase as a receptor for vascular endothelial cell growth factor. BBRC 187: 1579-86 (1992); **Vaisman N et al** Characterization of the receptors for vascular endothelial growth factor. JBC 265: 19461-6 (1990); **Quinn TP et al** Fetal liver kinase 1 is a receptor for vascular endothelial growth factor and is selectively expressed in vascular endothelium. PNAS 90: 7533-7 (1993)

● **Biological Activities: Alvarez JA et al** Localization of basic fibroblast growth factor and vascular endothelial growth factor in human glial neoplasms. Mod. Pathol. 5: 303-7 (1992); **Bikfalvi A et al** Interaction of vasculotropin/vascular endothelial cell growth factor with human umbilical vein endothelial cells: Binding, internalization, degradation, and biological effects. JCP 149: 50-9 (1991); **Breier G et al** Expression of vascular endothelial growth factor during embryonic angiogenesis and endothelial cell differentiation. Development 114: 521-32 (1992); **Brock TA et al** Tumor-secreted vascular permeability factor increases cytosolic Ca^{2+} and von Willebrand factor release in human endothelial cells. Am. J. Pathol. 138: 213-21 (1991); **Brown LF et al** Expression of vascular permeability factor (vascular endothelial growth factor) by epidermal keratinocytes during wound healing. JEM 176: 1375-9 (1992); **Charnock-Jones DS et al** Identification and localization of alternately spliced mRNAs for vascular endothelial growth factor in human uterus and estrogen regulation in endometrial carcinoma cell lines. Biol. Reprod. 48: 1120-8 (1993); **Claffey KP et al** Vascular endothelial growth factor. Regulation by cell differentiation and activated second messenger pathways. JBC 267: 16317-22 (1992); **Clauss M et al** Vascular permeability factor: a tumor-derived polypeptide that induces endothelial cell and monocyte procoagulant activity, and promotes monocyte migration. JEM 172: 1535-45 (1990); **Dvorak HF et al** Distribution of vascular permeability factor (vascular endothelial growth factor) in tumors: concentration in tumor blood vessels. JEM 174: 1275-8 (1991); **Ferrara N et al** Aortic smooth muscle cells express and secrete vascular endothelial growth factor. GF 5: 141-8 (1991); **Ferrara N et al** Expression of vascular endothelial growth factor does not promote transformation but confers a growth advantage *in vivo* to Chinese hamster ovary cells. JCI 91: 160-170 (1993); **Goldman CK et al** Epidermal growth factor stimulates vascular endothelial growth factor production by human malignant glioma cells: a model of glioblastoma multiforme pathophysiology. Mol. Biol. Cell. 4: 121-33 (1993; **Jakeman LB et al** Developmental expression of binding sites and messenger ribonucleic acid for vascular endothelial growth factor suggests a role for this protein in vasculogenesis and angiogenesis. Endocrinology 133: 848-59 (1993); **Leung DW et al** Vascular endothelial growth factor is a secreted angiogenic mitogen. S 246: 1306-9 (1989); **Pepper MS et al** Vascular endothelial growth factor (VEGF) induces plasminogen activators and plasminogen activator inhibitor-1 in microvascular endothelial cells. BBRC181: 902-6 (1991); **Pepper MS et al** Potent synergism between vascular endothelial growth factor and basic fibroblast growth factor in the induction of angiogenesis *in vitro*. BBRC 189; 824-31 (1992); **Peretz D et al** Glycosylation of vascular endothelial growth factor is not required for its mitogenic activity. BBRC 182: 1340-7 (1992); **Phillips H et al** Vascular endothelial growth factor is expressed in rat corpus luteum. E 127: 965-7 (1990); **Ravindranath N et al** Vascular endothelial growth factor messenger ribonucleic acid expression in the primate ovary. E 131: 254-60 (1992); **Soker S et al** Vascular endothelial growth factor is inactivated by binding to α 2-macroglobulin and the binding is inhibited by heparin. JBC 268: 7685-91 (1993); **Shweiki D et al** Patterns of expression of vascular endothelial growth factor (VEGF) and VEGF receptors in mice suggest a role in hormonally regulated angiogenesis. JCI 91: 2235-43 (1993); **Unemori EN et al** Vascular endothelial growth factor induces interstitial collagenase expression in human endothelial cells. JCP 153: 557-62 (1992)

● **Assays: Yeo KT et al** Development of time-resolved immunofluorometric assay of vascular permeability factor. Clin. Chem. 38: 71-5 (1992); **Yeo KT et al** Vascular permeability factor (vascular endothelial growth factor) in guinea pig and human tumor and inflammatory effusions. CR 53: 2912-8 (1993)

● **Clinical Use & Significance: Adamis AP et al** Synthesis and secretion of vascular permeability factor/vascular endothelial growth factor by human retinal pigment epithelial cells. BBRC 193: 631-8 (1993); **Berkman RA et al** Expression of the vascular permeability factor/vascular endothelial growth factor gene in central nervous system neoplasms. JCI 91: 153-9 (1993); **Goldman CK et al** Epidermal growth factor stimulates vascular endothelial growth factor production by human malignant glioma cells: a model of glioblastoma multiforme pathophysiology. Mol. Biol. Cell. 4: 121-33 (1993); **Kim KJ et al** Inhibition of vascular endothelial growth factor-induced angiogenesis suppresses tumor growth *in vivo*. N 362: 841-4 (1993); **Kondo S et al** Significance of vascular endothelial growth factor/vascular permeability factor for solid tumor growth, and its inhibition by the antibody. BBRC 194: 1234-41 (1993); **Morii K et al** Expression of vascular endothelial growth factor in capillary hemangioblastoma. BBRC 194: 749-55 (1993); **Plate KH et al** Vascular endothelial growth factor is a potential tumor angiogenesis factor in human gliomas *in vivo*. N 359: 845-8 (1992); **Shweiki D et al** Vascular endothelial growth factor induced by hypoxia may mediate hypoxia-initiated angiogenesis. N 359: 843-5 (1992)

VEG/PF: *vascular endothelial growth factor/vascular permeability factor* see: VEGF.

***Vg*-1:** *vegetalising factor* This gene has been identified in the genome of » *Xenopus laevis*. In *Xenopus* oocytes it is involved in the induction of mesodermal development. The factor is secreted by *Xenopus* cells cultured *in vitro*. *Vg*-1 also functions as an erythroid differentiation factor and therefore shows the same activities as » Activin A that belongs to the same protein family. The effects of *Vg*-1 are synergised by » EGF and » TGF-β.
Vg-1 is a *Xenopus* homologue of human » TGF-β. *Vg*-1 protein displays marked homology with a protein identified in *Drosophila*, called » *dpp* (*decapentaplegic*), the osteogenic protein » OP-1, and murine » GDF-1. ***Vgr-1*** and ***Vgr-2*** (*Vg*-related) are two related proteins with homology to *Vg*-1. *Vgr*-1 (= DVR-6 = BMP-6) appears to be constitutively expressed in the central nervous system. *Vgr*-2 is expressed at highest levels during midgestation mouse development, and transcripts are found in the osteogenic zone of developing bone. *Vgr*-2 is expressed in F9 teratocarcinoma cells, and its RNA levels are down-regulated within 24 h after differentiation with retinoic acid. Vgr-2 is identical with GDF-3 (see: GDF-1, growth differentiation factor).

Asashima M et al The vegetalising factor from chicken embryos: Its EDF (activin A)-like activity. Mech. Dev. 34: 135-41 (1991); **Dickinson ME et al** Chromosomal localization of seven members of the murine TGF-β superfamily suggests close linkage to several morphogenetic mutant loci. Genomics 6: 505-20 (1990); **Jones MC et al** Involvement of bone morphogenetic protein-4 (BMP-4) and *Vgr*-1 in morphogenesis and neurogenesis in the mouse. Development 111: 531-42 (1991); **Jones CM et al** Isolation of *Vgr*-2, a novel member of the transforming growth factor-β-related gene family. Mol. Endocrinol. 6: 1961-8 (1992); **Lyons K** *Vgr*-1, a mammalian gene related to *Xenopus Vg*-1, is a member of the transforming growth factor β gene superfamily. PNAS 86: 4554-8 (1989); **Sauermann U et al** Cloning of a novel TGF-β related cytokine, the *vgr*, from rat brain: cloning of and comparison to homologous human cytokines. J. Neurosci. Res. 33: 142-7 (1992); **Tiedemann H et al** The vegetalising factor. A member of the evolutionary highly conserved activin family. FL 300: 123-6 (1992); **Wall NA et al** Biosynthesis and *in vivo* localization of the *decapentaplegic-Vg*-related protein, DVR-6 (bone morphogenetic protein-6). JCB 120: 493-502 (1993); **Weeks DL & Melton DA** A maternal mRNA localized to the vegetal hemisphere in *Xenopus* eggs codes for a growth factor related to TGF-β. Cell 51: 861-7 (1987)

VGF: *vascular endothelial growth factor(s)* A generic name for growth factors supporting the growth and proliferation of vascular endothelial cells. See also: ECGF (endothelial cell growth factor) and VEGF (vascular endothelial cell growth factor).

Chen SC & Chen CH Vascular endothelial cell effectors in fetal calf retina, vitreous, and serum. Invest. Ophthalmol. Vis. Sci. 23: 340-50 (1982)

VGF: *vaccinia growth factor* VGF has a length of 77 amino acids. It is a kDa glycoprotein with a high content of mannose residues. The protein is encoded by the VV gene of Vaccinia virus and is expressed and secreted in virus-infected cells early during the infectious cycle as a 19 kDa protein.
The gene is localized in the ITR region (inverted terminal repetition) of the viral genome and hence occurs in two copies.
The factor is identical with » VVGF (vaccinia virus growth factor). It also binds to the receptor for » EGF and functions as an EGF receptor antagonist. The protein is not related immunologically to EGF but it belongs to the family of EGF-like proteins.
VGF-negative mutants of vaccinia virus show a similar behavior *in vitro* with respect to plaque forming efficiency and virus yields. These mutants are less neurovirulent *in vivo*. Viral mutants not expressing VGF are replication-deficient *in vivo*. In human » A431 cells which overexpress the EGF receptor infectious viral foci can be prevented by treatment of the cells with monoclonal antibodies directed against the EGF receptor. See also: Viroceptor.

Brown JP et al Vaccinia virus encodes a polypeptide homologous to epidermal growth factor and transforming growth factor-α. N 313: 491-2 (1985); **Buller RM et al** Cell proliferative response to vaccinia virus is mediated by VGF. Virology 164: 182-92 (1988); **Buller RM et al** Deletion of the vaccinia virus growth factor gene reduced virus virulence. J. Virol. 62: 866-74 (1988); **Chang W et al** Characterization of vaccinia virus growth factor biosynthetic pathway with an antipeptide antiserum. J. Virol. 62: 1080-3 (1988); **Lai AC et al** Attenuated deletion mutants of vaccinia virus lacking vaccinia growth factor are defective in replication *in vivo*. Microb. Pathog. 6: 219-26 (1989); **Lin YZ et al** Growth inhibition by vaccinia virus growth factor. JBC 265: 18884-90 (1990); **Twardzik DR et al** Vaccinia virus-infected cells release a novel polypeptide functionally related to transforming and epidermal growth factors. PNAS 82: 5300-4 (1985); **Twardzik DR et al** Vaccinia growth factor: newest member of the family of growth modulators which utilize the membrane receptor for EGF. Acta Neurochir. Suppl. Wien. 41: 104-9 (1987)

Vgr-1: *Vg-1-related* see: *Vg*-1 (vegetalising factor).

Vgr-2: see: *Vg*-1 (vegetalising factor).

viable motheaten: see: motheaten.

VIC: *vasoactive intestinal contractor* This protein causes a strong contraction of the ileum. Its gene is exclusively expressed in intestinal tissues. VIC is identical with Endothelin β (see: ET).

Bloch KD et al cDNA cloning and chromosomal assignment of the endothelin 2 gene: vasoactive intestinal contractor peptide is rat endothelin 2. Genomics 10: 236-42 (1991); **Saida K et al** A novel peptide, vasoactive intestinal contractor, of a new (endothelin) peptide family. JBC 264: 14613-6 (1989); **Saida K & Mitsui Y** Structure of the precursor for vasoactive intestinal contractor (VIC): its comparison with those of endothelin-1 and endothelin-3. J. Cardiovasc. Pharmacol. 17: S55-8 (1991)

vIL10: *viral IL10* see: BCRF-1.

VIP: *vasoactive intestinal peptide* VIP is a peptide of 28 amino acids which belongs to the glucagon/secretin family of neuropeptides. It is derived from a precursor protein of 170 amino acids by proteolytic cleavage. It is found in large amounts in the central and peripheral nervous system. VIP is synthesized and secreted by the hypothalamus and anterior pituitary and participates in the regulation of pituitary functions, stimulating, among other things, the release of » prolactin, » growth hormone, ACTH (see also: POMC, proopiomelanocortin), and » vasopressin and inhibiting the release of » Somatostatin by the hypothalamus. VIP is also found in the gastrointestinal mucosa, salivary glands, pancreas and the urogenital tract. The release of VIP is promoted by histamine liberators.

VIP influences the mineralisation of bones and stimulates the growth of human keratinocytes *in vitro*. This effect is enhanced by » EGF. In sensory neurons the synthesis of VIP is induced by » NGF. Glucocorticoids increase the VIP receptor density in mononuclear leukocytes. In general, the expression of the VIP gene is regulated by increased levels of cAMP and by activation of protein kinase C.

Apart from the classical role of neurohumoral hormone VIP also acts as a growth factor and therefore shows cytokine-like activities although it is not classified as a cytokine due to its small size. In the presence of insulin and inhibitors of cAMP-phosphodiesterase VIP stimulates the proliferation of murine 3T3 fibroblasts.

In vitro VIP reduces the mitogen-induced proliferation of leukocytes and enhances the synthesis of immunoglobulins, in particular IgA. VIP also has inhibitory effects on the proliferation of normal human thymocytes. *In vivo* VIP influences the activation of lymphocytes and the synthesis of immunoglobulins, influences the natural killer activity of cells, reduces the migration of lymphocytes into lymph nodes, and inhibits the production of reactive oxygen radicals in activated monocytes. VIP inhibits the tumoricidal activity of human NK cells at very low concentrations. It potentiates the suppressive effects of noradrenaline on macrophage-mediated tumoricidal activity although it does not affect tumor cell killing on its own.

VIP inhibits IgE production without affecting IgM or IgA production by mononuclear cells and also differentially modulates IgG subclass production. It also stimulates immunoglobulin production and growth of human B cells. VIP stimulates T lymphocytes to release IL5 in murine *Schistosomiasis mansoni* infections. VIP has been shown to inhibit the generation of » IL2 and » IFN-γ from mitogen-stimulated human and murine lymphocytes.

In some adenocarcinoma cells VIP induces the synthesis of interferons (see: IFN) and thus increases their resistance to viral infections. The expression of interferons » IFN-α and » IFN-β is also induced by VIP in glial cells, but not in neurons. The growth of non-small cell lung cancers transplanted into nude mice (see also: Immunodeficient mice), including adenocarcinomas, large cell carcinomas, and squamous cell carcinomas, can be inhibited by treatment with a VIP derivative that acts as a receptor antagonist. VIP increases the secretion rates of » bombesin-like peptides in small cell lung cancers.

VIP is an » autocrine growth factor for some human neuroblastoma cells lines. Also, VIP is a secretagogue for neuron survival-promoting activities and functions as an astroglial mitogen, thus maintaining survival of spinal cord neurons in primary culture during a critical period of development. VIP stimulates the synthesis of » IL6 in the anterior lobe of the pituitary. The synthesis of VIP in chromaffin cells is modulated by » IL1, » TNF-α, » LIF (see: CNDF, cholinergic neuronal differentiation factor) and by » Oncostatin M or » CNTF (ciliary neurotrophic factor) in human neuroblastoma cells.

Azzari C et al VIP restores natural killer cell activity depressed by hepatitis B surface antigen. Viral. Immunol. 5: 195-200 (1992); **Bondesson L et al** Dual effects of vasoactive intestinal peptide (VIP) on leukocyte migration. Acta Physiol. Scand. 141: 477-81 (1991); **Boudard F & Bastide M** Inhibition of mouse T cell proliferation by CGRP and VIP: effects of these neuropeptides on IL2 production and cAMP synthesis. J. Neurosci. Res. 29: 29-41 (1991); **Brenneman DE et al** Vasoactive intestinal peptide: a neurotrophic releasing agent and an astroglial mitogen. J. Neurosci. Res. 25: 386-94 (1990); **Chelbi-Alix MK et al** VIP induces in HT-29 cells 2'5'-oligoadenylate synthetase and antiviral state via interferon β/α synthesis. Peptides 12: 1085-93 (1991); **Chelbi-Alix MK et al** Induction by vasoactive intestinal peptide of interferon α/β synthesis in glial cells but not in neurons. JCP 158: 47-54 (1994); **Cutz E et al** Release of vasoactive intestinal peptide in mast cells by histamine liberators. N 275: 661-2 (1978); **Eskay RL et al** Interleukin-1α and tumor necrosis factor-α differentially regulate enkephalin, vasoactive intestinal polypeptide, neurotensin, and substance P biosynthesis in chromaffin cells. E 130: 2252-8 (1992); **Fahrenkrug J & Emson PC** Vasoactive intestinal peptide: functional aspects. Br. Med. Bull. 38: 265-70 (1982); **Fink JS et al** Cyclic AMP- and phorbol ester-induced transcriptional activation are mediated by the same enhancer element in the human vasoactive intestinal peptide gene. JBC 266: 3883-7 (1991); **Frawley LS et al** Stimulation of prolactin secretion in rhesus monkeys by vasoactive intestinal polypeptide. Neuroendocrinol. 33: 79-83 (1981); **Goetzl EJ et al** Generation and recognition of vasoactive intestinal peptide by cells of the immune system. ANY 594: 34-44 (1990); **Hohmann EL et al** Innervation of periosteum and bone by sympathetic vasoactive intestinal peptide-containing nerve fibres. S 232: 867-8 (1986); **Haegerstrand A et al** Vasoactive intestinal polypeptide stimulates cell proliferation and adenylate cyclase activity in cultured human keratinocytes. PNAS 86: 5993-6 (1989); **Ishioka C et al** Vasoactive intestinal peptide stimulates immunoglobulin production and growth of human B cells. Clin. Exp. Immunol. 87: 504-8 (1992); **Kimata H et al** Differential effect of vasoactive intestinal peptide, somatostatin, and substance P on human IgE and IgG subclass production. CI 144: 429-42 (1992); **Koff WC et al** Modulation of macrophage-mediated tumoricidal activity by neuropeptides and neurohormones. JI 135: 350-4 (1985); **Korman LY et al** Secretin/vasoactive intestinal peptide-stimulated secretion of bombesin/gastrin releasing peptide from human small cell carcinoma of the lung. CR 46: 1214-8 (1986); **Lamperti ED et al** Characterization of the gene and messages for vasoactive intestinal polypeptide (VIP) in rat and mouse. Brain Res. Mol. Brain. Res. 9: 217-31 (1991); **Lygren I et al** Vasoactive intestinal peptide and somatostatin in leukocytes. Scand. J. Clin. Lab. Invest. 44: 347-51 (1984); **Mathew RC et al** Vasoactive intestinal peptide stimulates T lymphocytes to release IL5 in murine schistosomiasis mansoni infection. JI 148: 3572-7 (1992); **Moody TW et al** A vasoactive intestinal peptide antagonist inhibits non-small cell lung cancer. PNAS 90: 4345-9 (1993); **Moore TC** Modification of lymphocyte traffic by vasoactive neurotransmitter substances. Immunology 52: 511-8 (1984); **O'Dorisio MS et al** Vasoactive intestinal peptide as a biochemical marker for polymorphonuclear leukocytes. J. Lab. Clin. Med. 96: 666-70 (1980); **O'Dorisio MS et al** Vasoactive intestinal peptide: autocrine growth factor in neuroblastoma. Regulatory Peptides 37: 213-26 (1992); **Ottaway CA** Vasoactive intestinal peptide as a modulator of lymphocytes and immune function. ANY 527: 486-500 (1988); **Pence JC & Shorter NA** Autoregulation of neuroblastoma growth by vasoactive intestinal peptide. J. Pediatr. Surg. 27: 935-43 (1992); **Rao MS et al** Oncostatin M regulates VIP expression in a human neuroblastoma cell line. Neuroreport 3: 865-8 (1992); **Rola-Pleszczynski M et al** The effects of vasoactive intestinal peptide on human natural killer cell function. JI 135: 2569-73 (1985); **Scholar EM & Paul S** Stimulation of tumor cell growth by vasoactive intestinal peptide. Cancer 67: 1561-4 (1991); **Sirianni MC et al** Modulation of human natural killer activity by vasoactive intestinal peptide (VIP) family. VIP, glucagon and GHRF specifically inhibit NK activity. Regul. Pept. 38: 79-87 (1992); **Soder O & Hellström PM** Neuropeptide regulation of human thymocyte, guinea pig T lymphocyte and rat B lymphocyte mitogenesis. Int. Arch. Allergy Appl. Immunol. 84: 205-11 (1987); **Spangelo BL et al** Production of interleukin 6 by anterior pituitary cells is stimulated by increases intracellular adenosine 3′,5′ monophosphate and vasoactive intestinal peptide. E 127: 403-9 (1990); **Stanisz AM et al** Differential effects of vasoactive intestinal peptide, substance P, and somatostatin on immunoglobulin synthesis and proliferation by lymphocytes from Peyer's patches, mesenteric lymph nodes, and spleen. JI 136: 152-6 (1986); **Tsukada T et al** Identification of a region in the human vasoactive intestinal polypeptide gene responsible for regulation by cyclic AMP. JBC 262: 8743-7 (1987); **van Tol EA et al** Modulatory effects of VIP and related peptides from the gastrointestinal tract on cell mediated cytotoxicity against tumor cells *in vitro*. Immunol. Invest. 20: 257-67 (1991); **Wollina U et al** Vasoactive intestinal peptide (VIP) acting as a growth factor for human keratinocytes. Neuroendocrinol. Lett. 14: 21-31 (1992); **Zurier RB et al** Vasoactive intestinal peptide synergistically stimulates DNA synthesis in mouse 3T3 cells: role of cAMP, Ca^{2+}, and protein kinase C. ECR 176: 155-61 (1988)

● TRANSGENIC/KNOCK-OUT/ANTISENSE STUDIES: **Agoston DV et al** Expression of a chimeric VIP gene is targeted to the intestine in transgenic mice. J. Neurosci. Res. 27: 479-86 (1990)

viral IL10: abbrev. vIL10. See: BCRF-1.

Viroceptor: A term proposed to describe virus-encoded homologues of cellular cytokine receptors the function of which is to intercept the activity of the cognate cytokine in order to short circuit the host immune response to the viral infection. Viroceptors in this sense are » T2 and » M-T7, encoded by myxoma viruses, cytomegalovirus » US28, » ECRF3 encoded by Herpesvirus saimiri, and » B15R encoded by vaccinia virus.

For other viral strategies allowing subversion of host immune responses see also: » A20, » BCRF-1 (viral IL10) of Epstein-Barr virus, VGF (vaccinia growth factor), SFGF (Shope fibroma growth factor), MGF (myxoma growth factor), and IL1β Convertase.

Gooding LR Virus proteins that counteract host immune defenses. Cell 71: 5-7 (1992); **Mallet S & Barcley AN** A new superfamily of cell surface proteins related to the nerve growth factor receptor IT 7: 220-223 (1991)

Vitronectin: abbrev. VN. This protein is also called ***Epibolin*** and ***serum spreading factor*** (abbrev. SF). It is a 75 kDa glycoprotein circulating in the serum. A 65 kDa form of this protein which is linked to a 10 kDa fragment by disulfide bridges has

also been described. Either form and also equal amounts of both forms may be prevalent in an individual. The human gene has a length of 5.3 kb and contains 8 exons. It maps to chromosome 17q11.
Vitronectin is a multifunctional protein. It is identical with the ***membrane attack complex-Inhibitor*** of the complement system, ***S protein*** (=***protein X, complement S protein***) and ***PAI-BP*** (plasminogen activator inhibitor 1 (PAI) binding protein), a protein circulating in human serum that prevents the interaction of PAI with the extracellular matrix. *In vitro* vitronectin promotes the adherence, spreading, and migration of many different cell types including tumor cells in serum-free medium (see also: SFM). Under some conditions vitronectin also promotes the proliferation and differentiation of some cell types. The protein also exists in a form associated with the cellular membrane. It binds collagen and forms complexes with thrombin/antithrombin III.
Vitronectin functions as a heparin-neutralizing factor and protects thrombin and factor Xa from inactivation by antithrombin II. Vitronectin also binds PAI-2 (plasminogen activator inhibitor 2), forming covalent disulfide-bonded complexes. Vitronectin also binds plasminogen and inhibits the fibrinogen-induced activation of plasminogen by tissue plasminogen activator. Both forms of vitronectin also specifically bind human β-endorphin.
The aminoterminal end of vitronectin contains the complete sequence of somatomedin B (44 amino acids) and is probably the precursor of this protein. Vitronectin contains heparin-binding domains that bind complement components C7, C8, and C9. They also bind » perforins produced by cytolytic T cells, thus effectively inhibiting their lytic activity. The vitronectin receptor consists of a 150 kDa (α) and a 115 kDa (β) subunit. The α subunit consists of a 125 kDa and a 25 kDa protein. The 25 kDa protein is obtained by proteolytic cleavage of the α subunit precursor. The human gene for this precursor maps to chromosome 2q31-q32. On murine T lymphocytes the α subunit of the vitronectin receptor is expressed as a T cell activation antigen (see also: Cell activation).
The receptor belongs to the integrins, a group of related membrane receptors that recognize an arg-gly-asp (RGD) sequence in their ligand protein. The α-subunit shows considerable sequence homology with the α subunit of the fibronectin receptor and the platelet membrane glycoprotein IIb. Its expression is stimulated by » TGF-β and inhibited by » TNF-α and » IFN-γ. Vitronectin also competes with fibronectin for binding to the platelet glycoprotein IIb/IIIa.
Vitronectin plays an important role in platelet aggregation and is probably also involved in the interaction of platelets with the vascular cell walls. Vitronectin thus is an important mediator of physiological and pathophysiological processes of hemostasis and thrombosis.

● REVIEWS: **Preissner KT** The role of vitronectin as multifunctional regulator in the hemostatic and immune system. Blut 59: 419-31 (1989); **Preissner KT** Structure of vitronectin and its biological role in hemostasis. Thromb. Haemost. 66: 123-32 (1991); **Tomasini BR & Mosher DF** Vitronectin. Prog. Hemost. Thromb. 10: 269-305 (1991)

● BIOCHEMISTRY & MOLECULAR BIOLOGY: **Barnes DW et al** Characterization of human serum spreading factor with monoclonal antibody. PNAS 80: 1362-6 (1983); **Barnes DW et al** Isolation of human serum spreading factor. JBC 258: 12548-52 (1983); **Barnes DW et al** Human serum spreading factor; relationship to somatomedin B. J. Clin. Endocrinol. Metab. 59: 1019-21 (1984); **Dahlback B & Podack ER** Characterization of human S protein, an inhibitor of the membrane attack complex of complement. Demonstration of a free reactive thiol group. B 24: 2368-74 (1985); **Declerck PJ et al** Purification and characterization of a plasminogen activator inhibitor 1 binding protein from human plasma. Identification as a multimeric form of S protein (vitronectin). JBC 263: 15454-61 (1988); **Fink TM et al** The human vitronectin (complement S-protein) gene maps to the centromeric region of 17q. Hum. Genet. 88: 569-72 (1992); **Jenne D & Stanley KK** Molecular cloning of S-protein, a link between complement, coagulation, and cell-substrate adhesion. EJ 4: 3153-7 (1985); **Jenne D & Stanley KK** Nucleotide sequence and organization of the human S-protein gene: repeating peptide motifs in the "pexin" family and a model for their evolution. B. 26: 6735-42 (1987); **Kubota K et al** Polymorphism of the human vitronectin gene causes vitronectin blood type. BBRC 167: 1355-60 (1990); **Poncz M et al** Structure of the platelet membrane glycoprotein IIb. Homology to the α subunits of the vitronectin and fibronectin membrane receptors. JBC 262: 8476-82 (1987); **Preissner KT et al** Physicochemical, immunochemical, and functional comparison of human S-protein and vitronectin. Evidence for the identity of both plasma proteins. BBRC 134: 951-6 (1986); **Radtke KP et al** Isolation of plasminogen activator inhibitor-2 (PAI-2) from human placenta. Evidence for vitronectin/PAI-2 complexes in human placenta extract. Biol. Chem. Hoppe Seyler 371: 1119-27 (1990); **Ruoslahti E et al** Purification and characterization of vitronectin. MiE 144: 430-7 (1987); **Stenn KS et al** Epibolin: a protein of human plasma that supports epithelial cell movement. PNAS 78: 6907-11 (1981); **Sosnoski DM et al** Chromosomal localization of the genes for the vitronectin and fibronectin receptors α subunits and for platelet glycoproteins IIb and IIIa. JCI 81: 1993-8 (1988); **Suzuki S et al** Complete amino acid sequence of human vitronectin deduced from cDNA. Similarity of cell attachment sites *in vitro*nectin and fibronectin. EJ 4: 2519-24 (1985); **Suzuki S et al** cDNA and amino acid sequences of the cell adhesion protein receptor recognizing vitronectin reveal a transmembrane domain and homologies with other adhesion protein receptors. PNAS 83: 8614-8 (1986); **Suzuki S et al** Amino acid sequence of the vitronectin receptor α subunit and comparative expression of adhesion receptor mRNAs. JBC 262: 14080-5 (1987); **Tollefsen DM et al** The presence of methionine or threonine at position 381 *in vitro*nectin is correlated with proteolytic cleavage at arginine 379. JBC 265: 9778-81 (1990); **Tomasini BR & Mosher DF** On the identity of vitronectin and S-protein: immunological crossreactivity and functional studies. Blood 68: 737-42 (1986); **Tschopp J et al** The heparin binding domain of S-protein/vitronectin binds to complement components C7, C8, and C9 and perforin from cytolytic T cells and inhibits their lytic activities. B 27: 4103-9 (1988)

● **Receptors: Fernandez-Ruiz E et al** Regional localization of the human vitronectin receptor α-subunit gene (VNRA) to chromosome 2q31-. q32. Cytogenet. Cell Genet. 62: 26-28 (1993); **Fitzgerald LA et al** Comparison of cDNA-derived protein sequences of the human fibronectin and vitronectin receptor α subunits and platelet glycoprotein IIb. B 26: 8158-8165 (1987); **Moulder K et al** The mouse vitronectin receptor is a T cell activation antigen. JEM 173: 343-7 (1991); **Suzuki S et al** cDNA and amino acid sequences of the cell adhesion protein receptor recognizing vitronectin reveal a transmembrane domain and homologies with other adhesion protein receptors. PNAS 83: 8614-8618 (1986)

● **Biological Activities: Asch E & Podack E** Vitronectin binds to activated human platelets and plays a role in platelet aggregation. JCI 85: 1372-8 (1990); **Delfilippi P et al** Tumor necrosis factor α and interferon γ modulate the expression of the vitronectin receptor (integrin β 3) in human endothelial cells. JBC 266: 7638-45 (1991); **Ehrlich HJ et al** Alteration of serpin specificity by a protein cofactor. Vitronectin endows plasminogen activator inhibitor 1 with thrombin inhibitory properties. JBC 265: 13029-35 (1990); **Hayman EG et al** Serum spreading factor (vitronectin) is present at the cell surface and in tissues. PNAS 80: 4003-7 (1983); **Hildebrand A et al** A novel β-endorphin binding protein. JBC 264: 15429-34 (1989); **Ignotz RA et al** Regulation of cell adhesion receptors by transforming growth factor-β. Regulation of vitronectin receptor and LFA-1. JBC 264: 389-92 (1989); **Mohri H & Ohkubo T** How vitronectin binds to activated glycoprotein IIb-IIIa complex and its function in platelet aggregation. Am. J. Clin. Pathol. 96: 605-9 (1991); **Podack ER et al** Interaction of S-protein of complement with thrombin and antithrombin III during coagulation. Protection of thrombin by S-protein from antithrombin III inactivation. JBC 261: 7387-92 (1986); **Preissner KT** Specific binding of plasminogen to vitronectin. Evidence for a modulatory role of vitronectin on fibrin(ogen)-induced plasmin formation by tissue plasminogen activator. BBRC 168: 966-71 (1990); **Thiagarajan P & Kelly K** Interaction of thrombin-stimulated platelets with vitronectin (S-protein of complement) substrate: inhibition by a monoclonal antibody to glycoprotein IIb-IIIa complex. Thromb. Haemost. 60: 514-7 (1988); **Wiman B et al** Plasminogen activator inhibitor 1 (PAI) is bound to vitronectin in plasma. FL 242: 125-8 (1988)

● **Assays: Korc-Grodzicki B et al** An enzymatic assay for vitronectin based on its selective phosphorylation by protein kinase A. AB 188: 288-94 (1990)

VN: abbrev. for » Vitronectin.

VP: abbrev. for » Vasopressin.

VPF: *vascular permeability factor* This factor is also called ***tumor vascular permeability factor***. It is a protein of 189 amino acids that is specifically produced by tumor cells (see also: U937). It is identical with one of the splice variants of the human » VEGF (vascular endothelial cell growth factor). A protein related to VPF is » PlGF (placenta growth factor).

The intradermal injection of VPF leads to an increased leakage of fluids at the site of injection and to the leakage of proteins from the capillaries. VPF also increases glucose transport and regulates the expression of tissue thromboplastin and of the glucose transporter. In terms of molar concentrations VPF is ≈ 50.000 times more active than histamine.

VPF is a strong mitogen for endothelial cells (see: VEGF, vascular endothelial growth factor). It is probably also involved in tumor angiogenesis (see also: Angiogenesis factors) and » wound healing. VPF is chemotactic for monocytes.

Clauss M et al Vascular permeability factor: a tumor-derived polypeptide that induces endothelial cell and monocyte procoagulant activity, and promotes monocyte migration. JEM 172: 1535-45 (1990); **Connolly DT et al** Tumor vascular permeability factor stimulates endothelial cell growth and angiogenesis. JCI 84: 1470-8 (1989); **Connolly DT** Vascular permeability factor: a unique regulator of blood vessel function. JCBc 47: 219-23 (1991); **Keck PJ et al** Vascular permeability factor, an endothelial cell mitogen related to PDGF. S 246: 1309-12 (1989); **Olander JV et al** Specific binding of vascular permeability factor to endothelial cells. BBRC 175: 68-76 (1991); **Senger DR et al** Tumor cells secrete a vascular permeability factor that promotes accumulation of ascites fluid. S 219: 983-5 (1983); **Senger DR et al** A highly conserved vascular permeability factor secreted by a variety of human and rodent cell lines. CR 46: 5629-32 (1986); **Yeo KT et al** Development of time-resolved immunofluorometric assay of vascular permeability factor. Clin. Chem. 38: 71-5 (1992)

VVGF: *vaccinia virus growth factor* VVGF, also known as Vaccinia 19 kDa protein (see: VGF, vaccinia growth factor), is detected in the medium of vaccinia-virus-infected cells. The factor is a processed form of a polypeptide encoded in the vaccinia virus genome which is related to » EGF and » TGF-α. VVGF, unlike EGF or TGF, is glycosylated. VVGF binds to the EGF receptor and stimulates its » autophosphorylation. and may, therefore allow the virus to subvert EGF receptor-dependent functions. VVGF can functionally replace » MGF (myxoma growth factor).

Reisner AH Similarity between the vaccinia virus 19K early protein and epidermal growth factor. N 313: 801-3 (1985); **Stroobant P et al** Purification and characterization of vaccinia virus growth factor. Cell 42: 383-93 (1985)

W: The murine gene locus *W* (white color) is identical with the c-*kit* proto-oncogene (see: *kit*) which encodes the receptor of » SCF (stem cell factor).

***waved*-1:** see: TGF-α, subentry Gene structure.

WBS: abbrev. for whole blood serum. See also: PDS (plasma-derived serum).

WEHI-231 BCGF 2: see: BCGF 2 (B cell growth factor 2).

WEHI-3B: A macrophage-like myelomonocytic leukemia cell line established from inbred BALB/c mice. This cell line is used as a standard cell line because it does not require exogenous growth factors.

Due to the integration of a retroviral genome in the vicinity of the IL3 gene this cell line constitutively produces » IL3 (see: IAP, intracisternal A particle). Constitutive production of IL3 by WEHI-3 cells is enhanced by sodium butyrate and is accompanied by the growth arrest of WEHI-3 cells in the G1 phase of the » cell cycle. Conditioned medium (see also: CM) from this cell line is frequently used as an undefined source of IL3 in a variety of assays. The cells also secrete » Activin A (see: WEHI-MIF), an unknown cytostatic factor (see: CF) and a poorly characterized ***differentiation autoinducing activity*** of ≈ 10-20 kDa that is different from » GM-CSF and » IL3.

Several factors, including » IL6 and » G-CSF lead to the differentiation of WEHI cells and this cell line is therefore also used to assay these factors.

A WEHI subclone, designated ***WEHI-164***, is used to assay » TNF. This assay is ≈ 5-10-fold more sensitive than another assay employing » L929 cells (see also: Bioassays). The assay is carried out in the presence of the RNA polymerase inhibitor actinomycin D, which increases the sensitivity of the cell line to TNF. An alternative and entirely different detection method for these factors is » Message amplification phenotyping.

Eskandari MK et al WEHI 164 subclone 13 assay for TNF: sensitivity, specificity, and reliability. Immunol. Invest. 19: 69-79 (1990); **Espevik T & Nissen-Meyer J** A highly sensitive cell line, WEHI 164 clone 13, for measuring cytotoxic factor/tumor necrosis factor from human monocytes. JIM 95: 99-105 (1986); **Ihle JN et al** Biological properties of homogenous interleukin-3. I. Demonstration of WEHI-3 growth factor activity, mast cell growth factor activity, P cell-stimulating factor activity, colony-stimulating factor activity, and histamine-producing cell-stimulating factor activity. JI 131: 282-7 (1983); **Kajigaya Y et al** The production of differentiation autoinducing activity by WEHI-3B D+ leukemia cells. EH 17: 368-73 (1989); **Kajigaya Y et al** Growth and differentiation of a murine interleukin-3-producing myelomonocytic leukemia cell line in a protein-free chemically defined medium. Leukemia 4: 712-6 (1990); **Lee JC et al** Constitutive production of a unique lymphokine (IL3) by the WEHI-3 cell line. JI 128: 2393-8 (1982); **Le Gros GS et al** The effects of sodium butyrate on lymphokine production. LR 4: 221-7 (1985); **Meager A et al** Assays for tumor necrosis factor and related cytokines. JIM 116: 1-17 (1989); **Metcalf D** Actions and interactions of G-CSF, LIF, and IL6 on normal and leukemic murine cells. Leukemia 3: 349-55 (1989); **Morgan CD et al** An improved colorimetric assay for tumor necrosis factor using WEHI 164 cells cultured on novel microtiter plates. JIM 145: 259-62 (1992); **Ymer S et al** Constitutive synthesis of interleukin-3 by leukemia cell line WEHI-3B is due to retroviral insertion near the gene. N 317: 255-8 (1985)

WEHI-3B differentiation inducing activity: This factor is found in conditioned culture media of Krebs II ascites cells. Some of the differentiation-inducing activity is due to » G-CSF and » GM-CSF.

Hilton DJ et al Resolution and purification of three distinct factors produced by Krebs ascites cells which have differentiation-inducing activity on murine myeloid leukemic cell lines. JBC 263: 9238-43 (1988)

WEHI 3 growth factor: abbrev. ***WGF***. This factor is secreted constitutively by the myelomonocytic leukemia cell line » WEHI-3B cells. It is identical with » IL3. For other leukemia-associated growth factors see also: LDA, LDGF, LGF, LGPF.

Greenberger JS et al *In vitro* induction of continuous acute promyelocytic leukemia cell lines by Friend and Abelson murine leukemia viruses. Blood 53: 987-1001 (1979); **Ihle JN et al** Biological properties of homogenous interleukin-3. I. Demonstration of WEHI-3 growth factor activity, mast cell growth factor activity, P cell-stimulating factor activity, colony-stimulating factor activity, and histamine-producing cell-stimulating factor activity. JI 131: 282-7 (1983); **Lee JC et al** Constitutive production of a unique lymphokine (IL 3) by the WEHI-3 cell line. JI 128: 2393-8 (1982); **Nicola NA et al** Purification of a factor

***waved*-2:** see addendum.

inducing differentiation in murine myelomonocytic leukemia cells. Identification as granulocyte colony-stimulating factor. JBC 258: 9017-23 (1983)

WEHI 3 hematopoietic growth factor: This factor is secreted constitutively by » WEHI-3B cells. It is identical with » IL3. See also: WEHI 3 growth factor.
Dexter TM et al Growth of factor-dependent hemopoietic precursor cell lines. JEM 152: 1036-47 (1980)

WEHI-164: see: WEHI-3B.

WEHI-MIF: *WEHI mesoderm-inducing factor* This factor, secreted by » WEHI-3B cells, was initially identified as an activity influencing mesodermal development in amphibians. It is identical with » » Activin A (see also: XTC-MIF).
Albano RM et al A mesoderm-inducing factor produced by WEHI-3 murine myelomonocytic leukemia cells is activin A. Development 110: 435-43 (1990)

Wellferon: Trademark (Wellcome) for » IFN-α.

WGF: see: WEHI 3 growth factor.

Whitlock-Witte cultures: see: LTBMC (long-term bone marrow culture).

Whole blood serum: see: WBS.

wingless: see: *Wnt*.

WI-26-VA4 factor: This factor was obtained from the culture supernatants of human fibroblasts. It induces the differentiation of the human monocytic cell line » U937. This factor is identical with » IL6.
Noda M et al Purification and characterization of human fibroblast derived differentiation inducing factor for human monoblastic leukemia cells identical to interleukin 6. ACR 11: 961-8 (1991)

Wilms' tumor susceptibility gene WT1: see: Egr-1.

WISH: A transformed human cell line with epitheloid morphology originally established from amnion epithelium. This cell line is used to assay human » IFN-α and » IFN-γ. The interferons are quantified by performing a cytopathic effect inhibition assay of vesicular stomatitis virus (VSV). The cells are significantly more sensitive for both IFN-α and IFN-γ when EMCV (encephalomyocarditis virus) is used instead of VSV.
The antiviral state of these cells is enhanced in the presence of mixtures of human interferons. IFN-γ inhibits the rise in » transferrin receptor mRNA level which is normally observed when stationary WISH cells are stimulated to proliferate.
Chloroquine, which raises the pH in lysosomes and endosomes, counteracts the induction by interferon of the antiviral state but has no effects on the induction of 2',5'-oligoadenylate synthetase.
Pretreatment of WISH cells with » TNF-α, » IL1, and TNF-β induces 2-5A synthetase activity, protects them from vesicular stomatitis virus cytopathic effect, and markedly reduces virus yields. Inclusion of polyclonal antibodies to IFN-β during cytokine pretreatment abrogates the antiviral state elicited by these cytokines. The antiviral activity of TNF can be attributed to the induction of IFN-β.
Treatment of WISH cells with » EGF, » TGF-α but not TGF–β, » TNF, » IL1 (IL1β), phorbol 12,13-dibutyrate, and phorbol 12-myristate 13-acetate (see also: Phorbol esters) stimulates the production of prostaglandin E2 in a concentration-dependent manner, while » IL6, » IL8, and » GM-CSF are inactive. Dexamethasone treatment results in a concentration-dependent inhibition of prostaglandin E2 production. Preexposure of WISH cells to epinephrine, norepinephrine, or dopamine inhibits EGF-induced prostaglandin E2 production.
Pretreatment of WISH cells with IFN-γ in the presence of 12-O-tetradecanoylphorbol 13-acetate (TPA) results in the down-modulation of » EGF receptors with respect to both receptor number and affinity.
WISH cells produce tumors in nude mice (see also: Immunodeficient mice) and induce large islands of bone with focal areas of cartilage immediately adjacent to the tumors.
Abu-Khabar KS et al Analysis and examination of cytokine interactions by the median-effect model: an example with antiviral action of tumor necrosis factor and interferon-γ. J. Interferon Res. 12: 161-5 (1992); **Bourgeade MF et al** Post-transcriptional regulation of transferrin receptor mRNA by IFN γ. NAR 20: 2997-3003 (1992); **Boyan BD et al** Epithelial cell lines that induce bone formation *in vivo* produce alkaline phosphatase-enriched matrix vesicles in culture. Clin. Orthop. April(277) 266-76 (1992); **Chelbi-Alix MK & Thang MN** Multiple molecular forms of interferon display different specific activities in the induction of the antiviral state and 2'5' oligoadenylate synthetase. BBRC 141: 1042-50 (1986); **Chelbi-Alix MK & Thang MN** Chloroquine impairs the interferon-induced antiviral state without affecting the 2',5'-oligoadenylate synthetase. JBC 260: 7960-4 (1985); **Harris AN et al** Characterization of prostaglandin production in amnion-derived WISH cells. Am. J. Obstet. Gynecol. 159: 1385-9 (1988); **Karasaki Y et al** Phorbol ester and interferon-γ modulation of epidermal growth factor

receptors on human amniotic (WISH) cells. JBC 264: 6158-63 (1989); **Kniss DA et al** Proinflammatory cytokines interact synergistically with epidermal growth factor to stimulate PGE2 production in amnion-derived cells. Prostaglandins 44: 237-44 (1992); **Kramer MJ et al** Cell and virus sensitivity studies with recombinant human α interferons. J. Interferon Res. 3: 425-35 (1983); **Meek WD & Davis WL** Fine structure and immunofluorescent studies of the WISH cell line. In Vitro Cell. Dev. Biol. 22: 716-24 (1986); **Oleszak E & Stewart WE 2nd** Potentiation of the antiviral and anticellular activities of interferons by mixtures of HuIFN-γ and HuIFN-α or HuIFN-β. J. Interferon Res. 5: 361-71 (1985); **Overall JC Jr et al** Activity of human recombinant and lymphoblastoid interferons in human and heterologous cell lines. J. Interferon Res. 4: 529-33 (1984); **Ruggiero V et al** Comparative study on the antiviral activity of tumor necrosis factor (TNF)-α, lymphotoxin/TNF-β, and IL1 in WISH cells. Immunol. Lett. 21: 165-9 (1989); **Ruggiero V et al** The *in vitro* antiviral activity of tumor necrosis factor (TNF) in WISH cells is mediated by IFN-β induction. Antiviral. Res. 11: 77-88 (1989); **Stebbing N & May L** Comparisons of dose-response data for various standard and recombinant DNA-derived human interferons. J. Virol. Methods 5: 309-15 (1982); **Su HC et al** Catecholamines modulate epidermal growth factor-induced prostaglandin E2 production in amnion-like (WISH) cells by means of a cyclic adenosine monophosphate-dependent pathway. Am. J. Obstet. Gynecol. 166: 236-41 (1992); **Tonew M et al** [Sensitivity of different cell lines to interferons: the relative antiviral activity as a function of the interferon subtype] J. Basic Microbiol. 29: 537-45 (1989); **Yousefi S et al** A practical cytopathic effect/dye-uptake interferon assay for routine use in the clinical laboratory. Am. J. Clin. Pathol. 83: 735-40 (1985); **Zoon KC et al** Purification and characterization of multiple components of human lymphoblastoid interferon-α. JBC 267: 15210-6 (1992)

W locus: This murine gene encodes the receptor for » SCF (stem cell factor).

***Wnt*:** *Wnt* genes encode a number of secreted proteins which are associated with the extracellular matrix (see: ECM) of cells. They are involved in the dorso-ventricular development of the mesoderm during embryogenesis in combination with other factors.

Wnt-1 has a highly specific temporal and spatial pattern of expression in fetal brain and spinal cord from 9- to 10-day-old mouse embryos. It is also expressed in the spermatids of the testes. The amounts of secreted proteins are considerably increased by treatment of cells with heparin (see also: HBGF, heparin-binding growth factors) or » suramin.

The *Wnt*-1 gene encodes a glycosylated cysteine-rich protein of 370 amino acids. The non-glycosylated protein has a molecular mass of 36 kDa; glycosylated variants of 38, 40, 42 and 44 kDa have been described. The protein is secreted. An intracellular form of the protein is bound to another protein, called BiP (= GRP78) which also functions to retain other secretory proteins within the endoplasmic reticulum.

Wnt-1 was initially identified as an » oncogene activated in various murine mammary carcinomas by the integration of the viral genome (mouse mammary tumor virus) and found to be ectopically expressed in tissues that normally do not produce this protein (hence its former designation ***int-1***). *Wnt*-1 plays an essential role in fetal brain development. It has » paracrine functions in the transformation of mammary epithelial cells. The human *Wnt*-1 gene maps to chromosome 12q13; the murine homologue to chromosome 15.

Fibroblasts expressing and secreting the *Wnt*-1 gene, that themselves are not transformed, can induce transformation of epithelial cells in cocultures (see also: Paracrine).

A strongly conserved *Drosophila* homologue of vertebrate *Wnt*-1 is ***wingless***, a segment-polarity gene (hence the new name *Wnt*-1 which is a hybrid of the symbols for *wingless* and *int*-1). The protein encoded by wingless regulates the expression of another gene, called ***engrailed***. Both genes control the development of segments in the fly embryo (see: Homeotic genes).

■ **TRANSGENIC/KNOCK-OUT/ANTISENSE STUDIES:** Mice homozygous for a *Wnt*-1 null mutation (created by targeted disruption of the gene in mouse embryo- derived stem cells and subsequent generation of chimaeric mice; see also: transgenic animals) exhibit a range of phenotypes from death before birth to survival with severe ataxia. Severe abnormalities in the development of the mesencephalon and metencephalon during embryogenesis suggest a prominent role for the *Wnt*-1 protein in the induction of the mesencephalon and cerebellum. It has also been possible to create » transgenic animals in which *Wnt*-1 levels were reduced by 98 %, using » antisense oligonucleotide technology. Male mice in which testicular expression of *Wnt*-1 was reduced to these levels were fertile and showed normal testicular histology, suggesting that the *Wnt*-1 gene product may not be essential for spermatogenesis.

The introduction of the *Wnt*-1 gene into mammary epithelial cell lines leads to cell transformation. In » transgenic animals (mice) expressing the *Wnt* gene in mammary tissues a marked hyperplasia of the mammary tissues and a subsequent development of carcinomas is observed in either sex. Mouse mammary tumor virus (MMTV) accelerates mammary tumorigenesis in *Wnt*-1-transgenic animals by insertional activation of two fibroblast growth factor genes, » *int*-2 (= FGF-3; unrelated to

Wnt-1; similarity in nomenclature is based on the criterion of being a target for MMTV insertion mutation), and » hst (= FGF-4).

Arheden K et al Chromosome localization of the human oncogene INT1 to 12q13 by *in situ* hybridization. Cytogenet. Cell Genet. 47: 86-7 (1988); **Baker N** Molecular cloning of sequences from *wingless*, a segment polarity gene in *Drosophila*: the spatial distribution of a transcript in embryos. EJ 6: 1765-73 (1987); **Bradley RS & Brown AMC** The proto-oncogene *int*-1 encodes a secreted protein associated with the extracellular matrix. EJ 9: 1569-75 (1990); **Christian JL et al** *Xwnt*-8 modifies the character of mesoderm induced by bFGF in isolated *Xenopus* ectoderm. EJ 11: 33-41 (1992); **Fung YK et al** Nucleotide sequence and expression *in vitro* of cDNA derived from mRNA of *int*-1, a provirally activated mouse mammary oncogene. MCB 5: 3337-44 (1985); **Gavin BJ et al** Expression of multiple novel *Wnt*/*int*-1-related genes during fetal and adult mouse development. GD 4: 2319-32 (1990); **Jue SF et al** The mouse *Wnt*-1 gene can act via a paracrine mechanism in transformation of mammary epithelial cells. MCB 12: 321-8 (1992); **Kitajewski J et al** Interaction of *Wnt*-1 protein with the binding protein BiP. MCB 12: 784-90 (1992); **McMahon AP** The *Wnt* family of developmental regulators. TIG 8: 236-42 (1992); **McMahon JA & McMahon AP** Nucleotide sequence, chromosomal localization, and developmental expression of the mouse *int*-1-related gene. Development 107: 643-50 (1989); **Nusse R et al** A new nomenclature for *int*-1 and related genes. Cell 64: 231 (1991); **Papkoff J & Schryver B** Secreted *int*-1 protein is associated with the cell surface. MCB 10: 2723-30 (1990); **Rijsewijk F et al** The *Drosophila* homologue of the mouse mammary oncogene *int*-1 is identical to the segment polarity gene *wingless*. Cell 50: 649-57 (1987); **Roelink H & Nusse R** Expression of two members of the *Wnt* family during mouse development – restricted temporal and spatial patterns in the developing neural tube. GD 5: 381-8 (1991); **Van Ooyen A & Nusse R** Structure and nucleotide sequence of the putative mammary oncogene *int*-1: proviral insertions leave the protein coding domain intact. Cell 39: 233-40 (1984); **van't Veer LJ et al** Molecular cloning and chromosomal assignment of the human homologue of *int*-1, a mouse gene implicated in mammary tumorigenesis. MCB 4: 2532-4 (1984); **Wainwright BJ et al** Isolation of a human gene with protein sequence similarity to human and murine *int*-1 and the *Drosophila* segment polarity mutant wingless. EMBO J. 7: 1743-8 (1988)

● **Transgenic/Knock-out/Antisense Studies: Erickson RP et al** Creating a conditional mutation of *Wnt*-1 by antisense transgenesis provides evidence that *Wnt*-1 is not essential for spermatogenesis. Dev. Gene. 14: 274-81 (1993); **Kwan H et al** Transgenes expressing the *Wnt*-1 and *int*-2 proto-oncogenes cooperate during mammary carcinogenesis in doubly transgenic mice. MCB 12: 147-54 (1992); **Lin TP et al** Role of endocrine, autocrine, and paracrine interactions in the development of mammary hyperplasia in *Wnt*-1 transgenic mice. CR 52: 4413-9 (1992); **McMahon JA & Bradley A** The *Wnt*-1 (*int*-1) proto-oncogene is required for development of a large region of the mouse brain. Cell 62: 1073-85 (1990); **Shackleford GM et al** Mouse mammary tumor virus infection accelerates mammary carcinogenesis in *Wnt*-1 transgenic mice by insertional activation of *int*-2/Fgf-3 and hst/Fgf-4. PNAS 90: 740-4 (1993); **Thomas KR & Capecchi MR** Targeted disruption of the murine *int*-1 proto-oncogene resulting in severe abnormalities in midbrain and cerebellar development. N 346: 847-50 (1990); **Tsukamoto AS et al** Expression of the *int*-1 gene in transgenic mice is associated with mammary gland hyperplasia and adenocarcinomas in male and female mice. Cell 55: 619-25 (1988)

Wound healing: Wound healing is a complex and protracted process of tissue repair and remodeling involving many different cell types. It requires a finely tuned control of various biochemical reaction cascades to balance degradative and regenerative processes.

Among other things the process comprises the migration of different cell types into the wound region, growth stimulation of epithelial cells and fibroblasts, formation of new blood vessels, and the generation of a new extracellular matrix (see also: ECM). At all phases the correct functioning critically depends on the biological activities of various » cytokines, including » chemokines, » EGF, » aFGF, » bFGF, » IGF, » PDGF, » TGF. These factors act in a controlled manner. They activate cells (see also: Cell activation), act as chemoattractants (see also: Chemotaxis), function as growth factors, or general mediators of neuroimmune activation (see also: Neuroimmune network).

The first step in wound healing is the activation of the blood clotting cascaded, initiated by the contact of plasma with tissue and basal membranes of cells.

The formation of a fibrin gel serves to fix plasma proteins and blood cells. This process leads to hemostasis and thus effectively produces a matrix that can be colonized by inflammatory cells such as neutrophils, monocytes, and macrophages (see also: Inflammation). Thrombin inside this plasma clot then induces platelets to degranulate, i. e. to release the contents of their α-granules. These contain locally active growth factors such as » TGF-β, » PDGF and » bFGF and a plethora of other chemotactically active mediators.

This initial reaction subsides with the synthesis of protein C which inactivates the coagulation factors V and VII, and the release of plasminogen activator which initiates the dissolution of the thrombus. Smooth muscle cells release prostaglandins which, in turn, prevent platelet aggregation and inhibit the synthesis of » PDGF. Several chemotactic cytokines mediate the migration of neutrophils and monocytes into the injured areas.

The differentiation of macrophages is initiated by several specific cytokines. These cells are major agents in counteracting bacterial infections and in the transportation of cell debris. Activated macrophages secrete more cytokines including » PDGF and » TGF-β which is a strong chemoattractant for monocytes and fibroblasts. Many macrophage-derived cytokines favor the further migration of in-

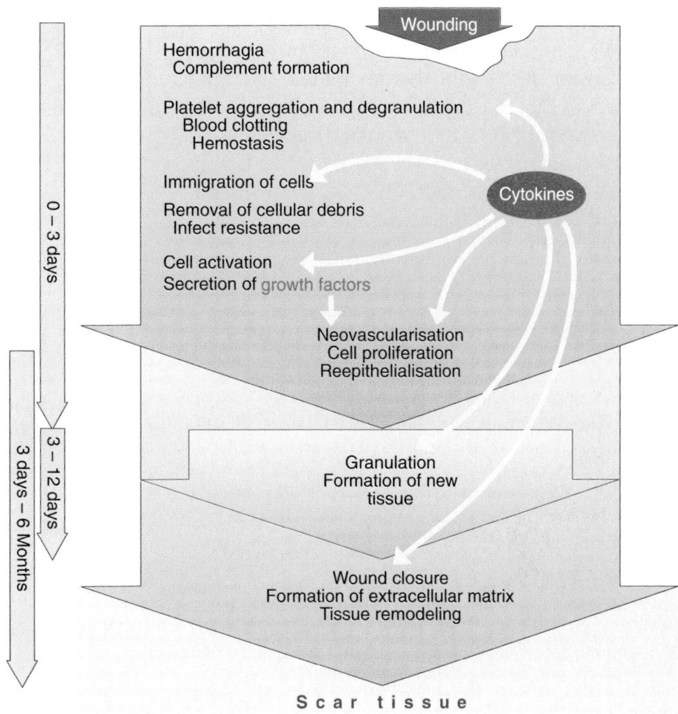

Schematic representation of processes involved in wound healing.
See text for details.

flammatory cells into the wound area (see also: Chemokines).

TGF-β appears to be the major factor responsible for the formation of granulation tissue and the synthesis of proteins of the extracellular matrix. The formation of new blood vessels within the wounded area is initiated and maintained as long as required by various » angiogenesis factors. Re-epithelialization is mediated by chemotactic and mitogenic growth factors of the » EGF family of growth factors.

The final phase of wound healing is characterized by the gradual replacement of granulation tissues by connective tissue. This process also requires locally acting cytokines. The synthesis of collagen and proteinase inhibitors is stimulated, among other things, by » TGF-β and related factors.

Animal experiments and also clinical experience have demonstrated that the topical administration of various cytokines, including » bFGF, » EGF, » KGF, » PDGF, » TGF-β, either alone or in combination, considerably accelerates wound healing (see: individual factors for details; subentry Clinical use & significance). Some of these factors even allow a complete healing of wounds that were previously refractory to conventional treatment. Treatment of incisional wounds in a rat healing model with TGF-β has shown that this factor enhances wound healing even if given up to 24 hours *before* wounding. The mechanism of action is unclear but it has been suggested that this effect may be due to the ability of TGF-β to induce its own synthesis. A single dose of TGF-β given before wounding is envisaged to prime cells throughout the body for further synthesis of TGF-β to respond more effectively to a future injury.

The main indications for the pharmacological use of cytokines involved in wound healing processes are at present the treatment of large wound areas and burn injuries, the regeneration of the cornea following abrasion or cataract treatment, and the treatment of chronic non-healing ulcers (see also: PDWHF, platelet-derived wound healing formula). At present the small number of trials carried out demonstrate that cytokine treatment can increase healing rates although proof of healing versus nonhealing under cytokine treatment is still lacking. One of the reasons for poor improvements in healing under cytokine therapy may be the rapid digestion of factors due to the high levels of proteolytic enzymes in chronic wounds. Some factors, such as » bFGF, have been shown to be highly effective in non-ischemic situations while being without effects in a hypoxic dermal ulcer models.

Bennett NT & Schultz GS Growth factors and wound healing: biochemical properties of growth factors and their receptors. Am. J. Surg. 165: 728-37 (1993); **Bhartiya D et al** Enhanced wound healing in animal models by interferon and an interferon inducer. JCP 150: 312-9 (1992); **Blitstein-Willinger E** The role of growth factors in wound healing. Skin Pharmacol. 4: 175-82 (1991); **Clark RAF** Wound repair. COCB 1: 1000-8 (1989); **Davidson JM & Broadley KN** Manipulation of the wound-healing process with basic fibroblast growth factor. ANY 638: 306-15 (1991); **Deuel TF et al** Growth factors and wound healing: Platelet-derived growth factor as a model cytokine. ARM 42: 567-84 (1991); **Eckersley JR & Dudley HA** Wounds and wound healing. BMB 44: 423-36 (1988); **Fiddes JC et al** Preclinical wound-healing studies with recombinant human basic fibroblast growth factor. ANY 638: 316-28 (1991); **Grotendorst GR** Growth factors as regulators of wound repair. Int. J. Tissue React. 10: 337-44 (1988); **Hammar H** Wound healing. Int. J. Dermatol. 32: 6-15 (1993); **Howell JM** Current and future trends in wound healing. Emerg. Med. Clin. North Am. 10: 655-63 (1992); **Hudson-Goodman P et al** Wound repair and the potential use of growth factors. Heart Lung 19: 379-84 (1990); **Joyce ME et al** Role of growth factors in fracture healing. PCBR 365: 391-416 (1991); **Kingsnorth AN & Slavin J** Peptide growth factors and wound healing. Br. J. Surg. 78: 1286-90 (1991); **Leibovitch SJ et al** Macrophages, wound repair and angiogenesis. PCBR 266: 131-45 (1988); **Lynch SE** Interactions of growth factors in tissue repair. PCBR 365: 341-57 (1991); **Leibovitch SJ et al** Macrophages, wound repair and angiogenesis. PCBR 266: 131-45 (1988); **McKay IA & Leigh IM** Epidermal cytokines and their roles in cutaneous wound healing. Brit. J. Dermatol. 124: 513-8 (1991); **Meyer-Ingold W** Wound therapy: growth factors as agents to promote healing. TibTech 11: 387-392 (1993); **Rollins BJ & Stiles CD** Molecular regulation of growth control: implications for wound healing. PCBR 266: 19-22 (1988); **Schultz G et al** EGF and TGF-α in wound healing and repair. JCBc 45: 346-52 (1991); **Skokan SJ & Davis RH** Principles of wound healing and growth factor considerations. J. Am. Podiatr. Med. Assoc. 83: 223-7 (1993); **Sprugel KH et al** Platelet-derived growth factor and impaired wound healing. PCBR 365: 327-40 (1991); **ten Dijke P & Iwata KK** Growth factors for wound healing. BT 7: 793-8 (1989); **Turk CW et al** Immunological mediators of wound healing and fibrosis. JCP 5: s89-93 (1987); **Wahl SM et al** Role of growth factors in inflammation and repair. JCB 40: 193-9 (1989)

Wound healing formula: see: PDWHF (platelet-derived wound healing formula); see also: wound healing.

WSXWS motif: see: Cytokine receptor superfamily.

WT1: Wilms' tumor susceptibility gene. See: Egr-1.

X: also: Xe. a prefix used in some abbreviations, e. g. XFGFR (fibroblast growth factor receptor) to indicate that the protein in question comes from » *Xenopus laevis*.

Xe: see: X.

XeFGF: *Xenopus embryonic fibroblast growth factor* This factor was recognized as a cDNA isolated from » *Xenopus laevis* embryos. The cDNA sequence reveals that it is closely related to fibroblast growth factor-like proteins of mammalians, called FGF-4 (product of the » *hst* oncogene) and » FGF-6. Unlike » bFGF (FGF-2) XeFGF is a secreted factor. XeFGF expression is highest in the gastrula stage of development and probably functions as a mesoderm-inducing factor (see also: XTC-MIF).

Godsave SF & Shiurba RA *Xenopus* blastulae show regional differences in competence for mesoderm induction: correlation with endogenous basic fibroblast growth factor levels. Dev. Biol. 151: 506-15 (1992); **Isaacs HV et al** Expression of a novel FGF in the *Xenopus* embryo. A new candidate inducing factor for mesoderm formation and anteroposterior specification. Development 114: 711-20 (1992)

***Xenopus* embryonic fibroblast growth factor:** see: XeFGF.

***Xenopus laevis*:** An African clawed toad that has been a favorite organism for developmental biologists and molecular biologists for many years. One of the reasons is the size of the fertilized egg. These are spheres of 1-2 mm in diameter with a dark brown pigmented upper hemisphere (animal hemisphere) and a light colored lower hemisphere (vegetal hemisphere). Embryogenesis occurs entirely outside the body of the female. In addition, development from egg laying to larval hatching only takes ≈ 48 hours. The transplantation of nuclei from intestinal cells of tadpoles to enucleated embryos and the demonstration that adult frogs are produced were a key experiment that demonstrated genetic pluripotency of the somatic cell nucleus. In addition *Xenopus* oocytes and unfertilized eggs were the first eukaryotic cells successfully manipulated by introducing foreign DNA. The analysis of *Xenopus* 5S RNA expression in *Xenopus* oocytes also produced the first definition of a eukaryotic promoter (see also: Gene expression). *Xenopus* oocytes are now an invaluable aid in the analysis of mechanisms controlling embryonic development and cell proliferation because the cells are amenable to microdissection and interference to follow the fates of cells during embryogenesis. *Xenopus* oocytes are also frequently used for analyzing the transcription and translation of cloned genes which are easily introduced by micro-injection. *Xenopus* cells have also been the source of some growth factors which have turned out to be homologues of known mammalian and non-mammalian cytokines or cytokine-like factors (see: XeFGF, Xotch, XTC-MIF).

Clemens MJ Translation of eukaryotic messenger RNA in cell-free extracts. In: Hames BD & Higgins SJ (eds) Transcription and translation, a practical approach. IRL Press, Oxford 1984; **Colman A** Expression of exogenous DNA in *Xenopus* oocytes. In: Hames BD & Higgins SJ (eds) Transcription and translation, a practical approach. IRL Press, Oxford 1984; **De Robertis EM & Mertz J** Coupled transcription-translation of DNA injected into *Xenopus* oocytes. Cell 12: 175-82 (1977); **Gurdon JB & Wickens MP** The use of *Xenopus* oocytes for the expression of cloned genes. MiE 101: 370-86 (1983); **Gurdon JB & Melton DA** Gene transfer in amphibian eggs and oocytes. ARG 15: 189-218 (1981); **Maller JL** *Xenopus* oocytes and the biochemistry of cell division. B 29: 3157-66 (1990); **Melton DA** Translation of messenger RNA in injected frog oocytes. MiE 152: 288-96 (1987); Slack JMW Embryological assays for growth factors. In: McKay I & Leigh I (Edts). Growth factors, a practical approach, IRL Press, pp. 73-84, Oxford (1992)

XG-1: A myeloma cell line that depends on » IL6 for its growth. It has been shown that GM-CSF stimulates IL6 responsiveness. GM-CSF does not induce any IL6 production in XG-1 cells, nor is it able to sustain their growth alone.

Zhang XG et al Granulocyte-macrophage colony-stimulating factor synergizes with interleukin-6 in supporting the proliferation of human myeloma cells. Blood 76: 2599-605 (1990)

Xid mouse: see: Immunodeficient mice.

X-linked hyper IgM syndrome: see: TRAP (TNF-related activation protein).

X-linked severe combined immunodeficiency: abbrev. XSCID. See: IL2 (subentry receptors). See also: Immunodeficient mice.

X-linked immunodeficiency mouse: see: Immunodeficient mice.

***Xotch*:** This » *Xenopus laevis* gene is a homologue of the *Drosophila* » *Notch*, which in turn is a homologue of human » EGF receptor. *Xotch* is involved in cell fate decisions and can functionally replace *Notch* in *Drosophila* development.
Coffman C et al *Xotch*, the *Xenopus* homologue of *Drosophila Notch*. S 249: 1438-1441 (1990); **Coffman CR et al** Expression of an extracellular deletion of Xotch diverts cell fate in *Xenopus* embryos. Cell 73: 659-71 (1993)

XSCID: *X-linked severe combined immunodeficiency* abbrev. XSCID. See: IL2 (subentry receptors). See also: Immunodeficient mice.

XTC: abbrev. *Xenopus* tissue culture. A cell line established from tissues of » *Xenopus laevis*. *Xenopus* cell lines have been used for a variety of studies in cell biology and molecular biology, one of their major advantages being their ability to growth at room temperature, usually without the need for a 5 % CO_2 atmosphere. Several *Xenopus* cell lines have recently been used intensively for transfection studies, involving the transfer of cytokine or cytokine receptor genes, and as a source of cytokines functioning as mesoderm-inducing factors.
Smith JC & Tata JR *Xenopus* cell lines. Methods in cell Biology Vol. 36: 635-54 (1991)

XTC-MIF: *XTC mesoderm inducing factor*
This heterodimeric factor, isolated from » *Xenopus laevis* cells, is a member of the transforming growth factor β family (see: TGF-β). Antibodies directed against TGF-β2 block the muscle-inducing activity of » XTC- cell-conditioned medium (see also: CM).
The first inductive interaction in amphibian development is mesoderm induction, when a signal from the vegetal hemisphere of the blastula induces mesoderm from overlying equatorial cells. Mesoderm-inducing factors cause isolated *Xenopus* animal caps to form mesodermal cell types such as muscle, instead of their normal fate of epidermis. XTC-MIF induces the development of the mesoderm in fertilized *Xenopus* oocytes. Induction of mesoderm during early amphibian embryogenesis can be mimicked *in vitro* by adding TGF-β to isolated ectoderm explants from *Xenopus* laevis embryos.
The mammalian homologue of XTC-MIF is » Activin A, the biological activity of which can be replaced by XTC-MIF in functional assays. XTC-MIF is, however, incapable of replacing TGF-β. The biological activities of XTC-MIF are concentration-dependent; three different biological effects on the development of the animal pole of the embryo are observed at three different concentrations.
Asashima M et al Presence of activin (erythroid differentiation factor) in unfertilized eggs and blastulae of *Xenopus* laevis. PNAS 88: 6511-4 (1991); **Chertov O et al** Mesoderm-inducing factor from bovine amniotic fluid: purification and N-terminal amino acid sequence determination. Biomed. Sci. 1: 499-506 (1990); **Friesel R & Dawid IG** cDNA cloning and developmental expression of fibroblast growth factor receptors from *Xenopus laevis*. MCB 11: 2481-8 (1991); **Green JBA et al** The biological effects of XTC-MIF: quantitative comparison with *Xenopus* bFGF. Development 108: 173-83 (1990); **Green JBA et al** Graded changes in dose of a *Xenopus* activin A homologue elicit stepwise transitions in embryonic cell fate. N 347: 391-4 (1990); **Rosa F et al** Mesoderm induction in amphibians: the role of TGFβ2-like factors. S 239: 783-5 (1988); **Smith JC et al** Purification, partial characterization, and biological effects of the XTC mesoderm-inducing factor. Development 103: 591-600 (1988); **Smith JC et al** Identification of a potent *Xenopus* mesoderm-inducing factor as a homologue of activin A. N 345: 729-31 (1990); **Smith JC** Purifying and assaying mesoderm-inducing factors from vertebrate embryos. In: Hartley DA (ed.) Cellular interactions in development. Practical Approach Series, IRL Press, Oxford 1993, pp181-204; **van den Eijnden-van Raaij AJM et al** Activin-like factor from a *Xenopus laevis* cell line responsible for mesoderm induction. N 345: 732-4 (1990)

XTT: Sodium 3'-[1-[(phenylamino)-carbonyl]-3,4-tetrazolium]-bis(4-methoxy-6-nitro)benzene-sulfonic acid hydrate. see: MTT assay.

Y

Y16: A mouse » IL5-dependent pre-B cell line. It constitutively expresses AIC2B (see: AIC2A), one of the components of the IL5 receptor. The cells can be used in » bioassays to detect murine IL5.
Takaki S et al Identification of the second subunit of the murine interleukin-5 receptor: interleukin-3 receptor-like protein, AIC2B is a component of the high affinity interleukin-5 receptor. EJ 10: 2833-8 (1991); **Tominaga A et al** Establishment of IL5-dependent early B cell lines by long-term bone marrow cultures. GF 1: 135-46 (1989)

Y73 avian sarcoma virus oncogene: see: *yes*.

YAC-1: A murine NK-sensitive T lymphoma cell line established from a tumor induced by Moloney sarcoma virus in A/Sn mice. The cells do not express Fc receptors. Among other things these cells are used to study the cytotoxic activities of naturally occurring killer cells, » LAK cells (lymphokine-activated killer cells), and factors influencing the cytotoxic activities of these and other lymphocytes.
Kiessling R et al Natural killer cells in the mouse. I. Cytotoxic cells with specificity for mouse Moloney leukemia cells. Specificity and distribution according to genotype. EJI 5: 112-117 (1975); **Bean P et al** Differential lysis of tumor target cells displayed by lymphokine activated killer (LAK) cell clones. IJCC 10: 190-5 (1992); **Pollack SB et al** Production and differentiation of NK lineage cells in long-term bone marrow cultures in the absence of exogenous growth factors. CI 139: 352-62 (1992); **Winter BK et al** Renal cell carcinoma and natural killer cells: studies in a novel rat model *in vitro* and *in vivo*. CR 52: 6279-86 (1992)

Yam 1B immunoregulatory factor: This poorly characterized factor of ≈ 43 and 65 kDa is secreted by Yam 1B cells, a human B lymphoblastoid cell line derived from the peripheral blood of an adult T cell leukemia patient. The factor suppresses blastogenesis and antibody formation by human lymphocytes. The factor also inhibits the expression of transferrin receptor, but not the expression of » IL2 receptors or the production of IL2 by lymphocytes stimulated with phytohemagglutin. The factor also inhibits DNA synthesis by human T and B lymphocytes and immunoglobulin generation by normal B cells and Epstein-Barr virus-transformed human B lymphoblastoid cell lines. The inhibitory activity of the factor is not blocked by antibodies directed against » TGF-β.
Morikawa K et al Regulation of lymphocyte blastogenesis and antibody production by soluble factor released by a human B lymphoblastoid cell line. Cytokine. 3: 609-10 (1991)

Yamaguchi sarcoma virus: see: *yes*.

Yeasts: see: *Saccharomyces cerevisiae*.

***yes*:** *yes* is a non-receptor cytoplasmic protein tyrosine kinase (62 kDa) of unknown function belonging to the *src* family of tyrosine kinases. The human gene, which is the cellular counterpart of the transforming » oncogene of the Y73 avian sarcoma virus (Yamaguchi sarcoma virus), is a proto-oncogene (see: Oncogenes) that maps to chromosome 18p11.32 less than 50 kb apart from the gene encoding thymidylate synthase. The *yes* gene at chromosome 18p is referred to as *yes*-1; a pseudogene, designated *yes*-2, is located at chromosome 22q11.2. In addition there are other *yes*-related sequences in the human genome that have not been mapped.
p62c-*yes* is highly expressed in a variety of mammalian cell types, including neurons, spermatozoa, platelets, and epithelial cells and is likely to have a general role in growth control.
yes tyrosine kinase is present both in the cell body region and in the axonal and nerve terminal region of the striatonigral neurons of rats, suggesting that it may be involved in both presynaptic and postsynaptic functions in mammalian forebrain neurons.
Transformation of cells induced by expression of *yes* can be reverted by treatment of cells with » Herbimycin A. *yes* has been shown to become associated after binding of » PDGF to its receptor. It has been shown that » oncostatin M utilizes a tyrosine phosphorylation signal transduction pathway in human endothelial cells involving the activation of the *yes* tyrosine kinase. *yes* also becomes activated after binding of » M-CSF to its membrane receptor.
Activation of platelets by thrombin and other phy-

siological agonists leads to a dramatic increase in tyrosine phosphorylation of multiple cellular proteins, and one of the tyrosine kinases involved in this process has been shown to be *yes*. *yes* is associated with the major platelet membrane glycoprotein IV(GPIV, CD36), together with » *fyn* and » *lyn*.

Cichowski K et al p21*ras*GAP association with *fyn*, *lyn*, and *yes* in thrombin-activated platelets. JBC 267: 5025-8 (1992); **Courtneidge SA et al** Activation of *src* family kinases by colony stimulating factor-1, and their association with its receptor. EJ 12: 943-50 (1993); **Huang MM et al** Membrane glycoprotein IV (CD36) is physically associated with the *fyn*, *lyn*, and *yes* protein-tyrosine kinases in human platelets. PNAS 88: 7844-8 (1991); **Kamps MP & Sefton BM** Identification of multiple novel polypeptide substrates of the v-*src*, v-*yes*, v-*fps*, v-ros, and v-*erb*-B oncogenic tyrosine protein kinases utilizing antisera against phosphotyrosine. O 2: 305-15 (1988); **Klages S et al** Molecular cloning and analysis of cDNA encoding the murine c-*yes* tyrosine protein kinase. O 8: 713-9 (1993); **Kypta RM et al** Association between the PDGF receptor and members of the *src* family of tyrosine kinases. Cell 62: 481-92 (1990); **Matsuzawa Y et al** Characterization of the promoter region of the c-*yes* proto-oncogene: the importance of the GC boxes on its promoter activity. O 6: 1561-7 (1991); **Schieven GL et al** Oncostatin M induces tyrosine phosphorylation in endothelial cells and activation of p62*yes* tyrosine kinase. JI 149: 1676-82 (1992); **Semba, K et al** Location of the c-*yes* gene on the human chromosome and its expression in various tissues. S 227: 1038-40, (1985); **Semba, K et al** Nucleotide sequence and chromosomal mapping of the human c-*yes*-2 gene. Jpn. J. Cancer Res. 79: 710-17 (1988); **Silverman GA et al** Chromosomal reassignment: YACs containing both *yes*1 and thymidylate synthase map to the short arm of chromosome 18. Genomics 15: 442-5 (1993); **Sugawara K et al** Distribution of c-*yes*-1 gene product in various cells and tissues. Br. J. Cancer 63: 508-13 (1991); **Sukegawa J et al** Characterization of cDNA clones for the human c-*yes* gene. MCB 7: 41-7 (1987); **Toyoshima K et al** Viral oncogenes, v-*yes* and v-*erb*B, and their cellular counterparts. Adv. Virus Res. 32: 97-127 (1987); **Twamley GM et al** Association of *fyn* with the activated platelet-derived growth factor receptor: requirements for binding and phosphorylation. O 7: 1893-901 (1992); **Yoshida MC et al** Regional mapping of the human proto- oncogene c-*yes*-1 to chromosome 18 at band q21.3. Jpn. J. Cancer Res. 76: 559-62 (1985)

YY-1: A transcription factor binding to sequence motifs of the serum response element (see: SRE). It is identical with » GATA-1.

Gualberto A et al Functional antagonism between YY1 and the serum response factor. MCB 12: 4209-14 (1992); **Gumucio DL et al** Phylogenetic footprinting reveals a nuclear protein which binds to silencer sequences in the human γ and ε globin genes. MCB 12: 4919-29 (1992); **Lee JS et al** Evidence for physical interaction between the zinc-finger transcription factors YY1 and Sp1. PNAS 90: 6145-9 (1993); **Safrany G & Perry RP** Characterization of the mouse gene that encodes the δ/YY1/NF-E1/UCRBP transcription factor. PNAS 90: 5559-63 (1993)

Z

ZENK: This gene has been shown to be identical with » Egr-1.

Zfp: abbrev. for » Zink finger protein.

zfp6: *zinc finger protein 6* This early response gene (see: ERG) has been shown to be identical with » Egr-1.
Darland T et al Regulation of Egr-1 (Zfp-6) and c-*fos* expression in differentiating embryonal carcinoma cells. O 6: 1367-76 (1991)

Zfp36: This » ERG (early response gene) is identical with » nup475 (nuclear protein 475).

zif268: *zinc finger protein 268* This » ERG (early response gene) expressed rapidly and transiently in murine » 3T3 fibroblasts after stimulation with serum, » Phorbol esters, » PDGF, or » FGF. zif268 has been shown to be identical with » Egr-1.
Christy BA et al A gene activated in mouse 3T3 cells by serum growth factors encodes a protein with "zinc finger" sequences. PNAS 85: 7857-61 (1988); **Christy BA & Nathans D** Functional serum response elements upstream of the growth factor-inducible gene zif268. MCB 9: 4889-95 (1989); **Christy BA & Nathans D** DNA binding site of the growth factor-inducible protein Zif268. PNAS 86: 8737-41 (1989)

Zinc finger proteins: Zinc finger proteins (abbrev. Zfp) are members of a multigene family encoding zinc-mediated nucleic acid binding proteins. They have been isolated from various organisms including yeast, *Drosophila*, *Xenopus* mouse and humans and some of its members have been highly conserved between species (see, for example: Egr-1). Zinc fingers are protein structural motifs of 28-40 amino acids that serve as common DNA-binding domains in nuclear proteins acting as regulatory proteins that exert positive or negative regulation of » Gene expression. Many Zinc finger proteins have been identified as » ERGs (early response genes) and thus play an important role in the response of cells to environmental stimuli, including cytokines (see, for example: Egr-1).
Berg JM Sp1 and the subfamily of zinc finger proteins with guanine-rich binding sites. PNAS 89: 11109-10 (1992); **Berg JM** Zinc finger domains: hypotheses and current knowledge. Annu. Rev. Biophys. Biophys. Chem. 19: 405-21 (1990); **Desjarles JR & Berg JM** Toward rules relating zinc finger protein sequences and DNA binding site preferences. PNAS 89: 7345-9 (1992); **Freedman LP** Anatomy of the steroid receptor zinc finger region. Endocrin. Rev. 13: 129-45 (1992); **Freemont PS** The RING finger. A novel protein sequence motif related to the zinc finger. ANY 684: 174-92 (1993); **Hoovers JM et al** High-resolution localization of 69 potential human zinc finger protein genes: a number are clustered. Genomics 12: 254-63 (1992); **Madden SL & Rauscher FJ 3rd** Positive and negative regulation of transcription and cell growth mediated by the EGR family of zinc-finger gene products. ANY 684: 75-84 (1993); **Reddy BA et al** A novel zinc finger coiled-coil domain in a family of nuclear proteins. TIBS 17: 344-5 (1992); Sluyser M et al (eds) Zinc finger proteins in oncogenesis. ARI entire Vol. 684 (1993); **Thiesen HJ & Bach C** DNA recognition of C2H2 zinc-finger proteins. Evidence for a zinc-finger-specific DNA recognition code. ANY 684: 246-9 (1993)

ZL5 autocrine activity: ZL5 is a pleural human mesothelioma cell line that grows in chemically defined serum-free medium and can be maintained over 3 weeks in protein-free RPMI basal medium alone. This cell line produces an uncharacterized » autocrine mitogenic activity that also affects the DNA synthesis of a variety of other lung-derived cells. The active principle is not identical with » EGF, » PDGF, or TGF-β, known to stimulate the growth of normal human mesothelial cells.
Lauber B et al An autocrine mitogenic activity produced by a pleural human mesothelioma cell line. IJC 50: 943-50 (1992)

Zn-FTS: *Zn-containing facteur thymique serique* Serum thymic factor is now called thymulin. See: Thymic hormones.

ZAP-70: see addendum. **Zeta-associated protein 70:** see addendum: ZAP-70.

Addendum

To be as up-to-date as possible the publishers have offered the opportunity to include this addendum. The information herein has come to my attention during the time between the submission of the manuscript and resubmission of the galley proofs (August 1994). It pertains mainly to novel cytokines, receptors, cytokine-regulated signaling components and transcription factors, and gene knock-out experiments.

I have added cross references pointing to information in this addendum as footnotes at appropriate places of the main body of the dictionary to facilitate easy access.

abundant *Src* homology: see: ASH (this addendum).

Activin A: The biological consequences of a activin/inhibin βB subunit gene disruption have been studied in mice generated from » ES cells carrying a targeted deletion of the gene. Mice homozygous for a targeted disruption of the activin/inhibin βB gene are viable, indicating that the βB subunit is not essential for mesoderm formation in the mouse.

Mutant mice show developmental and reproductive defects. Eyelid fusion during late embryonic development does not take place. Male mutant mice breed normally. Mutant female mice show a profoundly impaired reproductive ability, characterized by perinatal lethality of their offspring. In addition, the expression of the related βA subunit of activin is highly upregulated in ovaries of mutant females. For further information (inhibin-deficient mice) see: Inhibins.

Vassalli A et al Activin/inhibin βB subunit gene disruption leads to defects in eyelid development and female reproduction. Genes & Dev. 8: 414-27 (1994)

ANF: *atrial natriuretic factor* The circulatory effects associated with lifelong plasma ANF elevation have been examined in transgenic mice constitutively expressing a fusion gene consisting of the transthyretin promoter and the ANF structural gene. Mean arterial pressure is reduced by 24 mm Hg, associated with a 27% reduction in total heart weight. This chronic reduction in blood pressure is due to a 21% reduction in total peripheral resistance, whereas cardiac output, stroke volume, and heart rate are not significantly altered, despite a 15% elevation in plasma volume. Transgenic mice display reductions of 35%, 33%, 32%, and 19% in muscle, skin, brain, and renal vascular resistance, respectively, whereas coronary and splanchnic resistances are not significantly altered.

Barbee RW et al Hemodynamics in transgenic mice with overexpression of atrial natriuretic factor. Circ. Res. 74: 747-51 (1994)

ASH: *abundant src homology* The 25 kDa protein ASH contains one SH2 and two SH3 domains (see: *src* homology domains). It was cloned by screening human and rat cDNA libraries with an oligonucleotide probe directed to a consensus sequence of the SH2 domains. The amino acid sequence of ASH is strikingly similar to ***Sem-5***, the product of a nematode cell-signaling gene, and ASH is most probably a mammalian homologue of Sem-5.

In vitro ASH binds to phosphotyrosine-containing proteins, including activated » EGF receptor, through its SH2 domain (see also: GRB, this addendum) and is also involved in signaling through the insulin receptor. Induced expression of an antisense ASH cDNA leads to a reduction in cell growth.

Matuoka K et al Cloning of ASH, a ubiquitous protein composed of one *src* homology region (SH) 2 and two SH3 domains, from human and rat cDNA libraries. PNAS 89: 9015-9 (1992); **Tobe K et al** Insulin stimulates association of insulin receptor substrate-1 with the protein abundant *Src* homology/growth factor receptor-bound protein 2. JBC 268: 11167-71 (1993)

BDNF: *brain-derived neurotrophic factor* The biological consequences of a BDNF gene disruption have been studied in mice generated from » ES cells carrying a targeted deletion of the gene. Mice homozygous for a targeted disruption of the BDNF gene die within 2 days after birth although a fraction of these animals live for 2-4 weeks. These animals develop symptoms of nervous system dysfunction, including ataxia. BDNF-deficient mice have substantially reduced numbers of cranial and spinal sensory neurons. Although their central nervous systems show no gross structural abnormalities, expression of neuropeptide Y and calcium-binding proteins is altered in many neurons, suggesting that BDNF is essential for normal differentiation of the central nervous system. These animals do not have obvious deficits in motor neuron populations (see also: *trk* for information of BDNF receptor null mutations).

Jones KR et al Targeted disruption of the BDNF gene perturbs brain and sensory neuron development but not motor neuron development. Cell 76: 989-99 (1994); **Snider WD** Functions of the neurotrophins during nervous system development: what the knockouts are teaching us. Cell 77: 627-38 (1994)

BNP: *brain natriuretic peptide* BNP is a member is a member of the natriuretic peptide family of proteins (see also: ANF, atrial natriuretic factor). BNP has central and peripheral actions similar to those of ANP. The ventricle has been shown to be the major site of BNP production. The biological consequences of BNP overexpression have been studied in transgenic mice overexpressing BNP. These mice show a 10- to 100-fold increase in plasma BNP concentrations and have significantly lower blood pressure than their nontransgenic littermates, suggesting a potential usefulness of BNP as a long-term therapeutic agent.

Ogawa Y et al Molecular cloning of the complementary DNA and gene that encode mouse brain natriuretic peptide and generation of transgenic mice that overexpress the brain natriuretic peptide gene. JCI 93: 1911-21 (1994)

CD23: The biological consequences of a CD23 gene disruption have been studied in mice generated from » ES cells carrying a targeted deletion of the gene. Mice homozygous for a targeted disruption of the CD23 gene show a normal development of B and T cells. The formation of germinal centers is normal. Immune responses to the helminth *Nippostrongylus brasiliensis* are unaffected. Immunization with thymus-dependent antigens leads to increased and sustained specific IgE antibody titers compared with controls. CD23 thus appears to act as a negative feedback component of IgE regulation. It may play a part in the etiology of IgE-mediated diseases such as atopy, allergy, and asthma.

Yu P et al Negative feedback regulation of IgE synthesis by murine CD23. N 369: 753-6 (1994)

CD40: Gene and surface expression of the CD40 ligand (gp39; see: TRAP) by activated T cells has been shown to be depressed in a subgroup of patients with common variable immunodeficiency (CVI). These findings suggest that inefficient signaling via CD40 may be responsible, at least in part, for the failure of B cell differentiation in these patients.

The genomic structure of the CD40 ligand gene has been determined. The gene contains five exons. Mutations in this gene causally related to the X chromosome-linked hyper-IgM syndrome have been found to consist of clustered deletions. These arise by splice donor mutations with exon skipping, splice acceptor mutations with utilization of a cryptic splice site, and deletion/insertion events with the creation of a new splice acceptor site.

Farrington M et al CD40 ligand expression is defective in a subset of patients with common variable immunodeficiency. PNAS 1099-1103 (1994); **Villa A et al** Organization of the human CD40L gene: implications for molecular defects in X chromosome-linked hyper-IgM syndrome and prenatal diagnosis. PNAS 91: 2110-4 (1994)

CD40 ligand: see: CD40 (this addendum).

CD70: The gene encoding the surface antigen CD70 found on activated but not resting T and B lymphocytes has been cloned. It is identical with the ligand for » CD27 (see also: TRAP).

Bowman MR et al The cloning of CD70 and its identification as the ligand for CD27. JI 152: 1756-61 (1994)

Cell growth-inhibitory factor: see: CGIF (this addendum).

CGIF: *cell growth-inhibitory factor* This poorly characterized factor of ≈ 23-25 kDa is secreted by rat peritoneal exudate cells. CGIF completely inhibits Con A-induced thymocyte mitogenesis and also completely suppresses the growth of splenocytes stimulated by either Con A or bacterial lipopolysaccharides. Several findings suggest that CGIF is distinct from » IL1ra (IL1 receptor antagonist) or » TGF-β.

Yui S et al Characterization of cell growth-inhibitory factor in inflammatory peritoneal exudate cells of rats. Microbiol. Immunol. 37: 961-9 (1993)

Chromogranin A: see: Granins (this addendum).

Chromogranin B: see: Granins (this addendum).

Chromogranin C: see: Granins (this addendum).

Chromostatin: see: Granins (this addendum).

c-*mpl* ligand: see: thrombopoietin (this addendum).

CNTF: *ciliary neuronotrophic factor* The inactivation of the CNTF gene in mice causes motor neuron degeneration, indicating an essential role of CNTF in the survival of motor neurons. In contrast, mutations in the human CNTF gene that cause aberrant RNA splicing and abolish expression of CNTF protein do not appear to be causally related to any neurological disease in humans of various age groups. CNTF deficiency may, however, affect the manifestation, course and prognosis of neurological diseases.
Takahashi R et al A null mutation in the human CNTF gene is not causally related to neurological diseases. Nature Genetics 7: 79-84 (1994)

CREB: *cyclic AMP-responsive element binding protein* The biological consequences of a CREB gene disruption have been studied in mice generated from » ES cells carrying a targeted deletion of the gene. Mice homozygous for a targeted disruption of the CREB gene are healthy and exhibit no impairment of growth and development. An analysis of other CREB-related proteins in these mice shows a marked upregulated expression of CREM (CRE modulator protein) whereas there are no changes in the expression of ATF1, another CREB-related transcription factor. It appears that CREM and ATF1, which both can form heterodimers with CREB, can compensate for the loss of CREB (for further information see also: CRE (cyclic AMP-responsive element)).
Hummler E et al Targeted mutation of the CREB gene: compensation within the CREB/ATF family of transcription factors. PNAS 91: 3647-51 (1994)

CRF: *corticotropin releasing factor* The biological consequences of a CRF gene disruption have been studied in mice generated from » ES cells carrying a targeted deletion of the gene. Mice homozygous for a targeted disruption of the CRF gene are viable. For further information see also: POMC (proopiomelanocortin).
Muglia LJ et al Expression of the mouse corticotropin-releasing hormone gene in vivo and targeted inactivation in embryonic stem cells. JCI 93: 2066-72 (1994)

CSK: This cytoplasmic tyrosine kinase downregulates the tyrosine kinase activity of the c-src protein through tyrosine phosphorylation. Since cell transformation by src is caused by various mechanisms that interfere with this phosphorylation, the CSK gene might function as an anti-oncogene. The CSK gene maps to human chromosome 15q23-q25. It is ubiquitously expressed in human tissues as 2 mRNA species of 2.6 and 3.4 kb, although in some tissues and cell lines, only the larger mRNA is detected.
Armstrong E et al The c-src tyrosine kinase (CSK) gene, a potential anti-oncogene, localizes to human chromosome region 15q23-q25. Cytogenet. Cell Genet. 60: 119-20 (1992) **Partanen J et al** Cyl encodes a putative cytoplasmic tyrosine kinase lacking the conserved tyrosine autophosphorylation site (Y416-src). O 6: 2013-8 (1991)

EEK: *eph- and elk-related kinase* This novel tyrosine kinase is most similar to 2 putative receptor protein- tyrosine kinases, ELK and EPH, showing 69 and 57% identity, respectively. The EEK gene maps to human chromosome 1.
Chan J & Watt VM Eek and erk, new members of the eph subclass of receptor protein-tyrosine kinases. O 6: 1057-61 (1991)

Embryonic receptor kinase: see: EmRK2 (this addendum).

EmRK2: *embryonic receptor kinase* The gene encoding this kinase has been isolated from embryoid bodies generated by differentiating embryonic stem cells (see: ES cells). Sequence analysis of the cDNA predicts a receptor with a 755 amino acid extracellular region with seven immunoglobulin-like domains, a transmembrane region, and a 552 amino acid cytoplasmic region containing the kinase domain. The kinase domain is interrupted by a stretch of hydrophilic amino acids, the kinase insert. EmRK2 is expressed in embryoid bodies, in whole embryos at day 10 and 12 of gestation, and in the embryonic yolk sac and the fetal liver. EmRK2 appears to be the mouse homologue of human » flt receptor, a receptor for » VEGF (vascular endothelial growth factor).
Choi K et al Isolation of a gene encoding a novel receptor tyrosine kinase from differentiated embryonic stem cells. O 9: 1261-6 (1994)

EMT: *expressed mainly in T cells* EMT is expressed mainly in T lymphocytes and natural killer (NK) cells but not in other cell lineages and may play a role in thymic ontogeny and growth regulation of mature T cells. The cDNA predicts an open reading frame encoding a protein of 72 kDa. Sequence comparisons suggest that the protein is probably the human homologue of a murine IL2-inducible T-cell kinase, designated ITK. EMT is a member of a new family of intracellular kinases

that includes BPK (B cell progenitor kinase; see: Immunodeficient mice, subentry xid).
Gibson S et al Identification, cloning, and characterization of a novel human T-cell-specific tyrosine kinase located at the hematopoietin complex on chromosome 5q. Blood 82: 1561-72 (1993)

Eotaxin: This protein of 73 amino acids has been found in the bronchoalveolar lavage of guinea pigs used as a model of allergic inflammation. The factor belongs to the platelet factor 4 family of » chemokines. It exhibits homology of 53 % with human » MCP-1, 44 % with guinea pig MCP-1, 31 % with human » MIP-1α, and 26 % with human » RANTES.
Eotaxin induces substantial eosinophil accumulation at a 1-2 pmol dose in the skin without significantly affecting the accumulation of neutrophils. Eotaxin is a potent stimulator of both guinea pig and human eosinophils *in vitro*. The factor appears to share a binding site with RANTES on guinea pig eosinophils.
Jose PJ et al Eotaxin: a potent eosinophil chemoattractant cytokine detected in a guinea pig model of allergic airways inflammation. JEM 179: 881-7 (1994)

EPH: The EPH gene encodes a protein tyrosine kinase expressed in liver, lung, kidney, and testes of rat. Screening of various human cancers shows preferential expression in cells of epithelial origin. Overexpression of EPH mRNA has been found in a hepatoma and a lung cancer without gene amplification. The EPH locus maps to human chromosome 7q32-q36.
Many different tyrosine kinases related to eph (EPH family) have been isolated and most of them are believed to be putative growth factor receptors.
Maru Y et al Evolution, expression, and chromosomal location of a novel receptor tyrosine kinase gene, eph. MCB 8: 3770-3776 (1988); **Tuzi NL & Gullick WJ** eph, the largest known family of putative growth factor receptors. Br. J. Cancer 69: 417-21 (1994); **Yoshida M et al** Chromosomal location of a novel receptor tyrosine kinase gene, EPH, on chromosome 7. Cytogenet. Cell Genet. 51: 1113 (1989)

Eph- and elk-related kinase: see: EEK (this addendum).

Fas ligand: abbr. FasL. The ligand binding to the Fas antigen (see: Apo-1) has been purified from a cytotoxic rat T lymphocyte cell line as a membrane glycoprotein of 40 kDa.
The gene encoding the murine Fas ligand has been localized to the gld region of mouse chromosome 1. Mice homozygous for gld (generalized lymphoproliferative disease) are characterized by abnormal T cell populations in the spleen and lymph nodes and suffer from autoimmune disease-like systemic lupus erythematosus. The Fas ligand from gld mice has been shown to carry a point mutation in the C-terminal region. gld-derived Fas ligand cannot induce apoptosis in cells expressing the Fas antigen (see also: lpr, lymphoproliferative disease).
Suda T & Nagata S Purification and characterization of the Fas ligand that induces apoptosis. JEM 179: 873-9 (1994); **Takahashi T et al** Generalized lymphoproliferative disease in mice, caused by a point mutation in the Fas ligand. Cell 76: 969-76 (1994)

***fgr*:** see: *hck* (hematopoietic cell kinase) (this addendum).

FRK: *fyn-related kinase* FRK is a novel tyrosine kinase (505 amino acids) cloned from the human hepatoma cell line Hep3B. The predicted amino acid sequence contains characteristic tyrosine kinase motifs without a transmembrane region, suggesting an intracellular localization. There is a 49% amino acid sequence identity with human » *fyn* and 47% with human *src*.
Lee J et al Cloning of FRK, a novel human intracellular *src*-like tyrosine kinase-encoding gene. Gene 138: 247-51 (1994)

***Fyn*-related kinase:** see: FRK (this addendum).

GATA-2: Experiments with embryonic stem cells (see: ES cells) carrying a targeted disruption of the GATA-1 gene demonstrate that these cells fail to generate primitive erythroid precursors but retain their ability to differentiate normal numbers of definitive erythroid precursors. However, these cells undergo developmental arrest and death at the proerythroblast stage. These cells also express normal levels of GATA target genes. GATA-2 expression is markedly upregulated in these cells. These observations suggest that GATA-1 normally serves to repress GATA-2 during erythropoiesis and that GATA-2 can compensate at least partially for the loss of GATA-1 until the late stage of erythroid development.
Weiss MJ et al Novel insights into erythroid development revealed through *in vitro* differentiation of GATA-1^- embryonic stem cells. Genes Dev. (: 1184-97 (1994)

G-CSF: Severe congenital neutropenia (Kostmann syndrome) is characterized by profound absolute neutropenia and a maturation arrest of marrow progenitor cells at the promyelocyte-myelocyte stage (see also: Hemopoiesis). It has been

shown that a somatic point mutation in one allele of the G-CSF receptor gene in a patient with severe congenital neutropenia results in a cytoplasmic truncation of the receptor. The mutant receptor chain still transduces a strong growth signal but this signal is unable to trigger maturation. The mutant receptor chain probably acts in a dominant negative manner to block granulocytic maturation.
Dong F et al Identification of a nonsense mutation in the granulocyte colony-stimulating factor receptor in severe congenital neutropenia. PNAS 91: 4480-4 (1994);

GDF-5: *growth/differentiation factor 5* see also: GDF-1. GDF-5 is a new member of the » TGF-β family of proteins. GDF5 and the closely related new factors ***GDF-6*** and ***GDF-7*** define a new subgroup of factors related to known bone- and cartilage-inducing molecules, the bone morphogenetic proteins (see: BMP). Mutations in the gene encoding GDF-5 have been shown to be responsible for the mutation brachypodism, which alters the length and number of bones in the limbs of mice but spares the axial skeleton.
Storm EE et al Limb alterations in brachypodism mice due to mutations in a new member of the TGF β superfamily. N 368: 639-43 (1994)

GDF-6: *growth/differentiation factor 6* see: GDF-5 (this addendum).

GDF-7: *growth/differentiation factor 7* see: GDF-5 (this addendum).

generalized lymphoproliferative disease: abbr. *gld*. See: Fas ligand (this addendum).

GM-CSF: The biological consequences of an » GM-CSF gene disruption have been studied in mice generated from » ES cells carrying a targeted deletion of the gene. Mice homozygous for a targeted disruption of the GM-CSF gene are characterized by an unimpaired steady-state » hemopoiesis, demonstrating that GM-CSF is not essential for maintaining normal levels of the major types of mature hematopoietic cells and their precursors in blood, marrow, and spleen.
Most GM-CSF-deficient mice are superficially healthy and fertile but develop abnormal lungs. GM-CSF-deficient mice develop a progressive accumulation of surfactant lipids and proteins in the alveolar space, the defining characteristics of the idiopathic human disorder pulmonary alveolar proteinosis. Extensive lymphoid hyperplasia associated with lung airways and blood vessels is also found. These results demonstrate an unexpected, critical role for GM-CSF in pulmonary homeostasis.
Dranoff G et al Involvement of granulocyte-macrophage colony-stimulating factor in pulmonary homeostasis. S 264: 713-6 (1994); **Stanley E et al** Granulocyte/macrophage colony-stimulating factor-deficient mice show no major perturbation of hematopoiesis but develop a characteristic pulmonary pathology. PNAS 91: 5592-6 (1994)

gld: *generalized lymphoproliferative disease* see: Fas ligand (this addendum).

gonadotrope polypeptide: see: Granins (this addendum).

GP 87: see: Granins (this addendum).

Granins: Granins comprise a family of acidic soluble proteins known as Chromogranins A, B and C. They are acidic proteins of a molecular mass of 48, 76 and 67 kDa, respectively, found in the secretory granules of a wide variety of endocrine cells (collectively named APUD cells; amine precursor uptake and decarboxylation) and neurons, being stored together with many different peptide hormones and neuropeptides. Proteolytic cleavage of granins gives rise to a variety of biologically active peptides. The granins have been used as markers for normal and neoplastic endocrine and neuronal cells, as well as model proteins to understand the sorting mechanism involved in the formation of secretory granules.
Chromogranin A is also known as ***parathyroid secretory protein I***, a protein of unknown function costored and coreleased with parathyroid hormone from storage granules in the parathyroid gland in response to hypocalcemia. Chromogranin A is a low-affinity, high-capacity Ca^{2+} binding protein, postulated to be responsible for the Ca^{2+} buffering role of secretory vesicles. This protein is the major phosphorylated protein released by the parathyroid gland. The secretion of secretory protein I is inversely proportional to extracellular calcium concentration. In addition, the degree of phosphorylation of this protein is also inversely proportional to serum calcium. It has been suggested that chromogranin A may play an autocrine role as a glucocorticoid responsive inhibitor regulating the secretion of peptides derived from » POMC (proopiomelanocortin).
Chromogranin A is a protein of 439 amino acids costored and coreleased with catecholamines from storage granules in the adrenal medulla. The

human chromogranin A gene has 8 exons and 7 introns spanning ≈ 15 kb. The human gene maps to chromosome 14q32. Chromogranin A is secreted by a great variety of peptide- producing endocrine neoplasms: pheochromocytoma, parathyroid adenoma, medullary thyroid carcinoma, carcinoids, oat-cell lung cancer, pancreatic islet-cell tumors, and aortic-body tumor. A cleavage product of Chromogranin A with a length of 49 amino acids is ***pancreastatin***, which may be important for the physiologic homeostasis of blood insulin levels as well as pathologic aberrations such as diabetes mellitus. Pancreastatin inhibits insulin secretion, exocrine pancreatic secretion and gastric acid secretion, and stimulates glucagon secretion. Pancreastatin has been shown to inhibit transcription of the parathyroid hormone and chromogranin A genes and to decrease the stability of the respective messenger ribonucleic acids in parathyroid cells in culture. The pancreastatin sequence is contained within a variable domain encoded by exon VI of the chromogranin A gene.

Vasostatin is derived from the N-terminus of chromogranin A. It has vasoinhibitory activity. ***Parastatin***, which corresponds to residues 347-419 of chromogranin A, inhibits secretion of both parathyroid hormone and chromogranin A.

Another cleavage product of chromogranin A, encoded by exon VII of the gene, is ***chromostatin***, a bioactive peptide of 20 amino acids that may be involved in pancreatic beta cell functions. Chromostatin has been shown also to inhibit chromaffin cell secretion, probably functioning as an endocrine modulator of catecholamine-associated responses.

Chromogranin B is also known as ***secretogranin I***. Secretogranin II appears to be the same as the ***gonadotrope polypeptide*** (***GP 87***) released from pituitary cells under luteinizing hormone-releasing hormone stimulation. It is a tyrosine-sulfated secretory protein (657 amino acids, 76 kDa) found in a wide variety of peptidergic endocrine cells. The human chromogranin B maps to chromosome 20pter-p12. Chromogranin B has been found to be the major heparin-binding protein secreted by PC12 cells In vitro experiments demonstrate that Chromogranin B effectively promotes cell-substratum adhesion of NIH 3T3 and PC12 cells and also supports neurite outgrowth in primary hippocampal neurons.

Chromogranin C is also known as ***secretogranin II***. It is a protein of ≈ 86 kDa which was first detected in anterior pituitary. Subsequently, it was shown to be present also in adrenal medulla. It is synthesized by most neuroendocrine cells but occurs in greatest abundance in the anterior pituitary gland where it is localized primarily in gonadotrophs.

Secretoneurin is a 33 amino acid peptide derived from chromogranin C that is found in sensory afferent C-fibers. Stimulation of the hypothalamic magnocellular neurons can lead to the release of this neuropeptide. Secretoneurin has been found to trigger the selective migration of human monocytes *in vitro* and *in vivo*. Combinations of secretoneurin with the sensory neuropeptides, » SP (substance P) or somatostatin, synergistically stimulate such migration. In addition secretoneurin appears to trigger the release of dopamine.

Aardal S Helle KB The vasoinhibitory activity of bovine chromogranin A fragment (vasostatin) and its independence of extracellular calcium in isolated segments of human blood vessels. Regul. Pept. 41: 9-18 (1992); **Benedum UM et al** The primary structure of human secretogranin I (chromogranin B): comparison with chromogranin A reveals homologous terminal domains and a large intervening variable region. EJ 6: 1203-11 (1987); **Bhargava G et al** Phosphorylation of parathyroid secretory protein. PNAS 80: 878-81 (1983); **Cetin Y et al** Chromostatin, a chromogranin A-derived bioactive peptide, is present in human pancreatic insulin (beta) cells. PNAS 90: 2360-4 (1993); **Chanat E et al** The gonadotrope polypeptide (GP 87) released from pituitary cells under luteinizing hormone-releasing hormone stimulation is a secretogranin II form. Biochimie 70: 1361-8 (1988); **Chen M et al** Secretogranin I/chromogranin B is a heparin-binding adhesive protein. J. Neurochem. 58: 1691-8 (1992); **Cohn DV et al** Similarity of secretory protein I from parathyroid gland to chromogranin A from the adrenal medulla. PNAS 79: 6056-9 (1982); **Craig SP et al** Localization of the human gene for secretogranin 1 (chromogranin B) to chromosome 20. Cytogenet. Cell Genet. 46: 600 (1987); **Deftos LJ et al** A cloned chromogranin A (CgA) cDNA detects a 2.3kb mRNA in diverse neuroendocrine tissues. BBRC 137: 418-23 (1986); **Fasciotto BH et al** Parastatin (porcine chromogranin A347-419), a novel chromogranin A-derived peptide, inhibits parathyroid cell secretion. Endocrinology 133: 461-6 (1993); **Fischer-Colbrie R et al** Chromogranins A, B, and C: widespread constituents of secretory vesicles. ANY 493: 120-34 (1987); **Galindo E et al** Chromostatin, a 20-amino acid peptide derived from chromogranin A, inhibits chromaffin cell secretion. PNAS 88: 1426-30 (1991); **Gerdes HH et al** The primary structure of human secretogranin II, a widespread tyrosine-sulfated secretory granule protein that exhibits low pH- and calcium-induced aggregation. JBC 264: 12009-15 (1990); **Helman LJ et al** Molecular cloning and primary structure of human chromogranin A (secretory protein I) cDNA. JBC 263: 11559-63 (1988); **Huttner WB et al** The granin (chromogranin/secretogranin) family. TIBS 16: 27-30 (1991); **Kirchmair R et al** Secretoneurin - a neuropeptide generated in brain, adrenal medulla and other endocrine tissues by proteolytic processing of secretogranin II (chromogranin C). Neuroscience 53: 359-65 (1993); **Konecki DS et al** The primary structure of human chromogranin A and pancreastatin. JBC 262: 17026-30 (1987); **Kruggel W et al** The amino terminal sequences of bovine and human chromogranin A and secretory protein I are identical. BBRC 127: 380-3 (1985); **Metz-Boutigue MH**

et al Intracellular and extracellular processing of chromogranin A. Determination of cleavage sites. Eur. J. Biochem. 217: 247-57 (1993); **Modi WS et al** The human chromogranin A gene: chromosome assignment and RFLP analysis. Am. J. Hum. Genet. 45: 814-8 (1989); **Mouland AJ et al** Human chromogranin A gene. Molecular cloning, structural analysis, and neuroendocrine cell-specific expression. JBC 269: 6918-26 (1994); **Murray SS et al** The gene for human chromogranin A (CgA) is located on chromosome 14. BBRC 142: 141-6 (1987); **O'Connor DT & Deftos LJ** Secretion of chromogranin A by peptide-producing endocrine neoplasms. NEJM 314: 1145-51, 1986); **Reinisch N et al** Attraction of human monocytes by the neuropeptide secretoneurin. FL 334: 41-4 (1993); **Rosa P & Zanini A** Purification of a sulfated secretory protein from the adenohypophysis: immunochemical evidence that similar macromolecules are present in other glands. Eur. J. Cell Biol. 31: 94-8 (1983); **Sanchez-Margalet Vet al** Pancreastatin and its 33-49 C-terminal fragment inhibit glucagon-stimulated insulin in vivo. Gen. Pharmacol. 23: 637-8 (1992); **Saria A et al** Secretoneurin releases dopamine from rat striatal slices: a biological effect of a peptide derived from secretogranin II (chromogranin C). Neuroscience 54: 1-4 (1993); **Wand S et al** A proposed role for chromogranin A as a glucocorticoid-responsive autocrine inhibitor of proopiomelanocortin secretion. Endocrinology 128: 1345-51 (1991); **Wu HJ et al** Structure and function of the chromogranin A gene: clues to evolution and tissue-specific expression. JBC 266: 13130-4 (1991); **Zhang JX et al** Pancreastatin, a chromogranin A-derived peptide, inhibits transcription of the parathyroid hormone and chromogranin A genes and decreases the stability of the respective messenger ribonucleic acids in parathyroid cells in culture. Endocrinology 134: 1310-6 (1994)

GRB: *growth factor receptor-bound protein* A generic name for murine proteins containing SH2 and SH3 domains (see: *src* homology domains). These proteins bind to some tyrosine-phosphorylated cytokine receptors, either directly or by associating with tyrosine-phosphorylated receptor/adaptor protein complexes, and activate the *ras* signaling pathway.

GRB-1 encodes phosphatidylinositol 3-kinase-associated p85-α, which associates with activated growth factor receptors and modulates the interaction between PI3 kinase and the receptor for » PDGF. The human gene maps to chromosome 5q13.

The ***GRB-2*** protein is highly related to the *Caenorhabditis elegans* ***sem-5*** gene product (see: ASH, this addendum) and the human GRB2 protein and displays the same SH3-SH2-SH3 structural motifs. GRB-2 protein has been found to associate with tyrosine-phosphorylated receptors for » EGF and » PDGF via its SH2 domain. GRB-2 itself has been shown to interact with another signaling molecule designated » SOS1 (this addendum).

Overexpression of Grb2 has been shown to potentiate the EGF-induced activation of *Ras* and mitogen-activated protein kinase by enhancing the rate of guanine nucleotide exchange on *Ras*. Cellular GRB2 appears to form a complex with a guanine-nucleotide-exchange factor for *Ras*, which binds to the ligand-activated EGF receptor, allowing the tyrosine kinase to modulate *Ras* activity.

GRB-2 is also involved in mediating intracellular responses mediated by insulin binding to its receptor. GRB-2 has been implicated also in signaling through the receptor for » M-CSF.

The activation of *Ras* in T cells via the T cell receptor-CD3 complex is probably also controlled, at least in part, by mechanisms involving the formation of a complex of GRB2 and other phosphoproteins.

GRB-7 maps to mouse chromosome 11 to a region which also contains the tyrosine kinase receptor, HER2/erbB-2 (see: *neu* oncogene). The analogous chromosomal locus in man is often amplified in human breast cancer leading to overexpression of *neu*. GRB-7 has been shown to be amplified and overexpressed in concert with *neu* in several breast cancer cell lines. GRB-7, through its SH2 domain, binds tightly to *neu* such that a large fraction of the tyrosine phosphorylated *neu* is bound to GRB-7.

Buday L & Downward J Epidermal growth factor regulates p21ras through the formation of a complex of receptor, Grb2 adapter protein, and Sos nucleotide exchange factor. Cell 73: 611-20 (1993); **Buday L et al** A complex of Grb2 adaptor protein, Sos exchange factor, and a 36-kDa membrane-bound tyrosine phosphoprotein is implicated in ras activation in T cells. JBC 269: 9019-23 (1994); **Cannizzaro LA et al** The human gene encoding phosphatidylinositol 3-kinase associated p85-α is at chromosome region 5q12-13. CR 51: 3818-20 (1991); **Downward J** The GRB2/Sem-5 adaptor protein. FL 338(2): 113-7 (1994); **Egan SE et al** Association of Sos Ras exchange protein with Grb2 is implicated in tyrosine kinase signal transduction and transformation. N 363: 45-51 (1993); **Gale NW et al** Grb2 mediates the EGF-dependent activation of guanine nucleotide exchange on *Ras*. N 363: 88-92 (1993); **Li N et al** Guanine-nucleotide-releasing factor hSos1 binds to Grb2 and links receptor tyrosine kinases to Ras signaling. N 363: 85-8 (1993); **Li W et al** A new function for a phosphotyrosine phosphatase: linking GRB2-Sos to a receptor tyrosine kinase. MCB 14: 509-17 (1994); **Margolis B et al** High-efficiency expression/cloning of epidermal growth factor-receptor-binding proteins with *Src* homology 2 domains. PNAS 89: 8894-8 (1992); **Rozakis-Adcock M et al** The SH2 and SH3 domains of mammalian Grb2 couple the EGF receptor to the Ras activator mSos1. N 363: 83-5 (1993); **Skolnik EY et al** Cloning of PI3- kinase associated p85 utilizing a novel method for expression/cloning of target proteins for receptor tyrosine kinases. Cell 65: 83-90 (1991) ; **Skolnik EY et al** The function of GRB2 in linking the insulin receptor to *Ras* signaling pathways. S 260: 1953-5 (1993); **Stein D et al** The SH2 domain protein GRB-7 is co-amplified, overexpressed and in a tight complex with HER2 in breast cancer. EJ 13: 1331-40 (1994); **Suen KL et al** Molecular cloning of the mouse grb2 gene: differential interaction of the Grb2 adaptor protein with epidermal growth factor and nerve growth factor receptors. MCB 13: 5500-12 (1993); **van der Geer P & Hunter T** Mutation of Tyr697, a GRB2-binding site, and Tyr721, a PI 3-kinase binding site, abrogates signal transduction by the murine CSF-1 receptor expressed in Rat-2 fibroblasts. EJ 12: 5161-72 (1993)

GRB-2: *growth factor receptor-bound protein* see: GRB (this addendum).

GRB-7: *growth factor receptor-bound protein* see: GRB (this addendum).

Growth factor receptor-bound protein: see: GRB (this addendum).

Guanine nucleotide exchange factor: see: SOS1 (this addendum).

HARP: *heparin affin regulatory peptide* The human cDNA encoding HARP has been cloned. The recombinant protein is mitogenic for capillary endothelial cells and induces angiogenesis in an in vitro model. The biologically recombinant protein was shown to be an N-terminally extended form (three amino acids), suggesting the existence of naturally occurring protein variants. At present it is unknown whether these putative extended forms of HARP have physiological significance. For further information see: » HBNF (heparin-binding neurite-promoting factor).
Courty J et al Mitogenic properties of a new endothelial cell growth factor related to pleiotrophin. BBRC 151: 1312-8 (1991); **Laaroubi K et al** Mitogenic and in vitro angiogenic activity of human recombinant heparin affin regulatory peptide. GF 10: 89-98 (1994)

hck: *hematopoietic cell kinase* The biological consequences of an » *hck* gene disruption have been studied in mice generated from » ES cells carrying a targeted deletion of the gene. Hematopoiesis appears to proceed normally in these animals. Phagocytosis is impaired in *hck*-deficient mice. It has been suggested that the *src* family member » *lyn* may be able to compensate the *hck* deficiency.
Mice carrying a targeted deletion of the » *fgr* kinase gene also display normal hematopoiesis, as do animals carrying disruptions of both genes. Doubly mutant mice show a novel immunodeficiency characterized by an increased susceptibility to infection with *Listeria monocytogenes*, indicating that either *hck* or *fgr* is required to maintain a normal natural immune response.
Lowell CA et al Functional overlap in the *src* gene family: inactivation of hck and fgr impairs natural immunity. Genes Dev. 8: 387-98 (1994)

HEK: *human eph/elk-like tyrosine kinase* HEK is a new member of a family of tyrosine kinases with a high degree of homology to eph and elk tyrosine kinases. It is also known as ***human embryo kinase 1, EPH-like tyrosine kinase, ETK***.
The expression of HEK appears to be restricted to lymphoid tumor cell lines. HEK may thus play a role in some human lymphoid malignancies and also in normal lymphoid function and differentiation. The human gene maps to chromosome 3p11.2.
Boyd AW et al Isolation and characterization of a novel receptor-type protein tyrosine kinase (hek) from a human pre-B cell line. JBC 267: 3262-7 (1992); **Wicks IP et al** Localization of a human receptor tyrosine kinase (ETK1) to chromosome region 3p11.2. Genomics 19: 38-41 (1994); **Wicks IP et al** Molecular cloning of HEK, the gene encoding a receptor tyrosine kinase expressed by human lymphoid tumor cell lines. PNAS 89: 1611-5 (1992)

Hematopoietic cell kinase: see: hck (this addendum).

Hematopoietic consensus tyrosine-lacking kinase: see: HYL (this addendum).

Heparin affin regulatory peptide: see: HARP (this addendum).

HGF: *hepatocyte growth factor* The biological activities of HGF have been studied in transgenic mice expressing the factor under the control of albumin regulatory sequences. Hepatocytes of HGF-transgenic mice express increased levels of HGF as an » autocrine growth factor. The experiments also show that HGF increases growth and repair processes when expressed for long periods in the liver: the livers of HGF-transgenic mice recover completely in half the time needed for their normal siblings after partial hepatectomy. Levels of cellular » *myc* and » *jun* mRNA were increased in HGF-transgenic mice.
Shiota G et al Hepatocyte growth factor in transgenic mice: effects on hepatocyte growth, liver regeneration and gene expression. Hepatology 19: 962-72 (1994)

HYL: *hematopoietic consensus tyrosine-lacking kinase* HYL is a novel non-receptor tyrosine kinase identified in a human megakaryoblastic cell line, UT-7. The HYL gene contains a SH2 and SH3 domain (see also: *src* homology domains) and a tyrosine kinase catalytic domain. The deduced amino acid sequence of the protein encoded by this gene is most homologous to CSK (c-*src* kinase). Northern blot analysis reveals a 2.2 kb HYL transcript in various myeloid cell lines but not in adult tissues except for the brain and the lung. The expression of HYL is upregulated when these myeloid cells differentiate in response to » phorbol esters such as phorbol myristate acetate. The HYL gene maps to chromosome 19p13.

Sakano S et al Molecular cloning of a novel non-receptor tyrosine kinase, HYL (hematopoietic consensus tyrosine-lacking kinase). O 9: 1155-61 (1994)

Interleukin 15: see: IL15 (this addendum).

IL6: The biological consequences of an » IL6 gene disruption have been studied in mice generated from » ES cells carrying a targeted deletion of the gene. Mice homozygous for a targeted disruption of the IL6 gene develop normally. The development of bone marrow and spleen B cells in IL6-deficient mice appears to be normal. Numbers of thymocytes and peripheral T cells are consistently reduced by 20-40 % as compared to controls, suggesting that IL6 is involved in T cell proliferation. Thymocytes and T cells show a normal expression pattern of T cell receptor α, β, γ, and δ chains, » CD4, CD8, and some other markers.
IL6-deficient mice fail to control efficiently vaccinia virus and infection with Listeria monocytogenes. The T cell-dependent antibody response against vesicular stomatitis virus is impaired. IL6-deficient mice also show gross abnormalities with respect to the » acute phase response in response to tissue damage or infection. This response, however, is only moderately affected after challenge with bacterial lipopolysaccharides.

Kopf M et al Impaired immune and acute phase responses in interleukin-6-deficient mice. N 368: 339-42 (1994)

IL15: *interleukin 15* This factor is found in culture supernatants of a monkey kidney epithelial cell line, CV-1/EBNA. IL15 is a glycoprotein of 14-15 kDa. Human IL15 has been cloned from a human stromal cell line, IMTLH. It shares 97 % amino acid homology with the simian protein. High levels of IL15 mRNA are found in muscle and placenta, and a kidney epithelial line whereas T cells, the major source for IL2, do not express IL15 mRNA as determined by Northern blot analysis.
Some of the biological activities of IL15 resemble those of » IL2. IL15 stimulates proliferation of the established T cell line CTLL.2 as well as phytohemagglutinin-stimulated peripheral blood mononuclear cells. In addition, IL15 is also able to induce generation of cytolytic cells and » LAK cell (lymphokine-activated killer cell) activity *in vitro*. IL2 and IL5 do not share significant sequence homology, but modeling studies predict that IL1 and IL15 belong to the same family of four helix bundle type cytokines.
IL15 binds to the β and γ chains of the IL2 receptor, which are known to be required for ligand internalization and signal transduction. It does not bind to the α subunit of the IL2 receptor. High affinity IL15 binding has been observed on many lymphoid cell types, including peripheral blood monocytes, NK cells and PHA-activated peripheral blood mononuclear cells. CTLL.2 cells appear to express a single class of high affinity receptors for IL15 as determined by the analysis of » Scatchard plots. Receptors for IL15 have been observed on fresh human venous endothelial cells, which do not bind IL2. High affinity IL15 receptors are expressed also on stromal cells from bone marrow, fetal liver, and thymic epithelium. Some observations with cell lines transfected to express individual IL2 receptor subunits suggest the existence of other as yet undiscovered proteins that modulate binding of IL15 as they respond to IL15 but not to IL2.

Giri JG et al Utilization of the β and γ chains of the IL2 receptor by the novel cytokine IL15. EJ 13: 2822-30 (1994); **Grabstein et al** Cloning of a T cell growth factor that interacts with the β chain of the interleukin 2 receptor. S 264: 965-8 (1994)

IL-T: *interleukin T* This factor of 14 kDa is secreted by the adult T cell leukemia cell line HuT-102. It stimulates T cell proliferation and the induction of » LAK (lymphokine activated killer) cells. IL-T is produced by leukemic cells during the late phase of T cell lymphotropic virus I-induced adult T cell leukemia (ATL), which is characterized by a cessation of IL2 production although the cells continue to express high affinity IL2 receptors.
IL-T stimulates proliferation of the » CTLL-2 cell line in the presence of neutralizing antibodies directed against » IL2. IL-T also stimulates proliferation of the human cytokine-dependent T cell line » Kit225.
IL-T appears to be different from » IL1, » IL2, » IL3, » IL4, » IL5, » IL6, » IL7, » IL9, » IL10, » IL12, » IL13, or » GM-CSF. IL-T-mediated stimulation of T cell and lymphokine-activated killer cell activation has been shown to require the expression of the β subunit of the » IL2 receptor.

Bamford RN et al The interleukin (IL) 2 receptor β chain is shared by IL2 and a cytokine, provisionally designated IL-T, that stimulates T cell proliferation and the induction of lymphokine-activated killer cells. PNAS 91: 4940-4 (1994); **Burton JD et al** A lymphokine, provisionally designated interleukin T and produced by a human adult T cell leukemia line, stimulates T cell proliferation and the induction of lymphokine-activated killer cells. PNAS 91: 4935-9 (1994)

Interleukin T: see: IL-T (this addendum).

Invasion stimulating factor: see: ISF (this addendum).

IRF-1: *interferon regulatory factor 1* Embryonic fibroblasts derived from mice carrying a null mutation of the IRF-1 gene have been found to be transformable by expression of an activated c-Ha-*ras* oncogene. This property is also observed in embryonic fibroblasts derived from IRF-1/IRF-2 double negative animals but not in wild-type or IRF-2-negative embryonic fibroblasts.
The transformed phenotype of *ras*-expressing IRF-1-negative embryonic fibroblasts can be suppressed by the expression of the IRF-1 cDNA. IRF-1 thus appears to function as a tumor suppressor gene (see also: oncogenes). In addition, expression of c-Ha-*ras* oncogene causes wildtype but not IRF-1-negative embryonic fibroblasts to undergo » apoptosis when combined with a block to cell proliferation or treated with anticancer drugs or ionizing radiation. IRF-1 may thus be a critical determinant of oncogene-induced cell transformation or apoptosis. For further information see also: IRS (interferon response sequence).
The inhibition of encephalomyocarditis virus replication by » IFN-α and » IFN-γ has been found to be impaired in cells from mice carrying a null mutation of the IRF-1 gene. These mice are less resistant than normal mice to EMCV infection. The absence of IRF-1 does not affect replication of two other viruses (vesicular stomatitis virus, herpes simplex virus).
Kimura T et al Involvement of the IRF-1 transcription factor in antiviral responses to interferons. S 264: 1921-4 (1994); **Tanaka N et al** Cellular commitment to oncogene-induced transformation or apoptosis is dependent on the transcription factor IRF-1. Cell 77: 829-39 (1994)

ISF: *invasion stimulating factor* This poorly characterized factor of 78 kDa has been purified from the conditioned medium of a bone metastasizing human prostatic PC-3 ML clone. The producer PC-3 ML cells also express a receptor of ≈ 115 kDa, whereas the ISF non-producing, noninvasive PC-3 clones fail to express the ISF receptor.
Wang M et al Identification of the receptor for a novel M(r) 78,000 "invasion stimulating factor" from metastatic human prostatic PC-3 ML clones. CR 54: 2492-5 (1994)

Janus kinases: The human gene encoding JAK1 (see: Janus kinases) maps to chromosome 1p31.3. The gene encoding JAK2 maps to chromosome 9p24.
JAK2 has been shown to associate with the receptor subunit that is shared with the receptors for » GM-CSF, » IL3, and » IL5.
Pritchard MA et al Two members of the JAK family of protein tyrosine kinases map to chromosomes 1p31.3 and 9p24. Mammalian Genome 3: 36- 8 (1992); **Quelle FW et al** JAK2 associates with the chain of the receptor for granulocyte-macrophage colony-stimulating factor, and its activation requires the membrane-proximal region. MCB 14: 4335-41 (1994)

JAK3: The cDNA for rat JAK3 encodes a novel member of the JAK family of protein tyrosine kinases (see: Janus kinases). JAK3 is phylogenetically most closely related to JAK2 among the previously known JAK family members, JAK1, JAK2 and Tyk2. Southern analysis reveals that JAK3 is a single copy gene and well conserved in the vertebral genome. Northern analysis indicates that the 4.0 kb mRNA is transcribed in a variety of tissues including spleen, lung, kidney and intestine.
The human L-JAK (leukocyte janus kinase) has been renamed JAK3. It is expressed in NK and NK-like cells but not in resting T cells or in other tissues. Stimulated and transformed T cells express L-JAK, suggesting a role in lymphocyte activation. JAK3 has been shown to become phosphorylated and activated in response to » IL2 and » IL4 in human peripheral blood T cells and natural killer cells.
Johnston JA et al Phosphorylation and activation of the Jak-3 Janus kinase in response to interleukin-2. N 370: 151-3 (1994); **Kawamura M et al** Molecular cloning of L-JAK, a Janus family protein tyrosine kinase expressed in natural killer cells and activated leukocytes. PNAS 91: 6374-8 (1994); **Takahashi T & Shirasawa T** Molecular cloning of rat JAK3, a novel member of the JAK family of protein tyrosine kinases. FL 342: 124-8 (1994); **Witthun BA et al** Involvement of the Jak-3 Janus kinase in signaling by interleukins 2 and 4 in lymphoid and myeloid cells. N 370: 153-7 (1994)

Kostmann syndrome: see: G-CSF (this addendum).

LCF: *lymphocyte chemoattractant factor* The human gene encoding LCF has been cloned. LCF is a protein of 130 amino acids (13.4 kDa) that shows no significant homology to any known cytokine. The amino acid sequence predicted from the cDNA clone does not contain a consensus hydrophobic signal sequence or a potential transmembrane domain. The 3´ untranslated region of the mRNA contains several sequence motifs known to be involved in determining mRNA stability of certain other cytokines (see: ARE, AU-

rich element). Under physiological conditions LCF is found as a multimer of 55-60 kDa. Bioactivity is associated with the multimeric form and not with the monomeric form of LCF.
LCF has been shown to bind to » CD4 and to induce T lymphocyte expression of the » IL2 receptor. Cells expressing truncated forms of CD4 lacking the cytoplasmic part of the molecule do not respond to LCF, suggesting that LCF signaling involves the CD4-associated tyrosine-specific protein kinase » *lck*.
Cruikshank WW et al Molecular and functional analysis of a lymphocyte chemoattractant factor: association of biological function with CD4 expression. PNAS 91: 5109-13 (1994)

LESTR: *leukocyte-derived 7-transmembrane domain receptor* This novel receptor for CC-type » chemokines has been isolated from a human blood monocyte cDNA library. The cDNA clone encodes a protein of 352 amino acids belonging to a receptor of the 7-transmembrane domain, GTP-binding protein-coupled type.
The receptor shows 92.6% identity with a bovine neuropeptide Y receptor. Although the ligand for LESTR could not be identified among a large number of chemotactic cytokines, the high expression in white blood cells and the marked sequence relation to receptors for » IL8 suggest that LESTR may function in the activation of inflammatory cells.
Loetscher M et al Cloning of a human seven-transmembrane domain receptor, LESTR, that is highly expressed in leukocytes. JBC 269: 232-7 (1994)

Leukocyte-derived 7-transmembrane domain receptor: see: LESTR (this addendum).

Leukocyte janus kinase: abbr. L-JAK. See: JAK3 (this addendum).

Leukocyte tyrosine kinase: see: LTK (this addendum).

LIF: *leukemia inhibitory factor* The study of transgenic mice specifically overexpressing LIF in T cells show that LIF plays an important role in maintaining a functional thymic epithelium that will support proper T cell maturation. These animals display B cell hyperplasia, polyclonal hypergammaglobulinemia, and mesangial proliferative glomerulonephritis. These animals also show thymic and lymph node abnormalities. Cortical CD4/CD8-positive lymphocytes are lost in the thymus while numerous B cell follicles develop. Peripheral lymph nodes contain a vastly expanded CD4/CD8 double-positive lymphocyte population. The thymic epithelium is profoundly disorganized. Transplantation studies indicate that lymph node abnormalities can be rescued by wild type bone marrow but that the thymic defect cannot be rescued. This indicates that the thymic epithelium is irreversibly altered.
Shen MM et al Expression of LIF in transgenic mice results in altered thymic epithelium and apparent interconversion of thymic and lymph node morphologies. EJ 13: 1375-85 (1994)

L-JAK: *leukocyte janus kinase* see: JAK3 (this addendum).

LT-β: *lymphotoxin β* A membrane receptor specific for human LT-β has been described. This finding suggests that cell surface LT-b may have functions that are distinct from those of secreted LT-α (= TNF-β).
Crowe PD et al A lymphotoxin-β-specific receptor. S 264: 707-10 (1994)

LTK: *leukocyte tyrosine kinase* LTK, also known as ***Tyk1*** (tyrosine kinase 1) is a tyrosine kinase that has been suggested to be specific for hematopoietic cells (see also: Hematopoiesis) and neuronal cells. By using monoclonal antibodies directed against LTK the protein has been found to be expressed in human placenta.
Sequence analysis of the gene reveals similarities with several tyrosine kinase receptor genes of the insulin receptor family. The human Tyk1 maps to chromosome 15. Differential splicing of the LTK mRNA has been shown to yield various proteins including the putative receptor tyrosine kinase for an unknown ligand.
Ben-Neriah Y & Bauskin AR Leukocytes express a novel gene encoding a putative transmembrane protein-kinase devoid of an extracellular domain. N 333: 672-6 (1988). **Kozutsumi H et al** Identification of the human ltk gene product in placenta and hematopoietic cell lines. BBRC 190: 674-9 (1993); **Krolewski JJ et al** Identification and chromosomal mapping of new human tyrosine kinase genes. O 5: 277-82 (1990); **Toyoshima H et al** Differently spliced cDNAs of human leukocyte tyrosine kinase receptor tyrosine kinase predict receptor proteins with and without a tyrosine kinase domain and a soluble receptor protein. PNAS 90: 5404-8 (1993)

MCP-1: *monocyte chemoattractant protein 1* Two MCP-1-specific receptors have been cloned which signal in response to nanomolar concentrations of MCP-1. The two receptors differ in their carboxyl tails as a result of alternative splicing. They are closely related to the receptor for the » chemokines » MIP1α and » RANTES.

Charo IF et al Molecular cloning and functional expression of two monocyte chemoattractant protein 1 receptors reveals alternative splicing of the carboxyl-terminal tails. PNAS 91: 2752-6 (1994)

Megakaryocyte colony stimulating factor: see: thrombopoietin (this addendum).

Megakaryocyte maturation factor: see: thrombopoietin (this addendum).

Megakaryocyte growth and development factor: see: MGDF (this addendum).

MGDF: *megakaryocyte growth and development factor* This factor has been identified in aplastic canine plasma and the corresponding gene has been cloned from canine, murine, and human sources. MGDF bioactivity resides in two protein species of 25 kDa and 31 kDa that share the same N-terminal amino acids. Human, canine, and murine cDNAs for MGDF are highly conserved and encode open reading frames for proteins of 353, 352, and 356 amino acids, respectively, including secretory » signal sequences.
MGDF supports the development of megakaryocytes (see also: Hematopoiesis) from human CD34-positive hematopoietic progenitor cell populations in liquid culture. It also promotes survival of a factor-dependent murine cell line engineered to express the *mpl* oncogene (see: MPLV). MGDF has been identified as a ligand for the *mpl* cytokine receptor and is identical with an *mpl* ligand cloned and identified independently as » Thrombopoietin (this addendum).
Bartley TD et al Identification and cloning of a megakaryocyte growth and development factor that is a ligand for the cytokine receptor mpl. Cell 77: 1117-24 (1994)

MGF: *myxoma growth factor* Transgenic mice overexpressing the EGF-like growth factor MGF under the control of a metallothionein promoter have been generated. The transgene is expressed in epithelia and stroma of breast, lungs, liver and stomach. These mice display proliferation and arborization of breast ducts and ductules, with slight intraductal proliferation in virgin mice. They also show gastric epithelial hyperplasia. Stromal and epithelial hyperplasia are found in several organs.
Strayer DS et al Epidermal growth factor-like growth factors. I. Breast malignancies and other epithelial proliferations in transgenic mice. Lab. Invest. 69: 660-73 (1993)

Mk-CSF: *megakaryocyte colony stimulating factor* see: thrombopoietin (this addendum).

ML: *c-Mpl ligand* see: thrombopoietin (this addendum).

***mpl* ligand:** see: thrombopoietin (this addendum).

Neurotrophins: The article reviews the various effects of neurotrophins and » BDNF on the development of the nervous system and discusses results obtained with various neurotrophin/neurotrophin receptor gene null mutations.
Snider WD Functions of the neurotrophins during nervous system development: what the knockouts are teaching us. Cell 77: 627-38 (1994)

NGF: *nerve growth factor* The biological consequences of an NGF gene disruption have been studied in mice generated from » ES cells carrying a targeted deletion of the gene. Mice homozygous for a targeted disruption of the NGF gene survive for some days but all animals die by the age of 4 weeks. NGF-deficient homozygous animals show a profound cell loss in both sensory and sympathetic ganglia. The effects within the dorsal root ganglia appears to be restricted to small and medium peptidergic neurons. Examination of the central nervous system reveals that basal forebrain cholinergic neurons differentiate and continue to express phenotypic markers for the life span of the mutant mice, showing that differentiation and initial survival of central NGF-responsive neurons can occur in the absence of NGF.
An epidermal-specific gene promoter has been used to produce transgenic mice that overexpress the mouse NGF cDNA in the epidermis and associated hair follicles of the skin. The increase in NGF mRNA correlates with a hypertrophy of peripheral sensory and sympathetic nerves. Increased numbers of nerve processes in the transgenic skin display immunoreactivity for calcitonin gene-related peptide and tyrosine hydroxylase, indicating that both the sensory and sympathetic systems are hypertrophied. The trigeminal and superior cervical ganglia are greatly enlarged. Cell counts of trigeminal ganglia of control and transgenic mice show a 26-117% increase in the number of neurons in the transgenics, indicating a reduction or total prevention of the program of naturally occurring cell death (see also: Apoptosis). These results demonstrate that NGF production by the epidermal target tissue controls neuronal survival, and in so doing, establishes the level of innervation.
Mice overexpressing NGF in skin have been found to display a profound hyperalgesia to noxious

mechanical stimulation. Mice overexpressing an NGF antisense message (see also: Antisense oligonucleotides) display a profound hypoalgesia to the same stimuli.

Albers KM et al Overexpression of nerve growth factor in epidermis of transgenic mice causes hypertrophy of the peripheral nervous system. J. Neurosci. 14: 1422-32 (1994); **Crowley C et al** Mice lacking nerve growth factor display perinatal loss of sensory and sympathetic neurons yet develop basal forebrain cholinergic neurons. Cell 76: 1001-11 (1994); **Davis BM et al** Altered expression of nerve growth factor in the skin of transgenic mice leads to changes in response to mechanical stimuli. Neuroscience 56: 789-92 (1993); **Snider WD** Functions of the neurotrophins during nervous system development: what the knockouts are teaching us. Cell 77: 627-38 (1994)

NT-3: *neurotrophin 3* The biological consequences of an NT-3 gene disruption have been studied in mice generated from » ES cells carrying a targeted deletion of the gene. Mice homozygous for a targeted disruption of the NT-3 gene show severe deficits in sensory and sympathetic peripheral neuron populations. Most animals die shortly after birth.

NT-3-deficient mice lack muscle spindles and show abnormal limb positions. Spinal proprioceptive afferents are completely absent in homozygous mutant animals. Motor neurons, the enteric nervous system, and the major anatomical regions of the central nervous system seem to develop normally. In heterozygous mice the number of muscle spindles is half of that in control mice, indicating that NT-3 is present at limiting concentrations in the embryo

Ernfors P et al Lack of neurotrophin-3 leads to deficiencies in the peripheral nervous system and loss of limb proprioceptive afferents. Cell 77: 503-12 (1994); **Farinas I et al** Severe sensory and sympathetic deficits in mice lacking neurotrophin-3. N 369: 658-61 (1994); **Snider WD** Functions of the neurotrophins during nervous system development: what the knockouts are teaching us. Cell 77: 627-38 (1994)

Pancreastatin: see: Granins (this addendum).

Parastatin: see: Granins (this addendum).

Parathyroid secretory protein I: see: Granins (this addendum).

PBSF: *pre-B cell growth stimulating factor* This factor of 89 amino acids has been found to be secreted by the stromal cell line PA6 (see: LTBMC, long-term bone marrow culture). PBSF is a member of the » chemokine family of cytokines (α chemokines). It is distinct from » SCF (stem cell factor) and » IL7 and supports the proliferation of a stromal cell-dependent pre-B cell line » DW34. PBSF synergistically augments the growth of DW34 cells and bone marrow B cell progenitors in the presence of IL7.

Nagasawa T et al Molecular cloning and structure of a pre-B-cell growth-stimulating factor. PNAS 91: 2305-9 (1994)

Pre-B cell growth stimulating factor: see: PBSF (this addendum).

Receptor-mediated label-transfer assay: see: RELAY (this addendum).

RELAY: *receptor-mediated label-transfer assay* This assay exploits the extremely high binding specificity of the 55 kDa human » TNF receptor and allows detection of TNF with sensitivity limited only by limits in the detection of 131I. As little as 50 fg of active TNF (600,000 trimers) can be detected in a 5 ml sample of plasma, corresponding to the detection of TNF at a 200 aM concentration.

Poltorak A et al Receptor-mediated label-transfer assay (RELAY): a novel method for the detection of plasma tumor necrosis factor at attomolar concentrations. JIM 169: 93-9 (1994)

Renotropin: Renotropin is a collective name for poorly characterized organ-specific substance(s) found in the serum and in urine that mediate compensatory renal growth after partial nephrectomy. Renotropin may be more than a growth factor and is probably made up of a inhibitory as well as a stimulatory substance. In addition, some data suggest that renotropin also possesses vasomotor and natriuretic properties. Renotropin does not appear to be identical with » EGF, nor does it work by potentiating some activities of EGF.

Areas J et al The rabbit renotropic system. Am. J. Hypertens. 1: 152-7 (1988); **Memon S et al** Rabbit and human renotropin are not epidermal growth factor. J. Urol. 149: 1186-9 (1993); **Snow BW et al** Compensatory renal growth: interactions of nephrectomy serum and urine antisera leading to a new theory of renal growth regulation. Urol. Res. 15: 1-4 (1987)

Sarcolectin: The endolectin Sarcolectin is a protein of 65 kDa that is present in a great variety of conjunctival tissues (muscles, cartilage, sarcomas), but also in brain or placental extracts of vertebrates, including primates. Sarcolectin agglutinates cells and has an affinity for simple sugars. In addition, it functions as an interferon antagonists. It inhibits the synthesis of interferon-dependent secondary proteins and restores virus sensitivity 4-6 hrs after the establishment of antiviral protection (see also: IFN). Since there are no direct interac-

tions between sarcolectins and interferon or its receptors, it can be postulated that sarcolectins exert their effect through these interferon-dependent proteins. In a great variety of animal cells, sarcolectin can also initiate growth after it has been blocked by IFN.
The major cellular sarcolectin-binding protein in human placenta has been found to be identical with human macrophage migration inhibitory factor, MMIF (see: MIF, migration inhibition factor). The sarcolectin-binding protein reduces macrophage migration at a concentration of 100 ng/ml in MIF assays. Recombinant migration inhibitory factor and purified sarcolectin-binding protein react equally well with anti-MIF antibodies in immunoblot analyses.
Chany-Fournier F et al Sarcolectin and interferon in the regulation of cell growth. JCP 145: 173-80 (1990); **Jiang PH et al** Interferon- and sarcolectin-dependent cellular regulatory interactions. JBC 263: 19154-8 (1988); **Zeng FY et al** The major binding protein of the interferon antagonist sarcolectin in human placenta is a macrophage migration inhibitory factor. ABB 303: 74-80 (1993)

Scatter factor-like: see: SFL (this addendum).

Secretogranin I: see: Granins (this addendum).

Secretogranin II: see: Granins (this addendum).

Secretoneurin: see: Granins (this addendum).

Secretory protein I: see: Granins (this addendum).

Sem-5: see: ASH (this addendum).

SFGF: *Shope fibroma growth factor* Transgenic mice overexpressing the EGF-like growth factor SFGF (= SGF, shope growth factor) under the control of a metallothionein promoter have been generated. The transgene is expressed in epithelia and stroma of breast, lungs, liver and stomach. These mice display proliferation and arborization of breast ducts and ductules, with slight intraductal proliferation in virgin mice. They also show gastric epithelial hyperplasia. Stromal and epithelial hyperplasia are found in several organs.
Transgenic mice overexpressing SFGF under the control of a Rous sarcoma virus long terminal repeat develop atypical preneoplastic mammary ductal proliferations in both virgin and nonvirgin females by 6 months of age. In 1/3 of 8-month-old females, invasive secretory adenocarcinoma develops. These mice also develop severe epithelial atypia in the stomach, and papillary gastric tumors.
Strayer DS et al Epidermal growth factor-like growth factors. I. Breast malignancies and other epithelial proliferations in transgenic mice. Lab. Invest. 69: 660-73 (1993)

SFL: *scatter factor-like* This factor has been found to be secreted by a metastatic clone of the rat bladder carcinoma epithelial NBT-II cell line. The factor is able to dissociate epithelial clusters of NBT-II or MDCK cells and may be involved in tumor progression.
The activities of SFL resemble those of hepatocyte growth factor (see: HGF) on these two cell lines. SFL appears to be a factor that is distinct from HGF/Scatter factor since blocking antibodies against rat HGF do not inhibit SFL activities and SFL does not induce MDCK tubulogenesis, a biological assay that is specific for HGF activity.
Bellusci S et al A scatter factor-like factor is produced by a metastatic variant of a rat bladder carcinoma cell line. JCS 107: 1277-87 (1994)

SGF: *sweat gland factor* This poorly characterized factor is found in soluble extracts of rat footpad tissues. SGF induces cholinergic and reduces noradrenergic properties in the sweat gland innervation. In primary cultures of sweat gland cells production of SGF requires neurons. At present it is not known whether SGF is identical with any of the cytokines acting as cholinergic differentiation factors (see, for example, CDF, BDNF, CNTF, LIF).
Habecker BA & Landis SC Noradrenergic regulation of cholinergic differentiation. S 264: 1602-4 (1994)

Signal transducers and activators of transcription: see: STAT proteins (this addendum).

SOS1: SOS (son of sevenless) is a *Drosophila* gene which encodes a positive regulator of ras oncogene function by promoting guanine nucleotide exchange. The activation of the sevenless (sev) receptor tyrosine kinase is required for the specification of cell fate during the development of photoreceptor cells.
ras genes encode membrane-bound guanine nucleotide binding proteins. Binding of GTP activates *ras* proteins, and subsequent hydrolysis of the bound GTP to GDP inactivates signaling by these proteins. GTP binding can be catalyzed by guanine nucleotide exchange factors for *ras*, and GTP hydrolysis can be accelerated by GTPase-activating proteins (GAPs). Coupling of receptor

tyrosine kinases to ras signaling is mediated by a molecular complex consisting of growth factor receptor-bound protein 2 (see: GRB) and SOS1, the human homologue of *Drosophila* SOS. SOS1 and a related gene product (SOS2) have been implicated in the activation of *ras* by both the insulin and epidermal growth factor signal transduction pathways. Human SOS1 and SOS2 genes map to chromosomes 2p21-2p2 and 14q21, respectively.

Chardin P et al Human Sos1: a guanine nucleotide exchange factor for *Ras* that binds to GRB2. S 260: 1338-43 (1993); **Cherniack AD et al** Phosphorylation of the *Ras* nucleotide exchange factor son of sevenless by mitogen-activated protein kinase. JBC 269(7): 4717-20 (1994)>; **Webb GC et al** Mammalian homologues of the *Drosophila* Son of sevenless gene map to murine chromosomes 17 and 12 and to human chromosomes 2 and 14, respectively. Genomics 18: 14-9 (1993)

STAT proteins: *signal transducers and activators of transcription* STAT proteins are a family of unrelated cytoplasmic signaling proteins that function as latent cytoplasmic transcriptional activators that become activated by tyrosine phosphorylation in response to the engagement of various cytokine receptors. ***STAT1α*** and ***STAT1β*** are identical with p91 and p84, signaling components activated in response to » IFN (interferon) receptor engagement (see: IRS, interferon response sequence). ***STAT2*** is identical with p114, another component of the IFN signaling cascade (see: IRS, interferon response sequence). STAT proteins have been implicated in the signal transduction pathways mediated by a wide variety of cytokines which act through families of related receptors (see also: Cytokine receptor superfamily). **STAT3** has been shown to become activated by » IFN-α, » IFN-β, » EGF, and » CNTF. The ligand responsible for the activation of ***STAT4*** is not yet known. ***STAT5*** appears to mediate responses to » prolactin.

Darnell JE jr et al Jak-STAT pathways and transcriptional activation in response to IFNs and other extracellular signaling proteins. S 264: 1415-21 (1994); **Müller M et al** Interferon response pathways - a paradigm for cytokine signaling? J. Viral Hepatitis, in press (1994); **Yamamoto K et al** STAT4, a novel gamma interferon activation site-binding protein expressed in early myeloid differentiation. MCB 14: 4342-9 (1994)

Stem cell tyrosine kinase 1: see: STK1 (this addendum).

STK1: *stem cell tyrosine kinase 1* STK1, cloned from a CD34+ hematopoietic stem cell-enriched library, is the human homologue of murine » flk-2/flt-3. The cDNA encodes a protein of 993 amino acids with 85% identity and 92% similarity to the murine homologue. STK1 is a member of the type III receptor tyrosine kinase family that includes » kit, » fms, and the receptor for » PDGF. STK1 expression in human blood and marrow is restricted to » CD34-positive cells. The use of » antisense oligonucleotides directed against STK1 sequences shows that the inhibition of STK1 expression blocks hematopoietic colony formation, most strongly in long-term bone marrow cultures (see also: LTBMC).

Small D et al STK-1, the human homologue of Flk-2/Flt-3, is selectively expressed in CD34+ human bone marrow cells and is involved in the proliferation of early progenitor/stem cells. PNAS 91: 459-6 (1994)

Sweat gland factor: see: SGF (this addendum).

TGF-α: *transforming growth factor α* Neonatal mice overexpressing TGF-α in the epidermis have been generated. These mice are often smaller than normal littermates and have precocious eyelid opening and wrinkled, scaly skin with diffuse alopecia. Juvenile transgenic mouse epidermis is uniformly hyperkeratotic. Spontaneous, squamous papillomas occur at sites of wounding in adult mice expressing high levels of TGF-α; however, most are prone to regression. Expression of the » EGF receptor is down-regulated at the sites of increased TGF-α expression.

Transgenic mice overexpressing TGF-α have been used to establish lines of hepatocytes which can be maintained in long-term culture as replicating, differentiated cells while remaining nontumorigenic.

Dominey AM et al Targeted overexpression of transforming growth factor alpha in the epidermis of transgenic mice elicits hyperplasia, hyperkeratosis, and spontaneous, squamous papillomas. Cell Growth Differ. 4: 1071-82 (1993); **Wu JC et al** Establishment and characterization of differentiated, nontransformed hepatocyte cell lines derived from mice transgenic for transforming growth factor alpha. PNAS 91: 674-8 (1994)

Thrombopoietin: A factor, designated ML, that appears to stimulate both proliferation and maturation of megakaryocyte progenitor cells has been identified as the ligand for the c-*mpl* oncogene (see also: MPLV). ML is a protein of 332 amino acids that shows significant homology with » Epo (erythropoietin) in its N-terminal domain (152 amino acids). A comparison of human and porcine ML sequences shows 83 % identity between the erythropoietin-like domains but only 67 % between the C-terminal domains. Low levels of ML transcripts have been found in both fetal and adult liver.

c-*mpl* expression appears to be restricted to primitive stem cells, megakaryocytes, and platelets, and » antisense oligonucleotides directed against *mpl* selectively inhibit megakaryocytic colony formation without affecting the growth of erythroid and granulomacrophage colonies (see also: Hematopoiesis). Intraperitoneal injections of recombinant ML into mice increases circulating platelet levels by greater than fourfold after seven days. In vivo, ML also expands marrow and splenic megakaryocytes and their progenitors and shifts the distribution of megakaryocyte ploidy to higher values. It has been suggested that the factors described as » thrombopoietin and » MEG-CSF (megakaryocyte colony-stimulating factors are identical with the cloned thrombopoietin ML. ML is also identical with an independently cloned factor named » MGDF (megakaryocyte growth and development factor).

de Sauvage FJ et al Stimulation of megakaryocytopoiesis and thrombopoiesis by the c-*Mpl* ligand. N 369: 533-8 (1994); **Kaushansky K et al** Promotion of megakaryocyte progenitor expansion and differentiation by the c-*Mpl* ligand thrombopoietin. N 369: 568-71 (1994); **Lok S et al** Cloning and expression of murine thrombopoietin cDNA and stimulation of platelet production *in vivo*. N 369: 565-8 (1994); **Wendling F et al** c-*Mpl* ligand is a humoral regulator of megakaryocytopoiesis. N 369: 571-4 (1994)

Thymic stroma-derived T-cell inhibitory factor: see: TSTIF (this addendum).

TIMP-3: *tissue inhibitor of metalloproteases* Proteins of the » TIMP family bind and inactivate matrix metalloproteinases such as collagenases and gelatinases. TIMP-3 is the third member of the human TIMP family. The precursor protein deduced from the cloned cDNA includes twelve Cys and 27 other amino acids that are invariant in the TIMP family. The predicted amino acid sequence is 89, 39 and 46% identical to those of ChIMP-3, human TIMP-1 and TIMP-2, respectively.

Silbiger SM et al Cloning of cDNAs encoding human TIMP-3, a novel member of the tissue inhibitor of metalloproteinase family. Gene 141: 293-7 (1994)

TNF-β: *tumor necrosis factor β* The biological consequences of a » TNF-β gene disruption have been studied in mice generated from » ES cells carrying a targeted deletion of the gene. Mice homozygous for a targeted disruption of the TNF-β gene have no morphologically detectable lymph nodes or Peyer's patches. The development of the thymus appears normal. Within the white pulp of the spleen there is a failure of normal segregation of B and T cells. Spleen and peripheral blood contain $CD4^+CD8^-$ and $CD4^-CD8^+$ T cell in a normal ration, and both T cell subsets have an apparently normal lytic function. Lymphocytes positive for IgM are present in increased numbers in both the spleen and peripheral blood. These data suggest an essential role of TNF-β in the normal development of peripheral lymphoid organs. However, the mechanism by which TNF-bcontributes to normal lymphoid development remains unknown.

de Togni P et al Abnormal development of peripheral lymphoid organs in mice deficient in lymphotoxin. S 264: 703-7 (1994)

trk: The article reviews the various effects of neurotrophins, » BDNF, and their receptors on the development of the nervous system as revealed by studies with various neurotrophin/neurotrophin receptor gene null mutations.

Snider WD Functions of the neurotrophins during nervous system development: what the knockouts are teaching us. Cell 77: 627-38 (1994)

tyk1: see: LTK (this addendum).

TSTIF: *thymic stroma-derived T-cell inhibitory factor* This poorly characterized factor or ≈ 20-25 kDa has been isolated from culture supernatants of the MRL104.8a thymic stromal cell line. The supernatant alone induces proliferation of helper T cells because it contains » IL7. The addition of the supernatant to cultures of helper T cells stimulated with antigen plus antigen-presenting cells inhibits T cell proliferation. TSTIF exerts its inhibitory effect on the antigen-stimulated T-cell proliferation by acting on the antigen-presenting cells. Several experiments demonstrate that the effects of TSTIF are not mediated by » IL7, » IL10, or TGF-β.

Kita Y et al Thymic stroma-derived T-cell inhibitory factor (TSTIF) 1. TSTIF induces inhibition of antigen-stimulated T-cell proliferation. Thymus 21: 159-75 (1993); **Tai XG et al** Thymic stroma-derived T-cell inhibitory factor (TSTIF). 2: TSTIF acts on the antigen-presenting cell to inhibit antigen-stimulated T-cell proliferation. Thymus 21: 247-58 (1993)

Vasostatin: see: Granins (this addendum).

***waved*-2:** The wa-2 mutation (mouse chromosome 11) exhibits skin and eye abnormalities that are strikingly similar to those of TGF-α-deficient mice. The wa-2 phenotype has been shown to result from a single-nucleotide transversion in the » EGF receptor gene, substituting a glycine for a conserved valine residue near the amino terminus of the tyrosine kinase domain. The mutated receptor still binds its ligand but ligand-dependent » autophosphorylation of the wa-2 EGF receptor is greatly diminished.

Luetteke NC et al The mouse waved-2 phenotype results from a point mutation in the EGF receptor tyrosine kinase. Genes Dev. 8: 399-413 (1994)

ZAP-70: This 70 kDa tyrosine phosphoprotein associates with the zeta chain of the T cell receptor and undergoes tyrosine phosphorylation following receptor stimulation. The ZAP70 gene is expressed in T- and natural killer cells. The ZAP70 gene maps to human chromosome 2q12.

Mutations in the ZAP70 gene have been shown to occur in patients with severe combined immunodeficiency (SCID). A mutation of ZAP70 resulting in loss of the activity of this kinase has been shown also to be the underlying cause of a novel type of human immunodeficiency characterized by a selective T-cell defect (STD). Peripheral circulating T cells from STD patients exclusively express CD4, CD3, and T-cell receptor-α/β, but not CD8 molecules on their surface.

Arpaia E et al Defective T cell receptor signaling and CD8(+) thymic selection in humans lacking Zap-70 kinase. Cell 76: 947-58 (1994); **Chan AC et al** ZAP-70: a 70 kd protein-tyrosine kinase that associates with the TCR zeta chain. Cell 71: 649-62 (1992); **Chan AC et al** ZAP-70 deficiency in an autosomal recessive form of severe combined immunodeficiency. S 264: 1559-601 (1994); **Elder ME et al** Human severe combined immunodeficiency due to a defect in ZAP-70, a T cell tyrosine kinase. S 264: 1596-9 (1994); **Iwashima M et al** Sequential interactions of the TCR with two distinct cytoplasmic tyrosine kinases. S 263: 1136-9 (1994); **Roifman CM et al** Depletion of CD8(+) cells in human thymic medulla results in selective immune deficiency. JEM 170: 2177-82 (1989); **Tsygankov AY et al** Diminished tyrosine protein kinase activity in T cells unresponsive to TCR stimulation. J. Leukoc. Biol. 55: 289-98 (1994)

Zeta-associated protein 70: see: ZAP-70 (this addendum).